CALIFORNIA AND THE WORLD OCEAN '97

Taking a Look at California's Ocean Resources: An Agenda for the Future

VOLUME 1

CONFERENCE PROCEEDINGS

March 24-27, 1997
Town & Country Hotel
San Diego, California U.S.A.

EDITED BY
Orville T. Magoon
Hugh Converse
Brian Baird
Melissa Miller-Henson

ASCE American Society of Civil Engineers
1801 ALEXANDER BELL DRIVE
RESTON, VIRGINIA 20191

Abstract: The conference, California and the World Ocean '97 (CWO '97), was organized by the Coastal Zone Foundation and Resources Agency of California. CWO '97 was an opportunity to bring together the leading experts on ocean and coastal resource management issues from around the world to help California determine how best to begin approaching the recommendations in the strategy, *California's Ocean Resources: An Agenda for the Future.* The papers in this proceedings were presented in San Diego, March 24-27, 1997, and offer suggestions on integrating and making more effective the protection, enhancement, and management of California's extraordinary ocean and coastal resources. Topics covered are varied and include coastal economics, shoreline processes, ports and vessel traffic, oil and gas development, oil spill prevention and response, marine managed areas, habitats and ecosystems, fisheries management, water quality, governance and public policy, alternative and emerging technologies, education, and public participation.

Library of Congress Cataloging-in-Publication Data

California and the world ocean '97: taking a look at California's ocean resources: an agenda for the future: conference proceedings / edited by Orville T. Magoon ... [et al.].
p. cm.
"March 24-27, 1997, Town & Country Hotel, San Diego, California, U.S.A."
Includes indexes.
ISBN 0-7844-0297-3 (set)
1. Marine resources–California–Congresses. I. Magoon, Orville T.
GC1023.13.C2C25 1997 97-48327
333.91'64'09794–dc21 CIP

Library of Congress Catalog Card No: 97-48327
ISBN 0-7844-0297-3
Manufactured in the United States of America.

ACKNOWLEDGEMENTS

CWO '97 Executive Committee
Douglas P. Wheeler, Honorary Chair, Resources Agency of California
Orville T. Magoon, Conference Chair, Coastal Zone Foundation
Hugh Converse, Executive Officer, Coastal Zone Foundation
Brian E. Baird, Vice Chair, Resources Agency of California
Melissa Miller-Henson, Vice Chair, Resources Agency of California

Organizers
Coastal Zone Foundation
Resources Agency of California

Gold Sponsors
David and Lucile Packard Foundation
National Ocean Service, NOAA

Silver Sponsors
U.S. Minerals Management Service - Pacific OCS Region
University of California

Bronze Sponsors
California Department of Boating and Waterways
National Sea Grant College Program
Southern California Marine Institute
University of Southern California Sea Grant Institutional Program

Co-Sponsors
American Oceans Campaign
American Shore & Beach Preservation Association
American Society of Civil Engineers
California Aquaculture Association
California Coastal Commission
California Conservation Corps
California Department of Fish & Game
California Department of Parks & Recreation
California Environmental Protection Agency
California Fisheries & Seafood Institute
California Marine Affairs & Navigation Conference
California Shore & Beach Preservation Association
California State Association of Counties
California State Coastal Conservancy
California State Lands Commission
Center for Marine Conservation

Council of State Governments
Guenoc & Langtry Estate Vineyards and Winery
League for Coastal Protection
League of California Cities
League of Women Voters of California
Oregon Coastal/Ocean Management Program
Port of San Diego
San Diego Association of Governments
San Francisco Bay Conservation & Development Commission
Southern California Coastal Water Research Project
Surfrider Foundation
Texas A&M University - Ocean Engineering Program
U.S. Army Corps of Engineers - South Pacific Division
U.S. Environmental Protection Agency - Region IX
U.S. Fish & Wildlife Service - San Francisco Bay Program
U.S. Section - P.I.A.N.C.
Universidad Autónoma de Baja California

Data Base Managers
Sheila Robertson
Belinda Young

Over 30 years ago the State of California held a conference entitled, *California and the World Ocean*, which brought preeminent scientists together to address California's ocean and coastal management issues and needs. The Resources Agency of California's strategy, *California's Ocean Resources: An Agenda for the Future*, built upon that conference and many intervening events to identify the issues and concerns facing the State of California as we enter the next century.

California and the World Ocean '97 was the next step in continuing and proposing innovations across California to integrate and make more effective the protection, enhancement and management of California's extraordinary ocean and coastal resources. Envisioned as an international conference, CWO '97 would bring together the leading experts on ocean and coastal resource management issues from around the world to help California determine how best to begin approaching the recommendations in the Ocean Agenda.

We believe *California and the World Ocean '97* was a tremendous success! Over 800 individuals attended from California, other States, and nations. Participants included representatives from federal, state, and local governments, legislators and members of Congress, academia, the private sector, various non-profit groups, and members of the public who simply wanted to know more about managing California's ocean resources. Already the conference deliberations have been invaluable as California moves forward to implement the many recommendations in the Ocean Agenda, with numerous administrative and legislative proposals currently being considered.

We would like to thank the many sponsors and co-sponsors who helped make the conference and these proceedings possible. An undertaking of this magnitude is difficult and time-consuming, even under the best of circumstances.

We hope you find these proceedings of great value, whether for education, research, business, or decision-making purposes, and that you take an active role in helping find even more ways to better manage our precious ocean and coastal resources.

Douglas P. Wheeler
Honorary Chair

Orville T. Magoon
Conference Chair

Douglas P. Wheeler, Honorary Chair
Secretary for Resources, State of California

Orville T. Magoon, Conference Chair
President, Coastal Zone Foundation

P.O. Box 279, 21000 Butts Canyon Road • Middletown, California 95461 U.S.A.
Telephone: (707) 987-0114 • Facsimile: (707) 987-9351 • Electronic Mail: otmagoon@aol.com

California and the World Ocean '97:
An Overview of Major Themes

California and the World Ocean '97 was initially envisioned as a relatively small, focused event bringing together leading experts on ocean and coastal resource management issues to help California evaluate the findings and recommendations of the strategy titled, *California's Ocean Resources: An Agenda for the Future*. Response to the call for papers was overwhelming, leading the executive committee to re-evaluate the size and breadth of the conference. Eventually the concurrent sessions committee accepted over 270 papers and posters for presentation at CWO '97.

Outstanding financial and in-kind support from over 40 organizations substantially reduced the conference registration fees, and workshops and field trips were offered at little or no cost to conference attendees. As a result, over 800 people were in San Diego, California for CWO '97, many from the general public, not-for-profit organizations and local governments.

With such a large number of papers presented in six concurrent tracks during the three-day conference, it was impossible for any one individual to track the great diversity of topics. To help identify and summarize the conference themes, the National Sea Grant College Program provided funding to bring together marine policy graduate students from the Universities of Washington and California. These students joined University of California and University of Southern California Sea Grant staff to form a "theme team", which provided an overview of conference themes at the closing session and many of the observations for this summary of cross-cutting themes.

Major Topics

The document upon which this conference was based, *California's Ocean Resources: An Agenda for the Future* (Ocean Agenda), addresses a wide variety of issues regarding the spectacular marine resources that exist off the state's 1,100 mile shoreline. These issues were a foundation for the six tracks and almost 50 sessions of CWO '97, including coastal economics, coastal processes, ports and vessel traffic safety, oil and gas development, oil spill prevention and response, marine managed areas, habitats and ecosystems, fisheries management, water quality monitoring, governance and public policy, technology development, education, and public participation. Many suggestions were made regarding new ways to address these issues and how the recommendations made in the Ocean Agenda could be implemented. Substantial attention was given to the need for processes to address water quality, marine habitat and fisheries health, shoreline erosion, and economic sustainability.

Common Themes

Despite the multi-disciplinary nature of the conference and large number of participants, several common themes surfaced across most of the major subject area issues and approaches. The aforementioned "theme team" met each day to discuss observations made during the sessions they individually attended, eventually identifying a group of what they labeled "C words" and "I words" that were used frequently in session presentations. First identified were collaboration and communication. Existing systems need to be better coordinated through additional collaboration, and increased attention given to improving the way we communicate with others to achieve common resource protection goals or sustainable economic objectives. Two other "C words" repeated across subject areas were community and consensus, while three "I words" used by speakers were integration, investigation, and implementation. Each of these major themes are described here in more detail.

> ***Communication.*** Agencies, industries and the public must do a better job of communicating with one another. Lack of communication leads to misunderstandings, duplication of efforts, and sometimes conflicting outcomes. Communication must be both vertical and horizontal, with multiple channels available for communicating; horizontal communication is in greatest need of improvement.
>
> As information technology has improved and become less expensive (facsimile machines, electronic mail, the world wide web), communication has increased substantially, especially horizontally (within organizations) and between organizations. Improved information technology is enabling greater communication, collaboration and education.
>
> Communication (and coordination) is especially important for ocean and coastal resource management issues that tend to overlap political boundaries, such as populations or marine organisms, accretion and erosion of sand along the shoreline, and shipping safety. Political boundaries include not only local government jurisdictions along the coast, but also the many jurisdictions offshore such as State Tidelands, the territorial sea, the exclusive economic zone, and the high seas. The variety of jurisdictional, ownership, or user interests along and offshore the coast emphasizes the need for maintaining good communication among all parties.
>
> ***Collaboration.*** A prevalent theme was the need to create and, more importantly, sustain cooperative partnerships between and among government agencies, universities, conservation groups, and private interests, especially at the local level. Collaborative approaches allow

organizations to do more with less. However, collaboration is often impeded by institutional constraints, where an organization's design, policies or mission and goals conflict with those of potential partners. Organizations and their representatives must be willing and able to adapt to different circumstances and partners.

Participants also identified the need for collaborative permitting and funding. In one method for collaborative permitting, the agencies of jurisdiction come together at the beginning of a process to discuss their concerns, allowing issues to be addressed early, and then proceed together (or having one agency take the lead) in a coordinated effort to complete necessary environmental documents and permits.

Marine managed areas illustrate the need for more collaboration; California's system of marine managed areas is extremely complex, with multiple classifications and hundreds of local, state, federal and international designations along the coast. Management and protection measures vary widely from one classification to another. With little information easily accessible to the public and even managers themselves, it is uncertain that California's marine managed areas are achieving the goals intended. Partnerships are necessary to share expertise, achieve wider coverage of the issues being faced, and develop a more effective and efficient system of resource management and protection.

Community. Many speakers emphasized the need to work closely with the public and to take a more local or regional focus on issues in order to promote a strong sense of community. Concepts of shared governance or co-management were often discussed, with an emphasis on being proactive rather than reactive.

A more regional community focus is also necessary to sustain economic viability. Members of a community are more willing to help sustain and protect ocean and coastal resources when they understand the importance of those resources to both the environment and the economy. This takes on even more importance as population pressures continue to build along the California coast. Key examples of community based approaches include watershed management programs, coastal cleanups, and fishery enhancement projects.

Consensus. A number of speakers indicated that it is critical to build consensus among stakeholders, which encourages collaborative partnerships, improved compliance with regulations, and improved resource management. While this method of decision-making can require a substantial time commitment from participants, consensus-based decisions tend to have much greater support and involvement in implementation from the participating parties, improving the odds of a successful endeavor.

An example of stakeholder consensus-building in California is in vessel traffic safety through the harbor safety committees, where multiple stakeholders work together to find consensus-based solutions to safety issues. The result has been improved communication between interested parties; better information exchange between pilots, industry and interest groups; increased safety awareness; and development of workable regulations.

Integration. Ocean and coastal resource management issues are often addressed on a case-by-case or "urgency" basis by a multitude of organizations and agencies, impeding a comprehensive approach. Speakers repeatedly urged integrative strategies for resource management that take ecosystem or multi-species approaches, are multi-disciplinary, and integrate science as well as social values. Integrative strategies and programs must also be able to adapt to changing needs, shifting ecological conditions, and improved scientific understanding of natural processes.

Integration plays an important role in managing nonpoint source pollution. The number of participants can be rather large (in one case over 80 jurisdictions) and the time to implementation lengthy (18 months to several years) when using an integrative strategy, but the results can include region-wide cooperation, improved monitoring capabilities, long-term public education, attainment of performance standards, and reduced duplication of effort.

Investigation. Increased research at the state and federal levels to achieve more science-based policy decisions was consistently urged, including the need for continued investigation of nearshore oceanographic processes, harvest Refugia, wetlands restoration, fisheries management, and alternative energy technologies. While there needs to be adequate scientific information, there must also be an acknowledgement that there are currently flaws in monitoring, research, and current knowledge, requiring a certain amount of flexibility over time as new methods and information become available. This is especially true in the case of water quality monitoring where multiple sources of data across the State often cannot be integrated because monitoring methods and data are not standardized.

Inadequate scientific information and flawed methods also require a rather precautionary approach to management decisions. Consequences of management decisions are often difficult to predict in highly complex systems. A case given was fisheries management, where inadequate information and lack of a precautionary approach can lead to long term or permanent damage to the resource.

Investigation must also be closely accompanied by accessibility of the information gathered. Relatively new technologies, such as geographic information systems, global positioning systems and the world wide web, can help in decision-making and education by providing the public, resource managers and policy-makers with enhanced access to data. However, the data must also be in a usable format, with some standards for consistency and integration.

Implementation. It is not enough to simply identify problems and potential solutions; implementing changes in ocean resource management is necessary through legislation, restructuring of agencies and relationships, and funding of high-priority programs. Those involved in the implementation process need to also learn another "i" word, innovation, as they learn that the old ways of thinking and working together may not be appropriate for the challenges of today and tomorrow.

New regulatory approaches, which include a mix of both regulation and incentives, should be more closely examined and considered. This will require new ways of operating or "governing" for government agencies, as well as new mechanisms for collaboration between and among government agencies, industry and private interests. The very presence of so many different interests and government representatives at CWO '97 suggests that the climate may be right for reform.

A wide range of funding mechanisms currently exists for improved governance, but some are wasted or missed as a result of incomplete or ineffective management. Government budgeting processes frequently do not encourage innovative or proactive pursuit of funding for ocean and coastal resource management. Joint agency expenditures and budget combinations for those who share jurisdictions should be considered, with barriers to such efforts removed. These innovations will be important, but many speakers also observed that ocean and coastal resource management programs are a remarkably small fraction of government budgets. Participants commented that consideration must be given to providing additional funding for initiatives that represent effective and efficient approaches to managing ocean and coastal resources.

Conclusion

Papers presented at CWO '97 provide a wealth of information about a wide variety of issues affecting ocean and coastal management. Many issue-specific challenges exist and are described in detail in the papers presented. However, the seven themes identified in this summary (Communication, Collaboration, Community, Consensus, Integration, Investigation, and Implementation) are all process-oriented. An important lesson learned through this conference has been that while the specific challenges in each issue may be different, the need for similar processes to resolve them is not.

One method for improving these processes is through a coordinating entity, a proven tool in a number of processes, including watershed management, oil spill prevention and response planning, and air quality management. The need for enhanced coordination of fragmented ocean and coastal resource management in California is documented in the Ocean Agenda, which identifies the need for a forum to help resolve issues at the State level and a process of cooperating with and soliciting advice from other levels of government, the public, and the private sector. The Ocean Agenda recommends increasing accountability and coordination in matters affecting California's ocean and coastal resources in the following way:

> "Convene a State cabinet-level ocean resources management coordinating council, composed of agency and department directors with ocean resource management responsibilities, to help integrate the multiple agencies and programs of ocean and coastal jurisdiction. The effectiveness of this council will depend on its ability to work with public and private organizations to identify and develop solutions to ocean and coastal resource management issues of concern to the State of California."

The proposed council is one of many approaches that can help improve ocean and coastal resource management. However, this type of cabinet-level approach can only be successful if based on participation at the local level. CWO '97 participants have sent a clear message that the mission and goals identified for ocean management can only be achieved by using processes that provide for enhanced communication, collaboration, community (involvement), consensus, integration, investigation and implementation. Hopefully these lessons will be heard by all levels of government, academic institutions, public interest groups, and the private sector. Together, we must now seek new ways incorporate these suggestions in efforts to implement innovative approaches to managing California's ocean and coastal resources.

Acknowledgements

The conference organizers would like to thank the theme team for their hard work over the three days of CWO '97, making this summary possible. These individuals include the theme team leader, Christopher M. Dewees, Marine Fisheries Specialist and Coordinator of the UC Sea Grant Cooperative Extension Program, and theme team members Lillian Ferguson, Michael Kelly and Eli Weissman, graduate students at the University of Washington School of Marine Affairs; Mark Evans, graduate student at the University of California at Davis Department of Agricultural and Resource Economics; Sea Grant Marine Advisor Emeritus John B. Richards; and Susan E. Yoder, leader of the University of Southern California Marine Advisory Program.

SPECIAL SESSIONS AND KEYNOTE PRESENTATIONS

CWO '97 included three days of concurrent sessions where participants addressed a wide range of ocean and coastal management concerns. The papers submitted by those participants are provided in these conference proceedings. In addition to the concurrent sessions the conference held opening and closing plenary sessions, a special session on local government concerns, and other special luncheon presentations. The following is a listing of those important sessions and their participants.

Opening Session

Perspectives on California's Ocean Resources: Maximizing this Historic Opportunity

Honorable Sam Farr, US House of Representatives, 17th Congressional District of California

Orville Magoon, Conference Chair and President, Coastal Zone Foundation

Douglas Wheeler, Honorary Conference Chair and Secretary for Resources, State of California

Fred Hansen, Deputy Administrator, U.S. Environmental Protection Agency

Rolland A. Schmitten, Assistant Administrator, National Marine Fisheries Service, National Oceanic and Atmospheric Administration

Closing Session

Moving Forward with Ocean Protection and Management in California: Major Themes and Future Actions

Stanley Wilson, Assistant Administrator, National Ocean Service, National Oceanic and Atmospheric Administration

Richard Morrison, Senior Vice President, Environmental Policies and Programs, Bank of America

Robert Sulnick, Executive Director, American Oceans Campaign

Christopher Dewees, Coordinator of the University of California Sea Grant

Cooperative Extension Program

Peter Rooney, UnderSecretary, California Environmental Protection Agency

Brian E. Baird, Ocean Program Manager, California Resources Agency

Luncheon Speakers

Honorable John H. Chafee, US Senate, Rhode Island

Donald Murphy, Director, California Department of Parks and Recreation

Jacqueline E. Schafer, Director, California Department of Fish and Game

Local Government Roundtable

What is the Local Agenda for California's Ocean Resources?

Elliot Parks, Chair, San Diego Association of Governments

Carol Whiteside, Director of Intergovernmental Affairs, Governor's Office of Planning and Research

Ann J. Kulchin, City Council Member, City of Carlsbad

Laurence Laurent, County Supervisor, San Luis Obispo County

Contents

VOLUME I

* Manuscript not available at time of printing.

* Manuscript not available at time of printing.

* Manuscript not available at time of printing.

* Manuscript not available at time of printing.

* Manuscript not available at time of printing.

* Manuscript not available at time of printing.

* Manuscript not available at time of printing.

* Manuscript not available at time of printing.

* Manuscript not available at time of printing.

* Manuscript not available at time of printing.

Wetlands and Intertidal Planning

Sediments and Dredging

Challenges in Ocean Resource Management:
A Case Study of the San Francisco Bay Area

* Manuscript not available at time of printing.

* Manuscript not available at time of printing.

* Manuscript not available at time of printing.

* Manuscript not available at time of printing.

* Manuscript not available at time of printing.

* Manuscript not available at time of printing.

* Manuscript not available at time of printing.

* Manuscript not available at time of printing.

CHAPTER 1

An Ocean of Love

By
Orville T. Magoon

Conference Chair, *California and the World Ocean '97*
President, Coastal Zone Foundation
PO Box 279, Middletown CA 95461

Speaking version
March 25, 1997

Welcome to ***California and the World Ocean '97***! With over 800 registrants, I think the person who traveled the farthest is Dr. Ruben Kos'yan representing the Russian Academy of Sciences. Over dinner last night Ruben told me about fascinating changes in the political structure of Russia relating to science and technology, and directions in Russian Coastal Zone Management and about a conference on coastal processes being organized in Moscow Sept 98. I hope you all get a chance to chat with him during the conference.

Perhaps you might ask what is the Coastal Zone Foundation and how did this conference get started? The Coastal Zone Foundation is a small group of volunteers who are dedicated to providing a forum for resolution of the complex and multifaceted coastal decisions facing us now and in the future. When it was released almost two years, ago I purchased a copy of California's Ocean Resources: An Agenda for the Future and realized that this document contained a good outline on how to understand, maintain and hopefully improve and better manage California's Coasts and Oceans.

I then wrote California Resources Agency Secretary Doug Wheeler indicating that although the public hearings held along the coast provided a good vehicle for public comment, the Agenda for the Future would benefit from a conference that would help us to learn about different approaches to ocean management and to exchange ideas with professionals and individuals from California, other States, and nations. Our focus will be on California's new strategy " California's Ocean Resources: An Agenda for the Future." Secretary Wheeler will be describing the new State strategy and other State initiatives in his remarks this morning.

From my perspective, I've seen some major changes in the way the American public looks at government. Rather than thinking that our leaders in Washington D.C. will formulate every policy and dictate every role, I believe that the people in this room and the responsible population of California will be making the real decision for the future of our coastlines. Thus the Federal, State, County, City, and Local government roles are changing and it is up to us here at CWO '97 to define what those roles are.

Do you realize the first California and the World Ocean conference was held over 30 years ago? We are here in San Diego to celebrate a new beginning, our second conference on California and the World Oceans.

Our world is good, our lives are abundant with possibilities in this mysterious universe within which we sit on our ship, the Earth. What defines our own living space is the scraggly ribbon of our coast and shores. That ribbon of sand and rock, dunes and mangroves, is a place of prayer, of gathering, a place to take your children to build sand castles, investigate tide pools and climb rocks. It is a place to dream, to fall in love, to marry. It is a place to celebrate joy and harmony, to heal, to mend. In our Coastal Zone conferences in Japan, the Japanese have described the deeply spiritual meaning of Nagista, the triple line, where the land, the sea, and the air meet.

When I was a boy, growing up in Hawaii, I built a boat with my father called the Manukai, meaning, "where sea and seabird meet". I learned about life in and on the sea from that boat. I suppose that if I was to imagine a home for my soul, that is where it would be. Perhaps that's why it took some convincing for my father to persuade me to leave the Manukai and go to college.

On the final days I spent on the boat before trudging off to the scholastic world of books, papers, tests and deadlines, a hurricane built up over the ocean. As the waves were too high to get into a harbor on the Kona Coast of Hawaii I stayed on board trying to keep sediment-filled fuel tanks running the engine amidst the storm and huge breaking waves. Around midnight I saw a flashing light on shore a series of single dots, double dots and triple dots of light which was simple code of Hawaiian Fisherman, and knew it was my father. In the morning, the storm subsided, I made toward the harbor and safety. Reaching my father, I embraced him and then added, coolly, "Dad, we really should learn Morse code."

A true story, all of it. But it's the last part I want to talk about, the Morse Code of life, the reason we are here today, seeking solutions to problems centered on the world ocean—and in this instance, our great coastal state, California.

Nagista is easy. We didn't do it. Out there is the sky and ocean and we are on the shore, here in San Diego, close to the triple line. It was here before us and it will always be here, as long as earth endures. The coordinates may shift, we will change, the earth may get hotter or cooler. The ocean may rise or fall due to a Greenhouse effect or an Icehouse effect or other meteorological causes. But sky and sea and shore remain. *Nagista.*

What is not always easy is what I am calling our "Morse code", its translation and deciphering. In short, what's difficult is figuring out how to translate our responsibility to use our resources appropriately and creatively and at the same time protect the air and sea and land, so that its beauty and purpose may be enjoyed by everyone.

My wife and I were recently in India in January, where a beautiful people seem to have an entirely different idea from ours as to what is vital to earthen existence. It was in India that Maharajah Shah Jahan built the Taj Mahal after his wife died in childbirth in 1631. The child would have been their fourteenth offspring. Mumtaz Mahal, his beloved wife, had always accompanied the Maharajah to the battlefield. When she felt this last child die within her, she told her husband she too would be dying and to build a beautiful tomb to remember her. The Taj Mahal is the fulfillment of his promise and in my opinion the most exquisite work ever created by man.

The Taj Mahal, a last Beethoven string quartet, a Michelangelo David, a Mona Lisa, the inventions of human beings from Morse Code to the Internet, are, I believe, our humble attempt to translate or take responsibility for the gift of life, the power of creation given to us from a source which is our own beginning, to learn that which we too must teach those who come after us. I ask you, I ask myself, are we learning and imparting the right lessons? Are we good students? Competent teachers?

In Europe, I enjoyed having coffee in a little French city in Provence in one of the many squares that seem always to sprawl around the city halls of European cities, large and small. Here was a real meeting place where street musicians performed; children played, old folks or perhaps senior citizens like me enjoyed their coffee and ice cream or perhaps a glass of wine. Later that afternoon, there would be a concert there attended by thousands.

In America, although we don't have the city square, we do have our meeting places where we relax and play and listen to music. Those places are our coastlines and our boardwalks on the rivers and lakes of our great and small cities. There by the water's edge are social meeting places where violence seldom occurs. Rather, those meeting places by the water tend to encourage peaceful, harmonious public gatherings, essential to the life of our communities.

On the waterways and ocean coastlines, culture can spawn, theaters and musicians find stages and audiences where plays and music can be performed in a beautiful environment. Granville in Vancouver is an industrial island turned into an *Art Deco* cultural mecca. Five theaters, an open market, boat rides and street entertainment bring crowds everyday to mingle and relax. They also come to study poetry and listen to lectures on history and art. They come to view the art galleries and attend workshops.

These boardwalks by the sea and rivers are a way for all of us to get back to the community centered on the water. I believe every individual has an inalienable right to enjoy the coastlines and waterways of the world. With that right, as we know, comes also the responsibility to preserve a healthy environment centered around our sea, sky and shore. That is why we are here today. We are in the privileged position of being able to gather and disseminate knowledge. We are in the privileged position of having a mandate to learn about the coast and ocean and to use that knowledge to improve our world and teach others.

We are in a unique position. Because of that, it falls to us to find and impart the data we accumulate to those entrusted to us as well as to our elected officials. It is our responsibility to procure and disseminate correct and useful information to help the public and their elected representatives make the right decisions related to our environment and help to resolve the conflicts that arise generated by the tough decisions we are required to make concerning our coasts.

It lies with us to discover the message. We can surely find in our own coastlines the drama of a Hubble Telescope or an Odessey to the Moon interesting enough to make people listen. We must create a message so compelling that the public will respond with passion to their own innate love of the coast and teach their children how to translate that love into stewardship.

But it is incumbent upon us first to listen. We talk of erosion, we talk of sealevel rise, and we talk of pollution. Perhaps we talk too much. We need to listen and translate. We need to learn the "Morse code of the sea" so that we can understand our own beginnings and know where we are going. If you don't know where you're going, any road will take you there. If you do know, you can make miracles happen.

We can do it. We can make miracles. Let me ask you to define success. Is success an idea that has found its time or a time that demands an idea? Were the pyramids successful? That all depends on whether the pharaohs attained eternal life. Were Frank Lloyd Wright's revolutionary architectural designs successful? I'm told they all leak, but they presaged a whole new focus on organic architecture dedicated to building in harmony with the environment, building for families and small communities, building to improve the personal quality of life. The purpose of creating harmonic community units, attempted again by Soleri in Arizona, eventually led to a city, which is virtually deserted today. Success? Failure?

Success and Failure are two heads of what I like to call the *Complexity Dragon* which I've spoken about in many of our past Coastal Zone Conferences They may even be twin heads, for each failure contains the kernel of the next possible success. Today's jet aircraft and rockets are built on the remains of people's first and second and third attempts to fly. Similarly, today's problems with pollution and sand displacement, crumbling edifices and spiritual malaise have within them the blueprint for their own solutions. Every failure contains the blueprint of success, the perfect DNA.

In 1910, the beautiful, free-running Elwha River on the Olympic Peninsula which joins the Columbia River near it's mouth in the Pacific in Washington State was strangled. Two then economically needed hydroelectric dams interrupted the migration of the fish and the downstream movement of sand and gravel needed for spawning and habitat replenishment. These dams are no longer economically viable. Congress, however, responded to the controversy in October, 1992, by enacting Public Law 102-395, the Elwha River Ecosystem and Fisheries Restoration Act, which required the Secretary of the Interior to investigate how to restore the native anadromous fisheries and ecosystem of the Elwha. The Secretary recommended that the removal of both dams was both feasible and necessary to restore the fisheries and ecosystem and uphold the federal trust responsibility for affected Native American tribes, and to allow the river born sands to travel to the coast and to nourish the downcast beaches. In that report is the germ of another idea we will discuss in a Florida Shore and Beach Preservation Association Conference on Sand Rights next February - The right of a sand grain to go from the mountains to the sea and along the coast. A version of this concept "Sand Rights" by my wife, Karen Melander Magoon, and intended for children is enclosed in many of your registration packets.

Every time we fool with the ecostructure of our oceans, rivers, and shores, we also impinge upon the rights of naturally occurring phenomena to realize their destiny. When we block the flow of sand or gravel we subsequently infringe its right to move in balance with nature's demands. This question is of great consequence today as we seek to resolve controversies hinging on what may actually be *illegal* human intervention in the acts of nature.

However, as we learn to regulate ourselves better, we still need to move forward with new ideas and solve those problems, which have gratuitously appeared in the last few decades.

We here today enjoy the challenge of discovering new formulae to create entirely new blueprints for success. We have all read of Dr. Ho's research to find a solution or cure for AIDS. Each failure contained the germ of a new answer. So it must be with our search for solutions to problems found in our oceans, on our shores and at our coastlines. Although Pogo told us "We have met the enemy, and he is us", so are our solutions to be found within us, individually and collectively. And we should not forget that every failure contains a possible key to its own success.

Each one of you has an idea within you that can rock the world or redefine our lives. Each one of you is an artist or a poet or a scientist on the verge of discovering something extraordinary. And each one of you fuels your creative energy with your great love of the shores and beaches of the world.

This is an exciting moment, an exciting gathering. 1997 is officially The Year of the Reef, which is recognition of part of that vast and mighty network which composes our shores and near shores. Our conference, *California and the World Ocean '97*, recognizes how one fragment of our entire world coastline is related intrinsically to the whole. How much smaller than the fragment of a great state is that bit of earth we call a sand grain, and yet how powerful, how essential to the entire eco-structure of our planet. Everything we can touch or feel or see or smell contains a kernel of inalienability, which is the intrinsic right of everything to enjoy and be part of the earth, the sky and the shore.

We are in this world together. Few know the Morse code, as I do, which once was a miracle of communication. Today we have the Internet, the information super highway and, yes, gridlock! Gridlock on the highways, gridlock at the computer. The oceans wash and roll seemingly eternally, and we sit behind our desks to find solutions to problems of our own making, hoping to find the perfect equation to balance nature and our human needs. If perfect chromosomes of success are locked up in their own quagmires, then each one of you is capable of finding those keys, those answers, and those successes.

As for the balance, maybe there is an equation composed of a spoonful of knowledge, a pinch of humility, a dollop of humor, and an ocean of love.

I would like to acknowledge Dr. Craig Everts for his ideas which have been incorporated in this presentation, Sheila Robertson who worked tirelessly on CWO '97 and most importantly, my wonderful wife, Karen Melander Magoon for putting my concepts in this presentation.

CHAPTER 2

Beyond Whale Watching - the Future of Ecotourism

Bruce Monroe[1]
John Poimoroo[2]
Bill Wilson[3]

Abstract

The majority of the world's population lives within 60 km of a coast and, with the exception of China, this is certainly the case throughout the Pacific Rim. California, with over 20 million people in its Coastal Zone and 1100 miles of varied and often magnificent coastline, typifies the tension between development and preservation of the amenity that will become an increasingly significant segment of our economy. With tourism worldwide valued at over six trillion dollars and demand for "green tourism", "adventure tourism", "nature-based tourism", and "outdoor adventure" the fastest growing segment in that market, ecotourism will take on added dimensions as government and the private sector find it increasingly worthwhile to protect and restore our valuable coastal and marine assets. Watershed-based planning, solving problems of pollution and runoff, new research, innovative solutions to problems, and greater involvement of residents and visitors are just some of the tools that can be used to expand sustainable ecotourism opportunities. A nine step strategy and a planning template are proposed.

Background

Earth is the Ocean Planet, the pale blue dot. Humans have a water orientation with most of the world's six billion population

---

[1]Director, Algalita Marine Research Foundation
[2]Deputy Secretary for Tourism, CA Trade and Commerce Agency
[3]President, Algalita Marine Research Foundation

living within 60 km of a coast. California has 1100 miles of magnificently varied coast that will accommodate a wide range of tourism activities. Beckoning Pacific Rim tourists and ecotourists from the United States and abroad are fishing, SCUBA or snorkeling, sailing, sail training, charter cruising, cruise ship shore excursions, kayaking, whale and other marine mammal watching, catch and release fishing, watchable wildlife viewing, nature photography, wetland and marsh interpretations, coastal and island camping, hiking and coastal birding.

WATCHING WILDLIFE

Bird Festivals, Counts, Nesting Surveys

Habitat Stewardship - SCUBA, wetlands, beaches, hiking trails

Visiting coastal wildlife refuges

Map, monitor & guard coastal locations

Guides & Directories for Watchable Wildlife destinations

COASTAL BIRDING

- **Festivals**
  - **Morro Bay, Arcata**
- **Guided tours**
  - **Shearwater Tours - Monterey & Bodaga Bay**
  - **Pajara Dunes**
- **Self-guided tours**
  - **Coastal wetlands (Ramsar sites)**

<u>Proposals:</u>

California's population is projected to grow to 46.2 million by the year 2017 with additional tens of millions of tourists visiting the state annually. Will we have enough tourist spaces? Will these numbers stress the habitats? California's southern coast is losing more biodiversity today than any area in North America. Will there be any wildlife left to view?

Many terms are used as synonyms for ecotourism including:

"adventure tourism", "nature-based tourism"
"outside adventures" and "green tourism".

Each is defined later in this paper. Aquariums, marine theme parks, maritime museums, day nature cruises and tall ships offer shoreside and off-shore excursions. We have to use <u>all</u> our coastal areas to provide a variety of marine destinations:

wetlands, floating vs. land-based aquariums, submarine viewing craft, research vessels, overwater whale/marine mammal watching, island trips, recreational boating, tall ship passages, etc.

With the world's population growing at the rate of one million every four days, and with 99% of all habitat being marine habitat, coastal and off-shore are the only logical areas in which to expand sustainable ecotourism opportunities. A nine step strategy to do so is explained later in this paper.

Coastal wetlands are the ideal perspective from which to view the watershed and off-shore. Twenty-five years ago, the United States and 89 other nations signed the Ramsar Convention, a conservation treaty to preserve wetlands worldwide. Today, the Southern California Coastal Wetlands Complex of twenty wetlands from the Tijuana Slough (black arrow) north to Magu Lagoon (white arrow) and each of the 20 wetlands have the designation *Wetlands of International Importance*, the only such designation on the west coast of the United States and Canada.

WETLANDS OF INTERNATIONAL IMPORTANCE

20 Coastal Wetlands from Magu Lagoon near Oxnard to the Tijuana Slough near Baja California

Properly promoted, interpretive tours of coastal wetlands will draw ecotourists who may be uncomfortable with off-shore boating activities.

More and more tourism will be ecotourism as current trends whet the world's appetite for such travel experiences.

TRENDS FEEDING THE WORLD'S APPETITE FOR ECOTOURISM

- **Imax Theater**
- **TV programs - Discovery, PBS**
- **Travel magazines -Ecotraveler, Outside**
- **Competition -lower prices, more options**
- **Earthwatch & hands on research expeditions**
- **Internet - on line to adventure expeditions**

California, sitting at the center of tourism, a six trillion dollar global enterprise, should be preparing to meet the needs of eco-tourists, tourists vitally concerned about the environment. The state must be working in partnership with other governments, agencies, associations, organizations, and the tourist industry itself to solve threats from runoff and trash contamination. When California budgets for and is doing the best research, protection and conservation of coastal resources, ecotourists will come in large numbers. We propose as a starting place that the fine limits for littering in California be substantially increased and enforced with rewards for whistle blowers. If we do not invest in keeping our coast and ocean pristine, ecotourists will go elsewhere.

There are many innovative research ecotourism adventures available to ecotourism volunteers up and down the coast.

RESEARCH OPPORTUNITIES

US Fish & Wildlife Volunteers
SCUBA surveys
Endangered species recovery
Oceanic Society Expeditions
University expeditions
Revegetation with native plants
Earthwatch Expeditions & Exploration

THE ECOTOURIST & RESEARCH

- ▲ **Kelp forests**
- ▲ **Aquaculture**
- ▲ **Pickleweed**

An example is video diving on coastal vessels such as the Algita, a coastal restoration research catamaran operated by the Algalita Marine Research Foundation (AMRF) which gives ecotourists an opportunity to experience a dry, virtually seasick-free view of kelp forests, aquaculture and soft-edge pickleweed farms. Through the international cooperation of *proesteros*, the bi-national (Mexico and the US) estuary conservation group linked to AMRF, this research adventure is available, not only in California but ,also, extending down into Baja California and the Sea of Cortez.

To reduce competition and duplication, it is proposed that floating ecotourism facilities be constructed through memorandums of agreement in which areas share equipment and services that are readily moveable. A planning policy template described later weighs opportunities such as floating vs. land based aquariums.

SPECIAL INTEREST GROUPS

Naturalists, scientists, ecologists, nature photographers, research organizations using volunteers, SCUBA divers & snorkelers, learned societies, tour companies, whale watchers, surfer competitions, hikers, birders, kayakers, the physically challenged, boaters, etc.

The California state & local governments must issue regulations requiring <u>sustainable</u> ecotourism operations; encourage catering to special interest groups; offer tax incentives, grants, loans, and restoration scholarships and internships; provide amendments to the coastal plan, marine and estuary sanctuaries; support entrepreneurial centers, joint ventures, festivals, business incubators, & sport competitions that have a coastal orientation if the <u>sustainable</u> ecotourism industry is to dominate in this state.

Watershed-Based Planning--Protecting the Amenity

The coastal zone can be viewed as the ultimate destination of runoff from the series of inverted funnels that constitute the various drainages extending inland, in some cases virtually all the way to the continental divide. Under natural conditions, inputs into this system consist of rainfall and springs, to which we have added large amounts of piped imported water and pumped groundwater.

Outputs under natural conditions consist of rivers and creeks made up of water that has been filtered through the vegetative cover, soil biota, and geology of the watershed, to which we have added quantities of sewage and contaminated urban and agricultural runoff, artifacts of the rapid pace of development and a public works mindset that has regarded these discharges as something to be gotten rid of as fast as possible rather than from a resource recovery perspective.

An interesting way to look at the dynamic between the watershed and the receiving body of water at the coast is in terms of phosphorus, the essential mineral nutrient and energy driver in all living cells. Well-vegetated terrestrial systems, elaborate with their overstory/understory, groundcover, litter and mulch, roots and associated soil organisms, naturally retain phosphorus. When the landscape becomes stripped and eroded, phosphorus is easily leached and discharged to the receiving waters, where it kicks off an algal boom-and-bust cycle ending in large amounts of decaying matter, oxygen depletion, and a subsequent die-off of aquatic life.

Through activities associated with development, modern lifestyles, and agriculture, we discharge large amounts of phosphorus. As environmental planner Jim Bell of San Diego points out, food, representing the concentrated fertility of our agricultural regions, grown with massive amounts of chemicals and mined mineral reserves, travels great distances to get to our plates. When we're finished with it, it is flushed out through ocean outfalls after brief settling and biological interaction in a wastewater treatment facility. Additional volumes of phosphorous found in the wide range of consumer products and industrial processes are also part of this wastewater flow. From this perspective, ocean outfalls can be viewed as a continous form of erosion!

Unlike the more transitory nutrient nitrogen, phosphorus has no typical gaseous phase, and on a shallow continental shelf can cycle many times, building up and producing algal blooms that extend for longer durations until they are extremely damaging. There is good evidence that this has occurred on the East Coast of the United States, and is beginning to occur on the coast of Southern California. In the past this has resulted in 'red tides' consisting of algal blooms of familiar species such as the dinoflagellate *Gonyaulax*, but shifting the inputs into the ecology can have unpredictable results.

On the East Coast, this nutrient-based shift has produced blooms of a rare species of microalgae known as the 'fishkill algae' because of its production of potent neurotoxins that kill fish. The microalgae then produces exoenzymes which dissolve the flesh of the victim to produce nutrients that can be absorbed by the attacking organism. Since the toxins can volatilize, they pose a threat to beach-goers as well, and cases of waders going into convulsions have been documented.

Phosphorus removal at wastewater treatment facilities is a costly, tertiary process, and since marine systems have been regarded as *nitrogen* limited, has not been practiced at the treatment facilities that discharge through ocean outfalls. Thus, this vital nutrient contributes to ecological upsets when discharged, while being essential and subject to loss in the terrestrial ecology. According to UN estimates at the beginning of the decade, world mineable phosphate (and potassium) reserves will be exhausted sometime around 2040. A more aware resource recovery approach to sewage and runoff would seem mandatory.

A second subtlety that is beginning to manifest is the persistent presence of chemical hormone mimics in the ecology of receiving waters, the result of consumer product and industrial process residues in sewage and runoff. Present in the marine environment in the parts per *billion* range, these chemicals are implicated in reproductive and sexual abnormalities across the range of coastal populations, including fish, birds, mammals, and reptiles.

The third deleterious effect of our sewage and surface runoff discharge practices is the increasing presence of disease-causing pathogens along our beaches. Currently, in Southern California, indicator coliform bacteria counts are dangerously high from somewhat south of Rosarita (Baja, Mexico) to Latigo headland, just north of Malibu (Los Angeles), most of the year.

The impact of these three 'externalities' in our developmental accounting process are just beginning to be appreciated, but the potential for damage to an amenity-based economy such as that of Southern California and other coastal areas is massive. The solution is a shift in awareness to one that acknowledges our place in the watershed and accounts for all of the inputs and outputs, recovers valuable resources, and protects valuable assets. The solution is watershed-based planning.

To most public works engineers, the concept of watershed-based planning involves revegetation of vacant spaces, urban tree planting programs and reforestation of slopes, erosion control, and augmentation of porous and non-paved surfaces wherever possible. Indeed, an analysis of the percentage of paved and built-upon versus non-paved and open spaces in the urban region often yields an unexpectedly high number for unpaved surfaces that would indicate that there is great potential for reducing runoff and pollution to be gained from incorporating the practices of watershed enhancement into the public works program.

A vegetated and porous urban landscape mimics a natural watershed, accomplishing both quantitative and qualitative improvements in stormwater runoff via the processes of **1) deceleration, 2) sedimentation, 3) filtration, 4) infiltration, and 5) biological interaction.** Runoff is detained and absorbed by groundcover, roots, and soil; groundwater is recharged; and a natural hydrology is established that releases rainfall and other inputs over time to riparian habitats, wetlands, and finally to lakes and bays. In the process, sediments, nutrients, and pollutants are removed, and bioremediation can take place.

The application of watershed-based planning to the urban hardscape or 'sewershed' can involve many of these same precepts and interactions. For instance, the city of Denver found that by incorporating rooftop rainfall retention, storing runoff in parks and parking lots that had been designed as temporary spreading and retention basins, and desynchronizing release of runoff, annual flooding from the North Platt River that had plagued the downtown has been virtually eliminated. An added benefit has been a reduction in pollution levels.

Though the impacts of runoff are typically borne by the receiving body of water, lowland areas prone to flooding, and structures affected by unstable slopes and sinkholes, much of the activity of runoff in the urban hardscape takes place underground, in storm drains, sewers, and combined sewers. Watershed planning involves the modification of planning in the public works process to ask the questions: a) Why is there all of this runoff; b) Why is there all of this pollution; and c) How can we affect it in terms of the 5 processes outlined above?

In the urban environment, characterized by a high percentage of paved and built-upon surface areas, traditional public works projects are usually found to have focused on the speedy and efficient delivery of stormwater to the receiving body. In the case of combined sewers, the treatment plant at the bottom of the watershed acts as the final intermediary. When the rainfall intensity exceeds the capacity of the system, flooding and uncontrolled discharges result.

Today's large coastal sewage treatment facilities were located at the bottom of the watershed to take advantage of gravity and their convenient discharge location during the 'solution to pollution is dilution' era leading up to the present condition. Collection systems extend miles inland. In order to begin reducing wastewater discharges and make recycled water available, this locating of facilities has resulted in the necessity of building whole new distribution systems for reclaimed water and pumping this water back upgradient for reuse. The tremendous expense of the redistribution systems (as well as the required tertiary treatment) means that only large users of recycled water (and those sites lucky enough to be located near a recycled water main connecting the large re-user to the centralized treatment plant) will ever receive recycled water, thereby greatly limiting the use of this valuable commodity, and perpetuating the despoliation of the invaluable coastal amenity.

Watershed-based planning in the urban hardscape involves mapping the area involved and identifying existing structures, dividing the area into individual drainages, computing the acreages of each drainage, applying rainfall intensity-duration/frequency data to each drainage to accurately predict runoff, identifying the capacities of receiving conveyances and treatment facilities, constructing a hydrologic model, and designing for the deceleration of runoff and the removal of particulates and their associated pollutants, including metals, toxics, pathogens, and nutrients. And, it involves 'kicking' water out of the collection system for reuse whenever possible, including restoring and maintaining watershed landscapes.

Under an innovation being developed by the third author as an element of an integrated plan, the existing sewer system could be viewed as both a collection, treatment, and distribution system, since it is already in place under all parts of the urban landscape. Wastewater could be drawn from the collection system where needed and treated onsite for reuse, in the amount required, at low additional cost, thereby enabling reuse sites anywhere to enjoy the benefits of inexpensive recycled water and reducing the pressure on the potable water supply. Public works could even use onsite recycled water delivered in this manner to restore and enhance watersheds, green belts, freeway plantings, buffer strips, riparian corridors, habitats, and wetlands, and reduce the pollution from runoff.

A review of the planning documents from large urban sanitation and public works departments from the turn of the century through the '60's reveals that we were developing a decentralized wastewater treatment and reuse system, strictly for economic reasons. But with the advent in 1972 of the Clean Water Act, large public works engineering firms going after the tens of billions of dollars that the federal government made available instituted the current era of large, centralized treatment facilities and extensive collection systems resulting in the present bottleneck in water reuse. It is ironic that the CWA can be seen in this light as actually contributing to the degradation of the coastal environment.

The result is that large, expensive projects are now institutionalized, projects are put into motion that are too narrow in focus, and there is no incentive for the diversified, site-specific types of projects of various sizes and reduced costs that would restore the fabric of the natural and urban watershed, protect the coastal economy, restore rivers and streams, and return the seashore to healthy a condition. In this endeavor we are all stakeholders - we are all ecotourists.

To break the institutional gridlock in the planning and development community, an informed and involved public is necessary. A basic tenant of planning, taught in all university planning and policy programs, is to assess the values of the stakeholders at the outset, develop alternative proposals, and get feedback from past projects to inform new directions. A template for the planning process that upsets the practice of 'top-down' delivery now common consists of the following loop:

TENANTS OF THE PLANNING PROCESS

1.	**Assess community values**
2.	**Determine goals and objectives**
3.	**Collect data and information**
4.	**Design criteria and standards**
5.	**Develop alternatives**
6.	**Choose an alternative or mix of approaches**
7.	**Detailed design of the operational plan**
8.	**Modification/approval of the operational plan**
9.	**Implementation**
10.	**Monitoring and feedback**

after Melville C. Branch, Comprehensive Planning: Introduction & Explanation, American Planning Assoc., Chicago, 1985

As more and more people become involved in the interaction with the environment at all levels of intensity encompassed in the description of activities presented earlier in this discussion, including scientific excursions and visiting restored watershed elements like the Arcata wastewater treatment wetlands (which attracts over 100,000 visitors per year in a remote location), ecotourism can become a significant force in reclaiming environmental quality neighborhood by neighborhood and watershed by watershed, as it must be done.

A Strategy for Expansion of Sustainable Tourism

One of the basic problems in a discussion of sustainable tourism is that people often aren't speaking about the same thing.

Sustainable tourism is often described as being synonymous with ecotourism. However, the authors prefer to believe that any type of tourism can be sustainable, not just ecotourism. Furthermore, all tourism should be sustainable whether it is a visit to an amusement park or a trek in the wilderness.

The term 'sustainable tourism' has been too limited in describing only those types of tourism that involve nature. Further confusing the discussion is that the differences between nature-based tourism, adventure travel, and ecotourism are often blurred. Likewise, heritage tourism and cultural and multi-cultural tourism are similarly mistaken as being one and the same. To clarify, the following definitions are offered:

Traditional travel and tourism occurs by car, boat, plane, train, or bus, for purposes of recreation, education, business, leisure, enrichment, amusement, reunion, or any other number of purposes. It is not out of the ordinary, though it can be extraordinary.

All travel and tourism can be environmentally sensitive (including traditional travel and tourism), depending upon the destination and how it is conducted by the traveler and the travel company. Environmentally sensitive tourism considers its effect upon nature. And, culturally sensitive tourism considers its effect upon people and societies. Sustainable tourism, is, then, both environmentally and culturally sensitive.

Nature-based tourism is travel with an interest in wildlife or scenery. It can be anything from a busload of travelers getting a view of Yosemite Valley at a five-minute stop at Tunnel View, to cross-country skiing through Yellowstone, to bird watching in a Costa Rican rain forest. The term does not, of itself, convey environmental appropriateness or inappropriateness. It only describes travel in which nature has importance to the traveler, whether or not the traveler acts appropriately.

Adventure travel is "experiential." The experience, though it may be dependent upon the outdoors (such as white water rafting), may not necessarily be presented in an environmentally sensitive manner, though in my experience, most adventure-oriented tourism providers (scuba diving companies, ski areas, rock climbing schools, ocean kayak firms, etc.) understand the importance of protecting the environment that provides their bread and butter.

Ecotourism is always environmentally and culturally sensitive travel, and therefore - to be called ecotourism - it must be sustainable. Its foremost objective is to operate with minimal impact. Because of this, ecotourism tends to involve extremely small groups staying at small and specially designed facilities. Studies note that ecotourists believe ecotourism can exist only in untrammeled areas. They are only lightly interested in ecotourist trips to highly visited places. Because ecotourism involves small numbers of travelers going to mostly unvisited places, ecotourism tours, facilities, lodges, etc., tend to be remote, in areas of extraordinary environmental quality, and oriented toward affluent, highly committed travelers who seek one-of-a-kind places with high standards of service. Because this is a very small market, it limits ecotourism's ability to provide economic solutions.

This poses a dilemma for proponents of ecotourism in that it cannot be used as an economic panacea. That is not to say that ecotourism isn't commendable or preferable - it is, and it can be. However, in resource-based communities that are transiting from extractive uses to sustainable ones, it may not provide the mass of economic activity that the community demands to survive at its previous size. To do that, such communities need to consider a balanced approach to attracting different types of tourism, not all of which will be pure "ecotourism".

For the California Coast, it means committing to attracting sustainable tourism, but recognizing that not all of it will be ecotourism. A balance of sustainable traditional tourism, adventure tourism, nature-based tourism, and ecotourism may be best.

Here is a strategy for expansion of sustainable tourism:

1. **Encourage ecotourism development with a realistic expectation of what it can produce in economic benefit. It is not an economic panacea, but it can be the best solution in certain situations.**
2. **Encourage traditional tourism facilities and programs to re-engineer and adopt environmentally sensitive practices.**
3. **Encourage alliances between travel and tourism facilities and business groups in environmental preservation efforts.**
4. **Promote nature-based tourism as an inherent benefit and expectation in visiting California. Get businesses and destinations to promote nearby nature-based tourism activities.**

5. **Inform travelers of California nature-based tourism opportunities in advance of arrival.**
6. **Attract visitors that will seek out and support nature-based tourism activities during their stays.**
7. **Recognize good environmental practices and programs.**
8. **Encourage entrepreneurs and community leaders to explore ways in which environmental preservation can attract tourism and its economic benefits.**
9. **Seek solutions that will allow tourism growth in impacted areas by attracting types of visitation that do not worsen the impact.**

In coastal communities that have traffic congestion created by private vehicle visitation, seek to change travel patterns by encouraging travel by alternative means. That is... attract cruise ship port calls (solution: Reform the Passenger Services Act to allow foreign-flagged cruise ships to call at successive US ports); encourage the use of ferries, improve rail service, provide preferred parking to tour buses, and construct outlying parking areas and utilize shuttles or light rail to transport visitors to impacted areas.

Conclusion

Ecotourism, in all of its forms and varieties, forms a vital part of the coastal economy, whether in highly urbanized zones or more sparsely populated and pristine areas. In either case, the draw is environmental quality, which also impacts other major segments of the economy, such as resource availability, health, energy, citizenship, and the quality of life. As such, ecotourism provides an ever-increasing economic incentive to bridge the gap between the natural world, which provides us with a sense of connectedness and inspiration, and the built environment, enhancing the lifestyles of residents and visitors alike.

CHAPTER 3

The Blow Hole "Fideicomiso"

Oc. Claudia Avendaño .

The Blow Hole " Fideicomiso " is an example of the cooperative management policies among the three levels of government and the community.

Ensenada is a small city that has grown over the past 50 years (Since 1930) owing mainly to the tuna fisheries. Due to the international restrictions on this activity, Ensenada had to Search for a new means of income in order to survive the local economic crisis. For this reason, tourism is becoming more important for the city and has become the main income for the economy at this time. Most important to the tourist industry is the place known as the Blow Hole.

The Blow Hole receive more than 3,000 visitors weekly. Due to the volume of visitors and the cultural , economical and environmental value, the local government has promoted some actions in order to preserve the value of the Blow Hole in addition to solving some of the water pollution, waste management, safe hygienic conditions of food and security concerns, as well as community user and property conflicts.

In order to reach these goals the " Fideicomiso" was created. This is an association of the three levels of government (Federal, State, and local), in adition to the ecological and tourism agencies ,as well as community members, land owners and shop owners directly involved.

The Fideicomiso Authorizes a decision making committee that is in charge of planning and implementing the development for the Blow Hole in accord with the national and local government policies and strategies. This is done in order to increase the local economy and preserving the natural phenomenon within a sustainable development.

This document will explain how the "fideicomiso" works and gives an overview of the general problems and possible solutions.

CHAPTER 4

Managing Marine Wildlife and Eco-Visitors at Año Nuevo State Reserve

Gary Strachan

Abstract

This session will present operational and management techniques utilized at Año Nuevo State Reserve relative to the public viewing program of the Northern Elephant Seal rookery. The 4,000 acre state reserve is located on the central California coast adjacent to the San Francisco Metropolitan Area. The educational and interpretive programs at the Reserve are developed around the seal rookery located on the mainland portion. The primary programs include a Guided Walk Program into the rookery, a Roving Interpretive Program, an Equal Access Program for the physically challenged, and an Outdoor Education Program for school age children.

During the Northern Elephant Seal Breeding season from December through March, Guided Walks are offered to the general public. This Guided Walk Program is the only way that visitors may have the opportunity to view the seal rookery due to safety and wildlife protection aspects. These walks occur every 15 minutes. Trained Community volunteers present the majority of the Guided Walks.

The theme of this eco-visitor interpretation includes numerous aspects of the natural and cultural history of the Reserve. The State Park System has developed regulations and operational techniques to enable the general public to have opportunity to view and to learn about unique marine mammals. Management techniques are based primarily on the protection of the resource. Operational structure is organized to ensure visitor safety, enhancement of visitor education, research opportunities, and revenue generation.

Management considerations and techniques include; limiting and monitoring access points, setting carrying capacities through trail and parking design, developing visitor improvements based on visitor surveys, offering diversified interpretive levels an opportunities. utilizing community members as docent naturalists, provide adequate orientation to visitors before entering the wildlife protection area, and consistent enforcement of the regulations to ensure protection the visitors and the resources.

Gary Strachan, Supervising Ranger, California Department of Parks and Recreation, Año Nueve State Reserve, Pescadero, CA 94060

CHAPTER 5

MANAGING CUMULATIVE IMPACTS TO PUBLIC ACCESS

Tania Pollak[1]

Abstract:

Ensuring that public access opportunities to and along the coast are maintained is a primary mandate for the Coastal Commission. However, increasing demand for recreational opportunities and ongoing development along the coast have the potential to impact available access. Without addressing the wide range of impacts to access that occur as a result of increased use and development, California will face a decline in coastal access opportunities.

As part of a multi-year Regional Cumulative Assessment Project (ReCAP), Coastal Commission staff developed a methodology to assess the cumulative impacts to public access and to evaluate the effectiveness of California's coastal management program in maintaining public access opportunities. ReCAP documented significant impacts to access opportunities, and determined that many of the impacts could be better managed on a regional basis.

California's coastline and beaches have long been world-renowned for the numerous recreational opportunities they provide. The resulting tourism is a critical component to many local governments' economies and to the overall state's economy: as noted in *California's Ocean Resources: An Agenda for the Future*, ocean and coastal tourism contributed $9.9 billion to the state's economy in 1992 (California Resources Agency, 1997). The recreational opportunities and scenic beauty of the coastline also draw many people to reside there. In California, approximately 88% of the population resides in coastal counties (NOAA, 1997). Projections indicate an absolute population increase of 19.2 million people by 2010, representing a 122% increase from 1960 (NOAA, 1992). It is likely that coastal areas will continue to support much of the population growth and the corresponding development.

[1] Coastal Program Analyst, California Coastal Commission. 45 Fremont St. #2000; San Francisco, CA 94105. (415) 904-5297. The views expressed in this paper are those of the author and do not necessarily reflect those of the California Coastal Commission.

Over the past several decades, recreational use of the coastline has grown significantly. For example, between 1987 and 1993, the estimated household participation days for beach activities rose from 69 million to 105.3 million; this estimate excludes activities such as surfing, kayaking, and boating (DPR, 1987; DPR, 1994). This growth in use can occur both from increased tourism and from a growing residential population. As use of an area increases, the potential for negatively impacting the resource that initially drew tourists and residents alike also increases. It is likely that demand for coastal recreation will continue to grow, leaving managers with the increasingly difficult task of maintaining access to accommodate increasing use and protecting recreational areas from overuse.

The Coastal Management Framework

A number of agencies share responsibility in managing the recreational opportunities along the coast, including various state, local, and special district agencies. While these agencies generally manage specific areas of the shoreline, the California Coastal Commission plays a unique role in managing public coastal access *throughout* California's 1,100 mile long coastline. California Public Resources Code Section 30210 of the Coastal Act[2] requires that the Commission *maximize* public access and recreational opportunities "for all people consistent with public safety needs and the need to protect public rights, rights of private property owners, and natural resources from overuse." Maximizing public access includes both protecting existing access opportunities *and* assuring for the future needs of a growing population along the entire stretch of coast.

The Commission has two primary tools for meeting its mandate: 1) review of development projects in the coastal zone, guided by the policies in the Coastal Act; and 2) review and certification of Local Coastal Plans (LCPs). Each local jurisdiction within the coastal zone must develop its own LCP which includes policies and implementation tools to carry out the objectives of the Coastal Act. Once the Commission certifies an LCP as consistent with the Coastal Act, the local government assumes primary responsibility for regulatory review of development in the coastal zone.[3] Until an LCP is certified, the Commission retains jurisdiction over review of projects in the coastal zone. This regulatory review of development proposals seeks to ensure that where development would lead to negative impacts on access, those impacts are avoided or mitigated so access opportunities are protected. Often, mitigation measures are implemented through conditions on a coastal development permit.

---

[2] All future references to Coastal Act Sections refer to the California Resources Code.

[3] The Commission retains jurisdiction where development is "proposed or undertaken on any tidelands, submerged lands, or on public trust lands ... lying within the coastal zone (Coastal Act Section 30519). The Commission also hears appeals of local government decisions. All local government decisions between the first public road and the sea are appealable (Coastal Act Section 30603).

The ReCAP Project

Commission staff recently completed a pilot project in the Monterey Bay area, the Regional Cumulative Assessment Project (ReCAP), to evaluate the cumulative impacts of growth and development on coastal access opportunities and the effectiveness of the Commission's and local governments' programs in managing those impacts.[4] The evaluation focused on the following elements: 1) identifying the status of access over time; 2) identifying the problems or impacts to access (often derived from the analysis in step 1); 3) identifying the major factors or causal links contributing to cumulative impacts; and 4) evaluating the role and effectiveness of Coastal Act and LCP policies and procedures in managing these cumulative impacts. ReCAP also developed recommendations to address the problems identified.[5]

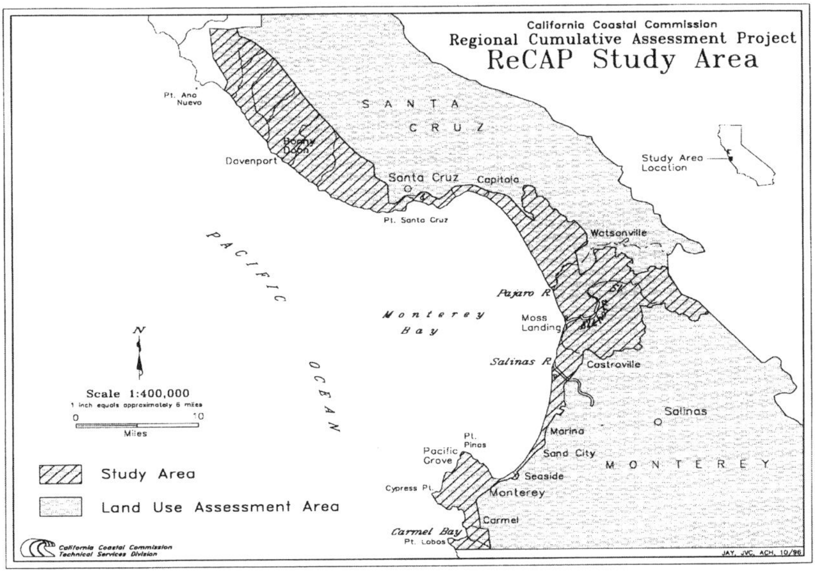

Figure 1: ReCAP Study Area

Cumulative impacts are often regional in nature and difficult to document through a project-by-project review of development. Coastal access is also a regional issue: a contiguous beach area can cross several political jurisdictions, as

---

[4] Cumulative impacts are the combined effects of a series of actions, development activities, or natural events over a geographic area and/or over time. Although a single action may not greatly affect the natural or human environment, the cumulative impacts created by many different actions over time may significantly alter environments.

[5] For a complete discussion of the ReCAP findings and recommendations, see *ReCAP Pilot Project Findings and Recommendations: Monterey Bay Region.* September, 1995.

can a person trying to reach an access site. In addition, actions taken in one area or by one managing entity can affect opportunities at adjacent sites. Therefore, the ReCAP analysis focused on a regional approach. Although there is no commonly accepted or scientifically derived boundary for an access-related analysis (analogous to a watershed for a wetland analysis or a littoral cell for analysis of shoreline erosion), Commission staff sought to define a region based on the resource, rather than political boundaries.[6] ReCAP defined the boundary for its analysis from the San Mateo/Santa Cruz County line through Point Lobos State Reserve in Monterey County, as shown in Figure 1 (previous page).

To analyze cumulative impacts to coastal access, Commission staff first needed to define both coastal access and "maximum" access. Coastal access is generally viewed as an issue of physical supply. The supply of access can be described as lateral access (along a beach), vertical access (access from the upland street or bluff to the beach), trails above a beach (for example, along a blufftop), or inland trails that lead to the coast. In many areas along the coast, protection of views and the provision of scenic turnouts are also a critical component, particularly in areas where the geography of the coastline prohibits access down to the actual beach area. However, for its analysis of supply, ReCAP focused primarily on lateral and vertical access immediately along the shoreline.

While it is a primary factor in assuring access opportunities, the issue of supply cannot be isolated from other factors which greatly affect a user's ability to reach and use a beach area. An early step in the ReCAP analysis was therefore identifying all the variables or factors that affect access opportunities.

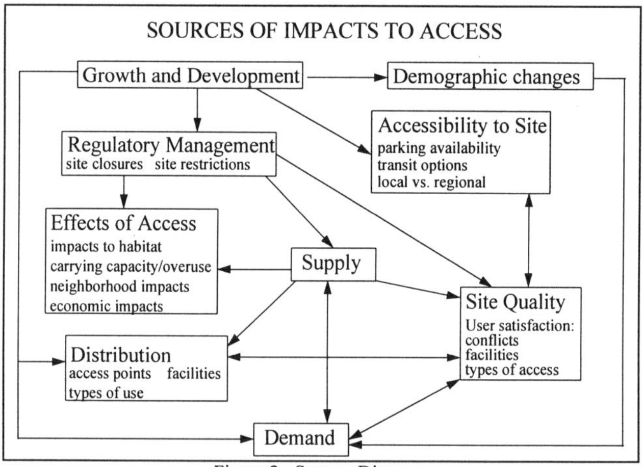

Figure 2: System Diagram

[6] Additional information on determining the ReCAP boundary for access is discussed in *Working Paper #2: Regional Cumulative Assessment Project Regional Designation and Preliminary Database Design.* April, 1993. California Coastal Commission. See also *Working Paper #1: Developing a Regional Cumulative Assessment Process for the California Coastal Zone: Issues and Concerns.* February 13, 1994. California Coastal Commission.

The system diagram in Figure 2 identifies these variables and their relationships to each other. Impacts to any one of these variables may affect the availability and use of a site, and will often lead to additional impacts to other variables. For example, the availability or lack of support facilities often affects the level and types of use at a specific site. A lack of support facilities at one site may lead to a perceived loss of access, and force users to other sites. This increase in use elsewhere may in turn increase congestion and the affect the quality of access at those other sites.

Data and Analysis

Based on the variables identified and the relationships developed in the system diagram, ReCAP developed an extensive matrix of analytical questions for its analysis. The matrix detailed methodologies for answering the questions, as well as the type, source, and time period of data needed (California Coastal Commission, 1993). Although a complete cumulative impact assessment would assess all factors that affect a resource, given the extent of resources available to undertake its analysis, ReCAP needed to focus the scope of its research to the factors most critical to the region. This prioritization was achieved through interviews with managers of access areas, local government officials, and environmental groups. In some cases, ReCAP was unable to analyze an issue or specific question due to a lack of necessary data.

Coastal development permits issued by both the Commission and, in areas with a certified LCP, by local governments were a primary source of data. ReCAP reviewed permits from 1983-1993, identifying which actions involved access issues and what, if any, mitigation was required to assure protection of access opportunities. ReCAP also extensively analyzed secondary data, including general land use trends and population changes to document areas of growth; use patterns for beaches; parking data; beach maintenance budgets and schedules; and changes in types and extent of facilities such as restrooms and coastal camping units in the pilot region. For some questions, ReCAP needed to rely on observational data and interviews with managers of access sites.

Summary of ReCAP Findings

General Land Use and Population Trends

Within the Santa Cruz and Monterey Counties, population and growth trends parallel the general trends for the state. Between 1930 and 1990, population in the two counties increased from 91,138 to 585,394 people, an increase of 542% (California State Senate, 1958; California Health and Welfare Division, 1993). If population trends hold true, by 2010, the population will have more than doubled since the passage of the Coastal Act (California Coastal Commission, 1995; AMBAG, 1994). This growth brings with it not only increased urbanization, but increased pressure on existing resources, including recreational areas.

The Monterey Bay area has also experienced an increase in recreational use; such coastal tourism has been an important component of the region's economy since the 1800s. Available data indicate that beach use between 1981-1992 has increased at about twice the rate of the population growth in Santa Cruz and Monterey Counties (California Coastal Commission, 1995). Interviews with managers of major beach areas throughout the region confirm a significant increase in use over the past decades (Mitchell, 1993). Coinciding with this increase in use is a rise in specialized sports, including jet-skiing, equestrian riding, hang-gliding, and mountain biking on trails and beaches.

Supply of Access

An increasing use generally leads to demand for additional supply, although an increase in supply may in itself generate more use by drawing more people to the area. Much of the Commission's and local governments' review of access opportunities focuses on the physical supply of access to and along the beach. The ReCAP analysis found significant increases in the supply of access in the Monterey Bay area through the development of new stairways, opening of new beach areas, and developing trails along the shoreline for walking and biking. A number of these improvements resulted from conditions on permits. In addition, the Commission and local governments often required an offer to dedicate a lateral or vertical easement as a condition of development. These offers provide additional *potential* access, as they become available to the public only after an agency agrees to accept and maintain the easement. As of 1994, of the 48 lateral and 31 vertical easements required under the Coastal Act and LCPs in the ReCAP area, only six had been accepted by a managing entity, and only one vertical access had been opened for public use.

In some areas in the pilot region, the potential for additional physical access is limited due to geographic constraints or conflicts with other uses, including sensitive habitat. In other areas, the supply of vertical access to the coast is at or close to identified standards.[7] In these areas, an additional vertical accessway will usually not significantly increase access opportunities.

In spite of the increase in supply, ReCAP found significant impacts to access opportunities. These impacts often occurred from a general increase in use of the coastline and development in the region, which contributed to a decline in site quality in some areas. As use continues to grow, placing additional burdens on access areas, managing beaches to assure site quality will play an increasingly vital role in assuring maximum access.

---

[7] ReCAP was unable to find a generally accepted standard for beach access. Therefore, ReCAP used the standard identified in the Santa Cruz County LCP for the entire ReCAP area; the Santa Cruz County LCP requires vertical access every 1/4 mile in rural areas and every 650 feet in urban areas. It is assumed that there should be no limit on lateral access.

Accessibility to the Coast

Beyond the issue of lateral and vertical access, one of the most apparent and significant factors to coastal recreation is being able to actually reach the coast; if a person cannot easily reach a desired site, access has not been provided. Availability of parking for beach areas and alternative transit serving access sites are both necessary to ensure accessibility to the coast.

Parking

The Coastal Act seeks to support recreational use of the coast by reserving upland areas "necessary to support coastal recreation" (Section 30223). In addition, the Commission and local governments often require that new development provide parking as part of the proposed development. As overall use and development in an area grows, competition for existing parking will also increase. By assuring that adequate parking is available to accommodate users of the *new* development, the Commission and local governments can help reduce competition for the existing supply, helping to assure its availability for recreational uses.

However, requiring parking for new development cannot address either historic shortages of parking or a growing demand for parking due to *recreational* uses. The concern over inadequate parking was an issue during LCP planning in many jurisdictions in the ReCAP area. Today, many of the areas formerly suitable for support facilities are developed with other uses. As demand continues to grow, the problem of inadequate support facilities becomes more of a problem.

In other areas, ReCAP found evidence that in spite of an increase in parking supply, access to the coast may be more difficult. Studies for Santa Cruz City indicate that while parking in the beach area expanded by approximately 2,400 spaces between 1980 and 1987, demand for parking still exceeded the supply (City of Santa Cruz, 1980; DKS, 1987). In 1980, parking deficiencies were noted only on summer weekends; by 1987, deficiencies also occurred during the week. It is likely that parking shortages are occurring, or will soon occur, in many other areas of the region.

In addition to a limited supply of parking, ReCAP also identified an increase in parking restrictions through red-curbing, no parking signs, and preferential parking programs. These actions further hinder a user's ability to reach the coast, as street parking is often vital to ensure beach access. In the Monterey Bay area, parking restrictions were often implemented due to conflicts between beach users and residents living near access locations. In some cases, signs restricting parking were placed by residents, or by a local government's Department of Public Works at the request of residents. Restrictions implemented by the Department of Public Works were not always coordinated with the local governments' planning and/or recreation departments; therefore, the loss of parking for recreational use was not analyzed or mitigated.

Implementation of such restrictions can often have an effect on access in areas beyond the immediate location of the restriction. Parking restrictions can force use to areas without restrictions, leading to increased congestion and potential conflicts in those areas as well.

Alternative Transportation

Even if the parking supply in an area increases, it is unlikely to ever keep pace with growing recreational use and demand; often an increase in parking will draw more visitors. Further, developing parking areas may not be the best use for the land, particularly as it may negatively affect the character of a community (City of Capitola, 1981). In many areas, this community character is as important to tourism as is easy access to the coast.

With a growing demand for coastal access and a limited availability of parking, alternative transportation becomes an increasingly important tool for providing access to the coast. A number of beach shuttles operate in the Monterey Bay region, although due to insecure funding, several shuttles have recently suspended service. While beach shuttles can help provide access to the coast and help relieve congestion along the coast, shuttles tend to run within a single city's jurisdiction. Much of the need for transit begins well inland of coastal cities, as there is little incentive for a user who has driven from an inland area to use a shuttle for the last several miles of a trip.

While the Coastal Act encourages transit options (Section 30252), planning for and implementing a transit system is usually impossible under a structure of reviewing individual projects for impacts on access.[8] Further, review of individual projects in the coastal zone will often not reveal the need for transit: while development in the coastal zone may *add* to congestion, much of the need for transit comes not from a development in the coastal zone, but from increasing use, often from beyond the political boundary of the coastal zone. To be effective, transit systems often must cross the political boundaries of the coastal zone, cities, and, often, counties, and service areas where users originate from. Transit should also be inexpensive, run frequently, stop at convenient locations, and accommodate various recreational equipment, such as surfboards. However, due to its jurisdictional constraints, the Commission is limited in its ability to pursue such systems.

Indirect Impacts due to Management Measures

As use of coastal areas increases, conflicts among users often result. While a certain amount of competing use is probably inevitable at most sites, left unaddressed, many of these conflicts can degrade the recreational experience. In other situations, managers of a beach area may take an action or restrict a use due to other needs or goals of the local jurisdiction. The result in either situation can be to

---

[8] In one case, for the construction of the Monterey Bay Aquarium, the Commission did require a shuttle as mitigation for impacts to traffic and parking.

transfer use to another area, leading to other secondary, and often unintended, effects that also affect access opportunities. Three actions that lead to these secondary impacts are the use of curfews, implementation of fees, and restrictions on various activities to manage conflicting uses in a beach area.

Fees and Curfews

While data are often incomplete and difficult to obtain, ReCAP saw an increase in the use of curfews and fees in the Monterey Bay region. Between 1981 and 1993, the Department of State Parks and Recreation (DPR) added curfews to 12 units; an additional 18 beach areas not managed by DPR also restricted hours of access. Once a manager at one beach area imposes a curfew, managers at neighboring beaches may follow: for example, after DPR imposed a curfew at Twin Lakes State Beach in Santa Cruz, the County supervisors imposed a curfew at their adjoining beaches. As the imposition of curfews spreads, access opportunities are lost not only at specific beaches, but over a wider, regional area. Overall, approximately 30% of the sites identified through the ReCAP study restricted hours of use.

ReCAP documented a similar trend with the imposition of fees: between 1981 and 1993, DPR imposed fees at four units which previously did not require payment, and increased fees in 1983, 1987, and 1990. Seven other beaches not managed by DPR also imposed fees for use, although two of these sites had previously been closed to the public. Because many factors affect the use of sites, ReCAP's review of available data could not specifically correlate a change in use with fees (California Coastal Commission, 1995). However, Commission staff has observed a transfer of use from fee to non-fee beaches. The receiving sites often had inadequate facilities to support the increase in use, leading to conflicts with public safety, parking, and other uses of the site, and/or had sensitive resources that were impacted by the increase in use.

Use Conflicts

As use at a site grows, congestion increases, often leading to conflicts between users and a perceived decline in access opportunities. New technologies, which often lead to new types of recreational activities, also may lead to conflicts with other users. An increasingly common example of conflicts evident in the Monterey Bay area is between jet-ski users and swimmers or surfers.

Managing for these use conflicts often leads to restricting at least one activity, either by completely prohibiting it or restricting it to certain hours or locations. While these actions may be necessary for safety concerns, and may enhance the quality of the area for some users, they severely restrict or preclude use by others. Similar to the issue of fees or curfews, simply imposing these restrictions may serve to only transfer the problem: users will find another site which may lead to similar conflicts there.

Another type of conflict becoming more evident is between access and impacts to sensitive habitat and wildlife. Many access sites currently open to public use in the ReCAP area have sensitive resources; many of these sites may not be able to support significant increases in use. Other areas of the region do *not* provide access due to conflicts with sensitive resources. However, it is likely that as use continues to increase, more people will seek less crowded beaches, many of which are those with sensitive resources. A number of areas in the ReCAP region already show evidence of impacts: erosion and vegetation loss are evident from vehicle intrusion, equestrian use, and from informal trails being cut through habitat to beach areas. Human activities may also be a factor in impacts to snowy plover nests (Page, 1994).

Quantifying and defining the impacts to sensitive resources is often difficult as information, when available, is based on informal observations. Impacts leading to a species abandoning a site due to human intrusion are particularly difficult to quantify. A number of techniques have been at least partially effective in the Monterey Bay area in reducing impacts to species: developing boardwalk to keep use to one trail; fencing or other barriers around certain areas; educational displays; restricting use to low-intensity activities; and in some areas, limiting the numbers of users. However, even with these techniques, impacts to resources are occurring, often due to inadequate enforcement of restrictions, inadequate monitoring, and staff with inadequate training in resource management. The priority of recreational use over resource protection is also a factor. Without a strong, on-going resource management program, growing demand for additional access will make adequate and effective resource protection more difficult.

Managing Cumulative Impacts

Although the regulatory program encompassed by the Coastal Act and LCPs provides a necessary framework to protect public access, many of the impacts resulting to access are difficult to manage through the existing program. Since many of the impacts occur from an overall increase in demand for recreational opportunities, they often are not attributable to a specific development project, and therefore cannot be mitigated through conditions on a permit. As discussed above, management actions taken at one location also may impact access opportunities. Addressing many of these impacts often requires coordination across jurisdictions as well as between departments within a local jurisdiction.

To address the cumulative impacts to public access that are not being fully addressed, different planning and management mechanisms may be necessary. Regional and site specific management plans can help address the day-to-day management of access that LCPs and project-by-project review of development cannot. As part of improving management of cumulative impacts to access, ReCAP developed a guidance manual for developing such plans; the manual identifies a variety of management measures for many of the issues raised in its analysis (California Coastal Commission, 1996).

A regional strategy for addressing access issues can often better provide for a diversity of sites and activities, better coordinate management of many of the issues, and reduce the spillover impacts resulting from management of one particular site. For example, as previously noted, transit systems to support recreational use will be most effective when developed on a regional basis. Even addressing parking issues has regional implications, as seen from the effects of parking restrictions in the Monterey Bay area. Without addressing the *regional* nature of access issues, impacts and problems are often simply transferred from one jurisdiction to another; as a result, access opportunities are not protected. Through regional strategies, local governments can also share resources, and may be able to reduce the need for many of the actions that can restrict access: for example, addressing curfews through a regional strategy may allow jurisdictions to share the costs for alternatives, such as night patrols, and negate the need for a curfew.

This regional management does not necessarily entail a new regional government. Enhancing coordination among jurisdictions, and often within a jurisdiction, can be a highly effective tool to enhance beach and recreational management; such coordination can ensure that all departments with a potential impact on access and recreational related activities are aware of those impacts and the need to respond to them. This coordination and information exchange can range from check-off lists for each department regarding policies and actions affecting access, to more structured taskforces.

Conclusion

Coastal recreation has increased significantly throughout much of California, as has development along the coastline. Both trends can significantly affect coastal access opportunities. Although ReCAP's analysis of impacts to access focused in the Monterey Bay region, it is likely that other areas of the coast are experiencing similar trends. Without adequately addressing the wide range of impacts to access that occur from an increase in use and development of the coast, California will face a decline in coastal recreation opportunities and the quality of those opportunities. In doing so, much of the treasure of California's coastline will be lost, as will a significant economic part of the local economies of coastal cities and counties. Adopting a regional perspective to planning for and managing coastal access can better assure for the long-term protection of access opportunities.

References

AMBAG, 1994. *1994 Regional Population & Employment Forecast.*

California Coastal Commission. 1993. *Cumulative Impacts Assessment Conceptual Framework and Analytic Matrix.* Working Paper #4. San Francisco.

California Coastal Commission. 1995. *ReCAP Pilot Project Findings and Recommendations: Monterey Bay Region.* San Francisco.

California Coastal Commission. 1996. *Beach Management: Issues and Solutions.* San Francisco.

California Health and Welfare Agency, Employment Development Department, Labor Market Information Division. 1993. U.S. Census of Population and Housing as found in *Annual Planning Information Santa Cruz Metropolitan Statistical Area 1993* and *Annual Planning Information Salinas-Seaside-Monterey Statistical Area 1993.*

California Public Resources Code, Division 20. Sections 30500-30600. January, 1997.

California Resources Agency. State of California. 1997. *California's Ocean Resources: An Agenda for the Future.* Sacramento.

California State Senate. 1958, 1970, and 1971. U.S. Census of Population and Housing in *California Statistical Abstract.*

City of Capitola. 1981. Land Use Plan of the *Local Coastal Plan.* Capitola.

City of Santa Cruz. 1980. *City of Santa Cruz Beach Area Plan.* Santa Cruz.

Department of State Parks and Recreation. 1987. *Public Opinions and Attitudes on Outdoor Recreational in California.* State of California, Resources Agency. Sacramento.

Department of State Parks and Recreation. 1994. *California Outdoor Recreational Plan: 1993.* State of California, Resources Agency. Sacramento.

DKS Associates. *Beach Area Transportation Study for Santa Cruz.* 1987.

Mitchell, Dave and Gretchen Iliff. Santa Cruz County; Parks, Open Space and Cultural Services. Personal Communication. September 9, 1993. Also: Doug Healey, State Department Parks and Recreation. Sacramento Personal Communication. October 6, 1993. Paula Jones. State Department of Parks and Recreation. Monterey County. Personal Communication. December 15, 1993. Anthony Lobay. City of Pacific Grove. Personal Communication. September 10, 1993.

National Ocean and Atmospheric Administration. 1992. *50 Years of Population Changes Along the National Coasts 1960-2010.* National Ocean Service. U.S. Department of Commerce. Rockville.

National Ocean and Atmospheric Administration. Office of Ocean Resources Conservation and Assessment. 1997. *Population and Development in Coastal Areas.* http://seaserver.nos.noaa.gov/projects/population/population.html (April 8, 1997).

Page, Gary. Point Reyes Bird Observatory. Personal Communication. May 13, 1994.

CHAPTER 6

ECONOMIC VALUE AND IMPACTS ON COASTAL ZONE THE GOLD COAST CASE STUDY

Mr L.A. (Angus)JACKSON [1]

Mr A.W.(Sam) SMITH [2]

The Gold Coast of Australia has developed because of its natural surfing beaches and the local economy is now very dependent on the correct management and protection of these assets. The severe erosion of 1967 showed the vulnerability of the area to temporary loss of the beaches due to storm cut. This erosion resulted in a 20-30% reduction in tourist revenue over 3 years. The cost of restoration by nourishment was only $1.5M ($1974) which if spent prior to the erosion would have prevented the loss of the beaches. There are now over three million visitors to the coast who stay at least 1 night with the average being ~5 nights. The number of day trippers is far greater. As these tourists spend over $1billion per year even a small reduction in tourist revenue would be significant.

Numerous manufacturing, retail and tourist support industries have developed in the area and these directly provide a large number of jobs and prosperity to the area. Surfing related businesses alone contribute to the employment of over 1200persons and a far greater number are employed in other areas such as accommodation, hotels, restaurants, tourist attractions etc.

Considerable ongoing management and infrastructure is required to provide safe and environmentally sustainable use of the coastal zone. The increasing use of the Coastal Zone for recreation and special events places considerable pressure on Coastal Managers and Engineers. The importance of activities such as surfing has now been recognised in the design of structures. The Council and State Government now recognise to some extent revenue generated by the beaches and a reasonable budget is provided for protection and improvement of the Gold Coast beaches and waterways. This budget also allows some practical reseach to be carried out. If properly managed the returns are high but if not, the risks are very high.

[1] Managing Director, International Coastal Management, P.O.Box 7196, Gold Coast MC 9726
AUSTRALIA PH/FAX 61-7-5592 9147 EMAIL; Jackson@on the net com.au
[2] Consulting Engineer, 5 Ilkinia Ave, Florida Gardens, Gold Coast 4217 AUSTRALIA
Ph 61-55381856

CHAPTER 7

The Port of Ensenada, B.C., Mexico: Problems and Prospects

Concepción Arredondo-García and David W. Fischer
Faculty of Marine Sciences, Autonomous University of Baja California, Ensenada, B.C. Mexico.

This paper summarizes the history and present conditions surrounding the Port of Ensenada. Included are tables describing the principal cargo and users of the port. The results of a survey are reported that focussed on the problems facing the port as it approaches the 21st Century. Suggestions are made to assist in a more sustainable and integrated plan for the port's development and growth.

I. Introduction

The Port of Ensenada is located in Baja California, Mexico, on the coast of the Pacific Ocean some 100 Km. south of the frontier with the United States of America. Currently, the Port of Ensenada is considered one of the high-priority ports for the modernization and privatization of its services, including ship repair, commerce, fishing and cruise ships. The Port occupies a strategic position in the Pacific Basin adjacent to the United States and Canada. This location gives Mexico the opportunity of forming one of its most dynamic commercial zones. To be incorporated into the world's commercial activity and contribute to the economic development of the country depends to a large extent on the functional characteristics of this transportation node (1).

The Secretariat of Communications and Transportation (SCT) recognizes that current port activity in Mexico does not yet correspond with the development needs of the country. The port system is not as competitive and efficient as demanded by world commerce. Meaningful changes in areas that complement the port such as improvements in the highway net, railroad capacity and greater participation of the private sector are needed (2).

Recently, the federal government accomplished important investments in port infrastructure that promotes and deregulates the activities of this sector such as establishing a new legal framework for restructuring and privatizing of its ports. During 1993 the SCT published the Master Plan for the Port of Ensenada which outlined the actions to be accomplished such as the construction of a new dock with two positions for cruise ships or commercial ships, the modernization of the port's equipment, the rehabilitation of the electrical system to operate 24 hours a day, new dredging that permitted four new positions for the tuna fleet, and new facilities for fuel supply were constructed. However, in general terms, it is observed that the port requires resolution of its administrative and planning problems as well as certain technical efforts to complete its modernization (3).

II. Development and Growth of the Port of Ensenada

The Port of Ensenada was opened to trade officially in the year 1877 and in 1888 was declared part of the municipality of Ensenada City. In 1933 it was granted a tariff exemption as part of the free zone (4). During the 1950's the breakwater and key infrastructure were built. This growth of the port facilities coincided with the population growth of the city (5) (Fig.1).

The port's activities began as support for the distribution of agricultural products from Mexicali, the capital of the state, located some 200 Km inland from Ensenada. In addition, fishing activities were served.

In the decade of the 1970's growth of port facilities increased to enhance fishing activities and repair vessels. In 1979 a dock for fishing use and the navy was built. During 1981 a floating repair and fishing docks were added along with the "El Gallo" dock and sport fishing area (6). Thereafter in 1987 new dredging occurred to capture tourist activity, especially the arrival of cruise ships with a designated passenger module called the "International Cruise Terminal" .

In more recent years, new commercial activity has included minerals in bulk as well as the distribution of fruit and automotive parts (4). Presently, activities in the Port of Ensenada include commercial, industrial, tourism, fishing and military uses. All these activities have been mixed over time without a planned foundation for integrating them. This unplanned growth has hindered the port's development and generated various environmental problems as well as cutting off the city from its waterfront.

This paper has as its objective to identify key port activities, actors and problems that have prevented its further progress toward the 21st Century.

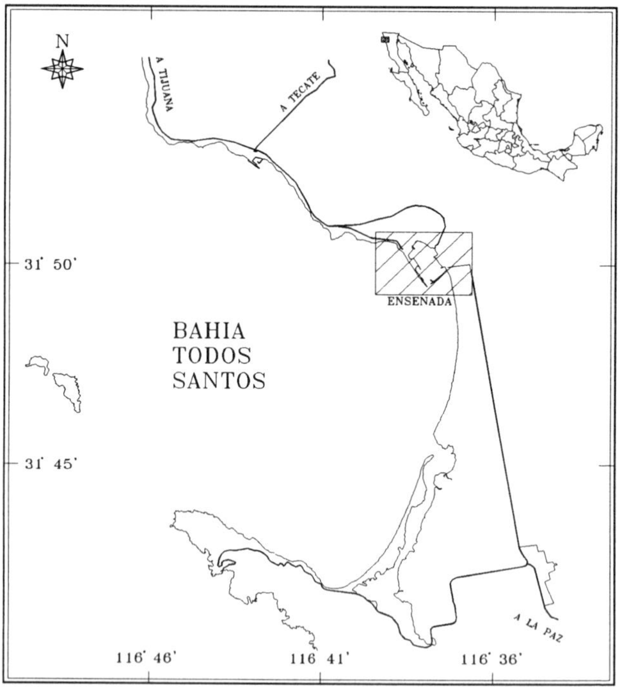

Figure 1.- Geographical Location of the Port of Ensenada, B.C.

III. Port Diagnosis

3.1 Load structure

Traditionally, the Port of Ensenada handled perishable items (tuna, anchovies) and bulk minerals, and at a smaller scale, agricultural items. The cargo containers began in 1990 as an important step. The composition of these loads during the period 1989-1995 is shown in Figure 2. During 1989-1996 the volume fluctuated with an annual mean rate of 14%; in particular, the minerals and perishables dropped to 20% and 19 % respectively, recovering in 1994 at almost 900,000 tons. (7, 8)

With respect to cargo movement an increase between 1989-1992 with an annual mean rate of 27% was observed. The following year it fell by 42%, recovering in 1994 to almost the level of 1992 with some 900,000 tons. This decrease was generated in part due to the decrease in cement from the CEMEX Plant.

Containers began to arrive at the Port of Ensenada in 1990 and increased 58% until 1992. The following year containers dropped 42% and fell drastically afterward because the principal clients, Hyundai, changed its production line, and the other client, Ford, changed shipping lines. In 1994 some containers began to arrive again; however, as of this year, the previous scale has not been achieved (9).

3.2 Cruise Ships

Ensenada is the second most visited port at the domestic level for cruise ships, representing a significant economic impact (Fig. 3). While in 1990 this impact represented USD $ 5.9 mill, for 1996 it is expected to be USD$15.9. (10). During 1993 an average of 29,000 passengers arrived on 53 ships monthly. However, the passengers only stay in the city approximately 8 hours. More arrivals are expected when casinos are permitted on cruise ships that touch an international port. However, this increase may be temporary since it could be modified at any time (7).

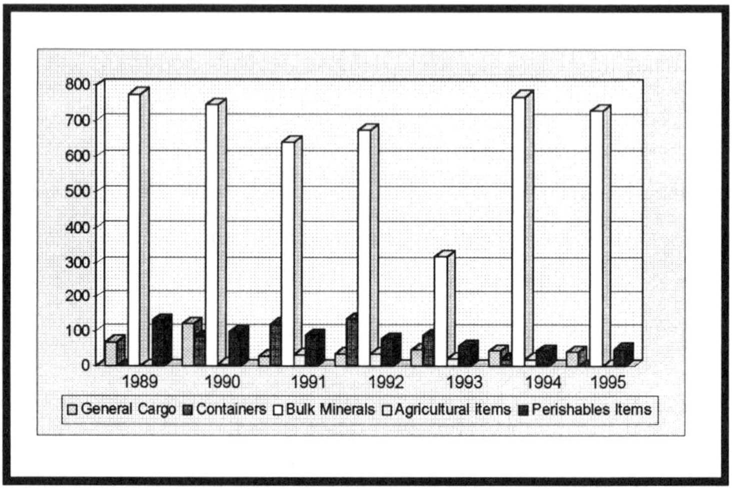

Figure 2.- Cargo Compostition of Ensenada Port between 1989-1996

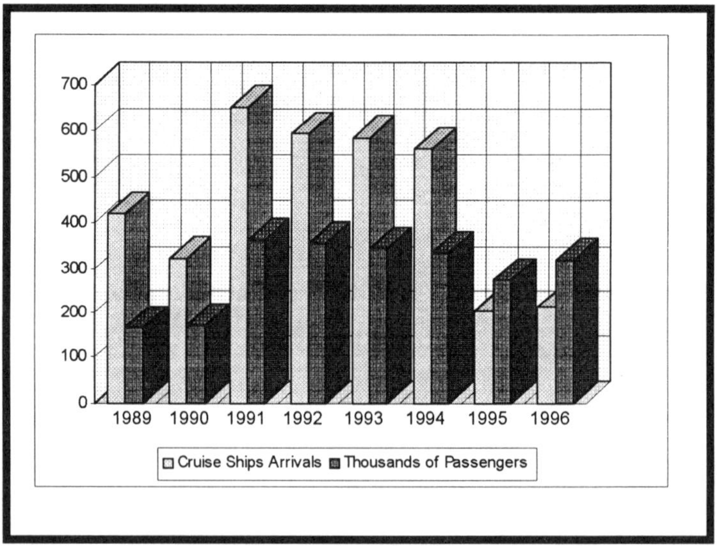

Figure 3.- Arrivals of cruise ships and numbers of passengers between 1990-1996

The State Economic Development Secretariat (SEDECO) noted that the port's income depends to a large extent on cruise ships since they are regular arrivals; therefore, the terminal for cruise ships is projected to be moved to a more accessible place adjacent to the tourist zone of the city (9). Furthermore, the report suggested a search for greater tourist attractions since only 60% of the passengers disembarked during 1994, the last year sucj a survey was completed (7).

IV. Environmental Diagnosis

Ensenada is considered a port that requires particular attention to water pollution control. At present population and industrial growth are deteriorating the natural conditions of the coastal zone. The principle types of pollutants poured into the port zone include solid tailings, organic matter, industrial residual waters, heavy metals, and hydrocarbons. Table II shows some of the principle concessions of the port and the types of pollutants generated and observed by practically all the users and concessionaires of the port. The deterioration of the water and air quality includes strong smells due to sulfuric acid from decomposition of organic matter.

Studies of pollution of the port and bay began in the decade of the 1970's when Lizárraga-Partida found that pollution in the port had not yet reached alarming levels, even though some zones were devoid of life (11). With respect to the bacteriological quality of the interior waters of the port, Orozco-Borbon and Gutierrez-Galindo reported that 100% of their samples exceeded the maximum values for coliforms established by Mexican Law (CT 1000 org/100 ml and C.F. of 200/100 org/ml for coastal waters of recreational use with primary contact (12). Romero-Vargas accomplished pollution studies of heavy metals in sediments and reported that the Port of Ensenada contained high copper and zinc levels attributed to the maintenance activities for vessels and dredging operations (13).

Table II . The principle types of pollutants in the port zone.

TYPE OF POLLUTANTS	SOURCES OF POLLUTERS									
	Administrative Offices	Arjona Agency	Seafood Market	Seafood and Fishing Plant	Navy Hospital	Port Services Shipyard	CEMEX Plant	Hotels, Shopping Centers	Little Boats and Sport Fishing Boats	Tuna and Commercial Ships
Chemicals , Pesticides, Acids		■				■				■
Heavy Metals		■				■				■
Particulate and Soluble Materials			■	■	■	■	■	■	■	■
Nitrogenous and Phosphorous Components			■	■	■				■	■
Pathogenic Microbes				■	■					■
Combustion Gas and Particulate Materials		■					■		■	■
Detritus	■	■	■	■	■	■	■	■	■	■
hydrocarbons		■							■	■

*** The shaded cells indicate the particular port actor contributes to wich types of pollutants.

V. Port Actors

Following the actor classification system developed by Fischer and Kieth (14), port actors were classified into different groups in relation to the development of the Port of Ensenada (Table I). The developer actors include the Port's Administration (API) and the three levels of government who are united in intensifying the port's development.

The allied actors are those that favor port development and are represented mainly by chambers of commerce and enterprises such as National Chamber Trade (CANACO), National Chamber of Industry and Transformation (CANACINTRA), National Chamber of Fishing Industry (CANAINPES), and Mexican Cement Company (CEMEX), as well as those providing port services such as ship repair and fuel provision.

Opposed to port development are some irregular concessionaires who are considered as rival actors. These concessionaires are opposed because their illegal status would be clarified and they would be excluded from the port. Included are some ship repair groups and small boat owners. Environmental actors include Secretariat of Environment, Natural Resources and Fishing (SEMARNAP), Federal Office of the Protection to the Environment (PROFEPA), State Ecology Office , Municipal Ecology Office, the Group Pro-Esteros, and the Sciences Museum (Museo de Ciencias).

In addition, the academic institutions, environmental consulting companies, and professional associations oppose development without rectifying the port's pollution and related problems. The Navy Secretariat (SM), Secretariat of Health and Public Assistance (SSA) are considered to be neutral to port development. The population as a whole would be affected both positively and negatively by any proposed growth.

The port authorities of Guaymas and Topolobampo, Sonora, and the Ports of San Diego, Long Beach and Los Angeles, as well as some financial groups, Bank of Foreing Trade (BANCOMEX) and Secretariat of Trade and Industrial Promotion (SECOFI) are considered exogenous agents who exercise some influence over port development strategies.

Table II. - Classification of actors in the Port of Ensenada, B.C.

Type of Actors	Representative Associations in the Port of Ensenada, involved in its development
Development	• Port Administration, S.A. (API) • Secretariat of Communications and Transportation (SCT) • State Government • Municipal Government • Terminal Operators
Environment	• Federal Office of the Protection to the Environment (PROFEPA) • National Commission of Water (CONAGUA) • State Ecology Office • Municipal Ecology Office • Science Museum (Museo de Ciencias) • Pro-Esteros, a local group
Allied	• National Chamber of Industry and Transformation (CANACINTRA) • National Chamber of Fishing Industry (CANAINPES) • National Trade Chamber (CANACO)
Rivals	• Irregular Concessionaires
Independent - Neutral	• Governance Secretariat (SG) • Navy Secretariat (SM) • Secretariat of Health and Public Assistance (SSA) • Associations of Professionals, Consulting Companies
Impactee	• Services Sector, casual users and city population
Exogenous	• Port authorities of San Diego, Long Beach, Los Angeles. Bank of Foreng Trade (BANCOMEX), Secretariat of Trade and Industrial Promotion (SECOFI)

This actor evaluation system characterizes the actors or social groups that form the core of a given project. These groups are involved to different degrees in the formulation of port policies, and they can be favorable or unfavorable to any development policies proposed.

The actors identified in the development of the Port of Ensenada have participated or had knowledge of the Port's Master Development Plan. These different groups were surveyed for their knowledge of the problems affecting the port and its future.

VI. Port Problems

A total of 40 surveys were distributed among the identified actors within the Port of Ensenada as shown in the Table II. Included were the three levels of government, cooperative groups, various public institutions and civil associations. On the part of the private sector were the different representative chambers of commerce, concessions and those providing services in the port. Of 40 distributed surveys, 24 actors responded. Table III shows the actors that answered the survey. Of the 24 answers, six interviews at the federal level and three each at the state and municipal levels were obtained. The social and private sectors returned six and six surveys.

Table III presents the problems of the port as mentioned and weighted by the participants. Twelve problems were identified and included lack of infrastructure , lack of public services, lack of definition for the port, lack of governmental organization and planning continuity, lack of public access and visual attractiveness, pollution problems, lack of control of residual waters, incompatibility among activities, lack of space for growth, lack of integration with the city, the loss of activities from the US Tuna Embargo, and the high operating costs in the port.

75% of the interviewees identified industrial and domestic residual waters into the port enclosure. The lack of infrastructure was the second most important problem, mentioned by 54% of the actors.

50% of the participants noted the insufficiency of public services, such as water supply, lighting and drainage. 33% identified the lack of definition in the use of the port, and the lack of organization and planning continuity.

VII. The Port's Future

46% of the participants mentioned that the responsibility for the problems rest with the governmental sector and the governing organization of the port. The federal level in particular recognizes that the port is its responsibility, even though 31% of those polled assigned responsibility to all three levels of government, as well as to the concessionaires and users of the port. 23% mentioned that joint efforts between local citizens, port users and concessionaires, as well as the three levels of government, were necessary to develop the port's future.

The federal level considers the solution of the port's problems to depend on privatization, in addition to requiring greater participation and conciliation between the participating sectors. The elaboration of a new integrated plan with new investment projects, as well as a simplified administrative structure would be indispensable to solving the port's problems.

Table III.- Actors Interviwed and the Problems of Ensenada Port.

Number of Actors / PROBLEMS	PUBLIC SECTOR												SOCIAL SECTOR						PRIVATE SECTOR						Number and Percent of Actors	
	Federal						State			Municipal																
	1	2	3	4	5	6	7	8	9	10	11	12	13	14	15	16	17	18	19	20	21	22	23	24	TOT	(%)
1.-Lack of Infrastructure		●	⊙	●		●	●				●			○	○		⊙		●	●		●	●	⊙	13	54.0
2.-Lack of Public Services			⊙		○	⊙	●				⊙			⊙	○		⊙	⊙	●		○			⊙	12	50.0
3.-Lack of Definition for the Port's Uses				●	○			●	⊙	●		●			●							⊙			8	33.3
4.- Lack of Organization and planning continuity	●				⊙				●	⊙			⊙		●					⊙	●				8	33.3
5.- Lack of public access and visual attractiveness			●		○					⊙			●	⊙					⊙						6	25.0
6.-Problems with communications		●	○				○		⊙	○												●			6	25.0
7.- Lack of control of residual waters		⊙	●	⊙	⊙		○	●	○	○	●	●		●	⊙	⊙	●	●	○		⊙		●		18	75.0
8.-Incompatibility among activities					●	●				○			⊙			●			⊙					●	7	29.1
9.- Lack of space for growth						⊙	⊙						●			●					⊙		⊙		6	25.0
10.- Lack of integration with the City	●						⊙	⊙		●					○			●			○			●	7	29.1
11.- Loss of activities from the US Tuna Embargo and the high operation cost in the port					●				●				○				●	●		●	●				7	29.1
12.- Lack of Administrative Simplification, Operations.	⊙					○								●				●		⊙		⊙		●	7	29.1
Number of Problems noted by Actor	3	3	5	3	7	4	7	3	5	7	2	2	5	5	6	3	4	5	5	4	5	5	3	4	x= 4.3 Prob./ actor	

Note: The numbers signify the numer of times actors noted a problems; The more times a problem is noted the more important the problem.
Importance of Problem: High = ● Medium = ⊙ Low = ○

With respect to the port's pollution problems, it is recommended that to implementation of existing laws and ordinances be made which in turn depends on the participation of local citizens. The application of sanctions and fines on violators of environmental laws are viewed as necessary. On the other hand, the municipal level mentioned that it is fundamental to clean up the Ensenada Arroyo, since it is considered one of the main sources of pollution into the bay, particularly in the rainy season. The relocation of fishing boats to the Port El Sauzal and the installation of a domestic and industrial sewage treatment plant also would assist in reducing water pollution.

The polled actors mentioned in general terms the critical aspects necessary for the future of the development of the port. The most important was the ability to obtain financing and to generate greater trade. Secondarily, it is necessary to solve the operational problems for public services and air and water quality. A critical point is the difficulty of obtaining financing for integrating the port's administration with the regional economic situation. Representatives of the municipal government stress the reordering and regulation of the port's activities as well as the endowment of more public services.

In addition, it is fundamental to give greater promotion to the port to increase the movement of operations by regional entrepreneurs. An administrative simplification of the port's services would lead to increased competitiveness of the port and to new projects. Private entrepreneurs cannot be expected to solve the port's problems without the governmental sector simplifying its governance and providing greater support services. One key issue is the integration of the city with any new port growth plan. The city is intimately involved with the port's pollution problems and to date is completely cut off from access to its own waterfront. Without the city's support, the port cannot succed.

Indeed, each actor involved in the port's development is important to its sucess as a growth enterprise in the 21st Century.

VIII. REFERENCES

1. Puertos Mexicanos, Secretaría de Comunicaciones y Transportes. Puertos Mexicanos. Abril 1991. Año II.Num 14. México, D.F.

2. Secretaría de Comunicaciones y Transportes.1994. Los Puertos Mexicanos. Inversión hacia el Futuro. Proceso de Reestructuración y Privatización (s.n.t.)

3. --------------.1994b. Puerto de Ensenada, México. México. D.F. Tríptico.

4. Secretaria de Comunicaciones y Transportes y Puertos Mexicanos (SCT-PM). 1993. Puerto de Ensenada. Oficinas Centrales, México, D.F. Delegación Ensenada, B.C. Plano.

5. Victoria-Prado, R. 1991. Investigación de las Actividades de Mayor Relevancia, sus Efectos y Medidas de Mitigación del Impacto Ambiental para la Bahía de Todos Santos y Baja California, México. Pregunta Problema para obtener el Título de Oceanólogo. F.C.M.-U.A.B.C. Ensenada, B.C. 56 pp.

6. González-Villagrán, A. 1992. Reevaluación de los Estudios Utilizados para el Diseño del Puerto de Ensenada, B.C. Tésis Profesional de Licenciatura. F.C.M., U.A.B.C. Ensenada, B.C. 126 pp.

7. Administración Portuaria Integral de Ensenada, S.A. de C.V. 1994. Programa Maestro de Desarrollo. s.p.

8. Secretaría de Comunicaciones y Transportes- Puertos Mexicanos, 1993

9. Secretaría de Desarrollo Económico del Estado. 1995. Puerto de Ensenada: Análisis de las Estadísticas del Movimiento Portuario. 20 pp.

10. Administración Portuaria Integral de Ensenada, S.A. de C.V. 1996. Puerto de Ensenda, Baja California, México. El Puerto de Ensenda en el Sistema Portuario Nacional. 10 pp.

11. Lizárraga -Partida, L. 1973. Contribución al Estudio de los Vermes Anélidos Poliquetos como Indicadores de Contaminación Orgánica. Tésis Profesional de Licenciatura. E.S.C.M., U.A.B.C. Ensenada, B.C. 45 pp.

12. Orozco-Borbón, V. y E.A. Gutiérrez-Galindo. 1983. Contamiación Fecla Costera en la Zona del Puerto de Ensenada, Baja California. Ciencias Marinas. Vol. 9 (1):27-34.

13. Romero-Vargas, I.P. 1995. Metales Pesados y su Fraccionación Química en Sedimentos de la Bahía de Todos Santos, Baja California, México. Tésis de Maestría en Ciencias en Oceanografía Costera. F.C.M.-U.A.B.C. Ensenada, B.C. 86 pp.

14. Fischer, D.W. and R.F. Kieth. 1977. Assessing the Development Decision-Making Process: A Holistic Framework. The American Journal of Economics and Sociology. Vol 36(1):1-17

CHAPTER 8

Balancing Revenue Needs with Public Open Space Needs

Donald E. Hillman, Jr.

Abstract

In 1996, in response to the proposed expansion of the San Diego Convention Center, the San Diego Unified Port District received unsolicited proposals for development of four new hotels located near the Convention Center. The District did not approve any of those proposals and instead embarked on a planning effort for the area which was specifically designed to encourage public input and accommodate public open space needs with the District's revenue needs. The masterplanning process, which took approximately four months to complete, resulted in the Board of Port Commissioners adoption of a development plan allowing for 2,300 new hotel rooms, including major expansion of two existing convention-related hotels, and development of a new major hotel and marina complex. The adopted plan, which provides for development was supported by the community and meets the District's revenue needs and the need for public open space and enhanced public access to the waterfront.

CHAPTER 9

Using Public/Private Partnerships to Enhance the Resource Value of San Francisco Bay Wetlands: The Role of Ports

Jim McGrath, Environmental Manager, Port of Oakland

INTRODUCTION

I'd like to start with a comment about innovative efforts like the restoration agreement for Bolsa Chica reached in October of 1996. I worked for the Coastal Commission for 14 years before I took this job. I did the staff work on Bolsa Chica for six years and I remember some intense moments before Commission meetings when Mel Nutter held the votes to be able to preserve Bolsa Chica from substantially more fill. I'd like to recognize the kind of innovative work by the agencies and by the ports that gets us to a situation where the Bolsa Chica wetland system will be restored, and there won't be a marina and a deep ocean channel in the middle of the restored system. The original proposal would have had all of the restoration funded by the state, and would have included a substantial loss of value and acreage to development of homes and a marina in the existing wetlands. I think that this agreement represents the kind of innovative cooperation between the public and private sector that is the direction that we want to go in, and that is the thesis of my talk.

I want to tell you about three things that the Port of Oakland is doing that involve wetland restoration, and then challenge you all to help improve this partnership. The first thing I'm going to tell you about is Sonoma Baylands, a completed project that used dredged material to restore a 290 acre tidal wetland system. The second thing I'm going to tell you about is Middle Harbor, where we're looking at 190 acres that was once damaged by dredging. We're looking at a concept very similar to one that's already been implemented by the Port of Los Angeles. Third, I'm going to talk a little about mitigation as a whole, and how it exists in the San Francisco Bay system.

SONOMA BAYLANDS

Figure 1 shows Sonoma Baylands. This was a former tidal wetland that was diked in the 19th century, and was very efficiently drained by the land owner. The owners farmed the site for decades, and in that process, the land settled and subsided. When the Coastal Conservancy and the Sonoma Land Trust acquired the land, it was three or four feet below the level of a tidal marsh plain. So the Corps of Engineers (Corps) and the Coastal Conservancy built a series of peninsulas to break up wave energy, and the Corps, the Port of Oakland and the Coastal Conservancy pumped two million cubic yards of dredged material into the site, then breached the outboard dike. This dredged material was clean, muddy material that was suitable for aquatic disposal but not of much structural value. This is what the site looks like now; it's very shallow and has been opened up to tidal action. It will be five to ten years before it's fully vegetated, but it's now working fairly effectively as a shallow mud flat and starting to vegetate well around the sides. So that's a success, in that it accelerates the natural sedimentation process and increases the acreage of tidal wetlands. The project seems to be performing in accordance with the restoration objectives.

MIDDLE HARBOR CHANNEL HABITAT RESTORATION

Figure 5 shows the Port of Oakland, and the three navigation channels: Inner Harbor, Outer Harbor, and Middle Harbor. Our business is shipping, and to compete with the Ports of Los Angeles and Long Beach we need to modernize our facilities and deepen our channels. We have an opportunity to do that now by redeveloping the Fleet Industrial Supply Center, which is slated to close and will be returned to the Port. We're now in the planning stage for redeveloping that site and deepening our channels to 50 feet.

If we are going to be competitive, we are going to need to deepen the harbor beyond the current project which is deepening it to 42 feet, and go down another 8 feet. That would generate about 20 million cubic yards of material. We plan to use some of that in construction, but we're looking at what we think is a unique opportunity, and that's restoration of some of the aquatic habitat values that used to exist in Middle Harbor before it was dredged to create a military shipping center.

The maps shown in Figures 2 through 5 are portions of the Coast and Geodetic surveys and navigation charts that show the changes that have occurred in Middle Harbor over time. Originally, the area was a series of tidal and subtidal flats, grading from intertidal mudflats to shallow mudflats, to sand flats, and the first map shows the 1857 shoreline. Over time, portions of the area were dredged, and the dredged materials were placed as fill to create the military port. The 1857 shoreline has been traced on the later maps so you can see what has happened over time. We're looking at about 190 acres that could be restored to shallow water - it's now about forty feet deep - and in the process restore some of the original habitat. (Figure 6) In addition, public access could be extended around this basin—currently, there's a four acre park at the end of Seventh Street, but no other substantial public access in Outer or Middle Harbors. So we are trying to find ways to provide the community of West Oakland with a new shoreline park, save money, and enhance habitat with dredged material. As you view the maps, you can see that the Middle Harbor area was shallow through 1921, even though the navigational channels had already been deepened to 34 feet. Indeed, Middle Harbor remained shallow until World War II began and the military port was developed. The final maps show further dredging and fill which has created the present, deep basin. The cross sections illustrate the incremental deepening that has occurred, and the concept of restoring shallow water depths that we have in mind.

This concept was really stimulated by what's happening across the channel in Alameda. There, the Naval Air Station is being closed and 525 acres is going to become a national wildlife refuge. That refuge was established primarily to protect habitat for the least tern which nests on and between the runways. The least terns feed, as Ralph Appy's presentation demonstrated, in shallow water. So we are looking at a plan to at least partially restore the natural conditions of the area. Similar projects have been implemented in the Port of Los Angeles. Indeed, the Port of Los Angeles has gotten mitigation credits for this type of project. In contrast, we're being told by our coastal management agency that "filling the Bay", even for habitat, is a non-starter. Since this portion of the Bay will become "filled" with mud over the next 50 to 75 years once maintenance dredging stops, we're hopeful that we can work with stakeholders and resource agencies to direct that inevitable sedimentation to enhance habitat more quickly. We've initiated an interactive planning process that includes the resource agencies and environmental groups. We hope to develop habitat enhancement plans over the next year and a half to two years that will reach some consensus on restoration goals for this area.

MITIGATION BANKING

Well, what's the context for this? We've talked a little bit about 500 acres of wetland restoration work that we're doing. How does this fit into the larger context? When I prepared an abstract for this presentation, I discovered that one of the statistics that I've used myself, that we've lost 90% of our wetlands, is not really true in San Francisco Bay. Instead, what we've done is alter over 90% of our wetlands, mainly by converting tidal wetlands to non-tidal wetlands. According to the Estuary newsletter, in 1987 there were 630,000 acres of wetlands - jurisdictional wetlands - remaining in San Francisco Bay. Figure 7 is a map of those areas, and comparing that map to the earliest maps of the areas shows that over 2/3 of the historic wetlands remain—but they have been highly altered. Nearly 400,000 acres of the remaining wetlands are farmed wetlands. Most of those areas were historically tidal wetlands, like Sonoma Baylands, which were diked for agricultural purposes. Other areas, particularly along the Napa River and in the South Bay were diked for salt production. So we're now looking at a system where there are about 600,000 acres of wetlands, of which over two thirds are privately owned. If we are going to be able to manage those areas for greater habitat value, we are going to have to either acquire the property, or reach some kind of private/public partnership with the landowners.

Progress towards public acquisition of those wetlands is slow, only about 19,000 acres have been acquired since completion of the Estuary Project's CCMP. This is a tiny fragment of the land that is in private ownership, and I believe that we are going to have to explore other options such as mitigation banking if we are going to succeed in enhancing (or recapturing) the historic values of the Bay area's wetlands.

So I'd like to pose the larger questions of how well wetland mitigation is working, and whether or not wetland mitigation should become a larger element of our wetland restoration tool kit. Let's start with how well wetland mitigation is working. In 1985, Wendy Eliot from the Coastal Conservancy looked at the mitigation projects for 58 wetland fill permits, and concluded that "Current mitigation policies have been largely unsuccessful in San Francisco Bay." In 1988, BCDC looked at 14 restoration projects, and concluded that 6 were successful, 5 were partially successful, and three were unsuccessful. That report cited a lack of clarity in restoration objectives, and lack of work, as major problems.

Josselyn and Zedler, prominent academics in the restoration field, took a more systematic look at restoration projects, and the developing professional field of restoration. Their conclusion: "...the majority of completed projects did create some wetland habitat...[but] a significant percentage of projects evaluated did not meet the expectations of wetland function..." The professionals working on wetland regulation and mitigation have responded to the criticism of their peers. Yet problems persist. A recent article in the National Wetlands Newsletter, "Mitigation's role in Wetland Loss", by Julie M. Sibbing of the National Audubon Society indicated that less than 1% of wetland permits are denied. Their investigation of 32 individual permits showed that actual mitigation involved a net loss of over 67 acres on a total filled of 371 acres, despite policies that call for a 1.5:1 mitigation ratio.

Thus, each wetland fill project involves a major regulatory decision over whether the fill should be allowed, and less concern and less scrutiny on how well the mitigation occurs when you get to the other end. We often see a fierce pitched battle over wetland fill, and if the fill is approved we have some very expensive mitigation measures that are not always successful. The Port of Oakland is an owner of wetlands, and our experiences illustrate the difficulties of creating an effective private-party/public agency partnership.

The last map, Figure 8, shows the jurisdictional wetlands at Oakland Airport, about 637 acres in all. The airport was created by placing hydraulic fill behind dikes and over the mudflats of the Bay. In that process, portions of the site settled, and other portions that were left lower for drainage continued to pond water and we have a very large wetland. If you look, particularly up in North Field, you see a patchwork of measles, as sort of little tiny scraps of what are jurisdictional wetlands, but don't have a tremendous amount of habitat value year round. One of the things we've discovered by looking at this system is that because of the oversight by the FAA, we're in a quandary. We cannot enhance these wetlands if that enhancement increases bird use. The FAA just has a fit about anything that increases bird habitat within two miles of a runway . On the other hand, environmental groups and wildlife agencies prefer on-site mitigation. At the airport, we have a system where the turf battle about mitigation between the federal agencies; the Environmental Protection Agency, the Corps and the FAA is not resolved and is not likely to be resolved in the short term. Tha system includes many hundreds of acres of wetlands, many of them scattered around in places that don't provide the best quality habitat. In mnaging that system, we're piddling around the edges, as we've done out there in two acre and three acre pieces, at tremendous cost.

The Port has completed a 4.1 acre mitigation project, at a 1.4:1 mitigation ratio, for placing dredged material at Galbraith, at a cost of about $212,000 for each acre filled. We are nearly ready to go to construction on a 71 acre wetland restoration project in San Leandro Bay that will resolve a decade long lawsuit. That project will cost $2.5 million, or $35,000 an acre. We are preparing a final EIS for airport expansion that would fill 5 acres of wetland, and realize that on-site restoration might involve wetlands with diminished habitat for birds. We are going to consider off-site restoration, possibly at a higher ratio to save money and hopefully create a superior quality habitat.

None of this work, or anything else going on around the Bay, is as bold as the mitigation project that has been completed at Batiquitos Lagoon in San Diego County, or the proposals that are moving forward for Bolsa Chica and San Dieguito River wetlands. Some innovative effort is needed to improve the quality and cost-effectiveness of mitigation, perhaps along the lines suggested by Will Travis and Jeff Blanchfield in their paper "San Francisco Bay Virtual Mitigation Bank." We're going to have to figure out different ways to get restoration projects funded or we're going to have to come to grips with some of our institutional constraints, particularly at the federal agency level, between the federal agencies that share responsibility but have no current incentive to try and resolve their conflicting goals.

REFERENCES

1. Eliot, Wendy, 1985, "Implementing Mitigation Policies in San Francisco Bay: A Critique"

2. San Francisco Bay Conservation and Development Commission, 1988 "Mitigation: An Analysis of Tideland Restoration Projects in San Francisco Bay."

3. Josselyn, Michael; Zedler, Joy; and Griswold, Theodore; 1990 "Wetland Mitigation along the Pacific Coast of the United States", in "Wetland Creation and Restoration: The status of the Science", edited by Jon Kusler and Mary Kentula

4. Sibbing, Julie M., "Mitigation's role in Wetland Loss", in the National Wetland Newsletter, January-February, 1997

5. San Francisco Estuary Project, "Comprehensive Conservation and Management Plan for the Bay-Delta, Implementation Progress 1993-1996", June 1996

PHOTOS: LAUREL MARCUS

FIGURE 1

San Antonio Creek
1857

FIGURE 2

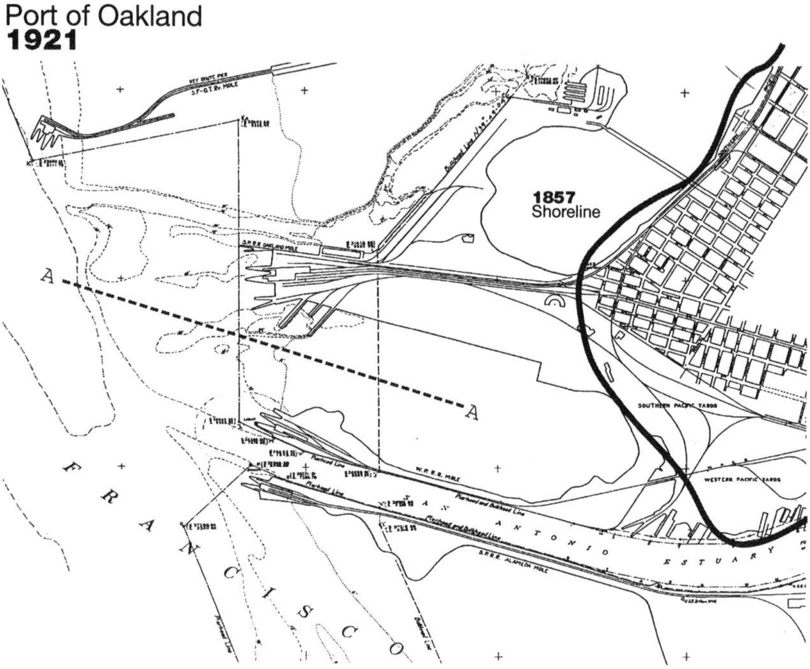
Port of Oakland
1921
1857
Shoreline
A
A
SOUTHERN PACIFIC YARDS
WESTERN PACIFIC YARDS
W.P.R.R. MOLE
S.P.R.R. ALAMEDA MOLE
KEY ROUTE PIER
S A N F R A N C I S C O
A N T O N I O
E S T U A R Y

FIGURE 3

Port of Oakland
1941

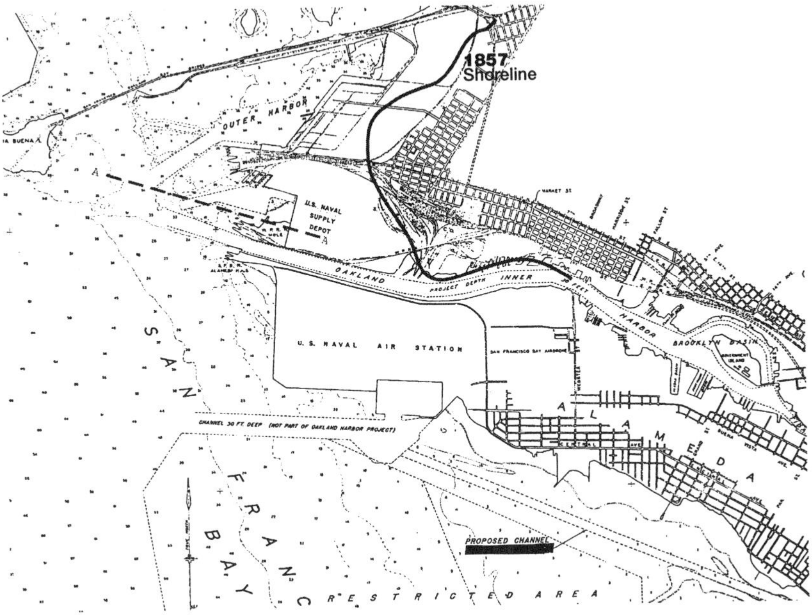

FIGURE 4

Port of Oakland
Present

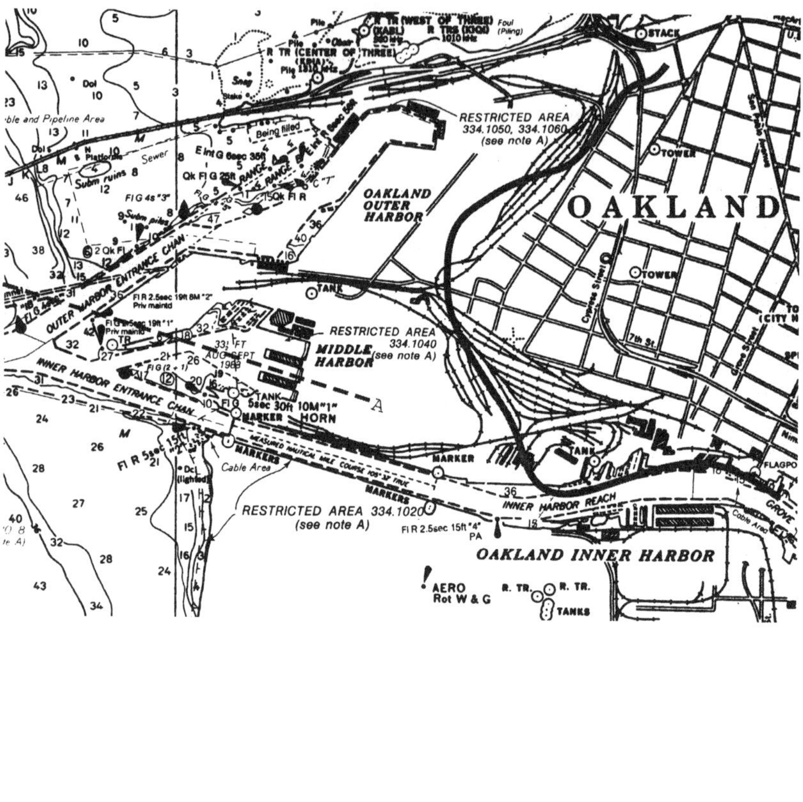

FIGURE 5

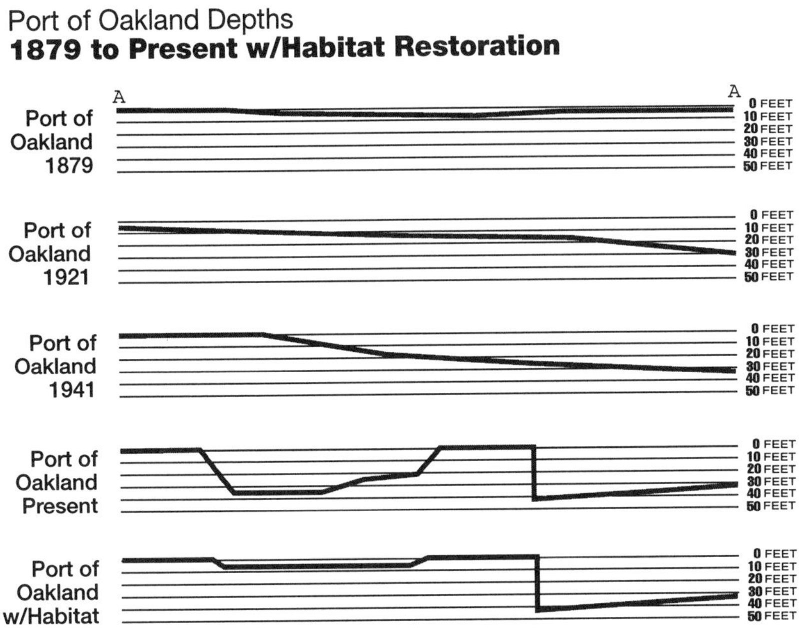
Port of Oakland Depths
1879 to Present w/Habitat Restoration
A
A
Port of Oakland 1879
Port of Oakland 1921
Port of Oakland 1941
Port of Oakland Present
Port of Oakland w/Habitat
0 FEET
10 FEET
20 FEET
30 FEET
40 FEET
50 FEET

FIGURE 6

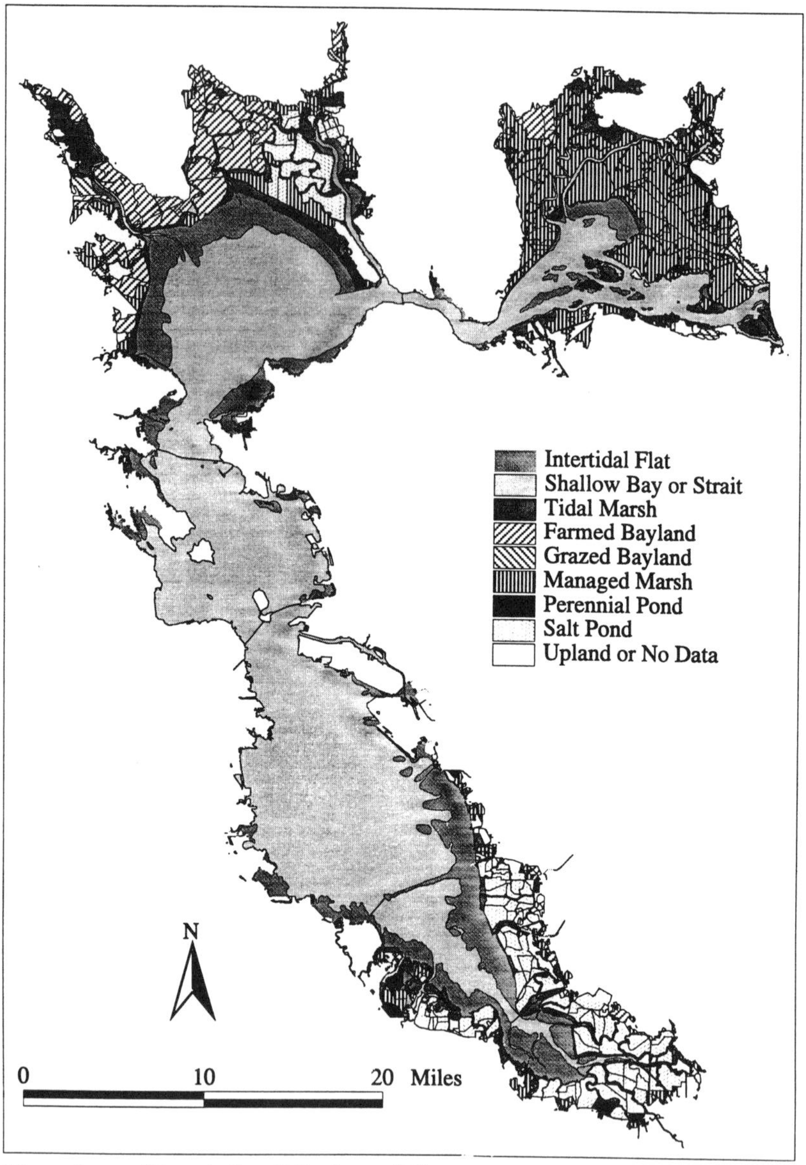

Bay Area EcoAtlas Modern View
(source San Francisco Estuary Institute)
FIGURE 7

Port of Oakland
Existing Juristictional Wetlands

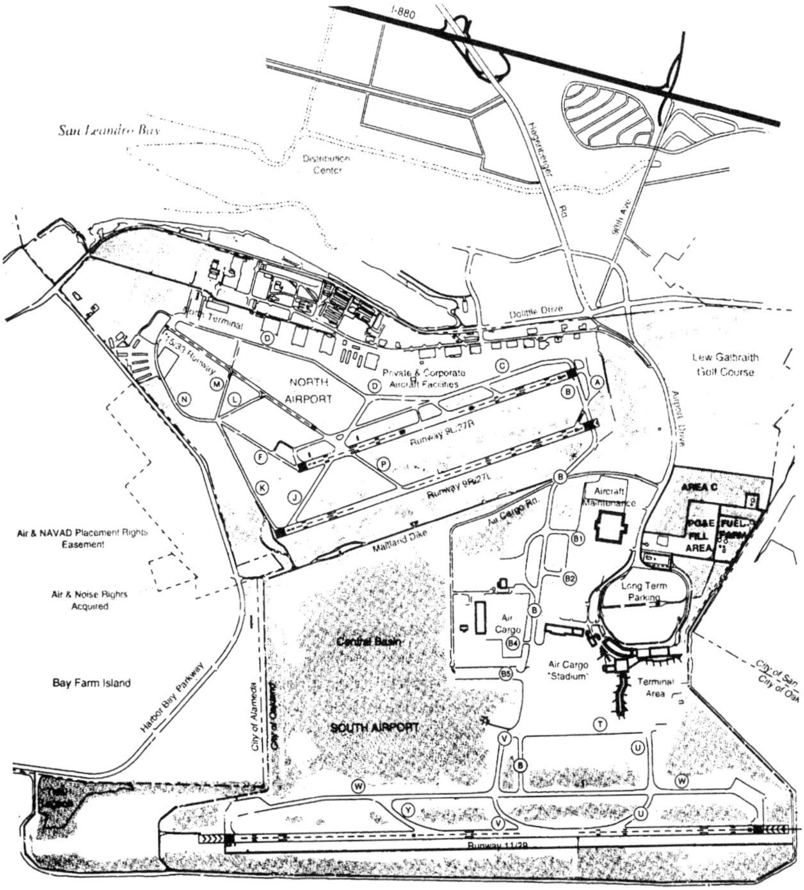

FIGURE 8

CHAPTER 10

Marine Protected Areas, Ecosystem Management, and Ocean Governance: Making pieces of different puzzles fit together (without a hammer).

Bradley W. Barr[1] and Stephanie R. Thornton[2]

Abstract

Considerable effort has been directed over the past few years at trying to find a new paradigm for the management of ocean resources and uses both in the US EEZ and elsewhere, one that moves us well away from largely unsuccessful past practices of single species management in a compartmentalized ecosystem. The management community, academic researchers, and others, spurred on by a growing body of concerned ocean resource users, have begun to grapple with the notion of ocean governance based on the principles of ecosystem management.

Marine protected areas (MPAs) are one key element in this growing era of bigger thinking and in the context of national ocean governance discussions. The inclusion of marine protected areas initiatives in the ocean governance discussions is a positive step toward integrating ecosystem principles within this governance structure, and a number of valuable lessons can be gleaned from taking a look at some of these efforts.

This paper focuses on one such program, NOAA's National Marine Sanctuary Program (NMSP), and provides some insights into its evolution and innovative efforts to foster ecosystem management in and around these coastal and ocean areas. This discussion will profile new ideas being formulated by the NMSP which address how MPAs such as these sanctuaries can provide opportunities for catalyzing broader ecosystem management initiatives and how they can serve as laboratories for developing innovative and alternative ocean governance mechanisms.

Finally, the paper will examine the critical role of forging effective partnerships to make this essential coordination and collaboration a reality for successful ocean governance.

---

[1] Sanctuary Manager, NOAA/Stellwagen Bank National Marine Sanctuary, 14 Union Street, Plymouth, Massachusetts 02360

[2] Chief, NOAA/Sanctuaries and Reserves Division, 1305 East-West Highway, Silver Spring, Maryland 20910

Introduction

At a recent meeting on marine protected areas (MPAs), a keynote speaker put up on the screen a picture of the Earth from space. He mused that a visiting extraterrestrial might wonder mightily about our having named this largely blue planet "Earth." Bringing it back to MPAs, his point was that we have focused our attention on the management of the human use of the terrestrial environment, and left the issues surrounding the use of the oceans largely unattended.

However, this situation is changing. There more attention being paid to ocean governance, ecosystem-based management of marine areas, and the development and implementation of marine protected areas programs. Learning from previous less than successful approaches to managing ocean resources, new directions and innovative approaches are being discussed and developed. Federal and state agencies, conservation organizations, the academic community, and user groups are seeking better ways to manage and govern ocean resources.

Like most times of transition, no common sense of direction, no shared vision has yet to surface. While there are many different initiatives actively being pursued, the point has not been reached where the ocean community has come together to seek consensus on where we need to go and how we might get there. We are getting close however, particularly as a result of the recent efforts of the National Academy of Sciences Marine Board regarding marine area governance, but all would agree that we remain well short of a collective vision. We are also seeing the principles of ecosystem management creep into Federal resource management efforts, most recently in the developing program of NOAA's National Marine Fisheries Service to conserve essential fish habitats. MPA and ocean governance are also hot topics in the international arena. Recent actions such as the passage of the Oceans Act in Canada have resulted in the initiation of an ecosystem-based ocean governance initiative for Canadian waters, and the development of what will ultimately be a comprehensive and collaborative MPA program at the Department of Fisheries and Oceans. Heritage Canada and Environment Canada.

The non-governmental conservation organizations are playing an important advocacy role, and the message is being heard in the halls of government. One place this message is resonating particularly loudly is in NOAA's National Marine Sanctuary Program, where ocean management is a principle focus, and where, since 1972, innovative approaches to MPA management and ocean governance have been and continue to be quietly developed and implemented. We believe that the NMSP is in a unique position to provide leadership in developing and implementing a national vision for collaborative, ecosystem-based ocean governance, calling upon the experience and insight that has come from 25 years of grappling with this issue.

Working Out The Bugs

One of the programs spawned during the period of the early 1970's, when the Federal government was attempting to respond to the public demand for more effective environmental laws, was the National Marine Sanctuary Program. Its goals were to protect areas of the marine environment determined to be of "special national significance." Whether intended or not, it was this program that largely formed the basis for anything approaching a comprehensive Federal ocean management program for most of the two and a half decades the NMSP has been on the scene.

The emphasis of the Sanctuary program has changed over the years, starting out as discrete areas where resources are provided significant sanctuary-specific regulatory protection. Today, sanctuaries are designated with management plans that "facilitate to the extent compatible with the primary objective of resource protection, all public and private uses of the resources of these areas not prohibited pursuant to other authorities."[3] Throughout this evolutionary process, it has consistently been a primary goal of the NMSP to provide "comprehensive and coordinated conservation and management" for these areas. While this was language developed before the term "ecosystem management" became popular, this approach to conserving ocean resources clearly shares many of the resource management concepts underlying this "new" approach.[4] However, the big problem has always been how do you draw the box around an area to insure effective management when you're not sure who's map you're drawing on.

When the Federal government was developing an approach to providing more effective management of ocean resources, focusing attention on those areas that are somehow special, that have high biodiversity, that are more sensitive to disturbance, that are essential habitats for certain key species, was, and is, quite appropriate. It is also not too surprising that such a prioritization strategy would be embraced given the available terrestrial protected area models, such as national parks and national wildlife refuges. These protected areas are generally owned by the government, function under a "command and control" management process, and involve terrestrial areas where functional ecosystems are easier to capture in reasonable-sized and politically acceptable boxes. However, this approach works most effectively protecting resources only if the site designated has high ecosystem integrity, i.e. it includes most if not all of the ecosystem being managed. Unfortunately, none of the MPA sites within the US NMSP (and some would say that none around the world) have sufficiently high ecosystem integrity to optimally protect resources in the designated boxes.

The critical flaw in this approach of relying on protecting discrete areas as the focus of an ocean management strategy is that almost all marine systems are open and function on multiple scales of space and time. Their function and structure is sometimes greatly affected by activities occurring a considerable distance away, and even the species composition of any MPA may change significantly in time, and sometimes in space within the designated area. Capturing entire functional

---

[3] Section 301(b)(5) of the NMSA, as amended.

[4] In fact, the recently adopted NMS strategic plan formally defines and adopts "ecosystem management" as an underlying principle of the program. While no consensus has yet been reached on a specific definition for ecosystem management as it applies to the sanctuary program, sources such as Kaufmann, et al. (1993), Grumbine (1994), Grumbine (1997), GAO (1994), and the recently published Keystone Center (1996) National Policy Dialogue on Ecosystem Management are some of the documents being used to frame these discussions.

ecosystems within a box delineating an MPA is neither practical nor politically achievable.

Is also questionable whether the sorts of "command and control management" used in many terrestrial protected areas can be as effective in marine systems, where ownership is not conveyed with a designation as it would be in a national park or wildlife refuge. While the waters of the US Exclusive Economic Zone (EEZ) are owned in common by the people of the US, there is perceived "private" ownership of many of the resources of the sea by a kind of claim of adverse possession...if a resource was exploited by a group for generations, who has the right to say that traditional use cannot continue in a certain area? While this "ownership" has little or no basis in law, it is nonetheless a large issue for the ocean resource manager. Thus, terrestrial "command and control" models are exceedingly difficult to apply effectively in these more fluid ocean systems..

Protecting critical habitats or special areas of the ocean is essential, but can only be effective when done within a context of an overarching ocean management framework. The lessons learned from 25 years of the NMSP is that if a box must be drawn, it must be drawn within another larger box, or within an ordered universe where the box is part of a larger management system that recognizes the box and embraces it as part of its governance system. Context is critical, and almost always absent.

Desperately Seeking Context

Since its inception in 1972, the National Marine Sanctuary Program has been wrestling with all of these issues as it evolved from a few "command and control" sites managed largely from Washington, DC, to a system of a dozen sites increasingly larger in size and complexity, managed by trained resource managers, specialists in ocean management, located both at the Headquarters in Silver Spring, Maryland, and at the sites. The evolution of this program has provided opportunities for field testing many successful, and a few unsuccessful, innovative management initiatives. This experience has placed the national marine sanctuaries in a unique position to play a leadership role in the development of an effective ocean management system for the waters of the US. The NMSP hopes to provide this leadership through catalyzing the development of a national vision for ocean governance, being the engine within the nation's ocean agency, NOAA, advocating for the implementation of that vision, and providing natural laboratories for new and innovative management initiatives and governance mechanisms.

As we approach the new millennium, the Sanctuary Program has adopted the strategic vision of demonstrating global leadership in managing marine protected areas. The NMSP is attempting to realize this vision, in part, by beginning to take small but determined steps to promote this vital ocean management context for the US EEZ . Embracing the collaborative management approach we have found to be most effective in ocean management initiatives, we are exerting this leadership not by attempting to somehow seize jurisdictional control over ocean management, but to show the "value-added" we can bring to others who share our desire to achieve effective ocean governance in US waters. We intend to lead by example.

The NMSP has, over the years, built a tremendous database of information regarding ocean resource management, dealing with systems from offshore banks in both the Atlantic and Pacific, to coral reefs in Florida, the Gulf of Mexico and the South Pacific, and the thin but biologically rich shelf edge along the Pacific coast. From these experiences, and what we learn everyday in our ongoing management activities, we have achieved some insight into effective ocean management. Our goal is to continue and expand our work in partnership with other Federal agencies, states, territories and tribal governments, as well as stakeholders and the public to develop successful collaborative programs of ocean management for areas that include NMS sites.

Partnerships have been critical to whatever successes the NMS Program has achieved. Where we have joint management authority with the states, as in the Florida Keys, innovative programs in areas such as enforcement and resource protection have been very effective. At the NMS site in American Samoa, the territorial government has been responsible for managing that site since it was designated in 1986.

Linking with other Federal agencies to address issues of joint concern have yielded very positive results. We have developed close working relationships with the US Coast Guard at the Stellwagen Bank and Olympic Coast Sanctuaries. The joint work of the Flower Garden Banks Sanctuary and the Minerals Management Service has provided mapping tools and technical analyses to help better understand the deep water reef ecosystem at risk there. The US Geological Survey and National Undersea Research Program's work in sanctuaries such as Stellwagen Bank, Monterey Bay, and the Florida Keys has also resulted in significantly broadening our understanding of these ecosystems. In all these cases, we have been able to enhance protection and conservation programs, and reduce redundancy, by taking advantage of economies of scale, existing technical expertise, and the natural enthusiasm of government scientists and managers who want to make a positive contribution to the protection of these exceptional areas of the marine environment.

Partnerships with academic institutions, non-governmental organizations, and the public have also been a key element in the success of sanctuary programs. Whether the collaborations at the Gray's Reef NMS with the University of Georgia and the Skidaway Institute of Oceanography (where the site's offices are located), the highly successful relationships of the Florida Keys Sanctuary with the Nature Conservancy and the Channel Islands site with the Santa Barbara Museum of Natural History, or the exceptional "Beach Watch" volunteer program at the Gulf of the Farallones NMS, what is achieved in partnership, this synergism, is considerably more than the additive efforts of each entity working separately.

This vision of leadership, with an emphasis on partnership, is already being put into practice at all of the sanctuaries, where we are seeking out partners, participating in regional interagency collaborations as well as national and international coordinating organizations, bringing NOAA and agency-sponsored science to the table to help support good management decisionmaking, and sharing our knowledge of these issues wherever we can find a willing audience. While examples abound, a sampling includes:

• Marine Zoning System developed for the Florida Keys NMS: Without precedent in US waters, the management plan for the FKNMS includes a comprehensive and detailed plan establishing use zones within the Sanctuary. Based on the best available scientific information, these zones establish permitted uses within each area, attempting to strike a balance between protection and sustainable use. Focused on protecting and enhancing the biological diversity of the Sanctuary, (which establishes the ecosystem context) the zoning plan provides protection for the most sensitive resource areas, while allowing less critical areas to be used more intensely.

This is by far and away the best example of an NMS creating its own regional management context within which the more sensitive habitats (which could be interpreted as the real MPAs, consistent with the biosphere reserve model) can be protected, conserved, or used in a sustainable manner. It establishes an overarching governance mechanism for the region, implemented jointly by the State and the Sanctuary, employing the principles of ecosystem management. What is somewhat unique about this approach is that while there are other examples of zones or areas being set aside for protection based on some special attribute of that area (such as the NMS program itself), this zoning system seeks to protect not only the special areas, but addresses the overall management of the defined natural system, the Florida Keys.

Sites as large and geographically well defined as the Keys may have considerable ecosystem integrity, but they represent only portions of a fully functioning ecosystem. In this case, the South Florida Ecosystem Project, discussed below, is attempting to address broader ecosystem issues.

• The "Rivers to Reefs" Program at the Gray's Reef NMS: While Gray's Reef NMS encompasses only 21 square miles of offshore live bottom reef some 17 miles off the coast of Savannah, Georgia, one of is most successful programs has been its "Rivers to Reefs" program. This outreach and education program, done in cooperation with The Sapelo Island National Estuarine Research Reserve and the Georgia Department of Natural Resources, begins the process of educating the public on ecosystems, and especially how activities in one part are likely to affect the whole system.

Such education and outreach programs are essential to the development of effective ocean governance through constituency building and decisionmaker education. The "bigger" thinking required in ecosystem management approaches, and in the establishment of regional and national ocean governance mechanisms requires something of a change in perspective by participants. Ecosystem management, collaborative by definition, requires that partner agencies be ready to share their sometimes autonomous authority over some aspect of that system. They need to see clearly the "value-added" of adopting this approach to accept the fact that they will have to give up some of their autonomy and controlling order to participate.

• Sanctuary Advisory Councils: At nearly half of the NMS Sites, formal Sanctuary Advisory Councils (SACs) have been formed to provide a link to the stakeholder and user communities, to be a two-way line of communication between sanctuary managers and users groups regarding major issues of policy, program development, and to address emerging issues affecting the Sanctuary.

The NMSP is committed to inclusive management, and has established these SACs as a formalized mechanism for coordination with these important constituent groups and the public.

There is the perception of an intangible wall between the "government" and the public, a perception based, in part, upon decades of "command and control" governance. While advisory groups are not new by any means, the notion of actually giving them the ability to play a substantive role in shaping policy is. SAC's are the NMS Program's initiative to make sanctuary management truly inclusive, consistent with the precepts of ecosystem management, and necessary to develop constituencies to support the implementation of effective ocean governance.

• Northern Right Whale Conservation Programs at Stellwagen Bank NMS: This is an example of looking at the bigger picture and working, in this case, well outside the designated boundary, The Stellwagen Bank NMS staff is involved in a multitude of projects and regional efforts to recover this critically endangered whale species, a seasonal visitor to the Sanctuary. The Sanctuary Manager serves as the vice-chair of the New England Whale Recovery Plan Implementation Team, and is the Chair of the sub-committee responsible for the reducing serious injury or mortality to right whales from collisions with ships, the greatest single threat to the survival of this species. The Sanctuary played a leadership role in developing and implementing an action plan to address this problem, and has had significant success in meeting the goals of that plan, including helping to bring to life an Early Warning System (warning passing ships to the presence of right whales), and significant progress on possible regulatory measures to allow more effective management of shipping activity in high-use areas. Sanctuary staff was also involved in an effort to develop an East-coast-wide plan to reduce entanglements of whales in fishing gear, and was a driving force behind the development of an offshore rapid-response disentanglement team, bringing together trained personnel from the Center for Coastal Studies and the Coast Guard to undertake this effort. In cooperation with the Grays Reef NMS, Sanctuary staff co-authored a Right Whale Curriculum Guide, and have nearly completed a film on right whales for general distribution. Sanctuary staff also were instrumental in developing a collaboration with the Internet-based education program "Journey North", charting the northward migration of right whales with over 2000 schools across the country.

While focused on a single species, this initiative represents an effort to establish working relationships with other management authorities, to promote ecosystem management principles in the decisionmaking that is a part of the effort to rebuild right whale stocks, and to work toward the development of a governance context within which areas like Stellwagen Bank can focus attention on better understanding, and thereby better protecting, essential habitats. Many times, high-profile species like the right whale (in its role as "charismatic megafauna") can be the vehicle for more comprehensive collaboration and coordination.

• Monterey NMS Integrated Water Quality Management Program: In order to establish a more coordinated program of water quality protection, the Sanctuary has been working collaboratively with the State of California, the US Environmental Protection Agency, and other state, regional, and local agencies with water quality protection responsibilities to devise a collaborative comprehensive program for water quality protection in the Sanctuary. Now in its fifth year, the collaborative management program is running successfully, and the Sanctuary has now expanded this effort to deal with another specific water quality problem, non-point source runoff from agricultural activities.

Ocean governance, and ecosystem management operate most effectively when there is an atmosphere of trust among the participants. Projects like this help to forge the working relationships essential to effective collaboration. Successes have many proud parents, and successful collaborations tend to pave the way for subsequent projects.

• South Florida Ecosystem Project: The Florida Keys NMS is a major participant in the multi-agency effort to deal with the severe water quality and quantity issues in South Florida, the Everglades, Florida Bay and the Florida Keys. One of the largest ecosystem-based management programs currently in operation, this initiative is critical to the overall health of the Keys, particularly the northern part of the reef around Key Largo, and expands the geographic area of concern of FKNMS Staff to the entire South Florida ecosystem from Lake Okeechobee to the most southern reaches of the Keys System.

There are some significant lessons being learned here in this ecosystem-based management initiative. In terms of ocean management, it is helping us to better understand, and clearly articulate, the linkage between onshore activities and offshore impacts. The partnership-building and development of working relationships among agencies will help to sustain and grow the collaborative environment in the region.

While there are many other examples of how the Sanctuaries are stepping back to view the "big picture" and taking partners along with them to help them adopt the same viewpoint, these give some sense of the kinds of initiatives currently ongoing in the Sanctuaries.

Toward the Development of a New Ocean Management Paradigm

There are lots of folks talking about ocean management these days.

There are the academic thinkers, who are spending time sitting around the table talking and listening. This is a good thing. Somebody has to think the deep thoughts.

There are the coastal states, who have worked very hard to develop good ocean management plans for their offshore waters, but generally have fallen short implementing those plans, not because of a lack of follow-through or good intentions but because of a lack of resources to support implementation. It is a good thing they are out there on the front lines slugging away, but many have walked away frustrated (and that's not good).

There are the fishery managers, who face declining fish stocks, a fishing industry scrambling to stay afloat, and tremendous pressures from all quarters to effectively address these problems. Historically, fisheries management has been done independently of most other resource protection efforts. However, fisheries managers need to be onboard and actively participating in any collaborative ocean governance mechanism that might be developed if it is to be successful. Experience is clear on this issue...you cannot effectively manage marine ecosystems if fish are dealt with separately.

There are the ocean users...the fishermen, shipping industry, and others...who are worrying about what they can do to stay in business. This is not good, because these folks are needed to help develop effective solutions to the problems we are wrestling with. Hopeless, frustrated people rarely become part of the solution. The inclusive nature of ecosystem-based ocean governance, as implemented in the national marine sanctuaries, may provide one model for effectively incorporating these user groups into the ocean policy development process.

There is the public, who, when given the choice, would support effective ocean governance but needs to be built into a constituency advocating ecosystem management through effective outreach and education programs.

Finally, there is NOAA's Ocean and Coastal Management Office, the home of the NMSP, who has been talking for a number of years about the need for better and more comprehensive ocean management, and has generally done a good job setting the stage for some initiative, like this one, that might happen along.

The strategy being proposed is two-fold. The first part of the initiative, as discussed above, is one involving the NMSP demonstrating leadership in ocean management through implementation of strategic plan goals at the sites. The NMSP will be seeking opportunities to continue to foster the development of ocean management initiatives that are consistent with an ocean management paradigm based upon:

- Collaboration: When most folks look at the ocean, they see water. Federal and state agency representatives may also see water, but it is covered in a rich and colorful tapestry of jurisdictions and agency mandates, each in splendid isolation. A successful ocean governance mechanism will be one that is as interconnected as marine ecosystems are. There must be buy-in at a decisionmaking level, a mechanism that respects the power structure underlying the tapestry and uses it to attract rather than repel, and one that respects the sovereignty of governments. We don't have to try to create new authorities, just better utilize the existing ones.

- Flexibility: It is critical that any mechanism be flexible, as what is being managed is itself very variable in space and time. Elemental changes must be made in the way business is done by managers to empower adaptive management. To use this adaptive process, we must also agree that failure is just another unexpected outcome that will help direct subsequent actions. This will not be easy, but it is clearly necessary given the level of understanding about the ocean systems we are attempting to manage.

- Inclusion: Stakeholders, user groups, and the public must be substantively engaged in this process. We also need to recognize that an important part of this engagement is effective communication with these groups (Weeks and Packard 1997). This participation is necessary not only to have the benefit of their knowledge of the activity in which their involved (to avoid the too often encountered rule written by someone who obviously knew little about how the thing being managed is done), and their support as a constituency, but because their buy-in will help to reduce the resources necessary for enforcement to insure compliance. We have much to learn about how to do this effectively.

• Ecosystem Management Principles: We absolutely must look for the big picture when managing such systems. Focusing on maintenance of biodiversity and sustainable use must be done at appropriate scales of space and time. We need to be able to discern the difference between inter-annual variability (noise) and actual effects of human activities. This is a type of knowledge that can only be acquired by rigorous long-term ecosystem monitoring programs.

• Precautionary Approaches: We should consider doing many innovative things, but only actually do the what we are sure will not cause more harm than potential good. Existing activities must also meet this same standard. Taking a phrase out of the Hippocratic Oath, we should "first, do no harm...)

• "Best Science": Science is an extraordinary tool, and while we never have enough money to support what need to be done, if we pool resources effectively, we may be able to get considerably more done for less. Scientists need to understand what managers need, and managers need to be able to articulate that need clearly.

• Integrity: In order for any collaborative management system to be effective, the participants must trust one another. This puts the responsibility on all involved to communicate honestly and follow-though on commitments. This is not a trivial issue...in fact, it is probably the most critical factor in the success or failure of any collaborative program.

Many of these efforts are underway or being initiated. There is no timeline for implementation, as the success or failure of this initiative will be, in large measure, dependent on how successful we are at catalyzing these initiatives, and how willing we find partner agencies to be to come to the table with us.

The second part of this initiative is a longer term, more speculative proposal involving the modification of the structure of OCRM to create an "Ocean Programs" division within the office. While the "Ocean and Coastal Resource Management Office" has a Coastal Programs Division, which administers the national Coastal Zone Management Program, it does not have a sister division to administer ocean programs like the National Marine Sanctuary Program. The mission of this Division would be to coordinate ocean programs, like NMSP, provide technical assistance to agencies and others who are addressing ocean management issues, and would otherwise seek out opportunities to develop programs supporting ocean management initiatives.

To avoid creating too much of our own jurisdictional isolation, we would also propose that a "ombudsman" team be created within the OCRM Director's Office to seek out opportunities for CPD and OPD to work collaboratively, exchanging information between CPD and OPD staff regarding ongoing programs and opportunities. With such a structure, the visibility of NOAA Ocean Programs would be elevated, and it would establish a dedicated staff at OCRM to address ocean issues more directly, and hopefully, more effectively.

Conclusion

The National Marine Sanctuary Program has tremendous potential for being the driving force behind better and more effective ocean governance. In addition to the extensive experience we have with ocean management, we also have the one thing that usually leads to success, a vested interest in seeing something happen. The context provided by comprehensive ocean governance will allow sanctuary managers to focus attention on the protection of those special places, and release us from this continuing quest for context.

The NMSP is not the appropriate venue for a nationwide program of ocean governance, as the focus of this program is the protection and conservation of our most significant, threatened, and unique areas of the coastal oceans of the US. This is an essential job, and we hope to do it even more effectively in the future as we continue to look for new and better ways to provide comprehensive and coordinated conservation and management for these special marine areas.

However, an equally necessary task is to provide the same sort of comprehensive and coordinated conservation and management for the other areas of the marine environment not designated as national marine sanctuaries. To achieve this most efficiently, we need to build upon what we have learned over the past 25 years instead going back to the beginning and starting over.

The Sanctuary Program has been working on models of ecosystem-based ocean governance mechanisms for a quarter of a century, and has learned quite a bit along the way about what works and what doesn't. The NMSP has also established strong and effective partnerships with many of the key agencies and organizations that will be critical players in the ocean governance issue, bringing that momentum to the table as well. We continue to learn from our successes and failures, and constantly look for ways of doing the job just a little better.

The NMSP hopes to make a significant contribution to the effort to develop a new paradigm for ecosystem-based ocean governance, to help to apply that paradigm in the formulation of a national vision for ocean governance, and continue to actively seek out opportunities to use the sanctuaries as field laboratories for innovative and alternative ecosystem management and ocean governance initiatives.

References

GAO 1994. Ecosystem Management: Additional actions needed to adequately test a promising approach. US General Accounting Office Report GAO/RCED-94-111, 87pp.

Grumbine, R.E. 1994. What is Ecosystem Management. *Conservation Biology* 8(1): 27-38.

Grumbine, R.E. 1997. Reflections on "What is Ecosystem Management?" *Conservation Biology* 11(1): 41-47.

Kaufmann, M.R., R.T. Graham, D.A. Boyce, Jr., W. H. Moir, L. Perry, R.T. Reynolds, R.L. Bassett, P. Mehlhop, C.B. Edminster, W.M. Block and P.S. Corn 1994. An Ecological Basis for Ecosystem Management. USDA Forest Service General Technical Report RM-246, 22 pp.

Keystone Center 1996. The Keystone National Policy Dialogue on Ecosystem Management. The Keystone Center, Keystone, Colorado, 43 pp.

CHAPTER 11

The Status of California Marine Protected Areas

Deborah A. McArdle

Abstract

In California there are 104 Marine Protected Areas (MPAs) that have been historically designated on a case by case basis causing information about them to be confusing and complex. This process has resulted in many MPAs being managed in isolation and not as a component of the statewide MPA system of which they are a part. MPAs in California are also listed under numerous classifications ranging from small highly protected reserves to larger multiple use sanctuaries. The objective of this project was to create a database that will provide researchers, resource managers, and resource users with comprehensive and centralized information on California MPAs. The database includes information on MPA classification type, county, geographic region, date established, responsible agencies, fishing-related regulations, official regulations and official boundaries. A Publication entitled, "California Marine Protected Areas" has also been published and includes the aforementioned information as well as a series of maps showing each MPA location, area and an MPA contact list. This database has been compiled to assist in the improvement of the methods of MPA site selection and to help to begin to integrate existing MPAs into a statewide system.

Introduction

Alternative management strategies are being considered by resource managers throughout the world. One alternative management strategy that is being considered is the use of Marine Protected Areas (MPAs). As opposed to or as a complement to traditional single species management strategies, MPAs (e.g., sanctuaries, parks and reserves) could represent a way of managing marine resources at an ecosystem level. While the establishment of well-designed new MPAs could potentially benefit marine resources, in many areas there are also numerous existing MPAs that could be improved and further studied.

MPAs were often initially developed as small extensions of terrestrial protected areas (e.g., California Park System underwater parks), but over time they have become designated worldwide in their own right. A 1994 estimate, by Kelleher (as stated in Gubbay, 1995) listed 1111 MPAs worldwide. Areas that have high numbers of MPAs include, but are not limited to, Australia which has ~300 MPAs, the United States which has at least 300 MPAs and the Caribbean which has ~135 MPAs. Just the Pacific Coast of the United States has many MPAs including

---

UC Sea Grant Cooperative Extension, 105 E Anapamu, #5, Santa Barbara, CA 93101

~100 in Washington (Mills, 1996) and 104 in California (McArdle, 1997).

Though not without controversy, MPAs may potentially offer many benefits including habitat protection, maintenance of genetic diversity, undisturbed research baseline areas, and locations for educational and recreational activities. Although the concept of putting a financial figure on the marine environment is controversial, MPAs may also offer financial benefits through ecotourism (Dixon and Sherman, 1990). For example, in 1988, Caribbean marine-related tourism income was estimated at $8 billion (Bloomstein, 1985). MPAs may also serve as sources of fisheries replenishment for surrounding harvested zones. However, the actual benefits of MPAs to fisheries are still being studied (Dugan, 1992). MPAs have the potential to increase fishery catches in two ways, through emigration of large fish across the border of a reserve and into local fishing areas (spillover) and through the export of larvae from a reserve, which may enhance recruitment into regional fishery stocks (Rowley, 1994). Many organizations and individuals (e.g., Man and Biosphere Program) are also actively studying how to incorporate resource use into the establishment and management of MPAs (Price, 1993).

In California the MPAs have primarily been established on a case-by-case basis. This is a result of many MPAs being initiated by "crisis" (some level of threat to a site that is considered important) rather than through a sound site evaluation process (Salm and Price, 1995). The MPA designation proposal is often initiated by a political action rather than scientific or management recommendation. As a result MPAs are scattered throughout the state (McArdle, 1997) and often managed in isolation. This situation is certainly not unique to California, but mirrors the status of many MPAs worldwide. MPAs are selected at a local level and not integrated into the statewide MPA system of which they are a component. Sites protected as a result of these initiatives have high local significance, but may contribute relatively little to the state conservation objectives. These areas may enjoy local support, but the danger is that these areas may be considered as substitutes for more valuable sites that really do contribute to state objectives for conservation and development (Salm and Price, 1995). The establishment of numerous ineffective MPAs may actually be counteractive to the establishment of useful MPA sites in the future.

As more MPAs are being considered (currently two additional MPAs are being proposed in California), it has become increasingly important to determine how to more effectively design and place them. It has also become more important to take a holistic look at the existing process of site selection and management and evaluate how it contributes to the overall MPA program. Although MPAs have varying objectives, the examination of the role of existing MPAs in more detail could help to guide future programs of site selection and management (Gubbay, 1995). The first step in this process is to collect data on all of the existing MPA sites (Salm and Clark, 1994).

Methods

MPA Definition

The commonly used definition for Marine Protected Areas, developed at the IVth World Wilderness Congress and adopted by the World Conservation Union

(IUCN) at its 17th General Assembly in 1988 is, "Any area of intertidal or subtidal terrain, together with its overlying water and associated flora, fauna, historical and cultural features, which has been reserved by law or other effective means to protect part or all of the enclosed environment." For this study, the MPA definition has been altered to include only protected areas with coastal subtidal components. This study does not include protected areas that contain only estuarine components or only intertidal components.

Objectives

There were three major objectives to this study over a two year period.

- The first objective was to compile the legal site description data including the year the MPA was established, the agency that established the site, the agencies responsible for the MPA site's management and enforcement, and a contact representative for each site.
- The second objective of the study was to locate and interpret all of the fishing-related regulations that applied to each MPA designation.
- The third objective was to compile the legal boundaries of each MPA in order to create a spatial database and a Geographic Information System (GIS).

Site Description Data

The information was collected by searching through legal public, state and federal code, locating and reviewing available reference material, and by personally contacting numerous individuals, public agencies, and non-governmental organizations responsible for or knowledgeable about the MPA sites. While compiling this data we realized that there was no document that explained the formation process of the particular MPA designations, so this information was also gathered and summarized (McArdle, 1997).

Fishing-Related Regulations

The information was compiled, in part, because of the numerous requests by MPA managers, fishermen and enforcement personnel that could not understand the complex matrix of fishing regulations that accompany each MPA. The data was also collected so that a "level of protection" for each site could be determined by future analysis. The information was collected by searching through legal public, state and federal code, locating and reviewing available reference material, and by personally contacting numerous individuals, public agencies, and non-governmental organizations responsible for or knowledgeable about the MPA sites. The regulations that were compiled are only those given to the site because of its MPA designation. For example, general fishing regulations for specific species, gear types, or areas were not included as they could have easily taken an additional several years to compile and interpret. Once obtained, we conducted extensive personal conversations with MPA managers to be certain that the legal regulations were interpreted correctly. In many cases management and/or enforcement officials (because of the complexity of the system) were unable to interpret the legal regulatory information. In addition, in large agencies, where we consulted individuals at the local, regional, and state levels, we often received conflicting opinions. In such cases, the information received from the "highest" management level was used. The data was entered onto FileMaker Pro. which enables the data to be sorted by any category of data.

Spatial Database and Geographic Information System (GIS) Database

The MPA legal boundary information was collected by searching through legal public, state and federal code, locating and reviewing available reference material, locating existing GIS databases, and by personally contacting numerous individuals, public agencies, and non-governmental organizations responsible for or knowledgeable about the MPA sites. The GIS created that depicts MPA boundaries was made from the most accurate information available. The emphasis here is on "available." The inconsistency in the availability of spatial data related to existing MPAs leads us to believe that some original information pertaining to MPA boundaries or goals may be lost or forgotten. We had to locate much of the data ourselves, as many of the individuals or agencies responsible for governing or managing the MPAs either did not have the boundary information we required, including accurate base maps of the MPA boundaries. The data compiled was entered into ArcInfo, by Gavin McGhie and Jeff Hempill at the Department of Geography, Remote Sensing Research Unit, UCSB to create the statewide MPA boundary GIS. From the GIS, map layouts for each MPA were made in ArcView. The layouts were then copied into Canvas and labeled with attributes (e.g., points of interest, major waterways, navigational points, roadways, etc.). Attribute information was obtained from USGS maps, NOAA charts, reference books and legal descriptions of the MPA boundaries. Latitude lines were determined in ArcInfo and placed on the layouts at the MPA northern and southern boundaries (to the nearest minute). The layouts were also labeled with overlapping MPA boundaries.

Determination of Overlapping MPA Boundaries

When all of the information was collected, it was analyzed to determine which MPAs had overlapping boundaries with other MPAs. This information was needed particularly to determine the fishing-related regulations for each site and to label the MPA site layouts. This step was particularly time consuming as many of the overlapping MPA designation regulations conflicted with each other. For this study, each legally designated MPA has been counted as one MPA. For example, Point Lobos had three MPA designations that have the exact same boundaries. It has been designated as an Area of Special Biological Significance, Ecological Reserve and Reserve. For an MPA designation to be counted (e.g., ecological reserve) it had to have a legal mission or goal (e.g., 14 California Code of Regulations 630.0) and a set of legal rules and/or regulations (e.g., 14 California Code of Regulations 630.0) that accompanied the designation. For Point Lobos, three MPAs were counted and are a part of the total of 104 MPAs. In future spatial analysis, the evaluation of overlapping boundaries will need to be tabulated differently.

Results

California has 104 MPAs with twelve designation types (Table 1). State MPAs have been established by the California Water Resources Control Board (Areas of Special Biological Significance), California Fish and Game Commission (Clam Refuges, Ecological Reserves, MRPA Ecological Reserves and Reserves), California State Legislature (Coastal Sanctuaries, and Fish, Game, and Marine Life Refuges), California Department of Parks and Recreation (Parks) and U.C. Regents (U.C. Natural Reserves). Federal MPAs have been designated by the

National Park Service (National Parks), National Oceanic and Atmospheric Association (National Marine Sanctuaries) and by Executive Order (Ecological Preserve). The International UNESCO Man and Biosphere Program has designated two Biosphere Reserves.

Table 1. California MPA Designation Categories

I. State Designations
1. Areas of Special Biological Significance (34)
2. Coastal Sanctuaries (1)
3. Ecological Reserves (10)
4. Marine Resources Protection Act (MRPA) Ecological Reserves (4)
5. Reserves (7)
6. Parks (beaches, historic areas, natural preserves, reserves, and underwater parks) (14) * 1 City Underwater Park included
7. Refuges (clam, fish, game, and marine life) (20)
8. U.C. Natural Reserves (1)
II. Federal Designations
1. Ecological Preserves (1)
2. National Marine Sanctuaries (4)
3. National Parks (national park, national seashore, national recreation area, national research natural area, national monument)(6)
III. International Designations
1. Biosphere Reserves (2)

Most California MPAs have been designated on a case-by-case basis, as a result of which they range greatly in size and in level of protection. The smallest MPA, is the six-acre Gerstle Cove Reserve in Sonoma county. At the opposite extreme are the largest MPAs, the million-acre Central California Coast Biosphere Reserve which extends from the coasts of Sonoma, Marin, San Francisco and San Mateo counties and the Channel Islands Man and Biosphere Reserve and National Marine Sanctuary in Santa Barbara and Ventura counties.

To determine geographical distribution of the MPAs, we placed the 104 MPAs into five Geographical Region categories created by the California Coastal Commission (1996) (Table 2). The majority of MPAs are inshore and adjacent to the coastline. The South Region (Santa Barbara, Ventura, Los Angeles, Orange, and San Diego counties) has the most MPAs, with forty-four percent. Twenty six percent of the MPAs are located in the North Central Region (Somona, Marin, San Francisco, and San Mateo Counties). Seventeen percent of the MPAs have been established in the Central Region (Santa Cruz, Monterey, and San Luis Obispo Counties) and twelve percent of the MPAs have been designated in the North Region (Del Norte, Humboldt, and Mendocino counties).

The level of protection varies among MPA sites. Some are dedicated to the protection of all species within their boundaries (e.g., the four Marine Resources Protection Act Ecological Reserves); others, such as the four National Marine Sanctuaries, allow multiple uses. Most of the California MPAs provide an intermediate level of protection and restrict only specific species from being taken. In ninety three MPAs, some level of recreational fishing is permitted and in ninety

Table 2. California MPA Distribution (North to South)

I. North Region (13 MPAs)
1. Redwood National Park ASBS
2. Redwood National Park
3. Kelp Beds at Trinidad Head ASBS,
4. King Range (Punta Gorda) Marine Resources Protection Act Ecological Reserve
5. Kings Range National Conservation Area ASBS
6. MacKerricher State Park (Underwater Park)
7. Pygmy Forest Ecological Staircase ASBS
8. Point Cabrillo Reserve
9. Russian Gulch State Park (Underwater Park)
10. Van Damme State Park (Underwater Park)
11. Manchester State Park (Underwater Park)
12. Arena Rock Marine Natural Preserve
13. Kelp Beds at Saunders Reef ASBS

II. North Central Region (27 MPAs)
14. Del Mar Landing Ecological Reserve ASBS
15. Del Mar Landing Ecological Reserve
16. Salt Point State Park (Underwater Park)
17. Gerstle Cove ASBS
18. Gerstle Cove Reserve
19. Fort Ross State Historic Park (Underwater Park)
20. Sonoma Coast State Beach (Underwater Park)
21. Bodega Marine Life Refuge ASBS
22. Bodega Marine Life Refuge
23. Cordell Bank National Marine Sanctuary
24. Central California Coast Man and Biosphere Reserve
25. Gulf of Farallones National Marine Sanctuary
26. Point Reyes National Seashore
27. Bird Rock ASBS
28. Point Reyes Headland Reserve and Extension ASBS,
29. Point Reyes Headlands Reserve,
30. Point Reyes Headlands National Research Natural Area
31. Double Point ASBS
32. Duxbury Reef Reserve and Extension ASBS
33.Duxbury Reef Reserve
34. Farallon Island ASBS
35. Farallon Islands Game Refuge
36. Monterey Bay National Marine Sanctuary
37. Golden Gate National Recreation Area
38. James V. Fitzgerald Marine Reserve ASBS
39. James V. Fitzgerald Marine Reserve (Marine Life Refuge)
40. Ano Nuevo Point and Island ASBS

III. Central Region (18)
41. Pacific Grove Marine Gardens Fish Refuge and Hopkins Marine Life Refuge ASBS
42. Hopkins Marine Life Refuge
43. Pacific Grove Marine Gardens Fish Refuge
44. Carmel Bay ASBS
45. Carmel Bay Ecological Reserve
46. California Sea Otter Game Refuge
47. Point Lobos Ecological Reserve ASBS
48. Point Lobos Ecological Reserve
49. Point Lobos Reserve
50. Julia Pfeiffer Burns Underwater Park ASBS
51. Julia Pfeiffer Burns State Park (Underwater Park)
52. Big Creek Marine Resources Protection Act Ecological Reserve
53. Ocean Area Surrounding the Mouth of Salmon Creek ASBS
54. Atascadero Beach Pismo Clam Preserve (Clam Refuge)
55. Morro Beach Pismo Clam Preserve (Clam Refuge)
56. Pismo Invertebrate Reserve
57. Pismo-Oceano Beach Pismo Clam Preserve (Clam Refuge)

III. South Region (46)
58. Vandenberg Marine Resources Protection Act Ecological Reserve
59. Channel Islands Biosphere Reserve
60. Channel Islands National Marine Sanctuary
61. Channel Islands National Park
62. Santa Barbara Channel Ecological Preserve
63. San Miguel Island Ecological Reserve
64. San Miguel, Santa Rosa, and Santa Cruz Islands ASBS
65. Anacapa Island Ecological Reserve, Ventura County
66. Santa Barbara Island Ecological Reserve
67. Santa Barbara Island and Anacapa Island ASBS
68. San Nicolas Island and Begg Rock ASBS
69. Big Sycamore Canyon Marine Resources Protection Act Ecological Reserve
70. Mugu Lagoon to Latigo Point ASBS
71. Abalone Cove Ecological Reserve
72. Point Fermin Marine Lite Refuge
73. Santa Catalina Island- Subarea One ASBS
74. Catalina Marine Science Center Marine Life Refuge
75. Santa Catalina Island- Subarea Two ASBS
76. Farnsworth Bank Ecological Reserve
77. Santa Catalina Island- Subarea Three ASBS
78. Lover's Cove Reserve, Santa Catalina Island
79. Santa Catalina Island- Subarea Four, ASBS
80. San Clemente Island ASBS
81. Newport Beach Marine Life Refuge
82. Newport Beach Marine Life Refuge ASBS
83. Crystal Cove State Park (Underwater Park)
84. Irvine Coast Marine Life Refuge
85. Irvine Coast Marine Life Refuge ASBS
86. Laguna Beach Marine Life Refuge
87. Heisler Park Ecological Reserve
88. Heisler Park Ecological Reserve ASBS
89. South Laguna Beach Marine Life Refuge
90. Niguel Marine Life Refuge
91. Dana Point Marine Life Refuge
92. Doheny State Beach (Underwater Park)
93. Doheny Beach Marine Life Refuge
94. City of Encinitas Marine Life Refuge
95. Cardiff and San Elijo State Beaches (Underwater Park)
96. San Diego La Jolla City Underwater Park
97. San Diego Marine Life Refuge
98. Scripps Coastal Reserve
99. San Diego Marine Life Refuge ASBS
100. San Diego-La Jolla Ecological Reserve
101. San Diego-La Jolla Ecological Reserve ASBS
102. Cabrillo National Monument
103. Point Loma Reserve

IV. All Coastal Regions
104. California Coastal Sanctuary

four of the MPAs, some level of commercial fishing is permitted. Nine MPAs currently restrict all species from being taken (Table 3). The nine "no-take" MPAs include ~6000 acres and 20 miles of California coastline. (It should be noted, however, that King Range and Big Sycamore Canyon MRPA Ecological Reserves do not actually include the coastline as they begin offshore at the 18' and 30' Isobath contour lines, respectively).

Table 3. California "No-Take" MPAs

1. King Range MRPA Ecological Reserve, Humboldt County
2. Hopkins Marine Life Refuge, Monterey County
3. Point Lobos Ecological Reserve, Monterey County
4. Big Creek MRPA Ecological Reserve, Monterey County
5. Vandenberg MRPA Ecological Reserve, Santa Barbara County
6. Catalina Marine Science Center Marine Life Refuge, Los Angeles County
7. Big Sycamore Canyon MRPA Ecological Reserve, Ventura County
8. Heisler Park Ecological Reserve, Los Angeles County
9. Scripps U.C. Natural Reserve, San Diego County

In many cases, two or more MPAs overlap and contain separate MPAs within their boundaries. Many of the MPA sites have more than one MPA designation, each with its own set of regulations. These overlapping regulations often conflict as shown in the example the Point Reyes Headlands in Sonoma County which has four overlapping MPA designations (Table 4, Figure 1).

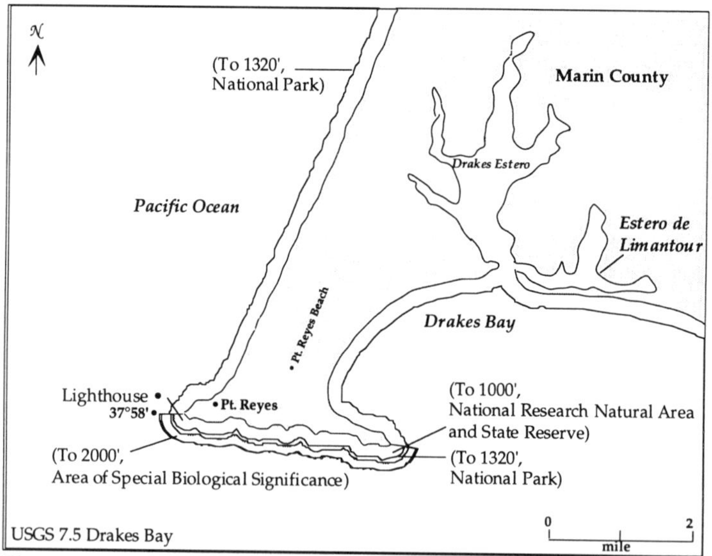

Figure 1. Point Reyes Headlands four MPA designations including Area of Special Biological Significance, State Reserve, National Park and National Research Natural Area.

Table 4. Point Reyes Headlands Summary of Designations and Regulations

a) 1962 Designated as a National Seashore, by the National Park Service. The offshore boundary goes to a distance of 1320'. Both commercial and recreational fishing are allowed, but limited by the designation.

[Recreational: To 1000' only the following invertebrates may be taken: abalones, chiones, clams, cockles, rock scallops, native oysters, crabs, lobsters, ghost shrimp, and sea urchins. Finfish may be taken (14 CCR 29.05)].

[Commercial: To 1000' only the following invertebrates may be taken: crabs, ghost shrimp, jackknife clams, sea urchins, squid, lobsters, abalones, and worms. Finfish may be taken (14 CCR 123 (d)].

a) 1972 Designated as a State Reserve, by the Fish and Game Commission. The offshore boundary goes to a distance of 1000'. Recreational fishing is prohibited by the designation and commercial fishing is allowed, but limited.

[Recreational: No form of marine life may be taken (14 CCR 27.30)].

[Commercial: To 1000' only the following invertebrates may be taken: crabs, ghost shrimp, jackknife clams, sea urchins, squid, lobsters, abalones, and worms. Finfish may be taken (14 CCR 123 (d)].

b) 1972 Designated as a National Research Natural Area, by the Point Reyes National Seashore. The offshore boundary goes to a distance of 1000'. Although management policy states that all consumptive activities shall be prohibited by this designation, there are no legal regulations to accompany the policy so therefore, both commercial and fishing is permitted by the designation.

d) 1974 Designated as an Area of Special Biological Significance, by the State Water Resources Control Board. The offshore boundary goes to a distance of 2000'. The mission of the ASBS designation is to protect the area from negative water quality conditions so therefore both commercial and fishing is permitted by the designation.

Discussion

The complexity of the 104 California MPA designations, regulations, overlapping boundaries and jurisdictions suggest that the site selection, designation process, and management of MPAs should be improved and integrated. Many California MPAs do not have clearly defined management goals and often even MPAs that initially had clearly defined goals saw political pressures cause those goals to be altered during the legislative designation process. Historically reserves established by the State Legislature and the Fish and Game Commission have been created independently and without guidelines to set either standard terminology in their description and use, or standards regarding restrictions on their various uses (Smith, 1989).

Selection of Sites

While it is preferable to choose MPAs through a system planning exercise, in the real world selection is often driven by crisis-some high level of threat to a site that is considered important for any reason (Salm and Price, 1995). This is certainly true in California where the establishment of MPAs has often been reactive rather than planned. For example, generally both legislative and Fish and Game reserves, until recently, have been established at the request of organizations outside the agency responsible for managing living marine resources, the California Department of Fish and Game (CDFG).

Selection of sites according to a well-laid plan that includes an overall and clearly understood goal and focused, practical criteria is preferable to any *ad hoc* means, such as crisis. Crisis are likely to arise at intervals, but managers need to get ahead of them if selection is to proceed along orderly and rational lines (Salm and Price 1995). The California MPA system could be improved by developing an integrated statewide plan that would include sound criteria for site selection. Methods on how to select MPA sites has been well reviewed in the literature (e.g., Ray 1976; Salm and Clark, 1984; Kelleher and Kenchington, 1992). A few questions that need to be furthered studied may be whether the existing systems of MPAs represent the diversity of ecosystems, habitats and species assemblages adequately; whether all of these individual MPAs, large or small, contribute to the maintenance of ecological processes and function at a regional and at larger scales; and whether proposed future MPAs are efficient and scientifically dependable (Brunkworst, 1995). An example of guidelines for establishing MPAs by focusing on biodiversity protection is in Norse (1993). A set of guidelines for establishing a representative system of MPAs is given by Ray and McCormick-Ray (1992).

Management of Sites

Well developed and practical management plans are critical to MPA success. Without them, a great deal of effort may be put into the designation of a MPA, with limited commitment to the long-term management of the area. Lack of finance is usually a major factor preventing their preparation, as funding for consultation and the designation of sites is often simpler to obtain than funding for day-to-day management and research (Gubbay, 1995). Most of the MPAs of California do not legally come with appropriations for management or research. The management of living resources in state MPA designated sites is the responsibility of the CDFG, who has limited resources. As a result, many of these sites do not have management plans or site managers at all. Some state and federal designations such as the National Marine Sanctuaries and National Parks often have managers and appropriations, but they are not allowed to use any federal laws to manage most commercial activities (including fishing) because living marine resources fall under the jurisdiction of the CDFG. As a result, federal agencies often cannot create or enforce the regulations that would support their management plans. This is also a problem with the State Parks established by the Department of Parks and Recreation. The management of existing California MPAs needs to be improved and the management of small marine protected areas needs to be integrated into the larger region of which they are a component. A compilation of information on developing management plans for MPAs can be found in Kenchington and Kelleher (1995).

Multi-Agency Cooperation

All steps that need to be taken to improve the California MPA system will need multi-agency cooperation because of the extreme complexity of the system. Agencies legally involved in the current designation and management of MPAs includes, but is not limited to, California Department of Fish and Game, California Coastal Commission, California State Lands Commission, California Minerals Management Service, California Water Resources Control Board, California Department of Parks and Recreation, California Fish and Game Commission, California State Legislature, University of California Regents, U.S. Fish and Wildlife Service, National Oceanic and Atmospheric Association, National Park Service, and the United Nations Environmental, Scientific, and Cultural Organization. In addition, the research community, numerous non-profit organizations and the general public are also often involved in the process of initiating, implementing, and managing MPAs. The Monterey Bay National Marine Sanctuary, in California, is an example of an MPA that has tried extensively to involve all agencies in all phases of the sanctuary development.

Conclusions

This database, represents our best effort to compile and organize available information on California's complex system of MPAs (McArdle, 1997). Although this database will not simplify the complexity of the system, we hope it will eliminate some confusion about California MPAs. It thus may provide a foundation upon which to build a more rational, manageable MPA system; one that is more successful in achieving the objective of better protecting marine resources. Finally, we hope that this database will help all interested individuals to better understand the system as it presently exists and to begin to develop one that is more integrated and more effective in conserving our valuable ocean resources.

Acknowledgments

The development of the California Marine Protected Area database was supported by a grant from the David and Lucile Packard Foundation and by the National Sea Grant College Program, National Oceanic and Atmospheric Administration, U.S. Department of Commerce under grant number NA36RG0537, Project number A/P-1. The spatial database could not have been completed without the creation of the GIS, created by Jeff Hempill and Gavin R. McGhie at the UCSB Department of Geography, Remote Sensing Research Unit, under the guidance of Joseph Scepan and Dr. John E. Estes. Thanks go to Jonathan Van Coops and Allyson Hitt at the California Coastal Commission for the creation of the MPA Regional maps. Information and interpretation of fishing-related regulations were provided, in part, by Kristine Barsky, Robson Collins, Greg Laret, Phil Nelms and Don Schultze of the California Department of Fish and Game. The original development of funding for the creation of the database was significantly supported by Brian Baird of the California Resources Agency. I would also like to thank Rosemary Amidei, Patricia Demmina, Rebecca Ditmore, Kathey Mooney and Richard M. Starr for their critical input. For more information contact the author at (805) 882-1889; damcardle@ucdavis.edu

References

Bloomstein, E. 1985. Tourism and the Environment: An Overview of the Eastern Caribbean. Port of Spain, Trinidad and Tobago: Economic Commission for Latin America and the Caribbean.

Brunkhorst, D.J. and Bridgewater, P.B. 1995. Marine Bioregional Planning: A Strategic Framework for Identifying Marine Reserve Networks, and Planning Sustainable Use and Management. Marine Protected Areas and Sustainable Use. (eds) N.L. Shackell and J.H.M. Willison. Science and Management of Protected Areas Association. ISBN 0-9699338-0-0. 300 p.

California Department of Fish and Game. 1996. Fish and Game Code of California. Gould Publications, Inc., Florida. 712 p.

California Department of Fish and Game. 1994. California Code of Regulations. CA Office of State Printing. 395 p.

Dixon, J.A. and Sherman, P.B. 1990. Economics of Protected Areas: approaches and applications. Island Press, Washington, DC., 234 p.

Dugan, J.E. and Davis, G.E. 1992. Applications of marine refugia to coastal fisheries management. Canadian Journal of Fisheries and Aquatic Science. Vol. 50, pp. 2029-2042.

Gubbay, S. 1995. Marine Protected Areas. Chapman & Hall, 2-6 Boundary Row, London SE18HN. pp. 1-14.

Kelleher, G. and Kenchington, R. 1992. Guidelines for Establishing Marine Protected Areas. A Marine Conservation and Development Report. IUCN, Gland, Switzerland, 79 p.

McArdle, D.A. 1997. California Marine Protected Areas. California Sea Grant Publication, San Diego, CA. Publication # T-039, ISBN 1-888-691-03-4. 268 p.

Mills, M.L. 1996. Washington Department of Fish and Wildlife. Personal Communication.

Norse. E. (ed) 1993. Global Marine Biodiversity. A Strategy for Building Conservation into Decision Making. Center for Marine Conservation, Washington DC., 383 p.

Price, A. and Humphrey, S. (eds). 1993. Application of the Biosphere Reserve Concept to Coastal Marine Areas. Papers presented at the UNESCO/IUCN San Francisco Workshop of 14-20 August, 1989. A Marine Conservation and Development Report. IUCN, Gland, Switzerland, 114 p.

Ray, G.C. 1976. Critical Marine Habitats, in Proceedings of an International Conference on Marine Parks and Reserves, IUCN Publication (n.s.) No. 37, Gland, Switzerland, pp. 15-64.

Ray, G.C. and McCormick-Ray, M. 1992. Critical Habitats and Representative Systems in Marine Environments: Concepts and Procedures. Paper for Workshop II.11. The Science of Conservation in the Coastal Zone. IV World Congress on National Parks and Protected Areas, Caracas, Venezuela.

Rowley, R.J. 1994. Case Studies and Reviews, Marine reserves in fisheries management. Marine and Freshwater Ecosystems, Vol. 4, pp. 233-254. Salm, R. and Clark, J.R. 1984. Marine and Coastal Protected Areas: A Guide For Planners and Managers. IUCN, Gland, Switzerland, 302 pp. Salm, R. and Price, A. 1995. Selection of Marine Protected Areas. In: Marine Protected Areas., ed. S. Gubbay, Chapman & Hall, 2-6 Boundary Row, London SE18HN. pp. 15-31.

Smith Jr., E. J. and Johnson, T. H. 1989. The Marine Life Refuges & Reserves of California. California Department of Fish and Game Marine Resources Information Bull. No. 1. 63 p.

CHAPTER 12

California Underwater Parks and Reserves Planning and Management

W. James Barry[1] and John W. Foster[2]

Abstract

Gaps in the underwater component of the California State Park System are evident in several regions of the State. To assess deficiencies in biological diversity we have divided the state into "Ecological Regions". Ecological regions are the basis for natural heritage planning of the State Park System (Sec. 5019.53 of the California Public Resources Code). The marine ecological regions are the focus of this plan; however terrestrial ecological regions are utilized for non-marine underwater analysis.

Offshore cultural heritage features include both prehistoric sites and artifacts as well as historical features such as shipwrecks. Many coastal archeological sites were inundated as the sea level rose after the last ice age. However very few of these sites have been identified due to the difficulties in relocation.

Recreational features often relate to areas of high natural diversity as these areas tend to be most attractive. Shipwreck sites are also popular with divers. Safe diving access is an issue to be addressed in this plan; facility development for skin and SCUBA divers is also important to the diving public. Interpretation of cultural and natural features is important to achieving the missions of the Department. Interpretation is an important product of research; as it informs the public of their California Heritage.

---

[1]Senior State Park Ecologist, California Department of Parks and Recreation, P.O. Box 942896, Sacramento, CA 94296-0001

[2]Senior State Archeologist, California Department of Parks and Recreation, P.O. Box 942896, Sacramento, CA 94296-0001

Introduction

Point Lobos State Reserve became the first marine protected area in the Nation on July 1, 1960. Seven hundred and fifty acres of submerged lands were annexed to the terrestrial portion of the reserve. John Pennecamp Coral Reef State Park, in the Florida Keys, was established in 1962. Later Buck Reef National Monument was established in the Virgin Islands.

California Department of Parks and Recreation (DPR) Underwater Parks Program was established by Director William Penn Mott, Jr. in 1968. He appointed a scientific and industry represented panel - the California Advisory Board for Underwater Parks and Reserves. The Board's mission was to identify outstanding and representative examples of underwater ecosystems (both marine and inland waters) and to recommend management methods for both preservation and recreational use.

The Board has investigated numerous sites and has made recommendations for inclusion into the State Park System (SPS) of suitable areas. Early additions include underwater extensions of Russian Gulch, MacKerricher, Van Damme, Salt Point, and Julia Pfeiffer Burns State Parks, Fort Ross State Historic Park, Sonoma Coast and Doheny State Beaches (1970). Currently 15 of the 264 units of the SPS include contiguous underwater portions totalling around 54,000 acres. Most are managed by DPR with 49 year leases from State Lands Commission. DPR has avoided stand - alone underwater park units due to difficulty in staffing and protection.

Gaps in the underwater component of the California State Park System are evident in several regions of the State. To access deficiencies in biological diversity we have divided the State into "Ecological Regions".

Offshore cultural heritage features include both prehistoric sites and artifacts as well as historical features such as shipwrecks. Many coastal archeological sites have been inundated as the sea level rose after the last ice age. However very few of these sites have been identified due to the difficulties in relocation.

Recreational features often relate to areas of high natural diversity, as these areas tend to be most attractive. Shipwreck sites are also popular with divers. Safe diving access is an issue to be addressed in this plan; facility development for skin and SCUBA divers is also important to the diving public. Interpretation of cultural and natural features is important to achieving the missions of the Department. Interpretation is an important product of research; as it informs the public of their California Heritage.

More than 25 marine additions and several freshwater additions to current units of the SPS are currently under investigation which are of high ecological and /or cultural importance. The SPS includes more than 250 miles of coastline with ten State Sea Shores.

Authority

The California Public Resources Code (PRC) authorizes DPR to establish State Park System Units in the underwater environs of the state. Specifically these include State Seashores (Sec. 5019.62), State Parks (Sec. 5019.53), State Reserves (Section 5019.65) and Underwater Recreation Areas (Sec. 5019.56(b). "The rules and regulations adapted by DPR also apply on any granted or ungranted tidelands or submerged lands abutting property of the department and used for recreational purposes of the general public in conjunction with their use of the department's property between the boundary of the lands under the jurisdiction of the department and a line running parallel to and 1,000 feet waterward of the ordinary high watermark, so long as the rule or regulation being applied is not inconsistent with any rule or regulation of any other public agency which is applicable to those tide or submerged lands" (Sec. 5003.05). "Commercial exploitation of resources in units of the state park system is prohibited." (Sec. 5001.65).

Natural Heritage Features

Ecological regions are the basis for natural heritage planning of the State Park System (Sec. 5019.53 of the California Public Resources Code). The marine ecological regions are the focus of this paper; however terrestrial ecological regions are utilized for non-marine underwater analysis. For more information on ecological regions refer to "Ecological Regions of California" (Barry, 1991) a document which was prepared for the Interagency Natural Areas Coordinating Council and to the 1:1 million scale map available the California Natural Diversity Data Base Geographic Information System. The four marine ecological regions are delineated in **Figure 1**. Unfortunately, the Channel Islands of the Southern-Baja Californian are not represented in the State Park System. Some of the worlds best diving sites are adjacent to the Channel Islands.

The establishment of underwater units in the State Park System requires petitioning the State Lands Commission for long term leases of tracts of submerged lands. These leases require some utilization of the land, which usually implies recreational use, i.e. SCUBA diving opportunities or interpretive use; the Commission has been somewhat reluctant to lease land strictly for preservation purposes. DPR has managed leased lands in partnership with the State Lands Commission; the public is informed of this partnership through interpretive programs. Fee title transfer has only occurred at Point Lobos State Reserve and legislatively at Mono Lake Tufa State Reserve. All other underwater areas are leased from State Lands Commission. These leases usually are for fifty years; if

DPR does not show active management by the end of the lease, it may not be renewed. Leases may allow State Lands Commission to utilize the resources of the submerged areas (i.e. offshore oil and mining) are still possible under the terms of some leases.

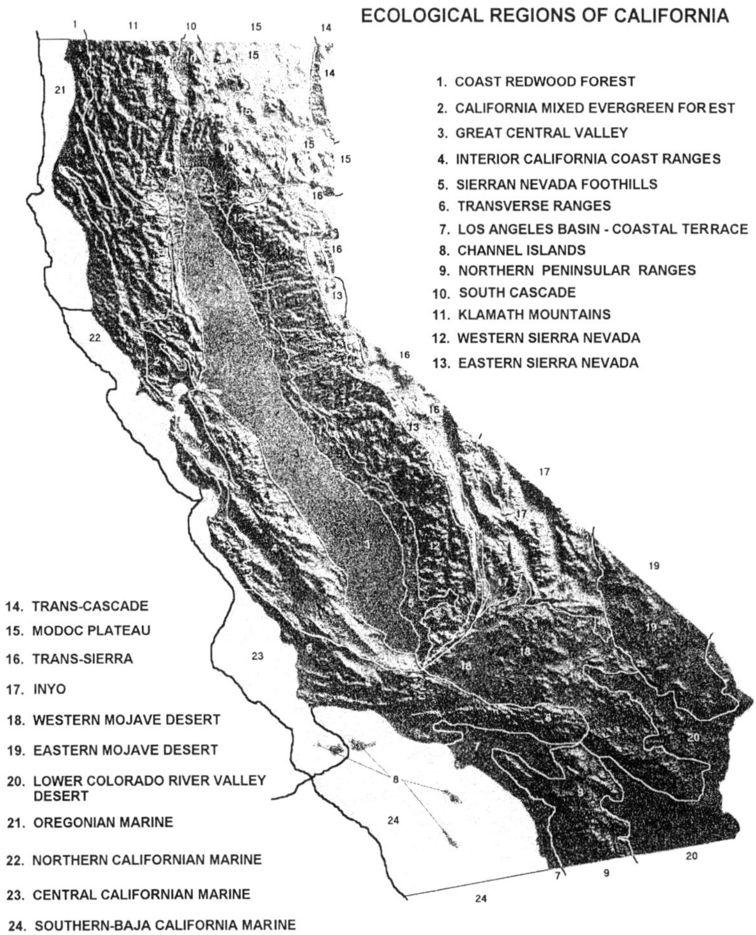

Figure 1. Terrestrial and marine ecological regions of California based on Barry, 1992 1:1,000,000 GIS map for the California Interagency Natural Areas Committee. Note that a major break occurs between the Central California Marine and the Southern-Baja California Marine Ecological Regions. The Boreal Marine Ecological Division is to the north and the Temperate Nearshore Marine Division is to the south.

The protection of natural features in marine environments is more complex than on land. The California Department of Fish and Game has statutory authority over living marine resources; therefore the Fish and Game Commission must approve any protection of living marine resources that DPR may propose. Regulations of take are specific for each "Ecological Reserve" in the Fish and Game Code; these regulations are set by the commission. Commercial fishing is regulated by the Department of Fish and Game, however the state legislature establishes commercial regulations, such as kelp cutting. If DPR wishes to protect living marine resources found in underwater units, "ecological reserve" classification must be superimposed on the DPR classification. The question of authority is further complicated by the State Water Quality Control Board's jurisdiction over the water column; a number of years ago they established "Areas of Special Biological Significance" (ASBS) along the California coast - many are adjacent to coastal units and most of DPR underwater units fall into this category. Basically the ASBS classification protects against point source pollution. With all the overlaying classifications necessary to protect the biological diversity of marine ecosystems, it appears that the most efficient and timely process for protection of the ocean's biological diversity is a comprehensive legislative mandate.

Cultural Heritage Features

Although formal exploration has only recently begun, two basic types of cultural sites have been located in the underwater environment of California. They are Native American archaeological sites (and objects), and historic-era shipwrecks, cargo spills or landing sites.

Our state has a rich and varied maritime history. Native California peoples lived in large, settled villages along the Pacific coastline for many centuries before European contact. In some areas, such as along the Santa Barbara Channel and San Francisco Bay, these villages developed a highly sophisticated technology, with widespread trade using native watercraft. Boat construction reached its highest development in California among the Chumash. Their plank canoe, called a *tomol*, impressed early explorers of the California coast for its versatility and seaworthiness. Recent dating of middens on San Clemente Island (some 60 miles offshore) has documented an ancient maritime culture dating back some 8,000 years. California coastal groups were technologically advanced and highly organized. Trading expeditions from the mainland were common to the Channel Islands to obtain steatite for soapstone bowls and effigy figurines. The result of this prehistoric seafaring is being recorded by underwater archaeologists in quantities of artifacts recovered and preserved from the offshore areas. At least 25 individual sites have been reported between Ventura Beach and Point Conception. A recent exploration of Goleta State Beach yielded six stone bowls in one dive. Many other similar sites no doubt await discovery.

The lakes and rivers of California also contain valuable prehistoric sites. Fish traps at Ahjumawi Lava Springs State Park demonstrate the systematic management and harvest of suckers by native people. Submerged bedrock mortars in Emerald Bay await dating and interpretation. Other submerged sites and preserved prehistoric watercraft have been reported in rivers and lakes.

The rise in sea level over the past 10,000 years has resulted in submergence of many archeological sites. One well known area at the edge of the La Jolla submarine canyon has yielded more than 2,000 stone bowls, from depths ranging up to 80 feet. This site is a submerged village, dated by archaeologists at 4,000 to 5,000 years old. A great deal of information on sea level change and the ancient cultures that first colonized California can be learned at such sites.

The first recorded European vessels to visit California were Cabrillo's caravels, *Victoria* and *San Salvador*, which cautiously made their way up the Pacific coast in 1542. Twenty-three years later, the first galleon made a trans-Pacific crossing, and opened the Manila route to be followed for the next 250 years. California's first recorded shipwreck is that of the *San Augustin*, which was driven ashore in 1595 at Drake's Bay. Since then, thousands of vessels have been lost in our waters. These ships, their cargoes, and the mooring systems which restrained them are the physical remains of a rich maritime legacy in California.

Discovery of the remains of an 1850 Baltimore clipper has sparked tremendous interest in the early history of Mendocino County. The *Frolic*, wrecked near Caspar, was carrying a cargo of trade goods to Gold Rush California. The artifacts provide a glimpse of the flurry of activity that brought people and goods from all over the world to the golden state. Silver tinder boxes, oyster shell window glass and Chinese porcelains have been recovered and exhibited. The wreck site remains an impressive historical feature.

Gold rush shipwrecks have also been recently documented in the Sacramento River within Old Sacramento State Historic Park. The *La Grange* and *Sterling*, former sailing vessels that carried pioneers around Cape Horn to the California gold fields in 1849, ended their days tied to the Sacramento levee. They sank into the mud only to reappear a century and a half later at the foot of the present flood wall. Examples of their timbers and contents have been preserved by underwater archaeologists.

Historic shipwreck sites, artifacts, and associated maritime features will be evaluated on an individual basis to determine their significance to California history. Important resources of statewide significance will be proposed for underwater park status in order to protect and interpret them. In this way, a record of California's maritime heritage can be made available for the enjoyment of present and future generations.

Recreational Features

The recreational aspect of diving is almost as diverse as the backgrounds of the divers. Most divers never have been, nor do they intend to become, professional divers. Recreation is the main interest of the estimated three million active self-contained underwater breathing apparatus (SCUBA) divers in the nation, and the 400,000 in California (Collier, 1984).

Some people dive just for the enjoyment of sightseeing and entering a new environment, or the sheer pleasure of a new experience. However, this attraction usually wears off, and divers find new underwater interests. Divers may branch into photography or behavioral studies of marine organisms. Underwater "fishing" for clams, lobster, abalone, or fish may become a recreational pursuit. Divers may collect shells or geological specimens, investigate caves or wrecks, or study underwater archeology.

The array of underwater activities by scuba divers is expanding. Providing divers and non-divers access to and interpretation of underwater environments will give them a better understanding of their importance, fragile nature, and beauty, and is a major part of DPR's underwater program.

Recreational diving is managed dependent on the fragility of the underwater environment of a given SPS unit. A "land carrying capacity survey" is required before any general plan is made (Sec. 5019.5). Attendance limits are then established (Sec. 5001.96). For example, at Point Lobos State Reserve 10 dive teams are allowed in the water at any one time and diving is restricted to Whaler's and Bluefish coves. This "allowable use intensity" for the marine environment was based on a subtidal survey conducted by De Martini and Barry (1977).

Recreational and commercial fishing have associated ecosystem management problems. At Arena Rock Marine Natural Preserve fishing has caused decline in rockfish breeding stock (Barry, 1974). Anchor damage from party boats as been noted on reefs at Crystal Cove State Park; commercial lobster fishing has conflicted with the DPR regulations.

Summary

The current status of State Park System marine underwater park units and proposed units is listed in **Appendix A**. A coarse filtered gap analysis of deficiencies in the marine ecological region has been completed. This matrix includes environmental factors such as depth, substrate types, salinity and shore line types. Cultural historic eras include Native American, Hispanic, and American (Barry, et. al., 1995).

<u>**Appendix A.**</u> Hierarchical Marine Ecological Classification of California with Listings of Underwater State Park System Units and Proposed Additions.

Pacific Marine Ecological Realm - Total Pacific Ocean (Biogeographical Realms of Udvardy, 1975)

North Pacific Basin Ecological Domain - North Pacific Ocean (Domains of Bailey, 1981; and/or Biogeographical Provinces of Udvardy, 1975; Temperate North Pacific of Ray, 1975)

Boreal Nearshore Marine Ecological Regime - Aleutian Is. to Point Conception (Regimes of Bailey, 1981; and/or Biogeographical Provinces of Udvardy, 1975; Boreal Zone of Hedgpeth, 1957; Columbian Biotic Province of Ray, 1975) According to Abbott and Hollenberg (1976), the most dramatic temperature and biological change occurs at Point Conception. The temperature usually remains above 18 degrees to the south and below 18 degrees to the north.

Oregonian Marine Ecological Province - Point Conception to Cape Flattery, Washington (Provinces of Bailey, 1981; and/or Biogeographical Provinces of Udvardy, 1975)

Oregonian Marine Ecological Region (Ecological Regions of the California Public Resources Code, Sec. 5019.53 and/or Ecoregions of Bailey, 1978; Oregonian Subdivision of Hedgepth, 1957)

The Oregonian Marine Ecological Region is bound on the north by Cape Flattery, Washington. Seaward it includes the continental shelf which generally extends to the south. The coastal boundary is the supra littoral zone of the Washington, Oregon and Northern California Coast from Cape Flattery southward to Point Arena, Mendocino County. The 10 meter depth temperatures range from around 8.5 degrees C in the north, during January through March, to a high of 16 degrees in the south in October through December. The entire region is 12.5 to 17.5 degrees in July, August, and September.

Areas with Underwater Designations in the Oregonian Marine Ecological Region

Underwater Units

Redding Rock (Proposed)
Point St. George (Proposed)
Trinidad State Beach (Proposed)
Sinkyone Wilderness State Park (Proposed)

Westport Union Landing State Beach (Proposed)
MacKerricher State Park
Jug Handle State Reserve (Proposed)
Van Damme State Park
Russian Gulch State Park
Greenwood State Beach (Proposed)
Manchester State Park and Arena Rock Marine Natural Preserve

State Seashores

Del Norte State Seashore - Pyramid Point to Point Saint George
Clem Miller State Seashore - Eel River mouth to Pudding Creek
Mendocino Coast State Seashore - Jughandle Creek to Gualala River

Californian Marine Ecological Province - Point Conception to Point Arena (Californian Faunistic Province of Hedgepeth, 1957).

Northern Californian Marine Ecological Region

The Northern Californian Marine Ecological Region is bound on the north by Point Arena and on the south by the Monterey Submarine Canyon to the Ascencion Canyon and landward at Año Nuevo Point. Seaward it includes the continental shelf which generally extends to the crest of the Farallon Escarpment. The coastal boundary is the supra littoral zone of the California Coast from Point Arena southward to Año Nuevo Point at the north end of Monterey Bay. The 10 meter depth temperatures range from around 11 degrees C in the north, during January through March, to a high of 15 degrees in the south in October through December. The entire region is 12.5 to 15 degrees in July, August, and September.

Areas with Underwater Designations in the Northern Californian Marine Ecological Region

Underwater Units

Salt Point State Park
Stillwater Cove (Proposed)
Fort Ross State Historic Park
Sonoma Coast State Beach
Mount Tamalpias State Park (Proposed)
Wilder Ranch (Proposed)
Año Nuevo State Reserve (Proposed)

State Seashores

Sonoma Coast State Seashore - Gualala River to Bodega Head
Año Nuevo State Seashore - Pillar Point to the City of Santa Cruz

Central Californian Marine Ecological Region

The Central Californian Marine Ecological Region is bound on the north by the Monterey Submarine Canyon and on the south by Point Conception. Seaward it includes the continental shelf. The coastal boundary is the supra littoral zone of the California Coast from Año Nuevo Point, Santa Cruz County southward to Point Conception. From Point Conception the southern seaward boundary cuts between Santa Rosa and Santa Cruz Islands (to the shore line of the latter). Santa Rosa and San Miguel Islands receive colder northern currents (the California Current) and therefore are included in the Central California Marine Ecological Region while all the other Channel Islands are within the Southern - Baja California Marine Ecological Region. The 10 meter depth temperatures range from around 12 degrees C in the north, during January through March, to a high of 15 degrees in the south in October through December. The entire region is 15 degrees in July, August, and September.

Areas with Underwater Designations in the Central Californian Marine Ecological Region

Underwater Units

Asilamar State Beach (Proposed)
Cannery Row (Proposed)
Point Lobos State Reserve
Garrapata State Park (Proposed)
Point Sur (Proposed)
Julia Pfeiffer Burns State Park
John Little State Reserve (Proposed)
Hearst San Simeon State Beach (Proposed)
Morro Bay State Park (Proposed)
Montana de Oro State Park (Proposed)
Dinosaur Point (Proposed)

State Seashores

Big Sur State Seashore (Proposed)
San Luis Obispo State Seashore - Cayucos to Lion's Head

Temperate Marine Ecological Division - Cabo San Lucus to Point Conception (Temperate Zone of Hedgpeth, 1957)

Baja Californian Marine Ecological Province - Punta

Eugenia to Point Conception (Lower Californian Faunal Province of Hedgepth, 1957; Californian Biotic Province of Ray, 1975).

Southern-Baja Californian Marine Ecological Region (Southern California Bite of SCCWRP, 1973).

The Southern-Baja Californian Marine Ecological Reserve extends southward from Point Conception, but it extends south east, across the Santa Barbara Channel then southward between Santa Rosa and Santa Cruz Islands (at the west shore of the latter). It then extends southwest to the edge of the continental shelf where it extends southward along the outer edge of the continental shelf to Cabo Colnett where the California current meets the land at 90 degrees and is deflected both north and southward along the coastline. Warm currents from off Baja California move northward and meet the California Current at Point Conception. The 10 meter temperatures vary from a low of around 14 degrees C in the north , during January through March, to a high of 20 degrees for the whole region in April, May and June.

Areas with Underwater Designations in the Southern-Baja Californian Marine Ecological Region

Underwater Units

Gaviota State Park (Proposed)
Naples Reef (Proposed)
Leo Carrilo State Park (Proposed)
Point Dume State Reserve (Proposed)
Royal Palms State Beach (Proposed)
Santa Rosa-Santa Cruz Islands (Proposed one mile perimeter which is State Lands)
Santa Catalina Island (Proposed)
Crystal Cove State Park
Doheny State Beach
San Onofre State Beach (Proposed)
San Elijo State Beach
Cardiff State Beach
San Nicolas Island (Proposed)
San Clemente Island (Proposed)

State Seashores

Santa Barbara Coast State Seashore - Gaviota to Las Llagas Canyon
Point Mugu State Seashore - Ormond Beach to San Nicholas Canyon
Capistrano Coast State Seashore - Newport Beach to San Mateo Point
San Diego Coast State Seashore - San Onofre State Beach to La Jolla

References

Abbott, Isabella A. and George J. Hollenberg. 1976. Marine algae of California. Stanford University Press. 827 p.

Bailey, Robert G. 1978. Description of the ecoregions of the United States. U.S. Department of Agriculture, Misc. Publ. No. 1391. 77 p.

Bailey, Robert G. 1981. (Assisted by Charles T. Cushwa) Ecoregions of North America. 1:12,000,000 Map. U.S. Forest Service and U.S. Fish and Wildlife Service. Washington D. C.

Barry, W. James. 1974. Ecological evaluation of the proposed Point Arena unit of the State Park System. State of California, The Resources Agency, Department of Parks and Recreation. Sacramento. 17 p.

Barry, W. James. 1991. Ecological Regions of California. Interagency Natural Areas Coordinating Committee, State of California, The Resources Agency, Department of Parks and Recreation. Sacramento. 67 p.

Barry, W. James, John W. Foster and Kenneth W. Collier. 1995. California underwater parks and reserves action plan. State of California, The Resources Agency, Department of Parks and Recreation. Sacramento. 34 p.

Collier, Ken. 1984. California State Park System underwater parks master plan. State of California, The Resources Agency, Department of Parks and Recreation. Sacramento 32 p.

DeMartini, John D. and W. James Barry. 1977. A subtidal resource inventory for Point Lobos State Reserve. State of California, The Resources Agency, Department of Parks and Recreation. Sacramento. 51 p.

Hedgpeth, Joel W. 1957. Marine biogeography. Memoir 67, Vol. 1, pp. 358-382. Geological Society of America.

SCCWRP. 1973. The ecology of southern California bight; Implications for water quality management. Three-year report of the Southern California Coastal Water Research Project, El Segundo, California. 531 p.

Udvardy, Miklos, D. F. 1975. A classification of the biogeographical provinces of the world. Internat. Union Conserv. Nature and Natural Resources (ICUN). Morges, Switzerland. Occas. Pap. 18:1-48.

Ray, G. Carleton, 1975. A preliminary classification of coastal and marine environments. Internat. Union Conserv. Nature and Natural Resources (ICUN). Morges, Switzerland. Occas. Pap. 14.

CHAPTER 13

International Recognition and Protection for United States National Marine Sanctuaries

Edward R. Cassano[1]

The United States National Marine Sanctuary Program (Marine Sanctuary Program) was mandated under Title III of the Marine Protection, Research and Sanctuaries Act (MPSRA) of 1972 as amended. The Marine Sanctuary Program was created to identify and manage areas of the marine environment that "posses conservation, recreational, ecological, historical, research and esthetic qualities which give them national and in some instances international significance". The size and scope of many of the designated United States Marine Sanctuaries include ocean areas that cross several jurisdictional zones. The Marine Sanctuary Program recognizes that maritime activities, in the form of vessel traffic constitute one of the major threats to sanctuary resources. This threat is in the form of marine pollution from vessels as, either operational or accidental discharge of pollutants, or physical groundings. This paper describes the existing national and international legal framework that limits the United States ability to fully protect the resources of a sanctuary in the Territorial Sea and the Exclusive Economic Zone (EEZ). The MPSRA is examined to see how sanctuary regulations may be implemented to protect resources by monitoring and controlling foreign vessel movement and discharges.

The authors of the MPSRA recognized the limits of the Unites States to effectively integrate sanctuary regulation across different jurisdictional zones. The clear intent of the MPSRA mandates the Marine Sanctuary Program to "develop and implement coordinated plans for protection and management of these areas with....internationally organizations... concerned with the health and resilience of these marine areas". The concept of Particular Sensitive Sea Areas (PSSAs), as defined by the International Maritime Organizations Resolution A. 720 (17), is introduced as an international recognized procedure that could be used to identify and impose protection measures for marine sanctuaries that exist in the EEZ. By seeking designation of marine sanctuaries as PSSAs, the United States fulfills the intent of MPSRA and major goals of the program: international recognition of marine sanctuaries, integration of sanctuary regulations through existing international law, and adequate protection of sanctuary resources.

---

[1] Manager, Channel Islands National Marine Sanctuary, 113 Harbor Way, Santa Barbara, CA 93109

CHAPTER 14

A review of research programs in central California coastal managed areas with a suggestion for improved collaboration

Andrew P. De Vogelaere[1] and Richard C. Green

Abstract

Between Pt. Reyes, located north of San Francisco, and Cambria, at the southern end of the Big Sur coast, there are over 30 managed coastal areas. These areas are designated as sanctuaries, reserves, preserves, refuges, wildlife areas, and parks. A review of these coastal programs found they have similar goals and overlapping jurisdictions. However, programs that incorporated scientific advisory groups had more current management plans, a broader view of resource management alternatives, larger research budgets, and a wider variety of research activities. These programs were developed, in large part, independently of each other and based on local history more than a comprehensive resource management vision. There is great opportunity for collaboration between these programs to improve their individual and common goals. The Monterey Bay National Marine Sanctuary (MBNMS), encompassing most of the coastline in central California, has a programmatic structure of working groups and advisory councils that take advantage of insight from academic scientists, multiple resource management agencies, user group representatives, and education specialists. The MBNMS may be the forum for improved collaboration between programs and an avenue to academic scientists in the development of science plans and enacting necessary research.

Introduction

One of the main purposes the California and the World Ocean '97 conference is to develop ways to improve management of coastal resources. Research is a critical component of resource management, as it provides the

[1] Research Coordinator/ Senior Scientist, Monterey Bay National Marine Sanctuary 299 Foam Street, Suite D. Monterey CA 93940

technical information needed for fact based decision making, a way to find solutions to resource management problems, a source of information for education programs, and, with monitoring, a way to assess "health" of the environment. The central California coast is a region with world-renowned research programs (e.g., Griggs 1995; Guerrero and Kvitek 1996) and boasts numerous protected and managed areas (Figure 1). The purpose of this review was to assess the role that research played in resource management programs and to suggest opportunities for improved coastal stewardship, if necessary.

Methods

The methods of this review were to develop a list of coastal managed areas, then to gather information on the role of research at each location. Several reliable lists of coastal management designation are available (Research Planning Inc. 1994, California Coastal Commission 1997). For each location, information was compiled on the program goals, the size and location of the managed area, the status of management plans, the status of research priorities, the use of scientific advisory committees, accomplishments in research, regulatory authority, staff size and budget. These information assessment categories were evaluated using written questionnaires, telephone interviews, site visits, existing public documents, and extensive personal knowledge of coastal programs that the authors have for the central California region.

Results

The numerous coastal programs were found to have similar goals and overlapping jurisdictions. All of the programs focused on resource protection and most had significant components for public education. Some research was occurring at almost all of the sites. Interestingly, the programs, in large part, operate independently of each other (even if within the same agency but at a different locations) and developed more as an artifact of local history than because of a comprehensive coastal management vision.

There was a distinct difference in programs that used scientific advisors (either with designated committees in the management structure or through numerous informal contacts) versus those that did not. Programs with science advisors (PWSA) had more general research goals, focusing on applied problems, than programs without science advisors (PW/OSA) which had specific research efforts on mostly exotic species removal and directing human traffic. PWSA had

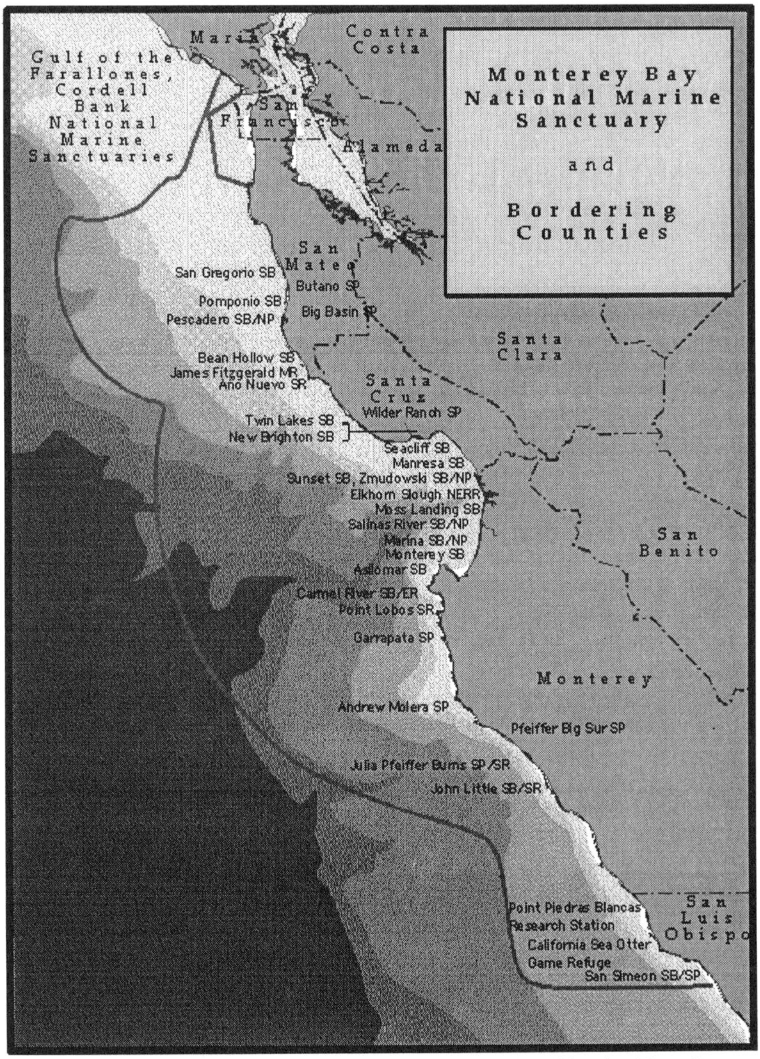

Figure 1. Protected and managed coastal areas in central California. Note that most are within or adjacent to the boundaries of the Monterey Bay National Marine Sanctuary. SB = State Beach, SP = State Park, NERR = National Estuarine Research Reserve, MR = Marine Reserve, ER = Ecological Reserve, SR = State Reserve, NP = Natural Preserve

more frequent updates of management plans and research priorities, while maintaining larger research budgets and leveraging outside grants. PW/OSA used mostly staff time and forms of match other than funds for research. Common research activities at PWSA included developing comprehensive plans, publishing results, monitoring habitats, and addressing specific applied problems while studying topics in general science. PW/OSA focused largely on monitoring particular species of concern, controlling exotic species, restoration and water quality monitoring.

Discussion

It is clear that coastal management programs benefit from having scientific advisors, and this asset is available with limited investment. Some programs have frequent contacts with academic scientists interested in specific resource management issues, without formal routes for input. Taking advantage of this resource simply involves an openness to outside input on a management program, and the initiative to contact regional scientific expertise and explain the benefits of their involvement. Some benefits to the scientist may include an avenue to serve the public, an interesting problem for them or their students, volunteers to assist in their research, secure research sites, equipment, or funds.

Programs which desire a more formal process for scientific council, or those not familiar with their regional scientific resources, may want to collaborate with the Monterey Bay National Marine Sanctuary (MBNMS) Research Activities Panel (RAP). This group of representatives, from 20 research organizations (see Table 1), meets monthly to address issues of concern to the MBNMS and to collaborate on research endeavors. Many of the issues of concern to the MBNMS are likely relevant to other central California coastal managed areas, and the MBNMS is adjacent to or overlaps the boundaries of at least 20 other programs. The meetings are open to the public, rotate in location between institutions, and agendas are announced by electronic mail.

The MBNMS RAP benefits both the Sanctuary Program and the scientists who serve on the panel. The MBNMS RAP wrote most of the Sanctuary Science Plan (MBNMSRAC, 1993), has provided reviews on numerous management issues (e.g., analyzing potential impacts of coastal development projects and management options for biological resources), contributed extensively to the text and editing of a comprehensive review of sanctuary resources (Guerrero and Kvitek, 1996), acted as a source for specific technical information (e.g., sea conditions related to vessel grounding events), trained volunteers, developed monitoring programs, and served as "ambassadors" for the program goals. The

Table 1. Membership organizations of the Monterey Bay National Marine Sanctuary Research Activities Panel.

MONTEREY BAY NATIONAL MARINE SANCTUARY RESEARCH ACTIVITIES PANEL

Moss Landing Marine Laboratories
California State University Monterey Bay
University of California at Santa Cruz
Hopkins Marine Station
Naval Postgraduate School
Monterey Bay Aquarium Research Institute
Pacific Fisheries Environmental Group, NMFS
United States Geological Survey
Elkhorn Slough National Estuarine Research Reserve
Gulf of the Farallones and Cordell Bank National Marine Sanctuaries
California Department of Fish and Game (CDFG)
Marine Pollution Laboratories, CDFG
University of California Sea Grant Extension Program
Monterey Bay Aquarium
National Undersea Research Program
Department of the Interior Office of Environmental Affairs
Morro Bay Foundation

At Large - Coastal Development Regulations
At Large - Social Sciences
At Large - Birds and Mammals

RAP members benefit from the MBNMS in several ways, including administrative support for monthly meetings that facilitates inter-institution collaboration, organized presentations on topics of interest, access to MBNMS resources such as ship time, access to a network of electronic mail lists and regional scientific directories, updates on regional and national initiatives and grants, some small MBNMS research grants, opportunities for their students who are interested in applied research problems, and opportunities to influence public policy. The MBNMS RAP is linked programmatically with similar working groups in education and conservation, a political council of with representative from numerous interest groups (e.g., industry, agriculture, fishing, other government agencies), and Sanctuaries and Reserved Headquarters in Washington DC (see a related paper in this same volume by T. Jackson).

More extensive use of the already successful MBNMS RAP would help achieve the California and the World Ocean '97 conference goal of comprehensive and integrated coastal management, and this is also a model for other regions on how to involve science in their programs.

Acknowledgments

We gratefully appreciate the invitation from Deborah McArdle and Richard Starr to participate in the Marine Managed Areas session. Michele Jacobi provided refreshing comments to the manuscript and formatted the text for publication.

Literature Cited

California Coastal Commission. 1997. Coastal Access Guide. University of California Press.

Griggs, G. 1995, Monterey Bay, national center for marine science. Sea Technology. May.

Guerrero, J. and R. Kvitek. Editors. 1996. Monterey Bay National Marine Sanctuary Site Characterization. (homepage address http//205.155.38.2/sitechar/index.html

MBNMSRAC, Monterey Bay National Marine Sanctuary Research Advisory Committee. 1993. Scientific research plan for the Monterey Bay National Marine Sanctuary.; National Ocean Service, Office of Ocean and Coastal Resource Management, Sanctuaries and Reserves Division, National Oceanic and Atmospheric Administration. 55pp.

Research Planning Inc. 1994. Sensitivity of coastal environment and wildlife to spilled oil: central California. Maps prepared for Office of Oil Spill Prevention and Response, California Department of Fish and Game; Hazardous Materials Response and Assessment Division, National Oceanic and Atmospheric Administration.

CHAPTER 15

Cooperative Management of the State's Marine Ecological Reserves: Preliminary Evidence from Big Creek

Caroline Pomeroy[1]
and
Julie Beck[2]

Abstract

The passage of Proposition 132 in 1990 banned the use of gillnets in California's nearshore areas, and mandated the establishment of four Marine Ecological Reserves (MERs) along the state's coast. To determine the MERs' effectiveness in protecting and enhancing fishery resources - their stated goal - it is essential to allocate resources and develop mechanisms for monitoring them over time. This, however, requires financial and personnel resources that may or may not be readily available from the state. An alternative or complement lies in the possibility of involving local fishermen in the collection of data and other, related management tasks. Big Creek Ecological Reserve, is unique among the four MERs in that such a system of cooperative management had emerged well before its formal establishment. Since 1991, the manager of the adjacent terrestrial reserve and local fishermen have mapped and systematically collected data on rockfish catches at sites near the marine reserve in an attempt to monitor the effects of local fishing on the MER and *vice verse*. We explore the Big Creek arrangement, its history and evolution, as well as current efforts to refine and enhance the system so that it may better serve research and management needs. We discuss the cooperative arrangement in connection with its benefits for both local fishers and managers including 1) the provision of low cost, high quality fishery-dependent data to complement fishery-independent data for the evaluation of the relationship between the MER and local fisheries, and 2) the maintenance of ongoing communication among local fishermen, researchers and managers. Taken together, these elements may enhance the effectiveness of the MER.

Introduction

Marine Protected Areas (MPAs) are gaining currency in fishery and marine resource management debates as new and potentially effective management tools. MPAs have been shown to protect and enhance within-reserve diversity and

[1] Institute of Marine Sciences, Earth & Marine Sciences Bldg. A316, UC Santa Cruz, Santa Cruz, CA 95064.
[2] Sociology Board, UC Santa Cruz, Santa Cruz, CA 95064.

abundances of stocks, and to have spillover effects in adjacent fishing areas (e.g., Bennett and Attwood 1991, Cole et al. 1990). Several challenges associated with the establishment and maintenance of effective MPAs, however, can limit these positive effects, and must be addressed in reserve design and management policy. In particular, it is critical that MPAs include monitoring mechanisms, both for the reserve itself, and for the fishing and other activities occurring adjacent to it (Roberts and Polunin 1993, Rowley 1994). Without such mechanisms, it is impossible to determine the effectiveness of the MPA and to insure that the benefits it produces are not offset by an increase in activity at its perimeter (Russ and Alcala 1996).

In 1990, California voters passed Proposition 132, banning the use of gillnets in nearshore waters and mandating the establishment of four Marine Ecological Reserves (MERs) along the state's coast by 1994. One of the sites, Big Creek Marine Ecological Reserve, located about 40 miles south of Monterey on the Big Sur coast, has been noted for the ease with which it was designated. In contrast to the other three MERs, it had broad local support, including that of local, small-scale commercial fishermen. Local fishers' support for the reserve was linked strongly to their cooperative arrangement with the reserve manager. As part of that arrangement, they had been maintaining an informal no-take zone and collecting fishery-dependent data in the area for several years.

This paper explores that arrangement as an experiment in cooperative management of local fisheries with important implications for the establishment and maintenance of the Big Creek Ecological Reserve and adjacent fisheries. We begin with background on the research we have conducted to date, then turn to the topic of cooperative management of local fisheries. We then focus on the Big Creek case, providing an overview of the fishery and the cooperative arrangement, followed by a preliminary evaluation of its strengths and weaknesses. We close with questions raised by the research to date, and a brief description of how our future work seeks to address those questions.

Background of the Study

This paper is based on exploratory research conducted during Spring and Summer 1996 as a fisheries social science demonstration project of the Monterey Bay Regional Studies Program (MBRS) at UC Santa Cruz.[3] We decided to focus on Big Creek because we had heard about the cooperative arrangement from several sources,[4] and because such an investigation would complement the work of MBRS graduate student Michelle Paddack on rockfish at the reserve.[5] Our aim in this exploratory research was to initiate MBRS-related research through an institutional analysis of the arrangement that would contribute to empirical and theoretical understanding of the situation and identify opportunities for future research. Our

---

[3] This work was conducted with MBRS graduate student support, funded by UCSC's Office of the Vice Chancellor for Research.

[4] John Smiley, Michelle Paddack, and CDFG fisheries biologists Paul Reilly and David VenTresca alerted Pomeroy to the situation. We are grateful to them, and to Big Sur fishermen, for their encouragement and input.

[5] Paddack (1996) conducted a comparative analysis of rockfish populations within and outside three Monterey Bay area marine reserves (Hopkins Marine Station, Pt. Lobos, and Big Creek) to determine reserve impacts in temperate waters.

objectives were: 1) to describe and analyze the fishery adjacent to the Big Creek MER, 2) to describe and analyze (preliminarily) the cooperative (i.e., institutional) arrangement between the reserve manager and local fishers, 3) to determine the strengths and weaknesses of the arrangement as a form of cooperative management with implications for the fishery and the reserve, and 4) to specify future research questions regarding the fishery and resource management system associated with the reserve. To address these objectives, we used a combination of in-depth interviews, observation, and archival research. Interviews with Big Creek fishers focused on five themes: biographical information, involvement in the Big Creek fishery and the reserve, cooperation between resource users and managers, the interface between resource use and scientific research, and attitudes toward conservation and marine reserves. We conducted informal interviews with the reserve manager and California Department of Fish and Game (CDFG) personnel. Archival research included the collection of data (e.g., landings data, regulatory information) on the fishery, as well as review of relevant literatures (e.g., cooperative management of local fisheries, the Big Creek/Big Sur area).

Cooperative Management of Local Fisheries

Pinkerton (1989:4) defines cooperative (or co-) management as negotiated agreements and other legal or informal arrangements made between groups or communities of fishermen and various levels of government responsible for fisheries management. Kearney (1989) notes that co-management may take a variety of forms which range in the scope of authority granted or left to local resource users (see Table 1). On one end of this co-management continuum is a consultative process directed by government whereby fishermen and others provide systematic information, but government officials retain sole decision-making authority. A 'middle ground' is an arrangement whereby fishers (rather than government authorities alone) implement and enforce government policy. On the other end of this continuum is comprehensive participation of fishermen in policy design, implementation and enforcement.

	policy design	implementation	enforcement
consultative	g,f	g	g
user-implemented	g	g,f	g,f
comprehensive	g,f	g,f	g,f

Table 1. Government (g) and fisherman (f) roles in variants of co-management (based on Kearney 1989).

Co-management is attractive as an alternative or complement to traditional, top-down resource management, the limitations of which are increasingly being recognized (see, e.g., McGoodwin 1990, Pinkerton 1989). Among the reasons for the failure of centralized, top-down management are limited financial and personnel resources for monitoring and enforcement, limited information available to resource managers, and lack of communication between resource managers and resource users (McGoodwin 1990). The lack of information and communication in fishery management, in particular, often leads to misunderstanding and mistrust between fishers and managers. These, in turn, contribute to ill-fitting regulations and fishermen's flouting of rules that make no biological, economic or social sense to them (Pollnac and Littlefield 1983). In sum, traditional, centralized resource management is costly.

Co-management makes an explicit link between resource managers and resource users, whereby it can overcome many of the limitations (or pitfalls) associated with centralized resource management. Among the possible benefits of co-management (in its various forms) are 1) better communication between fishermen and managers, 2) increased mutual understanding and trust, 3) improved and more detailed information on resource and fishery conditions, and 4) better fitting rules. These, along with resource users' participation in the management process, can enhance compliance with the rules and thus increase the likelihood of achieving management goals. In short, co-management can provide a cost-effective alternative or complement to 'traditional' resource management.

Examples of effective co-management systems are found in Iceland (Durrenberger and Palsson 1987), Norway (Jentoft and Kristofferson 1989), Japan (McKean 1992) and for a time, in the New York Bight (McCay 1980). Native Americans and First Nations have been involved in co-management with U.S. and state resource managers and with the Canadian government, respectively (see, e.g., Cohen 1986, Pinkerton 1989). Native American tribes in Washington State have worked out co-management of fisheries between their subsistence and commercial, small- and large-scale, and in-river and sound/ocean fishers (see, e.g., Cohen 1986, Pinkerton and Keitlah 1990). The State of Maine has recently formally devolved lobster fishery management to coastal communities, where local, traditional management has occurred for many years (see, e.g., Acheson 1988).

Co-management, however, is not a panacea for all fishery management situations, and certain caveats regarding its application are in order. McGoodwin (1990) cites both internal and external reasons for the failure of some co-management efforts. For example, while British Columbia empowers some communities to manage their own fishing activities, it pursues other policies that conflict with these efforts or otherwise create resource problems (Pinkerton 1988 in McGoodwin 1990). Others may fail for lack of technical competence (McGoodwin 1990). Or, they may be resisted by entrenched fishery management professionals, or thwarted by the historic hostility between state officials and fishermen, as happened in the New York Bight case described by McCay (1989). Moreover, co-management, like other forms of resource management, may best be done by fishermen and state managers muddling through (Lindblom 1959) - which at its best may be called 'adaptive management' (Walters 1986, Lee 1992).

A Brief History of the Big Creek Reserve

The Big Creek arrangement caught our attention as an experiment in co-management, in which local fishers are involved in data-collection, monitoring of resource conditions and use, and, to a lesser (and as yet not fully determined) extent in decision-making (with the reserve manager). At present, this arrangement has two key features: 1) the marine protected area now known as the Big Creek Marine Ecological Reserve, and 2) a fishery-dependent data collection system known as the Hook-and-Line Kelp Bed Survey. Importantly, both of these features were initiated locally and prior to the designation of the big Creek MER.

The Big Creek site, located about 40 miles south of Monterey in Monterey County, consists of two related elements: 1) the Landels-Hill Big Creek Reserve, a terrestrial UC Natural Reserve established in 1978 (see Georgette 1981 for a history of the reserve), and 2) the Big Creek Marine Ecological Reserve, established (legally) in 1994 pursuant to Proposition 132 (1990). The marine reserve covers an area of about 6.86 km2 adjacent to the Landels-Hill Big Creek Reserve.[6] Access to the reserves is available by land via a single dirt road off Highway 1 and by sea at Big Creek cove beach.

Eight to 12 fishermen participate in the Big Creek fishery, all but three of whom are full-time fishers.[7] They use 12-foot aluminum skiffs with small outboard motors, and hook-and-line to target shallow- and mid-water rockfishes including vermillion (*S. miniatus*), kelp (*S. atrovirens*), black and yellow (*S. chrysomelas*), gopher (*S. carnatus*), copper (*S. caurinus*), olive (*S. serranoides*), and black (*S. melanops*) rockfish, and cabezon (*Scorpaenithys marmoratus*) and lingcod (*Ophiodon elongatus*).[8] Fishers land an average of 30 to 40 fish per day (totalling 100-150 pounds), which they sell 'dead' (or barely alive) to local restaurants and/or transport to nearby cities (e.g., Monterey) for sale to restaurants and retail stores. In early 1995, they began a transition to the 'live fish' fishery, in response to a booming, primarily Asian market. For this fishery, they use 'Portuguese rigs' or 'trees' to target shallow-water rockfishes, which sell for $4.00 to $5.00 per pound, while 'dead' fish sell for $.25 to $.40 per pound. This opportunity notwithstanding, several of them are contemplating a return to the hook-and-line fishery for 'dead' fish, which they view as more biologically and economically sustainable.

The fishermen say Big Creek provides the best fishing compared to other sites (most of which are along the Big Sur coast to the south); in winter, Big Creek and Mill Creek (about 7 miles to the south) are the only viable launch sites on the Big Sur coast because of dangerous weather and seas. Any time of year, launching and landing even from Big Creek and Mill Creek is difficult in small skiffs, which easily capsize or are mangled, or have engines destroyed when hit from behind by waves.

In addition to the fishermen involved in the hook-and-line survey, other fishing takes place in the area. In the early 1990s, commercial passenger fishing vessels (CPFVs) based in Monterey and Morro Bay ran day trips to the Big Sur coast because of its relatively pristine and less crowded waters. The area's rough seas and long transit times, however, make this impractical for most CPFVs, and fewer are doing this now. A second type of fishing in the area is that by 'mother ships' from Morro Bay that motor to the Big Sur fishing grounds and launch

---

[6] The marine reserve is bounded by the 50 fathom contour to the west, and at 252 degrees magnetic north and south, respectively, of the terrestrial reserve's northern and southern boundaries.

[7] The fishermen are 'Anglos,' who range in age from 22 to their mid 50s and have worked as carpenters, handymen, mushroom hunters, and growers.

[8] Rockfish are robust, sessile, long-lived fish, distinguished as deep-, mid- and shallow-water species. Rockfish may live as long as 95 years, and mature between the ages of 5 and 15 years, depending on the species (VenTresca pers. comm.). Fifty-one species of rockfish (*Sebastes* spp.) have been identified along the 110-mile coastline that stretches from Monterey south to Cambria. This is the most diverse and abundant population of rockfish found anywhere in the world (Paddack pers. comm.).

several small skiffs at once to catch rockfish. Preliminary evidence suggests that this activity had sufficient negative impact on rockfish stocks to make it economically non-viable, and most of these operations have ceased. Clearly, these fisheries that have occurred (and in some cases continue) near Big Creek need to be investigated to ascertain their nature and implications for the reserve and nearshore fisheries.

Evidence of Co-Management at Big Creek

Two features of the Big Creek arrangement caught our attention as evidence of co-management of the local fishery. The first was the establishment of an informal marine reserve in 1988 by several Big Creek fishermen and Landels-Hill Big Creek Reserve manager Smiley, *prior to* the passage of Proposition 132 in 1990 and the legal designation of the Marine Ecological Reserve in 1994. The second was the collaborative design and implementation, by reserve manager Smiley and local fishermen of the hook-and-line kelp bed survey to collect data on and monitor fishing activities in the area.

Prior to 1988, a small group of local commercial fishers had been launching their skiffs from Mill Creek, and using hook-and-line gear to fish the kelp beds along the Big Sur coast. In October of that year, two fishermen approached John Smiley, who was then manager of the (terrestrial) Landels-Hill-Big Creek Reserve, in the interest of gaining permission to launch their boats from Big Creek in order to more easily reach preferred fishing spots to the north (e.g., Esalen). Smiley offered them access in the fall and winter when the terrestrial reserve was little used and harbor seals were not pupping; the fishermen agreed to a no-take zone in the kelp beds adjacent to the reserve.

The following season (1989-90), Smiley, CDFG biologist VenTresca, and fisherman and buyer John Peck discussed the possibility of research collaboration among Big Creek fishermen, CDFG and the University of California. They also further defined the boundaries of the informal marine reserve as a one square mile offshore area adjacent to Big Creek. In Fall 1991, Peck, Smiley and VenTresca developed a research protocol for the fishermen, and Peck trained them in its use. All of the fishermen who launched from Big Creek were to use hook-and-line gear to sample fish populations in conjunction with their is fishing activities, then upon returning to shore, record data on those fishing activities and their catch.

At present, the Big Creek fishing season runs from September through March, so as to minimize potential conflict with other visitors to the area, including humans and pupping sea lions. The season is further limited by a set number of 'sorties,' or one-day fishing trips made from the site, in accordance with the design of the hook-and-line survey (see Figure 1). One year (1993-94), the fishermen postponed the start of their season until December in order to be able to preserve opportunities to launch from the safer Big Creek site during rough winter weather.

Over time, the survey has been refined to address data collection problems and needs. The procedure is as follows. On each trip, fishers take their first five fish caught at each of three sites: a 'lightly fished,' a 'medium-fished,' and a 'heavily fished' kelp bed and place them in separately marked buckets. When they return to Big Creek, they record how long they fished and the conditions (e.g., current, cloud cover, water depth) at each site, along with the species, length and weight of each fish sampled. Fishermen weigh and measure their sampled fish at a central station set up by Smiley, and record their data on waterproof data sheets in the project notebook Smiley maintains. The resulting data consists of a large number of samples caught over the season, in an effort to gain broad coverage of the variation in fishing conditions and outcomes. In addition to providing data of potential use to resource managers, this procedure has enabled fishers to monitor their harvest and guard against overexploitation (Smiley unpub.). However, the data has not been quality-checked of thoroughly analyzed due to a lack of resources.

Two Studies
- Santa Lucia and Lime Creek (twin) kelp beds study
- Slate Rock study

Procedure
- sorties per season
  - 80 to twin kelp beds
  - 40 to Slate Rock
- sampling
  - first 5 fish caught at experimental sites(s) within 30 minutes
  - first 5 fish caught at preferred fishing site(s)
- data collected / recorded
  - date
  - fisher name
  - location, depth fished
  - conditions (cloud cover, temperature, currents)
  - species, weight, length

Figure. 1. Design of the hook-and-line survey.

The survey is one part of the cooperative arrangement at Big Creek, which is a type of local institution for common pool resource management (see Ostrom 1990). This institutional arrangement includes sets of rules that define participation. The rules of access to and launching at Big Creek are that fishermen may *not* fish in the marine reserve (this applied to the pre-reserve no-take zone as well), must collect and record their catch data, and must abide by all the rules of the Big Creek terrestrial reserve while on the property. Violations of the rules result in loss of Big Creek launching privileges. More than once, a fisher's Big Creek access right has been suspended because he broke the rules. However, infractions are rare, and the other fishermen have supported and participated in enforcement of the rules among themselves and with 'outsiders.'

As an example of co-management, the Big Creek arrangement appears to be a mix of Kearney's middle (user-implemented) and most participatory (comprehensive) forms. Government authorities, per se, have little direct role in the arrangement or its day-to-day operation. Rather, the reserve manager, who is not affiliated with the California Department of Fish and Game, has primary decision-making authority, and engages and negotiates with fishermen in rule-making, monitoring, enforcement and other aspects of local resource management.

Strengths and Weaknesses of the Arrangement

Preliminary evaluation of the Big Creek arrangement indicates a number of strengths and weaknesses, some of which we discuss here. Among the strengths of the arrangement, or perhaps elements that give it strength, are its informality and flexibility, its participants' shared experience, the good fit between the rules and the arrangement more generally and local conditions, and its capacity for and provision of ongoing resource monitoring. The use of an informal 'handshake' agreement, rather than a formal contract between the reserve manager and the fishermen to establish this arrangement fits well with local practices, and suggests a degree of informality and flexibility that fishers appear to appreciate. That flexibility has been instrumental in allowing the arrangement to adapt to changing conditions (e.g., in the fishery, its participants) and to help insure the rules of the arrangement fit local conditions. Fishers' and Smiley's experience together over time has helped to build shared understanding and mutual trust (which varies, of course, among participants).

An additional strength - and benefit - of the arrangement is its utility as a mechanism for monitoring and evaluating the fishery and its relationship with the reserve. The hook-and-line survey provides much needed and otherwise unavailable fishery-dependent data at virtually no cost to the state. Moreover, fishers' participation in this effort calls to attention their relevance to the reserve system. The MERs are intended to enhance fisheries, in both biological and human use terms. Their success is contingent upon fishers' respecting reserve boundaries and the nature and extent of the fisheries carried out in nearby waters.The arrangement also demonstrates that fishermen can and will participate in resource management (again, this varies across individuals). And, fishers' direct experience monitoring the fishery adjacent to the reserve invests them in resource management. This, together with the mutual understanding and trust mentioned above, contributed to fishers' support for the BC marine reserve. As Smiley notes,

> "As time went on, *because* we had the backing of the fishermen, we were a 'shoe-in' for a marine reserve, for none of the other marine reserves had the real backing of fishermen. Many of the reserves were placed there by default, that is, they were there because there was no opposition -- usually because there were no fish in them! [Big Creek] is one reserve that was put in place *because people really wanted it there.*"

This speaks of a local connection that has played an important role in the cooperative arrangement and the establishment of an MER at Big Creek. In particular, Smiley's role as a liaison between local interests and CDFG throughout this period appears to be a critical link. He built his credibility as such a liaison over time as *resident* manager of the terrestrial reserve, and through his and his family's involvement in the local community.

> "I think locals' fear (of CDFG management of the reserve) is the reason people were opposed to Big Creek becoming a marine reserve. [One of the large landholders] thought that CDFG would just take over, and it would be run remotely from Monterey, and probably not very well, because it's hard to run a marine reserve remotely. You kinda need to be there to interact with the neighbors,

> and to make happen things that need to happen. So by making me manager, I think it helped to overcome a lot of peoples' concerns about Cal Fish and Game making a reserve."

More generally, strengths of the arrangement lie in the sets of interdependencies it both hinges and capitalizes upon at once. Resource managers (Smiley, the UC Reserve System, and CDFG) and resource users have shared interests in and motivations for continuing to engage in the research. Although their material interests differ, all would like to see effective regulation and conservation of the resource. CDFG wants conservation via the reserve mechanism to increase (sport as well as commercial) fishing to assure future funding of agency research and management. Smiley wants the reserve to be used to monitor and conserve the natural environment in general and rockfish in particular. The fishers want the reserve because it provides a secure and reliable source of rockfish. (They say that rockfish are larger and more plentiful in the nearshore kelp beds compared to other places they fish). Corollary to this is the mutual reliance on and complementarity of the know-how and skills that each party has to offer. CDFG researchers depend on fishermen to catch these often elusive rockfish in the kelp beds, and to share their local understanding of the fishes' habits and characteristics. The fishermen, in turn, depend upon CDFG and Smiley to have (and use) their foresight and management expertise to interpret the data on their 'hidden under the water' resource.

These strengths notwithstanding, the Big Creek arrangement has weaknesses that limit its usefulness to resource management and raise questions regarding its sustainability over the long run. The first is most fishers' denial that they have any role in depletion of the resource. They point to other fishers (especially the mother ship operations mentioned above) as the source of any declines in fish stocks - a problem not unusual in fisheries (or any other shared resource use setting). In fact, researchers have suggested that fishing pressure in the early 1990s from the mother ship operations mentioned above may have had severe and lasting impacts on the resource. But Big Creek fishers' failure to recognize their own impacts on the resource suggests a weak link. Should fishing conditions take a turn for the worse, fishers might abandon the system rather than work through an adaptive solution in which they 'give a little' in the short term to insure the fishery over the longer term. Their economic self-interest in this, as they often cite their dependence on the resource -- specifically the small-scale fishing opportunity at Big Creek -- as an important (if not their only) source of income. They say they depend on Smiley to help them keep track of the resource and constrain their effort to some extent. His centrality to this arrangement, both in his relationship and work with the fishermen, and in his role as liaison with government authorities raises concerns as to the viability of the arrangement should he not be able to continue in these roles.

A second problem is resource managers' distrust of the fishermen's capacity to collect data scientifically, and to report the data reliably. Smiley trusts most of the data, and sees the fishermen as appropriately skilled to collect it because of their fishing experience and knowledge of the area. However, other resource managers and biologists question the accuracy and reliability of the data, its collection and recording. Fishermen, they argue, might lie about fish counts and sizes, for example, in order to insure they may continue to launch from and fish near Big Creek. (The data collection system, however, is structured to limit this problem.) This mistrust by outsiders can exacerbate the problem of fishers' commitment to the arrangement, as they come to feel their efforts to contribute to resource management are not valued.

Third, there is a lack of resources for data evaluation and analysis. To date, Smiley has taken primary responsibility for collating and analyzing the data, and sponsors an annual meeting of fishers and other Big Creek researchers to share his and others' results. But his time and resources (e.g., access to good database software and rockfish expertise) are quite limited, and have hindered full evaluation and analysis of the information collected to date. This has two implications. First, it means that information from the survey is not being used to help evaluate the fishery and its relationship with the reserve. Second, it may suggest to fishermen that their efforts are for naught, and could diminish their interest and investment in the effort.

Conclusions

As California considers the use of marine ecological reserves as a fishery enhancement and management tool, it is imperative to develop mechanisms to explore and evaluate the relationship between the reserves and adjacent fisheries, especially in terms of those fisheries' impacts of the reserves and its resources. Fishery-independent and fishery-dependent data constitute two important sources of information for monitoring this relationship. Increasing scarcity of resources available for fishery-independent research, together with its limitations in addressing the nexus between fisheries and resource conditions bolsters the need for fishery-dependent data.

The Big Creek hook-and-line survey is an unusual experiment in fishery-dependent data collection that may provide insights into the twofold need for of useable fishery-dependent data and marine reserve evaluation. The system is the product of a longstanding cooperative arrangement between the Reserve manager and local fishermen to monitor resource conditions and the local fishery to insure its sustainability. In addition to affording potentially useful data, the cooperative arrangement appears to have enhanced fishermen's understanding of the relevance of science to fishery management and encouraged their involvement in the data collection and fishery monitoring process.

Much fishery-dependent (e.g., logbook) data, however, are highly problematic in two ways: 1) they are (or are assumed to be) unreliable, and 2) they are not systematically collected, and therefore are plagued by multiple and unknown sources of variation. The Big Creek survey data may be no exception. Our preliminary work at Big Creek, however, has raised many questions that suggest a need for further study of the fishery, the cooperative arrangement, and its potential contribution to the management of both the Big Creek Marine Ecological Reserve and the fishery.

Nonetheless, our preliminary research suggests that this voluntary, cooperative arrangement between local fishermen and the Reserve manager provides an excellent opportunity for obtaining much needed fishery-dependent data on resource conditions adjacent to the reserve, and a possible model for analogous situations.

We have proposed (and received preliminary approval of) a three-year study at big Creek, the goals of which are to analyze the hook-and-line fishery adjacent to the reserve, to evaluate and analyze the hook-and-line survey data collected, and to assess and optimize the voluntary data collection system as a source of useable fishery-dependent data for management of both the reserve and the fishery. The proposed work will benefit California Department of Fish and Game, Big Creek researchers and fishermen, and the people of the state. CDFG, researchers, and the state will benefit from the optimization of the fishery-dependent data collection system at virtually no cost. The resultant data and analyses will 1) complement fishery-independent data from the region, and thereby enhance resource management, 2) afford opportunities for further collaborative and comparative research on Big Creek, marine reserves in general, and cooperative management of fisheries, and 3) provide fishermen with a more complete, longitudinal assessment of the fishery, and validation of their past and continued cooperation in the management of the marine reserve and nearby fisheries.

References

Acheson, J.M. (1988). *The lobster gangs of Maine.*, University Press of New England, Hanover.

Attwood, C.G., and Bennett, B.A. (1994). "Variation in dispersal of Galjoen (*Coracinus capensis*) (Teleostei: Coracinidae) from a marine reserve." *Canadian journal of fisheries and aquatic science*, 51, 1247-1257.

Bennett, B.A., and Attwood, C.G. (1991). "Evidence for recovery of a surf-zone fish assemblage following the establishment of a marine reserve on the southern coast of South Africa." *Marine ecology progress series,* 75, 173-181.

Cole, R.G., Ayling, A.M., and Creese, R.G. (1990). "Effects of marine reserve protection at Goat Island, northern New Zealand." *New Zealand journal of marine and freshwater research.* 24, 97-210.

Cohen, F.G. (1986). *Treaties on Trial*, University of Washington Press, Seattle.

Durrenberger, E.P., and Palsson, G., eds. (1989). *The Anthropology of Iceland*, University of Iowa Press, Iowa City.

Georgette, S.E. (1981). "In the rough land to the south : an oral history of the lives and events at Big Creek, Big Sur, California." Environmental Field Program, University of California, Santa Cruz.

Jentoft, S., and Kristoffersen, L. (1989). "Fishermen's co-management: The case of the Lofoten fishery." *Human organization*, 48, 355-65.

Kearney, J.F. (1989). Co-management or co-optation?: The ambiguities of lobster fishery management in southwest Nova Scotia. *Co-operative management of local fisheries,* E. Pinkerton, ed., University of British Columbia (UBC) Press, Vancouver, 85-102.

Lee, K.N. (1993). *Compass and gyroscope : integrating science and politics for the environment.* Island Press, Washington, DC.

Lindblom, C.E. (1959). "The science of 'muddling through.'" Bobbs, Merrill, Indianapolis. Reprinted from *Public administration review*, 19, 79-88

McCay, B.J. (1989). "Co-management of a clam revitalization project: The New Jersey 'spawner sanctuary' program." *Co-operative management of local fisheries*, E. Pinkerton, ed., UBC Press, Vancouver, 103-124.

McGoodwin, J.R. (1990). *Crisis in the worlds' fisheries.* Stanford University Press, Stanford.

McKean, M.A. (1992). "Management of tradtional common lands (iriaichi) in

Japan." *Making the commons work*, D. Bromley, ed., San Francisco, ICS Press, 63-98.

Ostrom, E. (1990). *Governing the commons: The evolution of institutions for collective action.* Cambridge University Press, New York.

Paddack, M.J. (1996). "The influence of marine reserves upon rockfish populations in central California kelp forests." M.S. thesis, UC Santa Cruz.

Pinkerton, E., ed. (1989). *Co-operative management of local fisheries*. UBC Press, Vancouver.

Pinkerton, E., and Keitlah, N. (1990). *The Point No Point Treaty Council: Innovations by an inter-tribal fisheries management cooperative*. Discussion Paper No.26, UBC Planning Papers, UBC Press, Vancouver.

Pollnac, R.B., and Littlefiled, S. (1983). "Sociocultural aspects of fisheries management." *Ocean development and international law journal*, 12, 209-46.

Roberts, C.M., and Polunin, N.V.C. (1993). "Marine reserves: Simple solutions to managing complex fisheries?" *Ambio,* 22, 363-368.

Rowley, R.J. (1994). "Marine reserves in fisheries management." *Aquatic conservation: marine and freshwater ecosystems,* 4, 233-254.

Russ, G.R., and Alcala, A.C. (1996). "Marine reserves: Rates and patterns of recovery and decline of large predatory fish." *Ecological applications,* 6, 947-961.

Smiley, J. (unpub.). "History of the Big Creek hook-and-line kelp bed survey."

Walters , C.J. (1986). *Adaptive management of renewable resources*. MacMillan, New York.

CHAPTER 16

The Pacific Fisheries Legislative Task Force

Mary A. Morgan, Executive Director

The Pacific Fisheries Legislative Task Force was founded in 1985 as a mechanism for dealing with Pacific fisheries, marine, aquaculture and seafood issues. California, Oregon, Washington, Alaska, Idaho and Hawaii have passed resolutions through their State Legislatures to become formal Task Force Members. Each state appoints two Senators and two Representatives or Assembly Members as their official Task Force Members.

The Task Force led the West Coast efforts to ban the use of tributyltin in vessel bottom antifouling paints. Members of the Task Force introduced legislation in their respective states on the same day. Additionally, we strongly supported the congressional efforts to ban TBT nationally. The Task Force was also very active in banning the use of high seas drift nets, as well as ballast water issues.

This kind of state government interactions, regarding the ocean resources is unique. The only other Legislative Task Force that addresses resource issues is the Western States Legislative Forestry Task Force.

Fish know no borders or boundaries. Thus, any state or territory of the United States, bordering or with streams tributary to the Pacific Ocean, may become a participating member of the Pacific Fisheries Legislative Task Force.

The Task Force objectives include, but are not limited to:

1) Advocating actions needed for healthy fisheries in the Pacific states for present and future generations.

2) Cooperating with, and recommending to federal, interstate and international fishery agencies with jurisdiction affecting participating

states, including regional compacts, regional fishery management councils, fishery and seafood development foundations, and the Pacific States Marine Fisheries Commission.

3) Monitoring congressional actions of importance to the Pacific states' fisheries.

4) Promoting cooperation among the Pacific states on legislation and administrative matters affecting common fishery concerns.

5) Serving as a clearinghouse for information and opinions from all the interests involved in Pacific States' fisheries.

6) Reporting to the legislatures of the participating states, and their congressional delegations, on means of protecting and enhancing Pacific fisheries.

The Task Force looks for fisheries, marine, aquaculture and seafood issues of concern that can be addressed in a coordinated, multi-state effort.

If you would like additional information, please do not hesitate to contact the Task Force at p.o. box 849, Garden Valley, CA 95633.

CHAPTER 17

SHIFTING BASELINES AND THE PROBLEM OF REDUCED EXPECTATIONS IN NEARSHORE FISHERIES

Mia J. Tegner[1] and Paul K. Dayton[1]

Abstract

Optimal management of marine ecosystems and their fisheries should be based on a well defined natural benchmark against which human impacts can be evaluated in relation to natural variation. A difficult proposition for all ecosystems, this is particularly problematic for marine systems which are difficult to sample, especially sensitive to nutrient availability, and where natural variation is often extreme. The long history of human intervention in nearshore regions exacerbates the problem of shifting baselines; each new generation of observers accepts the species composition and stock size at the beginning of their careers as baseline. As the baseline shifts with time, there are increasingly reduced expectations of the resource which constitute inappropriate reference points for evaluating overfishing or other disturbances, as well as for establishing objectives for restoration. A framework for incorporation of historical information into ecosystem management would help in part to overcome the shifting baseline problem.

Here we use the Point Loma kelp forest near San Diego, a well-studied and productive ecosystem, to illustrate the problems of shifting baselines and reduced expectations. The kelps themselves are strongly affected by large-scale, low-frequency variability in oceanographic conditions. Humans have been part of this system for thousands of years, and some effects, such as the cascading relationship between sea otters, sea urchins, and kelps, are well understood. Nevertheless, a meaningful baseline is virtually impossible to define because so many large animals have either disappeared, or their numbers and sizes have so decreased that they cannot exert their former ecological roles, many never defined by science. Historical observations of once common black sea basses, large sharks and rays, huge schools of yellowtail and white seabasses, and abalones stacked several deep suggest the historical framework we need to develop. Our challenge is to evaluate anthropogenic impacts on the kelp forest community in relation to short term oceanographic variability and longer-term change. Much has been learned about the role of ocean climate on the kelps themselves, but we may be left with a community of plants and only ghosts of the missing animals.

----

[1]Scripps Institution of Oceanography, University of California, San Diego, La Jolla, California 92093-0201

Introduction

The ideal of sustainable use fisheries has seldom been achieved in California; indeed the only real constant in the history of fisheries seems to be that resources are almost inevitably overexploited, often to the point of collapse (Ludwig et al. 1993). The reasons, which vary with the fishery and region, include poor management, overcapitalization of the fishery, pollution, loss of habitat, freshwater diversion, and excessive bycatch. The evolving lesson from this sad history is that the single species approach to management, in which the target species is considered in isolation from the community in which it lives, doesn't work. Single species management ignores the role of target species in food webs and rarely takes into account environmental variability affecting productivity. A growing consensus in fisheries management points towards the need for sustained ecosystems to support sustained fisheries, an approach that recognizes interactions among species and their changing environment.

An ecosystem approach to management should be based on a well defined community baseline against which anthropogenic impacts can be evaluated in relation to natural variation (Pauly 1995, Dayton et al. in press). This should include information on the important populations of plants and animals, their interactions, and how these vary with changing environmental conditions. A difficult proposition for all ecosystems, this is particularly problematic for marine systems where logistics and expense constrain data, population variability is high, and rare, episodic events have long-term consequences. The scientific problems of adequate baselines are compounded by a human observer effect; each new generation of observers tends to accept conditions at the beginning of their career as baseline. Thus, as the baseline shifts with time, there are increasingly reduced expectations of the resource which constitute inappropriate reference points for evaluating overfishing or other disturbances, as well as for establishing objectives for restoration. A framework for incorporation of historical information into ecosystem managment would help in part to overcome the shifting baseline problem (Pauly 1995, Dayton et al. in press).

Here we consider the problems of shifting baselines and reduced expectations for the Point Loma kelp forest community offshore of San Diego, California. Like other kelp forest communities, this multi-use resource has supported a variety of valuable sport and commercial fisheries and kelp harvest, as well as non-consumptive recreational use, for many decades. Accessible to direct observations by diving and under nearly continuous study since the 1950s, this community is one of the better known marine ecosystems in the state (e.g., North 1971, Tegner et al. 1996), yet we have little idea of the true baseline condition. We discuss the plant and animal communities, the long history of human intervention, and the role of ocean climate to argue that incorporation of historical information into an ecosystem approach to managment would better protect resources.

The Kelp Forest Community

Centered around the giant kelp, Macrocystis pyrifera, which provides most of the structure and primary productivity, kelp forest communities are highly dynamic ecosystems (reviewed by Foster & Schiel 1985, Murray & Bray 1993, North 1994). Strong seasonal and interannual variability in plant abundance is related to physical and biological factors; over the last century, the surface canopy of the Point Loma forest has varied from essentially zero to almost 10 km^2 (Tegner et al. 1996). Intense storms cause canopy loss and extensive mortality of kelps, but typically create conditions ideal for recruitment of new plants. Recent examples of virtually complete canopy loss include the winter of 1982-83 series of storms, and the single storm of January 1988 (Dayton et al. 1992). The importance of temperature is due to the strong inverse relationship between temperature and the availability of nitrate, the nutrient which limits kelp growth; periods of warm, nutrient-stressed conditions can lead to cessation of kelp growth, deterioration of the canopy, and mortality. A major lesson of the last two decades has been the importance of large-scale, low-frequency changes in ocean climate to local kelp forests. El Niño-Southern Oscillation (ENSO) events affect plant growth rates, successional processes, and community structure; El Niño events lead to intense stress for giant kelp populations and strong understory kelp development, whereas La Niña events favor domination by Macrocystis. There has also been a secular rise in sea surface temperatures since the mid 1970s which has been associated with a 2/3 decrease in the average size of giant kelp plants as measured by the number of stipes per plant, a substantial decrease in standing biomass and thus drift available for grazers, as well as export to other communities (Tegner et al. 1996, 1997). Whether this is an interdecadal shift or the harbinger of global warming will take years to determine, but the ecosystem implications are very serious.

Important biological factors affecting the distribution and abundance of kelps include competition among kelp guilds (although this is strongly mediated by water temperature) and grazing, especially by sea urchins. Many animals eat kelps, but in terms of the frequency and magnitude of destructive grazing episodes, sea urchins are the most serious problem. Normally sustained by detrital kelp, when food becomes limiting urchins attack attached plants, in extreme cases clearing large areas of all macroalgae. Called urchin barrens, these altered states of much reduced productivity may persist for months or years, and were especially prevalent in coastal kelp forests of Southern Califonia in the 1950s and 1960s (Leighton et al. 1966).

The kelps have also been subjected to anthropogenic disturbances. Poor strategies of sewage discharge in the 1950s and 1960s were associated with the decimation of kelp forests around the major metropolitan areas of Southern California, the Palos Verdes forest off Los Angeles, as well as Point Loma (Tegner & Dayton 1991). However, improved treatment and deeper, offshore discharges led to the recovery of both forests and there are neglible, if any, effects today. The Macrocystis canopy is harvested for the production of alginates. Harvest is limited to the upper four feet of the water column, and has minimal impacts on the plants. Indeed, while there is considerable interannual variability due to the physical and biological factors described above, the long-term stability of the kelp harvest suggests that it is the best managed marine harvest of wild populations in California.

Kelp Forest Fisheries

A wide variety of fishes and invertebrates is harvested from this community, and presumably each species plays a role in the structure and stability of the ecosystem. The effects of top level predator removal by fisheries may include replacement of one predator by others, the replacement of piscivores by planktivores, and, in extreme cases, ecosystem collapse and replacement by a different suite of species in an altered stable state (Parsons 1992). For Southern California kelp forests, the important predators clearly or likely to have exerted "top-down" control of community structure include sea otters, large fishes, and elasmobranchs, the sharks and rays, all of whose populations have undergone major decreases in this region (Dayton et al. in press). Sea otters (Enhydra lutris) were hunted out of Southern California in the middle of the last century, probably by the 1830s; our extensive knowledge of their dramatic ecosystem role comes from other parts of the otter range. Lacking the refuges afforded to the charismatic marine mammals, there are no significant remnant populations of the other predators left to study, and most of these populations were heavily impacted before science could properly evaluate their community roles.

Two very large fishes (up to hundreds of kg), black sea bass (Stereolepis gigas) and broomtail grouper (Mycteroperca xenarcha), were once common residents of San Diego kelp forests. Trophies for both anglers and spear fishermen as well as subject to commercial fishing, these became very rare in the middle of this century; in thousands of dives since 1970, our research group has never observed either at Point Loma. We have no measures of their former densities, but divers from the 1940s/early 1950s report that it was typical to see several of these large fishes on a dive (Dayton et al. in press). We also know very little about their diet but they certainly prey on fishes; the Scripps Aquarium curator reports that it is difficult to maintain populations of planktivorous blacksmith (Chromis punctipinnis) in the tank with a black sea bass (R. Burhans, pers. comm.). Thus, it is quite reasonable to surmise that populations of this presently abundant planktivore, hypothesized to prey on invertebrate larvae, play a different role today than when the large piscivores were more common. Divers in the early 1950s also reported seeing schools of hundreds of white seabass (Atractoscion noblis) and yellowtail (Seriola lalandi), pelagic species once common along the outer edges of kelp forests during the summer. Subjected to intense commercial and sport fishing, landings of both species underwent major declines by the 1960s. Smaller than the black sea bass, fishes are nevertheless important to the diet of these more pelagic species (Leet et al. 1992).

Some species of sharks and rays have been fished directly, but for many species, the more serious population impacts have been due to extensive incidental capture associated with gill net fishing. In most cases, fisheries statistics are the only indicators of abundance as bycatch is poorly documented. Targeted elasmobranch fisheries exploded in the 1980s with declining availability of other species, low prices, and new public acceptance. Many of the targeted species underwent rapid increases in landings and equally rapid declines. Even with the 1992 banning of nearshore gillnets, recovery will take decades due to slow growth rates, late ages of sexual maturation, low fecundities, and relatively long lives (Leet et al. 1992). Kelp forest elasmobranchs include a variety of species with different roles, some of which are partially known despite generally poor or non-existent abundance data. Several species common at Point Loma in the 1970s are now rare or absent.

Fisheries and Sea Urchins

Because of their potential for destructive grazing episodes, the most clear cut impacts of fishing in kelp forest communities involve animals interacting with sea urchins, both predators and putative competitors. Sea otters are voracious predators of sea urchins and other benthic invertebrates. Within established sea otter territory, urchins are scarce and confined to protective habitat; outbreaks of destructive urchin grazing in these kelp forests are rare. Thus, it was not surprising that the absence of sea otters was blamed for the destructive sea urchin grazing events that were common at Point Loma and elsewhere, especially in coastal forests, in the 1950s and 1960s (e.g., Leighton et al. 1966). The persistence of kelp forests for more than a century after the otters were hunted out of Southern California, however, suggested the importance of other predators, notably spiny lobsters (Panulirus interruptus) and sheephead (Semicossyphus pulcher), whose exploitation patterns in this century provided better explanations for the apparent increases in urchin populations (Tegner & Dayton 1981). Both predators are subject to commercial and sport fishing, including by divers who put a premium on size. After the demise of black sea bass and broomtail groupers, male sheephead as the largest reef fish in the forest were sought after by spear fishermen. In 1887, the average size lobster in the statewide commercial catch was about 150 mm carapace length; by 1955 this had fallen to 119 mm and from 1976-1980, it varied from 86-90 mm. Larger predators have higher feeding rates and consume larger urchin prey; thus the functional role of these predators is very different from the turn of the century. While we have not located useful historical data on sheephead size-distributions or abundance, the change in lobster catch per trap suggests dramatic decreases in density of this crustacean. In 1888, 260 traps yielded 231,060 lb; by 1975, 19,000 traps were required to harvest almost the same number of pounds, 233,179 (Tegner & Levin 1983).

A new threat to sheephead and other kelp forest fishes began in 1988 with the creation and rapid growth of the live fish fishery (California Department of Fish and Game 1996). This fishery focuses on animals sized for a single entree (1-3 lb), visually attractive, and hardy enough to survive the rigors of capture and transportation. Statewide the sheephead landings in this fishery jumped from 16,203 in 1989 to 194,942 metric tons in 1995; over the same time period, angler landings of this species on commercial passenger fishing vessels have dropped by an order of magnitude (Dayton et al. in press). This virtually unregulated fishery is having a huge impact on many kelp forest fish species, but the effects on sheephead populations are especially problematic. Sheephead are sequential hermaphrodites and the fishery only takes small females, many of which may be prereproductive. The ecosystem implications for this fishery include increased outbreaks of destructive grazing pressure by urchins, especially by minimally-exploited purple sea urchins (Strongylocentrotus purpuratus).

Abalones and sea urchins are the two major groups of large invertebrate grazers in kelp forest communities. The groups share similar microhabitat, food preferences, life-history characteristics, behavior patterns, and many of the same predators, including man--although abalone fisheries greatly predate those for sea urchins. Both are favored prey of sea otters, and stocks sufficient to support fisheries were clearly a product of the demise of these mammals in Southern California--yet abalones have been found in Native American middens dating back thousands of years. Of considerable relevance to kelp forest community structure is the impact that these two grazer groups had on each other's abundances in terms of competition for food and habitat. Five species of abalones in Southern California supported lucrative commercial and sport fisheries that date as far back as the 1850s. Densities are unknown but one can infer from landings data that they must have been very high; in 1879 Chinese immigrants, working intertidally and in shallow waters with rowboats and gaffs, landed 4.1 million lbs of black (Haliotis cracherodii) and green (H. fulgens) abalones in a fishery centered at San Diego (Cox 1962). Edwards (1913) observed hardhat diving fisheries for green and pink (H. corrugata) abalones on Santa Catalina and San Clemente Islands, and reported that a diver could gather 1.5 to 2 tons in a four hour shift. Observers in the San Diego region in the early 1950s reported green abalones so abundant that they were stacked in shallow water, and pavements of red (H. rufescens) abalones in the Point Loma kelp forest (Dayton et al. in press). Of less economic interest, we have found few observations of sea urchins until important coastal kelp forests near the major metropolitan centers of Southern California began to decline in the 1940s and 1950s. The apparent sea urchin population explosion was attributed to the demise of sea otters, the depletion of abalone competitors, and sewage enrichment of sea urchin nutrition (Leighton et al. 1966, North & Pearse 1970).

The history of the California abalone fishery integrates so many law changes, area shifts, and gear improvements that it is very difficult to infer population changes from fisheries statistics, but there was a sequential depletion of species to levels less than 10% of their maximum yields. Combined-species landings peaked in 1957 (Leet et al. 1992). As of May 1997, all five of the commercially exploited abalone species have been closed to sport and commercial fishing, and stocks of the white (H. sorenseni) abalone are so depleted that recovery without massive human intervention is unlikely (Davis et al. 1996). Adult abalones are rare and it is clear that they no longer exert their former ecological roles at Point Loma. While there is little question that abalones and sea urchins compete for food and space, their respective influence on each other's population sizes is a complicated issue affected by many factors including relative abundances, food availability, and predation pressure (Tegner & Levin 1982). In Northern California, where there is no commercial abalone fishing and sport take is restricted to free diving and shore picking, red abalone stocks have remained healthy. The intense sea urchin fishery in that region that began in the late 1980s resulted in an increase in the number and size of kelp forests (Leet et al. 1992), which was associated with increases in abalone abundance and growth rates (K. Karpov, pers. comm.), suggesting a release from competition with urchins.

The demise of some stocks and opening of new markets, especially Asian, has led to the creation of new fisheries on formerly unexploited groups such as sea urchins, snails, and sea cucumbers. The sea urchin fishery, which began in the early 1970s, focuses almost exclusively on the red sea urchin (S. franciscanus). While this is the largest and most important grazer, two other urchin species are still capable of generating barren grounds. In the Point Loma kelp forest, the red urchin fishery has substantially altered the distribution, abundance, and size-frequency distribution of this species. Once common throughout the forest, red urchins are now rare in much of the bed (Tegner & Dayton 1991). As the spine canopy of red urchins can be important nursery habitat for juvenile abalones, this change too may have ecosystem implications. The ecosystem consequences of newer fisheries for snails (Kelletia kelleti and Lithopoma undosa) and sea cucumbers (Parastichopus parvimensis) are totally unknown.

Ghosts and Shadows

We do not have space to go into more kelp forest ecosystem impacts of single species approaches to managment, but Table 1 lists important exploited animals in the Point Loma kelp forest whose ecological roles are gone, those which exert mere shadows of their former roles, and those currently under assault. Groups classifed as shadows are rare relative to their former abundances. Groups classified as under assault are still relatively common, but have greatly altered spatial distributions, abundances, and/or size-frequency distributions. Some were target species, others were discarded as bycatch.

In contrast with the poorly defined community roles of so many of the animal members of this ecosystem, the biology of the kelps themselves is comparatively well understood (reviewed by Foster & Schiel 1985, Murray & Bray 1993, North 1994). The forests vary in size with storms, ocean climate, and the occasional episode of destructive sea urchin grazing (Dayton et al. 1992, Tegner et al. 1996, 1997), with little hint of the loss of so much animal biomass (Dayton et al. in press).

Table 1. Effects of Fishing on Community Roles of Kelp Forest Species

GHOSTS	SHADOWS	UNDER ASSAULT
Giant Sea Bass	Sharks	Kelp and Sand Basses
Broomtail Grouper	Rays	Sheephead
White Abalones	Cabezon	Rockfishes
	Lingcod	Spiny Lobsters
	Scorpionfish	Sea Urchins
	White Seabass	Sea Cucumbers
	Yellowtail	Snails
	Other Abalones	Octopuses
		Rock and Spider Crabs

Reduced Expectations

> The writer has been observing the fisheries of Southern California for nearly thirty years. In that time, the supply has dropped off to a menacing extent, due to lack of laws, lack of protection, and over fishing where fishes should be protected.
>
> C. F. Holder (1914)

Holder, a professor at Throop College of Technology (which became the California Institute of Technology) and president of the Wildlife Protective League of America, founded the Tuna Club at Avalon in 1898 in a successful attempt to reduce 'wastage' by sportsmen by awarding prizes for the largest fish caught on the lightest tackle. He was also instumental in establishing a reserve from net fishing in the waters inshore of three miles around Santa Catalina Island to protect spawning grounds in 1913. One can only imagine what Holder, who based his actions on his personal experience dating to 1886, would have to say about the very different status of marine fisheries in 1997. We have no meaningful marine reserves where today's managers can study natural systems under their full potential. Expectations of what a resource could or should be underlie all management actions, and if those expectations are reduced by acceptance of a new status quo, management actions will not support the full potential of the ecosystem. The reality of the problem of reduced expectations was driven home when one of us (MJT) served on a white seabass advisory panel. Some members of that panel felt that because landing statistics had been stable for the last decade--despite being less than 10% of previous yields--there was no cause to take advantage of an opportunity for major corrective action.

Conclusions

Most marine ecosystems have no known baseline. Southern California kelp forests are one of the better known coastal communities yet, with a history of human exploitation dating back thousands of years, we will never fully appreciate what pristine conditions were. In addition to fishing pressure, many exploited populations, both plants and animals, are strongly affected by ocean climate which produces cycles of abundance on a variety of time scales from ENSOs of one to three years, to regime shifts of 50 to 70 years (e.g., MacCall 1996, Tegner et al. 1996). With the direct and indirect or ecosystem effects of fishing interacting with ocean climate and exacerbated by the problem of reduced expectations in managers, a useful baseline seems like an elusive goal. Nevertheless, for both the protection of ecosystems and realization of their productivity, it is critical that we recognize this limitation for both fisheries and environmental decisions. Shifting baselines are inappropriate reference points for evaluating overfishing or other anthropogenic disturbances ranging from sewage disposal to oil spills. Similarly enhancement goals based on recent baselines may severely underestimate the potential productivity of the ecosystem.

We submit that a framework for incorporation of historical information into ecosystem management would help overcome the shifting baseline problem, and allow an evaluation of the true social and ecological costs of fishing. An ecosystem based approach to management, one that recognizes that viable fisheries require sustained ecosystems, must be based on ecological assessments before new species are exploited and a more risk-averse management scheme for traditional fisheries. The history of failed fisheries in kelp forests and the economic return that we know that these fisheries are capable of providing demand that we do better.

It is not too late to rectify this situation because, except for the white abalone (Davis et al. 1996), none of these species is close to extinction. If the political will can be mustered, it is still possible to establish meaningful reserves or sanctuaries that are large enough to support at least the non-migratory species (Dayton et al. 1995). Unfortunately, most (or all) existing reserves are far too small to allow most such populations to become established, and we utterly lack the political will to enforce the law on even small reserves (e.g., Tegner 1993). Until there is the political interest and will power to radically change our management of marine resources, the future of coastal environments seems bleak indeed.

Acknowledgements

We thank P. B. Edwards and K. L. Riser for their many contributions to this research, and the National Science Foundation and the City of San Diego for their support.

References

California Department of Fish and Game. (1996). "Review of some California fisheries for 1995." Calif. Coop. Ocean. Fish. Invest. Rep. 37, 7-21.

Cox, K.W. (1962). "California abalones, family Haliotidae." Calif. Dept. Fish Game Fish Bull. 118, 1-133.

Davis, G.E., Haaker, P.L., and Richards, D.V. (1996). "Status and trends of white abalone at the California Channel Islands." Trans. Am. Fish. Soc. 125, 42-48.

Dayton, P.K., Tegner, M.J., Parnell, P.E., and Edwards, P.B. (1992). "Temporal and spatial patterns of disturbance and recovery in a kelp forest community." Ecol. Monogr. 62, 421-445.

Dayton, P.K., Thrush, S.F., Agardy, M.T., and Hofman, R.J. (1995). "Environmental effects of marine fishing." Aq. Cons. Mar. Freshw. Ecos. 5, 205-232.

Dayton, P.K., Tegner, M.J., Edwards, P.B., and Riser, K.L. "Ghost Communities and the problem of reduced expectation in kelp forests." Ecol. Appl. (in press).

Edwards, C.L. (1913). "The abalones of California." Popular Science Monthly, June 1913, 533-550.

Foster, M.S., and Schiel, D.R. (1985). *The ecology of giant kelp forests in California: A community profile.* U. S. Fish Wildl. Serv. Biol. Rep. 85(7.2).

Holder, C.F. (1914). "Attempts to protect the sea fisheries of Southern California." Calif. Fish and Game 1(1), 9-19.

Leet, W.S., Dewees, C.M., and Haugen, C.W. (1992). *California's living marine resources and their utilization.* Univ. Calif. Sea Grant Extension Publ. 92-12.

Leighton, D.L., Jones, L.G., and North, W.J. (1966). "Ecological relationships between giant kelp and sea urhins in Southern California." In: Proc. Fifth Ann. Seaweed Symp., E.G. Young and J.L. McLachlan (eds.). Pergamon Press, Oxford, pp. 141-153.

Ludwig, D., Hilborn, R., and Walters, C. (1993). "Uncertainty, resource exploitation, and conservation: lessons from history." Science 260, 17, 36.

MacCall, A.D. (1996). "Patterns of low-frequency variability in fish populations of the California Current." Calif. Coop. Ocean. Fish. Invest. Rep. 37, 100-110.

Murray, S.N. and Bray, R.N. (1993). "Benthic macrophytes." In: *Ecology of the Southern California Bight*, M.D. Dailey, D.J. Reish, and J.W. Anderson (eds.). University of California Press, Berkeley, pp. 304-368.

North, W.J. (1971). *The biology of giant kelp beds (Macrocystis) in California.* Beihefte Nova Hedwegia 32, 1-600.

North, W.J. (1994). "Review of Macrocystis biology." In: *Biology of economic algae*, I. Akatsuka (ed.). Netherlands, The Hague, pp. 447-527.

North, W.J. and Pearse, J.S. (1970). "Sea urchin population explosion in Southern California coastal waters." Science 167, 209.

Parsons, T.R. (1992). "The removal of marine predators by fisheries and the impact of trophic structure." Mar. Poll. Bull. 25, 51-53.

Pauly, D. (1995). "Anecdotes and the shifting baseline syndrome of fisheries." Trends Ecol. Evol. 10, 430.

Tegner, M.J. (1993). "Southern California abalones: can stocks be rebuilt using marine harvest refugia?" Can. J. Fish. Aq. Sci. 50, 2010-2018.

Tegner, M.J. and Levin, L.A. (1982). "Do sea urchins and abalones compete in California kelp forest communities?" In: *International echinoderm conference, Tampa Bay*, J.M. Lawrence (ed.). A.A. Balkema, Rotterdam, pp. 265-271.

Tegner, M.J. and Levin, L.A. (1983). "Spiny lobsters and sea urchins: analysis of a predator-prey interaction." J. Exp. Mar. Biol. Ecol. 73, 125-150.

Tegner, M.J. and Dayton, P.K. (1981). "Population structure, recruitment, and mortality of two sea urchins (Strongylocentrotus franciscanus and S. purpuratus) in a kelp forest near San Diego, California." Mar. Ecol. Prog. Ser. 5, 255-268.

Tegner, M.J. and Dayton, P.K. (1991). "Sea urchins, El Niños, and the long term stability of Southern California kelp forest communities." Mar. Ecol. Prog. Ser. 77, 49-63.

Tegner, M.J., Dayton, P.K., Edwards, P.B., and Riser, K.L. (1996). "Is there evidence for long-term change in Southern California kelp forest communities?" Calif. Coop. Ocean. Fish. Invest. Reps. 37, 111-126.

Tegner, M.J., Dayton, P.K., Edwards, P.B., and Riser, K.L. (1997). "Large-scale, low-frequency oceanographic effects on kelp forest succession: a tale of two cohorts." Mar. Ecol. Prog. Ser. 146, 117-134.

CHAPTER 18

California and International Conservation of Large Ocean Fishes of the Pacific Ocean

Michael L. Weber

Abstract

Many species of highly migratory, large ocean fishes utilize coastal and ocean waters off California. Some such as swordfish, marlins, sailfish, tunas, and sharks are caught in commercial and recreational fisheries. Some also are incidentally captured and discarded in fisheries. The conservation status and trends of many species are poorly known.

A pervious analysis of existing international fisheries management programs revealed enormous gaps in the conservation of large ocean fishes that at some point in their lives utilize coastal and ocean waters off California. These gaps are of two general types. Geographical gaps result form incomplete coverage of fisheries for large ocean fishes by existing management institutions. Functional gaps result from the lack of authority or capability in an international regime to carry out some key element in conservation, such as enforcement of data collection and analysis.

Internationally, the Inter-American Tropical Tuna Commission has limited authority for the management of tunas, and has exercised still less, especially in recent years. Nor does the IATTC exercise management responsibility over several species of tunas as well as billfishes. No international organization has undertaken study, much less management of oceanic sharks in the Pacific Region.

This paper expands the analysis above to include the Regional Fishery Management Councils established under the Magnuson Fishery Conservation and Management Act, as well as programs of the State of California. In analyzing Federal and state programs, this paper uses as a benchmark the recently concluded United Nations Agreement on Straddling Stocks and Highly Migratory Stocks.

The paper concludes with recommendations for ways in which the State of California might promote conservation of large ocean fishes in its waters through promotion of sound standards of conservation and management at national and international levels.

CHAPTER 19

LANDOWNER PERSPECTIVE ON ENVIRONMENTAL PLANNING

Sat Tamaribuchi

Abstract

Large scale comprehensive planning to address environmental issues has been a widely supported objective for many years. There are increasing efforts to encourage more comprehensive planning at the regional, subregional, and watershed level. However, many efforts in the past have not led to workable and effective plans. The multi-species, multi-habitat Natural Community Conservation Plan for the Orange County Central and Coastal Subregion and Section 208 Water Quality Plan for the Newport Bay Watershed have produced workable and effective programs. This paper provides a landowner perspective on making comprehensive environmental planning workable and effective.

CHAPTER 20

ORANGE COUNTY CENTRAL AND COASTAL PROGRAM

BY

Tim Neely **Rod Meade**

I. NEED FOR THE PROPOSED PROJECT

The County of Orange Environmental Management Agency (EMA) has prepared a Natural Community Conservation Plan and Habitat Conservation Plan (NCCP/HCP) for the Orange County Central and Coastal Subregion of the Southern California NCCP Coastal Sage Scrub (CSS) program. The County EMA and U.S. Fish and Wildlife Service (USFWS) are the lead agencies for this joint Environmental Impact Report/Environmental Impact Statement (EIR/EIS) prepared to facilitate environmental review of the proposed Central/Coastal NCCP/HCP.

Tim Neely, Administrator
Planning & Zoning
County of Orange
P.O. Box 4048
Santa Ana, CA 92702-4048

Rod Meade
R.J. Meade & Associates
7910 Ivanhoe Ave. Ste. 40
La Jolla, CA 92037

The need for the proposed subregional CSS NCCP/HCP has been established over recent years by a combination of cumulative impacts on habitat resources and the legislative and regulatory responses to those impacts. The listing of the California gnatcatcher as "threatened," and the proposed listing of several other species that rely upon the coastal sage scrub habitat have signaled the need to shift the conservation panning focus from single species, project by project efforts to conservation planning at the natural community level. The Southern California NCCP CSS Program was developed to address this need, facilitating regional protection of a range of species inhabiting a designated natural community - in this case CSS and its associated mosaic of habitat types - while allowing compatible land uses and appropriate growth and economic development.

The "take" (*i.e.*, killing, harm, harassment, including significant habitat loss) of threatened and endangered species is prohibited under Section 9 of FESA and its implementation regulations. The USFWS, however, may issue permits to incidentally take listed species in connection with otherwise lawful activities. In a manner similar to Section 10(a)(1)(B) of FESA, the NCCP Act contains statutory provision authorizing the take of species that would be prohibited or limited by CESA.

II. PUBLIC REVIEW PROCESS AND PROGRAMMATIC REVISIONS IN RESPONSE TO PUBLIC COMMENTS AND FURTHER AGENCY REVIEW

A. Documents Distributed/Available for Public Review

The NCCP/HCP, EIR/EIS and NCCP/HCP Implementation Agreement were prepared in cooperation with the California Department of Fish and Game (CDFG) and U.S. Fish and Wildlife Service (USFWS). These two resource agencies are responsible for implementing the California Endangered Species Act (CESA) and Federal Endangered Species Act (FESA). The Orange County EMA is the lead agency responsible for the preparation of the NCCP/HCP and the EIR while the USFWS is the lead agency responsible for managing preparation of the EIS.

The NCCP/HCP, Joint EIR/EIS and Implementation Agreement were distributed under a single cover by the County of Orange to facilitate public review of the project. The overall project documentation is presented in seven parts:

- an Introduction to the NCCP that provides planning and regulatory background information and the NCCP/HCP that contains the substance of the proposed subregional conservation strategies;

- the Joint EIR/EIS that evaluates environmental consequences and alternatives;

- the Implementation Agreement that outlines the specific enforceable measures and mechanisms that will be required to effectively implemented the NCCP/HCP;

- maps and other figures (termed "figures") referred to by the NCCP/HCP, the EIR/EIS and the Implementation Agreement (collected in one volume for ease of reference and economy of color reproductions); and

- appendices containing materials cited in or used in the preparation of the NCCP/HCP, EIR/EIS and Implementation Agreement.

B. Public Comments, Responses to Comments and Revisions Reflecting Responses

As part of the public review process pursuant to CEQA, NEPA, the NCCP Act and FESA, extensive public comments were received. The written public comments have been reproduced in the accompanying document title "County of Orange Central & Coastal Subregional Part III; Join Programmatic EIR/EIS - Comment Letters." Public testimony was also presented in oral form at public hearings before the County of Orange Planning Commission and the County of Orange Board of Supervisors.

Extensive written comments were prepared and are set forth in the accompanying document titled "County of Orange Central and Coastal Subregion Part III: Joint Programmatic EIR/EIS - Response to Comments." In broad terms, the responses are grouped under two categories: (1) "General Responses" presenting issues raised in a number of comments and/or which appeared to be of general interest and (2) "Specific Responses" which comprise responses to specific issues raised by only one party or a few parties.

In addition to providing substantive responses to specific comments, the Response to Comments document also indicates whether changes were to be made to the NCCP/HCP, the Implementation Agreement and/or the EIR/EIS in response to the comment and further lead agency assessment by the County of Orange, CDFG and USFWS. In addition to clarifications and corrections of specific matters in the documents, the responses included revisions that may be categorized as follows:

- ***Species-Related Considerations***
  - Three species (the golden eagle, the prairie falcon and the foothill mariposa lilly) were shifted to the category of "conditionally covered species" due to the particular needs of those species.
  - Three species which are particularly associated with grasslands (the white-tailed kite, the loggerhead shrike and the California horned lark) were deleted from the list of Identified species until such time as the NCCP Non-Profit is able to proceed with grasslands management programs, NCCP surveys and other information gathering.
  - The Implementation Agreement provisions relating to species dependent upon or associated with CSS and covered habitats were modified to provide for species-specific assessments at the time of any future listing of such species and to deal with regulatory contingencies involving assurances to *participating landowners* and requirements for the future issues of section 10(a)(1)(B) permits.

- Revisions were made to the provisions for the purchase of the Pacific pocket mouse site on Dana Point Headlands and construction-related minimization measures were incorporated into the EIR/EIS Mitigation measures and the Implementation Agreement.

- The Implementation Agreement has been modified to provide that an assessment will be made, within one year of the hiring of the NCCP Non-Profit Executive Director, as to whether special management measures should be undertaken for the cactus wren

- ***Future Public Involvement Considerations***

  - Membership of the Board of the Directors of the NCCP Non-Profit was modified to specifically provide for three "public" members.

  - A provision was made in the Implementation Agreement for the appointment of a Technical Advisory Committee to provide scientific input into policy matters considered by the NCCP Non-Profit.

  - Although likely required by existing law, the Implementation Agreement was modified to specifically commit the NCCP Non-Profit to open public meetings.

- ***Reserve System Configuration and Commitments***

  - Some revisions were made to the Reserve System boundaries - most revisions generally involved shifting areas proposed for inclusion in the Reserve System into the "Existing Use" category of lands subject to ongoing USFWS listed species jurisdiction (including the prohibitions against take in Section 9 of FESA).

- The Implementation Agreement was modified to provide for commitments of signatory jurisdiction to manage Reserve System lands in a manner consistent with the purposes of the NCCP/HCP after the expiration of the 75-year term of the Implementation Agreement/permits. Special Linkage provisions for signatory public agencies were also made subject to this commitment (private lands in Special Linkage areas were already subject to conservation easement requirements).

Other revisions involved the addition of some other lands to the "Existing Use" category, revisions to the MCAS El Toro re-use provisions and revisions to Implementation Agreement provisions relating to future CDFG actions in conjunction with state CESA listings.

Where applicable, the text of this EIR/EIS has been modified to reflect revisions, clarifications and corrections made in the Response to Comments including corresponding revisions to the NCCP/HCP, the Implementation Agreement and the NCCP/HCP Map Book.

III. PROJECT OVERVIEW

A. Regulatory Framework

Approval of the EIR/EIS and implementation of the NCCP/HCP would allow the conservation of large, diverse areas of natural habitat, including coastal sage scrub (CSS), the habitat of the federally threatened coastal California gnatcatcher. These actions, in turn, would provide the mitigation basis for allowing impacts to some CSS habitat and certain other habitat types designated as "covered habitats" caused by the construction of various private and public projects that conform to the NCCP/HCP.

The regulatory framework within which the NCCP/HCP and EIR/EIS were prepared is described in detail in both documents, and includes:

- *The California Natural Community Conservation Planning Act of 1991: The purpose of this act is to provide long-term regional protection of natural vegetation and wildlife diversity while allowing compatible land uses and appropriate development and growth. The NCCP process is intended to shift the focus of conservation efforts from single species to a range of species that inhabit a designated natural community, such as CSS. Subsequent to the passage of the NCCP Act, Process and Conservation Guidelines were prepared and adopted for the Southern California NCCP Coastal Sage Scrub program to assist in the preparation of NCCPs at the subregional level.*

- *The listing (March 30, 1993) of the coastal California gnatcatcher as a "threatened" species under the provisions of the Federal Endangered Species Act.*

- *The Special 4(d) Rule: This regulation was enacted by the Department of the Interior to encourage the preparation of NCCPs by establishing the NCCP Act as the substantive program for addressing the federal listing of the gnatcatcher.*

- *The "Region 1 Guidelines for Determining Covered species Lists and Assurances Relative to Habitat Conservation Planning (USFWS August 1, 1995)*

The County's Central and Coastal Subregion is one of eleven NCCP subregions within the five-county southern California area identified by the State of California's Southern California Coastal Sage Scrub NCCP program (Figure 2). This NCCP pilot program focuses on the protection of coastal sage scrub habitat (CSS), and adjacent habitats. By formulating conservation strategies for entire habitat systems, the state's NCCP program attempts to address long-term biological protection and management of multiple species and associated habitats at a subregional level.

Under the NCCP approach, the focus changes from protecting individual species to *conserving* natural communities and *accommodating* compatible land uses. The NCCP program is designed to provide incentives that will attract landowners, government agencies, and public interest to become stakeholders in a collaborative partnership. Conservation principles are applied at the natural community level, rather than focusing on new listings and regulating individual species. This shift is in focus toward protection of multiple species within a mosaic of natural communities is intended to enhance the ability of local, state and federal agencies to provide long-term protection for a broad range of species that are dependent on the natural communities. Accordingly, the NCCP/HCP focuses on creating a multiple-species, multiple-habitat subregional Reserve System and implementing a long-term "adaptive management" program that will protect coastal sage scrub (CSS) and other habitats and species located within the CSS habitat mosaic, while providing for economic uses that will meet the social and economic needs of the residents and businesses of the subregion.

B. Study Area

The Central and Coastal Subregion includes the central portion of Orange County from the coast inland to the Riverside Freeway (SR-91) (See Figure 1). Along the coast, the subregion extends from the mouth of the Santa Ana River in the City of

Costa Mesa to the mouth of San Juan Creek, in the City of Dana Point. The Central and Coastal Subregion covers approximately 208,000 acres. Approximately 104,000 acres are natural habitat. A total of 34,500 acres of CSS exists in the study area, embedded within a mosaic of non-CSS vegetation communities. CSS is a naturally fragmented and dispersed vegetation type. As a result, conservation planning efforts for CSS generally include significant non-CSS areas, such as chaparral and grasslands and their resident species.

C. Alternatives Reviewed

Four alternatives (see NCCP/HCP Alternatives Matrix at p. xx) have been selected for initial review and screening for further consideration in this EIR/EIS:

- the *Proposed Project* Alternative;
- the *No Project* Alternative (the "*No Action Alternative*" for NEPA purposes);
- the *No Take* Alternative; and
- a *Programmatic* Alternative.

The Section 10(a) FESA regulations require that permit applicants identify "what alternative actions to such taking the applicant considered and the reasons why such alternatives are not proposed to be utilized ..." CEQA and NEPA similarly require a review of a reasonable range of alternatives to the Proposed Project.

Two of the alternatives considered in the EIR/EIS, the No Project and No Take Alternatives, are fundamentally different from the Proposed Project because they focus on a subregional strategy of project-by-project review and regulation instead of formulating and implementing a subregional conservation strategy that defines a reserve system and management program at one point in time. The No Project Alternative would rely on the application of FESA Section 7 consultation and Section permit processes to protect the coastal California gnatcatcher, while the No Take Alternative would rely on the prohibitions of take included in Section 9 of the FESA to protect the gnatcatcher. Neither the No Project nor the No Take Alternatives would necessarily restrict or limit impacts on the cactus wren, the orange-throated whiptail (two of the three "target species") or the "identified" species recommended for regulatory coverage under the Proposed Project (although the No Project Alternative has potential for addressing unlisted species).

A fourth alternative, the Programmatic Alternative, would formulate a subregional Reserve System program, but involves a different approach to assembling the subregional Reserve System. Under the Programmatic Alternative, the subregional Reserve System would be assembled incrementally over time as specific projects requiring mitigation move forward and contribute mitigation fees or dedication lands to a management entity. This approach provides for more flexibility, but less certainty than the Proposed Project, in defining specific reserve boundaries and allows for a longer time period for accumulating scientific understanding regarding reserve design than is the case with the Proposed Project. The EIR/EIS addresses the primary substantive considerations of the Programmatic Alternative in the context of the review of alternative reserve design considerations in Chapter 5.

IV. PROPOSED PROJECT

A. Proposed Actions

The state and local government actions to be analyzed under CEQA are: (a) County approval of the NCCP/HCP as the CEQA lead agency, (b) CDFG approval as a CEQA responsible agency, including the approval of the NCCP/HCP and Implementation Agreement pursuant to NCCP Section 2810, 2820, 2825, 2830 and 2835 of the NCCP Act - the Implementation Agreement would serve as the management authorization for the immediate issuance of Section 2081 permits, and other appropriate take authorization pursuant to the NCCP Act, for all species treated as "identified species" in the NCCP and the Headlands plant species and future Section 2081 permits for species subject to the "covered habitat" provisions of the NCCP/HCP, and (c) Local government approval as signatories to the Implementation Agreement.

The federal actions analyzed under NEPA in this EIR/EIS are: (1) approval of the NCCP/HCP; (2) issuance of incidental take permits pursuant to Section 10(a)(1)(B) of FESA and amendment of the USFWS's FESA Section 10(a)(1)(A) permit for scientific study and recovery of the Pacific pocket mouse; and (3) execution of an Implementation Agreement (to be signed by USFWS, CDFG, participating local governments and *participating landowners* in conjunction with final state and federal approval of the NCCP/HCP) which provides for specific measures and assurances required to carry out the NCCP/HCP.

B. Results of the Proposed Actions

One result of the proposed actions would be the establishment of a subregional Reserve System that would include more than 37,000 acres of land consisting of two reserves - the "Coastal" reserve and the "Central" reserve (see Figure 12). Other areas, called Special Linkage Areas that enhance biological connectivity within the Reserve System and subregion and protect remnant populations of "identified species" and/or important habitat are also designated for specific habitat protection measures but not included in the NCCP Reserve System. Finally, certain areas are defined as "Existing Use Areas" where current endangered species regulation would remain the same as it presently exists (*i.e.*, anyone wishing to undertake an activity resulting in take of a state or federally listed species would have to obtain permits under CESA and FESA).

The proposed Reserve System is intended to provide the mitigation basis for allowing proposed CSS habitat conversion and impacts on "covered habitats" by meeting two fundamental requirements:

- assuring "the maintenance of net habitat value on a long-term basis" through the creation of a subregional Reserve System and Adaptive Management Program (addressing the provisions of the NCCP Act as applied through the NCCP Conservation Guidelines); and

- assuring that permitted activities will not "not appreciably reduce the likelihood of survival and recovery of the species in the wild" and will satisfy the other permit issuance requirements of Section 10(a)(1)(B) of the FESA.

The key elements of the NCCP/HCP conservation strategy are:

- *Reserve System* - creation of a publicly-owned habitat Reserve System that includes CSS and other habitat types representative of all but one of the major habitat types currently existing within the subregion (see Figure 4);

- *Special Linkages* - designation of "Special Linkages" to enhance biological connectivity within the subregion;

- *Existing Use Areas* - designation of areas containing significant populations of Target Species where existing uses are compatible with habitat protection and that would remain subject to currently existing CESA and FESA regulation;

- *Adaptive Management* - the creation of an institutional basis and funding program for undertaking management actions necessary to sustain populations over the long term, and in so doing, to adapt management actions to new information and changing habitat needs.

- *Interim Management* - provisions for extensive "interim" management of designated reserve lands prior to the time of the actual transfer of these lands to public ownership;

- *Funding* - establishment of a funding program to pay for the "Adaptive Management Program," additions to the Reserve System, and other mitigation measures; and

- *Mitigation Fee Option for "Non-Participating Landowners"* - provisions for an optional approach for mitigation of CSS impacts proposed to be authorized on lands located within the subregion but outside the Reserve System and owned by landowners who have not participated in the assemblage and management of the Reserve System through the contribution of reserve lands or planning/implementation funding.

Each of these elements is described in detail in the NCCP/HCP and reviewed in the EIR/EIS. Summary descriptions of each are included below:

1. NCCP/HCP Reserve System

The Reserve System encompasses more than 37,000 acres of land (see Figure 12). While NCCP/HCP reserve planning efforts focused on protecting significant coastal sage scrub habitat land, other habitat areas were also considered to be very important. Such areas were included in the reserve to assure that animal movement and genetic exchange continue to occur and to provide for bio-diversity within the Reserve System. The proposed Reserve System is comprised of two geographic "subareas" - The *Coastal Subarea Reserve* and the *Central Subarea Reserve*.

-- Description of the Coastal Subarea Reserve

The Coastal Subarea Reserve is proposed to contain 17,201 acres (more than 27 square miles) located primarily in and surrounding the San Joaquin Hills (Figure 16). The proposed reserve extends from the shoreline in Crystal Cove State Park inland almost 7.5 miles to a point close to the I-405 Freeway. Starting from the Upper Newport Bay reserve, the reserve extends southeast approximately 16 miles to the confluence of Oso Creek and Trabuco Creek, adjacent to the Southern NCCP subregion. Most of the proposed reserve is located within the unincorporated jurisdiction of the County; however, significant portions of the reserve are within the cities of Irvine, Laguna Beach, and San Juan Capistrano. Smaller portions of the reserve also are located within the cities of Costa Mesa and Newport Beach. The dominant physiographic features within the proposed subarea reserve include Upper Newport Bay, the various drainages and ridges contained within the coastal and inland slopes of the San Joaquin Hills, the Aliso-Woods canyons, and the Oso/Trabuco creek corridor (the latter in the City of San Juan Capistrano).

-- Description of the Central Subarea Reserve

The Central Subarea Reserve is proposed to comprise a 20,177-acre system (more than 31 square miles) locate south and west of the Cleveland National Forest in the foothills and frontal slopes of the Santa Ana Mountains (Figure 15). More than 92 percent of the Reserve System is located in the existing unincorporated County jurisdiction, but small areas on the western edge of the reserve are included in two County regional parks located within the cities of Anaheim, Orange, and Tustin. On the west, the subarea Reserve System extends from Santiago Oaks Regional Park, in the City of Orange, about 14 miles southeast to El Toro Road, the boundary with the adjacent South NCCP subregion. From its northernmost point in the Coal Canyon Preserve adjacent to the Cleveland National Forest boundary, the reserve extends about 7.5 miles southwest to the southern edge of the frontal slopes of the Lomas de Santiago. Major physiographic features contained within the Central Subarea Reserve include Windy Ridge, Weir Canyon, Irvine Lake, the frontal slopes of the Lomas de Santiago, and Limestone Canyon.

In addition to the CSS/gnatcatcher impacts discusses above, the NCCP/HCP creates a temporary 22-acre preserve on the Dana Point Headlands site for the federally-endangered Pacific pocket mouse. This temporary preserve is not a part of the subregional habitat Reserve System. It is created and funding is provided ($700,000 over and above the NCCP Endowment fund) to study the pocket mouse and determine the feasibility of alternative population conservation enhancement techniques.

-- Ownership of Reserve Lands and Coordination of Management

When fully assembled, the entire reserve would be owned and managed by public agencies, with management coordination provided by a non-profit corporation that would be created and would consist of representatives of individual public agency reserve owners, the CDFG and USFWS. This non-profit corporation would coordinate activities within the Reserve System, receive and disburse funds to the reserve owners, hire staff and biologists and prepare annual reports for public review.

The following local governments own lands, are responsible for dedication programs or are designated as potential recipients of future dedication of lands proposed to be included in the NCCP Reserve System:

- City of Anaheim;
- City of Irvine;
- City of Laguna Beach;
- City of Newport Beach;
- City of Orange;
- City of San Juan Capistrano;
- Unincorporated County of Orange.

The NCCP/HCP proposes that local government jurisdictions, along with other local governments within the subregion that would rely on the NCCP/HCP for mitigation for development activities affecting occupied CSS habitat, will be asked to become signatories of the Central and Coastal Subregion NCCP/HCP Implementation Agreement and participate in the implementation of the NCCP/HCP. The NCCP/HCP and the Implementation Agreement explain what participation in the NCCP/HCP would mean for local government signatories to the Implementation Agreement (a summary outline of local government roles and commitments is set forth in Section 2.3 of this EIR/EIS).

-- Protected Plan and Animal Species - "Identified Species"

Under the NCCP Conservation Guidelines, the subregional reserve design process focuses on protecting the habitat of three designated "*target species:*" the coastal California gnatcatcher, the coastal cactus wren and the orange-throated whiptail lizard. As envisioned by the NCCP Conservation Guidelines, the Reserve System designed for the three "target species" is intended to provide significant levels of protection for a much broader range of habitats and species than just CSS and the three target species. Accordingly, the NCCP/HCP proposes that target species and 36 additional species, most of which are not presently "listed" under state or federal endangered species law. The term "regulatory coverage" means that the species are threatened "as if listed" under NCCP Act and the state and Federal Endangered Species Acts, thereby allowing specifed conversion of CSS habitat in return for mitigation measures assured through the NCCP/HCP Implementation Agreement. The species proposed to receive "coverage" under the NCCP/HCP are termed "Identified Species" and include:

Target Species (3)
- * Coastal California gnatcatcher
- coastal cactus wren
- orange-throated whiptail

Mammals (3)
- San Diego desert woodrat
- coyote
- gray fox

Birds (6)
- northern harrier
- sharp-shinned hawk
- * peregrine falcon
- red-shouldered hawk
- rough-legged hawk
- southern California rufous-sparrow

Reptiles (6)
- coastal western whiptail
- San Bernardino ringneck snake
- red diamondback rattlesnake
- San Diego horned lizard
- Coronado skink
- coastal rosy boa

Amphibians (3)
- arboreal salamander
- western spadefoot toad
- black-bellied slender salamander

Plants (8)
- Catalina mariposa lily
- ** Laguna beach Dudleya (PE)
- ** Santa Monica Mts Dudleya (PT)
- Nuttal's scrub oak
- small-flowered mountain mahogany
- heart-leaved pitcher sage
- Coulter's matilija poppy
- Tecate cypress

Conditionally Covered Species (10)
- * least Bell's vireo
- * southwestern willow flycatcher
- * southwestern arroyo toad
- ** Quino (Wright's) checkerspot (PE)
- * Riverside Fairy shrimp
- ** San Diego fairy shrimp (PT)
- * Pacific pocket mouse
- golden eagle
- prairie falcon
- foothill mariposa lily

* *Species that currently are on the federal list of "threatened or endangered" species.*

** *Species that are proposed for federal listing of threatened or endangered species.*

For some of the above species, the NCCP/HCP proposes certain conditions that must be met in order for the species to be covered.

-- <u>"Covered Habitats"</u>

In addition to the regulatory coverage for incidental take of CSS habitat and the thirty-nine "Identified Species" cited above, the NCCP/HCP contains assurances to *participating landowners* relating to future impacts on other species located within specified habitats outside the proposed habitat Reserve System. The USFWS and CDFG have determined that the programmatic elements of the NCCP/HCP further the protection of certain habitats in a manner comparable to the protection provided for CSS habitat. These habitat types are referred to as "covered habitats" and include (Figure 69):

	ACRES OF "COVERED HABITATS" OUTSIDE THE RESERVE SYSTEMINSIDE	**ACRES OF "COVERED HABITATS" THE**
RESERVE SYSTEM		
• oakland woodlands	205	940
• Tecate cypress forest	3	191
• cliff and rock; and	28	74
• within the Coastal Subarea only; chaparral	260	3,337
TOTAL	496	4,542

For these habitats, CDFG and USFWS will assume the responsibility for assuring, to the extent allowed by law and consistent with the provisions of Section 8.3.4(d) of the Implementation Agreement, that all statutory and regulatory requirements necessary to issue Section 10(a)(1)(B) and/or Section 2081 permits to *participating landowners* for listed species found in these habitats that are affected by planned activities. USFWS and CDFG will issue Section 10/2081 permits to *participating landowners* concurrent with the listing of any currently unlisted covered species.

-- <u>Headlands Plant Species</u>

The NCCP/HCP proposes regulatory coverage for five plant species on the site of the Headlands Property only.

2. Adaptive Management

Adaptive management is a central theme of the NCCP Conservation Guidelines and is defined as a flexible, iterative approach to long-term management of biotic resources that is directed over time by the results of ongoing monitoring activities and other information. Under this approach, biological management techniques and specific objectives are regularly evaluated and adjusted, in light of monitoring results and other new information. The purpose of adaptive management within the framework of the NCCP/HCP Reserve System is to help maintain long-term, net habitat value within the subregion. The NCCP Conservation Guidelines recognize the need for such management to counter the effects of "benign neglect" under pre-NCCP conditions and to offset impacts on habitat lands proposed for conversion.

The Adaptive Management Program is described in detail in the NCCP/HCP. Key elements of this program are:

- monitoring and associated adaptive management of the biological resources located within the Reserve System;
- restoration and enhancement actions such as eradication of invasive, non-native plant species, predator control and grazing management plans;
- consideration of species enhancement, propagation and re-introduction within the reserve.
- Adaptive management carried out by means of short-term and long-term fire management programs within the Reserve System;
- adaptive management of public access and recreational uses within the Reserve System;
- adaptive management related to uses within the Reserve System that existed prior to approval of the Subregional NCCP/HCP;

- assurance that future permitted infrastructure uses and other public facilities within the Reserve System proceed in a manner provided for in the NCCP/HCP in order to minimize impacts of allowed uses;

- interim management of privately-owned lands for all of the above adaptive management elements prior to transfer of legal title to permanent public or non-profit ownership within the Reserve System (see discussion in Subsection "3" below);

- restoration and enhancement through the acquisition of existing CSS habitat or the creation of new CSS habitat to offset potential loss of net long-term habitat value due to development of CSS habitat outside the Reserve System on the part of *non-participating landowners (i.e.*, landowners who have not contributed significant lands or lands to the NCCP Reserve System - see discussion in Section "4" below); and

- restoration and enhancement of non-CSS "covered habitat" types.

3, <u>Interim Management Program</u>

Approximately 15,000 acres of the proposed Reserve System are currently publicly owned and would be available for inclusion in the reserve immediately following approval of the NCCP/HCP and signing of the Implementation Agreement by participants. However, because more than 21,000 acres of the proposed reserve are currently privately owned, and because most of the private ownership is subject to phased dedication commitments that preceded the NCCP/HCP, it would take many years to complete these open space dedication programs. To address the need for managing these lands prior to dedication, "*participating landowners*" would be required to allow the non-profit management entity to implement "interim" habitat management measures during the time following approval of the NCCP/HCP and the actual future transfer of lands from private to public ownership. The purpose of this interim management is to maintain and improve habitat values on lands designated for inclusion within the reserve.

4. Funding

The NCCP/HCP identifies three different sources of funding:

- An endowment to refund the Adaptive Management Program within the reserve lands over the life of the Reserve System, contributed by landowners within the subregion who were willing to fund the preparation of the NCCP/HCP plan and transfer land to the Reserve System at no cost. At this time, the NCCP/HCP projects the ultimate endowment fund to total more than $10.6 million.

- State/federal contribution to fund research, management, focused species inventories and future acquisitions of designated lands for the proposed Reserve System.

- A mitigation fee mechanism that gives landowners who are not contributing directly to creation/management of the reserve (*i.e.*, *non-participating landowners*) a choice of how to meet currently statutory requirements by mitigating proposed conversion of CSS habitat occupied by state or federally listed species located outside the Reserve System; mitigation funds provided by such landowners would be used to support restoration/enhancement activities within the Reserve System or to acquire lands to be added to the Reserves or Special Linkages. This mitigation option would involve payment of a fee in return for incidental take of CSS and gnatcatchers outside the Reserve System. Landowners who choose not to pay such a fee continue to have the option of separate Section 7 and 10(a) permits under applicable law (see discussion in Section "C.4" below).

C. Potential Impacts and Proposed Mitigation under the NCCP/HCP

1. Habitat Proposed to be Protected

The NCCP/HCP would establish a Reserve System that would protect more than 35,000 acres of wildlands, including more than 18,500 acres of CSS (the Reserve also contains about 2,200 acres of disturbed, agricultural and developed lands). In addition, almost 3,900 acres of non-reserve public open space is located within the subregion, and more than 5,700 acres are included within the "supplemental" non-reserve habitat areas (*i.e.*, the Special Linkage Areas and Existing Use Areas). In all, almost 47,000 acres of natural habitat would be included within the proposed Reserve System, other permanent public open space and the "supplemental" non-reserve habitat areas. These areas contain 487 of the surveyed gnatcatcher sites (81 percent), and 774 of the surveyed cactus wren sites (78 percent) identified during the NCCP field surveys.

2. Potential Impacts on Lands Located Inside the Habitat Reserve System or Supplemental Non-Reserve Habitat Areas

The NCCP/HCP would also authorize the incidental take of 512 acres of CSS supporting nine surveyed gnatcatcher sites located within the proposed Reserve System and the Special Linkage Areas (four surveyed sites, 106 acres of CSS). All of this incidental take is related to future activities proposed (primarily recreational facilities, arterial roads, water reservoirs and other water supply facilities and other utility facilities) by "*participating landowners*" (see subsection "4" below).

3. Potential Impacts on Lands Located Outside the Habitat Reserve System

Impacts on occupied habitat of "target/identified" species would be permitted outside the Reserve System subject to the terms of the NCCP/HCP, Implementation Agreement and applicable local, state and federal lows (*e.g.*, general plans, zoning, the federal Clean Water Act). These non-reserve areas contain approximately 7,000 acres of CSS habitat, of which approximately 1,100 acres are occupied by 108 surveyed gnatcatcher sites and 206 surveyed cactus wren sites. The NCCP/HCP proposes to authorize incidental take (*i.e.*, conversion of CSS habitat) within these lands for the coastal California gnatcatcher, and for identified species listed in the future under the terms of the NCCP/HĊP. Of the 108 gnatcatcher sites that would be impacted, 97 sites are located on *lands owned by "participating landowners, " and*

11 sites are on lands owned by "non-participating landowners" (see discussion in the following section).

When all of the potential impacts are considered, the total authorized incidental take proposed by the NCCP/HCP would include an estimate 1,217 acres of CSS habitat occupied by the federally-listed gnatcatcher containing 121 current gnatcatcher sites. Total "conversion" of CSS proposed to be permitted under the NCCP/HCP, would be 7,444 acres. The 7,444 acres amounts to 24 percent of the remaining CSS habitat within the subregion.

4. Mitigation Options for Landowners in the Subregion

Two categories of landowners are identified by the NCCP/HCP: *participating landowners* and *non-participating landowners*. Each of these landowner categories would be offered different endangered species habitat mitigation options (see discussion below and attached NCCP/HCP Mitigation Measures Summary).

-- *"Participating Landowners"*

Participating landowners are those contributing significant land and/or funding toward implementation of the Reserve System and Adaptive Management Program. The *participating landowners* includes:

- Southern California Edison;
- Metropolitan Water District of southern California;
- Irvine Ranch Water District;
- Santiago County Water District;
- Transportation Corridor Agencies;
- The Irvine Company;
- University Of California, Irvine;
- County of Orange; and
- Chandis-Sherman.

For these landowners, development activities and uses that are addressed by the NCCP/HCP would be considered fully mitigated under the NCCP Act and the state and federal ESAs for impacts to habitat occupied by listed and other "identified" species designated in the NCCP/HCP and for impacts to species dependent upon or associated with the "covered habitats". The "incidental take" proposed to be authorized by the NCCP/HCP is framed in terms of CSS acreage, not in terms of specific occupied habitat because the population levels and locations of occupied habitat vary considerably over time. Satisfactory implementation of the NCCP/HCP and term of the Implementation Agreement would mean that no additional mitigation will be required of *participating landowners*.

-- *"Non-Participating Landowners*

For purposes of the NCCP/HCP, other landowners within the subregional that are not contributing either significant land to the Reserve System or funding for the Adaptive Management Program are termed "*non-participating landowners*." The NCCP/HCP provides for a different mitigation approach for these "*non-participating landowners*" to assure that impacts to occupied listed species habitat resulting from activities on their lands maintain "net habitat value" within the subregion consistent with the NCCP Act, CESA and FESA (see Chapter 7 and 8). *"Non-participating landowners"* may satisfy the requirements of FESA and CESA with respect to listed species in any of the following ways: (1) avoidance of conversion of habitat resulting in "Take" under CESA or FESA through project site design or other actions; (2) satisfaction of applicable FESA and CESA provisions under the consultation and permit provisions of these statutes independently of the NCCP/HCP; or (3) payment of a mitigation fee, for mitigation of impacts to CSS Identified Species, to the non-profit mitigation fee, for mitigation of impacts to CSS Identified Species, to the non-profit management corporation as provided for in the NCCP/HCP and Implementation Agreement. Thus, if a "*non-participating landowner*" could not avoid impacting CSS habitat occupied by a state or federal listed CSS species designated as an NCCP "Identified Species," the landowner could decide either to work with the regulatory agencies to mitigate impacts on-site or off-site (pursuant to the federal Endangered Species Act Section 7 or Section 10 processes and the state Endangered Species Act 2081 and 2084 processes) or to choose the option of payment of a mitigation fee to the NCCP program.

Finally, some *non-participating landowners* are located in "Existing Use Areas" lands containing populations of CSS Target species for which the NCCP/HCP proposes maintaining the regulatory status quo (*i.e.*, no take of CSS habitat occupied by CSS Identified Species is proposed to be authorized, thereby maintaining USFWS and/or CDFG direct authority over habitat protected by applicable law).

NCCP/HCP IMPACTS/MITIGATION MEASURES PROGRAMMATIC SUMMARY

Project Impacts	Mitigation Measures	Unavoidable Impacts
Biology **Section 5.2.1.3**		
Approximately 4,700 acres of CSS, 97 gnatcatcher sites will be impacted outside the Reserve System by *participating landowners.*	Prior to the occurrence of incidental take, each *participating landowner* shall dedicate land/ provide fund for adaptive management as agreed to in the Implementation Agreement	The loss of CSS habitat, and other biological resources is unavoidable. Impacts are mitigated to a level below significant.
Approximately 512 acres of CSS, 13 gnatcatcher sites will be impacted within the Reserve System and Special Linkages by *participating landowners.*	Mitigation for *participating landowners* consists of the measures already undertaken or agreed to in funding the planning and creation of the Reserve System, including land contributions as provided for in the Implementation Agreement. Further minimization are specified in the Adaptive Management Program.	The loss of target species and CSS habitat is unavoidable. Impacts are mitigated to a level below significant.
Approximately 2,100 acres of CSS, and 11 gnatcatcher sites will be impacted outside of the Reserve System, Special Linkages, Existing Use Areas, non-reserve open space and North Ranch area by *non-participating landowners.*	Mitigation for impacts to habitat occupied by CESA or FESA listed species will be provided either through standard CESA 2081 and/or FESA Section 7/10 processes or through payment of the optional NCCP/HCP habitat mitigation fee.	The loss of CSS habitat occupied by listed species is required by FESA and CESA to be mitigated to a level of less than cumulative significant. "Net habitat value" is maintained under the optional NCCP mitigation fee and thus impacts are mitigated to a level below significant.
Incidental take beyond that authorized may be permitted at the discretion of USFWS	Mitigation for such take would be through the Section 7 or 10 process of through payment of the optional mitigation fee.	USFWS has the authority to reduce impacts to a level below significant.

Table ES 1
Central & Coastal Subregion NCCP
Vegetation, Target Species, and Proposed Habitat Reserve

Vegetation	Reserve	Special Linkage	Existing Use	Non Reserve Open Space	Policy Plan Area	National Forest OS	National Forest Private	Other Non Reserve	Total
	Area in Acres								
Dunes						9	8	2	18
Scrub	18,527	449	1,103	283	3,006	1,733	1,835	7,456	34,392
Chaparral	6.950	23	735	79	5,251	13,114	6,510	2,556	35,218
Grassland	5,732	518	1,053	1,402	694	105	346	12,025	21,874
Vernal Pools	9	2		0				42	53
Marsh	343		29	234				52	657
Riparian	1,770	116	116	379	240	804	497	1,204	5,126
Woodlands	940	16	33	52	157	253	179	291	1,920
Forest	191				2	563	43	5	804
Cliff and Rock	74	7	1	1	14	29	12	35	173
Marine & Coastal	362		15	0				1,553	1,930
Lakes, Reservoirs, Basins	99	10	1	790			0	456	1,357
Water Courses	182	1	22	8	0		9	563	784
Agriculture	577	90	5	83			21	12,489	13,265
Developed	694	199	415	324	23	12	254	81,210	83,131
Disturbed	929	475	269	195	68	10	59	6,004	8,008
Total	**37,378**	**1,906**	**3,796**	**3,831**	**9,456**	**16,632**	**9.772**	**125,942**	**203,713**

		Reserve	Special Linkage	Existing Use	Non Reserve Open Space	Policy Plan Area	National Forest OS	National Forest Private	Other Non Reserve	Total
Gnatcatcher	Total Sightings	370	20	87	10	5			108	600
Cactus Wren	Total Sightings	671	39	64		14			206	994
Total Sightings		1041	59	1511	10	19			314	1594

		Reserve	Special Linkage	Existing Use	Non Reserve Open Space	Policy Plan Area	National Forest OS	National Forest Private	Other Non Reserve	Total
CSS	Total Acres	18,527	449	1,103	283	3,006	1,733	1,835	7,456	34,392
OW	Total Acres	16,651	693	2,004	2,946	6,358	14,877	7,603	18,784	69,915
DDA	Total Acres	2,200	764	689	602	92	22	334	99,702	104,405

CSS - Coastal Sage Scrub Habitat

OW - Other Wildland Habitat

DDA - Developed, Distrubed and Agriculture

Notes:

1) *Target Species Sites in the National Forest are excluded from this analysis.

2) Target Species Sites impacted by Corridor Projects are excluded from this analysis.

CHAPTER 21

Closing the Loop on Natural Resources: Governance Lessons from the Sierra Nevada Ecosystem Project

Gregory B. Greenwood

Abstract

Created at the behest of Congress, the Sierra Nevada Ecosystem Project studied the entire Sierra Nevada in the context of developing strategies to maintain the health and sustainability of Sierra Nevada ecosystem while providing resources to meet human needs. The study cost nearly $7 million dollars, took three years to complete and in many ways epitomizes the on-going transformation of the Sierra from hinterland to core human habitat. SNEP considered the Sierra Nevada, including both public and private lands, as a whole system with physical, biological, social and institutional components. The institutional components bind together resources and beneficiaries, many of whom live outside of the region. The institutional arrangements, based on a technocratic, progressive vision, created piece-meal over time and weighted by bureaucracy inertia, seldom ensure sufficient reinvestment in the Sierra and often reflect poorly the evolving preferences of Californians for the Sierra Nevada. Improving institutional arrangements will mean going beyond improving technocracy to creating new institutions and linkages that better reflect the increasing complexity of the relationship between people and the ecosystems of the Sierra. These steps include increasing re-investment in Sierra Nevada ecosystems, restructuring relationships between existing agencies in the Sierra, involving local knowledge in management, reorienting and rescaling arrangements to reflect regions within the Sierra, and legal, financial and administrative reforms to support these new institutional arrangements.

CHAPTER 22

Undersea Pumped Storage for Load Leveling

Richard J. Seymour[1]

Abstract

The demand for electrical energy varies substantially over a 24 hour period. The supply of energy must be exactly matched to these changing demands.. Taking generating facilities on and off line and operating them at less than optimum performance levels is uneconomical. Therefore, when it is feasible, pumped storage is used to achieve load leveling. Water is pumped uphill and stored in a reservoir during low demand periods and allowed to flow down through turbogenerators when demand is high. In Southern California there are no locations where demand, water supply and environmentally acceptable reservoir sites coexist. Compressed Air Energy Storage (CAES) is a potential candidate for electrical energy load leveling in the absence of conventional pumped storage capability. In CAES, air is mechanically pressurized using surplus electrical power into man-made subterranean caverns. It is released during peak demand periods and either powers a turbine directly. or is mixed with natural gas and the mixture burned in the turbine. None of the suitable sites identified for CAES exist in coastal areas within the United States. Offshore Compressed Air Energy Storage (OCAES) is a conceptual modification of CAES in which the very high ambient pressures at depth in the ocean serves to reduce the storage vessel structural requirements.. Air is pumped into the system, displacing the seawater, until the storage section is filled. This air is maintained at the water pressure appropriate to the depth. Air can be withdrawn from the storage vessel at a constant pressure until water starts to fill the outfall. There are several sites along the California coast that offer close proximity to deep water and existing access to grid transmission lines, transformers, switching and other infrastructure. The paper discusses a 230 MW OCAES system sited at Carlsbad as an example.

---

[1] Center for Coastal Studies, Scripps Institution of Oceanography, University of California San Diego, La Jolla, CA 92093-0214

Introduction

The regional demands for electric power vary significantly over short time spans. In addition to seasonal and work-week-related cycles, the variation during any one day can be very large. Figure 1 shows the average daily demand during the summer time for San Diego Gas & Electric, a utility serving San Diego County. As the figure indicates, the demand varies by a factor of two in slightly over 10 hours. This varying demand must be precisely matched instant by instant through varying the power production rate. Excess power can sometimes be exported to another system connected to the same grid and shortages can be overcome to some extent by importing from other utilities. However, the bulk of the load matching must be accomplished by incremental changes to the number of generators on line as well as fine tuning through running facilities at less than full rated power. Any of the methods for load matching increase the costs of power generation when compared to base load facilities running continuously at optimum capacity.

Therefore, methods have been sought and developed to reduce the swings in demand, a process referred to as load leveling. One method has been reducing the price to the consumer during off-peak intervals. Some industries can take advantage of these savings through proper scheduling. Rebates to certain domestic users, such as those whose electric hot water heaters are disabled during peak demand times, have been employed by electric utilities. However, the most widely used approach is pumped storage. This method, employs pump-motor units which can operate as turbine-generators when flows through them are reversed. Sited at the base of a storage dam, they are used to push water uphill during periods of low demand and recover substantially all of the invested energy during peak periods by letting the impounded water return to the pump/generator station below. This concept is illustrated in Figure 2. Although the process is quite efficient and reduces the overall costs of power generation when the demand varies substantially, there are difficulties. The concept requires either a large river or aqueduct to supply water as needed and carry away the outflow, or the site must be adjacent to a lake which is large enough to accommodate the amount of water exchanged with the elevated reservoir. The large daily rise and fall of the reservoir level precludes or greatly restricts multiple uses such as boating and fishing. To reduce the construction costs of the reservoir, a valley with a narrow opening that is geologically competent to anchor a high dam is required. In short, most locations in the United States are either too arid, too flat or too far from large energy demand centers to permit economical pumped storage.

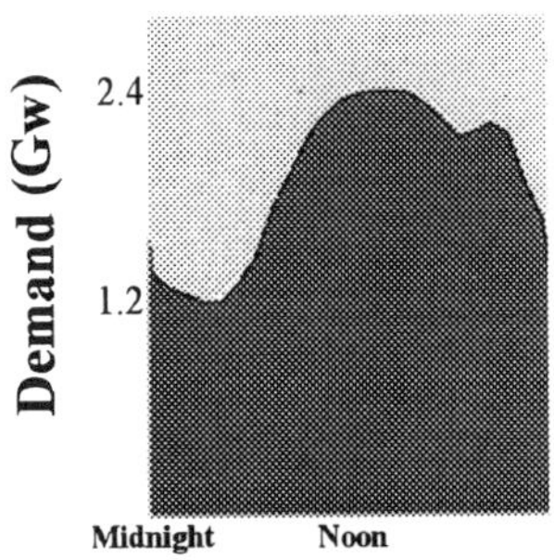

Figure 1: Average SDG&E summer weekday demand

Compressed Air Energy Storage

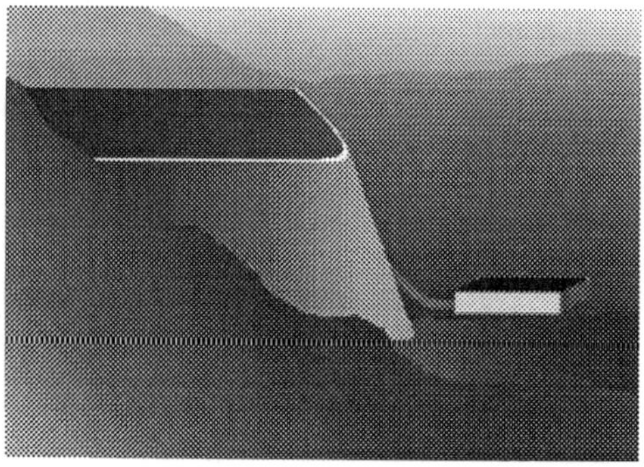

Figure 2: Conceptual arrangement of pumped storage system.

Compressed Air Energy Storage (CAES) is a promising alternative to pumped storage where the geological conditions allow. As shown schematically in Figure 3, the CAES system consists of a motor/generator which alternately drives an air compressor or is driven by a turbine. When in the energy storage mode, air is compressed and stored in an underground reservoir. During the generation phase the air flows back from the reservoir through a gas turbine, driving the generator. The turbine may use only the energy in the compressed air, or natural gas or other fuel may be added to increase the power. The air is heated as it is compressed and some of this heat may be extracted and stored temporarily. The cool compressed air can then be preheated before entering the turbine. The air storage chamber may be a natural cavity or it may be constructed. In either case, it requires non-permeable competent rock or salt. Along most of the Atlantic and Pacific coasts, where much of the nation's population and energy demand exists, the appropriate formations do not exist. In Southern California there are few if any suitable pumped storage sites and very few promising locations for CAES.

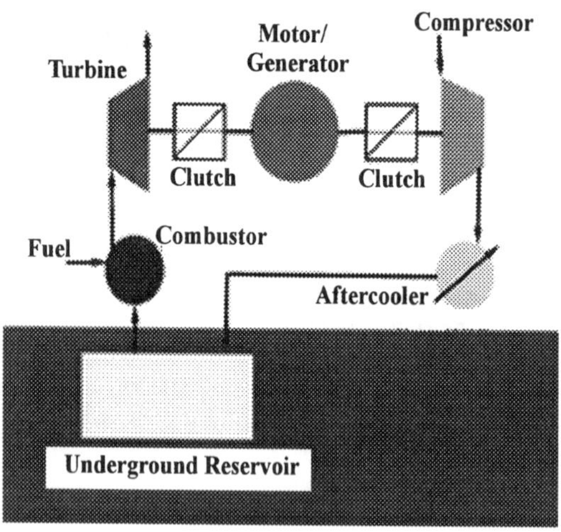

Figure 3 : Schematic of a typical CAES system.

Offshore Compressed Air Storage

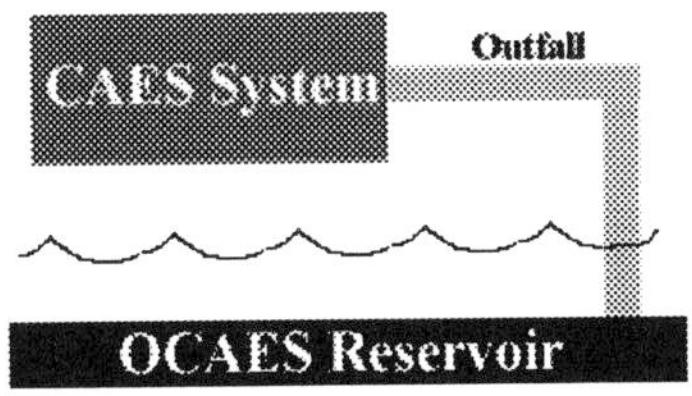

Figure 4: Schematic of OCAES system.

Offshore Compressed Air Storage (OCAES) is a proposed modification to a conventional CAES system in which the great hydrostatic pressure available at depth in the ocean is utilized to reduce the structural requirements for the air storage reservoir. In OCAES, air is stored in an open-ended reservoir at the local hydrostatic pressure. Because there is essentially no pressure differential across the walls of the reservoir, it may be constructed of any dense, inexpensive material -- such as unreinforced concrete. Other than installation loads, the only load on the reservoir are the buoyancy forces that exist when it is filled with air. The reservoir may use a flexible bladder to separate the air and the water, however it does not appear that this mechanism is a necessity. Figure 4 shows a schematic arrangement of an OCAES system with the conventional CAES machinery on the shore connected to the underwater reservoir by an outfall pipeline. This pipeline will see the full operating pressure as a differential across its walls at and above the sea surface, decreasing to zero differential at its base. Further, the upper portion of the outfall pipe must be protected from wave forces by burial, tunneling or ballasting.

Air is pumped into the reservoir until all of the water has been displaced. If the reservoir is at a near constant depth, hydrostatic pressure will assure that air is delivered at a constant pressure during the generation cycle. This results in the highest efficiency operation of the turbine. If no bladder is employed, it will be possible to further blow down the system and remove air from the outfall pipe as well, with a falling pressure.

OCAES Site Selection Considerations

To be economically attractive, the cost of construction of the ocean reservoir subsystem must be kept to a minimum. The cost of the storage reservoir is largely independent of the site so that the length of the outfall line -- which is set by the distance required to reach water of an appropriate depth -- is the most significant variable in selecting a site. In addition, considerations of aesthetics, competition for high-value coastal sites close to the energy demand and the necessity to be connected with high voltage transmission lines to the local grid argues for siting the OCAES unit at an existing coastal power plant. By coincidence, there are at least three power plants in California that have been built in very close proximity to submarine canyons. These sites are Carlsbad (near San Diego), Redondo Beach (near Los Angeles) and Moss Landing (near Monterey). The submarine canyons, incised over geologic time by turbidity flows of sand, provide the shortest paths to deep water. The pipeline could

Figure 5: Possible route for OCAES pipeline in Carlsbad Submarine Canyon.

follow the general route of the canyon while avoiding the active flow area at the bottom. Figure 5 shows a potential path along the inactive canyon at Carlsbad, very close to an existing fossil fuel generating plant, shown in Figure 6..

An OCAES System Example

To illustrate the application of OCAES technology to a coastal region with no pumped storage or conventional CAES potential, an OCAES system has been conceptually designed for the Carlsbad site, using the canyon route described above. CAES turbines have been designed to operate efficiently at about 60 atmospheres which would require a storage reservoir installed along a depth contour of about 650m. At the Carlsbad site, the length of the outfall pipe to reach this depth would be about 16km. A conceptual sketch of such a system in shown in Figure 7.

One CAES system that has been studied develops 230 Mw capacity. Assuming that the system would be required to generate at this rate for 10 hrs per day, the required volume of air stored at 60 atmospheres could be contained in a pipe 3.6 m in diameter and 12.8 km in length. To accommodate the flow rate, the outfall pipe requires a diameter of about 1.2 m. The reservoir could be fabricated and installed using essentially the identical technology employed for sewer outfalls -- rubber-gasketed concrete pipe lowered by crane from a moored surface vessel. The outfall pipe is comparable to large diameter welded steel oil and gas collection lines routinely installed at these depths.

Figure 6: Fossil fueled powerplant at Carlsbad, CA.

The San Diego Gas & Electric demand curve shown in Figure 1 can be used to illustrate the effects of using this OCAES system for load leveling. As illustrated in Figure 8, operating the 230 Mw system for 10 hours per day will reduce the load variation range required of the rest of the system from 1.2 Gw down to 0.75 Gw, a 40% reduction.

Acknowledgment

The author was introduced to the OCAES concept by James Wight of San Diego Gas & Electric, who also provided information on demand variation and load leveling requirements which were extremely valuable. His contribution is gratefully acknowledged. This work utilizes the analyses of Anton Rowe, which were performed as part of a student independent research project under the author's supervision while Mr. Rowe was an undergraduate at the University of California, San Diego. Mr. Rowe's enthusiastically executed and thoroughly performed preliminary designs were essential in establishing OCAES as a promising load leveling option for coastal regions and the author is indebted to him for his significant contributions.

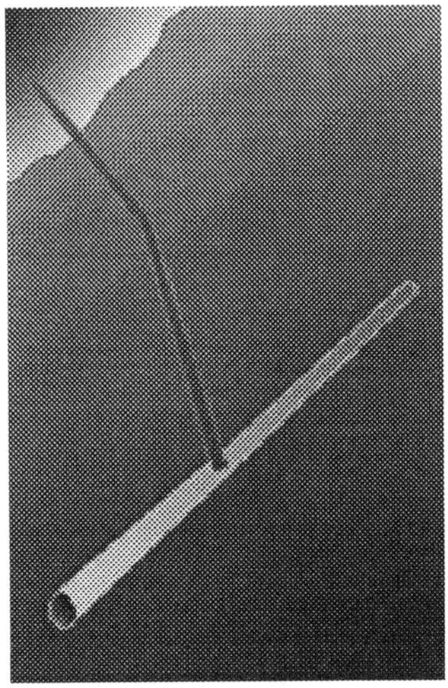

Figure 7: Schematic of OCAES outfall and storage reservoir.

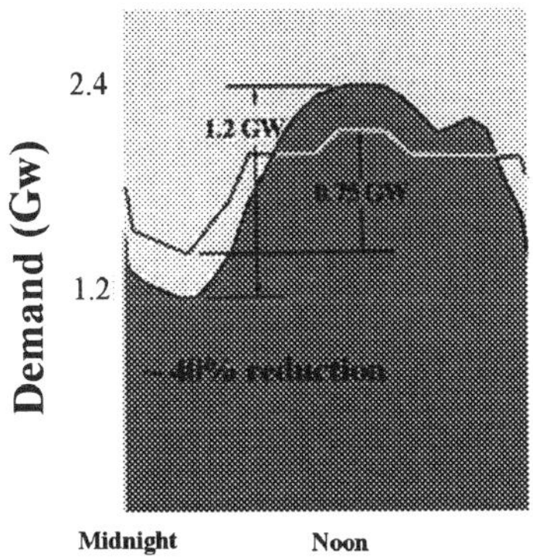

Figure 8: Demand curve as modified by a 230 Mw OCAES system generating 10 hours/day.

CHAPTER 23

The Environmental Issues of Seawater Desalination in California: An Overview of the 1990s

Timothy J. Cohen

Abstract

An overview is provided of environmental and regulatory issues related to seawater desalination in California during the 1990s. During this decade, several coastal jurisdictions in California have studied, and some have implemented, the use of desalted seawater as a potable water supply to supplement dwindling or unreliable surface and groundwater supplies.

As the decade continues, environmental issues related to seawater desalination are becoming more focused. State and local regulators had little guidance regarding the permitting of desalination facilities in the early '90s. Now, agencies have developed guidelines and the analysis and approval of desalination permit applications can rely on a greater body of existing information. These environmental issues and regulatory responses are outlined, including the key issues and policies related to desalination concentrate disposal in California.

Specific case studies are used to provide examples of the evolution in planning and environmental/regulatory analysis that has occurred during the decade, from the earlier drought-response planning to the later long-term use of desalinated seawater as a component of a community's water supply plan. Finally, a forecast and outlook is presented for the continued evolution of desalination planning and use in California.

Timothy J. Cohen, Vice-President, Woodward-Clyde Consultants, 130 Robin Hill Road, Suite 100, Santa Barbara, CA 93117

CHAPTER 24

Biodiesel Fuels for Marine Applications

Jack Whittier[1] and Harold M. Thomas[2]

Abstract

Biodiesel fuels are emerging as preferred alternative fuels in select market niches. Biodiesel fuel has several unique benefits which may be of important and significant value in certain engine markets, particularly in the marine environment. In general, regulatory pressures are instrumental in the decision to adopt biodiesel. The Oil Pollution Act of 1990 has enacted specific and potentially costly burdens on the marine market in connection with fuel spills. Biodiesel fuel characteristics may make it an attractive alternative to conventional diesel fuels to avoid damage and tort liability.

Biodiesel is typically derived from vegetable oils, although animal fats are also considered a feedstock, and is viewed as a renewable, indigenous resource. Biodiesel may be used as a neat fuel or blended with petrodiesel and used directly in diesel engines with little or no modifications. Blends of biodiesel and petrodiesel are also less toxic than petrodiesel. Biodiesel fuel is also biodegradable, and in the event of leakage or fuel spills, is rapidly degraded. Other biodiesel environmental advantages include virtually no SOx emissions, reduced particulate emissions, reduced exposure to carcinogens released from combustion of petrodiesel, increased lubricity, ease of cleaning spills, "less noxious" odors associated with combustion, and perception as a "green" product.

Widespread adoption of biodiesel is not anticipated in the near future because of multiple difficulties involved in introducing alternative fuels. Market barriers include higher price relative to conventional fuels, lack of a supply infrastructure, regulatory uncertainty, and negligible consumer awareness. Visible, practical actions by public and private organizations will assist market development for biodiesel fuels. Regulatory agencies require information regarding biodiesel market and technical characteristics to foster decision making processes that affect public policy. Private firms desire information addressing consumer acceptance and market demand for biodiesel. Consumers have limited knowledge of biodiesel as an alternative fuel for engines and want access to information regarding the advantages and disadvantages of biodiesel.

California recreational and commercial harbors are a logical market for the introduction of biodiesel. Fragile ecological systems coupled with a considerable and diverse boat population creates pressures that are posing increasing challenges for those who fish and swim in coastal waters. Regulatory agencies are empowered to find solutions to address environmental and other social issues. Private sector initiatives to provide environmentally-compatible fuels require cooperation from public sector entities to achieve broad societal goals. A broad-based trial of biodiesel in select ports has merit to gather information that will help to shape public and private actions.

---

[1] NEOS Corporation, 165 South Union Boulevard, Suite 260, Lakewood, CO 80228
[2] California Department of Fish and Game, Office of Oil Spill Prevention and Response, 1700 K Street, Suite 250, P.O. Box 944209, Sacramento, CA 94244-2090

Introduction

Biodiesel fuels are emerging as preferred alternative fuels in select market niches. In general, regulatory pressures are instrumental in the decision to adopt biodiesel fuels. Urban bus fleets, underground mining, regulated fleets, and the marine markets are the initial key sectors for use of biodiesel. Widespread adoption of biodiesel is not anticipated in the near future because of multiple difficulties involved in introducing alternative fuels. Market barriers include higher price relative to conventional fuels, product specification and quality control issues, lack of a mature supply infrastructure, regulatory uncertainty, and negligible consumer awareness.

The intent of this paper is to introduce a novel fuel with important marine applications. The paper addresses technical, economic and regulatory considerations involved in bringing biodiesel fuels into greater acceptance and usage in the marine environment.

Description of Biodiesel Fuels

Biodiesel Specification

Biodiesel is defined as the mono alkyl esters of long chain fatty acids derived from renewable lipid sources. This definition does not include cellulose-based materials such as tree oil derivatives, other woody products, or biological slurries. Biodiesel, as defined, is widely recognized in the alternative fuels industry as well as by the US Department of Energy (DOE), US Environmental Protection Agency (EPA) and the American Society of Testing and Materials (ASTM). Biodiesel is registered with the US EPA as a fuel and fuel additive. EPA is presently considering biodiesel for inclusion under the environmentally preferable products program established by Executive Order 12873. DOE has formally recognized biodiesel as an alternative fuel for the purposes of the Energy Policy Act as well as for the Clean Cities Program.

The first generation of a quality standard for biodiesel is now under consideration by the American Society for Testing and Materials (ASTM). ASTM is comprised of fuel producers, engine manufacturers, and third party interests which develop consensus testing methods and standards for petroleum fuels and a variety of other products for the United States. Table 1 provides candidate specifications for biodiesel that have been submitted to ASTM for consideration (*Biodiesel Report*). Note the specifications are for pure, 100 percent biodiesel. The proposed specification is a performance, not feedstock, based standard. The standard is being developed from the perspective of what qualities are needed for the fuel to perform in an engine.

Table 1. Specifications for Pure (100%) Biodiesel

Property	ASTM	Limits	Unit
Flash Point (Open Cup)	D93	100.0 min.	°C
Water and Sediment	D1796	0.050 max.	Vol. %
Carbon Residue	D4530	0.050 max.	wt. %
Sulfated Ash	D874	0.020 max.	wt. %
Kinematic Viscosity @ 40°C	D445	1.9-6.0	mm^2/sec.
Sulfur	D2622	0.05 max.	wt. %
Cetane Number	D613	40 min.	-
Cloud Point	D2500	by customer	°C
Copper Strip Corrosion	D130	No. 3b max.	
Acid Number	D664	0.80 max.	mg KOH/gm
Free Glycerine	G.C.*	0.020 max.	%
Total Glycerine	G.C.*	0.24 max.	%

NOTE: This specification is in the process of being evaluated by the American Society for Testing and Materials (ASTM). The ASTM designation is the test procedure recommended by the organization.
*Test method by Christiana Planc (Austria).

Neat biodiesel is registered as both a fuel and fuel additive with EPA. Biodiesel blends are under review for certification by EPA under the Urban Bus Retrofit/Rebuild Program of the Clean Air Act. EPA has not issued a determination of whether biodiesel fuel is "substantially similar" to diesel fuel. The purpose of a substantially similar designation is to insure that the engine's emissions with a given fuel are similar to the emissions from the engine when it was certified (engines are certified, not fuels). It is unlikely that biodiesel or blends above 10 percent will be considered substantially similar. Rather, it is likely that EPA will utilize the current 211(b) fuel registration process in place of the substantially similar status. The 211(b) process requires exhaust speciation.

Biodiesel Production

Biodiesel is produced by mixing a naturally occurring oil (typically, refined, degummed soybean oil) with a short chain alcohol (typically methanol) in an agitating reactor. Potassium hydroxide (KOH), sodium methoxide, or sodium hydroxide (NaOH) are used to catalyze the reaction. The catalyst is removed from the ester, the excess methanol is removed, and the ester is dried to remove water added in processing. With current esterification processes, slightly more than 7 pounds of soybean oil are required to make one gallon of biodiesel. About half a pound of crude glycerin is extracted as a co-product for every gallon of biodiesel. The extracted glycerin can be further refined to meet a 96 percent USP grade and therefore can be sold to the oleochemicals (i.e., fat-based compounds used in soaps, cosmetics, and detergents) industry. Biodiesel can also be manufactured by adding relatively inexpensive transesterification equipment to existing oil seed crushing and refining facilities.

Feedstocks

Biodiesel may be produced from a variety of vegetable and animal feedstocks. The primary feedstock in the commercial market today is soybean, which is processed into soy oil prior to use as in biodiesel. Other vegetable feedstocks include rapeseed, canola, cottonseed, and sunflower. Soybean is predominantly produced in the Midwest. Iowa is the leading producer followed by Illinois and Minnesota. Other major producers include Ohio, Missouri, and Indiana. Annual U.S. production of soy oil is approximately 2 billion gallons per year.

"Yellow or brown grease" are alternative biodiesel feedstocks that are anticipated to supply increasing portions of the market. Yellow grease is represented by restaurant grease and brown refers to grease recovered from traps. Animal feedstocks are another resource for production of biodiesel. Beef tallow or pork lard may be refined to produce biodiesel.

Producers and Suppliers of Biodiesel

Presently there are four producers of methyl esters which are sold as biodiesel: (1) Ag Processing Inc, Omaha, Nebraska (2) NOPEC, Lakeland, Florida, (3) Twin Rivers Technology, Quincy, Massachusetts, and (4) Columbus Foods, Chicago, Illinois.

The biodiesel industry is young, loosely organized, and actively working to develop the technical and institutional framework necessary for commercial production and distribution. Suppliers are small organizations, actively seeking to develop markets, who are dependent upon the producers for the supply of the raw product. New organizations, such as NOPEC, are entering the marketplace to supply product without reliance upon the established oleochemical industry. Supply infrastructure is similar to the ethanol industry (a several billion gallon per year fuel sector). Suppliers rely upon "specialty" deliveries such as rail tank cars to satisfy their customer needs. Established channels of distribution do not exist. For example, fuel bunkers that are designated for biodiesel are not readily available and many fuel brokers are not knowledgeable about biodiesel. The industry has established a supporting organizational structure to help foster biodiesel development, the National Biodiesel Board.

In marked contrast, the European biodiesel industry is considerably more mature and of more significant commercial status. It is estimated that annual production of biodiesel in various European countries exceeds 300 million gallons per year. France is the largest consumer of biodiesel fuels, followed by Germany and Austria. The underlying reasons for greater commercial acceptance of biodiesel fuels in Europe is both a larger tax burden for conventional fuels, which makes biodiesel price competitive, and greater regulatory insistence to support local farming interests and environmentally compatible fuels.

Potential U.S. Biodiesel Production Capability

Combined methyl ester production capacity is presently in excess of 25 million gallons per year, of which approximately one million gallons were sold as biodiesel in 1996. The United States consumes approximately 25 billion gallons of diesel fuel for transportation annually. The current U.S. production of all vegetable oils and animal fats is nearly 23 billion pounds per year. At a conversion rate of seven and a half pounds of oil/fat per gallon of biodiesel, the current U.S. production of oils and animal fats is only sufficient to produce some 3.1 billion gallons of biodiesel, about 12 percent of the transportation diesel market.

These figures are somewhat misleading, both because most of this oil already has a food or feed market and because it does not represent the full production potential. Over a billion pounds of soybean oil is in excess, capable of producing about 133 million gallons of biodiesel. Using land idled by the U.S. commodity programs (about 35 million acres) would permit additional production of about 3 billion gallons per year of biodiesel and not interrupt any current use of oils.

Approximately 2.5 billion pounds of waste cooking fats are collected annually from restaurants in the United States (Gaur and Graboski, 1992). In addition, 500 million pounds of fatty material are removed at municipal waste facilities nationwide and represents approximately 3.7 percent of all municipal solid waste recycled in the United States. Regulatory restrictions specify that none of this waste may be landfilled and municipal wastewater treatment plants have difficulty treating the material since it obstructs filters. Thus, an important option for the material is to be recycled for reuse. Currently, the spent greases are collected by various companies who process the waste for a feed supplement for cattle, hogs, and poultry.

Given these factors, it is reasonable to assume the plausible U.S. capability of producing biodiesel fuel is considerable, and economics will largely determine the size of biodiesel market. Economic analyses regarding the quantification of biodiesel benefits (e.g., no net CO2 emissions, reduced mutagencity) and subsequent information dissemination should also lead to greater production of biodiesel. Should incentives become available or regulatory mandates occur, markets will expand considerably.

Non-Fuel Markets for Methyl Esters

Methyl esters may be used for alternative applications such as industrial solvents, oil spill cleanup, drilling fluids, hydraulic fluids, and agricultural adjuvants ("spreader - stickers"). These markets are emerging concurrent with the fuel market and for many similar reasons (i.e., environmental regulations). It is likely that the fuel producer / supplier in the future will offer a range of products based on the esters available from vegetable and animal feedstocks.

Biodiesel Price

With the present feedstocks, the selling price of biodiesel depends greatly on the cost of the soybean oil. It is reported that soy oil price accounts for 75 percent of the final price of the methyl ester product (Peters). Furthermore, transportation costs can significantly influence the price of the delivered product. Transportation and delivery of biodiesel is presently accomplished in 55 gallon barrels, 300 gallon tanks, and 5,000 gallon over the road tankers, and 18,000 - 22,000 rail tank cars. Current prices, FOB the fuel supplier range from \$2.75 to over \$3.50 per gallon, depending upon the volume of the purchase.

Opportunities for reductions in the final delivered price of biodiesel are somewhat limited. It is unlikely that soy oil feedstock will be greatly reduced in price. Farmers desire higher prices and significant increases in soybean production efficiency are not foreseen. Alternative feedstocks such as yellow grease are less expensive to acquire but are more costly to process to meet fuel grade standards, leading to nearly equivalent prices as to soy oil (price reductions of \$0.25 - \$0.50/gallon may be possible). Animal fats may be obtained and processed less expensively than vegetable oils but there are alternative markets for these fats that establish minimum market values for which fuel producers must compete. If demand increases for animal fats then the price is likely to rise.

Production costs may be reduced with process improvements. The dedication of plants to the production of biodiesel will lead this trend and will also eliminate the increased prices due to the current dependency on the toll market. Price reductions of as much as \$0.75/gallon may be achievable with a dedicated biodiesel plant. However, the dominance of the feedstock costs relative to the total costs precludes significant opportunity for total cost reduction.

Biodiesel Operational Considerations

The major fuel characteristics include ignition quality commonly expressed by the cetane number; fuel stability, indicated by carbon residue and accelerated stability; low temperature flow performance, indicated by fuel cloud point; and fuel energy content from lower heating value in BTU/gallon.

Ignition Quality

One of the primary advantages of diesel is its high cetane rating which allows the fuel to autoignite when sufficiently compressed. The cetane rating is a number indicating a fuel's ability to self ignite, the higher the number, the easier it self-ignites. Reported cetane numbers for biodiesel range from 46 to greater than 60, depending on the feedstock used to make the ester. Whereas, a typical cetane number for No. 2 diesel is 51, and for No. 1 diesel, 48-52.

Fuel Stability

The stability of petroleum-based diesel fuels is characterized by carbon residue and accelerated stability. Lipid product stability can be described using the Peroxide Value (measured by American Oil Chemists Society (AOCS) method Cd 8-53) and Acid Value (measured by AOCS method Cd 3a-63). The Acid Value using the AOCS methods is similar as the Acid Number measured with ASTM D664, but there is nothing in the ASTM specification to characterize the extent of biodiesel oxidation. The Peroxide Value is capable of characterizing the extent of biodiesel oxidation. In practice, blends up to 30 percent are still within the satisfactory range for fuel stability.

Low Temperature Performance

The cold flow properties of biodiesel are worse than for petrodiesel fuels, especially for neat biodiesel. A 20 percent blend of biodiesel will increase the cold flow properties (cold filter plugging point, cloud point (the point at which crystals begin to appear in the fuel when cooled at a controlled rate), pour point (the coldest temperature at which no movement of the fuel is observed when cooled at a controlled rate)) of petrodiesel approximately 1 to 3 degrees Celsius. Thus far, no precautions have been needed for fueling with 20 percent blends, however, operation of neat (100 percent) biodiesel in cold weather will experience gelling faster than petrodiesel. The solutions for this potential issue are much the same as that with low-sulfur diesel (i.e. blending with No. 1 diesel, utilization of fuel heaters and storage of the vehicle in or near a building). While neat biodiesel appears to be unaffected by conventional pour point depressants, the depressants do work on the petrodiesel portion of blends and have been used by several customers to improve the cold flow performance of blends to mirror performance of non-blended petrodiesel fuel.

Fuel Energy Content

Neat methyl soyate has a lower heating value (BTU/pound) than No. 2 diesel fuel (12.5 percent lower) but higher than No. 1 diesel fuel. The heating value of a biodiesel blend is proportional to the methyl soyate concentration, because the effect on the heating value is a straight line blending function,. However, the density of methyl soyate is greater than diesel fuel, so the effect on energy density (BTU/gallon) is not as great. Therefore, because diesel engines inject fuel by volume, not mass, the impact of the lower heating value of methyl soyate will not be as great. Tests with urban buses with BD20, and compared with petrodiesel, showed approximately one-third had fuel economy decrease, one-third had the same fuel economy, and one-third had fuel economy increase. The difference has been attributed to the amount of No. 1 diesel used in the bus (transit operators often blend petrodiesel No. 1 and 2). It is highly probable that any change in the fuel energy content characteristics of 20 or 30 percent biodiesel blend compared to the base petrodiesel will be (with the exception of the oxygen content) within the chemical range of petrodiesel (Merrill, 1994).

Biodiesel Environmental Considerations

The environmental advantages of biodiesel, coupled with the renewable, indigenous resource base, are the major attractions for use of biodiesel. There are three key areas of environmental gain from the use of biodiesel fuel:

- Neat biodiesel is rapidly biodegradable and reported to be non-toxic to fish.
- Biodiesel use results in a reduction in greenhouse gas production.
- Biodiesel produces significantly lower levels of most exhaust emissions.

Biodegradability

Biodiesel is reported to be biodegradable to CO2 and water vapor, the by-products of bacterial activity. Over 98 percent of biodiesel degrades biologically in three weeks whereas in the same time only approximately 50 percent of fossil diesel degrades, after which degradation is much slower (Walker and Korbitz, 1994). Risks of ground water pollution would be significantly reduced if biodiesel were introduced as a primary fuel in key transportation corridors. Blends of biodiesel also degrade significantly faster than petrodiesel alone. Bacteria that are either present in the soil or that are introduced in response to an oil spill, are attracted to the esters which apparently accelerate the fuel degradation process.

A study conducted at the University of Idaho tested the "Biodegradability of Biodiesel in the Aquatic Environment" by the CO2 evolution method and gas chromatography (GC) and compared the results with diesel fuel. The biodegradability of various biodiesel fuels tested included Neat Rapeseed oil (NR) and Neat Soybean oil (NS) as well as their modified products Rape Ethyl Ester (REE), Rape Methyl Ester (RME), Soyate Ethyl Ester (SEE), and Soyate Methyl Ester (SME), and the blends of REE/diesel at different ratios in the aquatic environment.

Under aerobic conditions and nutrient supply (N, P), microorganisms will metabolize a substance to two final products, CO2 and water (University of Idaho). Therefore, CO2 is presumed to be the prevalent indicator of organic substance breakdown. If the substrate is the only carbon source, the amount of CO2 evolved will be proportional to the carbons consumed by microorganisms from the test substrate. Thus, the percentage of CO2 evolution is proportional to the percentage of substrate degradation. In effect, the biodiesel fraction creates a suitable environment for microbes to live. Thus the microbes are able to help degrade the diesel fuel fraction where in other conditions the microbes would have died.

Peterson reports that methyl and ethyl esters have high Chemical Oxygen Demand (COD) and Biochemical Oxygen Demand (BOD) values, necessary characteristics for a material to break down quickly (Peterson, 1994). Twenty-eight day biodegradation tests were conducted in water for rapeseed methyl ester, rapeseed ethyl ester, diesel, and dextrose. The esters degraded more rapidly than the dextrose control. At 28 days the esters were 95 percent biodegraded and the diesel fuel was 40 percent biodegraded.

Drilling fluids used to lubricate drill bits have been generally diesel oil based and have been documented to exhibit adverse environmental effects in the aquatic environment. Drilling mud research indicates that esters from vegetables oils exhibit high lubricating performance and are non-toxic and biodegradable. Biodegradation studies were conducted by the Baroid Company to compare the degradation rates of the ester to that of two mineral oils under standardized aerobic and anaerobic conditions. The aerobic test was designed to measure whether a substance should be classified as "readily biodegradable." To achieve this classification, a substance under the conditions of the test must reach 60 percent biodegradation within ten days of achieving an initial 10 percent biodegradation. Only the ester and the control substance - sodium benzoate - achieved the biodegradable threshold.

The aerobic tests were conducted over a thirty-five day period. The ester was 82.5 percent biodegraded compared to 3.9 percent for the petrodiesel. A field mud sample from a Norwegian test well was also evaluated for aerobic biodegradability using procedures specified by the State Pollution Control Agency. Over a twenty-eight day test period, the ester component of the mud was reduced by 81 percent, compared to an 85 percent biodegradation achieved by sodium acetate, the control substance.

Toxicity

Acute aquatic toxicity tests have been performed with biodiesel produced from rapeseed methyl ester (RME) and ethyl ester (REE) according to EPA/600/4-90/027 protocols (Reece and Peterson, 1995). The tests were conducted with both rainbow trout and Daphnia Magna by CH2MHill under contract to the University of Idaho. The LC50 (point at which 50 percent have died and 50 percent are still alive determined by interpolation) values for each of the substances tested with Daphnia Magna in parts per million were as follows: control (table salt (NaCl)) = 3.7, low-sulfur diesel control fuel = 1.43, RME = 23, REE = 99, Soy methyl ester (SME) (biodiesel from 100 percent soy oil), = 332. Duplicate tests were run with 10 organisms per replicate. LC50 numbers were not reported because of the failure to kill a sufficient number of fish at the concentrations tested, even with the diesel control fuel. The 20 percent and 50 percent blends had scattered losses of fish but none of the tests had less than 85 percent survival at any concentration after 96 hours. When compared to number two diesel fuel, RME is 16 times less toxic, REE is 69 times less toxic, and SME was 237 times less toxic.

Reece and Peterson report that biodiesel is not as toxic to Daphnia Magna as NaCl. When compared to the reference toxicant (sodium chloride), diesel fuel was 2.6 times more toxic, RME was 6.2 times less toxic, REE was 26 times less, and SME 89 less toxic.

Large scale spillage of biodiesel would cause temporary deoxygenation of water, but this is still significantly less damaging than an equivalent fossil diesel spill. These characteristics make it a valuable fuel, particularly in environmentally sensitive transportation and coastal corridors. It is thought that biodiesel will improve the biodegradability and reduce the toxicity of petrodiesel. The effect on biodegradability when biodiesel is blended with petrodiesel in varying percentages is being studied.

Toxicity tests have been conducted on vegetable esters in the North Sea for five years by Baroid Drilling Fluids. Testing included acute and chronic toxicity assays. Test organisms used in the toxicity studies were: the algae Skeletonema costatum, the barnacle Balanus improvisus, and the mussel Mytilus edulis. The results of the studies showed the formulated mud to be well within the limits of all government guidelines.

Biodiesel Marine Applications

A recent market study performed by Arthur D. Little, Inc. (ADL) for the National Biodiesel Board estimates that there are approximately 1.1 million diesel-propelled vessels consuming approximately 2.9 billion gallons of fuel per year (see Table 2). The great majority of the diesel fuel consumed is in the Inland and Harbor classification and includes vessels most likely to be active in environmentally sensitive waters.

Table 2. Marine Diesel Fuel Market Conditions

Market Segment	No. of Diesel Engine Vessels	Consumption of # 2 Diesel Fuel (million gal/yr.)	% of Fuel Market
Recreational	1,100,000	95	3%
Inland and Harbor	33,000	2,700	94%
Government-affiliated	2,100	30	1%
Research	100	40	1%
TOTAL	1,135,200	2,865	

Policies to Promote Biodiesel Use

California's water quality laws seek to control and limit pollution using two quite different theories of governance. The first body of law found primarily in both the Fish and Game Code and the Health and Safety Code seeks to prohibit the introduction of pollution into waters of the state. The codes provide for both criminal penalties as well as tort like damage remedies to discourage violations which degrade water quality and use. The prohibitions are wide ranging and protect both public health as well as fish and wildlife. Waters under California law are owned in fee by all the people by and through the various agencies of government. Pollution of the people's water is viewed a legal wrong against both property and public health. Public agencies and private parties have both injunctive and monetary remedies against illegal discharges and may collect damages for the adverse biological effects of such discharges. The utility of biodiesel in this context is predominantly an alternative fuel which avoids pollution by discharging a relatively benign fuel substance. Vegetable-based oils do not have the inherent toxicity to aquatic life that is found in petro diesel. The quantities of biodiesel typically discharged as the result of normal operations of marine diesel operations are sufficiently low impact as be not be deleterious to fish and wildlife. Even in the extreme situation when large fuel spills occur the effects of vegetable based fuels are significantly less impacting than petro diesel both on an acute time frame as well as a continuing or incremental basis. Damage awards from spill events involving vegetable based diesel are projected under our preliminary research to be law to very low far sills to fast moving streams with biological damages increasing for slower moving waters.

California damage and spill cleanup costs of marine oil spills are available from industry sources and total costs can range from \$1,000 to \$2,600 per gallon of oil spilled. (Oil Spill Intelligence Report). US East Coast clean up cost may be less as shown by the Casco Bay State of Maine spill which cost approximately \$300 per gallon. Depending on spill frequency and use patterns alternative fuels can be cost effective nationally when environmental externalities are accounted as internal or with in the institutional cost of a business.

The second body of California law impacting fuel use patterns is the water: quality regulatory program found in the Porter Cologne Act of 1969 as amended. These laws are administered through the regional and state water quality control agencies. Water quality is regulated through state and delegated federal permits which allocate pollution discharges depending on the assimilative capacity of the receiving waters. Environmentally sensitive receiving waters receive greater protections than water not used beneficially. The current water quality regulatory program is permit and compliance intensive because it seeks to protect a finite number of dischargers that each effect the common pool of the receiving waters. As water quality degrades due to incremental, outlaw, or unregulated pollution all dischargers suffer increasing regulatory compliance burdens.

Biodiesel users in and adjacent to enclosed aquatic ecosystems are absent spills, not significant dischargers and fall below the regulatory threshold thus escaping the shared costs of maintaining the common pool of water quality. While biodiesel has not received formal regulatory exemptions, the factual basis for such exemption is clear and available to users who seek such relief.

Conclusion and Recommendations

The marine industry has a significant direct and indirect financial interest in avoiding long term environmental liability from water pollution. In environmentally sensitive river corridors and enclosed marine ecosystems biodiesel is a fuel whose cost effectiveness is evident to those that understand the macro economics of pollution. California's large oil companies and specialty chemical industry has spent several years and extensive financial resources seeking to "reduce regulatory burdens" through legislative initiatives. It is now time to reduce regulatory burdens by reducing pollution through the use of alternative fuels. This paper has demonstrated that the biodiesel industry has a product whose market is maturing and will soon take it's place with the many "green" or low environmental impact products that wilt be useful in maintaining our water as " fishable and swimmable" as we enter the 21st century.

References

Ahouissoussi, N.B.C. and Wetzstein, M. E., The Economics of Engine Replacement/Repair for Biodiesel Fuels, Final Report, University of Georgia, Department of Agricultural and Applied Economics, March 1995.

Biodiesel Report, March 1996

"Esters from Vegetable Oil as Diesel Fuels." *Federation Francaise des Producteurs d'Oleagineux (FOP), AFPP-AGPO.* 12 Avenue George V. 75008 Paris.

Koch, D. from Ahouissooussi and Wetzstein, Novamont, Milan, Italy, 1992.

Long, E. "Spring Rape Rise." *Farmers Weekly*, March 1992.

Marshall, B. "Summary Reports on Soydiesel." National Institute for Petroleum and Energy Research (NIPER), Bartlesville, OK: NIPER Report No. B06874, 1993.

Merrill, P.N., *Biodiesel Alert*, Vol. 2, Number 5, March 1994

Muschalek, I., and K. Scharmer. "The Global Ecological Balance for Engine Fuel Production from Vegetable Oils." 7th European Conference on Biomass for Energy and Environment, Agriculture and Industry, Florence, 1992.

Oil Spill Intelligence Report "The Financial Costs of Oil Spills" Cutter Information Corp. 1994. See Avila Beach case study.

Pachter, H. and G. Hohl. "Rapeseed Oil Methyl Ester (RME) as an Alternative Diesel Fuel: Fleet Trials in the Austrian Army." AVL Conference, Engine Environment (1991):279-292.

Peters, R. A., *Methyl Ester Supply*, Proctor and Gamble, in Biodiesel '94, conference proceedings, Western Regional Biomass Energy Program,.

Reece, Daryl M.S., Engineer; Charles Peterson Ph.D., Acute Toxicity of Biodiesel to Freshwater and Marine Organisms, in Proceedings, Second Biomass Conference of the Americas: Energy, Environment, Agriculture, and Industry, Portland, OR 1995, NREL/CP-200-8098 DE95009230.

Sharp, C. A., Transient Emissions Testing of Biodiesel and Other Additives in a DDC Series 60 Engine, Southwest Research Institute, SwRI 6601, December 1994.

Wade, R. and R. Irving. "Biodiesel - Technical Note." MAFF - unpublished.

Walker K. C. and W. Korbitz. "Rationale and Economics of a British Biodiesel Industry." *British Association for Biofuels and Oils* (BABFO), April 1994.

Whitelegg, J., Gatrell A., and P. Naumann. "Traffic and Health." *Lancaster University*.

"Esters from Vegetable oil as Diesel Fuels." *Federation Francaises des Producteurs d'Oleagomeux (FOP). AFPP-AGPO.* 12 Avenue George V 75008 Paris

Zhang, X., Peterson, C. L., Reece, D., Möller, G., and Haws, R., Biodegradability of Biodiesel in the Aquatic Environment, ASAE 956742, Chicago, Illinois, 1995.

CHAPTER 25

Assessment of California's Ocean Wave Energy Recovery

Young C. Kim[1]

Abstract

With the prospect of an increasing shortage of energy resources, there has been a growing interest in renewable alternative sources of energy. An increasing effort is being directed towards resolving the problems of extracting energy from the world's oceans, as they represent a vast potential source of renewable energy. This paper summarizes the availability of California's coastal and ocean wave energy resources and examines the potentials and limitations of existing conversion technology for energy development in California. It was intended to provide a description of the various ocean wave energy technology and an assessment of the potential benefits that might be expected and the potential problems that could be encountered.

In the last two decades, considerable research and developments have taken place in ocean wave energy conversion technology. Suitable conversion technology for California include: (1) Wave power extracting caisson breakwater, (2) Backward Bent Duct Barge, and (3) Wave energy extraction combined with desalination. The wave power extracting caisson breakwater is an oscillating water column wave power converter, and the caisson functions as a breakwater and a wave power device at the same time. The Backward Bent Duct Buoy has an air chamber bent backward from the wave direction. Tests show that it has excellent conversion efficiency with low mooring force. The wave energy extraction combined with power desalination is a two point-absorber system, and the earlier test shows that it can produce 350 gpd of water as byproduct. Recommendations based on the engineering, environmental, and economic evaluations were made for future California's coastal and ocean wave energy conversion technology.

[1] Professor and Chairman, Department of Civil Engineering, California State University, Los Angeles, California 90032-8151

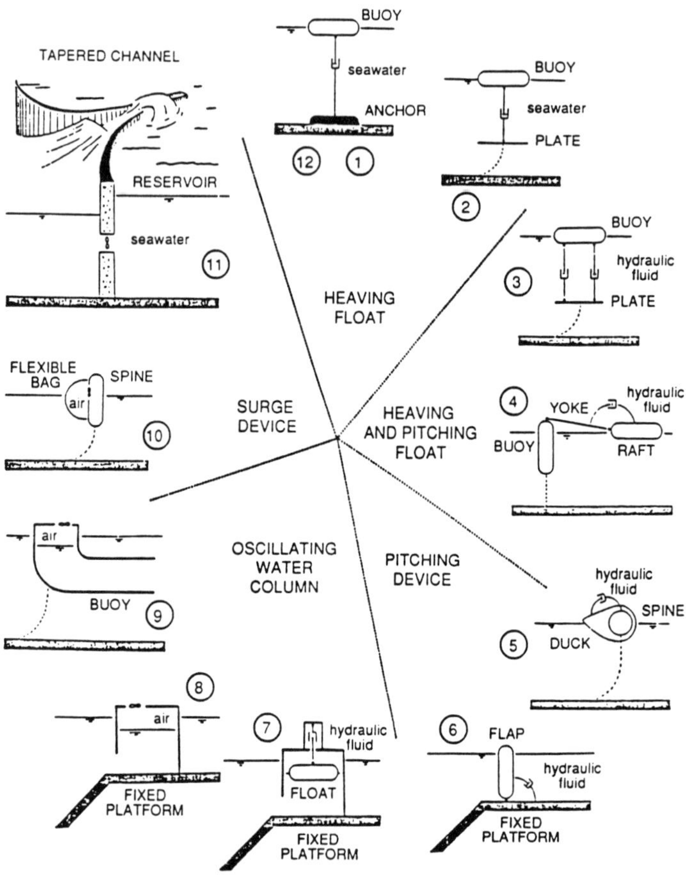

Figure 1. Stylized representation of the wave energy devices (after G. Hagerman 1990)

Development of Ocean Wave Energy Conversion Technology - Global Overview

Classification of Ocean Wave Energy Conversion

There are more than a thousand different proposals for the utilization of wave energy in the patent literature (McCormick, 1981), and there are many ways to describe and classify them. Figure 1 shows the classification of ocean wave energy conversion systems adopted by G. Hagerman. These devices can be classified into the following five categories:

1. Heaving
2. Heaving and Pitching
3. Pitching
4. Oscillating Water Column
5. Surging

Development of Ocean Wave Energy Conversion Technology

For heaving float devices, the KN System by Danish Wave Power was conducted with 45 kW prototype in the North Sea. The Gotaverken's Hose Pump Wave Energy Conversion which has been conducted since 1980, is based on the "hose pump", a hose with a helical pattern of steel wire reinforcement, which causes the hose to constrict when stretched. DELBUOY, two point absorber systems for wave energy extraction, has been developed by the International Science & Technology Institute, Inc. in the U.S. This system is comprised of a lightweight, shallow draft cylindrical buoy driving a submerged, single-acting positive displacement pump tethered to the sea floor. The motions of the buoy open the pump for the pressure stroke with energy stored in the natural rubber return springs used to refill the system. Water power desalination can be achieved by reverse osmosis. Full scale system tests were performed at the University of Puerto Rico in 1980-85, and a 350 gpd pilot commercial unit has been in operation at Coffin Island in Puerto Rico.

A Parallel Disk Wave Energy Module developed by the U.S. Wave Energy, Inc. combines the heaving and pitching motion. It consists of two parallel discs connected with hydraulic piston pumps. One disc is buoyant and rides on the ocean surface while the other is suspended at a depth where motion due to waves is small. Hydraulic fluid is transferred from the piston pumps to a high-pressure accumulator and is then fed to a hydraulic motor connected to a generator. 1 kW, 1:10 scale model tests were performed on Lake Champlain, Vermont. The Contouring-Raft Wave Energy Conversion System by Sea Energy Corporation uses the heaving and pitching motion. Test of a prototype 500 kW "one-raft" system has been conducted for regular and random waves.

For pitching devices, Salter from the University of Edinburgh has developed a very efficient wave energy conversion device known as the Salter Duck. The Duck is a cam shaped device that oscillates about a spine, connecting many devices. The spine is long enough to span several waves and provides a steady frame of reference for power extraction. 1:10 scale model tests were performed at Loch Ness. The Tandem Flap Wave Energy Converter by Q Corporation uses two flaps which are hinged at the bottom and allowed to move in the direction of wave. Because of its ability to capture both incident and a portion of the transmitted wave energy, it is significantly more efficient than a single flap device. A 10 kW Tandem Flap Wave Energy Converter system was tested in Lake Michigan in 1987. A Pendulor system was developed by Professor H. Kondo at the Muroran Institute of Technology in Japan. The pendulor is a pendulum gate which swings backwards and forwards in the opening at the front of a chamber facing the incident waves. The back of the chamber is a fixed wall, located a quarter wavelength behind the axis of the pendulum pivot. A 5kW system has been in operation in the bay of Uchiura near Muroran in Hokkaido, Japan.

Oscillating Water Column incorporates a wave-excited oscillating water column and a pneumatic turbine. The air above the water column is alternately compressed and expanded, resulting in an alternating pressure difference across the turbine. A one-meter Pneumatic Wave Energy Conversion System was tested in the second deployment of the KAIMEI in the Sea of Japan in 1985 (JAMSTEC, 1987). A 500 kW prototype, Multi-resonant Oscillating Water Column by Kvaener, Norway, was constructed in 1980. In Japan, Shore-Fixed 40 kW Wave Power Plant was constructed on the shore of Sanze, Sea of Japan. This plant consists of a caisson and air turbine generator, and Tandem Wells turbine was selected. The Wave Power Extracting Caisson Breakwaters developed by Y. Goda of Japan while he was with the Port and Harbor Research Institute has been built at Sakado, Japan. Caissons function as a breakwater and a wave power device at the same time. Another concept, Backward Bent Duct Buoy, developed by Ryokuseisha Corporation has an air chamber bent backward from the wave direction. Its originator, Y. Masuda, reported that this configuration of floating platform has a superior performance and that it has excellent conversion efficiency with low mooring force. A 2.4 m scale model was tested at Yura in 1987 and at Mikawa Bay in 1988. Recently, Professor M. McCormick at Johns Hopkins University conducted tests for optimum design of the Backward Bent Duct Buoy system.

For surging device, Tapered Channel known as TAPCHAN has been in operation since 1986 at Toftestallen, Norway. This Norwegian device converts the wave energy into potential energy in a reservoir on the shore.

Assessment of California's Ocean Wave Energy Recovery

Resource Characteristics

Monthly average wave power can be computed as $J = 5.7\ H^2$ where J is the monthly average wave power in kW/m of coastline and H is the significant wave height which is the average height of the highest 1/3 waves. Thus the wave energy is directly proportional to the square of the significant wave height. For example, at Santa Cruz, California, the average significant wave height is about 2.4m which can be expressed as J of 26 kW/m of coastline.

As waves generated in deep water travel shoreward across the continental shelf, they lose energy due to bottom friction and refraction. Furthermore, the incident wave power at the coast is much lower than offshore. A study by Hagerman shows that the energy at deep water north of Point Reyes, California is about 30-35 kW/m. This will be reduced about 25-30 kW/m as waves approach the Continental Shelf and reduced to about 5-15 kW/m at the water depth of 10-20 m at the coast. At the South of Point Reyes, California, the energy level is 25-30 kW/m and it reduces to 20-25 kW/m at the Continental Shelf. The annual average significant wave height at Central California is to be about 2.0m while at Southern California it drops to 80-90%. That means the power of 15-20 kW/m at Central California will be lowered to about 10 kW/m at Southern California. Therefore, it can be concluded that there is a reduction of about 5 kW/m of energy as waves travel from deep water to continental shelf another drop of 10 kW/m at the coast. A drop of 5 kW/m of energy is expected as we move from Central California to Southern California.

Not only there are geographical variations, but also seasonal and year-to-year variations. It is well known fact that the incident wave power on the west coast during the winter months is the highest and the summer months is the lowest. Data also show that between 1981 and 1985, there is a variation of annual wave power. The frequency of low-wave events, that is when the significant wave height is less than 1m, increases as we move from Northern California to Southern California. Conversely, the frequency of high-wave events increases as we move from Southern California to Northern California.

Economics of Wave Power

Land-based Tapered Channel technology, developed in Norway, is on the verge of commercialization and Japanese caisson-based systems are expected to enter the Pacific island market during the next five years. G. Hagerman's earlier study on the cost and performance of caisson-based wave power plants shows the following:

Device Double-Acting Double-Acting
Single-Acting Pivoting
MOWC MOWC OWC
Flap

Developer Kvaemer Brug Kvaemer Brug Takenaka
Muroran
A/S A/S Corporation
Institute of

Technology

Location Toftestallen Tongatapu Kashima Port
Muroran Port
(Norway) (Tonga) (Japan)
(Japan)

Cost of Energy 16-17 c/kWh 19-21 c/kWh 33 c/kWh
20 c/kWh

This data indicate that the energy cost for the caisson-based wave power plants is around 20 c/kWh.

There has been much less full-scale experience with offshore based wave energy devices than with caisson-based systems. Data from Bellamy, Hagerman, and Rasmussen for the cost and performance of offshore wave power plants show that:

Device Circular Swedish Swedish
Danish
SEA Heaving Heaving
Heaving
Clam Buoy Buoy
Buoy

Location Lewis Half Moon Bay Makapuu Point
Hanstholm
(Scotland) (California) (Ohau, Hawaii)
(Denmark)

Cost of energy 15 c/kWh 13 c/kWh 8.6 c/kWh
22-26 c/kWh

Comparing the energy cost for caisson-based and offshore based wave energy, the energy cost for offshore based is certainly cheaper than that of the land or caisson-based.

Various co-products can be obtained with wave energy production. These are:

1. Seawater renewal for closed-pond aquaculture
2. Fresh water for household use, crop irrigation
3. Breakwater protection for ports and harbors

A heaving buoy device (DELBUOY) was developed by Hicks for the desalination of sea water by reverse osmosis. Since 1982, a full-scale prototypes has been deployed off the southwest coast of Puerto Rico, producing fresh water at a rate of 950 liters per day from a single buoy. Combing wave energy generation with breakwater protection at coastal harbors will be an attractive alternative.

Data from Hagerman, Vega, Hemphill, and Doyle for energy cost comparison with other options show that when Wave, OTEC and Coal options were considered, the Wave option is cheaper than OTEC while Coal is the cheapest among three. When Wave, Wind, and Coal options were compared, the Wave option is the most expensive among the three.

California's Development Opportunities

The annual average wave power of the California coastline is said to be about 4,900 MW. Assuming that about half of this power can be extracted, it is still a large power resource along the California coast. The most attractive revenue-generating projects would be those with wave power plants for electricity with co-products. Examples of such projects are ones with electricity with breakwater protection and electricity with fresh water as a by-product.

Construction of caisson-type breakwaters with oscillating water column devices will be the ideal solution for coastal communities, specifically home port to a large commercial fishing fleet. Ideal candidates for this project will be Noyo Harbor in Fort Bragg and projected harbor construction at Santa Monica.

High-pressure sea water could be tapped to fresh water by reverse osmosis and electricity by passing the reject brine through a Pelton turbine. DELBUOY, heaving buoy systems for offshore use, have been proven to be a success in getting fresh water in island communities. Ideal candidates for this project are Santa Cruz, Santa Rosa, San Miguel, Santa Catalina, and San Clemente Islands.

Conclusions

1. Annual average wave power of California coastline is said to be 4,900 MW. Wave power in Northern California is around 15-20 kW/m while the wave power in Southern California is 10 kW/m.

2. For plant sizes of 30 MW or less, the cost of wave-generated electricity is lower than for other ocean energy technologies. The wave power is also less expensive than small hydropower or diesel generation.

3. Co-products enhance wave power's economic appeal. To construct breakwater in the coastal communities, a caisson-type breakwater with oscillating water column is recommended. To obtain fresh water in the island communities, heaving buoy system is recommended.

References

McCormick, Michael E., Ocean Wave Energy Conversion, Wiley Interscience, New York, New York, 1981.

McCormick, Michael E. and Kim, Young C. (editors), Utilization of Ocean Waves - Wave to Energy Conversion, Proceedings of the International Symposium, American Society of Civil Engineers, 1987.

Krock, Hans-Jurgen (editor), Ocean Energy Recovery, Proceedings of the International Conference on Ocean Energy Recovery, American Society of Civil Engineers, 1989.

Seymour, Richard J. (editor), Ocean Energy Recovery - The State of the Art, American Society of Civil Engineers, 1992.

Behrens, David W. and Champ, Michael A. (editors), A Global Review of the Development of Wave Energy Technologies, Proceedings of the Ocean '89 Special International Symposium, 1989.

CHAPTER 26

"The Economic Value of Coastal Resources and Coastal Conservation"

Beth Millemann

Abstract

The nation's coasts are an economic, as well as environmental, hotspot. Vitally important sectors of the American economy depend on healthy, clean and biologically diverse coastal resources. Water quality protection and improvement, habitat conservation and restoration, and fisheries and wildlife preservation are not ecological luxuries: they are economic imperatives. Federal coastal management programs - such as the Coastal Zone Management Act and the Coastal Nonpoint Runoff Control Program - are wise investments in economic prosperity and biological integrity. They benefit the nation's economy of California.

Economic Values

- According to the National Oceanic and Atmospheric Administration (NOAA), the U.S. coasts support 34% of national employment, providing more than 28 million jobs and $1.3 trillion.
- An estimated 94 million Americans spend time boating and fishing along the coast each year. In fact, the average American spends 10 recreational days a year on the coast.
- The U.S. coast produces 70% of the U.S commercial and recreational fisheries harvest.

Cost of Pollution and Habitat Loss

- In 1990, harvesting in 37% of the nation's shellfish beds was restricted because of pollution, compared to just 13% in 1966. Leading causes for these restrictions include urban and agricultural runoff, failed septic systems and boating.
- In 1993, 23 states issued nearly 2,500 beach advisories and closings, some lasting weeks, due to contamination of coastal waters with disease-causing

Beth Millemann, Executive Director, Coast Alliance

bacteria and viruses. Since 1988, more than 10,000 such advisories and closings have been issued. Major sources include street runoff collected by storm drains and combined with sewage and septic systems.

- When slicks of debris washed ashore on Long Island and New Jersey beaches in 1987 and 1988, beach attendance dropped as much as 90%, devastating local businesses.

Value of California's Coastal Resources

- Economic productivity in California's coastal zone makes up 56% of the Gross State Product, or more than one out of every two dollars.
- More than $30 billion is annually spent on tourism in California, with 86% spent in coastal counties.
- Birdwatchers contribute more than $27 million to the state's economy, while recreational boating brings in another $600 million a year, with commercial fishing, mariculture and kelp harvesting generating another $554 million annually.

Federal Programs That Help Protect The U.S. And California Coast

- The Coastal Nonpoint Pollution Control Program requires coastal states to develop plans to trap and prevent nonpoint runoff into coastal waters. This Program needs additional funding to help states complete the process of addressing polluted coastal runoff.

CHAPTER 27

The Economic Valuation of Coastal Resources: Methods and Applications

Rodney Weiher
Chief Economist
National Oceanic and Atmospheric Administration
Washington, DC, 20230

Abstract

Professionals who are responsible for coastal environmental and natural resource planning and management have a need to become conversant with new concepts designed to provide quantitative measures of the environmental benefits of natural resources.

The most widely used organizing tool for dealing with natural and environmental resource choices is benefit-cost analysis, which offers a convenient way to carefully identify and array, quantitatively if possible, the major costs, benefits, and consequences of a proposed policy or regulation.

However, a major shortfall of benefit-cost analysis has been the difficulty of quantifying both benefits and costs of actions that impact environmental assets not normally or regularly bought and sold in markets.

Economists and other social scientists have put a great deal of effort into addressing this shortcoming by developing techniques to quantify these non-market benefits.

To make that link, NOAA has undertaken a series of Regional Environmental Valuation Workshops to introduce and illustrate concepts of environmental valuation to state and local coastal planners and managers, fishery managers, and those concerned with the broad range of coastal sustainability issues.

This paper briefly summarizes the Handbook developed for these Workshops. The Handbook is intended for non-economists and addresses basic concepts of economic value such as willingness-to-pay and other tools often used in decision making such as cost-effectiveness analysis, economic impact analysis, and sustainable development.

Background

Professionals who are responsible for coastal environmental and natural resource planning and management have a need to become conversant with new concepts designed to provide quantitative measures of the environmental benefits of natural resources. These amenities range from beaches to wetlands to clean water and other assets that normally are not bought and sold in every-day markets.

At all levels of government--from federal agencies to townships and counties--decision-makers are being asked to account for the costs and benefits of proposed actions. Unfortunately, to non-specialists, the tools of professional economists are often poorly understood and sometimes inappropriate for the problem at hand.

The most widely used organizing tool for dealing with natural and environmental resource choices is benefit-cost analysis, which offers a convenient way to carefully identify and array, quantitatively if possible, the major costs, benefits, and consequences of a proposed policy or regulation. The major strength of benefit-cost analysis is not necessarily the predicted outcome, which depends upon assumptions and techniques, but the process itself, which forces an approach to decision-making that is based largely on rigorous and quantitative reasoning. However, a major shortfall of benefit-cost analysis has been the difficulty of quantifying both benefits and costs of actions that impact environmental assets not normally or regularly bought and sold in markets. Failure to account for these assets-- to omit them from the benefit-cost equation-- could seriously bias decisionmaking, often to the detriment of the environment. Economists and other social scientists have put a great deal of effort into addressing this shortcoming by developing techniques to quantify these non-market benefits.

The National Oceanic and Atmospheric Administration (NOAA) is charged with stewardship of our country's coastal ocean resources through rigorous scientific research, monitoring, and management oversight. Economic valuation and analysis is a major link between coastal stewarts and monitoring the policy/management process.

To make that link, NOAA has undertaken a series of Regional Environmental Valuation Workshops to introduce and illustrate concepts of environmental valuation to state and local coastal planners and managers, fishery managers, and those concerned with the broad range of coastal sustainability issues.

This paper briefly summarizes the Handbook developed for these Workshops. The Handbook is intended for non-economists and addresses basic concepts of economic value such as willingness-to-pay and other tools often used in decision making such as cost-effectiveness analysis, economic impact analysis, and sustainable development.

CONCEPTS IN ENVIRONMENTAL VALUATION

The term *value* in economics has a precise definition---it is the price individuals are willing to pay in order to obtain a good or service. However, the term *value*, in the context of coastal issues, can have different meanings to those with different interests. To an ecologist, the value of a salt marsh might mean the significance or importance of the marsh to the reproductive capacity of certain species of fish. To a coastal engineer, the value of a salt marsh may be associated with its contribution to shoreline stabilization. In general, these values are mathematical and functional: mathematical, meaning magnitude, and functional, meaning the physical or biological relationships of one entity to another. These values exist whether or not humans prefer them or are even aware of them.

ECONOMIC VALUE

A fundamental distinction between the way economics and other disciplines such as ecology use the term value·is the economic emphasis on human preferences. Thus, the functionality of economic value is between one entity and a set of human preferences. If a coastal area is degraded so it supports a lower abundance of organisms, an ecologist would characterize this degraded area as less valuable for those organisms than a non-degraded area. In economic terms, however, a polluted area only has less value than an otherwise equivalent non-polluted area if some individual members of society prefer non-polluted to polluted areas. If no one cares that there are fewer organisms in the polluted area, then there is no difference in economic value. Typically, some members of society will display a preference for an environment that is less degraded.

Economic value is a measure of the maximum amount an individual is willing to forego in other goods and services in order to obtain some good, service, or state of the world. This measure of welfare is formally expressed in a concept called willingness-to-pay (WTP). Thus, the lost value from the degraded environment is the maximum amount individuals are willing to pay to have a state where that same area is free of pollution.

In assessing the value of some policy or management plan, the economist is interested in estimating how much an individual's (or society's) well-being would change: how much it will decrease if a natural resource were lost or increase if a natural resource or resource service were better managed or its quality improved. In other words, when economists try to estimate the economic value of a coastal resource or resource service, they attempt to answer two questions:

- How much are people willing to trade (give up of other goods and services) to have some natural resource or resource service?
- How much better off would people be if a policy or management plan action were implemented and the amount or quality of a resource or resource service were improved?

SCARCE RESOURCES, LARGE DEMANDS

The economic definition of value is rooted in a simple idea: all resources are scarce, but the demands for those resources are large relative to their availability. There is never enough labor or land or water to do all the things that all individuals might wish. Because re-sources are scarce, it is necessary to make choices about how society will use what is available. We make choices about the amount of money to devote to schools, roads, libraries, and natural resource protection programs individually and collectively. These choices are often based on complex tradeoffs; thus, value is revealed in decisions about how individuals and society collectively choose to allocate these resources. People may recoil at the notion of placing a value on the natural environment, but there are other uses or alterations of that environment that might be proposed. Society always has to compromise, giving up something to get something else.

ENVIRONMENTAL VALUATION

Environmental Valuation is a series of techniques that economists use to assess the economic value of market and non-market goods, namely natural resources and resource services. It applies the welfare economics concepts of producer and consumer surplus to issues involving natural resources and the state of the environment. Welfare economics tries to answer the question "Is society better off?" Environmental valuation is the application of welfare economics when the differences in circumstances relate to the uses or states of natural resources or the quality of the environment.

When economists refer to evaluating societal benefits, it is necessary to recognize two "states at the world": *with* and *without*. *Without* is the base state if an activity circumstance and policy does not change. *With* is the state when the change occurs. A distinction is made between *with* and *without* and *before* and *after*. *Before* and *after* does not control for changes in the state of the world that do not result from the action or policy in question. Economists try, for example to weigh social benefits associated with a commercial development project against environmental benefits that would be lost should the project be implemented. Such a *social accounting* analysis tallies all real costs associated with an activity, including the cost of lost or damaged environmental assets and quality of life. Desirable characteristics of this social accounting scheme are these: it is internally consistent (i.e., the underlying theory does not change with circumstance), usually intuitively appealing, and acceptable in major courts of law.

MEASURING THE VALUE OF GOODS AND SERVICES TRADED IN MARKETS

If goods or services are traded in the market, there are well established and accepted empirical techniques for measuring welfare changes. These conventional techniques of measuring changes in value serve as a springboard for understanding non-market techniques of economic valuation.

To estimate so-called "use values", economists employ market resource valuation methodologies. For those resources for which markets exist, economists typically rely on directly observable behavior in the form of market transactions to reveal preferences or the value that individuals place on goods and services and their willingness to pay to avoid loss of such goods and services. The standard method for measuring the use value of resources traded in the marketplace is the estimation of producer and consumer surplus using market price and quantity data.

MEASURING THE VALUE OF NON MARKET GOODS AND SERVICES

Without the observable price and quantity data that are available when goods or services are traded in the market, economists have devised innovative techniques for measuring changes in value for natural resources and the environment. Three of the techniques, travel cost, random utility (not discussed in this paper) and hedonics use information to indirectly determine what a market might reveal in value if it did exist. The contingent value technique attempts to measure the change in value directly.

Some goods and services like recreational fishing and wildlife viewing are not traded in a well functioning, traditional market. That is, they are not supplied by private firms and consumers do not pay market prices. Nonetheless, individuals benefit from their use and, therefore, the loss of such environmentally related goods signifies welfare losses to these individuals. Conceptually, the same measure of benefit applies to market and non-market goods, that is, the maximum amount an individual would pay to avoid losing, or gaining, access to the good. Since these are non-market benefits, typically, there is no producer, or the consumer is both the producer and consumer. Thus, measures of non-market benefits are concerned with estimates of consumer demand and consumer surplus. There are a variety of methods that have been developed to measure this value concept in the absence of markets. These include:

- Travel cost (and random utility) models, which are based on expenditures and travel behavior for recreational opportunities
- Hedonic methods of decomposing prices of market goods to extract embedded values for related environmental attributes
- Experimental methods for eliciting preferences, either by using hypothetical settings, called contingent valuation, or by constructing a market where none existed

Travel cost models and hedonic methods are indirect measures based on observable behavior. Experimental methods, or contingent valuation, are based on direct surveys of individuals.

INDIRECT MEASUREMENT TECHNIQUES

Indirect techniques rely on observable behavior to deduce how much something is worth to an individual even though it is not traded in markets. These methods produce value estimates that are conceptually identical to market values, but they must be measured more creatively since market data are not available. Indirect techniques include travel cost models, random utility models, (not discussed here) and the hedonic pricing method.

- TRAVEL COST MODELS

The travel cost method is, in general, employed to estimate recreational values. This technique assumes that visitors to a particular site incur economic costs, in the form of outlays of time and travel expenses, to visit the site. In effect, these economic expenditures reflect the "price" (albeit implicit) of the goods and services provided by the site, and are an indirectly observable indication of the minimum amount that a visitor is willing to pay to use the site (with all its associated attributes).

By observing the characteristics of individuals visiting the site -- for example, the specific attributes of their trip to and from the site as well as the total number of visits -- economists are able to estimate the "derived demand" for the site. That is, for any given or implicit price, the derived demand relationship will determine the number of visits consumers will "purchase" at that site. The travel method technique has a number of applications -- it can be consumption used, for example, to measure the effects on a proper account of substitute sites and their characteristics because of changes in access costs to a recreational a site, or changes in it environmental quality.

The advantages of the travel cost technique are that it is relatively uncontroversial because it mimics empirical techniques used elsewhere in economics. Economists generally tend to prefer techniques of this sort because they are based on actual behavior rather than verbal responses to hypothetical scenarios. In the travel cost model, individuals are actually observed spending money and time, and their economic values are deduced from their behavior. In appropriate circumstances, this model can often be applied without enormous expense.

The greatest disadvantage of travel cost and other indirect techniques is that they can not be employed unless there is some easily observable behavior that can be used to reveal values. Thus, in the case of measurement of "nonuse" values these methods are inappropriate. In the case of nonuse values, there is no observable interaction between the individual and the resource in question.

· HEDONIC PRICING METHODS

The hedonic pricing method is another technique to determine environmental value. In its earliest applications, these techniques were intended to capture the willingness-to-pay measures associated with variations in property values that result from the presence or absence of specific environmental attributes, for instance, air pollution, noise, or water views. By comparing the market value of two properties which differ only with respect to a specific environmental attributes economists may assess the implicit price of that amenity (or its cost when undesirable) by observing the behavior of buyers and sellers.

The hedonic techniques, like travel cost models, depend on observable data resulting from the actual behavior of individuals. Market data on property sales and characteristics are available through real estate services and municipal sources and can be readily linked with other secondary data sources.

However, most environmental incidents will have only small, if any, effects on housing prices. Even where effects do exist, it may be difficult to estimate them using econometric methods because many factors, many of which are correlated, influence housing prices. For example, a house located near a factory with emissions that reduce air quality may be in a poorer section of town where schools are not as good and there are few other amenities like parks. Even when implicit prices for environmental amenities can be estimated, it is usually very difficult to obtain measures of value from these models. The connection between the implicit prices and value measures is technically very complex and sometimes empirically unobtainable.

DIRECT TECHNIQUES

CONTINGENT VALUATION METHOD (CVM)

The most obvious way to measure nonmarket values is through directly questioning individuals on their willingness-to-pay for a good or service. Called the contingent valuation method, it is a survey or questionnaire-based approach to the valuation of non-market goods and services. The dollar values obtained for the good or service are said to be contingent upon the nature of the constructed (hypothetical or simulated) market and the good or service described in the survey scenario.

The contingent valuation technique has great flexibility, allowing valuation of a wider variety of non-market goods and services than is possible with any of the indirect techniques. It is, in fact, the only method currently available for estimating nonuse values. In natural resources, contingent valuation studies generally derive values through the elicitation of respondents' willingness-to-pay to prevent injuries to natural resources or to restore injured natural resources. Since the first published contingent valuation study on valuing outdoor recreation appeared in 1963, more than 1,400 related documented papers, reports, and books have been published.

Contingent valuation studies are conducted as face-to-face interviews, telephone interviews, or mail surveys. The face-to-face is the most expensive survey administration format but is generally considered the best, especially if visual material needs to be present-ed. Non-response bias is always a concern in all sampling frames. In other words, people who do nor respond have, on average, different values than people who do respond.

In principle, contingent valuation methods can be used to estimate the economic value of any-thing, even if there is no observable behavior available to deduce values through other means. It is the only method that has any hope of measuring "existence values," i.e., the value that individuals place on simply knowing the natural resource exists in an improved state. This is because since existence values are not connected with use and all other methods depend on observing actual behavior associated with the resource.

Though the technique requires competent survey analysts to achieve defensible estimates, it is not difficult to understand. The responses must be statistically analyzed, but require no more than the understanding of a mean or median value.

When conducted to the exacting standards of the profession, contingent valuation methods can be very expensive because of the extensive pre-testing and survey work. In addition, while this technique appears easy, its application is fraught with problems, for example, the possibility of strategic bias by respondents or structural problems in questionnaire design. Moreover, question framing, mode of administration, payment formats, and interviewer interactions can all affect results.

In addition, many questions have been raised about the reliability of the contingent valuation method for the calculation of nonuse values, particularly in regard to natural resource damage assessment under OPA.

THEORY AND APPLICATION: RECONCILING DIFFERENCES

In the practical application of environmental valuation, issues such as choosing a discount rate, dealing with intergenerational transfers and equity, and decision-making under risk and uncertainty can become important to the outcome and interpretation of the analysis. For example, arguments about the appropriate discount rate can unduly obscure the underlying message that there is an economic value to natural resources and the environment.

Further, valuation of natural resources and environmental amenities can meet with difficulty under certain conditions. For example, if the use of a particular resource is impossible to reverse, the economic and social impacts over a long period of time must be considered. Such a consideration in turn raises the question of discounting or, more generally, the efficiency and equity of resource use in the long run. Moreover, where information about the costs and benefits of alternative uses is particularly poor, perhaps because of the long period to which it must apply and the non-market character of some of the uses, decisions should take this uncertainty into account.

Unfortunately, theory does not spell out the precise quantitative adjustments that would be required in applying these issues to estimate benefits and costs in empirical work. The major point is this: the traditional benefit analysis of resource use and allocation as a basis for public decision-making is only one part of the decision process which must be accompanied by subjective notions of risk-taking and equity. A benefit-cost analysis in isolation should not be the sole basis for decision-making.

CASE STUDY APPLICATIONS

A number of case study applications of environmental valuation techniques have been developed and presented in the Handbook that mimic real-world situations. They include such diverse situations as oyster restoration in Chesapeake Bay, salmon habitat restoration in Alaska, red snapper fishery management in the Gulf of Mexico, wetlands restoration in Louisiana, coastal barrier island protection in North Carolina, and nonpoint source pollution in California.

Reference: D. Lipton and K. Wellman, *Economic Valuation of Natural Resources: A Handbook for Coastal Resource Policymakers, NOAA Coastal Ocean Program,* Decision Analysis Series No. 5, National Oceanic and Atmospheric Administration, June 1995

CHAPTER 28

ECONOMIC ASSESSMENT OF SEVEN OCEAN-DEPENDENT ACTIVITIES

by Rosa Maria Moller, Ph.D., Senior Economist

INTRODUCTION

The purpose of this analysis is to quantify some of the economic contributions of the sea to the state economy, in order to provide a better understanding of the value of ocean resources to California. Instead of assessing the value of ocean resources themselves, this study focused on the economic benefits generated by those industries that could not exist without ocean access. These industries are: commercial fishing; mariculture; kelp harvesting; offshore oil and gas; coastal mineral production; and port activity. The economic benefit from coastal tourism and recreation was estimated in a separate paper.

The economic benefits of these selected industries are part of the economic contribution of the ocean to the state economy, since these activities necessarily require or depend on ocean access. For example, although oil refining occurs along the California coast, only the economic contribution of offshore oil and gas production and closely related activities, such as oil transportation from these locations, were included in this assessment as ocean-dependent activities. This decision was based on the fact that not all coastal refining operations require access to the ocean. However, to extract oil from beneath ocean waters, production operations and off-shore related activities must have ocean access.

Specific objectives of this study were to:

- provide a brief overview of these ocean-related activities

California Research Bureau, California State Library
P.O. Box 942837, Sacramento, California 94237-0001
(916) 653-6972, FAX (916) 654-5829, rmoller@library.ca.gov

- assess regional and statewide direct and indirect employment generated by these activities
- assess the direct and indirect income added by these ocean-related sectors to the state economy

METHODOLOGICAL STRATEGY

Employment, payroll employment, and output data were desegregated in 5 coastal regions, including 20 coastal counties. To preserve the confidentiality of employment and payroll information in cases where there is only one or two major employers in a given county, the information was presented by regions rather than counties. The selected 5 regions are:

1. North Coast includes the following three counties: Del Norte, Humboldt, and Mendocino.
2. North Central Coast, includes the following 8 counties: Alameda, Contra Costa, Marin, Napa, San Francisco, Santa Clara, Solano, and Sonoma.
3. Central Coast includes the San Mateo, Santa Cruz, and Monterey counties.
4. South Central Coast includes the San Luis Obispo, Santa Barbara, and Ventura counties.
5. South Coast includes the Los Angeles, Orange, and San Diego counties.

To evaluate the economic contribution of ocean-related industrial activities, the value added (or direct income) by industry was calculated. Value added differs from gross output which is the value of total production. Value added is the portion of gross output accruing to primary inputs. Indirect income or secondary effects derived from each ocean-related economic activities were approximated by using statewide average multipliers calculated from the RIMS II model, built by the U.S. Department of Commerce.[1] Total income was calculated as the estimated value added times the multiplier. Indirect income is defined as total income minus value added.

Employment data at the county level together with employment multipliers to income from the RIMS II model were used to calculate employment effects. The methodology followed in the calculation of value added varied for each of the ocean-related activities analyzed and depended largely on the availability and quality of pertinent data.[2]

---

[1] Based on RIMS II Regional Input-output model. To assess secondary effects, we have used California average multipliers throughout this analysis. These multipliers represent the total income generated in an economic region (in our case the State of California) for each dollar produced in a given industry. Indirect effects, and consequently regional multipliers, vary from region to region. Since regional impacts are calculated using the statewide average multipliers, regional impacts are just rough approximations.

[2] For a detail of the methodology followed in this study, please see Methodological Appendix attached to the Report "Economic Assessment of Ocean-Dependent Activities", included in the Resources Agency's publication: "California's Ocean Resources: An Agenda for the Future."

COMMERCIAL FISHERIES

The California Fishing Industry. California's fishing fleet consists of more than 4,000 vessels. *In 1992, California commercial fishermen landed approximately 294 million pounds of fish and invertebrates worth over $131 million*. Since 1970 commercial fishing has declined in the state. The California's share of the U.S. harvest has declined from 14 percent in 1970 to 4 percent in 1990. Ex-vessel revenues followed a similar pattern.[3]

The main reason for the drop in share is that total finfish and shellfish landings in California declined by more than half from 1970 to 1990, while total U.S. landings almost doubled over the same period. This decline was primarily a result of the dramatic drop in tuna landings. Salmon, abalone, and anchovy landings have also fallen.

The drop in tuna landings was due to increasing competition from imports of canned tuna and the movement of tuna processing operations from the state to American Samoa and Puerto Rico. In 1970 the proportion of ex-vessel revenues from tuna was 74 percent, while in 1990 it was only 17 percent. Despite the decline, tuna is still important. In 1992, 19.2 million pounds of tuna were landed into California, with a value of $10 million.

Other factors that have affected the state fisheries are: interest rate changes, fluctuations in species abundance and availability, weather, and changes in the availability of support facilities (as other uses such as recreational boating, oil and gas exploration, cargo handling, tourism, and urbanization compete for waterfront space).

The California fishing industry today is less dependent on tuna fisheries and has evolved into a mix of smaller fisheries. For instance, mackerel landings have increased dramatically since the 70's together with squid landings. A number of major fisheries are driven by the demand from markets in the Pacific Rim (especially Japan). Asia is the major market for red sea urchins, pacific herring, thornyheads, sablefish, and other species. Fresh sea urchin was California's major exported seafood product in 1990, accounting for 23 percent of the value of all seafood exported from the state. Main fisheries in the north of California are salmon, ground fish (cod, snapper, halibut, i.e.) albacore, crab, herring, and shrimp. Main fisheries in Southern California are tuna and mackerel. The per pound value of landings by the fisheries of the north of California is considerably higher than that for the south (especially South Coast region).

State Demand for Seafood. There is a large demand for seafood in California. Assuming that per capita seafood consumption in California is equal to the national

---

[3] Ex-vessel revenues are revenues obtained from the sale of finfish right off the boat.

average (15.5 pounds in 1990), 4.5 billion pounds of seafood (edible meat) were consumed in the state during 1990. This estimate is likely to be low since Californians have a strong preference for seafood (reportedly due to health-consciousness and the large proportion of Asian immigrants in the state).[4]

State's Seafood-Trade Balance. It is difficult to determine exactly how much of the seafood imported into the United States ends up in the state as well as how much of the imports that enter the state are actually consumed in California. Similarly, exports from California are not necessarily produced within the state. Despite the difficulty of determining exactly how much seafood California exports and imports, research have shown that the state's seafood trade deficit is substantial. According to one estimate, less than 10 percent of the seafood sold in California is derived from California fisheries and approximately 13 percent comes from other domestic markets (for instance Alaska and Pacific Northwest).[5] That would leave the overwhelming majority of seafood—about 75 percent—coming from foreign sources.

Economic Impact of Commercial Fishing. The commercial fishing industry is very important in the economy of many coastal communities as a source of food and as a source of employment. Some activities dependent on this industry are boat construction and repair yards, brokerage, dockhandling, trucking and other transportation, gear and rigging stores, fish processing, and commercial trade.

Table 1 presents the estimated impact on regional total income and employment brought about by commercial fisheries and fish hatcheries. Value added from commercial fisheries was calculated using the fact that payrolls represent approximately 40 percent of the total cost of commercial fisheries and assuming a ratio of costs to value added of 0.91.[6]

---

4 Source: Leet, William S., Dewees, Christopher M. and Haugen, Charles W. Editors. California's Living Marine Resources and their Utilization. California Sea Grant Extension Publication UCSGEP-92-12. 1992.

5 This estimate is also taken from: Leet, William S., Dewees, Christopher M. and Haugen, Charles W. Editors. California's Living Marine Resources and their Utilization. Op. cit.

6 The ratio of costs to value added was obtained from Hans Rradtke, Natural Resource Economist. This estimate is based on previous research. Using payroll data as a proxy for labor costs understates the actual labor costs of commercial fishing because there is a substantial amount of small boats operated by self-employed fishermen that are not included in these statistics. The ratio of total costs to value added was estimated from the ratio of costs to value added for all industries as published in the Bureau of Census: Census of Manufactures 1987.

Table 1

Commercial Fisheries and Fish Hatcheries
1992 Employment and Income Effects by Region

Region	Indirect Income Effect *(Dollars)*	Total Income Effect *(Dollars)*	Indirect Employment	Total Employment (Direct + Indirect)
North Coast	$431,062,919	$57,387,427	1,113	1,430
North Central Coast	7,397,815	13,667,150	265	475
Central Coast	7,753,333	14,323,954	278	424
South Central Coast	5,892,667	10,886,453	211	318
South Coast	93,200,443	172,183,869	3,340	4,329
All Coastal Counties	**$145,307,178**	**$268,448,854**	**5,208**	**6,977**

Table 2 summarizes statewide total benefits to the state from commercial fishing, fish processing, and marine fish oil production. Regional data for fish processing and marine fish oil production is not available. Value added for fish processing was calculated as the total cost of commercial fishing plus processor's margin (mark up). Since the production cost of commercial fishing is included in the cost of fishing processing, the value added of commercial fishing is thus subtracted from this total.[7]

Table 2

1992 Economic Impact of Commercial Fishing and Marine Fish Oil Production

	Indirect Income Effect	Total Income Effect	Indirect Employment	Total Employment (Direct + Indirect)
Commercial Fishing	$148,382,900	$274,131,120	5,246	6,836
Fish Processing	103,868,030	191,891,784	3,795	7,997
Marine Fish Oil	54,547,376	554,241,500	1,800	2,148
Total	**$306,798,306**	**$1,020,264,404**	**10,841**	**16,981**

The total effect of the fishing industry (commercial fishing and fish processing) on California income is estimated as $442,498,111. The income effect of the marine fish oil industry is estimated as $88,218,596. These industries provide employment (directly and indirectly) to at least 16,801 workers.

MARICULTURE

Mariculture, or the farming of marine species has increased sharply since 1979, as indicated by the doubling of the areas used for this purpose. Currently there are at least 4,872 acres leased for mariculture. The three most common mariculture crops in California are oysters, mussels, and abalone. Water quality is a primary concern in this industry, since the consumption of contaminated shell fish can pose a serious threat to human health.

[7] As a rule of thumb the mark up is 1.5 times the original price paid to the fisherman for the raw product. Here we used a ratio of 1.7 since a ratio 1.5 was low based on inferences from payroll data. Source: Dewees Chris, Radtke, Hans and Smith, Frederick. "The Fishing Industry and Pacific Coastal Communities: Understanding the Assessment of Economic Impacts."

Preliminary data for 1992 indicates that *157,900 pounds of abalone valued at $1,976,000* were produced in the state marine farming. Nearly 91 percent of abalone production takes place in the Central coast of California.

The table below shows regional data on oyster and mussel amounts and value of production during 1991. The production of single oysters for the half-shell trade is increasing each year in all state-growing areas. Tomales Bay is the leader in single oyster production, with over $1 million produced in 1991. Statewide mussel production decreased substantially during 1991 because a major southern California grower ceased operations due to bacterial contamination of his mussels and consequent financial difficulties. The bulk of statewide mussel production (85 percent) continues to come from offshore oil production platforms in southern California waters.

Table 3

1991 Value of Oyster and Mussel Production

Region	Oyster Production	Mussel Production
North Coast (Humboldt Bay)	$2,640,297	$12,981
North Central (Tomales, Drake Bay)	1,725,196	22
Central Coast (Morro Bay)	56,305	57,345
South Central (Santa Barbara)	8,205	70,348
Total	**$4,430,003**	**$140,696**

Source: Department of Fish and Game

Employment and payroll data for fish hatcheries were presented on Table 1 (together with commercial fisheries). Consequently, the impact of mariculture on the California economy was already included in the economic assessment of commercial fisheries in Tables 1, 2, 4, and 5.

KELP AND SEA VEGETABLE HARVESTING

Most of the kelp harvest is used for the extraction of algin, a product used as a binder, stabilizer, emulsifier, and molding material in pharmaceuticals, cosmetics and soaps, dental and food technology and a wide variety of food and industrial products. *In 1992, 91,251.25 tons were harvested, with a value of $2,933,721.*[8] Employment and payroll data for kelp harvesting were also included in the commercial fishing classification presented on Tables 1 and 2.

[8] The value of kelp harvested was estimated using the value of imported kelp from Mexico (the closest competitor).

OFFSHORE OIL AND NATURAL GAS PRODUCTION

The California's coastal zone has 11 sedimentary basins with favorable geologic structures for the accumulation of oil and gas deposits. These basins are the Southern California Shelf; the San Diego Offshore Area; the Los Angeles Basin; the Santa Barbara-Ventura Basin; the Santa Maria Basin; the Salinas Basin; the Santa Cruz Basin; the Bodega Basin; the Point Arena Basin; the Eel River Basin and the Bear-Mattole Offshore Area. Though all of the basins may contain oil and gas deposits, up until now only the Los Angeles, Santa Barbara-Ventura and Santa Maria basins have actually produced commercial quantities of oil and gas.

Oil Production. California's crude oil production decreased significantly from 1987 to 1990, when it leveled off at approximately 350 million barrels per year through 1992. *In 1992, the state's oil production was 348 million barrels* (a 0.8 percent decrease from the previous year).[9] *Offshore oil production in 1992 was 65 million barrels*. California's crude oil production has been declining due to continuing instability in crude oil prices, coupled with a general depletion of aging reserves as well as government restrictions on areas that can be explored or used for production. Low oil prices along with reduced production per well have resulted in the shut-in of many wells. Since 1991 the share of offshore oil production has increased significantly due to Point Arguello Offshore new field, which commenced production in June 1991. Since overall oil and gas production are not ocean-dependent economic activities, this economic assessment of the ocean-related industries in the state will include only state offshore oil and gas production.

Table 4 shows offshore oil production data and the gross value of output evaluated at the average 1992 price of posted Kern River 13 degree API gravity crude oil,[10] per county. The Table also shows the 1992 percentage of offshore oil production in total county oil production.

Table 4
1992 Offshore Oil

Region	Oil and Condensate *(Barrels)*	% Of County Total Oil Production	Daily Production per Well *(Barrels)*	Price of Oil (Average) (per Barrel)	Gross Value of Output
Santa Barbara	2,780,714	36.8	75.4	$12.8	35,593,139
Ventura	336,681	16.8	20.1	12.8	4,309,517
Los Angeles	17,062,923	43.4	51.2	12.8	218,405,414
Orange	3,063,014	32.4	37.5	12.8	39,206,579
All Coastal Counties	23,243,332	40	49.6	12.8	297,514,650
Federal OCS	31,613,651		223.7	12.8	404,654,733
Total	**54,856,983**			**$12.8**	**702,169,382**

Source: Department of Conservation. Division of Oil and Gas.

[9] Source: Department of Conservation. Division of Oil and Gas. Data based on ten months of production.
[10] Average price of Kern River 13 degree API gravity crude oil was 12.8

Total Income Effect of Offshore Oil Production in the State. The income portion (or value added) of the 1992 offshore oil production output of $702,169,382 is estimated as $407,258,243.[11] *The total effect on California income is estimated as $610,887,365.*[12] *Offshore oil production indirectly creates approximately 3,971 jobs.* Table 5 shows the value added, indirect income effect, and total income effect on the state, and indirect employment generated by offshore oil production by region.[13]

Table 5

1992 Value Added, Income and Indirect Employment Effects of Offshore Oil Production

Region	Value Added	Indirect Income Effect	Total Income Effect	Indirect Employment
Santa Barbara	$20,644,021	$10,322,010	$30,966,031	201
Ventura	2,499,522	1,249,761	3,749,282	24
Los Angeles	126,675,140	63,337,570	190,012,710	1,235
Orange	22,739,816	11,369,908	34,109,724	222
All Coastal Counties	172,558,498	86,278,249	258,837,747	1,682
Federal OCS	234,699,745	117,349,873	352,049,618	2,288
Total Offshore	**$407,258,243**	**$203,629,122**	**$610,887,365**	**3,971**

Oil Reserves. As 12/31/1991, *state total reserves are estimated as 4,126 billion barrels and offshore reserves as 655 billion barrels.*

Natural Gas Production. California's natural gas production in 1991 amounted to 383.6 billion cubic feet. Offshore production of natural gas represented 14.5 percent of the state total natural gas production. *State offshore production was 7.6 billion cubic feet and federal offshore production was 47.6 billion cubic feet. California's total net gas production for 1992 was 353 billion cubic feet*, 8.0 percent lower than the 1991 level.

Table 6 shows 1992 gas production data and the proportion of offshore natural gas in total natural gas production per county. *The total value of offshore natural gas production (gross output) is estimated as $87,758,229.*

---

[11] To roughly estimate the income portion of the 1992 offshore oil gross production we have used the direct income coefficient for crude petroleum (0.58) that was estimated in the input-output model developed by the State Department of Water Resources.

[12] Total income and employment effects were estimated using multipliers derived from the RIMS II model developed by the U.S. Department of Commerce. The California average employment multiplier for crude petroleum derived from the same model was used to estimate indirect employment.

[13] Regional effects have been approximated using the statewide average multipliers

Table 6

1992 Offshore Natural Gas

Region	Net Gas Production (Thousand Cubic Feet)	Percentage of County Total Net Gas Production	Average Price (1)	Total Value of Production
Santa Barbara	2,825,135	36.5	1.59	$4,491,965
Ventura	254,418	1.8	1.59	404,525
Los Angeles	3,769,955	25.7	1.59	5,994,228
Orange	727,868	29	1.59	1,157,310
All Coastal Counties	7,577,376	18.8	1.59	12,048,028
Federal	47,616,479			75,710,202
Total	55,193,855			**$87,758,229**

Source: Division of Oil and Gas.

The total effect of offshore natural gas production on California income is grossly estimated as $ 47,173,444.[14] Table 7 shows value added, total and indirect effects on income, and indirect employment generated by natural gas production.[15]

Table 7

1992 Value Added Income and Indirect Employment Effects of Natural Gas Production

Region	Value Added	Indirect Income Effect	Total Income Effect	Total Employment
Santa Barbara	$1,617,107	$808,554	$2,425,661	16
Ventura	1,629	815	2,444	0
Los Angeles	2,157,922	1,078,961	3,236,883	21
Orange	416,632	208,316	624,947	4
All Coastal Counties	4,193,290	2,096,645	6,289,935	41
Federal	27,255,673	13,627,836	40,883,509	266
Total	**$31,448,963**	**$15,724,481**	**$47,173,444**	**307**

Gas Reserves. As of 12/31/91, California gas reserves were estimated as *4.1 trillion cubic feet and offshore gas reserves as 0.7 trillion cubic feet.* Total federal gas reserves were estimated as *0.3 trillion cubic feet.* All federal reserves are offshore.

MINERAL PRODUCTION

A wide variety of minerals exist in California's coastal zone and waters. Present production includes sand and gravel for construction, sand for filtration, abrasives for glass making and other specialty uses, and shells as source for lime and magnesium compounds. Other resources waiting for markets and adequate

14 The income portion of the 1992 offshore natural gas production was estimated using the direct income coefficient for natural gas (0.36) from the input-output model developed by the State Department of Water Resources.

15 Indirect employment effects have been calculated using the RIMS II model California average employment multipliers for natural gas.

technology for mineral extraction are gold, phosphorite, an essential fertilizer material found offshore from Baja to San Francisco, and polymetallic sulphide nodules which may be recovered by dredging.

Salt Production. Two areas of the California coast are sites of commercial salt production: San Diego Bay and San Francisco Bay. The Western Salt Company is located at the south end of San Diego Bay and produces 40 to 80 thousand tons of salt per year. Salt production at this site is for industrial uses only. The Leslie Salt Company currently owns property in San Pablo Bay and in San Francisco Bay. The annual production of this firm is about 1.2 million tons. Only a small percentage of Leslie's salt production is sold as table salt, most of the production is for commercial and industrial uses.[16] As an approximation, we can use the average price f.o.b. mine and plant for salt from brine to assess the value of total salt output in these plants (5 dollars per ton). Therefore, *the total value of salt production (gross output) in the state is approximately $6,400,000. The income portion of the total value of salt production (value added) is roughly estimated at $3,904,000.*[17] *Total income (value added plus indirect income effects) added to the state by this sector is approximately $7,417,600.*[18]

PORT ACTIVITY: WATER TRANSPORTATION AND SHIP AND BOAT BUILDING AND REPAIR

There are six major commercial seaports in California handling over 90 percent of the state's inbound and outbound waterborne cargo. Port activities include cargo documentation, shipping lines, storage services, marine insurance, freight forwarding. Ports are a basic part of the infrastructure of the U.S. market system.

In addition to the transportation of cargo, California's major commercial ports support foreign trade, ship bunkering and repair activities, commercial fishing, cruise ships and passengers, and recreation. Major repair services with dry docks for above and below waterline repairs are available at Los Angeles, Oakland, Richmond, San Diego, and San Francisco ports.

Passenger and cruise ship services are concentrated at the ports of Los Angeles, San Diego, and San Francisco. In addition to the port of Oakland, these ports contain the most important state recreational facilities, for instance, Long Beach's Queen Mary complex and San Francisco's Fisherman's Wharf area.

---

16 Source: California State Lands Commission. (Op. cit.).

17 Gross output was estimated as $6,400,000. The estimation of value added or the income portion of salt production was approximated by using the direct income coefficients of .61 derived from the input output model build by the State Department of Water Resources. See: State of California: The Resources Agency. Department of Water Resources. "Measuring Economic Impacts. The Application of Input-output Analysis to California Water Resources Problems." Bulletin 210. March 1990.

18 Total income effect on the state includes secondary effects. The income effect was calculated using RIMS II multipliers.

Water Transportation. Table 8 shows direct and indirect income and employment effects by region derived from domestic and foreign transportation of freight, water transportation of passengers, and services incidental to water transportation. Direct income has been calculated using an average of $56,470 of annual earnings per employee.[19] *The total income generated directly and indirectly by water transportation was estimated as $3,4 billion.*

Table 8

Water Transportation: 1992 Direct and Indirect Income and Employment Effects

Region	Direct Income*	Indirect Income Effect	Total Income Effect	Indirect Employment	Total Employment
North Coast	$5,223,475	$8,968,707	$14,192,182	316	390
North Central Coast	522,841,613	897,719,050	1,420,560,663	31,679	39,086
Central Coast	15,105,725	25,936,530	41,042,255	915	1,129
South Central Coast	47,293,625	81,203,154	128,496,779	2,865	3,535
South Coast	$675,169,438	$1,159,265,925	$1,834,435,363	40,908	50,473

*(SIC codes 441, 442, 444, 448, 449)

Ship and Boat Building and Repair. The cuts in defense represents a challenge for the shipbuilding industry. The Navy's proposed fiscal year 1993-1997 shipbuilding program averages 7 ships per year, compared to a previous average of 19 ships per year. No commercial orders for new ships have been received in the state for the last two years. Closure of navy shipyards will probably reduce the size of this industry.

The market for the construction and repair of smaller type vessels, barges, and small military and government owned vessels has also decreased. In 1990 these industries had a surplus capacity. Employment and income benefits derived from ship and boat building and repair during 1992 are presented in Table 9. A proxy for income was estimated using the 1990 average earnings per worker employed in transportation equipment (approximately $45,000 dollars).[20] *The total income effect (direct and indirect income) of this industry is roughly estimated as $2,546,671,876. This industry provides (directly and indirectly) about 83,800 jobs in California.*

[19] This number is the 1991 California average earnings for water transportation. Data from Survey of Current Business. August 1992. Volume 72, #8. Direct income was estimated using the ratio of total earnings to direct income of .80, based on the observed relationship between these two variables, as reported by the Survey of Current Business. vol. 72, #8. August 1992. Total effects were estimated using RIMS II multipliers.

[20] This is the 1991 California average earnings for transportation equipment as published in Survey of Current Business, August 1992, #78. Direct income was approximated using the ratio of earnings to income. Indirect effects shown on Table 20 are calculated using income and employment multipliers from the Rims II model developed by the U.S. Department of Commerce.

Table 9
Ship and Boat Building and Repair:
1992 Employment, Payroll, Earnings & Direct Income

Region	Direct Income	Indirect Income	Total Income Effect	Indirect Employment	Total Employment
North Coast	$1,293,750	$1,293,750	$2,587,500	62	85
North Central Coast	453,825,000	453,825,000	907,650,000	21,784	29,852
Central Coast	3,262,500	3,262,500	6,525,000	157	215
South Central Coast	34,996,875	34,996,875	69,993,750	1,680	2,302
South Coast	$779,957,813	$779,957,813	$1,559,915,626	37,438	51,304

TOURISM

The economic impact of ocean-related tourism was estimated by Joe Fitz. The economic assessment of tourism was based on two reports provided by the California Trade and Commerce Agency: the California Travel Impacts by County, 1992, prepared by Dean Runyan Associates, and "Domestic Travel to California: Visitor Origins, Characteristics, and Attitudes," prepared by D.K. Shifflet and Associates. Once coastal counties, the coastal portion of these counties, and the purpose of traveling was defined, the proportion of county travel spending for coastal tourism and recreation was estimated based on total county travel spending.

The value of tourism and recreation along the California coast was $9.9 billion in 1992, and nearly 150,000 jobs were dependent on this activity. These figures include indirect effects.[21]

[21] For a detail of the methodology followed in this study, please see Methodological Appendix attached to the Report "Economic Assessment of Ocean-Dependent Activities", included in the Resources Agency's publication: "California's Ocean Resources: An Agenda for the Future."

CHAPTER 29

An Economic Evaluation of Beaches in California

Philip G. King, Ph.D.
Michael Potepan, Ph.D.[1]

Abstract

This study presents the results of a contingent valuation study of the economic valuation of beaches in California. 600 people were surveyed and asked their willingness to pay for beaches. A demand curve was constructed. Using standard methodology and techniques, we estimated the value of California's beaches to be $17.5 billion. As incomes in California rise the value of beaches can also be expected to increase at roughly three times the rate of the income increase. The study also examines spending patterns for California residents and for visitors. We breakdown these spending patterns into different components and estimate the spending multipliers. Finally, the study examines the value of beaches in preventing storm damage to public and private property.

Introduction

California's beaches are eroding. This is a basic geological fact that no one disputes. The Pacific ocean carries off millions of tons of sand every year from California's coastline. In particular, severe storms, such as the storm of the winter of 1994, wear away at beaches. Before humans built damns, the sand was replaced by silt carried by rivers, streams and other natural waterways. No precise estimate exists of the amount of sand which has eroded from California's beaches each year, but the amount is significant and over the next 50 years a significant percentage of California's beaches will have eroded away without some sort of prevention. To compound the problem the erosion is uneven and invisible; there are no media events and no photo opportunities.

California spends little maintaining and preserving its shoreline. As with most government spending, beach preservation must compete with other social needs and political lobbies. The view from Sacramento is that beaches have no real constituency

[1] Philip King and Michael Potepan are both Associate Professors of Economics at San Francisco State University, San Francisco, CA 94132.

and bring in few dollars. Unlike Florida, which spends far more on its beaches, California's beaches are not generally considered to be a tourist draw.

This paper will present the methodology for and summarize some of the results from research that was performed under a grant from the State of California supervised by the Department of Boating and Waterways. Dr. Philip King was the Principle investigator for the grant and Dr. Michael Potepan was the project director. The survey was conducted by the Public Research Institute at San Francisco State University.

The purpose of the grant was to give an economic analysis of the value of California's beaches. The grant announcement specifically states that the research must: (1) estimate the expenditures at California's beaches by natives and by foreign and domestic tourists; (2) estimate the value of California's beaches in serving as a buffer for storm damage prevention; (3) provide other appropriate valuations.

Survey Results

During October and November 1995, a telephone survey was conducted by the Public Research Institute (PRI) at San Francisco State University under contract with the California Department of Boating and Waterways. Over 600 California residents throughout the state were contacted and asked a series of questions regarding their household's beach-going activities during the previous year. The facsimile of the actual survey questionnaire appears as Appendix A of this report. Included also is a breakdown of the responses to each question.

The survey contained two different sets of questions that were used for two different purposes in this study. The first set of survey questions were concerned with how often people went to beaches, how much money they spend when they went to beaches, and what attitudes they held toward beaches. This set of questions asked households how many times they went to beaches during the previous year, how much their household typically spent on certain items when they went to beaches, and how they felt about the state's beaches. The results from this set of beach attendance, beach expenditure, and beach attitude questions are summarized in Table 2.1 below. These results also form the raw data used in formulating the estimates of the spending impact of beaches on California's state economy.

The survey also contained a second set of questions concerned with determining how much intrinsic value Californians place of the presence of beaches in the state. Since California's beaches are a "free" natural resource, people do not directly pay money to enjoy them. Nonetheless, the public receives enjoyment and economic benefit from their use of beaches. Since beaches are free, we cannot simply count how much money was spent on them to calculate their "intrinsic" economic value to the public. The conventional method used by economists to estimate intrinsic value in cases like this is called the contingent value method. The second set of survey questions closely conformed to conventional contingent valuation methodology. The results from this second set of questions concerning the intrinsic value of beaches to Californians are analyzed and presented in Section 4.

The Economic Impact of Beaches on California's Economy

This section of the report uses the survey information on beach visits and beach collected in the survey together with additional information about California's economy to estimate the total economic impact of beach spending on the state's output, employment, and sales tax revenue.

Table 3.1 takes survey information on the mean number of day trips taken, and expenditures made on day trips per household, and expands it to estimate attendance and expenditures for the entire California population of households. In part A. of Table 3.1, the mean number of day trips is presented for just those taking day trips as well as all those contacted in the sample. In part B. of Table 3.1, similar estimates are made for the total amount spent by each household on average for day trips.

Table 3.1 Estimates On Beach Attendance and Spending For Californian Households On Day Trips To The Beach

	Sub-Sample of 409 Taking Day Trips	Full Sample of 615 Total Households Surveyed
A. Attendance Days From Survey		
Mean Number of Day Trips	15.24	10.13
Mean Number of Persons Per Day Trip	4.00	2.66
Mean Annual Person Attendance Days	48.14	32.02
B. Spending From Survey		
Mean Household Spending Per Trip	$54.87	$36.49
Mean Per Person Spending Per Trip	$16.45	$10.94
Mean Annual Household Spending	$518.40	$344.75
Mean Annual Per Person Spending	$171.57	$114.10
C. Statewide Attendance Projections		
Mean Annual Person Attendance Days per Household		32.02
Total California Households (millions)		x 10.8
Total Person Attendance Days (millions		**345.78**
D. Total Direct Statewide Spending on Day Trips		
Mean Annual Household Spending per Household		$ 344.75
Total California Households (millions)		x 10.8
Total Statewide Spending (millions)		**$ 3,723.35**

Part C. of Table 3.1 presents the projections for total person attendance days for all Californians. It takes the mean per household from the sample (32.02 annual person attendance days) and multiplies by the number of California households in 1995 (10.2

million). This indicates that there were 345.78 million total person attendance days generated by California households taking day trips to the beach.

Part D. of Table 3.1 presents the projections for total expenditures for all Californians on day trips. The $344.75 spent on average by each household in California is multiplied by 10.2 million households. Therefore, an estimated $3,723.35 million was spent on day trips to beaches by California households in 1995.

Table 3.2 Estimates On Beach Attendance and Spending For Californian Households On Overnight Trips To The Beach

	Sub-Sample of 234 Taking Overnight Trips	Full Sample of 615 Total Households Surveyed
A. Attendance Days From Survey		
Mean Number of Overnight Trips	4.60	1.75
Mean Number of Days Per Overnight Trip	2.65	1.01
Mean Number of Persons Per Overnight Trip	4.34	1.65
Mean Annual Person Attendance Days	33.10	12.59
B. Spending From Survey		
Mean Household Spending Per Trip	$ 246.83	$ 93.92
Mean Per Person Spending Per Trip	$ 82.09	$ 31.24
Mean Annual Household Spending	$ 907.79	$ 345.40
Mean Annual Per Person Spending	$ 345.24	$ 131.36
C. Statewide Attendance Projections		
Mean Annual Person Attendance Person Days Per Household		12.59
Total California Households (millions)		x 10.8
Total Person Attendance Days (millions		**135.972**
D. Statewide Spending Projections		
Mean Annual Household Spending Per Household		$ 345.40
Total California Households		x 10.8
Total Statewide Spending (millions)		**$ 3,730.36**

Table 3.2 is similar to Table 3.1, but focuses instead on overnight trips to beaches taken by California households. Part A. indicates that for all California households on average, the mean number of overnight trips taken per year was 1.75. Part B. provides comparable spending information for California households taking overnight trips to the beach. It indicates that for all California households, the mean expenditure per overnight trip was $93.92, and the mean annual expenditure was $345.40.

Part C. projects the total person attendance days for Californians on overnight trips to the beach by multiplying the 12.59 mean annual person attendance days per household times the entire number of households in California. Almost 136 million person attendance days were generated by California households on overnight trips to the state's beaches. Part D. presents projections for total statewide spending on overnight trips to the beach. The mean annual expenditure on these trips per household ($345.40) was multiplied by the entire number of California households. Over $3,730 million was spent by California households taking overnight trips to the state's beaches.

Table 3.3 Estimates On Beach Attendance and Spending For Out-of-State Tourists Taking Trips To California's Beaches

A. Statewide Attendance Estimates	
Total Attendance Days (Tourists 15% of Total) (1)	566.76
California Resident's Total Attendance Person Days (2)	481.75
Out-Of-State Tourist Person Attendance Days (millions)	**85.01**

(1) Source: DBW survey of beach visitors, July 1996
(2) Source: Tables 3.1 & 3.2

B. Converting Attendance Days To Out-of-State Tourist Trips	
Out-Of-State Tourist Attendance Days	85.01
Mean Trip Length for Out-Of-State Tourists (days) (3)	/2.65
Out-Of-State Tourists Visiting State's Beaches (millions)	32.08
Mean Household Size of Out-Of-State Tourists (persons) (4)	/2.5
Total Out-Of-State Tourist Trips To Beach (millions)	**12.83**

(3) Source: Table 3.2
(4) Source: California Dept. of Tourism Survey

C. Statewide Spending Projections	
Household Spending Per Trip (3)	$ 246.83
Out-Of-State Tourist Trips (millions)	x 12.83
Total Statewide Spending (millions)	**$ 3,166.87**

Table 3.3 estimates beach attendance and spending by out-of-state tourists visiting California's beaches. Part A. of Table 3.3 presents an estimate of out-of-state tourist attendance days comparable to that shown in Tables 3.1 and 3.2 for California

households. The estimate was arrived at by using additional information from another survey conducted by the California Department of Boating and Waterways in July 1996. In that site survey of visitors to California beaches, 13,279 beach visitors were simply asked whether they were residents of California or were visiting from out-of-state. Tables 3.1 and 3.2 showed that 481.75 million person days of beach visits were generated by California households. Using the conservative assumption that these in-state visitors represented 85% of the total, and that the out-of-state tourists constituted the other 15%, this would indicate a total for statewide person attendance days at 566.76 million, out of which 85.01 million (15%) were from out-of-state tourists.

Part B. of Table 3.3 converts out-of-state tourist person attendance days into an estimate for the number of out-of-state tourist trips annually that included a beach visit. We conservatively assume that out-of-state tourists took only one trip to California's beaches during the year. We further assume that their spending pattern on these trips was similar to that of California households taking overnight trips to the beach. In Part B., the number of out-of-state tourist attendance days is divided by the mean trip length for out-of-state tourist visits. Part C. of Table 3.3 estimates the beach spending made by these out-of-state tourists. Assuming that out-of-state tourists spend $246.83 per trip (the same as California households who took overnight beach trips), this would translate to a total of $3,167 million in spending made by out-of-state tourists during visits to the state's beaches.

Table 3.4 Total Economic Impact of Beach Spending On Gross State Product

A. Total Direct Statewide Spending	
Spending By California Households On Day Trips	$ 3,723,348,293
Spending By California Households On Overnight Trips	$ 3,730,360,255
Spending By Out-Of-State Tourists	$ 3,166,870,000
Total Direct Statewide Spending	**$ 10,620,578,548**
B. Total Indirect and Induced Statewide Spending	
Total Direct Statewide Spending	$ 10,620,578,548
IMPLAN GSP Multiplier	x 1.5654
Total Indirect Statewide Spending	**$ 16,625,453,659**
C. The Combined Economic Impact of Beach Spending on Gross State Product	
Total Statewide Spending	**$ 27,246,032,206**

Table 3.4 presents estimates of the total economic impact of beach spending on California's gross state product. Part A. calculates total direct statewide beach spending at over $10,620 million, the sum of beach spending by California households and out-of-state tourists. Part B. shows estimates for the indirect and induced spending that results from direct beach spending. This estimate reflects the idea that direct spending on goods and services during visits to beaches stimulates demand for additional goods and services produced in service of industries (induced spending) serving beach visitors. Economists refer to this as the "multiplier process": each dollar of initial economic activity is re-spent a number of times and generates additional economic activity with each new round of spending. The multipliers used in this study to estimate this additional indirect and induced spending in California's economy were derived from the 1993 IMPLAN multi-regional input-output model of the U.S. economy prepared for the Minnesota Impact Group. To summarize, beach spending in 1995 contributed a combined total of over $27,246 million in economic activity for the state's economy, which constituted about 2.98% of the state's total economic activity that year.

Table 3.5 shows the estimated number of jobs that were generated as a result of the direct, indirect, and induced beach spending shown in Table 3.4. The combined effect of all this spending from all these sources resulted in the creation of over 516,000 jobs in the state economy in 1995, which constituted about 3.64% of total state employment that year.

Table 3.5 Total Economic Impact of Beach Spending On State Employment

Total Direct Statewide Spending	$ 10,620,578,548
IMPLAN Employment Multiplier	x 48.618
Total Employment Effect (total jobs)	**516,323**

The amount of economic activity generated by beach spending has a positive impact on the amount of tax revenues collected by California state and local government. The major tax revenue categories affected by beach spending would include sales tax revenue, personal income tax revenue from employment, and state gasoline tax revenue. These are the only tax revenues sources that are considered in this study. Besides these, however, additional tax revenues would undoubtedly have been collected from corporate income taxes, property taxes and business licenses. However, it was not possible, given the limited scope of this study, to provide estimates for revenues collected from these sources. Thus, if anything, the estimates of tax revenues generated by beach spending presented here are conservative, and are likely to understate the true total.

Table 3.8 estimates the total tax revenues generated from all three revenue streams: state personal income taxes, state sales tax, and state gasoline taxes. The tax base for personal income taxes was estimated by taking the 516,323 jobs shown in Table 3.5 and multiplying it by the 1995 average payroll per worker in California ($19,974). The estimated personal income tax rate used was 6.0%. The sales tax base came from Table 3.6, and the gasoline tax base came from Table 3.7. The statewide average of the sales tax was 7.25% in 1995. The gasoline tax rate was 14.0%. As shown in Table 3.8, an estimated $1,069 million in tax revenues were generated as a result of beach spending in 1995.

Table 3.8 Estimated Tax Revenue Generated From Beach Spending in California, 1995

Type of Tax	Tax Base	Tax Rate	Revenues
Personal Income Tax	$ 10,327,025,757	6.00%	$ 619,621,545
Sales Tax	$ 4,551,993,588	7.25%	$ 330,019,535
Gasoline Tax	$ 850,260,688	14.00%	$ 119,036,496
Estimated Total Tax Revenues Generated			**$ 1,068,677,577**

Contingent valuation Study

In addition to expenditure patterns, we decided to do a "proper" valuation of beaches using contingent valuation techniques (hereafter referred to as CV). Such a measure, most economists would agree, (though not necessarily most legislators) would give a far more reliable estimate of the true value of beaches to the state of California.

The literature on contingent valuation is quite large and it would be foolish to repeat it here. The reader is instead referred to the reference section at the end of the paper, particularly the excellent symposium in the *Journal of Economic Perspectives*. To be brief, as in all areas of human endeavor, CV has its proponents and detractors. By now, however, a consensus is emerging as to how best to conduct a contingent valuation study. Some of this consensus emerged in the wake a CV study conducted after the EXXON Valdez oil spill. In particular many economists have criticized the methodology of this study. The next section of the paper will detail how this literature was applied to estimating the economic value of beaches in a telephone survey.

According to the literature (and common sense) one must give the respondent an appropriate sense of the nature of the resource that is meant to be valued. For a variety of reasons, the conventional wisdom states that in-person interviews are to be preferred to other types of surveys. When valuing resources which many respondents may not be familiar with, pictures and other detailed information can be presented easily and concisely. Respondents have more time to discuss their answer at length and may be likely to reveal more information.

However, in-person interviews also have several disadvantages. In-person interviewers establish a more personal relationship with the interviewee and hence have more of an opportunity to impart any personal bias that they might have. Second, in-person interviews are expensive and, for a variety of reasons, it is far more difficult to get a truly random sample.

Fortunately, beaches are a natural resource that is widely understood. Consequently there is little need for interviewers to show pictures are explained what the beach is used for. Indeed early drafts of our survey contained excessive description of beaches such as "beaches are used for swimming, sunbathing and other recreational purposes." Common responses in early tests were: "I know what a beach is. Don't insult my intelligence." This proved to be an enormous advantage since it eliminated one of the major disadvantages of telephone surveys. The process of erosion was somewhat more difficult to get across in a telephone interview but it is not difficult to get across the notion that beaches are indeed disappearing and that some beaches literally disappear after severe storms. In our survey we were able to use respondents familiarity with beaches and build upon it relatively quickly.

The literature on CV also devotes a great deal of attention to the so-called warm-glow effect. Respondents are often likely to consider survey questions as a general referendum on the environment rather than on the specific resources being dealt with at hand. Thus any estimate is likely to be biased upwards. In addition, respondents often are not as familiar with the concept of thinking on the margin as we might like to think and questions involving paying more money to increase the amount of preservation often show little willingness to pay beyond an initial amount, implying a very inelastic demand curve.

Similarly, respondents in previous studies sometimes tend to overestimate their willingness to pay if they aren't reminded that these are real choices that have consequences on their household budgets. Respondents should be reminded of their budget constraint. They are likely to be far more parsimonious (and hence more realistic) when they consider competing alternatives. In particular, much of the literature suggests that questions should be posed as a proposition on a ballot.

Closed-ended questions are preferred to open-ended questions for a number of reasons. First of all, closed end questions are likely to reduce the warm-glow effect. Second, open ended questions often confuse respondents. The ballot format is ideal for closed ended questions, so the telephone survey presented respondents with a proposition and asked them if they would be willing to vote for it.

At this point, two problems presented themselves. First of all, closed ended questions are subject to a starting point bias. Whatever the initial amount is on the ballot proposition, respondents are likely to resist higher or lower amounts. In the literature on contingent valuation, one technique to eliminate this problem is to present respondents with a chart listing a range of options. They can choose, multiple choice style, what their willingness to pay is. In a telephone survey the problem with this approach is that respondents can only remember so many numbers so the number of choices is quite limited. In addition multiple choice does not lend itself readily to ballots. In the end we proposed use a hybrid approach starting with a few multiple choice options and the using a common technique in computer learning models--branching.

Respondents were asked their maximum willingness to pay in the form of a ballot proposition with several discrete options. Next, the questions branched up or down depending on their initial choice. If they voted no, the amount was decreased. The main constraint here is the fatigue of the respondent and two branches after the initial question was considered to be the maximum attainable. What this technique may loose in accuracy, it more than gains in unbiasedness, so we employed the technique. To our knowledge these technique has not been used elsewhere and represents a minor innovation.

The final issue is of course the issue of quantity. Contingent valuation necessarily means estimating a demand curve. The "price" or willingness to pay is simply the amount on the ballot. However varying Q is somewhat problematic. The grant specifically addressed the issue of beach erosion. Respondents were given a scenario-California's beaches are eroding and asked how much they would be willing to pay to prevent this erosion. Respondents were given two scenarios, one with 20% erosion, and one with 50% erosion. Erosion was measured by the amount of beach miles lost from the total of 600 beach miles in the state of California. After several discussion with geologists, I discovered that this was not as much as an oversimplification as one might think. Erosion is indeed a very uneven process and often engulfs entire beaches in a short period while having relatively little impact on other beaches.

Casual comments by surveyors indicated that survey respondents took the survey very seriously and gave very considered answers.

Results of Contingent Valuation Study

From the survey results a demand curve was estimated and the area under the curve, the consumers surplus was calculated using E-views. 462 respondents answered all of the survey questions.

The total consumer surplus was $40.320.

Dividing this by the number of respondents, we get an average consumer surplus per household of $87.27.

To get the full valuation of beaches we need to adjust for the number of households in California, 10.8 million.□

Californians as a whole value beaches per year at:

$87.27 x10.8 million households = $942 million per year.

What are beaches actually worth? If Californians are willing to pay $942 million per year we can calculate the present value of California's beaches. Using the 5 year Treasury bill rate as a discount factor, we get a total valuation of:

$942 million/.0536= **$17.5 billion**

The final number should be corrected for the income effect, however in this case the effect is less than 0.1 per cent and so can safely be ignored.

The survey results also indicate an income elasticity of 2.96. In other words, as Californian's household income grows, they value beaches at an even higher rate. For example, a 10% increase in household income will lead to a 29.6% increase in the amount they value beaches.

Dollar Value that Californians are Willing to Pay per Year to Prevent Beach Erosion

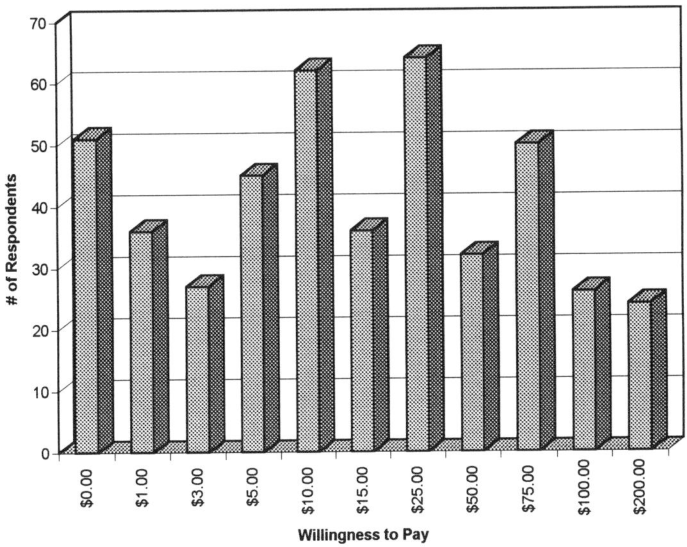

Conclusion

The state's beaches contributed over $27 billion to California's economy when spending from all sources is measured and the multiplier effects of the spending are accounted for. This represented almost 3% of total economic activity in the state. In addition, spending by people going to the beach allowed for the creation of over 500 thousand jobs, constituting over 3.5% of all jobs in the state. These figures clearly demonstrate that the state's beaches are a significant contributor to its economy.

The study also measured the intrinsic value to Californians of beaches in their state. This was done by asking state residents how much they would be willing to pay in additional taxes each year if the revenues were specifically earmarked for preserving state beaches. Using this technique, we estimate that Californians place an annual value of almost $1 billion on the presence of the state's beaches. This annual stream of almost $1 billion, projected each year into the future, has a present value today of over $17.5 billion. Our results are summarized in Table 4.1.

Table 4.1 Economic Impact of Beaches in California, 1995

A. Spending Impacts

	Gross State Product		Employment		State Tax Revenue	
Impact of Beaches	$27,246,032,206	(1)	516,323	(2)	$1,068,677,577	(3)
State Total	$913,700,000,000	(4)	14,206,000	(5)		
Impact of Beaches (%)	2.98%		3.63%			

(1) Source: Table 3.4
(2) Source: Table 3.5
(3) Source: Table 3.8
(4) Source: UCLA Business Forecast For Gross State Product 1995
(5) Source: Statistical Abstract of the U.S., Civilian Employment in California 1995

B. Intrinsic Value Placed On Beaches

Annual	$942,545,450
Present Value of Future	$17,584,803,000

References

Diamond, P. and Hausman, J., "Contingent Valuation: Is some number better than no number?" *Journal of Economic Perspectives*, Fall, 1994, v.8, no. 4. pp., 45-64.

Hanemann, M., "Contingent Valuation and Economics," Working Paper #697, Dept. of Agricultural Economics and Policy, U.C. Berkeley, March 1994.

Hanemann, M., "Valuing the Environment through Contingent Valuation," *Journal of Economic Perspectives*, Fall, 1994, v.8, no. 4. pp., 19-34.

Portney, P., "The Contingent Valuation debate: Why Economists should Care," *Journal of Economic Perspectives*, Fall, 1994, v.8, no. 4. pp., 3-17.

Schadke, D., and Payne, J., "How People respond to contingent valuation questions: A verbal protocal analysis of willingness to pay for an environmental regulation, " Journal of Environmental Economics and Management, v. 26., pp. 88-109, (1994).

CHAPTER 30

The Economic Benefits of a Major Urban Beach: A Case Study of Broward County, Florida

William B. Stronge and Ronald R. Schultz[1]

This paper[2] provides an overview of an economic study of the beaches of Broward County, Florida, one of the largest urban beaches in the U.S. Broward County is in the center of the three county southeast Florida region. Dade County (Miami) and Palm Beach County (West Palm Beach) are south and north, respectively. Broward is the twenty-fourth largest county in population in the U.S. and combined the three counties have more than four million residents. The official value of Broward's real estate property exceeds $60 billion. The principal cities along the coast are Fort Lauderdale, Hollywood, Pompano Beach, and Deerfield Beach. Broward's sandy beaches stretch continuously along twenty-five miles of narrow barrier island except for several maintained inlets to the Intracoastal Waterway. The barrier islands are occupied by mid to high rise condominiums and hotels, smaller one and two story motels, and single family residences, along with associated commercial businesses. There are two state parks that break up the continuous built-up nature of the coastal landscape. Because of Broward's subtropical location, its beaches are heavily used year-round. The combined effect of the large urban population, the heavily built-up nature of the barrier islands, and the

[1] Professor of Economics and Professor of Geography, respectively , Florida Atlantic University, 777 Glades Road, Boca Raton, Florida 33431

[2] The paper is drawn form an unpublished report, **Broward County Beaches: An Economic Study, 1995-96**, prepared for the Broward County, Department of Natural Resource Protection, Biological Resources Division. The report contains a fuller discussion than can be included in this paper.

subtropical location makes Broward's beaches an exceptionally valuable resource that increasingly demands public attention and evaluation.

In this paper we provide estimates of the economic importance of the beaches of Broward County to the economies of the county, the region, and the State. Beaches make three types of contribution to the local economy: they add value to properties; they create sales, incomes and jobs as a result of increased resident and nonresident spending; and they increase the local and state tax base. We estimate that Broward's beaches add $1.351 billion to local property values; that local economic production is higher by $547.9 million because of the beaches; and that the beaches cause the revenues of local governments to be higher by $29.2 million (Table A).

Beaches and Property Values

Conventional studies of the economic benefits of beaches identify certain benefits of beaches and provide estimates of each benefit. These studies, often conducted by or prepared for the U.S. Army Corps of Engineers, focus on the projection of property protection and recreational beach use benefits. Procedures for the forecast of project benefits for projects not yet constructed have been developed over many years. For this study of Broward County's beaches, benefits need to be determined for a beach system already in existence.

We approach the problem of determining the benefits of beaches by means of an analysis of private property values because the direct benefits from beaches will result in increased values for private properties. For example, if a home on the beach is vulnerable to storm damage, its value will be discounted by the expected property losses that it will incur. If the beach in front of the home is widened, there will be a reduction in the expected property losses and, therefore, a reduction in the discount buyers will demand to compensate for expected storm damage.

The protection accorded to public infrastructures by beaches will also show up in increased private property values.

If a section of public road used to access a coastal property suffers from periodic storm damage, the coastal property will be less attractive in the market and the result will be a discount in the property's value. If the beach adjacent to the road is widened, the extent of periodic damage to the road will be reduced and, once again, the discount applied to adjacent private properties will be reduced.

Beaches also provide recreational opportunities and the value of this opportunity will also show up in increased property values. Golf course communities are a familiar sight in Florida and other sun belt states, and nobody doubts that homes which are adjacent to golf courses (other things equal) are more valuable than those which are not beside the golf courses. Similarly, homes that are adjacent to beaches (other things being equal) are more valuable than those which are not.

Beaches may also enhance private properties in ways that are not allowed for in the conventional analysis. Beaches provide attractive views, opportunities to appreciate nature, and a generally attractive environment for residents and tourists. The value of these benefits will show up in property values.

Beaches may even create 'disbenefits' which reduce property values. For example, some beaches may increase traffic congestion or attract people who litter or commit crimes. The real estate market will assess the extent to which there are disbenefits and incorporate discounts into property values to allow for these disbenefits.

The point is that property values will reflect the direct economic benefits of beaches as well as the disbenefits. The total net benefit of the beaches can be found directly in property values, obviating the need to estimate the size of the itemized benefits and disbenefits themselves.

Most of the impact of beaches on property values will be found in their immediate neighborhood. For this study, the immediate neighborhood of beaches is defined to encompass the barrier islands along the coastline of Broward County. These islands are situated between the Intracoastal Waterway and the

TABLE A
Economic Benefits of Broward County Beaches
Impact on Broward County, 1995-96 in Millions of Dollars

Benefit	Amount
Impact on Property Values:	
Barrier Island Property Values	$ 1,048.7
Of Which: Recreational Value	60.6
Storm Protection & Other	988.1
Off Island Property Values	302.4
Of Which: Recreational Value	302.4
Total Impact on Property Values	**$1,351.1**
Impact on Broward County Economy:	
Direct Spending by Barrier Island Residents	$ 117.7
Direct Spending by Other Broward Residents	19.2
Total Direct Spending by Broward Residents	136.9
Direct Spending by Nonresidents of Broward in Broward	285.3
Total Direct Spending in Broward	422.2
Indirect Spending in Broward	125.7
Total Impact on Broward Economy	**$ 547.9**
Jobs Created in Broward	17,710 jobs
Impact on Local Taxes Annually	
Property Taxes	$ 28.2
Of Which: Barrier Island Properties	26.5
Of Which: Nine Cities + Unincorporated County	5.7
Broward County	8.0
Broward School District	9.9
Broward Hospital Districts	2.3
South Florida Water Management	0.6
District	1.7
Mainland Properties	1.7
State Revenue Sharing (e.g., gas taxes returned)	**$ 29.9**
Total Impact on Local Taxes Annually	

Note: The impacts on several small local taxing authorities such as the Hillsboro Inlet taxing district are estimated to be less than $100,000 annually and are not included here.

Atlantic Ocean. A study of real estate values on the islands compared to corresponding values immediately west of the Intracoastal Waterway was used as a basis to determine the contribution of the beaches to property values.

Although the bulk of the impact of beaches on property values occurs on the barrier islands, a relatively small impact on mainland property values will also occur. This is because the beaches provide recreational values to people residing or staying in mainland properties, as well as to people residing or staying in barrier island properties. The impact of recreational benefits on mainland property values is estimated by means of a survey of recreational beach use. The methodology follows the conventional study of the economic benefits of beaches as used by the U.S. Army Corps of Engineers.

The beaches in Broward County contribute $1.35 billion to property values in the county. This amounts to about 2.25 percent of the property value in Broward County. A little more than $1 billion of the contribution of beaches to property values occurs on the barrier islands, and $302.4 million occurs on the mainland.

Spending and Employment Resulting from the Beaches

Beaches make additional contributions to the economy beyond the direct benefits that show up in property values. They contribute to the economy as a result of the spending of residents occupying beachfront properties and as a result of the spending of nonresidents in connection with the recreational use of the beaches. Spending by either group, of course, results in increased sales by local businesses and, therefore, the creation of jobs and incomes for Broward County residents.

Spending by Broward Residents

One of the most obvious characteristics of homeowners is that high income homeowners usually live in expensive homes, and low income homeowners usually live in less expensive homes. In the absence of the beaches, beach area residential properties would lose value, and the occupants of the property would gradually be replaced by people of lower incomes. People with lower incomes spend less and so they have a lesser impact on the economy.

As a result, the loss of property values that would occur if Broward's beaches eroded away would be accompanied by a reduction in the amount of spending in the county. Likewise, expenditures by beach users in connection with their beach visits would also be lost.

The loss of Broward's beaches would result in a reduction in the amount of $136.9 million annually in resident expenditures in the county.[3]

Nonresident Spending

Additionally, if the beaches were to erode completely, nonresidents who are attracted to the county because of the beaches would not make expenditures in the county and this would result in a further negative impact on the county economy.

It can be argued that the loss of spending in the beach area of Broward County does not necessarily mean that it will all be lost to the county, since some beach area residents may locate to another part of the county or tourists may choose an alternative recreational experience in the county (such as golf). Similarly, not all beach-generated expenditure will be lost to Florida, since some beach area residents and tourists will move to the beach area of another county.

Indeed, from a national perspective, movement to other states will contain much of the lost expenditure. But even from that perspective there will be some loss to the extent that Canadians and other international tourists who currently visit Florida may choose alternative locations in the Caribbean or elsewhere for their winter vacations. To account for these effects, spending by nonresidents was estimated as an average between minimum and maximum values. The maximum value is

---

[3] Estimated using published data in the US Census of Housing and the Survey of Current Business. Ratios of housing value to personal income and personal disposable income for Florida and the ratio of national personal consumption expenditure to national personal disposable income were used. An adjustment for non-residential occupancy of barrier island residential properties was applied to the total expenditure generated by the barrier island residential properties. An estimate of Broward's mainland residents spending was derived from the interview survey of beach users discussed below.

the total expenditure in Broward County by nonresidents who visited the beaches in 1995-96. These expenditures include beach-related goods and services and 'tourist' expenditures (lodging, food, entertainment, local transportation, etc.). The minimum economic impact of the loss of Broward's beaches includes only the tourist expenditures of those who said they would not come to Broward County (55.4 percent) if there were no beaches.

Broward's beaches attract $285.3 million annually in nonresident expenditures to the county. Taking the increased expenditures of residents and nonresidents into account, total sales in the county are directly increased by $422.2 million on an annual basis directly as a result of the beaches.

Indirect Spending

The economic impact of beaches spreads beyond the immediate beach community. Spending in the beach community results in supplies being purchased by beach area businesses from businesses that are located away from the beach -- some on the mainland off the barrier island, some from neighboring counties in Florida, some from other states and, even, some from other countries around the world. Additionally, employees of beach area businesses (and their suppliers away from the area) spend their wages elsewhere in the county and in the region and this leads to further impacts (often called 'ripple' effects).

Technically, the spending on supplies and services that is stimulated as a result of direct spending is called 'indirect' spending and the purchases (spending) generated by the salaries and wages paid to employees is called 'induced' spending. Some of this indirect and induced spending occurs in Broward (an estimated 33 percent for this study) and most of the remaining indirect spending occurs in the rest of the Southeast Florida economic region[4].

The indirect spending in Broward resulting from the beaches was $125.7 million in 1995-96. Adding direct spending

[4] Broward County had 33 percent of the personal income on a place of residence basis for the Broward, Dade, and Palm Beach county region.

and indirect spending gives an estimate of the total impact on the Broward County economy of $547.9 million. The spending made as a result of the beaches creates 17,710 jobs in the county.

The Contribution of the Beaches to the Local Tax Base

Beaches also make a contribution to the tax bases of various taxing authorities. The protection of property values makes an obvious contribution to the revenues of those taxing authorities that levy ***ad valorem*** property taxes. These include municipalities, the school district, special taxing districts and the county government.

The population in the beach communities will place some demand on the services of the local governments, but this is usually much less than their contribution to tax revenues. This is because ocean front property is very valuable and does not generate its proportionate share of public school children or its proportionate share of the demand for other social services

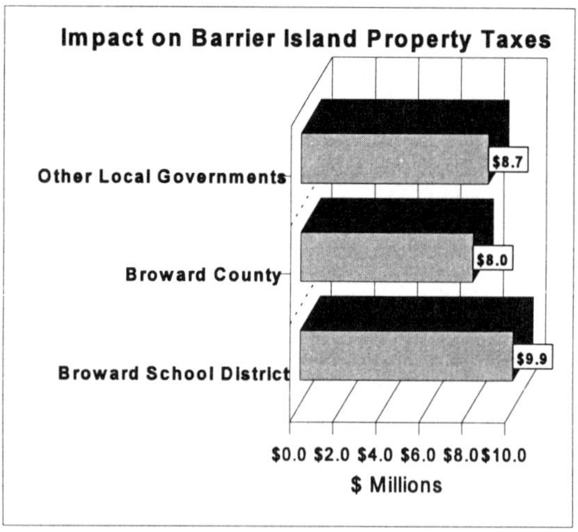

We estimate that property tax collections are increased in Broward County by $28.2 million annually, as a result of the beaches. The local taxing authority which receives the largest increase in revenues resulting from the beaches is the School Board, followed by the County Government. Most of the remaining revenues were received by the municipalities with jurisdiction on the barrier islands.

Impact of Broward Beaches on Southeast Florida

As noted previously, about two-thirds of the indirect spending due to Broward's beaches occurs elsewhere in the Southeast Florida region[5]. Estimates of this expenditure and its economic impact are given in Table B.

TABLE B
Economic Impact of Broward Beaches On the Southeast Florida Region, 1995-96 in Millions of Dollars

Benefit	Amount
Total Impact on Broward County	$ 547.9
Jobs Created in Broward County	17,710 jobs
Indirect Spending Elsewhere in Region	$ 255.2
Jobs Created	8,276 jobs
Total Impact on Southeast Florida	$ 803.1
Total Jobs Created in Southeast Florida	25,986 jobs

In 1995-96, Broward's beaches resulted in indirect expenditures of $255.2 million elsewhere in the region, resulting in a total impact in the Southeast Florida region from Broward's beaches of $803.1 million. The spending elsewhere in the region resulted in the creation of 8,276 jobs so that the total number of jobs created in Southeast Florida by Broward's beaches was 25,986.

Impact of Broward Beaches on the State Economy and State Revenues

Broward's beaches benefit the State of Florida by attracting out of state visitors to the state. The spending of these visitors creates sales and jobs in the state. Additionally, these visitors contribute to state tax revenues. The main

[5]Some of the indirect spending may spill out of the region to other parts of Florida, but given the large size of the Southeast Florida economy, these losses are likely to be negligible.

impacts of Broward's beaches are given in Table C.

Out of state visitors to Broward's beaches spent a total of $314.4 million in the state in 1995-96. Of this, $268.9 was spent in Broward, and $45.5 million was spent elsewhere in the region[6]. These expenditures stimulated indirect expenditures of $283.6 million, of which $80.0 million was expended in Broward and $203.6 million was expended elsewhere in Florida. As a result, the total impact on the state economy of out of state visitors to Broward County was $598 million. Of this impact, $348.9 million occurred in Broward, and $249.1 million occurred in other parts of Florida. Out of state Broward beach visitors created 19,395 jobs in Florida-- 11,342 in Broward and 8,053 in other parts of the state.

State Government also receives tax revenues resulting from Broward's beaches. Chief among these are sales taxes. There are other revenue sources, however, including car rental surcharges which are substantially affected by the attraction of Broward's beaches for tourism.

The contribution of Broward's beaches to state revenues is $18.5 million annually. Most of this is accounted for by sales taxes ($15.8 million) and car rental surcharges (2.1 million).

Recreational Use of Broward's Beaches

A major part of the economic study of Broward's beaches consisted of a personal interview survey of 4,556 beach users conducted between May 1995 and April 1996. Some results of that survey are presented in Table D. A total of 1,661,233 persons made 7,169,446 visits to Broward's beaches during 1995-96. Beach use in Broward is relatively constant year-round, with 56.6 percent of the visits in the winter season (November through April) and 43.4 percent of the visits in the summer (May through October). Nonresidents of the county, who account for the majority of winter visits (57.9 percent), are replaced in the summer season by residents of the county, who

[6] The estimates above allow for the fact that not all out of state beach visitors are attracted to the state by Broward's beaches.

account for the majority of summer visits (56.2 percent).

Broward's beaches appeal to a wide variety of people: the old, the middle aged and the young; students, employed people and the retired; people of modest income, middle incomes and high income. Summer visitors are younger than winter visitors,

TABLE C
Economic Benefits of Broward County Beaches
Impact on State of Florida, 1995-96 in Millions of Dollars

Benefit	Amount
Impact on State Economy:	
Direct Spending by Out of State Residents in Broward County	$ 268.9
Direct Spending by Out of State Residents in Other Counties	$ 45.5
Total Direct Spending by Out of State Residents in the State	$ 314.4
Indirect Spending in Broward	$ 80.0
Indirect Spending in Other Counties	203.6
Total Indirect Spending in Florida	$ 283.6
Total Impact on Broward Economy	$ 348.9
Total Impact on Other Counties	249.1
Total Impact on State Economy	$ 598.0
Jobs Created in Broward	11,342 jobs
Jobs Created in Other Counties	8,053
Total Jobs Created in Florida	19,395
Impact on State Taxes Annually	$ 18.5
Sales Taxes	15.8
Car Rental Taxes	2.1
Other State Revenues	0.6

and winter visitors are more likely to be retired than summer beach users. The heaviest use of Broward's beaches is in the Fort Lauderdale Beach area (about two million visits annually).

There is also heavy use in the Hollywood and Pompano Beach areas. The beach segments with the lightest use are in Hillsboro Beach and John U. Lloyd State Park.

TABLE D
Recreational Use of Broward's Beaches, 1995-96

Annual Visits	7,169,446
Annual Visitors	1,661,233
Visits Per Visitor	4.3
Summer Visits	43.4 %
Winter Visits	56.6 %
Visits by Broward Residents	48.2 %
Visits by Non-Residents of Broward	51.8 %
Broward Resident Visits in Summer	56.2 %
Broward Resident Visits in the Winter	42.1 %
Median Age of Adult Beach Visitor in Summer	37.6
Median Age of Adult Beach Visitor in Winter	40.5
Retired Adult Beach Visitors in Summer (Percent of All Visitors)	9.6 %
Retired Adult Beach Visitors in Winter (Percent of All Visitors)	19.3 %
Median Income of Adult Beach Visitor - Summer	$44,900
Median Income of Adult Beach Visitor - Winter	$42,700
Visits to Broward's Beaches by Beach Segment:	
Deerfield/Hillsboro	761,698
Pompano/Sea Ranch Lakes	1,091,445
Lauderdale-by-the-Sea/Galt Ocean Mile	692,024
Fort Lauderdale	2,002,059
John U. Lloyd/Dania	636,202
Hollywood	1,420,774
Hallandale	565,245

Characteristics of Beach Tourists

The personal interview survey of beach users enabled data to be collected on overnight tourists who used Broward's beaches in 1995-96.

A total of 725,521 overnight tourists visited the beaches of Broward County during 1995-96, with 30.3 percent visiting during the summer and 69.7 percent visiting in the winter season. Almost two-thirds of the tourists were from other states and one in four were international visitors. About one in six tourists was visiting friends or relatives and one in three were on their first visit to the county. The average size of a tourist party was about three persons. The average tourist party stayed about one week in the summer season and about ten days in the winter season. More than 50 percent of the tourists paid for their lodging in the county, spending $87 million on lodging during 1995-96. Spending per party averaged $925.39 in the summer and $1,407.13 in the winter. Much of the difference between these two figures is explained by the longer stay in the winter. Spending per person-day was $44.26 in the summer and $52.72 in the winter. Broward beach overnight tourists were asked whether they would come to Broward County if there were no beaches. About 60 percent of the tourists said they would not come and a further 12 percent said they would come less frequently, if there were no beaches in the county.

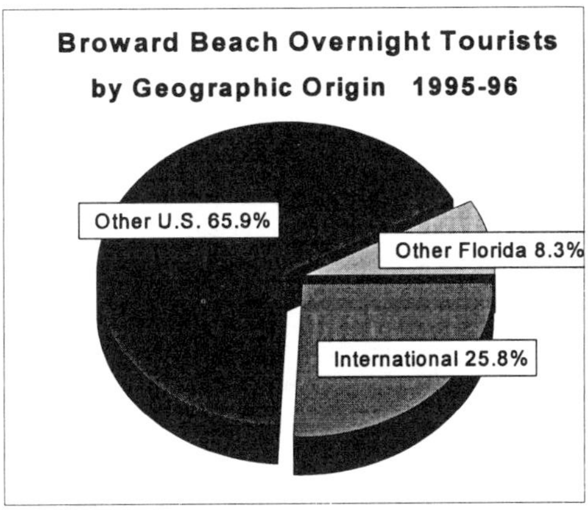

Conclusion

The existence, accessibility, and attractiveness of coastal marine beaches is a public/private affair. The economic evaluation of our beach resources is increasingly of concern to local and state communities. The estimates provided here suggest that urban beaches can be significant contributors to the local, regional, and state economy. Thus we would argue that they deserve careful evaluation as both a public and private resource.

TABLE E
Characteristics of Beach Tourists in Broward County, 1995-96

Number of Tourist at Broward's Beaches	725,521 Persons
Summer Tourists	30.3 %
Winter Tourists	69.7 %
Tourists from Elsewhere in Florida	8.3 %
Tourists from other States and Territories	65.9 %
International Tourists	25.8 %
Percent Visiting Friends or Relatives	16.9 %
Percent on First Visit	32.2 %
Average Party Size in Summer	3.1 Persons
Average Party Size in Winter	2.9 Persons
Average Length of Stay in Summer	6.7 Nights
Average Length of Stay in Winter	9.2 Nights
Person Nights in County in Summer	1,467,801 Prsn. N.
Person Nights in the Winter	4,666,998 Prsn. N.
Tourist Parties Paying for Lodging in Summer	59.8 %
Tourist Parties Paying for Lodging in Winter	53.2 %
Lodging Expenditures	$86,951,737
Spending Per Summer Party	$ 925.39
Spending Per Winter Party	$ 1,407.13
Spending Per Summer Person-Day	$ 44.26
Spending Per Winter Person-Day	$ 52.72
Percent of Tourists Who Would Not Come to Broward in the Absence of the Beaches	60.8 %

CHAPTER 31

PRIVATIZATION IN VESSEL TRAFFIC MANAGEMENT

Captain Richard B. McKenna[1]

Abstract

In a world of severe budget cutbacks, systems that have been supported by Governmental subsidy may no longer be affordable. Vessel Traffic Systems, which are becoming increasingly more desirable, are unfortunately finding themselves in this fiscal pinch. Privatization is the alternative being looked at more frequently as the solution. The Vessel Traffic Information Service (VTIS) at Los Angeles - Long Beach, the nation's first privatized service, serves as an example of what Public - Private partnership can achieve.

Operated by The Marine Exchange of Los Angeles - Long Beach Harbor, Incorporated, VTIS LA/LB has successfully demonstrated the feasibility of government partnership with the maritime industry for the past eighteen months. Using US Coast Guard Captain of The Port Authority, and the latest in maritime management technology, over fifty thousand vessel transits have been covered safely, with several lives saves and the environment protected. The maritime industry and boating public of Southern California have been well served at no expense to the taxpayer. This paper/presentation will discuss the genesis of this system, complete with facts and figures that went into its establishment. Organizational relationships, operational procedures and future prospects can also be addressed.

In a world of severe budget cutbacks, systems that have been supportable by Governmental subsidy may no longer be affordable. Vessel Traffic Systems, which are becoming increasingly more desirable, are unfortunately finding themselves in this fiscal pinch. Privatization is the alternative being looked at more frequently as the solution. The Vessel Traffic

[1]Deputy Executive Director, The Marine Exchange of Los Angeles - Long Beach Harbor, Inc., P. O. Box 1949, San Pedro, CA 90733

Information Service (VTIS) at Los Angeles - Long Beach, the nation's first privatized service, serves as an example of what Public - Private partnership can achieve.

Operated by The Marine Exchange of Los Angeles - Long Beach Harbor, Incorporated, VTIS LA/LB has successfully demonstrated the feasibility of government partnership with the maritime industry for the past three years. Using U.S. Coast Guard Captain-of-The-Port Authority, and the latest in maritime management technology, over 65,000 vessel transits have been covered safely, with several lives saved and the environment protected. The maritime industry and boating public of Southern California have been well served at no expense to the taxpayer.

The Marine Exchange has been operating the VTIS since 1 March 1994. This relatively recent date can sometimes be misleading, conveying the impression that this Vessel Traffic Service sprang from nowhere. In actuality, by that date, the Marine Exchange had been providing Vessel Traffic Service in an advisory capacity for over 12 years. This "Vessel Traffic Advisory Service" commenced in 1981, atop Warehouse #1 in the Port of Los Angeles. A modest ($50.00) user fee was charged, and the service achieved a subscription rate in excess of 90%. In 1989, because of the geographic limitations of that site for radar and communications, the Marine Exchange was relocated to its present position in Angel's Gate Park. This move, if not the exact beginning, is certainly to be considered as the seminal event in the development of the presently existing Public-Private Partnership. The Ports of Los Angeles and Long Beach, recognizing the importance of this move, jointly helped in the financing of the present building. The U.S. Coast Guard provided the actual site under a unique no-fee license agreement, with a lease which extends to the year 2007. Shortly after this event, the State of California also initiated legislation requiring vessel traffic services for their ports.

Under the then existing VTAS, advice and information was offered to all arriving and departing ships, identifying ship traffic in the area, or anticipated traffic, thereby alerting ship Masters of the situation around their particular vessel. Although this system was unregulated, informal and voluntary, the Marine Exchange still enjoyed virtual 100% participation by the regular commercial ship traffic involved. While this unique service greatly enhanced safe navigation, there were inevitably "close calls" and "misunderstandings" between ships as they made their arrivals and departures; something that had to be dealt with eventually. This was substantiated by numerous semi-formal safety meetings sponsored by the Marine Exchange which indicated wide spread support in the Maritime community for a more formalized system.

Spurred on by the Exxon Valdez disaster in Alaska, and driven by both State and Federal oil spill prevention legislation, a concerted effort was launched to establish a more formalized system. Various studies had been done for the Ports of Los Angeles and Long Beach, showing a definite need for a regulated, mandatory, pro-active Vessel Traffic System (VTS). Traditionally, the U.S. Coast Guard had provided VTS facilities at U.S. ports. However, with the budget cuts coming out of Washington during the early 1990's -- particularly hard on the U.S. Coast Guard -- plans for developing their VTS project here in Southern California were deferred annually through each budget process.

Meantime, the State of California, using the Oil Spill Prevention Act (OPA) of 1990, as a basis, created its own legislation requiring that the Ports of Los Angeles and Long Beach have a formal, mandatory, regulated VTS in operation by December of 1992. This deadline was missed. After that, the State's office of Oil Spill Prevention and Response (OSPR) authorized the Marine Exchange to develop a new Vessel Traffic Service in cooperation with the U.S. Coast Guard. Agreements were made, contracts were cut, and Memorandums of Understanding (MOU's) exchanged. This created a new system that was self-supporting, with Users' Fees mandated by State law, authorized by Port Tariffs, and endorsed by the U.S. Coast Guard. The State legislation also provided for the Marine Exchange to have a limitation of liability protection in rendering this new service. Thus, the VTIS became the first-ever, Government / Private Sector VTS operation in the country, funded entirely by industry, with no tax-payer money involved. This new concept captured the interest of Congress in Washington, DC -- who, in the Summer of 1995, declared VTIS to be a national model for other ports to study, evaluate and possibly emulate in their quest to create VTS programs for their areas. Today, the Marine Exchange is believed to be setting the pace and establishing procedures for future policy of VTS operations in the United States, and elsewhere around the World.

Marine Exchange and U.S. Coast Guard staff have all been trained and certified on state-of-the-art radar tracking equipment -- including VTS Simulator Training at the California Maritime Academy -- as well as training for navigation procedures, local geography, aids to navigation, rules-of-the-road, communications techniques, and more. This training continues, and has been recognized as an excellent program of high standard, on a parallel with the Coast Guard National Training Standard for VTS.

The Coast Guard has assigned 6 staff members, paid by the Marine Exchange as an operating expense. There is no taxpayer money involved. These personnel are all highly trained and motivated, and well integrated with the Marine Exchange staff in this new concept of providing service to the maritime industry. The Coast Guard has played a key role in the establishment of the VTIS. We've enjoyed strong support and participation from the Commandant's Office on down, making VTIS organizational training, equipment, personnel, and operational procedures second-to-none in America today. This unique partnership is succeeding admirably in its intended objective -- serving to make our harbors safer, cleaner, and more efficient in the management of ship traffic in and out of our two great ports. Since its inception, the VTIS has proven itself in the professional services rendered to all vessels calling into the Ports of Los Angeles and Long Beach.

For equipment, VTIS uses Racal Decca radars which supply ARPA input to a NORCONTROL VOC-5000 video display unit. The NORCONTROL VOC-5000 provides a radar overlay on electronic chart displays. The map system is a vector-based chart, digitized from official Defense Mapping Agency Charts, with a built-in editor to allow updates and changes through a NORCONTROL VMC unit. The NORCONTROL PAN-5000 system offers broad operational flexibility using a Function Key Menu divided into 10 dedicated areas.

Each dedicated area has a series of sub-menus for further definition. An extensive alarm system is built in, providing numerous target alarms to alert the operator of developing hazardous situations. Individual warning criteria can be defined for each target or for a specific geographic area as necessary. These alarms can be visual and / or audible and include:

(1) New Targets;
(2) Lost Targets;
(3) Navigation Channel Alarms;
  - high and low speed
  - course and position
  - encounter
(4) Anchor Alarms;
  - position and area alarms
  - anchor prohibited
(5) Collision Survey;
(6) Grounding Prediction;
(7) Striking Object;
(8) Restricted Area, and;
(9) Domain.

System warning alarms are also built into the NORCONTROL VOC-5000 to alert the Vessel Traffic Control Specialist (VTCS) of a system malfunction.

Degradation of the system caused by the effects of rain, snow, and wind (sea state) is adjusted accordingly using the same basic system functions found on all radar indicators. The VTIS presently has 100 percent coverage of its AOR, both by radar and, when meteorological conditions permit, eyesight.

Presently there are five radars capable of providing input to the NORCONTROL display. They are linked between the Long Beach Pilot Station and the Marine Exchange, with each station having the capability of selecting its radar input independently. The radar site locations are as follows:

(1) Long Beach Pilot Station radar "A";
(2) Long Beach Pilot Station radar "B";
(3) Pier- J radar;
(4) Marine Exchange radar "A";
(5) Marine Exchange radar "B".

A second hookup is anticipated in several months and will be located at the Chevron Oil Refinery in El Segundo. This will provide coverage of their offshore oil terminal as well as a significant part of Santa Monica Bay. This in turn will be followed by similar equipment at the Los Angeles Pilot Station, creating the capability for seamless coverage both inside and outside the port complex.

VTIS uses three SEA-156 VHF/FM Marine Band Radios. Each radio is a compact, fully frequency synthesized, VHF/FM Marine Transceiver providing 25-watt & 1-watt output. All USA, Canadian, and International channels, as well as 10 weather and 10 expansion channels are provided. Each radio has a six scan mode capability with a programmable priority channel selection. This allows the VTCS to monitor the VTIS working frequency, CH-14 VHF/FM, on one radio, and Channels 13/16 VHF/FM with priority 16 on the other radio at the console. A separate watch position, the desk watch, monitors the third radio, which is scanning the various harbor frequencies (pilot stations, tugs, Bridge-to-Bridge, etc.), with priority 16 VHF/FM. All VHF/FM communications are local sites with antennas located on the roof of the Marine Exchange. This configuration provides excellent VHF/FM coverage throughout the VTIS Area of Responsibility (AOR). There has been no degradation of service due to bleed over.

The VTIS console is also equipped with a TAIYO SIMRAD TD-L1550A Radio Direction Finder (RDF) receiver. This unit is capable of programming 100 channels into memory and scanning up to ten selected channels at one time. This capability can be programmed into the NORCONTROL display, and is being considered as a future upgrade.

The VTIS does not have CCTW equipment installed, nor is it considered necessary given the present AOR. There is however an Automated Dependent Surveillance (ADS) capability. Using VHF Channel-70 for polling, presentation on the tracker display console has been demonstrated to 105 miles.

The NORCONTROL VOC-5000 provides two tracker units capable of operating in conjunction with or independent of each other. The computer continually upgrades each contact and automatically provides information based on the best radar signal available. For example, if Pilot Station radar "A" is receiving a signal from a surface contact more strongly than Marine Exchange radar "A", the computer will track the contact using the Pilot Station radar "A", and transmit that signal to all systems linked. The NORCONTROL VOC-5000 is capable of predicting future position for up to 400 targets within an operator-selected, time-based computed course and speed. Changes in the course and speed of the real target will effect the prediction position. Each tracked contact will also display a course and speed vector set to a VTCS determined time. Once this vector time length has been established in the main computer, all contact vector lengths will be determined by their course and speed. The NORCONTROL VOC-5000 has a user friendly keyboard with many different fields and sub-fields available to the VTCS to assist in managing traffic. Acquired contacts are stored in the data base for up to three months in the hard disk. Raw radar video is stored for up to three days. All retrievable data can be stored on floppy disk for further archiving. VTIS also keeps a paper file of all vessels that participated in the system.

All primary equipment for operation is local and requires no signal relay from remote sensors. The Long Beach Pilot Station and Pier J radars are linked to the VTIS via special capacity phone lines. The VTIS can still function independently without access to, or computer input from, these radars through the local site radars, although the ability to link in with the other radars provides an enhanced capability.

Our computer database for reporting on ship specifics and port activities was designed and produced with off-the-shelf hardware and software, using existing technology and equipment in an open architecture environment. As such, we've been able to make direct on-line, real-time extension of our database to the U.S. Coast Guard's Marine Safety Office in Long Beach, which has greatly improved VTIS' participation in regulatory and law enforcement capabilities, as well as augmenting search-and-rescue efforts. Additionally, we are currently expanding our net to include the Ports' Wharfingers and Harbor Pilots. Potential exists to include the State's Oil Spill Prevention and Response (OSPR) Office, as well as other government agencies such as Customs. This will give us a closed loop information circuit throughout our harbor complex, whereby all interested parties will have immediate, first hand, real-time information on all vessel and port activities and; circumstances.

VTS operations are governed by a User's Manual, developed by the Ports' State mandated Harbor Safety Committee. It is an addendum to the Harbor Safety Plan and is reviewed annually.

This VTIS User's Manual has been distributed to the entire local maritime community, including Agents, Pilots, Ship Owners / Operators, the Coast Guard, and others. It outlines all the procedures, checkpoints, regulations / restrictions, and other important data. Approaching vessels are acquired twenty-five miles from Point Fermin. They report their destinations and ETA's, and in turn are advised of traffic in the area. Speed Limits are imposed in the Precautionary Area outside the breakwater, and safe separations have been mandated between vessels. We do not attempt to exercise ship control as a general routine. However, there may be that rare occasion when there is a requirement to exercise USCG "Captain-of-the-Port Authority", ordering a ship to slow down, stop, go to anchorage, take evasive action, stay at the dock, or return to port. This is only done in the event of some hazardous situation, or resulting from a direct non-compliance with regulations that creates an unsafe situation for ships in the area. Vessels are invited to make passing arrangements with other ships in their vicinity; and Vessel Traffic Control Specialists (both civilian and USCG watchstanders) also make hourly general broadcasts on VHF/FM CH-14, advising all listeners of current traffic in the 20-mile Area of Responsibility (AOR). A chartlet with a summary of procedures, as well as sample check-off lists for arriving / departing vessels, are provided as attachments to this paper (see App. I - App. III).

Although the VTIS has regulatory and enforcement functions to execute, it has come to be regarded as a value-added service for our two ports, a second set of eyes and ears for ship Masters to use, and a valuable safety tool for them to depend on.

To keep a finger on the pulse, Vessel Traffic Control Specialists - both civilians and USCG - are required to make trips on commercial ships to see what transpires on the bridge of a typical Merchant Vessel, and to gain an appreciation for what Masters and Pilots need from us. Conversely, the industry - both afloat and ashore - has been invited to come visit the facility. To date, hundreds of interested parties have toured the facilities.

Regarding finances, this project was started with a budget of $1.25 million, but actually came in under budget, with only $964,000 expended. The ports of Los Angeles and Long Beach both recognized the importance of this service, and contributed a grant of $250,000 each to help get the VTIS up and running in 1994. A low-interest loan agreement with the State of California was negotiated, with a one-million dollar line-of-credit. Less than half of that was used in this process.

MX-VTIS' annual budget is about $1.66 million, which is supported entirely by subscriptions and Users' Fees. The VTIS fee structure is based on a length-over-all formula, with the largest ships paying $340 per arrival, and the smallest ships paying $180 per call. The average fee for vessels calling into the Ports of Los Angeles and Long Beach is around $238. These rates represent a 10% decrease that the Marine Exchange was able to offer after the VTIS had been in operation for ten months. Moreover, any surplus revenues at the end of the year are considered for credits to the user in the next fiscal year. In 1996, the Marine Exchange returned 10% of its revenues in credits to its users.

The LA/Long Beach VTIS has proven its worth to the local maritime community. Through a history of incident reports and subsequent evaluation, much has been learned of ship operations in the approaches to the Harbor Complex. These lessons have provided the basis for significant improvement of the safety posture through Harbor Safety Committee deliberations, pilot coordination meetings and in-house reviews with the Captain-of-the-Port. These in turn have lead to a significant reduction in incidents, as well as the gravity of those which have occurred. Search and Rescue has been enhanced, with several lives saved as a direct result of VTIS participation. Law Enforcement and Port State Control operations have also been enhanced by the Marine Exchange linkage to the Captain-of-the-Port with a more accurate and available data base. There has been a synergy to this Public - Private Partnership that is gratifying to all who are involved.

Future plans call for this Partnership to continue. Having proven its worth to the community and the Coast Guard, the Marine Exchange was recently advised by Coast Guard Headquarters that any near term plans for a Coast Guard operated VTS in LA/Long Beach, or inclusion in VTS 2000, have been tabled for the foreseeable future. The Area of Responsibility is now extended twenty-five miles and into Santa Monica Bay. Traffic Regulations are identical to the Coast Guard's own National Regulations. The objective is consistency in regulations for the Mariner from port to port, whether Federally or privately operated. This is a significant statement in itself, in that it recognizes that privately operated Vessel Traffic Services can work as an alternative under the right circumstances, to the benefit of all users. At the same time, privately operated VTS's have latitude which allow them to pursue ancillary activities such as data base management and distribution, weather, tide and current information, and local port operations, which would not normally be within the reach of a single Federal organization.

In summary, we feel that by any available criteria, the Public-Private Partnership manifested in the LA/Long Beach VTIS has been an unqualified success. We feel equally strong that the reason for this success lies heavily in the "Partnership" aspect of that statement. Using the strong points of each of the partners has been the key factor in this successful endeavor. Thus, when Privatization is mentioned, we feel it important to qualify the concept with other such appropriate elements as governmental (local or otherwise) legislation, regulatory authority and tariff support. This blending of support, in our case from the Municipal, State and Federal level, has allowed us as the Private Operator to succeed in the manner intended. So much so, that we cannot envision a successful venture without some form of representation from each element. This need not connote excessive legislation or overbearing rules - in our case, we believe that it does not. However, it is most reassuring to have such backing on those infrequent occasions when clout is necessary and credibility is on the line. Thereafter, the strong private flavor of an industry operated, self regulating Vessel Traffic Service fares well with the local users, who can often feel that the particular needs of the local maritime community tend to go unrecognized when elevated to a State or National level (read government) operation. This local independence has proven well founded, as demonstrated in the high level of compliance to date. Other ports may well benefit from a similar arrangement. We at the Marine Exchange of Los Angeles / Long Beach Harbor would encourage the exploration of the concept as it serves a port's particular need. We are happy with what our community has explored, developed and achieved through such a process.

■ CHARTLET

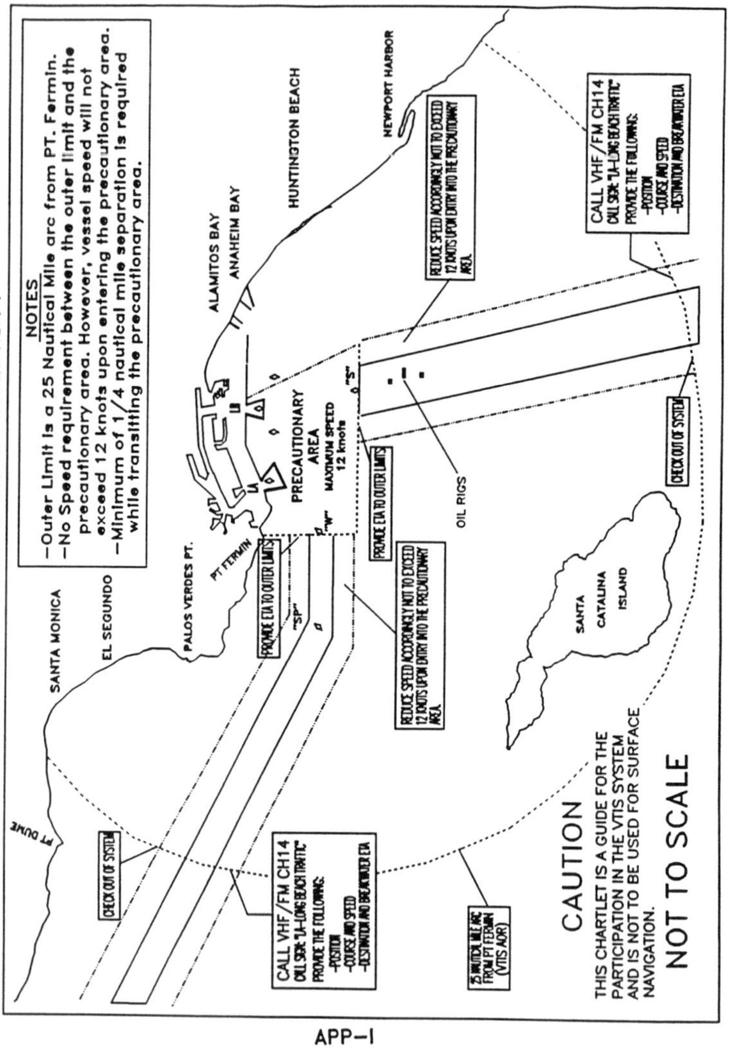

APP—I

CHECK OFF LIST FOR ARRIVING VESSELS

Upon Entering the 25 Mile Outer Limit:

Call Vessel Traffic Center (VTC, VHF channel 14, call Sign "LA—Long Beach Traffic.) Provide following information:

- Vessel name/call sign ________
- Position, course and speed ________
- Vessel destination ________
- State whether or not taking a pilot ________
- ETA to breakwater ________
- Reduce speed to be at 12 knots upon entry to precautionary area ________
- Advise of any known mechanical difficulty ________

Upon Entering Precautionary Area:

Call Vessel Traffic Center. Provide following information:

- Confirm vessel speed at 12 knots or below ________
- Confirm master on the bridge ________
- Confirm vessel in hand steering ________
- Contact pilots, channel 73 (LA) or 12 (LB) as appropriate ________

Notes of Interest

- The 25 mile outer limit is defined by a 25 mile arc from Pt. Fermin LAT 33°42'.3N, LONG 118°17'.6W.
- There is **no prescribed speed restriction** between the 25 mile limit and the Precautionary Area. All vessels shall proceed at a safe speed. However ships are required to be at 12 knots or less upon entering and transiting the Precautionary Area.
- A minimum vessel separation of 1/4 nautical mile is required in the Precautionary Area.
- Code of Federal Regulations, CFR 33, part 165, subsection 165.1109 identifies the Precautionary Area as a Regulated Navigation Area. The specific regulation is provided in Appendix IV of this User's Manual.

CHECK OFF LIST FOR DEPARTING VESSELS

Departing from Inside the Breakwater:

Call Vessel Traffic Center (VTC) (VHF channel 14, call Sign "LA—Long Beach Traffic".) Call should be made 15 minutes prior to breakwater entrance, or as practicable. Breakwater entrances include Angel's Gate (Los Angeles) Queen's Gate (Long Beach) and Anaheim Bay (Naval Weapons Station Seal Beach). Provide following information:

- Vessel name/call sign/present location ________
- Destination ________
- Acknowledge VTIS traffic report ________
- Maintain speed at 12 knots or less through Precautionary Area ________
- ETA to 25 nm limit from Point Fermin when practicable ________
- Report departure from Precautionary Area to VTC ________
- Report departure from VTIS at 25 nm limit ________

Departing from Anchorages Outside the Breakwater:

Call Vessel Traffic Center, VHF channel 14, fifteen minutes prior to heaving anchor. Provide following information:

- Vessel name/call sign ________
- ETA to sea buoy (if inbound) or destination (if outbound) ________
- Acknowledge VTIS traffic report ________
- Maintain speed at 12 knots or less through Precautionary Area ________
- Report departure from Precautionary Area to VTC ________
- If outbound, ETA to 25 nautical mile limit from Point Fermin ________
- Report departure from VTIS at 25 nautical mile limit ________

Notes of Interest

- The 25 mile outer limit is defined by a 25 mile arc from Pt. Fermin LAT 33°42'.3N, LONG 118°17'.6W.
- There is **no prescribed speed restriction** between the 25 mile limit and the Precautionary Area. However ships are required to be at 12 knots or less upon entering the Precautionary Area and throughout.
- A minimum vessel separation of 1/4 nautical mile is required in the Precautionary Area.
- Code of Federal Regulations, CFR 33, part 165, subsection 165.1109 identifies the Precautionary Area as a Regulated Navigation Area. The specific regulation is provided in Appendix IV of this User's Manual.

CHAPTER 32

Vessel Traffic Service Systems as Government-Private Sector Partnerships

Al Storm

Abstract

The Lempert-Keene-Seastrand Oil Spill Prevention and Response Act (Act) of 1990 grants the Administrator of the Office of Oil Spill Prevention and Response (OSPR) the authority and responsibility to provide for vessel traffic service (VTS) systems to protect California's harbors. The purpose of a VTS system is to furnish information, through radar tracking and radio communications, on vessel traffic in order to prevent collisions and groundings. Prior to the passage of the Act, VTS was solely a federal function under the jurisdiction of the Coast Guard.

In the Los Angeles/Long Beach (LA/LB) Harbor area, the Administrator has established one of the nation's first privately-funded vessel traffic information service (VTIS; so-named to set it apart from a Coast Guard VTS) systems. This endeavor required state and federal legislation (for start-up funding and Coast Guard participation) and extensive negotiations with the Coast Guard, as well as the local maritime community. The resultant VTIS has operated very effectively since March 1, 1994, meeting the highest federal VTS standards and significantly enhancing navigational safety in the LA/LB Harbor area. VTIS LA/LB exists as a partnership between the LA/LB Marine Exchange (MX), U.S. Coast Guard, OSPR, Port of Los Angeles (POLA), Port of Long Beach (POLB), and LA/LB Harbor Safety Committee (HSC).

VTIS LA/LB operates under state authority, with VTIS fees being collected locally, as authorized by state law, under the tariffs of the POLA and POLB. These fees cover all VTIS expenses, including Coast Guard personnel costs. VTIS operations are conducted jointly by MX and Coast Guard personnel on a 24-hour basis. The equipment and facilities are owned by the LA/LB MX, a nonprofit corporation. The OSPR oversees operations with the assistance of the LA/LB HSC, which provides recommendations to the Administrator and conducts periodic public hearings on VTIS.

CHAPTER 33

California's Harbor Safety Committees

Joan Lundstrom and Suzanne Rogalin[1]

Preface

During the past six years, the authors, Suzanne Rogalin and Joan Lundstrom, have served as voting representatives on all of the Harbor Safety Committees in the State of California, representing the California Coastal Commission and the San Francisco Bay Conservation and Development Commission respectively. To our knowledge, the California Harbor Safety Committees are the only ongoing committees in the United States focusing on harbor safety. This led to the question: How Effective are harbor safety committees in California?

Background

By way of background, in 1990, in response to the catastrophic Exxon Valdez oil tanker spill, the California Legislature passed the Oil Spill Prevention and Response Act. Those who have experienced even a small oil spill on the water realize there is no 'quick fix' to get the oily stuff off the water. So it's best to prevent the spill in the first place. The focus of the State Oil Spill Act is therefore twofold — to prevent oil spills from occurring in California's marine waters and to provide more effective oil spill response should a spill nevertheless occur. The Office of Oil Spill Prevention and Response (OSPR) was established to administer the state program, which is funded by a special surtax on crude oil transported by vessel and pipeline into California.

To prevent shipping accidents which could cause oil spills in our coastal ports, Harbor Safety Committees were created and appointed by OSPR in the five major harbors of the state — Humboldt Bay, San Francisco, Port Hueneme, Los Angeles/Long Beach, and San Diego. The committees represent a broad

[1] Joan Lundstrom, Coastal Analyst, San Francisco Bay Conservation and Development Commission and Suzanne Rogalin, Coastal Analyst, California Coastal Commission, c/o California Coastal Commission, 45 Fremont Street, Suite 2000, San Francisco, CA 94105-2219.

range of interests: port authorities, tanker and dry cargo companies, pilots, tug and barge companies, environmental, labor, fishermen and recreational boating interests are voting members of the committees. In addition, various nonvoting members included are the U.S. Coast Guard, U.S. Navy, U.S. Army Corps of Engineers, and vessel traffic services.

By law, each committee is directed to:

> "[plan] for the safe navigation and operation of tankers, barges, and other vessels within each harbor...[by preparing]...a harbor safety plan, encompassing all vessel traffic within the harbor."

By 1991 all five Harbor Safety Committees were functioning and meeting on a regular basis. The committees adopted Harbor Safety Plans within a year. As required by regulation, the plans were to be comprehensive documents which describe harbor conditions such as weather, tides and currents; the volume and mix of vessel traffic; aids to navigation; vessel routing and anchorages; tug escorting; communication; vessel traffic service; bridge management and other relevant topics. The plans addressed various safety problems in each harbor, making recommendations to prevent vessel accidents. The adopted plans were submitted to OSPR, for approval or disapproval of the plans based upon their consistency with state regulations. The OSPR Administrator is also charged with implementing the recommendations. After the initial Harbor Safety Plans were adopted and approved by OSPR, the Harbor Safety Committees did not sunset and dissolve, but continue to meet regularly, taking action on a variety of issues in each harbor. The Harbor Safety Plans are updated on an annual basis.

Example of a Harbor Safety Committee Issue: Tug Escorting

A good example of an issue that was successfully addressed by a Harbor Safety Committee is tug escort regulations for San Francisco Bay.

California was the first state to require that tankers and barges loaded with oil must have tug escorts in its harbors for quick stopping or maneuvering in case of an emergency. The state delegated that Harbor Safety Committees produce a set of tug escort guidelines, tailored to the conditions of each harbor. By law, the OSPR Administrator was guided by the committees' recommendations as the basis for tug escort regulations. In practice, the Administrator fundamentally followed the tug escort recommendations ultimately submitted by each harbor committee.

Because of the many hazards to navigation and the large volume of tankers and barges carrying oil in San Francisco Bay, the Oil Spill law mandated tug escort regulations be adopted *expeditiously* in the Bay. Starting in 1991, the San Francisco Harbor Safety Committee began extensive debate and research on how to match tug capabilities with various tanker sizes and speeds. While interim regulations took effect in 1993, the tug escort controversy continued over a number of years. Meetings — all public — often had forty to fifty people attending. At one point committee members were threatened by a lawsuit. Then the Committee's first set of permanent guidelines was rejected by the OSPR Administrator on the basis of the lack of rationale for the formula of matching tugs to tankers and barges. Subsequently, a naval architectural firm was hired to prepare a study analyzing the specifics of tug escorting in San Francisco Bay and to analyze failure probability and waterway specific standards. Though the final Harbor Safety Committee vote on tug escort guidelines was <u>not</u> unanimous, permanent regulations finally went into effect January 1st of this year.

Why did the entire process take almost six years? First of all, there was little regulatory history to refer to, since the California state regulations were a 'first' in the U.S. While the federal government mandated tug escorting in Prince William Sound and Puget Sound, it left the policing of the basic regulations up to industry. And secondly, the 'science' of how to effectively stop or retard loaded tankers was just emerging.

While the process was lengthy and often contentious, in the opinion of those involved many positive outcomes emerged. A spokesperson for the state oil spill office stated: "OSPR couldn't have gotten to the end product of tug escort regulations without the Harbor Safety Committee process. The tug escort regulations gained value because of the broad makeup of the committee." Industry spokespersons made such comments as: "This made a 'level playing field' for all tug and tanker companies"; "The regulations adopted best industry practices which could be monitored"; and "As a small company we felt our point of view was heard". It was also an established arena where the interests of environmentalists were heard — the Center for Marine Conservation was represented on the San Francisco Harbor Safety Committee. And finally, it is noteworthy that one Harbor Safety committee member felt that San Francisco tug escort regulations adopted a higher standard than any other port and suggested the San Francisco Bay regulations should be adopted elsewhere.

Survey

To gain a sense of whether the Harbor Safety Committees are effective, we conducted our own unscientific sampling of key players in the five harbors in California. The results don't reflect the opinions of every representative to every Harbor Safety Committee, but clear themes emerged.

We asked two questions: First, do you believe the Harbor Safety Committee has made a difference in this harbor, and if so, how. Secondly, have you used the Harbor Safety Committee as a forum for addressing relevant issues in this harbor, and, if so, what issues.

There was unanimous agreement that the committees were a positive asset to the harbor communities. We won't say we were shocked by this response, but we were surprised by the unanimity of it. Or, in the words of one environmental representative, "In spite of all the headaches and misery, it's been worth it. Because we now know each other, and have a forum." That was the first of the major themes we found, emerging in interview after interview: the value of having a forum in which to address problems, set up lines of communication, and exchange information. The problems addressed included such major issues as tug escorting and underwater rocks near Alcatraz Island in San Francisco Bay that pose a danger to vessel traffic. Tug escorting of tankers and barges carrying oil is now dealt with in all the harbors by at least interim regulations. And the Alcatraz underwater rocks are being addressed by both a subcommittee of the Harbor Safety Committee and in a congressional bill. Examples of more immediate, "nitty gritty" problems that were tackled and solved in this forum included placing channel markers during construction projects, installing racons on bridges, and securing a second radio channel for pilot/tug communications and a dedicated channel for the Vessel Traffic Information System (VTIS). The Marine Exchange used the Committee to address key issues in the establishment of the VTIS in Los Angeles/Long Beach Harbor.

The forum of the Harbor safety Committee has resulted in an improved level of communication among disparate groups. Recreational boating representatives and tanker vessel operators both believe the committees have led to better relations and a decrease in the number of small recreational boats interfering with large vessels. Representatives of a large west coast tug company and a small local tug company both agreed the Harbor Safety Committee had a positive impact as a forum for bringing interested parties together.

An environmental representative believed that the committee sensitized maritime people sitting around the table to environmental issues in the area. Another committee member said, "Speaking as a pilot, we tend to be one-dimensional. We want to get the ship from Point A to Point B. The Harbor Safety Committee has forced us to look at the big picture. We now see that concerns of non-port users must also be addressed." This increased forum for communication has resulted not only in an increased sensitivity to other interests, but to better information exchanges. An environmental representative said that having such on-the-water expertise from pilots and industry has helped her group in protecting the environment.

A second theme emergent from the interviews was increased safety awareness. A pilot and former chair of a Harbor Safety Committee believes the main difference resulting from the committees has been a stronger emphasis on safety concerns in the decision-making process. For him personally, prior to the establishment of the Harbor Safety Committee, it was typical to look at the record and say, "See, no problem." Now he says there is the idea that one should look forward to "what ifs."

The third theme is the use of the Harbor Safety Committees to formalize existing procedures and relationships. Some of the examples given by respondents were procedures for assist and escort tugs, pilotage training and disembarcation points, limited visibility guidelines, and VTIS. One tug representative said such formalization or institutionalization, for instance of bollard pull criteria, helped level the playing field for the tug industry.

The fourth and final theme was the use of the committees for the development of regulations. An Office of Oil Spill Prevention and Response respondent said she couldn't have produced tug escort, oil transfer, or Vessel Traffic Information System regulations without the committee. This theme of using the committee as a forum for producing workable regulation packages was echoed by other state government representatives and by industry.

In addition, respondents believed the Coast Guard used the Harbor Safety Committees to help develop and further its program for navigational safety. A Captain Of The Port confirmed this perception and said that on numerous occasions his organization had brought issues to the committee for comment and to help with implementation of its program. An excellent example is the current attempt by the Coast Guard to develop national underkeel clearance regulations. A pilot representative pointed out that this issue is a real balancing act of safety factors versus commerce. In his words, "Every foot is big money." Harbor Safety Committees had already developed local guidelines for underkeel clearance, and the Coast Guard had this ready made forum to work on the issue. A Captain Of The Port pointed out that his counter parts in other states are having to put such forums together to deal with the development of these regulations.

Conclusion

In conclusion, we think Harbor Safety Committees work. And we would suggest four reasons why.

First, these are ongoing committees. To come together twice a year to discuss problems isn't effective, as the Coast Guard discovered when it tried to have such advisory panels. In contrast, the Harbor Safety Committees are ongoing bodies with established lines of communication among disparate groups in the harbors.

Secondly, all the players are involved: the designated representatives of the ports, the harbor users, environmental representatives, and state and federal agencies. These individual committee members serve as channels for input from their wider constituencies. These different interests and viewpoints act as a reality check on committee proposals, helping to give the committee the big picture.

Thirdly, the committees have support staff. To effectively run committees composed mainly of unpaid volunteers with busy professional lives, such support staff is essential. Whether marine exchanges, ports, or environmental organizations staff the committees, we don't believe the concept would work without such support.

And finally, regulatory and enforcement agencies are represented and are required to consider the recommendations of the committee, whether by statute in the case of the state, or by committee peer pressure and political reality in the case of the federal and local representatives. Several respondents believed it was essential to have the Coast Guard and the state in the same room with the local players. And for their part, all three California Coast Guard Captains of the Port are strong supporters of the Harbor Safety Committees.

Our conclusion is that Harbor Safety Committees are effective and we believe such organizations would be useful in the harbors of other states. Another California concept that sweeps the nation? Why not.

CHAPTER 34

A Captain of the Port's View of Port Safety in California

James A. Watson IV
Commander, USCG

Abstract

The National Research Council's report on piloting entitled, "Minding the Helm" suggests that navigation safety in the nation's ports and waterways can best be improved by building on the inherent authority, experience and obligation of pilots, ports, state governments, and the federal government. The U.S. Coast Guard, the State of California, and the various port authorities in California are working to achieve a new standard for port safety by participating cooperatively in local Harbor Safety Committees to maintain Harbor Safety Plans, and by creating a Memorandum of Agreement for administration and oversight of California pilotage. This paper provides background, current status, and a prediction for port safety in California from a Coast Guard Captain of the Port's perspective.

Background

The responsibilities of shipping companies, California OSPR and the local port commission are very closely linked with mine as Captain of the Port. We share the task of ensuring a safe, clean San Diego Bay by monitoring quality shipboard and port navigation features, professional piloting and tug service for shipping from the time it reaches the sea buoy until the time it leaves it at the end of a port stay.

Shipboard features include modern steering and propulsion, navigational equipment such as gyro compasses and collision avoidance radar systems, and bridge radio and telecommunications equipment. The port's navigation features are its depth and width and aids to navigation such as buoys, ranges, racons, and lighthouses. Specific terminal features for mooring, fendering, fueling, and cargo handling are also important. San Diego does not have a vessel traffic service.

As Captain of the Port, I have the authority to direct any ship movements or to close the port if necessary. I'm responsible for safe navigation, environmental protection, port safety and the professional conduct of Coast Guard licensed mariners. If something goes awry, I am equipped to respond, enforce, investigate, penalize, cleanup or do whatever is necessary to ensure the port is protected and working safely again as quickly as possible. These authorities and responsibilities have rested with Coast Guard Captains of the Port since 1965. In ports where a federally supported VTS operates, the Captain of the Port's authority is used during daily traffic management communications to direct vessel movements. However, Captains of the Port do not manage pilotage of foreign vessels calling on the port.

Local management and control of pilotage in the United States dates back to 1789, one year before the predecessors of the Coast Guard: the Revenue Cutter Service and the Lighthouse Service were founded. State and local pilot commissions in ports throughout the country have guarded that control carefully through the years despite the continual evolution of federal management of foreign commerce, navigation and shipboard safety. Today there are volumes of federal and international treaty regulations for ship management and navigation, but local pilotage in the United States, except for the St. Lawrence Seaway, is preempted.

Local control of pilotage has served us well. Local pilot's associations throughout the country determine and enforce the requirements for foreign trade vessels calling on their ports. This includes location of embarkation and debarkation of the pilot, minimum ship size limitation for carrying a pilot, offshore anchoring conditions, and pilotage fees. Equally as important they determine the requirements for a person to be a pilot and they ensure that working pilots are properly equipped and supported. This includes pilot qualifications, experience and recency requirements, apprentice and training programs, physical standards, and a disciplinary program. Equipment and support includes a suitable pilot boat, radios, pilot's office, tugboat support, administrative support and adequate pay. Each commission accomplishes the tasks necessary to ensure safety in its own way.

Minding the Helm

The recent report from the Marine Board of the National Academy of

Sciences entitled "Minding the Helm" encourages Coast Guard to participate in the advancement of state pilotage systems and waterways management systems from both the local and national level. It recommends a fundamental change in the federal and state pilotage systems, drawing on the positive aspects of each to ultimately lead to a single pilotage system for each port. Additionally, it recommends risk based port safety analysis and application of new technologies to improve operational safety.

The Marine Board made 37 recommendations altogether. The following critical areas of pilotage administration were recommended for improvement: Pilot training, Qualification, Pilotage boards, Investigation of casualties involving pilot performance, and Gaps in pilotage coverage. The report argues that each port's pilotage system should be accredited nationally, but overseen locally by a nonfederal public organization with balanced membership representing state, pilot, marine industry, and public interests.

The Board looked at the loosely integrated management of the U.S. waterway system and turned to the national airspace system for ideas. The report encourages accredited vessel traffic systems that extend seaward of the pilot boarding areas and include real time environmental data. Additional air traffic control features like universal protocols, adherence to pre-planned routes, and near-miss reporting are also recommended. The report recognizes the same potential safety benefit from either exclusively federal, joint federal-private, or all private VTS-like services as long as they are authorized and accredited by the Department of Transportation.

Port of San Diego

In San Diego, a visiting ship embarks a pilot before it passes buoys 5 & 6 near Pt Loma. Four pilots have been authorized by the port to pilot vessels into and out of the harbor. A commercial tugboat company provides assistance maneuvering ships under the direction of the pilot assigned. Commercial ship movements average about 1000 each year. A tariff supports the pilots. Pilot and tug revenue is dependent on the number of ship visits.

The Coast Guard works very closely with the pilots in San Diego. While pilots control freight and cruise ships, and direct tugs in the bay, the Coast Guard manages and controls (when necessary) the rest of the activity out there. The Coast Guard oversees sport fishing and harbor excursion vessels, recreational boating events, commercial fishing vessels, maintenance of aids to navigation, dredging, dry docking, waterfront construction, oil transfers, water side port security, and coordination with the U.S. Navy.

The Navy recorded 7,157 large vessel and barge movements last year. Naval Station San Diego maintains its own pilots, tugs and vessel control center. Often the San Diego pilots communicate directly with the Navy pilots and the Navy control center. If a navigation problem or concern is known, the Coast Guard will initiate a broadcast notice to mariners and try to make direct contact with the Navy, the pilots, and anyone else that might be affected. When the pilots come across a problem with either the ship or navigation, they contact the Coast Guard.

I view the port as a system of interconnected people, plans and hardware. San Diego pilots are at the heart of this system. Shipping is set in motion through the efforts of business and market needs. The Coast Guard, the tugs, the ship agents, the terminal operators, the ship's crew and all the other elements of maritime commerce make up the rest of the system.

Harbor Safety Committee

The Port of San Diego has a Harbor Safety Committee, attended by representatives of the port's users including the pilots. Its goal is to keep the system working properly. The HSC has already completed projects that resulted in improving navigational aids, Coast Pilot information, anchorage limitations, publicizing of navigational safety, and a more reliable pilot boat. A tanker escort regulation that was drafted by the San Diego HSC and issued by California OSPR will go into full effect this year.

This year the HSC has an ambitious agenda to promote safe, reliable marine transportation in San Diego. One of the most important items on the HSC agenda is under-keel clearance. A critical feature for any port is its advertised minimum under-keel clearance. Deep draft ships must be assured that the prescribed limit really takes everything into consideration. That includes: shallow spots in the maneuvering areas, extra low tides, channel entrance swells that can cause ships to roll and heave, and an allowance for squat (the amount of aft trim that occurs when a ship is underway).

Another key item will be "Rule 9" signs, decals, flyers, and boater awareness training. Rule 9 is the provision of the navigation rules that gives large ships in restricted waterways the right of way over small craft. It's often ignored by small craft in San Diego Bay, so the HSC has applied for and received a grant from Boats U.S. to educate boaters. A special effort will be made to provide information at the boat rental companies.

The HSC also plans to continue to advise the Coast Guard and the California OSPR regarding regulations for anchorages, precautionary areas, and tug escorts for tankers.

Finally, a special challenge will be for the HSC to conduct a risk management study for shipping in the port. It will consider issues such as fog, radio interference, sharp turns in the channel, tug and barge traffic, pilotage, peak vessel movement periods, and foreign navy ship visits. The goal will be to prioritize the risks and reduce or eliminate them through non-regulatory means.

The California Pilotage Agreement

About a year ago the Journal of Commerce quoted Mr. Pete Bontadelli, director of the California Office of Spill Prevention and Response as saying, "California has virtually every type of system and non-system of pilotage that exists in this country".

For example, San Diego pilotage is different from the systems with state commissions that I had become familiar with in Puget Sound and Coastal Georgia. The pilots are good pilots, but they're not public servants; they're businessmen. There's no state license, no pilotage commission, no particular qualification process or apprentice program, and no law that requires a foreign registry vessel to take a pilot. The port authority simply authorizes certain federally licensed captains to provide piloting services in the port for a set fee or tariff. The Coast Guard currently recognizes the pilots as "state pilots" because California sanctions San Diego's port authorization system.

In Los Angeles the pilots are public servants. They are city employees. In Long Beach the pilots are employed by a pilot company that provides the service contractually to the port authority. San Francisco has a more conventional system with state licensed pilots and a commission.

I doubt there will ever be a uniform pilotage system in California. Uniformity is an unrealistic and unnecessary requirement. But there should be a performance criteria for the pilot systems that specifies qualifications, accountability, and discipline.

The Coast Guard is playing an important role in California's pilotage systems without taking sides for either the marine industry or the state's pilots. The recently signed pilotage Memorandum of Understanding between the State of California and the U.S. Coast Guard strengthens pilotage safety without new laws or regulations. It allocates the responsibilities for (1) initial competency and qualifications of pilots, (2) standards for maintaining proficiency and professional growth, and (3) enforcement. The agreement also describes a local pilotage advisory council that will provide recommendations for the pilotage system in each port.

Initial competency requirements will include federal licensing as a condition of employment, a period of apprenticeship, and other port specific prerequisites to be established by the local pilotage advisory council. The apprenticeship program will be vested with the port authority, not the pilots themselves. It will consist of a minimum period or target number of movements for various ship types and transits, a method for evaluating apprentices, and a standard for graduation to independent pilot status.

The port authority will also establish standards to maintain proficiency and professional growth. The standards must include periodic evaluations specific to ship types and sizes, routes, and operating conditions.

The Coast Guard will investigate any actual incident that may be indicative of a pilot's competency. The Coast Guard may take appropriate federal administrative action, or refer cases to the U.S. Attorney for possible criminal prosecution.

The State of California will enact legislation to regulate and license pilots in any port that does not implement a program in accordance with this agreement within twelve months of the effective date.

Port Safety in the Future

Looking into the future, port safety will be more than a pilot's advice, rules of the road, and an aids-to-navigation system. These are important parts of the system, but we must look beyond them and make better use of the technology available and the lessons learned from marine casualties like the Brightfield and Julie N accidents which occurred last year in other ports around the country.

We must develop standards for the use of automatic identification transponders that can provide voiceless ship-to-ship and ship-to-shore communications. We must develop vessel traffic system facilities and procedures that are cost effective and focused on both the human element and shipboard equipment failure contingencies. We must develop a waterways management system that integrates data from a host of sources including: floating buoy sensors that measure current, wave, tide and wind; Electronic Chart Displays (ECDIS) that alarm ships that stray from marked channels and anchorages; and risk analysis programs that aid in decision making.

I'm encouraged by the initiatives of public and private organizations from the federal forums down to the local level. The Interagency Committee on Waterways Management is coordinating the federal agencies in evaluating navigation safety and efficiency. The California Resources Agency is monitoring it's Agenda for the Future. The pilotage agreement for California ports exemplifies "Minding the Helm", and the local Harbor Safety Committee continues to revise and improve the plan for a safer San Diego Bay.

CHAPTER 35

MARINE PROTECTED AREAS AND BIOLOGICALDISTRIBUTIONAL DATA: LARGE AND SMALL SCALE PERSPECTIVES

R. Glenn Ford and Michael L. Bonnell

Abstract

Two studies involving extensive mapping of marine biological resources were compared to evaluate the data needs of marine resource managers working at the local or at the regional level. The first study was a compilation of GIS data for the Farallones National Marine Sanctuary; the second study was a delineation of marine ecoregions for the Pacific continental shelf region of the continental USA carried out for the World Wildlife Fund.

At the local level, a wide variety of mappable data were available, most already in digital form. Distributional information included density or sightings data for seabirds, cetaceans, pinnipeds, and fisheries. Human use data included coastal access, commercial use, shipwrecks, and shark attacks. Shoreline data were available showing pinniped haulouts, seabird colonies, coastal typing, and marine mammal strandings. Numerous versions of bathymetric and shoreline data were available at differing scales and covering differing geographic areas.

At the regional level, we classified pelagic habitats so that we could examine the level of protection available to the various habitat classes. While many of the same data used for management of the Gulf of Farallones were also available at the regional level, only a small subset of these data was of use in habitat classification at this scale. In general, the distribution of marine vertebrates was found to be a good basis for classification of pelagic habitats.

We examined the scale of variation of several general themes: vertebrate distribution, abiotic factors, and human use. Scales of variation were different across and along the shelf. The magnitude of the long shelf variation suggests that an integrated program of fisheries management at the continental scale and site specific management at the local scale are necessary for protection many or most temperate marine vertebrates.

R. Glenn Ford and Michael L. Bonnell, Ecological Consulting, Inc., 2735 N.E. Weidler Street, Portland, OR 97232

CHAPTER 36

Considerations for Marine Protected Areas: Sizing and Spacing

James F. Quinn

Abstract

Theoretical considerations and studies of terrestrial protected areas suggest that multiple protected areas, some of which may be small, are needed to effectively capture overall biodiversity within a region. Typically, capturing local uniqueness ("beta-diversity") is at least as important as identifying local diversity "hotspots" ("alpha-diversity") in representing the overall species pool. Existing broad-scale surveys of benthic marine communities have not typically yielded enough systematic data to convincingly distinguish among alternative MPA strategies as a means for protecting overall regional or global biodiversity, but new information, on both a regional and a global scale, promise to improve our knowledge for the potential role of MPAs. New initiatives form the international Man and Biosphere Program provide one model on how to increase the effectiveness of such efforts through cooperation among conservation organizations and international programs

James F. Quinn, Division of Environmental Studies. University of California, Davis, CA 95616

CHAPTER 37

Issues Relevant to the Design of Marine Reserves

Mark H. Carr[1] and Daniel Reed[1]

Abstract

The two primary reasons for establishing marine reserves are to: (1) protect the natural resources within a particular site or habitat, or (2) act as a fishery management tool for optimizing sustainable yields outside of the reserve. Regardless of purpose, the ultimate success of a reserve depends on its long term sustainability. In order to be sustainable, populations within a reserve must be replenished at a rate that is equal to or greater than the rate at which individuals die or emigrate form the reserve. The two types of marine reserves require different designs in order for them to effectively achieve their respective goals. In reserves designed for protection, sustainability will be maximized if replenishment is internally derived rather than being dependent on unprotected areas located outside the reserve. Because many marine organisms are highly dispersive at some stage in their life, larger reserves are more likely to be self-replenished than smaller reserves. Therefore, reserves established for purposes of protection need to relatively large in order to be effective.

In contrast to reserves established for protection, the goal of reserves established for fishery management (i.e., harvest refugia) is to enhance the resources outside the reserve. The idea here is that the reserve provides a protected area for organisms to grow and reproduce and that most of the production from within the reserve is exported (in the form of larvae or adults) to areas where fishing is allowed. Hence, self-replenishment is generally not desirable for harvest refugia since their primary goal is to replenish populations located outside the reserve. Consequently, the most effective harvest refugia design might consist of many relatively small protected areas that not only replenish the fishery, but replenish each other as well.

Determining the optimum size of a marine reserve will depend on the dispersal ranges of the organisms inhabiting the reserve. Dispersal ranges will vary not only among species, but also among locations as a function of oceanographic conditions. Information on sources of population replenishment, as determined by the biological and physical determinants of dispersal, is critical in the design and siting of marine reserves. However, to date, this has not been considered.

[1] Coastal Research Center, Marine Science Institute, University of California, Santa Barbara, CA 93106

CHAPTER 38

Goal Oriented Marine Reserves: A Zoogeographic Approach

Richard M. Starr[1] and Korie A. Johnson[1]

Abstract

The worldwide decline in the harvest from marine fisheries has resulted in a concern for both the economic viability of fisheries and the conservation status of selected species. These concerns have resulted in an increased interest in alternative methods for managing marine fisheries. One management option that is receiving increasing support is the use of marine protected areas (MPAs). MPAs theoretically have many intrinsic values, but also are characterized by a number of possible problems that can delay or prevent their projected benefits, especially if clear goals and objectives are not delineated prior to MPA development. More than 100 marine parks, reserves, and sanctuaries have been developed along the coast of California by agencies of different political levels. Each MPA has a different set of stated or implied goals. In many cases, MPAs with different goals, objectives, or regulations spatially co-occur. This has led to a disjointed MPA system in California that has several crucial problems. We are proposing a zoogeographic approach to the design and implementation of MPAs in California. The approach we propose acknowledges regional biotic differences and facilitates the development of goal oriented reserves. We provide an example of a way to develop a system of MPAs that has the goal of sustaining recreational and commercial fishery resources. A MPA system that is designed using this method will maximize the potential benefits of fish reserves while providing an environment for scientific research that can increase our understanding of marine ecosystems.

----

[1]University of California Sea Grant Extension Program, P.O. Box 440, Moss Landing, CA 95039.

Introduction

Catch data compiled by the United Nations Food and Agriculture Organization indicate that the world catch of marine fishes increased rapidly in the 1960-70's, then peaked in 1989 (FAO 1992). Since that time, the harvest of marine fishes has declined in many of the world's most productive fishing grounds. Throughout these marine areas, fish populations have experienced not only diminished abundance and reduced production, but also decreased lengths, weights, and ages at first maturity. The decline in these and other biological parameters indicate a severely negative response of fish populations to excessive harvest. These changes within fished populations have also led to changes in species composition, species distributions, and ecological relationships.

The dramatic worldwide decline in fisheries has resulted in an increased interest in fishery issues, not only by fishery management agencies, but by the general public as well. Currently, fishery managers are exploring management options beyond those traditionally used. Increasingly, marine protected areas (MPAs) are being considered worldwide as tools that may be used to help sustain the ocean's resources (Davis 1989, Rowley 1994, Bohnsack 1993). Although MPAs are rapidly being established, their effectiveness as viable management alternatives are largely untested and poorly understood (DeMartini 1993, Dugan and Davis 1993).

Marine protected areas theoretically have many intrinsic values, and benefits can accrue to marine species and habitats, and to several different human community sectors. Properly designed MPAs can increase the diversity and abundance of species in an area, protect critical habitats, and protect species at critical stages in their life history. Societal groups expecting to receive benefits of marine reserves include resource managers, commercial and recreational harvesters, conservation groups, non-extractive recreationalists, scientists, and the general public.

The expected benefits to marine environments and human communities can only be realized if goals and objectives of proposed MPAs are clearly defined. For example, the most obvious goal in fisheries management is to maintain a sustained harvest of a resource. A MPA designed with the goal of increasing protection of spawning or rearing areas for specific species can help fishery managers realize their goal of maintaining harvest. Some MPAs have been developed on the US east coast with the goal of improving fisheries. In those locations, area and seasonal closures have been used to protect adult stocks, restrict harvest of spawning adults or reduce bycatch and subsequent mortality of juveniles in nursery areas (Auster and Malatesta 1994).

MPAs are often advocated by conservationist groups as a means of wilderness protection. Included in the benefits of wilderness protection include the preservation of species diversity, genetic diversity, natural habitats, and large scale environmental processes. For these benefits to accrue, however, a MPA designed for wilderness protection should prevent potential ecosystem alteration, such as might occur from kelp harvest or from high intensity recreation.

Other examples of MPAs that would require specific goals include the development of marine research areas or marine recreation areas. In each case, the specific objectives for the reserve would include an acknowledgment that some potential benefits may be foregone at the expense of the chosen objective. Clearly, the scientific community would benefit from a system of marine research areas. Such a system of MPAs could provide relatively undisturbed areas for scientific research and education. The increasing level of marine recreation and tourism suggests that a system of MPAs for recreational activities may bolster local economies, and be of great recreational value as well. Goals for yet other types of MPAs may include the protection of cultural features and traditions that might not otherwise be sustained.

Unfortunately, many of the projected goals and benefits for different types of MPAs are mutually exclusive. For example, the allowance of recreational fishing in order to increase tourism could negate efforts towards increasing fish abundance or species diversity. Similarly, the selection of a MPA in a shallow, narrow strip along the coast may do little to increase the abundance of commercially harvested fishes in deepwater. Therefore, goals and objectives must be clearly defined before the establishment of a MPA. Once these goals are outlined, then MPAs need to be properly designed and implemented with those specific goals in mind.

There are a number of possible problems associated with MPAs that can delay or inhibit the projected benefits. Often, there is substantial political or social controversy surrounding the designation of a reserve that can slow or prevent the development of an effective MPA. For example, in order for some types of MPAs to effectively accomplish the stated goals, it is necessary to restrict a large resource area from many user groups, making public support even harder to achieve. In those cases, if specific goals and objectives are not predetermined and clearly articulated, public opinion can result in a MPA that is improperly designed or located, and for which the proposed benefits will not be realized. Another social problem inherent to MPAs is the need for enforcement. Even when goals and objectives for a MPA are clearly set, enforcement of rules and regulations is often lacking, due frequently to personnel and money constraints.

Once the milestones of public support, proper design, and sufficient enforcement are crossed, the benefits of an MPA are often slow to accrue. Continued support from the public during this transitional period is difficult to maintain. It is also important to remember that MPAs cannot be the entire solution to large scale marine resource management issues such as fisheries or wildlife management. The use of a MPA for management of many migratory species, for example, would require an MPA that is much larger than is feasible or practical to establish. Thus, traditional methods of resource management in concert with MPAs are still necessary.

The California coast provides good examples of MPAs that were established without associated goals and objectives being fully considered prior to implementation. There are more than 100 marine parks, reserves, and sanctuaries along the coast of California that have been implemented by agencies of different political levels (Table 1). These MPAs have a variety of names which usually stem from the specific mission of the implementing agency (Table 2). They are similar, however, in that many of these MPAs were established arbitrarily without the identification of specific purposes. Locations of the MPAs were usually selected haphazardly by public hearings and in the absence of scientific evidence to support the claims of proponents. Few of the existing MPAs have clearly articulated goals or objectives, thus the benefits accrued will be different than expected. Some California MPAs also have co-occurring or overlapping boundaries, have conflicting designations for use, and have conflicting rules and regulations (McArdle 1997). This disjointed system has created confusion and disorganization in the use of MPAs in California.

Table 1. Local, state, national and international agencies responsible for establishing Marine Protected Areas in California.
Source: McArdle 1997.

Organizing Agencies
Local
City Councils
State
State Water Resource Control Board
Fish and Game Commission
State Legislature
Parks and Recreation Commission
University of California
National
Minerals Management Service
NOAA Dept of Commerce
National Park Service
International
Man and Biosphere Program

Table 2. Specific designations of Marine Protected Areas in California established by local, state, national and international agencies.
Source: McArdle 1997

Reserve Designations
State
Areas of Special Biological Significance (34)
Coastal Sanctuaries (1)
Ecological Reserves (10)
Marine Resources Protection Act Ecological Reserves (4)
Parks (13)
Refuges (20)
Reserves (7)
UC Natural Reserves (1)
National
Ecological Preserves (1)
National Marine Sanctuaries (4)
National parks (6)
International
Biosphere Reserves (2)

The historically haphazard process of selecting MPAs in California has led to a system replete with crucial problems (Table 3). These problems prevent the current system as a whole from achieving regional goals and objectives, and has important future ramifications. The existing system may limit opportunities for future implementation of properly designed MPAs. We believe there is a high potential for public backlash, for example, if the current MPAs fail to solve the problem of declining fisheries. If this happens, scientists and resource managers would lose credibility in the public eye, and already limited funding for resource management could be decreased. The establishment of improperly designed reserves could also unnecessarily limit recreational or commercial use of marine areas.

Table 3. Problems associated with the current system of Marine Protected Areas in California.

Problems with the current system of MPAs in California:

1. No master plan for MPAs in California, there are too many categories and organizations involved
2. MPAs do not span range of species or habitats in California
3. MPAs are located haphazardly; they are not:
   -distributed evenly with respect to depth clines
   -distributed evenly with respect to latitudinal clines
   -based on known distribution of species or abundance
   -based on a habitat classification system, thus habitats are not representative
4. Few MPAs have set goals or objectives
5. Often reserve design is improper to accrue implied benefits
6. MPAs often have multiple, mutually exclusive objectives (e.g. recreational fishing in fish protection reserve)
7. Activities are allowed that are contrary to implied objective (e.g. intensive recreation in ecological reserve)
8. Enforcement is poorly funded or lacking in most MPAs
9. Many exceptions are made in MPA regulations

Proposed Zoogeographic Approach

The first step towards improving the potentially counterproductive MPA system in California is to develop a comprehensive plan for MPAs. A characteristic of the current confusing MPA system is that few MPAs have a defined set of goals and objectives. Thus, an important second step is to define a need and purpose for each type of MPA. The third step in the formation of a rational MPA system for California is to properly design individual reserves to meet the stated goals and objectives proposed for each type of reserve.

For this paper, we are proposing a model for the development of a logical system of MPAs in California, and thus will limit our discussion to the use of MPAs to help maintain viable commercial and recreational fisheries. There are many intricate aspects of fishery management to consider when establishing an MPA for the purpose of sustaining fisheries. We do not have the space needed to cover all the components of fishery management in this document, however, and will just focus on a few primary considerations for designing MPAs to enhance fisheries.

Selection of species or species groups intended to receive protection is a primary consideration for the design of MPAs for fisheries. This is complicated by the fact that commercial and recreational fisheries target different species which occupy different habitats. Most commercial fisheries for vertebrates occur in deep water, whereas most recreational fisheries occur in shallow coastal areas. Another important consideration is that fisheries operate on mixed groups of species. More than 200 species are harvested in the commercial fisheries, and over 150 species are harvested in the recreational fisheries for central California alone. A rational approach to developing MPAs for fisheries would include the establishment of criteria used to select species or species groups for which we would design MPAs. An example of a criterion would be to select economically valuable species. Another example criterion may be to select species at risk of overharvest or experiencing declining catches and abundance. In either case, an appropriate MPA could be designed for the selected species or species group.

Once these target species or species groups have been selected, we suggest a zoogeographic approach to the design and implementation of MPAs in California. Terrestrial scientists and managers have been using a zoogeographic approach to resource management for several decades (Wheeler 1996). The basic tenant of this approach is to define bioregions, define representative and key habitats in each bioregion, identify species-habitat associations within each representative habitat and bioregion, and then work to preserve representative habitats in each bioregion (Dethier 1992).

Three distinct zoogeographic regions, or marine bioregions, exist within California coastal waters. The boundaries of these bioregions are not explicit, but may be broadly defined by the major westward promontories at Cape Mendocino and Point Conception. South of Point Conception, species and habitats are typical of the less productive southern California bight waters; warmer water species dominate the fauna. Between Point Conception and Cape Mendocino, species and habitats are influenced by a mixture of California current water, strong seasonal upwelling, and periodic incursions of warmer water from the south. As a result, the fauna is typified by a mixture of species at both the northern and southern ends of their range. North of Cape Mendocino, the ocean environment is more uniformly cool with strong summer upwelling and strong winter downwelling. Species are much more typical of northern, colder water environments.

The steps involved in setting up a zoogeographic based system of MPAs within each bioregion include mapping the distribution of representative habitats, defining species-habitat associations, estimating fish and invertebrate densities, and estimating mean lengths of species within each of these three areas. Habitat distribution and species-habitat relationship information can be used to select appropriate locations for MPAs, and fish density and mean length data are important to provide information on fecundity and potential reproductive output as well as overall fish and invertebrate abundance.

Some broad scale information about bottom types is available and can be used as a first cut at defining habitat types. Additional information about bottom types can be collected by oceanographic sampling equipment that collects swath bathymetry. Broad surveys employing that equipment can be used to fill in gaps in our understanding of coastal bottom types. More detailed maps of the ocean bottom, such as those provided by sidescan sonar, can also help define habitats (Yoklavich et al. 1995). In addition to maps of bottom types and habitats, information is available with which to identify concentrations of important fish and invertebrate species. Fox and Starr (1996) describe methods that can be used to map locations of high fish density.

On a more finite scale, scientists continue to collect information about species-habitat associations. A multitude of scientific techniques are available to estimate the relative abundance of species. These techniques include both surface based sampling using acoustic equipment or fishing gear (e.g. nets, traps, and hook and line gear), and underwater sampling using submersibles, towed cameras, remotely operated vehicles, or scuba gear. Both the surface and underwater sampling techniques, however, have limitations with respect to area covered and habitats sampled, or have biases in the size of species sampled. Fishery scientists thus use a combination of techniques to estimate relative abundance of species in a region. The precision of those estimates increases with sampling, so for the purposes of developing a system of MPAs, it is advisable to focus sampling efforts on locations which are proposed MPAs.

Once representative habitats and species are delineated, major physical processes that will affect the MPAs also need to be identified. For example, current regimes need to be described and mapped throughout each bioregion to understand juvenile emigration from a reserve that may increase catches in nearby fisheries. A knowledge of currents also is important for assessing the value of a MPA in providing larvae for trasport out of the reserve (i.e. the objective of providing a source of progeny for surrounding areas). This understanding of large scale physical processes will allow managers to locate MPAs in areas that will best protect representative habitats and key species for each bioregion. It will also be helpful in estimating expected benefits from each type of MPA.

Once proper geographic locations have been designated, the next step is to maximize the chances of accomplishing the selected goals and objectives though appropriate reserve design. The size, shape, and location of a reserve will directly influence the effectiveness of the MPA. For example, assume a goal of a fishery reserve is to protect spawning fish. Obviously, the reserve must be placed in a location in which the species spawns. Next, a properly designed MPA for a spawning group will be of a sufficient size and shape to encompass the daily movements of the adults of that species. The reserve will be less effective if spawning fish consistently move outside the reserve.

Size, shape, and location of the MPA is extremely important when evaluating the differences between reserves designed for recreational and commercial fisheries. Most recreational fish species live in nearshore habitats, primarily shallow water kelp beds and sand or rocky habitats. For these species, a long narrow MPA encompassing those nearshore habitats may be appropriate for the objective of increasing species composition or relative abundance in the MPA. If properly located, that shape would also be effective for protecting important commercial invertebrate fisheries. It would not, however, effectively protect deep water commercial fish or invertebrate species. A MPA that spanned only a small portion of the coast, but extended well offshore may provide more protection for the commercially harvested species in deep water.

In all cases, the design and setting of MPAs should account for the habitat type and quality both inside and outside the reserve. It makes little sense to protect an area from fishing, for example, if the water quality inside and adjacent to the reserve is poor. Similarly, if an objective of the MPA is to improve fishing outside the reserve, it is important that suitable habitat is available within and adjacent to the reserve for emigration of juvenile and adult fishes as densities increase (Carr and Reed 1993, DeMartini 1993).

A structured system of MPAs such as we propose will facilitate scientific research to increase our understanding of the specific intrinsic values and characteristics of marine reserves and encompassed habitats. The approach we propose will enable studies to be conducted within and adjacent to the MPAs to estimate species composition, densities and sizes, habitat types, and currents to a finer degree of detail. The zoogeographic approach will also enable scientists to compare ecological processes occurring in different regions of our coast, thus providing a better understanding of the population dynamics of the wide variety of species in California.

In addition to increasing the diversity and abundance of natural resources, MPAs that have been developed and implemented using a zoogeographic approach will provide a rational system of reserves in California. A rational, goal oriented system of marine reserves in California will greatly increase the chances of MPAs succeeding as a viable resource management tool.

References

Agardy, M.T. 1994. Advances in marine conservation: the role of marine protected areas. TREE 9(7): 267-270.

Auster, P.J. and R.J. Malatesta. 1994. Assessing the role of non-extractive reserves for enhancing harvested populations in temperate and boreal marine systems. In: J. Lien, N. Shackell, and J.H.M. Willison (eds.), Marine Protected Areas and Sustainable Fisheries. Science and Management of Protected Areas Association, Wolfville, Nova Scotia.

Bohnsack, J.A. 1993. Marine reserves: they enhance fisheries, reduce conflicts, and protect resources. Oceanus, Fall. pp. 63-71.

Davis, G.E. 1989. Designated harvest refugia: the next stage of marine fishery management in California. CalCOFI Rep 30: 53-58.

DeMartini, E.E. 1993. Modeling the potential of fishery reserves for managing pacific coral reef reserves. US Fish Bull 91: 414-427.

Dethier, M.N. 1992. Classifying marine and estuarine natural communities: an alternative to the Cowardian system. Nat Areas Journal 12(2): 91-100.

Dugan, J.E. and G.E. Davis. 1993. Application of marine refugia to coastal fisheries management. Can J Fish Aquat Sci 50: 1-14.

FAO. 1992. Review of the State of World Fishery Resources Part I. The Marine Resources. Food and Agricultural Organization of the United Nations Fisheries Circular No. 710. Rome, Italy.

Fox, D.S. and R.M. Starr. 1996. Comparison of commercial fishery and research catch data. Can J Fish Aquat Sci 53: 2681-2694.

McArdle, D.A. 1997. California Marine Protected Areas. California Sea Grant College System, Publication No. T-039.

Quinn, J.F., Wing, S.R. and L.W. Botsford. 1993. Harvest refugia in marine invertebrate fisheries: models and applications to the red sea urchin, *Strongylocentrotus franciscanus*. Amer Zool 33: 537-550.

Rowley, R.J. 1994. Marine reserves in fisheries management. Aquatic Conservation: Marine and Freshwater Ecosystems. Vol 4, pp 233-254.

Sobel, J. 1993. Conserving biological diversity through marine protected areas. Oceanus 36: 19-26.

Wheeler, D.P. 1996. Ecosystem Management: An Organizing Principle for Land Use. In: Land Use in America. H. L. Diamond and P.F. Noonan (eds). Island Press: Washington D.C.

Willison, M. and N. Munro. 1994. Ecosystem monitoring and protected areas. Abstracts from the Second International Conference on Science and the Management of Protected Areas. May 16-20, Dalhousie University, Halifax, Nova Scotia, Canada. 69 pp.

Yoklavich, M. M., G.M. Cailliet, H.G. Greene, and D. Sullivan. 1995. Interpretation of sidescan sonar records for rockfish habitat analysis: examples from Monterey Bay. Alaska Dept. Fish and Game Special Publication No. 9:11-15.

CHAPTER 39

The Human Dimension:
The Missing Link in California's Fishery Management?

Caroline Pomeroy[1]

Abstract

The California Department of Fish and Game's mandate to manage the state's fishery resources places primary emphasis on the biology of the resource, to the near exclusion of the 'human dimensions' of its use and management. As the latest battle over managing the fishery for California market squid (*Loligo opalescens*) demonstrates, the Department's limited access to and constraints in considering the economic, social, cultural and political dimensions of fisheries have made it difficult to respond effectively to the issues at hand. Recent events in the fishery have raised questions regarding its sustainability, and have led to discussions of management alternatives for this, one of the last open access fisheries on the West Coast. As the debate indicates, the concerns about the fishery are social and economic, as well as biological. This paper argues for the importance of social science to effective fishery management. I begin by discussing the role of social science in fishery management, controversies surrounding its use, and examples of its contribution to fishery management elsewhere in the U.S. I then explore the case of the squid fishery, and concerns about its sustainability as both a natural resource system and a human enterprise. I then describe recent initiatives within and outside the legislature as they highlight the relevance of the social and economic, as well as the biological, issues at hand, and provide an overview of a recently approved Sea Grant project that will provide information to help address those issues and better inform impending resource management decisions. My concluding remarks include a call for establishing mechanisms to enable the state to more systematically engage the human dimensions in fishery management.

Introduction

The human dimensions - the social, cultural, economic and political aspects - of resource use are gaining recognition as critical to more effective management of public resources such as fisheries (Buck 1995, Orbach 1978). In the U.S., the 1969 National Environmental Protection Act (42 U.S.C. 4321-4347, P.L. 91-910) first mandated the investigation of social and economic, as well as environmental impacts, for any federally sponsored project or action. For fisheries, the 1976

[1] Institute of Marine Sciences, Earth & Marine Sciences Bldg. A316, UC Santa Cruz, Santa Cruz, CA 95064.

Magnuson Fisheries Conservation and Management Act (16 U.S.C. 1801 et seq., P.L. 94-265)[2] mandates the consideration of social and economic, as well as biological, information in the development and implementation of federal fishery management plans. In California, however, this mandate is not so clear. The California Environmental Quality Act (Pub. Res. Code 21000 et seq.) recommends consideration, but requires no systematic assessment of, social and/or economic impacts of proposed projects or actions. The California Fish and Game Code (California Code of Regulations Title 14) defines the Department's mission as one of protecting the resource, not to be compromised by social or economic considerations.

When it comes to fisheries, two challenges face resource managers. The first is determination of the biological condition of the resource and rate at which the resource can be used without threatening its reproductive potential. The second is the design and implementation of management measures that will insure the conservation of the resource while allowing human use. Because these measures are used to control human behavior, design and implementation require knowledge of both the fishery as a resource and as a human activity. In short, fishery management is people management (Fiske 1990). As such, effective resource management requires knowledge of peoples' values, preferences and behaviors, and the ability to use that knowledge in the resource management context (Clay and McGoodwin 1994, Orbach 1978). This is not to say that this information should be used to inhibit resource protection; rather, information on resource users is critical to its success.

This paper draws attention to the human dimensions of fisheries as a key, yet not fully reckoned with, element of fishery management in California through an exploration of recent changes in the squid fishery and discussions regarding its management. I begin by defining and providing examples of research and application of the human dimensions of fisheries in the U.S. I then give an overview of the squid fishery, with attention to dramatic changes in the industry in recent years. Next, I discuss emerging concerns about the fishery as manifest in recent legislative and other actions. I then present an overview of a socio-economic analysis of the fishery to be conducted with support from California Sea Grant, and will contribute information to assist resource management decisions. My closing remarks focus on the need for more systematic incorporation of the human dimension in the state's fishery management.

Fisheries Social Science: Focusing on the Human Dimension

The Magnuson Act defines a fishery as a stock fished or the act of fishing for such a stock; it may also be defined as the combination of fish and fishermen in a region, fishing for similar or the same species with similar or the same gear types (Fowle 1993:85). Whereas fisheries biology focuses on a fish stock or complex and its ecosystem), fisheries social science focuses on human activity, interaction, and organization relative to the resource and its use.

---

[2] The MFCMA was amended and reauthorized in 1996 (P.L. 104-297) as the Sustainable Fisheries Act.

Social science disciplines, ranging from anthropology to economics, have increasingly been brought to bear on fisheries as human systems, and as prime examples of human-environment interactions. Researchers have used ethnographic, survey, and a range of other research techniques to describe, explain and predict human behavior in fisheries. In general, fisheries social science produces two types of information - contextual and actionable (AFS Socioeconomic Section 1993). Contextual information provides a backdrop for managing a fishery, whereas actionable information bears directly on the development and implementation of regulations.

A growing body of literature demonstrates the relevance, and the effective and balanced use, of social science in fisheries management. In California, for example, Orbach and Beckwith (1980) conducted an ethnographic study of differences between Vietnamese (refugee) and European-American fishermen's values, beliefs and practices. This study provided information that described and explained both groups' practices in cultural context, and the conflicts between them, to enhance understanding between the two groups and to assist resource managers' efforts to work with them. Maiolo and Johnson (1989) analyzed communication networks between fishermen and managers in North Carolina to assist fishery managers' efforts to improve relations and enhance the effectiveness of their communications with fishermen. Hanna and Smith (1994) conducted a survey to determine Oregon trawl fishermen's attitudes and activities associated with work, resource use and management. Their work demonstrated the variation among fishermen - who are often assumed to be homogeneous - and the implications of their diverse attitudes for fishery management.[3]

The rationale for investigating the human dimensions of fisheries for resource management is threefold. First, as noted above, resource management is people management. We manage resources because people value them, and seek to derive benefits from them. Often, the nature and extent of resource use, be it consumptive or non-consumptive, varies considerably among (and within) groups. Certain uses may be incompatible, or in direct conflict, with others. When resources are limited, unbridled use can lead to resource destruction and conflict among users.[4] Resource destruction can be diminished or eliminated by applying regulations appropriate to the resource use activities of each group. Conflict can be averted or diminished by allocating the resource among user groups (although allocation itself is often a source of conflict). A second reason for the conduct and use of social science research for fishery management lies in the benefits afforded by the information it provides. This information on peoples' attitudes, values and beliefs and their social, cultural and economic underpinnings, enables managers to anticipate problems and opportunities in resource management. As one CDFG staffer put it, "managing with information [about those who fish] is a heck of a lot easier than without it" (anon. pers. comm.). Third, federal legislation mandates it:

---

3 See Acheson (1981), AFS (1993), Orbach (1978, 1980), Pollnac and Littlefield (1983), among others, for additional examples of social science research and application.

4 This problem is known as the 'tragedy of the commons' (Hardin 1968).

> The Secretary (of Commerce) shall initiate and maintain...a comprehensive program of fishery research...designed to acquire knowledge and information...on fishery conservation and management and on the economics and social characteristics of the fisheries (P.L. 101-627 SEC.404(a))

These include "the social, cultural, and economic relationships among fishing vessel owners, crew, United States fish processors, associated shoreside labor, seafood markets and fishing communities" (P.L. 101-627 SEC.404(c)). Where state and federal jurisdictions overlap (e.g., in the coastal pelagic species fishery), or federal funds are involved in resource management, this requirement will hold.

This rationale notwithstanding, there is valid concern regarding the appropriateness and efficacy of using social science information in fishery management. As one participant in California's fishery management debates noted recently,

> There appears to be no requirement for including social science information under...a number of [the state's] natural resource statutes, including those covering the protection of fish and shellfish. That is not to say, however, that social impacts are ignored; indeed, they are sometimes considered at the expense of sound resource protection (anon. pers. comm.).

However, social scientific information can play a critical role at various stages of resource management. For example, social and economic impact assessment can be used to help prepare those who will be affected by a change in the system/rules, and to better enable managers to anticipate potential resistance or other problems associated with implementing sound resource management policy (ICGP 1994). Moreover, in some cases, the biological condition of the resource is unknown or not clearly an issue (yet?), but there are social and economic concerns associated with its use - including concerns that continued growth in that use could threaten both the industry and the resource. The fishery for California market squid is a case in point.

Squid: A Changing Fishery

The fishery for California market squid, *Loligo opalescens*, was initiated in the 1860s at Monterey by the Chinese. They used torches to attract the squid to the surface and caught them using small purse seines (Dewees and Price 1983, Lydon 1985). In the early 1900s, Monterey's Italian fishermen introduced lampara nets into the fishery. The fishery remained limited to the Monterey Bay area until the 1960s, when the Southern California fishery developed on spawning aggregations near the Channel Islands (Dewees and Price 1983). Since then, landings in southern California have been on the rise, while those at Monterey have remained about the same (LMR 1995).

Historically, the squid fishery has been closely related to the fisheries for northern anchovy, Pacific sardine, and Pacific and jack mackerel. These 'wetfish' species comprise the Coastal Pelagic Species (CPS) fishery, the subject of the recently canceled federal CPS Fishery Management Plan. The CPS project team noted the squid fishery's close ecological, economic and social ties to those fisheries, especially because 1) it is an integral part of the CPS fishery, 2) much squid fishing is done by round haul gear used also in the CPS fishery, 3) many of the squid fleet were sardine fishermen and may resume participation in that fishery, and 4) squid boats are socially and culturally part of the CPS fishery (summarized from CPS FMP Development Team 1993a:55, 1993b:7).[5]

Through the early 1980s, squid landings stayed below 20,000 mt, except for 1981, when they reached just over 25,000 mt (LMR 1995). Following the record low catches associated with the 1982-83 El Niño, however, the fishery began to take off. By 1995, squid ranked first among California fisheries in volume and second in value, when boats landed over 77,000 mt worth more than $20 million. Landings topped 86,000 mt in 1996 (Vojkovich 1996).

This phenomenal growth in the squid fishery is the result of developments in markets, processing, and fishing strategies over the past thirty years. Many of these were stimulated by research on squid biology, processing, and market development by California Sea Grant and other state and federal agencies through the early 1980s (e.g., Brooks 1977, Berntsen 1988, Evans 1976, Fields 1965, Kato and Hardwick 1975, Wormuth 1976). Another important factor, however, has been the change in both domestic and international markets. Domestic markets have grown as consumers have come to value the nutritional benefits of seafood in general, and have developed a taste for products such as fried calamari and other specialty dishes. Growth in external demand has included that of traditional consumers such as Greece and Italy, especially as squid stocks elsewhere have declined. In addition, the emergence of a booming Chinese market for California squid over the past 2 to 3 years has created new opportunities for the state's fishermen and processors (Redmayne 1996).

Along with this growing demand have come changes in fishing practices that have resulted in both increased catch and fundamental changes in the structure and spatial distribution of the fishery. Purse seiners using round haul gear are now the dominant type of operation, although a number of lampara boats still operate. Some squid fishermen use spotter planes, as well as depth sounders and sonar, to locate fishable aggregations. Whereas lights used to be prohibited in Monterey Bay, they are now allowed, and widely used there as in Southern California. The central California fishery has spread from the inner waters of Monterey Bay to outer bay waters, with most of the catch landed at Monterey and Moss Landing; a few boats operate out of Half Moon Bay to the north. The fishery begin in late April or early may and runs as late as October. The southern California fishery focuses primarily on squid aggregations around the Channel Islands, with landings occurring at Port Hueneme and Oxnard, Ventura, and San Pedro. Although this used to begin in late November (after Thanksgiving), in recent years its start date has moved up to as early as mid to late October (Deluca pers. comm.).

---

[5] Barnes and Hanan note that "During periods of good market squid fishing, the (CPS) fleet routinely diverts fishing effort from sardine and targets squid instead...because they receive an ex-vessel price of approximately $140-300 per ton" (1996:3). Apart from the CPS fishery, some squid fishermen participate in other fisheries such as herring and salmon (in California and Alaska).

The fishery remains constrained, however, by squid's high perishability. It must be landed and processed within hours of capture, so boats prefer to land at the port nearest to the fishing grounds. With most squid fishing now occurring in southern California, processors from central California (who work with boats from that area) have developed receiving and transport capabilities to be able to receive fish from these boats, then deliver it to processing facilities.

These changes have in turn led to an increase in the number and capacity of vessels engaged in the fishery over the past three years, especially in the southern California fishery. CDFG estimates the fleet's maximum capacity in 1995 was 4,520 net tons, compared to 3,640 net tons in 1982 (1996:3).[6] Both CDFG and fishermen's records show there has been an increase in the number of out-of-state vessels participating in the southern California fishery. Of the 137 vessels that landed squid in 1982, six were from out of state and landed 6% of the catch; by 1995, 22 of 117 vessels in the fishery were from out of state, and accounted for 27% of landings (CDFG 1996: 2,3). A list compiled recently by fishermen shows that the number of vessels fishing squid on the southern California grounds doubled between 1994 and 1996, from 18 to 36.[7] This increase in out-of-state vessels participating in the fishery is attributed to limited entry and/or declines in other fisheries in California, Oregon, Washington and Alaska together with the strong demand (and good prices) for squid.

Perhaps the most notable change in practice, however, involves the wide use of lightboats - small motorized vessels equipped with halogen lamps to attract the squid - in the fishery. Lightboats scout for fishable aggregations of squid, then set their lights to attract and hold the squid for a seiner to net and haul on board.[8] This procedure has greatly increased the efficiency and the scope of squid fishing operations. Sources estimate that there are at least as many lightboats as there are squid fishing vessels.

Recent Management Debates

The fishery for California market squid is one of the last open access fisheries on the Pacific Coast. At present, the State of California has management responsibility for the fishery, following the recent cancellation of the federal Coastal

---

[6] Using the same data, I calculated a slightly different result: 1995 and 1982 maximum fleet vessel capacity was, respectively, 4,640 and 3,320 net tons. These calculations are based on size class information for vessels, using the maximum net tonnage for each size class (e.g., 20 tons, for vessels in the 11 to 20 ton category).

[7] CDFG data shows little change in the number of boats landing squid over the past several years, but CDFG personnel note that this does not account for changes in vessel size, gear, or practices such as the use of lightboats, all of which influence the fleet's catching power.

[8] These lightboats are part of a seiner operation (e.g., owned by the vessel owner) or operate as free-lancers. They locate an aggregation of squid, then radio seiners in the area to motor over and load the squid; the lightboats receive a percentage of the catch for their services.

Pelagic Species (CPS) Fishery Management Plan. Although the plan would not have managed squid directly, the fishery was included in the planning process because of its close links with the CPS fishery.

Much of the state's commercial fishery regulation is done by legislative initiative. Historically, squid fishing has been regulated at times with measures that restrict the use of lights to attract squid, limit days or times when fishing is allowed, and for several years, prohibited the use of purse seines in Monterey Bay (see, Dewees and Price 1983). Current regulation of the fishery includes a prohibition on the use of lightboats in District 10 (Half Moon Bay), and a closure of the fishery from noon Friday until midnight Sunday in Monterey Bay (California Fish and Game Code Div. 6, Part 3, Ch. 2, Art. 9, Sec. 8399.1 and 8399). Many of these regulations have been prompted by industry (i.e., harvester or processor) action.

Members of the industry, concerned about the implications of recent changes in the fishery, have spurred consideration of more comprehensive management for squid. A small group of central California fishermen, backed by the Pacific Coast Federation of Fishermen's Associations (PCFFA), has been calling for scientific, regulatory and managerial attention to the fishery for the past four years. They say they are concerned that the fishery may not be sustainable, either biologically or economically. Opponents argue that there is no evidence of a resource problem, and that there is no need for regulating the fishery. Initially, CDFG held that there was no evidence of a resource problem; it now adds that there is insufficient evidence to make such a detminreation.[9]

At present, the California Legislature is considering two bills regarding the squid fishery.[10] SB 364 (as amended April 28, 1997, Sher) would institute a squid vessel license and fee, authorize CDFG to establish annual catch quotas and reduce the hours or days of the week during which squid may be taken, and create a Commercial Squid Advisory Committee. AB 1204 (as amended May 1, 1997, Papan) is quite similar to SB 364, especially after its amendment following introduction and discussion of SB 364.

The two bills differ, however, in three fundamental ways. First, whereas SB 364 would create a non-transferable permit, with qualification based on a window between 1984 and 1998, AB 1204 establishes a transferable permit with qualification based on squid fishing activity between April 15 1996 and April 15, 1997. These differences will manifest themselves in how access to the fishery is allocated over the next three years (and likely beyond), and in the extent to which individuals and vessels within the fishery can change. Second, the Commercial Squid Advisory Committee as described in SB 364 would be responsible for making recommendations to the Legislature regarding conservation and management of the fishery, after seeing that appropriate research is conducted by the Department, other agencies, and academic institutions. By contrast, AB 1204

---

9 These positions and arguments are more complex, but are beyond the scope of this paper.

10 A third bill (SB 879), introduced by Karnette, was withdrawn at the April 15 hearing; its provisions were to be incorporated into SB 364.

grants more limited authority to the Advisory Committee, allowing it to sponsor meetings and make recommendations to the Department, rather than to the Legislature. In addition, the Department would have the authority to establish an annual catch quota, regulate squid lightboats, and conduct research to determine the status of the fishery.

These differences aside, both bills acknowledge and demonstrate the need for information on the structure of the fishery as a human enterprise, as well as biological information on the condition of the resource. The recommendations to be made after a period of study by the proposed Commercial Squid Advisory Committee under SB 364 highlight this mix of information needs. They include:

1. Whether a limited access plan is necessary, and if so, what criteria should be used for participation, and what the optimum number of vessels should be
2. Whether it is necessary or advisable to reduce the number of days that squid may be taken in the commercial fishery to protect the resource.
3. Whether harvest refugia for squid should be declared.
4. A research and monitoring program.
5. The biological feasibility of expansion of squid harvest north of Point Reyes, and measures for the inclusion of participants in the fishery.
6. The regulation of squid lightboats.
7. Coordination that may be necessary with a federal CPS management plan, should one be adopted.

These issues were part of a similar bill (SB 2122, Sher) introduced but tabled last year. According to the bill's supporters, the Department provided an inflated estimate of the costs of implementing the bill and thereby jeopardized its passage (PCFFA 1996). Soon afterward, CDFG's Marine Resources Division (MRD) initiated a dialogue with fishermen and processors, beginning with two 'ad hoc committee' meetings in Sacramento in late 1996. Participants included commercial fishermen, sport fishers, processors and MRD personnel.[11]

MRD framed these meetings as informal gatherings to discuss current knowledge of the resource and the fishery, participant interests and concerns, and potential management options. At the first ad hoc committee meeting (November 1996), an MRD staffer serving as facilitator opened the meeting by clarifying the Department's limited authority to regulate the fishery, and its interest in facilitating a discussion of the issues and options as viewed by the various participants. Next, MRD personnel made brief presentations on the biology of the resource, the structure and performance of the industry, squid fishery management elsewhere, and their assessment of information needs and constraints. They noted the limited knowledge of the squid life history and population dynamics, the limited utility of standard stock assessment procedures because of the specie's short lifespan (12 to 18 months; Spratt in Recksiek and Frey 1978), and the importance of environmental factors (especially temperature) in spawning, distribution and abundance of the squid.

---

[11] Lightboat operators were absent from the ad hoc committee meetings, but did participate in the town hall meetings.

Next, participants were encouraged to articulate their interests and concerns about the fishery. Although there was some overlap among groups' interests, there were important differences in perspectives on the resource, the industry, and management. For example, whereas harvesters emphasized their interest in the continued biological and economic viability of the fishery, processors said they wanted the fishery to remain open so as to maximize employment opportunities and the economic benefit from the fishery. Like harvesters, MRD personnel spoke of resource concerns, but their orientation toward the biological condition of the resource per se differed from harvesters' concerns about it as a long-term source of livelihood. (Not all fishermen shared the 'harvester' views, nor did all processors share the 'processor' views, as presented here.) Then discussion then turned to viable management options, that is, management measures that people could "live with and support for now." 'Naysayers' were obligated to explain why they objected to a given option, rather than being allowed to hinder the process without defined cause. The meeting closed with a rough set of management options, and agreement among participants to meet again to refine these, then take them to industry-wide discussion.

The second ad hoc committee meeting (December 1996) addressed questions left unanswered and revisited the management options discussed at the previous meeting. Participants had wanted to know, for example, the capacity (i.e., net tonnage) of the squid fleet, and how has this changed over time, what the catch of out-of-state boats was compared to the total fleet, and how long it takes for squid to spawn. MRD personnel had answers to many of these questions, but were unable, for example, to specify the capacity of the fleet. Then the group reviewed, clarified and narrowed the management options. Those options were: 1) area closures / reserves, 2) in-season (temporal) closures, 3) an effort (vessel) moratorium[12], 4) gear restrictions (e.g., depth of net), 5) restricted number of lightboats, 6) do nothing (status quo)[13] and 7) management legislation that allows flexibility. At the close of the meeting, MRD agreed to schedule two 'town hall meetings,' one in Monterey, the other in Oxnard in January 1997, to present the management options and solicit broader input.

The town hall meetings were similar to, and built upon, the ad hoc committee meetings. The agenda included welcome and overview of meeting objectives by MRD staff, followed by their presentations on squid biology and stock assessment, the history and status of the fishery, and the options that were generated by the ad hoc meetings.

Although the agenda was nominally the same for the two town hall meetings, however, the outcomes differed starkly. The Monterey meeting, held on

---

12 A vessel moratorium - one that limits the entry and number of vessels - does not necessarily limit effort, and may even lead to an increase in effort (Huppert 1987).

13 Participants cited a "need to get answers on economic, biological and sociological issues to cut the 'rumors' that are being discussed in the fishing community" before regulating the fishery. The need for socio-economic and other research on the fishery notwithstanding, this argument is strikingly similar to that offered in discussions of the sardine fishery several decades ago (Radovich 1982).

a Friday evening, was attended by about 20 fishermen, a couple of processors and a half dozen CDFG personnel. Most of the local industry members sat in the back of the room, and offered little comment. Only one fisherman signed up at the beginning of the meeting to offer public comment, although some additional comments were allowed after he finished. The Monterey meeting ended, though, half an hour ahead of schedule, having achieved little in the way of dialogue. By contrast, the Oxnard meeting, held on a Friday at noon, was attended by some 80 fishermen, processors, and others, and turned into a lively discussion that ran the full allotted two hours.

The difference between the two meetings was the result of differences between the central and southern California fisheries, and between the approaches used by MRD to engage members of the industry in the dialogue. Following the disappointing outcome of the Monterey meeting, MRD personnel ran this meeting differently. First, the meeting room at the Oxnard Harbor Master's office was better set up for discussion among participants (CDFG personnel, industry and others) compared to the Monterey City Council chambers, which had the appearance of a court room over which MRD personnel presided. Second, whereas public comment had been solicited through a formal sign-up at the beginning of the Monterey meeting, it was sought more informally by the MRD chief following MRD staff presentations.[14] The result was a much clearer message to participants that their input was expected - even welcomed - in sharp contrast to the Monterey meeting. Finally, it was clear that the two groups - the industry in Monterey and that in southern California - differ remarkably on several dimensions, including their socio-cultural and political characters.[15]

Human Dimensions Issues in the Squid Fishery

Clearly the 'squid issue' is both a socio-economic and a biological one. Participants are concerned about the viability of the fishery both in economic and biological terms, and about their continued opportunity to derive benefits from it. The biological questions to be answered focus on the current condition of the resource, its strengths and vulnerabilities, and specifically, the extent to which it can sustain present, or increased, levels of exploitation. The socio-economic questions that have arisen are, likewise, a mix of the straightforward and the complex. The most fundamental questions pertain to the current structure of the fishery and its qualitative and quantitative differences from the fishery of two, five and ten years ago. Studies such as those by Dewees and Price (1983) for the fishery as a whole, and by the CPS project team as it developed the CPS FMP in the early 1990s provide baseline information against which to compare today's industry (1993a,b). More complex questions, made evident in the legislative debate and the informal dialogue described above, relate to the optimum size, structure and

[14] The Monterey agenda did not include public comment on the agenda, but allocated 45 minutes for a "Summary" by the Department, and five minutes for discussion of "Where do we go from here?" by "All." On the agenda distributed at the Oxnard meeting, the summary had been crossed out, and replaced with "Input."

[15] This is a topic of particular interest to fisheries social scientists, and is to be explored in the recently approved Sea Grant project described below.

function of the industry, within the limits of the natural resource system. To some extent, these are questions of social and economic values (and socio-economic engineering) that often seem inappropriate to resource management. However, to the extent the biological limits of the resource suggest an appropriate level and nature of fishing activity, these questions are germane to resource management decisions. That is, if the resource is to be used over the long-term (i.e., conserved), it is essential that management measures be designed and implemented with knowledge of the structure of the industry, how it has changed, and how it is likely to change. As the management debate proceeds toward action - whether it be to maintain the status quo or to adopt a new management strategy - two questions arise:

1. How might the fishery's current structure affect the implementation and effectiveness of management measures?

2. How might such measures affect the fishery's organization?

A project entitled "Socio-Economic Organization of the California Market Squid Fishery: Assessment for Optimal Resource Management" (Co-PIs Caroline Pomeroy and Margaret FitzSimmons), recently approved for funding by California Sea Grant, is intended to provide information to help answer these questions. Using the combined approaches and methodologies of fisheries sociology (Pomeroy) and economic geography (FitzSimmons), the project will analyze the changing structure of the California market squid fishery to:

1. to characterize the changing practices and institutions in the California market squid fishery,
2. to determine the relationship between fishermen' and processors' dependence on the fishery and their fishing and processing strategies, and
3. to ascertain the spatial patterns, linkages, and developing relationships between the Central and Southern California fisheries.

In the two-year project, we will examine the changing (social, cultural and economic) relations of production, adaptation and flexibility within the industry, and resource dependence. We will collect and analyze data through a combination of archival (i.e., background documentary) research, observation of fishing, processing, and other industry activities, and interviews with industry participants and others knowledgeable of the fishery. Using this information, we will compare: (1) the present fishery with that of the past, (2) the social and economic institutions which organize the fishery within and among fishing areas and landing ports, and (3) this fishery with other fisheries. The proposed research will provide resource managers and industry participants with systematic documentation of the fishery as a human enterprise. This information, integrated with existing and new information on the nature and condition of the resource, can then be used to inform the design and implementation of effective resource management.

Concluding Remarks

As recent debates over the management of the fishery for California market squid demonstrate, fishery management has both biological and human dimensions. As such, it requires information on both the resource and those who use it or otherwise value it. Limiting the purview of a resource management agency such as the California Department of Fish and Game to biological considerations can inhibit effective resource management. In order to effectively manage the resource, the Department must effectively manage people.

A recognized pitfall of much of fishery (and other natural resource) management is the use of social and /or economic information to stand in the way of sound resource management policy. However, to ignore the socio-economic context of a fishery can be just as harmful to achieving resource conservation goals. Moreover, as the squid fishery suggests, the emergence of socio-economic issues may indicate a need to revisit knowledge of a resource and its use. Conditions of resource use change over time, especially as market, research, and management forces converge. As in the case of the squid, a previously 'underutilized' resource, these forces have led to phenomenal expansion of the fishery. Renewable resource are not limitless,

The challenge is to integrate social and natural science information to best inform the fishery management process, with safeguards against the 'improper' use of either type of information to the detriment of the resource. Although it is difficult to specify how that would be done, it is not so difficult to specify how social science can be used to inform the resource management process. First, the mandate for biological considerations to be the driving force behind resource management should be maintained, but this should not be taken as a mandate against systematic collection and analysis of social and economic information. As has been argued above, such information is essential to effective fishery management; without it, the implementation of management measures may well (and often does) go awry. Second, there is a need for mechanisms for collecting, analyzing and incorporating social scientific information into specific fishery management efforts. To accomplish this, the Department could develop a social science staff (analogous to MRD biologists) to carry out this research and/or develop and institutionalize cooperative arrangements with social scientists in other agencies, academic and research institutions. Then it would be necessary to develop mechanisms to facilitate collaboration among the 'natural' and 'social' scientists, and agency policymakers, to enable them bring this information to bear constructively on resource management design and implementation. The third, and perhaps most critical, component of meeting this challenge is the development and maintenance of open minds to the complexities and opportunities, as well as the potential pitfalls, of bringing social science to bear on fishery management.

References

Acheson, J.M. (1981). "The anthropology of fishing." *Annual review of anthropology* 10, 275-316.

American Fisheries Society Socioeconomic Section (AFS). (1993). Handbook of sources and applications of fisheries social science information. The Section, Bethesda, MD.

Barnes, Y., and Hanan, D. (1996). Status of the Pacific sardine (*Sardinops sagax*) resource and fishery. Marine Resources Division Administrative Report 96-1. State of California, The Resources Agency, Dept. of Fish and Game, Sacramento.

Berntsen, S.E. (1988). "Development of a restructured seafood product from squid (*Loligo opalescens*)." M.S. thesis, Oregon State University.

Brooks, L.A. (1977). "Development and testing of a device to automatically clean California market squid (*Loligo opalescens*)." PhD. thesis, University of California, Davis.

Buck, E.H. (1995). Social aspects of federal fishery management. CRS report 95-553 ENR, Congressional Research Service, Library of Congress, Washington, DC.

Clay, P., and McGoodwin, J.R. (1994). "Utilizing social science information in fishery management." *IIFET proceedings*.

Coastal Pelagic Species Fishery Management Plan (CPS FMP) Development Team and Contributing Experts. (1993a). Draft fishery description for the Pacific coast pelagic species FMP. October 26. PFMC, Portland, OR.

CPS FMP Development Team. (1993b). Draft SEIS, RIR and IRFA: Limited entry in the CPS fishery. October 28. PFMC, Portland, OR.

Dewees, C.M. and Price, R.J. (1983). The California squid fishery. Cooperative Extension Leaflet No. 21330. Cooperative Extension, Division of Agricultural Sciences, University of California. University of California, Berkeley.

Evans, R.G. (1976). "Aspects of the population biology of the California Market Squid (*Loligo opalescens*, Berry)." Ph.D. thesis, Moss Landing Marine Labs.

Fields, W.G. (1965). "The Structure, Development, Food Relations, Reproduction, and Life History of the Squid *Loligo Opalescens* Berry." *Fish Bulletin 131*. California State Resources Agency, Dept. of Fish and Game, Sacramento.

Fiske, S.J. (1990). Resource management as people management: Anthropology and renewable resources. *Renewable resources journal*, winter, 16-20.

Fowle, S. (1993). *Fish for the future: A citizen's guide to federal marine fisheries management.*. Center for Marine Conservation, Washington, DC.

Hanna, S.S. and Smith, C.L. (1994). Attitudes of trawl captains about work, resource use and fishery management. OSU-94-101. USDA Cooperative State Research Service Joint Research Project, University of Rhode Isalnd and Oregon State University.

Hardin, G. (1968). "The tragedy of the commons." *Science*, 162, 1243-1248.

Huppert, D., ed. (1987). Limited access alternatives for the Pacific groundfish fishery. NOAA technical report NMFS 52. NOAA, NMFS, Washington, DC.

Interorganizational Committee on Guidelines and Principles for Social Impact Assessment (ICGP). (1994). Guidelines and principles for social impact assessment. NMFS-F/SPO 16, U.S. Dept. of Commerce, NOAA, NMFS, Washington, DC.

Kato, S., and Hardwick, J.E. (1975). The California squid fishery. In *Expert consultation on fishing for squid and other cephalopods, Tokyo and Hakodate, Japan, 9-13 September 1975*. Fisheries Report 170 Suppl., 107-127. FAO, Rome.

LMR Fisheries Research, Inc. (1995). A review of the market squid industry of California. Prepared for the California Seafood Council. LMR Fisheries Research, San Diego.

Lydon, S. (1985). *Chinese Gold: The Chinese in the Monterey Bay regionn*. Capitola Book Co., Capitola, CA.

Maiolo, J.R., and Johnson, J.C. (1989). Discovering communication networks in marine fisheries: Implications for management. *Marine resource utilization: Proceedings of a conference on social science issues*, MASGP-88-039, J.S.

Thomas, L. Maril, and E.P. Durrenberger, eds., University of South Alabama and the Mississippi-Alabama Sea Grant Consortium, Mobile, 69-80.

Orbach, M.K. (1978). Social and cultural aspects of limited entry. *Limited entry as a fishery management tool,* R.B. Rettig and J.C. Ginter, eds., University of Washington Press, Seattle, 211-229.

Orbach, M.K., and Beckwith, J. (1980). *A report on the adaptation of Indochinese refugees to the fishing industry and community of Monterey Bay.* Center for Coastal and Marine Studies, UCSC, Santa Cruz, CA.

PCFFA. (1996). "Lesson of sardine collapse not forgotten - legislation introduced to protect California squid fishery." *PCFFA Friday* 24, 1-4.

Pollnac, R.B., and Littlefield, S.J. (1983). "Sociocultural aspects of fisheries management." *Ocean development and international law journal* 12, 209-246.

Radovich, J. (1982). "The collapse of the california sardine fishery: What have we learned?" *CalCOFI Report,* XXIII.

Recksiek, C., and Frey, H.W., eds. (1978). Biological, oceanographic, and acoustic aspects of the market squid, *Loligo opalescens* Berry. *Fish bulletin 169.* State of California, Resources Agency, Dept. of Fish and Game, Sacramento.

Redmayne, P. (1996). "The hungry giant: China's quest for seafood goes global." *Seafood leader*, May/June, 38-51.

Vojkovich, M. (1996). materials presented at the ad hoc committee and town hall meetings on the Calfornia market squid fishery, November 1996 through January 1997.

Wormuth, J.H. (1976). The biogeography and numerical taxonomy of the oegopsid squid family Ommastrephidae in the Pacific Ocean. Bulletin of the Scripps Institution of Oceanography, University of California, San Diego, v. 23. University of California Press, Berkeley.

CHAPTER 40

Cooperative Fisheries Research in Developing Management Strategies: Case Studies of Three California Fisheries

John B. Richards[1], John S. Sunada[2], Christopher M. Dewees[3]

Abstract

Fisheries management plans usually are developed under less than optimum conditions, often when an emerging fishery begins to show fluctuations in landings, increases in value, an increase in new participants, and a decrease in catch-per-unit effort. Biological and socioeconomic information on which to base regulatory decisions is, as a rule, inadequate or lacking.

In several recent cases, members of the fishing industry have shown a willingness to assist both resource managers and university researchers in acquiring much needed biological and life history data used in developing new management strategies. They have also been participants in formulating often stringent regulations when necessary.

This cooperative approach, which has been assisted by Sea Grant research and encouraged by Sea Grant Extension advisors, specialists, and Department of Fish and Game resource managers, is described for the ridgeback and spot prawn, angel shark, and sea urchin fisheries. In the case of the sea urchin fishery, industry members have voluntarily assessed themselves to provide the State with supplemental funding for research on resource enhancement and life history studies of the species. Information from this research is being utilized to cooperatively develop a fisheries management plan.

---

[1] Marine Advisor Emeritus, Sea Grant Extension Program, Marine Science Institute, University of California, Santa Barbara, CA 93106

[2] Associate Fisheries Biologist, California Department of Fish and Game, Region 5, Inland Fisheries Branch, 330 Golden Shore, Suite 50, Long Beach, CA 90802

[3] Extension Marine Fisheries Specialist and Coordinator, Sea Grant Extension Program, Dept. of Wildlife, Fish, and Conservation Biology, University of California, Davis, CA 95616-8751

Introduction

A core question being considered by the participants of the *California and the World's Oceans '97 Conference* and by resource managers for the past 25 years, is: "How, with an increasing human population, a resultant increase in the demand for seafood, and a diminishing tax base for State and federal programs, can our fisheries be managed to be both productive and sustainable?" In this paper we present a process which may help the State reach the desirable goal of productive, sustainable fisheries.

This process, "**cooperative fisheries research coupled with a modified system of co-management**," requires a great deal of commitment, hard work, cooperation, understanding, and trust between the resource users (fishermen), the resource managers (State or federal agency biologists), university researchers, and Sea Grant extension advisors and specialists. It requires at least a modest amount of funding through Sea Grant, resource management agencies, or other sources (County funding, State oil royalty revenues, fish and game fine monies) and relies on the financial or "in-kind" support from the fishing industry. A means of maintaining effective communication is an important component of the process. This is often the role of the extension advisor or specialist.

In these cases, the fishing industry has actually initiated or encouraged the management process and has contributed significantly to the development of biological and life history data needed to formulate a management strategy. A majority of the industry participants have been willingly involved in supporting the process even though in each case, they have anticipated often stringent regulations limiting effort or catch.

Development of three California fisheries management strategies are examined for: 1) the prawn trawl fisheries (ridgeback prawn and spot prawn), 2) the gill net angel shark fishery, and 3) the sea urchin fishery.

I. Prawn Trawl Fisheries: South-Central Coast

Spot prawns, ***Pandalus platyceros***, were being trapped in a small local fishery off Monterey over 65 years ago, but it was not until the early 1970's that a small trawl fishery began for these large pandalids in the Santa Barbara Channel. Finfish trawlers had taken spot prawns incidentally in previous decades, but no significant market developed until the 1973-74 season. An initial price of \$0.50/lb encouraged several skippers to explore for productive grounds, and in 1974 landings had jumped to 190,000 pounds. Between 1974 and 1977 the catch began to decline, and the price increased four-fold to \$2.00/lb. The spike and decline in landings drew the interest of California

Department of Fish and Game (CDFG) biologists and in 1976 a small catch-sampling effort began in Santa Barbara (David Ono, CDFG, pers. com.). A fair amount of biological knowledge had been developed on spot prawn populations in the Pacific Northwest, but little was known about this southern population.

Shifting Effort. The decline in spot prawn catch also prompted trawlers to shift their efforts to ridgeback prawns, ***Sicyonia ingentis***, a species fishermen and buyers had experimented with since 1966. These prawns have a desirable sweet flavor, but marketing was difficult due to a very short shelf-life when sold whole and fresh. The "heads-on" product became discolored within a few days after being caught. This discoloration developed even when it was chilled or frozen. A very limited fishery with local markets crawled along between 1974 and 1977, until two Santa Barbara processors renewed marketing efforts, one selling shell-on tails and the other, a peeled and deveined product. The expansion in sales to other areas stimulated the shift in trawl effort from spot prawns to ridgebacks and in 1978/79 landings jumped to 356,000 pounds (Sunada and Richards 1992).

Industry Requests Biological Information. News of the growing prawn fisheries had reached the Pacific Northwest and between 1979 and 1980, seven double-rigged trawlers (capable of towing two nets) joined the local fleet of 20 single-rig trawl vessels. This put additional pressure on both fisheries, and the ridgeback landings began a steep decline.

At this point, in 1979, representatives of the trawl fleet approached the local Sea Grant marine advisor to ask for information on the life histories and biology of these species. They also expressed concern about over-harvesting of the populations due to the increase in effort, the resulting drop in ridgeback catch, and the instability of spot prawn landings. Very little was known about the basic biology and life history of the California *Sicyonia*, and virtually no scientific literature had been published since the species was described in the 1930's.

Commitment from Resource Agency. The marine advisor had been in communication with the local DFG fisheries biologist and with the biologist's assistance, made contact with the new DFG Ocean Unit Manager (OUM) who worked out of Long Beach. The OUM had management responsibility for a wide-range of commercially harvested marine fisheries in a four-county area. The fishermen's concerns were discussed and a meeting was later convened with industry members.

The meeting was the start of a cooperative effort between the Santa Barbara trawl fleet, a seafood processor, and DFG biologists to exchange fisheries and biological information on the prawn fisheries. The marine advisor served as a continuing communications link with the participants between subsequent meetings.

With increasing demand for both species of prawns, the documented need for more biological information, and strong support from the industry, the OUM was able to secure a commitment from the DFG administration to begin a fisheries-independent research/sampling program. Initially, he was able to use the Department's research vessel, the Kelp Bass. Later he obtained funding to contract with a commercial trawl skipper who knew the fishing grounds and fished with state-of-the art equipment.

University Research Link. It was also in 1979 that a professor working on a Sea Grant funded crustacean aquaculture project at the University of California Bodega Marine Laboratory contacted the marine advisor with a request for assistance in finding a source of live ridgeback prawns for his graduate student who was beginning a study of the molt cycle and reproduction in *Sicyonia.* The marine advisor introduced the student to several trawl skippers and two processors in Santa Barbara, helped set up an arrangement allowing the student to accompany the trawlers during their regular fishing operations, and suggested adding a fisheries component to her aquaculture research project.

Industry Responds In-kind. The fishermen were willing to take the student at no cost, but eventually an arrangement was made with one of the skippers to make a special short tow at a fixed location one day each month for which he was reimbursed for fuel. Samples were also taken at other locations during regular trawl fishing operations.

The student traveled to Santa Barbara and went to sea consistently, once each month, for two years to collect the prawns and to compile field data on growth, recruitment, molt and reproduction cycles, and habitat. We estimate that the savings to the Sea Grant aquaculture project through the fishermen's donation of vessel time was somewhere between $11,000 and $14,000 (the cost of contracting a university or private vessel for the sample collections). The value of the fishermen's experience and knowledge of the species would be difficult to estimate, but certainly a benefit to the project. The three skippers most involved with the project had over 80 years of combined fishing experience to contribute. A Santa Barbara processor also contributed significantly to the project, both with information and material support.

Essential Communication. A critical element in the success of the cooperative arrangement was the feedback on the fisheries aspects of *Sicyonia* molt and reproduction that the student provided directly to trawl skippers with whom she worked and through written updates on her findings given to the marine advisor to distribute to the rest of the fleet, the processors, and DFG biologists.

Support for a Management Plan. By late 1981, the Ocean Unit Manager had enough information from his own investigations on both spot prawns and ridgebacks, from the preliminary Sea Grant data, and discussions and work with the fishermen, to develop management regulations for both species. The regulations proposed trawl fishing closures during the molt and egg bearing periods.

Subsequent meetings with the trawl skippers, processors, and DFG biologists were productive. Because nearly everyone had the same information, the industry fully supported the Unit Manager's proposal for seasonal closures. Within one year, the proposed regulations had moved through the legal system and became implemented in 1983 for the ridgeback prawns and in 1984 for spot prawn.

These regulations have helped stabilize the trawl catch of both prawn species. However, a new deep water trap fishery for spot prawns was beginning to develop in southern California and the seasonal trawl closure enticed more trappers to enter the fishery to fill the demand for prawns during the closed season. This served to point out the dynamic nature of these fisheries and the continuing need for developing communication links with new entrants. The trap fishery initially developed outside the Santa Barbara Channel fishing community.

Diminishing Resources for Research and Monitoring. The DFG Unit Manager was able to continue to develop life history data and monitor the health of the populations for a few years until State funding cuts limited the Department's ability to conduct fisheries research and monitoring. Only recently has an environmental document and management plan been completed for the spot prawn which considers both trawling and trapping methods (Mary Larson, CDFG, pers. com.). The ridgeback fishery continues to operate under the 1982 regulations as it remains a trawl-only fishery.

Advances in Applied and Basic Science. The U.C. graduate student completed her doctoral thesis in 1983 and published several papers contributing to the growing field of crustacean aquaculture as well as much needed life history information on *Sicyonia* to the field of fishery biology (Anderson, Botsford, and Clark, 1985).

II. <u>Angel Shark Gill Net Fishery: Santa Barbara Channel</u>

Angel Sharks, ***Squatina californica***, were taken and discarded as a nuisance species by halibut gill net fishermen and trawlers for several decades until seafood consumers began to accept sharks as high quality food fish.

Marketing of this species began in California in 1976 through the innovative work of a Santa Barbara-based processor with the help and encouragement of local fishermen. The fishery was conducted principally around the northern Santa Barbara Channel Islands and along the Santa Barbara-Ventura coast (Richards 1992).

Emerging Fisheries to meet Consumer Demand. The popularity of angel shark with health-conscious consumers rose in the late 1970's, closely following the rising demand for thresher shark, a larger, but seasonal, pelagic species. Between 1977 and 1981, recorded landings of *Squatina* rose from 328 pounds to 258,000 pounds, and by 1984 the catch reached 610,000 pounds.

Remembering Historical Declines. As demand and ex-vessel price increased, several fishermen began a directed fishery for the species by increasing the mesh size of their nets which yielded better catches than the traditional halibut gill net. At this time, they also began to express a concern about the lack of population and life history information on *S. californica*. Senior fishermen, who in the 1940's, had experienced the precipitous decline of soupfin shark (taken for the vitamin A contained in their livers), were especially wary.

The marine advisor had been working with the seafood processors and several fishermen on utilization of seafood waste and became involved with documenting the development of the angel shark fishery and market. He was also in contact with a university researcher at Moss Landing Marine Laboratory (MLML) who had an interest in elasmobranch biology and had submitted a proposal to Sea Grant for funding to study shark life histories, especially age, growth, and reproduction of various California species.

Blueprint for Cooperation. In 1979, a cooperative fisheries research effort was initiated involving Santa Barbara Channel angel shark fishermen, processors, DFG biologists, university researchers, and Sea Grant extension personnel. It was an ambitious undertaking to obtain information on angel shark distributions, migrations, growth rates, and eventually reproductive rates.

Directed Field Research: Graduate Trainees Gain Experience in Two Worlds. The MLML researcher was awarded Sea Grant funding and began work with the marine advisor, a processor, and the Santa Barbara fishermen, who in turn, provided vessel time, expertise, and specimens to the researcher and his graduate students over a four year period. One student directed her masters thesis research toward questions of age, growth, and reproduction in angel shark (Natanson and Cailliet 1986).

During the four year project, six graduate students from MLML, participated directly in the field research onboard the commercial fishing vessels, gaining valuable knowledge of the harvesting process and an understanding and appreciation of the fishermen's life at sea. Three of the students went on to obtain doctorate degrees, two are working in state and federal fisheries management agencies and one is a university professor in biological sciences. The others are all working in related biological fields. (Gregor Cailliet, MLML, pers. com.).

The student who had primary responsibility on the angel shark project is now a principal investigator studying sharks with the National Marine Fisheries Service on the east coast. She recently told one of the authors that her field experiences have been of significant value in her present job, especially in working with the industry and understanding the fishermen's point-of-view in regard to research and management issues.

Long-term, Low-cost Population Monitoring. A cooperative tag-recapture project was also begun with the help of several commercial fishermen, a processor, and the participation of the DFG, the marine advisor, and university researchers at MLML and U.C. Santa Barbara.. This followed and expanded upon an earlier study of *S. californica* movements, distribution, feeding, and growth at Catalina Island (Cailliet, et.al. 1992).

Integrating Local Fisheries Knowledge into a Management Plan. As in the case of the prawn fisheries, the commercial fishermen gained much information and insight by interacting with the researcher and his students. Being an integral part of the project, they were also communicating more frequently (with the advantage of having up-to-date life history information) with the DFG biologists. They were also able to contribute valuable suggestions on possible management strategies as more became known about the species.

In 1985-86, due to a rapid increase in landings, a major effort was launched by DFG biologists, with the support of the industry, to develop management regulations for *S. californica*. A management plan was written using much of the cooperative industry/Sea Grant/DFG research data, and in 1987, a size limit regulation was proposed by DFG. It became law in 1989.

It is not known if the regulation was effective in stabilizing the production, as the inshore commercial fishery for angel sharks essentially ended in 1994 with passage and implementation of Proposition 132, a citizens initiative which banned the use of set gill nets in State waters. The cooperative tag-recapture project and biological investigations also ended.

Biological Knowledge Utilized. Several scientific papers have been published by the researcher, his students, and DFG biologists adding to the knowledge-base of shark biology and fisheries science It is doubtful that this work could have been undertaken without the cooperation and support of the fishing industry as the cost of the field investigations would have been prohibitive.

Most participants in this effort agree that the biological information should have been integrated into a management plan prior to the major increase in catch in 1985-86, but these changes are hard to predict. The fishery was regarded by some in State and federal government as a minor localized phenomena that did not justify additional attention until landings exceeded those of thresher shark for two years. Seafood wholesalers and retailers now obtain most of their angel shark fillets from Mexico. (Michael Wagner, Seafood Buyer and Retailer, pers. com.).

III. Sea Urchin Fishery

Sea urchins, the red, ***Strongylocentrotus franciscanus***, and purple, ***S. purpuratus***, have been utilized as a source of human food by coastal Native Americans for thousands of years in California, but in the past 150 years a relatively small percentage of the State's population harvested urchins. They were, in fact, considered pests due to their destructive grazing on kelp (Parker and Kalvass 1992). In 1971, a small commercial fishery began with another cooperative effort involving a fisheries biologist with the National Marine Fisheries Service (NMFS), processors, and fishermen in Avila and San Diego.

The NMFS biologist had worked for several years on various fisheries development projects and had built an excellent rapport with the industry (from the authors' perspective, he made great progress in this area by combining good fisheries science with the principles of extension education). He also had several associates and colleagues in Japan where urchin gonad or "uni" was a highly regarded delicacy. By linking Japanese processing technologists and buyers with the California fishermen, a market and fishery quickly developed for the large red sea urchin (Kato and Schroeter, 1985). Many abalone divers found a supplemental income by diving for urchins and a significant number began to focus most of their efforts on this new fishery (Bruce Steele, Commercial Urchin Diver, pers. com.).

Rapid Expansion Points to Need for Research. By the mid-1970's, urchin landings had increased to nearly 10 million pounds, with most of the harvest from southern California waters, mainly around the Santa Barbara Channel Islands. The rapid expansion of the fishery and the lack of fishery related life history and population data stimulated the initiation of several Sea Grant funded research projects in the 1970's and 1980's (Stephen Schroeter, Ecometrics, pers. com.). Many of these take into consideration broader ecological aspects of the urchins' role in the marine environment (Tegner and Dayton, 1981).

A Move to the North. The fishery continued to expand in southern California, reaching a peak of nearly 25 million pounds in 1981, and eventually divers and processors began to further develop the previously limited northern California fishery. Landings in the north grew to 30.4 million pounds in 1988, overtaking the production of the southern California fishery. Combined, both the northern and southern California fisheries became the most valuable commercial living marine resource in the state, exceeding landings of 50 million pounds worth over $25 million, in 1988.

Management Void a Challenge to Industry. In a comprehensive review of red sea urchin biology and the fishery in California, Kato and Schroeter (1985) warned of an impending decline in the fishery, which had already begun in southern California after the peak 1981 season. In the absence of any State mandated regulations or management plan, they urged the members of the industry to impose self-regulatory measures to ensure the future of the fishery. The industry had already become active in a move to institute a limited entry program However, even though many of the leaders helped draft legislation in the early 1980's, a few members opposed it and the move failed. Limited entry was later instituted 1n 1989, but the delay had stimulated a rush of new divers and their numbers tripled from 300 to 900.

A Catalyst to Stimulate Research. The industry pulled together in 1986 and '87 forming a statewide organization that began to work with the DFG and legislature (Rep. Dan Hauser) to develop a management structure for the fishery. This included a limited entry/moratorium on permits for divers and a self-imposed landing tax which generates $150 to $250 thousand per year to do fishery monitoring , research and projects related to resource enhancement.

Approaching Co-management. The legislation also created the CDFG Director's Sea Urchin Advisory Committee (DSUAC) to advise the Department on spending of the landing tax and advise on management issues related to the fishery. The DSUAC consists of divers, processors, and a University of California representative. Funds have been used to examine the feasibility of enhancement, test out-plants of juvenile urchins, evaluate culture techniques and education, studies of recruitment/settlement, monitor landings and provide peer review of management options. Currently DSUAC is attempting to refine the process through which it defines and funds research activities.

Regulatory changes have been negotiated between DSUAC and CDFG. These include minimum size limits, reduction in number of fishing days, and fine tuning of the limited entry system. With the fishery now extending over a wide geographic range involving different ecosystems and different fishing communities, this process of defining appropriate management strategies has become much more complex. A DFG call for more restrictive management measures, especially in northern California, met with strong industry resistance.

A peer review was conducted of the proposed regulatory changes and none have currently been adopted. The industry has hired a scientific advisor to examine resource related issues that continue to surface.

University Research and Extension Encourages Cooperative Effort. Sea Grant researchers, extension advisors, and specialists have been involved in the urchin fishery development and management process since the early 1970's. University researchers and agency biologists have worked together to provide much of the data used in developing management strategies for the fishery with funding from both Sea Grant and the industry. Marine advisors have worked with the industry in both southern and northern California and have cooperated on many research projects. The UC-SGEP Marine Fisheries Specialist sits on DSUAC and entire staff of SGEP have helped in communications between the industry, the agencies, and the public through the extension of research results through workshops and other media.

The industry continues to pay a significant portion of the costs of research and monitoring of landings. They are also involved jointly with DFG in examining management issues. Even though it is at times a difficult exercise in the democratic process, it is definitely a good working model of the process of **cooperative fisheries research coupled with a modified system of co-management**.

Summary

The concept of cooperative research and co-management of fisheries is not new, either in the U.S. or the rest of the world. However, it is only recently that a multidisciplinary, international dialogue and exchange of information on the theory, practice, pros, and cons of this approach has been initiated. In a review of a number of presentations on fisheries co-management scenarios in the U.S. and Canada, a central thread was found to run through each of the examples; "---- fishery managers can and should involve local communities and groups of fishermen in the management process in a meaningful way" (Pinkerton1989).

In California and many other coastal states, Sea Grant extension and research is often available to help foster this process. In the three cases described in this paper, it is evident that the fishing industry in each case saw the need to develop management strategies to sustain the fisheries. Good communication was essential to the process. Involving the resource users in the research in a meaningful way is, thus, another part of the equation.

Pinkerton (1989) outlines a set of "preconditions" which are favorable in setting up a co-management or modified co-management process. We have selected several of these criteria that relate to the California experience and have included them with our own observations that relate to successful cooperative fisheries research projects. These are as follows:

Cooperative fisheries research and a modified form of co-management works best when:

1) A fishery is **local** or **regional** and has well defined boundaries. This facilitates easy communications between the stakeholders (resource users, managers, researchers).
2) When there are **no allocation** problems (i.e. few interactions with other users: sport, other commercial gear types, birds/mammals, etc.).
3) When **researchers** and **resource managers** are **willing to work with industry** to understand their ways of thinking and working --- to appreciate the hard earned knowledge of the fishermen and processors and able to **cooperate and share information in a common cause (sustainable fisheries).**
4) When **research** or a **call for management** is **initiated by industry.** Pinkerton notes that ço-management is most likely to develop out of a real or imagined crisis in stock depletion, or a problem of comparable magnitude (e.g. knowledge of historical declines in shark catches).
5) When fishermen show a **willingness to contribute financially** (or recruit other sources of support) to the rehabilitation of the resource, and/or contribute to other management functions. These conditions existed in all three cases.
6) Where **external support can be recruited** (university, non-government scientists, credible organizations). In the cases discussed above, Sea Grant served as the external support mechanism providing **extension, research,** and **"trainee" funding** to support graduate student participation in the field and laboratory research.

Conclusion

Experience shows, that to be effective the "process" needs to begin earlier. Strengthening the mechanisms which lead to communication and cooperation with the industry is essential. The process is viable when funding, resources, and information are available and shared. Cooperative fisheries research and a modified system of co-management has been shown to encourage stewardship of living marine resources.

References

Anderson, S.L., L.W. Botsford, and W.H. Clark, Jr. 1985 Size distributions and sex ratios of ridgeback prawns (*Sicyonia ingentis*) in the Santa Barbara Channel (1979-1981). Calif. Coop. Oceanic Fish. Invest. Rep. 26: 169-174.

Cailliet, G.M., H.F. Mollet, G. Pittenger, D. Bedford and L.J. Natanson. 1992. Growth and demography of the Pacific angel shark (*Squatina californica*) based upon tag returns off California. Aust. J. Mar. Freshwater Res. 43: 1313-1330.

Kato, S. and S.C. Schroeter. 1985. Biology of the red sea urchin, *Strongylocentrotus franciscanus*, and its fishery in California. Mar. Fish. Rev. 47(3):1-20.

Natanson, L.J. and G.M. Cailliet. 1986. Reproduction and development of the Pacific angel shark, *Squatina californica,* off Santa Barbara, California. Copeia 1986(4): 987-994.

Parker, D.O. and P. Kalvass. 1992. Other invertebrates: sea urchins. *In*: California's living marine resources and their utilization. W.S. Leet, C.M. Dewees, and C.W. Haugen (eds.). UCSGEP-92-12. Sea Grant Extension Program. University of California, Davis, CA.

Pinkerton, E. 1989. Co-operative management of local fisheries: new directions for improved management and community development. University of British Columbia Press, Vancouver, B.C. 320 pp.

Richards, J.B. 1992. Fish resources: Pacific angel shark. *In*: California's living marine resources and their utilization. W.S. Leet, C.M. Dewees, and C.W. Haugen (eds.). UCSGEP-92-12. Sea Grant Extension Program. University of California, Davis, CA.

Sunada, J.S. and J.B. Richards. 1992. Crustacean resources: ridgeback prawn. *In*: California's living marine resources and their utilization. W.S. Leet, C.M. Dewees, and C.W. Haugen (eds.). UCSGEP-92-12. Sea Grant Extension Program. University of California, Davis, CA.

Tegner, M.J. 1981. Population structure, recruitment and mortality of two sea urchins (*Strongylocentrotus franciscanus* and *S. purpuratus*) in a kelp forest. Mar. Ecol. Prog. Ser. 5: 255-268.

CHAPTER 41

Pacific Rim Experiences with Individual Transferable Quotas: Implications for Fisheries Policy in California

Christopher M. Dewees[1]

Abstract

The concept of individual transferable quotas (ITQs) in fisheries grew out of papers by natural resource economists F. Scott Gordon, Francis Christy, Peter Pearse and others between 1954 and 1979. Individual transferable quotas generally involve the assignment of a portion of the overall total allowable catch to individual fishers, companies, or communities. These quasi-property rights usually are transferable. In theory, this management technique provides incentives that will lead to reduction in the overcapitalization common in fisheries and lead to a more orderly, economically productive fishery.

Individual quota systems have been implemented around the Pacific Rim in New Zealand, Australia, Canada, and Alaska. In this paper I will report on my field studies on individual quota programs in New Zealand and British Columbia. I examined the effects of these quota management systems on individual fishers, processors, tribes, and agency personnel.

In general, the quota systems have ended the "race for fish" that has long plagued fisheries. Harvesters who own quota shares have usually received higher prices for their catch, often due to their motivation to develop innovative marketing and added value products. Crew employment has changed. In New Zealand, ITQs have accelerated fishery claims of Maori and they now control about 40 percent of all fishery quotas. Primary issues that have surfaced in the first decade of ITQ experience include methods for initial quota allocation, transferability, quota aggregation limits, research and management cost recovery, and resource conservation.

---

[1] Marine Fisheries Specialist and Sea Grant Extension Program Coordinator, Department of Wildlife, Fish and Conservation Biology, University of California, Davis, CA 95616-8751. email: cmdewees@ucdavis.edu

Individual transferable quota systems are just one in a suite of management tools available to Californians and should be examined relative to, and in combination with, other options. The current level of understanding of ITQ systems appears to be low among California fishery managers, fishing industry and other resource stakeholders.

Introduction

California is currently struggling to develop policies to restrict access to fisheries in an attempt to slow the "race for fish." H.S. Gordon pointed out the problems caused by the common property nature of fishery resources in his landmark 1954 paper. Lack of property rights has motivated fishers to race each other for the fish. This race often leads to overcapitalization and overexploitation of fishery resources. Others extended Gordon's concepts and have advocated the use of individual transferable quotas (ITQs) to promote economically rational fisheries exploitation (Christy, 1973; Moloney and Pearse, 1979). Interest in ITQs intensified during the past two decades as fishing power of fleets grew significantly and limited entry licensing often failed to slow the race for fish (Wilen, 1988).

Individual transferable quotas involve the allocation of a share of the total allowable catch (TAC) to individual operators. ITQs are a quasi-property right that can be a fixed quantity or a percentage of a variable TAC and can be allocated for specified time periods or allocated in perpetuity. For background discussion, see Christy (1973), Maloney and Pearse (1979), Grafton (1996). The details of ITQ systems vary widely depending on the fishery, economic goals, political realities, and industry structure and history. Major components of ITQs include initial quota allocation methods, degree of transferability, cost recovery or resource rentals, aggregation limits, and setting and adjusting total allowable catches. System details are important because they affect outcomes.

Implementation of ITQ systems in marine fisheries has accelerated since 1986. Experience to date suggests that individual transferable quotas do perform much as economic theory suggests; fishery participants are seeking to maximize profits within the constraint of an ITQ (Boyd and Dewees, 1992; Casey, et al., 1995; Arnason, 1996). The role of individual quotas in sustaining marine fisheries continues to be hotly debated by government and industry. Initial allocation methods, aggregation of quota shares, conservation of fish stocks, commercial/recreational allocation of TACs, and cost recovery by government are major policy issues.

This paper summarizes empirical work I have been doing since 1986 in New Zealand's individual transferable quota (ITQ) and British Columbia's halibut individual vessel quota (IVQ) systems. The details of these two systems differ considerably, but examining them together provides useful information about this type of fisheries management. The primary objective of each study was to document what happens when individual quota systems are implemented.

Methods

The New Zealand field work took place over two nine-month periods in 1986-87 and 1995. In 1987 I completed in depth personal interviews with 62 commercial fishermen and fishing company managers from an unstratified random sample chosen from the Ministry of Agriculture and Fisheries list of Auckland Region provisional quota owners. This included both people who stayed in the fishery and those who sold out. During 1995 I was able to re-interview 52 of the original 62 case study participants to obtain longitudinal data. For details on the 1987 interviews, see Dewees (1989) and Boyd and Dewees (1992). To gain a broader perspective on the effects of New Zealand's ITQ system in 1995, I conducted an additional 40 interviews throughout New Zealand with fishing company managers, individual fishers, Maori-owned fishing business staff, Ministry of Fisheries staff, and leaders from environmental and recreational organizations. In addition, I was able to gather some Ministry of Fisheries data to look at national changes over time (Annala 1996, New Zealand Ministry of Fisheries 1995).

The field work in British Columbia, Canada, was conducted over a six-month period during 1993-94. We completed in-depth interviews with 12 of the 14 primary halibut processors, and we surveyed all 435 licensed halibut fishers by mail (31% return) rate. Copies of all survey instruments are available from the author.

Results and Discussion

New Zealand

New Zealand implemented ITQs for most of its commercial fisheries in 1986. This followed (1) a 1982 moratorium on new fishery entrants in 1982, (2) implementation of enterprise allocations in deep water fish such as orange roughy in 1983, and (3) a general perception in the management agency and industry of a crisis due to overfishing and overcapitalization. In addition, New Zealand was beginning a major economic change from government-subsidized or -owned businesses toward privatization, removal of subsidies, reliance on market forces and recovery of costs (Boyd and Dewees 1992). The coinciding perceptions of a fishing crisis and a government economic policy favoring privatization facilitated a rapid and major transition to ITQs.

New Zealand implemented a system nearly identical to that described by Moloney and Pearse (1979) in their pioneering argument for the benefits of ITQs. See Dewees (1989) and Boyd and Dewees (1992) for a detailed description of the system and its early modifications. Basics of the New Zealand system include:

(1) Allocations made to fishing vessel owners based on 1982-84 catch histories.

(2) Ten fishery management areas (FMAs) set up and TACs established for species within these FMAs.

(3) Government used buyback and across-the-board cuts to reduce the historical catches to meet TAC levels at a cost of NZ$45 million (in 1986 NZ$1 ≈ US$.55).

(4) The government would enter the market to buy back or auction off quota to adjust TACs. By April 1990 government switched to a proportional ITQ system to remove the need for government to enter the market.

(5) Aggregation maximums were set at 20% of TAC (in each FMA for each species) for inshore species and 35% for deep water fisheries. When rock lobster came into the ITQ system in 1990, a 10% aggregation limit was set for that species.

(6) A low initial resource rental was charged ($3/ton). This rental was later raised on a species by species basis. Resource rentals were then frozen after the government switched to proportional quotas in 1990 and the resource rental funds were used to compensate industry for cuts in total allowable commercial catches (TACC) during a transitional period ending in 1994. In 1994 resource rentals were terminated and government started collecting fees to recover the costs of research, management and conservation (including non-ITQ species).

Data from the 1987 and 1995 interviews are presented in Table 1. See Dewees (1989) and Boyd and Dewees (1992) for details of the 1987 interviews. Numerically this sample was dominated by small-scale individual owner-operators. Five of the six vertically-integrated companies in the 1995 interviews had very positive responses to the ITQ system. These companies stated that their firms' relatively secure fish supplies resulting from the ITQ system enabled them to do longterm planning and value-added product development. The small-scale operators interviewed had mixed feelings. Many of these operators depend on the extremely valuable snapper fishery in the Auckland region which has undergone significant declines in TAC since 1986. Currently the Minister of Fisheries is attempting to reallocate 40% of the commercial catch to the recreational sector in the Auckland region. Participants in the snapper fishery perceived of holding snapper quota as very risky because of likely future cuts and reallocation of TACs. Cuts could significantly reduce the value of individual quota holdings.

Table 1. Responses of Auckland, New Zealand region fishers and fishing company managers to questions about individual transferable quotas, 1987 and 1995.

	1987	1995
Number who own quota (N = 62)	49	34
Positive ITQ effects (% of interviewees mentioning)		
Conserves fish stocks	53	50
Provides asset/security	42	27
Reduces effort	23	13
Improved quality	0	19
ITQ problems (% of interviewees mentioning)		
Highgrading	66	25
Enforcement	40	21
Company control	26	46
Resource allocation conflicts	6	33
Complexity	0	35
% who agree with statement		
NZ fishing industry better off	58	76
ITQ compatible with beliefs	56	73
ITQs conserve stocks	56	68
Fishing safer with ITQs	39	50
My economic situation improved	48	45
More secure about retirement	73	64
Difficult for young to enter	95	98

Data on national changes since 1986-87 are found in Table 2. Several factors have influenced these changes. There has been a continued trend toward New Zealandization of the fishery industry with more New Zealand harvesting and processing. Processing employment has grown as vertically-integrated companies and others continue to expand processing of value-added products. Seven more species groups have been added to the quota management system since 1986. In addition, non-ITQ fisheries such as tuna and shrimp have grown dramatically. According to a recent report, New Zealand catches have grown 40% since 1986 (New Zealand Ministry of Fisheries 1995).

Vertically-integrated companies landed two-thirds of the fish before ITQs and by 1989 the top ten companies held 82% of the quota (Bevin et al. 1990). In recent years, several large companies have left the fishing industry and their holdings flowed to other companies including those involved with settlement of Maori claims. This reduced the holdings of the ten largest vertically-integrated companies to 68%.

Table 2. Changes in New Zealand fishing industry 1986 to 1995[1]

Characteristic	1986-87	1994-95
Fishing industry employment	7900	9838
Number of quota owners nationally	1356	1733
Number of quota owners--Auckland region	902	855
Holdings of top three quota owners	28%	44%
Holdings of top ten quota owners	67%	68%
Number of fishing vessels	2331	2768
Seafood exports (1986 dollars)	$657 million	$896 million
Resource rentals/cost recovery	$25 million	$37 million
Number of species groups under ITQs	25	32

[1]Compiled from Annala 1996 and New Zealand Ministry of Fisheries 1995

Revenue generated for government has been quite variable, first, due to disputes about what is reasonable profit and, second, compromises due to the settlement of Maori claims. Resource rentals were replaced by cost recovery in 1994 which generated about 80% of the cost of research, management and enforcement in 1994-95 (Annala 1996). There has been considerable change in the New Zealand fisheries since 1986. Several changes and issues are summarized below.

Maori fishery claims related to the Treaty of Waitangi have led to settlements that have increased Maori quota assets from near zero to approximately 40% of New Zealand's fishery quota. Most of these assets are currently held in trust by the Treaty of Waitangi Fisheries Commission (TOWFC). This includes quota which is leased out by TOWFC to Maori *iwi* (tribes) at a discount from market price and quota owned by companies such as Sealord Products (50% Maori owned). The tribes are using a variety of strategies to utilize this quota including: (1) leasing to established fishing businesses, (2) leasing to Maori fishers, and (3) forming Maori-owned fishing businesses to manage quota assets. The primary issue now is how, and if, the Maori quota will be permanently allocated to the tribes.

The involvement and influence of recreational groups and environmental organizations on New Zealand fishery management has increased. The recreational fishery and support industries such as charterboats have grown as the New Zealand economy has improved since 1986. Recreational interests have formed organizations and they now participate in fishery policy negotiations. Environmental organizations have also become very active and have filed several suits against the Minister of Fisheries. Participation by non-commercial stakeholders now significantly influences the setting of total allowable catches (TAC) and total allowable commercial catches (TACC).

Major comprehensive fisheries legislation was passed by Parliament in 1996 and it includes several significant changes. See Dewees (in press) for a summary of these changes.

British Columbia halibut fishery

British Columbia (Canada) implemented individual vessel quotas in their Pacific halibut fishery (*Hippoglossus stenolepis*) in 1991. For a complete description of that fishery and the results of our study, see Bell 1981, Turris 1994, Casey et al. 1995.

This longline fishery had evolved into a derby fishery despite implementation of limited entry in 1979 (435 vessels). The season declined from 60 days in 1982 to 6 days in 1990, even with a larger harvest quota. After considerable government-industry consultation, an individual quota system was implemented.

(1) An 8-month season was established.

(2) Each licensed vessel received a share of the TAC based (70%) on catch history (1986-89) and the rest on vessel characteristics.

(3) A self funded monitoring and enforcement program was established.

(4) During the first two years the shares were non-transferable. In 1993 limited transferability was allowed. Each vessel's initial allocation was divided into two equal shares and quota owners could lease out or lease in up to two shares (maximum aggregation of four shares). The limits on transferability grew out of concerns about aggregation of quota shares. This gradual transition to transferability also allowed fishery participants time to adjust to the new system and think through their participation decisions.

Major findings of our interviews and mail survey field work are summarized below and in Table 3.

Table 3. Halibut fishers responses about IVQ attributes

Statement	% Agree or Strongly Agree[a]
IVQs conserve halibut resources	80
IVQs make fishing safer	72
I am better off under IVQs	80
I feel more secure about my retirement	84
Difficult for young to enter	77

[a]All values are different from 50% (binomial chi square test ($P \leq .05$)

(1) The low volume of halibut landed throughout the year allowed processors to sell 94% of the product fresh compared to 42% before IVQs.

(2) There was evidence of a 55% ex-vessel price gain associated with IVQs compared to the adjacent Alaska open-access derby fishing.

(3) Interviewees felt, and Department of Fisheries and Oceans data confirms, that there was a transfer of market share from large traditional processing firms to smaller firms specializing in halibut. The number of firms processing halibut grew from 57 to 69 and the locations of landings became less concentrated. These changes may be due to the reduced need for large freezing capacity to handle the glut of landings in the previous derby fishery.

(4) Seventy percent of the quota owners fished their entire allocation. Seventeen percent leased additional shares and 13% leased out one or both of their shares.

(5) No fishers increased their crew number and 44% reduced crew (usually by one person). Among those who reduced crew size, 59% increased remuneration to their crew. Those reporting maintaining the same crew size either kept crew remuneration the same (76%) or decreased it. We estimated that there was a 32% reduction in the number of crew in the fishery, but those crew still in the fishery seem to be better off.

(6) On average, the initial quota allocation of those leasing quota in 1993 was significantly higher than those leasing out quota to other halibut vessel owners.

The unique characteristic of the British Columbia halibut IVQ has been the gradual relaxation of transferability. There was great concern about aggregation, so the industry and government restricted transferability. Future issues in the fishery will revolve around the market effects of the 1995 implementation of individual quotas in the much larger Alaska halibut fishery and future relaxation of the transferability restriction in the B.C. fishery.

Policy Implications for Fisheries Management in California

There appears to be three generalizable outcomes across ITQ systems.

(1) The race for fish ends because each quota owner is assured access to a portion of the total allowable catch. This race (in terms of effort and technology) for fish is a characteristic of many fully or over-exploited fisheries. In California, the race for fish is particularly intense in the

Dungeness crab, herring, salmon and sablefish fisheries. ITQs would be likely to slow these derbies.

(2) Industry participants will seek to maximize the return from their quota holdings. This can take the form of value-added products, innovative marketing, quality improvements, market timing, and in some cases highgrading (discard of lower valued catch). California fisheries which have variation in market prices through the year (e.g., sea urchins), size (e.g., sablefish), or product form (e.g., live rockfish) are likely to see increased emphasis on price maximization under individual quotas.

(3) Pre-ITQ industry structure and the design details of each ITQ system strongly influence outcomes. These details include initial allocation methods, aggregation limits, transferability and cost recovery. Before 1986, vertically-integrated companies in New Zealand landed 67% of the fish. With relatively high aggregation limits (10 to 35%), full transferability/leasing, government economic policy encouraging export and market forces, vertically-integrated companies continue to land the majority of fish. In contrast, the north Pacific halibut fishery was made up primarily of small-scale owner-operators. The Canadian and Alaskan halibut IQ programs have low aggregation limits and restrictions on transferability and leasing. There has been some consolidation, but the fishing industry still consists primarily of small-scale owner-operators. California's fishing industry consists primarily of individual owner-operators with little of the company vertical integration found in New Zealand. It is most likely that any ITQs implemented in California would resemble the Canadian or Alaska model.

In terms of sustainability, ITQs appear to provide some stability on the important economic regulation side of fisheries management. However, they must be linked to good stock assessment work and well defined objective methodologies for setting total allowable catches (Sissenwine and Mace 1992). In New Zealand, the Fisheries Minister (a member of Parliament) sets the TACs using advice from the Ministry of Fisheries, the fishing industry and other stakeholders. In the long run, it might be better to base the TAC decisions on management plans rather than on the judgement of an elected official. Special attention will also have to be paid to management of bycatch in California's multispecies fisheries with trawl and longline gear.

Allocation of the TAC between user groups needs to be addressed before ITQ implementation, and there should be flexibility if unexpected changes occur. In the Auckland area snapper fishery, both the recreational landings and political influence of anglers have increased significantly since 1986. In 1995, the Minister reduced the total allowable commercial catch (TACC) by approximately

40%. The owners of snapper quota felt that the decision to reallocate TAC to the recreational industry was political. The issue is now in court. Rockfish, halibut, salmon, lobster, abalone and white seabass are examples of California fisheries that would need to address TAC allocation issues under ITQs.

The other allocation issue that New Zealand had to face after ITQ implementation was the allocation of quota to Maori based on the Treaty of Waitangi. This has been a long and difficult process. Currently most of the Maori fishery assets are held by the Treaty of Waitangi Fisheries Commission and leased out to tribes in a process somewhat analogous to Alaska's community development quotas (Ginter, 1995).

Probably the most important part of ITQ implementation involves the stakeholders deciding what they want the fishery to look like. Methods of initial quota allocation, aggregation limits and transferability are the key issues that will affect outcomes. For example, if the goal is to sustain communities, some quota could be allocated to the community rather than individuals. If the goal includes local management and enhancement of sedentary shellfish resources, smaller management areas with self regulated access to the shellfish beds (with recreational, subsistence participation) may be appropriate. If a goal was to encourage development of an offshore resource, aggregation limits could be set high enough to encourage long-term investment in product development and fishing technology.

Cost recovery, rather than resource rentals, is a big part of both the New Zealand and British Columbia cases. In both cases the industry has become intensely interested in how their funds are used for management and research. The quota owners appear willing to pay reasonable costs. The primary issue in this situation appears to be who controls how the funds are spent and how the costs are calculated. In California, some industries (e.g., sea urchin) help pay management and research costs. It seems likely that willingness to pay a larger portion of costs will be proportional to the security and exclusivity of fishers' resource harvest rights.

These two cases of ITQ implementation are very different, but the majority of the participants and managers in both cases think their fisheries are better off. The decade of New Zealand experience has shown that ITQs change fishing industry behavior significantly and create some economic stability. As world experience with ITQs grows, managers, industry and communities are learning how to address issues that arise with this very different management technique. ITQs, linked with good population assessment and TAC setting, can be a useful sustainable fisheries management tool. Recent papers on the Nova Scotia sea scallop fishery (Brander and Burke 1995) and Pacific halibut (Casey et al. 1995), species each subject to both open access and individual quota management, provide a snapshot of the different outcomes under different management regimes.

In California, the level of understanding of ITQs among resource managers, industry and other stakeholders is quite low. Current management discussions are revolving around limited access programs, sharing of research and management costs, and population assessment. It seems most likely that serious consideration of ITQs as a management option will not come until these other policy issues are resolved in California. Meanwhile, we should learn from the ongoing ITQ experience in the United States, Canada, Iceland, Australia and New Zealand.

References

Annala, J. H. 1996. New Zealand's ITQ system: have the first eight years been a success or failure. Reviews in Fish Biology and Fisheries 6:43-62.

Arnason, R. 1996. On the ITQ fisheries management system in Iceland. Reviews in Fish Biology and Fisheries 6:63-90.

Bell, F. H. 1981. The Pacific halibut: the resource and the fishery. Anchorage Alaska Northwest Publishing Company.

Bevin, G., P. Maloney, P. Roberts, and N. Redzwan. 1990. Economic Review of the New Zealand Fishing Industry 1988-89. Wellington, New Zealand: New Zealand Fishing Industry Board.

Boyd, R. O., and C. M. Dewees. 1992. Putting theory into practice: individual transferable quotas in New Zealand's Fisheries. Society and Natural Resources 5:179-198.

Brander, L., and D. L. Burke. 1995. Rights-based vs. competitive fishing of sea scallops *Placopecten magellanicus* in Nova Scotia. Aquatic Living Resource 8:279-288.

Casey, K. E., C. M. Dewees, B. R. Turris, and J. E. Wilen. 1995. The effects of individual vessel quotas in the British Columbia halibut fishery. Marine Resource Economics 10:211-230.

Christy, F. T. 1973. Fisherman quotas: A tentative suggestion for domestic management. Occasional Paper 19, University of Rhode Island, Law of Sea Institute, Kingston, RI.

Dewees, C. M. 1989. Assessment of the implementation of individual transferable quotas in New Zealand's inshore fishery. North American Journal of Fisheries Management 9:131-139.

Dewees, C. M. (in press) Effects of individual quota systems on New Zealand and British Columbia fisheries. Ecological Applications.

Ginter, J. C. (1995) The Alaskan community development quota fisheries management program. Ocean and Coastal Management 28:147-163.

Gordon, H. S. 1954. The economic theory of a common-property resource: The fishery. Journal of Political Economy 62:124-142.

Grafton, R. Q. 1996. Individual transferable quotas: theory and practice. Reviews in Fish Biology and Fisheries 6:5-20.

Moloney, D. G., and P. H. Pearse. 1979. Quantitative rights as an instrument for regulation of commercial fisheries. Journal of the Fisheries Research Board of Canada 36:859-866.

New Zealand Ministry of Fisheries. 1995. Country Report-New Zealand. Organization for Economic Cooperation and Development Committee for Fisheries. 55 p.

Sissenwine, M. P., and P. M. Mace. 1992. ITQs in New Zealand: the era of fixed quota in perpetuity. Fishery Bulletin 90:147-160.

Turris, B. R. 1994. Canada's Pacific Halibut Fishery: a case study of an individual quota fishery, in Karyn L. Gimbel, (ed.), Limiting Access to Marine Fisheries: Keeping the Focus on Conservation, Center for Marine Conservation, Washington, D.C.

Wilen, J. E. 1988. Limited entry licensing: a retrospective assessment. Marine Resource Economics 5:313-324.

CHAPTER 42

"Farming the California Seas"

Justin Malan and Jeff Young

Abstract

California is blessed with and extensive, rich and varied coast and ocean. Many ocean-dependent industries compete for the bountiful resources -- challenging local, State and federal governments to balance the often competing uses.

Increasing local and global demand for seafood strains the long-term sustainability of virtually all commercial fisheries and has forced many nations to turn to aquaculture as a source of seafood products.

While California has a long history of shellfish production in its estuaries and bays, it does not raise finfish in the ocean on a commercial scale.

This paper will explore the historical development of coastal and ocean aquaculture of mariculture in California, provide an outline of the opportunities and constraints to promoting mariculture, and explore some options for establishing a strong sustainable mariculture industry in California.

Justin Malan, Executive Director, California Aquaculture Association, 3700 Chaney Court, Carmichael, CA 93101

Jeffrey S. Young, Esq., General Counsel, California Aquaculture Association. 1307 State Street, Santa Barbara, CA 93101

CHAPTER 43

World Fisheries and California Fisheries Management

By Tana McHale, Fisheries Consultant

Abstract

In light of the growing world population and increasing seafood needs, it is this generation of resource managers, scientists, fishermen and conservationists who must work to stabilize and enhance fisheries harvests. As the world eighth largest economy with growing public interest in marine conservation issues, California has the opportunity to lead the world in designing and implementing innovative fishery management programs for continued, reasonable consumer access to publicly owned fisheries. However, problems with harvest records and their use, marine mammal issues and resource allocation disputes are crippling California's ability to efficiently manage fisheries resources. These problems are common in nearshore fisheries worldwide, impede efforts to accurately assess the status of fisheries stocks, and degrade the capacity to enhance fisheries productivity and harvests. This paper presents an overview of these issues with examples and discussions.

Background: The Role of Fisheries

Global

The United Nations Food and Agriculture Organization (FAO) 1996 "State of World Fisheries and Aquaculture" (SOFIA) report identifies and discusses issues of major importance for the sector's future contribution to food security and economic growth. Key statistics and projections, unavailable through any other source, are summarized here to provide perspective for addressing our own fisheries issues.

The SOFIA report points out that commercial and subsistence fisheries are a source of livelihood, they provide food, and demand for seafood is growing rapidly. Population increase is likely to bring worldwide seafood demand (97.85 million metric tons in 1990) to between 110 and 120 million metric tons (MT) by the year 2010. This increasing demand will have to be met by a combination of aquaculture and capture fisheries. A summary of fisheries production over the past five years, with two scenarios of projected fish and shellfish supplies for the year 2010:

	Millions of metric tons						Projections, year 2010	
	1990	1991	1992	1993	1994	1995*	Pessimistic	Optimistic
Capture fisheries	85.46	84.59	84.93	85.70	91.02	91.00	80	105
Aquaculture	12.39	13.17	14.42	16.48	18.56	21.30	27	39
TOTAL	97.85	97.76	99.35	102.18	109.58	112.30	107	144

** figures for 1995 are preliminary*

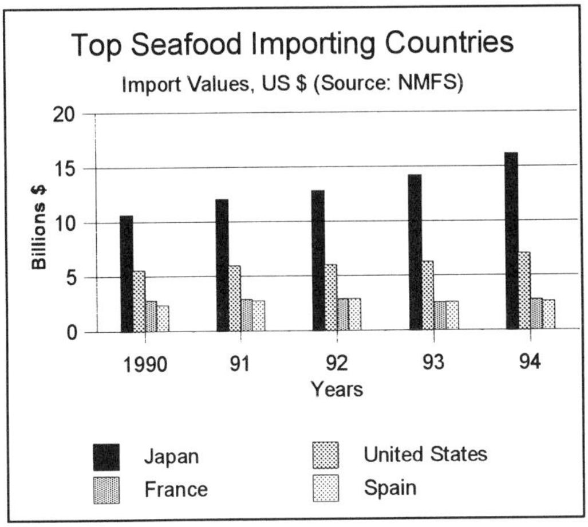

Figure 1: Top Seafood Importing Countries.

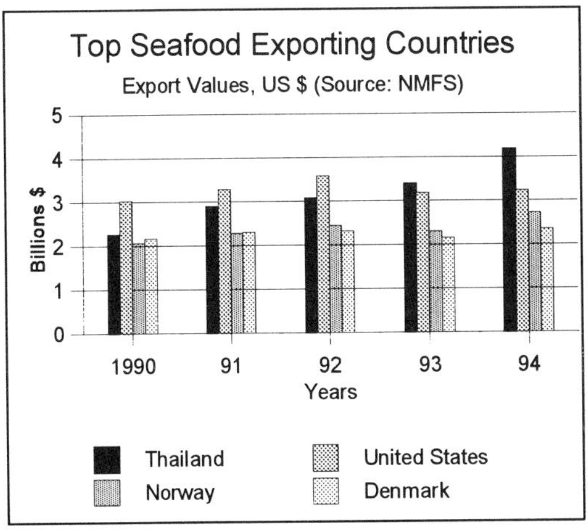

Figure 2: Top Seafood Exporting Countries.

The FAO reports that fish production is increasing the most in low-income food-deficit countries (LIFDCs), where harvests have increased by 50% since 1988 and grown an average of 6.9% annually during the period 1988-1994. On the American continent, Latin American and Caribbean fish production reached a record 24 MT in 1994, representing 22% of the total world catch. North American countries (the US and Canada) produced 7.1 MT and contributed roughly 6% of the global fisheries harvests. In order to meet the requirements of the next century, aquaculture production must increase while capture fisheries must be stabilized, or possibly increased, through improved management. Management know-how and capacity must be upgraded to meet these needs, and this means costs must be incurred.

Total world seafood imports and exports in 1994 were reported by the National Marine Fisheries Service[1] as $51.5 and $47.0 billion US dollars respectively. Figures 1 and 2 show the top four seafood importing and exporting countries worldwide.

United States

US harvests have remained relatively stable as indicated by 1990-1994 landings data published by the National Marine Fisheries Service (4.3, 4.3, 4.4, 4.7, and 4.7 MT respectively) and FAO (5.9, 5.5, 5.6, 5.9, and 5.9 MT respectively). However, there are problems with the landings data since the numbers and trends published by two accepted authorities are different.

NMFS reports that the United States is a major importer of seafood products, second only to Japan. While seafood imports continue to increase (growing from $5.5 billion in 1990 to $7 billion in 1994), seafood exports have remained stable. Figure 3 shows the relationship between US seafood imports and exports. The FAO also reports that an estimated 70% of US seafood expenditures is associated with food service purchases. Higher value seafood appears to be more frequently consumed in restaurants, and this trend of away-from-home consumption is expected to continue. US fisheries clearly represent a valued asset, and source of food, both domestically and abroad.

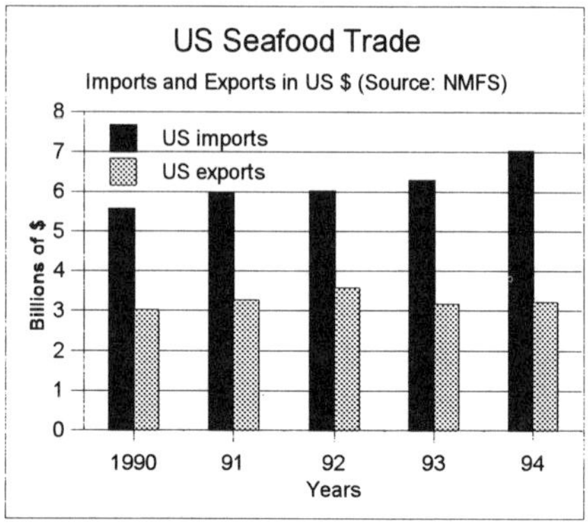

Figure 3: Value of US Seafood Trade.

US fisheries are administered nationally by the National Oceanic and Atmospheric Administration through NMFS. The Magnuson Fishery Conservation and Management Act (Public Law 94-265) serves as the primary conservation and management authority for all fishery resources with the US Exclusive Economic Zone (EEZ), continental shelf resources and anadromous species beyond the EEZ, although management authority of some fisheries is delegated to the individual states. The US has also recognizes and works within a series of international fisheries agreements and treaties (including the Law of the Sea, the Code of Conduct for Responsible Fisheries, the Straddling Fish Stocks and Highly Migratory Fish Stocks and the High Seas Fishing Compliance Act) and participates in the various intergovernmental fisheries management entities (including the Northwest Atlantic Fisheries Organization, the North Atlantic Salmon Conservation Organization, the Pacific Salmon Commission, the North Pacific Anadromous Fish Commission, the International Pacific Halibut Commission, the International Commission for the Conservation of Atlantic Tunas, and the Inter-American Tropical Tuna Commission).

There have been both successes and failures in the management of fisheries in the United States. The ability to understand the factors influencing the effect of management (what works, what doesn't and why) is limited by the lack of adequate, relevant, standardized information. In general, more information is needed about individual fish species, their distribution and abundance over time, response to environmental conditions, and relationship to and interactions with other marine species. The factors influencing harvests (the most reliable and consistent source of available fisheries data) must be defined, analyzed, documented, and formally considered if fisheries management capacity may be improved.

California's Fisheries

Fisheries are clearly an important source of food and employment worldwide, in the United States, and in California. A wide variety of species are harvested in this state to provide our own citizens, and consumers elsewhere, with nearly 300 varieties of seafood[2]. California fisheries of note include the abalone, crab, groundfish, herring, prawn, rockfish, salmon, sardine, sea urchin, shark, shrimp, squid, swordfish and tuna fisheries; some of these fisheries are managed by the state, others by the federal government (through the Magnuson Act fisheries council system), and still others fall under the jurisdiction of various international agreements.

While NMFS records show that California fisheries ranked fourth in the nation in 1994 and 1995, in terms of pounds landed, it is not possible to describe the state's seafood import and export trade balance, as noted in the California Resources Agency "California's Ocean Resources: An Agenda for the Future":

> "... It is difficult to determine exactly how much of the seafood imported into the United States ends up in the state as well as how much of the imports that enter the state are actually consumed in California. Similarly, exports from California are not necessarily produced within the state. Despite the difficulty of determining exactly how much seafood California exports and imports, research have shown that the state's seafood trade deficit is substantial..."

It is also difficult to make any general statements regarding the current status of California's fisheries due to the lack of fisheries independent data, gaps in the landing records and unresolved problems with interpreting landings records. These and other issues blocking our ability to efficiently manage fisheries to their full potential are described in the sections following.

A significant gap in landing record: recreational harvests

The California Resources Agency 1997 "California's Ocean Resources: An Agenda for the Future" acknowledges, "Managing California's ocean and coastal fishery stocks at sustainable levels is complex and depends on such things as accurately assessing wild stocks and judging the impacts that multiple harvesters have on the resource." However, while both commercial and recreational fishermen participate in California fisheries harvests, only commercial landings data is recorded in terms of species and pounds harvested. Only a small percentage of recreational landings are recorded via party-boat fish counts and telephone surveys. Even the most basic information about the recreational fishing community is difficult to estimate, as noted by the California Research Bureau[3]:

> "Over 2.2 million sport fishing licenses were issued by the California Department of Fish and Game in 1992. About two-thirds of the total 2.2 million licenses allow both ocean and inland fishing. Since these license holders can fish in both the ocean and inland, without further information it is difficult to estimate how

> many of them fish in the ocean. A total of 471,500 licenses issued are valid only for fishing in the Pacific Ocean... In 1991, there were approximately 270 (party boats) providing open ocean access to sport fishermen. A total of 660,000 anglers caught 4.2 million fish while aboard these commercial fishing vessels. Rockfish and bass were the most popular types of species caught..."

Managers are thus left without the most basic information needed to accurately assess harvest levels, and must make management decisions on the basis of a series of estimates rather than formally recorded data. Without knowing how many sportsmen participate in a specific fishery, the number of targeted species taken, and the average effort (harvest days) invested by individual sportsmen, recreational bag limits and other management tools must be set on the basis of guesses.

The lack of recreational fisheries information is a common problem shared by nations around the world. This is evidenced primarily by the general lack of references to, and discussion of, recreational fisheries in the UN Food and Agriculture Organization circulars and reports. The UN annual SOFIA reports, for example, provide detailed information and analysis of commercial fisheries and aquaculture -- without more than superficial mention of recreational harvests and their impacts, whether significant or otherwise.

Problems with interpreting harvest records

In the absence of adequate fisheries independent data, commercial landing statistics and trends are commonly used in attempts to assess the condition of marine fish and shellfish populations. However, it is important to recognize landing records only provide information about harvests, not necessarily information about the status of stocks left in the water or the impact of fishing on stocks.

To demonstrate this point, consider that the overall condition of California fisheries is often described through references to and comparisons of combined harvest landing highs and lows over time. The general downward trend over time, particularly since 1980s, has prompted widespread concern over the impacts of harvests on fishery stocks, and a widespread belief that fish and shellfish populations must be in decline since landings have declined.

However, California's overall commercial landings have been strongly influenced by the rise and fall of sardine and tuna landings (Figure 4), and landings for these two fisheries have been strongly influenced by factors unrelated to fishing pressures.

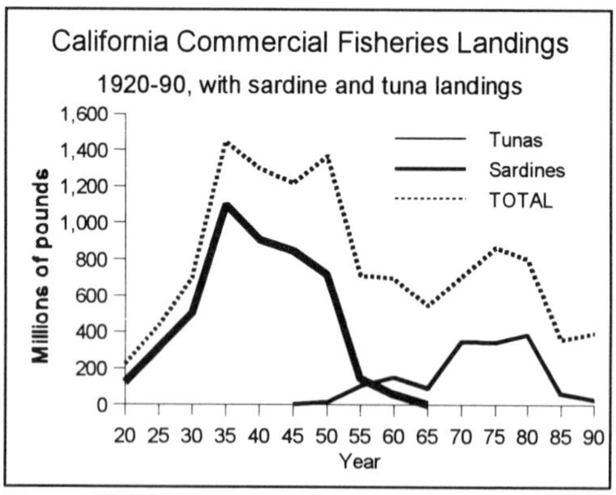

Figure 4 Calif. Commercial Fisheries Landings, 1920-90

In the case of the sardines, core samples taken from the Santa Barbara Channel documents extreme, natural fluctuations in sardine populations over hundreds of years (Figure 5). In 1992 researchers described these fluctuations as: "The scale-deposition records shows nine major recoveries and subsequent collapses of the sardine population over 1700 years. The average time for a recovery is defined as an increase from less than one to over four million metric tons of biomass..."[4]

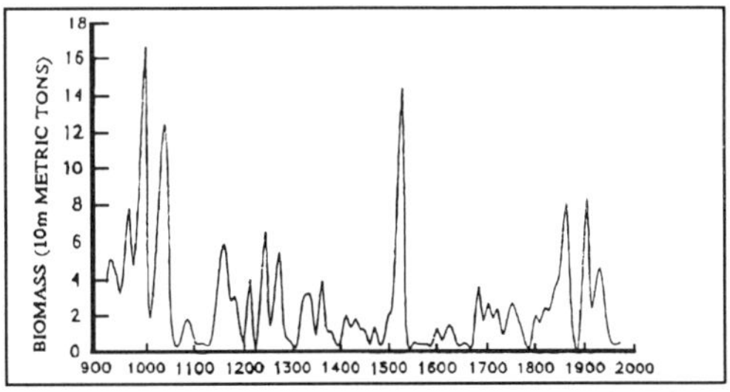

Figure 5 Pacific sardine, year 900 to the present (graph using Tim Baumgartner's data provided by Ocean Trust).

As for the tunas, California's landings crashed after the tuna canneries left the state, even though tuna harvests increased and were landed elsewhere. California Sea Grant Extension described the cannery exodus from this state:

> "... California's tuna canning industry had recovered from a post-World War II slump brought on by competition from imports to become one of the largest, most profitable fisheries in the world. By 1971, a modern fishing fleet of purse seiners based in San Diego and San Pedro fished throughout the eastern Pacific Ocean and was beginning to expand onto new fishing grounds in the western Pacific and Atlantic Oceans. Tropical tunas landed at California's canneries comprised almost half of the state's marine fish landings. During the early 1980's, California's tuna industry faced rising domestic labor costs, ageing processing facilities, more stringent waste discharge regulation, and increased competition from abroad. By 1985 the industry left California in favor of facilities located in American Samoa and Puerto Rico..."[5]

The sardine and tuna examples are not unique cases: all fisheries landings are influenced, to some degree or another, by any combination of the following factors: increased predation, pollution, loss of fishing grounds, natural oceanic cycles, increased recreational harvests, natural oceanic cycles, overfishing and/or regulations.

In particular, note that many regulations are adopted for the specific purpose of reducing harvests. Resource managers, biologists, ecologists, environmentalists and fishermen have invested thousands of hours, and an enormous amount of effort, into designing and implementing harvest regulations to purposely reduce commercial harvest rates. For example, size limits (for individual fish or shellfish) and minimum mesh size requirements (for net and trap gear) are used to protect juvenile and/or broodstock populations of commercially valuable species, and these populations are therefore not represented in harvest landing reports. Other regulations, such as seasonal closures, reduce or eliminate harvest opportunities, and limited entry programs are used to reduce the number of fishery participants over time.

It has been pointed out that harvests should stabilize in the years following the adoption of fishery regulations, yet this is likely to happen only when the following remain constant:

1. The same regulations have been in effect (without additional regulatory actions);
2. The regulations have been consistently enforced, and rates of compliance are stable;
3. The number of fishery participants remains the stable;
4. Water quality, availability of fishing grounds, weather conditions have been similar; and
5. Market conditions (including supply, demand, and exchange rates for those fisheries engaged in export sales) have been consistent.

Information for gauging the impact of these factors is often available, and the value of fishery landing reports in stock assessments will undoubtedly increase when these factors, and recreational landings, can be formally accounted for. However, until landings can be adjusted with regard to the various influencing factors, it is apparent that landing reports alone cannot be considered as a reliable source of information for judging the overall status of either individual fisheries or combined state, national or international fisheries.

Marine Mammal Issues

The federal Marine Mammal Protection Act (MMPA) provides complete protection to marine mammals in US waters regardless of their environmental impact or effect on other marine populations. The MMPA specifically prohibits any marine mammal from being killed, captured, hunted, harassed, pursued, tormented, or annoyed, effectively granting each individual marine mammal with "sacred cow" status. These protections far exceed what is needed to maintain healthy populations of these animals, or for recovering the relatively few threatened or endangered marine mammal populations that would otherwise fall under the jurisdiction of the Endangered Species Act.

Since the MMPA first passed in 1972, many California marine mammal populations have increased dramatically. For example, there were roughly 30,000 sea lions in California in 1972. Today their numbers have grown to an estimated 190,000, and, according to the National Marine Fisheries Service, the population continues to climb at an average rate of 8% per year. These sea lions are having a significant (and increasing) impact on the productivity of many fisheries, and they are hampering the recovery of threatened salmon and steelhead stocks. Other populations of fish-eating pinnipeds (i.e. harbor seals and elephant seals) are also growing, and also have an impact on fisheries.

Further, one trailblazing, key effort to resolve a well-recognized marine mammal/fisheries conflict has essentially died due to a lack of funding. This program implemented zonal management of recovering sea otter populations to protect abalone and other shellfish resources, from Point Conception to the Mexican border, following an experimental translocation of otters to San Nicolas Island (in 1987). Federal law mandates the capture and removal of sea otters from this zone, yet, as noted in the CDFG March 1997 Sea Otter Research Project newsletter:

> "The most significant change (in ending the Sea Otter Research Project) will be the loss of staff time focused on the sea otter-shellfish conflict and its resolution, which includes the loss of a field team trained in sea otter capture techniques.

> Delegated responsibility could not be accepted for implementing the containment management provisions of Public Law 99-625 without secure funding to support the effort. While the Department has maintained the capability for some time, the lack of federal funds to support the team has resulted in no containment related efforts since 1993."

The implications of losing the California's Sea Otter Research Project and containment capabilities are disheartening: if we lack the financial resources to *prevent* conflicts between shellfisheries and a relatively small, recovering population of sea otters, how will *ongoing* conflicts between fisheries and well-established, significantly larger populations of marine mammals be resolved? If California lacks the political will to comply with an existing federal mandate for protecting shell fisheries from excessive predation, how can fishermen, managers and conservationists hope to maintain healthy fish populations while the numbers of marine mammals continue to skyrocket?

These challenges are not restricted to California: similar marine mammal management issues occur in virtually all countries. For example, Britain passed the Grey Seals Protection Act in 1914 and the Conservation of Seals Act in 1970; the number of grey seals in British waters has doubled since 1960 to over 80,000 animals[6]. Their impact on fisheries is said to be significant, and increased numbers of breeding grey seals have also caused soil erosion on the Farne Islands while displacing other wildlife native to the islands. In Ireland, researcher Ciaran Crummery of the Irish Sea Fisheries Department found grey seals consumed 5.5 times the Irish white fish commercial harvest quota and damaged 92% of salmon harvested by fishermen[7]. In the northwest Atlantic the number of harp seals has risen to 4.7 million animals, and these stocks eat more than a billion cod a year (88,000 mt in the Gulf of Saint Lawrence and 54,000 mt off the New Foundland coast)[8]. Robust populations of Cape fur seals in South Africa and harp seals in Canada remain a matter of serious concern to fisheries managers[9].

The issue is not one of blame, or creating scapegoats of convenience when fisheries stocks have declined, but of objectively reviewing the predation factors influencing fisheries and making appropriate management decisions to ensure both productive fisheries and healthy marine mammal populations. As George Blichfeldt, Secretary of the High North Alliance, summarized:

> "The very simple message I will try to convey is that you can't have it both ways. You can not at the one side aim at the optimum sustainable yield from the living resources of a marine ecosystem, and on the other side totally protect an important group of top predators in that ecosystem... the development of the multispecies management is needed to assure sustainability and to consider the objective to have an optimum sustainable yield from the sum of resources."[10]

Clearly, providing favored "sacred cow" status to selected species at the expense of others makes management a nearly impossible task. The ability of managers to secure healthy and productive fisheries depends, in part, on their ability to manage other segments of the marine community, including marine mammals.

Allocation Disputes

The allocation of renewable (yet still limited) fisheries resources between different segments of the fishing community can wreck havoc on the already complex task of managing fisheries. Allocation disputes between competing groups of fishermen often involve passionate pleas to authorities to "save our resources" by disallowing the use of a specific type of gear or fishing method. These disputes often result in decisions made in reaction to upset people, and not necessarily fishery or seafood consumer needs.

For example, in response to concerns about the accidental take of sea otters by gillnets, CDFG, California Sea Grant and a group of fishermen worked to develop an alternative gear experimental fishery in Central California during the mid-1980s. Six Morro Bay fishermen participated in the development of trawls to reduce or eliminate sea otter bycatch; this experimental fishery was also subject to area restrictions, quotas and both weekend and seasonal closures. Participants caught no sea otters during the test period; the experiment successfully met the original program objectives.

Yet arguments between experimental fishery permit holders and other commercial fishermen who were not included in the program, as well as some sportsmen, led to unexpected opposition against the program. Claims about failing halibut resources followed. While CDFG biologists attributed fluctuations in halibut resources at that time to the effects of El Nino, the California Fish and Game Commission seriously considered ending the program anyway. The experimental fishery participants sued to save the program, successfully arguing there was no evidence of negative impacts on fishery resources. However, a recreational fishing group then sponsored legislation that eliminated the use of the experimental fishery trawls once and for all.

Many other California resource allocation disputes have occurred between competing groups of commercial harvesters, commercial and recreational fishermen, and fishermen and non-fishermen. Some of these conflicts have been resolved (for example, the Santa Barbara County -administered Fisheries Enhancement Fund program, designed to mitigate the impacts of oil development on commercial fisheries, has defused most conflicts between commercial fishermen and oil companies), while others disputes continue unabated year after year.

Resource allocation disputes are not unique to California and regularly complicate fisheries management throughout the United States and around the world, both on the ocean and inland. In south America, for instance, the FAO reports the following inland fisheries situation:

> "...In the south (Argentina, Chile and parts of Brazil), food fisheries are unimportant and have virtually become insignificant in certain cases, such as the Argentinean 'sabalerias'. In these areas, growing pressure from environmental lobbies and from urban-based recreational fishing interests has closed the fisheries to commercial exploitation, and they are now reserved mainly for recreational and subsistence fishing despite severe protein malnutrition. This area includes Pantanal of Brazil, and Bolivia, where there is a commercial fishery potential of over 100,000 mt per year..."[11]

In Africa, as another example, conflicts between gillnet fishermen and sport anglers on Lake Kariba have reached crisis proportions, as noted by Kefasi Nyikahadzoi of the University of Zimbabwe:

> "Anglers complain that poaching and illegal use of nets in closed areas indiscriminately deplete the resource and tangle boat motors. House boats cruising on the lake often have their boat engines snarled in gillnets. In both cases, anglers and house boat owners often confiscate and destroy the nets. The pattern of illegal setting of nets by fishers in prohibited areas of National Parks and the trespassing by tour operators and sport fishermen into designated commercial fishing grounds remains a problem. Fishers do not agree with the notion of 'a wilderness for tourists.' On the other hand tour operators and sport fishers continue to trespass into fishing grounds because these areas provide good catch... The safari operators argue that fishermen disturb wildlife, poach animals or scare them away, thus making their business less lucrative. Fishermen fear that they are going to loose their fishing grounds with time, as the number of safari operators increases in the area..."[12]

Note that allocation disputes focus on who should have access to a fishery rather than problem identification and resolution, and these disputes can block the interactive fisheries management process. Community participation in fisheries conservation programs and trust/support of resource management agencies may disintegrate under such circumstances, and attention typically shifts away from scientifically-based management towards politically-based manipulation of resources.

CONCLUSION

In California and other parts of the world, fisheries stocks are regularly judged on the basis of commercial landings; reliable recreational harvest records do not exist; marine mammals often proliferate at the expense of fish populations; and resource allocation disputes continue to complicate the already complex task of fisheries management. These problems are not only crippling our ability to manage fisheries resources to their fully potential, but also prevent managers from accurately assessing the past and current status of individual fisheries.

California's contribution to revitalizing both state and world fisheries could be directly related to taking a leadership role in addressing and resolving these common fisheries management challenges.

However, state government has yet to formally acknowledge the overall value of fisheries as they relate food security needs, the realities of a growing human population or international trade. The significance of California's fisheries is often described in terms of their intrinsic and recreational values alone, while the role of commercial harvests is often dismissed as mere exploitation. The capabilities of the California Department of Fish and Game have been severely limited by budget constraints, and existing state data collection and reporting systems fail to meet basic information and planning needs. These problems must be addressed at the policy level in the coming years so that programs to address and resolve the common fisheries management challenges, as discussed, can be properly funded, designed and implemented.

Notes

1. National Marine Fisheries Service "Fisheries of the United States, 1995."
2. California Seafood Council winter 1995/96 newsletter "Seafood Sense."
3. "Economic Assessment of Ocean-Dependent Activities" by Rosa Maria Moller, Ph.D and Joe Fitz, as published in the Resource Agency's 1997 "California's Ocean Resources: An Agenda for the Future."
4. "Reconstruction of the history of Pacific sardine and northern anchovy populations over the past two millennia from sediments of the Santa Barbara Basin, California," CalCOFI Reports 33:24-40, 1992, by T.R. Baumgartner, A. Soutar and V. Ferreira-Bartrina.
5. "California's Living Marine Resources and Their Utilization," Sea Grant Extension Publication UCSGEP-92-12, 1992, by William Leet, Christopher Dewees and Charles Haugen.
6. The Mammal Society (UK), Fact Sheet #9.
7. As reported by The High North News, no. 7, April 10, 1994 article "*Grey Seals and Fisheries Interaction.*"
8. "Report on the Status of Harp Seals in the Northwest Atlantic," Northwest Atlantic Fisheries Centre, Department of Fisheries and Oceans, St. John's, Newfoundland, 1996.
9. International Council for the Exploration of the Sea (ICES) summary of the September 1995 "Role of Marine Mammals in the Ecosystem" symposium, Nova Scotia, Canada.
10. Presentation before the EU Parliament Fisheries Committee public hearing on multispecies resource management in Brussels, September 27-28 1995 by Georg Blichfeldt, Secretary of the High North Alliance (Norway).
11. FAO Fisheries Circular No. 921 FIPP/C921.
12. "Multi User Conflicts at Lake Kariba Inshore Fishery," ALCOM News No. 21, 1996, by Kefasi Nyikahadzoi, Socio-economist, University of Zimbabwe.

CHAPTER 44

Ocean Conservation and the Public: An Issue Waiting to be Made SeaWeb: A Project of The Pew Charitable Trusts

Vikki N. Spruill and Mark Mellman

Abstract

In the summer of 1996, The Pew Charitable Trusts, a Philadelphia-based philanthropy, established SeaWeb, a multi-media educational initiative for ocean issues. Designed to raise awareness of the world ocean and the life within it, SeaWeb aims to provide information from a broad spectrum of sources to help all become more connected and involved in the life of the sea. SeaWeb's approach is objective but not neutral -- its bias is to protect the living ocean.

As an early component of this effort, SeaWeb commissioned a public opinion poll, the first of its kind, to gauge public attitudes towards the ocean. It is important to note that the SeaWeb campaign is not a direct response to the results of the public opinion poll. Although critically important, it is only one of several tools used to determine the salience of SeaWeb's communications. The research has given SeaWeb a strong sense of what will work to engage the public, and can therefore be useful as a starting point for ways to frame and discuss other, sometimes more significant issues.

The survey of 1300 adults nationwide (with a margin of error of 3.3%), conducted by telephone by the Washington D.C. based polling company The Mellman Group in May, 1996, yielded some surprising results. In particular, it found that deterioration of our ocean is an issue that can be embraced by the American public, but as yet has not been. Even though public awareness of the condition of the ocean and its great importance to us is generally high, their understanding of the specific nature of the problem is quite low. Nevertheless. people are anxious to learn about the ocean and they care if something is being done to harm it. Indeed, the data indicate that the ocean has the potential to become a powerful political issue.

Most Americans (67%) believe the ocean is in trouble. Not even one percent

Vikki N. Spruill, Executive Director, SeaWeb, 1731 Connecticut Ave. N'V, 4th Floor, Washington, DC 20009
Mark Mellman, The Mellman Group, Inc., 1000 Thomas Jefferson St. NW, Suite 520, Washington, DC 20007

of those surveyed report the condition of the ocean as excellent, and only 20% say it is in good condition. Over half (52%) view the deterioration of the ocean as a very serious threat ten years from now.

Most of the public attitudes required to create a major political issue are in place. A plurality of Americans (49%) say the condition of the ocean is very important to them personally. (In Coastal communities the number is 64%.) Another 29% say it is somewhat important, only 5% say not too important, and 1% not important at all. Understanding of the ocean's significance transcends traditional environmental categories, with 45% of those who do not consider themselves environmentalists and who do not live on the coast saying the ocean's condition is very important.

Americans are nearly unanimous (82%) in believing that the threat to the ocean stems from human activity. Only 10% believe the ocean is so vast and plentiful there is little humans can do to destroy it. In a country where the majority of the public often wants the federal government out of their lives, 85% believe that the government needs to do more to help protect the ocean. However, scientists have not communicated well with the public -- 61% of those surveyed believe that citizens and scientists disagree on which are the most important problems facing the ocean, and the government should focus on citizens concerns (42%) rather than scientists (37%). That does not mean that interest in ocean research is not high -- 72% believe funding for ocean exploration is a more important priority than funding for space exploration (17%).

In brief, the field is ripe for the ocean community to take action. Americans are ready and willing to accept the fact that the ocean is in danger. They are also ready and willing to do something about it. (Of all the potential actions presented, those that involved personal action were rated as something respondents were "almost certain" to do.) Concern is high, but information is low.

CHAPTER 45

POLITIC POLITICS AND OTHER REALITIES OF COASTAL MANAGEMENT

by
Peter M. Douglas
(California and the World Ocean)
(San Diego, CA)
(March 25, 1997)

INTRODUCTION

The California Coastal Commission celebrates its 25th anniversary this year. Established in 1972 by a citizen's initiative (Prop. 20) approved by voters despite well financed opposition, the Commission was mandated to prepare a coastal plan and to regulate development along the state's 1100 mile coastal margin. The Legislature adopted the Coastal Act in 1976 to implement most of the Plan's recommendations.

Born in controversy, the California coastal program, like the coast itself, continues to attract conflict and debate. Notwithstanding the controversy, California's coastal protection law, more than any other I know, is the public's law and whenever threatened, people of all political affiliations step forward in its defense. That was true in 1972 and again in 1976 when the Coastal Act made the Commission a permanent, powerful actor on California's environmental front-line. It was also true last year when the survival of the coastal program hung in the balance.

I should add, the overwhelming outpouring of public, media, legislative and local government support for the coastal program last summer was awesome and inspirational. The coast and integrity in public service emerged from that political firestorm as the real winners.

The public's spirited and successful defense of the coastal protection program last year carried through the November elections in California and played a significant role in rearranging the political landscape in Sacramento.

Public support was again demonstrated New Year's eve when the Commission sold the last whale tail coast and ocean protection automobile license plate needed to meet the midnight statutory deadline of 5000 plates sold. As the deadline neared and the public heard that the whale tail plate was in trouble, we sold over 2000 plates in two days! Proceeds from the sale of these coast and ocean protection plates will be used to expand the Commission's public education and involvement program.

The greatest threats to the coast and coastal ocean are public apathy and ignorance.

Because the public is anything but indifferent about the well-being of coastal environments, I am very optimistic about the future of coastal management. The challenge is to heighten public awareness about the problems facing the coast and ocean and to provide the information the public needs to be constructively and effectively engaged in efforts to address these problems.

The California public has forged an emotional and enduring bond with the coast and coastal ocean. That bond is both mystical and pragmatic. It is rooted in the power of land's end to inspire spiritual renewal, a rainbow of passions, and to give voice to an expansive score of artistic expression. It also derives from its earthly venue as a place to live, work and play.

The coast is many things to many people. And all the things it means to different people converge and compete on this relatively narrow reach of geography for their own place in the sun.

The coast, like any coveted place on earth, is never finally saved. Coasts always need saving and are always being saved.

Consequently and into the foreseeable future, coastal management must necessarily be an important fixture on the institutional and political landscape of global environmental stewardship. Accordingly, the principal questions should focus on the vitality and effectiveness of existing and emerging programs. And effectiveness will, in turn, depend on the politics of coastal management when a program is created and during implementation.

The politics of coastal management in most locations will, to some significant degree, be influenced by the extent and quality of public participation. Informed public involvement and active public support are vital to the success of any coastal protection program. Without it, newly emerging programs are doomed to fail and existing ones will atrophy and become irrelevant. With it, programs are continually renewed and refined, reinvigorated and reinforced.

THE POLITICS OF COASTAL MANAGEMENT

The economic, environmental, societal and ethical stakes in the evolving arena of coastal and ocean stewardship are profound and span generations. How we manage this critical geography, or more precisely, how we manage human use of it - what factors guide the value choices, who makes the judgments, and how decisions are informed and made - will involve questions as controversial and consequential tomorrow as they are today.

The politics of coastal management will always be important, intense, complex, volatile, sharp, confrontational and ethically challenged. How these politics are played out, and more specifically, how coastal managers chose to engage them will determine the coastal legacy we leave future generations.

Although the core conflicts in coastal management will remain largely unchanged over time, the specifics of venue and detail will differ. The competition among and collision of human values, interests, needs and desires in setting the agenda and allocating uses of coastal resources in the future will be as great, intense and explosive, perhaps even more so, as in times past.

Many factors will significantly influence the politics and dynamics of coastal management in coming years. These include population growth, an insatiable appetite for economic expansion, increasing freedom to work at home in remote locations as a result of rapidly evolving information technologies, the desire of people to leave developed neighborhoods to live in previously shunned outlying rural areas, the widening gap between groupings of varying economic means, increasing urban congestion, concerns about public safety, the deterioration of civic discourse, and the fragmentation and polarization of communities of interest. As inland recreational areas are displaced by development, crowding to beaches and coastal parks will exacerbate tensions between seaside residents and visitors. The desperate search by fiscally strapped local governments for new revenue streams threatens coastal resources and will impact coastal politics.

Another and perhaps dominating factor of coastal politics on the landward side of the coastal continuum involves the clash between public values, interests and rights and private property rights and values. One of the greatest challenges facing public sector coastal managers, and indeed, environmental resource managers everywhere, is how, in the face of evolving "takings" law, to protect the environmental commons of this country against the loss of significant natural and human community values

Environmental commons are those land and water environs that exist independently or as an integral part of a dynamic whole (i.e., eco-system, watershed, waterbody, coastal bight, public park, neighborhood, village, town, city) and that support features whose functional viability and integrity are essential to the survival of natural or human community values deemed to be of vital importance to public health, well-being and quality of life. Examples include, wilderness areas, wetlands and other sensitive habitats, lakes and watercourses, the ocean, and communities where people live, work and play. The environmental commons of a residential neighborhood, for example, are those places in it, whether on public or private property, the uses of which directly or indirectly impact the quality, functionality and safety of that human community. One important way to protect environmental commons is through reasonable land use controls.

Recent court rulings involving "takings" law have made it extremely difficult for public sector land use managers to protect significant public resources and values. In a quest for procedural fairness and a desire to preserve the private property domain of the individual, the courts appear to have lost sight of the importance of environmental commons to the health and integrity of life in natural and human communities. Individual interests and benefits relating to the use of private real property are being redefined at the expense of the collective best interests and well-being of living generations and those who follow and have to live with what we leave behind. The politics of this "public interest-private property rights" debate and the outcome will have a major impact on the future shape and effectiveness of coastal management programs.

POLITIC POLITICS

To be effective on the ground (and in the water), coastal program leaders must embrace and practice the art of prudent, tactful, wise, and judicious politics that I call "smart" or "politic" politics of coastal management. Politic politics means doing the public's business in a highly ethical, positive, deliberative, forthright, professional and knowledgeable manner that appeals to a wide cross-section of political partisans. It is a politics free of the shrill, divisive, negative, opportunistic, ego-dominated, sharply partisan electoral rhetoric so typical of the contemporary political scene. Smart coastal politics is about hope and the promise of a healthier, higher quality and more biologically productive coastal environmental future. Politic coastal politics are about promoting the public good and not about advancing any individual's career, nor any special interest group's particular agenda, nor the power of a political party.

The essential ingredients of "smart" or politic coastal politics are fairly logical and straight-ahead. I do not think there is anything new here. The trick is to talk the talk and then to walk that talk.

What is described here is based on the California experience but is, to a significant degree, transferable.

1. Public support: Dance with who brought you to the ball - and in California, that is the public. Public support cannot be taken for granted. Active support requires public good will toward the program and trust and confidence in its integrity. And these are attributes, attitudes and perceptions that must be earned and nourished. Public participation must be encouraged, welcomed and facilitated. Accordingly, the processes of program implementation must be designed and maintained to maximize public participation and involvement.

2. Keeping the mission clearly in sight: It is important that the program's mission and vision of what its purpose is and where it wants to be several decades from now are clear and well understood by those implementing it and the public. Then it is essential to set and maintain a steady course aimed by the mission at that vision and the goals and objectives that must be met to get there.

3. Set and maintain high standards: The program must set and maintain the highest standards of professionalism, integrity, accountability, propriety, quality work, openness, productivity, efficiency, objectivity and fairness. Because strong coastal programs inevitably attract controversy, staff must conduct itself in a manner that avoids even the appearance of impropriety. A test I use is to ask what impression a particular action or treatment of an issue would make on the front page of the Los Angeles Times. It is also important to be forthright and not to shy away from admitting when a mistake was made nor from setting a new course if a chosen approach is not working.

4. Attract, recruit and retain dedicated and skilled staff: The most valuable resource of any program is its staff. High motivation, dedication to the program's mission, commitment to public service and fine job skills are essential to the success of any program and critical in generating vital public good will and support.

5. Be always aware of the long-view: Sound coastal and ocean management are long-term propositions requiring patience and persistence. The delay between action and results is most often considerable. There is a natural tendency to reach for short-term gains, often at the expense of long-term achievement. Coastal managers must take the long-view and consider the consequences of their actions from this perspective. Politicians and policy decision-makers come and go, while the coast and ocean will always need protection. This makes it all the more important to strive for continuity and consistency, and a modicum of program stability. The program must not be

driven erratically on and off the road toward its vision depending on which political agenda happens to be in the driver's seat in government at the moment.

6. Avoid partisan politics: Certainly coastal protection is a political issue and may become a high profile issue in an election. Coastal managers should encourage all candidates to be supportive of good coastal management. After all, environmental protection can only succeed in the long run if it is seen and practiced on a non-partisan basis. Managers should resist the temptation to endorse or join the ranks of the most outspoken coastal protection candidate, especially at the state and congressional level. Managers must work with every candidate who becomes the incumbent once elected. It only hurts the program if a manager has difficulty working with an elected official because of involvement with the candidacy of an opponent. Obviously, it is important to understand the dynamics and implications of elections and outcomes. But being sensitive to electoral politics does not mean becoming directly involved.

7. Good media relations: It is essential that good working relations with the media are established and maintained. The print and electronic media are vital to getting the word out to the public about coastal program issues, problems and accomplishments. Managers must stress substantive responsiveness to media inquiries. Obviously, media coverage is a two-edged thing. Biased coverage or stories that misrepresent the facts can do considerable harm. And, of course, it is always wise to engage the brain before putting the mouth in gear. Environmental reporters and editorial writers are among the most influential players in coastal politics. Like good relations with the public, media relations should never be taken for granted.

8. Know who the key stakeholders are and keep them involved: The clients or constituents of coastal programs constitute a wide range of individuals, interest groups, organizations, agencies, and institutions. Anyone and everyone doing business with the program must be treated fairly and courteously. It is important to be civil and civilized in the conduct of program affairs. However, some stakeholders play a more pivotal role than others. For example, local governments are clients, stakeholders and partners in the program. More attention must be given to their needs and interests than those of many other stakeholders whose interests may be more peripheral or narrowly focused. In any event, it is essential to know who the key stakeholders are and to involve them in appropriate ways and phases of the program on a continuing basis.

9. Take good care of the public's business and the public's money: Scrupulous care and highest priority must be given to the proper expenditure of public funds. Steps must be taken to ensure programmatic and administrative efficiencies and accountability so that the public can be assured it is getting solid returns from its investment of public funds in the program. It takes effort to build a reputation for effectiveness, efficiency and good quality work. But that is what the public expects and what it deserves. Recognizing that public service is a privilege and not a right also helps build public confidence, good will and support. Program staff must be serious about doing more with less and doing it better. If a program is creative about finding appropriate ways and means for getting the job done better and at less cost, the public will respect the effort.

10. Good relations with elected officials at all levels: It is essential to establish and maintain good relations and open channels of communications with elected officials of all parties, at all levels of government. This takes considerable effort because and based on nearly three decades of experience in the arena, the egos of elected officials

usually tend to be somewhat larger than those of most people. It is true hell knows no fury like that of love spurned, but a close second may be that of some politicians ignored or slighted.

11. Avoid client-capture: Public trust and confidence requires knowing that program staff and decision-makers are not captured by the special monied interests doing business with the agency, especially by the regulated community. Safeguards must be built into every program to guard against the development of cozy relations with any special interest group that may affect sound judgment and that can erode program credibility.

12. Eschew extreme positions or those that are ideologically driven: Extremes have a way of coming home to roost. Programs lose credibility if seen to be ideologically driven. The distorted use of the phrase "seeking balance" is both a reaction to what was seen as environmental extremism and has lately often been used as a smoke screen behind which an agenda extreme in its own right is pushed in another direction. Managers should remember that no coastal law banned the use of sound judgment and common sense.

13. Carry out the law: In addition to common sense, another key guide for program implementation must be the letter and spirit of the law. Accordingly, program staff must ensure that policy and administrative decisions and actions are firmly grounded in law and principles of fundamental fairness and common sense.

14. Communicate clearly what is being done and why: It is good policy to explain clearly and simply the rationale for actions taken and decisions made. Others may not agree with a decision, but they have a right to know what it is and what the legal rationale underlying the decision was.

15. Stake out the strategic and philosophical high ground: One of the most helpful lessons to learn is not to take personally the verbal spears, stones and muck hurled in the heat of controversy and to not allow individual or institutional ego to get in the way of sound judgment. Vain ambition, opportunism in the hope of career advancement, and involvement in gutter-politics and political intrigues are slippery slopes to be assiduously avoided. Head for the philosophical and strategic high ground and stay there. That means adhering to strong principles of integrity and professionalism without being arrogant or self-righteous. It means being deliberate and adept without being clever or devious. Sticking with one's principles is good policy and politic coastal politics. It also ensures better sleep and renders more satisfying and less threatening the advance of the "golden years."

16. Beware the fads and traps of "new thinking" about management and environmental protection: Rhetoric means little if not backed by results. Similarly, management fads come and go with few delivering advertised results. It is wise and good practice for agencies and programs to regularly initiate and conduct a self-appraisal and to look for ways to improve performance and efficiencies. Institutional self-renewal ensures continuing program vitality and effectiveness and prevents programmatic stagnation and stultification.

At the same time, while it is necessary to be responsive to new policy and management shifts and directions, it is not unreasonable to ask whether these new management or environmental policy implementation paradigms will, in fact, work - that is, will they actually improve performance, save public funds, increase efficiencies, effectively maximize protection of our environmental commons and improve effectiveness of coastal management? For example, a systems approach to environmental protection and natural resources management is sensible and long overdue. One such example is the watershed approach to dealing with polluted runoff into coastal waters. On the other hand, some "systems" concepts are not well defined or understood and may not be suitable for certain purposes in certain areas. An example of the latter is "eco-system" management. In coastal management, coastal resources must be protected where they exist and where they are found. Examples include public access, wetlands and other environmentally sensitive habitat, agricultural lands, important public views, and coastal-dependent uses.

New management paradigms and reforms also often promise more than they deliver. Examples include "total quality management", "performance-based budgeting", "regulatory reform", "cutting red-tape" and "integrated coastal management." It is important to take and apply from these "paradigms" and "reforms" what works well and to resist spending scarce human and fiscal support resources on costly, "make-work" activities and "fire drills" that cannot reasonable be expected to produce tangible public benefits, such as increased program efficiencies and effectiveness without eroding implementation of the core program prescribed by law. The public is adept at distinguishing between public relations and actions that actually improve the way the public's business is conducted. The public cares most about results that affect their lives and cares little about how they were achieved, so long as the means to the ends are legal, ethical and of minimal cost to the taxpayer.

CONCLUSION

As I stated before, the coast, like any coveted reach of geography anywhere, is never finally saved. It is always being saved. That means there will always be work for those of us dedicated to environmental stewardship.

The challenge is to be effective and politically wise in the ways we chose to go about doing the public's business. We must, from time to time, renew our dedication to a strong coastal and ocean stewardship agenda by reminding ourselves what we do, why we do it, and how we do it. We must strive to cut through populist rhetoric and keep our mind's eye on the mission. We need to be clear about who the true beneficiaries of our labors are, what is actually at stake, and whether what we do will make a meaningful and enduring difference on the ground.

The politics of coastal and ocean management will not be simple or free from controversy in the years ahead. The major distinction from twenty-five years ago is that most participants in the debate today agree that coastal management is important and necessary. Accordingly, attention will focus on the ways and means to achieve improved results and increase program efficiencies. Pressures resulting from population and economic growth, changing quality-of-life needs in society, advances in technology, and evolving doctrines in environmental and property law represent enormous new challenges for coastal management programs. Whether coastal program leaders succeed in meeting these challenges will depend primarily on the political strategies and style they bring to the arena.

Thank you.

CHAPTER 46

Will California's Ocean Governance in the Year 2020 Be Myopic?

Ed Ueber

Abstract

Over the last 50 years ocean management has generally moved from "res nullius" (no ownership), to "res communis" (community ownership) and even to some private ownership (transfer of community ownership via lease). As the need for protecting our ocean resources increases the forces behind these changes often have narrow scope and utility. Increased resource vulnerability, human mobility and resource valve is inevitable in the Twenty First Century. Are present management goals, procedures and trends headed in the right directions? Are we moving to Nineteenth Century Management, maintaining Century policies which have not been effective or adopting new paradigm for the Twenty First Century?

Ed Uber, Sanctuary Manager, Cordell Bank, Gulf of the Farallones, Monterey Bay (northern portion) National Marine Sanctuaries, Fort Mason, Bldg. 201, San Francisco, CA 94123

CHAPTER 47

MANAGING OREGON'S OCEAN RESOURCES

Robert Bailey[1]
Don Oswalt[2]

Key Words

Oregon, Territorial Sea, Rocky Shores, Effects Assessment, Ocean Management, Ocean Resources

Introduction

The State of Oregon has instituted a comprehensive Ocean Resources Management Program within the state's overall Coastal Management Program. The Ocean Program, enacted by the state legislation in 1987 and amended in 1991, reflects long-standing public interests in protecting coastal resources and a commitment to a comprehensive planning approach to coastal management. The program, which includes policy, planning, administration, coordination, and research elements, has matured through four phases.

Development Phases

During <u>Phase One</u> the state relied on agency-based administrative processes and existing state policies to address federal OCS oil, gas, and marine mineral exploration and development proposals and other ocean-related activities during the late 1970s and early 1980s. This phase gave state agencies, led by the Coastal Management Program in the Department of Land Conservation and Development, experience with policy and management of ocean resource issues and with the limitations of scientific information and existing state capabilities to protect and

---

[1] Ocean Program Administrator, Oregon Dept. of Land Conservation & Development, 800 NE Oregon St., #18, Portland, Oregon 97232

[2] Coastal Policy Specialist, Oregon Dept. of Land Conservation & Development, 1175 Court St., NE, Salem, Oregon 97310

manage ocean resources, even if development of oil, gas, or mineral resources did not occur.

Phase Two, initiated by 1987 state legislation, was a broad-scale ocean planning assessment of resources and policy or management issues that was completed by a special 21-member task force assisted by the Coastal Management Program and various state and federal agencies. The result was a comprehensive, policy-oriented Ocean Plan for the US Exclusive Economic Zone off Oregon with attention to the continental margin, which was declared an Ocean Stewardship Area. The task force recommended to the legislature that a permanent ocean management program be established in the Office of the Governor and that a plan be developed for state territorial waters to address a number of management issues.

Phase Three, begun by 1991 legislation, established a permanent 23-member Ocean Policy Advisory Council (OPAC) in the Office of the Governor and specified the elements of a coordinated program. The principal initial task of the OPAC was developing a management plan for the state's territorial sea, which was completed in 1994 and approved as a part of the state's coastal management program in early 1995. The major elements of the plan are described below.

In the current Phase Four, several elements of the ocean program are being carried out. These include implementing the Territorial Sea Plan by state and federal agencies, refining and adding plan elements, coordinating ocean activities of several agencies, and promoting marine research to meet management needs.

Key Parts of the Territorial Sea Plan

The management-oriented Territorial Sea Plan adopted by the OPAC is unlike the broad Ocean Plan with its array of policy recommendations for the continental margin. The Territorial Sea Plan provides state and federal agencies with detailed guidance for managing resources or activities in the state's 3 mile-wide territorial sea and rocky shoreline. It is an "initial" plan and will be added to over time. As adopted in 1994 the Territorial Sea Plan includes three principal parts:

Part One, Ocean Management Framework is a compilation and brief description of all relevant state and federal statutes, agencies, and programs for managing ocean resources in state waters.

Part Two, Making Resource Use Decisions clarifies requirements for inventory, analysis, and coordination found in Statewide Planning Goal 19, Ocean Resources, part of the state's CZM program since 1977. Part Two lists requirements for state and federal agencies and local governments to provide information on and conduct evaluations of potential effects from their ocean-related activities. Provisions for small-scale "pilot-scale" projects allow for ocean-related activities to proceed in order to obtain information previously unavailable so that agencies can assess the effects of full-scale activities.

Part Three, Rocky Shores Strategy, is a multi-faceted program to manage rocky shore resources and areas within an ecosystem context. It is supported by an inventory and assessment of all coastal rocky shore areas and features, including cliffs, intertidal sites, associated rocks and subtidal reefs, and offshore rocks and reefs. The strategy contains overall goals, policies, and management objectives as a basis for an ecosystem-sensitive rocky shores management program. All rocky shore sites are designated into one of six management categories, each with certain management objectives and characteristics: Marine Garden, Habitat Refuge, Research Reserve, Marine Shore, Priority Rock and Reef, and Not Yet Designated. In addition to this "management-oriented" strategy, a companion communication strategy was prepared to coordinate interpretive and educational activities of various agencies so that overall rocky shore management goals and site-management prescriptions are supported and reinforced.

Implementing the Plan

The Territorial Sea Plan is being implemented in three tracks. One track is applying the requirements of Part Two to preparing inventory information and making effects-analyses prior to ocean-development activities. Activities of state and federal agencies, such as ocean disposal of dredged materials by the US Army Corps of Engineers s and creation of a "developmental fisheries" program by the Oregon Department of Fish and Wildlife are utilizing these requirements. Future agency actions, such as issuing permits for laying of submarine telecommunication cables, will follow these requirements.

A second track is enacting regulations prescribed in the rocky shores strategy through state and federal agency programs. Oregon's ocean legislation requires state agencies to carry out the plan while federal agencies are required to act "consistent" with the plan under the federal Coastal Zone Management Act. In fact, key state and federal agencies participated in developing the plan and thus have an interest in applying it to their rocky shore programs. For instance, the Oregon Fish and Wildlife Commission has adopted regulations for collecting or harvesting of intertidal organisms at certain Marine Garden, Research Reserve, and Habitat Refuge sites as prescribed in the plan. The Oregon Parks and Recreation Department will regulate the collecting or harvest of seaweeds in Marine Garden, Research Reserve, and other sites. The Oregon State Marine Board has enacted a seasonal 500-foot wide boating closure around Three Arch Rocks National Wildlife Refuge, an important seabird colony and marine mammal haul-out area. The US Bureau of Land Management and US Fish and Wildlife Service programs for rocky shores and wildlife resources are also carrying out the strategy.

The third track emphasizes interpretation, public awareness, and personal stewardship as primary management tools for rocky shores. The Rocky Shores Communication Strategy sketches out interpretive and other educational opportunities that relate to site-specific management prescriptions in the Rocky Shore Strategy. The Oregon Parks and Recreation Department has prepared and installed resource-interpretion panels at three key sites on the central coast; others are planned. Several state and federal agencies and Oregon Extension Sea Grant are collaborating to establish and provide assistance to community-based on-site interpretive programs for Marine Garden and other high-use rocky shore sites. The Oregon Department of Fish and Wildlife (ODFW) has prepared and distributed brochures and a guidebook for rocky shores.

Continuing Work

Amending the Plan. Amendments to the Territorial Sea Plan are being prepared by the OPAC. A primary amendment will be clarification and articulation of Oregon's overall ocean-resource goals and policies. This section is based on several existing sets of ocean-resource goals, policies, and recommended policies. Another amendment will likely be more detailed site-management plans for clusters of rocky shore areas that need additional work through a Special Management Area Plan process. Other plan policies and management prescriptions may be needed for activities such as artificial reefs, seafloor telecommunication cables, offshore disposal of dredged material, and oil, gas, or marine mineral exploration.

Research and Data Development. In addition to plan implementation and amendment, Oregon is continuing to pursue marine research to support management needs. One effort is a multi-year, modestly-scaled program focusing on nearshore kelp-reefs and other rocky shores that is funded by state and special federal CZM "program enhancement." This kelp-reef study is providing critical information previously unavailable that will support management of a complex mix of marine resources in these nearshore reef areas. In addition, because this program is being organized and conducted primarily by scientists in the Oregon Department of Fish Wildlife, that agency is increasing its capacity to understand and address complex ocean management issues. In addition, the Oregon Ocean Program has led formation of a regional, multi-year coastal ecosystem research program funded by the NOAA Coastal Ocean Program. Entitled the Pacific Northwest Coastal Ecosystem Regional Study, this program is a broad-scale research effort by a multi-disciplinary team of academic and agency scientists to assess the effects of natural variability in the ocean and atmosphere and human impacts on estuarine and nearshore ecosystem processes. This program will continue through 2002.

Conclusion

Oregon has successfully developed and implemented a program to manage ocean resources as an extension of its Coastal Management Program. The ocean program is maturing as an interagency institution and will continue to do so as plans and policies are amended to meet new needs and conditions. Positive results are being realized in the field from a coordinated, interagency approach to ocean resource-management. State agencies have incorporated an ocean-management conciousness into their programs and have come to appreciate the benefits of interagency planning and management. The public concerned with rocky shore protection at a local level has an institutional framework for carrying out their local activities with some assurance that the larger coastal environment is being considered. These program successes build a solid basis for addressing additional ocean resource issues in the future.

Oregon Ocean Program Development Phases

ONE: AGENCY ADMINISTRATIVE PROCESSES 1977-87

- OCS oil and gas leasing
- Marine mineral development
- Ocean sewerage outfalls

TWO: OCEAN PLANNING 1987-1990

- Legislative policies & program structure
- Ocean Resources Task Force
- Broad Ocean Policy Plan
  - Ocean Stewardship Area
  - Ocean resources policies & recommended actions
  - Legislative recommendations

THREE: MANAGEMENT PLANNING 1991-1994

- Legislation of permanent program
- Ocean Policy Advisory Council
  - 23 members
  - Office of the Governor
- Territorial Sea Plan
  - 3-mile territorial sea & shoreline
  - Management focus
  - Amends Coastal Management Program
  - Mandatory state implementation

FOUR: PLAN IMPLEMENTATION & UPDATE 1995-

- State & federal agency regulations & programs
- Ocean Policy Advisory Council plan update
- Research & information

Oregon Territorial Sea Plan KEY PARTS

PART ONE: OCEAN MANAGEMENT FRAMEWORK

- State & federal ocean resource laws
- State & federal agency programs & activities
- Plan mechanics & procedures

PART TWO: MAKING RESOURCE USE DECISIONS

- Inventory & effects evaluation requirements
  - Small-scale projects
- Joint Review Panels
- Consultation with local governments & tribes

PART THREE: ROCKY SHORES STRATEGY

- Ecosystem-orientation
  - coast-wide context
  - all areas & features
- Goals, Policies, Objectives
- Coastwide site inventory & analysis
- Site management categories
  - Marine Garden
  - Research Reserve
  - Priority Rock & Reef
  - Habitat Refuge
  - Marine Shore
  - Not Yet Designated
- Site management prescriptions
  - Regulations
  - Interpretation/Awareness

ROCKY SHORES COMMUNICATION STRATEGY

- Linked to overall resource-management goals
- Coastal network themes
- Site-based media prescriptions

IMPLEMENTING the Territorial Sea Plan

INVENTORY & EFFECTS ANALYSIS

- State agency actions/permits
- Federal agency actions/permits

ROCKY SHORE REGULATIONS

- Oregon Dept Fish and Wildlife
  - Collecting/harvesting marine Invertebrates
  - Permits for research or educational collecting
- Oregon Parks and Recreation Department
  - Marine algae collecting
- Oregon State Marine Board
  - Seasonal closures at Habitat Refuges

COMMUNICATION/INTERPRETATION

- On site interpreter programs during summer
  - Community groups
  - State/federal agencies
- On-site interpretive panels
  - Management theme links
  - Site-specific
- Print materials
- Curriculum materials for teachers/educators

CONTINUING WORK

AMENDING THE PLAN

- **OCEAN POLICY UPDATE**
  - Resolve/clarify existing
  - Meet new needs
- **SPECIAL MANAGEMENT AREA PLANS**
  - Complex shoreline intertidal/subtidal areas
  - Offshore kelp-reefs

RESEARCH & INFORMATION

- **ANALYSIS OF ROCKY SHORE INVENTORIES**
  - Fine-scale habitat usage
- **KELP- REEF ECOSYSTEM STUDY**
  - South coast kelp-reefs
  - Management-oriented
  - Basic inventory and mapping
  - Comprehsensive habitat assessment
- **PACIFIC NW COASTAL ECOSYSTEMS REGIONAL STUDY (PNCERS)**
  - Regional cooperative approach
  - NOAA Coastal Ocean Program
  - Multi-year, imulti-discipline
  - Management oriented
  - Regional ecosystem functions

CHAPTER 48

The Development of an Australian National Oceans Policy

Geoff Wescott
Associate Professor of Environmental Management
Deakin University
662 Blackburn Road
Clayton. 3168
Victoria. AUSTRALIA.
email: wescott@deakin.edu.au.

Introduction

Sometimes it appears that the full impact of the United Nations Convention on the Law of the Sea (UNCLOS) has yet to be realised.

The following two diagrams from the South Pacific should show its impact clearly.

Figure I is the territorial areas of the South Pacific prior to UNCLOS being ratified.

Figure 2 is the territorial areas claimed by these South Pacific nations after UNCLOS came into force in November 1994.

An area once thought of as being open sea with no restrictions now 'belongs' or is claimed by many nations.

The position for the island nation of Australia is equally dramatic. Australia has more than doubled its total territorial area, including making claims over Antarctic waters . In so doing it has also gone from a country without any bordering countries to one which now shares a border with Indonesia, Papua New Guinea and France (through New Caledonia). The seas concerned stretch from the tropical waters of the Great Barrier Reef to the cold waters of the Antarctic around Heard Island.

Origins of Australia's Ocean Policy

Australia's EEZ (Exclusive Economic Zone) contains an unknown value of natural resources scattered over 11 million square kilometres from the tropics to the Antarctica. The current economic value of use of this marine area is over A$19 billion ($US 14.63 billion) with over half of this coming from marine tourism mainly, but not exclusively based around the Great Barrier Reef (AMSIC, 1996).

In biological terms there is very high diversity in the tropical waters but just as importantly there is a staggering degree of unique species in the southern waters where about 90% of all species in the major taxa are endemic (Zann, 1995).

The ratification of the Law of the Sea and declaration of this massive EEZ has stimulated interest in marine affairs in Australia.

As well though a Federal Government program called 'Ocean Rescue 2000', which resulted from a major conservation conference in 1991, has played a significant role in increasing public awareness and compelling Australia towards the development of an integrated and comprehensive Oceans Policy.

Ocean Rescue 2000 consists of five inter-related programs (Figure 3).

The initial aim of the program was to prepare a national marine conservation plan but in light of the economic significance of the EEZ (described above) the derivation of a more comprehensive Ocean Policy has been promised, using the State of Marine Environment Report (Zann, 1995) as an initial data base and the Marine and Coastal Community Network as a disseminator of information.

In fact at the Australian Federal election in March 1996 both the major parties promised to prepare an Ocean Policy for Australia.

The current Government specifically promised:

> 'In consultation with all major stakeholders, we will develop a comprehensive and integrated Oceans Policy. Issues to be examined in formulating this Oceans Policy will include the future sustainable use of our marine environment, the protection of representative marine systems and the

development of effective marine management strategies' (Coalition Electoral Platform).

Before describing the process by which this Ocean Policy is to be prepared and the issues it may cover it is worth reflecting on the advantages and disadvantages Australia possesses in preparing such a policy.

<u>Impediments and stimulants to an Australian National Oceans Policy.</u>

Australia is in an excellent position to produce an Oceans Policy because:

- as an island nation we lay claim to an ocean territory which, with the exception of small common boundaries with PNG, France and Indonesia, is not complicated by interactions with other sovereign states,

- Australia possesses a National Strategy for Ecologically Sustainable Development which provides the conceptual framework for the preparation of a comprehensive Oceans Policy, and

- there is bipartisan support for the preparation of an Oceans Policy.

There are also some impediments to Australian development and implementation of an Oceans Policy. These are:

- the risk of domination by one, or a few, sectoral groups in the development of a policy;

- inter-agency rivalries and territoriality dominating the development of a policy to the exclusion of the crucial issues;

  the risk of a bureaucratically controlled process, based in Canberra, acting to the exclusion of user-groups and the community;

  friction between the various levels of Government in Australia; and the risk that the final policy is a series of platitudes with no suggestion of how they would be interpreted and implemented or brought into practice.

Clearly the advantages that Australia possesses in this area need to be capitalised on whilst the impediments expressed above have to be counteracted if a comprehensive and useful Oceans Policy is to be produced.

The **process** by which the policy is developed, the **content** and nature of the policy document itself and the method of **implementation** hence are critical to the eventual success of the Oceans Policy itself.

The Process for development of the Australia Oceans Policy.

After election in March 1996 the Federal Government asked the Department of Environment, Sports and Territories (DEST, the Department has been renamed "Environment Australia") to oversee the development of an Oceans Policy. They have been assisted by an inter-departmental committee made up of the major Federal Government agencies involved in marine affairs Wescott (1996a and b, 1997) describes these features in more detail.

On the 3 March 1997 the Prime Minister of Australia launched a 'consultation paper' on Australia's Ocean Policy (Howard 1997). The material that follows originates from that paper.

Draft Goal of an Oceans Policy

The draft goal is:

"To promote efficient, ecologically sustainable use of Australia's ocean resources while conserving the biological base and maintaining the underlying ocean processes".

Draft Objectives of an Oceans Policy

The draft objectives are:

"a. to understand, monitor, conserve and sustainably use the ocean environment and its resources;

b. to promote economic development through ecologically sustainable ocean industries;

c. to improve and use our expertise and capabilities in ocean related science, technology, research and engineering;

d. to accommodate identified and agreed community interests and responsibilities; and

e. to exercise and protect Australian rights and jurisdiction over offshore areas, including resources."

The paper says that:

"The Oceans Policy will provide a strategic framework for the planning, management and ecologically sustainable development of Australia's ocean industries, including fisheries, shipping, petroleum, gas and seabed resources. It will also enable the integration of other ocean related uses and services such as defence and climate and weather. prediction".

The Australian community has been asked to comment on the scope of an Ocean Policy before the end of April 1997 and have been promised a second round of consultation during 1997. The final policy is to be produced by the end of 1997.

The Issues

The issues to be considered in the policy fall under ten major headings (a full list of the preliminary issues is given in Appendix 1):

1. Planning and Management
2. Conservation and management of marine biodiversity
3. Marine pollution
4. Ocean Industries, Employment and Economic opportunities
5. Scientific understanding
6. Indigenous issues
7. Surveillance and enforcement
8. Shipping and sea safety
9. International considerations
10. Community participation, public awareness and understanding

Critique of the Australian approach with possible implications for other nations.

If one considers Australia's potential advantages and disadvantages in preparing an Ocean Policy (see previous pages), then in light of the consultation paper and proposed process the potential flaws are clearly revealed.

Australia's advantages speak for themselves and although some apply equally to the USA, what Australia does possess and other countries may not is the framework of a national strategy for Ecologically Sustainable Development (Commonwealth of Australia, 1992). This provides a 'balanced' framework in which to couch an integrated oceans policy and is a major theoretical advantage - a lot of the groundwork has been done. Despite this the principles enunciated in the national strategy are not immediately evident in the consultation paper except to use the phrase "ecologically sustainable use" on occasions.

Bipartisan support in Australia means that proponents of an Ocean Policy can debate its contents and direction rather dissipating energy on arguing for its very existence. But some disagreement has already occurred with the Opposition claiming the consultation paper is too development oriented.

On the negative side the policy could be 'hijacked' (or 'white anted') by sectoral interests. In the Australian context this is most probably by the mining sector and its technological support groups. Only a strong public education campaign and heavy lobbying from conservation groups will prevent this.

Inter-agency rivalries are also still a real risk to development of the final policy and the lack of an independent coordinating group, as I have recommended (Wescott, 1996b, 1997), with NGOs in the majority makes this more probable.

But perhaps the most alarming aspect so far in Australia's ocean policy development is that the process has been dominated so far, and in the absence of the independent coordinating group, will continue to be dominated by the Canberra (read Washington) based bureaucracy.

Nothing will diffuse enthusiasms and alienate the general community more than a continuation of this approach.

Friction between the various levels of Government (Federal / State/ Local) will always be an issue in Australia but in the context of the Federal Government controlling the great bulk of the area concerned (from 3 to 200 nautical miles, the States from high water to 3 nautical miles) this will be a less significant issue than usual.

Finally the risk that the final policy will be so general or so platitudinous that it will be impractical and not be able to be implemented is high, and may actually serve the wishes of some sectoral groups, if they are unsuccessful in 'capturing' the policy in the first place.

I believe the way to overcome these problems is to wrest the control of the development and implementation of the process away from the bureaucrats and give it to an independent coordinating group responsible to the Environment Minister. A starting model for this may be the approach adopted in Oregon with the Oceans Advisory Council (see Bailey's paper in these proceedings). Simultaneously a broad educational program and the building of a strong constituency in support of a comprehensive non-sectorally biased policy (possibly through the Marine network mentioned early) would provide Australia and other countries with the best chance for a truly visionary and practical integrated Oceans Policy.

Finally a fundamental question of how detailed the first Oceans Policy should be on completion needs to be debated. Does the Ocean Policy establish the overall objectives and principles for the management of human use in the Australian EEZ and leave the sectors to prepare separate detailed strategies consistent with these objectives and principle or is it more detailed and attempt to deal with the specific issues raised? The first would be achievable within the time-frame proposed. The latter would probably not be. There is much to be said, I believe, in getting something completed while the momentum (and political will) is there. Then the Coordinating Group could oversee the preparation of more detailed strategies over a longer period.

References:

Australian Marine Industries and Sciences Council, 1996. Draft Marine Industry Development Strategy. Dept. of Industry, Science and Tourism. Canberra.

Bergin, A. & Michaelis, F.B., 1996. Australia and the South Pacific: implementing the UNCED oceans agenda. Marine Policy 20(1): 47-62.

Commonwealth of Australia, 1992. A National Strategy for Sustainable Development. AGPS, Canberra.

Howard, J. (Commonwealth of Australia), 1997. Australia's Oceans. New Horizons. Oceans Policy Consultation Paper. Canberra. Available on: http:/www.erin.gov.au/portfolio/esd/coast-marine/oceans/consult.htnl.

Wescott, G., 1996(a). The Next Step towards developing an Australian Oceans Policy. School of Aquatic Science and Natural Resources Management. Technical Paper 1996/3. Deakin Univ., Melb.

Wescott, G., 1996(b). Recommendation on the Development of an Australian Oceans Policy. Discussion paper, Deakin University, Melbourne.

Wescott, G., 1997. Time to get serious about our oceans. Search 28(2): 50-53.

Zann, L., 1995. Our Sea, Our Future. Major Findings of the State of the Marine Environment Report for Australia. DEST, Canberra.

Appendix 1: Some Issues listed in the Consultation Paper.

Planning and Management

- a marine planning framework and associated institutional mechanisms; co-ordination between and within all relevant government agencies and advisory bodies;

- consistency in the application of multiple and sequential use principles;

- the adequacy and consistency of marine-related legislation; responsibilities for monitoring, data acquisition and supporting research;

- access to, and sharing of the benefits of, genetic resources; the management of highly migratory fish and straddling fish stocks;

- the management of emerging industries, eg. deep sea bed mining;

- limitations and/or inconsistencies in marine legislation applying to the Australian Antarctic Territory.

Conservation and Management of Australia's Marine Biodiversity

- marine biodiversity and the introduction of exotic marine pests and disease pathogens;

- the effectiveness of endangered/threatened and vulnerable species recovery efforts;

- marine and coastal habitat loss and degradation;

- the role of marine protected areas and the relationship to multiple and sequential use principles;

- impacts of fishing and bycatch on the marine environment; understanding the impacts or marine industries and recreation activities;

- understanding the causes and rates of degradation and recovery of the environment;

- existing and proposed oceanic World Heritage properties; impacts of climate variability and change;

  the existing framework for, and future development of, State of Environment reporting.

Marine Pollution

- the wide range of sources of marine pollution (land, sea and atmospheric);

- nutrient flows from river catchments;

- the environmental performance of sewage and stormwater systems;

- national water quality standards;

- ballast water strategies and hull fouling; oil spill response capacities;

- gaps in management regimes for marine pollution for example chemical, mineral and bunker fuel spills and shipping and barging operations;

- marine debris including impacts from vessel (inside and outside EEZ) recreational and urban sources.

Ocean Industries, Employment and Economic Opportunities

- current problems and ways to improve government regulatory and management regimes for marine industries;

- development of multiple use principles;

- collection of performance measures and information about future industry trends;

- the breadth of marine industries and their likely future significance, regionally, nationally and internationally;

- current and projected economic values and employment opportunities;

- mechanisms to encourage investment in capital intensive industries, eg. oil and mineral exploration and production;

- the supplier chain serving these industries and ways to improve the chain to minimise imports and develop domestic enterprises and technologies.

Our Scientific Understanding

- articulation of Australia's marine science and technology needs and how such needs can be funded over the longer term;

- institutional and infrastructure arrangements;

- information to support decision making and marine management;

- integration of marine-related data;

- the adequacy of existing databases;

- national mapping use and resources survey;

- national monitoring and data collection;

- development of reliable indicators;

- a multidisciplinary approach to marine science and other areas; Australia's expertise and capabilities in ocean related science, technology, research and engineering;

- access for foreign research programs in Australia's EEZ.

Indigenous Issues

- Aboriginal and Torres Strait Islander marine management and conservation aspirations;

- indigenous fishing.

Surveillance and Enforcement

- the roles of relevant Commonwealth and State/Territory agencies;
- ballast water management strategies;
- strategies on ship sourced pollution;
- marine weather forecasting services;
- marine areas sensitive to shipping and appropriate sea lanes; port waste reception facilities;
- interactions with marine mammals;

antifouling issues.

International Considerations

- regional co-operation including science and management and common resource management issues;
- conservation of migratory wildlife;
- conservation and sustainable use of straddling and highly migratory fish stocks;
- promotion and export of Australian expertise in marine science, management and technology;
- global dimension of unregulated high seas fishing, climate change, ozone depletion, ballast water waste etc.;
- need for international co-operation to implement an ocean observing system;

need for a clearly defined strategic international role for Australia in co-operation and maritime affairs.

Community Participation, Public Awareness and Understanding

- increased awareness and understanding of the marine environment, including costs, benefits and impacts of the use of the marine environment;

- the need for a stewardship ethic;

- capacity building of individuals and organisations.

Figure 1: The South Pacific prior to ratification of UNCLOS.

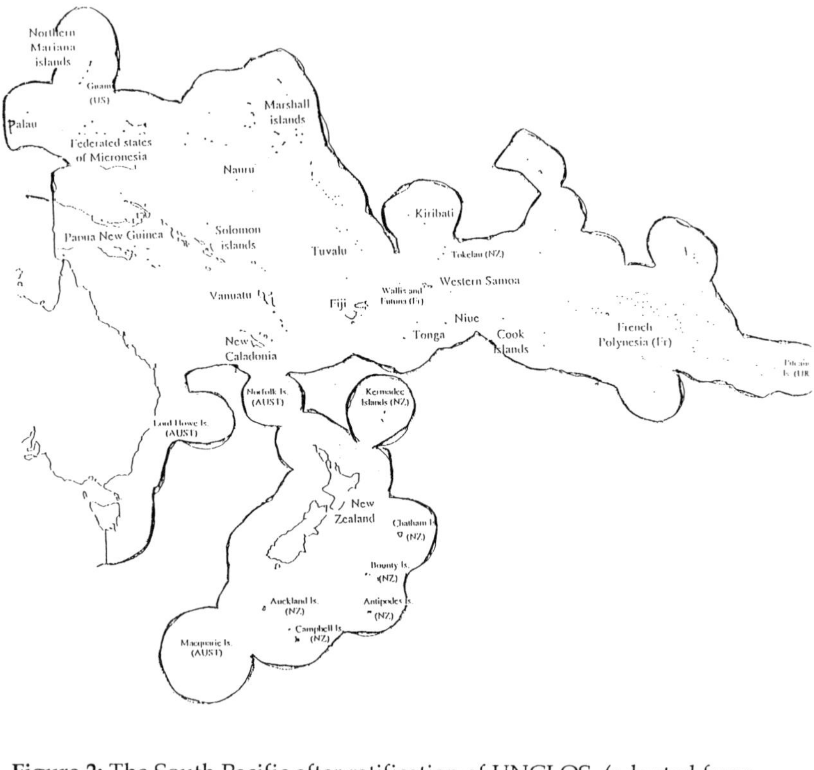

Figure 2: The South Pacific after ratification of UNCLOS (adapted from Bergin and Michaelis, 1996).

Figure 3:

Ocean Rescue 2000 Programs:

Launched August 1991, 10 year program, A$1.7 mill. in first year.

Objective: to develop and implement a *national marine conservation plan* based on the principles of ecologically sustainable development.

Five inter-related programs to contribute to the plan:

State of the Marine Environment Report (SOMER, published February 1995)

Marine and Coastal Community Network (established 1993)

National Marine Education Program

Marine Protected Area Program

National Marine Information System

CHAPTER 49

GIS Data Integration: Build It and They Will Come

By John Ellison and Paul Veisze
Technical Services Branch
California Department of Fish and Game

INTRODUCTION

When the Exxon Valdez ran around on Bligh Reef in Prince William Sound, Alaska in the spring of 1989, world governments were reminded once again about the enormous risks associated with the transportation and handling of petroleum products near areas with sensitive biological resources. California with its more than 1,100 miles of marine coast, thousands of species of native plants and animals, and its large petroleum industry was just such a region. The California Legislature responded with the passage of Assembly Bill 2040, the Lempert-Keene-Seastrand Oil Spill Prevention and Response Act, in 1991. AB 2040 established the Office of Oil Spill Prevention and Response (OSPR) in the California Department of Fish and Game (Department).

AB 2040 presented the Department with both a challenge and an opportunity. The challenge was to inventory and map the many sensitive biological, physical, and cultural resources at risk from marine oil spills, and to use this information to plan for their protection. The opportunity came from the funds associated with this new legislative mandate. Thus, the Department contracted with the consulting firm of Ernst and Young in 1992 to conduct a Department wide assessment of the feasibility of implementing an information management system to meet the needs of this new mandate and the Department=s many existing responsibilities. Because information about resource location (e.g., spatial or geographic data) was a critical element of most Department data, geographic information system (GIS) technology was to become the foundation for this integrated solution.

The Department=s experience with GIS dates back to the formation of the California Natural Diversity Data Base (CNDDB) in the early 1980s in what was to eventually become its Natural Heritage Division. At that time, The Department leased time on a Digital Equipment Corporation PDP 11/70 computer owned and operated by the California Department of Water Resources. This system used Computer Aided Design (CAD) software from the Intergraph Corporation and custom programing to produce rudimentary GIS-like functionality. In 1985, the Department contracted with BSI Consultants to do a feasibility study for a GIS to meet the data management needs of the CNDDB. This resulted in the procurement and implementation of a four seat turnkey GIS using Hewlett-Packard Unix workstations and Genamap, a GIS software product from Genasys.

The end result of the Department's most recent GIS feasibility study was a five year project to implement a Department-wide GIS starting in June of 1993. Facilities and staff to collect, process, manage, and analyze GIS data were placed in key Department programs. Most germane to the California and World Ocean '97 Conference was the GIS service unit established in the Department=s Technical Services Branch, and the GIS facilities and staff in the Department=s Office of Oil Spill Prevention and Response. These staff and facilities became involved in the mapping of ocean resources susceptible to damage from marine oil spills.

These activities soon began to draw the attention of other state agencies that were interested in GIS but had not yet launched their own GIS projects. Resource Agency staff were also interested in this technology for solving large area resource management problems and for tracking existing efforts in this regard. Thus, the Department became involved in two projects specifically having to do with coastal and ocean resources. The first was a request from the Resources Agency to concentrate on the assembly of data for the Monterey Bay region. The second was a contract from the State Water Resources Control Board (SWRCB) to support initial implementation of GIS in their Sacramento headquarters (see CASE STUDY section below).

The purpose of this paper is to share the experiences gained by Department GIS staff from responding to requests from biologists, managers, and policy makers. In particular, we want to share our perceptions about the availability and state of biological data sets of potential use in a GIS. We hope that this will lead to a better understanding of what this powerful technology can and can=t offer. In the end we are seeking a more productive relationship between those seeking GIS products and services and those that provide them.

LESSONS LEARNED

1. Do you expect that the data that you need has already been collected?

    It seems that many higher level managers, policy makers, politicians and others assume that enough data have already been collected, and all that remains is to affect a mechanism for its cataloging and conveyance. After all, haven=t government organizations and others been collecting environmental data for decades now? This seems to be associated with a desire to substitute traditional data collection efforts that scientists have longed relied on to support objective analysis and sound decision making with more politically expedient (and less expensive) processes. Efforts to conduct so called "professional judgement assessments" are examples of this. It is vital to recognize that opinion polls results cannot take the place of solid data objectively collected using repeatable and defendable methods.

Regrettably, our experience suggests that there is no vast, untapped "Mother Lode" of natural resource data awaiting discovery. A more suitable analogy is that of ore with a very high overburden-to-metal ratio. Data are often found to be outdated, incomplete, poorly attributed with locational data (e.g., no geocodes), often unsuitable for the intended purpose, unrecoverable (e.g., lost), or just plan missing. Data cataloging efforts, in spite of their obvious appeal as the logical first step, frequently become painful exercises in documenting our ignorance, data gaps, and the inadequacies of most previous data collection efforts. Most data were and continue to be collected for very narrow purposes or application. Its not unusual to find in those cases where data were collected with the thought that they might one day prove useful, that they are lacking in light of current demands and needs.

This is not to suggest that all is doom and gloom with respect to GIS data sets. There is a good supply of so called "framework" data sets important for general cartographic purposes thanks to the good efforts of the United States Geological Survey and other data publishing agencies and organizations. There is also a large amount of satellite and other remotely sensed data. Much of this raster imagery, however, is of too course a resolution (e.g., 10- to 25-meter pixels) to be useful for many large-scale, site specific natural resource management applications. This is changing as new satellites are launched with higher resolution sensors (e.g., 1.0- to 5-meter pixels). NASA is also a source of excellent data collected by use of sensors mounted on high altitude aircraft. These high resolution raster data tend to be expensive to acquire, store and process, however, due to their large file size and the technology used in their capture.

Department staff had an educational experience in the early nineties regarding GIS data availability when the Natural Community Conservation Plan (NCCP) process was first initiated for a region in Southern California. This regional conservation planning effort sought to protect from excessive development the Coastal Sage Scrub community and its associated species, especially the California gnat catcher. It was obvious from early meetings of interested parties representing both the private and public sectors that there had been numerous efforts to map this community. The Department was tasked with assembling the resulting data to see if it was possible to merge them into a coverage depicting the extent and condition of Coastal Sage Scrub community in Southern California. Unfortunately, few of the organizations used the same methods, community classification, or mapping procedures for their Coastal Sage Scrub mapping projects. In the end, the Department had to settle for being a central repository for these data sets without being able to merge them into a single comprehensive coverage of the targeted theme, principally due to the utter lack of documentation for the disparate data sets.

2. What problem or problems are you trying to solve?

Beware the hardware or software "bigot." Emotions can run high when discussions turn to what computer system are "best," often taking on the semblance of a holy crusade. Those that claim undeniable superiority for their platform of choice should be viewed with suspicion. Concentrate first on the business or operational problems that you are trying to address. Pay particular attention to solutions used by your partners and potential data sources. Use this information to derive a set of functional requirements, and then look around for the appropriate technology to answer your needs.

3. How good are the data?

The following points should be considered when planning your GIS data development efforts:

A. What can you find out about existing data sets (e.g., is there adequate database documentation or metadata)?

The reluctance to take the time to document one=s efforts seems to be a basic element of human nature judging by the general dearth of metadata (e.g., data about data) for many data collections. All too often we find that there is little documentation for where, why and how data were collected. Data dictionaries are seldom provided (e.g., look up tables for coded entries are not given). The absence of this documentation greatly hamper the already daunting task of coming to terms with extra- and sometimes inter-organizational data sets, and often is the chief contributing factor to the decision to risk duplication of effort by undertaking a new data collection effort.

B. Do you have good (adequate) locational data?

Although commercial GIS has been around since the early 1970s, many traditional data sets were not developed with this technology in mind. Consequently, attribution of location, the so called geocode, is frequently missing or so vague as to be useless. You and your staff can spend considerable time translating narrative descriptors about location into specific geocodes required for GIS. Even then the resulting locational references may be too imprecise to suit your needs (e.g., a zipcode instead of the street and house address that you really need).

C. Are your data consistent?

Ralph Waldo Emerson may have been right when he said that "a foolish consistency is the hobgoblin of little minds," but being consistent in the use and application of data entry rules certainly makes life easier for those that seek to make good use of databases. Data inconsistently entered or coded can confound database queries. Users will lose confidence that your efforts will yield valid and viable results. This may seem like an esoteric consideration, but your staff can spend many hours cleaning data and tuning a database to make it perform properly.

D. Are available data suitable to your needs?

You need to have a firm grasp on your specific needs when assessing the usefulness of other=s data to your situation. How accurate do you need the data to be (e.g., what resolution do you require)? Is a mapping specificity of 100 meters adequate? For large region planning this would be perfectly adequate, but consider the problems that could arise if you were using these data to decide where it was safe to trench for a pipeline. You would want to know within a few inches where that gas line was before your backhoe operator started breaking ground. It is reasonable to consider if you should ever rely on digital information where the consequence of failure or error is so high without first doing a detailed site survey.

4. What is your technology environment?

In planning the Department=s current GIS, we had to consider what barriers we needed to overcome before we could deliver GIS data and tools into the hands of our biologists throughout the Department. Of principal concern was the technology they already had available to them. It would do little good to implement a powerful system reliant on exotic, and costly technology, that we could not afford to give our biologists, could not support, or that would require extensive training to use.

We also had to consider where we were going to get our data from and what technologies our partners had. Given that about 80% of the on going cost of GIS is in data acquisition, we had ample motivation to select a system that optimized compatibility with systems used by our partners.

5. What is your corporate culture?

Does your organization have a good track record of intra-departmental cooperation and coordination? Do you have confidence that the hardware, software and procedural standards that are so important to the successful implementation of GIS will be followed by your organization? These are important considerations when planning where within your organization to place GIS technicians.

No single best answer can be given to this question. Much will depend on organizational procedures and political realities (e.g., is it required that all information technology related projects be placed within the Management Information Systems group?). Whatever is decided, it is very important to recognize that the setting and following of standards for data, application design, and collection protocols are critical to the success of GIS. If your organization has a history of programs and divisions operating in a largely autonomous manner, with little intra-agency cooperation, then serious consideration should be given to centralizing the authority and responsibility for standards development and enforcement. One measure of how much centralization of authority is needed is to determine to what extent your field staff are now using standardized field survey forms and procedures.

It is important that your GIS staff and equipment be physically located as close as possible to the staff that are to benefit from their support. The solution in the California Department of Fish and Game was to dedicate and locate GIS staff and equipment to specific programs while organizationally placing the GIS staff under a manager in our Technical Services Branch. This way program managers could be comfortable with having specific resources to call their own, and the communications needed to insure that standards were set and followed would occur.

CASE STUDY

The DFG - SWRCB INTERAGENCY GIS SUPPORT PROJECT

GOALS

The overall goal of this project was to conduct GIS operations of mutual interest to DFG and SWRCB, within the scope of non-point source water pollution analysis. Specifically for SWRCB this involved:

- Identification of the State's Hydrologic Units and the federal Cataloging Units encompassing impaired water bodies
- Production of digital maps of impaired water bodies and source waters
- Compilation, documentation, and registration to the above maps related themes such as jurisdictional boundaries, wetlands, landfills, groundwater basins, point source discharges, roads,

recreational areas, and marinas

- Coordination of GIS data acquisitions with US EPA, California Coastal Commission, Teale Data Center
- Consultation with SWRCB and Regional Water Quality Control Boards for presentation and interpretation of GIS data

BACKGROUND

DFG and SWRCB have complimentary regulatory roles. For example:

- Timber Harvest Plan (THP) reviews
- Streambed alteration agreements
- Water rights application reviews

An important precedent for interagency cooperation also existed in the form of "Lakes of California" (SWRCB 1978). Jointly produced by DFG and SWRCB, this electronic database described ecological and management aspects of over 4900 lakes in the state. In addition, both DFG and SWRCB had interest in applications of the United States Environmental Protection Agency's River Reach File (US EPA 1994), a national hydrographic addressing system.

PROBLEMS

The two main problems addressed in this project were inconsistent geocoding of water quality assessments and Reach File errors of omission and commission. Geocoding, in the SWRCB context of impaired water bodies, refers to the assignment of specific watershed codes to water quality assessments and problem statements. Reach File errors consisted of missing and incorrect water feature names and addresses throughout the state coverage. In pre-GIS operations, SWRCB and their Regional Boards were able to compensate for such errors by using local knowledge of water problem locations. However, when coordinating solutions to water quality problems, local databases would have to be shared among a variety of users, and the lack of consistent geocoding would prevent effective communication.

For example, the State code for the Klamath Hydrologic Unit (HU 105.00) encompasses seven federal Cataloging Units (CU 18010204 thru 18010210), whereas in southern California, the inverse situation occurs: a single CU (e.g. Salton Sea CU 18100200) can contain up to ten State HUs. These mixed subset/superset relationships are difficult to manage even under good adherence to the coding systems, let alone under less-disciplined location references.

SOLUTIONS

Given the above problems, the most viable approach was to

assemble digital maps of the different watershed coding systems into a single frame of registration. This would enable users to quickly identify the relevant codes and names of basins encompassing a water quality problem location. The basin coding would enable specifications of problem areas to a resolution of about 50,000 acres.

Regarding the River Reach File, the solution was to reconcile water feature names in the digital data with those on U.S. Geological Survey published maps, the de facto national standard--simple in concept, but labor-intensive and tedious in execution. Nevertheless, over the course of this project, over 40,000 records of water feature names were reviewed and/or corrected out of the approximately 238,000 total records in California. The most significant benefit of the Reach File work will be for support of SWRCB compliance with US EPA mandates to report water quality assessments using Reach File codes. Reach File codes will enable a greater level of specificity than the watershed coding described above by virtue of addressing individual water bodies and portions thereof, to a linear resolution up to about a tenth of a mile.

Table 1 presents a summary of the inputs and outputs of the project, and Figure 1 is a map produced from the compilation of digital data for one of this project's pilot study areas, Tomales Bay.

Table 1. DFG/SWRCB GIS Support Project Summary

INPUTS	
Funding:	$73,000
Time Period:	10/07/94 to 6/30/96
Personnel:	1 Research Program Specialist I (GIS)
	1 Research Analyst I (GIS)
	3 Technicians (Fish and Wildlife Scientific Aides)
Data:	Teale GIS Technology Center: hydrography, roads, etc
	CA Dept. Conservation: farmlands
	Counties: cadastral, agricultural
	CalPoly San LuisObispo: vegetation, land use
	CA Dept. Forestry: watershed boundaries
Equipment:	HP UNIX Workstations, IBM PCs
PROCESSING:	Data restructuring; GIS Integration; Documentation;Display
OUTPUTS	
GIS datasets:	Tomales Bay; San Luis Obispo Creek; Morro Bay
Databases:	Critical Coastal Areas; Beneficial Uses; Priority Watersheds; NOAA Water Quality Protection Program; 40,000+ corrected River Reach File records
Metadata:	58 text files
Live demos	
Trained personnel	

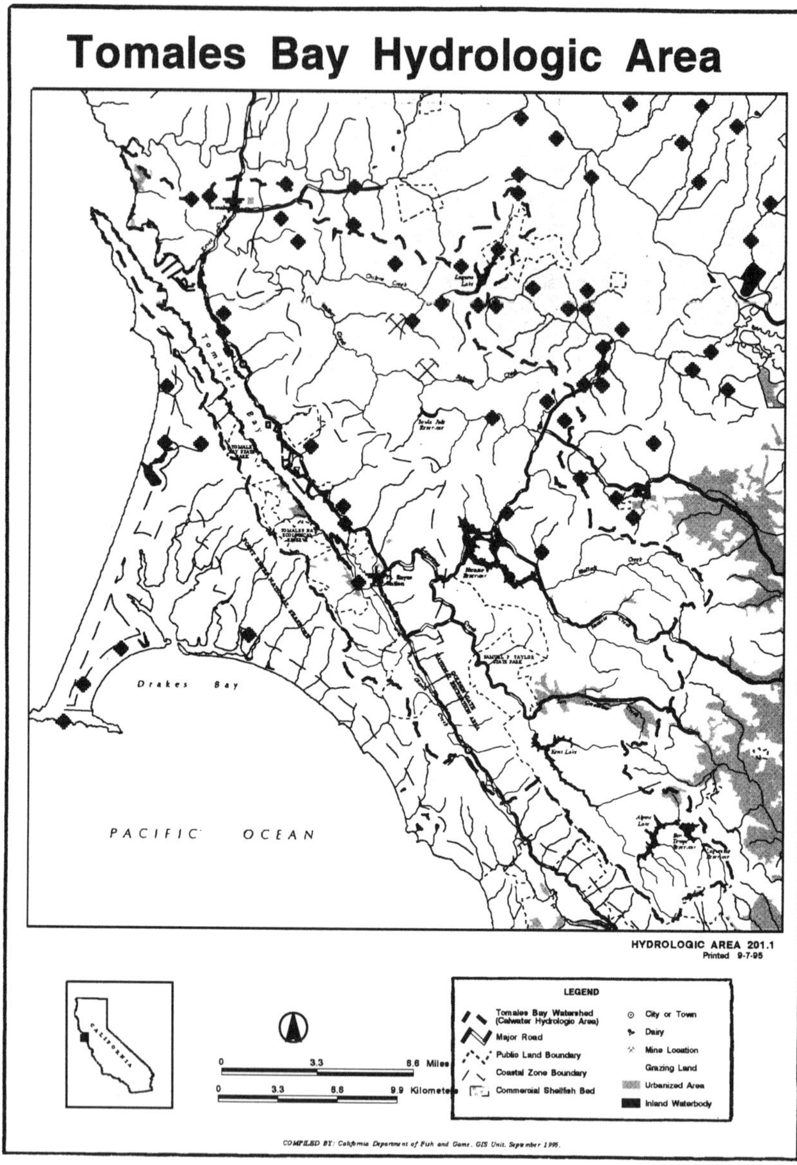

Figure 1. Map of Tomales Bay watershed created for SWRCB GIS project.

CONCLUSIONS

In general, and with only a few notable exceptions, it is our experience that state-wide or large region coverages of biological data sets appropriate to natural resource management in California are lacking, out-dated, fragmented, and poorly attributed and documented. In light of this, it is distressing to note what appears to be a tendency on the part of politicians, policy makers, and agency officials, to assume that enough data have already been collected, and that only minimal further investments need be made.

The emphasis appears to be on providing access to data and information resources without a commensurate investment in data or information substance. To be sure, developing technology infrastructure and mechanisms for data access and delivery are essential. However, we would like to see a more balanced and diverse investment strategy that addresses data quantity and quality as well.

Too often managers see data collection as a one time only exercise. The "snap shot" approach to data capture does not suffice with natural systems. Ecological entities are dynamic through time and space, and often exhibit a large natural variation even before anthropogenic effects are considered. Those that would work with these systems must realize that on going monitoring is required if we are to accurately assess the affects of societal actions on natural system health and viability.

More disturbing than this, however, is the under funding of agencies with public trust responsibilities so that they must rely on data from the very industries that they seek to regulate. This "fox in the hen house" approach to data sourcing has deficiencies too obvious to mention. Outsourcing may be valid when building cars or even in the design and the construction of our highways, but its role in data development for public trust matters must be carefully considered. Public agencies must be funded, staffed and equipped to define what is required of natural resource data and certify data sets from private sources that may be offered for use in public sector decision making.

GIS has much to offer biological resource management and conservation. GIS not only helps us better visualize and understand natural systems and the results of our actions upon them, but it is an excellent technology for integrating data, and a wide variety of planning and management efforts. GIS provides a sound technological foundation for a comprehensive, integrated, and holistic approach to ecosystem management.

Ecology is not an exact science and our understanding of natural system functions and mechanisms will probably always be imperfect. We must then approach ecosystem management with a profound sense of humility and appreciation for our own inadequacies. The benefit of a doubt should always be with mother nature. We must shift the burden of proof from those that seek to conserve natural resources to those that would benefit from their exploitation.

Collaboration is an effective and essential element of crafting information technology solutions. We strongly recommend that government agencies take the lead in assembling working groups around well defined resource issues to do a thorough assessment of data needs to as high a level of detail as can be tolerated. These need assessments would then form a template against which the adequacy of existing data may be judged and upon which future data development efforts would be designed.

It is expensive to place a qualified observer in the field and build and maintain the technological systems that facilitate efficient data capture, distribution, and use. About 80% of the on going costs of a GIS are in data collection and maintenance. However, the loss of our precious and irreplaceable natural biological heritage from our making the wrong decisions today would be a far greater price to pay. Consider how much more tragic and reprehensible this would be if we acted in wilful ignorance, unwilling to properly inform ourselves as to the consequences of our actions because the bottom line was threatened.

REFERENCES

U.S. Environmental Protection Agency. 1994. The U.S. EPA reach file version 3.0 alpha release (RF3-Alpha) technical reference. EPA contract number 68-C3-0303. First Edition December 1994. http://www.epa.gov/OWOW/NPS/rf/techref.html

California State Water Resources Control Board. 1978. Lakes of California, an electronically processed file. Division of Planning and Research, Surveillance and Monitoring Section. Sacramento, CA. February 1978.

California Department of Fish and Game. 1980. Lakes of California. Inland Fisheries Division Administrative Report 80-5. Sacramento.

CHAPTER 50

TECHNICAL AND HUMAN ISSUES IN GIS DATA INTEGRATION: THE TIJUANA RIVER WATERSHED PROJECT

Dr. Richard Wright, Department of Geography, San Diego State University
Dr. Alain Winckell, El Colegio de la Frontera Norte / ORSTOM

ABSTRACT

The Tijuana River Watershed (TRW) supports some of the most diverse ecosystems in North America and the largest urban population along the U.S. Mexican border. Because of the enormous development pressures in the San Diego/Tijuana urban core, which are becoming even more pronounced as a result of NAFTA, the human and natural environments have been unable to co-exist in a sustainable way. The huge demand for fresh water in this semi-arid region has had devastating effects on the Tijuana Estuary -- one of the last remaining functioning wetlands in southern California. Industrial, agricultural, and domestic sewage releases into the river have also had unacceptable consequences on the human populations in the form of outbreaks of gastro-intestinal diseases and other medical conditions.

Disparities between political, economic, and educational systems, and access to information have hampered efforts by both countries to jointly address the interrelated issues of environmental quality and human development within the watershed. Since 1994, funding from the National Oceanic and Atmospheric Administration (NOAA), augmented by support from other U.S. Federal agencies, has supported the development of a bi-national geographic information system (GIS) for the TRW to serve as an analysis and planning tool to facilitate transboundary watershed planning. The GIS has been developed jointly by San Diego State University (SDSU) and El Colegio de la Frontera Norte (COLEF). The agreement between SDSU and COLEF provides for data sharing, coordinated data development and scientific research, and joint use of the data for education and research. The project has five complimentary components: GIS database development, community outreach, education (university and K-12), scientific studies, and watershed management. The GIS data layers were developed in consultation with numerous individuals from city, county, and state governments of California and Baja California, as well as GIS experts, teachers, planners, and ecologists. A strong community input ensured that the information and scales were relevant to management-related applications on both sides of the border.

In this paper the authors describe the geographical characteristics of the watershed, major issues in the watershed, project goals, and the technical and human issues involved in creating and implementing a binational transborder GIS. Human factors discussed include legal ownership of data, institutional politics and priorities, and socio-economic characteristics. Technological factors include interoperability of GIS hardware and software systems, technological capability, data documentation, map coordinate systems and datums, and the accuracy, completeness, and currency of locational and attribute information.

CHAPTER 51

Developing Integrated Access to Contemporary Aerial Photography and Water Quality-Related Data for a Portion of the Elkhorn Slough and Vicinity, North Monterey County

Jonathan F. Van Coops[1]
Jayson A. Yap[2]

Abstract

In carrying out the day-to-day operations of their regulatory and planning programs for managing water quality, California's Coastal Zone planners and managers must routinely make use of spatial information in the form of maps, georeferenced tabular data, and aerial photography. In the central coast region of California, a wealth of such geographic information has been collected and developed during the last several decades. However, challenges to the integrated use of these data sets, by both staff level personnel and decision-making bodies, have persisted and increased due to the number of public agencies, academic institutions, and other research efforts collecting and generating data. Additionally, data from different agencies and jurisdictions are often initially incompatible due to variations in hardware, software, map scale, coordinate systems, data structure and other factors affecting the gathering and storage of data. In particular, effective management of water quality and non-point source pollution in the vicinity of the Elkhorn Slough National Estuarine Sanctuary (NES) in Monterey County has been hampered by lack of, or difficult access to, information required for implementation of adopted strategies and procedures, status and trends analysis, evaluation of permitting issues and impacts, and monitoring of mitigation sites and permit condition compliance. *This paper describes a project undertaken by the Technical Services Division of the California Coastal Commission to develop access to contemporary aerial photography and a variety of water quality-required information for seven critical subwatersheds in the vicinity of Elkhorn Slough NES, and to provide these essential elements of the material required for effective stewardship and balanced decision-making to all levels of resource managers and interested persons in the region.*

---

[1]Mapping/GIS Program Manager, California Coastal Commission, 45 Fremont Street, Suite 2000, San Francisco, CA 94105-2219
[2]Environmental Information Systems Specialist, California Coastal Commission

Introduction

Effective management of water quality and polluted runoff in California's coastal zone requires the ability to quickly access maps, georeferenced tabular data sets, and aerial photography. In order to analyze cumulative impacts and other issues that cross political and administrative boundaries, these data must be integrated and presented in a form that can be easily used by managers, decision-makers, and others who are not technical specialists in geographic information systems (GIS). Whether intended for developing finished cartographic products or custom software applications, this integration process can sometimes require levels of funding and staff expertise beyond the capacity of management entities, particularly local agencies that are responsible for carrying out many of the regulatory activities related to control of polluted runoff.

This project is one of several recent Coastal Commission efforts that will ultimately help move the central coast region towards ecosystem and watershed-based environmental management. It was undertaken in order to provide some of the integrated data products needed to facilitate basic access to aerial photography and water quality-related data for a portion of the Elkhorn Slough and vicinity, shown in Figure 1. In the Monterey Bay region, water quality, sediment, and tissue sampling data needed to improve management of polluted runoff are being collected by public agencies, academic institutions and other researchers. Contemporary aerial photography of the entire region was acquired by the National Ocean Service in 1994.[3] During the first part of 1996, after acquiring computer software capable of displaying linked map, image, and tabular data sets, the Coastal Commission's Technical Services Division designed and completed tasks that would allow managers to view a digital base map, showing sampling station locations and flight lines linked to scanned images of corresponding aerial photography. In addition, the product provides the ability to view and query the water, sediment, and tissue sampling data associated with any of the stations.

Project Description

The goal of this project was to create an ArcView "project" file, consisting of a base map of the four Elkhorn Slough area USGS 7.5 minute quadrangles (Watsonville West, Watsonville East, Moss Landing, and Prunedale), 61 "hot-linked" aerial photograph images (150 dpi tif files), and several hundred water,

---

[3] The Elkhorn Slough and vicinity was also chosen as the project area because of the availability of digital base maps, the national significance of the resources located in and around the slough, and the specific recognition of the need for improved information sharing by the Coastal Commission's Regional Cumulative Assessment Project (ReCAP), which focused on the Monterey Bay area for its pilot assessment.

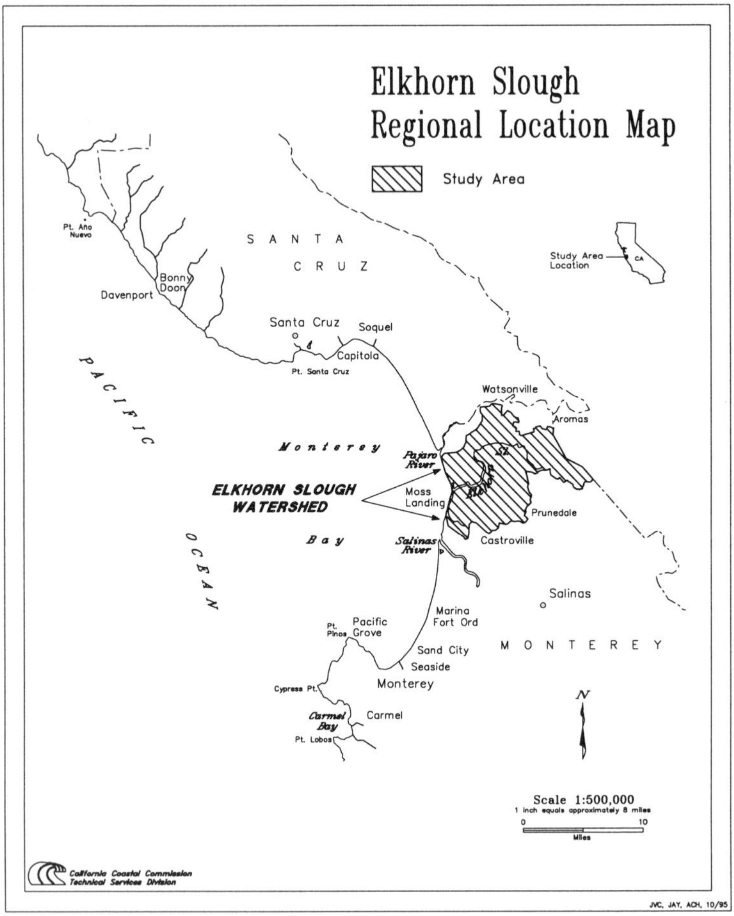

Figure 1. Project Area

sediment, and pesticide sampling records located by their decimal latitude and longitude values. The tasks required to complete this work included acquisition, preparation, and geocoding of the Monterey Bay National Marine Sanctuary (MBNMS) data sets, transformation of AutoCAD base maps from Cartesian to real-world coordinates, acquisition, scanning, and linking of aerial photography, and creation of point coverages for both sampling station locations and aerial photo frame center points. Preparation of written user materials was also required, and the final version of the project (approximately 150 MB in size) was planned for installation in the Coastal Commission's Santa Cruz office, the Department of Fish and Game's office in Sacramento, the MBNMS office in Monterey, the Association of Monterey Bay Area Governments (AMBAG) office in Marina, the Monterey County Planning Department's office in Salinas, and at the State Marine Laboratory in Moss Landing.

Methodology

The methodology employed to complete the required tasks was as follows:

1. *Project Design.* The decision was made to focus on recent aerial photography, water, tissue, and sediment sampling data. Test data sets were examined and evaluated. Scanning resolutions for the aerial photography were evaluated. Project limitations were identified. User interface and display requirements were identified. A schedule for completion of work was laid out.

2. *Acquisition of Aerial Photography.* Figure 2 is a flight line index map for 1994 color aerial photography, acquired by the National Ocean Service Photogrammetry Division during February and March of that year. The Commission was fortunate to obtain 9" by 9" paper copies of this high resolution 1:24,000 scale coverage for the six flight lines over Moss Landing and the Elkhorn Slough for this project. The Photogrammetry Division worked closely with the Coastal Commission and others in planning their 1994 California mission, and the Technical Services Division was able to successfully recommend extended coverage be flown inland in many critical coastal zone resource areas including Elkhorn Slough.

3. *Acquisition of Water, Tissue, and Sediment Sampling Data.* Figure 3 shows an example of the spreadsheet pages received from the Monterey Bay National Marine Sanctuary staff. The data sets were for varying sampling station locations and were collected by different persons, at different times, for differing purposes. The example shown is data from the Elkhorn Slough National Estuarine Research Reserve (NERR). Other data sets were from PG&E, National Refractories, the U.S. Army Corps of Engineers and other educational institutions. The Coastal Commission was very fortunate to have a Marine Sanctuary staffperson also working part-time with the Commission, and was able to take advantage of the opportunity to obtain copies of this data.

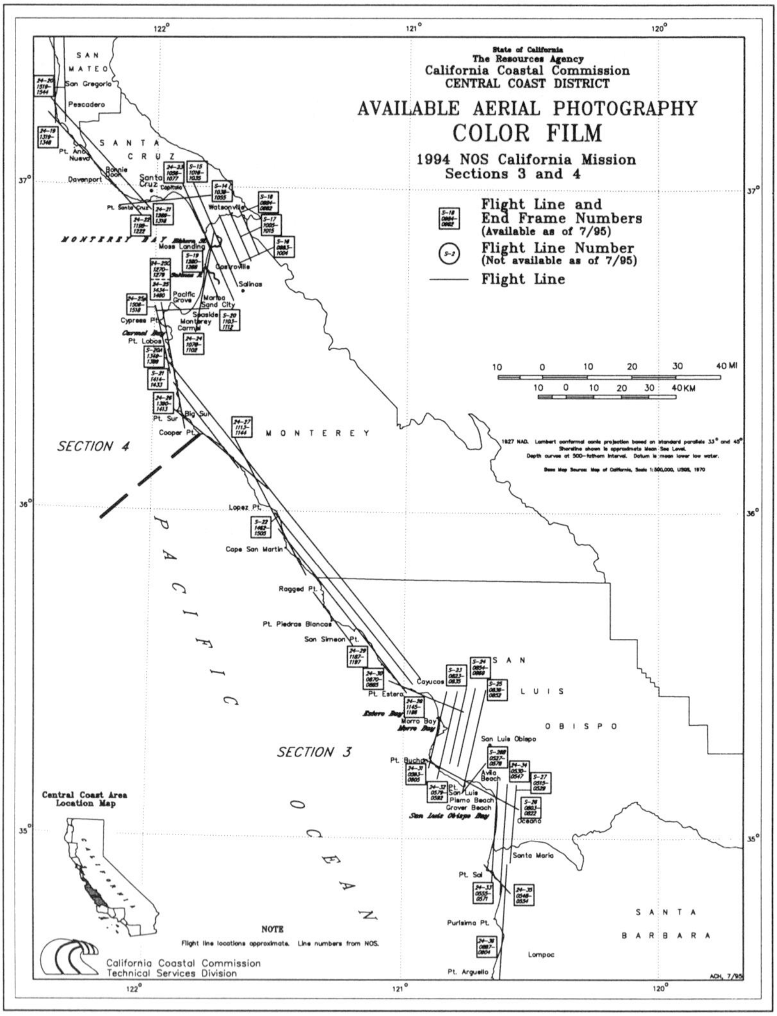

Figure 2. Central Coast Aerial Photograph Index

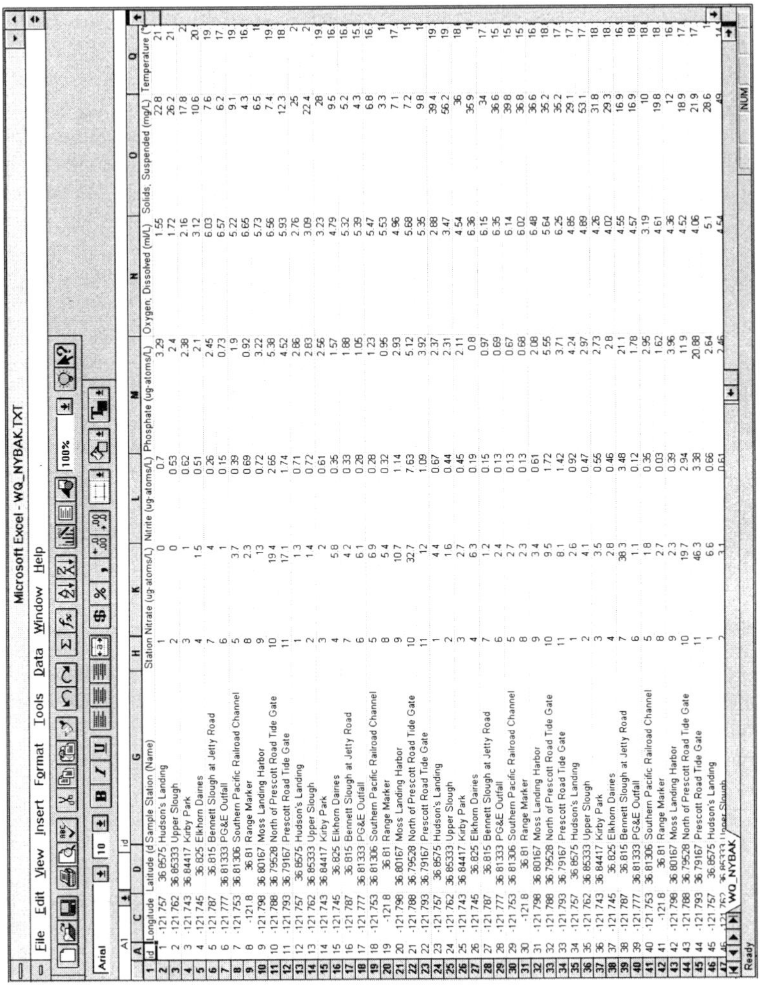

	A	C	D	G	H	K	L	M	N	O	Q
1	id	Longitude	Latitude (d	Sample Station (Name)	Station	Nitrate (ug-atoms/L)	Nitrite (ug-atoms/L)	Phosphate (ug-atoms/L)	Oxygen, Dissolved (ml/L)	Solids, Suspended (mg/L)	Temperature (°
2	1	-121.757	36.8575	Hudson's Landing	1	0	0.7	3.29	1.55	22.8	21
3	2	-121.762	36.85333	Upper Slough	2	0	0.53	2.4	1.72	26.2	21
4	3	-121.743	36.84417	Kirby Park	3	1	0.62	2.38	2.16	17.8	2
5	4	-121.745	36.825	Elkhorn Dairies	4	1.5	0.51	2.1	3.12	10.6	20.
6	5	-121.787	36.815	Bennett Slough at Jetty Road	7	4	0.26	2.45	6.03	7.6	19.
7	6	-121.777	36.81333	PG&E Outfall	6	1	0.15	0.73	6.57	6.2	17.
8	7	-121.753	36.81306	Southern Pacific Railroad Channel	5	3.7	0.39	1.9	5.22	9.1	19.
9	8	-121.8	36.81	Range Marker	8	2.3	0.69	0.92	6.65	4.3	16.
10	9	-121.798	36.80167	Moss Landing Harbor	9	13	0.72	3.22	5.73	6.5	1
11	10	-121.788	36.79528	North of Prescott Road Tide Gate	10	19.4	2.65	5.38	6.56	7.4	19.
12	11	-121.793	36.79167	Prescott Road Tide Gate	11	17.1	1.74	4.52	5.93	12.3	18.
13	12	-121.757	36.8575	Hudson's Landing	1	1.3	0.71	2.86	2.76	25	2
14	13	-121.762	36.85333	Upper Slough	2	1.4	0.72	2.83	3.09	22.4	2
15	14	-121.743	36.84417	Kirby Park	3	2	0.61	2.56	3.23	28	19
16	15	-121.745	36.825	Elkhorn Dairies	4	5.8	0.35	1.57	4.79	9.5	16.
17	16	-121.787	36.815	Bennett Slough at Jetty Road	7	4.2	0.33	1.88	5.32	5.2	16.
18	17	-121.777	36.81333	PG&E Outfall	6	6.1	0.28	1.05	5.39	4.3	15.
19	18	-121.753	36.81306	Southern Pacific Railroad Channel	5	6.9	0.28	1.23	5.47	6.8	16.
20	19	-121.8	36.81	Range Marker	8	5.4	0.32	0.95	5.53	3.3	1
21	20	-121.798	36.80167	Moss Landing Harbor	9	10.7	1.14	2.93	4.96	7.1	17.
22	21	-121.788	36.79528	North of Prescott Road Tide Gate	10	32.7	7.63	5.12	5.68	7.2	1
23	22	-121.793	36.79167	Prescott Road Tide Gate	11	12	1.09	3.92	5.35	9.8	1
24	23	-121.757	36.8575	Hudson's Landing	1	4.4	0.67	2.37	2.88	39.4	19.
25	24	-121.762	36.85333	Upper Slough	2	1.6	0.44	2.31	3.47	56.2	19.
26	25	-121.743	36.84417	Kirby Park	3	2.7	0.45	2.11	4.54	36	18.
27	26	-121.745	36.825	Elkhorn Dairies	4	6.3	0.19	0.8	6.36	35.9	1
28	27	-121.787	36.815	Bennett Slough at Jetty Road	7	1.2	0.15	0.97	6.15	34	17.
29	28	-121.777	36.81333	PG&E Outfall	6	2.4	0.13	0.69	6.35	36.6	15.
30	29	-121.753	36.81306	Southern Pacific Railroad Channel	5	2.7	0.13	0.67	6.14	39.8	15.
31	30	-121.8	36.81	Range Marker	8	2.3	0.13	0.68	6.02	36.8	15.
32	31	-121.798	36.80167	Moss Landing Harbor	9	3.4	0.61	2.08	6.48	36.6	16.
33	32	-121.788	36.79528	North of Prescott Road Tide Gate	10	9.5	1.72	5.55	5.64	35.2	18.
34	33	-121.793	36.79167	Prescott Road Tide Gate	11	8.1	1.42	3.71	6.25	35.2	17.
35	34	-121.757	36.8575	Hudson's Landing	1	2.6	0.92	4.24	4.85	29.1	17
36	35	-121.762	36.85333	Upper Slough	2	4.1	0.47	2.97	4.89	53.1	17
37	36	-121.743	36.84417	Kirby Park	3	3.5	0.55	2.73	4.26	31.8	18
38	37	-121.745	36.825	Elkhorn Dairies	4	2.8	0.46	2.8	4.02	29.3	18.
39	38	-121.787	36.815	Bennett Slough at Jetty Road	7	38.3	3.48	21.1	4.55	16.9	16.
40	39	-121.777	36.81333	PG&E Outfall	6	1.1	0.12	1.78	4.57	16.9	18.
41	40	-121.753	36.81306	Southern Pacific Railroad Channel	5	1.8	0.35	2.95	3.19	10	18.
42	41	-121.8	36.81	Range Marker	8	2.7	0.03	1.62	4.61	19.8	18.
43	42	-121.798	36.80167	Moss Landing Harbor	9	2.3	0.39	3.96	4.36	12	16.
44	43	-121.788	36.79528	North of Prescott Road Tide Gate	10	19.7	2.94	11.9	4.52	18.9	17.
45	44	-121.793	36.79167	Prescott Road Tide Gate	11	46.3	3.38	20.88	4.06	21.9	17.
46	45	-121.757	36.8575	Hudson's Landing	1	6.6	0.66	2.64	5.1	28.6	1
47	46	-121.762	36.85333	Upper Slough	2	3.1	0.61	2.46	4.54	49	14.

Figure 3. Excel Spreadsheet Showing Water Quality Sampling Data

The tabular data did present some challenges involving verification of geocoding and format of location references. After detailed analysis of the data sets, it became apparent that it would not be possible to stay on schedule and also rectify the location of each record's latitude and longitude. Instead, the decision was made to use a partial data set without major geocoding work needed, and to recommend to sanctuary staff that geocoding changes be made to the balance of the data.

4. *Data Manipulation and Integration.* This step in the project had many tasks, and included most of the time dedicated to actually developing the end product. The photography was scanned. An AutoCAD base map was transformed in ArcCAD and imported into an ArcView project. Point coverages were created for the sampling stations and aerial photo frame center points. A user interface was designed and created. Links were established between scanned photo images, geocoded tabular data, and the georeferenced base map. Testing the operation of the project, and preparation of the written user materials also took place during this step, along with planning the training sessions for two levels of users, managers and analysts.

5. *Installation and Training.* The final step was installing the project and training Commission staff in Santa Cruz and San Francisco, followed by installation and training for various users outside the Commission such as the California Department of Fish and Game, AMBAG, and the Monterey County Planning Department.

Project Results

The project was completed successfully and on time. Training sessions went smoothly, and comments indicate a relative ease of use for both analysts and managers. Figures 4 through 7 are intended to illustrate how the final version of the ArcView project works and samples of several possible results. The user interface shown in figure 4 has a selection of buttons operated by a "hot-link" tool, the use of which is illustrated with examples from the aerial photography and water-quality data in Figures 5, 6, and 7.

Figure 5 illustrates how the aerials are accessed. Selecting the aerial photography button from Figure 4 causes a georeferenced map of the four-quad area to display, with the flight lines and individual frame centers shown using the parallel lines and lightning bolt symbols. Selecting any of the lightning bolt photo center symbols displays the corresponding color aerial photo image. Both a table with flight line information and the scanned image are displayed. The image is *not* yet georeferenced to the map, nor is it even oriented true north. One of the decisions made in the project design phase mentioned above was to acknowledge the limitations of the high-quality, but still uncorrected photo images available to work with, and the images were simply oriented nominally north. This could be changed

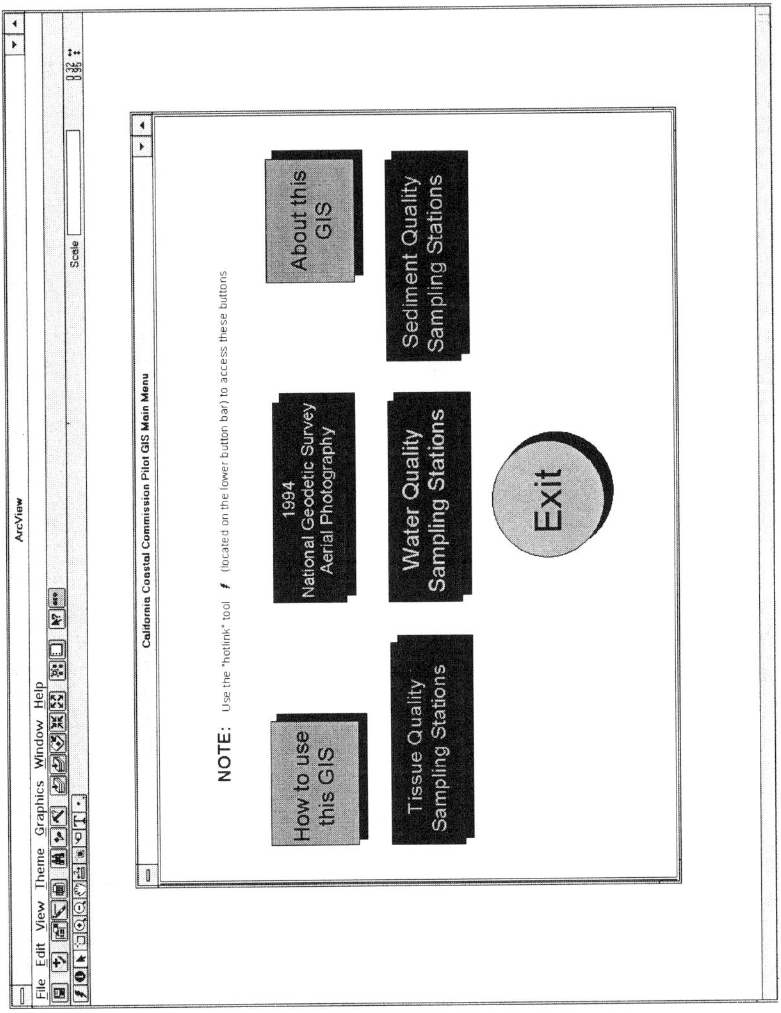

Figure 4. ArcView Project Interface

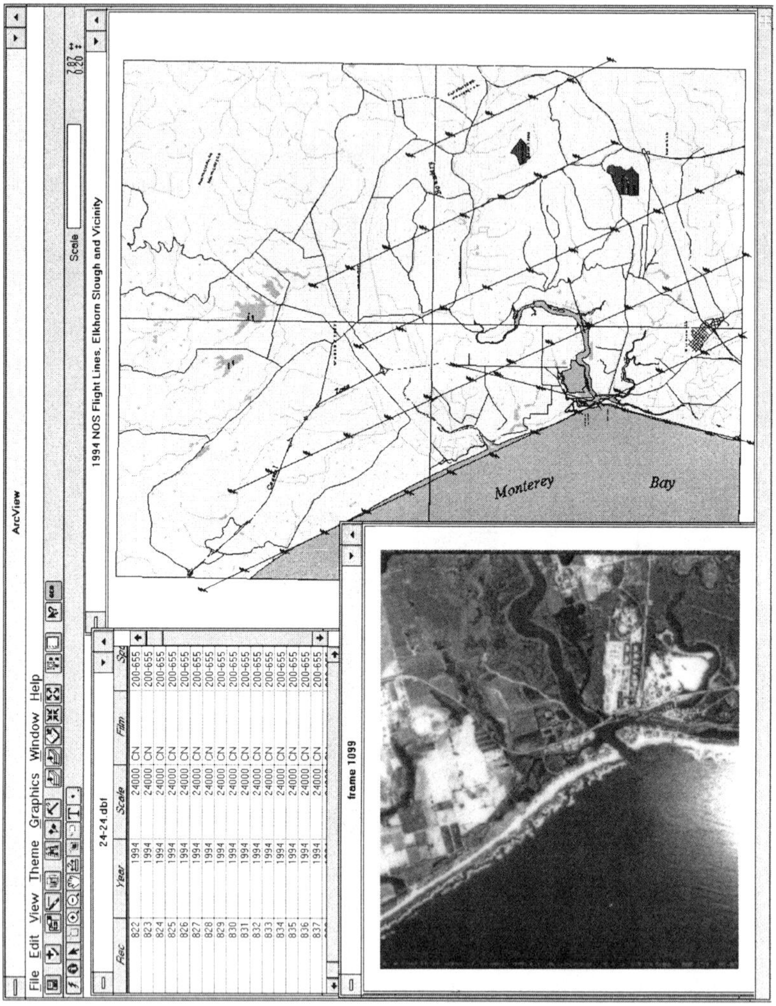

Figure 5. Example Display of Flight Line Map and Aerial Photography

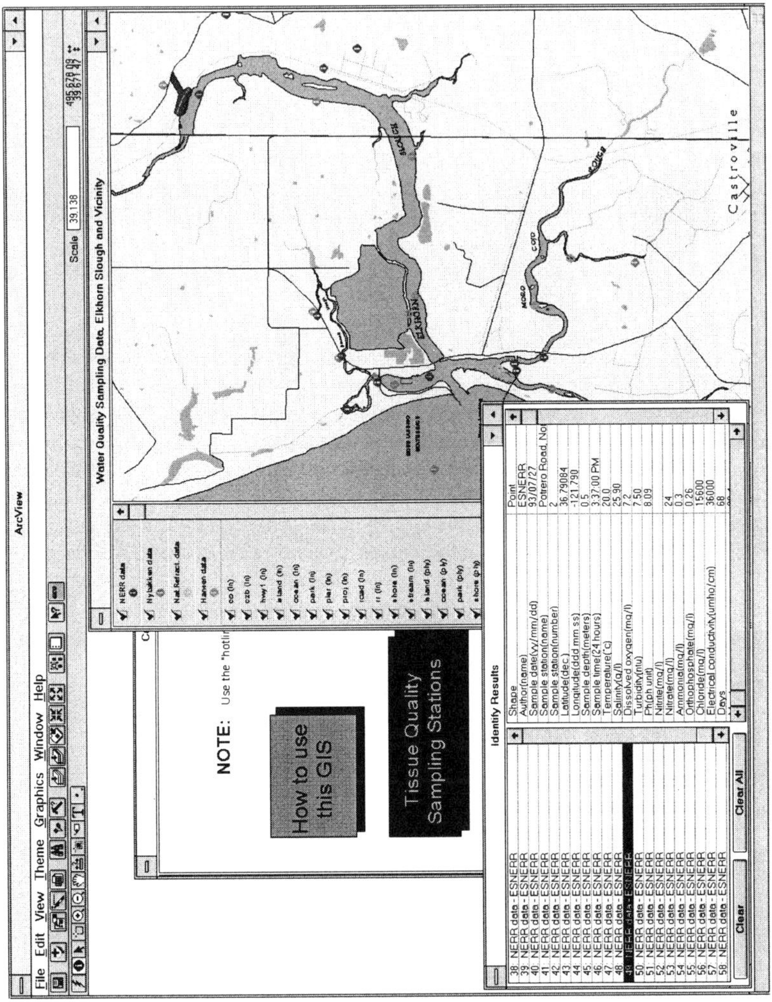

Figure 6. Example Display of Water Quality Sampling Data

in a future version of the project. Any of the standard viewing options can be used (e.g., zooming, sizing up window, etc.). Photo images can be annotated and plotted.

Figure 6 shows the results of selecting the water-quality sampling stations button from the user interface shown in Figure 4. The four-quad base map again displays, but instead of flight lines and photo centers, it depicts a series of dot symbols representing sampling stations located in and around Elkhorn Slough and Moro Cojo Slough. Selecting any of the dot symbols with the "identify" tool displays a table with that sampling station's data, which can also be sized up, scrolled through, and otherwise examined. Figure 7 illustrates the use of a simple query, where a search for nitrate levels greater than or equal to 80.5 mg/l on the NERR data displays a number of stations with samples at or above that level, each shown here in gray. The samples meeting the query can also be promoted to the top of the table for further examination, comparison, and printing.

Despite the successful completion of the project, there are several technical points which will ultimately require consideration beyond what was given in this effort, if a wider development of this application is undertaken. The most significant involves the introduction of small amounts of locational error during the process of transforming the AutoCAD base map into real-world coordinates. This was due to a limitation in the software used which relates to gridspace transformations and the rubber-sheeting of maps. *[Note: At this writing, the Commission is using a different software package capable of transforming and rubber-sheeting maps.]*

In addition, the quality of the datasets could be improved by identifying and adding the ungeoreferenced data, converting the data to consistent units of measurement, and integrating the resulting tables.

A third logical area for improvement in the project is the photo orientation. Since the direction of the flight lines varies, it might be best to add north arrows for each series of photos oriented similarly, which could then be moved away if it obscured any of the image being viewed.

Conclusion

Elkhorn Slough is a marvelous, rich natural resource, one that presents many challenges for its stewards and managers. The effects of industrial development and polluted runoff from intensive agriculture and residential development are serious threats to the ecological health of the slough and estuarine sanctuary These environmental challenges require decisions that can be improved by making better information available to the entire coastal community. The Coastal Commission has been successful in helping to do that with this project by providing integrated access to aerial photography and water quality-related data. Though modest in scope, the project was designed to be expanded and updated, and is already being used as a model for similar work in other coastal areas.

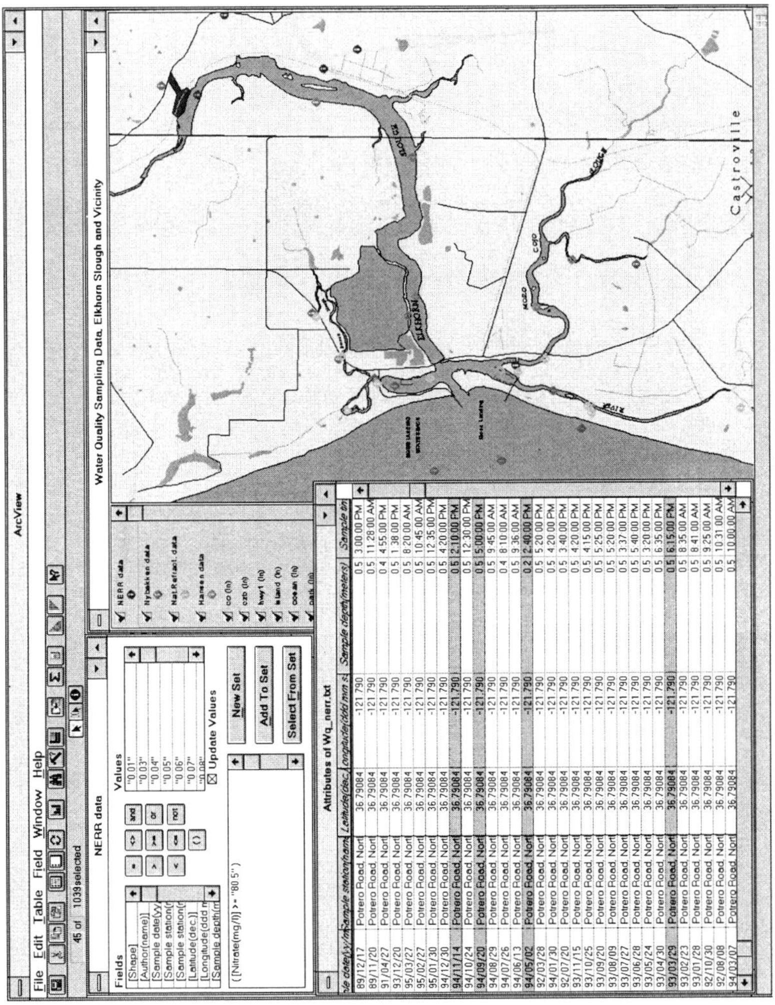

Figure 7. Example Display Showing Results of Water Quality Data Query

CHAPTER 52

Florida's Ocean GIS: Protecting Blue and Green Waters

Courtney Westlake[1]
Robert Hudson[1]
Christopher Friel[1]

Abstract

Florida's ocean policy should be based on the best available information, and where data gaps exist, a systematic approach to filling these information voids should be implemented. The Statewide Ocean Resource Inventory project addresses this need by developing a data assessment, information base, and distribution mechanism that can be used to facilitate a state ocean policy and provide state, regional, and local managers with much-needed marine resource information. The Coastal and Marine Resource Assessment section at the Florida Marine Research Institute is actively working to identify, collect, and package pertinent coastal and ocean information for use with a user-friendly desktop information system accessible to federal, state, and local decision-makers. The ArcView desktop application digitally links the Florida Administrative Code and Management Plans to GIS data layers. The application provides marine resource GIS data layers along with unique management tools through interactive mapping on the World Wide

Introduction

Environmental protection and conservation on the one hand, and the increasing pressure of human development activity on the other, lead inevitably to conflict between many different and often opposing uses of the coastal zone. In 1989, a report entitled Florida's Ocean Future: Towards a State Ocean Policy, produced by the Governor's Office of Planning and Budgeting, identified

[1]Westlake, C., R. Hudson, C. Friel, Florida Marine Research Institute, 100 Eighth Avenue, SE, St. Petersburg, Florida 33701

the need for a comprehensive ocean policy addressing the wise management of the state's offshore lands, coastal zone, and marine resources. However, a comprehensive knowledge of the resources and uses of Florida's coastal and marine environments is required before a comprehensive ocean policy can be developed.

Statewide Ocean Resource Inventory

The Statewide Ocean Resource Inventory (SORI) contributes to a project funded by the Florida Coastal Management Program. This project identifies, collects, and packages pertinent coastal and ocean information for use with a user-friendly desktop information system accessible by federal, state, and local decision makers. In order accomplish our goal of providing targeted and packaged Geographic Information System (GIS) data to marine resource managers in the state of Florida, we extended an invitation to the coastal and ocean management communities to enter into a partnership with the Coastal And Marine Resource Assessment section (CAMRA). CAMRA is a collection of scientists with backgrounds in geography, remote sensing, Geographic Information Systems, ecology, and biology at the Florida Marine Research Institute inside the Division of Marine Resources of the Florida Department of Environmental Protection (FDEP). The partnership spanned approximately one and one half years. The results of the project are available on the World Wide Web (WWW) site www.fmri.usf.edu/sori as of July 1997.

Marine Management Issues

Many of Florida's marine resource managers recognize that a great portion of the state's culture and economy are rooted in its valuable coastal and ocean assets. The peninsula of Florida is bordered by approximately eight thousand, one hundred miles of shoreline. Even the inland portions of the state are considered to be influenced by the marine environment due to the inherent connectivity of Florida's aquatic ecosystems. Florida lies between two distinct waterbodies, the Gulf of Mexico and the Atlantic Ocean. Each of these marine environments have been designated as Large Marine Ecosystems by the United Nations Environmental Programme. Florida also boasts of internationally significant ecosystems such as the Everglades, Florida Bay and the Florida Keys as well as world renown beaches and fishing grounds. Likewise, there are ten major ports, a large tourist industry, development pressures and military activities that Florida houses within these ecosystems. The diversity of resources coupled with the wide array of human pressures makes coastal zone management in Florida a challenging endeavor.

The partners were employed as our expert panel to insure that the GIS efforts aligned with the needs of Florida's marine management community. The partners developed a list of issues that are considered priority in managing Florida's waters. While Florida is faced with a multitude of marine management issues, the partner focused on a short-list of priority issues in the Statewide Ocean Resource Inventory project:

- Fisheries Management
- Marine Commerce
- Resource Quality/Critical Habitat
- Threatened/Endangered Species
- Disaster Response
- Law Enforcement

Nested within each of the management issues are more targeted issues. After defining the priority issues, we determined what information would be most useful in addressing pertinent management questions within a particular issue to support ocean policy.

Policy Information Support

Florida's ocean policy should be based on the best information available, and where data gaps exist, a systematic approach to filling these information voids should be implemented. The SORI project addresses this need by developing a data assessment and information base to facilitate a state ocean policy and provide state, regional, and local managers with much-needed local marine resource information.

Data availability and data management problems have deterred an approach to managing Florida's coastal and ocean waters in a holistic and self-sustaining way. Unfortunately, needed information is seldom readily available and rarely in the possession of decision-makers. Much information is in the possession of individuals; however, it is often scattered between different federal, state, and municipal government departments, research institutes, and universities in various formats.

The criteria for becoming a partner in the project included the ability to supply geographic information and the need to support management decisions in the coastal zone.

Multi-Agency Cooperation

The project was sculpted by our partners and the needs of each respective agency. Aligning a user community for the final product allowed CAMRA to optimize the information provided. The intent to create a product that alleviates the need for arcane information mediums also laid the foundation for Geographic Information Systems within each partner agency.

The project partners are a collection of representatives from the following agencies:

Florida Geological Survey
FDEP-Division of Law Enforcement
Apalachicola Bay National Estuarine Research Reserve
FDEP-Bureau of Aquatic Preserves
Florida Keys National Marine Sanctuary
Tampa Bay Regional Planning Council
Environmental Protection Commission of Hillsborough Co.
FDEP-Division of Historical Resources
FDEP-Bureau of Beaches & Coastal Systems
South Florida Regional Planning Council
Florida Coastal Management Program
Florida State University-College of Law
University of Florida-Geography Program
University of Miami-Marine Affairs Program
Center For Marine Conservation
FDEP-Division of Law Enforcement
FDEP-Florida Marine Research Institute
St. Johns River Water Management District

The multi-agency cooperation formed a consortium of GIS and natural resource managers, planners, policy makers, and scientists to transition to the management of Florida's waters with the best-available information. The many steps required in our partnership approach included the following:

- Assess information needed by local, state, and federal agencies to develop an ocean policy and implement local marine resources policy;

- Compile an inventory and catalogue of existing data and synthesize data and information, where appropriate,

- Enable statewide assessment and analysis and provide data access with a multiple user-friendly PC ArcView application;

- Develop an Internet-based tool for immediate access to synthesized data and regulatory information;

- Identify data gaps and create a strategy for multi-agency coordination to maintain, enhance, and augment existing information.

These six steps were carried out over the course of a year and one half. The efforts combined to create a resource information base and analysis tool upon which to base ocean policy. The methodology we employed to achieve our goal is described in this paper.

Information Assessment

The statewide data repository of marine resource information in Florida's waters resides at the Florida Marine Research Institute. The Marine Resource Geographic Information System (MRGIS) is a collection of approximately 300 GIS resource data layers. It houses digital resource information from collective efforts of state agencies in Florida.

Through an exercise of matching datasets with issues, we siphoned the MRGIS datasets into the issue categories. The results were a compilation of 15-25 targeted datasets per issue. The compilation was comprised to minimize the time a manager expends to browse the necessary information by optimizing the available information in a centralized location. The infusion of GIS with a management issue facilitates immediate and targeted information access. The issue-focused information provides managers access to digital datasets that were not previously available in one location.

The partners consist of a resource management community that ranges from state agency directors to GIS Managers. The initial phase was intended to determine the information needs of the managers. CAMRA presented a list of proposed coastal and ocean management GIS layers. The partners consolidated the information into a manageable number of layers.

The partners scrutinized each layer in two manners; categorized according to the number of management issues to which it applied and categorized by display speed. In the first exercise, the layers were matrixed with the management issues. The layers were then prioritized based on the number of issues to which a specific data layer applied. The layers that were applicable to the most number of issues were considered high priority. The subsequent layers were categorized as medium and low priority.

The second stage of prioritization accounted for the most important consideration for desktop GIS applications. An application design should allow users expedient access to browse all relevant information. GIS layers that have a high spatial resolution such as soils or topography layers, require a high level of processing time to display to the screen thereby shrinking the window of opportunity to browse and further analyze the information. Layers that have low spatial resolution received a rank of one (i.e. Shipping Lanes and ports, major roads, Aids to Navigation). The second category was mostly dense point data, line data and polygonal datasets (i.e. seagrass communities, propeller scaring, emergent vegetation). The GIS layers that are highly resolved (i.e. Benthic Resources, National Wetlands Inventory, Landuse/Landcover) received a rank of three.

The partners calculated the sum of the ranks for the applicable layers within each management issue. A large ranked sum indicated that the data layers for that issue were highly computer intensive therefore requiring a significant amount of time to display. A low sum indicated that the view would require less time to display. Therefore if soils was initially considered a highly desirable layer for the Threatened/Endangered species, but that layer substantially increased the drawing speed in ArcView and was not critical to the assessment within the management issue, it was removed from the list. The partners debated the applicability of each dataset in the high sum issues and narrowed the focus to more targeted data layers.

Statewide Inventory

Geographic information systems are known for their ability to house large, complex datasets, provide analytical tools to manipulate the information, and present the data in the form of a map. This is powerful because data can be manipulated to address specific questions and static atlas series can be modified to fit the needs of the user. Yet, in many cases, this is time consuming and planners do not have access to equipment, software and the technical expertise to manipulate the information. Therefore, the best available information may not be used when making decisions or plans regarding marine resource protection. Maps or corporate databases created in parallel with a planning effort assist in some decision support but, may not alter the courses of action based upon the available information or data gaps. This may be due to the fact that GIS skills are not part of the planner's training or the fact that the equipment needed to support spatial analysis is too expensive. Therefore, the role of spatial analysis is under-utilized throughout the planning process. The deliberate method of using GIS can be intertwined into the planning process. This will allow the data, or data gap, to become the rate determining step when making decisions.

Technology is helping move spatial data and the associated analyses from expensive Unix platform workstations to the planner's personal computer. The gap between the technical GIS analyst and the end-user who needs to browse or query data is becoming less of an impetus to using GIS in the planning process. There will always need to be an entity that maintains the corporate datasets and perform scphisticated spatial analyses but, the age of querying, browsing and displaying spatial information is here for the planning community. Hardware and software enhancements are providing GIS experts with an opportunity to create desktop information systems that address targeted planning scenarios. In addition, these systems can be integrated under an umbrella of similar applications that address several management concerns. This suite of modified software and hardware applications can be used across a wide variety of planning efforts. A change in planning methods, incorporation of spatial analysis within the planning process, and taking advantage of technology can add validity to the planning process.

Ocean GIS Application

Advancing technologies provide the means to execute the needs of marine resource managers in Florida. While 75% of the Floridian population resides on or near the coast, coastal managers are faced with constant change in the marine environment. The changing environment results in a continual shift in focus for the state's resource managers. Resource managers must continuously adapt to the problems that follow the influx of population. Population pressures on the coast are multiplied by two different climates. Florida's coastline migrates through a temperate climate in Northern Florida to a sub-tropical climate in South Florida. Two distinct climates present a varying set of circumstances surrounding each management issue.

The scope of the SORI project revolved around an ArcView (AV) desktop application for user-friendly with targeted datasets to support ocean policy. The application development required a two-step process. The first step involved creating a "prototype" application to offer the partners a proof of concept. The prototype arranged GIS data layers into interactive, issue-driven "views". Each view entertained customized ArcView functionality. The customized AV interface provided a toolbox of functionality for accessing and analyzing GIS layers paired with state and federal regulatory information. The toolbox allowed for cursory analysis and site assessment within each management issue. The partners' evaluation of the prototype revealed the shortcomings of the application and guided modifications for the final application with a release date of July 1997.

The table below examines the design considerations for the prototype application in contrast with enhancements to the final application.

Design Considerations	Prototype Application	Final Application
ArcView Compatible	Yes	Yes
Issue-Driven Views	Yes	Yes
Manager-Friendly	Yes	Yes
Expert-Friendly	No	Yes
ArcView Extensions	No	Yes
Internet Accessible	No	Yes
Dynamic Data Updates	No	Yes
Visual "Thumbnail" Metadata	No	Yes
On-Line Data Gap Analysis	No	Yes
Customized ArcView Interface	Yes	No
User-Customization	No	Yes
Spatially-Linked Regulatory Information	Yes	Yes
Florida Administrative Code	No	Yes
Management Plan Information	No	Yes
Florida Statutes	Yes	No
CD-ROM Dependent	Yes	No

Enhancements for the final application, based upon the partner the evaluations, steered the project to utilize advancing technologies. Florida's marine resource managers are concerned with the most current information when working with geographic data. Faced with only a limited time frame in which to make decisions, quick-access to the best-available information is vital. At the onset of an event requiring management decisions, valuable time is lost in decisions on what information is most pertinent, the search for the information, translation into a compatible format, in addition to familiarizing staff with new application software to view and analyze the geographic information. The continual evolution of information needs and management issues, paves the path towards quick, distributed access mechanisms inside the World Wide Web (WWW).

Internet Delivery

The WWW is revolutionizing how government agencies communicate, access, and utilize information. Until recently the only geographic information available on the Web were static maps that were limited in providing only a snapshot of information with no means for analysis. Recent breakthroughs in Web-based, interactive mapping are changing the way geographical information is accessed and used.

Prior to the new technology, the SORI application was available only through a customized ArcView interface drawing the static information from a CD-ROM. The application demanded that users have ArcView software, CD-ROM readers, 32MB RAM for optimal viewing, and 400 MB of free disk space. The demand for the geographic information originally exceeded the need for timely access to dynamic information. Due to the technological limitations early in the project development, users were forced to trade speed and continuous updates to the information for fingertip access.

Interactive Web-mapping removes the sacrifices that traditionally surrounds a "user-friendly, desktop information system". Resource managers can easily browse current information by theme or management issue through the SORI Web page. It is the easiest and most cost effective way for publishing the most up-to-date geographic information to a broad audience.

The project lineage cut a course through changing technology. ArcView 2.0 was targeted for application prototype development and distributed to the partners in the beginning of the project. The prototype was completed at the time of the ArcView version upgrade to

3.0. With enhanced functionality offered in ArcView 3.0, the final application is being designed to take advantage of stock programming. ESRI designed ArcView 3.0 as an open-architecture desktop development platform with the ability to "extend" the functionality. The added functionality of ArcView 3.0 extensions, allowed the SORI project to venture into the interactive Web-mapping environment. Taking full advantage of ESRI's new product, the ArcView Internet Map Server extension, the SORI Issue-driven views are available not only to project partners, but also to education facilities, the public, and other governmental and non-govermental organizations. While sustaining the nuances of today's technology, the SORI application prepares for tommorow's technological agenda.

Gap Analysis

Inherent to the data identification and prioritization effort is a data gap analysis. This analysis may alter decisions regarding courses of action based upon insufficient information. This analysis will also aid fiscal prioritization when deciding what types of studies to fund with limited financial resources. For example, if a dataset has been determined to be a priority for two management questions within one or more management issues, and is not available, then that could be a candidate for scientific study. This process removes the haphazardness of decision-making when information may not be available to the resource managers. This same scenario could go so far as to defer an irreversible management option in favor for an interim solution until the information is available for true analysis.

There is a vast amount of information that may not seem appropriate for a desktop GIS application but, is vital in making sound decisions. This may include regulatory information, financial information, and management plan information associated with the project area or management objectives. For example, regulatory information can be tagged to the appropriate managed area for users to query what locations are affected by the current regulatory requirements in their area of interest. Land acquisition priorities are another form of information that directly affects planning efforts as well as locations of management plans already in place for these locations. Advances in technology has made adding temporal data such as this feasible for GIS application developers. However, the identification and prioritization of non-spatial data must go through the same deliberate process as the spatial information.

Conclusion

To support marine resource protection activities, CAMRA is developing a suite of desktop information tools to address management challenges ranging from oil spill response to South Florida ecosystem restoration to statewide coastal/ocean resource inventorying and management. These tools have been created within the Coastal Ocean Management Planning and Assessment System (COMPAS) framework, an initiative evolving out of NOAA's Strategic Assessment Division (SEA). The COMPAS framework combines software development endeavors and data packaging efforts to support policy analysis within the marine resource management and emergency response communities.

CAMRA utilizes the COMPAS framework as an evolving process for desktop information tool development to ensure useful applications are produced and the best available information is synthesized and provided to Florida's marine resource managers. Future COMPAS efforts include partnering with NOAA's SEA Division and Coastal Services Center as well as, state agencies from around the Southeastern United States to develop the South Atlantic Ocean GIS for managing offshore marine resources at a regional scale. The flexibility of this process lends itself to management efforts at several scales, in many locations, for a variety of management issues.

Acknowledgements

The paper is based on work funded by the Florida Coastal Management Program. Geographic Planning Collaborative, Inc. and Metroscape Inc. provided technical expertise on the Internet mapping.

The authors thank Chris Johnson, Jim Poehlman, Henry Norris, Jennifer Bexley and Douglas Wilder for their invaluable contributions to the project.

CHAPTER 53

Marine fishery refugia and coastal fishery management

Jenifer E. Dugan[1] and Gary E. Davis[2]

Our exploited marine ecosystems are exposed to increasing alterations of natural population and community dynamics. Despite increased management efforts, marine fisheries and their related communities are collapsing in California and all over the world under traditional fishery management. Fisheries management regimes based on single species which use maximum sustainable yields, size and bag limits, seasonal closures and limited entry approaches appear to offer few prospects for the restoration of impacted fisheries and marine ecosystems.

New approaches to fishery management are critically needed. Marine fishery refugia, unharvested areas that serve as sources of replenishment, offer some hope for the restoration and stabilization of coastal fisheries and marine communities. Management which incorporates ecologically independent fishery refugia can provide more than another way to limit fishing effort for coastal species. The use of fishery refugia can potentially restore sustainable fishery yields by protecting spawning stock biomass, maintaining the genetic diversity of stocks, preserving natural population structures and densities of target species, enhancing ecosystem resiliency and stabilizing community structure.

The ability of marine refugia to serve as tools for fishery management and enhancement remains poorly understood, particularly in California coastal waters. Our review of studies of marine reserves worldwide indicated that effects which could lead to fishery and ecosystem restoration, such as increased abundance, individual size, reproductive output and species diversity, were reported in a variety of marine fishery species and communities in a number of relatively small reserves located in habitats ranging from coral reefs to temperate kelp forests. In the few published studies of reserves where spillover and fishery harvest were examined, areas surrounding reserves had increased yields of target species. Although none of those studies investigated reserves and fishery yields in temperate reef ecosystems, limiting the direct relevance of their results to California's fisheries, some applicable ideas emerged from our review. We found that positive effects of marine reserves on fishery yields and spillover were reported primarily in reserves which consisted of a portion of a relatively continuous habitat, e.g. a fraction of a fringing coral reef, rather than isolated reefs or entire habitat patches. This reserve characteristic could serve to facilitate exchange and spillover from the reserve area into adjoining fished areas and could be an important design consideration for fishery refugia in California and elsewhere. Further investigation of existing marine reserves, particularly in California and on other heavily populated coasts, and the establishment and investigation of experimental refugia over appropriate time spans are needed to allow scientists, fishery managers and legislators evaluate their potential application to fishery management for the next century.

[1]Marine Science Institute, University of California, Santa Barbara, CA 93106
[2]Channel Islands National Park, 1901 Spinnaker Dr., Ventura, CA 93001

CHAPTER 54

Results of a temperate reef marine reserve study and implications for use of marine reserves in nearshore fisheries management

Michelle J. Paddack

Abstract

Within the Monterey Bay National Marine Sanctuary, there are several marine reserves, yet none have been studied extensively for their effects on the local populations of benthic fishes. This study addresses three main questions: Does protection from fishing allow higher densities of fishes to persist in an area? Does fishing create a size selective pressure on a population? What is the relevance of these effects to fisheries management? Three marine reserves in central California were compared to adjacent, fished areas and ten species of benthic reef fish were assessed (most in the genus *Sebastes*). Slightly higher numbers of benthic fishes were observed in all of these marine reserves. Additionally, two of the three reserves had significant differences in both mean length and length frequency distributions, with larger fish inside marine reserves. The lengths were converted to biomass and to egg production to compare the possible contribution of these increased populations to the local fishery. The findings indicate differences among populations of benthic reef fishes along our coast, and that these differences may be either created or augmented by fishing pressure. These results are relevant to fisheries managers, providing partial support for the use of marine reserves as an alternative management strategy for management of nearshore fishes. This paper is a summary, more detailed analysis and discussion of this work can be found in Paddack (1996).

Introduction

In response to concerns over local marine resource declines, the number of established and proposed marine reserves in California has risen dramatically (e.g., Proposition 132 of 1990 established four new marine reserves, and a current bid to establish a new reserve in Monterey). Most marine reserves are established for general conservation, rather than fisheries management purposes, however, and often the reserves are established with no specified monitoring or management goals (Potter 1994). The recent surge in interest in setting up such areas raises the question of the value of protected areas and their usefulness to nearshore fisheries management. There is a wealth of evidence of marine reserve effects from many systems throughout the world (an excellent review of studies and discussion of marine reserve effects can be found in Dugan and Davis 1993); these effects include increased density, size, and species diversity of target organisms within reserves, and increases in fishery yields in adjacent areas. Unfortunately, however, few of these studies have been done along the California coast and when considering reserve effects along our coast, we must look toward evidence from similar systems in South Africa or New Zealand (see Cole et al. 1990, Buxton and Smale 1989).

While the work of establishing more and/or larger (and possibly connected) marine reserves is important, of perhaps equal weight is evaluating the reserves already in existence to determine the effects we may expect in this system, and how they may serve in fisheries management. Carr and Reed (1993) highlight the fact that a reserve may not necessarily contribute to fishery enhancement, and that a primary goal of a monitoring protocol should be to determine whether reserves are capable of maintaining a high level of larval production. I report here one of the first studies to look at this parameter for rockfishes along the California coast.

Rockfishes (*Sebastes spp.*) are among the most highly utilized of the west coast. Most near shore species of rockfish are reef-associated and are vulnerable to overfishing because of their late maturation, slow growth, and residential nature (Leaman 1991). Declines and closures of other fisheries, such as Washington's salmon fishery, have led to increased fishing pressure on California rockfish (P. Reilly, CDFG, pers. comm.), and rockfish landings have declined since the late 1940s (Dugan & Davis 1993), with some species showing precipitous declines. With growing concerns about the numbers, species, and sizes of fish being caught, marine reserves need to be more carefully considered for their use as part of an alternative strategy to maintain a viable rockfish fishery.

I surveyed three temperate reef, coastal marine reserves along the central California coast to look at the effects of the reserves on density and size of the reef fishes. The reserves included were: Hopkins Marine Life Refuge (Monterey Bay), Pt. Lobos State Reserve (Carmel Bay), and Big Creek Marine Ecological Reserve (Big Sur). The reserves ranged in age from 37 to 3 years old, and in size from 0.8 km^2 to 5.2 km^2. Because no data were available for these areas either prior to or since reserve establishment, a reserve/ non- reserve study was conducted, with each reserve being compared to adjacent, non-protected areas of similar habitat type. This study focused primarily upon the rockfish, of the genus *Sebastes,* including 6 shallow water, kelp forest species (*Sebastes atrovirens, S. carnatus, S. chrysomelas, S. caurinus, S. miniatus, S. nebulosus).* Four other reef fish were also included (*Scorpaenichthys marmoratus, Ophiodon elongatus, Hexagrammus decagrammos,* and *Semicossyphus pulcher*). All data were collected on scuba.

Results/Discussion

I found that while there were higher average densities of fishes in all of the reserves compared to the adjacent non-reserve areas, these differences were not statistically significant (2 Way ANOVA, p value = 0.06). The lack of statistical significance is likely due to the patchy distribution of the fish, leading to a high variance and a demand for a unwieldy sample size in order to overcome this error. However, the differences seen are close to significant and should not be discounted.

Size frequency information was collected by directly measuring the fish underwater, to the nearest centimeter (Paddack, 1996). One reserve, Big Creek Marine Ecological Reserve was established at the onset of this study (January 1994), and no differences were found in the size frequency in the three years following the reserve establishment. This is not surprising given the remoteness and lack of access along that coastline, resulting in very little historical or current fishing pressure in this area. For the other two reserves, however, the fish within the reserves were significantly larger than those in non-reserve areas (t test, $p \ll 0.001$). Additionally, the size frequency histograms of the fish had a significantly different distributions, with a greater proportion of large sizes classes inside the reserves, and relatively few in small size classes (Kolmogorov-Smirnoff, $p \ll 0.001$). This indicates that the fishing in the non-reserve areas is size selective.

The differences in size frequency distribution may have important implications about the reproductive capacity of these populations. Outside the reserves, most of the fish were below the minimum size at maturity known for these species (lengths from Lea et al., in press). This may indicate that the populations are becoming recruitment overfished, and the fished areas rely on other sources of larvae to replenish the populations. Size-specific fecundity for two species (*Sebastes atrovirens*, *S. chrysomelas*) were used to calculate an estimate of the reproductive potential of each reserve / non-reserve area. I found that the reserves were capable of producing up to three times more larvae per unit area compared to adjacent non-reserve areas.

Conclusion

The greater potential reproductive output of the reserve areas compared to adjacent non-reserve areas indicates that, despite small density differences, the size frequency differences between the exploited and non-exploited populations may translate into large differences in larval output. Reserve areas could provide the majority of recruits to outlying areas and thus be important to sustaining fished populations. Rockfish larvae are carried offshore for several months (Kendall & Lenarz 1987) and appear to return to nearshore regions via upwelling fronts (R. Larsen, S.F. State University, pers. comm.). These nearshore species tend to have shorter larval stages than deeper water species (Larsen, pers. comm.), which could decrease their dispersal and lead to more localized recruitment. Because of this, nearshore rockfish species can be classified in a "Limited Distance" model of larval recruitment (Carr & Reed 1993), meaning that neighboring populations would exchange more larvae than distant populations. If so, then the reserves would be of particular importance in restocking fishing grounds in nearby areas. However, it is not known whether these larvae are retained in localized currents and there is evidence that they are at least occasionally carried far offshore (Larsen, pers. comm.). More information on this aspect of rockfish life history is badly needed.

This study supports the widely held notion that marine reserves have a strong potential to enhance nearby fisheries and indicates that existing marine reserves along the California coast may already be playing a role in local fisheries management. The most striking evidence for a reserve effect is the differences found in size frequency distributions between fished and non-fished areas. Differences in density may be difficult to detect due to high variances and lack or time and/or resources to sample sufficiently to detect the changes. The size frequency information, however, was relatively easy to collect for these benthic reef fish and quickly showed that fishing pressure is likely to be lowering the mean size of these reef fish as well as shifting the size distributions downward. This information, coupled with size-specific fecundity for each species, should be included in monitoring programs to assess the effects of a marine reserve.

References

Buxton, C.D., and M.J. Smale. 1989. Abundance and distribution patterns of three temperate marine reef fish (Teleostei: Sparidae) in exploited and unexploited areas off the southern Cape coast. Jou. App. Eco. 26:441-451

Carr, M.H., and D.C. Reed. 1993. Conceptual Issues Relevant to Marine Harvest Refuges: Examples from Temperate Reef Fishes. Can. J. Fish. Aquat. Sci. 50:2019-2028

Cole, R.G., T.M. Ayling, and R.G. Creese. 1990. Effects of marine reserve protection at Goat Island, northern New Zealand. New Zealand Journal of Marine and Freshwater Research 24:197-210

Dugan, J.E., and G.E. Davis. 1993. Applications of marine refugia to coastal fisheries management. Can. J. Fish. Aquat. Sci. 50:2029-2042

Kendall, J., AW, and W.H. Lenarz. 1987. Status of early life history studies of northeast Pacific rockfishes. In: Proceedings of the International Rockfish Symposium, Anchorage, Alaska, Alaska Sea Grant Report 87-2. University of Alaska, Sea Grant Program, Fairbanks, AK. 128 p.

Lea, R.N., R.D. McAllister, and D.A. VenTresca. in press. Biological Aspects of Nearshore Rockfishes of the Genus *Sebastes* from Central California. Fish Bulletin 177

Leaman, B.M. 1991. Reproductive styles and life history variables relative to exploitation and management of *Sebastes* stocks. Env. Biol. Fish 30:253-271

Paddack, Michelle. 1996. The influence of marine reserves upon rockfish populations in central California kelp forests. Master of Science, Marine Sciences Thesis. University of California at Santa Cruz, Santa Cruz.

Potter, F.M. August 1994. What are coastal and marine protected areas supposed to do? D. J. Brunckhorst (ed.) Marine Protected Areas and Biosphere Reserves:' Towards a New Paradigm'. Proceedings of the 1st International Workshop on Marine and Coastal Protected Areas, Canberra, Australia.

CHAPTER 55

The Potential of Marine Reserves to Enhance Fisheries

David A. VenTresca[1], Marty L. Gingras, John Ugoretz, Andy Voss, Steve Blair, Josh Plant, Robert Hornady, and Cory Yoshiyama

Abstract

Rockfishes (*Sebastes spp.*) are an important and heavily exploited component of sport and commercial fisheries in central California. The success and quality of central California's nearshore rockfish sport fisheries have declined over the past decades, particularly in areas close to ports. Recognizing that a change in the current resource management approach is warranted, alternative management and enhancement strategies are of paramount importance. Our project, the Central California Marine Sport Fish Survey, has monitored nearshore recreational fisheries and documented life history characteristics of many rockfish species since 1956.

Marine reserves have been reported to enhance fisheries in other parts of the nation and world, but limited information is available to evaluate their effectiveness relative to California's sport and commercial rockfish fisheries. The establishment of the Big Creek Ecological Reserve (BCER), approximately 50 miles south of Monterey, California, presents a unique opportunity to evaluate the effects of a reserve on the rockfish resource. Obtaining baseline information on species composition, densities, and length frequencies of rockfish populations within and adjacent to BCER is the crucial first step for determining change in population parameters and future benefits to adjacent and distant fisheries. This information will allow resource managers to evaluate marine reserves as an alternative management tool for rockfish populations. Present population parameters of selected fish populations in nearshore habitats within and adjacent to BCER have been assessed visually (densities) by scuba divers and recorded with an underwater video camera equipped with paired lasers (length frequencies) during permanent and random transects.

[1] Calif. Dept. Fish and Game, 20 Lower Ragsdale Drive Suite 100 Monterey, CA 93940.

Introduction

Marine reserves established to enhance fisheries are designed to manage exploited stocks by protecting populations of sexually mature species vulnerable to overfishing, thus insuring a continual source of new recruits. Reserves have been reported to enhance fisheries in other parts of the nation and world; however, limited information is available to evaluate their effectiveness on California's sport and commercial rockfish fisheries. Rockfish (*Sebastes spp.*) are an important and heavily exploited component of sport and commercial fisheries in central California; however, because they are residential, slow-growing, and long-lived their numbers are quickly reduced by overfishing.

The success and quality of central California's nearshore rockfish sport fisheries have declined over the past decades, particularly in areas close to ports (Miller and Gotshall 1965, VenTresca and Lea 1984, Reilly et al. 1993, Karpov et al. 1995, Mason 1995). In the northern and central California sport fishery, between the late 1950s and early 1980s, the average weight per rockfish declined 13 percent, or 0.43 pounds per fish; average weight decreased in 12 of 16 major rockfish species (Karpov et al. 1995). In Monterey Bay, from 1987 to 1991, the average lengths of several rockfish species sampled in the Commercial Passenger Fishing Vessel (CPFV) fishery were below the average size at 50% sexual maturity (Reilly et al. 1993). Fishery managers are now concerned with catch and effort increases in the nearshore commercial hook-and-line fishery which is targeting on live rockfish, cabezon, lingcod and kelp greenling. The rapid expansion of this fishery is of greatest concern in the limited shallow rocky habitats along the coast. Many of these nearshore areas, especially along the Big Sur coast, had experienced little fishing pressure until recently and functioned as *ad hoc* reserves.

Recognizing that a change in the current resource management approach is warranted, alternative management and enhancement strategies are of paramount importance. Rockfish appear to be an excellent candidate for enhancement using the reserve management approach because in addition to being residential and long-lived they are extremely fecund and have a lengthy larval stage. Thus a reserve utilized as a rockfish management tool would protect critical spawning stock biomass and ensure a continual recruitment supply to fished areas via larval dispersal and emigration. However, to determine the effectiveness of a reserve to enhance rockfish fisheries, baseline information is first needed to demonstrate that densities and sizes of fish populations within the reserve are greater than adjacent fished areas and that absolute numbers of species of concern within the reserve are sufficient to provide a reasonable/significant source of larvae.

Marine fisheries reserves offer potential benefits to coastal fisheries and marine resource management, including: the enhancement and restoration of fishery yields through larval/egg transport; "spillover" of juveniles and adults from reserves to surrounding areas; protection of reproductive potential of targeted species; maintenance of biological diversity; and increased social and economic benefits to local communities (Russ 1985, Davis 1989, Bohnsack 1990, Roberts and Polunin 1991, Rowley 1994, Roberts et al. 1995). Reserves may also increase the resiliency of heavily exploited fisheries during times of environmental extremes by serving as a buffer against drastic declines and hasten the recovery of these fisheries by supplying harvested stocks with new recruits (Carr and Reed 1993). Although reserves provide an excellent opportunity for *in situ* testing of concepts related to the protection and enhancement of fish populations, there are few studies demonstrating a significant increase to fish population densities and average length within a reserve (Rowley 1992, Dugan and Davis 1993, Roberts 1995, Russ and Alcala 1996). Studies showing enhancement or "spillover" are rare but are probably most critical in convincing public user groups of the value of reserves (Rougharden and Iwaṣa 1986, Battershill et al. 1993, Rowley 1992).

Although marine reserves have been reported to enhance fisheries in other parts of the nation and world, limited information is available to evaluate their effectiveness on California's sport and commercial rockfish fisheries. Because improperly designed reserves may endanger a fishery by providing a false sense of protection, determining the effectiveness of a reserve is very important (Carr and Reed 1993). The establishment of BCER in Monterey County in January 1994 provided an excellent opportunity to evaluate beneficial effects of a reserve on the nearshore rockfish resource; however, without first obtaining baseline information on species composition, densities, and length frequencies of rockfish populations within and adjacent to BCER, determining change in population parameters or future benefits to adjacent and distant fisheries will be difficult at best. Furthermore, assessing the status of nearshore fish populations/stocks in the vicinity of BCER is timely and relevant due to the recent dramatic increase in commercial hook-and-line landings of nearshore species throughout California. The extensive kelp forests along the Big Sur coast, one of the last remaining unexploited marine habitats in California, are now being intensely fished to provide for the increasing demand in the lucrative live fish market. As fishing increases on these formerly pristine stocks, it becomes imperative to evaluate the effectiveness of a central California coast reserve to protect and enhance rockfish stocks.

Rockfishes are an excellent candidate species group to test. They are long-lived, residential, fecund, and have an open ended reproductive system that distributes offspring spatially. Resident populations of large sexually mature rockfishes within the reserve would be a continual source of larvae which in turn would be carried via nearshore currents to other sections of the coastline (Chelton et al. 1982, Wyllie-Echeverria 1987, VenTresca et al. 1996).

Methods

A central theme in our efforts to evaluate subtidal fish populations within and outside reserves is the use of fish counting and measuring procedures using scuba. Many methods have been used to quantify the abundance and composition of fish assemblages (Limbaugh 1955, Miller and Geibel 1973, Larson and DeMartini 1984, Bohnsack and Bannerot 1986, Roberts 1995, Russ and Alcala 1996). These studies have used both random and permanent transects to quantify fish assemblages. In general, *in situ* counts of fish are highly variable (Davis and Anderson 1989). As a result, a high number of replicate counts (*n*) must be made for statistically valid population estimates. We have developed a nondestructive sampling technique for obtaining fish lengths utilizing paired lasers mounted on an underwater housing containing a video camera, using a similar design to those used on submersibles (Caimi and Tusting 1987, Davis and Tusting 1991). This technology provides a method to obtain more accurate measurements of fish lengths than the method of estimating lengths visually underwater and allows us to gather subadult and adult fish length frequencies and species composition data within and adjacent to BCER without physically capturing fish.

Data from our transects will allow us to measure annual fluctuations of relative abundance at permanent stations and to make statistical inferences when comparing fish densities and length frequencies within and adjacent to BCER. Limited data using visual and video permanent and random transects have been collected during three cruises from 1993 to 1996 using the California Department of Fish and Game's (CDFG) **R/V MAKO**. These data will complement findings of other researchers working in this area (Crane and Paddack 1994, Paddack unpubl. data). We have also collected nearshore sea surface temperature intermittently at permanent stations in BCER with data loggers.

Population parameters of selected fish species in nearshore habitats within and adjacent to BCER are being assessed utilizing four related approaches: 1) fish population data are being collected using standardized scuba techniques; 2) portions of these data will be combined with a bathymetric/habitat map (produced from side-scan sonar and bathymetric data collected during spring 1998) and incorporated into a Geographical Information System (GIS) to estimate habitat associations, and biomass and spawning potential of fish populations within BCER; 3) species composition, length frequencies and catch-per-unit-effort (CPUE) of fishes taken in commercial skiff and CPFV fisheries in the areas adjacent to BCER will be collected from summer through fall in 1997 and spring through fall in 1998; 4) fish population data collected at BCER during this project will be compared to limited data collected at BCER in 1993 (prior to designation as a marine protected area) and during 1994-96 (first three years as a marine protected area). Additionally, data collected at BCER will be compared to data collected at Point Lobos State Reserve, a well established marine protected area, in 1989 and during this project (1997-98).

Subtidal Research Utilizing Scuba

The subtidal study site is divided into three areas, each encompassing approximately three miles of coastline: 1) within BCER, 2) south of BCER, and 3) north of BCER. Scuba transects are deployed in 7-25 m depths in kelp forest and non-kelp forest habitats. Target species (those susceptible to hook-and-line fishing) are predominately observed in the kelp forest strata; however, we are also conducting limited surveys in non-kelp forest strata. Nearshore subadult and adult fishes of interest and susceptible to fishing pressure include: *Sebastes atrovirens* (kelp rockfish), *Sebastes carnatus* (gopher rockfish), *Sebastes chrysomelas* (black-and-yellow rockfish), *Sebastes caurinus* (copper rockfish), *Sebastes nebulosus* (China rockfish), *Sebastes melanops* (black rockfish), *Sebastes miniatus* (vermilion rockfish), *Sebastes serranoides* (olive rockfish), *Hexagrammos decagrammus* (kelp greenling), *Ophiodon elongatus* (lingcod), *and Scorpaenichthys marmoratus* (cabezon). Densities and length frequencies of these species are readily quantifiable using described methods. Rare and unusual species are also recorded.

Population parameters of selected fishes are assessed visually by scuba divers and recorded with an underwater video camera, equipped with paired lasers for size reference, during permanent and random transects. To obtain species length estimates, the video camera is equipped with two parallel lasers, mounted on the underwater video camera housing. The parallel laser beams, which are 50 mm apart, produce paired dots on fish in the video frames. The video image of each fish with the paired red dots will subsequently be analyzed using computer technology by first capturing images from the video tapes, and then digitizing and integrating the images into computer files. Each fish can then be precisely measured and the data will be entered into a database to obtain estimates of mean lengths for each species. This method provides a nondestructive sampling technique for obtaining fish lengths which is more accurate than estimating lengths visually underwater.

Four permanent 20-m transects, deployed as sides of a square, are presently established at nine stations within and adjacent to BCER (three to the north, three within, and three to the south). To obtain visual counts for species composition and density estimates, dive partners will descend along a surface buoy line to the first corner of the square transect and position themselves on either side of the transect line. Both divers will observe an area 2 m high, 2 m wide and 20 m long on their respective sides of the transect line, starting 2 m from the beginning corner and ending 2 m short of the terminal corner. During random visual transects, each diver will survey 2x2x10 m swaths and record numbers and age classes of selected fishes encountered, as well as information on depth, substrate, relief, crevices and algal cover. Distance covered is determined by a 10-m retractable transect which consists of 10 m of line attached to a clip and small piece of wire at one end and a modified retractable dog leash unit at the other end.

Data from permanent transects, at three stations each within, south, and north of BCER, will be used as indices of annual changes of fish densities and length frequencies of targeted fish species. Random transects in each of the three areas will be used to correct for bias in the permanent transects and to determine if the permanent transects adequately reflect temporal changes taking place within and adjacent to BCER. The advantage of permanent transects should be lower within-sample variance, allowing detection of smaller differences in mean density within an area through time.

In conjunction with a bathymetric/habitat map, fish densities will be used to calculate habitat associations, biomass and spawning potential for selected fish populations within and adjacent to BCER. It will also be of interest to statistically compare species density and mean length estimates between permanent and random transects. This will provide an assessment on the effectiveness of the permanent stations as monitoring sites for these fish population parameters.

Bathymetric/Habitat Mapping

Nearshore (<35 m) habitat within BCER will be mapped and classified by ABA Consultants of Capitola, California. ABA Consultants specializes in description of biological and geological components of the sea floor. ABA will use state-of-the-art hydrographic survey technology to produce detailed charts of sea floor depth contours, substrate type and geomorphology. All hydrographic data will be digitally stored in real time with 2-m position accuracies using a Differential Global Positioning System (DGPS). Equipment includes: EG&G 260-TH dual frequency (100/500KHz) sidescan sonar with digital recorder, Innerspace 448 (208 KHz) digital depth sounder, Roxann Sea Floor Substrate Classification System, Trimble 4000 series DGPS, HYPACK navigation and data logging software and ArcInfo compatible GIS database.

Sea floor mapping will be conducted during late winter and early spring of 1998, to take advantage of sparse kelp canopy cover. To avoid kelp entanglement, acoustic transducers will be mounted on a floating platform on the surface from a boom mounted on the front of the survey vessel to avoid kelp entanglement.

Monitoring Catch of CPFV Fishery

The CPFV fishery in Morro Bay occasionally conducts 2-day, long range sport fishing trips to the area from Lopez Point north to Point Sur and encompassing the vicinity of BCER. An observer will be placed on these trips to record sport fishes caught, using established sampling methodology (Reilly et al. 1993). This will provide a cost-efficient supplementary database of CPUE, mean length, and length frequency for important nearshore sport fishes in the vicinity of BCER, an area that historically little information on sport or commercial fishing pressure. Trips originating in Morro Bay and fishing from Lopez Point to Point Sur will be sampled during 1997-1999.

Monitoring and Sampling Catch of Commercial Skiff Fishery

Commercial skiff fishery landings along the Big Sur coast will be monitored and sampled during 1997-1999. Sampling procedures will be based on criteria developed for CDFG/National Marine Fishery Service groundfish surveys. Data will be compared to CDFG's Marine Fisheries Statistical System database and examined for trends in the central California hook-and-line rockfish fishery. Historical analysis encompassed the years 1987 through 1995.

Results

A preliminary analysis of fish count/transect data shows that similar trends were documented by both transecting methods (Figure 1). The mean count of subadult and adult rockfish/*measured transect* in kelp forests within grouped locations tended to be similar. The means count of subadult and adult rockfish/*measured transect* was significantly ($p < 0.05$) greater in BCER and in the north control area than the south control area. The mean count of subadult and adult rockfish/*permanent transect* was significantly ($p < 0.05$) greater in BCER than in the north and south control areas (Figure2). Additional analyses will compare data from both types of transects to correct for possible biases from permanent transects.

The paired laser/underwater video system has proven to be a practical method for collecting fish length distribution data. The preliminary data presented here are from single measurements per fish with number of fish measured per species at each location group ranging from 1 to 166. Figure 3 is a depiction of subadult and adult fish length distributions (box plots) south of BCER, within BCER, and north of BCER. The mean length of gopher rockfish, for which we have the highest n, was significantly ($p < 0.05$) greater in BCER than control areas to the north and south. To make the best use of images, we are considering collecting several length measurements per fish (using successive frames) and comparing mean length among areas with repeated-measures ANOVA. These additional measurements will allow us to account for variation among measurements per fish, thereby increasing statistical power.

Conclusion

Rockfish are an important and heavily exploited component of sport and commercial fisheries in central California. The success and quality of central California's nearshore rockfish sport fisheries have declined over the past decades, particularly in areas close to ports. Recognizing that a change in the current resource management approach is warranted, alternative management and

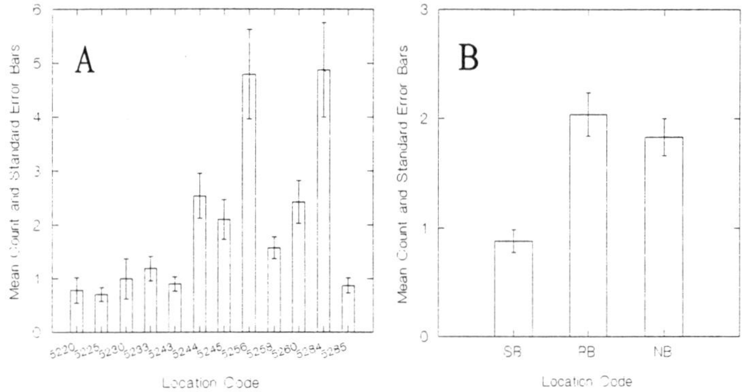

Figure 1. Mean count of subadult and adult fishes and standard error bars using measured transects. Numeric location codes (A)<= 5233 are south of BCER, from 5243 through 5256 are within BCER, and >= 5258 are north of BCER. Alphabetic location codes (B) denote grouped location codes south of BCER (SB), within BCER (RB) and north of BCER (NB).

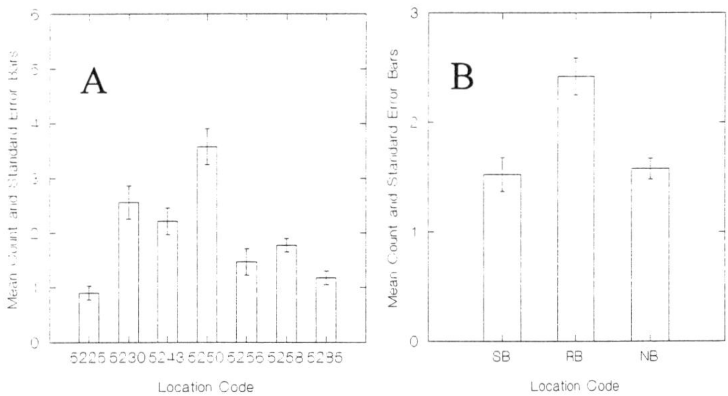

Figure 2. Mean count of subadult and adult fishes and standard error bars using permanent transects. Numeric location codes (A) <= 5233 are south of BCER, from 5243 through 5256 are within BCER, and >= 5258 are north of BCER. Alphabetic location codes (B) denote grouped location codes south of BCER (SB), within BCER (RB) and north of BCER (NB).

enhancement strategies are of paramount importance. Marine reserves have been reported to enhance fisheries in other parts of the nation and world; however, limited information is available to evaluate their effectiveness on California's sport and commercial rockfish fisheries. The establishment of BCER presents a

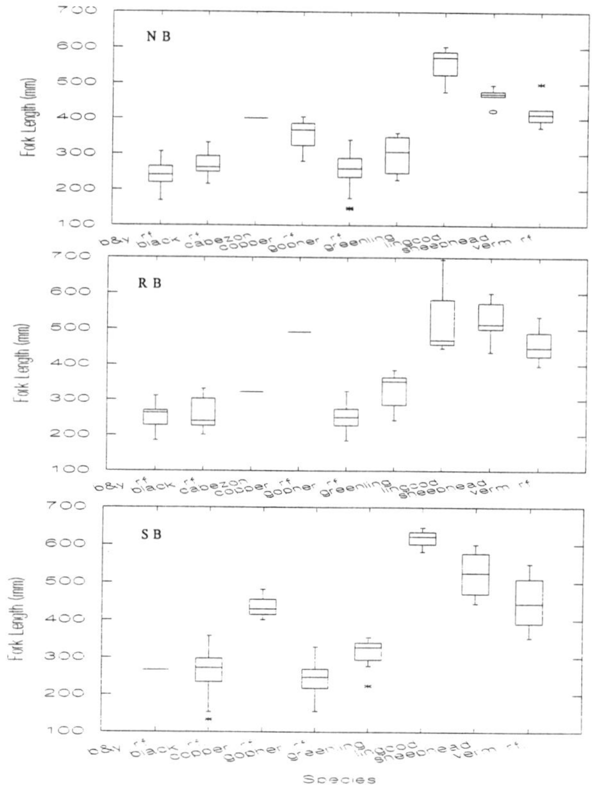

Figure 3. Preliminary depiction of length distribution (box plot) of fishes south of BCER (SB), within BCER (RB) and north of BCER (NB). All length measurements were obtained from recorded images of fish, scaled using fiducial marks projected by paired lasers.

unique opportunity to evaluate beneficial effects of a reserve on the rockfish resource. It is crucial to obtain baseline information on species composition, densities, and size frequencies of rockfish populations within and adjacent to BCER. This project proposes to estimate habitat associations, biomass and spawning potential of fish populations within BCER from fish population information collected using scuba and a bathymetric/habitat map, and incorporate this into a GIS. With this baseline data, we will determine changes in population parameters within and adjacent to BCER. Additionally, our sample design, which incorporates both random and permanent transects, will provide an opportunity to determine if permanent transects adequately reflect temporal changes in nearshore fish populations.

Resource managers will benefit by having a recently established marine reserve thoroughly evaluated as an alternative management tool, one which has been proven effective in enhancing fish populations in other parts of the world. If BCER enhances local fish populations, recreational anglers, the CPFV industry, the commercial skiff hook-and-line fishery, and associated local businesses will benefit from the long-term effects of a reserve. For the many important nearshore rockfishes, the reserve should insure a source of recruitment via larval transport to adjacent and distant fished areas. Additionally, the more direct benefits from the reserve will be emigration of sport fishes from areas of higher density within the reserve to fished areas of lower density. By providing evidence that marine reserves will enhance local sport and commercial fisheries, this project could further benefit these fisheries by recommending additional reserves in other fished areas. Also, the establishment and monitoring of underwater reserves and/or parks could provide positive benefits to local tourism, particularly recreational, non-consumptive divers.

References

Battershill, C.N., D.R. Schiel, G.P. Jones, R.G. Creese, and A.B. MacDiarmid (eds). 1993. Conference abstracts: Proceedings of the second international temperate reef symposium, January 7-10, 1992, Auckland, NZ. NIWA Marine.

Bohnsack, J.A. and S.P. Bannerot. 1986. A stationary visual census technique for quantitatively assessing community structures of coral reef fishes. NOAA, NMFS, SEFC. Tech. Rep. 41.

----------1990. The potential of marine fishery reserves for reef fish management in the U.S. Southern Atlantic. Plan Development Team, Reef Fish Management Plan, South Atlantic Fishery Management Council, Coastal Resource Division. CRD/89-90/04.

Caimi, F.M. and R.F. Tusting. 1987. Application of lasers to ocean research and image recording systems. In: Proceedings of the international conference on LASERS '86, Society for optical and quantum electronics. pp. 518-524.

Carr, M.H. and D.C Reed. 1993. Conceptual issues relevant to marine harvest refuges: Examples from temperate reef fishes. Can. J. Fish. Aq. Sci. 50:2019-2028.

Chelton, D.B., P.A. Bernal, and McGowan. 1982. Large-scale interannual physical and biological interaction in the California current. Jour. Mar. Res. 40:1095-1125.

Crane, N. and M. Paddack. 1994. Oceanic Society Expeditions: Monterey Bay National Marine Foundation Grant 94-7210, 41p.

Davis, D.L. and R.F. Tusting. 1991. Quantitative benthic photography using laser calibrations.In:Proceedings of undersea world, '91,San Diego, CA.

Davis, G.E. 1989. Designated harvest refugia: the next stage of marine fishery management in California. CalCOFI Reps. 30:53-58.

---------- and T.W. Anderson. 1989. Population estimates of four kelp forest fishes and an evaluation of three *in situ* assessment techniques. Bull. Mar. Sci. 44(1), 21p.

Dugan, J.E. and G.E. Davis. 1993. Introduction to the international symposium on marine harvest refugia, 1991-1992. Can. J. Fish. Aq. Sci. 50.

Karpov, K.A., D.P. Albin, and W.H. Van Buskirk. 1995. The marine recreational fishery in northern and central California: A historical comparison 1958-86, status of stocks 1980-86, and effects of changes in the California current. Calif. Fish and Game, Fish Bull. 176.

Larson, R.J. and E.E. DeMartini. 1984. Abundance and vertical distribution of fishes in a cobble-bottom kelp forest off San Onofre, California. Calif. Fish and Game. Fish Bull. 821. pp. 37-53.

Limbaugh, C. 1955. Fish life in the kelp beds and effects of harvesting. Univ. Calif. Inst. Mar. Res. IMR Ref. 55-9:1-156.

Mason, J.E. 1995. Species trends in sport fisheries, Monterey Bay, California. 1959-86. Mar. Fish. Rev. 57:1-16.

Miller, D.J. and D.W. Gotshall. 1965. Ocean sportfish catch and effort from Oregon to Point Arguello, California, July 1, 1957-June 30, 1961. Calif. Fish and Game, Fish. Bull. 130.

-------- and J.J. Geibel. 1973. Summary of blue rockfish and lingcod life histories; a reef ecology study; and giant kelp Macrocystis pyrifera, experiments in Monterey Bay, California. Calif. Fish and Game, Fish. Bull.:158.

Reilly, P.N., D. Wilson-Vandenberg, D.L. Watters, J.E. Hardwick, and D. Short. 1993. On board sampling of the rockfish and lingcod commercial passenger fishing vessel industry in northern and central California, May 1987 to December 1991. Calif. Fish and Game. Admin. Rep. 93-4.

Roberts, C.M. and N.V.C. Polunin. 1991. Are marine reserves effective in management of reef fisheries? Rev. Fish Biol. and Fish. 1:65-91.

--------. 1995. Rapid build-up of fish biomass in a Caribbean marine reserve. Conservation Biology. 94:815-826.

--------, W.J. Ballantine, C.D. Buxton, P. Dayton, L.B. Crowder, W. Milon, M.K. Orbach, D.Pauly, J. Trexler, and C.J. Walters. 1995. Review of the use of marine fishery reserves in the U.S. southeastern Atlantic. NOAA Tech. Memo. NMFS-SEFSC-376.

Roughgarden, J. and Y. Iwasa. 1986. Dynamics of a metapopulation with space-limited subpopulations. Theor. Pop. Biol. 29:235-261.

---------. 1992. Impacts of marine reserves on fisheries: A report and review of the literature. Sci. & Res. Series 50-51.

--------. 1994. Case studies and reviews: Marine reserves in fisheries management. Aq.Cons.: Mar. and Fresh. Ecosys. 4:233-254.

Russ, G.R. and A.C Alcala. 1996. Marine reserves: Rates and patterns of recovery and decline of large predatory fish. Ecological Applications. 63:947-961.

--------. 1985. Effects of protective management on coral reef fishes in the Central Philippines. Proc. 5th Int. Coral Reef Cong. Tahiti. 4:219-224.

VenTresca, D.A. and R.N. Lea. 1984. Feasibility of zonal management research delineation of nearshore sportfish stocks, July 1, 1983-June 30, 1984. Annual Job Performance Report Federal Aid in Sport Fish Restoration Act. Project Number:DJ-625-17:VII-2.

---------, J.L. Houk, M.J. Paddack, M.L. Gingras, N.L. Crane, and S.D. Short. 1996. Early life history studies of nearshore rockfishes and lingcod off central California, 1987-92. Calif. Fish and Game. Admin. Rep. 96-4.

Wyllie- Echeverria, T.E. 1987. Thirty-four species of California rockfishes: maturity and seasonality of reproduction. Calif. Dept. Fish and Game, Fish. Bull. 852:229-250.

CHAPTER 56

Marine Protected Areas and the Red Urchin Fishery

Laura Rogers-Bennett[1]

Abstract

There is a growing consensus among fishery scientists that marine protected areas (MPAs) offer a promising approach for managing California's fished populations. California's leading fishery the red urchin fishery has been declining for almost 10 years despite traditional management efforts. The biological characteristics of red urchins including their long life, slow growth, and sporadic recruitment suggest that previous landings came from an accumulation of old animals. MPAs can be used to ensure a portion of the catch remains to reproduce and replenish fished habitats. The most effective MPAs will encompass reproductively important "source" habitats. Shallow habitats in northern California appear to be "source" habitats because they harbor high densities of both brood stock and juvenile urchins. There is now a large body of evidence supporting the establishment of MPAs as a fishery management tool for benthic invertebrates.

Introduction

Many of California's former leading fisheries are currently depressed and the current largest, the red sea urchin fishery is in decline. Fisheries are in trouble worldwide with landings declining at an alarming rate. Two basic approaches have been employed to combat declining fish and invertebrate stocks. The first is the reduction of catch using gear restrictions, limited entry of fishers, season closures, and total allowable catch (TAC) quotas. There appears to be ample evidence from many fisheries that these traditional management strategies are not serving us well. For example, despite considerable management

[1]Postdoctoral Fellow, Friday Harbor Laboratories, Univ. of Washington, 620 Univ. Rd., Friday Harbor, Wa. 98250

efforts the north Atlantic cod fishery collapsed in 1991 (Hutchings and Myers, 1994). The second approach has been to enhance fished populations using hatchery reared juveniles released into the wild. Since 1984 California Department of Fish and Game (CDFG) has administered an enhancement program (OREHP) examining the feasibility of stocking juvenile white sea bass and California halibut in southern California (Kent, 1994). Currently, there is concern over the "success" of previous CDFG stocking programs because a wild red abalone population now appears to have diminished genetic diversity (Gaffney et al., 1997). Cases such as this have contributed to the widely accepted idea that enhancement programs offer limited success and can potentially have negative impacts on the behavioral and genetic diversity of wild animals (Stickney, 1994). Furthermore, enhancement programs typically have a poor track record following the fate of stocked juveniles and an even poorer record of determining if stocked juveniles displace wild individuals. Marine populations however, are able to rebuild themselves and this concept has been coined the natural hatchery. The classic example of this is the dramatic rebound of fisheries in the north sea during World War II. MPAs represent an attractive alternative for managing California's nearshore fisheries (Davis, 1989).

Metapopulations and MPAs

The metapopulation concept can be readily applied to MPAs used for either fishery management or marine conservation. Ecologists and resource managers alike recognize that vital rates (e.g. reproduction) can vary spatially. Reproductively important patches are known as "source" habitats, and occur where local reproduction exceeds local mortality, whereas "sink" habitats occur where local mortality exceeds local reproduction (Pulliam 1988, Dias 1996). Knowledge of metapopulation dynamics will aid fishery managers in determining which areas to preserve in MPAs and which areas are most suitable for harvest. The Tehuelche scallop population in Argentina is an example of an invertebrate fishery with metapopulation dynamics in which discrete beds are interconnected by larval dispersal. In this population "source" beds have low stranding mortality and are located where oceanographic currents will transport larvae to neighboring beds (Orensanz, 1986). Unfortunately, despite knowledge of the population dynamics, higher prices and more fishers recently increased fishing pressures on small scallops and "source" beds which precipitated fishery collapse

(Orensanz pers. comm.)

The most successful MPAs will supply fished habitats with surplus recruits. Many marine organisms have complex life histories with pelagic eggs and larvae followed by benthic juvenile and adult stages. Early life history stages in the water column have the potential to disperse 10's, 100's, and 1,000's of kilometers prior to settling. Indeed, many marine populations are "open" such that a larval pool is available for recruitment (Roughgarden et al., 1985). In addition, juveniles and adults may "spillover" from MPAs into neighboring fished habitats (Rowley, 1994). Dispersal of invertebrates into fished habitats has been demonstrated by tag and recapture studies for snow crabs (Yamasaki and Kuwahara, 1990) and pink shrimp (Gitschlag, 1986). This suggests that for many fished populations MPAs can be a viable management strategy.

There are few barriers to dispersal in the marine environment and few would argue that eggs, larvae, juveniles or adults do not disperse outside MPAs (D. Pauley pers. comm.). Debates do arise however over the quantity of fishable organisms which are produced and subsequently exported into fished habitats. In addition, there are questions about whether the surplus exported into fished areas equals or exceeds the quantity excluded from the fishery within the MPAs. It is likely that these issues will have to be examined on a case by case basis. There may be however cases where the number of fish exported is low, yet without the MPA the fishery could not persist. This appears to be the case for the seahorse fishery in the Philippines (A. Vincent pers. comm.). MPA may also have emergent properties which help sustain fisheries but which may be subtle and difficult to quantify. For example, ECOPATH trophic models consistently indicate that detritus is an important component in mature marine habitats and may be critical for sustaining marine communities (Pauley and Christensen, 1995). MPA may function to help rebuild detritus (Pauley pers. comm.) and may be analogous to terrestrial restoration efforts to maintain leaf litter in forests.

Design of Marine Protected Areas

The most successful MPAs will encompass important breeding populations which can serve as larval "sources". Skeptics worry that there is not enough biology known to appropriately identify the location, size, and configuration of MPAs. For many free-spawning species (e.g. sea urchins, abalone) these decisions may be easier than they appear. First, these animals require high densities to ensure reproductive success such that density appears to be a good predictor of fertilization success (Pennington, 1985; Levitan et al., 1992). Second, fecundity typically scales as a function of body size and so the larger the fish or invertebrate often the more gametes that animal can produce. Therefore, areas with more, and larger fish or invertebrates will have enhanced reproductive output compared with low density fished areas. There is now ample evidence from many regions that MPAs harbor higher densities of target organisms as well as individuals of larger size than comparable fished habitats (Roberts and Polunin, 1991; Dugan and Davis, 1993, Rowley, 1994; Palsson and Pacunski, 1995).

Local oceanographic patterns should be selected to ensure larval transport and "spillover" from MPAs to fished habitats. Areas with intense upwelling such as major headlands may receive little recruitment (Ebert and Russell, 1988) except in years when upwelling is relaxed (e.g. El Niño events). In contrast, gyres such as in the southern California byte appear to have good larval retention and as a result good settlement. For species with long lived larvae (>3 weeks) such as sea urchins (Strathmann, 1978) local oceanographic features may not be as important for MPA design since these larvae have the ability to disperse great distances. Abalone have short lived planktonic larvae and apparently limited larval dispersal (Tegner and Butler, 1985), however even brooded juveniles can disperse long distances (Johannesson, 1988; Helmuth et al., 1994). While biologists do not know all the important dispersal parameters for designing MPAs, areas can be adjusted as more information becomes available. Ballantine (1995) and others conclude that we now have enough evidence documenting the value of MPAs and therefore should proceed with their implementation.

California's Red Sea Urchin Fishery

California's largest fishery, the red sea urchin fishery landed over 20 million pounds worth approximately 20 million dollars in 1996, however, landings have been declining steadily since 1988. Invertebrates now dominate the state's fisheries with squid and crab being the second and third most valuable worth 20 and 14 million dollars, respectively (CDFG data). Fishers in California however, appear to be depleting one species after another as stocks decline following harvest. For example as abalone landings have declined urchin landings have taken their place (Dugan and Davis, 1993); a common phenomenon in fisheries.

Urchin landings in northern California have undergone an initial boom of 30 million pounds in 1988 followed by a bust of less than 4 million pounds in 1996. This dramatic decline has occurred despite fishery management efforts. Currently, three regulations are in place for the urchin fishery: (1) limited entry, (2) lower size limits, and (3) season closures. Limited entry was instituted to reduce the increase in the number of fishers, however this was enacted only after a large influx of fishers entered the fishery. Lower size limits are designed to allow individuals one or more spawning seasons before adults are harvested. The largest urchins however, appear to be the most important spawners and they are not protected (Ebert pers. comm). Season closures are intended to reduce catch. Red sea urchins, however are long lived, slow growing (Ebert and Russell, 1992) and slow moving animals which can easily be harvested after sites are identified during closed dates. Clearly this strategy has not been effective at maintaining sustainable urchin landings.

The red sea urchin fishery in northern California appears to be a classic example of being "fished down". Fishing long lived organisms is problematic since the initial harvest comes from an accumulation of old and unproductive subpopulations (Ricker, 1973; Hilborn et al., 1995). This problem occurs in forests and results in a decline in yield (or "fall down") after an initial high yield (Ludwig, 1993). Urchin harvesting in the north began in 1986 and peaked two years latter. Effort was high in part due to a shift in harvester effort from southern to northern California and the northern population has not sustained these high landings. Sustained landings may also be hampered by notoriously sporadic urchin recruitment (Pearse and Hines, 1987; Ebert, 1983) in combination with low survival rates of newly settled urchins (Rowley, 1989). In northern California urchin settlement on artificial substrates (brushes) is lower and more irregular than in southern California (Ebert et al., 1994). These biological features coupled with their sedentary behavior make red urchins particularly vulnerable to overharvesting.

Matrix population models indicates that the loss of the largest urchins may have serious negative consequences for the population (T. Ebert pers. comm.). Large adults have higher gonad mass than intermediate sized urchins, since gonad mass scales exponentially as a function of body size. High urchin densities ($> 3/m^2$) are required for fertilization success (Levitan et al., 1992). Protection of the largest urchins could be achieved using upper size limits, however slow growing animals may take years or decades to grow into the protected upper size during which time they are vulnerable to harvest (Breen, 1984). In addition, upper size limits would allow divers to thin urchin beds removing legal size urchins however this thinning can depress fertilization success (Levitan et al., 1992). The good news is that large individuals "boomers", with gonad weights in excess of 100 g, are not the most valuable to the sushi market in Japan. Therefore, given their importance to reproduction the largest urchins should be excluded from the fishery.

Studies have shown that shallow habitats in northern California harbor large urchins at high densities such that these areas may make ideal candidates for MPAs (Rogers-Bennett et al., 1995). Seasonally large amounts of drift algae are available in shallow habitats and urchin gonad indices there are 4 times greater than urchins in other habitats suggesting they are important spawners (Rogers-Bennett et al., 1995). Urchins in shallow habitats also have high gonad indices in Alaska (Carney, 1991) and Newfoundland (Keats et al., 1984) suggesting this pattern is widespread. In addition, in many areas adult urchins harbor juveniles under their spine canopy (Tegner and Dayton, 1977; Breen et al., 1985; Rogers-Bennett, 1989). During a four year study in northern California, juvenile urchin density was 10 times greater in shallow habitats than intermediate and deep habitats suggesting that shallow areas not only harbor brood stock but also function as critical nursery habitats (Rogers-Bennett et al., 1995). The implications of this can be seen in fished habitats where low adult density is coincident with lower juvenile densities than unfished habitats (Tegner and Dayton, 1977; Sloan et al., 1987; Rogers-Bennett et al., 1997). In northern California excluding harvesting in shallow habitats (<7m) would impact less than 10% of the catch (P. Kalvass, pers. comm.). Models of various MPA sizes and numbers support the strategy of establishing multiple high density patches (with large individuals) within larval dispersal range (Quinn et al., 1993). For these reasons fishery managers and policy makers should consider the value of MPAs when designing red urchin management strategies.

Caveats and Future Action

The establishment of MPA is not the "lazy mans" management strategy. It must be accompanied by rigorous long-term monitoring to determine it's effectiveness. Areas should be replicated and parameters such as density, individual size, and recruitment should be examined. Several analytical methods can be employed to examine the impacts of MPAs, even when replicate MPAs can not be established (e.g BACI-PS)(Schmitt and Osenberg, 1996). Caution must be exercised to ensure the establishment of MPAs does not simply intensify fishing effort in open areas. Similarly, all life history stages of the target species must be examined to ensure they are not fished before they disperse into the MPA. Highly migratory species, such as sharks, will not be protected in MPA. Many of the root causes of fishery problems like explosive human population growth, and excessive resource use (Ludwig et al., 1993) are not addressed by MPAs. In fact, the notion of sustainable yields has come under criticism in light of human short comings including greed and short sightedness (Ludwig et al., 1993). Poaching and pollution are major concerns particularly for management systems incorporating MPAs. While MPAs are not a panacea, they are a wonderful tool that we can use to allow the ocean to replenish itself. Fishery scientists and ecologists are in agreement and calling for the use of MPA for fishery management (Davis, 1989; Roberts et al., 1995; Bohnsack, 1997; Roberts, 1997; Lubchenko, pers. comm.).

In California a concerted effort is underway to investigate the adoption of MPAs for nearshore fishery management. The necessary first step, documenting the locations and regulations governing all the state's MPAs has been completed (McArdle, 1997). Of the 104 MPAs in California only 9 are "no take" (no fishing) areas occupying less than 20 miles of California's coastline. McArdle (1997) also reveals the confusion of bills and regulations that govern MPA's in the state. MPAs established in nearshore rocky habitats could be used as fishery management tools for a host of benthic species in California (Davis, 1989) including red sea urchins (Rogers-Bennett et al., 1995). MPAs are ideally suited for multispecies management. A comprehensive marine resource management plan is called for and being considered as evidenced by this conference. In addition, a "flotilla" of new legislation aimed at centralizing marine resource management is also currently being debated in Sacramento. Clearly, management strategies that incorporate the use of marine protected areas ("no take") to maintain California's valuable renewable marine resources must be seriously considered.

CALIFORNIA ANNUAL RED SEA URCHIN LANDINGS
IN THOUSANDS OF POUNDS ROUND WEIGHT

YEAR	NORTHERN CA.	SOUTHERN CA.	TOTAL
1980	103	21,100	21,200
1981	194	24,700	24,900
1982	92	19,300	19,400
1983	61	17,200	17,300
1984	59	14,900	15,000
1985	1,900	18,000	19,900
1986	10,200	24,000	34,200
1987	23,600	22,500	46,100
1988	30,400	21,400	51,800
1989	26,800	24,400	51,200
1990	18,600	27,000	45,600
1991	16,800	25,100	41,900
1992	12,200	20,100	32,300
1993	7,200	19,300	26,500
1994	5,800	17,800	23,600
1995	4,700	17,500	22,200
1996*	3,800	16,200	20,000

* Preliminary Figures

References

Ballantine, W.J. 1995. The New Zealand experience with "no-take" marine reserves. In Review of the Use of Marine Fishery Reserves in the U.S. Southeastern Atlantic. NOAA Tech. Memo. NMFS-SEFSC-376. Pp. c15-c31.

Bohnsack, J.E. 1997. Concensus development and the use of marine reserves in the Florida Keys National Marine Sanctuary. In Proceed. 8th Intern. Coral Reef Sympos. Panama. In press.

Breen, P.A. 1984. Sea urchins: Suitability of the present minimum size limit. Can. Manuscr. Rep. Fish. Aquat. Sci. 1174:25-47.

Breen, P.A., W. Carolsfield, and K.L. Yamanaka 1985. Social behavior of juvenile red sea urchins *Strongylocentrotus franciscanus* (Agassiz). J. Exp. Mar. Biol. Ecol. 92:45-61.

Carney, D. 1991. A comparison of densities, size distribution, gonad and total-gut indices, and the relative movements of red sea urchins *Strongylocentrotus franciscanus,* in two depth regimes. M.S. Thesis Univ. Ca., Santa Cruz, Santa Cruz, California, USA.

Davis, G.E. 1989. Designated harvest refugia: the next stage of marine fishery management in California. California Cooperative Oceanic Fisheries Investigations Report 30:53-58.

Diaz, P.C. 1996. Sources and sinks in population biology. TREE 11:326-330.

Dugan, J.E. and G.E. Davis, 1993. Applications of marine refugia to coastal fisheries management. Can. J. Fish. Aquat. Sci. 50:2029-2042.

Ebert, T.A. 1983. Recruitment in echinoderms. Echinoderm Studies 1:169-203.

Ebert, T.A., S.C. Schroeter, J.D. Dixon, and P. Kalvass 1994. Settlement patterns of red and purple sea urchins (*Strongylocentrotus franciscanus* and *S. purpuratus*) in California USA. Mar. Ecol. Prog. Ser. 111:41-52.

Ebert, T.A. and M.P. Russell 1992. Growth and mortality estimates for red sea urchins *Strongylocentrotus franciscanus* from San Nicolas Island, California. Mar. Ecol. Prog. Ser. 81:31-41.

Ebert, T.A., and M.P. Russell 1988. Latitudinal variation in size structure of the west coast purple sea urchin: a correlation with headlands. Limnol. Oceanogr. 32:286-294.

Gaffney, P.M. et al. 1997. Genetic effects of propagation: signals from wilds and hatchery populations of red abalone in California. Aquaculture In press.
Gitschlag, G.R. 1986. Movement of pink shrimp in relation to Tortugas sanctuary. N. Amer. J. Fish. Manage. 6:328-338.
Helmuth, B., Veit, R.R. and R. Holberton 1994. Long-distance dispersal by a brooding bivalve (*Gaimardia trapesina*) by kelp-rafting. Mar. Biol. 120:421-426.
Hilborn, R., Walters, C.J. and D. Ludwig 1995. Sustainable exploitation of renewable resources. Annu. Rev. Ecol. Syst. 26:45-67.
Hutchings, J.A. and R.A. Myers 1994. What can be learned from the collapse of a renewable resource? Atlantic cod *Gadus morhua*, of Newfoundland and Labrador. Can. J. Fish. Aquat. Sci. 51: 2126-2146.
Johannesson, K. 1988. The paradox of Rockall: why a brooding gastropod (*Littorina saxatilis*) is more widespread than one having a planktonic larval stage (*L. littorea*)?. Mar. Biol. 99:507-513.
Keats, D.W., Steele, D.H. and G.R. South 1984. Depth-dependent reproductive output of the green sea urchin, *Strongylocentrotus droebachiensis* (O.F. Muller), in relation to the nature and availability of food. J. Exp. Mar. Biol. Ecol. 80:77-91.
Kent, D.B. 1994. Evaluating the use of hatchery-reared juveniles to enhance depleted marine fisheries in southern California. Marine Fish Culture and Enhancement Conf. Proceed. Seattle, Wa. USA. Washington Sea Grant Program. Oct. 1993 pp.27-28.
Levitan, D.R., Sewell, M.A., and F.-S. Chia 1992. How distribution and abundance influence fertilization success in the sea urchin *Strongylocentrotus franciscanus*. Ecology 73:248-254.
Ludwig, D. 1993. Forest management strategies that account for short-term and long-term consequences. Can. J. For. Res. 23:563-572.
Ludwig, D., Hilborn, R. and C. Walters 1993. Uncertainty, resource exploitation, and conservation: Lessons from history. Sci. 260:17-18.
McArdle, D. 1997. California Marine Protected Areas. Ca. Sea Grant Coll. Sys., Univ. Ca. La Jolla, Ca.

Orensanz, J.M. 1986. Size, environment and density: The regulation of a scallop stock and its management implications. In G. Jamieson and N. Bourne (eds.) North Pacific Workshop on stock assessment and management of Invertebrates. Can. Spec. Publ. Fish. Aquat. Sci. 92:195-227.

Pauley, D. and V. Christensen 1995. Primary production required to sustain global fisheries. Nature 374: 255-257.

Palsson, W. And R.E. Pacunski 1995. Response of rocky reef fishes to harvest refugia in Puget Sound. Proceed. Puget Sound Research 1:224-234

Pearse, J.S. and A.H. Hines 1987. Long-term population dynamics of sea urchins in a central California kelp forest: rare recruitment and rapid decline. Mar. Ecol. Prog. Ser. 39:275-283.

Pennington, J.T. 1985. The ecology of fertilization of echinoid eggs: the consequence of sperm dilution, adult aggregation, and synchronous spawning. Biol. Bull. 169:417-430.

Pulliam, H.R. 1988. Sources, sinks, and population regulation. Amer. Nat. 132:652-661.

Quinn, J.F., Wing, S.R. and L.W. Botsford 1993. Harvest refugia in marine invertebrate fisheries: Models and applications to the red sea urchin, *Strongylocentrotus franciscanus*. Amer. Zool. 33:537-550.

Ricker, W.E. 1973. Two mechanism that make it impossible to maintain peak-period yields from stocks of Pacific Salmon and other fishes. J. Fish. Res. Bd. Can. 30: 1275-1286.

Roberts, C.M. 1997. Ecological advice for the global fishereis crisis. TREE 12:35-38.

Roberts, C. et al., 1995. Review of the use of marine fishery reserves in the southeastern Atlantic, NOAA Tech. Memo. NMFS-SEFSC-376

Roberts, C.M., and N.V.C. Polunin 1991. Are marine reserves effective in management of reef fisheries? Review of Fisheries Biology 1:65-91.

Rogers-Bennett, L. 1989. The spatial association of juvenile green sea urchins. M.S. Thesis University of Massachusetts, Boston, Massachusetts, USA.

Rogers-Bennett, L., Fastenau, H.C. and C.M. Dewees. 1997. Spatial patterns in the recovery of red sea urchins beds following experimental harvest. Echinoderms: San Francisco, 9th Intern. Echinod. Conf. Eds. R.Mooi and M. Telford.

Rogers-Bennett, L., Bennett, W.A., Fastenau, H.C. and C.M. Dewees 1995. Spatial variation in reproduction and morphology of red sea urchins *Strongylocentrotus franciscanus:* implications for harvest refugia. Ecol. Apps. 5:1171-1180.

Roughgarden, J., Iwasa, J. and C. Baxter 1985. Demographic theory for an open marine population with space-limited recruitment. Ecology 66:54-67.

Rowley, R.J. 1994. Marine reserves in fisheries management. Aquat. Conserv.: Mar. Freshwater Ecosystems 4:233-254.

Rowley, R.J. 1989. Settlement and recruitment of sea urchins Strongylocentrotus spp. in a sea-urchin barren ground and a kelp bed: are populations regulated by settlement or post- settlement processes? Mar. Biol. 100:485-494.

Schmitt, R. J. And C.W. Osenberg 1996. Detecting Ecological Impacts: Concepts an applications in coastal habitats. Academic Press. San Diego, Ca.

Sloan, N.A., Lauridsen, C.P., and R.M. Harbo 1987. Recruitment characteristics of the commercially harvested red sea urchin *Strongylocentrotus franciscanus* in southern British Columbia, Canada. Fish. Res. 5:55-69.

Stickney, R.R. 1994. The use of hatchery fish in enhancement programs. Fisheries 19:6-13.

Strathmann, R.R. 1978. Length of pelagic period in echinoderms with feeding larvae from the northeast Pacific. J. Exp. Mar. Biol. Ecol. 34:23-27.

Tegner, M.J. and R.A. Butler 1985. The survival and mortality of seeded native red abalones, *Haliotis rufescens*, on the Palos Verdes Peninsula. Calif. Dept. Fish Game 71:150-163.

Tegner M.J. and Dayton P.K. 1977. Sea urchin recruitment patterns and implications of commercial fishing. Sci. 196:324-326.

Yamasaki, A. and A. Kuwahara 1990. Preserved area to effect recovery of overfished Zuwai crab stocks off Kyoto Prefecture, p. 575-585. In Proceedings of the International Symposium on King and Tanner Crabs. 28-30 November 1989, Ancorage Alaska. Alaska Sea Grant College Program, University of Alaska, Fairbanks, Alaska.

CHAPTER 57

In Search of A Sea Urchin Fishery Management Framework: The Case of California Red Sea Urchin (RSU) Fishery Development, 1985~1996

I. Charles Igawa, Ph.D.

California's rsu fishery remained below the 20 million lbs. level until 1985. But, in 1986, sudden and unprecedentedly high demands for *uni* occurred in Japan, spurring the development of a "gold rush" like effect on California's rsu fishery. The California rsu landings rose to 34 million lbs. in 1986, 46 million lbs. in 1987, 51 million lbs. in both 1988 and 1989! The increase came almost totally from the North Coast where the rsu fishery had begun in 1985 and the North Coast virgin rsu stock was mercilessly exploited in the ensuing three-year period.

Urged strongly by the rsu industry voice, the CDFG placed a moratorium in the issuance of rsu fishery license (1987) and the State Legislature passed a bill (the "Hauser bill") to bring an order to California's run-away rsu industry. The CDFG was authorized to do the following: (1) to levy a special rsu tax for funding research projects toward rsu enhancement-directed activities, (2) to appoint a 12-member Director's Sea Urchin Advisory Committee (DSUAC), (3) to apportion at 40% of the special rsu levy for gathering fishery-dependent data - to be used for developing plans for rsu fishery management; (4) to collect a special tax from the rsu fishing permit holders for the purpose of funding a state-wide diver organization to gather and disseminate the rsu fishery-related data and information - for preservation and enhancement of California's rsu resources.

Both dive community and processing industry organized their respective state-wide organization in order to facilitate the development of intra-industry consensus on various issues to be discussed within the DSUAC structure. Under the leadership of the CDFG Marine Resource Division Chief, the DSUAC functioned cooperatively with the CDFG and generated some tangible management measures such as (1) size regulations, (2) the limited entry regimen - moratorium on permit issuance, (3) the seasonal and the area closure, (4) implementation of the "log book" requirement.

During the 1989~1995 period, the DSUAC approved funding of over $620,000 for enhancement-related research projects such as: (1) larval distribution and settlement, (2) test out plants and juvenile urchins, (3) recruitment, growth, and mortality, and (4) feedlot experiments, and (5) culture techniques. The Hauser bill was approved for extension in 1995 and the special levy for funding the stipulated activities was thus continued. DSUAC's most noteworthy development occurred in 1994 when the CDFG attempt to institute a new rsu regulatory regimen was challenged on the basis that it lacked scientifically sound basis. Accepting the industry concerns, the DSUAC recommended a "peer review" by the panel of sea urchin research specialists to determine the appropriate course of actions to pursue. The findings of the peer review process included the following: (1) CDFG's recommended management scheme was not based on adequate or sufficient data-backed knowledge, (2) the status of the rsu fishery in California warranted a more comprehensive management approach urgently.

Given that conclusion, the newly constituted DSUAC is trying to develop a more comprehensive management schema by stream-lining the deliberation process and enlisting the help of a "science advisor." While the DSUAC structure has shown to work effectively, much more collaborative effort must prevail among all participants in order for it to remain an effective body.

CHAPTER 58

An analysis of the importance of Allee effects in management of red sea urchins: Differences between Oregon/Washington and southern California

Thomas A. Ebert, Stephen C. Schroeter, and John D. Dixon

Abstract

Positive density-dependent responses, Allee effects, have been proposed for the red sea urchin, Strongylocentrotus franciscanus, as density-dependent fertilization rate and density-dependent survival of juveniles because of protection of adult spine canopies. A model of population growth was assembled using size categories with growth based on tagged animals and survival estimated from individual growth and size-frequency distributions: annual survival was estimated to be 0.95 yr^{-1} in Oregon/Washington and 0.79 yr^{-1} in southern California. Large changes in egg production and first year survival, which includes the Allee effects, would have a very small effect on population growth in the north. A 99% reduction in fertilization rate would change the population growth rate by only 3%. In southern California, a 99% reduction in fertilization rate would change population growth rate by nearly 10%. Most of the changes in population growth rate in both Oregon/Washington and southern California are determined by changes in stages that are older than 1 year.

Allee effects probably should not be given prominence in the development of management models for northern populations. In southern California, Allee effects appear to be more important and possibly should be considered by management. In northern populations. protection of large individuals would be most important for the long-term maintenance of populations whereas very large animals are less important in southern California.

CHAPTER 59

Influence of Local Productivity and Recruitment Patterns On Design of Reserves, the California Red Sea Urchin Population as an Example

Lance E. Morgan[1]
Louis W. Botsford
Stephen R. Wing

Abstract

Following rapid growth of the red sea urchin fishery in northern California there is now concern over its long-term viability. Catch per unit effort has declined, and fishery independent surveys have shown a substantial decrease in densities to near the minimum believed necessary for successful reproduction by a broadcast spawner. Management of coastal invertebrate fisheries is hampered by a lack of knowledge of spatial variation in productivity and demographic rates for managed populations. Attempts to design management policies based on temporary or permanent closed areas depend critically on knowing growth rates, mortality rates, the rate of entry of new individuals, and how these vary along the coast. Using a maximum likelihood procedure incorporating size distributions and growth increments based on the von Foerster equation, we have examined along-shore spatial variation in growth, natural mortality, harvest mortality and recruitment. Growth and mortality rates from fishery reserve sites show variation, but with no consistent along-shore pattern. Estimates of the von Bertalanffy K parameter range from 0.17- 0.35, while estimates of L infinity, the asymptotic test size, range from 108-127 mm. Instantaneous natural mortality rates are approximately 0.1. Instantaneous fishing mortality rates range from 0.12-1.6. The number of individuals in lower size classes indicates irregular, episodic recruitment in northern California during the last several years in a identifiable spatial pattern. Future work will incorporate these spatial patterns of productivity and recruitment in the design of management plans for sustaining the red sea urchin fishery in northern California.

Introduction

The red sea urchin (*Strongylocentrotus franciscanus*) has drawn considerable attention in the ecological and fisheries literature due to its proposed role as a keystone species (Harold and Pearse 1987, Estes and Duggins 1995) and the great monetary value of the fishery ($34 million ex-vessel value for California in 1991,

---

[1] Department of Wildlife, Fish, and Conservation Biology, University of California, Davis, CA 95616

Kalvass and Hendrix 1997). The California fishery developed rapidly in the late 1980s to become the state's most valuable fishery by 1990. Since 1988, catch and catch per unit effort (CPUE) have steadily declined in northern California and there is now concern that the population has been harvested to near the point of recruitment failure (Figure 1). Additionally, unknown but significant impacts to the highly productive nearshore kelp forest community could occur should the urchin population be significantly reduced (Harrold and Pearse 1987). The rapid decline of the fishery in northern California has provided increasing impetus to explore alternate methods of management (spatial management and/ or individual transferable quotas, ITQs).

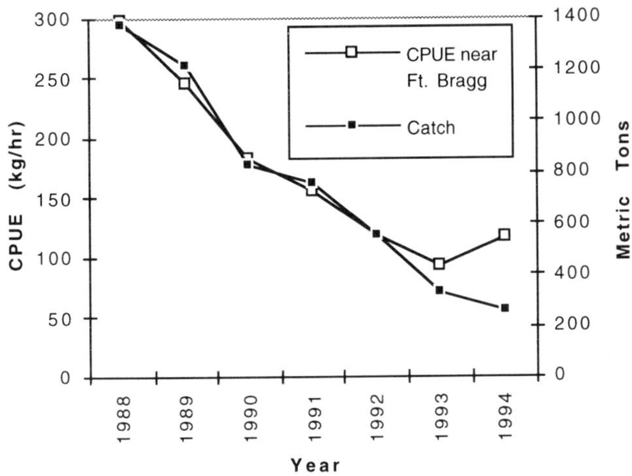

Figure 1. Decline in catch and CPUE in northern California red sea urchin fishery, 1988-1994.

Several authors have suggested using spatial management to prevent recruitment overfishing (Tegner 1989, Carr and Reed 1993, Dugan and Davis 1993). Others have found that spatial management can maintain sustainable catch for the red sea urchin when Allee mechanisms are a factor (Pfister and Bradbury 1996, Botsford et al. 1993, Quinn et al. 1993). Two Allee mechanisms have been proposed for the red sea urchin. The first concerns the protection of juveniles from predation under the spine canopy of adults (Tegner and Dayton 1977), while the second predicts recruitment failure as densities of adults fall below the threshold level necessary for successful fertilization of gametes (Penington 1985, Denny and Shibata 1989, Levitan 1991, Levitan et al. 1992). Both mechanisms may be significant factors at present. Fishery dependent data have shown a potential "fishing up" effect where by the biomass of large individuals (115-120mm) has decreased by approximately 90% in northern California from 1988-1994 (Kalvass and Hendrix 1997). This loss of adults results in the loss of protected nursery habitat for juvenile urchins via the spine canopy mechanism. Similarly, fishery independent surveys (Kalvass and Hendrix 1997) from the Mendocino County coast have measured densities below 0.7 individuals per square meter since 1990, a point at which fertilization declines to near 0 (Levitan et al. 1992). Spatial management offers amelioration from low density by affording either permanent areas of high density (harvest refugia) or shifting areas of increased density (rotating harvest) (see Botsford and Quinn this volume for a discussion of rotating spatial harvest and harvest refugia).

The basic theory underlying spatial management is to protect a flow of surplus individuals from areas with net production (sources) to areas without local production (sinks) (Pulliam, 1988). With regard to sea urchins, sedentary adult populations are linked together by planktonic larvae dispersing through coastal currents. Spatial management represents the special case of protecting areas with sufficient reproductive output to supply harvested areas of the coast with recruits through dispersal. The spatial variability in the fecundity of adults and the production of larvae needs to be considered side by side with the circulation of coastal waters which carry the larvae from one place to another. In order to successfully design a spatial management plan two factors need to be understood:

1) growth and mortality (productivity)

2) pattern and scale of larval dispersal (recruitment)

Here we describe ongoing work to estimate the pattern of spatial variation in red sea urchin growth and mortality from northern California (Morgan et al. 1997), and previous work on settlement of meroplankton in northern California (Wing et al. 1995a, 1995b) to interpret patterns of recruitment.

Productivity

Accounting for spatial variability in productivity directly in harvest strategies has been shown to increase fisheries yield (MacCall 1990). We have collected growth increment data from 3 sites (Salt Pt., Bodega Marine Reserve (BMR), and the Caspar Closure area), and size distribution data from 11 sites to estimate the rates of growth and mortality along the Sonoma and Mendocino County coasts of northern California. Growth increments were collected from individually marked urchins using passive interrogatable transponders (PIT) tags. Size distributions were collected along the coast from Bodega Head to Caspar Closure, spaced according to Department of Fish and Game management sectors (approximately 10° latitude, Figure 2).

A maximum likelihood procedure was used to estimate parameters from the von Bertalanffy growth equation and mortality rates (natural and fishing) (Smith et al. 1997). By incorporating growth increment and size distribution data in the procedure we avoided the statistical confounding of growth and mortality parameters. Following Botsford et al. (1994) variation in asymptotic test size (L_∞ from the von Bertalanffy eq.) was used to account for population variability in adult size. Model estimates for asymptotic test size (L_∞), standard deviation of L_∞, growth rate parameter K and natural mortality were calculated. The model was evaluated for sensitivty of results to sample size, recruitment variability, measurement error and sampling selectivity. Estimates of von Bertalanffy growth equation parameters and mortality were found to be robust with regard to these factors (Smith et al. 1997). Sample sizes of 300 individuals in size distributions and 30 growth increments gave statistically useful estimates of these parameters.

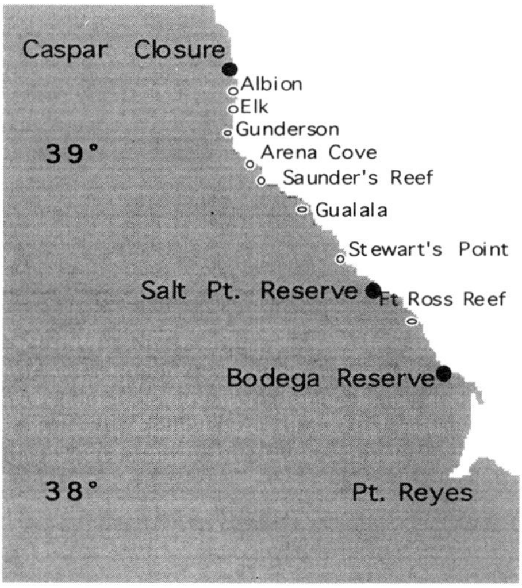

Figure 2. Coastline of Mendocino and Sonoma counties, California, showing sampling sites for size distribution (open circles) and growth increment (solid circles) data (from Morgan et al. 1997).

We have used the size distributions from reserves, along with growth increments to calculate a range of estimates for these population parameters (details can be found in Morgan et al. 1997). Estimates of L_∞ from Salt Pt. are not representative of the population due to harvesting at this site and were not used. Size distribution data from Pt. Cabrillo Marine Reserve were used to estimate L_∞ because they were collected near the beginning of the fishery in northern California (1989) and were less likely to be biased by recent changes in population dynamics related to the fishery. These estimates do not indicate a consistent pattern of variation in population parameters along-shore, but do suggest some variation in growth rates between sites. Growth rates at Salt Pt. were slightly higher than those from BMR. This range of growth rates indicates it takes a minimum of 6 and a maximum of 10 years for urchins to grow to the minimum legal size in northern California. A 1992 settlement pulse (see Wing et al. 1995a) was observed as a separate mode in each of the years sampled at Salt Pt. and BMR. The movement of this mode supplies independent validation of the population parameters we estimated. In 1994, 2 years post-settlement, the mode had a median of 32.5 mm at Salt Pt. and 30 mm at BMR, in 1995 it was 42-45 mm, and in 1996 approximately 55-60 mm. This corresponds well to the range of L_∞ and K values calculated from northern California.

Our estimated growth rates are higher than other estimates from northern California. Ebert et al. (1993) estimated that it may take up to 15 years for red sea urchins to reach the legal size of 89 mm. The difference is primarily in the slow growth of very small individuals. Their estimates also suggest large variation in growth rates between sites. At two sites less than 1.6 km apart estimates ranged from 3.2 to 7.2 years for urchins to grow from a size of 30 to 89 mm, although different growth functions were used in these calculations. The variation in growth rates between sub-populations may be influenced by the interaction of density, food quality and availability, temperature, physical disturbance and wave exposure, and competition (Bennet and Giese 1955, Swan 1961, Ebert 1967, 1968, 1977, Baker 1973, Edwards and Ebert 1991). Kato and Schroeter (1985) reported annual growth increments of 13 to 25 mm, with an average urchin being 4 to 5 years when harvested at the minimum legal size in southern California. Ebert and Russell (1992) estimated growth and mortality rates for *S. franciscanus* at 2 sites in southern California. Test diameters of tetracycline marked urchins at 10 years of age were estimated to vary from 67 to 79 mm.

Instantaneous natural mortality rates estimated by Morgan et al. (1997) and Ebert and Russel (1992) are similar for northern California sites. In northern California both groups calculated rates ranging from 0.05 - 0.1 for urchins at Pt. Cabrillo Marine Reserve (Ebert and Russel 1993, Morgan et al. 1997). Instantaneous mortality rate estimates ranged from 0.075 - 0.14 for sites in southern California in one study (Ebert and Russel 1992) and 0.19 to 0.2 in another (Ebert and Russel 1993).

Fishing mortality was calculated from size distributions collected at harvested sites (Morgan et al. 1997). This required a two step process of independently estimating the population parameters from the unharvested reserves, fixing these parameters, and measuring size specific mortality for urchins greater than 89 mm (minimum legal size in northern California). This was done for size distribution data collected in 1995 and 1996. Estimates of fishing mortality varied along the coast, increasing in a northward direction above Pt. Arena and above Pt. Reyes. The estimates from the area between Stewart's Pt. and Pt. Arena indicate substantial harvesting pressure relative to other sites sampled, as great as 80% annual mortality (F= 1.6). This pattern was consistent between years. The break at Pt. Arena is significant and may reflect differences in the coastal circulation and supply of larvae to the coast.

Recruitment

One of the obstacles to understanding the dispersal dynamics of sea urchins is a lack of knowledge of the physical mechanisms by which larvae are transported in the plankton and returned to the adult population. In northern California, red sea urchin larvae are released into the plankton during the period of the spring transition to upwelling season and are found to settle into the adult population one to several months later during the upwelling season (Wing et al. 1995a). Thus these larvae are subject to strong offshore, southwesterly transport in upwelling jets typically found off headlands in the California current system. Ebert and Russel (1988) proposed spatially varying purple sea urchin (*S. purpuratus*) recruitment associated with headlands along the Pacific coast, although the underlying mechanism was not clear.

Wing et al. (1995a, 1995b) described a physical mechanism for the transport of plankton in the intermittent upwelling and relaxation conditions of northern California. They combined a time series of settlement data along the coast north of Pt. Reyes, California with temperature data that suggested settlement of crab species and sea urchins occurred simultaneously with the arrival of warmer water in a northward moving pattern. They proposed that during upwelling larvae accumulate in the warm water that collects between the Farallon islands and Pt. Reyes, and more generally in the upwelling shadows of capes. When the winds which drive upwelling "relax", this larvae rich water moves northward in coastally trapped currents. Larvae competent to settle are thus delivered back to the coast. Further work by Wing et al. (1997a) sampling plankton in the Gulf of the Farallones has verified that larvae are abundant in these waters. This mechanism leads to intermittent settlement of meroplanktonic larvae and therefore irregular recruitment to areas north of headlands. As a consequence, within each upwelling cell sites furthest north of the point receive the lowest levels of recruitment.

Ebert et al. (1994) monitored settlement of juvenile red sea urchins and also found high inter-annual and seasonal variation in settlement. Infrequent recruitment appears to be a common pattern for red sea urchin populations from northern California to Britsih Columbia (Ebert 1983, Sloan et al. 1987, Pearse and Hines 1987, Ebert et al. 1994). There is also indirect support for irregular recruitment from observations of the shift in size of a recent lower mode in the size distributions from 1994 - 1996 at Salt Pt. Analysis of size distributions in this region has shown that areas to the north of prominent headlands (Pt. Arena and Pt. Reyes) have more young urchins than areas to the immediate south (Wing et al. 1997b).

Discussion

A major tenet of marine ecology is that the distribution and abundance of species is explained by the interaction of biotic and physical mechanisms. For meroplanktonic species (those with planktonic larvae and sedentary adults) knowledge of factors influencing adult dynamics and larval dispersal must be integrated to understand abundance (Tegner and Dayton 1981, Connell 1985, Roughgarden et al. 1988, Rowley 1989, Bertness et al. 1992, Botsford 1995, Osman and Whitlach 1996). Thus the applied view of spatial management represents a special case of integration over life history phases. Our analysis of factors influencing population dynamics of the red sea urchin in northern California suggests that the pattern of recruitment, along-shore spatial variation combined with irregular recent recruitment, explains spatial variation in population dynamics better than local variation in productivity. Whether this is due to impacts related to potential overfishing of adults or environmentally forced patterns of recruitment is unexplained.

Conservation of invertebrate fisheries is gaining more interest with increasing pressure on smaller populations and fewer species available for exploitation (Jamieson 1993). Tegner et al. (1996) discuss the probable extinction of white abalone in California, an event which was until recently thought to be impossible for invertebrates. The generally high fecundity of invertebrates was believed to be a safeguard which allowed for very high harvest rates. However irregular episodic recruitment coupled with high market value and intensive harvest pressure has doomed this species to extinction without immediate aquaculture to support stocks.

Our work suggests that intensive harvest with impacts to recruitment may be at play and spatial management could circumvent this population collapse.

The pattern of fishing mortality along-shore also lends support to the idea that variation in the pattern of recruitment is primarily reponsible for differences in local population dynamics. In addition to reflecting additional mortality to legal size urchins, fishing rate can also represent a lack of recent recruitment to the fishable size classes, as well as spatially varying natural mortality. Tegner and Dayton (1981) suggested that size distributions of red urchins in a southern Caifornia kelp forest were influenced predominately by predators. Rowley (1989) also suggested early mortality from predation affected populations in southern California. Size distribution data alone cannot distinguish between these alternative hypotheses, however the similarity of the spatial pattern of fishing mortality rate with the proposed pattern of larval dispersal supports the influence of recruitment on the former pattern. At sites north of Pt. Arena, the 1995 and 1996 fishing mortality estimates increase to the north. This pattern of fishing mortality does not fit the observed fishing patterns recorded in log books, or distance to fishing ports reported by Kalvass and Hendrix (1997).

Other authors have assessed spatial patterns of red sea urchin productivity in northern California with discussion of harvest refugia (Rogers-Bennet et al 1996). This work focused on depth as an independent variable influencing urchin dynamics. Greater gonad weight, size of urchins and increased numbers of small individuals in shallow areas (5m), led these authors to suggest protecting shallow areas as harvest refugia. While these are all beneficial aspects of shallow areas, there are no data to suggest that shallow urchins are the source of the larvae that settle there. In addition shallow areas are the preferred areas to fish, as well as the areas most subject to a fertilization Allee effect due to turbulence. Our approach to spatial management is to set aside along-shore sections of the coast so that no assumptions regarding depth are required, taking advantage of along-shore variation in recruitment rather than spatial patterns of gamete production.

The work discussed here provides a first attempt at providing information needed to implement a red sea urchin spatial management plan. It provides the basis for design of spatially explicit management that takes advantage of variation in population dynamics in northern California.

Acknowledgments

The authors wish to thank B. Smith, S. Bennett and P. Kalvass for assistance in preparing this manuscipt. This paper is funded by a grant from the National Sea Grant College Program, National Oceanic and Atmospheric Administration, U. S. Department of Commerce, under grant number NA66RG0477, project number R/F-169 through the California Sea Grant College System, and in part by the California Sate Resources Agency. The views expressed herein are those of the authors and do not necessarily reflect the views of NOAA or any of its sub-agencies. The U.S. Government is authorized to reproduce and distribute for governmental purposes.

Literature Cited

Baker, S. L. 1973. Growth of the red sea urchin, *Strongylocentrotus franciscanus* (Agassiz) in two natural habitats. M.S. Thesis, San Diego State. Univ., 83 p.

Bennet, J. and A. C. Geise. 1955. The annual reproductive and nutritional cycles in two western sea urchins. Biological Bulletin 109: 226-237.

Botsford, L.W. 1995. Population dynamics of spatially distributed, meroplanktonic, exploited marine invertebrates. ICES Marine Science Symposium 199: 118-128.

Botsford, L.W., J.F. Quinn, S.R. Wing and J.G. Brittnacher. 1993. Rotating spatial harvest of a benthic invertebrate, the Red Sea Urchin *Strongylocentrotus franciscanus*. In: Kruse, G., D.M. Eggers, R.J. Marasco, C. Pautzke and T.J. Quinn II (Editors). 1993. Proceedings International Symposium Management Strategies Exploited Fish Populations. Alaska Sea Grant College Program Rep. No. 93-02, University of Alaska, Fairbanks, AK.

Botsford, L.W., Smith, B.D. and J.F. Quinn. 1994. Bimodality in population size distributions: the red sea urchin, *Strongylocentrotus franciscanus* as an example. Ecological Applications 4: 42-50.

Carr, M.H. and D.C. Reed. 1993. Conceptual issues relevant to marine harvest refuges: examples from temperate reef fishes. Canadian Journal of Fisheries and Aquatic Sciences 50: 2019-2028.

Connell, J.H. 1985. The consequences of varitaion in inital settlement vs. post-settlement mortality in rocky intertidal communities. Journal Experimental Marine Biology and Ecology 93: 11-45.

Denny, M.W. and M.F. Shibata. 1989. Consequences of surf-zone turbulence for settlement and external fertilization. American Naturalist 117: 838-840.

Dugan, J.E. and G.E. Davis. 1993. Applications of marine refugia to coastal fisheries management. Canadian Journal of Fisheries and Aquatic Sciences 50: 2029-2042.

Ebert, T. A. 1967. Negative growth and longevity in the purple sea urchin *Strongylocentrotus purpuratus* (Stimpson) Science 157: 557-558.

Ebert, T. A. 1968. Growth rates of the sea urchin *Strongylocentrotus purpuratus* related to food availability and spine abrasion. Ecology 49: 1075-1091.

Ebert, T.A. 1977. An experimental analysis of sea urchin dynamics and community interactions on a rocky jetty. J. Exper. Marine Biology and Ecology 27: 1-22.

Ebert, T.A. 1983 Recruitment in Echinoderms. Echinoderm Studies 1: 169-203.

Ebert, T.A. and M.P. Russell. 1988. Latitudinal variation in size structure of the west coast purple urchin: a correlation with headlands. Limnology and Oceanography 33: 286-294.

Ebert, T.A. and M.P. Russel. 1992. Growth and mortality estimates for red sea urchin *Strongylocentrotus franciscanus* from San Nicolas Island, California. Marine Ecology - Progress Series 81: 31-41.

Ebert, T.A. and M.P. Russel. 1993. Growth and mortality of subtidal red sea urchins (*Strongylocentrotus-franciscanus*) at San Nicolas Island, California, USA- problems with models. Marine Biology 117: 79-89.

Ebert T.A., S.C. Schroeter, J.D. Dixon and P. Kalvass. 1994. Settlement patterns of red and purple sea urchins (*Strongylocentrotus franciscanus* and *S.-purpuratus*) in California, USA. Marine Ecology-Progress Series 111: 41-52.

Edwards, P.B. and T.A Ebert. 1991. Plastic responses to limited food availability and spine damage in the sea urchin *Strongylocentrotus purpuratus* (Stimpson). Journal Experimental Marine Biology and Ecology 145: 205-220.

Estes, J.A. and D.O. Duggins. 1995. Sea otters and kelp forests in Alaska: generality and variation in a community ecological paradigm. Ecological Monographs 65: 75-100.

Fogarty, M.J., M.P. Sissenwine and E.B. Cohen. 1991. Recruitment variability and the dynamics of exploited marine populations.Trends in Evolution & Ecology 8: 732-33.

Gaines, S, and J. Roughgarden. 1985. Larval settlement rates: a leading determinant of structure in an ecological community of the marine intertidal zone. Proceedings National Academy of Sciences, USA 82: 3707-3711.

Harrold, C. and J.S. Pearse. 1987. The ecological role of echinoderms in kelp forests. Echinoderm Studies 2: 137-233.

Kalvass, P. and J. Hendrix. 1997. The California res sea urchin (*Strongylocentrotus franciscanus*) fishery: catch, effort and management trends. (submitted)

Kato, S. and S.C. Schroeter. 1985. Biology of the red sea urchin, *Strongylocentrotus franciscanus*, and its fishery in California. Marine Fisheries Review 47: 1-20.

Levitan, D.R. 1991. Influence of body size and population density on fertilization success and reproductive output in a free-spawning invertebrate. Biological Bulletin 181: 261-268.

Levitan, D.R., M.A. Sewell and F.S. Chia. 1992. How distribution and abundance influence fertilization success in the sea urchin *Strongylocentrotus franciscanus*. Ecology 73: 248-254.

Osman, R.W and R.B. Whitlach, 1996. Processes affecting newly settled juveniles and the consequences to subsequent community development. Invertebrate Reproduction and Development 30: 217-225.

MacCall, A. 1990. Dynamic geography of marine fish populations. Washington Sea Grant Program, University of Washington Press, Seattle, WA.

Morgan, L.E., S.R. Wing and L.W. Botsford. 1997. Along-shore variation in the population dynamics of the red sea urchin, *Stongylocentrotus franciscanus* , in northern California. (in prep.).

Pearse, J.S. and A.H. Hines. 1987. Long-term population dynamics of sea urchins in a central California kelp forest: rare recruitment and rapid decline. Marine Ecology - Progress Series 39: 275-283.

Penington, J.T. 1985. The ecology of fertilization of echinoid eggs: the consequence of sperm dilution, adult aggregation and synchronous spawning. Biological Bulletin 169: 417-430.

Pfister C.A. and A. Bradbury. 1996. Harvesting red sea urchins - Recent effects and future predictions. Ecological Applications 6: 298-310.

Pulliam, H.R. 1988. Sources, sinks, and population regulation. American Naturalist 132: 652-661.

Quinn, J. F., S.R. Wing and L. W. Botsford. 1993. Harvest refugia in marine invertebrate fisheries: models and applications to the Red Sea Urchin *Strongylocentrotus franciscanus*. American Zoologist 33: 537-550.

Rogers-Bennett, L., W.A. Bennett, H.C. Fastenau and C.M. Dewees. 1995. Spatial variation in red sea urchin reproduction and morphology: implications for harvest refugia. Ecological Applications 5: 1171-1180.

Rowley, R.J. 1989. Settlement and recruitment of sea urchins (*Strongylocentrotus spp.*) in a sea urchin barren ground and a kelp bed: are populations regulated by settlement or post-settlement processes? Marine Biology 100: 485-494.

Sloan, N.A., C.P. Lauridsen and R.M. Harbo. 1987. Recruitment characteristics of the commercially harvested red sea urchin, *Strongylocentrotus franciscanus* in southern British Columbia, Canada. Fisheries Research 5: 55-69.

Smith, B. D., Botsford, L.W. and S.R. Wing. 1997. Estimation of growth and mortality paramters from size frequency distributions lacking age patterns: the red sea urchin (*Strongylocentrotus franciscanus*) as an example. (submitted).

Swan, E. F. 1961. Some observations on the growth rate of sea urchins in the genus *Strongylocentrotus*. Biological Bulletin 120: 420-427.

Tegner, M.J. 1989. The feasibility of enhancing red sea urchin, *Strongylocentrotus franciscanus*, stocks in California: An analysis of the options. Marine Fisheries Review 51: 1-22.

Tegner, M.J. and P.K. Dayton. 1977. Sea urchin recruitment patterns and implications for commercial fishing. Science 196: 324-326.

Tegner, M.J. and P.K. Dayton. 1981. Population structure, recruitment and mortality of two sea urchins (*Strongylocentrotus franciscanus* and *S. purpuratus*) in a kelp forest. Marine Ecology - Progress Series 5: 255-268.

Tegner, M.J., L.V. Basch and P.K. Dayton. 1996. Near extinction of an exploited marine invertebrate. Trends in Ecology & Evolution 11: 278-280.

Wing, S. R., J.L. Largier, L.W. Botsford and J.F. Quinn. 1995a. Settlement and transport of benthic invertebrates in an intermittent upwelling region. Limnology and Oceanography 40: 316-329.

Wing, S. R., L.W. Botsford, J.L. Largier, and L.E. Morgan. 1995b. Spatial variability in settlement of benthic invertebrates in a northern California upwelling system. Marine Ecology - Progress Series 128: 199-211.

Wing, S. R., L.W. Botsford, S.V. Ralston and J.L. Largier. 1997a. Meroplanktonic distribution and circulation assoicated with a coastal retention zone in the northern California upwelling system. (in prep.)

Wing, S. R., L.W. Botsford, L.E. Morgan, J.M. Diehl and C.J. Lundquist. 1997b. Spatial patterns of red sea urchin recruitment in northern California. (in prep.)

CHAPTER 60

A proposed method for red sea urchin (Strongylocentrotus franciscanus) resource assessment in California.

S.C. Schroeter [1], D.C. Barilotti [2], J.D. Dixon [3],
T.A. Ebert [4], and P.E. Kalvass [5]

Abstract

The red sea urchin is the basis for the largest nearshore fishery in California. In 1995 there were landings of 22 million pounds with an ex-vessel value in excess of $20 million. Unfortunately, the patterns of distribution and abundance of red sea urchins make an estimate of stock size prohibitively expensive. Nevertheless, concern for maintaining a sustainable fishery has made it necessary to formulate management regulations (such as size limits).

In the absence of a stock assessment, we suggest a three step method for estimating sustainability. The first is to increase the spatial resolution at which fishery-dependent data are collected. This will improve catch and effort statistics, reduce variability due to spatial factors, and allow a better understanding of how the fishery-dependent data is influenced by factors such as changes in kelp standing crops that in turn effect uni (sea urchin gonad) yields.

The second step is to use the improved fishery-dependent data along with information on kelp abundance to pick sites at which fishery independent data will be collected. The latter will consist of size measurements of all red sea urchins collected from samples of known area and gonadal yield of harvestable urchins. These indices can be used to track population dynamics and temporal changes in gonad yield and to estimate sustainability of the harvestable population.

Finally, data on jaw and body size will be combined with a geographically extensive database on individual growth which we have developed over the past 7 years. Combined with size-specific growth rate and the ratio of jaw size to body size, the size frequency data can provide population specific information on growth, size structure and mortality, This will provide the basis for estimates of harvest sustainability several years into the future. Applications of this method will be discussed.

e-mail addresses; [1]schroete@inetworld.net, [2]seafoam@pacbell.net, [3]jdixon@inetworld.net, [4]tebert@sunstroke.sdsu.edu and [5]pkalvass@mcn.org

CHAPTER 61

The Role of Divers' Associations in the Development and Stewardship of the California Sea Urchin Fishery

Peter Halmay[1], Kevin Crisp[2]

Abstract

The California sea urchin fishery began in 1972. For the first decade of its existence it was largely unregulated. Very few divers stayed in the fishery more than a few years.

As the sea urchin fishery developed into one of California's foremost commercial fisheries, the need for sea urchin divers' associations became apparent. They were needed to educate divers in order to form consensus on industry related issues. They were also needed to provide feedback to regulatory agencies for policy decisions.

In this paper we will discuss the formation, goals and short-term efficiency of local divers' associations. The development of statewide associations CUPA (California Urchin Producers' Association) in 1991 and SUHAC (Sea Urchin Harvesters' Association California) in 1995 with a legislatively mandated funding system is presented.

The advantages of marketing associations and cooperatives are discussed. The SUHAC Business Plan for 1997-1998 and a comprehensive list of the functions of a statewide sea urchin divers' association are also presented.

---

[1] Secretary SUHAC, PO Box 1607 Bodega Bay, CA 94923

[2] Vice President SUHAC, PO Box 1607 Bodega Bay, CA 94923

INTRODUCTION

The fishing community is composed of many parts: resource managers, scientific investigators, processors, consumers and fishermen. In order to have valid results from the interaction of the various components, valid processes must be established.

We will not attempt to find answers to specific problems and issues. In this paper we will attempt to make a strong case for the involvement of sea urchin divers' associations in the decision making process. The argument will be directed inward at sea urchin divers and outward to the other members of the community.

A good management scheme must take into account the health of sea urchins and the divers. The purpose of a sea urchin divers' association is to increase sea urchin fisheries' productivity and to improve the economic and social status of sea urchin divers. (Zengyoren 1987)

Productivity can be doubled while killing less sea urchins. This can be accomplished by selective harvesting, improved freshness, coordinating harvest with marketing and husbandry.

REGIONAL ASSOCIATIONS

The existing sea urchin fishery management was developed over the last 20 years. It is the product of 6 pieces of legislation and nine separate visits by the Fish and Game Commission. More than 60 regulations and modifications have been enacted. (J. Duffy 1997) Regional dive associations were by and large the force behind these regulations.

Several sea urchin regional associations were established during the period from 1979 to 1993. The reasons for organizing were very specific and once the goals were attained, the level of activity diminished radically. This is shown in Table I "Effectiveness of Local Associations" on the following page.

It can be seen from Table I that the regional diver associations were very effective in achieving their goals. The enforcement of the regulations was made more effective because it had the backing of the divers who helped to draft the regulations. Resource managers learned to respect the political power of the industry but remain intensely skeptical of the validity of their cognitive reality. (Finlayson 1994)

TABLE I
EFFECTIVENESS OF LOCAL ASSOCIATIONS

ASSOCIATION	GOAL	RESULT
San Diego Fishermen's Association (S.F.D.A.) Est. 1979 San Diego	Stop the use of quicklime to kill sea urchins	1980 Fish and Game Commission stopped the use of quicklime
California Urchin Divers Association Est.. 1985 Santa Barbara	• Reseeding programs • Limited entry • Management of sea otters • Minimum size regulation	• AB 972 (1987) Hauser - funding enhancement • 1988 Fish & Game Commission established limited entry • PL 99-625A (1986) Established zone for sea otters south of Pt. Conception • 1989 Fish & Game Commission established minimum size limit
Independent Urchin Divers Association Est. 1986 Ft. Bragg	• Limited entry • Reseeding programs	• 1988 Fish & Game Commission (see above) • AB 972 (1987) Hauser - funding enhancement
United Divers of San Pedro Est. 1986 San Pedro	• Limited entry • Apprentice program	• 1988 Fish & Game Commission (see above) • 1988 Fish & Game Commission establishes an apprentice program
Ventura County Commercial Fishermen's Association Est. 1989 Ventura	• Ventura to have representation on DSUAC • Establish a program to give preference to line tenders for new permits.	• AB 718 Hauser (1995) establishes Ventura representation on DSUAC • 1993 Fish & Game Commission establishes permits for line tenders and preferencefor lottery

Sea urchin divers make daily stock assessments and evaluations. It is called "reading the bottom". This knowledge must be considered by other investigators. How do we do this?

STATEWIDE ASSOCIATIONS

Once the regional associations were established it was observed that there was a large divergence of opinions between associations on issues related to the sea urchin fishery. Rather than argue, these differences in public it was preferable to establish a mechanism for trying to achieve consensus on issues.

Other problems such as lack of continuity once immediate objectives were attained and the funding required to hire outside professionals (scientific advisors, legal accounting, public relations help) had to be established.

An all-state meeting was held on January 20, 1990, in Ventura. Present at this meeting were representatives of divers' associations from San Diego, Los Angeles, Ventura, Santa Barbara, Ft. Bragg, and Bodega Bay.

Consensus was achieved in recommendations to the Department of Fish and Game regarding minimum size limits and tolerances to facilitate enforcement.

In addition a process was established to meet with the processing sector of the industry before the Director's Sea Urchin Advisory Committee (DSUAC) meetings held three times a year. In this manner the DSUAC meetings were a little less unruly and the length of the meetings was shortened.

CALIFORNIA URCHIN PRODUCERS ASSOCIATION (CUPA)

In 1991 the management company M.J. Johnson presented the sea urchin diving community a proposal to establish a statewide association. With their contacts in the governor's office, Department of Fish and Game and PCCFA (a federation of fishing associations) they would establish a law mandating a tax on all sea urchin divers to fund the activities of the association. Although never stated, it was understood that the newly established sea urchin association would hire the firm of M.J. Johnson to manage its affairs.

AB 642 was authored by Assemblyman Dan Hauser and signed into law in 1991.

According to the law, the divers' association would perform the following functions:

(1) To establish a communications network among sea urchin divers, through a periodic newsletter and such other means as are deemed necessary and appropriate by the organization, providing divers with information on policies, procedures, statutes, and regulations affecting the sea urchin fishery, meeting announcements, and for other information the department reasonably requests to be transmitted to sea urchin divers.

(2) To establish an education program for sea urchin divers and the public on the conservation and utilization of sea urchins.

(3) To work with the director's Sea Urchin Advisory Committee in developing an industry consensus on measurements.

(4) To establish a sea urchin diver safety program.

(5) To identify areas for recommendation to the director's Sea Urchin Advisory Committee for sea urchin enhancement.

(6) To develop a program to improve compliance with regulations and assistance to the department in the prevention of poaching.

When the tax was reauthorized by AB14 (Hauser) in 1993, the tax was extended for two years, and the association's functions were reduced to the following:

(1) To establish a communications network among sea urchin divers, through a periodic newsletter and such other means as are deemed necessary and appropriate by the organization, providing divers with information on policies, procedures, statutes, and regulations affecting the sea urchin fishery, meeting announcements, and for other information the department reasonably requests to be transmitted to sea urchin divers.

(2) To establish an education program on the conservation and utilization of sea urchins.

(3) To convene statewide conferences for members of the industry to meet for purposes of strengthening the industry and benefiting industry goals.

This plan went awry because of irregularities by M.J. Johnson Company. The California Department of Fish and Game froze the funding in 1993 and conducted an audit (CDFG 1994). The results of this audit indicated that CUPA invoiced the Department $289,560, of which $139,507 was ineligible for reimbursement.

SEA URCHIN HARVESTERS' ASSOCIATION, CALIFORNIA (SUHAC)

While the funds distributed to the divers association was frozen, the mandated tax was extended in 1993 by AB 14 (Hauser) and continued to be deducted from divers' paychecks.

A grass roots effort by divers began in 1994 to stop this mandatory tax and use the funds that had accumulated in the CDFG coffers for the benefit of sea urchin divers.

Sea Urchin Harvesters' Association, California (SUHAC) was established in 1995 consisting of more than 200 divers. It was established as a statewide association with directors domiciled in the counties of Mendocino, Sonoma, Santa Barbara, Ventura, Los Angeles, and San Diego. Officers and directors were selected by the membership at the annual meeting in June 1995.

According to the divers mandate, the tax was stopped by AB 612 (Hauser) in 1995.

The directors of SUHAC worked with the Department of Fish and Game representatives for two years to have the tax money returned to the divers.

In February 1997, a contract between CDFG and SUHAC was signed and approximately $160,000 was returned for SUHAC to perform the functions outlined in the Fish and Game Code.

The SUHAC Board of Directors have prepared a plan for 1997 and 1998 that proposes to do the following:

SUHAC Business Plan 1997 - 1998

Program Element [A] Educational Products:

A.1. Produce and Distribute Educational Video on California's Sea Urchin Industry from existing and new video footage

Background and Purpose: California's sea urchin industry is California's most valuable fishery, yet the public, the media, and policy makers are largely unaware of this fact. The purpose of this video is to illustrate the operation of California's sea urchin industry and present factual information on the importance of this industry to California. Distribution targets include the media and policy makers whose decisions may affect the industry.

A.2. Produce and Distribute CD-ROM on California's Sea Urchin Industry

Background and Purpose: Production and distribution of information on DC-ROM format is designed to complement and enhance the education video with updated graphic and statistical data. Distribution list targets public schools, media and policy makers.

A.3. Produce and Distribute Newsletter

Background and Purpose: A newsletter is essential to inform the sea urchin industry about the issues of importance. The newsletter, while listed as an educational product, serves a dual role by disseminating information in coordination with the communications element of this proposal. In addition to the members of the sea urchin industry, the newsletter is a timely vehicle to disseminate topics of importance re: the sea urchin industry to policy makers and the media.

A.4. Produce and Distribute Short Reports / Facts in Brief about the Sea Urchin Industry

Background and Purpose: In this element, SUHAC will develop and disseminate education materials designed to build a broad understanding of California's sea urchin harvesting industry. These reports will focus on providing accurate information to interested members of the public, scientists, the media, and decision-makers determining policy that will affect the industry.

Program Element [B] Industry Communications:

B.1. Establish statewide communications network within sea urchin diving industry

Background and Purpose: In addition to periodic news summaries such as those provided in a newsletter, a rapid-response communications network is essential to inform divers in a timely fashion about matters of importance and issues of concern to the sea urchin industry. Issues may include safety regulations, pending legislation affecting the industry, conservation/enhancement news or other topics of interest.

B.2. Coordinate ongoing industry education program, coordinate networking efforts with other organizations; coordinate / host annual issues forum

Background and Purpose: Developing and implementing public and industry relations strategies are important to best utilize the educational and research products produced in other elements. This element will monitor issues and trends potential affecting the sea urchin industry, educate divers on communications skills, refine industry position papers and assist in development of educational materials as needed, coordinate networking with other organizations, and facilitate meetings, culminating in an annual issues forum. The forum will bring licensed sea urchin divers together for the exchange of ideas and information. Guest speakers will be invited to address topics of concern.

Program Element [C] Administration

Background and Purpose: The SUHAC Board of Directors will be responsible for making decisions and implementing activities. SUHAC does not have an office, staff or executive director. Therefore, secretarial and accounting services are necessary to maintain correspondence and accurate records of financial transactions. An administrator is essential to review and/or prepare contracts, oversee projects and maintain accurate records of meetings. SUHAC will provide for efficient administration of the various program elements detailed in this proposal, in addition to its internal affairs, by subcontracting with the following:

MARKETING ASSOCIATIONS AND COOPERATIVES

General

The statewide association is a nonprofit mutual benefit corporation. These corporations are prevented by law from participating in certain activities. Divers may develop collective agreements for setting urchin prices only through a marketing association or cooperative. (The Fish Marketing Act, Part 3, Division 3, Title 1 of the California Corporations Code.) Marketing agreements may include, in addition to a price structure, arrangements to deduct a certain percentage of each load to be remitted directly to the association.

Existing Marketing Associations

Two sea urchin associations have been established as either a marketing association or a cooperative:

- Urchin Producers Marketing Association, San Diego (est. 1993)
- NOR PAC Divers Coop (est. 1996)

Types of Marketing Associations

There are four types of marketing associations:

1. Bargaining association
2. Vertically integrated associations
3. Processing cooperative
4. Marketing cooperative

1. Bargaining Association

In a bargaining association processors are contacted or an auction is set up in order to have the buyers competing for the limited supply of urchins.

2. Vertically Integrated Association

In a vertically integrated association, the association signs an agreement with a processor that would stipulate the supply of sea urchins for a certain period. The responsibilities of the producer and processor would be stipulated. The marketing of urchins would be an open book. The division of profits would be negotiated. The producers then would concentrate on the improvement of the quality and freshness of the product. The processor would concentrate on processing, packaging and marketing the product. A constant supply could be established.

3. Processing Cooperative

A processing cooperative would combine diving, delivery, processing and marketing into one unit. This would require a large capital investment and a good knowledge of the Japanese market.

4. Marketing Cooperative

A marketing cooperative would market the finished product from either vertically integrated associations or processing cooperatives. This would allow very small groups to achieve economies of scale in marketing their product.

Considerations for Marketing Associations

There are many services that marketing associations and cooperatives can provide their members. The services crucial to the profitability and sustainability of the sea urchin fishery are shown on the following page:

- Establish education programs
- Obtain marketing information on uni demand and price
- Negotiate satisfactory selling prices
- Capture economies of scale for transportation, processing and purchasing equipment and supplies
- Vertically integrate the industry through sales to wholesale distributors
- Promote the quality of California uni
- Protect against low cost foreign imports
- Manage the California resource so it is sustainable
- Identify consensus between divers and implement policy changes that will support this consensus
- Improve the sea urchin industry image with governmental and environmental organization

- Have a greater voice in regulatory decisions
- Provide services for divers not currently available/affordable (e.g. life insurance, loans)
- Improve communications between divers

Educational Programs

The following programs should be included in the services of a sea urchin association:

◆ **DEVELOP INFORMATION EXCHANGE NETWORK**
- Regulatory agencies (CDFG, NFS, NMFS, USFWS, OSHA, CG)
- Fishing and Environmental organizations
- Commission (F&G, Coastal)
- Sea urchin processors
- Legislators (State and Federal)
- Japanese markets
- Scientific community
- Local municipalities

◆ **COLLECT INFORMATION**
- Marketing
- Real-time fishery data
- Scientific literature
- Develop new technologies

◆ **DEVELOP CONSENSUS IN DIVING COMMUNITY**
- Leadership training
- Prepare white paper on issues
- Virtual meetings
- Develop new technologies

◆ **HOST CONFERENCES ON ISSUES**
- Closed areas
- ITQ
- Sea otter
- Diver safety

◆ **PROVIDE SUPPORT FOR HUSBANDRY**
- Experimental manipulations
- Enhancement

◆ **PROMOTION OF CALIFORNIA SEA URCHINS**
- Improve quality
- Define role in global marketplace
- Trademark product

◆ **DEVELOP & MAINTAIN WATERFRONT INFRASTRUCTURE (Denisoff 1991)**
- Offloading, docks, shoreside cold storage, shoreside processing
- Compatibility of fishing and tourism

◆ **ESTABLISH FUNDING MECHANISM**
- Check off system
- Grants

CONCLUSION

This report enumerates the functions and services provided by various types of sea urchin diver associations. It is clear that diver associations wish to be a part of the decision making process.

The concerns of divers as represented by a diving association, should be developed in the context of open communication rather than rebuttal. (Finlayson) In this manner true solutions can be achieved.

In the words of John Costlow, "The realms of science and politics need to collectively abandon their "ivory towers" and establish better mechanisms for technology transfer as it relates to intelligent management and conservation of our coastal environment. The 'partnership' however, urgently needs to be expanded to include aspects of a third dominant segment: e.g. industry... this third segment can only operate efficiently and profitably when it reflects the consensus of those it serves..." (Costlow 1989)

REFERENCES

Center for Cooperatives (1994), *Starting an Agricultural Marketing Cooperative*, J.B. Siebert, L. Ruth, D. Zollinger, U.C. Davis.

Denisoff C.R., Rote J.W. (1991), *Future of Commercial Fisheries in California Proceedings*, Coastal Zone 91 Conference.

Department of Fish and Game (1994), Audit of contract FG1232 (CUPA) report No. 30 audits BR May.

Finlayson A.C. (1994), *Fishing for Truth Social and Economic Studies A52*, Institute of Sical & Economic Research, Memorial University of Newfoundland.

Jentoft S. (1989), *Marine Policy*, April pp. 137-154.

Zengyoren (1987), *The Fisheries Law*, National Federation of Fisheries Cooperative Associations.

Zengyoren (1987), *The Fisheries Cooperative Association Law*, National Federation of Fisheries Cooperative Associations.

Costlow J. 1989, Our coastal environment: Science vs. Politics. Proceedings of the Internal Symposium on King and Tanner Crabs.

Duffy J. (1997) California Sea Urchin regulation synopsis compiled March 17, 1997.

Fish Marketing Act

CHAPTER 62

Learning to Cooperate: Joining Science and Policy for Ecosystems Governance

Rob Wilder, J.D., Ph.D.
Ocean and Coastal Policy Center
Marine Science Institute
University of California
Santa Barbara, CA 93106

phone: 805-683-6672
E-mail: robw@msi.ucsb.edu

There is growing agreement that conserving marine biological diversity requires far greater integration across policy/science/law perspectives, and better combined governance of both land/sea than practiced in the past (Ray 1996; Fiorino 1995; Norse 1993; Percival 1992). Fresh recognition is paid to fragmented ocean and coastal policies that are sub-optimal for conservation. Policies for land-based sources of marine pollution, or offshore oil, or governing coastal resources (Beatley 1994:195) or fisheries (Lee 1993) still fall short of integrated governance reflecting highly interconnected marine processes (Cicin-Sain 1994). A synthesis of science with policy/ law--along with integration of land and sea governance is needed (Wilder 1993*b*). Experience within academia tells us on the other hand that building these linkages between ecology, policy, and law is not a simple task. Notably the emerging and forward-looking ideas of the 'precautionary principle,' pollution prevention, clean production strategies and industrial ecology do point the way forward towards a useful synthesis.

Emergence of The Precautionary Principle

Despite the inevitable presence of scientific uncertainty in natural resources management, and scientists' (understandable) cultural reluctance to be engaged in politics in the policy arena, there are pathways to integrate science with policy in defense of biodiversity. One may begin by closely examining the 'default position' implicit in marine regimes. For instance the Third United Nations Convention on the Law of the Sea (UNCLOS) has established a global regime for, *inter alia,* both fisheries and ocean dumping. Careful examination reveals that language within this same Treaty will often rely upon the assimilative capacity of the marine environment as a key means to absorb harm. That is not an obscure matter of law but rather an observation with major consequences. Placing the 'burden of proof' on the seas in essence makes preservating marine biodiversity difficult. UNCLOS subtly uses the assimilative capacity of coastal seas, or relies upon hoped-for resiliency as seminal means to absorb harms (Wilder 1996). A troubling result for conservation is anthropogenic inputs are arguably understudied by science and so ecological impacts can go underappreciated by policymakers.

By placing the burden of proof upon the marine environment, advocacy is made more difficult for the biologists who by training speak in guarded terms of probabilities and uncertain data. Extraction is the assumption. It is difficult for scientists speaking in reasonably cautious terms to resist ecological deterioration or rally support for strong conservation measures. Consider the frightening crash seen in marine fisheries for cod, haddock and flounder off New England. Uncertain data made accurate modeling of stocks onerous. Nevertheless many fishery scientists saw the present over-fishing crisis coming a full decade ago. But these scientists discovered that "when they presented their data to the management councils, their penchant for speaking in terms of probabilities and confidence intervals often served them--and the fish--poorly" (Holmes 1994:1252). Given permissive marine regimes, a tendency is to err towards over-exploitation. Hence it is all the more important for biologists to understand the policy implications of this permissiveness, and to be able to speak to the propriety of it--given what are our complex and cascading anthropogenic inputs into the marine environment.

Hence the emerging Precautionary Principle is notable. This Principle has lately encouraged new thinking within international law and policy through its emphasis on clean production strategies, and the prevention of contamination such as of the coastal oceans. The concept stems from a belief that it is cheaper, easier and more practical to prevent pollution in the first place--than to try to clean contaminated systems later-on (Wilder 1994). Precaution also arises from emerging recognition in policy that scientific understanding of marine ecosystems is made exceptionally difficult by factors including uncertainty introduced by naturally chaotic population dynamics. In light of obstacles to predicting assimilative capacity of marine environments, this principle urges a conservative approach as often the most sensible tack (Wilder 1996). Far from marine law as usual, this new principle would reverse the burden of proof in most cases.

Unsurprisingly, this new Principle is both controversial, and evolving. While a generally accepted definition has so far evaded any international legal consensus (Birnie and Boyle 1994:95), this Principle is increasingly being cited in international environmental treaties. What is clear abou: this concept is it provides useful service to conserve marine biological diversity.

It can shift the burden of proof from ecosystems to industry, such as when a proposal is made to dump large amounts of waste that would be toxic, persistent, and likely to bioaccumulate within coastal seas. Thus an effect of this new Principle is to provide industry with incentives to pursue alternatives to dumping. It presumes the best options are to look upstream to achieve source reduction, waste minimization, and pollution prevention in land-based activities. Arguably this sort of upstream action is ultimately necessary to conserve marine biological diversity.

Looking Ahead: Industrial Ecology and Design for the Environment

Biological sciences are not the only relevant discipline for conserving marine ecosystems: as this Precautionary Principle implies, robust biodiversity is not maintained by one perspective alone. Rather the effort must include all media (land, air, sea) bringing together perspectives not normally enlisted in this enterprise. In fact conservation is well served by Industrial Ecology, and Design for the Environment (DfE). These fresh notions founded in engineering are helping to force realization that product design is crucial, and waste matters. Design influences the product's impacts from before cradle through after the grave, and hence significantly impacts biodiversity because so much follows the decisions made there. The most flexibility to determine benign materials is the design stage. Here is the best opportunity to choose 'green' manufacturing techniques, to consider ecological risks and the eventual characteristics of waste streams. Yet choices made in the design stage of manufacturing have been typically overlooked by academics such as conservation biologists, and the scholars of environmental policy and law. The utility of Industrial Ecology, and Design for the Environment for preservation of biodiversity are seen with needed holistic reflection.

Industrial ecology was recently introduced (Frosh and Gallopoulos 1989) with the aim of advancing a more cyclical approach to manufacturing design--where natural resources are used, and reused, over and over again as efficiently as possible (Graedel and Allenby 1995). The contrasts with linear models that still dominate manufacturing in which the cheapest material is extracted, used in production, and discarded without a significant regard for all the environmental burdens. Industrial ecology instead strives to emulate the highly interconnected cyclical ecosystems found in nature--since natural systems will abhor waste. The technological advantages in imitating designs in nature have been long recognized; dragonflies once served as a model for helicopter designs.

A manifestation of this ecosystems approach is Design for the Environment (DfE), adapted from 'Design for X' (DfX) in manufacturing. The idea of DfE is itself founded on interconnected biological systems, and so like nature it can suitably have many faces. In Germany for instance new legislative initiatives have led to new product designs that better incorporate ease of disassembly, recyclability and pollution prevention into original specifications. Pharmaceutical firms there emphasize pollution prevention while more broadly, German law is beginning to require that a spectrum of companies check production processes to detect the byproducts (waste) for which there might be some kind of use by other industries (Edgington 1993:33). This effort can bring about useful technology forcing. For instance simply changing one enzymatic process has reduced resulting wastes by about 90% while also making remaining waste nontoxic.

On a larger scale DfE may be applied through the 'eco-industrial park' which seeks to make industrial systems akin to biological systems using biologically-aware concepts like materials flows, carrying capacity, resilience, and connectivity (Kirschner 1995). An objective is networked interrelationships that take unwanted effluent from one plant or 'organism' (i.e, the heat, waste water, surplus gas, steam, organic debris) and use it as input for processes in other plants. By design the byproducts of metabolism can be coordinated so effluents are used in ways to support a much larger system. Though the biological metaphor is not intended to be perfect, our present state of industrial production can be roughly seen as akin to primitive life on earth when there was little recycling of materials and so toxic wastes may have accumulated to a point of presenting problems for life. Microorganisms evolved in response to become consumers of other's wastes. Once they evolved from primitive anaerobic to aerobic metabolism, and able to exploit what was poisonous oxygen--the web of life grew stronger (Edgington 1993).

A good example of industrial symbiosis is the biotech company Novo Nordisk and interrelationships seen among industrial plants in the small city of Kalundborg, Denmark. This complex interrelated system cost an estimated $60 million and returns roughly $10-12 million per year--with a payback of some $120 million so far (Edgington 1993:32). Interrelationships there work like this: A local oil refinery sells waste (cooling water, waste water, surplus gas and steam) to the nearby power plant. As that power plant operates, it sells heated water (its waste) to some 5000 homes and to a fish farm along with gypsum to a plaster board factory, and steam is sold to adjacent Novo Nordisk Biotech. Meanwhile Novo Nordisk sells nitrogen-rich biomass waste from enzyme-fermentation vats to farmers for plant fertilizer. Surplus yeast from Novo Nordisk's insulin production goes to farmers as fodder for their animals.

The U.S. Environmental Protection Agency has a voluntary DfE program that provides similar information to select industries on preventing pollution and increasing profits by designing their products or processes along ecological end-points (Hart, Boger and Kerr 1995). This program is beginning to establish cooperative government-industry partnerships for prevention within specific areas such as the dry (potentially wet) cleaning industry where an aim is to reduce perchloroethylene-use by demonstrating alternative greener processes. Through new green tools like 'cleaner technology substitutes assessment,' and 'life-cycle assessment', the Design for the Environment (DfE) program has a potential for headway towards preventing waste in sophisticated, multi-media settings. Voluntary integrated programs such as DfE that consider science, policy, law, technology, ecosystems, and profit, and that conserve resources as well ought to receive attention in the future. Each pound of pollutant avoided upstream is one less eventually ending within the ultimate receiving medium--the marine environment.

'Holism' and 'Governance': When Words Matter

Implementing this Precautionary Principle notably means overcoming considerable inertia in ways we view the world (Wilder 1996). This requires an overt recognition of values--and more precisely the critical meaning of words. Take Barry and Oelschlaeger's goal of a building a more 'holistic' paradigm (1996:910) in conservation biology. This word 'holistic' was invented by Jan Smuts over seventy years ago to convey the idea a single living organism or collection of organisms is interconnected, and the whole is more than the sum of its parts (Thomas 1992:72). Regrettably this word has itself been hijacked in recent years so that it now refers to a gamut of new-age ideals.

Perhaps a trendiness attaching to holism was inevitable given the that way this word alone expresses the idea that life is interconnected. As Lewis Thomas (Thomas 1992: 72) has wisely observed: "I wish holism could remain a respectable term for scientific usage, but alas, it has fallen in bad company. Science itself is really a holistic enterprise, and no other word would serve quite as well to describe it. . . the word is becoming trendy . . . . We need another word, a word to distinguish a system from the components of a system, and I cannot think of one." While holism remains the best way to describe grand biological interconnectedness, it is a word now too often discredited.

Likewise the word 'governance' while perhaps it is one seldom used by most conservation biologists, deserves brief reflection here. To 'govern' is defined by Oxford English Dictionary as to rule with authority. There is nothing in this definition that refers to the objective of acting wisely, or knowledgeably. Such a sterile meaning seems deeply unsatisfying to me however when thinking about the most proper goals and aspirations in governing the environment. Looking back in time a more appealing imagery for governance can be found. Earlier roots are the Latin *gubernare* which meant to helpfully direct or guide. That implies a sense of stewardship and indeed, the Greek root *kybernan* meant to steer wisely--as with a ship. Centuries ago then the idea of 'governance' meant to attend to, care for, or look after. These older meanings attaching to governance better reflect the wisdom, concern for biodiversity, and sense of ecological stewardship sought here. Governance in this sense merits greater use because like the emerging field of conservation biology, this values the presence of biodiversity.

Jargon as an Obstacle to Cross-Disciplinary Efforts

If synthesis across disciplines like with DfE can arguably assist to improve the conservation of marine biodiversity (Norse 1993), how might interested scientists and engineers become more aware of the relevant scholarship in political science (and law)? Here a frustrating difficulty not limited to political science and law begs review. An uncomfortable criticism of political science is that it is highly reductionist. The discipline is divided into at least seven sub-fields with little crossover. There is no political science analogy to the broad-ranging naturalist of an earlier era, who studied encompassing living systems across many academic boundaries. One consequence is that political models of decisionmaking are in effect less authoritative by going unread by the conservation biologists, or even by actual policymakers. A telling criticism is that not enough attention is paid to translating often salient theories into something accessible.

The jargon in political science like in other disciplines can describe a complex world in short hand. Technical terms may help to explain what is deemed reality among specialists. But consider that jargon can also constrain thinking. It can remove useful concepts from broader understanding and application in outside fields. This ought to be a more discussed matter in environmental policy and law--two sub-fields that should be intentionally more accessible to advance conservation of marine biodiversity. Importantly they need not be so hampered by inaccessibility. Through new recognition of this issue, and thus attention and care, the fields could become more accessible over a range of academic inquiry. In fact a model that I suggest is the integrative discipline of conservation biology--which strives to be both expansive and inclusive in content.

As for some advantages that accessibility can bring, consider that Hardin's (1968) 'Tragedy of the Commons' now carries a seminal influence across disciplines. Maybe because Hardin's article first appeared in the broadly-read *Science* and so sought to be understandable to a wide audience, technical terms were used sparingly. Indeed one has difficulty there in pinpointing Hardin's 'home discipline.' It is not easy for academics to write that way. An explosion of specialized policy journals, the 'reductionist' thinking in academia and the particularly wordy social science writing conspires against access and legibility. Certainly there are times where simplicity will be inappropriate; some theories are incapable of broad interest or resist accessible language. But with attention I believe policy and law studies could often be stated in more understandable fashion, and so link ecological policy/ law models to the goal of conserving biodiversity. Much can be done to demystify policy theory and thus to put this knowledge to use across disciplines.

The Cultural Divide

Such transdisciplinary efforts to conserve marine biodiversity will be a formidable proposition for conservation biologists as well, but for different reasons. As an accepted orthodoxy it is seldom discussed, yet reluctance to pursue interdisciplinary research stems in part from how the study of human institutions--and natural processes, was long ago pulled apart into distinct disciplines. At a fundamental level natural sciences are regarded as different from the social sciences/ humanities (Caldwell 1992; Hardin 1985). In what are customarily seen as the 'value-free' natural sciences, innumerable impacts humanity have on nature are studied, but issues of competing human needs are removed from analysis. Science might investigate various consequences of human-action for natural systems--yet intentionally ignore attached human values that drove action in the first place. The result is a divide. Conflicts over resource use are usually regarded by scientists as distasteful politics--a value laden world they deign to enter.

Some discomfort arguably arises from a perception deep inside academia distancing 'hard sciences' from what is often seen as a less rigorous culture within social sciences (which I often shorthand as 'soft science' for lack of better term). Social sciences are admittedly different as they do strive to comprehend values--a nettlesome proposition. Research is a debatable world dominated by shades of gray where human actions are the object of study. The result is an academic but a no less real gap between two cultures.

Note that these disciplines learn in different ways. In 'hard' sciences, typically a hypothesis is put forth and attempts are made to find instances where theory fails to fit all the facts. By rigorous efforts to refute the hypothesis (basically a 'what if?') a model is tested and new levels of scientific consensus are achieved. On the other hand policy theories are often slippery enough to be incapable of objective disproof. Knowledge is not built in a way that latest works are clearly superior to the past nor empirically testable with a rigor of 'harder' sciences, since models of human behavior can only approximate real-world events (Rosovsky 1990:103). Rather than helping to build governance regimes that will use biological concepts to preserve the integrity of marine ecosystem processes--the study tends to be upon <u>processes</u> of <u>human decisionmaking</u>, rather than the consequences for biological diversity. Lack of focus on ecosystems has led to subtle but quite thorny issues for environmental law and policy, which has often lacked a strong iterative capacity (Lee 1993:142; *see* Murphy 1991).

Some polarization pulling the natural sciences from social sciences is rational and inevitable; it is also a heritage that predates ecological concern. Yet from the standpoint of advancing resources governance--as well as the goal of effectively conserving marine biodiversity--there is a sizable penalty for this elementary splintering. Overviews of relational linkages tying human activities to marine ecosystems can be obscured. Complex feedback mechanisms that link complex anthropogenic inputs to ecological change can be easily overlooked. It is that old dilemma of not seeing the forest due to the trees. Because of inertia the more integrative aims of conservation biology can offer much in the future. Barry and Oelschlaeger urge conservation biologists to continue "extensive cross-disciplinary work--in essence, practical work toward a truly holistic paradigm--in order to fully realize the goals our science is attempting to attain." (Barry and Oelschlaeger 1996:910). Indeed conservation biologists have begun to bridge the academic divisions that stymie holism through their welcomed attention to many policy concerns (i.e., Wilson 1993; Earle 1995; Davis 1993; Woodwell 1992).

Conversely were more of the social scientists to address the ecological implications of biodiversity loss--in short to expand their transdisciplinary attentions towards the preservation of ecosystems--the results can be bountiful. I would argue that this is already happening, but just not quickly enough based on the broadening losses to marine biodiversity that we are facing. The two (potentially) closely related fields of environmental policy and environmental law are rightly starting to engage in newly transdisciplinary efforts for conserving biodiversity. Nonetheless that there has continued to be such a deep splintering of academic disciplines into dramatically separate perspectives is more than ever significant. It is a call to reassess goals and to consider possible contributions of synthesis. Finally a troubling obstacle ahead within conservation biology itself is the discipline has paid relatively little attention to the oceans and coasts (Ray 1988). This unfortunate fact poses a vexing dilemma for that discipline which is devoted to all biodiversity. Such an error can and must be cured by future recognition of the importance of broadly conserving marine ecosystems integrity.

Conclusion

In sum emerging tools like the precautionary principle, pollution prevention, industrial ecology, and design for the environment deserve greater attention. Applied robustly, these recent tools should ultimately strengthen the conservation of marine ecosystems integrity and preserve a robust marine biological diversity. Yet building a needed integration across academic disciplines also requires attention to traditional, practical obstacles. These include the existing deep divisions within academia stemming from non-interactive literature in countless sub-fields, differing modes of inquiry, splintering incentives for research, and the reductionist thinking common to academic efforts. Nonetheless by casting a wide net and efforts at synthesis, academic disciplines can cooperatively fashion a fresh ecosystems governance for the marine environment.

References

Barry, D., and M. Oelschlaeger. 1996. A science for survival: values and conservation biology. Conservation Biology 10:905-911.
Beatley, T., D. J. Brower, and A. K. Schwab. 1994. An introduction to coastal zone management. Island Press, Washington D.C.
Birnie, P. W., and A. E. Boyle. 1994. International law and the environment. Oxford University Press, New York.
Bormann, F. H., and S. R. Kellert. editors. 1991. Ecology, economics, ethics: the broken circle. Yale University Press. New Haven, Connecticut.
Caldwell, L. K. 1963. Environment: A new focus for public policy? Public Administration Review 23:132-139.
Caldwell, L. K. 1992. Between two worlds: Science, the environmental movement, and policy choice. Cambridge University Press, New York.
Carson, R. 1962. Silent spring. Houghton-Mifflin, Boston.
Cicin-Sain, B. 1994. Essay: a national ocean governance strategy for the United States is needed now. Coastal Management 22:171-176.
Davis, W. J. 1993. The need for a new global ocean governance system. Pages 147-170 in ed. J. Van Dyke, D. Zaelke, and G. Hewison, editors. Freedom for the seas in the 21st century: Ocean governance and environmental harmony. Island Press Washington D.C.
Edgington, S. M. 1995. Industrial ecology: Biotech's role in sustainable development. Bio/Technology 13:31-34.
Ehrenfeld, D. 1995. Conservation biology. Pages 147-148 in R. Paehlke, editor. Conservation and environmentalism: an encyclopedia. Garland Publishing Company, New York.
Fiorino, D. J. 1995. Making environmental policy. University of California Press, Berkeley, California.
Frosh, R. and N. Gallopoulos. 1989. Strategies for manufacturing. Scientific American 261:144-152.
Graedel, T. E., and B. R. Allenby. 1995. Industrial ecology. Prentice Hall, Englewood Cliffs, New Jersey.
Grobe, K. 1994. Composter links up With food processor. Biocycle. 35:40-47.
Haas, P. M. 1990. Saving the Mediterranean: the politics of international environmental cooperation. Columbia University Press, New York.
Haas, P. M., R. O. Keohane, and M. A. Levy. 1993. Institutions for the earth: Sources of effective international environmental cooperation. MIT Press, Cambridge, Massachusetts.
Hardin, G. 1968. The tragedy of the commons. Science 168:1243-48.
Hardin, G. 1985. Filters against folly: how to survive despite economists, ecologists, and the merely eloquent. Maple Press, Binghamton, New York.
Hart, K. M., D. L. Boger, and M. A. Kerr. 1995. Design for the environment: a partnership for a cleaner future. Printed Circuit Fabrication 18:16.
Hennessey, T. M. 1994. Governance and adaptive management for estuarine ecosystems: the case of Chesapeake bay. Coastal Management 22:119-145.
Holmes, B. 1994. Biologists sort the lessons of fisheries collapse. Science 264:1252-1253.

Imperial, M. T., T. M. Hennessey, and D. Robadue, Jr. 1993. The evolution of adaptive management for estuarine ecosystems: the national estuary program and its precursors. Ocean & Coastal Management 20:147-180.

Jasanoff, S. 1990. The fifth branch: Science advisors as policy makers. Harvard University Press, Cambridge, Massachusetts.

Kenchington, R. 1990. Managing marine environments. Taylor & Francis, New York.

Kirschner, E. 1995. Eco-industrial plants find growing acceptance. Chemical & Engineering News 73:15.

Lee, K. N. 1993. Compass and gyroscope: Integrating science and policy for the environment. Island Press, Washington D.C.

Lovejoy, T. 1989. The obligations of a biologist. Conservation Biology 3:329-330.

Lovelock, J. 1988. The ages of Gaia: a biography of our living earth. Oxford University Press, New York.

Maguire, L. A. 1996. Conservation values, conservation science: a healthy tension. Conservation Biology 10:916-917.

McCoy, E. D. 1996. Advocacy as part of conservation biology. Conservation Biology 10:919-920.

Murphy, D. D., and B. D. Noon. 1991. Coping with uncertainty in wildlife biology. Journal of Wildlife Management 55:773-82.

Norse, E. editor. 1993. Global marine biological diversity: a strategy for building conservation into decision making. Island Press. Washington D.C.

Norton, B. G. 1991. Ecological health and sustainable resource management. Pages 102-117 in R. Costanza, editor. Ecological economics: the science and management of sustainability, Columbia University Press, New York.

Noss, R. F. 1993. Whither conservation biology? Conservation Biology 7:215-217.

Percival, R., A. S. Miller, C. H. Schroeder, and J. P. Leape. 1992. Environmental regulation: Law, science, and policy. Little, Brown, Boston.

Ray, G. C. 1988. Ecological diversity in coastal zones and oceans. Pages 36-50 in E. O. Wilson, editor. Biodiversity. National Academy Press, Washington, D.C.

Ray, G. C. 1996. Coastal-marine discontinuities and synergisms: implications for biodiversity conservation. Biodiversity and Conservation 5: 1095-1108.

Rosovsky, H. 1990. The university: an owner's manual. Norton, New York.

Shrader-Frechette, K. 1996. Throwing out the bathwater of positivism, keeping the baby of objectivity: Relativism and advocacy in conservation biology. Conservation Biology 10:912-914.

Thomas, L. 1979. The medusa and the snail: more notes of a biology watcher. Viking, New York.

Thomas, L. 1992. The fragile species. Collier, New York.

Tracy, C. R., and Brussard, P. F. 1996. The importance of science in conservation biology. Conservation Biology 10:918-919.

Tuohy, W. S. 1993. Characterizing the San Francisco estuary: a case study of science management in the National Estuary Program. Coastal Management 21:113-129.

Vig, N. J., and M. E. Kraft, editors. 1994. Environmental policy in the 1990's: Towards a new agenda. Congressional Quarterly Press, Washington D.C.

Wilder, R. J. 1993*a*. Sea-change from Bush to Clinton: Setting a new course for offshore oil development and U.S. energy policy. UCLA Journal of Environmental Law and Policy 11:131-173.

Wilder, R. J. 1993*b*. Is this holistic ecology or just muddling through? the theory and practice of marine policy. Coastal Management 21:209-24.

Wilder, R. J. 1994. Prevention rather than cure. Nature 369:700.

Wilder, R.J. 1996. Law of the Sea Convention as stimulus for robust environmental policy: the case for precautionary action. Pages 207-222 in E. Mann-Borgese, N. Ginsberg, and J. Morgan, editors. Ocean Yearbook 12. University of Chicago Press, Chicago, Illinois.

Wilson, E. O. 1994. Naturalist. Island Press, Washington D.C.

Woodwell, G. M. 1992. How does the world work? Great issues of life and government hinge on the answer. Pages 31-37 in R. L. Peters, and T. E. Lovejoy, editors. Global warming and biological diversity. Yale University Press. New Haven, Connecticut.

CHAPTER 63

A Systems Approach to Enhancing Resource Utilization in the Coastal Zone

Charles B. Chesnutt[1] and Martin C. Miller[2]

Introduction

Though the U.S. Army Corps of Engineers is an arm of the Federal Government and a member of the defense community, it is much more than an organization for military, or even for Federal engineering interest. The Corps serves as the Nation's coastal engineer, supporting the national interest in coastal engineering by planning, designing, constructing, operating, and maintaining deep- and shallow-draft harbors and shore protection projects, by developing technology and design guidance and making it available to the public, and by maintaining expertise in the wide array of scientific and engineering disciplines necessary for the wise management of coastal resources.

The Corps maintains the largest, most comprehensive federal facility and staff for coastal engineering research in the country, and is committed to working with its non-federal partners to develop projects that are environmentally sound and economically beneficial and that promote the public safety. The Corps acknowledges its obligation to make its coastal expertise available for all projects, not just federal projects.

The State of California has also been a leader in the development of guidance for coastal management. The Resources Agency of California has gone to a great deal of effort to produce the strategic planning document, "*California's Ocean Resources:*

---

1. U.S. Army Corps of Engineers (CECW-PF), 20 Massachusetts Avenue, NW, Washington, D.C., 20314-1000. charles.chesnutt@inet.hq.usace.army.mil

2. Chief, Coastal Oceanography Branch, USAE Waterways Experiment Station (CEWES-CR-O), 3909 Halls Ferry Road, Vicksburg, MS 39180. m.miller@cerc.wes.army.mil

An Agenda for the Future." This visionary document sets a standard for other coastal states in the evaluation of important issues and in the establishment of direction for management guidance. The Corps of Engineers is committed to working with the state and local agencies, in all ways that we are able, to implement the *California Agenda.*

Toward a Systems Approach

The topic of this presentation, *"A Systems Approach to Enhancing Resource Utilization in the Coastal Zone,"* echoes themes that are played in the *Agenda* document. Ecologists have long recognized the interdependence of species and ecosystems. This interdependence has not always been acknowledged, however, in the way the ecological resources have been managed. This point is made explicit in recommendation A-1 of the *Agenda* in which future ecosystem management strategies are to be developed in consultation with the California Biodiversity Council. It is our contention that a similar philosophy (e.g., the systems approach) should be applied to non-living coastal resources such as sand.

The application of the systems approach was advocated in the recent National Research Council publication on Beach Nourishment and Restoration, in which it was pointed out that coastal managers and engineers spend too much time managing inlet projects and beach projects and too little (if any) time managing the sand that is moving through the inlet/beach system. This situation arises, not out of total ignorance of the system (though we are still a long way from total understanding), but because of the compartmentalized way of authorizing projects and the heavily bureaucratic way of budgeting and appropriating funds for those projects.

Though some projects have been successful, the approach has also led to some notable failures. The Santa Monica offshore breakwater which trapped sand in its wave shadow and the uncoordinated use of shore perpendicular groins to control longshore transport on New Jersey beaches are only two examples. The latter is a classic example of applying a solution to a local problem without consideration for the system. A groin built by a property owner to stop the longshore flow of sand on his beach deprives the downdrift beach of its source of sand. Downdrift property owners then felt compelled to build additional groins and the chain reaction of construction has led to the structuring of long sections of shoreline. Thus a local "solution," without consideration of the system, can lead to a regional problem.

Ignorance of the system can also lead to unwise practices and "one size fits all" policies. These may be encouraged by political and economic expediency. A policy that results in the offshore disposal of all dredged material might fall into this category. Though offshore disposal is justified in many circumstances, there are situations in which other alternatives could be considered, even if they are more expensive for the first or second cycle. Clean dredged material can be considered a resource that can be used to replenish beaches deprived of longshore sources of sand.

Though sometimes more expensive in the short term, systems considerations may reveal that such action is justified in the long term.

California has been a leader in the beneficial use of dredged material for building bird habitat in the Sonoma Baylands Project in North San Pablo Bay. This program is a showcase for "thinking outside the box." The Corps participated in this project by constructing levees and by moving utilities within the disposal area. Design efforts for the pilot project were initiated under the Coastal America Partnership and utilized the Corps of Engineer authority under Section 106 of WRDA 1992. The project's goal is to restore 348 acres of tidal wetland to the bay. This is being done using material dredged from the Petaluma River and the Oakland Harbor deepening project. The authorizing legislation states the following project purposes: "...to restore tidal wetlands, provide habitat for endangered species, expand the feeding and nesting areas for waterfowl along the Pacific Flyway and demonstrate the use of suitable dredged sediments as a resource, facilitating the completion of the San Francisco Bay Area dredging projects in an environmentally sound manner." The Corps has been responsible for design, construction and monitoring of the project and is providing 75 percent of the required funding. The California Coastal Conservancy is providing the balance of the needed funding.

The Corps supports other activities for environmental restoration through its association with Coastal America and is supporting this partnership at the highest levels.

Changing Modes of Operation

In order to implement the systems approach, we, the Federal, state and local agencies, must change the way we do business. Again, the visionary approach of the *Agenda* points this out and calls for removal of institutional barriers through the establishment of coordinating committees at the highest levels. Experience has shown that the successful implementation of the systems approach can only be achieved through the coordinated efforts of the public, all levels of government, the private sector and the academic community.

In order to achieve these levels of success, some of the institutions must reevaluate their methods of doing business, set new goals and resolve to work smarter. This is particularly true in today's climate of reduced budgets and restricted manpower ceilings. The Corps of Engineers is committed to carrying out its general coastal mission, which is to plan, design, construct and in some cases operate some of the infrastructure necessary to support the national economy. In order to fulfill this mission, the Corps is also committed to working in partnership with state and local governments, and with the public, public interest groups and with those in the private sector. The Corps is willing to take a leadership role in reaching out to others to define our relative roles and relationships.

The Corps' Strategic Plan

One of the Corps goalsin its strategic plan is to revolutionize. To meet this goal, we have developed eleven initiatives.

1. **Policy Review:** The Corps, at the recommendation of the Committee on Beach Nourishment and Protection of the National Research Council, is reviewing its technical practices and policy requirements with the goal of eliminating inconsistencies and inefficiencies in operations.

2. **Customer Outreach:** The Corps has begun a broad, generic outreach program aimed at all of its customers. Corps participation in this conference is an example.

3. **Public Involvement:** The NRC report also indicated that much of the public does not understand how beach nourishment projects are designed and recommended that the Corps seek to better educate the public about how our projects work and how success is defined. The Corps was quick to develop and implement a broad strategic approach to public education.

4. **Developing Critical Linkages and Partnerships:** The Corps has developed linkages with other organizations through, for example, Coastal America. This group of 15 Federal partners and more than 140 non-Federal partners seeks to improve the coastal environment by leveraging resources and expertise through cooperative efforts. The Corps is actively seeking to partner with organizations and states on broad, common interests in addition to our traditional role with cost-sharing sponsors of site-specific projects.

5. **Technology Transfer:** The Corps has long made its tecnology available to the general public through publications, workshops, and general access. *The Shore Protection Manual*, last published in 1984, is the world standard for coastal engineering design and is presently undergoing complete revision. In addition, the Corps has established sites on the World Wide Web to supply data and information to its customers.

6. **Coastal Engineering and Science Education:** In partnership with academia, the Corps is funding another study by the NRC of the coastal engineering education needs of the Nation. The Corps, in cooperation with Texas A&M University, conducts a one-year masters program in coastal engineering.

7. **Environmental Stewardship:** Achieve a new focus on environmental stewardship with an aim to restoring and enhancing the environment in a way that also stimulates the economic utilization of these valuable resources.

8. **Systems Approach to Sediment Management:** Take a systems approach to

managing sediments in the coastal zone.

9. **Re-targeting the Research and Development Program:** Retarget the Corps coastal engineering research and development program to support future needs of the Nation

10. **Hazard Mitigation:** Reduce and manage the impact of coastal hazards on the increasing numbers of people inhabiting the coastal zone.

11. **Developing New Opportunities:** Maintain openness to new opportunities to serve the people of the nation and promote our national economy within the world market.

So many of the Corps initiatives overlay and mesh with those expressed in the *California Agenda* that we must both be on the right track. The Corps can assist in virtually all of the eleven specific ocean resource and management issues addressed by the *Agenda*. Five of them are particularly suitable areas to use Corps coastal expertise, if not programmatic involvement.

Shoreline Erosion: As states and the Federal government work together on a national strategy for dealing with shoreline erosion, significant synergy can be achieved. The data collected and information generated by the Corps California Storm and Tidal Wave Study has already contributed to your information base. Regional studies of shoreline erosion, as authorized in the recent Water Resources Development Act, could support California's efforts enormously.

Ports and Harbors: The Corps re-targeted coastal engineering research program includes major new initiatives to improve the design of coastal waterways, channels, and harbors. This will focus on increasing benefits through more efficient vessel operation as well as decreasing costs in construction and maintenance of the channels. Environmental impacts of these changes in design will be examined generically during the research phase with the hope of reducing costs of environmental studies during the planning phase or mitigation during the construction phase.

Tourism and Recreation: While the Corps has not placed a high priority on supporting tourism and recreation development, the States, including California, have certainly recognized that beach-related tourism is the largest destination segment of the largest industry in the United States. That industry will have a great influence on the future of coastal zone management.

Research and Education: The recently initiated NRC study on Coastal Engineering Research and Education Needs will benefit states such as California as well as the Corps.

Information Technology: The Corps tremendous expertise in developing, cataloging, and retrieving data in GIS format should be considered available to the States as a national asset.

Summary

The Corps of Engineers is taking more care to work with our non-Federal partners as well as our cost-sharing local sponsors on a broad range of common issues and is adapting new strategies for working with others.

The Corps is seeking new ways to serve the people of this Nation and will work with a new array of partners in trying toserve the needs of the Nation. This certainly includes working with the State of California in trying to achieve the goals of its *Agenda for the Future*, as is appropriate.

CHAPTER 64

Scientific Data in Ocean and Coastal Resource Management Decisions: Pity the Poor Scientist

Frederick M. Piltz, Ph.D.
Chief, Environmental Studies Section
Pacific OCS Region
Minerals Management Service
770 Paseo Camarillo
Camarillo, CA 93010
e-mail Fred_Piltz@smtp.mms.gov
(805) 389-7850

The role of the scientist in ocean and coastal resource management and decision making has been the topic for several symposia and a few special studies by the National Academy of Sciences Ocean Studies Board. Among the diverse discussion topics in these fora, is usually one which deals with bridging the gap in communication between scientist and decision maker. A subset of this topic is the need for the scientist to understand what data and information is needed by decision makers and conversely the need for decision makers to understand the limits of scientific data and the proper use of that information. This paper addresses one aspect of that dialog: the need for all participants to cooperatively decide on the amount of change in the resource in question which is important for triggering decision making. This amount of change in a resource is critical to the proper role of science in decision making. The threshold change levels that trigger decisions need to be coupled with the levels of uncertainty in the data and information that the scientist brings to the decision making table. This paper argues that such discussion and agreement should proceed before experiments and studies are done in order to 1)assist the scientist in the proper design of research, 2)formulate a potential threshold of effect at which some decision will be made, and 3) avoid wasting time and effort in research that does not provide useful information. These points are illustrated with specific examples from a variety of ocean science-decision making efforts.

CHAPTER 65

Real Time Oceanographic Data from the Santa Barbara Channel-Santa Maria Basin Circulation Study Used for Oil Spill Response

David R. Browne[1]

Abstract

In 1991, the Minerals Management Service entered into a Cooperative Agreement with the State of California-Scripps Institution of Oceanography to conduct physical oceanographic studies that would be of mutual benefit in determining potential environmental impacts of offshore oil and gas activities in the Southern California Bight. Real time acquisition of currents, winds, satellite imagery, and surface drifter tracks from the resulting Santa Barbara Channel-Santa Maria Basin (SBC-SMB) Circulation study is proving helpful to oil spill response efforts in this area of active federal oil and gas leases. Characteristic circulation processes will be discussed in terms of their relative impact on oil spill transport. How near real time synoptic views of the oceanic circulation in the SBC-SMB area are obtained from this real time data, and how they facilitate oil spill trajectory predictions are described. Trajectory prediction performance as a result of using real time data in past oil spills and oil spill drills will be presented. Benefits of this program to local California communities, and ongoing MMS activities to enhance this program will be discussed.

Introduction

In 1991 the Minerals Management Service (MMS) entered into a Cooperative Agreement with the State of California to conduct physical oceanographic studies in the area of active OCS leases in the Santa Barbara Channel - Santa Maria Basin area. The research representative of the State of California in this Cooperative Agreement is the Scripps Institution of Oceanography (Scripps). The major objective of the MMS in this Cooperative Agreement was to obtain enough physical oceanographic information in specific areas offshore Southern California that would facilitate accurate oil spill risk analysis (OSRA) for strategic OCS oil and gas decision making.

What was originally intended to be a project that would determine and describe the oceanic circulation in the entire Southern California Bight was reduced in scope to the area of the Santa Barbara Channel (SBC or Channel) and Santa Maria Basin (SMB) due to MMS budget constraints.

1

Physical Scientist, Department of Interior, Minerals Management Service, Pacific OCS Region, 770 Paseo Camarillo, CA 93010

Thus, the title of the Cooperative Agreement became the Santa Barbara Channel-Santa Maria Basin (SBC-SMB) Circulation Study. Whatever research was done, it was (and still is) imperative that it be of high enough quality to comply with the National Research Council (NRC) recommendations. In regard to field work, this level has been achieved in the SBC with only some analysis work remaining to be done. At this time, however, two more years of field work is required in the SMB to achieve the same status of information in that area.

Early on in this effort, it was realized that expanded use could be made of the real-time environmental data that was routinely collected by Scripps researchers for their own analysis. The current and wind data that was being collected every two hours from the surface instruments, and the daily satellite imagery and free floating drifter data (3 - 4 location sites per day) could be used to estimate the movement of spilled oil as an additional service provided by the MMS during oil spill response operations. In November 1993, the MMS conducted an oil spill drill at Chevron's platform Gail, where the first use of this new real-time data resource was made to predict oil spill trajectories. Real-time current and wind data and satellite imagery was made available on the internet. The project was too young to have had the characteristic flow regimes described for the SBC, so there was much more dependency on an industry oil spill trajectory model. The real-time data was fed into the model which ultimately predicted the actual true path of floats on the water simulating spilled oil. The floats were moving directly north towards Santa Barbara Harbor at near right angles to a prevailing western current and an eastbound wind as predicted by the model. This was a very positive statement for the accuracy that can be achieved for oil spill models when they have real-time data as input. This was also a positive statement for the power of real-time data.

As the study progressed, it became apparent that the best way to realize the objectives associated with both higher quality OSRA and tactical oil spill trajectory prediction during an oil spill response would be to determine, and adequately define, the synoptic circulation flow patterns characteristic of the Channel. Understanding the state and trends of the transitions between these characteristic synoptic flow patterns would also be important to MMS objectives.

Scientific Results of the SBC-SMB Circulation Study:

Figures[2] 1 and 2 show the field measurement arrays for measurements from January 1993 to March 1996 primarily focussed in the SBC, and from March 1996 to September 1997 when measurements were primarily focussed in the SMB respectively. Both phases of this measurement program contain moored current meter, temperature, and bottom pressure measurements, NDBC buoy meteorological and ocean current measurements, meteorological stations deployed on the Channel Islands, oil and gas platforms, and along the mainland coast, free floating surface drifter buoy deployments (following the top 1 m of the water column), daily AVHRR satellite imagery, and ADCP, CDT, and XBT surveys. Figure 3 indicates the surface drifter release sites.

Results of the SBC-SMB Circulation Study indicate that there are two major forcing phenomena that determine (through their relative strength to each other) the type of flow regime that will exist in the SBC. They are:

2

Figures 1,2, and 3 were obtained from Dr. Guillermo Gutierrez of the Center for Coastal Studies, Scripps Institution of Oceanography.

(1) the northwest (direction from) upwelling favorable wind and its accompanying shear in the wind curl occurring in the western end of the Channel, and

(2) the alongshore pressure gradient due to the temperature differences between the water north and west of the Channel and the water in the Southern California Bight.

These opposing forcing mechanisms give rise to four of the six characteristic flow regimes in the Channel. They are the (1) Upwelling Flow regime, (2) Cyclonic Flow regime, (3) Relaxed flow regime, and the (4) Milling Flow regime. The fifth and sixth flow regimes, Flood East flow and Flood West flow, predominantly occur as the ocean's response to localized storms in the winter. All six flow regimes are pictorially characterized in Figure 4[3].

Upwelling Flow occurs when a strong northwest wind blows continuously along the mountainous California central coastline causing upwelling at the coastline and a south-south west flow along the coastal waters in that region. In the western entrance of the Channel, a tremendous wind curl and its accompanying shear at that location drives water south and eastward in the southern part of the Channel along the Islands. Western flow in the northern part of the Channel (along the mainland) is slowed, stopped and at times reversed. In this case, as may be obvious, the northwest wind is dominant over the alongshore pressure gradient.

Cyclonic Flow occurs when there is more of a balance between the northwest wind and the alongshore pressure gradient. In this case, western flow along the mainland of the Channel is strong enough to proceed to the western entrance, but not enough to round the corner and continue northward along the coast. Instead it cycles back into the channel to join the eastward flow along the Islands to form a cyclonic gyre at the western entrance of the Channel. A phenomena, dubbed the "Santa Rosa Squirt" by Scripps researchers, occurs most prominently in this flow regime. A combination of the eastern flow in the southern part of the Channel and the eastern edge of the western cyclonic gyre form a squirt or jet of water that transverses the Channel towards the northeast where it bifurcates or hammerheads against the western flow or the northeast coastline of the Channel. Part of the flow then joins the western current, and the other transverses the Channel once again toward the southeast where it typically causes surface drifters to beach in Chinese Harbor on Santa Cruz Island.

Relaxed Flow occurs when the northwest winds have sufficiently diminished or "relaxed" and the alongshore pressure gradient becomes the dominant forcing mechanism. Poleward flow from the Southern California Bight now enters the eastern entrance of the Channel, proceeds westward along the Mainland until it reaches the western entrance, upon which, it rounds the Points and proceeds poleward along the coast forming the Davidson Current. Eastern flow still exists in the southern part of the Channel, but in a diminished state.

Milling Flow mainly occurs due to the instabilities along the temperature front formed by the warm water flowing westward into the eastern entrance, and then along the northern section of the Channel from the Southern California Bight, and the colder water from the north and west

---

3

Obtained from "Circulation Induced by Winds and Pressure Gradients in the Santa Barbara Channel", a doctoral dissertation submitted by Ms. Sabine Harms, Scripps Institution of Oceanography.

flowing into the western entrance along the southern part of the Channel. The resultant cyclonic eddies advect towards the western entrance of the Channel typically escaping into the open ocean.

Flood East Flow mainly occurs in the winter due to local storm fronts passing through the southern and central California area. Cold water from the north and west enters the western Channel entrance and flows east through the eastern entrance and south through the island passes. Drifter deployments indicate that this southern flow continues past the southern extent of the Southern California Bight.

Flood West Flow is rare, and also primarily occurs in the winter due to local storm fronts passing through the southern and central California area. Relatively warm water from the Southern California Bight floods the Channel's eastern entrance and proceeds westward throughout the entire Channel, exits the Channel's western entrance where the flow typically bifurcates towards the northerly (poleward) and southerly (equatorward) directions.

The first three flow regimes occur as a direct response to the relative strengths of the poleward alongshore pressure gradient and the northwest upwelling favorable wind. These flow regimes generally occur in the sequence that they are discussed above. Figure 5[4] depicts twenty synoptic flow states (from hourly averaged 5m current velocities) as a function of the poleward alongshelf pressure gradient and the upwelling favorable wind stress. This figure is very useful for extended predictions of oil spill trajectories. The milling flow regime can occur as a result of instabilities along the temperature fronts existing in the first three flow regimes or during the transitions between them.

The characteristic flow patterns in the SMB are not yet so clearly defined. After one year of data collection from the larger field array deployed in that area, and analysis of the first half year of this data, it is seen that there is tremendous spatial and temporal variability in the oceanic circulation in the SMB. Beyond the typical seasonal upwelling (early spring to fall) along the central coast of California, a westward jet occurs just north of Pt. Arguello that seems to be the result of poleward flow coming out of the Channel meeting the southerly coastal flow off the central California coast. In April of 1997 an external review of the research completed in the MMS/Scripps Cooperative Agreement was conducted by a group of non-MMS, non-Scripps scientists simply named the "Scientific Review Panel." According to the Scientific Review Panel[5], reasonable dynamic and statistical descriptions of the westward jet phenomenon, as well as other events such as the cyclical occurrence of coastal eddies causing on- and offshore coastal flows in the SMB, will require a full three + years of field measurements. Understanding these phenomena are extremely important to future OCS oil and gas decision making concerning the SMB area.

---

[4] Obtained from "Circulation Induced by Winds and Pressure Gradients in the Santa Barbara Channel", a doctoral dissertation submitted by Dr. Sabine Harms, Scripps Institution of Oceanography.

[5] Scientific Review Panel members consisted of : Robert Beardsley (Chair), John Allen, Bill Boicourt, Cortis Cooper, Joe Niebauer, Robert Reed, and Robert Smith. The meeting was conducted March 31 to April 1, 1997 and was held at the Scripps Institution of Oceanography.

Oil Spill Trajectory Prediction

Once these characteristic synoptic flow patterns emerged from the increasing data base and the ensuing data analysis, it became clear that knowledge of these flow patterns could be used for tactical oil spill trajectory prediction during an actual oil spill with or without a model. Knowing the predominant currents in the area of the spill, and understanding the particular circulation phenomena that might be occuring in the path of the spilled oil has proven useful in two recent oil spills that occurred in the Channel. Accurate predictions were made without use of a model in both of the oil spills, but oil spill trajectory models were used as well. A model's output accuracy can be enhanced not only by the real-time data that can be used for input, but also simulating an eddy, squirt, or jet that the forecaster knows is present from his understanding of the total synoptic flow pattern, but is not specifically being picked up by the real-time moored data.

The Platform Hondo Oil Spill: Platform Hondo is located five miles offshore in the northwest corner of the Channel. On May 25, 1994 thirty barrels of oil spilled into the Channel due to a pump failure at platform Hondo. The Coast Guard contacted NOAA HAZMAT personnel who ultimately issued a prediction from their oil spill model that the oil would proceed westward out of the Channel. Scripps researchers on the SBC-SMB Circulation Study were also contacted early during the spill. They updated the real-time data and faxed to the MMS average currents, average winds, and satellite imagery for that day. The data indicated that while the oil would initially proceed westward, it would eventually be entrained in a cyclonic eddy (indicated by satellite imagery and supported by moored current data) at the western end of the Channel that would carry the oil cross-Channel to the south toward the Islands. In hindsight, we now know that the Channel was experiencing a **cyclonic flow condition**. Fortunately, Clean Seas was able to catch most of the oil in their containment booms before it ever reached the western eddy. However, a small sheen of oil from the spill was seen crossing the Channel, and a previously deployed Scripps free floating surface drifting buoy (surface drifter) that passed through the area where Platform Hondo is located during the spill actually crossed the Channel's western entrance, where the eddy was located, and ultimately beached on San Miguel Island.

The Platform Heritage Oil Spill: Platform Heritage is three miles further offshore than Platform Hondo and several miles further toward the west in the north west corner of the Channel. Near midnight on May 1, 1996 a pipeline control valve malfunctioned sending 150 barrels of oil out through the flare boom onto the ocean surface. Within minutes of notification, real-time data was obtained from the MMS/Scripps home page through the internet to determine the general flow regime and the specific wind and current velocities at the spill site. The daily averaged currents and winds, accompanied by satellite imagery, indicated either a **relaxation flow regime** or a **transition to a relaxation flow regime from a cyclonic flow regime** existed in the Channel. A strong westward flow throughout the northern part of the Channel (including the location of platform Heritage), coupled with only a weak indication of cyclonic action at the Channel's western entrance, led us to predict that the oil would continue westward outside the Channel. However, lack of spatially coherent current observations along the central California coast in the SMB, and mild southerly winds throughout the area, led us to believe that the oil would likely continue in a westward or south- westward direction outside the Channel. Normally, we would predict that the oil would go north along the Central California coast after exiting the western Channel entrance as would be the case in a fully developed relaxation flow regime.

An industry model made 8, 18, 24, and 36 hour oil spill trajectory predictions based on the first day of real-time data. The model predicted the spilled oil would first move westward outside the Channel. Once outside the Channel it would move continuously toward the southwest. Clean Seas collected most of the spilled oil at the Channel's western entrance. However, remnants of oil could be found miles west and south-west of the Channel. By coincidence, Scripps research team had deployed a number of surface drifters in the Channel, some of them close to Platform Heritage, five hours before the actual spill. The surface drifter tracks (also obtained in real-time during and after the spill response) confirmed all oil spill trajectory predictions, with the exception of one surface drifter which actually began to move around Pts. Conception and Arguello and proceed to the north (poleward) before it changed direction and headed westward. Apparently, the surface drifters followed the oil slick fairly well: field response team observers confirmed the locations of oil remnants which matched the locations of surface drifter tracts, including the one that rounded Pts. Conception and Arguello.

Future of Oil Spill Trajectory Capability in the SBC and the SMB

As stated earlier, while scientific research is near completion in the Channel, more field work is required in the SMB than the presently scheduled 1.5 years. According to the Scientific Review Panel, reasonable dynamic and statistical descriptions of the complex phenomenon observed in the SMB will require a full three + years of field measurements. Whether or not the SBC-SMB Circulation Study will receive the funding required to complete the field work in the SMB following NRC recommendations, as it did in the SBC, is yet to be seen. These decisions should be made by June 1997. If funding is approved, field work will continue until March 1999. If funding is not approved, field work will end in September 1997.

A monitoring program, after the larger field program is completed, would be extremely advantageous to the State of California. For very small funding each year, the information gathered in the larger \$11M to \$14M SBC-SMB CIRCULATION Study can be used together with real-time data obtained from a modest monitoring program to give very accurate surface pollutant trajectories in the areas of OCS oil and gas activities. The field array would consist of a total of four deployed current meter moorings (three in the SBC and one in the SMB), three NDBC meteorological buoys, daily satellite imagery, and a cache of 25 surface drifters to be deployed during a spill event would be administered by the Scripps Institution of Oceanography, University of California. The moored data would be continuously reporting in real time and satellite imagery would be processed daily, both of which would be made available on the Scripps Oil Spill Response web page. This data would not only be available to MMS and Scripps scientists, but anyone who has access to the Internet. In the event of a spill in either the SBC or the SMB, real-time data could be obtained within minutes of notification of the spill and surface drifters would be deployed near the spill site and other strategic locations within the SBC and the SMB. Initial predicted trajectories based on this accurate information could be made within an hour and transmitted to the appropriate Federal, State, and Industry agencies during oil spill response. State resources could then be more efficiently deployed with the knowledge of where expected landfall of an oil spill may occur. Predicted trajectories would be updated every few hours as new data is received from mooring transmitters, satellite imagery, and satellite transmitted surface drifter tracks.

Figure 1

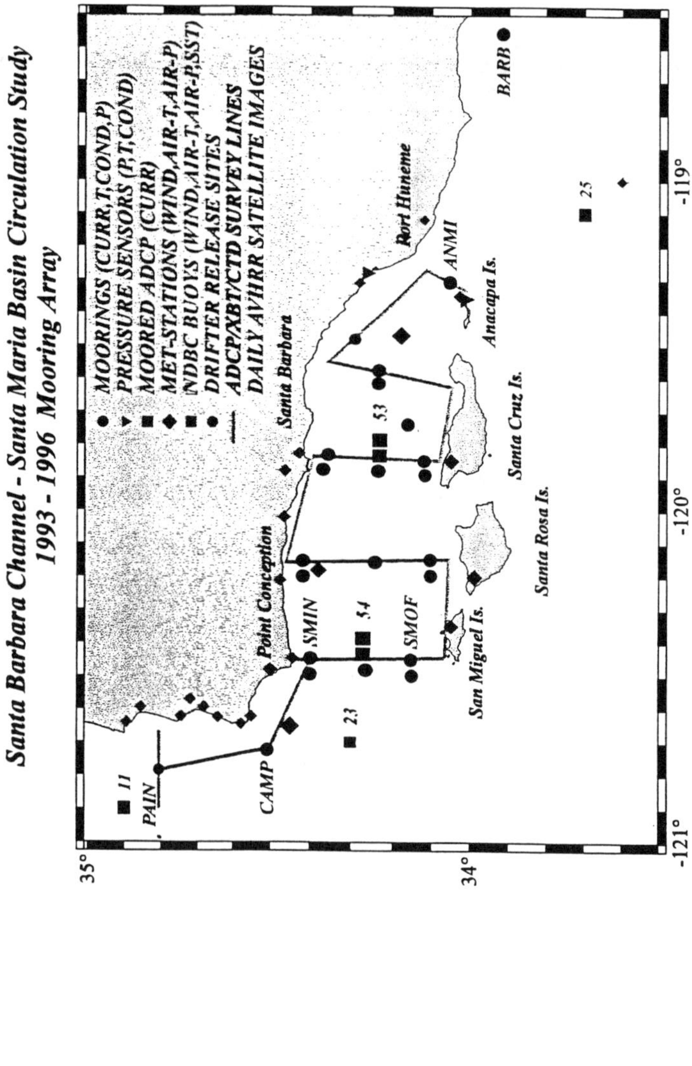
Santa Barbara Channel - Santa Maria Basin Circulation Study
1993 - 1996 Mooring Array
MOORINGS (CURR,T,COND,P)
PRESSURE SENSORS (P,T,COND)
MOORED ADCP (CURR)
MET-STATIONS (WIND,AIR-T,AIR-P)
NDBC BUOYS (WIND,AIR-T,AIR-P,SST)
DRIFTER RELEASE SITES
ADCP/XBT/CTD SURVEY LINES
DAILY AVHRR SATELLITE IMAGES
PAIN
11
CAMP
23
Point Conception
SMIN
54
SMOF
San Miguel Is.
Santa Rosa Is.
Santa Barbara
53
Santa Cruz Is.
Anacapa Is.
ANMI
Port Hueneme
25
BARB
35°
34°
-121°
-120°
-119°

Figure 2

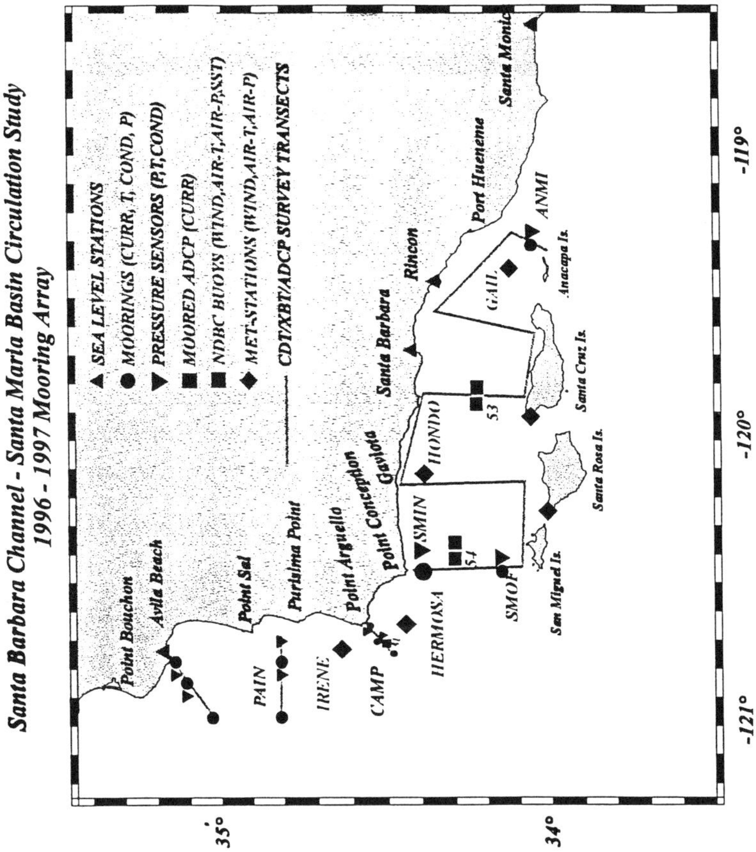
Santa Barbara Channel - Santa Maria Basin Circulation Study
1996 - 1997 Mooring Array
SEA LEVEL STATIONS
MOORINGS (CURR, T, COND, P)
PRESSURE SENSORS (P,T,COND)
MOORED ADCP (CURR)
NDBC BUOYS (WIND,AIR-T,AIR-P,SST)
MET-STATIONS (WIND,AIR-T,AIR-P)
CDT/XBT/ADCP SURVEY TRANSECTS
Point Bouchon
Avila Beach
Point Sal
Purisima Point
Point Arguello
Point Conception
Gaviota
Santa Barbara
Rincon
Port Hueneme
Santa Monica
PAIN
IRENE
CAMP
HERMOSA
SMIN
HONDO
SMOF
GAIL
ANMI
54
53
San Miguel Is.
Santa Rosa Is.
Santa Cruz Is.
Anacapa Is.
35°
34°
-121°
-120°
-119°

Figure 3

Santa Barbara Channel - Santa Maria Basin Circulation Study

Surface Drifter Release Sites

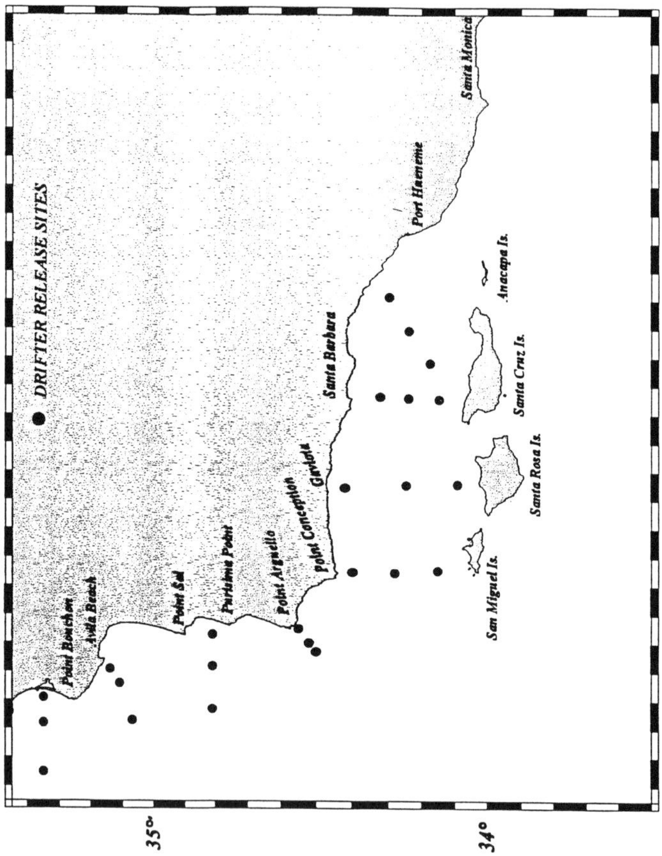

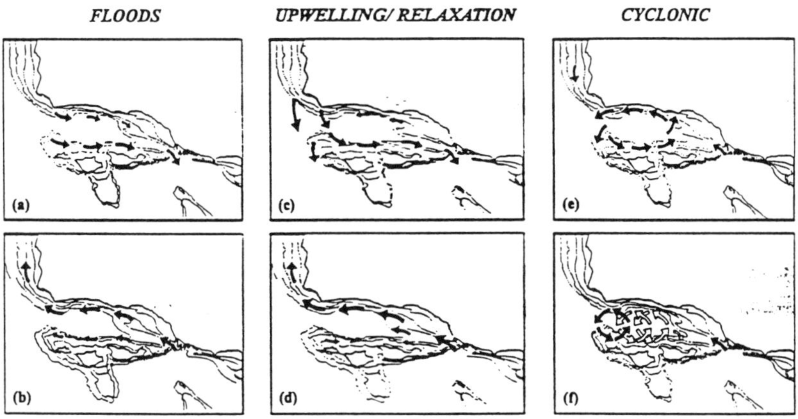

Figure 4 Synoptic views of the circulation in the Santa Barbara Channel. (a) Flood East, (b) Flood West, (c) Upwelling, (d) Relaxation, (e) Cyclonic, (f) Milling.

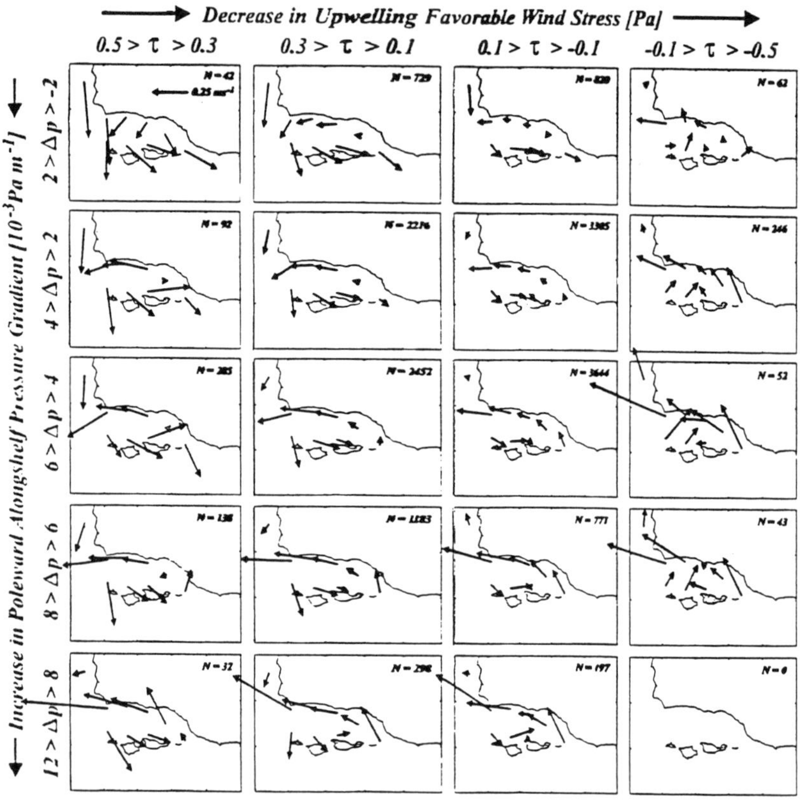

Figure 5 Averaged 5-m current velocity [m s^{-1}] as a function of equatorward wind stress (NDBC 54) [Pa] and alongshelf poleward pressure gradient (GOIN-PAIN) [10^{-3} Pa m^{-1}]. The analysis period is January 1994 to December 1995. The number of realizations of each combination of forcing is shown in the upper right corner of each panel.

References:

1. Harms, Sabine, "Circulation Induced by Winds and Pressure Gradients in the Santa Barbara Channel", Thesis Dissertation, Scripps Institution of Oceanography, October 1996.

2. Personal notes and discussions with Scripps Institution of Oceanography researchers: Dr. Clinton Winant et al.

CHAPTER 66

The Development and Use of a Computerized System to Examine Trends in the Components of Produced Water Discharges from Outer Continental Shelf Oil and Gas Platforms

David Panzer[1]
and
Philip Weisbecker[2]

Abstract

The Minerals Management Service's Pacific OCS Region (POCSR) has had an historic need for data on the types, volumes, and trends in the quantities of components of produced water and the toxicity of drilling muds discharged by the offshore oil and gas platforms. These data are important in working with local agencies to address environmental concerns, as well as addressing the National Environmental Policy Act (NEPA) requirements that decisions be made based on analysis of environmental impacts. This project could provide information to the State of California in its efforts to develop a comprehensive program to monitor the quality of ocean waters and coastal streams.

To analyze these data, the POCSR has developed a computerized system that enables the user to examine any produced water or drilling fluid discharge over time on a single or multiple platform, component, or mass-emission basis. QuattroPro, a spreadsheet program, was used for initial data entry and to perform primary analyses. The basic spreadsheet format is discussed, along with examples for both produced water

---

[1]Oceanographer
Department of the Interior,
Minerals Management Service, Pacific OCS Region,
770 Paseo Camarillo,
Camarillo, CA 93010

[2] Student Intern, University of California, Santa Barbara

and drilling fluid discharges. To further facilitate examination of statistical trends, export to a database such as Paradox can be performed. This information can be used for NEPA analyses, to indicate facilities where more or less sampling is needed, and to answer general questions from the public or other agencies concerning discharge characteristics and impacts on water quality.

Introduction

As the Federal bureau responsible for offshore oil and gas development, the Minerals Management Service's Pacific OCS Region (Pacific Region) has oversight for all aspects of offshore oil and gas platform and pipeline operations. Included in these operations are the separation, treatment, and discharge of produced water from crude oil and the discharge of drilling fluids (i.e., muds and cuttings).

The Pacific Region has worked closely with the Environmental Protection Agency (EPA), Region 9, located in San Francisco, California, to develop a Memorandum of Agreement (MOA) so that the effluents from the platforms could be more closely monitored. The MOA, which is based on a 1984 Memorandum of Understanding between EPA and the Department of the Interior, describes the specific needs and co-responsibilities of EPA and the Pacific Region. These responsibilities include, first, the development of an annual workplan detailing the location and schedule of that year's Pacific Region inspections of a platform operator's required National Pollution Discharge Elimination System (NPDES) self-monitoring performance (i.e., sampling, visual observations, record keeping, etc.). Second, the Pacific Region may assist in EPA enforcement actions according to any needs that EPA might have, particularly regarding operational and safety questions. Third, EPA and the Pacific Region are notified when final drilling mud discharges occur. Operators are required to perform this notification 48 hours prior to the commencement of a final mud dump. When these dumps occur, EPA may authorize the Pacific Region to collect a sample, which is sent to the EPA laboratory for analysis in order to determine compliance with the operator's permit. Lastly, the Pacific Region and EPA have worked together to conduct produced water compliance sampling. These are collected as stated in the annual workplan. No samples have been collected during the last two years due to technical issues that EPA and the Pacific Region are cooperatively striving to overcome.

The Pacific Region has had an historic need for data on produced water and drilling mud discharges. This includes such information as volumes, the key produced water components, including any changes in those components in the produced water stream, any permit exceedances that might have occurred, basic statistical trends that would show how an operator's equipment was working, and indications of facilities where more or less sampling (oversight) might be needed. This need arises from several origins and primarily stems from the fact that none of these data were readily accessible. Although oil and gas operators routinely submitted Discharge Monitoring Reports (DMRs) to EPA, as the terms of their permits specified, there was no ability to examine the data and perform NEPA analyses, answer questions from State and local agencies and members of the concerned public, make comparisons with other types of discharges, detect potential problem areas at facilities, or make the data available to the State of California for use in their efforts to characterize the quality of the State's waters. These topics are addressed in greater detail later in this paper.

Methods

To respond to these needs and questions, the Pacific Region developed a computer spreadsheet, using QuattroPro, that enables the user to examine any produced water or drilling fluid discharge over time on a single or multiple platform, component, or mass-emission basis. To assist in the designing of the spreadsheet and to perform the initial data entry, Philip Weisbecker, a student intern from the University of California, Santa Barbara, was hired. It was decided that the data in the spreadsheet should be able to be linked from one type of analysis to another. For example, from the raw data tables, formulas were inserted that calculated simple concentration averages and sums of volumes discharged. The averages and sums were then copied to other pages within the spreadsheet where mass emissions were calculated and graphs developed. To keep track of the many resulting pages and the addresses of key formulas, an index, contained in the first page of the spreadsheet, was developed. The index gives various types of information including the location (page name) of platform-specific data, the type of data contained on the page (e.g., produced water, drilling muds and cuttings or mass emissions), the cell locations of key formulas, the background seawater concentrations of various constituents, and the permit history for those platforms whose individual permits were newly issued or have been reissued.

Every page, regardless of the the specific contents, always contains the same basic type of information for a page type (e.g., produced water, drilling muds and cuttings, or mass emissions). The produced water and drilling muds and cuttings pages include the raw data as collected and submitted by the operator in their quarterly, semiannual or annual DMR, as the terms of their permit specify. For produced water, inserted within these data are formulas that allow the calculation of some basic statistics, such as the average concentration of components discharged with time, the overall (to date) average concentration and the total volume of produced water discharged (Table 1). A total of 15 pages in the spreadsheet contain produced water information.

Pages containing drilling muds and cuttings information give the number of days of discharge, the volume of drilling muds and cuttings discharged, the toxicity of the drilling mud according to the required LC_{50} test results, the number times sheen was noted according to the required static sheen test, and the amount of cadmium and mercury found in the barite (Table 2). This information, contained within eleven pages in the spreadsheet, is given on a well-by-well basis.

	Freq. ---> Limit (ppm)--->		1 / month 72		qrtly 0.032		qrtly 0.012		qrtly 0.008		qrtly 0.02		qrtly 0.02		qrtly 0.032		qrtly 0.00056		qrtly 0.08		qrtly 0.0018		qrtly 0.08		qrtly 0.12
Date	Flow (MGM)	Flow (MLM)	Oil & Grease		Arsenic		Cadmium		Total Chromium		Copper		Cyanide		Lead		Mercury		Nickel		Silver		Zinc		Phenol
Jan. 91	1.0754	4.07	17.94																						
Feb. 91	0.8781	3.32	51																						
Mar. 91	0.0245	0.09	41	<	0.0005	<	0.025		0.0237	<	0.1		0	<	0.25	<	0.0005	<	0.2	<	0.05		0.0906	<	0.1
Apr. 91	1.2177	4.61	24.33																						
May 91	0.9233	3.49	22.05																						
Jun. 91	0.7692	2.91	26.7																						
Jul. 91	0	0.00	67.73																						
Aug. 91	0	0.00	0																						
Sept. 91	0	0.00	0																						
Oct. 91	0.7048	2.67	0																						
Nov. 91	0.7025	2.66	27.6																						
Dec. 91	0.6444	2.44	26.15																						
subtotal 91	**6.9399**	**18.78**	**25.375**	<	**0.0005**	<	**0.025**		**0.0237**	<	**0.1**		**0**	<	**0.25**	<	**0.0005**	<	**0.2**	<	**0.05**		**0.0906**	<	**0.1**
Jan. 95	1.3283	5.03	54.75																						
Feb. 95	1.1888	4.50	23.25																						
Mar. 95	1.1209	4.24	23																						
Apr. 95	1.0659	4.03	6.06																						
May 95	1.4391	5.45	14.75	<	0.005	<	0.01	<	0.02	<	0.04		0	<	0.1		0.00069	<	0.08	<	0.02	<	0.04		0.21
Jun. 95	1.0859	4.11	47.5																						
Jul. 95	0.9497	3.59	0																						
Aug. 95	1.2834	4.86	18.75																						
Sept. 95	1.2667	4.79	20																						
Oct. 95	1.0476	3.97	no report																						
Nov. 95	1.159	4.39	27.75																						
Dec. 95	1.1545	4.37	20																						
subtotal 95'	**14.0898**	**39.56**	**21.43**	<	**0.005**	<	**0.01**	<	**0.02**	<	**0.04**		**0**	<	**0.1**		**0.00069**	<	**0.08**	<	**0.02**	<	**0.04**		**0.21**
Jan .96	1.168	4.42	22.5																						
Feb. 96	0.5041	1.91	22.5																						
Mar. 96	0.4433	1.68	27.5																						

Table 1. An example of a spreadsheet page for produced water showing the general layout and the years 1991, 1995 and 1996 (to date) for a platform under the current general permit.

	Mud Type (1-8)	Mud Toxicity	Volume Muds	Days Discharged	Volume Cuttings	Days Discharged	Sheen	Cadmium	Mercury	Well #
Units --->		(LC_{50})	(bbls)		(bbls)			(ppm)	(ppm)	
Limits --->		30,000 ppm					0	2 ppm	1 ppm	
Date										
Sept. 93	7	209,700	410	2	199	2	0			2
Oct. 93	7	209,700	3514	21	969	20	0	<.1	<.05	2
Nov. 93	7	209,700	2737	18	270	12	0	<.02	<.05	2
Dec. 93	7	>1E.06	4230	26	1238	20	0	<.1	<.05	3
Jan .94	7	>1E.06	1489	22	494	21	0			3
Feb. 94	7	>1E.06	1343	10	65	9	0	<.02	<.05	3
Mar. 94	7	>1E.06	2777	19	1864	18	0			4
Apr. 94	7	>1E.06	2190	20	277	16	0	<.08	<.05	4
May 94	7	>1E.06	1175	2	0	0	0	<.02	<.05	4
Oct. 94	7	>1E.06	225	2	60	1	0			7 (ST#1)
Nov. 94	7	>1E.06	5957	27	819	23	0	<.05	<.05	7 ST#1
Dec. 94	7	>1E.06	1752	7	18	4	0	<.05	<.09	7 ST#1
Feb. 95	7	>1E.06	3466	5	959	6	0	<.05	<.05	10
Mar. 95	7	>1E.06	5550	29	559	22	0	<.06	<.05	10
Apr. 95	7	>1E.06	6006	11	472	7	0			10&12
May 95	7	>1E.06	968	2	325	2	0	<.06	<.02	12
Jun. 95	7	>1E.06	12453	29	1759	24	0			12
Jul. 95	7	>1E.06	9737	22	1212	12	0	0.8	<.2	12&13
Aug. 95	7	>1E.06	9428	26	1137	17	0	<.1	<.05	13
Sept. 95	7	>1E.06	7118	22	1981	18	0			14
Oct. 95	7	>1E.06	3848	21	286	13	0	<.1	0.079	14
Nov. 95	7	>1E.06	10308	20	1572	12	0			14&15
Dec. 95	7	>1E.06	7766	31	949	24	0			-15
Jan. 96	7	>1E.06	6838	21	359	14	0	<.10	0.079	15&16
Feb. 96	7	>1E.06	11481	26	2297	18	0			16
Mar. 96	7	>1E.06	6090	19	542	14	0	<.10	0.079	16&17
Apr. 96	7	>1E.06	9238	23	941	13	0	<.08	<.05	17

Table 2. An example of a spreadsheet page for drilling muds and cuttings showing the general layout for a platform under an individual permit.

Twenty-three pages contain information in the form of mass-emissions of the components of produced water (Table 3). They reflect the total list of components required by EPA to be monitored by the OCS operators[3]. This information is in the form of the total amount of any produced water component discharged by a facility. It was arrived at by calculating the average concentration of an element that was monitored in a facility's produced water effluent for the period of time in question, usually annually, and multiplying by the volume of produced water discharged over the same interval. Each component of produced water required to be monitored, in either the General Permit or an Individual Permit, is contained in these series of pages.

Analysis

In addition to the simple, basic statistics that are performed on the pages containing the original, raw data, QuattroPro is capable of generating pictoral representations of the data via graphs of various types. For example, Figure 1 presents a graphic generated from a produced water page, which shows the total flow of produced water for the years 1995 and 1996 for all platforms. This can be done for any one or several components that are contained in the spreadsheet. An example of a graphic generated from a drilling mud page is given in Figure 2, which shows the total amount of drilling muds and cuttings discharges on a monthly basis. Mass emissions data show various relationships between any particular parameter (Figure 3), which shows cadmium mass emissions over time.

All of the data available can be exported from QuattroPro to a database such as Paradox. This would enable additional analysis to be conducted, such as more complex statistics (e.g., regression analysis, standard deviations, etc.). A database such as Paradox would also enable the user to compare in tabular format many other different combinations of time and parameter than would the more cumbersome spreadsheet methodology. At this writing, no use of a Paradox-type database has yet been attempted. These data were inputted during the summer and fall of 1996 and have not yet been quality checked. Thus, the spreadsheet data are presently considered to be in draft form. After a comprehensive check of the data, the formulas, and the graphics portion of the spreadsheet, the information will be used for various analyses.

3

An examination of the mass emissions pages by the reader might indicate an apparent inconsistency regarding numbers of entries on a per platform basis. Historically, not all platforms were required to monitor the same list of produced water components. This is due to the changes over time of the individual NPDES permits issued by EPA compared to the old General permit, which expired in 1984 and had been administratively extended. At this writing, a new General Permit is being developed that will bring all the offshore oil and gas platforms to the same standard of monitoring requirements andlimitations for discharges into Federal OCS waters.

Platform No. --->	1	2	3	4	5	6	7	8	9	10	11	12	13	14	15
Year															
1988	0				0.001		0	1.692							
1989	0	0	0	0	4.701		0	1.832			0				
1990	0	1.776	0	0	0.001		3.017	2.747			0.001		0.001		12.299
1991	0	0	0	0.009		0	1.026	2.568			0.001		0.148		9.989
1992	1.196	0.772	0	0.013		0	3.144	2.127					0.001		7.481
1993	0.451	0.212	0.032	0.156		0.316	28.270	0.624	0			0.320	0.001		0
1994	3.605	0.904	2.876	0.221		2.755	33.068	0	0.721	0		1.761	0.003	0.003	13.673
1995	5.704	0.107	0	0.197		0.227	11.011	0	0.917	0		2.470	0.001	0.003	12.087
1996	0	0.016	0.007	0		0.094	1.034	79.293	0	0.863		2.356			1.827
TOTAL	**10.756**	**3.787**	**2.915**	**0.596**	**4.703**	**3.392**	**80.590**	**90.873**	**1.638**	**0.863**	**0.002**	**6.907**	**0.156**	**0.006**	**57.356**

Table 3. An example of a mass emission spreadsheet page showing the emission of arsenic (in kilograms), by year on a per platform (numbers 1-15) basis.

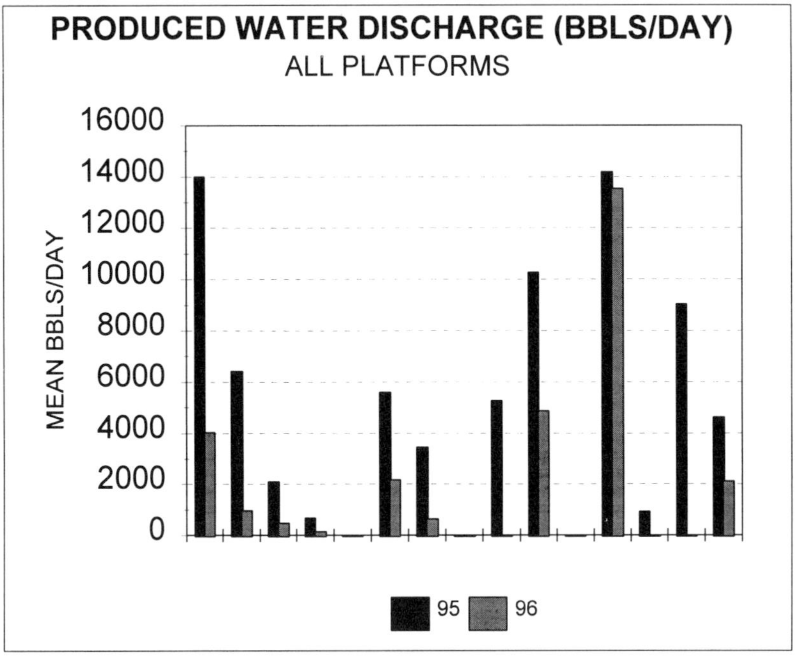

Figure 1. An example of a graphic generated using produced water data showing the total amount of produced water discharged for the years 1995 and 1996.

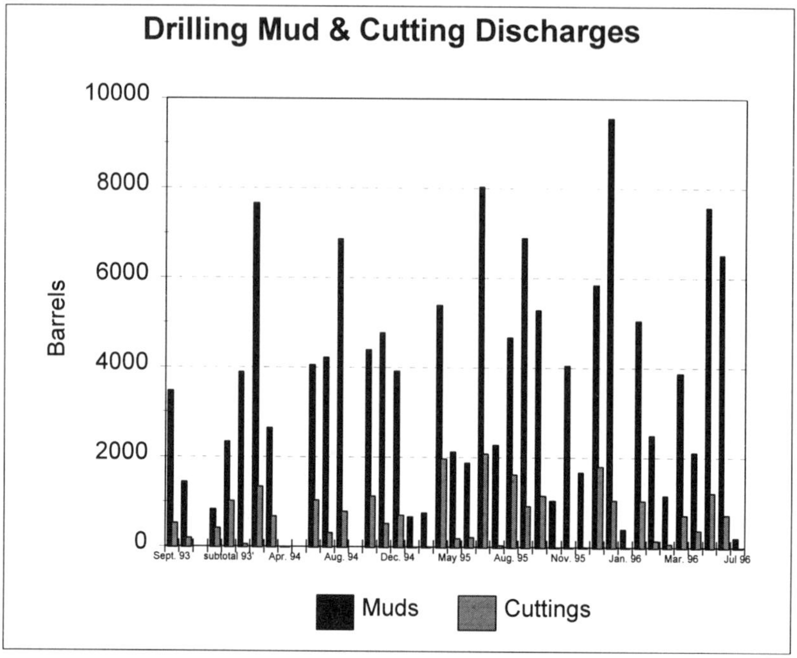

Figure 2. An example of a graphic generated from data contained within the spreadsheet showing drilling muds and cuttings discharged from all platforms from September 1993 to July 1996. Bars represent individual platforms.

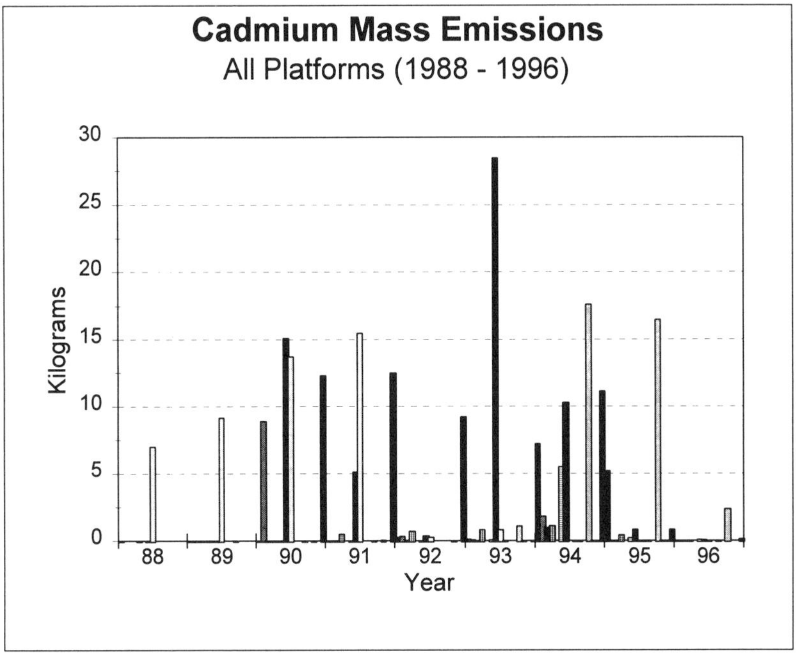

Figure 3. An example of a graphic generated from data contained within the spreadsheet showing cadmium mass emissions from all platforms from 1998 to 1996. Bars represent individual platforms.

Discussion

This information has several uses. First, the information in this spreadsheet will be useful for future NEPA analyses. Any Environmental Assessment or Environmental Impact Statement for a project that includes the exploration or development of offshore oil and gas resources must contain an analysis of the proposed project's impact on water quality. The information contained in this spreadsheet is from actual data collected by the operators according to their permit requirements. These real world data can allow an environmental impact analysis to be more specific, thus contributing to a more considered conclusion to be reached by Pacific Region decisionmakers.

Second, the Pacific Region has, in the past, been approached by members of the general public and local agencies with queries regarding the impacts that the effluents discharged by the offshore platforms might have on water quality and the organisms living in the waters of the Pacific Region. Some activitist groups have historically refered to these effluents as "toxic discharges". One purpose of this spreadsheet would be to allow a thorough examination of these NPDES data, showing that the monitored effluent components are, generally, well below the limits set by EPA and helping to alleviate some of these traditional concerns.

Third, specific trend analyses of the data may show problem areas. For example, the data might show that a particular facility is continually approaching the permit limitation for one or more parameters. Close consultation with the EPA may indicate the need for additional sampling, some adjustment of the operator's water treatment equipment or some other action that would prevent the operator from exceeding his permit limitation. The key here is that an analysis of historical data can show trends that might indicate circumstances under which an operator might be incurring problems.

Lastly, the State of California is one of the Federal government's customers. The information that is easily accessed through this spreadsheet could provide information to the State in its efforts to develop a comprehensive program to monitor the quality of ocean waters and coastal streams.

CHAPTER 67

A Coastal Monitor/Forecast System

Scott A. Jenkins[1],
and
Douglas L. Inman[2]

Abstract

The technology necessary for monitoring and forecasting civilian needs in the coastal environments of the world is now available and the design of a completely general, multipurpose monitor/forecast system for the coastal zone is a valid objective for the next decade. Proper management and maintenance of the waters of the coastal zone is essential to the health of the economy of coastal nations and states. Recreation, fishing, shipping, oil and mineral extraction, and military use each represent multibillion dollar industries to the United States, providing many thousands of jobs as well as attracting millions of visitors and billions of dollars in investments and expenditures. Protection of this vital asset will not only protect the existing industries, infrastructure and economy, but will also contribute to the establishment and growth of new industries, jobs, and to the revitalization of coastal economies.

The continual growth in world economies and population will place additional stresses on the coastal environment. High production farming and urban runoff from rapidly growing cities with aging infrastructure are releasing large amounts of particle effluents into the nearshore that combine with ocean disposal of treated sewage. These practices add further stress on the nearshore oceans as population increases. Recently, the National Research Council has identified the following issues (in approximate order of descending importance) as posing significant threats to the integrity of the State of California and the U. S. coastal ecosystems: (1) eutrophication; (2) habitat modification; (3) hydrologic and hydrodynamic disruption; (4) exploitation of resources; (5) toxic effects; (6) introduction of non-indigenous species; (7) global climate change

---

[1] Senior Engineer, and [2] Research Professor of Oceanography, both at Center for Coastal Studies, Scripps Institution of Oceanography, La Jolla, CA 92093-0209

and variability; (8) shoreline erosion and hazardous storms; and, (9) pathogens and toxins affecting human health. Problems associated with changes in the quantity and quality of inputs to coastal environments from runoff and atmospheric deposition are particularly important. A conceptual system architecture is outlined to address the need for continual monitoring and forecasting of these problems based upon existing sensor and modeling capabilities.

Introduction

The technology necessary for monitoring and forecasting civilian needs in the coastal environments of the world is nearly identical to that needed for military intelligence, amphibious operations, and mine detection. This report addresses an area where technology transfer will have the greatest combined civilian and military impact. An initial conceptual system is proposed here to address a dual-use set of monitoring and forecast goals. Because of the striking identity between civilian and military needs in the coastal zone, it has been argued that the most comprehensive design would result from solving the operational problem of the Special Warfare Groups (SEALS) and Amphibious Operations (AMPHIB).

The Coastal Warfare Operational Problem

The conduct of amphibious operations requires accurate, comprehensive means of determining the suitability of landing beaches and adjacent back-beach areas. Essential elements of intelligence include: 1) predicted surf and nearshore currents, 2) trafficability, 3) beach gradients, 4) bottom and back-beach composition and ground cover, 5) location of sandbars and other obstacles (natural and man-made). The required information is now obtained by initial study of charts and overhead photography and, for coverage of the vital area inward from the 3½ fathom line, by reconnaissance by SEAL Team swimmers. These methods, developed in the Pacific campaigns of WWII, require overt naval gun support and are inadequate in today's operational environment. Swimmer reconnaissance, inherently slow, limits the extent of coastline coverage in the preinvasion period. The resulting survey is both incomplete (obstacles in the intervals between swimmers are often missed) and static; it shows the state of the beach at that particular time, not the beach as it will be on some future day, altered by changing winds, waves, tides, and storms, and the longer term effects of sand movement, bar formation, and erosion. The feasibility of swimmer reconnaissance is further constrained by operational circumstances: in peacetime, by political or diplomatic prohibitions of such intrusion in other nation's territorial waters, and during conflicts by the unacceptable operational security, (OPSEC), and vulnerability, inevitably caused by preinvasion activity near candidate beaches.

Although the SPECWAR and AMPHIB operational problem focuses on five *inner variables* very near shore, they cannot be resolved without information about the mid- and outer shelf environments as well. Here, five additional *outer variables* must be resolved: 6) shelf waves, 7) visibility distance, 8) temperature profiles, 9) shelf currents, and 10) shelf bathymetry. The monitoring and forecasting of outer variables is also applicable to the operational problem requirements of Mine Countermeasures (MCM) and shallow water Anti-Submarine Warfare (ASW).

Subsets of the various SPECWAR, AMPHIB, MCM, and shallow water ASW operational requirements have applications to a variety of dual use needs in environmental monitoring and permitting. These include beach erosion, pollutant source detection, pollutant spill tracking and forecasting, littoral ecology assessment for a variety of EIR/EIS application, dredge disposal permitting and monitoring, salvage, news media diagnostics, graphics and animation, CD-ROM interactive teaching aids, et al.

Technical Solution

The operational problem presented above could be solved with a Coastal Monitor/Forecast System which has seven basic elements (Figure 1): 1) Sensor selector; 2) Overhead (remote) sensors; 3) In-the-Water sensors; 4) Data link for clandestine export of data to central processing; 5) Data analysis enhancement and fusion; 6) Nowcast/Forecast modeling; 7) Displays and decision aids.

The nearshore survey procedure for the nowcast of the five inner variables would be based largely on remote collection (primarily from satellites, but also manned and unmanned aircraft) of data by a variety of sensors, including imagery in both the visible and radar spectra. Also, tactical needs may require broad back-beach intelligence to accommodate initial penetration by helicopters, etc. Coastal geology would be applied to select the mix of sensors most appropriate for data collection and interpretation in a given coastal area. Multispectral scanning and image processing would be used to determine

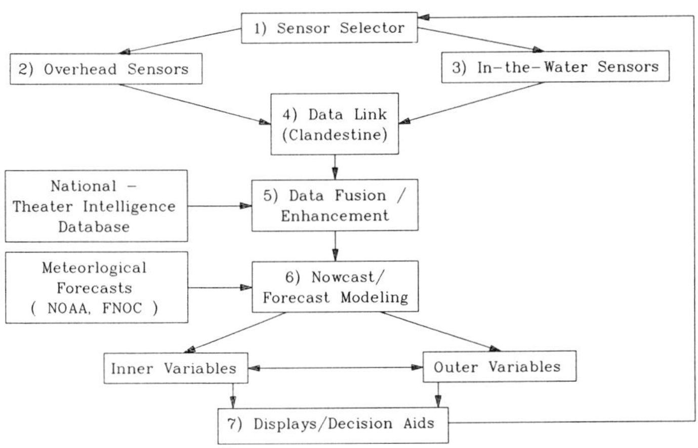

Figure 1. Flow chart for the coastal monitor/forecast system.

beach trafficability, to see the bottom, and to measure depths in shallow water. Synthetic aperture radar data would be processed for surface currents and wave patterns, analyzing in both the time and frequency domains to reveal bottom gradients and underwater obstacles. Direct detection LIDAR can provide a detailed mapping of the bottom topography. Integrating the results from each analysis method, the bathymetry of the entire nearshore area could be defined. Methods of remote sensing and analysis would be matched with differing sets of environmental conditions. Large amounts of data could be processed and integrated quickly by using an adaptive data fusion and assimilation system. The data links and data fusion components and communications can be coupled compatibly in the national-theater operational intelligence architecture. The intention is not to create a separate, stand-along coastal intelligence system, but instead to have the procedure draw data and processing capability from existing intelligence centers and architectures.

While overhead sensors give coverage of the large coastal areas involved with the computational domain of the forecast models, in-the-water sensors will provide control points for calibration of those models. The in-the-water sensors will also provide high resolution for certain subsurface observations for which the overhead sensors are not well suited, i.e., bottom composition, temperature profiles, waves and currents, and micro-bathymetry. While a total segregation of sensors among the inner and outer variables is neither possible nor desirable, the overhead sensors will likely provide the bulk of the measurements of the inner variables, while the in-the-water sensors will provide most of the direct measurements of the outer variables. The precise mix of sensor types and their domain of application will be determined by the Sensor Selector (element 1, Figure 1), according to the geomorphology of a particular coastal environment and the type of decision aids that are ultimately required. The Sensor Selector is envisioned as a front-end software-based decision aid, that has an in-stride adaptive menu structure which fully accounts for sensor compatibilities, dynamic range, resolution limits, sampling rates, and special calibration sequences and coastal geomorphology constraints. The purpose of the Sensor Selector is to allow the user to rapidly configure a sensor suite and to prevent the misapplication of sensors to particular coastal environments.

Because of a potential broad range in the mix of Overhead Sensors (2) vs. In-the-Water Sensors (3) as configured with the aid of the Sensor Selector (1), the Data Link (4) must be flexible and adaptive while remaining clandestine. For the non-DOD dual-use application, the clandestine criteria can be relaxed, with considerable cost-saving. The Data Link (4) will export the sensor data to a mobile command and control center which will house the remaining elements of the Coastal Monitor/Forecast System (Figure 1). The "junction box" for the data inflow is the Data Fusion/Enhancement element (5). It will sift, compress and enhance the incoming data streams to a manageable level containing only the requisite inputs in the proper spatial and temporal scales to produce the desired decision aids and graphics output from the Nowcast/Forecast models. The Data Fusion Enhancement element will also draw on the National Theater Intelligence databases to both augment the data stream and to eliminate time-consuming duplication of effort.

The throughput data stream from the Data Fusion/Enhancement element (5) will be used to initialize the Nowcast/Forecast models (6). These models consist of two generic types, inner variable and outer variable models. These two classes of models have inherently different spatial and temporal scales, and often very different gridding systems. The models Nowcasting/Forecasting inner variables typically use fine scale curvilinear grids with short time steps while the outer variable models tend to use longer, more averaged time steps with larger scale rectangular grids. The two classes of models are merged with the inner variable models nested inside the outer variable models, exchanging mass, momentum and energy across their overlap domains. Figure 2 gives an example of a data product which can be obtained from these nested grid models.

Figure 2 shows a comparison of the modeled suspended sediment vs. that which was actually observed in a LANDSAT RGB image of river plumes from the Santa Margarita and San Luis Rey Rivers at Oceanside Harbor, California at 10:00 a.m. PST for 23 January 1993. The modeled suspended particulate number concentration (#'s/ml) shown on the inner grid in this figure was driven by forcing functions specified by output from outer grid models which computed wave refraction/diffraction over the shelf, island sheltering effects, and river hydrology. The outer grid wave models encompassed the entire Southern California Bight and Channel Island arcs, while the river hydrology outer grid models included the system of drainage basins for the Santa Margarita and San Luis Rey Rivers which extended inland to Hot Springs Mountain, Mt. San Jacinto and the entire Elsinore Valley of southern California. These extended-domain outer grid models were necessary to accurately define the wave forcing functions and flood hydrographs which dominate the sediment loading within the inner grid shown in this figure. The inner grid simulation shows considerable replication of the details of the wave resuspension on the inner shelf, and the plume dispersion from these two rivers near Oceanside Harbor, CA. Also overlaid are the secondary and tertiary drainage basins of the minor streams along the coastal terrace. The discharge from these minor streams does not penetrate seaward of the surfzone, but merely enriches the local surfzone suspension. These simulated sediment dispersion features were hindcasted using only rainfall and deep water wave data derived from meteorological inputs. Given present forecasting accuracy of critical meteorological inputs, coastal forecasts out to 84 hours (3½ days) appear feasible. Longer forecasts for non-meteorological coupled variables, i.e., trafficability, bottom composition, also seem quite feasible.

Nowcast/Forecast modeled outputs of inner and outer variables shall then be assimilated, plotted, annotated or animated by the Display Decision Aids (7) element in Figure 1. This element shall contain a variety of graphics capabilities designed to provide rapid in-stride assessment of the present and near-future coastal environment. This element has a feedback loop to the Sensor Selector (element 1) to indicate the dependence of the mix of sensors upon the desired kinds of Nowcasts/Forecasts. This loop provides the user with

Comparison
Suspended Particulate - LANDSAT RGB Image
Modeled Concentration Contours (Log 10)

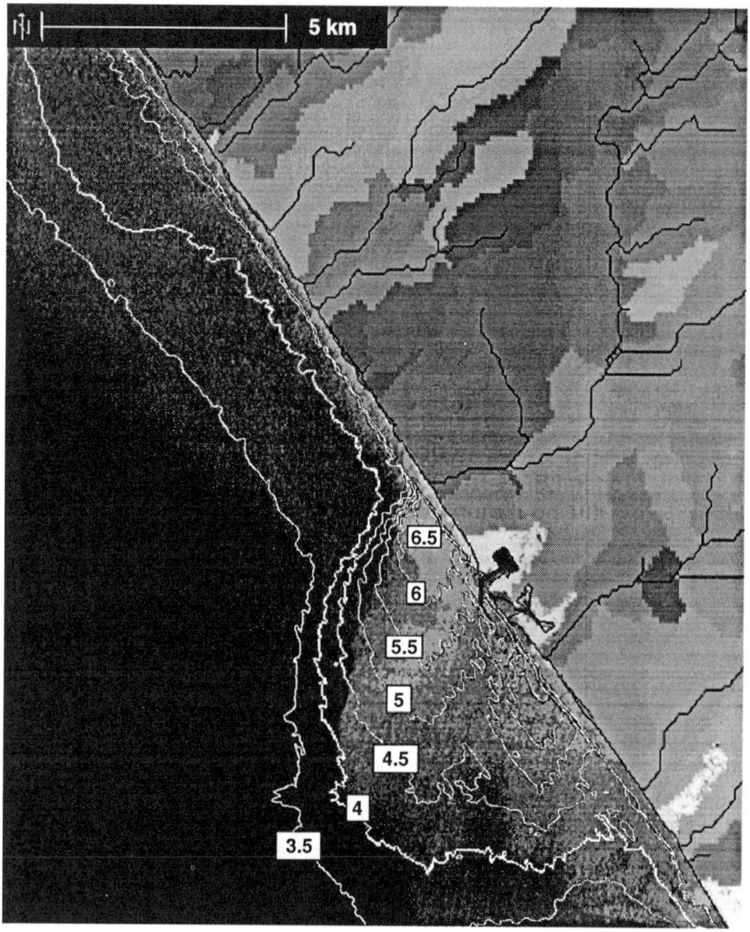

Figure 2

the system control that will optimize the display in accordance with the decision needs.

Sensor-Model Integration

Historic and Geologic Databases

Knowledge (state of the science) in coastal processes and coastal geology has improved markedly since WWII, spurred by environmental concern about increased use of fragile coastal areas and drastic erosion of beaches. Coastal processes, science, and geology can be applied initially to define the methodology (data collection including sensor mix and processing) most appropriate to a given coastal area. It can also be used as a cateloging tool to arrange data sets for most effective use by intelligence and operations staffs. At any particular beach, the database analysis can productively begin in the easily imaged area above the waterline, because beach features above the waterline given geotechnical clues about the frequency and severity of storm wave action, the range of extreme tides, the direction of longshore currents, the prevelence of rip channels and bars, and the structure of the nearshore circulation system. These details are usually absent from most digital databases derived from offshore soundings, yet they are vital details for the initialization of the nearshore hydrodynamic forcing and transport modeling portions of the system. These historic and geologic databases are also a primary criteria for the Sensor Selector in discriminating among the most appropriate mix of sensors.

Overhead Sensor-to-Model Integration

There are several above water sensor subsystems with which the data fusion and modeling integration efforts should be linked (see Figure 3). These sensor subsystems include:

1. MSS/HSS (Multi-Spectral Scanner/Hyper-Spectral Scanner)
2. Visible band
3. X, C, and L-Band SAR (X-Band Synthetic Aperture Radar)
4. LIDAR (Laser Imaging Doppler)

MSS/HSS. Multi- and Hyper-Spectral scanner data can be exploited to obtain estimates of beach trafficability. Multispectral sensing data from satellites or aircraft have proven capable, within the last 15 years, of delineating outcrop lithologies for mineral exploration. It has recently been demonstrated that these methods can be used to give estimates of mineralogy, grain size, and moisture content of beach sands. Once the multispectral scanner gives us estimated grain size and moisture content, it is possible to calculate the rated cone index. This index gives an indication of the particular types of vehicles that can traffic the beach. This procedure provides far better aerial coverage above the waterline than the necessarily cursory and error prone sampling of beach trafficability performed as an adjunct to nearshore swimmer reconnaissance. The MSS/HSS derived maps of rated cone index can

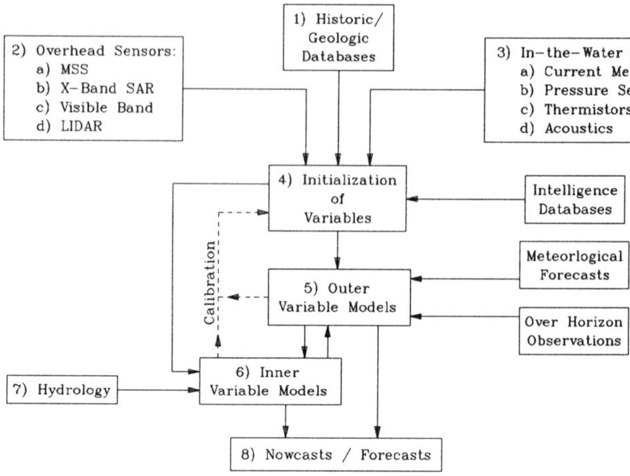

Figure 3. Sensor-to-model itegration scheme.

quantified for any mix of vehicles and throughput tonnage traveling above the beach waterline. Algorithms using band ratio techniques to calculate rated cone index have already been developed but these must be upgraded to fully utilize the capabilities of the MSS/HSS subsystems. Furthermore, a trafficability forecast model must be configured using these upgraded cone index algorithms, integrating them with the predicted beach profile changes forecasted by the hydrodynamics/transport models embedded in the undersea subsystems portion of the program.

MSS/HSS image processing can greatly increase the depth at which the bottom can be seen. Successive images of the same plot of ocean (and, therefore, the ocean bottom) are needed. Through processing, it is possible to reduce the large influence of surface reflection while enhancing the much smaller return off the ocean bottom. In this way, the relative visibility of the bottom, as compared to that of the surface, is increased.

The total signal that is returned from the ocean is made up of a time and spatially dependent surface component plus a spatially dependent bottom component. When this signal is integrated over time and Fourier-Series decomposed, we are left with only two components, the return from the bottom and the "DC" component of the surface. The surface component may then be subtracted-out leaving only the bottom return.

Such image processing does not give depth of water. It can, however, be adapted to any two frequency bands from a multispectral scanner to enhance bottom slope resolution. This adaptation relies on the assumption that a pair of wavelength bands can be found in the multispectral data output such that the ratio of the bottom reflectance in these two bands is the same for all bottom types within a given nearshore region. The water depth can then be calculated using two selected bands. Beach gradient variation follows directly.

Visible Band Images. Procedures for determining nearshore bathymetry from visible band images of shoaling wave patterns were developed as long ago as WWII by Sverdrup, Munk and others. The early calculations were based on a dispersion relation which is strictly valid only for flat bottoms, but they are reasonably accurate over sloping bottoms, provided that the bottom slope does not change appreciably over one wavelength. Recent developments in the parabolic mild slope equations for shallow water waves have greatly improved the accuracy of these calculations. When combined with high resolution overhead imagery of wave shoaling patterns, inverse solutions for the depth from the parabolic mild slope equations could greatly improve resolution of fine-scale transient nearshore bathymetry features such as bars and channels.

SAR Images. Synthetic Aperature Radar (SAR) images of wave refraction and diffraction patterns around obstacles such as reefs, bars, and rock outcroppings, and the new interferometric SAR which maps surface velocity components directly, are valuable tools for verifying and comparing estimated bathymetric variations and currents. The modulation of the surface wave roughness by water flowing over shallow bottom areas and associated wave refraction/diffraction patterns have been clearly demonstrated to depths of tens of meters, depending on current velocity.

SAR has also demonstrated its ability to locate anti-invasion obstacles in the nearshore area. In the 1985 Tactical Development and Evaluation (TACD&E) project, a brief test employing a Goodyear APD-10 SAR aboard a U. S. Marine Corps reconnaissance aircraft demonstrated the feasibility of locating such obstacles. The X-band SAR images can be processed to resolve waves having radial propagations, particularly the scattered waves from submerged objects in the nearshore. This domain is often too shallow and inaccessible in high-energy environments for undersea sensor systems. The SAR is also valuable in identifying boundaries of oily plumes, fronts and particulate-laden water masses which must be delineated and quantified for the initialization of the forecast models. Initialization algorithms for the model must be tailored to suite the particular SAR system capabilities in both temporal and spatial domains.

LIDAR. LIDAR systems are the most direct method for determining water depth and bottom profile. LIDAR can give the complete bathymetry of the nearshore as was shown by recent work completed by the Army Corps of Engineers. A LIDAR capable of imaging, by use of either an imaging receiver or by a scanning spot, can provide high resolution (< 1 meter) in both depth and horizontal size. In addition, these systems possess very favorable interrogation rates (> 10 square nautical miles per hour). LIDAR possesses the significant advantage of being able to operate at any time, 24 hours a day, under a variety of environmental conditions. LIDAR resolution is a factor which the inner variable models should exploit for reducing reliance on the in-the-water sensors and their inherently more vulnerable data links.

Application to Inner Variables. The desired accuracy and the particular application of the overhead vs. in-the-water sensors is another system integration criteria for the Sensor Selector. For example, a DOD/SPECWAR application for remote beach reconnaissance could expect the following relative performance for initialization of inner variables from the possible suite of overhead sensors:

	Depth Measurement	Obstacle Location	Beach Slope
Single Band Still Photography		2, 3	
Waterline Observation	1		1, 3
LIDAR Imager	1, 3	1, 3	1, 3
Single Band Image Integration		2, 3	
Multispectral Reflectance Ratios	1, 3		1, 3
Photographic Shoaling Transforms	1		1
Film Density Correlation	1	2, 3	
SAR Imaging	2, 3		

1 = Method will give quantitative measurements; 2 = Method will give qualitative information only; 3 = Method will reveal changes since last observation

The degree to which these overhead sensor data will resolve inner variables will dictate the numbers and deployment of in-the-water sensors necessary to refine these estimates.

Wet Sensor-to-Model Integration

If overhead sensors cannot adequately define initial values of inner variables, then in-the-water sensors must be used in conjunction with models to extrapolate these values from offshore in situ measurements. This procedure is close to conventional coastal processes methods, but the clandestine requirement is new. Fixed arrays and strings of pressure sensors, current meters and thermistor, will provide directional wave and current data and temperature profile data on the mid-to-inner shelf for initialization of the outer variable models (Figure 1, 3). The wet sensors will also provide ground truth at certain control points in the overhead images, for near real-time calibration of the data sets derived from overhead sensors. Additional water column and bottom details which cannot be ascertained by overhead sensors, e.g., acoustic properties, microbathymetry and sub-bottom profiles, may be collected by certain mobile platforms.

Initialization of Variables

This is one of the data fusion functions shown in Figure 3. Here, initialization data is reduced to calibrated values and enhanced by algorithms such as band ratio and curvilinear coordinate transformations, and subsequently formatted in structured files which are compatible with model requirements.

The real time estimates from overhead and in-the-water sensors together with stationary inputs from the historic databases will initialize the models for nowcast runs. Calibration nowcasts will be checked in-stride for fidelity with point-wise measurements at control points established by the in-the-water sensors. Interative loops of the models (see Figure 3) will adjust model calibration factors until nowcast calibrations have minimized variances with control point measurements. At this point, initialization of the models is complete and the models can then accept forecast values or over-the-horizon observations.

Over-The-Horizon Observations

Over the horizon ship observations of sea state and weather fronts offer a powerful means of predicting nearshore conditions. Forecasted sea state data and coastal winds are notoriously inaccurate. Over the horizon observations can improve upon this accuracy, although the range of the forecasts is limited by the offshore positioning of the observation platforms, usually Task Force assets. The approach requires accurate nearshore bathymetry (as provided by the remote methods described above), digitized and stored in a wave refraction program. Deep water wave parameters from over the horizon ship observations are shoaled numerically over the nearshore bathymetry to predict breaker height and angle, littoral currents, rip current spacing and strength of the candidate beach.

Outer Variable Models

These models are built to predict the outer variables on grid cell sizes 0 (100 m). The Coastal Water Clarity Model (CWC) developed under the sponsorship of the Office of Naval Research (ONR) is one such example that could be adapted to the system architecture and sensor capabilities represented in Figures 1 and 3. Such models need to have data assimilation capabilities such that they may be initialized at arbitrary times and grid locations by data from real time updates such as over-the-horizon observations. Outer variable models compute wave height, direction and periods, current vector components, tidal elevation, stratification and mixed layer depths as well as the full complement of optical properties including visibility distance.

Inner Variable Models

These models are nested inshore and inside the computational domain of the outer variable models. They will exchange momentum, mass, and energy with the outer domain models in their mutual overlap domain. They will compute the forecasts for inner variables in fine temporal and spatial scale resolution based upon forcing from the outer variable models and from the land-sea interface (see Figure 3). Land-sea interface forcing is primarily due to river hydrology inputs from either observation or forecasted flood hydrographs. The inner variable models give breaker heights, periods and directions, longshore currents, rip currents, beach profile evolution, trafficability, and position and dimensions of the breakpoint bar.

Conclusions

A conceptual plan for a dual-use coastal monitor/forecast system has been posed based upon available sensor and modeling capabilities. The primary challenge in implementing such a system is creating and providing funding for the team of industrial, military, and university partners who presently control the specialized sensor, data processing, and modeling skills. The Technology Reinvestment Program at one time had the potential for sponsoring such an effort, but has since been rendered disfunctional by budgetary cuts and a lack of political resolve. Nontheless, the concept of a coastal/monitor forecast system will age well, because the dual-use needs for such a system are likely to remain for the indefinite future: a future likely to be dominated by low intensity conflicts over access to raw materials, and retarded economic growth due to the population explosion and environmental degradation. In such a future a coastal monitor/forecase system will become a critical element in tracking global vital signs.

Acknowledgements

This study was supported by the Office of Naval Research, Ocean Modeling and Prediction, under Contract N00014-95-1-0005.

CHAPTER 68

The Marine Animal Database: Design, Methods and Use

Janet Clarke[1] and Sue Moore[1]

Environmental planners, representing oil companies, commerical fisheries and the U.S. military, must contend with the potential deleterious impact of offshore operations on marine animals. The National Environmental Policy Act (NEPA) requires preparation of environmental compliance documentation (i.e., Environmental Assessments or Environmental Impact Statements) for operations that may harm, or be perceived to harm, marine animals protected by the Marine Mammal Protection Act (MMPA) or Endangered Species Act (ESA). Unfortunately, there exists no standardized information base for preparation of these documents, nor can environmental planners easily access current data on marine animal populations for their geographic region of interest. Inaccurate and inadequate environmental documentation has led to litigation and costly program delays. A means to collate and disseminate up-to-date information on marine animal occurrence, migration, feeding/breeding habitats and bioacoustics is being developed to support the timely preparation of reliable and accurate environmental documentation. The best available data on marine animal populations, including marine mammals, sea turtles and seabirds, will be collated into a related database termed MAD (Marine Animal Database). The database will be developed on a PC-based workstation using off-the-shelf user-friendly software that can be tailored to the requirements of a variety of offshore operations. Sources of data will include published and unpublished literature, raw sighting data, and complementary databases developed by various agencies (e.g. NMFS, MMS, USFWS, state and local). Related databases will be organized by geographic region, season, marine sanctuaries and critical habitats, species-specific natural history, and audiometric/bioacoustic data. Once marine animal data are compiled, GIS software and data queries specifiying geographic area or time period will be used to derive various maps and summary tables. The end result will be a consolidated MAD that will enhance and streamline the environmental planning and environmental documentation process.

[1] SAIC, Maritime Services Division, 3990 Old Town Ave., #105A, San Diego, CA 92110

CHAPTER 69

Recycling Oil Industry Seismic Data for Geologic Hazards Analysis

Drew Mayerson[1]

During the 1970's and '80's, the oil industry collected hundreds of thousands of miles of marine geophysical data for use in the evaluation of offshore oil and gas deposits. Much of this data have lost their value for which they were originally collected because lease sales for oil and gas off the coast of California have been discontinued; with no plans for their resumption at least through 2002. The Minerals Management Service (MMS) and the academic community realize the value of these data for uses other than oil and gas exploration. Though these data were collected so that offshore structures favorable for the entrapment of oil and gas could be identified, their utility also extends to geologic hazard analysis, namely the identification of areas of high earthquake potential. The MMS has recently begun two separate endeavors that will aid in quantifying earthquake potential along the central and southern California coast.

The first undertaking analyzes proprietary industry collected geophysical data along the Hosgri Fault zone in south central offshore California. With academic collaboration from the University of California at Santa Barbara Institute for Crustal Studies (ICS) through the MMS-UC Southern California Educational Initiative, the MMS is helping fund a study entitled, "Structural Evolution of the Hosgri-Purisima-Lompoc Fault System, Santa Maria Basin, California, from 3-D Map Restoration: Economic and Earthquake Hazard Implications." The principle researchers for this study are Dr. Christopher Sorlien and Dr. Mark Kamerling of ICS. Expertise from geoscientists in the MMS is also being used to complete this project. The study is the most comprehensive study of the fault system to date including all public data, and for the first time, large quantities of industry collected proprietary data from MMS' data vault. When the results of the study are released, later this year, the public will have a document that will help resolve an ongoing debate within the geologic community; the movement and earthquake potential of this fault system and its implications for man-made structures in the vicinity of the fault system.

A second undertaking involves the acquisition of ownership rights to a proprietary marine geophysical data grid along the Newport-Inglewood Fault Zone in southern California. These data were collected in 1988 and are some of the best regional data collected in the offshore. MMS acquired full ownership rights to these data in September, 1996. The MMS plans to release these data to the public by April, 1997 via CD-ROM and the World Wide Web. Academia and the research community have expressed great interest in these data for their applicability to earthquake studies for the Los Angeles basin. By making these data available to the public in such an open format, MMS anticipates they will be utilized by professionals ranging from the National Earthquake Hazards Reduction Program seismologists to serving as the foundation for college theses, thereby bringing new ideas regarding earthquake potential into the debate over why, when, where, and how large future earthquakes might be in the southland.

To date, much of the earthquake hazards research in southern California has been limited to the analysis of land based data sets. The incorporation of larger volumes of offshore geophysical data into the earthquake hazards analysis puzzle is a welcome addition for researchers and planners alike.

---

[1] Geologist, U.S. Dept. of the Interior, Minerals Management Service, Camarillo, CA

CHAPTER 70

COASTAL EROSION: THE STATE OF THE PROBLEM AND THE PROBLEM OF THE STATE

*Benjamin T. Benumof *, Gary B. Griggs, Laura J. Moore*
Department of Earth Sciences and Institute of Marine Sciences
University of California, Santa Cruz

ABSTRACT

The coast of California is experiencing a conflict of increasing magnitude between the natural hazards which affect the coastline and the increased desire of the 32.4 million residents, 80% of which live within 50 km of the coast, to live virtually at the water's edge. Over 1500 km of the state's 1760 km of shoreline is actively eroding, yet development often continues in areas subject to coastal hazards such as shoreline erosion and coastal flooding. Careful investigation of the recent geologic history of oceanfront areas is relatively straightforward and is necessary to evaluate the hazards and risks present at a specific locality prior to any changes in land use, zoning, development/redevelopment, or permits.

In 1994, the National Flood Insurance Act (NFIRA) was enacted into law. Section 577 of the Act, *Evaluation of Erosion Hazards*, mandates that a state-by-state study of representative areas be conducted which focuses on developing 60-year shoreline hazard areas and comparing these to other FEMA coastal flood lines. The objective is to determine how projected or potential economic losses from shoreline erosion will impact the resources of the National Flood Insurance Program (NFIP). In California, these studies, funded by FEMA, are concentrated in San Diego and Santa Cruz counties and carried out at the University of California, Santa Cruz.

Although the purpose of these shoreline erosion studies is to determine long-term erosion rates in order to identify major problem areas and minimize potential 60-year economic losses, the changing federal position with respect to funding for shoreline protection will substantially change the economic feasibility of shoreline armoring projects. Currently, coastal protection in areas of active shoreline erosion is becoming a state issue with responsibility shifting away from federal investment in the shoreline.

State-level policies regarding cliff or bluff-top development are vague and an analysis of local governments policies reveals dramatic variations in both regulations and approaches to those regulations governing the siting of development and redevelopment in coastal bluff areas. "Long-term" coastal recession rates are usually the key factor in evaluating and conditioning oceanfront development projects and permits and also, in the case of FEMA, the key factor in evaluating 60 year potential shoreline losses. Those who rely on an "average annual erosion rate" in their decision making need to understand the uncertainties and limitations of these calculations. The representativeness of the time period of aerial photos or maps analyzed, the scale and clarity of the photos or maps, the methods or technology employed and skill of the investigator, and the effects of shoreline stabilization efforts, whether beach nourishment or coastal armoring, all significantly affect these erosion rates and, therefore, the decisions made based on these rates.

---

* Benjamin T. Benumof, Graduate Researcher, University of California at Santa Cruz, Earth Science Department, Santa Cruz, California, 95064

INTRODUCTION

The coast of California is experiencing a conflict of increasing magnitude between the natural hazards which affect the coastline and the increased desire of the 32.4 million residents, 80% of which live within 50 km of the coast, to live virtually at the water's edge. Over 1500 km of the state's 1760 km of shoreline is actively eroding, yet development/ redevelopment often continues in areas subject to coastal hazards such as shoreline erosion and coastal flooding.

In many local coastal jurisdictions, shoreline policies and land-use practices fail to correctly address the regionality and time-frame within which natural processes operate. For the most part, coastal natural processes such as littoral sand transport and sediment delivery to the coast by rivers tend to operate on regional, longer-term scales, whereas political practices and policies tend to reflect local concerns and shorter-term decisions. Furthermore, there exists a distinct lack of cohesiveness (in approaches to coastal management) between neighboring local coastal jurisdictions (often within the same natural regional systems) and between man and nature in general. As a result, the field of coastal management has become the battlefield among nature and human action, as coastal natural hazard issues evolve in both magnitude and complexity.

Figure 1. Coastal Erosion in Santa Cruz County, California.
Grand Avenue in the City of Capitola.
Photo taken by Strnad, L., 1980

Although the physical and economic impacts of large magnitude storms are the most widely publicized, the year-to-year public and private costs of intensive shoreline development and coastline modification are equally significant (Griggs, 1994). In many localities, coastal hazards issues have become increasingly magnified because pre-development impact analyses have been (and continue to be) far too narrow in scope and have consequently resulted in extreme cumulative impacts due to inaccurate assessments of true costs and benefits. Careful investigation of the recent geologic history of oceanfront areas and their regional setting are relatively straightforward and are necessary to evaluate the hazards and risks present at a specific locality prior to any changes in land-use, zoning, development/redevelopment, or permits.

Although coastal erosion is relatively easy to document qualitatively, and many photographic records of this process exist (figure 1), a need exists for quantitative documentation of the rates of cliff retreat (Griggs, 1994). Long-term erosion rates are generally the key factor in evaluating and conditioning oceanfront development and are necessary so that decisions based on the significance of erosion are sound. Over the past 25 years, numerous techniques have been developed to determine erosion rates so that the potential extent of future erosion may be assessed. The most common coastal mapping techniques employ the use of stereo aerial photography and range in complexity and accuracy from manual methods such as the photographic comparator and the stereo zoom transfer scope to state-of-the-art digital methods such as soft-copy photogrammetry.

FEMA CALIFORNIA *"EVALUATION OF EROSION HAZARDS"*

Beginning with the US Army Corps of Engineers National Shoreline Study (1971), there have been a number of efforts including Habel and Armstrong (1978) and Griggs and Savoy (1985) to try and quantify coastal hazards problems by mapping erosion hazard areas. These studies have had limited success in the transfer and use of information, regarding the impacts of erosion on coastal land-use planning, due to the lack of quantitative shoreline erosion measurements. Recently, the Federal Emergency Management Agency (FEMA) has initiated yet another erosion hazards study, however technological innovations have allowed for substantial improvement in the determination of erosion rates.

In 1994, the National Flood Insurance Act (NFIRA) was enacted into law. Section 577 of the Act, *Evaluation of Erosion Hazards*, mandates that a state-by-state study of representative coastal areas be conducted which focuses on developing 60-year shoreline hazard areas and comparing these to other coastal flood lines on FEMA Flood Insurance Rate Maps (FIRM'S). The objective of the erosion hazards study is to determine how projected or potential economic losses from shoreline erosion will impact the resources of the National Flood Insurance Program (NFIP) and the local coastal communities involved.

The Evaluation of Erosion Hazards study is to be conducted in three separate phases. The first phase of the study consists of (a) mapping 60-year erosion hazard areas using recent and historical aerial photography and (b) plotting current and 60-year projected flood zone boundaries onto base imagery (orthophotos). The first task is required in order to determine the number of structures located within the 60-year erosion hazard areas and the second task is required to determine how the landward migration of flood-zones might impact the NFIP and coastal communities if erosion continues. In California, the first phase, funded by FEMA, has been concentrated along the entire San Diego and Santa Cruz County shorelines and carried out at the University of California,

Santa Cruz (UCSC). For the purposes of this study, researchers at the UCSC Institute of Marine Sciences and Earth Science Department recently installed a "Coastal Imaging Lab" to employ the newest, state-of-the-art technique for measuring shoreline change (softcopy photogrammetry). A discussion of various photogrammetric techniques, including softcopy photogrammetry, is given by Moore (In press).

At the time the California *Evaluation of Erosion Hazards* study was initiated, the UCSC Coastal Imaging Lab was one of only three labs in the nation employing state-of-the-art softcopy photogrammetry technique for the purpose of shoreline mapping. The San Diego and Santa Cruz County studies are the first of their kind in California to employ such sophisticated, accurate mapping methods over such large portions of coastline. The UCSC Coastal Imaging Lab is equipped with a Sun Sparc20 workstation, Agfa Horizon Plus scanner, Hewlett Packard Design Jet 755 CM plotter, StereoGraphics Crystal Eyes, Pinnacle Micro Vertex 2.6 GB external drive, ESRI Arc-Info GIS software, Erdas Imagine software, and Trimble 4000 GPS receivers/GP Survey post-processing software. When combined, the equipment is designed to essentially eliminate shoreline mapping errors arising from distortion of imagery, ground control data, etc.

The objective of phases two and three, which will not be carried out at UCSC, are to inventory structures located within the mapped erosion hazard areas and conduct an economic analysis of erosion and erosion mapping on coastal communities and on the NFIP, respectively. The expected completion date of phase one of the California *Evaluation of Erosion Hazards* study is July, 1997.

In particular, San Diego County (figure 2) is a geologically interesting study area for a number reasons. First, the San Diego County shoreline is approximately 76.1 miles in length with lithologic and structural properties varying alongshore. Consequently, results of the *Evaluation of Erosion Hazards Study* (high resolution erosion rates) will provide valuable information regarding the long-term recession of the coast in differing rock types. Second, the San Diego County shoreline consists of multiple natural regional systems including the Oceanside, Mission Bay, and Silver Strand littoral cells. Third, large portions of the San Diego County shoreline are unarmored. As a result, this allows for accurate assessment of historic natural coastal erosion rates for many of San Diego's coastal areas.

Although inland water impoundment facilities, through reduction of beach sand supply, have definitely had an impact on the overall recession of the coastline, their effects are subsumed in most shoreline erosion studies which employ aerial photography as input data. In San Diego County, inland dams were built prior to the late 1920's; as pointed out by Anders and Byrnes (1991), quality stereo aerial photographs were not available until the 1930's.

Although the objective of the California erosion hazards study is to determine how projected or potential economic losses from shoreline erosion will impact both the resources of the NFIP and local coastal communities studied, by applying state-of-the-art coastal mapping methods and using longer-term data sets to obtain the greatest erosion rate accuracy possible, there remains a number of fundamental problems with the determination and use of erosion rate data. As a result, the focus of this paper is to clearly explain the limitations involved in the generation of erosion rate data despite the use of state-of-the-art technique. The significance or value of erosion rate information, to coastal scientists, engineers, decisionmakers, and planners, depends on whether a clear understanding of the limitations and appropriate use of average erosion rate data in coastal land-use planning and practice exists.

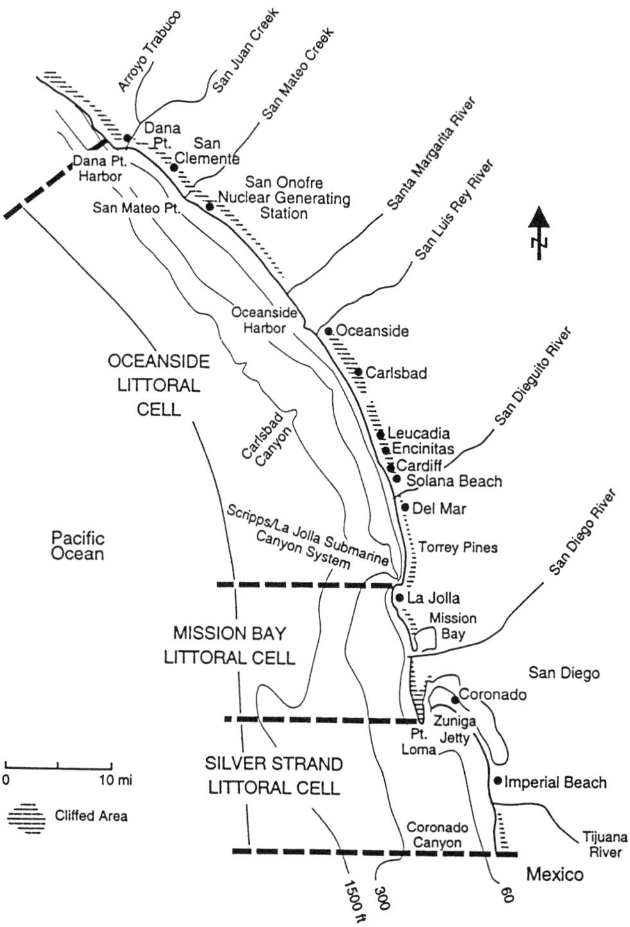

Figure 2. Location Map of the San Diego Region Showing Boundaries of the Three Major Littoral Cells *From Flick, 1994.*

It is important to note that the focus of FEMA's *Evaluation of Erosion Hazards* California study is concentrated on the landward migration of coastal bluff/cliff-tops (or dune-tops) and not the change in beach width or beach morphology. This is an extremely important point to note for two reasons. The first reason being that cliff/bluffs are distinctly different from beaches in geomorphic environments (figure 3). For eroding bluffs, there is no natural seasonal reconstruction as there is with beaches. Furthermore, cliff erosion results not only from waves undercutting the base of the cliff, but from landsliding, rockfalls, slumps, and saturation of cliff-tops from landscape irrigation, whereas beach erosion is mostly attributed to wave action. Secondly, the presence of shoreline armoring (seawalls, riprap, etc.) can be a major factor limiting the recession or landward migration of the shoreline. Coastal armoring serves to hold the shoreline in place, thereby disrupting the natural dynamic equilibrium of beach systems. Beach sand supply or beach morphology, on the other hand, is more often subject to the erosive power of waves due to the lack of lateral protection offered by seawalls and revetments (however beaches may be protected via groins, jetties, breakwaters, etc.). In addition, coastal erosion can be attributed to either marine or terrestrial processes or a combination of both marine and terrestrial processes. In many areas of San Diego County, Kuhn and Osborne (1987) have noted that coastal erosion is a product of both terrestrial erosion from groundwater as well as marine erosion from waves. The mechanisms of erosion along the San Diego and Santa Cruz County coastlines will not be investigated in the FEMA study, however Benumof and Griggs are investigating the main parameters responsible for differing erosion rates along the San Diego County coastline in a subsequent study.

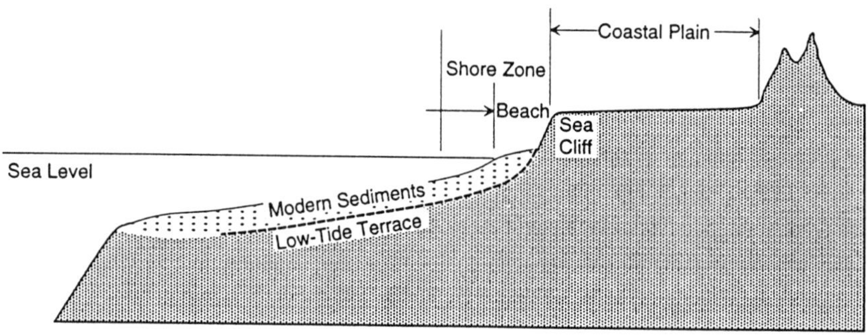

Figure 3. Cross Section Showing Differences in Geomorphic Environment with a Typical Sea Cliff and Beach.
Adapted from Flick, 1994.

As previously stated, long-term erosion rates are generally the key factor in evaluating and conditioning oceanfront development. Government officials are often confronted with questions of how to use erosion rate data for the guidance of setbacks, creation of zoning criteria, assessment of possible development/redevelopment proposals, or permits for shoreline armoring. As a result, it is the intention of this section of the paper to assist users of erosion rate data in their assessments of land-use planning, by specifically pointing out the main influences affecting the calculated values of erosion rates, when erosion rates are derived from aerial photographs.

THE VALUE OF EROSION RATES

Since erosion rates can only be as accurate as the data from which they are derived and the methods by which they are calculated, a thorough understanding of the errors inherent in the process of shoreline erosion rate determination is vital to the successful analysis and use of erosion rate data (Moore, In press). As a result of the episodic nature of the natural processes contributing to coastal erosion it is necessary, when determining erosion rates, to obtain quality stereo aerial photographs (or other sources of data) which cover the longest time-frame possible. In theory, this allows for sound determination of average erosion rates. However, as mentioned above, quality stereo aerial photographs were not available until the 1930's (Anders and Byrnes,1991). Thus, at best, shoreline mapping studies which utilize aerial photography are limited to approximately 65 years of data and may not be capable of accurately predicting true shoreline erosion patterns because of the episodic nature of erosion. Quite often in geological investigations, the key to understanding and predicting future outcomes is to carefully examine the past, although one can never be absolutely certain about predicting future natural processes. In any case, it is important to note that it may be possible to eliminate a significant degree of error resulting from the lack of a long-term data set, by choosing data sets which cover time periods including both above and below average rainfall/storm conditions as well as the expected moderate climatic conditions determined for a particular region. Combining the temporal variety of climatic conditions (over the long-term) in a given locality for a particular data set may help to account for variability in meteorological conditions and the episodicity of coastal natural processes. In the California FEMA *Evaluation of Erosion Hazards* study, both the San Diego and Santa Cruz County data sets cover relatively dry (1950's-1970's) and wet (1980's-present) periods, and are therefore considered to be adequately balanced.

Along much of the California coast, major shoreline erosion is attributable to extreme storm wave activity coincident with high tides. Storm waves are typically coincident with storm fronts which also generate significant rainfall. As illustrated in figure 4, if we assume that the cumulative residual precipitation/storm record for the South Coast Drainage Basin (from 1895 to 1990) is approximately correlated to the general pattern of coastal erosion, the importance of obtaining the longest possible data set for coastal mapping becomes very pronounced. In fact, the presence or absence of any particular data point has the potential to significantly affect the average of a group of data points. In other words, any particular time span will be unique, therefore it is recommended that the largest possible data set be employed.

In addition to the time-frame chosen for shoreline mapping studies, the degree of shoreline armoring in an area will considerably influence the value of erosion rates when rates are determined with the use of aerial photography. When analyzing erosion rate data in a specific locality or region, the investigator should evaluate the history of armoring. Depending on the structural design of the armoring and its capability of withstanding coastal wave attack, erosion rates may be significantly less than that of the natural recession of the shoreline because shoreline armoring tends to "hold" the shoreline in place (figure 5). It is recommended that when assessing erosion rates, that the investigator attempt to determine rates for both pre- and post-armoring time periods. This can be accomplished by using multiple sets of aerial photography from multiple years and calculating the recession of the shoreline between individual data sets. The end-point method (use of only two data sets- historical and recent) may be suitable if the study area is relatively deficient of shoreline armoring or if the area is very large.

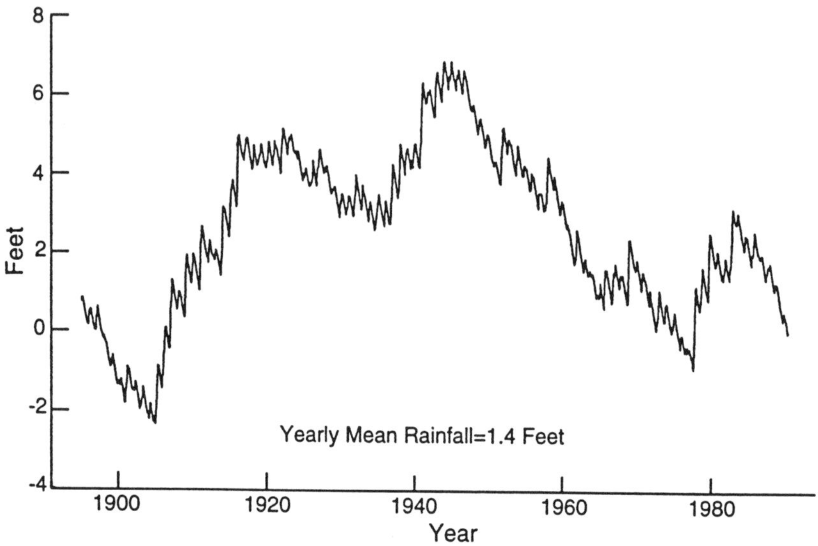

Figure 4. Cumulative Residual Precipitation for the South Coast Drainage Basin. Times of drought are indicated by the decreasing trend in this statistic, while periods with above-average rainfall show an upward trend. *From Flick, 1994.*

There are also technical and procedural factors which contribute to the quality and accuracy of erosion rate determination, including the scale of the photographs relative to the resolution demanded by the erosion study, clarity of the imagery (diapositives vs. contact prints from original negatives vs. contact prints from contact prints), scanning resolution of the imagery, and precision of ground control used to rectify aerial imagery. A detailed and thorough explanation of sources of error and analyses of different shoreline mapping techniques is given by Moore (In press).

CONCLUSION

In summary, it is the intention of this paper to inform users of erosion rate information about the uncertainties and limitations of "calculated average erosion rates" and the general value of erosion rate information. Often misleading, erosion rate information is used for a variety of purposes, chiefly to govern the siting of shoreline development. Users of erosion rate information should understand that erosion rate values may vary in quality depending on the input data from which they are derived and the methods by which they are determined. As a result, erosion rate information used to govern the siting of development should be analyzed for potential errors, to help ensure safety from coastal natural hazards and destruction of property.

Figure 5. Photo Showing Shoreline Armoring of Soquel Point, Santa Cruz County, California.
Photo taken by Benumof, 1996.

REFERENCES

Anders, F.J. and Byrnes, M.R., 1991. Accuracy of shoreline change rates as determined from maps and aerial photographs. *Shore and Beach*, 59(1), 17-26.

Griggs, G.B., 1994. California's Coastal Hazards. *Journal of Coastal Research, Special Issue No.12: Coastal Hazards,* 1-15.

Habel, J.S. and Armstrong, G.A., 1977. *Assessment and Atlas of Shoreline Erosion Along the California Coast State of California.* Sacramento, California: The Resources Agency and California Department of Navigation and Ocean Development.

Griggs, G. B. and Savoy, L.E., 1985. *Living with the California Coast.* Durham, North Carolina: Duke University Press.

Moore, L.J., In press. A Survey of Shoreline Mapping Techniques and Recommendations for Technique Selection. *Journal of Coastal Research.*

Flick, R.E., 1994. *Shoreline Erosion Assessment and Atlas of the San Diego Region.* Sacramento, California: The California Department of Boating and Waterways and the San Diego Association of Governments.

Kuhn, G.G. and Shepard, F.P., 1984. *Sea Cliffs, Beaches and Coastal Valleys of San Diego County.* Berkeley, California: University of California Press.

Kuhn, G.G. and Osborne, R.H., 1987. Sea Cliff Erosion In San Diego County, California. *Coastal Sediments,* ASCE, Vol. 2, 1839-1854.

CHAPTER 71

THE ARMORING OF CALIFORNIA'S COAST

Gary B. Griggs
Director-Institute of Marine Sciences
University of California, Santa Cruz

Abstract

Few issues along the coast of California are more complex, more poorly understood, and more divisive or controversial than the use of coastal protection structures. The extent of seawalls and revetments along the shoreline suggests that they are a necessity, but the numerous issues associated with them suggests much disagreement or differences of opinion about this necessity. The fact that approximately 130 miles of California's coast is now protected by armoring (~12% of the entire state's shoreline) is indicative of the magnitude of development in hazardous areas. That the extent of protection increased 400% in the 21 years between 1971 and 1992 suggests that this is a continuing issue.

Over the past fifty years, the principal response to coastal hazards in California, whether bluff erosion or wave attack, has been the construction of seawalls or rock revetments. These protective structures have usually been constructed only after existing shoreline development or infrastructure, whether public or private, has been threatened by wave attack or erosion. Rarely in the past has a protection strategy or armor plan preceded development. At present day costs of $5 - $15 million/mile for protective structures, this represents an enormous investment. Presently, some of California's coastal municipalities, primarily in the southern portion of the state, require a protective structure as a condition for development of oceanfront property. In striking contrast, other communities, particularly those along the central coast, will not allow oceanfront development in locations where a protective structure would be necessary to insure the survival of the property through the economic life of the structure (normally 50-100 years).

In recent years, considerable opposition has often arisen when new seawalls or revetments have been proposed because of the

In recent years, considerable opposition has often arisen when new seawalls or revetments have been proposed because of the perceived direct or indirect impacts of these structures. These potential impacts include visual impacts, restrictions on beach access, loss of beach due to placement of structure, public subsidy for private property owners, and reduction of sand supply from previously eroding coastal bluffs following armoring. An additional issue which has arisen with the proliferation of coastal armoring has been that of the potential impacts of a seawall on the beach itself. Does the emplacement of a seawall or revetment lead directly to beach loss? Long term field investigations of seawalls and adjacent beaches along the coastline of Monterey Bay have helped us answer this complex question.

The growing recognition of the limitations and impacts of "hard" protective structures has led to the consideration and implementation of "soft" strategies such as beach nourishment, which while soft, is neither free nor permanent. Moreover, the high capital and maintenance costs of seawalls and revetments, as well as the cutoff of federal funds for such projects, has led to the economic justification and reconsideration of relocating threatened oceanfront structures. The state of California needs to craft a consistent, comprehensive and long range policy on coastal armoring and protection for the entire coastline. Through a process of hazard recognition and evaluation, and then a standardized set of avoidance, mitigation, or hazard reduction policies, the private and public losses from future shoreline erosion, storm impact, and sea level rise can be significantly reduced.

INTRODUCTION

Coastal hazards are a function of the presence of human beings and their activities as they interact with naturally occurring coastal processes. With the exception of changes due to coastal erosion, the coastline of California has the same general configuration as it did in 1850 when the estimated population of California was 93,000. The state population grew steadily for the next 100 years, but following World War II, it virtually exploded; between 1950 and 1970 it nearly doubled, growing from 10.6 to 20 million (Figure 1; Griggs, Pepper and Jordan, 1992). The 1996 population of 32.4 million represents a tripling since 1950. California now is the most populous of the United States with approximately 80% of the population living within 50 km of the shoreline. Each resident now has about two inches of coastline to enjoy, and the population is growing at about 1500 people/day. As the population density increases, the competition for oceanfront property increases, along with the exposure to coastal geologic hazards. The population of California is projected to reach 50,000,000 by the year 2020.

In striking conflict with this population increase and rush to the coast in California is the loss and retreat of the shoreline. Of the state's 1100 miles of coastline, 85% is eroding (Griggs, 1996). Because of the historic land loss and potential for future losses, about 130 miles or 12% of the entire state's coastline has now been armored or protected with rip-rap, seawalls or other engineering structures. In response to the damaging coastal storms of the last two decades and the increased density of oceanfront construction, the extent of armoring increased 400% in the 21 years between 1971 and 1992 (Figure 2).

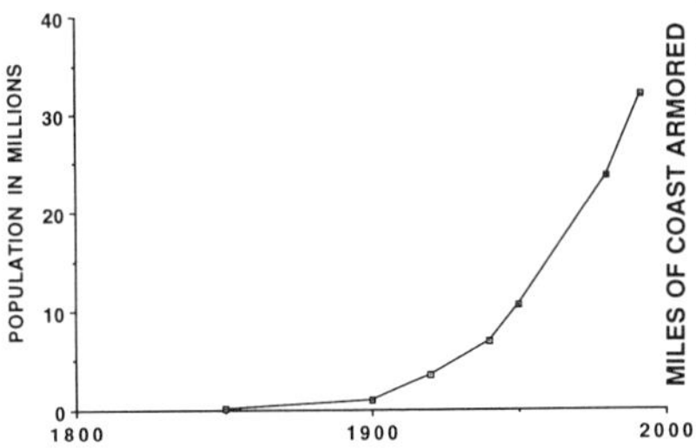

Figure 1. California's increasing population.

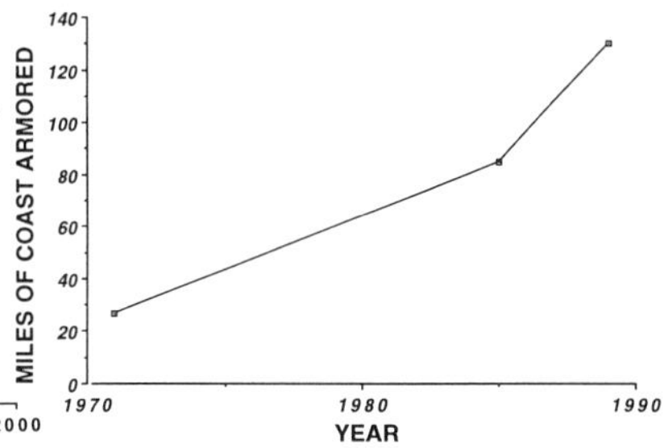

Figure 2. Increase in coastal armoring in California

HAZARDOUS COASTAL ENVIRONMENTS

Coastal geologic hazards in California occur most frequently in the form of shoreline erosion (both seacliff and beach) and coastal flooding (both wave impact and inundation). Human-induced interference with coastal processes (littoral drift and sand supply) and coastal bluff stability (altering the condition of coastal bluffs materials) can exacerbate hazard conditions.

Several factors combine to complicate shoreline erosion in California:

1] The process of cliff or bluff erosion tends to be an episodic one with the major erosion events taking place during the simultaneous occurrence of high tides and large storm waves. Thus while we tend to utilize "average" annual erosion rates (based on analysis of historic aerial photographs and/or maps) to describe the erosion risk or hazard at a particular location, the failure of coastal cliffs typically takes place in large incremental events during storm conditions.

2] The process of cliff retreat or erosion is very site specific and due to a combination of a) wave attack, b) rainfall and runoff which produce gullying, mass movements, and piping, c) and the lithology, stratigraphy and structure of the rock forming the seacliffs.

Along the California coast there are three coastal geomorphic environments where widespread development has taken place but that are potentially hazardous for any structure. These include: 1) eroding cliffs or bluffs, 2) the beach, and 3) coastal dunes.

Eroding Cliffs or Bluffs

Eroding cliffs and bluffs represents California's most extensive coastal hazard. The number of existing homes and "buildable" lots atop eroding seacliffs along the California coast is far greater than those either on the beach or on the frontal dune. In each case, the desire for unobstructed ocean views have induced many to invest in property literally within a stone's throw of the water. Because of its location along an active plate margin, tectonic uplift of the California coast has provided many square miles of flat marine terraces which are easily developed. As a result, terrace bluff tops throughout much of southern California have been intensively developed with homes, condominiums, apartments, hotels and restaurants. The Del Mar, Oceanside, Encinitas, and Santa Barbara areas provide clear examples of this phenomenon. Although the extent of cliff top construction along the central and northern California coast is considerably less, there has still been considerable development in places like Santa Cruz, Capitola, Rio del Mar and Pacifica.

Coastal cliff failure takes place as a result of several different processes: wave attack, subaerial erosion and mass movements, and seismic shaking. The degree to which a seacliff yields to wave attack depends primarily upon its exposure to wave energy and the physical properties or strength of the materials making up the cliff. The lithology or hardness of the cliff rock and the presence of internal weaknesses such as joints or faults all directly affect the resistance of the material to wave attack or to slope failure through subaerial processes.

Beach-Front Construction

Construction directly on the beach is common along the California coast. Homes have been built both elevated on pilings or piers, but also on slabs at beach level. The presence of beach sand surrounding a house is ordinarily clear evidence that the site is at least intermittently subject to high tides and storm waves.

Much of the $150 million in coastal storm damage over the past 20 years in California occurred when waves attacked beach-front homes in place like Stinson Beach, Rio Del Mar, Malibu, Redondo Beach, Del Mar, Oceanside and Imperial Beach. Storm damage took place during the winters of 1978, 1983 and 1988 when large storm waves arrived simultaneously with very high tides during a period of elevated sea level from El Nino events. Losses in 1983 included wave undermining of shallow piers, pilings or slabs so houses collapsed onto the beach, or direct wave impact on structures themselves. Waves also overtopped or destroyed low protective seawalls and revetments and then damaged or destroyed the house fronts facing the sea.

Sand Dunes

In contrast to the east and Gulf coasts of the United States, the coast of California is a geologically active continental margin, such that the coast is typically backed by steep cliffs and coastal mountains, only occasionally broken by alluvial valleys and river mouths. As a result, sand dunes, which require both a flat low-lying area on which to form, as well as an adequate supply of sand and a dominant wind direction, are restricted along the state's coastline. Extensive coastal dune fields have developed in Monterey Bay, Pismo Beach, and along portions of Los Angeles and Orange counties in southern California.

The most extensive area of oceanfront dune development along the central California coast is in Monterey Bay where numerous homes, condominiums, apartments, a hotel and other structures have been built directly on the crest of the frontal dune, overlooking the ocean. Additional high density dune-front development has been proposed. Concerns exist with these existing and proposed developments. Where structures have been built directly on the frontal or primary dune on conventional foundations or shallow piers, periodic dune erosion during episodes of high tides and storm waves can undermine the foundations and lead to collapse of these buildings.

RESPONSES TO COASTAL HAZARDS

Two different situations exist relative to coastal hazard exposure: areas which have already been developed and areas which are currently undeveloped. In regard to the latter, a comprehensive investigation of the recent geologic history of an oceanfront site prior to development is relatively straightforward, relatively inexpensive compared to property and home construction costs, and can determine hazards and risks present and make suitable recommendations so as to reduce future risk exposure. Dealing with existing development in hazardous locations is far more complex; several different options exist.

Do Nothing or Sell. The lowest cost and potentially highest risk approach is to do nothing. Depending upon a structure's particular location- its setback from the ocean, elevation, and past erosion or inundation problems- this approach may work for a limited period of time. There is no cost until a major storm finally does occur, and then either a rapid emergency response is necessary or losses may be very high. In many instances, threatened oceanfront homes have been sold to unsuspecting buyers, simply passing on the problem.

Relocate: Relocation of oceanfront structures or infrastructure is being given increasing consideration in a number of situations. Where a parcel is large enough, a threatened structure can be moved landward on the same parcel to extend its lifespan depending upon long term erosion rates. In many cases this will not be possible and relocation will require acquisition of a separate parcel. Recent examples of comparative costs of relocation or reconstruction compared to protection (Griggs, 1986) indicate that in the long run relocation is far less expensive. It is likely that this option has not been seriously considered by most threatened oceanfront property owners, simply because of a desire to protect their home and view at any cost.

Beach Nourishment: Beach nourishment or replenishment has emerged as an appealing soft approach to dealing with the problems of shoreline erosion. On the surface this strategy presents an attractive compromise to the extremes of abandoning the shoreline on the one hand and armoring it with concrete or rock on the other. The beach is nourished with sand from either an offshore or inland source. The goal is to increase the width of the beach so that more beach is available as a buffer to wave attack and also so that more beach is available for human use.

Although beach nourishment may offer significant benefits, it is a costly proposition with a number of limitations or concerns. Along the coast of California littoral drift rates are very high (200,000 - 1,000,000 cubic yards/yr) such that the lifespan of the sand added to a particular beach is likely to be fairly short. A recent study of west coast nourishment projects found that 18% survived less than a year, 55% lasted only 1-5 years, and 27% survived over 5 years (Leonard, et. al, 1990). A number of issues and questions have been raised as to what constitutes a successful nourishment project. At offshore dredging costs of $3-4/cubic yard, placing 300,000 cubic yards annually on a beach from adjacent offshore sources (were sand available of the appropriate grain size) would cost $900,000 to $1,200,000 annually. Inland sources of sand would have higher costs and more obvious environmental impacts. Availability of adequate volumes of appropriately sized sand, lifespan of a nourished beach, and who pays for the project are major issues. While the federal government has funded many Atlantic shoreline nourishment projects in the past, this practice is being significantly reduced as the federal government attempts to balance the budget.

Armoring Over the past 50 years, the typical response to coastal hazards in California has been the construction of seawalls and revetments intended to protect eroding or wave-impacted shorelines. At present, an astonishing 130 miles - approximately 12% of the state's entire 1100 mile shoreline - has been "protected" or armored (Figure 2). As expected, the percentage of shoreline armored is greater in the more densely populated southern California cities and counties than in the central and northern portion of the state (Figures 3 and 4). Protective structures have usually been constructed only after existing shoreline development has been threatened. Rarely did a protection strategy precede development.

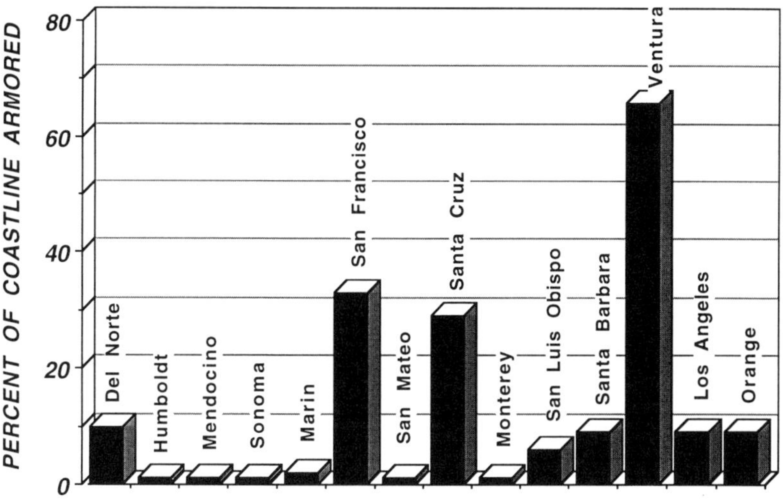

Figure 3. Percentage of shoreline of California's coastal counties armored

Protective structures can vary considerably in their cost, size, effectiveness, lifespan and impacts (Fulton-Bennett and Griggs, 1986). At one extreme, slabs of broken concrete or asphalt or other construction debris have simply been dumped at the base of the cliffs in an attempt to reduce wave impact. Most low-budget, unengineered efforts of this sort have been relatively futile or very short lived, often lasting less than a single winter. At the other extreme are massive, expensive, carefully engineered concrete seawalls, such as the O'Shaughnessy Seawall along Ocean Beach in San Francisco, which has functioned effectively for nearly 70 years. Costs for engineered coastal protection structures built over the past 15 years along the central California coast ranges today from perhaps \$750/foot at the low end to over \$3000/foot.

On a high energy and/or eroding shoreline, any protective structure built to withstand direct wave attack will probably eventually fail. Even a well-designed structure is likely to fail once its design life has been exceeded, especially if it has not been properly maintained. Engineers commonly think in terms of a 20 to 25 year life for a coastal

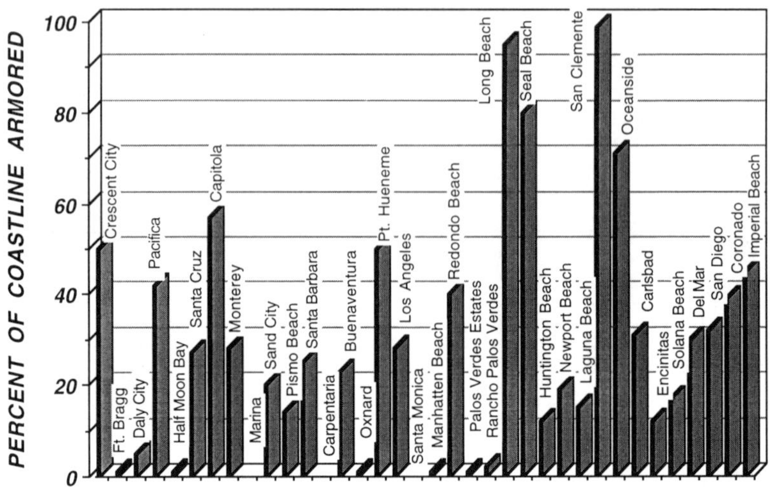

Figure 4. Percentage of shoreline of California's coastal cities armored

protection structure. The should be clearly understood by the homeowner or client, but often is not.

Spending large amounts of money on the installation of a coastal protection structure does not guarantee long-term, or in some cases, even short-term, protection for home and property (Fulton-Bennett and Griggs, 1985). The degree of exposure to wave attack, the foundation materials, as well as the specific design, construction and materials will all influence the lifespan and effectiveness of a structure. During prolonged or repeated wave attack under high tide conditions, such as occurred during the winter of 1983, protective structures which may have survived for many years may fail virtually overnight. Some structures have fared far better than others.

Whether rock revetments, concrete seawalls or timber bulkheads, these protective structures have historically failed by scour or undermining, outflanking, overtopping, or simply battering or wave impact or a combination of these. In a study along the central coast of California (Fulton-Bennett and Griggs, 1985), of the three major types of protection, concrete seawalls have been the most successful in reducing erosion and property damage, and have been the most durable over the long term. Concrete, in and of itself, however, does not guarantee survival. Rip rap typically fared less well than concrete walls, but better than wooden walls. The success of rip rap is a function of the foundation on which it is placed, the size, slope and height of the rock, and the internal layering. Wooden walls have been the least successful in preventing erosion and damage, and most are easily damaged during severe storms.

On the whole, few protective structures in the area studied by Fulton-Bennett and Griggs (1985) between San Francisco and Carmel, had stood the long-term tests of time, surviving unassisted and preventing damage and erosion, for more than twenty years or longer than their design life. There are some notable exceptions, such as the O'Shaughnessy seawall. Many structures have become structurally unsound, required considerable maintenance or repair, and/or failed to adequately reduce property damage for more than one severe storm period. Thus the effective lifespan of a structure often depends on how many mild winters pass before the next severe storm. However, most structures have reduced erosion rates, at least over the short term. Before any protective structure is built, its initial costs and maintenance costs and its probable lifespan should be carefully considered, as well as its design criteria relative to the site where it is to be emplaced and this history of other protective structures in the vicinity.

In contrast to the oceanfront homeowner's concern for the lifespan or effectiveness of a coastal protection structure, considerable opposition has arisen in recent years around proposals for new seawalls or revetments because of the perceived direct or indirect impacts of these structures. Many of the concerns, including aesthetic or visual impacts, restrictions on beach access, reduction of sand supply from previously eroding bluffs, loss of the beach beneath the rip rap or seawall, revolve around the issue of to what degree should private property owners be allowed to impact public beaches as they attempt to protect their own property, or in the case of government funded projects, how much taxpayer's money should be spent in efforts to stabilize the position of an otherwise eroding coastline.

Over the past decade concern has arisen regarding the direct impacts of seawalls or revetments on beaches. Seawalls are not designed to protect beaches but to reduce wave impact forces on uplands, whether dunes, bluff or cliffs. Pilkey and Wright (1988) and Griggs, Tait and Corona (1994) discuss three different potential impacts of armoring the shoreline.

Impoundment or Placement Loss: This effect is the most straightforward and predictable. When a structure is built seaward of the base of a bluff, cliff, or dune, well out on the beach profile, a known amount of beach will be covered (Figure 5). The effect is immediate beach loss; the extent of the loss being a function of how far seaward and alongshore the structure extends. The type of structure is also important. Where a vertical seawall is built against the base of a bluff or dune, there is essentially no beach loss. On the other hand, where a revetment is constructed to protect a bluff, it may reach a height of 15 to 20 feet or more, and extend seaward at a 1.5:1 or 2:1 slope, thus covering 30 to 40 feet of beach. This placement loss can easily be determined for

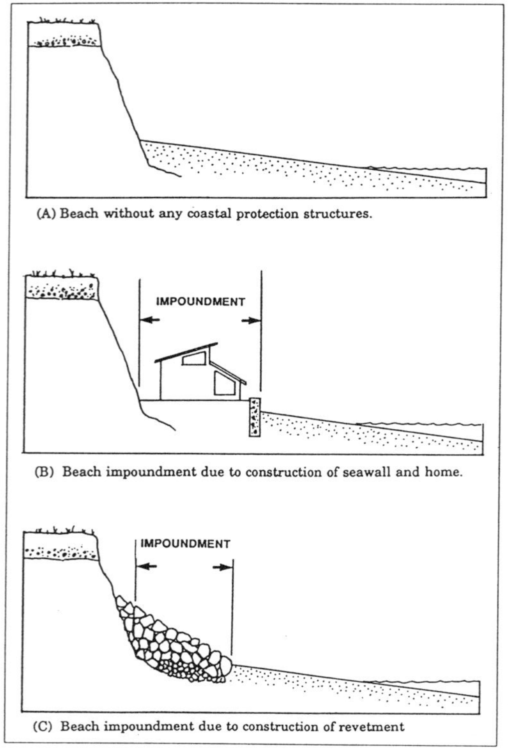

(A) Beach without any coastal protection structures.

(B) Beach impoundment due to construction of seawall and home.

(C) Beach impoundment due to construction of revetment

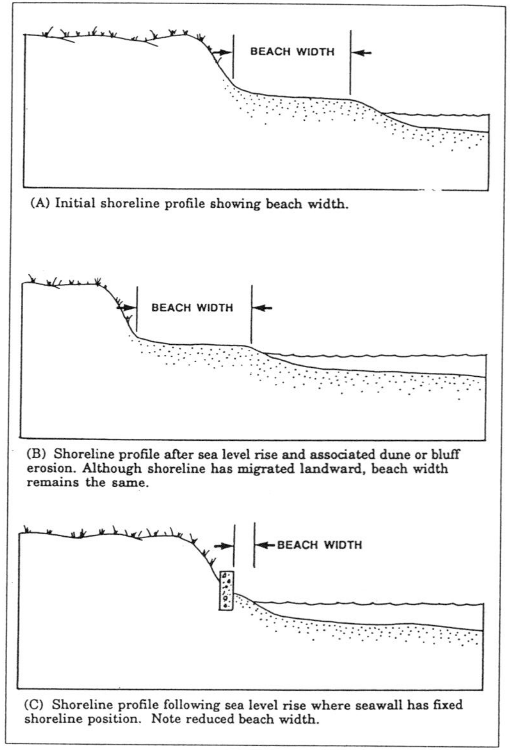

(A) Initial shoreline profile showing beach width.

(B) Shoreline profile after sea level rise and associated dune or bluff erosion. Although shoreline has migrated landward, beach width remains the same.

(C) Shoreline profile following sea level rise where seawall has fixed shoreline position. Note reduced beach width.

Figure 5. Examples of beach loss through emplacement. Figure 6. Examples of passive erosion.

any proposed revetment knowing the cross-sectional area and alongshore dimensions.

Passive Erosion: Whenever a hard structure is built along a coastline undergoing net long-term erosion, the shoreline will eventually migrate landward behind the structure (Figure 6). The effect will be gradual loss of the beach in front of the seawall or revetment as the water deepens and the shoreface profile migrates landward. This process is designated as passive erosion and appears to be the process which has been well documented along many of the barrier islands of the Atlantic coast, and now on Oahu (Fletcher, et.al., 1997). This process takes place regardless of the type of protective structure emplaced.

Active Erosion: The ability or potential for a seawall or revetment to induce or accelerate erosion, has been the source of most of the controversy over the past decade regarding the impacts of seawalls on beaches. Although differing opinions have been put forward regarding the impacts of these structures on adjacent beaches, until fairly recently there had been a noticeable lack of sustained or repeated field observations and measurements with which to resolve the conflicting claims.

A recently completed eight year study along the central coast of California was carried out to resolve some of the questions regarding the impacts of seawalls on beaches (Griggs, Tait and Corona, 1994). The study involved biweekly to monthly surveying of beaches adjacent to seawalls in Monterey Bay where beaches undergo significant seasonal erosion and accretion, but where there is no net long-term retreat. While the presence of a seawall or revetment has some local seasonal effects, comparison of time-averaged profiles for both seawall-backed and adjacent control beaches for both winter and summer months are virtually indistinguishable, indicating the lack of measurable or significant differences between these beach profiles. While rip rap or revetments have often been perceived and judged by permitting agencies to be more permeable and therefore expected to have less impact on beaches than "impermeable" seawalls, this is not supported by field observations and surveying (Griggs, Tait and Corona, 1994).

Conclusions

The most common response to coastal erosion on the shoreline of California in the past has been armoring, whether seawalls or revetments. Increasing pressure for oceanfront construction combined with an increased frequency of damaging coastal storms over the past decade has led to more requests for coastal protection structures such that 12 percent of the entire coastline of the state has now been armored. Opposition has developed in recent years when new seawalls or rip rap have been proposed due to a combination of actual and perceived environmental impacts including visual, access, sand supply reduction, and beach loss. Policies on approval of coastal protection structures varies widely along the state's shoreline due to a combination of local geologic conditions as well as local politics and economics. Given the past trends of increasing coastal populations and increasing dependence on armoring, the state needs to implement a consistent, comprehensive and long range policy on additional armoring which is informed by the environmental impacts as well as costs of additional armoring.

References

Fletcher, C.H., Mullane, R.A., and Richmond, B.M. 1997. Beach loss along armored shorelines on Oahu, Hawaiian Islands. Journal of Coastal Research 13: 209-215.

Fulton-Bennett, K.W. and Griggs, G.B., 1985. Coastal protection structures and their effectiveness. Joint publication of State of California, Dept. of Boating and Waterways, and the Institute of Marine Sciences, University of California, Santa Cruz: 48 p.

Griggs, G.B. 1986. Reconstruction or relocation: viable approaches for structures in areas of high coastal erosion. Shore and Beach 54: 8-16.

Griggs, G.B. 1996. California's coastal hazards. Journal of Coastal Research Special Issue No. 12: 1-15.

Griggs, G.B., Pepper, J.E. and Jordan, M.E., 1962. California's coastal hazards: a critical assessment of existing land-use policies and practices. California Policy Seminar Publication, 224 p. Berkeley, Ca.

Griggs, G.B., Tait, J.F., and Corona, W., 1994. The interaction of seawalls and beaches: seven years of monitoring, Monterey Bay, California. Shore and Beach v. 62:21-28.

Leonard, L., Dixon, K. and Pilkey, O.H., 1990. A comparison of beach replenishment on the U.S. Atlantic, Pacific and Gulf coasts. Journal of Coastal Research, Special Issue No. 6:127-140.

Pilkey, O.H. and Wright, H.L., 1988. Seawalls versus beaches. Journal of Coastal Research, Special Issue No. 4: 41-66.

CHAPTER 72

Littoral Cell Management Planning along the Oregon Coast

John J. Marra, PhD[1]
Paul Klarin[2]
Emily Toby[3]

Abstract

The Oregon Coastal Management Program is employing a new strategy to improve management of hazardous oceanfront areas. Littoral Cell Management Planning is a comprehensive, integrated, area-wide hazards management strategy adopted for the different physical and land use settings found in Oregon. A plan is composed of littoral cell inventories, a chronic hazards management strategy and local implementing mechanisms.

I. Introduction

The Oregon coast is subject to a variety of natural hazards associated with processes that occur across a broad range of spatial and temporal scales. Chronic hazards are those which are local in scale and scope. Dune-backed shoreline chronic hazards include; ocean flooding; beach and dune erosion; and sand inundation. Bluff-backed shoreline chronic hazards include bluff recession and landsliding/slumping.3,4,6

Twenty years ago Oregon began establishing policies to reduce risks to development from chronic coastal natural hazards. Recent reports evaluated the effectiveness of these policies in light of increased pressure for development and recent advances in scientific understanding and concluded that existing policies and

[1]Principal, Shoreland Solutions, P.O. Box 1046, Newport, Oregon 97365

[2]Coastal Specialist, Oregon Coastal Management Program, 1175 Court St. NE, Salem, Oregon 97310

[3]Policy Coordinator, Oregon Coastal Management Program, see address above

procedures need to be improved. It was recommended that oceanfront development and shore protection decision-making could be improved by addressing chronic hazard alleviation at the scale of individual littoral cells.[2,3,6,16]

Littoral Cells: Headlands divide Oregon's coast into compartments, or littoral cells. Approximately 45% of these sandy shores are dune-backed, the remainder being bluff-backed or occupied by inlets. Littoral cells or subcells define closed compartments in terms of 'sand supply', as the headlands restrict transfers of sediment to exchanges within individual littoral cells. These internal exchanges are typically described in terms of contributions to or losses from the littoral cell 'sediment budget', resulting in a negative or positive budget balance. Consequently, efforts to alleviate hazards along one segment of shoreline have the potential to adversely impact other segments. Also, an adequate littoral cell sand supply is important because it provides benefits of hazard alleviation, and also recreational, scenic, and economic opportunities.[3,6]

II. Littoral Cell Management Planning

Littoral cell management planning is designed to locate new development away from hazardous areas and to minimize threats to existing development, while improving both resource protection and economic opportunities. Its advantages include:

- Hazard assessment is more consistent in quality when it is carried out on an area- wide scale;

- Hazard alleviation is more effective and less costly when addressed at the scale at which factors affecting shoreline stability occur;

- The potential for adverse impacts and cumulative effects are minimized because there is an increased likelihood that these considerations will enter into the decision-making process; and

- Decision-making is more timely and predictable because interagency and inter-governmental coordination is enhanced.

Littoral cell management planning is designed to serve as a guide to local planners, state and federal resource managers, property owners, business leaders, community activists, scientists and other interested parties, and its framework is intended to be flexible enough to accommodate a broad range of community needs and capabilities. The resulting littoral cell management plan (LCMP) is a comprehensive, integrated, area-wide hazards management strategy unique to a specific physical and social setting. It is focused on the reduction of risk to new and existing oceanfront development from chronic coastal natural hazards and should include littoral cell inventories, chronic hazards management strategy and implementing mechanisms. The LCMP three step planning process is carried out in two phases. The 3 steps - inventory, analysis, and implementation - are repeated during both phases of the process and each step involves different activities and results in different products.

Littoral Cell Inventory: Information describing physical, biological, and cultural characteristics within a littoral cell\subcell. Inventory information such as maps, database and text formats, form the basis for desicion-making. The content of the various inventories follow below.

Chronic Hazards Management Strategy: preferred management measures and the policies and procedures needed to implement them. Two parts of an chronic hazards management strategy are the *hazard avoidance strategy* and the *beach and shore protection strategy*.

□ The *hazard avoidance strategy* identifies chronic coastal natural hazards within a littoral cell or subcell, assesses potential risks from these hazards, formulates policies to avoid unacceptable risks, and prescribes measures to implement these policies.

The strategy focuses on policies and procedures for the siting and design of new development. It includes a series of options or techniques which can be used to reduce potential risk by influencing the location, elevation, and design of structures and infrastructure. Though usually applied to new development, hazard avoidance options may also be applied the the rebuilding of existing development. 3,5

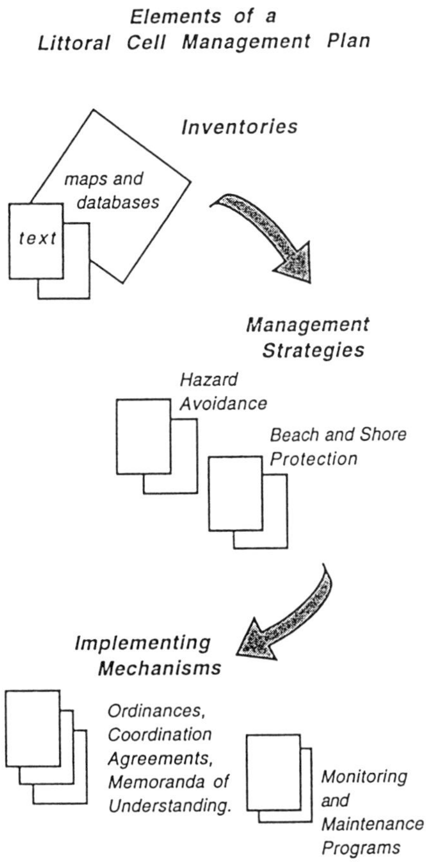

Hazard avoidance options include; (1) Siting, design & construction standards; (2) Oceanfront construction setbacks, and; (3) Relocation and land acquisition programs. Educational programs, environmental regulations, and zoning controls and infrastructure planning might also be considered.[5]

□ The *beach and shore protection strategy* identifies potential beach and shore protection techniques, assesses the positive and negative impacts of these techniques, formulates policies regarding preferred alternatives, and prescribes measures that can be taken to implement these policies. It focuses on policies and procedures pertaining to hazard alleviation for existing development and includes techniques that can be used to reduce potential risk by acting to retard or prevent the effects of wave attacks and/or mass wasting.[3,5] Such options are used for existing development, and include:

1. 'Soft' Options for processes of Wave Attack include:
   a) Dune and Bluff Enhancement
   b) Beach Nourishment
   c) Boulder berms (also called 'dynamic revetments')

2. 'Hard' Options for Processes of Wave Attack include:
   a) Revetments and seawalls
   b) Breakwaters
   c) Groins

3. Options for processes of Mass Wasting include:
   a) Vegetation Management
   b) Drainage Controls
   c) Slope Regrading
   d) Reinforcing Structures
   e) Surface Fixing

Implementing Mechanisms: include local ordinances, coordination agreements, memoranda of understanding, or other mechanisms which adopt policies and procedures prescribed in the management strategy. These materials, together with monitoring and maintenance programs, are needed to ensure the success of a LCMP.

III. Phase I : Hazard Avoidance

The first step in the planning process is to collect information on physical and human factors affecting shoreline stability within a littoral cell, called the *physical inventory*. The environmental, socioeconomic, and jurisdictional factors make up the *biological and cultural inventories*.[5]

Physical Inventory

A. *Regional Setting*: Major geographic and geologic features

B. *Factors Affecting Chronic Shoreline Stability*: Long term trends of shoreline change, episodic flooding and erosion events, episodic sliding and slumping events needed to assess the potential risk to development.[4]

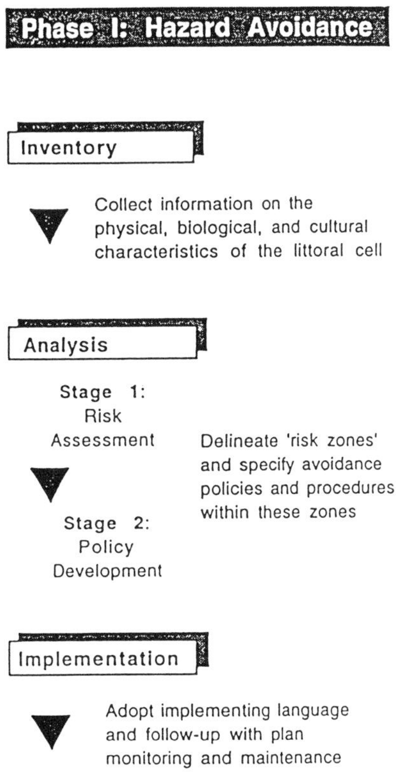

1. Long-term trend:
a) Historical dune/bluff retreat
b) Relative sea-level rise
c) Sediment budget and sources

2. Short-term trend from Episodic Flooding/Erosion:
a) Wave overtopping/ undercutting
b) Direct evidence from existing and antecedent conditions
c) Inlet dynamics
d) Dune Stability

3. Short-term trend from Episodic Sliding/Slumping
a) Surface features
b) Material & structural properties
c) Surface/Subsurface drainage
d) Wave attack

Biological and Cultural Inventories

A. *Environmental factors*
1. Wetlands and major marshes
2. Significant wildlife habitat
3. Exceptional aesthetic resources
4. Cultural resources

B. *Socioeconomic factors*
1. Recreational (water-dependent or not);
2. Residential (low or high density);
3. Commercial;
4. Industrial; and
5. Critical facilities and infrastructure (hospitals, schools, roads, power lines etc.)

C. *Jurisdictional factors*
1. Private property owners;
2. Local government (city, county, port districts other special districts)
3. State government regulatory and resources agencies
4. Federal agencies (ACOE, FEMA, BLM, EPA, USFW)

Analysis - Stage 1

Risk Assessment: The collected inventory information is analyzed to assess the potential risks attributable to the individual and cumulative effects of chronic hazards present within any given littoral cell or subcell. This step is a fundamental component of the LCMP process which provides information needed to make decisions at other steps in the planning process. Based on the analysis shoreline segments may be classified according to levels of risk.

- Active -Areas currently subject to or likely subject to wave attack or mass wasting within the next~10 years;

- High Potential - areas likely subject to wave attack or mass wasting within the next~10 to 50 years; or

- Moderate Potential - areas likely subject to wave attack or mass wasting within the next~50 to 100 years.

This classification system expresses levels of risk in terms of the relative potential for activity as used in the LCMP process. Other classification systems may be devised for this purpose.

Analysis - Stage 2

Policy Development : The next step in the planning process is to examine the results of the risk assessment in conjunction with biological and cultural inventory information in order to formulate hazard avoidance policies appropriate to different segments of littoral cell shoreline and establish procedures to implement these policies. There are a variety of hazard avoidance options that can be considered during this policy development stage of the littoral cell management planning process. One option is construction setbacks for various types and levels of new oceanfront development such as:

- Low density 'rural' residential development (e.g. single and multi-family dwellings or other structures < 5000 sq. ft. and/or those which are readily movable) is appropriate in areas with Moderate Potential for activity and may be appropriate in areas with High Potential for activity;

- High density "urban" residential development (e.g. residential dwellings > 5000 sq. ft.) and commercial and/or industrial development (e.g. hotels, motels, restaurants, factories, etc.) Is generally inappropriate in areas with High Potential for activity but may be appropriate in areas with Moderate Potential for activity; and

- All new habitable structures are generally inappropriate in Active areas. Similar standards can also be applied to critical facilities and infrastructure, or the repair or remodeling of existing structures. Exceptions may need to be made for unique situations where, for example, a lot is found to lack sufficient depth to meet the established standard. In those cases, hazard avoidance options such as relocation or land acquisition might be considered. Other hazard avoidance options to consider include zoning controls and infrastructure planning. Clustering, down-sizing or limiting the level of future services can also be appropriate means of reducing risks to new development in hazardous areas. Similarly, existing siting, design, and construction standards can be reviewed and updated.

Implementation: The next step in the process is to adopt the appropriate mechanisms for the LCMP, through formal or informal action at local, state, and federal levels of government. Plan monitoring and maintenance also need to be implemented. Various types of administrative actions or agreements can be taken to ensure that the policies and procedures which constitute a hazard avoidance strategy are applied consistently across intergovernmental boundaries.[6]

- Local government formally adopts relevant portions of the LCMP into their comprehensive plan and zoning ordinances including maps data-bases, avoidance policies and procedures. They form coordination agreements with other jurisdictions, port districts, or other special districts as needed.

- State and federal agencies implement relevant portions of the LCMP, first by being subject to consistency with local comprehensive plans, and second through policy letters or memoranda of understanding.

It is intended that littoral cell management plans be adopted voluntarily and implemented within the existing regulatory and land use planning programs. It is unlikely that new or revised statutes, goals, or rules will be needed. This is consistent with the purpose of littoral cell management planning - to simplify and coordinate the existing management system, not create a new one.

Post Adoption Measures: The monitoring and maintenance measures of the LCMP are vital components that need to be identified during the policy development stage of the planning process. These measures include:

- *Day-to-day review* and approval of building and beach improvement permits. For example, a process for coordinating the review of applications needs to be established. Preapplication meetings where affected interest groups are able to review proposals prior to submittal may be conducted as part of this process.

- *Annual review* of decisions and their cumulative effects under the plan. Regular meetings of the affected local jurisdictions and regulatory agencies may be needed to make minor adjustments and to maintain continuity.

- *Five year evaluation* of the plan's success relative to objectives identified during policy development stage. The evaluation should also determine if there has been a substantial change in circumstances since plan adoption, and if that requires it be amended. Additionally, the inventory information should be reviewed and updated at this time.

IV. Phase II: Beach and Shore Protection

Inventory

Much of the inventory information collected during the first phase of the LCMP process can be used to develop beach and shore protection strategies. However, additional inventory information such as littoral cell sediment budgets may be required to more accurately assess hazard alleviation needs and to better understand the positive and negative impacts of certain shore protection options. Additional biological and cultural inventory information may also be needed.

Analysis - Stage 1: *Alternatives and impacts Assessment*

At this stage the inventory information is analyzed to identify potential beach and shore protection techniques and their respective positive and negative impacts. This will lead to the identification of preferred alternatives. Typically, a combination of beach and shore protection techniques will be employed, reflecting the preferred alternatives and community perspectives.[5]

Beach and shore protection techniques can be compared through a cost-benefit analysis. An elaborate cost-benefit analysis may not be needed, simply structuring the alternatives and impacts assessment in a cost-benefit context can be sufficient to reveal preferred alternatives.

<u>Analysis - Stage 2:</u> *Policy Development.* The results of the alternatives and impacts assessment are used to formulate beach and shore protection policies appropriate to different segments of littoral cell shoreline and the implementing measures.
For example, during this policy development stage of the littoral cell management planning process it might be determined that:

- Dune and Bluff Enhancement (e.g. foredune - bluff grading, vegetative stabilization, drainage controls) is appropriate in Active areas and in areas with High Potential for activity, and may be appropriate in areas with Moderate Potential for activity; and

- Revetments and seawalls may be appropriate only in Active areas where it has been shown that non-structural solutions have failed to provide adequate hazard alleviation and only when constructed according to prescribed design standards. This is an 'imminent threat' criterion.

More elaborate strategies involving active management of the sediment budget may be feasible in some littoral cells. In some cells shallow-water disposal of material dredged from navigation channels could be allowed. In others, sand that has accumulated at the northern end of the cell may be back-passed to the southern end.

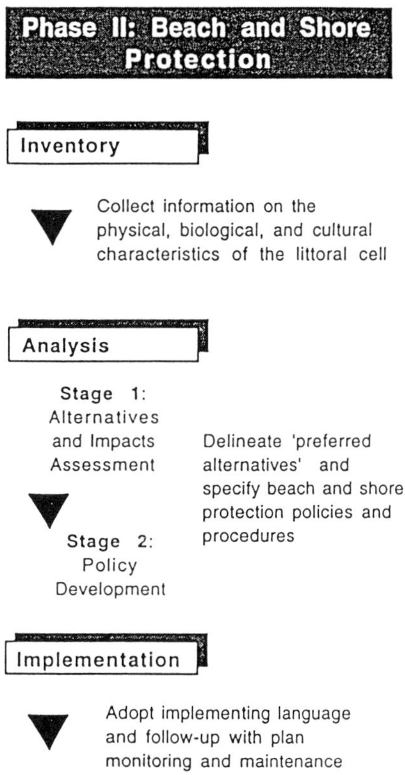

These measures have the advantage of providing hazard alleviation to segments of shoreline where flooding/erosion potential is greatest, and also where sand accumulation is a potential hazard. Also, they recognize the value of a wide sandy beach - from the perspective of hazard alleviation, and for economic, social, and environmental concerns.

Structural stabilization may be allowed on a lot by lot basis, provided it is built in accordance with applicable design standards and losses to the sediment budget are compensated. In some instances, such as along lightly developed dune-backed shores, structural stabilization may be prohibited, and foredune management measures may be applied. In developed or significantly altered shorelines, protection strategies may prescribe design standards for shore protection structures.

Implementation:

The beach and shore protection strategies need to be adopted into the LCMP either formally or informally by local, state, and federal agencies. Special elements for monitoring and maintenance may be needed for specific techniques such as beach nourishment or rip rap revetments.

V. Who does a Littoral Cell Management Plan?

Participants, Roles and Responsibilities:

A broad range of expertise and perspectives is needed to develop and implement a littoral cell management plan, requiring a variety of individuals assembled to perform various functions. Separate *facilitation, technical, and management groups* are needed.[1] The composition and function of these groups is described below though it may vary with community needs and resources.

The facilitation group will play the lead role in guiding a community through the LCMP process. It defines the scope and sets the agenda, establishes the boundaries of the planning area, and selects the members of the technical and management groups. It also compiles and disseminates information to participants and acts as a liaison between other groups in the planning process. The facilitation group should be established as a formal entity that convenes on a regular basis until the LCMP is implemented. It should be composed of city and county planners, or other local officials who initiated and oversee the process. State and federal resource and regulatory agency staff should also be members, since they provide the technical and financial support needed to develop and implement a LCMP. A community may use a consultant to lead them through the process or to provide assistance at specific stages. The use and selection of a consultant is one of the group's first decisions.

The technical group is composed of planners, geologists, geographers, oceanographers, engineers, vegetation specialists, economists, lawyers, and other individuals with specialized knowledge, skills, or experience. Although the group may be formally established, it is probably best viewed as a fluid entity that meets on an as needed basis. The technical group will play a lead role during the inventory stages of the process. Its members will be responsible for data collection and analysis needed to develop the inventories. This group will also have a key role during the risk assessment and alternatives and impacts assessment stages and will conduct much of these analyses. Some members of the technical group may also be needed during policy development.

The management group is composed of representatives and officials from city and county government, port or other special districts; landowners and developers, citizens, special interest groups, other potential stake-holders; and federal and state regulatory and resource agencies. The management group should be established as a formal entity that convenes on a regular basis. This group plays a limited role during the inventory stage, but becomes more prominent in later stages. Its members are responsible for regulatory and resource management decisions, and therefore play a lead role in the policy development stages. It formulates the specific policies and procedures which make-up a chronic hazards management strategy. Its members will assist in accomplishing plan adoption and will be responsible for conducting post-adoption plan monitoring and maintenance.

Appendix

1. Ames, S.C. editor. *A Guide to Community Visioning: Hands on information for Local Communities.* Oregon Visions Project, Oregon Chapter, APA, 1993.

2. Good, J.W. and Riddington, S.S. editors, *Ocean Shore Protection Policy and Practices in Oregon. Coastal Natural Hazards*: Science, Engineering, and Public Policy, Oregon Sea Grant ORESU-B-92-001:145162, 1992.

3. Good, J.W., Editor, *Improving Natural Hazards Management on the Oregon Coast: Recommendations of the Coastal Natural Hazards Policy Working Group.* Oregon Sea Grant, ORESU-T-94-002, 1994.

4. Marra, J. J., *All Chronic Hazards Mapping Pilot Project.* Report to the Oregon Department of Land Conservation and Development. Sept. 1993.

5. Marra, J.J., *Appraisal of Chronic Hazard Alleviation Techniques with special reference to the Oregon Coast.* Report to the Oregon DLCD, December 1994.

6. Oregon Department of Land Conservation and Development, *Coastal and Ocean Resources Planning: An Assessment of Oregon's Coastal and Ocean Resources Issues and Management Capability - Coastal Hazards.* DLCD, January 1992.

CHAPTER 73

Changing Wave Climate and Littoral Drift Along the California Coast

Douglas L. Inman[1], and Scott A. Jenkins[2]

Abstract

Changes in wave climate that occurred along the California coast during the El Niño of 1982/83 can have catastrophic effects on beaches and may reverse the prevailing direction of transport within littoral cells. Research during the past two decades has advanced our understanding of world climate and the abrupt change in wave climate associated with the *El Niño/Southern Oscillation (ENSO)* phenomenon. Although ENSO events are not clearly understood, it is now possible to forecast their occurrence several months in advance; and their frequency of occurrence requires that they be factored into existing estimates of wave climate and the yield of sediment from streams and cliff erosion.

Introduction

The El Niño event of 1977/78 followed by the strongest El Niño on record in 1982/83 appear to have been harbingers of a continuing period of changes in wave climate and rainfall along the coastlines of California and the world. Since then El Niño's have occurred with a frequency unmatched during the past century (Figure 1a). This change has been accompanied by global temperature increases and unusually warm water off the eastern boundaries of major oceans such as the west coast of the Americas. Four of the warmest years since global records began in 1860 occurred in 1990, 1991, 1995 and 1996 with 1995 being 0.4 C above the 30-year average from 1960 to 1990 (e.g., Monastersky, 1995; 1997). Also, the second largest number of Atlantic hurricanes (19) since 1800 occurred in 1995. These climate changes are basin-wide in the Pacific and have affected the ocean and terrestrial climates world-wide (e.g., Miller et al. 1994; Kleeman et al. 1996; Ridgway & Godfrey, 1996; Gu & Philander, 1997).

[1] Research Professor of Oceanography, and [2] Senior Engineer, both at Center for Coastal Studies, Scripps Institution of Oceanography, La Jolla, CA 92093-0209

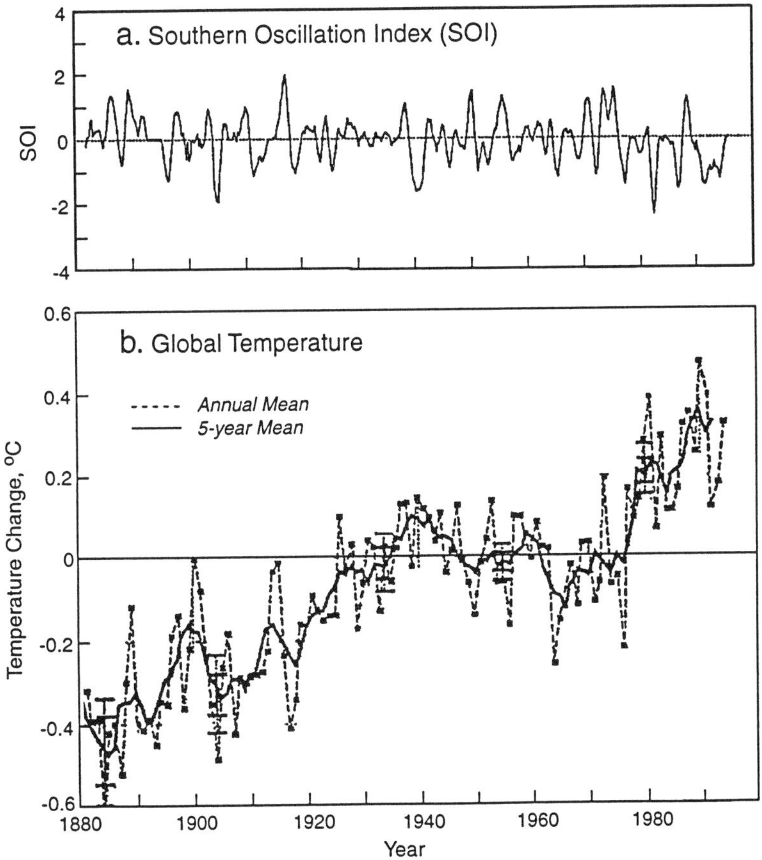

Figure 1. Southern Oscillation Index, SOI, (a) and global temperature change (b). [SOI is 1 year running mean of 3-month averages from NOAA; tempera tures from James Hansen, NASA Goddard Institute for Space Studies]

Paleo-time series obtained from Greenland ice cores, tree rings, coral reefs, and ocean sediments, all show that abrupt, fluctuating, changes in climate have occurred throughout the Holocene, and that these shifts may occur within a few years and last for several decades and longer (e.g. Meko, 1992; Cole et al, 1992; 1993; Heusser & Sirocko, 1997; Soutar & Isaacs, 1974). These paleo events and the presently occurring climate changes are generally thought to be related to the extreme phases (El Niño and La Niña) of the El Niño/Southern Oscillation (ENSO) phenomena. Hypotheses explaining the causes of unusual ENSO events are numerous and range from the extrusion of hot lava on the seafloor to variations in the intensity of the suns' radiations (e.g. Walker, 1995; Kerr, 1996; Haigh, 1996), with most climate modelers favoring a complex of changes in the atmosphere-ocean interaction that cause fluctuations in trade winds, monsoon intensity, and sea surface temperature (e.g. Latif & Barnett, 1994; Chen et al, 1995; Somerville, 1996). However, in spite of the disagreement on causes of climate change, a clear pattern of atmospheric-ocean events associated with ENSO phenomena has emerged. It is this pattern that gives rise to three month forecasts of ENSO events, even though the causes remain obscure. However, ongoing ENSO events are superimposed on a progressive rise in mean global temperature (Figure 1b); consequently, their effects are likely to be less predictable and perhaps more intense than previous events.

Pattern of ENSO Events

Seasonal variations in Earth's exposure to the sun produces interannual variations in atmospheric pressure fields which in turn cause the Earth's interannual seasons. Upon occassion, the typical seasonal weather cycles are abruptly and severely modified on a global scale. These intense global modifications are signaled by anomalies in the pressure fields between the tropical eastern Pacific Ocean and Australia/Malaysia known as the Southern Oscillation. The intensity of the oscillation is often measured in terms of the Southern Oscillation Index (SOI), defined as the monthly mean sea level pressure anomaly in mb normalized by the standard deviation of the monthly means for the period 1951-1980 at Tahiti minus that at Darwin, Australia (Figure 1a). ENSO events behave as enhancements and protractions of the interannual seasonal cycles, and their two extremes are referred to as El Niño (SOI negative) and La Niña (SOI positive), which apply to enhancements of Earth's January and July seasonal weather cycles, respectively.

Under prevailing, normal conditions, general ocean circulation and the westerly-blowing trade winds maintain a large pool of warm water in the tropical western Pacific which is ultimately regulated by the strength of the trade wind-driven component of gyre circulation. Associated with these prevailing conditions is a stationary area of low atmospheric pressure, centered on Darwin,

Australia, as compared with the relatively higher sea level pressure in the eastern Pacific from Tahiti to the coast of the Americas. El Niño events are associated with the release to the east of the warm water pool and a reversal of the relative pressure fields at Darwin as compared with Tahiti and the coast of the Americas, giving rise to a negative value of SOI. These changes in the pressure fields in the lower layers of the atmosphere redirect the upper level jet streams that steer the tracks of traveling storms and markedly alter the wave climate of the world's coastlines. The sequences within an El Niño event are illustrated in detail in Inman et al. (in press).

Examples of average atmospheric pressure anomalies and the storm-track enhancement (arrows) associated with La Niña (SOI positive) and El Niño (SOI negative) are shown in Figure 2 for the northern hemisphere. The strongest pressure gradients occur along the boundaries of the height anomalies, and these boundaries indicate the likely location of storm paths. Figure 2a (La Niña) shows large areas of high pressure over the North Pacific and eastern North Atlantic. These high pressure areas would enhance storm tracks as shown by the two arrows for waves approaching North America and Europe. The El Niño condition (Figure 2b) is characterized by a large area of negative height anomaly over the North Pacific Ocean and a relaxation in intensity but an increase in area of the positive height anomaly in the eastern North Atlantic. Thus, the La Niña/El Niño pattern over the North Pacific is bipolar and would be expected to give distinct changes in the storm track locations for the two conditions (Inman, et al. in press).

Along the Southern California coast a period of mild-stable weather occurred during the 30 years between the mid-1940's and mid-1970's when mild La Niña type weather prevailed. The average SOI for this period was +0.1, with strong La Niña events in 1950, (SOI = +1.4); 1955/56, (+1.2); 1970/71, (+1.0); 1973/74, (+1.0); and 1975/76 (+1.4). Winters were moderate with low rainfall, and winds were predominantly from the west-northwest. The principal wave energy was from Aleutian lows having storm tracks which usually did not reach southern California (Figure 2a). Summers were mild and dry with principal wave energy coming from the southern hemisphere. The wave climate in southern California changed, beginning with the El Niño years of 1978/79 and 1982/83 and extending at least until 1996. The average SOI for this period was -0.5, with the 1978/79 El Niño averaging -1.2, the 1982/83 El Niño averaging a record -1.7 and the 1993/94 El Niño recording a mean of -1.0. The prevailing northwesterly winter waves have been replaced by high energy waves approaching from the west (Figures 2b, 3), and the previous southern hemisphere swell waves of summer have been replaced by tropical storm waves from the waters off Central America. The net result appears to be a decrease in the southerly component of the longshore transport of sand that prevailed during the preceding thirty years (Inman, et al, in press). Previous southward drift rates of 200,000 m^3/yr in the vicinity of Oceanside have decreased to 50,000 m^3/yr and have reversed direction for protracted periods of time. Figure 3 shows a dramatic increase in both the numbers and intensity of extreme wave events during this recent flurry of ENSO

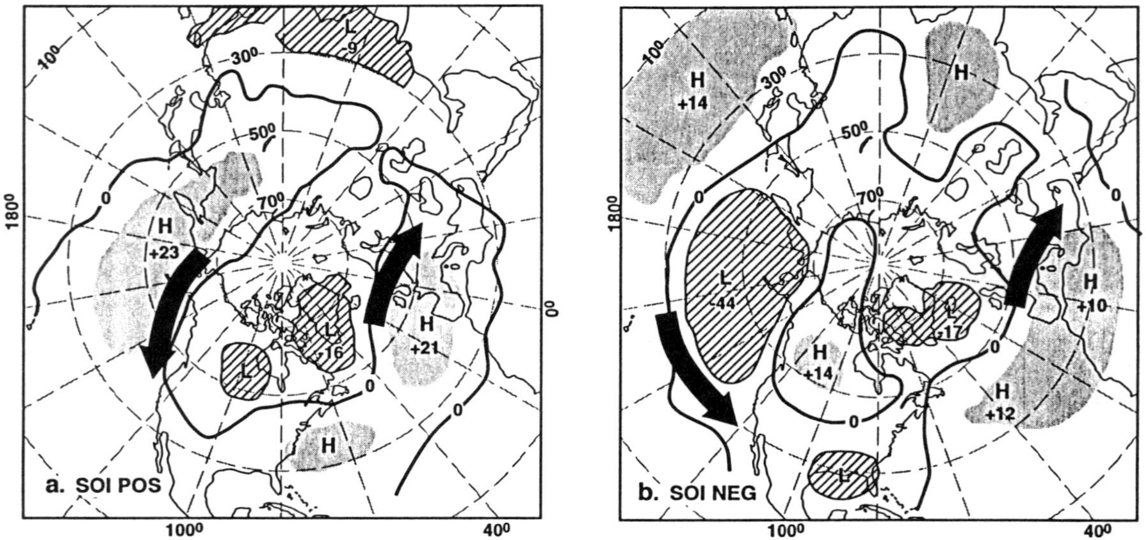

Figure 2. Height anomalies of the 700 mb atmospheric pressure surface derived from averaging winters (October-March) between 1947 and 1992 with preceding June-Nov. for 11 SOI of +0.5 or greater (a) and for 12 SOI of -0.5 or less (b). Shading and hatching indicate areas of negative (L) and positive (H) anomalies with maximum departure in meter. Arrows shows position of storm-track enhancement associated with the anomalies. [data from Redmond & Cayan, 1994]

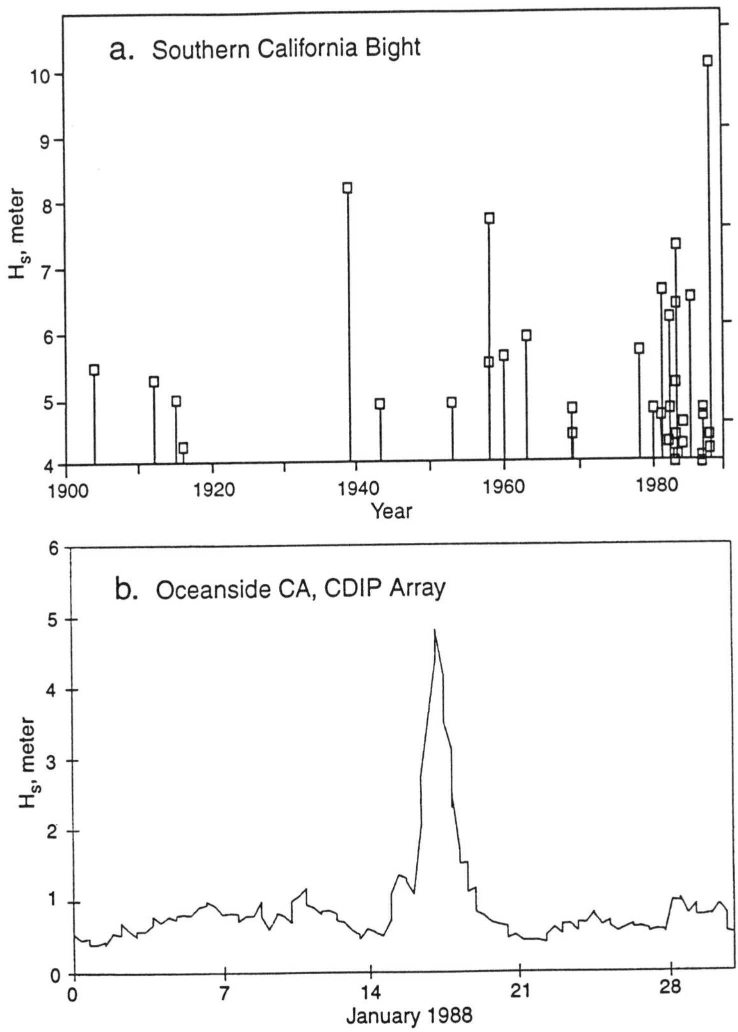

Figure 3. Occurrence in the Southern California Bight of wave events with deep-water significant heights $H_s \geq 4$ m (a) and measurements in 9 m depth from the CDIP array during January 1988 at Oceanside, CA (b). [data from Inman & Masters (1991) and Coastal Data Information Program, U. S. Army Corps. Eng. and California Department of Boating and Waterways]

events. In addition the modern record extreme wave event occurred in southern California on 18 January 1988, at which time the SOI was -0.50. This storm caused serious erosion and property damage throughout the Southern California Bight.

Other strong El Niño events of the past have also been accompanied by extreme wave events, although none of these have been as sustained as the succession of ENSO cycles from 1978 to 1995. The 1939/42 El Niñohad an average SOI of -1.3 and was associated with a series of destructive wave events in the Southern California Bight, the most intense being the 24/25 September 1939 storm which seriously damaged the breakwater system at Long Beach, CA. The El Niño of 1904/05 had a mean of -1.4 and was attended by a series of damaging west swells in March 1904 and again in March 1905 (Horrer, 1950; Marine Advisors, 1961).

The local effects of periods of mild-dry and warm-wet climate are manifest in the rainfall records of coastal cities and in the flow rate and sediment yield of adjacent rivers. The wet and dry periods become more visually apparent when the usual time history of rainfall is presented as a cumulative residual of the rainfall time series (compare Figure 4a, b). The graph in Figure 4b shows the cumulative plot of the summed residuals when the mean of the series is subtracted from each successive annual rainfall. In this representation periods of dry climate appear as intervals of decreasing residual (negative slope), while wet periods are represented by intervals of increasing residual (positive slope). The cumulative residual rainfall record for San Diego, California clearly shows the long period of relatively dry climate from 1944 to 1977, corresponding to the period of successive La Niña events. This is followed by the wet period beginning in 1978 and continuing to the present, which corresponds with the period of El Niño events discussed previously.

Along the coasts of California and Peru the La Niña dry periods and the El Niño wet periods are associated respectively with periods of relatively colder and warmer ocean water. These periods of colder and warmer water in the eastern boundary currents of the Pacific have had dramatic effect on the sardine and anchovy fisheries. The fishery flourished during the warm water periods of 1905-1944 (El Niño), collapsed in the colder waters during 1945-1978 (La Niña), and have reappeared during the warm El Niño water of recent years. Interestingly, the biomass of macrozooplankton in the waters off California has decreased by 80% since about 1975 (Roemmich & McGowan, 1995). The chronology of the fisheries off the California coast has been reconstructed from the fish scales preserved in the anoxic sediments of the Santa Barbara Basin in the Southern California Bight and from the Gulf of California. (Soutar & Isaacs, 1974; Baumgartner et al, 1992a;b; in preparation). The chronology shows that these fluctuating periods of fish abundance extend back in time for over two millenia.

The effects of the alternate series of La Niña (cool-dry) and El Niño (warm-wet) periods on sediment yield on three major rivers along the California coast is shown in Figure 5. However, there was a relatively weak El Niño

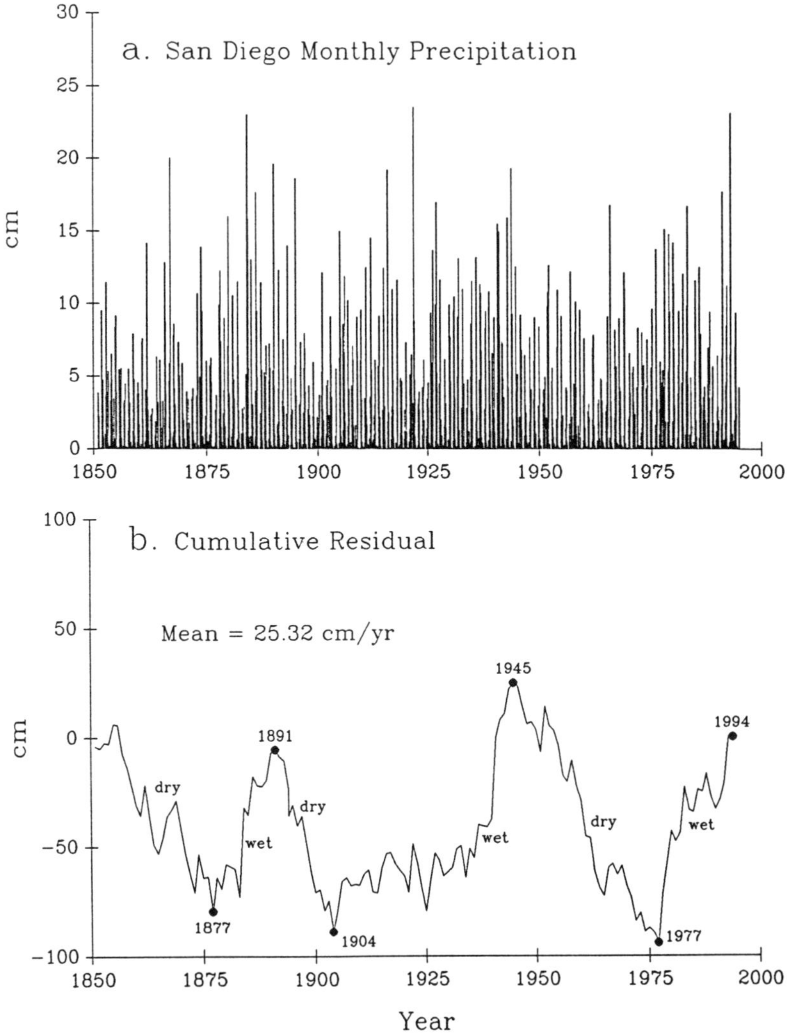

Figure 4. Monthly rainfall at San Diego, CA, 1851-1994 (a), and the cumulative residuals of annual rainfall for the same period (b).

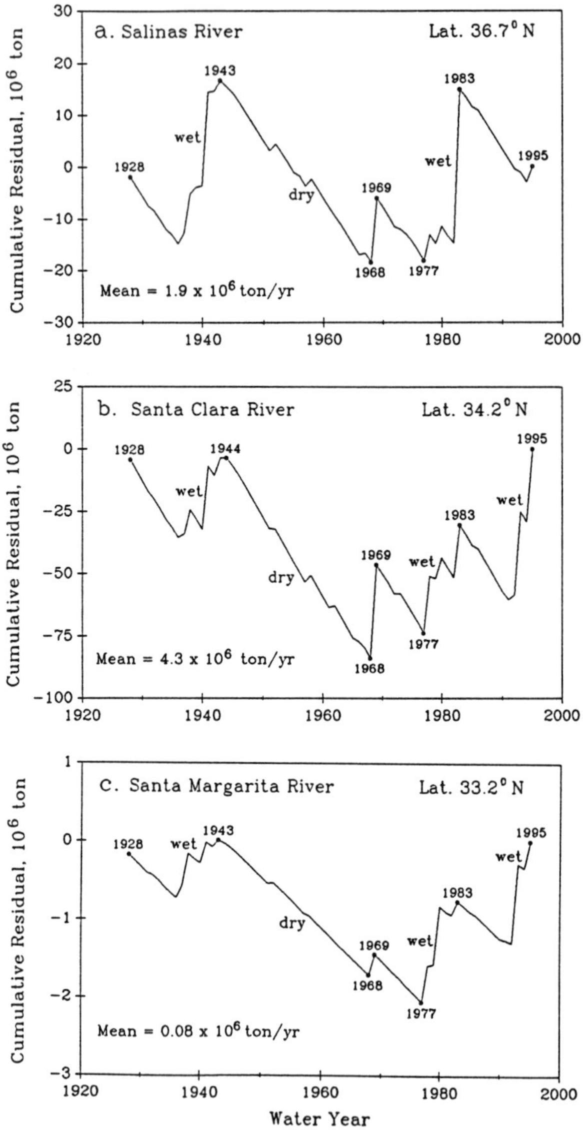

Figure 5. Cumulative residuals of annual sediment yield for (a) Salinas River, (b) Santa Clara River and (c) Santa Margarita River. [sediment yield following procedure of Brownlie & Taylor (1981) using stream flow data from the USGS]

embedded in this drought period in 1969, which had a peak SOI of -0.8. A storm associated with the El Niño caused significant flooding and high sediment yield on the Salinas and Santa Clara Rivers. Even more dramatic has been the increases in sediment yield which accompanied the powerful El Niño of 1982/83, and caused flooding on all rivers throughout California. The most recent, yet weaker, El Niño's of 1993/95 appear to have had the greatest increases in sediment yield in southern California, as shown in Figure 5 for the Santa Clara and Santa Margarita Rivers vs the Salinas River in central California. Perhaps this is an indication that the global warming effects evident in Figure 1 are intensifying the El Niño storm track patterns shown in Figure 2b to a greater degree than they have in the past, and thereby diverting rainfall away from the more northerly latitudes.

Conclusions

The Southern Oscillation Index (SOI) appears to be a useful diagnostic for identifying periods of anomalous wave climate and sediment yield along the coast of California. The occurrence of episodes of extreme event waves and enhanced sediment yield seems to correlate well in the modern record with protracted intervals of negative SOI, corresponding to the El Niño events of the ENSO cycles. The likelihood of extreme flooding and high energy waves apparently increases as the SOI becomes more negative for a greater length of time. Conversely, sustained periods of positive SOI, corresponding to the La Niña events of ENSO cycles, correlate with diminished sediment yield and mild wave climate along the southern California coastline. Because ENSO cycles are global in extent, quasi-systematic relationships between SOI and wave climate or sediment yield are likely elsewhere, particularly along the eastern boundaries of major ocean basins. The interaction between global mean warming and ENSO events is uncertain. However, comparison of Figures 1, 4b and 5 suggest that ENSO related wave climate and sediment yield along the coast of southern California intensify during periods of increased global temperature.

Acknowledgements

This study was supported by the Office of Naval Research, Ocean Modeling and Prediction, under Contract N00014-95-1-0005.

References

Baumgartner, T. R., V. Ferreira, J. Cowen & A. Soutar, 1992a, Reconstruction of a twentieth century wave chronology from the central Gulf of California, p. 606-616 in J. P. Dauphin & B. Simoneit, eds, *The Gulf and Penninsular Province of the Californias,* Amer. Assoc. Pet. Geol., Memoir 47,

Baumgartner, T. R., A. Soutar & V. Ferreira-Bartrina, 1992b, Reconstruction of the history of Pacific sardine and northern anchovy populations over the past two millennia from sediments of the Santa Barbara Basin, *CalCOFI* Reports, v. 33:24, 40 pp.

Baumgartner, T. R., & seven others, in preparation 1997, A sedimentary record of small pelagic fish abundances from the west coast of Vancouver Island, *Canadian Jour. Fisheries.*

Brownlie, W. R. & B. D. Taylor, 1981, Sediment management of southern California mountains, coastal plains and shoreline, Part C, Coastal Sediment Delivery by Major Rivers in southern California, *California Institute of Technology, Pasadena, CA,* Environ. Quality Lab. Report 17-C, 314 pp.

Chen, D., S. E. Zebiak, A. J. Busalacchi & M. A. Cane, 1995, An improved procedure for El Niño forecasting: implications for predictability, *Science,* v. 269, p. 1699-1702.

Cole, J. E., G. T. Shen, K. G. Fairbanks & M. Moore, 1992, Coral monitors of El Niño/Southern Oscillation dynamics across the equatorial Pacific, p. 349-375 in H. F. Diaz and V. Markgraf, eds., *El Niño, Historical and Paleoclimatic Aspects of the Southern Oscillation,* Cambridge Univ. Press, Cambridge, U. K., 476 pp.

Cole, J. E., R. G. Fairbanks & G. T. Shen, 1993, Recent variability in Southern Oscillation: isotopic results from a Tarawa Atoll coral, *Science,* v. 260, p. 1790-93.

Gu, D. & S. G. H. Philander, 1997, Interdecadal climate fluctuations that depend on exchanges between the tropics and extratropics, *Science,* v. 275, p. 805-807.

Haigh, J. D., 1996, The impact of solar variability on climate, *Science,* v. 272, p. 981-984.

Heusser, L. E. & F. Sirocks, 1997, Millennial pulsing of environmental change in southern California from the past 24 k.y.: a record of Indo-Pacific ENSO events, *Geology,* v. 25, n. 3, p. 243-246.

Horrer, P. L., 1950, Southern hemisphere swell and waves from a tropical storm at Long Beach, California, *Beach Erosion Bull.*, v. 4, n. 3, p. 1-18.

Inman, D. L. & P. M. Masters, 1991, Budget of sediment and prediction of the future state of the coast, chapter 9 (105 pp.) in State of the Coast Report,

San Diego Region, Coast of California Storm and Tidal Waves Study, U. S. Army Corps of Engineers, Los Angeles District, Chapters 1-10, Appen. A-Q.

Inman, D. L., S. A. Jenkins & M. H. S. Elwany, in press, Wave climate cycles and coastal engineering practice, *Coastal Eng., 1996, Proc. 25th Int. Conf., (Orlando),* Amer. Soc. Civil Eng., New York.

Kerr, R. A., 1996, A new dawn for sun-climate links, *Science,* v. 271, p. 1360-61.

Kleeman, R., R. A. Colman, N. R. Smith & S. B. Power, 1996, A recent change in the mean state of the Pacific basin climate: observational evidence and atmospheric and oceanic responses, *Jour. Geophys. Res.*, v. 101, n. C9, p. 20,483-99.

Latif, M. & T. P. Barnett, 1994, Causes of decadal climate variability over the North Pacific and North America, *Science,* v. 266, p. 634-637.

Marine Advisors, 1961, A statistical survey of ocean wave characteristics in southern California waters, prepared for the U. S. Army Corps of Eng., Los Angeles District by *Marine Advisors, Inc., La Jolla, CA*, 30 pp,.

Meko, D. M., 1992, Spectral properties of tree-ring data in the United States southwest as related to *El Niño, Historical and Paleoclimatic Aspects of the Southern Oscillation,* Cambridge Univ. Press, Cambridge, U. K., 476 pp.

Miller, A. J. D. R. Cayan, T. P. Barnett, N. E. Graham & J. M. Oberhuber, 1994, The 1976-77 climate shift of the Pacific Ocean, *Oceanography,* v. 7, n. 1, p. 21-26.

Monastersky, R., 1995, Tropical trouble, two decades of Pacific warmth have fired up the globe, *Science News,* v. 147, p. 154-155.

Monastersky, R., 1997, 1996: year of warmth and weather reversals, *Science News,* v. 151, p. 38.

Picaut, J., M. Ioualalen, C. Menkes, T. Delcroix & M. J. McPhaden, 1996, Mechanism of the zonal displacements of the Pacific warm pool: implications for ENSO, *Science,* v. 274, p. 1486-89.

Ridgway, K. R. & J. S. Godfrey, 1996, Long-term temperature and circulation changes off eastern Australia, *Jour. Geophys. Res.*, v. 101, n. C2, p. 3615-27.

Roemmich, D. & J. McGowan, 1995, Climatic warming and the decline of zooplankton in the California Current, *Science,* v. 267, p. 1324-26.

Sommerville, R. C. J., 1996, *The Forgiving Air, Understanding Enironmental Change,* University of California Press, Berkeley, 195 pp.

Soutar, A. & J. D. Isaacs, 1974, Abundance of pelagic fish during the 19th and 20th centuries as recorded in anaerobic sediment off the Californias, *Fishery Bull.*, v. 72, p. 257-294.

Walker, D. A., 1995, More evidence indicates link between El Niño, and seismicity, *EOS,* v. 76, n. 4, p. 33-35.

CHAPTER 74

Human Impacts on California's Coastal Sediment Supply

Douglas J. Sherman[1]

Abstract

Most of the beaches in California are eroding. Much of this erosion is at a rate fast enough to be termed "critical" in a recent study by Griggs. Chronic (as opposed to short-term cyclic) beach erosion represents a substantial threat to property, coastally dependent economic enterprise, and aspects of coastal ecosystems. Beach erosion, one manifestation of a negative local sediment budget (usually calculated at the scale of a littoral cell, or smaller reach) is often the result of natural processes. Frequently, however, human activities may alter rates of sand transport into or out of a littoral cell, thereby increasing erosion or sedimentation rates, or changing the sign of the net budget.

The purpose of this paper is to provide a general overview of the nature and magnitude of human impacts on California's coastal sediment supply. In particular, emphasis will be on those alterations that increase secular erosion rates. Direct impacts include the reduction of alongshore transport rates as a result of engineering structures such as groins, jettys, breakwaters, or cliff protection, and sand mining operations from beaches and adjacent dune systems. Indirect impacts include reduced sand supply to beaches as a result of dam construction and aggregate mining in coastal stream, and because of residential and transportation development at the foot of coastal cliff systems. In Southern California, the cumulative effects of these alterations have reduced sand supply in some littoral systems by up to 75% of its natural volume.

[1]Director, USC Sea Grant Program, and Professor of Geography, University of Southern California, Los Angeles, CA, 90089-0373.

This has greatly accelerated local erosion rates and caused the reduction of the recreational values of these beaches and increased the vulnerability of development to erosion and flooding. The magnitude of these impacts, and the impracticality of redesigning coastal land uses to minimize future impacts, requires the development of direct intervention policies as the only sure method of preserving and enhancing the value of California's beach resources.

Introduction

The beaches of California represent an important aesthetic, cultural, and economic resource. For many, these beaches are an integral part of the image of our state. Millions of bathers, surfers, fishermen, shell collectors and others have enjoyed the amenities provided by this vital infrastructure. Unfortunately, as a result of natural and human-induced changes, most of these beaches are eroding -- literally undermining this high-value resource base. The California Research Bureau (1993) estimates that approximately two-thirds of the state's $17.3 billion coastally-dependent economy derives from coastal recreation and tourism. The bulk of that portion of the economy reflects beach visitation. A study by Leeworthy and Wiley (1993, cited in Resources Agency 1997), for example found that the market value of the beaches at Santa Monica, Leo Carillo State Beach, and Cabrillo Pier/City of Long Beach was approximately $12 billion (1989 $), with a net, annual user value in excess of $357 million. Beaches also provide infrastructure for coastal development, especially residential real estate and related facilities, a buffer zone against coastal erosion and flooding, and may be critical habitat for a number of plant and animal species. To the extent that degradation of our beach resources can be ameliorated through improved management strategies, the economic (and other) value of our beach resources can be maintained or enhanced. This requires that we understand the processes that cause beach changes. The purpose of this paper is to address some of the challenges and issues associated with the identification and nature of human impacts on California's coastal sediment supply and the subsequent effects on beaches.

The coast of California is approximately 1100 miles (1760 km) long, and most of this coastline is fronted by at least a nominal width of sand or gravel beach. In a recent study, Griggs and Savoy (1985) suggested that 29% of the length of the coast, (about 500 km (or 315 miles), may be considered at "high risk" to erosion and flooding. Another 37% of the coast (650 km, or about 405 miles) is designated as requiring "caution" in terms of present risk and future development. Partly as a response to real or threatened loss, about 12% of the California coast has been protected with hard, engineered structures (Griggs et al., 1991). Along some of the highly developed reaches, most of the shoreline might be armored. Griggs et al. (1991) note that about 77% of the coastline between Carpenteria and Ventura has engineered structures. Much of the erosion and coastal flooding hazard can be directly attributed to inappropriate coastal development, especially from the period before the mid-1970's. Other components of the hazard result from natural variations in sediment supply to the beaches, and human-induced modifications to the littoral cell systems. For example, large scale erosion, both in spatial and temporal dimensions, has been attributed to the construction of the small craft harbor at Santa Barbara beginning in 1927. Other human impacts, especially beach nourishment projects, have had beneficial effects. Many of these impacts have received attention (e.g. Department of Navigation and Ocean Development, 1977; or Inman 1976).

Flick (1993, p. 3) has reviewed a diverse literature concerning the beaches of southern California from a sediment supply perspective , and notes that "...the widest sand beaches in southern California have been created and maintained by human intervention." The high, wide beaches along Santa Monica Bay, the Silver Strand (San Diego County), and at Ocean Beach (San Francisco County). Indeed, in many locations where natural sand delivery is minimal, recreational beaches are completely artificial. Such is the case at Avalon, on Santa Catalina Island.

From these simple examples, it can be seen that both beneficial and deleterious impacts may result from human influences on coastal sediment supply. But there are many instances where the magnitude and direction of human impact are difficult to recognize against the variability resulting from natural change. This represents one of the first challenges to estimating human impacts. The interactions between long and short term cyclicity, especially as a result of climate variability or secular environmental trends, present a puzzle difficult to unravel. Against this backdrop of causal uncertainty, human-induced change may be too subtle for recognition and definition. Further, there are few large beach environments that might be considered the pristine control site for comparative studies. This is true for California in general, and southern California in particular. It is difficult to imagine any beach south of Point Conception that does not reflect substantial human agency. Further, there are many complex interactions between human and natural systems, and between different human alterations. But we can designate a framework for the evaluation of human impacts, based upon the sediment budget and littoral cell concepts.

The Sediment Budget Concept

The sediment budget concept has been discussed extensively elsewhere (e.g., Bowen and Inman 1966), and is only reviewed briefly here. Figure 1 is a schematic designed to highlight the primary components of a typical coastal sediment budget. The volume of material contained in a beach represents the balance of sediment into and out of the system. The usual natural sources of sediment to a specific beach system include fluvial

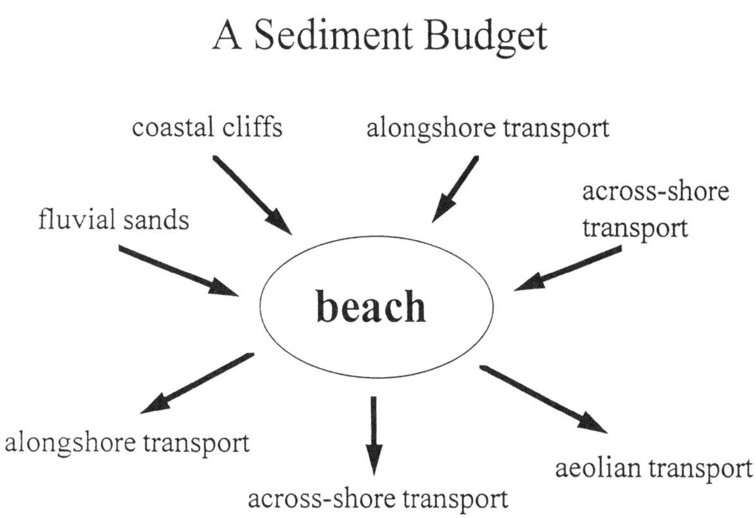

Figure 1: Components of a simplified coastal sediment budget

systems debouching to the shore, coastal cliff erosion, and alongshore and across-shore sediment transport systems. Alongshore and across-shore transport, and aeolian transport move sediment out of a particular beach. The ultimate sediment sinks are usually deep water repositories such as submarine canyons and spit termini, and, occasionally, coastal dune systems. The complete transport system has been defined as a littoral cell (e.g. Inman and Frautschy 1965), and presents an ideal organizational structure for coastal management and planning. Increased sediment supply to a cell, or decreased outflow, cause a positive sediment budget, and the beach volume will increase. Decreasing sediment supplies or increased outflow cause a negative sediment budget, and the beach will erode. Frequently, human activities alter rates of sand transport into or out of a littoral cell, thereby increasing erosion or sedimentation rates, or changing the sign of the net budget.

We can consider two basic categories of human influences on sediment budgets: direct human impacts and indirect human impacts. Direct human impacts include those that result from intentional intervention in the sediment budget. Examples of direct impacts include beach nourishment -- probably the single most important form of intervention -- and its counterpoint, sand mining from beaches and dunes, and the installation of coastal engineering structures, especially groins. Examples of indirect human impacts include those that result from human modification of the environment where sediment budget changes are incidental to the original impact of the action. Examples of indirect impacts include the reduction of fluvial sand delivery to the coastal zone as a result of dam construction or upstream sand mining, or the reduction of coastal cliff recession rates -- thereby reducing sand supply -- by cliff armoring, or changes in sediment delivery into fluvial systems as a result of land use changes.

Assessing Human Impacts

When considering a specific littoral cell along a developed coastline, or where coastal watersheds have substantial inland development, its sediment budget will usually reflect a suite of direct and indirect impacts. This often makes environmental interpretation, and the selection of appropriate management strategies difficult. Figure 2 illustrates coastal accretion along part of the southern end of the Santa Monica littoral cell -- the reach beginning at the Redondo Beach Breakwater and extending updrift for a distance of about 3,000 m. There are several notable aspects of these data (from LA County, 1960). First, this reach has a generally negative sediment budget, yet the data show substantial accretion over the 35 year period between 1935 and 1960. The accretion here represents the persistent effects of large scale beach nourishment projects, mainly in the vicinity of the Hyperion Treatment Plant at Dockweiler Beach -- a direct impact. Second, there is a pronounced increase in beach width (and accumulation rates) adjacent to the Redondo Beach Breakwater, constructed in 1939. In this instance, the breakwater functions as a very long groin, reducing alongshore transport out of the reach. If the structure had been intended as a groin, this impact would have been categorized as direct. As an incidental impact of harbor creation, the impact is better considered indirect (the distinction is made, although the net impact is identical). Third, there are morphological indications of an indirect impact associated with Hermosa Beach Municipal Pier, constructed in 1912. At this location, net accretion is at a minimum for this reach. The presence of the pier may alter local hydrodynamics in a manner that enhances sand transport away from the pier. Alternatively, the presence of the accretionary minimum at this site may be merely

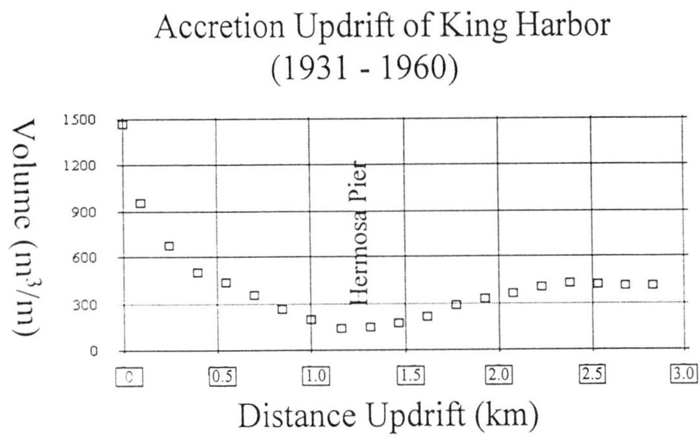

Figure 2: Sand accumulations updrift of the Redondo Beach Breakwater, 1931-1960.

coincidental. Without other, similar data for different time increments, the appropriate attribution is moot. Notably, these data represent conditions beginning more than 60 years ago. However, the large volumes of sand involved in the nourishment projects have left a residual imprint on this system, detectable to the present, and dominating natural controls..

The development of future management strategies must recognize that the beaches along most of Santa Monica Bay have been similarly altered through nourishment, and bear little resemblance to their natural condition (at least through historical times). Over the period 1941 - 1988, for example, average nourishment rates along this coast have been about 565,000 m^3/yr (Flick, 1993). Between 1939 and 1988, a total of about 24 million cubic meters of sand were emplaced along that stretch of shoreline (Liedersdorf et al. 1994). All of these nourishment effort represent direct human impacts. Because of the abundance of sand from these nourishment projects, the presence of cobbles along this coast (Shepard 1935) is largely a thing of the past, unless nourishment practices are ended.

Assessment of indirect impacts are more difficult. We can consider the behavior of fluvial sediment supply as an example. There are three primary factors that make the assessment of human impacts in such systems complicated: climate-induced hydrological variability, human-induced environmental changes, large-scale land use changes, and the construction of dams. First, there is a tremendous natural variability, associated largely with climate cycles and trends, and with the occasionally resulting high-discharge flood events. For many of the state's littoral systems, most sediments delivered over a hundred year period might be transported during only a few extreme flood events. Orme (1985) for example, notes that the average annual sediment discharge of the Santa Clara River is about 600,000 m^3/yr, although the median discharge is substantially less than that. The floods of 1969 caused a discharge of about 16.5 million m^3 from the Santa Clara, representing almost 30 years worth of the average. Note that the average, however, has been raised substantially because it includes the 1969 event, which in fact probably represented about 50 years worth of sediment delivery. This short-term event has long-term residual sediment-budget effects.

In contrast, we can consider the human-induced environmental changes associated with long term activities that lead to long-term, large spatial-scale effects on sediment budgets. One example is the persistent effects of the hydraulic gold-mining that took place in the Sierra Nevada Mountains between 1850 and 1914. Gilbert (1917) estimated that transport of the mining debris delivered about 1 billion m^3 of sediment to San Francisco Bay. He estimated further that about 20 million m^3 got to adjacent beaches. Anecdotal stories relate stories of placer mining along Ocean Beach as locals tried to recover gold from the former Sierra sediments. And the wedge of sediments sent downstream through the fluvial system is still working its way through the Sacramento/San Joaquin Rivers Delta region, to the bay and beaches. The sediment budget impacts continue more than 80 years after the cessation of hydraulic mining at the distant inland sites.

Third, land use changes, especially through the twentieth century, can have profound effects on sediment delivery to beaches. A dramatic example is presented by the growth of the city of Los Angeles across the largest coastal plain in the state. The extensive urbanization has covered the alluvial deposits created by the Los Angeles, San Gabriel, and Santa An Rivers. In some parts of the basin, sedimentary deposits are approximately 10,000 m thick -- in the vicinity of the Los Angeles River and Rio Hondo, for example (Norris and Webb 1990). Flooding through this plain was once able to mobilize surface sediments for delivery to the coastline. For example, the floods of 1862 enlarged the Santa Ana River to a width of about 3 miles in places (Norris and Webb 1990). In 1938, extensive flooding on the San Gabriel and Los Angeles Rivers covered wide portions of the basin, and also delivered large volumes of sediment to the coast. The surface sediments that once would have been available to transport through flooding have been removed from the modern system because of the anthropogenic armoring. There are many other, well-documented land use effects that increase sediment delivery, especially as a result of converting naturally vegetated systems to agriculture, or as the interim result of real estate development where raw earth is exposed to rainfall and runoff.

Finally, the development of water resources and increased need for flood protection has led to extensive damming of California's rivers and streams. This aspect of indirect human influences is perhaps the most studied and best documented for California. Reduction in sediment delivery occurs as a result of decreased flood flows -- both in terms of discharge and frequency -- and because of sediment trapping behind dams and in catch basins. Most of the dams in California are multiple use facilities, designed to enhance flood protection while storing water. The protective function acts as a filter to reduce the frequency and magnitude of runoff events. This may severely limit the ability of a particular fluvial system to push sediment through the coastal zone (including estuaries and lagoons). This will increase sediment storage in the lower reaches of stream.

In addition to reduced flood discharges, there is direct trapping of sands behind dams. Potter (1985) has estimated the volume of sand trapped in the San Gabriel system to have totaled approximately 30 million m^3 between 1938 and the early 1980's (although some of this sand has been removed). Wasyl et al. (1991) note that the sediments trapped in Lake Henshaw have caused subsidence in the Warner Valley of approximately 0.15 m due to static loading since the San Luis Rey River was dammed at that site. Reservoir sedimentation not only has negative impacts on coastal sediment supplies, but also reduces the capacity of reservoirs.

It is difficult to distinguish the magnitude of each of the paired impacts, but several studies have evaluated net effects. Brownlie and Taylor (1981), for example, have estimated present fluvial sediment delivery for all rivers and streams in southern California, and have made comparisons with delivery under "natural" conditions. Flick (1993) has gone a step further and, for the six southern-most littoral cells in California (Santa Barbara to Silver Strand), has compared the ranges of "natural" and "actual" river yields with annual nourishment budgets. His findings show that, for several of these cells, that beach nourishment totals approximate the reduction of natural sand delivery. This represents a shift in responsibility for rational sand management practices away from the water resource managers and onto coastal managers. More importantly, in several of the cells, even nourishment efforts leave the system underfed. Flick's (1993) work indicates that natural yields in the San Pedro cell have decreased by about 600,000 m^3 per year, with only about half of that volume replaced through nourishment projects. Even under conditions where it appears substantial maintenance efforts are being made, the beach resource continues to be devalued.

Conclusions

Much has been written about individual aspects of human impacts on coastal sediment supplies, and this paper has only highlighted a small portion of that literature. But the examples presented here serve to illustrate that complexity and importance associated with efforts to understand how we modify coastal systems -- intentionally and accidentally. Much research remains to be done before we can begin to confidently identify the magnitude of effect from particular management strategies. It is simple to conclude that unless we are prepared to limit inappropriate impacts, and invest substantially to correct past mistakes, the allure of California's beaches may become mythical indeed.

References

Brownlie, W.R., and B.D. Taylor (1981). *Sediment Management of Southern California Mountains, Coastal Plains and Shoreline -- Part C, Coastal*

Sediment Delivery by Major Rivers in Southern California. California Institute of Technology, Environmental Quality Lab Report No. 17-C, 314 pp.

Bowen, A.J., and D.A. Inman (1966). *Budget of Littoral Sands in the Vicinity of Point Arguello, California*. US Army Coastal Engineering Research Center, Technical Memo 19.

California Research Bureau (1993). *Economic Assessment of Ocean-Dependent Activities*. The Resources Agency, Sacramento, California.

Department of Navigation and Ocean Development (1977). *Study of Beach Nourishment along the Southern California Coast*. The Resources Agency, Sacramento, California, 151 pp.

Flick, R.E. (1993). The myth and reality of southern California beaches. *Shore and Beach*, Vol. 61: 3-13.

Gilbert, G.K. (1917). *Hydraulic Mining Debris in the Sierra Nevada*. US Geological Survey Professional Paper 105, Washington, D.C.

Griggs, G.B., J.E. Pepper, and M.E. Jordan (1991). California's coastal hazards policies: a critique. In Domurat, G.W., and T.H. Wakeman (eds.), *The California Coastal Experience*, A.S.C.E., New York, pp. 89-107.

Griggs, G.B., and L. Savoy (1985). *Living with the California Coast*. Duke University Press, North Carolina, 344 pp.

Inman, D.A. (1976). *Summary Report of Man's Impact on the California Coastal Zone*. Department of Navigation and Ocean Development, The Resources Agency, Sacramento, California, 150 pp.

Inman, D.A., and J.D. Frautschy (1965). Littoral processes and the development of shorelines. *Proceedings, 9th Coastal Engineering Conference*, A.S.C.E., New York, pp. 511-533.

LA County (1960). *Mass Diagram, Division 2: Redondo Breakwater to South Boundary, City of Los Angeles, 1935-1960*. Produced by Department of

Engineers, Los Angeles.

Leeworthy, V.R., and P.C. Wiley (1993). *Recreational Use Value for Three Southern California Beaches*. National Oceanic and Atmospheric Administration, Department of Commerce, Rockville, MD.

Leidersdorf, C. B., R.C.Hollar, and G. Woodell (1994). Human intervention with the beaches of Santa Monica Bay. *Shore and Beach*, Vol. 62: 29-38.

Norris R.M., and R.W. Webb (1990). *Geology of California, 2nd edition*. John Wiley and Sons, Inc., New York, 541 pp.

Orme, A.R. (1985). California. In Bird, E.C., and M.L. Schwartz (ed.s), *The World's Coastlines*. Van Nostrand-Reinholt, New York, 1071 pp.

Potter, D. (1985). Sand sluicing from dams on the San Gabriel River - is it feasible? In McGrath, J. (ed.), *California's Battered Coast; Proceedings from a Conference on Coastal Erosion,* San Diego, California Coastal Commission, pp. 251-259.

Resources Agency (1997) *California's Ocean Resources: An Agenda for the Future*. The Resources Agency, Sacramento, Ca.

Shepard, F.P. (1935) Gravel cusps on the California coast related to tides. *Science*, Vol. 82: 251-258.

Wasyl, J., S.A. Jenkins, and D.W. Skelly (1991). Sediment bypassing around dams - a potential beach erosion control mechanism. In Domurat, G.W., and T.H. Wakeman (eds.), *The California Coastal Experience*, A.S.C.E., New York, pp. 251-265.

CHAPTER 75

OIL SPILL PREVENTION AND RESPONSE: PARTNERSHIPS MAKE IT HAPPEN

Commander William J. Wetzel, USCG

Abstract

The Coast Guard has a long history of working within the marine community to improve oil spill prevention and response. The Oil Pollution Act of 1990 (OPA '90) gave added emphasis to these efforts. Today's approach to these problems is characterized by partnerships among those with a stake in their solution. This paper will outline recent Coast Guard initiatives, particularly within the Eleventh Coast Guard District (which encompasses California), that support our environmental goals.

It is increasingly recognized that human error is the greatest contributing factor to marine accidents, including oil spills. A new, non-regulatory initiative - Prevention Through People - is a nationwide program, involving partnerships with the marine community, to address the human factors, including organizational complacency and fatigue, which are the greatest obstacles to improved safety performance. This program builds upon port state control initiatives which have greatly reduced the numbers of "substandard" foreign-flag vessels calling at U.S. ports.

Within California, the Coast Guard is involved in a number of partnerships to improve maritime safety - with direct impacts on oil spill prevention and response - such as: the Vessel Traffic Information Service in Los Angeles/Long Beach; a recently-revised agreement with the State of California on cooperation in all aspects of marine safety; and a new agreement with the State and ports to improve ship pilotage.

Similarly, the Coast Guard helped create the Standard Oil Spill Response Management System (STORMS) task force, which has been instrumental in gaining acceptance of the Incident Command system (ICS) for oil spill response throughout the nation. This group also created the California Unified Oil Spill Control Course and developed the Field Operations Guide, and is now working to expand and improve ICS training throughout the response community.

Commander William J. Wetzel, USCG Chief, Pollution Response Branch, Eleventh Coast Guard District (Pm), Bldg 50-6, Coast Guard Island, Alameda CA 94501

The Coast guard also has a long-standing commitment to the Regional Response Team (RRT) for coordination of federal/State planning and response. The RRT is involved in projects to gain acceptance of alternative response technologies (e.g. in situ burning, use of dispersants) for offshore spills. Since 1991, the Coast Guard has spearheaded local Area Committees, which are responsible for developing unified response plans for oil spill contingencies. One of our more important achievements has been a joint U.S./Mexican oil spill contingency plan that improves inter-governmental cooperation.

CHAPTER 76

Oil Spill Prevention on the California Coast Facts, Failures and the Future

Warner Chabot

Abstract

Six years after the Exxon-Valdez oil spill, California's coast is still vulnerable to major risks from oil spills. Congressionally required Coast Guard studies on improving tanker safety languish, years overdue. Oil industry research and development budgets for new spill cleanup and prevention technology have been eliminated. Public vigilance and media interest are waning. Recent decisions to allow export of Alaskan crude oil to Japan, will mean an increase of west coast crude imports in foreign flagged tankers. That means more west coast tanker traffic with poorly maintained tankers and poorly trained crews.

In addition, recent Interior Department studies project California's offshore oil reserves may be twice as much as previously estimated. This may renew industry pressure to open the California's coast to offshore oil drilling. All of these issues indicate a need reconsider if California's coast is really protected from the future risk of oil spills.

Warner Chabot, Director of the Center for Marine Conservation's Pacific Regional Office, 580 Market Street, Suite 550, San Francisco, CA 94104

CHAPTER 77

The California State Lands Commission Marine Oil Terminal Pollution Prevention Program

Gary L. Gregory[1]

Abstract

The California legislature recognized the importance of prevention of oil pollution in its passage of the Lempert-Keene-Seastrand Oil Spill Prevention and Response Act of 1990. The Act placed responsibility for oil spill prevention at bulk liquid terminals with the California State Lands Commission. The Commission instituted a new program with the implementation of the Marine Facilities Division.

Since implementation, there has been a sharp decrease in volume of oil spilled at bulk liquid terminals in California. This paper will outline the systematic development of the Division and will discuss the effectiveness of this governmental program using analysis of data generated by the Commission and other entities.

Introduction

The California legislature recognized the importance of prevention of oil pollution in its passage of the Lempert-Keene-Seastrand Oil Spill Prevention and Response Act of 1990 (Act), when it declared, "Because of the inadequacy of existing cleanup and response measures and technology, the emphasis must be put on prevention, if the risk and consequences of oil spills are to be minimized." {Government Code § 8670.2. (f)} The Act placed responsibility for oil spill prevention at marine oil terminals with the California State Lands Commission. The Commission instituted a new program with the implementation of the Marine Facilities Division (Division).

The Commission's Responsibilities Under the Act

The Commission's historic role has been as a land management agency responsible for the sovereign lands of the state including coastal sea floor and river

[1]Chief, Marine Facilities Division, California State Lands Commission
330 Golden Shore, Suite 210, Long Beach, CA 90802

and stream beds throughout the state. In this role, the Commission has issued industrial leases for some twenty marine oil terminals and thus has had a proprietary interest in the safe operation of these facilities. The Act recognized the Commission's past expertise and expanded its responsibilities to all marine oil terminals in California state waters. Marine oil terminals are defined as those land based facilities that transfer oil to or from tankships and barges. There are over eighty such facilities in California. These terminals conduct almost 8000 transfer operations annually moving over 18.7 billion gallons of oil each year.

The Act specifically requires the Commission to:

1. "Inspect or cause to be inspected, on a regular basis, all marine facilities, along with associated equipment and shall monitor their operations and the effects on public health, safety, and the environment." P.R.C. §8757(a)

2. "...the commission shall adopt rules, regulations, guidelines, and commission leasing policies for reviewing the location, type, character, performance standards, size and operation of all existing and proposed marine terminals within the state..." P.R.C. §8755(a)

3. "Each operator of a marine facility shall prepare an operations manual describing equipment and procedures which the operator employs or will employ to protect public health and safety, an the environment and to prevent oil spills...The operations manual shall be submitted for approval to the commission." P.R.C.§8758(a)

Implementation

The Commission took extraordinary steps and care in creating the Marine Facilities Division. A Total Quality Management type environment was created to ensure the development of quality systems, a focus on service to the people of California, and the need for hard data for decision making. The Commission made a commitment to creating a field operations based, professional maritime organization.

First, a refined personnel classification (Marine Terminal Safety Series) was created to ensure that maritime expertise was available. Over two hundred individuals have applied and been examined for employment as a marine terminal safety professionals. The top candidates serve today as one the thirty-five professionals responsible for conducting daily monitorings of oil transfer operations and inspections of marine terminals as well as providing supervision and management. Division personnel have on average more than 19 years of oil transportation and environmental protection experience.

Secondly, the Commission focused on operational field enforcement. Field Offices were created, personnel were hired, trained, and field activities were commenced within three months of passage of the Act. In implementing its monitoring responsibilities, the commission believed that, much as the CHP Officer patrolling the highway causes drivers to be more careful, field inspectors cause marine transfer personnel to act in a more careful, safe and pollution free manner.

Field offices were established in Long Beach and Hercules, on Northern San Francisco Bay, as the home bases for inspector operations. Each field office operates seven day a week from 6:00 a.m. to 10:00 p.m. Special union agreements were made regarding rotating shift arrangements and special work hours to ensure almost 24 hour per day coverage in our port areas, while making efficient use of the limited funds available.

In order to establish a workable and comprehensive program, contact and coordination with other regulatory and safety agencies was essential. The Division identified over 20 agencies with some jurisdiction over marine terminals. The Division contacted those agencies and has worked closely with many of them. In January 1991, a Memorandum of Understanding was concluded with the U.S. Coast Guard to ensure coordination of activities to make the most of both organization's limited resources and to eliminate duplication of effort and needless or ineffective burdens on the industry.

A Memorandum of Understanding was completed in 1994 with the California State Fire Marshal. The MOU provides a procedure to ensure effective coordination and cooperation in carrying out both agency's responsibilities under the Act. In particular, the issue of jurisdiction over pipelines on marine terminals was coordinated by a joint physical inspection of each terminal with the terminal operator. Detailed jurisdictional determinations were made and documented to ensure effective regulatory compliance by terminal operators.

In 1995, detailed cooperative agreements and protocols were developed by the Division, the U.S. Coast Guard and the Department of Fish and Game's Office of Oil Spill Prevention and Response (OSPR). The Division works closely with other agencies, port authorities, and industry working groups.

Monitoring Program

Field operations remain the lifeblood of the Commission's program. An assessment is made daily as to the prioritization of monitoring individual transfer operations, based on a variety of factors such as the facility's oil spill and regulatory violation history as well as the vessel's frequency in California ports and its spill and violation history.

Inspectors monitor the operations of marine oil terminals to:

1. Ensure compliance with Commission regulations regarding safe and pollution free transfer operations;

2. Verify that terminal and vessel personnel are operating in accordance with the terminal's approved operations manual;

3. Verify vessel and terminal compliance with other applicable state and federal requirements as well as with industry accept practices.

Oil transfer operations are comprised of five separate portions or "events" which have different risk factors associated with them. Of the over 200,00 events in the years 1992 through 1996, Division inspectors have monitored over 50,000.

The Inspection Program

In spite of seeming to have a modern infrastructure, California's marine oil terminals are showing their age. The median design year for the terminals is 1949. Today's tankships using these terminals range in size from 30,000 deadweight tons (DWT) to over 200,000 DWT. In the late 1940's tankships were more typically in the 10,000 to 15,000 DWT category. Our terminals today resist and restrain forces much in excess of what was conceived during their design and construction.

This problem is further exacerbated by a lack of agreed upon industry or government standards for design, construction, maintenance, and repair of marine terminals. In the absence of a standard, the Division conducted inspections of terminals based upon the American Petroleum Institute's recommended practices for the inspection of marine structures such as oil drilling platforms.

The inspections were conducted jointly with the Coast Guard when possible and concentrated on the more critical structural areas of the facilities providing initial input for additional inspection if found to be necessary. Unfortunately, we found that most terminals had not had a comprehensive engineering inspection conducted in the near past, if at all.

Overall, 65% of the facilities inspected were found to be in good to excellent condition; 22% were far to good with the remainder being in poor condition. In all, 343 structural deficiencies were noted. Several terminals have been or are now being upgraded to ensure their safe operation. Some terminals however, are still in need of extensive engineering review and retrofit.

Under the Division's inspection program, a series of more comprehensive "audits" of terminal system safety has begun. A comprehensive "Audit Manual" is in the final stages of development. A facility audit will include underwater inspection of structural members and combines good engineering practices to provide a tool for facility operators to evaluate the terminal's fitness-for-purpose.

The Commission's future efforts will involve the continued development of structural design criteria for marine oil terminals and the study of seismic safety.

Regulatory Authorities

The Commission's responsibilities under the Act include regulation of all marine oil terminals in the state to ensure the "Best Achievable Protection" (BAP) of the public health and welfare and the environment using the "Best Achievable Technology" (BAT). Under the Act, BAP means the highest level of protection which can be achieved through both the use of the best achievable technology and those manpower levels, training procedures, and operational methods which provide the greatest degree of protection available {P.R.C. §8750 (c)}. BAT means the technology which provides the greatest degree of protection taking into consideration (1) processes which are being developed, or could feasibly be developed anywhere in the world, given overall reasonable expenditures on research and development, and (2) processes which are currently in use anywhere in the world {P.R.C. §8750 (d)}. The determination as to whether a measure constitutes BAP/BAT must give reasonable consideration to (1) the protection provided by the measure, (2) the technological achievability of the measure and (3) the cost of the measure.

At the outset, emergency rules were created that provided the regulatory base for the Division's activities. These regulations required compliance with the federal regulations that existed as well as providing notification to the Division of transfer operations. These regulations spanned the gap until more comprehensive requirements that incorporated BAP/BAT could be developed and moved through the legal process.

The Division's comprehensive regulations were based upon information gathered from around the world. Maritime nations worldwide and each coastal state in the Union were queried as to the existence of marine terminal standards. Maritime libraries were searched and international code making bodies were contacted in order to uncover the existing BAP/BAT standards. As a result, the division took the best of the standards and developed a draft set of requirements. The Division then convened a Technical Advisory Group (TAG) composed of a number of experts from various parts of the maritime and oil transportation community, state and federal agencies, engineering firms, environmental groups and academia. The TAG reviewed and revised the draft to ensure that BAP/BAT standards were being met. The draft then went though the regulatory review process to become a final regulation in December 1992. This comprehensive set of regulations is the foundation of the Division's regulatory program. Also, the use of a TAG has become the model for all rulemakings undertaken by the Division.

Additional rulemakings have been completed in response to actual and perceived risks. In 1994 the Commission adopted a program for the "Marine Terminal Personnel Training and Certification." This package was the first to address human and organizational error (HOE) as a major contributing factor in oil spill and non-compliance. The program requires a set of minimum standards for training of personnel and a certification by terminal management that personnel are qualified to perform the tasks assigned them. Also included are work hour restrictions that are meant to reduce or eliminate chronic fatigue as a contributing factor to oil spills. HOE continues to be a major concern of the Division and several projects are underway to develop tools for the analysis and elimination of high HOE risk factors at marine terminals.

Rulemakings have also included Vapor Control System Structural Requirements that look at installed vapor recovery systems used in loading tankvessels as part of the overall structure and transfer system at the facility. Marine terminal pipeline standards will be promulgated in June 1997 as required by the Act and in accordance with our MOU with the California State Fire Marshal.

As mentioned earlier, standards for the design, construction, maintenance, inspection and repair of terminals will require major efforts in the future.

Prevention First!

As part of its mandate to ensure the best achievable protection using the best achievable technology, the Division held two symposia entitle **Prevention First!** and **Prevention First! 96**. These symposia brought together people and organizations who believe that prevention is the key to protecting our resources.

Academia, government, environemtal groups and industry participated to work toward their mutual goal of preventing oil pollution incidents and their consequences. **Prevention First! 96** included the first ever prevention trade show of technologies for prevention of spills. Over 600 people attended the conferences and shared ideas on engineering of marine oil terminals and seismic risk, best achievable technologies, and human and organizational error. Watch for **Prevention First! 98**.

Oil Spill and Incident Investigation

Investigation of oil spills and incidents is a major tool in preventing marine oil pollution. The Division investigates incidents and spills at marine oil terminals to determine the cause of the event, not who is to blame. (Investigations for law enforcement are conducted by the Department of Fish and Game.) Inspectors are trained to determine not "what happened" but rather " why it happened". The root cause of the spill or incident may be hard to discern. But, by asking the question, "what could have been done to prevent this spill?" it is often possible to see the major cause in a complex chain of events. This information is a significant portion of our database and is used to analyze the program to assess shortfalls in the regulatory program.

This incident investigation methodology has reaped benefits when the data is entered into our database management systems. What was once held as "common wisdom" as to the riskier parts of oil transfer operations and oil spill causation is now know to be not entirely accurate. We now know that eighty per cent of spills are caused by human error. Almost half of the spills at marine terminals occur when the terminal is not actively involved in transfer operations. When terminals are idle and have a spill, the spill is most likely to be an equipment failure. Twenty five per cent of those spills resulted from testing and maintenance. We have also learned that during transfer operations, the "steady state" event which is believed to be the least risky part of a transfer operation accounts for almost thirty per cent of all spills.

It is clear that good investigatory practices, procedures and data collection can be invaluable in discerning the true causes of oil spills and other terminal incidents.

Program Measures of Effectiveness

As with any safety or prevention program, it is difficult to establish clear measures of program effectiveness. It is impossible to measure the events that "didn't occur." Data capture and proper analysis are the key to developing effective measurement tools.

Comprehensive data capture and analysis is another major building block of the Commission's program. Every marine terminal inspection, transfer monitoring activity, deficiency or non-compliance and oil spill investigation has been carefully documented in an electronic database. Trend analysis is conducted and is used for several purposes including advising industry officials and developing future program direction. The Division's comprehensive oil spill database is the only known database of small and large spills that effectively discerns spill causation. The database has been adopted for implementation in several other states and has been recommended as the model for database development by the States/British Columbia Oil Spill Task Force.

One measure of effectiveness is the measurement of spill events. While large spills are very rare events, smaller spills happen at a somewhat more frequent basis. Oil spill risk is examined as in any other risk based system. The risk associated with spills is the product of the probability that an event will occur and the consequences of the event. Small spills have a greater probability of occurrence with a smaller consequence. Large spills are very rare events with very large and unacceptable consequences. By having a spill database that includes all spills, we are able to get a more complete picture of risk management.

The following figures shows oil spill occurrence and the total amount spilled from 1992 through 1996. The data for 1992 shows spill events after one year of program activity. Our comprehensive regulations package went into effect in December 1992. If the regulations are effective, we would expect to see a change in spill events in the following year. As show, in 1993 there was a decrease of over 50% in the volume of oil spilled. Interestingly, the number of oil spill events did not change. So, the program appeared to change the consequences of oil spill incidents without changing the probability of the event's occurrence.

We believe that there are a number of issues at play here but the key issue is human and organizational error. Our training and certification for marine terminal is the start of a program to deal effectively with this issue. In fact, in 1996, spill numbers did decrease which could be the start of a trend toward fewer incidents. However, basing any conclusion on one data element is obviously a risky proposition.

Figure 1 also includes the amount of oil spilled at federal Department of Defense (DOD) terminals in California waters starting in 1995. Prior to then, these terminals and their spills were not under the jurisdiction of the Commission. Base on current federal law and executive orders. DOD terminals now comply with Commission regulations and the Division's prevention programs. There were 11 DOD spills in 1995.

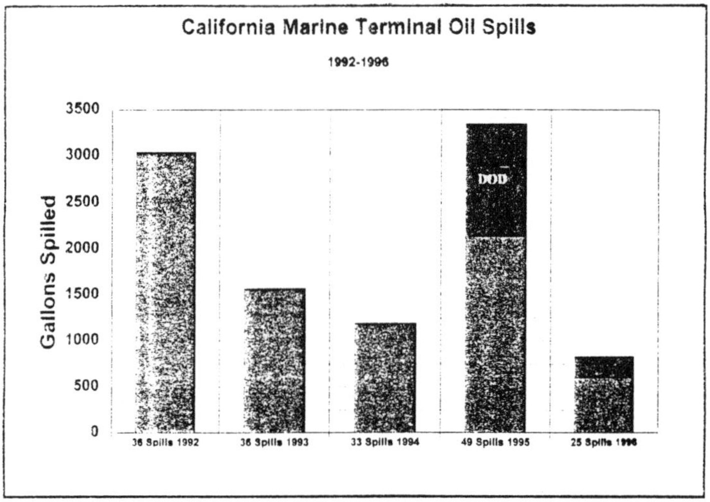

Figure 1

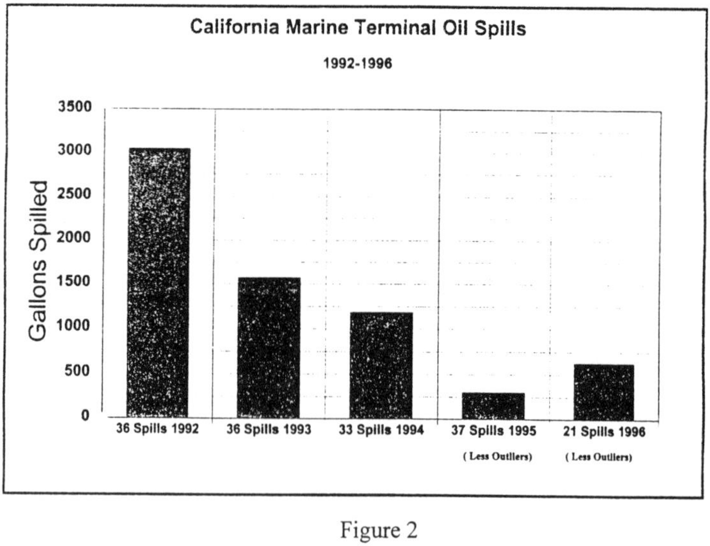

Figure 2

Figure 2 shows a similar view of incidents at California marine terminals but removes DOD terminals from the graph. One additional spill was removed in 1995 as a statistical outlier because its volume was much greater than all the other spills combined. What is left as Figure 2 gives a better presentation of the trends of oil spill incidents at marine terminals since the inception of the program. On average, the volume of spills is decreasing and the number of spill incidents is, at least, steady.

Another measure of effectiveness is compliance with the regulations adopted by the Commission and enforced by the Division. Deficiency histories show that non-compliance was at its highest in 1993. This should be no surprise as new comprehensive regulations went into effect in December 1992. In 1993 there were 972 deficiencies noted. In 1994 there were 91 deficiencies noted for the same set of requirements. Records for 1995 and 1996 show that compliance has been at about the same rate for the same set of deficiencies as in 1994. It appears that non-compliance is at a steady rate. A review of the deficiencies noted and their consequences are now being study to assess trends.

Conclusion

Under the Lempert-Keene-Seastrand Oil Spill Prevention and Response Act of 1990, the California State Lands Commission established a program to implement its oil spill prevention responsibilities. Through the Marine Facilities Division program, significant progress has been made in decreasing the risk of oil spills associated with marine oil terminals in California state waters. Through its database management program and quality based programs, the Division will ensure that spill rates and consequences are minimized.

CHAPTER 78

Joint Agency Cooperation to Ensure Safety and Protection of the Environment Through Pipeline Inspections

Theresa P. Bell[1], Catherine M. Hoffman[1], and James B. Hart[2]

Abstract

The U.S. Department of the Interior, Minerals Management Service (MMS), in reviewing recent offshore pipeline inspection surveys, recognized the need for refinements that would benefit both industry and government while ensuring safety and protection of the environment. Realizing that other Federal, State, and local agencies have different inspection requirements on the same pipelines, the MMS invited interested agencies to participate in a technical workgroup to review existing Federal and State agency requirements for inspections of pipelines on the Pacific Outer Continental Shelf (OCS). The Pipeline Inspection Quality Improvement Team (PIQIT), composed of representatives from the MMS, California State Lands Commission (CSLC), California State Fire Marshal (CSFM), and California Division of Oil, Gas, and Geothermal Resources (DOGGR), convened in spring of 1994. The CSFM acted as the representative for the U.S. Department of Transportation (DOT). After extensive evaluation of existing inspection requirements, the PIQIT concluded that the best approach to inspecting offshore pipelines is to critically examine each line individually and develop a performance-based inspection schedule for each based on the present condition of the line. The Offshore California Pipeline Inspection Survey (OCPIS) Plan was developed by the PIQIT to provide guidelines to improve the quality of pipeline inspection surveys.

The OCPIS Plan is a consensus-based, decisionmaking process intended to provide user agencies with an analytical framework for assessing the present condition and inspection needs of offshore pipelines. Implementation of the OCPIS Plan will permit pipeline operators to develop inspection strategies that are tailored to the needs of individual lines based on the actual condition of the pipeline. Requiring operators to conduct the most beneficial surveys based on the actual condition of the line will improve pipeline safety and

---

[1] Petroleum Engineer, U. S. Minerals Management Service, 770 Paseo Camarillo, Camarillo, CA 93010

[2] Petroleum Engineer, California State Lands Commission, 330 Golden Shore, Suite 210, Long Beach, CA 90802

reduce the risk of failure while affording industry an opportunity to reduce survey costs as a benefit of diligent and innovative inspection and maintenance.

The paper focuses on the OCPIS Plan, the implementation of the OCPIS Plan, the development of a Memorandum of Agreement between the PIQIT agencies, and updating external pipeline inspection requirements and guidelines. The paper also focuses on the cooperative nature of this effort and the advantages provided by this level of cooperation.

Introduction

In spring of 1994, the Minerals Management Service (MMS), Pacific OCS Region (POCSR), initiated an interagency review of pipeline inspection requirements. The Pipeline Inspection Quality Improvement Team (PIQIT), composed of representatives from the MMS, California State Lands Commission (CSLC), California State Fire Marshal (CSFM), and the California Division of Oil, Gas, and Geothermal Resources (DOGGR), was charged with the task of reviewing each agencies' pipeline inspection requirements and developing pipeline inspection guidelines that are consistent and non-duplicative. In particular, differing frequency requirements for some surveys are a source of concern for some pipeline operators who desire more uniformity in agency inspection requirements. The CSFM acted as the representative for the U.S. Department of Transportation (DOT).

After extensive evaluation of existing inspection requirements, the PIQIT concluded that the best approach to enhancing pipeline safety would be to abandon the traditional "one-size-fits-all" approach and adopt performance-based standards to ensure adequate inspection of pipelines. The Offshore California Pipeline Inspection Survey (OCPIS) was developed by the PIQIT to provide user agencies with an analytical framework for assessing the present condition and inspection needs of offshore pipelines. The OCPIS Plan facilitates coordination between agencies early in the process to identify issues and concerns and develop consensus on regulatory actions.

The OCPIS Plan encourages pipeline operators to develop pipeline inspection strategies specific to their individual pipelines and pipeline systems mindful that no offshore pipeline system can be comprehensively documented with any single inspection tool or technique.

PIQIT Agencies' Goals

At the outset, the PIQIT's goal was to standardize as much as possible the inspection requirements for offshore lines, which at that time varied considerably from agency to agency. After examining in great detail the many variables that influence survey methodology and frequency, the PIQIT determined that the principle factors influencing surveys tend to be line specific and do not conform well to prescriptive inspection requirements. Offshore pipeline operators will be asked to critically examine each of their lines and develop an inspection schedule based on the present condition and risk potential of the line for review and approval by the appropriate PIQIT agencies. In most cases, with the absence of an approved inspection schedule, the appropriate agencies' existing policy and requirements have been serving as a default plan for POCSR lines. The uniform process eliminates the need for uniform regulations and related statutory revisions that might be required for implementation by some agencies. This approach achieves the other PIQIT agencies' goals of cooperation and ensuring safety.

The Offshore California Pipeline Inspection Survey (OCPIS) Plan

The **OCPIS Plan** process and procedures (described below and detailed in the checklists in fig. 2-5) is a nine-step process to:

- streamline regulatory process for operators,
- identify agencies' issues and concerns,
- focus deliberations to resolve concerns,
- develop a partnership between agencies to exchange information and resolve differences through coordination with operators,
- develop alternative inspection or remediation proposals as needed,
- build consensus among agencies, and
- make appropriate recommendations.

The **OCPIS Plan** flowchart (figure 1) and the decision checklist (figure 2) are utilized when considering each of the following proposed actions:

(1) An operator's proposed survey plan or request for a waiver from existing survey requirements,

(2) An agency's request for a change in current survey requirements, or

(3) An agency's requirement for an unscheduled inspection following an offshore incident or accident.

Once an action is initiated, either by an agency or an operator, a lead agency is identified. The lead agency examines the jurisdictional issues and regulatory requirements of each affected Federal, State, and local agency and determines if a joint review or consultation is needed. If coordination is necessary, the lead agency will inform all affected agencies of the pending action.

Using Checklist I (fig. 3) as a guide, the lead agency gathers pertinent design, operational, inspection, repair, environmental, and other data and information from agencies' and operators' files. The evaluation of the pipeline through Checklist I is based on an analysis of eight general categories of information (containing 36 influential criteria) related to pipeline design, operation, inspection, maintenance, incident history, physical environment, and other factors.

Individual criteria on Checklist I are relatively weighted as "primary", "secondary", or "non-applicable" depending on how much weight (or influence) should be placed on the factor in assessing internal and external survey methods and frequencies. The purpose of defining and weighting criteria was to provide an analytical basis for evaluating the integrity and inspection needs of offshore pipelines to assist in regulatory decisionmaking. However, the weight a user places on an individual criterion may be, in many cases, line and setting dependent. The user, utilizing sound engineering practices, must decide how much weight to place on a given criterion for a given situation or environment in reaching a decision.

The lead agency uses Checklist II (fig. 4) to synthesize the information in Checklist I and assess the present condition of the line, the compliance history of the operator, and the potential for future pipeline failures.

Using Checklist III (fig. 5) as a guide, the lead agency initiates an evaluation of the proposed action, either independently or jointly with the affected agencies, as appropriate. The agencies identify and attempt to resolve concerns relating to the pipeline evaluation, the proposed action, or an alternative recommended action(s) through coordination with all parties including the operator. The agencies work towards achieving a consensus decision on the proposed action, if possible, and issue either joint or independent recommendations to their respective managements to approve or deny the proposed action (i.e, plan or waiver request) or to require an alternate inspection or remediation plan, as appropriate. The operator is subsequently notified in writing of the agency's(ies') decision(s).

Pipeline Evaluation Workshop

The PIQIT convened a table-top exercise in July and August 1995 to test and refine the **OCPIS Plan** process and procedures. The PIQIT used an actual pipeline inspection waiver request as a test-case to assess the adequacy and utility of the **OCPIS Plan**. As part of the exercise, participants made a detailed evaluation of the integrity of the pipelines using the draft process and checklists as a guide.

Overall, the exercise demonstrated the usefulness of the **OCPIS Plan** process in evaluating the integrity of offshore pipelines and illustrated some areas that needed refinement. Those refinements were made.

Implementation of the OCPIS Plan

The **OCPIS Plan** is currently being implemented by the PIQIT agencies. The PIQIT agencies and the DOT are in the process of finalizing a Memorandum of Agreement (MOA) to jointly implement the **OCPIS Plan**. When this new MOA is signed and implemented it will:

- officially adopt and implement the **OCPIS Plan**,
- have the agencies review and update the **OCPIS Plan**,
- strive toward achieving consensus decisions between agencies,
- establish guidelines for the lead agency, and
- share information on offshore pipeline and pipeline studies.

The MMS has met with the Counties of Santa Barbara and Ventura and the City of Carpinteria to discuss the local government's involvement in the process. There have been discussions on modifying the **OCPIS Plan** process to include those onshore portions of pipelines that come from Federal and State offshore facilities to onshore processing facilities. The MMS is currently reviewing this proposal.

Currently, the POCSR is in the process of updating its default inspection policy and finalizing its external survey requirements. The POCSR envisions that these requirements will be incorporated into a comprehensive Notice to Lessees on offshore pipeline inspections, as was recommended by the PIQIT.

One Case History

Recently, an operator in the POCSR requested to delay the internal pipeline inspections for 1 year. The operator wanted to synchronize the internal inspections for all the pipelines in the one pipeline system for their two platforms. The MMS acted as the lead agency and was able to use the **OCPIS Plan** to evaluate the pipeline condition and the waiver request with the other agencies involved. This pipeline system has permit conditions and regulatory requirements from five agencies: MMS, DOT, CSLC, County of Santa Barbara, and City of Carpinteria. Not all of these agencies were involved with the review of the waiver or the final decision because some of them were not impacted by way of permit conditions or regulatory requirements, but all were informed of the final decision.

The staffs of the agencies impacted by the waiver request were able to agree and presented their finding to management who then in-turn also agreed with the finding. Based on the finding, the operator was given two choices: (1) perform the inspections on schedule, or (2) delay and perform all the inspections for the pipeline system within 6 months. The operator chose the latter and was pleased to be given a choice.

Conclusion

The **OCPIS Plan** has been implemented for the last year and been a useful tool in making consensus decisions between the agencies and local governments. The successful implementation of the **OCPIS Plan** presents a systematic approach to inspecting and evaluating offshore pipelines. It facilitates the development of uniform agency inspection requirements for individual pipelines and is especially helpful for pipeline segments that fall under multi jurisdictions with the goal of all agencies of ensuring safety and protecting the environment.

Acknowledgments

The authors wish to thank the other members of the PIQIT: Richard Clingan (Team Leader) and John Smith, MMS; Greg Pelka, CSLC; Robert Gorham, CS FM; and Steve Fields, DOGGR for all their work and perseverance. The authors also wish to thank the management of our agencies for adopting the **OCPIS Plan** and our recommendations.

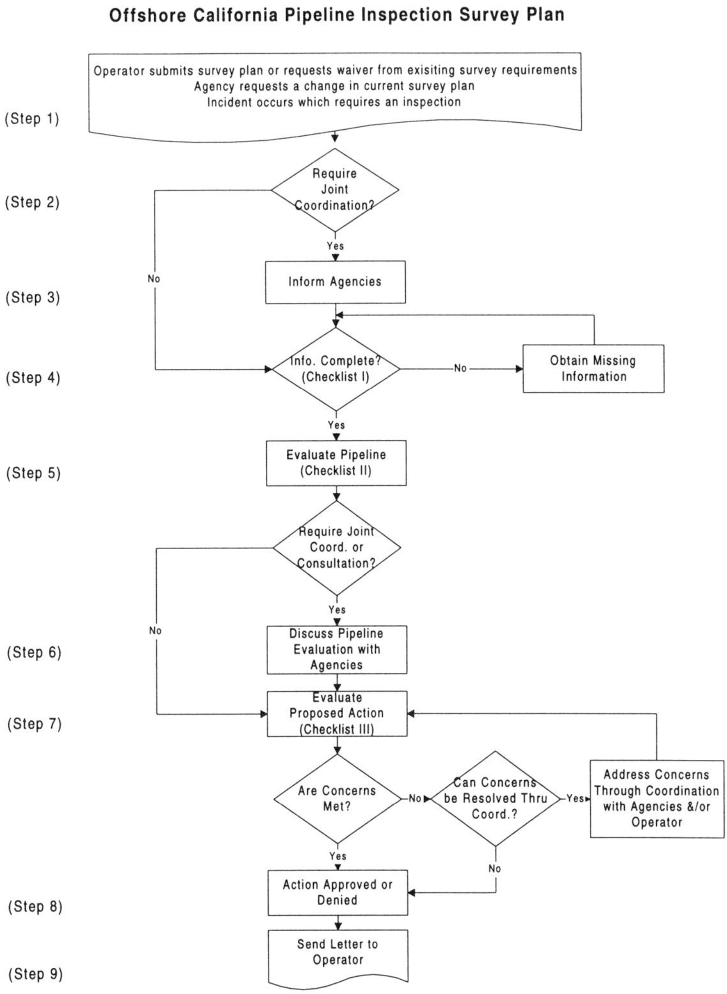

Figure. 1 Offshore California Pipeline Inspection Survey Plan flowchart

OFFSHORE CALIFORNIA PIPELINE INSPECTION SURVEY (OCPIS) PLAN DECISION CHECKLIST
(For fig. 1)

STEP 1. Action identified that requires a decision: (check one)

Operator submits a survey plan ____
Operator requests a waiver from current requirements ____
Agency requests a change in current survey plan ____
Pipeline incident requires an inspection ____
- Describe:________________________________

STEP 2. Agency receiving or initiating action examines jurisdictional issues and regulatory requirements and determines if a joint review or consultation is required.

Agencies with inspection authority over the pipelines and their requirements ________________
____
Conflicting jurisdictional requirements between agencies ________________
Lead agency(ies) ________________
Additional agencies needing notification ________________
Interagency agreements and conditions ________________

STEP 3. Lead agency informs affected agencies of pending action if decision process includes joint review or consultation.

Agencies contacted ________________
Agencies requiring joint review ________________

STEP 4. Lead agency compiles data and information for pipeline evaluation using "Offshore California Pipeline Evaluation Checklist I." (check)

Missing data and information obtained from operators ____
Checklist I completed ____

STEP 5. Lead agency evaluates pipeline using "Offshore California Pipeline Evaluation Checklists I & II." (check)

Checklist II completed ____

STEP 6. Lead agency discusses pipeline evaluation with affected agencies if decision process includes joint review or consultation.

Agencies consulted ________________

STEP 7. Lead agency evaluates the proposed action independently or jointly with affected agencies, as appropriate, using "Offshore California Pipeline Evaluation Checklist III." Agencies identify and attempt to resolve concerns through plan revisions. Plan revisions reevaluated to determine if concerns are adequately addressed. (check/explain)

Agency concerns ________________
____
- if concerns are resolved, go to Step 8 ____
- if concerns may be resolved through coordination, attempt to resolve concerns ____
- if concerns cannot be resolved through coordination, go to Step 8 ____
Unresolved concerns ________________
Checklist III completed ____
Final recommendation ________________

STEP 8. The proposed action is approved or denied.

Action approved ______Conditions ________________
Action denied _______ Explain ________________
Alternate plan or remediation ________________

STEP 9. Operator is notified in writing of agency(ies) decision. (check one)

Joint actions ____
Independent action(s) ____

Figure 2 Offshore California Pipeline Inspection Survey (OCPIS) Plan Decision Checklist

CHECKLIST I: OFFSHORE CALIFORNIA PIPELINE EVALUATION - DATA AND INFORMATION

	Criteria I/E
A. Pipe specifications:	
Diameter ______	P/N
Wall thickness ______	P/N
Process of Manufacture ______	P/N
Steel grade ______	P/N
Flange rating ______	P/N
Installation date ______	S/N
B. Operating conditions:	
Normal operating pressure ______	P/N
Maximum operating pressure (MAOP) ______	P/N
Flow rate ______	P/N
Product type and composition ______	P/S
% Water ______ % CO_2 ______ ppm H_2S ______ Other ______	
C. Environmental factors:	
Water depth ______	N/S
Geological/geotechnical conditions along route ______	N/S
Effects of currents on pipeline integrity ______	S/S
Proximity to environmentally sensitive habitats ______	S/P
D. Present pipeline characteristics:	
Is line smart piggable? ______	P/S
Types of internal corrosion controls ______	P/N
Type of cathodic protection ______	P/P
Type of external coating ______	S/P
Buried or exposed sections ______	S/P
Spanned sections ______	S/P
E. Inspection history:	
Date, results and quality of most recent:	
- internal inspection ______	P/S
- external inspection ______	N/P
- cathodic protection survey ______	N/P
- pressure test ______	P/N
Extent, location and rate of:	
- internal corrosion ______	P/N
- external corrosion ______	S/P
F. Maintenance history:	
Date, location and description of repairs:	
- leaks ______	P/P
- spans ______	P/P
- other safety deficiencies (specify) ______	P/P
- third party damage ______	P/P
What maintenance records are available? ______	P/P
Additional corrective and preventive maintenance ______	P/P
G. Recent incidents:	
Impacts on pipeline integrity from:	
- seismic loads ______	S/P
- storm loads ______	S/P
- third party damage ______	S/P
H. Waiver history (explain): ______	S/S

Criteria Key: P-primary, S-secondary, N-non-applicable, I-internal survey, E-external survey

Figure 3 Checklist I: Offshore California Pipeline Evaluation - Data and Information

CHECKLIST II: OFFSHORE CALIFORNIA PIPELINE EVALUATION - ANALYSIS AND CONCLUSIONS

A. **Pipeline evaluation:**
Present condition of the aggregate pipeline:
- Internal ____
- External ____
Present condition of the riser:
- Internal ____
- External ____
Problem areas identified in past inspections ____

B. **Compliance history (waiver requests):**
Operator's diligence in inspecting the line ____
Operator's diligence in maintaining the line ____

C. **Risk assessment:**
Identify potential for pipeline failure(s) due to each of the following (individually or collectively):
- internal corrosion ____
- external corrosion ____
- leaks ____
- spans ____
- third-party damage ____
- natural phenomena ____
- weight-coating damage ____
- operator non-compliance ____
- other: (specify) ____

Figure 4 Checklist II: Offshore California Pipeline Evaluation - Analysis and Conclusions

CHECKLIST III: OFFSHORE CALIFORNIA PIPELINE EVALUATION - RECOMMENDATIONS

A. **Identify action or incident:** ____

B. **Identify agency concerns:** ____
Can concerns be resolved through coordination with agencies or operator? ____ If no, explain: ____
Resolution: ____

C. **Alternative actions:**
Plan or waiver approved ____ Conditions of approval: ____

Plan or waiver disapproved/denied ____ Explain: ____

Alternate inspection plan recommended (explain): ____

- internal inspection (smart pig, other): ____
- pressure test: ____
- external inspection (SSS, diver/ROV visual search, other): ____
- cathodic protection: ____
- no survey required: ____

Remediation recommended (explain): ____
- replacement, upgrade, or improvements: ____
- reduce operating pressure: ____
- other (specify): ____

D. **Final Recommendation:** ____

Figure 5 Checklist III: Offshore California Pipeline Evaluation - Recommendations

CHAPTER 79

POLLUTION IMPACTS TO OCEAN AND COASTAL RESOURCES: THE COMPLEX PROCESS OF ASSESSMENT AND RESTORATION

Dr. Pierre duVair

Abstract

California's's Office of Oil Spill Prevention and Response (OSPR) has a comprehensive program designed to prevent and respond to oil spills that affect marine and coastal resources. This paper describes OSPR's approach to coordinate cleanup and response operations with the assessment of impacts to natural resources. Early identification of restoration alternatives is emphasized as part of injury determination and valuation of the spill damages. The benefits of coordination both injury assessment and restoration planning are described through case examples of: improves site access for sampling; improved documentation of ephemeral impacts; avoidance of redundant sampling; minimized collateral injuries from cleanup or response actions; expedited analysis of restoration options and determination of the types of injured resources that can be restored.

The resource agencies use a "team concept" for assessing oil spill impacts in California. Potential benefits of the team approach as well as the pros and cons of joint damage assessments and restoration planning with responsible parties are presented as learned by OSPR through recent spill cases in California. Benefits of pre-spill planning are described, including the need for response training and selection of acceptable and scientifically defensible methods to assess spill impacts. This paper concludes with a discussion of criteria or factors commonly used to select restoration alternatives to make the environment and public whole. The importance of regional restoration planning is identified with respect to improved marine resource conservation.

CHAPTER 80

Impact of Oil Spills on Commercial Fisheries: Lessons of Alaska for California

Timothy R. Holder

Abstract

California and Alaska each have valuable commercial fisheries in their adjoining ocean waters. This paper explores the opportunity to use predictive models prepared for Alaska commercial fisheries to reduce time spent and awards made in litigation. Litigation after the *Exxon Valdez* Oil Spill has resulted in multi-billion dollar damages. Use of predictive models could help reduce the litigated damages.

1. Estimated Damages to Commercial Fisheries from the Exxon Valdez Oil Spill

The *Exxon Valdez* Oil Spill (EVOS) reduced ex-vessel revenue for southcentral Alaska's commercial fishermen during 1989 by an amount estimated between \$6 million and \$42 million and in 1990 between \$11 and \$45 million. These fisheries are primarily in Prince William Sound and Cook Inlet. Many fishermen declared bankruptcy because they could not make loan payments as a result of an inability to fish caused by EVOS. (Cohen, Maurice J., "The Economic Impacts of the *Exxon Valdez* Oil Spill on Alaska's Commercial Fishing Industry" in "*Exxon Valdez* Oil Spill Symposium: Program and Abstracts," February 1993, Anchorage, Alaska.)

Actual damages to commercial fishermen were determined in court to be \$287 million. The punitive damages could be even larger. As of 1996, the court case against Exxon U.S.A. could result in payments to commercial fishermen and other associated entities for punitive damages totaling \$4.29 billion. The punitive damage amounts are according to a plan being considered by the court. Exxon U.S.A. is likely to appeal such a decision and this appeal could take several years (Anchorage Daily News, March 17, 1996, page B-3).

2. Value of California Commercial Fisheries

The value of California's commercial fisheries as a whole is on the same order of magnitude as Alaska's fisheries affected by EVOS. In 1992, California's commercial fishermen landed fish and invertebrates worth over \$131 million. The fishing industry sector of the California economy, which includes commercial fishing and fish processing, generated a total income estimated at $466 million, and the income generated by the marine fishing industry is estimated at approximately $88 million. These industries provide employment directly and indirectly to nearly 17,000 workers. In 1992 kelp and sea vegetable harvest had a value of $2.9 million ("California's Ocean Resources: An Agenda for the Future," Appendix A: Economic Assessment of Seven Ocean-Dependent Industries, pp. 1, 8, and 11).

3. Forecasting Value of Commercial Fisheries

While the risk of a significant oil spill is small in both California and Alaska, the EVOS experience has shown that damages to commercial fishing can be significant when a large spill does occur. Compensation for damages resulting from spills of oil produced on the federal Outer Continental Shelf is provided for in the Outer Continental Shelf Lands Act. Compensation for damages resulting from any oil spill in U.S. waters is provided for in the Oil Pollution Act of 1990. A prediction of the value of commercial fisheries could be of use in minimizing litigation after a spill. Predictive commercial fisheries models could provide a framework for anticipating potential damages to commercial fisheries. The variables used in models could provide a framework for anticipating potential damages to commercial fisheries. The variables used in models developed for the commercial fisheries in the Bering Sea and Gulf of Alaska by Minerals Management Service could be useful in the California situation.

CHAPTER 81

Conserving the Marine Resources of the Southern California Bight

Marc Beyeler, Elena Eger,[1] Vinicio Macias Zamora,[2] and Laura Durazo[3]

Abstract

In 1995, Canada, Mexico and the United States joined more than one hundred nations in adopting "A Global Programme of Action (GPA) for the Protection of the Marine Environment from Land-Based Sources." The Commission for Environmental Cooperation (CEC) is assisting the three North American countries in implementing the GPA by establishing cooperative partnerships which may form the basis for ongoing regional cooperation and joint actions in marine and coastal areas.

The binational marine ecosystem of the Southern California bight was chosen as one of two pilot projects to initiate the implementation of the GPA in North America. This effort within the Southern California bight is a unique attempt to promote collaboration to address common problems on both sides of the border. This project, while only in its initial stage, has sought to encourage cooperative partnerships, involving governmental institutions and organizations, communities, the private sector and non-governmental organizations. This effort is a model project of hemispheric and binational regional cooperation and coordination.

Introduction

The Global Programme of Action (GPA), adopted by over 100 nations in 1995, is aimed at facilitating national actions and regional cooperation in identification of problems, priorities and measures for the protection of marine habitats and for sustainable management of marine and coastal environments (United Nations, 1995). The GPA is intended to provide guidance on developing a process for management actions to

---

1 California State Coastal Conservancy, Oakland, Ca.;

2 Universidad Autonoma de Baja California, Ensenada, Baja Ca.;

3 Proyecto Fronterizo Educacion Ambiental, Tijuana, Baja Ca.

result in sustainable use of coastal and marine resources. The binational marine ecosystem of the Southern California bight was chosen as one of two pilot projects to initiate the implementation of the GPA in North America.

The Southern California Bight

The binational marine ecosystem of the Southern California bight was chosen as one of two pilot projects to initiate the implementation of the GPA in North America. The Southern California Bight extends from Point Conception west of Santa Barbara in Southern California, United States, to Cabo Colonett and Bahia de San Quintin in Baja California, Mexico. South of Point Conception, the coast of California angles sharply to the east and forms the Bight, with is characterized by unique oceanographic conditions and complex bethymetry (Southern California Coastal Water Research Project, 1989-1993; MBC Environmental Sciences, 1991; Daily, et al, 1993).

The Southern California bight is one of the most important marine ecosystems in North America, home to more than twenty million residents. The Southern California Bight coastal region contains marine and terrestrial ecosystems with extraordinary biological diversity, equaling that anywhere within North America (National Research Council, 1990). At the same time, the greater southern California region is beset with large, complex, and controversial planning, land use, resource and environmental challenges, in addition to significant regional and transboundary economic, social, environmental and political problems (California Environmental Protection Agency, 1995).

Organizing the Pilot Project

In 1996, staff of the California State Coastal Conservancy, working with staff of the Commission for Environmental Cooperation (CEC), staff from the Ensenada, Baja California campus of the Universidad Autonoma de Baja California (UABC), and staff of the Proyecto Fronterizo de Educacion Ambiental of Tijuana, Baja Califorrnia, convened key stakeholders from throughout the Southern California bight to address a broad range of issues related to marine pollution prevention from land-based activities.

The initial efforts have been focused on cataloguing the programs and projects already in place to address land-based sources of marine pollution within the bight (Coastal Conservancy, 1995). In this regard it is important to note that efforts to address land-based sources of marine degradation within the Southern California Bight go back several decades. These efforts, while only partially successful and still evolving to address the full range of land-based activities, form a complex backdrop to the effort to apply the GPA to the Southern Californis Bight. These efforts include comprehenisve management of coastal resources using a variety of planning, permitting, regulatory and non-regulatory mechanisms while relying on cooperation among key stakeholders.

The majority of these ongoing efforts were described in a working document prepared for the CEC by the Coastal Conservancy and UABC consultants, entitled, ***Conserving Marine Resources of the Southern California Bight***. This discussion paper describes more than two dozen ongoing programs and projects within the Bight, on both sides of the international border, which address one or more land-based sources of marine degradation.

Regional Workshop for the Implementation of the GPA

From September 9 through 11, 1996, approximately 60 representatives from all levels of government-federal, state, regional and local-as well as representaives of non-governmental organizations, academia, and the private sector attended a workshop in Tijuana, Baja California, Mexico with the objective of discussing the application of the GPA to ongoing efforts in the Bight.

Perhaps the most important action of the workshop was to identify and convene a large number of key stakeholders in the region involved in addressing land-based sources of marine degradation. Although several of the participatns greeted old friends and colleagues, this was the first opportunity for many of the participants to meet one another. In addition, many of the participants were hearing about the full range of ongoing programs and projects in the Bight for the first time.

After much discussion it was agreed that the workshop participants were not prepared to discuss priorities for action, but instead the discussion focused on the purpose and goals of the GPA and its application to the Bight. However, the group did discuss several suggestions for next steps in the process of facilitating the implementation of the GPA within the Bight and facilitating improved communication among key stakeholders. The result of these discussions is contained in a second document prepared by the consultants from the Coastal Conservancy, UABC and Proyecto Fronterizo de Educacion Ambiental, entitled ***Southern California Bight Cooperation Strategy***.

Cooperation Strategy: Improved Communication, Collaboration and Cooperation

The draft cooperation strategy is the result of the discussion at the pilot project workshop held in Tijuana, Baja California, Mexico and subsequent information supplied by numerous organizations and individuals. Among the many groups and individuals actively involved in addressing the conservation of marine resources within the Southern California Bight, there is strong agreement that there are growing needs, and many opportunities, for improved communication, collaboration and greatly expended cooperation.

Several specific actions have been suggested which could form the basis of improved communciation, collaboration and cooperation in this area. It is the conclusion of the diverse key stakeholders involved in the initial stage of the pilot project that ongoing efforts should be the building blocks of any new binational initiatives. There is a stong feeling among all those who have participated to date, that given the breadth and scope of existing efforts, it would be counter productive to try to start yet another project. Rather, the strong consensus of the participants was to build incrementally on work already being undertaken in order to maximize the usefulness of the pilot project. Described below are several actions which could be undertaken in the next two to three years which would advance the work of many existing initiatives in the region.

Facilitate Improved Communication:

Complete Survey of Key Stakeholders- A first follow-up activity is a survey of key stakeholders. The overview of current initiatives was cited by several participants as an essential component of improved information gathering and exchange. This initial inventory however was not an exhaustive list of programs and projects addressing marine degradation within the Southern California Bight. . In order to develop a complete inventory of programs and projects, it is necessary to compile a survey response from all known initiatives. The product will be a compendium of all current initiatives within the Bight. The compendium can then be widely circulated to interested parties as an initial effort to improved information exchange within the Bight.

Organize Southern California "Ad-Hoc Working Committee- A second suggestion is the establishment of an informal, binational "ad-hoc working committee" for the Southern California Bight. The purpose of the the working committee would be to take a limited number of initial projects that have already devleoped independently on both sides of the border and that have immediate potential for extension and cooperation across the border. Staff of the CEC, working with pilot project consultants, have already begun the organization of the ad-hoc committee and the initial meeting of the commitee took place in the Spring of 1997 in San Diego.

Establish Information Clearinghouse- Finally, the last suggestion for improving communication within the Bight was the proposal to design and establish some kind of electronic clearinghouse (home-page; listserve; electronic database, etc.). Discussions are currently underway among different stakeholders to establish a home-page focused on marine pollution policy within the Bight and separately to establish an "umbrella" electronic database of select marine water quality data sets that are now separately organized.

Promote Community Capacity Building:

Organize Sub-Regional Issues Meetings- One suggestion is to institute a series of sub-regional issue meetings. Already a caucus of stakeholders from Baja California has been meeting and exchanging information. Periodic meetings in Southern California and Baja California could be convened to identify the current "State of the Bight" and could be combined with sub-regional meetings of ad-hoc working committee members.

Training Community NGOs- Another suggestion is the proposal to train non-governmental organizations (NGOs) in marine pollution science and policy issues. It is clear that non-governmental organizations will play a crucial role in the future in addressing land-based sources of marine pollution. While their role is sure to increase, limits on their organizational skills and human and financial resource will be a constraining factor. Additional training in science, policy and organizational development is an immediate need.

Increase Technical Exchange and Technology Transfer:

Binational Science Meeting- Cooperation among scientists within the Southern California Bight is limited. A major advance in acientific cooperaation and communication could be made if a bight-wide scientific meeting were organized which brought together scientists from both countries to review the "State of the Bight" and to discuss joint scientific initiatives.

Regular Binational Scientific Exchanges- Another major advance could be made if there were regular opportunities for exchange between scientists from the California and Baja California. Over the next two to three yerrs, a regular series of scientific exchanges could be organized in order to faciliate better coordination and communication of scientific information. An initial exchange has already been organized by UABC and the Southern California Coastal Water Research Project (SCCWRP). An all-day seminar for scientists from Ensenada was hosted in early April, 1997 by SCCWRP.

Support Binational Demonstration Projects:

Regional Monitoring Program- As part of increasing technical and scientific exchange there could be an effort in the next yeasr to expand marine monitoring from Southern California into Baja California. The regional monitoring program organized by SCCWRP earlier this decade is proposed to be updated and repeated in 1998. Staff of SCCWRP have offered to help develop protocols for expanding the regional monitoring into Baja California so that a region-wide monitoring program can be achieved.

Binational Wetlands Inventory- Currently there are parallel efforts underway to devleop information regarding wetland resources in Southern California and Baja California. A demonstration project could be devleoped which creates an integrated and coordinated wetlands inventory for the entire Southern California Bight.

Next Steps

To follow through on the suggestions made at the September Tijuana workshop and in subsequent communications from participants, and to assist regional stakeholders in drafting a regional implementation plan consistent with the GPA, the Commission for Environmental Cooperation (CEC) is underwriting the establishment and the initial meetings an "Ad-Hoc Working Committee."

Over the next year, the Ad-Hoc Working Committee will hold a series of meeting to develop an implementation plan for the region. The focus of the committee's work will be devleoping a plan for implementing the GPA within the context of Southern California Bight. In the regard, the activities of the committee will, no doubt, be focused on current regional efforts at managing land-based sources of marine degradation in order to suggest approaches aimed at strengthening and broadening current initiatives and facilitating the implementation of new binational efforts aimed at addressing land-based sources of marine degraddation.

Conclusion

The pilot project in Southern California-Baja California seeks to develop a process which integrates coastal management with watershed and ecosystem planning in order to develop a comprehensive program aimed at preventing marine degradation from land-based activities. This project has sought to encourage cooperative and collaborative partnerships, involving governmental institutions and organizations, communities, the private sector and non-governmental organizations. This unique effort is potentially a model project of hemispheric and binational regional cooperation and coordination.

Appendix 1: References

Coastal Conservancy. 1995. Binational Coordination of Environmental Activities in the San Diego-Tijuana Border Region: Preliminary Survery Report. California State Coastal Conservancy: Oakland, CA.

California Environmental Protection Agency. 1995. Preliminary California-Mexico Border Environment Needs Assessment. California Environmental Protecction Agency: Sacramento, CA.

Dailey, Murray D., Donald J. Reish and Jack W. Anderson. 1993. Ecology of the Southern California Bight, A Synthesis and Interpretation. University of California Press: Berkeley, CA.

MBC Applied Environmental Sciences. 1991. Southern California Bight Physical Oceanography, Proceedings of a Workshop. US Department of Interior, Minerals Management Servicee: Camarillo, CA.

National Research Council. 1990. Monitoring Southern California's Coastal Waters. National Academy Press: Washington, D.C.

Southern California Coastal Water Research Project. Annual Reports 1989-90, 1990-91, 1991-92, and 1993-94. Southern California Coastal Water Research Project Authority: Westminster, CA.

United Nations Environment Program. 1995. Global Programme of Action for the Protection of the Marine Environment from Land-Based Activities. United Nations: New York, New York.

CHAPTER 82

Integrated Natural Resources Planning: Ecosystem Management on a Coastal Air Force Base

Christina McGinnis and Darcelle Pruitt

Abstract

Integrated Natural Resources Management Planning is currently being promoted by the Air Force as a comprehensive ecosystem management tool to preserve and enhance biodiversity. All Air Force bases are now required, under a Department of Defense (DOD) directive, to synthesize existing natural resource plans into one document using an ecosystem approach. Since these new plans require the integration of several environmental factors including wetlands, threatened and endangered species, fish and wildlife management, grounds maintenance, outdoor recreation, public access, agricultural outleasing, and numerous coastal issues, they present opportunities for overcoming past management problems due to a lack of synthesized management practices. The Integrated Natural Resources Management (INRMP) provides information on management activities, agencies responsible for management, and policy issues, and presents tangible, integrated management goals for a five-year period. Many of these goals were developed to reconcile previously conflicting goals. Goals are defined to provide specific performance criteria, and are implemented using an adaptive management approach.

This approach has broad-reaching application potential, and stresses the need for integration of existing information to properly manage diverse and sensitive natural systems.

CHAPTER 83

Integrated Coastal Management: a concept in practice in central California through the Monterey Bay National Marine Sanctuary

Terry Jackson[1]

Abstract

The Monterey Bay National Marine Sanctuary (MBNMS) encompasses over 5,000 square miles of ocean area along the California coast between San Francisco and Cambria. The overall goal of the program is to protect and enhance natural resources in a wide array of habitats within its boundaries. MBNMS achieves this goal through a responsive integrated coastal management approach that promotes cooperation between the Sanctuary office and other public agencies and capitalizes on the expertise of individuals and interest groups that share a common interest in preserving the natural resources of the Sanctuary. The structure of the program involves integrating a small Sanctuary staff with various management agencies and interest groups. An example of this integration includes coordination with the Sanctuary Advisory Council, made up of twenty-four representatives from public agencies and user interests plus specific working groups for research (22 marine institutions), education (20 educational institutions) and conservation (15 conservation organizations). The Sanctuary also is involved in developing a water quality protection program for the region that presently involves twenty-seven different agencies plus interest groups. Additionally, the public has opportunities for input into the program through council meetings, through representatives from working groups or council members, and through contact with Sanctuary staff. The MBNMS program is linked nationally with eleven other Sanctuaries and twenty-six Estuarine Research Reserves through the Sanctuaries and Reserves Division in Washington DC. This integrated approach has addressed numerous issues ranging from assessing former military exclusion zones and improving water quality to facilitating multiple resource use conflicts such as shark chumming. In this manuscript, I will outline the structure and process of the MBNMS integrated management approach and provide examples of resource management issues that have been addressed.

---

[1] Manager, Monterey Bay National Marine Sanctuary 299 Foam Street, Suite D. Monterey CA 93940

Introduction

The Monterey Bay National Marine Sanctuary (MBNMS) is part of a marine Sanctuary system run by the National Oceanic and Atmospheric Administration (NOAA). The Monterey Bay Sanctuary encompasses over 5,000 square miles of ocean area along the central California coast between San Francisco and Cambria. It is the largest Sanctuary in the national system and the second largest marine protected area in the world -- after the Great Barrier Reef in Australia. To put it in perspective, the MBNMS is about one-and-a-half times the size of our largest National Park, Yellowstone. Unfortunately, the resources of Yellowstone National Park are far greater than that of the entire National Marine Sanctuary program. It is for this reason that an integrated approach to marine resource management was established to ensure the success and survival of the Sanctuary. In this paper, I will highlight some of the extraordinary features of the Monterey Bay region, discuss goals and objectives of the Sanctuary program, and describe how the Sanctuary program is developing cooperative management strategies with the community and other agencies to establish an integrated coastal management program for the region.

Background

The MBNMS is a unique place! Designated in 1992 for its pristine waters and abundance of wildlife, the Monterey Bay region is fortunate to possess one of North American's largest submarine canyons -- an area so rich in diversity that it serves as a habitat to hundreds of marine species. Furthermore, several endangered marine mammal species inhabit the Monterey Bay region including the California gray whale, fin whale, humpback whale, sperm whale, and the California sea otter. In addition, all species of pinnipeds commonly located off the central and northern California coast are found in the Monterey Bay area. The area is also an important resting and staging area for migratory birds and contains one of our nation's most expansive kelp forests.

The MBNMS is one of our nation's most scenic places, making it an attractive destination for thousands of visitors each year. Not surprisingly, tourism is a major industry, supporting jobs and providing a substantial economic base for the surrounding community. This success is attributable, not only to the natural beauty of the area, but also to the relatively clean water which allows visitors and residents alike to enjoy such recreational activities as sailing, surfing, kayaking, scuba diving and snorkeling, sport fishing, whale watching, and many more. Local businesses depend upon the Sanctuary and other agencies to guard against pollution and contaminated waters. Unfortunately, as the human population increases, so do the demands on the Sanctuary's resources and the need for cooperative resource management. Programs such as the MBNMS Water Quality Protection Program (WQPP) have developed into world-class models of coordination amongst local, State, and Federal agencies to conserve and protect the resource.

SANCTUARY GOALS AND OBJECTIVES

The mission of the Sanctuary program is to manage and protect the natural resources within its boundaries so the region's ecological and cultural integrity is preserved for present and future generations. Specifically, the four goals of the Sanctuary program are:

A) Resource protection -- Assigned the highest priority for management, specific objectives of resource protection efforts include:

- coordinating policies and procedures among agencies sharing responsibility for the protection and management of resources;
- encouraging participation by interested agencies and organizations;
- promoting public awareness of the Sanctuary;
- assessing the water quality of the Sanctuary; and
- reducing threats to Sanctuary resources.

B) Research -- The Sanctuary's basic research goals are to improve understanding of the Monterey Bay environment, resources and qualities, to resolve specific management problems, and to coordinate and facilitate the flow of information between the numerous research institutions. Ever increasing knowledge about Sanctuary resources ensures responsible resource protection and informed decision making. The research program attempts to reach its goals by employing the following strategies:

- establishing a framework and procedures for administering scientific projects to ensure that research is responsive to management needs of the Sanctuary;
- ensuring that research results are incorporated into interpretive and educational programs about the Sanctuary and its resources;
- developing and coordinating data collection efforts so a comprehensive picture of Sanctuary characteristics can be documented and understood;
- initiating a monitoring program to assess environmental changes as they occur;
- identifying the range of environmental effects that would result from human activity and/or natural phenomena; and
- encouraging information exchange.

C) Education -- Sanctuary education goals focus on improving public awareness, understanding, and promoting stewardship of the Sanctuary through public interpretive and education programs. Management objectives to achieve this goal include:

- providing information about the Sanctuary to the general public;
- broadening support for the Sanctuary through cooperative programs; and
- encouraging public feedback on Sanctuary programs and policy decisions.

D) Manage Multiple Use -- The Sanctuary program recognizes the many different uses of the area's natural resources and the need to balance resource protection with multiple use. Goals for accomplishing this often challenging task include:

- educating the general public and encouraging them to respect Sanctuary resources;
- providing relevant information about Sanctuary regulations and policies to the public, special interest groups, and regulatory authorities;
- collaborating with public and private organizations to promote compatible use of Sanctuary resources; and
- monitoring the level of use to identify potential degradation of resources and minimizing the potential for user conflicts.

The Sanctuary strives to achieve these goals by working closely with the local communities and other agencies in an integrated way that builds communication and cooperation.

INTEGRATING SANCTUARY MANAGEMENT IN THE MBNMS

As I mentioned in the introduction, integrated management is a fundamental part of the MBNMS program. One of the primary reasons for this is that the Sanctuary encompasses over 360 miles of coastline which includes many communities and authorities. Along these boundaries, people establish businesses, participate in water sports and recreation, and seek answers to marine related phenomena. For the MBNMS, management issues do not so much involve the natural resources, as they require understanding and guiding human interactions. Through integrated management strategies employed by the MBNMS, the surrounding communities are active and responsible participants in achieving Sanctuary goals. This strong base of public participation in a Federal program has created a far stronger position from which Sanctuary management can operate. A twenty-four member Advisory Council and three working groups dedicated to research, education, and conservation advise and assist the Sanctuary in the integrated management task.

A. Sanctuary Advisory Council

The Sanctuary Advisory Council was established to advise and assist the Manager in making decisions that affect the Sanctuary. Composed of individuals representing various constituencies of the MBNMS, the Advisory Council serves as a forum for the public to provide input to the Sanctuary. The purpose of the Advisory Council is to:

- help strengthen and provide support for the growth of the Sanctuary;
- assist in protection of Sanctuary resources by helping identify research needed to rebuild or protect Sanctuary resources; and,
- assist in building community support through problem solving, consensus building, new constituency development, increasing opportunities for revenue enhancement, and increasing understanding about the Sanctuary.

Advisory Council members volunteer their time and expertise for a three-year term. They compete for their positions as representatives of local user groups and are ultimately appointed, per the recommendation of the Sanctuary Manager, by the Director of the Ocean Coastal Resource Management Division of the National Oceanic and Atmospheric Administration (NOAA). Also serving as active members of the Council are representatives of eight State and Federal government agencies with expertise in the management of natural resources in the jurisdiction of the Sanctuary. To ensure relevant information exchange and consistent management within the National Marine Sanctuary and National Estuarine Research Reserve programs, the Managers of the four California Sanctuaries and the Elkhorn Slough National Estuarine Research Reserve participate as non-voting members of the Council.

The Advisory Council meets regularly, usually six to eight time a year, to discuss issues of importance to the MBNMS. The Council Chair and the Sanctuary Manager work closely to draft agendas that appropriately address Sanctuary related matters. All of the meetings are open the public and time for public input is scheduled at every session. This ensures the public has a chance to bring matters to the Council's and Sanctuary's attention. Moreover, public participation in discussions during Council meetings provides valuable insight about public sentiments related to Sanctuary policies and practices.

Over the years, the Council has helped the Sanctuary pursue solutions to problems and issues such as shark chumming, vessel traffic safety, jade collection, enforcement, permit processing, and various aspects of water quality.

B. Advisory Council Working Groups

The Advisory Council working groups also assist in the decision making process. These groups, concerned with research, education, and conservation, meet on a monthly basis to coordinate their own efforts and offer their expertise to the Advisory Council and the Sanctuary Manager on certain issues.

The working groups continue to expand the range of their activities throughout the Sanctuary. In 1996, special emphasis was placed on efforts to improve communication and cooperation opportunities among the three groups, culminating in the highly successful *Sanctuary Currents Symposium*. The Symposium combined the interests of the research, education, and conservation communities into a conference that was well-attended by many aspects of the public at-large. The *1997 Sanctuary Symposium* held just a few weeks ago, was another success, making the *Symposium* sure to become an annual event.

C. Water Quality Protection Program (WQPP)

During the past three years, the Sanctuary office has spearheaded the development of a comprehensive Water Quality Protection Program for the Monterey Bay region. The effort is a partnership among twenty-seven Federal, State, and local agencies, public and private groups to enhance and protect water quality in the Sanctuary and the eleven watersheds which drain to it. It has served to improve coordination and pooling of resources among over 100 existing programs which address water quality issues. Using a consensus-based approach involving a broad array of public and private interests, plans have been completed addressing urban runoff, regional monitoring and data sharing, and marinas and boating activities. Local cities, state agencies and nonprofit foundations are currently working with the Sanctuary to carry out the plans, developing model pollution control programs for the region's small cities and installing convenient waste disposal and recycling facilities at local harbors. Planning has also been initiated with the agricultural community to develop practical, voluntary solutions to reduce sedimentation and polluted runoff from the region's farms and ranches. The program's extensive outreach and education efforts have succeeded in raising public awareness about the connection between land and sea, involving a diverse array of participants in the program, and promoting simple changes in everyday activities which can prevent polluted runoff to the sanctuary.

Conclusion

What has been presented here is merely the tip of the iceberg. The potential for protecting the natural resources of the MBNMS continues to expand exponentially as resources are discovered and shared throughout the region. The amount of positive results that the Monterey Bay Sanctuary has been able to accomplish in just four years is commendable. Of singular importance in the succession of events leading to this success is that the Monterey Bay NMS has fostered and earned the public trust. Although understandably wary in the beginning, the public has realized that the Sanctuary is here to work with them and that they are encouraged to be active participants.

The stage is set for the MBNMS to continue achieving milestones through a responsive integrated coastal management approach. The obvious success of the program exemplifies the strengths of integrated management theories. The MBNMS has worked to maintain that much of the burden of ownership for Sanctuary resources remains with the people and organizations directly connected to it. With the active participation of public agencies and interested citizens, the Sanctuary can persist in capitalizing on the region's abundance of expertise. Through management practices, which promote cooperative solutions to multiple use conflicts, the MBNMS is facilitating a powerful and responsive method of government; thus, ensuring the protection of precious Sanctuary resources for present and future generations.

CHAPTER 84

A METHODOLOGY FOR DEVELOPING A MANAGEMENT PROGRAM FOR PROTECTED AREAS: THE BIOSPHERE RESERVE OF THE UPPER CALIFORNIA GULF AND COLORADO RIVER DELTA, MEXICO.

[1]José Luis Fermán Almada, [2]Alejandro Varela Romero, [2]Antonio Cruz Varela, [1]Lorenzo Gómez Morin-Fuentes, [1]David W. Fischer, [3] Anamaria Escofet, [4] Ileana Espejel.

ABSTRACT.

The concept of the Biosphere Reserve is defined within UNESCO'S program for " Man and the Biosphere". In Mexico, the General Law of the Ecological Equilibrium and Environmental Protection, Article 48, defines this concept. The Biosphere Reserve of the Upper California Gulf and Delta of the Colorado River was decreed in July of 1993, and required the development of a management program to define the needs and opportunities found within the reserve. An objective of this work was to define a methodological design for managing the reserve, including its contributions to socioeconomic development, while at the same time preserving the natural processes of the area. This paper outlines a methodology which includes the participation of the inhabitants of the reserve from the formulation of its objectives through to proposal of management strategies. An important part of the methodology is the process of ecological regionalization which sets the stage for capacity for suitable uses, as well as the integration of social demands into the definition of management strategies. The management strategies suggested used the concept of sustaintability based on balanced actions related to the protection and the utilization of its natural resources.

[1]Facultad de Ciencias Marinas. Universidad Autónoma de Baja California, (UABC).
E-mail jlferman@faro.ens.uabc.mx
[2] Centro de Investigaciones Científicas y Tecnológicas de la Universidad de Sonora, (CICTUS).
[3] Centro de Investigaciones Científicas y Estudios Superiores de Ensenada, (CICESE).
[4]Facultad de Ciencias. Universidad Autónoma de Baja California, (UABC).

INTRODUCTION.

The Biosphere Reserve of the Upper California Gulf and Colorado River Delta is presented as basis for solution of the resource use conflicts generated by those living and visiting the region and for the protection of the ecosystems that sustain the diversity of its environmental and economic resources. The decree establishing the Biosphere Reserve indicates that a social perspective is necessary for creating an Integrated Management Program for the area, (Fig 1).

An integrated program for coastal management can be defined as "A governmental program established in order to use and preserve a resource or coastal environment. Implied is the existence of a unit of the State that distinguishes the coastal zone and frames it within a geographical area whose extension will depend on the type and distribution of the resources administered" (Sorensen, 1990).

Administration programs for the coastal zone can be defined by the following points (Fermán and Gómez-Morin, 1993):

1.- To identify the problems related to the coastal area such as: to define and characterize the coastal zone, to describe the types of demand that different sectors of society make on the coast, and at the same time to identify the conflicts that are generated by such demands.

2.- To define the policies undergirding any management program, including social, economic and environmental dimensions.

3.- To provide alternative solutions to the problems identified which are based on the outlined policies. These solutions should be based on integrated studies of the use capacities of the environmental resources complemented with a socioeconomic analysis.

4.- To analyze the legal structure and existing development plans and integrate them with policies and alternatives.

5.- To generate a social consensus that can guarantee the sustaintability of the program and gain social recognition and support for the solutions and policies outlined.

6.- If necessary, to define a legal authority entrusted with implementing the program of coastal zone administration based on an acceptable and workable legal structure.

The Biosphere Reserve of the Upper California Gulf and Colorado River Delta was originated through a series of activities that were begun formally in June 1992, at the Port of Mazatlán, Sinaloa. There a workshop was developed for the identification of high-priority marine areas for conservation, organized by the Secretaria de Desarrollo Urbano y Ecología (Mexico) and the World Wildlife Fund (USA). On February 19, 1993, at the request of Instituto de Ecologia de la Secretaria de Desarrollo Social (SEDESOL), the Delta of the Colorado River was declared to be an international reserve of the Hemispheric Program for Coastal Water Fowl Reserves.

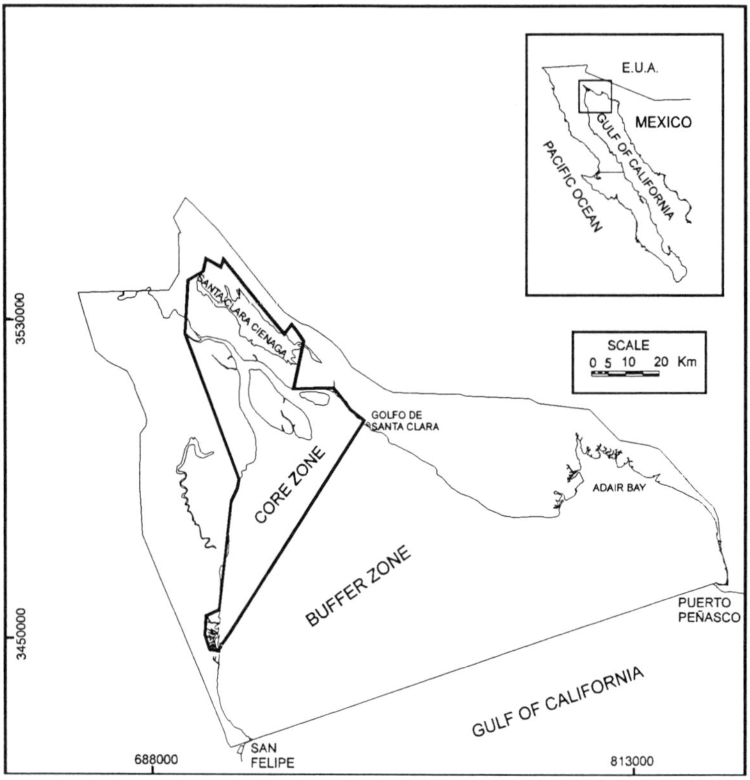

Fig. 1.- Study area localization.

These efforts emphasized the ecological importance of the area and the need to formalize a management program that would guarantee the protection of natural values while providing for the sustainable utilization of its natural resources. In March 1993, the Technical Committee for the Preserve of the Vaquita and the Totoaba in the Upper California Gulf presented to the Federal Government a document, "Proposal for the Biosphere Reserve of the Upper California Gulf and Colorado River Delta.

Based on these efforts, the President of Mexico, Carlos Salinas de Gortari, decreed on June 10, 1993, the Biosphere Reserve suggested. The general characteristics that motivated the declaration included:

- A desire to include different ecosystems of the Upper California Gulf and the Delta of the Colorado River, such as the falls zone, the tidelands and the channels of the outlet of the Cienaga of Santa Clara .
- The integration of the management of ecosystems along with various environmental problems into an administrative structure that must answer to the environmental and social needs of the area.
- The need to arrest the deterioration in the interior of the reserve and reveal the potential for conservation of environmental resources of the area.
- The need to preserve the genetic diversity of the area to include those species in danger of extinction, such as the vaquita (*Phocoena sinus)*, the totoaba (*Totoaba macdonaldi)*, the cachorrito of the desert (*Cyprinodon macularius macularius)*, the palmoteador of Yuma (*Rallus longirostris yumanensis)*, and endemic kinds such as the grouch (*Colpichthys hubbsi)*. Also to be preserved are species of economic importance such as the blue shrimp (*Penaeus stylirostris)*, brown shrimp (*P. californiensis)*, white shrimp (*P. vannamei)* and the curvina yellow mouth (*Cynoscion xanthulus)*, among others.

This Reserve occupies an area of 934,756 hectares (85% is a marine environment) constituted into two management zones: The core zone with 164,779 hectares and a buffer zone with 769,976 hectares which area is to have the following two goals:

- To maintain and to strengthen the economic activities of the region through the sustainable use of natural resources. Conservation in the Upper California Gulf will permit an orderly utilization of the different natural resources of economic and ecological importance which reduces negative impacts from chaotic use. The economic and social welfare of the inhabitants of the region will be promoted through the rational utilization of the area's resources by local inhabitants with technologies that conform with the sustained maintenance of natural populations.
- To preserve the biological values of the ecosystems of the Upper California Gulf and Colorado River Delta. However, very important species for preservation purposes that motivated the creation of the Reserve still need specific management programs to promote their survival.

After the decree of the Reserve, the negotiations for the elaboration of the management program were begun by the Scientific and Technological Research Center of the University of Sonora (CICTUS). Included were the Ecological Center of the State of Sonora and the Research and Development Center for the Natural Resources of the State of Sonora (CIDESON) which were to accomplish the inventory of the natural resources of the area.

In January 1994, other groups were added: Center for Study of the Desert and Ocean in Sonora, College of the Northern Frontier in Tijuana County and the Marine

Sciences Faculty of the Autonomous University of Baja California (UABC). With the administrative coordination of this Base Group in CICTUS, UABC participated through training the others on ecological regionalization and developing the methodological framework for the integrated management program for the Reserve. Given the three coastal reserves in Mexico, this Reserve is the first containing a large marine area and the participation of two States and the Federal Government (SEMARNAP-INE, 1996).

GOALS OF THE MANAGEMENT PROGRAM

To design a management program for the Biosphere Reserve of the Upper California Gulf and Colorado River Delta in order to promote the region's socioeconomic development and to preserve the natural processes of its ecosystems through rational utilization of their resources, sustained through activities of research and environmental education.

Specific Objectives of the Management Program:

1.- To preserve the ecosystems of the Sonoran Desert, the Upper California Gulf and Delta of the Colorado River.
2.- To establish permanent protection areas for endangered species such as totoaba, vaquita, the cachorrito of the desert and several kinds of water fowl;
3.- To regulate economic activities to safeguard the area's natural resources;
4.- To promote economic and administrative activities that can increase the quality of life in resident communities;
5.- To outline lines of research and environmental education for the region;
6.- To restore and to preserve the flora and fauna and the environmental quality of the ecosystems.

METHODOLOGICAL FRAMEWORK.

The methodology for evaluating the environment for the purpose of planning for the Reserve has two different approaches: the analytical and the synthetic (Cendrero, 1989). The first considers the nature of the land surface as an integration of elements or natural characteristics that can be represented and evaluated separately to determine their capacity for use (Bojórquez-Tapia, 1992). The synthetic approach, on the other hand, considers the surface of the land as a set of environmental units with homogeneous features (Cendrero, 1982; SEDUE, 1988; Cendrero, 1989). Recently, the need for including socioeconomic factors in the definition of environmental units has been outlined (Gómez-Morin, *et al.*, 1992).

Zoning.

Coastal areas frequently include population centers whose residents depend on its natural resources. This population impacts the area's resources and the conservation of its ecosystems. A conflict occurs in the difficulty of delimiting areas suitable for resource development that satisfies the needs of the residents, as well as the area's environmental requirements for achieving sustaintability. Zoning is a method widely accepted as a land management instrument which can be applied for resolving conflicts between incompatible activities of land use. Here the issue is to designate conservation areas that preserve the continuity of biological processes and the productivity of natural ecosystems (Slam and Clark, 1989).

To carry out this process an environmental evaluation is based on two key concepts: the ecological regionalization of the territory and the evaluation of the capacity of the territory to support an environmental activity or specific land use. The capacity implies an evaluation of the present use of resources without introducing modifications to the environment in order to adapt it to the environmental policy or use determined for the territory.

The system of environmental units developed to date depends not so much on the scale as the number of environmental factors that are used to define the units (Gómez-Morin and Fermán, 1991). Because of the dynamic nature and variability of natural systems on land as well as in marine areas, it is difficult to develop an applicable classification system. Existing classification systems must be adapted to the specific requirements of planning and management programs, including the particular characteristics of the area. The projects of Naveh and Lieberman (1984), Cendrero (1989), and Gómez-Morin *et al.* (1992), among others, present an adequate description of the methodologies using this synthetic approach.

The regionalization of the Reserve was based on the technique described in the Ecological Classification Handbook of the Territory (SEDUE, 1988; p.96-111). This technique uses this criteria for defining units in the territory. Each unit is represented on maps and superimposed one on the other to certify the degree of homogeneity. This system of regionalization presents a structure of different hierarchic levels (" environmental units") in different scales. Together they comprise the basic criteria used to define homogeneous environmental units of the territory. This classification system was modified and adapted for the particular requirements of the management program of the Reserve. The governing criteria for the definition of each hierarchic level were the following:

a) Environment: Defined as the components of a coastal region (land and marine zones.)

b) System and Subsystem: Defined based on criteria presented by SEDUE (1988) based on geomorphological and hydrological zones. The first modification to this classification is for the Marine Subsystem which is defined as a function of the depth and the degree of turbidity.
c) Landscape: Defined by the criteria presented by SEDUE (1988), these zones are geomorphological based on topography, type of sedimentary deposit, and type of vegetation.

Using these criteria and applying the technique of superimposed maps, environmental units of the territory are defined, (Fig 2). The application of this system of regionalization permits an integration of the environmental policies for a coastal region where compatible activities are assigned, such as aquaculture, fishing and natural area conservation. It will be possible, furthermore, to establish specific program proposals for managing each unit on the basis of their resource attributes or natural processes relevant to the zone. The system is flexible in that other criteria for unit definition can be used.

Capacity Evaluation.

The technique of territorial use capacity is described in Cendrero and Díaz de Terán (1987) and is known as a weight-valuation technique. The valuation of the factors or attributes of each environmental unit is accomplished using the following model:

$$C = \sum_{j=1}^{n} W * V$$

where:
C = Capacity of the environmental unit to sustain the activity or environmental policy "a";
n = Number of factors or parameters that influence the determination of the capacity;
W = Weight or weighting coefficient that reflects the importance of the factor or parameter "i" on the capacity for the activity or environmental policy "a"; and
V = Value that the factor or parameter "i" has in that unit in terms of its capacity to sustain the activity or environmental policy "a".

Each factor is weighted according to its relative importance for each use, and the weights are based on the Delphi method of conferencing with experts (Cendrero and Díaz de Terán, 1987). Experts were consulted including marine ecology, land ecology, flora and fauna, geomorphology, sociology, fishing, natural resources management, economics and fishing administration. Once the importance of each factor is weighted for the capacity of use, the condition that is represented by each factor is valued in each unit (i.e. presence of endemic kinds of flora) in terms of an assignment of an environmental policy and in terms of the impact expected on that factor. This valuation is carried out using a nominal scale of -1 (unfavorable condition for the development of the activity and unfavorable impact of the activity on the factor), 0 (indifferent condition and indifferent impact), or +1 (favorable condition, positive impact). The weight product for each value represents the contribution of each one of the factors to the capacity of use, and the sum of such products provides the total capacity for an environmental unit to sustain a proposed use.

The total capacity values are split into classes whose limits are defined by intervals of 0.60 standard deviation. Thus, the classes of use capacity include the following:

```
      LOW       MEDIUM       HIGH      VERY HIGH
<---------------+---------------+--------------+---------------->
             0.60 S          M           0.60 S

         M = Mean                      S = Standard Deviation
```

In this case the capacity values of the different intensities of use were transferred to regional maps where the units with equal capacity for different environmental policies were analyzed in the context of the objective of the management program. Based on this analysis and associated discussions, recommendations for environmental policies for each natural unit of the area were made.

Community Participation.

Participation of residents and users of the area must be given weight in the elaboration process of the management program. The union of community groups with the management process assures greater support and program acceptance, while providing local information for greater relevancy and helping merge agreed strategies with local needs. This process allows local perceptions, traditions and habits to be protected and the rights of local interests to be included in the decision making process, thereby facilitating agreements for management of the area (Fischer, 1981).

The biosphere reserve concept is based on the idea of protecting the natural resources of an area, but it often ignores the area's socioeconomic problems in the research and management of the reserve. A long list of failures has been experienced through managing protected areas as isolated entities from local societies and their utilization of its natural resources. It is important "to open an area's conservation" to the human beings living there and therefore to include them in the management of the reserves. This inclusion allows ecological criteria to achieve a balance between the utilization of the natural resources and the protection of the ecosystems sustaining the area (Slam and Clark, 1989).

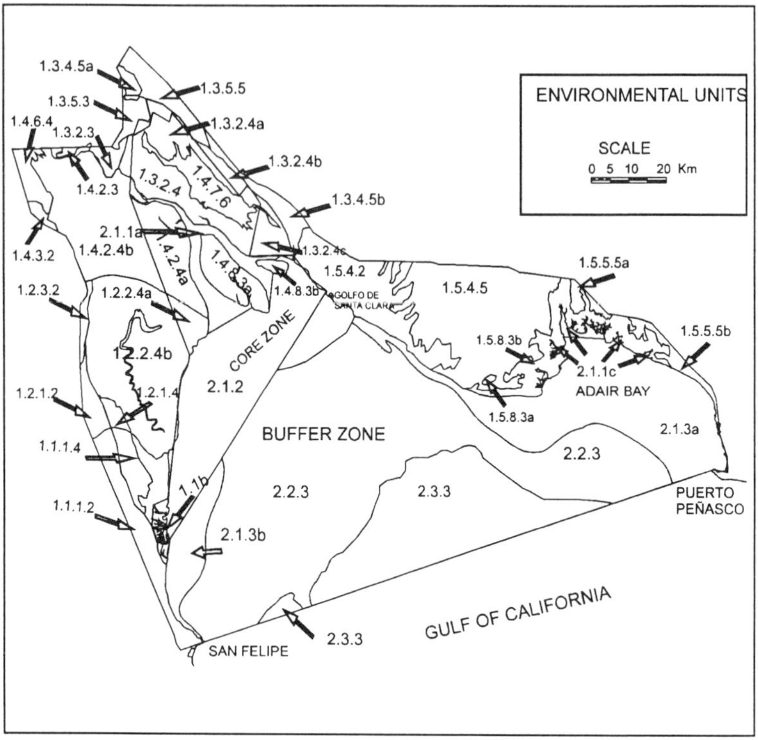

Fig 2.- Coastal and marine environmental units in the Reserve.

With this philosophy, a series of public meetings were held in the largest communities of the region, Puerto Peñasco and Golfo de Santa Clara in Sonora and San Felipe in Baja California. To ensure the objectives of the management program were known and to spread knowledge of the possibility of direct participation, a series of information forums these three communities of the Reserve were held, as well as direct interviews with in key groups.

The community conferences were envisaged at the beginning of the planning process, in order to learn of the anxieties, doubts, problems and proposals of different groups living in the Reserve. The need for forming a link with the natural or representative leaders of each sector or group in these communities was necessary to explain the reserve management program, as well as to obtain alternative solutions to management problems. To obtain a high degree of participation in the community forums invitations were sent directly to each community, personalized according to previous lists of cooperatives, industries and trades,. and to those interested persons who had previously indicated an interest during the declaration process. In addition, invitations to authorities at municipal, state and federal levels were made, as well as to representatives of academic and non-governmental organizations of the States of Sonora and Baja California.

RESULTS.

Participation of the Inhabitants.

In each community interviews were accomplished by use sector to understand the response of the community to the management proposals. A key issue was the regionalization of the reserve with different environmental management policies for each area. Representatives of the community groups suggested proposals for the management plan (Table I). The inhabitants of the communities displayed three types of attitudes during the process: positive, neutral and negative. These attitudes were demonstrated through degrees of participation specified as high, middle and low (Table II), (Fischer, 1981).

In three of the towns strong participation from the fishing sector was experienced with concern over shrimp fishing limits. In the tourist sector proposals of mega-projects were made. Aquaculture representatives were negative to the Reserve; both tourism and aquaculture projects were considered to be incompatible with the Reserve because of the large degree of landscape alteration and water pollution involved.

Table I.- Proposal for resources management by Puerto Peñasco's community.

Activity	Region	Issue/Application
Fishing and aquaculture	Buffer Zone	1.- Start trawl fishing in the buffer zone at the end of December.
		2.- Regulate Shrimp fishing during reproduction seasons.
		3.- Take research shrimp samples from the trawlers.
		4.- Minimal net mesh size for Escama fishing.
		5.- Regulate snail fishing.
		6.- Add two more areas for reproduction.
		7.- Apply regulation to sport fishing.
		8.- No hydrocarbon exploration.
Development projects (tourism, aquaculture, salt , agriculture)	Adair Bay (Buffer Zone)	1.- Aquaculture develompment (shrimp, crabs and oysters).
		2.- Sport fishing and salt exploitation.
		3.- Grow coconuts and dates on shoreline.
Multiproject development	East of Adair Bay	1.- Consider marina project.
		2.- Urban, tourism, ecotourism, agriculture (hydroponic), mining (salt) development.

Thirty-seven percent of the economically active population participated in the local forums. Positive comments were received from the commercial fishing sector, sports fishing and research groups. Aquaculture and tourism groups wanted to know more about the management proposals. Only the mining industry had a low participation rate. Federal, state, municipal, private industry and educational centers were generally neutral in attitude.

The indigenous communities of the Cucapa and Papagos (Tohono O′odham) had actively participated in the forums. Both groups were considered in the management program because of their traditional fishing activities in the Colorado River Delta as well as their traditions and religious ceremonies in the Desert. For the Papagos or "people of the desert" the region of the Pinacate is very important, since according to their traditions, the world emerged from the caves and craters of this region. Due to their belief system the Papagos have requested that sacred sites found in the region of the Pinacate and throughout the coast of the uninhabited area be respected, particularly those which are found near Bay Adair.

The proposals of the O′odham and the Cucapa were based on continuing their traditional economic activities, as well as to assure the maintenance of an environmental balance in the Reserve. They wanted the right to intervene in management activities and research since their impact on the environment has been minimal, and they wanted to continue their ability to preserve this delicate area.

"We have not elected you, the governments of Mexico and of the United States, to occupy our land, as it is all sacred, and you have assumed the same responsibility as we have supported. For this reason we are here to teach you these responsibilities and to help you learn to fulfill them" (Jesus Aguilla and Angello Mattia, from the community of Papagos Indians).

Table II.- Degree of Involvement of the inhabitants of the Communities and Sectors.

		ATTITUDE TO THE RESERVE		
		Positive	Neutral	Negative
INVOLVEMENT	High	Aquaculture Fishing Sector (ribereña and sport) Tourist Sector Research Groups (NGO's, Educational) Indigenous community (Papagos and Cucapa)	Commercial fishing, (Escama, Shark fishery)	Aquaculture Shrimp Fishing Fishing Industry (Ship construction)
	Middle	Research Groups (NGO's, Educational) Tourist Sector Recreation	Administration Official Research Commercial Fishing Maquiladora (Industry).	Residential Tourism
	Low		Mining Industry	

Table III. Classification system for the environmental units.

Environment	System	Subsystem	Landscape
1. Coastal Land.	.1. Hydrological Basin RH4A. .2. Hydrological Basin RH4B. .3. Hydrological Basin RH7A. .4. Hydrological Basin RH7B. .5. Hydrological Basin RH8C.	.1. Piamont .2. Coastal Plain .3. Hills .4. Dune Fields .5. Saline Plain .6. Lacustrine Plain .7. Cienaga .8. Island	.2 Desert Scrub .3 Halophytes Vegetation .4 Without Vegetation .5 Sand Desert Vegetation
2. Coastal Marine.	.1 0 - 10 m depth. .2 10 - 30 m depth. .3 > 30 m depth.	.1. Turbidity T < 0.5 m .2. Turbidity 0.5 m < T <1.0 m. .3. Turbidity T > 1.0 m	

Regionalization and Environmental Policies.

The classification system was used is shown in Tables III and IV. Eight marine environmental subsystems and 28 different types of landscapes were defined. These were classified by means of a four digit code, where the first number represents the environment, the second the system, the third the subsystem and the fourth the landscape. The environmental policies define the intensity of possible uses in the territory (Gómez Morin, *et al*, 1992). The definition of these policies for the Reserve is based on social demand for natural resources, the environmental capacity and the management norms applicable to the Reserve (Art. 48 of the General Law of the Ecological Balance and Protection to the Environment). To establish the management strategies of the Reserve the following environmental policies were set: (Ferman, 1994).

1. Protection.

The objective of this policy is to provide the technical measures necessary to regulate and prevent environmental deterioration, and if necessary, to restore deteriorated areas. It is applied to areas with ecosystems of delicate ecological balance and to areas of natural hazards that require active intervention to prevent environmental deterioration. Under this policy is an emphasis on maintenance of the natural environment where the degree of human intervention is limited. In these areas are uses that do not consume natural resources, but rather support their protection through are authorized such as research and environmental education. Potential uses include:

- Non-manipulative research;
- Environmental education;
- Passive recreation;
- Infrastructure construction to allow the permitted activities to occur;
- Traditional uses, particularly fishing by Cucapas Indians in the outlet of the Colorado River Delta and the digging for clams by Lugareños in a portion of Montague Island. Specific measures for these activities will have to be established by the SEMARNAP (the federal oversight agency for environmental matters).

2. Protection with Active Use.

This policy is applied in areas that require measures to control the uses of the natural resources of regional economic importance, as well as environmental restoration measures for ecosystems affected by previous development. Under this policy the maintenance of the natural environment is stressed but with an increased degree of human activities. Infrastructure construction to support artesian activities is permitted as well as for research activities, environmental education and ecotourism. Potential uses include:

- Environmental education activities;
- Infrastructure to mitigate hazards;
-Indigenous vegetation for stabilization of soils and water margins;

Table IV. Enviromental Units of the Management Program for the Biosphere Reserve of the Upper California Gulf and Colorado River Delta.

Coastal Environmetal Units	
1.1.1.2	Piamont with Desert Scrub, San Felipe.
1.1.1.4	Piamont without Vegetation, El Chinero.
1.1.2.4	Coastal Plain without Vegetation, El Chinero.
1.2.1.2	Piamont with Desert Scrub, El Chinero.
1.2. 1.4	Piamont without Vegetation, El Chinero.
1.2.2.4a	Coastal Plain without Vegetation, La Ventana.
1.2.2.4b	Coastal Plain without Vegetation, Montague.
1.2.3.2	Hills with Desert Scrub, La Ventana.
1.4.2.3	Coastal Plain with Halophyte Vegetation, Delta del Rio Colorado, B.C.
1.4.2.4a	Coastal Plain without Vegetation, Laguna Salada.
1.4.2.4b	Coastal Plain without Vegetation, Delta del Rio Colorado, B.C.
1.4.3.2	Hills with Desert Scrub, Laguna Salada.
1.4.6.4	Lacustrine Plain without Vegetation, Laguna Salada.
1.4.7.6	Cienaga of Santa Clara.
1.4.8.3a	Island with Vegetation, Montague.
1.4.8.3b	Island with Halophyte Vegetation, El Pelicano.
1.3.2.3	Coastal Plain with Halophyte Vegetation, Delta del Rio Colorado, Sonora.
1.3.2.4	Coastal Plain without Vegetation, Delta del Rio Colorado, Sonora.
1.3.4.5a	Dune Field with Sandy Desert Vegetation, Riito
1.3.4.5b	Dune Field with Sandy Desert Vegetation, El Doctor
1.3.5.3	Saline Plain with Halophyte Vegetation, Ej. Encinas Johnson
1.3.5.5	Saline Plain with Sandy Desert Vegetation, Riito
1.5.4.5	Coastal Dunes with Sandy Desert Vegetation, Santa Clara
1.5.4.5	Coastal Dunes with Sandy Desert Vegetation, El Tornillal
1.5.5.5a	Saline Plain with Sandy Desert Vegetation, Bahia Adair.
1.5.5.5b	Saline Plain with Sandy Desert Vegetation, Puerto Peñasco.
1.5.7.3a	Cienaga with Halophyte Vegetation , La Salina.
1.5.7.3b	Cienaga with Halophyte Vegetation, Bahia Adair.
Marine Environmental Units	
2.1.1a	Outlet of the Rio Colorado.
2.1.1b	Tideland La Ramada.
2.1.2	Machorro - La Ramada.
2.1.1.c.	Estuaries of Bahia Adair
2.1.3a	Don Abel Bay
2.1.3b	Adair Bay
2.2.3	Lower Depth Zone
2.3.3	Southern Zone

- Infrastructure for low density environmental education activities and ecotourism;
- Artesian or ribereña fishing, using small vessels and selective fishing arts;
- Aquaculture of low impact that complies with environmental impact regulations;
- Collection of selected flora and fauna, land and marine, under specific management program guidelines.

3. Utilization with Control.

The objective of this policy is to provide the technical measures necessary to regulate the utilization of natural resources in such a way that sustainable development of the region is promoted. It is applied to areas with current or potential productive uses, as well as to areas that require a reduction in the secondary impacts associated with productive activities. Through control of the growth and pace of these activities, under the strict application of regulatory procedures based on ecological criteria, it is expected that sustainable activities will take place. Potential uses include:

- Productive activities for the development of selected sectors under an integrated management plan for resources. The modification of rivers and their zones of discharge are to be avoided;
- Preventive measures to avoid the desertification process will be taken while protecting indigenous vegetation through recovery or repopulating activities;
- Fishing with large vessels and selective fishing arts under programs that regulate the extraction, season and volumes of fishing;
- Aquaculture of low impact that complies with environmental impact regulation;
- Sport fishing under a specific management program with particular limits for each area;
- Utilization of natural resources through specific management programs for each type of resource based on a previous evaluation of expected environmental impacts;
- Water treatment of residuals guaranteed in any work that requires it;
- Residual waters from domestic and productive activities are to be recycled;
- Specific handling of solid wastes from urban and rural zones is to be regulated;
- Any developments established in the Reserve are to observe a buffer zone of 30 additional meters from the Federal Maritime Zone of 20 meters for a total of 50 meters from mean high tide. Also, they must comply with environmental impact regulations.

Capacity Evaluation of Uses for Environmental Policies.

For evaluating the capacity of use of the soils of the environmental units a characterization of each was made in three classes: abiotic, biotic and socioeconomic (Table V). These classes were evaluated with respect to each environmental policy in order to define the factors that were favorable (F), indifferent (I) or unfavorable (U) for the establishment of each policy. The definition of the capacity of use for each natural unit is presented in tables VI and VII. The results of the evaluation are presented in the form of tables that contain the name and the key of each environmental unit, its capacity of use (environmental policy), and its relevant ecological characteristics (Table VIII).

The strategies for the management program of the Reserve are defined first as general management strategies which frame the limits and include ecological criteria for each of the environmental policies. Use intensity permitted under the application of each policy is emphasized. Uses are determined for more specific management of each marine system or landscape with the objective of integrating the environmental and socioeconomic elements in the defined areas.

The definition of limits and ecological criteria specific for each one of the landscapes and marine subsystems is made with the objective of considering the ecological and socioeconomic characteristics of each natural unit. To obtain this level of specificity two elements are necessary: an information base for the dynamic characteristics of the environments and an administrative and legal structure that facilitates the operation of the management program. Instrumentation for particular management strategies will show the need for outlining specific management programs to define uses of the resources, as well as for their administration.

Table V.- Description of the environmental units.

SYSTEM		LANDSCAPE	
		2.5 Natural Attractiveness	
Landscape Distribution	extensive/restricted	Vegetation	Presence/Absence
1. ABIOTIC FACTORS		Fauna	Presence/Absence
1.1 Geology		Hydrology	Presence/Absence
Rocky Type		Landscape	Presence/Absence
Relief	plain/undulate/rugged	**3.- SOCIO ECOMOMIC FACTORS**	
	(<10 %) (10-20%) (> 25%)	**3.1 Land**	
1.2 Underwater Hydrology		**3.1.1 Actual Land Use**	
Aquifer	underexploited	Recreation	presence/ absence
	overexploited	Tourism	presence/ absence
	polluted	Urban	presence/ absence
1.3 Overwater	**Hydrology**	Industrial	presence/ absence
Permanent	presence/ absence	Agriculture	presence/ absence
Seasonal	presence/ absence	Fisheries	commercial/artesan
1.4 Climatology		Aquiculture	semintensive/artesa
Atmospheric	Rain	Navigation	presence/ absence
Oceanic	Temperature	Natural Area	presence/ absence
	Salinity	**3.1.2 Degree of Transforma**	high/ middle / low
1.5 Natural Hazard		**3.1.3 Land Property**	
Flooding	fluvial	Public	presence/ absence
Erosion	coastal/terrestrial/absence	Private	presence/ absence
1.6 Soil Types		Ejido	presence/ absence
General classification		Conflict	presence/ absence
Texture	sand/mud/clay	**3.1.4 Equipment Level**	high/ middle / low
Aquaculture capacit	high/ middle / low	**3.1.5 Infraestructure Level**	high/ middle / low
2. BIOTIC FACTORS		**3.2 Anthropogenic Features**	
2.1 Flora		**3.2.1. Contamination**	
Vegetation type		Soil	presence/ absence
Quality	introduced/mixed/native	Aquifer	presence/ absence
Species Status		Ocean	presence/ absence
Endemics	presence/ absence	Water body	presence/ absence
In danger of extincion	presence/ absence	Rivers / Currents	presence/ absence
Threatened	presence/ absence	**3.2.2 Risk**	presence/ absence
Vulnerable	presence/ absence	**3.4 Heritage**	
Key	presence/ absence	Historic Attraction	presence/ absence
Useful	presence/ absence	Archeological Attraction	presence/ absence
2.2 Fauna		**OBSERVATIONS**	
Species Status			
Endemics	presence/ absence		
In danger of extincion	presence/ absence		
Threatened	presence/ absence		
Vulnerable	presence/ absence		
Key	presence/ absence		
Useful	presence/ absence		
2.3 Productivity	high/ middle / low		
2.4 Plankton	high/ middle / low		

Table VI.- Policy with respect to Value Factors (I= Indifferent, F= Favorable, U = Unfavorable).

	Protection	Protection with active use	Utilization with Control
ABIOTIC FACTORS			
1.- Landscape Distribution (extensive/restricted)	I /F	F/U	F/U
2.- Underground Hydrology (over//under/cont)	U/F/I	F/U/U	F/U/U
3.- Superficial Hydrology (permanent/seasonal)	F/I	F/U	F/U
4.- Natural Hazards (flood/erosion)	F/I	U/F	U/F
BIOTIC FACTORS			
FLORA			
5.- Quality (introduced/mixed/natural)	U/I/F	F/F/F	F/F/U
6.- Endemic (presence/absence)	F/U	U/I	U/F
7.- Threatened (presence/absence)	F/I	U/I	U/F
8.- Key (presence/absence)	F/I	U/I	U/F
9.- Useful (presence/absence)	F/I	F/I	U/I
FAUNA			
10.- Endemic (presence/absence)	F/I	U/I	U/F
11.-In danger of extintion (presence/absence)	F/I	U/I	U/F
12.- Threatened (presence/absence)	F/I	U/I	U/F
13.- Migrant (presence/absence)	F/I	U/I	U/F
14.- Key (presence/absence)	F/I	U/I	U/F
15.- Useful (presence/absence)	F/U	F/I	U/F
REPRODUCTION ZONES			
16.- Reproduction zones (presence/absence)	F/I	U/I	U/F
NATURAL ATTRACTIONS			
17.- Vegetation (presence/absence)	F/I	U/I	U/I
18.- Fauna (presence/absence)	F/I	U/I	U/I
19.- Hydrology (presence/absence)	F/I	I/I	I/I
20.- Landscape (presence/absence)	F/I	I/I	I/I
SOCIOECONOMIC FACTORS			
LAND USE	U/I	F/I	I/I
21.- Recreation (presence/absence)	U/I	F/I	I/I
22.- Tourism (presence/absence)	U/I	F/I	F/I
23.- Urban (presence/absence)	U/I	U/I	F/I
24.- Industrial (presence/absence)	U/I	U/I	F/I
25.- Agriculture (presence/absence)	U/I	I/I	I/I
26.- Fishing (commercial/artesan)	U/U	U/F	F/I
27.- Aquaculture (semintensive/artesan)	U/U	U/F	F/I
28.- Navigation (presence/absence)	U/I	F/I	F/I
29.- Core Zone (presence/absence)	F/I	U/F	U/F
30.- Degree of Transf. (high /middle/low)	U/I/F	U/I/I	I/I/I
LAND PROPERTY			
31.- Public (presence/absence)	F/I	U/I	U/I
32.- Private (presence/absence)	U/I	F/I	F/I
33.- Ejido (presence/absence)	U/I	U/I	U/I
34.- Conflict (presence/absence)	U/I	U/I	U/I
URBAN LEVEL			
35.- Equipment level (high /middle/low)	U/U/F	U/I/F	F/I/I
36.- Infrastructure level (hgh /middle/low)	U/U/F	F/F/I	F/F/U
RISKS AND ATTRACTIONS			
37.- Risks (presence/Absence)	U/I	U/I	U/I
38.- Historical Attraction (presence/Absence)	F/I	I/I	U/I
39.- Archeological Attraction (presence/Absence)	F/I	I/I	U/I

Table VII.- An example of the evaluation an environmental unit. (e = extensive, na = not apply, a = absence, p = presence, h=high, m = middle, l = low).

UNIT: 1.1.1.2			Protection		Protection with active use		Utilization with control	
ABIOTIC FACTORS	Factor	Weight	Value	V*W	Value	V*W	Value	V*W
1.- LANDSCAPE DISTRIBUTION	e	12	0	0	1	12	1	12
2.- UNDERGROUND HYDROLOGY	na	2	0	0	0	0	0	0
3.- SUPERFICIAL HYDROLOGY	p	7	1	7	1	7	1	7
4.- NATURAL DANGERS	p	3	1	3	-1	-3	-1	-3
BIOTIC FACTORS								
1. FLORA		9						
Quality	n		1	9	1	9	-1	-9
Endemic	p		1	9	-1	-9	-1	-9
Threaten	a		0	0	0	0	1	9
Key	p		1	9	-1	-9	-1	-9
Useful	p		1	9	1	9	-1	-9
2.- FAUNA		10						
Endemic	p		1	10	-1	-10	-1	-10
In danger of extintion	a		0	0	0	0	1	10
Threatened	a		0	0	0	0	1	10
Migrant	a		0	0	0	0	1	10
Key	p		1	10	-1	-10	-1	-10
Useful	a		-1	-10	0	0	1	10
3. REPRODUCTION ZONES	a	11	0	0	0	0	1	11
4.- NATURAL ATRACTION		4						
Vegetation	p		1	4	-1	-4	-1	-4
Fauna	a		0	0	0	0	0	0
Hydrology	a		0	0	0	0	0	0
Landscape	p		1	4	0	0	0	0
SOCIOECONOMIC FACTORS								
1.- LAND USE		7						
Recreation	p		-1	-7	1	7	0	0
Tourism	p		-1	-7	1	7	1	7
Urban	a		0	0	0	0	0	0
Industrial	a		0	0	0	0	0	0
Agriculture	a		0	0	0	0	0	0
Fishing	na		0	0	0	0	0	0
Aquaculture	a		-1	-7	1	7	0	0
Navigation	na		0	0	0	0	0	0
Core Zone	a		0	0	1	7	1	7
Transformation Degree	l		1	7	0	0	0	0
2.-LAND PROPERTY		6						
Public	p		1	6	-1	-6	-1	-6
Private	p		-1	-6	1	6	1	6
Ejido	p		-1	-6	-1	-6	-1	-6
Conflict	p		-1	-6	-1	-6	-1	-6
3.- URBAN LEVEL		1						
Equipment Level	l		1	1	1	1	0	0
Infrastructure Level	m		-1	-1	1	1	1	1
4.- RISKS AND ATTRACTIONS		5						
Risk of Disasters.	p		-1	-5	-1	-5	-1	-5
Historical Attraction	p		1	5	0	0	-1	-5
Archeological Attraction	a		0	0	0	0	0	0
				38		5		9

Based on the foregoing, the limits and specific ecological criteria are outlined in three periods: short term (one to two years), medium term (three to four years), and long term (more than 5 years). In table IX an example of the definition of specific strategies for each natural unit over time for each environmental policy is presented.

Table VIII. - Summary of the regionalization and evaluation of use capacity (environmental policy) and the most relevant ecological characteristics of the landscape by environmental unit and by hydrological region.

Environment:	**Coastal Land**			
System:	**1.1**	**Hydrologic Basin RH4A**		
Landscape	**Environmental Policy**	**Relevant Ecological Characteristic**	**Environmental Issue**	**Social Issue**
1.1.1.2.	Utilization with Control	Endemic vegetation (cactaceas), attractive desert vegetation.	None	Tourist , urban and aquaculture developments.
1.1.1.4	Protection with Active Use	Tidelands.	None	Mining salt development.
1.1.2.4.	Protection with Active Use	Tidelands. Rest area for migrant birds.	None	None
System:	**1.2**	**Hyrologic Basin RH4B**		
Landscape	**Environmental Policy**	**Relevant Ecological Characteristic**	**Environmental Issue**	**Social Issue**
1.2.1.2.	Utilization with Control	Endemic vegetation (cactaceas), attractive desert vegetation.	None	Tourist and aquaculture developments demand.
1.2.1.4	Protection with Active Use	Tidelands.	None	None
1.2.2.4.a.	Protection	Tidelands. Rest area for migrant birds in the Core Zone.	None	Traditional activity for the Cucapá community.
1.2.2.4.b.	Protection with Active Use	Tidelands. Rest area for migrant birds.	None	None
1.2.3.2.	Protection with Active Use	Endemic vegetation (cactaceas), attractive desert vegetation.	None	None

Table IX.- Protection Policy: limits and specific ecological criteria by environmental unit.

Coastal Landscape	Limits and Specific Ecological Criteria, to short term.	Limits and Specific Ecological Criteria, to medium term.	Limits and Specific Ecological Criteria, to long term.
1.2.2.4.a. 1.4.2.4.a. 1.3.2.4.a.	-. Reduction and control of road access to the area.	-. Establishment of meteorological station.	-. Follow-up of the environmental conditions of the delta.
1.4.7.6.	-. Development of an environmental education program. -. Creation of basic minimal infrastructure for the establishment of passive recreational activities. -. Any type hunting forbidden - Fauna apprehension and flora extraction or commercial use forbidden. - Use of all terrain vehicles and airboats forbidden. - Formal measures for fire prevention.	- Proposals and negotiation to assure water in sufficient quantity and quality to maintain the ecosystem of the Santa Clara Cienaga. - Program for the control of tular growth .	-. Systematic evaluation of environmental conditions, vegetation and species with designated status..
2.1.1.a 2.1.1.b 2.1.2	- Any type hunting forbidden - fishing activity forbidden, except clam and artesan fishing by the cucapa community (in the outlet of the delta), in the volume and terms that SEPESCA establishes.	-. Systematic evaluation of marine fauna.	-. Systematic evaluation of marine fauna.
1.4.8.3a. 1.4.8.3b.	- Any type hunting forbidden. - Unguided tourist groups forbidden. - Introduction of domestic animals or other exotics forbidden. - Access to islands during the migrant and resident water fowl reproduction limited.	-. Regulate ecotourism use of the islands.	-. Systematic evaluation of water fowl populations.

CONCLUSIONS.

The Upper California Gulf and Colorado River Delta is the first Biosphere Reserve of the three coastal reserves of Mexico that contains marine issues. This particular characteristic presents technical challenges, as well as social, political and management opportunities. The participation of local inhabitants of the Reserve in special forums indicated a clear resistance and dread toward the reserve, as well as little interest to participate within the formulation process of the management program. Nevertheless, they showed strong interest in developing a planning program for economic activities rather than the protection of species.

A methodology for creating an ecological classification program was presented. This ecological classification method permitted an integration of environmental and social factors into a geographical context with alternative management strategies defined for each administrative jurisdiction. It can be expected that administrative conflicts will occur, and such should be facilitated through a process of flexible agreements in the formulation and instrumentation of the program.

One of the first goals of the program will be to motivate greater participation of those living in the Reserve through presenting the objectives of the program as a base for an overall regional development plan. In this way the management strategies will be based on social demands and economic activities important to those living in the region.

REFERENCES.

Bojórquez-Tapia, L.A. and E. Ongay-Delhumeau, (1992). International Lending and Resource Development in Mexico: Can Environmental Quality Be Assured?. Ecological Economics. 5: 197-211.

Cendrero, A., (1982). Técnicas e Instrumentos de Análisis para la Evaluación, Planificación y Gestión del Medio Ambiente. CIFCA, Serie Opiniones, Política y Planificación Ambiental. Madrid, España. 6: 67 pp.

Cendrero, A., 1989. Mapping and Evaluation of Coastal Areas for Planning. Ocean and Shoreline Management, 12: 367-381.

Cendrero, A. and J.R. Díaz de Terán, (1987). The Environmental Map System of The University of Cantabria, Spain. In P. Arndt and G. Lüttig (eds.): Mineral Resource Extraction, Environmental Protection and Land-Use Planning in the Industrial and Developing Countries. E. Schweizerbart-Verlag, Stuttgart; 149-181.

Fermán J.L. and L. Gómez-Morin, (1993). Legal and Regulatory Aspects to Support a Mexican Coastal Zone Management Program. In Ferman, J.L., L.

Gómez-Morin and D. Fischer (eds.) Coastal Management in México: The Baja California Experience. Coastlines of the World Series, A.S.C.E., N.Y.; 7-13.

Fermán J.L., (1994). Programa de Manejo de la Reserva de la Biosfera del Alto Golfo de California y Delta del Rio Colorado. Tesis de Mestria en Ocenografia Costera. Facultad de Ciencias Marinas. Universidad Autonoma de Baja California. Ensenada, B.C. 78 pp.

Fischer, D.W., (1981). North Sea Oil: an Environment Interface. Institute of Industrial Economics, Bergen, Norway. 350p.

Gómez-Morin, L. and J.L. Fermán, (1991). Classification System of Environments Units for Land Use and Coastal Planning in Baja California, México. Coastal Zone '91. Proc. 7th. Symposium on Coastal and Ocean Management. American Society of Civil Engineers, New York, N.Y. 423-430.

Gómez-Morin, L., J.L. Fermán, A. Escofet, G. Torres, I. Espejel, L.G. Alvarez, G. Arámburo, N. Bringas, (1992). Ordenamiento Ecológico de la Microregión de Punta Banda-Estero de Punta Banda, B.C. Organización de Estados Americanos-Secretaría de Desarrollo Urbano y Ecología. Technical Report, 50 pp.

Naveh, Z. and A. S. Lieberman, (1984). Landscape Ecology. Spinger-Verlag. New York; 343 pp.

SEDUE (Secretaría de Desarrollo Urbano y Ecología), (1988). Manual de Ordenamiento Ecológico del Territorio. Subsecretaría de Ecología, Dirección General de Normatividad y Regulación Ecológica, México, D.F.; 356 pp.

SEMARNAP-INE, (1995). Programa de Manejo Areas Naturales Protegidas 1: Reserva de la Biósfera del Alto Golfo y Delta del Rio Colorado. 104 pp.

Slam, R.V. and J. R. Clark, (1989). Marine and Coastal Protected Areas: A Guide for Planners and Managers. IUCN, Gland; 302 pp.

Sorensen J.C., (1990). Coasts: Institutional Arrangements for Management of Coastal Resources. Research Planning Institute. Inc. Columbia, South Carolina; 265 pp.

CHAPTER 85

Local Issues in California Coastal Zone Planning: Case Studies in the Sacramento-San Joaquin Delta, Salinas River Basin, and San Francisco Bay

Gretchen Grani[1]

Abstract

This paper examines case studies of three current coastal zone planning programs in California that attempt to solve large-scale resource management problems. The case studies focus on different coastal regions of California and include Anadromous Fish Restoration in the Sacramento-San Joaquin Delta, Managing Seawater Intrusion in the Salinas River Basin, and the Long-term Management Strategy for Dredged Material from San Francisco Bay. Some of the local issues and stakeholders associated with these programs are highlighted, along with the opportunities and/or constraints they pose to the development of regional management programs. Examination of these case studies reveals the importance of local issues and has implications for the future governance of California's coastal resources.

Introduction

The term coastal zone planning has come to signify a comprehensive approach to considering the needs of an entire coastal region when formulating policy decisions. With the growing acceptance and practice of coastal zone planning, an opposite tension is increasingly becoming an issue: the disparity between the best interests of a coastal region as a whole and the interests of the individual communities that comprise the region. Although these interests are not necessarily in opposition, they are often perceived that way. This adds to the challenge of the planning process and requires that local participation and input be

[1] Environmental Planner, EDAW, Inc., 753 Davis St., San Francisco, CA 94111, Phone: 415-433-1484, Fax: 415-788-4875

incorporated, in order to ensure that regional management programs are accepted and effective in the long term.

Since the regional goals for California's ocean and coastal resources, as defined in *California's Ocean Resources: An Agenda for the Future (1997),* are to be achieved through "stakeholder partnerships" and "require cooperative efforts and understanding among all stakeholders," it may be useful for those implementing the Agenda to look at case studies where this has or has not worked. The three case studies presented in this paper provide insight into current coastal planning practices and the lessons learned from these practices, which are relevant to the future governance of coastal regions.

The case studies involve large-scale, resource management problems and solution strategies for the coastal regions of California shown in Figure 1. The case studies focus on Anadromous Fish Restoration in the Sacramento-San Joaquin Delta, Managing Seawater Intrusion in the Salinas River Basin, and the Long-term Management Strategy for Dredged Material from San Francisco Bay.

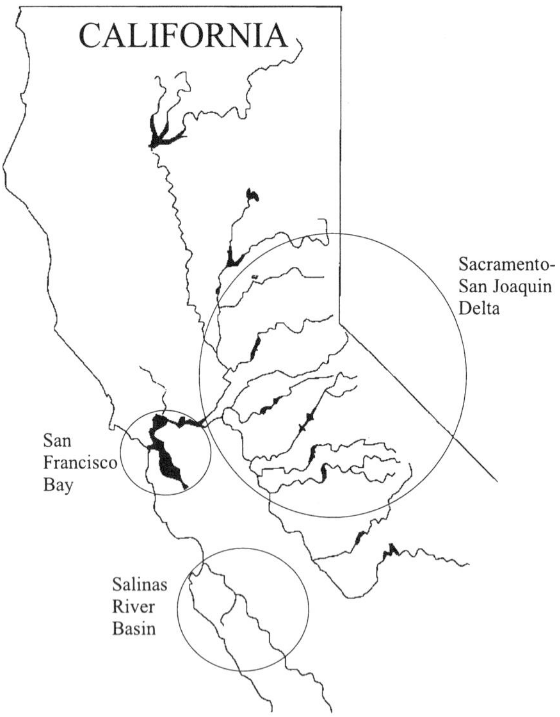

Figure 1. Regional Setting

Case Study 1: Anadromous Fish Restoration in the Sacramento-San Joaquin Delta

The Sacramento-San Joaquin Delta provides water supply for irrigating the Central Valley agriculture industry, as well as drinking water for two-thirds of California (including the San Francisco Bay Area and Central Valley residential communities). The Delta is also home to fish and other biological resources which depend on water as the medium in which they survive. The physical and ecological conditions of the Delta and its rivers have been significantly altered over time, and fish populations have severely declined. The Central Valley Project Improvement Act (CVPIA) of 1992 mandates an Anadromous Fish Restoration Program (AFRP) for the Central Valley rivers and streams that flow into the Sacramento-San Joaquin Delta. The goal of the AFRP is to double the natural production of threatened anadromous fish in that region.

In response to the CVPIA mandate, the U.S. Fish and Wildlife Service (USFWS) prepared as part of the AFRP a draft *Working Paper on Restoration Needs* (1995). This working paper suggests preliminary policies for water diversion operations, such as decreasing diversions to increase in-channel flows and coordinating the timing of diversions with fish migration. The paper also recommends river enhancement measures, such as restoring spawning habitat, decreasing predation, installing fish ladders at dams, and implementing water quality control measures. Other management measures such as increasing law enforcement and public awareness are also included.

In order to assist the USFWS with development of the AFRP, the California Urban Water Agencies sponsored a series of public workshops held in the fall of 1995 to discuss the *Working Paper on Restoration Needs* and facilitate a response to the paper from stakeholders. The objective of the workshops was to reach a consensus among all participants on restoration actions that could be submitted to the USFWS and various policy bodies for approval, and ultimately implemented as part of the AFRP. Another objective was to identify policy measures on which there was disagreement and/or uncertainty and should therefore be addressed in a longer-term process (CUWA 1995).

Local stakeholders attending the workshops represented various organizations with an interest in the Sacramento-San Joaquin Delta, including resource management agencies, urban and agricultural water use agencies, environmental groups, and fish biologists. Individuals and organizations with specific technical knowledge pertaining to the Central Valley rivers or aquatic biology discussed their views on the science of fish restoration and the unique conditions of the individual rivers. All parties participating in the workshops expressed an interest in identifying, as quickly as possible, specific immediate-term measures. Discussion of the more complex and contentious issues, such as the relationship between river flows and natural fish production and adequate flow amounts, was postponed.

The result of the workshops was consensus on sixty-one policy recommendations for fish restoration and management, which have since been incorporated into the Draft Anadromous Fish Restoration Plan (1995). Consensus was not reached, however, on the most contentious issue of flow allocation. The crux of this issue was the question of how the mandated reallocation of 800,000 acre-feet of river water for fisheries restoration would be utilized.

A lesson learned from this case study is the usefulness of reaching consensus on common stakeholder issues, even if conflicting issues are waiting to be resolved. The resulting consensus on issues held in common was a significant first step in the overall collaborative process and served to build partnerships and mutual understanding. This step also allowed for development of the regional management program to proceed with the identified immediate-term restoration measures.

Case Study 2: Managing Seawater Intrusion in the Salinas River Basin

The coastal aquifers of the Salinas River Basin in Monterey County are subject to significant seawater intrusion and aquifer system compaction, which has reduced groundwater storage and caused land subsidence. This condition threatens the sustainability of the aquifers and their ability to support regional demand for groundwater. Groundwater is not only the primary source of water for local communities, but also the crucial source of irrigation for the highly productive and fertile agricultural land of the Salinas Valley, the "nation's salad bowl" (Soil Conservation Service 1978).

The Monterey County Water Resources Agency (MCWRA) is responsible for developing the Salinas River Basin Management Plan (BMP). The goals of the BMP are to stop seawater intrusion, assure future water supplies to the year 2030, and balance the Basin aquifers (MCWRA 1994). Many complex local issues have complicated the MCWRA's efforts towards developing the BMP since 1992.

Local stakeholders have expressed a number of concerns, including the equitable allocation of project benefits and costs, project impacts affecting the local agriculture industry, community resistance to future growth, natural resource protection and enhancement, and the ability of the Basin aquifers to continue to provide local water supplies. The process of developing the BMP has required extensive and lengthy public involvement to address local stakeholder concerns and reach a consensus on key issues. This is due in part to the high level of community activism. Local stakeholders are well informed about the BMP and have continued to be vigilantly involved over many years.

Compounding the effects of the extensive planning process is the uncertainty that exists with regard to undefined scientific issues, relating to seawater intrusion, and undefined economic issues relating to the value of groundwater. The amount of water needed to stop seawater intrusion and balance the Basin is not precisely known, nor has the BMP project benefits and allocation of costs been given an exact economic value. The ambiguity and complexity surrounding these issues, combined with the volume of information and opinion generated by the planning process, has at times confused stakeholders and heightened their distrust of the planning process. The public voiced their distrust in 1994 when they disapproved of the initial set of BMP project alternatives, which were then in turn rejected by the Board of Directors.

One of the lessons learned from this case study is that managing the flow of public information is critical to gaining the confidence of stakeholders and facilitating consensus. It is important when releasing works in progress and interim reports to ensure that all parties are kept up to date with the most recently developed information. It is also important to present and explain information when it is released, and to clarify assumptions to minimize confusion and distrust of the problem solving process. In addition, it should sometimes be made clear that policymakers are searching not for the only answer but for the best answer they can find for a complex problem. If not adequately managed, the public participation process can pose a constraint to the successful development of regional management programs.

Case Study 3: Long-term Management Strategy for Dredged Material from San Francisco Bay

The dredging of San Francisco Bay for maritime trade, recreational boating, and other public trust uses supports billions of dollars of regional economic activity which depends on maintaining Bay shipping channels. This dredging creates an annual disposal requirement of approximately 4 million cubic yards of sediment (Resources Agency 1997). Current port expansion plans to deepen existing harbors would result in an even greater disposal requirement.

The primary and most economical method of dredge disposal to date has been to sites within the Bay. However, capacity constraints caused by severe mounding at the primary in-Bay site near Alcatraz Island, combined with concern over long-term environmental effects, limits the future use of in-Bay disposal and increases the need for an overall dredge management plan for San Francisco Bay.

The Long Term Management Strategy (LTMS) is a multi-participant program established to prepare a plan for disposal and/or reuse of dredged material from the Bay over the next 50 years. As part of this strategy, studies are being conducted and managed in a collaborative effort by a group of regulatory agencies consisting of the U.S. Army Corps of Engineers, San Francisco Bay Conservation and Development Commission, U.S. Environmental Protection Agency, and San Francisco Regional Water Quality Control Board. Local stakeholders are also actively involved, including the ports, local counties and communities that border the Bay, and public and private entities.

Local stakeholders have concerns relating to feasibility, regulatory, and ownership constraints, potential contamination and associated health issues, and the need for wetland restoration in the Bay. As part of the LTMS, opportunities and constraints for the reuse/upland disposal of dredge material were analyzed, with the goal of maximizing the use of dredged material as a resource. Existing land use planning, regulatory and permitting, and funding and implementation systems were reviewed. Review of these systems revealed that there is an absence of priority programs within key agencies to address the reuse or upland disposal of dredged material. Proponents of sediment reuse must deal individually with multiple local and regional regulatory agencies, since two or more agencies may have jurisdiction over the same aspects of projects. In addition, existing regulations are often applied without recognition of the beneficial aspects of many proposed projects (particularly wetland restoration). These projects are viewed most often as waste disposal projects (EDAW 1993).

As a consequence of these regulatory and planning practices, opportunities to combine sediment disposal needs with the protection or even enhancement of the environment have not been widely explored, much less pursued. One of the interim results of the LTMS collaborative process has been an opportunity to incorporate into the management strategy the use of dredged sediment for wetland restoration. The State Coastal Conservancy and Sonoma Land Trust's restoration plan for a 322-acre diked hayfield and former tidal marsh in the Sonoma Baylands fit well with the regional need for sediment disposal. This tidal marsh restoration project used more than 2.5 million cubic yards of clean dredged material from the Port of Oakland to raise a subsided hayfield to the elevation required for salt marshes to occur (Marcus 1995). In addition to restoring Bay wetlands, this restoration project allowed for the much needed deepening of the port's channels to accommodate larger container ships.

One of the lessons learned from this case study is that collaboration and building partnerships can result in creative and unexpected solutions to regional planning problems. The LTMS collaborative process brought together individuals from all the regulatory agencies involved with various aspects of dredging activity, in addition to other local stakeholders with an interest in the Bay. This allowed them an opportunity to view dredge management within a comprehensive regional framework and to refine inefficient regulatory and planning systems. With the involvement and encouragement of local stakeholders, especially those associated with the Baylands project, innovative solutions for combining sediment disposal needs with wetland restoration needs was explored. In part due to the success of the Baylands project, the regulatory agencies coordinating the LTMS have all endorsed wetland creation as an option for the future reuse of dredged material.

Conclusion

Examination of these case studies reveals the role that local issues and stakeholder groups can have in influencing the development of regional management strategies. These studies illustrate that the planning process, whether it attempts to incorporate the needs and concerns of a community or an entire region, is still in essence a political process. This process will be driven by the constituencies that have something at stake. Regardless of the geographic scope of a given planning program, its outcome will be determined by the stakeholder groups that have organized around the process, whether they consist of small or large numbers of people.

While the goals of regional management programs may be focused on large-scale, measurable scientific and engineering strategies (e.g. double the natural production of anadromous fish; stop seawater, balance the Basin and provide future water supplies; or reuse dredged material from the Bay), the achievement and implementation of these strategies occurs at the local stakeholder level. In other words, coastal zone policy can guide and limit coastal activities but cannot build local solutions and opportunities. Coastal zone planning, therefore, should consider not only the scientific and economic realities of an entire coastal region when formulating policy decisions, but also how a policy program will support local solutions and facilitate dispute resolution.

Appendix - References

California Urban Water Agencies, 1995. *Recommended Immediate and Near-term Restoration Actions for Doubling Anadromous Fish in the Central Valley*, October 30.

EDAW, 1993. *Long Term Management Strategy Reuse/Nonaquatic Studies*. Prepared for the U.S. Army Corps of Engineers, San Francisco, California.

Monterey County Water Resources Agency (MCWRA), 1994. *Salinas River Basin Management Plan Alternatives Analysis*. Monterey, California, August.

The Resources Agency, 1997. *California's Ocean Resources: An Agenda for the Future*. State of California, Sacramento, California, March.

Soil Conservation Service, 1978. *Soil Survey of Monterey County, California*. U.S. Department of Agriculture in cooperation with the U.S. Forest Service and University of California Agricultural Experiment Station.

U.S. Fish and Wildlife Service, 1995. *Working Paper on Restoration Needs*.

U.S. Fish and Wildlife Service, 1995. *Draft Anadromous Fish Restoration Plan*. Prepared with assistance from the Anadromous Fish Restoration Program Core Group under authority of the Central Valley Project Improvement Act, December.

CHAPTER 86

LOCAL GOVERNMENTS ON THE EDGE: PLANNING FOR THEIR OCEAN INTERFACE

David W. Fischer* and Ingrid Lenz**

Acknowledgements

The authors wish to thank their research assistants without whose efforts this paper would not have been possible. Our thanks to Veronica Dolleschel and Nanci Eastman.

Abstract

This paper describes a survey of southern California municipal planners to determine their degree of emphasis given to ocean-related concerns within their jurisdictions. With growing local impacts and issues associated with the coastal zones of southern California, the study was undertaken to assess the initiation and use of scientific information in planning for coastal land use decisions. In addition, the kinds of ocean-related problems of local planning concern were identified along with the measures local governments use to mitigate these problems. Also, the planning measures local governments were willing to consider were identified. An emphasis of this study was denoting the types of ocean-related hazards recognized in their local coastal plans, if any, and the types of measures used to avoid such hazards. Sea level rise was singled out and each municipality was requested to determine their expected actions, timing and payment strategy for this chronic, uneven event. The findings show that planners lack knowledge of their coastal zones and view the coastal zone as just one element among the many elements involved in the planning process. Their emphasis on day-to-day development permitting overshadows any proactive stance on strategic planning for their ocean interface.

Introduction

The twin driving forces of population growth and the need for economic development are the leading sources of stress on the global environment.[1] Nowhere are these forces more apparent than in the southern California coastal zone with its increasing traffic congestion and lack of parking at local beaches, frequent sewage spills and beach closures, infilling and expansion of existing coastal developments and continuing property losses from coastal hazards. A fifty year description of past and projected population growth shows the tremendous growth experienced by the California coastal zone (Table 1). Such growth, in part, generates the demand to live in beautiful, yet hazardous, places in the coastal zone. Indeed, California has the largest total population in coastal counties in the United States.[2]

Los Angeles, Orange and San Diego Counties comprise more than 200 miles of the total 1100 mile California shoreline and account for the popular image of California scenic beach areas. These three counties also contribute the majority of coastal residents and have an intense infilling of their coastal seeking places to stay. More residents and tourists demand more services and new businesses are opened to meet the need. Additional municipal services are required alike by residents, tourists and businesses.

---

*Facultad de Ciencias Marinas, Universidad Autonoma de Baja California, Mexico, and Graduate Center for Public Policy and Administration, California State University, Long Beach, USA.
** Administrative Analyst, Community Development, City of Anaheim, California, USA.

Such growth impacts negatively on the environment, but the mere presence of this population and infrastructure bear impact from the ocean interface as well. Coastal hazards are many in southern California. Winter storms along with torrential rains interacting with erosive soils have generated cliff slumping, mud slides and beach loss along with whatever structures were associated with these areas. For example, rains caused landslides, subsidence and debris in Orange County resulting in a loss of homes, highway, railroad and municipal services for a total of $75 million.[3] The City of Malibu suffered floods and mudslides from winter rains at an estimated coast of $22 million.[4] The City of Redondo Beach was hit three times by winter storms in 1988, resulting in a loss of structures with 18 businesses destroyed and 400 jobs lost for a total of $32 million lost.[5] These costs involve direct losses and do not include the costs of lost business, litigation and additional protective works to mitigate future storm damage. California suffers an average of $10 million in property losses annually due to winter storms.[6]

Not only are there recurring hazards of high probability, such as winter storms, but hazards of lower probability also exist. These hazards include tsunamis, earthquakes and sea level rise. Given the loosely consolidated soils underlying coastal bluffs and comprising beaches in southern California, these hazards can be termed significant, especially in conjunction with winter storms. While tsunamis or earthquakes have hit southern California's immediate coastal zone in historic times, the devastation recorded in tsunamis and quake events in Alaska was immense.[7] The recent quake in Kobe, Japan, with the epicenter in the coastal zone resulted in widespread loss of life, property and infrastructure.[8]

Sea level rise is a "rising" hazard of immense scope. The expected national coastal property loss just in wetlands has been compared to the loss of the entire state of Massachusetts.[9] While California losses are estimated to be less than elsewhere in the United States, such losses will still be major and include impacts on the entire economy, the state's water resources, wetland habitats, fisheries, endangered species, coastal bluffs and beach and coastal property.[10] San Francisco Bay would be irrevocably changed through the necessity of having to build seawalls and levees at an estimated cost of $1 billion with an annual maintenance of $100 million.[11] Coastal beach retreat in southern California has been estimated to be between 30 - 200 feet by the year 2050 with an even greater risk inland from winter storms and wave run-up.[12]

With the southern California population attracted to coastal activities in spite of coastal hazards, a planning and regulatory framework was created to account for impacts of coastal use and development.[13] The general policies of the California Coastal Act include:

1. Providing for maximum public access to and recreational use of the coast, consistent with private rights and environmental protections.

2. Protecting marine and land resources, including wetlands, rare and endangered habitat areas, environmentally sensitive areas, tidepools, and stream channels.

3. Maintaining productive coastal agricultural lands.

4. Directing new housing and other development to urbanized areas with adequate services rather than allowing a scattered, sprawling, wasteful pattern of subdivision.

5. Protecting the scenic beauty of the coastal landscape.

6. Locating any needed coastal energy and industrial facilities where they will have the least adverse impact.

It is of interest to note that coastal hazards do not appear as a general policy goal in the Coastal Act. The Act does not recognize coastal hazards; however, only one section notes that developments shall, "minimize risks to life and property in areas of high geologic, flood and fire hazard" (Section 30253). No coastal hazard requirements are set forth in either the Coastal Act or its implementing guidelines. Thus, local governments which must implement the California Coastal Act can be expected to have a wide variety of responses to coastal hazards.

Each coastal city is required to prepare and maintain a Local Coastal Plan (LCP). The LCP incorporates the policies outlined in the Coastal Act and must be approved by the California Coastal Commission. A LCP is the city's specific, long-term coastal management plan which includes a land use plan, zoning ordinances and other implementing actions. LCP's are drafted by coastal cities, submitted to the Coastal Commission for approval and, upon approval, are formally adopted by the City Council of the authorizing city.[14] Not all coastal cities have complied with this requirement.

Because municipal governments bear the brunt of coastal planning for environmental and hazard impacts, this paper focuses on this level of government. The objective is to identify the extent to which coastal municipalities in southern California plan for ocean-related issues and what mitigation measures, if any, are used or being considered. A further objective is to assess the extent of knowledge planning officials have concerning coastal problems, resources and hazards and what features are incorporated, if any, into their Local Coastal Plans.

Selected Previous Studies

Information about the role of local government in coastal planning is scarce. Recent case studies about California coastal cities are scant because most studies are conducted statewide. One such state-sponsored study focused on the coastal policies of locals governments in the Los Angeles area before and after the passage of the California Coastal Act of 1976.[15] The study looked at development patterns and the permit processes to obtain permission to develop coastal land. For example, before the Coastal Act, Redondo Beach followed a policy of extensive coastal development where the City was transformed from a small, "locals only" beach community to a larger municipality with King Harbor and an urban redevelopment project. After passage of the Coastal Act, the new permit process halted many proposed projects in Redondo Beach, including some projects with tremendous citizen support. The authors of this study concluded that the Coastal Commission cannot make decisions that are universal, but rather should take the time to review local government policies which reflect the historical patterns and preferences of the local citizenry.

The study of growth management without specific reference to coastal concerns was prominent in the U.S. in the 1980's. For example, both state and local levels of concern were addressed in surveys of development growth pressures and planning responses by the appropriate government agencies.[16] In general, these studies show that innovative yet aggressive approaches by state governments were successful in guiding local growth planning in order to preserve the physical environment. At the local government level, the use of mandated comprehensive planning fostered innovative approaches to balancing the demands for environmental quality with new development.[17]

One recent study tracked the "American Trader" oil spill in Huntington Beach.[18] The problem which surfaced during this incident was the lack of coordination between local governments that were affected by the spill. There was little communication between and among the five immediately affected cities and the 49 other federal, state and local agencies involved in the clean-up efforts. This study stressed the importance of taking a proactive approach by planning for ocean-related contingencies.

Two other studies concerning coastal hazards focused on increased coastal erosion resulting from sea level rise. The first study, conducted in Ocean Beach, California, near San Francisco, stated that by the year 2100, sea level rise will provoke a tremendous amount of erosion.[19] In Ocean Beach, a Sewer Transport project located in the coastal zone was approved by the California Coastal Commission and the participating cities without full knowledge of the effects of sea level rise on the project. This study showed that erosion caused by sea level rise would undermine the approved Sewer Transport project. The second study on sea level rise focused on planning for this hazard. The author states that, "planning for global (warming) is made difficult not only as a result of the diversity of agencies involved in producing country assessments and/or recommendations for action, but also because such assessments have been undertaken in an uncoordinated manner, as a crisis response to current concerns, and without clear definition of spatial and temporal boundaries."[20] This study concluded that many, 'local, regional and national studies have failed to define precisely the changed conditions or the time frame under which projected scenarios will occur and have often been based on general rather theoretical reviews of broad areas of impact which may or may not occur in any given location."[21]

Two studies of local government responses to coastal hazard based on surveys were recently completed. The first concentrated on California municipal efforts to develop and protect their coastal zones via municipal ordinances and regulations.[22] This study relied on a questionnaire and interviews noting the use of setback standards, technical study requirements, regulations of seawalls, and desired changes from state agencies. The second study used a telephone survey to determine Louisiana coastal resident and local officials' views on the impacts of sea level rise.[23] The California and Louisiana surveys showed the need for clearer policies from state governments to assist local land use planning in potentially hazardous coastal areas. Coastal hazard information was deemed lacking as well as the regulatory measures needed to reduce development in threatened areas. Surprisingly, only four out of the 48 California local governments surveyed had a specific ordinance dealing with geologic hazards. Even though the Louisiana study dealt with sea level rise and the California study dealt with coastal erosion and flooding, both studies showed that local officials felt they lacked the regulatory measures to address the problems they faced. While no official wanted to restrict development in response to coastal hazard, local governments seemed increasingly aware of the conflicts they faced between public and private concerns.

A survey study involving the authors was conducted among northern Spanish coastal municipalities. It showed that local officials tend to rely on personal observations, legal requirements and tourist demands for making coastal land use decisions.[24] Scientific information in the form of expert studies had not played a role in decision-making. The views of these officials with respect to the need for coastal protection and hazard avoidance were at variance with scientific studies of the same region.

These three survey studies, California, Louisiana and Cantabria, show that coastal municipalities desire clearer policies and regulatory measures from the next highest level of government to assist them in planning for coastal protection and hazard avoidance. As well, a gap seems to exist between what is known among scientists and what is being implemented locally by municipalities.[25]

Methodology

The data for this study was obtained through a survey administered by three research assistants, one each for Los Angeles, Orange and San Diego Counties. In Los Angeles County, the cities of Malibu, Santa Monica, Rancho Palos Verdes, Manhattan Beach and Redondo Beach were contacted. In Orange County, the cities of Seal Beach, Huntington Beach, Newport Beach, Laguna Beach, Dana Point and San Clemente were contacted. In San Diego County, the cities of Encinitas, Solana Beach, Del Mar, Coronado and Imperial Beach participated. The three large cities of Long Beach, Los Angeles and San Diego were excluded because of their size relative to all the other coastal cities.

It is recognized that coastal planning can be influenced by the national, regional and local levels of government as well as non-governmental organizations and the general public. However, this study focused on local government because it is the central arena where coastal plans are forged, interpreted and implemented. Local government officials integrate the requirements of other government levels with demands from their constituents to create the plans that shape the development of their respective coastal zones. Therefore, this study was directed solely to local governments of small to medium-sized cities in the three county region.

An advance copy of the questions was sent to the planning director of each municipality included in the study. Along with the questions a cover letter was enclosed to request that the questions be given to the municipal planner with responsibility for technical coastal considerations prior to the interview. Each respondent was asked each of the pre-determined, multiple-option questions in the order presented in the questionnaire and their responses were recorded by the interviewer.

The questions asked of these local officials included the topics of what coastal problems were being experienced, what coastal features protected, what coastal hazards were acknowledged, their knowledge of sea level rise, planning response to conflicts involving coastal protection and development, and their preferences for coastal scientific information. The questions used were similar to the California, Louisiana and Spanish studies previously described.

Survey Results And Discussion

Table 2 shows the recent and recurring coastal zone problems being experienced by each of the municipalities. Sixty percent of the cities reported problems with cliff slumping, storm water run-off and water pollution, making these three issues the most prevalent problems affecting the region. Congestion and beach erosion were each also cited as on-going concerns for the surveyed cities. Nearly all of the municipalities reported experiencing persistent coastal problems.

Table 2 also shows the reasons cited by the municipalities for their continuing coastal problems. Municipalities viewed urbanization pressures and lack of funding as the reasons behind their coastal problems. Municipalities cited inadequate public infrastructure, including sewage treatment, as a contributing cause of their coastal zone problems. This lack of funding may discourage cities from designating beach erosion and cliff slumping as fiscal priority items because in times of financial hardship voters tend to favor basic city services. Also reported was tourist growth, lack of local authority and conflict as recurring problems.

Table 3 shows the specific types of natural coastal features protected by each municipality. The table indicates that a majority of the municipalities actively protect beaches and open spaces, while few cities protect dunes, farms and rivers. The table also indicates the measures employed by municipalities to protect these natural features. Frequently-used protection measures include regulations, buffer zones, special use plans/zones, building codes and engineering structures. No municipalities reported buying-out owners as a coastal protection measure. As shown in Table 2, this may be due to lack of funding for local coastal cities. Few municipalities reported banning activities or providing tax incentives as coastal protection methods. No proactive stance toward ocean-related phenomena seem to exist in these municipalities.

Table 4 denotes the municipal actions taken or considered in the process of granting legal protection to the coastal zone. The table indicates the types of studies done to support coastal protection and the potential solutions that municipalities are willing to consider to protect the coastal zone. The types of studies done to support coastal protection include urban planning, vegetation and wildlife identification, engineering studies and environmental impact assessments. The least used were socio-economic studies. It was surprising that environmental impact studies were not used in one hundred percent of the cities due to the problems associated with urbanization and its destruction of natural habitats. Just one-half of the municipalities surveyed indicated that they were willing to consider land use plans based on environmental capacity as a solution to coastal protection. This was closely followed by buffer zones, density, educational programs, lot/building density restrictions, and restricting new public infrastructure. No municipality reported even considering leasing public land as a viable solution to coastal protection. The cities of Dana Point, Seal Beach and San Clemente were interested in protection, using studies of endangered species, environmental impacts and urban planning. Huntington Beach, with the second largest stretch of coastline in the surveyed municipalities, only cited urban planning as a coastal protection method. Surprisingly, educational programs were not considered as an option among San Diego County cities, except for Del Mar; however, buffer zones were considered important by all cities in this area. Again, no proactive movement toward ocean-related planning could be detected.

Table 5 shows preferences for new projects in the coastal zone and is based on the hypothetical question of how each municipality would spend $25 million dollars. Coastal and non-coastal choices were included in the list of spending alternatives. Interestingly, there was little agreement among the municipalities about the most desirable new projects. Preferences were for hazard mitigation and new parks followed by new parking lots, a museum or aquarium, a recreational pier and engineering structures. No municipalities favored new roads to the coast, a new public marina or enhanced sewage treatment. Most municipalities would recommend the same expenditures if their municipality had to provide 50% of the matching funds for the new projects, particularly for hazard mitigation. Orange and San Diego County cities appear more willing to provide matching funds than Los Angeles County cities.

Table 6 describes the types of ocean hazards experienced by local municipalities and the studies undertaken to assess and mitigate those hazards. Most municipalities identified storms and beach erosion as hazards, but interestingly, the majority of municipalities reported using urban planning as a means of hazard assessment and mitigation. Urban planning cannot solve the problems that arise from beach erosion and storms, but engineering studies, equally used, are more useful for this type of problem. Santa Monica and Encinitas are the only municipalities that use the types of studies identified in the survey to assess and mitigate identified coastal hazards.

Table 7 illustrates the measures used or considered by municipalities to avoid coastal hazards. Again, Santa Monica used all studies/measures noted in the survey. Most of the municipalities use, or are willing to use, land-use planning measures because of the California Coastal Commission mandate requiring municipalities to include a Land Use Element in their Local Coastal Plan. The City of Malibu is the only survey municipality that does not have a Local Coastal Plan and could not respond to this section of the survey.

Table 7 also described hazard avoidance measures that municipalities are willing to adopt. Most of the municipalities were willing to engage in erosion setbacks and land-use planning based on hazard potential and post storm/flood reconstruction restrictions. A lesser number of cities would use educational programs, remodel infrastructure, create special hazard study zones and require hazard insurance of residents in the coastal zone. Not one municipality was willing to pay part of the costs of residents to relocate or purchase hazardous areas as hazard avoidance measures. These results are similar to a survey of coastal municipalities done in North Carolina.[26]

Table 8 illustrates the commonality of a regional hazard facing all of the municipalities in the study area and how they may be affected by sea level rise. A majority of the municipalities reported that sea level rise would result, most significantly, in loss of beaches, loss of housing and flooding of inhabited areas. Only one municipality, Dana Point, believed that they would experience no effect from sea level rise, even though they have considerable beachfront development. Many of the municipalities reported that their cities would respond to sea level rise within one to five years and others thought their cities would respond in six to ten years. Although sea level rise is not perceived as an immediate hazard, all municipalities registered a more immediate response than was experienced in Spain.[27]

Sea level rise mitigation strategies preferred by a majority of the municipalities were building shoreline protection structures, accommodating existing structures and prohibiting new construction. Payment for these mitigation measures was reported to be primarily by owners of shoreline development, followed by federal and state governments at 50% each. As expected, only two municipalities reported that the entire city would pay for sea level rise mitigation measures. No proactive local government stance or joint efforts were reported.

Conclusion

The results of this study confirm that local coastal planners are aware of existing and potential coastal problems in their municipalities and are exploring these issues in accordance with prescribed mandates and their own perceptions of the magnitude of these problems in relationship to other city issues. One issue is the inherent difficulty in dealing with coastal zone issues separately from non-coastal zone issues. Although many coastal problems really need special use review and planning, the magnitude of the coastal planning equation in relation to the entire municipal planning process is often lost or denied. Although only one survey municipality, Manhattan Beach, reported employing a planner dedicated to coastal issues as opposed to just one or two specially-designated planners. Perhaps this lack of specialization illustrates best the municipal perspective on coastal zone issues as being just part of the usual mix of planning issues in general, even though each municipality with a coastal zone is responsible for a unique natural resource/hazard area.

Some coastal problems may receive less than proactive attention because of the lack of coastal information. Not one surveyed municipality indicated that they would be willing to seek expert advice about coastal issues. This finding may be based on the assumption that expert advice must be purchased and, since coastal issues are not considered a priority, funds are not forthcoming for ocean-related issues. Municipalities do, however, have access to experts conducting scientific research on coastal problems whose findings could assist coastal cities in identifying and mitigating potential coastal hazards. Often such research is conducted by local universities and is free to those interested in the information.

It is encouraging, however, that some long term coastal hazards are gaining increased attention. This is particularly evident in the sea level rise information obtained in this survey. The results of two previously-cited case studies differ from those in this survey in that in the previous studies, sea level rise was not considered a pressing issue. However, this survey indicates that sea level rise is on the "radar screens" of local coastal municipalities. Further, this study also shows that most surveyed municipalities would be significantly impacted by sea level rise. This indicates an increased perception of long term coastal hazards and could be the start of an encouraging exploration of potential long term mitigation measures.

Given the unique planning and mitigation issues associated with the ocean, the prevailing lack of information about it, and the municipal fiscal scarcity hampering the creation of such valuable information, perhaps it should be recognized that the coastal zone has a very special, untapped resource at its disposal...its wealthier-than-average residents. Coastal municipalities in southern California generally have a large population of upper and upper-middle class residents. As an example, few surveyed municipalities were willing to require hazard insurance or consider the leasing of public lands. Such revenue streams could help pay for coastal hazard identification and mitigation as well as fund other coastline preservation activities. The concept of having hazard zone residents bear the brunt of the cost of living in that hazard zone is not new, but must be revisited as we move into the era of sea level rise. This new era is characterized by a new understanding and acceptance of the unique short and long term issues and remedies associated with the coastal zone and the real costs of inhabiting, maintaining and preserving this unique area.

Although the California Coastal Commission works with local governments to protect the coastal zone, they do not appear to help local governments prepare for future, long term hazards. The Coastal Commission is largely responsible for the permit process in the coastal zone by granting and denying developers the right to build on certain coastal properties, even hough it often lacks local knowledge of municipal preferences. This situation can lead to Commission decisions being ignored. This permit process has become the focus of energy for the state and local coastal municipalities rather than the larger picture which deals with identification, protection and mitigation of marine environmental resources and hazards. Currently, governance of the coastal zone is administered reactively, ultimately costing more and achieving less per dollar expended than if a proactive governance approach were used in the coastal zone. Generally, the region's storm damage was far more expensive than the mitigation measures that could have prevented or minimized storm damage. A shift in the perception of coastal zone planners is important, especially moving the primary focus way from individual development projects and toward coastal zone management.

It is evident from the general results of this survey that not nearly enough is being done to protect the southern California coastline by the municipalities. Survey results indicate that coastal municipalities are taking inconsistent approaches to local coastal planning and protection. Local governments must be empowered and encouraged to identify broad coastal problems and mitigation strategies and work in concert with each other and the California Coastal Commission to review those strategies from a regional perspective. The more local coastal development issues should be left solely to the municipalities. This shift in governance and perspective, when combined with financial and legislative support for ocean planning, would provide a true foundation for cost-effective, long term, coordinated proactive coastal ocean planning.

Tables

Table 1 **Coastal Zone Population In California**

1960	13,073,000
1970	16,662,000
1980	19,238,000
1990	23,091,000
2000	25,605,000
2010	27,429,000

Source T J Culliton, M.A. Warren, et al , 50 Years Of Population Change Along The Nation's Coasts, 1960 - 2010 (Strategic Assessment Branch, National Oceanic and Atmoshperic Agency, Washington, D C 1989)

Table 2 Officially recognized coastal zone problems experienced by selected Southern California coastal municipalities in order from north to south

Coastal Zone Problems

Municipality	N	BE	WP	CS	WL	SC	LN	IU	SI	DS	F	RC	SR	OT	SP	WD	C	O	P
Malibu						X					X		X				X		
Santa Monica			X	X				X	X				X			X	X		
Rancho Palos Verdes		X	X	X									X		X			X	
Manhattan Beach								X				X					X		
Redondo Beach																			X
Seal Beach		X	X																
Huntington Beach				X'															
Newport Beach		X	X	X						X	X	X	X		X		X	X	
Dana Point			X	X									X	X			X		
San Clemente		X'		X'									X'					X'	
Encinitas		X	X	X							X		X						
Solana Beach		X		X							X								
Del Mar		X	X	X		X				X	X	X	X		X		X		
Coronado			X										X				X		
Imperial Beach		X	X																
Total Number	0	8	9	9	0	2	0	2	1	2	5	3	9	1	3	1	7	3	1
Percentage	0	54	60	60	0	14	0	14	7	14	34	20	60	7	20	7	47	20	7

' = recent problems, X persistent problems

Coastal Zone Problems N = none, BE = beach erosion, WP = water pollution, CS = cliff slumping, WL = wetland loss, SC = silting of channels, LN = loss of natural areas, IU = increased urbanization, SI = scarcity of infrastructure, DS = dredging/spoil disposal, F = flooding, RC = recreational use conflicts, SR = stormwater runoff, OT = oil/toxic spills, SP = scarcity of public services, WD = waste disposal, C = congestion, O = other, P = no response

Table 2 Officially recognized coastal zone problems experienced by selected Southern California coastal municipalities in order from north to sout (continued)

Reason Coastal Zone Problems Still Occurring

Municipality	TG	IP	UP	LI	PS	LS	LF	IL	LA	SA	AC	O	P	N
Malibu							X					X		
Santa Monica			X		X	X						X		
Rancho Palos Verdes			X				X		X					
Manhattan Beach	X		X											
Redondo Beach													X	
Seal Beach			X				X				X			
Huntington Beach														X
Newport Beach		X	X		X							X		
Dana Point					X									
San Clemente	X				X		X				X			
Encinitas							X		X		X			
Solana Beach												X		
Del Mar			X		X		X	X		X				
Coronado	X					X		X	X					
Imperial Beach						X						X		
Total Number	3	1	6	0	5	3	6	2	3	1	3	5	1	1
Percentage	20	7	40	0	34	20	40	14	20	7	20	34	7	7

Reasons For Continuing Problems TG = tourist growth, IP = industry practices, UP urbanization pressures, LI - lack of scientific information, PS = inadequate public infrastructure, LS - lack of complete sewage treatment, LF = lack of funding, IL - inadequate legislation, LA - lack of local authority, SA - lack of state authority, AC of agreement among contenting parties, O = other, P = no response, N = no continui problems

Table 3 Parts of the coastal zone legally protected by municipalities

Natural Features Protected

Municipality	A	B	C	D	E	F	G	H	I	J	K	M	N	O
Malibu													X	
Santa Monica													X	
Rancho Palos Verdes		X		X	X		X		X			X		X
Manhattan Beach		X							X					
Redondo Beach													X	
Seal Beach		X	X		X						X	X		
Huntington Beach	X	X	X	X	X	X	X	X	X	X	X	X		
Newport Beach	X	X	X		X		X				X	X		X
Dana Point		X	X	X			X		X	X		X		
San Clemente		X		X	X		X	X	X			X		
Encinitas		X	X	X	X	X	X		X			X		
Solana Beach		X		X			X	X				X		
Del Mar		X	X	X			X		X	X				
Coronado	X	X	X		X		X	X			X		X	
Imperial Beach	X	X	X				X							
Total Number	4	12	8	7	7	2	10	4	7	3	4	8	4	2
Percentage	27	80	54	47	47	14	67	27	47	20	27	54	27	14

Natural Features Protected A = dunes, B = beaches, C = wetlands, D = cliff tops, E = fauna, F = farms, G = open spaces, H = old buildings, I - hazardous areas, J = rivers, K = bays, M = vegetation, N = no response, O = other

Table 3 Parts of the coastal zone legally protected by municipalities (continued)

Measures Used To Protect

Municipality	B	R	SP	PS	BC	EP	TI	ES	BZ	IS	BO	O	N	P
Malibu														X
Santa Monica		X												
Rancho Palos Verdes		X	X		X				X	X				
Manhattan Beach													X	
Redondo Beach								X						
Seal Beach			X		X			X				X		
Huntington Beach												X		
Newport Beach		X			X	X		X		X				
Dana Point		X	X	X		X		X	X					
San Clemente	X	X	X	X	X	X	X		X	X				
Encinitas		X	X	X				X	X					
Solana Beach		X							X					
Del Mar		X	X						X					
Coronado								X	X	X				
Imperial Beach					X			X	X		X			
Total Number	1	8	6	3	5	3	1	7	8	4	1	2	1	1
Percentage	7	54	40	20	34	20	7	47	54	27	7	14	7	7

Measures Used To Protect B = ban on activity, R = regulation (ban), SP = special use plans/zones, PS = performance standards, BC = building code, EP = educational programs, TI = tax incentive, ES = engineering structures, BZ = buffer zones/setbacks, IS = impact studies, BO = buy-out owner(s), O = other, N = none, P = no response

Table 4 Actions taken or considered in the process of granting legal protection to the coastal zone by municipalities

Studies Done To Support Protection

Municipality	E	ES	EI	NR	HA	VW	SS	UP	O	N	P
Malibu											X
Santa Monica			X					X	X		
Rancho Palos Verdes									X		
Manhattan Beach								X			
Redondo Beach	X										
Seal Beach	X	X	X	X	X	X					
Huntington Beach								X			
Newport Beach									X		
Dana Point	X	X	X	X	X	X	X	X			
San Clemente	X	X	X	X		X		X			
Encinitas	X				X	X		X			
Solana Beach			X			X					
Del Mar				X							
Coronado											X
Imperial Beach					X	X	X				
Total Number	5	3	5	4	4	6	2	6	3	0	2
Percentage	34	20	34	27	27	40	14	40	20	0	14

Studies Done To Support Protection E = engineering, ES = endangered species, EI = environmental impacts, NR = natural resource assessments, HA = hazard assessments, VW = vegetation/wildlife, SS = socio-economic studies, UP = urban planning, O = other, N = none, P = no response

Table 4 Actions taken or considered in the process of granting legal protection to the coastal zone by municipalities (continued)

Potential Solutions Willing To Consider

Municipality	DB	BC	LL	LD	RP	EP	TC	HE	BZ	LR	US	LU	PD	O	N	P
Malibu		X		X	X				X	X		X				
Santa Monica	X				X	X				X		X	X	X		
Rancho Palos Verdes				X		X			X	X	X	X	X			
Manhattan Beach	X			X	X					X						
Redondo Beach																X
Seal Beach		X				X						X				
Huntington Beach															X	
Newport Beach				X		X		X	X	X		X	X			
Dana Point		X		X	X	X		X	X	X		X				
San Clemente	X					X					X	X				
Encinitas		X		X	X			X	X	X						
Solana Beach		X							X							
Del Mar						X			X							
Coronado				X					X			X				
Imperial Beach				X					X							
Total Number	3	5	0	8	5	7	0	3	9	7	2	8	3	1	1	1
Percentage	20	34	0	54	34	47	0	20	60	47	14	54	20	7	7	7

Potential Solutions Willing To Consider DB = development ban, BC = building code, LL = leasing public land, LD = lot/building density restriction, RP = restrict new public infrastructure, EP = educational programs, TC = raise/lower taxes for compliance, HE = higher engineering standards for new infrastructure, BZ = buffer zones around fragile areas, LR = lot/building size restriction, US = university studies, LU = land use plan based on environmental capacity, PD = private deed restrictions, O = other, N = none, P = no response

Table 5 Municipal preferences for new coastal zone projects and planning measures

If Had $25 Million Dollars/
If Had To Match By 50%

Municipality	RC	NP	P	MA	PC	HM	RP	M	ST	WP	ES	O	N	P	YM	NM
Malibu						X	X			X						X
Santa Monica			X				X					X			X	
Rancho Palos Verdes	9					X					X					X
Manhattan Beach		X	X				X									X
Redondo Beach				X		X										X
Seal Beach											X	X				X
Huntington Beach		X		X						X	X				X	
Newport Beach												X			X	
Dana Point			X	X	X										X	
San Clemente		X			X										X	
Encinitas			X			X					X					X
Solana Beach		X	X												X	
Del Mar						X				X		X			X	
Coronado					X		X								X	
Imperial Beach					X	X									X	
Total Number	0	4	5	3	4	6	4	0	0	3	4	4	0	0	9	6
Percentage	0	27	34	20	27	40	27	0	0	20	27	27	0	0	60	40

If Had $25 Million Dollars RC = new roads to coast, NP = new parking lots on coast, P = new parks in city, MA = museum/aquarium, PC = promenade along coast, HM = hazard mitigation on coast, RP = recreational pier, M = public marina on coast, ST = enhanced sewage treatment, WP = wetland protection, ES = engineering structures, O = other, N = none , P = no response

Matching 50% Funding YM = yes - would match funding by 50%, NM = no - would not match funding by 50%

Table 6 Types of natural and technological hazards applicable to coastal municipalities (continued)

Hazards Studies Undertaken

Municipality	E	G	UP	VA	SE	0	N	P
Malibu								X
Santa Monica	X	X	X	X	X			
Rancho Palos Verdes			X		X			
Manhattan Beach						X		
Redondo Beach	X		X					
Seal Beach	X							
Huntington Beach					X			
Newport Beach	X	X	X					
Dana Point	X	X	X					
San Clemente	X	X	X		X			
Encinitas	X	X	X	X				
Solana Beach							X	
Del Mar	X		X					
Coronado								X
Imperial Beach	X		X	X				
Total Number	9	5	9	3	4	1	1	2
Percentage	60	34	60	20	27	7	7	14

Hazard Studies Undertaken E = engineering, G = geological, UP = urban planning, VA = vegetation assessment, SE - socio-economic, O = other, N = none, P = no response

Table 7 Type of measures used and willing to use to avoid hazards (continued)

Measures Willing To Consider

Municipality	ES	EP	LU	RI	PR	BZ	SL	DR	DS	BS	DB	RI	CR	HZ	BH	RI	BN	O	P
Malibu	X		X		X	X								X		X	X		
Santa Monica	X	X	X		X	X	X		X			X		X		X			
Rancho Palos Verdes	X			X						X		X		X					
Manhattan Beach		X										X							
Redondo Beach											X								
Seal Beach			X		X					X							X		
Huntington Beach		X	X		X											X			
Newport Beach																		X	
Dana Point	X	X	X	X	X					X	X	X		X		X	X		
San Clemente	X									X	X								
Encinitas	X		X			X								X			X		
Solana Beach	X		X																
Del Mar	X		X							X	X			X				X	
Coronado	X		X		X	X				X		X							
Imperial Beach								X		X						X	X		
Total Number	60	4	9	2	6	4	1	1	1	7	4	5	0	6	0	5	5	2	0
Percentage	50	27	60	14	40	27	7	7	7	47	27	34	0	40	0	34	34	14	0

Measures Willing To Consider ES = erosion setback, EP = educational program, LU = land-use planning based on hazard potential, RI = remodel infrastructure, PR = post storm/flood reconstruction restrictions, BZ = buffer zone around hazardous areas, SL = stop leasing public land, DR = dune revegation, DS = destroy sea walls/groins, BS = build sea walls groins, DB = development ban, RI = restrict new public infrastructure, CR = pay part of costs to relocate, HZ = create special hazard study zones, BH = buy hazardous

Table 6 Types of natural and technological hazards applicable to coastal municipalities

Hazards Officially Recognized

Municipality	S	F	CS	LM	BE	SL	T	W	WF	P	AS	E	O	N	P
Malibu															X
Santa Monica															X
Rancho Palos Verdes	X	X	X	X	X		X		X	X			X		
Manhattan Beach										X	X				
Redondo Beach	X	X			X	X	X	X		X					
Seal Beach	X	X			X						X		X		
Huntington Beach	X	X	X	X	X	X	X	X	X	X	X	X			
Newport Beach	X	X	X	X	X		X		X	X	X	X	X		
Dana Point	X	X	X	X	X	X	X		X	X	X				
San Clemente			X		X					X					
Encinitas	X	X	X	X	X			X							
Solana Beach														X	
Del Mar	X	X	X	X	X		X								
Coronado					X										
Imperial Beach		X		X	X		X						X		
Total Number	8	9	7	7	11	3	7	3	4	7	5	2	4	1	2
Percentage	54	60	47	47	74	20	47	20	27	47	34	14	27	7	14

Hazards Officially Recognized S = storms, F = floods, CS = cliff slumping, LM = landslides/mudflows, BE = beach erosion, SL = sea level rise, T = tsunamis, W = winds, WF - wildfires, P = pollution, AS = accidental spills, E = explosions, O = other, N = none, P = no response

Table 7 Type of measures used and willing to use to avoid hazards

Measures Used To Avoid Hazards

Municipality	BA	PS	BC	EP	RL	BO	DH	ES	HI	EP	LU	BZ	O	N	P
Malibu															X
Santa Monica	X	X	X	X	X	X	X	X	X	X	X	X			
Rancho Palos Verdes			X				X				X	X			
Manhattan Beach													X		
Redondo Beach			X					X			X				
Seal Beach			X	X				X	X	X	X		X		
Huntington Beach			X				X	X			X				
Newport Beach			X				X	X		X	X	X			
Dana Point		X		X				X		X	X	X			
San Clemente	X	X	X				X			X	X	X			
Encinitas		X					X	X			X	X			
Solana Beach											X				
Del Mar		X						X			X	X			
Coronado		X						X			X	X			
Total Number	2	6	7	3	1	1	6	9	2	5	12	8	2	0	1
Percentage	14	40	47	20	7	7	40	60	14	34	80	54	14	0	7

Measures Used To Avoid Hazards BA = ban activity, PS = performance standards, BC = building code, EP = educational program, RL = reimbursement for loss, BO = buy-out owner(s), DH = designation of hazard zones, ES = engineering structures, HI = require hazard insurance, EP = evacuation plan, LU = Land Use Planning, BZ = buffer zones, O = other, N = none, P = no response

Table 8 The expected actions by municipalities to sea level rise

Expected Sea Level Rise To Affect

Municipality	NE	F	LB	NB	FW	SC	LI	LH	DK	O
Malibu		X	X	X	X	X		X		
Santa Monica		X	X	X		X		X		
Rancho Palos Verdes		X	X			X	X	X		X
Manhattan Beach			X							X
Redondo Beach		X	X							
Seal Beach		X	X	X	X		X	X		
Huntington Beach		X	X	X	X	X		X		
Newport Beach		X	X	X	X		X	X		
Dana Point	X									
San Clemente			X			X	X	X		X
Encinitas			X	X	X	X		X		
Solana Beach			X	X		X				
Del Mar				X	X	X				
Coronado				X			X			
Imperial Beach		X		X	X			X		
Total Number	1	8	11	10	7	8	5	9	0	3
Percentage	7	54	74	67	47	54	34	60	0	20

Expected Sea Level Rise To Affect NE = no effect, F = flooding of inhabited areas, LB = loss of beaches, NB = narrowing of beaches, FW = flooding of wetlands, SC = slumping of cliffs, LI - loss of infrastructure, LH = loss of housing, DK = don't know, O = other

Table 8 The expected actions by municipalities to sea level rise (continued)

Years To React

Municipality	1 - 5 YR	6 - 10 YR	11 - 20 YR	> 20 YR	UNKNOWN
Malibu					X
Santa Monica	X				
Rancho Palos Verdes	X				
Manhattan Beach	X				
Redondo Beach			X		
Seal Beach		X			
Huntington Beach		X			
Newport Beach	X				
Dana Point		X			
San Clemente		X			
Encinitas	X				
Solana Beach			X		
Del Mar					X
Coronado		X			
Imperial Beach		X			
Total Number	5	6	2	0	2
Percentage	34	40	14	0	14

Table 8 The expected actions by municipalities to sea level rise (continued)

Strategy To Combat

Municipality	PC	AS	RS	BS	O	N	P
Malibu	X	X					
Santa Monica	X	X		X			
Rancho Palos Verdes	X	X		X			
Manhattan Beach		X		X			
Redondo Beach				X			
Seal Beach				X			
Huntington Beach		X		X			
Newport Beach					X		
Dana Point				X			
San Clemente			X				
Encinitas		X		X			
Solana Beach		X		X			
Del Mar	X	X		X			
Coronado				X			
Imperial Beach				X			
Total Number	4	8	1	12	1	0	0
Percentage	27	54	7	80	7	0	0

Strategy To Combat PC = prohibit new construction, AS = accommmodate existing structures, N = nothing, RS = retreat from/abandon shoreline, BS = build shoreline protection structures, O = other, P = no response

Table 8 The expected actions by municipalities to sea level rise (continued)

Paying For Strategy

Municipality	OS	EM	CG	SG	FG	IC	D	O	P
Malibu	X						X		
Santa Monica					X				
Rancho Palos Verdes	X						X	X	
Manhattan Beach			X						
Redondo Beach			X	X	X				
Seal Beach	X		X	X					
Huntington Beach	X			X	X	X			
Newport Beach	X	X	X	X	X		X		
Dana Point	X								
San Clemente	X	X	X	X	X				
Encinitas	X			X	X			X	
Solana Beach	X								
Del Mar	X								
Coronado				X	X				
Imperial Beach	X			X	X		X		
Total Number	11	2	5	8	8	1	4	1	0
Percentage	74	14	34	54	54	7	27	7	0

Paying For Strategy OS = owners on shoreline, EM = entire municipality, CG = county government, SG = state government, FG = federal government, IC = insurance companies, D = developer, O = other, P = no response

References

1. J.F. Coates, "Factors Shaping and Shaped by the Environment: 1990-2010" Futures Research Quarterly 7, 1-51.
2. R. Warren, L.F. Weschler and M.S. Rosentraub, "Local-Regional Interaction in the Development of Coastal Land Use Policies: A Case Study of Metropolitan Los Angeles" Coastal Zone Management 3, 331-360 (1977).
3. T. Walker and T. Berg, "Storm Damage Estimate Rises" Orange County Register 3 (March 2, 1993).
4. B. Pool, "The Slide Toward The Tide" Los Angeles Times B1 (February 9, 1995).
5. D.W. Fischer, "Local Coastal Storm Response: The 1988 Redondo Beach Experience" International Journal of Mass Emergencies and Disasters 8, 49-59 (1990).
6. G.B. Griggs, J.E. Pepper and M.E. Jordan, "California Coastal Hazards: A Critical Assessment of Existing Land Use Policies" California Policy Seminar Report (University of California, Berkeley, 1992).
7. D.L. Brown, Tsunami Activity Accompanying the Alaskan Earthquake of 27 March 1964 (U.S. Army Engineer District, Anchorage, 1964).
8. T.R. Reid, "Kobe Wakes to a Nightmare" National Geographic 188, 112-136 (1995).
9. J.G. Titus, "Greenhouse Effects and Coastal Wetland Policy: How Americans Could Abandon an Area the Size of Massachusetts at Minimum Cost" Environmental Management 15, 39-58 (1991).
10. California Energy Commission, The Impacts of Global Warming on California (California Intergovernmental Relations Committee, Sacramento, 1989).
11. M.A. Stein, "Havoc to San Francisco Bay Possible Due to Big Rise in Seas" Los Angeles Times, A3 (April 16, 1990).
12. R. Gustaitus, "Cliffs and Beaches Will Go" California Waterfront Age 5, 29 (Fall 1989).
13. California Coastal Act of 1976, Public Resources Code, Division 20 (San Francisco, 1988).
14. California Coastal Commission, Statewide Interpretive Guidelines (San Francisco, 1981).
15. R. Warren, op.cit.
16. D. Browner, D. Godschalk and D. Porter, Understanding Growth Management (Urban Land Institute, Washington, D.C. 1989); J. De Grove and N. Stroud, "New Developments and Future Trends in Local Government Comprehensive Planning," Stetson Law Review, 17 573-605 (1988).
17. J. De Grove, Balanced Growth: A Planning Guide for Local Government (International City Management Association, Washington D.C. 1991).
18. D.W. Fischer and L. Martinet, "Local Government Response to the 'American Trader' Oil Spill of 1990," Ocean and Coastal Management, 19, 59-73 (1993).
19. P.J. Wilcoxen, "Coastal Erosion and Sea Level Rise: Application for Ocean Beach," Coastal Zone Management Journal, 14, 173-185 (1986).
20. J.C. Pernetta and D.L. Elder, "Climate, Sea Level Rise and the Coastal Zone: Management and Planning for Global Changes," Ocean and Coastal Management, 18, 113-160 (1992).
21. Ibid., 123.
22. G.B. Griggs, J.E. Pepper, M.E. Jordan, op.cit.
23. S. Lascha and R. Emmer, Resident and Public Official Perceptions of the Effects of Coastal Erosion and Sea Level Rise on Coastal Louisiana (Environmental Social Science Research Institute, University of New Orleans, 1992).
24. D.W. Fischer, V. Rivas and A. Cendrero, "Local Government Planning for Coastal Protection: A Case Study of Cantabrian Municipalities, Spain," Journal of Coastal Research 11, 135-152 (1995).
25. V. Rivas, D.W. Fischer, A. Cendrero, "Perception of Indicators of Coastal Environmental Quality by Municipal Officials in Northern Spain," International Journal of Environmental Studies 45, 217-225 (1994).
26. D.R. Godschalk, D.J. Brower and T. Beatley, Catastrophic Coastal Storms: Hazard Mitigation and Development Management (Duke University Press, Durham, NC. 1989).
27. D.W. Fischer, V. Rivas and A. Cendrero, op.cit.

CHAPTER 87

STRATEGIES FOR CONDUCTING PERIODIC REVIEWS OF COASTAL PLAN IMPLEMENTATION

Rick Hyman[1]

ABSTRACT

The California Coastal Act requires the state Coastal Commission to periodically review a local government's implementation of its certified coastal program. An evaluation of Sand City's plan implementation was completed by the Commission. This small city's mostly undeveloped coastline is part of the diminishing Monterey Bay dunes system. The evaluation examined permit and other actions that the City took, initiatives of regional agencies, and changed circumstances since certification. Certain deficiencies were found both in Sand City's plan and in the manner in which it had been implemented, resulting in a report recommending 59 corrective actions.

Preparing the report for Sand City proved to be a complex and time-consuming task. The experience gained in this pioneering evaluation has value in establishing better procedures for conducting periodic reviews. The process and schedule, as well as the penalties for not abiding by the recommendations, need to be institutionalized. Conducting a thorough periodic review relies on diverse, dispersed information, thus requiring a variety of staff skills and support to collect and analyze. It triggers performing follow-up work, such as preparing and approving plan amendments.

The approach used for Sand City of comprehensively evaluating all coastal issues has its limitations in terms of staffing, ability to address regional concerns, and political realities. The experience suggests effectiveness and efficiency criteria that may lead to other approaches of conducting periodic reviews. Exercises that are frequent, integrated with general plan updates, regional, supplementary-staffed, or of limited focus may each satisfy some criteria. The Coastal Commission decided that a regional focus on a few key issues was an approach subsequently worth trying.

---

[1] Coastal Program Analyst, California Coastal Commission, 725 Front Street, Santa Cruz, CA 95060. The views expressed in this paper are those of the author and do not necessarily reflect those of the Coastal Commission.

INTRODUCTION

The California Coastal Act requires the state Coastal Commission to review a local government's implementation of its certified coastal program. Such a periodic review is critical for assuring that a plan is enforced and remains timely, but is potentially complex. Therefore, it is important to carefully structure the review process. In 1990 the Commission reviewed Sand City's implementation of its local coastal program (California Coastal Commission, October 1990). This paper evaluates that experience as a basis for suggesting refinements for future reviews.

CONTEXTS OF PERIODIC REVIEW

The State-Local Collaborative Planning Process Context

The California Coastal Act of 1976 is one of many state programs based on a state-local collaborative planning model. Typical steps include: state promulgates policies, local governments prepare plans based on these policies, state certifies these plans, local governments regulate based on these plans with some state oversight, such as appeal authority (Dickert and Sorensen, 1978). Periodic review is a subsequent evaluation step.

Under the Act each coastal jurisdiction must prepare a Coastal Commission-approved local coastal program to provide the blueprint for regulating development. Furthermore, the Coastal Act mandates that the Commission conduct periodic reviews to determine if local governments are implementing their programs in conformance with the Act. The review process provides an opportunity for public input and to respond to new information and legislation and changed circumstances.

The Procedural Context

Despite incorporation of periodic review provisions in the 1976 Act, few follow-up institutional initiatives to render the process operational had occurred by 1990. Only three documents were available to guide preparation of the Sand City report: the law, a manual, and an earlier report. The enabling act, consisting of just two paragraphs, outlined the basic approach. The review is undertaken at least every five years, it analyzes whether the local program is being effectively implemented, and it makes recommendations for (but does not impose) corrective actions (Coastal Act). An unadopted manual prepared by consultants for the Coastal Commission suggested a series of steps for performing periodic review (Sedway-Cooke Associates, 1986). A written report on the first periodic review provided a format that could be followed. (California Coastal Commission, 1989).

The Geographic Context: The City and Region

Sand City, population 200, is among the smallest cities along the California coast. It contains 1.4 miles of largely undeveloped sand dune shoreline on the Monterey Peninsula. Its local coastal program was certified in 1984, with the land use plan portion based on working papers developed in 1980-1 (Environmental Management Consultants, 1981 and 1984). The documents had not been amended in any substantial way through 1989. Large hotel and residential projects were allowed along the entire shoreline. No extensive development had occurred, but some major permits were pending. Thus, not only was periodic review due, it was appropriate before these projects (one being a reapplication of a hotel previously denied by the Commission) were individually considered by the Commission (Brown, 1989).

Sand City is one of six contiguous jurisdictions containing remnants of the South Monterey Bay sand dunes complex, home to the endangered Smith's blue butterfly. In addition to dune/open space protection, regional issues affecting the City and beyond its independent authority to solve included housing needs, wastewater allocation, water supply constraints, and air quality. Regional and state agency directives on these matters have a major effect on the City's future development.

Political and Philosophical Context

The official posture of Sand City at the time could be characterized as favoring development of their coastline as their certified local coastal program allowed. Attempts for public open space purchase had been opposed by City officials (California Coastal Commission, October 1990). The City recognized the Coastal Commission's authority to perform a periodic review (Goblirsch, January 1990), but expressed concerns that there were no regulations and that a response would cost substantial money. The City speculated that the review would result in lower property values making landowners suffer and making it easier for the State to purchase land in opposition to City desires (Goblirsch, February 1990). The City maintained and wanted assurances to the effect that the review not be an attempt to redo its local coastal program, but only examine how well it had been implemented. Noting its sparse permit activity and the Commission's completion of only one other periodic review, the City questioned the review's timing, especially in light of the pending hotel project (Pendergrass, 1989 and February 1990). The City also questioned whether Commission staff could be "fair, objective, and independent."

Some Coastal Commissioners harbored similar feelings about ensuring that the scope of periodic reviews was limited. With the Commission experiencing a heavy workload and budget cutbacks in the 1980's, some Commissioners felt that performing periodic review was a low priority (California Coastal Commission, January and February 1990). Indeed by 1990, 27 jurisdictions had certified plans from five to nine years old whose implementation awaited the required evaluation.

THE SAND CITY REVIEW EXPERIENCE

Review Report Preparation and Adoption

The Commission's review of Sand City's plan implementation began in December 1989 and was completed at an October 10, 1990 meeting (California Coastal Commission, October 1990). The three-hour public hearing featured testimony alternately praising and faulting the City. The review report, as presented by staff, was adopted by the Commission on a 7 to 4 vote. The final report consists of 131 pages plus bibliography and cover letter summary. The report contains 59 recommendations to the City. Some examples are that the City: pursue some immediate access improvements and not oppose public acquisitions for parks; not support additional sand mining; only permit seawalls to protect existing structures; limit overall site coverage to 15% to protect the dune habitat; provide a habitat corridor for the Smith's Blue butterfly; revise policies to limit intrusions into the public viewshed; limit development to that which could be served by available water; scale back maximum allowed development intensities. Recommendations take many forms: local coastal program revisions, calls for more closely following existing policies in official decisions, new guidelines, improved procedures, better enforcement, concluding a habitat planning process, and providing public access.

The recommendations were derived from an extensive evaluation of the City's activities, new information and regional circumstances that occurred since certification. In addition to the permits which it issued, City documents examined included a Redevelopment Plan, two sand mining reclamation plans, lawsuit papers against the Commission for denying the hotel project, bike path and street alignment choices, new no parking zones and street closures, oppositions to public acquisitions by recreational agencies, plans for City acquisitions, initiation of a Specific Plan, and other Council resolutions. The author, a generalist planner, was primarily responsible for this exercise, but received assistance from staff versed in geology (staff engineer, student intern geologist) and visual analysis.

Several conclusions were drawn for each coastal issue areas (e.g., public access, recreation, shoreline processes and structures, sensitive habitats, visual resources, water supply, other public services, land use locations, types and densities, and procedures). In some cases the City's policies were found adequate, but not always being fully implemented in the permit process nor followed in non-permit actions. In some other cases, the plan provisions themselves were found to be deficient. A draft report was released in April 1990, followed by a City public hearing in May, subsequent discussions between Coastal Commission and City staff, and release of the final report in September.

Follow-Up

The Coastal Act gave Sand City one year to either take the recommended actions or report to the Commission why it declined to do so. Instead, the City filed suit against the Commission for allegedly exceeding its legal authority, by causing an entire plan rewrite (City of Sand City, 1990, 1993). The courts ultimately upheld the Commission's action. Sand City is currently preparing a formal response.

In the meantime, largely as the result of continuing pressure from the regional parks district, Sand City's plan has been amended to achieve some of the review recommendations, and public agencies have acquired some beach and dunes areas.

DIRECTIONS FOR FUTURE PERIODIC REVIEW

In some aspects the Sand City experience proved useful to emulate; in other aspects, deficiencies are apparent that warrant correction. Some deficiencies point to a need to reevaluate the basic approach taken. The resultant procedural and approach recommendations should prove generally applicable to any review component of a state-local collaborative planning process.

Procedural Recommendations

First, periodic review should be institutionalized. The Sand City experience of having to obtain buy-in to prepare the report should not have to be repeated for each subsequent review. Everyone -- the public, decision-makers, local governments -- should have knowledge of what will occur.

Periodic review should be a task incorporated into the oversight agency's ongoing work and pronouncements. In documents prepared for the initial certification of a local program mention could be made that periodic review will occur. Subsequent contacts with the local government could contain similar reminders. A schedule of when periodic reviews are to occur should be established. Public comment procedures should be set in advance in order that citizens and local officials will know what to expect. When the state agency considers appeals of local permit decisions, issues that emerge to potentially be examined in periodic review should be flagged. For example, a project on appeal may be found consistent with the local coastal program and hence merit approval, but still raise resource protection concerns. A data base should be compiled of other important state agency decisions and initiatives on matters that would be applicable to local jurisdictions.

Second, the project scope needs to be in sync with available support. The Sand City experience demonstrated that substantial staffing is needed to produce the type of comprehensive product envisioned. If resources are constrained, then the review's scope should be commensurately scaled back to a doable level.

Review staff should be governed by a work program, have adequate supervision, and be freed of competing duties. A team with relevant experience should be assigned. Where desired technical assistance is unavailable (in Sand City's case expertise in dune habitat issues, for example), the resultant recommendations may have to be more process-oriented and contingent on future conclusions.

Third, on-going monitoring and documentation of a local government's activities should always be occurring. The experience of having kept fairly close tabs on Sand City over the years proved valuable in focusing the review process and determining the material to be evaluated.

Although the collaborative process relies on local government to daily implement the coastal program, state staff should be monitoring local actions. Local permits and other items received should be maintained in organized files. Tracking sheets, including lists of documents received and comments made, communications, local hearings, field reviews, and problems detected, should be kept. Both the state and local agencies should have up-to-date versions of the certified local coastal program as it is amended. If monitoring assignments change, the new lead planner should be fully briefed by his or her predecessor to ensure continuity and historical memory.

Fourth, an extensive data collection phase should initiate the actual review. The Sand City experience of consulting a wide-range of sources resulted in a thorough, defensible product, but the reliance on professional planning staff to collect most documents was not very efficient.

Prior to writing the review report, the state agency's own generated files related to the subject jurisdiction, such as for appeal and enforcement cases, should be gathered. In some instances these agency's files might need perfection (e.g., solving an outstanding violation). Interns, librarians, and other support staff could be assigned to collect and organize information. Any special background studies necessary to derive recommendations should be completed. Where resources are unavailable, the review report will have to recommend the future preparation of studies necessary have a full understanding of the issue in question. It may be desirable, although a little extra work, to prepare a separate background data report first that hopefully all sides accept, before work on the analysis starts.

Fifth, follow-up procedures should be established and committed to. Sand City's response of a lawsuit provided an unintended respite as the Commission was thin on resources (e.g., staffing, financial assistance) and procedural guidance (it had only the Act's requirement for a city to respond within a year) for effective follow-up immediately after the review report was adopted.

Implementing recommendations is a process in itself that should be guided by a mutually-agreed upon work program. The action taken on the review report should clearly lay out what follow-up is expected. Where the review requires various or complex responses, benchmark points should be established. State staff needs to be available to explain its recommendations, offer guidance, and promptly process local responses. If the state agency controls a source of funds to distribute, it should make some available to local government to use to effectuate the recommendations, along with establishing guidelines for use of the money. For example, support could be targeted to making program changes that arise out of new legislation or changed circumstances, rather than those that arise out of locality's failures to abide by their own policies. The local jurisdiction should be required to document how it has addressed each recommendation. At the deadline, there should be a public hearing on the complete response.

Sixth, remedies for inactions or disputes over acceptable responses need to be in place. In the Sand City case the somewhat vague Coastal Act provision of a follow-up Coastal Commission report to the Legislature if a local government declines to take recommended actions may be partly responsible for a slow response.

The available remedies and how they will be selected should be clear from the outset of the review process. A variety of incentives and penalties should be available, given that there may be different levels and kinds of disagreements between the state oversight agency and the local government (e.g., policy dispute, philosophical agreement but no funding to implement). Possible remedies include:

- a moratorium on issuance of permits until local plan revisions are made;
- decertification of an obsolete or inappropriate local plan or plan component;
- retracting local authority to issue its own coastal permits;
- funding for acquisitions of threatened resources that the local government has been unable to protect;
- funding for other non-regulatory programs (e.g., litter control) found necessary to fully protect resources;
- grantings of additional authorities (e.g., regulatory or fund-raising) to local or regional governments to address discovered deficiencies.

Towards New Review Approaches

The Sand City experience highlighted three limitations that suggest a reevaluation of the basic approach taken: staff availability, jurisdictional boundaries, and political realities.

Staff Workload Implications: The Sand City review involved substantial staff time. Approximately eight total months of professional time and one month of clerical support were spent on preparing the report. A volunteer assistant prepared the bibliography, consisting of over 200 items. The time spent would have been greater had state staff not been monitoring City coastal issues over the years and participating in their habitat task force. The review also resulted in discovery and pursuit of related enforcement and monitoring work in Sand City which was the Commission's responsibility, but had lingered.

Release of the review report did not bring an end to the workload. State staff reviewed a City request for funding to amend its plan and responded to the lawsuit. As the City resumes acting on the report's recommendations, state staff will be expected to offer planning assistance. The resulting local plan amendments, as well as submittal of outstanding revisions, will need State staff evaluation. Local staff spent some time meeting with and providing information to coastal staff and could be expected to put in considerable time to respond to the report. The City estimated its cost would be $135,000 to do so (Goblirsch, August 1990).

The staffing implications of performing similar individual periodic reviews every five years are staggering. The California coastline is over 700 times longer than Sand City's alone. Devoting the same staffing level (.8 staff years every 5 years per 1.4 mile of coastline) projects to needing at least 100 coastal planners devoted solely to preparing periodic reviews. And Sand City's permit and plan amendment output is not as great per mile of shoreline as other jurisdictions'. Therefore, a statewide projection based on review time spent per item could yield an even larger staffing requirement. This is exclusive of necessary follow-up time.

<u>Resource Protection Implications</u>: The Sand City review attempted to address resource and access issues that extended beyond City boundaries, revealing the limitations of evaluations confined to single jurisdictions. Some extralocal issues were found to have been already addressed by regional agencies; e.g., Sand City had water and wastewater allocations. Had problems been discovered with these, the report would have not be able to recommend binding corrective actions, as it was not directed to these responsible agencies. For another extralocal issue -- sand supply and coastal erosion -- previous studies of the South Monterey Bay littoral cell were informative enough to enable recommendations to be offered about sand mining and shoreline structures directed to Sand City. On the other hand, there was neither adequate information nor a comprehensive interjurisdictional forum for addressing the South Monterey Bay sand dunes complex. For this case the review both outlined what would be necessary to complete such a study and recommended that in the absence of one, sand dune coverage in Sand City be limited to 15%.

The implications of conducting periodic reviews of individual jurisdictions are that addressing some issues will be fragmented and recommendations will be less than optimal. Even where extra effort is taken to understand the entire resource system, the need to tailor recommendations to what the local government has jurisdiction over will be limiting (e.g., where pollution is emanating from an adjacent city not under review, the best that could be recommended to the subject jurisdiction would be: coordinate with your neighbor). Where the information is unavailable or there is no mechanism to effectuate a regional agreement, a recommendation to the local government may have to be more conservative and less equitable.

Local Government Implications. The Sand City review, being conducted entirely by state staff, needed acceptance by the City Council in order to be successful. State staff attempted to achieve rapport in various ways, such as consulting with city officials, eliciting public comments, concentrating on the shoreline part of the City, noting the City's accomplishments, being professional and objective, and trying to fit recommendations into the City's existing habitat planning program.

Nevertheless, while the City was very cooperative in providing information, it was displeased with the review process. In additional to philosophical differences with the Commission over the appropriate scope of the review process and whether development or open space had priority, City officials complained about inadequate response time (Pendergrass, May 1990; Goblirsch, September 1990). After the Commission decision, the City tried a variety of maneuvers: requesting grant money to basically redo their local plan; by considering, but not necessarily following, the review's recommendations (Goblirsch, August 1990); challenging the review's legitimacy in a lawsuit; and offering selective and limited local coastal program amendments only partially addressing some recommendations. No direct penalties have been imposed on the City for its slow response, although it was compelled to accept one amendment through the initiative of the regional parks district.

The implication of a state agency unilaterally performing individual jurisdiction reviews is that the subject locality may be recalcitrant in responding for political, financial, or other reasons. While the above-suggested follow-up and remedial procedural improvements may help smooth the process, the basic approach may still alienate local officials. Since they control the response phase, full and timely compliance may be difficult to achieve.

Parameters for a Different Approach: Elements of the approach followed for Sand City involved examining that single jurisdiction alone, waiting for the maximum review period (actually even extending a little beyond the five year mandate), relying exclusively on the state agency staff to produce the report, comprehensively reviewing City activities and issues, and presenting the City with a multitude of recommendations to take in the allotted year response time. As this approach, meritorious in many respects, nevertheless taxes staff resources, incompletely addresses regional resource issues, and potentially alienates local governments, more effective and efficient variations deserve consideration. In order for periodic review to be effective, offering timely recommendations and ones acceptable to local governments is important. Recommendations must also be tailored to solving identified resource problems. In order for periodic review to be performed efficiently, staffing must fit budgetary limitations. Another measure of efficiency is to ensure that the review is manageable.

Examples of Alternative Approaches: An approach which may be effective is to perform on-going monitoring and take action each time a problem is discovered. Not only would state staff be kept informed of City activities and at times participate in them, but would have an established mechanism of immediately transmitting concerns to the jurisdiction and receiving their responses.

A similar approach would involve frequent reviews. Since under the Coastal Act the time interval must be no more than five years, the period chosen could vary, depending on the amount of activities and degrees of controversy in each locality. For example, annual information exchange meetings with local governments, as recommended in Sedway Cooke's manual, could serve as periodic reviews (1986).

Another approach is to impart state evaluation concerns into local governments' general plan or related update processes. A review could be timed to offer recommendations when a local government is beginning background work on a plan overhaul or undertaking a specific plan covering an important area. Since there would already be a presumption that a plan needs preparing or amending, the local government may be more receptive to the state's input.

At times resource protection may benefit from a mulitjurisdictional approach. For example, an existing regional entity (e.g., council of governments, CRMP - coordinated resource management program) could be made responsible for acting on recommendations. Another scenario would be for the reviewing agency to facilitate the creation of a regional framework to act on review recommendations.

A more efficient approach would be to rely on others than just the state staff, such as local planners or consultants, to perform at least a portion of the review. The state agency would set guidelines for what information to collect and when and then, later on, become active in formulating review recommendations. Including the subject locality in the review process may foster better acceptance of the results.

Another approach is to focus only on the results of permit actions that the local government took. These are the primary means of implementing the Coastal Act.

Another narrower scope approach is to focus on smaller geographical units within large cities or counties. Periodic review can also be segmented by subject matter, concentrating on one or a small group of issues at a time and enabling that one staff person with expertise on the subject to conduct the review.

A related approach would be to couple statewide issue initiatives with local reviews for that issue. Periodically, the Coastal Commission, for example, devotes staff to focus on a major issue, such as wetlands or shoreline erosion. Usually, to prepare statewide guidance, individual local policies and procedures are examined.

Extending such an exercise to requiring a response from local governments to discovered deficiencies would be an efficient review approach.

CONCLUSION AND FOLLOW-UP

Although the Sand City review was an official work item with legal standing, it was somewhat of an experiment, being only the second one completed. As a follow-up, the author wrote a memo to management outlining many of the concerns expressed in this paper (Hyman, 1990). While this retrospective look results in some logical procedural recommendations, the experience with one single jurisdiction is not sufficient to draw conclusions as to the best approach to take for all future reviews. And in deriving the alternatives, it becomes apparent that there is not likely to be one single approach that satisfies all concerns. For example, a regional evaluation holds the promise of more comprehensively addressing resource issues, but might even be more costly to undertake and less effective at producing regulatory changes, which are the purview of local government.

This analysis suggests that conducting additional reviews of a pilot nature, experimenting with different approaches, can elicit answers about workability and help shape priorities. As experience is gained with various approaches, the positives and negatives can be judged against a set of prioritized criteria. For example, budgetary limitations dictate that efficient use of staff must be a top Coastal Commission priority. An approach that scores poorly on this point will have to be rejected in California, at least until circumstances change. This suggests continuing flexibility in performing the periodic review function. Ultimately a combination of approaches may be found optimal. For example, annual reports could be produced, concentrating on a single issue each year. Single issue reviews can be performed for several jurisdictions in the region or even statewide, possibly as part as an overall annual agency focus (e.g., year of access, of wetlands). More limited single jurisdictional reviews could still occur as a summarizing exercise every few years, examining local procedures and important issues not subject of the larger focus.

Coastal Commission managers recognized these dynamics in deciding that the subsequent review would be a pilot project. And they accepted the conclusions of the Sand City experience by deciding to try another approach; one on a regional, instead of single jurisdictional basis and one focusing on a few key issues, rather than on all of those present. The story of that effort is told in a companion conference paper by Elizabeth Fuchs (Fuchs, 1997).

ACKNOWLEDGMENTS

Thanks to Liz Fuchs, David Loomis, and Les Strnad of the Coastal Commission staff and Dr. Herbert Hyman, Hunter College for their most helpful comments.

REFERENCES

Brown, Edward, District Director to All [California Coastal] Commissioners, Memorandum re: Sterling Hotel Project, November 28, 1989.

California Coastal Commission. *Report to the City of Trinidad on the Implementation of the Local Coastal Program*, July 28, 1989.

California Coastal Commission. Comments of Commissioners Wornum and MacElvaine, Transcript of January 10, 1990 public hearing.

California Coastal Commission. Comments of Commissioner Wright, Audiotape of February 14, 1990 hearing.

California Coastal Commission. (Adopted) Staff Recommendation. *Report to the City of Sand City on the Implementation of its Local Coastal Program,* October 10, 1990.

City of Sand City vs. California Coastal Commission. Monterey County Superior Court Case No. M 22716. Filed December 7, 1990; California State Court of Appeal, Sixth District Case No. H009748. Decided June 3, 1993.

Coastal Act of 1976. California Public Resources Code, Section 30519.5.

Dickert, Thomas and Jens Sorensen. *Collaborative Land-Use Planning for the Coastal Zone. Volume 1 A Process For Local Program Development*. Berkeley: Institute of Urban and Regional Development, University of California, March 1978.

Environmental Management Consultants. *Sand City Local Coastal Program: Land Use Plan*, December 1982.

Environmental Management Consultants. *Sand City Local Coastal Program: Implementation Plan*, March 1984.

Fuchs, Elizabeth, "Managing Cumulative Impacts in the California Coastal Zone: Lessons of the Regional Cumulative Assessment Project," presentation at California and the World Ocean '97 Conference, march 28-30, San Diego, California.

Goblirsch, Dick (City Manager, Sand City) to David Loomis, Letter of February 2, 1990.

Goblirsch to Douglas, Cover Letter and Grant Application, August 31, 1990.

Goblirsch. Comment on periodic review. Transcript of Coastal Commission hearing of January 10, 1990

Goblirsch to Thomas Gwyn, Chairman and [Coastal] Commissioners, Letter of September 26, 1990.

Hyman, Rick to Liz Fuchs, "Lessons of Periodic Review," memorandum, May 18, 1990.

Pendergrass, David (Mayor, Sand City). Letter to Chairman and Members of the Coastal Commission re: Sand City Five Year Review, February 1, 1990.

Pendergrass. Comment on periodic review. Transcript of Coastal Commission hearing of December 12, 1990

Pendergrass, Letter to Thomas Gwyn, May 2, 1990.

Sedway Cooke Associates for California Coastal Commission. *LCP Review Program*, January 1986.

CHAPTER 88

Historical Review of Planned Land Uses for Coastal Areas in the Counties of Santa Cruz, Monterey, and San Luis Obispo, 1965 - 1995

Keith Woodcock

Abstract

The General Plan is the Constitution of development for local government. It establishes the desired location and intensity of physical development and the use of natural resources within each jurisdiction. The use of Coastal Resources as expressed in the General Plans, often differs between neighboring jurisdictions and frequently changes over time.

This session examines the General Plans adopted by local jurisdictions in three contiguous counties, Santa Cruz, Monterey, and San Luis Obispo, over a 30 year period (1965-1995) to see how land use planning in the coastal area has changed over time with respect to both the individual jurisdictions and the three county area as a whole.

How Accomplished:

Compare the adopted Land Use Plans and relevant policies from the General Plans at the midpoint for the following decades: 1960s, 1970s, 1980s, and 1990s for the cities and counties of the area including Santa Cruz, Monterey, and San Luis Obispo counties.

Keith Woodcock, Planning and Environmental Consultant, Woodcock Planning and Environmental services, 6345 N. Spalding Ave., Fresno, CA 93710

CHAPTER 89

Managing Cumulative Impacts in the California Coastal Zone: Lessons of the Regional Cumulative Assessment Project

by Elizabeth A. Fuchs, AICP[1]

Abstract

The California Coastal Commission's Regional Cumulative Assessment Project (ReCAP) in the Monterey Bay area is the Commission's most recent effort to improve program evaluation and the management of cumulative impacts. A summary of the key findings and recommendations are outlined. ReCAP provides valuable lessons for future evaluation and management of coastal resources, including: the need to define the "region" or "management area" to reflect multiple issues; the need to address the interrelatedness of coastal issues; the importance of targeting key issues; the need to initiate evaluation from identification of impacts to the resources rather than policy intent; the importance of improved monitoring and information management in addressing data gaps in key areas; the increasing importance of methods to improve spatial analysis and presentation; the need to focus more resources on non-regulatory management efforts; and the continued importance of the California Coastal Program in managing cumulative impacts.

I. Introduction

The California Coastal Act (Act) establishes a unique planning partnership between state and local government for management of coastal resources. Local governments develop a Local Coastal Program (LCP) to conform with and carry out the Act. After certification of the LCP by the Commission, most permitting authority is delegated to the local government to implement the LCP. The Coastal Commission is required to monitor the implementation of the program and review amendments to the certified LCP. In addition, the Act requires that the Commission, at least once every five years, review the implementation of each certified LCP and determine if it is being effectively implemented consistent with the policies of the Coastal Act (California Public Resources Code, 1997).

[1] Coastal Program Manager, California Coastal Commission, 45 Fremont Street, Suite 2000, San Francisco, CA. 94105. The views expressed in this article are those of the author and do not necessarily reflect those of the California Coastal Commission.

While significant progress has been made in the development and certification of LCPs - over 85% of the geographic area of California's coastal zone is governed by a certified LCP - the Commission evaluation of the implementation of the local plans has been limited. Since 1981 only two such reviews have been completed. Monitoring and evaluating the effects of policy application in decisions is the bridge connecting long range plans with regulatory decisions and is a way to keep plans current (Stollman, 1988). The Commission's limited ability to monitor and evaluate local coastal program implementation and to recommend revisions and updates has undermined the ability of the California Coastal Management Program (CCMP) and LCPs to continue to provide current and effective mechanisms for carrying out state policy (California Coastal Commission, 1993a). The Regional Cumulative Assessment Project (ReCAP) is the Commission's most recent effort to address this lack of effective policy and program evaluation and to improve the ability of the Commission and local governments to manage coastal and ocean resources by addressing the cumulative impacts of growth and development.

II. The ReCAP Project

A. Background: In Section 309 of the 1990 amendments to the Coastal Zone Management Act (CZMA), the U.S. Congress identified cumulative and secondary impacts as a priority area for improving state coastal management programs. The federal grant program provides for states to assess their program in eight priority enhancement areas and adopt strategies to improve their coastal management programs in selected priority areas. The Commission adopted a strategy in 1992 and identified hazards, public access, wetlands and the management of cumulative impacts as the priority areas for program improvement.

The ReCAP pilot project conducted in the Monterey Bay region of California comprised the core of the 1992 strategy. The purpose of the project was to develop new ways to: 1) strengthen program evaluation and monitoring and, 2) improve comprehensive management of cumulative impacts of growth and development to multiple coastal resources across jurisdictional lines.

Cumulative impacts management was increasingly important to the Coastal Commission for a number of reasons. Significant population growth and development had occurred in the coastal zone. The state experienced a 62% increase in population growth since 1970 and from 1980-1990, 10 of 15 coastal counties grew by a rate of at least 20%. Projections were that such rates of growth would continue (California Department of Finance, 1983). Nationally, California, along with Florida, led the nation in coastal permitting from 1970-89 (NOAA, 1992). Since 1981, the Commission had reviewed over 30,000 permit applications.

At the same time as coastal zone population and development was growing, certified LCPs were aging. Of all LCPs certified, three-fourths were approved prior to 1989 and amended numerous times. The Commission's previous attempts at evaluation and program review were less than successful. The reviews tried to evaluate implementation of all policies in an LCP and taxed the Commission's limited staff resources. Only two were ever done (Hyman, 1997). Without ongoing evaluation and feedback, there was limited assessment of the cumulative changes under an LCP in relation to Coastal Act policies and, except in these two LCPs, no determination of whether the LCPs were effectively managing impacts to coastal resources. In addition, judicial decisions were increasing requirements on public agencies to improve the level of documentation and quantification of impacts of development proposals in order to require mitigation measures.

These factors combined to give rise to the Commission's concern that the integrity of the state coastal management program was eroding and resources were being lost as a result of numerous incremental regulatory decisions on both permit applications and LCP amendments (California Coastal Commission, 1992b). The Commission initiated the ReCAP project to develop a better way to evaluate the effects of these numerous incremental decisions and to evaluate the ability of the LCPs to continue to effectively manage coastal resources.

B. The Pilot Process: The ReCAP pilot project in the Monterey Bay area set out to:

- assess the cumulative impacts of development on hazards, access and wetlands that might not otherwise be evident through project by project permit review;
- analyze major factors contributing to impacts and identify broad coastal resource trends;
- evaluate the effectiveness of LCP and CCMP policies and procedures in responding to cumulative impacts; and,
- develop a new regional framework for undertaking program monitoring and evaluation.

The first step was to determine the project boundaries for analyzing cumulative impacts to hazards, access and wetlands. Setting the project boundary proved to be a complex but critical task because it served to focus the scope and nature of the evaluation (California Coastal Commission, 1993a). Designating a regional boundary that would allow assessment of a variety of natural and cultural resource issues also presented a challenge.

The regional project area for evaluating cumulative impacts of shoreline and bluff retreat was based on the northern and southern boundaries of three adjacent littoral cells. Since many of the policy questions identified in the project area focused on the immediate shoreline, it was reasonable to establish the landward boundary at the first inland public road and seaward at the seaward extent of the littoral cells (California Coastal Commission, 1993b).

For public access assessment, something comparable to a littoral cell or watershed was not available. Depending on the access policy question evaluated, impact assessment and policy evaluation could be based on a single shoreline area or on a regional transportation network used by visitors coming to the coast. To address this, ReCAP sought to define a geographic area that crossed jurisdictional lines, included a variety of different types of access resources and facilities, and included sensitive resource areas where access conflicts were evident. The access boundary extended from the Santa Cruz County line on the north to Point Lobos State Reserve on the south and inland and seaward to the coastal zone boundary. The access boundary did not extend into the Big Sur area of Monterey County because the access issues in that part of the coastline were distinctly different than issues encountered in the more developed north Monterey County and Santa Cruz County.

For wetland policy evaluation the project identified an area that included a variety of wetlands. The inland boundary for wetland analysis was drawn to include two entire watersheds for significant resource areas, Watsonville Slough and Elkhorn Slough, in order to evaluate water quality and land use issues (California Coastal Commission, 1993b).

These three separate issue boundaries were combined and differences reconciled by evaluating how the three areas overlapped and estimating the amount of data collection and extent of analysis required for evaluating each geographic area. The final boundary for the project used the littoral cell boundaries as the northern and southern project limits and the coastal zone boundary for the seaward and inland project limits. It extended from the Santa Cruz/San Mateo county line on the north through Pt. Lobos State Reserve in Monterey County to the south, a total of about 83 miles. It included all or portions of 11 jurisdictions, 7 of which were implementing certified LCPs which had been amended over 58 times. The remaining 4 jurisdictions were not yet certified. In defining the boundary, it was decided that for purposes of regional population growth and land use change analysis, a land use assessment area would extend beyond the coastal zone and include data for the two counties as a whole because of the impacts of such growth on the resources within the coastal zone.

After identifying the project boundary, focus shifted to identifying the key evaluation questions and setting priorities for analysis. Outreach to the public and technical experts helped identify what aspects of, or problems related to, hazards, public access, and wetlands appeared to be most important in managing the resource. Two tools were developed to assist in focusing the review: a systems diagram which identified all the variables in the particular resource system and how they interact in order to fully understand the cumulative impacts on the resource;

and an evaluation matrix which identified the range of possible policy questions to evaluate, the possible methods and measures to use in the evaluation, and potential data sources.

During the assessment phase of the ReCAP project, staff collected data from numerous sources and built a database containing key information on all coastal permits issued from 1982-1992.

C. Summary of Key Findings and Recommendations: While the final ReCAP report contains more extensive analysis and recommendations, some of the key findings and recommendations report are summarized below (California Coastal Commission, 1995).

Overall, regional growth and development in the two county area mirrored that of the state as a whole, increasing by 65%. Urban and suburban land uses increased by roughly 43%. Of the development permits reviewed in the coastal zone between 1973-1992 in the Monterey Bay area, roughly 40% were for single family residences which added an estimated 558 acres of impervious surface to the area (Lester, 1996). Overall, ReCAP found that while the California Coastal Management Program (CCMP), including LCPs, was fairly successful at managing the effects of this development and protecting significant coastal resources within the pilot area, cumulative impacts problems still existed.

<u>Shoreline Hazards:</u> By using aerial photointerpretation and information from permit application files and by estimating the dimensions of approved projects, ReCAP found that since 1978, the length of shoreline armoring increased by 2.4 miles to a total of 12 miles in the project area. ReCAP projected, based on land use patterns and shoreline characteristics, that up to 27.7 miles could eventually be armored if current policy implementation continued, thus more than doubling the length of shoreline armoring that existed in the project area (California Coastal Commission, 1995). ReCAP also estimated that the existing 12 miles of armoring covered about 25 acres of sandy beach and projected that up to 65 acres of beach could be covered by future armoring (California Coastal Commission, 1995).

The ReCAP evaluation also found that many of the permits for armoring were issued as emergency authorizations, some without issuance of follow up permits, which limited the ability of the Commission and local government to consider other non-structural alternatives or to mitigate impacts of the structures on public access.

The range of recommendations to address the hazards management problems evaluated were aimed at assuring that future development avoid as much as possible continued armoring of the bluffs and shoreline in the project area. For example, ReCAP recommended improvements to LCP policies to revise how setbacks from bluff edges were calculated and to expand the extent of the setbacks;

improvements to emergency permitting procedures to encourage consideration of more temporary measures during emergency response and to ensure follow up consideration of alternatives; steps to develop mechanisms to manage shoreline resources regionally to minimize the need for future protective structures; and possible legislative responses to minimize future armoring.

Public Access: ReCAP found that in evaluating public access, physical access had increased in the project area since 1982 through new acquisitions and stairways opened. About 7000 acres of new parkland was acquired, new stairways were opened, and new parking spaces added.

At the same time, while physical access increased, other factors contributed to impacting public access. Beach use increased by 50% and the demand for a variety of recreational activities grew. Complete site mitigation for impacts of growth and development on access was not always realized because mitigation through offers to dedicate public access easements (OTDs) required as conditions on coastal permits were not always recorded, and only one had been accepted and opened for public use. As noted before, sandy beach areas were being lost to seawalls, including about 5 acres under implementation of the Coastal Act.

ReCAP also found that maximizing access was increasingly dependent on the provision of adequate upland facilities such as parking and support facilities. For example, parking had increased in the project area but demand was still not met and impacts resulted from actions such as approvals of preferential parking programs. Attention to these secondary accessibility issues is key to minimizing cumulative impacts and maximizing access in the project area in the future and recommendations were developed to address this finding. For example, project recommendations included suggested policy and procedural changes to: improve monitoring and recording of OTDs to achieve compliance with mitigation; reduce armoring; increase upland support and balance parking resources with transit alternatives; and improve regional management of access and recreation facilities to address spillover impacts and regional accessibility issues between jurisdictions.

Wetland Resources: ReCAP found that most major losses to wetlands occurred prior to 1970. No appreciable changes to wetland acreage had occurred from 1982-1992 since losses had generally been offset by gains through restoration. However, the habitat of a number of wetlands and lagoons in the project area were degraded from changes in hydrology and water quality. Wetland edges were being "hardened," thus affecting the transitional habitat and overall wetland system. Buffer policies in LCPs were inconsistently applied and inadequate in restricting uses. Impacts were occurring from agricultural activities expanding into wetlands. In response, ReCAP recommended a number of measures, including: improvements to identification and monitoring of wetlands; revisions in policies to implement a minimum 100 ft. buffer in LCPs and to improve management of buffer areas; and, suggestions to improve wetland and watershed management.

The final recommendations of the project included numerous short and longer term implementation mechanisms ranging from changes in LCP policies and in Commission procedures, to possible legislative initiatives, new regional intergovernmental coordination efforts, and educational measures. Efforts to implement the recommendations from the pilot project through amendments to LCPs and other Commission efforts continues in the Central Coast District.

III. The Lessons Learned

In completing this project, the Commission gained valuable lessons in carrying out monitoring and program evaluation.

A. Regional Definition: Defining the area of analysis is a critical element in program evaluation. No single boundary or landscape unit can serve all purposes for monitoring and evaluation. The project found that "region" can mean many different things depending on the institutional level and the resource issues being evaluated. Regional analysis can be based on watersheds, air basins, or political jurisdictions, among other things. For example, the evaluation of wetland issues can rely on a boundary of analysis as small as a wetland or as large as the Pacific Flyway, depending on the focus of the analysis.

Defining "region" or "management area" must reflect multiple and interacting issues. Assessing impacts based on a watershed may allow adequate analysis of wetland impacts or water quality issues but will not address regional transportation impacts on public access nor regional shoreline effects of seawall construction. Selection of an area of analysis for monitoring and evaluation may need to be flexible and vary depending on the resource focus of the evaluation. Defining the project area requires making choices about: 1) priority resource values to evaluate (Williamson, 1993); 2) the methods and measures that will be used to assess impacts and evaluate change; and 3) the available data and technical expertise (California Coastal Commission, 1993a).

Choosing how to evaluate an issue in large part helps define the boundary. For example, deciding to evaluate the impacts of shoreline structures as a priority hazards management issues obviated the need to define a boundary which considered inland earthquake, geologic or flooding hazards, among others. There may not be one set boundary. Project boundaries may need to vary and be flexible depending on what policies are being evaluated.

B. Integration Among Multiple Issues: It is necessary to address the interrelatedness of many coastal issues in order to effectively address cumulative impacts. For example, the ReCAP analysis determined that one of the most important ways to address cumulative impacts on public access in the project area was to modify hazard management policies on seawall construction to minimize effects on recreational beach areas. Developing management measures which are integrated to address multiple resources concerns is important.

C. Need to Focus Issues: Recognizing limits in time, staffing and available data early in the project and scoping the analysis accordingly is important to successful program evaluation. Complete cumulative impact analysis and detailed, quantified evaluation of the results of policy implementation can be daunting if attempting to analyze impacts to every aspect of a resource system. The Commission's past efforts at program evaluation tried to look at the implementation of all policies of a single LCP. It was found to be more effective to start with critical resource issues then look regionally at several LCPs and their ability to address those one or two key resource issues that seem most important to the management of the resource.

Identifying appropriate policy questions to analyze and the indicators to quantify results of policy implementation are useful ways to help focus the program review. The process of focusing the review can proceed along two tracks: one track of inquiry determines what data currently exist and are available related to the issue areas; another track develops the ideal policy questions needed to fully evaluate the implementation of the program and its effects. For example, with ReCAP, a survey of available data related to wetlands and water quality was occurring at the same time the project team was identifying policy questions such as those related to acreage lost, biodiversity modification, and hydrology changes that would be answered in a complete program evaluation. The two tracks then merged to help focus the review by identifying which policy questions could reasonably be answered with available data. As a result of this approach, however, the availability of data turned out to be a major factor and determinant in scoping and designing the review. In addition, the identification of existing data gaps and recommendations for ways to improve monitoring to address these gaps became an important result of the review.

D. Resource Based Policy Evaluation: The Commission ReCAP found it important to start the policy evaluation with determining the condition of the resource on the ground and identifying the problem rather than beginning from the point of asking what was intended to occur by implementation of a particular policy and whether the objective was achieved. Starting analysis from the effects is an effective means to assess cumulative impacts as well (Williamson, 1993).

A cause of impacts to resources may not be the policy statement itself but rather how it is applied through the local processes or some other factors. For example, LCP policies in the project area required that adequate parking in new development be provided and data indicated that thousands of new parking spaces had been provided. If just the policy intent was evaluated a reasonable conclusion would be that the LCP was implemented consistent with the access policies of the Coastal Act. However, evidence of parking impacts in the project area were examined and included evidence of reduced accessibility to the coastal area. Analysis indicated that it was other measures such as placement of red curbing, no parking signs and preferential parking programs that had further constrained public access, even though parking was provided in new development. The inquiry which focused on how the existing policies were being implemented through the local process yielded a different conclusion. The public safety actions of other departments pursuant to other ordinances and spillover effects from preferential parking programs approved for neighborhoods were contributing to impacts. Thus, it remains important to begin the evaluation process by examining the problems to the resources evident on the ground.

E. Working Within Data Constraints: Another factor important to consider when undertaking regional program evaluation is the need to establish limits to data collection and recognize that there will never be enough time or data for a complete analysis. Large gaps in available scientific or monitoring data for some resource factors existed which were beyond the Commission's ability to resolve. For example, while the project had adequate tools and data to measure changes in overall wetland acreage, it quickly became apparent that sound data on the changes to wetland biodiversity in the project area were lacking. The limited data that were available was sketchy and incomplete and could not be used to forecast trends with any reliability. The presence of large gaps in data required that explicit assumptions be made about data limitations. In some cases the only data available was observational and the reliability and limitations of the data to support policy revisions and recommendations had to be acknowledged. From the ReCAP experience, it appears that developing interjurisdictional mechanisms to address gaps in data by sharing existing data and coordinating research and data collection efforts will be increasingly important in assessing and managing cumulative impacts.

F. Importance of Mapping/GIS: Related to the lessons learned about data collection, it is increasingly apparent that with advances in mapping and geographic information systems (GIS), these tools are becoming more powerful and necessary for effective cumulative impacts assessment and management. As a result of years of budget cutbacks (Douglas, 1996), the Commission was less advanced in developing GIS tools. The ReCAP project helped expose the Commission's limitations in this area and focused attention on the importance of these tools and of the need to increase the Commission's capabilities in this areas. As a result, the Commission is currently undertaking new projects to expand the agency's capability in this area. Future program evaluation will look toward GIS as an efficient mechanism for monitoring, analyzing and describing resource changes.

G. Considering Alternative Implementation Techniques: In carrying out recommendations resulting from a program evaluation, it is necessary to increase emphasis on alternative implementation techniques to address solutions to cumulative impacts and regional problems. The most effective solution for the cause of a cumulative impacts problem may not be in a regulatory action. Rather, the best solution may lie within a range of possible alternative responses including revisions to internal procedures for carrying out existing policies, interjurisdictional coordination mechanisms, new public or local government educational efforts or possible new legislation. However, such non regulatory solutions are often more complex and time consuming to effectuate, especially on a regional basis. State and local coastal managers will need to reallocate staff time and resources to non-regulatory approaches to cumulative impacts management and will need to accept a long term commitment to solutions as opposed to the short term response usually available through the regulatory process.

H. Importance of Ongoing Monitoring: Completing a cumulative assessment and regional program review allows coastal managers to document the current conditions of the resources. Once developed, it is critical to maintain this baseline data not only to help coastal analysts have easy access to information to incorporate findings and trends into day to day decisions but also to continue to track achievements and results of implementation of the program review recommendations. Pressure continues to increase on coastal managers at all levels of government to document results on the ground of recommended management measures. Consistent and reliable monitoring, and ongoing update of baseline data and information is necessary to respond to this growing pressure.

I. LCPs Remain An Effective Framework: In California, there are many examples of organizations attempting to respond to regional problems, including for example, regional efforts to manage beach nourishment or efforts to manage natural biological communities and biodiversity. But, these still tend toward single issue problem solving, whether it is beach erosion or habitat management. The coastal management program in California remains a viable mechanism for integrated and comprehensive management of varied physical, natural and cultural coastal resource systems. With stronger monitoring and program evaluation components developed through ReCAP, the California Coastal Management Program and the LCPs will continue to provide an effective management framework for balancing multiple issues and for developing comprehensive measures to address cumulative impacts across policy issue areas and across political jurisdictions.

IV. Conclusion

In conclusion, after completing this pilot project, the Commission found the ReCAP framework to be effective in assessing and managing cumulative impacts and in reviewing policy implementation. Lessons learned from the pilot project effort in the Monterey Bay region will assist the Commission in transferring the process to other coastal areas through procedural guidance to staff and making future improvements in the Commission's coastal management program.

References

California Coastal Commission. 1992a. *Final Assessment of the California Coastal Management Program, (CCMP), performed under Section 309 of the Coastal Zone Management Act (the Coastal Zone Enhancement Grants Program),* San Francisco.

California Coastal Commission 1992b. *The California Coastal Commission's Enhancement Grants Strategy, prepared pursuant to Section 309 of the Coastal Zone Management Act*, San Francisco.

California Coastal Commission. 1993a. *ReCAP Working Paper No. 1, Developing a Regional Cumulative Assessment Project for the California Coastal Zone: Issues and Concerns*, San Francisco. pp.ii-6.

California Coastal Commission. 1993b. *ReCAP Working Paper No. 2, Region Designation and Preliminary Database Design*, San Francisco. pp.3-4..

California Coastal Commission. 1995. *Regional Cumulative Assessment Project, Pilot Project, Findings and Recommendations: Monterey Bay Region*, San Francisco.

California Department of Finance. 1983. *Population Projections for California Counties 1980-2020 with Age/Sex Detail to 2020: Baseline 83*, Population Research Unit, Sacramento.

California Public Resources Code, Division 20, Sections 30500- 30600, January, 1997.

Douglas, Peter M. 1996. *Memo to Douglas P. Wheeler, Secretary for Resources from Peter M. Douglas, Executive Director on FY 1997/98 Budget Changes Proposals,* San Francisco.

Hyman, Rick. 1997. "Strategies for Conducting Periodic Reviews of Coastal Plan Implementation", *presented at the California and the World Ocean '97 Conference, March 28-10, 1997, San Diego.*

Lester, Charles. 1996. "Cumulative Impact Management in the Coastal Zone: Regulation and Planning in the Monterey Bay Region of California", *California Sea Grant Technical Report*, *Working Draft*, University of Colorado, Boulder, pg. 24.

National Oceanic and Atmospheric Administration. 1992. *Building Along America's Coasts: 20 Years of Building Permits, 1970-1989.* National Ocean Services. Rockville.

Stollman, Israel. 1988. "The Values of the Planner", in *The Practice of Local Government Planning*, Frank S. So and Judith Getzels, eds., International City Management Association, Washington D.C., p509.

Williamson, S.C. 1993. "Cumulative Impacts Assessment and Management Planning", in *Environmental Analysis: the NEPA Experience*, S.G. Hildebrand and J.B. Cannon, eds., Lewis Publishers, Boca Raton, pp.391-407.

CHAPTER 90

Permit Streamlining Efforts In Alaska

Kerry Megan Howard[1]

Abstract

Permitting is often labeled as a regulatory area in need of improvement, and the current administration in Alaska has emphasized permit streamlining and administrative efficiency as high priorities. Permit streamlining first began in Alaska in the late 1970s, and various piecemeal improvements have been made since that time. However, as budgets decrease and staff are reduced, interest in streamlining has increased once again. A number of collaborative efforts are currently underway to improve the permitting process in Alaska. This paper describes past and current efforts.

Introduction

Alaska's economy depends on the development of its natural resources. This dependence requires a balancing of resource development with environmental protection to ensure a viable economy yet retain a high quality of life. Striking the appropriate balance between these sometimes competing needs means resource management is challenging and complex. This complexity is complicated by the number of state agencies with resource management and permitting responsibilities.

Permitting is often labeled as an area in need of improvement. Not only is the array of permits large, and the regulations governing their issuance complex, permit processes are often perceived as duplicative and cumbersome. A recurring theme in government these days is to seek improved efficiency and effectiveness of operations. With regard to permitting, a continuing goal is often to simplify and streamline.

---

[1]Senior Project Analyst, Division of Governmental Coordination, P.O. Box 110030, Juneau, AK 99811-0030.

With the release of his proposed FY 98 budget in December, the Governor of Alaska emphasized permit streamlining and efficiency as one of his priorities. The current administration is committed to budget discipline and customer-based government, while meeting its constitutional and statutory responsibilities to protect the public trust and maintain renewable resources for future generations of Alaskans. This paper provides a history of past permit streamlining efforts in Alaska, describes current permit streamlining efforts, and then concludes with some summary observations.

A History of Permit Streamlining in Alaska

The earliest permit streamlining effort in Alaska commenced with the passage of the Environmental Procedures Coordination Act (Alaska Statute 46.35) in 1977. This legislative mandate required a simplified permit application procedure, better opportunities to obtain public comment, more certainty for applicants, increased coordination between federal, state, and local administrators, and the creation of permit information centers.

Through the Department of Environmental Conservation (DEC), this Act created a voluntary, coordinated review process available to any applicant who required multiple state environmental permits. This process included a master application, coordinated public notice, joint hearings, and joint appeals. The State also established permit information centers in several locations to disseminate information about permits to both applicants and members of the public. In addition, the State published a Directory of Permits which summarized general information about permits, their requirements, review procedures, and authorities.

Although the Coordination Act is still in place, the process has received limited use, particularly within the last decade. In part, this may be due to its voluntary nature and to the 1984 enactment of the coordinated consistency review process, which is discussed below. Also, although the DEC Permit Information Centers no longer exist and the Directory of Permits is no longer routinely updated, they were both used extensively by the public for a number of years. The State Department of Natural Resources has since established Public Information Centers that serve some of the functions of the earlier centers.

The next major permit reform effort came through an Administrative Order (AO) implemented by Governor Hammond in 1979 that affected the state's coastal management program. With its "networked" structure, the Alaska Coastal Management Program (ACMP) provided a logical home for the state's next major regulatory streamlining effort. "Networked" refers to a program that uses existing permitting authorities as a mechanism to implement and enforce coastal consistency requirements for projects. To be most effective, a networked program requires a coordinated review process to ensure a consistent approach to permitting.

Interestingly, the original ACMP approved in 1979 did not include a coordinated process for state consistency review of permitted uses and activities. Instead, state agencies independently conducted consistency reviews for projects which often resulted in different findings of consistency for the same project. As such, the AO, among other things, created a clearinghouse within the Governor's Office to coordinate the state review of federal actions and directed state agencies to develop a process for the review of projects within the coastal zone. The AO also noted that the ACMP required the establishment of a system and procedures to ensure uniform, coherent, and aggressive implementation of the program.

Further reform to the ACMP came in 1983, when legislation was passed that created the Office of Management and Budget (OMB). Among other things, this legislation required the OMB to render a "conclusive state consistency determination" and to coordinate the state review of projects within Alaska's coastal zone that required a permit, lease, or authorization from a federal agency or two or more state resource agencies.

Efforts to establish review time lines and to define a role for state agencies, applicants, and coastal districts in the consistency review process were further expanded under the Sheffield administration in the early 1980s. Specifically, in September 1983, the resource agency commissioners and the director of OMB signed a Memorandum of Understanding (MOU) which set out various objectives of permit reform, including streamline and expedite state reviews and decisions on development activities, establish uniformity in state comments and decisions on federal activities and federal permits, and eliminate repetitive state reviews and decisions on the same project.

This initial MOU was followed by an AO and eventually regulations that established a coordinated interagency consistency review process. These regulations, which were implemented in 1984 and are still in place today, established a review process that is project-oriented rather than based on individual permits. This coordinated review provides a multi-party, interactive process whose intent is to eliminate duplicative reviews and establish a consensus-based decision making process. The regulations also provide definitive milestones to ensure the review process is timely and efficient.

Other, more focussed streamlining efforts have also emerged over the last decade. These include the establishment of joint offices, joint application forms, and expedited review lists. Specifically, a joint resource agency office on the North Slope was created to provide a greater agency presence, primarily for monitoring and compliance of oil industry development around Prudhoe Bay. The office was staffed on a rotating basis by representatives of the three resource agencies, and the costs of maintaining the joint office were shared. The office operated successfully from 1984 until about 1989, when it was closed with the development of another joint office, the State Pipeline Coordinator's office. This latter office, now officially known as the Joint Pipeline Office, was officially created by an AO in 1990. This joint state/federal office serves as oversight coordinator for certain common carrier pipelines, and is specifically responsible for processing and adjudicating applications, and for monitoring design, construction, operations, and closeout of designated pipeline facilities.

Streamlining also looked at consolidated sectoral permit applications. A joint placer mining application, originally known as the "Tri-Agency Placer Mining Application" was created in 1981. That application is still in use, although its name has been changed to the "Alaska Placer Mining Application." A joint application form was also created for use by potential aquatic farmers. That form was first put into place in 1989, and revised versions are still being used. The purpose of these sectoral applications is to ease and consolidate the review process for applicants by combining all necessary permits into one application form. Another consolidated application form, known as the Coastal Project Questionnaire, provides a tool to assist applicants and agencies in identifying state and federal permits that are needed for activities in the coastal zone. This questionnaire is regularly updated and has been in continual use since 1984.

Finally, applicants for projects in the coastal zone have also benefitted from a document known as the "Classification of State Agency Approvals." Using this document, the state classifies permits and activities by their potential effects on the coastal zone and, by doing so, has been able to predetermine that a number of permits and activities do not need further consistency review. In particular, permits on the "Categorically Approved" list have de minimus effects on coastal resources and do not need further review. In addition, routine projects on the "General Concurrence" list do not need further review if they can meet specified terms and conditions. The General Concurrence list began modestly in 1984 with 10 projects; today the list contains 50 routine projects.

Current Permit Streamlining Efforts

Many variables in Alaska have changed profoundly since earlier streamlining efforts were initiated. Two significant factors include shrinking budgets and major staff reductions. Also, given the multitude of environmental permitting requirements, there is an increased perception between agencies and applicants that the various processes are duplicative, lengthy, disjointed, and confusing. In addition, there is an expanding awareness that if real gains are to be made, permit streamlining efforts can not focus on just one aspect or one agency. All elements of, and participants in, the permitting arena must be fully examined to achieve real gains. All of these variables--both real and perceived--have sparked renewed interest in pursuing further permit streamlining in Alaska.

Recognizing that effective permit streamlining must involve all agencies with permitting responsibilities, a Resource Agency Coordinating and Streamlining Taskforce (RACST) was convened by the Knowles' administration in spring 1996 to evaluate options to simplify and streamline permitting in Alaska. A specific goal of this group was to improve statewide review across agency lines for projects that require multiple agency permits. A subgroup consisting of senior staff from the three state resource agencies and the Division of Governmental Coordination (DGC) was initially tasked with drafting recommendations to achieve this goal.

Twelve draft subgroup recommendations were forwarded to the RACST for consideration. The 12 recommendations covered a wide array of options, and acknowledged that permit streamlining efforts could vary considerably depending on available resources and level of commitment. After review, three initiatives were highlighted by the administration for further collective action including, from low to high level of effort: 1) develop a standardized application form; 2) develop several multi-agency general permits; and 3) standardize administrative appeal processes. To augment these efforts, an additional initiative was added later to consider efficiencies that could be gained if a master project review process, which would bypass most existing statutory and regulatory requirements, could be developed.

Standardized Application Form

The goal of this task is to simplify paperwork for permitting by reformatting a "project-based" application so that applicants can use one form, supplemented by attachments, to apply for all required approvals. Specific instructions for this task were to develop an application for statewide use, with consideration given to electronic distribution of project information, including scanned maps, diagrams, and blueprints.

An initial task was to identify the most common permits applicants needed for multi-permit projects. The DGC initially collected 34 permits from four agencies. These permits or application forms were then reviewed to identify common questions that could potentially lend themselves toward consolidation onto a single form.

Not surprisingly, DGC determined that many applications have similar questions regarding such topics as applicant information, location and description of the proposed project, and identification of necessary state and federal permits. Also not surprising was the vast array of dissimilar questions, many of which were very technical. This was most notable for applications from the DEC which regulates various discharges to the air, water, and land.

Armed with this information, the State determined action was appropriate to develop a common cover sheet of basic questions that could be used by any applicant for any project. This common cover sheet would then be supplemented by specific application forms that contain agency-specific questions which regulators need to process their permits. Advantages of this approach are that an applicant need only respond to basic questions one time, agency coordination is increased, and a common application form becomes more easily recognizable to all users.

As of this writing, a draft form has been developed that an applicant could use to apply for up to 12 different state permits, and possibly three common federal permits. There is a consensus that the draft application form needs to be field tested before a decision is made to adopt it in final form.

In addition to this broad-based look at common multi-party permits, an additional joint sectoral application has also recently been developed for seafood processors. Other industry-specific applications may be developed in the future.

General Permits

General permits are a type of permit typically used to authorize categories of activities where applicants can meet specified criteria without the need for individual agency review. Activities authorized by general permits are usually routine and have minimal environmental impacts. The primary advantage of general permits, from a permit streamlining perspective, is that they reduce the administrative burden on regulators and also relieve qualified applicants from the need for time-consuming permit reviews.

To be most effective, all agency permits needed for a routine activity must be covered by a general permit. If only a portion of the needed approvals are covered by a general permit, certain administrators and applicants must still expend resources for routine activities, which reduces overall efficiency. In Alaska, a single general permit that covers all agency approvals is not possible. Rather, each affected agency must issue its own general permit and then "package" the various general permits together to collectively cover all necessary authorizations.

General permits have been used successfully by individual agencies in Alaska for a number of years. However, the concept of "packaging" individual permits to collectively cover all authorizations needed for an entire activity is relatively new. It is this "packaging" concept that the administration wanted to explore further.

Based on input from industry and permitting staff, three categories of activities were initially targeted for development of general permits: winter North Slope oil and gas exploration wells, exploration mining camps, and tourism guide and outfitter activities. This assignment was given with the understanding that the development of general permits was the goal, but if general permits were found to be impractical, other streamlining options would be considered.

To date, the following recommendations have emerged for each of these sectors. For oil and gas, a series of general permits could be created to streamline permitting and reporting requirements for oil and gas exploration activities on the North Slope and adjacent state waters. For specific project proposals, reviewers would have seven days to request an individual review if a project would have site-specific effects not covered by stipulations on the general permits. The proposal also recommends simplifying reporting by use of an electronic system accessible by all agencies.

For mining camps, a similar series of general permits, with an abbreviated initial review to identify projects with site-specific impacts, has been proposed. However, if site-specific concerns are identified, these general permits may contain a provision that additional site-specific conditions could be added without full review, if acceptable to both the applicant and the agencies. The general permits would be designed to address the parameters typically needed by temporary mining exploration camps in terms of size and location.

The proposal for tourism and outfitter guides has taken a different tact. Upon examination, it was determined that there are no major state permitting issues for these activities that general permits could rectify. As such, other actions could be taken including developing comprehensive information for applicants on how to apply for necessary permits, and developing a consolidated database to track any permits that are issued.

Standardized Administrative Appeal Processes

In recent years, the State of Alaska has become increasingly aware that a single project could be administratively appealed multiple times. This opportunity exists because, even if state agencies with permitting authority collectively review a project, each agency has its own appeal process that could be invoked. For example, a project in the coastal zone that requires permits from three state agencies could potentially be appealed six times, and if the applicant were still disgruntled, judicial appeal could follow. Granted, this is a worst case scenario that rarely, if ever, occurs, but the potential does exist. For controversial projects, it is becoming increasingly common to have two or three administrative appeals to the state.

Multiple appeals challenge limited state resources and are highly redundant. In spite of these concerns, the State recognizes that applicants and the public do need adequate access to government to dispute decisions that may be flawed. As such, the streamlining challenge is to provide adequate options for administrative appeal to preserve the public's due process while not overtaxing increasingly limited agency resources with unnecessary reviews. Also, because administrative appeals are legally provided through either statute or regulation, this streamlining task is more complicated because it requires legal revisions.

A group of senior-level agency staff and a representative from the Department of Law were convened to investigate this streamlining challenge. Their specific goal was to simplify and standardize appeals between agencies by providing one informal, internal process to reach agency agreement on permit decisions, followed by one administrative appeal by any party on any aspect of the project decision. If a party were still aggrieved following the administrative appeal, they could appeal the decision to superior court.

A draft proposal that meets this streamlining goal has been developed. The proposal also discusses at some length who can file and hear appeals, and under what circumstances, but a detailed discussion of these parameters is beyond the scope of this paper.

Bypass Project Review Process

As mentioned in the history of streamlining, examining only one permitting element or one permitting agency will result in limited streamlining improvements. As such, the streamlining initiatives discussed thus far acknowledge this broader perspective and are incorporating a multi-agency perspective. However, with the exception of consolidating appeals, these initiatives are still being developed within the context of existing agency permit requirements and structures.

Although the entire history of environmental permitting in Alaska cannot be overlooked, this initiative acknowledges that permitting has evolved in a piecemeal fashion, over time, and that broader efficiencies may be gained if actions are recommended that go beyond current organizational requirements and structures. The initiative acknowledges that permitting agencies each have their own permit review requirements which, among others, may include public notice, evaluation procedures, and decision documents. These varying and sometimes conflicting requirements have never been fully harmonized and are frequently perceived as redundant and disjointed.

A group of senior agency staff and the Department of Law was convened to develop the "ideal" project review process that, if legislatively implemented, would bypass most existing, disjointed statutory and regulatory review requirements. The group's overall goal was to improve and streamline the review process for projects that require multiple state resource approvals, and the group identified two specific tasks:

- Task 1: Develop statutory changes that create a mandatory multi-agency review process for coastal and non-coastal areas, that bypasses current organizational requirements and structures; and

- Task 2: Determine agency responsibilities under this multi-agency permit review process, recognizing that some consolidation of agency functions and changes to interactions among the state, the public, and applicants may be required to achieve the greatest efficiencies.

At the time the group was convened, elements of the first task were already being acted on by the previously discussed streamlining initiatives. Specifically, actions were being taken to develop a joint application form and to streamline appeals. Therefore, the group decided to integrate these efforts into the broader outline of a master project review process. As of this writing, a draft master process has been developed to meet Task 1 and is undergoing internal review. While the process superficially appears similar to the master process developed in 1977, there are key differences. Specifically, the new process would be mandatory for projects that require multiple permits, pre-application meetings are a key element, review time lines are negotiable, and the process fully integrates coastal management requirements and other permit processes.

Regarding Task 2, the process currently recommends that the DGC, which has over 12 years experience in coordinating projects, serve as coordinator for all projects. An exception is that the overall coordinator for large, complex projects would be negotiable.

Conclusion

Beginning in the 1970s, a series of environmental laws and regulations began to be passed at the state and federal levels to ensure protection of important natural resources. Over time, this legal framework has continued to grow in size and complexity.

In addition, because this legal framework was developed incrementally, with many laws created to address discreet problems, their various requirements have not been harmonized. Specific procedures have been developed to address specific problems, and are often not integrated with other existing regulatory requirements. The implementation of these varying requirements has become increasingly frustrating to applicants who confront a plethora of permits and challenging to regulators whose operating budgets and staffing levels continue to decline.

As long as natural resources continue to be developed, existing environmental laws continue to be implemented, and new laws continue to be enacted, permit streamlining will remain a concern for state and federal regulators. And, as any regulator will agree, there are rarely quick fixes when it comes to permit streamlining. Also, truly effective streamlining requires a thoughtful and collaborative effort on the part of all affected regulators to ensure that the public interest is well served.

Early permit streamlining efforts in Alaska focussed on permit information centers, joint applications, joint review processes, and joint agency offices. Each effort met with some success. More current reform efforts are focussing on improvements through joint applications, general permits, consolidated appeals, and development of a bypass project review process.

These efforts encompass an array of changes ranging from those that can be implemented administratively to those that require involvement of both executive and legislative branches of government for statutory revisions. While sweeping statutory reform may be desirable, statutory remedies are often not achievable. Efforts that require statutory changes can be increasingly difficult to achieve as political agendas influence outcomes. Absent the ability to change statutes, many bureaucracies seek efficiency through incremental changes to smaller components of the permitting process.

Regardless of their scope, the purpose of most streamlining efforts is to improve some aspect of the permitting process by making the process more efficient or more effective. However, these parameters and the overall success of permit streamlining are often difficult to measure. Many efforts result in improvements--communication, satisfaction, joint interactions--that are hard to quantify and even more difficult to translate into actual cost savings. Also, permit streamlining does not always equate with measured cost savings, although reducing costs is often an objective.

In spite of these limitations and difficulties, permit streamlining remains a laudable goal for most bureaucracies. A complete evaluation of Alaska's current efforts will not be possible until they are fully implemented. However, the administration's commitment to improve permitting will remain an impetus to help ensure these efforts are successful.

CHAPTER 91

Educating Industry, Government and the Public About Maintaining a Healthy Ecosystem

Stephanie McDonald[1]

ABSTRACT
In 1993 the Santa Monica Bay Restoration Project (SMBRP) established the Public Involvement and Education (PIE) Fund to support programs that creatively and effectively educate Los Angeles County residents about ways to 1) restore and protect Santa Monica Bay, and 2) maintain a health ecosystem for the region. The SMBRP funds numerous outreach programs developed both by and for government and other public agencies, educators, environmental organizations, trade groups, business professionals, artists, etc. These projects represent the efforts of those who are committed to protecting coastal and inland resources such as aquatic, marine and coastal habitats; drinking water sources; wetlands; beaches/intertidal zones; etc. A specific project funded through the PIE program is highlighted in this paper -- it addresses the need for providing a tool to help cities and industries educate themselves about storm water-related issues. The tool, a *workshop module*, was developed to do just this; it's a step-by-step manual that outlines the procedures to successfully host a storm water educational workshop.

Introduction

Where does the responsibility lie in educating industry, government and the public about maintaining a healthy ecosystem? The responsibility lies with everyone -- everyone who in some way benefits from maintaining the health of our local habitats, whether economically, recreationally or for sustainability. A majority of California's residents live near the coast. Not coincidentally, numerous industrial and

[1] PIE Program Coordinator, Santa Monica Bay Restoration Project, 101 Centre Plaza Drive, Monterey Park, CA 91754

commercial businesses are also located in the coastal regions. Therefore, it is vitally important that all of these groups understand the value of maintaining a healthy ecosystem as it relates not only to them, but to the entire region.

As a way to encourage industry, government and the public to become involved in pollution prevention educational efforts and to become environmental advocates among their peers, the Santa Monica Bay Restoration Project has developed the Public Involvement and Education (PIE) Fund. The intent of the PIE Fund is to empower these groups to educate their constituents about the many sources and effects of pollution, how it affects the surrounding ecosystem and what can be done to prevent pollution from adversely impacting valuable resources. The SMBRP feels that by providing grant funds to members of government, industry and the general public, these groups will conduct educational outreach to their constituents in a way that proves most effective for changing behaviors and reducing pollution.

Grant funds are provided through a competitive process to anyone (except State and federal agencies) who submits a proposal in response to the SMBRP's "Request for Proposal" released each year. Awards are given to universities, teachers, businesses, artists, non-profit organizations, environmental groups and others. Approximately $260,000 has already been awarded to 42 programs over the past three years, reaching an estimated 200,000 people throughout the Los Angeles area about restoring and protecting Santa Monica Bay. In many cases, the funds awarded have been the seed money needed by an organization to get a program started. By providing seed money, the SMBRP helps a PIE recipient establish a track record which can then be used to lobby for other sources of funding. Several projects supported by the SMBRP have launched ongoing programs using PIE funds. In supporting these projects, the SMBRP supports diverse, creative and innovative educational approaches to environmental awareness and protection.

One of the 1994 PIE projects developed was a "workshop module" aimed at educating municipalities and businesses (both industrial and commercial) about the impacts of contaminated storm water, the importance of pollution prevention, the roles of storm water regulatory agencies, and the effectiveness of implementing Best Management Practices (BMP). The module has been developed for (1) municipalities to use in their city-wide storm water educational outreach efforts and (2) industry representatives to use in peer-to-peer educational outreach.

With passage of the new Storm Water National Pollution Discharge Elimination System (NPDES) permit, cities are now required to conduct educational outreach to industrial and commercial establishments regarding storm water pollution control. However, many cities do not have the appropriate personnel or sufficient funds to carry out educational activities. This module attempts to help meet the storm water-related educational objectives required by cities. More specifically, the module has been designed to help everyone, even those with little or no experience, present a successful storm water workshop. The module recommends that the workshop last approximately two hours and allows time for guest speakers to present real-world successes/strategies in implementing storm water best management and pollution prevention practices. Components of the module include:

- An outline of all the preliminary steps that need to be taken to successfully host a workshop, including how to request and obtain guest speakers;
- Script suggestions for storm water-related presentations;
- Attendee and Presenter Agendas;
- Sample overheads that can be used during the presentation;
- A list of storm water reference materials;
- Contact agency names and phone numbers where more information can be obtained; and
- Samples of the types of handouts that can be obtained and given attendees.

The initial goal of the module is to help individuals determine the *Who, What, Why, When, Where, and Cost* of hosting a workshop (Figure 1). It presents questions and offers suggestions to help planners focus on important pre-workshop issues, such as:

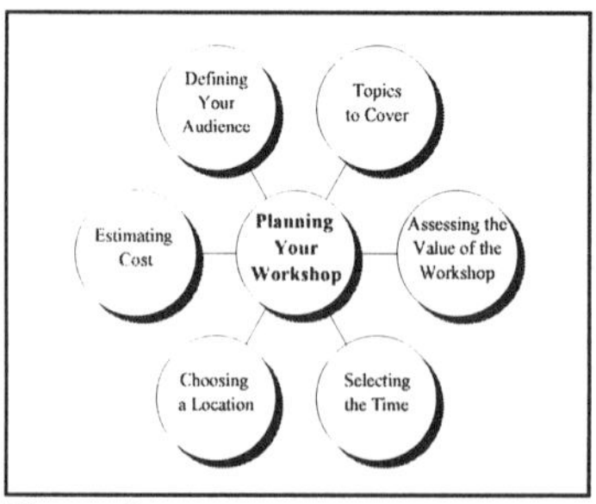

Figure 1

- Where do I begin?
- Who should be invited?
- What storm water pollution issues should be addressed?
- How do I ensure that people will attend?
- When is the best date and time to host the workshop?
- Is there a free facility that could be reserved for the workshop?
- How much will it cost to host and will attendees have to pay?

Workshops are interactive approaches to education and can be an effective tool to reach a large audience with a clear, concise storm water pollution prevention message. However, the module recommends targeting a specific audience (e.g., restaurants, service stations and auto repair shops, etc.) for outreach; if topics are too broad or diverse, the perceived value of the workshop diminishes and turnout may be low. Workshops also provide a neutral/informal setting where business owners and operators can ask questions about pollution issues/regulations without having to worry about penalties and/or potential site visits from regulatory agencies. The module strongly encourages requesting guest speakers that will showcase real examples of implementing Best Management Practices (BMPs) at similar sites and discuss obstacles they may have encountered.

Education and awareness are worthwhile source control measures and should be conducted in a manner that encourages both the sharing of information and the willingness to make changes to prevent storm water pollution. If carefully planned and if vendors are secured to help reduce out-of-pocket expenses, workshops can be a very cost-effective method of meeting outreach objectives. The workshop presented here is but one example of how education and outreach objectives can be creatively accomplished.

CHAPTER 92

Enhancing Education -- Partnerships That Work!

Liz Love[1] and Laura Gorodezky[2]

The Monterey Bay and the Channel Islands National Marine Sanctuaries are two of twelve marine sanctuaries nationwide under direction of the National Oceanic and Atmospheric Administration (NOAA). The mission of the National Marine Sanctuary Program is to manage marine areas of special national significance to protect their ecological and cultural integrity for the benefit of current and future generations. Mission objectives include conducting research, education and stewardship programs that foster public understanding, support and participation, and promote the ecologically sustainable use of the nation's marine resources. The Sanctuary Program strives to provide leadership and link the assets of governmental and non-governmental organizations to focus people's attention on the need to protect marine resources, and depends for its success upon extensive participation by other agencies and organizations, as well as the public.

Participants will learn how the Monterey Bay and the Channel Islands National Marine Sanctuaries, in collaboration with a variety of organizations and agencies, have implemented educational programs that foster public stewardship of the marine environment. One program that will be featured is the Great American Fish Count, a fish censusing activity for volunteer divers that increases public awareness about the diversity and condition of marine fish populations. This annual event provides divers with a goal-oriented activity focused on non-extractive resource use.

[1] Education Coordinator, Monterey Bay National Marine Sancutary, 299 Foam Street, Suite D, Monterey, CA 93940

[2] Education Coordinator, Channel Islands National Marine Sanctuary, 113 Harbor Way, Santa Barbara, CA 93109

CHAPTER 93

Adopt-A-Beach: A Model for Marine Debris Education

Amy Windrope[1] and Chris Parry[2]

Introduction

The marine environment is being seriously harmed by plastic debris that is being dumped on land and in the ocean (Joyner and Frew, 1991). But the average person has no idea how each of these impacts are directly related to her life and therefore is at a loss to mitigate her individual impact of the marine environment. The Coastal Commission has endeavored to create a program that offers every Californian a way to improve the coastal environment through the direct action of cleaning the beaches. The trash on our shores is indicative of the trash in our oceans; by cleaning our local beaches we are cleaning the oceans that cover our watery planet.

The California Coastal Commission's Adopt-A-Beach Program encompasses several public outreach and education programs, including the Annual California Coastal Cleanup Day, the Year 'Round Adopt-A-Beach Program, and the Save Our Seas curriculum. The Annual Coastal Cleanup Day is the most visible event and enjoys statewide participation of over 40,000 people annually. A number of school-based Adopt-A-Beach programs have been established as an outgrowth of the success of Coastal Cleanup Day. These programs integrate classroom and experiential learning by combining the use of the Save Our Seas curriculum with a field trip to the beach for a clean up. This paper focuses on several educational partnerships that can serve as a model for marine debris education. The programs we highlight in this paper demonstrate the power of engaging students in issues immediate to their world and then offering them ways to effect positive change.

The programs we highlight can be adapted to other communities that are interested in offering a hands-on, issue based, stewardship project for students and teachers. Strong partnerships with local groups and individuals have been key to the success of the Adopt-A-Beach model. The paper focuses on three programs: Adopt-A-Beach School Assembly

---

[1] Save San Francisco Bay Association, 1736 Franklin Street, Oakland, CA 94612
[2] California Coastal Commission, 45 Fremont Street, Suite 2000, San Francisco, CA 94105

Program, the Assembly Program Teacher Trainings, and the Golden Gate National Recreation Area Adopt-A-Beach Program.

The Adopt-A-Beach School Assembly Program in Los Angeles County

The Adopt-A-Beach School Assembly program began in 1991 as the brainchild of Michael Klubock, a sailor turned environmentalist. Mr. Klubock's love for the beaches and oceans inspired him to create a school assembly program to get more kids to the beach and in touch with nature. The Adopt-A-Beach School Assembly Program was designed to motivate students to care about the marine environment and then give them a way to make the environment safer. The students are engaged in the beach cleanup process and learn why it is of critical importance to marine animals that the ocean environment be clean and safe. One of the program's participants, Allen Martinez, a student who had never been to the beach before, was quoted in the L A Times as saying "cleaning the beach with my classmates made me feel hopeful." The Assembly program is about creating hope in children's lives by empowering them to make a difference.

The Assembly program has three distinct components: an awareness raising assembly, fundraising through recycling, and a beach cleanup field trip. Students are exposed to the harm that everyday refuse has on the ocean. The students learn how the ocean is connected to their communities through the storm drain system, and that most of the garbage found on our shores is from inland sources. Most students are unaware that the plastic blowing down the streets of Los Angeles County could eventually end up in the ocean where it poses a serious threat to cetaceans, pinnipeds, turtles, fish and seabirds.

The assembly begins with a slide show. The slides open with images of the wonders of the oceans - beautiful fish, giant sea anemones, swaying kelp forests; and then proceeds to show pictures of trash on the shores, trash in the stomach of birds, and a whale trapped in a ghost net. The sequence of images are designed to engender love for the ocean and its creatures and a sense of empathy for them. The images turn from awe inspiring to tragic as beautiful seascapes are followed by the reality of entangled birds, pinnipeds, even whales. But, instead of just conveying a negative message, the program focuses on solutions to the marine debris problem: recycling, reuse, and proper disposal.

Throughout the assembly, Klubock explains about recycling and the cost effectiveness of reusing. His hat is made of recycled bottles and his shirt from reclaimed cotton. He challenges the students to think about being a consumer. What does it mean when you buy a recycled product? How does that change the marine environment? He teaches the connection between consumer decisions and environmental degradation with marine debris as his central focus. He draws his students into frank discussions about recycling in the context of reusing plastics in order to reduce the amount of plastic in the oceans. He supplies the students with striking statistics: Americans throw away 2.5 million plastic bottles an hour, and for the last 13 years, plastic is the single largest type of debris found on the beaches. After the slide show, the students understand how their individual actions, when taken as an aggregate, affect the animals they love - seals, dolphins and whales. He successfully engages his students minds and hearts.

After the slide show and the discussion of the need to reuse and recycle, Klubock offers the students a challenge. If they can raise enough money by recycling aluminum cans to pay for half the cost of leasing a school bus, he will pay the other half for a trip to the beach for a cleanup and a picnic. This fund-raiser raises not just funds but the acknowledgment that recycling doesn't just feel good, it is a wise economic decision.

The students then complete the Adopt-A-Beach experience by cleaning a beach near their community. The students have raised money for the bus, learned about why they are cleaning and finally get to do something to improve their community. The students are given trash bags and data cards on which they record exactly what they have picked up. Teachers can then use this data in the classroom to continue the learning process by integrating the data cards into future science, math and social studies lessons. This program has reached more than one hundred thousand students in the last five years.

The key to implementing this program is having a source of funding that can match the funds raised by the students. The beach trip isn't just about picking up garbage, it is about working with their classmates to raise money and educating their parents about the necessity of recycling. In addition, the program needs a dynamic speaker to engage the students in why they should care, why they should want to go to a beach cleanup. The entire program is about engendering students with a sense of personal environmental responsibility: for the coast and ocean, for their communities, and for making a positive impact.

Save Our Seas Teacher Trainings in Los Angeles County

The second model of marine debris education is also coordinated by the Malibu Foundation. The Coastal Commission was interested in a way to reach more teachers and classrooms than was possible through the Malibu Foundation's Adopt-A-Beach Assembly Program.

The Coastal Commission and the Center for Marine Conservation had previously developed a marine debris curriculum named Save Our Seas or SOS. SOS is an issue-based curriculum which focuses on the impact of marine debris on the marine environment. The curriculum is designed for grades K-12 in accordance with the California Science Framework. It is designed to be used by teachers who are interested in engaging their students in a unit focused on cleaning a beach or waterway. It offers activities for both before and after a beach cleanup and stresses the connection between inland sources and the marine environment.

The Coastal Commission developed a teacher education component to complement the Assembly program. The Malibu Foundation, in partnership with the Coastal Commission, hired Dr. Marilyn Kronmal to spearhead an intensive teacher training program over the summer of 1996. As part of this program, the SOS handouts were translated into Spanish. The goal was to reach 400 teachers in the greater Los Angeles area, with a focus on the inner-cities, through workshops and seminars. Marilyn Kronmal went far beyond this goal and was able to reach 670 teachers through 15 workshops ranging in size from eight teachers to 125 teachers per workshop.

The reason for the success of this intensive summer training program was largely due to Dr. Kronmal's understanding of the Los Angeles educational community. She was able to tap into already existing workshops throughout Los Angeles county to present the curriculum. Dr. Kronmal used a team teaching approach to present the curriculum. She also enhanced the curriculum by providing the teachers with an extensive array of additional related resources, including materials from the Department of Conservation, the US Coast Guard, the City of Los Angeles, and Heal the Bay.

Each workshop began with the Marine Debris Slide show created by the Center for Marine Conservation which graphically portrays the problem of marine debris. The slide show is available to all teachers on a check-out basis from both the California Coastal Commission and the Los Angeles County Office of Education. The slide show familiarizes teachers with the issue of marine debris and the effect it has on marine life. Save Our Seas is then introduced with explanations of the different sections. Save Our Seas is relevant in the context of a cleanup whether the cleanup happens at the beach, in the shoreline of a local stream, or in the school yard. The activities are designed to assist the students in interpreting the experience of cleaning an environment.

Similar to the assemblies, the curriculum is about motivating students to take responsibility for the environment. The program armed teachers with activities and resources for educating their students and then offered them a way to get to their classrooms to the beach on Coastal Cleanup Day. Dr. Kronmal worked with Heal the Bay and Los Angeles County to obtain free buses to the beach on Coastal Cleanup Day 1996; 1500 students participated. By integrating the Los Angeles Assembly program, the teacher training program and Coastal Cleanup Day, the students of Los Angeles County have several opportunities to learn about their role in marine debris and to take action to help solve the problem.

The San Francisco County Adopt-A-Beach Program

The San Francisco Adopt-A-Beach program began in 1986. In 1992, the program was greatly expanded and became a cooperative effort between the San Francisco Conservation Corps (SFCC), the National Park Service (NPS), and the California Coastal Commission. Rebecca Ramos, a NPS Ranger, began the current incarnation of the San Francisco County Adopt-A-Beach program with funding from the National Oceanic and Atmospheric Administration (NOAA). The program centers on Ocean Beach in San Francisco, which is part of the Golden Gate National Recreation Area, and is managed by the National Park Service. Ocean Beach, one of the most accessible urban beaches in the

Bay Area, which makes it one of the most polluted beaches and also an excellent staging ground for educational programs.

The program has three components, similar to the Los Angeles SchoolAssembly program: classroom presentation, teacher training, and a beach cleanup. But, instead of doing the classroom presentation herself, Ramos has mobilized the SFCC to go into the classroom for a one hour orientation with the children.

The SFCC has provided Corp members to help facilitate the program since 1986, but was reconsidering its involvement in 1992. Ramos realized that the corps involvement could be greatly enhanced if they were better trained in the issues of marine debris. She developed a one day training that took the corps members out to a garbage transfer station at Candlestick Point and educated them about garbage issues in relation to Ocean Beach. This one day training then expanded to two days to allow the corps to be trained in interpretive skills, and how to deal with children. The corps members emerge from the training well-prepared for the program and have proved vital to the program's success.

Teachers apply to this program for thirty slots both spring and fall. The teachers are required to attend a one day training with Rebecca Ramos that will prepare them to prepare their students for the beach cleanup and gives them tools to use in the classroom. Training the teachers increases the impact of the beach cleaning experience. Teachers receive the Save Our Schools curriculum, the San Francisco Recycling Center's materials and other materials that will enhance student learning and curiosity about the marine environment.

As with the Malibu Foundation program, the training begins with the Center for Marine Conservation's Marine Debris Slide Show. She then has the teachers participate in several activities from the Save Our Seas curriculum to familiarize the teachers with the curriculum and engage them in the same experience that their students will have. The teachers are also prepared for the actual logistics of the beach cleanup and schedules are reviewed. Finally, the training ends with a presentation by a SCRAPPS educator. SCRAPPS is a non-profit organization that teaches all about recycled materials and how to use those materials for art projects.

Once the teachers have returned to the classroom, Ramos provides additional support. Prior to the cleanup, the students are visited by Conservation Corps members who lead them through a series of activities designed to help the students understand why they are doing a beach cleanup. They lead them through art projects that ask the students to draw the beach, they ask students to write about their experiences on the beach. They also engage the students with activities from the Save Our Seas curriculum. A sample activity from SOS is called: Early Bird Gets the Plastic? The game encourages the students to discover what happens when a bird eats plastic instead of food. The game teaches about feeding adaptations, feeding strategies and the harm of eating plastic. Students are encouraged to relate the experience to their own community and are asked what kind of food they see on the ground in their own neighborhood for birds.

When the students finally get to the beach they are met by a ranger who gives them a small presentation on safety and what to watch for while on the beach. The students are then broken into small groups of five to seven children for beach cleaning and beach discovery. Each small group is assigned to one of the same Corps members who visited the students in the classroom. The students already know their leader and the beach cleaning can begin at once with the Corps member serving as naturalist and guide. The arrangement provides an opportunity for the students to develop a relationship with a young adult who is modeling environmental stewardship. After several hours, the classroom returns to the starting point and the ranger facilitates a closing discussion.

Conclusion

The Adopt-A-Beach program has provided a model, materials and a curriculum that supports organizations or individuals who want to start a program that engages students in a beach cleanup that extends back into the classroom. The programs described here all have common elements: the Marine Debris Slide Show, classroom materials, teacher training, and a service project and field trip. These elements combine to form a powerful and comprehensive learning experience for students. They learn that they can make a difference.

Joyner, Christopher C. and Scot Frew. "Plastic Pollution in the Marine Environment" Ocean Development and International Law, 1991 Volume 22, pp. 33-69.

Parker, Patricia. "The Plastics Threat." Sea Frontiers, March-April 1990.

CHAPTER 94

CLEAN BOATING EDUCATION PROGRAMS OF CALIFORNIA

Kevin Atkinson[1]

California is one of the nation's top three boating states with more than 840,000 registered boats. This large boating population represents eleven billion dollars of economic activity in California each year, yet it has the potential to severely impact the marine ecosystem that it depends on.

Over the years increased environmental awareness has shown that boaters and other recreational marine constituents do not exercise the same environmental conscienceness on the water that they normally would use on land. Typical areas of concern are trash/garbage disposal, oil and fuel spill prevention, and sewage disposal. Many boater education programs are in place that address these issues. However, due to California's large geographical area, education groups are often unaware that others throughout the state have already produced materials that cover the same subject matter and consequently end up "reinventing the wheel."

The following list of organizations represents some of the active campaigns currently targeting clean boating issues. Due to space limitations many organizations are not listed, however, those listed are a good starting place and can also lead you to other contacts.

[1]California Clean Vessel Act Grant and Education Coordinator
California Department of Boating and Waterways
1629 S Street, Sacramento, CA 95831

Name:	California Department of Boating and Waterways 1629 S Street Sacramento, CA 95814 (916) 445-2616
Affiliations:	None.
Funding Sources:	State of California, Federal grants.
Target Area:	Statewide.
Audience:	Recreational boaters, marina operators.
Methods of Outreach:	Brochures, films, videos, home study course.
Potential Sources of Funding for Other Programs?	None.
Available Materials:	The Department has over 40 different brochures, films and videos available. These items cover a wide variety of subjects such as boating safety, boating trail guides, marine pollution laws, and environmental issues. The Department also has education programs available for grades K-12. Contact the Department for a complete list of available materials and programs.
Additional Information:	The Department is administering the Clean Vessel Act Grants (CVA) for the U.S. Fish and Wildlife Service. This grant program is targeted at increasing the number of pumpout and dump station facilities for sewage disposal available to boaters. The program also includes an education component for boaters. The Department has granted funds (CVA) to the San Francisco Estuary Project, Santa Monica Bay Restoration Project and the Coast Guard Auxiliary, Division 16 to assist in this education component.

. . . .

Name:	San Francisco Estuary Project 2101 Webster Street, Suite 500 Oakland, CA 94612 (510) 286-0460

Affiliation:	Association of Bay Area Governments.
Funding Sources:	National Estuary Program, State funds, grants, donations, subscriptions.
Target Area:	12 counties of the San Francisco Bay - Delta Region.
Audience:	Recreational boaters, marina operators/owners, government agencies, general public.
Methods of Outreach:	Presentations, brochures, slide shows, videos, radio PSA's, newsletters, marina visits, training sessions. Many of the Estuary Projects materials are distributed by other organizations such as BayKeeper, U.S. Coast Guard, Power Squadrons, Sea Partners, boat supply shops and marina operators. The Estuary Project participates in a variety of pollution prevention programs and uses these events to distribute their materials also.
Potential Sources of Funding for Other Programs:	None.
Available Materials:	Brochures: *Ship Shape Sanitation, Clean Vessel Act Program, How Vessel Sewage Discharges Affect the Environment, Clean Boating Guide to San Francisco Bay, Clean Boating Guide to the Sacramento-San Joaquin Delta.*

. . . .

Name:	Santa Monica Bay Restoration Project 101 Centre Plaza Drive Monterey Park, CA 91754 213/266-7516
Affiliations:	Santa Monica Bay Restoration Foundation (non-profit).
Funding Sources:	State of California, Federal (EPA), California Department of Boating and Waterways (CVA Grant).
Target Area:	Los Angeles, Orange, Ventura and Santa Barbara Counties.

Audience:	Boaters, marine supply stores, marinas, harbor patrol, boat retailers, environmental groups, public/government agencies and other interested parties.
Methods of Outreach:	Publications (booklets, brochures); newsletters, establishing personal contacts, attending boat shows, forming partnerships, and submitting articles to boating magazines, newspapers, etc.
Potential Sources of Funding for Other Programs?	The SMBRP provides some seed money to projects that address pollution issues in and around Santa Monica Bay. Pilot programs that address polluted marina waters and boating wastes are eligible to receive SMBRP funding through a competitive process.
Available Materials:	*Ship Shape Sanitation* booklets, a list of available pumpout locations in the four counties mentioned above, and newsletters (starting in April 1997).
Additional Information:	The SMBRP is currently concentrating their education program on vessel sewage waste problems in their area. The SMBRP also hosted the second regional California Clean Boating Network (CCBN) meeting in February, 1997, in an effort to strengthen outreach efforts for the upcoming boating season.

. . . .

Name:	U.S. Coast Guard Auxiliary, Division 16 (a volunteer organization) c/o U.S. Coast Guard Marine Safety Officer 2716 North Harbor Drive San Diego, CA 92101 (619) 683-6500
Affiliations:	U.S. Coast Guard (active duty), Activity Center San Diego and the National Coast Guard Auxiliary, under the direction of the Commandant of the Coast Guard.
Funding Sources:	Membership dues, California Department of Boating and Waterways (CVA Grant)

Target Area:	San Diego County.
Audience:	The boating public and those interested in boating and water safety, and covers all sectors from public, private, and government, including the military. The Auxiliary has programs for audiences as young as age 5.
Methods of Outreach:	Boating education classes and public appearances at boat shows, civic organizations, yacht clubs etc. Courtesy Marine Examinations: A one-on-one education and check of a boats safety equipment (with the owner). Marine Dealer Visitations: Distribution of boating literature through the boating dealers, marinas, yacht clubs and retail stores. In addition to the above items, the Auxiliary participates in functions organized by others, such as *the Bay Clean Up* in San Diego Bay, sponsored by the Marine Dealers Association.
Potential Sources of Funding for Other Programs?	None
Available Materials:	There are a number of brochures available from the U.S. Coast Guard concerning boating safety, pollution and general regulations. Contact your local Coast Guard station for these brochures.

. . . .

Name:	American Oceans Campaign, (non-profit organization) 725 Arizona Avenue, Suite 102 Santa Monica, CA 90401 (310) 576-6162
Affiliation:	American Oceans Campaign (AOC), a national non-profit organization.
Funding Sources:	Private donors.
Target Area:	Coastal states.

Audience:	Public, private entities, and government agencies.
Methods of Outreach:	Audiences are reached through videos, television, public service announcements, educational materials, legislative action, etc. AOC also hosts beach clean-ups, distributes environmental information, holds press conferences, and sponsors environmentally sound products.
Potential Sources of Funding for Other Programs:	None.
Available Materials:	Videos: *Help Save Planet Earth, AIR NEW ZEALAND FEATURETTES on Driftnets, Marine Mammal Reserves, Whale Watching, Penguins, Heaven on Earth, Coral, Dolphins, Turtles.*
	PSA's: 3 different Used Motor Oil tapes, Estuaries, Wetlands, U.S. Coast Guard, disappearing fish, and others.
	Films and Documentaries: a series of Dinner Films from 1993 to 1996, and *Danger at the Beach-a National Audubon Special with Ted Danson.*

. . . .

Name:	United States Power Squadrons, (non-profit organization)
Affiliations:	United States Power Squadrons, a nationwide non-profit organization 1-888-FORUSPS.
Funding Sources:	Membership.
Target Area:	Nationwide.
Audience:	The United State Power Squadrons is a non profit, private membership organization that is dedicated to safe boating, boating education for the public as well as continuing education for USPS membership.

Methods of Outreach:	There are 450 squadrons across the nation, that provide as a public service, the *USPS Boating Course*. This is a six to eight week program and is instructed by experienced, highly trained members of local squadrons. The course is tuition free with only a small charge for the materials. Other educational programs that are available from USPS are designed for self instruction and are provided in the form of the USPS Video Boating Course and a computer based program called *BoatPro*. Two new class room instructed programs will soon be made available. They are *Boat Smart*, targeted to the general population and *Jet Smart*, designed for the PWC boaters. These instructed and self taught courses all have elements that teach applicable parts of the Federal Boating Law, specifically the Federal Water Control Act of 1972. Topics of instruction also include proper fueling procedures, waste disposal, as well as operational sensitivity for the environment.
Potential Sources of Funding for Other Programs?	None
Available Materials:	Within California there are 35 squadrons that provide these educational programs to the boating public several times each year. The video and computer programs are available at a small charge from any of the squadrons or may be obtained by calling 1-800-SEASKIL, a statewide toll free number specifically for California and Southern Nevada boaters. The National Headquarters of the United State Power Squadrons may be contacted by calling 1-888-FORUSPS.

. . . .

Name:	San Mateo County Used Oil / Marina Public Outreach Program 590 Hamilton Street Redwood City, CA 94063 (415) 599-1514

Affiliations:	None.
Funding Sources:	San Mateo County , California Integrated Waste Management Board (Grant).
Target Area:	San Mateo County.
Audience:	Recreational boaters and marina operators.
Methods of Outreach:	Marina newsletters, brochures and public displays.
Potential Sources of Funding for Other Programs?	None.
Available Materials:	Call for examples of outreach materials.
Additional Information:	San Mateo County Used Oil Program has been working closely with county marinas to promote used oil recycling and reduce marine pollution. The program has purchased oil absorbent pads for boaters to remove oil from their bilge water and has installed collection drums at 5 marinas for disposal of the used pads. In addition, the Used Oil Program has provided Pillar Point Harbor in Half Moon Bay with a 55 gallon drum for disposal of used oil filters, and has participated in a pilot project to supply bilge pump-out systems for use by resident boaters. The Used Oil Program is also currently working with the Monterey Bay National Marine Sanctuary and Save Our Shores who just received a grant from California Integrated Waste Management Board as a result of the success of San Mateo County's Used Oil Program.

. . . .

Name:	California Coastal Commission Used Oil / Public Outreach Program 45 Fremont Street, Suite 2000 San Francisco, CA 94105 (415) 904 -5200
Affiliations:	None.

Funding Sources:	State of California, California Integrated Waste Management Board (Grant).
Target Area:	State coastal areas.
Audience:	Recreational boaters and marina operators.
Methods of Outreach:	Working with boating groups and other organizations.
Potential Sources of Funding for Other Programs?	None.
Available Materials:	Materials will be available in the future.
Additional Information:	Starting in mid 1997, the California Coastal Commission will conduct a three year program designed to educate California boaters about ways to minimize oil discharges and other types of pollution associated with recreational boating.

The primary objectives of this project are to provide state-wide education and to facilitate the installation of services needed by California's boaters in order to reduce the discharges of pollutants associated with boating.

The Public Outreach Program, in addition to producing some new boater education materials, will be continue the California Clean Boating Network (CCBN) that was started several years ago. The CCBN is a consortium of environmentally conscience individuals from public agencies, industry, and environmental groups. Their purpose is to improve the communication and coordination between marina operators, general boaters, and boating pollution educators in order to help boaters pursue their interests with minimal adverse impacts.

Author: The CCBN has compiled a manual for its members called "Pollution Solutions - A Compendium of Marina and Recreational Boater Pollution Education Materials. " This manual is a must see for anyone interested in boater education. Chances are materials covering your area of interest can be found here.

Name:	Save Our Shores (SOS), (non-profit organization) 222 East Cliff Drive #5A Santa Cruz, CA 95062 (408) 462-5660
Affiliations:	None.
Funding Sources:	Save Our Shores is currently using funding received under the 1996/98 Nonprofit Organization - Used Oil Grant from the California Integrated Waste management Board.
Target Area:	Monterey Bay National Marine Sanctuary.
Audience:	General public.
Methods of Outreach:	SOS will have brochures and flyers available in the near future.
Potential Sources of Funding for Other Programs?	None.
Available Materials:	Using the grant funds, SOS will implement a two-year pilot project to recycle used oil in Monterey Bay harbors. The grant will enable SOS to purchase absorbent oil pads and collection drums to receive the used oil pads. The pads and disposal containers will be placed in selected sites within Monterey, Moss Landing and Santa Cruz harbors. The project will also entail an education program in the harbors to promote the proper disposal of waste oil.
Additional Information:	SOS is collaborating with Monterey Bay National Marine Sanctuary Water Quality Protection Program staff, and harbor staff to implement the Used Oil Grant.

. . . .

Name:	Monterey Bay National Marine Sanctuary Water Quality Protection Program (WQPP) 299 Foam Street, Suite D Monterey, CA 93940 (408) 647-4217

Affiliations:	National Marine Sanctuary Program Office of Ocean and Coastal Resource Management National Oceanic & Atmospheric Administration Washington, D.C. (202) 673-5126
Funding Sources:	U.S. Department of Commerce, National Oceanic & Atmospheric Administration (Sanctuaries and Reserves Division).
Target Area:	Area encompassed by the Monterey Bay National Marine Sanctuary (MBNMS).
Audience:	All users of the Monterey Bay National Marine Sanctuary.
Methods of Outreach:	The WQPP is working with four harbors within the MBNMS to promote good harbor practices. Each year the MBNMS helps support an underwater clean-up in Monterey during Coastal Clean-Up Day. WQPP is also generating funding support for bilge water and oil/water collection facilities. Articles are also submitted to local marina newsletters concerning best management practices.
Potential Sources of Funding for Other Programs?	The WQPP has established a grant data base and is aggressively seeking funding to implement strategies developed during workshops devoted to harbor and marina issues. The WQPP also works with other agencies and non-profit groups to help solicit additional programs and support for outreach to the boating community.
Available Materials:	Publications: *Prevent Boating Pollution, Boating in the Monterey Bay National Marine Sanctuary*. Also available *Action Plan III: Marinas and Boating, May 1996*, which outlines a set of strategies proposed to address potential water quality problems resulting from activities in and around marinas and boating.

CHAPTER 95

Out and About: Learning about marine sanctuaries through field-based education

Maria Brown[1] and Candice Brown[2]

Abstract

The Farallones Marine Sanctuary Association (FMSA) is developing an education program that integrates experiential learning in the Gulf of the Farallones National Marine Sanctuary and focuses on a systems approach to education. The education program is based on the principles of environmental literacy: understanding the relationship between healthy ecosystems and human well-being.

Research in the field of education has shown that programs which involve teachers and students in meaningful, experiential education activities foster more positive attitudes towards science, interest more diverse populations of students, promote stewardship of resources, and develop critical-thinking skills necessary to solve complex problems. This program will educate future decision-makers about the need for national marine sanctuaries and the role sanctuaries play in conserving local biodiversity for global ecosystem health. "An informed citizenry opts for sustainability."

Introduction

Education programs can contribute to habitat protection by encouraging the public to develop habits of conservation and wise use in the marine environment. By developing public awareness and appreciation for the complexities of the marine world, education programs can increase public support and commitment for the marine environment and the National Marine Sanctuary Program.

As land resources dwindle and coastal populations swell, there is a need for greater awareness of human actions and their impacts on the oceans. This is not only desirable and useful—it is an obligation (Norse, 1993; Kaza, 1985). Ocean awareness is underrepresented in almost all current educational curricula. Schools do not traditionally teach about our fundamental dependence on water and the oceans for food,

[1] Maria Brown, Executive Director, Farallones Marine Sanctuary Association, The Presidio, P.O. Box 29386, San Francisco, CA 94129. Mariabrown@aol.com. Telephone (415) 561-6625.
[2] Candice Brown, Ph.D. candidate in Educational Psychology, 214 A West Arrellaga, Santa Barbara, CA 93101. cbrown2@education.ucsb.edu. Telephone: (805) 957-1029.

transportation, mineral resources, recreation, and for life itself. Except for sudden disasters such as oil spills, damage to ocean and estuary environments often goes unnoticed, because it happens gradually in a slow accumulation of "insults, degradations, and poor management" (Goodwin and Schaadt, 1978). Though knowledge about oceans and estuaries is limited, marine education programs can begin to offer basic understanding of the principles of life in ocean systems (Norse, 1993).

The Goal of FMSA's Education Program: Marine Environmental Literacy

The goal of Farallones Marine Sanctuary Association's (FMSA) education program is to expand the learner's perspective and integrated understanding of the marine biological and physical environment and its relationship with the social environment in order to produce a citizenry that lives in a sustainable manner.

In order to achieve a sustainable society, the majority of the literature suggests that environmental education needs to produce environmentally literate citizens. In this context, the term literacy means the integration of thinking, talking, interacting, and valuing, in addition to possessing reading and writing skills (Disinger and Roth, 1992). A distinguishing characteristic of environmental literacy has been its "action" perspective (Cherif, 1988). It requires interpreting the knowledge and becoming motivated to take action.

Environmental literacy is having the awareness, desire, and skills to choose and promote sustainable practices at the individual and collective level, based upon a knowledge and understanding of the interrelationships between natural and social systems on a local, regional, and global scale.

Environmentally literate individuals routinely evaluate the impacts and consequences of his or her actions, gather and synthesize pertinent information, choose among alternatives, advocate action, and take actions that work to sustain or enhance a healthy environment (Roth, 1992). This behavior should be reinforced throughout a student's education so that the behavior becomes second nature.

The Objectives of FMSA's Education Program

It is generally recognized that to achieve environmental literacy, environmental programs must provide students the opportunities to acquire knowledge, skills, and attitudes necessary to make the choices needed to protect the environment (UNESCO, 1977). Hungerfold et al. (1981) have suggested four levels of education needed to achieve environmental literacy. FMSA has developed the following set objectives based on these four levels:

1) to develop a marine ecology foundation in students;
2) to develop a conceptual awareness of coastal and marine environments in students;
3) to develop investigation and evaluation skills in students; and
4) to develop environmental action skills in students.

A Framework for Achieving Marine Environmental Literacy

Developing a Marine Ecology Foundation

The first step towards achieving sustainability is educating a marine ecologically literate populace. All life on Earth is based upon productive ecosystems, and the citizens of the Earth must understand how these systems function. This knowledge should be included among the basic skills of reading, writing, and arithmetic, for these skills are irrelevant if the Earth can no longer sustain healthy ecosystems.

Before we can expect students to behave in a sustainable manner and promote sustainable practices, they must understand the rationale behind their behavior. "Knowledge is the necessary precondition of thoughtful behavior and action (Disinger and Roth, 1992)." In addition, experiencing the natural world is essential to developing an awareness of the natural environment and understanding how ecosystem's function. Cherif (1988) asks "How can a country that does not have a sufficient ecological understanding of its flora and fauna find solutions to environmental problems and sustainably use its natural resources?"

Experiencing the Marine Environment

The first step towards understanding the natural world lies in developing a sense of belonging. Through direct experience with the marine environment an individual begins to value the ocean, and the importance of his or her own connection with the marine environment. By taking students outside of the classroom to the beach, tidepools, or estuary they begin to develop an appreciation for the marine environment. This first step is important in developing an environmental ethic that reflects a commitment and sense of responsibility for sustaining the health of marine and estuarine environments.

Direct experience is believed to result in more attitude-behavior consistency than indirect experience (Jordan, Hungerfold, and Tomera, 1986). For instance by taking students to a productive wetland and then to a wetland that has been developed, they can begin to appreciate the loss of habitat. Living systems can be powerful and awe inspiring. Humans are continually trying to control living systems through dredging, building sea walls and other barriers, and dyking wetlands instead of learning how to live within these natural systems. Understanding how natural systems function is a necessary precursor to living sustainably within those systems. Through guided experiences students can begin to explore how marine systems function.

While learning about marine systems, it is important for students to have fun—to enjoy being outside. Students must associate the marine environment with pleasure, and not surprisingly students will become more interested in learning. Also, associating pleasant memories with the natural environment can be a motivating factor in helping to protect natural areas.

Learning about Marine and Human Systems

It is important for students to understand the basic concepts of ecology. Students must develop the ability to think about the Earth as a system in which everything is ultimately connected to everything else. They need to be taught that the Earth is a relatively closed system with a limited supply of resources that must be shared by all living things and that people depend on other living things for survival. Students must also understand that people are an integral part of the system and are subject to the rules of nature. Our social systems directly and indirectly impact the Earth and consequently marine systems, and human success lies more in cooperating with the forces of nature, rather than trying to dominate them (Chiras, 1993).

Developing a Conceptual Awareness of Coastal and Marine Environments

Understanding Human Social Systems

Many of us do not know where our clean water supply comes from, or where our wastes go when they disappear down the drain. If we do not trust the quality of our tap water, then we may buy bottled water instead of addressing the root of our concerns. By developing an ecological and conceptual awareness we reconnect ourselves to the planet and begin to understand where things come from, where they go, and how much energy and material is used along the way (McClaren, 1989).

Understanding that Human Systems and Marine Systems Interact

As students begin to understand the foundations of ecology, educators can begin developing a conceptual awareness of how people interact with marine systems. This can be demonstrated by outlining any basic human action, for example drinking a glass of water. The educator can begin by asking where the water came from and ask students to trace the drinking water back to its source. The class can discuss the impacts of removing the water from its place of origin. Next the educator can trace the path of the water after it has been consumed to the local waste water treatment plant and then to the ocean. The class can then discuss the water cycle and consequences of removing and disposing of the water. The educator can illustrate how humans fit into the cycle by demonstrating that what we take out of the Earth must again return to the Earth teaching the basic concepts of sustainability. No species, including humans, should take more resources than can be replaced.

Incorporating Investigation and Evaluation into Education

After students have developed a conceptual awareness of marine ecology and the role of humans and human social systems, students will learn to investigate how individual and group actions effect marine ecosystems. By learning how to investigate an issue students will learn the skills necessary to make informed decisions.

Learning How to Investigate

As part of the investigation process students must be able to apply their knowledge and skills to the "real" world. The campus provides an excellent opportunity for students to investigate "real" world environmental issues that effect them. Students can examine water use, food policies, energy use, materials purchased, and waste flows as they relate to marine environments and to their lives.

These types of projects are excellent group projects and provide an opportunity for students to identify which issue their group will investigate. The projects will include:

- gathering information from primary and secondary sources;
- comparing and analyzing the information and information sources;
- evaluating the information;
- summarizing the environmental issues; and
- proposing solutions.

There is evidence that these kinds of experiences influence an individual's participation in environmentally responsible action (Hungerfold et al., 1981).

Learning How to Analyze Information

By using primary investigation methods such as surveys, opinionnaries, and questionnaires students learn how to gather information and record data. Students will also learn how to interpret the data, produce and interpret graphs, draw conclusions and inferences, and formulate recommendations based on their research.

Understanding the Role of Beliefs and Values

During the investigation process, class discussions will examine the impact that beliefs and values have on the marine environment; thereby exposing students to both the analytical and emotional side of environmental issues. Students will also begin to examine how their own beliefs and values effect their behavior and examine how their habits contribute to problems in the marine environment and solutions. This will give students insight into the complexity of marine issues.

Teaching Responsible Behavior

Once students understand how marine ecosystems function, conceptualize how humans fit within marine ecosystems, and learn how to investigate and evaluate problems, they are ready to learn how to be leaders in solving marine problems. Instead of externalizing the problem, educators can demonstrate how the school as well as the students and their families can contribute to solving problems in marine environments. Most global problems originate close to home and must be tackled there to provide global solutions (Lakshmi, 1991).

Educators can use the campus project to work toward larger principles, making the connection between the students' behavior and the marine environment. People need to see the connection between their own lives and marine issues such as water pollution, global warming, and solid waste (Van Matre, 1990). Citizens must take responsibility for their actions in order to begin to solve environmental problems.

Applying Problem Solving Skills

Often educators transfer knowledge and teach skills, but do not provide the opportunity for students to apply those skills to real life problems (Gigilotti, 1990). By taking a real situation and having the students investigate the problem, students will learn how to apply their investigation and evaluation skills to real life situations and begin to solve real problems.

Taking Action

Students should be given the opportunity to implement their recommendations. Based on their research, students can begin to develop recommendations and take action. By taking action, students will learn how difficult it is to influence change as well as learn the gratification of success. Students will learn the processes of working out problems. If the first recommendation does not work, then the students can analyze the reasons why, and modify the recommendation. Solving problems is a process of trial and error.

Individual and societal environmental behavior does not change from increased knowledge and skills alone (Iozzi, 1989). Research has shown that students that learned about environmental issues and action strategies exhibit a greater number of environmentally appropriate behaviors (Jorden et al., 1986). Becoming an environmentally responsible citizen requires knowledge about how the environment functions, how humans effect the environment, skills to investigate and evaluate problems, and experience applying the above knowledge and skills.

<u>Evaluation Tool</u>

FMSA recognizes the need to evaluate the process by which individuals change. Based on the preceding objectives an evaluation tool will be developed to measure achievements in marine environmentally literacy. An evaluation tool will allow FMSA to compare different types of education programs and their effectiveness in achieving our objectives. The tool will concentrate on evaluating the skills, thought processes, content, and behavior change. Based on the results of the evaluation, FMSA will modify curricula and program content to better meet our objectives.

Conclusion

Developing knowledge about the marine environment as well as critical thinking, problem-solving skills obtained through meaningful minds-on, outdoor experiences are a necessary precursor to responsible behavior and action. However, often marine education programs neglect the critical thinking, problem-solving component in outdoor experiences. FMSA's program is a pioneer in this sense.

FMSA's education programs will guide individuals through the stages of environmental literacy and evaluate the process. Understanding the proper mix of experiences, skill development, and concept building is challenging. The evaluation tool will help FMSA understand program strengths and deficiencies in meeting our goal and objectives. Information emerging from FMSA's research will help FMSA and others design and develop programs that produce environmentally literate citizens that are committed to taking action.

Appendix

Cherif, Abour, H. (1988). "History and Ecological Education." *American Biology Teacher*, 50(1) 33-38.

Chiras, Daniel D. (1993). "Eco-Logic: Teaching the Biological Principles of Sustainability." *American Biology Teacher*, 55(2) 71-76.

Disinger, John F., and Charles Roth. (1992) "Environmental Education Research News." *Environmentalist*, 12(3) 165-168.

Gigliotti, L. M. (1990). Environmental Education: What went wrong? What can be done? *Journal of Environmental Education*. 22(1).

Goodwin, Harold L. and James G. Schaadt. (1978). *The need for marine and aquatic education*. Sea Grant Publication, University of Deleware.

Hungerford, H., Peyton, R., & Wilke, R. (1981).Goals for curriculum development in environmental education. *Journal of Environmental Education*, 11(3), 42-47.

Iozzi, Luis A. (1989). What research says to the educator: Part one, environment education and the affective domain. *Journal of Environmental Education*, 20(3), 3-9.

Jordan, J. R., H. Hungerfold, and A. Tomera. (1986). Effects of two residential workshops on high school students. *Journal of Environmental Education*, 18(2): 1-8.

Kaza, Stephanie. (1985). Point Reyes-Farallon Islands National Marine Sanctuary Education and Interpretation Plan. Washington, D.C.: National Marine Sanctuary Program, National Oceanic and Atmospheric Administration.

Lakshmi, Ahana. (1991). Approaches to Environmental Literacy. *Indian Jouranl of Environmental Education*, 11(7).

McClaren, M. (1989). Environmental literacy: A critical element of liberal education for the 21st century. Manitoba: Education Manitoba.

Newhouse, Nancy. (1990). Implications of attitude and behavior research for environmental conservation. *Journal of Environmental Education* , 22(1) 26-32.

Norse, Elliot, ed.(1993). *Global Marine Biological Diversity*. Washington, D.C.: Island Press.

Roth, Charles E. (1992). "Environmental Literacy: Its Roots, Evolution and Directions in the 1990s." *ERIC CSMEE Digest. ERIC* Document Reproduction Service No. EJ 300318.

UNESCO. (1977). Belgarade Charter (Paris).

Van Matre, Steven. (1990). Earth Education: a new beginning. Warrenville, IL: The Institute for Earth Education.

CHAPTER 96

Annual variation in effects of UV radiation on plankton

Lynn J. Rothschild[1] and Karen F. Grimmer[2]

Abstract

Environmental effects on coastal communities, especially when they can be monitored remotely, are a critical component of NASA's Mission to Planet Earth. Some environmental variables that have a profound biological effect are changing; for example, surficial fluxes of UVB radiation are rising. What is not known for aquatic systems is the natural annual variability in the effect of UV radiation on plankton. An exciting opportunity to obtain these data has recently become available. The Marine Sciences Institute (MSI) of Redwood City, a science educational organization that runs educational cruises for teachers and students, has begun a collaboration with NASA/Ames to monitor annual variability in UV radiation and its effect on the biota of the San Francisco Bay Estuary. The students and teachers at MSI obtain weekly samples of surface water, make basic physical and chemical measurements of key environmental parameters, and survey fish. High school and college students based at NASA/Ames collect the samples from MSI and take them to NASA for analysis of UV-absorbing pigments and biomass. We have hypothesized that phytoplankton will produce UV-absorbing pigments in direct response to surficial fluxes of UV radiation, and the early results appear to support the hypothesis. Additional components will be added to this project in the coming year, including analyses of UV-induced DNA damage in the plankton and the acquisition of remotely-sensed data. The educational value of this work will be enhanced through the formation of a Web Site with some data along with information from NASA researchers to assist participating students and teachers in analyzing and interpreting their data, and anyone worldwide in using this study as a classroom exercise. The total data set will be published in the scientific literature. The result will be a critical environmental study on annual variation within this important aquatic community that could not be done without the student contribution.

Background

San Francisco Bay Estuary is one of the most economically and ecologically important estuaries in the United States. It consists of three sections: the Delta, North Bay and South Bay (Fig. 1). The three areas differ physically, chemically, biologically

[1]Research Scientist, MS 239-12, NASA/Ames Research Center, Moffett Field, CA 94035
[2]Director of Education, Marine Science Institute, 500 Discover Parkway, Redwood City, CA 94063

and in their importance to coastal processes. At Golden Gate, water from the North and South Bay feeds into the Pacific Ocean and oceanic water enters the Bay. The SF Bay is surrounded by a large urban population. During the last 150 years, 95% of the tidal wetlands have been levied and filled, the flow of freshwater has been greatly reduced by water diversion, and the harbor and the channel have been dredged altering water flow and salinity. Contaminants enter the water as sewage, and urban and agricultural runoff. Exotic species have been introduced into the Bay (Carlton, 1995), including 212 non-native animals and plants with an arrival rate of about one new organism every 12 weeks (Cohen and Carleton, 1995). Since most other estuaries are also becoming urbanized[3], SF Bay Estuary is a key area to study to understand the future of estuaries in this country. There is heavy recreational use of the Bay, including sport fishing (e.g., sturgeon, bass, sharks), windsurfing, jet skiing and sailing. For these reasons, there is great interest in understanding what goes on in the SF Bay by both government and industry. Questions include, what is the effect of discharges (i.e., sewage in the South Bay and agricultural run-off from the Delta) on the Bay biota? What is the effect of varying salinity on phytoplankton? How can the remaining 125 km^2 of wetlands be preserved and new wetlands created?

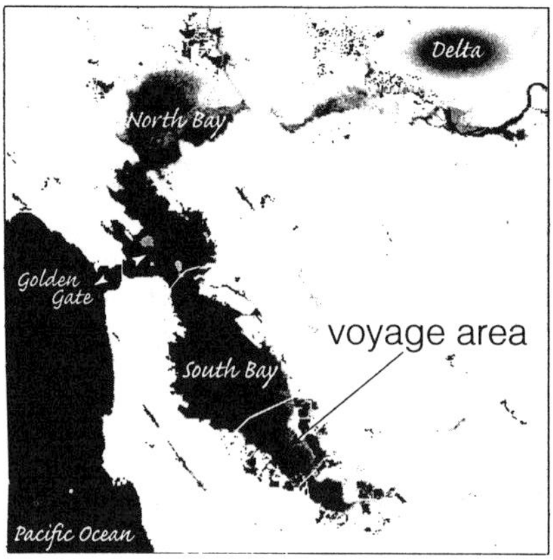

Figure 1. The San Francisco Bay Estuary system. Currently, MSI conducts daily voyages in the South Bay.

The Marine Science Institute (MSI), a nonprofit, marine science educational organization, was created in 1970 to educate and acquaint the community with the ecology of the San Francisco Bay Estuary. During the past twenty-seven years, MSI has taken over 300,000 teachers and students aboard the research vessel, Inland Seas (Fig. 2). The shipboard programs serve 13,000 grade four through university students per year, 80% of whom are fifth and sixth graders. Students in the Discovery Voyage field activities set and pull a trawl net to catch fish, deploy a mud grab to sample the bottom for benthic invertebrates, collect plankton at the water's surface with a plankton tow net,

[3]Today few, if any, estuaries are considered "pristine". By the year 2010 it is estimated that 75% of the US population will live within 75 km of a coast (Williams et al., 1991) thereby exacerbating the problem.

and analyze water samples for physical properties such as salinity, temperature, and oxygen. Another 20,000 students per year, ranging in age from pre-school to high school, participate in MSI's land-based programs. The Shoreside Program and Ocean Lab take place on MSI's dock, on the beach and in the waterfront classroom. The programs introduce younger students to the use of real oceanographic equipment and experiments. Through an outreach program, the Inland Voyage, instructors bring live fish and invertebrates to the actual school site in a trailer-mounted aquarium that can accommodate larger specimens such as sharks.

Figure 2. MSI's research vessel, the Inland Seas.

The strength and success of MSI's programs lie in reinforcing the natural inquisitiveness of children and in helping to build their critical thinking and problem solving skills. These skills will extend beyond science to other disciplines and, of course, to the students' lives.

With a mission to provide affordable K-12 environmental science programs to the public schools, MSI raises funds every year to help offset the cost of the programs. While some schools can afford to pay full price, the majority of public schools can not. In most cases, MSI works to match funders with schools and teachers based on financial need.

In the autumn of 1995, MSI contacted the U.S. Geological Survey (USGS) in Menlo Park and NASA/Ames Research Center (NASA/ARC) to see if there were ways that they could integrate their activities with scientific studies of the Bay. Since 1968, USGS has conducted an intensive applied research effort on the Bay, including monitoring salinity, phytoplankton and clam abundance, and toxins in the Bay (e.g., Cloern, 1996). One of us (Rothschild), in consultation with MSI and USGS, has proposed a research activity that is oriented towards basic scientific questions that are of interest to NASA (i.e., the effect of global change on ecosystem functioning) and exploits the biochemical, oceanographic and educational expertise of ARC, and ultimately remote sensing as well.

The scientific objective of our collaborative project is to monitor annual variation in UV radiation and other physical and chemical parameters, to determine annual variation in the health and composition of the planktonic community, and to analyze the results to elucidate causal links. Of particular interest is to determine if a correlation exists between annual variations in UV flux and DNA damage in the plankton. Aquatic communities, particularly estuaries and coastal waters, are critical to the economic, recreational and quality of life of our country. Plankton are the base of the food web and are known to be particularly sensitive to UVB radiation (rev. in Häder, 1993; Vincent and Roy, 1993). Thus, changes in UV radiation may have profound effects on the aquatic community as a whole. UVB radiation can effect aquatic organisms up to depths of meters depending on the water chemistry and turbidity (Gieskes and Kraay, 1990; Larkum & Wood, 1993). While UV penetration in the South Bay is not known, the South Bay is shallow (1-2 meters) except for a central channel (~10 meters). With UV fluxes currently rising seasonally as the result of depletion of the stratospheric ozone layer, there is concern about the effect of UV radiation on biological systems.

It is very difficult logistically and financially to understand current annual variability in the effect of UV radiation on plankton because of the labor and ship time necessary. Also, there are few workers with the molecular and biochemical expertise who have the logistic interest or support for such a study. Thus, scientifically the results of this work will be an invaluable source of data about current annual variability in UV effects and thus can be used for predictive work (including modeling and remote sensing) in the face of a continued depletion of stratospheric ozone. MSI conducts over 310 field-study Voyages a year (two trips per day, five and six days per week) making the ship an ideal platform to conduct such experiments (see Fig. 1). Specifically, MSI is responsible for taking readings of physical and chemical parameters. Measurements have begun of chlorophyll, temperature, photosynthetically active radiation, UV radiation, salinity, pH and turbidity). MSI collects weekly plankton samples and monitors the presence of major fish species. At NASA/Ames, further analyses of the plankton sample are done to assay for UV-absorbing pigments and chlorophylls, and plankton species composition. The possible use of remotely sensed data on phytoplankton distribution and abundance is being explored at ARC. We hope that as this project develops, the results will compiled and interpreted at ARC, the physical measurements will be returned to the students in the form of an on-going Web Page and the physical measurements and biological experimental results will be made available to the scientific community through peer-reviewed publications and ultimately through the Web Page.

There is a collaborative nature to the educational aspects of this project. MSI, as a teaching institute, involves students in all phases of their participation. Additionally, NASA supports many educational opportunities for students and teachers to learn of the Agency's mission and projects. Participants typically are involved at a superficial level (i.e., they receive information in a publication, attend a workshop or seminar, or in some other way learn what NASA is doing). Some programs, such as the Space Biology Intern Program pairs outstanding Bay Area High School students with a scientific mentor at NASA/Ames. These students spend approximately 6 hours per week conducting research in the host lab. Two students have already begun work on this project through the Space Biology Intern Program. Sarah Kreps, a junior at Harvard University majoring in Envionrmental Studies, spent the summer of 1996 acting as a scientific and educational liaison between MSI and NASA/Ames. Thus, through this work we are involving students and teachers at the very heart of a crucial scientific investigation - one that bears both immediate significance for the participants and long term importance to our understanding of this important aquatic environment.

The research described here supports NASA's recently released Strategic Plan for Education in terms of goals ("develop out-of-school programs that support formal school instruction") and partnerships ("expand the impact of NASA's education program by strengthening partnerships with external constituencies.") and even in principle ("... programs will emphasize student-centered pedagogies... problem-solving and cooperative endeavors) (from NASA's Strategic Plan for Education - A Strategy for Change). The use of the World Wide Web to involve students and teachers as they return to their schools aligns well with NASA's emphasis on next-generation educational technologies. Ames is presently the agency lead in the use of electronic networking for K-12 education, this project provides another excellent demonstration of the effective use of networking for education.

In summary, the main scientific objective of this research is to measure annual variability in UV and biological responses to UV damage on the plankton community in the SF Bay. These data will lead to an understanding of the role of UV-induced damage on aquatic communities, particularly wetland and coastal communities. The main educational objective is to involve a large number of Bay Area students and teachers in a novel environmental project that would not be practical without their support. Eventually we hope that the educational benefits of this study will extend beyond the participating students by making educational information and data available on a web site, and possibly creating exhibits for local science museums.

Methods

During the summer of 1996 and spring of 1997, we collected samples of water from the South Bay. The South Bay receives millions of gallons of treated sewage daily. Personal communications with USGS, MSI, South Bayside System Authority and the Regional Water Quality Control Board suggest that the discharge does not prevent typical annual phytoplankton blooms (Cloern, 1996). In addition, the quality and quantity of the discharge should not affect DNA damage. MSI will be purchasing a new ship that will be used for educational programs in the North Bay, including Rio Vista, Richmond, and Golden Gate. These areas of the North Bay are deeper and are physically and chemically closer to coastal oceanic waters; in fact, there is tidal forcing across the Golden Gate (indicated by the arrow in Fig. 1). Thus, in future years there will be the potential to extend our studies to the North Bay as a way to start to understand oceanic processes.

Surface samples are being be taken by MSI at channel marker 8 (37° 32.5' N, 122° 11.6' W) and a location north of marker 8 and closer to the shoreline in SF Bay. The positional data are obtained using Global Positioning System units donated by Trimble Navigation. The samples (7 x 1 liter) are collected with a Van Doren sample bottle in the morning, between 900 and 1200 h. The samples are collected in this manner rather than with plankton tows since the majority of the phytoplankton are smaller (2-8 μm) than the mesh size of a very fine net. The samples are placed in sterile polypropylene bottles. Environmental measurements are made at that time. These measurements currently include temperature of the sample as determined using a temperature probe, and salinity as determined using a YSI meter (Yellow Spring, Incorporated). Water turbidity is measured by Secchi depth. Fish trawls are conducted with a 19 foot otter trawl, 15 minutes per trawl, and six times per day.

The samples are returned to shore in a dark ice chest and kept cool and in the dark at MSI. Within a few hours of collection, samples are transported to NASA/ARC. Samples (0.5 l) are collected on a Whatman GF/F glass microfiber filter (retention size 0.7 μm) by vacuum filtration. Triplicate samples are extracted in the dark on ice in 2 mL acetone overnight. The acetone extract is analyzed spectrophotometrically 1) to assay for UV-protective pigments by measuring absorbance in the UV range, and 2) to obtain an estimate of biomass by assessing chlorophyll a concentrations. In the future, a complete chlorophyll profile (chl. a, b, and c - Jeffrey and Humphrey, 1975) will be done as an assessment of community composition.

Results

Regular sampling has occurred since January 1997. During that time (January to March) , there was neither a strong secular change in surface or bottom temperature nor salinity (Fig. 3). There was an increase in the average number of fish caught per

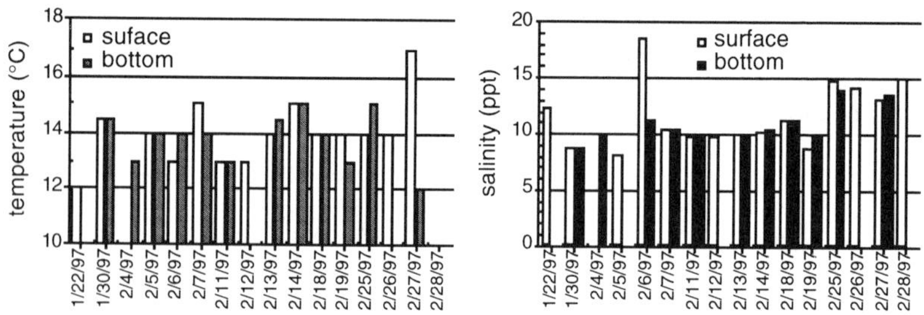

Figure 3. Temperature and salinity readings in the South Bay, January to March 1997.

trawl from 23 January until 13 February, a decrease until the 21 February followed by another increase (Fig. 4). The error bars suggest that the only change that is significant is the increase from late February through the end of the data series in late March. Pigmentation data are presented in Fig. 5, and are expressed as absorbance of an acetone extract of a 500 mL grab sample. Data are given for absorbance at 665 nm. The absorption maxima for chlorophyll a in 100% acetone are 430 nm and approximately 662 nm (Rowan, 1989). Chlorophyll a is used as a rough estimate of phytoplankton biomass, and thus the 665 nm readings are an estimate of phytoplankton biomass. The 665 nm absorbace data suggest an increase in phytoplankton biomass from early to late February. Similarly, the absorption spectrum for the samples in the UV region of the spectrum as measured from 275 to 400 nm, shows a significant increase in UV absorption from early to late February. These preliminary data suggest that there is an increase in accumulation of UVA absorbing pigments in the phytoplankton during the month of February. We would like to stress that, while the pigmentation results are suggestive, these data are very preliminary.

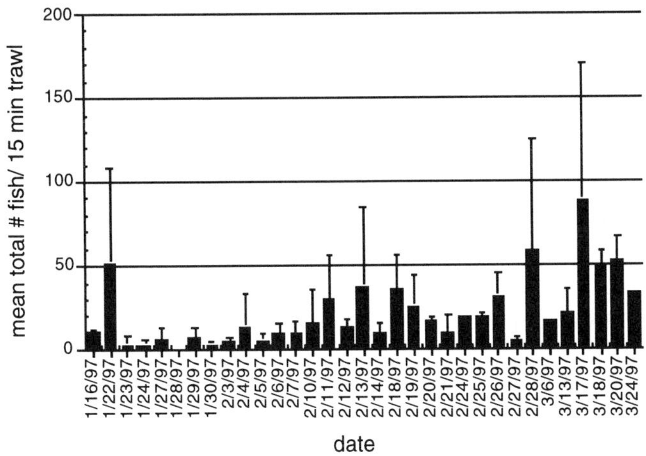

Figure 4. Average fish caught in 15 minute trawl in the South Bay, January to March 1997.

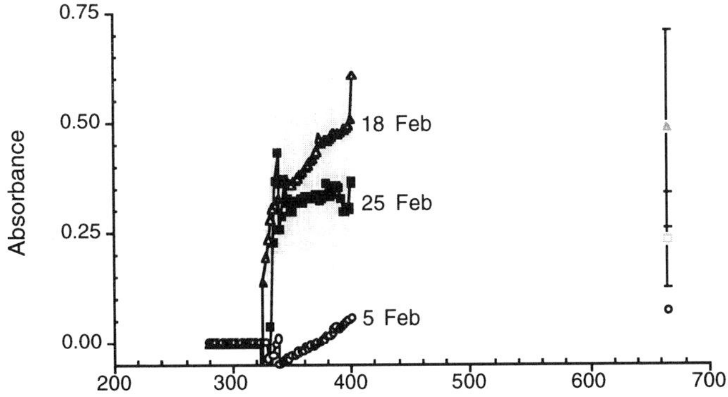

Figure 5. Pigment data from grab samples, February 1997. Values are the mean of 2-4 samples.

Future Plans

Scientifically, our highest priority is to continue the data set through several years. Additional measurements will be made in conjunction with the water sampling when we have obtained the equipment. Most importantly, we will measure radiation, both photosynthetically-active radiation and UV radiation. Measurements of the pH of the water will be made. In situ chlorophyll a estimates (a standard estimate of total primary productivity) will be made using a flow-through fluorometer (donated by Turner Designs, Sunnyvale) that was used last year on the ship. These in situ measurements will supplement the chlorophyll measurements taken on the samples, and provide an estimate of the patchiness of the phytoplankton in the voyage area. General weather conditions are be recorded. Total hours of daylight are calculated from sunrise and sunset data from the National Weather Service. As the project develops, pCO_2, pO_2, and concentrations of inorganic nitrogen and phosphate will be determined on ship using probes.

Currently, the pigment results are expressed per volume water, and per unit of chlorophyll a. By the summer, protein content will be analyzed spectrophotometrically using bicinchoninic acid reagent (BCA assay; Pierce, Rockford, IL).

Detailed plankton analyses are being done by USGS and their contractors so will not be replicated here. The students at NASA/Ames do periodic microscopy on the samples on a Zeiss Axioscope compound microscope, using phase contrast optics.

Within the year, we hope to extend the research to assay for UV-induced DNA damage. DNA will be purified from triplicate filtered samples using either standard techniques (Wilson, 1989) or a commercially-available kit. We have successfully extracted high molecular weight genomic DNA (as assessed by agarose gel electrophoresis) from Bay plankton tows using the Invitrogen Easy DNA kit (Invitrogen, Palo Alto, CA). A sample of the DNA will then be sent to a commercial lab (Dermigen, Inc., TX) for detection of photoproducts per unit DNA, a standard way to assess UV-induced DNA damage (Karenz, 1994).

The data will be graphed as a function of Julian day of the year to monitor annual variability. Second, the environmental data will be compared to the biological data to look for correlations. For example, the chlorophyll content is an estimate of community primary productivity as well as taxonomic distribution of photosynthesizers. We will determine if in SF Bay this is best correlated with radiation, temperature, or nutrient levels. However, the major focus will be to look for a correlation between UV radiation and 1) the presence and amount of UV-absorbing pigments and 2) DNA damage as assessed by photoproduct content. We predict that these would be greatest at peak UV fluxes, but it is possible that at that time of year protective mechanisms such as changes in taxonomic distribution of protective pigments would uncouple these processes. Third, our data will be compared to that obtained by USGS and additional data from MSI (e.g., fish abundance) to see if other correlations are possible.

Educational materials will be developed by NASA/Ames scientific and educational personnel in collaboration with the staff of MSI. The materials will then be evaluated and revised according to recommendations made by the educational consulting board of MSI which consists of curriculum consultants from Bay Area school districts. The materials will include written matter and possibly lectures to acquaint teachers and students with the purpose of the research in advance of the voyage, materials to use in data collection on the ship, and follow-up activities in the classroom. Some of these materials will, of necessity, be in the form of handouts. Most material will be available on a Web page. This will allow the materials to be updated as data are collected and make it widely available. It is anticipated that the Web page will be an important educational tool for not only the participating students and teachers, but anyone who wishes to use these data in classroom exercises.

In summary, this project has several unique aspects to it. It is a scientific project that would not be practical without the educational aspect, and it is an educational project that involves students and teachers in important primary research. This project is developing into an exciting collaborative research effort involving two government agencies (NASA and USGS), a private educational institute (MSI) and dozens of Bay Area schools. Outside collaborators in academia as far away as Florida and Canada have indicated an interest in participating. In the meantime, it has begun to yield data regarding the response of phytoplankton to the annual variability in surficial UV fluxes in the San Francisco Bay.

References

Cloern, J.E. 1996. Phytoplankton bloom dynamics in coastal ecosystems: A review with some general lessons from sustained investigation of San Francisco Bay, California. Rev. of Geophys. 34: 127-168.

Cohen, A.N. and J. T. Carlton. 1995. Biological Study. Nonindigenous Aquatic Species in a United States Estuary: A Case Study of the Biological Invasions of the San Francisco Bay and Delta. A Report for the United States Fish and Wildlife Service, Washington, D.C., and The National Sea Grant College Program, Connecticut Sea Grant NTIS Report Number PB96-166525 246 pp. + Appendices.

Gieskes, W.C. and G.W. Kraay. 1990. Transmission of ultraviolet light in the Weddell Sea. Report on the first measurements made in Antarctic. Biomass Newsletter 12: 12-14.

Häder, D.-P. 1993. Effects of enhanced solar ultraviolet radiation on aquatic ecosystems. in UV-B Radiation and Ozone Depletion (M. Tevini, ed). Lewis Publishers, Boca Raton. pp. 155-192.

Jeffrey, S. W. and Humphrey, G. F. 1975. New spectrophotometric equations for determining chlorophylls a,b,c1, and c2 in higher plants, algae, and natural phytoplankton. Biochem. Physiol. Pflanzen. 167: 191.

Karenz, D. 1994. Ultraviolet tolerance mechanisms in Antarctic marine organisms. Ultraviolet Radiation in Antarctica: Measurements and Biological Effect. Antarctic Research Series 62: 93-110.

Lloyd, S.A. 1993. Stratospheric ozone depletion. The Lancet 342: 1156-8.

Rowan, K.S. 1989. Photosynthetic Pigments of Algae. Cambridge University Press, Cambridge. 334 pp.

Vincent, W.F. and S. Roy. 1993. Solar ultraviolet-B radiation and aquatic primary production: damage, protection, and recovery. Environ. Rev. 1: 1-12.

Williams, S.J., K. Dodd & K.K. Gohn. 1991. Coasts in Crisis. U.S. Geol. Surv. Circ.,

1075, 32 pp.

Wilson, K. Preparation of genomic DNA from bacteria. in Current Protocols in Molecular Biology (eds Ausubel, F.M. et al.) 2.4.1-2.4.5 (Greene Publishing Associates and Wiley-Interscience, New York, 1989).

CHAPTER 97

Measuring Shoreline Erosion Rates: Strategy, Techniques and Accuracy

[1]*Laura J. Moore and* [2]*Gary B. Griggs*

ABSTRACT

When coastal change in populated areas manifests itself as a landward migration of the shoreline occurring on a human time scale, quantification of erosion rates becomes important. Numerous techniques to measure shoreline change have been developed over the last 25 years (Stafford, 1971; Dolan, et al., 1978; Clow and Leatherman, 1984; McBride *et al.,* 1991; Thieler and Danforth, 1994, Fisher and Overton, Per. Com.). These methods have been used to study beach erosion, barrier island migration and dune/bluff erosion. Although applied to the investigation of similar problems, these techniques vary considerably in approach, accuracy and cost. They range from qualitative approaches such as the measurement of shoreline position from ground or oblique photos to the very precise quantitative approach of "softcopy photogrammetry." Along with an increase in accuracy provided by the more quantitative techniques is an increase in cost and project duration. This, among other factors, has likely prevented the emergence of a standard technique for measuring shoreline change.

While many coastal mapping studies would benefit from application of more advanced and accurate techniques, not all studies of the changing coastal zone require the high resolution these techniques offer. Thus, choosing appropriately from among the available methods is challenging. The many factors to be considered when selecting a shoreline mapping technique are outlined in this paper along with a framework for technique selection. Two applications of this framework are presented using studies of the California coast.

INTRODUCTION

With a large and growing percentage of our nation's population residing along the coast, study of our changing shorelines has become more than a topic of scientific curiosity. All coastal areas are dynamic with changes occurring over many time scales. However, when this change occurs as a landward movement of the shoreline on a human time scale, quantification of erosion rates becomes important for many reasons. No longer are erosion rates used only to study sediment budgets or the role of natural processes in shoreline alteration, they are now also used to determine

----

[1]*Graduate Student Researcher*, Earth Sciences Department
University of California Santa Cruz, 1156 High Street, Santa Cruz, CA 95064.
[2]*Professor* , Earth Sciences Department and *Director* , Institute of Marine Sciences,
University of California Santa Cruz, 1156 High Street, Santa Cruz, CA 95064.

safe construction setbacks, settle property ownership disputes, study the effectiveness of shoreline protection structures and to make land use changes.

The range of purposes for which shoreline erosion is studied, the variety of equipment and funding available to each scientist, engineer or planner, and the varying expertise of professionals studying the coastal zone, has lead to the lack of a standard method for analyzing shoreline change.

Existing shoreline mapping techniques vary from simple measurements made directly from uncorrected aerial photographs to very precise measurements made from computer rectified aerial orthophotographs in digital format. In an ideal world of unlimited time, money and expertise, the standard technique for calculating shoreline change would be the latter of these two. In the real world, project requirements and available resources must be assessed and not all studies of the changing coastal zone require the highest resolution available.

When beginning a coastal mapping project or choosing to upgrade laboratory facilities, researching established coastal mapping techniques in order to choose from among them requires extensive literature review and a significant time investment. To assist researchers, engineers and planners who wish to undertake a coastal mapping project, this paper provides a brief overview of the errors associated with shoreline mapping, an introduction to the different methods available and a short discussion of factors to be considered when selecting a technique.

Numerous manuscripts have been published on the topic of shoreline mapping and erosion rate determination. In each of these manuscripts, the authors have included some combination of a discussion of errors inherent in shoreline mapping, an analysis of source information, an introduction to a new technique and/or a description of existing techniques. Since this paper is intended to server as an overview, the authors refer you to the bibliography and to Moore (In Press) for additional or more specific information.

It is worth mentioning that in addition to selecting an appropriate shoreline mapping method, it is critical to consider how representative average shoreline erosion rates are of long term trends. Due to the episodic nature of erosion, the most representative rates for a retreating coast, will be generated from a data set which encompasses the longest time interval possible. For coastal areas which experience both accretion and erosion over time, an average erosion rate will likely not be useful. In addition, when using past erosion rates to predict future coastal erosion, an implicit assumption that erosion will continue at the same rate is often made. In some locations, changing sea level as well as changing wave and storm conditions may invalidate this assumption.

SOURCES OF ERROR IN SHORELINE MAPPING

Since erosion rates can only be as accurate as the data from which they are derived and the methods by which they are calculated, a thorough understanding of the errors inherent in the process of shoreline erosion rate determination is vital to the successful completion of a project. A brief explanation of the errors involved in shoreline mapping is provided here. For the purpose of the following discussion, errors are divided into two categories: 1) errors introduced by data sources and 2) errors introduced by measurement methods.

Potential Data Source Errors

Historical Maps

If a study requires reconstruction of shoreline position before the aerial photographic record begins, or if a particular location has poor photographic coverage, the use of historical maps in conjunction with aerial photography will be necessary. USGS Quadrangles and NOS Topographic (T) sheets are the most commonly used maps for shoreline mapping. NOS T sheets date back to the 1830s while USGS Quadrangles date back to the formation of the USGS in 1879 (Ellis, 1978). The use of historical maps introduces several potential errors to the coastal mapping process. The severity of these errors depends on the accuracy standards met by each map and on physical changes in each map since publication (Anders and Byrnes, 1991).

Both USGS Quadrangles and NOS T sheets meet or exceed the National Map Accuracy Standards of 1941 (Ellis, 1978, p. 60). Under these standards, the maximum allowable error for 90 percent of points on a 1:24,000 USGS Quadrangle is 12.2 m while the maximum allowable error for 90 percent of points on a 1:10,000 and 1:20,000 NOS T sheet, is 8.5 m and 10.2 m, respectively (Ellis, 1978).

When dealing with maps constructed prior to the Map Accuracy Standards of 1941, and especially when dealing with maps created prior to the first T sheets in the 1830s, reliability is a serious issue. In these cases, map accuracy depends on the standards set forth by the chief surveyor of each individual survey party (Anders and Byrnes, 1991). Shalowitz, (1964), evaluated the accuracy of early surveys from which T sheets were constructed and found that the intention of NOS surveyors was to determine the location of the HWL for delineation on maps. Therefore, despite the lack of general accuracy standards, surveys of the HWL were taken seriously (Shalowitz, 1964). As long as errors are considered, the use of older maps seems reasonable, but only to obtain estimates of general shoreline trends (Anders and Byrnes, 1991). For studies requiring a high degree of accuracy, it is prudent to assess the accuracy of each map used. Crowell et al., (1991) discuss the assessment of map accuracy and present a detailed accuracy analysis of 232 T-sheets from the state of Massachusetts.

Map accuracy is also influenced by displacements of feature position resulting from changes in the horizontal reference datum. Before shorelines from maps with different projections, ellipsoids and datums can be compared, they must be converted to a common projection, ellipsoid and datum.

Finally, map shrinkage, stretch and other defects also add error to shoreline change analysis. For example, typical map paper can expand by greater than 1 percent with a humidity increase of 60 percent (Snyder, 1987). The replacement of paper maps with maps printed on mylar, will significantly reduce these errors.

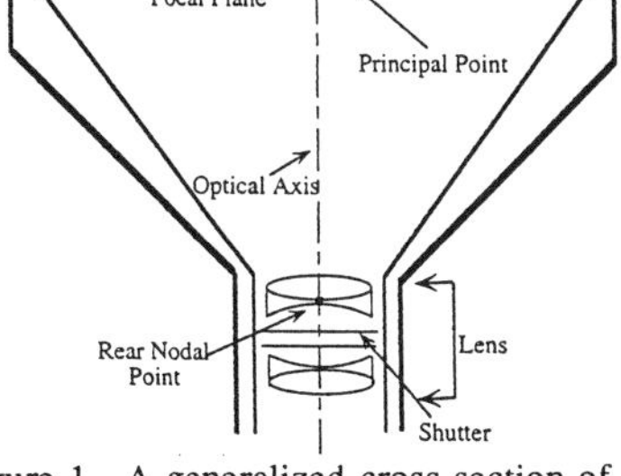

Figure 1. A generalized cross section of an aerial camera illustrating the position of the principal point (after Wolf, 1983).

Aerial Photography

Aerial photographs are the most commonly used data source in shoreline mapping. Because many coastal areas of the United States have extensive aerial photo coverage and because aerial photographs are taken fairly often, they provide a valuable record of shoreline position. Black and white vertical aerial photographs date back to the late 1920's but quality stereo aerial photographs were not available until the 1930's (Anders and Byrnes, 1991).

Aerial photographs cannot be successfully interpreted as maps because various distortions and displacements are introduced throughout the photographic process. These distortions and displacements are perturbations of the geometric relationship between image space and object space (Slama, 1980).

Image space refers to a three dimensional, rectangular Cartesian coordinate system defined inside the camera with the principal point (Figure 1) as the origin. In this coordinate system, the z axis corresponds to the optical axis and the x axis typically corresponds to the flight direction (Thieler and Danforth, 1994). Object space refers to the real-world geographic coordinates of the photographed area outside the camera.

Image space distortions are the result of lens distortion and film deformation. As an image passes through a camera lens, two types of distortion, radial and tangential, can occur. Radial distortion, caused by imperfections in the lens elements, distorts image points along radial lines from the principal point. Tangential distortion, caused by faulty centering of the camera lens, distorts image points at right angles to radial lines from the principal point.

The second cause of image space distortion is film deformation which may occur during an aerial survey due to buckling of film in the camera, during processing of original negatives, or in subsequent processing of each generation of prints. Deformation may also occur due to the instability of photographic media once the image has been printed.

If film deformation and camera lens distortion are present in a photograph, measurements made directly from that photograph will be significantly in error. In some cases, making measurements only at the center of a photograph may reduce the effect of these errors to a tolerable level. For more recent photography, camera calibration test results can be obtained, or in the case of older photography, initial camera parameters can be estimated. Modern photogrammetric techniques utilize this information to remove most image space distortions by either correcting the image itself or by applying a correction to measurements made from an image.

Object space displacements are generated by conditions outside of the camera, such as ground relief and aerial camera tilt, which cause objects on the photo image to be displaced from their true ground positions. Relief displacement, for example, causes objects above ground level to be displaced outward from the center of the photograph and objects below ground level to be displaced inward toward the center of the photograph. When the camera axis is unintentionally tilted from vertical at the time of exposure, a nearly vertical, instead of a truly vertical photograph results. The scale of this image will be smaller on the upward side of the tilt axis and larger on the downward side of the tilt axis (Figure 2).

Most shoreline mapping methods involve techniques to minimize the effect of geometric perturbations on measurements. Since the magnitude of perturbations, and thus the magnitude of displacements and distortions, increase with scale, it is advisable to use the largest scale photography available. In fact, most coastal researchers consider 1:20,000 the upper limit usable for shoreline mapping (Tanner, 1978; Byrnes et al., 1991; Thieler and Danforth, 1994). For a discussion of photogrammetry as it

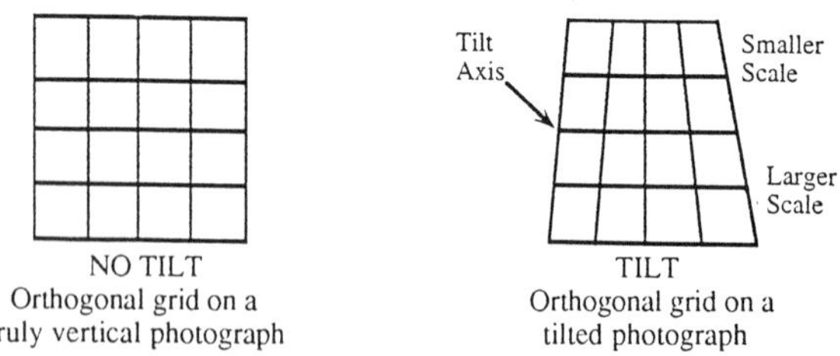

Figure 2. On an untilted photograph, the scale will be uniform across the image. On a tilted aerial photograph the scale is smaller on the upward side of the tilt axis and larger on the downward side (after Leatherman, 1983).

relates to coastal mapping beyond what appears above, see Moore (In Press) or Thieler and Danforth (1994) and for detailed explanations of photogrammetric principles refer to Wolf, 1983.

Ground Control

Ground control points are used to rectify aerial photographs and to transfer older maps from obsolete to currently used datums. Depending on the application used, photorectification (removal of previously discussed distortions and displacements) typically requires four to preferably six or nine control points evenly distributed throughout each photograph (Thieler and Danforth, 1994). Control is typically gathered in one of two ways: 1) by locating stable points on current NOS T sheets or USGS Quadrangles or 2) by locating the latitude and longitude of stable points by field survey, for example, using Global Positioning System (GPS) technology. GPS field surveys undoubtedly provide the most reliable control.

The accuracy of shoreline mapping is not only affected by the quality of ground control but also by the availability of ground control. Availability will depend on both the time period of the study and on the extent to which the study site has been developed. The best control points are road intersections or low relief buildings.

Finally, regardless of how accurately control point locations are known, some error will remain due to the difficulty of precisely locating each control point on a photograph. For example, the geographic coordinates of the center of a road intersection may be very accurately known, but the precision with which the center of an intersection can be identified on a photograph is limited.

Potential Measurement Errors

Interpretation

The line between wet sand and dry sand, which can usually be clearly seen as a significant tonal change on aerial photographs, has traditionally been the most commonly used proxy for shoreline position (Dolan et al., 1980; Smith and Zarillo, 1990; Anders and Byrnes, 1991; and Shoshany and Degani, 1992). The wet/dry line closely approximates the high water line (HWL) which in turn approximates the mean HWL (Dolan, 1980). There are, however, many errors associated with using the wet/dry line to represent the shoreline. These errors result from the natural migration of the HWL, interpretation of the wet/dry line on a photograph and measurement of the interpreted line position. These errors have been investigated by Smith and Zarillo (1990), Dolan et al., (1980) and Crowell et al., (1980). In some areas, where beach morphology permits, the bluff edge, bluff toe or a vegetation line may be a better choice for shoreline analysis (Thieler and Danforth, 1994). When a proxy other than the HWL cannot be used, errors can be minimized by using photos taken during the same season, correcting for tidal differences and using photos of high quality on which the HWL can be clearly identified.

Annotation

Finally, errors arise due to the difficulty of measuring a single line on a photograph. The magnitude of these errors will vary according to the shoreline mapping technique used. When digitizing the wet/dry line instead of annotating with a drafting pen, an investigator must account for digitizer and digitizer operator error. Operator error may be significant and will vary according to working conditions, steadiness of hand and the operator's level of experience.

DESCRIPTION OF MAPPING TECHNIQUES

Numerous shoreline mapping techniques have been developed throughout the last 25 years. The progression of techniques from manual, to partially automated, to fully automated is consistent with decreases in the cost of personal computers and workstations as well as improvements in data processing and storage capabilities. With such quickly changing technology, it is not surprising that a standard technique for shoreline analysis has not emerged. The length of this paper restricts the discussion below to a brief description of each method. For more complete descriptions including capabilities and requirements, see Moore (In Press) or the cited material.

Point Measurements

Point measurement, a widely used method to measure shoreline change, can be performed fairly inexpensively, quickly and simply. This technique, developed by Stafford (1971) to study erosion on the outer banks of North Carolina, involves computing the scale of each individual photograph and then measuring the distance between a fixed reference object and the shoreline. This procedure is repeated for as many sets of photographs as desired. An erosion rate for each point is then calculated by dividing the difference between distances to the shoreline from a reference point by the time interval between photographs.

Orthogonal Grid Mapping System (OGMS)

Drawing upon Stafford's work, Dolan et al. (1978) developed the OGMS. The first step in this procedure is the generation of base maps of the study site by photo enlargement of 7.5 minute USGS maps (scale 1:24,000). The long axis of each base map is oriented parallel to the shoreline and the oceanward map edge serves as the baseline from which all measurements are made. Historical aerial photographs are then enlarged to the exact scale of the base map and superimposed onto the base map using a projecting light table. Once the high water lines from several sets of aerial photography have been superimposed and traced onto an overlay, shoreline change is determined using a rectilinear grid to take measurements at 100 meter intervals along the shoreline. These measurements are then entered into a computer and shoreline maps are generated.

Stereo Zoom Transfer Scope

For a period of time, the stereo zoom transfer scope (ZTS) was considered state-of-the-art in coastal mapping (Fisher, 1979). This technique or a variation of it, is still used in coastal mapping today (Smith and Zarillo, 1990; McBride et al., 1991). The ZTS can be used to bring photography of different scales to the same scale by superimposition onto a base map and rectification to ground control points. Once a photograph has been rectified and adjusted in scale, shoreline features can be traced. By tracing shoreline features, such as the HWL, from several sets of aerial photography, shoreline changes can be recognized and quantified.

Metric Mapping

Metric Mapping, a semi-automated technique introduced by Leatherman (1983) and developed by Clow and Leatherman (1984), was the first shoreline mapping technique developed to utilize a computer to apply an analytical treatment of photogrammetry premised on mathematical models and numerical solutions. Clow and Leatherman (1984) used NOS T-sheets as a source of both primary control points (geographic coordinates of labeled grid marks on maps) and secondary control points (features on photos which can be identified on a map and assigned geographic coordinates using primary control points). Once control has been selected, the shoreline and control points are manually annotated on the aerial photographs. Primary and secondary control points on the NOS T-sheets are then digitized and primary control coordinates are used to obtain the state plane coordinates of the secondary control points. The annotated shoreline and secondary control points are then digitized from the air photos and run through a Fortran IV space resection program. This program, developed by Clow and Leatherman (1984), removes radial distortion and tilt displacement from each individual photograph and adjusts all of the photos to the same scale. A MESH program adjusts the junctions between adjacent photographs and a shoreline map is plotted as the final product.

Zoom Transfer Scope and GIS

This technique, developed by McBride et al. (1991) to map barrier island changes in Louisiana, combines ZTS, computer-aided design and drafting (CADD), computer cartography and geographic information system (GIS) technology to improve accuracy. McBride et al. (1991) used both aerial photographs and NOS T-sheets as sources of shoreline information (McBride (1989) used the same technique solely for cartographic data).

The first step in this procedure is to register aerial photographs to USGS 7.5 minute quadrangles using a ZTS and to annotate the shoreline. Following photoregistration and annotation, aerial photo and NOS T-sheet shoreline data are digitized in original projection, ellipsoid and datum using computer mapping hardware and software. Data are then converted to a common projection, coordinate system, horizontal datum and ellipsoid. After conversion, shore-normal transects are established and average erosion rates are determined by dividing measurements by elapsed time.

GIS Strategy

This technique, presented by Hiland et al. (1993) and Byrnes and Hiland (1994), is similar to the one described above. Both techniques use CADD, computer cartography and GIS software packages, begin with compiling and evaluating the data sources, and use similar procedures for map digitization and transformation. However, in addition to using the ZTS to rectify photographs, Hiland et al. (1993), present a second method which involves scanning a photograph, rectifying it using computer software and control points from a previously digitized map, and then digitizing the shoreline as it appears on the computer screen.

Digital Shoreline Mapping System (DSMS) and Digital Shoreline Analysis System (DSAS)

Thieler and Danforth (1994), designed the DSMS/DSAS to provide a highly accurate coastal mapping solution which can be carried out by one person in a small laboratory with a digitizing tablet, workstation, DSMS/DSAS public domain software written by Thieler and Danforth (1994), GIS software and a plotting device. The first step in this process is to establish a network of fully known control points and secondary or supplemental control points. Secondly, the control points, shoreline and fiducial marks (marks in the corner or along the edges of photos) are digitized using the Digital Shoreline Mapping System. Camera calibration data are then used to correct for errors in the image space coordinate system. The digitized photo information, tied together by the ground control network, is then put through a simultaneous adjustment program. This program performs an absolute orientation which solves for camera position at the time of each exposure.

Once the photographs and maps have been digitized and oriented with respect to the ground, the DSMS uses the camera position and image space coordinates of the digitized shoreline to compute the geographic position of the shoreline. The digital shoreline analysis program can then be used to determine rates of shoreline change. Final shoreline maps can be generated by importing shoreline position data from the DSAS to a GIS system for output on a plotting device.

Softcopy Photogrammetry

Softcopy photogrammetry is the newest method to be used in the measurement of shoreline change. It utilizes photogrammetric techniques of the traditional stereoplotter (a manual instrument typically used to generate topographic maps), but is automated and can be performed by non-photogrammetrists who have been trained to use a softcopy system. Softcopy technology has been applied to coastal mapping by Fisher and Overton (pers. com, 1994 and 1996), Stumpf (pers. com, 1996), and is currently also being applied by the coastal group at the University of California Santa Cruz.

This technique is the most automated, accurate, and thorough but easily the most expensive of the techniques currently applied to the study of shoreline change. The many softcopy systems on the market, varying in user friendliness and capabilities, range in price from $25 K to $100+ K (Saleh and Scarpace, 1994). A lab equipped to use this technique will contain the following: softcopy photogrammetric software and the corresponding computer platform, a GIS system or computer program to measure distance between shorelines, a scanner, significant storage devices and a large-format printer.

The first step in the mapping process is to transfer the photographic image to digital format by obtaining an accurate, high-resolution scan of the image. Once scans have been obtained, the group of images is imported to the softcopy system and triangulation is performed. During triangulation, camera calibration information, fiducial point locations, ground control points, and tie points, are used to solve for camera position at the time of each exposure. Once triangulation is complete, digital stereo pairs are created for each stereo pair in the block and a digital elevation model (DEM) is created, mosaicked and then combined with the triangulated images to generate an orthophoto mosaic. From this orthophoto mosaic, the shoreline is digitized on-screen to create a vector file of shoreline position. Once these steps have been repeated for one or more sets of photographs, the vector shoreline files are imported to a GIS or a stand-alone program which takes measurements between two shorelines at specified intervals and from this information, calculates erosion/accretion rates.

Selecting a Technique

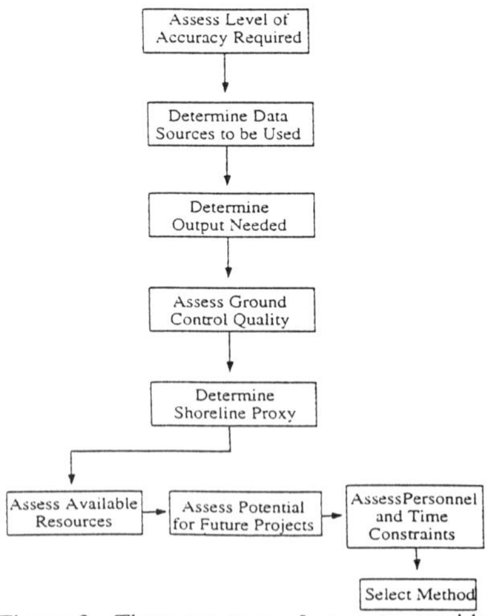

Figure 3. There are many factors to consider before selecting a shoreline mapping technique. This flow chart summarizes the discussion provided above.

TECHNIQUE SELECTION

Selecting from among the established techniques described above can pose a significant challenge. Throughout the following discussion, several factors are recommended for consideration including: the level of accuracy required, type of output desired, method of ground control point collection, availability of funding and/or equipment, and the potential for similar future projects. Consideration of these factors will aid in preventing selection of a technique which will ultimately not meet project needs or which unnecessarily (and perhaps wastefully) goes beyond project needs. Figure 3 is a generalized flowchart which illustrates the considerations discussed below as a framework for technique selection. Figures 4 and 5 provide examples of how this framework may be applied to aid in technique selection.

Determining Accuracy Requirements

When determining which technique will be best suited for a particular project, the first step is to assess the resolution needed. This will then assist in determining the error reduction capabilities required to achieve the desired resolution. For example, the following questions should be addressed: 1) What are the estimated/suspected erosion or accretion rates for the study site? 2) How long of a time period is being investigated? and 3) How precisely do erosion or accretion rates need to be known, i.e. is the project goal simply to identify whether or not change has occurred, or is the goal to provide the most accurate rates of change possible?

Figure 3 is a generalized flow chart designed to serve as a starting point for determining the accuracy requirements of a particular project. Some projects may unrealistically require a level of accuracy which cannot be achieved. For example, it may be difficult to resolve an average annual erosion/accretion rate for a very slowly eroding coast over a 20 year time span, especially since HWL migration alone could account for enough error to prevent resolution of rates under these circumstances. If, however, rates of change are expected to be high, the time period of study is long, and only a qualitative analysis is required, it would be difficult to justify the use of a more time consuming method, such as softcopy photogrammetry, since point measurements would likely satisfy project requirements.

Data Input and Desired Output

Before undertaking a project, it is wise to consider the data sources to be used. If the only source of shoreline and control point information is aerial photography,

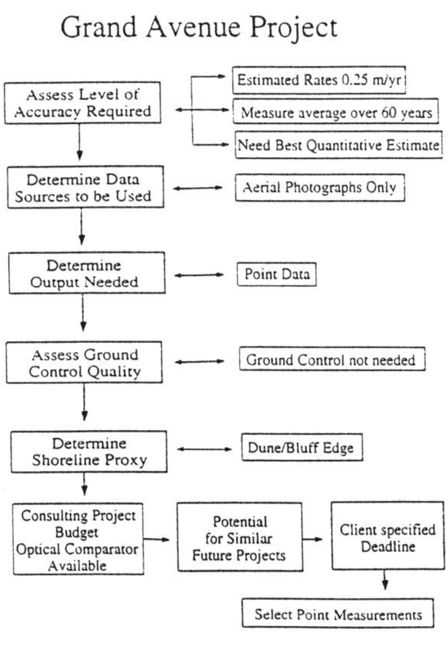

Figure 4. This project was carried out by a consultant within a limited budget and time frame. Aerial photographs were the only data source and point measurements were the required output.

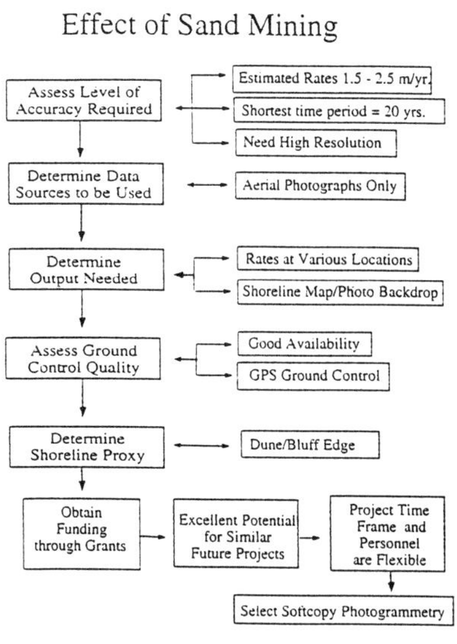

Figure 5. This project requires high resolution of erosion rates. Good control and a softcopy laboratory were available and influenced technique selection.

technique selection will not be limited by this factor. If however, the project is to incorporate information from maps in addition to aerial photography, it would be most prudent to choose from among the methods which allow for map transformation.

In addition to considering the inputs, the output to be generated should also be taken into account. If the project goal is to generate average annual erosion/accretion rates for selected points along the shoreline, technique selection will not be limited by output requirements because all of the methods discussed can generate data in this format. However, if the project goal is to provide a shoreline map or a photo backdrop, selection will be more limited.

Ground Control Considerations

Once the level of accuracy and type of input/output required have been determined, the availability of ground control points and the method of control point collection must be considered. In undeveloped areas, it may be difficult to establish a control network with a sufficient number of points to use an approach which requires a great number of control points. In such cases, it may only be possible to achieve results using the DSMS/DSAS or Softcopy Photogrammetry for which six to eight high-quality, fully known points are generally sufficient to control a group of ten aerial photographs.

Since USGS Quadrangles are not of sufficient accuracy to provide reliable control points, it is not justifiable to use an automated method of shoreline analysis for projects in which control cannot be collected from NOS T-sheets or by GPS survey. To benefit from the error reduction capabilities of the more robust automated methods of shoreline analysis, high-quality control is essential. For this reason, although results may be achieved using control from NOS T-sheets, the time and monetary commitments required to generate results using the more recently developed methods, can only be justified when using control collected by GPS survey.

Shoreline Proxy

The feature or features chosen as proxy for shoreline position may also affect technique selection. As discussed in earlier sections, migration and interpretation of the HWL alone can be a source of considerable error. Depending on the level of accuracy required, if the HWL must be used as a proxy for shoreline position, it may be prudent to select a method which significantly reduces error from other potential sources. On the other hand, if the potential error due to the migration of the HWL is too great, it may not be worth the extra time involved to utilize the techniques which significantly reduce other errors since they will likely be insignificant when compared to errors in HWL position. When using the edge of a dune, bluff or cliff, the choice of shoreline proxy will not likely influence technique selection.

Monetary and Time Constraints

The following questions summarize additional critical factors recommended for consideration when selecting a shoreline mapping technique: 1) What equipment, or hardware and software is already available?, 2) How much funding is designated for this project?, 3) How many future projects of a similar nature will be carried out, i.e. can the cost of a more expensive method be justified?, 4) What is the expertise of personnel responsible for completing the project?, and 5) Are there time constraints on completion of the project?.

CONCLUSIONS

There are many errors inherent in the process of analyzing shoreline change and determining erosion/accretion rates. Though project results are ultimately limited by the quality of available inputs, numerous techniques have been developed over the last 25 years, each in an attempt to improve upon previous methods by reducing the potential for error. With advances in computer technology and decreases in the cost of advanced technology, the more recently developed methods of shoreline mapping are not only quite successful in reducing error, they are also becoming more accessible. With techniques ranging from ruler measurements on uncorrected photographs to high-tech generation of orthophotos and measurement in a GIS, the methods and the numerous publications stemming from their development can be a source of considerable frustration when selecting a shoreline mapping technique.

This paper has been crafted to provide an introduction to the shoreline mapping process and to present a summary of considerations to be made when selecting a technique. I refer you to the bibliography and the bibliography of Moore (In Press), which contains at least a representative sampling of articles which address the topic of shoreline mapping.

ACKNOWLEDGMENTS

We extend thanks to Robert Thieler, Frank Scarpace, John Fisher, Margerie Overton and Richard Stumpf for their discussions with Laura Moore as she worked to determine which mapping method would best serve the requirements of her dissertation project.

LITERATURE CITED

Anders, F.J. and Byrnes, M.R., 1991. Accuracy of shoreline change rates as determined form maps and aerial photographs. Shore and Beach, 59(1), 17-26.

Byrnes, M.R. and Hiland, M.W., 1994. Shoreline position and nearshore bathymetric change. In Kraus, N.C. Gorman, L.T. and Pope, J.(ed.), Kings Bay Coastal and Estuarine Physical Monitoring and Evaluation Program: Coastal Studies, Army Corps of Engineers Technical Report CERC-94-9.

Clow, J.B. and Leatherman, S.P., 1984. Metric mapping: An automated technique of shoreline mapping. Proceedings, 44th American Congress on Surveying and Mapping. Falls Church, Virginia: American Society of Photogrammetry, pp. 309-318.

Crowell, M., Leatherman, S.P and Buckley, M.K., 1991. Historical shoreline change: Error analysis and mapping accuracy. Journal of Coastal Research, 7(3), 839-852.

Dolan, R.; Hayden, B. and Heywood, J., 1978. A new photogrammetric method for determining shoreline erosion. Coastal Engineering, 2, 21-39.

Ellis, M.Y., (ed.), 1978. Coastal Mapping Handbook. Washington: U.S. Department of Commerce, U.S. Department of the Interior, 199p.

Fisher, J., and Overton, M., Personal Communication, 1996. The use of softcopy photogrammetry to measure shoreline change in North Carolina for the Federal Emergency Management Agency Hazard Mapping Project. Erosion Hazard Mapping Workshop, June 14.

Fisher, J., Personal Communication, 1994. The use of softcopy photogrammetry to measure shoreline change in North Carolina.

Fisher, J.J., and Simpson, E.J., 1979. Washover and tidal sedimentation rates as environmental factors in development of a transgressive barrier shoreline. In Leatherman, S.P., (ed.), Barrier Islands, New York: Academic Press pp. 127-148.

Hiland, M.W., Byrnes, M.R., McBride, R.A., and Jones, F.W., 1993. Change analysis and spatial information management for coastal environments. Microstation Manager, March, 58-61.

Leatherman, S.P., 1983. Shoreline mapping: A comparison of techniques. Shore and Beach, 51, 28-33.

McBride, R.A., 1989. Accurate computer mapping of coastal change: Bayou Lafourche shoreline, Louisiana, USA. In: Magoon, O.T. (ed.), Coastal Zone '89, New York: ASCE, pp. 707-719.

McBride, R.A.; Hiland, M.W.; Penland, S.; Williams, S.J.; Byrnes, M.R.; Westphal, K.A.; Jaffe, B.E. And Sallenger, A.H., 1991. Mapping barrier island changes in Louisiana: Techniques, accuracy and results. In: Kraus, N.C. (ed.), Coastal Sediments '91, New York: ASCE, pp. 1011-1026.

Moore, L.J., In Press. A survey of shoreline mapping techniques and recommendations for technique selection. Journal of Coastal Research.

Saleh, R.A., and Scarpace, F.L., 1994. Softcopy photogrammetry: The concepts and the technology. Washington DC: Workshop #2, Mapping and Remote Sensing Tools for the 21st Century, August 26, 86p.

Shalowitz A.L., 1964. Shoreline and Sea Boundaries. V1. Washington, DC: U.S. Department of Commerce, Coast and Geodetic Survey, U.S. Government Printing Office, 420p.

Slama, C.C., (ed.), 1980. Manual of Photogrammetry. Falls Church, Virginia: American Society of Photogrammetry, 4th ed., 1056p.

Smith, G.L. and Zarillo, G.A., 1990. Calculating long-term shoreline recession rates using aerial photographic and beach profiling techniques. Journal of Coastal Research, 1, 111-120.

Snyder, J.P., 1987. Map projections - A working manual. Washington, DC: U.S. Geological Society professional Paper 1395, 383p.

Stafford, D.B., 1971. An Aerial Photographic Technique for Beach Erosion Surveys in North Carolina. Washington DC: U.S. Army Corps of Engineers, Coastal Engineering Research Center, Technical Memorandum No. 36, 115p.

Stumpf, R., 1996, Personal Communication. The use of softcopy photogrammetry in determining shoreline erosion rates, March 6.

Tanner, W.F. (ed.), 1978. Standards for Measuring Shoreline Changes: A Study of The Precision Obtainable, and Needed, in Making Measurements of Changes (Erosion and Accretion). Tallahassee, Florida: Coastal Research, Florida State University, 87p.

Thieler, E.R. and Danforth, W.W., 1994. Historical shoreline mapping (I): Improving techniques and reducing positioning errors. Journal of Coastal Research, 10(3), 549-563.

Wolf, P.R., 1983. Elements of Photogrammetry, 3rd ed., New York: McGraw-Hill, 562p.

CHAPTER 98

GENERAL OVERVIEW OF THE COASTLINE IN TODOS SANTOS BAY

Oc. Rigoberto Guardado France
Facultad de Ciencias Marinas
Universidad Autónoma de Baja California
Ensenada, B.C. México.

Introduction

Todos Santos Bay, located on the West coast of Baja California and 100 km South of the international border México- USA, has experimented severe erosion (Renaul y García, 1994) due to natural and human related events. Natural events include waves, tides, currents, sea level rise and climatic events. Superimposed on this are factors due to human interference such as beach front development and river damming. The protection alternative used to confront the erosion problem is the construction of hard protection structures however this protection devices have different degree of effectiveness and in general have produced an increment on the erosion rate which in term produce the loss of sandy beaches. Due to the fact that the beaches of the bay are an attraction to the tourism and hence a form of income to the city, the purpose of this paper is to provide a genera overview of the coastline of Todos Santos Bay.

Erosion

Renaul and García (1994) compiled information of several authors that have reported erosion rates in Todos Santos Bay coastline and concluded that it is eroding at a rate of about 1 - 1.5 m/yr., however, this is the only study that has been made on the subject. With the purpose of developing a coastal erosion risk map we gather information about coastal erosion rates, density of coastal protection devices along the coast as well as information about erosion resistance of the material composing the coastline, wave energy distribution along the coast, and sea level risk zones. Once the information was obtained it was superimposed on a map to obtain different erosion risk zones along the coast. The results show that considering the total length of the coastline of the bay which is 18 km, seven of them are considered as very high erosion risk, five as high erosion risk, four as medium erosion risk and only two as low erosion risk.

Coastal protection

The results obtained by the erosion analyses are very important because it explain the fact that right now there are a lot of coastal protection structures being constructed along the coast. By December, 1996 there were 43 protection devices and by January, 1997 the number was increased to 45. Among the structures we can find breakwaters, groins, seawalls and Rip-rap. As in the case of coastal erosion, there were no integral studies concerning coastal protection devices so a recoinance field trip was made to detect and visually evaluate the condition of the structures. As a result we find that about 90% of all the structures had some kind of problem. The seawalls have problems of erosion at the base and erosion at the sides as well as inadequate drainage system and cracks. The Rip-rap has problems of base and side erosion, rock sliding, rock settlement and wave overtoping during storm season. The breakwaters have problems of overtoping during storm season, rock splitting and sliding meanwhile the groins are working quite well. Because much of this problems can be prevented or reduced during the design process, an interview with the proper authorities was carried on to find out if a rigorous design process was required to obtain a construction permit. As a result of the interview we find out that right now the aren't any specific guide used to design a coastal protection structure. Because the design process is of such importance on the success of a coastal protection device it is suggested that the design process proposed by Edward et al. (1995) can be used to serve as a basis for the design process in an attempt to consider all the information required to make a proper design which take us to another problem, that is, all the information about oceanography, geology, meteorology and so on is very disperse and in this condition cannot support a good design so we are almost finish with the elaboration of a database which contains information about geomorphology, sediment transport, wave statistics, currents, wind, wave refraction, storms, run off, precipitation, and sea level rise. The next step is incorporate this information to a geographical information system. The use of soft coastal protection such as coastal sand dunes must also be implemented in the bay because in some parts of the coastline which are considered as very high risk erosion zones exist dune fields that have been substantially altered by human action mainly by the circulation of motorcycles and all terrain vehicles (ATV's) and the reconstruction of this dune fields can help to avoid further damage to the area behind the dune field. Because the length of the damaged dune fields is about 5 km long it is considered necessary to initiate a program to reconstruct it so we elaborated a dune field restoration manual that is easy to understand and gives all the information required to take on a project of this magnitude.

Land use

Land use is another important issue concerning coastal protection in Todos Santos Bay because as in other places, it has to support decision making concerning the type of coastal structure device if any can be constructed on the coastline based on his actual or future use. Land use strategies such as requiring a certain distance from the high tide mark to construct infrastructure on high erosion areas to act a buffer zone are highly recommended as a mean of avoiding the excessive construction of coastal structures and thus preventing serious damage to the beaches which in term are one of the main attractions of the bay. The erosion risk map definitely can be used to implement land use strategies according to the degree of risk. Right now the municipal government has a map on which it represents the actual and future land uses of the city but almost all of the coastline is refereed as touristic use but it doesn't show specific uses, that is, where are located the Marinas, residential areas, Hotels and so on. Because of this, we elaborated a map specifying actual uses which combined with the map that they already have and shows potential uses can serve as the basis on which land use strategies can be implemented as well as the basis for deciding which coastal protection device, if any can be constructed.

Conclusion

Coastal erosion is a problem in Todos Santos Bay and the current approach to deal with the problem is the use of hard protection devices which in term make the problem worse. Right now the government has to consider the coastal erosion risk map and the coastal land use map as a basis to establish coastal use strategies in a attempt to avoid the construction of new coastal protection devices and if it is necessary to construct them, they must be sure that it is properly design.

References

Edward, F.T., Donald, L.W., Domurat, G.W., Pirie, D.M. and Oliver, J. 1995. Process for Coastal Projects. Journal of the American Shore and Beach Preservation Association, Vol. 11 (2)

Renaul, G. W. y García, G.C. 1994. Estudio de razones de retroceso en la Bahía de Todos Santos. No publicado

Profile Variation Under Storm Conditions in the Area of Playas de Tijuana, B.C., Mexico

Héctor Lazcano V.[1] and Róman Lizárraga A.[2]

Abstract

This paper describes simulations of storm-induced beach erosion performed with a numerical model of beach profile evolution **(SBEACH)**. The numeric model was applied to beach profile backed by a bluff. To identify storm events wave data was analyzed from November through March of 1983 to 1995 from Imperial Beach, CA. array. Using criteria of wave height and persistence during winter seasons near 120 storm events where detected and from those, selection of the average storm event, the maximum wave height storm event and a consecutive storm event where obtained. Simulations where done for each condition show that the erosion of beach profile and exposure of the bluff to wave action is more severe when consecutive storm events occur.

Introduction

In 1963 began the residential and commercial development known as Fraccionamiento Playas de Tijuana situated in a coastal area west of the city of Tijuana, B.C., México. In the north section of Playa de Tijuana, atop of the bluff it was constructed a boulevard (malecon) settled over a land fill in which nearly 400,000m^3 of much finer that original beach material was used, the dimensions of the land fill where 1.5km long, 50m wide and a average height of 5m (Gonzales-Calvillo and Cupul-Magaña,1986). Since the construction of the land fill the area did not suffer any erosional event until the winter season of 1977-78 when a series of severe storm attacked the area causing a mayor erosion of the land fill and the destruction of houses and building situated near the edge of the bluff. Cupul-Magaña(1983), conclude that the major causes responsible for the beach erosion in Playas de Tijuana where: a)Occurrence of high period waves; b)Mayor run-up coinciding with high tide levels acting on the base of the bluff; c)Textural

[1] U.A.B.C. Student, Facultad de Ciencias Marinas, Km. 107 Carretera Tij-Ens, Ensenada,B.C., Mexico Address: Av. Hidalgo # 33 int.1 Zona Centro Ensenda, B.C., Mexico, 22860.

[2] M.Sc., Researcher, Intituto de Investigaciones Oceanologicas. Km 103 Carretera Tij-Ens., P.O. Box 453,Admon 1, Ensenada,B.C., Mexico, e-mail: roman@bahia.ens.uabc.mx.

incompatibility of land fill material with original beach material; d)Effect of the Gral. Abelardo Rodriguez dam and channeling of the Tijuana River acting as sediment trap.

Over an interval of just days, storms can produce serious damage and life-threatening situations to the coast by rapid erosion of the beach, upland inundation, and generate scour to cliffs that leads to collapse over time and cause slumps of the unconsolidated overlying strata (Larson and Kraus, 1989; Harker and Flick, 1991). Because of its narrow beach, Playas de Tijuana is subjected to severe beach profile erosion, exposing the bluff to wave action generated by storm events.

Purpose of the Study

To analyze the variation of the beach profile and possible effect on the bluff produced by the occurrence of high energy waves generated by winter storms in the region, by means of a numerical model named SBEACH Numerical Model for Simulating Storm-Induced Beach Change (Larson and Kraus 1989; Larson et al. 1990; Rosati et al. 1993, Wise et al. 1996).

Location

The residential development Playas de Tijuana is situated 9km west of the city of Tijuana, B.C. México, and 3.2 km. south of Imperial Beach, CA. located at 32°32',32°28'lat.N and 117°08',117°06'long.W, the area is limited to the north with a metal structure situated on the international Mexico-U.S. border (fig.1), and to the south by a rock headland known as Punta Costa Azul. On the shore a narrow beach starting on the border, backed by averaging 5m high bluffs extends 2km long to the south with median to fine grain sand.

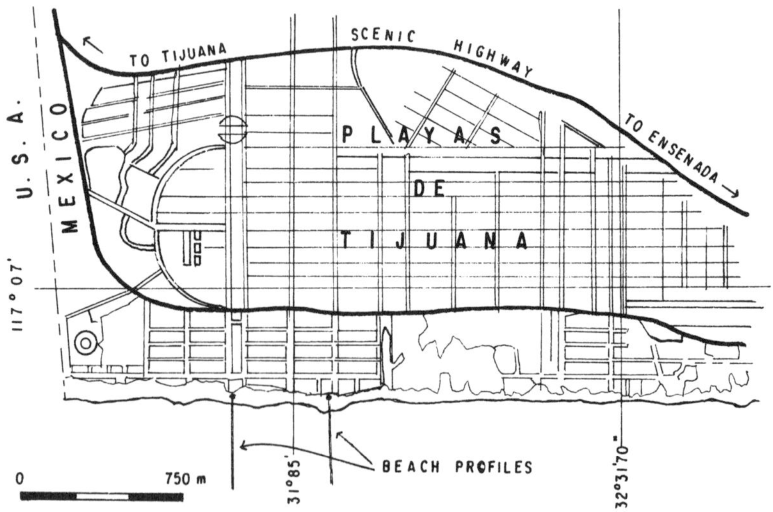

Fig. 1 Location of study area.

Data Analysis

To evaluate beach erosion and detect the maximum changes of the beach profile under a variety of storm conditions in the area of Playa de Tijuana, B.C. steps indicated by Larson and Kraus, (1989) for the use of the numerical model SBEACH where followed. The execution of the model requires of structuring the information in individual input files.

Wave Data

Due to the proximity of the study area and the continuity of their records it was selected the wave information registered by a sensor array located in Imperial Beach, CA. (32°35.0'N,117°08.2'W) at a depth of 10.2m (shallow water) operated by the Coastal Data Information Program (CDIP) monthly reports (Dormurat, 1981). To detect the occurrence and persistency of storm events based on wave height approaching the coast, wave data was analyzed every 6h. for the winter seasons of November throw March of the years 1983 to 1995.

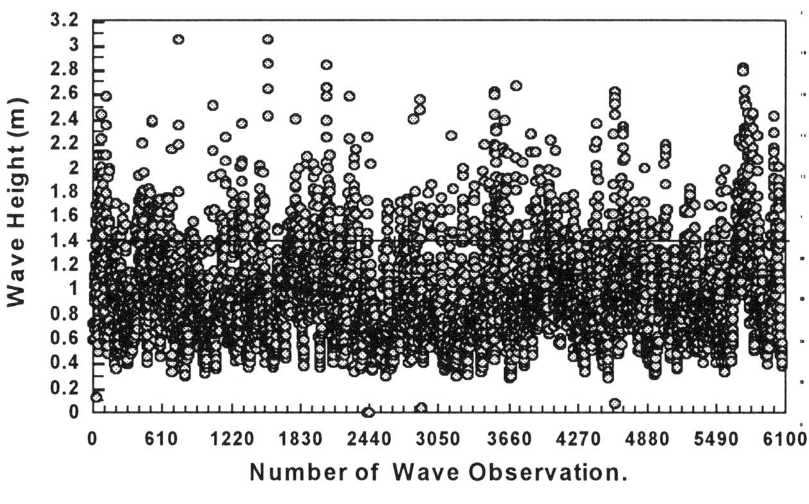

Figure 2. Winter season (1983-1995) wave height observations in Imperial Beach, CA. array.

The determination of the total storm events for each winter season was made considering the average wave height of all the registered wave data (0.96m) and for defining a storm event one-standard deviation was added, therefore a storm event is identified when a wave height is equal o grater than 1.40m (fig.2). Since the wave height records of 1.40m is observed with duration of 6 an 12 hours and the wave records includes wave generated by local and distant storms, to eliminate isolated wave occurrence referred to this wave height, the selection of records with wave height of 1.40m and persistence 18hr or more in a day were selected. This criterion is based according on the prediction of storm events with extreme waves for 1990 to 2040 accomplish by (US Army Corps of Engineers, 1991a) for the coast of California which presents a duration of about 1 day storm event and less than 2 days. From these, 120 storm events in the region were identified.

Considering that variability in intensity, duration and storm periodicity influences differently on the form of the beach profile, the 120 events were analyzed to obtain:

a) The average storm event: determined by averaging duration of each of 120 storm events which resulted. From all events of 2 day duration it was selected the one with grater wave height corresponding to the event occurred on March/21/91 to March/22/91 with an average wave height 2m.

b) The maximum storm event: the storm event with the largest wave height, occurred on Feb/01/1986 with a registered average wave height of 2.32m of the event and a duration of one day.

c) A consecutive storm event: identifying a period of a maximum number of consecutive storm events during a lapse of 8 days and given a maximum of 2 intermediate days between storm event, selecting the storm (fig.3) that occurred on February/28/91 to March/07/91 with an average wave height of 1.74m.

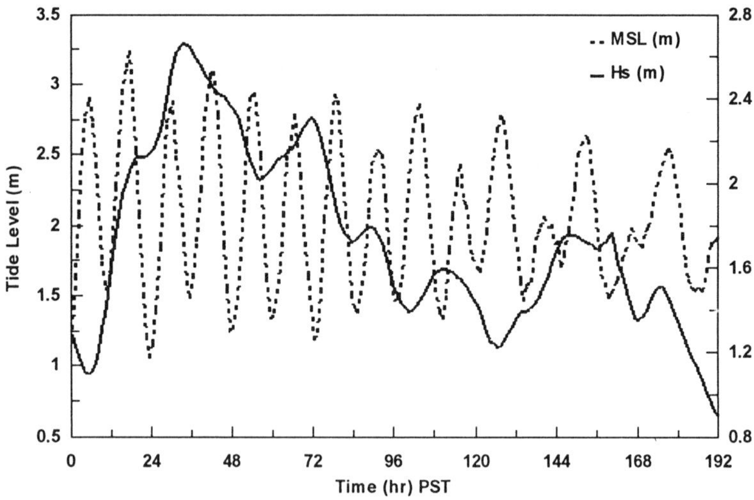

Figure 3. Selected Consecutive Storm Event February/27/1991 to March/07/1991

Tide Data

Tide levels where analyzed from the tide station located in Scripps Pier, CA. (32°52.0'N,117°15.4'W.) (NOAA, 1988). A series of hourly records where obtained from NOAA and CICESE archives of the years of 1986 to 1992, data are referred originally to "0" of the rule. Data was corrected to the MSL datum of Scripps Pier,CA. (2.091m). The position of tide level referred to MSL was obtained for each hour in which storm days where established.

Beach Profile Data

Corresponding beach profile of pre-storm and post-storm conditions were obtained. Pre-storm beach profile was obtained from US Army Corps of Engineers (1991b) measured on Oct/83 [SS-0003A] in the area immediately north of Playas de Tijuana. Post-storm beach profile was obtained by measurement of the beach and back beach by simple topographic leveling referred to a bench mark 15.19m above MSL, installed by SEDUE (State Secretary of Urban Development and Ecology). Two beach profiles 500m apart where measured, at the time of measurement, breaker height and breaker depth was estimated using the McCowan criterion (Komar, 1976); on this basis, beach profile was extended 500m offshore from the bench mark.

Sediment Data

This variable was obtained by sand samples collected on the surf zone and berm, from this samples was determined the sand size (0.25mm) by means of a sediment analysis in agreement to the method described by (Folk and Ward, 1957).

Results and Discussion

The calibration of the numerical model **SBEACH** was accomplished trough trial and error approach until finding the calibration parameters that better reproduce changes on the post-storm beach profile (measured) were defined (Appedini-Albrechtsen, 1995). Pre-storm and post-storm beach profiles used in the calibration where taken north of the study area [SS-0003A] on Oct/83 pre-storm profile and Feb/84 post-storm profile (US Army Corps of Engineers, 1991b) with this information, calibration of the model was performed simulating the storm event that occurred on Feb/09/84 to Feb/11/84 with a duration of 3 days with an average wave height 1.41m . Calibration begun with the default calibration parameters values propose by SBEACH: $K=1.75x10^{-6}m^4/N$, $Eps=0.002m^2/seg$, LAMM=0.2 (Rosati et al., 1993), several combinations between the sediment transport coefficients where accomplished, determining the best simulation with a transport rate coefficient $K=1.6x10^{-6}m^4/N$, $Eps=0.002m^2/seg$, and LAMM=0.2.

Verification for these SBEACH simulations was performed with the same beach profiles used in calibration, but simulating another set of storm event data which occurred on Jan/13/86 to Jan/15/86 with a duration of 3 days and a wave height average of 1.66m, within verification the same calibration parameters were used and simulation was similar as performed calibration.

Wave conditions generally were measured every 6hr by the Imperial Beach array, about 6000 wave heights observations were analyzed of the 11 winter seasons obtained, 15% of all wave heights were equal o grater than 1.40m wave height. Since the time step used in the numerical model (typically 5 min.) was shorter than the wave measurements, cubic spline interpolation was used to obtain intermediate wave heights between measurement points of every storm event selected. A time step of 5 min, and a length step of 3m were used for all simulations.

SBEACH numerical model reproduce reasonably well the beach profile response to the three storm events selected the transport rate coefficient obtained in Playas de Tijuana ($K=1.6x10^{-6}m^4/N$) is similar to those from large wave tank simulations ($K=1.6x10^{-6}m^4/N$) and from Torrey Pine Beach, California ($K=1.4x10^{-6}m^4/N$) (Larson and Kraus,1989) and from Mar de Ajo, Argentina ($K=1.5x10^{-6}m^4/N$) (Caviglia and Pousa,1996

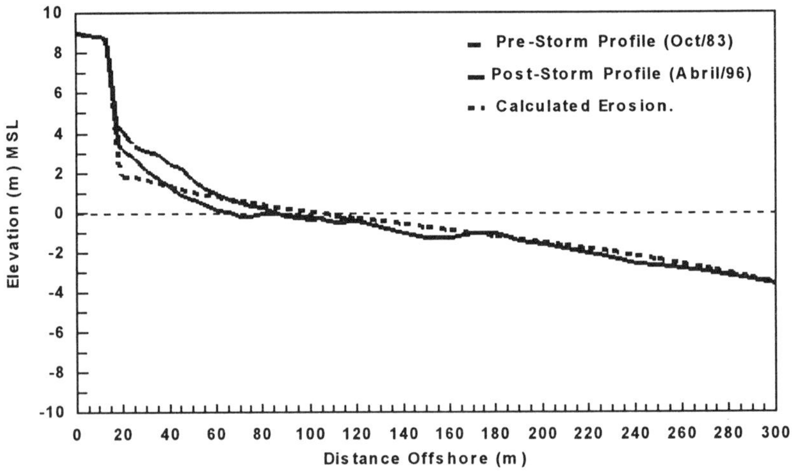

Figure 4. Simulated profile under consecutive storm condition.

Difference in total volume between the post- and pre-storm beach profile is $55m^3/m$ (fig.4) shows beach profile 300m distance offshore even though total distance is 500m. Calculated changes on all three simulations showed maximum recession of the -1 to +2m elevation contour registered the maximum recession. The simulated consecutive storm event registered the maximum calculated erosion difference in total volume between initial and calculated profile ($0.8m^3/m$) with a maximum recession on the -1m elevation contour of 39.1m and a calculated erosion depth of 1.50m between +1 and +2m elevation contour.

Conclusions

In spite of limited beach profile information, SBEACH Numerical Model simulated reasonably well the erosion on beach profile in short period of time. Simulation of the three storm selected, register on the beach profile a maximum recession on -1 to

+2 elevation contours exposing the backbeach and bluff to high energy waves. Effect of a one day maximum storm event has a minor erosive influence on the base of the bluff than a consecutive storm event, even if they produce lower wave height arrival to the coast.

References

Appendini-Albrechtsen, C.M.(1995) Aplicación de un modelo numérico para la estimación de volúmenes de azolve en la central termoeléctrica de Rosarito, B.C. Tesis de Licenciatura. Escuela Superior de Ciencias Marinas, U.A.B.C., 93pp.

Caviglia, F.J. y Pousa, J.L.(1996) Beach Profile Response Simulation to Storm on the Argentine Coast. Shore and Beach.64(2):11-15.

Cupul-Magaña, L.A.(1983) Causas de Erosión en Playas de Tijuana durante el invierno de (1977-78). Tesis de Licenciatura. Escuela Superior de Ciencias Marinas, U.A.B.C. 56pp.

Domurat, G.W.(1981) The Coastal Data Information Program. Shore and Beach. 49(4):12-18.

Folk, R.L. and Ward W.C.(1957) Brazos River Bar: a study in the significance of grain size parameters.Jour.Sed; Petrology, Vol.27:3-26.

González Calvillo, A. and Cupul Magaña, L.A.(1986) Causas de erosión en Playas de Tijuana B.C., durante el período 1975-1985. Ciencias Marinas. 12(3):82-93.

Harker, A.H. and Flick. R.E.(1991)Beach and Cliff Erosion at Solana Beach, California, 1984-1990. Proceeding of Coastal Zone '91, American Society Of Civil Engineers, pp. 2123-2135.

Komar, P.D.(1976)Beach Processes and Sedimentation. Printece-Hall, 1st. Edit. Englewood Cliffs, N.Y. 429pp.

Larson, M., and Kraus, N.C.(1989)Prediction of Beach Fill Response to Varying Waves and Water Level, Proceeding of Coastal Zone '89, American Society of Civil Engineers, pp 607-621.

Larson, M., Kraus, N.C. y Byrnes, M.R.(1990) SBEACH: Numerical model for simulating storm-induced beach change; Report 2, Numerical formulation and model test, Technical Report CERC-89-9, U.S. Army Engineer Waterways Experiment Station, Viksburg, MS,115pp.

Larson, M., and Kraus, N.C.(1989) "SBEACH: Numerical Model for Simulating Storm-Induced Beach Change", Report 1, Empirical Foundation and Model Development, Technical Report CERC-89-9, US Army Engineer Waterways Experiment Station, Coastal Engineering Research Center, Viksburg, MS, 256pp.

NOAA (1988) U.S. Department of Commerce. National Oceanic and Atmospheric Administration, National Ocean Services. Tidal Bench Marks, La Jolla, CA. Pacific Ocean. http://www.olld.noaa.gov/bench/9410230.txt.5pp

.

U.S. Army Corps of Engineers (1991a). Coast of California Storm and Tidal Waves Study, State of the Coast Report San Diego Region. USACE.,Los Angeles, District Vol. I. Main Report CCSTWS 91-9. 350pp.

U.S. Army Corps of Engineers(1991b) Coast of California Storm and Tidal Waves Study, State of the Coast Report San Diego Region. USACE.,Los Angeles, District: . Main Report CCSTWS 91-9. Vol II Appendices.55pp.

Rosati, J.D. and R.A. Wise, Kraus, N.C., Larson, M.(1993) SBEACH: Numerical Model for Simulating Storm-Induced Beach Changes. Report 3 Users Manual. Coastal Engineering Research Program, 54pp.

Wise, R.A., Smith, S.J., y Larson, M.(1996) "SBEACH: Numerical model for simulating storm-induced beach change"; Report 4, Cross-shore transport under random waves and model validation with SUPERTANK and field data, Technical Report CERC-89-9, U.S. Army Engineer Waterways Experiment Station, Vicksburg, MS, 139pp.

CHAPTER 100

Coastal Process Cyclicity

Vladimir Peshkov

Science Department, SIA Krasnodarberegozaschita, 68, Krasnaya str. Krasnodar, 350000 Russia

Abstract.

This work considers the influence of the climatic changes on the dynamics of the Caucasian Black Sea coast. The main attention is paid to middle periodic cycles about 20 years duration, which are caused fluvial sediments yield fluctuation and wave climate. Every such cycle is characterized by accumulative or abrasion principal of development.

Introduction.

The cyclicity phenomenon in sea coast dynamics was marked by number of researchers in connection with atmospheric season processes or with periodic solar activity fluctuations /Shepard, Wanless, 1969; Foster, Stone, 1965; Fox, Davis, 1978; Hayden, 1975; Hume, Shalk, Hume 1972; Marabini, Veggiani, 1993 and other/. The researches of last years testify that coastal process cyclicity has more complex character. Some cycles have not precisely expressed temporary duration even at strict factor periodicity causing them. This is explained by that a mechanism of the first course influence on coastal processes is rather complete one and the back reactions to the influence of external environment are not unequivocal. First, an atmosphere perceives such influence and transfers them to a climate and so on, thus causing the sort of chain reaction.

Thus, the separate components do not react at once to external changes because of certain inertia, therefore the shift of phases happens, i.e. the time variety or heterogeneity in coastal cyclicity of the processes of the coastal origin.

The most usual cyclicity displays in beach dynamics in phase of development, stabilization and damping of storm. This process specificity allows to allocate storm spatial - temporary coastal zone system or storm cycle. It is important to note the following. If irreversible changes in sediment balance structure do not occur during the storm cycle, then the beach comes back to initial condition or close to it. Cyclic character of storm process is one of sea coast dynamics display of steady mode.

Coastal Process Cyclicity

The seasonal changes of wave regime in the Black Sea are connected with its presence in interaction zone of various air flows. In cold period an area of low pressure, connected with anticyclones, is established above basin. It promotes high atmospheric pressure gradients formation, often intrusion of strong winds and storm excitement development. At land heating in warm period there is the alignment of atmospheric pressure gradient and consequently, the air flows become less steady. Strong storms happen seldom in this period, their total energy decreases. On Caucasian Black Sea coast the processes of lithodynamic are submitted to the seasonal cycles of near rivers mouth, capacity of the longshore drifts, the beach material losses in submarine canyons and other.

The analysis of cartographic surveys shows, that the maximum storm and seasonal cycles intensity is marked at the moment of the greatest contradiction between wave regime and coastal zone relief. When the conflict is removed, the speed of relief transformation is reduced. Observations of wave processing of artificial filling testifies, that this process proceeds more intensive in initial period, and then fades sharply.

In modern sea coast dynamics the important role is played by cycles, in a basis of which fluctuations of climate coinciding with "by cycles of solar activity" in 11, 22, and 44 years. On the eastern Black Sea coast the short periodic duration of climate oscillations are approximately 20 years.

The presence of long-term changes of wave regime with inner century fluctuations of climate proves to be true by researches of many authors (Balabanov, 1984; Lamb, 1982; Marabini, Veggiani, 1993 and other). General climate cooling in 50's has resulted, for example, in sharp decrease of wave activity on the Black Sea. In this case direct connection between year average repeatability of storm and average temperature of air in Northern Hemisphere is observed. Such dependence is revealed for Sochi region since 1916. Balabanov /1984/ made the analysis of the dependence of wave activity on inner century climate fluctuations on the Black Sea basin and came to the conclusion, that: a) strong and very strong storm repeatability is connected by a certain way with deep cyclones penetration (pressure in center 990 mb and below); b) deep cyclone repeatability has direct connection with year average temperature changes of Northern Hemisphere. For the last 60 years in this large region well expressed dependence is established between repeatability of deep cyclones and a deviation of average year temperature from norm.

Researches made during last decades show that general trend in Holocene is directed to the climate improvement in comparison with the last glacial stage (Wurm). Inside this trend in Holocene it is possible to distinguish secondary climate changes, that have strongly influenced the environment, in particular, on physical and biological processes. Using materials of geological, geomorphologic, glacialogical, paleobotanical, archaeological researches, it is possible to reconstruct with high accuracy the sequence of climatic variations, especially for the historical times. Ice distribution study is one of typical examples, used in paleoclimatology. Front changes observation of Fernau ice in Tirol for the last 3 thousand years have allowed to identified 5 humid and cold periods. They induced in Europe and Mediterranean basin some hydrogeological calamities. These periods had the following duration: 1400-1300 and 900-300 BC, 450-750, 1150-1300 and 1550-1850 AD (Mayr, 1964). With in these large scale cycles, the shorter periods (10-35 years) of cold rainy weather, alternated with warm dry and weather periods, which are known as Brukner cycles.

On Adriatic coast of Italy from river Po up to Gabicce for periods of cold and humid weather the coast advancing is characteristic (Marabini, Veggiani, 1993). This process is correlated with sharp advancing of river Po delta with 1600 on 1812 during little ice age. The reason of advancing there was powerful fluvial sediment yield, exceeding ability of waves action has moved material shorewards. During warm and dry weather volume of floods supply sediment becomes less and simultaneously, the repeatability of strong storms is reduced. From the end of 70's dry and warm weather without the plentiful rain, floods and storm, that is quiet situation in comparison with previous period of cold and humid weather was established mainly. Then, the strong storms prevail over natural fluvial sediments from the rivers. Regression of the shoreline was the leading action in during 20 years. The present period is characterized by stabilization and even advancing of the coastal line. It occurs in areas, where coast protection works have been done and where they have not been carried out. It does not mean, that abrasion is not dangerous. According to climate fluctuations cold and humid weather conditions are expected to happen. The general storm waves attack can cause new coastal destruction. This process, alongside with the natural preconditions, can be strengthened by technogenic factor imposing.

The influence of climate oscillations has an effect not only for coastal processes. During cold and humid weather the plentiful deposits are combined with strong storms, that provoke landslides activity. Thus, the large blocks of ground can be advanced into the sea as large tongue. During dry and warm weather condition a volume of firm sediment yield is reduced and simultaneously wave activity decreases. Cliffs become stable as well as beaches. Artificial stabilization of cliffed coast has resulted in that the beaches from Gabicce up to Punta Penna have remained without former power supplies material. Therefore during period of cold and humid weather they will be subjected strong erosion (Marabini, Veggiani, 1993).

According to observations in Pitsunda region (the Black Sea coast), the total energy of storms changes considerably roughly every 20 years. So, from the middle of 20s to the beginning of 40s it has exceeded 66 thousand energy units (ceu). In 1939 an absolute maximum - 8400 ceu was traced, that exceeded the norm thrice. (2660 ceu). The next cycle (1945-1963) differed by general decrease of wave activity (about 30 thousand ceu). From middle of 60s cycle of high activity of excitements began (about 65 thousand ceu by 1985). Following the original maps, photographs and archival data since 1903, first 20 years of our century differed by rather small total energy of storms (about 35-40 thousand ceu). The modern cycle should be characterized by the decrease of general storm activity. It does not exclude, however, the passage of separate extreme storms as in 1949 and 1954 during the previous similar cycle.

Calculation technique of temporary numbers for the storm activity forecast distribution was offered on the basis of observation of wave energy in Sochi region. Thus, the distribution of total annual energy well coincides with other regions of eastern part of the Black sea, in particular, with Pitsunda and has average duration of 20 years. According to forecast of wave activity estimation the reduction happens at last 20 years of century, that coincides with trend of the author.

The existence of long-term wave regime changes is revealed in many regions of World Ocean. Foster and Stone (1965), on the basis of cartographic surveys, locia and photo documents have established cycles in dynamics of a beach Gronula (Australia). From 1901 to 1924 sediments accumulation has prevailed on this beach, then erosion cycle has followed, and since 1951 the tendency to accumulation again began. Coast in Barrow on Alaska region testify about cycle abrasion course and accumulation (Hume, Shalk, Hume, 1972). The detailed analysis of wind - wave regime for Baltic 5 sectors from 1880 to 1972 testifies about storm activity cyclicity with period 24-26 years. Shuisky (1986) demonstrate data about abrasion cyclicity on Burnassky region of the Black

Sea. The increase of speed cliff regression was marked here at the end of 50s and in the beginning of 70s. In the beginning of 60s and in the middle of 70s the speed of cliff cutting was minimum, and in the second half of 60s and in the beginning of 80s abrasion was again activated. The specified phenomena have touched also beaches and accumulative forms. Modern erosion amplification of USA Atlantic coast is connected with cyclic growth of storms and its duration (Hume, Shalk, Hume 1972). Abrasion cyclicity in connection with non-equal energy distribution of storms is marked in Crimea. The number of researches in Klemskerke-Vosseslag region (Netherlands) testifies that the beach erosion phase is replaced here by relative balance and accumulation phase. Construction of coast protection structures in erosion phase (1912) is the confirmation to that, which have appeared to be buried under sediments in accumulation phase.

The alternation of erosion cycles and accumulation is marked on Bretan peninsula coast in places of sand-gravel beaches development. Accumulative coasts here are rather dynamic: the erosion cycles are replaced by advancing. During separate cycles the coast is advanced in the sea up to 250 m (1961-1971); up to 800 m from 1963 to 1980. Coast retreat on some sectors occurs to speed up to 140 m (1975-1981). Cyclic beach dynamics changes with period about 10 years are observed in region of Freemantle port (western coast of Australia) /Eliot, Clark, Rhodes, 1982/.

The existence of periods with low and high storm activity is revealed for a number of Baltic sectors, and Caspian Sea. The average duration of storm cycles activity is defined in 24-28 years. The maximum of such activity is dated to 1902-1910, 1925-1932, 1948-1952; minimum -1914-1920, 1940-1944, 1955-1970. During high storm activity influence of waves on coast and hydraulic engineering structures sharply grows. Sedimentation is amplifies of ports. The cycles are rather steady, that enables the chance of qualitative forecast growth and decrease of the sea wave activity. Shepard, Wanless (1969) mark the existence of coast sea cycles by average duration of 35 years (inner century). In storm activity periods the outlets from Mexican gulf lagoons are displaced much faster in a direction of sediment flow, than during general storm repeatability decrease. In Fresendorf region on the Baltic Sea the growth and coast abrasion speed decrease with an interval of 25-45 years are witnessed. The variability of wave regime renders large influence beach material moving and quantitative parameters longshore sediments flows.

Simultaneously with long-term fluctuations of wave activity, a fluvial sediment supply of the mountain rivers in western Georgia is undergone to cyclic changes. Fluctuations of amount of floods supply sediments of some rivers have here determined connection with energy distribution of excitements. This explains, in particular, characteristic relief changes near rivers mouth and sea coasts adjacent to them. The periods of high total wave activity coincide usually in common fluvial sediment yield amount of rivers decrease. As a consequence - the coast is washing out, especially near rivers mouth. Accordingly, when values of sediment runoff and wave activity are large there occurs the decrease of a general coast advancing.

Data comparison of longshore drift on capacity in Pitsunda region with amount of sediment from the river Bzib shows, that near river mouth sharp sediments deficit or their surplus occurs periodically. So, from the middle of 20s to the first half of 40s longshore sediment flow reached 250 thousand cubic meters and more in some years. In these conditions south-western Pitsunda coast has tested strong erosion, is especial from river Bzib mouth up to Inkitsky cape (erosion volume is near 1.5 million cubic meters). The coast about 2 km length has receded on the average of 40-50 m. About 0.9 million cubic meters of sediments were washed out from a beach to the south of Inkitskaya bay. The similar situation has arisen at that period on Adzaria, where the coast, according to an exposition to western rumbas prevailing excitements, represents Pitsunda mirror display. The erosion amount estimation according to cartographic dates shows, that the longshore drift of beach material from Chorochi river mouth up to Batumi cape reached 250-300 thousand cubic meters a year.

From 1945 till 1963 accumulative processes prevail on Pitsunda. Total Bzib river fluvial sediment yield amount has exceeded about 0.8 million cubic meters total capacity of longshore flow (year average volume is 30 thousand cubic meters). In conditions of wave activity decrease the maximum of accumulation is at the near river mouth area; river Bzib promontory is advanced seaward almost for 100 m. Since 1964 the tendency to river Bzib firm drain decrease was planned at simultaneous increase "capacities" of sediments flow. As a result the Southern-Western Pitsunda coast has tested a new erosion cycle, which is 0.7 million cubic meters volume only for a beach. To the beginning of 80s the coast SE from river Bzib has receded on the 45-50m average. Erosion of Inkitskaya bay beaches activated appreciably. Summary of beach sediments balance during some periods and as a whole from 1903 to 1985 allows to assume, that of the first 20 years of XX century were characterized by accumulative processes prevalence of (amount of floods supply sediments excess over longshore sediment drift volume is estimated by 1.5-1.7

million cubic meters). The modern cycle is characterized by general wave activity decrease (1986-2005).

Variability of river Bzib sediment runoff and wave regime has a certain influence on beach material losses in submarine canyons. The heads of canyons in front of rivers mouth of the Caucasian Black Sea coast there are in 150-200m from the coast, (8-15m depth). The most active canyon heads cross a migration zone not only sand, but also gravel. Researches of the Institute of Oceonology of the Russian Academy of Sciences (1967-1969) have shown, that the part of beach material really fills canyons. Sand and shingle in canyon "Shark" were found out up to 420m depth. Detailed repeated measuring (1970-72) gave the author an opportunity to estimate such losses quantitatively. The considered period differed by high storm activity. In the subsequent period the heavy storm repeatability has much decreased and, accordingly, sediments losses on depth have decreased. This is established by careful supervision with application of special methods of underwater geomorphologic researches. The subsequent year works show, that at storm activity strengthening of the submarine canyons influence on sediments balance of a coastal zone again grows, So, in 1983 and 1985, sand and shingle volume in canyon in front of river Bzib has a few tens thousand of cubic meters. The beach material was marked also in the head of canyon "Shark" 1983, 1985, 1989. Thus, basic reason of different opinions concerning beach materials losses into submarine canyons are explained by that they are received in periods with various hydrodynamic conditions.

When considering of coastal processes cyclicity as the important peculiarity of dynamics and sea coast development, it is necessary to state some preliminary conclusions. Each of cycles, irrespectively of character and duration, reflects the occurrence of the certain contradiction between various coastal zone components elements and external environment. Therefore basic process way within the framework of that or other cycle is connected to an establishment of new balance. For storm cycle it is expressed in consecutive relief changes and sediments so that to supply maximum dispersion of waves mechanical energy at the least capacity of the process. It is reached by manufacture of dynamic balance structure or by aspiration to such condition. Under condition of balance preservation of substance and energy, storm cycle does not cause unidirectional relief changes and coastal zone sediments structure. The cycles of the seasonal character have the same dynamics. The basic distinction that the last is covered with more long intervals of time and comprise a set of storm cycles.

Though the seasonal cycles are not always identical each other in quantitative expression, preservation of qualitative character of changes is their basic attribute, i.e they are made in one variant limits. The seasonal cycles are rather steady, though in some years can considerably change both volume of a firm drain, and energy of excitements. Accordingly, the relief near river month changes have as though pulsing character rather of some average statistic condition. Seasonal and storm cycles unite and one more attribute: the intensity of processes is maximum in an initial phase at the greatest relief contradiction in relation to the wave factor and is reduced on a measure consecrated of such contradiction. One more important attribute is convertibility or stability of the basic changes at quality preservation.

We shall address to the analysis of long-term cycles, which reflect the certain tendency in dynamics and coast development in some temporary interval. And though thus an opportunity to return to former or close to it condition is admitted, within the framework of such cycle qualitative changes are possible. So, during general depletion sand-gravel beach, it can be turned into gravel one. The outcrops of relict sediments into surf zone can be the reason of a qualitatively new stage in the coast development. Thus, major attribute of long-term cycles an opportunity of transition of quantitative changes in qualitative is, that is accumulative on genesis the coast turns in typically abrasion. At that time, the cyclicity concept means an opportunity of coast return to former qualitative condition. Or else, the accumulative coast within the framework of a cycle can become abrasive, and then again to be transformed in accumulative one. Integration of materials on coastal processes cyclicity allows to allocate a number of characteristic stages, namely: a) relative coast relief stabilization (structure, coastal line, grain structure), which comes in accordance with contradiction deleting in relation to the power characteristics of excitement; b) gradual intensity attenuation relief reformation down to a condition, at which the wave energy completely dissipates and does not cause unidirectional changes during this or that temporary cycle. It concerns to storm, seasonal and long-term cycles. It is possible to assume, that between cycles of a different rank there are no basic distinctions. Zenkovich (1967) considers, that in regressive-transregressive phases, when a wave regime in geological time scale is established, faster relief are developed, then sediments balance and in last turn - grain structure. Or else, the development proceeds from youth to old age, when the basic contradictions between the various factors are removed. High abrasion and accumulation speeds, bad sediments sorting - attributes of development proceeding on a way of balance achievement. Accumulative process periodicity in Holocene transgression time specifies of relict accumulative generation forms on shelf, and about accumulative cycles plurality it is possible to judge to one age generations ridges arrangement on different bathimetric shelf levels. Modern accumulative form degrade off testifies about the accumulative cycles change on erosion.

Conclusions

The allocation of cycles in dynamics and sea coast development in a general case is the reliable tool for understanding of occurring changes. Each cycle is characterized inherent only to it by peculiarities, but each of them is rather independent system. All this allows to speak about two aspects of the structural - system analysis, about mutual relation of system with external environment, and also about systems hierarchy and universal involve of any system as a subsystem in more complex. Thus, the behavior of separate system is defined not only by internal structure, but also structure of its external connections.

The relative stability of cycles allows to expect possible changes of sea coast and to use thus their peculiarity for the forecast. This is especially important for acceptance of the practical decisions. Not taking into consideration the natural erosion processes change and accumulation can become the reason of the serious miscalculations and large unproductive expenses on measures on protection of sea coasts. The example of that the coast Pitsunda resort could be served, which has tested catastrophic erosion in 1969. For its protection the projects of cost in many millions rubles were developed. And only a duly cyclic dynamics character prediction of Pitsunda cape has allowed to accept the correct decision. To middle of 70s the beaches of the resort were completely restored at the expense of NW longshore sediment flow in complete volume. One more example. At development of coast protection measures in Gagra (the Black Sea coast) the fact of sediments moving is accepted for basis in 50-60s years from river Bzib mouth up to Gagra. After the first groynes construction (70s) the beach material has become to move in a direction of the basic sediment flow on SE. As a result, the abrasion has begun and for its elimination it was necessary to build new groynes.

Coastal process cyclicity data are too sketchy yet for any strict integration. It is much so unclear concerning the influence of the mechanism of the primary factors and back reactions to them on the part of a coast remains. There is some uncertainty concerning possible displacement in time of various cycles, and also duration of their action. In a number of uncertainty cases of estimations is caused by absence of sufficient long-duration period of supervision. Extreme insufficiency is given for integration about influence of technogenous factors on natural processes cyclicity of a coastal nature. The stated material is thus preliminary integration of existing nature phenomena, which require further study.

References

Balabanov, I.P., 1984. The Black Sea wave regime changes in the late Holocene, Izvestiya of the USSR Academy of Science, Geography Series, N 5, pp. 70-81.

Eliot, J.G., Clark D.J., Rhodes A., Beach-width variation at Scarborough, Western Australia. J. Roy. Soc. W. Austria, 65, 1982, pp.153-158.

Foster, D.N., Stone, D.M., 1965. Historical evidence of erosion at Gronulla. J. Inst. Eng. Australia, 37, N 10, pp.275-286.

Fox, W.T., Davis, R.A., 1978. Seasonal variation in beach erosion and sedimentation on the Oregon coast. Bull. Geol. Soc. America, 89, N 10, pp.1541-1549.

Hayden, B.P., 1975. Storm wave climates at Cape Hatteras, North America recent secular variations. Sciences, 190, N4218, pp.981-983.

Hume, J.D., Shalk, M., Hume, P.W., 1972. Short term climate changes and coastal erosion, Barrow, Alaska. Arctic, 25, 4, pp.177-184.

Lamb, H.H., 1982. Climate, history and the modern world. Methuen, London, 432 pp.

Marabini, F., Veggiani, A., 1993. The evolution of the northern Adriatic coastal zone (Italy): the climatic changes and the environment. In: Coastlines of the Mediterranean. N 4, ASCE, pp.1-15.

Mayr, F., 1964. Untersuchungen uber Ausmass und Folgen der Klima Gletcherschwankungen seit dem Beginn der postglazialen Warmezeit. Ausgewahlte Beispiele aus dem Stubaier Alpen in Tirol. Zeitschr. P. Geomorphol. Wien. pp.257-286.

Shepard, F.P., Wanless, H.R., 1971. Our changing coastlines. N.Y. Mc Grow-Hill Book Co., 579 pp.

Shuisky, Yu.D., Problems of sediment balance in the coastal zone. Leningrad, Hydrometeoizdat Publ., 1986, 240 pp. (In Russian)

Zenkovich, V.P., Processes of Coastal Development. Oliver and Boyd Publ. Co., Edinburgh, 1967, 738 pp.

CHAPTER 101

FIGHT AGAINST EROSION OF THE CAUCASUS COAST

Kos'yan R.D., Yesin N.V.

The Southern Branch of the P.P.Shirshov Institute of Oceanology , Russian Academy of Sciences. Gelendzhik-7, 353470 Russia. Tel. 7-86141-23261. Fax. 7-86141-31631. E-mail:kosyan@sdios.sea.ru

Introduction

The enormous wind energy, absorbed by the waves is mostly spent in shallow areas. Here, the dissipation of wave energy generates quite a few specific hydro-, litho- and morphodynamical processes, which lead to the development of strong and highly variable currents and complex patterns of intensive water exchange, the formation and active displacement of underwater bars and migration of the shoreline. A specific peculiarity of coastal zone is that the dynamic al processes are complex here, and individual processes are involved in this complex in the very different combinations.

Depending on the local conditions waves and wave flows either erode the original rocks in the coastal zone moving the coast landwards, or build up it by means of sediment accumulation along the shore, or keep a stable outline and profile of the coast manifesting a mobile equilibrium between acting factors.

But the process of the bottom profile and a stable coastline evolution cannot be researched isolately, since it depends greatly on an economic activity of a man.

Really, the practical considerations make researchers and engineers concentrate their attention on this part of the sea area. Coastal zone had long become the field of intensive economic expansion, which has considerably accelerated during the past decades and it will grow up even further in the near future. Of course, the economic development today is being extended to depths, which can no longer be regarded as coastal zone. It is obvious, nevertheless, that coastal zone will retain its special economic importance to mankind for a long time (Kos'yan, Magoon, 1993).

The importance of coastal zone in human life is by no means exhausted by the above examples. They only serve to illustrate how momentous and manifold are our interests in this part of the Black Sea. At the same time, the anthropogenic factor becomes to be one of the basic in the coastal zone dynamics. The coastal zone represents the part of the sea, which is especially sensitive to any kind of exposure. Here the problem of environmental protection and conservation takes a very acute form. Therefore, all other decisions concerning feasibility and ways of any technical expansion must be preceded by a forecast of its consequences. A prediction of this kind is only possible on the basis of exact knowledge and careful analysis of dynamical processes occurring in coastal zone.

Processes leading to coastal erosion

Among the mechanisms of anthropogenous transformation of physical processes in the Black Sea coast zone we can single out the following: influence of hydrotechnical erections upon the coastal processes; influence of the mineral mining upon ecosystems of the coastal zone and upon coast; consequences of the river run-off regulation (Kos'yan, Magoon, 1993).

On the Coast of the Black Sea the first ones negative factors are transformations of the coastal processes resulting in the change of features of the cross profile and the shore outline evolution by means of hydraulic constructions of various economic importance and different direction to the shoreline (longitudinal constructions - mooring and baffle walls, quays, slope shoring and breakwaters; cross ones - port moles, piers, groins)(Aibulatov, Artiukhin, 1993).

According to available data (Kiknadze, Kos'yan, Yesin, 1994), about 2 billion US dollars have been spent during the last three decades on the protection of the Caucasian sea coast. Only within the Big Sochi coast 80 km breakwaters, and 50 km walls and more than 900 jetties have been built, which only aggravated the problem of coast protection. These constructions restricted the natural circulation of nearshore water and turned these waters and beaches into highly polluted zone.

For a long time engineers aspired to reduce negative influence of sea walls upon the bottom and beach sediment accumulation by means of building of curvilinear walls. But in spite of all their diligence, undermining of such constructions almost everywhere took place. The analysis of the efficiency of different walls between the Tuapse and Batumi proves this. Breakwaters became one of an essential factors of the destruction of gravel beaches on the Caucasian Black Sea coast.

A serious shortcoming of the majority of above mentioned constructions is an emergence of regional washout, provoked by a concentration of wave energy on the junctions of those constructions with a natural coast. Regional washouts occur both on the abrasive and accumulative coasts.

Longitudinal constructions of different types have similarities in negative impact upon the shore: (Aibulatov,1993)

1. When protecting a slope from a destruction they exclude a possibility of additional sediment supply from precipices.

2. Walls, while breaking waves favor the destruction of beach and intensify a bottom erosion.

Submerged and above water breakwaters and similar constructions redistribute sediments along shore artificially, i.e. they deform a space structure of sediment balance. If they have provoked a local beach erosion, then sediment deficit on that section of the coastal zone will not be more than 10-20 percent from the value of longitudinal sediment discharge. It could be compensated naturally rather soon, if its reason would be removed. Anomalous zones of accumulation are more serious danger for a coastal stability. Constructions with parallel port jetties have a long direct and indirect influence upon the evolution of large coastal sections (1-2 or 5-10 km long, and even more).

Table 1. Value of technogenous impact upon the Sea coast.
Extracts from (Aibulatov, Artiukhin, 1993)

Sea coast	Length of coast	Length of constructions, km				Coefficient of technogenous impact K km/lin.km
		groynes jetties	submerged breakwaters	walls	fills	
Caucasian						
Anapa-Tuapse	225.0	4.5	0.2	8.9	-	0.06
Tuapse-Adler	106.4	35.0	3.8	80.6	-	1.12
Adjaria	112.5	3.5	-	17.6	-	0.19

Latitudinal constructions on the Black Sea coast are represented by different types of constructions, including crib retaining walls, groynes and barrier flows. Their principal purpose is to accumulate beach sediment by means of artificial breaking of longitudinal sediment transport. The length of a zone of a coastal transformation by latitudinal constructions depends on the dimensions of longitudinal sediment flows. In the paper of Aibulatov, Artiukhin (1993), a coefficient K was considered, which determines ratio between the length of engineering constructions (jetties, groynes, walls etc.) and a total length of a coast, for the estimation of technogenous impact on the coast. This coefficient has the largest value on the Caucasian Black Sea coast.

From the table 1 one can see that on the coast between Tuapse and Adler the groynes, walls and breakwaters are longer than the length of a shore line. But it does not means, that there is no natural coasts. In some places there is an extraordinary density of engineering constructions, the others (Lazarev Cape, Dagomys etc.) have not them.

As it follows from above mentioned, coastal and marine hydrotechnical erections are able to influence mechanically upon the sediment dynamics and the coast relief. As a result of serious disturbance of the structure of the moving water mass, the development of the whole coastal zone has changed. Anomalous phenomena have being formed, which are not peculiar to a natural interaction of two media, i.e. water and sediments. And dynamic balance of this complex nearshore system can be achieved only due to such an interaction.

In some areas the transport of sediment to the coast by rivers form the biggest source of material to the coastal zone. Dams constructed on these rivers form sediment traps and reduce the sediment supply to the coast which results in coastal erosion.

Excavation of beach material is undertaken in many parts of the sea. This material is sometimes mined for the mineral it contains; in other places it is used for construction purposes. Such actions are direct losses of available supply of material for sediment transport in the nearshore zone.

The dredging of nearshore bars and shoals can change the pattern of energy dissipation on a beach face. If the change increases the wave energy acting on a given section of beach, erosion will likely result at that section.

Other things being equal, the larger an energy flow comes from the open sea, the greater technogenous impact upon the coast. Therefore, the main feature of the technogenous impact upon a coastal process is a distortion of energy and the matter shift.

Measuring of wave erosion

Russian coast of the Black Sea stretches from the Kerch Strait to the Psou River mouth. A total length of the coastline is about 460 km. Here are sections of accumulative, abrasive, abrasion - landslided and abrasion - denudated shore. (Fig.1).

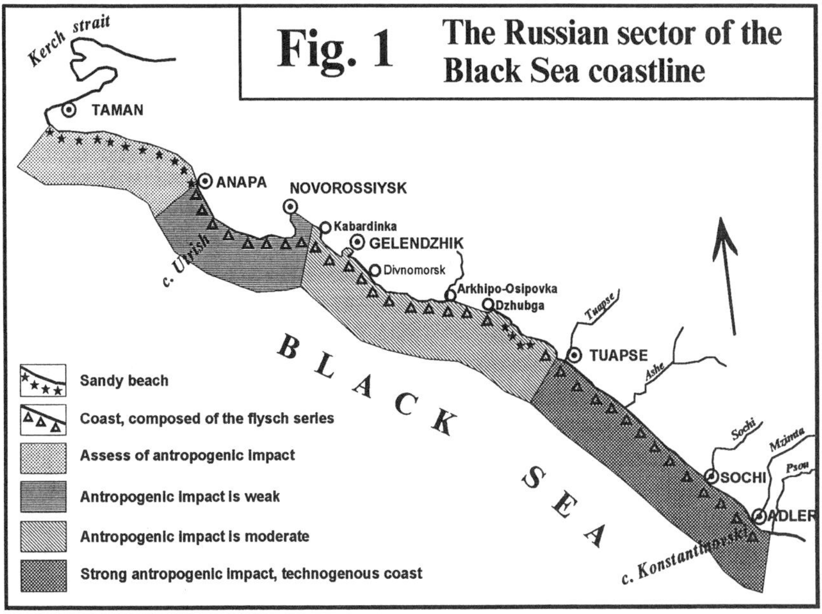

Fig. 1 The Russian sector of the Black Sea coastline

As it was said above, sandy coast from Taman peninsula to Anapa Town had been formed by sediments brought here by southern arm of the Kuban River, which then became shallow one and ceased in the end of the last century. Thus, today there is no a source of sediments that has formed the shore here. Though small amount of sand is brought here from eroded cliff of the Taman peninsula and from sandy dunes being on the shore, none the less, the coast here is developed in the conditions of a definite deficiency of sediments, as far as sand is gradually grated and left the nearshore zone for a large depth. This results in very slow recession of the coastline. Since the longitudinal sedimentary flux is directed from North-West to South-East, towards Anapa, and do not practically reach the latter, just here a pronounce process of the coast recession has begun. It was intensified by a wrong strategy of the coast protection by means of baffle wall. The coastline, 350 m long, has retreated average for 10 m. And buildings constructed on the shore have turned out to be under the threat of destruction.

Between Anapa and Sochi there are the following types of the shore: abrasion-denudated, abrasion-landslided and accumulative ones. Abrasion is the basic process in the formation of the coastline shape. Denudation and land-sliding processes are the consequence of abrasion, the latter one when eroding the cliff basement creates the conditions for the land-sliding and intensifies the denudation as far as the rate of the slope retreat as a result of denudation is proportional to its gradient. Clastic material that appears in the coastal zone as a result of abrasion accumulates in some areas (mainly in small bays). Thus, it forms accumulative sections of the coast. The formation of that or another type of the coast depends on the definite conditions, i.e. the cliff height, the resistance of rocks, orientation of the shoreline.

Peculiarities of the coastal processes here are the following. Since the flysch strata are rather resistant ones, the waves cannot destruct them. The basic destructive influence upon the submerged coastal slope and the cliff foot is exerted by clastic material, i.e. pebble, boulders, coarse sand. In the course of storm clastic material grates the bottom, strikes the cliff and forms surf wave niches. Besides attrition, clastic material, when moving, breaks off small pieces of advancing flysch strata.

In examined region the coast is composed of alternating layers of fine easily eroded shells and solid sandstones and marls of a high strength. During storms firstly loose rocks are eroded and then more resistant ones. As a result a ridged bench is formed. In the near edge zone where pebble is always abundant the apexes of jutting beds are being grated and hence the height of the ridges here is not large. Deeper where pebble lies as one thin layer or as separate lenses, it erodes loose layers and resistant ones remain as underwater relics of ridges.

Pebble is attrited in the course of storm. Very small dusty particles transported sea off is the product of this process. They do not participate in coastal processes. During one year from 4 to 10% of fragmentary material forming a beach are grated (Zenkovitch, Shuisky, 1988).

A detail observation of the shore abrasion on the coast section between Novorossiisk and Arkhipo-Osipovka settlement has been conducted. Research was performed by drilling of the bench marks, i.e. hollows on the coastal slope with the next measuring of their depth (Yesin, Savin, Zhilyaev, 1980). As it can be seen from this work, the rate of attrition varies from zero to 20 mm/yr. Zero rate was observed in places where there was no clastic material.

The rate of abrasion is sometimes more than velocity of attrition because bench marks have been drilled in layers of solid rocks. And the rate of erosion of loose layers is many times more than the others. An average rate of erosion is 1-2cm/yr. Such a conclusion appears from the results of field study and calculation.

Withdrawal of sediments from the coast and river mouths for construction has become one of the reason of intensification of abrasion on the section between Anapa and Adler. As a result of this the width of beaches between Anapa and Novorossiisk has reduced twice, in Divnomorsk - for 75%, in Arkhipo-Osipovka - for 50%. Sediment removal here is hundred thousand m^3 per year.

On the section between Tuapse and Sochi there have been removed 10 million m^3 of gravel since 1945 till 1955. As a result, the beach there stopped to be a coast protective factor. In addition, being scanty on the submerged slope, pebble has become an active factor of abrasion that breaks the foot of cliff and grates the underwater slope.

On some sections abrasion became more intensive one after the construction of groins and moles which interrupted sedimentary flux. So, after the building of the Sochi port in 1936-1939 the process of the coast erosion became more intensive one on the section between Sochi and Adler. Besides, in 50ies about 5 million m^3 of gravel were removed from the Adler Cape that resulted in the reducing of beach for 35000 m^2. And this, in its turn, has made more active the process of wave erosion.

The total length of Georgian coast of the Black Sea is 313 km (Fig.2). The length of the eroded coast of Georgia was 155 km in 1961; 183 km in 1971; 220 km in 1981, when the total length of the coastline in Georgia is 313 km. The basic reasons of the intensification of the erosion here was withdrawal of sediments from the coastal zone, construction of moles which obstruct longitudinal sedimentary flux, construction of hydrotechnical erections in beach zone and adjusting of river runoff.

According to georgian scientists papers (Kiknadze, 1991; 1993; Salukvadze, Elieshvili, 1993; Dzhaoshvili,1986; Saf'yanov,1993) a total amount of sediment withdrawal out of the river channels and coastal zone in Georgia since 1945 till 1990 was 40 million m^3. And in the end of 80ies it increased for 1.5-2 million m^3 per year. On some coast sections it provoked an intensive coast destruction.

An intensive erosion of the coast along the town of Gagra and washing away of a considerable area of a seaside park were caused by the construction of a mole in 1914-1917. Similar situation was observed in Ochamchire. There in the end of 30ies a mole was built which obstructed the sedimentary flux. This intensified the process of the coast erosion southward from the port.

In 70ies the rate of the coast recession near the port became 9 m/yr (Yesin, Savin, Zhilyaev, 1980). Approaching the town it reduced to 3 m/yr. A total area of erosion in Ochamchire from 1933 to 1969 is 32 hectares. A belt of the town area, 40 m wide, has been washed away.

The same situation occurred after the construction of Sukhumi, Poti and Batumi ports. They interrupted longitudinal sedimentary fluxes.

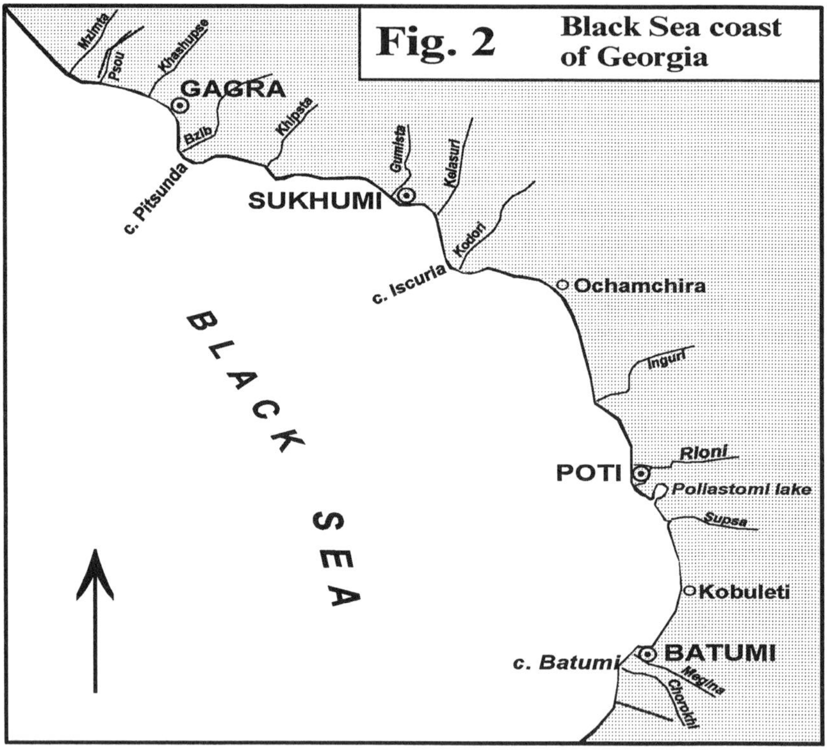

Fig. 2 Black Sea coast of Georgia

Real size of the coast recession is: in Kodori region + 49 m (1872-1933); - 27 m (1934-1954); - 42 m (1955-1976); - 11 m (1977-1980); - 24 m (1981-1985); - 8.4 m (1986-1987). A coast which till 1933 was protruding to sea with the velocity of + 0.8 m/yr, since 1933 it began to retreat firstly with the velocity of -1.3 m/yr then - with the velocity of - 4.2 m/yr.

Roughly in the same region (between Inguri and Chobi rivers) the coast dynamics was the following: (1872-1933) +61 m; (1934-1954) - 10.5 m; (1955-1976) - 22 m; (1977-1980) - 6.4 m; (1981-1985) - 16 m; (1986-1987) - 6.8 m. Thus the coast which was protruding to sea with the velocity of + 1 m/yr started to be eroded with the rate that increased in time from - 0.5 m/yr to - 3.4 m/yr.

Fight against erosion of the Russian coast

The fight against the wave erosion of the Black Sea coast has been combating since the middle of the last century. By the present a material has been obtained that affords to make conclusions about the most expedient and acceptable methods of the coast protection in either hydrophysical situation.

The practice of protection of sea coasts from erosion has shown that there is no universal method of averting of the coast destruction. Methods of coast protection depended on a specific physical-geographical situation and on its geological structure. But a general strategy of the coast protection has been already formed by nowadays and it will be described in this chapter. One-and-a half-century history of "the struggle" with sea has shown also, that wrong method of the coast protection creates more complex problems that the initial problem of the strengthening of a concrete coast section.

The coast between the Taman Peninsula and Anapa was composed of sand that had been brought there by the Kuban River as early as the last century. Now a stable coastline is formed here, a small (as compared with the volume of beach-forming material on shore) sandy flux is shifting along it from north to south. There are two sources of its nourishment. It is an erodible cliff of the Taman peninsula containing sand fractions and sand lying on shore as dunes. When heavy northern and north-eastern wind is blowing (up to 20 m/sec.) this sand is blown off to the coastal zone and hence it compensates to a certain extent that amount of sand which is attrited and washed away to a large depth.

The coastline of the described shore sections reacts very vividly upon the deficiency of sand that is blown off the shore. Thus, in the beginning of 80ies the coastline on one coast section, about 300 m long, has retreated for several tens metres in the course of one storm. The reason of this phenomenon was the construction of the children's playground on the beach. This erection has blocked the sand shift to sea under the action of wind (a dune has been formed around the playground, which has buried the playground). So, this structure caused deficiency of sediments. The reaction of the coastline resulted in its recession. When the playground was pulled down and sand got a possibility to shift freely off shore, the coastline was restored approximately within early existed boundaries.

It should be noted that at present the coast of this region is intensively covered with hotels and health resorts. As a result, the amount of sand transporting off shore reduces year after year, and in future it will result in deficiency of sediments and in progressive recession of the coastline.

On the south a sandy beach ends within the Anapa Town. Early in 90ies the construction of the embankment was planned. For this purpose a ferro-concrete baffle-wall was built across the sandy beach, it ran along the shore. It provoked the erosion and recession of the shore for 10-15 m at the length of 300 m. As a result, the sea has approached the buildings being on shore. To prevent their destruction, sand brought from quarry has been poured into water and a small artificial island has been made opposite the eroding coast section.

Slightly southward from the sandy beach within the Anapa Town in 1993-1995 an artificial gravel beach, roughly 1 km long, was built. Its destination was recreation and coast protection from erosion. The beach is 30-40 m wide here.

The coast between Anapa and Novorossiisk has not received an economic development and has remained in its original natural view. There is no any coast protective constructions. But lately the building of holiday homes here has started. Very likely that new beaches and groins will be built here.

In the Tsemess Bay groins were built in its southern part near the Kabardinka settlement. Their destination is to retain a gravel beach. There are several tens of groins and some new ones are being built. At present the beach has been washed away from the areas between groins, and groins have ceased to perform coastprotective functions.

Between Tsemess and Gelendzhik Bays there is no coast protective erections.

In 80ies in Gelendzhik Bay there were built 44 groins but spaces between groins were not filled with beach material. Moreover, in some places groins were built very close to each other (50 m or less). They hamper waterexchange between the surf and deep zones in the bay. As a result a rubbish is accumulated between groins and pollutants are accumulated in water. All this reduces greatly a recreative potential of the bay. Thus, the construction of the groins in Gelendzhik Bay resulted in a number of negative consequences, i.e. worsening of ecological situation and the lowering of aesthetic qualities of the coast. For this motive the construction of a gravel beach near the southern cape has resulted in the improvement of sea water quality and creation of a picturesque marine view.

Between the Gelendzhik Bay and Novo-Mikhailovka settlements there are several coast sections protected by groins and beaches. Groins were built in the latest decade. By nowadays there is scanty gravel material between groins, several m^3 per one metre of the shoreline.

Between the Novo-Mikhailovka Settlement and the Tuapse Town ferro-concrete baffle-plates and groins are everywhere used for the coast protection and retaining of beach material. It should be noted that the efficiency of their use is not large. The thing is that in time pebble is being washed away from the areas between groins, the beach becomes narrow and it stops to protect the shore. On the contrary, pebble begins to play a part of an abrasive material and precipitates the destruction both of the baffle-plate and of groins. In the course of storm it bombards the baffle-plate and groins, makes holes in them during the first year of their exploitation.

So, in 1994 a beach was built for the guests of the BELAYA RUS sanatorium. It was protected by two groins and baffle wall. In 1995 12000 m^3 of pebble were filled between groins. But in the beginning of 1996 pebble was already washed away from this area, and a hole was made in one groin. The baffle-wall began to ruin too.

A quick destruction of ferro-concrete constructions on shore suggested an idea to some planners to build more heavy coast protective constructions. It should be noted that those constructions deteriorate the view of the coast greatly, they make it technogenous one and uncomfortable for bathing.

A total length of the coast between Anapa and Tuapse is 225 km. The length of the coast protective constructions is the following: groins and moles - 4.5 km, breakwaters - 0.2 km, baffle-walls - 8.9 km.

On the section **between Tuapse and Adler** there runs a railway. The whole length of this section is protected by various coast protective erections.

In 1914 a railway running along the coast and connecting Tuapse and Adler was built. When building the railway the gravel material was withdrawn from beaches for ballast and for coast protective walls. A long section of the coast which suffered from the deficiency of beach material earlier lost beaches at all and began to be eroded more intensively and to retreat more rapidly. On some sections the land sliding became more intensive one as a result of the beach destruction.

To protect the railway from the land sliding and abrasion more and more new coast protective constructions were built, beach sediments were used in them as an inert concrete aggregate. The most important factor which influenced negatively upon the coast evolution was the withdrawal of material from beaches and river beds. The withdrawal reached several million m^3. Only during 1945-1955 more than 10 million m^3 of pebble were removed from there. In addition, in 1936-1939 the flux of the coastal sediments was broken off by moles of the Sochi sea port that was built in this period.

At present the coast is protected by baffle-plates and groins stretching for a long distance. In some places artificial beaches have been built in areas between groins, they need a periodical supply of the sediment loss for attrition. The use of passive methods of the coast protection is not effective one. During 1914-1955 an average width of the beach belt has reduced from 46 to 23 m on the coast section between Tuapse and Adler. And some coast sections turned out to be without beach at all by the beginning of 50ies.

The coast between Tuapse and Adler, near Sochi in particular, always was an abrasive one. But the processes of the coast erosion and recession were more slow ones due to the existence of the beach belt. Beaches were supplied with alluvium from numerous rivers and temporary water flows, and they were bound by a single longitudinal sedimentary flux directed from the north-west to south-east. But, above mentioned manifestation of a man activity has sharply intensified the rate of abrasion all length of the coast and necessitated the coast protection on the most dangerous coast sections. The discharge of the sedimentary flux has sharply reduced after the coast strengthening. At present the section between Tuapse and Adler is "the hottest point" on the Black Sea coast within the Russia. Though, it should be noted, that in spite of rather unfavorable conditions created to the south from the Port of Sochi russian engineers managed "to patch up" the breach and to solve successfully the problem of the coast protection when sedimentary flux is broken by moles of the Sochi Port.

The section between Sochi and the Konstantinovsky Cape is a margin of the Adler accumulative body formed by alluvium of the most powerful river in this region Mzimta and Psou. Submerged slope here is very steep one, therefore for formation of a broad beach and its nourishment in this region demanded a large amount of sediments and a long period of their accumulation.

In 50ies a disastrous erosion of the beach began owing to the reducing of the nourishment from the north-west due to a general coast protective arrangement on the coast section between Sochi and Adler and the construction of the Sochi Port and due to the withdrawal of clastic material for the construction purposes More than 5 million m^3 of the material have been removed from the beach at the Adler Cape, the area of the beach has reduced for 35 000 m^2. In 1965-1970 groins were built and a man-made restoration of the eroded beach belt began. During 1965-1970 400 thousand m^3 of clastic material were added. After the construction of groins the intensity of the longitudinal drift of the material on the section near Adler has reduced twice. Simultaneously with the beach erosion the deepening of the submerged slope took place here: here was revealed a steady approaching of the 5 m isobath to the coastal wall. It is impossible to restore in the nearest future the natural conditions of the Adler accumulative form balance.

The construction of the Sochi Port, abundant withdrawal of clastic material from beaches and river beds without preliminary study of the coastal processes on the coast section between Tuapse and Adler have changed the course of the coast evolution. The scale of mentioned transformations cannot be compared with the scale of the storm influence upon the shore in this region. A man activity in this case turned out to be the most important one in the whole complex of the coast forming factors.

Coast section from the mouth of the Mzimta River to the mouth of the Psou River. A total length of the coastline here is 8 km. The outline of the coast consists of two capes, i.e. Konstantinovsky and Psou, and two bays. From the geological point of view they are advanced parts of the deltas of two rivers. As far as about 2000000 m^3 of gravel were removed from the mouth of the Mzimta River for the construction, it caused a considerable deficiency of sediments. It will be appreciable for 100 years. 10 groins were built in this section. By the present almost all of them have been already eroded.

The length of the coastline on the section between Tuapse and Adler is 106 km. The ratio of the coastline length to the length of the coast protective constructions is 1.12, i.e. the length of the latter is more than the length of the shoreline.

Longer than one century practice of the use of traditional coast protective constructions, i.e. baffle-walls, groins, breakwaters, ripraps of blocks of concrete or clastic material, has shown their small efficiency. Firstly, those constructions are completely or partially broken off during the first 10-20 years of their exploitation. Secondly, they protect only a coastal cliff but they do not protect the bottom from the erosion. In the surf zone the bottom becomes deeper one and larger waves approach the coast protective constructions, their attacks become stronger ones. Thirdly, groins block a longitudinal sedimentary flux and cause so called down-stream erosion. Fourthly, groins and breakwaters deteriorate the waterexchange. Pollutants are accumulated in the surf zone, i.e. zone destinated for bathing, an ecological situation here becomes worse. Therefore every year beaches are closed for some short period. Fifthly, ferro-concrete constructions mar the view of the shore landscape. When the constructions are dilapidated the coast looks ugly. There are another negative consequences of the use of ferro-concrete constructions for the coast protection.

Fight against erosion of the Georgian coast

Since 1980 another non-traditional strategy of the coast protection has been used on the coast of Georgia. Its essence is to restore a morphodynamic balance in the coastal zone by means of the management of the lithodynamic processes (Kiknadze, Kos'yan, Yesin, 1994). The following arrangements can be a controlling factor: the creation of artificial beaches, ripraps of clastic material to the coastal zone, the cape cutting off, destruction of groins and other erections hampering the sediment shift, the transferring of the river mouth to the another coast section, artificial transportation of sediments from one section to another one and so on. All these actions are of natural character and are destinated for more even distribution of the river and other sediments on the protected coast.

Ferro-concrete coast protective constructions almost are not used here.

The study of the mechanism of the coastal processes, which are supported by ripraps of clastic material, has afforded to formulate a conclusion about the possibility of the most optimum and the most economically expedient management of the coastal processes.

By the present a considerable work has been done for the redistribution of solid river run-off along the shore and for the decrease of the amount of sediments drifted to canyons. Such a work was fulfilled on the following coast sections: between the mouth of Bzib and the

Pitzunda Gulf; between mouths of Gumista and Kelasuri; between mouths of Kodori and Inguri.

Every year about 50000 m^3 of sediments are discharged to sea from windward side in Kelasuri and Chorokhi regions in order to satiate a sedimentary flux. As a result, the sedimentary flux was restored on 60% of a total length of the coast. And the beach was restored on the sections where it was absent during some latest decades.

Near the Sukhumi and Ochamchire ports sediments have been transferred from the windward sides of the moles to the lee ones. Artificial gravel beaches were built in Gagra, Pitzunda, Kabuleti and other towns.

In order to stop the coast erosion in the mouth of Rioni, sand from its mouth is transported by pipeline,8.5 km long, to the coastal zone. It is supposed that a land belt, 100 m wide, will be restored.

Arrangements performed in 1982-1990 resulted in the reducing of the length of the eroded coast from 220 to 80 km. 110 hectares of a new land have been created during the same period. A receipt budget of sediments has grown for 18.1 million m^3. One kilometre of protected coast saves 4500 tons of cement and 70 tons of reinforced steel.

Calculation has shown that for the beach building, 50 m wide and 15 km long, near Gagra it needs about one million m^3 of clastic material. It would result in the coast stabilization since 1982 till 1995.

1-1.2 million m^3 of clastic material are necessary for the same purpose near Pitzunda. And 20 thousand m^3 are necessary to be added to the beach every year.

To protect coast of the Western Abkhazia, there needs 6.5-6.8 million m^3 of clastic material. About 30 thousand m^3 must be added every year for the compensation of grated material.

Having used above described strategy, there were restored beaches, and the wave erosion was stopped near Gagra and Pitzunda. The coast of Kolkhida was strengthened by gravel beaches. A broad gravel beach was restored on the coast section, 30 km long, between the mouth of the Chorokhi River and Batumi.

Unfortunately, observations for the coast state here were stopped in 1992, the creation of new beaches and the support of stability of earlier built ones were broken off too. If such works are being performed here in a smaller scale now, we know nothing about it. In this paper we have to limit ourselves to the description and analysis of the work that has been made by Georgian scientists and engineers before.

Built beaches are both of great coast protective and recreative importance. Besides, aesthetic qualities of the coastal zone has improved. Millions tons of cement and steel have been saved. The cost of the coast protection by stimulation of natural coast protective processes is 2 -2.5 times less than that of traditional methods used earlier.

Conclusions

The practice of the Black Sea coast protection from the wave erosion in Russia and Georgia has shown that strict methods of the coast protection by construction of ferro-concrete erections are rather expensive and non-effective ones. In 10-20 years waves erode ferro-concrete, and the coast begins to go to ruin more rapidly. At the same time an artificial intensification of natural coast protective processes including the creation of artificial gravel beaches, sediment nourishment of existing longitudinal sedimentary fluxes, destruction of barriers hampering the longitudinal sediment shift, transference of the river mouths far from canyons absorbing sediments, etc., gives highly large positive effect. Natural picturesque coast fit for recreation and the other kinds of economic activity is reconstructed.

As a specific research shows (Yesin, Dmitriev, 1988), the coast protection by the creation of broad gravel beaches will be effective one in conditions of expected ocean-level-rise under "greenhouse effect" too.

An experience of the coast protection in Russia and Georgia by creation of beaches, which are of great coast protective and recreative importance, can be used by other Black Sea countries.

Acknowledgment

This work was fulfilled under the support of the Global Environment Facility Black Sea Environmental Program and the State Committee of Sciences and Technology of Russian Federation.

References

Aibulatov N.A. 1993. Geoecology of the Black Sea coastal zone. In "Coastlines of the Black Sea". (edited by R.D.Kos'yan, O.T.Magoon). Publ.ASCE, N.Y., p.103-124.

Aibulatov N.A., Artiukhin. 1993. Geoecology of the shelf and coasts of the World ocean. S-Peterburg. Gidrometeoizdat Publ. 304 p. (in Russian).

Coastlines of the Black Sea (edited by R.D.Kos'yan, O.T.Magoon). 1993. New York. Publ. by the American Society of Civil Engineers. 573p.

Dzhaoshvili Sh.V. 1986. Alluvial sediment and beach formation on the Georgian Black Sea coast. Tbilisi. Sabchota Sakartvelo Publ., 158 p. (in Russian).

Kiknadze A.G. 1991. Morphodynamics of the nearshore zone and optimization of its using. Thesis of Doctor of Sciences Dissertation. Tbilisi. 66p (in Russian).

Kiknadze A.G. 1993. Scientific basis of regulation of coastal processes (Case study: coast of the Black Sea). In "Coastlines of the Black Sea". (edited by R.D.Kos'yan, O.T.Magoon). Publ.ASCE, N.Y., p.201-213.

Kiknadze A.G., Kos'yan R.D., Yesin N.V. 1994. Problems of the Black Sea coast protection. Proc. of the Black Sea Symposium. Istanbul,.p.85-88.

Kos'yan R., Magoon O. 1993. A man on the Black Sea Coast. In "Coastlines of the Black Sea". (edited by R.D.Kos'yan, O.T.Magoon). Publ.ASCE, N.Y., p.1-13.

Kos'yan R.D., Pykhov N.V. Hydrogenous Sediment Shift in the Coastal Zone. Moscow, "Science", 1991. 280 p. (in Russian).

Kos'yan R.D., Yesin N.V. 1995. An anthropogenic impact upon the Black sea coast of Russia. Proc. of the International Conference "Coastal Change'95". Bordo. France, p.615-623.

Peshkov V.M. 1974. Observation of relief dynamics of the top of the underwater canyon "Akula". Geomorphology, N3, p.82-87 (in Russian).

Peshkov V.M. 1994. Pebble beaches of tideless seas. Thesis of Doctor of Sciences Dissertation. Moscow State University. 62p (in Russian).

Saf'yanov G.A. 1993. Pitzunda peninsula coast dynamics. In "Coastlines of the Black Sea". (edited by R.D.Kos'yan, O.T.Magoon). Publ.ASCE, N.Y., p.246-263.

Salukvadze N.I., Elieshvili B.A. 1993. New structures used for protection of the Black Sea coasts. In "Coastlines of the Black Sea". (edited by R.D.Kos'yan, O.T.Magoon). Publ.ASCE, N.Y., p.264-277.

Yesin N.V., Dmitriev V.A. 1988. Control of intensity of abrasion process in coastal area. Oceanology, V. XXVIII, N1, p.115-121 (in Russian).

Yesin N.V., Savin M.T., Zhilyaev A.P. 1980. Abrasion process on a sea shore. Gidrometeoizdat Publ. Leningrad. 200p. (In Russian).

Zenkovitch, V.P. and Shuisky, Yu.D. 1988. USSR: The Black, Azov, Caspian and Aral sea coasts. In Artificial structures and shorelines (edited by J.H.Walker). Amsterdam. Kluwer Acad. Publ. Co., p.223-240.

CHAPTER 102

The U.S. Minerals Management Service Oil Spill Research Program

Joseph V. Mullin[1]

Abstract

The U.S. Minerals Management Service (MMS) is the principal U.S. Government agency sponsoring offshore oil spill response research. MMS and its predecessor organization, the Conservation Division of the U.S. Geological Survey, have sponsored oil spill research since the late 1970's. For the past 11 years, MMS has maintained a comprehensive international research program to improve oil spill response technologies and procedures, enhancing the capabilities to respond to an open ocean oil spill.

Efforts have focused on improving: capabilities to burn oil on the waters surface; modeling the dispersion pattern of smoke emissions from in situ burns; updating the performance database for new and improved booms and skimmers by reopening Ohmsett - The National Oil Spill Research Test Facility, located in Leonardo, New Jersey; remote sensing and measurement of spilled oil; oil spill chemical treating agents including dispersants; sorbent research; understanding the properties and behavior of spilled oil in the marine environment; and innovative shoreline cleanup strategies.

Significant advances in oil spill response have been made, as a result of the MMS research program in the last two years. This paper presents an overview of the results associated with these projects.

[1] Research Program Manager, U.S. Minerals Management Service, Engineering and Research Branch, 381 Elden Street, (MS-4700), Herndon, Virginia, USA 20170-4817. Telephone (703) 787-1556, fax (703) 787-1555, Internet: joseph_mullin@mms.gov

I. In Situ Burn Research

Research results from the mesoscale burns in Mobile, AL (1991-94), the Newfoundland Offshore Burn Experiment (NOBE), and the Alaska Clean Seas, Emulsion Burn Experiments, continue to indicate that burning is an rapid, effective and environmentally safe means for removing large quantities of oil from the surface of the water. In public and government forums, burning has become accepted as a first response method. However, questions remain about both water and air quality during burning. In addition, equipment to conduct in situ burns (fire resistant booms and igniters) is lacking. MMS is participating in the following research projects that will advance the understanding of burning of oil spills and foster public acceptance of the method through experiments conducted in the laboratory, at mesoscale, and at sea.

1. Burning of Weathered Oils and Emulsions: Laboratory and Mesoscale Studies

Research at the National Institute of Standards and Technology (NIST) and elsewhere has established that in most cases fresh crude oil spills on water can be ignited and burned. In the case where the oil is confined either by natural barriers such as ice or mechanical fire resistant booms, greater than 90 percent of the contained oil can be expected to be removed by burning. In the case of accidental spills, response to the incident can be expected to take at least one-day. During this time the oil may weather and emulsify with the water.

The limiting conditions under which weathered and emulsified oils can be ignited are largely unknown. A systematic study of ignition of weathered and emulsified oils over a range of properties was recently started at NIST. Analytical means will be sought to relate laboratory scale measurements to ignition of fuels under wind and wave conditions. This is a challenging objective that has not been undertaken by any other investigators. We can expect to make progress over the next year with relative measurements of ignition resistance based on the laboratory scale tests.

2. Improvements to the ALOFT (A Large Outdoor Fire Trajectory) Smoke Plume Dispersion Model

At present, the ALOFT model of smoke transport, is capable of predicting time-averaged downwind concentrations of particulate matter from a large fire which is blown by a uniform ambient wind over relatively flat terrain. The model has been expanded to include the effect of wind velocity variations with altitude and add the ability to calculate smoke plume concentrations as a function of time. The model has recently been further expanded to calculate the soot production from multiple smoke plumes over flat terrain.

3. ALOFT- FT-PC - A Smoke Plume Trajectory Model for Personal Computers

In situ burning of spilled oil continues to gain acceptance as an oil spill response tool as we gain a better understanding its capabilities and limitations. One widely imposed criteria for the use of burning is limiting the exposure of downwind population to smoke particulates. NIST has developed ALOFT a smoke plume trajectory model to predict the downwind, ground level concentrations of smoke particulates. The ALOFT model was refined over a period of years using a computer workstation and the results have compared favorably with the limited data available from experimental burns.

ALOFT-FT-PC is a public domain version of the model for high-end, windows based personal computers. The model inputs include wind speed and variability, atmospheric temperature profile, and fire parameters. The output is the average concentration of smoke downwind of the burn for flat terrain. ALOFT is designed to aid in the in situ burn planning stage. The ALOFT-FT-PC model is currently available in a beta test version only. Project participants include: State of Alaska, Department of Environmental Conservation

4. Testing of Fire Resistant Boom in Waves and Flames.

This objective of this project is to develop a system and protocol to test fire resistant containment booms in waves, while exposing it to full-scale heat loads. Currently, when a fire boom is tested to determine its ability to contain and hold oil, there is no follow-on mesoscale tests to determine the booms ability to withstand the intense heat and mechanical flexure from waves. Full-scale testing a fire resistant boom at-sea, with oil, is very expensive, requires permits, and is not possible in U.S. waters. A mesoscale-scale test system and protocol is required to evaluate candidate fire booms, before expensive full-scale testing.

In this project, full-scale heat loads on the boom would be produced by burning propane in the boom pocket. The burn test protocol to be followed involves a cycle of exposure to flames for a period of one hour, followed by a period of no flame, over a period of days. Booms that survive this mesoscale fire test without structural degradation, would be tested at full-scale for their abilities to contain thick oil slicks and survive extended exposure to more energetic wave conditions. Testing in both waves and high-temperature flames for extended time periods simulating proposed operating conditions is the only way to evaluate a particular fire boom system's capabilities and limitations.

In the first phase of this project, a series of propane burn experiments were conducted, November 5-12, 1996, in the 120m x 60m x 3m deep Outdoor Ship Maneuvering Basin, at the National Research Council, in Ottawa, Ontario. These experiments were larger in scale than the Mobile, Al, propane burns conducted by

NIST in September 1996 and the results proved encouraging. Further development of the test protocol will be conducted, particularly in the areas of increasing the heat flux, improved heat flux and temperature monitoring, and pre-tensioning of the boom during the outdoor tank tests.

5. Laboratory Testing to Determine Operational Parameters for In Situ Burning of United States Outer Continental Shelf (OCS) Crude Oil Spills.

To determine the efficacy of in situ burning with fire booms, in response to crude oil spills, it is vital that specific oils be tested. Data from each oil must be collected to document the effects of slick weathering on ignitability, burn rate and oil removal efficiency. A laboratory test program is proposed to determine the following parameters with respect to the in situ burning of six OCS crudes:

- Natural limits to ignition via gelled gasoline imposed by evaporation and emulsification.
- The ability of commercially available oil spill emulsion breakers and alternative gelled fuel igniters to extend the window of opportunity for igniting stable emulsions.
- The feasibility of applying emulsion breakers simultaneously with igniter by premixing it with gelled fuel.
- The effects of wave action on the combustion of emulsion slicks.
- The likelihood of the residues sinking after efficient burns of thick slicks of crude oils.
- Soot yield measurements

6. Soot Production from In Situ Oil Fires: Review of the Literature and Calculations of Values from Experimental Spills

Recently, the generally accepted soot yield of oil fires on water has been questioned. A thorough review of the literature was undertaken and experimentally sound values of soot yield were collected. The valid soot yield data was plotted against fire diameter with fuel type and soot particle diameters sampled as variables.

7. Fire Resistant Containment Boom Tests at Ohmsett - The National Oil Spill Response Test Facility

Six commercially available offshore oil spill booms produced by five manufacturers for use a fire booms were tested Ohmsett in Leonardo, New Jersey. The tests were conducted between July 16 and October 4, 1996. The U.S. Coast Guard Research and Development Center U.S. Coast Guard, Minerals Management Service, Canadian Coast Guard, and Environmental Protection Agency sponsored the tests which evaluated the sea keeping and oil containment ability of the booms.

Phase II to be conducted in the Fall 1997, will be to evaluate the six booms tested in Phase I for thermal stress and mechanical performance in burning oil. These tests will be conducted at the USCG Fire and Safety Test Detachment, Mobile, AL. If new prototype booms are available from U.S. Navy (SUPSALV) they will be included in the tests.

II. Physical Behavior of Spilled Oil

The objective of these projects are to develop an accurate understanding behavior and fate of spilled oil. This project will continue to study spilled oil to acquire understanding of the long-term behavior and to develop models to predict this behavior. The physical and chemical properties of oils will be measured and included in a database. Elements to be addressed include: oil weathering, evaporation, water-in-oil emulsification, dispersion, dissolution, and photooxidation.

1. Catalog of Crude Oil and Oil Product Properties

The oil properties catalog was first compiled by Environment Canada (EC) in 1984. MMS has jointly funded the catalog program since 1989. The catalog was started to provide in one place, the physical and chemical data about oils, relevant to oil spills. No catalog of this type existed and data were found only by random searching of literature. The current (December 1996) catalog contains information on over 380 different types of crude oils and petroleum products, including many Outer Continental Shelf (OCS) crude oils. The catalog is available in electronic format and has been distributed to all OCS regions and other Federal Agencies. The data is also available on a public bulletin board (EC's SPILLS bulletin board). EC also has an Internet Homepage that will be on line in February, 1997. The catalog has been immensely popular because it satisfies a need that had not previously been recognized. The catalog contains many new data points including adhesion measurements, evaporation equations, new form of distillation data, BTEX and C3 analysis. It is now possible to generate over 800 data points for each oil.

2. BOSS - Behavior of Oil SpillS

The BOSS project is designed to provide a comprehensive collection and review of data and concepts related to oil spill behavior. Topics include behavior of oil spilled at sea but will also include the lesser-documented topics of oil on land, on freshwater and in the ground. It will combine into one source, the literature on oil spill behavior and findings from previous joint research projects No literature collection or review of this type exists now, over 5,500 papers have been collected and initially reviewed to date. Completion of this project will result in a series of volumes combining not only the review of literature, but also data tables and unpublished results. This will enable future reevaluation of processes and use of data for model inclusion.

3. Evaporation of Oil

This joint project with Environment Canada will study the fundamental relationships involved in the evaporation of crude oils. The relationship between evaporation and temperature was investigated. In FY 1996, it was found that the evaporation rate was linear with temperature change. Empirical equation were developed to accurately predict evaporative loss. The evaporation equations for 60 oils and petroleum products were completed. The evaporation equations were provided to NOAA for inclusion into their oil spill response models. In FY 1997, evaporation equations for additional oils and petroleum products will be developed.

4. Emulsion Formation and Stability Studies

Emulsion formation is a major factor complicating oil spill cleanup. A critical aspect to the study of oil spills is the assignment of stability to emulsions and a means to measure this. Studies in the past year have focussed on stability as caused by asphaltenes and resins. The purpose of the present study is to move the emulsion studies onwards to focus on the classes of compounds responsible for their formation: asphaltenes and resins, and particularly on the migration of these compounds from an oil solution to the oil-water interface.

Asphaltenes are the primary agent responsible for the formation and stability of water-in-oil emulsions. These large compounds are interfacially active and behave like surfactants. Like surfactants, they can stabilize droplets of oil and water within each other, in this case water-in-oil. The role of resins may be important, however the experimental results did not encompass resins to the same extent as asphaltenes.

Asphaltenes migrate to the oil-water interface from solution in oil. This process can continue over weeks. Experiments conducted as long as 3 months after emulsion formation, indicate that the asphaltene migration may still continue. This migration may explain the observation that many emulsions increase in stability and visco-elasticity after sitting or periods of times It was found that a class of relatively unstable emulsions, named meso-stable emulsions, exist and degrade into stable emulsions, oil and water or into oil and water. Further work discriminating between the 2 and 3 part meso-stable emulsions are necessary. Tests for emulsion breakers will also be improved.

5. Oil Analysis

Work will continue at the Environment Canada's, River Road Environmental Technology Center, in Ottawa to improve oil measurement techniques. In 1996, a new "total analysis" technique, a BTEX (benzene, xylene, toluene) method, a standard weathering method, and a method for analyzing biodegradation were developed. Validation testing is underway. Methods to analyze the polar fractions will continue.

6. Oil Adhesion Test

Researchers studying the penetration of oil into shoreline sediments have identified oil adhesiveness as a possible influencing factor. To provide a measure of this property, a test was developed by Environment Canada. In this test, oil adhesion is defined as the mass per unit area, remaining on a standard test surface, after a defined time, under prescribed conditions. In FY 1996, six different surface materials (stainless steel, glass, plastic, Teflon, wood, and ceramic) were tested with five different oils. The results indicate that the relative adhesiveness of the oils is unaffected by the use of different materials. In 1997, work will investigate the basics of oil-to-substrate adhesion process, controlling variables such as oil properties and the influence on oil retention and persistence in beach sediments. This is a joint project with Environment Canada.

7. Basics of Oil Spill Cleanup

The current "Basics of Oil Spill Cleanup" manual was published in 1983 by Environment Canada (EC). The manual explains what happens to oil when it is accidentally spilled on water or land and the specific cleanup strategies which are possible under varying environmental conditions. Work has been completed on a "new " manual. The draft chapters are undergoing final review. Over 200 high quality photographs have been collected for potential use and the line graphics are complete. This manual will be printed in FY 1997.

III. Remote Sensing and Surveillance

1. Laser Fluorosensor Technology

The objectives of this project include; laser fluorosensing technology development for oil spill applications including oil on water, oil on shoreline, oil among debris, ice or weeds and oil on ice; develop prototype instruments; test these instruments and develop a new generation instrument which provides scanning capability.

a. Laser Environmental Airborne Fluorosensor (LEAF)

The Laser Environmental Airborne Fluorosensor (LEAF) continues to be developed into an operational system. In FY 1996, the LEAF was tested in several short test flights, including demonstration flights over the Chesapeake Bay. It was used extensively during the raising of the IRVING WHALE barge in the Gulf of St. Lawrence in July-August 1996. A series of test flights were conducted January, 1997 over the natural oil seeps offshore Santa Barbara, CA. The flight mission was designed to test the LEAF system over the natural oil seeps in day/night conditions. EC will provide their remote sensing plane, a modified DC-3, with the operational LEAF and photographic systems installed.

b. Scanning Laser Environmental Airborne Fluorosensor (SLEAF)

The Scanning Laser Environmental Airborne Fluorosensor (SLEAF) development is nearing completion with delivery and aircraft installation set for June, 1997. The SLEAF system will employ a state-of-the-art laser optimized for fast deployment and the detection of oil contamination on shorelines, land, snow, ice, and the marine environment. It will have an adjustable scanning capability that will allow for selection of the optimum swath to respond to various spill scenarios. This capability will maximize coverage on shorelines, ice-infested waters and other complicated surfaces.

The SLEAF has been designed to provide a real-time annotated map output product. This geo-referenced map, will be faxed or down-linked to oil spill response personnel. The timely information provided by the SLEAF sensor will help mitigate the harmful effects of oil spill by ensuring fast ad effective response. The SLEAF will detect and classify oil in real-time. Geo-referenced oil contamination locations will be easily visible on the hard-copy map output. Once the SLEAF has been installed in Environment Canada's aircraft it will undergo a period of flight testing to verify the proper operation of the system under airborne conditions and ensure the system complies to all design specifications.

2. Frequency Scanning Radiometer (FSR) for Measuring Oil Slick Thickness

A major problem associated with oil spills on the ocean is determining the thick-oil regions so appropriate countermeasures can be applied to those areas. Although oil slicks can cover very large areas, more than 80 percent of the slick is contained in less than 20 percent of the slick area. Therefore, it is crucial that an accurate spatial-oil-thickness map is constructed so spill countermeasures can be employed efficiently. MMS has contracted the Massachusetts Institute of Technology's Lincoln Laboratories to develop an all-weather, airborne instrument capable of measuring and mapping oil thickness over an oil slick in real-time and transmitting images to the response vessels. In FY 1996, the long-term project plan, detailing the technological advances expected and milestones to be met to create a dual-band, airborne, radiometric sensor for remote sensing oil slick thickness was completed. The existing 26-40 Ghz (Ka-Band) FSR was upgraded to a multichannel instrument (faster measurement), and validated during tests at at Ohmsett - The National Oil Spill Response Test Facility, in Leonardo, New Jersey.

In FY 1997, the existing 26-40 Ghz (Ka-Band) FSR will be further evaluated during tests at Ohmsett. The design for the 75-110 Ghz (W-Band) FSR will be completed and a prototype developed. Additional tests are scheduled following each major hardware development milestone. Project participants include the U.S. Coast Guard Research and Development Center.

IV. Oil Spill Chemical Treating Agents

This objectives of this joint MMS/Environment Canada project are to: develop repeatable tests for oil spill treating agents; test commercial products; understand dispersant action mechanisms; and test new product ideas. Work will continue to develop standard test protocols for evaluation of performance for ten categories of oil spill chemical treating agents. Work will also continue to evaluate existing agents including dispersants for efficacy and toxicity, and to apply lessons learned to develop new oil spill chemical treating agents.

1. Oil Spill Chemical Dispersants

Environment Canada will continue to study the chemistry and physics of dispersants. In FY 1996, work focused on filling gaps of dispersibility in the Catalog of Crude Oil and Oil Product Properties. Over 400 new entries were added to the oil catalog. Five new dispersant products were tested. None passed the minimum acceptability criteria. In FY 1997, work will include study of several new dispersants formulations. Work will also focus on the filling additional data gaps for dispersibility in the oil catalog .

2. Other Spill-Treating Agents

Work will continue on the development of tests for treating agents. Tests will ultimately be developed for the effectiveness measurement of nine classes of treating agents. To date, tests have been developed for dispersants, solidifies, and surface-washing agents. Preliminary tests have been developed for biodegradation agents (both fertilizers and organisms), emulsion breakers, recovery enhances, and emulsion preventers. Work this year will focus on bringing the latter tests to a final form of standardization and to test more products currently on the market. Work will also include developing tests for remaining agents and testing new agents. Tests will be conducted on the variation of effectiveness results (dispersants, solidifiers, and surface washing agents) with various simple parameters such as time, temperature and apparatus set up.

V. Mechanical Containment/Recovery

1. Testing and Evaluation of Oil Spill Sorbents

Standards developed by this project, based largely on American Society of Testing and Materials (ASTM) and Environment Canada's (EC) methods, have been widely adopted in North America. EC continues to work with the ASTM F-20 committee in methods development and the testing of sorbent booms, and in the development of new test protocols for sorbents used to combat chemical spills. In FY, 1997, work includes testing of new products and final development of the sorbent database, for inclusion on EC's Internet Homepage. Project participants include: Environment Canada, Canadian Coast Guard, U.S. Coast Guard.

2. Svalbard Shoreline Project

The Svalbard Shoreline Field Trials are part of a series of studies in a long-term program to better understand the behavior of oil on shorelines and apply appropriate response options. The Svalbard Shoreline Field Trials, in combination with its partner, the Oil and Fines Interaction (OFI) Basin Trials, will investigate both the effectiveness of mainstream in situ shoreline cleanup techniques and the natural processes for oil removal from shorelines, in particular by oil and fine particle interaction. The implementation of shoreline cleanup following a spill in a remote area is limited by the constraints of available equipment and personnel, and the desire to minimize waste materials that require transport and disposal. In such cases, the preferred option is to treat the oil in situ so that environmental recovery is accelerated without the requirements for labor intensive effort. The FY, 1996 field season of the Svalbard Shoreline Field Trials Project concluded on August 20, 1996, at experimental sites near Sveagruva, Svalbard, Norway. Baseline field work and field protocol development were successfully completed. The 1996 field activities met all of their objectives and will be used to design the final full-scale field trials scheduled for Summer 1997.

The Svalbard Shoreline Field Trials are being supported by an international partnership of spill response and spill research agencies which include: Minerals Management Service, Environment Canada, Exxon Research , Imperial Oil Canada, Marine Pollution Control Authority (U.K.), Texas General Land Office, Norwegian Pollution Control Authority, and the Swedish Rescue Service

VI. Ohmsett - The National Oil Spill Response Test Facility

Ohmsett is The National Oil Spill Response Test Facility, located in Leonardo, New Jersey, on the grounds of Naval Weapons Station Earle. Ohmsett is the only facility in the world where clients can conduct full-scale oil spill response equipment tests with a variety of crude oils and refined petroleum products. Located an hour south of New York City, the facility features a test tank 203 meters (667 feet) long by 20 meters (65 feet) wide and with a water depth of 3.4 meters (11 feet) capable of reproducing various sea states, a fully computerized data collection system, above and below waterline video, and a complete oil storage and handling system. The tank is filled with 9.84 million liters of crystal clear water.

Ohmsett is available on a reimbursable basis for use by Government, private industry and academia as a research center to meet a variety of objectives involving marine oil spills, such as evaluation of new equipment or techniques (skimmers, booms, storage devices, remote sensors), hands-on training of response personnel.

In FY'95, six tests, totaling 81 days were completed. In FY'96, eight tests totaling 114 days were completed. In FY'97 ten tests are currently scheduled with several more in the planning stage. Concerted efforts are being made to improve utilization and reduce testing and operational costs. MMS is actively researching and developing new ways to detect, understand, and ultimately clean up spills. Encouraging research and development projects, along with evaluating new or existing oil spill response equipment has successfully increased the number of tests at Ohmsett.

CHAPTER 103

OIL SPILL CLEANUP TECHNOLOGIES AND PLANNING

RAYMOND NOTTINGHAM

When the drilling from an offshore oil production platform in the waters off of Santa Barbara caused a massive oil spill in 1969, the age of innocence with regard to oil spills in California waters was over. It became woefully clear that even though some technology for oil spill clean up existed, the tools for clean up were mostly primitive and not in place where they might be needed.

Within a year after the Santa Barbara incident, the California State Lands Commission set down guidelines that would change the direction of spill response in California dramatically. The State Lands Commission mandated that before any more drilling and production activities could take place in California waters, adequate means to clean up an oil spill must be in place. By 1976, the same kind of regulations were in place for federal waters. The wheels were turning toward a more responsible attitude regarding oil spills.

In response to the federal and state mandates, companies like the one for which I work, Clean Coastal Waters, Inc., were formed. These companies were a series of non-profit, oil spill clean up cooperatives, funded by the major oil companies, designed to protect California and federal waters and the California coastline. Other privately held companies existed that did spill response as an adjunct to their principal lines of work, like industrial cleaning, vacuum truck hauling, etc., but the co-ops were the first companies formed solely to combat oil spills. The co-ops purchased or leased spill response vessels, containment booms and oil skimmers. Some of the equipment was designed for use in shallow or protected waters while other equipment was designed to recover spilled oil on the open sea.

To a large degree, the type of equipment chosen to fight oil spills twenty-five years ago is the type of equipment that is still in use today. Many of the original concepts involved in oil containment and recovery have proven to be successful but refinements on those concepts have been on-going. The key tools involved with oil spill response on water, besides well trained response personnel, continue to be boats, containment booms and skimmers.

Boats that were initially used for spill response activities were usually adapted from some other service, i.e.: offshore supply, water taxi, equipment transport, etc. Today, although many vessels of that type are still used, there has been a greater emphasis on boats designed specifically for oil spill response and recovery. Co-ops in California have fast response boats that are specifically designed to get to the scene of an oil spill as quickly as possible and attempt to contain the oil before it has a chance to spread. The co-ops also have skimming vessels positioned all along the coast, designed or re-designed for the express purpose of recovering spilled oil from our offshore or near shore waters. Clean Coastal Waters' boats have highly selective, brush type skimming systems that have high recovery rates and our vessels have on board storage tanks for the recovered oil. They can stay offshore for long periods of time, if necessary, and continue to recover oil, pumping large accumulations to oil storage barges that are also pre-identified and located along our coast.

Containment booms vary in size and type and also have many different functions. Some booms are rigid and designed to stay in the water at all times and might be found in the waters around a marine terminal or surrounding a ship that is taking on fuel or discharging oil. Clean Coastal Waters primarily uses two different types of boom to fight oil spills; boom that uses air as the flotation medium and boom that uses foam for the flotation.

The first air flotation booms were prone to problems with how the boom actually got inflated, which caused the boom to be ineffective or even to sink. Improvements in technology have made today's air flotation booms more reliable with more sophisticated means of delivering the air to the flotation chambers. These booms have proven to work well containing oil on open seas because they have good wave following characteristics, they can usually be deployed easily and they have an ability to be stowed compactly when not in use.

The other type of boom typically used by CCW is foam flotation boom. Foam flotation booms tend to be more rugged than air flotation booms and might be used near rocky shorelines or in other areas where chafing from pilings, ship hulls or other structures would be expected. CCW also uses foam flotation boom in instances where the boom is expected to be in the water for extended periods of time since it is not susceptible to water seeping in to the flotation chambers. Foam flotation booms have improved over time with regard to the types of foam used as flotation, the length of the boom segments, the durability of the boom fabric, and the types of connectors.

The third tool we will mention is the oil skimmer. ASTM identifies eleven different types of oil skimmers on the market, each with an intended application. Skimmers might be designed to work in refineries, protected waters, rough waters, or any number of other applications. Basic concepts of how to skim and recover oil have not changed to a great degree, but over the last several years refinements have been made that have allowed skimmers to become more effective, much more efficient and frequently easier to operate.

As with containment boom, CCW tends to utilize two main types of skimmers; the weir type and the oleophyllic brush type. A weir skimmer attempts to collect the oil floating on the surface of the water by allowing or encouraging it to spill into a receiving unit floating in the water. Refinements have been made over time to allow weir skimmers to collect a higher percentage of oil and less water than the first models allowed. This type of skimmer is particularly useful when recovering hydrocarbon in calm water as less water will typically enter the receiving unit.

Oleophyllic skimmers operate under the premise that oil will stick to a surface if that surface is made available. Oleophyllic skimmers include rotating discs, belts and brushes, which, when presented to the oil, tend to get coated by the oil. While the oil is coating these surfaces, water that comes in contact with these same surfaces simply runs off. The oil on the rotating discs, belts or brushes is wrung out or wiped off and channeled to a reservoir where it is pumped to storage. CCW has three ocean going skimming vessels, each of which carry two oleophyllic brush skimming systems. That gives us an approved recovery rate of nearly 420,000 gallons per boat per day. Use of this type of skimmer allows for an extremely high oil to water ratio in the recovered product and can frequently mean the need for less storage capacity on scene.

In addition to the tried and true technologies, newer alternatives have begun to gain a small amount of acceptance. In-situ burning, a topic which will be addressed by Yvonne Adassi in today's session, is one that has undergone extensive testing and has been found to be effective, but is a hard sell because of the fear of air borne contaminants. Dispersant use is a second viable alternative technology which has been around for quite awhile but unused in the waters off California's coastline because of the fear that the dispersant liquid itself might cause more harm to the environment than the oil. Regulatory bodies have come to the realization, however, that in certain situations dispersant is the best answer and Quick Approval Zones have been set up off the coast which will allow the approval process to proceed quickly (as the name implies), for dispersant use. Clean Coastal Waters feels that these two alternative technologies are not mutually exclusive. On certain spills, burning, dispersants and mechanical means might all be used to reach a successful conclusion. We support using any appropriate approved means that might cause an oil spill to be cleaned up as fast as possible.

Although the equipment used for oil spill response and recovery has not greatly changed over the last 25 years, what has changed is the philosophy involved. Whereas all of the component parts I have discussed are critical to the effectiveness of the clean up effort, historically each part has been thought of as an individual item. Today all of the component parts involved in an oil spill response are thought of as part of a system. The realization is that any part of the system by itself is not going to be very successful but the union of those component parts into a functioning system can get the job done.

The philosophy of oil spill response has also taken a dramatic turn into the realm of pro-activity. Whereas in the 70's and 80's the goal was to have people and equipment ready to react to an incident, in the 90's the vision has focused heavily on prevention and on advance preparation. Advance preparation means many things but it mostly means planning ahead to decide who will do certain things, what they will use, and where they will use them in the event of an incident.

As part of this advance preparation, Clean Coastal Waters has teamed with the Coast Guard and the Department of Fish and Game, Office of Oil Spill Prevention and Response (OSPR), to develop strategies to best protect California's sensitive southern coastal areas from the threat of an oil spill. These strategies are contained in the Area Contingency Plans, plans that were developed by the Coast Guard as a result of OPA 90 Legislation. Area contingency plans contain specific information on California's coast, how best to protect it and what resources are available to provide that protection.

With the OSPR providing guidance on which areas of our coastline would suffer the most severe impact from an oil spill incident, a priority on protection and a system for providing that protection has been established and included in the Area Contingency Plans. Clean Coastal Waters participates in the development of the protection strategies and volunteers the people and equipment to test those strategies to ensure their feasibility and the reality of what would actually be required to put the strategy in place.

As part of this program, Clean Coastal Waters develops an annual plan showing existing, sensitive, coastal locations whose protection strategies need verification. The proposed annual plan is agreed to by the Coast Guard and the OSPR representatives having jurisdiction in those geographical areas and the strategy verifications begin. One area is exercised each month and some of the most sensitive areas, like Anaheim Bay, may be visited two or three times in a year.

Anaheim Bay is the entrance to the Bolsa Chica Wetlands and the Seal Beach National Wildlife Refuge. These two sanctuaries comprise the most extensive and valuable wetlands in southern California. They are utilized by over 150 different species of waterfowl, shore birds, and sea birds of which four are threatened or endangered. In addition, this wetland area is inhabited by various types of juvenile fish and provides nourishment to a wide variety of vegetation. The strategy for protection as contained in the Area Contingency Plan is extensive and actually employs four lines of defense that would have to be breached for oil to enter from the sea.

Clean Coastal Waters works with personnel from the Orange County Sheriff's Department and the Seal Beach Naval Weapons Station on how to use and deploy the booms in this area. Both groups realize the potential danger to these wetlands areas and have trained with CCW, volunteering their own people and boats, to deploy the containment boom that CCW has pre-positioned on the shorelines of the Naval Weapons Station property. Since the Naval Weapons Station has workboats located on scene at all times and since the Sheriff's Department has patrol boats located in the vicinity at all times, and since they are both willing to pull the booms out in the event of an emergency, they are an invaluable asset.

Other groups who work with CCW deploying booms in sensitive areas include the Long Beach Fire Department, Long Beach Lifeguards, and private companies such as Vessel Assist, Sea Tow, Foss Environmental and Advanced Cleanup Technology. Chevron USA participates with a very well trained spill response team from their El Segundo refinery. Chevron personnel practice regularly using their own boats and equipment and are actually the first line of defense for the entire Santa Monica Bay.

Strategies for other environmentally sensitive areas including San Diego Bay, Batiquitos and Agua Hedionda Lagoons, Newport Bay, Cabrillo Beach Wetlands, Palos Verdes Peninsula, Ballona Creek Wetlands and Malibu Lagoon are also worked with on a regular basis. Protection strategies even exist for the islands off of California's coast. The sea otter breeding areas of San Nicolas Island, and the seal and sea lion pupping areas of San Clemente and Santa Catalina Island are planned for in the Area Contingency Plan.

Environmentally sensitive areas nearly always take precedence over economically sensitive sites but the strategies for economically sensitive sites are also exercised each year. Economically sensitive sites as defined in the ACP can involve many different activities. Examples of economically sensitive sites include locations like the Occidental College Fish Hatchery, the Alamitos Bay Marina, Leo Carillo State Beach Campgrounds, Los Angeles Harbor's commercial fishing boats, Long Beach Harbor's cargo terminals and San Diego Electric's cooling water intakes. The most recent site strategy verifications for economically sensitive areas have been at Oceanside Harbor, Dana Point Harbor and Marina Del Rey.

Oil spills always pose a threat to California and the world ocean. However, the laissez faire attitude of years ago is long gone. The pro-active attitude of the regulatory community coupled with the willingness of the oil producers and transporters to plan ahead and act responsibly has made the California waters and coastline safer from the potential of oil spill pollution than ever before.

CHAPTER 104

PREVENTION FIRST OR BUSINESS AS USUAL

Richard W. Holly/Megan Thomas

ABSTRACT:

This paper details the process used by two states, California and Washington, to develop common reporting of oil spill causes. Many different agencies use databases to store oil spill information, but there is a general lack of standardization in reporting requirements, rendering difficulty in the sharing of information. Many of these databases only reflect initial spill reports and few contain information on spill causes that could be utilized to prevent the recurrence of oil spills.

Washington State's Office of Marine Safety has developed an extensive database including spill and casualty information from vessels transiting Washington state waters. California State Lands Commission, Marine Facilities Division, has similarly developed an oil spill database for marine terminals in the state of California.

What makes the above-mentioned systems unique is how they are geared to integrate reliable, accurate information on incident causes, with specific regard to human error. Both agencies have joined together to create in a pilot program to share oil[1] spill causal information using common terms. This paper addresses the approach used to establish common data dictionaries, oil spill report formats, and the methodology to determine causal information. Currently, California and Washington are working conjointly with other states, including British Columbia, to develop common reporting techniques throughout the West Coast. This vital information collected and shared can be used by each jurisdiction to create and develop appropriate prevention strategies.

---

[1]Richard W. Holly, California State Lands Commission, Marine Facilities Division, 330 Golden Shore Suite 210, Long Beach CA 90802-4246; Megan Thomas, Washington Office of Marine Safety, 711 State Avenue N.E., P.O.Box 42407, Olympia, WA 98504-2407

INTRODUCTION

As the Oil Pollution Act of 1990 (OPA '90) approaches its seventh anniversary, the impact of stricter regulations is becoming the focus of many data collection projects and reports. Prevention activities have been given new emphasis, in an industry that has traditionally focused on response. Coordinating spill response is difficult in the field, clean-up is marginal at best, and costs can become exponential as time spent on clean-up is extended and natural resource damages are assessed.

Collection and analysis of oil spill data is an essential element of many spill prevention programs. Valuable prevention lessons are gained when investigators are able to identify the causes that contribute to oil spills. To prevent oil spills, meaningful risk data and oil spill causal information need to be available, in a user-friendly format, to agencies in the oil spill prevention business. Most public and private organizations agree that prevention is better than cure and recognize the importance of collecting meaningful spill data (Lee, 1995) . Prevention of oil spills is a preferred approach, but little has been published on prevention methods, compared to the amount of material on spill response. Comprehensive analysis of oil spill data has been hindered in the past by three problems.

Firstly, previous attempts at analysis of oil spills were sometimes limited by the infrequency of high impact spills, which are usually the focus of investigation efforts. Conclusions about spill causes can be called into question, when they are based on a small number of incidents. Due to the limited information on oil spill causes, it is crucial that agencies coordinate data collection efforts and share findings.

Secondly, many agencies collect spill data independently, with differing authorities and with different goals. For example, attempts have been made to identify high risk entities, such as tank vessels (carrying large volumes of petroleum products) and marine oil terminals (where large spills could occur). Many organizations, such as the Tanker Advisory Center, Lloyd's of London, and the U.S. Coast Guard, have databases with information on oil spills. However, most of these systems are not compatible with one another, nor do they have the capability to share real-time data with other systems in a timely manner.

Thirdly, causal information that has been collected often does not delve far enough into the incident to identify the "root cause". Research by the U.S. Coast Guard and the international maritime community indicates that eighty percent of all marine accidents are caused by human error. (U.S. Coast Guard, 1995). Unfortunately, too often oil spill data indicates the cause of a spill as a mechanical or equipment failure, instead of addressing the human error factor.

To address these problems, the Washington State Office of Marine Safety (OMS) and California State Lands Commission, Marine Facilities division (MFD) have initiated a pilot program to share oil spill data. Both agencies have developed incident data collection and analysis programs that concentrate on human and organizational errors contributing to spills or marine casualties. This paper outlines the progress the agencies have made toward developing a system to ensure that the shared data are compatible with the same reference points, quality, and meaning.

THE FIRST STEPS

The Washington State Office of Marine Safety (OMS) and California State Lands Commission, Marine Facilities Division (MFD) have independently developed similar oil spill prevention organizations, one with primary jurisdiction over vessels and the other with primary jurisdiction over marine oil terminals. The first step in this project was to gain an understanding of each agency's regulatory authority and how collection of oil spill information fit into their respective prevention programs.

California State Lands Commission, Marine Facilities Division

The California State Legislature recognized the importance of oil pollution prevention in its passage of the Lemper-Keene-Seastrand Oil Spill Prevention and Response Act of 1990(Act), declaring, "Because of the inadequacy of existing clean-up and response measures and technology, the emphasis must be put on prevention, if the risk and consequences of oil spills are to be minimized" (Lempert et al.1996). The Act placed responsibility for oil spill prevention at marine oil terminals with the California State lands Commission (Commission), which developed a new program and created the Marine Facilities Division (Division).

The Commission's responsibilities under the Act include regulation of all marine oil terminals in the state to ensure that the "*best achievable technology*" is being used to provide the "*best achievable protection*" of the public health and safety, and the environment. The Commission's program is based on comprehensive regulations developed from world-wide sources and in conjunction with technical advisory groups comprised of industry, governmental and environmental experts. Since implementation of the program, the volume of oil spilled at marine terminals in California has declined sharply.

Data collection and analysis became a major building block of the Commission's program. Every marine terminal inspection, transfer monitoring activity and oil spill investigation has been carefully documented in an electronic database. Trend analysis is conducted and used for several purposes including advising industry officials and developing future program direction. The Division's comprehensive oil spill database is the only database in the state that contains comprehensive information on the causes of spills that occur at marine terminals. Information in the database has been shared with several other states and the database itself has been offered as a model for a States/British Columbia Oil Spill Task Force (States/BC) Project[2].

Washington State Office of Marine Safety

After the *Exxon Valdez* spill in Alaska in 1989, and two major spills along the Washington Coast in 1988 and 1991, the Washington State Legislature passed the Oil Spill Prevention and Response Act of 1991, a comprehensive law increasing state involvement in oil spill prevention activities. This legislation established the Office of Marine Safety (OMS) to implement oil spill prevention and maritime safety programs in the state. The mission of OMS is to reduce the risk of oil spills in Washington waters by promoting safe marine transportation.

OMS implements its oil spill prevention mission through a variety of innovative marine safety programs. Risk identification and management are the foundations of OMS prevention strategies. Similar in nature to California's program, one of OMS's programs also focuses on "best achievable protection" standards, by requiring tank vessels owner/operators to submit "Prevention Plans" (Chapter 317-21 Washington Administrative Code). OMS also uses accepted industry standards to evaluate the risk that cargo and passenger ships pose to the public health and safety and the environment (Chapter 317-31 Washington Administrative Code). OMS prevention programs focus on the human element and improving human performance in maritime practices.

Like California's Marine Facility Division (MFD), OMS has a core of

[2]States/British Columbia Oil Spill Task Force (States/BC). The States/British Columbia Oil Spill Task Force is comprised of state, and provincial representatives from Alaska, Washington, Oregon, California and British Columbia organized in 1989 to strengthen efforts to prevent, prepare for, and respond to marine oil spills on the West Coast.

experienced mariners that conduct vessel inspections and investigations. The investigations program focuses on identifying the human and organizational factors that cause marine accidents or oil spills. Investigation findings assist in evaluating risk, but more importantly, the information is used to develop technical and educational materials that can prevent future accidents or spills.

OMS and MFD have worked together on similar projects in the past. In April, 1994, an information-sharing agreement was reached between OMS and MFD to share information on tank vessels which received violation notices during inspections or monitoring in either state. MFD initiated the reports and OMS agreed to reciprocate when their tank vessel inspection program was in place. Without enough personnel to inspect every vessel, information shared between the states could be considered when deciding which vessels to board. The shared information could also be used to develop a vessel history, which is an integral part of evaluating risk.

In September, 1995, further initiative was taken to enter into a regional oil spill data base project being sponsored by the States/British Columbia Oil Spill Task Force. Through this effort, OMS and MFD started an information exchange which began to show the similarity of their programs, philosophy of prevention, and their oil spill analysis, with particular attention to human error. In April, 1996, MFD and OMS agreed in principal to jointly pursue a pilot program which expanded on the shared information on tank vessels. The agencies began the process of developing a program to share oil spill causal information using common terms.

THE PROCESS

OMS and MFD began the data sharing process by discussing what they hoped to accomplish and determining project goals. Both agencies agreed that quantifying reliable information on oil spill causes, from their combined databases, was a desirable outcome. Sharing investigation strategies and data collection systems to get at oil spill causes would also be beneficial. Both OMS and MFD have developed programs aimed at collecting causal information; the challenge was finding a way to assimilate the data.

The primary goal of the project was to develop a data dictionary that would ensure that the shared data be comparable. This data dictionary would reduce both state's oil spill terminology into common language and permit the individual states to integrate the information. Before the data dictionary could be developed, it was necessary to determine the scope of each states' authority and their respective areas of oil spill investigation.

OMS has regulatory authority over cargo and passenger vessels 300 gross tons and larger, and all tank vessels. Part of the agency's database stores oil spill, marine casualty, and "near miss" information for each vessel operating in state waters. The database contains information for incidents world-wide, but investigation efforts focus on those that occur in state waters. OMS does not have the resources to fully investigate each incident to determine root cause, so each case is evaluated to determine if a valuable prevention lesson could be learned. For those cases that are investigated, information on primary and contributing factors are stored in the database.

MFD has regulatory authority over marine terminals in general, and specifically, during oil transfer operations to or from vessels or barges with an oil carrying capacity of 250 barrels or more. MFD's database contains oil spills, vessel and terminal deficiencies during oil transfer operations and terminal deficiencies recorded during periodic terminal inspections. This information is used, in part, to determine allocation of inspector resources by assessing risk factors associated with terminals and vessels. Like OMS, MFD does not have the personnel to monitor all petroleum transfer operations. However, all oil spills at marine terminals are investigated.

The agencies were able to identify basic information (database fields) that both spill databases contained. At this point, defining the standardized entries for each field became extremely important to ensure the data would be comparable. The Table One below summarizes the common data fields.

Table One

COMMON DATA FIELDS		
Elements	**OMS**	**MFD**
Person making the report	YES	YES
Date and time of report	YES	YES
Date and time of incident	YES	YES
Location name (where the spill occurred)	YES	YES
Source of the spill (where the spill originated, the name of the vessel or facility)	YES	YES

COMMON DATA FIELDS		
Elements	**OMS**	**MFD**
Material spilled (type of petroleum product)	YES	YES
Quantity spilled (in gallons)	YES	YES
Activity at time of the incident (what was occurring immediately prior to incident)	YES	YES
Immediate cause (actual action or error that led to the incident)	YES	YES
Contributing factor (conditions or occurrences that led to the immediate cause)	YES	YES

It became clear that both agencies collected oil spill data, but their databases were organized differently. OMS collects information on vessels that have oil spills; MFD collects data for spills that occur at marine terminals. The agencies continued to compare database elements. Tables Two and Three show the common areas that both states focused on as they tried to assimilate data. The data could be segregated by either of these categories. For example, MFD could segregate those spills involving vessels that OMS has authority over and OMS could segregate those spills involving a petroleum product transfer. At this time the agencies decided to focus on spill causes, but more detailed approaches will be considered in the future.

Table Two

OIL SPILL DATA COLLECTION BY VESSEL TYPE		
VESSELS (> 300 T)	**OMS**	**MFD**
Tank ships	YES	YES*
Passenger vessels	YES	YES*
Fishing vessels	YES	YES*
Cargo vessels**	YES	YES*
Bulk carriers	YES	YES*
Tug boats	NO	YES*
Public vessels	NO	YES *
Tank barges***	YES	YES*
Marine terminals	NO****	**YES***
Mobile marine terminals	NO****	**YES***

* MFD if the spill occurred at a marine terminal.
** OMS breaks this category down into general cargo ships, refrigerated carriers, purecar carriers, and container ships.
*** OMS information includes petroleum and chemical barges.
**** The OMS database does not gather information on terminals per se, but if a regulated vessel has a spill at a terminal, the spill data would be collected.

Table Three

ACTIVITY AT THE TIME OF THE SPILL		
ACTIVITY	**OMS**	**MFD**
Bunkering	YES	YES*
Lightering	YES	YES*
Vessel underway	YES	NO
Vessel at anchor	YES	NO
Ballasting/deballasting	YES	YES*
Bilge pumping	YES	YES*
Discharging cargo (specify petroleum or non-petroleum)	YES	YES*
Loading cargo (specify petroleum or non-petroleum)	YES	YES*
Internal transfer	YES	YES*
Other		

* MFD investigates all oil spills occurring at marine terminals and further categorizes the spills into discretely defined evolutions of the oil transfer process (hook-up, start, steady state, top-off/stripping, and disconnect).

Thorough incident investigations are the most reliable way to obtain accurate causal information. In reviewing historic spill data there is a great deal of variance in reported spill causes. OMS and MFD have agreed that the cause of the spill should not be confused with "what happened?" but rather, "why did that happen?" For example, pump failure may be reported as a spill cause, but poor maintenance on the pump is what caused the pump to fail. When trying to identify spill causes, the investigator needs to focus on two primary questions: Firstly, they must ask "why did that happen?", until they reach the root cause. Secondly, by answering the question "what could have been done to prevent this spill?", they may arrive at the root cause.

To help distinguish the "what " from the "why," OMS and MFD carefully differentiated between data elements. They agreed to approach spill causal analysis by using the terminology: Spill source, activity at the time of the spill, immediate cause, contributing factor, and primary or root cause. Table Four defines these terms and provides a sample of how spill information might be categorized.

Table Four

SPILL TERMINOLOGY		
ELEMENT	**DEFINITION**	**SAMPLE DATA**
spill source	where the spill originated, was it a vessel tank, a facility hose, etc.	Vessel fuel tank
activity at the time of the spill	what was occurring immediately prior to the incident; was fuel being transferred was maintenance being performed, was the vessel underway	bunkering
immediate cause	the actual action or error that led to the incident, such as pump failure	tank overflow
contributing factor	the conditions or occurrences that led to the immediate cause, such as lack of training or poor maintenance	lack of communication between vessel and facility personnel
primary or root cause	the lack of control or management and safety systems that allowed conditions to exist that led to the immediate cause	failure of company to set procedures for "topping off"

Finally, OMS and MFD had to agree on how to classify spill causes, once they were determined. OMS had three main categories: external, equipment/material, and human error; while MFD had seven categories:

mechanical, organizational, personnel, seepage, unknown, vessel distress and weather. After discussing the meaning of each entry, the agencies were able to "match-up" the spill causes.

This list of spill cause terminology is subject to change as each agency continues to develop and improve it spill cause classification system. For this reason, creating and maintaining a "dictionary" of each term is extremely important. This "data dictionary" should be provided along with data transfers to ensure that the data is not misinterpreted. The agencies tried to keep the list of terms short enough to be easy to use, but long enough to be inclusive of the many circumstances that could exist. Both agencies agreed that a narrative was the best way to clarify what a simplified one-word cause meant.

Table Five shows the areas where OMS and MFD's spill cause terminology was easily comparable. Both agencies had general categories with varying levels of detail in sub-categories. At this stage of the project, the data was segregated into the major categories. The level of detail of sill causes will be refined as the project continues.

Table Five

INTEGRATION OF SPILL CAUSE TERMINOLOGY	
OMS	**MFD**
EXTERNAL OMS has 10 subcategories	WEATHER
EQUIPMENT/MATERIAL FAILURE Mechanical/equipment failure Structural failure OMS has 4 additional subcategories	MECHANICAL Equipment Structural MFD has 16 additional subcategories
HUMAN ERROR Communication Drugs/Alcohol Improper maintenance Inattention Insufficient personnel Procedural error Lack of training Other OMS has 7 additional subcategories	ORGANIZATIONAL & PERSONNEL* Communication Intoxication Maintenance program Inattention Insufficient personnel Bad procedure/Procedures not followed Training Other MFD has 6 additional subcategories

INTEGRATION OF SPILL CAUSE TERMINOLOGY	
OMS	**MFD**
UNKNOWN	UNKNOWN
OMS has no equivalent	VESSEL DISTRESS
OMS has no equivalent	SEEPAGE

Once the elements were agreed upon, actual data transfer was possible. The pie charts (Charts I, II, and III) that follow reflect a comparison of the data generated in both states individually, followed by a combination of the two. Through an analysis of the shared data, some common root causes can be identified. Each state can use the information to develop appropriate prevention strategies. Both agencies will continue to work on this pilot program, which will be evaluated in June, 1997. The combined data supports the conclusion that most oil spills are caused by human error. Since both agencies are interested in reducing human error, the next phase of the project will be to further define and segregate human error by type.

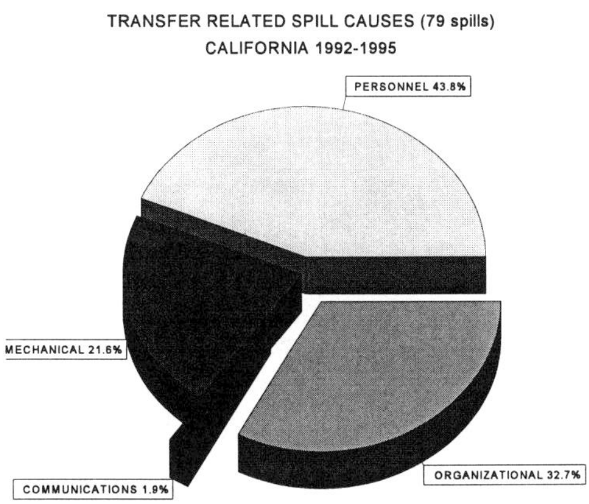

Chart I

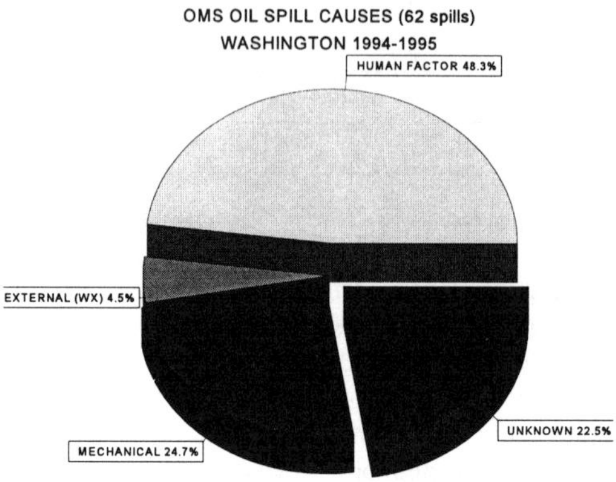

Chart II

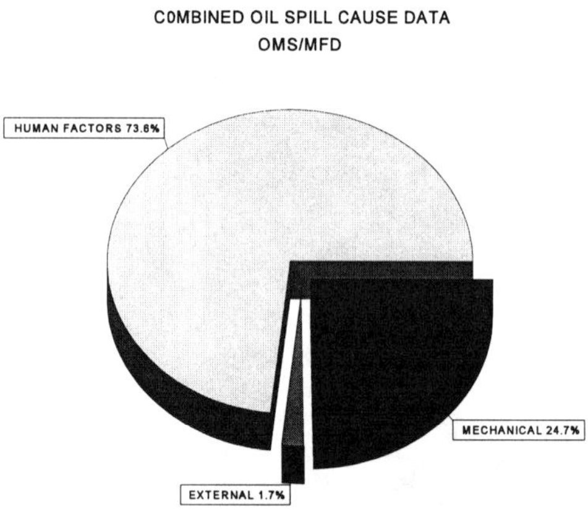

Chart III

THE FUTURE

As the development of a common data dictionary of oil spill cause terminology continues, the agencies also want to examine the training that investigators receive, in order to be able to determine spill causes. They will try to determine how a baseline training program could be structured to insure consistent oil spill cause reporting. Both agencies have developed investigator training programs and Table Six lists elements that could be used to build a common baseline of training to improve the collection of causal information.

Table Six

INVESTIGATOR TRAINING
Spill Investigation reporting formats
Definition of spill cause terminology
Oil types and estimating spill quantities
Responsible party responsibilities
Root cause analysis and human error

CONCLUSION

The work of Washington State Office of Marine Safety (OMS) and the California State Lands Commission, Marine Facilities Division (MFD) demonstrates the benefits of sharing oil spill data. Agencies and organizations interested in sharing data must be willing to take the time to clearly define data elements, to ensure that shared data is not misinterpreted. OMS and MFD found the best way to accomplish this was to create a data dictionary of spill terminology.

The lesson learned by OMS and MFD will be shared with the States/British Columbia Oil Spill Task Force, to assist in the development of common oil spill reporting formats throughout the West Coast.

BIOGRAPHY

Megan Thomas is a transportation specialist for the Washington Office of Marine Safety. Prior to OMS, Megan gained significant transportation experience both managing cargo shipments for federal agencies and as a U.S. Army logistics officer in Saudi Arabia during the Gulf War. She holds a Masters degree in Environmental Studies.

Gary Gregory has been the Chief, California State Lands Commission, Marine Facilities Division since its inception in 1990. Gary has extensive knowledge in port safety and environmental protection gained through 20 years of U.S. Coast Guard experience. He holds a Masters of Business Administration from George Washington University.

Rick Holly is a master mariner with a Masters degree in Oceanography. He heads the Planning Branch of the California State Lands Commission, Marine Facilities Division. He is currently a member of the American Council of Master Mariners, and the United States Naval Academy Alumni Association.

APPENDIX

Chapter 317-21 Washington Administrative Code, "Tank Vessel Oil Spill Prevention Plans," December 1994

Chapter 317-31 Washington Administrative Code, "Cargo and Passenger Vessel Substantial Risk Rules," July 1996

Lee, N. Robin, 1995. Operational Pollution Prevention Programs for Loaded Tankers
Transiting Coastal Waters. 14th biennial International Oil Spill Conference Proceedings
U.S. Coast Guard et al.,Long Beach, California, pp 311-314

Lempert-Keene-Seastrand Oil Spill Prevention Act (As amended as of January 1996) California Resources Code, Sections 8574-8760

U.S. Coast Guard. Prevention Through People Quality Action Team Report. Washington D.C.: U.S. Department of Transportation, U.S. Coast Guard, Office of Marine Safety, Security and Environmental Protection, Office of Navigation, Safety and Waterways Services, July 15, 1995.

CHAPTER 105

IN-SITU BURNING POLICY DEVELOPMENT FOR CALIFORNIA

by
Yvonne Najah Addassi
California Department of Fish and Game
P.O. Box 944209
Sacramento, CA 94244-2090

Abstract

To satisfy California's large petroleum requirements, hundreds of millions of barrels of crude oil and its refined products are shipped annually through its coastal waters and into ports. In addition, tens of million of barrels of petroleum are produced each year from offshore oil wells and transported ashore by subsea pipelines. Consequently, the California marine environment, including sensitive biological resources and amenity resources, is under continual threat of damage as a result of accidents that lead to spills of oil. The use of burning as an available oil spill response method can play an important role in reducing the possible widespread impact of oil spills to marine resources. The use of this technology, however, often requires the approval of many different oversight agencies and its use has often been very controversial.

This paper outlines the current efforts being spearheaded by the Office of Oil Spill Prevention and Response (OSPR) in the California Department of Fish and Game, to develop an in-situ burning policy to address oil spills in the open-water marine environments of the State. It describes the importance of early coordination and cooperation amongst the stakeholders in an in-situ burning decision, the formation of a working-group, and a discussion of the consensus approach integral to current policy development efforts.

The author concludes with a description of the pertinent sections of the draft policy, a discussion of current projected timeline for completion of the finalized in-situ burning policy, plans for public review and comment on the draft policy and procedures for policy adoption and integration into appropriate

State and Federal Plans. The implications of "negative public perception" regarding the use of in-situ burning will be discussed and future plans for a public outreach and education program will be outlined.

Introduction

There are presently two commonly recognized approaches to remove significant quantities of spilled petroleum from marine surface waters. The most common technique involves mechanical skimming devices which typically remove less than 20% of the spilled petroleum (National Research Council, 1989). The second and more controversial method is the use of chemical agents (e.g. dispersants) to disperse oil into the water column. The effectiveness of chemical dispersants has been reported to range from zero to 100 percent depending on the type of petroleum spilled, the dispersant used, and the approach employed to estimate effectiveness (National Research Council, 1989).

There is a third approach, in-situ burning, which is beginning to gain greater acceptance as a viable oil spill response option. In-situ burning has distinct advantages over other oil spill countermeasures. It offers the potential to rapidly convert large quantities of oil into its primary combustion products with a small percentage of other unburned and residue byproducts (Evans et. al, 1992). Until recently, this response technology has not been regularly used, due largely to the lack of understanding of the combustion products and the principles governing the combustibility of oil-on-water (Evans, et. al., 1992) as well as the lack of the equipment necessary to carry out a burn within the window of opportunity. Much of the renewed interest in in-situ burning has resulted from years of study of both the dynamics of burning oil on the water and the combustion products produced during an in-situ burn.

In-situ burning removes the surface oil by driving much of it into the atmosphere in the form of combustion gases and soot. As such, in-situ burning reduces the environmental threat and impacts posed by on-water spills but only at the cost of increasing the potential threat posed by the airborne plume. In-situ burning, however, does have the potential to accelerate cleanup of spilled petroleum on the surface of the water and at the same time reduce the risk of petroleum-related impacts on environmentally sensitive areas. In the case of California, environmentally sensitive areas include the productive intertidal regions, tidal inlets, tidal marshes and other wetland areas of the coastal islands and mainland and the surface waters where endangered marine mammals and large concentrations of sea birds might exist. The problem for decision makers is to compare the effects of burning versus not-burning and choose the option that provides the greatest net benefit to the environment, without causing undue public health impacts.

In the wake of the *Exxon Valdez* and the *American Trader* oil spills, the Lempert-Keene-Seastrand Oil Spill Prevention and Response Act (Act) was enacted to protect waters of the State of California from oil pollution. This Act created a new and comprehensive statewide program which consolidated the primary authority for prevention and response efforts under the authority of the Administrator of the OSPR. Government Code Section 8670.7(g) requires the Administrator to develop a decision-making document for the use of in-situ burning to address oil spills within marine waters of the State. In furtherance of these requirements, the OSPR established an in-situ burning workgroup April of 1995 tasked with the development of a statewide in-situ burning policy. The workgroup consisted of representatives from diverse interests including industry, environmental organization, response organizations, federal, state and local government agencies and was tasked with the development of recommendations for the use of in-situ burning to address marine oil spills. The workgroup adopted a consensus-based, decision-making approach for all-phases of policy development and is currently reviewing its first internal draft in-situ burning policy document. Once the in-situ burn policy is finalized, a public education and risk communication program will be undertaken. Such a program is considered an integral component in the successful use of alternative response technologies.

The Use of Alternative Response Technologies in California

The use of any oil spill alternative response technology, whether dispersants or in-situ burning, will probably always be controversial. The is especially true for a State like California, which is characterized by a large, multi-cultural population, diversified business interests and a strong environmental movement. Although the response to every oil spill is unique, some situations and dynamics hold true for most, if not all, oil spills. An example is this is that virtually all catastrophic oil spills occur under extreme and/or unusual circumstances, either through negligence, unforeseeable events or an "Act of God." With the proverbial "genie out of the bottle," responders are at the mercy of the given circumstances and must quickly develop a plan to cleanup the spill.

This situation is further complicated by the fact that the system usually established to combat an oil spill, the Incident Command System (ICS) does not correspond to the democratic tenets espoused by the greater society. The public has come to expect government to react in particular ways in response to catastrophes which adds another layer of complexity to response operations (Ott et.al., 1997). In these types of scenarios, the ICS must respond to both the oil spill and to some extent, the expectations of the public. This includes concepts of public-right-to-know, the ability for individuals to comment on decisions within a public forum, the ability to express outrage or dissatisfaction with decision-makers and have these concerns heard and/or addressed in a timely, efficient manner (Sabatier, et.al. 1993). If these areas of potential conflict are not addressed early during a spill response, the public may begin to distrust any and all actions taken during oil spill cleanup. Unfortunately, this is often the very arena in which the use of alternative response technologies are decided.

It is important to point out that although oil spill are not the only emergency situations which require a multi-agency response (floods, forest fires, earthquakes), the public views the occurrence and therefore the resulting consequences of an oil spill differently from these other disasters. An earthquake is considered an "Act of God" and as such, has no response party by definition. The public may come to question government agencies regarding the appropriateness of safety standards for buildings or freeway structures after the incident is over, but this is different than having a specific target in which to focus outrage and blame for causing the incident. Even in the case of forest fires being caused by arson, the public usually does not have a Responsible Party with potentially deep pockets in which to focus its aggression and sense of loss, as the arsonist may be a reckless teenager, a homeless person trying to stay warm or a family camping in which a small fire got out of control. Oil spills are different. Unlike the forest fire or flood, oil spills are rarely considered unavoidable circumstances by the general public and as a result, the oil industry is always to blame, negligently, for the spill. Indeed oil companies are and should be responsible for the cleanup and damages associated with spills; but the dynamics associated with public frustration and outrage surrounding oil spill response are quite distinct from those of many other emergency response operations. These difference should not be overlooked for they can easily overwhelm an Incident Command System focused on response and not structured to deal with these issues directly, if at all.

The Workgroup Dynamic and Draft Policy Proposals

Although Government Code Section 8670.7(f) gives the Administrator of the OSPR the State authority over the use of all response methods, including, but not limited to in-situ burning and dispersant use, many other local, state and federal regulatory agencies as well as environmental groups, industry and response community representatives are stakeholders in an in-situ burn use decision. These organizations have both differing mandates, missions and jurisdictions as well as expertise. For example, the Administrator of the OSPR is responsible for overseeing cleanup operations and can also serve as the State trustee for wildlife resources for the Department of Fish and Game; while the local air pollution control districts are responsible for maintaining containment zones standards as required by the Federal Clean Air Act (standards vary among the air basins within the state). During an oil spill response, all the "stakeholders" need to come together and make a decision that is the best "for the State of California" but this may not necessarily "be the best" for their individual interests. It is with this general concept of the "greater good" that the in-situ burn working group was established.

From the beginning, workgroup members agreed to utilize a "consensus-based" approach to policy development. Additionally, members also agreed to try and move beyond the specific confines of individual and agency perspectives and try to work towards a policy that would provide the greatest net benefit to the environment, without causing undo risk to public health and safety. The work group had to address specific question regarding the impact an in-situ burn would have on the environmental resources of the State, the public health and safety, and develop criteria by which these could be safeguarded. Currently, the workgroup is reviewing the following policy options for in-situ burning along the coast: 1) segmenting the California coast along air basin designations, to allow flexibility in burn/no burn decisions; 2) utilizing a combination of federal and state air quality standards for air pollutants of concern, such as PM10 (particulate materials 10 microns or smaller); designations of quick approval zones in areas where the weather and prevailing wind direction will ensure that the public health is not threatened; and, 4) undertaking modeling efforts to provide further information for areas not designated quick approval zones.

Finally, throughout the process, members of the workgroup felt strongly that once the draft policy was completed, it must be followed by a public education and outreach program which would help provide a framewcrk in which the public could understand a decision favoring the use of an alternative response technology. Without this type of outreach effort, the general public would have to rely primarily on mass media for information in which to form their opinions. Given that the mass media is not a scientifically rigorous source of information and given the nature of news coverage during emergency response situations, it is likely that "fear" and "outrage" could be the foundation for many of the decisions made by the general public. Such an outcome would serve neither the public nor those tasked with emergency response operations. A brief summary of the concepts for a proposed public outreach and risk communication program are discussed below.

Current Timelines

At the time this paper was published, the workgroup was reviewing its first draft of the in-situ burn policy. It is estimated that after revisions to the policy are made and internal reviews are completed, a draft policy will go out for public review and comment in late Fall, 1997. Timeframes for finalization of the policy will be greatly dependant on the nature and scope of the public comments but it is hoped that the policy will be incorporated in the California State Oil Spill Contingency Plan and the Coast Guard Area Plans by Spring of 1998. Once this is completed, the OSPR will begin work on a public outreach and risk communication program.

Public Outreach and Risk Communication

Any policies that are established regarding the use of alternative response technologies such as in-situ burning, will only be as good as our ability to implement the policies and use them to address a real spill scenario. Although this seems obvious, how such implementation is accomplished is not so obvious. The development of an in-situ burn policy for use in California is the first important step towards the use of this technology, but it is by no means, the end of the story.

"People's perceptions are their realities. . . . " and the public's perceptions of in-situ burning are critical to the use of this technology during any spill event. It is not enough to have a policy which was developed by experts, open for public input, finalized and then put on a shelf until needed for a spill event. The general public needs to be educated on response to oil spills. Groundwork needs to be laid for people to gain a reasonable understanding of the nature of oil spill response - what is being done and why. It is for these reasons that the OSPR is proposing a public outreach and risk communication program whose primary goal will be to facilitate a dialog with the public regarding oil spill response options, prevention & concepts of risk, and the concepts of trade-offs and the inherent trade-offs associated with different decisions made during response operations. The program will be centered around three specific points:

1. <u>Prevention is the first and best line of defense, but accidents do happen.</u> As with so many aspects of modern society, the general public only becomes aware of an issue when something goes wrong. The public becomes concerned with the problems associated with moving oil on or through the marine waters when a spill occurs and it impacts things that are of value to them (gasoline prices, beaches, cancelled fishing or whale-watching trips, impacted wildlife). Professionals in the field however, (whether they be from industry, response organizations or government regulators) are keenly aware of what is being done to prevent spills and how these measures greatly reduce the risk of spills. The risk of spills, however, approaches but <u>never</u> reaches zero. It is often stated that the general public does not have a good understanding of risk nor the risks associated with the transportation of oil. However, the author maintain that the general public deals with risk everyday of their lives (the risk of speeding or not putting money in a parking meter; the risk of

floating checks a couple of days before payday; or the risk of a pregnancy even when utilizing birth control)(Sandman, 1995). If the public is able to weigh risk and the consequences of particular behaviors or activities, then perhaps government agencies and others involved in oil spill response are beholden to educated the public on the risks associated with oil transportation and production and what is being done to reduce this risk. It seem ineffectual to hold on to an outdated idea that the general public wants the impossible (zero risk) or that they can't understand the complexities of an ever decreasing risk.

2. "Once oil is spilled, there will be damage." Oil spill specialist must dispel the public perception that in response to a catastrophic spill, the oil can be completely removed from the environment by traditional cleanup methods (if only the responsible party were to employ enough resources). For example, the public must be made to understand that no amount of mechanical cleanup equipment could have completely cleaned-up the *Exxon Valdez* oil spill within the first few days of good weather following the initial release. *In-situ* burning, however, had the potential to eliminate a significant amount of oil during this timeframe of good weather and calm sea conditions (Allen, personal communication) had the option been exercised.

3. "The question becomes how to minimize damage not how to eliminate them?" Unfortunately, trade-offs must be made during response to a catastrophic oil spill (since point #2 is true) and it becomes the job of the response planners within the ICS to determine where the environment can best afford the impacts of oil. Critical to this discussion is also the recognition that if alternative response technologies, such as dispersants and *in-situ* burning, are not allowed, a de facto decision is being made that on-water and on-shore resources can best afford the impacts of an oil spill. These on-shore/on-water resources include not only the "charismatic megafauna" such as otters and puffins, but also the very beaches and shorelines that the public values so highly. It must be made clear that all actions taken during an oil spill response, including the use of alternative countermeasures, are consistent with society's concerns: to protect the public health and safety, to protect environmentally sensitive areas; and to protect economic resources. If an alternative response technology is used, it is because their use is consistent with these goals, and not simply because their use makes the cleanup easier or cheaper for a responsible party (although this may be a beneficial side-effect).

Summary

The movement of hundreds of millions of barrels of oil over and through the marine waters of California poses a threat to the public heath and environmental resources from oil spills resulting from such transport. Although prevention is the key to reducing the risk of oil spills, these risks can not be reduced to zero (save not transporting oil on or through the water). As a result, "pre-planning" for oil spill response becomes critical in ameliorating the adverse effects of spills once they do occur. Alternative response technologies, such as dispersants and in-situ burning, are simply more tools that should be available for use at the time of a spill. Given the political and logistical constraints associated with the use of in-situ burning, a predesignated approval process, agreed upon by all the stakeholders in advance of a spill, would greatly facilitate the use of this technology. This paper outlined the approach taken by the State of California in developing a statewide in-situ burning policy, including: establishment of a workgroup; consensus approach to decision making adopted by the group; questions addressed and resolved by the group; proposed policy directions adopted by the group; finalization of the policy which includes public review and comment. The OSPR sees the finalization of an in-situ burn policy as a first step in the use of alternative response technologies. As stated previously, any policy that is established regarding the use of alternative response technologies will only be as good as ones ability to implement the policies and use them to address a real spill response scenario. To address this, the OSPR is proposing a public outreach and risk communication program whose primary goals will be to facilitate a dialog with the public regarding oil spill response options and the trade-offs associated with different decisions made during response operations.

Acknowledgments

The author would like to acknowledge the members of the *in-situ* burn work group without whose tireless work and efforts, none of this would have been possible. Barbara Lee, Sonoma Air Pollution Control District (APCD); Mohsen Nazemi, South Coast APCD; Kathleen Shimmin, United States Environmental Protection Agency; Karen Brooks, San Luis Obispo APCD; Glen Long, Bay Area APCD; Kent Field, Ventura APCD; Ron Tan, Santa Barbara APCD; Ed Ueber, Terry Jackson and Ed Cassano, National Marine Sanctuary; Chip Demerest, Department of the Interior; Scott Stoltz, National Oceanographic and Atmospheric Administration; Christopher Gregory, Clean Coastal Waters; Bela James, Shell Oil; Ann-Hayward Walker, SEA, Inc.; Scott Robertson; ARCO; Steve Ricks, Clean Bay; Darryl Waldron, Clean Coastal Waters; Kathy Gregory, U.S. Coast Guard; Ike Ikerd, Marine Spill Response Corporation; Warner Chabot, Center for Marine Conservation; Gary Zimmerman, California Air Resources Board.

Special thanks goes to Kevin McGrattan of the National Institute of Standards and Technology for his willingness to honor our modeling requests (and those still to come); the Marine Preservation Association and Scientific and Environmental Associates, Inc. for their assistance in coordination of the *In-Situ* Burn Symposium; and to Joe Mullin for all the support from Minerals Management Service.

References

Allen, A.A. Personnal Communication. *In-Situ* Burning Symposium, Concord California, October 1-2, 1996.

Allen, A.A. "Contained Controlled Burning of Spilled Oil During the *Exxon Valdez* Oil Spill. *In Proceedings of the Thirteenth Arctic and Marine Oil Spill Program Technical Seminar*; June 6-8, Edmonton, Alberta, pp. 305-313, 1990.

ATSDR (Agency for Toxic Substances and Disease Registry). Preliminary Health Advisory Related to Burning Oil Wells in Kuwait. In-situ Burning Workshop, 1991.

Campagna, P.R. & Humphrey, A. 1992. Air sampling and monitoring at the Kuwait oil well fires. In *Proceedings of the Fifteenth Arctic and Maine Oil Spill Program Technical Seminar,* pp. 221-232, 1992.

Covello, Vincent, *Environmental Risk Communication and Public Dialogue.* Center for Risk Communications. New York, 1996.

Evans, D.D., W.D. Walton, H.R. Baum, K.A. Notarianni, J.R. Lawson, H.c. Tang, K.R. Keydel, R.G. Rehm, D. Madrzykowski, R.H. Zile, H. Koseki, and E.J. Tennyson. 1992. *In-situ* Burning of Oil Spills: Mesoscale Experiments. In *Proceedings of the Fifteenth Arctic and Maine Oil Spill Program Technical Seminar*, June 10-12, Edmonton, Alberta, pp. 593-657. 1986.

Ferek, et. al. Chemical Composition of Emissions from the Kuwait Oil Fires. Journal of Geophysical Research; 97: pp. 14483-14489, 1992.

National Research Council. Using Oil Spill Dispersants on the Sea. National Academy Press, Washington D.C., 335 p., 1989.

Ott, Gary. and Stalfort, David. Planning and Exercising for Success: The Four Step, Scenario-Based Process. *Proceedings of the 1997 International Oil Spill Conference.* American Petroleum Institute, Washington, D.C., pp. 3-8, 1997.

Sabatier, Paul and Jenkins-Smith, Hank. *Policy Change and Learning: An Advocacy Coalition Approach.* Westview Press. Boulder Colorado, 290p., 1993.

Sandman, Peter. The Nature of Outrage, Consulting, Training and Research in Risk Communication, Newton Centre, Massashuetts, 123 p., 1995.

S.L. Ross Environmental Research LTD. "Evaluation of Capabilities to Respond to Large Oil Spills in California Marine Waters." Prepared for the California State Interagency Oil Spill Committee, 308 p., 1990.

CHAPTER 106

A Quantitative Basis for Optimal Spill Response

J. A. Galt and D. L. Payton[1]

Abstract

Decision-making in the increasingly complex oil spill response and planning community is based on "best-available" information. This "best-available" information involves analysis from a number of experts. The decision-maker must compile these discrete pieces of information to make decisions on protection strategies and priorities. In response situations, it can sometimes be obvious when less-than-optimum decisions have been made. In planning or drill situations, the evaluation of these decisions can be very difficult.

This paper proposes a quantitative approach to integrate this information into a "game" that can be played, in a planning phase, to examine alternate scenarios. Game theory is used to design the "game", and linear programming is used to quantify and optimize the results. Potential players in this planning game could include:

1) cleanup or response contractors who wish to check the adequacy of equipment inventories and determine the differential cost benefit of individual pieces of equipment either to establish pricing policy or justify future acquisitions,

2) resource trustees who would like to investigate alternate evaluation schemes used to characterize coastal resources or damage assessment tables,

3) insurance underwriters or potentially responsible parties who wish to check alternate response plans against local damage assessment tables and/or potential future Natural Resource Damage Assessment (NRDA) claims, either during training exercises or actual responses.

[1]NOAA/HAZMAT, 7600 Sand Point Way, NE, Seattle, Washington 98115 USA

Coastal Watershed Policy and Management in California: Notes on the Development of Cooperative Strategies

Michael V. McGinnis, Ph.D.[1]
and
John T. Woolley, Ph.D.[2]

Abstract

In light of recent emphasis in California and elsewhere on managing natural resources on a watershed basis, the successes of such efforts to date should be evaluated. This paper has three objectives: 1) To generally characterize watershed management processes and activities in California, 2) To report results of our study of the place of values and scientific information in watershed management and restoration, and 3) To describe the techniques policy-makers can use to resolve value-based conflicts between watershed managers, planners and participants.

Introduction

California's watersheds supply water for drinking, recreation, industry and farming and at the same time provide critical habitat for a wide variety of animals and plants. In California, rapid population growth in watersheds has led to: (a) increased conflict between human use and natural resources; (b) dramatic loss of native diversity; and (c) general decline in the health of ecosystems. Because of the decline in the health of the state's watersheds, federal and state governments have turned to watershed-based organizations to develop and implement environmental policy (California State Resources Agency,

---

[1] Assistant Research Political Scientist II, Ocean and Coastal Policy Center and Resident Fellow, Interdisciplinary Humanities Center, UC-Santa Barbara 93106
[2] Professor, Department of Political Science, UC-Santa Barbara 93106

1991; Jensen, 1994; Kier et al., 1995). California's watershed-based organizations include among their members (in varying proportions) community activists, private property owners, representatives from local, state and federal governments, industrial interests and environmentalists. Despite the burgeoning emphasis in watershed management in California, the short and long-term implications of these activities are not well understood by scientists, policymakers or the general public.

The human impacts on watersheds and the associated biological or natural communities are forcing resource agencies and scientists to rethink how ecosystems are managed. Scientists have documented a sharp decline in aquatic species diversity, such as salmon, in the state (Nehlsen, 1991). Among the 50 states, California ranks second in numbers of freshwater fish species that are declining (Williams et al., 1989).

To protect and restore the biodiversity in California, a number of agreements have been signed that are based on a watershed-based approach. In 1991, twenty four state and federal agencies signed a Memorandum of Understanding on Biological Diversity (MOU) (State of California Resources Agency, 1991) to promote regional approaches to conflict resolution, local participation, inter-agency cooperation and ecosystem management in the state (Jensen, 1994). Also, a Framework Agreement between the Governor's Water Policy Council of the State of California and the Federal Ecosystem Directorate was signed in 1994 to establish a comprehensive program for coordination and communication between the Council and the federal government with respect to environmental protection and water supply dependability in the San Francisco Bay, Sacramento-San Joaquin Delta Estuary and its watershed.

These MOUs and other agreements have accelerated the formation of watershed associations, councils and conservancies. The California Watershed Project Inventory lists more than 200 watershed projects statewide. Several of these organizations are developing and implementing specific cooperative projects to restore and manage entire watershed ecosystems. Several of these watershed organizations are playing an important role in the management of California's natural resources (Coastal Conservancy, 1996). Yet, several organizations are stagnant or moribund, and have failed to successfully reach agreement on a plan of action. Watershed organizations have obtained funds to begin a range of resource related activities, such as environmental education, flood and pollution control, general ecosystem management and restoration, and the enhancement of water quality.

Despite these recent initiatives and agreements, it remains unclear how policy-makers can increase the probability of successful watershed management. To date, there exist no dominant policy, statute or agency in California responsible for watershed management and planning. Rather, there exists a number of government and non-government actors that influence public policy in this area. Despite the increasing polarization over issues such as preserving and protecting wild steelhead or coho salmon, the prevailing view in watershed planning and management is that decisions should be reached by consensus. State and federal resource agencies believe that consensus-based strategies are the best way to gain cooperation of private property owners and water users. The assumption is that it is better to address conflicts related to private property, for example, through cooperation rather than regulation. Also, because of the rural character of many watershed communities, resource agencies are seeking management approaches that support local control over local resources. Reconciling the disputes and conflicts that emerge in local, watershed-based management and planning, however, is difficult because of the range of values, users, preferences and perceptions of stakeholders that are associated with a common watershed or bioregion.

With respect to watershed management in the state, resource managers have proposed and initiated consensus-based approaches rather than moved toward imposing decisions notwithstanding the opposition from other organization (Jensen, 1994; Western Water, 1995; Montgomery et al., 1995; State of California Resources Agency, 1995). Consensus-based decision making is a great idea, but will it work with watershed organizations? The answer to this question lies in the place of science and values in the watershed planning process.

The Place of Science and Values in Watershed Management

Watershed management and planning exists in the larger context of values, or what we refer to as a restoration concourse. A concourse is a variety of existing opinions bearing on the issues associated with a particular activity, such as watershed management. A concourse is made up of discourses that consist of shared values. Each discourse rests on a shared set of assumptions regarding the purpose and values of watershed management. Participants may be committed to more than one discourse, some of which may be incompatible (McGinnis and

Woolley, under review). If these discourses are recognized and discussed, we have argued that the development of watershed-based strategies and plans will be more successfully both ecologically and culturally (McGinnis and Woolley, 1997).

Watershed management should be approached as a trans-scientific enterprise involving the intermingling of scientific facts and values. In a trans-scientific enterprise, burdens of proof, values, politics, and other qualitative aspects of decision-making play an influential role. Watershed restoration and management occurs in spite of, or in light of, the absence of reasonable scientific consensus (Lee, 1993). Even at a purely technical level, there are social and institutional conditions that influence the practice of watershed management. Therefore, we believe that one of the most important determinants of watershed management success susceptible of influence by policy-makers is the nature of the value-based conflict in watershed management.

Consequently, we believe that policy makers and other watershed managers would benefit from a clearer understanding of the range of discourses that exist in different contexts. One primary goal of watershed management and planning should be to map the range of discourses that exist in a particular decision making situation. This mapping is a first step toward reconciling disputes rooted in different perceptions and values regarding science, the watershed and the landscape.

Our study of watershed restoration focuses on the ways in which values and perceptions affect the process of the planning and implementation of restoration projects. To explore this we investigated the range of values and beliefs held by participants and stakeholders involved in three large-scale river restoration projects (Washington's Yakima River, California's Upper Sacramento River and the Santa Ynez River). To sample the values and beliefs of those involved in these projects, we prepared a mail survey that included sixty five statements or questions related to restoration efforts, and asked recipients to indicate their response to each of these on a five point numerical scale from 1 for "strongly agree) to 5 for "strongly disagree". We sent these surveys to 442 persons members of various watershed groups that exist or existed within the river system. For a more complete characterization of the survey results and the survey sample, see Woolley and McGinnis (under review). We received 221 completed surveys, or a 50% return.

Because of the limited space available, we report a sample of the survey results here and focus on the failure of the Santa Ynez Watershed Enhancement Project.

The Case of the Santa Ynez River Project

With technical, financial and administrative support from the State Coastal Conservancy ($100,000), the Land Trust for Santa Barbara County, the EPA ($100,000), Coastal Resource Enhancement Funds (CREF) from Santa Barbara County ($100,000) and watershed planning process began for the Santa Ynez River, the longest river in Central California. Early on, this watershed process was a highly contentious one, reflecting the fundamental divisiveness between northern and southern Santa Barbara County, government and nongovernment interests and up-river versus down-river inhabitants. Debates over what the ultimate goals and objectives of the plan continued for approximately one year. In 1995, the Land Trust distributed 850 surveys to stakeholders and asked questions related to the future direction the planning process should take. Based on the surveys, the Land Trust decided to continue to facilitate the watershed planning process.

One early result of the survey was the consideration for considering "parts" of the watershed rather than the entire watershed with its tributaries. For example, the Los Padres National Forest, which exist upriver, was no longer considered part of the watershed for planning purposes. Unfortunately, steelhead trout depend on this part of the watershed. Indeed, the Los Padres Wilderness Act of 1992 permanently sets aside 31.5 miles of the Los Padres for steelhead trout protection and designates the Sespe Creek as a "wild trout stream". What happens downriver continues to impact the threatened steelhead runs.

It was a valuable coincidence for us that the organization devoted to restoring the Santa Ynez River completely disintegrated during the period of our study. If we take this as a bottom-line indicator of the level of conflict, we can show with our current data that the Santa Ynez project appears to be distinctive in several ways. Figure 1 shows that the Santa Ynez project has (at least as reflected in our survey responses) the highest proportion of self-identified career civil servants, and the lowest proportion of self-identified academics and representatives of private industry.

Also, the Santa Ynez project is an unique case with the strongest respondent sentiment that scientists involved agreed on their findings on the health of the river (see Figure 2) and the strongest respondent sentiment that there were not formal channels for scientists to communicate their findings to policy makers. There is no statistically significant difference between the cases in the extent of agreement with the statement that "scientific and technical information has been available to make good decisions about restoration". Moreover, scientists involved in the Santa Ynez were no more likely that scientists in the other projects to agree that such information has been available. Santa Ynez respondents do not score particularly high on a factor that measures the extent to which participants have faith in science to solve restoration problems. Of equal or greater importance, the results suggest that there was an internal preponderance of anti-restoration sentiment among respondents in the Santa Ynez case--many private property owners who live near the river preferred the term "enhancement" to restoration. In the other two cases, there was more of a balance between pro- and anti-restoration respondents. While local control and concern for property values are important issues that separate participants, that does not appear to be a dimension on which the Santa Ynez case is especially distinguished.

There exist a potential lesson from the vignette described above. Generally, each watershed organization deals with complex issues related to the development of appropriate boundaries, in other word, <u>boundary identification</u> (McGinnis, forthcoming). Each group struggles over defining watershed and planning boundaries. Each boundary definition translated into different priority actions and issues. In the Santa Ynez bioregion, for example, the watershed group initially debated and defined the boundary as the 500 year flood plain. Later, the steering committee redefined the boundary to include the 100 flood plain. Six months later, with landowner support, the boundary was again redefined to parts of the watershed from Bradbury dam, up river, to the ocean, to primarily deal with pressing problems related to flood control near developing cities. Each <u>boundary shift</u> translates into a different scope of conflict, and different priorities. Broadening the scope of conflict entails many more issues than a narrow boundary definition. A focus on the entire watershed boundary, for instance, may promote the needs of wild salmon. A refocus on part of the river reflects an interest in flood control not salmon restoration. Each boundary redefinition supports one value over another.

Membership of watershed organizations indicate there primary goals and common interests--generally, to protect and sustain

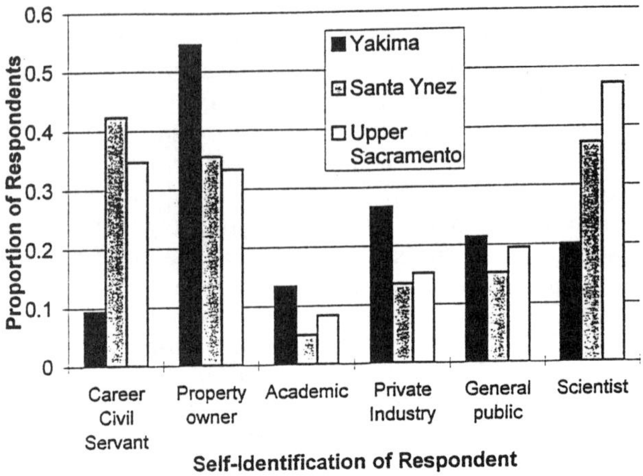

Figure 1. River Basin Survey Respondents by Self-Identification

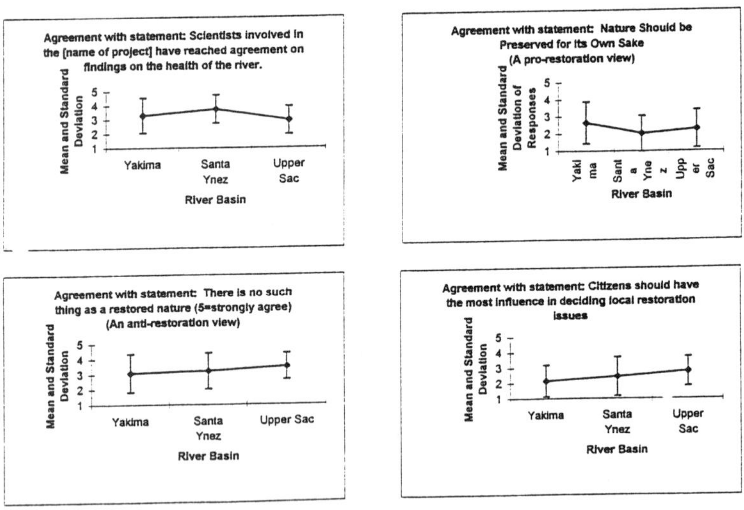

Figure 2. Selected Survey Responses With Implications for Scientists and Policy Makers Involved in Restoration Activities

economic activities associated with private property rights and interests for development. This is often represented by the fact that members is based on the level of private property one own; the more property one owns, the more "representation" that owner has in watershed planning. Restoration of watershed health may be a secondary goal of many watershed organizations. As a member of a watershed group interviewed for this paper stated, "I represent private landowners on the Watershed Board. Our goal is better use of water. We recognize that fish survival is a major priority, but while establishing habitat for fish, we do not want to deteriorate the economic benefits that are based on water use. We see a need for timber production, agricultural irrigation, and flood control. Hopefully, we can do these things, and improve fish habitat."

Watershed organizations may also be created to fend off the influence of the scientific community and government. Landowners, for example, may create a watershed conservancy with a basic philosophy "dedicated to protecting natural resources, private property rights, and responsible land stewardship". Government officials and environmental interests may not be represented on the conservancy's board. Those who do not own property are not formal stakeholders and are not granted voting privileges. In addition, voting is based on the amount of property one owns; the more property one owns the more votes one receives in the planning process. Moreover, the public fears and risks associated with scientific evidence and biological data lead to conflict between contending interests.

Watershed-Based Conflict Resolution

There are huge obstacles to success in watershed management in California, and there is no guarantee that a consensus-based approach will work, for example, to deal with pressing problems related to endangered species.

There are three major factors that influence the process of conflict resolution: 1) organizational mandate, 2) ratio of private to public land in the watershed, and 3) the age of the organization. As previously noted, California's MOU, federal, state and local government organizations have fostered the creation of several watershed organizations. Top-down organizations are those based on formal state government authority while grass-roots organizations are created by local citizens. Age is an important indicator of the degree to which many conflicts have been resolved in an organization and the degree to which a distinctive "organizational culture" may have developed (Wilson, 1989). Finally, the patterns of land ownership interests are important not just for affecting the nature of conflict, but also in the way that scientific information is used by the participants.

We propose five preconditions for consensus-based decision making: 1. Sustaining Cultural Values: Watershed boundaries are social constructions and therefore necessarily include social values and concerns. The public decides what values to place on issues surrounding watershed-based approaches. Watershed management should, therefore, attempt to preserve and respect diverse cultural values and interests. 2. Participation: Watershed planning can only be successful if it involves all of the stakeholders who have an interest in the outcome. The partnership that is created in a watershed organization is about issues and relationships. Relationship-building is the most significant aspect of the entire process. 3. Representation and Assistance for Weaker Parties: Stakeholders should be adequately and equally represented. Less resourceful parties need to be included in the process, and may need technical and financial assistance. 4. Legitimacy: Participants in the process need to believe that they are engaged in a valid and genuine process. To gain legitimacy, the process should: (a) account for attitudes and perceptions of the participants, (b) represent the interests of all stakeholders, and (c) be shown what they can gain from participation. 5. Creating Adaptive Decision-Making: Watershed management should be sensitive to the changing character of culture and ecology. Organizations and management strategies should therefore be flexible and experimental, in order to adapt to changing social values, perceptions, political pressures, available data and ecological conditions. Watershed management implies relearning lost community and social skills.

Ultimately, a consensus-based approach may not be enough. There are certain environmental issues that inevitably defy a consensus-based approach. To better understand the diversity of California's watershed management activities, and the cooperative nature of these organizations, we propose that policy maker evaluate the status of watershed-based planning in the state. This will require more than an inventory of watershed-based initiatives in the state or merely an appropriation of funds to study watershed systems.

Several important questions should guide the evaluation: 1) how to strengthen cooperative decision-making in watershed management, 2) how to enhance the capacity of the public to play a constructive role in resource planning, 3) whether agency performance could best be enhanced by investing in scientific information, by working to deal with the nature of interest conflicts within agencies, or by altering the objective structure of legal incentives and mandates that agencies face.

Acknowledgment

This essay is part of a larger work and research project funded by the National Science Foundation's Ethics and Values Studies Program under award SBR 9511599. Any opinions, findings, and conclusions or recommendations expressed in this essay are those of the authors and do not necessarily reflect the views of NSF. With a second grant from NSF (SBR 96-16958), the authors are continuing to investigate the place of values and science in California watersheds.

References

California State Resources Agency. 1991. Memorandum of Understanding: California's Coordinated Regional strategy to Conserve Biodiversity. "Agreement on Biological Diversity." Sacramento, California.

Coastal Conservancy. 1996. Watershed Summit. Oakland, California. May 9.

Jensen, D.B. 1994. Conservation through Coordination: California's Experiment in Bioregional Councils. In Environmental Policy and Biodiversity. R.E. Grumbine. ed. Island Press.

Kier, William et al. 1995. Watershed Restoration: A Guide for Citizen Involvement in California. NOAA Coastal Ocean Program. Decision Analysis series No. 8. U.S. Department of Commerce. Washington, D.C. December.

McGinnis, M.V. (ed) forthcoming. Bioregionalism: The Tug and Pull of Place. London and New York: Routledge.

McGinnis, M.V. and J.T. Woolley. 1997. The Discourses of Restoration. Restoration and Management Notes 15,1 (Summer): 76-79.

McGinnis, M.V. and J.T. Woolley. under review. Understanding the Concourse of Ecological Restoration.

Lee, Kai N. 1993. Compass and Gyroscope. Covello, CA: Island Press.

Nehlsen, W., J.E. Williams, and J.A. Lichatowich. 1991. Pacific salmon at the crossroads: stocks at risk from California, Oregon, Idaho, and Washington. Fisheries 16:4-21.

Press, D. 1995. Environmental Regionalism and the Struggle for California. Society and Natural Resources 8: 289-306.

Western Water. 1995. Watershed Management. Published by the Water Education Foundation. November/December.

Wilson, J. 1989. Bureaucracy. New York, Basic.

Woolley J. and McGinnis, M. Under review. Values and Science in River Restoration.

CHAPTER 108

Coastal Links--Partnerships for Watershed and Marine Protection

Holly Price[1]

Abstract

The Water Quality Protection Program (WQPP) is an interagency-public-private partnership to enhance and protect the physical, chemical and biological conditions in the Monterey Bay National Marine Sanctuary and its 11 adjacent watersheds. Currently, 27 federal, state and local agencies and organizations are working with businesses, landowners, scientific and public representatives to develop and carry out strategies for a range of water quality and watershed issues. Issues addressed by the program include urban and agricultural runoff, marinas and boating activities, wetland/riparian degradation, point sources of pollution, and regional monitoring. A prime focus is to more efficiently integrate the large number of existing programs and projects related to these issues (over 100!), eliminate gaps and redundancies, and pool interagency resources to address problems.

Plans have been completed and are underway for urban runoff, marinas and boating, and regional monitoring, in collaboration with agencies, cities, harbors, and volunteers. Workshops sponsored by the WQPP and local Farm Bureaus have also resulted in a variety of approaches to addressing water quality issues related to agriculture, including permit streamlining to encourage landowners to adopt conservation practices. The WQPP has been successful in pooling the resources of various agencies and groups to address marine and watershed issues, and in raising public awareness about the connection between everyday activities in the watersheds and the ocean's health.

[1] Water Quality Protection Program Director, Monterey Bay National Marine Sanctuary, 299 Foam St., Suite D, Monterey, CA 93940

Introduction

The Monterey Bay National Marine Sanctuary includes over 5000 square miles of marine waters, and stretches over 360 miles along California's Central Coast. During the designation of the Sanctuary, water quality was recognized as a key element linking all the natural resources of the area. The eleven major watershed areas adjacent to the Sanctuary (Figure 1) contain a diverse array of land uses and habitats, including dense urban areas, rural communities, irrigated agriculture, grazing land, forests, rivers and wetlands. Because of this proximity to land, protecting water quality in the Sanctuary requires not only addressing the marine waters, but also water quality issues in the various watersheds that drain to it.

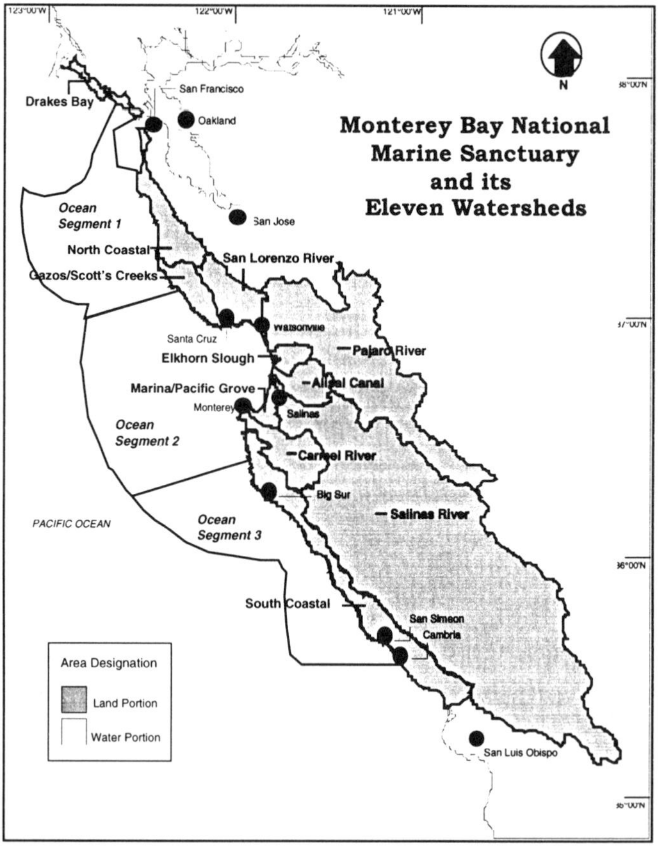

Figure 1: Sanctuary and its 11 major watershed areas

The Water Quality Protection Program (WQPP) for Monterey Bay National Marine Sanctuary is a partnership among federal, state and local agencies, public and private groups to enhance and protect water quality in the Sanctuary and its watersheds. The program implements a key element of a Memorandum of Agreement signed by a variety of water quality agencies during Sanctuary designation in 1992. Currently 27 federal, state and local agencies and organizations are working with businesses, landowners, scientists and the public to develop and carry out strategies addressing a range of water quality and watershed issues.

Problem assessment

The Sanctuary's watersheds receive rainfall and irrigation water, which runs downhill to streams, rivers, wetlands, and ultimately out to the Sanctuary. While it is moving downhill, it picks up a variety of pollutants which can potentially be carried to the Sanctuary and to the surrounding wetland areas which sustain many Sanctuary species. In addition, marinas and harbors at the shoreline contribute a variety of contaminants from boating activities.

Fortunately, water quality within the Sanctuary itself is in relatively good condition compared to many other areas of the country. The WQPP thus has the opportunity to proactively address existing water quality issues, preventing them from becoming the expensive crises which have arisen elsewhere in the country. Water quality problems and sources within the Sanctuary region are summarized below and in Fig. 2. Maps of the State's Water Quality assessment can also be found in NOAA (1995a), indicating levels of impairment of water bodies within the Sanctuary and its watersheds due to ongoing water quality problems. The State Water Resources Control Board and Regional Water Quality Control Boards designate waters as "impaired" if they do not attain or maintain applicable water quality standards. Impaired water bodies in the region are primarily in the lower reaches of the rivers, and along the coastline of Monterey Bay.

Elevated levels of pollutants such as tributyl tin, copper and zinc have been found in Monterey, Moss Landing and Santa Cruz Harbors (SWRCB, 1995a). Elevated zinc levels have also been reported in runoff from cities adjacent to the Sanctuary (Camp Dresser & McKee et al, 1993).

Contamination from persistent pesticides such as DDT, chlordane, dieldrin, toxaphene, endosulfan, etc. have been reported for the Old Salinas River Channel, Salinas River Lagoon, Moss Landing Harbor, Watsonville

Slough and Elkhorn Slough (SWRCB, 1990, 1995a,b; Stephenson et al., 1995). Although most of these persistent pesticides are currently banned, they remain present in the soil and are continually transported to the water bodies and tributaries surrounding the Sanctuary (Kleinfelder, 1993; SWRCB, 1986).

Elevated levels of coliform bacteria have been reported in the Sanctuary and its watersheds, including locations near the San Lorenzo River mouth, Aptos and Soquel Creeks (Ricker, 1995), and near the Pajaro River mouth. Elevated levels have also been reported in Elkhorn Slough (Cotter and Strnad, 1995), attributable to natural wildlife populations and to runoff from dairy operations. Coliform elevations have occurred due to discharges from Neary Lagoon in Santa Cruz, and remain a problem in the Half Moon Bay region due to periodic sewer system overflows. Coliform contamination along the coastline can potentially lead to health impacts for contact recreational activities such as surfing and swimming, and to closure of beaches.

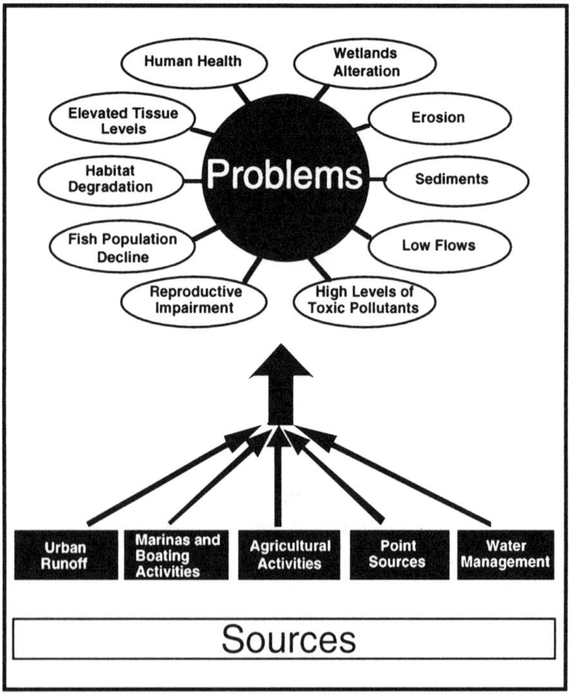

Figure 2: Water quality problems and sources

Sedimentation is an ongoing problem for the rivers, wetlands and harbors which border the Sanctuary (NOAA, 1995a; USGS, 1992). During flood conditions, sediment plumes from tributaries may extend several miles into the Sanctuary itself. Sedimentation may smother riverine, estuarine and marine habitats which can kill benthic organisms, reduce colonization, interfere with filter-feeding, and cover important spawning grounds for anadramous species such as steelhead and salmon. Sedimentation is also an important transport mechanism for persistent pesticides which are adsorbed on particle surfaces (Kleinfelder, 1993).

Elevated nutrient levels are also a potential problem in some of the areas surrounding the Sanctuary. Elevated levels have been reported in the Elkhorn Slough and Salinas River watersheds (Caffrey, 1996; Habitat Restoration Group, 1991), and in Pajaro River and Watsonville Sloughs (Questa, 1995).

Water quality issues are also related to total water flows in over 60 rivers, streams and estuaries which drain into the Sanctuary. Water flow in many of these tributaries has been greatly reduced due to increasing demands from agricultural and urban users over the past 50 years (USGS, 1992). Low flows and increased sedimentation have impacted anadramous species such as salmon and steelhead which relied on these tributaries as migratory routes to their upstream spawning grounds. Continuing habitat degradation, in combination with overfishing, has left only remnant populations of these species within the Sanctuary's coastal watersheds (Brown et al., 1994). Historical increases in water demands have also greatly reduced the amount of wetlands and riparian areas which act as natural filters for pollutants.

Issue areas and program approach

Due to the large geographic area and complexity of the issues to be addressed, the WQPP committee adopted a watershed approach which focuses sequentially on a variety of issues. This phased approach has allowed the group to maximize its limited staff and funding resources, complete plans for individual issue areas, and begin implementation while initiating planning for additional issues. Key issue areas identified by the group include a) regional monitoring, data sharing and interagency coordination, b) urban runoff, c) marinas and boating; d) agricultural activities, e) wetlands/riparian degradation and f) point sources of pollution.

The program takes a "bottoms-up" stewardship approach, focusing heavily on public outreach and developing consensus among a wide range of stakeholders for each issue, including government representatives, businesses, landowners and environmental groups. Priority strategies were identified and refined for each topic by working with these diverse stakeholders to evaluate the feasibility of recommendations, environmental benefits, and economic impacts. This "Integrated Coastal Management" approach emphasizing stewardship contrasts with the more traditional regulatory approaches typical for addressing point sources of pollution (Fig. 3).

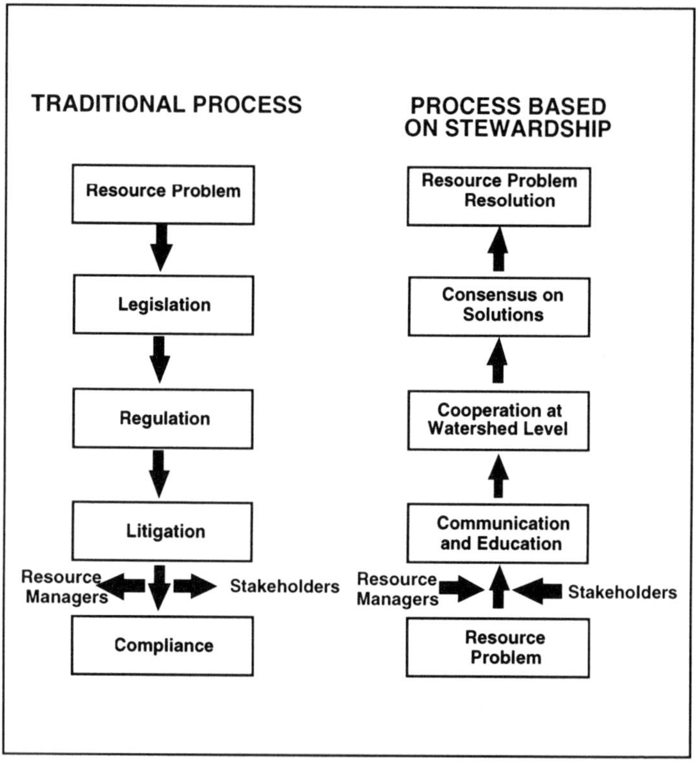

Figure 3: WQPP stewardship approach vs traditional approaches

A main focus of the program is to more efficiently coordinate the large number of existing programs and projects related to these issues, eliminate gaps and redundancies, and pool resources to address problems. These existing programs are then tied in to plans that are developed, focusing on how they might be modified, coordinated or their resources pooled to carry out WQPP goals and activities.

Planning and Implementation

Program participants have worked jointly on a variety of tasks. Key accomplishments in the initial three years include:

1) assembling an ongoing team of agencies, public and private groups to oversee problem assessment, planning and implementation;
2) evaluating and prioritizing water quality and watershed issues, including assessment worskhops, compilation of existing data, and mapping of areas of water quality impairments;
3) identification and assessment of over 100 existing federal, state and local programs and plans related to water quality, to lay the groundwork for improved integration and coordination;
4) identification and assessment of over 20 federal, state and local government water quality monitoring programs and 15 volunteer programs, to develop a basis for coordinated regional monitoring programs;
5) developing consensus on effective solutions and detailed implementation steps for three completed action plans addressing urban runoff, marinas and boating, and regional monitoring/data sharing;
6) pooling existing resources and attracting new resources to begin carrying out these three initial plans in collaboration with a variety of agencies and groups;
7) initiating development of an agricultural plan through agricultural workshops and outreach with growers and ranchers;
8) development of extensive outreach and education efforts to support the planning process and focus public attention on healthy watersheds and marine protection.

In addition, WQPP participants have initiated development of an ongoing Water Quality Coordinating Council which will ensure continuous collaboration and oversee implementation of the issues areas outlined in more detail below.

Monitoring, Data Sharing and Interagency Coordination

Although the studies noted above broadly define problem areas, existing water quality data are inadequate to thoroughly characterize the health of the waters, sediments and biota of the Sanctuary, its adjacent wetlands and riparian areas. WQPP participants agreed that development of a coordinated regional water quality monitoring program is needed to adequately characterize the health of the Sanctuary, monitor hot spots and long-term trends, and evaluate the success of water quality management measures. Similarly, the group recognized that although there are numerous databases containing water quality data, there is little or no coordination among these various efforts.

To address these issues, a detailed action plan has been completed and implementation begun on regional monitoring, data sharing, and development of a water quality council to ensure continued collaboration (WQPP, 1996b). EPA assisted by funding an assessment of twenty existing federal, state and local monitoring efforts in the Sanctuary region (Tetra Tech, 1995), adding details to an initial survey conducted by the WQPP (NOAA, 1995). Building upon this information, the state's Regional Water Quality Control Board recently hired staff to begin implementation of a regional monitoring strategy to coordinate and strengthen these monitoring programs. Concurrently, coordination and standardized training for over 20 volunteer monitoring efforts scattered throughout the Sanctuary's watersheds has been initiated with the Center for Marine Conservation, the Coastal Watershed Council and other local nonprofit groups. Implementation has also begun on several other steps in this plan, including compilation of GIS data layers for the Sanctuary's watersheds, and development of an ongoing water quality council.

Urban Runoff

Runoff from the cities surrounding the Sanctuary carries a variety of contaminants into the watersheds and the Sanctuary itself (Camp Dresser & McKee et al., 1993). These include oil, grease, metals from automobiles, pesticides and fertilizers from lawns, sediment, pet droppings, and human fecal contamination from septic tank leakages, leaky sewer lines, etc.

A detailed action plan outlining steps to implementation and multiagency responsibilities has been completed to address these urban runoff issues (WQPP, 1996a). Strategies include: regional public education programs, regional technical training, storm drain inspections and mapping, structural and nonstructural controls, sedimentation and erosion control programs, and modification of the California Environmental Quality Act (CEQA) checklist to address urban runoff. Development of a regional stormwater management program is also included to allow cost-sharing and division of labor among the region's municipalities.

Implementation of the education, CEQA and regional stormwater strategies has begun using existing multi-agency resources. Additional funding from the State Water Resources Control Board to the cities of Monterey and Santa Cruz, the California Coastal Commission, Monterey Bay Sanctuary and the Association of Monterey Bay Area Governments has begun to carry out many of the other WQPP strategies in the 2 cities, and is serving as a pilot for a larger regional effort. Efforts to date include:

1) Joint development and distribution of a revised CEQA checklist and training package with the County of Monterey, and checklist adoption by three coastal counties.

2) Mapping of critical biological resources, storm drains and pollution "hot spots" in Monterey and Santa Cruz.

3) Initiation of a regional urban runoff education and volunteer monitoring program, including posters, flyers, displays, a hands-on watershed runoff model for events and classroom use, teacher training workshops, stenciling and monitoring efforts, media campaign, etc.

Marinas and Boating

Marinas and boating activities also generate a variety of pollutants which can enter the harbors and adjacent Sanctuary waters. These include TBT, copper, and other toxins from hull paints, waste oil, bilge oils, spilled fuels, toxic materials such as paint thinner, batteries, etc. These materials are contributed in small amounts from a variety of commercial and recreational vessels.

An action plan has been completed addressing marinas and boating activities, working with the region's harbormasters, boatyard operators and boaters (WQPP, 1996c). Seven priority strategies include: regional public education programs, regional technical training, waste oil/bilge waste recovery and disposal, hazardous and toxic materials management, BMPs for vessel maintenance and underwater hull cleaning, and a pollution reduction progress review for harbors. Implementation of the education and bilge waste strategies is underway in a collaborative effort with the nonprofit organization Save Our Shores, local harbormasters and the Sanctuary, using funding from the Integrated Waste Management Board.

Agriculture:

Sediment, pesticides and nutrients are drained from agricultural fields into adjacent rivers and wetlands, and may ultimately reach the Sanctuary itself. This includes contaminants such as DDT, chlordane, toxaphene, dieldrin, endosulfan, etc. Heavy sedimentation from the Salinas Valley and Elkhorn Slough watersheds also increases the necessity for annual dredging operations for Moss Landing Harbor. These operations have recently been complicated by problems with limited disposal options for dredge spoils contaminated with persistent pesticides. Agricultural activities, along with urban growth, have also greatly decreased the amount of wetlands and riparian areas which function as natural filters for sediment and associated pollutants.

The WQPP has begun working with the agricultural community on the development of a water quality plan to address these issues. Initial workshops in three counties were cosponsored with the local Farm Bureaus and Resource Conservation Districts, and attracted approximately 160 participants. These workshops focused on generating interest in watershed stewardship and working with the growers and ranchers to identify problems and initial solutions. Workshops have been accompanied by extensive outreach to agricultural groups and committees to involve them in the planning processs. Initial recommendations from this process are currently being developed by the WQPP committee members and workshop participants into draft strategies.

The team has also begun early implementation on one of the key workshop recommendations--streamlining of the permit process for landowners who wish to carry out water quality conservation practices. Currently a landowner who wishes to adopt conservation practices such as a sediment basin, vegetated buffer or berm near a riparian zone, or a grade stabilization structure faces a confusing and sometimes contradictory array of federal, state and local permits. Agencies participating in the WQPP are working together with the Natural Resources Conservation Service and the nonprofit group Sustainable Conservation to develop a streamlined permitting system to work with strawberry growers in the Elkhorn Slough watershed. This effort will serve as a pilot for regional permit coordination throughout the Sanctuary's watersheds.

Outreach and Education

Efforts in the above specific issue areas have been accompanied by general outreach efforts to agencies, businesses, landowners, and public groups. These efforts have focused on raising awareness about the link between healthy watersheds and the marine environment, inviting participation in the planning process, and identifying simple steps which can be taken to enhance and protect water quality. Outreach efforts include approximately 50 presentations per year, a program newsletter, citizen's guides to the technical plans, teacher training workshops, watershed stewardship training workshops, and a watershed video. A watershed educator's committee has also been established to coordinate and build on existing education programs in the Sanctuary region.

Future Needs

The WQPP has been successful in pooling the resources and expertise of various agencies, public and private groups to evaluate and address water quality issues, and in raising public awareness about the connection between everyday activities in the watersheds and the ocean's health. Despite these successes, much remains to be done. Future work includes completing the plans for agriculture, wetlands/ riparian issues and point sources, achieving full implementation of all plans, and completing establishment of an ongoing Water Quality Coordinating Council which will replace the existing planning committees. In addition, the program requires continuous management to ensure ongoing cooperation of the participating groups, evaluate the success of management efforts, modify them where necessary, and address new problems as they arise. The partnerships which have been established provide a foundation for continued stewardship of the Monterey Bay National Marine Sanctuary and the diverse watersheds of California's Central Coast.

References:

Brown, L.R., P.B. Moyle and R. M. Yoshiyama. 1994. Historical decline and current status of coho salmon in California. North American J. Fish. Man. Vol 14. pp. 237-261.

Caffrey, J.M. 1996. The Effect of Land-use Practices on Benthic Processes in a West Coast Estuary, Elkhorn Slough, CA. Monterey Bay National Marine Sanctuary Currents Symposium Poster Session.

Camp Dresser & McKee, Inc., Kinnetic Laboratories, Inc. and Larry Walker Associates. 1993. Urban runoff water quality management plan for Monterey Bay region. Marina, CA: Association of Monterey Bay Area Governments. 141 pp.

Cotter, P.J. and L. Strnad. 1995. Data Evaluation Report for the Elkhorn Slough Watershed. California Coastal Commission. 63 pp.

Habitat Restoration Group, 1991. Salinas River Lagoon Management and Enhancement Plan. Monterey County Water Resources Agency.

Kleinfelder, Inc. 1993. Elkhorn Slough Uplands Water Quality Management Plan. Association of Monterey Bay Area Governments.

NOAA. 1995a. Monterey Bay National Marine Sanctuary Water Quality Protection Program Working Document: Comparison of the California Water Quality Assessment and the January 1994 workshop results. Silver Spring, MD 76 pp.

NOAA. 1995b. Initial assessment of existing water quality monitoring efforts in the Monterey Bay Region. Silver Spring, MD 23 pp.

Questa Engineering Corporation. 1995. Water Resources Management Plan for Watsonville Sloughs Santa Cruz County. Association of Monterey Bay Area Governments.

Ricker, J. 1995. Salinas River Lagoon Management and Enhancement Plan. Santa Cruz County Environmental Health Services.

SWRCB. 1986. DDT in the Salinas Valley; A special report on the probable source of technical grade DDT found in the Blanco Drain near Salinas, California. Sacramento, CA.

SWRCB. 1990. Toxic Substances Monitoring Program: Ten Year Summary Report 1978-1987. Water Quality Report no 90-1-WQ.

SWRCB. 1995a. California State Mussel Watch, Water Quality Monitoring Report, Summary 1987-1993 Data. Sacramento, CA

SWRCB. 1995b. Region 3: Bay Protection Toxic Cleanup Program Summary, Station Information, Chemistry Data, Toxicity Data. 15 May 1995.

Stephenson, M.D., M. Martin, and R.S. Tjeerdema. 1995. Long-term trends in DDT, polychlorinated biphenyls, and chlordane in California mussels. Arch. Environmental Contamination and Toxicology. Vol. 28: pp 443-450.

Tetra Tech, Inc. 1995. Assessment of ongoing water quality monitoring in the Monterey Bay National Marine Sanctuary and its watersheds. Lafayette, CA.

USGS. 1992. Water Resources Data California. Vol 2.

WQPP for MBNMS. 1996a. Action Plan 1: Implementing Solutions to Urban Runoff. Monterey, CA. 71 pp.

WQPP for MBNMS. 1996b. Action Plan 2: Regional Monitoring, Data Access, and Interagency Coordination. Monterey, CA. 32 pp.

WQPP for MBNMS. 1996c. Action Plan 3: Marinas and Boating. Monterey, CA 52 pp.

CHAPTER 109

Land Management Practices and Endangered species in the Watershed Are they compatible? A case history

Jim Steele[1]

Abstract

California Endangered Species Act protection was granted to the coho salmon south of San Francisco by the California Fish and Game Commission effective in 1996. Among those affected by this needed protection are forest land owners of Santa Cruz and San Mateo counties. Projects such as timber harvest, stream crossings and road maintenance must contain measures that protect this vulnerable specie before permit approvals can be granted. To develop protective measures, the California Department of Fish and Game met with professional foresters and engaged in unique problem solving workshops and training. Foresters and biologists were able to develop protection measures and monitoring methods for timber plans guided by four land management principles.

Introduction

The coho salmon (*Oncorhynchus kisutch)* was listed as endangered south of San Francisco under the California Endangered Species Act (CESA) by the California Fish and Game Commission on January 1, 1996. Testimony that resulted in the listing indicated that coho populations were in decline, many streams no longer had populations of coho and habitats for the remaining fish

---

1

Program Manager, California Department of Fish and Game, 1416 9th street, Sacramento, Ca 95814

were affected by logging, agriculture and housing development. The protection afforded by the listing added to the myriad water pollution laws, planning laws and regulations already in place in California watersheds. Advantage given to the coho by the listing is a requirement for focus on fish problems when applying other laws and to consult with the California Department of Fish and Game (DFG) before project approvals.

Coho populations south of San Francisco are in the counties of San Mateo and Santa Cruz. Logging practiced in the coastal mountains of these counties is governed by extensive State and local rules and requires a permit from the California Department of Forestry and Fire Protection (CDF). Consultation with the CDF regarding their permit process by DFG was necessary to comply with the CESA. This was accomplished through a series of meetings by DFG with both the regulators (CDF) and Registered Professional Foresters (RPF) who represented the regulated timber growing community. Public meetings were held to keep the public informed and an agreement was signed between the two Departments in April 1996 (DFG, 1996) that defined the focused protection measures needed to avoid further harm to the coho from logging.

Two training workshops were conducted with RPFs and CDF Foresters which explained elements of the protection needed for coho from land management activities. Four principles of land management were used to develop protection measures focused on protecting the biological needs of anadromous fish, particularly coho salmon. To develop the principles, the biology of coho needed to be understood as well as the regulatory process for reviewing proposed timber harvest activities.

Coho Taxonomy

The coho salmon is one of seven species of Pacific salmon belonging to the genus *Oncorhynchus*, and one of five such species found in California. It presently ranges in freshwater basins from Hokkaido, Japan and the Russian Far East, around the Bering Sea and Aleutian Islands to mainland Alaska, and south along the North American coast to Monterey Bay, California. Coho salmon have been successfully introduced into more non endemic lakes and streams than any other Pacific salmon species, including all of the Great Lakes. It is a member of the family Salmonidae (salmon, trout and char), of the order Salmoniformes (salmon-like fishes) in the class Osteichthyes (bony fishes). The coho salmon was first described as a species, *Salmo kisutch*, by Walbaum in 1792, from specimens taken from the rivers and lakes of the Kamchatka Peninsula, Russia. It was eventually described as *Oncorhynchus kisutch* by Jordan and Evermann in 1896-1900. *Oncorhynchus* means "hooked snout" and *kisutch* is Walbaum's interpretation of the local name for this species used in Kamchatka. Coho salmon is the accepted common name as adopted by the American Fisheries Society and Federal and State agencies. The name "coho" comes from an American Indian name for this salmon. Other common names include: silver salmon, blueback and sea trout (DFG 2, 1994).

General Life History Cycle

The life history of the coho salmon in California has been well documented by Shapovalov and Taft (1954) and Hassler (1987). Coho salmon generally return to their natal streams to spawn after spending two years in the ocean but some males called "jacks" may return after one growing season in the ocean. Spawning migrations begin after heavy late-fall or winter rains breach the sand bars at the mouths of coastal streams, allowing the fish to move into them. However, migration typically occurs when stream flows are either rising or falling, not necessarily when streams are at a peak flood. Early part of the run is dominated by males, with females returning in greater numbers during the latter part of the run.

The timing of their return varies considerably, but usually they return earlier in the season in more northern areas and in the larger river systems (Baker and Reynolds 1986). In the short, coastal streams of California, most coho return during mid-November through mid-January (Baker and Reynolds 1986). For example, in Waddell Creek, Santa Crus County spawning migrations often do not occur until November or December (Shapovalov and Taft 1954).

Females choose the spawning sites (redds) usually near the head of a riffle, just below a pool, where the water changes from laminar to turbulent flow and there is a medium to small gravel substrate. The flow characteristics of the location of the redd usually ensure good aeration of eggs and embryos, and flushing of waste products. The water circulation in these areas helps fry emergence from the gravel. Each female builds a series of redds, moving upstream as she does so, and deposits a few hundred eggs in each. Thus, spawning may take about a week to complete and a female can lay between 1,400-7,000 eggs. A positive correlation exists between fecundity and size of females. Hassler (1987) noted a dominant male accompanies a female during spawning, but one or more subordinate males also may engage in spawning. He also found both males and females die after spawning, although the female may guard a nest for up to two weeks.

Embryos hatch after 8-12 weeks of incubation, the time being inversely related to water temperature. Hatchlings remain in the gravel until their yolk sacs have been absorbed, 4-10 weeks after hatching. According to Baker and Reynolds (1986), under optimum conditions, mortality during this period can be as low as 10 percent; under adverse conditions of high scouring flows or heavy siltation, mortality may be close to 100 percent. Upon emerging, they seek out shallow water, usually along stream margins. Initially they form schools, but as they grow bigger the schools break up and the juveniles (parr) set up individual territories. An optimal habitat is in deep pools created by large woody debris (LWD) and boulders in heavily shaded sections of stream.

As water temperatures decrease into the fall and winter months, fish stop or reduce feeding due to lack of food or in response to the colder water and growth rates slow down. Toward the end of March and the beginning of April they begin to migrate downstream and into the ocean. Storms that flush fish out of the system before they are ready can result in heavy mortality. An out-migration in California streams typically peaks in mid-April and to mid-May, if conditions are favorable. Migratory behavior is related to rising or falling water levels, size of fish, day length, water temperature, food densities, and dissolved oxygen levels. At this point, the out migrants are about one year old and 10-13 cm in length. The fish migrate in small schools of about 10-50 individuals. Parr marks are still prominent in the early migrants, but the later migrants are silvery, having transformed into smolts.

Coho spawn in many substrate conditions. Usually the available spawning habitat is limited to that with less than optimal small gravel, containing high sand and silt content and moderate to high embeddedness. Coastal streams are characterized by large quantities of highly mobile sediment bedload. Sometimes, juveniles must survive in pools that exhibit elevated summer and fall water temperatures at the margin of acceptability. The ability of coho to use marginal spawning and rearing habitats is an expression of their wide range of tolerance compared with other salmonids. Coho may be the least particular of all Pacific salmon in choice of spawning sites and are opportunistic in the use of a wide range of spawning substrates (Sandercock 1991). However, a reduction in substrate suitability that is already marginal can have a considerable impact to hatching success. Although juveniles prefer cold water, they have tolerance for diurnal temperature peaks that can reach the low 2O's C (70's F), providing nocturnal temperatures drop back into the 15-19 C (60-66 F), range or less.

Population Declines

The DFG determined that coho salmon population declines or their inability to recover in streams in the "listed" area was due substantially to in-stream habitat degradation caused by many factors such as:

- insufficient stream flow at critical life cycle stages due to water appropriations and low rainfall;
- water temperature increases;
- natural and human-induced sedimentation of spawning gravel and rearing pools;
- lack of large woody debris (LWD) or dense canopy from side-stream trees and other vegetation that supports in-stream habitats;
- changes in the stream fluvial geomorphology due to flood control projects and public structures; and
- blockages to upstream migration due to dams, culverts and other watershed alterations.

Other factors contributing to the decline of the coho salmon include catastrophic flood events, changes in ocean conditions, disease, over harvest of fisheries and hatchery influences. Reversing as many of the adverse conditions for coho as possible is necessary for recovery of the species. Despite the present problems with coho in-stream habitats, the DFG recognized that current timber operations and other land management practices occurring on private timberlands within coho inhabited watersheds contributed only to part of the continuing risk to in-stream habitat conditions and thus coho salmon population declines in streams south of San Francisco. In fact, appropriate stream-side management during logging such as certain tree thinning practices may increase habitat recovery.

Timber Harvest Project Review process

The CDF conducts their review and approval process under authority granted by the California Forest Practices Act and extensive rules promulgated by the Board of Forestry. The CDF is a state lead agency under the California Environmental Quality Act ("CEQA") when administering the review and approval of timber harvest plans (THP). The THP is reviewed by a multi-agency Review Team chaired by CDF and includes members from the DFG and other agencies or groups.

Land Management Principles that Protect Salmonid Species

The consultation process with CDF lead to agreement that a strategy was needed to develop protection measures for fish using practices common to the timber industry. It was agreed that an effective strategy to protect salmonid resources began with an understanding of watershed processes and attributes that shape the in-stream habitats essential to salmonid life cycle (life history) needs. The protection measures had to be certain that seven life history needs were met:

1) adults must be able to enter the channel or estuary;
2) migration up the channel must not be blocked;
3) spawning activities should be in clean gravel;
4) eggs must successfully reside in gravel until they hatch;
5) fry must be able to escape the gravel;
6) juvenile rearing habitat must be available; and
7) smolts must be able to safely migrate down stream when they are ready, through estuary waters and out to sea.

Foresters and biologists determined that in order to manage land compatibly with anadromous fish streams, fish needs must be understood in terms of problems possible from land management activities. Project changes and mitigation measures can then be developed. The following four basic land use principles describe the physical characteristics making up essential habitat and stream functions for anadromous fish (Flosi and Reynolds, 1994) which must be protected:

1. The key elements necessary for functioning salmonid habitats must be protected/retained. For land managers, it is not necessarily what is taken, but what is left that is important. So, essential elements must be recognized and left.

Life history requirements of a given species determine what physical elements are necessary to make up suitable habitats. For example, some birds need nest trees, snags, etc. and for salmonids it is appropriately sized spawning gravel, wood debris, bottom roughness and clean cold water. The composition of these physical attributes present in or near the channel, shape and influence the degree and quality of habitats available to salmonids.

The quality and quantity of a habitat are measured by its ability to support all of the life history stages of the species in question. Each species may have different requirements and for a community of species the sum of all life history needs in a watershed can result in a large inventory of important elements forming a complex matrix. The restoration of these elements is the center piece of many restoration projects. Some of these elements may be more essential to a species than others. An important element for coho is a canopy that shades and overhangs the channel. This physical attribute not only provides shade that maintains cool water temperatures and food (through insect and leaf drop) but also overhead cover that darkens the water as a competitive advantage over some species (e.g., steelhead trout). If this habitat element is missing, the advantage in survival may shift to a competing species that can cope with its loss. Understanding the importance of key elements is essential to recognizing or avoiding watershed problems. Therefore, a strategy for retaining important elements should be part of any landuse practice.

Inventory of important elements for Foresters to leave salmonids following a THP for each coho life history stage:

- Migration; leave large logs in the channel and large trees in the side stream area (that will recruit to the stream) to provide in-stream hiding from flood flow currents.
- Spawning; needed are clean (not-embedded with fine sediment) gravel of suitable size. THP's should include erosion protection in stream-side zones such as gravel surfaces on roads to reduce silt production. Ground disturbance close to streams should be as little as possible. Water runoff from roads should not carry sediment.
- Egg and Fry survival in gravel and hiding space under cobble and wood; The same as above, in addition to seasonal restrictions for roads crossing streams.
- Juvenile rearing; Boulders, logs and roots provide escape from flows and predators. Side stream trees that provide habitat shading and maintain low water temperatures should not be harvested. Trees that have a high potential to recruit to the channel should be protected and competing trees thinned to promote fast growth.
- Smolts; As above, elements that provide escape during downstream migration should be protected.

2. <u>The basic structure necessary for functioning habitats must be protected/maintained. To some extent, it is not so important to chronicle what is removed from or changed in a functioning system, as is the</u>

structure of what must remain to function properly as habitat.

A drought may influence the fish production of a stream system because of low flows that decrease the habitat area available. However, if enough structural features of the stream, such as pool depths, are protected, life history needs can be met and the cycle completed for surviving fish. Fish production can then quickly return in years with adequate flows.

Important structural elements for salmon are pools, undercut banks, side channels or woody debris large enough and situated to provide an escape from flow or predators. The influence of structure on coho life history is determined by; shape, location, amount and size. The more nooks and pockets available increases escape opportunities, locations throughout the stream may accommodate more individuals and larger size may influence channel flow characteristics to benefit fish. LWD transverse to the flow can hold spawning gravels in place and form plunge pools for summer survival. Removing or changing structural attributes can severely diminish habitat quality.

The value of some structure is difficult to appreciate. Consider a situation where the appropriate channel depth and all the appropriate elements exist for fish migration, spawning and juvenile out-migrant passage. However, the basic structure of the channel is altered by a project that removes in-channel objects, thus creating a laminar flow without flow shadows for fish to hide behind. In this situation, the channel capacity is improved from the stand point of flood control, but the structural changes can effectively remove scour pools and escape areas and increase flow velocities. Even though the channel appears to allow fish use because the water is deep and large surface area is visible, storm flow velocities may deter effective upstream migration for adult fish that cannot overcome the higher velocities or find hiding areas. Spawning gravels can be swept away and banks can collapse making the channel shallower and wider. Storms later in the year can cause eggs, fry and juveniles to be flushed out of the system prematurely and not survive.

Bank stabilization projects or even culvert placement for steam crossings often changes the channel's structure but rarely improves it for fish. Culverts in a fish passage area must be designed to allow fish passage. Bank stabilization projects should result in reducing harmful sediment levels and not result in decreasing side channel hiding areas or increased flow velocities.

The amount of structure is important. Woody debris in a channel can alter flow characteristics in a way to produce scour pools or plunge pools used by fish. Debris upstream of habitat areas can also hold banks together and act as small check dams to meter sediment evenly through the system rather than let it flush immediately to the low gradient areas. Evenly spaced woody structure throughout the system usually keeps the channel bed at a base level that slows side slope sediment movement. However, too much wood debris in one location can plug the channel and block fish passage. This situation commonly occurred in early logging practices. The resulting jams led to several decades of clearance projects that are still going on in some streams. Therefore, log loading sites should not be placed in areas where large amounts of abandoned logs, limbs and debris can enter the stream during a storm.

Removing channel structure is a less obvious change to a system but can result in negative impacts to habitat quality if done incorrectly. For example, channel straightening projects that remove channel meander can increase flow velocity, and remove slow current areas and side scour pools.

Also, removing debris that is jamming the flow of sediment or water is an important project to do only if fish passage is blocked, surface water habitat area is reduced or human dwellings could be flooded, etc. If necessary to do so, debris alteration should be accomplished so that escape habitat or channel altering functions are maintained. Remove too much debris and important structure can be lost. Strategies for restoring and maintaining watershed structure should be included in land management activities.

Inventory of important structures to maintain:

- Side-stream trees providing temperature blocking shade as well as habitat shade and can later recruit to the stream as LWD structure.
- Channel shapes with bottom irregularities that form pools and side channel areas and allow escape from high flow velocities. The channel should be deep enough compared with its width to maintain flow velocities that clean and sort gravel and reform irregular channel shapes.
- Fish ladders should be available to provide passage over dams.
- Undercut banks held by roots provide an escape habitat and vegetation should be protected .

3. The rate of recovery in an ecosystem must be measurably greater than its rate of degradation. A system in balance with respect to fish will have disturbance regimes less than or equal in time and spatial scope to recovery processes.

When habitat recovery is slow with respect to repeated human caused disturbances, affected fish populations may respond slowly and even disappear. This is particularly evident where change is locked into place by projects that also alter natural systems. Land conversion to roads, buildings, dams, or other permanent fixtures that produce sediment, block migration or change channels for flood control will lower the recovery ability of fish populations and their habitat.

Important watershed processes that are rates driven are; delivery of water flow, sediment, fine and course woody debris, solar radiation (water temperature) and dissolved substances to the channel. Too much of these inputs can cause a negative result and conversely too little can result in depleted habitats. Sometimes the recovery rate can be increased through enhancement projects that stabilize bank and place wood structures in the channel). Strategies for improving and measuring recovery rates for habitats (monitoring) should be included in land management plans.

4. The functional integrity of natural systems must be protected and maintained as measured by their biological outputs (e.g., fish production). When these are hard to measure, important physical characteristics can sometimes be substitute indicators.

Of course, measurements of physical characteristics cannot replace biological outputs as a true measure of habitat health. Nevertheless, with this in mind, physical products normally associated with salmonid production can be considered indicative of a healthy riverine system. For example, a sufficient canopy can often be quantified through temperature measurements, and woody debris can be counted and its function assessed. Conversely, low numbers of these elements can also be used to estimate the level that a system is impaired. Because in-stream habitats are so important to fish survival, physical channel attributes such as LWD and gravel cleanliness are used in many monitoring schemes. The best water quality measurements for fish are dissolved oxygen, temperature, and pH.

A watershed may have recovered to the point where important physical elements are in place and situated in ways to provide optimum structure for species life history. However, if the continuity of the life history has been broken such that the population is to low to recover to the habitat, population recovery might not occur. This argues for identifying and providing even higher levels of protection for critical pockets of biological health or fish "refugia" within the watershed so that recovery into renewed habitats can occur.

The best land use protection for salmonid populations is protection of a habitat's functional integrity. For example, sediment introduced to channels at a rate above a level that the system can redistribute it through storm flows could result in reduced available wetted habitat area where the fish need it. The stream can literally flow entirely below a gravel surface during the summer. As pools are filled and channels widen, it is possible to reach a point where the functional integrity of the habitat is compromised and biological use (no surface water) for fish is lost. Loss of habitat integrity has happened in cases of sedimentation, temperature increases, water diversions or chemical changes in the worst case scenario of dams, channelization, outfall pollution or watershed disturbance.

For the forestland manager, knowledge about the existing watershed condition and the optimum condition that will promote the most positive output is key to compatible land management. As a practice, timber operations that have the least risk and narrowest margin of error are called for when sensitive species such as coho could be affected. A land management plan should seek a high level of protection for key elements and their arrangement, and for channel structure (DFG 3,1994). The plan should ensure proper habitat function by determining the functional integrity of the system (habitat complexity) and enhance recovery rates through restoration wherever possible. All timber harvest operations should have a monitoring plan to assess the effectiveness of protection measures.

Strategies for establishing baseline bench marks and monitoring the continued functional integrity of the watershed is key to understanding the effectiveness of watershed protection and restoration measures. To date, the coho and their habitats south of San Francisco are still in a recovery mode, but the Foresters are using land management principles compatible with coho to make sure their projects are part of the solution and not the problem.

References

Baker, P. and F. Reynolds. 1986. Life history, habitat requirements, and status of coho salmon in California. Report to the Calif. Fish and Game Commission. Sacramento, Calif. 37 p.

DFG 2. 1994. Petition to the Board of Forestry to list coho salmon as a sensitive species. Prepared by the Dept. Fish and Game. Sacramento, Ca. 110 p.

DFG 3. 1994. Qualitative Assessment Technique for Registered Professional Foresters. Prepared for the Board of Forestry by Dept. Fish and Game. California Board of Forestry. 31 p.

DFG. 1996. California Endangered Species Act, Biological Opinion issued by the California Department of Fish and Game to the Department of Forestry and Fire Protection. Signed by Schafer April 1996.

Flosi, Gary and Forest Reynolds. 1994. California Salmonid Stream Habitat Restoration Manual. California Dept Fish and Game. Sacramento Ca.

Hassler, T. J. 1987. Species profiles: life histories and environmental requirements of coastal fishes and invertebrates (Pacific Northwest)--coho salmon. U. S. Fish and Wildlife Service Biological Report 82 (11.70). U. S. Army Corps of Engineers, TR EL-82-4. 19 p

Sandercock, F. K. 1991. Life history of coho salmon, *Oncorhynchus kisutch*. Pages 397-445 *in* C. Groot and L. Margolis, eds. Pacif salmon life histories. Univ. Of British Columbia Press, Vancouver, B. C.

Shapovalov, L. and A. C. Taft. 1954. The life histories of the steelhead rainbow trout (*Salmo gairdneri)* and silver salmon (*Oncorhynchus kisutch*) with special reference to Waddell Creek, California, and recommendations regarding their management. Calif. Dep. Fish and Game, Bull. 98. 375 p.

CHAPTER 110

SAN FRANCISCO WATERSHED MANAGEMENT PLANS ACHIEVING WATER QUALITY AND BALANCING MULTIPLE USES

Darcey C. Rosenblatt
Tina Stott

Abstract

The San Francisco Water Department (SFWD) is in the process of preparing Management Plans for the 40,000 acre Alameda Watershed and the 23,000 acre Peninsula Watershed. On daily basis, these lands store and provide water for 2.3 million homes and businesses in San Francisco, as well as in portions of San Mateo, Santa Clara and Alameda Counties. SFWD's mission statement for watershed management is "to provide the best environment for the production, collection and storage of the highest quality water for the City of San Francisco and suburban consumers". Existing SFWD policies and plans do not address the management of these two Watersheds in a comprehensive manner, making it difficult to evaluate proposals for new Watershed activities and their effects on Watershed resources. The Management Plans establish a framework for decision making by providing a comprehensive set of Watershed management policies and procedures reflecting the unique qualities of each Watershed.

As part of the process developing Management Plans, SFWD established a primary goal and secondary goals to drive the creation of management policies. These goals were intended to provide direction for plan development, guidelines for decision making and a basis for the evaluation of plan implementation. The primary goal for the Management Plans is to maintain and improve source water quality to public health and safety. Secondary goals are as follows:

- Maximize the contribution of the Watersheds to the total water supply;
- Preserve and enhance the ecological and cultural resources of the Watersheds;
- Protect the Watersheds, adjacent urban areas and the public from fire and other hazards;
- Continue existing compatible uses and provide opportunities for new compatible uses on Watershed lands (including educational, recreational and scientific uses);

- Provide a balance between financial resources, revenue generation activities, and overall benefits; and
- Enhance public awareness of water quality water supply, conservation.

The proposed Peninsula Watershed Management Plan includes the following general directives for Watershed management.

- Establishment of baseline monitoring and ongoing water quality monitoring;
- Moderate level of ecological resource monitoring;
- Significant reduction in the fuel load;
- Moderate increase in staff to support ecological resource protection and some increased access
- Establishment of Best Management Practices for SFWD operations and maintenance activities, and;
- Acquisition of critical Watershed acreage to protect source water supplies.

This paper will report on the processes of developing a comprehensive watershed management plan, including the development of policy implementation strategies and assessing public input.

CHAPTER 111

Where the Tern Meets the Tide: Public Participation in Bay Water Quality Issues for Morro Bay

Charmaine Marie Gallagher[1]

Abstract

Now that the Morro Bay Estuary has received both State and National Recognition through inclusion to these management programs, a management plan must be constructed which achieves the goals of both directives. Morro Bay is the marine tourist industry's, ocean activist's and fisherfolk's lifeblood of the Central Coast of California. The City of Morro Bay gives priority to projects which support commercial fishing and use on the waterfront. The management plan must address trends in the estuary's water quality, pollutant loadings and causes of environmental problems in the estuary. It is clear that multiple agencies, businesses and individuals have a vested interest in the water quality of the estuary and bay. A specific work-group to address bay water quality and the issues, fears and preconceptions surrounding the effects of management are presented. Dredging, tidal flushing, oil spill response and storm water runoff are foremost in the mind of the public. Solutions to address concerns must rely on existing users cooperation and support in the development of the plan.

Introduction

Public participation in the formulation of coastal management plans provides needed support for agency implementation. The entire process surrounding the establishment of Morro Bay as a State and National Estuary has been predicated on grass roots and citizen involvement. Although state, regional and local agencies have facilitated the public process, citizen initiative was required to assure acceptance in to the National Estuary Program. Acceptance now mandates that management plans be formulated within two years to address trends in the estuary's water quality, natural resources and uses. These plans require identification of causes of environmental problems through research, assessment of pollutant loading in the estuary and observation of changes in water quality and natural resources. Results of the plan will be to recommend and schedule priority actions to restore and maintain the estuary.

[1] former Chair, Bay Water Quality Work Group, Morro Bay Watershed Management Council, P.O. Box R Corvallis OR, 97339

The focus of this paper is to stress continuance of public participation in the formulation of specific actions to be included in the Morro Bay Management Plan. More specifically, to use the information provided by the public work-group on bay water quality to find reasonable and workable solutions to improving shore based and marine practices affecting water quality on Morro Bay. This is where ' the rubber meets the road' or in this case where the tern meets the tide. Only with public participation in the formulation of realistic action plans to improve water quality will the implementation of the federal mandate be achieved. State and National Estuary Process as well as additional public work-groups on specific topics related to the bay and its watershed will be the subject of future reports.

Morro Bay Physical Description

Morro Bay is one of the most dangerous harbors on the west coast with an average of 55 days per year declared hazardous by the coast guard. The mouth of the bay is subject to shoaling, sand spit movement and 4 knot tides. Morro Bay was once a lagoon with the mouth distinguishable by a large rock which is a remnant of a volcanic plug. Road construction in the early part of this century made use of the rock as a quarry and created a causeway from land to sea thereby shutting off the northerly opening of the lagoon. Morro Bay as it is now called has been estimated to flush 11,000 acre feet of water per day but is presently operating at 6,500 acre feet per day (Haltiner et. al 1991). This reduction in tidal flushing has grave consequences pertaining to health of the bay and bay water quality. The central portion of the bay has both intertidal areas mudflats and tidal channels that drain the mudflats during low tide to the mouth of the bay. The back bay is characterized by mudflats and shallow channels with both Chorro and Los Osos creek deltas creating marshlands.

Economics and Demographics

The City of Morro Bay supports a variety of commercial fisheries, processors and restaurants. In addition, it is one of only two commercial harbors between Monterey and Santa Barbara. The commercial coldwater fishery of dungeness crab, groundfish and seasonal catch. Recreational fisheries of California halibut within the bay, clams, baitshrimp and offshore rockfish species. The commercial sector accounts for substantial revenues to waterfront businesses. According to Dr. William Fox, head of National Marine Fisheries Service "Over 70 percent of the U. S Commercial and recreational landing that provide over $30 billion to the U. S. Economy are composed of species associated with estuaries at some time in their life history. Therefore, seventy percent of our production is dependent upon the preservation of nearshore habitats.

The estuary supports 200 species of migrating birds on the Pacific Flyway important for tourism which is vital to the region providing over $745 million annually. The mudflats provide a home for juvenile crab and eelgrass beds a home for juvenile fish nurseries needed for commercial and recreational fishing. Commercial fishing and its supportive processing and distribution elements constitute a substantial portion of the Morro Bay economy. Therefore, Morro Bay supports elemental plans, projects, and regulation which give priority to ocean dependent uses. The estuary is also the backyard of approximately 20,000 people in the unincorporated town of Los Osos which has had no storm drain system (until 1996), no sewage system, and no provisions for curb gutters and sidewalks. As a result sediment and possible nutrient loading from this region impacts the back bay.

Legislation on Water Quality

Under the Federal Coastal Zone Act Reauthorization, non-point sources came to the forefront in addressing nearshore and coastal waters with a requirement to upgrade the State's Non Point Source Plan to provide direction for the application of federal and state water quality regulations. The State Water Resources Control Board is the authoritative agency to address compliance and enforcement of regulatory measures. In Morro Bay jurisdiction ranges from the city to state parks to private enterprise. Application of federal requirements is unique at best.
State mandates to address water quality are by the State Water Resource Control Board administered through the Central Coast Regional Water Quality Control Board who also carries out provisions of EPA for this region.

The National Estuary Program has its own set of guidelines to address water quality and within these activities required are two which address public participation. The first requires transfer of scientific and management information, experience and expertise to program participants. The second, pertaining to public buy-in, is to increase public awareness of pollution problems and ensure public participation in consensus building. The experience with the early task force and water quality work-group is that participation is heavily laden with agency personnel. These participants are just as concerned as stakeholders, and are every bit aware that a management plan can and may affect their own job description as well as affect their recreation or families pleasure of the Morro Bay estuary.

Water Quality Public Work-group

Having stated that the composition of the public work-group is largely agency personnel, this mixture provide benefits of its own. For here constitutes the best available collection of scientific, management information, experience, and expertise on the issue of bay water quality. Yes where the rubber meets the road, with Harbor District, City Public Works, state parks marina, local mariculturists. These people have an invested interest in addressing water quality for they have been forced in the past by both regulation and administration. Yet there continue to be areas which plague managers in the quest for improved water quality. Getting the user to take responsibility for his own share in the degradation of bay water quality is the biggest challenge. It is this next step and the non-point sources of pollution from all areas and shore based and ocean activities which need to be addressed if water quality is to improve.

People whose career functions are to ensure the safety and continuance of uses in and adjacent to the bay are well versed in applicable provisions to protecting water quality. These represent some of the vital supporters in developing management programs which will be understood and applicable to every situation on the bay. Addressing concerns, fears and preconceptions pertaining to water quality is a large part of this public work-group. Identifying where research is needed, potential action programs and solutions are a logical next step in the formulation of management regimes. What follows is a selection of major issues pertaining to water quality in Morro Bay and the business activities both dependent on and designed to address management and improvement of the bay.

Morro Bay Concerns/Activities to be Protected

Sediment

Beneficial uses of bay 11,000 acre feet/day should flush according to Army Corps of Engineers the bay is flushing 8500-9000 acre feet per day. This bay is nearly 10 times natural conditions in the rate it is filling with sediment (Worcester, K. personal communication).
Fears/ perceptions

Sediment enters the bay via three sources 1) littoral transport throughout the lagoon mouth, fluvial transport of sediment from the watershed and aeolian sand transport from the sand spit (Haltiner et. al, 1991). Volumetric changes in the bay indicate that in the last century Morro Bay has lost between 25 and 30 percent of its volume below +4 ft and in deeper parts of the bay below mean low low water 35-40% of volume has been filled in. The change in mean tidal prism (difference between mean high and mean low water) shows an average of 37,000 cubic yards per year are accumulating in the bay with most of it coming from the central and south bay areas.

Dredging

Dredging is always believed to be a problem with water quality but in truth dredging supports commercial efforts, bay hydrology, fishery nurseries and recreational sites. Dredging operations are designed to reduce the numbers of hazardous days in the mouth of Morro Bay from 55 to 10-15 per year and reduce thermal effects of the bay affecting power generation and dissolved oxygen levels. Dredging in the back bay supports recreational use of the marina and improved tidal flushing. Major dredging which is scheduled every three(minor) seven years (major) requires Environmental Impact Reports and Statements respectively. Eel grass mitigation plan from dredging come about as a result of these efforts. Harbor Mouth dredging in addition to channel mouth clearing constitutes $3.9 million (R. Algert, personal communication).

If large dredging efforts are curtailed, then pirate (prop) dredging is common for both commercial and recreational uses for it is difficult to obtain permits and expensive for boatowners ($175,000) with no guarantee of a permit to dredge. Dredging has always been completed with funds from Army Corps of Engineers; now that this funding is limited there is likely to be less legal means of dredging going on. Adverse affects from dredging include loss of nursery habitat from eelgrass encroachment, suspension of fine particles, re-suspension of heavy metals, bacteria, pesticides or other sediment toxics which when suspended affect aquaculture activities. Additional concerns arise with the disposal of dredge spoils and the content of dredged materials. The work group identified areas needed for research including underestimating hydrological dynamics and the link between dredging and economics of the commercial fishing industry.

Power Generation

A Major power generator is located at the mouth of Morro Bay which uses seawater for cooling and discharges outside the bay. Thermal pollution is not clearly delineated but raises concern for a one degree increase in ambient temperature constitutes a loss of thermal effects of bay. an increase in back bay limits function of plant effects efficiency of power generation

Aquaculture Activities and Bacterial Pollution

Morro Bay is the cleanest estuary along the entire pacific coast, yet it is subject to as much as 36 closed days per year to shellfish collection and distribution. Certainly there is room for improvement. Some of the best indicators of what is actually happening in the bay are the aquaculture farming practices which have been ongoing in Morro Bay for decades. At present, a single oyster farm provides living indicators of the water quality conditions in the bay. proof that the bay has improved over time but subject to all forms of bacterial pollution.

This single active grower in Morro Bay is one of 12 active oyster growers in California with over 2 million Pacific oysters in 275 of 754 leased acres in the back bay. This aquaculture venture is subject to temperature fluctuations from reduced tidal flushing, freshwater flooding, siltation, agency mandates for bay closures, and any storm situation where over one half inch falls in a 24 hour period. And given that oysters are filter feeders, the farmer is first to detect any water quality changes and consistently monitors health of the bay.

In 1995, 3 0,000 oysters were lost due to sediment from storms. Bacterial pollution can shut down operations for days constituting a loss to the farmer. Bacteria come from effluent sources from areas surrounding the bay as well as from upstream creek discharges. In addition, live-aboards may contribute when discharging or dumping with un-enforced ordinances.

Oil Spills/Leaks/Launch Ramps

Much effort on catastrophic oil spills has stimulated the U. S. Coast Guard and California Department of Fish and Game to devise an Oil spill contingency plan with assistance from San Luis Obispo County Planning. The Oil spill Contingency Plan includes the San Luis Obispo Natural, Economic and Public Resource Inventory Maps which provide all needed contacts, responders, equipment, ownership information and specific concerns related to an ocean based spill. What is not included are the small "Sunday morning spills" which are inherently difficult to identify the source and contain before the strong tide takes them away. In addition, boaters have taken to spreading Dawn dish-washing liquid to dispense of spillage. According to marine ecologist, John Hardy, oil from power boat engine use on a daily basis settles in the sea-surface microlayer and continually photooxidized (Mele, 1993).

Public launch ramps and marinas have become focus locations to assist in information dissemination, visitor awareness, and toxic collection stations. Initial projects such as oil recycling, oil/water separators from builge water, port-a-potty dump stations , and interpretive signage. In addition, preventative measures to suggest natural antifoulants for boats, prevention of scraping/painting boats on mudflats, mobile pump out service, waste recycling and discard stations and additional behavioral measures to encourage the public at large to comply.

Urban Influent

This area is the most concentrated source of non-point source pollution from residences and urban areas. Few regulations are imposed on the household and this sector of the public is harder to reach on bay water quality issues. Each households contribution is incremental compared to the plethora of chemicals funneling through the storm drains and into the bay. Concentrations of hydrocarbons from parking lots and streets, paint and construction debris, yard and pet wastes, pesticides and litter constitute the bulk of urban runoff into storm drains. Efforts to stencil storm drains were completed for the entire City of Morro Bay in attempt to educate in the simplest of terms how public facilities are constructed.

Scientific/Monitoring efforts

Baseline information to address pollution parameters is established throughout the watershed by the Environmental Protection Agency Section 319 National Monitoring Program. In it the Non point source monitoring program was developed and is co-managed by the Central Coast Regional Water Quality Control Board and California Polytechnic State University. EPA also has a Quality Assurance Plan to establish guidelines for water quality sampling and analysis.. These demonstrated high pH levels, elevated nutrients, depressed oxygen levels and elevated fecal cloakroom levels in dryer years and high sedimentation loads and fecal coliform levels in storm years. Relative biological and chemical data sets are to be collected to demonstrate the effectiveness of best management practices.

Despite concerns the bay is actually doing better with water quality than it had many years prior. This bay has seen as many as 17 leases for oyster cultivation dating back to the 1950's but was always tempered by pollution for the city of Morro Bay had no treatment for sewage. By 1986 the City began treatment of sewage and provided a cleaner wastewater discharge to the ocean.

This supports newer concepts pertaining to bio-inicators of water quality. Oysters are obdurate in their demand for pollution free water, incapable of compromise (Rauber, 1995). They are the coastal waters coal mine canaries for they are deemed inedible when water quality is reduced. Oysters are bio-accumulators and should be treated as indicator species in Morro Bay for these are the only species monitored on a regular basis. The California Ocean Plan supports the use of Indicator Organisms for water contact bacterial standards and will broaden this use for water contact fecal coliform standards (1997). Clearly the indicator organisms exist and can be used. Mussels present in the bay are the best at accumulating toxins at a rapid rate. Further support of mussel watch programs is needed as mussels are quicker at picking up toxins and can indicate contaminants sooner. Funding for this program is in jeopardy and little is being done to investigate the results of this information. Oysters and the associated aquacluture venture are excellent sources to determine contamination levels. The benefits are twofold; continuous monitoring of oysters provide the most consistent source of information

Conclusions

The only realistic way to expect residences, business owners and all waterfront activities to agree that the State and National Estuary Management Programs are designed to assist in in-going efforts to maintain and enhance water quality in Morro Bay is by inclusion. Agencies should recognize the wealth of expertise and consistency in parceled efforts to assure water quality through existing oyster cultivation, power plant influent/effluent systems, fishery logbook data as well as volunteer monitoring efforts of creek deltas into Morro Bay. Implementation of any management Plan will require tie-in not additional mandates to add to the list of regulations. Much can be incorporated through the beneficial participation of pubic and private groups interested in water quality.

References

Algert, Rick. 1995. Morro Bay Harbor Director.

Bell, J. A 1993. Water quality analysis of Chorro creek using benthic invertebrates as indicators. California Polytechnic State University. San Luis Obispo, California.

California Department of Health Services, S. B. D. (1985). Sanitary Investigation of Shellfish and Water Quality, Morro Bay. September 1984 and 1985. California Regional Water Quality Control Board, Central Coast Region.

Haltiner, J. P., M. ASCE and Devin Thor. 1991. Sedimentation Processes in Morro Bay, California. Coastal Sediments '91 Proceedings Seattle, WA 25-27 June 1991.

McCray, J. 1994. Bay is his Oyster. San Luis Obispo County *Telegram Tribune* 20 Jan. 1994.

Mele, A. 1995. Ban the Putt-Putt. Audubon. May-June 1995.

Rauber, P. 1995. The oyster is our world. *Sierra*. 16 September 1995.

State Water Resources Control Board. 1997. California Ocean Plan. California Environmental Protection Agency.

USDA Water Quality Hydrologic Unit Area. 1995. Morro Bay Watershed, San Luis Obispo, California. 1995 Annual Report: Accomplishments and Impacts.

U. S. Army Corps of Engineers. 1986. Unpublished historical dredging data.

Worcester, K. 1996. Biologist, Regional Water Quality Control Board

CALIFORNIA AND THE WORLD OCEAN '97

Taking a Look at California's Ocean Resources: An Agenda for the Future

VOLUME 2

CONFERENCE PROCEEDINGS

March 24-27, 1997
Town & Country Hotel
San Diego, California U.S.A.

EDITED BY
Orville T. Magoon
Hugh Converse
Brian Baird
Melissa Miller-Henson

ASCE American Society of Civil Engineers
1801 ALEXANDER BELL DRIVE
RESTON, VIRGINIA 20191

Abstract: The conference, California and the World Ocean '97 (CWO '97), was organized by the Coastal Zone Foundation and Resources Agency of California. CWO '97 was an opportunity to bring together the leading experts on ocean and coastal resource management issues from around the world to help California determine how best to begin approaching the recommendations in the strategy, *California's Ocean Resources: An Agenda for the Future.* The papers in this proceedings were presented in San Diego, March 24-27, 1997, and offer suggestions on integrating and making more effective the protection, enhancement, and management of California's extraordinary ocean and coastal resources. Topics covered are varied and include coastal economics, shoreline processes, ports and vessel traffic, oil and gas development, oil spill prevention and response, marine managed areas, habitats and ecosystems, fisheries management, water quality, governance and public policy, alternative and emerging technologies, education, and public participation.

Library of Congress Cataloging-in-Publication Data

California and the world ocean '97: taking a look at California's ocean resources: an agenda for the future: conference proceedings / edited by Orville T. Magoon ... [et al.].
p. cm.
"March 24-27, 1997, Town & Country Hotel, San Diego, California, U.S.A."
Includes indexes.
ISBN 0-7844-0297-3 (set)
1. Marine resources–California–Congresses. I. Magoon, Orville T.
GC1023.13.C2C25 1997 97-48327
333.91'64'09794–dc21 CIP

Library of Congress Catalog Card No: 97-48327
ISBN 0-7844-0297-3
Manufactured in the United States of America.

Contents

VOLUME I

* Manuscript not available at time of printing.

Regional and International Fisheries Management

Governance Lessons from Innovative Ecosystem Management Strategies

* Manuscript not available at time of printing.

* Manuscript not available at time of printing.

* Manuscript not available at time of printing.

* Manuscript not available at time of printing.

* Manuscript not available at time of printing.

* Manuscript not available at time of printing.

* Manuscript not available at time of printing.

* Manuscript not available at time of printing.

* Manuscript not available at time of printing.

* Manuscript not available at time of printing.

* Manuscript not available at time of printing.

* Manuscript not available at time of printing.

Coastline Processes and Management

Oil and Gas Facilities Decommissioning

Biodiversity and Threats

* Manuscript not available at time of printing.

* Manuscript not available at time of printing.

* Manuscript not available at time of printing.

* Manuscript not available at time of printing.

* Manuscript not available at time of printing.

CHAPTER 112

SURFACE WATER MONITORING UNDER THE COUNTRY'S LARGEST MUNICIPAL STORMWATER PERMIT

Bill DePoto[1], P.E., M.ASCE; Mert Ramos[2], P.E.; Tim Smith[2], P.E.

Abstract

The Los Angeles County Department of Public Works and its predecessor Flood Control District have been actively engaged in stormwater quality monitoring for many years. As the principal permittee of the new Los Angeles County Municipal Stormwater NPDES Permit, the Department is undertaking one of the most comprehensive water quality monitoring programs of any municipal stormwater program in the country. The program features water quality monitoring of storm and dry weather flows from both mass emission and land-use specific drainage areas, an illegal connection elimination program, a critical source monitoring program, and a receiving waters impact study. The list of constituents sampled is extensive, including metals, hydrocarbons, pesticides, solids, nutrients, semi-volatile organics, and selected minerals.

The information collected from the monitoring program will be used to: track water quality status, trends, and loads and identify pollutants of concern; monitor and assess land use and watershed pollutant loads; identify and assess significant water quality problems in watersheds; identify sources of pollutants in runoff; identify and eliminate illicit connections and discharges; evaluate the effectiveness of best management practices, and; assess impacts on receiving waters.

Introduction

Los Angeles County is one of the largest governmental jurisdictions in the country. At more than 10,200 square km (4000 square miles) with a population of over 8.8 million, it is larger than two states in size and 42 states in population. Roughly three fourths of the land surface drains to the Pacific Ocean; only the northeast corner of the county drains to the Mojave Desert at the edge of the Great Basin. The downtown Los Angeles

---

[1]Supervising Civil Engineer II, L. A. County Public Works, 900 So. Fremont Ave., Alhambra, CA 91803

[2]Supervising Civil Engineer I, L. A. County Public Works, 900 So. Fremont Ave., Alhambra, CA 91803

area averages about 380 mm (15") of rainfall a year, most of it falling in the six months between October and April. Higher elevations in the San Gabriel Mountains can expect up to 1143 mm (45") of rain a year.

Ever since its inception in 1915, the Los Angeles County Flood Control District has been concerned with inland surface water quality. Because of its mandate to recharge stormwater runoff into underground aquifers, the District has always been concerned with conventional water quality parameters like turbidity and suspended solids. In 1967, the District expanded water quality monitoring into a regularly scheduled activity of dry weather and wet weather grab sampling. In 1985, the Los Angeles County Flood Control District was integrated, along with the County's Road and Engineering Departments, into the newly formed Los Angeles County Department of Public Works. Over the years, the water quality monitoring program has been revised and modified a number of times to support infrastructure and operational needs and meet regulatory requirements.

The latest regulatory requirements derive from the 1990 and 1996 Municipal Stormwater NPDES permits issued by the State Regional Water Quality Control Board under the federal Clean Water Act amendments of 1987. Recognizing that nonpoint source pollution was at least as much a threat to water quality as point source pollution, the amendments focused on controlling pollutants associated with stormwater and urban runoff. The regulations apply to urban areas with a population of at least 100,000. Since the Department of Public Works is the major owner and operator of the storm drain system, it became the "principal permittee." The other 85 incorporated cities within the portion of the county that drains to the Pacific Ocean, including the City of Los Angeles, became "copermittees."

The total drainage area covered by the stormwater permit comprises some 7900 square km (3100 square miles) and includes tributary area from three adjacent counties (see map). The population is distributed on some 2 million parcels served by roughly 4000 km (2500 miles) of storm drains, channels, and rivers. The permit area also contains some 150,000 licensed businesses.

The 1990 NPDES Municipal Stormwater Permit

The 1990 permit, one of the first in the nation under the newly revised federal Clean Water Act, called for the development of a comprehensive monitoring plan, an illegal connection and discharge elimination program, and implementation of in-stream monitoring in three distinct phases. The plan that the Department developed departed from the historical program already in place. The largest departure was in characterizing runoff quality in terms of the land use of the drainage area. Under this approach, the Department would focus on runoff from seven land use types developed from information compiled by the Southern California Association of Governments (SCAG), namely:

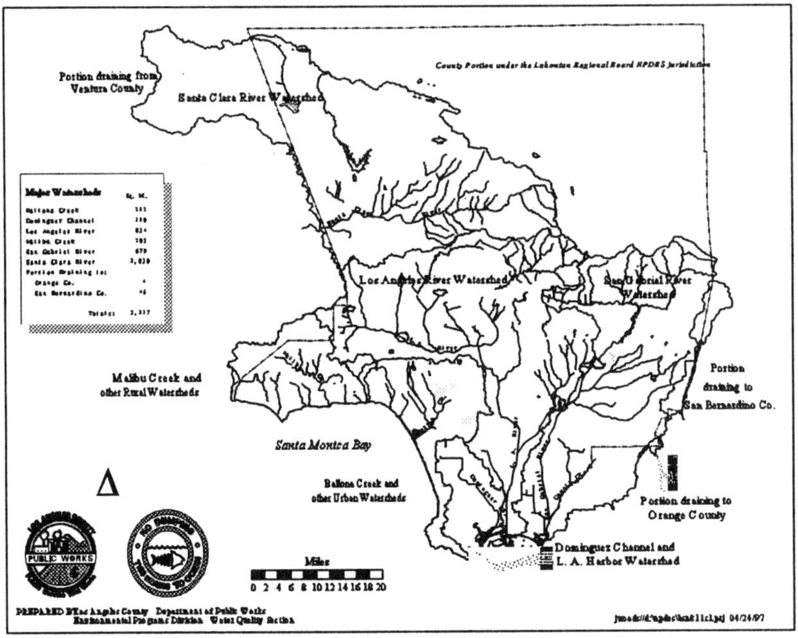

Figure 1. Map of Los Angeles County and Watershed Management Areas

- open space/recreational
- low density residential
- high density residential
- multifamily
- commercial
- industrial
- transportation

The seven land use types, aggregated according to similar imperviousness characteristics, closely resembled the types already in use by the Department's peak rainfall/runoff model. In addition, the plan included monitoring "mass emission" runoff from large drainage areas that have a wide variation of land use types.

To begin implementing the in-stream monitoring phases of the plan, Department staff manually overlaid land use transparencies over watershed and drain alignment maps to identify the best candidates for nonpoint source monitoring. Siting was accomplished by answering the following questions:

- Is the chosen land use type predominant and homogeneous throughout the tributary area?

- Are the design flow rate and drain hydraulics known for the tributary area so that the drain will not flow surcharged in a moderate storm? Are there tidal or backwater influences that would affect flow rate measurements?

- Is there usually water in the drain during periods of no rain? If so, what is the flow rate?

- Is the site near a source of electrical power? For underground drains, is there a nearby manhole?

- Is the site in a relatively safe neighborhood?

- Does additional right of way need to be purchased?

Once these questions were answered, and the individual designs and construction permits completed, the installation of Phase I stations began in the Spring of 1994. Phase I covered all of the surface drainage area tributary to Santa Monica Bay, some 1100 square km (414 square miles). Through the land use analysis process, the Department chose five land use sites monitoring a total of 26 square km (10 square miles) of drainage area, and four mass emission monitoring sites covering 530 square km (209 square miles) of drainage area. Fifty three percent of the monitored area would be undeveloped while the balance would be urbanized coastal basin. By Jan. 15, 1995, all nine stations were installed.

Included as a requirement in the 1990 plan was the estimation of total pollutant loading to the ocean. The estimation would be accomplished by the application of a spreadsheet model that converts rainfall to runoff via the use of imperviousness values for each of the seven land use types. The Department had been using such impervious values for years in its peak rainfall/runoff modeling. The volume of runoff derived from recorded rainfall amounts would then be multiplied by an "event mean concentration" for a number of "pollutants of concern" identified in the permit. The result, multiplied by an appropriate conversion factor, would represent total loading, in pounds per event, discharging to the ocean.

Event mean concentrations would be derived from flow-weighted compositing of samples. The scope of the program meant that the only feasible method of collecting composited flows was with automated refrigerated sampling equipment. The samplers, with 38 liter (10 gallon) water collection capacity, could be programmed in advance to cover a wide range of storm sizes without requiring one or more bottle changes during a rainfall event. A pressure transducer in the bottom of the storm drain would sense the depth of flow once a predetermined triggering level is exceeded. Converting the depth to a flow rate for the drain's size and slope, the processor would start measuring the volume of flow going past the station. After a predetermined volume of flow ensued, the sampler would begin its cycle of purge and collection. Programming for each station would take into account the size and overall imperviousness of the drainage area so that runoff from a storm of 10 mm (0.4") rainfall would fill one 10 liter (2-1/2 gallon) sample bottle in increments of 400 ml aliquots. Similarly, runoff from a storm of 40 mm (1.6") would develop 38 liters (10 gallons) of sample.

The 1990 monitoring plan identified a wide range of constituents to be tested, including:

- bacteria
- general minerals
- biochemical oxygen demand
- total organic carbon
- total petroleum hydrocarbons
- volatile organic compounds
- semi-volatile organic compounds

Those constituents requiring short holding times or not amenable to compositing would be collected via manual grab samples. If safe and convenient, open channels would be sampled from nearby bridges. Underground drains would have to be sampled manually with the automatic sampler's pump already in place.

Finally, the plan called for monitoring five storms per year. Dry weather sampling would also be conducted every other month. For the dry weather case, compositing would be time weighted.

The 1994-95 storm sampling season, the first under the new stormwater permit plan, was the last season of the historic sampling program. By the advent of the next storm season, 1995-96, the Department had identified and installed an additional 15 land use and mass emission stormwater monitoring stations under Phases II and III of the 1990 NPDES permit, bringing the total to 24 sites in the permit area. These additional stations monitored runoff in the Los Angeles River, San Gabriel River, and Santa Clarita River watersheds. All told, the Department was monitoring runoff from 14 land use sites and 10 mass emission sites; 11 were open channel and 13 were underground drains. Mass emission monitoring covered 4240 square km (1657 square miles), and land use monitoring 59 square km (23 square miles).

The need for collecting meaningful storm grab samples meant arriving at the sampling site before the peak flow occurred, any time of day. This requirement introduced a logistical task of mobilizing a squad of up to 26 people into teams, each with a dozen bottles and sample collection paraphernalia. A command and coordination system was thus set up to train personnel, assign people to collection sites, follow weather forecasts, notify the contract laboratory, and send out samplers in the event of rain. Storm event coordinators were able to track weather forecasts and actual rainfall in real time through the Department's real time remote sensing system, otherwise known as ALERT. At the first sign of sustained rainfall, the coordinator would activate the sampling teams to fan out across the county to collect grab samples at the predetermined locations. By virtue of the refrigerated units, composited samples could be kept at each site until personnel could retrieve the samples during normal working hours.

An extra monitoring requirement imposed by the Regional Water Quality Control Board was a toxics runoff sampling program. The program called for sampling runoff from a municipal corporation yard and two industrial sites to detect the presence of toxics of concern as set forth in EPA's NPDES Stormwater Sampling Guidance Document. In both the 1994-95 and 1995-95 storm seasons, runoff was sampled at the Department's road maintenance yard in Westchester. In the 1995-96 season, the program expanded to include sampling of stormwater runoff at an auto dismantling site and a heavy industrial site.

The 1996 NPDES Municipal Stormwater Permit

As the original stormwater permit was drawing to a close in 1995, negotiations began on the succeeding permit. The new document issued in 1996 varied in a number of ways from its predecessor. One of the differences was a statement of objectives to be attained by the monitoring program. Those objectives are to:

- track water quality status, trends, and loads and identify pollutants of concern;
- monitor and assess land use and watershed pollutant loads;
- identify and assess significant water quality problems in watersheds;
- identify sources of pollutants in runoff;
- identify and eliminate illicit connections and discharges
- evaluate effectiveness of Best Management Practices;
- assess impacts on receiving waters;

To achieve these goals, the new permit separated the monitoring program into five basic elements:

- Continued land use monitoring
- Continued mass emission monitoring
- Continued storm drain inspection
- New "critical source" monitoring
- New receiving waters study

The permit further divided the 7900 square km (3100 square mile) drainage area into six management areas based on natural watershed boundaries. Those management areas are:

- Malibu Creek and neighboring rural watersheds
- Ballona Creek and neighboring urban watersheds
- Los Angeles River watershed
- San Gabriel River watershed
- Santa Clara River watershed
- Dominguez Channel and Los Angeles Harbor watershed

While stormwater monitoring under the original permit was designed to collect runoff after an initial rainfall of 10 mm (0.4"), the new permit called for collection of runoff from as little as 6 mm (0.25") of rainfall. In addition, a pilot study would be set up on one station to evaluate the feasibility of collecting runoff from a storm of 2.5 mm (0.1") rainfall. The basic list of constituents would stay the same, but volatile organic compounds would be dropped in favor of the herbicides diazinon, chlorpyrifos, diuron, and malathion.

Land Use Monitoring

While keeping with the primary objective of "developing and supporting effective programs towards reduction of pollutants to the maximum extent practicable," the new permit refocussed the land use monitoring program. It described in detail an 8 step process by which the Department would reassess its ongoing program. The object of the reassessment was to identify and then monitor types of land uses believed to cause the greatest mass loading of pollutants. The loading calculation would take into account not only land use type but also areal extent and rainfall distribution.

Starting with an initial list of 104 land use types arranged in 45 groups identified by the Southern California Association of Governments (SCAG), Department staff aggregated and split the land use types into a list of 37 categories. Of these, the top 12 urban uses based on total area were chosen for field survey. For each of the 12 types, 8 representative areas no larger than a city block were chosen for field survey during the spring of 1996.

With field-derived estimates of imperviousness, a loading model was run for four hypothetical constituents (copper, phosphorus, COD, and TSS) and the results ranked according to a marginal benefit analysis. The top seven land use types that had marginal benefits above or equal to the maximum and were chosen for monitoring. They were:

- Vacant
- High Density Single Family Residential
- Light Industrial
- Transportation
- Retail/Commercial
- Multifamily Residential
- Educational Facilities

As it turned out, the first 5 land use types listed above were already being monitored under the 1990 stormwater permit. The Department would be able to keep one site from each land use for continued sampling; the excess sites could be abandoned under the terms of the 1996 permit. New stations to monitor the last two land use types, multifamily and educational, would be installed by the end of the 1996-97 storm season.

In addition to the loading analysis, land use was also ranked by total area in each of the six individual watershed management areas. Four land use types not already on the list were then identified, namely:

- heavy industrial
- rural residential
- utility facilities
- mixed residential

By the start of the 1997-98 storm season, two of these four land uses will also be chosen for monitoring, bringing the total to nine land use monitoring stations.

Lastly, in the 1996-97 storm season, the permit requires the Department to sample up to 100 "station events." That is, the number of sampling stations multiplied by the number of sampled storms would not exceed 100. In 1997-98 and subsequent years of the permit's duration, the Department would sample up to 200 station events.

The constituents of concern to be monitored under the land use program are:

Total PAH	Mercury
Copper	TSS
Chromium	Malathion
Selenium	Total PCB's
Total phosphorus	Cadmium
Chlorpyrifos	Lead
Total DDT	Zinc
Chlordane	Total nitrogen
Nickel	Diazinon
Silver	Simazine

Elimination of the time dependent constituents has allowed the Department a respite in responding personnel to land use stations to collect grab samples during storms.

Mass Emission Monitoring

In addition to monitoring the runoff from specific land uses, the Department of Public Works is also monitoring the major drainage areas near their outfall to the ocean. Five of the mass emission monitoring stations installed under the original 1990 permit would be retained under the new permit, namely Los Angeles River, San Gabriel River, Coyote Creek, Ballona Creek, and Malibu Creek. Four of the original mass emission sites would be abandoned, while one would be kept to continue the calibration of the SWMM model begun under the 1990 permit. Amounting to some 4200 square km (1654 square miles) of drainage area, the monitoring stations would produce information used to calculate total loading to the ocean.

Constituents tested for mass emissions include those under the land use program plus bacteria, oil and grease, total phenols, cyanide, and TPH. Because these additional constituents could only be collected by grab, the storm response coordinating task continued, although at a reduced scale.

The new permit also allowed sampling at mass emission stations to total five events per year--dry weather, storm, or a combination of both. In addition, the Ballona Creek station is the site of a one-year "wide channel" study designed to determine if there is any variation of pollutant concentrations across the width of a large channel. Results of this study will define what measures, if any, need to be taken in modifying results in this situation.

Critical Source Monitoring

The toxics evaluation of runoff from the three industrial sites under the original stormwater permit expanded into an enhanced "critical source" monitoring program under the 1996 permit. Similar to the land use monitoring evaluation process described above, the Department of Public Works undertook a five step process to identify and prioritize a list of critical industries within the county that may contribute significant pollutants to stormwater runoff. Standard Industrial Codes, or SIC's, played a major role in the selection process. Once selected, appropriate sites would be monitored over a two year period for the duration of the permit to measure runoff quality with and without remedial cleanup actions. These remedial actions are called Best Management Practices, or BMP's.

The first step was to develop an initial list of candidate industries. This list contained industries both included and excluded under the State's General Industrial Activities Stormwater permit process. Initial candidate selection was based on prevalence in the county and the extent of outdoor activities. The resulting list yielded a group of 30 candidate industries ranked by the number of facilities.

The next step involved developing a set of criteria to prioritize the list. A number of empirical factors were used to assign levels of significance to each SIC category. Loading (Q) would be addressed by the number of sources at a site and the likelihood of release. Imperviousness (R) of a site would be represented by the percent of paved area. Pollutant toxicity (T) would be denoted by the number of toxic pollutants and inherent toxicity of the mix. An exposure factor (E) signifies if activities are exposed to rainfall. And finally, number (N) would represent the total number of sites in the county. Each variable would be assigned a qualitative number from 1 to 10, with 10 representing the worst condition. The pollutant potential (P) used to rank the results would thus be the product of all the factors, or

$$P = Q \times R \times T \times E \times N$$

Based on this ranking scheme, the top five "critical source" industries were:

- Wholesale Trade (scrap and auto dismantling)
- Automotive Repair/Parking
- Fabricated Metal Products
- Motor Freight

Chemical Manufacturing

A literature search was simultaneously conducted to identify what "critical source" industries, if any, have already been analyzed. The search reveled that similar stormwater studies have yet to be performed.

After the identification and prioritization, the Department then had the task of finding six companies of any one of the top five industries to enlist for monitoring runoff from five storms during the 1996-97 storm season. For the subsequent season, half of the chosen sites would be fitted with the same structural or nonstructural BMP at the Department's expense. The other half would remain as controls in order to evaluate BMP effectiveness. At the same time, stormwater monitoring of six companies of the next critical source industry would begin. This scenario would continue for six years until five critical source industries and remedial BMP's are tested and evaluated, or until another search reveals similar studies underway in some other part of the country.

Constituents tested would include:

- Total petroleum hydrocarbons (diesel and gasoline)
- pH
- Electrical conductivity
- Oil and grease
- Semivolatile organics
- Chemical oxygen demand
- MBAS
- Total suspended solids
- Total dissolved solids
- Total organic carbon
- Total and dissolved aluminum
- Total and dissolved cadmium
- Total and dissolved chromium
- Total and dissolved copper
- Total and dissolved iron
- Total and dissolved lead
- Total and dissolved nickel
- Total and dissolved zinc

Illegal Connection Program

Begun under the 1990 permit, the illegal connection and discharge elimination program was instituted by the Department to identify, investigate, and eliminate illegal connections and discharges to the municipal storm drain system. The program is centered around a geographic data base and a field inspection operation.

Compilation of the geographic data base began in 1990 with the monumental task of inventorying the entire County 4000 km (2500 mile) storm drain system. The foundation of the inventory is a CADD depiction of all of the Department drain alignments. Department staff from the beginning have been superimposing over each CADD alignment information extracted from the drain's construction drawings. The data collected relate to each storm drain's name, size, material, and stationing. In addition, all catch basin and manhole locations are entered on the CADD drawing. Along with the CADD entry, staff is entering the identical data into a tabular data base.

A second tabular data base has also been compiled that enumerates all of the connection permits granted by the Department to other agencies and private parties over the years allowing their connection to the Department's storm drains. Permitted connection information includes receiving drain name, permit applicant, drain material, and station. To date, more than 86,000 connection records have been entered and approximately 3500 km (2200 mi.) of storm drain attribute data have been plotted and tabulated. A geographic information system soon to be installed in-house for the NPDES Stormwater Permit program will greatly enable staff to compile and analyze these kinds of data.

The field inspection operation is conducted by the Department's storm drain maintenance division. The maintenance division develops its own inspection schedule and requests the storm drain connection maps and data tables in advance. Since the beginning of the inspection program in January 1996, 75 drains have been inspected, yielding a total of 984 undocumented connections. Office staff will then determine if the undocumented connections are legal or illegal and issue permits or orders to seal the outlet accordingly.

Receiving Waters Monitoring

Perhaps the most eagerly awaited study is the receiving waters study being performed in Santa Monica Bay off the mouths of rural Malibu Creek and urban Ballona Creek. The Department of Public Works, pulling together ongoing works in progress, is funding a half-million dollar 3-phased ocean monitoring program. The participants are the University of Southern California Sea Grant Program, Southern California Coastal Waters Research Project, and U.C. Santa Barbara. The work plan is comprised of:

- a plume dispersion study that will deal with stormwater spreading, mixing, and impacts at the ocean interface;
- a benthics study that will assess stormwater impacts on the sediments and benthic invertebrate community;
- a toxics effects study that will determine if water column or sediment toxicity results from stormwater discharge;

The plan calls for the collection of data for the ocean studies to end after the 1996-97 storm season. However, if either the benthics or toxicity studies need further investigation, they will be extended into the 1997-98 storm season.

On the fresh water side, three samples will be collected at the Los Angeles River and San Gabriel River monitoring sites in 1997-98 and 1998-99 to conduct a sea urchin toxicity assay to be funded by the Department.

Ever since the first nonpoint source regulations were proposed, stormwater management agencies nationwide have argued that not enough scientific evidence exists to quantify a cause-and-effect relationship between stormwater runoff and receiving water impacts. Already in its second year, this study will go a long way in defining those impacts.

Conclusion

The Los Angeles County Department of Public Works is actively executing the terms of the latest NPDES Municipal Stormwater permit, which includes a comprehensive monitoring program. The permit applies to more than 7900 square km (3100 square miles) of area that eventually drains to the Pacific Ocean. The drainage area is the largest such permitted area in the nation.

In 1990, the Department modified its historic sampling program under the terms of one of the nation's first stormwater permits. The new program included in-stream monitoring of runoff from specific land use types, an illicit connection elimination program, an industrial runoff toxics study, and an estimation of total loads to the Ocean.

In 1996, under the terms of the succeeding stormwater permit, the Department added a critical source monitoring element and a study of possible stormwater impacts on ocean receiving waters.

The Department of Public Works has committed and will continue to commit significant resources to develop and support effective programs towards the reduction of stormwater pollutants to the receiving waters in and adjacent to Los Angeles County.

References

State of California Regional Water Quality Control Board, Los Angeles Region; Order No. 90-079; NPDES No. CA0061654 (CI 6948); Waste Discharge Requirements, Stormwater/Urban Runoff Discharge for Los Angeles County and Co-permittees; 1990.

State of California Regional Water Quality Control Board, Los Angeles Region; Order No. 96-054 (NPDES No. CAS614001); Waste Discharge Requirements for Municipal Storm Water and Urban Runoff Discharges with the County of Los Angeles; 1996.

World Almanac, 1993.

CHAPTER 113

An Epidemiological Study of Possible Adverse Health Effects of Swimming in Santa Monica Bay

Dr. Guang-yu Wang[1]

Abstract

An epidemiological study of possible adverse health effects of swimming was conducted in Santa Monica Bay during the summer of 1995. The study was conducted in response to wide public perception and evidence that there may be health risks associated with swimming in beach areas contaminated by runoff. Higher risks of a broad range of symptoms (gastrointestinal and upper respiratory) were observed for subjects swimming close to storm drains (0 versus. 400 yards). Higher risks of several symptoms were also observed for subjects swimming in water with high levels of single bacterial indicators and in water where enteric viruses were detected. Low ratio of total to fecal coliforms was found to be associated with higher risks of a broad range of symptoms. In response to the study results, an action agenda has been adopted and implemented by the local community to reduce the exposure to the risk and the sources of contamination.

Introduction

As one of the 28 National Estuary Programs, the Santa Monica Bay Restoration Project (SMBRP) is charged with assessing the Bay's pollution problems and with producing a Bay Restoration Plan to serve as a blueprint for the Bay's recovery. Since its genesis in 1988,, a primary focus of energy has been to find the answer to a fundamental human health question: "How safe is it to swim in Santa Monica Bay?" Nearly fifty million tourists and local residents come to Santa Monica Bay's public beaches each year to enjoy its recreational resources, but there has been wide public perception and some scientific evidence that there may be health risks associated with swimming in beach areas contaminated by runoff.

---

[1] Staff Scientist, Santa Monica Bay Restoration Project, 101 Centre Plaza Dr. Monterey Park, CA 91754

In previous investigations conducted by the SMBRP (SMBRP, 1992, 1991, 1990), human pathogens were detected in summer runoff, an unexpected result since sewer and storm drain systems in Los Angeles are completely separate. Possible sources of pathogens contamination into the storm drain system include illegal sewer connections, leaking sewer lines, malfunctioning septic systems, illegal dumping from recreational vehicles, or direct human sources such as campers or transients. Other potential sources of human pathogens in near shore areas include sewage spills into storm drains, small boat waste discharges and swimmers themselves.

The members of the SMBRP therefore decided that the definitive step necessary to answer this question of swimming-related health risks was an epidemiological study. The study was launched in the summer of 1995 under the auspices of the SMBRP, with a wide support of governmental agencies, dischargers, and environmental communities. It was hoped that the study would provide a scientific basis for addressing the wide public perception about the health risks associated with swimming in urban runoff-impacted beach areas, for developing pollutant control measures if necessary, and ultimately for promoting the local tourism-based economy.

Study Overview

During the course of the study (June to September 1995), 15,492 beachgoers who swam at three Santa Monica Bay beaches located near flowing storm drain outlets were interviewed. Nine to 14 days after the beach interviews, 13,278 follow-up telephone interviews were conducted to ascertain the occurrence of 16 symptoms (including fever, chills, eye discharge, earache, ear discharge, skin rash, infected cut, nausea, vomiting, diarrhea, diarrhea with blood, stomach pain, coughing, coughing with phlegm, nasal congestion, sore throat), and a group of symptoms indicative of "highly credible gastrointestinal illness" (HCGI)[2] and significant respiratory disease" (SRD)[3].

Water samples were collected daily in ankle depth at various distances from the drains (0, 100 yards north and south, and at 400 yards[4]) and analyzed for

[2]Two definitions of HCGI were used in this study and grouped as HCGI-1 (vomiting, diarrhea and fever, stomach pain and fever) or HCGI-2 (vomiting and fever).

[3] Symptoms including fever and nasal congestion, fever and sore throat, and cough with sputum.

[4] Previous studies have showed that indicator bacteria levels at 400 yards are generally low, therefore comparisons could be made between rates of illness in swimmers at this distance and at 0 yards.

total and fecal coliforms, enterococci, and *E. coli*. In addition, water samples were collected at storm drain sites every Friday, Saturday, and Sunday and analyzed for enteric viruses.

Persons who bathed and immersed their heads in the ocean water were potential subjects for this study. There were no restrictions based on age, sex or race. Persons who had bathed at the study beaches within seven days of the survey date (before and after) were excluded, as were subjects who bathed on multiple days. Since a primary research question was whether the risk of illness was associated with levels of particular indicator organisms in the water (which could vary from day to day), it would have been impossible to link subjects' experiences with specific counts on a given day if they were in the water on numerous days.

Summary of Findings[5]

Fifty-five percent of the subjects surveyed were male, 45 percent female. Forty-eight percent of the subjects were children (under 12 years of age); 13-to-25 year-olds comprised 26 percent of the survey population and the remaining 26 percent were aged 26 and over. The ethnicity of the survey population was 45 percent white, 43 percent Latino, 3 percent black, 3 percent Asian, 3 percent multi-ethnic, and 2 percent "other." Children and Latino subjects tended to swim closer to the drain. Sixty-three percent of subjects swimming at the drain were children under 12. Eighty-eight percent of the surveyed subjects were residents of California.

The analyses conducted in this study addressed two questions: a) What are the risks of illness relative to the distance one swims from a flowing storm drain? and b) Are the risks of illness associated with measures of water quality? The major findings resulting from these analyses are as follows:

1. **There is an increased risk of illness associated with swimming near flowing storm drain outlets in Santa Monica Bay.** Statistically significant increases in risks for a broad range of adverse health effects were found for subjects that swam in front of storm drains in comparison to those who swam over 400 yards away (Table 1). These increases in risk appeared to be limited to the 0 yard distance, as a significant drop-off effect was observed at other distances from the drain (Figure 1).

   The estimated number of excess cases of illness attributable to swimming at the drain reached into the 100's per 10,000 exposed subjects (greater than 1 percent, Table 1) suggesting that significant numbers of beachgoers swimming near storm drain outlets are subject to increased health risks.

[5] See SMBRP 1996 for detailed summary.

Table 1. Comparative health outcomes for swimming in front of drains versus 400+ yards away.

Health Outcome	Relative Risk (0 vs. 400+ Yds)	Estimated No. of Excess Cases per 10,000 Persons
Fever	57%	259
Chills	58%	138
Ear Discharge	127%	88
Vomiting	61%	115
Coughing w/ Phlegm	59%	175
Any of the above symptoms	44%	373
HCGI-2	111%	95
SRD	66%	303
HCGI-2 or SRD	53%	314

The results did not change when adjusted for age, beach, gender, race, California versus out-of-state resident, socioeconomic status, or worry about potential health hazards at the beach. Distance results also did not change substantially when controlled for each bacterial indicator.

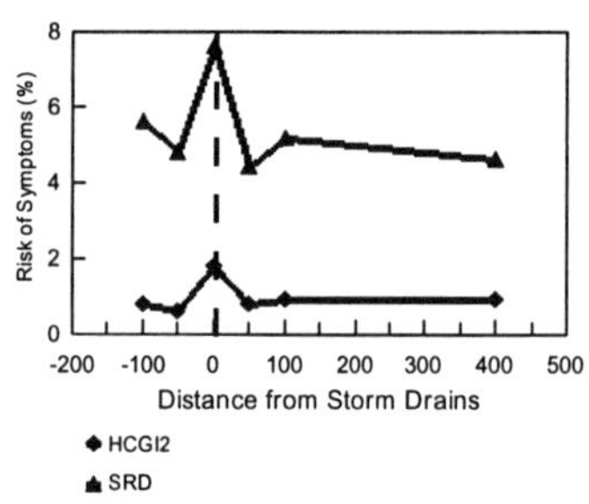

Figure 1. Reports of HCGI-2 and SRD relative to distance from drains.

2. **There is an increased risk of illness associated with swimming in areas with high densities of bacterial indicators.** "Cutoff points" were used to determine whether there were differences in the incidence of illness for those who swam in waters with bacterial densities "greater than" versus "less than" certain cutoff levels. Symptoms were found to be associated with swimming in areas where bacterial indicator counts were greater than the cutoff points that are used as part of federal and state water quality standards (Table 2). However, effects were noted for only few symptoms for the three bacterial indicators analyzed in this study.

Table 2. Health outcomes associated with swimming in areas with high bacterial indicator counts.

Indicator (cutoff)	Health Outcomes	Increased Risk
E. coli (>320 cfu)	Earache Nasal congestion	46% 24%
Enterococcus (>106 cfu)	Diarrhea w/ blood HCGI-1	323% 44%
Total coliform (>10,000 cfu)	Skin rash	200%
Fecal coliform (>400 cfu)	Skin rash	88%

cfu: colony forming unit.

3. **The total coliform to fecal coliform ratio was found to be one of the better indicators for predicting health risks.** When analyses were restricted to time when total coliforms exceeded 1,000 cfu, significant associations were observed, with incidence of illness generally increasing as the ratio of densities of total coliforms to fecal coliforms decreased (Figure 2). The strongest effects were generally observed when the total-to-fecal coliform ratio of 2:1 was used for comparison.

None of the bacterial results changed when adjusted for age, beach, gender, race, California versus out-of-state resident, socioeconomic status or worry about potential health hazards at the beach.

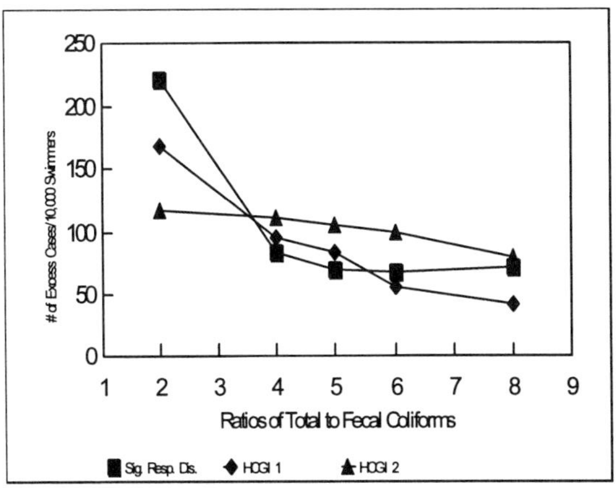

Figure 2. Relationship of excess cases of illness and total-to-fecal coliform ratios (when total coliform exceeded 1,000 cfu).

4. **Illnesses were reported more often on days when the samples were positive for enteric viruses.** Seventeen samples were positive for enteric viruses. Although based on small numbers, a comparison of subjects who were swimming within 50 yards of the drain on days when samples were tested for viruses indicates that a number of outcomes were reported more often on days when the samples were positive for viruses versus days when samples were negative. Symptoms for which increased risks were noted include: fever (53% increase), vomiting (89% increase), HCGI-1 (74% increase), and HCGI-2 (126% increase). Results remained essentially unchanged when adjusted for covariates or for each bacterial indicator.

5. **High densities of bacterial indicators were measured on a significant number of survey days, particularly in front of drains.** A great deal of day-to-day variability in bacterial indicator counts was recorded, however, high bacterial densities in water samples were detected most frequently directly in front of drains (at 0 yards). High densities of *E. coli*, fecal coliforms, and enterococcus occurred on over 25 percent of survey days. Total coliform levels were exceeded less frequently (8.6 percent of days). Total-to-fecal coliform ratios of less than 5 occurred on 12 percent of survey days.

Follow-Up Actions

This epidemiological study was the first large-scale study in the nation to investigate possible adverse health effects associated with swimming in ocean waters contaminated by urban runoff. The results of this health risk investigation provided both good news and cause for concern. The good news is that, of the Bay's 50-plus mile coastline, less than 2 miles are problematic. However, the study has also confirmed that there is a risk of illness associated with swimming immediately adjacent to flowing storm drains. Although it is not yet known what specific pathogens cause illness, the study confirms that the distance from storm drains and the bacterial indicators that are being monitored do help to predict risk. In addition, the total-to-fecal coliform ratio has been found to be a useful predictor of illness and could be used as a new analytical tool.

The scientific findings documented through this study laid the foundation to develop new policies and actions that will improve our ability to protect the public's health. The SMBRP has identified an "Epi Study Action Agenda" to respond to the findings of this study. This Action Agenda included steps to be taken in five areas: education and advisories to the public, implementation of source control measures, identification and prevention of pathogen sources, modification of water quality standards and monitoring programs, and financing. Since adoption of the Action Agenda in May 1996, significant progress has been made in its implementation. New warning signs with stronger language have been posted near flowing storm drains in Santa Monica Bay. Dry-weather flows of eight problematic drains have been scheduled for diversion or on-site treatment. In addition, given the results of this study, the SMBRP recommends that the US EPA and the State health agencies develop national and state-wide bathing beach water quality standards/criteria.

References

Santa Monica Bay Restoration Project. 1996. An epidemiological study of possible adverse health effects of swimming in Santa Monica Bay. Santa Monica Bay Restor. Proj., Monterey Park, CA . 211 p. plus appendices.

Santa Monica Bay Restoration Project. 1992. Pathogens and indicators in storm drains within the Santa Monica Bay Watershed. Santa Monica Bay Restor. Proj., Monterey Park. CA . 64 p.

Santa Monica Bay Restoration Project. 1991. Storm drains as a source of surf zone bacterial indicators and human enteric viruses to Santa Monica Bay. Santa Monica Bay Restor. Proj., Monterey Park, CA . 32 p.

Santa Monica Bay Restoration Project. 1990. An assessment of inputs of fecal indicator organisms and human enteric viruses from two Santa Monica storm drains. Santa Monica Bay Restor. Proj., Monterey Park, CA . 39 p. plus appendices.

CHAPTER 114

AN INTEGRATED STRATEGY FOR MANAGING URBAN RUNOFF POLLUTION IN LOS ANGELES COUNTY

Dr. Xavier Swamikannu[1]

Abstract

Contaminated urban runoff can harm important water resources, including estuaries, coastal waters, and wetlands. Urban runoff typically contains the same types of pollutants that are found in wastewater and industrial discharges. The enormous volume of runoff discharges transport significant mass loads of pollutants.

California issued the first urban runoff permit program to eighty-four municipalities (now eighty-six) in Los Angeles County in 1990. A countywide approach was used because of the complexity and networking of the urban drainage system. The permit program was revised in July 1996 to promote watershed management. Staff utilized consensus-building techniques and worked with a committee of stakeholders (the "negotiating group") comprised of representatives of selected municipalities and environmental groups. Separate discussions were also held with other interested parties to resolve areas of disagreement.

The municipal storm water and urban runoff program in Los Angeles County offers a unique challenge to protecting the coastal environment because of the large number of municipal jurisdictions participating to implement an innovative and cost-effective approach. This paper describes the process followed to reach agreement on this landmark effort, the progress to date, and imminent challenges that lie ahead.

---

[1]Storm Water Program, CalEPA, California Regional Water Quality Control Board - Los Angeles Region, 101 Centre Plaza Drive, Monterey Park, CA 91754-2156

Introduction

Storm water and urban dry weather runoff pollution can damage important water resources, including streams, lakes, estuaries and wetlands, and ground water (USEPA 1994). Many recent studies have shown that runoff from urban areas typically contains significant quantities of the same general types of pollutants that are found in wastewater and industrial discharges and often causes similar water quality problems (CRWQCB-LA 1988; Pitt and Field 1990; SCCWRP 1993). These pollutants include heavy metals (e.g., chromium, cadmium, copper, lead, mercury, nickel, zinc), pesticides, herbicides, nutrients, bacteria, and synthetic organic compounds such as fuels, waste oils, solvents, lubricants, and grease.

In addition, large impervious surfaces in urban areas increase the quantity and peak flows of runoff, which in turn cause hydrologic impacts such as scoured streambed channels, instream sedimentation, and loss of habitat (TI 1994; CWPR 1996; Arnold and Gibbons 1996). Furthermore, because of the enormous volume of runoff discharges mass loads of pollutants in storm water can be significant (SMBRP 1994). There are multiple sources of pollution that contaminate storm water, including land use activities, operation and maintenance activities, illicit discharges and spills, atmospheric deposition, and vehicular traffic conditions (USEPA 1992A). While the impacts on the beneficial uses of water bodies are site-specific and vary due to differences in local land use conditions, geography, geology, hydrologic conditions, and the type of receiving water, the characteristics of urban storm water are similar.

Amendments to the Clean Water Act (CWA) in 1987 established new statutory requirements to control industrial and municipal storm water discharges to waters of the United States [CWA Section 402 (p)]. Storm water from municipal separate storm sewer systems (MS4s) are required to be managed with, "controls to reduce the discharge of pollutants to the maximum extent practicable, including management practices, control techniques and systems, design and engineering.." (40 CFR Pt. 122.26)

Urban Runoff Pollution in Southern California

Urban runoff and storm water contamination in the Southern California Bight, which includes the Los Angeles region, was identified as a regionwide problem by the National Research Council (NRC 1990). Further, water quality assessments conducted by the California Regional Water Quality Control Board - Los Angeles Region (CRWQCB-LA) identified urban runoff as a leading cause of impairment of a number of water bodies in Los Angeles County (CRWQCB-LA 1993). Pollutants associated with the

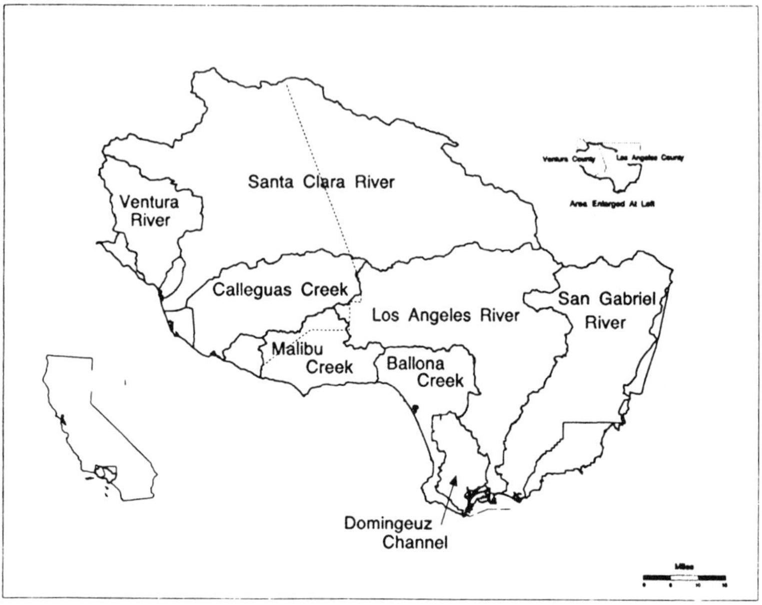

Figure 1. Map of the watershed management areas in Los Angeles and Ventura counties that are under the regulatory authority of the California Regional Water Quality Control Board - Los Angeles Region.

impairment include: heavy metals, coliform, enteric viruses, pesticides, nutrients, polycyclic aromatic hydrocarbons, polychlorinated biphenyls, organic solvents, sediments, trash, debris, algae, scum, and odor.

More recently, an epidemiological study demonstrated the increased risk of acute illness associated with swimming near flowing storm drain outlets in Santa Monica Bay (SMBRP 1996B). Human pathogens have also been identified in summer urban runoff (SMBRP 1992). Possible sources of human pathogen contamination include pet and livestock feces, illicit sewer connections to the storm drains, leaking sewer lines, malfunctioning septic systems, improper waste disposal by recreational vehicles, campers or transients. Additional potential sources of human pathogens in nearshore waters include sewage overflows into storm drains, small boats waste discharges, and bathers themselves.

The Los Angeles County Storm Water Program

Los Angeles County, California, is the most densely populated county in the State of California. The storm water program area includes eighty-five cities as well as the unincorporated areas of the County within the jurisdiction of CRWQCB-LA. The storm drain system serves a population of about 9 million who live on approximately 7,000 km^2 (3,100 mi^2) of land area. Nearly a third of statewide industry is included within the program boundary. There are six major watershed management areas (Figure 1).

The CRWQCB-LA is the National Pollutant Discharge Elimination System (NPDES) permitting authority for the Los Angeles region under delegation provisions of the Federal Clean Water Act and the Porter-Cologne Act (California Water Code). This delegation authorizes it to regulate and control the discharge of pollutants into waters of the U.S. and the State of California.

Pursuant to changes to the CWA, the CRWQCB-LA issued the first storm water permit (Order No. 90-079) in June 1990, to the municipalities in Los Angeles County. Because of the complexity and networking of the storm drain system and drainage facilities within and tributary to the County of Los Angeles, a countywide approach in permitting storm water and urban runoff discharges was adopted. The County of Los Angeles was designated as Principal Permittee under that permit. As Principal Permittee, the County of Los Angeles is responsible for the general administration of the permit and facilitates cooperation among municipalities. The first permit term expired in 1995.

Program Restructuring

The first step in restructuring the storm water permit program was to convene a stakeholder group to work with CRWQCB-LA staff to develop permit requirements and implementing language. Since direct discussions with eighty jurisdictions was not practical, the CRWQCB-LA asked the County of Los Angeles to assemble a representative group of city delegates. Besides the County, three small city representatives and the City of Los Angeles were chosen from the membership of the Storm Water Executive Advisory Committee, a pre-existing group that occasionally met to coordinate implementation of the first storm water permit. These parties comprised the stakeholder group, along with Heal the Bay, an advocacy organization which was asked to participate and provide the environmental perspective on this issue. This organization's involvement was important in order to reduce the likelihood of third party citizen lawsuits or appeals. This advisory committee, often called the "negotiating group," met approximately

60 times over an 18-month period to identify areas of agreement and narrow areas of disagreement.

In addition to holding meetings of the negotiating group, CRWQCB-LA staff held numerous meetings with the cities, watershed committees, stakeholder attorneys, industry spokespeople and individual city representatives. Working with the negotiating group's input, staff sent out a first partial draft permit in September, 1995. A first full draft was distributed for comments in December, 1995. A Public Service Announcement video was made available to elected officials and decision makers to facilitate understanding of the objectives and the benefits of the effort (SMBRP 1996A). In large part due to the involvement and buy-in of the majority of participants, the permit was ultimately approved by the CRWQCB-LA Board in July, 1996 (CRWQCB-LA 1996).

Storm Water Program Elements

The permit requires municipalities to implement comprehensive pollution prevention and management programs to reduce storm water pollution. The Storm Water Management Program (SWMP) must include controls necessary to reduce the discharge of pollutants from the MS4 to the Maximum Extent Practicable (MEP). These controls consist of a combination of best management practices, control techniques, system design and engineering methods.

Specifically, municipalities are required to develop and implement programs to control storm water pollution in the following areas: (1) Illicit connections and illicit discharges elimination; (2) Development planning and construction controls; (3) Public agency activities management; and, (4) Public information and participation initiatives. The development and implementation are scheduled over a period of thirty six months.

The requirements of Section 6217(g) of the Coastal Zone Act Reauthorization Amendments of 1990 (CZARA) were also reviewed to ensure consistency with guidelines for urban runoff pollution management. CZARA requires coastal states with approved coastal zone management programs to address nonpoint pollution impacting or threatening coastal water quality (USEPA 1993)

Monitoring is performed by Los Angeles County to characterize sources, evaluate water resource impacts, and to assess the effectiveness of pollution control methods. Preliminary results on the impact of storm water pollution on selected receiving waters are reported in the Conference Proceedings (Bay 1997; Schiff 1997).

No	Land Use	Cu Load	Cum Pcent
0			0
1	High Density SFR	7912	0.31
2	Vacant	5405	0.52
3	Light Industrial	2415	0.61
4	MF Residential	1678	0.68
5	Transportation	2416	0.77
6	Retail and Commercial	1311	0.83
7	Low Density SFR	557	0.85
8	Mixed Residential	507	0.87
9	Educational Facilities	494	0.89
10	Nat Resource Extraction	338	0.90
11	Heavy Industrial	307	0.91
12	General Office Use	245	0.92
13	Open Space and Rec	219	0.93
14	Institutions	217	0.94
15	Utility Facilities	191	0.95
16	Mobile Homes	173	0.95
17	Other Commercial	165	0.96
18	Mixed Transport/ Utility	158	0.96
19	Floodways and Structures	117	0.97
20	Under Construction	114	0.97
21	Remaining (21-37)	676	1.00
	Total	25615	1

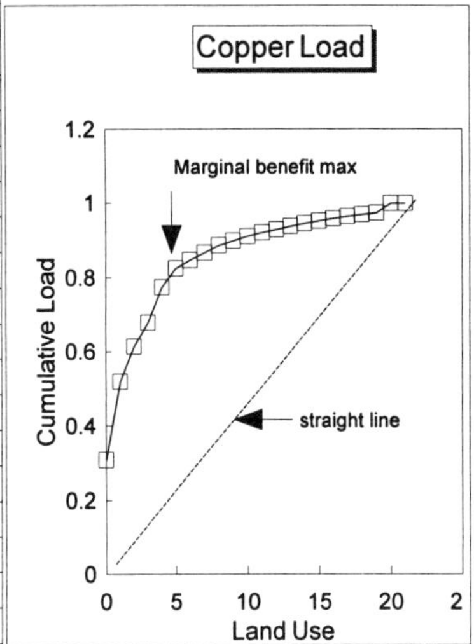

Figure 2. Modelled pollutant loads for copper (Cu) based on thirty seven landuse types in Los Angeles County. The inflexion point of the cumulative load curve indicates the cut-off point for diminishing marginal benefit. Additional landuse type sampling for the pollutant beyond the cut-off point is of marginal benefit. The copper EMC for the transportation landuse was calculated using storm water data provided by the California Department of Transportation.

Prioritizing Actions

A scientifically defensible approach is essential to prioritizing pollution control efforts. Consequently, predominant landuses that contribute significant pollutant loads, and the numerical dominance of certain types of industrial and commercial activities in Los Angeles County have been selected for further study.

Landuse selection: Landuse for special monitoring were selected in a multi-step process as follows, (i) selection of eight representative locations from selected landuse categories identified from an initial landuse list compiled from landuse overlays and landuse data from the Southern California Association of Governments (SCAG); (ii) performing a site survey at representative locations; (iii) evaluation of the use of aerial photographs as an alternative measure of imperviousness for each survey location; (iv) modelling of mass emissions based on landuse categories using Event Mean Concentrations (EMCs) from past landuse monitoring, and (vi) performing a marginal benefit analysis to select landuses for monitoring. The marginal benefit analysis was conducted for representative pollutant categories: solids (total suspended solids, heavy metal (total copper); nutrient (total phosphorous; and organics (chemical oxygen demand), for all Los Angeles County and also by watersheds.

Seven landuses out of thirty seven were selected for monitoring and characterization because their landuse-mass emissions were equal to or greater than the maximum marginal benefit for at least one pollutant category. The seven are: (i) High density single family residential; (ii) vacant; (iii) light industrial; (iv) transportation; (vi) retail and commercial; multi-family residential; and (vii) educational facilities. The marginal benefit analysis for Cu for Los Angeles County is shown in Figure 2. Just seven landuses contribute 85% of the Cu pollutant load in storm water. Cu is a priority pollutant in storm water and Cu concentrations in many estuaries exceed water quality criteria (Moran 1997).

Figure 3 illustrates modelled pollutant loads for Cu by dominant landuses in the two watersheds that drain to Santa Monica Bay: (i) Ballona Creek and urban; and, (ii) Malibu Creek and rural. The Source Significance Index (SSI) which is the ratio of pollutant load percent to land use percent is also shown. The SSI is influenced directly by pollutant concentrations and inversely by landuse area percent, and thus may be utilized as an indicator of transient in-stream effects such as toxicity or to evaluate cost benefit of control measures implementation. Currently, there is no evidence that Cu in storm water has any measurable impact on Santa Monica Bay (Schiff 1997).

Critical Pollution Sources Selection: The County identified a list of candidate industrial sources based on the numerical prevalence countywide, designation under USEPA Phase I storm water regulations, and storm water pollution characteristics (LACDPW 1996). These industrial facilities were then ranked based on a semi-quantitative measure. The Pollution Potential (P) was computed as a product of several criteria and normalized to fall between 0 and 100.

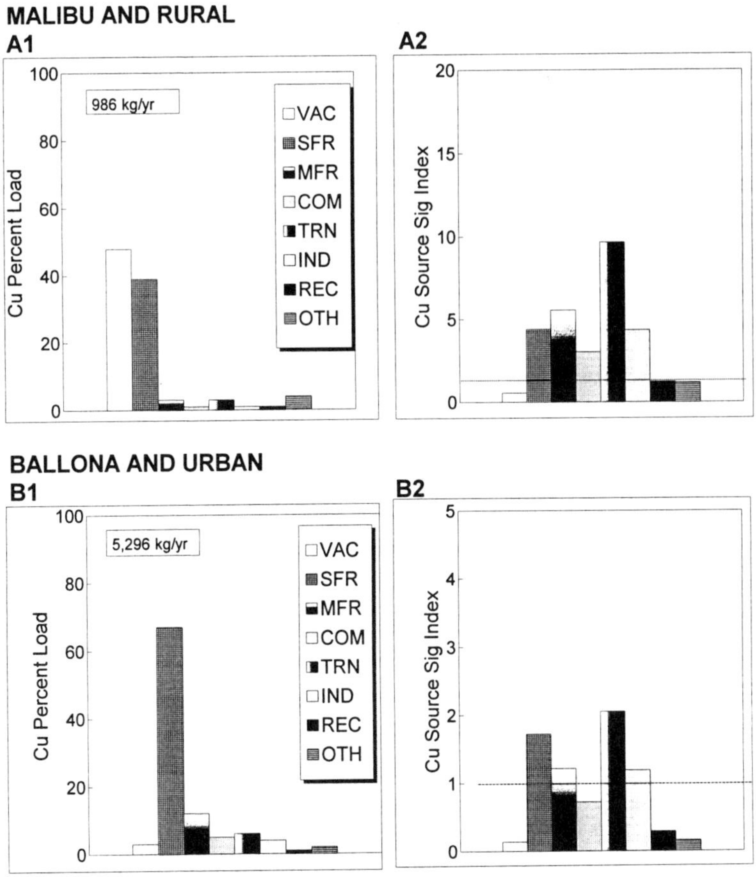

Figure 3. Modelled percent load for copper (Cu) in two watersheds which differ in the dominant type of landuses. A1- Malibu and rural; B1- Ballona and urban. A2 and B2 are their respective Source Significance Index (SSI) which is the ratio of percent load to percent area for a particular landuse. The dashed line indicates SSI value of 1, the theoretical value were pollutant concentrations all equal. (VAC - Vacant; SFR - Single Family Residential; MFR - Multi Family Residential; COM - Commercial; TRN - Transportation; IND - Industrial; REC - Recreational; and OTH - Other)

Pollution Potential (P) = Q x R x T x E x N,
where Q is pollutant quantity; R is the runoff amount; N is number of facilities; T is pollutant toxicity, and E is the rain exposure factor.

Table 1. Ten highest ranking industrial facility types selected for storm water characterization and pollution control assessment in Los Angeles County. (SIC - Standard Industrial Classification Major Group number)

SIC (1)	Industry (2)	Number of Facilities (3)	Pollution Rank (4)
50	Scrap / Auto Dismantling	587	1
75	Automotive Repair / Parking	6,067	2
35	Fabricated metal Products	3,283	3
42	Motor Freight	872	4
28	Chemicals / Allied Products	1,069	5
55	Automotive Dealers / Gas Retail	2,744	6
33	Primary Metal Products	703	7
49	Electric / Gas / Sanitary	2,001	8
45	Air Transportation	319	9
30	Rubber / Miscellaneous	1,034	10

The top ten ranks from the thirty industrial types that were analyzed for Pollution Potential (P) are shown in Table 1. Los Angeles County selected the first five ranking industry types for further study to characterize storm water pollution and evaluate potential best management practices (BMPs) to control pollution. In each industry type, six facilities will be evaluated over a five year period. In the first year of the study, a baseline characterization of each industry type will be completed by sampling storm water discharges from all six facilities. Three of the facilities in each industry type will then be augmented with appropriate BMPs and monitored in subsequent years to measure reductions in storm water pollution.

Challenges Ahead

Numerous challenges lie ahead both for the municipalities and the CRWQCB-LA. The large size of the Los Angeles Storm Water Program ensures that the gamut of environmental participation - from pro-activism and passivity to active resistance - is well represented among municipal participants. Any inherent inertia on the other hand is adequately counterbalanced by a vocal and vigilant environmental community represented by nationally active groups. The CRWQCB-LA must reconcile competing interests and be fully engaged in program implementation. This will be difficult to do if staff resources become limited by State budget constraints. Similarly, municipalities face competing budget priorities in an era of voter and legislative fiscal conservatism.

Los Angeles County is in the process of preparing countywide model components . These models will be submitted to CRWQCB-LA for approval over the next three years. The integrated cooperative approach offers the opportunity to cut costs from economies of scale, cost sharing, pooling of resources and coordination of implementation actions. No doubt some of the municipalities will make modifications to the models to reflect conditions within their municipal areas. Balancing municipal flexibility with countywide consistency will offer a challenge to the CRWQCB-LA.

The Watershed Area Management Plans (WMAPs) are the long-term blue prints to aggressively protect water resources, human health, and aquatic habitats in the proximate geographic subunit of the cities. Watershed committees are required to develop WMAPS in order to focus priority actions once countywide baseline programs have been established. The critical financial mechanism to fund development of the WMAPs has not be identified.

Los Angeles County is required to identify and develop performance standards for SWMP components by the year 20001 to quantify level of effort, program effectiveness and provide BMP baselines. The candidate parameters for the performance standards must be identified and evaluated during model development in order to ascertain meaningful indicators of storm water pollution reduction.

The monitoring program conducted by Los Angeles County must be sustained long-term and integrated with other regional monitoring programs to track the condition of water resources and assess the effectiveness of BMPs. In addition, watershed monitoring programs may need to be diversified to assess the needs and status of other watersheds beyond the present two. Such undertakings will need the financial participation and resources support from the other eighty-five municipalities.

Conclusion

The efforts to control urban storm water pollution is a huge undertaking. The current implementation agreement in Los Angeles County reflects the benefits of taking a consensus seeking approach to protecting coastal environments when there are a large number of stakeholders. In order for the storm water management program to be successful, the full participation of divergent interests - the municipalities, business, environmental advocates and the public - is essential to ensure effective implementation of a storm water programs to achieve a healthier coastal environment. In addition, both the State and the Federal governments must expend adequate resources to ensure continuous engagement with the implementing community. In this new environment of openness and responsiveness any perceived governmental apathy will erode public trust.

References

Arnold, Jr., C.L. and Gibbons, J.G. (1996). "Impervious Surface Coverage", *APA Journal* 62(2), 243-258.

Bay, S. (1997). "Storm Water Runoff into Santa Monica Bay: Effects on Surface Water and Sediment Toxicity", (In this Conference Proceedings).

CRWQCB-LA (1988). *Storm Runoff in Los Angeles and Ventura Counties, Final Report*, Los Angeles Regional Water Quality Control Board.

CRWQCB-LA (1993A). "Waterbodies and their Beneficial Uses", *Waterbodies, Wetlands and their Beneficial Uses in the Los Angeles Region, Volume 1.* Los Angeles Regional Water Quality Control Board.

CRWQCB-LA (1996). Waste Discharge Requirements for Municipal Storm Water and Urban Runoff Pollution within the County of Los Angeles. Order No. 96-054 (NPDES No. CAS614001).

CWPR (1994). "The Importance of Imperviousness", *Watershed Protection Techniques* 1(3), 100-111.

LACDPW (1996A). *Evaluation of Landuse Monitoring Stations* and *Critical Source Selection and Monitoring Report*, Los Angeles County Department of Public Works, September 1996, Supplemented October 1996.

Moran, K.D. (1997). "Solving the Copper Problem: The Brake Pad Partnership", *CWEA Bulletin* 33(2), 14-16.

NRC (1990). *Monitoring Southern California's Coastal Waters*, National Research Council, Washington D.C.

Pitt and Field (1990). "Hazardous and Toxic Wastes Associated with Urban Storm Water Runoff", *Proceedings of the Sixteenth Annual RREL Hazardous Waste Reduction Symposium, EPA 600-9-90-037*(1990);

SCCWRP (1993A). "Surface Runoff to the Southern California Bight", and "Characteristics of Effluents from Large Municipal Wastewater Treatment Facilities in 1990 and 1991". *SCCWRP Annual Report 1990-1991 and 1991-1992*, Southern California Coastal Water Research Project, Huntington Beach, CA.

Schiff, K. (1997). "Stormwater Runoff into Santa Monica Bay: Benthic Impacts in the Nearshore Coastal Zone", (In this Conference Proceedings).

SMBRP (1992). *Pathogens and Indicators in Storm Drains within the Santa Monica Bay Watershed*, Santa Monica Bay Restoration Project, Monterey Park, CA.

SMBRP (1994). *State of the Bay 1993: Characterization study of the Santa Monica Bay Restoration Plan*, Santa Monica Bay Restoration Project.

SMBRP (1996A). *Los Angeles County Municipal Storm Water Permit Public Service Announcement*. Santa Monica Bay Restoration Project, June 1996, 5:04 min.

SMBRP (1996B). *An Epidemiological Study of Possible Adverse Health Effects of Swimming in Santa Monica Bay*, Santa Monica Bay Restoration Project.

USEPA (1992A). *Guidance Manual for the Preparation of Part 2 of the NPDES Permit Applications for Discharges from Municipal Separate Storm Sewer Systems,* Document No. EPA 833-B-92-002.

USEPA (1992B). *Investigation of Inappropriate Pollutant Entries into Storm Drainage Systems,* Document No/600/R-92/238.

USEPA (1993). *Guidance Specifying Management Measures For Sources of Nonpoint Pollution In Coastal Waters*, Document No. EPA-840-B-92-002.

TI (1994). *Fundamentals of Urban Storm Water Management*, Terrene Institute, Washington D.C.

CHAPTER 115

Storm water Runoff into Santa Monica Bay: Identification, Impact and Dispersion

Burton H. Jones[1] and Libe Washburn[2]

Abstract:

Storm water runoff into the coastal ocean can have major impacts on a variety of oceanographic properties and processes that include coastal currents, coastal nutrient and optical characteristics, and contamination from heavy metals, organic compounds and microbial input. Off southern California these inputs are low for most of the year, but during the winter season, large rainstorms result in large inputs of freshwater and the associated contaminants from storm water runoff. These inputs have both event and seasonal scale effects on Santa Monica Bay. Towyo and surface mapping were used to map the three-dimensional distribution of dissolved and particulate components associated with storm water runoff into Santa Monica Bay during the winter of 1996. Results from this study indicate that the plume can extend offshore approximately 5 km from the coast and alongshore at least 5 km. The freshwater generally forms a thin surface layer that is initially less than 5 m thick, although the influence of the plume (salinity < 33.0 pss) can be detected to nearly 10 m depth. Three major particle groups are observed during these periods: 1) particles within the storm water runoff, 2) phytoplankton in the water column, and 3) resuspended sediments. These particles have a large effect on the water column optics, and may carry a measurable load of absorbed contaminants including heavy metals and organic compounds which may be toxic to marine organisms. Following the plume over the course of 2-3 days at the end of a storm indicates that the plume

----

[1]Research Associate Professor, Department of Biological Sciences, University of Southern California, Los Angeles, CA 90089-0371

[2]Associate Professor, ICESS, Department of Geography, University of California, Santa Barbara, CA 93106-4060

remains in the coastal area for at least 3 days with gradual dilution during this period. Dinoflagellate blooms can develop after the rain event providing a secondary particulate field associated with the plume. It is likely that the stratification and nutrient input provided by the freshwater inflow contribute to the presence and development of this dinoflagellate population.

Introduction:

The storm water component of urban runoff into the coastal ocean can have major impacts on a variety of oceanographic properties and processes that include coastal currents, coastal nutrient and optical characteristics, and the presence of toxic materials such as heavy metals and organic compounds. Storm water runoff is a small part of urban runoff in southern California during most of the year. But during the winter season, November through March, large rainstorms result in large inputs of freshwater and the associated contaminants from storm water runoff (SCCWRP, 1992). At the least, these inputs have both event and seasonal scale effects on Santa Monica Bay. Two major drainage systems, the Ballona Creek and Malibu Creek watersheds, contribute nearly 60 percent of the freshwater inflow into Santa Monica Bay (Table 1). The Ballona Creek watershed is highly urbanized with only 17 percent open land. In contrast, the Malibu Creek watershed is comparatively rural with 88 percent open land.

Table 1. Comparison of land use characteristics for the two watershed systems: Ballona Creek and Malibu Creek

Land Use	Ballona Creek	Malibu Creek
Single Family	46%	8%
Multiple Family	18%	1%
Commercial	8%	1%
Public	4%	0%
Industrial - Light	4%	1%
Urban - other	4%	2%
Open	17 %	88%
Total Percent	100%	100%
Total Acres ($*10^3$)	83.3	70.3
Pct of FW Inflow	31%	27%

The impact of these storm water inputs on the coastal receiving waters has been poorly understood until recently, because there has been little research on the freshwater impacts on the coastal zone in southern California. In contrast, other inputs, such as treated sewage outfalls, have been well studied, partly because of environmental regulation, engineering interests, and community concern. In other regions where freshwater input from rivers is a continous process there have been studies of the physics (e.g. Garvine, 1982; Muchow and Garvine, 1992), nutrient chemistry, and biological impacts on the coastal region (e.g. Malone, 1982; Lohrenz et al., 1990).

Methods:

Field sampling consists of several types of measurements to study the distribution of freshwater inputs from river sources during winter storms. Continuous mapping of physical and optical properties was performed with a towyo system that we have developed and used successfully for sewage outfalls (Washburn et al., 1992; Jones et al., 1993; Wu et al., 1994). The towed platform carries a conductivity-temperature-depth instrument (CTD) that is used to measure temperature, salinity, and density of seawater. Seawater turbidity is measured with a transmissometer having a path length if 0.25 m and a 660 nanometer (nm) light source and detector. The presence of chlorophyll in seawater, which can derive from ocean and terrestrial sources, is quantified with a fluorometer. The light distribution in the water column is measured with a Photosynthetically Available Radiation (PAR) sensor. The platform is moved through the water column along a zigzag tow path by winching the platform up and down as the ship steams ahead at about 4 knots. Vertical sections obtained in this way are referred to as towyo sections. The upper limit of the towyo sections is the sea surface; we tried to break through the sea surface in order to sample to the top of the water column. The lower limit is generally within about 5 m of the sea floor, although on a few occasions the platform made contact with the bottom.

Continuous surface sampling is accomplished using the same set of instruments through which near-surface water is pumped continuously while towyo sampling is occurring. An instrument which measures spectral absorption and attenuation at 9 visible wavelengths is also incorporated in the near-surface measurements. Ship's position is recorded at 10 second intervals from a GPS receiver.

Ocean currents are measured with an electromagnetic current sensor that also measures temperature and instrument depth were also measured by the current meter. The current meter, deployed by SCCWRP, was mounted on a mooring that was located at 33° 56.172' N, 118° 30.103' W in a water depth of 40 m. The depth of the current meter was about 15 m from the surface. The current meter was configured to sample as follows: 1) raw current speed and direction were obtained at a rate of 2 samples per second; 2) raw current samples were averaged over 10 seconds and 30 of these 10-second averages were stored in memory every 30 minutes. That is a 5-minute segment of 10-second averages was recorded every 1/2 hour.

Observations:

The study region for the impact of storm water runoff on the receiving waters of Santa Monica Bay is shown in Figure 1. The two major focus areas are the regions off of the mouths of Ballona Creek and Malibu Creeks. Not all track lines are sampled each time. Rather, the track lines are chosen according to the direction that the plume is flowing and the wave and wind patterns.

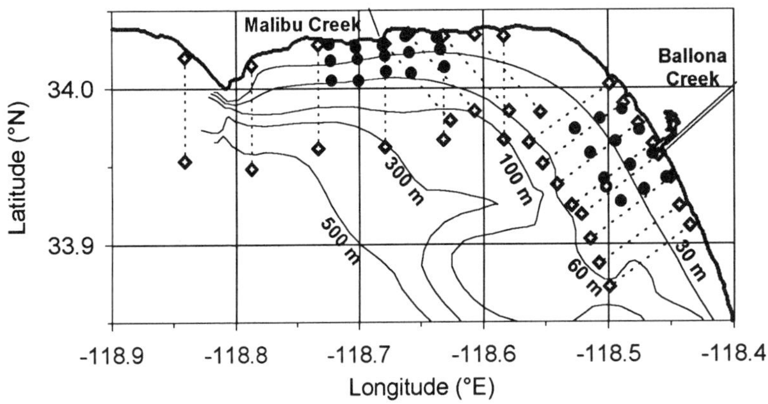

Figure 1. Map of northern Santa Monica Bay showing the way points (diamonds) and sampling tracks (dashed lines) for towyo mapping and benthic sampling sites (black circles used by SCCWRP for studying the benthic impacts from the storm water runoff.

The storm water study that is described below occurred during the period of March 5-7, 1996. The rainfall began on March 4 and ended early on March 5, yielding more than 0.8 inches of precipitation (Figure 2). The integrated flow from Ballona Creek for the period of the storm was about $1.2x10^6$ m^3 over the period of 24 hours between 0730 on March 4 and 0730 on March 5. We sampled for 3 days sequentially during this study, permitting a longer study of the temporal evolution of the coastal ocean following the storm event.

On March 5, we obtained four towyo transects near Malibu Creek and five towyo sections near the mouth of Ballona Creek. From the towyo tracks near the mouth of Ballona Creek we were able to construct the near-surface maps of temperature, salinity , chlorophyll fluorescence and beam attenuation coefficient (Figure 3). It is evident in these maps that there is a significant freshwater plume that extends between 2 and 4 km offshore from the beach. The plume is evident from the warmer (lighter) and fresher (darker) water that was present near the mouth of Ballona Creek. Both suspended particulate matter, as indicated by the beam attenuation coefficient, and chlorophyll fluorescence were high within the Ballona Creek plume and decreased seaward.

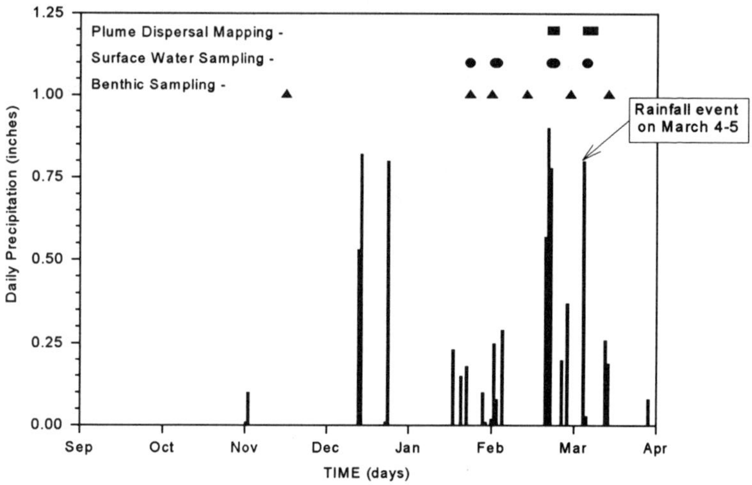

Figure 2. Rainfall events during the winter of 1995-96. Sampling events are indicated across the top of the figure: squares indicate plume mapping, circles indicate water column toxicity sampling, and triangles indicate benthic sampling.

Towyo 9 extended seaward from the south end of the Marina Del Rey breakwater and reveals the subsurface structure associated with the plume across the shelf. When the sampling began, the flow from the storm had already subsided significantly. As seen in the surface map, the plume characteristics of fresher, warmer, and more turbid water were evident in the nearshore, near-surface layer (Figure 3). This layer was less than 5 meters near the coast and thinned offshore extending to a distance of about 4 km from the coast. The subsurface water on the shelf was cool (<14 °C) and saline (>33.4 pss) as shown in Figure 4. High particle concentrations, indicated by beam c, were observed in the near-surface layer associated with the storm water plume and in the bottom boundary layer (Figure 4). In the nearshore 0.5-1 km of the transect the particle concentration was high (beam c > 1 m^{-1}) throughout the water column, perhaps resulting from a combination of bottom resuspension and sinking from the runoff plume. The near-bottom layer is similar to near-bottom layers that we have often observed off of Palos Verdes (e.g. Washburn et al., 1992). Deeper water appeared to intrude up the shelf near the bottom as evidenced by the cool (<12.5°C), salty (>33.45), low chlorophyll water.

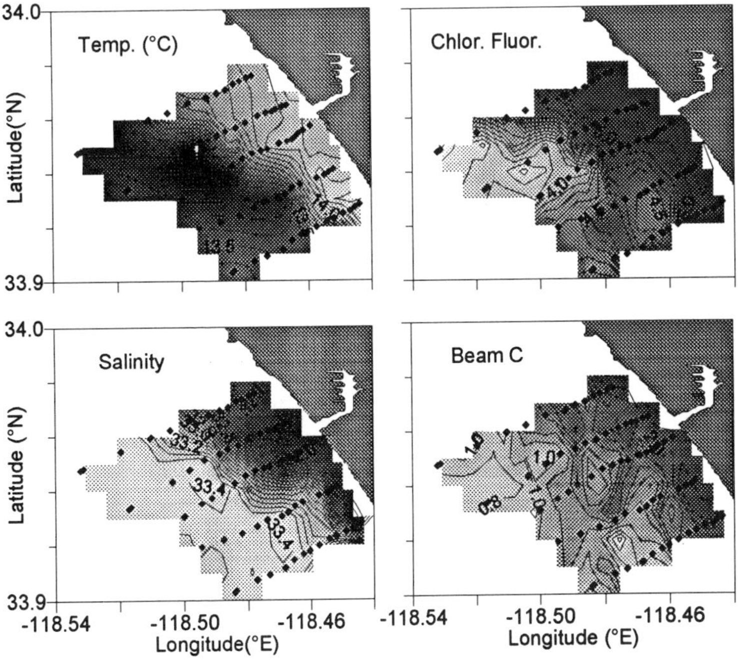

Figure 3. Surface maps constructed from the near-surface observations of the towyo indicated by the diamonds, which indicate locations where the towyo was within 2-3 m of the surface. This map is from March 5, 1996. Chlorophyll fluorescence is in relative units (the voltage output of the sensor, where 1 volt corresponds to about 0.6 μg/l chlorophyll); beam attenuation is in the units of m^{-1}; salinity is expressed in the units of practical salinity scale (pss), essentially parts per thousand.

Subsurface chlorophyll fluorescence at ~20 dbar between 2.5 and 4.5 km is likely to be evidence of phytoplankton rather than terrigenous plant material.

Two days later on March 7, a larger scale map of the area was obtained with transects extending from south of Marina del Rey to Point Dume, west of Malibu Creek. Low salinity water, evidence of the storm water runoff, was still present nearshore throughout the area (Figure 5). However, salinities were not as low as observed on March 5. Nearsurface particle concentrations were highest in the low salinity water and corresponded somewhat with higher chlorophyll fluorescence in the surface layer. The highest chlorophyll fluorescence values were observed both nearshore and south of Malibu Creek. The largest values found between Point Dume

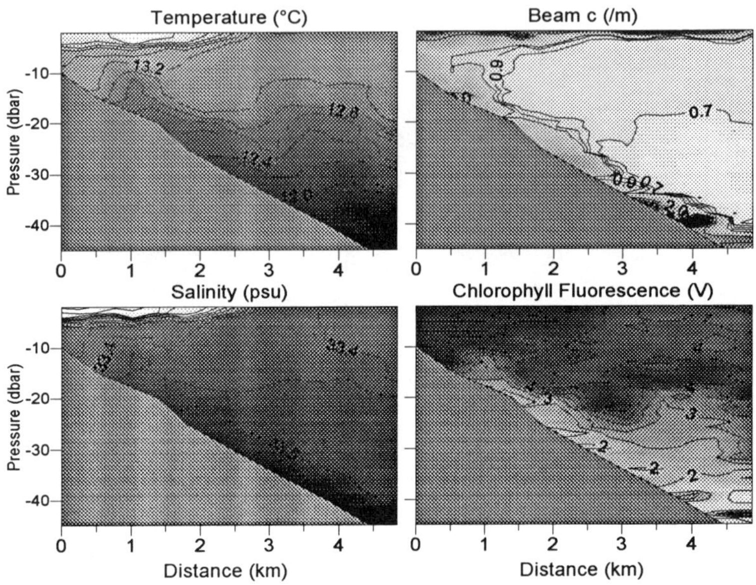

Figure 4. Towyo section extending southwestward from the southern end of the Marina del Rey breakwater. Corresponds to the central transect shown in Figure 3. Units are the same as in Figure 3.

and Malibu Creek were associated with reddish brown water typical of dinoflagellate blooms.

The transect extending offshore from the southern end of the Marina del Rey breakwater was again profiled with the towyo on March 7 (Figure 6). As seen in the surface map, there was still evidence for the presence of runoff in the inner 2.5 km of the transect where salinities of <33.4 penetrated to >10 dbar. Particle concentrations, as indicated by beam c, were still high in the upper layer, but did not show the extreme values observed on previous transects. Chlorophyll fluorescence were no longer high the low salinity, turbid water near the surface, but showed subsurface maxima that were independent of the higher particle concentrations in the surface layer and near the bottom.

An unusual feature was observed within this transect. Between 2.5 and 4 km there was a dynamic feature that was apparent in all of the variables (Figure 6). It is most noticeable in the elevation of the contour lines of temperature and salinity. But there is also evidence in chlorophyll fluorescence with the bifurcation of the subsurface chlorophyll maximum and in beam c with the elevation of particles to more than 15 dbar above the bottom. This narrow uplifting is probably the result of an internal wave passing through the region. Examining the surface map in Figure 5,

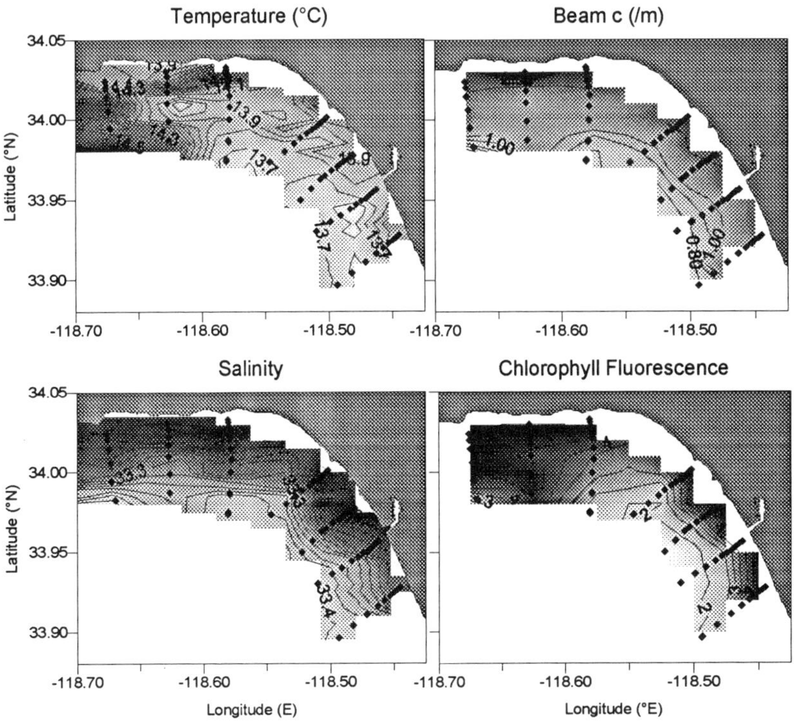

Figure 5. Surface maps from the towyo mapping on March 7, 1996. Units are the same as in Figure 3.

it appears that the cool spot in the center of this transect may be the surface expression of this feature and provides no indication of this feature in any of the other transects.

Moored sampling

The current time series at 15 m reveals that ocean currents near the Ballona Creek outflow are highly variable. A summary of some basic statistics of the current data are presented in Table 2. For both the eastward and northward components of current, the standard deviation exceeds the mean which confirms the variable nature of currents in this region. The mean current speed is 8.9 ± 6.2 cm s^{-1} in the direction of $162^{o} \pm 83^{o}$ true. Thus the mean current direction is toward the southwest and is oriented alongshore, more or less parallel to lines of constant depth (isobaths).

Because these data are effectively sampled at a rate of 2 points per hour, they resolve fluctuations in currents on a minimum time-scale of about 1 hour. The total

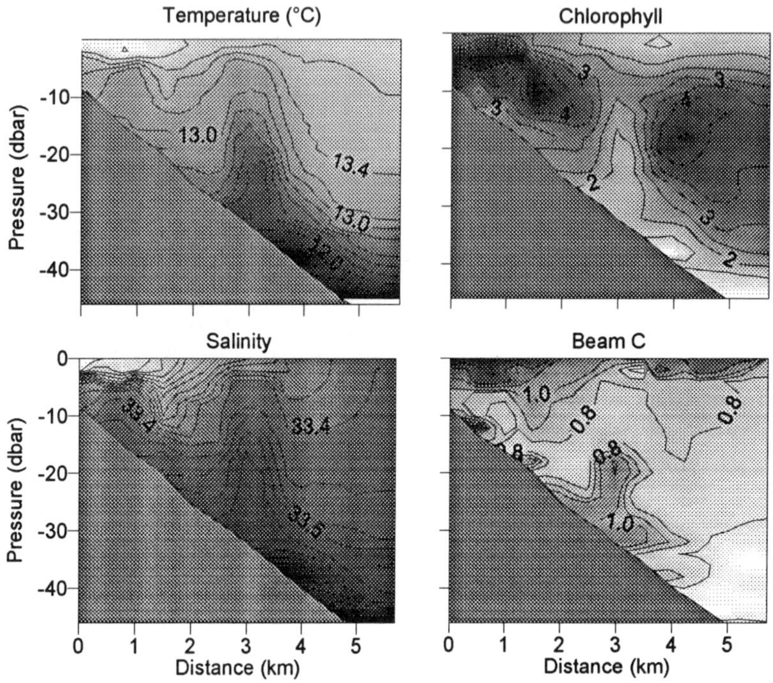

Figure 6. Towyo section obtained on March 7, 1996. As in Figure 4, this section extends southwestward from the southern end of the Marina del Rey breakwater. This transect corresponds to the second line from the south in Figure 5. Units are the same as in Figure 3.

Table 2: Basic current statistics

Parameter	Mean	Standard deviation
eastward velocity (cm s^{-1})	3.0	6.8
northward velocity (cm s^{-1})	-1.7	7.7
speed (cm s^{-1})	8.9	6.2
direction (degrees true)	162	83

length of the time series is about 70.9 days, so the maximum resolved time scale of current fluctuations is about 2 months. Several physical oceanographic processes produce variability over this range of time scales including tides, wind forcing, coastal-trapped waves, and eddies. All of these processes may be important for dispersing storm water runoff plumes from Ballona and Malibu Creeks.

It is important to determine the dominant time scales of currents during runoff events to understand the dynamic processes which disperse the storm water runoff plume. To aid in identifying these time scales, power spectra are computed for the eastward (u) and northward (v) velocity components. A clear peak is found in the spectra at frequencies corresponding to the lunar and solar semi-diurnal tides (periods in the range 11.97 to 12.42 hours). A smaller peak is associated with the lunar and solar diurnal tides (periods in the range 23.93 to 26.87 hours). In addition to the tidal peaks in the spectra, current fluctuations occur over a continuum of time scales from 12 hours to about 11 days, the longest period which is resolved by the spectra.

Other important quantities measured by the moored current meter are temperature and pressure. The temperature time series shows a 2.5^{o} temperature oscillation over a period of about 30 days. A full period of the oscillation was observed, but it is not known it continued beyond the observation period. High frequency temperature fluctuations with a period of about 12 hr are clearly due to the diurnal tides. The pressure time series measures the water depth with respect to the current meter. Peak-to-trough variations of 2 m in water depth occur on time scales of 12 hours and are due to the semi-diurnal tide. The low frequency "envelope" of fluctuations with a period of about 14 days results from the spring-neap cycle of the tides. Clearly tides are a dominant source of current fluctuations are likely to be very important in transporting the storm water plumes over periods of several hours.

Discussion:

Towyo mapping along with nearsurface continuous pumping has been used to successfully map the distribution of the storm water runoff into Santa Monica during several storm events. Results from a single smaller event in March 1996, demonstrate several aspects of the impact of storm water runoff into the coastal ocean.

Initially the plume spreads as a thin layer over the surface extending offshore 2-5 km and alongshore for larger extents depending on the alongshore advection. During the initial stages, when there is still significant discharge from the creek, the plume is demarcated at its offshore boundary by a sharp front which is apparent visibly (i.e., optical properties), in salinity and temperature. During this period the plume is often less than 5 m deep, and the most intense part of the plume is probably less than 2 m deep, as seen in Figure 4.

Over time the plume becomes more dilute and the particle concentrations decrease in the upper layer. This can be seen by comparing the surface maps (Figures 3 and 5) and the towyo sections (Figures 4 and 6) from March 5 and 7, 1996. Other factors indicate that the impact of the plume is greater than just the initial input of dissolved and particulate material. The development of the dinoflagellate bloom seen on March 7 near Malibu indicates that the nutrients and stratification provided by the runoff plume contribute to the biological productivity of the region. In other areas, river plumes are a significant factor in coastal primary and secondary production through the transport of buoyancy and nutrients into the coastal ocean (Malone, 1982, 1984; Lohrenz et al., 1990). Initially, primary production may be inhibited or limited in the region of the storm water due to the rapid attenuation of light by the high paritcle concentrations in the storm water, and perhaps by the toxic components that may be present (e.g. SCCWRP, 1992).

There are several deleterious effects that can result from runoff plumes. The most obvious factors are the introduction of pathogenic bacteria and viruses which affect the human population, and toxic elements and compounds which can affect marine life. It is likely that these effects are intially directly correlated with the proportion of storm water in the sea water. Bay et al. (this volume), as part of this project, have shown that toxicity to indicator organisms correlates well with salinity. They found that toxic metals are most likely to be the toxic component in the storm water runoff. These toxic effects can be apparent when runoff concentrations are as little as 5% (salinity $\leq\sim 32$) in the surface seawater. Therefore, the distribution of the storm water in the surface layer, as measured by salinity, is an important tool for mapping the dispersion of toxicity effects in the coastal ocean.

Coastal currents are highly variable in this region and therefore prediction of the direction and speed of dispersion of the plume will be difficult. The mean current speed during the current meter deployment was about 3.4 $cm \cdot s^{-1}$, or <0.1 knots, suggesting very low mean advection. However, because of the strong tidal variation of currents, the actual dispersion can be more than 10 km alongshore during a single phase of the tidal cycle. The circulation within Santa Monica Bay is complex (Hickey, 1993) and the maps that we have obtained indicate that the frontal boundaries are spatially variable (e.g. Figure 3). Therefore, although the predominant direction of advection will be alongshore, eddy systems within the Santa Monica Bight could result in significant offshore transport of storm water runoff.

The results presented here provide an initial look at the impacts of storm water runoff on the receiving waters of Santa Monica Bay. Plume mapping, combined with toxicity and benthic sampling, provide a powerful technique for assessing the spatial impact of storm water and for evaluating the influence of physical dynamics on the dispersion of the storm water. More detailed analysis of the data set is ongoing and is expected to yield additional insight into the interaction of storm water runoff with the coastal ocean off of southern California.

Acknowledgments:

This research has been supported by the USC Sea Grant Program and by the County of Los Angeles, Department of Public Works. Ship time was generously provided by the City of Los Angeles. We thank the crew of the Research Vessel La Mer for their helpful support during field sampling, and for the scientists of the City of Los Angeles Environmental Monitoring Division who allowed us to interrupt their schedules when storms occurred. Technical support during field sampling was provided by Zhihong Zheng and Matthew Ragan at USC, Brian Emery at UCSB, and Harold Stubbs and Dario Diehl at SCCWRP.

References:

Bay, S., D. Greenstein, A. Jirik, and A. Zellers. 1997. Storm water runoff into Santa Monica Bay: surface water and sediment toxicity. This volume.

Garvine, R.W. 1982. A steady state model for buoyant surface plume hydrodynamics in coastal waters. Tellus 34:293-306.

Hickey, B.M. 1993. Physical Oceanography, in: M.D. Dailey, D.J. Reish, and J.W. Anderson (eds.), Ecology of the Southern California Bight: A Synthesis and Interpretation, University of California Press, pp. 19-69.

Jones, B.H., T.D. Dickey, L. Washburn, and D. Manov. 1993. Physical and biological dynamics of sewage outfall plumes in the coastal region: an integrated observational approach. In: L.C. Wrobel and C.A. Brebbia (Eds.), Water Pollution II: Modeling, measuring and prediction, Computational Mechanics Publications, Southampton, pp. 527-534

Lohrenz, S. E., M.J. Dagg, and T.E. Whitledge. 1990. Enhanced primary production at the plume/oceanic interface of the Mississippi River. Continental Shelf Research 10:639-664.

Malone, T. C. 1982. Factors influencing the fate of sewage derived nutrients in the lower Hudson estuary and New York Bight. In: G.F. Mayer (Ed.), Ecological Stress in the New York Bight: Science and Management. Estuarine Research Foundation, Coliumbia, South Carolina, , pp. 301-320.

Malone, 1984

Munchow, A. and R.W. Garvine, 1993, Dynamical properties of a buoyancy-driven coastal current. Journal of Geophysical Research 98:20063-20077.

Southern California Coastal Water Research Project, 1992, Surface runoff to the Southern California Bight, In: Annual Report 1990-91 and 1991-92, J.N. Cross and C. Francisco (eds.), pp. 19-28.

Washburn, L., B.H. Jones, A. Bratkovich, T.D. Dickey, and M.S. Chen. 1992. Mixing, dispersion, and resuspension in the vicinity of an ocean wastewater plume. J. Hydraulic Engineering, 118: 38-58.

Wu, Y., L. Washburn, and B.H. Jones. 1994. Buoyant plume dispersion in a coastal environment: Observations of plume structure and dynamics. Continental Shelf Research 14:1001-1023.

CHAPTER 116

Stormwater Runoff Effects on Santa Monica Bay: Toxicity, Sediment Quality, and Benthic Community Impacts

Steven Bay[1], Kenneth Schiff[1], Darrin Greenstein[1], and Liesl Tiefenthaler[1]

ABSTRACT

Results from the initial year of a three year study of the effects of stormwater in Santa Monica Bay are described. Surface water and sediment samples were collected for analysis following four significantly-sized storm events in January through March, 1996. Toxicity was present in water samples offshore of Ballona Creek and was proportional to the concentration of runoff in the plume. Changes in sediment characteristics, such as grain size and TOC, were evident offshore of both Ballona and Malibu Creeks. Not only were the changes in sediment characteristics temporally stable (similar patterns observed after storms and during dry weather periods), but there was a gradient of change decreasing both upcoast and downcoast away from each creek mouth. Sediment contaminants, such as lead, total DDT, total PCB, and total PAH, were elevated at stations directly offshore Ballona Creek compared to sediments at similar depths offshore Malibu Creek. The first year's results have not detected significant stormwater-related changes in benthic infaunal community assemblages or sediment toxicity in the vicinity offshore the discharge of either creek.

INTRODUCTION

Urban runoff has been shown to discharge large quantities of contaminants (Schiff and Stevenson 1996; SCCWRP 1990, 1994a) and can be toxic to marine organisms (Bay *et al.* 1996). Unlike municipal wastewater, stormwater runoff enters the nearshore marine environment, often through estuaries or wetlands, wholly untreated. New regulations and increased monitoring efficiency are

---

[1] Southern California Coastal Water Research Project, 7171 Fenwick Lane, Westminster, CA 92683 (www.sccwrp.org)

enhancing in-channel measurements, yielding better information on characterization of wet weather inputs and effectiveness of best management practices (LACDPW 1996). However, virtually no information exists on contaminant fates and their biological effects once wet weather discharges enter the marine environment.

The fate and effects of contaminants on the receiving environment cannot be predicted from in-channel measurements alone. The mixing of the freshwater plume with ambient seawater alters the chemical state and solubility of some contaminants; particle aggregation and settling are also affected in complex ways. Furthermore, the nearshore environment is very dynamic with waves and currents having a strong influence on the deposition and distribution of stormwater contaminants. Moreover, much of what we know about the effects of contaminants in the benthic environment has been learned from studies of ocean wastewater outfalls. These systems differ greatly from stormwater discharges in terms of discharge composition and variability, mixing and dispersion, as well as the receiving environment characteristics. Therefore, directed studies of stormwater discharge are needed in order to identify the contaminants of concern and their biological effects.

The research described in this report represents the initial results of a three year program to investigate the linkage between stormwater discharge and environmental effects. This study, conducted in collaboration with the University of Southern California (USC), USC Sea Grant, and the University of California at Santa Barbara (UCSB), had three principal objectives. The first objective was to measure the dispersion and mixing of stormwater plumes in Santa Monica Bay; this work was conducted by our collaborators and is presented in an accompanying article. The second objective was to examine the magnitude and characteristics of water column and sediment toxicity near stormwater discharges. This work element examines potential contaminant effects and provides an important link with similar data from measurements within the watershed. The final objective of this project was to measure the impacts to benthic communities in the immediate vicinity of the discharge. This article presents the results from the toxicity and benthic studies for the first wet season of sampling.

METHODS

The Ballona Creek and Malibu Creek watersheds were selected for evaluation in this study. Both watersheds are roughly similar in size (Figure 1), and together they encompass over half of the entire Santa Monica Bay drainage area (Stenstrom and Strecker 1993). The Ballona Creek drainage basin is highly urbanized; 83% of the watershed is developed and comprised of predominantly residential land use. Almost the entire channel is concrete-lined. Conversely,

Malibu Creek is predominantly undeveloped; 88% of the watershed is open land and the channel is almost entirely earthen. These differences in watershed characteristics, combined with localized diversity in rainfall, lead to large variations in flow and pollutant loading to the ocean, even for the same storm event (LACDPW 1996). By comparing impacts associated with each watershed type, we hope to distinguish between effects arising from urban and non-urban stormwater runoff.

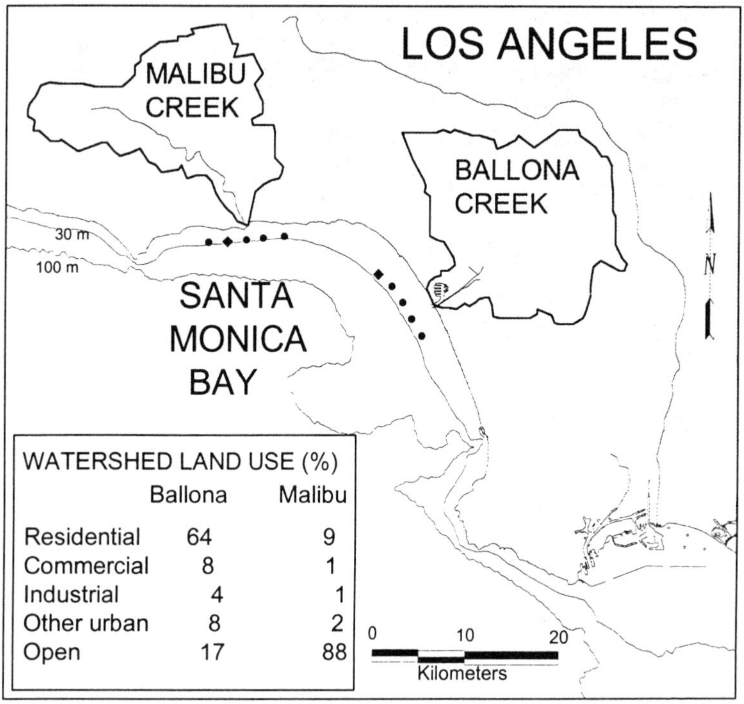

Figure 1. Watershed characteristics and sediment sampling sites for the two study locations. All stations were located along the 25 m isobath. Circles (•) indicate stations sampled after multiple storms for chemistry and infaunal community analyses. Sediment from all 10 stations were collected for toxicity tests following the February 1996 storm event.

Sampling and Sample Handling

Wet season sampling was accomplished after four 1996 storm events of variable magnitude: January 21 (1.52 cm rain at Los Angeles Civic Center), January 31 (0.51 cm), February 19-21 (10.16 cm), and March 4 (1.78 cm). Wet season sampling was coordinated with plume dispersion studies conducted by USC and UCSB. Water samples for toxicity were collected as soon as possible following a storm (8-48 hours), while sediment samples were collected up to 14 days following target storms in order to allow suspended particles to settle.

Surface water samples (upper meter) were collected during each of the four storm events. Five water samples were obtained from the Ballona Creek area during each event. Fewer samples were collected off Malibu Creek, due to difficulties in accessing the study area and the relatively small amount of runoff found in the surface water. Water sampling methods varied because of the necessity to use different boats for some sampling. Surface water samples were either collected by dipping a glass jar attached to the end of an aluminum pole or from a submersible pump deployed from a boom amidships of the boat. Samples were stored under refrigeration at SCCWRP and tested within 48 hours.

Locations of the water sampling stations were determined during each cruise and varied between events. Salinity measurements were used to select locations that represented a gradient of runoff concentration, usually aligned along a transect running between the creek mouth and a reference station containing no measurable runoff, located up to 6 km offshore.

Four stations offshore of each creek mouth were targeted for post-storm sediment chemistry, toxicity, and infauna measurements (Figure 1). All eight stations were located at roughly 25 m, a depth indicated by a pre-season spatial survey to show strong gradients of change across the area of stormwater influence (Bay and Schiff 1997). One additional sediment sample at each site was collected for sediment toxicity and chemistry following the February storm.

Sediments were collected using a 0.1 m^2 modified Van Veen grab. For contaminant analysis, only surficial sediments (top 2 cm) from undisturbed, representative grabs were collected. Sediment samples were placed in separate containers for grain size, TOC/TN, trace organics, and metals analyses. Samples were either stored under refrigeration (grain size) or frozen until analyzed. Samples for sediment toxicity tests were taken from replicate grabs and stored under refrigeration.

For benthic invertebrate community (infaunal) analysis, entire sediment grab samples were gently washed through a 1.0 mm mesh stainless steel screen on the boat. The organisms retained on the screen were "relaxed" using $MgSO_4$ (Epsom salts) in seawater. After 30 minutes the sample was fixed with 10% borax buffered formalin and returned to the laboratory. After 24 hours, samples were rinsed with freshwater to remove formalin and preserved in 70% ethanol.

Analytical Chemistry

Grain Size Analysis. Sediment grain size was measured using a Horiba Model LA-900 laser scattering particle size distribution analyzer. Sediment samples were first homogenized, then a representative aliquot was passed through the instrument and the particle sizes determined by detection of scattered (refracted and reflected) laser light.

Total Organic Carbon, Total Nitrogen and Total Volatile Solids Analysis. Total organic carbon and total nitrogen (TOC/TN) measurements were conducted using a Carlo Erba 1108 CHN Elemental Analyzer, according to methods developed by SCCWRP (1993b). Sediment samples were homogenized, dried, and then digested with acid to remove inorganic carbon. Samples were then oxidized by combustion in the analyzer and the evolved carbon and nitrogen quantified using a thermal conductivity detector.

Total Volatile Solids (TVS) was measured using a Thermolyne Model 62700 muffle furnace. Sediments were dried at 60 °C overnight, combusted at 500 °C, and then weighed after cooling. TVS was determined from the net loss in weight after combustion. While not as specific a measure as TOC, TVS has been shown to be significantly correlated with TOC measurements in reference areas of the Southern California Bight (Thompson *et al.* 1993).

Metal Analysis. Samples were prepared for metal analysis (aluminum, antimony, arsenic, beryllium, cadmium, chromium, copper, iron, lead, mercury, nickel, selenium, silver, and zinc) in accordance with EPA Method 3051 (EPA 1996). Dried sediment samples were digested using a nitric acid:hydrochloric acid mixture. Metal concentrations were determined using a Hewlett Packard Model 4500 inductively coupled plasma-mass spectrometer (ICP-MS) according to EPA Method 200.8 (EPA 1991).

Pesticides and Polychlorinated Biphenyls (PCB). Analytical methods for chlorinated pesticides and polychlorinated biphenyls followed EPA protocols (EPA 1986 or EPA 1983). Six DDT isomers and metabolites (o,p'-DDT, p,p'-DDT, o,p'-DDD, p,p'-DDD, o,p'-DDE, p,p'-DDE) and 27 individual PCB congeners (congeners 8, 18, 28, 29, 44, 50, 52, 66, 77, 87, 101, 104, 105, 118, 126, 128, 138, 153, 154, 170, 180, 187, 188, 195, 201, 206, 209) were quantified. Samples were also examined for twelve additional chlorinated pesticides (isomers of chlordane and lindane, hexachlorobenzene, and derivatives of endosulfan), but none of these compounds were detected for any sample in this study.

Specific methodological details for the analyses can be found in SCCWRP (1994b) or Zeng and Khan (1995), but a general description of the procedure follows. Samples for DDT and PCB analysis were homogenized and then centrifuged to remove pore water. Following sediment extraction by methylene chloride, the extracts were cleaned of interfering compounds using activated copper addition and preparative columns of alumina and silica. Extracts were concentrated to 1 mL and injected into a Hewlett Packard Model 5890 II gas chromatograph equipped with a 60m x 0.25 mm ID (0.25 μm film thickness) DB-5 fused silica capillary column and a ^{63}Ni electron capture detector (GC-ECD) for analyte measurement.

Polynuclear Aromatic Hydrocarbons (PAH). PAH analyses were conducted using EPA protocols (EPA 1986 or EPA 1983) which quantify 28 different PAHs. Specific methodological details can be found in Zeng and Khan (1995) or SCCWRP (1995a). Analysis was accomplished by injecting a portion of the same solvent extract used for the chlorinated hydrocarbon measurements into a Hewlett Packard Model 5890 II gas chromatograph equipped with a DB-5 column (60m x 0.25 mm ID x 0.25 μm film thickness) and a Hewlett Packard Model 5870 Mass Selective Detector in electron impact ionization mode.

Infaunal Community Analysis

Each infaunal sample was sorted into six different taxonomic groups - annelids, molluscs, arthropods, ophiuroids, miscellaneous echinoderms, and "other phyla" A minimum of 10% of each sample was resorted by another person to detect missed organisms. If sorting efficiency was less than 95%, then the entire sample was resorted. Biomass measurements were obtained by weighing each group of organisms to the nearest 0.01 g (wet weight).

Each organism was identified to the lowest taxon possible, using standardized nomenclature developed for the Southern California Bight (SCAMIT 1996). Species level identifications were assigned by scientists who were experts in their taxonomic group and were active members of the Southern California Association of Marine Invertebrate Taxonomists (an interagency quality assurance group). Ten percent of all samples were re-identified and enumerated by a second taxonomist for quality assurance. All new species encountered were maintained in a voucher collection which is located at SCCWRP.

Toxicity Measurement

Three types of environmental samples were tested for toxicity: surface water, sediment interstitial water, and whole sediment. Surface water samples were not filtered or centrifuged before testing. Brine (prepared by the partial freezing of seawater) was added to samples with a salinity below 30 g/kg to adjust the salinity to 34 g/kg. Each water sample was tested at a single concentration, 100% sample or the maximum concentration after salinity adjustment. Four replicates of each sample were tested. The percentage of runoff present in each toxicity test sample was calculated from the initial salinity value and included dilution resulting from salinity adjustment. This calculation assumed that the percent of runoff present in the original sample was inversely proportional to the relative salinity, expressed as a percentage of the background value (outside of plume).

Interstitial water was extracted from the sediment samples by centrifugation twice at 3,000 x g for 30 minutes. Laboratory seawater was added to the samples to produce three test concentrations containing 100, 50, and 25% interstitial water. Three replicates of each concentration were tested.

Water quality measurements conducted during each toxicity tested consisted of salinity, dissolved oxygen, pH, and total ammonia content. Measurements were made using electrodes that were calibrated daily. Measurements were made at the start of each test and also at the end of the 10-day amphipod survival test. Electronic thermometers were used to measure water temperature continuously during each experiment.

Sea Urchin Fertilization. All samples of surface water and interstitial water were tested for toxicity using a sea urchin fertilization test (Chapman *et al.* 1995). The test consisted of a 20 minute exposure of sperm to the samples at 15 °C. Eggs were then added and given 20 minutes for fertilization to occur. The eggs were preserved and examined later with a microscope to assess the percent fertilized. Toxic effects are expressed as a reduction in fertilization percentage.

Purple sea urchins (*Strongylocentrotus purpuratus*) used in the tests were collected from intertidal areas in northern Santa Monica Bay. The tests were conducted in glass vials containing 10 mL of solution. A negative control (0.45 μm and activated carbon filtered natural seawater from Redondo Beach) and a brine control (distilled water containing 50% brine) were included in each test series for quality assurance purposes.

Amphipod Survival. The toxicity of sediment samples was assessed by measuring the survival of amphipods following a 10 day exposure period. Test methods followed standard guidelines (ASTM 1991). A one liter sediment sample was removed from storage and homogenized with a plastic spoon. A 2 cm layer of sediment was added to five replicate one-quart glass canning jars for each station. Approximately 750 mL of lab seawater, adjusted to a salinity of 30 g/kg, was added to each jar. The jars were fitted with aeration tubes and allowed to equilibrate overnight before addition of the amphipods.

Twenty amphipods (*Rhepoxynius abronius*) were added to each jar. The test animals were collected from Puget Sound (Washington). A sample of collection site sediment was also included in the test, as a negative control. The test was conducted at 15°C, under constant illumination. Surviving animals were removed from the sediment at the end of the exposure by sieving and counted to determine the percent survival. Reburial success was not determined.

RESULTS

Surface Water Toxicity

Toxicity was present in surface water samples collected near the mouth of Ballona Creek during three of the four sampling events. No toxicity was detected in water samples collected one day following the January 21 storm. Water samples from this event contained less than 4% runoff, below the concentration range of Ballona Creek runoff (>6 %) shown to produce toxicity in related studies (Bay et al. 1997). The relatively small amount of runoff present in the January 22 samples may have been due to a delay in collection following the storm.

A map of the results for water samples collected following the January 31 and February 19-21 storms shows that toxicity was usually present near the mouth of Ballona Creek. Toxic water samples were restricted to a relatively small area (≤1 km offshore of Ballona Creek) following the January 31 (Figure 2) and March 4 storms (Bay and Schiff 1997). A greater area of toxicity appeared to be present on February 21, with toxic water present up to 4 km from the mouth of Ballona Creek. No toxicity was detected in water samples collected near the mouth of Malibu Lagoon in January and March. Toxicity was not expected in the Malibu samples, since all samples from the area contained less than 2% runoff. Too few samples were collected to examine the longshore distribution of toxic surface water.

The magnitude of toxicity was proportional to the amount of Ballona Creek runoff present in the sample (Figure 3). This relationship was also similar to the dose response pattern measured for Ballona Creek stormwater samples collected during the same storm (Figure 3).

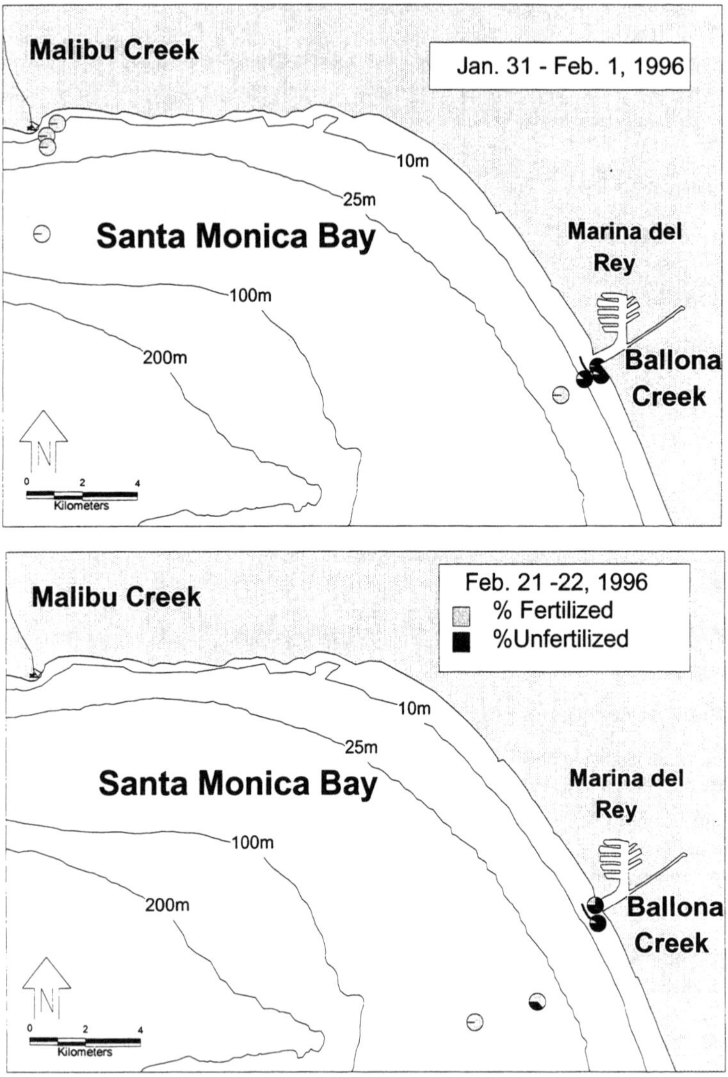

Figure 2. Surface water toxicity results for stations sampled after two storms. Pie diagrams show results of sea urchin fertilization test.

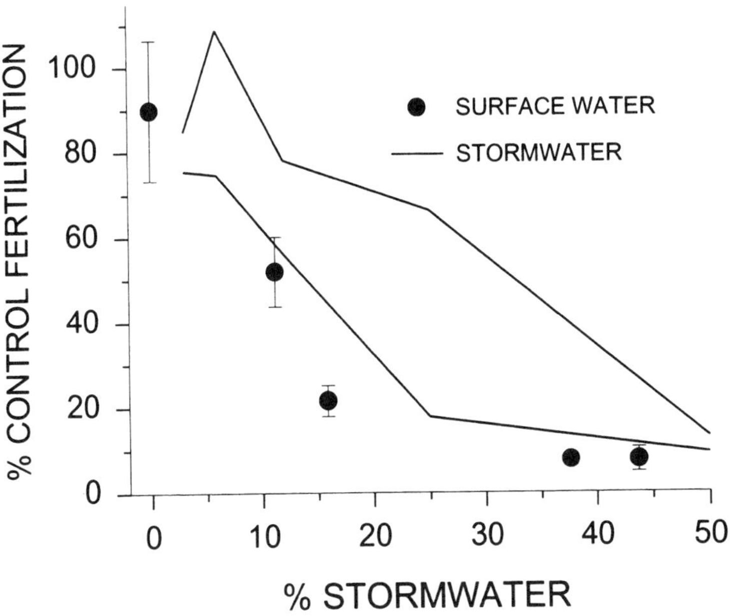

Figure 3. Comparison of sea urchin fertilization test results for February 21-22, 1996 samples of Ballona Creek stormwater (upstream) and nearby surface water (mean and standard deviation). Stormwater dose response lines are for two samples that represent the range of toxicity measured during the storm (Bay *et al.* 1997). All values have been normalized to the control response.

Effects on Sediment

Sediment Concentrations. The analytical results from the four sets of post-storm sediment samples were used to determine the wet season average concentrations of sediment characteristics and contaminants directly offshore the mouths of Ballona and Malibu Creeks. Sediments sampled at Malibu Creek contained twice the fines, 50% more TOC, and 25% greater TN than Ballona Creek (Table 1). Of the 14 different inorganic/metal constituents, seven were substantially greater in sediments offshore of Malibu Creek (Al, Be, Cd, Cr, Fe, Ni, Se), three were substantially greater in sediments offshore of Ballona Creek

(Pb, Hg, Ag), and the remaining three constituents were roughly similar in the sediments offshore of the two drainages (As, Cu, Zn). Ballona Creek was significantly greater in sediment concentrations of total DDT, total PCB, and total PAH than Malibu Creek.

Table 1. Mean (± 95% confidence intervals) sediment characteristics and pollutant concentrations offshore a highly urbanized (Ballona Ck) and less urbanized (Malibu Ck) watershed. Sediment samples were taken directly offshore each creek mouth at roughly 25 m depth following storm events.

	Ballona Creek (n=4)		Malibu Creek (n=3)	
	Mean	95% CI	Mean	95% CI
General Characteristics (% dry wt)				
Fines	30.5	1.2	53.6	8.2
TOC	0.662	0.263	0.912	0.106
TN	0.064	0.012	0.080	0.008
Inorganic Contaminants (ug/dry g)				
Aluminum	14075	258	21500	898
Arsenic	4.9	0.7	5.0	1.3
Beryllium	0.39	0.03	0.57	0.10
Cadmium	0.45	0.07	0.68	0.15
Chromium	41.8	3.1	57.5	13.7
Copper	11.2	2.0	13.0	2.6
Iron	15575	123	22933	2103
Lead	26.8	2.2	10.3	1.2
Mercury	0.17	0.02	0.10	0.01
Nickel	14	1	29	4
Selenium	0.49	0.05	0.69	0.04
Silver	0.93	0.13	0.31	0.05
Zinc	56	3	58	7
Organic Contaminants (ng/dry g)				
Total DDT	26.5	6.7	17.5	3.3
Total PCB	26.0	11.0	3.6	2.1
Total PAH	289.3	185.3	73.5	2.2

Gradients of Stormwater Influence. Sediment characteristics offshore of Ballona Creek followed a distinct pattern across the gradient of stormwater influence (Figure 4). The proportion of sedimentary fine-grained materials (silt + clay) was greatest directly offshore the creek mouth and declined both upcoast and downcoast. Grain size was significantly higher directly offshore Ballona Creek (31% fines) relative to sediments 4 km upcoast (19% fines). Similarly, a spatial pattern in sediment TOC and TN content was associated with Ballona Creek discharges. Sediment organic carbon content doubled directly offshore Ballona Creek (0.66% TOC) compared to sediments 4 km upcoast (0.30% TOC).

The spatial pattern in sediment characteristics offshore Malibu Creek was less distinct than at Ballona Creek, but was still discernible (Figure 4). Fine-grained sediments at 25 m depth directly offshore the creek mouth were 15% greater than in sediments collected at similar depths 4 km upcoast. In general, background conditions at Malibu Creek, as indicated by sediment characteristics such as grain size, were substantially different compared to background conditions offshore Ballona Creek.

Inorganic contaminants offshore of Ballona Creek followed a pattern across the gradient of stormwater influence which was similar to the pattern observed in sediment characteristics. Stormwater associated metals such as lead, copper, and zinc were significantly higher in sediments sampled directly offshore of the creek mouth and concentrations declined both upcoast and downcoast (Figure 4). Wet season averages at distant stations 4 km upcoast were between 41% (for copper) and 70% (for zinc) of average concentrations directly offshore of Ballona Creek mouth.

The pattern of inorganic sediment contamination across the gradient of stormwater influence offshore Malibu Creek was small or non-existent compared to the patterns observed offshore Ballona Creek (Figure 4). As with the sediment characteristics, where patterns did exist the magnitude of changes were smaller than at Ballona Creek. Wet season averages at distant stations 4 km upcoast were between 58% (for copper) and 80% (for zinc) of values measured directly offshore of the Malibu Creek mouth. Average metal concentrations were highest in sediments sampled 2 to 4 km downcoast from the Malibu Creek mouth.

Organic contaminants offshore of Ballona Creek followed a distinct pattern across the gradient of stormwater influence (Figure 4). Total DDT, total PCB, and total PAH were highest in sediments directly offshore of the creek mouth and concentrations declined both upcoast and downcoast. Wet season averages at distant stations 4 km upcoast were between 6% (for total PAH) and 41% (for total DDT) of average concentrations directly offshore of Ballona Creek mouth.

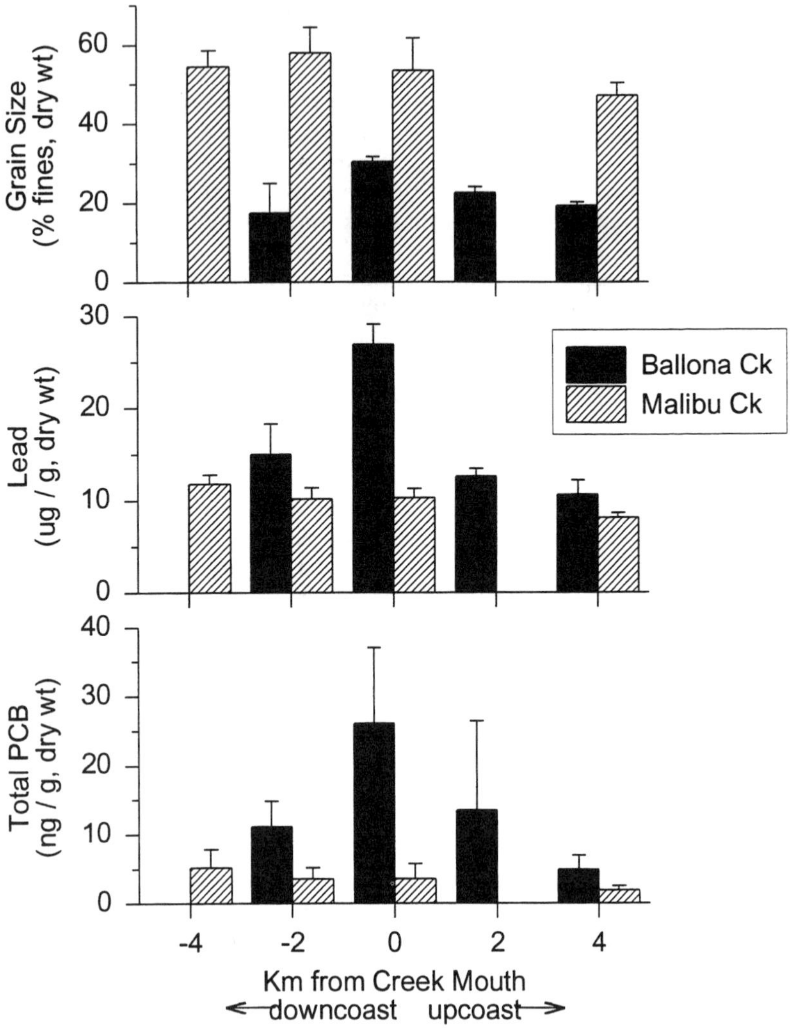

Figure 4. Grain size, lead, and total PCB concentrations in surficial sediments offshore of Ballona and Malibu Creeks. Values are mean (+95% CI) of 3-4 storms during the 1995-96 wet season. Distance from creek mouth refers to the upcoast (+) or downcoast (-) direction at a depth of 25 m.

Organic contaminants offshore of Malibu Creek generally did not follow a pattern across the gradient of stormwater influence (Figure 4). There was no consistent trend to the data and organic contaminant concentrations were very low overall. The greatest wet season concentrations of total PCB and total PAH were not directly offshore of Malibu Creek, but 4 km downcoast instead. However, the lowest values for all three organic compound classes were regularly observed 4 km upcoast from the Malibu Creek mouth.

Toxicity. Amphipod survival was high (89-98%) and indicative of no toxicity in all sediment samples (Table 2). The concentration of ammonia in the water overlying the sediment was slightly higher for stations within 2 km of either creek, possibly reflecting the organic enrichment identified by chemical analyses. These ammonia concentrations were not toxic and were within the range typically found in sediment toxicity tests.

Interstitial water from nine of ten sediment samples was nontoxic to sea urchin sperm (Table 2). Only interstitial water from the Ballona Creek station located 6 km upcoast was toxic, reducing fertilization by about 60% (relative to the control).

Table 2. Summary of toxicity test results for whole sediment (amphipod survival) and interstitial water (sea urchin fertilization). Sediment samples were collected February 28, 1996. Distances refer to the upcoast (+) or downcoast (-) direction at a depth of 25 m.

	Amphipod survival		Sea urchin fertilization	
Distance from creek (km)	% Survival mean (SD)	Ammonia[a] (mg/L)	% Fertilized mean (SD)	Ammonia [b] (mg/L)
Ballona Creek				
-2	93 (4)	2.0	98 (2)	4.2
0	95 (6)	3.1	98 (2)	6.2
2	89 (11)	2.6	95 (4)	4.1
4	97 (4)	1.5	96 (1)	3.4
6	98 (3)	1.7	42 (5)	9.8
Malibu Creek				
-4	93 (6)	1.1	98 (1)	2.1
-2	91 (8)	1.6	99 (1)	2.0
0	90 (7)	2.8	95 (1)	2.7
2	93 (6)	2.0	99 (1)	1.5
4	89 (10)	0.6	99 (2)	1.7

Infaunal Community Structure. A total of 30 samples were sieved, sorted, weighed and identified for infaunal community structure analysis. A total of 8,531 individuals were identified comprising 389 different taxa. About 90% of the total abundance at each creek represented species common to both sites. The dominant species recorded at each creek site included the polychaetes, *Spiophanes missionensis* and *Paraprionospio pinnata*, the mollusc, *Tellina modesta*, and the amphipods, *Amphideutopus oculatus* and *Ampelisca brevisimulata*. Interestingly, the Malibu Creek site also contained some organisms (e.g., *Amphiodia urtica*) which are typical of fine-grained habitats common in deeper water (Bergen 1995).

For the entire wet season, stations directly offshore of Ballona Creek and Malibu Creek at 25 m depth had similar diversity (Shannon-Wiener H'), evenness (Pielou's J), and species richness (Table 3). Abundance was slightly reduced offshore of Ballona Creek relative to Malibu Creek.

Table 3. Summary of infaunal results for the 1995-96 wet season. Distances refer to the upcoast (+) or downcoast (-) direction at a depth of 25 m.

Distance from creek mouth (km)	Mean (95% Confidence Interval)			
	Abundance [a]	Species Richness	Diversity [b]	Eveness [c]
Ballona Creek (n=4)				
-2	324 (46)	94.5 (12.4)	1.73 (0.08)	0.88 (0.03)
0	244 (18)	80.0 (4.4)	1.70 (0.04)	0.89 (0.01)
2	216 (33)	66.5 (4.4)	1.60 (0.05)	0.88 (0.02)
4	145 (47)	67.3 (13.1)	1.68 (0.08)	0.93 (0.02)
Malibu Creek (n=3)				
-4	374 (66)	94.0 (10.4)	1.70 (0.10)	0.86 (0.03)
-2	378 (325)	85.3 (57.1)	1.60 (0.30)	0.90 (0.08)
0	333 (38)	95.7 (9.1)	1.74 (0.02)	0.88 (0.02)
4	303 (52)	91.3 (10.4)	1.71 (0.08)	0.87 (0.04)

[a] Number of individuals/0.1 m^2

[b] Shannon-Wiener H'

[c] Pielou's J

Examination of trends across the gradient of stormwater influence did not reveal any significant relationships to stormwater discharges (Table 3). Instead, mean abundance decreased moving upcoast offshore of both creeks. Similarly, species richness was highest downcoast and lowest upcoast of the Ballona Creek mouth. Species richness was fairly constant across the gradient of stormwater influence at Malibu Creek, but the station 2 km downcoast showed high variability. Diversity and eveness measures showed no strong trends between stations.

DISCUSSION

Results from the first year of this study, though preliminary, have already helped address several important questions regarding runoff effects in receiving waters. Stormwater runoff in Southern California is extremely variable between storms and years. We cannot be sure that the patterns or magnitudes of effects discussed below are representative without conducting similar measurements over a longer time span, as planned for the subsequent years of this study.

- **Do stormwater discharge plumes contain toxic materials?**

The results indicate that surface water toxicity in the Ballona Creek area may occur when discharge plumes contain greater than about 10% runoff. Malibu Creek discharge plumes did not contain toxic concentrations of runoff, primarily because of greater dilution. Differences in runoff concentrations in the surface water samples prevent a direct comparison of the results for Malibu and Ballona plumes. Less concentrated runoff plumes off Malibu are probably the result of several factors, including much lower flow rates (up to two orders of magnitude difference) due to the more permeable watershed and additional dispersion resulting from delays in obtaining samples from the Malibu area.

The surface water toxicity results are consistent with tests of stormwater (sampled upstream of the creek mouth) conducted as part of a project funded by the Santa Monica Bay Restoration Project (Bay *et al.* 1997). Toxicity of Malibu Creek stormwater is usually less than similar samples from Ballona Creek, with concentrations of ≥ 25% usually needed to produce toxic effects. Extending these results to surface waters, it is likely that less toxicity will be present offshore of Malibu Creek.

- **Do stormwater discharges produce long-lasting alterations in Santa Monica Bay sediments?**

Discharges from Ballona and Malibu Creeks appeared to alter the general characteristics (e.g. grain size and organic content) of offshore sediments. The spatial pattern of altered sediment characteristics was persistent (present in dry weather) and could be observed at least 2 km distant. Some alterations in sediment characteristics were observed at depths of 40 m, but were not seen in depths of less than 10 m (Bay and Schiff 1997). These changes are not unexpected sine the combined loads of suspended solids from these two channels were estimated to be greater than 21 x 10^3 mt during the 1994-95 water year (LACDPW 1996). Moreover, these results are consistent with other studies which observed increases in sediment fines at distances $\geq$ 2 km offshore of the Santa Clara River following large winter storms (Kolpack and Drake 1985).

Runoff from urbanized watersheds appear to have some effect on the receiving water benthic environment. Sediments offshore of Ballona Creek were higher in concentrations of organic contaminants such as total DDT, total PCB, and total PAH as well as lead, a stormwater-associated metal, compared to sediments offshore of the less urbanized Malibu Creek watershed. This contamination covaried with sediment characteristics across the gradient of Ballona Creek stormwater influence. Sediment concentrations of organic and inorganic pollutants were highest offshore of the Ballona Creek mouth and then decreased upcoast and downcoast. Similarly, other researchers have shown that runoff from watersheds with as little as 15 to 25% urbanized land use(i.e. imperviouseness) have altered freshwater ecosystems (Yoder and Rankin 1996, Schueler 1994).

It is unlikely that the changes in sediment contaminants measured near Ballona and Malibu Creeks represent most of the many tons of contaminants entering Santa Monica Bay from stormwater each year. More research is needed to define the range of influence and eventual deposition of storm discharged particles in Santa Monica Bay. This work is continuing at SCCWRP, and is part of the effort being conducted with our collaborators at USC and UCSB.

- **Is sediment toxicity affected?**

Results from the sediment toxicity tests indicate that the altered sediment characteristics and increased contaminants resulting from stormwater discharge did not result in increased sediment toxicity. Amphipod survival was not reduced by exposure to sediment from any station. These results are similar to sediment toxicity data from the Southern California Bight Pilot Project (SCBPP), a regional study of coastal sediment quality conducted in 1994 (SCCWRP 1996). Sediment from 72 stations in Southern California (in depths of 10-200 meters), including 13 sites in Santa Monica Bay, were collected and tested for toxicity using a similar amphipod survival test. No significant amphipod mortality was found at any of the SCBPP stations.

Amphipod survival tests do not always provide a highly sensitive measure of sediment toxicity. Recent amphipod survival tests on much more highly contaminated sediments located near Southern California wastewater outfalls failed to detect toxicity in samples that caused sublethal effects on other toxicity test species (SCCWRP 1995b) or contained altered benthic communities (SCCWRP 1993c). The 10 day amphipod survival test was used in this program because it is a reliable method commonly used in sediment quality studies, and toxic effects observed with this test are often associated with adverse benthic community changes.

Interstitial water toxicity measurements were included in this program to provide a more sensitive measure of sediment quality. Previous studies have detected interstitial water toxicity in Southern California sediments that did not affect amphipod survival (SCCWRP 1995b and 1996). The sea urchin fertilization test of interstitial water did detect toxicity in one sample at the northern end of the Ballona Creek study area. Sediment chemistry results indicate that this station is outside the area most influenced by the Creek (Figure 4). As interstitial water toxicity did not correspond to variations in sediment contamination, it is unlikely that the toxicity was related to Ballona Creek stormwater. A cause for the interstitial water toxicity cannot yet be identified. This toxicity may indicate the presence of unidentified contaminants from an unknown source, or it may reflect temporary variations in sediment quality caused by natural factors or sediment storage. Additional research is needed to clarify the significance of the results. Additional toxicity tests of sediments from the Ballona Creek study area are in progress to assess the sublethal effects and bioavailability of the sediment-associated contaminants.

- **Are infaunal communities impaired?**

No dramatic biological effects in the benthic community structure were evident from the first wet season of sampling. There were no indications of a strongly degraded environment. Differences in community composition between sites offshore of the two creeks were most likely a result of variations in sediment characteristics (e.g., grain size) rather than differences in sediment-associated contaminants.

Due to relatively small sample sizes ($n \leq 4$), it is premature to reach conclusions regarding community disturbance at this time. However, these preliminary results are similar to previous surveys of reference areas which have reported 273-358 individuals and 78-91 species per grab at depths of 30 m during the summer months (Thompson *et al.* 1987, 1993). Completion of the second year of infaunal analysis will provide a greater ability to detect subtle differences in communities across gradients of stormwater influence and determine whether the results are consistent over time.

ACKNOWLEDGMENTS

This project would not be possible without the dedication and effort by many outstanding SCCWRP employees. These individuals include: Dario Diehl, Andrew Jirik, Harold Stubbs, Kim Tran, David Tsukada, Cherrie Vista, Charlie Yu, and Ann Zellers. Grain size analyses were performed by the City of San Diego Environmental Monitoring and Technical Services Department. Metal analyses were performed by CRG Environmental Laboratories (Los Angeles, CA). MEC Analytical Systems, Inc. (Carlsbad, CA) conducted the infaunal taxonomic analyses. We also wish to acknowledge our co-investigators Patrick Hartney (USC Seagrant), Burton Jones (USC), and Libe Washburn (UCSB). This work was funded by the Los Angeles County Department of Public Works and we are grateful for the cooperative interaction by their staff, including Jim Noyes, Gary Hildebrand, Bill DePoto, Tim Smith, and Mert Ramos. Additional review of this project was provided by Richard Gersberg (San Diego State University).

REFERENCES

American Society for Testing and Materials, 1991. E 1367, Guide for conducting 10-d static sediment toxicity tests with estuarine and marine amphipods. pp 1052-1075 *in*: Annual Book of ASTM Standards, Volume 11.04, American Society of Testing and Materials, Philadelphia, PA.

ASTM see American Society of Testing and Materials

Bay, S.M. and K. Schiff. 1997. Impacts of stormwater discharges on the nearshore environment of Santa Monica Bay. Southern California Coastal Water Research Project Annual Report 1995-96. Westminster, CA. *in press*.

Bay, S.M, D.J. Greenstein, S.-L. Lau, M.K. Stenstrom, and C.G. Kelly. 1996. Toxicity of dry weather flow from the Santa Monica Bay watershed. *Bulletin of the Southern California Academy of Sciences* 95:33-45.

Bay, S.M., D.J. Greenstein, A. Jirik and A. Zellers. 1997. Toxicity of stormwater from Ballona and Malibu creeks. Southern California Coastal Water Research Project Annual Report 1995-96. Westminster, CA. *in press*.

Bergen, M. 1995. Distribution of the brittlestar *Amphiodia* spp. in the Southern California Bight in 1956 to 1959. *Bulletin of the Southern California Academy of Sciences* 94:190-203.

Chapman, G., D. Denton and J. Lasorchak. 1995. Short-term methods for estimating the chronic toxicity of effluents and receiving waters to west coast marine and estuarine species. National Exposure Research Laboratory, U.S. Environmental Protection Agency, Cincinnati, OH. EPA/600/R-95/136.

EPA see U.S. Environmental Protection Agency

Kolpack, R. and D. Drake. 1985. Transport of clays in the eastern part of Santa Barbara Channel, California. *Geo-Marine Letters* 4:191-196.

LACDPW see Los Angeles County Department of Public Works

Los Angeles County Department of Public Works. 1996. Report of stormwater monitoring; Winter of 1994-95. Los Angeles County Department of Public Works, Environmental Programs Division, Alhambra, CA. 376 pp.

SCAMIT see Southern California Association of Marine Invertebrate Taxonomist

SCCWRP see Southern California Coastal Water Research Project

Schiff, K. and M. Stevenson. 1996. The San Diego Regional Stormwater Monitoring Program: Contaminant inputs to wetlands and bays. *Bulletin of the Southern California Academy of Sciences* 95(1): 7-16.

Schueler, T. 1994. The importance of imperviousness. *Watershed Protection Techniques* 1(3): 100-111.

Southern California Association of Marine Invertebrate Taxonomists. 1996. A taxonomic listing of soft bottom macro- and mega-invertebrates from infaunal and epibenthic monitoring programs in the Southern California Bight. Edition 2. The Southern California Association of Marine Invertebrate Taxonomists. San Pedro, CA. 86 pp.

Southern California Coastal Water Research Project (K. Schiff, J. Cross and H. Schafer). 1990. Mass emissions for selected constituents from the Los Angeles River. pp. 25-37 *in:* J.N. Cross and D.M. Wiley (eds.), Southern California Coastal Water Research Project Annual Report 1989-90. Long Beach, CA.

Southern California Coastal Water Research Project (J. Cross, K. Schiff and H. Schafer). 1993a. Surface Runoff to the Southern California Bight. pp. 19-28 *in*: J.N. Cross and C. Francisco (eds.), Southern California Coastal Water Research Project Annual Report 1990-91 and 1991-92. Long Beach, CA.

Southern California Coastal Water Research Project (M. Castillo and A. Khan). 1993b. Total organic carbon and total nitrogen in marine sediments, sediment traps, wastewater effluents, and surface runoff. pp. 50-60 *in*: J.N. Cross and C. Francisco (eds.), Southern California Coastal Water Research Project Annual Report 1990-91 and 1991-92. Long Beach, CA.

Southern California Coastal Water Research Project (S. Bay and D. Greenstein). 1993c. Temporal and spatial changes in sediment toxicity in Santa Monica Bay. pp. 81-87 *in*: J.N. Cross and C. Francisco (eds.), Southern California Coastal Water Research Project Annual Report 1990-91 and 1991-92. Long Beach, CA.

Southern California Coastal Water Research Project (S. Bay and J. Brown). 1994a. Preliminary Toxicity Identification Evaluation (TIE) of dry weather urban discharge. pp. 142-148 *in*: J.N. Cross, C. Francisco and D. Hallock (eds.), Southern California Coastal Water Research Project Annual Report 1992-93. Westminster, CA.

Southern California Coastal Water Research Project (E. Zeng, M. Castillo and A. Khan). 1994b. Congener specific analysis of polychlorinated biphenyls by GC/ECD and GC/MS: Characterization of an Aroclor mixture. pp. 35-53 *in:* J.N. Cross, C. Francisco and D. Hallock (eds.), Southern California Coastal Water Research Project Annual Report 1992-93. Westminster, CA.

Southern California Coastal Water Research Project (A. Khan and E. Zeng). 1995a. Extraction of hydrophobic organics from aqueous samples with 90-mm C-18 bonded disks. pp. 25-34 *in*: J.N. Cross, C. Francisco and D. Hallock (eds.), Southern California Coastal Water Research Project Annual Report 1993-94. Westminster, CA.

Southern California Coastal Water Research Project, 1995b. Toxicity of sediments on the Palos Verdes shelf. pp. 79-90 *in*: J.N. Cross, C. Francisco and D. Hallock (eds.), Southern California Coastal Water Research Project 1993-94 Annual Report. Westminster, CA.

Southern California Coastal Water Research Project (S.M. Bay) 1996. Sediment toxicity on the mainland shelf of the southern California Bight in 1994. pp. 128-136 *in*: M.J. Allen, C. Francisco and D. Hallock (eds.), Southern California Coastal Water Research Project Annual Report 1994-95. Westminster, CA.

Stenstrom, M. and E. Strecker. 1993. Assessment of Storm Drain Sources of Contaminants to Santa Monica Bay: Volume 1-Annual Pollutant Loadings to Santa Monica Bay from Stormwater Runoff. UCLA-Engr. Report 93-62, University of California, Los Angeles, CA.

Thompson, B., J. Laughlin and D. Tsukada. 1987. 1985 Reference Site Survey. Southern California Coastal Water Research Project, Long Beach, CA. 50 p.

Thompson, B., D. Tsukada and D. O'Donohue, 1993. 1990 Reference Site Survey. Southern California Coastal Water Research Project, Long Beach, CA. 105 p.

U.S. Environmental Protection Agency. 1983. Analysis of water and wastes. U.S. Environmental Protection Agency, Environmental Monitoring and Systems Laboratory. Cincinnati, OH. EPA-600/4-79-020.

U.S. Environmental Protection Agency. 1986. Test Methods for evaluating solid waste, physical/chemical methods, 3rd edition. United States Environmental Protection Agency, Office of Solid Waste and Emergency Response, Washington, DC SW-846.

U.S. Environmental Protection Agency. 1991. Methods for the determination of metals in environmental samples. Office of Research and development, Washington, DC EPA/600/4-91/010.

U.S. Environmental Protection Agency. 1996. Addendum to SW-846: Test Methods for evaluating solid waste, physical/chemical methods, 3rd edition. United States Environmental Protection Agency, Office of Solid Waste and Emergency Response, Washington, DC

Yoder, C. and E. Rankin, 1996. Assessing the condition and status of aquatic life designated uses in urban and suburban watersheds, *In*: Effects of Watershed Development and Management on Aquatic Ecosystems. Engineering Foundation Conference Proceedings, August 5-7, 1996, Salt Lake City, Utah (*in press*).

Zeng, E. and A. Khan. 1995. Extraction of municipal wastewater effluent using 90-mm C-18 bonded disks. *Journal of Microcolumn Separations* 7: 529-539.

CHAPTER 117

Striking a Balance: Improving Stewardship of Marine Areas

Edward P. Ames, Robert L. Bendick, Brock B. Bernstein,
Leo R. Brien, William M. Eichbaum, Charles G. Groat,
Marc J. Hershman, Michael F. Hirshfield, Eldon Hout,
Robert L. Howard, Robert W. Knecht, Robert C. Repetto,
Alison Reiser, Katharnie F. Wellman, and George M. Woodwell

Abstract

The National Research Council will release in mid-1997 the results of a major study on the governance and management of coastal and ocean resources and space, entitle *Striking a Balance: Improving Stewardship of Marine Areas*. A committee of experts was assembled by the Marine Board to develop guiding principles and effective processes for strengthening governance and management for both environmental stewardship and development of ocean resources. The study was sponsored by the National Oceanic and Atmospheric Administration, the Minerals Management Service, and the Environmental Protection Agency.

As part of the project, three case studies* were undertaken to provide representative examples of marine governance and management issues (e.g. marine sanctuaries and parks, areas where multiple uses are in conflict, regions where living marine resources are threatened, and regional planning and management efforts). Additional existing federal and state coastal and marine programs were also reviewed during the course of the study, including the Coastal Zone Management program, the National Estuary Program, fisheries management activities, the Outer Continental Shelf oil and gas leasing program, and state ocean and coastal programs. Base on the finding from these in-depth examinations, recommendations are made for improved processes and structures for long-term stewardship of marine areas and resources.

The report presents an overview of the importance of the marine sphere to the nation and a definition of principles and criteria upon which to base improved governance and management of this vital area. An assessment is provided of the value to the marine environment and resources in economic and other terms. The report then describes a number of critical problems that need addressing and the role of existing management and governance institutions and processes. The lessons learned from the case studies and the examination of existing institutions and programs are presented in terms of an analytical structure based on the principles and criteria for achieving and improved governance and management process. A number of options are proposed for improving marine governance along with a broad array of tools for improving management.

The report will be of interest to the general public, but is designed for use by federal and state agencies with management and/or regulatory responsibility for coastal and marine areas and resources; scientists and policy analysts with a particular interest in marine and coastal species, habitats, and ecosystems and associated economic, social, and political issues; individuals, organizations, and companies with an interest in utilization and/or preservation of marine and coastal resources and space; international organizations that seek to establish marine reserves and parks; public interest groups focused on environmental and natural resource issues; and decisionmakers in state and federal legislative bodies.

*the following areas were selected for the case studies:

1. **The Gulf of Maine/Massachusetts Bay/N.W. Atlantic** (including Georges Bank and Stellwagen Bank National Marine Sanctuary) [conducted by the Marine Policy Center, Woods Hole Oceanographic Institution]
2. **Florida Keys National Marine Sanctuary/Florida Bay Ecosystem** [conducted by the National Academy of Public Administration}
3. **Southern California Coast** (offshore and coastal region form the San Luis Obispo/Monterey County Line in the north to the U.S./Mexico border in the South, including the Monterey Bay and Channel Islands National Marine Sanctuaries)[conducted by the National Academy of Public Administration]

CHAPTER 118

MANAGING THE OCEANS: INTO THE MURKY WATERS OF JURISDICTION

John A. Saurenman
Deputy Attorney General
California Department of Justice

One of the most difficult problems which policy makers face when dealing with managing the oceans is that a wide variety of federal, state and local governments all have some jurisdiction over the oceans. The federal government has the greatest jurisdiction over the oceans adjacent to our coast generally allowing it to make decisions on the use of the oceans and its resources within 200 miles of shore. The federal government has ceded a significant portion of its authority to California for those waters within three miles of the shore, and California has in turn ceded some of its authority to local governments. The presentation examines the jurisdictional overlaps which exist in general and in the context of several specific situations disputes.

The specific situations examined are the following:

1. EPA is now considering a CERCLA response action on the Palos Verdes shelf offshore Palos Verdes, California to address a major deposit of DDT and PCBs which exists on the shelf. Because EPA is acting pursuant to CERCLA, some jurisdictional problems are eliminated, e.g., EPA need not apply for many local of governmental permits. However, a real issue remains of the extent to which EPA must comply with the procedural mandates of the Coastal Zone Management Act.

2. The Secretary of Commerce has closed the commercial salmon fishery offshore Northern California in order to preserve and restore the fishery. However, certain tribes are allowed to continue fishing for these salmon. In so doing, the Secretary rejected the recommendations of the Pacific Regional Fishery Management Council and thereby engendered litigation.

3. Throughout the 1980s, the federal government sought to lease Outer Continental Shelf lands for oil and gas development. Such leasing and development implicates a variety of environmental concerns including air and water quality as well as the more general concern of aesthetics. Such leasing and development also implicates federal, state and local jurisdictions. That is, there are federal, state and local permits, consistency review pursuant to the federal Coastal Zone Management Act, and local governments exercising jurisdiction delegated from EPA.

Each of these situations is reviewed to determine the federal, state, and local players and the manner in which they interact, coordinate and conflict with comments on how the current system of overlapping jurisdictions is performing and with suggestions about how the system might be improved.

CHAPTER 119

America's Unmanaged Territories: Its Extended Territorial Sea and its Exclusive Economic Zone

John Briscoe[1]

Setting out to sea from one of the coastal States of the United States, a vessel passes out of that State three nautical miles from the coast line. (That is true of all the coastal States except Texas, and Florida on that State's Gulf Coast. In those two cases the seaward limit of the States is not reached until one reaches nine nautical miles.) Beyond three nautical miles, however, the vessel is still within the territorial boundaries of the United States, odd as that may be, until it passes 12 nautical miles from the coast. At that point the vessel is in international waters, which is to say "high seas." Well, it is except for the fact that the United States claims exclusive resource rights out to 200 miles, and in the case of the Continental Shelf, to distances far beyond that in many places.

Put another way, the seaward limit of the United States off the coast of California is 12 nautical miles from the coast, though the seaward limit of the State of California is but three nautical miles. So there is a belt or band of American territory nine miles in width lying seaward of the State of California. Too, there is an additional band at least 188 miles wide lying seaward of that 12-mile belt, which is not claimed as sovereign territory, but whose resources are wholly the province of the United States. Call it "quasi-sovereignty."

That band, the "Exclusive Economic Zone" (often called the 200-mile zone), was proclaimed by President Reagan in 1983. At that time--as had been the case for many years--the width of the American territorial sea was three miles. In 1988, however, in one of his final acts as President, Mr. Reagan proclaimed a 12-mile territorial sea for the United States. Putting aside the merits of these two proclamations—that is, their validity under international law or their prudence as international policy—the fact remains that the United States has done virtually nothing to provide for the management of the extended territorial sea and the new EEZ. With the exception of fisheries and oil, there is virtually no management of the myriad of resources of the EEZ. What is worse (or at least more egregiously illogical), no effort has been made to extend the boundaries of the coastal States from the former American territorial limit of three miles to the new one at 12 miles.

It is textbook American civics that the United States consists of the 50 States, the District of Columbia and five territories (Guam, American Samoa, the Commonwealth of the Northern Marianas Islands, the Commonwealth of Puerto Rico, and the U.S. Virgin Islands). The nine-mile band of extended territorial sea, being as much the territory of the United States as, say the District of Columbia (but for one minor qualification), must stand as a sixth territory. And the 188-mile-wide band of Exclusive Economic Zone, being so plenary a jurisdiction, might arguably qualify as a seventh.

The burden of this paper is to suggest two things: (1) that the seaward boundaries of the coastal States ought to be made coextensive with the seaward boundary of the nation of which they are a part, and (2) that the United States ought to enact legislation providing plenary management over the resources of the EEZ. To articulate this two-fold burden, it is unhumbly suggested, is to carry it.

It is useful, first, to describe the several maritime zones the United States presently claims.

A. The Territorial Sea.

The Territorial Sea may be the maritime zone most often mentioned in the literature. On December 27, 1988, President Reagan by Proclamation extended the American Territorial Sea from three to twelve miles.[2] The Territorial Sea is that zone over which a coastal nation exercises sovereignty as fully as over its land territory, subject only to the rights of foreign vessels to innocent passage.[3] Most of the other coastal nations of the world now also claim twelve miles as the breadth of their territorial seas. Functionaries of the United States federal government have long maintained that, until 1988, the United States had steadfastly adhered to a claim of a mere three-mile breadth of territorial sea. Notwithstanding those assertions, the United States probably had a twelve-mile territorial sea, or at least a nine-mile territorial sea, for some time in the past, though it was not actually called that then.

In 1782 the United States asserted that nine miles was a reasonable breadth for its territorial sea.[4] In 1790, Congress extended American authority over smuggling—what must have been a "territorial sea" claim—to a distance of twelve nautical miles from the coast.[5] In February of 1793

Congress established customs jurisdiction to a distance of nine nautical miles.[6]

On April 22, 1793, to be sure, Secretary of State Jefferson declared a three-mile American territorial sea for neutrality purposes.[7] But several months later, on November 8, Jefferson wrote both the Spanish and French ministers to the United States that America was entitled to "as broad a margin . . . as any nation," and reserved "the ultimate extent of the [territorial sea] for future deliberations."[8] Jefferson later explained that the United States had been forced to accept three miles, and in 1805 suggested that the Gulf Stream—many miles to sea—would make a good outer limit of the American territorial sea.[9]

During the mid-nineteenth century, foreign governments were not at all beguiled by American pretensions to a mere three-mile territorial sea. The clear-headedness of these governments is illustrated by a diplomatic incident in the 1860s precipitated by America's protest of Spain's claim to a six-mile territorial sea off Cuba. In reply to a letter from Secretary of State Seward, Spanish Minister Tassara wrote on December 30, 1862 that the United States' claim to a "much more extensive" territorial sea, one having a breadth of twelve miles, was quite notorious in the international community.[10]

Nevertheless the United States persisted in maintaining that it claimed but a three-mile territorial sea. It maintained that position in its correspondence with Minister Tassara in 1862. It maintained it during World War II, when it declared defense zones of several hundred miles[11]; it maintained it in 1945 when it broke with the rest of the world and proclaimed jurisdiction over the Continental Shelf[12]; it maintained it in 1958 and 1960 when it sought international agreement on a six-mile territorial sea[13]; it maintained it when, in 1970, it proposed that all the coastal nations of the world adopt a twelve-mile territorial sea[14]; it maintained it in 1976 when it declared a 200-mile Fishery Conservation Zone[15] (when the International Court of Justice had just written that twelve miles was the maximum permitted breadth for such a zone[16]); and it maintained it when, in 1983, having refused to sign the 1982 Convention on the Law of the Sea, it unilaterally proclaimed a 200-mile Exclusive Economic Zone, which of course was then provided for only by the Convention.[17] The United States' may have been revisionist, but unarguably consistent in its revisionism.

The venerable territorial sea is the most long-lived of the maritime zones in practice today but, with the coming of claims to Continental Shelves and Exclusive Economic Zones, its importance to coastal nations is far less than what it once was. The other American maritime zones are:

B. The Contiguous Zone

The Contiguous Zone, which lies adjacent to a state's Territorial Sea, is a maritime zone in which the coastal state may exercise powers necessary to

> (a) prevent infringement of its customs, fiscal, immigration or sanitary laws and regulations with its territory or territorial sea;
>
> (b) punish infringement of the above laws and regulations committed within its territory or territorial sea.[18]

The 1958 Convention on the Territorial Sea and the Contiguous Zone, in Article 24.2, permitted a maximum breadth of twelve miles for the contiguous zone; Article 33.2 of the 1982 Convention allows 24 nautical miles. The President's 1988 Proclamation did not extend the breadth of the United States' Contiguous Zone, and so it maintains its former breadth of twelve miles, which makes it precisely coextensive with the American Territorial Sea.

C. The Fishery Conservation Zone (now called The Exclusive Economic Zone.)

Until its amendment in 1986, the Fishery Conservation and Management Act of 1976 ("FCMA") established the Fishery Conservation Zone, which began at the seaward boundary of each coastal state[19] and extended to 200 miles from the baseline from which the territorial sea was measured.[20] Congress in 1986 renamed this zone, in honor of the President's 1983 Proclamation, the Exclusive Economic Zone.

D. The Continental Shelf

The United States has claimed jurisdiction over the resources of the Continental Shelf since 1945. The 1958 Convention on the Continental Shelf defined the Continental Shelf by what is known as the "exploitability" criterion. That is to say, the Continental Shelf extended seaward to a depth of 200 meters and, beyond that, to whatever depth would still permit the exploration of the resources of the seabed and the subsoil of the Shelf. This criterion has been supplanted in Article 76 of the 1982 Convention. The new formulation provides: "The Continental Shelf of a coastal state comprises the sea-bed and subsoil of the submarine areas that extend beyond its territorial sea throughout the natural prolongation of its land territory to the outer edge of the continental margin, or to a distance of 200 nautical miles from the baseline from which the breadth of the Territorial Sea is measured where the outer edge of the continental margin does not extend up to that distance."

The geologic definition of the continental margin contained in the 1982 Convention is exceedingly complicated, but suffice it to say that, in certain geological circumstances, the 1982 Convention permits states to claim continental shelves to distances as great as 350 nautical miles from the baseline from which the territorial sea is measured, or even farther. Two such claims were made early, by Chile and by Ecuador, each in September of 1985. On September 12, Chile proclaimed jurisdiction over the Continental Shelves of the Easter and Sala y Gomez Islands extending to a distance of 350 nautical miles from the territorial-sea baseline. On September 19, Ecuador issued a similar proclamation asserting its jurisdiction over the Continental Shelf extending between the mainland of Ecuador and the Galapagos Islands. The United States contested these claims on the ground the geologic facts did not satisfy the criteria of Article 76.4 of the 1982 convention.[21]

Whether the United States, which has not signed the 1982 Convention, is permitted to claim a continental shelf in accordance with Article 76 is a nice question.[22]

By 1980, it was generally thought that the seaward limit of the American Continental Shelf stood at the 1000-fathom isobath (20 to 50 miles offshore in the case of California)--the then-existing limit of exploitability. The President's Proclamation of March 10, 1983, declaring a 200-mile Exclusive Economic Zone (EEZ) produced an unusual claim by the Department of Interior relating to the Outer Continental Shelf Lands Act. The Solicitor of the Department of Interior on May 30, 1985, issued an opinion on the question of the seaward extent of the Department's jurisdiction under the Outer Continental Shelf Lands Act (OCSLA) in light of the EEZ Proclamation.

The Solicitor's Opinion was that on March 10, 1983, by no act of Congress but by virtue of the Proclamation (which made no mention of the OCSLA), the outer limit of the Department's jurisdiction over the "outer *Continental Shelf*" leapt from its location on March 9, at approximately the 1000-fathom contour, to a new limit precisely 200 miles from the territorial sea baseline. (The landward limit of the jurisdiction of the Department of Commerce under the Deep Seabed Hard Minerals Resources Act, would, if the Solicitor is correct, have made a corresponding leap seaward. *See* 30 U.S.C. § 1403(4).) The new limit is not related to any geologic concept of continental shelf, nor to the legal concept of "continental shelf" that had been employed in American law from 1945 until, at least, March 9, 1983.

The author of the Opinion fills 52 printed pages in *Interior Decisions*, but its essence is found in two sentences on its third page. The Solicitor first notes that in section 2(a) of the OCSLA, Congress defined the expression "outer Continental Shelf" to mean

> . . . all submerged lands lying seaward and outside of the area of lands beneath navigable waters as defined in section 1301 of this title, and of which the subsoil and seabed appertain to the United States and are subject to its jurisdiction and control.

43 U.S.C. § 1331(a).

The Opinion then writes that the question presented is "when may the subsoil and seabed of its submerged lands be said to 'appertain' to the U.S. and to be 'subject to its jurisdiction and control.'" 92 I.D. at 461. The critical two sentences are then stated:

> On its face, the area described by this statutory definition is an expanding one: if a given area of subsoil and seabed becomes subject to the U.S. jurisdiction and control and appertains to the U.S., then that area falls within the definition. The plain meaning of section 2(a) must be followed, unless some unstated limitations must be inferred.

92 I.D. at 461 (footnote omitted).

Without citation of authority, the Solicitor concludes that the plain-meaning rule holds sway, and not any of the several other, often more preferred rules of statutory construction, such as those suggesting a construction in accordance with contemporaneous circumstances. *See generally Train v. Colorado Public Interest Research Group*, 426 U.S. 1, 9-11, 23-24 (1976).

Even so, the "plain meaning" attributed by the Solicitor to the OCSLA's definition of "outer Continental Shelf" is not so plain at all. The Solicitor takes great lengths to show that, ever since 1953, the area of the American Continental Shelf had been thought to be expanding. 92 I.D. at 465-68. That is true enough. But the Shelf expanded not because its definition continually changed, but because its definition, fixed in its terms, spoke of the limits of exploitability--and those limits changed. The Truman Proclamation of September 28, 1945 made emphatic reference to the exploitability concept,[23] and the concept, as is well known, was subsequently codified in Article 1 of the 1958 Convention on the Continental Shelf. Substantial areas of the seabed and subsoil that were unexploitable in 1953 were, by 1980, exploitable. In 1953, 100 fathoms of depth was considered the limits of exploitability; by 1980 it was 1000 fathoms.

Regardless of the soundness of the Solicitor's logic, his and his Department's assertion of extended jurisdiction, having gone unchallenged by foreign states, represents one, albeit minor, American assertion of resource management over the resources of the American Exclusive Economic Zone. I say "minor" because the authority of the Outer Continental Shelf Lands Act extends only to the seabed and subsoil—*i.e.*, practically speaking, petroleum—of the zone in which it operates.

E. Exclusive Economic Zone.

The 1982 Convention on the Law of the Sea provides in Articles 55 and 57 that every coastal state enjoys an Exclusive Economic Zone ("EEZ") adjacent to its coast, extending as far as 200 nautical miles from the territorial sea baselines. The President's Proclamation of March 10, 1983 proclaimed a 200-mile EEZ for the United States, notwithstanding that we have refused to ratify the 1982 Convention.[24]

This has been a somewhat tongue-in-cheek overview of American claims to offshore jurisdictions. The silliness aside, the facts remain that since 1983 the United States has claimed a 200-mile EEZ, and since 1988 a 12-mile territorial sea. Yet it has enacted no legislation to provide for the plenary management of either of these expansive zones. Worse, while the territory of the United States has been extended to 12 miles offshore, the seaward boundaries of the coastal States have been left at their former limit of three miles. This leaves, next to those States' seaward limits, a bizarre, anomalous nine-mile-wide band of American territory—as much the territory of the United States as the City of Omaha (save for the right of innocent passage of foreign vessels)—unincorporated into the adjacent coastal State, and mostly unlegislated for.

California state legislators in 1997 have introduced a plethora of coastal and marine regulation bills. As well-intentioned as such local legislation may be, wise and effective coastal and marine management for America won't be in the offing until Congress enacts thoughtful legislation providing for the plenary management of the Territorial Sea and the Exclusive Economic Zone. Making the seaward boundaries of the coastal States coextensive with the seaward boundaries of the United States is but an absurdly self-evident first step.

Appendix

[1] Mr. Briscoe has been a member of the San Francisco law firm of Washburn, Briscoe & McCarthy since 1980. He has represented a number of coastal States in ocean and coastal matters, among them California, Alaska, Hawaii, Georgia, and as well the Territory of Guam.

[2]Proclamation No. 5928; 54 Fed. Reg. 777 (1989).

[3]*See* Article 1 of the 1958 Convention on the Territorial Sea and Contiguous Zone, and Articles 2 and 17 of the 1982 Convention on the Law of the Sea (New York: United Nations, 1983), Sales No. E83.V.5.

[4]Crocker, *Extent of the Marginal Sea* (1919), p. 630; Jessup, *The Law of Territorial Waters and Maritime Jurisdiction* (1927), p. 50; 7 North Atlantic Coast Fisheries Arbitration, p. 46.

[5]Act of August 4, 1790, ch. 35, §§ 12-13, 31, 64, 1 Stat. 157-58, 164-65, 175 (1845).

[6]1 Stat. § 21,305, 313-14.

[7]Proclamation of Neutrality, April 22, 1793. See Fulton, *The Sovereignty of the Sea* (1911), 572-574.

[8]Letter from Secretary of State Jefferson to George Hammond, British Minister to the United States, November 8, 1793, reprinted in Swartztrauber, The Three-Mile Limit of Territorial Sea (1972), p. 57.

[9]Conversation between President Jefferson, John Quincy Adams, and Mr. Gaillard, as related in Memoirs of J. Q. Adams, Vol. 1, p. 375-376.

[10]This letter is in evidence in *United States v. Alaska*, United States Supreme Court, No. 84 Original, as Alaska's Exhibit 85-027.

[11]*See, e.g.*, Department of State Bulletin, 1, No. 15 (October 7, 1939), 5 Foreign Relations of the United States, pp. 36-37.

[12]Proclamation No. 2667, September 28, 1945, 10 Fed.Reg. 12303, 59 Stat. 884.

[13]This we sought at the First and Second United Nations Conferences on the Law of the Sea held in Geneva those years. *See generally*, 1 Shalowitz, *Shore and Sea Boundaries* (1962) pp. 269-270.

[14]1970 Public Papers of the Presidents--Richard Nixon; Statement about

United States Oceans Policy, p. 454, May 23, 1970.

[15]Fishery Conservation and Management Act of 1976, 16 U.S.C. §§ 1801 *et seq.*

[16]*United Kingdom v. Iceland* (Fisheries Jurisdiction) [1974] I.C.J. 3, 130. C F. Magnuson, The Fishery Conservation and Management Act of 1976: First Step Toward Improved Management of Marine Fisheries, 52 Wash. L. Rev. 427, 438-441 and Esp. 441 fn. 46 (1977).

[17]Proclamation 5030, Exclusive Economic Zone of the United States of America, 1 Public Papers of the President--Ronald Reagan, p. 380, March 10, 1983. Of course, this Proclamation asserted that the EEZ had become customary international law.

[18]Article 33.1 of the 1982 Convention on the Law of the Sea; see also Article 24.1 of the 1958 Convention on the Territorial Sea and the Contiguous Zone.

[19]By "seaward boundary" Congress probably meant the seaward boundaries of the lands restored to the states by the Submerged Lands Act, which is discussed below, rather than the seaward limit of the territorial sea. A Legislative History of the Fishery Conservation and Management Act of 1976, Senate Committee on Commerce & Nat'l Ocean Policy Study, 94th Cong., 2d Sess. (Gov't Printing Office 73-982) (1976), p. 678. The Submerged Lands Act boundaries, most pronouncedly in the cases of Texas and Florida, on Florida's gulf coast, may extend seaward of the territorial sea boundary.

[20]16 U.S.C. § 1811 (pre-1986). Selecting as the inner boundary of the Zone the seaward boundary of the coastal State, rather than the seaward boundary of the territorial sea, was done intentionally. If the United States, as many then expected, changed its claim of territorial waters from three to twelve miles, the boundaries of the fishery conservation zone would not have to be shifted. See A Legislative History, etc., *supra*, at 1051, 1101. 16 U.S.C. § 1811.

[21]*See* Kilaparti Ramakrishna, Robert E. Bowen, and Jack H. Archer, "Outer Limits of the Continental Shelf: A Legal Analysis of Chilean and Ecuadorian Island Claims and U.S. Response," 11 Marine Policy 56-68 (Jan. 1987).

[22]The United States ratified the four 1958 Geneva Conventions, which had been sent to the Senate for ratification in 1959 by President Eisenhower, in 1961. The Continental Shelf Convention, like the Territorial Sea Convention, went into force in 1964. Hollick, U.S. Foreign Policy and the Law of the Sea (1981) 159 and 442. U.S. T.I.A.S. 5578; 15 U.S.T. 471. The United States, of course, has not ratified the 1982 Convention.

[23]Proclamation No. 267, 59 Stat. 884.

[24]Proclamation 5030, 1 Public Papers of the President--Ronald Reagan 380 (March 10, 1983).

CHAPTER 120

FISHERIES COMPLIANCE AND ENFORCEMENT FOR THE 21ST CENTURY

Christopher J. Carr*

Abstract

Illegal fishing bears a large share of the responsibility for the crisis in marine fisheries world-wide. Fishery regulators are increasingly looking to satellite-based vessel monitoring systems (VMS) to enhance compliance and enforcement. VMS also offers management benefits and can assist fishermen in a variety of ways. California fishery management and enforcement officials should give serious consideration to VMS.

Introduction

The press abounds with reports of the crisis in fisheries around the world; fishery resources are being exploited at levels that cannot long be sustained. For example, The Economist in 1994 decried on its cover "The Tragedy of the Oceans," and National Geographic devoted most of its November 1995 issue to "Exploiting the Ocean's Bounty: Diminishing Returns." In California, a recent series in the Sacramento Bee documented the unsustainable exploitation of many of the state's marine fisheries. While excessive fishing effort is primarily responsible for this crisis, inadequate compliance with and enforcement of fishing regulations shares much of the blame. Off the coast of California, an estimated 43 to 64 million pounds of sea life is illegally taken from the Pacific Ocean every year -- the largest natural resource theft in California. (Knudson and Vogel 1996).

Recent international agreements and changes in national fishery laws signal a world-wide commitment to make the transition to sustainable fisheries. Examples of recent international fishery agreements include the

---

* Associate, Washburn, Briscoe & McCarthy, 55 Francisco Street, Suite 600, San Francisco, CA 94133 and Ph.D. Student, Jurisprudence and Social Policy Program, University of California, Berkeley, CA 94720

1993 FAO Agreement to Promote Compliance With International Conservation and Management Measures by Fishing Vessels on the High Seas, the 1995 United Nations Agreement on the Conservation and Management of Straddling Fish Stocks and Highly Migratory Fish Stocks, and the FAO Code of Conduct for Responsible Fisheries, also adopted in 1995. Each of these recognizes the critical role of effective compliance and enforcement to achieving sustainable fisheries and strengthens the basis in international law for states to take the necessary actions.

At the national level, states are re-orienting the focus of their framework fisheries laws to sustainable use of fishery resources. For example, in Fall 1996, the Magnuson-Stevens Fishery Conservation and Management Act (the framework law for management of U.S. marine fisheries) was amended by the Sustainable Fisheries Act to ensure greater attention to preventing overfishing, rebuilding depleted fisheries, reducing waste, protecting habitat, eliminating excess fishing capacity, and strengthening the scientific basis for fishery management. Similarly, New Zealand enacted a new Fisheries Act, also in 1996, which altered the Ministry of Fisheries' focus to supporting the goal of sustainable fisheries management. Numerous other countries as well have recently modified, or are considering modifying their fisheries laws, to emphasize sustainable fisheries management. National fisheries administrators are turning to new and innovative compliance and enforcement tools in their efforts to respond to these new mandates for sustainable fisheries.

<u>New Technologies - Vessel Monitoring Systems</u>

In seeking to enhance fisheries compliance and enforcement and achieve more effective management, administrators are looking to new technologies. One new tool is the satellite-based vessel monitoring system (VMS). On the most basic level, a VMS unit consists of a transceiver which receives a position-fixing signal from the Global Positioning System (GPS), a satellite-based navigation system operated and maintained by the United States Department of Defense. The VMS transceiver is placed on a fishing vessel so the location of the vessel can be monitored. After receiving the vessel position information from the GPS satellite, the transceiver forwards the data to a land earth station, which in turn sends the information to a base station. The base station then monitors the vessel, which appears on a geographic display. The vessel's location, movements, and speed can be tracked. All VMS systems allow for shore-to-ship messaging, while the more advanced systems allow two-way messaging. The use of VMS in two Hawaii fisheries demonstrates this new technology holds advantages for both fishery administrators and fishermen.

The Hawaii Longline and Lobster VMS Programs - A Case Study

Under international law, as reflected in the 1982 United Nations Convention on the Law of the Sea, states have sovereign rights over living marine resources to a distance of 200 miles from their coasts. The Law of the Sea Convention also places upon states obligations to conserve and manage fisheries occurring in their 200 mile zones. The Magnuson-Stevens Act implements the United States' rights and responsibilities with respect to marine fisheries. It establishes regional fishery management councils to develop management measures for the marine fisheries from 3 to 200 miles from the U.S. coasts. These councils are federal agencies comprised of federal and state officials, commercial and recreational fishermen, environmental representatives, and others concerned with fisheries management.

The Western Pacific Regional Fishery Management Council ("Council") has jurisdiction over the marine fisheries of Hawaii, American Samoa, Guam, and the Commonwealth of the Northern Mariana Islands. The Council has implemented a VMS program for the longline tuna fishery and has been conducting trials for VMS in the Northwestern Hawaiian Islands lobster fishery. Under Council regulations, which have the force of law, longline fishing vessels are prohibited from fishing in certain areas around the Hawaiian islands in order to protect endangered monk seals and avoid gear conflicts with recreational fishermen. These closed areas extend along the Hawaiian Islands for a distance of some 1,500 nautical miles. Since late 1994 U.S. officials have enforced these closed area restrictions through VMS. The 100 plus vessels of the longline fleet have had to carry operational VMS units so their positions and movements can be remotely tracked by enforcement personnel based in Honolulu.

For the past two years VMS has been used on an experimental basis in the Northwestern Hawaiian Islands lobster fishery. While not subject to closed area restrictions to prevent harm to endangered species, lobster vessels may not begin fishing before the opening day for the fishery and must cease fishing on the day managers announce as the closing day for the fishery. VMS permits remote monitoring of vessel compliance with these requirements.

Benefits of VMS for Fishery Administrators

VMS enhances fisheries enforcement and management. Under the longline VMS regime, patrol vessels or aircraft need only be dispatched to intercept vessels detected by remote monitoring to be fishing in a closed area. Recent Coast Guard projections place the costs of achieving a level of compliance through conventional vessel and aircraft patrols commensurate with that achieved through VMS at almost $23 million annually. The cost to enforcement agencies of operating VMS over the same period of time was less than $250,000.

Not only does VMS deliver greater fisheries enforcement presence at less cost than conventional enforcement assets, VMS also provides a state-of-the-art tool for fishery managers. This can be most readily appreciated in the use of VMS to enhance quota management. Many fisheries, like the Northwestern Hawaiian lobster fishery, are managed on the basis of an annual catch quota. If they are following sustainable fisheries practices, fishery scientists and managers will set the annual quota, or amount of fish that can be caught in a year, at a level that will ensure that the stock will reproduce at a level which will permit a comparable harvest the following season.

Traditionally, quota management has been conducted through the transmission of daily radio reports from vessels on the fishing grounds to management officials on shore. Radio, however, is not a very reliable means of communication with fishing vessels. Among other limitations, radio communication requires coordination on both ends of the connection. As a result, some catch reports are not communicated in a timely fashion and fishing vessel operators may not receive from managers on shore the closing date of the fishery. VMS is far more reliable. With VMS, the fishery catch can be determined more precisely and the closure date set more precisely. Because VMS communication systems have a store and forward capability, vessels need only send their catch data and managers need only transmit the closure date of the fishery, and the information will be received.

<u>Benefits of VMS for the Private Sector</u>

It is not surprising that fisheries enforcement officials and managers would aspire to use this new, powerful, and cost-effective technology. What is surprising, however, is that, after many years of opposition to VMS as "electronic cowbells" abrogating their "freedom of fishing," increasing numbers of fishermen are coming to appreciate VMS for its potential benefits to them. These include more reliable and secure communication among vessels and greater control over vessel operations by vessel owners, enhanced fishing opportunities, proof that fishermen are in compliance with fishery regulations, and increased safety of operations at sea. Thus, VMS presents a happy convergence of the interests of fishery administrators and fishermen.

Improve Communication and Control

Most fishing vessels communicate among each other, and from ship to shore, by radio. Radio, however, is neither a secure nor reliable medium for communication. Because radio communication is not secure, when vessels working in concert communicate with each other about favorable fishing grounds, captains speak in coded language and often lie about their vessel positions to protect the information from their competitors. Communication from vessel to vessel by VMS, in contrast, is secure and confidential. Since each vessel has a unique VMS registration number, only the intended vessel can receive the communication. This makes it much easier for fishermen to communicate with each other at sea and coordinate their activities.

In addition, radio is not a reliable means of communication for vessels. Among other limitations, radio communication requires coordination on both ends of the connection. Captains have to maintain a radio schedule so that, for example, one is not off the bridge tinkering with the engine when called by another captain over the radio. With VMS there is no need for such coordination since the VMS stores and forwards information. This makes communication far more reliable.

In areas where there are a number of ports, such as the North Pacific, captains can use VMS to determine which port to put in, based on the price being offered for the vessel's catch. More generally, vessel owners can monitor and control their vessels if they have their own VMS base station computer. For example, in the Pacific shark fins are highly valued. Under industry custom the captain and crew of a fishing vessel, not the owner, own fins that are removed from sharks caught through fishing operations; shark fins are not regarded as part of the "catch." Shark finning, however, is dangerous to crew and costly to vessel owners.

Crew have been mauled by sharks while finning, injuring the crew member and visiting financial liability upon the vessel owner. At the same time, vessels have been known to leave fishing grounds to target shark for purposes of finning. Even if a vessel has not targeted sharks but, as is usually the case, catches them incidentally in the course of fishing for tuna, a vessel will often leave the fishing grounds to rendezvous with a vessel to which it transships shark fins. When this happens, the vessel owner loses fuel and time off the fishing ground and incurs myriad other costs. Finally, many species of sharks have extraordinarily low reproduction rates. With VMS, a vessel owner can monitor if a vessel has left the fishing grounds and thus control finning operations.

Enhance Fishing Opportunities

VMS can lead to enhanced fishing opportunities for vessel operators. In the Northwest Hawaiian Island lobster fishery, lobster vessels cannot be west of 161 degrees longitude 24 hours before the opening of the fishery. As a result, vessels are hovering at 161 degrees for the dash to the nearest bank supporting lobster. Because it is a derby fishery (fishermen fish for all they can catch until the quota is reached and the fishery closed), fishermen have reason to stay on one bank, rather than spend fishing time moving to more distant banks while their competitors are busy fishing. This concentrates all the effort on one bank, so the pressure on the resource is uneven. This is undesirable from a conservation standpoint. With VMS, the rule preventing vessels from venturing west of 161 degrees before the opening could be eliminated since VMS could be used to monitor vessel activity; vessels would permitted west of 161 degrees so long as they were not fishing for lobster.

Moreover, under the current regime, fishermen have to be in port in Honolulu within 48 hours after the closure of the fishery. Therefore, vessels will not venture into the outer banks (further west). With VMS, this requirement could be eliminated and fishermen could instead be required to be moving off the grounds within some period of time following the closure. This would enable vessels to fish the outer banks, providing new and possibly better fishing opportunities for the fishermen, and more evenly distributing the pressure on the resource.

Demonstrate Compliance

With increasing pressure from environmental organizations, commercial fishermen throughout the world feel under siege. Many responsible fishermen feel that they have been shouldered with the burden of proving they are complying with fishing regulations, rather than enjoying a presumption of innocence. In Australia and New Zealand large segments of the fishing industry, hoping to quiet the complaints of environmental organizations, support VMS to prove that they are not violating closed area requirements. In Hawaii, the longline VMS has demonstrated to the satisfaction of environmental organizations that longline fishermen are not harming endangered monk seals.

Recently several longline vessels from the Hawaii fleet were detained in port by California enforcement officials who suspected that swordfish on board the vessels had been caught inside the 200-mile zone off California, contrary to regulations. The vessels, which had VMS on board and operating under the Hawaii longline program, were able to use the VMS position and signature records for their vessels to demonstrate that they had not been fishing inside 200 miles, but rather had legally caught the swordfish on the high seas.

Safety

Most fishing vessels of industrialized countries carry devices called EPIRBs -- electronic position indicating radio beacons -- that send out a radio signal when a vessel is in distress. But these often fail and do not permit precise location of the vessel. VMS enhances safety because it automatically transmits vessel position reports at certain time intervals. For example, recently a vessel far to the east of New Zealand was struck by a rogue wave. The EPIRB failed and though the VMS was destroyed, the last position report enabled rescuers to locate the vessel with relative ease, thus speeding up the rescue process.

While all of these benefits to fishermen are important in aiding the transition to sustainable fisheries, most important of all, of course, is the fact that VMS will ensure the conservation of the resource through more effective compliance and enforcement. After overcoming their traditional aversion to surveillance felt to impinge their long-held freedom of fishing, fishermen are coming to appreciate this. With a VMS requrement, fishermen can be secure in the knowledge that their competitors are complying with regulations: that the playing field is level. Incentives for non-compliance are diminished and responsible, sustainable fishing is fostered.

Conclusion

Fishery management and enforcement officials responsible for overseeing the 220,000 square mile state and federal ocean that is California's inheritance should afford serious consideration to requiring vessels subject to their jurisdiction to use VMS. VMS would provide a cost-effective supplement to limited agency resources. If closed areas or "replenishment" areas are implemented for certain fisheries, VMS would help ensure that such areas serve their purpose. As shown by the Hawaii experience, VMS can also enhance the effectiveness of quota management. Among the many benefits VMS provides to fishermen, it would allow those of California to take the lead in demonstrating their compliance with fishery rules and regulations. In the 21st Century VMS will be standard equipment required of fishing vessels. California regulators and fishermen should recognize this and embrace VMS as an indispensable tool for responsible fisheries management and enforcement.

References

Knudson, Tom and Nancy Vogel, "Secret Slaughter by Poachers Takes Huge Toll on Wildlife," Sacramento Bee, Dec. 24, 1996, p. A1.

CHAPTER 121

Tide and Beach Fluctuations and the Mean High Water Line

Reinhard E. Flick, Ph.D.
California Department of Boating and Waterways
Scripps Institution of Oceanography
La Jolla, CA 92093-0209
ref@coast.ucsd.edu (619) 534-3234

M. Hany S. Elwany, Ph.D.
Center for Coastal Studies
Scripps Institution of Oceanography
La Jolla, CA 92093-0209
hany@coast.ucsd.edu (619) 534-6119

Abstract

This paper examines how tide elevation and beach width fluctuations in southern California influence the location of the Mean High Water Line (MHWL). Tide height changes on time scales ranging from daily to the lunar node cycle of 18.6 years duration are reviewed. Beach profile data from Del Mar, California measured between 1978 and 1996 are used to illustrate beach width changes. The conclusion is that the position of either a fluctuating or an average MHWL can be determined on a typical southern California beach, provided that sufficient profile data exists.

Introduction

The Mean High Water Line is defined as "the line on a chart or map which represents the intersection of land with the water surface at the elevation of mean high water" (Hicks, 1989). The MHWL has in the past been accepted as the common law boundary between the uplands, which can be held as private property, and the tidelands, and so also the submerged lands, both of which are held by the state in trust for the public. This definition was settled upon in 17th century England, adopted by the sovereign States after the Revolution, and subsequently by the United States (Shalowitz, 1962).

In recent years, confusion has arisen in the courts about whether the MHWL is or should be the boundary, and if so, how its location can be calculated (see other papers in this session). Notwithstanding this confusion, the MHWL remains an important concept that still has enormous practical consequences pending a clear definition of an alternative boundary.

Similarities and differences between tide elevation fluctuations and beach width changes are discussed below. While there are some similarities, the chief practical difference is that there are usually far less beach width data than there are water level data. Since beach changes are also less predictable than the tides, the lack of measurements represents the most serious drawback to any effort at rationalization of shoreline boundaries in terms of the MHWL.

Mean High Water and Mean High Water Line

Mean High Water (MHW) is a tidal datum. It is defined as the average height of all high water elevations observed over (or adjusted to) a specified 19-year tidal epoch (Hicks, 1989). The definition of the MHWL states that the water level datum represents an average (*mean* high water), but implies no averaging of the horizontal position of its intersection with the beach. If the beach changes width, as most open coast beaches do, this line moves with the beach.

Beach width is usually defined as the distance from a (fixed) back shore location to the intersection of the beach profile with a given elevation plane, and is therefore closely related to the definition of the MHWL itself. This means that the MHWL is an ambulatory boundary that changes with beach width, conceivably over very short time spans. This is less than ideal, especially for a public-private boundary. In some areas, the boundary has been determined and fixed, either by agreement or court decision. But, depending upon the local shoreline processes, this may not be an ideal, long-term solution either, although boundaries can and are occasionally re-litigated.

The definition of MHW is relatively clear and generally not disputed, at least on coasts where the tide dominates the water level fluctuations. In contrast, very little seems to be understood, determined or accepted about how to account for changes in the location of the MHWL resulting from beach width changes and the resulting uncertainty in property boundaries, or even if anything should be done to improve this situation. Even where beach width data are available, there are no specific, published methods to calculate an "average" MHWL on the basis of, for example, a mean high water beach width, much less an agreed upon standard that is in any way analogous to methods developed to compute tidal datum elevations.

Water Level Fluctuations

Ocean water level changes on the coast of California are mainly due to the tide, which has a mean range of about 5.5 ft, and an extreme range of about 9 ft. Strictly defined, tides are the periodic rise and fall of waters of the earth in response to the gravitational forces of the moon and the sun. Effects of wind and barometric pressure (storm surge) reach up to about 2 ft (Flick, 1991), excluding the influence of wave set-up and surf beat in shallow water. Seasonal changes in water level are about 0.5 ft (Reid and Mantyla, 1976). In addition, climate changes and geologic processes such as subsidence, tectonic uplift and earthquake related movements produce long-term changes in relative sea level. Tide gauges indicate that relative sea level in southern California has risen about 0.7 ft over the past century (USACOE, 1989).

Tide Fluctuations

Since the tide is the dominant portion of the sea level fluctuation on most open coasts, it is logical that water level datums be computed with consideration to the inherent astronomical time scales of tidal periodicities. This is true even though the datum computations use the total water level record, not just the part attributable to the tide. In order of increasing span, the periodicities of practical importance are: 2 cpd, 2 cpm, 2 cpy, 1 cpy, 1 cp 4.4 y and 1 cp 18.6 y (where cp stands for cycle per, and d, m and y stand for day, month and year, respectively).

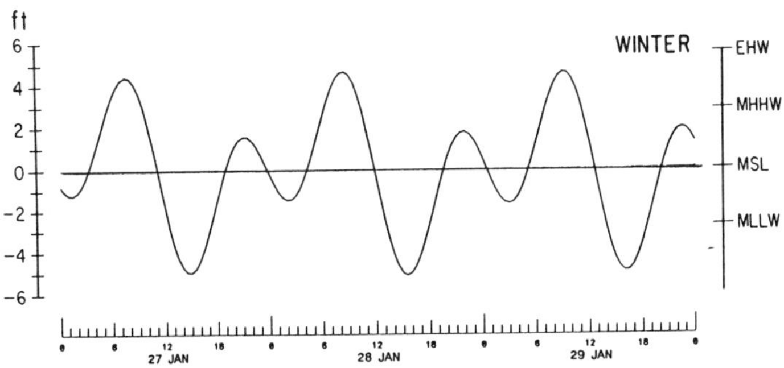

Figure 1
West-coast (San Diego) tide curve for time of peak tide range in winter.

In order to accommodate the 18.6 y lunar node cycle and the annual cycles, tidal datums are calculated over a 19-year period, the nearest whole number of years. To account for long-term sea level change, a specific 19-year period is chosen and periodically updated. This is the called the National Tidal Datum Epoch, and is currently set at 1960 - 1978 (Hicks, 1989).

Some of the important characteristics of west coast tide fluctuations are illustrated in Figure 1. The tide is mixed with nearly equal semi-daily and daily components (Zetler and Flick, 1985). The two high tides and two low tides each day are, respectively, unequal in amplitude. Monthly changes are dominated by the spring-neap cycle formed by the interference of semidiurnal constituents. This produces two periods of relatively high tides (spring) around full and new moon, and two of lower ranges (neap) around the times of half-moon (not shown). One spring tide range per month is usually higher than the other, because of changes in the moon's distance and declination. The declination dependence is called the tropic cycle and is caused by interference of diurnal constituents.

Longer period variations also occur. There is a 4.4 y cycle caused by the precession of the lunar perigee that results in higher peak monthly tides of about 0.5 ft, compared with years in between. This cycle crested in 1982-83, 1986-87, 1990-91, and 1995-96, and will peak again in 1999-2000, and periodically thereafter. The 18.6 y regression of the lunar node causes an 0.3 ft variation in peak tide heights with a maximum in 1986-87, a minimum in 1997, and so on (Zetler and Flick, 1985).

Beach Profile Changes

An 18-year long record, extending from 1978 to 1996 with gaps, of beach profile data from Del Mar, California has been accumulated at Scripps Institution of Oceanography. The data were collected under the sponsorship of various agencies, including the City of Del Mar, the State of California, the U. S. Army Corps of Engineers and Southern California Edison Company. Beach profiles represent an elevation cross-section of the beach. They are surveyed using standard surveying instruments to measure the elevation of the sand surface as a function of distance relative to a fixed benchmark in the back shore.

Figure 2 shows three beach profile traces taken in October and December 1982, and in January 1983. The October 1982 profile shows a moderately wide, flat beach berm extending up to about 150 ft from the benchmark, and a concave beach face decreasing in slope offshore (Inman, *et al.*, 1993). By December, winter storm waves had transported sand from the berm offshore, narrowing the beach width and generally flattening the beach profile. By late January 1983, the beach had been greatly reduced in width, with essentially all the sand being removed from the upper beach and moved offshore. Note that because of the concave shape of the profiles, the respective changes in beach width depend upon the elevation under consideration. At the elevation of MLLW (-2.56 ft below NGVD), the decrease in width (295 ft) between October and January was almost twice the loss (157 ft) at the elevation of MHW (2.07 ft above NGVD).

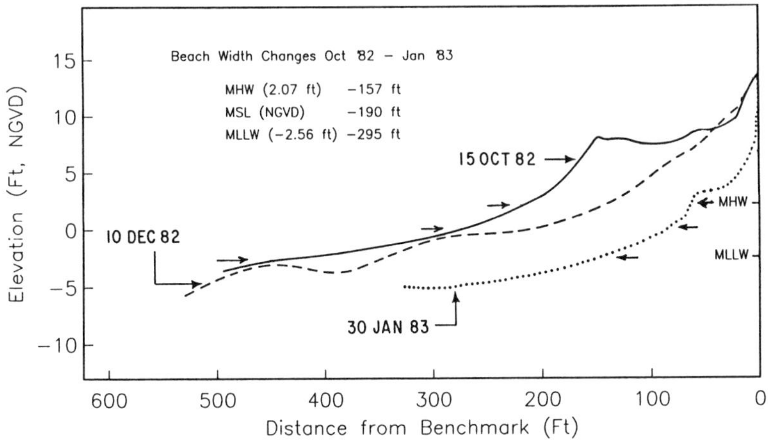

Figure 2
Beach profiles at Del Mar showing beach width decrease at elevations of MHW, MSL and MLLW during the severe winter of 1982-83.

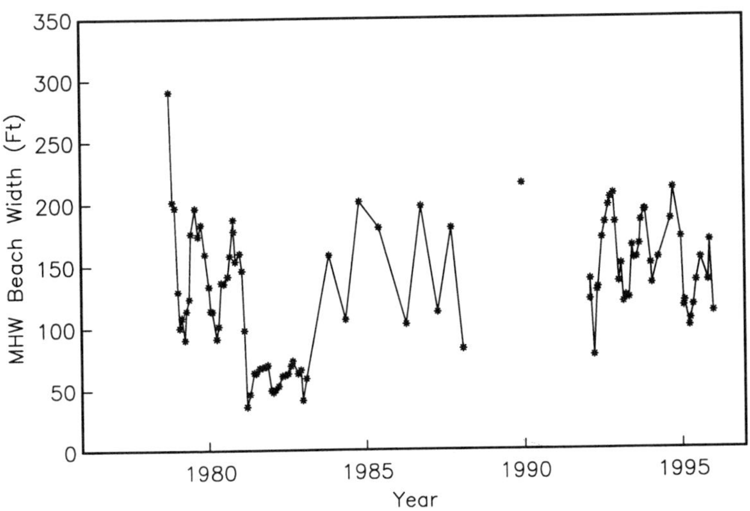

Figure 3
Time history of beach width at Del Mar, CA from 1978 to 1996.

Beach Width Changes

Figure 3 shows the time history of beach width measured at the elevation of NGVD at Del Mar from 1978 to 1996. Del Mar Beach is a 1-mile long, moderately wide, low-lying, sand and cobble spit that spans the historical range of the San Dieguito River mouth. The beach has a moderate wave climate, typical for southern California. It is heavily used for recreation, and is backed by a nearly continuous line of seawalls and revetments protecting residential development.

Figure 3 shows that the average beach width at the MHW elevation at Del Mar is about 150 feet, with little or no upper beach erosion evident over the past 18 years. The typical seasonal range of beach width change in Del Mar is about 100 ft, from a seasonal maximum of 200 ft wide in the autumn, to a minimum of about 100 ft in the spring. Several episodes of above average width changes are also evident in Figure 3. In 1978 - 1979 there was a 150 ft drop, in 1982 - 1983 the beach retreated as described above. The data shown in Figure 3 are useful to illustrate the changes in the location of the MHWL over time, and to explore various options for utilizing these data to determine an average MHWL.

The seasonal changes in beach width are analogous to the periodic fluctuations in tide range. Although Del Mar beach shows little or no sign of any trends in beach width, such trends would be present in areas of long-term beach erosion or accretion, and would be analogous to the observed rise in sea level. Carrying the analogy further, it is clear from Figure 3 that, at least in Del Mar, a reasonable estimate of the average position of the MHWL can be made with confidence. This would be at a location of 150 ft from the benchmark. Similarly, a reasonable estimate of the average maximum and average minimum position of the MHWL could be made. These would be located at distances of 200 ft and 100 ft from the benchmark, respectively.

Conclusion

In beach areas with adequate profile data, such as Del Mar, CA, the position of either a fluctuating or an average MHWL can be determined.

Acknowledgements and Disclaimer

The authors are grateful to the agencies who have sponsored beach profile data gathering at Del Mar since 1978. Ms. Claudia Avendano processed much of the beach profile data. Any opinions expressed in this paper are those of the authors, and do not necessarily represent official positions of the State of California.

References

Flick, R.E., 1991, "Joint Occurrence of High Tide and Storm Surge in California," *Proc. World Marina '91*, Amer. Soc. Civil Eng., p. 52-60.

Hicks, S.D., 1989, *Tide and Current Glossary*, National Ocean Service, NOAA, Washington, D.C., 28 pp.

Inman, D.L., M.H.S. Elwany and S.A. Jenkins, 1993, "Shorerise and Bar-Berm Profiles on Ocean Beaches," *Jour. Geophys. Res.*, v. 98, n. C10, p. 18,181-18,199.

Reid and Mantyla, 1976, "The Effect of the Geostrophic Flow Upon Coastal Sea Elevations in the Northern North Pacific Ocean," *Jour. Geophys. Res.*, v. 81, n. 18, p. 3100-3110.

Shalowitz, A.L., 1962, *Shore and Sea Boundaries*, Coast and Geodetic Survey, U.S. Dept. Commerce, Washington, D.C., two vols.

USACOE, 1989, *Historic Wave and Sea Level Data Report, San Diego Region,* U.S. Army Corps of Engineers Los Angeles Dist., CCSTWS 88-6.

Zetler, B.D. and R.E. Flick, 1985, "Predicted Extreme High Tides for Mixed-Tide Regimes," *Jour. Phys. Oceanog.*, v. 15, n. 3, p. 357-359.

CHAPTER 122

PUBLIC BEACH OR PRIVATE BACKYARD: A CASE STUDY OF DEL MAR BEACH AND HOW THE SAND WAS DIVIDED UP.

By Jamee Jordan Patterson[1/]

Public and private rights to the beach come from a variety of sources and can often conflict. Nowhere was this more apparent than on the beach in Del Mar, California, where the public's right to use the beach clashed with the private property owners' right to protect their homes and their backyards from the ocean. The dispute about who owned what on the sandy beach involved the California Coastal Commission, the City of Del Mar, the State of California acting through the State Lands Commission, and the citizens of Del Mar, including beachfront property owners and inland residents. Out of this conflict has come a truce and a solution which may provide a model for other beachfront cities to emulate. While the solution was reached with a great deal of difficulty, the result has proven quite satisfactory.

HISTORY OF THE BEACH

Del Mar is a small beachfront community in northern San Diego County located between the San Dieguito River on the north and the Penasquitos Lagoon on the south. From the river mouth south approximately a mile and a half lies a sandy beach backed by coastal dunes. Now developed with luxury homes, this beachfront was originally subdivided in 1885, then resubdivided from 1910 to 1922 by the South Coast Land Company. The South

---

1. Ms. Patterson is a Deputy Attorney General for the State of California. The views expressed here are her own and do not necessarily reflect the views of her office or client agencies. Ms. Patterson wishes to acknowledge and thank the City of Del Mar, particularly its planning staff, for valuable assistance in researching this paper.

Coast Land Company was formed by William Kerckhoff and Henry Huntington, owner of the Los Angeles Red Car system, and two others with Colonel Ed Fletcher as general manager. Kerckhoff and Huntington also founded the Pacific Light and Power Company to build dams, railroads and provide electricity. The primary mode of transportation at the time was the railroad which had come to Del Mar in 1882. A railroad depot was built which provided ready access to the beach. The South Coast Land Company also developed a large, luxury hotel on a hill next to the beach, a bathing plunge, and a 1,000 foot pier. The resubdivisions were more in tune with the natural contours of the beach, creating individual lots and realigning the streets almost due east-west ending at the sandy beach. A fifteen to twenty foot wide public right of way ran north along the beach from 17th Street.

Resubdivision No. 2 in 1910 created lots between 17th Street and 21st Street, created a sidewalk labeled lot 22 fifteen feet wide running between the lots and the beach and designated the sandy beach as lot 21. Resubdivision No. 3 in 1912 continued the pattern of beachfront lots separated from the sandy beach by a public right of way. This map covered the area from 21st Street to 25th Street. The right of way was clearly labeled a public sidewalk on this resubdivision map. The North Strand Map in 1922 created lots from 25th Street through 27th Street and showed the right of way as an alley twenty feet wide. As a part of its subdivision sales program, the South Coast Land Company built a concrete sidewalk seven and a half feet wide, running from the train depot north to the end of the developed portion of the beach. Electric lights were installed along the sidewalk, apparently by the County of San Diego. Because the sidewalk connected the beachfront lots to the train depot, it provided the primary access to the beach until the automobile replaced the train as the primary mode of transportation to the area.

As houses began to be built along the beach, the surrounding area also underwent development. Between 1916 and 1918, the Lake Hodges dam was built on the San Dieguito River in response to disastrous floods in 1916. This dam holds some twenty-two billion gallons of water and forms an eight plus mile lake quite some ways east of Del Mar. The construction of the dam decreased the flow of the San Dieguito River and prevented the natural transport of river sand to the beach. Other impacts on the natural condition of the beach over time included the construction of the pier in 1913 (later demolished in the early 1960's), the increase in sand mining in local rivers, the construction of a boat basin at Camp Pendleton during World War II which interrupted the littoral drift of sand along the coast and the construction of shoreline protective devices on the Del Mar beach itself. In 1927 heavy rains and storms washed out much of the concrete sidewalk on the beach and in 1928/1929 the first seawall was built on the beach. In 1929 the Fletcher family constructed a seawall across their five lots between 18th and 19th Streets. This wall was built thirty-five feet out onto the sandy beach and enclosed the sidewalk and a large portion of the beach.

In 1938 the beach west of the public right of way was conveyed to the Del Mar Civic Association by the William Kerckhoff Company. This deed of gift provided that the property shall be used exclusively as a bathing beach by the residents and property owners of Del Mar, their tenants, friends and guests and prohibited the construction, erection or placement of any building of any kind upon the beach. Unfortunately, this prohibition was honored in the breach and over time nearly every lot on the beach was protected by some form of shoreline protective device or had some structure encroaching on the beach.

The beachfront now comprises 85 lots and is developed with approximately 82 single family homes, two restaurants and four lifeguard stations located at an elevation slightly above the beach. Many of these properties are subject to flooding and wave action during storms. According to the City of Del Mar, between 1973 and 1983, Scripps Institution of Oceanography conducted a series of beach profiles which proved statistically that the beach at Del Mar was eroding faster than it was accreting and losing approximately 250,000 cubic meters of sand per year.

SHORELINE PROTECTIVE EFFORTS

Over time, encroachments extended over the public right of way or farther out on to the beach on all 85 beachfront lots. In the 1970s, temporary sand berms were constructed to provide some protection and sand bags and rip rap were installed by property owners to protect against wave action. In some cases, the temporary sand berm was planted with vegetation or otherwise improved and made into a more permanent encroachment. In 1980 a test section of Longard Tube was installed between 27th Street and 29th Street, intended as a first line of defense for the beach backshore. Initially this appeared to work well but severe storms in 1982-1983 proved otherwise. The Longard Tube was simply not a substantial enough barrier to dissipate or reflect wave energy and prevent beach erosion. In 1979 the City of Del Mar began efforts to remove private encroachments from the sandy beach. These efforts included a number of reports and studies and resulted in a variety of proposals. Ultimately, the voters took the matter into their hands and adopted the Beach Protective Initiative Ordinance.

PUBLIC VERSUS PRIVATE INTERESTS

The recitation of property rights above was not universally accepted. Many beachfront owners disputed the public's ownership rights in the right of way and the sandy beach. An early legal challenge occurred in 1951 when owners of a lot between 18th Street and 19th Street, the Goods, extended the porch of their house over the sidewalk area and destroyed the sidewalk crossing seventy feet of their property. The owners of neighboring property, including descendants of Colonel Fletcher, sued the Goods for destruction of the sidewalk. In a December 1951 decision in Crabtree v. Good, San Diego Superior Court number 166161, the trial court ruled in favor of plaintiffs. The judge held the sidewalk had been dedicated to the public and that the neighboring property owners also had an easement right to the sidewalk. Judgment pursuant to stipulation was entered in March 1952 providing that plaintiffs together with the general public were the owners of and entitled to an easement for sidewalk purposes over the Goods' property and extending south to the bath house, pier and railroad station. The Goods were enjoined from interfering with plaintiffs' or the public's use of the easement area and were ordered to remove the porch extending over the easement as well as to replace the sidewalk which had been destroyed.

Perhaps the turning point for this beach occurred in 1983. The United States Weather Service predicted severe storms and high tides for fall of 1983, following on the heels of a severe winter in 1982-1983. In response, a group of property owners between 24th and 26th Streets began unauthorized construction of a 480 foot long seawall located fifteen feet out on the sandy beach. When the California Coastal Commission became aware of the construction, its staff notified the property owners that a coastal development permit was required. The property owners continued construction but applied for an emergency permit. The Commission issued an emergency permit with several provisos, including the requirement that a regular permit be applied for and, if a permit were not received, the seawall would be removed. Commission staff expressly warned the homeowners that the location of the seawall was very controversial and there was a strong likelihood that staff would recommend relocation of the seawall. Staff urged the homeowners not to spend too much money on the wall.

Ignoring staff's advice, the property owners spent over $300,000 on constructing a seawall some nineteen to twenty five feet from their residences and fifteen feet on to the public beach. The wall displaced over 7,200 square feet of sandy beach. Behind the wall, the homeowners installed unpermitted non-structural privacy walls and brick pavers, creating private patios on the public beach. The homeowners applied for a regular permit to make the wall permanent. In December 1983, the Coastal Commission approved the application conditioned upon the seawall being reconstructed and relocated so that it encroached no more than five feet from the westerly property lines, payment for the encroachment upon the public beach based on the fair market rent for the land, and restoration of public access at the street ends. The Commission required the property owners to remove the temporary wall, privacy walls and brick pavers by June 1984. After numerous administrative extensions of the deadlines to allow the City of Del Mar to complete its planning for the area, the Commission ultimately rejected an indefinite extension and required the wall be removed.

Predictably, the homeowners sued. Equally predictably, they lost. In a 1987 published appellate decision which constitutes binding precedent, the Court of Appeal upheld the Commission's decision. (Barrie v. California Coastal Commission (1987) 196 Cal.App.3d 8.) The Court found the homeowners did not have a vested right to retain the seawall, refused to estop the Commission from requiring relocation and basically found the emergency faced by the homeowners was one of their own making. By commencing construction prior to applying for a permit and continuing in light of staff's warning that the location was problematic, the homeowners attempted to coerce the Commission into approving the wall's placement, a result the Court would not countenance. The Court found the Commission's decision to relocate the wall was based upon substantial evidence and was consistent with applicable law. Ultimately, the seawall was removed and a new seawall was built on the homeowners' property lines.

THE BEACH PRESERVATION INITIATIVE

In April 1988, after the Barrie decision, the voters of the City of Del Mar approved Measure D, the Beach Preservation Initiative. The initiative created a Beach Overlay Zone to regulate the use of the beach and established a Shore Protection Area zone (SPA zone) and a Shore Protection Area line (SPA line) which is comparable to a set back line. The initiative also established permit requirements for protective structures in the SPA zone. Only privately owned protective structures, publicly owned protective structures and publicly owned development may be permitted in the SPA zone. The initiative has specific criteria for protective structures in order for them to be permitted. Perhaps the most important component of the initiative is its treatment of existing private development in the SPA zone. All non-conforming private development in the SPA zone must be abated unless a permit has been obtained which establishes an amortization period for the development. The initiative declared all privately owned non-conforming development to be a public nuisance if no amortizing permit is obtained, whether such development occurred before or after the effective date of the initiative. A property owner is entitled to a public hearing before the City Council to determine whether a development is non-conforming and whether abatement should be ordered. Failure to comply with a properly issued abatement order may result in the city abating the nuisance and charging the property owner for its efforts.

Following the enactment of the initiative, a flurry of litigation was filed. A number of lawsuits resulted in the initiative being upheld while others resulted in settlements which balanced private and public rights. One lawsuit involved a 1985 city permit issued for replacement of the Fletcher seawall which was originally constructed in 1928 to protect five lots and was destroyed by a storm. (Lang v. City of Del Mar, et al. San Diego Superior Court Case No. N314780.) Because the seawall replaced a prior wall in the same location, some thirty five feet from the property line, the Coastal Commission did not have jurisdiction over the new wall. However, Joe Lang, a resident opposed to the seawall, sued the City of Del Mar and the State Lands Commission, alleging that the wall encroached on public beach or public tidelands. After conducting a boundary investigation, the State Lands Commission concluded that the wall was likely constructed on historic public tidelands. Comparisons of United States Coast and Geodetic Surveys from 1889 and 1934 along with other historic information buttressed this conclusion, prompting the State Lands Commission and the City of Del Mar to realign themselves in the litigation with Lang against the Fletchers. The case was settled after much negotiation. The Fletchers agreed to remove the seawall and the State agreed not to assert a boundary interest landward of the SPA line so long as the Fletchers complied with the initiative. The seawall was removed and over 12,000 square feet of sandy beach became available for public use. The settlement in this case became a blueprint for settlements in a number of the other cases.

One final case warrants a brief discussion. The Scott and Lynch families owned two adjacent houses within Resubdivision No. 3. The Lynches' predecessors had erected a seawall with a patio behind it in 1928 over the right of way designated as public sidewalk and out onto the beach. The Scotts' predecessor erected a seawall and patio in 1946 over the right of way and onto the beach. Rip rap was subsequently placed on both properties, extending the encroachments farther seaward. Neither family applied for permits to legalize the encroachments under the initiative. When the City of Del Mar sent them abatement notices, they ignored the notices. Ultimately, the City exercised its power of abatement and removed the illegal encroachments. Scott and Lynch both sued, claiming the sidewalk was not public or had been abandoned, they had acquired a prescriptive right to build and retain their structures on the beach and the City's removal of those structures required the City to pay compensation. The trial court rejected their arguments, holding that the sidewalk was indeed public having been dedicated when the resubdivision map was accepted and recorded in 1912, the sidewalk had not been abandoned, the private property owners had no right to build or retain the encroachments and the City's removal of the illegal encroachments was a lawful exercise of the police power which did not require compensation be paid. The court also found that any vulnerability to storms the property owners suffered when their seawalls were removed was caused by their refusal to comply with the initiative. Having failed to avail themselves of the opportunity to build seawalls consistent with the initiative, they could not recover damages. They had a duty to protect themselves and did not take any action to do so. (Statement of Decision, Scott v. City of Del Mar, San Diego Superior Court No. 666702.)

SUCCESS OF THE INITIATIVE

Although the initiative was objectionable to many beachfront property owners, it has nevertheless been resoundingly successful. To date, the City of Del Mar reports that of the 85 properties affected by the initiative, 64 (75 percent) have either voluntarily removed non-complying structures, applied for permits, processed an amortization schedule or been subject to an abatement hearing. Approximately 66,740 square feet of usable beach has been returned to the public. The beachfront owners benefit from using this property along with the public as well as from having a greater degree of certainty about what methods may be legally employed to protect their valuable investments. A beautiful, usable and wider beach is now available to residents and public alike.

BIBLIOGRAPHY

In addition to information developed through litigation, the following resources were relied upon in this paper:

A Beach Proposal, Councilman Harvey Shapiro, City of Del Mar, September 1983

Autobiography of Colonel Edward Fletcher, privately published

Beach Encroachment Committee Report, City of Del Mar, July 13, 1979

Del Mar Looking Back, Nancy Hanks Ewing, The Del Mar History Foundation, 1988.

Performance Documentation of the Longard Tube at Del Mar, California 1980-1983, Reinhard Flick and Walton Waldorf, Coastal Engineering 8 (1984) 199-217.

Update and Recommendations: A Beach Proposal, Councilman Harvey Shapiro, City of Del Mar November 1983

CHAPTER 123

Locating the Landward Limit of the Mean High Water Line

F. D. Bud Uzes[1]

Abstract

The seasonal landward limit reached by the ambulatory mean high water line is now an important element in regulating public property interests in the California coastal zone. Because the mean high water line has been generally treated as the line separating public tidelands from upland property, it has special significance to state agencies concerned with protecting the public's right to use the tidelands.

Historic surveys seldom show the mean high water line at its seasonal landward limit because they were not made at the time that condition existed. Some past use has been made of historic vertical aerial photographs to determine an approximate location of the line. These efforts have usually considered the border of wet sand as representative of the approximate mean high water line.

The method presented here also relies upon historic aerial photographs. It differs from the former procedure by applying corrections for tide height and wave uprush to increase the accuracy of the result.

Introduction

Along much of California's coastal shoreline the littoral processes are not directly affected by works of man. In these areas public agencies often need to know both the present and landward seasonal positions of the mean high water line to better oversee the public

---

[1] Principal and owner, Boundaries Unlimited, 6840 Pine Cone Rd., Loomis, CA 95650

tidelands. The present location is needed for everyday property boundary purposes. The landward limit of the mean high water line is needed to identify property that is tidelands for part of the time. This is to insure that development does not interfere with public use of tidelands during any portion of the year. Typically the landward limit is found during the winter season after larger waves have eroded the beach. The landward limit of the mean high water line is virtually impossible to determine without a field survey made when the seasonal event occurs, but photographic evidence can be used to show its approximate position.

Locating the Mean High Water Line directly by survey

The present location of the mean high water line can be readily determined by field survey. This involves locating an invisible line on the beach surface that has the same elevation as the tidal height of mean high water. This procedure involves working with (1) measured tide heights, (2) points having known horizontal positions relative to a reference system, and (3) bench marks of known elevation relative to tide heights.

Tidal measurements were first conducted in California in 1852 although the earliest work was not continuous. Tide records now used for boundary purposes are based upon 19 years of measurement. If the nearby tide stations lack that duration of measurement, data from those stations is extrapolated to 19-year mean values by comparison with long-term stations. Tide height information is usually obtained from the National Ocean Survey in Silver Spring, Maryland.

Tide heights along the California coast vary in elevation depending upon location. In determining the height of mean high water at a site located some distance from a tide station, the common practice is to interpolate between the nearest upcoast and downcoast stations.

Points of known horizontal position are typically found in the form of subdivision boundary markers or other like monuments. It is often a good idea when surveying large areas, and sometimes required by regulatory agencies, that these points be connected by survey to the national geodetic control network.

Bench marks are survey monuments of known elevations that are often located along highways and railroad lines. They are used by surveyors to transfer elevations to the beaches being surveyed. This enables surveyors to find points on the beaches that correspond in height to the elevation of the mean high water line.

Locating the landward limit of the mean high water line is performed when the tide line can be expected to be at that position. At other times one could expect to wait several months before performing a survey of the landward line. It should also be borne in mind that the landward limit of the mean high water line may vary from year to year. When the information is needed immediately, one can use past surveys to aid in the determination if they are available and show the tide line in a landward position. Oftentimes, however, there are no adequate prior surveys to show the location of the landward mean high water line. In such instances, resort must be had to alternative techniques that offer a reasonably accurate approximation of the line.

Locating the Mean High Water Line other than by survey

This paper describes a technique that was recently used to determine a number of approximate locations of the mean high water line along a 1,000 feet-long coastal beach in Southern California. The procedure involved using historic single-frame aerial photographs gathered from several government and commercial sources. Measurements were made using the photographs that incorporated uprush and backwash limits, beach slope, and tide heights.

The locations of the approximate mean high water lines developed by this technique were in this instance upheld under both state administrative review and a challenge in Superior Court. The case is presently on appeal.

The setting

A proposal to build homes on a one-thousand feet long stretch of sand beach in western Los Angeles County resulted in several coastal studies. These included design work for shore protection and determining the landward reach of the mean high water line. Early studies furnished by the builder concluded that all of the proposed homes were going to be located upland of the landward locations of the tide line. This finding was based upon statistical information derived from summer surveys as no winter surveys showing a landward position of the line were available. Regulatory agencies were being asked to act based upon this projected information.

Prior to action by the agencies, one additional study was requested that would focus on determining the landward location of the tide line. As a result, further research was conducted for existing surveys showing the feature. When this effort failed, resort was made to obtaining the desired information from secondary sources.

The use of ancillary data always raises a concern about accuracy. While data meeting modern survey standards is highly desirable, courts have readily accepted historic shoreline surveys prepared by the U.S. Coast & Geodetic Survey. These late-19th and early-20th century surveys are stated by the government to be accurate to within 10 meters, or about 33 feet. Because of its prior acceptance in shoreline boundary determination, that figure was adopted as the accuracy limit for this study.

The only available resource information other than surveys are aerial photographs. These materials are commonly used by oceanographic researchers for locating approximate mean high water lines, but concern was felt about the accuracy of the methods employed in those situations. They typically adopt the wet sand line as the mean high water line without correcting for tide height or wave uprush. It was decided to pursue the use of aerial photographs to determine if suitable results could be obtained by using more sophisticated techniques.

Use of aerial photographs

The project then focused upon getting photographs of the site for all seasons, but particularly including winter views. A total of 29 aerials dating between 1946 and 1989 were collected from a combination of government and private sources. The types of prints obtained included black and white, natural color, and color infrared. The scales of the various prints also differed, ranging from about 1" = 200 feet to about 1" = 1,000 feet, although including one print at about 1" = 2,000 feet. Enlargement prints were made from the original copies from time to time as needs arose.

The next order of business involved making measurements from the prints. Single-frame photographs can be difficult to measure on because they have no single scale. Differences in ground elevations of the various features can cause havoc in scaling. The saving factor in this instance is working along a relatively flat sandy beach where scales are more stable. One added strength in this case was the existence of ample survey control points that permitted many cross-checks for scale uniformity. In this instance a very detailed large-scale photogrammetric survey had previously been done that showed the positions of many shoreline rocks on the beach and just offshore.

These rocks were identifiable on the photographs and were used to determine the working scales of the photographs.

Once the framework for scale control was set up, the work continued to the interpretation of features on the photographs. The particular ones needed were the limits of uprush and backwash. These features bracket the section of the shore called the beach face. They were used to locate the theoretical intersection of the beach surface with the ocean at rest at the time of the photograph. The ocean at rest corresponds to the instantaneous tide height of the sea. The intersection line of the beach surface with the ocean at rest was concluded to be midway between the two limits of uprush and backwash. This decision on where the ocean at rest would most likely intersect the beach was made on the basis of earlier personal observations and recent consultation with an expert in oceanography. It was learned from studies conducted at the U.S. Army Corps of Engineers' Waterways Experiment Station in Vicksburg that the actual intersection ranged within the middle 40% of the beach face, but the mid-line was a good single working position. Error analysis would consider the entire range of possible intersections, however. Later in the study other researchers contended that the ocean at rest would intersect the beach face within the middle 56% of the beach face, so this range was also included in the error analysis study.

One other element in data preparation was determining the tide height at the instant the photographs were taken. This involved two factors, (1) the date and time of the photograph, and (2) the tide height corresponding to that instant of time. The hour is sometimes shown directly on the margin of the photograph, typically in the form of a clock face with hands. One must be careful to check that the clock appears to be running, as an unlikely time might indicate it was stopped. If the time is unknown and cannot be found from the source of the photograph, resort must be made to estimating time. This can be done by using shadows from vertical features on the photograph such as power poles, flag poles, edges of buildings, etc.

The directions of shadow edges relative to true north can be used to estimate time. Care must again be used considering the scale distortion of photographs. Actual calculation of the time of the photograph is done using standard formulas in navigation and surveying astronomy. These will not be further detailed here. The estimation of time by shadows was similarly done on photographs of known time to determine the possible inaccuracy of the procedure. Comparisons showed that all estimates were within one hour of being correct. This disparity was then used in conducting the error analysis.

Knowing the time each photograph was taken leads directly to finding the tide height. Because of the delay in getting tide height data from the government, initial measurements used estimated heights. The official tide tables published by the National Ocean Survey provide formulas and tables for making these estimates. Once the actual measured heights were known, they were substituted for the estimates.

The technique used for estimating the mean high water line location from photographs involves making offset corrections from the invisible ocean height line. After positioning that feature as described above, corrections are made for the differences between the tide height of mean high water (4.7' above mean lower low water at this location) and the individual ocean heights determined for each photograph. This is done by relating the vertical differences in height to their horizontal components based upon the slope of the beach. Deciding on an appropriate slope value is another of the tasks of this procedure. The slope value decided upon would directly impact the accuracy of the result.

The problem is knowing that the slopes on all the photographs could differ depending upon oceanographic factors. Furthermore, there would be no way to determine what the actual slope on each photograph was. The only solution was to estimate a working slope as closely as possible, and consider the amount of error introduced depending on the range of possible slopes for this beach.

It would have helped to first measure a number of slopes at the project site but there was no provision in this study for doing a field survey. Values were therefore determined using measurements previously made on nearby beaches of similar configuration. These sources yielded slopes ranging from 1 in 7 to 1 in 19. A working slope of 1 in 12 was finally adopted for this work, with error analysis considering the full range of slopes that were encountered. Later field measurements determined an average beach slope at the site of 1 in 13, confirming the reasonableness of the value used in the photographic study.

The initial solution of mean high water line location used the (1) ocean height at rest as being midway between the lines of uprush and backwash, (2) tide heights for individual photographs based upon times either estimated or known, (3) mean high water elevation of 4.7 feet above mean lower low water, and (4) beach slopes of 1 in 12. With this data it was found that the approximate mean high water line overlapped into the proposed home building area as determined from 7 of the 29 aerial photographs. In some instances the overlap exceeded 40 feet. The focus then shifted to these 7 instances for additional study and testing of the results.

All the error estimates were brought into consideration in the continuing analysis of the 7 photographs showing overlap. Ocean height intersections with the beach corresponding to the 40% and 56% mid-zones of the beach face were alternatively used, as were slopes ranging from 1 in 7 to 1 in 19. The one hour differences in estimating time from shadows were also applied for differing tide heights. Eventually all photographs with estimated time were withdrawn from consideration because of outside contentions they were unsuitable for use. Even when applying all of the possible error factors in combination so as to be most favorable to the developer, the tide line still overlapped into the project area in two instances. Based upon these results, a conclusion was reached that the mean high water line has at times been located inside the area proposed for building. Eventually this and other factors resulted in the denial of building permits to the developer.

Accuracy of results

It would be helpful if there were some objective means for determining the accuracy of the results of this study. In constantly making checks of the scaled measurements, and considering the possible sources of error and their effect, it was felt that this method gave results consistent with the old government shoreline surveys that were set as a standard. For individual photographs it is believed that most determinations were better than the 10-meter limit although a few may have approached it. While not a true indicator of individual accuracy, one test was devised to provide some measure of insight into the quality of the result. This involved comparative results with field surveys.

Many field surveys of the mean high water line were made following the photographic study. In all, about 33 shoreline surveys were performed over the next few years. These confirmed the finding by this study that the mean high water line at times overlapped well into the proposed building area. The overall survey results not only upheld the conclusions reached in the photographic study, but also showed its results to be generally conservative. Two-thirds of the field surveys showed the mean high water line extending into the proposed building area while only about one-fourth of the photographs showed this outcome. The results of individual surveys, however, cannot be compared with corresponding photographic determinations because there is no way to match beach conditions.

The sole accuracy test consisted of comparing extreme results produced by the two systems. Four composite lines were drawn showing the most landward and seaward positions of the mean high water line based upon both systems of measurement. These lines are combinations of mean high water line segments taken from various years, and everywhere depict the local extreme positions known for that feature. On the average the landward and seaward limits were about 100 feet apart. Using measurements along the beach at intervals of 60 feet, it was found that the difference between the photographic and surveyed locations of the mean high water line on the landward extreme averaged about 3 feet. On the seaward extreme the average difference is about 7 feet. This is a surprisingly close agreement between the surveyed and estimated results. It also supports the belief that the results fell within the self-imposed accuracy standard of 10 meters, or 33 feet.

Conclusion

The photographic technique was devised only when it became clear that needed information was not readily available and could not be developed by survey in a timely manner. Due care was exercised to control possible sources of error and limit their effect. The disputed conclusion that the mean high water line overlapped into the building area was later confirmed by numerous reliable field surveys. This photographic technique for determining an approximate location of the mean high water line is not of the same reliability as a field survey, but provides reasonable results. If adequate survey information does not exist this can be the only feasible alternative when seeking to determine historic locations of the mean high water line.

CHAPTER 124

LIMITATIONS OF "AVERAGING" COASTAL SURVEY DATA IN THE DETERMINATION OF COASTAL BOUNDARIES

Peter E. Gadd[1]

Introduction

The location of the coastal boundary in California separates upland private land ownership from publicly owned tidelands.

California law defines the private land – public tidelands boundary as follows:

> ". . . the owner of the upland, when it borders on tidewater, takes to ordinary high-water mark . . .".
>
> *California Civil Code, Section 830*

The term "ordinary high water mark" has been declared to be the "mean high tide line" (MHTL). The MHTL is the intersection of the plane of Mean High Tide elevation and the shore.

To precisely establish the elevation of mean high tide, all high tides must be measured over a Tidal Epoch, an 18.6 year period during which all the complex interactions of the moon, sun, and earth that effect the tides occur. The elevation of Mean High Tide during the most recent Tidal Epoch (1960-1978) has been determined by the U.S. Department of Commerce, NOAA, using tide gauge measurements located along the Southern California coast. The coastal boundary at a given time may be established by surveying the elevation of Mean High Tide at a given location.

On a sandy beach, the location of the Mean High Tide Line (MHTL) can move significantly in response to the configuration of sand on the beach. Given this condition, the coastal boundary is not fixed, but fluctuates with changes in sand profile.

[1] Principal, Coastal Frontiers Corporation

> " . . . a boundary marked by a water line is a shifting boundary, going landward with erosion and waterward with accretion . . . "
>
> *City of Berkeley vs. Buteau (1919)*

Figure 1 indicates the fluctuation of the MHTL (and the coastal boundary) as a result of coastal erosion (typically a winter phenomenon) and coastal sand accretion (typically a summer process). In many areas of the Southern California beach environment, the coastal boundary can shift horizontally by 100 feet or more in response to the shifting sands of the beach.

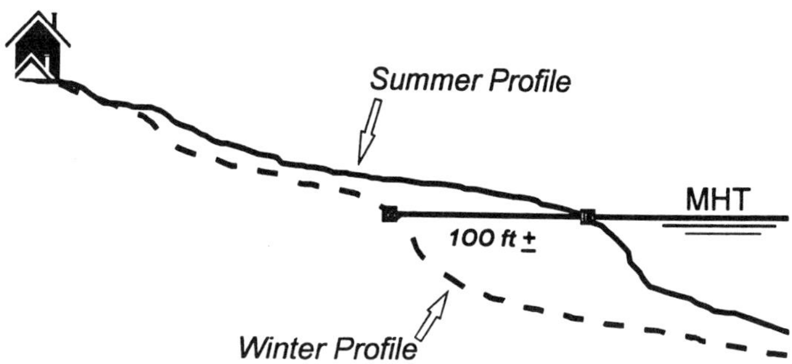

Figure 1: MHTL Boundary Fluctuation Caused by Erosion/Accretion

Recently, however, a decision by the Los Angeles County Superior Court sought to fix the position of a coastal boundary in a case related to the development of residential properties on Lechuza Beach, within the City of Malibu. One finding of the Court provided that:

> " . . . to remediate future disputes regarding boundaries, . . . survey to determine the average MHTL for the previous 2 years, encompassing at least one survey for each previous season (winter, spring, summer, fall) . . ."
>
> *Case No. BC0768555, Superior Court County of Los Angeles, 11 June 1996*

The goal of this presentation is to illustrate the consequences of the recent Superior Court verdict (currently under appeal) and to indicate the limitations of averaging coastal survey data to determine the boundary between private and public land.

Coastal Setting

Lechuza Beach is a narrow sand beach backed by a coastal bluff, located north of Zuma Beach, in the City of Malibu. An established residential community was developed at Lechuza Beach beginning in the early 1930's. The coastal road at this beach, Sea Level Drive, originally ran the length of the beach but the western portion of the road was destroyed by a storm in the late 1930's.

Presently, a developer has proposed to reconstruct the formerly washed-out portion of Sea Level Drive and construct a total of at least 12 new homes at the site. In addition, a 1,000 foot long rock revetment is planned to protect the homes from storm wave impacts.

When the plan of development was presented to local and State permitting authorities, a determination of the location of the seaward private property boundary was required. The State Lands Commission believed that the MHTL moved sufficiently landward during eroded beach periods that the development project, as proposed, would encroach on State tidelands. Historical survey data indicating the fluctuating location of the MHTL at the site included only three surveys conducted in 1928, 1932, and 1951. To augment this data, frequent surveys of the MHTL were undertaken along the proposed project frontage during the 1990-1997 period yielding a total of 43 MHTL surveys of the beach. This data served to develop a previously unavailable knowledge of the fluctuations of the location of the MHTL along the project frontage.

Survey Data

The surveys of the MHTL at Lechuza Beach span the 1928-1997 period, with 40 of the 43 total surveys having been performed since 1990. As would be expected, there exists a substantial distance (100+ feet) between the most landward MHTL and the most seaward MHTL. Figure 2 presents a time history of the fluctuating position of the MHTL at a given transect within the project area during the 1990-1997 time period. The vertical scale is the distance in feet from the fixed northern lot line while the horizontal axis displays the time period. As is evident, the summer months typically experience a seaward migration of the MHTL, although that is not the case every summer (note the summers of 1994 and 1995). Likewise, the winter months typically experience beach recession, although this is not the case every winter (the winter of 1990, for example). Previous California law would indicate that the MHTL fluctuations illustrated in Figure 2 also represent the fluctuations in the public-private property boundary throughout this period.

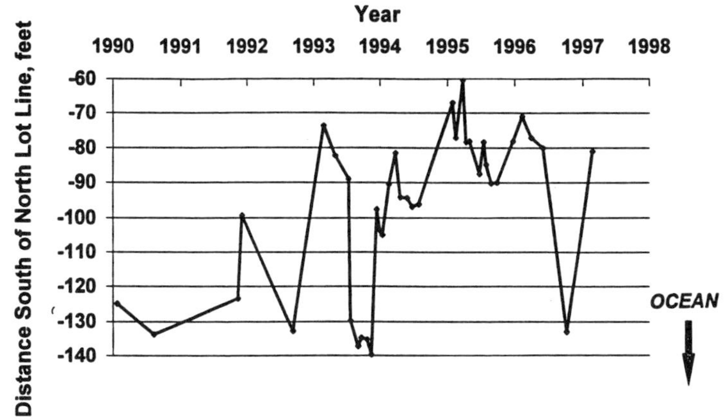

Figure 2: Historical MHTL Positions, Lechuza Beach, 1990-1997

Under its view of the law, the trial court determined that it must establish a permanently fixed "average" boundary line. The decision of the Superior Court was to average all surveys available (at the time, 38 surveys had been conducted), to determine the coastal boundary within the project area. The position of this average MHTL is identified as the "Court" line in Figure 3. In this figure, the vertical axis is the distance (in feet) from the north lot line and the horizontal axis is the distance from the west end of the project site. Subsequent to the Court decision, five additional MHTL surveys were performed. These surveys were averaged with the original 38 surveys used by the Court and the resulting "average" MHTL moved as much as three feet. This second "average" MHTL encompassing 43 surveys is identified as "Ave #1" in Figure 3.

The decision of the Superior Court acknowledged that not all coastal properties would be able to draw upon such a prolific survey data base as was used at Lechuza Beach. In fact, historical MHTL surveys are conducted very infrequently in California. This project area had been surveyed only three previous times prior to the survey activities associated with this litigation. Therefore, to determine the boundaries of coastal property in the future, the Court required that there be a total of eight conducted during each season (winter, spring, summer, fall) of the two previous years.

In Figure 3, an effort was made to establish average MHTL positions using the eight survey protocol of the Court. Because there are a total of 43 surveys spanning each of the four seasons, multiple combinations of eight surveys spanning two consecutive years are possible. These are indicated as "Ave 2, Ave 3, and Ave 4" in Figure 3. It is clear at a given site that the selection of eight MHTL surveys can result in substantial discrepancies in the "average" MHTL boundary—at this site, variation of the average MHTL location measures 20 feet or more. It is apparent that use of a MHTL "average" position to establish a coastal boundary is arbitrary—based entirely upon the conditions that exist when the survey data is collected. By the proper timing of survey activities, the boundary can be skewed in favor of the State (by concentrating the survey effort during landward extremes of the MHTL locations) or the private property owner (by surveying entirely at times of seaward extremes of the MHTL). Figure 2 indicates that requiring seasonal (fall, winter, spring, summer) surveys will not create consistent "average" MHTL positions.

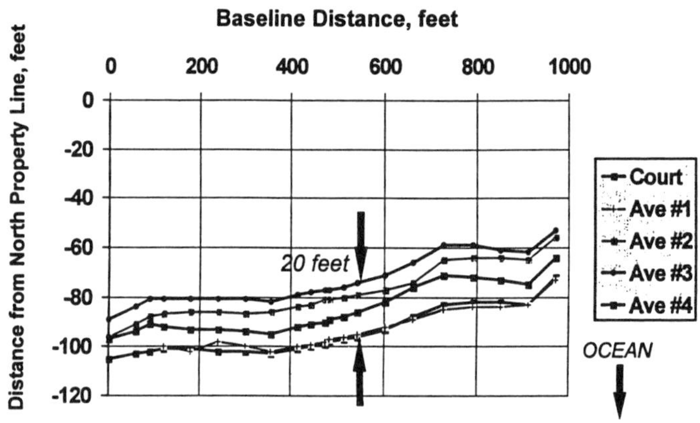

Figure 3: Boundary Fluctuation With Choice of "Average" MHTL Position

The long-term MHTL fluctuations at Lechuza Beach are shown in Figure 4, for the 1928-1997 time period. The wealth of data collected during the 1990's is evident. Also evident is the erosional trend of the beach that, when subjected to linear regression analysis, indicates that the beach along this transect is experiencing long-term erosion of about one foot per year. Any proposed "averaging" technique to establish a coastal property line must take into account the frailty of establishing such an average on a beach experiencing long-term erosion. The "average" MHTL position of the 1920's is clearly not the "average" position of the 1990's. Diminishment of the coastal sediment supply and rising sea level indicate that the determination of an "average" MHTL is a time dependent process, and not subject to long-term stability.

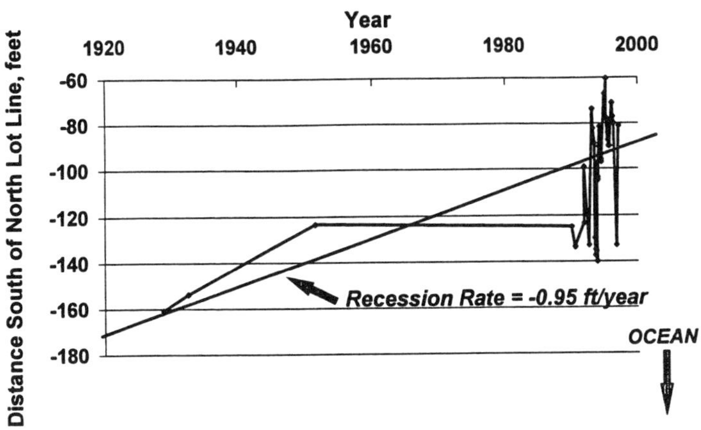

Figure 4: Historical MHTL Positions, Lechuza Beach, 1928-1997

The averaging methodology proposed by the Court would allow individual property owners to create their own seaward property lines. At Lechuza Beach, this is now being proposed for a 1,000 foot beach frontage. However, consider the possibility of 30 individual property owners having adjacent 30 foot wide lots establishing 30 different coastal property lines based on the "average" MHTL determined by each owner's eight surveys. The resulting boundary would be an alongshore saw-tooth pattern, as shown in Figure 5. A fixed coastal boundary could possibly be fenced or improvements (decks, barbecue pits, etc.) could be placed on the beach in a haphazard configuration conforming to such a disjointed property boundary.

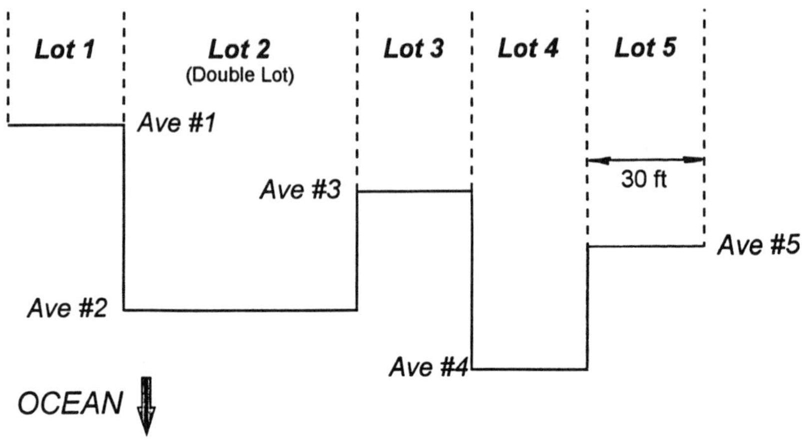

Figure 5: Disjointed Coastal Boundary Based on Individual Selection of "Average" MHTL Location

Conclusions

The rule that the MHTL is the seaward boundary of coastal property in California has been challenged in the Superior Court of Los Angeles County. An alternate boundary determination method utilizes the "average" position of the MHTL based on (at the least) eight MHTL surveys collected during each of the four seasons within the previous two year period. Averaging the MHTL data in this way invites either the private property owner or the State to select survey times that will be advantageous to them (the landwardmost positions from the State standpoint, and the seawardmost positions from the property owners point of view). Substantial variation in the "average" MHTL boundary can result from the selection of different survey data bases. A disjointed boundary, changing at each lot line (lot widths in Malibu can be as narrow as 30 feet) results from this boundary determination method.

Selection of an "average" MHTL boundary is particularly disruptive of beach use during periods of eroded beach conditions. Such periods, which typically occur during the winter months, can cause the public beach access to be inundated by high tide and wave run-up. Under severe conditions, the lateral beach access that is confined seaward of an arbitrary "average"

MHTL may be impassable to all but those who would choose to swim along the beach. Should private property development encroach outward onto the beach environment (onto beaches that in many areas of California are experiencing both long-term erosional trends and the effects of sea level rise), the residential properties will be exposed to the damaging effects of energetic storm wave impacts. An arbitrary selection of the coastal property boundary by averaging positions of the MHTL based on limited survey data can adversely affect the public right to safe coastal access and recreational opportunities along the sandy shore.

CHAPTER 125

THE MMS OPA '90 IMPLEMENTATION PLAN FOR THE PACIFIC OCS REGION

J. Manuel Saenz, Craig K. Ogawa, James R. Grant[1]

Abstract

The Exxon Valdez oil spill in March 1989 was the catalyst for new environmental legislation which included the Oil Pollution Act of 1990 (OPA '90). Prior to the passage of OPA '90, the Minerals Management Service (MMS) had jurisdiction for spill response plans and prevention for oil and gas facilities on the Outer Continental Shelf (OCS), but not in State waters. Under OPA '90 and Executive Order (EO) 12777, the MMS authority for oil spill prevention and response has expanded to include all offshore facilities except for those associated with deepwater ports. MMS was also given authority for financial responsibility of offshore facilities under OPA '90 and its 1996 amendments.

To ensure that the responsibilities delegated to the MMS under OPA '90 for all offshore oil and gas facilities are implemented, the Pacific OCS Region (POCSR) of the MMS has embarked on the development of an implementation plan. Additionally, the OPA '90 Implementation Plan emphasizes components of several Memorandums of Agreement and Understanding between the MMS and other agencies to ensure a coordinated oil spill prevention and response program. This coordination provides for, among other measures, the sharing of technical expertise in drilling, production, pollution prevention, and other related "specialty" areas of offshore operations and safety.

This paper discusses several critical efforts undertaken by the POCSR to develop a well-coordinated oil spill prevention and response program under OPA '90 for offshore California. Some of these efforts include a joint review of oil spill

---

[1]U. S. Department of the Interior, Mineral Management Service, Plans, Platforms, and Pipelines Sections, 770 Paseo Camarillo , Camarillo, CA 93010

contingency plans, the initiation of a series of systemic inspections entitled "Focused Facility Inspections (FFI)," coordinated oil-spill drills with other agencies, and dissemination on the latest oil spill response research and technologies.

Introduction

The POCSR OPA '90 Implementation Plan is designed to ensure that the POCSR oil spill prevention program includes all regulatory responsibilities mandated to the MMS by OPA '90 and subsequent delegation from EO 12777 for offshore oil and gas facilities. The Plan augments the POCSR's present program by integrating MMS's new responsibilities under OPA '90, and Memorandum of Understanding and/or Agreement with other agencies, with existing programs and activities in the POCSR. The Implementation Plan will also provide the POCSR (District and Region) personnel with guidelines for fulfilling the requirements and responsibilities delegated to MMS under OPA '90.

The Implementation Plan outlines MMS's current requirements for spill prevention and containment, response plans, drills and inspections, training and issues regarding oil-spill prevention and response research, financial responsibility, and civil penalties. The Implementation Plan also discusses new requirements to prevent and contain discharges of oil and hazardous substances from offshore facilities, including associated pipelines.

Federal Prevention, Containment, and Removal Authority

Requirements:

Section 4201(c)(1)(A) of OPA '90 states that the Federal Removal Authority mandated by "the President shall in accordance with the National Contingency Plan and any appropriate Area Contingency Plan, ensure effective and immediate removal of a discharge, and mitigation or prevention of a substantial threat of a discharge of oil or a hazardous substance--

(I) into or on the navigable waters;

(ii) on the adjoining shorelines to the navigable waters;

(iii) into or on the waters of the exclusive economic zone; or

(iv) that may effect natural resources belonging to, or appertaining to or under the exclusive management authority of the United States."

Section 4201(c)(2)(A) entitled Discharge Posing Substantial Threat To Public Health Or Welfare - states that "if a discharge or substantial threat of a discharge, of oil or a hazardous substance from a vessel, offshore facility, or onshore facility is of such size or character as to be a substantial threat to the public health or welfare of the United States, the President shall direct all Federal, State, and private actions to remove the discharge or to mitigate or prevent the threat of the discharge."

Responsibilities:
Section 2(b)(3) of the EO 12777 states that "the functions vested in the President by Section 311(j)(1)(C) of the Federal Water Pollution Control Act (FWPCA), respecting the establishment of procedures, methods, and equipment and other requirements for equipment to prevent and contain discharges of oil and hazardous substances discharge from offshore facilities, including associated pipelines, other than deepwater ports subject to the Deepwater Ports Act (DPA), are delegated to the Secretary of the Interior."

Implementation:
The MMS has some 41 regulations (or parts of) that address and govern prevention, containment, and removal of discharges from OCS oil and gas facilities. Examples of regulations that address pollution prevention, inspections, training and drills, and oil spill contingency plan (OSCP) requirements can be found at the following quotations:

• 30 CFR 250.40 states that "all hydrocarbon-handling equipment ...such as separators, tanks, and treaters shall be designed, installed, operated and maintained to prevent pollution. Additionally, curbs, gutters, drip pans, and drains shall be installed in decks areas...to collect all contaminants not authorized for discharge."

• 30 CFR 250.41 entitled Inspections and Reports states that "drilling and production facilities shall be inspected daily... to determine if pollution is occurring. It also states that pollution response equipment shall be inspected at least monthly. Additionally, all spills of oil and liquid pollution shall be reported."

• 30 CFR 250.43 entitled Training and Drills "requires OCS lessees to provide oil spill response teams with hands-on training annually. Additionally drills for familiarization of pollution-control equipment and operational procedures shall be held once every 12 months."

•30 CFR 250.56 entitled Blowout Preventer Systems and Systems Components states that "BOP systems... shall be designed, installed, used, maintained, and tested to assure well control."

• 30 CFR 250.121 states that "all tubing installations open to hydrocarbon-bearing zones shall be equipped with subsurface safety devices that will shut off the flow from the well."

• 30 CFR 254 requires that "all offshore facilities shall have submitted a spill-response plan. This section states that for operators of offshore facilities in Federal waters, OSCP's are to be submitted in accordance with 30 CFR 250.42. For offshore facilities located in State waters, this section gives clear direction for submittal of spill-response plans in accordance with this part."

The MMS has in place a number of programs to further fulfill Federal pollution prevention requirements:

• The MMS is meeting pollution prevention and containment requirements and responsibilities by implementing a new inspection program entitled FFI at platforms in the POCSR. The FFI covers all aspects of platform operations and management from a systemic perspective concentrating on the areas of facility condition, safety systems, documents, training, hazardous materials, electrical, policies/performance, and environmental aspects.

• The MMS Engineering and Research (E&R) Branch conducts research for equipment to prevent and contain discharges of oil and hazardous substances from offshore facilities. The E&R program implements and manages research to ensure that the Best Available and Safest Technologies are used by industry.

• The MMS is continuously evaluating its existing spill prevention regulations to ensure that the requirements of OPA '90 and applicable regulations and laws are fulfilled. MMS also reviews other Federal and State regulations to ensure consistency with the requirements of OPA '90 and other applicable laws and regulations.

• MMS personnel are currently participating in the Area Contingency Plan (ACP) committee with the USCG, the California Office of Oil Spill Prevention and Response (OSPR), other Federal and State agencies, industry, and other interested parties. Participation in the ACP committee allows for the information sharing regarding the availability of oil spill response, resources at risk, and ensures consistency between facility OSCPs and the ACPs.

Response Plans

Requirements:
Section 4202(b)(4) of the OPA states that the Secretary of the Interior is required to issue regulations requiring the owners or operators of offshore facilities, including associated pipelines, to prepare and submit response plans that ensure the availability of private spill response personnel and equipment and to permit the operation of offshore facilities, including associated pipelines, without approved response plans if certain conditions are met.

Responsibilities:
Section 2(d)(3) of EO 12777 states that "the functions vested in the President by Section 311(j)(1)(5) of the FWPCA and Section 4202(b)(4) of OPA requiring the owners or operators of offshore facilities, including pipelines, other than deepwater ports subject to the DPA, to prepare and submit response plans, the approval of means to ensure the availability of private personnel and equipment, the review and approval of such response plans and the authorization of offshore facilities, including

associated pipelines, other than deepwater ports subject to the DPA to operate without approved response plans, are delegated to the Secretary of the Interior."

Implementation:
Examples of the currently established MMS requirements for owners or operators of offshore facilities to prepare and submit response plans, the approval of means to ensure the availability of private personnel and equipment, and the review and approval of such response plans can be found in regulation and internal practices.

• 30 CFR 254 describes oil-spill response plan requirements for offshore facilities in Federal and State waters. 30 CFR 254 states that "Spill-response plans submitted to the MMS shall be reviewed and updated annually, with all modifications submitted to the MMS office of original submission. The spill-response plans originally submitted to a state shall be updated in accordance with the requirements of the State."

• The MMS, POCSR has developed guidelines to aid owners and operators of facilities in the preparation of oil spill contingency plans. All OSCP's submitted for approval to the MMS must conform to the guidelines.

• Some OPA '90 requirements for state water facilities are being met by implementing the Memorandum of Agreement (MOA) between the MMS and OSPR. This MOA states that the agencies shall cooperate in the coordination and implementation of research and other informational programs of mutual benefit to both organizations. The agencies will work to ensure State plans and policies for marine environmental protection are consistent with the NCP; the parties agree to consult with each other to enhance Area Contingency Plans; to streamline and avoid duplication of efforts during the OSCP review process; and to conduct joint inspection and oil spill drills.

• The MMS and USCG MOU allows for the joint review of exploration plans, development and production plans, oil spill contingency plans, and applications for permits to drill as necessary for prompt and efficient exploration, development, and production of a lease area.

• Currently, under the MMS and OSPR MOA, the agencies are finalizing processes for joint review of OSCP for state and federal facilities.

• The MOU between the MMS and USCG is currently being revised to meet oil pollution prevention responsibilities so as to ensure a technical review of the OSCPs under OPA '90.

Drills and Inspections

Requirements:
Section 4202(a)(6)(A) of OPA '90 requires that "periodic inspection of containment booms and skimmers, vessels, and other major equipment used to remove discharges."

Responsibilities:
Section 2(e)(3) of the EO 12777 states that "the functions vested in the President by "Section 311(j)(6)(A) of the FWPCA, respecting the requirements for periodic inspections of containment booms and equipment used to remove discharges at offshore facilities, including associated pipelines, other than deepwater ports subject to the DPA, are delegated to the Secretary of the Interior."

Implementation:
Below are some examples of the currently established the MMS requirements for periodic inspection of containment booms and equipment used to remove discharges at offshore facilities:

• Exercise requirements are set forth by the National Preparedness for Response Exercise Program (PREP) guidelines. The MMS has determined that the proposed requirements for tabletop drills for spill management team satisfy the purpose and goal of the Act's requirement that the response plan describe the periodic unannounced exercise to be carried out under the plan. The tabletop exercises will involve owner or operator personnel who make decisions and organize the response to a spill. These personnel must be exercised using a spill scenario that is unannounced prior to the exercise. The MMS will also periodically initiate unannounced drills to test the preparedness of owners and operators.

• Two unannounced major oil-spill drills are conducted annually by the MMS involving a simulated spill of over 100 barrels of oil. These major spill drills are designed to test the operator's response at the facility and command center which are established by the operator for large oil spills. Representatives from the MMS and other Federal and State agencies are at the platform and the command center to observe response to the spill scenario.

• MMS initiated 22 unannounced oil-spill drills during the period of September 1995 through September 1996 in the Pacific OCS Region. These unannounced drills involved simulated spills up to 15 barrels of oil or less. The oil-spill drills are complemented with an unannounced pollution prevention inspection in which oil-spill equipment (i.e., booms and skimmers) and records (i.e., exercise and pollution records) are reviewed and inspected.

• The MOU between the MMS and USCG is currently being revised to meet oil pollution prevention responsibilities under OPA '90. Each agency will provide the other agency with assistance and technical support, including transportation. Such exchange of services and use of personnel will benefit the application and enforcement of safety regulations. This will enable the MMS and USCG to conduct scheduled and unannounced inspections, in a cooperative effort, to ensure compliance with Federal requirements.

• The MOA between the MMS and OSPR is currently being implemented to meet oil pollution prevention responsibilities under OPA '90. Each agency will provide the other agency with assistance and technical support, including transportation. Such exchange of services and use of personnel will benefit the application and enforcement of safety regulations. This will enable the MMS and OSPR to conduct scheduled and unannounced oil spill drills and inspections, in a cooperative effort, to ensure compliance with Federal and State requirements, and to avoid and minimize duplication of efforts.

Oil Pollution Research and Training

Requirements:
Section 7001(a)(3) of OPA '90 mandates that "the Interagency Committee shall include representatives from... the Department of the Interior (including the Minerals Management Service...)..."

Section 7001(c)(1) of OPA '90 states that "the Interagency Committee shall coordinate the establishment, by the agencies represented on the Interagency Committee, a program for conducting oil pollution research and development, as provided in this subsection."

Responsibilities:
• Oil Spill Research -- MMS currently operates the Offshore Hazardous Material Simulated Environmental Test Tank (OHMSETT) test facility. The MMS POCSR role in oil spill research has been to evaluate potential projects voicing regional needs for research and disseminating oil-spill response research and data to the user community in the POCSR. For some proposals, the process will enable the research projects selected for MMS funding to gain acceptance and to be evaluated at OHMSETT.

Implementation:
The MMS facilitates or assists in the comprehensive programs for oil pollution research and technology in a number of ways. In addition, examples of the newly established or proposed MMS programs for oil pollution research, technology and training include:

• Laser Environmental Airborne Fluorosensor (LEAF) -- The MMS funded research has developed and successfully flight tested prototypes of new generation oil spill detection technologies to remotely sense, detect, and quantify spilled oil in day/night conditions. The LEAF system incorporation of new measures to remotely measure the thickness of oil slicks will allow rapid and efficient deployment of response assets to the optimum locations for clean up (i.e., 80% of the volume of an oil slick is contained in 20% of the total surface area). Operational tests of the LEAF system were conducted in January 1997 over the oil seeps, offshore Santa Barbara, CA. Future operational tests are operational in Santa Barbara County for early 1998.

• Dispersants -- The MMS research on the use of chemical treating agents (e.g. Dispersants) is concentrated on improving understanding of how these products interact chemically with different types of oil.

• *In-situ* Burns -- Results of the MMS funded research on *in-situ* burning (the controlled burning of oil slicks on the waters surface) has shown this technique to be an effective and efficient response option. *In-situ* burning can remove large quantities of oil from the waters surface quickly, before the oil can impact coastal resources. The States of California, Oregon, and Washington are developing pre-approval plans for the use of in situ burning as an option to respond to mechanical cleanup of offshore oil-spills.

•The MMS E&R Branch conducts research for equipment to prevent and contain discharges of oil and hazardous substances from offshore facilities. The E&R Program implements and manages research to ensure that the Best Available and Safest Technologies are used by industry. The E&R Program is divided into four subsets: operational safety; structures and pipelines integrity; air pollution control; and oil spill response.

• Dissemination of Oil Spill Research Information -- The MMS has held several oil-spill information dissemination meetings in Camarillo, CA and at the Regional Response Team meetings. The information was used to update attendees on MMS Oil Spill Response Research and to elicit feedback for current and future research from other agencies. The POCSR has been disseminating MMS oil spill research journals and articles by mail, by E-mail, and by the Internet access.

• Training Programs -- The MMS has a number of regulations that address and govern training performance standards for operators from OCS oil and gas facilities. Examples of regulations that address training and drills for pollution prevention, and inspections requirements can be found at the following quotations:

• 30 CFR 250.210 requires that the "Lessees and contractor employees engaged in drilling, well-completion, well-workover, or production operations in the OCS shall be trained in the proper operation of equipment, methods of operation, and techniques to avoid hazards to people and property and to prevent pollution to the environment ."

• 30 CFR 250.43 (a) requires that Lessees conducting oil, gas, oil and gas, sulphur operations in the OCS shall ensure that the oil spill response team is provided with hands-on training classes at least annually in the deployment and operation of pollution control equipment to which they are assigned."

The MMS has in place a number of programs to further fulfill Federal training requirements:

• MMS recognizes the potential hazards to health and safety associated with field inspections and possible oil spill activities at the POCSR platforms. It is the intent of the MMS to establish a health and safety program to supply MMS personnel engaged in on-site activities associated with field activities a level of health and safety training consistent with an employees job functions and responsibilities.

• The MMS POCSR has conducted 40-Hour, 24-Hour HAZWOPER certification, and 8-Hour recertification courses for its personnel to fulfill Federal requirements. The MMS will also conduct a HAZCOM course in accordance with Federal regulations for Fiscal Year 1997. In addition, MMS personnel attend the oil-spill response training provided by the operator through the co-ops (i.e., Clean Seas and Clean Coastal Waters).

Financial Responsibility

Requirements:
Section 1016 (e) of OPA '90 mandates that "the financial responsibility under this section may be established by any one, or by any combination, of the following methods which the Secretary or the President determines to be acceptable: evidence of insurance, surety bond, guarantee, letter of credit, qualification as a self-insurer, or other evidence of financial responsibility."

Responsibilities:
Section 5 (a)(1) of the EO 12777 discusses that "the establishment of the issuance of regulations concerning financial responsibility, the determination of acceptable methods of financial responsibility, and the specification of necessary or unacceptable terms, conditions, or defenses."

Implementation:
• Currently, COFR regulations reside in 33 CFR 135. The MMS is developing new COFR regulations which meet the requirements of OPA '90. Recently, MMS published advanced notice of proposed rule making which is a precursor to the development of the new Certificate Of Financial Responsibility (COFR) regulations.

• The MMS has assumed COFR authority under OPA '90 for facilities from the USCG. In the POCSR, 16 parties hold COFR's for 26 leases, and 7 parties hold COFRs for specific facilities. The MMS is maintaining and managing a COFR data base of POCSR facilities. The POCSR ensures that operators or owners for facilities in the region have COFRs which are current. The POCSR has reviewed and commented on the preparation of draft OPA '90 COFR regulation prior to the promulgation of a final rule.

Civil Penalties

Requirements:
Section 8201 of OPA '90 amends Section 24(b)(2) of the Outer Continental Shelf Lands Act to (OCSLA) ensure civil penalties reflect current consumer prices and to ensure civil penalties can be assessed for violations that result in "a threat of serious irreparable, or immediate harm or damage to life (including fish or other aquatic life), property, any mineral deposit, or the marine, coastal, or the human environment, a civil penalty may be assessed without regard to the requirement of expiration of a period allowed for corrective action."

Responsibilities:
Section 5(b)(2)of the EO 12777 states that "the functions vested in the President by Section 4303 of OPA '90, respecting the assessment of civil penalties, the compromising, modification or remission, with or without condition, and the referral for collection of such imposed penalties, and requests to the Attorney General to secure necessary judicial relief are delegated to the Secretary of the Interior."

This section strengthened Section 24(b) of the OCSLA amendments of 1978. Subtitle B of OPA '90, entitled "Penalties," by increasing the amount of the civil penalty from a maximum of 10,000 to a maximum of 20,000 for each day of noncompliance. More importantly, in cases where a failure to comply with applicable regulations constitutes or constituted a threat of serious, irreparable, or immediate harm or damage to life (including fish or other aquatic life), property, any mineral deposit, or the marine, coastal, or the human environment, OPA '90 provided the Secretary of the Interior with the authority to assess a civil penalty without regard to the requirement of expiration of a period of time allowed for corrective action.

Implementation:
• 30 CFR 250.200 states that the Regional Director may appoint a reviewing officer to determine if civil penalties should be assessed for actions that may constitute a threat of serious, irreparable, or immediate harm or damage to life, property, any mineral deposit, or the marine, coastal, or human environment.

• The POCSR civil penalties program helps assure safe and clean operations through the pursuit, assessment, and collection of civil penalties. Six civil penalties were assessed in the POCSR since 1990, with four cases involving oil discharged into the ocean.

• The MMS maintains a nationwide data base of OCS accidents reported by lessees, operators, and permit holders which are due to oil, gas, and sulphur operations. The nationwide data base is called Offshore Pollution and Accident Compilation (OPAC). The Events File contains records of accidents occurring before January 1, 1990, and OPAC will contain records of accidents occurring on or after that date. All reported accidents are included in OPAC except minor injuries, i.e., those involving less than 16 hours lost time.

Conclusion

The goal to the MMS OPA '90 Implementation Plan is ensure that the POCSR implements the requirements under OPA '90 and EO 12777, and MMS regulations pertaining to oil spill prevention, preparedness, and response. To implement these new requirements, the POCSR is developing programs to meet these new responsibilities and has integrated them into existing programs and activities. The POCSR is also working closely with other Federal and State agencies. The partnerships with Federal and State agencies are essential to the Implementation Plan and MMS's POCSR's commitment to ensure safe and environmentally sound oil and gas operations in California Coastal waters.

References

Code of Federal Regulations, 30 CFR, Parts 200 to 699, Office of the Federal Register National Archives and Records Administration, revised 1996.

Executive Order 12777, Implementation of Section 311 of the Federal Water Pollution Control Act of October 18, 1972, as Amended, and the Oil Pollution Act of 1990, 1991.

Oil Pollution Act of 1990, Hundred First Congress of the United States of America, An Act, 1990.

CHAPTER 126

Navy Oil Spill Contingency Planning

John B. Locke[1]

Introduction

The United States Navy is committed to protecting the environment through various programs and policies that are integral to all operations. Maintaining an oil spill response contingency capable of providing a complete response to all oil spills which could originate from Navy ships and facilities worldwide is one of many environmental programs that have been established to protect the environment in which the Navy operates. The city of San Diego is located near a large natural harbor and is a major U.S. Navy port. San Diego is homeport for 69 Navy ships, representing nearly one-sixth of the Navy's entire fleet and one-third of the U.S. Pacific Fleet. Local coordination of Navy oil spill response assets and planning efforts are essential to the protection of San Diego Bay. Commander Naval Base San Diego (CNB-SD) assumes responsibility for oil spill response and planning as the Navy On-Scene Coordinator (NOSC) for an area which includes Mainland EPA Region IX (California, Arizona and Nevada) and territorial waters out to 12 nautical miles.

---

[1]Environmental Engineer, Navy On-Scene Coordinator - Planning, Commander Naval Base San Diego, 937 North Harbor Drive, San Diego, CA 92132

Navy On-Scene Coordinator

The Navy On-Scene Coordinator (NOSC) assumes the position of Responsible Party within the Unified Incident Command System for pollution responses which originate from Navy vessels or facilities. The NOSC acts as the Federal On-Scene Coordinator for hazardous substance releases occurring on Navy property. The NOSC assumes the position of Responsible Party's Qualified Individual possessing authority to commit Navy and contractual assets. There are fourteen NOSCs located throughout the country covering areas which are similar to the U.S. Environmental Protect Administration (EPA) Regions (Figure 1). The NOSC also develops policies and provides guidance to naval activities within their area of responsibility in the form of an Oil and Hazardous Substances (OHS) Contingency Plan (COMNAVBASESANDIEGO INST 5090.1B).

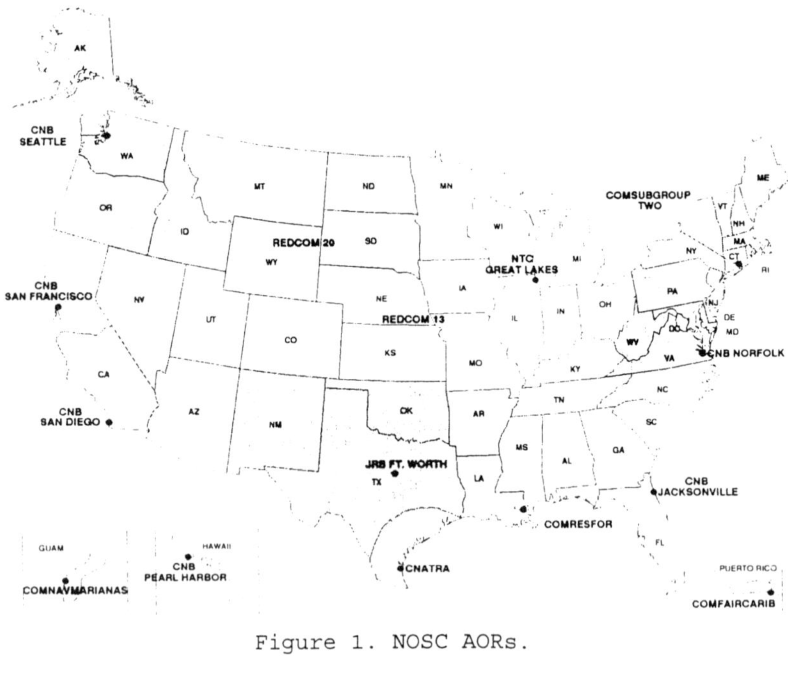

Figure 1. NOSC AORs.

The OHS Contingency Plan has become a standardized document for consistency throughout the naval community, while providing additional policies and information that is locally specific to the NOSC area of responsibility. The plan provides internal and external reporting requirements, local points of contact, Navy policies, training and drills documentation and appropriate jurisdictional boundaries.

The Navy's oil spill response strategy has been developed as a two tiered approach. Pre-established Oil Spill Operation Teams (OSOTs) fulfilling Tier I response are typically located within the base confines of predesignated Facility Incident Commanders (FICs). While Emergency Ship Salvage Material (ESSM) Bases, managed by Naval Sea Systems Command, Supervisor of Salvage, located at central locations throughout the world provide Tier II response.

Facility Incident Commander

The Facility Incident Commander (FIC), usually the Commanding Officer of the individual naval station, maintains an OSOT on call 24 hours a day with equipment and personnel capable of containing most probable spill volumes within their assigned areas of responsibility. Each OSOT maintains a full array of containment and recovery equipment which includes skimmers, large quantities of boom, various work boats and storage containers. The OSOTs are typically managed through the Ports Services Department of the station, performing reoccurring tasks that include booming of vessels in-port and offloading of oily waste. Facility Incident Commanders within the San Diego area include Naval Station San Diego, Naval Amphibious Base Coronado, Naval Air Station North Island, SUBASE and Fleet Industrial Supply Center San Diego.

Assets of each OSOT may be combined when responding to a spill that is beyond the capability of a single OSOT or is a time critical cleanup. Oil spill response operations which require the concerted effort of two or more OSOTs are coordinated by the designated Navy On-Scene Coordinator (NOSC) for the region. In the event that additional response assets are required and not locally available, the NOSC is empowered to activate SUPSALV assets and civilian contractor services through U.S. Coast Guard Basic Ordering Agreements. Naval Facilities Engineering Command provides equipment, equipment evaluation for procurement and planning support to the individual OSOTs.

Naval Facilities Engineering Service Center

Naval Facilities Engineering Service Center (NFESC) located in Port Hueneme, California a division of Naval Facilities Engineering Command (NAVFAC) has been created to provide engineering support to the Naval community. Within the discipline of oil spill response NFESC provides oil spill response training, compiles Navy oil and hazardous substance spill data, and maintains on-site OSOT equipment inventories. As a training provider, NFESC conducts Oil Spill Operations Teams Training and Spill Response Equipment Maintenance Training for OSOTs worldwide. This type of training gives OSOT personnel the required knowledge to maintain existing response equipment and respond to oil spills on the water in an efficient and safe manner. Maintaining world wide Navy oil spill data is a function of NFESC. Reduced data is published annually in the Naval Oil and Hazardous Substance Spills Annual Report.

The Annual Allowance and Requirements Review (AARR) Program was established to research pollution response equipment performance and design for issuance to the Oil Spill Operations Teams. NFESC manages the AARR Program by identifying OSOT equipment short falls. Shortfalls may be a requirement for additional skimmers, workboats or boom. Upon shortfall identification, NFESC may conduct research for procurement of existing equipment to fulfill the requirement or may have to design the equipment in-house. The equipment is then purchased or contracted for manufacture through NFESC and issued directly to the OSOTs.

Engineering Field Divisions

Naval Facilities Engineering Command, Southwest Division (SWDIV) provide contract services for access to United Coast Guard (USCG), Basic Ordering Agreements (BOAs). The BOA contracts are typically utilized for shoreline cleanup activities. Navy utilization of USCG BOAs has been made possible through a Memorandum of Understanding with the USCG. SWDIV also provides contractual support in the development of Facility Response Plans that are submitted for approval by various federal and state regulators.

Supervisor of Salvage

Naval Sea Systems Command, Supervisor of Salvage (SUPSALV) provides technical support, training and equipment for spills which exceed local (OSOT) response assets. Technical support includes the development, testing, and procurement of new equipment and systems; diving system inspection and certification; oil spill contingency planning; and other related functions. SUPSALV maintains inventories of pollution response equipment that are maintained in constant readiness for immediate air or surface shipment to spill respons/operations sites. This equipment is centrally located at Emergency Ship Salvage Material (ESSM) Bases for rapid deployment to pollution sites worldwide. ESSM Bases are located near major fleet operating areas at the following sites: Williamsburg, Virginia; Stockton, California; Pearl Harbor, Hawaii; Anchorage, Alaska; Aberdeen, Scotland; Livorno, Italy; Sasebo, Japan; Bahrain; and Singapore.

SUPSALV also provides training which consists of the Navy On-Scenes Commander (NOSC)/Facility Incident Commander (FIC) Course, providing NOSCs and FICs the knowledge required to utilize the Navy's pollution response resources effectively and keep abreast of new policies and regulations that may effect spill response and planning activity.

Summary

The following is diagram illustrates the organizational structure of the support groups which form the Navy's oil and hazardous substance spill response network.

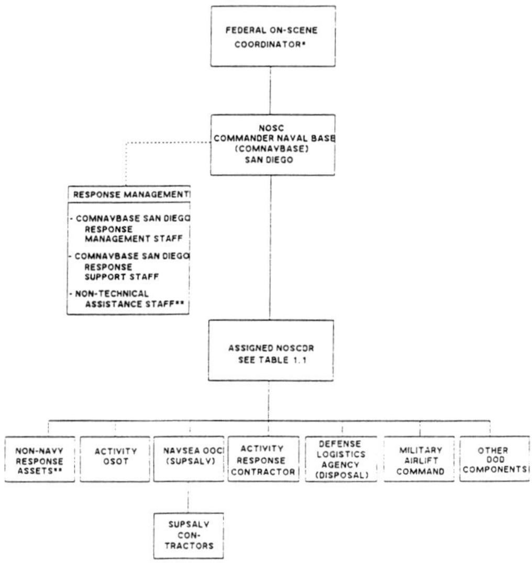

Figure 2. Organizational Structure Navy's Oil Spill Response Contingency.

The Navy's oil and hazardous substance spill response network is a well established set of specialized working groups. The working groups or commands described above provide the Navy with a spill contingency capable of efficiently responding to a pollution incident originating from naval operations. The Navy identified the need for effective response and mitigation of oil and hazardous substances releases over 20 years ago. Since that time the programs supporting the response contingencies have been improved and expanded to their present form.

CHAPTER 127

Reducing Duplicative Efforts in Oil Spill Contingency Planning

Craig Ogawa[1], Barbara Foster[2], and J. Manuel Saenz[1]

Abstract

Passage of the Oil Pollution Act of 1990 (OPA), subsequent delegation of authority under Executive Order 12777, and the California Senate Bill 2040 gave the Minerals Management Service (MMS) Pacific OCS Region (POCSR) and the California Office of Oil Spill Prevention and Response (OSPR), respectively, overlapping oil spill regulatory responsibility for offshore facilities in state waters. As a result, the MMS and OSPR have worked together closely on issues related to oil spill preparedness and response to avoid duplicating efforts for the facility operators. In 1993, the MMS and OSPR agreed that a Memorandum of Agreement (MOA) would be the most effective method of facilitating coordination between the two organizations.

The MOA, signed in April 1995, was designed to allow for the parties to exercise their respective authorities regarding oil spill planning, prevention, and response in a manner that ensures the best achievable protection for the State of California. A major focus of the MOA is coordinating the review of oil spill contingency plans (OSCP).

In the spirit of the MOA, the MMS and OSPR have developed a joint OSCP review and approval process, which meets the regulatory needs of both agencies, and enhances protection of the environment. The review and approval process was designed to provide a coordinated review of the OSCPs, eliminate duplication of efforts, simplify and streamline the process, and address concerns of industry by allowing operators of offshore facilities to submit a single OSCP which meets the needs of both agencies. This paper details the process designed by MMS and OSPR to review and approve the OSCPs and the MOA that facilitated the development of a coordinated

----

[1]U.S. Department of the Interior, Minerals Management Service, Office of Development, Operations and Safety, 770 Paseo Camarillo, Camarillo, CA 93010
[2]California Department of Fish and Game, Office of Oil Spill Prevention and Response, 1700 K Street, Sacramento, CA 95814

process.

Introduction

In response to major oil spill incidents, Federal and State legislators have enacted laws that were designed to augment and strengthen existing legislation. In some cases, this legislation has created regulatory overlaps which the agencies have had to contend with. To cope with these overlapping authorities, agencies such as the U.S. Department of the Interior, Minerals Management Service (MMS) and the California Department of Fish and Game, Office of Oil Spill Prevention and Response (OSPR) have developed a Memorandum of Agreement (MOA). The MOA ensures that there is a coordinated and cooperative approach to resolving these overlaps. The MOA also forms the framework for guidelines and processes developed to fully implement issues covered.

One major area of overlap is the responsibility for the review and approval of the Oil Spill Contingency Plans (OSCPs) for offshore facilities in state waters. The MMS and OSPR have developed an OSCP review approval process that is streamlined, efficient and reduces duplicative efforts while meeting the regulatory needs of both agencies. The process also benefits industry by reducing the amount of paperwork and burden on the facility operator.

Background

On March 23, 1989, the Exxon Valdez went aground on Bligh Reef in Prince William Sound, Alaska creating one of the largest oil spills in American history. This spill exposed deficiencies in the Clean Water Act, the applicable Federal Law at the time of the spill. Federal Legislators quickly enacted a law that amended the Clean Water Act called the Oil Pollution Act of 1990 (OPA).

The OPA toughened existing oil spill prevention and created new requirements for operators of vessels and offshore facilities while giving new authorities and responsibilities to Federal Agencies such as the MMS. The MMS is the federal agency that manages the Nation's natural gas, oil and other mineral resources on the Outer Continental Shelf and collects revenues from offshore federal mineral leases and from onshore mineral leases on federal and Indian Lands. Prior to OPA, the MMS had oil spill preparedness and planning responsibilities for facilities in Federal waters. The OPA gave the MMS new authority for oil spill preparedness and planning for offshore facilities in state waters.

On February 7, 1990, the American Trader drifted over its anchor causing a major oil spill off of Huntington Beach, California. In response to this incident, California State legislators saw deficiencies in existing state laws and enacted California Senate Bill 2040 (SB 2040) or the Lempert-Keene-Seastrand Oil Spill

Prevention and Response Act. SB 2040 created a new agency under the California Department of Fish and Game called the Office of Oil Spill Prevention and Response. The OSPR is the lead state agency charged with oil spill prevention, preparedness and response within California's marine environment. The OSPR Administrator has substantial authority to direct spill response, cleanup, and natural damage assessment. The OSPR's prevention programs include oversight of the safe operation of marine vessels, terminals, facilities and pipeline transportation.

Offshore Facilities and Contingency Plans

Under the OPA, the MMS was given responsibility for the review and approval for OSCPs for all offshore facilities that produce, store or transport petroleum products. This includes offshore platforms, islands, piers, pipelines and mobile drilling units both in Federal and State waters. Under SB 2040, OSPR has OSCP review and approval responsibilities for all offshore facilities in State waters, their pipelines, and any onshore facilities which can spill oil that can impact state waters.

Currently there are 10 offshore facilities in State waters and 23 offshore facilities that are required to maintain OSCPs under OPA. These facilities include offshore platforms, piers, and manmade islands. Each of these facilities and their associated pipelines must be covered by an OSCP. OSCPs are also required for mobile drilling units that will be used in any future abandonment or drilling operations in State or Federal waters.

The OSPR has always recognized the need for the elimination of duplicate reviews by regulatory agencies. In its contingency plan regulations developed pursuant to SB 2040 and effective November 1993, OSPR allows plans prepared for other agencies to be submitted in substitution for the California contingency plans, if a response manual and an adequate cross-reference are included. The OSPR also allows platforms that fall under the jurisdiction of both the MMS and OSPR, with a reasonable worst case spill of less than 250 barrels, to apply for an exemption form the California oil spill contingency plan requirements if the platform has an OSCP approved by the MMS.

The MMS also recognizes the need to eliminate duplicative plans. The MMS, in its Interim Final Rule for oil spill response plans allow operators of offshore facilities in State Waters to provide a copy of the same plan submitted to a State agency for approval, to the agency. As the responsibility for approval of response plans was delegated to the MMS by the Secretary of the Interior, the MMS cannot further delegate this responsibility to another Federal or State agency.

OSCP Review and Approval Process

On April of 1995, MMS and OSPR signed the MOA on *Oil Pollution*

Prevention and Response. Very soon after, agency officials met to discuss ways to implement the MOA. Four areas that the MOA focused on were oil spill response, preparedness, prevention of spills, response, and enforcement. It was decided that a new joint review and approval process for OSCPs needed to be developed.

In the spirit of the MOA, OSPR and the MMS, after a year of meetings and discussions, have developed an OSCP joint review and approval process. This process meets the regulatory needs of each agency without compromising protection of the environment. The MOA allows for each of the parties to exercise their respective authorities regarding oil spill planning, prevention, and response in a manner that ensures the best achievable protection from the impacts of oil pollution incidents for the State of California.

The joint review and approval process was designed to provide a coordinated review of the OSCPs, eliminate duplication of efforts, and simplify and streamline the entire process for both the agencies and industry. This process also addresses the concerns of industry by allowing operators of offshore facilities to develop a single OSCP which meets the needs of both agencies. The MOA has provisions for updating guidelines agreed upon so as to ensure that the MMS and OSPR are able to freely address current and future issues and situations as they arise.

To simplify the OSCP review and approval process, a flow diagram was created (fig. 1). This process will be used for all OSCPs for offshore facilities that both the MMS and OSPR must review and approve under current law and statue. The process will be used for new OSCPs as well as updates to the plans. Plans submitted to the MMS and to OSPR are updated biennially.

The MMS and OSPR review process has three major phases; the review of the plan or update, reviews of the revisions to the plan or update (if any are required), and final approval of the plan or the update. In the review stage (Fig. 2), both the MMS and OSPR have 45 days in which to review the plan. One agency will act the lead agency for the review and approval of the plan. For offshore facilities in Federal Waters, MMS will be the lead agency. For facilities in State Waters, OSPR is the lead agency. Should one agency not receive a copy of the plan, the lead agency would contact the plan preparer and request that a copy of the plan be sent to the other agency.

It is during this phase that agencies other than MMS or OSPR review and provides comments to the plan. The MMS sends a copy of the OSCP to the U.S. Coast Guard for review and comment. These comments, as appropriate are then added to MMS's comments to the OSCP or update. Provisions for this review of the OSCP are found in the Memorandum of Understanding between the MMS and U.S. Coast Guard. The OSPR receives additional comments from two State agencies that have statutory authority to review the OSCPs, the State Lands Commission and the California Coastal Commission/San Francisco Bay Conservation and Development Commission. The

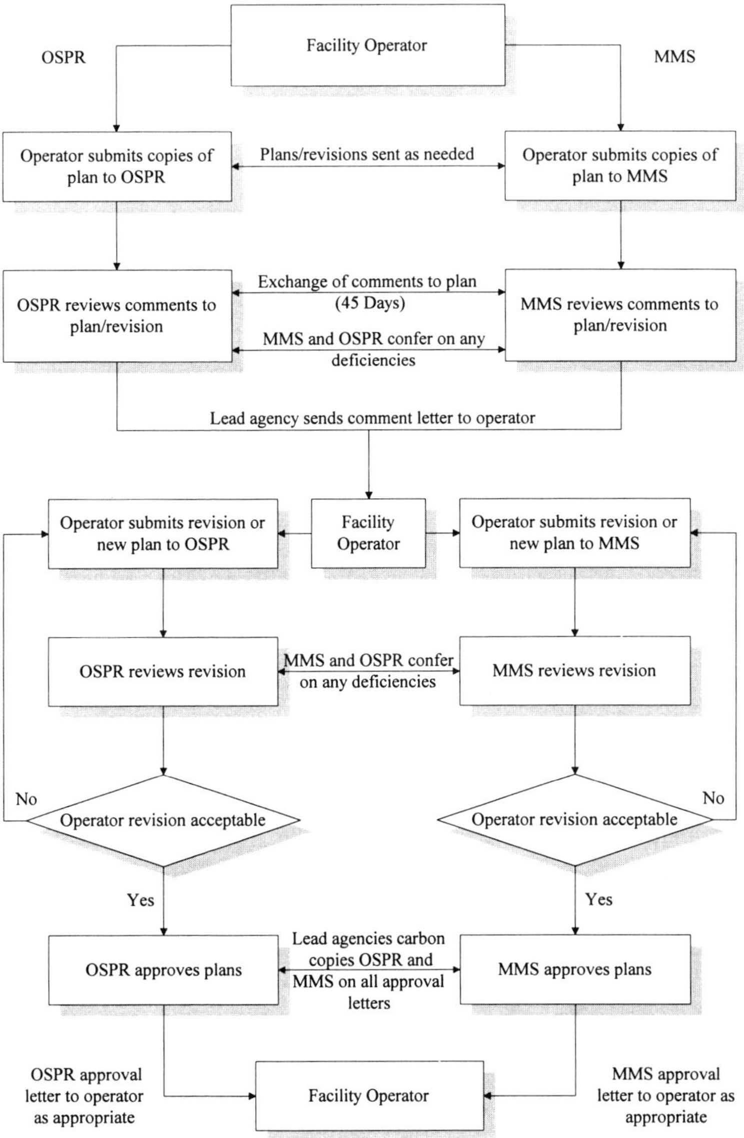

Figure 1. The MMS/OSPR Joint Oil Spill Contingency Plan Review Process.

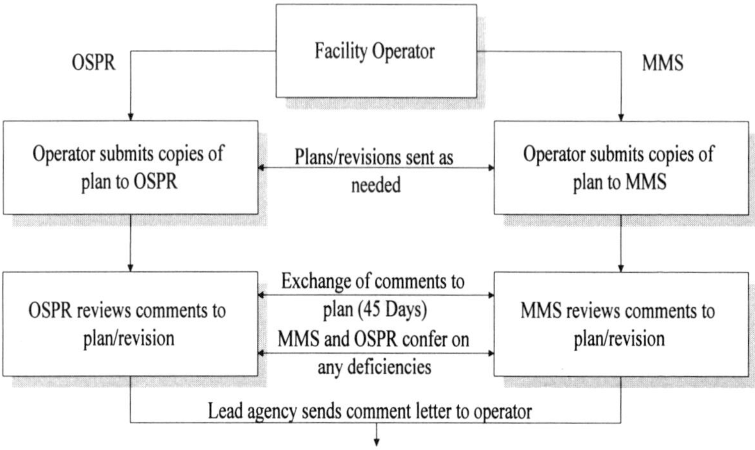

Figure 2. The review phase of the MMS/OSPR Joint OSCP Review Process.

OSPR considers all comments from other agencies in approving or disproving the OSCP.

Once the OSCP is approved, the MMS provides copies of the plan to the U.S. Environmental Protection Agency, National Marine Fisheries Service, National Oceanographic and Atmospheric Administration, U.S. Fish and Wildlife Service, National Park Service and sends additional copies to the U.S. Coast Guard. These agencies can then review the plan and send comments to the MMS for consideration during the next update of the plan.

To reduce duplicative efforts, the review of the OSCP by the secondary agency will be cursory, while still meeting the minimum requirements of the agency and relying heavily on the lead agency for the bulk of the review. After the agencies complete their review of the plan, comments are exchanged within 45 days. If there are deficiencies or conflicts in the agencies' comments to the plan, the agencies discuss and resolve these issues. After resolving any conflicts and agreeing on the deficiencies, the lead agency then sends a comment letter to the OSCP submitter.

By using this process, the agencies reduce duplication of efforts by relying on the lead agency for the bulk of the plan review and sending of a comment letter to the plan submitter. The OSCP submitter benefits from this process by having to respond to one set of comments and making one a set of revisions to the plan.

The second phase of the review process (Fig. 3) involves the review of revisions made to the OSCP in response to the agencies' comments. The plan submitter, in

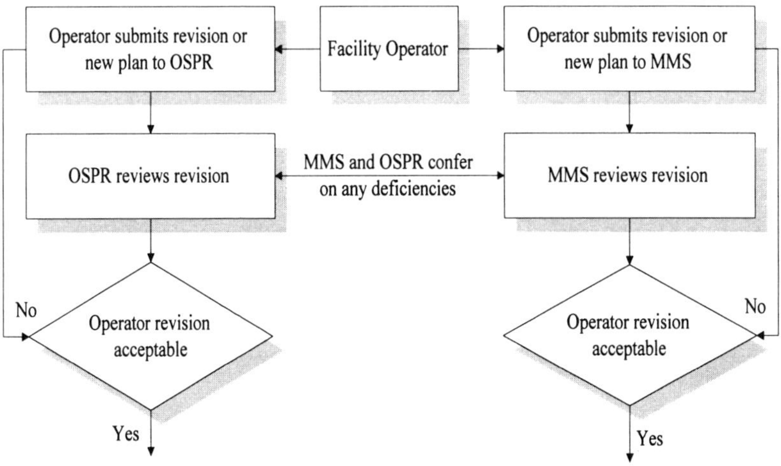

Figure 3. The revision phase of the MMS/OSPR Joint OSCP Review Process.

response to the comment letter from the lead agency, makes the requested changes to the plan, then resubmits copies of the revision to the agencies for review. The agencies then determine if the revision is acceptable or has any additional deficiencies. This phase ends when both agencies find the revisions to the plan acceptable.

This phase ensures that the best achievable protection is maintained at the offshore facility as both agencies must agree that the plan meets the regulatory needs of the agencies during a joint review. Should additional revisions be required, the OSCP is remanded to the plan submitter for additional changes and a revision is again resubmitted for review.

The last phase of the plan is the approval phase (fig. 4). Once a plan is deemed acceptable, an approval letter is sent to the OSCP plan holder. For offshore facilities in State waters, both OSPR and MMS will send approval letters to the operator and also carbon copy the letter to one another. For OSCPs that cover only facilities in the Federal waters, the MMS will send an approval letter to the plan holder and carbon copy the approval letter to OSPR.

Conclusion

The OPA, SB 2040 and other Federal and California State Oil Spill Prevention and Response legislation have created regulatory overlaps which agencies such as the MMS and OSPR have had to cope with. This overlap was surmounted by creating a MOA between the MMS and the OSPR. Through the development and implementation

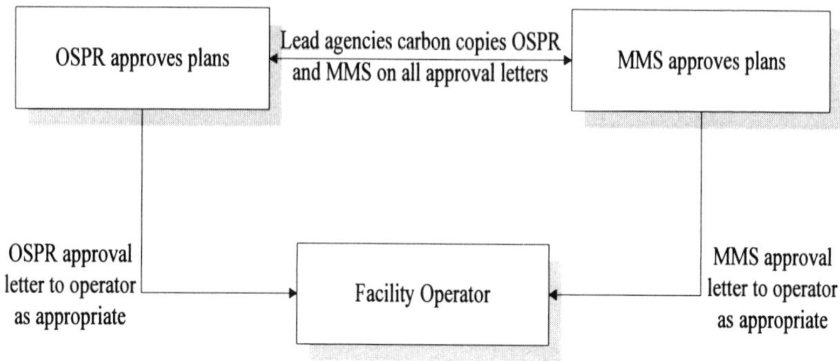

Figure 4. The final approval phase of the MMS/OSPR Joint OSCP Review Process.

of the MOA between the two agencies, a spirit of cooperation and coordination has been created by MMS and OSPR.

This cooperation has lead to the development of an OSCP review process that provides a coordinated and organized review while reducing duplicative efforts both for the agencies and industry, fosters trust between the agencies, and provides a thorough review of the plans. This process provides the State of California with the best achievable protection from spills occurring at offshore facilities through the efficient use of Federal and State resources.

CHAPTER 128

SHARING RESOURCES FOR EFFECTIVE OIL-SPILL RESPONSE: DEVELOPMENT AND TESTING OF AN AGENCY OIL-SPILL RESPONSE PLAN

Mark O. Pierson, David E. Panzer, Craig K. Ogawa, David L. Gebauer, Catherine Hoffman, Dan Knowlson, John Romero, and Khaleeq Siddiqui[1]

Abstract. As the Federal agency responsible for offshore oil and gas development, the Minerals Management Service (MMS) Pacific OCS Region (POCSR) has drafted a comprehensive plan to respond to oil spills. The Regional spill response action plan documents procedures and actions to be taken by POCSR personnel during a spill response. The plan, which is consistent with the National Contingency Plan and local Area Contingency Plans, serves to provide a structure within which MMS POCSR personnel may respond to a spill within the greater framework of the response community. It is, fundamentally, a preparedness document, serving as a constant reminder to be prepared for oil-spill drills, exercises, and actual events, and as a starting point from which timely response can be made. The primary sections of the plan include: Introduction, Guidelines, Exercises And Training, Command Structure, Notification, Response, Logistics, Environmental Support, and Investigation. The plan has been tested during oil-spill exercises at offshore facilities regulated by MMS and by actual spills.

Introduction

The Minerals Management Service is the agency in the Department of the Interior responsible for managing oil and gas exploration and development on the Federal Outer Continental Shelf (OCS). Under the OCS Lands Act, MMS's primary

[1]U.S. Department of the Interior, Minerals Management Service, Pacific OCS Region, 770 Paseo Camarillo, Camarillo, CA 93010

responsibilities for oil-spill response consist of spill abatement and investigation. Once a spill occurs, the agency plays a secondary response role, in coordination with the primary spill responders, the U.S. Coast Guard and lead State agency. However, with the advent of recent legislation such as the Oil Pollution Act of 1990 and increased public awareness of oil spills, the Pacific Region determined that a more proactive response role was necessary.

As a result, the MMS Pacific Region (Figure 1) has drafted a comprehensive plan to respond to oil spills. Development of the Regional Spill Response Action Plan (RSRAP) began in 1993, in part due to the Region's participation in a major oil-spill drill. As the results of the drill were analyzed, it became apparent that actions taken by Pacific Regional personnel during a spill response needed to be formalized and documented. To accomplish this and to incorporate additional improvements to current procedures, a team consisting of eight members representing the key offices within the Pacific Region was assembled. These eight team members are listed as the authors of this paper.

There are nine core sections in the plan, plus appendices. We will describe each section briefly and discuss the testing and validation of the plan through drills and actual events.

Section I. Introduction

The introduction describes the purpose of the plan, which is to define and document the actions, roles, and responsibilities of Regional personnel during an oil-spill response. That is, to tell Regional responders the who, what, where, when, and why of any action. As discussed in Section VI, Response, the location, size, type, and duration of a spill will determine the actions taken by Regional responders. For example, a large oil spill from an OCS platform could generate a major response, while an onshore spill that threatened to impact ocean waters might trigger a minor or no response.

In addition to its value in organizing and coordinating the Region's internal response, the plan identifies available resources and outlines the means of providing these resources to the spill- response community. These resources include communications equipment; transportation (primarily helicopters); near real-time ocean and atmospheric circulation data; wildlife and fisheries databases; oil-spill databases; as-built diagrams of offshore facilities and pipelines; and a broad range of expertise, from engineers and geologists to wildlife biologists.

Figure 1. The Pacific OCS Region of the Minerals Management Service. Current OCS activities occur within the Southern California Planning Area.

Section II. Guidelines

This section provides guidelines to Pacific Regional personnel for coordinating an integrated oil-spill response. The guidelines include important information such as to which spills Regional personnel would respond; how large a spill would have to be to mount a response, given the location of the spill (i.e., Federal facility versus one in State waters); the relationships between Regional personnel and their counterparts at facilities not under Federal jurisdiction; and a general list of personnel, equipment, communication, and training that would be required to mount a fully prepared response. The guidelines served as aids to the drafting of the original plan and will retain this function during future revisions.

Section III. Exercises and Training

This section describes how oil-spill response exercises conducted with facility operators are used to test both the agency's preparedness for responding to an oil spill in the Region and the effectiveness of the plan. The Pacific Region conducts an intensive program of drills and exercises, consisting of both equipment-deployment and table-top exercises. Equipment-deployment exercises can be major or minor, depending on the scenario.

Minor equipment-deployment exercises require the successful deployment and operation of primary response equipment (e.g., booms and skimmers), are unannounced, and are scheduled to exercise every offshore facility at least once per year. The Pacific Region also conducts one major equipment-deployment exercise each year. Major equipment-deployment exercises call for full implementation of an operator's oil spill contingency plan (OSCP), including the establishment of an onshore incident command center and the successful deployment and operation of primary and some secondary response equipment.

The Pacific Region uses major drills for exercising the plan. This includes setting up the command structure, deploying personnel to the field and the responsible party's command center, conducting notifications, and making decisions on what resources available in the Region would be needed in a real situation corresponding to the drill scenario.

The entire plan does not need to be exercised at every opportunity. As in the U.S. Coast Guard's National Preparedness for Response Exercises Program (NPREP), only critical portions may need to be exercised at any one time. Since various parts of the plan, including notification, command structure, and logistics, are all part of a spill response, any portion of the plan can be exercised more frequently. Minor drills also are used for such partial testing. Finally, other drills and exercises of opportunity, such as those conducted by the Coast Guard via the NPREP are used to exercise some part or all of the plan.

Section IV. Command Structure

The Pacific Region has adopted the Unified Command System/Incident Command System, which is becoming the standard for both industry and government agencies. The Unified Command System (or UCS) is intended to coordinate the command and control of a response, allowing input from the responders to reach the decisionmakers so that appropriate actions can be taken. The Incident Command System (or ICS) organization is composed of four functional areas: 1) operations section; 2) planning section; 3) logistics section; and 4) finance section.

The role of the command level within the UCS is: 1) to set the strategy for the spill response; 2) to coordinate the Pacific Region's response with the responsible party and Federal and State on-scene commanders and corresponding ICS sections; and 3) to coordinate public affairs issues with the media and other Federal, State, and local agencies.

Section V. Notification

This section of the plan identifies the process used to mobilize Pacific Regional resources and coordinate response efforts in support of the responsible party, the response organizations, and Federal and State lead agencies.

Notification is a vital step in the critical early response to an oil spill. It is, therefore, essential that the notification sequence occurs efficiently. Once affected, or potentially affected, parties are alerted and provided with basic information on the spill, they can more easily assess the need to mobilize their resources and orchestrate any response efforts in coordination with the Unified Command.

In general, notifications are made to internal and external parties. Internal parties include the Regional Supervisor of the Pacific Region's Office of Development, Operations, and Safety, who initially oversees the response, the Department of Interior or MMS headquarters duty officer, the Pacific Regional Director's office, the Regional Supervisors of the Offices of Environmental Evaluation and Resource Evaluation, and the Public Affairs Officer.

Externally, notification is made to seven Federal agencies, five State agencies, six county agencies, and one fisheries liaison group. A standard notification report is used to ensure that the information conveyed to all parties is timely, accurate, and consistent.

Section VI. Response

This section describes the actions to be taken by Pacific Regional personnel in the event of an oil spill. Since spills can occur in all situations and sizes, can consist of various types of petroleum, and may occur in nearly any location at any time, the plan must be flexible, while still providing the responders with guidelines for appropriate decisions.

Pacific Regional oil-spill response has four aspects: equipment and resources, coordination and communication, response time, and the degree of response.

Equipment and Resources. Equipment and resources available to the Regional spill response team include transportation (vehicles and helicopters), cellular phones, still and video cameras, spill reporting forms, and personal safety equipment. Personnel safety is of paramount importance to the team. This equipment includes hard hats, steel-toed work boots, safety glasses, flight helmets, hearing protection, and personal respirators. In addition, 24- and 40-hour HAZWOPER training has been provided to all identified Regional responders.

Regional Coordination and Communication. Coordination and communication go hand-in-hand throughout a spill response. Communication is maintained via telephone, fax, e-mail, and, whenever possible, face-to-face conversations. The importance of communication cannot be understated--the philosophy of the Regional response team is: "communicate every chance you get."

Response Time. Response time is defined as the amount of time it takes for Regional personnel to arrive at a spill site (or at a command center) and be ready to act. This time includes the time it takes to be notified, travel to the assigned location, interact with other responders, and begin assigned duties.

Degree of Response. Four scenarios were developed to determine the degree of Regional response appropriate to a given spill situation. The scenarios are based upon those developed for spill drills conducted by the Region and on experience with actual spills that have occurred at OCS facilities during the past 5 years. The scenarios describe the initial actions of Regional personnel and are defined by the size and source of the oil spill: Scenario 1 is for a small spill (5 bbl) from an OCS platform; Scenario 2 is for a medium spill (100 bbl) from an OCS platform; Scenario 3 is for a large spill (3,000 bbl over 3 days) from an OCS platform; and Scenario 4 is for a non-OCS oil spill of any size.

Pacific Regional personnel generally respond to three places during an oil spill: the spill site, the responsible party's command center, and the Pacific Regional command center. The actual decision as to whether and how Regional personnel respond is made on a case-by-case basis. The decisionmaker for the Pacific Region is the Regional Supervisor, Office of Development, Operations, and Safety, in close consultation with his counterpart in the Office of Environmental Evaluation and the Regional Director.

Section VII. Logistics

The logistics function is normally one of the four operational roles included in the ICS, and as such is covered generally in Section IV, Command Structure. However, for the purposes of this plan, it was felt that logistics should be discussed in additional detail, so that personnel would know where equipment is kept and who is responsible for maintaining it and assuring its accessability during a drill or a spill.

A pre-established logistical plan is necessary to bring equipment and services into a response effort in a timely manner. This section addresses facilities, transportation, supplies, equipment maintenance, communication equipment, and all other resources that are available to the Region during a response effort. This section also includes an up-dated inventory of all equipment and supplies. Additional logistically related lists of equipment, including the persons responsible for specific items, are presented in the appendix.

Section VIII. Environmental Support

Section VIII contains two parts: a guide to environmental sensitivity maps, and a discussion of Pacific Regional contributions to post-spill natural resource damage assessment efforts.

Environmental Sensitivity Maps. This section of the plan provides a guide to the maps of environmentally sensitive sections of the California coastline that are contained in the U.S. Coast Guard Area Contingency Plans (ACPs). The ACP sensitive areas appendices are intended to provide basic identification of the environmental, economic, and cultural considerations and sites within a plan area.

Under their current organization, there are 5 ACPs applicable to the Pacific Region's Southern California Planning Area. They range from Monterey County in central California to the Mexican border. These maps are found in Annex E, Appendix V, Tab A of each plan.

Natural Resource Damage Assessment (NRDA). Section 1006(e)(1) of the Oil Pollution Act of 1990 requires the development of regulations establishing procedures for natural resource trustees to use in the assessment of damages for injury to, destruction of, loss of, or loss of use of natural resources covered by the Act. Section 1006(b) of the Act provides for the designation of Federal, State, Indian tribal, and foreign resource trustees. In the Department of the Interior, the designated trustee agencies are the U.S. Fish and Wildlife Service and the National Park Service. For the State of California, the Department of Fish and Game, Office of Oil Spill Prevention and Response (OSPR) is the trustee agency.

The MMS is not a designated trustee agency under the Oil Pollution Act of 1990. However, the need exists for damage assessment efforts as part of Pacific Regional response to oil spills. These efforts could take two forms: 1) preliminary damage assessment in support of Regional needs for information on spill impacts; and 2) contribution to a post-spill NRDA effort by Federal and State trustee agencies.

Experience has indicated a need for some preliminary damage assessment when oil spills occur at MMS-regulated facilities. This preliminary assessment is of particular importance for small spills, where Federal and State trustee agencies may not become involved, but where the Pacific Region may have responsibility for determining impacts (as in the case of possible civil penalties).

The Pacific Region can contribute to a post-spill NRDA effort in several ways: 1) through the application of existing MMS databases, such as the Marine Mammal and Seabird Database or the California Sea Otter Population Model (OTPOP); 2) through pre- and post-contact intertidal and shorebird surveys conducted by the Regional biologists in conjunction with ongoing in-house and cooperative studies; and 4) through participation by Regional biologists in OSPR's seabird and marine mammal damage assessment survey flights.

Section IX. Investigation

This section addresses the general process involved in investigating an oil spill occurring at an OCS facility within the Pacific Region. The goals of an investigation are to determine the root cause of the oil spill and to develop procedures to prevent recurrence where appropriate.

Investigations are conducted by Pacific Regional District personnel under the authority and direction of the appropriate District Supervisor, with assistance from other offices within the Region. Coordination with the U.S. Coast Guard, OSPR, the responsible party, major oil-spill cooperatives, and other organizations may be required. The Coast Guard and OSPR can be requested to participate in the investigation as discussed in the respective Memoranda of Understanding/Agreement between the Pacific Region and these two agencies. An investigation may result in the issuing of Incidents of Non-Compliance (INCs), as well as the application of civil and/or criminal penalties.

Conclusion

Throughout its development, the Regional oil-spill response plan has been reviewed internally and tested via drills and spills. In its current form, the plan reflects the lessons learned from several events, including two major oil-spill drills and two moderate (30- to 100-bbl) spills at OCS facilities. The plan is a "living" document, and it will be updated and fine-tuned as circumstances and personnel change.

One major aspect of the plan that has yet to be fully tested is how it allows the Pacific Region to fit into an overall response to a major spill. We have tested this during the development of the plan, but, fortunately, no large oil spill has occurred that would demonstrate how the plan would integrate into the larger response community. The primary accomplishment during development of the plan to this point has been the formalization of the substantial resources that the Pacific Region could bring to an oil-spill response, regardless of the source of the spill.

While not a trustee agency, MMS has mitigation, planning, and prevention responsibilities for marine facilities and spills in both Federal and State waters in the Pacific OCS Region, which it coordinates through a Memorandum of Agreement with the State of California. Given this unique position of responsibility for offshore platforms and pipelines, the Pacific Region expects to continue to assist the overall oil-spill response community with resources and expertise.

CHAPTER 129

Oiled Wildlife Care for Sea Otters in California: A Government, Private Sector, Not-For-Profit Partnership

David A. Jessup and Jonna A. K. Mazet

Although State wildlife agencies are the trustees for wildlife resources, few wildlife agencies actually provide care for wildlife compromised by oil spills. Legislation enacted in 1990 established the Office of Oil Spill Prevention and Response (OSPR) within the California Department of Fish and Game (CDFG). A $.04 per barrel fee was placed on oil transported or processed in California, the proceeds of which are to be used, among many other things, for care of wildlife affected by oil spills and required the establishment of rescue and rehabilitation stations for sea birds, sea otters, and other marine mammals.

OSPR is building one such facility at the University of California, Santa Cruz which, when completed in the summer of 1997 at a cost of approximately 5 million dollars, will be capable of caring for 125 sea otters, but flexible enough to care for other species of marine animals and house ongoing joint and cooperative research projects. Additional legislation in 1993 and 1995 allowed OSPR to approximate 9 million dollars over a 4 year period to help establish an Oiled Wildlife Care Network (OWCN) to serve all marine species for the entire California coast. These additional centers are being developed in San Diego, Orange County, Los Angeles, Santa Barbara, the San Francisco Bay Area and Humboldt County. Each of these centers is being developed in conjunction with an existing marine park, marine wildlife rehabilitation program, educational institution, or combination of compatible programs. Each program has separate and additional sources of financial, political and personnel (volunteer) support. This approach assures that all OWCN centers will provide useful and appreciated services to the people of California when not in use for oil spill response.

Operations and maintenance of the OWCN, and support of appropriate technology development and research via competitive grants, will be accomplished through the Wildlife Health Center at U.C. Davis. Current OSPR funded research at California universities and Hubbs Institute is addressing the effects of oil on sea otters, harbor seals and key marine bird species. All research is designed to improve our ability to care for oiled marine wildlife, and to improve our ability to determine both the immediate and the sublethal effects of oil pollution on marine animal populations.

The OWCN is one of the first government, private sector, non-profit, educational institution partnerships of its kind. It allows the petroleum industry to meet legal requirements and represents a State government commitment to integrate wildlife care with resource management and ongoing research.

CHAPTER 130

Investigations and Governmental Cooperation on the Coastal and Marine Impact of Oil Spill Responsiveness to Recent Shore Based Spills in Central California

Charmaine Marie Gallagher[1]

Abstract

Coastal related spills in California are subject to the National Oceanic and Atmospheric Administration's proposed rulings under the Oil Pollution Act of 1990. The local response to shore based spills on the central coast is examined in light of participation of local governments and the public. Compensation and on-site restoration issues are addressed given the criticism surrounding no use values and contingent valuation. Suggestions for greater public participation and clearer guidelines for trustee agencies are presented.

Introduction

The Oil Pollution Act (OPA 90) provides for the prevention, liability, removal and compensation from a discharge or threat of discharge to the ocean (NOAA, 1993). The primary state trustee is given the authority to ensure these mandates are completed and that local trustee agencies are identified for compensation and to oversee restoration. This paper addresses the process by which compensation is implemented through mandate in the Department of Commerce's rulemaking intended to be used by the trustees. Descriptions on the result of these procedures for three oil spills on the central coast of California and the on-going progress of both compensation and restoration are presented. Specifically it will address the successes, frustrations and areas of improvement for 3 land based spills in San Luis Obispo County; one surface spill and two subsurface leaks and present the local government response and implementation of addressing clean up, compensation and restoration. The involvement of the Natural Resource Damage Assessment (NRDA) process along with the perceived adherence to the changes in NOAA's proposed rulemaking is described by its local impact. Suggested measures are included to improve this relationship to deter future events and to address the public concerns for on-site restoration. Direct measurement techniques to address compensation of non market values of damaged coastal sites are also included.

---

[1] former Chair, Ocean Sanctuary Coalition and San Luis Obispo County Planning Commissioner, P.O. Box R Corvallis, OR 97339

Oil Pollution Act Ruling

In 1995 NOAA sent out an additional proposed ruling as an update of the 1994 Rules with greater emphasis on restoration. Under the Oil Pollution Act, a Natural Resource Damage Assessment is designed to supplement response actions by providing a process for making the public whole for injury to natural resources as a result of oil pollution (NOAA. 1995). A particular goal of the Oil Pollution Act is to invoke the NRDA Process, which is designed to meet four procedural elements along and act as guidance documents to assist with performing these assessments. The procedural elements are as follows: 1) Identification and evaluation of injuries to natural resources, 2) Employment of assessment methods relevant to circumstances of the incident, 3) Identification and evaluation of restoration alternatives, and 4) Involvement of the public in the process of selecting appropriate restoration actions.

Oil Incidents

The local spill scenarios in San Luis Obispo County began in 1992 with the pipe leakage at Whales Cave. This stretch of shoreline is owned by the oil company, the Harbor District, a private landowner with a public easement to a recreational nude beach which is under County jurisdiction. A major stakeholder is the Native American population of Chumash who value this living cultural resource site as second only to Point Conception. The spill caused damage to the rocky shores, intertidal marine and ocean life, local fishing fleet, and the tidal estuary. It caused the closure of public beaches and losses to the adjacent community and fisheries. Clean up efforts damaged the archaeological remains of this important Chumash site. Primary in the efforts for on-site clean-up and restoration were oil spill trained Chumash elders and public citizens. These people continued with the process of oil spill contingency planning and involvement in additional training to address future oil incidents. Local government was quick to respond with necessary emergency permits but was then barred from further involvement as the NRDA process took precedence. This then excluded both government and locals as the primary agency collected vital information on damage which would be used to determine compensatory information pertaining to the damages rendered at this site. A big issue are the non-market use values here which can be under represented by agencies but are extremely significant to residents and the public.

A second incident involving leaking diluent into dunes and groundwater aquifers became apparent in 1993. Previous investigations in 1990 led to the installment of a bentonite wall, but citizen initiative prompted the U.S. Coast Guard and Fish and Game to assess jurisdictional uncertainties with EPA and to halt the free flow of product to the ocean. Citizen participation again was active, informed and present. Once the incident command was established, citizen participation was curtailed as the NRDA process took precedence. Multiple land use and restoration decisions were needed and a constrained local government, non profit landowner and non-trustee state agencies had little input into the decision making process for remedial action

This area is comprised of the oceano dunes which are the highest dune complexes on the pacific coast with 26 miles of undeveloped coastline and ocean recreational activities. In May 1974, it was designated under the National Natural Landmarks (NPS, 1994).

The third spill in progress is the Avila Beach Spill. Leaking pipelines which run from an oil dock to tank farms under a frontal street through the town of Avila have so polluted beaches and homeowner parcels that the town has become hostage to this event. Failure to properly acknowledge and address clean up efforts has resulted in losses to beach front businesses, property values and the ability to obtain loans to build improve or sell. The issue here is the oil company's unwillingness to include social effects in the Environmental Assessment Report, claiming the costs would be prohibitive. Here the public has its greatest opportunity to address true compensatory values for proper restoration.

Local Actions

The fourth procedural element described above, involvement of the public in the process of selecting appropriate restoration actions, is where the breakdown has occurred in San Luis Obispo County. It has been the trustee agencies with their own priorities for the region which have selected restoration sites and actions, thereby excluding the public. If the public has no access to the NRDA reports, then it is powerless in assisting selection of appropriate actions. This has resulted in restoration efforts being concentrated on off-site areas with the exclusion of public input. Understanding the criteria necessary to justify off-site restoration vs on-site is a necessary element in the selection process. Public involvement in the decision making process is necessary for appropriate actions, restoration on-site or other relevant site selectivity. It is imperative that compensation and restoration requirements do not become the opportunity to create pet projects of agencies and politicians and those of regional interest prioritized over the marine, landowner, cultural and recreational resources damaged. There is much administrative discretion in restoration when as many as ten factors are to be considered for determining alternatives for restoration, rehabilitation, replacement and/or acquisition of equivalent resources (Castle et. al 1994).

Having observed and participated in the public involvement of these spills, it has been citizen initiative which both motivated and assisted agencies and government into action in the Whales Cave and Guadalupe spills. But what transpires once the spill falls under the jurisdiction of the primary agency is then bound by law in the NRDA process. Agency personnel and natural resource damage consultants abound but the public is shut out of this informational gathering process, relying on the ability of local government to understand and explain a process by which even they are excluded from and relying on the honesty and integrity of the oil company to provide vital information. With this exclusion from information relevant to the real on-site damages the public looses interest. The whole inclusion of civil law has caused an exclusion of information for a minimum of two years until the judicial process is complete and the money as been exchanged from the oil company to the trustee agencies. That is not where I think NOAA proposed the system to end. Instead, I argue that this is where the process begins.
Perhaps the public is worn out by the process, but I suggest it is this information barrier given that the NRDA information is not available for two years. This violates any freedom of information and given that this is public information rests uneasy with the implementation of this procedure.

This argument is further strengthened by the major goals of NOAA outlined in the rulemaking: 1) Involve the public in the decision of what actions will make them whole, 2) Ensure appropriate scientific procedures and methods to determine restoration actions, 3) reduce transaction costs. In this region, the public has worked diligently with the Coast Guard and the Office of Oil Spill Prevention and Response in the Development of the Oil Spill Contingency Plan. It was unfortunate that once we established rapport with the Monterey contingent, San Luis Obispo was transferred to the Long Beach sector which included Santa Barbara and San Luis Obispo lost much of the established identity of this central coast region in the transition. For San Luis Obispo County is much more aligned with the interests of Monterey than that of Santa Barbara to the south in off shore and onshore issues relating to oil. There has been no lack of public interest and involvement in both oil and coastal issues in this region.

A credible concern is that the trustee agencies rely heavily on what NOAA has identified as the Restoration Alternatives and Regional Restoration planning where the desired objectives come from public entities not necessarily the public. These are used for small incidents and may be an ad-hoc solution as a result of the presentation preceding this paper. As of 20 March 1997 there did not exist an Avial Beach Trustee Council or if it did it surely was not a public forum (Sneed, 1997) . Nearly five years after the spill and three years after $370,000 were provided to a local land conservation trustee, the newspaper runs an article claiming this Avial Beach Trustee Council exists and invites the public to a forum to discuss the $$950,000 watershed and $100,000 sea otter restoration plans. Whose plans are these? And the public was told that this is a part of the $1.5 million settlement between the state and oil company. Where was public involvement until now? NOAA's provisions make itself and the primary trustee agencies responsible to the rulings but something is lost when it compensates local trustee agencies with private interests outside of the concerns for on-site restoration. The local trustee agency is not even a stakeholder in the resources damaged, merely a convenient trustworthy agency to handle big funds with other environmental priorities. The public and local Chumash received only what government was able to convince the oil company to restore.

Restoration and Compensation

Under the restoration Planning Phase where trustees evaluate injuries there are two components: Primary and Compensatory. An additional breakdown is the compensatory restoration which are the actions to make the environment and public whole for resources lost from incident until recovery. Under the Comprehensive Environmental Response, Compensation and Liability Act of 1980 (CERCLA) the public must be compensated for natural resources and services that are injured or lost in order to make them as well off as they would have been with no spill (Lipton et. al, 1995). In the development of the initial damage claim the trustee agencies are depended on to use valid and reliable methods to determine the type and scale of restoration efforts. Questions concerning whether the trustee agencies have sufficiently identified public social damages without the involvement of the public arise. If indirect techniques are used, they may not sufficiently cover the non market values of these regions. The less developed the area, the less likely that use values, even outdoor recreation use values will be sufficient to justify needed restoration efforts (Castle et. al, 1994). Perhaps this is why the public is always in despair when agencies display their restoration plans and the agencies believe they have covered all necessary inputs. Whereas direct techniques would seek out losses and injuries to that portion of the public not obviously affected and would not undervalue the need to restore on-site resources.

This is clearly a problem in this region where those affected have not been a part of the compensation process. Instead it has been left up to the trustee agencies. Section 1006.5 requires that the restoration process be open to the public before final decisions made and actions taken (NOAA, 1995). Here is where the system seems to exclude interested citizens for once a suit is filed the NRDA is unavailable to the public for two years. How then can citizen participants be involved prior to final decisions. Does this require lobbying our local trustee agencies? Do the trustee agencies compete for these compensatory resources with specific projects which are evaluated on a peer review level? It is questionable if this system is serving the intent of the open process or if the concerns if this paper are merely regional. NOAA suggests public involvement should be through pre-incident planning and regional restoration plan development, but I have yet to experience any restoration plan development. In addition NOAA states that selective solicitation of comments from interested members of the public as a supplement to agency peer review is sufficient to satisfy NEPA requirements.

Perhaps the breakdown is in the trustee agencies responsibilities to the public, therefore I propose that those agencies obtaining funds from settlements as a result of Fish and Games litigation be required to address restoration with interested parties prior to any off-site restoration pans being constructed. This is not to delay the process, this is to restore on site injury to coastal and marine resources. The rulemaking designates trustees to determine injury or destruction to assess natural resource damages, present a claim, recover damages and develop and implement a plan for the restoration with much wiggle room as stated before. What was never available to the public or to local government is the Natural Resource Damage Assessment prior to compensation from litigation.

Results of Proposed Rulings

In light of the NOAA Blue Ribbon Panel's response to contingent valuation and that damage assessments have the status of rebuttable presumption of validity in judicial proceedings, the oil company would be expected to balk in the case of the Avila spill. More specifically the 1995 version eliminated the need for the determination of Compensable values as a separate component of a natural resource damage claim The results of the Panel created a procedure for valuing non-use values which was prohibitively expensive. But local governments trying to assess the damages both environmental and social must take into account losses in the form of lost revenue, lost pleasure, lost non-market values which are challenging to address but without them the effects on the community are misunderstood. The Harbor may obtain needed updates, the fisherfolks compensated, the creek cleaned up, the hotels and businesses compensated, the homeowners given loans but the losses to non-valued marine life, fisheries and environmental amenities after 50 years of continuing pollutants will never be recovered.

What has not been done in this region has been called for in the Blue Ribbon Panel's 1993 document which requested the federal government to produce standard damage assessments for a few specific reference oil spills ether hypothetical or actual ranging from small to large. The only one is that completed by Hawaii Sea Grant in a Risk assessment but not as the Panel suggested where is conducts CV studies to obtain benchmarks and reference points. Much skepticism has surrounded the use of Contingent Valuation as it applies to non-use values. The possible inflation of values from the spill in Prince William Sound caused the compensatory damages values to be extreme. Non use values ranged in the billions while use values only in the millions of dollars. There is considerable skepticism concerning the validity and reliability of contingent values expressed by people who have never been to a site (Whitehead et. al, 1995). But relative prices and income have predictable effects on willingness to pay for both on-site and off-site users of a resource. Willingness to pay can be validly applied from respondents with information for both on-site and off-site use.

Conclusion

It is clear that local government has had to address damage as the oil company has been a neighbor and participant of this community for over 70-years. It is only in the last 15 years that local governments must address the costs associated with this type of activity with its associated impact on the adjacent ocean and coastal resources. Yes, the learning curve has been very steep but the most affective approach has been the multi-agency cooperation where all environmental aspects can be addressed cohesively and without the oil company orchestrating who will receive what information and which agency is more effective at getting response. The unclear and uncomfortable area is how the state agencies limit the local agencies. It was uncomfortable in the case of Guadalupe. Where local government and associated agencies were to play backfield to the state agency not for operation command but for information assessment and decisions. The public is resigned to believe that trustees are to seek on behalf of the public, restoration of what was lost. The proposed rulings authorize trustees to determine appropriate assessment method on incident specific events ensuring that assessment methods be reliable, valid and cost effective.

References

Castle, E. M., R. P. Berrens, and R. M. Adams. 1994. *Land Economics*. 70(3): 378-85.

National Oceanic and Atmospheric Administration (NOAA). 1993. Natural Resource Damage Assessment Under the Oil Pollution Act of 1990. Department of Commerce, 15 CFR Chapter IX. Vol. 58 No. 10.

National Oceanic and Atmospheric Administration (NOAA). 1995. Natural Resource Damage Assessments. Department of Commerce, 15 CFR Part 990. Vol. No. 149.

National Park Service. 1994. Damaged and Threatened National Natural Landmarks. U. S. Department of the Interior 29pp.

Levine & Fricke. 1994. Remedial Action Plan Beach Project Guadalupe Oil Field, San Luis Obispo County, California 32pp.

Lipton, D. W., K. F. Wellman, I.C. Shefifer and R. F. Weiher. 1995. Economic Valuation of Natural Resources -- A Handbook for Coastal Resource Policy Makers. NOAA Coastal Ocean Program Decision Analysis Series No. 5. NOAA Coastal Ocean Office, Silver Spring, MD 131pp.

Sneed, D. 1997. New Life for the Creek: San Luis Obispo Creek Watershed Will be Renovated. *Telegram Tribune* 28 March 1997.

Unocal Energy Resources Division. 1994. Report on Options Considered for Remediation of the Beach Plume at the Guadalupe Oilfield San Luis Obispo County, California: 22 pp.

Unocal Energy Resources Division. 1994. Incident Action Plan: Guadalupe Beach Project Beach Excavation. 200pp.

Whitehead, J. C., G. C. Blomquist, T. J. Hoban and W. B. Clifford. 1995. Assessing the Validity and reliability of contingent values: a comparison of on-site users, off-site users, and non-users. *Journal of Environmental Economics and Management*, 29 (2): 238-251.

Wilder, R. J. 1994. Prevention rather than cure. *Nature* Vol. 369. June 1994.

CHAPTER 131

New Tools for Assessing Coastal Habitats

J. B. Zedler, J. Desmond, S. Phinn, B. Nyden, G. Sullivan, G. D. Williams, J. Callaway, K.E. Boyer[1], A. N. Powell[2]

Introduction

Efforts to restore and create coastal wetlands in southern California need to be assessed with extra care and precision, because the resource is so sensitive and highly valued. It is not sufficient for restored or created wetlands simply to be green or to achieve a predetermined species list. The region's natural wetlands are so depleted in area and functioning, that restoration and enhancement efforts must aim to provide self-sustaining ecosystems that are functionally equivalent with the natural wetlands that are no longer in existence. Criteria and standards for "successful projects" need to be very clear (e.g., FWS 1988, CCC 1991) and the assessment very detailed (e.g., PERL 1990). At the same time, sampling needs to be nondestructive or minimally intrusive.

At the Pacific Estuarine Research Laboratory, we have pioneered the development of several tools that are designed to provide high-precision data on the condition of restored and created wetlands, to serve as surrogates for measures of function and to have low impact on the restored or created site. In this paper, we describe the use of: remote sensing to quantify habitat types, a global positioning system to map rare plant distributions, a geographic information system to analyze tidal creek networks, sampling to clarify fish use of different channel types, dataloggers to characterize water quality, soil cores to assess marsh development, height distributions to characterize plant canopy architecture, and a reproductive index to assess the value of habitats to nesting birds.

Quantifying habitat types with remote sensing (high resolution airborne multi-spectral image data)

Remote sensing provides a non-intrusive and repetitive approach for obtaining information on the composition, configuration and condition of coastal habitats from micro to regional scales. To derive accurate data from coastal environments, attention should be paid to the spatial, spectral and temporal resolutions of remotely sensed data. Spatial resolution concerns the ground resolution element (GRE) size and image extent required to detect certain landscape elements. Spectral resolution pertains to the

---

[1] Pacific Estuarine Research Laboratory, San Diego State University, San Diego, CA 92182-4625.

[2] USGS-Biological Resources Division, San Diego Field Station, San Diego State University, San Diego, CA 92182.

number and location of bandwidths for which reflected or emitted electromagnetic radiation is measured (e.g., red and near-infrared wavelengths enable discrimination of vegetation cover types). Specifications for the time of year, day or tidal stage, along with the required revisit cycle for monitoring dynamics are the temporal dimensions. Phinn and Stow (1996a, 1996b) identify characteristics of southern Californian wetland environments to consider when selecting appropriate remotely sensed data. The techniques described below rely upon high spatial resolution (GRE < 2.0m) and multi- or hyper-spectral image data. These data are particularly suited to the fragmented and species-diverse southern California wetlands (Zedler 1982). The next 5-10 years (1997-2007) will see an increase in the public/commercial availability of remotely sensed data from both airborne and satellite platforms (Johnson-Freese 1995; Fritz 1996).

High-resolution multi-spectral image data have been successfully collected from Pacific Coast wetlands and other coastal environments for: (1) mapping the extent and composition of coastal wetlands; (2) assessing the configuration or structure of wetlands and their component vegetation types; and (3) detecting and analyzing changes in (1) and (2) over time. Mapping the composition of wetlands requires preliminary knowledge of the existing vegetation assemblages and image data acquired at a time when vegetation and surface cover types can be differentiated. Image classification approaches have been successfully applied by Phinn et al. (1996) in southern California wetlands and Jensen et al. (1986, 1993) at slightly larger scales on the Atlantic Coast. Improvements in the accuracy of vegetation composition maps can be achieved by incorporating hierarchically structured classification schemes (e.g., Ferren et al. 1995) and digital elevation data (e.g., Bradshaw et al. 1996). Maps of vegetation cover derived from georeferenced images also provide a basis for examining the spatial distribution of ground survey data on the locations of invasive plants and bird habitat (Nyden et al. 1996; Brewster et al. 1996).

The first step in assessment is developing an accurate map delimiting the extent of vegetation cover types. These maps are then subject to analyses by landscape structure metrics or spatial statistical functions. Landscape structure metrics summarize information on the area, boundary features and topology of each vegetation, substrate or channel cover type (Haines-Young and Chopping 1996). Analyses of wetland vegetation and hydrogeomorphic configuration from remotely sensed data have been limited, although extremely promising results were provided by Phinn and Stow (1996a) from southern California wetlands and Mertes et al. (1995) from the Amazon. Biophysical properties of wetland vegetation have also been successfully estimated from remotely sensed data (Phinn and Stow 1996b). Maps of vegetation cover types and hydrogeomorphic features can be established from remotely sensed data obtained at seasonal or yearly periods, allowing analysis of wetland change in relation to the function of the wetland environment (Phinn et al. 1995; Mertes et al. 1995; Stow et al. 1996).

Remotely sensed analyses of composition, configuration and change are not intended to replace ecological assessments or field-work, but to act as a complementary source of data, enabling more spatially explicit and multi-temporal analyses. Ecological field work and analyses should be conducted in association with remote sensing analyses to maximize the accuracy and utility of information extracted.

Characterizing topographic complexity of marshes using GIS

Coastal wetlands exhibit a great deal of topographic complexity, with intricate networks of creeks, which vary in elevation and size. This complexity is rarely seen in restored marshes, because small intertidal creeks and rivulets are generally not included in restoration plans. Recent studies have linked topographic complexity to habitat function in riverine systems (Hatfield and Milne 1996; Kirchofer 1995; Jungwirth et al. 1995), and we recommend that the complexity of natural systems be mimicked at restored sites. In order to characterize natural complexity, we have obtained quantitative data on the configuration of creek networks in natural systems (Desmond 1996; Desmond et al., in review).

We used remote sensing images along with Geographic Information Systems (GIS) software to obtain quantitative information on the creek network of the Tijuana Estuary, which is the largest remaining natural wetland in San Diego County. Using a remotely sensed image as a template, the Estuary's creek system was manually digitized with the GIS software. Creeks were then classified into stream orders, and the digitized image was queried for information on total tidal creek length, tidal creek density, and other geomorphic parameters.

Using the vegetation classification from the ADAR image, we were further able to characterize the creek systems of different marsh zones (i.e., low-, mid-, and high-marsh). Determining the distribution of creeks with respect to marsh zone is important because the relative proportion of marsh zones will vary depending on the site to be restored, and a recommendation based solely on the area of a site may not be sufficient. We found that at Tijuana Estuary, over 50% of the total creek length was composed of small (first and second-order) creeks, and that nearly all of the tidal creek length was contained within the low-marsh zone. Information on the distribution and density of different creek types readily leads to designs for creek networks at individual restoration sites.

GPS Mapping

Global Positioning System (GPS) Mapping is a technique that is being used by the Pacific Estuarine Research Laboratory to provide very high resolution spatial data ($\pm$ 5 cm) about vegetation, channel topography and other wetlands attributes. L1 carrier phase GPS receivers and kinematic surveys are used to achieve maximum positional accuracy. A typical survey utilizes multiple carrier phase receivers. One unit is static over a known benchmark, and the others serve as roving units for rapid field measurements. Approximately 20 seconds of occupation time is required for each data point. Data are post-processed using differential correction with a pre-determined network of reference points, and then fine-tuned with geoid modeling (Geoid 96). The results provide spatial coordinates that are accurate to within 2-3 cm in latitude/longitude and 3-5 cm in elevation with optimal satellite constellations. This method can effectively pinpoint a position to within a baseball sized sphere of error from 12,000 miles in space.

The accuracy of data obtained from pocket-sized portable GPS receivers differs by several orders of magnitude. Inexpensive GPS receivers are commercially available for as little as $150. However, their primary shortcoming for scientific application is the large circular error probability, which is approximately 100 m. Most low-cost GPS units calculate distance to satellites and triangulate positions by monitoring coded signals broadcast by the U.S. Department of Defense (DOD) from the 24+ satellites that are available. Carrier phase GPS receivers take maximum advantage of the GPS satellite network by recording the transmitted code information as well as monitoring the subtle shifts (Doppler Effect) in the actual carrier wave frequency as satellites move closer and then farther away from the receiver.

Carrier phase receivers are unaffected by the periodic and deliberate degradation of the DOD's coded signals (selective availability and anti-spoofing). The primary drawbacks to the carrier phase receivers are their high cost ($20,000-$80,000) and requirement for continuous lock on the satellite signals. Even a brief interruption of a satellite's signal by a tree branch is enough to cause loss of lock. For obvious reasons, these receivers are well suited for short-canopy marshes but not very practical for rainforest monitoring.

Mapping rare plant distributions using GPS

Degraded salt marsh habitats are characterized by low diversity of native plants and dominance by weedy or stress-tolerant species (PERL 1990). Attempts to restore or create salt marshes are hampered by a lack of knowledge of the population and physiological requirements of many component species, especially those restricted to narrow or ephemeral environments (Skinner et al. 1995). Accurately assessing the range and physical parameters of rare populations or those occurring in restricted environments becomes more important with increasing degradation and loss of natural habitat. Further, understanding what limits the distribution of rare species and what is required for their growth and reproduction is critical for constructing fully-functional, self-sustaining salt marsh ecosystems.

Adequate assessment of rare plant populations can be facilitated by utilizing multiple techniques (Menges et al. 1996). We now use GPS to investigate and monitor the spatial and vertical distributions of rare plants of coastal salt marshes, such as *Lasthenia glabrata* at Los Peñasquitos Lagoon and *Cordylanthus maritimus* at Tijuana Estuary and Sweetwater Marsh. Both species are narrowly distributed annuals: *L. glabrata* is found on the margins of salt pans, and *C. maritimus*, a hemiparasite, is restricted to its hosts' ranges. Growth and reproduction of these species vary a great deal both within and between years in response to the quantity and timing of rainfall. GPS not only defines species' spatial distributions, but also their elevational limits. By superimposing data from year to year, we also monitor the temporal dynamics of population growth over time. Temporal differences in distribution can be related to elevation of occurrence, which in turn suggests specific limiting conditions such as soil moisture and salinity. Simultaneous data on covariate factors such as plant density, the identity of co-occurring species, and specific microhabitat conditions can be coded into each three-dimensional position recorded. This type of information is invaluable for planning restoration projects that are designed to support a diverse assemblage of species.

Characterizing fish use of different channel types

Physical structure is an important determinant of fish habitat in coastal wetlands. Channel morphometry, proximity and type of channel vegetation, channel location, flow rate, and sediment type are all known to influence the type and abundance of fishes in Atlantic and Gulf Coast wetlands (e.g., McIvor and Odum 1988, Baltz et al. 1993). We have found correlations between relative fish species composition and the morphology and hydrology of a variety of channel habitats in Sweetwater Marsh National Wildlife Refuge on San Diego Bay (Williams and Zedler, in prep.). Seine samples from narrow, steep-banked channels with clay sediments, elevated salinities, and low dissolved oxygen levels were generally dominated by longjaw mudsuckers (*Gillichthys mirabilis*), a species tolerant of low dissolved oxygen levels. Broad, shallow channels with low slopes, often fringed by emergent vegetation, were dominated by California killifish (*Fundulus parvipinnis*). Finally, samples from broad, deep channels (mean depth of 1.0 m) were generally dominated by topsmelt (*Atherinops affinis*), a schooling pelagic species. Associated research has also shown that shallow, vegetated, first-order creeks may be important habitat for juvenile killifish, which are found in higher proportion here (Desmond 1996).

Although the existence of species-habitat associations has been well-established in other regions, these habitat requirements or preferences have not been previously quantified in southern California.

In southern California, coastal wetlands have been constructed to compensate for the loss of deep-water habitat (e.g., MEC 1995) or to provide habitat and food for endangered birds (Haltiner et al. 1997). Such sites have been designed to provide deep subtidal channels with primarily sandy sediments, low sloping banks, and few shallow creeks. While several studies have shown that fish are quick to use constructed tidal habitats (Minello and Zimmerman 1992; Vose and Bell 1994; Havens et al. 1995; Simenstad and Thom 1996), it is important to consider which species of fish are desired and what habitats can attract and sustain them. If the objective is restore habitat for endangered birds (e.g., terns that feed on fish), then we need to mimic as closely as possible the natural systems that still support feeding by the target species. If the objective is to replace fish species that occur in deepwater habitats, then the fish assemblage and its habitat requirements need to be understood. An early understanding of project goals and an intimate knowledge of the system's ecology are imperative in any restoration project.

Characterizing water quality using dataloggers

Like channel morphology, water quality has a strong influence on fish use of wetland habitats, and is an important consideration in designing and monitoring constructed wetlands. At various wetlands in San Diego County, we have used dataloggers to characterize variability in physical-chemical conditions at reference sites as well as to monitor restoration sites.

Data on water quality (salinity, dissolved oxygen, pH, depth) have been collected (with measurements taken at 30-minute intervals) since 1995 from the main channel at Tijuana Estuary. Long-term monitoring of this site, which serves as a reference for the region (Zedler et al. 1992), will eventually allow us to develop water quality guidelines for future restoration projects. At Los Peñasquitos Lagoon, we have employed dataloggers to evaluate the effects of periodic mouth closures on lagoon water quality (Williams 1996). And in an evaluation of fish use of small intertidal creeks, dataloggers were used to compare tidal attenuation at two study sites in San Diego County. Greater tidal attenuation in creeks at one site, and the resulting decreased habitat availability, may have partially explained lower fish use (Desmond et al., in review).

Because human activities have negatively impacted the hydrology of many estuaries in southern California, improved tidal flushing is a main goal of several restoration projects. We are using dataloggers to evaluate one such restoration site at Tijuana Estuary; water depth data measured from either end of a newly excavated channel (prior to and after the excavation) will allow us to determine whether the channel is bringing about the intended changes in tidal amplitude and duration of tidal flushing.

Dataloggers allow analysis of fluctuations in water quality on a finer scale than was previously feasible. Detailed data from both restored and natural systems improves our ability to monitor current restoration projects and allows us to develop water quality guidelines for future projects.

Assessing marsh development with soil cores

Slow development of soils is a problem at many coastal wetland restoration projects. Because restoration projects are often completed using dredged material or other coarse-grain sediments, restoration sites frequently have sandier soils and lower organic matter and nutrient concentrations (especially total Kjeldahl nitrogen, TKN) than natural wetlands (Langis et al. 1991). It is anticipated that, as material accumulates within the restoration site, soil conditions will improve, both through the accumulation of organic matter and nutrients in the existing soil, and through the accretion of new sediment on the surface of the marsh (this method has been proposed for many restoration projects in sediment-rich San Francisco Bay). However, there are few data to determine if the desired changes actually occur at restored wetlands.

We have collected soil cores periodically from 1988-1996 to assess soil development at a restoration site in San Diego Bay. Despite predictions that restoration sites will follow a "trajectory" or predictable path of improvement, the data from San Diego Bay fail to indicate consistent improvements in any soil characteristics. On the contrary, annual changes in sediment organic content and TKN have been similar to changes at a nearby natural marsh, indicating that interannual variability is greater than any trend in accumulation of organic matter or nitrogen at this site. At present both sediment organic matter and TKN remain less than 70% of that in the natural marsh. Sediment texture is very coarse at this restored marsh and is the likely factor limiting nutrient accumulation. Even with additions of fertilizer there have been little retention of nutrients or accumulation of organic matter (Gibson et al. 1994; Boyer and Zedler 1996a, 1996b, in review).

Sedimentation rates are low at this site (Haltiner et al. 1997), so there is unlikely to be significant accumulation of fine sediment. We have documented a thin layer of new sediment on the surface of the marsh that is significantly higher in clay content and is likely to have more organic matter and nutrients. However, this thin layer has not significantly affected plant growth in the marsh. In addition, the marsh was designed with "target" elevations in place; if substantial amounts of new sediment were to accumulate, this would result in changes in marsh elevation and in the plant community (Haltiner et al. 1997).

The long-term comparison of soil texture, soil organic matter concentration and soil TKN is a useful tool for tracking the development of wetland soil. If these attributes follow a trajectory, one can predict when the site will achieve functional equivalency with natural wetlands and thus indicate when mitigation projects should be judged in compliance with reference wetlands. If they fail to show progressive development, the factors responsible need to be explored and corrective measures undertaken or alternative mitigation measures identified.

Determining plant responses to nitrogen fertilization through experimentation

Experimental manipulations are useful tools for assessing problems at restoration sites. We have used nitrogen fertilization to increase cordgrass height growth in sandy constructed marsh soils (Gibson et al. 1994; Boyer and Zedler 1996b; Boyer and Zedler in review), and we have used several methods to monitor the response of plants to increased nitrogen (N). Along with height histograms (see below), aboveground tissue nitrogen (TKN) and biomass (as estimated by total stem length; Zedler 1983) have proved useful in assessing the development of cordgrass in these constructed marshes (Boyer and Zedler, in review). The standing crop of N (which incorporates both N concentration and biomass) is higher in natural cordgrass stands and remains similar before and after fertilization, suggesting adequate nutrient supply from soils and internal recycling. In contrast, cordgrass in constructed marshes increases its N standing crop significantly with fertilization, matching that of reference natural marshes, but returns to pre-fertilization conditions in the next year (Boyer and Zedler, in review). As a result, the functions provided by taller, more robust and more N-rich cordgrass may also be lacking; these functions include habitat support for arthropods and nesting birds (Boyer and Zedler 1996a; Boyer and Zedler, in review).

When measured in the winter after aboveground tissue has senesced, N crop in the belowground tissues suggests the potential for growth in the next year. In constructed marshes fertilized for one growing season, belowground N crop of cordgrass approaches that of natural marshes; however, we know that this N retention response does not result in sustained high aboveground biomass (see above). We conclude that recycling in the tissues is inadequate to support optimal growth without sufficient soil N pools to augment supply. Thus, through experimentation, we have learned how important adequate soil N is to the structure and function of salt marshes. In other words, larger soil N pools are needed to sustain tall vegetation. Continued N additions over multiple years might eventually improve nutrient supply through increased organic matter deposition, building of N pools, and recycling within tissues.

While fertilizers may improve growth of salt marsh plants, and plant response to N additions may be used to compare nutrient sufficiency in constructed and natural marshes, fertilizers should be used with some caution. First, herbivory can increase with fertilization, although herbivorous insects did not increase following our experimental N additions (Boyer and Zedler 1996a). Second, fertilizing mixed-species stands can lead to changes in community composition (Boyer and Zedler, in preparation). Finally, fertilizers can lead to eutrophication and associated problems. Opportunistic green algae may be useful as an *in situ* indicator of changes in water column nutrients if N pollution of open waters is a concern (Fong et al., in review).

Characterizing marsh canopy architecture

Marsh vegetation varies in composition and vertical structure, and both attributes help determine the value of the site for nesting birds. A comparison of the canopy architecture of cordgrass stands that are and are not used by light-footed clapper rails for nesting indicates that suitable nesting habitat can be identified using height histograms (Zedler 1993). Light-footed clapper rails were found to nest where there were $\geq$100 stems m^{-2} and where at least 90 stems were taller then 60 cm and at least 30 of those stems were taller than 90 cm. Tall stems are woven into a canopy over the nest, which presumably protects the eggs and hatchlings from aerial predators, e.g., raptors. Tall stems are also presumed important for preventing the flotable nest from being washed away by a high tide.

Sampling for suitable nesting habitat is accomplished in September, after the nesting season. Sampling sites of 10-m radius or 20x20-m dimension should be selected within areas of about 2 ha, each of which might be a suitable home range for one pair of birds (based on densities of 0.52 nests/ha, Jorgensen 1975). A stratified random procedure is recommended to locate 10 circular quadrats per sampling site, with each quadrat being 0.25 m^2 in area. A random starting point can be identified and the 10 quadrats placed at 2-m intervals in a regular pattern that coincides with the shape of the patch of the most robust cordgrass. Plant heights are measured for every stem that is present within the quadrat; the tallest leaf is extended along a meter stick to record height. Flowering stalks are not included in height measurements, as they are not present during the nesting season; the above height standards relate only to leaf extension. It is not necessary to establish permanent quadrats for multi-year comparisons, as the tallest stand of cordgrass may occur in a different location the following year.

In our studies of constructed marshes at San Diego Bay, we have found few areas of cordgrass tall enough to support clapper rail nesting, and the rails have not nested in the transplanted cordgrass. The problem appears to be soil quality (texture too coarse, organic content too low, nitrogen concentration too low; Langis et al. 1991). Thus, special attention should be paid to the sediment provided. Nitrogen additions can increase cordgrass height, but the effect is not sustainable in coarse soils (Boyer and Zedler, in review).

Vertical structure is important to a variety of insects, spiders, and other birds. Belding's Savannah sparrows perch on the tallest plants and build their nests at elevations where tidal inundation is less likely (Brewster 1996). Pickleweed (*Salicornia virginica*) appears to be the preferred vegetation, but we do not yet have standards for pickleweed canopy architecture. Tall, dense vegetation may not be sufficient for sparrow nesting to succeed, however. In the next section, we indicate the need to protect marsh habitats from predators.

Monitoring avian use in coastal marshes using a reproductive index

Monitoring wetland birds in this region is often focused on rare or endangered species. For example, Belding's Savannah sparrows (*Passerculus sandwichensis beldingi*) are endemic to coastal wetlands of southern California and Baja California, Mexico. The subspecies is a metapopulation, with each marsh representing a distinct local population. Statewide totals of Belding's Savannah sparrows have been compiled by individual researchers and the USFWS since 1973. Based on these results, it appears that the metapopulation increased dramatically from 1973 to 1987, then experienced a decline. However, any assumptions about population growth or decline based on these numbers are tenuous because of inconsistent and inefficient monitoring efforts (Bradley 1973, Massey 1979, Zembal et al. 1988, James and Stadtlander 1991).

We (the San Diego Field Station, USGS-Biological Resources Division) conducted intensive research on Savannah sparrows at Sweetwater Marsh from 1994-1996. This area is highly disturbed, and several of its associated wetlands are distinct fragments surrounded by an urban landscape. Past surveys used walk-through census techniques and revealed high variability in local population size. We used a combination of walk-through, line-transect and spot mapping techniques to determine nesting densities. Line-transects and spot-mapping showed that walk-through censuses grossly underestimated the number of sparrows. Densities (i.e., number of birds/ha), varied among seasons, tides, and vegetation zones within the marshes. In spring, densities were highest in the high-marsh zones, which was consistent with preferred nesting habitat (Powell 1993). Interestingly, densities were higher during high tides in all zones of the marsh. This may be because birds were more easily seen when perched high in the vegetation. In addition, line-transect data also showed that detectability of birds decreased dramatically with distance from the transect. For example, at low tide, detectability was less than 50% only 10 m from the transect line.

We also used mark-recapture and spot-mapping to examine productivity of Belding's Savannah sparrows within the Sweetwater Marsh Complex. We used a reproductive index to evaluate reproductive success that was based on specific observed behaviors as well as the presence of fledglings within territories (Vickery et al. 1992a and 1992b). The high-marsh zones of Sweetwater Marsh appeared to be population sources because they produced an excess of young during the breeding season. Mid-marsh habitats within Sweetwater Marsh and at a connected marsh fragment produced enough young to maintain their populations. However, a small marsh isolated from the marsh complex produced no fledglings, and it was therefore a population sink. We saw little evidence of movement of birds between marshes.

Presence/absence data can establish basic information on extirpation and colonization of avian populations in coastal marshes. However, information on population size should be determined more accurately with density estimates. Unfortunately, we are currently lacking accurate information on actual areas of suitable habitat that is available to wildlife. Focusing on avian productivity may be the most useful information for managing the remaining marsh remnants and planning restoration. Management efforts should then key in on source and sink habitats to ensure sustainability of metapopulations. It is essential that clear objectives for monitoring are established in order to obtain the best information for effective management of these species.

Appendix 1. References.

Baltz, D., C. Rakocinski, and J. Fleeger. (1993). "Microhabitat use by marsh-edge fishes in a Louisiana estuary." *Environmental Biology of Fishes* 36, 109-126.

Boyer, K. E., and J. B. Zedler. (1996a). "Damage to cordgrass by scale insects in a constructed salt marsh: Effects of nitrogen additions." *Estuaries* 19, 1-12.

Boyer, K. E., and J. B. Zedler. (1996b). "Producing tall cordgrass vegetation with multiple additions of nitrogen." *In*: J. Zedler, principal author. *Restoring tidal wetlands: A scientific perspective and southern California focus.* California Sea Grant College System, University of California, La Jolla, CA. Report No. T-038.

Boyer, K. E., and J. B. Zedler. (In review). "Effects of nitrogen additions on the vertical structure of a constructed salt marsh."

Boyer, K. E., and J. B. Zedler. (In preparation). "Restoration of cordgrass at a California salt marsh: Species composition shifts with nitrogen enrichment."

Bradley, R. A. (1973). "A population census of the Belding's Savannah sparrow." *Western Bird Bander* 48, 40-43.

Bradshaw, B. K., D. Stow, and A. Brewster. (1996). "High resolution digital imagery and digital terrain data for mapping southern California coastal salt marshes." *In: Proceedings, ERIM Second International Airborne Remote Sensing Conference and Exhibition, San Francisco*, II:662-671.

Brewster, A. E. (1996). "Utilizing geographic technologies to analyze the nesting habitat preferences of the Belding's Savannah sparrow." M.S. Thesis, San Diego State University. 84 p.

Brewster, A. E., D. Stow, and L. Deysher. (1996). "Integrating airborne multispectral image data, kinematic GPS and GIS to analyze the habitat preferences of the Belding's Savannah Sparrow." *In* : *Proceedings, ERIM Second International Airborne Remote Sensing Conference and Exhibition, San Francisco*, II:653-661.

CCC (California Coastal Commission). (1991). "Staff Recommendation to Further Condition Permit No. 183-73, San Onofre Nuclear Generating Station Units 2 and 3." California Coastal Commission, San Francisco. 60 p. plus appendices.

Desmond, J. (1996). "Species composition and size structure of fish assemblages in relation to tidal creek size in southern California coastal wetlands." M.S. Thesis, San Diego State University. 71 p.

Desmond, J., J. B. Zedler, and G. D. Williams. (In review). "The influence of tidal creek size on fish assemblage structure in southern California coastal wetlands."

Ferren, W. R. Jr., P. Fiedler, R. Leidy, K. Lafferty, and L. Mertes. (1995). "Classification and description of wetlands of the central and southern California coast and coastal watersheds." *Madroño* 43, 125-182.

Fong, P., K. E. Boyer, and J. B. Zedler. (In review). "Developing an in situ bioassay of nutrient enrichment in coastal estuaries and lagoons using tissue nitrogen content of the opportunistic alga, *Enteromorpha intestinalis."*

Fritz, W. (1996). "The era of commercial earth observation satellites." *Photogrammetric Engineering and Remote Sensing* 62(1), 39-45.

FWS (United States Department of Interior Fish and Wildlife Service). (1988). "Biological Opinion 1-1-78-F-14-R2, The combined Sweetwater River Flood control and highway project, San Diego County, California." Los Angeles District, Corps of Engineers, March 30, 1988. 28 p plus appendices.

Gibson, K. D., J. B. Zedler, and R. Langis. (1994). "Limited response of cordgrass (*Spartina foliosa*) to soil amendments in a constructed marsh." *Ecological Applications* 4, 757-767.

Haines-Young, R. H., and M. Chopping. (1996). "Quantifying landscape structure: A review of landscape indices and their application to forested environment." *Progress in Physical Geography* 20, 418- 445.

Haltiner, J., J. B. Zedler, K. E. Boyer, G. D. Williams, and J. Callaway. (1997). "Influence of physical processes on the design, functioning and evolution of restored tidal wetlands in California, USA." *Wetlands Ecology and Management* 4, 73-91.

Hatfield, C. A., and B.T. Milne. (1996). "Patterns of vegetation response to creek network geometry." Abstract, 1996 Annual meeting, Ecological Society of America.

Havens, K. J., L. M. Varnell, and J. G. Bradshaw. (1995). "An assessment of ecological conditions in a constructed tidal marsh and two natural reference tidal marshes in coastal Virginia." *Ecological Engineering* 4, 117-141.

James, R., and D. Stadtlander. (1991). "A survey of the Belding's Savannah sparrow (*Passerculus sandwichensis beldingi*) in California, 1991." U.S. Fish and Wildlife Service, Southern California Field Office, Laguna Nigel, CA.

Jensen, J. R., M. Hodgson, E. Christensen, H. Mackey, and L. Tinney. (1986). "Remote sensing of inland wetlands: A multi- spectral approach." *Photogrammetric Engineering and Remote Sensing* 52, 87-100.

Jensen, J. R., D. J. Cowen, J. D. Althausen, S. Narumalani, and O. Weatherbee. (1993). "An evaluation of CoastWatch change detection protocol in South Carolina." *Photogrammetric Engineering and Remote Sensing* 59, 1039-1046.

Johnson-Freese, J. (1995). "Expanding remote sensing utilization: Issue and initiatives." *Earth Observation Magazine* April, 45-48.

Jorgensen, P. D. (1975). "Habitat preference of the Light-footed Clapper Rail in Tijuana Marsh, California." M. S. Thesis, San Diego State University. 115 p.

Jungwirth, M., S. Muhar, and S. Schmutz. (1995). "The effects of recreated increek and ecotone structures on the fish fauna of an epipotomal river." *Hydrobiologia* 303, 195-206.

Kirchofer, A. (1995). "Morphological variability in the ecotone: An important factor for the conservation of fish species in Swiss rivers." *Hydrobiologia* 303, 103-110.
Langis, R., M. Zalejko, and J. B. Zedler. (1991). "Nitrogen assessments in a constructed and a natural salt marsh of San Diego Bay, California." *Ecological Applications* 1, 40-51.
Massey, B. W. (1979). "The Belding's Savannah sparrow." U.S. Army Corps of Engineers, Work Order No. 6, Los Angeles, CA.
McIvor, C. C., and W. E. Odum. (1988). "Food, predation risk, and microhabitat selection in a marsh fish assemblage." *Ecology* 69, 1341-1351.
MEC Analytical Systems, Inc. (1995). "Anaheim Bay biological monitoring project, volume 1." Report to Port of Long Beach, Long Beach.
Menges, E. S., and D. R. Gordon. (1996). "Three levels of monitoring intensity for rare plant species." *Natural Areas Journal* 16, 227-236.
Mertes, L. A. K., D. Daniel, J. Melack, B. Nelson, L. Martinelli, and B. Fosberg. (1995). "Spatial patterns of hydrology, geomorphology and vegetation on the floodplain of the Amazon River in Brazil from a remote sensing perspective." *Geomorphology* 13, 215-232.
Minello, T. J., and R. J. Zimmerman. (1992). "Utilization of natural and transplanted Texas salt marshes by fish and decapod crustaceans." *Marine Ecology Progress Series* 90, 273-285.
Nyden, B., D. Stow, and J. B. Zedler. (1996). "Tracking an invasive plant species in a southern California wetland using high resolution digital multispectral imagery." *In: Proceedings, ERIM Second International Airborne Remote Sensing Conference, San Francisco*, II: 646-652.
PERL (Pacific Estuarine Research Laboratory). (1990). "A manual for assessing restored and natural coastal wetlands with examples from southern California." California Sea Grant Report No. T-CSGCP-021. La Jolla, California.
Phinn, S. R., D. A. Stow, and J. B. Zedler. (1995). "Monitoring wetland habitat restoration using airborne digital multi-spectral video data in southern California." *In: Proceedings, ERIM Third Thematic Conference on Remote Sensing for Coastal and Marine Environments, Washington*, I: 322-333.
Phinn, S. R., and D. A. Stow. (1996a). "Spatial, spectral, radiometric and temporal dimensions of remotely sensed data for monitoring wetland vegetation in southern California." *In: Proceedings, ERIM Second International Airborne Remote Sensing Conference, San Francisco*, I:64-73.
Phinn, S. R., and D. A. Stow. (1996b). "New techniques for assessing restoration and mitigation sites: Use of remote sensing to monitor vegetation properties." *In*: J. Zedler, principal author. *Tidal Wetland Restoration: A Scientific Perspective and Southern California Focus*. California Sea Grant College, La Jolla, p. 88-98.
Phinn, S. R., D. A. Stow, and J. B. Zedler. (1996). "Monitoring wetland habitat restoration using airborne multi-spectral video data in southern California." *Restoration Ecology* 4, 412-422.
Powell, A. N. (1993). "Nesting habitat of Belding's Savannah sparrows in coastal salt marshes." *Wetlands* 13, 129-133.
Simenstad, C. H., and R. Thom. (1996). "Functional equivalency trajectories of the restored Gog-Le-Hi-Te estuarine wetland." *Ecological Applications* 6, 38-56.
Skinner, M. W., D. P. Tibor, R. L. Bittman, B. Ertter, T. S. Ross, S. Boyd, A. C. Sanders, J. R. Shevock, and D.W. Taylor. (1995). "Research needs for conserving California's rare plants." *Madroño* 42, 211-241.

Stow, D., A. Hope, A. Nguyen, S. Phinn, and C. Benkelman. (1996). "Monitoring detailed land surface changes from an airborne multispectral digital camera system." *IEEE Transactions on Geosciences and Remote Sensing* 34, 1191-1203.

Vickery, P. D., M. L. Hunter Jr., and J. V. Wells. (1992a). "Use of a new reproductive index to evaluate relationship between habitat quality and breeding success." *Auk* 109, 697-705.

Vickery, P. D., M. L. Hunter Jr., and J. V. Wells. (1992b). "Is density an indicator of breeding success?" *Auk* 109, 706-710.

Vose, F. E., and S. S. Bell. (1994). "Resident fishes and macrobenthos in mangrove-rimmed habitats: evaluation of habitat restoration by hydrologic modification." *Estuaries* 17, 585-596.

Williams, G. D. (1996). "The physical, chemical, and biological monitoring of Los Peñasquitos Lagoon." Final Report prepared for the Los Peñasquitos Lagoon Foundation, Cardiff, CA 92007.

Williams, G. D., and J. B. Zedler. (In preparation). "Fish assemblage composition in constructed and natural tidal channels of a southern California salt marsh: a comparative evaluation of channel morphology, hydrology, and alteration history."

Zedler, J. B. (1982). "The ecology of southern California coastal salt marshes: A community profile." FWS/OBS-81/54, U.S.D.I., Fish and Wildlife Service, Biological Services Program, Washington D.C.

Zedler, J. B. (1983). "Freshwater impacts in normally hypersaline marshes." *Estuaries* 6, 346-355.

Zedler, J. B. (1993). "Canopy architecture of natural and planted cordgrass marshes: selecting habitat evaluation criteria." *Ecological Applications* 3, 123-138.

Zedler, J. B., C. Nordby, and B. Kus. (1992). "The ecology of Tijuana Estuary, California: A National Estuarine Research Reserve." NOAA Sanctuaries and Reserves Division, Washington, D.C.

Zembal, R., K. J. Kramer, R. J. Bransfield, and N. Gilbert. (1988). "A survey of Belding's Savannah sparrows in California." *American Birds* 42, 1233-1236.

CHAPTER 132

Regional Wetlands Planning: *A Case Study of Coastal Wetlands Planning in the San Francisco Bay Area & Southern California*

Craig Denisoff[1]
Maziar Movassaghi[2]

I. Introduction

Historically, wetland habitats were often seen only as a breeding ground for disease-carrying mosquitoes. From approximately the mid-18th century through the middle of the 20th century, the vast majority of wetlands in the United States were drained and converted into agricultural land through policies of the federal and state governments for what were then considered more "productive" uses. For example, the Federal Swamp Land Acts -National Swamp and Overflowed Land Act- of the 1800's gave 65 million acres of wetlands to 15 states, including California, for reclamation (Green 1996). In 1866, the California Legislature Commissioned the Board of Swamp and Overflowed Land to manage these properties. In turn, by 1870, the majority of wetlands had been transferred to private ownership. Between 1850 and 1920, about 70 percent of California's wetlands were destroyed, largely by levee and drainage projects (Dennis et al 1984). These projects where in some cases subsidized to aid private developers in reclaiming swamplands for agricultural purposes, helping to make California the leading agricultural state in the Nation by 1887 (CA Department of Water Resources 1993). Estimates of wetlands that historically existed in California range from 3 to 5 million acres. The current estimate of wetland acreage in California is approximately 454,000 acres; this represents an 85 to 90 percent reduction, the greatest percentage loss in the nation (Dahl 1990).

The majority of government efforts in response to these significant losses have come in the form of legal restrictions and regulations on the uses of wetlands. These restrictions are in part based on knowledge that 43% of endangered and threatened species rely directly or indirectly on wetlands for their survival. Also considering that 75% of existing wetlands are on privately owned lands, wetland protection efforts

[1] Deputy Assistant Secretary/ Wetlands Program Coordinator, California Resources Agency, 1416 9th Street, Suite 1311, Sacramento, CA 95814
[2] Executive Fellow, California Resources Agency.

have led to combative and antagonistic approaches between private landowners and those who wish to protect these fragile ecosystems through cumbersome regulatory controls. (US EPA 1993).

Recognizing the loss of these habitats, in 1993 California Governor Pete Wilson established the State Wetlands Conservation Policy (SWCP). The three central goals of this state-wide policy are to:

- Ensure no overall net loss and achieve a long-term net gain in the quantity, quality, and permanence of wetlands acreage and values in California in a manner that fosters creativity, stewardship and respect for private property;
- Reduce procedural complexity in the administration of State and Federal wetlands conservation programs;
- Encourage partnerships to make landowner incentive programs and cooperative planning efforts the Primary focus of wetlands conservation and restoration.

A primary focus in achieving these stated goals is to promote regional wetlands planning efforts. This regional approach was designed to permit State government to tailor statewide policies and programs to local conditions. In addition, these regional planning efforts are intended to help the State learn what works and what does not, and what type of wetland ecosystem management programs can be implemented in those parts of the State with a high likelihood of success. Two such efforts addressing coastal wetland habitats are currently underway in the San Francisco Bay Area and Southern California. In this paper we will focus on a comparative case study of these regions, detailing resource issues, regional issues, and conflicts surrounding coastal wetlands.

II. State Wetlands Conservation Policy

The first step in addressing the wetlands conservation in California, as called for in SWCP, was the development of a plan to address regional conservation of these habitats. This regional conservation approach was intended to help develop a wetlands conservation plan by having local entities incorporate wetlands into their planning process. In order to promote the comprehensive ecosystem planning approach it is important to develop baseline data on the extent of regional wetland resources and their interrelationship. Thus, data is collected from a number of sources including the U.S. Fish & Wildlife Service's National Wetlands Inventory, historical data from U.S. Geological Service maps, and other relevant environmental reports . These types of data are intended to help resource managers in identifying high value wetland habitats for regional restoration and enhancement activities.

The SWCP specifically called for a "no net-loss" of wetlands and a long-term gain in the quality, quantity and permanence of wetlands. Secretary for Resources, Douglas Wheeler, set "*a goal of a 30 to 50 percent increase in the quantity, quality and permanence of wetlands by the year 2010.*" That is equal to approximately 14,397 acres per year in California, using the 454,000 National Wetland Inventory figure as a base. This 30 to 50 percent increase is to correlate with the quality and diversity of wetland habitats, not just a pure increase in total acreage of wetlands. Since 1993, over 112,000 acres of wetlands have been restored and enhanced in California, while, an additional 50,000 have been acquired (see figure A). These gains include restoration and enhancement activities mostly in the Central Valley. The information does not account for cumulative loss from Nationwide Permit's (notably Nation Wide Permit 26's). However, contrary to assertions by some environmental organizations, recent comparative wetland losses from approved wetlands permits issued by the U.S. Army Corps of Engineers are relatively low. The current regulatory laws, both state and federal, require mitigation for all individual permits and even some Nationwide Permits (reporting). Thus only non-reporting Nationwide Permits go unmitigated and while very little data exists on the extent of those impacts, they do not come close to matching the wetland restoration and enhancement gains made over the past 3 to 5 years.

III. Regional Planning Efforts

The SWCP called for the development of regional strategies and efforts to achieve wetland conservation. However, the level of detail and direction for the two case study areas differed markedly. In the San Francisco Bay Area, the SWCP called for:

- -- an inventory of wetlands in the San Francisco Bay region;
- -- incorporating the resulting wetland inventory into a broader planning effort;
- -- promoting wetland acquisition, restoration and enhancement activities;
- -- developing landowner incentives; and
- -- incremental assumption of the Clean Water Act § 404 program, beginning with a pilot project by the San Francisco Bay Regional Water Quality Control Board.

In the Southern California region, the SWCP was more vague. The policy called for:

- -- develop a regional strategy to initiate better coordination and communication among diverse interests;
- -- involve principle stakeholders;
- -- long-term goals and priorities for conservation; and

FIGURE A

Long-term net Gain in the Quality, Quantity, & Permanence of Wetland Acreage & Values in California 1993 - 1997

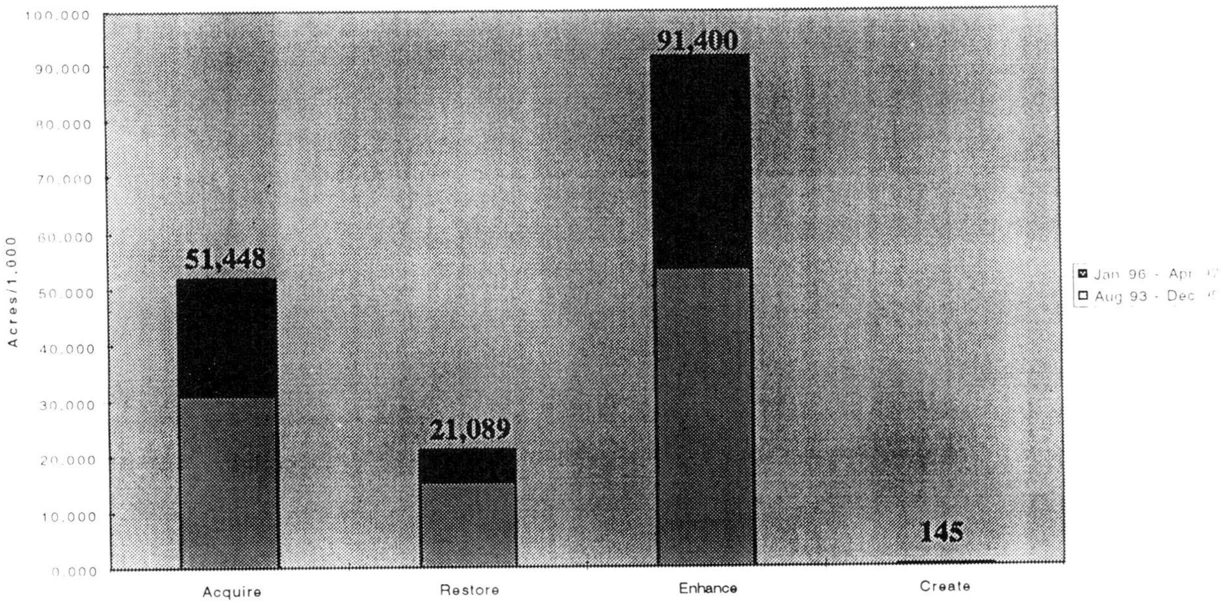

-- creation of a "joint venture" type effort.

The different level of specificity in the two regions could account for the differences in the institutional structures and existing planning activities in the San Francisco Bay Area and Southern California.

A. San Francisco Bay Area

The coastal area of the San Francisco Bay stretches from the mouth of the Golden Gate in the west to the confluence of the Sacramento & San Joaquin Rivers in the east, with the southern portion of the Bay extending to San Jose, and to Sonoma & Napa Counties in the north.

The wetlands of the San Francisco Bay Area have been greatly altered, due mainly to human activities. Recorded data from the late 19th century indicate that the tidal marshes of the San Francisco Bay Area numbered approximately 545,375 acres, which have been reduced to an estimated 44,371 acres (SFEP Status and Trends Report 1992). Impacts to these wetland habitats date back to the Gold Rush and the hydraulic mining in the Sierra Nevada Mountains, farming activities and, more recently, urban development.

The importance of these wetland ecosystems were recognized with the passage of one of the first coastal protection mandates in the United States, the 1965 McAteer-Petris Act, which created the San Francisco Bay Conservation and Development Commission. This body was designed to provide regional planning for the tidal and wetland resources surrounding the Bay.

Recently, a number of wetland initiatives, such as U.S. EPA's North Bay Initiative, the San Francisco Estuary Project (SFEP), the Long Term Management Strategy (primarily focused on dredging activities), the Regional Wetlands Management Plan, and a number of other efforts have began to address regional wetland planning in the Bay Area. All of these efforts recognize the need to develop a shared resource vision for wetlands in the Bay ecosystem.

Following the direction of SFEP's Comprehensive Conservation and Management Plan (CCMP) and Governor Wilson's SWCP, a number of state and federal agencies developed the San Francisco Bay Ecosystem Goals Project, in cooperation with the San Francisco Estuary Institute (SFEI). The primary purpose of this effort was to determine the number and types of wetlands necessary to sustain a healthy wetland ecosystem in the San Francisco Bay. With financial support from state and federal agencies, SFEI used historic and current maps, along with species and other habitat information in developing habitat mosaics depicting various wetland landscapes. Using these data, the agency scientists, along with academic and consulting wetland experts, are in the process of setting habitat goals for wetlands in the Bay Area. One of the major issues intended to be resolved by the Ecosystem Gaols project revolves around which type of wetland habitats should be restored in the Bay Area. For example, should wetlands be restored to tidal or seasonal wetland conditions? The Bay Area as previously mentioned has suffered a tremendous loss of tidal wetlands from sedimentation and the diking and draining of wetlands for agricultural and urban development. This has resulted in a number of species associated with tidal wetlands being listed as endangered (clapper rail, salt water harvest mouse). However, concerns have been raised in regards to where to locate these projects.

The San Francisco Bay Area also has many different, yet complimentary regional wetland planning efforts underway. While each one of these efforts may address a different aspect of the planning spectrum, they all share the need for overarching goals for achieving a self-sustaining and healthy wetlands ecosystem in the region. For example, while the newly created San Francisco Bay Joint Venture is focusing on implementing goals for wetland restoration and enhancement, the Bay Area Wetlands Planning Group is focusing on improving interagency coordination on planning and permitting issues. Both of the above mentioned groups are cooperating in their respective efforts to use the best available science to protect high-value wetlands habitats in the region.

B. Southern California Coastal Wetlands

The Southern California coastal area stretches from the Mexico-United States border at Tijuana to Point Conception, and is often referred to as the Southern California Bight. The actual acreage of historic coastal wetlands in the Southern California Bight is uncertain. It is thought that the coastal wetland habitats were mostly shallow water habitats (salt marshes and intertidal flats) rather than sub-tidal fish habitats (Zedler 1996). Similar to the San Francisco Bay Area, this area has also suffered a substantial alteration and loss of wetland habitats. Estimates of habitat loss and/or alteration range from 60 to 90 percent.

Except for the State Wetlands Conservation Plan, which is admittedly vague, their were no existing policies discussing methods to develop regional wetlands planning in Southern California. In addition, with notable exception of the initial campaign by local environmental groups, *Save California Wetlands* , there were no existing private or public regional comprehensive wetland planning efforts underway. The majority of wetland restoration or planning efforts in Southern California have been associated with site-specific wetland projects such as the local efforts to save Ormond Beach, Ballona Lagoon, and Bolsa Chica. This site-specific wetlands approach, however, can be somewhat problematic since entities focused on site-specific projects often do not take into account what is best for the overall coastal wetland resources in the region. The primary focus of such activities is only on what is best for their particular site. These site-specific efforts have resulted in conflicts between areas and even litigation regarding the preferred type of restoration at a project site.

Unlike the San Francisco Bay Area, there has not been a clear mandate for how to begin a comprehensive wetlands planning effort in Southern California. Based on the recommendations of the scientific community, federal and state wetland planning and regulatory agencies have worked together to develop a coastal wetlands inventory. With funding from the U.S. EPA and California Resources Agency, the State Coastal Conservancy (in cooperation with the California Coastal Commission and the U.S. Fish and Wildlife Service) undertook a comprehensive study of the coastal wetlands in Southern California to determine both their historic and current extent and to compile existing habitat and resource data on the over 40 coastal wetlands sites. The study was designed to serve as a basis for setting comprehensive regional wetland restoration and enhancement goals.

In order to help facilitate comprehensive wetlands planning, the Wilson administration has introduced a $5.75 million funding measure to promote the Southern California Wetlands Clearinghouse. The primary rationale behind the Clearinghouse was, similar to the Bay Area, to develop a shared vision for the restoration and enhancement of coastal wetlands in Southern California. Unfortunately, the early proposal met with substantial opposition due to the initial focus on restoration primarily through mitigation banking. However, given the limited future demand for coastal wetland mitigation (due mainly to the acquisition of Bolsa Chica lowlands by mitigation monies from the Ports of Los Angeles & Long Beach), the Clearinghouse effort has been re-focused on restoration and enhancement, with mitigation banking being only one component of the effort.

IV. Comparison of Regional Wetlands Planning Efforts

Due to a number of factors, the regional coastal wetland planning efforts are further along in the San Francisco Bay Area than in Southern California. First, The San Francisco Bay Area is more focused in its effort to develop a shared resource vision for the area in geographic terms. Second, the Bay Area region is generally comprised of one major watershed, while Southern California has many different watersheds spread-out in a much larger geographic area. Thirdly, The Bay Area has many different, yet cooperative and coordinated planning efforts underway. As compared to an unsuccessful attempt to create a "joint venture" in Southern California, the San Francisco Bay has recently created a joint venture mechanism to allow for early implementation of wetland restoration and enhancement efforts. In addition, there are two existing and relatively detailed guidance documents to help guide wetland restoration activities in the Bay Area (SFEP's CCMP and SWCP).

In Southern California, there is very little ties between the wetland stakeholder groups in different portions of the area. For example, the Santa Barbara wetland advocates are usually not sitting at the same table as advocates from the San Diego area, and thus may not share the same vision for comprehensive restoration for the region. These local restoration activities in effect then begin to compete with one another for a limited pool of resources, creating further fracture between these local groups. Public entities in this area are also competing with one another. For example there are four separate Regional Water Quality Control Boards (RWQCB) that have regulatory authority along the Southern California coastal area. Each RWQCB may have its own set of plans focused to meet the needs of their respective region. This fracturing of planning and restoration activities has led to very little overall guidance in terms of developing comprehensive wetlands planning for the entire region and has predominantly centred around strong local and site-specific wetland protection efforts.

In order to address this issue, the Clearinghouse concept was initiated to help set habitat goals using the Coastal Conservancy's Southern California coastal wetlands inventory. In addition, the Coastal Conservancy will be convening a set of scientific workshops for noted wetland ecologists and biologists to provide input to state and federal planners regarding wetlands restoration and enhancement goals. The Coastal Conservancy will also hold public forums and workshops to solicit input from stakeholders on how best to begin wetlands restoration in the area. However, without more public support, "how and when" wetlands conservation will be achieved in Southern California is in question.

V. Conclusion

Each region has its unique and varied characteristics, and thus transposing a set of regional plans to another is doomed for failure. However, there are some basis precepts that are vital to the success of any effort:

1. Importance of a shared vision: identifying and setting habitat restoration goals within a given region provides a road-map for both the public and private sectors to follow and reduces inter-agency conflicts (both federal-state and local-state-federal conflicts) while increasing constituency support;

2. Geographic focus: it is important to shape regional planning efforts so that all parties understand the benefits of comprehensive regional

ecosystem restoration over a site-specific focus. Thus it is important to educate stakeholders to the habitat benefits from restoration in adjacent areas not just their backyard;

3. Flexible mechanisms: no one entity should be responsible for implementing the shared vision. In the San Francisco Bay Area there are many different groups and entities working to restore wetlands, and as long as these different entities share the same goals and cooperate in attaining said goals, then no one person or group must shoulder complete responsibility. This flexible and cooperative mechanism leads to more active participation and greater "buy-in" to the implementation process from all participating entities.

4. Different methods to achieve overall wetlands conservation. a combination of methods such as restoration, enhancement, and acquisition (less than full fee easements), along with mitigation can be used to achieve ones goals. For example, the acquisition and restoration of the Bolsa Chica wetlands was finally realized with the use of mitigation monies from the Ports of Los Angeles and Long Beach.

5. Stakeholder participation and coordination: governmental entities can assist and provide the tools (such as inventories or initial funding) for regional planning, but the involved stakeholders must see the benefit in moving forward in a similar direction in order for any planning process to be successful. For example, in the case of the Natural Conservation and Community Planning in San Diego, cooperation from both the main-stream environmental and business entities helped to move the process forward in order set and implement habitat goals.

It is fair to say that in both the San Francisco Bay Area and Southern California it is too early to determine if these initial regional wetlands planning efforts will be completely successful. However, it is evident that efforts that attempt to find common ground and work toward similar resource goals are more likely to be successful, which regardless of the potential vehicles, will lead to improved wetland habitats.

Bibliography

California Department of Water Resources. 1993. *Sacramento-San Joaquin Delta Report.* Sacramento, California.

Dahl, T.E. 1990. *Wetlands-Losses in the United States, 1780's to 1980's.* US Fish & Wildlife Service Report to Congress, Washington D.C.

Dennis, N.B., M.L. Marcus, & H. Hill. 1984. *Status and Trends of California Wetlands.* Report to the Resources Sub-Committee, the California State Assembly. Sacramento, California.

Green, Collen. 1996. *The Wetlands Restoration Revolution.* Erosion Control. Nov/Dec 1996. pp. 24-37.

San Francisco Estuary Project. 1992. *Status and Trends Report on Wetlands & Related Habitats in the San Francisco Estuary.* SFEP, Oakland, California.

Zedler, Joy B., Principal Author. 1996. *Tidal Wetland Restoration: A Scientific Perspective and Southern California Focus.* California Sea Grant College System, University of California. La Jolla, California.

CHAPTER 133

Wetland Reference Sites in San Francisco Bay (Regional Water Quality Control Board, Region 2)

Andree Breaux[1], Feride Serefiddin[1], and Michael Carlin[2]

Abstract

The San Francisco Bay Regional Water Quality Control Board has received funding from the U.S. Environmental Protection Agency to establish guidelines for wetland mitigation and monitoring activities within the Regional Board's jurisdiction. What follows here is the first of several commentaries based on review of approximately 116 mitigation and monitoring reports between 1988-1995. This review deals with the use of reference systems for parallel monitoring of mitigation projects in the San Francisco Bay area. Results indicate that less than 25% of the compensatory mitigation projects reviewed have used, or have planned to use, reference systems to guide the determination of success. While the monitoring of reference sites is not a requirement for mitigation projects and, in many cases, may not even be necessary, references are undoubtedly important, along with the scientific literature and information from other projects in the so-called gray literature, for establishing overall standards to measure the no net loss policy. When reference sites are used as measures of success, it is important that appropriate sites be selected for standards, and that records are kept of the range of those measures.

We suggest that resources applied to wetland reference systems for mitigation and monitoring projects might be better spent by pooling wetland types within the same watersheds and subwatersheds. The systematic observation and review of these reference systems will increase the likelihood that these sites will be evaluated by scientists and the interested public, instead of receiving little or no review by overloaded permit reviewers in agencies with

[1] Regional Water Quality Control Board, San Francisco Bay Region, 2101 Webster Street, Oakland, CA 94612; (510) 286-1277.

[2] San Francisco Public Utilities Commission, 1155 Market Street, San Francisco, CA 94103.

Disclaimer: the following report reflects the opinions of the authors, and not of the Regional Board members, the Executive Officer, or the general staff.

relatively high turnover rates. Consolidating wetland reference systems and the information derived from them in the same watershed or subwatershed will benefit wetlands and other natural resources in the region, first, by encouraging a landscape perspective in viewing the interdependence of wetlands on other resources and, second, by allowing the public to visualize existing wetland resources and to participate in the prioritization of those resources. This prioritization could extend to the location, type, number, and methods of monitoring reference sites in general, in addition to those that might be selected as mitigation banks.

Introduction

As interest in restoration ecology has continued to grow nationally during the 1990's, so has the number of wetland project sites in the San Francisco Bay Region. Among the large-scale efforts to establish wetland study sites or to increase wetland acreage are: the Long Term Management Strategy to restore wetlands with dredged sediment; the Montezuma project, which will be a case study under this strategy; the establishment of five or six National Estuarine Research Reserve sites to preserve the best wetlands in the area; and the Wetland Ecosystem Goals Project, which is attempting to set goals by type and amount. There are many other projects and studies as well. In addition, considerable attention is being paid to the evaluation of wetland impacts through the Hydrogeomorphic Approach, which relies on the use of numerous reference sites to provide standards for measurement, as well as to the establishment of mitigation banks to facilitate the review of small projects and to encourage mitigation for those projects that might otherwise be dismissed as inconsequential. For alongside these large, expensive, and slow-moving efforts to establish long-term study sites or simply to increase wetland acreage, many small-scale projects continue to pour into inadequately staffed resource agencies for review and approval.

Reference sites are defined here as areas to be monitored in conjunction with a given project for purposes of comparison to help in the formulation of standards for mitigation projects. In the future, results from such references may actually become standards for mitigation projects, but at this time our information is too fragmentary and incomplete. This paper covers projects using references for parallel monitoring of ecosystems, and leaves the design aspect to consultants or engineers. Unfortunately reference systems sometimes are given the status of controls in laboratory experiments which they should never hold, because outdoor, large-scale projects cannot be controlled for climate, temperature, edaphic conditions, hydrology, nutrient cycling, and other natural and unnatural processes. Furthermore "large-scale" might refer to anything over one square foot, depending on the scale of one's subject of study. Reference systems should serve only for general guidelines. And in the broadest sense, the information from all monitoring projects can constitute a reference for other projects. Therefore, they should be viewed with flexibility, and not as inflexible standards that projects must meet or else be classified a failure. Moreover, the use of reference sites is not always necessary or even appropriate, for there may be ample information already available to defend a hypothesis of expected growth rates and patterns for creation or restoration projects based on the literature or on reliable data from other study sites. Where reference sites are used as a standard of success, however, it is important that they be appropriately assessed to avoid potentially large losses of both function and acreage.

Three additional cautions emerge when reviewing wetland mitigation sites, whether or not parallel monitoring of reference sites is planned: (1) avoidance -- the first element of mitigation sequencing -- should be emphasized to obviate the risk of losing what we may not be able to replace; (2) the scientific literature should complement both the field experience of the wetland consultant and the use of references; and (3) the more, the larger, and the longer the monitoring of wetland reference systems, the more reliable the information will be to make scientific and management decisions dealing with wetland systems. This last caution holds, however, only insofar as the sites are studied by means of reliable and consistent scientific methods. Otherwise, the references used as standards may do more harm than good.

The Role of the San Francisco Bay Regional Water Quality Board in Regulating Wetlands within Watersheds

The San Francisco Bay Regional Water Quality Control Board (hereafter referred to as Region 2), is in a favorable position to contribute to the protection and restoration of wetlands, due to its authority over wetland permits through both the Clean Water Act and the Porter-Cologne Act. This authority requires wetland permittees to submit applications, no matter how small the activity, and to compensate, through mitigation, for losses that cannot be avoided or minimized. The Regional Board can, then, determine or direct mitigation either on a case-by-case basis, or for mitigation sites established in watersheds or subwatersheds. Our efforts can be directed towards identifying reference sites within watersheds of the San Francisco Bay, which may then be reviewed by the community and either dismissed as inappropriate or selected as a site that merits continued future monitoring as a reference. Only by encouraging long-term study sites will we begin to recognize trends in wetland science that will prove useful for wetland policy and management. Many sites are monitored briefly and then abandoned because they serve no immediate purpose to the permit applicant. These reference sites should be built upon rather than discarded. We present the Suisun Basin in this report as a model of how information on reference sites typically contained in mitigation reports might be usefully presented. The information is not complete, and further demonstrations of watersheds and subwatersheds will depend on funding for future efforts..

As it is, operating under a "no net loss policy", the Regional Board requires permittees to compensate for wetland loss by at least a one-to-one ratio of acres lost to acres created or restored. In addition, in following the lead of the U.S. Army Corps of Engineers for mitigation projects, the Regional Board directs its permit applicants "to provide a summary of the functions and values of the wetland to be filled" (San Francisco District Corps of Engineers 1991). More often than not, neither functions nor values are measured by the permittee or wetland consultant, and never are all the functions or values of a site determined. Agency staff are dependent, in large part, on the opinions and determinations of the wetland consultants which are sometimes excellent, and sometimes not. Directing financial resources toward the development of a few well monitored reference sites in each watershed or subwatershed, will inevitably lead to a greater understanding of exactly what we are gaining and losing in terms of wetland functions. Use of the Hydrogeomorphic Approach (HGM) to analyze the loss and gain of wetland functions may prove to be a useful tool in the future. The jury is still out, however, as to the usefulness of this relatively new and untested approach, which will be reviewed as the next paper under this grant. A statement from the proponents of HGM is worth noting here to emphasize the importance of reference systems, whether they are used in large numbers or as single systems:

> ...the use of reference forces one to constantly revisit and calibrate on reference standards. In so doing, reference standards and reference wetlands take the lead in guiding mitigation, and all parties involved in mitigation are fully informed of the process. This potentially leads to greater consistency and efficiency in achieving goals of the Clean Water Act Section 404." (Brinson & Rheinhardt 1996)

Again it is important to note that, whatever the number of reference sites selected, if the wrong ones are used, substantial functional losses may occur.

In the absence of agreed upon techniques for measuring wetland functions, Regional Board staff have attempted to sort the wetland projects within Region 2 on a strictly acreage basis, particularly for 401 certifications and the projects reviewed for the 404 pilot program. We have begun to locate all wetland projects by watershed in a General Wetland Project Inventory (GWPI). Also, we are compiling detailed records of past mitigation projects in Region 2 with a view to compiling both the losses and gains by wetland type, and the technical information used to monitor these projects in a Mitigation Database (MD). The purpose of such project tracking is three-fold:

(1) to determine what data is useful or lacking in permit applications, mitigation plans, and monitoring reports,
(2) to account for wetland losses and gains by wetland type in Region 2, and
(3) to direct future projects toward watersheds or subwatersheds that are deemed to have both a need for increased wetland acreage and a chance of success for wetland creation, restoration, enhancement, or preservation of uplands or wetlands.

This report deals with a subset of the two major projects described above (GWPI and the MD), specifically what information has been collected and analyzed for reference sites in the San Francisco Bay Area that falls under the jurisdiction of Region 2.

As the agencies reviewing 401 applications (and, for Region 2, 404 applications under the pilot program), the nine Regional Board offices in the state of California are in a strategic position to receive information on wetlands that are being lost, restored, and used as standards for comparison for creation, restoration, or enhancement projects. It is crucial that the information passing through the regions be recorded, analyzed, and steered toward some regionally agreed upon goal. We hope to have provided a start to this project.

Methods

General Methods used for General Wetland Project Inventory and Mitigation Database

A major assumption in compiling a GWPI and MD is that many wetland projects are being designed and carried out by scientists and individuals who know and understand how to create and restore wetlands, and where appropriate reference models for those wetlands exist. We have attempted to compile this information in the hope of building on it, and of avoiding the loss of valuable monitoring data in the future. We have relied largely on state funds and the existing NWI as basemaps for starting a General Wetland Project Inventory (GWPI), which currently contains over 220 wetland projects in Region 2 including mitigation, restoration, enhancement, reference, scientific, and wastewater treatment (Figure 1). The projects can be viewed on GIS using the EcoAtlas as well. The purpose of compiling an inventory of wetland projects is threefold: first, to establish a baseline for study sites or future mitigation banks; second, to assist with the general accounting of wetland losses and gains in Region 2; and, finally, to assist with developing a strategy for establishing reference sites for watersheds.

In addition to the GWPI, our Mitigation Database (MD) contains another 116 mitigation sites, 27 of which have monitored, or expressed an intention to monitor, reference sites as standards of comparison for their mitigation projects. Precise locations for approximately 100 of the 116 mitigation sites have been determined to date (Figure 2). Locations for the 27 proposed reference sites from the MD are depicted on Figure 2(a). We hope eventually to provide maps of all of these reference sites, by watershed and wetland type. It should be noted that the data derived from the MD is based on both proposed mitigation sites and those that have actually begun construction. A preliminary review of the data in the MD indicates that less than ten of the total 116 projects entered in the Mitigation Database have been determined to be completed and successful. It follows, then, that any results based on the information contained in the mitigation projects should also be viewed as preliminary.

For the MD and the GWPI, the approximate center of each wetland project was first located based on review of project descriptions contained in reports or described over the phone. The following guidelines were generally used:

(1) if the shape of the mitigation site is square or rectangular, the intersection of bisecting lines was taken
(2) if the shape is circular, the midpoint of the diameter is taken. This is also estimated and precise to hundreds of feet
(3) if the project site is linear, as in riparian corridors, the midway point along the length of the project site was taken. If the corridor curved, the same method was used.
(4) after the center point was chosen, the file was reviewed for a description of the project location, and any maps depicting the site. The corresponding USGS 1:24,000 quad map was then used too relocate the site for those in the MD. For the GWPI, project locations were placed on NWI maps, which are also on a scale of 1:24,000.
(5) the site was marked and coordinates determined using "The Coordinator". This tool is used to determine precisely the North Latitude and West Longitude of any point on all United States Coast & Geodetic Survey Maps at several scales.

For projects located for the MD, we estimate that the precision of the center point is in hundreds of feet, and the latitudes/longitudes are also accurate to hundreds of feet. For projects located in the GWPI, the locations may not be as accurate since our methodology evolved as the Inventory grew. We urge caution, then, in relying too heavily on the GWPI for precise locational information. As always, we urge readers with more precise information, to notify us so that we can make the necessary changes.

Files were reviewed and information gathered from the Department of Fish and Game at Yountville, and also from the Contra Costa Mosquito Abatement District. Most other information was gathered from the *Ecological Restoration*

Directory for the San Francisco Bay Area (1990) (approximately 47 of the original projects for the GWPI came from this directory), project descriptions and maps supplied by Ducks Unlimited, files housed at the U.S. Army Corps of Engineers in San Francisco, and the gray literature on wetland permits and monitoring plans contained at the Regional Board Office in Oakland. Overall, more than 45 different agencies and private consulting firms were contacted for new or revised data. The GWPI provides a generalized certainty ranking of "1" for locations mapped from reliable narratives or graphic data and a "2" for locations determined from less reliable data. Hardcopies of the data used to compile the GWPI are kept at the Regional Board office in Oakland.

Reference Sites in GWPI and MD

A list of reference sites used in the mitigation projects reviewed for this study are contained in Appendix 1 (total reference sites = 27 of approximately 116 mitigation projects). A supplemental list of reference sites described in the Regional Board's GWPI is contained in Appendix 2 (total reference sites = 29 of approximately 220 wetland projects). The GWPI was sent out in its entirety (including all project types, not simply reference sites) to approximately 20 reviewers in April 1996, and their comments were addressed. The information contained in the MD (Appendix 1) is much more extensive than that in the GWPI (Appendix 2) because the latter is meant to be simply a list of wetland project locations, while the former is designed to answer numerous and specific questions on the type of monitoring being conducted at impacted wetland sites and their associated reference sites.

Appendix 3 contains a list of the wetland types used in both the previous appendices based on the MD and the GWPI. The types were recommended by Molly Martindale (U.S. Army Corps of Engineers, San Francisco District) who has had extensive experience in reviewing Mitigation and Monitoring Plans in the region. The selected wetland types were also based on experience with the Wetland Ecosystem Goals Project. A final list of habitat types has not yet been finalized by the group that sets policy for the project. It is our belief, however, that the wetland types used for the Mitigation Database and GWPI can be cross-referenced with the wetland types eventually decided on. This will help form the basis for a uniform accounting system for use by agencies and individuals interested in determining wetland loss and gain by type in the region. Appendix 4 lists projects in the first two appendices, sorted by the Suisun Basin Hydrounit or watershed. The fifth and final appendix contains a list of potential reference sites listed by the Plant Focus Team of the Regional Wetland Goals.

Again, an important feature of collecting the kind of information presented here is to encourage overall review by outside parties. The more reviewers, the greater the chance that future suggestions will be based on the agreement of many, rather than the opinions of one or a few. We urge readers to notify us of any mistakes that are found so we can rectify them in the databases.

Results

Results from the projects reviewed and entered in the Mitigation Database (Appendix 1) indicate that only 27 of the 116 projects (23%) reviewed used or considered using reference sites as a means of determining success of the proposed mitigation project. Of those 27 projects, very few, in their mitigation reports, describe the actual or proposed reference sites by name, location, or attributes to be measured at the sites. Descriptions are often limited to "adjacent wetland sites" for references, with no indication of where the reference wetland is or what will be measured there. Most sites using references rely on only one site (70%) which is usually described as adjacent to the site. One project, however, used over 50 reference sites which were combined under the Hydrogeomorphic Approach to wetland analysis, which aims to give great emphasis to the reference condition. In addition to the attempts to use adjacent sites, presumably in the same watershed, most projects reviewed sought to select a reference of the same type as the wetland to be created or restored. While this may seem like an obvious requirement for site selection, scientists and consultants have complained about the difficulty in finding suitable reference sites in the same watershed.

Measurements planned or actually taken included the following for the 27 projects (totals in parentheses): 44% proposed vegetation monitoring (12); 26% hydrology (7); 22% wildlife (6); 15% water quality (4); and 4% benthos (1). One of the projects also measured soil salinity. The variation in type and frequency of measurements was large. One project, for example, monitored three major site features (hydrology, vegetation, and wildlife), under which ten measurable attributes were listed: tidal monitoring, surface water levels, inundation regime, channel geometry/stability, salinity, percent cover, plant species diversity/richness, productivity, density and wildlife species diversity/richness. A different project, on the other hand, measured only percent cover and species diversity of vegetation. Forty-eight percent of the 27 projects (13) proposed no performance criteria at all, but simply stated a plan to use reference sites.

In our final guidelines due in the Fall of 1997, we hope to suggest an adequate number of attributes to measure per site, in addition to ranges for some of the major site attributes monitored for wetland studies. What is evident from this first look at the reference sites used in wetland studies in Region 2 is that the data proposed or collected for these relatively small projects is sparse and highly variable. These two features alone suggest that where reference sites are being used as standards for success, we might be losing more wetlands than we are gaining. This is to say that case-by-case reference sites may be too small and too short-term to merit the kind of technical review that would make a strong and reliable reference site. If, however, knowing the amount of percent cover after five years is virtually all that is required in a compensatory mitigation project, then these case-by-case reference sites may be adequate. Nonetheless, to begin to understand wetland functional losses and gains, long-term reference sites would provide a more solid basis for evaluating the success of mitigation projects because they would begin to account for the variation that natural wetland systems are subjected to. We should, at least, be tracking the location and overall performance of reference sites. Those that are abandoned will at least have a recorded performance record, and those that continue to be monitored will shed light on succession and on the reaction of created or restored systems to future changes in climate and hydrology.

For larger wetland restoration and creation projects which have not yet been permitted, or do not necessarily require compensatory mitigation, reference sites are proposed on a much larger scale, both as a model and as a standard for success. The Arrowhead Marsh Restoration project (approximately 71 acres), for example, has relied on information from six different references systems in its design phase. The use of so many is necessary, in large part, because no one wetland system is completely unaltered in the San Francisco Bay region. Other large-scale projects such as Montezuma (approximately 1,800 acres of tidal wetland proposed for restoration), Sonoma Baylands (320 acres of restored wetland area), and Ora Loma have also proposed the simultaneous monitoring of project areas with reference sites. The criteria being set for these extremely large sites may well serve to guide the smaller sites. At the very least, methods should be similar for both large and small sites. Moreover, it may be that funds put toward small projects with questionable chances of success might be better spent as contributions toward more extensive and thorough wetland study sites -- either established references, or creation or restoration sites -- where the proper study design, statistical analysis, and large-scale monitoring can be designed for both the scientific study of wetlands and for the increase of wetland acreage and quality.

In sum, based on the mitigation reports covering the period from approximately 1988 to 1996, it appears that a relatively small number of mitigation plans have even considered using reference systems to compare the mitigation wetland to existing wetlands of the same type. Those that have used references have selected a minimum of attributes to measure, and very few do frequent or extensive monitoring. It is generally up to the wetland consultant to select an appropriate site for analysis, which in itself is controversial, given the extensive artificial changes in wetlands in the San Francisco Bay area over the last 150 years. Compiling data on the location, on the performance criteria and, eventually, on the success of the mitigation site associated with the reference sites, should provide a framework in which knowledgeable scientists and resource managers can review the pool of references.

Policy options to consider include:

(1) allowing the status quo to continue with each mitigation project selecting its own methods of selection and review;

(2) encouraging the centralization of information on mitigation sites and their associated references either by Regional Board staff or by staff at the San Francisco Estuary Institute (the body expected to oversee monitoring of the San Francisco Bay and its wetlands). This would involve completing the compilation of wetland reference sites in the remaining 6 watersheds using the Suisun Hydrounit as a model (Appendix 4); ranking existing references by the feasibility of preserving, acquiring, and expanding the area; and finally, encouraging large-scale and long-term studies by local universities and colleges. While still in draft form and not yet depicted by wetland type, Figures 3 and 4 give a preliminary idea of the location of some reference and mitigation sites in the Suisun watershed: Figure 3 depicts the 12 reference sites from the GWPI, and Figure 4 depicts the 24 mitigation sites from the MD. An additional 5 reference sites from the MD remain to be mapped on GIS in the Suisun Hydrounit. Though incomplete, the GIS maps suggest how we might begin to group wetlands for mitigation and monitoring: first by watershed (shown here for Suisun), and later by subwatersheds and wetland types. Future goals can then be based on the current need for wetlands in a subwatershed determined by the interested public (e.g., larger riparian buffers along stream channels to purify water vs. more perennial wetland storage areas for winter floodwaters), or determined by scientists and resource managers for the Wetland Ecosystem Goals project based on the habitat needs of important wetland species.

(3) as an extension of number 2, adopting mitigation bank sites that could serve as tightly monitored models for created or restored systems, in addition to requiring appropriate and well-monitored reference sites for each bank. If well-designed mitigation banks and associated reference sites were established for each wetland type in each watershed or subwatersheds, the pool of wetland study sites in Region 2 would be substantially increased. At the same time, extensive review of proposed study designs and monitoring programs for the mitigation banks by experts from different agencies and disciplines, would be more likely to prevent the loss of wetland functions than the current style of one person from each agency (if that), struggling under time clocks for scores of permits requiring review. It is this extensive review of each small mitigation project that is currently lacking.

To the extent that reference sites continue to be used on a project by project basis, being selected and monitored exclusively by the wetland consultants with no overall guidance from resource agencies or the interested public, the following are some necessary components of mitigation reports. First, clear and precise locations of both the project site and all reference sites used should be provided. USGS quad maps are invaluable, but should be accompanied by the latitude/longitudes as well, so that input into the GWPI or MD would then be a simple data-entry step before GIS input. Secondly, reference sites should be the same wetland type and be located within the same watershed or subwatershed as the mitigation site. Finally, a scientifically and statistically defensible study design should be provided that specifies criteria to be monitored and defines the goals to be achieved. As stated earlier, reference sites are not required components of mitigation plans, and their use as a standard is not always appropriate. To the extent that the consultant or permit applicant wishes to rely on reference sites as a standard for success, however, they need to be reviewed for their suitability and adequately monitored.

Future Directions

Along with other agencies and scientists in the San Francisco Bay Area, Regional Board staff are involved with the Wetlands Ecosystem Goals Project which is establishing recommendations for types and amounts of wetlands in the region. We expect that these recommendations will form the basis for directing future mitigation projects to meet specific goals. With or without the Goals, however, we believe that our efforts to locate and describe wetland projects -- from reference systems to mitigation sites -- can be a useful tool in determining where future study sites should be directed. It is this author's opinion that more effort should be directed toward developing permanent, long-term wetland study sites in each watershed as these will provide more reliable answers to scientific questions than short-term studies. Small-scale mitigation projects are expensive for the permittee and for the tax payer who must pay for staff review. And, if success of compensatory mitigation is based on the selection of an inappropriate reference site, or on inadequate scientific methods, then the small-scale projects may be costing us more than mere money.

As we move into the next phase of our EPA grant -- an attempt to establish technical guidelines for mitigation and monitoring of wetland projects -- our efforts will need to be focused on that task, and less on maintaining the general wetland inventory for the San Francisco Bay area. Our resources have been cut back in the Wetland unit and we cannot maintain both the Inventory and the MD. We can, however, attempt to continue with the MD in terms of the latitudes and longitudes of site locations, so that regional planning can be assisted by GIS used elsewhere. It may be most efficient to turn over the maintenance of our GWPI to the San Francisco Estuary Institute (SFEI), which is currently coordinating the Wetland Goals Project and will be attempting regional analysis of those Goals on GIS. The Regional Board would continue to have authority over its mitigation and monitoring projects, but would be working with the resource community in determining which wetlands should be increased or improved, where those increases or improvements should take place, and how wetlands should be monitored to provide the best scientific and administrative information . We have proposed such a plan in our 1998 grant to the U.S. EPA. As interest in wetlands grows in the lay and scientific communities, it behooves us to continue what we have started in terms of a regional wetland program that will streamline the processing of 401 certifications and 404 permits, and increase the quality and quantity of wetlands in the San Francisco Bay Area.

In sum, this report describes our approach for organizing wetlands information and for analyzing wetlands mitigation projects. If wetland projects are visualized on wetland maps, by type and in each watershed or subwatershed, then we can begin to direct the isolated wetland parcels toward a broader system of connected and larger units that are favored by most species.

Acknowledgments

This work has been partially funded by a grant from the U.S. Environmental Protection Agency. Specific thanks is directed to Molly Martindale, Janet Meth, Carrie Salazar, Rachel Bonnefil, and Lynn Suer.

References

Brinson, M.A. and R. Rheinhardt. 1996. The role of reference wetlands in functional assessment and mitigation. Ecological Applications, Vol. 6, No. 1: 69-76.

San Francisco District Corps of Engineers. 1991. Habitat Mitigation and Monitoring Proposal Guidelines. Department of the Army, San Francisco, CA.

Appendix 1: Reference Sites Used for Mitigation Projects, SF Bay, Region 2

Appendix 1: Reference Sites used for Mitigation Projects (Total mitigation projects reviewed = 116)

Count	Project name	# in Mit. D.B.	Permit type	COE #	Project size (ac)	Hydro-unit	# of ref sites*	Monitoring Veg	Hydrol	WQ	W/life	Benthos	Adjacent to site?
1	Redwood Landfill Levee	1	NWP 33	17463N83	3.00	206.30	1	Y	Y		Y		Y
2	Anderson Drive Extension	2	NG	18893N83	0.9	203.20	1	Y					Y
3	Sheldon North Subdivision	4	IP	19900023	19.6	NG	1	Y	Y	Y			Y
4	Big Lagoon	6	IP	18568N36	6.3	201.30	TBD	Y		Y			NA
5	Pacifica Wastewater Treat.	8	NWP 26	20147536	0	202.21	56	Y	Y	Y			Y
6	Keller Canyon	9	NWP 33	18888E76	8.6	207.31	6	Y	Y		Y		Y
7	Hamilton AFB	10	NG	NG	12.5	206.27	2	Y	Y	Y	Y		Y
8	Moss Landing	11	IP	17659S27	2.5	NG	1						Y
9	Joe Blaney, Vallejo	14	IP	15439E24	0.2	206.50	1						Y
11	Jordan Smith/Gallinas Cree	21	IP	17418N17	0.10	206.20	1						Y
12	Caltrans Automall Pkwy	37	IP	19005E75	8.1	205.20	1						Y
13	Ruby Hill	41	NWP 26	19853E75	2.20	204.30	1						Y
14	Ponderosa Homes	42	NWP 26	17114E80	2.1	204.30	1						Y
15	Baylands Park	45	IP	15334549	6.6	205.00	1						Y
16	EBRPD-Wildcat Creek	53	NWP 26	19896E80	0.10	206.60	1						N
17	Berlex Biosciences	55	NWP 26	20000E76	0.10	NG	1						Y
18	ABA Energy Corp	57	NWP 33	20150E95	0.60	207.24	1						Y
19	Sonoma Valley Sanitation	62	IP	16087N96	20.3	206.00	1						Y
20	Fairfield Suisun Sewer	63	IP	17114E95	3.0	207.23	1						Y
21	Caltrans Hwy 237	64	IP	16686549	22.5	205.30	2				Y		N*
22	Chevron, Van Sickle Island	68	IP	18385E62	0.45	207.10	1						Y
23	Allied Signal Bay Point	89	NWP 38	19237E63	1.4	207.10	1	Y					II
24	Willow Pass Grade,Hwy 4	101	NWP 26	17000E80	2.7	207.33	2	Y					II
25	Calabazas Creek	106	IP	16833592	2.0	205.50	2	Y	Y		Y	Y	II
26	Mayhews Landing	112	NG	13199E75	5.9	204.00	1	Y					Y
27	Route 87 Freeway	116	IP	19713518	1.1	205.00	1	Y			Y		II
*TBD=to be determined						**TOTAL**		**12**	**7**	**4**	**6**	**1**	**19**
*NG=not given						**% of 27**		44%	26%	15%	22%	4%	70%
*NA = not appropriate													
*II = Insufficient Information													
*size refers to acres created or restored for compensatory mitigation													

*Pacifica (#5 above), used the following references (see report for more details):
San Mateo County north drainages including Pilarcitos, Tunitas, San Pedro, Denniston, and Purisima
San Mateo So. Creeks: San Gregorio, Pescadero, Butano Creeks (multiple sites for each creek)
Santa Cruz County Creeks: Waddell + Scott Creeks
Multiple sites at remaining creeks located on central coast: Arroyo de los Frijoles; Bradley, Gazos, Liddell Creeks, Rockaway.

Appendix 1: Reference Sites Used for Mitigation Projects, SF Bay, Region 2

Same w/l type?	Is location described?	Are Lat/Longs provided?	Is Reference name provided?	Comments
Y	N	N	N	seasonal, diked brackish
Y	Y	N	N	along Mahon Creek, San Rafael
Y	Y	N	N	
TBD	NA	NA	N	Very lilttle information
Y	Y	N	Y (see below*)	Hydrogeomorphic study; riparian systems in South Bay.
Y (?)	Y	N	recheck	impacted seeps but refs are perennial fresh??
Y	Y	N	N	refs are 800 ft outside of watershed
Y	Y	N	N	not specified?
NG	N	N	N	simply states "adjacent marsh"
Y	Y	N	N	
Y	N	N	N	adjacent seasonal w/l
N	N	N	N	nearby undisturbed ponds
Y	N	N	N	other creeks in immediate area
Y	N	N	N	adjacent to existing pickleweed habitat
Y	N	N	N	upstream locations
ll	N	N	N	adjacent wetlands
ll	N	N	N	adjacent undistrubed sites
N	N	N	N	adjacent salt marsh
ll	N	N	N	adjacent area, not specified
Y (?)	Y	N	Alviso Marina + New Chicago Marsh	monitoring for birds; refs are near but not touching;type suitability unc
ll	N	N	N	adjacent managed w/l w/in a duck club
ll	N	N	Hastings Slough adjacent to SP line	seasonal, diked, salt for impact project, not clear for ref.
Y and N	Y	N	N	Seasonal and riparian refs for riparian impacts (?)
Y	Y	N	downstream of confluence +toe of levee	
Y	N	N	N	
Y	Y	N	(between 2 avenues in San Jose)	
17	**11**	**0**		
59%	37%			

Appendix 2: Reference Sites Listed in Region 2's General Wetland Project Inventory (GWPI)

Num	On Wall Map/ Project Name	Acres	Loc. Cert.	W/L Type	Lost	Created	Type RMSFEWO	Status	Status Date	Lat	Long	Contact	Ctc Phone	Start	End	Hydro Unit	Moni-tored	Yrs	Comments
78b	Arrowhead Marsh	45	1	A1a			000F00	E	Jul-95	37 44 43	122 12 45	Levine-Fricke	652-4500	1996	1996	4.2	N		Ref.for 78a
184	Browns Island	50	1	A1a			000F00	E	Mar-96	38 02 03	121 52 50	Mike Vasey	415/338-1957	1996		7.1			SFSU/NERRS and Levine-Fricke Montezuma Ref.
199	Carquinez Strait	340	2	A1a			000F00	E	Mar-96	38 02 50	122 04 45	Levine-Frickes	652-4500	1996		7.31			Pot. ref. for Montezuma
186	Chain Island	64	1	A1a			000F00	E	Mar-96	38 04 18	121 51 22	Levine-Fricke	652-4500	1996		7.1			Pot. ref. for Montezuma
51	Corte Madera Ecological Reserve	88	1	A1a			00SF000	E	May-95	37 56 26	122 30 56	Barbara Salzman	415-924-6057	1982	1987	3.2			Heerdt Marsh Never diked. CCR Study. Xmas Bird Count
149	Coyote Creek Marsh	89.4	1	A2b			00SF00	E	May-96	37 28 01	121 57 09			1990		5.3			Salt Marsh Conversion Study
120	Damon Marsh	13	1	A1a			000F00	E	Mar-96	37 45 13	122 12 44	EBRPD	635-0138		1995	4.2	N		Arrowhead Ref., Levine-Fricke
68b	Dumbarton (ref.)	0.7	2	A1a			000F00	C	Jun-95	37 30 30	122 05 23	Jenkins-Sanders	818-710-8545	1994	1994	5.2	N		1990 Bay Ref.
195	Dutton Island	49.5	1	A1a			000F00	E	May-96	38 04 48	121 58 37	Levine-Fricke		1996		7.1			Pot. ref. for Montezuma
9	Faber Tract Marsh	107.1	1	A1a			R00FE0	C	Apr-95	37 27 57	122 07 30	USFWS	792-0222	1971	1989	5.5	Y	2	City of Palo Alto, LTMS
191	Grizzly Bay	402.2	1	A1a			000F00	E	Mar-96	38 07 42	121 59 28	Levine-Fricke		1996		7.24			Pot. ref. for Montezuma
190	Hill Slough	410.5	1	A1a			000F00	E	May-96	38 13 43	122 00 00	Levine-Fricke		1996		7.23			Pot. ref. for Montezuma
100	Joice Island	970	1	A2a			000FE0	C	Oct-95	38 07 32	122 04 44	D. U.	916-852-2000	1995	1995	7.23	N		Private hunting club/Ref. for Montezuma
204	Jones Sand Marsh	15	1	A1a			000F00	E	May-96	38 01 57	122 07 24	CCMAD				7.33			
52	Laumeister Tract	87.6	1	A1a			000F00	E	May-95	37 28 19	122 07 30	USFWS	792-0222	1991	1992	5.5	N		Mgt. w/City of Palo Alto, LTMS

Appendix 2: Reference Sites Listed in Region 2's General Wetland Project Inventory (GWPI)

125	Marina Marsh	20	1	A1a			000F00	C	May-96	37 26 20	121 58 54	Dan Stephens	408-263-1814	1993	1996	5.3	Y		Ref. for bird use for Hwy 237
16	Muzzi Salt Marsh	130	1	A1a	80	125	RM0F00	C	Apr-95	37 55 49	122 30 18	P. Faber	415-388-6002	1976	1981	3.2	Y		Nat. rest./no dredge, LTMS
17	New Chicago Marsh	340	1	B4b			RM0FE0	C	Apr-95	37 25 58	121 58 35	USFWS	792-0222	1985	1994	6.3	Y		Hwy 237 ref. POST
132	Petaluma Marsh Wildlife Area	2000	1	A2a			000F00	E		38 10 50	122 33 07	CDFG				6.3	N		Hunting, Arrowhead ref.
143	Peytonia Slough Ecol. Reserve	326	1	A2a			000F000	E		38 13 40	122 02 38	CDFG		1996		7.33			Pot. ref. Levine-Fricke
106	Point Edith Marsh	411	1	A1a			R0SF00	P		38 02 40	122 03 56	Karl M-Roam	685-9301	1996		7.31			Karl-Plant study,Aud. Soc. -Xmas bird count
67	Ravenswood (ref.)	0.7	2	A1a			000F000	C		37 30 04	122 08 38	Jenkins Sanders	818-710-8545	1994	1994	4.4			For 1990 Bay Rd.
177	Roe Island	91.3	1	A1a			000F00	E		38 04 21	122 02 17	Levine-Fricke		1996		7.1			Potential Ref.
75b	Rush Ranch	1050	1	A1a			000F00	C		38 11 49	122 01 46	Sonoma	707-547-1925	1988	1994	7.23	Y	5	Pristine, Montezuma ref.
113a	Skaggs Island	3000	1	B4a			R00F00	P		38 10 35	122 23 32	US Navy	415/877-7608	1995		6.4			LTMS rec.
113b	Skaggs Island	?	1	B4a			000F00	E		38 08 43	122 23 15	PRC	415/543-4880	1994		6.4	N		PRC ref.
208	Warm Springs Seasonal Marsh (Caruff Property)		1	B00			000F00	E		37 28 29	122 56 49			1970	1985	5.2			Salt Marsh Conversion Study
202a	Whittel Marsh (Pt. Pinole #3)	5	1	A2a			R00FE0	U		38 00 33	122 21 54	EBRPD		1993	1996	6.6			Ref. for Selby spill

TYPE: R=Restoration, M=Mitigation, S=Scientific, F=Reference, E=Enhancement, W=Wastewater

STATUS: C=Complete, P=Proposed, U=Under Construction, N=Not Done, E=Existing

Legend for Appendix 2

Appendix 2, Legend: Explanation of Column Headings

Column Heading	Explanation
Count	Nominal list
Num	Number assigned to project on existing NWI maps (hardcopies),
	and in binder containing project descriptions located at Regional Water Board, Oakland CA
Project name	The most recent and widely-accepted title given the project.
Acres	Number of acres
Location Certainty	Degree of certainty of researcher in mapping the project area; 1 = certain, 2 = moderately certain, and
	3 = not very certain of the accuracy of the data
Wetland type	See "Appendix 2, Legend 2" below
Lost	Acres lost (primarily for mitigation projects)
Created	Acres created (primarily for mitigation projects)
Type RMSFEWO	The 29 projects listed here are all "F" which refers to References. Some fall under other classifications reserved,
	primarily, for R = Restoration; M = Mitigation; S = Scientific, E = Enhancement; W = Wastewater
Corps #	Permit number assigned to project by U.S. Army Corps of Engineers
Status	C = Complete; P = Postponed; U = Under Construction; N = Not done; E = Existing (e.g., wetlands on a reserve)
Status Date	Date that previous column (status) was entered.
Lat	Latitude
Long	Longitude
Contact	Person or organization to contact
Ctc. phone	Phone number of contact
Start	Start date of project
End	End date of project
Hydro Unit	Watershed unit used by Regional Board. 1=Marin Basin; 2=San Mateo Basin; 3=Central Bay Basin; 4=South
	Bay Basin; 5=Santa Clara Basin; 6= San Pablo Basin; 7=Suisun Basin
Monitored	Is the project monitored
Yrs.	If monitored, for how many years?
Comments	Any pertinent information not included above. An example might be the type of scientific study conducted at
	a wetland site.

Appendix 3: Wetland Types Used in Mitigation Database and General Wetland Project Inventory

Appendix 3: List of Wetland Types			
A00	Perennial Wetlands		
	A10. Perennial-tidal		
		A1a. Perennial-tidal-salt	
		A1b. Perennial-tidal-brackish	
		A1c. Perennial-tidal-fresh	
	A20. Perennial-muted tidal		
		A2a. Perennial-muted tidal-salt	
		A2b. Perennial-muted tidal-brackish	
		A2c. Perennial-muted tidal-fresh	
	A30. Perennial-nontidal		
		A3a. Perennial-nontidal-salt	
		A3b. Perennial-nontidal-brackish	
		A3c. Perennial-nontidal-fresh	
BOC	Seasonal Wetlands		
	B10. Seasonal-vernal pool/swale		
	B20. Seasonal-dune swale		
	B30. Seasonal-no dike		
		B3a. Seasonal-no dike-salt	
		B3b. Seasonal-no dike-brackish	
		B3c. Seasonal-no dike-fresh	
	B40. Seasonal-diked		
		B4a. Seasonal-diked-salt	
		B4b. Seasonal-diked-brackish	
		B4d. Seasonal-diked-fresh	
	B50. Seasonal-saline flat (inland)		
	B60. Seasonal-managed (e.g., duck ponds)		
	B70. Seasonal-palustrine-farmed (i.e., ponds on farms)		
	B80. Seasonal-inactive salt ponds		
C00	Open Water (non-channel)		
	C10. Open Water (non-channel)-ocean		
	C20. Open Water (non-channel)-estuary		

Appendix 3: Wetland Types Used in Mitigation Database and General Wetland Project Inventory

	C30. Open Water (non-channel)-bay		
	C40. Open Water (non-channel)-lake		
	C50. Open Water (non-channel)-natural pond		
	C60. Open Water (non-channel)-salt pond		
D00	Mudflat (no vascular plants)		
	D10. Mudflat (no vascular plants)-salt		
	D20. Mudflat (no vascular plants)-brackish		
	D30. Mudflat (no vascular plants)-fresh		
E00.	Rocky Intertidal		
F00.	Eelgrass Bed		
G00	Seep/Spring		
H00.	Riparian		
	H10. Riparian-intermittent/seasonal freshwater		
		H1a. Riparian-intermittent/seasonal freshwater-free-flowing	
		H1b. Riparian-intermittent/seasonal freshwater-controlled flow	
	H20. Riparian-permanent-freshwater		
		H2a. Riparian-permanent-freshwater-free-flowing	
		H2b. Riparian-permanent-freshwater-controlled flow	
	H30. Riparian-tidal		
		H3a. Riparian-tidal-fully tidal	
		H3b. Riparian-tidal-muted tidal	
I00.	Upland		
J00.	Other		

Appendix 4: Reference Sites in Hydrounit 7, Suisun Basin

Appendix 4: Reference Sites in Hydrounit 7, Suisun Basin					
Information from General Wetland Project Inventory:					
Count	**Map #**	**Project Name**	**Acres***	**Wetland Type**	**ydro Unit**
1	184	Browns Island	50.00	A1a = Perennial-tidal-salt	207.10
2	199	Carquinez Strait	340.00	A1a = Perennial-tidal-salt	207.31
3	186	Chain Island	63.60	A1a = Perennial-tidal-salt	207.10
4	195	Dutton Island	49.50	A1a = Perennial-tidal-salt	207.10
5	191	Grizzly Bay	402.20	A1a = Perennial-tidal-salt	207.24
6	190	Hill Slough	410.50	A1a = Perennial-tidal-salt	207.23
7	204	Jones Sand Marsh	15.00	A1a = Perennial-tidal-salt	207.33
8	106	Point Edith Marsh	411.00	A1a = Perennial-tidal-salt	207.31
9	177	Roe Island	91.30	A1a = Perennial-tidal-salt	207.10
10	75b	Rush Ranch	1050.00	A1a = Perennial-tidal-salt	207.23
11	100	Joice Island	970.00	A2a = Perennial-tidal-brackish	207.23
12	143	Peytonia Slough Ecol. Reserv	326.00	A2a = Perennial-tidal-brackish	207.33
*acreages are only estimates					
Information from Mitigation Database:**					
1	NA	Keller Canyon	NA	Perennial-nontidal-fresh; seasonal freshwater; seeps	207.31
2	NA	ABA Energy Corp	NA	I.I.	207.24
3	NA	Fairfield Suisun Sewer	NA	I.I.	207.23
4	NA	Chevron, Van Sickle Island	NA	I.I.	207.10
5	NA	Allied Signal Bay Point	NA	I.I.	207.10
6	NA	Willow Pass Grade, Hwy	NA	Sesonal, Riparian	207.33

**Acreage available only for size of restoration or creation project, not for reference site(s) used

I.I. = Insufficient Information in Mitigation Report

Appendix 5: Suggested Reference Sites from Ecosystem Goals Project

Appendix 5: Reference Marshes Suggested by Plant Focus Team for Wetlands Ecosystem Goals Project (7/96) (Reprinted with permission of Project Chairman)

ECOTONE	REFERENCE LOCATION
Tidal Fresh to Tidal Brackish	n.a.
Brackish Tidal	Browns Island
	Rush Ranch
	Hill Slough
	Petonia (/)
	Roe Island
	Chabot Creek (Vallejo)
	Guadalupe/Coyote Creek/Alviso (South Bay)
Tidal Brackish to Salt	n.a.
Tidal Salt	Mowry
	Bird Isle
	China Camp
	Herdt (Corte Madera)
	Greco Island
	Fagan Slough
	Highway 37 marshes
	Hamilton
	Petaluma Marsh
Intertidal/Subtidal	n.a.
Riparian to Bayland*	Green Valley Creek
	Suisun Creek
	Carneros Creek
	American Canyon
	Scott Creek
	Coyote Creek

Appendix 5: Suggested Reference Sites from Ecosystem Goals Project

	San Antonia Creek		
Salt Ponds	Alviso Pond B-2 and A-1		
	Baumberg 10 and 11		
	Plant 226 (hypersaline)		
Diked Salt (check Shellhamme	Montezuma		
	Fremont Airport		
	CDFG property (North San Pablo Bay)		
	Agricultural areas: Leonard Ranch		
Adjacent Uplands			
Valley Grassland	Rush Ranch		
	Coyote Hills		
Oak Savannah	South Sonoma		
	Palo Alto		
Wet Shrublands	South Marin at north terminus of GGBridge		
	Angel Island		
Mixed Evergreen	Marin, around Richardson Bay		
	China Camp		
Coastal Prarie	Point Richmond		
	Angel Island (coastal analogs)		
Serpentine Grassland	Golden Gate		
	Ring Mountain		
Vernal Pools	Potrero/Hill Slough		
	Coyote Hills		
	Denverton (Jepson Prairie)		
*Note: unknown whether any of these areas are unaltered			

Figure 1: General Wetlands Project Inventory Sites in Region 2

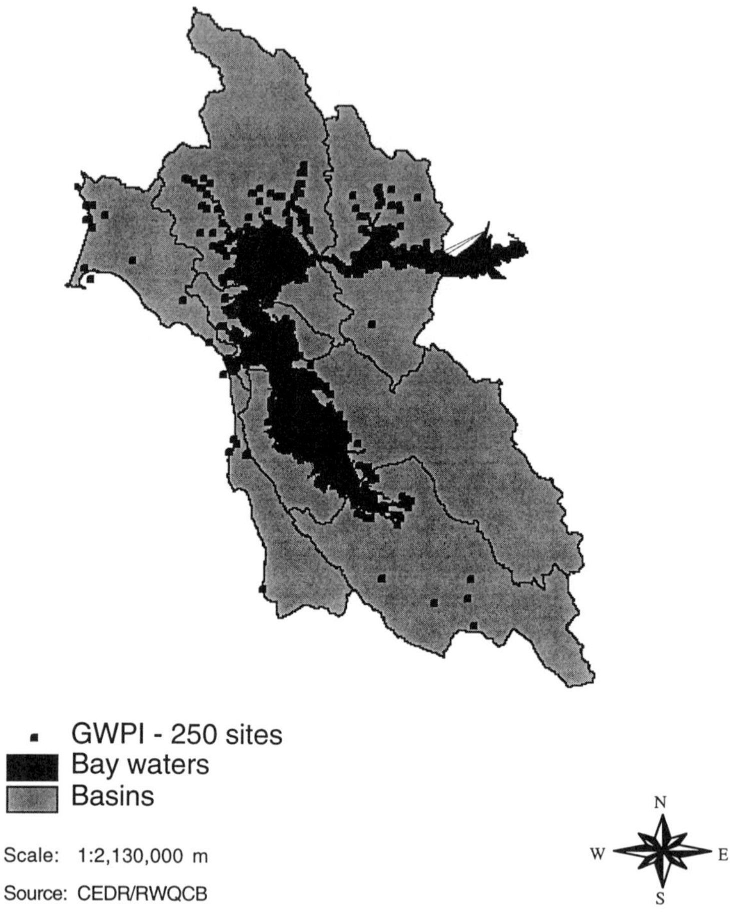

Figure 2: Mitigation Database Sites in Region 2

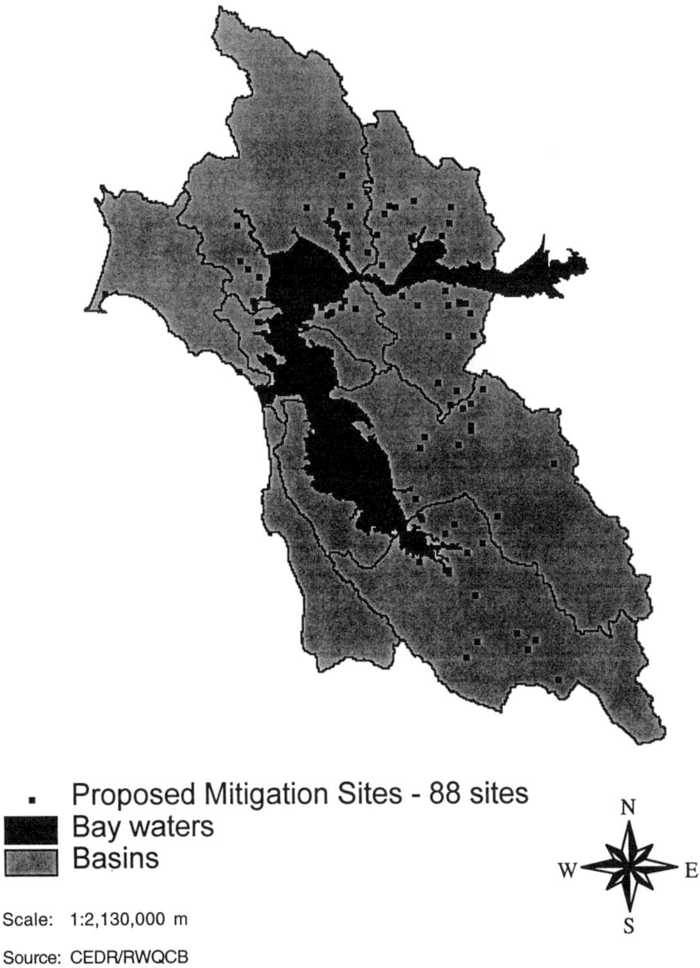

Figure 2a - Proposed Reference Sites for Mitigation Database Sites

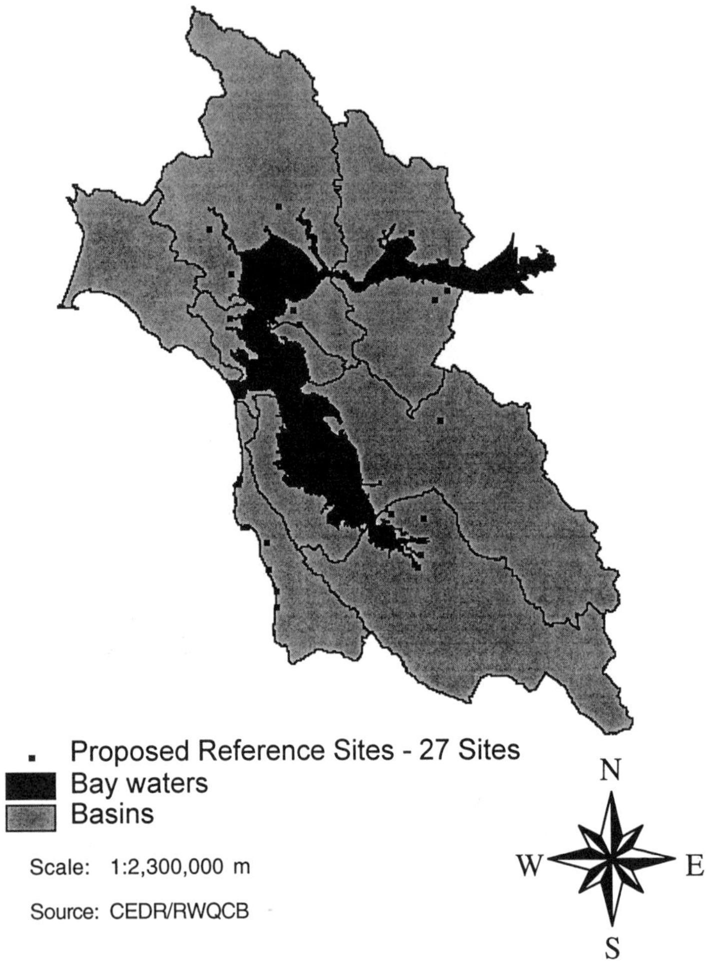

Figure 3: Suisun Basin General Wetlands Project Inventory

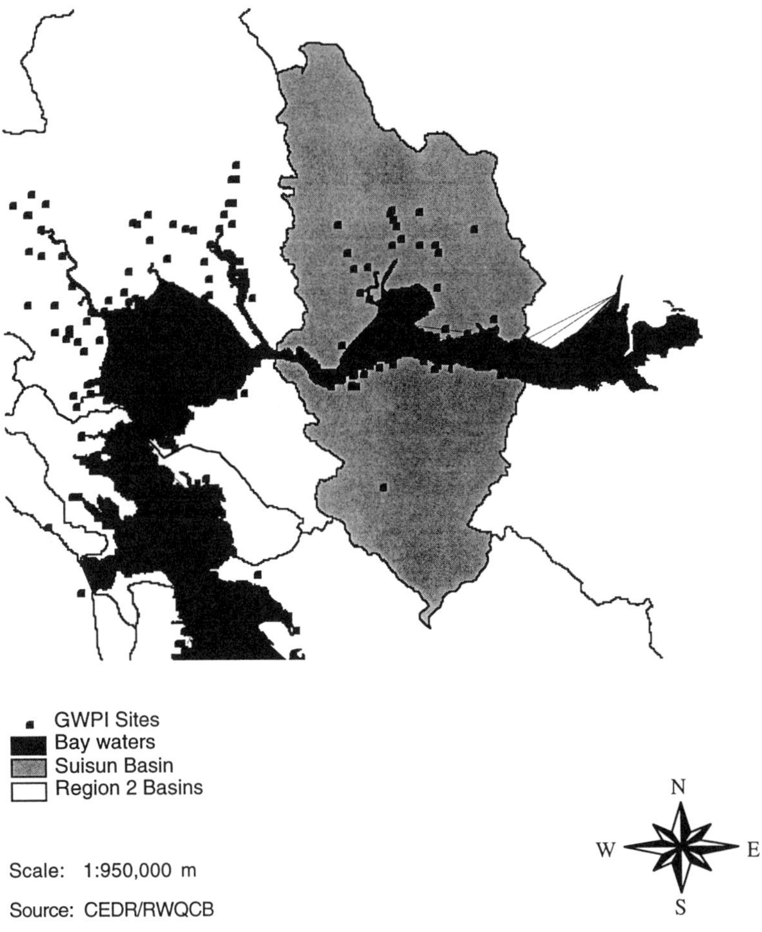

Figure 4: Suisun Basin Mitigation Database Sites

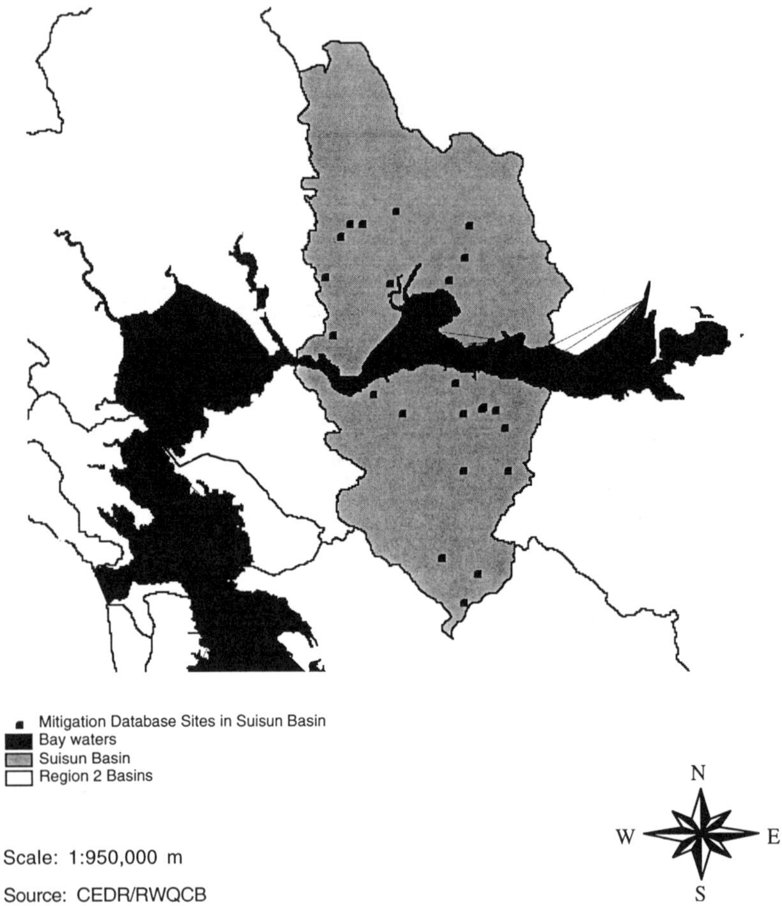

Southern California Coastal Wetlands Inventory: Centralizing Existing Information

Patricia Heasly[1], Reed Holderman, William Martin, Jonathan Van Coops, Allyson Hitt, Susan Wynn, Deborah Fleischer

Abstract

The State Coastal Conservancy, Coastal Commission, and U. S. Fish and Wildlife Service have developed a database of maps and texts summarizing recent data on biological and physical conditions at 41 of Southern California's coastal wetlands. It was designed with guidance from local resource managers to be a widely accessible tool for wetlands conservation planning in the region. It will be posted on the worldwide web and available in a modified form on an ArcView-generated relational database.

Introduction

The Southern California Coastal Wetlands Inventory (the Inventory) is a database of existing information on 41 coastal wetlands that lie between the Mexican border and Pt. Conception in northern Santa Barbara County (Figure 1). Three types of information are provided for each site: 1) a map showing the historical extent of the wetlands; 2) a map showing recent distributions of habitat types; and 3) a "profile" or brief summary of ecological conditions, land use and enhancement histories. The profile also includes an annotated list of sources. All of the information in the Inventory, both mapped and written, is derived from data that were generated previously.

---

[1]Project Manager, Enhancement Program, State Coastal Conservancy, 1330 Broadway Suite 1100, Oakland, CA 94612

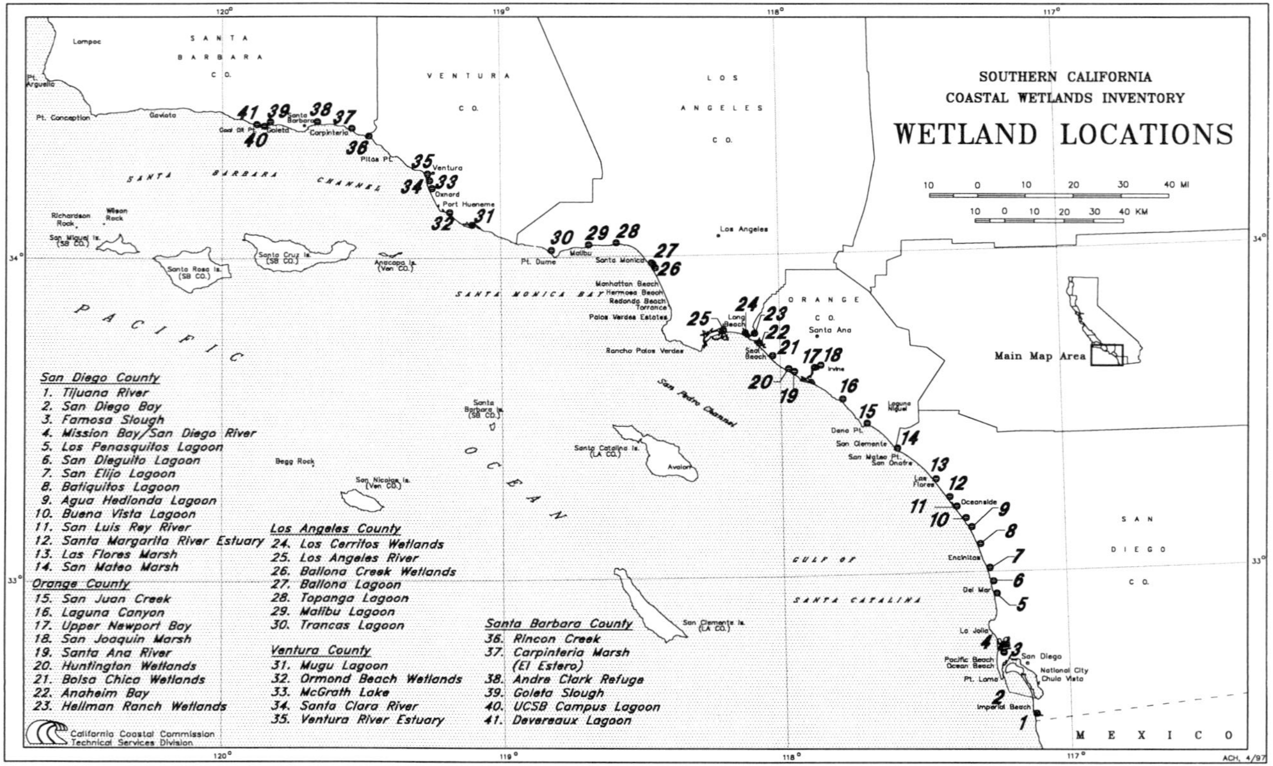

Figure 1. Wetlands described in the Inventory

The Inventory is a component of the Governor's 1993 Wetland Conservation Strategy, which recommended development of a statewide wetland inventory to support identification of regional and statewide restoration and enhancement goals. The Inventory implements this recommendation for the Southern California region. It was prepared jointly by the State Coastal Conservancy, the California Coastal Commission and the U.S. Fish and Wildlife Service. Overall guidance for the project came from the Southern California Wetlands Group, a committee representing regional, state, and federal resource agencies whose purpose is to improve coordination of wetlands protection and regulatory activities in the region.

The purpose of the Inventory is to enable a regional perspective on coastal wetland resources and habitats. Providing basic descriptions of wetlands in the entire area, it emphasizes breadth rather than depth. It will be used to answer in broad terms such questions as: Where are the wetlands now ? Where were they historically ? What are some of the constraints to land use, restoration, enhancement, and protection in and around those areas ? The Inventory also constitutes a foundation upon which new and other types of data can be added.

Database Elements

Historic Maps

The historic wetland perimeter for each site is shown on United States Geological Survey 7.5′ quadrangles. These were digitized from the U.S. Coast and Geodetic Survey (currently NGS) Topographic Maps, prepared between 1851 and 1893.

Habitat Maps

The habitat maps show vegetative communities at each wetland. Their purpose is to give a general indication of the location, extent and subhabitats within Southern California wetlands. They also include the historic wetland boundary and, to provide a sense of the current degree of continuity between wetland habitats and adjacent uplands, they depict vegetative communities of surrounding uplands. A legend identifies the communities represented by each color on the map with habitat nomenclature following that developed by Holland (1986)[2].

Wetland Profiles

The profiles are distillations of the most recent studies documenting physical and biological characteristics of each wetland area (see excerpt in Table 1). Presented in tabular form they highlight hydrology, vegetative communities, animal use, species of concern, ownership, onsite and surrounding land use, and the history of enhancement efforts. Sources are referenced next to each entry and annotated citations are provided at the end of each profile. The profiles are intended to convey overall pictures of ecological conditions at each site.

Methods

Capturing the Historic Wetland Data

The Geodetic Survey data were available in two formats, the first being paper reproductions of the original topographic maps, or T-sheets, at the original scale (usually 1:10,000), and the second being mylar copies of USGS 7.5' quadrangles onto which the wetland boundaries had been transferred by the California Department of Navigation and Ocean Development (currently Department of Boating and Waterways). To minimize the amount of error introduced by DNOD's boundary transfer, historical wetland boundaries were digitized directly from the paper USC&GS T-sheets whenever possible. The following information was digitized from the historic maps:

-Historical shoreline
-Landward boundary of wetland symbols, including marsh, salt ponds and alkali flats
-T-sheet number and year of survey
-Scales of capture 1:10,000, 1:20,000, and 1:24,000

In order to facilitate registration of the historical wetland extent with the USGS base map, the 1927 North American Datum (NAD) was used for control. The 1927 NAD was available for all but three T-sheets used in the Inventory. For these map sheets it was generated from adjacent T-sheets.

---

[2] Holland, Robert F. 1986. Preliminary descriptions of the terrestrial natural communities of California. State of California, The Resources Agency, Department of Fish and Game. 146 pp.

Tijuana Estuary

ANIMAL USE		Source
Birds	Winter 1988/89 survey - 90 species total, 63 waterbird, 9 raptor, 18 upland; 9 species of special concern. 1995 - 61 pr light-footed clapper rail#; 1994 - 5 active snowy plover+ nests; 1991 - 303 pr Belding's Savannah sparrow#.	6, 16, 17, 18
Fish	Sep 1986 - Dec 1989 sampling (gear not specified) collected 21 species from 15 families; arrow goby dominant, also topsmelt and California killifish; dominance shifted during the sampling period from co-dominance by topsmelt and arrow goby.	4
Benthic Invertebrates	Various studies since 1969 document changes in community structure after catastrophic events; Sep 1986 - June 1988 sampling of north channels collected 58 taxa of polychaetes and bivalves, also the California ghost shrimp; highest bivalve densities were at the deepest site also farthest from sewage inflow; Capitellids and Spionids were the dominant polychaetes.	4
Insects	Fall 88 - Spr 89 survey of dune, Salicornia marsh, transition and upland found 16 orders; beetles, springtails, flies, leafhoppers and planthoppers, bees and wasps, butterflies and moths were most abundant; salt marsh wandering skipper butterfly+ was present, also globose dune beetle+ on dunes.	6
Other Wildlife	Sep 88 - June 89 survey of reptiles and amphibians found 13 of 25 - 30 species expected according to the habitats onsite - 4 species of frogs, 6 of lizards, and 3 of snakes, including the African clawed frog* and the bullfrog*; also the San Diego coast horned lizard+ in dunes. Evidence of 16 mammal species was found Nov 1988-May 1989,10 of those on the marsh; no Pacific pocket mice+. Domestic dogs and long-tailed weasel are seen frequently.	6

* Non-native species
\+ Special status species
\# Nesting special status species

Table 1. Excerpt of the wetland profile for Tijuana Estuary. A comparable level of detail is provided for other categories including water quality, tidal influence, habitat types and dominant vegetation, ownership, current and historic landuse, and management efforts. The numbers in the source column refer to the source documents which are identified in a list of annotated citations at the end of each profile.

Creating the Base Map

Both the historical and recent habitat layers were added to a digital base map. It was created by digitizing the following data layers from USGS 7.5' quadrangles:

-Shoreline
-Hydrography
-Latitude/Longitude graticule
-County boundaries
-Coastal Zone Boundary
-Railroads
-NGS control points and benchmarks
-Place names and other text labels
-Scale of capture 1:24,000

Habitat Maps

The habitat maps were composed in ArcInfo using the most recent and highest resolution digital data available for each site. The sources of data vary in the collection and interpretation methods used, as well as in the date and time represented. Professional judgment of U.S. Fish and Wildlife Service biologists and field knowledge of a given area informed selection of the most appropriate data for each site. In general, if more than one dataset mapped at comparable levels of detail were available, then the most recent was used. However, decisions were made on a case by case basis. For example, two sources of data might have been available for a given wetland in San Diego County, one being the National Wetlands Inventory (NWI) from 1985 and the other, the 1993 Natural Communities Conservation Planning effort involving coastal sage scrub habitat. Although less recent, the NWI data would have been preferred because it was generated by an effort explicitly designed to capture wetland data as opposed to an effort focused on upland habitats but which also included wetlands. Actually, the resulting map in this case was a composite of the two datasets. Because of the way the two datasets were originally generated (using the same projection), they could be easily overlaid without creating cartographic discrepancies. By overlaying the data, both the detail of the NWI map and more recent alterations to the landscape, such as a new wetlands created since 1985, could be captured.

The data source(s) are indicated on the respective maps. The base map and historic wetland boundary were those developed by the Coastal Commission as described above. Updates will be prepared by USFWS as additional data are made available.

Wetland Profiles

The reservoir of information used to develop the wetland profiles encompasses a good deal of variation. Studies were carried out for different reasons at different scopes, at varying levels of detail on various dates. Some provide data describing the hydrology, water quality, and plant and animal species of the entire site while others focus on a particular subhabitat or species; some describe data collected over several years while, others report on a single survey. These differences were exhibited both within wetland sites and among them. The two main criteria used to determine which sources to include were recency and comprehensiveness, the relative weight afforded each depended on the nature of the available data. An effort was made to maintain internal consistency within a profile where a dataset existed that described several habitat characteristics collected concurrently, with broad coverage of the site and for the same project. If more recent data were available for selected parameters these were included in addition to the information from the broader dataset.

The heterogeneity of the source material also meant that the targeted information was not always available for every site and so consistency in the type of information provided for all the wetlands was not possible. A table accompanying the profiles identifies both the type of information sought for the entry and the general objective for the category.

The draft profiles and bibliography were reviewed for accuracy and completeness by resource managers, selected wetland scientists, and the landowners and/or other groups having management responsibility for a particular site.

Using the Inventory

The Inventory's primary function is as a planning tool. It describes recent conditions at each site - one area has recently supported nesting light-footed clapper rails; another recently had salt marsh but no cordgrass grew there; permanent structures have been built across this site; invasive exotics are among the dominant species here. The user can also learn about the availability or lack of information for each area - few sites may have data on hydrology, water quality, plant and animal species that were collected in the same year; sources for some sites compare data from several years of consistent sampling, while others report on a single survey, the only data collected in the last ten years. You can also ask basic questions about every site and get a sense of how things look from a regional perspective. How many of the sites' tributaries are free-flowing ? How many sites have some sort of industry in the immediate vicinity ?

How many sites have recently supported nesting endangered birds ? endangered plants ? How many are surrounded by development ?

In exploring potential applications for the data, it is important for users to be aware of some characteristics of the Inventory that impose limitations on its uses. Of singular importance is the dynamic nature of the resource being described; responding to numerous forces both external and internal, the measurable components of a wetland can exhibit tremendous variation. The Inventory's reliance on existing information means that the source materials will represent a wide range of dates, scope and purpose. The brevity of the wetland profiles allows for a mere glimpse of the system as a whole. In addition, while the Inventory was reviewed for completeness and accuracy by wetland scientists and resource managers, no independent evaluation of the sources used in the profiles was undertaken. Finally, with these characteristics in mind it is important to state that: 1) the information in the Inventory is of insufficient resolution to serve as the sole basis for regulatory decisions and is not intended for such use; and 2) the information in the Inventory is not meant to provide in-depth, definitive portrayals of the sites' current or potential condition or significance.

Database Formats

The Inventory can be viewed on the worldwide web at ceres.ca.gov/wetlands/geo_info/so_cal.html. To facilitate the kinds of analyses mentioned above, the data were also put onto a relational database, which enables the user to query a subset of the profile data (see Table 2) and relate the answers to the historic and habitat maps. It was developed using ArcView and is available from the Coastal Commission's mapping unit in San Francisco. The Inventory data will also ultimately be available through the Department of Fish and Game's Natural Heritage Division. Finally, a bound hard copy of the Inventory as well as a library of all the source documents, will be housed at both the Coastal Conservancy office in Oakland, and at an as yet undetermined university library in coastal Southern California.

ArcView Data Categories

Wetland name	Predominant and second most abundant wetland habitat types
USGS 7.5' quadrangle	Open Water
County	Salt Marsh
Approximate wetland acreage	Fresh/brackish marsh
Approximate historic wetland acreage	Riparian
Ownership/Acreage	Seasonal Wetland
Federal	Cordgrass present (yes/no)
State	Tidal Condition
Local	Full tidal
Quasipublic (i.e. utilities, ports)	Restricted
Private	Muted
Designated Land Use	Seasonal/intermittent
Onsite Land Use	Nontidal
Special Status Species	Included in Water Resources Control Board list of impaired water bodies
Nesting Special Status Species	

Table 2. Data categories adapted from the profiles for use in the ArcView relational database. Paring down and standardizing the profile data enables queries of the dataset.

Conclusion

The Inventory is the first effort to summarize in a standardized format, the existing mapped and field survey data describing coastal wetlands of the entire Southern California bight. It is our hope that it will lay the foundation for development of a regional wetlands conservation strategy, provide a baseline for a growing regional dataset, and through its wide availability, increase public awareness of the unique character and diversity of these coastal systems.

CHAPTER 135

Batiquitos Lagoon Enhancement Project: Concept to Construction.

Michael Josselyn[1], Ralph Appy[2], and Adam Whelchel[1].

***Abstract*:** The 593 acre Batiquitos Lagoon Enhancement Project is a joint restoration project of the City of Carlsbad and the Port of Los Angeles. The $55 million project is being funded by the Port of Los Angeles as mitigation for the loss of deep water marine habitat in the outer Los Angeles harbor area. The enhancement project involves the re-establishment of a tidal entrance to the blocked lagoon mouth and dredging approximately 3.1 million cubic yards of accumulated sediments from the lagoon basin to increase the tidal prism and restore coastal lagoon habitats. The project has been under construction since 1994 using electric dredges. A portion of the dredged material is being used for beach replenishment and a portion for the construction of nesting areas for the California least tern and western snowy plover. Use of the nesting sites by these birds has exceeded historic levels and contributes significantly to the coastal populations of these federally listed species. The project will be completed in mid-1997.

INTRODUCTION

The Batiquitos Lagoon Enhancement Project is the largest coastal wetland restoration project in California and provides an example of the effort and planning required to implement large scale projects. It is one of a series of six lagoons on the northern coast of San Diego County, several of which are also proposed for restoration (Figure 1). Most of the other proposed restoration projects involve restoration of tidal action as was done at Batiquitos.

The need for restoration at Batiquitos Lagoon was clearly recognized in 1976 when the Department of Fish and Game issued its report on the current conditions in the lagoon (Mudie et al 1976). The lagoon had been tidally influenced during historic times; however, the construction of road and railroad bridges and causeways restricted tidal flows. With increasing agricultural and urbanization pressures on the lagoon's

[1]Wetlands Research Associates, Inc., 2169 E Francisco Blvd Ste G. San Rafael, CA 94901, (415) 454-8868

[2]Port of Los Angeles, Environmental Management Division, PO Box 151, San Pedro, CA 90733-0151, (310) 732-3497

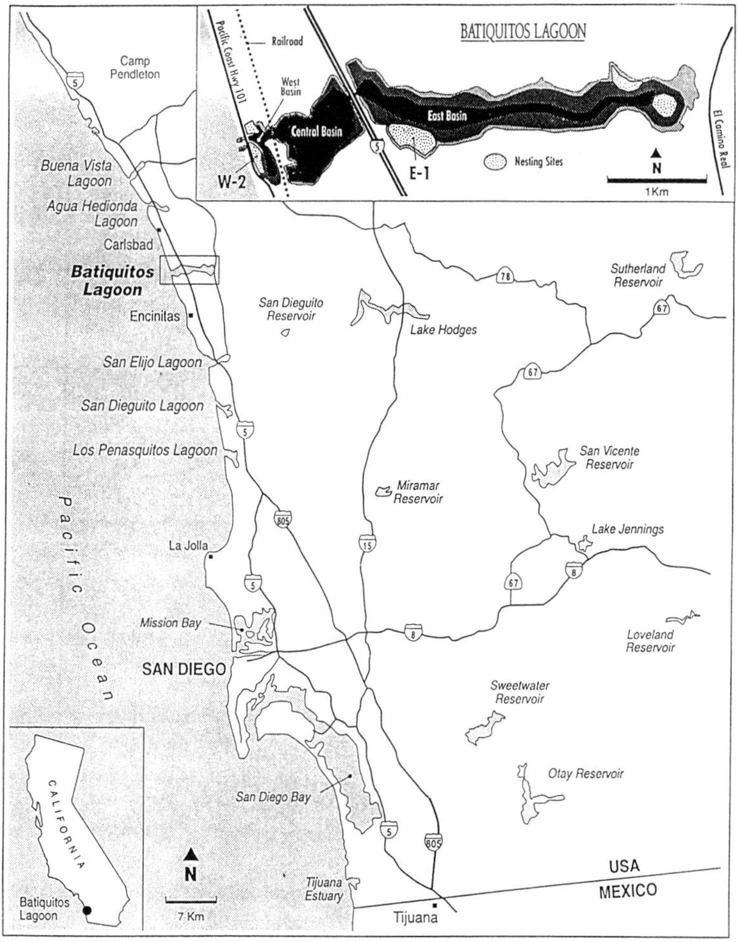

Figure 1. Batiquitos Lagoon is located in northern San Diego County in the City of Carlsbad, CA. The Lagoon Enhancement Project was undertaken jointly by the City of Carlsbad and the Port of Los Angeles. The lagoon consists of three basins: West, Central, and East. Each basin was dredged in order to create subtidal and intertidal habitat. The Central Basin was overdredged and the sand used for beach replenishment north of the lagoon mouth. Sand was also used to create five nesting sites (two of which are labelled W-2 and E-1). Silts from the East Basin were placed in the Central Basin and these silts were capped with sand from the West Basin. Jetties were constructed at the mouth to help maintain an open tidal entrance.

watershed, sediments filled the lagoon, decreasing its tidal prism. In addition, portions of the lagoon were filled for development, diked for solar salt production, and freshwater flows damned and diverted. These influences resulted in a lagoon environment that was frequently closed and sometimes completely fresh. In summer months, the closed system dried out and hypersaline mudflats were prevalent. In less than 150 years, the lagoon had been converted from a tidally influenced, marine community to one that altered annually between freshwater and hypersaline conditions.

The lagoon continued to support wildlife, particularly migratory birds (California Coastal Conservancy 1987). In addition, the federally listed, endangered California least tern (*Sterna antillarum browni*) and threatened western snowy plover (*Charadrius alexandrinus nivosus*) were observed and sometimes nested on the mudflats of the lagoon. The concern expressed by most observers was that despite the existing values, the frequent long-term closure of the lagoon coupled with the decreasing tidal prism due to siltation would eventually eliminate the existing wetland functions and values.

In 1987, the Port of Los Angeles, the City of Carlsbad, the California State Lands Commission, the California Department of Fish and Game, the US Fish and Wildlife Service, and the National Marine Fisheries Service entered into a formal Memorandum of Agreement to enhance Batiquitos Lagoon for the mitigation of Port fills in San Pedro Bay. The state and federal resource agencies determined that the enhancement of Batiquitos Lagoon would be an appropriate mitigation to offset the adverse environmental impacts that would result from filling activities in the shallow waters in Los Angeles Harbor. The mitigation was both off-site and out-of-kind, but was deemed valuable because "shallow, estuarine coastal embayment habitat in southern California, with its relatively high value to marine fishes and migratory birds, has been reduced in area at a greater rate than that of deep water habitat".

A joint EIR/EIS document was prepared for the enhancement of the lagoon (City of Carlsbad 1990). The EIR/EIS evaluated a number of alternatives for the enhancement project. These alternatives ranged from no action to large scale dredging of the lagoon. The City and the California Coastal Commission eventually determined that a project referred to as "Mitigated Alternative B" had the best opportunity to restore tidal action to the lagoon and to provide for improved habitat for marine and estuarine fish and wildlife.

Engineering studies and specifications were prepared by the City and the Port and construction on the project commenced in April of 1994. The purpose of this paper is to summarize the construction procedure and to document the use of newly constructed nesting areas by least terns and western snowy plovers.

CONSTRUCTION SUMMARY

The restoration of Batiquitos Lagoon consists of several major construction activities:

- Dredging the lagoon's central basin and use of the sandy material for beach nourishment and construction of shorebird nesting sites.
- Dredging the lagoon's east basin and the use of the silty material as fill within the central basin.
- Dredging of the west basin and use of the sandy material for nesting sites and beach nourishment.
- Construction of five California least tern and western snowy plover nesting sites totaling 38 acres.
- Construction of two 300 ft rock jetties extending into the ocean from the mouth of the lagoon.
- Replacement and/or rock revetment of bridges crossing the lagoon.
- Revegetation of salt marsh vegetation disturbed during construction of the project.

The enhancement project is a unique approach to disposal of bottom materials that are unsuitable for beach nourishment. Within the west and central basin, sandy materials are prevalent. However, in the east basin, the material is largely fine silt and unsuitable for placement on area beaches. As a result, the construction involves an over-excavation of sand from the central basin and the placement of the fine silts into the pit created. Approximately 1.6 million cubic yards of sand will be placed on area beaches at the conclusion of the project. The sand was also used to create the nesting areas for shorebirds. Five nesting sites range in size from 1 to 19 acres and were created to be above tidal and flood water influence. Once the fine materials from the east basin are placed within the central basin, it will be capped with sandy material from the dredging in the western basin. At the end of the project, approximately 3 million cubic yards of material will have been moved.

The project is also being constructed in a manner to avoid impacts to existing natural resources. Dredging is being conducted using electric dredges to assure quiet and reduce air pollution. All dredging is halted during the least tern/western snowy plover nesting period (April 1 to September 15). Construction activities must also avoid the state listed Belding's savannah sparrow habitat during the summer nesting period. All salt marsh vegetation disturbed by the project must be transplanted or planted on a one to one ratio. These measures and others are being supervised by an on-site environmental monitor throughout the construction.

UTILIZATION OF NESTING SITES DURING CONSTRUCTION

One of the conditions established in the permits issued for the project was that suitable nesting sites for California least tern and western snowy plover be constructed during the project and the utilization of these sites monitored. During the first year of construction, one 2 acre site (W-1) was constructed, managed, and monitored. During the second year of construction, two additional sites (W-2--5.2 acres and E-1--19.1 acres) were made available. Each of the sites was constructed with sand either tracked in or dredged from the lagoon. The sand contained shell materials. Two of the sites (W-1 and W-2) were surrounded on landward facing portions with chain-link fencing up to 7 ft in height. On the water facing sides, they were bounded with 3 foot high polyethylene mesh fencing to retain least tern chicks inside the nesting area and to provide some predator control. Site E-l was not fenced except at the access road.

To facilitate monitoring, a grid system of various dimensions was established on each of the three nesting sites. On W-1 and W-2, a 20x20 meter grid system was utilized and on E-1 a 60x60 meter grid system was used. The grid system assisted in locating and monitoring nest sites and chicks and was supplemented with marking each nest location with wooden slats (tongue depressors). Monitoring of the nesting areas was conducted daily either by observations outside the nesting area or by walk-throughs. Counts were kept on number of eggs laid, hatched, abandoned or depredated, and chicks fledged (Table 1).

Table 1. Summary of California least tern and Western snowy plover nesting activity at recreated nesting sites at Batiquitos Lagoon.

Year	Nesting pairs		Eggs laid		Fledgings	
	CLT	WSP	CLT	WSP	CLT	WSP
1994	72	0	147	0	68	0
1995	80	15	165	40	71	24
1996	182	39	434	113	228	45

During the 1994 breeding season, approximately 72 adult pairs of the California least tern were observed nesting at the Batiquitos Lagoon breeding colony (Whelchel and Keane 1994). Nesting at Batiquitos Lagoon in 1994 represented approximately 5 percent of the total nesting pairs in San Diego County and 3 percent of the State total. The 72 adult pairs initiated a total of 80 nests. A total of 147 eggs were laid; of these, 129 hatched (88% hatching success). Mean clutch size was 1.84. Based on fledgling counts, an estimated 68 chicks fledged from Batiquitos Lagoon.

During the 1995 nesting season, 82 adult pairs of the California least tern were observed at Batiquitos Lagoon (Whelchel and Keane 1995). These birds initiated a total of 93 nests. A total of 165 eggs were laid; of these 146 hatched (88% hatching success). The mean clutch size was 1.77. Based on fledgling counts and recaptures, an estimated 71 chicks fledged (0.87 fledgling per pair).

During the 1996 nesting season, the total number of nesting pairs increased to 182 adult pairs. These nesting pairs laid a total of 434 eggs. Based on fledging counts, an estimated 228 chicks fledged.

In addition to least tern, nesting activity for the western snowy plover was also recorded. This species did not utilize the constructed nest site in 1994, but did utilize E-1 during the 1995 nesting season (Whelchel and Keane 1995). Fifteen nesting attempts were documented between April 3 and July 25. A total of 40 eggs were laid and of these 32 hatched (80% hatching success rate). The modal clutch size was 3 eggs with a mean clutch size of 2.7. Based on fledgling surveys, an estimated 24 chicks fledged between June 28 and September 19. In 1996, 39 pairs of plovers laid a total of 113 eggs resulting in 45 fledgings.

Predation at the nesting sites was reduced through the efforts of an active trapping and monitoring program conducted by Animal Damage Control. Most predation attacks on least terns were from American kestrels, great horned owls, and a feral cat. Most predation attacks on western snowy plovers were due to common ravens.

DISCUSSION

The construction of the Batiquitos Lagoon Enhancement Project represents the culmination of a 20 year planning effort. Restoration of degraded systems is made difficult by the conflicts that arise due to surrounding land uses, recreational demands on beaches where inlets are proposed, infrastructure and transportation corridor constraints, protection of federally and state listed species, and implementation costs. The cooperation of federal and state agencies coupled with the economic resources of the Port of Los Angeles and the desire of the City of Carlsbad to restore its natural resources was the combination that made this project possible. The project benefits, especially to beach nourishment and habitat enhancement, provided the public incentives to accept the project.

The use of the nesting areas by federally listed shorebirds has far exceeded expectations. This may be due in part to the protection afforded these areas by both passive and active predator control methods. However, the wide-spread use of the site during construction suggests that the design and implementation of the project has been successful. Only long-term monitoring to be conducted after the completion of the construction will tell whether the restoration itself functions as designed.

REFERENCES

California State Coastal Conservancy. 1987. Batiquitos Lagoon Enhancement Plan. Revised Draft. Oakland, CA.

City of Carlsbad. 1990. Batiquitos Lagoon Enhancement Project. Final EIR/EIS.

Memorandum of Agreement. 1987. Agreement among the City of Los Angeles, the City of Carlsbad, the California Department of Fish and Game, the California State Lands Commission, the National Marine Fisheries Service, and the United States Fish and Wildlife Service to Establish a Project for Compensation of Marine Habitat Losses Incurred by Port Development Landfills within the Harbor District of the City of Los Angeles by Marine Habitat Enhancement at Batiquitos Lagoon.

Mudie, P.J., B.M. Browning, and J.W. Speth. 1976. The Natural Resources of San Dieguito and Batiquitos Lagoon. Coastal Wetland Series #12. Department of Fish and Game, State of California.

Whelchel, A.W. and K. Keane. 1994. California Least Tern Breeding Survey, Batiquitos Lagoon, San Diego County. 1994 season. Batiquitos Lagoon Environmental Monitor, City of Carlsbad.

Whelchel, A.W. and K. Keane. 1995. Western Snowy Plover Breeding Survey, Batiquitos Lagoon, San Diego County. 1995 season. Batiquitos Lagoon Environmental Monitor, City of Carlsbad.

Whelchel, A.W. and K. Keane. 1995. California Least Tern Breeding Survey, Batiquitos Lagoon. San Diego County. Batiquitos Lagoon Environmental Monitor, City of Carlsbad.

CHAPTER 136

The Long Term Management Strategy For San Francisco Bay Dredging

Steven Goldbeck

Abstract

In July 1990, the San Francisco Bay Conservation and Development Commission (Commission), U.S. Army Corps of Engineers (Corps), U.S. Environmental Protection Agency (USEPA), and the San Francisco Bay Regional Water Quality Control Board, joined together in a consensus-based program to develop a Long Term Management Strategy (LTMS) for San Francisco Bay dredging and dredged material disposal. The purpose of the LTMS program is to formulate a long-range management plan and implementation program to guide the dredging and disposal of materials from San Francisco Bay. The development of the LTMS is in its sixth year. Technical studies have been completed and a draft environmental document analyzing the impacts of alterative LTMS policies for dredging was released in April, 1996.

The LTMS involves crafting a program that will meet the dredging needs of the region while achieving the goals of the LTMS. The four main goals are to (1) maintain channels in an environmentally and economically sound manner; (2) dispose of dredged material in the most environmentally sound manner; (3) maximize the use of dredged material as a resource; and (4) establish a cooperative permitting framework for dredging applications. The Commission's LTMS efforts have focused on investigation and promoting the use of dredged material as a resource to: restore diked historic baylands to tidal wetlands, cap and line landfills, and bolster levees surrounding the Bay and Delta. The most notable example of this is the Sonoma Bayland Wetland Restoration Project sponsored by the State Coastal Conservancy and constructed by the Corps that used approximately 2.5 million cubic yards of dredged material to restore over 200 acres of tidal wetlands. However, several policy issues will need to be resolved before an LTMS can be implemented.

Steven Goldbeck, Coastal Program Manager, San Francisco Bay Conservation & Development Commission

For example, a critical concern is the need or regional planning, support, and financing for upland sites. The Commission's technical studies indicate that the direct costs of reuse and upland disposal projects will often be significantly higher than the historic cost of in-Bay disposal. Constructing sites that are sufficiently large to achieve economies of scale would reduce the cost per cubic yard of such projects. However, because of their significant costs and complex logistics, many of the otherwise feasible sites might become infeasible and might be used for other purposes unless a regional approach is taken. Therefore, the LTMS agencies must consider how to provide for a regional, state, and/or federal approach to provide financing, take needed actions, and address regional concerns, in order to increase the likelihood of upland disposal and reuse. Another major issue of the LTMS is how to provide a more streamlined permit process given the involvement of multiple permitting agencies mandated under several state and federal laws. Recently, the LTMS agencies have formed a pilot Dredged Material Management Office to provide the applicant with a single permit application process, while ensuring implementation of the mandates and laws of the regulatory agencies.

This paper will provide a background and status report on the LTMS program, discuss the alterative policies proposed for managing Bay dredging and disposal, and address considerations involved in implementing the program.

CHAPTER 137

STUDY OF POLLUTANT DISPERSION OF COASTAL DREDGING AND RECLAIMING

Hou, Ho-Shong Ph.D., P.E.
Director, Transportation Engineering Division
Institute of Transportation, MOTC
9F, 240 Tun-Hua N. Rd., Taipei, Taiwan, R.O.C.

ABSTRACT: Coastal dredging and reclamation will increase the turbidity of sea water on the one hand and cause sediment transport on the other hand and then affect the offshore environmental and ecological system. New reclaiming land changes its coastal morphology, while dredging area deepens its surrounding depth. Both of them heavily cause environmental impact of natural oceanographic condition and ecological system.

For pollutant dispersion problem, numerical simulation model of sediment transport is used to realize its diffusion phenomena. While the sea water quality measurement is conducted to analyze the environmental impact.

This study involves Hydrodynamic Module by using Alternating Direction Implicit Method to simulate hydraulic characteristics of coastal and harbor area.

1. INTRODUCTION:

Ministry of Transport and Communications intends to extend the Kaohsiung Port as a deep water harbor. Simultaneously it will work as a hub port by reclaiming coastal zone, and building the area covering from the south breakwater of the second entrance of Kaohsiung Port to the river mouth of Kao-Ping Hsi. The seaward direction of Kaohsiung Deep Water Port will be extended to the depth of -28 meter deep, about 6 km away from the coastline.

The total reclaiming area of this project is about 6177 hectares and the filling sand volume is about 1.3 billion cubic meter. It is expected to be completed within 20 years. Therefore, the filling sand quantity is about 65 million cubic meter per year. Since massive dredging and reclaiming works for building the new land, they will cause sediment pollutant transport and pollute around the offshore area.

For pre-calculating the environmental ecological effect due to dredging and reclaiming sand pollution, it is necessary to apply numerical model to predict the littoral sediment pollutant diffusion and its serious effect of the environmental ecological aspects.

2. CLIMATE AND OCEANOGRPHIC CONDITIONS

For Kaohsiung offshore wave data, summer waves are generated by south -westerly monsoon and typhoon (Maximum typhoon wave height is 6.9 meters), while autumn and winter waves are generated by typhoon (maximum typhoon wave height is 9.29 meters)and north-easterly monsoon.

Ocean tidal current occurs dominantly parallel to the coastline and flow direction is toward south-easterly direction. The maximum tidal range is 171 cm, and the mean tidal range is 74 cm.

Monthly spring wind speed variation is from 1.15m/s to 3.78m/s, summer season is from 2.19m/s to 3.97m/s. Autumn is from 1.28 m/s to 3.37 m/s, and winter wind speed variation is from 1.43 m/s to 4.11m/s. Wind speed in winter season is mostly produced due to north-easterly monsoon, while wind speed of summer is caused by south-westerly monsoon and typhoon action. Both show a little higher value.

3. GOVERNING EQUATION

Continuity Equation

$$\frac{\partial\xi}{\partial t}+\frac{\partial p}{\partial x}+\frac{\partial q}{\partial y}=s-e \qquad (1)$$

X-direction Momentum Equation

$$\frac{\partial q}{\partial t}+\frac{\partial}{\partial y}[\frac{p^2}{h}]+\frac{\partial}{\partial y}[\frac{p\cdot q}{h}]+gh\frac{\partial\xi}{\partial x}+\frac{\sqrt{\frac{p^2}{h^2}+\frac{q^2}{h^2}+\frac{p}{h}}}{c^2}-fvv_x$$
$$-\frac{h}{p_a}\cdot\frac{\partial p_a}{\partial x}-\Omega q-[\frac{\partial}{\partial x}[E_x\cdot h\cdot\frac{\partial u}{\partial x}]+\frac{\partial}{\partial x}[E_y\cdot h\cdot\frac{\partial u}{\partial y}]] \qquad (2)$$
$$=S_{ix}$$

Y-direction Momentum Equation

$$\frac{\partial q}{\partial t}+\frac{\partial}{\partial y}[\frac{p^2}{h}]+\frac{\partial}{\partial x}[\frac{p\cdot q}{h}]+gh\frac{\partial\xi}{\partial y}+\frac{\sqrt{\frac{p^2}{h^2}+\frac{q^2}{h^2}+\frac{p}{h}}}{c^2}-fvv_y$$
$$-\frac{h}{p_a}\cdot\frac{\partial p_a}{\partial y}-\Omega q-[\frac{\partial}{\partial x}[E_x\cdot h\cdot\frac{\partial u}{\partial x}]+\frac{\partial}{\partial y}[E_y\cdot h\cdot\frac{\partial u}{\partial y}]] \qquad (3)$$
$$=S_{iy}$$

h (x,y,z): Water Depth (m)
s (x,y,z): Water level (m)
p (x,y,z): Flux Density in x-direction (m^3/s/m)
q (x,y,t): Flux Density in y-direction(m^3/s/m)
e (x,y) : Momentum Coefficient(m^2/s)
s : Source Per unit plane(m^3/s/m)
g : Gravity acceleration
c : Bottom Resistance Coefficient($m^{1/2}$s)
f : Wind Resistance Coefficient
V, U_x, U_y: Wind speed, and its components of x and y direction
Ω : Corriolis Coefficient (s^{-1})
p_a : Pressure (kg/m/s)
x, y : Space Coordinate
u , v : Mean Velocity of x and p directions
t : Time, s_{ix}, s_{iy}: Source of x and y directions
Six, Siy : Source of x and y

4. BOUNDARY CONDITION

Tide Elevation at anytime at certain position

$$H_t=H_0+\sum^{p} f_i H_i\cos((V+u)_i+w_{it}-k_i)) \qquad (4)$$

H_0 : Mean Water Level

H_i : Amplitude of the ith component tide
W_i : Time lag of the ith component tide
K_i : Angular velocity of the ith component tide $\Delta t_{max}=\Delta x \bullet C_r/c$ (5)
$C_r=C\bullet \Delta t/\Delta x$

$$=\sqrt{g\cdot h}$$

Cr :Courant Number
C :Celerity
Δt :Time Increment
Δx :Grid Size
g :Gravity Acceleration
h :Water Depth

5. LITTORAL TRANSPORT MODEL

Diffusion Equation

$$\frac{\partial \bar{c}}{\partial t}+u\frac{\partial \bar{c}}{\partial x}+v\frac{\partial \bar{c}}{\partial y} = \frac{1}{h}\frac{\partial}{\partial x}(hD_x\frac{\partial \bar{c}}{\partial x})+\frac{1}{h}\frac{\partial}{\partial y}(hD_y\frac{\partial \bar{c}}{\partial y})+Q_1C_1\frac{1}{h}-S \tag{8}$$

$\bar{c}$:Concentration of Mean Water Depth
U,v :Velocity Component of Mean Water Depth
D_x, D_y:Diffusion Coefficient of x, y direction
h :Water Depth
S :Settling or Flushing terms
Q_1 :Source Flux per Unit Horizontal Area
C_1 :Concentration of Source

Diffusion and Settling Balance Equation

$$-\varepsilon\frac{dC}{dz}=wC \tag{9}$$

ε :Diffusion Coefficient
C :Concentration of"Z"function
Z :Vertical Axis of Cartesian Coordinate
W :Mean Settling Velocity of Littoral Transport

Chezy Number could be epressed as

$$=M\cdot h^{1/6}$$

C : Chezy Number
M : Manning Number
h : Water Depth

Smagorinsky Equation

$$E=\iota^2\{(\frac{\partial\mu}{\partial X})^2+(\frac{\partial v}{\partial Y})+\frac{1}{2}(\frac{\partial\mu}{\partial Y}+\frac{\partial v}{\partial X})^2\}^{\frac{1}{2}} \tag{6}$$

E :Eddy Diffusion Coefficient

u,V :Velocity Component of x,y Direction
L :Mixing Length
L :$C_s \bullet \Delta x$

Wind Stress is expressed by using Square Law:

$$C_w \bullet \rho_{air} / \rho_{water} \bullet W^2 \tag{7}$$

C_w :Wind Friction Coefficient
W :Measured Wind Speed of 10^m high above Mean Sea Level

Turbulent Eddy Viscosity is expressed as

$$\varepsilon = kU_f \cdot z[1-\frac{z}{h_1}] \tag{10}$$

k :Von Karman's universal constant

U_f :Friction Velocity $U_f = \sqrt{\frac{\tau_b}{\rho}}$]

ρ :Fluid Density]

Vertical Concentration Profile is expressed as

$$= C_a[\frac{a}{h-a}\cdot\frac{h-z}{z}]^R, a \le z \le h \tag{11}$$

C_a :Reference Concentration, when Z=a
a :Reference Level
Z :Distance of Sea Bottom
R :Rouse number $= \frac{w}{0.4U_f}$]

Mean Settling Depth of particle

$$\frac{h_a}{h} = \frac{\int_0^1 s(\frac{1}{s}-1)^R ds}{\int_0^1 (\frac{1}{s}-1)^R ds} \tag{12}$$

Where s=h/z
Settly velocity is expressed as

$$S = -\frac{w\bar{c}}{h_a}\cdot[\frac{\tau_b}{\tau_{cd}}-1], \tau_b \le \tau_{cd} \tag{13}$$

W :Settling velocity of suspended particle
h_a :Mean Settling Depth of a particle
τ_{cd}:Critical Shear Stress of Settling Material
τ_b :Bottom Shear Stress
W :from stokes law is obtained as

$$= \frac{(s-1)gd^2}{18\nu} \tag{14}$$

W :Settling Velocity of Settling Particle

S :Specific Weight of Settling Particle
g :Gravity Acceleration
d :Floc Diameter
v :Dynamic Viscosity of Water

Shear Stress is determined by Manning formula

$$\tau_b = \rho g \frac{v^2}{M^2 h^{1/3}} \qquad (15)$$

V :Mean Depth Concentration
M :Manning number

Settling velocity is expressed by combining the bed shear stress and settling rate

$$S = \frac{w\bar{c}}{h_a}[1-(\frac{v}{v_{cd}})^2], v \le v_{cd} \qquad (16)$$

Critical Settling Velocity

$$_{cd} = 1.25w \frac{Mh^{1/6}}{\sqrt{g}} \qquad (17)$$

Flushing Velocity of Littoral Sediment with Viscous

$$S = -\frac{E}{h}[\frac{\tau_b}{\tau_{ce}} - 1], \tau_b \ge \tau_{ce} \qquad ((18)$$

τ_{ce} :Critical Shear Stress before Flushing Occurring
E :Bottom Erodeability
Flushing Speed is Expressed by Current Velocity

$$S = \frac{E}{h}[1-(\frac{V}{V_{cd}})^2], V \ge V_{ce} \qquad (19)$$

V_{ce} :Critical Flushing velocity

6. SIMULATION RESULTS AND SUGGESTION

.A. For the condition of 1 km Seawall construction of the north end, the simulation results show that suspension load conceentration around the second entrance of the Kaohsiung port is about 1,500 ppm. For the case of no Seawall construction, suspension load concentration will be high above 2,000 ppm around the second entrance, as the filling sand source is 1.75 m^3/sec and it is taken as the 100% concentration of the filling sand location..

B. From the simulation results, it is found that whether extension construction reclaiming from north end or from south end, the better way is to build Seoul 6 km long at the first step. In this way, the littoral sediment shoaling the second entrance will reduce to the minimum..

C. After the first 6 km seawall construction is completed, filling sand position need to be chosen near the corner of the Seoul and the coastline, since there the current speed is lower. Littoral sediment easily settles and loss of filling sand quantity will be less.

D. It is suggested that during construction stage, the field measurement of turbidity is necessarily executed for the comparison with the simulation result of filling sand volume is processing. As both concentrations are coincident with each other, then the model is verified and confirmed. Therefore, the Hydrodynamic Module is fitted well and could be extended its use to the advection-dispersion module of coastal geomorphologic change vs wave and current interaction..

REFERENCE

[1] Ho-Shong Hou & CECI: Taiwan West Coast Climate and Oceanographic Data Analysis. Institute of Transportation, MOTC, Dec., 1987.

[2] Ho-Shong Hou & Tainan Hydraulic Lab:Kaohsiung Offshore climate and Oceanographic Survey. Institute of Transportation, MOTC, Jan., 1993. .

[3] Ho-Shong Hou & Robert Huang: Kaohsiung Deep Water Port Littoral Transport and Pollutant Transport Study. Institute of Transportation, MOTC, June, 1993. .

Acknowledgement

This research project is jointly carried out by Division of Transportation, IOT, MOTC, and Department of River & Harbor Engineering ,NTOU, financial support by Institute of Transportation, contributions from Prof. W.H.Yang and Deam Robert, Huang, especially Administration support from IOT'S Directoral General Dr. Chia-Juia Chang are greatly appreciated.

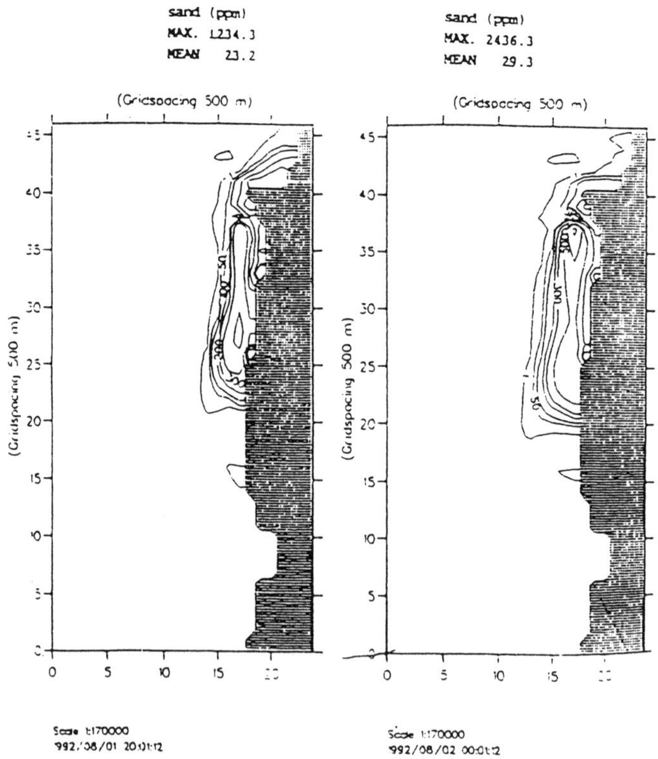

Fig. 1. Deepwater port original pattern of Littoral Sediment Diffusion

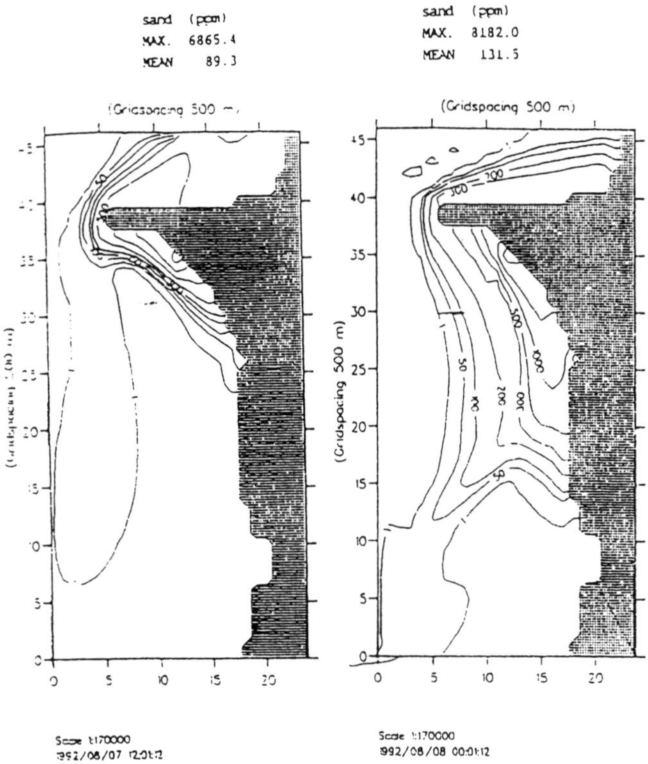

Fig. 2 Littoral Sediment Diffusion Pattern --

North End of Deep Water Port has been built 6 km seawall

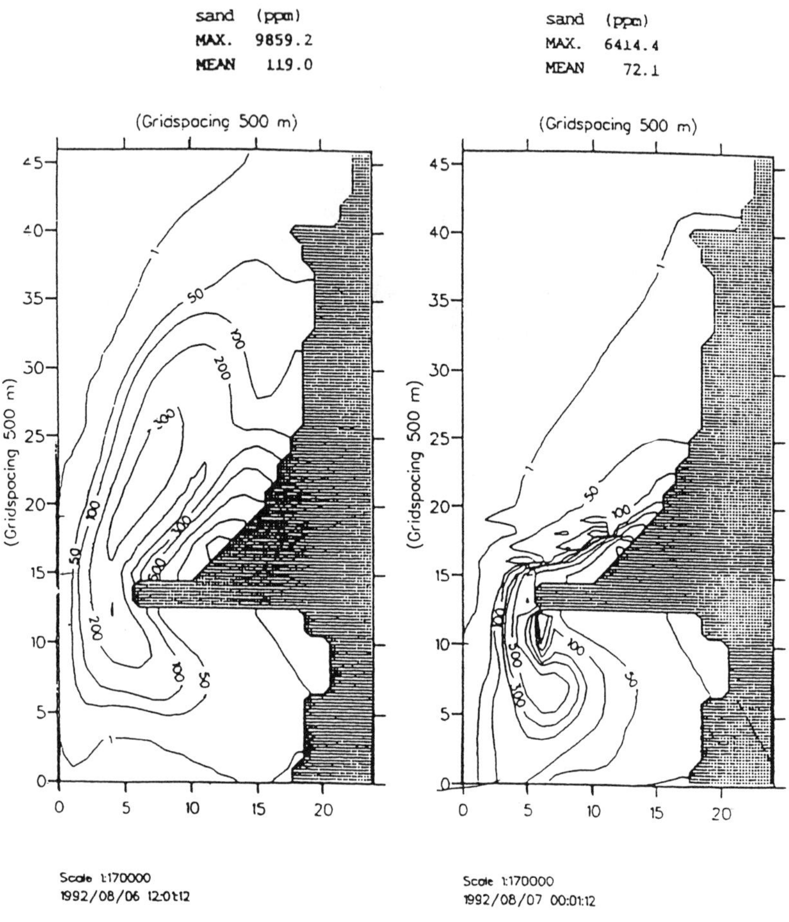

Fig. 3 Littoral Sediment Diffusion Pattern --

South End of Deep Water Port has been built 6 km seawall

CHAPTER 138

CONTAMINATED SEDIMENTS IN PORTS AND WATERWAYS: CLEANUP STRATEGIES AND TECHNOLOGIES[1]

Joseph L. Zelibor[2]
National Research Council
National Academy of Sciences

There is no simple solution to the problems created by contaminated marine sediments,[3] which are widespread in U.S. coastal waters and can pose risks to human health, the environment, and the nation's economy. Marine sediments are contaminated by chemicals that tend to sorb to fine-grained particles; contaminants of concern include trace metals and hydrophobic organics, such as dioxins, polychlorinated biphenyls (PCBs), and polyaromatic hydrocarbons. Contamination is sometimes concentrated in "hot spots" but is often diffuse, with low to moderate levels of chemicals extending no more than a meter into the seabed but covering wide areas. Approximately 14 to 28 million cubic yards of contaminated sediments must be managed annually, an estimated 5 to 10 percent of all sediments dredged in the United States.

---

[1]Reprinted with permission from Contaminated Sediments in Ports and Waterways: Cleanup Strategies and Technologies. Copyright 1997 by the National Academy of Sciences. Courtesy of the National Academy Press, Washington D.C.

[2]Study Director, Committee on Contaminated Marine Sediments, National Research Council, Marine Board, 2101 Constitution Avenue N.W., Washington D.C. 20418

[3]For purposes of this report, contaminated marine sediment is defined as containing chemical concentrations that pose a known or suspected threat to the environment or human health.

The many challenges to be overcome in managing contaminated sediments include an inadequate understanding of the natural processes governing sediment dispersion and the bioavailability of contaminants; a complex and sometimes inconsistent legal and regulatory framework; a highly charged political atmosphere surrounding environmental issues; and high costs and technical difficulties involved in sediment characterization, removal, containment, and treatment. The need to meet these challenges is urgent. The presence of contaminated sediments poses a barrier to essential waterway maintenance and construction in many ports, which support approximately 95 percent of U.S. foreign trade. The management of these sediments is also an issue in the remediation[4] of an estimated 100 marine sites targeted for cleanup under the Comprehensive Environmental Response, Cleanup, and Liability Act (CERCLA) (P.L. 96-510), commonly known as Superfund, as well as in the cleanup of many other near-shore contaminated sites.

The Committee on Contaminated Marine Sediments was established by the National Research Council under the auspices of the Marine Board to assess the nation's capability for remediating contaminated marine sediments and to chart a course for the development of management strategies. In the committee's view, cost-effective management of contaminated marine sediments will require a multifaceted campaign as well as a willingness to innovate. The committee determined that a systematic, risk-based approach incorporating improvements to current practice is essential for the cost-effective management of contaminated marine sediments. The committee identified opportunities for improvement in the areas of decision making, project implementation, and interim and long-term controls and technologies, as outlined in this summary. Although the study focused on evaluating management practices and technologies, the committee also found it essential to address a number of tangentially related topics (e.g., regulations, source control, site assessment) because problems in these areas can impede application of the best management practices and technologies.

As part of the three-year study, the committee compiled six case histories of recent or ongoing contaminated sediments projects, visited one of those sites, analyzed the relevant regulatory framework in depth, held separate workshops on

---

[4]For purposes of this report, sediment management is a broad term encompassing remediation technologies as well as nontechnical strategies. Remediation refers generally to technologies and controls designed to limit or reduce sediment contamination or its effects. Controls are practices, such as health advisories, that limit the exposure of contaminants to specific receptors. Technologies include containment, removal, and treatment approaches. Treatment refers to advanced technologies that remove a large percentage of the contamination from sediment.

interim controls and long-term technologies, and examined in detail how various decision-making approaches can be applied in the contaminated sediments context. The committee also examined the application of decision analysis in contaminated sediments management.

IMPROVING DECISION MAKING

Decision-Making Tools

Contaminated sediments can best be managed if the problem is viewed as a system composed of interrelated issues and tasks. Systems engineering and analysis are widely used in other fields but have not been applied rigorously to the management of contaminated sediments. The overall goal is to manage the system in such a way that the results are optimized. In particular, a systems approach is advisable with respect to the selection and optimization of interim and long-term controls and technologies. Although unlimited time and money would make remediation of any site feasible, resource limitations demand that trade-offs be made and that solutions be optimized.

A fundamental aspect of the committee's recommended approach is the delineation of the trade-offs among risks, costs, and benefits that must be made in choosing the best course of action among multiple management alternatives. A number of decision-making tools can be used in making these trade-offs. Available tools include risk analysis, cost-benefit analysis, and decision analysis.

Cost-effective contaminated sediments management requires the application of risk analysis—the combination of risk assessment, risk management, and risk communication. Contaminated sediments are considered a problem only if they pose a risk that exceeds a toxicological benchmark. In its most elemental form, risk assessment is intended to determine whether the chemical concentrations likely to be encountered by organisms are higher or lower than the level identified as causing an unacceptable effect. The "acceptable risk" needs to be identified, quantified, and communicated to decision makers, and the risk needs to be managed. First, management strategies need to be identified that can reduce risk to an acceptable level. Second, remediation technologies need to be identified that can reduce the risk associated with contaminants to acceptable levels within the constraints of applicable laws and regulations. Third, promising technologies need to be evaluated within the context of the trade-offs among risks, costs, and benefits, a difficult task given the uncertainties in risk and cost estimates. The next step is risk communication, when the trade-offs are communicated to the public.

At present, risk analysis is not applied comprehensively in contaminated sediments management. Risks are usually assessed only at the beginning of the decision-making process to determine the severity of the in-place contamination; the risks associated with removing and relocating the sediments or the risks remaining after the implementation of solutions are not evaluated. The expanded application of risk analysis would not only inform decision making in specific situations but would also provide data that could be used in the selection and evaluation of sediment management techniques and remediation technologies.

Cost-benefit analysis can also be useful for evaluating proposed sediment management strategies. Although risk assessments may provide information about the exposure, toxicity, and other aspects of the contamination, they may result in a less-than-optimum allocation of resources unless additional information is considered. For example, a given concentration of contaminants at a particular site might be toxic enough to induce mortality in a test species, but this information alone does not indicate the spending level that would be justified for cleanup. Cost-benefit analysis combines risk and cost information to determine the most efficient allocation of resources. The basic principle of cost-benefit analysis is that activities should be pursued as long as the overall benefit to society exceeds the social cost. The difficulty lies in the measurement of the benefits and costs, or, more to the point, the projection of what they will be, before a strategy is implemented.

Cost-benefit analysis is not applied widely in contaminated sediments management. It is generally carried out only for major new navigational dredging projects, and the analyses are usually narrow in scope. Cost-benefit analysis could be used in many cases to help identify the optimum solution in which the benefits outweigh the costs (i.e., to maximize benefits for a given cost or to minimize costs for a given level of benefits). The costs and benefits involved in contaminated sediments management are difficult to calculate and cannot be measured precisely, but cost-benefit analysis may be worth the effort; comprehensive cost-benefit analysis may be warranted in very expensive, or extensive projects. Informal estimates or cost-effectiveness[5] analyses may suffice in smaller projects.

---

[5]Cost effectiveness is defined here as a measure of tangible benefits for money spent.

As the demand for the remediation of contaminated sediments grows, and as costs and controversies multiply, decision makers need to be able to use information about risks, costs, and benefits that may be controversial and difficult to evaluate, compare, or reconcile. One approach that could help meet this need is decision analysis, a computational technique that makes use of both factual and subjective information in the evaluation of the relative merits of alternative courses of action. Decision analysis involves gathering certain types of information about a problem and selecting a set of alternative solutions to be evaluated. The evaluation is used to determine and assess possible outcomes for each alternative. The outcomes are rated, and the results are used to develop a strategy that offers the best odds for successful risk management.

Formal decision analysis is not yet widely used in the management of contaminated sediments. The committee examined this technique using a test case and determined that applications of decision analysis may be particularly timely now, because recent advances in computer hardware and software make it possible to perform such analyses in ways that are user friendly and interactive. Decision analysis could be especially valuable because it can accommodate more variables (including uncertainty) than techniques such as cost-benefit analysis that measure single outcomes. Decision analysis can also serve as a consensus-building tool by enabling stakeholders to explore various elements of the problem and, perhaps, find common ground. However, because decision analysis is technical in design and involves complex computations, it is probably worth the effort only in highly contentious situations in which stakeholders are willing to devote enough time to become confident of the usefulness of the approach.

Regulatory Framework

Few aspects of sediment handling, treatment, or containment are unregulated at the federal, state, or local level, but the regulatory approach is inconsistent, primarily because the applicable laws were originally written to address issues other than contaminated marine sediments. As a result, the current laws and regulations affecting contaminated sediments can impede efforts to implement the best management practices and achieve efficient, risk-based, and cost-effective solutions. This is a shortcoming of the governing statutes, not a criticism of regulatory agencies charged with implementing them. The timeliness of decision making is also an issue, given that it typically takes years to implement solutions to contaminated sediments problems. In the committee's case histories, the delay between the discovery of a problem and the implementation of a solution ranged from approximately 3 to 15 years.

At least six comprehensive acts of Congress, with implementation responsibilities spread over seven federal agencies, govern sediment remediation or dredging operations in settings that range from the open ocean to the freshwater reaches of estuaries and wetlands. When environmental cleanup is the driving force, the relevant federal laws include Superfund; the Resource Conservation and Recovery Act (RCRA) (P.L. 94-580); and Section 115 of the Clean Water Act (CWA) (formerly the Federal Water Pollution Control Act [P.L. 80-845]). When navigational dredging is the issue, the applicable statutes are likely to be the CWA; the Rivers and Harbors Act of 1899 (P.L. 55-525); the Marine Protection, Research and Sanctuaries Act (MPRSA, commonly known as the Ocean Dumping Act) (P.L. 92-532); and the Coastal Zone Management Act (P.L. 92-583). In addition, states also exercise important authority related to water quality certification and coastal zone management. In some cases, local laws may also apply. To complicate matters further, federal, state, and local authorities often overlap.

The principal federal agencies involved are the U.S. Environmental Protection Agency (EPA), which is responsible for implementing Superfund and has major site designation, regulation development, and veto responsibilities under the CWA and MPRSA; the National Oceanic and Atmospheric Administration, which assesses the potential threat of Superfund sites to coastal marine resources and exercises significant responsibilities for research, under the MPRSA, and review and comment, under CWA and MPRSA; and the U.S. Army Corps of Engineers (USACE), which assists in the design and implementation of remedial actions, under Superfund, and has responsibilities for dredged material, under the CWA, MPRSA, and Rivers and Harbors Act. The federal navigational dredging program is the joint responsibility of the EPA and USACE; the EPA regulates disposal, whereas USACE handles the dredging.

The committee identified several areas of the current regulatory framework in which changes might be beneficial. For example, the CWA, the MPRSA, and Superfund use different approaches for evaluating remedial alternatives, but none fully considers either the risks posed by contaminated marine sediments or the costs and benefits of various solutions. The MPRSA requires biological testing of dredged material to determine its inherent toxicity but does not fully consider site-specific factors that may influence the exposure of organisms in the receiving environment, meaning that, at best, risk is considered only indirectly and the actual impact is approximated. Although the CWA procedures, which consider chemical and physical as well as biological characteristics in assessing whether the discharge of dredged material will cause unacceptable adverse impacts, are not risk-based, at least they do not specify rigid pass-fail criteria. They are geared to identification of the least environmentally damaging, implementable alternative. The Superfund remedial action program addresses risks and costs to some degree—an exposure assessment (but not a full risk analysis) is required to assess in-place risks; remedial alternatives are identified based on their capability of reducing exposure risks to an acceptable level; and the final selection involves choosing the most cost-effective solution. However, there are no risk-based cleanup standards for underwater sediments. Insufficient attention to risks, costs, and benefits impedes efforts to reach technically sound decisions and manage sediments cost-effectively.

Similar inattention to risk is evident in the permitting processes for sediment disposal. It is currently necessary to secure different types of permits for the placement of sediments in navigation channels or ocean waters as part of the construction of land or containment facilities (under the Rivers and Harbors Act), the dumping of sediments in the ocean (under the MPRSA), the discharge of sediments in inland waters or wetlands (CWA), and the containment of contaminated sediments on land (RCRA). In addition, different regulations come into play depending on whether sediments are removed during navigational dredging (CWA or MPRSA) or are excavated for environmental remediation (Superfund). The committee can see little technical justification for the differential regulation of contaminated sediments, given that neither the location of the aquatic disposal site (freshwater versus saltwater) nor the reason for dredging (navigational dredging versus environmental remediation) necessarily affects the risk posed by the contamination. The regulatory regime does not adequately address risk; instead it focuses rigidly on the nature of the activities to be carried out. This problem has been eased in some instances by the interpretation of regulations based on the intent of the underlying statute(s).

Systematic, integrated decision making can also be undermined by dredging regulations governing cost allocation and cost-benefit analysis. The federal government pays for most new-work dredging and all maintenance dredging but not for sediment disposal, except in open water. The local sponsors of federal navigation projects bear the burden of identifying, constructing, operating, and maintaining dredged material disposal sites, under the "project cooperation agreement" of the Water Resources Development Act (WRDA) of 1986 (P.L. 99-662). Because project sponsors must pay for disposal on land, whereas open-water disposal is paid for by the federal government as a component of dredging costs, the WRDA provision creates a strong preference for open-water disposal. Furthermore, a local sponsor bearing the full burden of disposal costs has little incentive to seek out opportunities for the beneficial uses of dredged material (discussed in the next section). The cost of making use of dredge material adds to the project cost and may benefit only third parties. This inconsistent approach to cost sharing can lead to the economically irrational allocation of scarce societal resources. Additional inconsistencies are introduced in the area of cost-benefit analysis. As noted earlier, costs and benefits must be weighed for new dredging projects but not for the maintenance dredging of existing channels or for the disposal of dredged material.

IMPROVING PROJECT IMPLEMENTATION

Stakeholder Interests

Contaminated sediments are not managed in a political or social vacuum. Most contaminated sediments sites are located in highly populated areas near the Great Lakes or the oceans. The nature of these sites virtually ensures that complicated ecological situations and difficult technical problems will have to be accommodated along with complex political circumstances involving multiple resource users and interest groups. Stakeholders include port managers and transportation officials who have strong economic reasons for dredging; federal, state, and local regulators responsible for protecting natural resources and enforcing regulations; and environmental groups, local residents, fishermen, and other marine resource users who are concerned about public health and natural resources. The successful management of contaminated sediments must respond to all dimensions of the problem: ecological, technical, social, and political.

The committee determined that remediation and disposal projects need strong proponents and that the identification and timely implementation of effective solutions depend heavily on how project proponents interact with stakeholders, who often have different perspectives on the problem and proposed solutions. Because any participant in the decision-making process can block or delay remedial action, project proponents need to identify all stakeholders and build a consensus among them. The development of a consensus can be fostered by the use of various tools, including mediation, negotiated rule making, collaborative problem solving, and effective communication of risks.

Stakeholder acceptance of contaminated sediments management projects can be fostered by the reuse of dredged material. Dredged material has been used for many purposes, including the creation of thousands of islands for sea-bird nesting, landfills for urban development, and wetlands, as well as for beach nourishment and shoreline stabilization. The policy focus and most of the experience to date have concerned the use of clean materials, but some contaminated sediments can also be used safely for certain beneficial purposes. Reuse can provide alternatives to increasingly scarce disposal sites while also making management plans more attractive, or at least palatable, to stakeholders. Some contaminated sediment sites have been successfully transformed into wetlands, and productive USACE research is under way on the safe use of contaminated sediments for "manufacturing" topsoil and landfill covers. However, funding for this type of research is limited, and technical guidelines have yet to be developed. Other barriers include the USACE policy of selecting lowest-cost disposal options with little regard to the possibilities of beneficial use and the uncertainties about whether the incremental costs of beneficial use should be borne by the project proponent or the beneficiary.

Source Control

Because accumulations of sediments interfere with deep-draft navigation, ports have no alternative but to dredge periodically in order to remain economically viable. If the sediments to be dredged are contaminated, then ports become responsible for both sediment disposal and any necessary remediation, even though they have no control over the source of the contamination. Upstream generators of contaminants often cannot be identified or held accountable, leaving ports to manage a problem that is not of their making. This responsibility could be shared by states (when states do not already operate or oversee port agencies), which benefit economically from dredging and already engage in watershed management. Under the CWA (Section 303), the EPA and the states set total maximum daily loads for waterway segments and develop load allocations for pollution sources in an effort to control water pollution. This approach could be readily expanded to address sources of sediment contamination. In addition, government regulators and ports could use all available legal and enforcement tools for ensuring that polluters bear a fair share of cleanup costs.

Site Characterization

Accurate site characterization is essential to the cost-effective management of contaminated sediments. Site assessments need to be sufficiently comprehensive and accurate to ensure that the contamination is well defined both chemically and geographically. Inaccuracies and incompleteness can leave areas of unidentified contamination that pose continuing unmanaged risks. Another compelling argument for accurate site assessment is the need to control remediation costs;

precise site definition is necessary to facilitate removal of only those sediments that are contaminated, thus controlling the volume of material that requires expensive remediation. But the high cost of commonly used site characterization technologies (i.e., physical profiling and chemical testing) has limited the precise definition of either horizontal or vertical contaminant distributions, which may have led to the removal and "remediation" of large quantities of uncontaminated sediments at unnecessarily high costs.

Thus, the development and wide use of new or improved site characterization technologies that are less expensive than current methods would enhance the cost-effective management of contaminated sediment sites. One technology that may prove useful in the future is acoustic profiling,[6] which helps define the thickness and distribution of disparate sediment types. Because contaminants tend to be associated with fine-grained material, acoustic profiling may provide for cost-effective remote surveying of contaminated sediments, thereby increasing the precision and accuracy of site assessment. Additional research and development is needed, however. Sediment characterization may also be enhanced through the adaptation of chemical sensors now used in the assessment of soil and groundwater sites.

INTERIM AND LONG-TERM CONTROLS AND TECHNOLOGIES

The following is a brief assessment of the controls and technologies that are applicable to contaminated sediments. The section concludes with a comparative analysis reflecting the committee's overall judgments of the feasibility, effectiveness, practicality, and cost of each control and technology.

Interim Controls

Interim controls may prove helpful when sediment contamination poses an imminent hazard. Identification of an imminent hazard is usually a matter of judgment, but in general an imminent hazard exists when contamination levels exceed by a significant amount the sum of a defined threshold level plus the associated uncertainty. Administrative interim controls (e.g., signs, health advisories) have been used a number of times. Only two applications of structural interim approaches (e.g., thin caps) were identified by the committee, but

[6] Acoustic profiling involves high-resolution mapping of the acoustic reflectivity of sediments.

additional structural approaches, such as the use of confined disposal facilities (CDFs) for temporary storage, appear promising. Few data are available concerning the effectiveness of interim controls because to date they have not been used often or evaluated in detail.

Long-Term Controls and Technologies

Technologies for remediating contaminated sediments are at various stages of development. Sediment-handling technologies are the most advanced, although benefits can be realized from improvements in the precision of dredging (and, concurrently, site characterization). The state of practice for in situ controls ranges from immature (e.g., bioremediation) to evolving (e.g., capping). Ex situ containment is commonplace. A number of existing ex situ treatment technologies can probably be applied successfully to treating contaminated sediments, but full-scale demonstrations are needed to determine their effectiveness. But these technologies are expensive, and it is not clear whether unit costs would drop significantly in full-scale implementation.

The cost of cleanup depends on the number of steps involved—the more handling required, the higher the cost—and the type of approach used. The costs of removing and transporting contaminated sediments (generally less than \$15 to \$20/yd^3) tend to be higher than costs of conventional navigational dredging (seldom more than \$5/yd^3) but much lower than the costs of treatment (usually more than \$100/yd^3). Volume reduction (i.e., removing only sediments that require treatment and entraining as little water as possible) will mean greater cost savings than increased production rates; improved site characterization coupled with precision dredging techniques hold particular promise for reducing volume. Treatment costs may also be reduced through pretreatment.

In situ management offers the potential advantage of avoiding the costs and potential material losses associated with the excavation and relocation of sediments. Among the inherent disadvantages of in situ management is that they are seldom feasible in navigation channels that are subject to routine maintenance dredging. In addition, monitoring needs to be an integral part of any in situ approach to ensure effectiveness over the long term.

Natural recovery is a viable alternative under some circumstances and offers the advantages of low cost and, in certain situations, the lowest risk of human and ecosystem exposure to sediment contamination. Natural recovery is most likely to be effective where surficial concentrations of contaminants are low, where surface contamination is covered over rapidly by cleaner sediments, or where natural processes destroy or modify the contaminants, so that contaminant releases to the environment decrease over time. A disadvantage of natural recovery is that the sediment bed is subject to resuspension by storms or anthropogenic processes. For natural recovery to be pursued with confidence, the physical, chemical, and hydrological processes at a site need to be understood adequately; however, no capability currently exists for completely quantifying chemical movements. Extensive site-specific studies may be required.

In situ capping promotes chemical isolation and may protect the underlying contaminated sediments from resuspension until naturally occurring biological degradation of contaminants has occurred. The original bed must be able to support the cap, suitable capping materials must be available to create the cap, and suitable hydraulic conditions (including water depth) must exist to permit placement of the cap and to avoid compromising the integrity of the cap. Changes in the local substrate, the benthic community structure, or the bathymetry at a depositional site may subject the cap to erosion. Improved long-term monitoring methods are needed. A regulatory barrier to the use of capping is the language of Superfund legislation (Section 121[b]), which gives preference to "permanent" controls. Capping is not considered by regulators to be a permanent control, but available evidence suggests that properly managed caps can be effective.

Neither in situ immobilization nor chemical treatment of contaminated sediments has been demonstrated successfully in the marine environment, although both concepts are attractive because they do not require sediment removal. Their application would be complicated by the need to isolate sediments from the water column during treatment, by inaccuracies in reagent placement, and by the need for long-term follow-up monitoring. Other constituents (e.g., natural organic matter, oil and grease, metal sulfide precipitates) could interfere with chemical oxidation. Immobilization techniques may not be applicable to fine-grained sediments with a high water content.

Biodegradation has been observed in soils, in groundwater, and along shorelines contaminated by a variety of organic compounds (e.g., petroleum products, PCBs, polyaromatic hydrocarbons, pesticides). However, the use of biodegradation in subaqueous and especially marine environments presents unresolved microbial, geochemical, and hydrological issues and has yet to be demonstrated.

When sediments must be moved for ex situ remediation or confinement, efficient hydraulic and mechanical methods are available for removal and transportation. Most dredging technologies can be used successfully to remove contaminated sediments; however, they have been designed for large-volume navigational dredging rather than for the precise removal of hot spots. Promising technologies offering precision control include electronically positioned dredge heads and bottom-crawling hydraulic dredges. The latter may also have the capability to dredge in depths beyond the standard maximum operating capacity. The cost effectiveness of dredging innovations can best be judged by side-by-side comparisons to technologies in current use.

Containment technologies, particularly CDFs, have been used successfully in numerous projects. A CDF can be effective for long-term containment if it is well designed to contain sediment particles and contaminants and if a suitable site can be found. A CDF can also be a valuable treatment or interim storage facility, allowing the separation of sediments for varying levels of treatment and, in some cases, beneficial reuse. Costs are reasonable; in some parts of the country it may be cheaper to reuse CDFs than to build new ones. Disadvantages of this technology include the imperfect methods for controlling contaminant release pathways. There is also a need for improved long-term monitoring methods.

Contained aquatic disposal (CAD) is applicable particularly to contaminated sites in shallow waters where in situ capping is not possible and to the disposal and containment of slightly contaminated material from navigation dredging. Although the methodology has been developed, CAD has not been widely used. Among the advantages of CAD are that it can be performed with conventional dredging equipment and that the chemical environment surrounding the cap remains unchanged. Disadvantages include the possible loss of contaminated sediments during placement operations. Improved tools are needed for the design of sediment caps and armor layers and for the evaluation of their long-term stability and effectiveness.

Scores of ex situ treatment technologies have been bench tested and pilot tested, and some warrant larger-scale testing in marine systems, depending on their applicability to particular problems. Chemical separation, thermal desorption, and immobilization technologies have been used successfully but are expensive, complicated, and only effective for treating certain types of sediments. Similarly, because of extraordinarily high unit costs, thermal and chemical destruction techniques do not appear to be near-term, cost-effective approaches for the remediation of large volumes of contaminated dredged sediment.

Ex situ bioremediation, which is not as far along in development as are other ex situ treatment approaches, presents so many technical problems that its application to contaminated sediments would be expensive. If these technical problems can be resolved, however, ex situ bioremediation has the potential, over the long term, for the cost-effective remediation of large volumes of sediments. Ex situ bioremediation is much more promising than in situ bioremediation because conditions can be controlled more effectively in a contained facility. The approach has been demonstrated on a pilot scale with some success, but complex questions remain concerning how to engineer the system.

Comparative Analysis of Controls and Technologies

Table S-1 summarizes the committee's overall assessment of the feasibility, effectiveness, practicality, and costs of controls and technologies. For each control and technology, the four characteristics were rated separately on a scale of 0 to 4, with 4 representing the best available (not necessarily the best theoretically possible) features. The effectiveness rating is an estimate of contaminant reduction or isolation and removal efficiency; scores represent a range of less than 90 percent to nearly 100 percent. The feasibility rating represents the extent of technology development, with 0 for a concept that has not been verified experimentally and 4 for a technology that has been commercialized. The practicality ranking reflects public acceptance; 0 means no tolerance for an activity and 4 represents widespread acceptance. The cost ranking is inversely related to the cost of using the control or technology (not including expenses associated with monitoring, environmental resource damage, or the loss of use of public facilities).

The overall pattern of the ratings underscores the need for trade-offs in the selection of technologies. No single approach emerges with the highest scores across the board, and each control or technology has at least one low or moderate ranking. In general, interim controls and in situ approaches are feasible and low in cost but less effective than the most practical ex situ approaches, which tend to be high in cost and complexity. Decisions about which approach is the most appropriate must be made on a project by project basis.

TABLE S-1 Comparative Analysis of Technology Categories

Approach	Feasibility	Effective	Practicality	Cost
INTERIM CONTROL				
Administrative	0	4	2	4
Technological	1	3	1	3
LONG-TERM CONTROL				
In Situ				
Natural recovery	0	4	1	4
Capping	2	3	3	3
Treatment	1	1	2	2
Sediment Removal and Transport	2	4	3	2
Ex Situ Treatment				
Physical	1	4	4	1
Chemical	1	2	4	1
Thermal	4	4	3	0
Biological	0	1	4	1
Ex situ Containment	2	4	2	2
SCORING				
0	< 90%	Concept	Not acceptable, very uncertain	$1,000/yd^3$
1	90%	Bench		$100/yd^3$
2	99%	Pilot		$10/yd^3$
3	99.9%	Field		$1/yd^3$
4	99.99%	Commercial	Acceptable, certain	< $1/yd^3$

CHAPTER 139

Streamlining Contaminated Sediment Management: The Use of Sediment Treatability Profiles During Site Assessment

Sabine E. Apitz and Victoria J. Kirtay

Abstract

Contaminated marine sediments are highly complex systems. Sediments can be composed of naturally occurring organic matter (from fresh detrital material to highly humified residues), and one or many minerals, exhibiting varying properties such as surface chemistry, grain size and porosity. Optimum procedures for adequately assessing and cleaning up contaminated sediments prior to on-land or at-sea disposal have not yet been determined. A critical issue in contaminated marine sediments is that in almost all cases, they are impacted by more that one class of toxic chemical. Therefore it is necessary to determine how multiple contaminants (*e.g.*, toxic metals, PAHs, PCBs), interact with one another and with the sediment itself. Each component differs in its reactivity, solubility, volatility, mineral surface affinity, and biodegradability. Furthermore, interactions between components and the mode of introduction of the contaminant into the soil or sediment, postdepositional weathering and diverse mobility characteristics can drastically alter the composition of the bulk contaminant, as well as its bioavailabilty, risk, and potential treatabilty.

For all these reasons, simply knowing the geographic extent of sediment contamination within a bay or estuary does not provide sufficient insight to help a site manager or stakeholder to begin to assess how to manage sediments deemed either an ecological risk or regulatory violation. A better approach would be to assess contamination distribution and behavior at a micro-scale in parallel with macro-scale site assessments. In such an approach, a site investigation includes an assessment of contaminant/sediment biogeochemistry, as it will impact bioavailability, risk, and potential treatability. With such information in hand, a site manager can make a streamlined and informed decision about what remedial options are available, based upon the site specific sediment characteristics, allowing for rapid progress toward a decision and cleanup. If this is not done in parallel with

Sabine E. Apitz and Victoria J. Kirtay, Remediation Research Laboratory, NCCOSC RDTE DIV 361, 53475 Strothe Roam Rm. 258, San Diego, CA

sample collections for site assessment, there is an great increase in potential cost and time delay.

Such a sediment treatability profile has several parts, examining different aspects of contaminant/sediment interactions. 1) Contaminant distribution as a function of grain size and mass. This assessment determines how contaminants are distributed within the sediment itself. Such an assessment provides insight into whether sediment is amenable to particle separation for volume minimization, whether the contaminant load is subject to easy resuspension during disturbances, and, often, how highly sequestered the contaminants are. 2) Contamination leachibility. Such an assessment helps determine how easily contaminants are remobilized during disturbance, and how amenable sediments are to remediation by chemical leaching. 3) Bioavailability. Such assessments help determine how sequestered or available contaminants are to target biota.

4) Microbial health and capacity. This assessment determines the health and biodegradative potential of the native microbial community.

CHAPTER 140

Organochlorines: The Silent Killer

Maria Brown[1]

Abstract

An ever-growing body of evidence indicates that certain pollutants that build-up through the food chain are capable of entering animals and humans and tampering with vital body systems. The fact that some chemicals persist and bioaccumulate up the food chain has been known for some time now; however, what is new is the understanding that some of these chemicals interfere with operations of the body's endocrine system. Organochlorines as a class of chemicals have been implicated in human and wildlife reproductive, development, and immune disorders. This paper examines the threat of organochlorines to marine life and consequently to humans, and discusses regulating organochlorines from an ecosystem health-human health perspective.

Introduction

The rates of breast cancer incidence have risen sharply (Wolff, 1993); the sperm count of males in industrialized countries has fallen by fifty percent; increasingly male children are born with undescended testes and shrunken penises; and the cases of endometriosis are on the rise (National Wildlife Federation, 1994). At no other time in human existence have we witnessed such startling findings. Something in our environment is altering the reproductive potential of humans, and increasingly scientists are implicating human-made chemicals as the culprit—in particular organochlorine compounds (Wingspread Retreat, 1991). The chlorine industry argues that the only information we have regarding the effects of organochlorines is anecdotal (The Chlorine Chemistry Council, 1994); however, this anecdotal reporting of increased health risks from organochlorines may be the best indicator we have that the global reproductive health of ecosystems including human health is entering a period of unusual and drastic decline.

Organochlorines

Organochlorines are chemical compounds built upon a foundation of sturdy carbon-chlorine bonds. Chlorine has a high electron affinity drawing other electrons toward it. This electron affinity allows chlorine to stabilize some chemical structures making the chemicals inert and long-lasting.

[1] Maria Brown, Executive Director, Farallones Marine Sanctuary Association, The Presidio, P.O. Box 29386, San Francisco, CA 94129. Mariabrown@aol.com. Telephone (415) 561-6625.

Fifteen thousand different organochlorine compounds are known to exist. Organochlorines range from refrigerants such as chloroflorocarbons (CFC), to pesticides such as (DDT), to industrial chemicals such as polychlorinated biphenyls (PCBs). Organochlorines, such as polychlorinated dibenzodioxins (dioxins), are also the byproducts from a number of different processes such as bleaching paper and incineration.

Chemicals such as dioxin can be classified as a chemical family. Different members of a family are called congeners. PCBs and dioxin are not single chemicals; they are blanket terms for two families of chemicals that have many variations. There are 209 congeners of PCBs and 75 congeners of dioxin, some more dangerous than others.

Persistence

Scientists have discovered that the characteristics that make organochlorines an ideal chemical—inert and long-lasting—also create environmental problems. Scientists assumed implicitly or explicitly that organochlorines emitted into the environment would be assimilated by nature: either the natural system would convert them into harmless, naturally occurring substances, or the chemicals would be diluted to such an extent they would pose no threat to life. However, many organochlorine substances are not assimilated; they remain unaltered by light, water, air, or microorganisms for extremely long periods of time. For example toxaphene, an banned in 1982, was so persistent that lakes containing toxaphene runoff could not be successfully restocked for years (Baird, 1995).

Because chlorine creates long-lasting, persistent compounds that do not disperse evenly throughout the environment, they can build up to toxic concentrations. In addition, when chlorine compounds are broken down, for instance during incineration or chemical processes such as the bleaching of paper, they are converted into long-lasting molecules such as dioxins and furans.

Bioaccumulation and Biomagnification

Long-lasting chemicals that resist weathering and remain in the environment for years are more likely to enter the food web. Organochlorines enter the environment from numerous sources, such as farms, incinerators, power plants, pulp and paper mills, and factories, often reaching the same final destination—aquatic ecosystems. Most organochlorines are hydrophobic: they do not readily dissolve in water, but are soluble in oils or fatty tissue. Once they reach an aquatic environment, they attach to flecks of organic material or settle on bottom sediments, and aquatic organisms filter and/or ingest these chemicals (Baird, 1995).

Organochlorine's stable structure and fat solubility allows it to pass through digestive chambers chemically intact and to be stored in fat tissue where it can bioaccumulate, building up to toxic levels. For example, fish bioaccumulate organic chemicals from the food they eat. The chemicals are not metabolized, but accumulate in the fatty tissue, where the concentration increases with time. The concentration of DDT in trout from Lake Ontario increases almost linearly with the age of the fish. This linear increase also occurs up the food chain (Baird, 1995).

Biomagnification begins when phytoplankton absorb persistent chemicals along with the nutrients they filter from the water. An invertebrate will eat phytoplankton, and a fish will eat the invertebrate. Over its lifetime, a fish will eat many times its weight in food and retain rather than eliminate most organochlorine chemicals from these meals (Baird, 1995). From dilutions in the water or sediment measured in parts per quadrillion, a persistent, fat-soluble organochlorine can magnify to concentrations millions times higher in fish. As a result of biomagnification, a bald eagle, which subsists on fish, could have concentration of organochlorines million times higher than levels in water (Hileman, 1992). For example, the ratio of PCBs in the eggs of Herring Gulls in the Great Lakes is 50,000 times the levels in phytoplankton, and in the 1980s the levels of PCBs in eggs of Foster's tern in Green Bay, Wisconsin was about 180,000 times those in the water.

> As an example of biomagnification, consider that the DDT concentration in seawater in Long Island Sound and the protected waters of the southern shore is about 0.000003 ppm, but that it reaches 0.04 ppm in the plankton, 0.5 ppm in the fat of minnows and 2 ppm in the needlefish that swim in these waters, and 25 ppm in the fat of cormorants and osprey that feed on the fish, for a total biomagnification factor of about ten million. It is by such mechanisms that DDE [the degraded version of DDT] levels in some birds of prey became so great that their ability to reproduce successfully was impaired. The bioaccumulation of organochlorines in fish and other animals is the reason that most of the human daily intake of such chemicals enters via our food supply rather than from the water we drink (Baird, 1995).

For humans, the amount of organochlorines ingested by eating a single Great Lakes fish is generally greater that the total organochlorine content in a lifetime of drinking water from the same Great Lake. More than 90% of human exposure to dioxins is attributable to the food we eat, particularly meat, fish, and dairy products of which the largest source is ocean fish (Baird, 1995).

Biological Effects

Endocrine disrupters

Research has shown persistent chemicals such as PCBs, dioxin, atrazine, hexachlorobenzene, as well as other organochlorines to be strongly implicated in the disruption of endocrine systems, including estrogenic effects, in laboratory animals and in wildlife. The chemicals mimic estrogen hormones. Individual cells within an animal's body often cannot distinguish these imitators from natural hormones. This confounds the endocrine system's messages to turn crucial body functions on and off.

Laboratory experiments exposing fetuses to endocrine-disrupting chemicals have found that these chemicals disturb organ differentiation especially of the reproductive system. The organs effected in female fetuses include the mammary glands, fallopian tubes, uterus, cervix, and vagina, and in male fetuses it includes the prostrate, seminal vesicles, epididymides, and testes. In both sexes the external genitalia, brain, skeleton, thyroid, liver, kidney, and immune system are also potential targets for endocrine-disrupting chemicals (Colborn et al, 1993).

A group of twenty-one scientist, who met at Wingspread, Racine, Wisconsin to discuss chemically-induced alterations in sexual and functional development (1991), concurred that portions of a mother's lifetime load of persistent, bioaccumulative chemicals can pass to her young at a particularly vulnerable stage of development, either across the placenta in the case of mammals or into her eggs in the cases of fish, birds, amphibians, and reptiles. While some of the problems are plainly visible at birth, many others are not revealed until the exposed offspring become adults. Either prolonged exposure or a "quick hit" can "masculinize" females or "feminize" males resulting in smaller sex organs, testes that are delayed in descending, lowered sperm generation, and lessened interest in sex.

Effects on Wildlife

Exposure to endocrine-disrupting chemicals on wildlife has been associated with abnormal thyroid function in birds and fish; decreased fertility in birds, fish, shellfish, and mammals; decreased hatching success in fish, birds, and turtles; demasculinization and feminization of male fish, birds, and mammals; defeminization and masculinization of female fish, gastropods, and birds; and alteration of immune function in birds and mammals (Colborn et al, 1993). These effects can be observed throughout the food chain culminating with the most devastating effects on top predators.

For example, in the Great Lakes researchers from Guelph University report a 100% prevalence of thyroid enlargement in 2-4 year-old salmon, poor egg survival (<15%), and low egg thyroid hormone content. Consequently, bird species that feed on Great Lakes fish have been documented with multiple abnormalities, including behavioral changes, reproductive loss, and early mortality in offspring (Colborn et al, 1993).

In the 1970s the American public became aware of the devastating effect of DDT on the national symbol of freedom—bald eagles. DDT was passed from the mother to her eggs and caused eggshell thinning and cracking. Although, DDT use has been restricted for twenty-five years, embryonic and chick survival is not adequate to maintain stable populations. Bald eagles that migrate to the shoreline and consume fish and other food from the Great Lakes for two or more years have difficulty producing viable offspring. The eggs have as much as ten times the levels of DDT determined to maintain a stable population of eagles in addition to elevated levels of chlordane, dieldrin, and PCBs. Researchers are reporting similar findings in nesting bald eagles along the Columbia River in Washington State (Colborn et al, 1993).

Laboratory studies using DDT and other pesticides which remain in wide use today (dicofol, kethane, and methoxychlor) caused severe reproductive system malformities in kestrels, western gulls, and California gulls. These effects include oviducts and gonads resembling ovaries in male birds and abnormal development of oviductual system in female birds. These same effects occurred in herring-gull embryos and newly hatched chicks containing high levels of DDT in Lake Ontario during the 1970s (Colborn et al, 1993).

Human Risk

While most of the harmful effects of these chemicals have been chronicled in wildlife and in laboratory animals, researchers say that each person harbors levels of some of these same chemicals close to those found in impaired wildlife (Thomas and Colborn, 1992). Organochlorines are not considered acutely toxic to humans, because people do not immediately become deathly ill from ingesting them in small quantities nor suffer mutagenic effects. Although humans consume doses of these chemicals considered safe for an adult, they can cause irreparable harm to a fetus. The array of potential disorders range from infertility and birth defects to immune suppression and even gender confusion.

For example, DDT ingested by humans is slowly but eventually eliminated. Of more concern is the DDT that has been converted to *d*ichlorodiphenyl*d*ichloreo*e*thane (DDE) in the environment and is present in the food we eat. DDE is almost biologically nondegradable and extremely fat-soluble, so it remains in our bodies for a long time

(Baird, 1995). Although not traditionally considered to be a human carcinogen, Mary Wolff and her colleagues (1993) from Mount Sinai Hospital in New York have found that women with the highest exposure to DDE had four times the breast cancer risk of women with the lowest exposure.

Laboratory experiments have linked another organochlorine family, dioxins, to disrupting sexual development and impairing nervous system development, behavior, and learning ability. Disturbingly, dioxins have been identified in the breast milk of women, and scientists hypothesize that dioxins are linked to the fifty percent decline in sperm counts in men in industrialized nations.

PCBs are of particular concern when large quantities of fish have been consumed in which PCBs have accumulated in the fatty tissue. PCBs in particular have been associated with developmental effects in humans living in the Great Lakes region of the U.S. Mothers who ate moderate amounts of Lake Michigan fish, roughly two to three fish meals a month, and had moderate blood concentrations of PCBs gave birth to children with reduced psychomotor function and visual recognition, and their children at age four had impaired verbal and memory ability (Jacobson et al, 1990).

As a result of the widespread use of chlorine compounds and their persistent nature, people all over the world have measurable amounts of DDT and other organochlorine pesticides, PCBs, dioxins, and furans in their tissue (Thomas and Colborn, 1992)

Ecosystem Health as an Indicator of Human Health

Although there is a correlation between exposure to organochlorines and reproductive abnormalities and cancer, it is unlikely that scientist will be able to determine a causal relationship in humans: Scientists may not use humans in experiments and often the affects from exposure to organochlorines appear in offspring. Even though some of the hormonal and reproductive abnormalities we are witnessing in humans are simultaneously occurring in wildlife populations that are exposed to organochlorines, this data can only be extrapolated to humans and, therefore, does not prove causation.

Colburn and her colleagues (1993) point out that animal models using the estrogenic-mimicking hormone diethylstilbestrol (DES), a synthetic estrogen that was used by physicians to prevent spontaneous abortions in women from 1948 until 1971, corroborate clinical studies in humans and should be used as a human model for exposure during early life to any estrogenic chemical, including pollutants in the environment that mimic estrogen. Similar reproductive problems documented in wildlife exposed to high concentrations of organochlorines occur in daughters whose mothers took DES (about 1 million or more between 1960 and 1970). The reproductive problems include reproductive organ dysfunction, abnormal pregnancies, a reduction in fertility, and immune system disorders.

Organochlorines are a ubiquitous substance. They are in the tissue of people all over the world (Thomas and Colborn, 1992). As a result of global air circulation, PCBs can be found in at least trace amounts in almost every ecosystem around the world and at every level of the food chain from phytoplankton to top predators to humans (National Wildlife Federation, 1994). They travel around the globe in fresh water, salt water, and in the atmosphere. They are transported to other countries in the food and in migratory animals. And they persist.

Industrial countries have already begun to respond to the organochlorine issue. Regional agreements, such as the Paris Convention on the North Atlantic, Barcelona Convention on the Mediterranean, and the International Joint Commission on the Great Lakes have called for the reduction of discharge of organochlorines into shared water bodies, and the Montreal Protocol has called for the global phase-out of chlorofluorocarbons. However, domestic policies, treaties between industrial nations, and protocols on individual organochlorine compounds do not adequately address the organochlorine issue.

For example, DDT was banned in the United States twenty-two years ago; however, concentrations in the Great Lakes are actually on the rise again, after leveling off in the 1980s. Researchers suspect that DDT particles are being swept north on winds from Central and South America—where the chemical is still in use—and are being deposited in the lake ecosystem through rain and snow (Hileman, 1993). Researchers also estimated that 90% of the PCBs entering Lake Superior is derived from the atmosphere (Colborn et al, 1993). The only solution to the organochlorine issue is to internationalize the regulation of the production and use of organochlorines as a class of chemicals on a global scale. Industrial countries must work with developing countries to write an international agreement to regulate and phase-out the production and use of certain organochlorines. The international convention needs the support of local communities, states, regions, and countries to be effective. Together natural resource managers, policy-makers, researchers, universities, and the public can make an international convention a reality

Appendix

Baird, Colin. 1995. *Environmental Chemistry*. Witt Freeman and Company. New York, New York.

Colburn, Theo, Federick S. vom Saal, and Ana M. Soto. 1993. Developmental effects of endocrine-disrupting chemicals in wildlife and humans. Environmental Health Perspectives, 101:5, 378-384.

Hileman, Bette. 1992. Effects of organohalogens on marine animals to be investigated. Chemical and Engineering News. March 9, 1992, pp. 23-24.

Jacobson, J. L., S. W. Jacobson and H. E. B. Humphrey. 1990. Effects of in-utero exposure to polychlorinated biphenyls and related contaminants on cognitive functioning in young children. Journal of Pediatrics, 116:38-45.

Statement from the work session on Chemically-Induced Alterations in Sexual and Functional Development: The Wildlife/Human Connection (Wingspread Retreat, Racine, Wisconsin, July 26-28, 1991).

Thomas, Kristin B. and Theo Colborn. 1992. Organochlorine endocrine disrupters in human tissue. In: Chemically-induced alterations in sexual and functional development: The wildlife/human connection. Theo Colborn and C. Clement, eds. Princeton, NJ: Princeton Scientific Publishing Company, p. 365.

Wolff, Mary S. et al. 1993. Blood levels of organochlorine residues and risk of breast cancer. Journal of the National Cancer Institute, 8:648-52.

CHAPTER 141

Using Geography as a Management Tool: The National Ocean Service San Francisco Bay Project and Pacific Coast Program

David McKinnie

Abstract

The National Ocean Service is the primary civil agency within the federal government for maritime charting, geodesy, and sustainable coastal management. As such, the National ocean service goal is to help ensure that nation's vital marine-based economic activities can continue to expand while its invaluable ecological resources are protected and managed wisely for future generations. Through its charting, geodesy, coastal management, and oil spill prevention and response products and services, NOS supports decision makers in the coastal zone, whether vessel masters or resource managers.

NOS cannot implement these broad mission effectively from headquarters, however: most of the agency's are not regionalized and the geographic distance from the user community makes it difficult to identify and respond to regional needs effectively. Further, NOS charting and geodesy programs on the one hand, and coastal management and spill response programs on the other, often operate independently of each other with the result that opportunities To address this geographic limitations. which are inherent in programs such as those NOS operates, the agency launched the san Francisco Bay Project in October of 1995.

San Francisco Bay

San Francisco Bay and the Delta comprise a rich and diverse coastal ecosystem. Draining over 40% of California, they constitute the largest estuary on the West Coast. Unfortunately, development, polluted runoff, and the diversion of fresh water from the Delta have significantly reduced the quality and quantity of the Bay's aquatic habitat. These losses have important implications for the state's economy, as well as the health of its coastal and riverine resources.

The Bay is also home to six major ocean shipping ports, oil refineries, and petroleum blending facilities. Maritime activities are vital to the Bay area's economy and it is critical to the region economic health that maritime commerce continue to grow. At the same time, new, deeper-draft vessels require more accurate navigation information and, eventually, deeper, wider channels. Navigation in the Bay is made difficult by confined maneuvering areas, depth limitations, fog, and strong currents. These factors restrict vessel transits, increase the risk of an oil or hazardous materials spill, and complicate efforts that night be undertaken to mitigate a spill's effects.

Given this set of coastal management, spill prevention and response, and maritime safety and efficiency issues and community willingness to participate with NOS in developing a pilot project, NOS selected San Francisco Bay to demonstrate the integration of its maritime commerce and coastal management missions in a regional context.

Three Major principles underlie NOS' San Francisco Bay Project:

> *Local Focus*: Regional problems are too complex to address with broad programs implemented at the national level. To meet a particular area's needs, these programs must be tailored, with local community input, for that particular geographic area.
>
> *Local involvement*: The kinds of relationships needed to address regional navigation issues effectively are not possible at the national level. Consultation and collaboration with the local maritime community are essential improving the design and delivery of NOS, Coast Guard, and other agency products and services.
>
> *Efficiency*: Information must be accessible to a broad array of users. NOS and the Coast Guard are working with the maritime industry and community groups to identify and implement specific cost effective solutions for managing maritime safety and preventing pollution.

CHAPTER 142

San Francisco Bay Wetland Mitigation Banking System

Will Travis[1]
Jeffry Blanchfield[2]
Michael L. Fischer[3]

Wetland Mitigation Banks

The concept of wetland mitigation banking has been around for about 20 years. In its most elementary form, the idea involves restoring wetlands to produce mitigation "credits" which can be used to offset unavoidable damage to existing wetlands—the "debits"—that will result when an allowable development project is constructed.

Using a mitigation bank has several advantages over having a real estate developer initiate its own wetland mitigation project. First, mitigation banks are usually large and, thus, often ecologically superior to the "postage stamp" wetlands created to mitigate the damages of smaller individual projects. Second, real estate developers and other permit applicants do not always have the expertise or incentive to assure the long-term success of individual mitigation projects. Third, mitigation banks allow the cumulative impact of many small wetlands losses—too small to be mitigated individually—to be addressed. Fourth, by buying mitigation credits a project developer can avoid the occasional "double permit" dilemma where one permit is needed for the development project and a second one for the mitigation project. Finally, regulatory agencies may find it easier to monitor success and compliance of mitigation banks than of individual wetland restoration projects.

The value of mitigation banks was recognized in the Comprehensive Conservation and Management Plan for the San Francisco Estuary (CCMP) which was developed under the guidance of the U.S. Environmental Protection Agency in the early 1990s using a process involving all the major stakeholders in Bay use and protection. The CCMP was ultimately approved by California Governor Pete Wilson and U.S. Environmental Protection Agency Administrator Carol M. Browner. The proposal advanced in this report is consistent with the policies on mitigation and mitigation banking contained in the CCMP.

[1]Executive Director, San Francisco Bay Conservation and Development Commission
Thirty Van Ness Avenue, Suite 2011, San Francisco, California 94102

[2]Chief Planner, San Francisco Bay Conservation and Development Commission
Thirty Van Ness Avenue, Suite 2011, San Francisco, California 94102

[3]Executive Officer, Executive Officer, California State Coastal Conservancy
1330 Broadway, Suite 1100, Oakland, CA 94612-2530

Problems with Mitigation Banks

Despite the advantages of wetland mitigation banks, nationwide there are relatively few successful mitigation banks. On average there is less than one bank per state, and there are currently no banks available to provide mitigation for impacts to San Francisco Bay wetlands.

There are a number of reasons for this poor track record. Property in the Bay Area, and particularly land along the Bay shoreline that would be suitable for a mitigation bank, is both scarce and expensive. Start-up costs are high; an owner of a mitigation bank must incur substantial upfront costs in acquiring a suitable property, developing an acceptable restoration plan, implementing the plan, and monitoring its success. As a result, it may be years before a bank owner begins to see any return on its investment. Wetland restoration is an inexact science with a spotty success record making the commercial viability of a mitigation bank a risky venture. And mitigation bank developers face as much regulatory uncertainty as do other developers.

Regulatory agencies are also troubled by some aspects of mitigation banking. Quantifying mitigation credits is not as easy as measuring wetland losses. The productivity of an acre of wetland in a mitigation bank may not exactly equal the productivity of an acre of wetland being damaged by a development project. Furthermore, enhancing an acre of existing wetland is clearly not satisfactory mitigation for the total destruction of an acre of wetland by a development project. But what multiplier should be used when mitigating destruction with enhancement? The quantification issue is further compounded when a mitigation bank contains one type of habitat (e.g., tidal wetlands) and a development project will impact another type of resource (e.g., seasonal wetlands). Furthermore, even though the recently developed federal policy on mitigation banking calls for mitigation to be provided before any impacts to natural resource occur, some proponents of mitigation banks advocate collecting mitigation fees which can be pooled with other fees to later implement a mitigation project. This approach is particularly troublesome to critics who correctly note that it involves damaging the environment now, while the mitigation, which may or may not ultimately be effective, would occur sometime in the future and possibly at a location which is presently undetermined.

Finally, Bay Area environmental organizations have generally been reluctant to endorse the mitigation banking concept. The past failures in wetland mitigation have made these organizations hesitant to conclude that any amount of artificially-created, enhanced or restored wetlands can compensate for the permanent loss of natural wetlands. Furthermore, some skeptics view the purchase of mitigation credits as a way for a developer to "buy" a permit for a project that would not otherwise gain approval from regulatory agencies. And, if, as they contend, the availability of a mitigation bank allows developers to "buy" permits, they believe there is nothing cynical about them both opposing development projects which do not provide mitigation and also opposing the creation of mitigation banks. In their view the existence of the banks—and thus the availability of mitigation—fosters unwanted development.

San Francisco Bay Wetland Mitigation Banking System

In order to have any hope of being successful, every proposal for an individual mitigation bank in the Bay Area will have to overcome all of the obstacles cited above. Judging by the record thus far, few, if any, successful mitigation banks will be established in the Bay Area unless government takes a more active role in making mitigation banks work.

Government agencies can take a number of steps to advance the role mitigation banks can play in enhancing the quality of the natural environment and improving the efficiency of government regulatory programs. Public agencies can work with private bank operators to help them overcome the many problems they will face in operating public mitigation banks. Government agencies can, as has been proposed by the California Coastal Conservancy, establish and operate public mitigation banks that would serve as pilot projects where the bugs in the mitigation banking system can be worked out. But perhaps the most important step that government can take to facilitate the creation of individual mitigation banks would be to expand the idea of mitigation banking from one which involves a single wetland restoration project within a single bank that is explicitly linked to another project that results in the loss of specific wetland resources. In its place a mitigation banking system could be created to harness free market economic forces and use them to advance the restoration and enhancement of Bay wetland habitats.

Perhaps the best way to describe the proposal advanced in this report is to draw an analogy between mitigation and shoe making. Until the Industrial Revolution, if you wanted a pair of shoes you went to your local cobbler and had him measure your feet and hand-make the shoes from scratch. This situation is not unlike what happens when a developer initiates his own mitigation project to compensate for the loss of wetlands caused by the developer's project. The cobbler does his best to create a pair of shoes that will match the style you have in mind and will fit your feet. The developer tries his best to create wetlands that will match the type, amount and functionality of the wetland being destroyed. But in both instances the time taken to create the product wanted is slow, the cost high, and the quality of the results unpredictable.

With a mitigation bank in place the analogy changes from a cobbler to a shoe store. Now if you want a pair of shoes you go to a shoe store, choose a pair in the style and size you want and buy them. In theory, a mitigation bank works the same way. When a developer needs to provide mitigation for a certain type and size of wetland loss, he can go to a mitigation bank and buy the type and amount of the mitigation he needs.

Unfortunately, there are two gaps between the theory and reality in the mitigation banking business. First, with so few mitigation banks being operational (and, again, none around San Francisco Bay), the selection is limited. If you are shopping for dress shoes you won't find what you want in an athletic shoe store. Similarly, if a developer needs to mitigate the loss of tidal wetlands that provide habitat for Clapper Rails in the South Bay, a mitigation bank that contains only seasonal wetlands in the North Bay will be of little value to him.

Second, unlike the person who is shopping for shoes, the real estate developer cannot revise his preferences to accommodate what is on the market. If you want new dress shoes but the athletic shoe store is the only one open, you might be willing to buy the most stylish casual shoes in stock. But the developer can't make this sort of compromise. In fact, the decision isn't his to make at all. It is up to the regulatory agencies that require the mitigation to determine whether the mitigation provided by a particular mitigation bank would be adequate and suitable. This role for the regulatory agencies is entirely appropriate. Their job is to protect the wetland resources, not to help the real estate developer make a profit by allowing him to compromise the overall quality of the Bay ecosystem.

Eliminating this gap between theory and reality in mitigation banking requires a twofold challenge to be met: first, a means must be found to stimulate the creation of a lot of mitigation banks that would provide a wide variety of wetland resources so developers can find the type of mitigation credits they need; and second, regulatory agencies need to be provided with the assurance that, by accepting mitigation credits, they will be guaranteed that high quality wetlands of the type and functionality needed will be provided regardless of where the developer buys his mitigation credits or how much he pays for them.

To meet this challenge a new element needs to be introduced into the mitigation banking concept. Its purpose would be to create an alternative to the current direct linkage regulatory agencies establish between the wetlands being damaged and the restored wetlands used as mitigation. This new element would be a type of currency—a mitigation banking certificate—that would provide regulatory agencies with the assurance they need and encourage the creation of mitigation banks.

The analogy between shoes and mitigation has been used to explain how the delivery of products—whether shoes or mitigation credits—is improved when purchases are from a well-stocked warehouse instead of waiting for a made-to-order delivery. Another historical progression can be used to describe how the idea of a mitigation banking certificate would improve the delivery of mitigation. Currently, mitigation transactions are handled through much the same type of barter system that was used when a farmer traded a certain number of bushels of corn for a pair of shoes. A mitigation bank operator barters an area of restored wetland for an area of degraded wetland deemed by government regulators to be compensation of equal value. Regulatory agencies rely on this barter approach to ensure that the mitigation being provided is comparable in functionality with the wetland being destroyed.

With the introduction of currency, the farmer got cash for corn which he could then use to buy shoes, or whatever else he wanted, whenever he needed it, all at the lowest price he could find. Currency eliminated the need to directly link the value of corn with the value of shoes. But in order for the currency system to work, both the farmer and the shoemaker had to have confidence that in accepting currency they were getting something of equal value to the real commodities they were selling.

The same principle applies in mitigation banking. Regulatory agencies must have the assurance that the mitigation being provided is comparable in environmental value with the wetland being destroyed. They now gain this assurance by evaluating the specific restored wetland that is offered as mitigation. But what if the agencies could get this assurance in a different way? What if they had the authority to inspect all of the assets (i.e., the restored wetlands) in each and every mitigation bank and to establish the type, environmental value, and other relevant features of the wetlands? The agencies could describe this information, which is what is envisioned as being contained on a mitigation banking certificate, to fill in the supply side of the equation that is established by the current regulatory practice. The demand side would be filled in—again by the wetland regulatory agencies—when they issue permits requiring mitigation. At that point they establish how much of what kind wetland needs to be mitigated. Thus, a system would be created that delivers the kind and quality of mitigation needed far more efficiently than does the current barter system.

Regulatory agencies would retain control over the supply side of the mitigation equation by placing them in charge of the mitigation banking system that issues the mitigation banking certificates. Because the regulatory agencies would have control over the issuance of the currency in the banking system, they should have no reason to challenge the adequacy of the actual mitigation backing the certificates. As important as it is to put the regulatory agencies in charge of overseeing the mitigation banking system, it is even more essential that no regulatory agency be involved in the operation of a mitigation bank. An arms length relationship between regulatory agencies and mitigation bank operators is critical to avoiding challenges that developers can "buy" permits for a projects that would not otherwise gain approval by purchasing mitigation credits. If a regulatory agency is perceived as having an interest in the success of a mitigation bank, that agency's regulatory decisions which involve the use of mitigation credits will always be suspect.

Moreover, even with a mitigation banking system in place the current basic philosophy on wetland protection must be maintained. Regulatory agencies should continue to require that wetland destruction be avoided if at all possible. If it is necessary to damage wetlands to achieve an overriding public purpose, the wetland loss should be minimized. Ideally, mitigation to compensate for the justifiable loss of wetlands mitigation should be in-kind and on-site. And consideration of using off-site mitigation should come only after it is determined that off-site mitigation would provide more or better habitat than would result from on-site mitigation.

To put these ideas into effect a San Francisco Bay Wetland Mitigation Banking System needs to be established and operated as follows:

- The Wetland Mitigation Banking System would be managed by a board of governors composed of representatives of the principal regulatory and management agencies responsible for the protection of Bay Area wetlands. The Board of Governors would include at least the following: the Association of Bay Area Governments; the California Coastal Conservancy; the California Department of Fish and Game; the California Environmental Protection Agency; the California Resources Agency; the California State Lands Commission; the San Francisco Bay Conservation and Development Commission; the San Francisco Bay Regional Water Quality Control Board; the State Water Resources Control Board; the National Marine Fisheries Service; the U.S. Army Corps of Engineers; the U.S. Environmental Protection Agency; and the U.S. Fish and Wildlife Service.
- A public agency would be assigned the responsibility of administering and staffing the operation of the San Francisco Bay Wetland Mitigation Banking System. The California Coastal Conservancy has been selected for this role because of its experience in operating mitigation banks and its proposed role as the manager of the Southern California Wetlands Clearinghouse, a venture similar to the San Francisco Bay Wetland Mitigation Banking System.
- The Banking System would establish the specific types of habitats that would be encompassed by the Banking System. These would probably include at least subtidal and tidal waters, intertidal mudflats, tidal salt marshes, tidal brackish marshes, tidal freshwater marshes, diked seasonal and perennial wetlands, salt ponds, lakes and ponds, adjacent riparian woodland, and adjacent uplands. Because of the vast size of the San Francisco Bay estuary system, the region would then be subdivided into subregions which might include the South Bay, Central Bay, North Bay or San Pablo Bay, and Suisun Bay and Marsh.
- The Conservancy, working under the direction of the Wetland Mitigation Banking System Board of Governors, would be responsible for evaluating a proposed wetland restoration project; determining whether a mitigation bank operator could successfully restore a habitat area; establishing the type of habitat that would be restored; and establishing performance standards that the restored habitat would have to meet. The results of this evaluation would be presented at a public hearing to provide the mitigation bank operator, the affected local government, and the general public with the opportunity to comment on the accuracy and adequacy of the conclusions of the evaluation, as well as whether it would be appropriate to restore the wetlands as proposed and to operate a mitigation bank at the site.

- A new commodity—a mitigation credit certificate—would be created for issuance by the Board of Governors. A certificate would be issued when the Board of Governors determines by consensus, after holding a public hearing on the matter, that the wetland has been fully restored and meets the performance standards established for the restoration project. Each certificate would be backed by the full assurance that it represents a viable wetland of the size, type, and location specified on the certificate.
- To address concerns that restored areas may not be as productive as natural habitats, a multiplier of two or more to one would be utilized so that at least two acres of wetland would have to be created for each acre of mitigation credit certificate issued.
- The mitigation credit certificate would be issued to the owner of the property on which wetlands have been restored (the mitigation bank operator). The operator of a mitigation bank would be responsible for complying with local zoning, meeting the requirements of the National Environmental Policy Act and the California Environmental Quality Act, securing the necessary permits for the wetland restoration project, undertaking the project, and managing the site to ensure the permanent viability of the selected habitat type.
- The Coastal Conservancy would charge a fee for issuing a mitigation credit certificate. The purpose of the fee would be twofold. First, to provide the Conservancy with sufficient funds to remediate any failed wetland restoration projects. And second, to recover the public investment in getting the banking system operational so that public funds are not used either directly or indirectly to subsidize private development projects.
- If a regulatory agency determines that off-site mitigation is appropriate in a particular situation, a developer who needs mitigation for wetland impacts would be allowed to satisfy regulatory agency mitigation requirements by delivering to the agency the appropriate mitigation credit certificates. At least initially, the credit certificates could only be used for small amounts of wetland impacts. Regulatory agencies would be required to make their decisions on whether a project should or should not be approved based on the merits of the project and not on the mitigation that would be provided as a result of the project.
- If mitigation credits are used, they would have to be of the same type and from the same subregion as the impact. In other words, impacts to tidal wetlands in the South Bay would have to be mitigated with credits from tidal wetlands in the South Bay.
- Mitigation bank operators would be permitted to sell mitigation credit certificates on the open market at a price established by the dynamics of supply and demand.

As noted, regulators often tie specific development impacts to specific mitigation activities. For example, some government regulatory agencies incorporate mitigation credits into their regulatory decisions by specifying the location of the mitigation bank where the credit is generated, outlining performance standards or other criteria for the mitigation work, and imposing conditions that give the regulatory agencies the authority to monitor the effectiveness of the mitigation project. Moreover, the project developers may be held responsible for the long-term viability of the mitigation projects even though their involvement is limited to buying mitigation credits from a third party.

With the Wetland Mitigation Banking System in place, if a regulatory agency determines that the wetland impacts resulting from a development project cannot be mitigated on-site, the agency would determine the precise amount and type of wetland area that would be impacted by the project, and the project developer could choose to satisfy its mitigation responsibilities by delivering the necessary number and type of credit certificates to the regulatory agency. (The developer could also choose to avail itself of the two mitigation options currently available: it could undertake its own mitigation project; or it could agree to have the development project linked directly with a specific wetland restoration project.)

If a developer chooses to take advantage of the mitigation credit certificate option, a regulatory agency would not have to craft elaborate permit conditions or monitor mitigation project effectiveness. In fact, the agency might not even know precisely where the credit was generated. But the mitigation credit certificates would provide the agency with the assurance that somewhere within the same geographic sub-area a viable mitigation project has restored an adequate amount of the same type of habitat to mitigate the impacts of the development project. If the mitigation project fails, the project developer would have no responsibility for remediating the project. The developer's responsibility would end with delivering the credit certificates to the regulatory agency. At the same time, if the mitigation project fails, the regulatory agency would have assurance that the Coastal Conservancy would step in to assure the mitigation project is remediated. Under the worst of conditions, such as a mitigation bank operator's bankruptcy, the Coastal Conservancy could take possession of the property and bring the wetland back to acceptable standards.

The Mitigation Banking System would provide regulatory agencies with a tool they desperately need to deal with the cumulative impact of small projects that impact wetlands. Studies by BCDC and others have found that a large amount of loss of wetlands is occurring because of the cumulative impact of small projects, such as seawalls, riprap, and boat docks. For example, a 1988 study by BCDC found that between 1978 and 1987 BCDC had administratively authorized approximately 41 acres of Bay fill to control shoreline erosion without any required mitigation. Despite the cumulative impacts of practices like this, regulatory agencies often find it impossible to require mitigation for the impacts of small projects. The projects are often carried out by a single lot owner or small business which undertakes the project to protect its property from erosion or to conduct its business. The expense and difficulty of undertaking a small mitigation project—especially if it has to be off-site—is far out of proportion with the scale and cost of the development project. Faced with this dilemma, regulatory agencies often forego requiring mitigation for small projects.

In addition to providing regulatory agencies with a mechanism for mitigating the impacts of small projects, the Wetland Mitigation Banking System would simplify matters for regulatory agencies. Nevertheless, these agencies will probably not embrace the System unless they have complete faith that it will work better to protect and enhance wetland resources than does the current regulatory regime. Most importantly, the agencies have to agree to accept the mitigation credit certificates in lieu of embodying a specific mitigation project into their regulatory decision-making. Just as everyone is willing to accept that a small, rumpled piece of paper with no practical use does, in fact, have value when that piece of paper is a U.S. Federal Reserve Note, the regulatory agencies must be convinced that mitigation credit certificates are as good as the wetlands backing them. To gain this level of acceptance, regulatory agencies are going to have to be satisfied that they are making a good trade when they give up the control they currently have over mitigation projects through permit conditions in return for the control they would gain through serving on the Board of Governors of the Wetland Mitigation Banking System where decisions are made by consensus.

In addition to gaining support from regulatory agencies, the success of the Wetland Mitigation Banking System depends on consensus being reached among regulatory agencies, developers, and wetland protection advocates that mitigation banks can play a valuable role in advancing the goals of wetland protection laws. Taken at face value, the arguments for this consensus seem compelling. If the Wetland Mitigation Banking System were in full operation, wetlands would exist where they do not now exist. Overall, the stock of wetlands in the region should increase beyond what would exist were the current regulatory regime continued.

Despite these benefits, there will probably be strong skepticism to embracing a Wetland Mitigation Banking System whose success relies heavily on using market incentives to restore wetlands on private land. To date, private market forces have been the powerful engine driving the massive destruction of wetlands. Hoping that these same forces can be relied upon to create wetlands demands a giant leap of faith. However, it may be prudent to make this leap for two reasons. First, creating a market for enhanced privately-owned wetlands could effectively counter the argument that the primary economic effect of wetland protection laws is to deprive private property owners of all economic use of their lands. Second, there has been considerable success in reducing the cost of improving air quality by incorporating an emission trading system into the air quality regulatory program. The techniques that proved effective in that program could be incorporated into the wetland mitigation credits program.

Advantages of the San Francisco Bay Wetland Mitigation Banking System

The Wetland Mitigation Banking System would have the following advantages over the current approach:

- The overall quality of the regional wetland ecosystem would benefit from the requirement that the wetlands which generate mitigation credits be viable before the credit certificates are issued for sale and that the ratio between wetlands created and mitigation credits created be more than one to one. Thus, in terms of quantity and timing, the environment would always be "ahead of the curve." Further, even unsold mitigation credit would create an environmental benefit because they would represent wetlands created which were not offset by wetland losses from development impacts.
- Mitigation would finally be possible to address the cumulative impact from the loss of many small amounts of wetlands from numerous small projects such as seawalls, riprap, boat docks, and other seemingly benign projects.
- There would be an economic incentive for private property owners to restore wetlands on their properties.
- Mitigation for the impacts of development would be of the same type and general location as the impacts, would be of an amount greater than the area impacted, and would occur before project impacts.
- The more mitigation banks are used the more regulatory agencies are relieved of the burden of monitoring mitigation compliance in developer-sponsored wetland restoration projects and the more developers can focus their talents in the area of their expertise—property development.
- The regulatory process would be faster, more certain, and probably less costly.

- In contrast with the current situation where a private owner may fear that the presence of historic wetland conditions will limit development potential of his property and reduce the land's market value, property owners would be able to profit by restoring historic wetlands and selling mitigation credits.
- A market would be created for a new type of entrepreneur—a habitat restoration specialist who could either act as a consultant to property owners wishing to develop wetland mitigation credits, buy or lease property for restoration, or enter into a partnership with a property owner to share the costs, risks and profits of developing mitigation credits.
- Project developers needing mitigation credits would have the choice of undertaking their own mitigation projects, buying mitigation credits from a specific mitigation bank, or shopping for the least costly mitigation credit certificates on the open market.
- Public resource agencies could sell mitigation credits generated from their restoration projects and use the revenues to undertake additional wetland restoration projects, preserve existing wetlands by acquiring them, or undertake other environmental projects.
- Environmental organizations and public resource agencies could prevent restored wetlands from being used for mitigation by purchasing the mitigation credit certificates generated by the wetlands and taking the certificates off the market.
- The issue of the availability of mitigation would be largely eliminated from the debate over the pros and cons of a proposed development project in regulatory agency public hearings. Project proponents could no longer argue that mitigation is either infeasible or the cost unknown. Project opponents could no longer argue that the absence of mitigation dictates the project should not be approved.
- A unique public/private partnership would be created in which the private capital would be used to achieve public goals that could not be met with public funding alone.

Next Steps

The benefits of the concept described in this paper would exist only after the banking system is fully operational and a large number of individual mitigation banks with a wide variety of habitat types are in place. That end cannot be reached overnight. The first step in putting the Banking System in place involves gaining formal endorsement of the Wetland Banking System from affected and interested agencies and organizations, and particularly from the government regulatory agencies that would be expected to accept mitigation credit certificates. This would best be accomplished by having the agencies draft and ultimately approve a Memorandum of Agreement which lays out in detail how the Mitigation Banking System will be operated. The Bay Area Wetlands Planning Group, an informal coalition of government agencies involved in wetland protection and regulation organized by the California Resources Agency to coordinate wetland policy development and implementation in the San Francisco Bay Area, should be used as a forum for developing the MOA. Based on the experience gained from evaluating other successful mitigation banks, the following specific tasks need to be accomplished to complete the development of such a MOA:

- **Secure Support for the Wetlands Banking System**. The Wetland Mitigation Banking System concept needs to be further discussed with government policymakers, regulatory agencies, environmental interests, the development community, landowners, potential mitigation bank operators, and other interested parties with a goal of achieving a broad consensus that the region's environmental quality will improve and the development process will benefit by having the Mitigation Banking System in place.

- **Develop Mitigation Banking System Goals, Objectives and Policies**. Clear and concise goals and objectives for the establishment and operation of the Mitigation Banking System must be set to provide a clear direction for the System and to measure success. Further, operational policies for the Banking System must be established to guide its use and operation. To do this, literature should be reviewed, existing mitigation bank managers interviewed, and meetings held with government policy-makers, regulatory agencies, environmental interests, the development community, academic community, landowners, and other interested parties to develop consensus Mitigation Banking System goals and objectives.
- **Develop Mitigation Banking System Administrative Structure**. An essential step in drafting the Mitigation Banking System MOA is to reach agreement on the structure to administer and manage the System. To do this, the literature should be reviewed, existing mitigation bank managers interviewed, and meetings held with public agency and private organization representatives with experience in mitigation banking. The analysis should focus on the institutional structure of the Banking System, determine the administrator of the system, be it the Coastal Conservancy or another agency or individual, and establish a Board of Governors or similar body to oversee the operations of the Mitigation Banking System. In addition, agreement on whether a scientific technical review team that would advise the Board of Governors and operators of the mitigation banks should be made. More specifically, analysis and agreement on the following matters will be needed in formulating the MOA:

   1. **Powers and Authorities**. An analysis of the legal authority of agencies to participate in mitigation banking and to assume responsibilities for administering and operating mitigation banks should be evaluated. In addition, a determination should be made as to what, if any, additional authority needed to permit agency participation or assumption of duties and responsibilities for mitigation banking needs evaluation. Research is needed to determine which laws, regulations, policies or practices, if any, need to be amended to allow the Wetland Mitigation Banking System to fully succeed.
   2. **Governing Body**. An analysis should be made of the desirable membership, organization and role of a Board of Governors, or similar policy setting and oversight body. The management responsibilities of the Board of Governors should be identified, such as review and approval of mitigation bank sites, individual plans for mitigation banks, and mitigation bank debits and credits. The administrative and operating procedures of the Board of Governors also needs identification.
   3. **Administrator**. The administrator would provide the day-to-day administration of the Mitigation Banking System. The Coastal Conservancy has been identified as the most appropriate agency to serve as the administrator. The powers, duties and responsibilities of the administrator need to be defined. Further, the relationship of the administrator to the Board of Governors and the operators of the mitigation banks should be established.
   4. **Operators**. The operators of mitigation banks would be responsible for mitigation bank acquisition, development, maintenance, monitoring and reme-

diation. An analysis and recommendation of the duties, responsibilities and obligations of a mitigation bank operators should be made and the relationship of the operators to the system administrator and Board of Governors should be set.

5. **Technical Review Team**. An analysis should be made of establishment of a technical review team which be responsible for advising the Board of Governors, administrator and mitigation bank operators on the scientific aspects of mitigation banking, including bank site selection and approval, mitigation bank plan approval, and appropriate mitigation credit and debit approval. Agreement on the membership, role and responsibilities of the technical review team should be reached in developing the MOA.

- **Develop Mitigation Banking System Operation Procedures**. Development of the procedures, guidelines and standards for the creation and operation of the mitigation banks is an essential ingredient of the Mitigation Banking System. The following specific issues need to be addressed in establishing these procedures.

  1. **Service Area**. The geographic area of mitigation bank location and the area for project eligibility for using the Banking System should be determined. Options include using the same area as being used by the San Francisco Bay Area Wetlands Ecosystem Goals Project, using the extent of the jurisdiction of the Coastal Conservancy, or establishing boundaries unique to the Banking System.

  2. **Project Qualification**. The kinds of projects that would be eligible for participation in the Mitigation Banking System must be determined. This analysis should heavily rely on the goals, objectives and wetland definitions which would be incorporated into the MOA.

  3. **Bank Establishment Requirements**. An analysis of the criteria that must be met by a proposed mitigation bank to be eligible for participation in the Mitigation Banking System should be conducted. The criteria should include such concerns as bank purpose, project description, baseline conditions, work proposed, implementation timetable, financial assurances, and real estate provisions.

  4. **Performance Review System**. Clear benchmarks, a process of performance review, and an auditing program need to be developed so the effectiveness of the Mitigation Banking System can be measured and improved over time. An accounting system which tracks the credits and debits administered by the Banking System will be needed to evaluate the success of the system. The use of a geographic information system (GIS) with attribute tables to graphically show and link the mitigation sites with the development projects and the details of the individual mitigation projects participating in the Banking System offers great potential. Such a system could be developed and made available for review over the Internet. The San Francisco Estuary Institute's EcoAtlas would provide a good basis for the GIS mitigation project tracking system.

  5. **Maintenance and Monitoring System**. Requirements for maintenance of mitigation banks and an annual monitoring reporting program need to be developed as part of the MOA.

6. **Small Fill Definition.** The definition of a "small amount of wetland destruction" that is eligible for mitigation with the credit certificates needs to be established.

7. **Activities Eligible for Credits.** Further consideration is needed to determine whether mitigation credits should be allowed to be created through a spectrum of activities ranging from wetland enhancement to wetland creation, as well as by wetland restoration.

8. **Emissions Program Analysis.** A detailed evaluation of the air emissions trading program should be undertaken so the effective techniques from that program can be incorporated into the Mitigation Banking System, and the shortcomings of the emissions program avoided.

- **Identify Potential Mitigation Bank Sites.** The identification and description of potential regional wetlands habitat mitigation sites is an important step in establishing the Mitigation Banking System. The San Francisco Estuary Institute (SFEI) has completed its computer mapping of the San Francisco Bay Area wetland habitats (EcoAtlas) in cooperation with the San Francisco Wetland Ecosystem Goals Project. The EcoAtlas contains infrastructure and ownership data and aerial photography as well as habitat information. In addition the San Francisco Bay Conservation and Development Commission has developed land use planning, ownership, and wetland project data on its GIS system for the North San Francisco Bay wetlands habitats which is compatible with SFEI's EcoAtlas. The EcoAtlas provides an excellent tool for screening and evaluating potential mitigation sites. The EcoAtlas can be queried, for example, to find all sites meeting given criteria within a certain distance of a project site. The Atlas can also be used to evaluate how well additional acres of various habitat types in a given location would advance regional or subregional goals.

  Using the EcoAtlas and working with resource managers, a set of criteria for mitigation sites should be developed. The GIS could be queried to develop a data coverage of potential mitigation sites. These sites, and others identified through other planning processes should be evaluated as to how well the establishment of various configurations of habitat types in these locations advance regional wetland restoration goals.

- **Pilot Projects.** Several public wetland restoration projects need to be undertaken to serve as pilot projects and to generate an initial pool of mitigation credit certificates to test the feasibility of the mitigation banking concept.

Memorandum of Agreement Outline

Following is a general outline for the type of memorandum of agreement that would be needed to establish the San Francisco Bay Wetlands Mitigation Banking System:

Banking System Authority and Purpose

1. Findings and declarations of need for banking system
2. Statement of authorities to form banking system
3. Statement of purpose of banking system

Banking System Goals, Objectives and Policies

1. Establish banking system goals

2. Establish banking system objectives (measurable and obtainable)
3. Establish banking system operation policies

Definitions

1. Define words and terms used in the Memorandum of Agreement

Administrative Structure

1. Administrator
    a. Identification of administrator
    b. Responsible for day-to-day administration
    c. Reports to the Board of Governors
2. Funding
    a. Funding for administration
    b. Funding for operations (banking system projects, maintenance, monitoring and remediation)
2. Board of Governors
    a. Policy and management direction
    b. Approves mitigation bank sites and plans
    c. Issues mitigation credit certificates
    d. Prepares annual performance report
3. Technical Review Team
    a. Advisory to the Board of Governors and administrator
    b. Reviews proposed mitigation bank plans
    c. Reviews mitigation bank(s) maintenance and monitoring programs and annual performance reports

System Operation

1. Establish who may be a mitigation bank developer
    a. Board of Governors
    b. System administrator
    c. Public agencies, non-profit and private organizations and individuals
    d. Establish the Mitigation Banking credit system
    e. Establish unit of credit, e.g., acres
    f. Mitigation credit calculation methodology
    g. Credit certification process (e.g., based on the guidelines, criteria, performance standards and functional assessment determined by the Board of Governors)
2. Establish performance review system
    a. Maintain an accounting of the number of credits available for sale and sold to date and costs associated with the mitigation credits and program operation

b. Develop tracking system to identify mitigation bank and credit purchase with development program requiring mitigation integrated with a GIS tracking system accessible for review over the Internet

3. Establish criteria for preparation and approval of individual mitigation plans
   a. Project description
   b. Existing conditions
   c. Planned conditions
   d. Plan goals
   e. Plan specifications
   f. Maintenance plan
   g Monitoring plan
   h. Performance reporting requirements
4. Establish performance criteria for mitigation projects
   a. Plant regime
      (1) Plant species
      (2) Percent of coverage per plant species
      (3) Survival rate per time period (e.g., by five years, 10 years)
      (4) Plant diversity and structure by habitat type

Banking System Geographic Area

1. Identify geographic area covered by the Mitigation Banking System
   a. Limit development projects participation in the Banking System to the geographic area
   b. Limit mitigation banks participating in the Banking System to the geographic area
2. Identify and evaluate potential mitigation bank sites

Monitoring and Remedial Measures

1. Set time period for mitigation bank monitoring from date of bank completion (e.g., five years)
   a. Require annual inspection and report for the first few years
   b. Perform maintenance as needed
   c. If the bank fails to meet expectations, determine the cause of the failure; develop and implement a remediation plan
2. Completion of the monitoring period
   a. The Governing Board should certify the mitigation bank upon completion of the monitoring period and with evidence that the site has performed satisfactorily to the standards set in the approved mitigation plan
   b. Any financial guarantee funds or other encumbrances should be released to the bank creator

CHAPTER 143

RECAPTURING THE BAY WATERFRONT THROUGH MILITARY BASE CONVERSION

Louise Aiello
Executive Director
Bay Area Defense Conversion Action Team
(BADCAT)
A Partnership for Successful Military Base Conversion

ABSTRACT

Military base closures targeted for the Northern California Bay Area comprise almost 15% of the national total. The closing bases, situated on or fronting the San Francisco and San Pablo Bays as they are, comprise some of the region's more beautiful locations with spectacular scenic vistas. Moreover the bases, with over 10,000 acres of dry land becoming available, provide the Bay Area with an unprecedented circumstance in which the region may address a wide array of interests that together create livable communities and sustainable economic growth. To optimize these opportunities, the Bay Area Defense Conversion Action Team (BADCAT) focused on linking base conversion to existing strengths and demands within the regional economy rather than as primarily a land-use planning task.

Challenge

The Northern California Bay Area faces the greatest number of military base closures anywhere in the nation. Base closures targeted for the region comprise almost 15% of the national total. Additionally over 30% of the total national civilian job losses resulting from military base closures will hit the Bay Area in the upcoming years. While the base closures and downsizing of military operations will occur in 5 counties within the 9 county region comprising the Bay Area (namely, in Alameda, Contra Costa, Marin, San Francisco, Santa Clara, and Solano counties--*see Map 1 on page 9*), military and civilian workers as well as businesses which provided goods and services for the bases and employees are spread throughout the region.

Consequently, the unprecedented defense downsizing and closure of twelve significant military facilities in the same region creates enormous uncertainty for communities, employees and their families, and businesses.

The military bases, however, situated as they are, on or fronting the San Francisco and San Pablo Bays, comprise some of the region's most beautiful locations with spectacular scenic vistas. With over 10,000 acres of dry land becoming available, the bases provide the Bay Area with an unprecedented circumstance in which the region may address a wide array of interests that together create livable communities and sustainable economic growth: 1.) creating or restoring parks, open space, natural and wildlife habitat and recreational facilities; 2.) meeting the needs of low-income and homeless populations; 3.) addressing public trust requirements and purposes; and 4.) supporting business growth and job generation.

Additionally, the skills and experience of base and defense workers can benefit Bay Area businesses and industries. Thus, the challenge for the region is to avoid the historical pattern and get ahead of the curve with strategic economic planning and investment.

Response

Recognizing that defense conversion will be more successful if it is rooted in the comparative strengths of the regional economy, over 20 government agencies, businesses, and community and environmental organizations, in 1994, formed the **Bay Area Defense Conversion Action Team (BADCAT)** to systematically consider the best prospective economic opportunities for defense conversion and base reuse.

BADCAT is managed by the Bay Area Economic Forum (BAEF). **BADCAT** was funded from 1994-96 by the Department of Commerce Economic Development Administration (EDA Project Number: 07-49-04071) and the James Irvine Foundation. BAEF is sponsored through a partnership of the Bay Area Council (BAC) and the Association of Bay Area Governments (ABAG). BAC represents 250 major employers and ABAG represents 109 cities and counties.

The basic mission of **BADCAT** to enhance the success of local conversion efforts by: 1.) coordinating regional activities to minimize duplication; 2.) facilitating joint ventures to efficiently utilize and leverage resources; and 3.) providing economic information and analysis to improve local planning, increase the potential for job generation, and promote the greatest added-value to local and regional economies. The major purpose of **BADCAT** is to create, over a ten year period, an equal number of jobs at comparable salaries and to foster sufficient economic growth to generate an equal or greater contribution to the regional economy and local tax base as were contributed by the military bases. By improving the likelihood of economic success from defense conversion, **BADCAT** partners help generate greater resources to also address other human, community, and environmental needs.

In contrast to approaching base reuse or conversion as primarily a land-use planning task, the **BADCAT** partners focused on linking base conversion to the existing strengths and demands within the regional economy. Specifically, from 1994-96 **BADCAT** partners:

1. Analyzed constraints and growth opportunities for 7 key industry clusters for which the region has a competitive advantage--multimedia, movie and television production, telecommunications, environmental technologies, healthcare and biotechnology, banking and financial services, visitor and tourism; and prepared employment and growth data for 3 additional industry clusters--food processing and agriculture, petrochemical manufacturing, and computers and electronics;

2. Evaluated 7 infrastructure components critical in supporting economic competitiveness and developed strategies to address infrastructure needs;

3. Identified priority infrastructure requirements for reuse of each closing military base;

4. Considered the role of the universities and national laboratories in maintaining regional economic competitiveness;

5. Developed two new regional partnerships to support economic growth and environmental remediation of the closing bases, namely, the Bay Area Multimedia Partnership and the Bay Area Environmental Technology Partnership;

6. Promoted exports through linkages with BAYTRADE and the Greater Bay Area Regional Marketing Partnership;

7. Considered policy issues to improve the base closure process and accelerate economic conversion.

Industry Cluster Analyses

To support regional strategies for defense conversion and to attract private investment on closing military bases, **BADCAT** completed industry cluster analysis reports and held workshops involving industry leaders and local reuse officials for seven industry clusters in which the Bay Area shows a comparative economic advantage. An industry cluster is a grouping of companies that represents the entire value chain of a broadly defined industry from suppliers to end producers, including supporting services and specialized infrastructure. The seven industry clusters analyzed by **BADCAT** were: information technologies, including multimedia, telecommunications, and movie and television production; visitor and tourism; environmental technologies; biotechnology/healthcare; and banking and financial services.

One of the major opportunities for industry growth identified in all industry cluster workshops is the multicultural diversity of the Bay Area. Opportunities exist for businesses to draw upon this multicultural resource pool for their own development of products and services and in targeting ethnic and language specific markets. the Pacific Rim. Chart 1 on pages 10 and 11 summarizes employment and company information as well as the strengths, challenges, and infrastructure needs identified by industry leaders.

Information Technologies
(Telecommunications, Multimedia, and Movie and Television Production)

The advancement of information technologies is less a function of the development of new breakthrough technologies than of the integration and application of available, but until now, distinct technologies. The convergence of talent and technology is essential to the growth of information technologies as a whole, and in particular, for multimedia and movie and television production. The following specific actions should be taken to support the growth of Information Technologies:

1. Complete an inventory of available information technologies and telecommunications facilities on the closing bases.

2. Facilitate regional efforts to accelerate deployment of advanced telecommunications and assess feasibility of aggregating demand for use of closing military bases.
3. Identify facilities on closing bases and in host communities for sound studios and production and post production activities within the Movie and TV Production Industry Cluster and prepare location brochures to assist in marketing the facilities.

4. Link multimedia arts and digital technologies to support growth of Movie and TV Production and job generation within the Bay Area.

Environmental Technologies

BADCAT established a second regional partnership as part of its initial work program to support the use and commercialization of emerging environmental technologies. The Bay Area Environmental Technologies Partnership is comprised of the Bay Area Economic Forum, the Bay Area Regional Technology Alliance, the U.S. Navy, the U.S. Environmental Protection Agency, the California Environmental

Protection Agency, Chevron, and other private sector and community interests. The Environmental Technologies Partnership is assisting in testing environmental contaminant identification and remediation technologies on the closing bases and supporting the following actions:

1. Convene stakeholders for identification of regional clean-up priority needs based on the Final Report of the Federal Facilities Environmental Restoration Dialogue Committee, April 1996.

2. Collaborate with the Department of Defense, Local Reuse Authorities, real estate developers and other interested parties and stakeholders to ensure newly tested and identified emerging and innovative technologies are employed to cleanup parcels on the closing bases.

3. Complete environmental technology testing program.

4. Foster resource conservation technologies where data shows a competitive advantage exists, such as developing international markets and attracting "green businesses," especially recycling and remanufacturing companies, to the Bay Area's closing bases.

Biotechnology

The Bay Area is the birthplace of biotechnology and remains the world leader in core bioscience industries (drugs and diagnostics) as well as in newer sectors such as instrumentation, medical equipment, agriculture. To foster growth in biotechnology and healthcare, the following action should be taken:

1. Encourage development of biotechnology incubator(s) and expansion of biotechnology, pharmaceutical, and medical instrument manufacturing businesses on the closing bases or in the host communities of the closing bases.

Financial Services

San Francisco historically has been the financial center of the West Coast, and the Bay Area's prominence as a financial center is supported by its contributions to new ideas in finance such as passive investment, discount stock brokerage, and specialty investment banking. The Bay Area has become the center of the venture capital industry for the U.S., and an increasingly important investment banking source, particularly for technology based industries.

As a result of meetings with industry leaders it was determined that the most effective role for a Financial Services Roundtable is to develop a program of "Capital for Communities" concentrating on:

1. Developing new financial support to address infrastructure needs for military base conversion and for inner-city and "brownfields" development.

2. Attracting private equity capital for community development, infrastructure needs, and affordable housing.

Visitor and Tourism

The Bay Area is among the most popular tourist destinations in the world, attracting visitors from every continent. San Francisco hosts more than 13 million visitors annually and San Jose attracts another 4.7 million conventioneers and tourists each year. The prospects are tremendous for creating new jobs and enhancing the regional economy by supporting the visitor and tourism industry. Currently, $5 billion is spent annually in the Bay Area as a result of tourism. Three actions will support visitor and tourism throughout the region:

1. Foster region-wide consortia to support the Visitor and Tourism Industry by linking military history and ecotourism.

2. Collaborate in promoting attractions and natural assets within the region, nationally and internationally using a shared brochure, advertising and movie and television production in the Bay Area.

3. Promote opportunities to increase visitor and tourist use of the region's historical, cultural, and recreational attractions by assisting efforts to enhance water access and water shuttle services.

Infrastructure Component Evaluations and Strategies

Research Associates at the Institute of Urban and Regional Development (IURD) at UC Berkeley developed six infrastructure component evaluations: Workforce Education and Training; Transportation; Telecommunications; Housing; Environment and Natural Resources; and Government Operations and Regulations. **BADCAT** participants and industry leaders rated the infrastructure components and identified key strategies to support economic conversion of the closing military bases and growth in the regional economy.

Along with infrastructure improvements required for economic conversion of the closing bases, reuse options must take into consideration land which is designated for public trust. Additionally, over 6,000 acres of land on the closing military bases will be returned to wetlands as listed in Chart 2 on page 12. To effectively address these issues the work program for Phase II of **BADCAT** calls for the development of a methodology for evaluating the economic value of alternative uses, including the economic benefit of improved environmental quality of the San Francisco and San Pablo Bays resulting from wetland restoration, to assist local communities in making local land use decisions within the regional context.

The following actions are critical in supporting economic conversion:

1. Promote improvements which increase water access to the closing bases.

2. Foster environmental quality to "celebrate the majesty of the San Francisco and San Pablo Bays" by supporting effective methods to establish "livable communities."

3. Develop a methodology to address the economic benefits of public trust lands in themselves and in relation to other reuse options for the closing military bases.

4. Encourage the deployment of advanced telecommunications systems.

5. Advance restoration and reuse of housing stock on military bases to augment affordable housing units in the region.

6. Develop industry-led workforce education and training efforts to support key industry clusters by linking Private Industry Councils, community organizations, Local Reuse Authorities, Community Colleges, and other interested stakeholders.

7. Develop alternative funding and foster public-private partnerships to attract private sector investment in financing priority infrastructure requirements critical for successful economic conversion and reuse.

Policy Issues To Improve the Base Conversion Process and Accelerate Economic Conversion

In 1995 the **BADCAT** working with Local Reuse Coordinators, public and community organizations, and businesses prepared a policy paper outlining a series of issues impeding implementation of President Clinton's Five-point Program and the Pryor Amendment to ease the economic burden on local communities seeking to acquire former military base property. While the federal government, during 1996, adopted regulatory changes to implement the Pryor Amendment, three (3) issues continue to plague local communities in their efforts to expedite economic reinvestment through base reuse:

Issue #1: **The federal government must, as part of the federal budget, ensure and encumber sufficient funds as part of the federal budget to complete environmental clean-up.**

For the past three years, less than 1/2 of the projected costs needed for each year's clean-up activities was budgeted. As a result, the clean-up process is agonizingly slow and extending well beyond the originally estimated completion timeframes.

Issue #2: Streamline leasing procedures and federal review procedures to support local communities reuse activities as means of generating revenue to minimize economic impact on local communities.

Under current procedures reviews of the subleases have taken several months. Additionally, other reviews required by the federal government with regard to environmental documents, economic development conveyance requests and so forth have taken over a year to complete.

Issue #3: Address infrastructure requirements to support reuse of existing buildings and facilities and to assist local communities in upgrading to code all infrastructure necessary for successful economic conversion of the closed or closing military bases.

Leasing facilities during the interim and potential permanent reuse are very difficult because buildings and other structures, utilities and roads do not conform to state and local standards.

To address these issues effectively, **BADCAT** supports using public-private partnerships that would provide that the local community, federal government, and private sector share decision making, risk, and return. Implementation of a public-private partnership approach to military base conversion would encourage all parties to expedite closure, transfer, and economic conversion of the former military bases.

CONCLUSION

The strategies outlined by **BADCAT** support local decision makers in identifying the best prospective economic opportunities for military base reuse available within the context of the regional economy. The sheer magnitude of the base closures, however, combined with the effects on former base employees and their families, local communities and businesses requires regional collaboration to assure successful base conversion and economic reinvestment.

The opportunity to address the wide array of interests that together create livable communities and sustainable economic growth does not come often in a lifetime. The strategies developed by the **BADCAT** partners seek to turn the closing military bases into productive lands which allow, in their reuse, local communities and the entire region to "celebrate the majesty of the Bay."

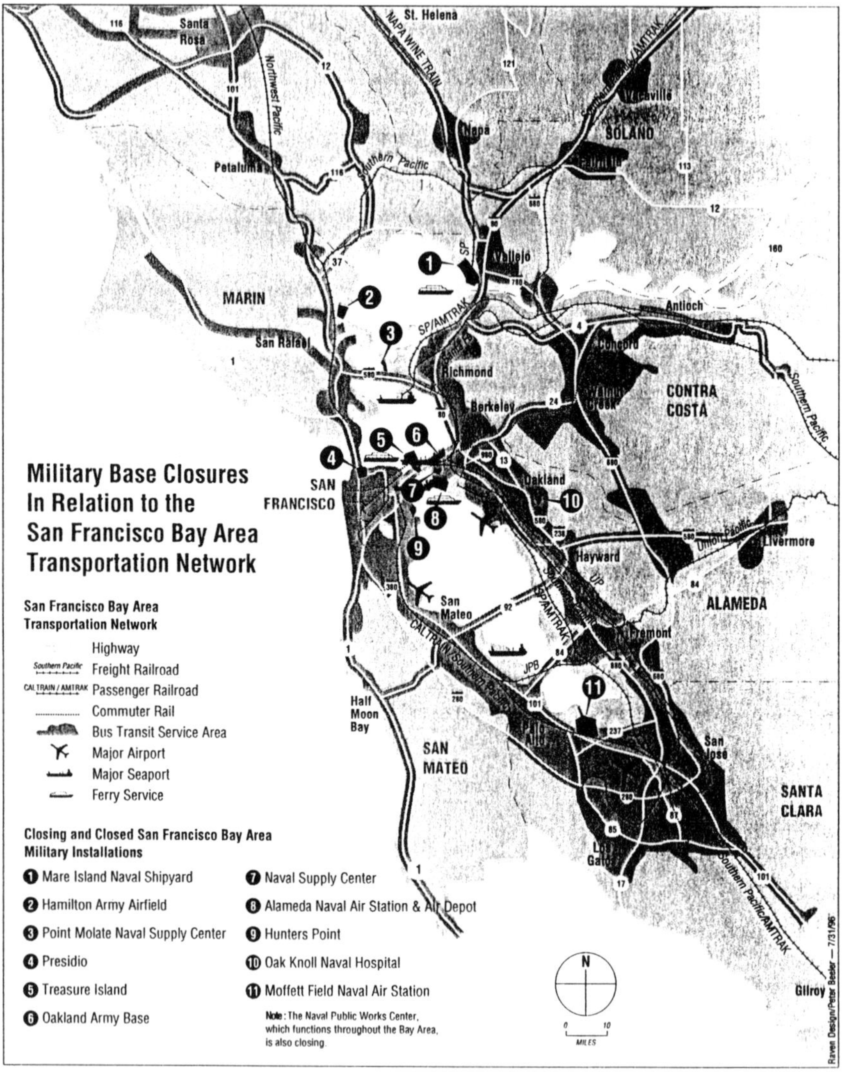
Military Base Closures
In Relation to the
San Francisco Bay Area
Transportation Network
San Francisco Bay Area
Transportation Network
Highway
Freight Railroad
Passenger Railroad
Commuter Rail
Bus Transit Service Area
Major Airport
Major Seaport
Ferry Service
Closing and Closed San Francisco Bay Area
Military Installations
1 Mare Island Naval Shipyard
2 Hamilton Army Airfield
3 Point Molate Naval Supply Center
4 Presidio
5 Treasure Island
6 Oakland Army Base
7 Naval Supply Center
8 Alameda Naval Air Station & Air Depot
9 Hunters Point
10 Oak Knoll Naval Hospital
11 Moffett Field Naval Air Station
Note: The Naval Public Works Center, which functions throughout the Bay Area, is also closing.
St. Helena
NAPA WINE TRAIN
Santa Rosa
Northwest Pacific
Petaluma
Southern Pacific
Napa
Vacaville
SOLANO
Vallejo
MARIN
San Rafael
SP/AMTRAK
Richmond
Berkeley
Antioch
Concord
CONTRA COSTA
SAN FRANCISCO
Oakland
Hayward
Livermore
Union Pacific
ALAMEDA
San Mateo
CALTRAIN/Southern Pacific
Fremont
Half Moon Bay
SAN MATEO
San Jose
SANTA CLARA
Gilroy
Southern Pacific/AMTRAK
N
0 10 MILES
Raven Design/Peter Beeler — 7/31/96

INDUSTRY CLUSTER ANALYSES SUMMARY

Industry Cluster	**Telecommunications**	**Multimedia**	**Movie and Television Production**	**Visitor and Tourism**
Number of Employees	79,499	62,063	35,986	88,211
Number of Firms	2,510	2,262	3,726	7,753
Employ. Concentration	2.56	2.49	2.08	1.01
Payroll Per Employee	$51,842	$46,374	$52,436	$15,519
% Payroll Growth '88	17.8%	NA	12.5%	6.7%
Strengths	Technology development. Market penetration. Consumer demand.	Digital technology. Critical mass of design, art & technology talent.	Digital technology. Cultural diversity.	Wide variety of resources and attractions. Moderate climate. Diverse population. World-class technology.
Challenges	Public official knowledge of technology. Fair policies for competitive market. Protection of community design and health. "Universal access."	Consumer knowledge, market definition, distribution mechanisms. High-speed broadband infrastructure. Incompatible platforms. Uncoordinated resources.	Production facilities. Cooperation from local communities.	Integrated transportation system. Additional destinations & attractions.
Infrastructure	Education & Training Telecommunications	Education & Training Telecommunications	Education & Training Environmental Resources Telecommunications	Transportation Environmental Resources Telecommunications
Action Strategies	Form partnership to educate local officials and accelerate deployment of advanced telecommunications.	Form partnership to leverage resources & integrate industry. Establish industry-led consortia for workforce preparation. Facilitate deployment of "Digital Diamond". Promote market development through fostering applications in education, health, business & gov't.	Foster gov't. support. Streamline permitting. Facilitate networking, link industry, & community. Develop production facilities and large-scale sound stage. Provide access to digital & broadband technology.	Establish regional program for integrated marketing. Integrate transportation systems and expand water shuttle service. Integrate tourism into base reuse plans. Promote eco-tourism, military history tourism & family-oriented activities. Develop Treasure Island to showcase region's assets. Profile Bay Area in movie & TV production.

INDUSTRY CLUSTER ANALYSES SUMMARY

Industry Cluster	**Bioscience and Health Technology**	**Banking and Financial Services**	**Environmental Technology**
Number of Employees	30,033	119,125	13,540
Number of Firms	920	11,017	1,200
Employ. Concentration	1.93	1.72	1.59
Payroll Per Employee	$49,734	$51,522	$48,726
% Payroll Growth '88		13.3%	NA
Strengths		Headquarters location. Electronic commerce, new software & multimedia applications. Entrepreneurial culture. Talent pool.	Research institutions. Entrepreneurial culture.
Challenges	Affordable wet lab space for R & D. Affordable land for manufacturing.	Maintain region as a major financial center.	Funding for demonstration of new technology. Certification. Performance-based regulations. Positive working relationship with gov't.
Infrastructure	Education & Training Capital	Education & Training Telecommunications	
Action Strategies	Facilitate establishment of an incubator. Link biotechnology companies to bases. Advocate for improved FDA procedures.	Maintain region as a major financial center in west. Form Finance Leadership Roundtable. Deploy advanced telecommunications infrastructure. Export services. Create "patient" capital pool for economic development & base conversion.	Form partnership to link industry to base clean-up: (a) use more regional firms; (b) test emerging technologies on bases. Facilitate positive working relationship with regulators. Link research to industry. Promote performance-based regulations and appropriate risk-based standards.

Bay Area Defense Conversion Action Team, March 1997

OPEN SPACE, WETLANDS, AND MARITIME FACILITIES ON THE BAY AREA'S CLOSING BASES

BASE/ TOTAL ACREAGE	WETLANDS	PLANNED OPEN SPACE	MARITIME FACILITIES
Hunter's Point Naval Annex 936 acres	443 acres submerged $2.5 million received to restore tidal marsh lands	134 acres	22 deep water berths 880 ton capacity crane Biggest dry-dock in the West
Treasure Island Naval Station 1,310 acres	314 acres submerged - TI 243 acres tide and submerged lands - YBI	74 acres	4 piers for large ship repair & berthing, recreational marina, & fishing sites
Presidio Army Base 1,480 acres	20 acres to be restored to tidal marsh - $3 million is allocated of $16 million project. Also $500,000 allocated to restore Mountain Lake	780 acres currently, removal of 276 buildings will result in 1000 restored acres	Exploring development of ferry access.
Mare Island Naval Shipyard 5,460 acres	3,810 acres 670 acres of nontidal wetlands to transfer to USFWS	438 acres total 200 acres of total are hills	Mare Is. Strait (dredged to 30 ft. depth) provides waterway from Pacific to Stockton & Sacramento
Alameda Naval Air Station 2,842 acres	1,027 acres of which 525 acres will be National Wildlife Refuge for CA least tern	1,794 acres	Deep piers & marina
FISC Alameda 176 acres			No public trust lands
FISC Pt. Molate 423 acres	133 acres below high tide line	156 acres Hillside Op. Spc. 14.4 Shoreline Open Space 20.9 acres Shoreline Park	1,450 sq. ft. Pier
FISC Oakland 533 acres	102 acres of submerged lands	102 submerged lands	392 acres transferring to Port of Oakland 14 Ship berths for loading container cargo
Oakland Army Base 422 acres	44 - 54 acres are submerged		189 acres transferring to Port of Oakland 100 ton crane -only one in Oakland
Hamilton AFB 1,624 acres	655 acres to be potentially returned to wetlands on Army parcel		
Oakland Naval Hospital 183 acres	0 acres	37 acres	No shoreline access

CHAPTER 144

A Comprehensive Conservation and Management Plan for the San Francisco Bay-Delta Estuary, With a Focus on Wetlands Ecosystem Goals

Marcia L. Brockbank[1]
Michael W. Monroe[2]

As one of the nation's most biologically diverse and productive estuaries, the San Francisco Bay-Sacramento/San Joaquin Rivers Delta also supports many other uses including commercial and sport fisheries, shipping, recreation and tourism. It is the largest estuary on the west coasts of North and South America; its watershed is over 60,000 square miles and drains forty percent of the state.

In 1987, Congress created the National Estuary Program in the amended Clean Water Act to help reverse the decline of the nation's estuaries. The establishment of the San Francisco Estuary Project, one of 28 programs around the country, signaled the beginning of a federal-state partnership to develop a Comprehensive Conservation and Management Plan (CCMP) for the Bay-Delta Estuary. Its overall goal has been to promote effective management of the estuary while restoring its water quality and natural resources, and maintaining its economic vitality.

The Clean Water Act required that the CCMP be developed by stakeholders and agencies involved with regulating the estuary's beneficial uses. Over one-hundred people, including representatives of local/state/federal government and diverse interest groups from the twelve Bay-Delta counties worked for five years to address this complex and vital challenge. Governor Pete Wilson and U.S. EPA Administrator Carol Browner approved the plan in late 1993. The San Francisco Bay Regional Water Quality Control Board is the lead agency for implementing the CCMP and the Association of Bay Area Governments provides technical and administrative support.

The Plan's 145 actions and nine program areas (Aquatic Resources, Wildlife, Wetlands, Water Use, Pollution Prevention and Reduction, Dredging and Waterway Modification, Land Use, Public Involvement and Education, and Research and Monitoring) recognize the estuary's environmental value, including the need to manage habitats and ecosystems within the sub-watersheds and without compromise to the overall economic diversity of the region. The CCMP is the only approved, completed ecosystem-wide plan

---

[1] Program Manager, San Francisco Estuary Project, 2101 Webster St., Ste. 500, Oakland, Ca 94612
[2] Environmental Scientist, U.S. EPA, Region 9, 75 Hawthorne St., San Francisco, CA 94105

for balancing environmental protection and beneficial use of the estuary's resources and waterways.

During the development of the CCMP, we brought together agency resource managers, and interest groups from throughout the region. We carry out restoration actions by coordinating activities, improving communication, promoting collaborative partnerships among the many diverse stakeholders and supporting community-based environmental stewardship.

The Estuary Project faces many challenges in managing the estuary's resources - the first of our continuing challenges is PUBLIC INVOLVEMENT - the Estuary Project recognizes the importance and need for public involvement in the decision-making process for managing the Estuary. We call this the "hook" - how do you get folks to care enough about the environment so that they will change their behavior? We need informed citizen activism, so we build awareness and appreciation for the estuary by getting people involved in outreach and restoration activities. One of our most effective projects works with volunteers to do a fish consumption survey on piers in the East Bay.

Another challenge for us is ENVIRONMENTAL JUSTICE - the merging of environmental concerns with social justice issues, and economic concerns. We need to do a better job of multicultural outreach so that all social, economic and ethnic groups participate in the decision-making process for managing our natural resources. One of our outreach projects works with inner-city high schools, neighborhood groups, and other agencies to provide enriched education and career opportunities to students.

ECONOMIC CONCERNS are also a challenge - the "jobs vs. environment" or "fish vs. people" debate has been with us for a long time. Our approach is to ask interested industry, agriculture, and other forces of economic development to work with us in restoring the estuary.

Another challenge for us is the role of LOCAL GOVERNMENT in protecting the estuary. With responsibilities ranging from regulating land use and stormwater runoff to promoting water reuse and preservation of open space - local government has significant power to influence the health of the estuary. Pressured by many interest groups, local government often has a difficult time in enforcing policies that protect the environment. We respond to this challenge by outreach and education programs that support local government in developing effective, long range plans for protection of our natural resources.

One of our most important challenges, is EVALUATING PROGRESS in carrying out the plan. We need to keep public and political pressure on the implementing agencies so that they will take action to protect and restore the estuary. We produced a workbook that evaluated the progress made on each of the 145 actions in the CCMP. This report card tracks the time and resources spent by agencies and the public in restoring the estuary. It puts the spotlight on agencies and entities that aren't taking an active role in restoration efforts. It informs the public, and facilitates adaptive management by providing an opportunity for agencies and stakeholders to re-evaluate priorities and set new goals.

An example of how we put all of this into action is our Delta in-channel islands workgroup. Workgroup members represent Delta agencies and stakeholders. They reached consensus on the need to create a management strategy for protection and restoration of the fast disappearing in-channel islands.

These are small remnant islands that remained after dredging took place, years ago to straighten out channels throughout the Delta. They provide habitat for many special status species and are an important fish and wildlife habitat resource - they also provide recreational, aesthetic and levee protection benefits. The management strategy is ecosystem based, uses consensus decision-making and provides for voluntary participation. Using different restoration techniques, the workgroup intends to carry out demonstration projects within the Delta that increase and enhance in-channel island habitat. And in order to coordinate better with CALFED and not duplicate efforts, we have become a subgroup of their Levee and Channel Technical Team.

In summary, our management challenges are: 1) increased attention to social justice on the one hand and economic constraints on the other; 2) increased focus on local government control; 3) the need to keep public and political pressure on the implementing entities and agencies so that they will protect and restore the estuary and; 4) the need to change our own individual behavior so that we can have a healthier and more productive Bay-Delta Estuary, and protect it for future generations.

Wetlands Ecosystem Goals

The San Francisco Estuary Project's Comprehensive Conservation and Management Plan recommends establishing regional wetlands habitat goals for use as the basis of a regional wetlands management plan to protect, enhance, restore, and create wetlands in the San Francisco Bay-Delta estuary. In 1994, representatives of state and federal resource agencies involved in wetlands regulation and management began meeting with staff of the San Francisco Estuary Institute to design a process for establishing these goals. These discussions led to the initiation, in mid-1995, of the San Francisco Bay Area Wetlands Ecosystem Goals Project.

The Goals Project is using available information to identify the kinds, amounts, and distribution of wetlands and related habitats needed to sustain diverse and healthy communities of fish and wildlife resources in the San Francisco Bay area. When completed in early 1998, the goals will provide a guide for private, local, state, and federal entities seeking to protect and improve wetlands.

The geographic scope of the project includes the four primary sub-regions of the estuary downstream of the Sacramento-San Joaquin Delta: Suisun Marsh and Bay, San Pablo Bay, Central Bay, and South Bay. The effort is focused primarily on the area's baylands, which include mudflats, tidal marsh, and seasonal wetlands within diked historical tidal marshlands. Adjacent riparian, subtidal, and upland areas are being considered to the extent necessary to develop ecological goals for the baylands.

The Goals Project is directed by the Resource Managers Group, representatives of eight state and federal agencies with major responsibilities for Bay area wetlands. Assisting the RMG are teams of species experts, physical scientists, and a science review group. The San Francisco Estuary Institute provides project participants with technical guidance and analytical support, and the San Francisco Bay Regional Water Quality Control Board assists with project administration. Funding comes from several local, state, and federal sources.

Establishing the habitat goals involves several steps. These include identifying the bayland's wetland habitats and representative species of plants, fish, and wildlife that inhabit them; assembling and analyzing data and other information regarding historical and current distributions of habitat and species; describing the ecological relationships between the habitats and species; and, finally, recommending the amounts and distribution of the various wetland habitat types. The goals will be expressed as quantitative and qualitative objectives, and will be described in narratives and in maps and other graphics.

In addition to establishing habitat goals, project participants will produce guidelines for implementing projects to attain them, a set of general principles to guide wetlands improvement efforts, and an adaptive management model for assessing progress towards achieving the goals and for revisiting and adjusting them. Drafts of these documents and the goals will be circulated to the public for review and comment in October 1997. Based on public input, final habitat goals will be prepared in early 1998.

The completion of the wetlands habitat goals will be followed by the development of a regional wetlands management plan for the Bay-Delta Estuary. This effort should involve a wide spectrum of stakeholders including landowners, environmental groups, and local, state and federal agencies. The regional wetlands management plan will implement many of the wetlands action items in the Estuary Project's Comprehensive Conservation and Management Plan.

CHAPTER 145

The California Coast: Of, By and For the People

Karin Strasser Kauffman

Abstract

In order to adequately protect our marine resources in California it is essential to continue to involve the public. Governmental and educational institutions cannot do that job alone, nor are there sufficient funds available in the public sector. Recognizing the wealth and variety of California's coastal and ocean resources, the complicated interplay of political jurisdictions, and the high expectations on the part of the public, this is our challenge: We must find ways to mobilize and utilize coastal advocates more effectively. We must be ready to channel energies and enthusiasm. We must develop formal structures to plug in willing volunteers and recruit more.

Some existing role models which already provide such structure are the Sanctuary Advisory Council of the Monterey Bay National Marine Sanctuary, the Big Sur Multi-Agency Advisory Council and the Monterey Bay Sanctuary's Water Quality Protection Program. Each serves as an example of integrated management and inter-agency coordination on a variety of governmental levels. Each insists on public involvement and a network of organizational and agency support. Each has a history of successes and recognized achievements.

Utilizing California's unique coastal and ocean resources--which include its people--we have the potential to provide global leadership in designing more productive patterns of governance.

CHAPTER 146

"Musical Chairs - Who's Minding the Coast?"

Ann Notthoff

Abstract

Increasingly frequent leadership changes in the State Capitol, due, in very large part, to the advent of term limits in California, have a destabilizing effect on the California Coastal Commission. Commitment to coastal protection and the institutional memory that informs sound decision making have waned in recent years.

The California Coastal Commission is carefully balanced body composed of twelve voting members, appointed evenly by the Governor, the Senate Rules Committee and the Speaker of the Assembly. Each appoints four members who serve for two year terms "at the pleasure" of their appointing authority. The absence of set terms also tends to destabilize Commission decision making. Commission members now routinely serve two to four years and some far shorter. In February 1997 the longest tenure of a sitting Commissioner was five years.

Given the complexity of the Coastal Act, the Commission's governing statute, and its detailed resource management policies, long term tenure tends to improve the ability of Commissioners to make sound policy decisions, consistent with state law. Charts tracking the votes of Commissioners show widely varying outcomes on development decisions, indicating unpredictability.

Sound protection of the California Coast requires consistent application of well understood policies. Despite increasing changes among appointed Commissioners, the professional staff of the Commission has provided virtually seemless consistency since California's coastal program began 24 years ago. This stable staffing has provided certainty for project applicants and the public, yet the confounding effects of high Commissioner turnover cannot be ignored. California's coast would be better protected if both Commission members and staff were left to the business of faithfully applying Coastal Act policies free from the fear that controversial decision could lead to political calls for their removal.

Ann Notthoff, Senior Planner, Natural Resources Defense Council

CHAPTER 147

25 Years Later -- Is There Still a Coastal Constituency?

Naomi Schwartz

Abstract

In 1972 the people of California passed Proposition 20. After years of unbridled coastal development, loss of public access to the beach and legislative inaction, from Sea Ranch to Mission Bay, from the Mexican to the Oregon borders, the voters took coastal protection into their own hands and established the California Coastal Commission.

Throughout the state, citizens re-organized from signature-gatherers to Commission watchdogs. Six Regional Commissions of the State Commission held meeting throughout the coastal zone for three years, giving unprecedented opportunity for citizen organizations and just plain "folks" to be a part of a state agency's work.

Reflecting the effectiveness and importance of this "coastal constituency," in enacting the 1996 law which made the Commission a permanent agency, the Legislature described the coastal zone as "belonging to all the people," and provided explicitly for the broadest possible public participation in Commission activities.

Twenty-five years later, the Regional Commissions have been eliminated, State Commission offices have been consolidated and the Commission meets primarily in just two geographic locations statewide.

What has happened to public access to the Commission. as well as public access to the coast?

Has the coast been "saved," and the public's concerns been satisfied? Or has public participation become so burdened that it has become a mere shadow of its former self?

CHAPTER 148

Building Alliances to Save Coastal Waters

Warner Chabot

Abstract

Successful campaigns to protection coastal habitats and sealife requires coalitions of leaders from the environmental community, commercial and sport fishing industry, local governments and commercial interests that depend on clean coastal waters.

Past models of successful campaigns include those that passed the California coastal initiative, blocked federal offshore oil drilling plans and created national marine sanctuaries.

An armada of proposed state legislation to protect fisheries and coastal water quality and reform coastal management, offer a clear challenge to renew and expand this coalition. A rare window of opportunity exists in 1997 to pass dramatic legislation to improve and protect the quality of California's coastal waters.

Warner Chabot, Director of the Center for Marine Conservation's Pacific Regional Office, 580 Market Street, Suite 550, San Francisco, CA 94104

CHAPTER 149

Bridging the Gap between Science and the Public Interest

Susan Jordan

Abstract

"Scientists are not generally activists and activists are not generally scientists and therein lies the rub."

Ms. Jordan addresses this dilemma and offers concrete strategies for bringing the best available science to the heart of the public debate over existing and proposed projects that impact the marine environment.

Using the Scripps's ATOC project as an example, Ms. Jordan argues that it is up to the public to get informed and that the best way to do this is by actively seeking out scientists who work in the field, asking them to review and evaluate the information at hand, and then bringing this information to the forefront of the public debate by sharing it within the context of the appropriate public forum, e.g. public hearings, newspaper articles, etc.

She argues that most scientists are interested in doing the research and are more than willing to share their work, but are reluctant to 'carry the banner' in support or opposition to controversial projects. Therefore, it is the responsibility of all interested parties, e.g. the environmental community, members of the public, etc. to seek them out and make their expert opinions available to the public debate.

CHAPTER 150

The Need for New Coastal Legislation

Howard Wayne

Abstract

In light of the reawakened coastal constituency in California in 1997, new coastal initiatives have a better chance of success than in the recent past. The California Legislature is currently considering a number of coastal legislative packages. This presentation will provide a summary of efforts underway in the California State Assembly, highlighting measures and solutions of particular concern to Assemblymember Wayne.

CHAPTER 151

The Coastal Boundary Line in California: From Certainty to Chaos

Howard D. Coleman,[1/]

The purpose of this paper is to describe why in boundary determinations in California between the land and the water in the coastal area, the law no longer works under current judicial thinking. Basically, the problem is associated with the fact that tides move, and this fact poses too much of a puzzle for our legal system given the methodology developed to date to confront this issue.

Historically, a visual inspection was generally all that was necessary to determine the location of a tidal boundary line: what was subject to water was in sovereign ownership and what was not, was in private ownership. This "what you see, what you have" approach to tidal boundary lines, was pragmatic and worked quite well in simpler times. The principle was addressed in a famous dissertation from the last century which analyzed hundreds of years of British jurisprudence covering coastal areas, which included the River Thames.

> ""[I]t is said:
>
> > 'the Prince [King] shall have all lands left by or gained from the sea. I have seen and perused a treatise thereon and therein many examples of Romney Marsh and Bromhill, in Kent, of which there are farmers to the King; and there is vouched a memorandum they came to an agreement with the Lord the

1/ Partner, Law Firm of Nossaman, Guthner, Knox & Elliott, LLP, 445 South Figueroa Street, Los Angeles, California 90071

> King.' Trin. 43 Ed. 3, Rot. 13, ex parte of the Treasurer's Remembrancer.'
>
> 'If the sea-marks are gone, so that it cannot be known if ever there was land there, the land gained from the sea belongs to the King. But if the sea cover the land at the flux of the sea, and retreat at the reflux so that the sea-marks are known, if *such* land be gained from the sea, it belongs to the owner.' 8th Eliz., Corporation of Rumney's case. Hall's 803-804.

This is an addition to Dyer, by the Editor from Dyer's own notes. So Lord Hale lays it down, (f) that 'if a subject hath land adjoining the sea, and the violence of the sea swallow it up, but so that there be yet reasonable marks to continue the notice of it, or though the marks be defaced, yet if by situation and extent of quantity, and bounding upon the firm land, the same can be known, though the sea leave the land again, or it be by art and industry regained, the subject doth not lose his property; and accordingly it was held by Cooke and Foster, M. 7 Jac. C.B. though the inundation continue forty years. for which he cites Dy. ub. sup. and a notable case of an overflowing by the Thames, Rot. Parl. 8 E 2 M. 23. Burnell v. the Bishop of Bath and Wells. (g) Callis (h) also, in his book on sewers, puts his case; 'the sea overflows a field where divers men's grounds lie' promiscuously, and there continueth so long, that the same is accounted parcel of the sea; and then after many years the sea goes back and leaves the same, but the grounds are so defaced as the bounds thereof be clean extinct, and grown out of knowledge, it may be that the King shall have those grounds; yet in histories I find that Nilus every year so overflows the grounds

> adjoining, that their bounds are defaced thereby, yet they are able to set them out by the art of geometry.' At this day it may be concluded that the former ownership may be identified by mensuration, so that if the sea suddenly swallow up ten acres, and after several years leave twenty acres dry, the ten acres may be reclaimed by admeasurement, but then the locality must be proved."

Hall's Essay on the Rights of the Crown and the Privileges of the Subject in the Sea-Shores of the Realm (3rd Edition) found in Moore, A History of the Foreshore and the Law Relating Thereto, pp. 803-804 (1888).

This very practical approach to defining coastal rights based on water-oriented purposes was also reflected in the Spanish law which governed property rights in California under both Spanish and Mexican rule prior to Calfornia's admission to the Union in 1850. Law 4 and 6 of Partida 3, Title 28 of the Spanish real estate law made clear that the seashore was protected for the building of galleys and other vessels, the mending of nets, and making use of trees to which to fasten a vessel. In Lux v. Haggin (1886) 69 Cal. 255 324, the California Supreme Court, in commenting on Mexican law affecting waters, stated that "the property of the nation is in the river and its bed, while it is the bed of the river; the common use continues while the water is the water of a river."

In California in its early years, coastal uplands, unless having immediate use for ocean-related commercial purposes such as piers, wharves or warehouses, were typically thought of as worthless. Certainly, it was not a place to build a home and raise a family; the salinity of the soil and the coastal breezes made agriculture use and the raising of livestock in such areas undesirable. The disenos (land grants) issued by the Mexican and Spanish governments for ranchos in California in coastal areas typically showed the rancho house (casa) was not adjacent to the ocean. (See R. Becker, Disenos of California Ranchos (1964).)

The late nineteenth and the twentieth centuries abruptly changed this view of the coast. California's coastal area became more utilized for commerce; people started to use the beaches as recreational areas; and even residential development on the coast became popular as the "modern" family no longer needed to raise their own food or pasture their own cattle for purposes of self-preservation. In addition, the coastline of California became subject to more artificial changes as the result of dredging and filling activities and the construction of wharves, revetments, seawalls and groins. Such work began to change the physical configuration between uplands and tidelands as sand became trapped in new harbors and blocked by seawalls and groins. Thus, the synergy between surging interest in the coastal areas coupled with resulting changes in the coastline worked together to heighten the importance and value of coastal lands.

As man-made physical developments continued to increase, California's coastal jurisprudence also blossomed magnifying the significance of the changing perceptions of the ocean zone and the continuing transformation of the boundary between land and water. Case law in California increasingly was called on to classify land as tidelands, submerged lands, swamp lands, overflowed lands or just "plain vanilla" uplands. Such classifications were of vital importance since they determined whether one's land was properly granted and if the answer was yes, what public rights should such grant be subject to under the law. In the latter half of this century, case law moreover greatly expanded the public trust rights residing in the State of California (as the surrogate sovereign replacing the King) and its municipal grantees to include environmental and scenic rights. (Marks v. Whitney (1971) 6 Cal.3d 251.) In Marks, the California Supreme Court stated that public trust uses have been traditionally defined in terms of navigation; commerce and fisheries. However, the "public uses to which tidelands are subject are sufficiently flexible to encompass changing public needs . . . [O]ne of the most public uses of the tidelands -- a use encompassed within the tideland trust -- is the preservation of those lands in their natural state, so that they may serve as ecological units for scientific study, as open space, and as environments which provide food and habitat for birds and marine life and which favorably affect the scenery and climate of the area." Thus, the requirement that the land contain "water" was eliminated as a factor in determining sovereign rights; dry, former ocean properties would now be subject to the public trust even though their use had no connection to traditional maritime and commercial purposes associated with such type of land. In fact, the open space public trust rights meant that the land need only look good, and the fact that it was near water need not be critical.

Approaching the issue of coastal boundaries from another end, California courts found that accretions formed by artificial means (that is by works of man) belonged to the State and not to the upland owner -- even accretions formed gradually and imperceptibly. (See Carpenter v. City of Santa Monica (1944) 63 Cal.App.2d 772; City of Los Angeles v. Anderson (1929) 206 Cal. 662.) This meant that dry lands created as a result of construction of groins or wharves, continued to be the State's property. Moreover, the State could justify such retention based on the ever broadening of the public trust.

Thus, the fact that tidal waters affected the property no longer necessarily mattered for boundary determinations, so long as the lands had been, at one time, tide or submerged land. This doctrine, coupled with the extensive types of public rights to be protected, meant that traditional concepts relating to the marine-oriented usage of the property were thrown out the window. No longer was the coastal area to be protected because nets could be mended or ships anchored; rather, it would be deemed "maritime-like" in terms of rights if anything of public benefit could be attributed to the property. The result was that "what you saw is what you had" could not operate, because tidal action no longer was the criterion. Taking water out of the equation of setting a coastal boundary line naturally would have significant and unanticipated consequences.

In addition, as time passed, technology and the accumulation of data also made the determination of an historic high tide boundary line (the line in its last natural condition before artificial change to the shoreline) more feasible regardless of the base criteria for calculating such line.

As a result of Muchenberger v. City of Santa Monica (1929) 206 Cal. 635, it became imperative to determine the true boundary -- that is the high tide line in its last natural condition -- in order to legally identify the tideland/upland property line simply just guessing or compromising was insufficient for California courts. With that, the different approaches to determine coastal boundary lines began to abound and run amok. One could literally go to the boundary line supermarket to choose from a selection of alternative approaches to apply to determining a boundary line. Chart I (on the next page) shows the basic selections which were potentially available since the California courts were unsure as to how to proceed, given that the presence or absence of actual tidal action was not necessarily determinative.

In approaching a boundary line determination, the first step is to pick out one's preferred approach as to type of tide to be used (whether it be winter, average, neap, etc.). The second step was to figure out the amount of data for such approach (e.g., one year, two years, a hundred years or the 18.6 year complete lunar cycle). An elevation calculated vertically in feet from mean sea level, based on one's preferred approach could then be utilized based on the applicable slope which might be chosen for the seashore. The problem with this last step is that the slope of the seashore typically also continually changes with appreciable differences generally apparent between winter and summer seasons.

CHART
Beach Profile and the Sea Surface Intercepts

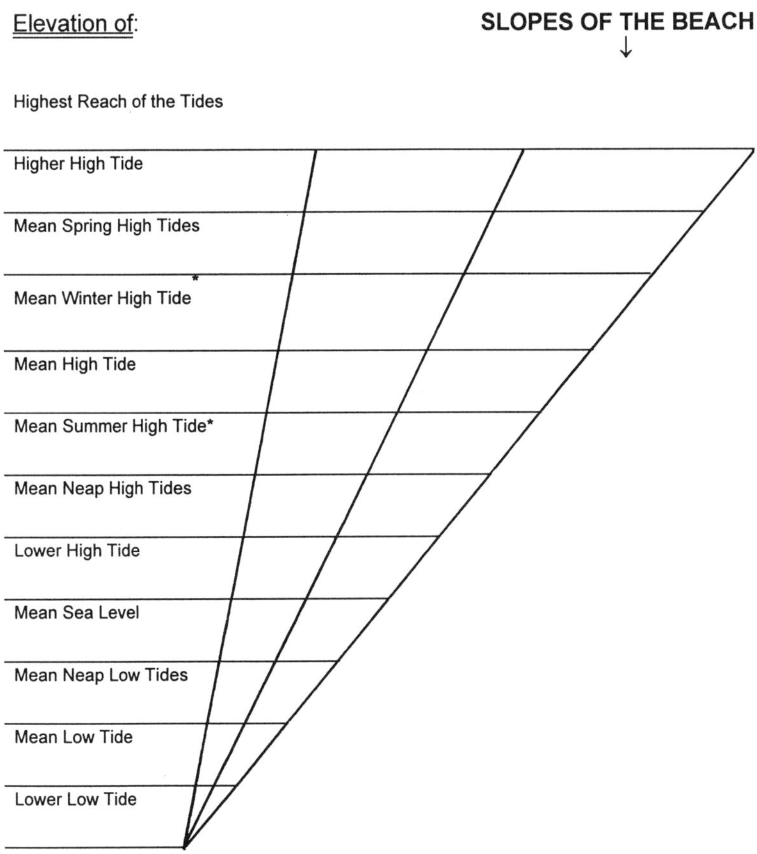

The Mean Winter High Tide intercept and the Mean Summer High Tide intercept are depicted on this Chart using the assumption that the beach profile for the Mean Winter High Tide intercept would be based on a winter beach profile and the beach profile for the mean Summer High Tide intercept would be based on a summer beach profile. A summer beach profile would usually be significantly different than a winter beach profile. Of course, if the same beach profile were used, both the mean Winter High Tide and Summer High Tide intercepts would touch the beach profile at the exact same point.

The following are some of the *approaches* to lines which are, could or will be appropriate in California:

- Average of All Mean High Tides. Since California has two tides a day of unequal reach, there would be an average calculated based on the lower high tide and the higher high tide for a period of time. Borax Consolidated v. City of Los Angeles (1935 296 U.S. 10.

- Neap High Tide Line. Neap tides typically extend inland and seaward to a lesser extent than spring tides. They occur when the sun and moon are at right angles to the earth and thus their respective gravitational forces are not working together. Teschemacher v. Thompson (1861) 19 Cal.11; People v. Wm. Kent Estate Co. (1966) 242 Cal.App. 156; Otay v. Carmel Sanitary District (1933) 219 Cal. 310.

- Mean of Winter and Summer High Tide Lines (for Beach Profiles). This is combining high tidelines as they intersect with the coast line during two different seasons. See Kent case, supra.

- Highest High Water Mark. This is the average of the higher of the high water marks or the highest extent of a typical winter wave. See Aptos Seascape Corp. v. County of Santa Cruz (1982) 138 Cal.App.3d 484.

- Spring High Tide Line. This is a tide which occurs when the moon and sun are in line on the same side of the earth, or opposite, which causes the tides to "spring" forward (landward) further than the average high tide and spring back (waterward) lower than the average low tide.

- Chance -- This generally means a California's court trying to make sense of this area and designing its own line (usually unsuccessfully), examples of which are described below.

So inconsistent and imprecise are the California cases that judges today do not know how to really set a coastal boundary line (if indeed such a line is really practically possible or only academically capable of being done). This entire process generally results in setting a high tide line which is not the real high tide line as it currently exists based on tidal action.

This chaos is best evidenced in two recent court cases which are thankfully not precedent except to those parties actually involved in such cases. In Lechuza Villas West v. California Coastal Commission et al. [Case No. BC076855, Superior Court, County of Los Angeles] the trial judge in his statement of decision filed on June 11, 1996 stated that the coastal boundary line would be decided based on the Kent case, supra, and that surveys will need to be prepared to "determine the average mean high tide line for the previous 2 years [and] the date for the computation shall encompass at least one survey for each previous season (the previous winter, spring, summer and fall)." This is certainly not Kent. First, there are four seasons to meld together. The judge gives no indication of how the seasonal data should be compiled nor does the judge indicate that the "neap" high tides will only be used for each of the four seasons which would be the methodology if Kent were to be followed. The two years is an oddity without precedent and is probably a result of the judge wanting to utilize a line closer to the continuing moving present high tide line.

In Antoine v. California Commission [No. B051709, issued July 31, 1992; modified August 26, 1992, which decision was thankfully de-certified on October 29, 1992 by the California Supreme Court], the California Court of Appeal attempted to provide "some guidance in resolving this question" on setting the coastal boundary. In fact, the court probably knew it was going to offer some pretty incoherent advice since it concluded its guidance by noting that the "parties should also keep in mind that a negotiated agreement is a worthwhile avenue to pursue." The Antoine court specifically rejected the Kent case neap tide approach in favor of the average of all tides. The Court of Appeal seemed to throw out the idea of a fixed boundary line because ["[g]iven the definition of mean high tide line and the physical characteristics of soft sand beaches, . . . the result is a continually changing boundary between public and private land." [At least the Court was cognizant of the basic problem.] However, the Court then concluded that using an ambulatory line was appropriate although "[a] moving boundary may therefore result in a development project encroaching on public land at some times of the year and not others." The Antoine decision is precedent for there being no boundary line at all given the absurdity of attempting to adjust land ownership based on each day's multiple changes in the line. Imagine the private owner continually taking away his improvements when the tide line surges forth and then installing them once again when the tides recede. Think of insurance, liability and trespass issues in this regard.

The current law in locating a coastal boundary line in California is that there is no applicable maxim which is both practical and realistic. In other words, the law is that there is no law. The reason for this is that any line changes as each wave impacting the coastline adds or subtracts new sand into the system. Fixing a line can certainly be done with scientific certainty based on whatever formula is chosen and then applied to a specific set beach profile. However, the line so set is arbitrary despite the fact that it was "scientifically" determined. Moreover, a line which is so changeable and based on many competing criteria does not provide a guaranty of establishing the real water/land boundary line.

It is the author's opinion that since the technological approach to establishing a high tide line does not provide a practical answer, the courts should return to the historic grounds of determining the boundary line -- use. It is fairly easy to establish the coastal boundary where water is the actual demarcation point. In this regard, there are two common situations. The first situation is where water is stopped from further inland intrusion as a result of a seawall, breakwater or other such improvement. The boundary line in this case should be the intersection of the water and such improvement. The second situation is where the water is not so rigidly confined (and defined), and to the extent there are no prior existing private improvements or use, the boundary line would be set along the present average vegetation or debris line, thereby preserving the beaches for recreational, net mending and other purposes.

More difficult are those accreted upland areas which were once tidelands by any definition. Continuing the thread of the "use" criterion, if such "historic" tidelands are not being used for any private purpose and are useful for the current public trust purposes, whatever they are, then the public land boundary should be pushed inland to the farthest high tide line as determined in Borax, the average of all tide lines over a lengthy period of time. However, if the ordinary high tide line at the time the property was first subject to artificial influences could be shown to be even further than the Borax line, then such further inland ordinary high tide line should be used. However, if private use of historic tidelands, no longer subject to water, is physically verifiable and under color of title then such lands should not be classified as tidelands for determination of property rights.

This suggested approach is contrary to current California jurisprudence. However, it represents a truer approach based on the origins of coastal boundaries in the first place. The use and purpose of what is happening presently should guide the setting of a coastal boundary line since practicality should be the byword. The alchemy of judicial and scientific approaches to this issue has certainly led to confusion and chaos; more importantly, it is leading to lines which may be supported by theories and data, but which have no relevance in feasibly dividing public and private ownership in the coastal zone.

CHAPTER 152

PRIVATE AND PUBLIC RIGHTS IN THE BEACH AND SHORE IN CALIFORNIA

Edgar B. Washburn, Esq. [1]
Bruce S. Flushman, Esq

INTRODUCTION

The respective rights of the public and private individuals in the beach and shore in California are governed by a legal framework grounded in the English common law, applied to the vestiges of Mexican law, and shaped by changing public policy concerns implemented by the state legislature. Recently, certain administrative excesses have been constrained by the due process provisions of the United States Constitution. The result has been a series of conflicts arising from fundamental differences in public policy that prevailed for a century—when private ownership and development was of paramount importance—in contrast to that existing today, which emphasizes public ownership and use. Implementation of today's policies have often been at the expense of real property concepts thought to be unassailable for the first 100 years of California's existence. While there might be a clear consensus that today's emphasis on public rights should be the rule if one were writing upon a clean slate, the Constitution places limits upon how far the state can go in destroying private rights previously vested in order to achieve that goal. The tension created by the now prevailing interest in preserving the beach and shore for the public and legal rules that originally favored private ownership will remain at the heart of conflicts concerning rights in the California shorezone.

THE SOURCE OF PRIVATE AND PUBLIC RIGHTS DERIVED FROM TITLE

Initially, private and public rights initially derived from title, as opposed to legislatively created regulatory schemes, and were grounded in the original physical condition of the property. In this regard, four distinct categories of land are recognized.[2] The first are areas that are dry uplands. This includes beaches and dunes lying above the ordinary high water line ("OHWL").[3] The second category consists of salt marshes and similar areas lying above the OHWL which are periodically covered by the tides.[4] These lands were officially classified as "swamp and overflowed" lands and are a category of uplands. A third category, tidelands, are those areas lying below the OHWL and above the ordinary low water line. The last category consists of submerged lands, i.e. those lands lying below the ordinary low water line.[5]

Rights Derived From Spanish and Mexican Law

California was occupied and settled by the Spanish in the 18th Century. While California was a part of Mexico, prior to the conquest by the United States, the Mexican governors of California granted into private ownership vast expanses of land. These grants—referred to as "ranchos" (grants to private individuals) and "pueblos" (grants to municipalities)—included much of the most valuable land in

California. With respect to the Pacific Coast, rancho grants covered all coastal property from Sonoma County south to the Mexican border, with the exception of a relatively small stretch along the Big Sur coastline in what is now Monterey County. Although the Mexican government did not regularly convey beaches, bays or navigable waters into private ownership, shallow lagoons, tidal sloughs and the like were often included within Mexican grants. In evaluating the extensive rancho grants that bounded upon tidewater, it is important to keep in mind what the Pacific Ocean was under Mexican law. In contrast to the common law applicable to California after it became a state,[6] Mexican law considered the border of the Pacific Ocean to extend to the highest wash of the waves in winter (rather than the OHWL). Thus, much of the beach and salt marsh bordering tidewater was included within the Mexican definition of the Pacific Ocean, although excluded from the common law definition. Mexican law considered the ocean to be owned by the sovereign and usable by the public.

In 1848, the United States acquired California from Mexico under the terms of the Treaty of Guadalupe Hidalgo. The treaty imposed upon the United States an obligation to recognize valid grants made by the government of Mexico to private persons during the time that California was a part of that country. With the exception of these prior grants, title to all other lands acquired from Mexico passed to the United States. This acquisition included the beds of navigable waters.[7] As to the beds of navigable waters not included within prior Mexican grants, the United States held title in trust for the future State of California.[8]

When California became a state on September 9, 1850, it received by reason of its sovereignty the beds of all navigable waterbodies under the Equal Footing Doctrine of the United States Constitution.[9] As noted above, an exception to California's acquisition exists with regard to the beds of navigable waters included within the private grants that had been made by the Mexican government. Therefore, if the bed of a navigable waterbody had been conveyed into private ownership by the Mexican government, the State of California did not acquire any sovereign interest in such lands and the state public trust for commerce, navigation and fishing does not encumber such waterbodies.[10]

The boundary of tidal waters (tidelands and submerged lands) that California received by reason of its sovereignty is the OHWL.[11] This boundary is determined by common law principles and not by the civil law applicable during the time California was a part of Mexico.[12] To the extent that the definition of the boundary of navigable waters differs under common law from that applicable under the civil law, the State of California only received lands up to the common law boundary as sovereign lands. This fact is particularly significant on tidal waters where there is a major difference between the two methods of locating sovereign boundaries.

In 1851, Congress, in fulfillment of obligations imposed upon the United States by the Treaty of Guadalupe Hidalgo, passed "An Act to Ascertain and Settle the Private Land Claims in the State of California."[13] The act's purpose was to provide a procedure to determine the validity of Mexican land claims in California and to locate the boundaries of Mexican grants made while California was a part of Mexico. The Act of 1851 established a means by which the United States could separate private land from the public domain over which the United States exercised ownership and control.[14]

The purpose of the 1851 Act and the confirmation proceedings it authorized was to establish once and for all the status of title held by the claimant to the Mexican rancho. As a result, once the patent was issued and the boundaries determined, those boundaries became fixed and the status of lands contained within the patent was established by the terms of the patent. Thus, if the patent conveyed fee simple to the lands and the patent included navigable waters within its boundaries, the recipient received absolute fee title free of the state public trust for commerce, navigation and fishing.[15]

Sovereign Rights of the State of California Effective at Statehood (September 9, 1850)

Since the United States Supreme Court's decision in *Phillips Petroleum Company, et al. v. Mississippi*,[16] it has been clear that those navigable waters that the state received by reason of is sovereignty upon its admission to the Union, encompassed *all* tidal waters below the OHWL. This includes small, non-navigable and shallow tidal arms, sloughs and creeks that are not navigable-in-fact. It also includes areas that are merely subject to tidal *influence*, as opposed to being tidewater themselves.

Although each state, including California, received title to all tidal land below the OHWL as an incident of its sovereignty, the rules of property to be applied to these lands is for the state to decide. California was free to adopt property rules governing boundaries and the extent to which, if any, the state would exercise its sovereign rights through the public trust.

The title of the state to the beds of navigable waterbodies is subject to the state public trust for commerce, navigation and fishing.[17] The beneficiaries of this trust are the citizens of the state. By virtue of the trust capacity in which California owns the beds of navigable waters, the state may not indiscriminately convey such lands into private ownership.[18] However, subject to statutory limitations that have existed from time to time and the state constitutional prohibition against the conveyance of tidelands within two miles of an incorporated city enacted in 1879, California may grant the beds of navigable waterbodies into private ownership if the grant does not destroy or substantially impair the public interest in the remaining lands or waters and the grant is made for the purpose of promoting navigation and commerce.[19] Unless an intent to the contrary is clearly manifested by the legislature, grants of lands below the OHWL implicitly reserve the public trust easement for the benefit of the public.[20]

FEDERAL RIGHTS AND INTERESTS IN NAVIGABLE WATERS

In addition to the servitude that exists over the beds of navigable waters in California represented by the public trust, there is a second servitude that benefits the public: the federal navigational servitude. The federal servitude is said to have arisen under the commerce clause of the United States Constitution and extends to all waters navigable-in-fact (which includes tidewaters) below the OHWL.[21] This servitude provides the right of free public passage over encumbered lands and subjects lands and improvements located below the OHWL to uncompensated takings by the United States.[22] The federal navigational servitude is akin to an easement and is to be distinguished from the federal government's right to regulate navigable waters—a right that also arises under the commerce clause.[23] The federal government's right to regulate and the existence of the federal navigational servitude are unrelated to current ownership of the underlying land.

In addition to the servitude just discussed, the acts admitting states into the Union generally provided that navigable waters within the state are to remain common highways and forever free. California's admission contains such a provision which has been interpreted as incorporating the common law right of passage and requires that waterways remain open without preference and that exaction of a toll for the privilege of navigation is prohibited.[24]

CALIFORNIA EXERCISES ITS SOVEREIGN RIGHTS OVER THE BEDS OF TIDEWATER

Less than one month after California became a state, it received a grant from the United States of all "swamp and overflowed lands" within the state under the Act of September 28, 1850, the Arkansas Swamp Act.[25] Swamp and overflowed lands passing under the 1850 Act were in fact two categories of land. "Swamplands" were considered to require drainage to make them fit for cultivation. "Overflowed lands," on the other hand, were those lands subject to periodic or frequent overflows so as to require the construction of levees or embankments to keep out the water in order to render them suitable for cultivation.[26]

Commencing in 1855 after the state received swamp and overflowed lands under the Arkansas Act of 1850, California adopted a series of acts authorizing the sale of tidelands, as well as swamp and overflowed lands. Because it was often difficult to distinguish between the two, the 1858 Act (as did subsequent statutes) included tidelands within the general statute authorizing the sale of swamp and overflowed lands. Although the swamp and overflowed lands sold under this statutory scheme are free of any state servitude, tidelands so sold are not because the general statutory scheme authorizing their sale did not manifest a specific intent to terminate the public trust.[27]

In addition to the general statutes authorizing the sale of swamp and overflowed lands and tidelands, the state early on adopted a series of what were referred to as special acts authorizing the sale of various submerged and tidelands to private owners as a part of waterfront development. Under these acts, large portions of the waterfronts of San Francisco, Eureka, Benicia, Oakland, Martinez, and Crescent City, and various areas within five miles of San Francisco were conveyed into private ownership.[28]

In 1915, the California Supreme Court, in reviewing the legislative intent and effect of the special acts for waterfront development, concluded that these grants were designed to facilitate the construction of waterfronts and, as a necessary part of that act, contemplated the elimination of navigable waters and their replacement with fill or a bulkhead essential to the existence and operation of the waterfront. Therefore, the legislation was viewed as being in furtherance of the state public trust for commerce, navigation and fishing and was held to effectuate a termination of that trust in the area to be filled or removed from navigation.[29] However, sixty-five years later, the California Supreme Court rewrote history and reversed the case of *Knudson v. Kearney* in *City of Berkeley v. Superior Court.*[30] Under the *City of Berkeley* decision, lands sold under certain of the special acts--which had not yet been filled--remained subject to the public trust.

CALIFORNIA EXERCISES ITS PUBLIC TRUST

The phrase "public trust" has been widely used, often in inappropriate circumstances. There are several basic, but different, concepts that relate to public rights in waterbodies, only one of which is the public trust. These concepts are: (1) the federal navigational servitude that attaches to the beds of all waterbodies that are navigable in fact for commercial purposes;[31] (2) the ability of the federal government to regulate navigable waters under the commerce clause of the United States Constitution;[32] (3) the ability of the state government to regulate land and water use under the police power;[33] and, lastly, (4) the public trust for commerce, navigation and fishing.[34] While the power of the federal government and the states to regulate under the commerce clause and police power are essentially unrestricted in the geographic sense,[35] the federal navigational servitude and the public trust have well defined geographic limits.[36] They apply only to the beds of navigable waters below the OHWL.

The "public trust" easement, as it has developed in California law, vests most of the attributes of property ownership in the State. The State may "enter upon and possess" trust-encumbered lands to improve them for trust-related uses.[37] It has virtually unfettered discretion in dictating the uses to which the lands may be put. The State's options range from prohibiting any use whatsoever to occupying the property to promote any of a vast array of public purposes. The uses for which the State may appropriate lands subject to the public trust have been held to be as varied as: (1) harbor development and hydrocarbon, geothermal and mineral exploration/ development;[38] (2) construction of a YMCA building;[39] and (3) development of public playgrounds and parks.[40] Under the public trust doctrine, the State can designate property as open space.[41] It has been held that the State retains its trust powers even after the lands have been filled and improved.[42]

The existence of the public trust is of fundamental importance when considering public rights in California's beach and shore. Within areas subject to the trust, the public may engage in many forms of recreation such as: walking, swimming, picnicking, and the like. Trust lands are also subject to virtual unrestricted public access. Conversely, the ability of the owners of private lands subject to the trust to restrict public use is very limited.

IMPLIED DEDICATION

Not all public rights in California's beaches and shores are derived from title to the underlying property. The public, under certain circumstances, can acquire prescriptive rights over beachfront property under the doctrine of implied dedication. This doctrine—akin to adverse possession or prescription—has its genesis in *Gion v. City of Santa Cruz* and *Dietz v. King* in 1970.[43]

Basically, if the public can show that large numbers of individuals used beachfront property believing that the public had a right to such use, without objection or interference by the landowner for more than five years, it is presumed that the use was adverse and that the landowner had impliedly dedicated the property to the public for such use. The doctrine applies only where there has been significant public use, as opposed to limited use by a definable number of persons. Again, the public must show that the area was used as if it were a public recreation area and it is helpful to point to the expenditure of public funds or other public actions facilitating such use. To negate a finding of intent to dedicate based on uninterrupted public use for more than five years, the landowner must either affirmatively prove the public was granted a license to use the property or demonstrate that the landowner made a bona fide attempt to prevent such use. Subsequent cases have taken a more restrictive view of the application of the doctrine and have required only that the landowner show a reasonable effort to curtail public use under the circumstances, as opposed to an actual, effective prevention of public use to defeat the presumption of dedication.[44]

The doctrine enunciated by the Gion-Dietz decision created an uproar, to say the least. The California legislature responded one year later by amending Civil Code section 813 and adding section 1009, towards the end of affording landowners a mechanism by which they can cut-off the ability of the public to acquire implied dedication rights while at the same time allowing public use. Consequently, the recording and publishing of appropriate notices, as dictated by these two code sections, will prevent the public from acquiring implied dedication or prescriptive rights in the beachfront property.

IMPACT OF THE CALIFORNIA COASTAL COMMISSION AND THE CALIFORNIA COASTAL ZONE ACT

One of the most significant developments in preserving and enhancing public access and rights along the California coast was the adoption of the California Coastal Zone Conservation Commission in 1972 (by Proposition 20). The Commission developed the California Coastal Plan, which formed the basis for the legislature's adoption of the California Coastal Act in 1976.[45] Proposition 20, California's local manifestation of concern about its 3400 miles of shoreline, was adopted the same year (1972) as the federal Coastal Zone Management Act,[46] which represents a national effort to facilitate coastal protection of managed coastal resources. California's coastal management program was approved by the federal government in 1977. That program consists of the California Coastal Act and the regulations of the California Coastal Commission, along with the authorizing legislation and regulations of the San Francisco Bay Conservation and Development Commission and the Coastal Conservancy. The San Francisco Bay Conservation and Development Commission, which served in large part as a model for the California Coastal Act, regulates development in and 100 feet from the shore of San Francisco Bay.

The California Coastal Act regulates the "coastal zone," which is the land and water area extending seaward to the State's outer limit of jurisdiction (the three-mile limit) and inward generally 1000 yards from the mean high tide line. However, where significant estuarine habitat and recreational areas exist, jurisdiction extends to the first major ridge line paralleling the sea, or five miles from the mean high tide line, whichever is less.[47] In addition to requiring the formulation of plans as to how development will occur within the coastal zone, the California Coastal Act establishes a rigid procedure for the issuing of permits for activities within that zone.

In granting or denying permits for activities within the area of its jurisdiction, local agencies (cities and counties) implementing the Coastal Act are directed to carry out the requirements of Article X, section 4 of the California Constitution, providing for maximum access for the public. The constitutional provision prohibits owners possessing the frontage on tidal lands of a harbor, bay, inlet, estuary or other navigable waters, from interfering with public access or excluding the right-of-way to the waters whenever it is required for public purposes and prohibits the obstruction of free navigation of such waters.[48]

The public access provisions of the Coastal Act have, in practice, been vigorously (and, some would say, excessively) enforced by the Commission. Until curtailed by the United States Supreme Court decision in *Nollan v. California Coastal Commission*,[49] the Commission required those agencies implementing the local coastal plans to exact public access to and along the shore as a condition for the issuance of permits for developments, both large and small. In fact, these exactions became so numerous that the agencies charged with administering dedications of public access did not have the resources or ability to do so. Many of the dedications were exacted as conditions for construction of structures or improvements which in no way adversely affected public access. This fact led the United States Supreme Court in *Nollan* to curtail the Commission's excessive implementation of the public mandate and to limit requiring public access to those circumstances where there was a nexus between the project for which the permit was sought and public access. Absent the nexus, the imposition of a public access condition was viewed as being a "taking" for which compensation must be paid. At the end of the day, however, the California Coastal Act must be viewed as a very effective tool in preserving California's coast and enhancing the public's ability to use the shore.

PROPERTY BOUNDARY DETERMINATION ALONG THE OPEN OCEAN COAST – THE ORDINARY HIGH WATER LINE

Early Use of Scientifically-Based Measurements As Indicia of the Physical Location of Ordinary High Water Line

The boundary between State and privately-held lands on tidewater is the OHWL. In the early days of California's statehood, the OHWL was determined by reference to physical features. In the first decade of the 20th century, interested professionals attempted to provide more precision to determination of the location of the littoral property boundary.[50] These attempts were due, in part, to several factors including ever-escalating development and land use pressure. These endeavors gained some recognition in the courts.

Although opinions consistently stated the OHWL was the line reached by the "ordinary" or "neap" tides, courts in the early years of this century had not formulated any physically quantifiable definition of what they meant. Pinpointing the physical location of that property boundary merely by using the written description provided by court opinions was not really possible.

Use of the term "neap tides" grew from an early California case, *Teschemacher v. Thompson.*[51] *Teschemacher* concerned location of the OHWL in an estuarine marsh. Other cases, however, adopted and applied its description of the OHWL property boundary in cases concerning the open coast shoreline property boundary of uplands and tidelands.

The opinion defined "usual" or "ordinary" high water mark as:

> [T]he limit reached by the neap tides; that is, those tides which happen between the full and the change of the moon, twice in every twenty-four hours.[52]

Nevertheless, "neap tides" do not occur as frequently as twice in every twenty-four hours. Technically, "neap tides" only occur twice a month when the forces of the sun and moon act opposite one another.[53] There is at least a question of whether one can term this "ordinary" in the dictionary sense of the word. Moreover, at least on the Pacific Coast, the two high waters are not the same height. Instead, the two high waters each day differ markedly in their height. Even if one accepts the "neap" tide test, deciding which of the two daily high tides supplies the limit reached by the "neap" tides is still not possible.

As can readily be seen, the use of the term "neap tides" as descriptive of the physical phenomenon defining "ordinary tides" is fraught with ambiguity.[54] The courts finally recognized and resolved this uncertainty with help from the government and the scientific community.

Background, Explanation and Definition of Tidal Datums

The United States government recognized it was important for waterborne commerce to understand and quantify tidal phenomena. Mariners required accurate navigational charts of the coastal and estuarine shorelines to protect waterborne resources and commerce destined for or travelling from such areas. To supply information for those purposes, the U.S. Coast Survey and its successors, the USC&GS and the NOS, operated a system of tide stations at harbors and particular coastal locations. At these places continuous tide observations were and have been taken and recorded over many years.

For accurate results, determination of tidal characteristics should be based on a series of observations or measurements systematically taken over many years.[55] Indeed, an 18.6-year (sometimes called a 19 year) period is considered a full tidal cycle. This is because the more important of the periodic tidal variations due to astronomic causes will have gone through complete cycles. In addition, we assume randomly recurring meteorologic variations balance out during this long a period.[56]

Through statistical and mathematical means, one may convert the observations from a tide station into various vertical planes of reference, known as tidal datums.[57] Tidal datums are quite simply defined and can be readily, accurately and certainly determined (assuming there is tidal data available). And it is not essential that tidal data be obtained for an entire 19-year tidal cycle. Through statistical and mathematical means, a 19-year mean can be derived from a shorter series of observations.[58]

The basic tidal datums, from the lowest to the highest in terms of relative elevation, are: mean lower low water,[59] mean low water,[60] mean sea level,[61] mean tide level,[62] mean high water[63] and mean higher high water.[64] There is nothing magical about these tidal datums. They are useful tools for measuring for different purposes water levels which are constantly in flux. Certain of these tidal datums were also used as reference points for the mapping and charting work of the USCS, the USC&GS and the NOS.

Adoption of the Mean High Water Line as the Physical Location of the Ordinary High Water Line—The Borax Cases

At least in California, it was during the first decades of the 20th century that use of the mean high water line[65] as the physical location of the legal property boundary between tidelands and coastal uplands first gained currency.[66] The origin of and impetus for the adoption of this methodology by the courts was most probably the development and use of tidal datum planes by the United States government and its coastal and shoreline mapping agencies.

During the early 20th century, the unsatisfactory state of court-crafted equivocal or obscure descriptions of the physical location of the legal property boundary along tidal shorelines was confirmed. A vigorous debate grew up between surveyors and coastal engineers over just how to translate the legal term "ordinary high water line" into a physical location. Legislative enactments and court decisions began using the term "mean high water line" or "line of mean high water" in describing the shoreward property boundary of sovereign lands.[67] All these developments led to a pressure on the courts to decide finally and definitively how to convert the legal term "ordinary high water line" into a physical location.

In *Borax Consolidated, Ltd. v. City of Los Angeles,*[68] the United States Supreme Court decided as a matter of federal law to locate the OHWL, the property boundary between federally patented uplands and tidelands owned by the state, at the mean high water line.[69] In *Borax Consolidated, Ltd. v. City of Los Angeles,*[70] the United States Supreme Court decided as a matter of federal law that the location of the OHWL littoral property boundary was not the physical mark made on the ground by the water. Instead, the Supreme Court held that, for a tidal regime, tidal observations would determine the physical location of the OHWL.[71] Thus, the OHWL property boundary of lands adjacent to or along tidal waters would be physically located by use of the mean high water line.[72]

The Open Coast Mean High Water Line – Fluctuation of the Land Form

As we now know, as far as open coast tidelands are concerned, one physically locates the OHWL property boundary of such lands at the line of mean high water. While the plane of mean high water is a fixed plane,[73] the land surface or land form which the water (the datum plane) intersects may be constantly in flux. This is because of the processes of erosion, deposition, subsidence, upheaval or any of the many other physical processes which may affect the terrestrial form. The dynamic property of the land surface is most dramatically evidenced in the case of sandy beaches.

The legal principles concerning property boundary movement are applicable along the open coast.[74] Thus, if land is added through the process of accretion[75] or washed away through the process of erosion[76], the property boundary follows the changing physical location of the mean high water line.

The definitional problem of what is and what is not gradual or imperceptible along the open coast is difficult. In addition, another variable is both pervasive and its effect on the location of the ocean shoreline property boundary seems to be both legally and technically unpredictable. Although we might wish it otherwise, man has modified the coastlines along the continent.[77]

In California and in some other states, alluvion deposited gradually and imperceptibly and attributable to the works of man in the immediate vicinity of the property does not benefit the upland owner; the coastal property boundary is fixed at its location prior to the occurrence of the artificial accretion.[78] This result is similar to the property boundary consequence of an avulsive change.

When the federal government is the upland owner, however, the highest court in the land established a different rule. In that case, the federal government unmistakably caused a gradual and imperceptible increase in the seaward extent of the Pacific Ocean shoreline through construction of a massive coastal jetty. The court held that even so, the United States, as the upland owner, receives the benefit of the artificially caused accretion, even in California.[79] The law in this area is in flux.[80] One can generally state, however, that what some have called the "California rule," is not the rule in the majority of states. Thus, in states other than California, the OHWL moves with accretion and erosion—even if it is the result of man's activities—so long as it was not effected by the owner in order to expand his lands.

#4668 v03

---

[1] Messrs. Washburn and Flushman are principals in the law firm of Washburn, Briscoe & McCarthy, 55 Francisco Street, Suite 600, San Francisco, California 94133.

[2] This discussion focuses only on beaches and shores in tidal areas.

[3] The OHWL boundary is determined by calculating the mean high water line in tidal areas as determined by the procedures endorsed by the United States Supreme Court in *Borax Consolidated, Ltd. v. City of Los Angeles*, 196 U.S. 10 (1935).

[4] Due to the fact that tides on the Pacific Coast are of two unequal heights each day and vary from day-to-day and month-to-month, extreme high tides (often referred to as "spring tides") occur during various times of the year which periodically cover the surface of the salt marsh.

[5] The means and concepts by which these various boundary locations are determined is discussed later in this paper.

[6] The boundary of tidewater and all other navigable waters under the common law is the OHWL.

[7] Knight v. United Land Ass'n, 142 U.S. 161 (1891); City & County of San Francisco v. LeRoy, 138 U.S. 656 (1890).

[8] Pollard's Lessee v. Hagan, 44 U.S. 212 (1845); Weber v. Board of Harbor Commissioners, 85 U.S. 57 (1873).

[9] 9 Stat. 452; Weber v. Board of Harbor Commissioners, *supra*, 85 U.S. 57; Knight v. United Land Ass'n, *supra*, 142 U.S. 161; Pollard's Lessee v. Hagan, *supra*, 44 U.S. 212; Shively v. Bowlby, 152 U.S. 1 (1894). When the original colonies ratified the Constitution, they succeeded to the Crown's title and interest in the beds of navigable waters within their respective borders. New states were admitted to the Union on an equal footing with the original colonies and were held to have acquired the same rights, sovereignty and jurisdiction as the original states possessed within their respective borders. Accordingly, under this "equal footing doctrine," title to lands beneath navigable waters passed from the federal government to the new states upon their admission to the Union. Pollard's Lessee, *supra*.

[10] Knight v. United Land Ass'n, *supra*, 142 U.S. 161; City & County of San Francisco v. LeRoy, 138 U.S. 656 (1890); Summa Corp. v. California *ex rel.* State Lands Comm'n, 466 U.S. 198 (1984), *rev'g.* City of Los Angeles v. Venice Peninsula Properties, 31 Cal. 3d 288 (1982).

[11] United States v. Pacheco, 69 U.S. 587, 590 (1864); Borax Consolidated, Ltd. v. City of Los Angeles, *supra*, 269 U.S. 10; Marks v. Whitney, 6 Cal. 3d. 251 (1971).

[12] United States v. Pacheco, *supra*, 69 U.S. 587; Borax Consolidated, Ltd. v. City of Los Angeles, *supra*, 269 U.S. 10.

[13] Act of March 3, 1851, 9 Stat. 631.

[14] S. Exec. Doc. No. 26, 32d Cong. 1st. Sess. 2 (1852); Rodriguez v. United States, 68 U.S. 582 (1863).

[15] Summa Corp. v. California *ex rel.* State Lands Comm'n, *supra,* 466 U.S. 198; United States v. Coronado Beach Co., 255 U.S. 472 (1921).

[16] 484 U.S. 468 (1988).

[17] City of Berkeley v. Superior Court, *supra,* 26 Cal. 3d 515 (1980); People v. California Fish Co., 166 Cal. 567 (1913).

[18] State v. Superior Court (Fogerty), 29 Cal. 3d 240 (1981); State v. Superior Court (Lyon), 29 Cal. 3d 210 (1981); City of Berkeley v. Superior Court, *supra,* 26 Cal. 3d 515; People v. California Fish Co., *supra,* 166 Cal. 567; Illinois Central R.R. v. Illinois, 146 U.S. 387 (1892).

[19] City of Berkeley v. Superior Court, *supra,* 26 Cal. 3d 515; State v. Superior Court (Lyon), *supra,* 29 Cal. 3d 210; Illinois Central R.R. v. Illinois, *supra,* 146 U.S. 387. In 1879, article XV, section 3 of the California Constitution was adopted. That provision prohibited the sale of tidelands within two miles of an incorporated city or town. In 1909, section 3443(a) of the Political Code was adopted. That provision forbade further sale of tidelands.

[20] City of Berkeley v. Superior Court, *supra,* 26 Cal. 3d 515; State v. Superior Court (Lyon), 29 Cal. 3d 210 (1981).

[21] United States v. Rands, 389 U.S. 121 (1967); Kaiser Aetna v. United States, 444 U.S. 164 (1979).

[22] Kaiser Aetna v. United States, *supra,* 444 U.S. 164; United States v. Rands, *supra,* 389 U.S. 121; United States v. Grand River Dam Authority, 363 U.S. 229 (1960).

[23] Kaiser Aetna v. United States, *supra,* 444 U.S. 164.

[24] 9 Stat. 452. *See also* Willamette Iron Bridge Co. v. Hatch, 125 U.S. 1 (1888); Cardwell v. American Bridge Co., 113 U.S. 205 (1885).

[25] 9 Stat. 519, 43 U.S.C. § 981, *et seq.*

[26] San Francisco Savings Union v. Irwin, 28 F. 708 (9th Cir. 1886), *aff'd.* 136 U.S. 578 (1890).

[27] People v. California Fish Co., *supra,* 166 Cal. 576.

[28] Stats. 1851, ch. 41 (San Francisco); Stats. 1851, ch. 44 (San Francisco); Stats. 1853, ch. 137 (San Francisco); Stats. 1851, ch. 37 (Martinez); Stats. 1851, ch. 83 (Benicia); Stats. 1855, ch. 187 (Benicia); Stats. 1857, ch. 82 (Eureka); Stats. 1863, ch. 299 (Crescent City); Stats. 1868, ch. 543 and Stats. 1870, ch. 388 (San Francisco Bay, Board of Tideland Commissioners).

[29] Knudson v. Kearney, 171 Cal. 250 (1915).

[30] 26 Cal. 3d 515 (1980).

[31] Kaiser Aetna v. United States, *supra,* 444 U.S. 164; United States v. Rands, *supra,* 389 U.S. 121; United States v. Grand River Dam Authority, *supra,* 363 U.S. 229; Gibbons v. Ogden, 22 U.S. (9 Wheat.) 1 (1824).

[32] Kaiser Aetna v. United States, *supra,* 444 U.S. 164.

[33] Hitchings v. Del Rio Woods Recreation & Park District, 55 Cal. App. 3d 560 (1976); People *ex rel.* Baker v. Mack, 19 Cal. App. 3d 1040 (1971).

[34] Illinois Central R.R. v. Illinois, *supra,* 146 U.S. 387.

[35] Kaiser Aetna v. United States, *supra,* 444 U.S. 164.

[36] United States v. Rands, *supra,* 389 U.S. 121; United States v. Grand River Dam Authority, *supra,* 363 U.S. 229; Shively v. Bowlby, 152 U.S. 1 (1894); Illinois Central R.R. v. Illinois, *supra,* 146 U.S. 387; Hardin v. Jordan, 140 U.S. 371 (1891).

[37] People v. California Fish Co., *supra,* 166 Cal. at 588.

[38] Boone v. Kingsbury, 206 Cal. 148 (1928); Mallon v. City of Long Beach, 44 Cal. 2d 199 (1955); Cal. Pub. Res. Code §§6501-7062.

[39] People v. City of Long Beach 51 Cal. 2d 875 (1959).

[40] Los Angeles Athletic Club v. City of Long Beach 128 Cal. App. 427 (1932).

[41] Marks v. Whitney, *supra,* 6 Cal. 3d at 259-260.

[42] Atwood v. Hammond, 4 Cal. 2d 31, 41 (1935); City of Long Beach v. Mansell, 3 Cal. 3d 462, 486-487 (1970).

[43] 2 Cal. 3d 29 (1970).

[44] *See* County of Orange v. Chandler-Sherman Corp., 54 Cal. App. 3d 561 (1976); Aptos Seascape Corp. v. County of Santa Cruz, 138 Cal. App. 3d 484 (1982).

[45] *See* Cal. Pub. Res. Code §§ 30,000-30,900.

[46] 16 U.S.C. §§ 1451-1464.

[47] Cal. Pub. Res. Code § 30103.

[48] Cal. Gov't Code § 39934.

[49] 483 U.S. 825 (1987).

[50] D.E. Hughes and O. Von Geldern, *The Determination of the Plane of Ordinary High Tide for Pacific Coast Harbors, with Particular Reference to San Diego Harbor California*, J. Ass'n Engineering Societies, vol. XLIV, no. 4 (1910) (hereinafter "Hughes & Von Geldern"); Eichelberger v. Mills Land, etc. Co., 9 Cal. App. 628 (1908).

[51] 18 Cal 11 (1861).

[52] Teschemacher v. Thompson, 18 Cal 11, 21-22 (1861).

[53] H.A. Marmer, *Tidal Datum Planes,* U.S. Coast & Geodetic Survey Spec. Pub. 135 (rev. ed. 1951) (hereinafter "Marmer") at p. 5 and R.S. Patton, *Relation of the Tide to Property Boundaries*, set out in 2 Shalowitz, Shore and Sea Boundaries, App. E, p. 669 (Dep't Comm. Pub. 10-1 1962) (hereinafter "Patton at 2 Shalowitz, p. __").

[54] This ambiguity has been the subject of much critical comment by commentators and courts. *E.g.,* Patton at 2 Shalowitz, p. 668; 1 Shalowitz at p. 93; J. Briscoe, *The Use of Tidal Datums and the Law,* (1983) 43 Surveying & Mapping (J. Am. Congress Surveying & Mapping 1983) No. 2, pp. 115, 116; Maloney & Ausness, *The Use and Legal Significance of the Mean High Water Line in Coastal Boundary Mapping,* 53 N.C. L. Rev. 185, 204 (1974); City of Los Angeles v. Borax Consolidated, Ltd., 74 F.2d 901, 905 (9th Cir. 1935), *aff'd,* 296 U.S. 10 (1936).

[55] Marmer at pp. 7-8; Patton at 2 Shalowitz, p. 675.

[56] Patton at 2 Shalowitz § 2311, p. 59.

[57] Shalowitz defines "datum plane" as "[a] surface used as a reference from which heights or depths are reckoned. The plane is called a Tidal Datum when defined by a phase of the tide . . ." Patton at 1 Shalowitz, p. 286.

[58] Marmer has a good explanation of the process. Marmer at p. 87-95.

[59] The average height of only the lower of the low waters over a 19-year period. Marmer at p. 113; Patton at 2 Shalowitz, p. 581. On the Pacific Coast this is the datum for soundings taken for hydrographic charts. Patton at 2 Shalowitz, p. 581.

[60] The average height of all the low waters over a 19-year period. Marmer at p. 104; Patton at 2 Shalowitz,

p. 581.

[61] The mean level of the sea at a particular location determined by averaging the height of the water levels for all stages of the tide. This is the primary tidal datum. Marmer at p. 45; Patton at 2 Shalowitz, p. 528.

[62] This datum is also known as half tide level. It is a tidal datum midway between mean high water and mean low water. Marmer at 69; 2 Shalowitz at 568.

[63] The average height of all the high waters at a location for a period of 19 years. Marmer at 86; 2 Shalowitz at 581.

[64] The average of only the higher of the high waters at a location over a 19-year period. Marmer at 86; 2 Shalowitz at 581.

[65] The mean high water line is the intersection of the tidal datum mean high water with the shore. 2 Shalowitz at 581; Swarzwald v. Cooley, 39 Cal. App. 2d 306, 313 (1940).

[66] Forgeus v. County of Santa Cruz, 24 Cal. App. 193, 195 (1914); Strand Improvement Co. v. City of Long Beach, 173 Cal. 765, 769 (1916).

[67] Los Angeles v. San Pedro, etc., R.R. Co., 182 Cal. 652, 663 (1920), *cert. denied* 254 U.S. 636 (1920); City of Los Angeles v. Anderson, 206 Cal. 662, 664 (1929).

[68] 296 U.S. 10 (1935).

[69] *Id.* at 26-27.

[70] 296 U.S. 10 (1935).

[71] Borax Consolidated, Ltd. v. City of Los Angeles, *supra,* 296 U.S. at 22 (". . . it means the line of high water as determined by the course of the tides.").

[72] *Id.* at 26-27.

[73] Swarzwald v. Cooley, 39 Cal. App. 2d 306, 313 (1940); People v. Wm. Kent Estate Co., 242 Cal. App. 2d 156, 159-160 (1966).

[74] Strand Improvement Co. v. City of Long Beach, 73 Cal. 765, 772-773 (1916).

[75] *E.g.*, City of Los Angeles v. Anderson, 206 Cal. 662, 666-667 (1929)

[76] Miramar Co. v. City of Santa Barbara, 23 Cal. 2d 170, 175 (1943).

[77] For a discussion of the impacts of a jetty on the coast and the problem of attempting to relate general statements about the effect of man's works on beach processes to a particular beach, see *Surfside Colony, Ltd.*, 226 Cal. App. 3d 1260, 1263-1264, 1268-1269 (1991).

[78] State of California *ex rel.* State Lands Comm'n v. Superior Court (Lovelace), 11 Cal. 4th 50, 76-80 (1995) (must be direct cause); Carpenter v. City of Santa Monica, 63 Cal. App. 2d 772, 794 (1944); People v. Hecker, 179 Cal. App. 2d 823, 837 (1960); Lorino v. Crawford Packing Co., 175 S.W. 2d 410, 414 (Tex. 1943). But in *Lovelace, supra,* 11 Cal. 4th at 74, the California Supreme Court observed federal law may apply to oceanfront titles derived from the federal government. The court also noted the choice of law was different for inland property over which the federal government has no interest. *Id.*

[79] California *ex rel.* State Lands Comm'n v. Superior Court (Humboldt Spit), 457 U.S. 273, 284-285 (1982).

[80] The California Supreme Court decided the scope of the application of the artificial accretion rule, but there are no other cases applying the rule.

CHAPTER 153

PUBLIC AND PRIVATE RIGHTS WHERE THE LAND MEETS THE SEA

By Joseph J. Barbieri* and Dennis M. Eagan**

Abstract

Various systems of law including the common law have recognized that the boundaries of real property along bodies of water are not like other boundaries. Such boundaries are not fixed in location, but rather move as the "natural monument" moves under the influence of natural processes, whether that "monument" be the bed or bank of a lake or river, or the shoreline of the ocean. Even this concept of a dynamic boundary, however, is not adequate to fully comprehend the range of public and private interests in the shorezone of navigable bodies of water. The fee boundary, although moving, is most often tied to some "ordinary" water level, which leaves unclear the existence, nature, and extent of the rights that obtain when the waters of the lake, river, or ocean are at some level other than their "ordinary" level. Accordingly, adjunct legal principles have been developed to more completely deal with these interests. As have the boundary principles themselves, these adjunct principles have taken different forms in different jurisdictions. This paper explores how different jurisdictions and different courts have dealt with the littoral boundary issue and related issues concerning rights in the shorezone, with particular emphasis on the ocean shoreline.

Introduction

Most parcels of property are bounded on each side by land. The boundaries of these properties are fixed by survey and may be easily relocated many years

---

*B.A., Northwestern University, 1973.
J.D., University of California, Berkeley, 1978.
**A.B., Harvard College, 1963.
LL.B., University of California, Berkeley, 1966.

Mr. Barbieri and Mr. Eagan are deputy attorneys general in the Land Law Section of the California Attorney General's Office, specializing in litigation involving coastal land use and public and private rights in tidal and inland navigable waterways. The views expressed are their own, and do not necessarily reflect the views of the California Attorney General's Office.

after their initial establishment. Property boundaries along waterbodies, particularly on the open coast, present a special set of problems. These include not only the difficulty of identifying the seaward boundary along the dynamic ocean shoreline, but the framing of principles to govern the competing interests of coastal landowners and the public in using shorelands, independent of property ownership. This paper explores these problems, identifies the ways in which the courts have addressed ownership issues and other rights in coastal properties, and discusses possible resolutions of the issues presented.

The Nature of the Problem

Most real property is landlocked, with readily definable boundaries. These boundaries have been located by surveys of generally unquestioned accuracy and the boundaries remain established as the parcel of property is conveyed from one owner to the next. Even when disputes arise as to the technical location of boundaries, the boundaries may be recovered using the metes and bounds description of the original survey. Of course, even in inland situations, disputes over ownership may sometimes arise. For example, an adjoining owner may claim that boundaries have been reestablished by a fence that serves as an "agreed boundary," or a person may claim a right in another's property based on adverse possession. In general, however, ownership rights are generally well-defined and landowners may use the property within their boundary lines to the exclusion of others.

On the open coast, however, boundaries are far less certain. Where the land meets the sea, the water itself is considered a natural monument that defines the boundary. See, e.g, Cal. Civ. Code, § 830 (boundary is the "ordinary high-water mark"). Although that concept sounds simple enough, the dynamic physical regime in the area where ocean waters intersect the shore profile makes the actual location of an ocean boundary a much more complicated proposition. Variables abound. The level of the sea itself rises and falls with the tide; a typical tidal elevation range over the course of a day may be four to five feet or more. Further, the ocean is not still water. Winds, swells, and storms supplement tide height as a determinant of how far inland waters from breaking waves extend. Further still, on a sandy beach, the configuration of the land itself changes as changing wave energy causes the seasonal erosion and accretion of the shore profile, further altering the point of intersection of land and water. Finally, beach slope, which can vary over time for a given beach, is also a factor; the flatter the slope, the farther inland the water extends. These variables in combination result in a changing intersection of land and water. In addition to all of these factors, there are long-term changes such as sea level rise and diminished sand supply. These, too, affect the boundary.

Thus, the complicated physical relationship of land and sea by itself presents obstacles to the creation of stable and permanent boundaries separating private uplands and public tidelands. The technical problems associated with using ocean waters to mark an identifiable boundary, however, are just the beginning of the problem. Competing legal principles produce further uncertainty. The ocean is a resource of unique importance, and all legal systems have recognized that members of the public generally share common rights to its use. The public makes use of the ocean and the shorezone for a variety of commercial, recreational and environmental purposes.[1] On the upland, in contrast, the law recognizes rights of exclusive use in the private owner. The upland owner has a special interest both in protecting this exclusive use and in maintaining proximity to the ocean. As a result, the intersection of land and sea results in a collision not just of physical forces but of different legal regimes whose competing principles must be reconciled in the course of sorting out public and private rights in the shorezone. Given the dynamic physical regime and the importance attached to public and private rights in coastal shorelands, it is not surprising that courts have struggled to accommodate the competing public and private interests in this area. The problem is far from trivial. In California, for instance, sandy beaches constitute over half of the 1,072 miles of ocean shoreline, and over half of those beaches border privately-owned uplands.

In fashioning rules of law, courts have generally seemed aware of these differences between the stable interior and the more dynamic world of the coastline, but they sometimes have expressed a yearning for the simpler rules that obtain in everyday property ownership where no waterbodies are involved. Some courts, to borrow a phrase from Professor Mount, Chairman of the Geology Department at the University of California at Davis, have seemed "obsessed with a static view of the Earth." This latter view is expressed in a quote from *People* v. *William Kent Estate Co.*, 242 Cal.App.2d 156, 51 Cal.Rptr. 215 (1966), a California court of appeal decision, concerning movements of the mean high tide line on a sandy beach:

> "We recognize that the 'almost mathematical line bounding Blackacre' . . . cannot be achieved, but somewhat greater certainty should be possible than the constantly moving line contemplated by the present decree." *Id.* at p. 161.

Unfortunately, such certainty is exceedingly elusive on the ocean shore, and probably not obtainable at all, at least to the degree that we have come to expect from boundaries in the interior, away from the meeting of land and water. In the *Kent Estate* case, the Court wondered if seasonal changes in beach width might be "constant, in offsetting pairs occurring annually" and whether beach movements might be "substantially the same distance each year, thus affording a basis for fixing an average, mean, or ordinary line of the shore against which the average plane of the water at high tide may be placed to determine a reason-

---

1. Coastal tide and submerged lands, as are other lands underlying navigable waters, are subject to a public trust for a variety of public uses including commerce, fishing, navigation, recreation, environmental study and open space. E.g., *Marks* v. *Whitney*, 6 Cal.3d 251, 257, 491 P.2d 374 (1971); *National Audubon Society* v. *Superior Court*, 33 Cal.3d 419, 435, 658 P.2d 709 (1983). Although bed ownership may reside with the government, use of the waters is shared in common by all members of the public.

ably definite boundary line." (*Id.* at pp. 160–161.) On retrial in that case, and again more recently in a beach boundary case in southern California, the evidence established that no such neatly predictable average line existed in nature.

Perhaps more to the point, even if such an average line could be identified, there would only be rare occasions during the year when the actual physical condition of the shoreline precisely conformed to the position of the "average line." During most of the year, the shoreline would be in a different location, in response to constantly changing physical conditions. This means that during much of the year land under water would be in private ownership—as the actual mean high tide line crept landward of the average line—thereby precluding the public from making use of the tidelands along the actual water's edge for recreational and other uses. At other times of the year, as the beach built out due to seasonal accretion, dry sandy beach that otherwise would belong to the upland owner would instead be owned by the public. In other words, the law can attempt to impose an average fixed boundary line, but the physical forces of nature are not bound to obey it. Such arbitrary lines will not satisfactorily resolve the competing needs of private and public interests over the use of this special resource.

The import of this paper is that, through one means or another, through some combination of legal principles—a combination that often varies from state to state—our courts have or will evolve a body of law that will deal effectively and fairly with the special circumstances that exist along the seashore. And the courts that best understand and accept the variable, unpredictable and powerful nature of the natural forces at work will be the most successful in articulating reasonable rules of law to govern public and private rights in this special area where more traditional rules must be bent or altered to fit very different physical circumstances.

A discussion of how courts have struggled to accommodate these competing interests involves both issues of land ownership and issues of use of the shore, independent of ownership of the underlying land.

Differing Locations of the Shoreline Boundary

Recognizing from the outset the special value of the ocean shoreline, legal systems other than the common law have generally defined the littoral boundary in a way that puts more emphasis on a commonality of shared interests than on exclusive private rights. In Roman law, the principle was that "the sea-shore extends as far as the greatest winter flood runs up." Sandars, *The Institutes of Justinian* (4th ed. 1867) 2.1.3. Similarly, under the French Civil Code (the Code Napoleon):

> "Ways, roads and streets kept up by the State, rivers and streams floatable and navigable, *beaches*, alluvion and derelictions of the sea, ports, harbors, anchorages, and generally all parts of French territory which cannot become private property, are considered as forming part of the Public Domain." (Italics added.) Art. 538, 1804 French Civil Code, from Cachard, *The*

French Civil Code (rev. ed. 1930), p. 173.

And under French law, the coast extends up to the *highest* regular tide line. Aubry & Rau, *Droit Civil Francois* (7th ed. 1961) (1966 translation by Mayda & Esmein), p. 51, n.16. Likewise, under Mexican law, those parts of the land "covered and uncovered by the water, up to the limits of the greatest ebbtide (reflujo)," plus the "terrestrial maritime zone," a strip of land 20 meters farther inland, are considered part of the public domain for the common use of the public. Wheless, *Compendium of the Laws of Mexico* (2d ed. 1938), p. 699 (Art. 1005, Lands of Public Domain for Common Use).

This expansive view of the inland extent of public ownership and rights is shared by those states whose law can be traced to sources other than the common law. Louisiana describes the upper boundary of its sovereign-owned tideland as being where "the waters of the sea spread in the highest tide during the winter season." La. Civ. Code Ann., art. 451. This description parallels that in *The Institutes of Justinian*, where it is stated that the boundary is "as far as the greatest wave extended itself in the winter." *The Institutes of Justinian*, Lib. 2, title 1, § 3, as reprinted in Angell, *A Treatise on Tide Waters* (1826), p. 65. In Hawaii, the boundary is the upper reach of the wash of the waves, *Application of Sanborn,* 562 P.2d 771 (Haw. 1977); *In re Ashford*, 440 P.2d 76, 77 (Haw. 1968), and the waters presumptively extend to the vegetation line. (*County of Hawaii* v. *Sotomura*, 517 P.2d 57 (Haw. 1973).) Texas has exhibited a split personality, reflecting both its Roman/civil law origins and its later decision to adopt the common law. Pre-1840 grants that are governed by the early precedents go to the highest tidal intrusion of ocean waters (now interpreted in Texas as the mean higher high water line). *City of Galveston* v. *Menard*, 23 Tex. 349, 399 (1859); *Heard* v. *Town of Refugio*, 103 S.W. 2d 728, 732–733 (Tex. 1977). After the adoption of the common law by Texas in 1840, subsequent grants are bounded by the ordinary high water mark, construed to be the mean high water line. See *Matcha* v. *Mattox*, 711 S.W.2d 95 (Tex.App. 1986).

Under the common law and in states such as California, the boundary on tide water is the "ordinary high-water mark." E.g., Cal. Civ. Code, §§ 670, 830; Cal. Code Civ. Proc., § 2077. Prior to developing systematic measurements of tide heights, the common law used a *mark*, such as a debris line or a vegetation line. Following the decision of the United States Supreme Court in *Borax Consolidated, Ltd.* v. *City of Los Angeles*, 296 U.S. 10, 26–27 (1935), however, the federal courts and most state courts including California have equated the ordinary high water mark with the mean high tide line. E.g., *Kruse* v. *Grukap*, 349 So.2d 788, 789 (Fla. 1977).

The mean high tide line is the intersection of the elevation of mean high tide and the shore. *Miller & Lux* v. *Secara*, 193 Cal. 755, 765, 227 P. 171 (1924). The elevation of mean high tide is obtained by averaging all the twice-daily high tides at a particular location over a 19-year "tidal epoch." *Borax Consolidated, Ltd.* v. *City of Los Angeles* (1935) 296 U.S. 10, 26–27. Although the *elevation* of mean high tide is a fixed plane, the mean high tide *line* is subject to change as the shore itself changes through the deposition and erosion of material.

Consequently, the mean high tide line is an "ambulatory" boundary that moves as the shore profile changes through accretion and erosion. *City of Oakland* v. *Buteau,* 180 Cal. 83, 87–89, 179 P. 170 (1919).

Although the mean high tide line can be measured with precision at any point in time, its use as a boundary line presents a number of practical problems. The mean high water component of the mean high tide line is a plane representing the still-water average elevation of all high tides over a 19-year period. Because this plane reflects only an *average* elevation of the high tides, the height of the tide in a still-water condition will often be above the plane of mean high tide; approximately 14 percent of the time, the tide height will exceed the height of this plane. Nor do measured tide heights reflect increases in water height due to phenomena such as swells and wave setup, because devices for measuring tide height dampen the effect of such influences. Finally, the waters of the ocean are of course not still; the water is constantly moving through wave runup or backrush as the waves crash upon the beach. This means that there will be only brief periods each day when the actual elevation of the water coincides with the mean high water elevation.

Moreover, the shoreline profile component of the mean high tide line is itself subject to change through the ongoing processes of accretion and erosion, both seasonal and long-term. These changes in the shore profile cause the mean high tide line to shift, so that the line may move horizontally up to 100 feet or more as the beach profile changes over the course of the year.

As a result, an approach that focuses only on boundary does not provide a complete answer to the issues raised when competing public and private uses meet in the shorezone. Although it can be precisely measured, the mean high tide line at any one instant in time may bear only a tenuous relationship to the actual intersection of land and sea. This means that at any one moment in time, it is possible that the mean high tide line—the boundary line—may be under water, seaward of the water's edge—the place where most recreational activity takes place.

That the mean high tide line cannot be visually observed but must be precisely surveyed tends to compound this problem. Short of performing daily mean high tide line surveys, neither the public nor the upland owner knows at any particular time where the boundary line is and where, accordingly, the respective ownership rights apply. More importantly, although the mean high tide line is capable of precise location at any one moment in time, its location rarely reflects the actual point of intersection of ocean waters with the shore. Consequently, although it may serve as a boundary for ownership purposes, it has limited utility as a reasonable point of division between public and private rights.

Public Rights Independent of Public Ownership

A legal regime that focuses solely on boundary, particularly a mean high tide line boundary, ignores a significant portion of the inland extent of the resource. Just as in the words of one legal commentator, "the fickle Missouri refused to be bound by [a] Supreme Court decree," *Nebraska* v. *Iowa*, 406 U.S. 117, 119

(1972), the ocean refuses to be bound by legal notions of "boundary" transplanted from the interior. As a result, various courts and legal systems have addressed the existence and extent of rights in the area inland of the formal coastal boundary line. Public rights to use shorelands have been recognized even though at any particular moment in time these rights may extend landward of the public-private boundary. Such rights have been variously rationalized and legitimized.

Some of these recognitions of public rights in private property may be characterized as "secondary easements"—rights that are necessary to fully and effectively enjoy the adjacent public resource. For example, concerning privately-owned river banks, Roman law provided that:

> "The public use of the banks of a river is part of the law of nations, just as is that of the river itself. All persons therefore are as much at liberty to bring their vessels to the bank, to fasten ropes to the trees growing there, and to place any part of their cargo there, as to navigate the river itself. But the banks of a river are the property of those whose land they adjoin; and consequently the trees growing on them are also the property of the same persons." Sandars, *The Institutes of Justinian*, *supra*, 2.1.4.

Concerning ocean beaches, Roman law further provided that "any person whose use of the sea or beach would be impaired by any construction or building could move for an interdict . . . to prevent such nuisance in public places." Hunter, *Roman Law* (4th ed. 1903, translations by J. Cross), p. 310, citing *Digest* of *Corpus Juris Civilis*, 43, 8, 2, 8; see also *Lux* v. *Haggin*, 69 Cal. 255, 320–21, 10 P. 674 (1886) (indicating that under Roman law the right to navigate rivers carried with it the right to moor ships to the banks).

Similar recognition of public rights prevailed under Mexican law, which derived from Las Siete Partidas, a compendium of Spanish law produced in the thirteenth century during the reign of King Alphonso the Wise. "The laws of the Partidas place the playa amongst the common things which all men can use . . . ." *Lux, supra*, 69 Cal. 255, 316. Mexican law specifically provided that:

> "All persons may, within the law and police regulations, make use of the shores and of the maritime-terrestrial zone [extending 20 meters farther inland], for passage, bathing, spreading clothing and nets, building and repairing vessels, tending cattle, gathering shells, plants and sea-food; also of the banks of rivers, lakes and lagoons and of the fluvial zone, for the needs of navigation, fishing, irrigation, for water for motive power, drinking purposes, and other uses of public interest . . . ." *Wheless, supra*, at pp. 699–700.

Individual states have also developed rules allowing public use of private property adjacent to public waters. Montana and Idaho, for instance, have recognized a number of such rights including the right to use private land to portage around obstacles. *Montana Coalition for Stream Access, Inc.* v. *Curran*, 682 P.2d 163, 172 (Mont. 1984); *Southern Idaho Fish & Game Ass'n* v. *Picabo Livestock Co.*, 528 P.2d 1295, 1297 (Idaho 1974). New Jersey recognizes access to the foreshore across the dry sand area, even if it is in private ownership, and also recognizes use of the dry sand area above the mean high tide line boundary for recreational uses incident to the use of the foreshore and ocean waters. *Matthews* v. *Bay Head Improvement Ass'n*, 471 A.2d 355, 365 (N.J. 1984). Pursuant to the doctrine of "custom," other states have recognized public use of the beach above the mean high tide line inland to the line of vegetation, irrespective of ownership of the underlying land. *Matcha* v. *Mattox*, 711 S.W. 2d 95 (Tex.App. 1986); *State* ex rel. *Thornton* v. *Hay*, 462 P.2d 671 (Or. 1969); see also *State Highway Commission* v. *Fultz,* 491 P.2d 1171 (Or. 1971).

In numerous states, the public has the right to use waters for recreational purposes, even where the water body is not navigable under the federal test for navigability and therefore the underlying bed is in private ownership. See, e.g., *People* ex rel. *Baker* v. *Mack*, 19 Cal.App.3d 1040, 1045–46, 97 Cal.Rptr. 448 (1971); *Hitchings* v. *Del Rio Woods Rec. & Park Dist.*, 55 Cal.App.3d 560, 566, 127 Cal.Rptr. 830 (1976); *Day* v. *Armstrong*, 362 P.2d 137, 143–147 (Wyo. 1961); *Montana Coalition for Stream Access, Inc.* v. *Hildreth*, 684 P.2d 1088, 1091 (Mont. 1984); Elder v. *Delcour*, 269 S.W.2d 17, 23–26 (Mo. 1954); *Lamprey* v. *State*, 53 N.W. 1139 (Minn. 1893); *Curry* v. *Hill*, 460 P.2d 933, 935–36 (Okla. 1969). And a number of state courts have grounded the public right of use on the "public" nature of the state's waters, choosing not to make the public right dependent on one or another definition of "navigability." In the words of the Wyoming Supreme Court, "[T]he exclusive control of waters is vested in the state, whether the waters are deemed navigable in the Federal sense or in any other sense." *Day*, *supra*, 362 P.2d at 143; accord, *Hitchings*, *supra*, 55 Cal.App.3d at 567. Moreover, the public's right to make recreational use of navigable waters, regardless of bed ownership, is not limited to navigation alone but includes uses that are simply "recreational" in nature. See, e.g., *National Audubon Society* v. *Superior Court*, 33 Cal.3d 419, 434, 658 P.2d 709 (1983); *Mack*, *supra*, 19 Cal.App.3d at pp. 1045–46; *Day, supra*, 362 P.2d at 147; *Montana Coalition for Stream Access, Inc.* v. *Curran*, 682 P.2d 163, 170–71 (Mont. 1984); *Luscher* v. *Reynolds*, 56 P.2d 1158 (Or. 1963).

The potential for conflict between private upland uses and public rights to use shoreline resources is acute where private owners seek to develop shoreline property that lies at the fluctuating intersection of land and water. The Supreme Court of Washington has addressed a related issue in a situation where a private property owner sought to develop a portion of a lake shore that was subject to periodic inundation as the level of the lake rose and fell at different times during the year. *Wilbour* v. *Gallagher*, 462 P.2d 232 (Wash. 1969). The court prohibited development within the zone of inundation because "the public has the right to go where the navigable waters go, even though the navigable waters lie over privately-owned lands." *Id.* at p. 238. Although it protected the public's right to be free from the impacts of development that would obstruct legal public use, the Washington court nevertheless gave effect to private rights during those periods when the lake receded and privately-owned dry land was exposed: "When the level of the lake is lowered so that the defendants' land is no longer submerged, then they are entitled to keep trespassers off their land, and may do with the land as they wish consistent with the right of navigation when it is submerged." *Id.* at p. 239.

Conflicts over the use of shoreline areas are likely to increase in the foreseeable future. Public need for the use of the shoreline is likely to increase as the population grows. The desire of upland landowners to use the shoreline for strictly private purposes such as residential housing may increase as the value of private property increases and as technology allows the construction of structures within areas that in other times might have been thought too perilous or costly to build upon. Given the relative absence of these conflicts in the historical past, it is not entirely surprising that legal rules applicable to the shorezone which define the extent of public ownership, and of public rights independent of ownership, have not yet been fully developed. There are still some gaps.

In California, for example, there is a well-defined body of law that establishes the mean high tide line as the legal boundary separating private uplands and public tidelands; defines sovereign interests in tidelands as including a wide variety of public trust uses, including recreation; and, as to inland navigable waters, establishes that the public has a right to use the waters for navigation and other recreational purposes regardless of the ownership of the underlying bed. The California appellate courts, however, have yet to directly consider whether the public may have additional rights in the ocean shorezone under the doctrine of custom or, by extension of the principle established on inland waters, a right to use the navigable waters of the ocean wherever they may reach, independent of underlying ownership. As conflicts increase, courts may be asked to define and resolve these issues, and the California courts can be expected to borrow from the collective wisdom of other courts, whose law is ultimately based on the same sources from which California law derives.

Possible Legislative Resolutions

Leaving the development of these principles to future development by the courts can be unsatisfying because of the slow, case-by-case development of court-made law. Legislative solutions are an alternative to leaving resolution of these issues to ad hoc judicial rulings. Legislative solutions may be useful in either of the two contexts where conflict between private and public use of the shorezone is most acute: (1) in the context of public use and access, where the question arises whether the public's use of a particular area of the shoreline conflicts with private ownership of the underlying land; and (2) in the context of development, where a proposal to construct permanent improvements within the zone periodically subjected to inundation by ocean waters, if permitted, might impair public rights to use the shore or ocean waters.

With regard to the development issue, for example, the legislature might enact a bill requiring that any development proposed along the shoreline be set back an appropriate distance (e.g., 25 feet) from the most landward surveyed location of the mean high tide line. To insure that the setback properly accounts for the most landward movements of the mean high tide line, the statute might require that the applicant finance a number of mean high tide line surveys, to be performed at the discretion of the State within a one-year period following the application. The most landward of these surveys then could be used as a base line for establishing the development setback line. A more ambitious legislative initiative would be establishment of a comprehensive statewide program for identifying and mapping the extent of public rights. See, e.g., Florida Coastal Mapping Act, Fla. Stats. 177.25.

With regard to access, the legislature might provide that the public has a right to use navigable and tidal waters wherever they are located. Or it could regulate access along the shore by establishing a rebuttable presumption that the public has a right to use the wet sand areas of the beach. And it might provide that a person seeking to establish that a member of the public was trespassing on private property could rebut this presumption only by showing that on the day in question the alleged trespasser was above a properly surveyed mean high tide line and intended to trespass on private property. See *State* v. *Ibbison,* 448 A.2d 728, 732 (R.I. 1982).

One risk of seeking to clarify or establish public rights through legislation is the possibility that landowners might claim that the legislative action divests them of vested property rights, thereby effecting an unconstitutional taking of property. Such a claim, however, would be difficult to establish. In order to establish a taking, it is necessary to show some interference with established private property rights. *Lucas* v. *South Carolina Coastal Council,* 505 U.S. 1003 (1992). None of the propositions suggested above, however, divests the upland owner of any clearly-established property rights landward of the mean high tide line boundary. Nor do any of the proposals unconstitutionally impair the ability of landowners to use their property. Persons owning property adjacent to navigable waters, especially along the ocean, have far more limited property rights than persons owning non-littoral property. Although the littoral owner has a general expectation of maintaining access to the water and enjoying other advantages of contiguity, 5A *Powell on Real Property* (1995) § 66.01[3], p. 66–13, that expectation is subject to numerous limitations. Persons acquiring coastal property know or should know that the boundary is the mean high tide line, a boundary that by its very nature is subject to change and that is uncertain in location in the absence of a contemporary survey. It has been recognized that the precise location of the shoreline is one of the hazards incident to littoral or riparian ownership. *County of Hawaii* v. *Sotomura,* 517 P.2d 57 (Haw. 1973). Title reports and title insurance policies for littoral property routinely contain exceptions for public rights in navigable waters and acknowledge that property ownership is subject to exceptions for claims of public ownership.

The expectations of littoral owners on coastal property are subject to other uncertainties and exceptions. In California, for example, the state can cut off littoral owners from access to the water without compensation. See *Henry Dalton & Sons Co.* v. *Oakland*, 168 Cal. 463, 465, 143 P. 721 (1914); *People* v. *Hecker*, 179 Cal.App.2d 823, 840–841, 4 Cal.Rptr. 334 (1960). And California's doctrine of artificial accretion—by which the State is entitled to ownership of accretions on state property directly caused by human activities in the vicinity, *State of Cal.* ex rel. *State Lands Com.* v. *Superior Court*, 11 Cal.4th 50, 79–80, 900 P.2d 648 (1995)—can likewise result in the termination of a littoral owner's right to contiguity with the ocean waters. That the waters of the ocean are encumbered with the public trust also serves to diminish any expectation of unfettered use of adjoining property. See *National Audubon Society* v. *Superior Court*, 33 Cal.3d 419, 434–435, 658 P.2d 709 (1983).

Further, since 1879, a provision of the California Constitution has acted to limit and qualify any expectation that upland owners are free to use their property without consideration of coexisting public rights in waterfront property. Article X, section 4 of the California Constitution provides:

> "No individual, partnership, or corporation, claiming or possessing the frontage or tidal lands of a harbor, bay, inlet, estuary, or other navigable water in this State, shall be permitted to exclude the right-of-way to such water whenever it is required for any public purpose, nor to destroy or obstruct the free navigation of such water; and the Legislature shall enact such laws as will give the most liberal construction to this provision, so that access to the navigable waters of this State shall be always obtainable for the people thereof."

Finally, added to all of these qualifications to a coastal property owner's expectations is the long history of public ownership, secondary easements, and other public rights that have been historically impressed on coastal shore land and coastal waters, as well as inland waters, as described in this paper. In short, ownership of coastal property is a highly qualified form of ownership, and any person owning such property must reasonably expect that ownership of that portion of the property that fronts upon the ocean will be subject to legal principles that protect public rights to the use of coastal waters and that will differ significantly from the principles that apply to inland property ownership. Legislative solutions that are consistent with long-established limitations on private property rights in the coastal shorezone should be immune from claims that they effect an unconstitutional taking of property.

Conclusion

Legal concepts concerning boundaries and property rights that have been developed to govern real property in the stable interior—where parcels are bounded on all sides by fast land—have limited application in the unique coastal realm of rising and falling tides, pounding surf, and vagrant, shifting sands. Throughout western civilization, legal regimes have recognized the inherent importance of paramount public rights to use coastal waters both for recreation and commerce. Conflicts between public and private rights are likely to increase over time in the fluctuating zone of periodic inundation where the land meets the sea. Courts and legislatures will need to respond to these conflicts in a manner that protects private property rights, but also recognizes that private property rights in coastal lands have long been qualified by the public right to make use not only of publicly-owned tidelands but also of public waters themselves, whose inland reach does not conveniently conform to the static concepts of real property boundary that apply to landlocked parcels.

CHAPTER 154

FROM HERE TO ETERNITY - WERE THEY TRESPASSING?

Curtis L. Fossum*

This paper seeks to expose the irrational and inconsistent state of the law relating to boundaries separating public and private rights along California's coastal shoreline. The premise I wish to put forward is not unique or new,[1] it is that the law of water boundaries in California and in much of the United States is in a state of disarray.[2] While the dilemma and confusion are not new, I wish to focus the criticism on what I believe is the primary roadblock to resolving the location and extent of public/private rights along open coastal shorelines. The roadblock, or wrong turn in the road, that is perpetuating an illogical aberration of Common Law principles is the erroneous and irrational decision in 1935 by the United States Supreme Court entitled *Borax Consolidated v City of Los Angeles*, 296 U.S. 10. In fact, several lines of cases in California are incompatible with the historical legal concepts that reflect natural law and the perceptions and expectations of reasonable people — an expectation that the Common Law should fulfill. Specifically, I will address the ill-conceived and unsupportable notion that the Common Law legal terms for a boundary between land and a waterway — the **High Water Mark** (**HWM**) or **Ordinary High Water Mark** (**OHWM**) and the principles associated with them — can in any fashion be equated with the so-called **Mean High Tide Line** (**MHTL**) described in the *Borax* case, which if it is a **Line** at all, is only so on a statistician's graph.

The paper's title was chosen to create a visual image in the readers' minds. The famous scene, from the film which won 8 Academy Awards (including the Oscar for Best Picture of 1953), serves to graphically portray what I believe is the incongruity between common expectations of public rights on the beach and the result of the *Borax* decision. The scene with Burt Lancaster and Deborah Kerr, lying on the sandy beach along Oahu's west shore, being soaked by the waters of the Pacific Ocean alternately flowing in and ebbing back from the shore, is considered to be among the most memorable scenes in

* Senior Staff Counsel
California State Lands Commission
100 Howe Ave., Suite 100 South
Sacramento, CA 95825

Phone - (916) 574-1828
Fax - (916) 574-1855
e-mail - fossumc@ca.ca.gov

The opinions expressed are those of the author and do not necessarily represent the position of the California State Lands Commission.

motion picture history. When one visualizes this couple being washed by the sea water, it is difficult to conceive of them as not having been lying in the ocean below the **OHWM**, the legal definition of the boundary along California's shoreline at least as early as its codification in the Civil Code in 1872[3] and arguably since adoption of the Common Law at statehood in 1850. Yet, a plausible argument can be asserted that, had they been in California rather than Hawaii[4] as they lay in the water, they were not really on public lands or in the ocean, but above the **MHTL**. This so-called "line" is in reality an amorphous and indistinguishable location on the beach,[5] more precisely defined as the theoretical intersection of the elevation of mean high tide and land (*terra firma* — or more properly on a sandy beach: *terra infirma*?). Blackstone in his *Commentaries* says that accretions which he calls "lands gained from the sea" or "alluvium" caused "by the washing up of sand and earth, so as in time to make *terra firma*" by small and imperceptible degrees, belong to the adjoining owner.[6]

The *Borax* decision involved the location of the boundary of Mormon Island, a small island in San Pedro Inner Bay. It was surveyed in 1880 as 18.88 acres, but thereafter was challenged as being as little as one acre. The issues included whether the survey fixed the boundary or could other facts be presented in court to prove where the **OHWM** was located. More details of the case will be discussed later; however, the impact of the US Supreme Court decision was that it adopted the theoretical intersection of the arithmetic average of the elevation of mean high water and land, without citation to authority (other than a recent USC&GS pamphlet)[7] as the boundary between State-owned sovereign public trust lands and privately owned uplands. To locate this "line" on the beach requires measuring, then averaging, 135,788 sequential high tide elevations (over 18.6 years),[8] followed by a controlled field survey from known elevation benchmarks to determine the location of that average elevation and hence the so-called "line" — at least its location during the time of the survey.[9] Should someone resurvey the elevation a year, month, week, day, or even a few hours later, they are likely to find someone or something has moved the line! This mystery is simply the law of nature showing a total lack of deference to the laws of man. The *Borax* court seems to assume a static and non-dynamic interface of land and water and a logical and locatable boundary between private upland and public waters. However, the ocean and the sands on a beach are in a constant state of flux. The other enigma is that to survey the elevation of mean high tide on a coastal beach the survey party must time the survey of the beach when the tide is at a minus tide, preferably three feet or more below mean high (see Photo # 1). That is because the ocean does not lie in a stable repose and simply rise as the tide comes in. Instead, the dynamic ocean, unlike inland estuaries and bays protected from the ocean, experiences breaking waves as they approach the shore (see Photo #2). This real-life anomaly resulted from the court equating the Common Law **HWM** with the then (1935) relatively recently obtained mathematical average of *offshore* measured tidal elevations. A dose of common sense seems needed here — laws should reflect broadly held and longstanding human perceptions of what is by its very nature the public's to use. The call to a *natural monument* (e.g. a tree, rock, river, bay or ocean) in deeds was for purposes of simplicity and easy identification by the average man; such simplicity would normally not require 19 years of scientific measurements and analysis followed by an, at least, second order level survey.

PHOTO # 1

Surveyor is standing on MHTLwhile projected tide is still 1 foot below Mean High Water.

PHOTO # 2

Wave wash leaves visible signs on shoreline. Mean High Tide Elevation indistinguishable and under water.

How did the current situation arise? Legal historians trace the roots of the Common Law from the ancient Roman Law of the 2nd Century C.E. and the codified Institutes of Emperor Justinian in the 6th Century,[10] then to the Magna Charta (1215)[11] and Bracton in 13th Century England[12] to Lord Hale in the 17th Century.[13] Under the ancient Roman Law and Civil Laws of Europe and America, the separation of private upland from public waters is the location of the reach of the highest tide in winter — the beaches were public. So, for nearly two millennia, Roman Law has passed on to the majority of European and American nations the public use and ownership of its beaches. And although California was a Spanish colony and Mexican territory, the courts in this country have, to date, not upheld the rights of California's citizens to public rights on the beaches of this state, even those lands subject to the Treaty of Guadalupe Hidalgo.[14]

Unlike the ancient Roman Law and Civil Law of Europe, the Common Law of England, as related by the reporters, at times seems as changeable and transitory as a sandy beach. And frankly, given the inconsistency of judicial decisions reported even today, it should not be a surprise that cases from hundreds of years ago sometimes were vague, ambiguous and contradictory. It does appear, however — if one gives weight to *De Jure Maris* attributed to Lord Hale — that at least by the 17th Century the English courts normally described the natural boundary as the **HWM** or **OHWM**[15] (ordinary being occasionally added to clarify that the extraordinary levels of floods on rivers or the heights that storm waters reached annually were not intended — as in the Civil Law countries). Although this terminology may seem somewhat ambiguous to those whose desire is to reduce it to statistical or scientific analysis and measurement, the term does reflect clear and distinct Common Law concepts. The **mark** left on the ground by **high water** was the natural boundary — something the common man could judge if not the common law judge.

The oft-cited 19th Century English decisions of *Blundell v Catterall*[16] (1821) and *Attorney General v Chambers*[17] (1854) are themselves internally inconsistent and they are cited for justification as to several decisions with likewise fallacious reasoning and results.[18] Furthermore, the attempts to decipher Lord Hale's 17th Century pronouncements by American courts has only further muddied the **"tidal"** waters.[19] Both the English and American cases and their wide range of terminology and explanations and descriptions of the legal boundary are too numerous to describe in this short paper. However, since the *Chambers* case is supposed to be strong support for the location of the boundary along the shore, I would like to include two quotes (the emphasis added is mine). The first quote is from Lord Chancellor Cranworth:

> "The principle which gives the shore to the Crown is that it is land not capable of ordinary cultivation or occupation, and so in the nature of unappropriated soil. Lord Hale gives, as his reason for thinking **that lands only covered by the high spring tides do not belong to the Crown, that such lands are for the most part dry and maniorable**; and taking this passage as the only authority at all capable of guiding us, the reasonable conclusion is that the Crown's right is limited to land which is for the most part, not dry or maniorable."[20]

And this quote from Baron Alderson discussing *Blundell v Catterall*:

> "Mr. Justice Holroyd, no mean authority, in his very elaborate judgment in the case of Blundell v Catterall, 5 B. & Ald. 268, 290, mentions this as one of the instances in which the Common Law differs from the Civil Law, and says that it is clear that according to our law it is not the limit of the highest tides of the year, but **the limit reached by the highest ordinary tides of the sea, which is the limit of the shore.**"

Note there is no mention of mean, average, elevation, or line, but the limit (mark) reached by the highest ordinary (not extraordinary) tides. It appears that later courts have misinterpreted this to be a measurement of the elevation of tides rather than the location of the limits reached by the sea on the shore. This is the crux of the problem.

CALIFORNIA — By statutory and judicial decrees and adoption of the Civil Code in California in 1872, the **OHWM** has been defined as the location of the landward extent of sovereign public trust lands in California, whether the location is at Lake Tahoe, the Colorado River, Tiburon, or Malibu. The 19th Century legislatures in California created considerable confusion in their dealings with public trust lands by adopting programs of giving away the fee title to lands between **OHWM** and Ordinary Low Water Mark on non-tidal waterways[21] and selling off lands designated as swamp and overflowed, salt marsh, and tidelands (generally at one dollar an acre, which often was credited rather than paid). The early judicial decrees interpreted the sale of these lands as applying only to inland waters that were generally non-navigable and reasonably capable of reclamation and agricultural development. The California Courts on the other hand, as early as 1864,[22] struck down the attempted sale of coastal tidelands which were unfit for reclamation and agriculture. In inland areas affected by the tides, a significant dilemma developed in trying to locate any discernable mark left by the **OHWM** or by any particular tide elevation.[23] Bays and estuaries being rich with nutrients and unaffected by the waves, surf, and wash of the ocean supported substantial vegetation in the waters. Various species of salt-tolerant grasses and other plants thrived in the muddy muck of the Bay of San Francisco and Inner Bay of San Pedro. This caused problematic application of the Common Law test. The maps drawn by the Coast Survey and later the US Coast and Geodetic Survey for the open coast, whether they be hydrologic maps for mariners or topographic maps for terrestrial use, depicted a **HWM** in the true Common Law sense.[24] The topographic surveyors were to locate high water or the shoreline **by visual observation not measurement of the elevation of the tides** and depict on their maps the **mark left by the ocean waters**[25] (i.e., Ordinary High Water Mark), excluding extraordinary events. Typically they used a debris line, and drew a heavy **high water** line on the map.[26] The **hydro maps**, drawn by those who measured tidal elevations, used the topo surveyor's line — **and not tidal data**.[27] Because vegetation grows both below and above mean high tide and ordinary high water, surveys involving bays and estuaries, where marsh vegetation was depicted, have led to long drawn-out and costly analysis of what is lawfully upland and what are public trust lands. This is often due to manmade changes that have occurred since the early mapping and the inability to find the natural location of the mean high tide line in a marsh. The only reason I bring up the bay and estuary issue — since I am focusing on open coast boundaries — is that the cases, such as *Borax*, which have set in motion this irrational application of water gauge level measurements, did not involve open coast areas, but rather inland waters protected from the seas. As to the logic or reasonableness of water gauge measurements in bays or estuaries I offer no opinion; however, as to open coast beaches their utility is overstated and misapplied to the extreme.

The difficulty in locating the mark left by ordinary high water in bays and the disarray of guidance by court decisions led to a hope that science and available data could *divine* a test that equated to the Common Law precept. Unfortunately, science had *not* been seeking to answer this inquiry. However, measurements of tides, even though taken for other purposes, and the use of statistics and available data (regardless of the reason gathered) gave the courts a seemingly appropriate and easy scientific solution. Surveyors and engineers had measured certain ocean phenomena and from them developed very useful information. The problem arose with the attempt to apply a statistical average to such a dynamic setting as a coastal beach and shoreline. The courts clearly understood less than the engineers and surveyors — and, after all, the **MHTL** *sounded like* **OHWM**, *sort of.*

The problem is that statistical averaging of scientific data is only that. The rationale for measuring the tides was **not** to determine the *onshore* boundary but to help mariners understand *offshore* depths and tide heights in order to prevent running aground. It retains its importance to navigation in certain shallow waterways and for other scientific use, but as a means of determining an elevation for a boundary on the shores of the open coast it is misleading and woefully inappropriate.

Rear Admiral Patton, head of the USC&GS from 1929 to 1937, discussing property boundaries and the tides, said it very well:

> "Statistical data, however, serve the ends of justice only to the extent that they are properly interpreted, and proper interpretation of tidal data is usually possible only when the tabulation is appraised against a background of pertinent scientific principles and physical facts pertaining to the situation under consideration."[28]

The Admiral goes on to state it is impossible to be certain of what Justice Field intended the rule to be in the 1861 California Supreme Court case of *Teschamacher v Thompson* (this case is the genesis of the "neap tide" line of cases in California). He then speaks of the custom of judicial precedent, saying, "It probably has some tendency to extend the application of imperfect decisions to subsequent cases" and a "similar tendency" to where the underlying body of facts "superficially seem identical, yet with a searching study of both cases would reveal as having important points of difference if only those concerned knew where to look for them."[29] These statements were made while the *Borax* case was pending.

Rear Admiral Patton's words were prophetic. Not only did the United States Supreme Court use the statistical approach for Mormon Island in the Inner Bay of San Pedro, but it has been applied in virtually every tide water case regardless of factual analysis since. Much has been said and written about the Common Law by the English courts and commentators as well as the American courts and treatise writers regarding whether the **OHWM** should be measured by the average of the neap tides or all the tides. Possibly too much analysis but not much common sense has gone into this exercise. An associate of mine (with over 30 years experience in surveying) when referring recently to the concept of the intersection of the plane or elevation of mean high tide as a boundary on the coast said, "it may be Common Law but it isn't Common Sense."[30] I believe it is neither and that the true Common Law rule of the **OHWM** does reflect common sense.

It's interesting to note that, apparently, the twelve-member jury of common folk in the *Teschamacher* case didn't understand how to find the **OHWM** — they found the boundary to be the **mark** left by the water on the ground![31] Respondents' counsel in defending the jury finding must have understood Chief Justice Stephen Field's politics even less. Twice in his brief submitted in **April 1861** he referred to the federal government as the **Confederacy**.[32] I'd like to think that Chief Justice Field, whom President Lincoln appointed to the US Supreme Court in 1863 (and who later authored the famous *Illinois Central Railroad*[33] opinion), didn't let the remark influence his decision that the lands (waters) at issue were not the state's sovereign lands, but had been "confirmed" by the United States to private parties (even though neither the Mexican grant or US patent was in the record)! Interestingly, the US Supreme Court in the 1984 *Summa* case, made a remarkable, but incongruous statement: "Patents issued pursuant to the 1851 Act were, of course, confirmatory patents that did not expand the title of the original Mexican grantee."[34] But just as Shalowitz criticizes *Teschamacher* as unclear and inaccurate,[35] I believe that *Borax* likewise has misapplied latter-day statistical analysis of offshore water levels to what the Common Law had adopted as the *onshore natural monument*.

The *Borax* case was one in a long and important series of court dealing with title in San Pedro Bay.[36] One important case in the *Borax* line was *Dominguez de Guyer v Banning*[37] in which Justice Harlan described Mormon Island as less than an acre. This case determined that, even though the Mexican land grant and the 1857 United States District Court decree had determined the bay and its islands to be within the Rancho which ran "thence along the sea coast," the federal survey and United States patent issued in 1858 excepted the inner bay and islands. In 1880 the US surveyed and in 1881 patented Mormon Island as public lands. The meander survey determined there were 18.88 acres on it. In 1886 Banning won a state court judgment that the island was not within the Rancho. An 1891 decision in the California Supreme Court initially went for Dominguez, with three votes plus one concurring and two for Banning. On rehearing, the vote was six for Banning and none for Dominguez. Dominguez petitioned the US Supreme Court, the case was argued March 20 and 23, 1896, and ordered reargued January 8, 1897. The confusion, exemplified by reversal on rehearing by the California Supreme Court and three days of hearings before the US Supreme Court, is well founded. The issues are complex. At trial Banning had testified:

> "This tract of land known as Mormon island is an island; at about half tide it is an island, and now an island at low water; at low water it is only partly surrounded by water; at low water it would not be surrounded with water; at mean tide there would be about 2 feet of water around it; at high tide it is almost all covered with water.... A very small portion of the island is above ordinary high water. At mean tide, I don't think there is an acre above water. The descriptive clause in the patent to me extends to mean low water. I think to include 18 acres would carry to mean low water. We occupied a portion of it that was covered with water. I have ship ways there and houses on piles. About an acre is covered that way. Another portion of the island we run lighters on and pile lumber on when the tide falls."[38]

By the turn of the century development of the harbor was well underway. The breakwater was begun in the 1890s. The City of Los Angeles was granted the state-owned tide and submerged lands of San Pedro Bay in trust in 1911.[39] Two years later the California Supreme Court rendered its famous decision in *California Fish.*[40] Los Angeles challenged the Borax Company's title to the federal patent of 18.88 acres. The US District Court in a 1-½ page decision found that the patent as surveyed and approved by the US Government Land Office was conclusive as to the 18.88 acres.[41] The City appealed and the Ninth Circuit Court of Appeal reversed. The appellate court quoted from *US* v *Boynton*, 53 F.2d 297, 298, "Meander lines, as used in land surveys, where they indicate a course along navigable waters, refer to the line **as the water itself may delineate it**, subject to changes that different tides or **water flows may work**"[42] (emphasis added). After stating this rule — that the water delineates the boundary — they proceed to throw away this test for *the promise of a new scientific approach.*

> "We will now consider the location of the shore line. It has been assumed by the appellant that the mean high tide line is that boundary....The appellant, however, contends for the rule that the boundary line between the tidelands and upland is determined "by definite mark upon the ground which has been left by the tide." This rule as to a definite mark is applicable to the high-water line of streams but not to a boundary line of tidewaters."[43] (***No authority cited.***)

The court goes on to criticize Justice Field's interpretation of *Hall on the Sea Shore* in *the Teschamacher* case and quotes from Hall, "But it has long been settled, that that portion only of the land adjacent to the sea, which is alternately covered and left dry by the ordinary flux and reflux of the tides is, in legal intendment the shore.... But in fact, no more of this unreclaimed tract is, at law, sea shore, than that portion which lies between high-water and low-water mark, at ordinary tides." The court continues, "It is evident that when we consider the above language from Hall on the Sea Shore, and, when we consider **the fact that the shore line is the boundary between tillable land or land available for agricultural purposes,**"[here's where the quantum leap and non-sequitur comes] "**that the average or mean of the high tides would be the appropriate line defining the boundary between tillable upland and the lands which were ordinarily submerged**

by the tides." The court then explains its rationale. "An increased **knowledge** of the effect of the moon and sun upon the tides and the **necessity** of more definite and certain delineation of the boundary line between tide lands owned by the state in its sovereign capacity, **has fixed the boundary line as the mean high tide line**" (emphasis added). That's right: "knowledge and necessity" have fixed the line, not the judges — how fortunate! For a definition the court cites USC&GS special publication N. 135. The court adds gratuitously and erroneously, "This mean high tide line is the one usually referred to by the United States government in its patents and in the work of its various departments delimiting the boundary between upland and tideland [again apparently taking judicial notice of an unsubstantiated fact]."[44]

The case was appealed by Borax. Both parties argued that state law should decide the case, not federal law. The city argued, "The 'line of mean high tide' is the boundary between tideland and upland."[45] The Court stated:

> "The tideland extends to high water mark. [Citations omitted.] This does not mean as petitioners contend, a physical mark made upon the ground by the waters; it means the line of high water as determined by the course of the tides. [No citation.] By the civil law, the shore extends as far as the highest wave in winter. Inst. lib. 2, title 1, section 3; Dig. lib. 50, title 16, section 112. But by the common law, the shore 'is confined to the flux and reflux of the sea at ordinary tides.' Blundell v. Catterall, 5 Barn. & A. 268, 292, 106 Eng. Reprint, 1190.[46]
>
> "The subject was thoroughly considered in the case of Atty. Gen. v. Chambers, 4 De G. M. & G. 206, 43 Eng. Reprint, 486. In that case Lord Chancellor Cranworth....thought that the authorities had left the question 'very much at large.' He found the principle to be that 'it is land not capable of ordinary cultivation or occupation, and so is in the nature of unappropriated soil...' and 'Lord Hale gives as his reason for thinking that the lands only covered by the high spring-tides do not belong to the Crown, that such lands are for the most part dry and maniorable;...."[47]

The Supreme Court then affirms the Court of Appeals decision. On remand the District Court did not even attempt to decide the location of the **MHTL**; it held the city was estopped by its actions (encouraging $1 million in improvements and actually buying some of the island) from challenging Borax's ownership of the island.[48]

The *Borax* courts frankly were not faced with the facts presented by an open coast shore. The dynamics of the tides on an island in an estuary are significantly different than on an ocean beach. However, we have no court (or rather no reported decision by a US or California court) that has adequately addressed this issue. At least two California appellate decisions sought to address the issue of the migratory nature of a shoreline boundary due to erosion and accretion, but neither dealt with the actual location of the water itself and the first upon remand and was mooted and the second depublished.[49] What is needed is to present the factual dynamics to a court in a manner that they can be understood. It is likewise necessary to challenge or at a minimum alert the court to the incongruity of the reasoning and conclusion reached in *Borax*. It is clear that the Common Law rule relating to water bodies as a natural monument of property boundaries has been lost in the misplaced desire to average and fix an ownership line. It should be remembered that Solomon *never really* cut the baby in half.

Arguments will be raised that *Borax* cannot be challenged at this late date and that state, federal and private parties have relied on it. What is likewise true is that the Common Law, which *Borax* was supposed to be applying, had never used an intersection of an average of measured tidal elevations to a shore as a boundary. One should not forget the United States Supreme Court reversing its 1973 *Bonelli* decision in the 1977 *Corvallis* case, nor the California Supreme Court in the *Alameda Conservation* case and its *Knudson v Kearney* decision in the *Berkeley* case.[50] All of these cases dealt with the issue of the boundary of public rights on inland waterways allegedly in private ownership. Where there has been reasonable and justifiable reliance, the courts typically find ways to protect private parties as in the *Borax* retrial, *Berkeley* decision and even *Illinois Central Railroad.*

Can it really be said that *Borax* is the definitive answer when at this late date the US Supreme Court is still making the following references to the boundary:

> (1) ordinary high-water mark, ordinary high water mark, mean high water line, line of mean high tide;(2) mean high water mark, high water mark; (3) lands submerged at high tide, line of high water, line of mean high water, high water line, line of high water, line of ordinary high water.[51] (4) all land lying under any waters influenced by the tide, mean high water mark, high-water mark, lands flowed by the tide water, tide waters and in the lands under them, soils under the tide waters, lands beneath tidal waters, all lands under waters subject to the ebb and flow of the tide, lands beneath waters under tidal influence, soils beneath waters affected by the tide, soils under tidal waters, title to all land under tidewater, *all lands beneath waters subject to the tide's influence*.[52]

Please note that neither of these recent Supreme Court decisions refer to the boundary as the intersection of the elevation of mean high water and the shore.

What are the alternatives? Initially one could challenge the conclusions reached by the Court, that the test adopted by the Ninth Circuit was appropriate. No legal precedent was cited for the conclusion that the **OHWM** was not a mark on the ground nor for the use of the statistical average of tide measurements as opposed to the actual location of the water on the shore. Irrespective of the Court's statement that the **MHTL** had been used by the US agencies, the record does not support that. The surveyors used the mark of the water on the ground. The Court's conclusions are as solid as the sand on the beach when the up rushing wave drains back around one's feet. Certainly this rule adopted for an estuary is not appropriate for an open coast shore. Even if the boundary location of fee title remains this amorphous line, what of public easements or rights of use in the ocean and on the beach above the line? What about Burt and Deborah? Lucky for them, they were in Hawaii where the line of terrestrial vegetation is the boundary between public beach and private upland, so they weren't trespassing — even on the **dry** sandy beach.[53]

Does California law protect the public's right to use a dry beach, wet beach or even be in the waters of the ocean, and does the law protect the common perceptions and reasonable expectations of its citizens? There is substantial California case law that establishes public easements on private lands, whether it be a Public Trust Easement, Implied Dedication Easement, or Public Easement in Navigable Waters. Furthermore, our Act of Admission and 1879 Constitution provide additional support.[54] Additionally, other states have adopted custom and usage of beaches for hundreds of years as sufficient to create a public right akin to the Civil Law.[55] Archaeology and written history support that California has had such customary and common use of its beaches for thousands of years: first by the Native Americans (California was the most densely populated area in North America),[56] then Spanish and Mexican colonists and later by early settlers.

As previously noted the State of Hawaii, plus the states of Oregon and Texas use the vegetation line as the boundary of public rights on the beach. In *Hughes v State*, the Supreme Court of Washington likewise ruled that private property stopped at the vegetation line; however, they ruled that it was the 1899 vegetation line and that no rights to natural accretion applied thereto. The facts showed the *surveyed* **MHTL** was located 130 feet seaward of the *mark or line reached by water when the ocean was at mean high tide* and 431 feet from the 1899 vegetation line. It appears that the beach had likely accreted 400 feet from 1859 to 1899, and another 175 feet by 1963. The US Supreme Court reversed the decision finding that the right to accretion was protected by federal common law.[57] Other Pacific Coast jurisdictions include British Columbia,[58] which provides that the beach to the vegetation line is public and Mexico,[59] where rights often go even further inland.

However, in a Rhode Island Supreme Court case which by rote followed the *Borax* decision, a citizen who was picking up trash in the water on the beach was arrested for trespassing above the **MHTL**. His defense was that he was in the ocean below the **HWM** he defined as "*a visible line on the shore indicated by the reach of an average high tide and further indicated by drifts and seaweed along the shore*." The Court found him in technical trespass, even though the **MHTL** was under water at the time (see Photo #1 as an example). They dismissed the criminal charges because "no man shall be held criminally responsible for conduct that he could not reasonably understand to be proscribed."[60] This points to the need for courts to return to a rule of reason.

Fifty years before William the Conqueror, King Canute II (the Great) ruled England and later also Denmark, Norway and Scotland. He is portrayed on the Bayeux Tapestry and his legend is told by the Chronicler Henry of Huntington. When at the height of his power, he ordered his throne placed on the sea-shore as the tide was coming in, seated himself, surrounded by his courtiers, and said to the tide, "You are within my jurisdiction and the land on which I sit is mine — no one has ever resisted my command with impunity. I therefore command you not to rise over my land nor presume to wet the limbs or clothes of your lord." As the tide apparently was unaware of his power — it proceeded to wet his robes — he jumped back from the shore and stated, "Be it known to all inhabitants of the realm — the power of Kings is empty and superficial and no one is worthy of the name of king except him whose will is obeyed by heaven, earth and sea in accordance with eternal laws." He then removed his golden crown and never put it back on.[61] This lesson in reality and humility should be learned by all who wear robes, sit on thrones and try to command the sea.

1. C. Corker. *WHERE DOES THE BEACH BEGIN, AND TO WHAT EXTENT IS THIS A FEDERAL QUESTION*, University of Washington Law Review, Vol. 42:33.
2. In 1933 the California Supreme Court in *Otey v Carmel Sanitary Dist.*, 219 Cal 310 referred to the Ordinary High Water Mark as "the limit reached by the neap or twice-a-day tides."
3. Civil Code Section 830 amended 1873. Civil Code Section 670.
4. *In re Ashford*, 50 Haw. 314 (1968), *Sotomura v County of Hawaii*, 402 F. Supp. 95 (D. Haw. 1975), *In re Sanborn*, 57 Haw. 585 (1977)
5. O. W. Swainson. *Topographic Manual*, USC&GS (1928) at 69.
6. Sir Wm Blackstone. 2 *Commentaries 262 (1766).*
7. *Borax Consolidated v City of Los Angeles*, 296 U.S. 10, 26 (1935).
8. Id. at 27.
9. R.S. Patton. *Relation of the Tide to Property Boundaries,* USC&GS (1940) at 10.
10. H. Althaus *Public Trust Rights*, US Fish & Wildlife Service, USGPO (1978), at 1-10; D. Slade. *Putting the Public Trust to Work* (1990) at xvii, 15.
11. Althaus, at 23-38.
12. Ibid; J. Gould. *A Treatise on the Law of Waters* (1883) at 6.
13. Althaus, at 23-38.
14. *Aptos Seascape Corp. v County of Santa Cruz*, 138 Cal. App. 3d 484, 503 (1982).
15. Gould, at 8 footnote 2 citing Hale's *De Jure Maris* c. 4.
16. *Blundell v Catterall*, 5 B. & Ald. 268 (1821) quoted in *Borax Consolidated v City of Los Angeles*, 296 U.S. 10, 22 (1935)
17. *Borax Consolidated v City of Los Angeles*, 296 U.S. 10, 24-25 (1935) quoting *Attorney General v Chambers*, 4 De Gex, Mac N. & G. 206 (1854).
18. *Borax Consolidated v City of Los Angeles*, 296 U.S. 10, 23-25 (1935).
19. Shalowitz, *Shore and Sea Boundaries, Vol. 1* (1962) at 89-97.
20. Gould, at 38 footnote 1;
21. *State v Superior Court* (Lyon), 29 Cal. 3d 210, 229 (1981).
22. *People v Morrill*, 26 Cal. 336 (1864).
23. Shalowitz, *Shore and Sea Boundaries, Vol 2* (1964) at 176-183.

24. Id. at 171-175.
25. Id. at 174 and 344-345.
26. Id. at 171; United States Coast and Geodetic Survey, Report of the Superintendent (1865) at 230.
27. Shalowitz, *Vol. 2* at 174.
28. Patton at 1.
29. Id. at 1-2.
30. J. Michael McKown. Senior Boundary Determination Officer, California State Lands Commission. PLS # 4443.
31. *Teschamacher v Thompson,* 11 Cal 10, 21 (1861).
32. Id. at 19.
33. *Illinois Central Railroad Co. v Illinois*, 146 U.S. 387 (1892).
34. *Summa Corp. v California ex rel. State Lands Comm.*, 466 U.S. 198 (1984).
35. Shalowitz, *Vol. 1* at 92-97.
36. See *People v California Fish Co.*, 166 Cal. 576 (1913) and companion cases.
37. *Dominguez de Guyer v Banning*, 167 U.S. 723 (1897).
38. Ibid..
39. Chapter 656, Statutes of 1911.
40. *People v California Fish Co.*, 166 Cal. 576 (1913).
41. *City of Los Angeles v Borax Consolidated*, 5 F. Supp. 281 (S.D. Cal. 1933).
42. *City of Los Angeles v Borax Consolidated*, 74 F.2d 901, 902 (9th Cir. 1935).
43. Id. at 904.
44. Id. at 905.
45. *Borax Consolidated v City of Los Angeles*, 296 U.S. 10, 12 (1935).
46. Id. at 22.
47. Id. at 24-25.
48. *City of Los Angeles v Borax Consolidated*, 20 F. Supp. 69 (S.D. Cal. 1937), affirmed by *City of Los Angeles v Borax Consolidated*, 102 F.2d 52 (9th Cir.), *cert. denied,* 307 U.S. 644 (1939).
49. *People v Wm Kent Estate Co*, 242 Cal.App.2d 156 (1966) — neap tide case; and the depublished opinion in *Antoine v California Coastal Commission*, [8 Cal.App.4th 641 (1992) **depublished**] — migratory boundary case.
50. *Bonelli Cattle Co. v Arizona*, 414 U.S. 313 (1973); *State Land Board v Corvallis Sand and Gravel Co.*, 429 U.S. 363 (1977); *Berkeley v Superior Court*, 26 Cal.3d 515 (1980); *Knudson v Kearney*, 171 Cal. 250 (1915); *Alameda Conservation v City of Alameda*, 264 Cal.App.2d 284 (1968).
51. *California ex rel. State Lands Comm. v United States*, 457 U.S. 273 (1982) : (1) opinion of the Court, (2) concurring opinion, (3) US Solicitor's brief.
52. *Phillips Petroleum Co. v Mississippi*, 484 U.S. 469 (1988): (4) from the opinion of the Court — the final italicized phrase describes the decision as to what "the States, upon entering the Union were given ownership over...."

53. See note 4.

54. An Act for the Adm. of Calif. into the Union, Sep. 9, 1850; Calif. Const. Art. X, Sec. 4.

55. *State ex rel. Thornton v Hay,* 254 Or. 584 (1969); Tex. Nat. Res. Code Ann. Sec. 61.011.

56. A. Josephy. *The Indian Heritage of America.* Bantam Books (1968) at 137-145.

57. *Hughes v State*, 67 Wash. 2d 787 (1966); *Hughes v Washington*, 389 U.S. 290 (1967); Corker at 46.

58. *Land Act, Province of British Columbia* [Consolidated January 17, 1989], Section 1.

59. J. Wheless. *Compendium on the Laws of Mexico*, 699 (1938) quoted in Althaus at 19.

60. *State v Ibbison*, 448 A.2d 728, 733 (1982).

61. M. Magnusson. *Tales from Viking Times, Canute and the Waves.* Music for Pleasure, Limited (1981).

CHAPTER 155

The Kobe, Japan Earthquake and Lessons for Port Facilities in California

[1]Martin L. Eskijian and [2]Gary L. Gregory

Abstract

The Marine Facilities Division of the California State Lands Commission was invited to inspect port damage in Kobe, Japan, following the magnitude 6.9 earthquake of January 17, 1995. The damage assessment served as an educational experience, due to similarities between Kobe and California's ports. The Kobe, Japan region is similar to major port areas of Northern and Southern California, both in faulting regimes and their relationship to waterfront and port facilities. The Kobe earthquake was centered about 12 miles southwest of the city and was a strike-slip fault event. This faulting structure is similar to the Hayward Fault in the San Francisco bay area and the Palos Verdes Fault and other nearby faults that extend into the ports of Los Angeles and Long Beach. Although construction methods are different in the two countries, the ground failure information is relevant, along with design criteria decisions. Selection of appropriate seismic criteria is important, even though public safety may not be at risk. Having Kobe's artificial islands designed to a significantly lower seismic hazard made the entire port inoperable following the earthquake.

Introduction

As a guest of the Port and Harbour Research Institute of Japan (PHRI), we were invited to Kobe, Japan about two weeks after the earthquake of January 17, 1995 (Ref. 1) to see first hand the damage and compare with possible scenarios for California. The inspection focused on Kobe's two artificial islands of Port and Rokko. Port Island was completed in 1981 after 15 years of construction; Rokko Island was completed in 1991 after 20 years of construction. The older mainland-type port structures such as pier or wharf-type, were built 20-50 years ago.

---

[1]Senior Engineer, Petroleum Structures, Marine Facilities Division, California State Lands Commission, 330 Golden Shore, Suite 210, Long Beach, CA 90802
[2]Chief, Marine Facilities Division, California State Lands Commission

The damage reported in this study focuses on the two artificial islands; more severe structural damage in the city has been adequately addressed by other investigations (Refs. 2, 3).

Seismic Records

PHRI provided time history records and areal values of peak ground motion. The seismic event has been characterized as follows:

Hypocenter:
Latitude 34.6
Longitude 135.0
Depth of 20.0 Km
Magnitude 6.9
Time: 5:46 AM, January 17, 1995

Figures 1 and 2 are acceleration records taken along a quay wall in the Port. Figure 1 indicates the strongest motion was from NE to SW, in line with the direction of faulting. Figure 2 provides time history records at the same location. From these records, representative ground motion at the port was determined:

Direction	Acceleration	
EW	683 Gals	0.695 g's
NS	394. "	0.40 "
Vert.	334. "	0.34 "

(Note: 1.0 g = 981.7 cm/sec**2 = 981.7 gals = 32.2 ft/sec**2)

Inspection and Observations of Port Facilities

The Port of Kobe has three major components, the near-shore islands of Sinko, Onohama, Maya Wharf, Nadahama-Higasimati, Mikage-Hammamati, Uozaki-Hammati, Hukae-Hammati, and the two larger islands of Port and Rokko.
At the time of our visit, about two weeks after the earthquake, the entire port area appeared to be shut-down, except for the hydrofoil ferry operating from the K-KAT terminal, a few Navy vessels at their wharves, and repair/crane barges at other facilities.

The following description and photographs summarize observations made during this brief visit:

A. Oil and Petrochemical Terminals and Tank Farms

Figures 3 and 4 show typical damage to oil and petrochemical terminals and tank farms. Marine oil terminals were shut-down, due at least in part to bent and deformed pipeline systems. Equipment had toppled and as indicated in Figure 4, the berm wall around a tank farm had lost functionality. Notice the soil dumped and ready to fill-in gaps in the berm, to prevent petroleum products from running into the ocean. Gaps of over 300 cm were frequently noted, indicating a modular caisson-type of construction, with a translation of the entire caisson. There was no evidence of large amounts of oil spilled anywhere in the harbor. Figure 5 shows the typical precast concrete caisson blocks used to form the perimeters of the offshore islands. Although this basic design had many variations, in general this type of precast concrete caisson member was filled with sand and used to construct Port and Rokko islands.

B. Container Wharves

Figures 6 and 7 show horizontal and vertical displacement of concrete caisson elements. The caissons didn't fail, they just displaced laterally and vertically. Figure 7 also shows the buckling of the outside box-section steel columns of the container cranes. Almost all of the cranes suffered an almost identical failure, consistent with a uniform design criteria. Crane manufacture data indicates that design criteria was 0.20g's; however, they were subjected to peak accelerations of 0.80 g's. Thus, it isn't surprising that the cranes started to buckle. Figure 8 shows another container crane wharf with the concrete grade beams in good condition and piles intact. Failure of the soil between the adjacent grade beams failed or liquefied; the wharf settled more than 1 meter.

C. Port Terminal Building and Freeway Access to Port Island

Figure 9 shows extensive soil failure of the parking lot of the new port building on the mainland harbor portion of Kobe. Figure 10 shows the new concrete arch bridge to the islands. Notice ground failure in the foreground of Figure 10 and the displacement of the concrete arch to the right of the concrete supporting bent. The port building itself was in excellent condition as viewed from the outside, but was totally unaccessible due to ground failure adjacent to the entrance and the parking lot failure.

D. Port Island K-KAT Building

Figure 11 shows the settlement of Port Island around the K-KAT terminal building. The ground settled more than a meter; the building itself was intact, without any broken glass or diagonal shear cracking. The K-KAT building, used for public hydrofoil transport to Kansai International Airport and the city of Osaka was semi-operational, without any utilities. Preliminary estimates indicate that the island settled about 1 meter; however, the pile-supported structure did not settle.

E. Pipeline Failures - Under Bridge

Figure 12 shows the failure of a pipeline connected to the underside of a suspension bridge. The pipeline is connected to the bridge, at about 1 meter intervals; the connections "unzipped" and the pipeline dropped into the ocean. It is believed that this was a water pipeline.

Structural Opinions and Assessment

The following conclusions and structural opinions were formulated after observing damage on Port and Rokko Islands:

1. The major cause of damage to the Port of Kobe was related to liquefaction, soil displacement or other type of soil failure. The concrete substructures, such as caissons or grade beams with piles performed quite well, but fill and underlying soil did not. Modern structures performed quite well; there was no evidence of failure of modern buildings on the two islands.

2. Pipelines and berms are at great risk, especially with major soil displacement or failure.

3. Although modern buildings performed quite well, utilities and vehicular access was cut-off, thereby reducing functionality to zero. Bridges were severely damaged.

Seismic Design Criteria at Kobe

Werner and Dickenson (Ref. 4) have investigated the recent design criteria used for Port and Rokko Islands. This information indicates that peak design accelerations were between 0.19 to 0.42 g's, with associated return periods of 75 to 200 years.

However, the January 17, 1995 Kobe earthquake appears to be a 500 to 1000 return period event. In 1596, the area experienced an estimated magnitude 7.2 event. Kobe was about due for another magnitude 7.0 event, yet facilities were designed for a much lower criteria.

Although modern commercial and industrial buildings performed as expected, having artificial islands designed to a significantly lower standard made the entire port inoperable following the earthquake.

Lessons and Application to California

There are a number of lessons and applications for California and other high seismic regions with port facilities.

1. Older, concrete vertical pile wharves and piers are fairly resistant to moderate earthquake damage. From the Kobe observations, there was no case where complete collapse of the concrete wharf/pier occurred.

2. Liquefaction in areas of poorly engineered fill material is a serious risk in a moderate earthquake in port areas. Damage was noted in terms of settlement in the Port of Los Angeles after the 1994 Northridge earthquake (about 50 miles away from the epicenter) (Ref. 4). This type of soil failure can seriously affect transportation and usability of structures, even though the basic concrete or steel structure has little or no damage. If roads, bridges, utilities and operational characteristics of a well-designed port structure are disabled, having the structure survive but not being able to perform its operations, may be a financial disaster for a port authority or terminal operator.

3. Selection of appropriate seismic criteria is extremely important, even though public safety may not be at risk. The financial losses from a moderate earthquake, in terms of shutting a major port facility down completely can be devastating to a regional or even state or national economy.

Acknowledgements

The Kobe visit was made possible with funding from the California State Lands Commission and with the cooperation and assistance of PHRI. A special thanks to Dr. T. Uwabe and Mr. Noza who provided maps, seismic records, transportation and translation for our inspection of Kobe.

Appendix

1. Eskijian, M. L. "Preliminary Report of the Hyogo Earthquake- Port of Kobe, Japan", February 1995, California State Lands Commission, unpublished.

2. "The January 17, 1995 Kobe Earthquake - An EQE Summary Report", April 1995.

3. "The Hyogo-Ken Nanbu Earthquake January 17, 1995, Preliminary Reconnaissance Report, Earthquake Engineering Research Institute, February 1995.

4. Werner, Stuart D. and Stephen E. Dickenson, "Hyogo-Ken Nanbu Earthquake of January 17, 1995: A Post-Earthquake Reconnaissance of Port Facilities", September 1995.

Figure 1 - Directional acceleration record from the quay wall (DAI8-G location). Note the maximum acceleration of 726.6 gals (0.74 g's).

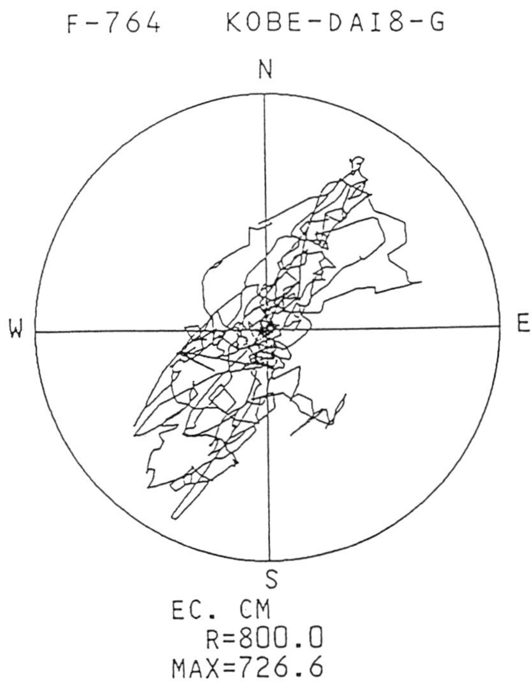

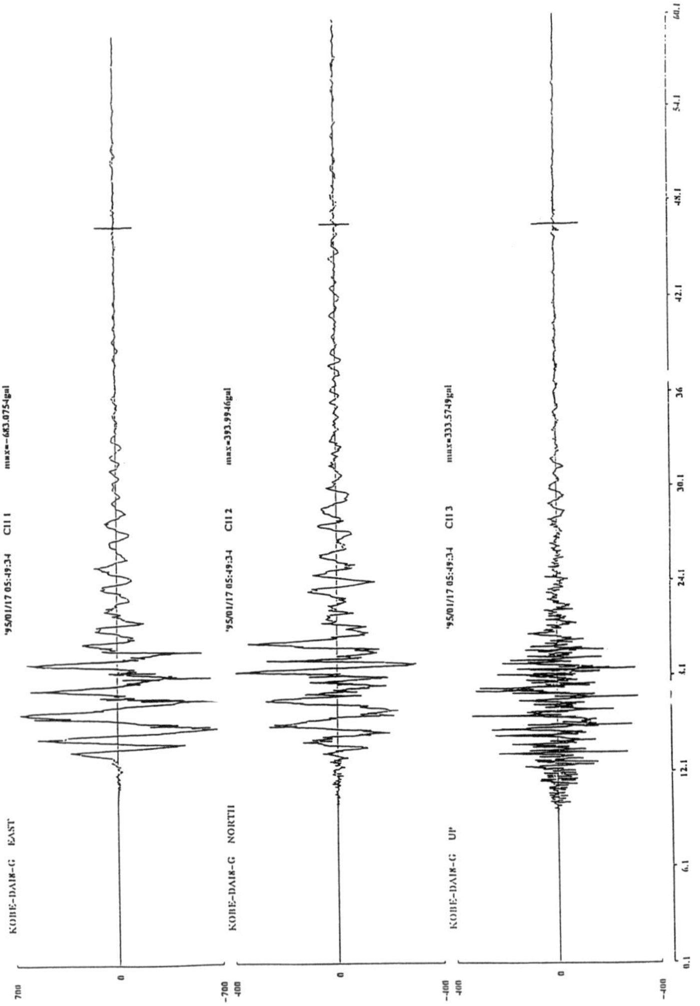

Figure 2 - Acceleration Time Histories, Two Horizontal and Vertical (Units are gals and seconds) (DA18-G)

Figure 3 - Marine oil terminal damaged, with equipment overturned and pipelines bent/distcrted.

Figure 4 - Tank farm and berm wall; notice large gaps in berm wall, and soil behind wall to keep petroleum products from reaching the water.

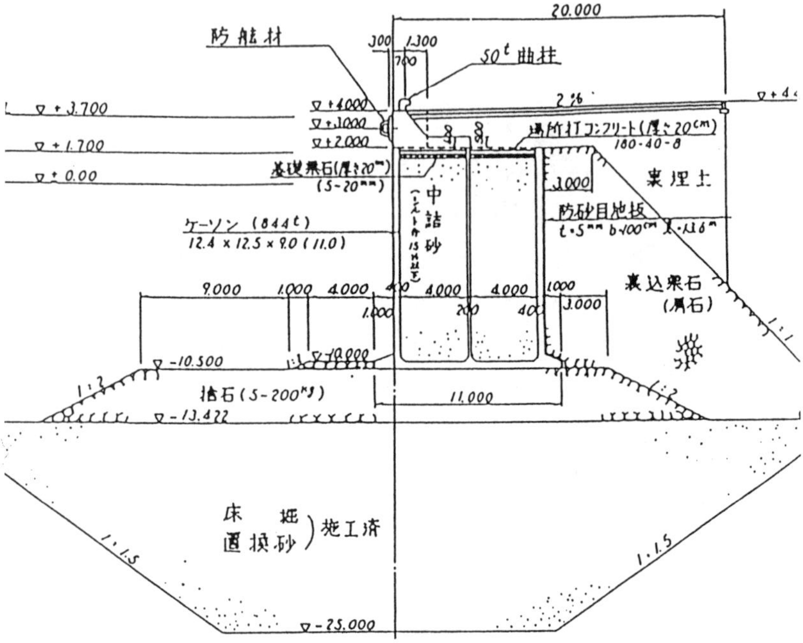

Figure 5 - Typical cross-section of the precast concrete caisson structural elements. Note that caisson is filled with sand.

Figure 6 - Horizontal displacement of adjacent caissons along outside face of container wharf.

Figure 7 - Vertical displacement of adjacent precast caisson elements at a container wharf.

Figure 8 - Crane rail supported by a concrete grade beam with circular concrete piles. Grade beam and pile were in good condition; soil failure between grade beams.

Figure 9 - Soil failure of parking lot under new terminal building.

Figure 10 - Displacement (horizontal) failure of concrete arch bridge, with failure of soil and wharf in the foreground.

Figure 11 - K-KAT Building at rear; notice ground level has displaced more than 1 meter.

Figure 12 - Failure of pipeline supported by bridge to islands.

CHAPTER 156

Installation of a Seafloor Earthquake Monitoring System

Catherine M. Hoffman[1]

Abstract

Very little is known about the earth's movement underwater during earthquakes. Data on the response of seafloor sediments to earthquake induced ground motion are scarce. In an effort to learn more about the response of seafloor sediments, the U.S. Minerals Management Service (MMS) installed a seafloor earthquake monitoring system (SEMS) network offshore southern California. The data gained will be used to help strengthen the safety standards as they relate to the structural aspects of offshore facilities. This SEMS is the only existing offshore seismic network in the United States.

Due to recommendations provided by the Marine Board of the National Academy of Sciences to obtain data on seafloor seismic ground motions, a series of prototype instruments known as SEMS were developed and installed by the Sandia National Laboratories under the sponsorship of the MMS and the U.S. Department of Energy. Due to the significance of the information received from these instruments, the MMS sponsored the development and installation of improved seismic data gathering networks offshore southern California.

Since 1979, the MMS has installed four generations of SEMS equipment, SEMS I, II, III and IV. Preliminary data collected by SEMS I, II and III suggested that there were major differences between onshore and offshore recordings of earthquake ground motions.

---

[1]Petroleum Engineer, U. S. Minerals Management Service, 770 Paseo Camarillo, Camarillo, CA 93010

A study was launched to identify and substantiate differences that might exist between onshore and offshore records and to compare the response of a typical mid-water depth offshore platform subject to both onshore and offshore recorded ground motions from the same earthquake. The study concluded that there are significant differences in platform responses when using the onshore and offshore earthquake records as a forcing function, but before specific recommendation can be put forth for using offshore recordings, as opposed to onshore recordings, additional offshore records would have to be obtained and analyzed (Smith, C. E. 1996).

In July of 1995, the MMS completed the installation of its fourth prototype of SEMS equipment, now called SEMS IV. The SEMS IV is a network of three seismic probes installed in locations of varying water depths. The probes are imbedded in the seafloor approximately 400 feet from a platform and hardwired up to the platform topsides, where it is connected to a telephone modem. The three probes were installed offshore Los Angeles County at CalResources LLC's Platform Eureka, Ventura County at Chevron Company, U.S.A's Platform Grace, and Santa Barbara County at Torch Operating Company's Platform Irene.

The SEMS IV area of coverage is quite extensive (Figure 1). A probe will record a 4.0 earthquake occurring within a 25-mile radius of its location and a 4.5 earthquake occurring within a 65-mile radius. With the distance from Platform Irene to Platform Eureka being 160 miles, the probes' ability to detect a moderately sized earthquake in the Southern California area is quite good.

On September 20, 1995, at 4:27 pm, a 5.8 earthquake ripped through the Ridgecrest area. With the earthquake's epicenter approximately 192 miles away from Platform Irene, the earthquake registered loud and clear on her computer. The data have not been compared to the onshore data so no conclusions can be made at this time as to whether the data are consistent with previous data collected offshore. In addition, Platform Grace has recorded several aftershocks to the 1994 Northridge earthquake.

The MMS and the California Division of Mines and Geology (CDM&G) are in the process of developing a joint agency work agreement for the maintenance and monitoring of our SEMS equipment. Through this work agreement, the CDM&G will:

- Periodically monitor and test the SEMS to ensure that the instruments are operational.
- Perform on-site maintenance and inspection of the equipment once a year or as needed.
- Troubleshoot and repair the system as necessary to assure reliability.

The data received from the SEMS will be incorporated into the CDM&G's onshore network and processed for public distribution to the engineering community. The CDM&G will be responsible for archiving all data received. Anyone can check out the data from the SEMS by logging onto their homepage at www.consrv.ca.gov/dmg/csmip.

In closing, the SEMS program has demonstrated that there is a difference between data collected from onshore and offshore instruments. Data from the SEMS will provide input to structural and soil response models used in the design and requalification of offshore platforms. Very importantly, the SEMS program has demonstrated that long-term underwater measurements are possible.

APPENDIX

Smith, C. E. 1996 OTC Proceedings. "Response of a Steel-Jacket Platform Subject to Measured Seafloor Seismic Ground Motions.

CHAPTER 157

CONSIDERATIONS FOR SEISMIC ASSESSMENT OF EXISTING PETROCHEMICAL AND TERMINAL FACILITIES

Gayle S. Johnson[1]

Abstract

This paper discusses methods for performing seismic assessments of existing petrochemical facilities, chemical plants, and marine terminals, where seismic events could result in unacceptable releases of hazardous or environmentally sensitive materials into the air and water.

There is no widely used and accepted standard applicable to evaluating existing facilities for earthquake loads. It is impractical, if not impossible, to require facilities to conform to standards for new design. Design codes change constantly in order to incorporate lessons learned from past earthquakes and ongoing research, and almost always become more restrictive and more conservative. Code requirements that are inexpensive to satisfy during the design phase may be prohibitively expensive to incorporate as a retrofit to an existing structure.

It should also be noted that new design codes do not necessarily adequately address conditions that are unique to existing facilities, such as deterioration and corrosion, minor damage, changes in weights and operating conditions, interaction effects (impact and differential displacement) and safety requirements for functionality as well as structural integrity. Building codes for new design do not necessarily handle all of the causes of damage experienced in past earthquakes.

This paper will discuss methods currently being used and documented for evaluating existing facilities. One primary method utilizes methodical field reviews by experienced engineers to act as a rapid screening and prioritization of further evaluations. This method is used throughout California to satisfy Risk Management and Prevention Program seismic requirements. Guidance is also provided in a soon-to-be-published document by the ASCE Petrochemical Energy Committee, written by a group of owners, architect engineers, and specialist seismic consultants.

The paper will discuss methodologies and show examples of the type of damage that has historically occurred and how it is addressed. The paper will address equipment and structures typically found in petrochemical facilities, chemical plants, and marine terminals. Recommendations will be given for ways to realize a significant risk reduction in a cost-effective manner that will be logical and defensible to owners and regulators.

---

[1] EQE International, Inc., 44 Montgomery Street, 32nd Floor, San Francisco, CA 94104

CHAPTER 158

An Update on Seismic Reassessment of Aging Platforms Offshore California

[1]Leslie A. Monahan, [2] Martin L. Eskijian

ABSTRACT

Several platforms offshore California, used for the production of oil and gas, are over 25 years old. The majority of these platforms were installed before industry-wide design methods and criteria were established. As a result of California's 1989 Loma Prieta earthquake, regulators have been developing a policy to address seismic reassessment of platforms offshore California. There are two questions about these aged platforms: (1) Are the structures fit-for-purpose? and (2) what types of monitoring and structural reassessment are required?

After numerous public workshops, a consensus of what would constitute an acceptable level of fitness-for-purpose was developed. The final document has been approved by the American Petroleum Institute (API) and is now available.

In addition to the guidelines, structural inspection and monitoring are an important aspect of maintaining the fitness-for-purpose of these aged structures. Structural inspection intervals are clearly specified in API RP 2A; damaged members can be accurately described in finite element models. So, the analyst can evaluate whether or not repair is mandated by a reduction in structural performance. In addition, the Minerals Management Service (MMS) has instituted a "Seafloor Earthquake Measurement System" (SEMS) to collect and analyze seismic data offshore. Many recently constructed platforms have accelerometers at various deck levels and having the free-field data available will provide the perfect calibration tool for post-event analyses.

---

[1]Petroleum Engineer, Minerals Management Service, 1201 Elmwood Park Blvd., New Orleans, LA 70123

[2]Senior Engineer, Petroleum Structures, Marine Facilities Division, California State Lands Commission, 330 Golden Shore, Suite 210, Long Beach, CA 90802

In order to verify that offshore operators have performed the reassessment adequately, the State and Federal regulators have instituted methods to check operators. The MMS has developed the "Certified Verification Agent" (CVA) program wherein a respected peer reviews the structural analysis/retrofit of the aged platform. At the State level, the California State Lands Commission has instituted an in-house review process, wherein the operator submits all structural information, including applied loading, configuration, and computer models, and the work is independently verified.

Background

After the Loma Prieta earthquake struck the San Francisco Bay area in October 1989 and caused a number of structural failures of bridges, highways, and other buildings, the MMS and the California State Lands Commission (CSLC) began evaluating seismic analyses that had been performed for existing offshore platforms, which had been in service for many years. A number of structures had been installed prior to the development of seismic criteria for new offshore structures. These structures had been designed using the Uniform Building Code (UBC) or other company-specific criteria developed for conventional onshore structures (see Table 1). Early UBC versions simply call for a 0.1g lateral load to be applied for seismic design.

Today, the API develops recommended practices for the entire oil industry. These publications commonly become "code" when adopted by regulatory bodies. API's publication for offshore structures is entitled "API Recommended Practice for Planning, Designing, and Constructing Fixed Offshore Platforms" (API RP 2A). API did not include seismic design criteria for new structures in API RP 2A until the 7th edition, published in 1976.

National codes and recommendations have been in place to address the issue of reassessment of existing conventional, onshore buildings; however, codes had not yet been developed that addressed reassessment of existing offshore structures. Therefore, after the Loma Prieta earthquake, the MMS initially required operators of Outer Continental Shelf (OCS) platforms installed before 1976 to conduct preliminary seismic analyses using API platform design criteria for new platforms. These analyses showed that two of the platforms passed while four others revealed potential problems. Experience gained from this process demonstrated the need for development of uniform seismic reassessment criteria that would be more appropriate for use with existing offshore structures.

The MMS encouraged API to become involved in the development of criteria for platform reassessment. API contracted with an independent panel consisting of four earthquake engineering experts (hereby referred to as the "Panel") to

Table 1

Platform Name	State or Outer Continental Shelf (OCS)	Water depth (feet)	Year Installed
Emmy	State	47	1963
Eva	State	57	1964
Esther	State	35	1985
Ellen	OCS	265	1980
Elly	OCS	255	1980
Edith	OCS	161	1983
Eureka	OCS	700	1984
Gina	OCS	95	1980
Gail	OCS	739	1987
Gilda	OCS	205	1981
Grace	OCS	318	1979
Habitat	OCS	290	1981
Heidi*	State	125	1966
Hope*	State	136	1965
Hogan	OCS	154	1967
Houchin	OCS	163	1968
Henry	OCS	173	1979
Hillhouse	OCS	190	1969
A	OCS	188	1968
B	OCS	190	1968
C	OCS	192	1977
Hazel*	State	100	1958
Hilda*	State	106	1960
Holly	State	211	1966
Hondo	OCS	842	1976
Harvest	OCS	675	1985
Hermosa	OCS	603	1985
Hidalgo	OCS	430	1986
Harmony	OCS	1200	1989
Heritage	OCS	1075	1989
Irene	OCS	242	1985

*removed in 1996

recommend a methodology for the seismic reassessment of offshore platforms. The Panel consisted of Professors G. Housner and W. Iwan of the California Institute of Technology, and Professors C. A. Cornell and C. Thiel of Stanford University. The Panel prepared a report entitled "Seismic Safety Requalification of Offshore Platforms" (Ref. 4). As a result of that report, API then formed an industry/MMS task group to formulate what has since become Supplement 1 to API RP 2A, 20th edition (Ref. 1)

The MMS sponsored three workshops for the purpose of involving industry, academia, environmental groups, and the general public in the development of reassessment guidelines. Both technical and policy issues were addressed. The first meeting, entitled "International Workshop on Seismic Design and Reassessment of Offshore Structures" (Ref. 6) was held in Pasadena, California, in December 1992. This workshop brought together structural and earthquake engineers, along with topside equipment experts and seismic hazard geotechnical experts, to review the Panel's findings, which were ultimately used as a framework in developing platform reassessment guidelines.

A second workshop, entitled "Public Policy Issues on the Seismic Requalification of Offshore Structures" (Ref. 7), was co-sponsored by the MMS and the CSLC and was held in Camarillo, California, in November 1993. This workshop bought together representatives from agencies involved in seismic safety, environmental groups, and academia to seek a consensus of what constitutes acceptable risk offshore. A consensus was reached that existing platforms should maintain their structural integrity when subjected to an earthquake with a median return period of 1,000 years. It was established that criteria should be more conservative than for conventional buildings onshore, because in the event of a sudden platform collapse (unlike hurricane scenarios in the Gulf of Mexico where timely warnings allow platform personnel to be evacuated), worker safety is at risk, and possible pollution could be catastrophic.

The third and final workshop, entitled "International Workshop on Assessment and Requalification of Offshore Production Structures" (Ref. 5), was held in New Orleans, Louisiana, in December 1994 and was sponsored by MMS, API, and others with participation from industry, academia, regulatory agencies, and the general public. This final workshop dealt with the reassessment of structures located in all areas of the United States. While developing guidelines for seismic reassessment, the API/MMS Task Group began looking into other loading conditions in different coastal areas of the U.S. For instance, ice loading is a major design factor for platforms located offshore Alaska, whereas in the Gulf of Mexico, wave forces from storms and hurricanes constitute a major design load. For platforms located offshore California, seismic loading conditions are of utmost concern and in most cases govern their design. However, there are circumstances where waves would reach lower deck areas, and in these cases it is possible that wave loads may be the dominant design criterion.

The results of the Camarillo, California, workshop were presented to the public policy work group in New Orleans, Louisiana, and there was consensus that the structure must maintain integrity when subjected to the median 1,000-year return period seismic event. This consensus was ultimately adopted by API.

Reassessment Criteria

API issued the Supplement to API RP 2A in December 1996. It provides technical criteria for use in reassessing existing structures and embraces a fitness-for-purpose evaluation coupled with the risk of structural failure and the consequence of that failure. It is very specific about applied environmental loads and requires that the platform maintain structural integrity without collapse when subjected to specified horizontal and vertical seismic, storm, and ice loads. A summary of the seismic reassessment criteria follows:

- Manned platforms or any platform whose collapse would produce a high consequence event must maintain global structural stability when subjected to the median 1,000-year return period seismic event.
- Unmanned platforms and platforms whose collapse would not produce a high consequence event must maintain global structural stability when subjected to the median 500-year return period seismic event.
- As an alternative to the use of analyses based on the guidance specified in the Supplement, the document also allows for the calculation of explicit probabilities of failure.
- Topside equipment and appurtenances must also withstand these seismic loads.

Four platforms referred to at the beginning of this paper showed potential problems in the crew quarters, crane pedestal areas, and deck legs. The crew quarter and crane pedestal areas were eventually reinforced. Subsequent analyses of these platforms revealed that they would maintain global structural stability when subjected to the median 1,000-year return period earthquake.

Reassessment Triggers

There are various reassessment triggers identified in the Supplement that require a seismic analysis be performed. They include such factors as significant structural damage, new fault or seismic hazard information, as well as an increase in platform loading that is significantly greater than that which was used in the original design (i.e., a new drilling program with a heavier rig, additional processing equipment, etc.). The definition of "significant" is important, in that it is the increase in loading and/or cumulative damage resulting in a change in capacity or applied loads of more than 10 percent of the original design. The regulatory agencies are moving ahead with a set of consistent triggering mechanisms in their rulemaking process.

In the absence of reassessment being initiated by these "active" triggers, the public policy working group at the New Orleans workshop proposed that a "passive" trigger requirement be invoked such that all platforms would be reassessed within five years of publication of reassessment regulations. The MMS is considering reducing this time period to three years.

Structural Verification of Analyses and Design

At the Federal level, the MMS has a Certified Verification Agent (CVA) program wherein a certified peer verifies design, fabrication, and installation of certain new facilities in Federal waters. Federal regulations (Title 30, Code of Federal Regulations 250.130) provide for CVA review of new construction and major modifications, but currently there are no regulations addressing reassessment of platforms. However, this program has been quite successful and is being performed for the reassessment process in an ad-hoc manner.

At the State level, the CSLC has instituted a review process, wherein the operator submits all structural information, including applied loadings, structural drawings, and computer models, and the work is reviewed. The goal of this review process is to independently verify that the operator has performed an adequate analysis and that the proposed strengthening measures, if any, are appropriate. The review also includes an assessment of damage, as reported in the underwater inspection report. The damaged capacity of the individual members is also verified, using the methodology developed by J. Ricles (Ref. 8). For example, seemingly small dents (i.e., one-tenth of the diameter) in tubular sections can result in major reductions in the load-carrying capacity of the member. Analytical simplified models have been derived and compare favorably with actual scale model behavior.

Currently, onshore data from seismic events are extrapolated for the offshore environment and used for design criteria for platforms. However, it should be clearly understood that offshore structures have several structural and environmental load differences from typical structures onshore (Smith, 1990). Structurally, offshore oil and gas facilities have longer periods of vibration, deeper pile foundations, a large number of redundant members, and most of the mass is located at the top of the structure. Differences in environmental loading include the dampening effects of the water column, the virtual mass of the water adjacent to the structure, and the geology/response of the subsea soils.

These factors have contributed to a decision by MMS to support a research project to develop accurate and reliable seafloor instruments to measure strong ground motion of the free-field area adjacent to offshore platforms. The present goals of this project are (1) to collect and analyze data from offshore seismograph stations, and (2) to verify or, if necessary, modify existing models for predicting the response of saturated sediments to earthquake-induced seismicity (Sleefe, 1990). Many recently constructed platforms have accelerometers at various deck levels, and having the free field data would provide the perfect calibration tool for post-event analyses.

The seafloor instrument is called the "Seafloor Earthquake Measurement System" or SEMS unit. The unit, designed by Sandia National Laboratories, consists of a sea probe that can be either hardwired to the platform with a cable, or self-contained with a battery pack. Prototypes of the units were first installed in the Santa Barbara Channel in 1979. First and second generation models detected only vertical accelerations. The third generation unit, installed offshore in 1989 and 1991, measured vertical and 90-degree horizontal accelerations. However, both of these latter generation units encountered problems that made them inoperable after recording several small local seismic events. In July 1995, three of the newer SEMS IV models were installed adjacent to three platforms located in Federal waters and have recorded several earthquakes throughout the Southern California area.

Structural Inspection and Monitoring

The API RP 2A 19th edition guidelines, which have been incorporated by reference into MMS regulations, provide specific intervals for performing various levels of structural inspection in order to monitor the structural health of platforms. These inspections or surveys require that structural members both above and below the water line receive visual examination, cleaning of critical locations, and an evaluation of the corrosion protection system. Surveys are also required after direct exposure to major environmental and accidental events, such as earthquakes, hurricanes, and vessel impact. Survey reports are used by both regulators and operators to monitor structural health. If the survey indicates potential problems, the operator will submit a plan for corrective action to the regulator.

Conclusions

As a result of the Loma Prieta event of 1989, structural reassessment of offshore platforms has become a matter of concern for both regulators and industry. Reassessment guidelines for offshore platforms have been developed, based on input from the oil industry, academia, regulatory agencies, and the general public.

Structural inspection requirements have been in place for several years; reported damage, if left unrepaired, must be included in the structural model for reassessment purposes. Damage assessment of individual components is based on analytical and experimental results. The structural verification process ranges from an independent, third-party CVA review, to regulatory personnel themselves independently verifying the design/analysis provided by the operator.

For technical reasons unrelated to platform reassessment issues, MMS has not yet adopted the 20th edition of API RP 2A into its regulations. Both MMS and CSLC have worked very closely with the API and others over the years in trying to formalize reassessment criteria. The regulators are moving forward with rulemaking that will be consistent with the seismic reassessment aspects of the Supplement to the 20th edition of API RP 2A. The MMS is considering adoption of the 20th edition when certain technical matters are resolved, in which case separate rulemaking will not be necessary.

Reference

1. API Supplement 1 to Recommended Practice for Planning, Designing, and Constructing Fixed Offshore Platforms - Working Stress Design. 20th Edition, December 1996.

2. Eskijan, M. L. and L.A. Monahan, "Monitoring and Structural Reassessment of Aging Offshore Platforms: The Regulator's Viewpoint," presented at the First World Conference on Structural Control, 3-5 August 1994, Los Angeles, California.

3. Dyhrkopp, F. G., "Minerals Management Service Perspective on Platform Reassessment," presented at the ASME Energy and Environmental Expo '95, January 29-February 1, 1995, Houston, Texas.

4. Iwan, W.D., C.C. Thiel, G.W. Housner, and C.A. Cornell, "Seismic Safety Requalification for Offshore Platforms," prepared for the American Petroleum Institute, 1992.

5. "Proceedings of the International Workshop on Assessment and Requalification of Offshore Production Structures (1993), New Orleans, Louisiana..

6. "Proceedings of the International Workshop on Seismic Design and Reassessment of Offshore Structures," 1992, Pasadena, California.

7. "Public Policy Issues on the Seismic Requalification of Offshore Structures," co-sponsored by the MMS and the CSLC, Camarillo, California, November 1993.

8. Ricles, J. M. and T. Gillium, "Residual Strength of Damaged Offshore Steel Tubular Bracing," Proceedings of the Offshore Technology Conference, OTC Paper No. 6938, 1992.

9. Smith, Charles E. (1990), "Seismic Design Considerations for Offshore Oil and Gas Structures," Proceedings of the 21st Joint Meeting of the U.S.-Japan Cooperative Program in Natural Resources Panel on Wind and Seismic Effects, Tsukauba, Japan, 1989. National Institute of Standards and Technology Special Publication 776.

10. Sleefe, G.E. (1990), "The Long Term Measurement of Strong Motion Earthquakes Offshore Southern California." Proceedings of the Offshore Technology Conference, OTC Paper 6336.

CHAPTER 159

EVALUATING THE TSUNAMI RISK IN CALIFORNIA

Costas Emmanuel Synolakis [1], Dick McCarthy[2], Vasily V. Titov [3], and Jose Borrero [4]

Abstract We present existing pre–1985 predictions on the tsunami risk in California, and we evaluate them given the more recent field results and advances in computational methods. We find that the existing predictions quantitatively only account for farfield events, they don't consider the possibility of a CSZ event, they do not consider tsunami generation from nearshore strike–slip faults, and they use threshold computations which have been shown to substantially underpredict inundation. Using a field–validated state–of–the–art methodology, we present preliminary results on current tsunami simulations in California, suggesting a method for addressing the problem of estimating the tsunami risk in the State.

1. Introduction Tidal waves or tsunamis are long water waves generated by impulsive geophysical events such as submarine earthquakes, coseismic coastal or submarine landslides, and volcanoes. In the deep ocean, tidal waves may travel at speeds up to $500mph$, and they can propagate rapidly across the world oceans and strike distant shorelines.

Up until 1992, the tsunami hazard in California was primarily attributed to teletsunamis, i.e., to tidal waves generated farfield; pre–1985 hazard predictions had only identified an overall small risk, subject to disclaimers. As a result, most of the tsunami risk reduction in the US concentrated to mitigating the hazard in Hawaii and Alaska. The Cape Mendocino tsunami triggered more comprehensive analyses of historic events in California, and now the risk from locally generated (nearshore) tsunamis is believed to be high along the coast from Crescent City to Cape Mendocino, moderate, south of the Cape to north of Monterey, high, south of Monterey to Palos Verdes, and moderate south of PV to Sand Diego (McCarthy et al, 1993).

In the period 1992–1996 and immediately following the Cape Mendocino event, eight large earthquakes, generated tsunamis with runup heights ranging from five to thirty meters around the Pacific; before these events, the last major tsunami of similar magnitude occured in 1983. These events caused extensive inundation and claimed the lifes of at least 2000 people. The post—event surveys produced field data, at exactly the time when inundation codes had started breaking the computational barriers of the notoriously difficult shoreline calculation, and as seismological interface models started producing seafloor deformation contours instead of average deformations. Also, the surveys identified previously unrecognized generic land and seafloor features which greatly increase the inundation potential, as well as unidentified generation

[1] Professor of Civil Engineering University of Southern California, Los Angeles, California 90089–2531, costas@usc.edu.

[2] Executive Director, Seismic Șafety Commission, 1900 K Street, Sacramento, Calif. 95814.

[3] Research Scientist, NOAA–PMEL, Seattle, Washington.

[4] Research Assistant,University of Southern California, Los Angeles, California 90089–2531.

mechanisms; before the 1994 Mindoro event, strike–slip faults were not believed to be capable of triggering tsunamis.

While the first inundation computations of the 1992–1994 period underestimated the inundation height, some inundation models have now proven capable of modeling extreme events accurately; for example, the $30m$ high runup values and $20m/sec$ overland currents measured and inferred during the Hokkaido–Nansei–Oki tsunami of 1993 have been modeled and explained thoroughly (Titov, 1997, Titov & Synolakis, 1997).

The new inundation models allow for quantitative evaluation of the risk from nearfield tsunamis. Also, they allow deductions as to how appropriate the use of older results may be. Here, we will discuss some preliminary results of tsunami inundation in Southern California and present a methodology for evaluating the tsunami risk in the entire State and elsewhere.

2. The existing analysis of tsunami hazard in California.

The most comprehensive calculation of tsunami hazards for California is the work of Houston and Garcia (1974) and of Houston (1980), both of which focused on the hazard in Southern California from farfield events. McCulloch (1985) also focused on the hazards in the Los Angeles region primarily from farfield events, but also considered several local events. Satake and Sommerville (1992) analysed the Lompoc 1927 earthquake and the associated local hazards. In a seminal review, McCarthy et al (1993) analysed the historic records of tsunamis in California and predicted qualitatively the hazard over the entire State.

As a preamble, we will define the terms runup and inundation, which are sometimes misused. Wave *runup* is the rush of water up a structure or a beach; it is called the uprush. The maximum runup is the vertical height above stillwater that the rush of water reaches as it climbs onshore. Specific knowledge of the maximum wave runup on a given beach is essential both in shore protection and in the design of coastal structures. Sometimes the term tsunami height is used to refer to the runup height, begging the question the tsunami height at what depth ? Inundation refers to the horizontal distance the wave penetrates inland. Depending on land use, either runup or inundation are relevant, and most often both. In addition the wave–front velocity as the wave strikes the shoreline is an important design parameter. An inundation computation includes predictions of runup heights, inundation distances and inundation currents.

2.1 The hazard from farfield events

Houston and Garcia (1974) used a combination finite–difference solution (FD) and analytic solution of the linearized shallow–water wave equations (LSW) to calculate tsunami propagation, except in the Santa Monica and San Diego bays where they used a finite–element solution (FE) to resolve possible local resonance effects. They argued that the only reliable data for defining source cahracteristics at that time were from the 1964 Alaskan and the 1960 Chilean earthuakes. Based on these data, they approximated the initial ground deformation by a hypothetical uplift mass of ellipsoidal shape, about $600miles$ long with an aspect ratio of $1:5$ and maximum vertical uplift of $8m(25ft)$ to $10m(30ft)$. They then divided the Aleutian trench into 12 segments and calculated the wave evolution from each segment, and repeated the procedure for tsunamis from the Peru–Chile trench. Some representative results for the 100–year (R_{100})

and 500–year (R_{500}) tsunami runup heights are summarized below.

Table 2
Houston and Garcia (1974) predictions for the 1964 Alaskan tsunami

Location	R_{100}	R_{500}
La Jolla	6.1ft	12.7ft
San Onofre	5.7ft	11.1ft
Newport Beach	6.1ft	10.8ft
Long Beach	7.0ft	9.7ft
Dockweiler Beach	9.6ft	15.3ft
Topanga	10.4ft	16.6ft
Ventura	10.5ft	21.7ft

It is important to note that for their time, the methodology used and and the calculations was groundbreaking, not only in its scope, but also in the combined use of analytical and numerical methods. Houston and Garcia (1974) first solved a linear form spherical long wave and they propagated the tsunami from the source to the edge of the continental shelf, by using a finite difference model; at the continental shelf they derived an analytic expression to match the outer and inner wave applitudes, and then they used that expression to derive a simple amplification factor for a sinusioidal tsunami. Even though they did not match the slope of the water surfaces in the inner and outer continental shelf regions, their results compared extremely well with measurements from tidal gage records, whenever suitable tidal gage records were available that did not need additional signal processing to filter harbor resonance effects. These comparisons are presented below.

Table 2
Houston and Garcia (1974) predictions for the 1964 Alaskan tsunami

Location	*Predicted*	*Observed*
Alamitos Bay	2.2ft	1.8ft
Santa Monica	2.8ft	2.6ft
Avila Beach	3.7ft	4.4ft
Crescent City	7.3ft	8.0ft

The excellent accuracy of these predictions of the 1964 Alaskan event allowed the extrapolation of the results for nearfield events, despite Houston and Garcia's (1974) disclaimers, thereby masking three important aspects of the inundation calculations used. One, nearfield events are quite three dimensional, and the same methodology used for farfield events may not be appropriate for nearshore quakes. Two, in the middle 1980's it became apparent that superposition of sinusoids is not as straightforward as previously assumed; the reason is that during the reflection process –not accounted in Houston and Garcia's calculation– there is a phase lag introduced which is frequency dependent (Synolakis, 1987, Liu et al, 1991). Three, comparisons of numerical model predictions with the field data from the field surveys of the 1992–1996 tsunamis suggested that even small scale nearshore features can influence inundation

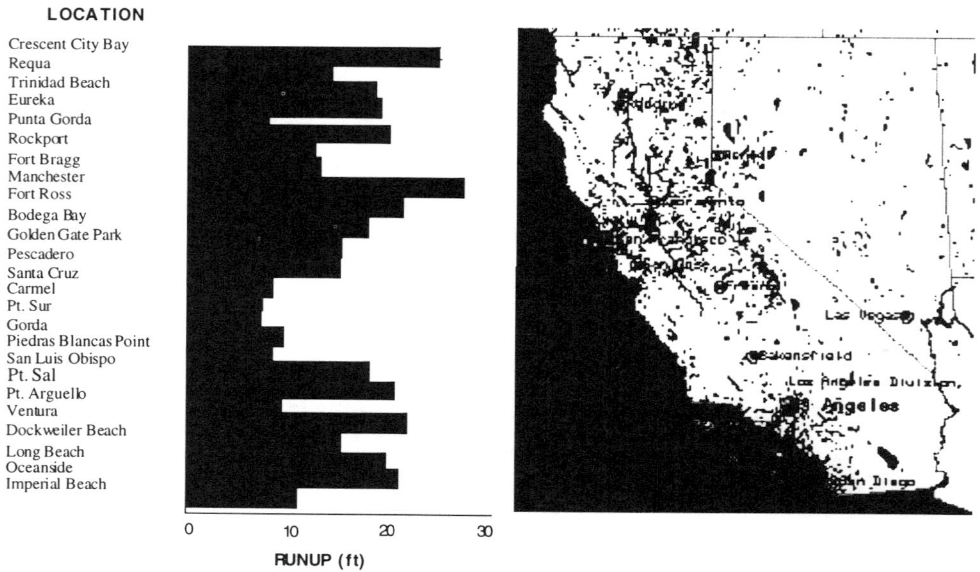

Figure 1. Houston and Garcia's (1978) predictions for the 500-year farfield runup heights along the California coastline.

to first order, casting doubt on predictions from coarse grid computations, because they may miss extreme events.

Houston (1980) performed another comprehensive study for tsunami predictions in California in a series of two reports, Houston and Garcia (1978), and Houston (1980). By this time, numerical solutions of the shallow water wave equations had started being available; they solved using finite difference algorithms the nonlinear form of the shallow–water wave equations (NSW) including frictional terms. The results from these computations are shown in figure 1.

Even though these newer computations were an improvement over the 1974 study, it is important again to revisit the computational assumptions with which these 1978 and 1980 predictions were arrived at.

- One, only farfield events from Alaska and South America were considered.
- Two, the Pacific ocean was modeled as a $500m$ constant depth basin with a $2mile$ square grid. In the nearshore region,the bathymetry was also modeled with a $2mile$ square grid, and the transition between the offshore and nearshore grids was at least "one and one half wavelength of a $30min$ wave.
- Three, the computational boundary was a vertical wall at the shoreline, i.e., there were no inundation computations. Houston (1980) noted that the runup elevations, i.e., the elevation of the maximum inland penetration of the tsunami may not equal shoreline elevations at locations where dunes prevent flooding, or if the land is flat, where inland flooding maybe extensive. Houston (1980) used the best available methodology of the time, i.e., threshold modeling which propagates the wave close to the shoreline, where a vertical wall is placed to facilitate the numerical computations. As shown later, wall–type calculations not only underpredict the runup substantially, but they may also miss extreme events. Although the degree of under prediction varies with the local topography, it is often a factor of two and sometimes it has been reported as large as five.

Nonetheless, even under the fairly restrictive assumptions and the 1970's technology with which the above predictions had been arrived at, these results on their own are a substantial cause of concern, a fact not fully recognized until the post 1992 tsunami field surveys.

2.2 The hazard from nearshore events.

As noted earlier all existing studies had focused on farfield events. Houston (1980) noted that the frequency of occurrence of locally generated tsunamis in southern California is not known and predictions of locally generated tsunami elevations wre beyond the scope of his reports. McCulloch's (1985) study is a seminal bwork on tsunami hazard potential in Southern California. McCulloch relied on Houston's results for farfield tsunamis and then used seismological data to make predictions for nearshore events. Since he did not use any hydrodynamic model, he relied on several empirical formulae developed in Japan. These formulae had been extensively used before 1992; since then the availability of high quality runup data from the 1992–1996 event for many areas around the Pacific have shown that these formulas are only applicable in Japan and that they substantially underpredict the runup elsewhere.

McCulloch (1985) tried to explain the fairly moderate predictions of Houston by arguing that in "California, the major tsunamis generated in the Pacific–Eurasian plate boundary that have repeatedly decimated the Japanese coast are reduced to small but detectable waves after crossing the Pacific." Referring to local tsunamis, he wrote that four have been observed along the southern California coast during the period of record from 1812 to 1975, for three of which events the wave heights were not indicated. He continued by arguing that in Southern California the displacement between the North American and Pacific Plates is accommodated in part by movements along strike–slip faults, some of which are in the offshore borderland, but that however, there was some suggestive evidence of episodes of vertical displacement capable of tsunami generation associated with the offshore extension in the Palos Verdes Hills reverse fault to the southwest in an area otherwise dominated by strike–slip displacement. McCulloch (1985) did not provide any runup estimates from nearshore events, that was beyond the scope of his work. However, his implication is clear that the hazard is low. It is interesting to examine why.

- One, before the 1994 Mindoro event (Imamura et al, 1995) where a strike–slip fault generated a moderate tsunami which caused spotty, but extensive inundation, the hazard from nearshore strike–slip faults was not recognized.
- Two, he relied on the relations between tsunami size and, earthquake magnitude and hypocentral depth established for Japanese earthquakes. Using them, he argued that a local seafloor earthquake having a magnitude of 7.5 and a hypocentral depth of $4km$ to $14km$ could produce a tsunami accompanied by a runup height of $4m(13ft)$ to $6m(20ft)$. In 1985, a $6m(20ft)$ tsunami may have appeared a marginal hazard, even though the tsunami height in the 1964 Alaskan tsunami in Crescent City was about $20ft$. The 1992–1996 post–event field surveys have shown that even a $4m(13ft)$ tsunami can cause extensive damage and flooding in flat coastlines, such as those in Santa Monica bay or in Orange and San Diego counties.

McCulloch (1985) also presented a rather interesting analysis of the cost associated with tsunami events. He compared the cost of damage in California from the 1964 Alaskan tsunami the cost of all 26 earthquakes in historic record, and he concluded that the cost was 0.2% in 1983 dollars. Although the cost of tsunami damage was 0.2% of earthquake damage, the bulk of the cost was incurred in Crescent City, where 11 of the 123 tsunami deaths occurred in California.

McCarthy et al (1993) performed a systematic analysis of all historic and possible tsunami hazards in California and they qualitatively calculated the tsunami hazard in California as *high* along the coast from Crescent City to Cape Mendocino, *moderate* south of the Cape to north of Monterey, *high*, south of Monterey to Palos Verdes, and moderate south of Palos Verdes to Sand Diego. Provocative as these results were, they remain to be quantified with studies of scenario events.

The most recent evaluation of earthquake hazards in California is by the Working Group on California Earthquake Probabilities (1995), working for the Southern California Earthquake Center (SCEC). Examining Southern California, they identify the Palos Verdes, Santa Cruz Island and Santa Rosa Island faults at type B, i.e., faults which are active and documented, and which may have characteristic earthquakes, but for which information about their behaviour is lacking. The most studied of these faults is the Palos Verdes fault, also identified by McCulloch (1985) for its tsunamigenic potential. The fault is said to be dominated by lateral strike–slip motion with a slip rate of $3mm/yr$, with vertical rate estimates ranging from $0.2 - 3.5mm/yr$; however, there are no data on earthquake reccurence intervals. and the SCEC analysis quoted above relies on standard probabilistic hazard assessment methods. The same analysis groups

Source Zone	Major Offshore Faults[1]	Major Submarine Canyons	Maximum Historical Earthquake[2]	Historical Taunaml
Crescent City to Cape Mendocino	Little Salmon Fault (T) Mad River Fault Zone (T) Mendocino Fault (S) Cascadia Subduction Zone (T)	Trinity, Eel, Mendocino, Mattole	M_S=7.4 (1923) M=7.2 (1923)	1.1-m (1992)
Cape Mendocino to San Francisco	San Andreas Fault (S) Point Reyes Fault (T)	Spanish, Delgada, Vizcaino, Noyo, Navarro, Arena, Bodega	M=7.7 (1906)	0.1-m (1906)
San Francisco to Monterey	San Gregorio Fault (S)	Pioneer, Ascension, Monterey	M=7.1 (1989)	0.3-m (1989)
Monterey to Point Arguello	Hosgri Fault Zone (RS) Santa Lucia Bank Fault (RS?)	Sur, Lucia	M_S=7.3 (1927)	3-4 m (1812)
Point Arguello to Los Angeles (Santa Barbara Channel and Santa Monica Bay)	Santa Barbara Channel faults (T) Anacapa-Dume Fault Zone (RS) Santa Monica Fault (T)	Arguello, Hueneme, Mugu, Dume, Santa Monica, Redondo	M_I=7-7½ (1812)	3-4 m (1812)
Los Angeles to San Diego (Inner Borderland)	San Clemente (R) Catalina – San Diego Trough (S, RS?) Palos Verdes(RS)—Coronado Bank (NS) Newport-Inglewood – Rose Canyon (S)	San Gabriel, Newport, Carlsbad, La Jolla, Coronado	M=6.25 1933	Uncertain (1862, 1933)
Northern Channel Islands to San Nicolas Island (Northern Outer Borderland)	East Santa Cruz Basin Fault Zone (S) Ferrelo Fault Zone (S) San Nicolas Island Escarpment	Santa Cruz	M_L=5.1 (1969)	?
San Nicolas Island to Mexican Border Southern Outer Borderland	East Santa Cruz Basin Fault Zone (S) Ferrelo Fault Zone (S)	unnamed	M_L=5.3 (1948)	

EXPLANATION

Fault Character: T = Thrust of Reverse; RS = Reverse-Oblique; S = Strike-Slip; NS = Normal-Oblique
Magnitude: M = Moment Magnitude (M_W); M_S = Surface Wave Magnitude; M_L = Local (Richter) Magnitude;
M_I Seismic Intensity Magnitude; ? = Magnitude Not Given or Unknown

Table 3. Local tsunami sources in California after McCarthy et al (1993).

all other offshore faults, i.e., the ones located in the vicinity of the Channel Islands, and the ones off the Santa Barbara channel as type C, i.e., as faults where evidence for a preferred characteristic earthquake is lacking, although the fault length suggests one.

Table 4

SCEC predictions for Southern California offshore faults

Fault Name	*LimitingEvent*	*FullRate*
	M_x	$10^{-3}/yr$
Palos Verdes	7.10	2.54
Santa Cruz Island	6.94	2.19
Santa Rosa Island	7.05	1.37
Offshore Channel	7.05	16.81
Offshore Islands	7.05	23.37

We emphasize that the SCEC results do not directly refer to tsunamigenic faults; yet the slip rates identified and the magnitude of characteristic events is such that, it is highly likely the characteristic earthquakes along these faults would trigger moderate to large local tsunamis.

3. Methodology for evaluating the hazard.

Clearly, the evaluation of the tsunami hazard to the California coastlines involves both the probabilistic hazard assessment of different sources and the hydrodynamic computation of the tsunami evolution from the source to the target coastline. In this section, we will focus of the hydrodynamic computations, and we show that numerical methods now exist that can model even extreme inundation flows.

Important as it is for hazard mitigation planning for great tsunamis, the calculation of three–dimensional tsunami inundation flows remains formidable. Up until recently (Shuto, 1991, Liu et al. , 1995, Synolakis , 1995 Takahashi et al. 1995), field–applicable inundation predictions were only possible with what is referred to as threshold–type models; the latter only propagate the tidal wave from its generation region up until the $10m$ depth contour, and then baptize the wave height at $10m$ as the runup height † The field surveys of the 1992–1995 tsunamis showed that threshold models sometimes produced predictions differing from field observations by factors ranging from five to ten. These differences have raised concern because threshold models have been used extensively in in civil defense applications, and they have fueled controversy as to whether the poor predictions were due to insufficient or incorrect ground deformation data affecting the initialization of the calculations, or to inherent limitations in the propagation models.

The $M_w = 7.8$ Hokkaido–Nansei–Oki (HNO) earthquake of July 12, 1993 allowed the measurement of high–quality runup data and the inference of fairly unambiguous ground deformation contours due to its proximity to seismic instrument arrays, in a locale where high–resolution bathymetry and topography data exist. Hydrodynamic computations reported to date (Satake and Tanioka, 1995, Takahashi et al., 1995) using the NSW approximation have reproduced to first–order the runup heights distribution along the coast of Hokkaido and to some extent around Okushiri, but neither the wave current velocities nor the extreme runup values observed in the front of the island; at best (Takahashi et al., 1995), predictions for the extreme measurements have differed by a factor of two from the field data, even though quantitatively correct elsewhere.

To explore the relative effects and magnitudes of inundation heights and inundation–flow velocities, and to determine whether the modeling difficulties are due to inherent limitations of the SW models or their application outside their range of validity or to splash, Titov and Synolakis (1997) investigated this event using a newly proposed inundation model known as VTCS–3. VTCS–3 differs from previous model in that it does not include any dissipation or artificial viscocity terms. These terms are often used to model the prototype friction, but their choice has been shown repeatedly not to affect the predictions, except possibly for extreme

---

† If the calculation threshold had been placed at the original shoreline, the tsunami rundown would almost always expose the seafloor necessitating moving boundary computations, whose avoidance is the whole purpose of not performing inundation computations. Therefore, it is necessary to place the threshold at a depth which would allow the wave to move up and down the wall without exposing the seafloor; for reference, this depth has historically been $10m$.

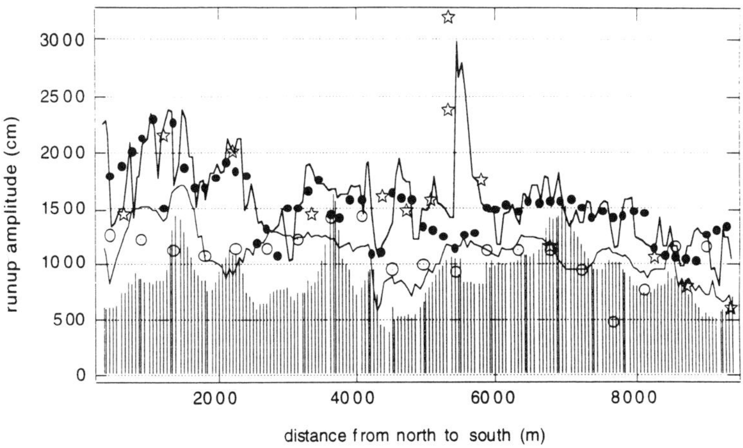

Figure 2. Comparisons of field measurements (stars) of runup around Okushiri island with the computed runup heights using different grid resolutions; heavy solid line shows computed results with $50m$ resolution, solid and empty circles with $150m$ and $450m$ grid resolutions respectively. The vertical bars are threshold model resuls with $50m$ grids. The light solid line is the inferred runup at $10m$ from the $50m$ inundation computations.

events. Since, at best these terms are ad–hoc and, at worst, chosen to simply match predictions with a specific data set, it is best to avoid using them whenever possible. VTCS–3 has been shown in excellent agreement with large–scale laboratory experiments (Titov and Synolakis, 1995, Titov, 1997), even for mildly plunging waves.

Figure 2 shows the distribution of computed runup heights around Okushiri using VTCS–3 and it compares them with the field measurements (HTSG, 1993, Shuto and Matsutomi, 1995). The computed distribution of the maximum runup heights reproduces most features of the field measurements well. Note the variation in the computed runup height over small longshore distances, verifying that the variation in the field data over the same areas is due to hydrodynamic effects and not to random observation scatter. More comprehensive information can be found in Titov and Synolakis (1997). Overall, for this extreme tsunami event (notice the $30m$ runup measurments) the SW model appears adequate for modeling runup heights, suggesting that wave breaking and bottom friction are apparently not important to first–order for predicting inundation.

The same figure also presents additional results to evaluate the effect of other aspects of numerical modeling, such as the grid resolution and the $10m$–depth threshold. Note that for identical grid resolutions, interrupting the computation at $10m$ using reflective "wall"–type boundaries underpredicts the runup by a factor of two (vertical lines in figure 2), as compared with the inundation computations (heavy solid line). Even the computed maxima at $10m$ monitored during the "full" inundation calculation (thin solid line) differ substantially from the computed wave height maxima from the $10m$–threshold calculations, suggesting that any tsunami "inundation" predictions done with threshold type models need to be re–evaluated.

4. Evaluating the hazard in California; what have we learned ?

Without detailed inundation calculations of both farfield and nearshore tsunamis it is clearly impossible to evaluate the hazard. As an example, to demonstrate the hazard from one particular event, we present in figure 3 predictions of the runup height distribution in Southern California from a magnitude 7.0 event along the San Clemente compression fault. Ward and Valensise (Pers. Comm.) have argued that the topographic data in this locale are best explained by the motion of two very shallow dipping blind faults. To perfom this simulation, we assumed a fault area of $100km \times 50km$ and vertical displacements ranging from $3m$ to $6m$.

Figure 3 also compares the predictions from threshold computations to inundation computations. Note that near the extrema of the runup distribution, the threshold computations underpredict the runup by a factor of two.

These preliminary results from inundation modeling in Southern California, and the experience from the 1992–1996 post–tsunami field surveys suggest following points have to be carefully considered in assessing the hazard from the existing pre–1985 simulations.

1. Threshold–type calculations such as those of Houston and Garcia (1974) or Houston (1980) for farfield events probably underestimate the inundation for some locales in Southern California. Even when threshold type calculations produce reasonable estimates for the wave height, they don't provide any data on the tsunami currents which may quite large. For example, overland flow as observed in Wuhring during the 1992 Flores event or in 1994 in South–East Java, even with moderate $3m(10ft)$ heights, this type of overland flow may produce extensive devastation due to sizable currents and strong rundown flows. Such flows may be possible in low lying areas around Malibu, Seal Beach, Huntington Beach and South Orange County.

2. Any prediction should be done with models which have been validated with both laboratory and field data, particularly when the models include ad–hoc additional frictional terms which even in high–resolution models tend to produce results which miss extreme runup heights and currents.

3. Strike–slip faults may be inefficient generators of tsunamis as compared to thrust faults, but nearshore strike–slip faults may cause moderate tsunamis and sizable inundation currents. Before the 1994 Mindoro earthquake, the tsunamigenic potential of strike-slip faults was unrecognized. Imamura et al (1995) wrote that, the strike-slip fault of the 1994 Mindoro event generated a $2m(6f)$ tsunami, which nonetheless caused a 6,000 ton diesel–power generation barge to break away from its mooring and be transported one mile inland. The Palos Verdes fault may produce a similar tsunami with enhanced risk to the Port of Los Angeles and Long Beach. Tsunamis generated by submarine landslided triggered by onshore strong ground shaking due to strike–slip faults need also to be carefully evaluated in assessing the hazard.

4. The changes in land use intensify the risk, as compared to the historic events, the last of which in California was over 30 years ago. If one conjectures that the CSZ could affect the

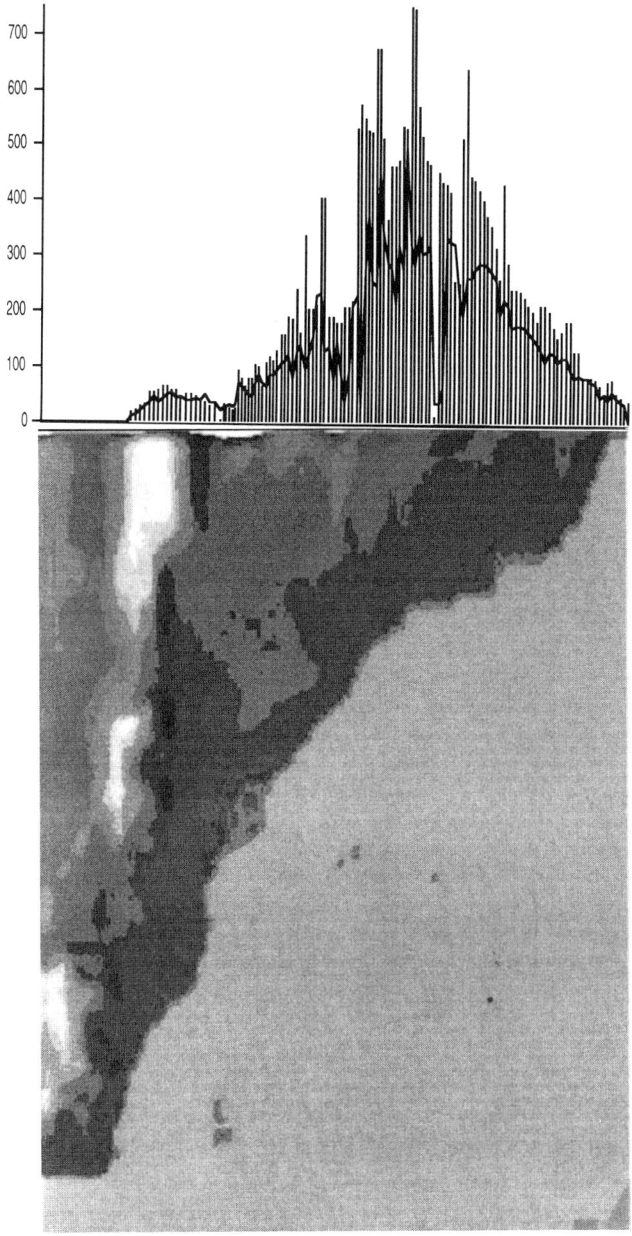

Figure 3. The maximum runup height distribution in Southern California for a magnitude 7.0 event along the San Clemente fault, using threshold type computations (solid line) and using inundation computations vertical lines. Clearly the threshold computation underpredicts the runup substantially.

Port of LA as the Alaska 1964 quake affected Crescent City, the damage would be far larger than the $32 million damage in Crescent City in 1971 dollars, and this does not even account for economic costs associated with loss of commercial use. Even in locales where the tsunami hazard may be small, the land use may render the overall risk high. The rapid onset of the hazard and its long duration also intensifies the risk from nearshore events.

5. The 500–year tsunami heights in California may be as high as $10m(30ft)$ in most locales, with the possibility that local topographic effects may produce local amplification which may result in $30m(90ft)$ heights, as those observed recently in Flores in 1992 and in Okushiri in 1994.

Acknowledgements This work was supported by a grant from the National Science Foundation. The assistance of Dr. Utku Kanoglu and Mrs. Irina Hoffman is greatly appreciated.

References

Imamura, F., Synolakis, C.E., Titov, V.V., Lee, S., *Pure and Applied Geophysics*, **144**, 3/4, 839–855, 1995.

Liu, P., L–F., M. J. Briggs, U. Kanoglu, C. E. Synolakis, *J. Fluid Mech.* **302**, 259–285, 1995.

Houston, J.R., and Garcia, A.W., Type 16 Flood Insurance Study, USACE, WES Report H–74–3, 1974.

Houston, J. R., Type 19 Flood Insurance Study, USACE, WES Report HL–80–18, 1980.

Geist, E. L., Local tsunamis and earthquake source parameters, *Advances in Geophysics*, **39**, 1997.

Liu, P. L..-F., Synolakis C.E.., Yeh, H.H., 1991, *J. Fluid Mech.,* **229**, 675–688.

McCarthy, R.J., Bernard, E.N., Legg, M.R., Coastal Zone93, Proc. Amer. Shore and Beach Preserv. Assoc., ASCE, New Orleans, Louisiana, 2812–2828, 1993.

McCulloch, D.S., Evaluating Tsunami Potential, USGS Prof. Paper 1360, 374–413, 1985.

Myers, E. P. and A. M. Baptista, *Pure and Applied Geophysics*, **144**, (3/4), 769–802, 1995.

Satake, K. and Y. Tanioka, *Pure and Applied Geophysics*, **144**, (3/4), 803–822, 1995.

Shuto, N., *Numerical simulation of tsunamis*, in Tsunami Hazard edited by E. Bernard, 171–191, Kluwer Academic Publishers, Doldrecht, The Netherlands, 1991.

Shuto N. and H. Matsutomi, *Pure and Applied Geophysics*, **144** (3/4), 649–664, 1995.

Synolakis, C.E., 1987, *J. Fluid Mech.,* **185**, 523–545.

Synolakis, C. E. et al.,*EOS, Trans. AGU* , **76** (26), 257, 261–262, 1995.

Synolakis, C.E., *Science* , **270**, 15, 1995.

Tadepalli, S. and C. E. Synolakis, *Physical Review Letters*, **77**, 2141–2144, 1996.

Takahashi, To. et al., *Pure and Applied Geophysics*,**144**, (3/4), 747–768, 1995.

Titov, V.V. and Synolakis, C.E., , *J. Wat. Ports Coast. Eng.,* **121**, 308–316, 1995.

Titov, V. V. and C. E. Synolakis, Extreme inundation flows during the Hokkaido– Nansei–Oki tsunami, *Geophysical Research Letters*, June 2, 1997.

Titov, V. V. Numerical modeling of long wave runup, Ph.D. thesis, 130pp., University of Southern California, Los Angeles, California, January 1997.

Working Group on California Erathquake Probabilities Seismic hazards in Southern California, 1994–2024, *Bull Seism. Soc. Amer.*, **85**, 379–439, 1995.

Yeh, H., Imamura, F., Synolakis, C., Tsuji, Y., Liu, P., Shi, S., *EOS, Trans. AGU*, **33**, 1993.

CHAPTER 160

TIDAL CREEK CHANGES AT THE SONOMA BAYLANDS RESTORATION SITE

John R. Dingler and David A. Cacchione[1]

Abstract

Over the past 150 years, human activity has had a major impact on tidal wetlands adjoining the San Francisco Bay-Delta estuary Growing concern about the effect of this change on the ecology of the estuary has prompted Bay area managers to attempt to reclaim tidal wetlands. The Sonoma Baylands Restoration Project is designed to use dredge material to convert 348 acres from farmland to wetland. This paper describes changes to a tidal creek that flows from that restoration site to San Pablo Bay (north San Francisco Bay) through an existing tidal wetland during different phases of the project. Hydrologic measurements near the bottom of the creek and cross-creek profiles show how the creek responded to non-tidal flow conditions introduced by filling the site with dredge materials. At the time of this study, the creek had deepened by approximately 40 cm but had not widened.

Introduction

Within two years after the discovery of gold in the Sierra Nevada foothills in 1848, San Francisco had changed from an isolated trading outpost to a bustling community. Subsequently, the entire San Francisco Bay-Delta estuary experienced such extensive population growth that most of the bay's shoreline has been modified (fig. 1). As a result, the estuary is considered the major United States estuary most modified by human activity with respect to diversity of change (Cross and Williams, 1981; Nichols *et al*, 1986). For example, Gilbert (1917) estimated that 1400 km^2 of freshwater tidal wetlands and 800 km^2 of saltwater tidal wetlands fringed the estuary before 1850. By the early 1920's, all of the freshwater tidal wetlands had been reclaimed, mostly for farmland (Kockelman *et al*, 1982), and, at present

[1] Oceanographers, U.S. Geological Survey, 345 Middlefield Road, Menlo Park, CA 94025

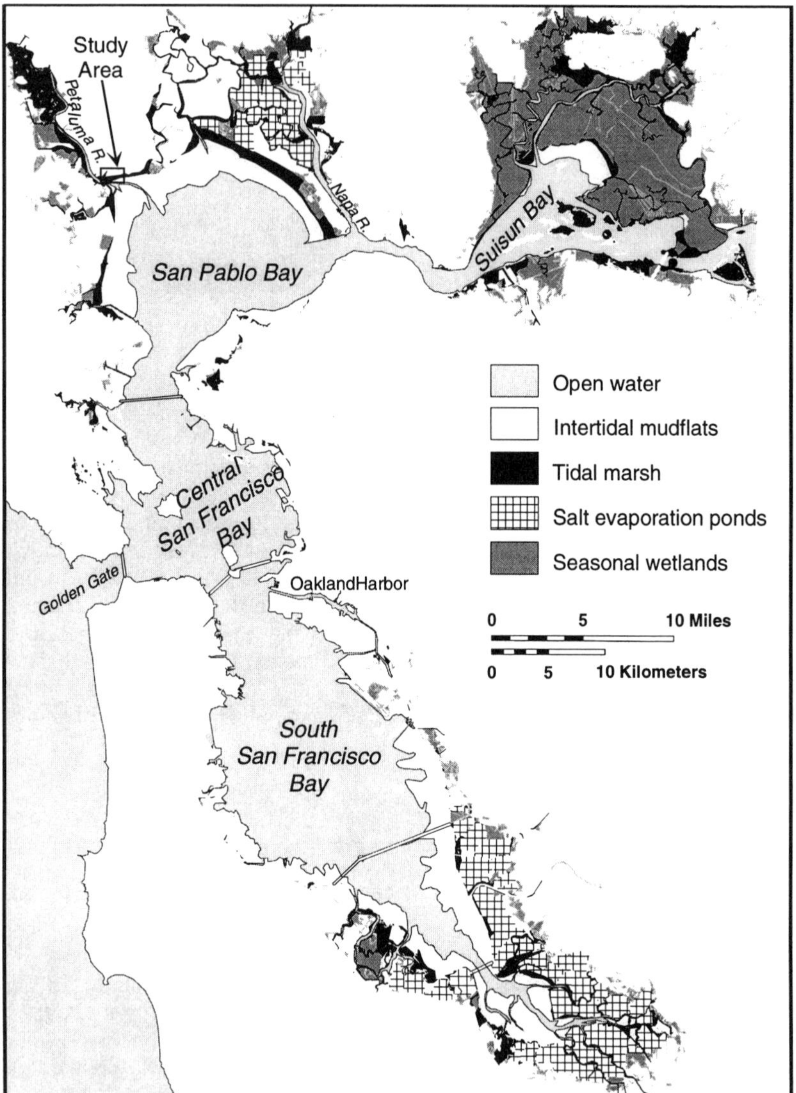

Figure 1. Present-day distribution of shoreline types in the San Francisco Bay-Delta estuary. Before the 1848 gold rush in the Sierra foothills, the intertidal mudflats were less extensive and the seasonal wetlands, salt evaporation ponds and tidal wetland were all tidal wetlands.

all but 125 km^2 of the saltwater tidal wetlands have been reclaimed as farmland, salt evaporation ponds, or residential and industrial land (Atwater *et al*, 1979). That change represents a huge loss of habitat for the bay's wildlife and those aquatic species that use the wetlands during breeding or maturing stages. Furthermore, the remaining tidal wetlands are still threatened by development, erosion, pollution, and sea-level rise.

To preserve a viable wetland habitat around the bay, management agencies have begun to identify the threats to existing wetlands and to determine viable strategies for wetland restoration. Those agencies, in coordination with other interested groups, have initiated a major wetland restoration endeavor, the Sonoma Baylands Restoration Project, at a site adjacent to the mouth of the Petaluma River (fig. 1; box in upper left). At the restoration site, material dredged from the mouth of the Petaluma River and from the Oakland Harbor was pumped as a slurry into an impoundment until the sediment surface was approximately at mean bay level. After letting the sediment consolidate for several months, part of the impounding levee was breached to allow tidally driven circulation in the tract.

The Sonoma Baylands Restoration Project, which bay managers hope will be a model for future wetland restoration around the bay, requires a detailed monitoring program to measure the changes to the project site and the wetland that exists between the site and San Pablo Bay. An important aspect of the restoration-project design is that existing tidal creeks change in size to handle the increased flow expected into and out of the site. The design model for the project predicts that creek cross-sectional area will increase through a combination of deepening and widening. This study is designed to provide information about changes in the cross-sectional area of one creek through measuring current, salinity, and suspended sediment patterns and tidal-channel development. Hydrologic measurements were made every 15 minutes at a bridge that crosses the tidal creek approximately 45 m bayward from the breach at weir 2, and creek profiles were taken along bridge.

Setting

The Sonoma Baylands Restoration Project covers 348 acres of diked farmland in southwestern Sonoma County. The site is adjacent to the mouth of the Petaluma River and the northern shoreline of San Pablo Bay (figs. 1 & 2). Tidal wetland lies between the site and both the river and bay. To create the restored wetland, dredged material from the mouth of the Petaluma River and from the Oakland Harbor was pumped as a slurry into the site to bring the elevation from approximately 2 m below Mean Higher High Water (the National Geodetic Vertical Datum of 1929) to slightly below the expected equilibrium elevation of 1.04 m (3.4 ft), which is the elevation of the adjacent tidal marsh. The fill elevation was chosen to be below the

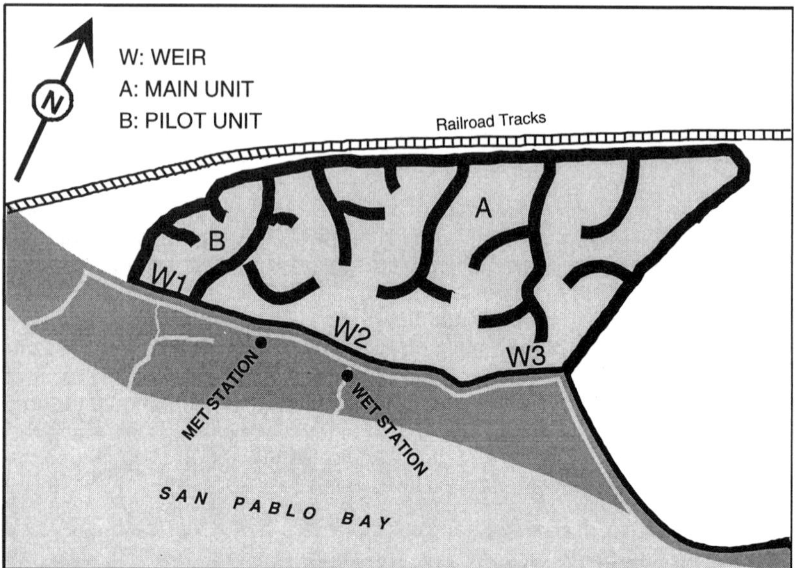

Figure 2. Sketch of the Sonoma Baylands Restoration Project. Wide black lines represent levees and berms. Dark grey represents existing tidal wetland. Light grey represents water in tidal creeks and bay and covering material pumped into the site.

equilibrium elevation to encourage proper channel formation and to allow for natural on-site sedimentation during the restoration process (U.S. Environmental Protection Agency, 1996).

Site preparation by the U. S. Army, Corps of Engineers included adding a levee on the landward side of the site, strengthening and adding weirs to the existing dikes, and adding several low, interior berms (fig. 2). The berms are designed to limit the length of wind-wave fetches across the site and to direct the formation of major tidal channels away from the toe of the landward-side levee. An internal levee divided the site into a 29-acre Pilot Unit, which received the Petaluma River material, and a 260-acre Main Unit, which received the Oakland Harbor material.

The Pilot Unit was filled during October-November 1994 and was exposed to tidal action on 24 January 1996 by breaching the levee at weir 1. The Main Unit was filled during May-November 1995, and was exposed to tidal action on 25 October 1996 by breaching the levee at weir 2. Between filling and breaching, the site sat with enough water to cover all the emplaced material.

Procedure

Instruments for this study and a companion study on wind dynamics were installed in the existing tidal wetland bayward of the site (fig. 2). A 10-m tall meteorological tower was installed at the 'Met' site, and a package of instruments to measure hydrologic conditions in the tidal creek flowing through the wetland was installed at the 'Wet' site. The 'Wet' station data consist of current, water pressure, elevation from an arbitrary elevation to the water surface or bottom, temperature, conductivity, and turbidity. The instruments were mounted from a bridge that spans the tidal creek with the altimeter above the bridge and the other instruments below. Throughout the study period, the altimeter elevation remained unchanged at 88 cm above an arbitrary bridge datum. Initially, the other instruments were 92 cm below the datum except for the current meter, which was 105 cm below. In October 1995, all the instruments were lowered to 108 cm, and in July 1996 to 121 cm below the datum.

Changes to the tidal creek were measured by surveying across the creek at the bridge with an infra-red distance meter. Surveys were conducted before filling the Main Unit, during filling, while the Main Unit was sitting quiet, and after the breach at weir 2 (fig. 3).

Results and Discussion

In May 1995, before material was pumped into the Main Unit, the Wet station instruments were almost always out of the water except during highest water (fig. 4a). Maximum flow velocity at flood slightly led the highest water level, which is consistent with the pattern produced by a progressive tide. When the Main Unit was being filled, the flow of slurry water out of the site was enough to keep water continuously in the creek, reduce the maximum flood velocity to approximately 0 cm/s, and produce a bayward flow of approximately 60 cm/s during much of the ebb (fig. 4b). After the pumping ended in November 1995, the current pattern was similar to the May 1995 pattern (figs. 4c,d). After the levee was breached, water remained in the tidal creek during the tidal cycle, and maximum velocities ranged from approximately 100 cm/s toward the site on the flood to approximately 80 cm/s bayward on the ebb (fig. 4e).

The currents shown for October 1995 and December 1996 (figs. 4b,e) are notably different from the currents at the other three periods (figs. 4a,c,d). During the latter times, there was little or no flow into and out of the site through weir 2, so that the flow in the creek was dominated by the tidal activity in the bay. During the former periods, there were strong flows through weir 2 or the breach that replaced it. In October 1995, slurry water that was pumped into the site flowed out of weir 2, which created a continuous bayward flow that was only reduced to approximately zero during

Figure 4. Schedule for surveys across the tidal creek at the bridge relative to the filling of the main unit and breaching of weir 2.

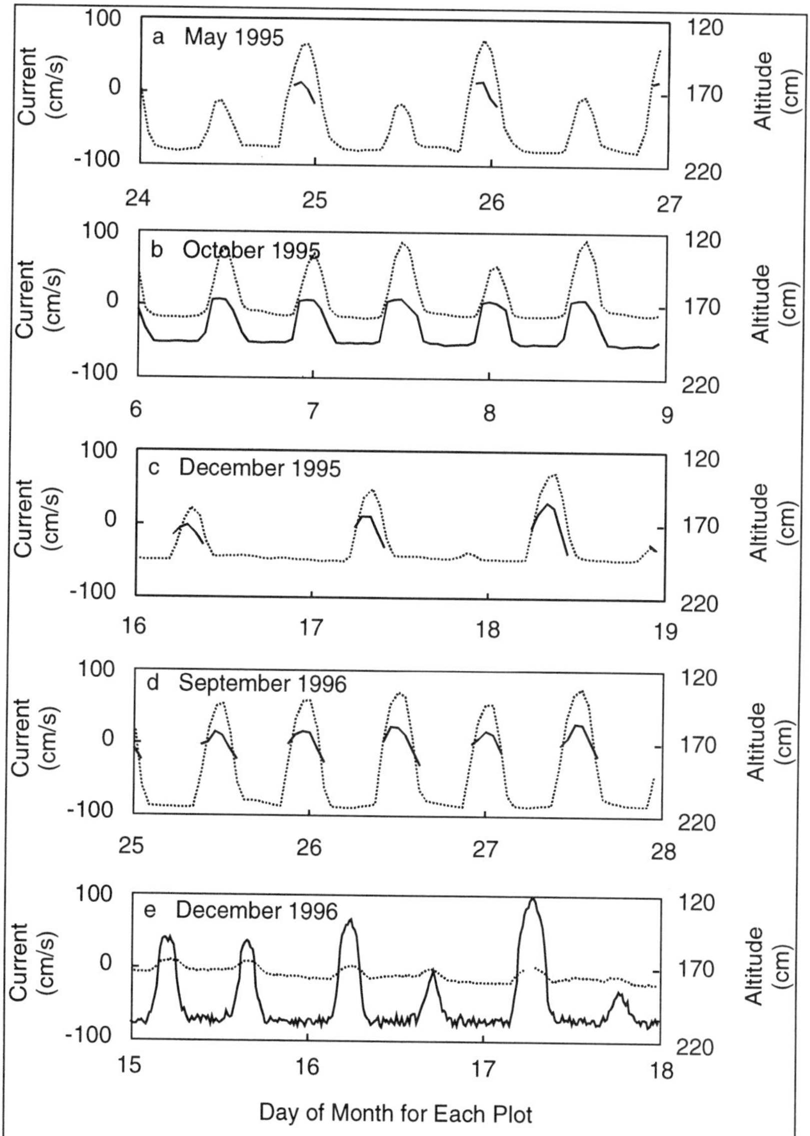

Figure 5. Channel current (solid line) and vertical distance from altimeter to water or bottom (dotted line) for various dates. Positive currents indicate flow from bay to site. Smaller altitudes indicate higher water. There is no current data whenever the sensor is out of the water.

the time of maximum flood current. During the quiescent period after the site was filled, the water level was kept above the sediment level. Therefore, after the breach there was a large head of water in the site that tried to drain bayward on lower tidal levels. This changed the hydrodynamics of the tidal flow from the bay as shown in figure 4e. Added to the man-made changes, heavy rainfall occurred in December 1996, which produced regional flooding that was capable of affecting local water levels.

Before the impoundment was filled, the creek was roughly 2.5 m wide and 0.8 m deep (fig. 5). After the pumping of the slurry ended, the creek had deepened by approximately 0.2 m. The creek deepened another 0.1 m during the consolidation period, presumably because the stop logs at the weir were set to allow some tidal flow. Four months after the levee was breached, the creek had deepened another 0.1 m. Throughout this period, the width of the creek remained constant.

Eventually, the creek will not be able to deepen further because it will reach a base level relative to the tidal range in the bay whence the creek gradient will be insufficient to produce further downcutting. Then, if the creek is to enlarge enough to handle the increased volume of water resulting from opening the restoration site to tidal flow, it will have to widen. Because the creek bed is mud and the sides are mud strengthened by roots, it seems reasonable that channel deepening precedes widening. Continued monitoring will document the process by which the creek reaches a new equilibrium shape. It is possible that the creek is already sufficiently wide to handle the increased water volume, but the flow profile (fig. 4e), whereby the currents are much stronger than before filling began (fig. 4a) suggests that is not the case.

Other factors may affect tidal flow in the creek being monitored. Other creeks, such as the one flowing east-west along the levee (fig. 2) could enlarge to carry some of the flow; other researchers are looking at this possibility with surveys and air photos. Ebb flow could occur primarily in one creek and flood flow in another. The breach could silt in, which would create a pond with little tidal action. At present, monitoring has not progressed to the point where the relative importance of each of these possibilities is known. When their importance is determined, remedial steps can be taken to ensure that the site reaches the desired final wetland state.

Conclusions

At this time, the tidal creek extending between breach at weir 2 and San Pablo Bay has deepened, but not widened at the survey site in response to increased flow during site filling and breaching of the levee. Currents in the creek also increased substantially after the site was opened to tidal action. The fact that water levels in the creek during December 1996

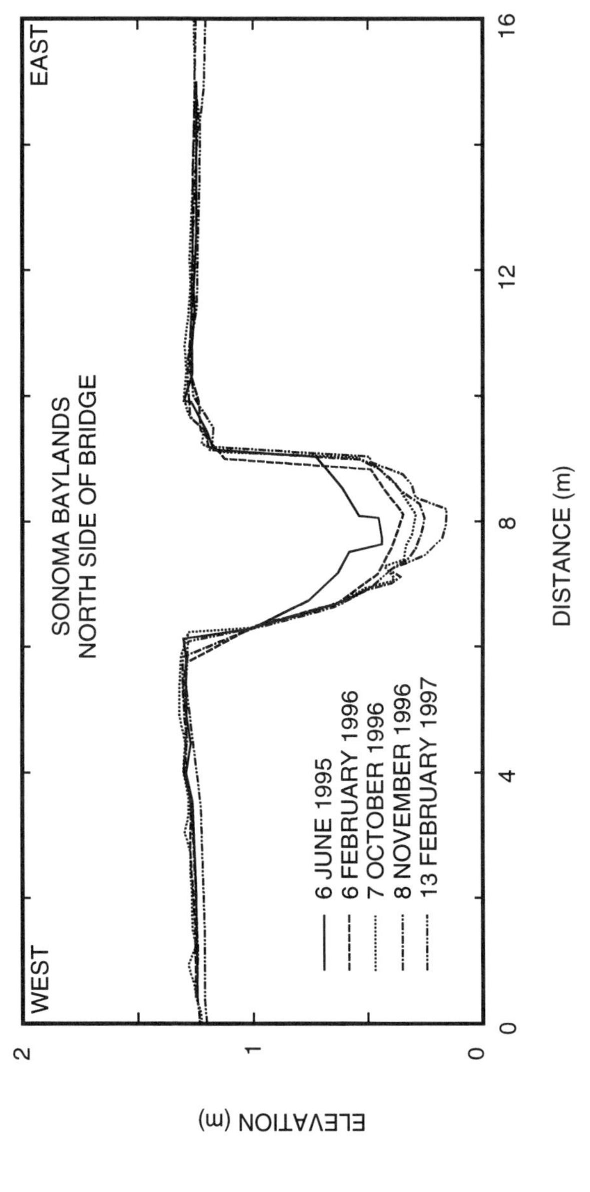

Figure 5. Profiles taken across the tidal creek that flows bayward from the breach at wier 2.

were lower than those expected to be produced by tidal fluctuations in the bay are not yet understood, but could be caused by the large volume of water in the site, or by the heavy flooding in December 1996, or by both.

Acknowledgments

Partial funding for this study came from the USGS Ecosystem Program. The U.S. Fish and Wildlife Service has supported this study by allowing us access to the existing wetland, a nesting area for the endangered clapper rail. The State of California Coastal Conservancy, which manages the project for the state, authorized our access to the site. The U.S. Army, Corps of Engineers, which managed the restoration work, gave us access to the work area and to their plans and data. Steve Goldbeck and Eric Larson of the San Francisco Bay Conservation and Development Commission have helped us in many ways to solve problems associated with conducting this project.

Bibliography

Atwater, B.F., Canard, S.G., Dowden, J.N., Hedel, C.W., MacDonald, R.L., and Savage, W., 1979. History, landforms, and vegetation of the estuary's tidal marshes. *In:* Conomos, T.J., Ed., San Francisco Bay: The Urbanized Estuary. Am. Assoc. Adv. Sci., San Francisco.

Cross, R. and Williams, D., 1981. Proceedings of the National Symposium on Freshwater Inflow to Estuaries. U.S. Fish and Wildlife Service Office of Biological Services, FYS/OBS-81/04.

Gilbert, G.K., 1917. U.S. Geological Survey Professional Paper 105.

Kockelman, W.J., Conomos, T.J., and Leviton, A.E., Eds., 1982. San Francisco Bay: Use and Protection. Am. Assoc. Adv. Sci., San Francisco.

Nichols, F.H., Cloern, J.E., Luoma, S.N., and Peterson, D.H., 1986. The modification of an estuary. Science, 231: 567-573.

U.S. Environmental Protection Agency, 1996. Long-Term Management Strategy for the placement of dredged material in the San Francisco Bay region. Draft regulation, 123 pp.

CHAPTER 161

Tidewater Goby *(Eucyclogobius newberryi)* Management in California Estuaries

Mark H. Capelli[1]

Abstract

The Tidewater goby is a small brackish water fish endemic to California estuaries. Loss and degradation of California estuaries has reduced historically suitable habitat and resulted in the extirpation of one quarter of the disjunct populations within the species' known range. Remaining habitat is threatened by a variety of on-going habitat modifications, including artificial breaching of sand and cobble berms which seasonally form at the mouths of most estuaries and create the brackish water regime requisite for the Tidewater goby. Breaching causes rapid fluctuations in water salinity levels, as well as in the amount of suitable habitat, which can adversely impact the reproductive and recruitment life phases of the species. The classification of the Tidewater goby as a federally endangered species creates challenges for the management of naturally restricted estuarine habitats along the California coast. Alternative approaches to controlling water levels and other factors affecting artificially managed estuaries are necessary to protect the small suite of endemic estuarine species, including the Tidewater goby.

Introduction

Coastal rivers and streams along the 1000 mile coast of California terminate in estuaries ranging in size from several square kilometers to a few square meters. Those seasonally closed off to the ocean with a sand and cobble berm provide a protected habitat of brackish water for a small estuarine species, the Tidewater goby, which is restricted to the coast of California. These estuaries lie within the California Coastal Zone established by the California Legislature in 1976, and are subject to the planning and regulatory authority of the California Coastal Commission pursuant to the California Coastal Act of 1976 (California Legislature 1976, 1997).

1. Coastal Program Analyst, California Coastal Commission; Lecturer, Environmental Studies Program, University of California, Santa Barbara, CA.

Because many California estuaries are surrounded by urbanized areas, or areas occupied by recreational or agricultural uses, elevated water levels which develop following the seasonal closure of the mouths of estuaries often create problems for adjacent landowners, including flooding, odors, and an increase in insect vectors. The pooling of polluted waters can also create problems for fish and wildlife and recreational activities involving water body contact. As a result, artificial breaching of the mouths of estuaries has become a common means of controlling the water levels and increasing tidal flushing in many California estuaries. Artificial breaching of closed estuaries by mechanical means can result in the sever alteration of the amount and quality of brackish water habitat, affecting both reproductive behavior and subsequent recruitment of Tidewater gobies. The California Coastal Commission has addressed the conflicting interests in the maintenance of estuarine water levels using a variety of management techniques.

Tidewater Goby Distribution and Habitat Characteristics

Tidewater gobies have been reported from estuaries in California ranging from Tillas Slough at the mouth of the Smith River (Del Norte Co.) to Agua Hedionda Lagoon (San Diego Co.). The distribution of the Tidewater goby corresponds to the distribution of sand deposition within the littoral cells along the California coast. These cells supply the sediment which facilitate the seasonal formation of sand and cobble berms which close off the mouths of most California estuaries for a majority of the year (California Department of Navigation and Ocean Development 1977; Cooper 1967; Swift 1975).

The distribution of Tidewater goby populations along the California coast is marked by several large geographical gaps. The geographical gaps north of Tomales Bay (Marin Co.), as well as between San Francisco Bay and San Gregorio Creek (San Mateo Co.), and between the Salinas River (Monterey Co.) and Arroyo del Oso (San Luis Obispo Co.) may be natural, due to the precipitous nature of the coastline and the absence of estuaries with extended periods of seasonal closure to the ocean. The gap south of Agua Hedionda Lagoon (San Diego Co.) may be the result of pre-World War II extirpations, though no historical occurrences of Tidewater gobies have been documented to date (Ballard and Swift 1996; Swift 1975; Swift et al. 1989; Swift, personal communication).

Estuaries which provide suitable Tidewater goby habitat exhibit a brackish water regime ranging from 5 to 20 ppt, with Tidewater gobies displaying a preference for salinity levels between 10 and 15 ppt. Extensive brackish water areas are generally associated with estuaries which develop a seasonal sand and cobble berm at their mouths. Estuaries with a more permanent connection with the ocean have a more saline water regime (20 < 33 ppt) and often do not support Tidewater gobies. Where the build up of a seasonal sand and cobble berm creates a brackish water regime throughout the estuary, Tidewater gobies may utilize the entire estuary, including vegetated backwater areas. In larger estuaries with prolonged exposure to tidal action Tidewater gobies are generally restricted to the upper reaches of the estuary near the freshwater-saltwater interface. Estuaries supporting Tidewater gobies vary in size from a few square meters to several square kilometers. Today most estuaries supporting Tidewater gobies are small, ranging from approximately 1 to 2 hectares (Ballard and Swift 1996; Swenson 1995; Swift et al. 1989).

Figure 1. Malibu Creek Estuary, Los Angeles County. Typical Tidewater goby estuarine habitat is closed to the ocean by a seasonal sand and cobble berm until naturally breached by winter run-off. Natural breaching pattern of this estuary has been altered by upstream wastewater discharges and mechanical breaching practices.

Tidewater goby habitat is naturally subject to considerable seasonal fluctuations. This variability is most strongly affected by the Mediterranean climate which is characterized by a short wet winter and a long dry summer. In coastal California a majority of the precipitation falls between December and February, and frequently during a few brief storms lasting three to five days. Little or no rain falls between March and November. As a result of this climatic regime, the freshwater flows into estuaries are episodic, and frequently of short duration. The mixed semi-diurnal tidal regime along the Pacific coast, with two unequal high and low tides per twenty-four hour period, results in frequent changes in the salinity levels of the estuary when the mouths remain open to the ocean.

Following winter rains, upstream flows decline sufficiently by April or May to allow the surf and littoral current to build up a sand and cobble berm at the mouths of most estuaries. The seasonal formation and natural breaching of the sand and cobble berms at the mouths of California estuaries is also influenced by the marked winter/summer seasonal variations in littoral sand movement along the California coast. During the winter months steep high energy waves erode sand from beaches and berms and facilitate the natural breaching of the sand and cobble berms at the mouths of estuaries. During the summer months shallow low energy waves move sand from off-shore sand bars landward and build up deposits of sand and larger sediments along beaches and berms at the mouths of estuaries (Davis 1985; Komar 1976).

Under natural conditions estuarine water levels during the late spring through early winter months remain relatively stable as a result of a balance between upstream freshwater inflow, evaporation, and percolation through the porous sand and cobble berm. Occasional surf washing over the berm during the summer and fall, along with percolation of marine water through the berm into the estuary, also helps sustain the mildly brackish water regime preferred by the Tidewater goby (Smith 1990; Swift et al. 1989).

These seasonal variations in freshwater inflow from rivers and streams during the spring, summer, fall, and winter months, coupled with changes in the wave climate and mixed semi-diurnal variations in tides, results in the prolonged closure of most California estuaries to the ocean. The closure of the estuary during the late spring, summer and fall months and the gradual conversion of the estuary from a marine to a low salinity brackish water regime creates a maximum amount of suitable spawning and rearing habitat for the Tidewater goby (Josselyn and Zedler 1990; Shapovalov and Taft 1954; Smith 1990; Swift et al. 1989).

Tidewater Goby Natural History

The Tidewater goby *(Eucyclogobius newberryi)* is one of seven species of gobies native to California estuaries, and one of only three fish species along the Pacific coast dependent upon a low salinity water regime. The other two species of estuarine fishes of California which are dependent upon low salinity water (Delta smelt *Hypomesus transpacificus* and Splittail *Pogonichthys macrolepidotus*) are known only from the fresh and brackish waters of the Sacramento-San Joaquin Delta of San Francisco Bay (Moyle 1976). As such, the Tidewater goby is an important indicator of the health of the unique brackish water conditions characteristic of California estuaries.

The other native gobies (Arrow goby *Clevelandia ios,* Cheekspot goby *Ilypnus gilberti,* Shadow goby *Quietula y-cauda,* Bay goby *Lepidogobius lepidus,* Longtail goby *Gobionellus sagittula,* and Longjaw mudsucker *Gillichthys mirabilis)* generally do not occupy the same niches as the Tidewater goby. Three of these goby species (Shadow, Cheekspot, and Bay gobies) are found in more marine intertidal habitats and have not been collected with the Tidewater goby. The Longjaw mudsucker is strongly associated with fresh water, and the Longtail goby is apparently extinct in California (Swift et al. 1989, 1993; Swift and Ballard 1996).

The Tidewater goby is a relatively small and inconspicuous species which may be confused with other commonly associated species such as the Longjaw mudsucker *(G. mirabilis).* Both males and females vary in color from translucent gray brown to olive with darker mottling, and are marked with a distinctive white or clear tipped spinous dorsal fin. Tidewater gobies reach a maximum length of 50 millimeters. Tidewater gobies generally occur in 1 to 2 meters of standing water over a sandy or mixed sandy/silty bottom. They are weak swimmers, and generally avoid swiftly moving waters which can act has a velocity barrier preventing movement upstream of the estuary (Swenson 1995; Swift et al. 1989).

Figure 2. Adult Tidewater goby *(Eucyclogobius newberryi)* from Aliso Creek Estuary, Orange County. Top: Female, c. 38 mm. Bottom: Male, c. 40 mm. Photograph courtesy Camm C. Swift.

The Tidewater goby is a short lived species, with a majority of individuals completing their life cycle within one year. The life cycle of the Tidewater goby is closely tied to the annual hydrologic cycle of California estuaries, with successful recruitment dependent on an adequate amount of spawning habitat and a suitable salinity regime during the reproductive and rearing period. Spawning is most prevalent from spring to mid-summer when most California estuaries are naturally closed to the ocean and exhibit low-salinity brackish water conditions. Spawning is generally halted by the onset of colder weather and the first winter rains which result in breaching of the sand and cobble berm and the temporary conversion of the estuary to fresh water. However, spawning activity may persist into early fall and winter if water temperatures remain warm, and freshwater inflow is not increased sufficiently to breach the sand and cobble berm at the mouth of the estuary (Swenson 1995, 1997; Swift et al. 1989).

Before spawning, males excavate breeding burrows in clean, unconsolidated sand. Females aggressively compete with other females for access to the males with burrows. Females lay between 400 and 600 eggs per clutch, which are attached to the sand grains in the ceiling and walls of the burrow. Tidewater gobies typically produce multiple clutches (4 to 6) during the spawning season. The eggs are tended by the males, which may occasionally support several clutches from different females. Tidewater goby eggs hatch in 7 to 9 days depending on water temperature. The young remain planktonic for several days before becoming benthic oriented. Tidewater gobies become sexually mature within three to six months, depending on the timing of hatching, and the rearing conditions (i.e., water temperature, salinity regime, and available food supply). Juveniles and adults feed on a wide variety of non-marine benthic invertebrates, including ostracods, amphipods, snails, and chironomid larvae and pupae (Swenson 1995, 1996, 1997; Swift et al. 1989).

The Tidewater goby spends its entire life cycle within the estuary, though there are records of fish moving several kilometers upstream in shallow gradient streams such as San Antonio Creek, Santa Barbara County (Ballard and Swift 1989; Swift, personal communication). The lack of a marine phase in the life history suggests that there is little likelihood that populations from distant estuaries form metapopulations. This isolation is supported by genetic investigations which have identified fixed allelic differences between the extreme northern and southern populations, and some variation in populations from the central portion of the species' range (Crabtree 1985). However, recent recordings of Tidewter gobies from sites from which gobies had apparently been extirpated during the 1987-92 drought suggests that there may be some recolonization and/or genetic exchange between populations from nearby estuaries (Ambrose 1995; Ballard, personal communication).

Habitat Modifications

Populations of Tidewater gobies have suffered declines in parallel with the degradation of California coastal wetlands. It has been estimated that approximately 75-90% of the original estuarine wetland acreage of California has been lost since 1850 (California Department of Fish and Game 1983; California Department of Parks and Recreation 1988). Of the 110 sites from which Tidewater gobies have historically been reported, 26 sites currently do not support Tidewater gobies, and 50 sites are currently not considered viable (Ballard and Swift 1996). Recent investigations, however, have reported Tidewater goby populations from estuaries where the species was previously thought to be extirpated (Ambrose 1995; Ballard personal communication).

Encroachment of development, channelization of coastal streams, diversion of surface flows, groundwater extraction, importation of point and non-point sources of pollution (including sedimentation), and the introduction of exotic species of plants and fishes have all contributed to the loss of estuarine habitat and the extirpation of individual populations of Tidewater gobies (Ballard and Swift 1996; Ferren et al. 1995, 1996).

Figure 3. Pismo Creek Estuary, San Luis Obispo County. Urban development encroaching on the coastline of California has exacerbated conflicts between habitat maintenance and adjacent land uses. The agricultural field adjacent to the mouth of the estuary has been converted to recreational uses since this photograph was taken in 1973.

The mouths of some estuaries have also been modified through the installation of jetties which have resulted in the permanent opening of the mouth and the conversion of the seasonal brackish water regime to a permanent marine water regime unsuitable for Tidewater gobies. In several instances Tidewater gobies in artificially created tidal conditions have become extirpated after a few years of exposure to continuous tidal flushing. Examples of estuaries which formerly supported Tidewater gobies before permanent artificial modification of the mouths include Morro Bay Estuary, Carpinteria Salt Marsh, Goleta Slough, Mugu Lagoon, and the Santa Margarita River Estuary. Other estuaries which have been subjected to regular and frequent artificial breaching and no longer support Tidewater gobies include Freshwater Lagoon, Ten Mile River, and the Elkhorn/Bennett Slough. Humboldt Bay is the only remaining tidal system which currently supports Tidewater gobies (Ballard and Swift 1996; Swift, personal communication).

Figure 4. Ormond Beach Estuary, Ventura County. Hand-dug channel between Ormond Beach Estuary and ocean, cut to lower lagoon water levels.

In addition to the widespread loss of estuarine habitat, the viability of remaining Tidewater goby habitat is threatened by the practice of artificially breaching the seasonal sand and cobble berms which form at the mouths of estuaries. The methods of breaching the sand bar can involve the use of mechanical equipment (e.g., bull-dozer or back-hoe), or can be accomplished with hand-tools (e.g., shovel, hoe, or stick). Because of the ease of digging through unconsolidated sand and the elevated water level of the estuary above the adjacent marine water, cutting a small channel several inches wide and deep can initiate a flow from the estuary to the ocean which can quickly expand the channel through erosion and result in the rapid draining of the estuary. Draining can be particularly catastrophic if the artificial breaching occurs during minus or low tide conditions. While the sand and cobble berms may reform relatively quickly (within a few days or weeks), if breaching occurs during the spring or summer months when freshwater inflow is substantially reduced, estuarine water levels may stabilize at a significantly lower level (Smith 1986, 1990; Smith and Robinson 1987; Swift et al. 1989).

Artificial breaching of estuaries through mechanical means has become an increasingly common practice as a result of development encroaching into periodically flooded areas. Artificial breaching of estuaries is undertaken for a variety of reasons: to lower water levels which inundate adjacent land uses (e.g., agricultural, residential, commercial, and recreational), to reduce surface water areas supporting mosquitoes or other insect vectors; to improve water quality in the estuary by increasing tidal circulation; to increase recreational birding activities by exposing mud-flat habitat; to alter the configuration of off-shore sand bars to improve recreational surfing conditions; and to induce the upstream movement of anadromous fishes for sport fishing. Some of these activities are conducted by state or local governmental authorities as part of routine maintenance activities, while others are carried out by private individuals without supervision or authorization.

The life-cycle of the Tidewater goby is adapted to the natural seasonal breaching cycles of estuaries but can be disrupted by artificial breaching, particularly during the spring and summer. Breaching results in immediate reduction in the depth and aerial extent of the estuary. Because the water level in closed California estuaries is generally several feet above mean sea-level, and as much as 6 feet above lower-low mean sea-level, breaching can result in the complete draining of the estuary. Sudden lowering of water levels can sweep Tidewater gobies into the marine environment, or strand fish in shallow pools or on exposed substrates, increasing their vulnerability to predation by shore and other water birds. The sudden artificial breaching of the estuary mouth, unlike natural breaching in response to rainfall and increased runoff, provides no natural cues which may allow Tidewater gobies to seek refuge in backwater or marginal areas of the estuary.

The artificial draining of a closed estuary in late spring or summer may result in the estuary being dominated by marine water for extended periods until winter run-off increases the inflow of freshwater into the estuary. Marine water is denser than freshwater and will dominate the lower depths of the estuary directly over the substrate where Tidewater gobies congregate. This marine water acts as a solar collector heating up and reducing oxygen levels, and thus providing sub-optimal or lethal conditions for the Tidewater goby. The sudden influx of marine water as a result of artificial breaching can also sharply reduce the abundance of non-marine invertebrates which provide the primary food source for the Tidewater goby.

Artificially breaching the estuary during the spring, summer and fall seasons can adversely affect the reproductive and rearing life-history phase of the Tidewater goby. Lowering of water levels can expose juveniles and incubating eggs in breeding burrows to the air, leading to desiccation and death (Smith 1990; Swenson 1995; Swift et al. 1989). Recent research provides evidence that vegetated areas at the margins of the estuary serve as refugia for rearing Tidewater gobies. These areas can become isolated from the main water body of an estuary as a result of artificial breaching and draining, thus depriving Tidewater gobies of productive rearing sites and effective protection from predators (Swenson 1995; Worcester 1992).

The natural pattern of breaching of estuaries (i.e., timing, frequency, and location of breaching) has been further modified as a result of changes in the watersheds of California estuaries. Upstream water withdrawals as a result of both surface diversions and groundwater extractions has resulted in lowering the input of freshwater into some estuaries, reducing estuarine water levels, and delaying the conversion from a marine to a low-salinity brackish water regime following the closure of the mouth. In some cases, summer base stream flows have been substantially increased as a result of upstream municipal wastewater discharges, particularly in southern California. These flows have modified the natural breaching regimes of estuaries by advancing the timing and increasing the frequency of spring and summer breachings (Ballard and Swift 1996).

California Coastal Commission Jurisdiction

The California Coastal Commission has both planning and regulatory authority within the Coastal Zone. Landward of the mean high-tide line this zone varies from as little as 100 yards, to over five miles; seaward of the mean high-tide line the zone extends three miles. The Coastal Zone includes every estuary within California, with the notable exception of San Francisco Bay, which is outside the Coastal Commission's jurisdiction, and is governed by the San Francisco Bay Conservation and Development Commission.

Under the California Coastal Act, the Coastal Commission retains permitting authority over historically tidally influenced lands, and appeals jurisdiction over those areas which are generally seaward of the first public road paralleling the coast. As a result, the Commission has in may cases exclusive, state land use regulatory responsibilities over the habitat of the endangered Tidewater goby. Specific policies of the California Coastal Act provide for the protection, and where feasible, the restoration of the biological productivity of coastal waters, including streams, wetlands, and estuaries. Additionally, the Coastal Act places stringent limits on the diking, filling, or dredging of coastal wetlands, including estuaries. Where there are conflicts between the protection of natural coastal resources and other Coastal Act objectives (e.g., protection of agriculture or recreational opportunities), the Coastal Act provides that "such conflicts be resolved in a manner which on balance is the most protective of significant coastal resources." (California Legislature 1976, 1997)

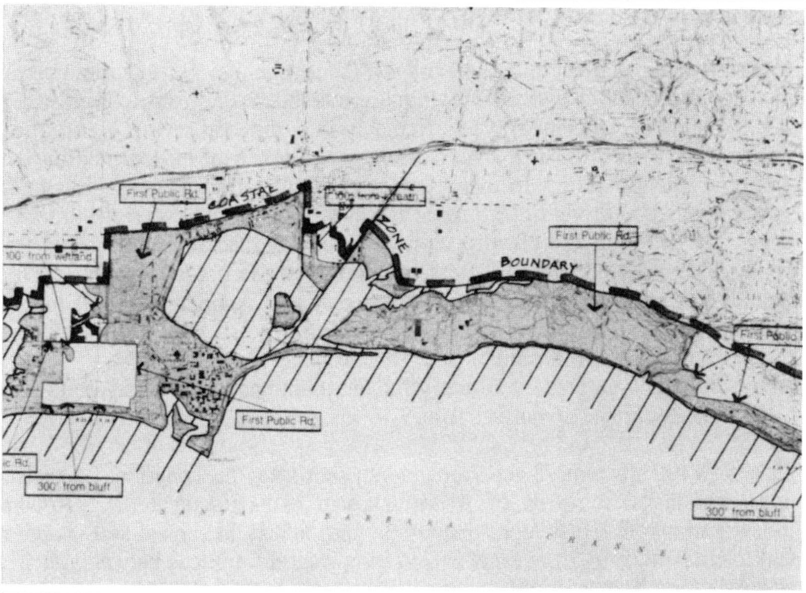

Figure 5. Coastal Zone along a portion of the California Coast from Coal Oil Point to Arroyo Burro, Santa Barbara County. Original permit jurisdiction (cross-hatched lines), and appeals jurisdiction (shaded area).

Tidewater Goby Management Issues

Management of Tidewater gobies is inseparably linked to the management of California's remaining estuaries. Maintenance of water quality (through regulation of point and non-point sources of pollution), prevention of the physical modification of habitat (through regulation of filling, ditching draining, and disturbance of native vegetation), regulation of consumptive and non-consumptive human uses (through regulation of fishing, claming, swimming, and boating), and the control of exotic species (through regulation of the release of fishes or amphibians, or the introduction of non-native plants), is critical for the maintenance of Tidewater gobies, as well as a suite of other native estuarine species. Control of fine sediments originating within the watershed is also necessary to ensure that the sandy substrate preferred by Tidewater gobies is not overlain with muds and silts.

Maintenance of the hydrological regime is recognized as the single most important objective in preserving the composition and ecological functions of estuarine habitat (Macdonald 1977; Mitsch and Gosselink 1993; Onuf 1987; Onuf and Zedler 1987). Of particular importance for the maintenance of Tidewater gobies is the regulation of water levels and salinity regimes in estuaries (Smith 1990; Swenson 1995; Swift et al. 1989).

One rationale for artificially breaching of the seasonal sand and cobble berms at the mouths of California estuaries has been the improvement and maintenance of water quality, particularly in southern California estuaries, where depressed dissolved oxygen levels as a result of eutrophication is a common phenomenon. In many estuaries regular tidal flushing has been accomplished by dredging extensive areas of the estuary to increase the tidal prism and through the construction of jetties at the mouths of estuaries. While the direct physical impacts of some of these practices on estuarine habitat have been questioned, the maintenance of regular tidal flushing in California estuaries, which naturally experience protracted seasonal closure of the mouth and only irregular tidal flushing, has frequently been viewed as an acceptable management strategy for the maintenance of water quality in California estuaries (Elwany et al. 1997).

This view reflects a more general tendency to portray the functions of Pacific coast estuaries in terms of Atlantic coast estuarine systems. Recent investigations of Pacific coast estuaries, particularly in central and southern California, however, have emphasized important differences between Pacific and Atlantic coast estuarine systems (Ferren et al. 1995, 1996; Josselyn and Zedler 1990; Lewis 1995; Zedler et al. 1992). Atlantic coast estuaries have developed on a tectonically divergent coast with a broad continental shelf, and a temperate climate. They are characterized by extensive intertidal floodplains which are subjected to both regular fresh-water inflow and a simple diurnal tidal regime. As a result, they are subjected to daily inter-change between the terrestrial and marine hydrologic systems. Pacific coast estuaries have developed on a tectonically convergent coast with a narrow continental shelf, and a Mediterranean climate. Many small California estuaries are formed at canyon mouths and do not have extensive intertidal floodplains. Almost all Pacific coast estuaries experience only episodic fresh-water inflow, and a short seasonal connection with the ocean (Davis

1985; Ferren et al. 1995; Macdonald 1986).

As a result of the recent listing of the Tidewater goby as an endangered species, the disruptive effects of artificial breaching of estuaries supporting populations of Tidewater gobies are being more carefully scrutinized, and alternative methods of controlling water levels are being explored (U.S. Fish and Wildlife Service 1994). Artificial breaching activities which have been previously undertaken without regulatory review are increasingly being subjected to the California Coastal Commission's regulatory process, and estuarine management plans are being developed to address the control of estuarine water levels (Habitat Restoration Group 1994; Swanson and Associates et al. 1993; Philip Williams and Associates 1993).

The California Coastal Commission has reviewed and provisionally approved a number of techniques for controlling estuarine water levels to protect Tidewater gobies from the adverse effects of artificial breaching. These include: regulating the frequency and timing of breaching (e.g., Russian River, Lake Earl and Lake Talawa, Salinas River, Carmel River), redirecting the location of breaching (e.g., Malibu Creek), installing a sill at the mouth of the estuary to limit the level of drainage (e.g., Soquel Creek), and installing temporary culverts in the sand berm at a set elevation to limit the level of drainage (e.g., San Lorenzo River, Batiquitos Lagoon). The Commission has also considered proposals for installing pumps to allow the regulated lowering of estuarine water levels (e.g., Ormond Beach Lagoon). None of these efforts have been monitored on a long time basis and their long-term cumulative effects is unknown (California Coastal Commission 1992, 1993a, 1993b, 1996a, 1996b, 1996c, 1997a, 1997b, and 1997c).

Discussion

The listing of the Tidewater goby has focused renewed attention on the management of California's estuaries, particularly the management of the hydrologic regime. Of particular note has been the attention paid to small estuaries - including those measuring only a few square hectares - which have frequently been over-looked because they do not provide habitat for significant numbers of estuarine species.

Inducing regular tidal flushing has often been viewed by estuarine managers as an effective, and in many cases necessary, strategy for maintaining the ecological functions of California estuaries. However, the effects of artificial breaching of naturally closed estuaries on federally or state protected species which are sensitive to sudden changes in the estuarine water regime will require a re-examination of this management technique. The regulation of water levels in estuaries through artificial breaching for flood

Figure 6. Ormond Beach Estuary, Ventura County. This estuary had been regularly breached for flood control and mosquito abatement before the discovery of Tidewater goby. With the cessation of breaching the estuary has expanded from less than 1 acre to over 20 acres. Expansion has impinged on flood control facilities and adjacent lands.

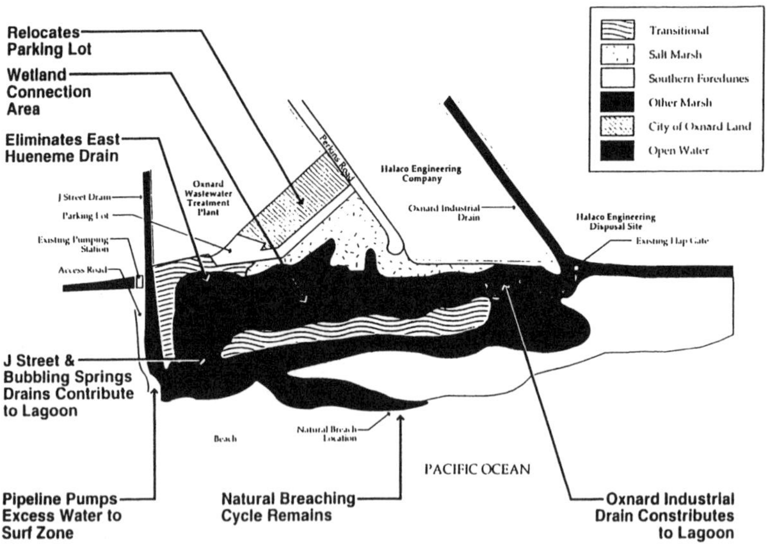

Figure 7. Ormond Beach Estuary, Ventura County. Conceptual proposal to regulate water levels in the estuary through the installation of water pumps to discharge estuarine water into the surf zone (Impact Sciences, Inc. 1995).

control, mosquito abatement, and various recreationally related purposes must be reconciled with the habitat requirements of the Tidewater goby and other estuarine species sensitive to changes in salinity and water levels.

Alternative means of addressing conflicts between Tidewater goby habitat requirements and adjacent land uses should be explored. These include controlling the amount and quality of upstream waste discharges, and modifying land uses to accommodate naturally fluctuating estuarine water levels. Where artificial breaching is necessary to address problems associated with the natural seasonal rise in estuarine water levels, artificial breaching patterns should mimic natural breaching patterns to the maximum extent practicable to minimize impacts to the spawning and rearing phases of the Tidewater goby. Mechanical breaching of the sand and gravel berm provides the least control over the ultimate reduction of estuarine water levels and should be used only when other methods are not practicable. Complete draining of the estuary, in particular, should be avoided; this can be controlled to some extent by breaching at the onset of a rising high-tide to minimize the elevational difference between the water level in the estuary and the adjacent ocean.

Each estuary presents a unique set of circumstances which must be evaluated in terms of over-all management objectives of the estuary. In assessing the effects of artificially lowering estuarine water levels on the Tidewater goby, the effects on other fish species must also be considered. Of particular importance are anadromous fishes such as Chinook salmon *Oncorhynchus tshawytscha,* Coho salmon *O. kisutch*, and Steelhead trout *O. mykiss* which utilize estuaries during portions of their life-cycle (e.g., rearing, acclimating to fresh or marine water). Finally, other species such as the California least tern *Sterna antillarum,* Belding's savannah sparrow *Passerculus sandwichensis beldingi,* California clapper rail *Rallus longerostris obsoletus,* and Light-footed clapper rail *R. longerostris levipes,* are also dependent on estuarine habitat and sensitive to sudden changes in estuarine water levels (e.g., post-hatch feeding of juvenile Least terns) and therefore must also be considered in the regulation of estuarine water levels.

As one of the few fish species in California that is restricted to a low salinity brackish water habitat, the Tidewater goby is a significant component of the estuarine community. Historically Tidewater gobies were one of the most abundant small fishes in California estuaries and have played important roles as both a secondary invertebrate predator and as prey for other estuarine species of fishes and birds. The Tidewater goby's decline can be an important indicator of the general health of individual estuaries, and its loss could have subtle but significant effects on the over-all ecological functioning of California's remaining estuaries.

References

Ambrose, Richard F. (ed.). 1995. Coastal Wetland Resources of the Santa Barbara County Mainland. Prepared for the County of Santa Barbara.

Ballard, Edgar D. and Camm C. Swift. 1996. Tidewater Goby *(Eucyclogobius newberryi).* Draft Recovery Plan. U.S. Fish and Wildlife Service, Region 1.

California Legislature. 1976, 1997. The California Coastal Act (Public Resources Code §§ 30000 et seq.) as amended through 1997. Sacramento, CA.

California Coastal Commission. 1992. Staff Report. Application No. 3-92-15. Installation of a water level control structure for San Lorenzo River Lagoon, Santa Cruz, CA.

_________________, 1993a. Staff Report. Application No. 3-90-41 A2. Erection of sand bar across Soquel Creek Lagoon, Capitola, CA.

_________________, 1993b. Staff Report. Application No. 6-93-41. Installation of pipeline in cobble berm at the mouth of Batiquitos Lagoon, Carlsbad, CA.

_________________, 1996a. Staff Report. Application No. 1-96-09. Periodic breaching of sand bar at the mouth of the Russian River, Jenner, CA.

_________________, 1996b. Staff Report. Application No. 1-94-49. Periodic breaching of the sand bar across Lake Earl and Lake Talawa, Crescent City, CA.

_________________, 1996c. Staff Report. Application No. 3-95-58. Periodic breaching of the sand bar at the mouth of the Salinas River, Monterey County, CA.

_________________, 1997a. Staff Report. Application No. CP-1-97. Carmel River Lagoon Enhancement Plan, Monterey County, CA.

_________________, 1997b. Staff Report. Application No. 4-96-192. Temporary installation of a submersible pump and pipeline to control water level of Ormond Beach Lagoon, Oxnard, CA.

_________________, 1997c. Staff Report. 4-97-076G. Emergency redirection of mouth of Malibu Creek Lagoon, Malibu, CA.

California Department of Fish and Game. 1983. A Proposal to Inventory California's Wetlands. The Resources Agency. Sacramento, CA.

California Department of Parks and Recreation. 1988. California Wetlands. An Element of the California Outdoor Recreation Planning Program. The Resources Agency. Sacramento. CA.

California Department of Navigation and Ocean Development. 1977. Assessment and Atlas of Shoreline Erosion Along the California Coast. The Resources Agency. Sacramento. CA.

Crabtree, C.B. 1985. Allozyme Variability in the Tidewater goby, *Eucyclogobius newberryi* (Pices: Gobidae) *Isozyme Bulletin,* 18:70.

Cooper, William S. 1967. Coastal Dunes of California. Memoir 104. The Geological Society of America.

Davis, Richard A., Jr. (ed.) 1985. Coastal Sedimentary Environments. Springer-Verlag.

Elwany, M. Hanny S., Saima Aijaz, Alan B. Thum, and Reinhard E. Flick. 1997. A Strategy to Maintain Tidal Flushing in Small Coastal Lagoons. Proceedings, California and the World Ocean Conference.

Ferren, Wayne R., Jr., Peggy L. Fiedler, and Robert A. Leidy. 1995. Wetlands of the Central and Southern California Coast and Coastal Watersheds: A Methodology for their Classification and Description. U.S. Environmental Protection Agency, Region IX.

Ferren, Wayne R., Jr., Peggy L. Fiedler, Robert A. Leidy, Keven D. Lafferty, and Leal A. K. Mertes. 1996. Wetlands of California: Part I. History and Habitat Classification; Part II. Classification and Description of Wetlands of the Central and Southern California Coast and Coastal Watersheds. *Madrono.* Vol. 43, Number 1, Supplement. California Botanical Society.

Habitat Restoration Group. 1994. Salinas River Lagoon Management and Enhancement Plan (Revised Draft). Prepared for Monterey County Water Resources Agency.

Impact Sciences, Inc. 1995. Draft North Ormond Beach Coastal Management Plan. Prepared for Ventura County Flood Control District.

Josselyn, Michael, Joy B. Zedler, and Theodore Griswold. 1990. "Wetland Mitigation Along the Pacific Coast of the United States" in Jon A. Kusler and Mary E. Kentula (eds.) Wetland Creation and Restoration: The Status of the Science. Island Press.

Komar, Paul D. 1976. Beach Processes and Sedimentation. Prentice Hall.

Lewis, William L. Jr. (ed.) 1995. Wetlands: Characteristics and Boundaries. Committee on Characterization of Wetlands. National Research Council. National Academy Press.

Macdonald, Keith B. (ed.) 1986. Pacific Coast Wetlands Function and Values: Some Regional Comparisons. Annual Meeting, Western Society of Naturalists, Long Beach, CA.

Mitchell Swanson and Associates. 1993. Pajaro River Lagoon Management Plan. Prepared for Santa Cruz County Public Works Department and the California Coastal Conservancy.

Mitsch, William J. and James G. Gosselink. 1993. Wetlands. Van Nostrand Reinhold Company.

Moyle, Peter B. 1976. Inland Fishes of California. University of California Press.

Onuf, Chris and Joy B. Zedler. 1987. "Pattern and process in arid-region salt marshes" in D.D. Hook et al. (eds.) The Ecology and Management of Wetlands. Timber Press.

Onuf, Chris. 1987. The Ecology of Mugu Lagoon, California: An Estuarine

Profile. Biological Report 85(7.15) U.S. Fish and Wildlife Service.

Philip Williams and Associates, Ltd. 1993. Russian River Estuary Study 1992-1993. Prepared for Sonoma County, Department of Planning.

Shapovalov, Leo and Alan C. Taft. 1954. The Life Histories of the Steelhead Rainbow Trout (*Salmo gairdneri gairdneri)* and Silver Salmon *(Oncorhynchus kisutch)* with Special Reference to Waddell Creek, California, and Recommendations Regarding their Management. Fish Bulletin No. 98. California Department of Fish and Game.

Smith, Jerry J. and Mark A. Robinson. 1986. Effects of Inflows on Water Quality and Habitat Conditions in Small Central California Estuary/Lagoon Systems. Department of Biology, San Jose State University.

Smith, Jerry J. 1987. Aquatic Habitat and Fish Utilization of Pescadero, San Gregorio, and Waddell and Pomponio Creek Estuary/Lagoon Systems. Department of Biology, San Jose State University.

Smith, Jerry J. 1990. The Effects of Sandbar Formation and Inflows on Aquatic Habitat and Fish Utilization in Pescadero, San Gregorio, Waddell and Pomponio Creek Estuary/Lagoon Systems, 1985-1989. Department of Biology, San Jose State University.

Swift, Camm, C. 1975. Distribution and Status of the Estuarine Lagoon Goby, *Eucyclogobius newberryi*. Prepared for the California Department of Fish and Game.

Swift, Camm, C., Jack L. Nelson, Carolyn Maslow, and Theodore Stein. 1989. Biology and Distribution of the Tidewater Goby, *Eucyclogobius newberryi*. Contributions in Science. Number 404. Natural History Museum of Los Angeles County.

Swenson, Ramona O. 1995. The Reproductive Behavior and Ecology of the Tidewater Goby *Eucyclogobius newberryi* (Pices: Gobiidae). Ph.D. Dissertation. Department of Integrative Biology, University of California, Berkeley.

Swenson, Ramona O. and Arja T. McCray. 1996. Feeding Ecology of the Tidewater Goby. *Transactions of the American Fisheries Society*. 125:956-970.

Swenson, Ramona O. 1997. Sex-Role Reversal in the Tidewater Goby. *Environmental Biology of Fishes* (in press).

U.S. Fish and Wildlife Service. 1994. Endangered and Threatened Wildlife and Plants: Determination of Endangered Status for the Tidewater Goby. Federal Register 59(24):5494-5499.

Worcester, K. R. 1992. Habitat Utilization in a Central California Coastal Lagoon by the Tidewater Goby *(Eucyclogobius newberryi)*. M.A. Thesis, California Polytechnic State University, San Luis Obispo.

Zedler, Joy B., Christopher S. Nordby, and Barbara E. Kus. 1992. The Ecology of Tijuana Estuary, California: A National Estuarine Research Reserve. NOAA Office of Coastal Resource Management.

CHAPTER 162

A STRATEGY TO MAINTAIN TIDAL FLUSHING IN SMALL COASTAL LAGOONS

M. Hany S. Elwany[1], Alan B. Thum[2], Saima Aijaz[1] ,
John M. Boland[3], and Reinhard E. Flick[4]

Abstract

A strategy of minimal, but frequent tidal inlet and channel maintenance dredging, coupled with seasonal flows of freshwater runoff, can be successfully harnessed to maintain tidal flushing in certain small southern California coastal lagoons. We have applied this technique to three lagoons in San Diego County. The results of our studies strongly suggest that it is feasible to maintain tidal flushing in many small lagoons without massive enlargement and deepening to increase the tidal prism, and without intrusive and costly structures.

Introduction

San Diego County's watersheds are drained by 29 major and minor creek and river systems (Simon, Li & Associates, 1988). These terminate in a variety of river mouths, estuaries, marshes, lagoons and bays (Boland and Zedler, 1996). There are 7 lagoons along the coast of San Diego County relevant to this study (Figure 1). Each receives varying amounts of ephemeral flows of freshwater, depending on the watershed size, which ranges from 18.8 to 1724 mi^2. The strong seasonality in stormwater runoff imposes unique hydrological and ecological conditions on these lagoons, which range in size from 246 to 1320 acres of wetlands.

Buena Vista Lagoon is a freshwater lagoon isolated from the ocean by a weir adjacent to the beach. Three lagoons, Batiquitos, Agua Hedionda, and Tijuana Estuary, are open to year-round tidal flushing. The ocean inlets at Batiquitos and Agua Hedionda lagoons are sustained by jetties and occasional maintenance dredging.

---

1 Research Associate and Postgraduate Researcher, Center for Coastal Studies, Scripps Institution of Oceanography, La Jolla, CA; 2 Senior Scientist, Coastal Environments, 810 Third Street, Suite A, Encinitas, CA 92024; 3 Ecologist, California Coastal Commission, 3111 Camino del Rio North, San Diego, CA 92108; 4 Oceanographer, California Department of Boating and Waterways, Center for Coastal Studies, Scripps Institution of Oceanography, La Jolla, CA 92093.

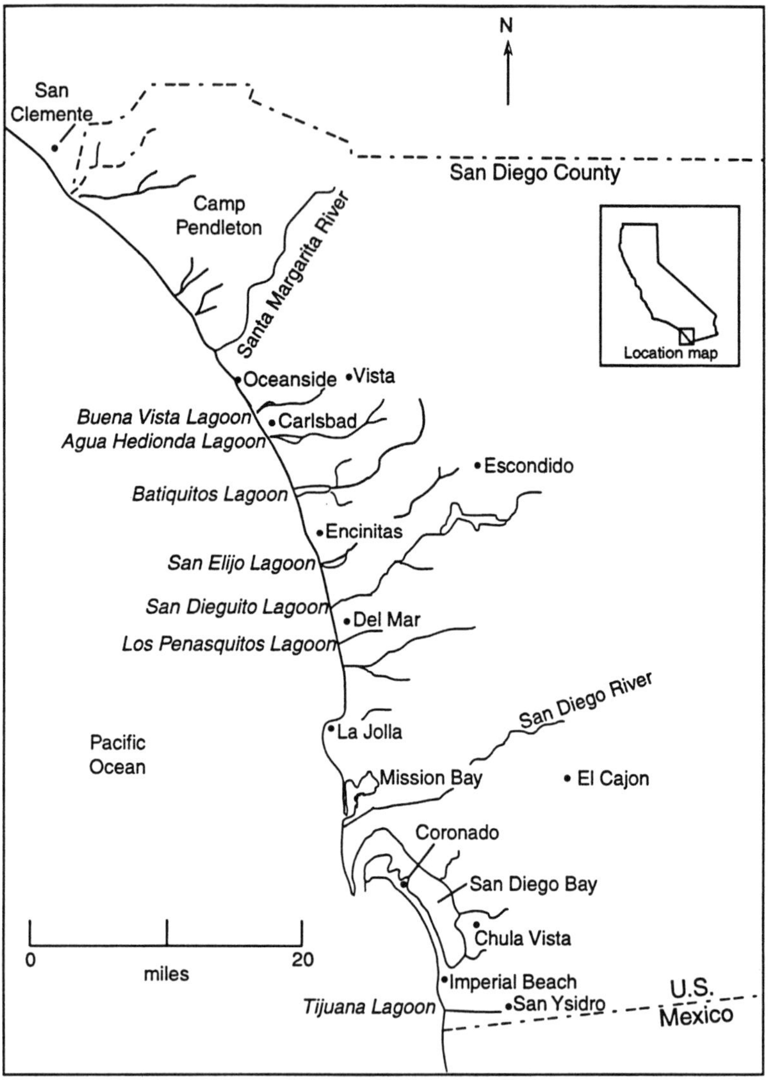

Figure 1. Map showing major lagoons and bays along San Diego County coast.

Tijuana Estuary is open naturally, due to its large tidal prism, configuration, and large watershed area.

The remaining lagoons, San Elijo, San Dieguito, and Los Penasquitos, are open only intermittently and require maintenance dredging of the inlet and main channel to sustain tidal flushing. This paper describes studies that have been conducted in these three lagoons.

Factors Controlling Lagoon Conditions

Three physical factors control the conditions in San Diego's (and many other) coastal lagoons: tidal flushing, seasonal fluvial runoff, and the deposition of sediments.

Tidal Flushing

Coastal lagoons will stay open to the ocean if the tidal prism is large enough to facilitate scouring of flood tide deposited marine sediments from the channel bed. Since most of San Diego's lagoons have relatively small tidal prisms and ebb flows are constricted by bridges, their inlets are dominated by flood tides, accumulate marine sands, and tend to close.

When tidal flushing ceases, seasonal stormwater runoff accumulates, fresh or brackish, eutrophic water conditions prevail, and plant biomass and sediments accumulate within the lagoon. If rainfall runoff is low, the lagoon inlet may open briefly or may remain closed. During summer, high water temperatures, low dissolved oxygen and large changes in lagoon water salinity (2 to 50 ppt) may occur. The long term outcome is that the marine communities degenerate and fresh/brackish water plants such as cattails (*Typha spp.*) and bulrushes (*Scirpus spp.*), colonize and expand.

On the other hand, when the lagoon inlets are open, the accumulated organic and fine sediment burden and associated eutrophic conditions will be reduced, the benthos, fisheries and salt marsh habitat will be enhanced, species abundance, diversity and trophic complexity will increase, human health conditions will improve and the potential impacts from any contaminant spills will be reduced. When tidal flushing is sustained, secondary productivity is enhanced and develops linkages with the nearshore ocean environment, e.g., production of juvenile fishes, like California halibut (*Paralichthys californicus*).

Stormwater Runoff

The historical record of rainfall from 1850 to 1994 recorded at Lindbergh Field (near downtown San Diego, CA) shows that there are large variations from year to year (Elwany et al., 1995). A plot of the cumulative residual rainfall, r_t, calculated for each successive year t shows that there are long term cycles of wet and dry periods Figure 2). The value r_t is a running departure from the mean annual rainfall calculated according to the following equation:

$$r_t = \sum_{i=t_o}^{i=t} (r_i - \bar{r}),$$

where $\bar{r}$ is the mean rainfall over the entire record.

The ascending parts of the curve reflect above average rainfall, and the descending parts reflect below average rainfall. For example, the years between 1945 and 1977 were the third period of semi-drought in the record, with an average yearly rainfall of about 8.5 in. We are currently experiencing a wet period. Between 1978 and 1994 the average annual rainfall was approximately 11.3 in, 14% above normal. During wet periods, stormwater runoff is more likely to assist in scouring the ocean inlet and facilitate tidal circulation than during dry periods. However, the recurrence rate of major stormwater runoff events is not high enough to keep the lagoon ocean inlets open for any prolonged period of time.

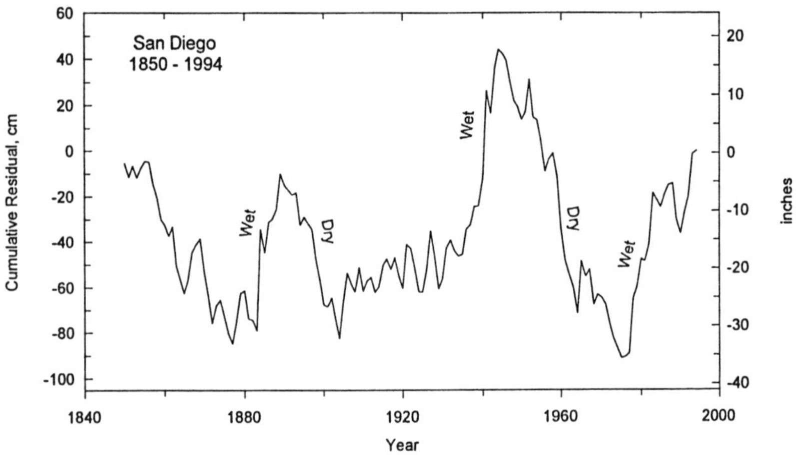

Figure 2. Cumulative residual rainfall in San Diego, CA, illustrating wet and dry periods.

Sedimentation

Since stormwater runoff is highly seasonal in southern California, river sediment discharge is also episodic. Discharge is limited by the presence of dams, urbanization, channelization, and soil type within the watershed. In fact, dams control 51.5% of the entire 4530 mi^2 of watershed that drain to the ocean in San Diego County. Simon, Li & Associates (1988) investigated the delivery of river sediments to the coast in the San Diego region and concluded that for the rivers that discharge into lagoons, nearly 100% of the sediments greater than 0.063 mm were deposited within the lagoons. Only suspended silts and clays transit the lagoons and are discharged to the ocean. Four lagoons, San Elijo, San Dieguito, Los Penasquitos, and Tijuana Estuary, showed the largest rates of sediment deposition.

Ocean Inlet Opening and Closure Processes

Inlet closure is a complex process that is poorly understood, because it involves numerous variables that may interact to different degrees depending on the time course of events. These variables include waves and beach characteristics, lagoon hydrology, inlet configuration, channel morphology, sediment characteristics, presence of cobbles, and adjacent shore morphology. Beach width is of particular importance because of its effect on the length of the inlet channel and volume of beach sediment available for resuspension and transport into the inlet and lagoon channels during flood tides. Consequently, changes in beach width may have an important influence on inlet dynamics. For example, a long inlet channel that crosses a broad sand beach during summer will be more sensitive to closure processes than a channel that crosses a narrow sand beach during winter. Deposits of beach cobbles may also contribute to inlet closure. Waves may cause the cobbles to migrate into the inlet, restrict tidal exchange, and reduce ebb flow scouring of the channel sand bed.

The duration of inlet closure events varies. Short term closures result from periods of high ocean wave activity that cause substantial sediment suspension, transport, and redeposition. A beach berm barrier can be established across a lagoon inlet in a few hours. Since minimal sediment is deposited within the adjacent lagoon channel, the inlet may reopen under the right conditions. Long term closure events lasting months to years that emerge from cumulative deposition of sediment in the inlet as well as the adjacent lagoon channel, are more difficult to reverse, because they typically require major stormwater runoff to blow out the beach berm and to scour the channel bed. As the long term closure process matures, there may be brief episodes of limited tidal exchange, for example, during spring tides.

When tidal flushing has ceased, seasonal stormwater runoff may accumulate and increase the lagoon water level and pressure head. Fresh or brackish water conditions may prevail for many months, or blow out the barrier beach berm, scour the channel bed, and restart tidal flushing. The extent to which the channel is scoured is a significant factor in the duration of tidal flushing. The tidal flushing may continue briefly or continue for many months until depositional processes close the lagoon inlet once again.

Lagoon Restoration

Several of the coastal lagoons in San Diego County have undergone significant long term degradation from construction of dams, urbanization, periods of drought, lack of tidal flushing, conversion of salt marsh habitat to brackish water marsh, and invasion of exotic species. As a result, there is renewed interest in restoring these lagoons (Boland and Zedler, 1996). However, these lagoons are complex ecosystems and our ability to successfully create, restore, or enhance them is limited and largely experimental.

Although the federal government recommends that restoration be given precedence over creation of wetlands habitat (US Government, 1995), California government agencies seem to recommend creation over restoration or enhancement projects due to the historical losses of wetland acreage and their desire to add new wetland acreage. This policy plus the desire for self-sustaining mitigation projects has led to applying hydrological solutions to lagoon problems. Hence, increasing the tidal prism and construction of jetties at the ocean inlet have dominated approaches to rehabilitating these lagoons. These two elements necessarily result in risky destruction/conversion of one habitat to another and incurring high costs for dredging and jetty construction. In turn, since the increased tidal flows may affect the stability of the older highway bridges and railroad trestles, these structures may also have to be replaced or protected; increasing the project cost even further. Also, depending upon the characteristics of lagoon sediments, the dredged materials are not always easily disposed.

Inlet Maintenance Strategy

Earlier plans for restoring coastal lagoons in southern California were based on the view of maximizing the tidal prism so that the inlets would sustain themselves through increased tidal flushing. The tidal prism needed was determined by using hydraulic modeling data and the results of classical studies on the stability of lagoon inlets by O'Brien (1931), Escoffier (1977), and Bruun (1978). These studies emphasized the roles of the tidal prism and longshore sediment transport in governing the stability of tidal inlets. The net result has been the assumption that massive enlargement and deepening of the lagoon basin by dredging was required to increase the tidal prism of the wetland enough for the inlet to become stable.

The main drawbacks to this approach are the high costs of dredging, disposal of sediments, the ecological disruptions during the construction phase, the fact that one system of habitats is often replaced by another, and the need for cross-beach jetty structures to help stabilize the inlet. In addition, the outcome of this approach is unpredictable, ecological success is not guaranteed, and the mitigation credit that should be awarded is difficult to estimate beforehand. Results of our studies suggest that it is feasible to maintain tidal flushing in small lagoons without resorting to large scale dredging (e.g., Boland, 1993; Boland and Elwany, 1994; Elwany et al., 1994). We propose an alternative strategy of inlet maintenance that is more economical, and less intrusive.

The proposed plan consists of one or two small scale tidal inlet and channel dredgings per year. The plan coupled with scouring from seasonal flows of stormwater runoff, can be successfully used to maintain tidal flushing in small coastal lagoons. Maintaining the inlet according to this plan will greatly reduce the frequency of inlet closures, but may not prevent rapid storm wave induced closures. The plan calls for prompt reopening if the lagoon does not reopen naturally within a few days. This approach is the result of several years of monitoring the inlets at

San Elijo, San Dieguito, and Los Penasquitos Lagoons located in San Diego County, California.

A one-time restorative dredging would be carried out to return the inlet and main channel to a near-equilibrium configuration. Subsequent maintenance dredging efforts would remove newly deposited marine sediments from the inlet channel before it migrates farther into the lagoon. Dredging with conventional earth-moving equipment, which can work in water depths up to three feet, or with a small barge-mounted pump and slurry pipeline is more economical and can be deployed more frequently than use of large scale dredgers. This approach does not require jetties to stabilize the inlet, and fosters continued recreational use and public access to the beach.

Inlet Maintenance Methodology

Three proven dredging methods were evaluated for their usefulness in maintaining inlets to small coastal lagoons as follows: conventional earth-moving equipment, cutterhead suction dredge, venturi pump dredge (Creek & Sagraves, 1995). Each dredge method was assessed for its ability to: 1) Complete entire scope of work, 2) Work within the constraints of the work site and on short notice, 3) Conduct both large and small maintenance jobs 4) Compete cost effectively. Other methods are discussed by Inman & Harris (1970) and Jenkins et al (1980).

The greatest constraint to dredging operations is the access problem created by presence of road and railroad bridges that traverse the lagoon inlet and main channel. The dredge method or combinations of methods must be able to dredge beneath these bridges and to transport sediments to the beach disposal site. Typical dredging rates for each method are summarized in Table 1.

Table 1. Estimated dredging capabilities of excavation equipment

Dredge type	Dry Capacity (yd^3)	Wet Capacity (yd^3)	Dredging Rate (yd^3/hr)
Front end loader	7	5	150-250
Scrapper	22	16	300
Backhoe excavator	3.5	3	250-350
Cutterhead Dredger (12 - 14 inch)	–	–	up to 875
Eddy Pump (12 - 14 inch)	–	–	330

Conventional Earth-moving Equipment

Most earth-moving vehicles are capable of working in 2-3 ft of water during periods of low tide. Earth movers, front end loaders, and tractor excavators have been used successfully to maintain the ocean inlets at Los Penasquitos, San Dieguito, and San Elijo Lagoons. Depending upon haul distance, a single earth mover can move about 150 - 250 yd^3 of wet beach sediment per hour. The mobilization and demobilization time for this equipment is minimal. There is no slurry pipeline involved. There is no de-watering runoff or turbidity plume generated at the disposal site and disposal incorporates grading.

Successful operation of these vehicles in the marine environment requires operators experienced in operating safely on sand beaches, in sea water, and under tidal conditions. These vehicles can work in lagoon channels that pass under most highway bridges, but not under railroad trestles. A disposal route must be accessible in order to be able to transport and dump the excavated sediments. The excavation effort must be scheduled around workable tide levels and ebb flows, otherwise flooding tides can rapidly undo previous progress. The inlet channel is opened temporarily and the lagoon is allowed to drain during ebb tide to lower the water level so that the vehicles can work in the channel. The inlet is then closed off again by construction of a temporary berm. The inlet is then reopened during a low spring tide, after the excavation is completed.

Cutterhead Suction Dredger and Slurry Pipeline

This is one of the most common dredgers and represents a proven technology. Typical dredger units are capable of dredging at rates of 150 yd^3 per hour and pumping a sediment slurry of 15% to 20% at 15 - 20 ft/sec. It operates by differentially winching forward by alternating lowering and raising it port and starboard spuds and by using anchors. The slurry pipeline is sensitive to wave motion and may have to be submerged or routed over land. The dredger should be operated in an upriver direction, particularly if there is a possibility of excessive stormwater runoff and may have to be redirected into the flooding tidal currents. A two knot current is thought to be maximal for safe operation since the spuds may slip in fast currents. Similarly, waves in excess of 2 - 3 ft can damage the ladder supporting the cutterhead. Consequently, this dredger would probably have to operate when the lagoon is closed. So opening of the inlet would have to be done by conventional earth-moving vehicles. Presence of cobbles may be a problem for a system that depends on pumping. Initial launching requires a body of deep water adjacent to the shore and several launchings may be required if it can not pass under certain bridges. The mobilization period may be lengthy.

Venturi Pump and Slurry Pipeline

The Eddy pump (Creek & Sagraves, 1995) is a relatively new high velocity, submersible pump dredging technology that offers a 50% energy savings and an ability to pump a sediment slurry of up to 70% solids. The pump is mounted on a small barge, lowered over the side to the dredge depth, and used to dredge the sediments by suction without the aid of a cutterhead. It has a maximum pumping rate of 10,000 gpm and can pump 250 ft of head. The 12 in pump can pump up to 560 yd^3 of sand per hour. Alternatively, the pump could be operated via a crane boom from shore. These attributes, including increased maneuverability and ability to pump cobbles, offers some significant advantages in terms of efficiency and in minimizing discharge of slurry water on the open beach. The pump requires operation of a slurry pipeline and would be difficult to operate in the ocean inlet.

Application of the Maintenance Plan

Three experimental dredgings were undertaken at San Elijo Lagoon between 1994 and 1997 (e.g., Boland and Elwany, 1994, Elwany et al., 1997). Conventional earth-moving equipment was used during the three experiments for dredging the inlet and the lagoon inlet channel. Figure 3 shows the equipment used for these experimental excavations. The first experiments was conducted in April 1994 and resulted in an open inlet for 4 months (22 April 1994 - 23 August 1994). The second experiment was conducted in June 1995, and the lagoon remained open until the end of October 1996 (4 months). The third experiment was conducted in winter 1996-97 and kept the lagoon open intermittently from the end of September until 1 March 1997 (5 months). Figure 4 (A) gives time periods of lagoon open or closed conditions at San Elijo Lagoon from 1986 to 1 March 1997.

At San Dieguito Lagoon our studies (Elwany et al., 1994) show that the typical inlet opening and closing sequence that is observed begins when a major flood scours the inlet channel bed below the equilibrium depth sustainable by the maximum available tidal prism. The inlet recharges with littoral sand, rapidly filling the entrance and exterior portions of the lagoon channels to a near-equilibrium depth. This condition has repeatedly been observed to remain semi-stable for up to several years. However, the inlet and channels slowly fill, building sills which choke off the inlet over a period of 2 to 5 years, eventually leading to inlet closure and deterioration in lagoon water quality. The lagoon remains closed until a large enough flood again scours the inlet and interior channel, or until the inlet is opened artificially. If a major flood occurs, the cycle starts over. This sequence of events implies that the inlet will remain open naturally as long as sufficiently large floods occur every 2 to 5 years. When there are no floods, the lagoon remains closed.

Based on our understanding of the physical and dynamical processes controlling San Dieguito Lagoon an inlet maintenance plan was developed. A one-time restorative dredging of about 25,000 yd^3 would be carried out to return the inlet channel to a near-equilibrium configuration. Subsequent maintenance dredging would remove an average yearly volume of about 15,000 yd^3 of sand from the inlet channel. Dredging with conventional earth-moving equipment, which can work in water depths up to three feet, or a small barge-mounted pump and slurry pipeline has been examined (Elwany et al., 1994), and found to be economical and far less disruptive than massive dredging. This approach requires no structures to stabilize the inlet, leaving recreation and lateral access along the beach area unaffected.

Figure 3. Conventional excavation equipment used to dredge San Elijo Lagoon.

A minimal dredging strategy was developed and applied with good success to Los Penasquitos Lagoon by the Los Penasquitos Lagoon Foundation from 1985 until present. The inlet was dredged frequently using earth-moving equipment. Figure 4 (B) shows time periods of lagoon opening and closing before and after adopting the maintenance plan. Between 1985 and 1991 the lagoon was open only 33% of the time, while after 1991 it has been open over 90% of the time. The Foundation believes that the improved success rate of their recent inlet maintenance efforts is due to their focussing on the dredging, during early spring, of the cobbles that were deposited in the inlet during winter storms (Boland 1993).

Conclusions

Southern California coastal lagoons are small, shallow, and subject to closure of the inlet. These lagoons are backed by rivers or streams. Episodic stormwater runoff contribute to scouring deposited marine sediments and restarts tidal flushing. Active dredging maintenance of the ocean inlet and main channel is required in order to sustain tidal flushing.

A site specific dredging maintenance plan should be designed on the basis of monitored marine sedimentation rates and changes in lagoon water quality conditions. On the basis of costs and flexibility in scheduling and deployment, conventional earth-moving vehicles are the best method for carrying out inlet maintenance programs in small coastal lagoons in San Diego County, California. The discussed maintenance plan was applied successfully to San Elijo and Los Penasquitos lagoons in San Diego County.

Acknowledgements

This study was supported by Southern California Edison Company, the San Elijo Lagoon Conservancy and the Los Penasquitos Lagoon Foundation.

References

Bruun, P. 1978. Stability of Tidal Inlets, Theory and Engineering. Elsevier Scientific Publishing Co., New York, 506 pp.

Boland, J. 1993. Maintaining the mouth at Los Penasquitos Lagoon 1991-1993: review and recommendations. Unpublished report to the Los Penasquitos Lagoon Foundation, P.O. Box 866, Cardiff, CA 92007, 10 pp.

Boland, J. and H. Elwany. 1994. Sand and cobble removal from the mouth of San Elijo Lagoon. Unpublished report to the San Elijo Lagoon Conservancy, P.O. Box 230634, Encinitas, CA 92023-0634, CE Ref No. 97-15, 17 pp.

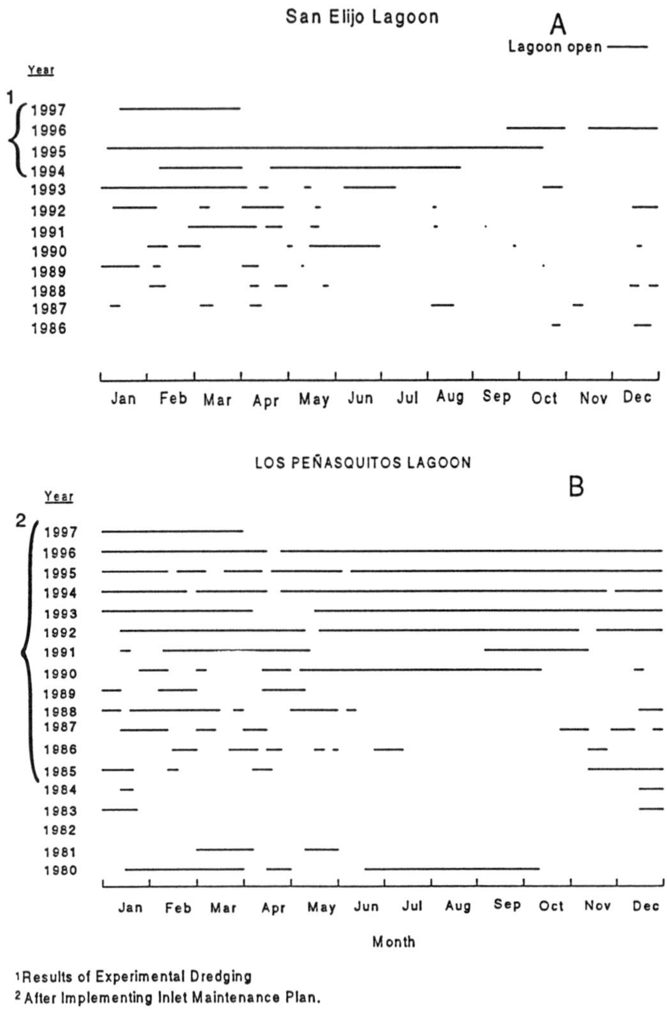

Figure 4. Inlet status (open or closed) at San Elijo and Los Penasquitos Lagoons to March 1, 1997. Data from San Elijo Lagoon Conservancy (unpublished), Los Penasquitos Lagoon Foundation (unpublished) and Boland (1993).

Boland, J. and J. Zedler, 1996. Maintaining tidal flows in southern California lagoons. In Tidal Wetland Restoration, A Scientific Perspective and Southern California Focus. CA Sea Grant College System, University of California, La Jolla, CA 92093-0232, Pub. No. T-039, Section 2.2, 11-25.

Creek, K. D., T. H. Sagraves, 1995. Eddy Pump Dredging: Does it reduce water Quality, Impacts? Proceedings of the International Conference on Hydropower, ASCE, Held July 25-28, 1995, San Francisco, CA., 2246-2252.

Elwany, M.H.S., R. Flick, J. Reitzel and S. Aijaz. 1994. Inlet Channel Maintenance Plan for San Dieguito Lagoon. Final Report, submitted to Southern California Edison, Rosemead, CA 91770, CE Ref. No. 94-11, 112 pp. and 3 appendices.

Elwany, M.H.S., R. E. Flick, and S. Aijaz. 1995. Long-Term Probability of Open Inlet at San Dieguito Lagoon. Final Report, submitted to Southern California Edison, Rosemead, CA 91770, CE Ref No. 95-8, 28 pp.

Elwany, M.H.S., C. Avendano, and R. Flick. 1997. Winter 1996-97 Experimental Dredging at San Elijo Lagoon. Final Report, submitted to San Elijo Lagoon Conservancy, P.O. Box 230634, Encinitas, CA 92023-0634, CE Ref No. 97-15.

Escoffier, F.F. 1977. Hydraulics and Stability of Tidal Inlets, GITI Report 13. U.S. Army Coastal Engineering Research Center, U.S. Army Engineer Waterways Experiment Station, Vicksburg, Mississippi, 72 pp.

Inman, D. L., and R. Harris, 1970. Cater-sink sand transfer system. Proc. 12th Conf. Coastal Eng., ASCE, New York, 2, 919-933.

Jenkins, S. A., D. L. Inman, and J. A. Bailard, 1980. Opening and Maintaining Tidal Lagoons & Estuaries. Proc. of the 17th International Coastal Engineering Conference, ASCE, Sydney, Australia, March 23-38, 1980, 1528-1546.

O'Brien, M.P. 1931. Estuary Tidal Prisms Related to Entrance Area. Civil Eng., Amer. Soc. Civil Eng., 1(8), 738-739.

Simon, Li & Associates, Inc. 1988. River Sediment Discharge Study San Diego Region. Coast of California Storm and Tidal Waves Study. Report No. CCSTWS 88-3, Tech. Rep. prepared for U.S. Army Corps of Engineers, Los Angeles, CA., 38 pp.

US Government, 1995. Federal Guidance for the Establishment, Use and Operation of Mitigation Banks. Federal Register, dated March 1995, 60 (43), 12-286 - 12-293.

CHAPTER 163

Establishing a Southern California Rocky Intertidal Monitoring Network

Dunaway, Ambrose, Campbell, Engle, Hill, Hymanson, and Richards[1]

Abstract

Continuous monitoring of both the nature of rocky intertidal habitats and the temporal dynamics of key species along the southern California coastline is critical to our ability to detect and predict change in the marine environment and to develop measures to anticipate and reduce harmful effects. Although a large part of the Southern California coastline is being actively monitored and informally coordinated by interested agencies, no coordinated program has been established over the region. We believe establishing a long-term monitoring network for rocky intertidal resources in the Southern California Bight will greatly improve our understanding of the changes to rocky intertidal communities by providing a regional context and by increasing communication among researchers. The network could serve as a model for future monitoring of other resources or in other areas.

Long-term monitoring is not a new idea in the region. The National Park Service's Channel Islands long-term monitoring program was established in the early 1980's and contains both intertidal and subtidal programs. The Shoreline Inventory of Resources of Santa Barbara County program co-sponsored by Santa Barbara County and the Minerals Management Service was patterned after the Park Service's program with the intention of maintaining a long-term monitoring program along the adjacent mainland. This mainland inventory, now in its 5th year, has been linked to several other monitoring programs along the coastline extending from San Luis

---

[1]Dunaway, Mary Elaine. U.S.Minerals Mgmt Ser, 770 Paseo Camarillo, Camarillo, CA 93010-6064
Ambrose, Richard. Environmental Science and Engineering Program, UCLA, CA 90095-1772
Campbell, Jackie. Santa Barbara County Energy Division, Santa Barbara, CA 93101
Engle, John. Marinc Science Institute, University of California, Santa Barbara, CA 93106
Hill, Maurice. U.S.Minerals Management Service, 770 Paseo Camarillo, Camarillo, CA 93010-6064
Hymanson, Zachary. California Coastal Commission, Energy Div. San Francisco, CA 94105-2219
Richards, Dan. Channel Islands National Park, Ventura, CA 93001-4354

Obispo to San Diego. These additional programs are currently sponsored by several agencies/institutions including the California Coastal Commission, National Park Service, University of California, and Minerals Management Service.

The success of these programs and the desire to examine marine resources from a regional perspective is leading agencies concerned with these resources to develop a coordinated, regional, long-term monitoring network. The network is presently focused on rocky intertidal resources and will provide for continued annual/semiannual monitoring of intertidal resources over the area, a joint data management system, an integrated research program and shared long-term monetary support.

The benefits of a regional monitoring program are numerous and somewhat obvious, yet obtaining consistent funding for such an effort is difficult and requires the dedication and teamwork of many agencies and researchers. Accordingly, this paper is co-authored by several agencies and researchers who have had key roles in developing the network and who are working toward its continued future. This paper will discuss the goals of this network, several of the ongoing monitoring programs in southern California, and the future plans of the network.

Introduction

The efforts of many scientists are required to set up a large multi-agency monitoring network. To obtain the input from scientists from academic, governmental and private interests, a workshop was held at the Cliff House on the University of California, Santa Barbara (UCSB) campus February 26th and 27th, 1997. Fifty-one invited scientists attended the workshop and provided their recommendations for the establishment of the Multi-Agency Rocky Intertidal Monitoring Network (MARINE). The workshop has led to renewed commitments from sponsoring agencies and new participation by a variety of groups. The specific results will be summarized at the end of this paper.

Monitoring has not always been valued by researchers, but this attitude is changing. In the mid- to-late 1970's, intensive monitoring of 22 sites in the Southern California Bight were funded by the Bureau of Land Management (BLM) (now the Minerals Management Service (MMS)) (Littler, 1978a, 1978b, 1979). This intensive three year study was part of a larger baseline study of the Southern California Bight which included the simultaneous monitoring of subtidal, benthic, and marine mammal and seabird resources, geochemical and physical oceanographic parameters. It was intended that the initial effort would provide the information needed to establish a long-term monitoring program. However, funding for the entire program was cut after the National Academy of Science reviewed the program and directed BLM/MMS to contract for more focused research. Prevailing thought among scientists was that conducting "monitoring" or collecting "baseline" data was too general to be a useful expenditure of resources. It is only fairly recently that scientists and managers have identified monitoring as a critical part of the resource management program, a philosophy which is changing the way National Parks are managed (Halvorson and Davis, 1996).

Fortunately, in the 1990's there has been a renewed appreciation for monitoring as a regulatory tool for detecting impacts to the environment. This is demonstrated by the large sections of the Southern California coastline currently being monitored by state, Federal and local agencies, private groups and institutions. We can now also realize the synergistic benefits of a network as technology advancements make communication among organizations and data synthesis comparatively easy. A network of coordinated groups conducting research and comparing data provides a wealth of information about the coastline, allowing scientists to understand changes in the marine environment in a regional context. This paper will describe the existing monitoring programs sponsored by the key agencies from a historical perspective and discuss significant findings from the monitoring to date. The paper will also discuss the issues presented at the workshop and summarize the recommendations made by the scientists present.

EXISTING MONITORING PROGRAMS

National Park Service - Channel Islands National Park and Cabrillo National Monument

The need to monitor the environment was recognized early in the establishment of the Channel Islands National Park and monitoring of the rocky intertidal and subtidal resources has been pursued continuously since the early 1980's. This ongoing program covers four of the Park's islands and is soon to be expanded to include Santa Cruz Island. They have been able to continue this monitoring by incorporating data collection as part of the ranger's normal activities in order to ensure the health of the resources under their purview.

The National Park Service's program was initially established at Anacapa Island to monitor visitor impacts. Low visitation control sites were established with visitor impact sites, and experimental plots tested the effects of trampling and clearing of organisms. Other concerns of the National Park Service include impacts from fishery harvest, oil spills, marine pollution, and global warming. Monitoring was expanded to Santa Barbara, San Miguel, and Santa Rosa islands in 1985 with a broader goal of collecting ecological baseline information. Sites were chosen to provide the full range of exposures and substrate types. Additional sites were added on Santa Rosa Island in 1986 and 1988 as access improved. Currently there are 14 sites monitored on four islands (Richards and Davis, 1988).

Three intertidal monitoring sites were established in high, medium, and low use areas of rocky reef at Cabrillo National Monument (NM) in 1990. Tidepools at Cabrillo NM are among the most accessible in San Diego County resulting in thousands of visitors exploring the reef every year.

Shoreline Inventory-Santa Barbara County/Minerals Management Service

The Shoreline Inventory Program was established by the Santa Barbara County Board of Supervisors in 1990 to create an inventory of subtidal, intertidal and wetland resources of Santa Barbara County's mainland coast to improve oil spill response and overall planning for habitat restoration efforts in the event of an oil spill. The program was co-sponsored by Santa Barbara County and MMS and has since been linked to other shoreline monitoring segments both to the north and south along the mainland.

The goal of the Shoreline Inventory Project is two-fold. First, the inventory is intended to provide a database to be used to establish a record of pre-existing conditions in the event of a marine oil spill and resulting damage to the coastline. This portion was funded by the offshore operators (Chevron, Exxon, Texaco, Unocal, and Clean Seas) and managed by Santa Barbara County. The database includes a Literature Search, species list, bibliographic and natural history databases, a set of maps for the County and an educational booklet entitled *Seashore Life of Santa Barbara County.* Agencies and the public alike are able to access and rely on these data for a variety of needs including biologic study, community planning and education. This information is provided to interested parties free of charge and has proven helpful in analyzing and monitoring potential impacts of development on coastal resources through the California Environmental Quality Act (CEQA) and the National Environmental Policy Act (NEPA) environmental review processes.

Second, the inventory provides responsible agencies and the public with increased information and a database of intertidal and subtidal habitats along the coastline, including ten wetland habitat locations in Santa Barbara County. As part of this effort, a three-year field monitoring study of the rocky intertidal shoreline in Santa Barbara County was funded by MMS and carried out by UCSB scientists in the early 1990's; MMS has committed to continued long-term monitoring of these nine rocky intertidal sites. Data reports and the database can be obtained by contacting the County Energy Division or MMS.

California Coastal Commission Shoreline Monitoring

The shoreline monitoring programs initiated in Ventura and Los Angeles Counties and the Northern Channel Islands were required through a regulatory action. In January 1993, the California Coastal Commission approved a coastal development permit to allow the Point Arguello Producers to transport up to 50 thousand barrels of crude oil per day from the Gaviota Marine Terminal in Santa Barbara County to Los Angeles for up to three years. A special condition of this permit required the Point Arguello Producers to pay $600,000 for the purpose of funding shoreline inventories of coastal ecological resources at the Northern Channel Islands and the mainland of Ventura and Los Angeles Counties. Thus, the available funds served to extend the geographic coverage of the existing Santa Barbara County mainland shoreline inventory.

The Coastal Commission required funding for a shoreline inventory as one of several advanced preparedness measures in the event of a tanker oil spill. The goal of the Northern Channel Islands and Ventura/Los Angeles Counties shoreline monitoring programs is to establish a baseline record of select biological resources at representative nearshore habitats potentially affected by an oil spill. In such an event, the resulting information will be used to provide strategies for the protection of resident biota and sensitive habitats, as well as for the clean-up and rehabilitation of oiled areas. Other objectives of the monitoring program, include: 1) establishment of monitoring sites and methods suitable for use in a long-term monitoring program; 2) expansion of the information base for the study regions through collection of baseline information in locations or habitats not previously sampled; and 3) provision of baseline information useful in completing natural resource damage assessments resulting from other catastrophes.

Minerals Management Service Mainland Shoreline Monitoring

The MMS is responsible for the regulation of offshore oil production in federal waters beyond three miles from the Southern California coastline. Pursuant to the Outer Continental Shelf Lands Act, the MMS is required to monitor the effects of offshore oil operations under their regulatory jurisdiction. Specifically, the Act states that MMS "shall conduct such additional studies to establish environmental information as [it] deems necessary and shall monitor the human, marine, and coastal environments of such area or region in a manner designed to provide time-series and data trend information which can be used for comparison with any previously collected data for the purpose of identifying any significant changes in the quality and productivity of such environments, for establishing trends in the areas studied and monitored, and for designing experiments to identify the causes of such changes." Furthermore, the National Environmental Policy Act also requires federal agencies to monitor the effects of the operations that they regulate. Of particular concern to the MMS is the effect of an oil spill from an offshore oil platform or subsea pipeline; or from other sources outside the regulatory authority of MMS, such as tankers, barges, or offshore oil platforms and pipelines that are located within state waters.

As a result of MMS's regulatory responsibilities and an increased public concern for oil spills following the EXXON VALDEZ spill in Alaska, the MMS funded a three-year study of rocky intertidal resources in Santa Barbara County (previously in the discussion of the Shoreline Inventory) which consists of sampling selected sites along the Santa Barbara County coastline to fill in gaps in our knowledge about these areas, and provide current data on the shoreline resources. A total of 8 rocky intertidal sampling sites were established in Santa Barbara County in 1992. The selection of these monitoring sites was based on a review of existing information at the time (Ambrose et al. 1992), a qualitative reconnaissance of many rocky intertidal sites in the County, and quantitative sampling of 11 potential sites (Ambrose et al. 1995). The monitoring sites are located from just south of Point Sal in north County to Carpinteria in the south County. A site located at Purisima Point was established in Fall 1993 to monitor the black abalone population for a total of nine sites in Santa Barbara County (Figure 1).

Through the UCSB Coastal Marine Institute funding, four new rocky intertidal monitoring sites were established in San Luis Obispo County in fall 1995 and four sites were established in Orange County in fall 1996 to provide monitoring data for areas adjacent to offshore oil production and transportation that were not being monitored (Figure 1).

Sampling Design

The monitoring program for long-term programs described in this paper consists of repeated sampling during spring and fall of each year at a series of permanent sites along the coast. These sites range from southern San Diego County in the south to northern San Luis Obispo County in the north, and include most of the Channel Islands. Currently, there are 59 sites being monitored, with 49 sites using similar monitoring methods. This similarity in the sampling design makes forming a network easier and logical.

With the exception of San Diego County, the sites were chosen to provide good geographic coverage of the area so that at least one site affected by a large oil spill would have baseline data. Along the mainland, four sites in San Luis Obispo County and four sites in Santa Barbara County north of Pt. Conception ensure that this section of the central coast, which is predominantly rocky, has good coverage (Figure 1). South of Pt.Conception there are four sites in Santa Barbara County, two in Ventura County, two in Los Angeles County, and four in Orange County. Although the density of sites is lower along this section of coast, this is appropriate because rocky areas are less common. Finally, there are five sites clustered in the southern area of San Diego County. Because these sites were established for a different purpose, the sites in San Diego County were not spaced throughout the entire County.

Seven of the eight Channel Islands are also included in the monitoring (Figure 2). There are two sites on Santa Catalina and Santa Barbara Island, four sites on Anacapa and San Miguel, five sites on Santa Rosa, and six sites on Santa Cruz Island. Five sites on San Nicolas Island are sampled by the U.S. Geological Survey, Biological Resources Division. The islands have a relatively high density of sites compared to the mainland, and coverage is fairly equal except for relatively few sites on Santa Catalina, the north side of Santa Rosa, and of course San Clemence Island, which currently has no sites.

All sites discussed here are rocky intertidal habitats, and most are rocky benches with a fairly consistent gradient from upper to lower intertidal. Where possible, we chose sites where historic data was available, particularly those sampled during the earlier baseline study funded by BLM (Littler 1978a, 1978b, 1979). At each site, a series of permanent plots have been established by epoxying stainless steel bolts into the rock substrate. The plots target key intertidal species. A number of criteria were used to select the target species (Ambrose et al. 1992), including ecological importance, abundance, susceptibility to human impacts (especially oil spills), and practicality. Examples of target species include barnacles (*Chthamalus* spp., *Balanus glandula*), mussels (*Mytilus californianus*), rockweed (*Pelvetia fastigiata*), turfweed (*Endocladia muricata*), owl limpets (*Lottia gigantea*), black abalone (*Haliotis cracherodii*), and surfgrass (*Phyllospadix* spp.).

Most target species are sampled using 50 cm x 75 cm (0.375 m^2) photoplots. Cover is estimated for these species by scoring the photographs taken in the field using a point-contact method (Table 1). For each species at a site, five photoplots were

Table 1. Summary of sampling techniques used for representative targeted taxa and methods of data collection. Refer to text for common names. NA = not applicable

Sampling Technique	Field	Laboratory
Quadrats: barnacles *Endocladia* *Pelvetia* *Mytilus*	Photographs	% cover by point-contact method
Transect line: *Phyllospadix*	% cover by point-intercept method	NA
Circular plots: *Lottia gigantea*	Counts/size measurements	NA
Irregular plots or band transects: sea stars black abalone	Counts/ size measurements (abalone)	NA

Figure 1.
Location of rocky intertidal monitoring sites along the central and southern California Coast.

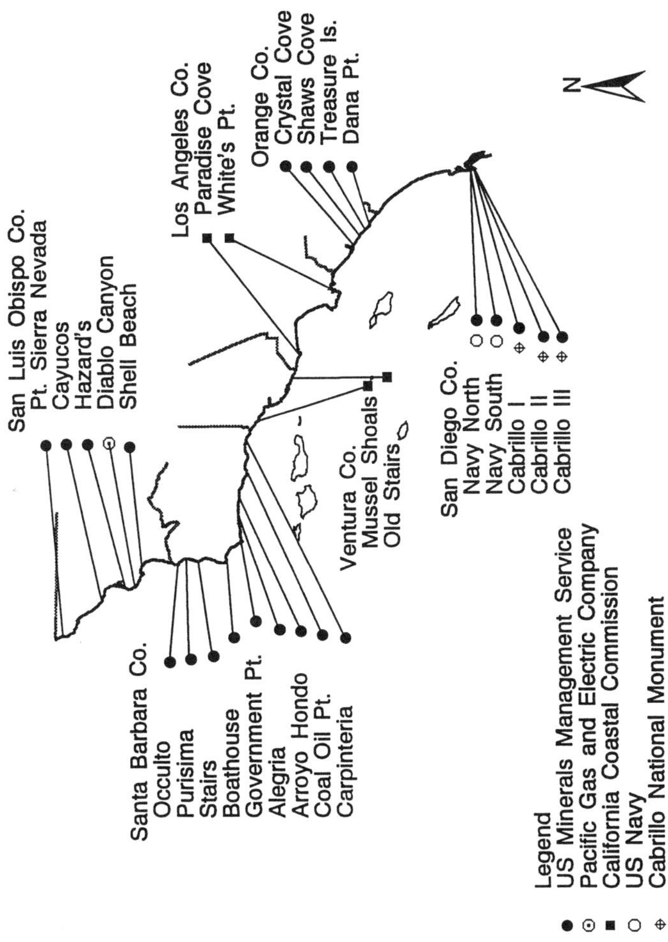

Figure 2. Location of rocky intertidal monitoring sites at the Channel Islands.

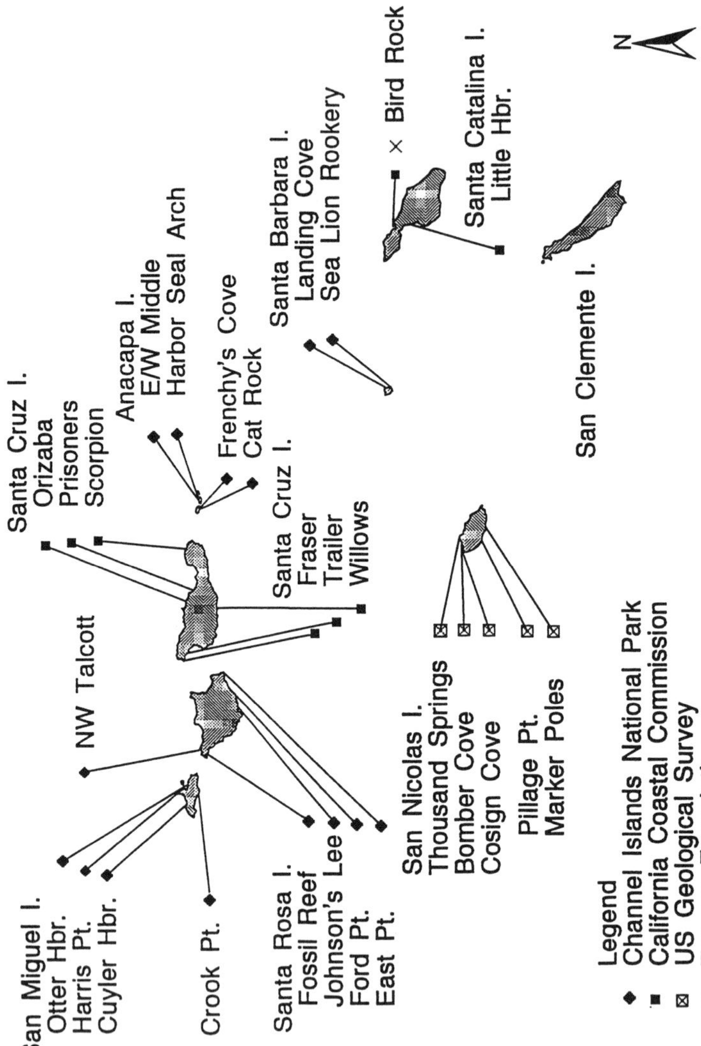

established; the photoplots were not randomly located, but rather placed in areas with high abundance of the species at that site. Surfgrass cannot be sampled adequately using photoplots, so three line transects, each 100 m long, are used. Other target species are counted, either in 2-m diameter circular plots (*Lottia*) or larger irregular plots or band transects (sea stars and abalone). The size and shape of the irregular plots or band transects depend on the abundance of the species and the configuration of the habitat at a particular site. At some sites, timed searches are used for species that are too rare for establishing permanent plots for estimating density.

The California Coastal Commission's shoreline monitoring program is collecting and analyzing information on intertidal, subtidal, and wetland resources during the study period (March 1994 through October 1998). Study emphasis is on those resources and habitats most vulnerable to oil spill impacts. One-time reconnaissance surveys to characterize biotic assemblages and habitat types over the study areas are included, as well as periodic surveys. The periodic surveys provide data on species dynamics to assess seasonal and long-term changes in community structure.

Surveys at the Northern Channel Islands, including Anacapa, Santa Cruz, Santa Rosa, and San Miguel Islands, emphasize rocky intertidal communities. Sandy intertidal communities are also surveyed at a reduced intensity. Subtidal surveys concentrate on reconnaissance surveys of shallow soft-bottom habitats, particularly eelgrass and clam beds. Surveys of subtidal rocky habitats are already ongoing by the Channel Islands National Park and other organizations.

Surveys of Ventura and Los Angeles Counties cover the mainland coast and Santa Catalina Island, and like the Northern Channel Islands program, include rocky intertidal and sandy intertidal communities. Coastal wetlands are also a primary focus of the Ventura and Los Angeles Counties program. This is because coastal wetlands are a habitat of concern in California, and are disproportionately susceptible to damage from oil spills. The utility of the wetland data collected is further increased by the sampling of comparable sites in Santa Barbara County.

Special Studies

Permanent survey sites are ideal locations for research and special studies that complement the monitoring program. Time-series data are necessary for many ecological investigations and raise questions that can be addressed by research projects. Research increases basic understanding of ecological structure and processes, resulting in enhanced impact predictability, improved means to detect change, and other insights benefiting long-term monitoring. Special studies conducted at the monitoring sites include sampling method evaluations, community studies, manipulative experiments, and additional monitoring for particular species and life history parameters. Methods-testing studies have emphasized evaluation of biotic cover estimates by various techniques, including video. Community studies include comparison of boulder and bedrock habitats in Los Angeles County and characterization of rockweed assemblages in Orange County. Few experimental studies have been done at the monitoring sites, in part because manipulations could affect "natural" monitoring. In a study at the northernmost site, mussels had not fully recovered eight years after plots were cleared (MMS 1996). At several other sites, habitat studies involve transplants of surfgrass, abalone, or mussels. The importance of lobster predation on mussels in determining intertidal community structure at Santa Catalina Island was determined using predator-exclusion cages and other field experiments (Robles et al. 1990; Dittman and Robles 1991; Robles and Robb 1993).

Many of the special studies focus on temporal and spatial dynamics of particular species or processes (e.g., recruitment, growth, mortality). This research requires additional monitoring efforts beyond the basic key species surveys. Rockweed, surfgrass, turban snails, abalone and mussels have been targeted for population and life history studies at various sites. Also, the type and extent of human recreational activities have been surveyed at some of the urban sites. Recruitment is a critical process about which surprisingly little is known. Colonization studies of barnacles, hermit crabs, abalone and mussels at representative island and mainland sites indicate unique local and regional patterns that likely are related to varying oceanographic regimes. For example, at 11 sites along 200 mi of coastline, barnacle settlement was consistently high at 2 sites, intermediate at 6, and low at the remaining 3. Settlement "hotspots" and "coldspots" were not in geographic clusters; hence, local chronic conditions prevailed.

Additional studies also help tie the biological data collected to physical measurements in order to better explain trends seen in the monitoring. Temperature loggers were recently placed in the intertidal zone on the northern Channel Islands. Coupled with the temperature loggers at subtidal sites, this will help in understanding local oceanographic conditions and possibly explain recruitment patterns.

Significant Findings

One of the more significant findings of the monitoring has been the marked decline of the black abalone, *Haliotis cracherodii*. Black abalone density declines at the Channels Islands were observed at some sites as early as 1986. The monitoring program identified that there was a serious problem early on and directed research towards identifying the cause (Haaker et al.1992). Monitoring documented the mass mortality of the species (over 90% parkwide), and guided research about the mass mortality. Monitoring results, including size-frequency measurements and growth studies, documented the magnitude and regional pattern of mortality. Associated special studies confirmed the presence and spread of "withering foot syndrome" and searched for a causative agent. It is believed to be caused by a microorganism which causes the foot to shrink and leads to mortality. The microorganism appears to be linked to warm water conditions though there is no clear understanding of how it is spread.

Sharp declines of the black abalone were first noticed on the mainland at Government Point in 1992. Monitoring programs provided quantitative data to document the mass mortality of black abalone from the beginning, thus attention was drawn to this issue early on. The decline on the mainland has since continued north to other MMS sites on Vandenburg Air Force Base at Boathouse and Stairs (Altstatt et al. 1996.) As of spring 1997, the number of abalone at Boathouse is holding at a fairly constant level (at approximately one-fourth of the original number of abalone). At Stairs the numbers continue to drop; currently the mean value is around a third of the original mean value (pers.comm., M.Wilson, UCSB). Fortunately, the Purisima Point site further north (established after the first decline was noted to document black abalone abundance) is still showing undiminished numbers of abalone and withered foot animals have not been specifically identified at that site.

Monitoring data from the islands and mainland provided the non-fishery data used to support closure of the black abalone fishery in 1993 to protect the species (Richards and Davis 1993). Other research is underway to evaluate community changes associated with black abalone loss, and determine if future black abalone recruitment will be affected by these community alterations.

Observations from the monitoring of Cabrillo National Monument have led to a management program of education and limited access to one portion of the reef to allow recovery. Dramatic declines in mussel density have occurred since at least 1990. While the cause is unknown, it is evident that the mussels have suffered from a lack of recruitment. The monitoring program was able to detect changes from the 1992 sewage spill off Point Loma. Though some species were temporarily impacted from the spill, they recovered and others actually benefited from closing the area to visitation for two months (Engle and Davis 1996b).

THE MULTI-AGENCY ROCKY INTERTIDAL NETWORK

The authors of this paper strongly believe that formation of a monitoring network is advantageous for all parties concerned. Some of these key advantages include:

- ✓Increasing the reliability, efficiency and cost-effectiveness of programs
- ✓Enhances long-term support to ensure continuity of sampling
- ✓Permits the evaluation of large scale patterns
- ✓Provides for better analysis synthesis and reporting of the data

Cliff House Workshop Findings

A workshop was held at the University of California, Santa Barbara to discuss formation of a network and develop recommendations for improving the monitoring of the rocky intertidal communities in the Southern California Bight. The concept of the network has been extremely well received and the number of participants continues to grow.

There was general consensus that establishing a network met the needs of both resource management agencies and public use agencies. Though the goals of agencies managing a specific resource differed from agencies that need to assess the impact of their program on or public use of these resources, the need for consistently funded long term monitoring programs were the same. Long-term monitoring is valued because it provides the information to gather information on ecosystem health, establish limits of variation diagnose abnormal conditions, and identify potential agents of change (Davis et al.1994). The formation of the network greatly enhances individual monitoring programs by providing for better communication among researchers, by permitting evaluation of large-scale spatial and temporal patterns, by providing more timely access to the data by all users, and by integrating information for efficient analysis, synthesis, and reporting. The network also provides a more certain funding mechanism through a formalized joint sharing of the process; a value which cannot be underestimated.

A synopsis of the findings from the workshop has been released and is available upon request from the MMS or UCSB (Engle et al. 1997). The four key recommendations drafted by the scientists attending the workshop were:

1. Continue monitoring, create a network, and secure stable funding.

It was recognized by all the workshop participants that the most critical part of the program was to continue the monitoring so that data gaps are not created. While aspects of the network will be aimed at improving sampling and developing a database, the highest priority for the network is to maintain sampling at the existing sites. Once basic monitoring needs for the Bight are met, the participants can focus on setting up the structure needed for the network and addressing network goals.

One of the highlighted benefits of the network is the opportunity to pool resources so that a stable funding base is established. Obtaining long-term funding is difficult for most agencies and organizations as most programs are funded on a year-to-year basis. Shared long-term funding or in-kind services are critical to ensure the stability of the network and to maximize its effectiveness.

2. Form a Steering Committee, Scientific Review Panel, and Data Review Panel.

Some structure was recommended to assure continuity and good communication among participants. A Steering Committee, made up of groups committing resources and involved in joint assessment of the data, would ensure long-term support of the monitoring and provide oversight of a Scientific Review Panel and Data Review Panel.

3. *Develop a management system for network data.*

The scientists believe there is an immediate need to develop a system capable of managing the data from all programs in the network. Initially, this system should provide an efficient means to access standardized data from each monitoring program. Eventually, all monitoring data should be placed in a central repository maintained by a data specialist. Members appointed to a Data Review Panel would oversee the design and development of the system, and report their findings to the Steering Committee.

4. *Review existing data from network programs.*

High priority should be placed on reviewing pooled data from all network programs. This review is needed before any recommendations for improving the monitoring can be made. This review would be conducted by the Scientific Review Panel and would serve as the basis for any decisions for future modifications.

A number of detailed recommendations were also made by the invited scientists. This input will be used to further delineate areas of improvement once the initial tasks are completed. Many of these recommendations relate to possible technical improvements in sampling protocol, species selection, and increasing the efficiency of the tasks. One of the recommendations which was discussed at length was identifying practical ways to improve the ability to extrapolate information within and among sites.

Participation and Support

Thirteen agencies and organizations have committed money, or in-kind support (staff field time, boat transport to the islands, computer support, etc.) for the network. Six of these organizations/agencies have committed support for long-term monitoring of specific sites along the mainland or islands; nine have indicated interest in participating on a Steering Committee. We anticipate that at least a dozen more groups will participate in the network for the purpose of data exchange.

The scientists at the workshop discussed the existing programs and anticipated funding needs. Several organizations including Federal agencies, local entities, and volunteer organizations came forward to fill identified funding gaps. By the end of the workshop, funding was identified to support almost all of the 62 sites being monitored in the Southern California Bight. The remaining gaps in LA County and San Clemente Island which will occur when the CCC funding expires spring of 1998, are of interest to several groups and are likely be funded.

Conclusion

Long-term monitoring of shoreline resources provides the data needed for knowledge-based decisionmaking and strong resource management programs. As has been seen in the monitoring conducted to date, consistent sampling provides resource managers with early warnings of abnormal conditions and allows for reduction of environmental effects. The network concept evolved from the natural process of working together, combining resources, and sharing a common goal. We believe that establishment of this network furthers the common needs of many organizations while advancing our understanding of the dynamics of the rocky intertidal communities in the Southern California Bight. We are excited about this opportunity and look forward to the progress we will make in the coming year.

REFERENCES

Altstatt, J.M., R.F. Ambrose, J.M. Engle, P.L. Haaker, K.D. Lafferty and P.T. Raimondi. 1996. *Recent declines of black abalone Haliotis cracherodii on the mainland coast of central California*. Mar. Ecol. Prog. Ser. 142: 185-192.

Ambrose, R.F., P.T. Raimondi and J.M. Engle. 1992. *Final study plan for Inventory of Intertidal Resources in Santa Barbara County.* U.S. Department of the Interior, Minerals Management Service.

Ambrose, R.F.J.M. Engle, P.T.Raimondi, M.Wilson, J. Altstatt. 1995. *Rocky Intertidal and Subtidal Resources: Santa Barbara County Mainland.* Final Report. U.S. Department of the Interior, Minerals Management Service. OCS Study MMS 95-0067. pp. 171.

Dittman, D.E. & C. Robles. 1991. *Effect of algal epiphytes on the mussel Mytilus californianus*. Ecology 72: 286-296.

Davis, G.E., K.R. Faulkner, and W.L. Halvorson. 1994. *Ecological Monitoring in Channel Islands National Park, California.* In: the Fourth California Islands Symposium: Update on the Status of Resources. Santa Barbara Museum of Natural Hisroty, Santa Barbara, California.

Engle, J.M. and G.E. Davis. 1996a. *Rocky Intertidal Resources Monitoring Handbook, Cabrillo National Monument, San Diego, California.* U.S. Department of the Interior, California Science Center, Davis, CA 38pp.

Engle, J.M. and G.E. Davis. 1996b. *Ecological condition and public use of the Cabrillo National Monument Intertidal Zone 1990-1995.* Technical Report. U.S. Department of the Interior, California Science Center, Davis, CA 183pp.

Engle, J.M., R.F. Ambrose, and P.T. Raimondi. 1997. *Synopsis of the Interagency Rocky Intertidal Monitoring Network Workshop. Final Report.* U.S. Department of the Interior, Minerals Managment Sevice. OCS Study MMS 97-0012.

Halvorson, W.L. and G.E Davis. 1996. *Science and Ecosystem Management in the National Parks.* University of Arizona Press.

Haaker, P.L., D.V. Richards, C.S. Friedman, G.E. Davis, D.O. Parker, and H.Togstad., 1992. *Mass Mortality and withering Syndrome in Black Abalone, Haliotis cracherodii in California,* In: Abalone of the World: Biology, Fisheries, and Culture. Proceedings of the first International Symposium on Abalone pp. 214-224. Blackwell Scientific Publications, Inc. Cambridge.

Haaker, P.L., D.O. Parker & C.Y. Chun. 1995. *Growth of black abalone, Haliotis cracherodii Leach, at San Miguel Island and Point Arguello, California.* J. Shellfish Res. 14: 519-525.

Littler, M.M., ed. 1978a. *Spatial and temporal variations in the distribution and abundance of rocky intertidal and tide pool biotas in the Southern California Bight.*, Year I - Southern California OCS.

Littler, M.M., ed. 1978b. *The annual and seasonal ecology of Southern California rocky intertidal, subtidal and tide pool biotas.* Year II - Southern California Outer Continental Shelf Program. Bureau of Land Management, Technical Report III-1.1. U.S. Government Printing Office, Washington, D.C..

Littler, M.M., ed. 1979. *The distribution, abundance and community structure of rocky intertidal and tide pool biotas in the Southern California Bight*, ed. M. M. Littler. Year III - Southern California Outer Continental Shelf Program. Bureau of Land Management, Technical Report III-1.0. U.S. Government Printing Office, Washington, D.C..

Minerals Management Service & Kinnetic Laboratories, Inc. 1996. *Mussel recovery and species dynamics at four California rocky intertidal sites: 1992 Data Report.* U.S.Department of the Interior, Minerals Management Service, Pacific OCS Region. OCS Study MMS 96-0009. 151 pp.

Richards, D.V. 1994. *Effects of disturbance on population dynamics of selected taxa in the rocky intertidal zone of Channel Islands National Park, California.* In the Fourth California Islands Symposium; update on the status of resources. Halvorson, W.L. and G.J. Maender, Eds.pp. 45-58. Santa Barbara Museum of Natural History, Santa Barbara, CA.

Richards, D.V. and G.E.Davis. 1988. *Rocky Intertidal Communiites Monitoring Handbook.* Channel Islands National Park.

Richards, D.V. & G.E. Davis. 1993. *Early warnings of modern population collapse in black abalone Haliotis cracherodii, Leack, 1814 at the California Channel Islands.* J. Shellfish Res. 12: 189-194.

Robles, C., D.A. Sweetnam & J. Eminike. 1990. *Lobster predation on mussels: shore-level differences in prey vulnerability and predator preference.* Ecology 71: 1564-1577.

Robles, C. and J. Robb, 1993. *Varied carnivore effects and the prevalence of Intertidal turfs.* J.Exp. Mar. Biol. Ecol. 166: 65-91.

CHAPTER 164

RECENT DEVELOPMENTS IN OCEAN OUTFALLS

Fredric L. Hettinger 1

ABSTRACT.

As one of the world's leading underwater service companies, Oceaneering International has been involved in a wide range of ocean and coastal projects since it was founded in 1969. Many of these have been major water quality improvement projects performed by Oceaneering's Western Region office located in Ventura, California. This paper addresses one important type of water quality management system - the ocean outfall, and specifically highlights recent developments in their design and construction.

A brief summary of the basic outfall designs is presented, including typical construction methods. The major elements of a typical marine outfall system are described in the paper, which generally include the main pipeline, diffuser section, diffuser ports, pipeline stabilization or burial means, seafloor preparation, cathodic protection, manholes and flapgate structures. This section also includes the results of a series of outfall material cost-benefit studies which highlight several economic and performance advantages of high density polyethylene (HDPE). This material has been widely used on onshore liquid transmission projects and is increasingly being considered and specified for offshore outfalls as well. During the past three years, Oceaneering has designed and successfully installed several ocean outfall systems on the West Coast and Australia using HDPE. As with all construction materials, HDPE has a range of advantages and disadvantages. While the relatively low overall cost of an HDPE outfall is compelling, several other important physical characteristics as well as site environmental conditions must be considered before selecting this material. Both technical and economic factors relating to outfall materials selection are discussed in the paper.

Included in this paper are several project descriptions of recent HDPE outfall installations made by Oceaneering offshore California and Australia. These descriptions include photographs taken during the construction phase which illustrate new developments in pipeline stabilization techniques.

In summary, recognizing two major longterm factors affecting California and the World Ocean - namely limited project funding and increased human coastal impact - the outfall designer must consider economic factors as well as technical performance. A major element of ocean outfall construction is the price of the pipeline material and the cost to install it. High density polyethylene (HDPE) is a pipeline material that possesses several attractive cost and technical features. The use of this material, where appropriate, can result in significant cost savings to the owner. During the past three years, Oceaneering has successfully designed and installed several HDPE outfall pipelines. We believe that this material will increasingly be the preferred choice for many outfall installations.

---

1 Project Manager, Oceaneering International, 2055 N. Ventura Ave. Ventura, CA 93001

1.0 INTRODUCTION.

As the State of California enters a new era of economic recovery and longterm growth, it will be faced with many challenges and areas of uncertainty. Certainly, ocean and coastal related subjects will be important issues during this exciting millennial period. Whatever the specific patterns of growth emerge, we can be fairly certain of two longterm factors that will affect California and the World Ocean - limited financial resources and the impact of increased human coastal contact.

This paper briefly describes recent developments in the design and construction of ocean outfalls. While the function and technical features of ocean outfalls are largely unknown by the general public, we are all affected to some degree by their presence and operation. In terms of watershed management, water quality and offshore pollution, the capabilities and performance of ocean outfalls will be of increased importance to us. As human activities and demands increasingly impact the coastal zone, it is anticipated that current and future legislation will mandate cleaner nearshore water. This will have the practical result of new outfall projects reflecting improved and more cost-effective designs that recognize both economic realities and California's population growth. It is the purpose of this paper to discuss current outfall designs which address these requirements.

2.0 OCEAN OUTFALL ELEMENTS.

General. As a simplistic description, an ocean outfall is a long pipe that carries wastewater or effluent from a land source out into the ocean far enough and under such design conditions that result in an acceptable degree of dispersion at the end. Several hundred ocean outfalls have been constructed in California marine waters during the past several decades. These range in size from small pipe diameters of 12 inches up to the massive 120 inch diameter reinforced concrete pipe outfalls used offshore San Diego and Orange County.

While there are no specific Federal design guidelines, all outfalls constructed offshore California must conform to numerous federal, city and county specifications in order to complete the requisite permitting process prior to construction. The Federal Water Pollution Control Act of 1972, as amended, prohibits any discharge of pollutants into a United States waterway unless the discharge is authorized by a permit. The fundamental purpose of the outfall is to ensure adequate mixing of the outfall effluent and the receiving volume (ocean) to standards specified by the EPA. Therefore, recognizing that the outfall is in essence a complex mixing *system*, the outfall designer must consider a number of critical factors such as effluent flow and composition, outfall length, diameter, number and size of diffuser ports, ocean current and water depth.

The construction work associated with these projects generally involves offshore vessels and underwater operations, particularly diving services. Since these operations are conducted in the marine environment, weather is always a significant risk. Hence, it is recognized within the industry that the construction of an ocean outrall involves considerably higher costs and risk than a comparable onshore pipeline project.

Outfall Elements. The major components of a marine outfall include the main pipeline, diffuser section, diffuser ports, manholes and flapgate structures. The main pipeline is the conduit that carries the effluent from the land source (e.g., wastewater treatment plant, pulp plant, etc.) to the disposal area, usually thousands of feet offshore in water depths typically from 30 ft. to over 300 ft. The effluent flowrate is generally delivered from an onshore pumping station, although gravity-head flow is sometimes used. In order to preserve its structural integrity during all weather and sea conditions, the main pipeline is often buried in a trench or covered with ballast rock or otherwise weighted to protect the pipe and prevent seabed movement.

The final length of pipe through which the effluent is discharged is called the diffuser and is typically several hundred feet in length. The effluent is discharged into the ocean through a series of small holes, or ports, spaced along the sides of the diffuser for its entire length. When the effluent enters the ocean through the diffuser ports, it is normally buoyant since the main constituent is fresh water. This dispersed effluent initially travels horizontally after being discharged thru the ports, but then gradually ascends in a more vertical trajectory. The strength and direction of the local bottom current also affects the movement of the effluent stream. This resulting discharge is referred to as a plume, and the diffuser design criteria generally specifies a minimum level of effluent concentration at a prescribed distance from the port. Hence, the plume becomes increasingly mixed and less concentrated as it travels out from the port.

A large gate valve structure called a flapgate is often installed at the extreme end of larger outfall diffusers. During normal operation this large gate is closed, causing the effluent to be discharged through the diffuser ports. However, during periodic cleaning operations, the flapgate is opened to permit the outfall diffuser to be hydraulically flushed. During this operation, a large flowrate is generated at the onshore source for a short period to maximize the flushing effect. This process is often required at scheduled intervals to clean out biological growth and sand ingress that would degrade outfall performance. Alternatively, special cleaning devices, called "pigs" are periodically run through the outfall to perform the same cleaning effect.

The designer should make provision in the outfall for internal inspection and/or maintenance during its operating life. Inspection and maintenance have historically been performed by divers who "penetrate" the outfall pipe up to several hundred feet. Current trends are to use remotely operated vehicles (ROVs) and "intelligent" pigs when possible and cost-effective. To provide reasonable and safe access for these operations, manholes are often installed at periodic distances and strategic locations along the outfall.

3.0 MATERIALS OF CONSTRUCTION.

The outfall engineer must consider a wide range of factors during the design phase of the project. These include the usual considerations necessary for marine construction such as environmental loads, seawater corrosion, constructability, specific gravity, reliability, maintainability, seabed stability and of course - cost! In terms of overall project significance, the primary cost drivers for the project are usually the pipe and the materials required to provide seabed stabilization.

Historically, outfall pipes have been constructed of a variety of materials including cast iron, wooden staves, vitrified clay, reinforced concrete pipe (RCP), steel, and plastic. Current trends make increasing use of plastics which include high-density polyethylene (HDPE), polyvinyl chloride (PVC), and fiberglass-reinforced pipe (FRP). At present, most new outfall designs specify RCP, steel or HDPE as primary pipeline materials. A more detailed comparison of these materials is made in Section 5.

A major task facing the designer is to ensure that the outfall pipe can withstand all environmental forces, namely waves, currents, earthquakes and storm surge. Generally accepted analytical techniques are available to calculate the combined effects of these forces. After the resulting forces and pipeline loads have been determined, the designer can then select the most appropriate method of stabilizing the pipeline on a specific seabed geology. These methods generally include some combination of concrete weights, tremie (poured) concrete, rock ballast, anchor piles, screw augers, trench and jetting-in. In addition to the materials required for stabilization, small bedding rock or sand is often required to establish a level and consistent base for the pipe.

Because such materials are usually not available locally in sufficient quantities, costly onshore and offshore transportation must be used. Hence, it is often the case that the materials used for pipeline stabilization represents a significant portion of the total project cost.

4.0 GENERAL CONSTRUCTION METHODS.

The construction of an ocean outfall generally consists of the same elements as the onshore construction of a pipeline with the added challenges of the marine environment. A wide variety of techniques have been developed in recent years to address these challenges - all having the intent of minimizing risk and cost. Nearly all ocean outfalls require the use of offshore vessels and underwater operations which are subject to adverse weather conditions and costly shutdowns. The following sections briefly describe current trends in outfall construction, several of which reflect the application of new technologies.

Initial Survey. The outfall designer requires specific details of the proposed outfall route bathymetry and geology before the detailed design work can commence. It is generally the practice to run detailed bathymetric and subbottom profile surveys, including core samples, at an early stage in the design phase. These details enable the designer to specify many features of the outfall design, such as optimum route, required seabed support, outfall grade and pipe stabilization method. The survey equipment generally used includes side-scan sonar, subbottom profiler and coring devices, supplemented with ROV and/or diver services.

Route Selection. The results of the survey are used by the designer and client to select the optimum outfall route. The final determination of the outfall corridor often requires consideration of existing seabed restrictions, state and federal code requirements and designated uses as well as the geophysical features obtained during the survey. The approved outfall route is then specifically described in the project plans and specifications.

Seabed Preparation. In most cases, some preparation of the seafloor is necessary before the outfall is installed to ensure a uniform grade and to avoid prominent obstructions. When the designer calls for the outfall to be buried, a trench is often dug using a dredge vessel or specialized trenching machine. In the past, underwater blasting was often used as an effective method to break up bedrock but this technique is now banned in most U.S. waters. A recent large outfall was installed in San Diego(1) in which the seabed was carefully graded and a thick layer of bedding rock deposited along the outfall route prior to installing the pipe by a purpose-built rock handling machine (Photo #1).

Pipe Installation. The method of installing the outfall pipe is generally dependent upon the pipe material and construction. Traditional methods of installing conventional RCP (reinforced concrete pipe) makes use of a laybarge and surface crane, installing the pipe section by section. Because this is usually a labor intensive and lengthy approach, methods have been developed to speed up this process. Photo #2 illustrates a hydraulically-operated "horse", or pipe handling machine, that is often used to speed up repetitive pipelaying operations. This method reduces the installation time and also requires less labor.

Another technique used for constructing shorter outfalls or beach crossings is the trestle, which is essentially a temporary pier extending out into the ocean along the outfall route. The trestle is also used in waters that are too shallow for pipelaying vessels or barges. By using the trestle method, much of the labor and pipe handling can be accomplished from the surface rather than underwater.

A very effective technique used for steel and HDPE pipelines is the beach pull, in which the pipeline is assembled onshore and pulled out over a set of roller frames by a large winch mounted on an offshore barge or tug (Photo #3). In this method, the pipeline can be pulled and subsequently towed to the installation area either floating on the surface or submerged (bottom pull). Photo #4 illustrates a surface tow in which a 3,000 foot length of HDPE pipeline is towed to location (2), then deballasted (flooded) and lowered to the bottom.

Stabilization. The means used to secure, anchor, and protect the outfall is a critical element of the project. As noted earlier, the outfall engineer must carefully identify and calculate all the environmental and operating forces that can be expected to affect the outfall. He must then determine the appropriate methods of providing stabilization to prevent any pipe movement during the life of the outfall. Common techniques of providing seabed stabilization include trenching and burial, anchor (ballast) rock, tremied concrete, concrete weights (Photo #5), and auger/screw anchors.

As-Built Survey. To ensure that the outfall is constructed in accordance with the designer's plans and specifications, an as-built survey is usually performed at the end of the construction phase. Typically, a diver or ROV (Photo #6) is used to "swim" the outfall to video record its position, stabilization, integrity and diffuser operation. It should also be noted that the client generally has a quality control inspector involved during all phases of the construction to ensure compliance with the contract design.

Areas of Risk. A significant cost element of an ocean outfall is the risk provision that the contractor must include in his bid. In addition to the usual risks associated with land construction, the marine contractor must also consider weather downtime, vessel and offshore equipment breakdown, material transportation delays, and underwater safety - particular during diving operations.

5.0 MATERIAL SELECTION CRITERIA.

The selection of the outfall materials is done during the design phase. Because the outfall materials alone usually represent a significant portion (30% - 60%) of the total project costs, the material selection is an important aspect of the project. During the design phase, the engineer must consider both the technical properties and economic factors associated with the candidate materials. The important technical characteristics to be evaluated include strength, durability, specific weight, effluent compatibility, seawater corrosion, dissimilar material corrosion and resistance to biological attack. Important economic factors include unit cost, availability and transportation costs - both land and marine.

For larger outfalls (62 inch diameter and greater), current trends favor RCP, which possesses good long term stability, corrosion resistance, ready supply and, when suitably lined, good corrosion resistance. Alternatively, for the more numerous small pipelines, HDPE has several favorable characteristics which include chemical resistance, durability, low unit weight and low cost. HDPE's low weight is a particularly important factor in that its use permits smaller offshore vessels and barges required during installation. This greatly reduces the cost of offshore operations which often range $20,000/day - $50,000/day, including crews. As an example, Oceaneering recently evaluated the use of 3 material types for a 48 inch outfall - RCP, steel and HDPE - and found the overall project costs using HDPE was 20% less than the other two materials. As a result of these favorable characteristics, Oceaneering has successfully designed and installed 6 ocean outfalls on the West Coast and Australia using HDPE. These overall cost savings using HDPE should be an important consideration in planning future outfall projects.

6.0 RECENT POLYETHYLENE (HDPE) OUTFALL INSTALLATIONS.

The following outfall design and/or construction projects have been performed by Oceaneering within the past three years. In all of these outfalls, HDPE was used as the primary pipe material or lining:

o Project: Wastewater Outfall Installation
Client: Chevron U.S.A.
Location: El Segundo, California
Outfall Length: 3,000 feet
Pipe Diameter: 60 inches

o Project: Wastewater Outfall Installation
Client: Lousiana Pacific Pulp Company
Location: Eureka, California
Outfall Length: 8,000 feet
Pipe Diameter: 48 inches

o Project: Desalination Plant Intake/Discharge Outfall
Client: Cambria Community Services District
Location: Cambria, California
Outfall Length: 2,500 feet
Pipe Diameter: 10 inches (diffuser)
10 inches (intake)

o Project: Wastewater Outfall Installation
Client: Australian Paper
Location: Burnie, Tasmania
Outfall Length: 4,000 feet
Pipe Diameter: 40 inches

o Project: Wastewater Outfall Installation
Client: Australian Paper
Location: Devonport, Tasmania
Outfall Length: 7,000 feet
Pipe Diameter: 25 inches

7.0 PHOTOGRAPHS.

1. ROCK HANDLING MACHINE

2. PIPE INSTALLATION HORSE

7.0 PHOTOGRAPHS (cont.)

3. PIPELINE LAUNCH FRAME

4. PIPELINE SURFACE TOW

7.0 PHOTOGRAPHS (cont.)

5. CONCRETE ANCHOR WEIGHT

6. REMOTELY OPERATED VEHICLE (ROV)

8.0 CONCLUSIONS.

As the result of successfully designing and installing several ocean outfalls offshore California during the past three years, Oceaneering International has pioneered the use of several improved techniques and materials. We believe these innovations are highly relevant to California and other coastal states because they offer the outfall owner - municipality, utility or commercial firm - the means to handle ocean deposited wastewater more cost-effectively. In particular, we have found the application of high density polyethylene (HDPE) pipe to offer several technical and economic advantages over conventional materials for smaller outfalls. We believe that the use of HDPE outfall pipe will result in more economic designs that address the combined realities of limited project funding and greater human use of the World Ocean.

CHAPTER 165

SATELLITE REMOTE SENSING AS PART OF AN OCEAN OUTFALL ENVIRONMENTAL MONITORING PROGRAM

Jan Svejkovsky[1] and Irwin Haydock[2]

Operational utilization of existing satellite sensors as part of a coastal sanitation outfall monitoring program were evaluated through a pilot project with the County Sanitation Districts of Orange County, California (CSDOC).

With the satellite sensors presently available, trade-offs exist between spatial resolution, spectral resolution and re-sampling (i.e. overflight) frequency. We used the AVHRR for daily monitoring of surface temperature patterns, upper layer advection, and patterns of high turbidity from river runoff, storm mixing or heavy algal blooms. Although the very high resolution TM or Spot sensors could best resolve water patterns around the outfall, their poor resampling frequency (10-16 days) make such data of very limited use for operational monitoring.

The CSDOC's monthly field sampling grid covers an area of 22 km longshore by 14 km offshore, centered on the outfall which terminates 8.2km offshore at a 60m depth. Even at the AVHRR's relatively coarse (1.1km) spatial resolution, the images provide sufficiently detailed information to add significant interpretive value to field data collected during regular monthly surveys. Multiple image time series covering the field sampling intervals are used to derive surface flow fields over the region surrounding the site. The AVHRR imagery also provides information on the existence of water mass boundaries, localized upwelling events, runoff, and other features with respect to the field data station grid. Such surface manifestations often can be related to subsurface characteristics observed on the CTD profiles. In some cases, turbidity maxima as deep as 10m could be resolved in the satellite images and their spatial pattern and possible source identified. Most recently, there is some indication that areas with high near-surface coliform bacteria concentrations could be associated with features resolved by the satellite imagery.

The satellite data also proved useful for extending the CSDOC's monitoring efficacy between the monthly field efforts. The AVHRR analyses are most useful for helping document events such as upwelling, coastal runoff or red tide blooms which cause rapid environmental changes in the vicinity of the outfall, but are not directly related to the effluent discharge. The analyses thus provide an aid in the interpretation of data obtained during the District's surveys as well as a means to extend the District's awareness of environmental variability between the survey periods. In isolated cases, primarily during winter and spring after rain storms, the effluent itself could be identified in the thermal and turbidity images.

The three greatest impediments of AVHRR imagery for sanitation outfall monitoring are coverage limitations imposed by cloud cover, the need for better spatial resolution to resolve smaller-scale features around the plume and the sensor's inability to provide subsurface information. However, despite its inherent limitations, present day satellite imaging can add valuable information to a coastal outfall monitoring program when used in conjunction with field sampling. The CSDOC project is providing opportunities to evaluate the utility of satellite remote sensing for this application and to establish user criteria for future sensors and data types.

---

[1]President, Ocean Imaging Corp. 201 Lomas Santa Fe Dr., Suite 370, Solana Beach Ca 92075

[2]Chief Scientist, County Sanitation Districts of Orange County. P.O. Box 8127, 10844 Ellis Ave., Fountain Valley, CA 92728-8127.

CHAPTER 166

An Ocean Outfall As an Artificial Reef: Impacts to the Benthic Environment and a Balanced Indigenous Population

Douglas Diener[1], Brian Riley[1], George Robertson[2], Don Maurer[3], Tom Gerlinger[2], and Irwin Haydock[2]

Abstract

A large, wastewater-outfall pipe and diffuser have been shown to produce two types of impacts to the surrounding environment: discharge effects and reef effects. To determine the relative significance of these two effects upon the nearby infaunal community, the infauna were sampled, and physical measurements were made at various distances from the outfall structure at two transects. One transect was conducted in the wastewater discharge area at 56 m (181 ft), and another transect was conducted along the 30 m (97 ft) depth contour, 2.6 km (1.6 miles) inshore from the wastewater discharge area. The results of this study found significant alteration in grain size and sediment organic carbon content at the inshore transect that extended several hundred meters downcurrent and less than 200 m (650 ft) upcurrent from the reef structure. While abundances of crustaceans and molluscs are usually enhanced near wastewater discharges, abundances of these groups decreased near the pipe structure at the inshore transect. Abundance of polychaetes, particularly *Melinna oculata*, increased near the pipe structure at the inshore transect. This increase in density was positively correlated with increase in fine sediments near the pipe. The brittlestar *Amphiodia urtica*, generally considered a sensitive indicator of wastewater discharges, had depressed abundances at the inshore transect up to 400 m (1300 ft) from the pipe structure. Discharge effects upon this species extended much further. While recovery of this species over large areas near the discharge correlates with improved treatment strategies, effective source control, and decreasing sediment contaminant concentrations, our results indicate that this species may never fully recover to historical abundance levels near the pipe structure because of reef effects. Compliance with wastewater discharge permits generally requires that the discharge does not adversely impact the environment and that a balanced indigenous population persists near the discharge. Thus, the purpose of this study was to distinguish changes in infaunal communities due to discharge effects from those associated with reef effects.

---

[1]MEC Analytical Systems, Inc., 2433 Impala Drive, Carlsbad, CA 92008

[2]County Sanitation Districts of Orange County, 10844 Ellis Avenue, Fountain Valley, CA 92708

[3]Biology Department, California State University Long Beach, 1250 Bellflower Blvd., CA 90840

Introduction

A large, wastewater-outfall pipe and diffuser located on the San Pedro Shelf off Huntington Beach, California, is operated by the County Sanitation Districts of Orange County (Districts) (Figure 1). Results from the Districts' Ocean Monitoring Plan have provided clear evidence that the ocean outfall produces two types of effects upon the environment: (1) chemical and water quality effects related to the discharge of over 250 million gallons per day of treated wastewater effluent (discharge effects), and (2) physical effects from the outfall pipe structure and associated rock ballast (reef effects) (e.g., CSDOC 1995, 1996). The discharge structure is a 10-ft diameter, concrete reinforced pipe and associated ballast rock laid on the ocean floor traversing a broad area of the subtidal and mainland shelf communities in depths ranging from the intertidal to 56 m. The structure has been photographed and videotaped during inspections (Undersea Graphics 1991). The pipeline is well-covered and supported by ballast rock along its entire length, resulting in a structure that resembles many artificial reefs built in southern California (Lewis and McKee 1989). The pipe structure extends 8.2 km offshore covering approximately 100,000 m² (23 acres) of soft bottom habitat. The terminal 1.2 km of the outfall structure, the diffuser area, contains 500 ports where the discharge of wastewater occurs. Thus, the outfall pipe structure, the diffuser area, and associated ballast rock represent one of the largest artificial reefs in the Southern California Bight.

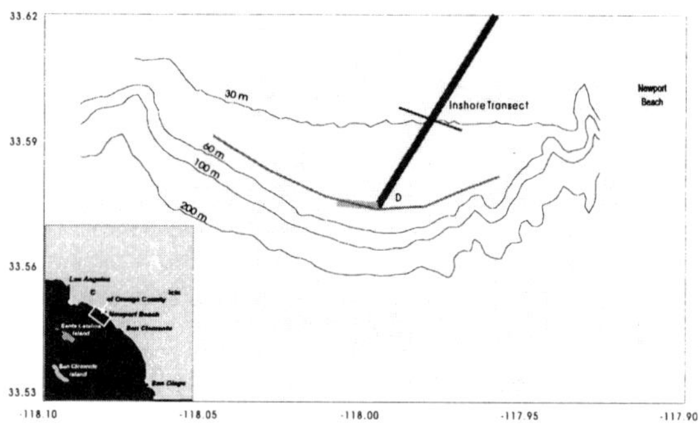

Figure 1. The County Sanitation Districts of Orange County outfall structure and study site transects.

Artificial reefs provide numerous habitats for biota because of their complex topography and hard surfaces that otherwise would not occur on a sand-mud bottom. Furthermore, the presence of a three-dimensional structure (e.g., an artificial reef) on a relatively featureless soft-bottom area is likely to attract many reef-associated fishes. Soft-bottom demersal species may also be attracted to areas adjacent to the reefs for food or refuge. In Puget Sound, there was a four- to five-fold increase in density and biomass of the five most prevalent species of flatfish after installation of artificial reef modules (Walton 1982). Fish move to reefs soon after construction (Turner et al. 1969), and maximum population sizes can be reached within months (Gascon and Miller 1981; Matthews 1985). Invertebrate and plant communities require longer time periods for reef maturity and for larger, and longer-lived biota, such as gorgonians, this can take decades. Since the outfall structure is now 25 years old, it appears to harbor a well-developed mature reef community (Undersea Graphics 1991). While some disagreement remains on whether artificial reefs actually increase biomass and fish standing stocks, other studies have shown increased biomass associated with the presence of reefs (e.g., MEC 1991). However, artificial reefs are generally considered to be beneficial for the marine environment, consequently much effort and expense has been expended in building reefs in the Southern California Bight.

The presence of the reef and associated organisms is likely to affect the surrounding community. This effect can be caused by the feeding habits of reef-associated animals and the physical effects of the reef structure on local currents. The feeding of mobile animals on the reef is not restricted to the reef itself and can change the densities of nearby biota. The present study provides a description of the infaunal community and physical measurements made along transects at various distances from the outfall pipe structure at two different locations. The physical structure of the outfall was similar in both locations. Other than different water depths, the main difference between these transect was the presence of the wastewater discharge at the deeper transect. Thus, the inshore transect represents changes in the adjacent soft-bottom community due to reef effects, while the deeper transects represent changes due to both reef and discharge effects. It is important to make this distinction, since in order for the Districts to be in compliance with their discharge permit they need to demonstrate that the discharge is not significantly degrading the biota. Consequently, understanding the spatial extent and significance of reef effects is necessary to properly evaluate effects upon the infaunal community due solely to discharge effects.

Materials and Methods

In July 1994 the Districts' outfall structure was sampled along two parallel transects at varying distances from the pipe (Figure 1). One transect followed the 56 m depth contour through the wastewater discharge area. This transect included the regular quarterly monitoring stations with stations located from 20 m up to 7.7 km from the discharge (Diener et al. 1995; CSDOC 1995, 1996). The second transect extended along the 30 m depth contour 2.6 km inshore from the wastewater discharge. Samples were taken both upcoast and downcoast along this transect at 6 m, 60 m (upcoast only), 200 m, 400 m, and 750 m. Infauna samples were collected with a 0.1 m² Van Veen grab sampler and processed through a 1.0 mm sieve. The same taxonomists counted and identified all the organisms from all the samples. Three replicate samples were collected at the inshore transect and five replicate samples were collected at the deeper transect. Three separate sediment samples were collected at each location for grain size analysis and total organic carbon content.

Results Sediment Grain Size and TOC

Sediment grain size and TOC are important sediment properties that can influence infaunal communities. Percent fine sediments (smaller than 64μ) and TOC were plotted against distance from the outfall pipe structure along the 30 m transect (Figure 2). Both percent fine sediments and TOC increased nearer the structure on the downcoast side. Percent fine sediments peaked at about 55% 6 m downcoast, while TOC peaked at about 0.7% 200 m downcoast

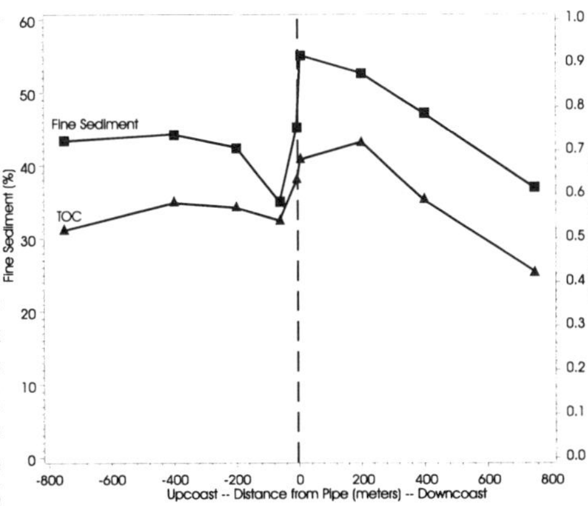

Figure 2. Percentage of fine sediments and TOC as a function of distance from outfall structure at 30 m.

from the pipe structure. Near the pipe structure on the upcoast side, there was a drop in percentages producing an asymmetrical distribution of these measures relative to the pipe structure. This reef effect extended at least 400 m downcoast and somewhere between 60-200 m in the upcoast direction. This distribution suggest that the general upcoast, flowing, bottom current has caused deposition on the downcurrent side and scour and erosion on the upcoast side of the structure. Alterations in local current patterns and changes in sediment properties might be expected to produce concomitant changes in infaunal organisms.

Results Infaunal Patterns

The mean number of infaunal species, total abundance, and abundance of major taxonomic groups (crustaceans, molluscs, and polychaetes) were plotted as a function of distance from the outfall structure along the 30 m transect (Figure 3) to assess the influence of the structure on the infaunal community. Gradients in infaunal densities indicate that the pipe structure alters infaunal communities. The number of species and total abundance tended to be lower on the downcoast side, area of deposition, and was enhanced on the upcoast side in the area of disturbed sediments. Because this transect is not subject to discharge effects, it was concluded that the pipe structure acts as an artificial reef and alters the infaunal community through reef effects. Crustacean densities were lower on the downcoast side and enhanced on the upcoast side extending about 200 m on each side of the structure. This distribution is consistent with many crustaceans being highly mobile and able to respond quickly to localized disturbances (Parr et al. 1978; VanBlaricom 1978). Changes in mollusc densities decreased near the structure but were enhanced at 200 m distances. Reduced densities of crustaceans and molluscs near the structure are typical of the "halo" reef effect associated with artificial (Davis et al. 1982; Fricke et al. 1986) and natural reefs (Posey and Ambrose 1994). Molluscan density increased from a mean of 42 individuals near the pipe (6 m) to over 90 individuals at 200 m from the structure.

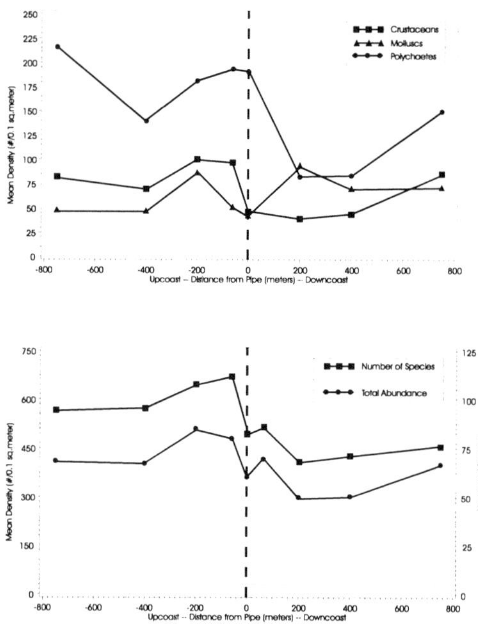

Figure 3. Mean infaunal measures as a function of distance from outfall structure at 30 m.

Polychaete density increased near the pipe structure and was enhanced up to 400 m from the structure. This increase was positively correlated with the increase in fine sediments near the structure ($r^2=0.65$, $p<0.05$) and was primarily due to the species Melinna oculata, which accounted for approximately 25% of the polychaetes found at the two near-pipe stations (Figure 4). The mean density of this species declined sharply with increasing distance from the structure, falling from 45 individuals at the two stations nearest the structure to an average of 6.8 individuals at all other stations. M. oculata is a tube-dwelling, surface deposit feeder and may be responding to the relatively high percentage of fine sediments and organic carbon found near the structure (Figure 2).

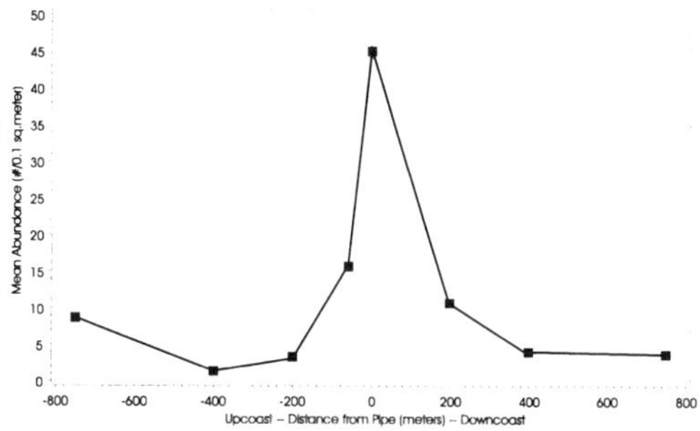

Figure 4. Density of Melinna oculata as a function of distance from outfall structure at 30 m.

The mean number of infaunal species, total abundance, and abundance of major taxonomic groups (crustaceans, molluscs, and polychaetes) were plotted as a function of distance from the outfall diffuser structure along the 56 m transect (Figure 5) to assess the influence of the structure and the wastewater discharge on the infaunal community. Gradients in infaunal densities extended over greater spatial scales indicating the effect of the wastewater discharge on these communities. The number of species per sample showed some fluctuations near the discharge, but the magnitude of these changes was comparable to the range of variability found for the farfield reference stations. Thus, as observed for the 30 m transect, there was some minor variability in the number of species but there were no significant depression of diversity near structures. However, there was some change in species composition as stations near the pipe structure included some species more representative of hard-bottom habitats (i.e., reefs), possibly representing displaced species or taxa preferring transitional habitats. Infaunal density exhibited the greatest gradient effect being enhanced several kilometers on either side of the diffuser structure,

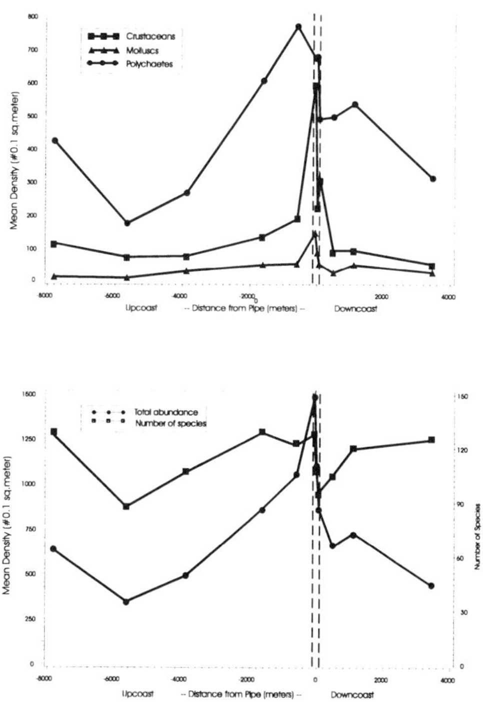

Figure 5. Mean infaunal measures as a function of distance from diffuser structure at 56 m.

but more so on the upcoast side than downcoast due to the prevailing bottom current patterns. This enhancement of density indicates the spatial extent of discharge effects influencing the infaunal community. Crustacean and molluscan densities were enhanced near the diffuser structure and there was no indication of a "halo" reef effect as observed for the 30 m transect. Thus, the enhancement of these groups by the wastewater discharge more than compensates for the reef "halo" effect. Polychaete densities accounted for more than half of the total abundance, and thus, their distribution largely determined the pattern for total density. The enhancement of polychaete densities due to discharge effects was about ten times that due to the reef effect.

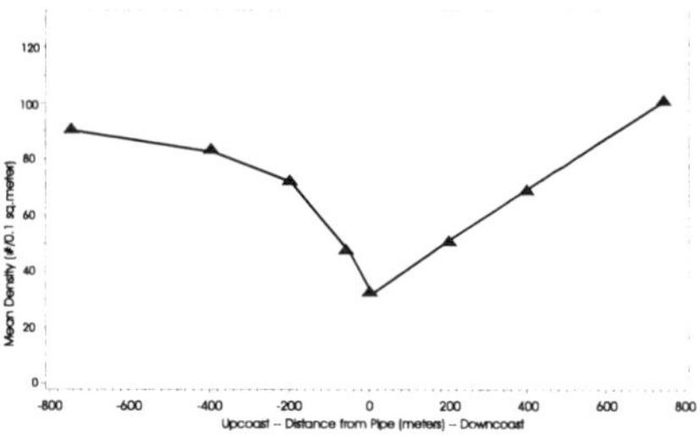

Figure 6. Density of Amphiodia urtica as a function of distance from outfall structure at 30 m.

While the densities of most major taxa and species generally have shown enhancement or localized decreases relative to the diffuser structure, Amphiodia urtica (a brittlestar) has decreased densities over a broader area. Amphiodia densities decreased significantly near the pipe structure at the 30 m transect, and this effect extended at least 400 m downcoast and 200 m upcoast (Figure 6). Densities near the pipe structure were reduced by about 70% compared to the most distant stations. Because the 30 m transect is not subject to wastewater discharge effects, the sharp decrease in the density of this species near the pipe structure most be due to reef effects. The decreases in density due to reef effects at the 30 m transect are in sharp contrast to the much broader decreases observed along the 56 m transect indicating the significant influence of discharge effects on this species (Figure 7).

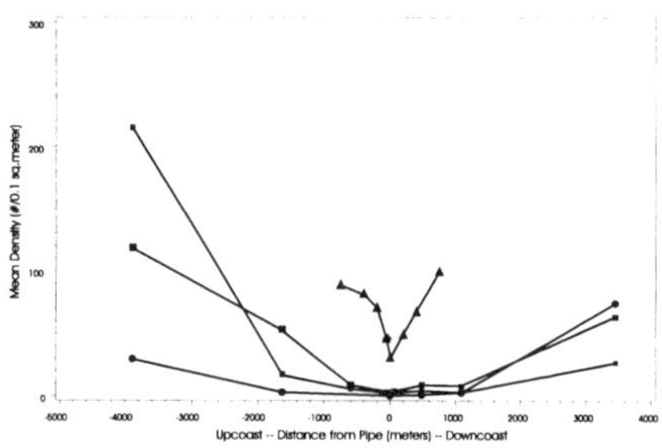

Figure 7. Density of Amphiodia urtica as a function of distance from outfall structures at 30 m (>) and 56 m (•S•S• 1985, P–P–P 1990, and #–#–# 1994).

Amphiodia has been considered a sensitive indicator of wastewater discharges as it has been found to have depressed densities near wastewater outfalls (e.g, Word 1978; Word and Mearns 1979). Densities of this species have shown a dramatic and continuous increase within the study area since 1985 (Figure 8). Historical impacts to this species relate to construction of the ocean outfall and its early operation. However, since 1985 improved treatment levels, reduced mass emissions, and effective source control programs have significantly improved the marine environment near the diffuser. A repeated measure ANOVA on stations grouped by distance from the diffuser structure indicates a positive trend of increasing density for this species at stations near the diffuser (Figure 9). Density trends for Within and Boundary stations (20 and 60 m from the structure and discharge, respectively) are almost identical. Nearfield stations (stations 0.5 to 1.5 km from the discharge) have a more positive slope and are recovering faster. Farfield stations (greater than 3.5 km from the discharge) have a slightly negative slope indicating a slight, long-term decline, representative of the region. Reasons for this regional decline are unknown, but a contributing factor may be increased commercial trawling in the reference area since 1990. Given these significant positive trends and the continuous recovery depicted in the spatial plots, there is reason to expect continued increases in the density of Amphiodia, although this study shows a significant reef effect for this species along the 30 m transect.

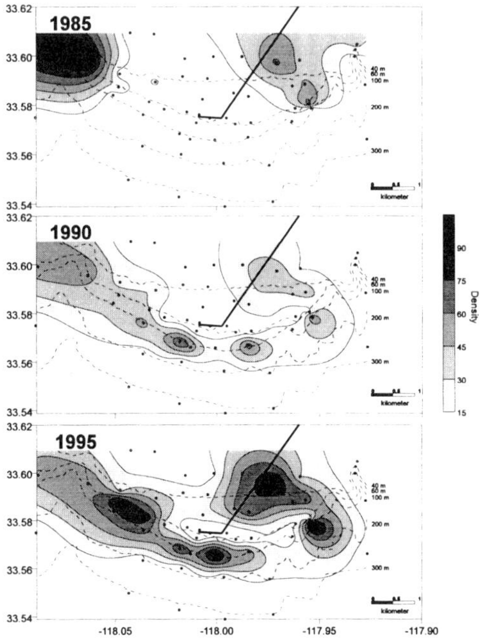

Figure 8. Spatial contour plot of Amphiodia urtica density (#/0.1m^2) for the study area by year. Not contoured for reef effect.

Discussion

The outfall pipe structure and associated ballast rock, acting like an artificial reef, have changed nearby infaunal communities. At the inshore 30 m transect this change appears attributable to feeding habitats of reef-associated animals and the physical effects of the reef structure on local currents and subsequent sediment modification. The feeding of mobile animals on the reef is not restricted to the reef itself and can change the densities of nearby biota. Several studies have concluded that the presence of a reef community creates a "halo" reef effect, in which organism abundances in neighboring soft-bottom regions are altered, due to predation effects and changes in sediment parameters (Ambrose and Anderson 1990; Posey and Ambrose 1994). For example, reef fish foraging eradicated seapens within 200 m of a newly constructed artificial reef in southern California (Davis et al. 1982). The "halo" reef effect was observed during an eight-year monitoring study of a small wastewater outfall in 45 m off Carlsbad, California (Diener and Fuller 1995). Abundances of the brittlestar, Amphiodia urtica, were found to be depressed near the wastewater discharge structure (MEC

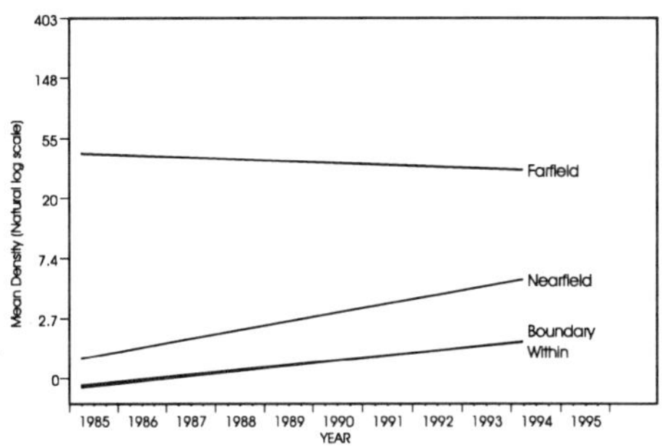

Figure 9. Repeated measures ANOVA of Amphiodia urtica densities by grouped stations as a function of distance from the diffuser structure. Within=20 m, Boundary=60 m, Nearfield=0.5 to 1.5 km, Farfield=>3.5 km from structure.

1992). However, during the study period the effluent quality was improved significantly as the treatment process was upgraded with no change in the "halo" reef effect. The only hypothesis consistent with these data is that the depression in brittlestar abundance was due to reef effects (i.e., probably predation), and not due to discharge-related effects.

Infauna communities have been shown to be sensitive to changes in sediment grain size (Gray 1974; Oakden 1984) and organic carbon content (Pearson and Rosenberg 1978; Oakden et al. 1984). Thus, some of the observed alterations of infaunal densities near pipe structures may have been due to alteration of bottom current flows that produced changes in the physical environment. Changes in grain size and organic carbon were observed up to 400 m from the pipe structure along the 30 m transect. This reef effect has been observed in the vicinity of oil rigs (Davis et al. 1982) and sunken boat reefs (Fricke et al. 1986; Baynes and Szmant 1989) and documented along the 56 m transect (CSDOC 1995, 1996) and in the vicinity of the diffuser structure (Gerlinger et al. 1993). Turbulence generated by swells, bottom currents, and secondary entrainment near diffuser ports causes sediments to be resuspended and sorted by size. In addition, shells from hard-bottom organisms growing on the reef fall onto the sediments and increase sediment coarseness. The sediment redistribution changes the suitability of the area for soft-bottom species and may influence their distribution.

The effects of the reef and associated biota on the soft-bottom community are likely to be most pronounced in the immediate vicinity of the reef, decreasing with distance from the outfall structure. This is the same pattern that would be expected from discharge effects related to organic carbon gradients, contaminants, or changes in grain size. The biological and physical effects of the reef on soft-bottom communities may thus resemble impacts from wastewater discharges. It is important to make this distinction, since in order for the Districts to be in compliance with their discharge permit they need to demonstrate that the wastewater discharge does not preclude a balanced indigenous population (BIP) within the Districts' study area. Determination of a BIP involves comparing various community metrics and indicator species abundance near the discharge with reference stations in the study area and throughout the Southern California Bight. Consequently, understanding the spatial extent and significance of reef effects is necessary to properly evaluate effects upon the infaunal community due solely to discharge effects.

Discharge effects are due to the discharge of an effluent composed of organic carbon particulates, absorbed contaminants, and dissolved contaminants (CSDOC 1995, 1996; Diener et al. 1996). Thus, gradients in the infaunal community along the 56 m transect represent biological accommodation to wastewater organic particulates as a food source, different sensitivities to contaminants, and reef effects. Thus, gradients in community measures and species density are indicative of natural response to their environment (Jones 1969; Nichols 1970; Pearson and Rosenberg 1978; Gray 1981; Weston 1990; Warwick and Clark 1994). Thus, community and species density gradients result from environmental gradients produced by both natural (e.g., reef effects) and anthropogenic influences (e.g., wastewater discharges).

Compliance assessments for determining the effects of wastewater discharges on infaunal communities usually involve evaluating community measures, species spatial distributions, and density patterns. However, these assessments often are subjective and even where statistical significance can be demonstrated, there still remains the question of whether these significant observations represent ecological significance. The use of an ecological model, e.g., community response to stress, might provide more objective evaluations. The response of individual species to environmental impacts (stressors) is well documented: however, few attempts have been made to infer or predict changes at the population level or on complete communities (Gray 1989). Environmental stress can be defined as a change that causes some response by a species or community. Numerous studies have shown reductions in diversity from a variety of stressors as one of the first indications of environmental stress (Pearson and Rosenberg 1978; Gray 1979, 1989). Thus, for compliance assessments, a decrease in the number of species is a good indicator of environmental stress and a community undergoing change. In the present study while densities were significantly enhanced, there was no significant decrease in the number of species near pipe structures along either transect. Thus, the observed gradients in the infaunal community are not representative of a degraded community, and the communities near the pipe structures appear to be representative of a BIP for those habitats.

The spatial scale of reduced Amphiodia densities along the 56 m transect suggests that historically the community was more highly stressed by the wastewater discharge and large areas may not have been representative of a BIP. Recovery for this species is slow because it is more dependent upon adult immigration rather than recruitment success (Scanland 1995), and recovery may take years even after the stress has been removed. The pattern of recovery for this species supports the contention that these historical conditions (greater environmental stress) no longer apply. While there is still a small area where the brittlestar density is lower than might be expected, full recovery to historic densities may not occur due to the reef effect. Future compliance assessments for infaunal communities and this indicator species need to consider the environmental stress associated with reef effects upon their density patterns which in this study extended at least 400 m from outfall structures. This is important as many compliance assessments test for statistical differences in density at the zone of initial dilution boundary, 60 m from the structure for this outfall, with reference areas located several kilometers distant. Clearly, in these situations reef effects need to be considered as significant contributing factors in determining brittlestar densities near pipe structures. Finally, it has been demonstrated that factors other than wastewater discharges can significantly alter the density patterns of Amphiodia within its natural habitat.

References

Ambrose, R.F. and T.W. Anderson. 1990. Influence of an artificial reef on the surrounding infaunal community. Mar. Biol. 107:41-52.

Baynes, T.W. and A.M. Szmant. 1989. Effect of current on the sessile benthic community structure of an artificial reef. Bull. Mar. Sci. 44(2):545-566.

CSDOC. 1995. Annual Report, July 1993-June 1994. Marine Monitoring, Fountain Valley, CA.

CSDOC. 1996. Science Report, Ten year Synthesis, 1985-1995. Marine Monitoring, Fountain Valley, CA.

Davis, N., G.R. Van Blaricom, and P.K. Dayton. 1982. Man-made structures on marine sediments: Effects on adjacent benthic communities. Mar. Biol. 70:295-303.

Diener, D.R. and C. Fuller. 1995. Infaunal patterns in the vicinity of a small coastal wastewater outfall and the lack of infaunal community response to secondary treatment. Bull. So. Calif. Acad. of Sci. 94:85-107.

Diener, D.R., S.C. Fuller, A. Lissner, I. Haydock, D. Maurer, G. Robertson, and T. Gerlinger. 1995. Spatial and temporal patterns of the infaunal community near a major ocean outfall in southern California. Mar. Poll. Bull. 30(12):861-878.

Fricke, A.H., K. Koop, and G. Cliff. 1986. Modification of sediment texture and enhancement of interstitial meiofauna by an artificial reef. Trans. Royal Soc. of S. Africa 46:27-34.

Gascon, D. and R.A. Miller. 1981. Colonization by nearshore fish on small artificial reefs in Barkley Sound, British Columbia. Can. J. Zool. 59:1635-1646.

Gerlinger, T., D. Maurer, and G. Robertson. 1993. Rapid-response benthic monitoring of a southern California ocean outfall. Sea Tech. 34:23-28.

Gray, J.S. 1974. Animal-sediment relationships. Oceanogr. Mar. Biol. Ann. Rev. 12:223-282.

Gray, J.S. 1979. Pollution-induced changes in populations. Phil. Trans. Roy. Soc. Ser. B, 286:545-561.

Gray, J.S. 1981. The Ecology of Marine Sediments. Cambridge, UK: Cambridge University Press.

Gray, J.S. 1989. Effects of environmental stress on species rich assemblages. Biol. J. Linnean. Soc. 37:19-32.

Jones, G.F. 1969. The benthic macrofauna of the mainland shelf of southern California. A. Hancock Monogr. in Mar. Biol. No. 4. 219 pp.

Lewis, R.d. and K.K. McKee. 1989. A Guide to the Artificial Reefs of Southern California. California Department of Fish and Game 73 p.

Matthews, K. 1985. Species similarity and movement of fishes on natural and artificial reefs in Monterey Bay, California. Bull. Mar. Sci. 37(1):252-270.

MEC (MEC Analytical Systems, Inc.). 1991. Production and Valuation Study of an Artificial Reef Off Southern California. Final Report. Port of Long Beach, Port of Los Angeles, National Marine Fisheries Service.

MEC. 1992. Encina Receiving Water Monitoring: Summary Report July 1985-June 1991. Carlsbad, CA.

Nichols, F.H. 1970. Benthic polychaete assemblages and their relationship to sediment in Port Madison, Washington. Mar. Biol. 6:48-57.

Oakden, J.M. 1984. Feeding and substrate preference in five species of phoxocephalid amphipods from Central California. J. Crustacean Biol. 4(2):233-247.

Oakden, J.M., J.S. Oliver, and A.R. Flegal. 1984. Behavioral responses of a phoxocephalid amphipod to organic enrichment and trace metals in sediment. Mar. Ecol. Prog. Ser. 14:253-257.

Parr, T., D.R. Diener, and S. Lacy. 1978. Effects of Beach Replenishment on the Nearshore Sand Fauna at Imperial Beach. Miscellaneous Report No. 78-4, U.S. Army, Corps of Engineers.

Pearson T.H. and R. Rosenberg. 1978. Macrobenthic succession in relation to organic enrichment and pollution of the marine environment. Oceanogr. Mar. Biol. Ann. Rev. 16:229-311.

Posey, M.H. and W.G. Ambrose Jr. 1994. Effects of proximity to an offshore hard-bottom reef on infaunal abundances. Mar. Biol. 118:745-753.

Scanland, T. 1995. Succession and the role of ophiuroids as it applies to the marine infaunal association off Palos Verdes, California. Bull. S. Cal. Acad. Sci. 94:103-116.

Turner, C.H., E.E. Ebert, and R.R. Given. 1969. Man-made reef ecology. Calif. Dept. Fish Game Fish. Bull. 146. 221 pp.

Undersea Graphics, Inc. 1991. Orange County Outfalls 1991 Inspection. Unpublished report, UGI.

Van Blaricom, G.R. 1978. Disturbance, Predation, and Resource Allocation in a High-Energy Sublittoral Sand-Bottom Ecosystem: Experimental Analyses of Critical Structuring Processes for the Infaunal Community. Dissertation, University of California, San Diego, CA.

Walton, J.M. 1982. The effects of an artificial reef on resident flatfish populations. Mar. Fish. Rev. 44(6-7):45-48.

Warwick, R.M. and K.R. Clarke. 1994. Relearning the ABC: Taxonomic changes and abundance/biomass relationships in disturbed benthic communities. Mar. Biol. 118:739-744.

Weston, D.P. 1990. Quantitative examination of macrobenthic community changes along an organic enrichment gradient. Mar. Ecol. Prog. Ser. 61:233-244.

Word, J. 1978. The infaunal trophic index. In: SCCWRP Annual Report, 1979. El Segundo, CA: SCCWRP.

Word, J. and A.J. Mearns. 1979. Sixty-Meter Control Survey off Southern California. SCCWRP TM 229, June, 1979. 58 pp.

CHAPTER 167

Currents of Santa Monica Bay

Barbara M. Hickey[1]

Abstract

Santa Monica Bay is a semi-enclosed shallow shelf located within the Southern California Bight. Several current/temperature studies of Santa Monica Bay took place in the period 1986-1988. The goal of these studies was to determine how (and why) currents and water properties change within the Bay on time scales from a few hours to several months. A primary result from these studies is that Santa Monica Bay cannot be considered in isolation from the adjacent shelf areas or from the entire Southern California Bright. Changes in water properties are primarily determined by large scale processes; i.e., processes such as intrusions of warm water from the south or coastal upwelling, that include the entire Bight region. With respect to these large scale processes, Santa Monica is merely a small indentation in the coastline. Tracking of water masses via drifting sensors illustrates that water from Santa Monica can move to the San Pedro shelf; and water from the Palos Verdes shelf can move into Santa Monica Bay.

Currents within Santa Monica Bay can be relatively strong (10-40 cm s^{-1}) in spite of the rather light winds common in the area. Currents change both direction and strength on time scales of 0.5-20 days. The patterns of currents within the Bay are spatially complex: the mean circulation pattern consists of two gyres, a counterclockwise gyre in the northern half of the Bay and a clockwise gyre in the southern half of the Bay. The fluctuating currents often have very short spatial scales; i.e., unlike water properties, currents are frequently unrelated at different sites within the Bay. Only during the spring season is there any evidence that currents within the Bay are forced directly by the local wind. However, even during spring much of the variability cannot be accounted for by local wind. In summer, currents are still strong and variable, but their forcing mechanisms are presently unknown. Much of the variability in currents in Santa Monica Bay appears to be caused by processes that occur seaward of the Bay. These results have important implications with respect to modeling of currents within the Bay for purposes such as sewage outfall management.

1. Introduction

Santa Monica Bay is a semi-enclosed shallow shelf about 50 km in length located offshore of Los Angeles within the Southern California Bight (Fig. 1). Several comprehensive current/temperature studies of Santa Monica Bay took place

[1] Professor, School of Oceanography, Box 357940, University of Washington, Seattle, WA 98195-7940

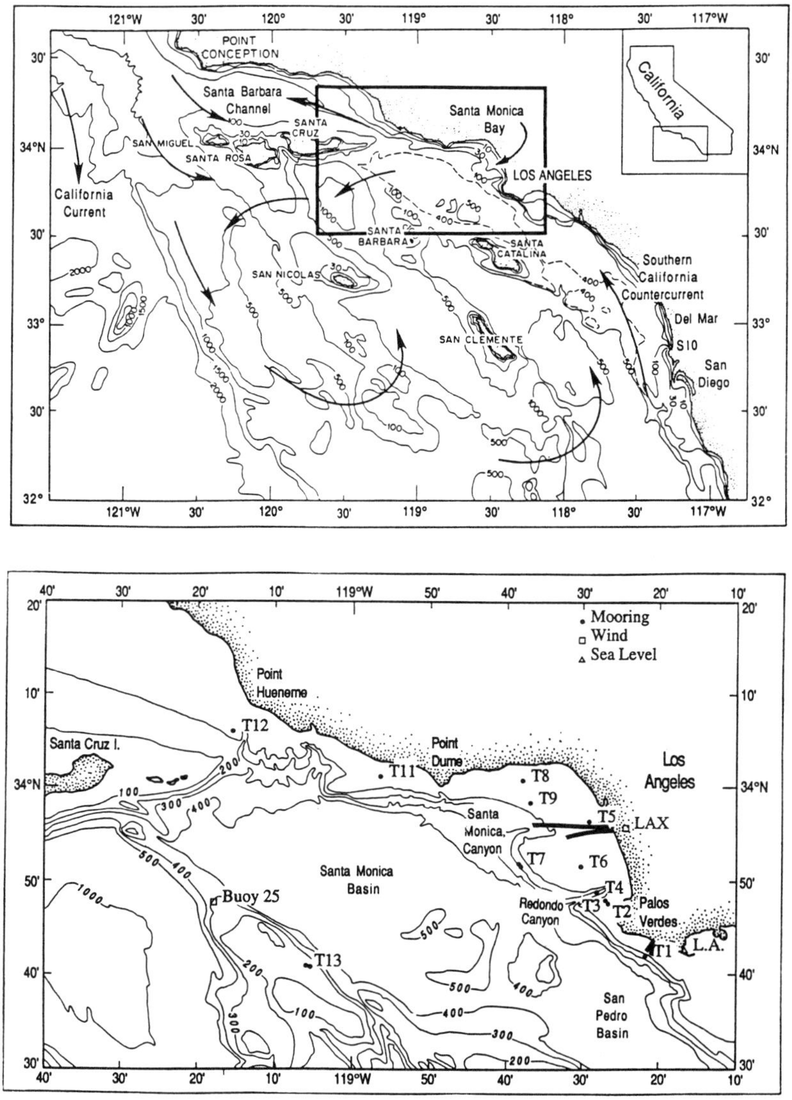

Figure 1. Maps of the Southern California Bight and the Santa Monica region showing bottom topography, landmarks and schematic mean circulation pattern in the Southern California Bight. Sewage outfalls in Santa Monica Bay and off Palos Verdes are shown with solid straight lines. Locations of sensor arrays (currents, wind, or sea level) are shown as dots with a label (e.g., T12). Depth contours in these and all subsequent figures are in fathoms.

during the period 1986-1988 as part of an interdisciplinary study (CaBS) of the Santa Monica/San Pedro region (Jackson et al., 1989). The study included as many as 34 instruments at 13 locations within the Bay and in the Santa Barbara and Santa Monica/San Pedro Channels. Two locations were within 1/2 mile of regional sewage outfalls (Fig. 1). The goal of the program was to determine how (and why) currents and water properties change within the Bay on time scales from a few hours to several months. The present paper will describe aspects of this study as well as related, larger scales studies, as they pertain to sewage and pollutant dispersal within the Bay and between the Bay and regions outside the Bay.

The topography of Santa Monica Bay is complicated. North and south of the Bay, the shelf is relatively narrow (~7 km); inside the Bay the shelf is wider (~20 km) but is incised by two submarine canyons. The larger of the canyons, Santa Monica, bisects the Bay; the smaller, Redondo, is located in the southern part of the Bay (Fig. 1). The topography of the adjacent landmass is generally mountainous, with the exception of the Los Angeles basin to the east. The mountainous topography blocks winds in the region, making them lighter and more variable than winds outside the Southern California Bight. In the Bay, winds with velocities higher than 15 m s^{-1} are rare. Such winds generally occur only during the winter and spring, or during the afternoons of the summer and fall when a strong sea breeze can occur (Stevenson et al., 1956).

Santa Monica Bay lies within the California Current system, shown schematically in Figure 1. The California Current is a large scale current system that controls much of the seasonal variability in currents and water properties along the entire U.S. West Coast (Hickey, 1979, 1997; Lynn and Simpson 1987). Briefly, the California Current flows equatorward offshore of the Bight and is most fully developed in late summer and early fall. South of Point Conception and inshore of the equatorward flowing California Current, the Southern California Counter current flows poleward through the Bight (Sverdrup and Fleming, 1941). The Countercurrent is strongest in summer and early fall, and weakest or even non-existent in spring. Equatorward flow has sometimes been observed inshore of the Southern California Countercurrent on narrow coastal shelves in the Southern California Bight (Tsuchiya, 1980; Hickey, 1992, 1993). Beneath the sea surface (~100-200 m core depth) the poleward flowing California Undercurrent is a dominant feature of circulation in the Bight.

In this paper, we will focus on three aspects of oceanic variability as it relates to Santa Monica Bay: 1) the physical connection in water masses between the Bay and the adjacent shelves and offshore ocean; 2) the variability in currents within the Bay; and 3) the relative lack of importance of local wind in driving currents within the Bay. Each of these topics has important consequences with respect to understanding processes controlling water properties and circulation within the Bay and in developing a working numerical model for physical processes within the Bay.

2. Connectivity of Santa Monica Bay with Other Oceanic Regions

Santa Monica Bay cannot be considered in isolation from the surrounding ocean. Maps of sea surface temperature illustrate that Santa Monica Bay is merely a small coastal indentation (~20 km wide) in a region in which temperature differences occur over very large distances along the coastline (~300 km) (Fig. 2). For example, on January 29, 1990, warmer water floods Santa Monica Bay as an intrusion pushes water northwestward through the whole coastal region. Several days later, colder water is observed all along the coast when upwelling occurs due to the southeastward winds over the region. Again, the process is large scale, involving Santa Monica as it does much of the coast within and outside the Southern California Bight.

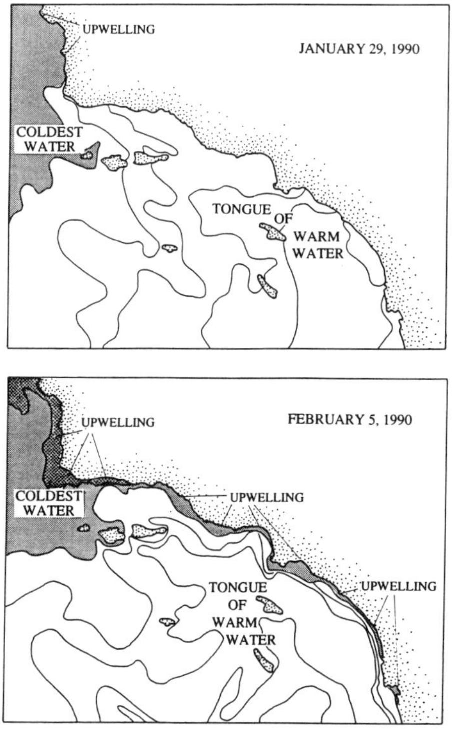

Figure 2. Temperature isotherms derived from a satellite image of sea surface temperature in the Southern California Bight region for January 29, 1990 (upper panel) and for February 5, 1990 (lower panel). Temperature generally increases from south to north except in the lower image in the nearshore region where temperature is colder all along the coastline than it is farther offshore. Original images are shown in Figure 34 in Hickey (1992).

Tracks of "drifters" (sensors that follow water masses at 15 or 40 m below the sea surface and transmit location data to satellite) illustrate another aspect of the connection between Santa Monica and the world ocean (Fig. 3). The drifter tracks were obtained during the same periods as the sea surface temperature maps shown in Figure 2. Drifters deployed just offshore of the Palos Verdes shelf move onshore onto the Palos Verdes shelf and then up into Santa Monica Bay. Drifters deployed on the edge of the Santa Monica Bay shelf move across the Santa Monica and San Pedro deep basins, then back onshore onto the San Pedro shelf during a strong upwelling event (see Hickey, 1992 for a more detailed discussion of these drifter pathways). Drifter tracks shown in Hickey (1992) demonstrate that water masses offshore of Santa Monica Bay shelf can get pulled up onto the Santa Monica shelf via Santa Monica Canyon. They can also circulate around the canyon and then return to the offshore ocean. This canyon is the site of a major sludge outfall. Thus, suspended and particulate materials from other regions might be brought to Santa Monica Bay; conversely, materials from Santa Monica might be exported to the world ocean.

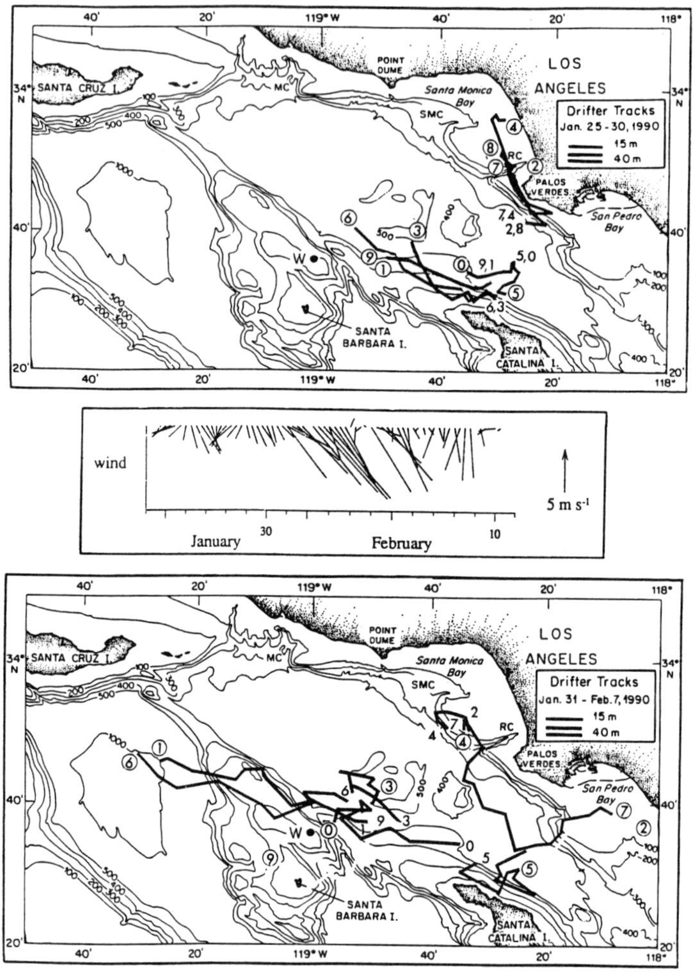

Figure 3. Tracks of satellite-tracked drifters for the period January 25-30, 1990 (upper panel) and January 31-February 7, 1990 (lower panel) and vector wind for the time period of the drifter deployments. Drifters were drogued to follow water at depths of 15 m or 40 m. Data suggest the ability of water to move into Santa Monica Bay from the Palos Verdes area (upper panel) or to move from the edge of Santa Monica Bay into San Pedro Bay (lower panel). The drifter ID is given at the beginning and end of each track; the circled ID indicates the end of the track. A pair of numbers indicates that a pair of drifters (surface, deep) were deployed together. The location of the wind buoy is indicated by a "W" on the maps. Important submarine canyons are indicated by the notation MC, SMC and RC (Magu, Santa Monica and Redondo, respectively). Adapted from Hickey (1992).

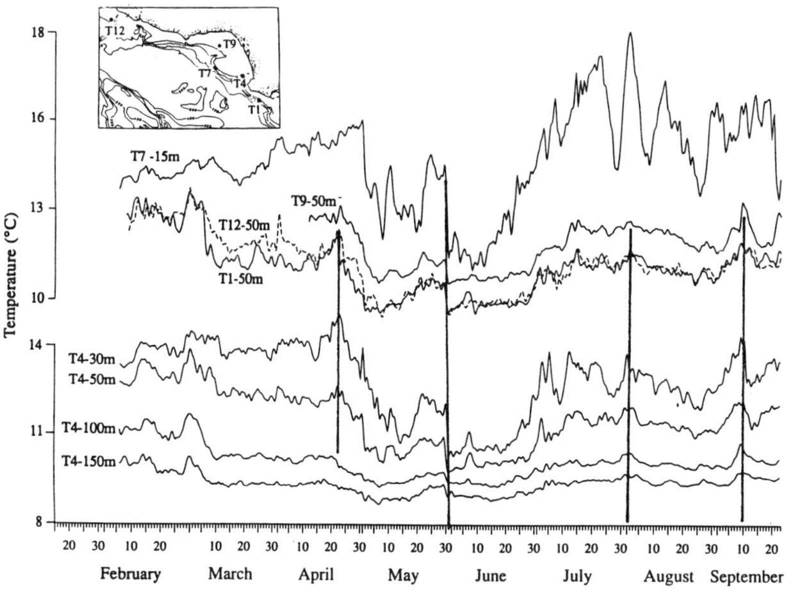

Figure 4. Time series of temperature measured hourly at selected locations throughout the study area, including several depths (30 m, 50 m, 100 m, and 150 m) at mooring T4 above Redondo Canyon. Mooring locations are shown on the inset map. Data show the variability of temperature from day to day as well as the similarity of temperature changes from site to site. Several events are marked with vertical lines for site to site comparison. Adapted from Dobbins and Hickey (1997).

The close correspondence among time series of temperature from moored sensors within Santa Monica Bay and on the shelves to the north and south of the Bay at similar depths (~50 m) illustrates that over both the course of a year and on scales of a few days, water properties are not unique to Santa Monica Bay (Fig. 4). The temperature data were obtained at hourly intervals for a period of about 8 months. Note the dramatic changes that occur over time periods of a few days. Whatever the processes that control the state of the water properties, they occur over a very large scale and are not determined by events that occur only in Santa Monica. We note that water properties near the sea surface (~15 m) differ from those at deeper depths. This is because heating and cooling are more important in determining water temperature near the sea surface than movement by currents, as is the case at deeper depths.

3. Variability of Current Patterns within Santa Monica Bay

Current patterns, unlike temperature patterns, have a high degree of spatial variability over Santa Monica Bay and the adjacent regions (Fig. 5). Seasonal mean currents, as shown in the figure for the spring season, show that a two gyre system

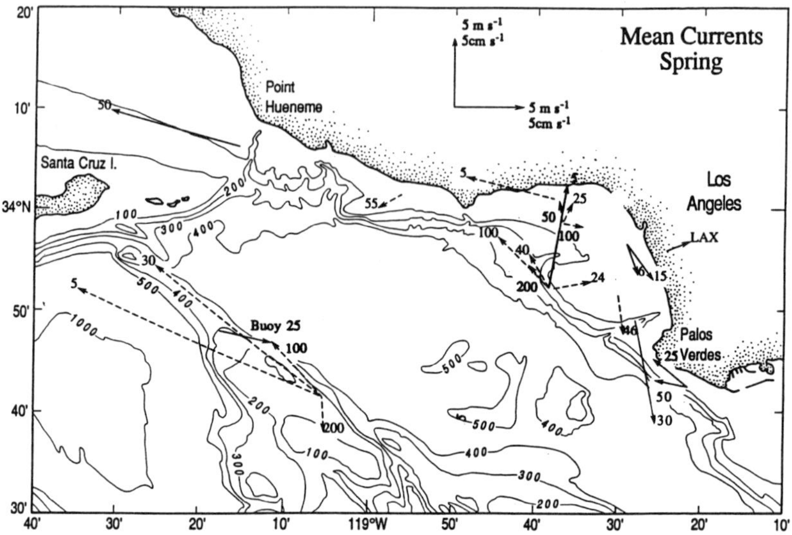

Figure 5. Mean currents from selected locations and mean wind velocities plotted as vectors for the spring season. Mean currents are given in cm s^{-1} and mean winds are given in m s^{-1}. The depth of the current data in meters is indicated by the number at the head of the arrow. Means calculated with shorter record lengths are indicated by dashed arrows. The figure illustrates the complex nature of mean flow patterns; in particular, the existence of a two gyre pattern within Santa Monica Bay.

occurs in Santa Monica Bay in both seasons. Water moves into the Bay at some locations along the shelf edge, and diverges into northward and southward branches as it near the coast. In general, mean currents are stronger in winter and spring than in summer. Currents over the adjacent shelves and in the Santa Monica Channel are northwestward at all sites and are largest in the mouth of the Santa Barbara Channel. At the southeastern end of the Bay, mean currents are strongly affected by Redondo Canyon. In particular, below some depth, currents are steered along the canyon walls rather than following the general trend of the Bay topography (not shown). This depth of canyon influence rises from about 100 m in spring to less than 30 m in summer (Dobbins and Hickey, 1997).

Time series displaying actual measured currents demonstrate the high degree of temporal variability that accompanies the spatial variability (Fig. 6). Much of the variability is due to tidal currents, with an amplitude of about 10 cm s^{-1}. These are displayed for the two dominant tidal frequencies (using band-pass filtering) in the panels beneath the raw data records. The tidal time series themselves illustrate a large degree of temporal variability: periods of weak tidal currents are followed by periods of stronger tidal currents. Some of these differences are due to the fortnightly tide (i.e., the tidal spring-neap cycle). However, much of the variability is likely due to changes in stratification of the coastal waters. Stratification (the ocean density) affects the pathways of waves generated as the tide passes over the continental slope. Current ellipses for the principal semi-diurnal tidal currents (M_2) at several locations

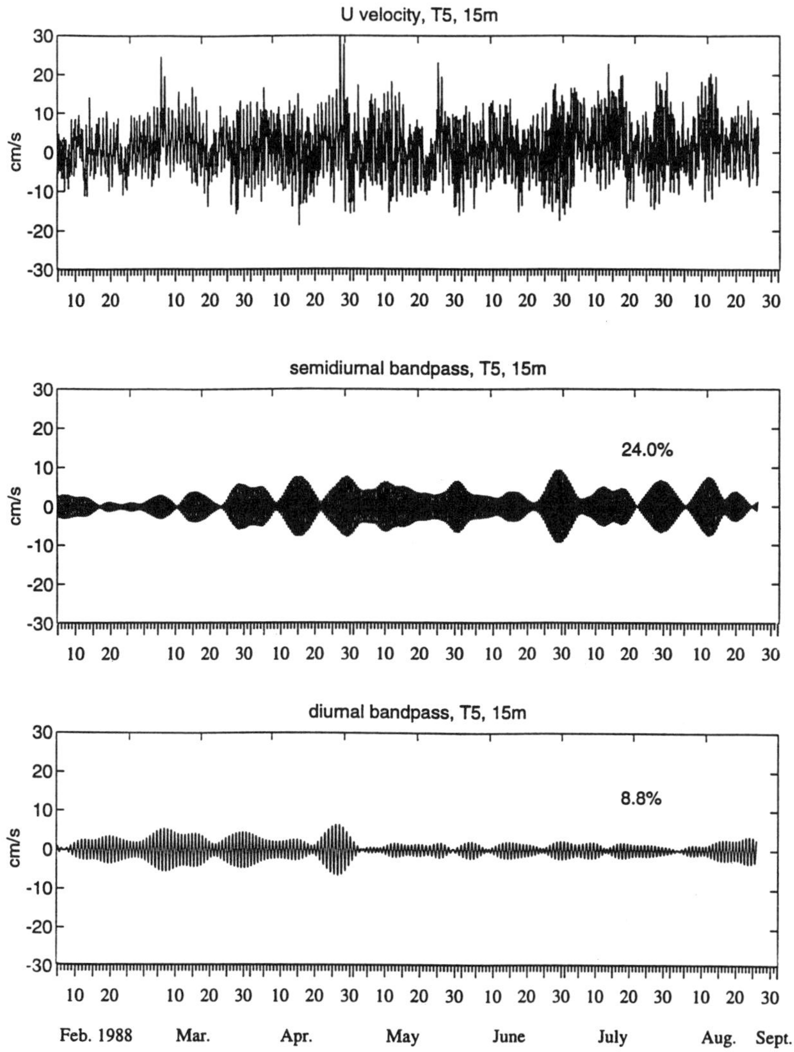

Figure 6. Time series of hourly east-west currents at a site on the Santa Monica shelf: original data (upper panel); band-passed diurnal and semi-diurnal records (mid and lower panels). "Band-passed" records show the currents associated with a specific period, such as the daily (diurnal) or twice daily (semidiurnal) tides. The figure illustrates the extreme variability in the direction and magnitude of measured currents on a day by day basis. The figure also shows that even tidal currents change significantly over several days.

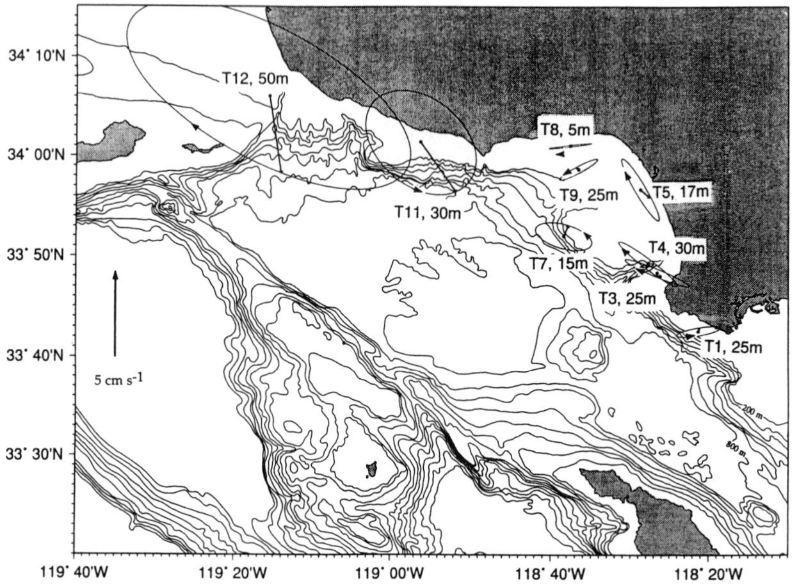

Figure 7. Spatial distribution of semi-diurnal tidal current amplitudes and phases in the upper water column, showing the high degree of spatial variability over the region. Measurement depth in meters is shown next to each ellipse. Tidal currents typically change direction during the day. The tidal ellipse shows the tip of the current vector (the direction of the current and its amplitude or speed) as it rotates through a 12.4 hr period. Direction of rotation is indicated by an arrow head on each ellipse.

at similar depths are shown in Figure 7. The direction of the tidal current rotates around the ellipse over the course of 12.4 hr and the amplitude of the current vector at any time is equal to the scaled distance between the ellipse center and the edge. The figure demonstrates that, like the mean currents, tidal currents have a high degree of site-to-site variability. The strength of M_2 tidal currents in the 30-50 m depth interval increases toward the Santa Barbara Channel.

At most locations, current energy occurs in three dominant time intervals: near 0.5 d (tidal origin), near 1.0 d (mostly of tidal origin) and in the range of 2-20 d (mostly not of tidal origin) (Dobbins and Hickey, 1997). Inspection of time series from which daily and shorter period tidal currents have been removed illustrates the extent of the variability on scales of 2 to 20 d: at most sites currents reverse direction every few days (Fig. 8). Speeds range from about 10 to 30 cm s^{-1} (or ~10 to 30 km d^{-1}). Reversals are similar to the abrupt changes observed in temperature records (Fig. 4). However, although the three major upwelling events (marked by a drop in water temperature) are associated with a strengthening of southward or southeastward flow at most sites) it is difficult to find a one to one correspondence between temperature changes and current reversals.

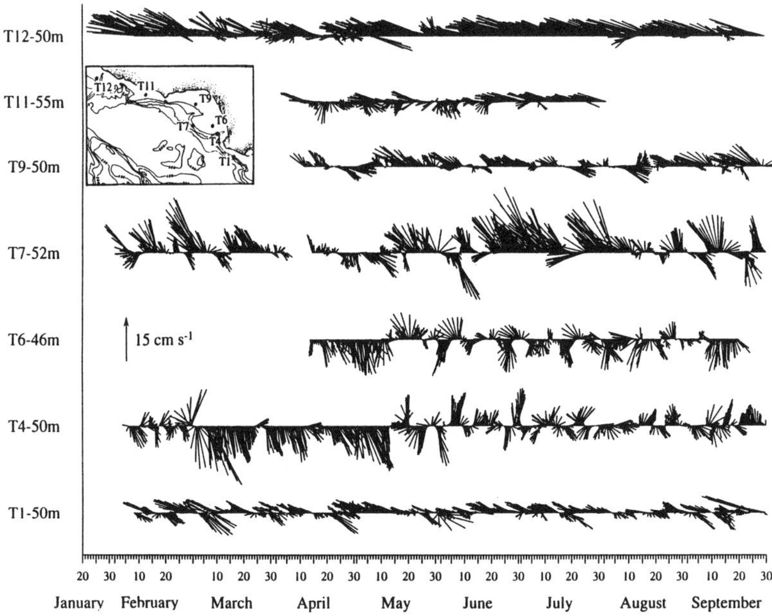

Figure 8. Vector time series of subtidal currents (current records from which tidal currents have been removed) measured near 50 m from the sea surface. Mooring locations are shown on the inset map. Time series are plotted with the most northern mooring at the top of the page, proceeding southward along the shelf. The figure illustrates the high degree of both temporal and spatial variability in current speed and direction within and near Santa Monica Bay. Unlike temperature records (see Figure 6), it is difficult to track changes between sites. Adapted from Dobbins and Hickey (1997).

Statistical analysis of the current measurements indicates that there is some degree of spatial coherence; i.e., currents are in the same direction at the same time at a number of sites (Dobbins and Hickey, 1997). During some periods in both spring and summer, currents tend to form a clockwise gyre within the Bay. During other periods, currents within the Bay and on the adjacent shelves are all in roughly the same direction.

4. The Weakness of Direct Wind Driving within the Bay

At most locations along the U.S. west coast, the alongshelf component of the wind is primarily responsible for alongshelf currents and their temporal changes (see review in Hickey, 1997). The relationship between wind and currents has been exploited in such regions to develop reasonably successful models of the coastal circulation (e.g., Chapman, 1987). Strong wind driving does not occur frequently in Santa Monica Bay. For example, although current patterns are similar in spring and summer, the relationship to the wind field is not. In summer, when winds are weak within the entire Bight, there is no correspondence with observed wind time series.

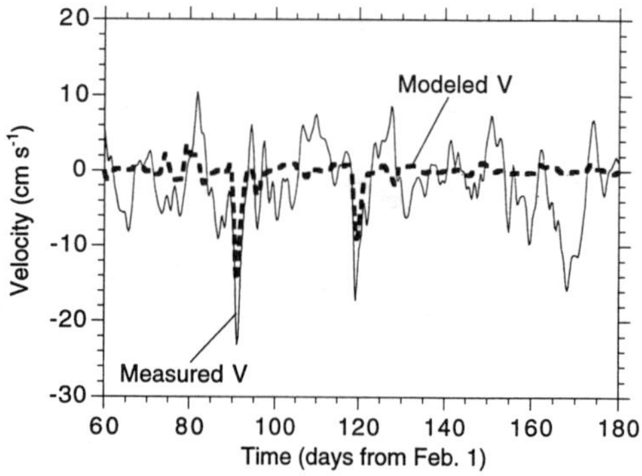

Figure 9. Comparison of depth-averaged alongshelf velocity as obtained from measurements in Santa Monica Bay with velocity predicted by a simple wind-driven model.

In spring, some relationship to the wind is observed, particularly during the several largest wind upwelling events.

If currents along the Bay are directly forced by local wind stress, then the acceleration of the vertical average of currents directed along the Bay would be balanced by wind stress minus bottom stress (both divided by water depth) (Hickey, 1992, and others). When this relationship is used to calculate vertically-averaged currents over time for Santa Monica Bay, comparison with the observed currents demonstrates that the wind/current relationship does not account for the majority of current reversals within the Bay: only 3-4 events over the 3 month record shown are clearly wind-driven (Fig. 9). The difference between Santa Monica and other coastal regions is likely a result of the semi-enclosed nature of the Bay (50 km vs. 1000 km long) as well as the relatively weak winds that occur over the Bay. Hickey (1992) presents evidence suggesting that much of the current variability is related to (driven by) current variability over the shelf edge.

5. Summary

Waters of Santa Monica Bay are controlled by and interchange freely with those of the Southern California Bight region; water within the Bay cools during upwelling events and warms during periods when warmer water pushes through the Bight from the south. Currents within Santa Monica Bay can be relatively strong (10-40 cm s^{-1}) in spite of the rather light winds common in the area. Currents change both direction and strength at intervals of 0.5-20 d. The spatial patterns of currents within the Bay are much more complex than temperature patterns: the mean circulation pattern consists of two gyres, a counterclockwise gyre in the northern half of the Bay and a clockwise one in the southern half of the Bay. The fluctuating currents often have very short spatial scales; i.e., currents are frequently unrelated at different sites within the Bay. Only during the spring season is there any evidence that currents within the Bay are forced directly by the local wind. Even during spring, however, much of the variability cannot be accounted for by local wind. In summer, currents are still strong and variable, but their forcing mechanisms are completely unknown. Much of the variability in currents in Santa Monca Bay appears to be caused by processes that occur seaward of the Bay. These results have, important implications with respect to modeling of currents within the Bay for purposes such as sewage outfall management.

An ideal model for oceanic processes in Santa Monica Bay would include, in order of priority, four components:

1) Time-dependent information on large scale (Bight-wide) processes;
2) Fine resolution topography that adequately resolves the two canyons as well as local headlands such as Point Dume (< 0.25. km);
3) Time-dependent stratification (density) information;
4) Time-dependent wind information for the entire Bay and the adjacent continental slope.

Acknowledgments

Preparation of this paper was funded by Contract No. 14-35-01-96-CT-30819 to B. Hickey from the Minerals Management Service, U.S. Department of the Interior. E. Dobbins, C. Moore, and N. Kachel were responsible for the illustrations.

References

Chapman, D.C. (1987) Application of wind-forced, long, coastal-trapped wave theory along the California coast. *Journal of Geophysical. Res.*, **92**, 1798-1816.

Dobbins, E. and B.M. Hickey (1997) Circulation and water properties in Santa Monica Bay. In preparation.

Hickey, B.M. (1979) The California current system - hypotheses and facts. *Progress in Oceanography*, **8**, 191-279.

Hickey, B.M. (1992) Circulation over the Santa Monica-San Pedro Basin and Shelf. *Progress in Oceanography*, **30**, 37-115.

Hickey, B.M. (1993) Physical Oceanography. In: *Ecology of the Southern California Bight: a synthesis and interpretation,* Eds. M. Dailey, D. Reish, and J. Anderson, 19-70. University of California Press, Berkeley, California.

Hickey, B.M (1997) Physical Oceanography of the West Coast of North America from the Strait of Juan de Fuca to the tip of Baja California. Chapter for a Volume of The Sea. In press.

Jackson, G., F. Azam, A. Carlucci, R. Eppley, B. Finney, D. Gorsline, B.M. Hickey, C.A. Huh, R. Jahnke, I. Kaplan, M. Landry, L. Small, M. Venkatesan, P. Williams and K. Wong (1989) Elemental cycling and fluxes off the coast of southern California. *Eos Transactions of the American Geophysical Union*, **70**.

Lynn, R.S. and J.J. Simpson (1987) The California current system - the seasonal variability of its physical characteristics. *Journal of Geophysical Research*, **92**(C12), 12,947-12,966.

Sverdrup, H.U. and R.H. Flemming (1941) The waters off the coast of southern California, March to July 1937. Scripps Institution of Oceanography, *University of California Press*, **4**(10), 261-378.

Tsuchiya, M. (1980) Inshore circulation in the Southern California Bight, 1974-1977. *Deep Sea Research*, **27a**, 99-118.

CHAPTER 168

STATE-OF-THE-ART DESIGN OF OCEAN OUTFALL DIFFUSER SYSTEMS

Joseph V. Perrone
Michael Dowling

Abstract

Even with modern advances in wastewater treatment technology, as well as increasingly stringent regulatory requirements, effluent from all sewage treatment plants contains significant concentrations of pollutants. The principal purpose of an ocean outfall is to discharge treated effluent to the ocean so as to minimize the environmental impact of the discharged effluent on the shoreline and other sensitive areas. The degree to which this can be achieved is dictated largely by the location of the outfall terminus, or diffuser, and the design of the diffuser system itself.

The location of the outfall diffuser is chosen to provide sufficient dilution, dispersion and time of decay of micro-organisms and other pollutants to meet the specified water quality objectives in bathing water areas, fishing areas and other designated areas (e.g., kelp beds).

The principal objectives in the design of the diffuser system itself are:

(a) To ensure that the initial dilution which is achieved is sufficient to ensure compliance with environmental water quality standards in the immediate vicinity of the diffuser.
(b) To ensure that head losses in the diffuser, together with head losses through the rest of the outfall system, do not exceed the available driving head, even under projected future peak flow conditions.
(c) To prevent intrusion of sea-water and sea-bed sediments.
(d) To ensure that flow velocities are adequate to remove deposited sediment, slime and other materials which could impact the long term operability of the outfall.
(e) To ensure that the system can fulfill its specified design life in the harsh marine environment.

The outfall diffuser design must achieve the above objectives while minimizing capital cost as well as long term operations and maintenance requirements.

Depending on the flow rates, level of treatment and regulatory requirements, the above objectives may be achieved by a diffuser system as simple as an open-ended steel or concrete pipe or as complex as a multi-port system several thousand feet long, constructed of corrosion resistant materials.

The objectives of this paper are to chronicle recent advances in outfall diffuser design, to discuss the characteristics and features of a number of advanced diffuser systems, and to discuss further potential advances in this area. Projects referenced include the Boston Ocean Outfall, the world's largest tunneled outfall, as well as several other outfalls which have been at the forefront of innovative technology and its application to outfall design and construction.

CHAPTER 169

Public Partipation in Managing California's Ocean Resources: "The Role of the Public in Governance Affecting Coastal and Offshore Resources"

By Linda Krop/[1]

Abstract

This paper describes suggestions postulated by Linda Krop, Senior Staff Attorney for the Environmental Defense Center (EDC), a non-profit public interest environmental law firm that represents community organizations in a variety of environmental and land use issues before public agencies and in court. As part of its activities, the EDC seeks to improve public education, participation, and influence in the decision-making process. Governance supported by an informed and participatory public is more effective and fosters an increased level of faith in governmental functions than governance that is inaccessible or unresponsive to the public.

EDC has been involved in many decision-making processes that raised the ire of public members because they did not believe they were given adequate notice, provided an adequate opportunity to review and comment on written materials, given an adequate opportunity to testify at public hearings, afforded public hearings at all, or were otherwise disenfranchised from the process. Without timely and appropriate opportunities for public participation, the result is often late intervention, enhanced distrust between the parties, and exacerbated levels of conflict. This result does not serve the best interests of the public, private or governmental parties.

In this paper, the EDC proposes solutions that address both procedural (e.g., noticing, commenting, hearings)

[1] Environmental Defense Center, 906 Garden Street, Santa Barbara, CA 93101.

and substantive (e.g., collaborative task forces, citizen advisory committees, dispute resolution mechanisms) aspects of governance that can enhance public participation.

The Role of the Public in Governance Affecting Coastal and Offshore Resources

History confirms the significance of the public's interest in coastal issues, especially in California where the voters established not only a statutory framework for regulating coastal development and resource protection, but also created a state-wide commission to oversee activities on the coast. The public's role in shaping coastal policy has been impressive and enduring. However, ensuring effective public participation in the implementation of such policy has been frustrating at times.

I am an attorney and community educator for the Environmental Defense Center, a non-profit public interest environmental law firm in Santa Barbara, California. We represent community organizations throughout the South-Central Coast (San Luis Obispo, Santa Barbara, and Ventura Counties) in a variety of coastal environmental and land use issues. Our representation includes enforcement of local, state and federal laws and policies.

By working closely with public groups, and joining them in their participation in the public review process, EDC has first-hand experience working with the agencies that implement and enforce our coastal regulations. Based upon this experience, we offer the following advice to ensure full compliance with public participation laws and rules, and to encourage continued public participation in issues of critical coastal importance.

First, I urge all agency staff and representatives to consider what it is like to be a member of the public with an interest in a coastal issue. Consider how important it is to receive timely notice of a hearing or comment opportunity. Consider what it would be like to receive hundreds of pages of technical documents to review a few days before a critical hearing. Consider what it is like to be told that a staff report will not be available until the day of the hearing. Finally, consider what it would be like to work full-time and find out that the hearing on your issue will be held hundreds of miles away on a Wednesday morning. Or imagine what it would be like to find out that a decision was already made on an important issue without your knowledge. All of these situations occur on a routine basis, and make it difficult for the public to provide meaningful input to the decision-making process, let alone participate at all.

The following sets forth specific recommendations for adoption by public agencies:

I. PUBLIC NOTICE

Provide adequate notice of public hearings, document availability, and comment opportunities. Public notice should be provided in a direct manner whenever possible. Federal register notices and newspaper ads are required, but are often not visible enough to notify the general public. Notices should also be directly provided to interested organizations and individuals. Agencies such as the California Coastal Commission, State Lands Commission, California Resources Agency, and local cities and counties often have mailing lists of organizations and individuals that are interested in coastal issues.

Public notice must also be timely. It takes time to request staff reports and other relevant documents, review them, consult experts, neighbors and other interested parties, and prepare a coherent response.

II. ACCESS TO INFORMATION

Again, access to information (e.g. staff reports, environmental review documents) must be provided in a timely manner to allow the public to respond in an intelligent and meaningful way. Some agencies do not provide staff reports and accompanying materials until the day of the hearing! Others provide documents only a couple days before a hearing. The Coastal Commission has one of the better policies, which requires that staff reports be available to the public at least ten days prior to a hearing.

The public must have access to documents at an affordable rate to ensure adequate distribution and review.

Finally, agencies must ensure independent and thorough review of applicant-supplied information. The burden should not be on the public to ensure an objective analysis.

III. MEANINGFUL PUBLIC INPUT

Everyone benefits if the public is involved in a meaningful manner. Agencies may receive additional information that is helpful in their review of a proposal; they will also have a better sense of public sentiment.

The public must be afforded an adequate time to provide both written and verbal comments. To provide effective written comments, the public must receive timely notice and access to information, as stated above. Written comments should be disseminated to decision-makers sufficiently in advance of a hearing to ensure that they have adequate time to review and digest all comments. At the hearings themselves, the public must be given an adequate time to comment on an issue. Many agencies provide applicants with lengthy or unlimited time to comment, but limit public speakers to 1 - 3 minutes. This protocol is unfair to the speaker as well as to the decision-makers, who receive an incomplete and biased presentation.

Location and timing of hearings are important. Hearings should be held in an accessible location. The Coastal Commission holds hearings at various locations around the state (although hearings are not always held in a proximate locale); it is important for all state and federal agencies to try to schedule hearings at appropriate locations that are accessible to interested members of the public.

EDC recommends the use of informal public workshops as a means to apprise the public (and decision-makers) of major upcoming issues. Workshops are useful tools for educating the public and identifying key areas of concern. In Santa Barbara County, we negotiated an arrangement with the Energy Division whereby the County held an early public workshop regarding a proposed "slant drilling" gas development project. The workshop was held as the Draft Environmental Impact Report was released, to inform the public about the project and the opportunity to comment on the DEIR. A second hearing was held towards the end of the review period to accept comments on the DEIR. The first workshop facilitated the public's review of the DEIR, and made for more effective comment on the report. The workshop also ensured that the County provided copies of the DEIR to all interested parties well in advance of the comment deadline. The County has used the workshop model for planning issues as well (e.g. when considering new or amended general plan policies or zoning ordinances).

IV. PUBLIC POLICY AND PLANNING

In addition to providing input regarding specific projects or development proposals, the public has a tremendous interest in the development of public policy and long-term planning. As stated above, public workshops can be an effective tool for providing public input in such processes. In addition, adequate public representation on agency task forces and committees is essential. The representatives should be carefully chosen to ensure that a broad range of interests are represented. The representatives should have the tools necessary to participate effectively: (1) funding should be made available to pay for the representatives' time and expenses; (2) meetings should be held in accessible locations and at appropriate times; and (3) independent expertise should be made available to review technical reports and documents. Where feasible, task force and committee meetings should be open to the public for observation and comment.

V. FACILITATION AND MEDIATION

More often, agencies, applicants and the public are willing to try to resolve differences through the use of facilitation or mediation. EDC is very supportive of such efforts, provided all interests are represented and all issues are addressed. Most important, the participating parties must commit to a solution that will address all interests. One essential role of the facilitator or mediator is to help identify all appropriate participants. The facilitator or mediator can also assist in the procurement of technical information or expertise that will support negotiations. In any event, agreements forged through facilitation or mediation should still be subject to a public review process.

CONCLUSION

Early and effective public participation is critical to the agency process. If the public feels left out of the decision-making process, last-minute objections may obstruct or delay the process. Extended deliberations and even litigation may result. It is better for all parties - the applicants, the decision-makers, and the public - if participation is extended to all throughout the process in a meaningful manner.

CHAPTER 170

Meaningful Marine Monitoring: the Three R's to Responsible Stewardship

Jan Roletto, Dan Howard, and Candice Brown

Gulf of the Farallones National Marine Sanctuary,
Fort Mason, Bldg. 201,
San Francisco, CA 94123
tel. no. (415) 561-6622, fax no. (415) 561-6616

Abstract

Research and Responsible Stewardship result in Resource Protection and are the basis of the management program for the Gulf of the Farallones (GFNMS), Cordell Bank (CBNMS) and the northern portion (Año Nuevo to Rocky Point) of Monterey Bay National Marine Sanctuaries (MBNMS), which are administered through the Sanctuary office, in San Francisco. Through the use of volunteers (citizen scientists) the Sanctuary directly educates hundreds of local constituents on marine resources and indirectly reaches 1000's of visitors to the Sanctuary each year. The volunteers collect data that will be used to protect the natural resources of the Sanctuary while they learn how to be responsible stewards of the Sanctuary. Through the use of volunteer monitors we can accomplish our research objectives: 1) to assess the gaps in the Sanctuary's information base that may affect our ability to manage the area; 2) conduct studies of species or marine communities to identify resources most at risk or in need of management attention; 3) promote the Sanctuaries as a site for management related marine research by providing financial and logistical support for scientific investigations that address critical marine resource protection issues; 4) design research projects that are responsive to management concerns and that contribute to improved management of the Sanctuary; 5) make effective use of research results by incorporating them into interpretive and resource protection programs; and 6) encourage information exchange and cooperation among all the organizations and agencies undertaking management related research in the Sanctuaries to promote more timely and informed management.

Introduction

Research and Responsible Stewardship result in Resource Protection and are the basis of the management program for the Gulf of the Farallones (GFNMS), Cordell Bank (CBNMS) and the northern portion (Año Nuevo to Rocky Point) of Monterey Bay National Marine Sanctuaries (MBNMS), which are administered through the Sanctuary office, in San Francisco. Through the use of volunteers (citizen scientists) the Sanctuary directly educates hundreds of local constituents on marine resources and indirectly reaches 1000's of visitors to the Sanctuary each year. The volunteers collect data that will be used to protect the natural resources of the Sanctuary while they learn how to be responsible stewards of the Sanctuary. Through the use of volunteer monitors we can accomplish our research objectives which are:

- Assess the Sanctuary's information base to identify gaps in knowledge that can affect our ability to manage the area;
- Conduct studies of species or marine communities to identify resources most at risk or in need of management attention;
- Promote the Sanctuaries as a site for management related marine research by providing financial and logistical support for scientific investigations that address critical marine resource protection issues;
- Design research projects that are responsive to management concerns and contribute to improved management of the Sanctuary;
- Make effective use of research results by incorporating them into interpretive and resource protection programs;
- Encourage information exchange and cooperation among all the organizations and agencies undertaking management related research in the Sanctuaries to promote more timely and informed management.

Five Year Sanctuary Research Plan, 1996-2001

Research within the Gulf of the Farallones National Marine Sanctuary (GFNMS), Cordell Bank National Marine Sanctuary (CBNMS) and the northern portion (Rocky Point, Marin County, south to the San Mateo-Santa Cruz County border) of the Monterey Bay National Marine Sanctuary (MBNMS) is administered through the GFNMS offices in San Francisco. Research is one of three primary management goals of the Sanctuaries. Scientific research is encouraged at the Sanctuaries, particularly where research results can help resolve key management issues. The GFNMS-CBNMS-MBNMS Research Plan outlines general priorities for research for the next five years. These include:

- Baseline studies for populations and habitats whose presence were critical in the Sanctuaries designation, yet whose distributions and other basic characteristics remain poorly understood;
- Directed monitoring studies focusing on indicator species and representative habitats and undertaken jointly with other Sanctuaries and agencies;
- Analytical studies aimed at determining the cause of a condition or impacts (i.e. in a specific population).

The Research Program includes additional procedures aimed at ensuring that research projects address management issues and are effectively integrated into the education and resource protection programs of the Sanctuary and other resource trustee agencies. Research should help in solving specific management problems, enhance resource protection efforts, and assist in the interpretation of the resources for the general public.

Research for Management

Research conducted and funded within the Sanctuaries will continue to focus on management issues that relate to protection of significant resources. Research will typically examine questions involving communities or whole ecosystems and occasionally it may be species specific. General directions and priorities for additional research are provided in this section as a guide for identifying and selecting future research projects. Research projects are directed to answer four basic management questions:

- What are the general characteristics and distribution of species within different marine communities (documentation of baseline characteristics);
- What are the short and long term changes in species' populations and marine communities (i.e. abundance and distribution) and what are the influences of natural variability and human-caused perturbations on these changes (detectable through monitoring);
- What are possible cause and effect relationships accounting for changes observed in resource distribution, abundance, and diversity (analysis of experimental and predictive studies);
- Which ecosystems are best studied, by students of all ages and education backgrounds, which allow for information exchange among researchers, resource managers and involves the public in conservation and protection of the marine environment and resources?

Baseline studies are designed to obtain a basic understanding of the Sanctuary's resources. These studies generally refer to abundance, distribution, and movement of species. Elementary characteristics of many important populations are not known (cetaceans, fish, subtidal and intertidal invertebrates, and marine algae). Inventories of selected species within these populations, particularly Endangered and Threatened Species, represent an important direction for research. Other baseline studies focus on habitat inventory. There is a need for detailed inventory of the subtidal, intertidal and coastal habitats in the Sanctuaries and these studies build on previous surveys conducted by other Federal and State agencies.

Monitoring studies are fundamental to the successful management of marine resources. By regularly recording the abundance of organisms or the diversity of a community, it is possible to detect ecological change and perhaps changes after human caused perturbations. We attempt to detect and relate unusual changes in natural cycles and trends, to sources of disturbance. Long term monitoring provides an indication of the health of the important populations within the Sanctuaries as well as health of the overall system.

Analytical and predictive studies are conducted to determine the causes of changes in the ecosystem or to predict the effects of changes. Examples of completed studies for the Sanctuaries include studies of premature pupping in Steller sea lions, an updated study is nearing completion, and human disturbance on seals in Tomales Bay and Bolinas Lagoon. A public education project has been developed to reduce disturbance. Such in-depth projects will continue to be a means of focusing on specific management related issues and provide follow-up studies and projects which will assess effectiveness of management related decisions.

We encourage information exchange and cooperation among the organizations and agencies undertaking management related research in the Sanctuaries to promote more informed management. In order to identify and assess the gaps in knowledge that can affect our ability to manage the area, the Sanctuary hosts a biennial research workshop. Researchers from all marine disciplines gather at this workshop to discussion current and ongoing project results and to discuss future projects and joint funding. The biennial workshop is one of several research workshops which the Sanctuary coordinates to encourage information exchange among Bay Area organizations. The Sanctuary also hosts smaller workshops concentrating on specific topics or areas of interest to various researchers in the Bay Area, the State, nationally and internationally. In the past these workshops include annual seabird monitoring, multi-agency geographic information systems, research on marine mammals in multiple Sanctuaries, quarterly coordination meetings with local resource trustee agencies such as the National Park Service, use of volunteers for collecting biological information, long term monitoring programs, and the biology of white sharks. These workshops allow resource managers to extend limited funds and eliminate duplicative research and monitoring efforts.

Current Management Concerns and Associated Research Priorities

The Sanctuaries are constantly influenced from upland decisions and coastal development. The task for all resource mangers is resource protection while allowing appropriate economic uses that do not damage or stress Sanctuary resources. In order to accomplish this, a Sanctuary Management Plan must include an adequate research and monitoring program to continuously ascertain the health of the Sanctuaries. Current and ongoing problems facing the GFNMS, CBNMS and northern portion of the MBNMS primarily come from the need of local agencies to discharge unwanted substances, disturbance to the seabed, and disturbance to marine resources from aircraft and thrill craft (i.e. jet skis).

Reduction of biodiversity and productivity is an underlying concern of Sanctuary Managers. Dredge spoils, pollutants, siltation, and physical disturbance can all lead to the eventual reduction in species numbers and populations. Management must have knowledge of Sanctuary resource population trends and indices, critical habitats, and predator-prey relationships in order to adequately assess how an activity may affect Sanctuary resources.

Specific management concerns include dredging for environmental enhancement without adequate assessment of the effects of dredging on the benthic infauna. Bolinas Lagoon and Tomales Bay are two areas in the GFNMS where environmental enhancement (dredging) has been proposed without adequately assessing all of the consequences of the proposed dredging and deposition of dredge materials. The Long Term Management Strategy of the Army Corps of Engineers, for the clean dredge spoils from major military and commercial ports in San Francisco Bay, stipulates that clean dredge spoils from major ports be deposited 5 miles west of the western boundary of the GFNMS, so as not to contaminate the GFNMS with excess siltation, scouring and burial of epifauna.

Disturbance to the seabed also comes in the form of physical damage via vessel grounding. Sanctuaries with coastal boundaries work with local, state and federal agencies to remove toxic substances from grounded vessels (hydrocarbons, batteries, and other caustic materials found on many boats) and the Sanctuary, land manager and vessel owner (if known) must remove the vessel before physical damage occurs to the seabed or intertidal zone. Often funding is available only for the removal and clean up of pollutants but not for the physical removal of vessels.

Cordell Bank is commercially fished for groundfish and shrimp. The fishing communities realize that the continued practice of trawling directly on Cordell Bank could destroy the brittle habitat and decrease the productivity of the area. The Pacific Fisheries Management Council, Habitat Committee has proposed evaluating trawling activities on the Bank to investigate the effects of fishing on the Bank habitat.

Outfall from the San Francisco sewage treatment facility is not within the Sanctuary but may effect transboundary resources and eventually drift into the MBNMS and GFNMS. There is a proposal to increase the amount of effluent and to include effluent from the Central Valley Water Project. This sewage contains concentrated waste, including high levels of organochlorides and heavy metals, from areas such as the Kesterson Wildlife Refuge. These pollutants are known to cause a variety of birth defects and reduced fecundity in wildlife and humans.

Disturbance to seabirds and marine mammals by aircraft and thrill craft is an ongoing problem. Fixed-wing aircraft without mufflers and helicopters create most disturbances. Military aircraft, commercial sightseeing aircraft and commercial filming also contribute to disturbances. The GFNMS and MBNMS regulations require aircraft to stay above 1000 feet in sensitive areas. Researchers interested in conducting investigations that will contribute to the management of Sanctuary resources, can apply for exemption from altitude restrictions by obtaining a research permit from the Sanctuary. Thrill craft zones, where individuals can safely ride water craft such as jet skis, have been established at several areas within the MBNMS and are being investigated for the GFNMS. The zones avoid areas such as seal haul-out and pupping areas, sea otter habitat, and reduce conflict with other recreational users. Conflicts occur between recreational users of the Sanctuaries. Disturbance has occurred by recreational users (kayakers, beach goers, people and dogs, clam diggers, jet skiers, etc.) to seals at their haul-outs. Disturbance to the seals can be particularly damaging to local populations during the seals' pupping and molting seasons. The Sanctuary has developed a monitoring and education program to reduce disturbance to the seals and still provide areas for recreational activities.

Meaningful Marine Monitoring for Responsible Stewardship

Marine resource monitoring programs are costly and can be complex. For these reasons we have chosen to monitor habitats and communities which relate to important management questions, that involve the use of volunteers, and are projects that work with other marine resource agencies, thus providing technical and in-kind support to multiple agencies.

Two of our research and monitoring programs involve the service of citizen scientists and achieve the three R's: Beach Watch and SEALS (Sanctuary Education Awareness and Long-term Stewardship). Both programs fulfills the three R's by: 1) educating the public about the coastal environment; 2) encouraging the public that they can make a difference in protecting their habitats and resources and develop a sense of stewardship for the Sanctuaries; 3) assisting the Sanctuary in the detection of natural and human-caused environmental perturbations; 4) providing a baseline of information on the indices of marine organisms; and 5) developing a network of local experts who can document and discuss the natural changes an area will undergo over a period of several years.

Beach Watch: The Beach Watch program is a long term, baseline monitoring project implemented by volunteers and administered through the GFNMS. The area of operation is from Bodega Head to the San Mateo-Santa Cruz County border and includes the northern portion of the MBNMS. The program goals are to: 1) educate the public about the coastal environment; 2) encourage the public that they can make a difference in protecting their beaches; 3) assist the Sanctuary in the early detection of natural and human-caused environmental perturbations such as warm or cold water events and oil spills; 4) provide a baseline of information on the average presence of live and beachcast marine organisms; and 5) develop a network of local experts who can document and discuss the natural changes a specific beach will undergo over a period of several years. Surveyors record live and dead marine organisms including: seabirds, shorebirds, marine mammals, humans, and large racks of invertebrates and marine plants. They collect samples of oil or oiled organisms and notify the Sanctuary in the event of a violation, stranded marine mammal or unusual sightings.

Beach monitors cover one or a series of pre-defined beach segments within seven regions of the coast; 86 beach segments have been designated to be monitored within this program. Surveys are conducted once every two to four weeks. A total of 58 beaches (120.6 km in length) were monitored during 1996. Currently (1996) 38 beach segments are monitored once every four weeks and 20 beach segments are monitored every two weeks. Data collected, is compatible with pre and post-spill assessment programs implemented by federal and state resource trustees. An annual report is distributed.

SEALS in Tomales Bay and Bolinas Lagoon: SEALS, Sanctuary Education, Awareness and Long-term Stewardship, is a joint program for both the Gulf of the Farallones National Marine Sanctuary and the Farallones Marine Sanctuary Association (FMSA). Occasionally human activities in Bolinas Lagoon and Tomales Bay disturb harbor seals at their haul-outs. In Tomales Bay, pupping season coincides with clamming season, which brings large numbers of people to two seal haul-outs. In Bolinas Lagoon, seals haul-out near Highway One where people are sometimes causing disturbance from unleashed dogs, kayaks, canoes, and clam digging near the seals' haul-out. A decline in the breeding population has been documented at Hog, Clam and Seal Islands in Tomales Bay. The SEALS program has been developed to educate the public on proper behavior near the seals so that recreational activities can occur without disturbing the seals. SEALS volunteers monitor human activities, assess the fecundity of the local seal population and provide information that allows better management for the seal colonies. Monitors collect data on daily and seasonal disturbance rates, haul-out behavior of the seals prior to and after a disturbance or a series of disturbances, population and distribution of the seals in the Lagoon. SEALS volunteers also conduct informational interpretation talks to reduce the level of disturbance to the seal colonies in Bolinas Lagoon and Tomales Bay. Both areas have high visitor use by several different recreational users: clam diggers, bird and marine mammal watchers, kayakers, sailing, boating, and jet skiers. Each user-group presents a unique potential disturbance to the seal colonies. The Sanctuary and FMSA have been working with several federal and local environmental organizations to develop a plan to allow compatible recreational uses of the Sanctuary and still protect natural resources. The interpretation program has shown that it's effective in educating the public visiting the haul-out areas and reducing disturbance to the seals.

Other Research Projects Within the GFNMS, CBNMS and Northern Portion of MBNMS

Each of the following research projects address at least one of the management questions about indicator species and habitats (such as long term monitoring studies), abundance and distribution (baseline studies), perturbated systems, and assess cause and effect (analytical and predictive studies), or has an element of information exchange among researchers or educates the public. The following list of current research and monitoring projects include lead and contributing organizations for each project. Lead organizations are those groups that significantly contribute personnel, equipment or funding for the project and contributing organizations contribute small amounts of funding, are advisory, or contribute small amounts of in-kind funding such as equipment or personnel time. For further information, contact the Sanctuary office.

Projects for Which the GFNMS is the Lead Organization

1. Development of a Geographic Information System (GIS) Data Library for the GFNMS, CBNMS, and northern portion of MBNMS.

- Lead Organizations: GFNMS and Ecological Consulting, Inc.
- Contributing Organizations: Point Reyes Bird Observatory (PRBO), Avian Systems, Cascadia Research Collective, National Marine Fisheries Service (NMFS), Environmental Protection Agency (EPA), Golden Gate National Recreation Area (GGNRA), National Park Service (NPS)-Western Region, United States Geological Survey (USGS), California Department of Fish and Game (CDFG)-Office of Oil Spill Prevention and Response (OSPR), California Academy of Sciences, NOAA Prediction Center, and CIRIOS.
- Management and Research Objectives Met by this Project: Baseline studies and information exchange.

2. Beach Watch.

- Lead Organization: GFNMS.
- Contributing Organizations: CDFG, OSPR, NPS, State Parks and Recreation, NMFS, Marine Mammal Stranding Network, USGS Biological Resources Division (BRD), Natural Resource Damage Assessment (NRDA), Farallones Marine Sanctuary Association (FMSA), and NOAA.
- Management and Research Objectives Met by this Project: Baseline studies and information exchange.

3. Sanctuary Education, Awareness and Long-term Stewardship (SEALS).

- Lead Organizations: GFNMS and FMSA.
- Contributing Organizations: CDFG, NPS, NMFS, Audubon Canyon Ranch-Cypress Grove Preserve, Marin County Open Space, and Lawson's Landing.
- Management and Research Objectives Met by this Project: Baseline studies, analytical and predictive studies, and information exchange.

4. Establishing a Biological Baseline Monitoring in the Estero de San Antonio.

- Lead Organization: GFNMS.
- Contributing Organizations: San Francisco State University-Romberg Tiburon Center for Environmental Studies (SFSU) and GGNRA.
- Management and Research Objectives Met by this Project: Baseline studies, monitoring studies, analytical and predictive studies, and information exchange.

5. Abalone Mortality, Mark and Recapture at the South Farallon Islands.

- Lead Organization: GFNMS.
- Contributing Organization: CDFG.
- Management and Research Objectives Met by this Project: Baseline studies.

6. Evaluation of the Radioactive Waste Dump Site.

- Lead Organizations: GFNMS, NOAA HazMat, USGS, United States Navy, EPA (1990-1993).
- Contributing Organizations: NOAA and Moss Landing Marine Laboratories (MLML).
- Management and Research Objectives Met by this Project: Analytical and predictive studies and information exchange.

7. Common Murre Restoration Project.
- Lead Organizations: United States Fish and Wildlife Service (USFWS), GFNMS, CDFG, and OSPR.
- Contributing Organizations: NPS and PRBO.
- Management and Research Objectives Met by this Project: Baseline studies, monitoring studies, analytical and predictive studies, and information exchange.

8. Assessment of the Ashy Storm-petrel Population on the South Farallon Islands.
- Lead Organizations: GFNMS and PRBO.
- Management and Research Objectives Met by this Project: Baseline studies and monitoring studies.

9. Non-point Source Pollutants Seabirds and the Steller Sea Lion.
- Lead Organizations: GFNMS and PRBO
- Contributing Organizations: University of California (UC), Santa Cruz, Chemical Ecotoxicology Group (CETOX), Canadian Wildlife Service, and Western Foundation of Vertebrate Zoology.
- Management and Research Objectives Met by this Project: Monitoring studies and analytical and predictive studies.

10. Monitoring Parasitic Infestation Rates in *Emerita analoga*.
- Lead Organizations: National Wildlife Health Lab (NWHL) and GFNMS.
- Contributing Organization: CDFG.
- Management and Research Objectives Met by this Project: Monitoring studies, analytical and predictive studies, and information exchange.

11. Rocky Intertidal Monitoring at the South Farallon Islands and Central California Coast.
- Lead Organizations: GFNMS and Cosentino Consulting.
- Contributing Organizations: Humboldt State University and UC, Berkeley.
- Management and Research Objectives Met by this Project: Baseline studies, monitoring studies, analytical and predictive studies, and information exchange.

Projects for Which the GFNMS is a Cooperating Organization

1. Land Margin Ecosystems Research (LMER) and Biochemical Reactions in Estuaries (BRIE).
- Lead Organizations: University of Hawaii and SFSU.
- Contributing Organizations: National Science Foundation (NSF), GFNMS, et al.
- Management and Research Objectives Met by this Project: Monitoring studies, analytical and predictive studies, and information exchange.

2. Quarterly Monitoring of Fish Populations in Bolinas Lagoon.
- Lead Organization: CDFG.
- Contributing Organizations: GFNMS and U.C. Davis, Bodega Marine Laboratory.
- Management and Research Objectives Met by this Project: Baseline studies and monitoring studies.

3. The Relationship Between Physical Oceanographic Conditions and Larval Distribution in and Around the Gulf of the Farallones.

- Lead Organization: U.C. Davis, Bodega Marine Laboratory.
- Contributing Organizations: California Sea Grant, GFNMS, and CBNMS.
- Management and Research Objectives Met by this Project: Baseline studies and monitoring studies.

4. Reproductive Development in Yellowtail Rockfish.
- Lead Organization: NMFS, Tiburon Laboratory.
- Contributing Organization: CBNMS.
- Management and Research Objectives Met by this Project: Baseline studies, analytical and predictive studies, and information exchange.

5. Quarterly Monitoring in Estero Americano.
- Lead Organization: CDFG.
- Contributing Organization: GFNMS.
- Management and Research Objectives Met by this Project: Baseline studies and monitoring studies.

6. Lone Tree Creek Slide and Highway One Repair Site.
- Lead Organizations: California Department of Transportation (CalTrans) and MLML.
- Contributing Organizations: GFNMS, MBNMS (north), NPS, State Lands Commission (SLC), CDFG, Army Corps of Engineers (ACE), Federal Highway Commission (FHC), Regional Water Quality Control Board (RWQCB), Audubon Canyon Ranch (ACR), et al.
- Management and Research Objectives Met by this Project: Monitoring studies, analytical and predictive studies, and information exchange.

7. Big Lagoon, Muir Beach Restoration.
- Lead Organizations: CalTrans and GGNRA.
- Contributing Organizations: GFNMS, MBNMS (north), NPS, SLC, CDFG, ACE, FHC, RWQCB, MLML, ACR, et al..
- Management and Research Objectives Met by this Project: Monitoring studies, analytical and predictive studies, and information exchange.

8. Bolinas Lagoon Mitigation Site.
- Lead Organizations: CalTrans, GGNRA, and Marin Open Space.
- Contributing Organizations: GFNMS, SLC, CDFG, ACE, FHC, RWQCB, ACR, et al.
- Management and Research Objectives Met by this Project: Analytical and predictive studies and information exchange.

9. Seafloor Mapping and Oceanographic Characterization, Gulf of the Farallones Region, Central California Atlas.
- Lead Organization: USGS.
- Contributing Organizations: GFNMS, NMFS, UC Davis, et al.
- Management and Research Objectives Met by this Project: Monitoring studies and information exchange.

10. Biotoxin Monitoring.
- Lead Organizations: California Department of Health Services (CDHS) and U.S. Food and Drug Administration.
- Contributing Organizations: County Health Departments, Bay Bottom Beds, Cove Mussel Company, Hog Island Oyster Company, GFNMS, et al.

- Management and Research Objectives Met by this Project: Baseline studies, monitoring studies, analytical and predictive studies, and information exchange.

11. Feeding Ecology and Population Assessment of Minke Whales in the Gulf of the Farallones.

- Lead Organization: Texas A & M.
- Contributing Organizations: GFNMS, CBNMS, et al.
- Management and Research Objectives Met by this Project: Baseline studies and monitoring studies.

12. Gray Whale Population Monitoring.

- Lead Organization: NMFS-Southwest Fisheries Science Center.
- Contributing Organizations: GFNMS et al.
- Management and Research Objectives Met by this Project: Baseline studies, monitoring studies, analytical and predictive studies, and information exchange.

13. Humpback and Blue Whale Population Assessment and Photo Identification.

- Lead Organizations: Cascadia Research Collective and NMFS.
- Contributing Organizations: GFNMS, CBNMS, MBNMS, Channel Islands National Marine Sanctuary, et al.
- Management and Research Objectives Met by this Project: Baseline studies, monitoring studies, analytical and predictive studies, and information exchange.

14. Photogrammetry and Population Monitoring of Steller Sea Lions.

- Lead Organization: NMFS.
- Contributing Organizations: UC, Santa Cruz, GFNMS, PRBO, USFWS, et al.
- Management and Research Objectives Met by this Project: Baseline studies, monitoring studies, and analytical and predictive studies.

CHAPTER 171

Surfrider Foundation Blue Water Task Force

Citizen Water Quality Monitoring And Education

By Gordon Labedz, M.D.

There are two great movements represented in this conference, the environmental movement and the conservation movement. The conservation movement started at the turn of century, perhaps the benchmark was when John Muir started the Sierra Club in 1895. It was a movement to preserve wild places for the recreational values they provided for humans. The environmental movement is quite different, its symbolic birth date was Earth Day, 1970. The environmental movement organized to protect human health. We deal with issues like garbage, sludge, transportation, food choices, air pollution, water pollution and recycling.

The Environmental Movement

In 1970, the environmental movement was a grass roots phenomenon. The meetings took place in living rooms and community centers. The slogan was, "Think globally. Act locally!" National environmental groups grew and prospered during this period. Their budgets swelled and they hired staff. The national groups became influential in politics and many environmental laws were passed. The air started to clear up. The water wasn't as dirty. Things got better.

"A Decade of "Compromise"

During the eighties, the large national environmental groups grew complacent. Compromise was the word of the day. Grass roots activism became an embarrassment to "environmentally friendly" politicians. For example, Greenpeace was a powerful group of media savvy activists with legendary and almost suicidal courage, putting themselves between whales and harpoon shooting whale ships. During the eighties it transformed into a huge, bureaucratic, fund-raising machine. Another example, The Sierra Club hired Doug Wheeler as executive director. Wheeler now works for Pete Wilson who recently tried to dismantle the Coastal Commission with his friends in the Assembly.

The Surfrider Foundation was started in 1985. It followed the grass roots model from the first day of its existence. The twentieth anniversary of Earth Day, 1990, marked another transformation of the environmental movement. A decade of compromise gave us Ronald Reagan and James Watt environmentalism. "You've seen one redwood, you've seen them all," Reagan proclaimed. Local people became active again.

1990 was the year the Surfrider Foundation started the Blue Water Task Force. We had a group of people who were passionately connected to the coast, but they were largely uneducated as to its problems. Our members were very aware that coastal water quality had declined, but they weren't sure of the whys or hows.

Our members felt that they were becoming ill from the water. We initiated a water testing program that was designed to evaluate human health risks from swimming and surfing in polluted water. We found an easy-to-use bacteria monitoring kit and we encouraged all our local chapters (now 35 in the U.S.) to check their favorite beaches and all the flowing creeks, rivers and storm drains. Creeks and rivermouths are popular surfing areas where well-shaped waves can be found.

The water testing program taught our activists that the problem with our beach water quality was urban run off. Thus sparked the quest for solutions. This launched our "storm drain stenciling program". Surfrider activists stenciled signs on street gutters to educate people that all street gutters (on the west coast) were connected directly to the beach. This created an army of activists that became educators for non point source pollution.

Conservation and Environmentalism Unified

An interesting thing emerged. The conservation movement and the environmental movement started to merge into one. Surfrider Foundation activists were finding that the reasons we had so much urban run off pollution were from things that happened far inland. Urban sprawl, so-called flood control, the filling of wetlands, the paving of our rivers and streams, the lack of forests, the lack of natural open space, cattle grazing, etc. All of sudden Surfrider activists became conservationists. The natural environment was the healthiest for humans. Nature was our best water purifier.

All over the country, our grass roots, community activists are fighting to preserve or restore our natural coastal areas. Some examples: The Santa Barbara chapter has fought a battle to save the natural coast of Gaviota from development; the Ventura chapter has become involved in watershed management and protection of the Santa Clara River; the Malibu chapter is interested in restoring Malibu Creek and Malibu lagoon; the South Bay chapter wants to restore Ballona wetlands; the Huntington Beach chapter helped save and restore 900 acres of wetlands from development in Bolsa Chica; the list goes on.

The Surfrider Foundation has become a model of an effective, national organization. So much so that the president of the Sierra Club has sought our advice on our grass roots empowerment programs. Educating local activists and orienting them on preserving our coasts has been our mission. We have evolved into, for our budget, one of the most effective environmental/conservation organizations in the U.S.

Gordon Labedz is a full time Family Physician at Kaiser-Permanente. He chairs the Surfrider Foundation Environmental Issues Team.

CHAPTER 172

Citizen Volunteerism in Coastal Education and Management Programs

Rachel T. Saunders[1]

The direct engagement of citizens in the protection of marine resources and the decisions that affect them is essential to good environmental stewardship. Citizen involvement is also key to the success of many government initiatives focused on marine resource management. Without public support and involvement, protection of marine habitats and wildlife are dependent on ever-shrinking government budgets. More often than not, public participation is also necessary to create the political will to carry out these much needed initiatives.

In 1995 the Center for Marine Conservation (CMC) launched an innovative volunteer-based outreach program known as **BAY NET** (the Monterey Bay National Marine Sanctuary Volunteer Network) where trained shore front and on-the-water volunteer educators act as on-site interpreters and sanctuary "ambassadors." The ultimate goal is to establish a cooperative sanctuary-wide regional network of citizen volunteers to provide education, field observation, and assistance in the promotion and enforcement of sanctuary policies.

BAY NET is one example of the direct and positive contribution the public can make. To link citizens with other hands-on volunteer-based programs in communities adjacent to the sanctuary, CMC has also produced a *Citizens Stewardship Guide*. These programs not only invite the public to enjoy the sea around them, but also to appreciate, understand and take care of it. It is a responsibility we all must share.

---

[1] Director, Pacific Ecosystems Protection Program, Center for Marine Conservation, 580 Market Street, Suite 550, San Francisco, CA 94104

CHAPTER 173

Advancing Pollution Prevention (P2) Through Advocacy and Education: A Combined Approach

Libby Lucas[1]

The Environmental Health Coalition (EHC) is a 17 year old community-based, non-profit organization. EHC is the only organization in the San Diego region that works to protect people and the environment by promoting the reduction of the use of toxic chemicals. At the foundation of all our work is **pollution prevention** (P2). Herein we discuss the need for broad application of P2 and describe EHC's current P2 efforts.

What is P2? P2 is also called waste prevention, toxics use reduction, or source reduction. Simply put, and consistent with federal and state law, P2 means reducing the volume of hazardous materials coming in the door, whether at home, work, or school while maintaining the same level of production, service or activity. It involves the use of materials, or practices that reduce or eliminate the generation of pollutants or wastes, thereby reducing the amount of any hazardous substance, pollutant or contaminant entering any waste stream or otherwise released into the environment prior to recycling, treatment or disposal.

What are the benefits of P2? The benefits of P2 are several and far-reaching. P2:

- averts adverse impacts on the environment and people throughout the life-cycle of a product or material
- is the most common sense, effective and least costly way to reduce these impacts
- reduces/eliminates direct net costs associated with toxic chemicals
- reduces/eliminates need for or/and indirect costs associated with: monitoring, control, treatment, cleanup, lost worker productivity, health care costs
- is a SOLUTION, not a band-aid
- provides a legitimate reason for regulatory relief
- is sustainable.

[1] Director, Watershed Protection Program; Environmental Health Coalition (619 235-0281; e-mail: ehcoalition@igc.apc.org)

Goals of Clean Water Act (CWA). In 1972, the CWA called for the elimination of the discharge of pollutants to navigable waters nationwide by 1985 and for the all such waters to be fishable and swimmable by 1983. Whether in the law or not, these goals are what we should be aiming for. What right do we have to contaminate the water, land or air? None. Why are we able to send a man to the moon or clone a sheep and yet we can't manage to stop polluting the environment? There is no acceptable reason.

Discharge prohibitions have been in place for years, yet we clearly have failed to reach both of these goals largely because of polluted runoff, noncompliance with discharge prohibitions, and a lack of enforcement against violations. Additionally, the approach to dealing with polluted runoff and traditional point source discharges has been for the most part reactive: monitoring, more studies, structural best management practices (BMPs), end-of-pipe treatment and controls. We need to ask "how can we justify continuing putting so many resources into such reactive endeavors, without dedicating commensurate or greater resources concurrently to the SOLUTIONs to the problem?"

There is already ample pier-reviewed scientific evidence that clearly indicates that we must err on the side of caution. Contaminants in concentrations that are so low that they cannot be measured or even detected using standard analytical procedures have been shown over and over again to be persistent in the environment and to concentrate in the tissue and accumulate exponentially as they move up the food chain. The testing and review done by manufacturers and government regulatory agencies have focused largely on whether a chemical might cause cancer. But, science indicates that persistent and bio-accumulative toxins wreak havoc in many other often more nefarious and far-reaching ways -- causing population declines, reproductive effects, tumors, immune suppression, and behavioral changes.

It is time to move beyond the traditional end-of-pipe pollution controls that are aimed to reduce discharges to a single medium. It is time to move beyond the reactive mode; to move on to the integration into laws, policies, permits and other regulatory mechanisms the visionary, proactive, and the most practical and economical approach of P2 to address the loading of toxic chemicals into the environment, including the bays, lagoons, rivers and oceans.

EHC's current efforts in P2. EHC's efforts to advance P2 are multi-faceted. But, all of them involve education in the sense that we inform the people with whom we work about P2. This includes the general public, students, city staff, agency staff, legislators, and employees of industry. Obviously, it's necessary for people to understand what P2 is and what its benefits are before they will take actions that will bring it about. So, education is the first step in effecting P2, the common sense solution to the loading of toxic chemicals into coastal waters. Here we describe EHC's current P2 efforts.

Watershed Education: In the more traditional vein of education, EHC has conducted a watershed education program for 6 years in a 25 square-mile sub-watershed of San Diego Bay, with funding primarily from US EPA. As just one aspect of this year's program, entitled *Student-To-Student*, we hired 19 high school students, provided them with 90 hours of instruction primarily on polluted runoff and how to give presentations about it. P2 was a theme throughout the training. So, when these student interns gave 45 50-minute presentations to over 1,500 3rd-6th graders in October, the elementary school students heard their role models telling them about the importance of things like:

- using bugs as natural pesticides instead of chemical pesticides (like ladybugs to keep aphids under control);
- using non-toxic antifreeze with propylene glycol instead of the highly toxic ethylene glycol (CA. condors have died from drinking antifreeze leaked from vehicles);
- recycling used oil and using re-refined oil to decrease our dependence on virgin oil (just 6 tablespoons of waste oil can contaminate 25,000 gallons of water, the capacity of a standard residential swimming pool);
- using alternatives to the family car to get around (mass transit, walking, biking, carpooling); and
- using water-based paint instead of oil-based paint.

It is gratifying that interviews conducted in March for a radio show with the student interns and the younger students who have experienced EHC's watershed education program both this year and years past demonstrated their understanding for the need to use non-toxic products.

Municipal P2:

1. First, EHC has participated in a local effort to integrate the consideration of protecting water quality and beneficial uses into the planning and environmental review process for new development and redevelopment. The San Diego Association of Government's (SANDAG) Regional Growth Management Strategy (RGMS) identifies actions that jurisdictions should take to protect and improve the quality of life in the San Diego Region. If implemented, the actions identified in the strategy should allow for growth in the region without compromising quality of life.

About 1½ years ago the Regional Water Quality Control Board and SANDAG launched a process to address the fact that the RGMS does not contain language related to water quality. EHC's participation in the development of the RGMS' *Water Quality Element* (WQE) was driven by our desire for regional compliance with the regulatory requirements[2] related to nonpoint source pollution from construction and post-construction activities. Our specific interest has been to make sure that the WQE:

- establishes P2 as a priority over pollution control, treatment or disposal;
- establishes P2 as fundamental to the state's watershed management approach; and
- institutionalizes P2 into the planning, environmental review, construction, and post-construction processes for development.

As the draft WQE goes through the process of review and approval, we will work with citizens to cultivate support for it.

2. The second of EHC's current municipal P2 ventures is the Municipal P2 Project. This work began in 1993 with funding from the Jessie Smith Noyes Foundation. If you are wondering about the need for such a project, consider that cities use a tremendous volume of toxic and hazardous materials. For instance, the City of San Diego with 1 million people, use over 4.2 million pounds of toxic or hazardous materials annually. The City of Chula Vista, home to 132,000 residents, uses approximately 200,000 pounds annually (with gasoline it's 2.2 million pounds). In Chula Vista, these are largely paints, pesticides, gasoline, pool chemicals, and miscellaneous other materials. The potential is high for these materials to become pollutants in the environment. One indication of this is the fact that toxic chemical spills or accidents occur in the US more than 20 times a day (according to a study released on 12/4/1996 by the US PIRG and the National Environmental Law Center).

   We chose to work with Chula Vista because:

   - the City had already demonstrated environmental leadership;
   - the City is smaller than San Diego and therefore more conducive to a pilot project of this scale and intent; and
   - the City is adjacent to San Diego Bay.

[2] These are the requirements under the CWA municipal and construction storm water permits and §6217 of the Coastal Zone Act Reauthorization Amendments of 1990.

The latter points to the pertinence of this work to the coastal environment. Obviously, as for any activities in the watershed of San Diego Bay, pollutants released to the environment by the City's activities will eventually drain into San Diego Bay untreated.

This project started with an audit of the City's use of hazardous materials. We then proceeded to consult with staff in the departments of the City using these materials to determine alternatives for them and to plan for field testing.

Concurrently, EHC developed a Municipal P2 Policy which the City Council adopted in August of 1996, memorializing the City's belief that *Local governments have the opportunity and obligation to act as regional leaders in pollution prevention.* The City has already made several major changes to use safer substitutes or different processes in many operations. For example, the City has:

- instituted integrated pest management
- replaced the lead-based paint used in the divider strips on the streets with non-leaded water-based paints
- at the City's two pools, replaced chlorine gas with chlorine bleach, and replaced muriatic acid with carbon dioxide.

Based on this work and previous experience, EHC has determined that the top 5 targets for municipal P2 are, in no special order:

- volatile organic compounds
- toxic pesticides
- halogenated organic compounds
- toxic gases, and
- heavy metals.

And, based on this project, previous experience, and additional research, EHC has developed a manual entitled *Toxic Turnaround.* It provides step-by-step guidance to reducing pollution and is for use by local government employees,and managers of cities, counties, school districts, other public agencies and for environmental and community organizations. The manual will be available in August of '97.

P2 at the Regional Water Quality Control Board: While we have advocated for P2 at the RWQCB since the inception of our Clean Bay Campaign for San Diego Bay in 1987, it was in 1994 that we made our first concerted attempt to integrate P2 into the fabric of the agency's regulatory activities. This was during the update in 1994 of the Basin Plan, which is the water quality control plan designed to preserve and enhance water quality and protect the beneficial uses of all the region's waters. We requested that the update incorporate a P2 Policy into the Plan. In September of that year we lost that fight, though the section of the Basin Plan on urban runoff emphasizes that P2 is the way to achieve compliance with the objectives of the urban runoff permits. Since then, because of our advocacy and staff's growing understanding of the full implications of P2, there have been some victories along the way, most involving language in NPDES permits. They include:

- the May 1995 draft of the San Diego municipal urban runoff permit which includes an emphasis on P2 in the requirements for education of municipal employees, business, and the general public -- this permit has not yet come before the Board for consideration;
- the general permit for boatyards adopted in October of 1995 requires that they *"identify any techniques and/or policies to eliminate or reduce the generation of pollutants;"*
- the SDG&E permit adopted in November, 1996 requires the discharger to submit a written report that identifies and evaluates activities including P2 *"which could prevent, or minimize the potential for, the discharge of pollutants and wastes..... to San Diego Bay;*
- just this month the RWQCB adopted an urban runoff permit for Caltrans that encourages Caltrans' maximum use of P2. Additionally, the permit requires that Caltrans *"shall consider the use and/or promotion of: integrated pest management......[and] the promotion of zero-emission vehicles, and electric vehicle charging stations."*

The P2 language in these permits represents a start. The RWQCB is prohibited from prescribing specific ways for dischargers to comply with the discharge prohibitions in their permits. The RWQCB can, however, through enforcement actions, require violators to apply P2. But, staff has to be aware of P2 to do that. The more that P2 is in the regulators' and dischargers' consciousnesses, the better chance it has of becoming a reality. We are also advocating for P2 in the context of watershed management.

P2 and Industry: In May, 1993 EHC commenced litigation to achieve P2 and protection of a small Latino community near the waterfront. The San Diego Unified School District joined the suit against the Port District over their plans to fumigate imported fruit with the gas methyl bromide at a dockside cold storage facility. The used gas was to be vented directly into the downwind community; a screening health risk assessment showed a potential acute health hazard to residents and school children.

Methyl bromide is a potent reproductive toxin, a suspected carcinogen, affects the central nervous system and respiratory systems, and is an ozone depletor.

The problem is that the USDA requires methyl bromide for certain imports, so material substitution or process change were not feasible P2 options. We reached a settlement with the Port which required them to install a methyl bromide recapture/recycle unit on the facility. No off-the-shelf technology existed; an experimental unit has been installed and received interim approval to operate. When fully permitted and operational if ever, it will reduce the volume of methyl bromide needed by over 26,000 lbs. annually, a 75% reduction.

The litigation on this continues. The recapture unit is not working but the Port continues to fumigate. Residents in both the surrounding community and in Coronado have been very closely involved in this struggle to protect their health and, in fact, have been crucial to the process.

Legislative efforts: EHC has really just begun to work on these efforts with the advent of the bills related to coastal protection. Our ideas for effecting P2 through legislation include:

- formalizing its integration into the State's watershed management efforts;
- requiring its integration into permit writing and enforcement actions;
- the development and implementation of a program to train the RWQCBs' and SWRCB's staffs and appropriate city staffs about P2;
- providing a P2 coordinator/expert to the RWQCB
- requiring that municipalities and counties make a transition to integrated pest management by a certain date;
- requiring that School Districts do the same;
- mandating a shift of funding priorities from reactive activities to P2 (as in the development of electric vehicle technology using fuel cells, not the conventional batteries);
- mandating a phase-out of biocidal anti-fouling paint for recreational boat hulls.[3]

---

[3] This paint is a major source of copper in San Diego Bay and likely for other waterbodies with recreational boats. This ban/phase-out would need some work because one of the challenges in making the transition from biocide-based anti-fouling paints to non-biocidal paints is that the latter currently available are very slick and many boatyards are not equipped to handle the hulls with the slick paints on them. To address this, a loan or grant program could provide seed money (or more) for boatyards to retrofit their facilities.

Conclusion. The Governor's *California's Coastal Resources: An Agenda For The Future* calls for economic sustainability. Economic sustainability is impossible without environmental sustainability. P2 is the most common sense and cost effective SOLUTION to achieving coastal protection from toxic pollutants. We challenge our legislators to take a bold step, to push P2 into the forefront, to go beyond lip-service about the need to protect the coastal environment and take definitive action to effect P2. EHC's hope is to move P2 into the mainstream approach to protecting the coastal environment and its inhabitants, including people, from toxic chemicals.

****

CHAPTER 174

THE STATE OF OUR COASTS

Stephen P. Leatherman

Abstract

There have been few attempts to survey the U.S. coasts in order to understand the extent and magnitude of the problems that are being faced by coastal zone managers The last national assessment of shore erosion and its economic implications was undertaken by the U.S. Army Corps of Engineers in 1971. Individual researchers have compiled more up-to-date information on erosional trends, but this information has not been uniform in quality nor has it been used to address important public policy issues, such as beach nourishment. FEAM has produced a series of Flood Insurance Rate Maps (FIRMs) that depict the V-zones and 100 year flood levels, but the topographic data were acquired several decades ago without updates, and beaches and dunes are very dynamic landforms. For areas that are undergoing high rates of erosion, these floods maps are especially dated; for example, some houses along the Outer Banks of North Carolina rated as in safe (non-flood zones) now stand at the water's edge. Coastal pollution is another area where national standards and monitoring are lacking. Each year the Natural Resources Defense Council (NRDC) puts out a report on beach quality as obtained from state data on swimming closures, but the data are incomplete. Finally, coastal areas are continuing to urbanize at a rapid rate, but the only information available is a historic compilation from the U.S. Census data and extrapolation of those rates of growth into the future by NOAA. While this document has helped fill a void, more analysis is needed to truly understand how we have arrived at the present situation and where we are headed in the future. This paper will review the above issues so that we can formulate a national approach to better understand our number one coastal resource for recreational purposes--the oceanic beach.

Stephen P. Leatherman, Laboratory for Coastal Research, University of Maryland, College Park, Maryland

CHAPTER 175

Implications for California of the Marine Board Study on Beach Nourishment and Shoreline Protection

Richard J. Seymour*

Abstract

The Marine Board of the National Research Council undertook a study, published in 1995, of the state of the art in shoreline protection with emphasis on the applicability of beach renourishment as a mechanism for accomplishing this function. In addition to providing a thorough review of the methodologies for the design, construction and performance analysis of beach nourishment -- including environmental impact assessments -- the study also included a number of findings related to the economics of the process. The study committee also made a number of policy recommendations. Some of these were directed primarily to Federal Agencies such as the Corps of Engineers and the Federal Emergency Management Agency. Because of regional differences between coasts, some of their findings do not apply to west coast shorelines. However a number of significant recommendations are applicable to California's beaches and of interest and value to local and state agencies involved in the permitting, financing and -- in some cases -- the construction of beach nourishment projects. The paper concentrates on those findings of the study which are of particular interest to California, although many of these are universally applicable. The study findings include a number of explicit recommendations for regional planning, building and maintaining public support for projects, the proper role for hard structures as adjuncts to nourishment, keeping the stake holders informed, assessing the economic and engineering performance of beach nourishment projects and the significance of planning for multiple renourishments over the lifetime of a project. Emphasis is placed upon the identification and qualification of adequate sources of nourishment sand and upon the environmental impacts of the removal of this material.

---

*Center for Coastal Studies, Scripps Institution of Oceanography, University of California San Diego, La Jolla, CA 92093-0214

Introduction

The National Research Council (NRC) is the operating organization for the National Academies of Science and Engineering. The NRC conducts studies for the federal government on a broad range of topics. The standing board of the NRC charged with providing advice to the federal government on engineering and technology issues related to the ocean is the Marine Board. The Marine Board was asked by the U.S. Army Corps of Engineers (USACE), the National Oceanic and Atmospheric Administration (NOAA), the U.S. Geological Survey (USGS) and the Federal Emergency Management Agency (FEMA) to provide an improved technical basis for decsiion making related to shoreline protection through exploring the engineering, environmental, economic, and public policy aspects of beach nourishment. The Marine Board nominated an eleven-person committee selected for their expertise and wide range of experience and viewpoints.

The Committee on Beach Nourishment and Protection held a series of meetings, utilized a number of fact-finding methodologies and drafted a final report. This report was then subjected to a rigorous external and internal review by the NRC to assure, balance, fairness and completeness. It was approved and published in late 1995 as "Beach Nourishment and Protection" (Marine Board, 1995.) In addition to the findings, conclusions and recommendations of the committee, the book contains several extensive appendices which provide a valuable assessment of the state of good practice in beach nourishment planning, performance prediction and assessment, design, economics, construction, and both environmental and physical monitoring.

Although many of the recommendations are directed to specific agencies in the federal government, a number of them are generic in character and can provide valuable advice and insights to state and local agencies as well. In particular, given the trends towards reduced federal participation in public works projects, non-federal bodies will be taking stronger roles in the planning and execution of beach nourishment projects. There are, of course, dramatic regional differences in characteristics among the various coasts, such that California shares on a subset of the problems dealt with in this report. The following treatment identifies those findings of the Committee on Beach Nourishment and Protection which are most critical to California.

Committee Findings, Conclusions and Recommendations

Basic Finding

The committee found that beach nourishment is a viable engineering alternative for shore protection, that it is the principal technique for beach restoration, and that it is suitable for some, but not all eroding beaches. The limit of applicability was found to be the rate of background (pre-nourishment) erosion. In almost every case, the beaches of California have very low background erosion rates because many of the beaches are underlain by rock and backed by cliffs. Therefore, under the conditions described in the report, eroding California beaches would almost universally be candidates for nourishment projects.

The report provided further caveats for assuring success in beach nourishment projects:

1. Local erosion processes must be understood. In California, this would include the effects of flood control measures and the role of cliff erosion in sediment supply.

2. Design uncertainties must be realistically accounted for. For California beaches, those uncertainties include sediment supply rates (because of river flow variations) as well as the highly variable wave climate.

3. State-of-the -art engineering must be employed.

Associated Hard Structures

The committee recommended that state and local agencies modify prescriptive regulations to allow the use of fixed structures in conjunction with beach nourishment where project performance would be significantly improved. In California, one result of such a change in regulatory policy would be to stop discouraging well-designed and placed seawalls at the back of the beach. A second effect might be the consideration of constructing artificial headlands to divide long, unbroken stretches of sandy coastline (e.g., Oceanside to La Jolla) into a series of short, stable pocket beaches.

Severe restrictions are recommended on the use of hard structures. Any structure that will interfere with the longshore transport of sand should be filled to capacity from sources external to the project after construction. If a project is abandoned (nourishment not continued) the structures should be required to be removed if they will cause environmental degradation in the absence of a viable nourishment program.

Public Information and Involvement

The committee found that erroneous expectations about beach nourishment has caused the perception of failure in some projects. Therefore, a strong series of recommendations were made to the project sponsors to establish, as integral parts of the beach nourishment project, public information and involvement programs starting with the design phase and continuing through subsequent maintenance renourishments.

The local citizens who are most closely identified with bearing the cost of the project should understand and anticipate certain important attributes.

1. Beach nourishment is a continuing process and periodic maintenance and renourishment will be required.

2. Typical construction methods result in unnaturally wide and steep beaches initially. These eventually reform by natural wave-driven processes to the stable and narrower form of a natural beach.

3. Sand loss to erosion may not be uniform over the whole project and may be significantly intensified at some locations.

Sources of Sand

The report identifies four significant recommendations concerning the source of sand for nourishment:

1. Sufficient sand should be identified at the initiation of the project to provide for all of the planned renourishment events over the life of the project, which is typically fifty years.

2. Sources for emergency repair operations should be identified and permits obtained in advance to allow rapid response following major erosion events.

3. Where there are sand sources within the project or restrictions on longshore transport, careful consideration should be given to the needs for bypassing or back-passing sand.

4. Source areas are the locations where there is the greatest likelihood for damage to the environment. Emphasis should be given to these areas in planning and executing the project.

Contingency Planning

The committee found that planning for contingencies was often overlooked in beach nourishment projects. Three major recommendations were made:

1. Contingency plans for repairing severe storm damage should be an integral part of the project.

2. The sponsors should investigate methods for expediting the procurement procedures for dredging services or other pacing elements of repair activities.

3. The beach configuration should be restored as quickly as possible following a severe storm (or season) to protect against subsequent damage.

Wave Measurement

The report included a recommendation that the collection, analysis and dissemination of directional wave data be a part of every major beach nourishment program.

Monitoring Projects

The report finds that funding monitoring programs is typically unpopular with local project sponsors. Further, monitoring programs are often inappropriate to the project. The recommendations on monitoring are:

1. Monitoring should be utilized to evaluate changes in the physical and environments conditions at the project.
2. The scope of the program should be appropriate to the size of the project.
3. Monitoring effort should be strictly limited to those parameters that will be used to make project-related decisions.
4. Data should be analyzed in a timely manner to support management decisions.

Relationship to Channel Dredging

The committee found that there was insufficient coordination between navigation projects and beach nourishment needs and that beach-quality sand was often inappropriately dumped in deep water when it would have provided significant value to local beaches. The Home Porting Program in San Diego Harbor, scheduled to begin in Summer 1997, is an excellent example of effective coordination. A very substantial amount of beach grade material, which would characteristically have been disposed of by ocean dumping, will be used to nourish eroded beaches within San Diego County. The report further recommends that the federal share of beach nourishment projects reflect the impacts of nearby navigational projects even though they may not be the primary cause of erosion.

Regional Planning

The committee recommended that planning for beach nourishment and protection be undertaken on a regional basis including together shoreline sections with common problem and solutions regardless of arbitrary political boundaries. California already has an outstanding example of the benefits of regional planning in the very successful San Diego Region Shoreline Preservation Strategy (SANDAG, 1993.)

APPENDIX

References

Marine Board, 1995. Beach Nourishment and Protection, National Research Council, National Academy Press, Washington, D.C., 334 p.

SANDAG, 1993. Shoreline Preservation Strategy for the San Diego Region, San Diego, CA, San Diego Association of Governments.

CHAPTER 176

BEACH NOURISHMENT IN THE 21ST CENTURY

Lesley Ewing[1]

Beach nourishment is a valuable activity for shoreline protection. Nourishment can reverse the loss of beach area resulting from erosion and protect the back beach development, provide ongoing recreational opportunities and provide some assurance that future Californians can enjoy sand beaches. As such, beach nourishment is a powerful tool in a local or regional shoreline management program.

Historically, beach nourishment has been a means to dispose of excess excavation or dredge material. More recently, some local or regional governments have begun to look for excess beach quality material and incorporate it into their shoreline management program. Such programs should become more routine as beach quality sand becomes a more scarce and valuable commodity. Just as we now find it unacceptable to dispose of beach quality material in landfills or through deep ocean disposal, we should find it unacceptable to place sand in the littoral zone without a management plan or understanding of transport characteristics and suitability of the site for nourishment.

The 1995 Marine Board report on beach nourishment (Marine Board, 1995) emphasizes that a beach nourishment <u>program</u> is a plan for conducting a long-term series of beach nourishment projects for the development and maintenance of a beach at a specified site, including technical studies, financial plans and identification of funding and sand sources for the duration of the program. Beach nourishment should switch from "one-time projects" to developed, long-term local or regional programs. As much as possible, future beach nourishment efforts should:

- *be targeted for areas with well-suited coastal conditions*
- *be placed in areas with good baseline data on existing beach conditions*
- *provide for post-nourishment monitoring*
- *provide for long-term maintenance needs, including identified funding sources and structural "fixes" to improve long-term beach retention.*

---

[1] Associate Civil Engineer with the California Coastal Commission, 45 Fremont Street Suite 2000, San Francisco, California 94105.

NOTE: The views expressed in the article are those of the author and do not necessarily reflect those of the California Coastal Commission.

BACKGROUND

With the approach of the new millennium -- ignoring that apparently there is some dispute about when that will begin -- there is a tendency to think about how this new century will be different from the one we are in now. There will be changes to some obvious things like tele-communications, the internet, and new car designs. But there will also be changes in how we will live and treat our environment -- the natural resources, urban resources and infrastructure which make up the twentieth century environment, and how, in the coming century, we will recreate and spend our free time. In this examination of the future, in combination with the theme of this conference, the condition of California's beaches in the 21st century and the future of beach nourishment must also be given some thought.

Before we can speculate on future changes to beach nourishment, we need to have a decent idea about the historic and current approaches to beach nourishment and how it has been applied in California.

HISTORICALLY DREDGE OR EXCAVATION PROJECTS CONTROLLED NOURISHMENT EFFORTS

California has a long history of beach nourishment through much of the 20th century -- especially when beach nourishment encompasses harbor bypassing, back passing and nearshore disposal of dredge material. Since the 1930's Santa Monica Bay has been the site of numerous beach nourishment projects placing collectively over 31.6 million cubic yards of sand on the beaches or in the near shore (Leidersdorf, Hollar and Woodell, 1994). Further south, harbor and lagoon dredging has placed over 58 million cubic yards of sand on San Diego beaches or in the nearshore from 1957 to 1992 (Wiegel, R. 1994 and Vallianos, L. 1990) and another 7.5 million cubic yards have been or will be placed in San Diego beaches or in the near shore environment from the current Navy Homeporting project, the Bataquitos lagoon restoration project and other efforts. To the north in Santa Barbara and Ventura, approximately 24.5 million cubic yards of sand had been placed or by-passed by the late '80's; at Morro Bay, approximately 2.7 million cubic yards have been placed on or near beaches in the past 18 years; at Santa Cruz, approximately 4 million cubic yards of sand has been by-passed since the mid '60's; in San Francisco, 5 million cubic yards of sand placed on the beaches from '87 to '96; and until recently 600,000 cubic yards of sand were annually placed in the nearshore area at Humboldt (Wiegel, R. 1994, Vallianos, L. 1990, Commission Federal Consistency project files for Morro Bay (CD-44-93) and Humboldt Bay (CD-111-94 and CD-69-94)). The combination of all these projects ultimately leads to a lot of experience with placing sand on the beach or near shore, and with moving it to different parts of the shoreline.

The bulk of these projects have not had beach building as their primary purpose. Most of these projects have been undertaken as opportunistic beach nourishment projects, or projects which take advantage of "opportunistic sand". Opportunistic sand is material which can be used for beach nourishment, but the primary reason this material is available is that it is excess material from another project and the beach is a convenient disposal location. The method is similar to someone placing a sign on a pile of dirt saying "Free Topsoil", knowing that in a short time someone who can use some extra dirt would come by with a pick-up and take away the dirt pile.

A similar general "sign on the sand pile" program has functioned quite well for beach nourishment in California. Someone digs a huge hole on or near the coast -- for a new or deepened harbor, for a building foundation, for a landfill, for additional flood capacity, or some other reason. They find that the easiest or best way to get rid their excess material is to put it on or near the beach. Such activities usually have a more formal arrangement that a "Free Topsoil" or, in the beach nourishment cases, "Free Sand" signs and usually need more than a spare pick-up truck for transport; nevertheless most beach nourishment in California has been from "opportunistic sand".

There has been a massive amount of construction along the California coast throughout the 20th century, and, because of this, there has been a massive amount of opportunistic sand available for beach nourishment. Since transportation has been the highest cost of these "opportunistic" beach nourishment projects, most of the opportunistic beach nourishment/sand disposal sites have been fairly close to the source sites. In general, the coastal areas nearest the excavation or dredge sites have been the ones to get the sand and the excavation or dredge project has dictated how much material will be available for placement.

Of course, there have been some exceptions to this general situation. For many years the Corps and the local government have done periodic nourishment at Surfside Sunset, with the sites for nourishment and the quantities both determined by need, in combination with availability of material. In 1991, Los Angeles County Department of Beaches and Harbors trucked sand several miles to Cabrillo City Beach as a targeted nourishment program. The current Navy Homeporting Project will be placing sand throughout San Diego County on or near targeted beaches, and while the total quantity of sand is limited by the channel and harbor deepening project, the nourishment locations and quantities have been allocated by a general determination of need.

BEACH NOURISHMENT IN THE FUTURE

New Ideas for "Opportunistic Sand Sources"

There will continue to be sources of "opportunistic sand"; however, the quantities are not likely to compare with the past 60 or so years. Many of the developable areas along the coast have already been used for construction. It is very unlikely that California will experience the same scale and amount of coastal construction that occurred during and after World War II. For example, most harbor construction is likely to be the maintenance and deepening of existing harbors rather than the wholesale construction of another five or six new harbors.

A possible new source of opportunistic sand which could replace the harbor construction of the 20th century could be the maintenance and renewal of flood control structures. We have enormous quantities of sand trapped behind dams and in sedimentation basins and these structures must be cleaned out to remain functional. It probably seems logical to many coastal supporters that the sand behind dams, in flood control channels or in sediment basins near the coast would eventually have gone to the ocean and thus should be used for beach nourishment. Unfortunately, much of this sand occurs in thin layers mixed in with organic debris, silts clays and other material. The still water environment behind a dam does not provide the natural sorting and mechanical separation which occurs when this material is carried naturally by river flow to the ocean and moved through the littoral cell by waves and currents. Even if material can be removed easily from the impoundment areas, it may be difficult and costly to replicate the natural sorting processes which separate out the beach quality material from the less desirable organic and fine material. It may be costly be get this material to the coast; there may be significant environmental disruption associated with this transport; and impoundment lakes have themselves become recreational areas which would be affected by dredging activities.

Beach quality material is and will remain a valuable commodity. Whenever there are sources of beach quality material or sources of material which can economically be cleaned of organic debris, silts, clays and other fines, there very likely will be a number of parties interested in acquiring this material for uses such as site grading, construction, and others. Recently in conjunction with channel maintenance in the Santa Marguerita River, the Marine Corps sold over three-quarters of the available beach quality sand in order to pay the cost for transporting the remaining 40,000 cubic yards of sand to a beach in Oceanside (SanDAG, 1996). Beach quality material will be available in the future; it just may be a bit harder to find and there may be more competition for available material.

Changes in the Treatment of "Opportunistic Sand"

The beach community will need to actively look for sources of new beach material and will have to be vocal in asserting the need for this material to go to the beaches. One example of a program which is attempting to actively seek opportunistic sand is being started in Carlsbad (Moffatt & Nichol, 1996b, also, Jantz, S. and Webb, C. in these proceedings). The City of Carlsbad is establishing a program to take advantage of inland construction projects and is attempting to get all excess, appropriate material from these construction projects to the beach. A critical concern for the City has been to determine what inland construction material is appropriate for beach use, or what tests should be done to test for appropriateness. The City wants to be sure that the nourished beaches will contain relatively clean sand or sand and fines; they do not want to place material on the beach which is contaminated by oil, agricultural chemicals, biological waste, etc. Once the City has a screening process developed to determine that the material is appropriate,, the City will try to take advantage of material which has not traditionally been considered for beach nourishment, but which the City can identify through its review of grading permit applications and inland construction activities. There may not be as many "Free Sand" signs in the future, but there will be available sand if the coastal communities look for it.

All of this leads to some general changes which may occur in beach nourishment in the next century. First, beach nourishment will increasingly be targeted to specific areas which are well-suited to receive and maintain nourishment. Nourishment will occur less and less at the closest site and more at the "best" close site. The two key conditions which would determine suitability would be (1) whether we can safely get the material to the beach and (2) once the sand it there, whether it will serve the intended purpose.

The two ways to get material to the beach are by land or by sea. Either way, it must be safe to the transport operators, the public and the environment. For example, in Humboldt Bay, the US Army Corps of Engineers annually dredges approximately 600,000 cubic yards of sand from the harbor. For many years the sand was placed in the nearshore environment, an area where it should have moved onto the spit. However, material movement from the site was slow, the disposal site shoaled and it became a navigation hazard. A second nearshore site was selected which was closer to shore and provided for more rapid onshore transport. This site was found to have safety problems and the new site was abandoned after two years due to concerns about barge safety and environmental degradation. The wave climate made nearshore disposal very tricky and barge operators were legitimately concerned about operating broadside to the waves. In addition, lobsters and other scavengers would congregate at the dump site to feed on the various animals in the dredge material or the animals that were suffocated by the dredge material. These scavengers would then be suffocated by the next barge load of material, attracting new scavengers and continuing the cycle. The disposal site was becoming an attractive death site for lobsters. Ultimately, nearshore disposal was abandoned in Humboldt and the Corps went to a safer, environmentally acceptable deep water site. Because of the vast quantities of sand in the Humboldt area, it was thought that the loss of this material would not have any noticeable effect on the littoral sand supply. The Humboldt spits are being carefully monitored for irreversible sand losses, so the disposal site may change again if problems are identified. For now, the deep water site is being used successfully (Moffatt & Nichol, 1996a).

Humboldt is very different from much of southern California where, in many locations, 600,000 cubic yards of sand could be the entire volume of material in the littoral cell. For these areas, the idea of losing 600,000 cubic yards each year may seem almost criminal. Maybe in the future, it will be cost effective to bring this material south to San Diego or Santa Monica, rather than dispose of it in deep water. As we begin to look for new sources of "opportunistic sand" we may turn to some very unlikely areas such as the Humboldt Bay, or even the Arizona desert.

Recently the coastal communities in California have begun to treat our beaches more like a protected resource or important coastal infrastructure, and this trend is one which should continue in the future. As nourishment material becomes more valued, we will also want to make sure we are making intelligent decisions about where and how to attempt beach nourishment. Southern California loses huge amounts of sand to submarine canyons. In past nourishment efforts, we have tried to avoid beach nourishment in areas where we know the material is very likely to be carried into the canyon during the next storm season. Similarly, harbor dredging projects have tried to place the nourishment material in a location down current from the harbor, to avoid continually redredging the same material. In many cases, this has amounted to site selection by elimination -- eliminate the bad sites and use transport distance to pick amongst the remaining "not-bad" sites. In the future we will want to pick sites because they are affirmatively good sites for beach nourishment, not because they aren't bad.

BEACH INFORMATION AND PROJECT MONITORING

Monitoring is one of the only ways to determine whether a beach nourishment project has accomplished its intended purpose. In the future, beach nourishment programs should be monitored -- not to collect esoteric data, but to anticipate changes and provide answers to questions directed at the functioning of that particular beach. The selection of good beach nourishment sites will require information about the beach sites -- what natural features exist to hold sand in place; whether there is sufficient waves energy to move material on to the beach if it is placed in the nearshore; if we can build a beach there, can people get to it and use it. Once a beach is built, the coastal community will want information on how it performs and what were the wave conditions to which it was exposed.

In Maryland, the state and the Corps of Engineers recently undertook a huge beach nourishment program. They built nice wide beaches and then the coast was hit by some of the worst storms of the century. That was it for the wide beaches and initially people were saying that the project was a failure. The Corps and the state however, had data on beach changes and storm conditions and were able to predict the amount of property damage that might have occurred if the "unsuccessful" beach had not been in place. The data on the project turned this supposed "failure" into a success. People were not talking about the millions of dollars that were thrown into the ocean on that "stupid" beach nourishment project. Instead, regular beach users whose normal beach conversation was the appropriate Skin Protection Factor to use while playing volleyball were talking about winter and summer beach profiles, extreme storm events, protective beaches, sacrificial beaches, and the benefits of beach nourishment. And people were willing to continue funding beach programs because there was information about the beach and how it would function if nourished.

The Ocean City Maryland project was monitored and this monitoring made it possible to distinguish between success and failure. Closer to home, Oceanside California in north San Diego county is participating right now in a sand for trash program. Trash from San Diego is being carried by truck to a landfill in La Paz County, Arizona. Rather than dead-haul the trucks back to San Diego, the excess sand from the landfill in La Paz is being trucked to Oceanside for an experimental nourishment program. That nourishment effort is being monitored to see if the desert sands are acceptable beach material (Coastal

Environments, 1996). Monitoring is really the only way to determine if a project is successful and if it should be continued, repeated, changed, modified, or abandoned. A business needs some type of business plan and follow-up financial accounting to get a bank loan; similarly coastal communities should establish an overall plan and follow-up monitoring for a beach nourishment program. The information from the plan and monitoring will be critical to future evaluations of the program, decisions by the community concerning the timing and design of future nourishment projects, as well as decisions on the benefits from continuing to invest sand resources in this area.

CONCLUSION

Beach nourishment will, in the future, become more and more an investment program -- support for that critically important coastal infrastructure we call our beaches. Many of these beaches will require long-term maintenance and a long-term financial commitment. With knowledge and data on the beach area, the financial program may emphasize the targetted needs of the specific beach area -- the addition of sand retaining structures; improved access for either nourishment or continued recreational use; scheduled periodic maintenance; exploration for innovative sources of sand. The final, but probably most critical element of beach nourishment in the 21st century, will be the recognition of our beaches as valuable resources which can be protected if there is the necessary long-term financial planning and stewardship responsibility that comes from this recognition. Future generations of Californians will be able to enjoy the many benefits which come from wide, stable beaches if there is a long-term commitment to beach nourishmetn and maintenance.

REFERENCES

Coastal Environments, Oceanside Beach Nourishment Demonstration Project, prepared for La Paz County Landfill, July 1996.

Leidersdorf, C., Hollar, R. and Woodell, G., "Human Intervention with the Beaches of Santa Monica Bay, California," *Shore and Beach*, American Shore and Beach Preservation Association, Vol. 62, No. 3, July 1994, pp. 29-38.

Marine Board, National Research Council, *Beach Nourishment and Protection*, National Academy of Sciences, Washington, DC, 1995.

Moffatt & Nichol, Engineers, Shoreline Mapping for Humboldt Entrance, prepared for Philip Williams & Associates, Ltd. under contract with US Army Corps of Engineers, August 1996a.

Moffatt & Nichol, Engineers, Final Technical Report: Opportunistic Beach Fill Criteria and Concept Design, prepared for the City of Carlsbad, M&N File: 3497, November 1996b.

SanDAG, Memo to the Shoreline Erosion Committee from the SanDAG Staff, entitled, "Update on Opportunistic Sand Projects", November, 26, 1996

Vallianos, L. "Beach and Nearshore Placement of Material Dredged from Federally Authorized Navigation Projects," US Army Corps of Engineers IWR - Policy Study, April 1990.

Wiegel, R. "Ocean Beach Nourishment on the USA Pacific Coast," *Shore and Beach*, American Shore and Beach Preservation Association, Vol. 62. No. 1, January 1994, pp. 11-36.

CHAPTER 177

WINNING SUPPORT AND FUNDING FOR BEACH MANAGEMENT

Stan Tait[1]

Abstract

Most coastal states have great difficulty in obtaining adequate funding for beach nourishment and other shore protection projects. Florida has been an exception. In the last two decades, more than 100 miles of eroded beach have been restored with a commitment to maintain each project with periodic renourishment for up to 50 years. This paper offers strategies, based on Florida's experience, to win state and local support for funding a comprehensive beach management program.

Introduction

The easiest part of beach preservation is engineering design. The hardest part is getting the money to do the job. So let's cut to the chase and talk about how to get the money.

[1]President, Florida Shore & Beach Preservation Association, 2952 Wellington Circle North, Tallahassee, FL 32308

Specifically, I am going to give you a three-step plan to get the money you need for implementing a comprehensive, statewide beach management program in your state. It is an approach that has worked well in Florida. It can work in any coastal state where beaches are important.

1. ORGANIZE A STRONG STATEWIDE ASSOCIATION OR ADVOCACY GROUP FOR BEACHES. THINK OF IT AS A STATE SIERRA CLUB FOR BEACHES.

Good things like a comprehensive, well-funded beach management program don't happen by themselves. They don't happen by wishful thinking. Someone or some organization has to provide the vision and the momentum to make things happen.

Think about it. We wouldn't have the good environmental laws we have today if it were not for groups such as the Audubon Society and Sierra Club. They carried the banner for pollution laws and endangered species laws and all the other environmental laws on the books today. Beach proponents need to do the same. It works.

The real impetus for Florida's beach management program began in 1957 when an informal group of 33 city and county government officials, university professors and consulting engineers met in Orlando to try to come up with a strategy to deal with Florida's growing beach erosion problem. They recognized that local governments, by themselves, couldn't cope with beach erosion. And they concluded that the solution was to create a state agency to combat beach erosion.

That meeting also led to the creation of an action organization to be known as the Florida Shore & Beach Preservation Association -- because the founders recognized that it would take a major lobbying effort to persuade the Florida Legislature to create a new beach agency. The new FSBPA differed from the American Shore & Beach Preservation Association in two important ways. (1) It was created as a "league of cities and counties" on beach preservation issues and only representatives of governmental agencies could serve on the Board of Directors; and (2) An important function of FSBPA was to lobby both the state and federal governments, something that ASBPA scrupulously avoided in those days.

The new FSBPA enjoyed almost instant success. Two months after its founding, FSBPA and its contingent of local government officials persuaded the 1957 Legislature to create the first state beach agency -- the State Division of Beaches and Shores. FSBPA was an organization with instant political clout because its membership represented the City of Miami Beach, the City of Fort Lauderdale, the City of Palm Beach and many other cities and counties.

Since those days, we have expanded our membership to include non-governmental members -- coastal engineers and geologists, hotel and resort owners, business leaders, university professors and just plain folks who live on the beach and want to keep it. That's not a bad nucleus for an action organization of any kind.

THE VALUE OF A PROFESSIONAL STAFF

Along the way, FSBPA learned another important lesson. It's not enough to simply charter a non-profit corporation to advocate for beaches. At some point, you need to hire a paid executive director to keep things moving. FSBPA foundered in the early days because it relied entirely on volunteers. Some did the job. Many didn't. The result was that the state's beach agency remained woefully underfunded and understaffed because it did not get the public support it needed in the Legislature.

In 1975, FSBPA hired its first executive director. His job was to rebuild the organization, lobby the Legislature and get Florida's beach program off the ground. And that's exactly what happened. Today, FSBPA has an annual budget of $500,000 and a staff of seven, including a top-notch lobbyist who is a former director of the state beach agency. It sponsors three major conferences annually, including the National Conference on Beach Preservation Technology (300 delegates) and the National Hurricane Conference (1,200 plus delegates).

2. REWRITE YOUR OWN STATE BEACH MANAGEMENT LAW TO PROVIDE A STRONG STATE BEACH AGENCY AND A STRONG STATUTORY MISSION FOR THAT AGENCY.

In fact, that's the first job of your state association as it was with FSBPA -- to persuade your legislature to do a major overhaul and strengthening of your state beach law. The idea is to elevate the status and importance and mission of the state beach program.

Why? Because the way the law is written sets the stage for project funding as well as agency staffing. Your goal should be to put beaches high in the pecking order of your state's environmental and natural resources program. And that is the real key to getting adequate funding for beach preservation projects.

I think that over the years we have developed a very good beach management law in Florida. And that has been a key to the successes we have had.

It's not accidental that Florida's law is entitled the ***Beach Management Act*** nor that it refers throughout to ***Comprehensive Beach Management***. Even the name of the state beach agency reflects the wide scope of its mission -- the BUREAU OF BEACHES AND COASTAL SYSTEMS. The agency's main mission is to preserve and restore the sandy beaches of Florida. But it does much more. Since most erosion in Florida is caused by sand loss at navigation inlets, the agency has powers to require navigation districts to bypass annually the amount of sand they block from the system. The beach agency also has an important environmental role, especially protecting endangered sea turtles and other species. In a nutshell, there are two basic parts to our very comprehensive beach law:

A. BEACH MANAGEMENT PLANNING AND CONSTRUCTION

The State Bureau of Beaches and Coastal Systems, with a staff of over 60, is responsible for planning and implementing the best management strategy for every mile of sandy beach in Florida. Authorized management strategies include but are not limited to:

- Beach nourishment
- Structures (few and far between)
- Dune reconstruction
- Beach and dune revegetation
- Environmental protection and enhancements
- Identifying land for acquisition for parks, including pristine areas.
- Inlet management, including sand bypassing and feeder beaches.
- Demonstration projects of experimental technology.
- A policy of retreat from the coast in areas of severe erosion.

By law, the Bureau is required to set spending priorities based on need as a guide to legislative appropriations. For example, the FY1997-98 budget request lists 29 projects of all kinds with a price tag of $25 million in state funding, $34 million federal and $30 million local --a work program totaling $89 million. (The Legislature actually appropriated $21 million in state funds, generating $80 million in projects for the coming year.)

B. REGULATION OF COASTAL CONSTRUCTION

FSBPA recognized early that many property owners were damaging the state's beach and dune system by building houses and hotels too close to the water. Thus an early priority was to set up within the state beach agency a coastal construction permitting program. In 1970, FSBPA persuaded the Legislature to establish the Coastal Construction Control Line program which restricts construction in the coastal area. The law does two things:

* it requires siting of new construction as far landward as practicable -- behind the beach and primary dune.

* it requires strict constructions standards, including storm surge elevations, so that new buildings have a good chance of surviving hurricanes.

This program has been highly successful in reducing hurricane damage. Hurricane Opal, which hit Northwest Florida in 1995, destroyed many old buildings. But not a single structure built under current CCCL standards was destroyed or seriously damaged.

3. **AGGRESSIVELY EDUCATE AND LOBBY THE PUBLIC AND YOUR LEGISLATURE ON THE IMPORTANCE OF YOUR BEACHES. KEEP FINE-TUNING YOUR BEACH LAW TO REFLECT CHANGING TIMES.**

Enacting a strong, comprehensive beach management law for your state is just the beginning. You've got to keep the pressure on for adequate funding. And you've got to keep an alert eye for those who would weaken your program.

Almost every session of the Florida Legislature someone introduces a bill to weaken the program. In our case, the assault is usually against the coastal construction permitting part of the beach law. And every time our lobbyists must work to kill these bills.

I want to emphasize that developing and keeping a good beach law is done incrementally over the years. In Florida's case, there were four landmark changes in our law -- all spearheaded by FSBPA -- to bring us to where we are today:

1957 -- Creation and gradual expansion of the Division of Beaches and Shores and providing authority for local government to serve as project sponsors, including a 50-50 state/local match of the non-federal share.

1970 -- Establishment of the coastal construction control line law to prevent construction on the dynamic beach and dune system and requiring strict building standards to withstand hurricanes.

1978 -- Start of major beach nourishment program statewide; designating nourishment as the preferred method of beach preservation; since then over 100 miles of beach have been restored with a commitment to keep them periodically renourished for up to 50 years.

1986 -- Enactment of the comprehensive beach management act which directs the Bureau of Beaches and Coastal Systems to identify an appropriate management strategy for every mile of sandy beach and which begins the first comprehensive effort at erosion prevention by requiring navigation districts to bypass annually an amount of sand that they block.

We've made a lot of progress in the past 50 years but we realize our job is far from over. The passage of time brings new conditions and new situations that we must respond to. The most important challenge we now face in Florida -- and nationally -- is the fact that federal funding of shore protection projects is going to decline significantly. Historically, about 50% of funding for our beach nourishment projects has come from the federal government. That may drop to 25%. It may even drop to zero, though that is unlikely.

How will we replace the lost money? Our strategy is to persuade the Florida Legislature to establish a dedicated state funding source for beach management of $35 million annually. With local matching funds, that will generate a $70 million construction program, even without federal money. The Bureau of Beaches and Coastal Systems calculates it will take $95 million annually to fully implement the beach management program.

In view of its "no tax" commitment, the 1997 Legislature did not levy new taxes for a beach trust fund. But they passed a bill stating legislative intent to appropriate $35 million annually for the beach trust fund and they set up a procedure to identify funding sources for this purpose. These will be considered by the 1998 Legislature.

CHAPTER 178

Platform Decommissioning: Chevron's Experience with the 4H State Platforms Abandonment Project Santa Barbara Channel, California

Lee Bafalon, Esq.*
Greg W. Sinclair **

Abstract

This paper describes the planning, permitting and execution of the decommissioning and removal of four oil and gas drilling and production platforms operated by Chevron U.S.A. Inc. The "4H" platforms--Hazel, Hilda, Hope and Heidi-- were located in California state waters in the Santa Barbara Channel, offshore Santa Barbara County.

The 4H platforms were installed between 1958 and 1965: platforms Hazel and Hilda, jointly owned by Chevron and Exxon, were installed in 1958 and 1960, respectively. on State Lease PRC 1824.1 to develop the Summerland Field: platforms Hope and Heidi, jointly owned by Chevron and Arco, were installed in 1965 on State Lease PRC 3150.1 to develop the Carpinteria Field. A total of 135 wells were drilled from the platforms that tapped reserves primarily form the Vaqueros and Repoetto formations. Between 1958 and 1992, when the platforms' wells were permanently shut-in, the 4H platforms produced approximately 62 million barrels of oil and 132 billion cubic feet of natural gas.

* Senior Land Representative, Chevron U.S.A. Inc., 646 County Square Drive, Ventura, CA 93003.
** Project Engineer, Chevron U.S.A. Inc., 935 Gravier Street, New Orleans, LA 70112.

CHAPTER 179

OFFSHORE FACILITY DECOMMISSIONING: ISSUES AND OPPORTUNITIES

Pete Johnson

Abstract

Offshore and nearshore California oil and gas production facilities are maturing; the decommissioning of many of these structures will occur during the next twenty years. Permitting, decommissioning, removal, disposal, or re-use of these facilities presents challenges for operators, construction contractors, agencies, other ocean user groups, and the public. The impacts of choices made during the decommissioning process on other ocean users may outlive the impacts of development. Central issues identified during recent projects are discussed, and opportunities for productive collaboration are identified.

Pete Johnson, Chief Petroleum Engineer, State Lands Commission

CHAPTER 180

Decommissioning of Offshore Oil and Gas Facilities: Contrasts Between Southern California and the Gulf of Mexico and Implications for Ecological Research.

Mark H. Carr[1], Graham E. Forrester[2] and Michael V. McGinnis[3]

Abstract

Decommissioning of offshore oil and gas production facilities is increasing at a rapid rate world-wide and in the coastal waters of southern California. Of the 30 platforms off southern California, four have just been removed from the Santa Barbara Channel (summer 1996), three more are slated to be removed in the next five years and most of the remaining platforms are predicted to be removed or abandoned during the next 20 years. Present policy for decommissioning in California requires entire removal of such structures from the ocean. However, in the Gulf of Mexico many platforms are decommissioned/abandoned at sea to create artificial reefs used extensively by recreational fisheries, some commercial fisheries and some mariculture projects. The Gulf of Mexico program suggests an alternative fate and policy for decommissioning but marked differences in geologic and oceanographic features between southern California and the Gulf have important consequences for how abandonment is approached, and particularly what kinds of environmental studies will be useful to regulatory agencies for developing and implementing decommissioning policy.

Critical to understanding the relative benefits and costs of decommissioning structures at sea is an assessment of the effects of abandoned structures on regional fish production. This assessment involves distinguishing the extent to which such structures enhance regional fish production or simply re-distribute (i.e. attract) fish from natural reefs. Determining the relative extent of "production vs. attraction" requires, among other things, knowledge of larval dispersal and survivorship and the relative demographic rates of fish populations on both artificial and natural reefs. Furthermore, a complete assessment of the potential value of structures abandoned at sea requires knowledge of their effects under different oceanographic regimes. We provide examples of many of these issues to demonstrate the complexity of the problem and their implications for research.

---

[1] Coastal Research Center, Marine Science Institute, University of California, Santa Barbara, CA 93106

[2] Dept. of Biological Sciences, University of California, Los Angeles, CA 90024

[3] Ocean and Coastal Policy Center, Marine Science Institute, University of California, Santa Barbara, CA 93106

CHAPTER 181

An Analysis of the Role of Ecological Science in Offshore Continental Shelf Decommissioning Policy

Michael Vincent McGinnis, Ph.D.[1]

Abstract

In California, several offshore oil and gas production facilities in the Outer Continent Shelf (OCS) are nearing the end of, or have ended, their "economic lives". Present international, federal and state law requires that offshore structures be completed removed. Complete removal requires significant economic costs to the oil industry, and may have an impact on the rock reef fish species which have been shown to inhabit the offshore oil and gas structures. In California, the ultimate cultural and ecological impacts of decommissioning has only recently received systematic study. Based on the history of the role of ecological science in the development and implementation of environmental policy, this paper focuses on the factors which will likely influence the role of scientific information in the politics of decommissioning policy making for California's OCS oil and gas structures.

History, Contingency, Ecology and Development

Scholars have examined the political dimensions of oil and gas leasing, exploration and production associated with the Outer Continental Shelf (OCS) (Cicin-Sain and Knecht 1987; Hershman et al. 1988; Farrow 1990; Lester 1991; Lima 1994; Freudenberg and Gramling 1994; Gramling 1996). Researchers have uncovered critical differences between OCS oil and gas regions; the politics of OCS oil and gas development is based on important local, regional, legal,

[1] Assistant Research Political Scientist II, Ocean and Coastal Policy Center, and Resident Fellow, Interdisciplinary Humanities Center, UC-Santa Barbara 93106.

technological, and socio-economic considerations and factors. These factors are unlikely to be generalizable across regions, places or locations. In other words, OCS oil and gas development is contingent upon socio-cultural factors that are endemic to particular places, regions and localities and not just the presence of OCS oil and gas reserves, government's willingness to create opportunities for the development of oil and gas, or the availability of modern technologies to develop oil and gas.

The history of OCS oil and gas development varies from one region to region (Cicin-Sain and Knecht 1987; Hershman et al. 1988; Farrow 1990; Lester 1991; Lima 1994; Freudenberg and Gramling 1994; Gramling 1996). From 1896 to 1970, Lima (1994) identifies six periods of technological development of Pacific OCS oil and gas activities. Lima (1994) shows that California and local government, for example, was generally supportive of development during the early stage of offshore drilling off Santa Barbara. After 1966, however, renewed jurisdictional disputes between the state and federal government and conflicts between offshore development and competing land uses heightened community concerns over development. Since 1970, however, public support has generally declined. The decline in public support is a byproduct of the public perceptions of risk associated with development and the "pull of local control" over proposals for offshore oil and gas development. In an examination of public opinion poll data in the state, Smith and Garcia (1994) show that public support for offshore oil development has declined since 1980. This decline in support for offshore oil and gas development is exemplified by the over 15 initiatives that have been passed by counties and cities to restrict the placement of onshore OCS support facilities along California's coast. The most recent initiative was Santa Barbara County's Measure A. These various place-based initiatives affect the politics of OCS oil and gas activity.

In general, the transition from exploration/development to decommissioning of OCS oil and gas facilities marks a fundamental change in the history of development. Rather than construction and operation of oil and gas facilities (or what one Minerals Management Service employee ironically referred to as the "oil and gas ecosystem"), the oil industry is mandated by international, federal and state law to completely remove, and thereby deconstruct offshore oil and gas structures. The ecological and cultural implications of which have received little systematic consideration by scientists or policy makers. The deconstruction of oil and gas facilities is likely to be based on the political culture of various regional areas and localities. For example, the role of scientific and technological information in the process of decommissioning policy making will vary from one region to another.

California has entered a new era of OCS oil and gas activity--decommissioning of offshore oil and gas facilities is rapidly becoming an issue of general concern. The NRC (1985b) estimates that the cumulative costs for removal of all platforms in the OCS could total $2.9 billion by 2005 and $9.9 billion by 2020 while the GAO (1994) estimates the total decommissioning costs at $4.4 billion. The GAO (1994) reviewed offshore structure removal operations and concluded that the MMS needs to better understand the risk of environmental damage posed by certain decommissioning practices, such as the use of explosives as a removal technique. What is certain is that the economic costs of decommissioning on the oil industry will be significant.

With a focus on the Gulf, the National Fishing Enhancement Act of 1984 called for the development of a National Artificial Reef Plan. Under the direction of the National Marine Fisheries Service, this plan: 1) established general guidelines for artificial reef materials, siting and design; 2) identified oil and gas structures as excellent material for developing offshore reefs of opportunity; 3) charged coastal states with the primary responsibility for developing and implementing site specific artificial reef plans; 4) assigned the Army Corps of Engineers the responsibility of developing artificial reef regulations and for permitting artificial reef projects in the waters of the US; and 5) limited liability of approved reef sponsors who complied with the terms and conditions of artificial reef permits.

From 1987-1995, of the over 941 platforms removed from Louisiana and Texas waters, 90 platforms or approximately 10% were transformed into artificial reefs. The primary purpose of the National Artificial Reef Plan is to enhance the socio-economic values associated with the Gulf's sports and commercial fisheries, and is not based on the various ecological values that may be associated with these artificial structures (e.g., the preservation of biodiversity, predation or succession). Indeed, the commercial and sports fishing industry lobbied the state and federal government to create the National Artificial Reef Plan.

The human and ecological character of the Gulf of Mexico is much different than that of California. For example, the commercial fishing industry's relationship with the oil industry in California is very different from that of the Gulf experience (Freudenberg and Gramling

1994). The conflict between commercial fishing interests and oil development interests are well documented. Learning from the Gulf experience is important, but one should recognize that the history of California's diverse cultural and social systems are qualitatively and quantitatively different than those that exist in the Gulf. These differences are as much ecological and geographical as they are political and cultural. The biogeography of the Gulf is substantially different than that of the southern California bite. The Gulf has less reef habitat in comparison to that of southern California. There are also important cultural differences. As the NRC (1992: 68) states, "[T]he definitions of OCS issues and problems--and solutions to them--are as much based on those persons' perceptions, biases, culture, and experience as are those of any affected person or community". Robert Gramling (1996: 149) points out, "Because of divergent processes over time and different socioeconomic factors...they are fundamentally different... social realities" and Freudenberg and Gramling (1994: 75) show that offshore drilling evolved in a social environment characterized by a low level of environmental concern. The social reality and environmental concern of California will likely shape the politics of decommissioning policy making in different ways.

In California, there exist a number of important ecological issues that will likely influence the politics of decommissioning policy making. Rock reef fish represent $3 Billion in sports fishing values and between $2-300 million/annually in commercial landings. Preliminary evidence from Love (1997) shows that between 90 to 95% of fish near rigs are rock reef fish. Some of the particular rigs also include juvenile Bocaccio, which may be listed as a threatened species. These fish also serve an important role in the diverse ecosystems off southern California as well. Moreover Love (1997) shows that "not all platforms are created equal -- each platform has a different ecological and socioeconomic worth" (see also the essay by Carr and Forrester in this volume). Also the ultimate impact of micro-climate and global climate change on these ecosystems is unknown.

To investigate the relationship between rock reef fisheries and OCS oil and gas structures in southern California, MMS and the US Geological Survey are: 1) synthesizing offshore fish data gathered by a number of different sources for southern/central California (which will be place on the internet); 2) analyzing the overall health of genetic stocks and species diversity of fish in southern/central California; 3) exploring the distribution and abundance of fish (specifically rock fish) in relation to ecosystems and OCS oil and gas structures; and 4) exploring the deep water community dynamics and fish assemblages near rigs and natural reefs.

With these recent ecological investigations in mind, the remainder of this paper deals with two primary questions: Given the history of the use of ecological information in environmental policy, what are the problems associated with relying on scientific information in decommissioning policy making? The current policy for complete removal of OCS oil and gas structures was not developed with due ecological considerations in mind. Rather, scientific information played little if any role in decommissioning policy making. With this in mind, what are the strategies that can be employed to integrate new ecological information in decommissioning policy making?

The Role of Science

The integration of scientific information in ocean and coastal zone policy is a political and cultural issue. Scientific investigation does not take place in a political vacuum, but involves important perceptual, value-based and ethical issues. In a comprehensive analysis of the history of California coastal and ocean studies from 1945-1973, Harry Scheiber (1995) describes several factors that influence the use of scientific knowledge to inform ocean policies: 1) the dilemma of incomplete information (policy makers often act without a comprehensive understanding of a policy area); 2) the frustration of political entanglement (most policy areas include a number of government and non government interest); 3) the minefield effects of multi-level government (ocean and coastal policy making is an intergovernmental process that involves a number of different actors who may hold different interests and values); and 4) the pitfalls of native faith in science.

The politics of integrating scientific information in the OCS decision making process takes place within the context of a highly charged and contentious intergovernmental and non governmental process. This process is often resolved by litigation (Lester 1991). In such a process, there is the likelihood that "unfavorable" information will be forgotten or remain "marginalized". In addition, "good" scientific information is that which supports one's position in the decision-making situation. Access to information is a function of how well information flows within institutions, among institutions, and between institutions and their environment. This has been a major problem for the MMS, the lead administrative agency for OCS activities (Lester 1991). Under conditions of high scientific uncertainty (such as the assessment of the ecological impacts of decommission in California's OCS), effective management requires that institutions do more than merely acquire information. Information must be communicated and shared. Access to information should be made open.

In a scientifically-oriented society like that of the West, there is the real problem that to resolve a dispute, government and industry will rely on primarily on scientific information or scientists. However, this naive faith in science inevitably leads to conflict. This is an issue that I now turn to.

The Pitfalls of a Naive Faith in Science

Shrader-Frechette (a philosopher) and McCoy (an ecologist) received funding from the National Science Foundation to evaluate the role of ecological theory and information in environmental policy making. Their findings and analysis are supported by recommendations made by the National Academy of Sciences and the National Research Council. Their findings are also relevant to the politics of decommissioning policy making.

Shrader-Frechette and McCoy (1994) showed the following: 1) Ecologists do not agree on what the basic principles or ecological laws ("community" and "stability") are; 2). All scientific facts are laden with epistemic or cognitive values. Values can be divided into three types: bias values, contextual values, and methodological values; 3) Ecology cannot dictate ends or goals (but can act as a guide to good environmental policy); 4) Ecologists and laypersons do not share the same uncontroversial and unambiguous goals; 5) Ecology cannot tell us what is "natural"; 6) Ecological applications arise when and because scientists have a great deal of knowledge about the (qualitative and quantitative) natural history of a specific organism or place (which is not the case in this policy area). This final point is especially relevant to the politics of decommissioning policy making. Scientists do not have a great deal of understanding of the natural history of the diverse ecosystems of southern California., nor are they likely to have such an understanding before offshore oil and gas platforms and structures are removed.

Decommissioning policy making should be understood as a "transcientific" process which involves the intermingling of facts and values, science and culture, society and nature. A focus on the ecology of OCS structures should necessarily include human beings, their values and perceptions. A view of scientific investigation in isolation from the human community will not suffice.

With these conditions in mind and given the high level of ecological uncertainty in California's OCS, it is unlikely that science alone will provide the foundation for a future decommissioning policy or program. People will act out of their perceptual and place-based understandings of the world.

Ecology or scientific information alone cannot provide policy makers with goals or values for policies, but can guide policy makers regarding the means to attain or the reasons to attain ends or goals. The ends should based on the relevant categorical judgment of values held by society, not merely the industry or government. In the case of decommissioning, place-based and regional values are likely influence the role that scientific information plays in the politics of decommissioning policy making. The politics of decommission will also vary from one region to another.

Policy Recommendations: Strategies to Integrate Ecology in Decommissioning Policy Making

Given the complexity and intergovernmental character of decommissioning policy making, I have identified three strategies for integrating ecological information in decommissioning policy making.

First, policy makers should adopt a case-specific approach to reviewing the ecological and socioeconomic importance of each platform and structure. Each platform is embedded in a particular ecological system. Removal of the platform may have an impact on this system. Yet, we should also recognize that it is unlikely that we will have "the facts" necessary to make a comprehensive decision regarding a particular rig's removal. Ecological information is most useful when it does not try to predict complex interactions among many species, but instead attempts to predict what will happen for only one or two taxa in a particular case (e.g., Bocaccio rock fish or widow rockfish). With this in mind, policy makers should avoid adoption of programmatic EIRs which would cover broad ecological and biogeographical areas. Those platforms northern of Point Conception, for instances, should be treated differently than those south of Point Conception because of the ecological character of these two distinct biogeographical areas.

Second, if a rigs to reef alternative is adopted for a particular reef, I propose the ideal of a living permit (personal correspondence with Lima 1994). The ideal of a "living permit" supports values of adaptive decision-making. Adaptive decision making views the permitting process as ongoing and open ended. An artificial reef's ability to enhance commercial/sports fishery, recreational and ecological values should be studied over time, and should not be viewed as a one-shot operation by the industry or government. Complete removal should remain an option if the ecological impacts of the artificial reef are shown to be significant.

Third, policy making should be based on cooperative approaches that can coordinate the values and interests of a broad range of government and nongoverment participants. Decommissioning planning should take place in a "neutral" forum, and mediated by professionals who are not members of government or the industry. At the outset of the planning process, technical advisory committees or steering committees should include interests and parties which have traditionally not been included in the oil and gas decision-making process. Such a process should be developed at the regional and local levels to reflect the needs of the public and community, and not merely the interests of the federal government or industry.

Acknowledgment

The author thanks Dr. James Lima, Troy State University (Alabama) and Dr. Mark Carr at UC-Santa Barbara's Coastal Marine Institute for their assistance. This paper is based on research supported by a grant from the CMI-MMS Program at UC-Santa Barbara. The opinions and recommendations expressed in this paper are those of the author and do not reflect the views or opinions of MMS, CMI or the University of California. For additional information, contact mcginnis@lifesci.lscf.ucsb.edu

References

Cicin-Sain, Biliana and Robert Knecht. 1987. Federalism under Stress: The Case of Offshore Oil and California." In Harry Scheiber, ed. Perspectives on Federalism. University of California: 149-17666.

Farrow, Scott R. 1990. Managing the Outer Continental Shelf Lands: Oceans of Controversy. New York: Taylor and Francis.

Freudenberg, William R. and Robert Gramling. 1994. Oil in Troubled Waters: Perceptions, Politics, and the Battle Over Offshore Drilling. State University of New York Press.

GAO. 1994. Offshore Oil and Gas Resources: Interior Can Improve Its Management of Lease Decommissioning. Washington, D.C.: U.S. Government Printing Office.

Gramling, Robert. 1996. Oil on the Edge: Offshore Development, Conflict, Gridlock. State University of New York.

Hershman, Marc J. David L. Fluharty and Scott L. Powell. 1988. State and Local Influence Over Offshore Oil Decisions. Seattle, WA: Washington Sea Grant Program.

Lester, Charles. 1991. The Search for Dialogue in the Administrative State: The Politics, Policy, and Law of Offshore Oil Development. Ph.D. Dissertation. University of California, Berkeley.

Lima, James T. 1994. The Politics of Offshore Energy Development. Ph.D. Dissertation. University of California, Santa Barbara.

Love, M. 1997. The Ecological Role of Natural Reefs and Oil and Gas Production Platforms on Rocky Reef Fishes in Southern California. Presentation at a MMS/USGS Environmental Study Briefing. March 19, Camarillo, CA.

National Research Council. 1996. An Assessment of Techniques for Removing Offshore Structures. Washington, D.C.: National Academy Press.

Scheiber, H.N. 1995. "Success and Failure in Science-Policy Interactions: Cases from the History of California Coastal and Ocean Studies, 1945-1973," In National Research Council, Improving Interactions Between Coastal Science and Policy. Washington, D.C.

Shrader-Frechette, K.S. and E.D. McCoy. 1994. Method in Ecology: Strategies for Conservation. Cambridge University Press.

Smith, Eric R.A.N. and Sonia R. Garcia. 1994. California's' Attitudes Toward Energy Policy. Paper delivered at the Annual Meeting of the Western Political Science Association. Albuquerque, New Mexico. March.

CHAPTER 182

Some Results of Fish Surveys Around California Offshore Production Platforms

Milton Love, Associate Research Biologist

Abstract

As part of a National Biological Service-funded study of southern and central California reef fisher, we are surveying fish populations around offshore oil and gas platforms. We have found substantial variation in the species composition and numbers of fishes inhabiting different platforms. Platforms in the Santa Maria Basin, off central California, harbor the largest populations, some with at least 25,000 individuals. These structures act as nursery grounds, as most of the fishes are juvenile rockfishes, many of them newly-settled young-of-the-year. Some other platforms, particularly those in the Santa Barbara Channel, have far fewer fishes around them. Some of these differences are probably due to long-term poor recruitment of rockfishes in southern California, apparently caused by adverse oceanographic conditions.

Fish populations around at least some of the platforms are quite dynamic. Monthly sampling around one platform has shown that its fish population varies with season. with peak diversity and numbers during the summer and a depauperate fish community in late winter and spring. Even species, such as seaperches., that are closely tied to structure seem to leave the platform when water conditions are unfavorable.

CHAPTER 183

What do we really know about marine biodiversity in central California?

Gregor M. Cailliet[1]

Abstract

This is a brief overview of what is known about marine biodiversity in the central California coastal region. The term biodiversity has become very popular, but detailed knowledge is often lacking, especially in the ocean. Although much information on the flora and fauna of this coast exists in the form of books, guides and the scientific literature, some habitats are still poorly understood. Marine ecosystems can be divided into: 1) well known, 2) moderately known, and 3) poorly known systems. These categories in general follow a gradient from shallow, coastal habitats to deeper, offshore areas. Marine biologists have an idea which factors are associated with high or low biodiversity, and those that act as threats to it. However, basic information on which species occupy specific habitats, especially those in deeper water, is often lacking. To better understand marine biodiversity, we need to develop and deploy non-destructive survey techniques, increase training of taxonomists, increase support for museums, produce more comprehensive taxonomic aids, and undertake more intensive and long-term surveys of marine ecosystems. Cooperative efforts among state and federal reserves, refuges, preserves, and sanctuaries, specifically the Gulf of the Farallones, Cordell Bank, Monterey Bay and Channel Island National Marine Sanctuaries, can result in major advances in our understanding and preservation of marine biodiversity.

Introduction

Biodiversity has become a very popular subject, mainly stimulated by our concern that human activities have caused serious and irreversible damage to the health and abundance of many species on earth (Soule 1986, Lubchenco et al.1990, Wilson 1988a, Ryan 1992, Ricklefs and Schluter 1993). Emphasis has

----

[1]Moss Landing Marine Laboratories, P.O. Box 450, Moss Landing, CA 95039

been on terrestrial systems, but this is now switching to marine habitats (Peterson 1992a, Yoon 1993). The first marine biodiversity studies concentrated on temperate coastal zone habitats (Ray 1985, 1988), followed by tropical coral reef ecosystems (e.g. Ogden et al. 1993). Recently, interest has expanded to other marine habitats (Peterson 1992a,b, Butman and Carlton 1993 & 1995, Norse 1993, National Research Council 1995).

Early efforts to understand marine diversity centered around biogeography (e.g. Briggs 1974, Angel 1992, McGowan 1992), evolutionary trends in diversity as recorded in the fossil record (Raup 1988), and theories about the processes of speciation occurring in the sea (e.g. Harding and Tebble 1963). More recently, efforts have focused on describing the patterns of biodiversity among different marine habitats (e.g. Rex 1973, Wolff 1977, Rex 1981, Foster and Schiel 1985, Emmett et al. 1991, Monaco et al. 1992, Angel 1992, Davis et al. 1994). Also of interest have been the effects of man's activities on diversity, such as global warming (Peters 1988, Peters and Lovejoy 1992, Ray 1992, Suchanek 1994), disturbance (Addessi 1994), oil pollution (Lissner et al. 1991, Suchanek 1993), Mariculture (Tseng 1992) and sewage discharge and fishing pressure (Omori et al. 1992, Grigg 1994, Boehlert 1996). Very little work has been done to determine what causes trends in biodiversity in the sea, but some interesting debates have occurred, for example, regarding the relatively high biodiversity in the deep sea (Hessler 1967, Dayton and Hessler 1972, Grassle and Sanders 1973, Thistle 1978, Grassle et al. 1990).

The need to better understand marine biodiversity generated a national research intitiative and international taxonomy, biogeography, and biodiversity initatives (Grassle et al. 1990, Lubchenco et al. 1991, Butman and Carlton 1993 & 1995, Systematics Agenda 2000 1994, NRC 1995), and one of the focuses of the National Marine Sanctuaries and Reserves Division (SRD; Norse 1993, Cava 1994, Sullivan 1996). In addition, many actions have been proposed to enhance or preserve the biodiversity of marine ecosystems. These include initiating monitoring studies (Burley 1988), designating state seashores (Emory 1992), establishing and enforcing marine protected areas (Sobel 1993, Gubbay 1994), undertaking restoration actions (Zedler 1988), and establishing national estuarine research reserves and marine sanctuaries (Norse 1993, Cava 1994).

What is Biodiversity?

Wilcox (1984) defined biological diversity (or its shortened form, biodiversity) as "the variety of life forms, the ecological roles they play and the genetic diversity they contain." Norse (1993) and Angel (1992), reviewed the history of the term as it pertains to the marine environment, and divided it into at least three levels: 1) **Genetic diversity** is the genetic heterogeneity within and among species; 2) **Species diversity** is number of species within certain larger taxonomic entities (e.g. genera, families, orders), often disregarding geographic or ecosystem features; and 3) **Ecosystem diversity** is the variety of taxa occupying and functioning within ecosystems. The latter includes the composition, structure and function of such systems as sandy beaches, salt marshes, coral reefs, the rocky intertidal or deep sea.

Ecologists also distinguish between ecological and evolutionary diversity (Ricklefs and Schluter 1993). Ecological diversity is the number of species occurring in certain ecosystems in the present tense. These can be assessed at two spatial scales: local or regional (Ricklefs 1987). Evolutionary diversity refers to the total number of species occupying a specific ecosystem, having accumulated these via speciation over geological or evolutionary time (Ricklefs and Schluter 1993). Diversity can be measured several ways, ranging from indices of richness (a simple measure, the number of species) to indices of evenness (the distribution of relative abundance over all of the species in a system) (Pielou 1975). Norse (1993) further defined diversity at higher taxonomic levels (i.e., the diversity of genera or families), and the functional diversity of an ecosystem (i.e. the variety of trophic guilds). These are even more difficult to interpret than the first definitions.

What are the Central California Marine Ecosystems and Habitats?

The lengthy central California coastline has many marine habitats in close proximity. The nearshore habitats, primarily due to their accessibility, are often the easiest to describe, are the best studied, and have suffered the greatest threats (Ray 1985, 1988). Parts of the coast are lined with sand dunes, resulting from onshore winds that accumulate sand and successional processes by which plants further anchor the developing dunes. Seaward of these dunes are sandy beaches, which are dynamic tide- and surf-influenced sedimentary habitats. Inserted at numerous areas of the coastline are bays and estuaries, some of which are connected to river systems or streams, others of which are lagoonal in nature.

Other parts of the central California coastline are lined with rocky habitats, both inter- and subtidal, where macroalgae and kelp forests occur (Foster and Schiel 1985). Offshore from both sandy and rocky shores, there are progressively deeper soft-bottom (demersal) habitats, which range to depths of several thousand meters, within miles of the coast (see Cailliet et al. 1992 and Nybakken et al. 1992). Above these demersal habitats are volumes of open water column, ranging from the epipelagic surface to the deeper meso- and bathypelagic zones.

What is Known about the Biodiveristy in these Habitats?

Any discussion of local or regional biodiversity must start with the perspective of global biodiversity. Although it is unknown exactly how many species there may be worldwide, some estimates do exist. For example, Barnes (1984) estimated that there are ~1,333,611 species of organisms in the world, organized into five kingdoms, at least 40 phyla and more than 400 orders. Wilson (1988b: pages 4-5) summarized several sources and arrived at a figure of

1,392,485 but guessed that the absolute number might be between 5 and 30 million. May (1986) and Roberts (1994) illustrated several reasons why assessing such an elusive figure is difficult.

In the marine environment, accurate estimates of local, regional and total diversity are likewise very difficult to derive and most recent texts do not even attempt an estimate (e.g. Norse 1993). After all, marine habitats are not easy for humans to visit, study and sample. Those which are in closest proximity to humans are the best studied. The proportion of global species that inhabit marine habitats is likewise unknown, but the best estimate is that 20% of all species known to science are marine and almost all phyla occur in the sea (Ogden et al. 1993).

Off central California, detailed knowledge of the biodiversity for many marine habitats is often incomplete. The extent of knowledge about the biodiversity of these systems decreases with distance offshore and therefore depth. For the sake of this broad synopsis, I have divided central California's marine ecosystems into three categories, those in which the biodiversity is: 1) well known; 2) moderately known; and 3) poorly known. Because it is beyond the scope of this paper to comprehensively cover all systems, I have picked typical examples of these three categories.

Well-Known Systems

Perhaps the best known ecosystems in central California are the beaches, bays and estuaries. Numerous studies have been done and are summarized in reports from NOAA's Estuarine Living Marine Resources (ELMR) reports (e.g. Monaco et al. 1990 and Emmett et al. 1991). One particularly good example is Elkhorn Slough. Because it is part of the National Estuarine Research Reserve System (NERRS), knowledge about its biodiversity is perhaps as complete as for any system. A cumulative list of organisms utilizing Elkhorn Slough and environs (Andrew DeVogelaere, former ESNERR Research Coordinator and now MBNMS Research Coordinator), reveals that there are more than 1,300 species, at least 600 of which are marine, ranging from fungi, plants, and protozoa to invertebrates, fishes, amphibians, reptiles, birds and mammals. Thus, reserve status has increased knowledge of the biodiversity of such a system, making it easier to consider management strategies to preserve this biodiversity.

Several specific studies have been done to accurately characterize the fauna of Elkhorn Slough. The best examples are surveys (Cailliet et al., 1977, Yoklavich et al. 1991, 1992) which characterized the fish fauna using net sampling for adult, juvenile and larval fishes. These baseline survey data, documenting more than 79 species of fishes in the slough, along with trophic information (Barry et al. 1996), are presently being compared to the recent ichthyofauna to evaluate the effect of erosional changes in the slough.

The other well-known marine habitats off central California are also coastal and accessible, including sand dunes, have been extensively studied (Foster and Dawson 1982). Many efforts are being made to restore the native plant species to these dunes, and to stimulate designation of reserve areas, such as the proposed Monterey Bay State Seashore (Emory 1992).

Another habitat, the rocky intertidal areas along the central California coast, has been long studied, with several faunal guides available, ranging from Between Pacific Tides (Ricketts et al. 1985) to the two classics, Light's Manual (Smith and Carlton 1975) and Intertidal Invertebrates of California (Morris et al. 1980). In addition, several spatial and temporal studies have been done in this habitat, stressing the variability in species composition (hence biodiversity). For example, Foster et al. (1988) and Yoshiyama et al. (1986) assessed the spatial variability in community structure of numerous rocky intertidal habitats along the central and northern Calfornia coasts. On a temporal scale, Barry et al. (1995) assessed long-term changes (1931 to 1994) in an intertidal study site near Hopkins Marine Station in Pacific Grove.

The rocky subtidal habitats, often dominated by the giant kelp, *Macrocystis pyrifera*, also have been extensively studied (see Foster and Schiel 1985). However, details on the diversity of all taxa are certainly not complete nor equivalent. The fishes are well known (Miller and Lea 1972, Eschmeyer et al. 1983). However, in examining the invertebrates occupying the holdfast of the giant kelp, one often encounters unidentifiable specimens.

Moderately Known Systems

Both the nearshore and offshore soft bottom habitats off central California are moderately known because they have been surveyed by many biologists at Moss Landing Marine Laboratories (and elsewhere) for decades (Oliver et al. 1980). The estimates of diversity of the invertebrate fauna in this habitat depend entirely upon sampling considerations (type of samples taken and seive size used to separate them from the sediment) and their environmental setting (depth, distance offshore, water movement, sedimentary characteristics, amount of predation, etc.). Knowledge about the fishes occupying this habitat is a bit more detailed (Heimann 1963, Day and Pearcy 1968, Kukowski 1973, Horn 1980, Engineering Science, Inc. 1976, Ecomar, Inc. 1981, Allen 1985) but much work needs to be done. Indeed, Butman and Carlton (1995) described these habitats as "the underdescribed coastal zone."

Poorly Known Systems

The poorly known systems are primarily offshore and deeper, including the deep-sea benthos, and the epi-, meso- and bathy-pelagic environments. The lack of knowledge is primarily due to the remote nature and difficulty in surveying these habitats, leading Burman and Carlton (1995) to term this "the undersampled deep sea." In one descriptive study, Cailliet et al. (1978) analysed sea survey data and described pelagic assemblages, especially those associated with the market squid, *Loligo opalescens*. Also, Morejohn et al. (1978) analyzed feeding habits of the main large open water predators and predicted trophic links which may help understand the biodiversity and functioning of this system. From many midwater trawling samples taken in the Monterey Submarine Canyon, we commonly encounter at least 70 species of meso- and bathypelagic fishes. However, the majority of open water, pelagic organisms are poorly known. Studies of their taxonomy, morphology and ecology are just beginning, especially using the Monterey Bay Aquarium Research Institute's (MBARI) Remotely Operated Vehicle (ROV) *Ventana* (Bruce Robison, Judith Connor, George Matsumoto and Mary Silver, Personal Communication).

Based upon experience with soft bottom, deep-water invertebrates over more than two decades, only a small portion (perhaps 50%) of the species occupying habitats from 15 to >1,000 m have been described and are known to science (James Nybakken, MLML, Personal Communication). This lack of knowledge is partly due to the difficulty of sampling these habitats and the use of relatively large seive sizes, but also due to the paucity of taxonomists actively working on virtually the entire list of invertebrate phyla (Winston 1987).

There have been several deep bottom-trawling surveys off the west coast that have provided fish diversity data (e.g. Allen 1982, Day and Pearcy 1973). Results to date indicate that there are about 33 nearshore, demersal fish species, while the deep benthic total is ~ 86 species. Recent sampling at depths from 2000 to 3,500 m off central California (SAIC 1991, Cailliet et al. 1992, Nybakken et al. 1992) has provided preliminary statements of deep-benthic biodiversity, but estimates vary among types of sampling gear, with camera sleds generally estimating more species (or taxa, since many of these are undescribed or unidentifiable by most zoologists) per unit area.

During recent submersible studies sponsored by NOAA's National Undersea Research Program (NURP), Yoklavich et al. (1993) investigated deep canyon habitats as refugia for otherwise exploited rockfish populations. A side project has been visually to compare species abundance and diversity estimates between deep, rocky outcrop areas where fishing has been heavy, light or very low. Preliminary results on fish diversity in rocky habitats between 65 and 100 m indicate that the same number species occupy heavily fished habitats, but that their relative frequencies differ. For example, 20 species or rockfishes occurred in both fished and unfished sites, but there were five species that did not overlap. Also, the abundance and average size of fishes seen in heavily fished areas is significantly less. Analysis of the invertebrates observed in these surveys has yet to be done, partly due to the inability to accurately identify many of them.

What are the Threats to Marine Biodiveristy?

In a very good review of threats to marine biodiversity, Suchanek (1994) has listed numerous human activities. These include fishing (see also Grigg 1994), oil and gas development (Lissner et al. 1991), pollution such as sewage discharge (Grigg 1994), physical habitat modifications such as trawling and anchoring (Auster et al. 1994, 1995), human recreational activities (Addessi 1994), dredge spoil disposal, alien species (Norse 1993), and global change (i.e. warming; e.g. Peters 1988, Peters and Lovejoy 1992, Ray et al. 1992, Barry et al. 1995). Although this list may not be complete, it is certainly a good starting point for discussing modifications of biodiversity in central California.

It is my contention, however, that to propose studies of the effects of these factors on biodiversity is premature. So little is known of the biodiversity of marine ecosystems that few conclusions could be drawn about possible threats. Rather, I would encourage more studies on the basic taxonomy of the groups in marine habitats before we would really know how to manage for it.

What is the Role of Marine Managed or Protected Areas?

Many recent mechanisms have been proposed and intitated to monitor, promote and manage for marine biodiversity (Burley 1988, Jenkins 1988, Bohnsack and Ault 1996, Eichbaum et al. 1996). Much confusion exists about the words sanctuary, refuge, and reserve. In California, attempts are underway to define accurately the differences between "Marine Protected" and "Marine Managed" areas (Resources Agency of California 1995, McArdle 1997).

National Marine Sanctuaries and Estuarine Research Reserves

Federal sanctuaries appear to serve both as protected and managed habitats (Eichbaum et al. 1996). Although biodiversity was only barely mentioned in the 1992 Final Environmental Impact Statement and Management Plan (SRD 1992) for the MBNMS, it has since become a major objective of marine sanctuaries (Cava 1994, page 66). This was generated by the "Biodiversity Intitative" which has goals of: 1) increasing NOAA's ability to manage and conserve marine species and ecosystems before they become endangered; 2) changing from species to ecosystem-level management; and 3) regulating factors threatening biodiversity and ecosystem functions. Biodiversity is specifically identified as a subject for study in the MBNMS Research Advisory Panel's (1993) Scientific Research Plan.

State Refuges and Reserves

In California, there are at least five different designations for the protection and/or management of marine life (SRD 1992: page II-57 to II-62, Resources Agency of California 1997). These are Ecological Reserves (e.g. Point Lobos and Carmel Bay), Game Refuges (e.g. California Sea Otter Game Refuge in Monterey and San Luis Obispo counties), Marine Life Refuges (e.g. the Hopkins Marine Life Refuge in Pacific Grove), Fish Refuges (e.g. the Pacific Grove Marine Gardens Fish Refuge), and Marine Reserves (e.g. Ano Nuevo Point and Island, on which elephant seals are protected, but also studied and observed by the paying public).

All of the above refuges and reserves tend to aid in the protection of biodiversity because they limit human activities in specific ways. However, none really specifically are designed for this purpose. Recent recommendations have been made that "ecosystem management strategies, developed in consultation with the California Biodiversity Council, should be applied for purposes of sustaining ocean resources" and "to safeguard the State's ocean and coastal biodiversity" (Resources Agency of California 1995: page ES-5).

Protected Areas (Harvest Refugia or Non-Extractive Reserves)

Recently, the establishment of marine protected areas has attracted interest, especially as they might relate to biodiversity (Gubbay 1994). It has been suggested that some marine areas be designated as protected from exploitation ("Non-Extracted Reserves" Auster and Malatesta, In Press). Sobel (1993) strongly endorsed establishing marine protected areas for conserving biological diversity. Indeed, preliminary descriptions and evaluations by Auster et al. (1994, 1995). Indeed, they have found that biodiversity is influenced by fishing and other extractive processes. Thus, we need to further examine whether or not fishing activities should be regulated in the marine environment to protect biodiversity.

Although National Marine Sanctuaries do not prohibit commercial or recreational fishing (SRD 1992), these activities are regulated by other state and federal agencies. However, the regulations under which fisheries are regulated are based upon single species management concepts, not biodiversity concepts (Boehlert 1996). In a recent review, Bohnsack (1993) found that tropical marine reserves enhance fisheries, reduce conflicts and protect resources. However, only later (Bohnsack and Ault 1996) did a similar review indicate that Marine Sanctuaries ought to also serve to protect biodiversity.

What is Needed to Better Understand Marine Biodiversity?

There are several ways we could improve our very limited knowledge of marine biodiversity in central California and elsewhere. The conclusions of several workshops (e.g. Grassle et al. 1990, Peterson 1990a, Ogden et al. 1993) and the NRC Committee on Biological Diversity in Marine Systems (Butman and Carlton 1993 & 1995, NRC 1995) provide several suggestions. Among those that I most strongly support are:

More comprehensive surveys in all marine habitats

This objective could be met with funding to support surveys in the habitats mentioned earlier. These surveys would have to follow specific guidelines and quality control measures to insure the utility of their results over a long period of time. It would also be very beneficial if several techniques were used, and compared, within each marine habitat. This is one of SRD's (1992) objectives.

Development of non-destructive survey techniques

In many marine habitats, surveys can be observational rather than extractive. These would include SCUBA, towed camera sleds and video cameras, remotely operated vehicles (ROVs), and submersibles. With new technology, these and other techniques should be developed, evaluated and deployed to add to our growing knowledge through extractive sampling. One example of this approach would bottom trawling versus ROV surveys (Adams et al. 1995).

Increased training of taxonomists

My review of marine biodiversity in central California indicates that we know very little, at least part of which is due to the lack of support for taxonomy - either in education or research support (see Winston 1987 , Vecchione and Collette 1996). Thus, both the federal and state governments should increase support for training of taxonomists and using them in biodiversity studies at all levels (species, genus, family and above, plus biogeographic analyses and genetic relatedness studies (see also Angel 1992, McGowan 1992).

Increased support for museums

To house specimens, data sheets, videos, photographs, and voucher collections from biodiversity studies, support for museums needs to increase. Museums also provide space and facilities for taxonomists to work on collections, thus increasing our knowledge of biodiversity (see also McGowan 1992).

Studies on biodiveristy in sanctuaries, refuges and reserves

Without qualitative and quantitative studies on the actual effect that managed or protected areas have on biodiversity, it does not make sense to establish more of these. Research in managed and protected areas is very important and should be foremost in their establishment and maintenance.

Literature Cited

Adams, P.B., J.L. Butler, C.H. Baxter, T.E. Laidig, K.A. Dahlin, and W.W. Wakefield. 1995. Population estimates of Pacific coast groundfishes from video transects and swept-area trawls. Fishery Bulletin 93(3):446-455.

Addessi, L. 1994. Human disturbance and long-term changes on a rocky intertidal community. Ecol. Appl. 4(4):786-797

Allen, L.G. 1985. A habitat analysis of the nearshore marine fishes from southern California. Bulletin, Southern California Academy of Science 84(3):133-155.

Allen, M.J. 1982. Functional structure of soft-bottom fish communities of the southern California shelf. Ph.D. Dissertation, University of California, San Diego, California.

Angel, M.V. 1992. Managing Biodiversity in the Oceans. Chapter 2, pages 23-59 In: Peterson, Melvin N.A. (Editor) 1992. Diversity of Oceanic Life: An Evaluative Review. Significant Issues Series, Volume XIV, Number 12, The Center for Strategic and International Studies, Washington, D.C.

Auster, P.J. and R.J. Malatesta. In Press. A framework for assessing the potential role of non-extractive reserves for management of groundfish in temperate and boreal marine systems. In: Shackell, N. and J.H. M. Willison (Editors). Marine Protected Areas and Sustainable Fisheries. Science and Management of Protected Areas Association, Wolfville, Nova Scotia.

Auster, P.J. R.J. Malatesta, and I.G. Babb. 1994. The impacts of mobile fishing gear on benthic habitat complexity. Gulf of Maine News, Reg. Assoc. Res. Gulf of Maine 2(4): 1-4.

Auster, P.J., R. J. Malatesta, R.W. Langton, L. Watling, P.C. Valentine, I.G. Babb, and L.B. Cahoon. 1995. Management implications of mobile fishing gear alterations to benthic habitats in the Gulf of Maine: Science Program Summary. NOAA's National Undersea Research Center, The Univ. of Conn., Avery Point, Groton, Conn. 19 pages.

Barnes, R.S.K. (Editor). 1984. A Synoptic Classification of Living Organisms. Blackwell Scientific Publication, London, England, 273 pages.

Barry, J.P., C.H. Baxter, R.D. Sagarin, and S.E. Gilman. 1995. Climate-related, long-term faunal changes in a California rocky intertidal community. Science 267:672-675.

Barry, J.P., M.M. Yoklavich, G.M. Cailliet, D.A. Ambrose and B.S. Antrim. 1996. Trophic ecology of the dominant fishes in Elkhorn Slough, California, 1974-1980. Estuaries 19(1):115-138.

Boehlert, G.W. 1996. Biodiversity and the sustainability of marine fisheries. Oceanography 9(1):28-35.

Bohnsack, J.A. 1993. Marine Reserves: They enhance fisheries, reduce conflicts, and protect resources. Oceanus 36(2):63-71.

Briggs, J.C. 1974. Marine Biogeography. McGraw-Hill Book Company, New York, 475 pages.

Burley, F. W. 1988. Monitoring biological diversity for setting priorities in conservation. Chapter 26, pages, 227-230, In: Wilson, E.O. (Editor). 1988. Biodiversity. National Academy Press, Washington, D.C. 521 pages.

Butman, C.A. and J.T. Carlton. 1993. Biological diversity in marine systems (BioMar): A Proposed National Research Initiative. National Science Foundation Workshop Report: 1-20 pages.

Butman, C.A. and J.T. Carlton. 1995. Marine biological diversity: some important issues, opportunities and critical research needs. U.S > National Report to International Union of Geodesy and Geophysics 1991-1994, Reviews of Geophysics, Supplement, pages 1201-1209.

Butman, C.A. and J.T. Carlton. 1995. Marine biological diversity: some important issues, opportunities and critical research needs. Reviews of Geophysics, Supplement, pages 1201-1209, U.S. National Report to International Union of Geodesy and Geophysics 1991-1994.

Cailliet, G.M., B. Antrim, D. Ambrose, S. Pace and M. Stevenson. 1977. Species composition, abundance and ecological studies of fishes, larval fishes and zooplankton in Elkhorn Slough. pp. 216-386. *In*: J. Nybakken, G. Cailliet and W. Broenkow. Ecologic and hydrographic studies of Elkhorn Slough, Moss Landing Harbor and nearshore coastal waters July 1974 to June 1976. Moss Landing Marine Laboratories Report, 465 pages.

Cailliet, G. M., W. W. Wakefield, G. Moreno, A. Andrews, and K. Rhodes. 1992. The deep-sea fish fauna from the proposed Navy Ocean Disposal Site, using trap, otter and beam trawl, and camera sled samples. Final Report on Navy CLEAN Contract No. N62474-88-D-5086, 69 pages.

Cava, F.M. (1994). An overview of the Sanctuaries and Reserves Division. March 1994 Congressional Briefing, Second Edition, NERRS and NMSP: page 66: Biodiversity Initiative.

Davis, S.M., J.C. Ogden, and W.A. Park (Editors). 1994. Everglades. The Ecosystem and its Restoration. St. Lucie Press, Delray Beach, FL, 826 pages.

Day, D. and W.G. Pearcy. 1968. Species association of benthic fishes on the continental shelf and slope off Oregon. J. Fish. Res. Bd. Canada 25:2665-2675.

Dayton, P.K. and R.R. Hessler. 1972. Role of disturbance in maintaining diversity in the deep sea. Deep-Sea Research 19:199-208.

Ecomar, Inc. 1981. Final Oceanographic Studies Report (November, 1979 to December, 1980) for James M. Montgomery, Consulting Engineers, Inc. C-06-1128. Goleta, CA.

Eichbaum, W.M., M.P. Crosby, M.T. Agardy and S.A. Laskin. 1996. The role of marine and coastal protected areas in the conservation and sustainable use of biological diversity. Oceanography 9(1):60-70.

Emmett, R.L., S.A. Hinton, S.L. Stone, and M.E. Monaco. 1991. Distribution and abundance of fishes and invertebrates in west coast estuaries, Volume II: Species life history summaries. ELMR Rep. No. 8. NOAA/NOS Strategic Environmental Assessments Division, Rockville, MD, 329 pp.

Emory, J. (Editor). 1992. The Monterey Bay State Seashore: A Study for the Preservation of the Monterey Bay Dunes. The Big Sur Land Trust. 42 pages.

Engineering-Science, Inc. 1976. Predischarge Oceanographic Investigations in Central Monterey Bay, 1977. Monterey, CA. Reports to the Monterey

Peninsula Water POllution Control Agency.
Eschmeyer, W.N., E.S. Herald, and H. Hammann. 1983. A field guide to Pacific coast fishes of North America. Peterson Field Fuide, Houghton Mifflin Company, Boston, 336 pages.
Foster, M.S. and E.Y. Dawson. 1982. Seashore plants of California. California Natural History Guides, University of California Press, Berkeley, California, (47):226 pages.
Foster, M.S. and D. R. Schiel. 1985. The ecology of giant kelp forests in California: A community profile. Department of the Interior, U.S. Fish and Wildlife Service, Biological Report 85(7.2):1-152.
Foster, M.S., A.P. DeVogelaere, C. Harrold, J.S. Pearse, and A.B. Thum. 1988. Causes of spatial and temporal patterns in rocky intertidal communities of central and northern California. Memoirs of the California Academy of Sciences 9:1-45.
Grassle, J.F., P. Lasserre, A.D. McIntyre and G.C. Ray. 1990. Marine Biodiveristy and Ecosystem Function: A proposal for an International Programme of Research. Biology International, Special Issue (23):1-19.
Grassle, J.R. and H.L. Sanders. 1973. Life histories and the role of disturbance. Deep-Sea Research 20:643-659.
Grigg, R.W. 1994. Effects of sewage discharge, fishing pressure and habitat complexity on coral ecosystems and reef fishes in Hawaii. Marine Ecology Progress Series 103:25-34.
Gubbay, S. (Editor). 1994. Marine Protected Areas, principles and techniques for management. Chapman and Hall, New York, 350 pages.
Harding, J.P. and N. Tebble (Editors). 1963. Speciation in the Sea. The Systematics Association, c/o British Museum (Natural History), Cromwell Road, London, S.W. 199 pages.
Heimann, R.F.G. 1963. Trawling in the Monterey Bay area, with special reference to catch composition. California FIsh and Game 49(3):152-173.
Hessler, R.R. 1967. Faunal diversity in the deep sea. Deep-Sea Research 14:65-78.
Horn, M. 1980. Diversity and ecological roles of non-commercial fishes in the California Current. California Cooperative Oceanic Fisheries Investigations, Report XXI.
Jenkins, R.E. Jr. 1988. Information management for the conservation of biodiversity. Chapter 27, pages, 231-239 In: Wilson, E.O. (Editor). 1988. Biodiversity. National Academy Press, Washington, D.C. 521 pages.
Kukowski, G.E. 1973. Results of the Sea Grant fishes sampling program for the 1971-1972 season. Annual Report, Part 6, Tech. Pub. MLML 73-6.
Lissner, A.L., G.L. Taghon, D.R. DIener, S.C. Schroeter, and J.D. Dixon. 1991. Recolonization of deep-water hard-substrate communities: potential impacts from oil and gas development. Ecol. Appl. 1(3):258-267
Lubchenco, J., A.M. Olson, L.B. Brubaker, S.R. Carpenter, M.M. Holland, S.P.

Hubbell, S.A. Levin, J.A. MacMahon, P.A. Matson, J.M. Melillo, H.A. Mooney, C.H. Peterson, H. R. Pulliam, L.A. Real, P.J. Regal, and P.G. Risser. 1991. The sustainable biosphere initiative: an ecological research agenda. Ecology 72(2):371-412.

May, R.M. 1986. How many species are there? Nature 324:514-515.

McArdle, D.A. 1997. California marine protected areas. California Sea Grant College System, Publication No. T-039:1-268.

McGowan, J. A. 1992. Oceanic Species Diversity. Chapter 3, pages 60-62. In: Peterson, Melvin N.A. (Editor) 1992. Diversity of Oceanic Life: An Evaluative Review. Significant Issues Series, Volume XIV, Number 12, The Center for Strategic and International Studies, Washington, D.C.

Miller, D.J. and R.N. Lea. 1972. Guide to the Coastal Marine Fishes of California. California Department of Fish and Game, Fish Bulletin (157):1-235.

Monaco, M.E., D.M. Nelson, R.L. Emmett, and S.A. Hinton. 1990. Distribution and abundance of fishes and invertebrates in west coast estuaries, Volume 1: data summaries. ELMR Rep. No. 4. NOAA/NOS Strategic Environmental Assessments Division, Rockville, MD, 240 pages.

Morejohn, G. V., J. T. Harvey and L.T. Krasnow. 1978. The importance of *Loligo opalescens* in the food web of marine vertebrates in Monterey Bay, California. California Fish and Game, Fish Bulletin (169):67-111.

Morris, R.H., D.P. Abbott, and E.C. Haderlie. 1980. Intertidal invertebrates of California. Stanford University Press, Stanford, California, 690 pages.

Monterey Bay National Marine Sanctuary Research Advisory Committee. 1993. Scientific Research Plan for the Monterey National Marine Sanctuary. U.S. Department of Commerce, National Oceanic and Atmospheric Administration, National Ocean Service, Office of Ocean and Coastal Resource Management, Sanctuaries and Reserves Division: 1-54.

National Research Council. 1995. Understanding Marine Biodiversity: A Research Agenda for the Nation. National Academy Press, Washington, D.C., 114 pages.

Norse, E.A. (Editor). 1993. Global Marine Biological Diversity Strategy: Building Conservation into Decision Making. Island Press, Washington, D.C. 383 pages.

Nybakken, J., L. Smith-Beasley, A. Summers, S. Craig, and L. Weetman. 1992. Invertebrate megafauna collected from the proposed U.S. Navy Ocean Disposal Site as determined from beam and otter trawls and camera sled samples, Final Report on Navy CLEAN Contract No. N622474-88-D-1991, 56 pages.

Ogden, J., T. Done and B. Salvat. 1993. Ecosystem function and biodiversity on coral reefs: Report of a workshop. International Council of Scientific Unions (ICSU)/Scientific Committee on Problems of the Environment (SCOPE) Workshop.

Oliver, J.S. P.N. Slattery, L.W. Hulberg, and J.W. Nybakken. 1980.

Relationships between wave disturbance and zonation of benthic invertebrate communities along a subtidal high energy beach in Monterey Bay, California. FIshery Bulletin 78:437-454.

Omori, M., C.P. Norman, and H. Yamakawa. 1992. Biodiversity: human impacts through fisheries and transportation. Chapter 4, pages 63-75 In: Peterson, Melvin N.A. (Editor) 1992. Diversity of Oceanic Life: An Evaluative Review. Significant Issues Series, Volume XIV, Number 12, The Center for Strategic and International Studies, Washington, D.C.

Peters, R.L. II. 1988. The effect of global climatic change on natural communities. Chapter 51, pages 450-461 In: Wilson, E.O. (Editor). 1988. Biodiversity. National Academy Press, Washington, D.C. 521 pages.

Peters, R.L. and T.E. Lovejoy (Editors). 1992. Global Warming and Biological Diversity. Yale, 336 pages.

Peterson, Melvin N.A. (Editor) 1992. Diversity of Oceanic Life: An Evaluative Review. Significant Issues Series, Volume XIV, Number 12, The Center for Strategic and International Studies, Washington, D.C. 108 pages.

Peterson, M.N.A. 1992. Diversity of Life: its meaning and significance in Oceanic realms. Chapter 1, pages 1-22 In: Peterson, Melvin N.A. (Editor) 1992. Diversity of Oceanic Life: An Evaluative Review. Significant Issues Series, Volume XIV, Number 12, The Center for Strategic and International Studies, Washington, D.C.

Pielou, E.C. 1975. Ecological Diversity. John Wiley and Sons, New York, 165 pages.

Raup, D.M. 1988. Diversity crises in the geological past. Chapter 4, pages 51-57 In: Wilson, E.O. (Editor). 1988. Biodiversity. National Academy Press, Washington, D.C. 521 pages.

Ray, G.C. 1985. Man and the sea - the ecological challenge. American Zoologist 25:451-468.

Ray, G.C. 1988. Ecological diversity in coastal zones and oceans. Chapter 4, pages 36-50. In: Wilson, E.O. (Editor). 1988. Biodiversity. National Academy Press, Washington, D.C. 521 pages.

Ray, G.C. et al. 1992. Chapter 7, Effects of global warming on the biodiversity of coastal-marine zones. pp. 91-104, In: R.L. Peters & T.E. Lovejoy (Editors), Global Warming and Biological Diversity. Yale University Press.

Resources Agency of California. 1997. California's Ocean Resources: An Agenda for the Future. Ocean Resources Management Program, The Resources Agency of California, pages i-viii, ES-1 to BB-9.

Rex, M.A. 1973. Deep-sea species diversity: decreased gastropod diversity at abyssal depths. Science 181:1051-1053.

Rex, M.A. 1981. Community structure ijn the deep-sea benthos. Annual Review of Ecology and Systematics 12:331-353.

Ricketts, E.F., J. Calvin, J.W. Hedgpeth, and D.W. Phillips. 1985. Between

Pacific Tides. Fifth Edition. Stanford University Press, Stanford, California, 652 pages.

Ricklefs, R.E. 1987. Community diversity: relative roles of local and regional processes. Science 235:167-171.

Ricklefs, R.E. and D. Schluter (eds.). 1993. Species diversity in ecological cojmmunities: historial and geographical perspectives. University of Chicago Press, Chicago, Illinois, 414 pp.

Roberts, W.E. 1994. How many students know, how many species are on earth? Bulletin, Ecological Society of America, 75(4):268-269.

Ryan, J.C. 1992. Life Support: Conserving Biological Diversity. Worldwatch Institute, Washington, D.C. Worldwatch Paper (108):62 pages.

Sanctuaries and Reserves Division. 1992. Final Environmental Impact Statement and Management Plan for the Proposed Monterey Bay National Marine Sanctuary. Volume 1, Parts 1-VII. Office of Ocean and Coastal Resource Management, National Ocean Service, NOAA.

Science Applications International Corporation (SAIC). 1991. Biological and sedimentological investigations of the sea floor at the proposed U.S. Navy 103 Ocean Disposal Site. Final Report to PRC Environmental Management, Inc.

Smith, R.I. and J.T. Carlton. 1975. Light's manual: Intertidal invertebrates of the central California coast. Third Edition. University of California Press, Berkeley, 716 pages.

Sobel, J. 1993. Conserving biological diversity through marine protected areas: A global challenge. Oceanus 36(3): 19-26.

Soule, M.E. (Editor). 1986. Conservation Biology: The Science of Scarcity and Diversity. Sinauer Associates, Sunderland, MA, 584 pages.

Soule, M.E. and K.A. Kohm (Editors). 1989. Research priorities for conservation biology. Island Press Critical Issues Series, 110 pages.

Suchanek, T.H. 1993. Oil impacts on marine invertebrate populations and communities. American Zoologist 33:510-523.

Suchanek, T.H. 1994. Temperate coastal marine communities: biodiversity and threats. American Zoologist 34:100-114.

Sullivan, K.D. (Special Editor). 1996. Marine biological diversity: a special issue commemorating 25 year of science and service by NOAA. Oceanography 9(1), 110 pages.

Systematic Agenda 2000. 1994. Charting the Biosphere: A Global Initiative to Discover, Describe and Classify the World's Species. Technical Report, 34 pages.

Thistle, D. 1983. The stability-time hypothesis as a predictor of diversity in deep-sea soft-bottom communities: a test. Deep-Sea Research 30:267-277.

Tseng, C.K. 1992. Mariculture on coastal China and biodiversity. Chapter 5, pages 76-98 In: Peterson, Melvin N.A. (Editor) 1992. Diversity of Oceanic Life: An Evaluative Review. Significant Issues Series, Volume XIV, Number 12, The Center for Strategic and International Studies,

Washington, D.C.

Vecchione, M. and B.B. Collette. 1996. The central role of systematics in marine biodiveristy problems. Oceanography 9(1):44-45.

Wilson, E.O. (Editor). 1988a. Biodiversity. National Academy Press, Washington, D.C. 521 pages.

Wilson, E.O. 1988b. The current state of biological diversity. Chapter 1, pages 3-20, In: Wilson, E.O. (Editor). Biodiversity. National Academy Press, Washington, D.C.

Winston, J.E. 1987 . The systematists' perspective. *In*: Fautin, D. (editor). Biomedical importance of marine organisms. Memoirs, California Academy of Sciences, Number 13.

Wolff, T. 1977. Diversity and faunal composition of the deep-sea benthos. Nature 267:780-785.

Yoklavich, M.M., G.M. Cailliet, J.P. Barry, D.A. Ambrose, and B.S. Antrim. 1991. Temporal and spatial patterns in abundance and diversity of fish assemblages in Elkhorn Slough, California. Estuaries 14(4):465-480.

Yoklavich, M.M., M. Stevenson, and G.M. Cailliet. 1992. Seasonal and spatial patterns of ichthyoplankton abundance in Elkhorn Slough, California. Estuarine, Coastal and Shelf Science 34: 109-126.

Yoklavich, M.M., G.M. Cailliet, and G. Moreno. 1993. Rocks and fishes: submersible observations in a submarine canyon. American Academy of Underwater Sciences, Diving for Science, pages 173-180.

Yoon, C.K. 1993. Counting creatures great and small. Science 260:620-622.

Yoshiyama, R.M., C. Sassaman and R.N. Lea. 1986. Rocky intertidal fish communities of California; temporal and spatial variation. Environmental Biology of Fishes 17(1):23-40.

Zedler, J.B. 1988. Restoring diversity in salt marshes... Can we do it? Chapter 36, pages. 317-325 In: Wilson, E.O. (Editor). 1988. Biodiversity. National Academy Press, Washington, D.C. 521 pages.

CHAPTER 184

Ecological and Geological impacts of exotic plants on coastal ecosystems and processes

Dr. David Chipping[1]

Introduction

Exotic plants are non-natives that have been introduced to native ecosystems during historic times. Much of the California landscape, such as annual grassland, is dominated by these plants. Relatively recently, several invasive exotic plants have seriously impacted and threatened coastal ecosystems. The following is a discussion of some of the most troublesome invasive plants, but the list of problem species is far longer and is constantly increasing. Many examples are taken from the central coast of California, an area most familiar to the author. At the present time the war we have been fighting against ecological weeds is one of progressive defeats due to the lack of funding and human resources. The prognosis for California's coastal ecosystem is poor.

Invaders Of Mudflats And Tidal Marshes

In the estuaries of the Pacific Northwest, European and Atlantic coast cordgrasses have started to occupy mud flats and convert them to tidal grasslands. With the change in substrate, free movement of animals and water is reduced, sedimentation rates increase, and the habitat for mud dwelling organisms such as clams is eliminated. The native *Spartina foliosa* is far less aggressive and provides habitat for the endangered California clapper rail and salt marsh harvest mouse (Strong and Daehler, 1995), but the Atlantic coast originated *Spartina alterniflora* extends into mudflats unsuitable for the native grass where it eliminates the open mudflat environment, and packs with much higher stem density the habitat of the native cordgrass. Other species of cordgrass, *S. anglica, S. patens*, and *S. densiflora* also pose a similar threat to the intertidal mudflats, and all are currently present in San Francisco Bay. Any tidal marsh capable of supporting the native cordgrass is a likely candidate for invasion by the exotics (Strong and Daehler, 1995). Ecological impacts include the filling and elimination of mudflat tidal channels, the loss of accessibility to mudflat-foraging wading birds, the loss of the mudflat habitat for marine invertebrates, and the loss of the open waters at high tide that serve as habitat for fish . The plants spread from fragments broken by tides and waves, or are transported by boats and equipment (Strong and Daehler, 1995).

----

[1] Physics Dept. CalPoly State University, San Luis Obispo, CA 93407

Control can be accomplished with glyphosate application in the spring, or by shading stands under plastic sheeting (Chamberlain, 1994). Biological control would not be possible in areas supporting the native cordgrass, and little is known of the potential for this type of control (Daehler and Strong, 1994). Coming to terms with a control method is urgent, as 31 estuaries on the west coast of North America may be succeptible to invasion (Strong and Daehler, 1995).

Invaders Of Coastal Dunes

Beach Grass And Veldt Grass

In coastal dune systems newly introduced sand-binding plants alter the sand dynamics and surface morphology European beach grass (*Ammophila arenaria*) was introduced to North America for this purpose. The plant spreads through rhizomes, or from fragments which blow downwind and root. The most profound effect is a fast upward growth of the foredunes as a rampart of sand along the rear of the beach, quite unlike the form of the beach that is defined by sand trapping native plants such as *Abronia*. Blowouts are rare, and the bare sand so vital to birds such as Snowy plover is eliminated. Back dune habitats are gradually converted to a dune grass monoculture as fragments of grass are windblown from the foredunes. The grass therefore can turn a naturally dynamic landscape into a frozen sea of grass.

However a new addition to the dunes, Veldt grass (*Ehrharta calycina*) from South Africa, may have greater impact on dune ecosystems. It was introduced to the dunes of central California as experimental forage, and the rare coastal dune scrub plant communities are being totally replaced by a monoculture of the wiry bunchgrass. The Veldt grass spreads from seed, and from the spread of small bulblets when the soil is disturbed. In the central coast of California, losses of dune ecosystems to Veldt grass are much higher than losses to any other cause. Unlike European beach grass, the Veldt grass will infest any sandy soil along the coast, and is destroying old dune sands with maturing soils that support some unique plants. In Montana de Oro State Park, approximately one quarter of the coastal dune scrubs have been destroyed within 10 years, and loss of bare sand eliminates habitat for the endangered snowy plover, the federally listed Morro Bay kangaroo rat, Morro Bay banded dune snail and Morro manzanita, as well as most of the other plants and animals that depended on the plants of the original scub community.

Control of these grasses is difficult, and can best be approached through a combination of chemical and mechanical means. European beach grass cannot easily be removed by mechanical means, as it will resprout from root fragments, and removal of the Veldt grass requires removal of all the bulblets, but can be successful. Veldt grass is a very prolific seeder, and control of young grasses solely by mechanical means is almost impossible. In monocultures of the grasses, direct spraying with any general herbicide works, but special grass-specific compounds must be used if the remaining native species are to be retained to reoccupy the niche after the invasive grasses are gone. Spraying in the vicinity of endangered species clearly complicates the removal process.

Ice Plants

Ice plant and other succulent species from South Africa play a similar role in coastal dunes. Three genera, *Carpobrotus, Conicosia,* and *Mesembryanthemum* have been imported as landscaping plants, and were commonly found as plantings along roadsides and railroad tracks to prevent erosion. (Albert 1995). The ice plants are particularly suited to pioneering loose dune sand , where they can form a continuous carpet from the salt sprayed foredunes to the more mature soils and vegetated of the back dune areas. The plants effectively trap sand, slowing or stopping the natural flow of sand to the rear of the dunes, and raising the height of the foredunes. The ice plants are not as severe a sand trop as the exotic dune grasses, but they the ability to form a continuous carpet in which there is little or no plant diversity. The mats of ice plant change the acidity of the sands and build up a loamy soil that would not otherwise develop , and which attracts other exotic plants not normally found in the dune sands (Albert, 1995).

Ice plants can be eliminated by repeated chemical spraying, and *Carpobrotus* and *Mesembryanthemum* can be removed mechanically. Conicosia, on the other hand, forms a deep root and cannot be mechanically removed with ease. A recent successful restoration of Asilomar State Beach on California's Monterey Peninsular used chemical control with glyphosate to eliminate large mats of Carpobrotus..

Yellow Bush Lupine

Lupinus arboreus is a native of the California central coast dunes around the Santa Maria river, but was introduced in 1908 to the North Spit on Humboldt Bay to stabilize the sand. From a small scattering of seed the lupine now occupies over 500 acres of the dunes, where it displaces several rare plants such as the federally endangered beach layia, and the Humboldt Bay wallflower (Miller 1995). This illustrates the ecological reality that the term 'native' should be used only in an extremely small geographical context when restoration projects are being designed. Removal is usually accomplished by mechanical means.

Invaders Of Coastal Riparian Systems

German Ivy

Riparian and wetland areas of California are being invaded by German ivy, which forms a monoculture that can carpet woodland floors and tree canopies. German Ivy was introduced to the United States from South Africa as an ornamental plant. The plant was well established as an escaped exotic plant thirty to forty years ago. It is generally known in California as *Senecio mikanoides* , although taxonomic revision suggests it be called *Delairea odorata.* The plant is generating an ecological crisis in coastal California through almost total displacement of other plants. This is achieved mainly by smothering competitors and cutting off light, but also through stem breakage of shrubs and vines generated by the biomass of the German Ivy. It appears to have no natural enemies, and is not grazed by animals due to the presence of alkaloids. The thick masses of the plant choke off all other growth, kill and fell trees, and block the passage of flood waters. In wetlands adjacent to Morro Bay, the effective floor of the flood plain has been raised several feet due to the blockage of flow from downed trees and the mass of the vine itself. The plant is expanding its range into Redwood forest, coastal dune scrub, coastal sage scrubs and coastal marshes.

The plant appears to have increased its range in three phases. The plant first appeared in central California within relatively frost-free, coastal riparian corridors, mainly where the water table was high and there was a shady overstory. The plant is described in the early 1960's from scattered coastal watersheds of central California, and in some of the cooler inland valleys of the Coast Range. This is still the most abundant habitat, where it has completely replaced the understory of many coastal streams. In the second phase it has been seen to expand its range into non riparian habitats. These include Monterey Pine forests at Cambria and Monterey, eucalyptus plantations such as the Presidio in San Francisco and other areas that offer some shade and are usually within the coastal summer fog belt. It has become a serious threat to the redwood ecosystems of the Big Sur Coast, such as Julia Pfeiffer State Park. The third phase is just now beginning to be reported, in which the plant occupies extensive areas on unshaded slopes, and even sunny, exposed sunny slopes close to the coast. It has been seen occupying coastal dunes at Ragged Point, in northern San Luis Obispo County and at San Simeon State Park. In Morro Bay State Park it grows into the margins of coastal salt marsh.

Control of the plant cannot be accomplished by mechanical means. Until recently it was though that the plant did not set viable seeds in California, and that the remarkable spreading rates were achieved vegetatively. Many sites in Central California appear to be associated with road work or vehicles, with the plant presumably being carried on dirty equipment. The plant starts from a small, rooting fragment which sends lateral rhizomes distances of up to 3 meters. Fragments of stem less than 1 mm in length will root to produce viable new plants. Multiple stems grow from the root mass, extending as vines across the ground surface. The vines lack tendrils, but will wrap around any hanging object and climb any available surface such as the bole of a tree, shrubs, wires etc., but they do not establish any adhesive attachments such as English Ivy. The vines crossing the ground can develop roots at any of the leaf nodes, which are commonly 10-15 centimeters apart. Each rooting can become a separate plant clone, and the original vine can either continue to act as a connector between plants or can be established as part of the root system for the new clone. In areas where the lower canopy is covered by German Ivy or other plants, the vines may extend very rapidly but without leaves, running for several meters in darkness below the canopy before breaking through to the light and leafing. In this manner a plant surrounded by shaded ground unsuitable for clone development can extend its influence beyond the shade. In areas where the plant roots can easily be removed for examination, as in the humus of an oak woodland, the rooting masses are very dense, and will develop a mat below the population. There is little quantitative data on the growth rate of individual plants, although rates of at least 30 cm a month is recorded for a dry year by Steven Hillis in the San Francisco area. Rates of 1 meter per month were measured in a high water table area of Los Osos (by author).

Control may best be achieved using a spray mixture of 0.5% triclopyr, 0.5% glyphosate, 0.1% silicone surfactant in water, sprayed at a concentration of 640 liters/Ha. (Bossard and Benefield, 1995).

Giant Reed

Giant reed, *Arundo donax* , is a bamboo-like plant from around the Mediterranean Sea. It was introduced to the U.S. for both erosion control and for horticultural purposes. (Perdue, 1958). The weed was plentiful in waterways of southern California in the 1820's, when it was gathered for roofing material. (Robbins, Bellue, and Ball, 1951). It spreads quickly vegetatively by rhizomes, and is spread throughout riparian areas by the transportation of fragments during floods, eventually displacing all other plants. Rerooting of sprouts takes place on the bare river bed exposed by flood events. Growth rates of 0.5 m/week are known (Hoshovsky, 1986) It prefers open areas with full sun, and therefore has been the fastest invader in shrub dominated areas.

Arundo donax is of concern because it displaces most other species and presents a serious fire hazard, especially as the plant grows to a height of 8 meters and clumps accumulate an extensive thatch of dead fuel. The introduction of intense fire into riparian corridors is an ecological factor that is not normally experienced, and while the *Arundo* resprouts, most other species are eliminated in the intense heat generated from the large biomass of the *Arundo.* Rieger and Krieger (1989) note that "the presence of giant reed essentially creates a zone devoid of wildlife" .

In the Santa Ana river it is calculated that *Arundo* evapotranspires 56,200 acre-feet of water / year, (which would otherwise support 280,000 people), and that the displaced native vegetation would have used only 18,700 acre-feet of water (Iverson 1995). The ecological problem of dewatering wetlands further increases the plant's impact. The plant therefore offers a serious obstacle to the restoration of steelhead runs in southern streams. The major central California drainages of the Salinas, Santa Maria, Santa Ynez, and Santa Clara rivers are all very heavily infested.

The serious impacts caused by giant reed resulted in a multi-county, multi-agency attack on the problem that enlisted the help of such diverse interests as the Monsanto Company, the Nature Conservancy, the California Native Plant Society and many others. A control experiment by 'Team Arundo' along the Santa Ana River successfuly controlled the plant through fire followed by chemical treatment of the resprouts (Jackson, 1993).

Pampas Grass

Pampas grass is a tall South American grass that was introduced to the west coast as an ornamental plant, and as erosion control. It spread rapidly in areas of the central coast of California that are summer-cool and generally frost-free. The distinctive white plumes of *Cortadaria jubata* , the most common of the two invasive species, contain thousands of fertile seeds that are produced by apomixis or cloning. They establish easily in wetlands and damp places, but also occupy bare ground with ease. In a classic case of how exotic invasions take place, a roadside planting at the north end of California's spectacular Big Sur coast resulted in tourists driving south with the plumes tied to their cars. Today there is hardly a road cut along 100 miles of highway that does not have the plant. The plant can be a major problem in coastal wetlands, where it can dominate interdune marshes, as at Oceano Dunes Campground and Oso Flaco Lake in San Luis Obispo County, California. California dune systems and coastal terraces north of San Francisco Bay are highly impacted, and it rapidly infests clearcuts and disturbances in coastal forests.

There is some evidence that another species of pampas grass, *C. selloana* , is starting to appear in coastal wetlands. This has been debated, as the nursery trade was supposed to deal only in female plants, but it is possible that male plants are present or that crosses have taken place with *C. jubata*.

Control efforts have largely been manual, but very large scale infestations such as several hundred acres around the Vandenberg Air Force Base runways cannot be controlled in this way. Glyphosates are effective chemical treatment, but may require several passes to kill the larger plants.

The 'Whitetops'

Whitetop, *Cardaria draba* has invaded the saltmarshes and flood plain in the delta of Chorro Creek, in Morro Bay State Park. This Eurasian member of the mustard family is a prodigious seeder, and forms mats of plants that displace all other life. The distinctive white flowers can be seen on portions of the floodplain marginally higher than the area occupied by Pickle Weed (*Salicornia virginica*). Elsewhere in the U.S. the weed is a major problem on grazing lands, but in the context of the coast, it is another monoculture that reduces the amount of Salicornia marsh by replacing the *Salicornia* occupying the upper end of it's elevation range. At the present time this seems to be a problem only in the coastal wetlands around Morro Bay.

A very similar plant with the same common name, *Lepidium latifolium* (tall whitetop, perennial pepperweed) is also beginning to show up in coastal marshes or other areas with salty soils (Young and Turner 1995). The pepperweed is a particular problem in Suisun Marsh in San Francisco Bay , where it had been held in check by cultivation until the land became a wildlife preserve. The end of cultivation resulted in the marsh being overrun by the weed, threatening its value as habitat for coastal wildfowl. Control appears to be limited to chemical methods, and elimination is not though to be possible. Broadleaf specific herbicides were used, as non -specific herbicided left bare ground that was rapidly reinvaded from the seed bank (Trumbo, 1994).

Invaders Of Forests And Scrubs

Coastal forests are linked to coastal processes though the provision of habitat (Marbled murrelet) and the production of water and sediment . There are several weeds invading the forest floor that might change the charactaristics of the forest. Of particular concern is Eupatory, (*Ageratina adenophora*), which is beginning to invade the sorrel-covered understory of coastal redwood forests and some shaded coastal wetlands in southern California, and various species of European brooms such as French broom (*Genista monsspessulana*) and Scotch broom (*Cytisus scoparius*) that invade clearings and repress the seeding of young trees and the native plant understory. The brooms are considered the major pest for northern California's coastal forests and woodlands. Mechanical removal, herbicides, and controlled fires are being used for control.

Monitoring And Control

Exotic pest plants are being tracked by the California Native Plant Society and the California Exotic Pest Plant Council. Control is dependent on the numbers and energy of volunteer labor, and small and decreasing amounts of state and federal funding. Except in a few very successful cases control is insufficient to halt the accelerating degradation of coastal ecosystems. The key to control is a dedicatedand knowledgable group of team leaders and a large number of volunteers who are prepared to work long hours against depressingly long odds. Successes against German ivy and the brooms has been won at Golden Gate Natural Recreation Area, and against *Arundo* along the Santa Ana river. Research is badly needed for biological control agents against massive weed infestations, and to support efforts that reach beyond volunteer labor.

References

Albert, M. (1995) "Portrait of an Invader II: The ecology and management of Carpobrotus edulis." CalEPPC News 3(2) pp. 4-7, California Exotic Pest Plant Council

Chamberlain, S. (1994) "Comparison of efforts to control an invasive cordgrass in San Francisco Bay." MA thesis. San Francisco Sate University

Daehler, C. and Strong, D. (1994) "Variable reproductibe output among clones of *Spartina alterniflora* (Poaceae) invading San francisco Bay, California: the influence of herbivory, pollination, and the establishment site" Amer. Journ. Bot. 81 pp.307-313

Hoshovsky, M. (1986), "Element Stewardship abstracts- Tamarisk and Arundo donax." The Nature Conservancy, San Francisco, California

Iverson, M (1995) "Effects of Arundo donax on Water Resources." CalEPPC News 3(1) p. 10, California Exotic Pest Plant Council

Jackson, N. (1993) "The story of Team Arundo." CalEPPC News 1(2) pp. 6-7, California Exotic Pest Plant Council

Miller, L (1995) "Yellow Bush Lupine." CalEPPC News 3(1) p. 6, California Exotic Pest Plant Council

Rieger, J. and Kreager, D. (1989). "Giant Reed (Arundo donax): a climax community of the riparian zone." pp 222-223 in Dana L. Abell, Tech. Coord., Proc. California Riparian Systems Conference: "Protection, Management and Restoration for the 1990's." USDA Forest Service Gen. Tech. Report PSW-110, Berkeley, CA 544pp.

Perdue, R. (1958). "Arundo donax - source of musical reeds and industrial cellulose." Econ. Bot. 12(4), 368-404.

Robbins, W., Bellue M., and Ball W. (1951). "Weeds of California. " California Dept. Agriculture, Sacramento, California

Strong, D. and Daehler, C. (1995). "Alien cordgrasses in Pacific estuaries." Proc., California Exotic Pest Plant Coucil Symposium '95, 25-28

Trumbo, J. (1994). "Perennial Pepperweed: A threat to wildland areas. " CalEPPC News 2(3) pp. 4-5 California Exotic Pest Plant Council

Young, J. and Charles, T. (1995) "Lepidium latifolium L. in California." CalEPPC News 3(1) pp. 4-5 California Exotic Pest Plant Council

CHAPTER 185

The Exotic Species Threat to California's Coastal Resources

Andrew N. Cohen[1]

Abstract

The potential scale and impact of biological invasions in coastal waters is demonstrated by invasions in the San Francisco Bay/Delta Estuary. Biodiversity, ecosystem processes and human activities have already been substantially affected, while new organisms continue to be introduced into the ecosystem. Protecting coastal biodiversity, and preventing new impacts on the human use of coastal resources, will require regulation of the various mechanisms by which exotic marine and freshwater organisms are transported across oceans and continents to California. The broad scope of the problem is known, as are some of the steps needed to reduce the rate of transport and invasion. The most critical vector now operating is undoubtedly the transport of organisms in ships' ballast water. To reduce the flood of invasions into California's coastal waters, there is an urgent need for the adoption and implementation of clear legal requirements for the high-seas exchange of ballast water.

Introduction

In 1852 Commodore John Sloat arrived on the Pacific coast with orders to find a site for a naval base that was "safe from attack by wind, wave, enemies, *and marine worms*" (Lott, 1954). The worm that worried the U. S. Navy was the Pacific shipworm *Bankia setacea*—which is not, taxonomically-speaking, a worm, but rather a clam with a

1 Environmental Scientist, San Francisco Estuary Institute, 1325 South 46th Street, Richmond, CA 94804

skinny, worm-like body and small shells modified for boring into wood. Shipworms were infamous for infesting, weakening and frequently destroying the hulls of ships and boats and the pilings that supported shore-side structures. Commodore Sloat eventually chose a site in the northern part of San Francisco Bay, where the waters were too fresh for the Pacific shipworm to survive. Thus the Mare Island Naval Shipyard was founded and operated for many decades free from attack by any of its feared enemies.

But then in 1913 an Atlantic shipworm, which had probably travelled from its native waters in the hull of a ship, was discovered in a piling at the shipyard (Hill 1927; Neily 1927). The Atlantic shipworm could tolerate fresher water than its Pacific cousin, and reproduced and spread so vigorously that it soon began to chew its way through the entire maritime infrastructure of northern San Francisco Bay. Over a two-year period, an average of one major wharf, pier or ferry slip collapsed into the water every two weeks, including the Union Oil Company railroad dock at Oleum, which took several loaded freight cars into the Bay along with it, the Benicia Municipal Wharf and Customs House, three grain warehouses, one highway and two railroad bridges and twelve ferry terminals. Mare Island lost several docks, the causeway and an 8,000-foot-long dike. The present value of the damage to these structures is estimated at several billion dollars, not including collateral damage (such as the freight cars), lost business, damage from the shipworm in other parts of the Bay, subsequent damage, and subsequent costs of treating structures to be resistant to the shipworm (Hill 1927; Neily 1927; Lemmon and Wichels 1977; Cohen, 1996).

Biological Invasions

Exotic organisms (also known as nonindigenous, introduced or alien organisms) not only endanger maritime activities, but may also constitute the largest single threat to the biological diversity of the world's coastal regions. In California, exotic organisms have invaded many coastal habitats, including beaches and dune areas, marshes, mudflats and open waters, in fresh, brackish and salt-water environments.

These invasions have been most intensively studied in the San Francisco Bay/Delta Estuary, which now hosts over 200 exotic species including plants, protists and invertebrate and vertebrate animals. More significant than the sheer number of exotic species is their dominance in many habitats, accounting for 40% to 100% of the common species at many sites. The organisms living on the muddy bottom of the Bay and on the sides of the docks are mostly exotic, most of the Delta's fish are native to the eastern United States, and the zooplankton in the northern part of the Estuary now consist primarily of several species of Asian copepod and Asian mysid shrimp. Moreover, the rate of invasion has been increasing, probably due in large part to the development of new or more active or more diverse mechanisms for transporting organisms across and between oceans, related to the rapid modern expansion of international trade and travel (Cohen and Carlton 1995).

Exotic species are not confined to the San Francisco Estuary, but are commonly reported in Los Angeles, San Diego and many smaller harbors and embayments in California, and in coastal regions around the world. Once established in one bay, organisms may readily invade another along the coast: the European green crab, first reported in 1989-90 from San Francisco Bay, had spread to Bodega Bay by 1993, Elkhorn Slough in Monterey Bay by 1994, Humboldt Bay by 1995, and Coos Bay, Oregon by 1997 (Cohen *et al.* 1995; Grosholtz and Ruiz 1995; Miller 1996; N. Richmond, pers. comm. 1997). Exotic species may also deploy out of the bays to invade the open coast. A predatory New Zealand sea slug that was collected in San Francisco Bay in 1992 and is now found from San Diego to Bodega Harbor, has become the most commonly collected sea slug in southern California's coastal waters (Gosliner 1995; D. Cadien, pers. comm. 1996).

Types of Impacts

In recent decades, negative impacts from accidentally-introduced exotic marine and freshwater organisms have attracted increasing attention in many parts of the world.

- The Atlantic comb jelly *Mnemiopsis leidyi*, introduced into the Black and Azov Seas by the early 1980s, became phenomenally abundant and ate up the zooplankton, contributing to the destruction of the region's anchovy and sprat fisheries and of the fishing fleets (from six countries) that depended on them (Travis 1993).
- European zebra mussels, appearing in the Great Lakes in the late 1980s, have become a massive nuisance, causing billions of dollars of damage by closing beaches that became fouled with sharp-edged mussel shells and rotting mussel flesh; agglomerating in enormous numbers on navigational buoys, and sinking them; and blocking the water intake systems of cities, factories and power plants. In a few year's time, zebra mussels spread across much of North America, from Canada to New Orleans and from the Hudson River to Oklahoma (Nalepa and Schloesser 1993)

- Toxic red tides thought by many to be caused by introduced dinoflagellates have appeared in many parts of the world (Hallegraeff et al. 1989; Hallegraeff and Bolch 1991; G. Lembeye, pers. comm. 1996). The toxins accumulate in clams or mussels, sickening and occasionally killing the people that eat them. In New Zealand, a recent red tide outbreak was so severe that people walking along the shore became ill (O'Hara 1993).
- In 1991 during the South American cholera epidemic, the cholera-causing bacterium *Vibrio cholerae* was discovered in oysters and fish in Mobile Bay, Alabama. The U. S. Food and Drug Administration subsequently sampled ships arriving from South America and found *Vibrio cholerae* in one third of them (*Federal* 1991). It was even possible that cholera initially reached South America via ballast water (Ditchfield 1993).

Table 1 lists some categories of effects resulting from the introduction of marine, estuarine and aquatic organisms. Ecological effects can alternately be sorted into two broad classes based on our assessment of their impacts. On the one hand, each introduction produces specific changes which may be judged to be positive, negative or neutral, depending on the system, the observer and the cultural context. On the other hand, every introduction of an exotic species into a region has a negative impact on the region's native diversity, and diminishes to some degree the distinctive faunal characteristics and ecological relationships that have developed or evolved there.

In a similar manner, economic effects and other effects on human activities may be sorted into a few broad classes. Again specific changes may be judged positive, negative or neutral, depending on the situation, but there are two general aspects of such change that are usually or invariably negative. First, human activities as they interface with natural resources are typically dependent on the stability of those resources—for example, when we invest in a fishing fleet and fish processing facilities we do so anticipating that the fishery will remain viable, at an adequate level of production, for some years. The introduction of exotic organisms brings change, and if the effect on existing resources is substantial there will be a cost in adapting to it, irrespective of whether the change in the long run is economically harmful or useful. From an economic perspective, other things being equal, we'd prefer to not have our ecosystems change on us. Second, many types of economically-important activities, such as the construction of dams and water diversions, the dredging of waterways for navigation, the creation or expansion of port facilities, and the development and maintenance of commercial fisheries, are increasingly subject to regulations that require a demonstration that the proposed activity will not unduly harm native populations of organisms or endangered species. To make a convincing demonstration requires some basic understanding of the affected ecosystem and some ability to predict its future response, and these requirements become harder to meet if the ecosystem is continually being altered by the addition of new species.

Table 1. Types of Effects from Exotic Species in Coastal Waters

Effects on Ecosystems	Effects on Human Activities
species-species interactions • predation • competition • hybridization • introduction of new parasites and diseases	*simple effects* • fouling of boats and marine structures • boring of wooden boats and structures • fouling of waterways and water systems • new fisheries • changes in populations that support existing fisheries • new human parasites or diseases
ecosystem effects • habitat alteration • changes in productivity • changes in trophic pathways • changes in nutrient or contaminant cycling • ecosystem instability	*systemic effects* • disruption of services • management uncertainty and failure
diversity impacts • reduction of populations and loss of native species • loss of genetic diversity • increase in local species number • loss of geographic diversity	
indirect effects • pollution from herbicides, pesticides and anti-fouling compounds • introduction of biocontrol organisms	*indirect effects* • costs of maintaining facilities and services against fouling and other damages • costs of monitoring, intercepting and controlling exotic species

This latter problem has been noted by the managers of California water systems whose diversions affect the size and timing of outflows from the Sacramento-San Joaquin River Delta into San Francisco Bay. Concern for many years has focussed on the level of productivity of the phytoplankton—the microscopic floating plants that are important contributors to the pelagic food web—in the northern part of the Estuary. Productivity was believed to be dependent on the size of the outflows, and years of effort were spent on scientific data collection and analysis in order to illuminate the nature of that dependence (Arthur and Ball 1978; Nichols 1985). Then, just about the time that relationship began to be reasonably understood, an Asian clam was introduced into the ecosystem and became so abundant, and was so effective at feeding on phytoplankton, that the dynamics of phytoplankton productivity were immediately and dramatically altered. Almost overnight it was as if a new ecosystem had appeared, and all that had been learned about predicting and managing the response of ecosystem productivity to flows was, in essence, lost (Nichols *et al.* 1990; Cohen 1990).

Finally, while exotic species that have been intentionally introduced to aquatic and marine ecosystems to support or enhance human activities have sometimes served their intended function, sometimes failed to do so, and sometime produced harmful results, only rarely has an accidentally introduced organism produced a positive economic result. The impacts of accidental introductions are more commonly negative, and occasionally disastrous. Thus, while there may be cause for lively debate about proposed intentional introductions, there should be no question about the value of reducing or eliminating accidental introductions.

Ballast water

Major pathways responsible for the introduction of exotic organisms into California waters include the transport of organisms in ships' ballast water, the introduction of organisms via aquaculture activities, and the release or escape of organisms imported in the aquarium and live food trades. Ballast water is by far the most important mechanism in terms of the number and variety of organisms transported and released.

Ballast water is water that ships pump into empty cargo holds or into dedicated ballast tanks at the start of a voyage in order to achieve proper trim and buoyancy, and later discharge on arrival at a new port prior to taking on cargo. The amount of water involved can be as much as tens of millions of gallons per ship. Water was first used for ballast as early as the 1880s when iron hulls and steam-driven pumps came into use, but the amount of ballast water transported around the world has been rapidly increasing in recent decades with the expansion of international trade.

Numerous studies have found that ballast water can carry an enormous number and variety of aquatic organisms (Carlton 1985; Carlton and Geller 1993), and in most parts of the world, including the entire California coast, these are freely discharged into coastal waters without regulation. Ballast water is credited with transporting the zebra mussel to North America, the comb jelly *Mnemiopsis* to the Black Sea, toxic red tides to many parts of the world, and the bacterium that causes cholera to coastal waters of the southeastern United States. Roughly 50 exotic organisms are thought to have been introduced into the San Francisco Estuary via ballast water in recent decades (Cohen and Carlton 1995).

In the long run it might be possible to stop the introduction of exotic organisms in ballast water by means of some technological fix—a treatment process to kill ballast water organisms that can be installed on-board ship—but such a fix will take at least several decades to develop and to install in the world's merchant shipping fleet. In the meantime, however, we could substantially and immediately reduce ballast water invasions by requiring ships to exchange their ballast water at sea, replacing the coastal water (and associated coastal organisms) in their ballast tanks with mid-ocean water (and organisms). Because of differing conditions in the coastal and mid-ocean environments, coastal organisms discharged into the middle of the ocean are unlikely to survive or compete effectively, and mid-ocean organisms discharged into coastal waters are also unlikely to survive and compete.

Federal regulations requiring such high-seas ballast water exchange have been in effect for several years for ships entering the Great Lakes from overseas ports. The shipping industry has apparently complied with these requirements without ill effect and at insignificant cost. However, despite last year's passage of much-ballyhooed federal legislation that purported to deal with aquatic invasions, these regulations have not been extended to the rest of the country. Thus untold numbers of living marine and freshwater organisms from other parts of the world, comprising probably thousands of different species, continue to be dumped into California waters every year, unrestricted by any law or regulation.

References

Arthur, J. F. and Ball, M. (1978). *Entrapment of suspended materials in the San Francisco Bay-Delta Estuary*. U. S. Department of the Interior, Bureau of Reclamation, Sacramento.

Carlton, J. T. (1985). "Transoceanic and interoceanic dispersal of coastal marine organisms: the biology of ballast water." *Oceanogr. Mar. Biol., Ann. Rev.* 23, 313-371.

Carlton, J. T. and Geller, J. B. (1993). "Ecological roulette: The global transport of nonindigenous marine organisms." *Science* 261, 78-82.

Cohen, A. N. (1990). *An introduction to the ecology of the San Francisco Estuary.* San Francisco Estuary Project, U. S. Environmental Protection Agency, Oakland.

Cohen, A. N. (1996). "Biological invasions in the san francisco estuary: a comprehensive regional analysis." Ph. D. dissertation, University of California at Berkeley, Calif.

Cohen, A. N. and Carlton, J. T. (1995). *Biological report. Nonindigenous aquatic species in a United States estuary: a case study of the biological invasions of the San Francisco Bay and Delta.* U. S. Fish & Wildlife Service, Washington DC.

Cohen, A. N., Carlton, J. T. and Fountain, M. (1995). "Introduction, dispersal and potential impacts of the green crab *Carcinus maenas* in San Francisco Bay, California." *Mar. Biol.* 122, 225-237.

Ditchfield, J. (1993). "Cholera, plankton blooms, and ballast water." *Global Biodiversity* (Canadian Museum of Nature) 3(3), 17-18.

Federal Register. (1991). 58(No. 239), 64831-64836.

Gosliner, T. (1995). "The introduction and spread of *Philine auriformis* (Gastropoda: Opisthobranchia) from New Zealand to San Francisco Bay and Bodega Harbor." *Mar. Biol.* 122:, 249-255.

Grosholz, E. D. and Ruiz, G. M. (1995). "Spread and potential impact of the recently introduced European green crab, *Carcinus maenas*, in central California." *Mar. Biol.* 122, 239-247.

Hallegraeff, G. M., Bolch, C. J., Bryan, J. and Koerbin, B. (1989). "Microalgal spores in ship's ballast water: a danger to aquaculture." Toxic Marine Phytoplankton, E. Granéli, B. Sundström, L. Endler and D. M. Anderson, eds., Elsevier, New York, 475-480.

Hallegraeff, G. M. and Bolch, C. J. (1991.) "Transport of toxic dinoflagellate cysts via ships' ballast water." *Mar. Poll. Bull.* 22(1), 27-30.

Hill, C. L. (1927). "Origin and development of the San Francisco Bay Marine Piling Survey Project." Marine Borers and their Relation to Marine Construction on the Pacific Coast, C. L. Hill, C. A. Kofoid, eds., final report of the San Francisco Bay Marine Piling Committee, San Francisco, 1-12.

Lemmon, S. and Wichels, E. D. (1977). *Sidewheelers to nuclear power: a pictorial essay covering 123 years at the Mare Island Naval Shipyard.* Leeward Publications, Annapolis.

Lott, A. S. (1954). *A long line of ships: Mare Island's century of naval activity in California.* U. S. Naval Institute, Annapolis.

Miller, T. W. (1996). "First record of the green crab, *Carcinus maenas*, in Humboldt Bay, California." *California Fish and Game* 82(2), 93-96.

Nalepa, T. F. and Schloesser, D. W. (1993). *Zebra mussels: biology, impacts, and controls.* Lewis Publishers, Boca Raton, FL.

Neily, R. M. (1927). "Historical development of marine structures in San Francisco Bay." Marine Borers and their Relation to Marine Construction on the Pacific Coast, C. L. Hill, C. A. Kofoid, eds., final report of the San Francisco Bay Marine Piling Committee, San Francisco, 13-32.

Nichols, F. H. (1985). "Increased benthic grazing: an alternative explanation for low phytoplankton biomass in northern San Francisco Bay during the 1976-1977 drought." *Estuar. Coastal Shelf Sci.* 21, 379-388.

Nichols, F. H., Thompson, J. K. and Schemel, L. E. (1990). "Remarkable invasion of San Francisco Bay (California, USA) by the Asian clam *Potamocorbula amurensis*. II. Displacement of a former community." *Mar. Ecol. Prog. Ser.* 66, 95-101.

O'Hara, P. (1993). "Overview of the marine biotoxin crisis in 1993." Marine toxins and New Zealand shellfish, J. A. Jasperse, ed., Royal Society of New Zealand, Misc. Ser. 24, 3-6.

Travis, J. (1996). "Invader threatens Black, Azov Seas." *Science* 262, 1366-1367.

CHAPTER 186

THE FISH FAUNA OF BALLONA MARSH, AN URBAN ESTUARY ON THE WESTERN LOS ANGELES BASIN

Camm C. Swift

Abstract

Ballona Marsh is a highly modified remnant marsh on the western edge of the Los Angeles Basin bordering Santa Monica Bay. Thirteen stations (collected with small seines) in the marsh are being sampled bimonthly in 1996-1997 to compare with identical monthly collections at the same stations in 1980-81. Seven stations are from the lower marsh (north of Culver Blvd) that is regularly tidal and with relatively high salinities; six stations are in the upper marsh (south of Culver Blvd) which is at higher elevation and usually has lower salinity. Recent samples were taken in October, December, and February. The numbers of fishes are comparable with the earlier collections, but fewer stations have fish and the species composition has changed. In 1980-81 the fish numbers and biomass peaked in the spring and fall and were low in the mid-winter and mid-summer. In October and December of 1996 all but one station in the upper marsh were dry. In 1980-81 only the uppermost two or three stations were dry in the summer and fall. Our October, 1996 collections also had relatively large numbers of fishes and very few fish were taken in December and February. Naive California killifish, topsmelt, arrow goby, mudsucker, and staghorn sculpin are common and mostly in the lower marsh as in 1980-81. Shadow and cheekspot gobies, California halibut, and diamond turbot have been rare and found only in the lowest portions of the marsh in both studies. The introduced mosquitofish are common and mostly in the upper marsh. The native striped mullet and introduced yellowfin goby were major components of the 1980 study and have yet t be collected in 1996-97. The introduced sailfin molly appeared in the marsh between 1981 and 198/6 and is now abundant. Much of the fish fauna is about the same as earlier, buy the wetted habitat is reduced, two species (one native and one introduced) no longer use the marsh for a nursery, and an additional introduce species has become dominant in much of the marsh.

Camm C. Swift, Department of Biology, Loyola Marymount University, 7900 Loyola Boulevard, Los Angeles, CA 90045-8220

CHAPTER 187

RESTORING SOUTHERN CALIFORNIA STEELHEAD: DAM. WHY CARE?

Jim Edmondson

Abstract

Arguably why spend attention, energy and funds on a handful of Steelhead in Southern California when Northern California salmon and Steelhead populations are at risk? Southern California Steelhead, those populations located south of San Francisco, represent important populations to the biodiversity of Steelhead along the Pacific Coast. Dr. Robert Benhke theorized western trout evolved from ancestral parents residing in the Gulf of California around the end of the last ice age. Cutthroats trout developed throughout the Great Basin using the Colorado River, while Steelhead advanced along the Pacific Coast to populate coastal steam form Baja California northward.

After examining the mtDNA of Southern California Steelhead, Dr. Jennifer Neilsen believes these population may well represent the most ancient of all rainbow trout. They persist in the harshest of all Steelhead habitat conditions. River flows are flashy and represent a fraction of annual runoff compared to Northern California river systems. Spawning migrations, in order to persist, need to be timed to late winter seasonal runoff patterns and be opportunistic in nature. Low steam flows create harsh water quality conditions including water temperatures once thought to be stressful of worse to steelhead.

Once numbering in the tens of thousands, Southern California steelhead populations have now declined to less than 1% of their 1850 abundance. Dozens of individual coastal steam populations are now extinct from Los Angeles south to San Diego. Those populations which endure are limited by the constructions of dams to support the 20th century urbanization of the region, where 62% of California's populations resides on a fraction of the state's landscape.

Dams without fish passage are present on all major Southern California river and steam systems. Most do not have passage facilities to allow free passage to upstream habitats which are largely in tact compared to the coastal plain developed

Jim Edmondson, Executive Director, California Trout, Inc.

areas. Because of the ecological processes of the region (highly erodible soils, rapidly building mountains, and frequency of periodic high flow events) sediment loads are overwhelming dam facilities. Already two dams are now obsolete because of sedimentation(Matilija Dam on the Ventura and Rindge Dam on Malibu Creek). Ironically these dams are also effecting beach formation and nourishment of the region's highly important and popular beaches.

If steelhead are to be restored in Southern California, obsolete dams must be removed, and operation dams must provide fish passage to high quality habitat reaches above.

CHAPTER 188

"ONE LOCAL VIEW OF THE CALIFORNIA BIODIVERSITY COUNCIL"

by
Bud Laurent, San Luis Obispo County Supervisor
Executive Board member, California Biodiversity Council

The California Biodiversity Council (CBC) was formed in 1991 largely as a result of leadership provided by Mr. Ed Hastey, State Director of the Bureau of Land Management, and Mr. Douglas Wheeler, Resources Secretary for the State of California, with the support of the Wilson Administration. The original primary purpose of the CBC was to improve communication between and among the myriad federal and state agencies having anything to do with wildlife and other natural resources.

When it was first organized, The California Biodiversity Council consisted only of state and federal resource agency executives, and local government representatives were conspicuously absent. After hearing concerns expressed by many local officials about this oversight, the CBC expanded its membership to include representatives of regional associations of county supervisors, resource conservation districts, and, in the case of Southern California, the San Diego Association of Governments (SANDAG) and the Southern California Association of Governments (SCAG). For the past four years, nine locally elected representatives have been fully-involved participants in the activities of the CBC with the 27 member state and federal agencies.

A question which might be asked is "why was the California Biodiversity Council formed since we have many (some say too many) agencies responsible for conserving natural resources?" In this observer's view, the basic answer is that the present system of resource conservation is not working, and, in fact, is substantially dysfunctional. To begin with, California is complicated biologically, possessing at least nine terrestrial bioregions, and at a minimum two marine provinces, yet the state is "managed" as one area with little regard or distinction for regional differences. The natural resources of the state are not managed ecologically, and, as the Greeks might say, an oikos divided against itself cannot stand.

Overlaying the bioregions, each with numerous watersheds and biomes, is a jumble of political jurisdictions of local government : 58 counties, about 415 cities, nearly 5,000 special districts and over 1,200 school districts, all of whom can (and do) exert effect on natural resources. Added to this maze of local jurisdictions, where there is little or no coordination or communication between and among them, are private and public land ownership patterns which spill over artificial boundaries of government.

Compounding the above problems created by governmental organization, is the fact that post-World War II patterns of human population growth, which have eaten up large amounts of agricultural lands and remaining wildlife habitat, show little sign of changing. The housing industry, which relies on a dependence for the private automobile, continues to consume land at rates 3-4 times faster than the rates of population growth. Recent Supreme Court decisions on property dedications sought in exchange for development entitlement have also tipped the balance toward "private property rights" over the rights of wildlife or the interests of community.

Along with the shift in the legal landscape, there has also been a seemingly growing antipathy to resource agency efforts to conserve threatened or endangered species. In many cases, it seems that the closer a listed species is to someone's backyard, the less likely it is to receive local sympathy and support if it stands in the way of 'progress' -- especially if it lacks fluffy fur and limpid brown eyes. Woe be unto a species if it can be characterized as being merely a snail, a salamander or a weed.

If human habits and attitudes are to change, they will most likely change only through civic discourse and education, rather than through imposition of mandates. It is unlikely that statewide or regional land use planning initiatives could succeed in the foreseeable future. Nor is it likely that protections offered through the Endangered Species Act, which has come under increasing attack, will be any more effective as threats of expanding human development increase. The greatest potential benefit of the California Biodiversity Council is its commitment to fostering education through civic (and hopefully civil) discourse so that there can be, at all levels in the state, an honest exchange of information. Destructive attitudes and habits, if they are to experience lasting change, must be re-examined and changed within each individual-- or at least enough individuals to make a difference.

The stated goals of the CBC are as follows:

- to develop a statewide strategy to conserve biological diversity
- to coordinate bidiversity-related science
- to improve communication about biodiversity-related activities
- to support Natural Communities Conservation Planning efforts, such as the project to conserve coastal sage scrub habitat in Southern California
- to support regional Geographic Information System projects, including Klamath Metadata Project, Trinity Community GIS Project, the Sierra Nevada Catalogue, and the Sierra Nevada Information Inventory Project
- to encourage volunteer-based, inclusive, grassroots watershed or bioregion projects

- to maintain and enhance an established 16 agency "Science Coordination Committee" for the coordination and sharing of natural resources information and data to support informed decision making throughout the state

- to continue to support the California Environmental Resources Evaluation System (CERES)

Given its rather ad hoc status, the California Biodiversity Council has accomplished a number of significant results in its 5-1/2 years of existence. It has held up to four forums throughout the state each year on local and statewide environmental issues since 1994 which have brought agency administrators and managers into much closer contact with California citizens. An educational video designed for local public and private sector interests on the uses of Geographic Information System technologies was produced in 1994. A first-ever "Regional Managers" meeting between state and federal resource agencies was held in 1993 to improve interagency communication. CBC sponsored the Sierra Summit in 1992 which began the process to create the Sierra Nevada Ecosystem Project. The CBC publishes a quarterly newsletter, the "California Biodiversity News," which serves as a valuable linkage between resource agencies and the public.

What the CBC is designed to be is neither another layer of government, nor to have any authority on its own, but to serve as a vehicle for "friendly persuasion" and information exchange among and between public and private sector interests. While this has been one of the primary reasons for its success to date, and its growing acceptability in the eyes of local officials and other citizens often apprehensive about resouce agencies, it has also been a focus of criticism by some observers anxious for faster action in conserving California's shrinking wildlife habitat. Given the degree of philosophic separation between participants in this debate, the question about an expanded role for a "more effective" CBC will not be answered easily or soon. Personally, I believe the CBC deserves to receive stable, adequate funding from both federal and state governments. It presently lives in a hand-to-mouth fashion, relying on "contributions" member agencies are able to scrape from existing budgets. As a consequence, what it can afford to do (and communication of information on broad scales takes signicant funding) is limited by the unpredictable amount of funding available from year to year.

I also believe that the CBC should conduct a "campaign for habitat", on the scale of the Smokey Bear advertising campaign, in order to tell the story of the real crisis natural resources face. Public awareness as to the needs for habitat conservation and restoration must be raised far above current levels, and public education efforts should be increased ten-fold, at a minimum. A possible vehicle for raising public awareness would be an annual "State of Wildlife" report to the people of California produced by state and federal resource agencies working together; this annual report could provide information at the level of a watershed and compiled to form a composite picture of the ecological health of the state.

In order to lessen some of the tension which exists between state and federal resource agencies with local governments of cities and counties, agencies' field staff must find ways of working more closely (and sympathetically) with city and county staff on environmental issues. Too often, local governments only hear from resource agencies when there are problems, usually violations alleged or real. If mutual understanding can be increased through improved communication and pro-active involvement, the number of violations should correspondingly decrease. It is tragic when suspicion and distrust become the prevalent characteristics in the relationships between local governments and resource agencies -- and this should be wholly preventable.

Finally, a question to which the CBC should direct its attention, in the opinion of many, is whether state and federal agencies can be better organized, possibly even reduced in number, to better and more cost-effectively accomplish their missions, and to achieve a truly comprehensive ecosystems approach to managing natural resources. These and other difficult questions will be subject to growing discussion, especially given the expected impacts of a rapidly increasing human population in California. Finding suitable answers won't happen without the political will of a developed consensus in what is presently a highly charged political atmosphere, and the California Biodiversity Council has established a foundation for civil discourse which can --and hopefully will-- be built upon.

CHAPTER 189

Coastal Zone Management & Point Source Monitoring: Direction for the Future

Irwin Haydock[1], George Robertson, Michael A. Rozengurt and Alex Steele

Abstract

The County Sanitation Districts of Orange County, California (Districts) has proposed a three-part ocean monitoring strategy incorporating: (1) temporal and spatial trend studies (including point-source compliance); (2) special process studies; and (3) regional monitoring studies. This new design addresses various methodic procedures and relevant questions and takes advantage of information gained over the past decades. Using a systems approach and developing partnerships between the regulators and the regulated community recognizes the regional and long term nature of potential impacts of a growing human population to the watershed-coastal zone relationship. Designed to answer both specific permit requirements and questions that the public perceives as important, this approach to water quality and living resources monitoring and evaluation (e.g., impaired runoff impacts and losses of habitats) integrates coastal resources management and consists of: (1) Core Monitoring, (2) Strategic Process Studies, and (3) Regional Monitoring.

Introduction

The role played by ocean outfalls in the changing southern California marine environment is clear in the historical monitoring data gathered by sanitation agencies from Santa Monica Bay, Los Angeles to the coastal waters off Point Loma, San Diego. The mission of receiving water monitoring and reporting programs is to protect public health and the marine environment by conducting scientific investigations that can assess the environmental

---

[1]Chief Scientist, County Sanitation Districts of Orange County, P.O. Box 8127, Fountain Valley, CA 92728-8127

significance of wastewater discharges. These results then have to be effectively communicated to regulators, decision-makers and the public.

Southern California sanitation agencies have carried out prescribed point-source monitoring programs for more than three decades. Regulators designed "end-of-pipe" monitoring to address concerns that were prevalent when National Pollution Discharge Elimination System (NPDES) permitted monitoring programs were initiated following amendments to the Clean Water Act in 1972. This led to a large amount of spatially limited data (Bergen et al., 1995). Note. The present conditions near major outfalls contrasts sharply with those described historically in the Southern California Bight (SCB). Today the impact of improved treated wastewater discharged is much more subtle.

The recoveries documented at these outfalls in the face of overall declining natural resources have made it apparent to practitioners in the SCB that factors other than point sources need to be incorporated into monitoring to manage this significant area of biological diversity for effectively. Rozengurt and Haydock (1993; 1994) evaluated the role of inland water development and freshwater flow diversions on the declining biodiversity of the coastal seas. This linkage between watersheds and the health of the coastal environment is apparent in southern California. The runoff from twenty-six rivers and more than 120 streams that collect runoff from the coastal mountain ranges have been modified by man. These rivers and streams historically had represented important, episodic sources of freshwater, suspended sediments, organic and inorganic materials to the southern California coastal environment (Rozengurt, 1993). In southern California, multiple sources of contaminant inputs and inland discharges are also important because they may influence the results and subsequent interpretation of data from coastal monitoring programs.

With the continued pressure to the coastal zone from increasing human population, the need for coastal recreational areas, industry and declining space for living natural resources justifies the concern regarding the quality of the coastal zone. However, currently mandated monitoring does not recognize the regional and long term nature of these potential impacts; indeed sometimes these programs, designed to answer specific permit requirements, cannot address the questions the public perceives as important. A new approach to monitor point-source ocean discharges and to evaluate regional changes (e.g., impaired runoff impacts and losses of habitats) is needed that can better integrate watershed and coastal resource management.

Since 1985 the County Sanitation Districts of Orange County, California (Districts) has discharged an average of 8.8×10^8 L (232 million gallons) of treated wastewater per day onto the San Pedro Shelf 8.2 km (5.1 mi) offshore through a 1.8 km (6000 ft) long, 500-port outfall diffuser at a depth of 60 meters.

The Districts' 1985 301(h)–modified NPDES discharge permit requires a multifaceted ocean monitoring program (i.e., water quality, sediment geochemistry and invertebrate infauna, epibenthic fish and macroinvertebrates) to study the effects of this discharge. A synthesis of the results of 10 years of monitoring, costing more than 20 million dollars, is presented in CSDOC (1996). Specific requirements, goals and objectives, and methods used can be found in various annual reports such as CSDOC (1996; 1997).

A vast amount of data has been collected and analyzed to demonstrate compliance with receiving water limitations. While this extensive monitoring program has documented conditions near the Districts' discharge, it has also raised several important questions that cannot be answered by standard, ongoing monitoring programs. These questions require special studies or regional monitoring to address. Programs can benefit from the flexibility and adaptability of special studies and regional monitoring while maintaining a balance of consistency derived from standard monitoring (Bernstein and Dorsey, 1991). Other authors have also addressed the shortcomings of present ocean monitoring and proposed new designs to improve them (Eichbaum and Bernstein, 1990; NRC, 1990).

Based upon our experience, the Districts has proposed a three-part monitoring strategy incorporating: (1) temporal and spatial trend studies (including point-source compliance); (2) special process studies; and (3) regional monitoring studies. This new design addresses various relevant questions and takes advantage of information gained over the past decades and formally solidifies a partnership between the regulators and the regulated community.

Concurrence by both groups should allow regulators to facilitate flexibility into the monitoring program during the permit period. In turn, the discharger shows its commitment by undertaking a monitoring program using the best available science using adaptive management strategies and undergoing periodic review by the interested parties (e.g., regulators, managers, the public and scientists). This proposed monitoring program addresses the need to practice environmentally sound and *integrated* monitoring to protect the coastal marine environment better (Gerges, 1994).

Program Approach

Ocean monitoring programs study the physical, chemical and biological components of the marine ecosystem (Figure 1). However, to be truely effective, a monitoring program needs to evaluate and understand how the three components interact with each other. The Districts' have proposed a balanced, integrative approach to monitoring as necessary to meet differing goals of: (1)

determining compliance with permit terms and conditions, and state water quality standards for regulators; (2) answering special questions relevant for operations management; and (3) developing a regional program important to the public.

The approach used is to conduct three types of studies: Core Monitoring, Strategic Process Studies and Regional Monitoring. This new integrated program is based upon a systems approach to environmental goals, and sets priorities based on results of existing studies and available funding. Each type of monitoring is designed to provide independent information plus supplement and enhance the information content gained separately (Figure 2).

The guiding principles for this program include monitoring only those pollutants that are important to the marine environment *and* can be quantified. Too often dischargers or monitoring agencies measure chemical compounds that are not detectable, and would not be expected to persist in the marine environment (e.g., volatile organic compounds). Instead efforts could be better applied to understanding the sources, fates and effects of important pollutants (e.g., heavy metals, pesticides).

Core Monitoring

Building and/or continuing time series used to assess the state of the coastal environment is a key component to any monitoring program. As proposed, Core Monitoring consists of ongoing field studies conducted for the full term of the permit. It measures long-term physical, chemical and biological temporal and spatial trends near the Districts' discharge and at "unaffected" reference sites. Core surveys include elements for all receiving water monitoring components including surfzone, nearshore and offshore water quality, bottom sediments, benthic infauna and epibenthic fish and

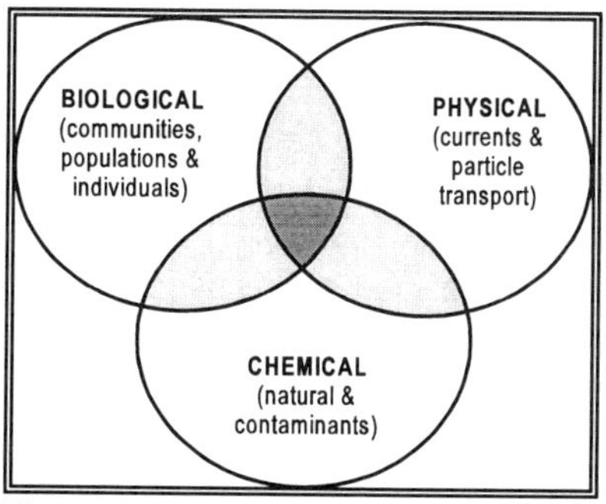

Figure 1. Model of Marine Ecosystem.

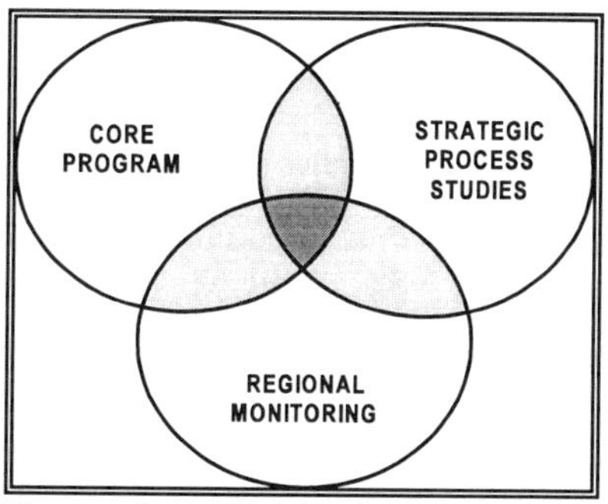

Figure 2. Model of the Districts' Proposed Ocean Monitoring Program.

macroinvertebrates. This program element is also designed to allow mandatory compliance evaluation with receiving water criteria and permit limits and conditions.

Strategic Process Studies

Strategic Process Studies are relevant to the ocean monitoring program's goals and objectives or the Districts' overall mission. Results will provide either clear direction for the Districts' treatment of wastewater and operation of its outfall or address issues of vital public concern (e.g., storm water runoff). These studies supplement Core and Regional Monitoring by addressing questions best answered by intensive, short-term sampling and analyses versus ongoing standard monitoring efforts or larger-scale studies. These investigations will be conducted within the set guidelines of an overall level of effort for the monitoring program.

As noted above, the scope, funding and timing of Strategic Process Studies will depend on the effort required for Core and Regional Monitoring. Designed with flexibility and adaptability in mind, these studies will primarily analyze the sources, quantity, pathways, effects and fate of marine pollutants. Their proposal, design and results will be subject to careful and critical review by not only the scientific and environmental communities, but also by regulators who are to be active partners in this endeavor and management who will be asked to fund it.

Regional Monitoring

Several authors have seen cooperative regional monitoring as a critical component to the management of the coastal zone (Gerges, 1994; Eichbaum and Bernstein, 1990; NRC, 1990). Using a Regional Monitoring approach to studying the marine environment off southern California allows development of better ways to address public health concerns, monitor the status of natural resources and nearshore habitats and to assess the impacts of a variety of contaminant sources.

Regional Monitoring involves cooperative efforts with regulators, researchers and dischargers in the SCB to obtain the ultimate goals of integrating the separate NPDES programs and developing a comprehensive regional program for resource management in the SCB. It is, therefore, important that the more localized work done by individual dischargers be comparable so that it may be applied to regional questions being asked by the public.

Discussion

The proposed program creates a comprehensive systems approach that is adaptable to changing needs yet is efficient, effective and of high quality. Efficiency comes from two general areas. The first is to measure only those parameters that are important to the marine environment and are expected to be found. The second is that the data and information generated can be used for more than one purpose; for example for comparison with the permit (compliance) and threshold effects (process) criteria. Effectiveness will come in the data allowing appropriate decisions to be made regarding the management of the coastal resources. High quality is inherent if research and monitoring are mixed and allows the evaluation and comparison of data arising from different sources.

To carry out the proposed permitted monitoring program, partnerships between regulators, dischargers, the scientific and academic communities, and the public needs to be developed and maintained. Periodic independent reviews and workshops by panels of experts will assure that the Districts is best applying its resources to protect the environment and public health for the residents of Orange County.

Benefits to this monitoring approach include the ability to make better management decisions and to assess the utility of pollution control measures. The improved understanding of the way the coastal environment functions, coupled with modeling and prediction of responses to human activities will be one key to ensure that the southern California coast is managed for its long term benefit. We believe that the proposed program addresses the practical needs and concerns of regulators and resource managers trying to balance conflicting public and environmental needs evidenced in watershed and regional management frameworks.

The Districts is now at a point where it can begin to build on the success of past monitoring activities. Previous special studies conducted by the Districts' ocean monitoring staff (CSDOC, 1995) and our participation in a 1994 regional pilot program (Bergen, et al., 1995) have shown that the three-part monitoring program is advantageous to all parties. It produces a better understanding and integration than one that mandates point-source monitoring alone (CSDOC, 1996). Results from these additional studies have allowed better assessment of the relative impacts of the Districts' point source discharge to the marine environment.

Currently we are working with the U.S. Environmental Protection Agency, Region 9 and the Regional Water Quality Control Board, Santa Ana Region in fulfilling the partnership by collaboratively developing a new ocean monitoring;

one that provides the flexibility to address regional and strategic process studies while maintaining required and necessary compliance monitoring. An integrated program of mandated monitoring, process research and modeling, and regional assessments, coupled with information distribution, promises to provide a sound basis for managing California's coastal waters.

References

Bergen, M., E. Zeng, and C. Vista. 1995. The Southern California Bight Pilot Project: An Experiment in Cooperative Regional Monitoring. In *Proceedings Oceans '95, Challenges of Our Changing Global Environment.* Marine Technology Society. San Diego, CA, October 9-12. Vol. 1. pp 526-536.

Bernstein, B.B. and J. Dorsey. 1991. Marine Monitoring in Heterogeneous Environments. Journal of Environmental Management. Vol. 32. Pp. 227-240.

County Sanitation Districts of Orange County (CSDOC). 1995. Marine Monitoring Annual Report, July 1993–June 1994. Fountain Valley, CA. 360 pages plus appendices.

County Sanitation Districts of Orange County (CSDOC). 1996. Marine Monitoring Annual Report: Science Report — Ten Year Synthesis, 1985-1995. Fountain Valley, CA. 377 pages plus appendices.

County Sanitation Districts of Orange County (CSDOC). 1997. Marine Monitoring Annual Report, July 1995–June 1996. Fountain Valley, CA. 161 pages plus appendices.

Eichbaum, W.M. and B.B. Bernstein. 1990. Current Issues in Environmental Management: A Case Study of Southern California's Marine Monitoring System. Coastal Management. Vol. 18. pp 433-445.

Gerges, M.A. 1994. Marine Pollution Monitoring, Assessment and Control: UNEP's Approach and Strategy. Marine Pollution Bulletin. Vol. 28. No. 4. pp 199- 210.

National Research Council (NRC). 1990. *Managing Troubled Waters. The Role of Marine Environmental Monitoring.* Washington, D.C. National Academy Press. 125 pages.

Rozengurt, M. 1993. Watershed Management: The Effects of Water Development on Estuarine and Coastal Zone Ecosystems in the United States and Other Countries. Presentation: Environmental and Energy Study Conference Briefing, United States Congress. March 22. Washington, D.C.

Rozengurt, M. and I. Haydock. 1993. Freshwater Flow Diversion and its Implications for Coastal Zone Ecosystems. In *Proceedings Transactions of the 58th North American Wildlife and Natural Resources Conferences*. Wildlife Management Institute. pp 287-295.

Rozengurt, M. and I. Haydock. 1994. The Role of Inland Water Development in the Systemic Alteration of the Coastal Zone Environment. In *Proceedings Watershed '93, A National Conference on Watershed Management*. Alexandria, Virginia. March 21-24. pp 755-760.

CHAPTER 190

Strategies for Creating a Comprehensive, Integrated Coastal Water Quality Monitoring Network

Linda Sheehan[1]

Abstract

Comprehensive, long-term monitoring of our coastal and ocean waters is essential to ensuring the health of those waters. However, despite federal and state mandates to conduct coastal water monitoring and the general need for reliable information, there still is little knowledge of the overall health of our waterways. Several steps must be taken to ensure that the quality of California's coastal waters is adequately and regularly assessed. First, current monitoring efforts must be inventoried to minimize duplication and identify gaps in monitoring efforts. Second, monitoring gaps must be filled on a priority basis through new programs or extensions of existing programs. Third, agencies must work to create consistent monitoring and reporting protocols wherever possible so that they may share data, and should post data in one area easily accessible by other agencies and the public. Finally, the role of the public in assessing their own waters must be elevated. Citizen monitoring groups throughout the state are becoming increasingly active in testing local waters. Citizen monitoring enhances stewardship of local waters, educates the public about the health of the waters that border their homes and workplaces, and serves as a potentially valuable source of information for agencies and other decisionmakers. Using the Monterey Bay Citizen Water Quality Monitoring Network as a model, this paper will explore strategies to enhance and expand the role of citizen monitors in creating a complete picture of the health of California's coastal waters.

[1] Pollution Programs Manager, Center for Marine Conservation, 580 Market Street, Suite 550, San Francisco, CA 94104

CHAPTER 191

Organizational Issues in Coastal Governance: A Regional Monitoring Example

Dr. Brock B. Bernstein

Abstract

There is increasing recognition that coastal governance systems (the network of laws, regulations, roles, responsibilities, reporting pathways, etc.) have not kept pace with increases in scientific understanding, the nature of problems facing the coast, or with expectations about what effective governance and management should accomplish. There are far too many instances of unresolved conflicts about resource use, inefficient and unproductive decision making, outmoded and rigid regulations, and high transaction costs, among other problems. In contrast, there are examples where existing governance and management systems have been successfully modified in order to address such problems. While every situation is unique is some respects, there are nevertheless useful, and more broadly applicable, lessons to be learned from such successes.

One such success is the recent redesign of core compliance monitoring programs in the Los Angeles, CA, USA region. The Santa Monica Bay Restoration Project, a part of the U.S. EPA's National Estuary Program, undertook to develop a comprehensive regional monitoring program. A primary goal was to modify existing compliance monitoring programs, rather than to create an additional layer of monitoring, thus doing better instead of doing more. Working through a series of structured planning steps and then with focused technical workgroups, the Project successfully reconfigured existing monitoring programs to:

- improve technical standardization
- focus on specific decision-maker's information needs
- enhance flexibility
- make better use of science
- fill gaps and reduce duplication of effort
- improve information management

Shortcomings in the existing monitoring system reflected and derived

Dr. Brock B. Bernstein, Vice President, EcoAnalysis, Inc., 221 E. Matilija St., Suite A, Ojai, CA 93023 from specific features of the existing governance and management system. Improving the monitoring system therefore required that changes be made to the governance system. As one example, making monitoring responsive to decision makers' needs (a common requirement in monitoring design guidance) required identifying decisions and decision makers, not always an easy task. It then sometimes necessitated integrating the relevant decision maker into the regional governance system, a step with the potential for changing existing roles and responsibilities. Thus, it proved impossible to improve the technical aspects of environmental monitoring without simultaneously addressing organizational issues in the governance system.

The Santa Monica Bay monitoring redesign example is instructive for two main reasons. First, it demonstrates that addressing governance issues from the "bottom up" can be extremely effective, since it provides a tangible rationale for suggesting governance changes. Second, the specific lessons learned about dealing with organizational and behavioral issues are applicable to a wide variety of other situations.

CHAPTER 192

Standard Southern California Monitoring Methods

M. James Allen[1]

Abstract

Although marine monitoring has been conducted in the Southern California Bight for more than 25 years, there is no overall coordination of monitoring programs. This lack of coordination means that different monitoring methods have been used in different programs. Because data collected in these programs cannot be easily integrated, a good regional assessment of conditions was not possible. However, this problem was resolved by the conduct of a regional survey in 1994 by the Southern California Bight Pilot Project (consisting of 13 federal, state, and local agencies and organizations). This survey required that participating agencies use standardized methods. This one-time standardization of methods is evolving toward a permanent standardization of receiving-water monitoring methods as the opportunity to incorporate these methods into discharge permits is discussed. Factors that led to the success of this effort included a review of Southern California monitoring programs by the National Research Council, the recognition of the need by both regulatory and regulated agencies, a regional survey to provide a focus for the standardization, cost-neutral participation of regulated agencies, and facilitation by a neutral organization administered by both regulatory and regulated agencies. An on-going Marine Monitoring Methods Committee provides for the periodic reevaluation and update of the methods.

Introduction

Marine monitoring has been conducted over much of the Southern California Bight (SCB) for more than 25 years. The National Research Council has noted that marine monitoring programs in Southern California cost more than $17 million annually (NRC 1990). Most monitoring is done to satisfy permit requirements that

----

[1] Principal Scientist, Southern California Coastal Water Research Project, 7171 Fenwick Lane, Westminster, CA 92683-5218

regulate discharge activities. Although there are many monitoring programs, there is no overall coordination of monitoring in the SCB. This lack of coordination means that existing data cannot be easily integrated. This is because different programs monitor different parameters, use different methods, and produce data that cannot be easily shared with other organizations. These short-comings prevent a regional assessment of conditions.

This problem was resolved when a regional survey of the SCB was conducted in 1994. The survey involved the participation of many organizations, with a cooperative field sampling effort among five agencies. The survey required a one-time standardization of methods. This survey led to the production of standardized tools, including a methods manual for a regional survey, a receiving-water methods manual, a standard taxonomic list of benthic invertebrates, a quality assurance plan, and an information management plan, all of which are currently available (SCBPP 1994a,b,c; SCBPP 1995). This paper provides information on the receiving-water methods manual, the process by which it was developed, and future plans for updates.

Regional Survey of the SCB

The regional survey of conditions on the mainland shelf of the SCB was conducted by the Southern California Bight Pilot Project (SCBPP) (Cross and Weisberg 1996). The SCBPP consisted of 13 federal, state, and local agencies or organizations (Table 1). The goal of this survey was to assess conditions on the mainland shelf of the Southern California Bight in summer of 1994. The survey was conducted on the mainland shelf from Point Conception, California, to the United States-Mexico border at depths of 10 to 200 m.

The regional survey was developed as an extension of the existing, local monitoring programs of the four largest municipal wastewater dischargers of Southern California. These dischargers routinely monitor the receiving waters near their ocean outfalls, which discharge waste at a depth of 60-100 m near the bottom. Most of these monitoring programs include characterization of the water column and assessment of the health of the benthos and fish. In these programs, these components of the environment are assessed or sampled using conductivity-temperature-depth (CTD) profilers, benthic grabs, and otter trawls. These sampling methods were also appropriate for sampling the mainland shelf of the SCB in the regional survey. By conducting a regional survey using these methods, we hoped to obtain data that would be comparable to those collected in the local, routine monitoring programs. However, although the monitoring methods used by the different dischargers were generally similar, they differed in many details of equipment specifications and sample collection and processing protocol. Thus there was a need to standardize sampling methods for the survey. Although previous efforts to standardize sampling methods among the agencies had not been successful, all agencies were willing to adjust their normal sampling protocol for a one-time regional survey.

TABLE 1. Agencies and Organizations Participating in the Southern California Bight Pilot Project.

- Southern California Coastal Water Research Project
- City of Los Angeles, Bureau of Sanitation
- County Sanitation Districts of Los Angeles County
- County Sanitation Districts of Orange County
- City of San Diego, Metropolitan Wastewater Department
- Regional Water Quality Control Board, Los Angeles Region
- Regional Water Quality Control Board, Santa Ana Region
- Regional Water Quality Control Board, San Diego Region
- California State Water Resources Control Board
- United States Environmental Protection Agency, Office of Research and Development
- United States Environmental Protection Agency, Region IX
- Santa Monica Bay Restoration Project
- California Stormwater Quality Task Force

Process for Developing the Marine Monitoring Methods Manual

The process for developing the marine monitoring methods manual was consensus-based and involved the participation of the 13 organizations comprising the SCBPP (Table 1). After the focus of the study was determined by the SCBPP Steering Committee, a Field Coordination Team was created, which consisted of representatives of the five agencies directly involved with marine monitoring. This team integrated existing sampling methods manuals of each of the participating agencies. The process focused the appropriate level of standardization and avoided methods that represented the lowest common denominator. It focused on methods appropriate for achieving a regional survey. In particular, methods were considered that provide good quality data but which could be used by all participating agencies. This resulted in a field operations manual for the SCBPP survey, which described sampling methods, protocol, and information specific to this survey (such as station locations, protocol for maintaining a coordinated sampling effort, etc.) (SCBPP 1994a). This manual was approved by the SCBPP Steering Committee which included representatives of all 13 agencies.

After the field sampling was successfully conducted by five agencies that had never worked together as a team on a single survey, it became apparent among all participants (both regulatory and regulated agencies) that the adjustments that were made to the field sampling protocol for this one-time survey could easily be extended to the routine monitoring programs. Thus the SCBPP field operations manual was modified by eliminating information specific to the SCBPP survey and adding some additional protocol related to the existing routine monitoring programs. This resulted in a receiving-water methods manual that could be used by all participating agencies in their routine monitoring programs or in regional surveys (SCBPP 1995). This manual was approved by the SCCWRP Commission, a nine member body with representatives of five regulatory agencies (United States Environmental Protection Agency, State Water Resources Control Board, and three Regional Water Quality Control Boards -- Los Angeles Region, Santa Ana Region, and San Diego Region) and four regulated agencies (City of Los Angeles, Bureau of Sanitation; County Sanitation Districts of Los Angeles County; County Sanitation Districts of Orange County; City of San Diego, Municipal Wastewater Department -- all municipal wastewater dischargers). Consensus was ultimately achieved because both regulators and regulated were involved in the process.

Receiving-water Methods Manual

The receiving-water methods manual provides information on navigation, conductivity-temperature-depth (CTD) profiler methods, benthic grab methods, and otter trawl methods (SCBPP 1995). It provides information on gear specifications, gear deployment protocol, field sample processing protocol, data recording, and quality assurance/quality control.

Navigation equipment and charts are standardized as well as the type and format of navigational information recorded. Standard CTD instruments and protocol are described, along with provisions for technician training, quality assurance and quality control (QA/QC) of instrument performance, deployment procedures, criteria for assessing cast acceptability, and type and format of recorded data. Grab equipment and protocol are standardized, with criteria for determining the acceptability of the grab sample. Methods for collecting and processing benthic infaunal samples and sediments samples are described in detail. Quality assurance procedures are also developed. Similarly, otter trawl specifications are standardized as are trawl deployment procedures, sample processing protocol, and sample preservation methods. Protocol for identification, enumeration, measurement, weighing, and examination for pathology of fish and invertebrates are standardized. QA/QC procedures are also delineated.

Factors Leading to Success

Although efforts had been made from time to time during the past 25 years to standardize marine monitoring methods, none had been successful up to the organization of the SCBPP regional survey. A number of factors led to the success of this project in standardizing the methods:

1) The National Research Council review of monitoring programs in Southern California (NRC 1990) was crucial to the success of this process. This study pointed out that although there were many marine monitoring programs in Southern California, there was no coordination among the programs and hence data produced by different programs are often not comparable or accessible. This resulted in a great inefficiency of effort.

2) The need for methods standardization was recognized by both the regulators and the regulated, and throughout the chain of command of both groups. In previous efforts, the need was often recognized at organizational levels that were not able to effect change. By involving agency participation at all levels, ranging from management to field personnel, the SCBPP was able to produce changes that were previously not possible.

3) The regional survey provided a framework and focus for methods standardization. It provided a goal for focusing efforts. All participating organizations agreed to the value of the conducting a regional survey and hence the need for standardizing the methods. The survey also provided a framework for organizing participation by different agencies. It provided an organizational framework with Program Managers, Quality Assurance Officers, and Field, Laboratory, Information Management, and Report Coordinators. Agency personnel with particular expertise contributed to the process at the appropriate level. The survey also provided a sense of urgency to achieve standardized field methods because a fixed time for the survey absolutely required that methods and protocol be standardized by a particular date.

4) Cost-neutral participation was another factor leading to the success of the project. The regulated community participated willingly since participation was cost-neutral due to one-time, regulator-approved, cost-neutral trade-offs of some normal monitoring activities for participation in activities crucial to the success of the regional survey.

5) The process was facilitated by a neutral organization (SCCWRP), which is jointly administered by regulator and regulated agencies. This was important because there are few opportunities for regulators and the regulated to work together. The nine member SCCWRP commission with representatives of five regulatory and four regulated agencies is a reporting body with the power to make decisions and hence to approve the use of the methods manual.

Recent Developments and Future Plans

A number of developments have occurred recently which will help to assure that standardized sampling will continue in Southern California. These include the following:

1) Incorporation of the receiving-water methods manual into receiving-water monitoring sections of discharge permits is now being discussed. If this occurs, then what was initially a one-time standardization of methods for the purpose of conducting a regional survey would then be a permanent standardization of methods.

2) To assure that the manual would evolve as technology or needs change, an on-going Marine Monitoring Methods Committee was established. This committee meets annually to consider updates to the manual. It is a body to evaluate the comparability of existing methods or the merit of new methods.

3) Plans are currently underway to conduct another regional survey in 1998. This survey may expand into new habitats and may include additional participants. If so, modifications of existing monitoring methods or incorporation of new methods will be considered as appropriate, with new agencies participating in the consensus-based standardization process.

In summary, standardization of the most commonly used marine monitoring methods had been elusive for more than 25 years until a National Research Council study focused attention on the issue. The conduct of a cooperative regional survey involving both regulatory and regulated agencies provided the urgency, need, and means to accomplish this standardization.

Acknowledgments

I thank Stephen B. Weisberg (SCCWRP) for insightful suggestions regarding this paper and providing leadership in the SCBPP project. I thank the SCBPP Field Coordination Team for their efforts in developing the monitoring methods manual, and Jeffrey N. Cross (now at National Marine Fisheries Service, Northeast Fisheries Science Center) and the SCBPP Steering Committee for their direction during the production of the manual.

Appendix I. References

Cross, J. N., and Weisberg, S. B. (1996). The Southern California Bight Pilot Project: an overview. M. J. Allen, C. Francisco, and D. Hallock, eds., Southern California Coastal Water Research Project annual report 1994-95. Southern California Coastal Water Research Project, Westminster, Calif. 104-108.

NRC (National Research Council). (1990). Monitoring Southern California's coastal waters. National Academy Press, Washington, DC. 154 p.

SCBPP (Southern California Bight Pilot Project). (1994a). Field operations manual for the Southern California Bight Pilot Project. Southern California Coastal Water Research Project, Westminster, Calif. 185 p.

SCBPP (Southern California Bight Pilot Project). (1994b). Information management plan for the Southern California Bight Pilot Project. Southern California Coastal Water Research Project, Westminster, Calif. 60 p.

SCBPP (Southern California Bight Pilot Project). (1994c). Quality assurance project plan for the Southern California Bight Pilot Project. Southern California Coastal Water Research Project, Westminster, Calif. 60 p.

SCBPP (Southern California Bight Pilot Project). (1995). Field operations manual for marine water-column, benthic, and trawl monitoring in Southern California. Southern California Coastal Water Research Project, Westminster, Calif. 53 p.

CHAPTER 193

Developing watershed Stewardship Through Improved Water Quality Monitoring

Heather Trim[1]

Years of existing water quality data were inadequate to perform waste load models and to completely assess water quality in the Los Angeles and Ventura County area. Although baseline data had been collected from as early as the 1930's, stations were spaced widely and monitoring did not focus on watershed-specific pollutants or constituents. Over the past several years, through the initiation of a watershed approach to managing water quality, the Regional Water Board and others have performed intensive surveys of watersheds in partnership with watershed stakeholders (e.g., agency staff, residents, environmentalists and surfers). These multi-station simultaneous sampling and habitat assessment surveys have provided valuable usable data as well as promoted cooperation between disparate groups for preserving and improving resources.

A Southern California Volunteer Monitoring Network was initiated in the fall of 1995. This group of agencies and environmental groups has developed a training program and a data collection system to provide data useful for water quality assessment and input into the environmental decision making process. New field observation protocols were developed that were specific to southern California. Some of these protocols focus on human health concerns, a high priority for user groups such as surfers, and on pollutants that are of great concern in urban and stormwater runoff, such as trash, that have not been adequately quantified by existing data collection protocols. Many of the volunteer monitoring projects are tied into ongoing watershed management efforts.

Both the intensive surveys and the volunteer monitoring program provide an excellent way to educate the public, provide feedback to public concerns, and provide valuable needed data for agency water quality and resource assessments.

[associate] Engineering Geologist, California Regional Water Quality Control Board, Los Angeles Region, 101 Centre Plaza Drive, Monterey Park, CA 91754, (213) 266-7553

CHAPTER 194

Effectiveness of Marine Life Refuges on Southern California Shores

Steven N. Murray[1]

Abstract

California's valuable coastal resources are used intensely for educational, scientific, commercial, and recreational purposes. Because of the intensity of this utilization, rocky intertidal systems throughout the State are being impacted by human activities, including the pulse disturbances of lawful and unlawful harvesting and the damaging effects of visitor foot traffic and human manipulation of organisms and substrata. In California and elsewhere, Marine Protected Areas (MPAs) serve as a primary tool for protecting and sustaining intertidal ecosystems. Throughout much of urban southern California, California Marine Life Refuges (CMLRs) are the principal form of MPA designed to protect intertidal organisms. However, CMLRs allow recreational fishing and, while placing restrictions on the collecting of most species of invertebrates and seaweeds, do not limit visitor access.

In a year-long study of eight sites in Orange County, California, four of which have been designated CMLRs since 1968-71, unlawful collecting was observed to be of widespread and frequent occurrence both in and outside historical CMLR boundaries. Human visitation is often extreme on southern California rocky shores and visitor foot traffic is known to damage intertidal organisms. Results of an on-going experimental study reveal that the canopy-forming rockweed *Pelvetia compressa* suffers substantial damage from trampling, including the loss of large amounts of potential reproductive material. It is concluded that CMLRs, as currently designed and patrolled, are ineffective in protecting coastal populations in regions of high visitor density. Knowledge of the current status of coastal ecosystems and the processes that affect them should be used to evaluate California's MPA system with the goal of designing truly effective refuges and sanctuaries for sustaining coastal populations.

Introduction

Natural shore and nearshore habitats and their biological resources are under constant pressure from anthropogenic activities as a consequence of the high population density and intense utilization of California's coastal zone. These pressures result in

----

[1]Professor of Biology, Department of Biological Science, California State University, Fullerton, P.O. Box 6850, Fullerton, CA 92834-6850

both press and pulse disturbances to coastal ecosystems. Press disturbances are defined as chronic perturbations, such as those resulting from continuous point and non-point discharges into nearshore waters, whereas pulse disturbances occur as periodic disruptions to coastal ecosystems such as those occurring during a major storm event or the spillage of oil or other toxicants into the environment (Bender et al., 1984, Underwood, 1989). Besides changes in coastal habitats and their living resources directly attributable to the press and pulse disturbances of regional human activity, California's coastal ecosystems are also experiencing change due to global processes that affect climate and local sea temperatures (Roemmich and McGowan, 1995; Hayward, 1996; Tegner, et al., 1996) and are being threatened by invasions of introduced species (see Carlton, 1989; 1992, 1996; Carlton and Geller, 1993).

Over the past three decades, sustained press disturbances (*sensu* Underwood, 1989), such as sewage and industrial discharges produced by the growing human population, are believed to have contributed to a long-term and widespread decline in the biodiversity of southern California's rocky shores (e.g., reviewed in Murray, 1974; Littler, 1980a,b; Litter et al., 1991; Murray and Bray, 1994). Unfortunately, quantitative documentation of this trend has often been difficult to obtain due to the absence of bench-mark data against which to compare the current status of coastal populations. However accumulating evidence indicates that southern California's heavily-used intertidal systems are being damaged by pulse disturbances resulting from human use in the form of lawful and unlawful harvesting and the damaging effects of visitor foot traffic and human manipulation of intertidal rocks and organisms.

Marine Protected Areas and The California Marine Life Refuge

In California and throughout the world, the establishment of what are broadly referred to as Marine Protected Areas (MPAs) serve as a primary tool for protecting and sustaining marine habitats and their bioresources (Ticco, 1995). As reviewed by Gubbay (1995), MPAs range in their objectives from 'no-take' zones designed to provide strict protection of marine life and habitat to those where emphasis is placed on multiple-use management. Due perhaps to the diversity of their objectives, MPAs have received a variety of names, including those often used for the designation of protected terrestrial areas such as 'reserves', 'sanctuaries' or 'parks' (Gubbay, 1995). In recognition of its rich and diverse coastline and the value of its natural coastal resources, California has been active historically in establishing MPAs and interest in creating new protected areas continues to grow. For example, according to a recent, extensive review by McArdle (1997), California currently has 104 MPAs, a figure that does not include MPAs designed to protect solely intertidal or estuarine habitats. As discussed by McArdle (1997), MPAs have arisen in California on a case by case basis by different mechanisms and often in response to a perceived threat, a process that has resulted in a complex and highly fragmented MPA system that includes many MPAs that lack clearly defined goals.

One of the most common types of MPA in California is the Refuge, particularly in the South Coast region (Santa Barbara, Ventura, Los Angeles, Orange, and San Diego Counties) where 11 of the State's 14 Marine Life Refuges are found (McArdle, 1997). California Marine Life Refuges (CMLRs) vary somewhat from site to site in the level and form of protection; however, except under special permitting conditions, it is unlawful to harvest most species of fish, invertebrates, and marine plants inside their boundaries (Smith and Johnson, 1989; McArdle, 1997). Throughout much of urban southern California, CMLRs are the sole form of MPA designed to protect most coastal organisms from pulse disturbances resulting from harvesting or collecting. However, CMLRs, and almost all other MPAs in California, do not include regulations that place limits and constraints on human access or on most forms of recreational activity other than the collecting and harvesting of organisms. Since many of California's MPAs have been created to provide some level of protection for coastal marine life, it is important to measure their success in light of the observed decline here and elsewhere in marine biodiversity (Suchanek, 1994; Norse, 1995; Murphy and Duffus, 1997). Yet, to our knowledge the effectiveness of California's MPAs, individually or collectively, has never been rigorously evaluated. The purpose of this paper is to address the effectiveness of the CMLR concept in conserving and protecting seashore life in urban regions subject to high visitor use. I discuss previous relevant work and present selected results from our on-going studies along the heavily-used Orange County, California, shoreline to raise questions concerning the performance and potential of CMLRs to serve as effective tools for protecting and conserving marine life in heavily urbanized areas.

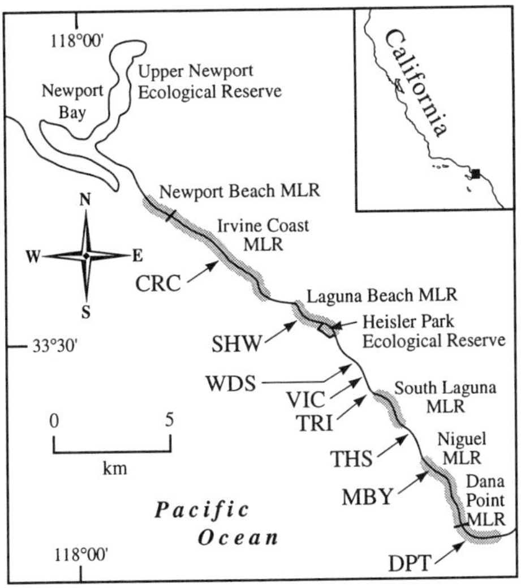

Figure 1. Map of the Orange County coastline between Newport Beach and Dana Point showing the eight study sites and the boundaries of MPAs and other sites discussed in the text. Shaded coastal areas are historical CMLRs. Historical CMLR study sites: Dana Point (DPT), Monarch Bay (MBY), Shaw's Cove (SHW), Crystal Cove (CRC); historical non-CMLR Sites: Thousand Steps (THS), Treasure Island (TRI), Victoria Beach (VIC), Wood's Cove (WDS).

Much of the rocky shore habitat of Orange County, California, lies within one of seven CMLRs whose boundaries were originally established between 1968 and 1971 (Fig. 1). Effective January 1, 1994, this shoreline area was expanded to include previously undesignated sections of the Laguna Beach coast which were accorded CMLR status as a result of SB-716. Since much of this heavily-used rocky coastline had received CMLR protection for more than 25 years, I hypothesized that differences in biological parameters and in the patterns of human activity should exist between these historical CMLRs and historical non-refuge sites if CMLR designation has been effective in protecting and conserving seashore life and deterring the intensity of unlawful collecting and other human activities. To evaluate the effectiveness of CMLRs, eight rocky intertidal study sites were established between Crystal Cove State Park and the Dana Point Marine Life Refuge along an ca. 18 km section of relatively uniform Orange County coastline (Fig. 1). Specific study sites were selected based upon the availability of substantial rocky substratum supporting well developed assemblages of the canopy-forming rockweed *Pelvetia compressa* (J. Ag.) DeToni. Here, this brown seaweed is an indicator species for protection from direct wave action and also provides important habitat for numerous species of invertebrates and algae

(Hill, 1980; Oates, 1981). Four of the eight study sites (Crystal Cove, Shaw's Cove, Monarch Bay, and Dana Point) have been located within CMLRs since 1968-1971; four sites (Wood's Cove, Victoria Beach, Treasure Island, Thousand Steps) were established outside of the historical boundaries of CMLRs or other MPAs and have only been located within CMLR boundaries since 1994. Considerable variation in accessibility based on distance from the beach entrance, the proximity of housing developments, and the availability of public parking exists among the eight sites.

The Intensity of Unlawful Collecting

As emphasized by Underwood and Kennelly (1990), protection of any biological system from anthropogenic pulse disturbance requires knowledge of the forms and intensity of impacting human activity. Unfortunately, quantitative studies of human activities when visiting rocky intertidal habitats are rare and, when performed, have usually been completed within only a few days. If CMLRs are effective in deterring unlawful activities, then collecting of algae and invertebrates should be a rare event in these MPAs and restricted to individuals with special permits. This hypothesis was tested by observing and quantifying human use, including collecting incidents, at each of our eight study sites over an entire year. Each site was visited four times during each monthly tidal series for a total of 48 visits between February 1995 and January 1996. For each bimonthly pair of assessments, one visit was performed during the week while the second occurred on a weekend day. All visits took place between sunrise and sunset over a two-hour period from one hour before to one hour after the predicted time of lower-low water. Throughout this study period, the public remained almost completely unaware of the 1994 inclusion of our historical non-refuge sites within CMLR boundaries and informative signage remained absent for all but one entrance to a beach near our Thousand Steps study area.

Unlawful collecting was a common occurrence at all of our sites, ranging from an annual monthly mean of 0.2 to 2.2 person-hours per two-hour low tide period (Fig. 2). Greatest collecting intensity occurred during the fall and winter when low tides occurred during the afternoon. Interestingly, flagrant unlawful harvesting of large quantities of organisms was rarely detected. CMLR designation and signage did not appear to deter unlawful collecting, which was most common at our most accessible study sites. Least collecting activity was observed at Monarch Bay and Thousand Steps, two of our most difficult sites for the public to reach. Greatest collecting intensity occurred at Dana Point, a well-signed CMLR, and Victoria Beach, two sites that are easily accessible to visitors. One of the most frequently observed forms of collecting was the removal and use of mussels (*Mytilus californianus* Conrad) for fishing bait. Since some form of recreational fishing is lawful inside all Orange County CMLRs, with the exception of a small area also contained inside the boundaries of the Heisler Park Ecological Reserve (Fig. 1), on-site bait removal may represent an important source of damage to mussels and other intertidal organisms at sites commonly visited by recreational fishers.

In corroboration of our own work, records of Ecological Advisements made by Laguna Beach Lifeguards during summer 1996 provide additional evidence for the widespread occurrence of unlawful collecting or ecologically damaging activity both in and outside of CMLRs and other MPAs. Despite generally arriving on duty following the occurrence of the early morning lower-low tides, during summer 1996 Laguna Beach Lifeguards issued a total of 12,269 Ecological Advisements over a three-month period (June, July and August), at 19 sites located in the northern portion of the City of Laguna Beach (Mark Klosterman, Marine Safety Chief, City of Laguna Beach, personal communication). The greatest number of Advisements (1,814) was recorded at Crescent Beach, a signed MPA location where CMLR protection has existed since

1968 and, with the exception of recreational fishing, where collecting or disturbing marine organisms is prohibited. Moreover, more than 2,278 Advisements were made by Lifeguards at three stations located inside the Heisler Park Ecological Reserve, a no-take MPA nested inside the boundaries of the Laguna Beach Marine Life Refuge (see McArdle, 1997). All collecting of marine organisms, including recreational fishing, is prohibited inside the Heisler Park Ecological Reserve without special permits.

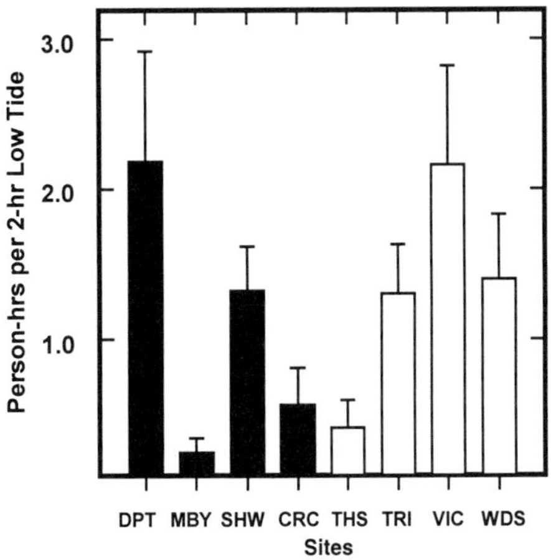

Figure 2. Human Collecting Activity at the eight study sites. Data represent the mean (+ 1 SE) of the monthly averages for number of person hours of collecting activity. See Figure 1 for site abbreviations. Historical CMLR sites are indicated by the solid bars. Person-hours of collecting were calculated for each daily visit based on the average numbers of individuals obtained over five 10-min observation periods completed at 30 min intervals during the 2-hr observational period.

Non-commercial human collecting of marine organisms has long been known to potentially impact the abundances and size structures of targeted coastal invertebrate populations (Chan, 1970; Zedler, 1978; Ghazanshahi, et al., 1983; Moreno, et al., 1984; Hockey and Bosman, 1986; Keough and Quinn, 1991; Keough, et al., 1993) and even seaweeds (Castilla and Bustamante, 1989). Although apparently unstudied along the Orange County coastline, recreational fishing pressure may also be strongly impacting local fish populations as apparently has occurred in northern New Zealand where Cole et al. (1990) found a reciprocal relationship between fish abundances and the intensity of site use. Since human harvesting usually results in greater mortality rates for larger individuals (Moreno, et al., 1984; Castilla and Durán, 1985; Hockey and Bosman, 1986; Oliva and Castilla, 1986; Catterall and Poiner, 1987; Ortega, 1987; Keough, et al., 1993; Newton, et al., 1993), collecting may result not only in declines in numerical abundance but also in significant reductions in mean individual size and shifts towards higher frequencies in smaller size classes. For many marine invertebrates, larger, older animals exhibit the greatest fecundity. Hence, continuous harvesting pressure on larger and often older individuals can reduce life expectancy, lower the reproductive value within each age class, and severely impact the reproductive potential of an entire population. And, for species like the owl limpet *Lottia gigantea* Sowerby, a sequential hermaphrodite that changes sex with age, harvesting of larger, and presumably older, individuals may provide the added impact of altering the gender makeup of the population.

Visitor Density and Damage to Intertidal Populations

Orange County's rocky shores have long served as an important resource for the southern California human population. Pre-college educational institutions throughout the region have historically made extensive use of these intertidal habitats to provide field experiences for their students. In addition, at Crystal Cove State Park and the Orange County Marine Institute at Dana Point, organized educational experiences for visitors are offered besides those designed for pre-college school classes. As examples of the total intensity of this use, in 1996 organized field trips for educational purposes brought 7,690 low-tide visitors to Crystal Cove State Park located in the Irvine Coast Marine Life Refuge (Mike Eaton and Gregory Scott, California Park Service, personal communication), 12-15,000 to a single site (Little Corona) in the Newport Beach Marine Life Refuge (Tony Melum, City of Newport Beach, personal communication), and 12,204 to the rocky intertidal zone of the Dana Point Marine Life Refuge (Helling and Goodson, Orange County Marine Institute, personal communication). Besides these visitations, numerous other groups, including college classes, make unrecorded organized visits to these and other rocky shores in the region. In addition to organized field trips, rocky intertidal habitats along the Orange County coast are also intensely used for recreational purposes on a day-to-day basis by drop-in visitors, including fishers, tidepoolers, collectors, and even walkers and sitters. At times, the number of simultaneous shore users can be extreme, as exemplified by the occasion during November 1993, when 1,443 persons visited the rocky intertidal zone at the Dana Point Marine Life Refuge during a single afternoon low tide (Helling, Orange County Marine Institute, personal communication).

Although the intensity with which southern California rocky shores receive human visitation is often extreme it is highly unusual for visitors to be denied or limited in their access to the shore. In fact, except for the three MPAs that generally must be accessed through the entrance to a military base (Vandenberg Marine Resources Protection Act Ecological Reserve), a State Park (Crystal Cove State Park), or a State Beach (Doheny State Beach), none of the 18 MPAs located along the southern California mainland from Santa Barbara through Orange Counties contain regulatory language that places limits on user access (McArdle, 1997). Even though research is limited, clear evidence has accumulated indicating that human use, and particularly the effects of visitor foot traffic and human manipulation of rocky substrata, can have significant negative impacts on vulnerable rocky intertidal populations (Boalch, et al., 1974; Zedler, 1978; Littler, 1979; Beauchamp and Gowing, 1982; Ghazanshahi, et al., 1983; Keough and Quinn, 1991; Povey and Keough, 1991; Brosnan, 1993; Brosnan and Crumrine, 1994; Addessi, 1995). Based on previously published studies and on-going experimental work (Gibson and Murray, unpublished data), it is clear that a wide variety of rocky intertidal organisms are susceptible to damage from visitor foot traffic including most species of fleshy seaweeds, geniculate coralline algae, certain tube forming polychaete worms, mussels and other bivalves, acorn barnacles, and limpets. Elimination or reduction in the cover of canopy-forming seaweeds or mussels due to trampling can have ecologically important effects not only on these populations but also on others dependent on algal canopies and mussel beds for habitat. Upper-shore brown seaweeds appear to be particularly vulnerable to human foot traffic (Boalch, et. al., 1974; Beauchamp and Gowing, 1982; Povey and Keough, 1991; Brosnan and Crumrine, 1994). These seaweeds include the rockweeds *Pelvetia compressa* and *Hesperophycus californicus* Silva whose fronds provide upper-shore habitat for numerous species of algae and invertebrates throughout southern California (Hill, 1980; Oates, 1981; Sapper and Murray, unpublished data).

Results of an on-going study (Gibson and Murray, unpublished data) indicate that southern California *P. compressa* populations suffer substantial damage from trampling and that at sites subjected to high visitor density, established rockweed thalli may loose less biomass when subjected to additional experimental foot traffic than at sites where visitor density is known to be low (Table 2). We randomly located 15 plots (0.5 m x 0.7 m) in representative *P. compressa* assemblages at each of three study sites. Based on the previously described human use surveys (Murray,

Table 2. *Pelvetia compressa* biomass dislodged from control (0 foot-steps) and experimental (150 or 300 foot-steps) plots at three study sites following the first application of trampling treatments during June and July 1996. Values shown represent mean ± 1 SE (n = 5) for the detached total and reproductive wet biomass, and the % reproductive biomass. Reproductive biomass is defined as receptacle tissue. Data were log10 transformed to achieve variance homogeneity and then analyzed by two-way ANOVA to test for significant differences among sites and treatments. Highly significant results were obtained for both site and treatment factors. *A posteriori* multiple comparison tests (Student-Newman-Keuls) revealed that all treatment means (0, 150, and 300 Steps) differed significantly and that the quantity of detached biomass (total and reproductive) recovered from Monarch Bay was greater than that collected from either of the other two sites whose means could not be differentiated from each other. (Source: Gibson and Murray, unpublished data).

Site (Visitor Density) Treatment	Total *P. compressa* Biomass (g)	Reproductive Biomass (g)	% Reproductive Biomass
Monarch Bay (Very Low)			
Controls: 0 Steps	0.9 ± 0.5	0.5 ± 0.3	52.2
150 Steps	129.6 ± 40.0	44.8 ± 10.8	34.6
300 Steps	240.6 ± 53.5	88.5 ± 13.4	36.8
Shaw's Cove (High)			
Controls: 0 Steps	0.5 ± 0.3	0.1 ± 0.1	23.4
150 Steps	41.6 ± 9.9	10.8 ± 3.1	25.9
300 Steps	78.4 ± 20.2	20.6 ± 6.6	26.2
Treasure Island (High)			
Controls: 0 Steps	0.2 ± 0.2	0.1 ± 0.1	44.0
150 Steps	43.7 ± 14.3	14.5 ± 3.3	33.1
300 Steps	89.7 ± 13.6	34.7 ± 4.3	38.7
Results: Two-Way ANOVA			
Site	$\rho < 0.001$	$\rho = 0.005$	
Treatment	$\rho < 0.001$	$\rho < 0.001$	
Site x Treatment Interaction	$\rho = 0.221$	$\rho = 0.316$	

unpublished data), we selected Monarch Bay, a site subjected to very little human use throughout the year, and two Laguna Beach study sites (Shaw's Cove and Treasure Island) where *P. compressa* assemblages are exposed to high visitor density (Fig. 1). At each site, plots were grouped into 5 blocks of 3 and, within each block, plots were randomly assigned to one of three treatments: 0 (Controls), 150, or 300 foot-steps.

We found that the initial application of the two levels of trampling resulted in the detachment of significantly more *P. compressa* frond biomass, and also frond parts (receptacles) required for reproduction, than was dislodged by wave action or other natural events from untrampled controls. Moreover, the amount of frond and receptacle biomass lost was greatest in plots receiving the 300-step trampling treatment (Table 2). These results provide evidence that visitor foot traffic can significantly damage rockweed fronds and result in the loss of large amounts of frond biomass.

Perhaps more important than simply the loss of vegetative fronds, however, is the loss of significant receptacle material. In *Pelvetia* and other rockweeds, reproduction is sexual and the reproductive organs are restricted to receptacles which develop almost exclusively on branch tips. Hence, trampling damage to receptacles can hypothetically reduce gamete production and affect reproductive success in *P. compressa*, a seaweed known (Hill, 1980; Murray and Littler, 1979) to recover slowly following disturbance.

The results of our trampling study also suggest that *P. compressa* thalli occurring in heavily-visited rocky intertidal habitats show effects of visitor pulse disturbance. At Shaw's Cove and Treasure Island, where human visitor density is high, *P. compressa* thalli are smaller, appear more ragged, and are more likely to be attached in small crevices or on the sides of small elevated surfaces that may offer protection from damaging human foot-steps. Comparisons of *P. compressa* thalli at these and our other study sites reveal that at Monarch Bay, where accessibility is difficult and visitor density is low, individual thalli are much larger and are attached on more open, flattened surfaces than at more accessible sites where human use is much higher. Interestingly, trampling treatments dislodged approximately three times more *P. compressa* biomass at Monarch Bay than at either of the other two, more heavily-used sites; a similar pattern was found for receptacle tissue where the quantity of material lost was also much greater at Monarch Bay (Table 2).

Conclusions

The results of our work, together with tallies of Ecological Advisements made by Laguna Beach Lifeguards, indicate that unlawful harvesting and damage to marine organisms takes place on a routine basis in and outside of MPAs, including signed CMLRs that have been established for > 25 years. These data suggest that widespread unlawful collecting is damaging rocky intertidal populations in urban southern California and, that despite their intent, MPA designations such as CMLRs are largely ineffective in preventing or reducing this form of anthropogenic pulse disturbance. This is in part because regulations of CMLRs and other MPAs, including the no-take Ecological Reserve at Heisler Park, are poorly enforced. As evidence of the absence of enforcement, patrolling officials of the California Department of Fish and Game, the State agency given primary responsibility for CMLRs, were never observed during 768 hours of prime low-tide observations made at our eight study sites (Murray, unpublished data). Only Park Rangers at Crystal Cove State Park and the seasonal presence of Lifeguards provided an official enforcement presence during the period of our research. Besides the absence of enforcement, often signs are not present at the entrance or inside many MPAs throughout much of urban southern California. Moreover, even when present, signage occurs only in English despite an increasing number of users lacking English proficiency, and often fails to effectively communicate MPA status and existing regulations. For example, throughout the City of Laguna Beach, numerous signs along the major coastal highway visibly advertise access to the beach and coast but, where posted, signs fail to emphasize the existence of MPAs and their regulations (Fig. 3). Through the efforts of State agencies and Legislators, attempts have been made to protect coastal marine life in California by creating CMLRs, Ecological Reserves, and other MPAs. However, it appears unlikely that the forms of protection offered by our existing MPA system, even if enforced and effectively communicated to the public, can successfully protect and conserve coastal and intertidal biodiversity in urban areas that experience extremely high levels of human visitation. In addition to the deleterious effects of unlawful harvesting, the immense volume of visitors appears to be causing ecologically important damage to populations inhabiting heavily-used rocky intertidal shores.

Throughout the State, conflicts are constantly arising between groups advocating development and access and those promoting conservation of natural resources. Unfortunately, in southern California conserving marine biodiversity has taken a back seat to terrestrial-based issues such as completing planned developments and developing transportation corridors. Securing and ensuring public access to the coast has been a high priority for many action-oriented groups and the California Coastal Commission. Yet, many of California's seashores are being subjected to high levels of human disturbance, particularly in southern California, where a highly mobile and recreation-oriented public is active year-round along a very accessible coastline. Because of its intensity, especially at certain locations that serve as attractions due to their organized programs, human visitation along the southern California coast clearly represents a significant source of damaging pulse disturbance to rocky intertidal populations and communities. To date, this form of disturbance has received little attention from governmental officials who have shown more interest in providing access and allowing use than in protecting California's coastal resources.

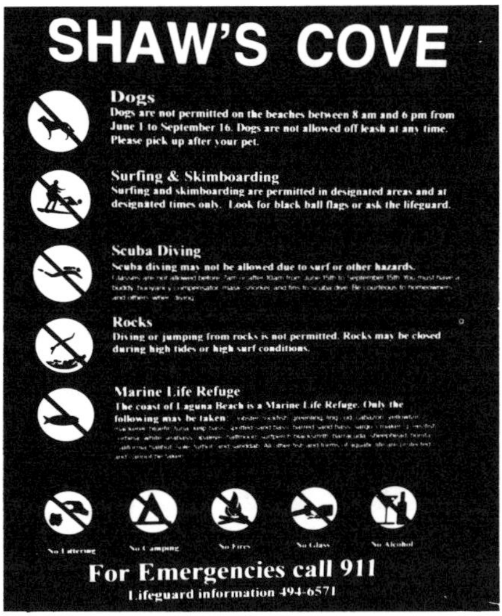

Figure 3. Sign posted at the entrance to Shaw's Cove containing easily accessible rocky intertidal habitat located inside the Laguna Beach Marine Life Refuge. Communication of CMLR status and restrictions on collecting and other unlawful activity occurs on the fifth line beneath warnings about walking dogs, surfing and skimboarding, SCUBA diving, and the dangers of diving and jumping from rocks.

Clearly, CMLRs, as currently designed and patrolled, are ineffective tools for protecting and conserving coastal populations in urban southern California. Hence, officials who seek to enable the sustainability of coastal ecosystems must identify rational policies that intercept degradative processes even if these, at times, conflict with desired forms and levels of human use. To accomplish this difficult goal, public awareness of the effects of unlawful collecting, visitor foot traffic, and other damaging activities on southern California's deteriorating coastal ecosystems must become more widely known. Scientific knowledge of the current status of coastal populations and communities and the processes that affect them must be used to evaluate the success of existing MPAs and MPA criteria with the goal of designing and obtaining truly effective marine refuges and sanctuaries if the long-term conservation of coastal resources is to be attained.

Acknowledgments

I would like to acknowledge the able and dedicated assistance of many students who have worked with me at CSU Fullerton over the past five years but particularly Teri Gibson, Julie Goodson, Shana Heid, Janine Kido, Jill Koehnke, Elizabeth Pearson, Oscar Rivas, Stephanie Sapper, Laurie Sato, Jayson Smith, and Robert Wakefield. In addition, I would especially like to acknowledge the assistance provided by Leslie Kelly and Karen Kubitz with the human use surveys and Paul Denis, Amanda Gerrard, and Sherri Stewart who made important contributions to the trampling study. I am also grateful to Kelly Donovan for preparing Figure 1 and to Steve Karl for assistance with Figure 3. This research was supported by NOAA Grant NA46RG0472 to USC Sea Grant and the California State Resources Agency.

References

Addessi, L. (1995). "Human disturbance and long-term changes on a rocky intertidal community." Ecol. Appl., 4, 786-797.

Beauchamp, K.A., and Gowing, M.M. (1982). "A quantitative assessment of human trampling effects on a rocky intertidal community." Mar. Envir. Res., 7, 279-293.

Bender, E.A., Case, T.J., and Gilpin, M.E. (1984). "Perturbation experiments in community ecology: theory and practice." Ecology, 65, 1-13.

Boalch, G.T., Holme, N.A., Jephson, N.A., and Sidwell, J.M. (1974). "A resurvey of Colman's intertidal traverses at Wembury, South Devon." J. Mar. Biol. Ass., U.K., 54, 551-553.

Brosnan, D.M. and Crumrine, L.L. (1994). "Effects of human trampling on marine rocky shore communities." J. Exp. Mar. Biol. Ecol., 177, 79-97.

Carlton, J.T. (1989). "Man's role in changing the face of the ocean: biological invasions and implications for conservation of near-shore environments." Conserv. Biol., 3, 265-273.

Carlton, J.T. (1992). "Introduced marine and estuarine mollusks of North America: an end-of-the 20th century perspective." J. Shell Res., 11, 489-505.

Carlton, J.T. (1996). "Pattern, process, and prediction in marine invasion ecology." Biol. Conserv., 78, 97-106.

Carlton, J.T. and Geller, J.B. (1993). "Ecological roulette: the global transport of nonindigenous marine organisms." Science, N.Y., 261, 78-82.

Castilla, C.P. and Bustamante, R.H. (1989). "Human exclusion from rocky intertidal of Las Cruces, central Chile: effects on *Durvillea antarctica* (Phaeophyta, Durvilleales)." Mar. Ecol. Prog. Ser., 50, 203-214.

Castilla, J.C. and Durán, L.R. (1985). "Human exclusion from the rocky intertidal zone of central Chile: effects on *Concholepas concholepas* (Gastropoda)." Oikos , 45, 391-399.

Catterall, C.P. and Poiner, I.R. (1987). "The potential impact of human gathering on shellfish populations, with reference to some N.E. Australian intertidal flats." Oikos, 50, 114-122.

Chan, G.L. (1970). "Analysis of the effects of public and educational school field trips on a marine environment, Duxbury Reef." Ph.D. Dissertation. University of California at Berkeley, California.

Cole, R.G., Ayling, T.M. and Creese, R.G. (1990). "Effects of marine reserve protection at Goat Island, northern New Zealand." N.Z. J. Mar. Freshwat. Res., 24, 197-210.

Ghazanshahi, J., Huchel, T.D., and Devinny, J.S. (1983). "Alteration of southern California rocky shore ecosystems by public recreational use." J. Envir. Management , 16, 379-394.

Gubbay, S. (1995). "Marine protected areas - past, present and future." Marine protected areas. principles and techniques for management, S. Gubbay, ed., Chapman and Hall, London, 1-14,

Hayward, T.L. (1996). "Long-term change in the North Pacific Ocean: a consideration of some important issues." CalCOFI Rep., 37, 41-44.

Hill, M.L. (1980). "Structure, organization, and persistence of the *Pelvetia fastigiata* (Phaeophyceae: Fucales) community on a rocky intertidal shoreline at Dana Point, Orange County, California." M. A. Thesis, California State University, Fullerton, California.

Hockey, P.A.R. and Bosman, A.L. (1986). "Man as an intertidal predator in Transjei: disturbance, community convergence and management of a natural food resource." Oikos, 46, 3-14.

Keough, M.J. and Quinn, G.P. (1991). "Causality and the choice of measurements for detecting human impacts in marine environments." Aust. J. Mar. Freshwat. Res., 42, 539-554.

Keough, M.J., Quinn, G.P. and King, A. (1993). "Correlations between human collecting and intertidal mollusc populations on rocky shores." Conserv. Biol., 7, 378-390.

Littler, M.M. (1979). "Assessments of the distribution, abundance, and community structure of rocky intertidal organisms at the South side of Frenchy's Cove, Anacapa Island." The distribution, abundance and community structure of rocky intertidal and tidepool biotas in the Southern California Bight, M.M. Littler, ed., Bureau of Land Management, U.S. Dept. Interior, Washington, D.C., 48 pp.

Littler, M.M. (1980a). "Overview of the rocky intertidal systems of southern California." The California Islands: proceedings of a multidisciplinary symposium, D.M. Power, ed., Santa Barbara Museum of Natural History, Santa Barbara, 265-306.

Littler, M.M. (1980b). "Southern California Rocky Intertidal Ecosystems: methods, community structure and variability." The shore environment. Vol. 2. Ecosystems, J.H. Price, D.E.G. Irvine, and W.H. Farnham, eds., Academic Press, London. 565-608.

Littler, M.M., Littler, D.S., Murray, S.N. and Seapy, R.R. (1991). "Southern California rocky intertidal ecosystems." Ecosystems of the world. Volume 24. Intertidal and littoral ecosystems, A.C. Mathieson and P.H. Nienhuis, eds., Elsevier, Amsterdam, 273-296.

McArdle, D.A. (1997). "*California marine protected areas*" California Sea Grant College System, University of California, La Jolla, Publication No. T-039, 268 pp.

Moreno, C.A., Sutherland, J.P., and Jara, F.H. (1984). "Man as a predator in the intertidal zone of central Chile." Oikos, 42, 155-160.

Murphy, D.D. and Duffus, D.A. (1997). "Conservation biology and marine biodiversity." Conserv. Biol., 10, 311-312.

Murray, S.N. (1974). "Benthic algae and grasses." A summary of knowledge of the southern California coastal zone and offshore areas, Vol. II, M. Dailey, B. Hill, and N. Lansing, eds, U.S. Department of the Interior, Bureau of Land Management, Washington, D.C., 9-1 to 9-61.

Murray, S.N. and Bray, R.N. (1994). "Benthic Macrophytes." Ecology of the Southern California Bight: a synthesis and interpretation, M.D. Dailey, D.J. Reish, and J. Anderson, eds., University of California Press, Berkeley, 125 pp.

Murray, S.N. and Littler, M.M. (1979). "Experimental studies of the recovery of populations of rocky intertidal macro-organisms following mechanical disturbance." The distribution, abundance and community structure of rocky intertidal and tidepool biotas in the Southern California Bight, M.M. Littler, ed., Tech. Pap., Bureau of Land Management, U. S. Dept. Interior, Washington, D.C., 170 pp.

Newton, L.C., Parkes, E.V.H. and Thompson, R.C. (1993). "The effects of shell collecting on the abundance of gastropods on Tanzanian shores." Biol. Conserv., 63, 241-245.

Norse, E.A. (1995). "Maintaining the world's marine biological diversity." Bull. Mar. Sci., 57, 10-13.

Oates, B.R. (1981). "Ecological and physiological studies concerning *Hesperophycus harveyanus* (Decne.) S. & G. and *Pelvetia fastigiata* f. *gracilis* S. & G., two intertidal fucoids from southern California." M.A. Thesis, California State University, Fullerton, California, 120 pp.

Oliva, D. and Castilla, J.C. (1986). "The effect of human exclusion on the population structure of key-hole limpets *Fissurella crassa* and *F. limbata* on the coast of central Chile." PSZNI Mar. Ecol. 7, 210-217.

Ortega, S. (1987). "The effect of human predation on the size distribution of *Siphonaria gigas* (Mollusca: Pulmonata) on the Pacific coast of Costa Rica." Veliger, 29, 251-255.

Povey, A. and Keough, M.J. (1991). "Effects of trampling on plant and animal populations on rocky shores." Oikos, 61, 355-368.

Roemmich, D and McGowan, J. (1995). "Climatic warming and the decline of zooplankton of the California Current." Science, 267, 1324-1326.

Smith, E.M., Jr. and Johnson, T.H. (1989). "The marine life refuges and reserves of California." *Mar. Res. Inform. Bull. No. 1, State of California. The Resources Agency,* 63 pp.

Suchanek, T.H. (1994). "Temperate coastal marine communities: biodiversity and threats." Amer. Zool., 34, 100-114.

Tegner, M.J., Dayton, P.K., Edwards, P.B and Riser, K.L. (1996). "Is there evidence for long-term climatic change in southern California kelp forests?" CalCOFI Rep., 37, 111-126.

Ticco, P.C. (1995). "The use of marine protected areas to reserve and enhance marine biological diversity." Coastal Management, 23, 309-314.

Underwood, A.J. (1989). "The analysis of stress in natural populations." Biol. J. Linnean Soc., 37, 51-78.

Underwood, A.J. and Kennelly, S.J. (1990). "Pilot studies for designs of surveys of human disturbance of intertidal habitats in New South Wales." Aust. J. Mar. Freshwat. Res., 41. 165-173.

Zedler, J.B. (1978). "*Public use effects in the Cabrillo National Monument intertidal zone.*" 1978 Project Report for the U.S. Department of Interior, National Park Service, Washington, D.C. 52 pp.

CHAPTER 195

AFTER THE PERMIT: RESOLVING CONFLICTS BETWEEN OCS DEVELOPMENT AND LOCAL LAND USES AND RESOURCES

Alice McCurdy and Nancy Minick[1]

BACKGROUND

Oil development has been a part of Santa Barbara County's coastal scene for more than a century. Between 1895 and 1900, 400 oil wells were constructed on the bluff and beaches of Summerland, on the County's south coast. Other coastal areas, such as Santa Barbara's Mesa and Goleta's Ellwood Mesa, were extensively developed as oil fields in the early part of this century. In the 1950s, the first State leases were issued for offshore drilling from platforms. The federal leasing program within the Outer Continental Shelf off of Santa Barbara County began in 1966.[2] In 1969 Platform A, south of Santa Barbara, spilled an estimated 70,000 barrels of crude oil into the Santa Barbara Channel, fouling over 100 miles of Channel beaches and killing wildlife. The effects of this significant spill were still reverberating in the late 1970s and early 1980s, when Santa Barbara County was faced with several proposals for major expansion of oil and gas production off its coast. A total of fourteen new platforms were proposed to be added to the seventeen which then existed. All of these new platform proposals were accompanied by proposals for major onshore oil and gas processing facilities. The County also was faced with applications for large-scale, oil and gas related developments such as marine terminals, pipelines, and supply bases.

With the effects of the 1969 spill etched in their memories, many of the residents of Santa Barbara County were vehemently opposed to additional oil development off of Santa Barbara's coast. However, two recent experiences may have made the County decisionmakers feel that they couldn't "just say no" to more oil. In the first instance, local permitting for a proposed Liquefied Natural Gas (LNG) Terminal at

---

[1] Alice McCurdy and Nancy Minick are Energy Specialists in the Energy Division of Santa Barbara County's Planning and Development Department.

[2] Today, much of the Santa Barbara Channel and Santa Maria Basin to the north have been leased but are not yet developed for oil and gas production. These leases account for the majority of known, developable reserves off the coast of California. Thus, current federal and state moratoria on oil and gas leasing do not preclude additional oil development in the Santa Barbara County area in the future.

Point Conception had been preempted by the State in the late 1970s. Also, a tanker converted to treat and store crude oil (Offshore Storage and Treating vessel; OS&T) was moored in federal waters just beyond the 3-mile State boundary, in place of an onshore processing facility which had been proposed for a coastal canyon. The onshore facility had been approved by the County and the State, but with air quality and oil transportation conditions that were more strict than those imposed by the federal government for the OS&T. However, the safety, air quality, and visual impacts of this offshore facility were greater than those of the original, onshore project. For the years that this facility was moored off our coast, it served as a reminder that local control over offshore oil and gas could be circumvented. With the unrest in the Middle East oil-exporting countries and the oil embargoes of the early 1970s and early 1980s, the potential for federal preemption due to national emergency was also raised by industry advocates. Both the preemption by the state for siting LNG facilities and the permitting of the oil treating facility in federal waters beyond the County's jurisdiction influenced the position the County took thereafter in considering proposals for additional oil and gas development. This position was that oil development could be allowed, provided that it was consistent with the rules set by the County. Key among these were four major requirements:

1. All oil must be transported to refineries by pipeline;
2. Oil and gas projects must reduce emissions as much as possible and provide a net benefit to local air quality (via emission reductions and pollutant offsets);
3. All environmental impacts must be mitigated to the maximum extent feasible; and,
4. Onshore support facilities for offshore oil and gas development must be consolidated to avoid proliferation of industrial development along the rural coastline.

As County planners, we faced a populace which was understandably concerned about additional oil and gas development. The people of Santa Barbara County made it clear that protection of our coast's tourism, fisheries, public safety, and natural resources was (and is) paramount. We worked for County leaders who were both supportive of the economic advantages of oil and gas development and yet adamant about the need for local control. We also had, with the California Environmental Quality Act, a strong Coastal Land Use Plan, and AB 3180 (a state requirement to monitor the effectiveness of imposed mitigations) in place, the popular support, the political will, and the planning tools to require maximum feasible mitigation of the adverse environmental effects and land use conflicts associated with oil and gas development.

RESPONSE TO IMPACTS--MITIGATION APPROACHES

Five new major oil and gas projects--Exxon's Santa Ynez Unit Expansion, Chevron's Point Arguello Project, Unocal's Point Pedernales Project, the Gaviota Interim Marine Terminal, and the All American Pipeline Project--were reviewed and permitted by Santa Barbara County in the 1980s.[3] Extensive Environmental Impact Reports/Statements were prepared to identify all environmental impacts and feasible mitigation measures for these projects. Based on these analyses, numerous measures and plans were required to mitigate the impacts of constructing and operating the new oil and gas projects. The County's intent was to eliminate, where feasible, or at least reduce, impacts and thus the conflicts among oil development and other local land use priorities and resources. The County's responses to the anticipated effects of these proposed projects took four basic forms:

1. Development of new policies and ordinances regulating activities associated with large-scale oil and gas production;
2. Mitigation programs that addressed project-specific impacts to minimize conflicts with other uses and/or natural resource protection;
3. Creation of new post-permit review committees to coordinate the examination of important details of the projects, such as design and safety features to ensure consistency with the project approvals and County requirements; and,
4. Compensatory programs to provide mitigation for significant residual and cumulative impacts that could not be fully mitigated through project-specific measures.

All four approaches have been applied to the oil and gas projects permitted by the County. Together they provide substantial mitigation of both the direct and indirect impacts of construction and operation of these large industrial projects. The following table summarizes the major mitigation approaches which the County has undertaken to reduce conflicts between oil and gas development and local land uses and resources. For each program, an assessment of its effectiveness is made.[4] Specific examples, taken from this table, of each of the four mitigation approaches are described in the remainder of this paper.

SANTA BARBARA COUNTY MITIGATION PROGRAMS
ADDRESSING OIL DEVELOPMENT AND COASTAL RESOURCE CONFLICTS

Problems	Response/Solutions	Effectiveness
Oil transportation	Policies and permit conditions requiring pipeline transportation	Moderate (almost all locally produced oil is transported to refineries by pipeline; some limited marine tankering has been requested and approved; some tankering has occurred without County approval)

[3] Several other proposed projects were reviewed by the County; some were approved and others denied. For this paper, we focus on these five projects in discussing mitigation approaches implemented by the County to reduce impacts, and thus land use and natural resource conflicts, of large-scale petroleum-related industrial development in the County.

[4] For any of these programs that may be applicable to another jurisdiction, we would be happy to provide additional information.

SANTA BARBARA COUNTY MITIGATION PROGRAMS ADDRESSING OIL DEVELOPMENT AND COASTAL RESOURCE CONFLICTS		
Problems	**Response/Solutions**	**Effectiveness**
Proliferation of industrial facilities along rural coastline	New policies requiring consolidation of oil and gas processing facilities at designated sites in the County	High
Hazards of gas liquids transportation	New policies requiring pipeline transport, truck safety measures, funding for CHP patrols	High
Hydrogen sulfide hazards	Permit conditions limiting H_2S concentration; safety review of pipeline design, facility inspection and maintenance plans	Moderate to high
Air pollution	Requirements for offsets and Best Available Control Technology	Moderate to high
Surfgrass restoration	Replanting and research	Moderate
Oak tree restoration	Replanting w/seed, seedlings, and large container-grown trees; research	Low to moderate
Marine biota impacts	Requirements to avoid and mitigate impacts to marine biota (e.g., kelp beds, surfgrass, hard bottom areas)	Moderate re: avoiding impacts; Low re: mitigating impacts that occur
Emergency access	Require new secondary access road	Moderate (should have been considered as part of the original project)
Operational noise nuisance	Noise monitoring and mitigation plans; surveys of sensitive receptors; investigation of noise-making equipment and implementation of noise reduction measures	Moderate re: avoiding impacts; Low re: mitigating impacts that occur
Wastewater discharge	Requirements for well discharge or deep-water discharge and discharge sampling	Moderate to high
Impacts to coastal recreation and viewsheds	Coastal Resource Enhancement Fund (CREF)	Moderate to high
Socioeconomic impacts (demand for housing, schools, etc.)	Socio-economic Monitoring and Mitigation Program (SEMP)	High; $9.5 million in payments to counties, cities, and special districts for impacts on public services, housing stock, and schools
Impacts to fisheries	Fisheries Enhancement Fund and Local Fishermen's Contingency Fund	Moderate to high

SANTA BARBARA COUNTY MITIGATION PROGRAMS ADDRESSING OIL DEVELOPMENT AND COASTAL RESOURCE CONFLICTS		
Problems	**Response/Solutions**	**Effectiveness**
Parking and transportation	New offsite parking lot and ridesharing required via permit conditions	High re: traffic, but may have permitted more offsite parking than necessary
Systems safety	System Safety and Reliability Review Committee; Areawide and site-specific Emergency Response Plans; Quantitative Risk Assessments	High
Coordination and tracking of multiple mitigation efforts	Environmental Quality Assurance Program (EQAP)	High
Assessing mitigation effectiveness and providing for technological advances in mitigation approaches	Periodic Condition Effectiveness Review	High
Overall permit compliance	Permit Compliance Committee and EQAP	Moderate to high

New Policies and Ordinances

Consolidation Policies

A prominent example of the County's use of policies and ordinances to reduce coastal conflicts is our adoption of strong consolidation policies. The South Coast Consolidation Policies require that new onshore processing facilities for offshore oil and gas production be located within one of two currently designated sites. The North County policies provide for the identification of up to three consolidated gas processing facility site, one of which is currently designated. On the South Coast, all existing oil and gas processing sites outside of the designated consolidation sites were rezoned to non-industrial uses. The consolidation policies have been successful in preventing a proliferation of industrial facilities related to offshore oil and gas along the County's coast and have effectively guided the location of new oil and gas development since their adoption. In March 1996, the County electorate approved an initiative (Measure A96) which requires that legislative approvals of oil and gas activities on the South Coast outside of the consolidation sites be subject to a vote of the people in a regular election. This initiative indicates continuing popular support for the County's consolidation policies.

Oil Transportation

During environmental review of several projects proposed in the 1980s, the County (and other regulatory agencies) determined that transporting crude oil by marine tanker poses significant risks to public health and safety, and to the environment. In approving the major oil and gas projects, the County found that mitigating these risks to the maximum extent feasible requires transporting the crude oil by pipeline to its refinery destinations outside the County. A major study of oil transportation conducted by the County in 1984 resulted in the County's adoption of policies and ordinance language requiring that all offshore oil processed in the County be transported to refinery destinations by pipeline. The pipeline transportation requirement described in our General and Local Coastal Plan policies was imposed as a permit condition on all major offshore oil and gas projects approved by the County. Pipelines constructed pursuant to the County's (and State's) pipeline policies now distribute locally produced oil to refineries in Los Angeles, San Francisco and Texas. These policy requirements and permit conditions have been tested several times and, for the most part, have been effective tools to mitigate against the potential for major oil spills along the County's and the State's coast by avoiding marine transport of crude oil.

Public Safety/NGL Transportation

The County also has established policies related to offshore oil and gas development by resolution, rather than adoption of amendments to the General Plan and zoning ordinances. For example, in 1985, the Board of Supervisors adopted a policy resolution addressing public safety impacts of natural gas liquids transportation. (Natural gas liquids, such as propane and butane, are removed from the oil and gas produced from the offshore platforms and are hazardous materials due to their flammability and volatility.) The Board's policy resolution requires that NGLs be blended to the maximum extent feasible into the crude oil that is processed and shipped out of the County via pipeline to refineries. For NGLs that cannot be blended into the "crude stream," the resolution originally required that the NGLs be transported out of the County via a regional NGL pipeline, if feasible, and if not, by rail or truck. The NGL policy resolution was applied through specific permit conditions to three offshore oil and gas projects (Chevron Point Arguello, Exxon Santa Ynez Unit, and Unocal Point Pedernales). In 1993, the County conducted further risk assessment of the transport of natural gas liquids and concluded that, with specific safeguards, the hazards of shipping natural gas liquids by tanker truck could be significantly reduced, and would be lower than the hazards associated with rail transport. Based on that analysis, the Board revised the policy resolution to require that those safety measures be implemented. Some of the required measures include limiting shipments to vehicles equipped with systems for governing or monitoring vehicle speed, requiring audits of carriers' safety records, requiring driver training programs, and funding increased enforcement by the California Highway Patrol on the route most commonly used by NGL carriers. By conducting a detailed risk assessment and adopting a policy resolution applying stringent safety measures to certain appropriate projects, the County was able to resolve a serious conflict between oil and gas development and public safety.

Mitigation Programs

Mitigation measures and integrated mitigation programs are one of the most common mechanisms the County has used to address the impacts and land use conflicts associated with oil and gas development on our coast. The mitigation programs initiated by the County to deal with oil's impacts and conflicts have met with varying degrees of success. Mitigation programs have been implemented to deal with habitat restoration, operational noise, socioeconomic impacts, and public safety, among others.

Habitat Restoration

To reduce the conflicts inherent between habitat preservation and oil and gas development, the County has required habitat restoration programs. For some habitats, this has proven to be relatively straightforward. As an example, a coastal site which contained a gas processing facility until its abandonment in 1992, now is a thriving coastal scrub habitat. Restoration of grassland, chaparral, and coastal sage scrub habitats has been successful in most instances along more than a hundred miles of new pipeline corridors within the County. Other biological impacts have proven to be more resistant to full mitigation, despite the efforts of both the County and the oil and gas industry. An example of a biological impact that has been very resistant to full mitigation involves oak habitat restoration. Oak trees grow very slowly, and even if they could be easily re-established, it would be decades before a seedling provided the value of a mature tree. Moreover, the operators have had limited success getting seedlings to become established and grow. The attempts to replace the thousands of oak trees (*Quercus agrifolia, Q. lobata*, and *Q. douglasii*) lost primarily to oil pipeline construction are still ongoing, ten years later. Different strategies of replanting (acorns, seedlings, and large container-grown plants) have all been attempted. One operator has funded a major oak research and mitigation project, in association with both Santa Barbara County and the University of California at Santa Barbara, in lieu of continuing with its own independent replanting efforts.

Other habitat restoration efforts have focused on restoring surfgrass (*Phyllospadix torreyi)* in oil and gas offshore pipeline corridors. As with the oaks, surfgrass removed during construction of the pipelines had to be actively restored at the site. For example, construction of pipelines and power cables through a nearshore area removed boulder and cobble that supported surfgrass. The pipeline developer originally agreed to mitigate lost surfgrass habitat through replanting efforts. However, after the impact occurred, experts hired by the developer claimed that restoration using seeds or seedlings was not feasible and that only adult plants could be moved to the site to replace the lost habitat. Similar to the oak restoration program, though of much smaller magnitude, a two-part mitigation program was developed. This program consists of a four-year restoration and monitoring effort, carried out by the operator, and a two-year laboratory research effort, carried out by the University of California. Both efforts focused on identifying effective methods for restoring surfgrass. The restoration effort tried such things as gluing surfgrass rhizomes onto rocks and clearing areas on rocks for surfgrass to spread vegetatively. This restoration and monitoring effort is scheduled to conclude this summer, at which time the overall effectiveness of the program will be assessed.

The operator provided the initial funding for UCSB's research efforts. Although indications throughout the four-year program are that restoration efforts have had only limited success, important information has been gained regarding the prospects for certain specific restoration techniques using adults, as well as seeds and seedlings.

The County, in a collaborative effort with project operators and the State Department of Fish and Game, developed another programmatic approach to mitigation for impacts to Gaviota tarplant (*Hemizonia increscens ssp. villosa)*, a protected species under the California Endangered Species Act. Essential features of this program include preservation of tarplant seed, seed bank, and soil where construction disturbs tarplant habitat; 1:1 replacement of removed plants at a "Mitigation Site;" and enhancement of tarplant (e.g., through removal of weeds/competing species) where it has not been disturbed.

Mitigation for Socioeconomic Impacts

Another generally successful mitigation program is the Tri-County Socioeconomic Monitoring and Mitigation Program, known as SEMP. This program involved monitoring the demographics of the work force related to offshore oil and gas. Offshore operators were required to make payments to documented claimants, such as school districts and providers of public services. SEMP payments were used to compensate the counties, cities, and special districts that were impacted by the increased population associated with offshore oil and gas development. A total of $9.5 million in SEMP payments were made to compensate for impacts to public services, affordable housing stock, and schools. As an example of how SEMP funding worked, a school district in Santa Maria was allocated $150,000 to accommodate its increased enrollment attributable to new offshore oil and gas development. The district used these funds to equip new temporary classrooms with desks, chairs, and other equipment.

We also have had to address chronic noise nuisance resulting from operation of one of the oil and gas processing facilities on the coast. Since the plant began operating, we have received complaints from residents about nighttime noise emanating from it. It appears that, under certain meteorological conditions, normal operational noise from the plant disturbs several residents to the northwest, often interrupting their sleep. Because a permit condition of approval and the operator's approved Noise Monitoring and Mitigation Plan require that the operator take action to reduce nighttime noise if complaints are received, we have been working with the operator to identify and mitigate the offending activities. This has been a painstaking process that, after about three years of intermittent investigation, has finally resulted in specific equipment modifications that all parties hope will eliminate the problem.

Condition Effectiveness Review

One key permit condition allows the County to continue to strive to minimize the impacts and conflicts associated with oil and gas development over time. This condition requires the periodic review of all permit conditions to determine their effectiveness in reducing project impacts. This condition also provides for the permit to be "reopened" and for new mitigations to be imposed if original conditions have proven to be inadequate in mitigating project impacts. Also, where recent proven technological advances capable of reducing a project's impacts are identified, additional reasonable conditions are required to be imposed.

New Review Committees

System Safety and Reliability Review Committee (SSRRC)

The SSRRC was established by the Board of Supervisors through specific permit conditions on the major oil and gas projects. During review of these large industrial projects, it became clear that the Board of Supervisors, as the permitting authority, could not review all the details of design and construction, and thus be able to require specific mitigation for all potential safety impacts, at the discretionary permitting stage. Companies were not willing to go forward with detailed design until they received County approval of the project but the County still had the obligation to ensure the projects were as safe as was feasible. The Board delegated the detailed safety review of the projects to a new committee comprising certain County department heads, the core members being the Fire Department/Office of Emergency Services, Building and Safety division (then in Public Works Department, now in P&D), Air Pollution Control District, and the Planning and Development Department. The SSRRC was set up as a "Brown Act" (open to the public) committee chaired by P&D. In practice, staff from each department, rather than the department heads, meet monthly (or more often if necessary) to ensure coordination among the departments in maintaining safe operations and in reviewing new projects or changes to existing projects for safety implications.

One of the primary goals for setting up the SSRRC was to provide "one-stop," integrated review by appropriate County departments for hazards and safety mitigation measures associated with the projects. The safe design and operation of each major project could be addressed collectively through this coordinated review, substantially reducing the total time required for the County to ensure its standards would be met. The SSRRC's mandate is to ensure that oil and gas projects are designed and built consistent with good engineering standards and practices, and consistent with approved safety-related permit conditions. Besides reviewing detailed construction drawings, the SSRRC also reviews operational safety and inspection plans and conducts periodic safety audits at the plants. For plants in operation, the SSRRC continues to review modifications to them to ensure safety features are adequate.

Permit Compliance Committee (PCC)

Another County response to addressing potential conflicts associated with oil and gas development was to establish a County-wide permit compliance program for facilities related to offshore oil and gas. An important component of this program was the establishment of the Permit Compliance Review Committee in 1986 by the Board of Supervisors. The PCC is composed of staff from the same departments as the SSRRC, with the addition of other departments to address specific issues of concern as needed. The PCC's role is to ensure coordination among the various departments with regulatory oversight of the oil and gas projects to avoid or resolve conflicts that can arise when regulatory mandates, permit conditions, and/or code requirements conflict with one another. The PCC now meets monthly to facilitate interdepartmental communications regarding the status of permit compliance and upcoming modifications to existing facilities, and to review new proposals. One of the goals of the Committee is to identify and resolve potential conflicts among the various requirements placed on oil and gas projects. (Enforcement of regulations and permit condition requirements is still the responsibility of each department.)

Compensatory Programs

Coastal Resource Enhancement Fund (CREF)

Some of the more innovative and far-reaching mitigations employed by Santa Barbara County to reduce the conflicts between oil and gas development and other coastal resources involved the development of compensatory programs. These new programs were intended to compensate for the cumulative and residual impacts that go beyond what can be mitigated on a case-by-case basis. As an example, the County established a Coastal Resource Enhancement Fund to help mitigate impacts to coastal recreation, coastal views, coastal tourism, and environmentally sensitive coastal resources. Each year, oil and gas operators are required to contribute to the CREF fund. In the nine years since its establishment, the County has collected approximately ten million dollars in this fund, and has allocated the funds to coastal enhancement projects as follows:

Santa Barbara County Coastal Resource Enhancement Fund

Type of project	Total allocation (1988 through 1997)
Acquisition of coastal property	$5,300,000
Capital improvements	$3,800,000
Planning and research	$ 175,000
Education/other	$ 329,000

CREF allocations this past year have funded the acquisition of land with a rare form of maritime chaparral, access improvements at a popular surfing beach, bikeway construction linking urban areas with the coast, appraisal of a large property being considered for future acquisition, and expansion of an existing marine science exhibit for school children. Ocean Beach Park is one of the best examples of coastal improvements funded through the County's CREF. Ten years ago, this county property contained only a small parking lot and a portable toilet. Recent CREF-funded improvements include a small playground with a coastal theme, picnic tables, landscaped berms to cut down on the ever-present wind, a wheelchair-accessible walkway to the beach, and permanent restrooms. The connection between this site and offshore oil development is that Platform Irene is visible off the coast from this park and its oil and gas pipelines come onshore nearby. The CREF program is one of the more successful mitigation programs which the County has undertaken to compensate for the residual impacts of oil development to reduce oil development's conflicts with local land uses and coastal resources.

Fisheries Funds

The potential for significant conflict between the oil and gas industry's use of ocean resources and commercial fishing was recognized during the County's review of the major oil and gas projects. Many conflicts, such as the exclusion of the fishing industry from large areas during construction of the offshore platforms and pipelines, could not be fully mitigated through avoidance or other conventional measures. Resolution of this type of conflict was achieved partly through the establishment of the Fisheries Enhancement Fund (FEF). The major oil and gas projects were required to make contributions to the fund, according to their contribution to the impact to commercial fisheries. In the past ten years, the Fisheries Enhancement Fund has funded eleven projects, totaling over $500,000. These projects include the construction and operation of an ice machine in the Santa Barbara Harbor for the fishermen's use, dedicating a weather radio broadcast for mariners, and funding the start-up costs for a fishermen's market at the Santa Barbara Harbor. By providing for other needs of the fishing industry, the FEF has helped compensate for some of the conflicts between these two industries. This program has been an effective tool to overcome difficulties created for one coastal zone user by another.

The County also established the Local Fishermen's Contingency Fund (LFCF) to compensate for oil and gas related gear losses suffered by commercial fishermen. When such losses occur, and the responsible oil company cannot be identified, the fisherman can apply to a federal Fishermen's Contingency Fund for reimbursement. However, this is typically a time-consuming process. The LFCF provides short-term loans that enable the fishermen to replace equipment and get back to work sooner than with the federal fund. Fishermen who ultimately are reimbursed by the federal government are required to repay the County fund. If the lost gear is found to have resulted from oil and gas development in State waters (from shore to the 3-mile limit), the LFCF payments disbursed by the County are made as grants and do not require repayment. In this case the LFCF addresses a previously unresolved conflict since the State has no fishermen's contingency fund similar to that of the federal government. The LFCF, though not widely used, has been very successful. The County has paid out claims of approximately $72,000 from the LFCF.

SUMMARY AND CONCLUSIONS

As the California County most affected by offshore oil an gas development, we have learned several significant lessons:

- There is much that can be done to reduce the conflicts and impacts associated with offshore oil and gas development;

- Visual impacts of these major facilities on our rural coastline are very hard to mitigate; facilities should be required to be located outside of public viewing corridors;

- More work needs to be done up front in planning for the abandonment of oil and gas projects. Only now are we coming to grips with the fact that abandonment of fixed deep-water (>600 ft. deep) platforms has never before been attempted.

- In assessing how well oil and gas development and coastal resources can co-exist, the risk associated with oil spills or other major industrial accidents is a wildcard. Santa Barbara County has weathered the effects of one major crude oil spill. As coastal resource managers, we are working to prevent future spills. Upholding County policies requiring pipeline transportation of crude oil is a primary means for avoiding a spill like the devastating 1969 experience.

One of the most helpful aspects of the County's permitting and compliance efforts regarding oil and gas projects has been the continuous feedback loops built in to the review and compliance tracking efforts. This communication between the permit review efforts and the permit compliance efforts is enhanced by having the same staff perform both functions, rather than having a stand-alone permit compliance unit, as is common in many regulatory agencies. For us, this feedback is invaluable since the Energy Division continues to review new and modified oil and gas projects, and because there could be considerably more offshore oil and gas development in this area in the future. We have been able to make adjustments to reduce the cost and effort of permit compliance when possible and still minimize impacts by constant attention to the need for and effectiveness of our methods for resolving land use conflicts.

In summary, Santa Barbara County has used a variety of mechanisms to reduce the impacts and thus the conflicts between oil and gas development and protection of our coastal resources. New policies and ordinances, mitigation programs, post-permit review committees, and compensatory programs have all helped to make it easier for Santa Barbara County to live with oil and gas development and to plan for future industrial development. Based on our experience, the County's answer to the question, "Multiple Use of the Coastal Zone: Conflict or Accommodation?", would perhaps be: it's undoubtedly both. It will always be hard mixing oil and water, but by diligently pursuing maximum feasible mitigation, most of the conflicts associated with accommodating oil and gas development can be minimized.

CHAPTER 196

Sport vs. Commercial Allocation Disputes in California

By Tana McHale, Fisheries Consultant

Abstract

In recent years some commercial fishermen have claimed fishery management tools have been used to exclude them from participating in harvests for the benefit of the sportfishing community. Some recreational fishermen, on the other hand, believe commercial fishermen have an unfair advantage in harvesting publicly owned resources, and assert the commercial fishing lobby has continually stalled attempts to protect California's fisheries. This paper presents the results of an investigation into "sport vs. commercial" allocation disputes, and describes common characteristics and components of such disputes so they may be identified and addressed by policy makers.

Background

Although fish and shellfish are renewable resources, even healthy populations of marine species are limited by the availability of suitable habitat, predator-prey relationships, environmental events, natural cycles and other factors. This means fisheries must be responsibly managed and shared by different groups within the harvest community, and this need to share sometimes results in resource allocation disputes.

Such disputes are particularly likely to occur in fisheries shared by both recreational and commercial fishermen. The number of recreational fishermen far exceeds the number of their commercial counterparts in California (see Figure 1); some, and possibly many, of these sportsmen are very concerned over commercial fishing gear and techniques designed to harvest large amounts of fish. While the average commercial fisherman provides tens of thousands of consumers with seafood each year and therefore harvests more than the average recreational fisherman, some sportsmen still believe that commercial industry takes "more than a fair share" of resources. This has lead to a number of allocation disputes in a range of fisheries (including the halibut, rockfish, abalone, lobster and white sea bass fisheries), with most disputes instigated by concerned sportsmen.
Allocation disputes are common not only in fisheries but in any number of other environmental resource situations, and they almost always involve some level of hostility,

posturing, politicking and outright attacks by one group against another. In his 1991 article "On The Threshold: Environmental Changes as Causes of Acute Conflict," Thomas Homer-Dixon, head of Peace and Conflict Studies Program at the University of Toronto, described components of civil strife and social unrest as related to competition for limited resources. While his work focuses primarily on the social consequences of catastrophic resource collapse, community response to simulated resource scarcity -- i.e., limiting harvests to a subset of the greater resource pool, through regulations and other harvest controls, for conservation purposes -- are similar. Homer-Dixon's findings are thus relevant for describing common responses to resource collapse and simulated scarcity situations as well:

> "...environmental degradation will 'ratchet up' the level of stress within national and international society, thus increasing the likelihood of many different kinds of conflict and impeding the development of cooperative solutions...
>
> "Individuals may have a need for a sense of camaraderie or 'we-ness' that can be satisfied in a group when it discriminates against or attacks another group; similarly, a person's sense of self-worth may be strengthened when his or her group's status is enhanced relative to that of other groups. By attacking outside groups, leaders may try to exploit these needs in order to increase their political power within their own groups, but this behavior makes divisions between groups deeper and more acrimonious..."

Conflicts such as those described by Homer-Dixon impair not only the cooperative processes but can also obstruct the application of science in resource management processes. In the context of fisheries, note that as long as fish managers control a fishery well, the fish are protected from harm whether they are chased by 500 men in 500 inefficient boats or 50 men in five efficient boats -- regardless of their commercial or recreational status. However, when fishermen begin campaigning over who should be allowed to participate in harvests, or who should have a larger or smaller share of available resources, control of the fishery is typically transferred away from managers and into the political arena. At that point the process of formally identifying,

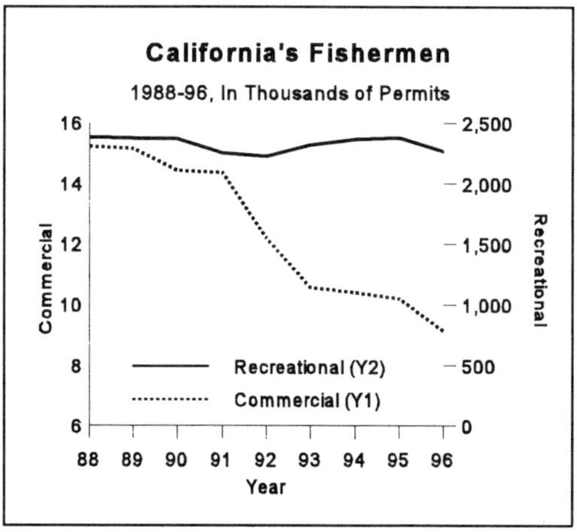

Figure 1: California fishing permits, using data reported by CDFG LRB on 02/28/97. Recreational permits include resident permits (annual and 1 day), non-resident (annual and 10 day), Pacific Ocean Only (annual, 1 day, 1 day with enhancement stamp) and reduced fee sport fishing permits, making it difficult to estimate exactly how many recreational permits were used for marine fishing during any one year.

describing, evaluating and systematically addressing specific problems is often abandoned as emotionally based arguments and accusations become the focus of attention. This encourages important decisions to be made in reaction to upset people rather than the objective analyses of facts, a fishery's overall needs or with consideration to the long-term strategic use of fisheries. As Teresa Platt of The Fishermen's Coalition observed in her 1996 article "Totems, Taboos, Sacred Cows and Tunafish,":

> "Ignorance and packaged hate are a lethal combination in the manipulation of public opinion and the mobilization of massive numbers of concerned people. The resulting policies are based on dark fantasies of the world very similar to the beliefs of primitive societies where the harvest and the earth were protected by ritual and rote, not by examination and understanding... One tribe is sent into battle against another with cries of "right and the gods are on our side!" If all the world's people were well fed, clothed and housed, a little dissent and dark entertainment would be of small consequence. But with 6 billion people and more coming soon to a planet near you, our policies need to be based on fact, not fantasy and fiction."

Likewise, public hysteria generated by some allocation disputes has preempted the role of science, the cooperative fisheries management approach and manager control over fishery resources, and also blocked California's ability to address the larger issues of seafood trade [1] and long-term food security needs[2]. The two case studies following have been selected as examples of how "sport vs. commercial" allocation disputes in California have generated such hysteria, encouraged poor decision making, and ultimately reflect the observations of Homer-Dixon and Platt.

Case Study: Proposition 132

The California nearshore gillnet fishery was heavily regulated, regularly scrutinized and subjected to at least 20 legislative actions during the 1970s and '80s[3]. By the late 1980s nearshore gillnet fishery management included an array of seasonal closures, mesh-size restrictions, species limitations and minimum size limits, gear modifications (to enable whales to break through the nets) and a series of area closures to protect seabirds, dolphins, sea otters and whales. While some sea lions were still accidentally taken in gillnets, sea lion populations were very healthy and growing.

Regardless, several legislative proposals to simply ban nearshore gillnets were introduced by then-Assemblywoman Doris Allen with sponsorship from United Anglers of California (UAC, a recreational fishing group). These bills were introduced on the premise that gillnets indiscriminately killed wildlife and threatened fisheries resources, particularly halibut stocks. At that time the public and media were already caught up in the debate over high-seas driftnetting and the Eastern Tropical Pacific (ETP) tuna/dolphin controversy; these distant fisheries and the local California gillnet fishery were often confused by the press and the pubic. However, commercial fishermens' representatives were able to address the gillnet ban proposals before legislative policy committees, and pointed out the differences between the high-seas driftnet fishery and the aggressively regulated nearshore gillnet fishery. The California Department of Fish and Game (CDFG) also released data showing healthy halibut populations, and the estimated total annual sport halibut landings were similar[4]. The bills therefore died in process.

Allen and UAC then decided to ask state voters to implement a southern California gillnet ban through Proposition 132 in 1990. This Proposition 132 "Marine Resources Protection Act" proposed to:

1. Eliminate the use of gill and trammel nets in southern California waters by January 1, 1994;
2. Increase permit fees and set up a temporary sportfishing "marine protection" stamp to finance a displaced gillnetters compensation program and research projects in ecological reserves (the Legislative Analyst estimated these fees and stamps would provide "approximately $5 million... compensation for fishermen who surrender gill and trammel nets could total up to $3.4 million if necessary legislation [is] enacted"[5]); and
3. Create four ecological reserves.

The Committee to Ban Gillnets (CBG, chaired by Allen) was established, eventually raising $603,014 to promote and run the gillnet ban campaign ("No More by '94!"). Records from the California Secretary of State's office show American Petition Consultants received $159,850 to help gather the necessary signatures for qualifying Proposition 132 for the ballot, and $320,000 for professional campaign management and consulting services. Other expenditures (including non-monetary donations) covered the costs of press conferences, radio and television airtime, mailers, hats, bumper stickers, posters, and a yacht vacation raffle prize[6].

Proponents claimed that the proposed gillnet ban would save "thousands of whales, dolphins, sea lions, harbor seals, sea otters and birds" from "indiscriminate slaughter"[7] while protecting fishery resources. The Marine Mammal Fund, the Dolphin Connection and Earth Island Institute promoted the initiative using the same claims, and the media latched onto the image of saving marine mammals and other sea life from "curtains of death."

In the meantime, the captains of small, family owned commercial fishing vessels were completely unprepared for the realities of the initiative process or the accompanying media smear campaign. They eventually raised $41,436 to oppose Proposition 132, yet the relatively small budget (less than 7% of the CBG budget) limited their ability to communicate with the voting public and deliver their message regarding the regulatory history of California's gillnet fishery, the status of fish stocks and the differences between nearshore gillnets, high seas drift nets and the ETP dolphin associated tuna fishery. In the end, Proposition 132 passed by a relatively narrow 56%-44% margin.

As a result of Proposition 132, the mandated license fee increases and stamp program generated a total of $2,655,000 by October 31, 1994. 557 gillnetters claimed to be impacted Proposition 132 by applying for compensation funds, although only 136 qualified under the final compensation program rules. 86 displaced gillnetters eventually shared a total of $1,203,421[8] (considerably less than the $3.4 million figure mentioned in the 1990 voter ballot pamphlet) with an average amount of $13,993 each[9]. CDFG records show the total number of gillnet permits dropped from 665 in 1989 to 289 in 1993 (the last year nearshore gillnetting was allowed). The number of gillnet permits dropped to 204 in the 1996 permit year[10].

Recreational and commercial catch patterns shifted in the years following implementation of the nearshore gillnet ban. CalCOFI[11] "*Fisheries Review: 1994*" summarized the change in commercial catch: "Since nearshore gillnetting has historically been the major method of take in southern California, the closure had a dramatic effect on California halibut (*Paralichthys californicus*) landings. In 1994, 232 MT [metric tons] of halibut were landed statewide, a 32% reduction from the 1993 catch. Landings in southern California decreased by 54% (80 MT landed); landings in northern California decreased by only 14% (149 MT landed); illustrating the effect of Proposition 132 on the fishery in the south..." Commercial white sea bass landings experienced a similar, 42% decline between 1990 and 1995.

While the precise effect of Proposition 132 on recreational harvests cannot be documented (recreational landings are not formally recorded), CDFG party boat survey data indicates: (1) in 1993, the last year in which inshore commercial gillnetting was permitted, recreational landings of halibut were estimated at 295,000 lbs.; (2) by 1995 their estimated catch nearly doubled, reaching 521,000 lbs. Recreational landings of white seabass nearly tripled from an estimated 36,000 lbs. in 1993 to 94,000 lbs. in 1995[12]. Finally, processors and wholesalers report they have been purchasing more seafood from sources outside of California (primarily Mexico) to compensate for reduced local harvests[13].

Proposition 132 Discussion:

This ballot initiative successfully eliminated one sector of the fishery harvest community using a combination of money, emotional appeals and by disguising an allocation dispute with conservation-oriented claims and promises. Proponents effectively converted public concern for marine mammals into "yes" votes, demonstrated apparent compassion for targetted gillnetters through the compensation program and appeared to sacrifice some of their own harvest privilages through the creation of four no-fishing ecological reserves.

However, sportsmen who promoted Proposition 132 using photographs of gillnet-entangled sea lions also support the concept of reducing California's sea lion populations through culling, birth control or other means. The average amount paid to 86 displaced gillnetters, through the compensation program, represented less than three months worth of income for the average fishermen's household. The four new no-fishing ecological reserves were eventually established in areas generally unused by recreational fishermen, and recreational harvests have apparently increased. What did Proposition 132 really accomplish? Beyond eliminating our own heavily regulated nearshore gillnet fleet, California consumer dependence on seafood imports grew, and the likelihood of eating fish harvested without the benefit of size limits, gear restrictions and other controls subsequently increased.

Case Study: Abalone Allocation Dispute

In the spring of 1995 the Abalone and Marine Resources Council (AMRC, a group of recreational abalone divers) sponsored legislation (AB 229) proposing to eliminate commercial abalone diving along the San Mateo County coastline and Farallon Islands (Fish and Game District 10, the only north-central California area open to commercial abalone diving) while allowing for continued recreational harvests in the same area. Recreational divers had been promoting variations of the plan for over six years as a solution to poaching and to address abalone resource concerns. As one AMRC representative explained in an 1990 Pacific Diver Magazine article, "Blatant abalone poaching is occurring along the north coast of California. The base cause is commercial fishing laws inspired by the commercial fishing lobby. The problem is so severe that quality recreational abalone diving on the north coast may soon be a thing of the past."

Abalone poaching had become an issue during the late 1980s, when hundreds of commercial sea urchin divers flooded into the north state. Abalone landings in District 10 increased without explanation just as northern sea urchin harvests peaked in 1987 and 1988; this seemed to indicate that abalone were being taken from areas closed to commercial abalone access in the deep north and "laundered" for sale through District 10. Furthermore, in 1989 one commercial abalone diver was sentenced to three years in state prison for poaching 196 abalone, and in 1990 CDFG concluded a two-year undercover sting operation resulting in the arrest of 12 abalone poaching suspects, four of whom were commercial abalone divers. The California Abalone Association (CAA, representing the state's 124 commercial divers[14]) responded in 1989 by sponsoring legislation (AB 3705) to (1) declare the illegal possession or transport of abalone by commercial divers north of Point Lobos as a special-circumstance crime and (2) mandating seizure and confiscation of the fishing vessel and a fine of either five times the market value of abalone possessed or $10,000, whichever is greater. This new law went into effect 1991 and reportedly served as an effective deterrent: by 1995, CDFG reported that there hadn't been any major abalone poaching arrests involving professional divers "in a period of years."[15]

Yet illegal abalone harvests continued to be a serious problem in the north state. For example, in 1990 CDFG found illegal abalone in 10% of cars stopped in one Sonoma County location; wardens wrote three dozen citations and issued 120 warnings in one day. In 1994 members of a poaching ring were arrested for smuggling an estimated 50,000 pounds of abalone to a San Diego dealer for sale on the black market; one CDFG biologist noted, "The amount taken represents 18% of the yearly legal sport take for [Sonoma] county."[16] While commercial divers were not involved in these incidents, continual references by the press to "commercial poaching" (all cases involving the sale of illegally harvested abalone) and "commercial poachers" (all individuals who sell illegally harvested abalone, regardless of their permit status) confused the situation considerably.

AMRC also continued to implicate the 15 commercial abalone divers from District 10[17] as the source of north state poaching problems and resource declines. As CDFG noted in their 1995 _Kelp Forest Newsletter_, "The question of who has the selfish attitude with regard to the remaining abalone stocks will probably pit recreational interests in the north (40,000 participating sportsmen annually generating an estimated $10,000,000 in economic activity) against the 'black market' interests and remnant commercial fishery of central and southern California." The same document also noted the northern sport fishery "removes almost two million pounds annually"; in comparison, District 10 commercial abalone harvests average roughly 80,000 pounds a year, and statewide commercial harvests declined to 264,266 pounds total by 1995.

The difference between District 10 commercial and recreational abalone regulations was also a matter of concern to sportsmen. Commercial divers are allowed to take up 84 red abalone each day using hooka (surface supplied air) gear, while recreational divers are restricted to a 4 abalone daily bag limit and breath-holding diving only in District 10 and the north state. While sportsmen assert this gives their commercial counterparts an unfair advantage, commercial divers claim the advantage is counterbalanced by difference in the recreational (7") and commercial (7¾") minimum size limits (providing sportsmen with an 7-year average "head start" on harvesting individual abalone).

Sportsmen were also enraged when AB 3705 granted commercial access to shallow waters (less than 20' deep) in District 10, while commercial divers pointed out that hundreds of miles in northern California are reserved for sport use only. These matters were debated on countless occasions in fishery advisory committee meetings, newspaper letters to the editor, newspaper and magazine articles, and before the California Fish and Game Commission over a period of years.

In 1995, when Assemblyman John Burton introduced AB 229 to "help maintain District 10 abalone populations and prevent commercial abalone divers from diving themselves out of business,"[18] both sport and commercial interests hired professional lobbyists; the six year old resource allocation dispute continued before legislative policy committees. The original proposal to eliminate District 10 commercial abalone diving eventually evolved into a package to reduce the commercial abalone bag limit from 7 dozen and 3½ dozen, reduced the recreational bag limit from 4 to 2, create a $5 Recreational Abalone Management Program stamp and corresponding advisory committee, and delete the moratorium on black abalone harvests. Yet AMRC, the bill's original sponsors, then reversed it's position and testified in opposition to AB 229, stating that anything less than eliminating commercial abalone divers from District 10 was not enough.[19] While the bill had other supporters, it eventually died for reasons that remain unclear.

The dispute continues to this day, although it recently changed form. On March 6th, 1997, AMRC representatives and divers from other recreational diving groups testified before the California Fish and Game Commission in support of a new proposal to implement an emergency closure of the entire abalone fishery, both sport and commercial, south of San Francisco. Sportsmen waved the morning newspaper (which praised the closure proposal and criticized the commercial abalone), read poetry, and dismissed CDFG's suggestion of developing an Environmental Document as too time consuming. A Natural Resources Defense Council (NRDC) representative testified AMRC had contacted and asked for assistance in the effort, and NRDC supported the emergency closure proposal.

Commercial diver representatives supported the development of an Environmental Document to formally identify, describe, evaluate and systematically investigate specific problems impacting abalone populations (including the impacts of commercial harvests, recreational harvests, sea otter predation, pollution, withering syndrome, and Sabellid worms). They also offered the use of their vessels to Department biologists for data collection purposes, and testified that newspaper articles are not a useful management tool. One commissioner noted that while sport divers appeared to be sacrificing their own harvest privileges out of concern for the resource, they would still be allowed to harvest the resource north of San Francisco.

Another stated that their decision on abalone would serve as a symbol of how they intended to address fisheries management problems. The Commission eventually voted to reconsider the emergency closure request at their May 1997 meeting; CDFG agreed to complete an environmental document by the end of the year.

Abalone Allocation Dispute Discussion:

Despite the time, money and effort invested into the eight year long Save-the-Abalone campaign, southern and central California abalone resources continue to suffer the effects of withering syndrome, increasing sea otter predation, pollution and poaching. This is hardly surprising since *allocation disputes focus on who should have access to a fishery, not problem identification and resolution.*

Further, it appears obvious that fishery management has undergone a breakdown at the expense of a once-abundant resource. The breakdown is almost certainly related to the lack of basic, fundamental information: who (and what) is impacting the resource, and to what degree? Information on human take is essentially unknown: although commercial harvest records have been formally documented for decades, recreational harvests (which began during the 1950s with the advent of scuba equipment) remain unknown. Telephone and party boat surveys designed to estimate recreational harvest levels were not implemented until 1972, and recreational abalone harvests were not restricted by either bag limits or size limits until the 1980s. The number of abalone scuba divers is also unknown (although one sport diver representative writes "There are approximately 500,000 active recreational divers in California. Each year, about 50,000 new divers are certified. An estimated 40,000 Californians harvest abalone recreationally [in the north state] each year... In southern California [there are] more than 10 times the active [sport] divers..."[20]). This information is needed not as ammunition to be used in disputes, but in recognition that fisheries cannot be properly managed on guesswork and gut-feelings.

While it is evident that managers have not controlled the abalone resource well, the press, the legislature, concerned citizens, protesters, warring factions within the harvest community and "*policies based on dark fantasies of the world*"[21] are a poor substitute for the science-based cooperative fisheries management approach.

FINDINGS

Identifying allocation disputes is not always easy

Allocation disputes are almost always presented in the context of caring for a resource through the severe restriction, or elimination, of harvest methods designated as destructive or overly efficient. Nearly everyone wants to care for a resource, and doing so normally involves an array of restrictions and prohibitions; therefore, detecting the differences between legitimate resource management and allocation dispute -based proposals can be difficult. However, after reviewing a number of alleged allocation dispute cases (involving lobster, white croaker, far-offshore longlining, a discontinued central California experimental trawl program, and the Proposition 132 and abalone cases reviewed in this paper) the following patterns emerged:

Allocation disputes are generally instigated by an active participant (or group) who:

1. Raises a resource concern relating to the activities of another fishery participant;
2. Promotes the adoption of a specific action to eliminate or severely restrict use of a gear type or fishing method;
3. Discourages the process of formally identifying, describing, evaluating and investigating various options to address the specific problem(s) or, if this process has already taken place, rejects the results;
4. Portrays the value of the targeted fishery, and it's participants, as insignificant;
5. Highlights any cases involving illegal activities in the targeted fishery;
6. Privatizes the resource through references to "our fish"; and
7. Promotes the specific action (often referred to as "the only real option" or "the right thing to do") using a combination of:
   - Campaign devices (petitions, letter drives, flyers, press releases, Letters to the Editor, and/or situation-specific slogans, banners, hats, bumper stickers or t-shirts);
   - Impassioned pleas to authorities to "save our resources" (if one authority level rejects or questions the proposed action, a higher authority is approached); and
   - Adjectives and verbs soliciting emotional responses and creating mental images (i.e., words such as "precious," "irreplaceable," "pristine," "slaughter," "devastating," "tragic," "disastrous," "doomed," "immoral").

Conversely, fishery participants who are *not* involved in an allocation dispute, or who are working to address one:

a) Focus on specific problems rather than specific solutions;
b) Advocate the collection and analyses of relevant information;
c) Show interest in investigating the potential use of more than one specific action;
d) Regularly communicate with managers, scientists and other harvesters; and
f) While conflicts, disagreements and emotional outbursts may still occur, the central focus of attention remains on identifying, describing and resolving specific resource and fishery problems.

State "preference" policies may encourage allocation disputes

"Sport vs. commercial" allocation disputes may continue to distort fishery management in California since they seem to be encouraged by (1) the availability of the initiative process and (2) policies granting apparent preference to sportsmen. These policies were developed in appreciation for the economic benefits of sportfishing and monies providing a significant portion of state fisheries agency budgets through license sales and federal Federal Aid in Sport Fish Restoration Act (SFRA) grants (see Table 1 for California fishing license and grant revenues).

CDFG Fishing License and SFRA Grant Revenues

	1992	1993	1994	1995	1996
Sport License sales	40,490,851	42,066,451	43,954,080	44,851,515	44,854,862
Sport SFRA grants	9,745,000	8,731,435	9,996,650	9,856,845	13,661,447
Sport licenses + grants	50,235,851	50,797,886	53,950,730	54,708,360	58,516,309
Commercial permits	2,659,791	2,725,780	2,849,335	2,872,885	2,895,550

Table 1: California Department of Fish and Game Fishing License and SFRA Grant Revenues (figures provided by CDFG LRB and the Inland Fisheries Division).

Funding for the SFRA program is generated through a series of national, sport fishing-oriented taxes (a 10% excise tax on fishing equipment, a 3% tax on electric trolling motors and sonar fish finders, taxes on motorboat fuels and import duties on fishing tackle and pleasure boats). These taxes raised $197.1 million in FY 1996 and another $273.2 for FY 1997 alone[22]. The funds are then allocated to individual states based on the land and water area and the number of fishing license holders for each state.

The financial burden placed on individual sportsman is relatively small, although large amounts of money are generated each year by millions of recreational fishermen who purchase taxed items for their personal use and enjoyment. A similar program for commercial fisheries does not exist, nor would it be possible for a small number of professional fishermen to pay taxes on behalf of the millions of seafood consumers who are the beneficiaries of commercial harvests. Due to general opposition to taxing food products in the United States, it is also unlikely that a seafood tax, which could reconcile the difference between sport and commercial fisheries funding, will be implemented anytime soon.

Appreciation for sport fishing related revenues has resulted in apparent preferential treatment to the recreational fishing sector. For example, the Preserving Hunting and Sport Fishing Opportunities Policy, as unanimously passed by the California Fish and Game Commission in 1994, recognizes "sportsmen's generous monetary contributions to the economy and protection of our fish and wildlife resources"[23] through the following:

> "... The Department shall emphasize programs that ensure and enhance hunting and fishing opportunities. These activities shall be integrated into all Department programs... the Department shall establish the goal of preventing loss of hunting and sport fishing opportunities."

Another policy document, the January 1966 California Fish and Wildlife Plan, enjoys wide circulation within the sportfishing community. It includes the following recommendation:

> "To give priority to recreational uses where a species or species-group under State jurisdiction is incapable of supporting both the reasonable requirements of the sport fishery and the existing or potential commercial

> harvest. Where the optimum sustainable harvest of a species or species-group is insufficient to support both the recreational and commercial demands, first priority should be given to satisfying the reasonable and legitimate demands of the recreational fishery; the commercial fishery should be encouraged to use any harvestable surplus remaining after the recreational demand is satisfied..."[24]

Whether or not this recommendation has been ratified by any legislative mandate is unclear, yet this document has been promoted by sportsmen during various Commission and legislative policy committee hearings.

CONCLUSIONS

The balance between conservation and utilization of marine resources is the signature environmental challenge of our times, requiring rigorous application of good science, a sense of commercial and environmental balance, and vigilant attention to negotiated settlements that reflect fair and equal access to available resources. Fisheries cannot be responsibly managed on the basis of preferences, hysteria, media smear campaigns, popular opinions, petitions and public votes: California's policies, laws and resource agency programs should therefore acknowledge, and discourage, "sport vs. commercial" allocation disputes that have encouraged such chaos, dissention, and reactionary decision making in the past.

Notes

1. The US seafood trade deficit is considerable; exports remaining relatively stable ($3-3.6 billion during 1990 and 1994) as imports increase steadily (growing from $5.5 billion in 1990 to $7 billion in 1994). Source: NMFS.
2. The United National Food and Agriculture 1996 report, "State of World Fisheries and Aquaculture," underscores the importance of fisheries in global food security planning. The report predicts worldwide seafood demand (109.5 million metric tons in 1994) will increase to between 110 and 120 million metric tons by 2010.
3. Title 14, California Code of Regulations, 1990.
4. Symposium on the State of California Halibut, July 1989.
5. From the 1990 Proposition 132 ballot pamphlet, Legislative Analyst Summary, provided to all registered California voters.
6. Figures provided by the California Secretary of State via copies of Committee to Ban Gillnets 1990 Form 420s.
7. From the 1990 Prop. 132 ballot pamphlet, "Argument in Favor of Proposition 132."
8. Another $1,277,000 was reserved for scientific research in the four new ecological reserves.

9. 1995 "Compensation of Displaced Gill Netters Pursuant to Proposition 132" report by J. Thomas Barnes, CDFG.

10. Permit statistics provided by the CDFG Licensing and Revenue Branch (02/28/97 printouts).

11. California Cooperative Oceanic Fisheries Investigations (CalCOFI) Report, Vol. 36, 1995.

12. Figures compiled from CDFG survey data by Santa Barbara -based fisheries consultant Mick Kronman.

13. 04/17/95 Santa Barbara News Press "*Fishing for Survival*"; 04/28/96 Ventura County Star "*Gill Nets Gone South.*"

14. CDFG Licensing and Revenue records show that there were 124 commercial abalone permits issued during 1989, the year this legislation was first drafted. The number of commercial abalone permits has since declined to 101 during the 1996/97 permit year.

15. The Waypoint Fisheries Newsletter, Spring 1995.

16. Point Reyes Light, 01/30/97 "Abalone Threatened Along California Coast."

17. California Abalone Association, pers. comm.; Letter to the Editor, 01/15/97 Sacramento Bee.

18. April 4, 1995 testimony before the Assembly Water, Parks and Wildlife Committee.

19. June 11, 1995 testimony before the Senate Natural Resources and Wildlife Committee.

20. "Financial Fact Sheet: Economic Benefits of the Recreational Abalone Fishery and Myths About the Value of the Commercial Abalone Fishery"distributed via electronic mail over the internet.

21. "Totems, Taboos, Sacred Cows and Tuna Fish" by Teresa Platt.

22. US Fish and Wildlife Service press release, "State Wildlife Agencies to Share Record $439 Million for Fish and Wildlife Projects," April 3, 1997.

23. Minutes of the June 1994 California Fish and Game Commission hearing, item 15, consideration of adoption of commission policy re: preserving hunting and sport fishing opportunities.

24. January 1966, California Fish and Wildlife Plan Volume 1, Chapter 1, major recommendations, item 4.

Keywords: conflict, allocation, dispute, abalone, Proposition 132, gillnet, policy, preference

CHAPTER 197

Governing A River of Sand--Developing Policies, Plans and Programs to Manage the San Diego Region's Beaches

Mayor Mike Bixler, City of Imperial Beach
Steve Sachs, Senior Regional Planner, SANDAG
Anne Kramer, Intern, SANDAG

The major storms of the late 1970's and early 1980's elevated shoreline erosion to the position of a major public issue in the San Diego region. During several preceding decades, people in the San Diego area had generally taken the shoreline and beaches for granted during a long period of relatively calm coastal weather. However, in the post-World War II period, human and natural events were occurring, the human ones being predominant, that would ultimately make preservation and enhancement of the region's shoreline a high priority.

Development of the region's coastal plain limited sediment flowing into streams and rivers that historically had been the largest source of beach sand. Construction of water storage reservoirs on the major streams and rivers provided a water supply and flood control for this new development; and trapped beach sediment behind dams.. As this development limited sand resources, coastal structures were put at risk. A slowly, but continually rising, sea level and more intense weather patterns resulted in more shoreline erosion and greater public concern. The growing economy also developed a strong tourism component (third largest in the region today), based to a large extent on the area's spectacular shoreline and usually wonderful weather.

When the storms of 1978 through 1983 dramatically brought the implications of all these factors to the public's attention, the local governments facing shoreline problems decided to try a new approach. They organized the Shoreline Erosion Committee, a voluntary group advising SANDAG, the regional council of governments. The Committee is made up of elected officials from the 9 coastal jurisdictions, the San Diego Unified Port District, the U.S. Navy and technical advisory members.

The group recognized that the shorelines of the various coastal jurisdictions were physically interconnected; and therefore, its mission emphasizes coordination of local actions and pursuit of cooperative regionwide efforts. With the organization of the Shoreline Erosion Committee, the region's approach to the shoreline gradually changed from piecemeal damage control, to a proactive, long-term commitment to looking for solutions to properly manage the shoreline before problems occur.

The region had several good examples of proactive shoreline solutions to draw on. The most prominent being the beachfill at Coronado during World War II which turned an almost non-existent beach into one of the widest and stablest in Southern California. While this project was a fortunate happenstance, and not the result of public policy initiatives, it does hold up a fine example of what an enormous difference shoreline management actions can make.

The problem the Shoreline Erosion Committee initially encountered in its work was a scientific and technical one. There was no real understanding of shoreline processes by the officials and decision-makers who were trying to respond to public concerns about erosion. There were also many important technical issues on which a strong consensus did not exist in the scientific and technical communities.

As a first step, the region enlisted the U.S. Army Corps of Engineers to complete a precedent setting study of the region's shoreline process and problems, the *Coast of California Study*. This $6 million, six-year study developed a detailed and comprehensive database describing shoreline sand sources and sinks, wind and wave patterns, and explaining the ebb and flow of sand along the coast.

While it did not resolve all of the scientific and technical issues, it helped the technical community and public agency planners and decision-makers agree on a core of assumptions and conclusions that allowed significant progress on policy-making and management actions. The Corps' work has recently been supplemented by the preparation of a valuable two-volume *Assessment and Atlas of Shoreline Erosion in the San Diego Region*, developed by the California Department of Boating and Waterways and SANDAG.

Using this information, SANDAG's Shoreline Erosion Committee was able to develop planning level analysis to identify critical erosion areas in the region and estimate future erosion rates to help define the extent of the problem. Potential sand sources for beach replenishment were identified, as well as beach stabilization strategies such as groins and breakwaters. The recreational and economic value of the region's beaches was estimated and goals, objectives, policies, and recommendations for shoreline management were developed. This led to the preparation of the *Shoreline Preservation Strategy*, the San Diego region's overall strategy for managing and preserving the shoreline.

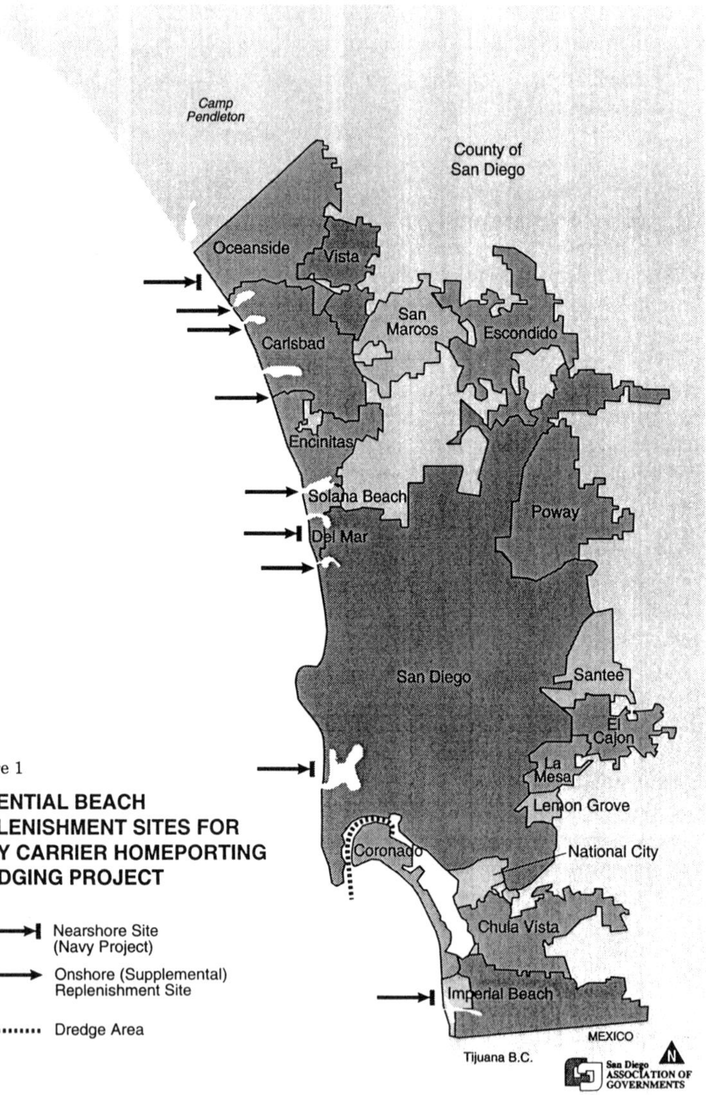

Figure 1

POTENTIAL BEACH REPLENISHMENT SITES FOR NAVY CARRIER HOMEPORTING DREDGING PROJECT

The primary tactic recommended in the *Shoreline Preservation Strategy* is "beach-building," defined as beach replenishment and nourishment. The Strategy recommends stabilization structures when they increase cost effectiveness, would not have negative impacts on down-coast beaches, and are acceptable to the local community in which they would be located.

Immediately after the Strategy was adopted in 1993, the Shoreline Erosion Committee started to work on its implementation. The state of the economy and the state of government finance, at the local, state, and federal levels, precluded a quick solution through massive public funding of the region's beach-building program.

The Shoreline Erosion Committee, while continuing to look for regionwide beach-building financing and projects, decided to devote an equal effort to promoting "opportunistic sand projects," projects that would be excavating large amounts of beach compatible material at or near the coastal zone. Historically, very few of these types of projects dispersed the excavated material on the beach; however, they provided a "target" source of beach replenishment material that might be leveraged with a modest amount of regulatory coordination and funds.

The most significant opportunistic sand project the Shoreline Erosion Committee has had the opportunity to be a part of is the Navy Homeporting Sand Replenishment project. This project involves the dredging of approximately 7 million cubic yards of clean, beach compatible sand from San Diego Bay to allow for the homeporting of a Navy carrier at North Island Naval Air Station. Sand from the Homeporting Project will be pumped onto the region's beaches starting in the summer of 1997. (Potential beach replenishment sites shown in Figure 1.) The project will go on for approximately one year, depositing enough sand to fill two times the stadium where the Padres and Chargers play.

Through the development of the *Shoreline Preservation Strategy* and the promotion of opportunistic sand projects, the San Diego region has taken great strides in developing a philosophy and program of stewardship of our beaches. This is the beginning of a very important undertaking in the preservation and management of our valuable coastal resources.

CHAPTER 198

Development of a Shoreline Preservation Strategy

Christian M. Appendini[1] and Román Lizárraga-Arciniega[2]

Abstract

Rosarito Beach, Baja California, is a small town with a tourism oriented economy based on its beaches. Erosion has been observed in the last years and it is considered a critical problem for the town. A shoreline preservation strategy that meets the interests of the community over its beaches, working in the context of coastal processes, is needed to maintain this important economic asset. This paper describes the steps towards the development of a shoreline preservation strategy for the area.

Introduction

Shoreline erosion is of great concern to coastal urbanized areas since it represents a threat to infrastructure, development, private property and to the sandy beaches used for recreation. Rosarito Beach, Baja California, is located 20 km south of the Mexico-US international Border. The primary economic activity in this zone is tourism, this activity has developed increasingly in recent years, with the long sandy beaches being one of its greatest attractions. Sandy beaches are now being threatened by erosional process, decreasing its recreational capacity and protective nature. Beach preservation is a high priority, however, there is no plan to preserve the shoreline nor any program for monitoring the shoreline evolution. The importance of the beaches as an economic asset pinpoints the need for the development of a shoreline preservation strategy.

---

[1] Graduate student, Facultad de Ciencias Marinas, Universidad Autónoma de Baja California. Suite ESE-1306, 4492 Camino de la Plaza, San Ysidro, CA 92173- 3097.

[2]Researcher, Instituto de Investigaciones Oceanológicas, Universidad Autónoma de Baja California. P.O. Box 453, Admon 1, Ensenada, B.C., Mexico.

Historical background

Rosarito Beach is a recently formed municipality in the Pacific coast of Baja California, 20 km south of the Mexico-US International Border. Rosarito Beach started as a cattle ranch known as El Rosario in the mid-eighteen hundreds. Since those days, it was a tourism destination for San Diegans, who went to enjoy its magnificent beaches. In the 1920's two major resorts where built: the Shore Acres Country Club, now known as the Rosarito Beach Hotel; and the Rosarito Beach Resort, now know as Renee's. With these resorts the economy of the area started to shift towards tourism, and in the 1930's tourism became the first economic activity of the town. In the second half of this century, this part of Mexico started to shift its economy towards the maquiladora industry, that became the most important activity for cities as Tijuana and Mexicali, but Rosarito maintained its tourism oriented economy (Piñeda-Ramirez, et al. 1989). Nowadays, tourism is the primary economic activity for Rosarito Beach, bringing more revenues to the city than any other activity and employment for 65% of the active economic population (EURA, 1992).

Problem statement

A fundamental piece of Rosarito Beach tourism vocation are its beaches, clearly being the natural asset that support its economy, as pointed out in several brochures that states phrases as: "20 miles of magnificent coastline, wide sandy beaches ..." (RTT), "pristine coastlines that are ideal for tourism related developments ... Rosarito is known for its lovely, wide beaches ..." (SECTUR). Although the importance of the beaches is significant for the region economy, development along the shore has been accelerated and lacking of coastal management, threatening this important asset.

The increased recreational demand has lead to the construction of new resorts, hotels and houses along the shores of Rosarito, augmenting the pressure over the beaches. The Rosarito beaches have undergone significant retreatment, as appreciated by the old local residents. The erosion problems along the shore have resulted in the construction of coastal structures, in particular seawalls, affecting the coastal processes of the area. Erosion is now considered a critical problem by the local government; despite this fact, there is not a management plan for the beaches, a policy to protect this resource or a monitoring program to address the problem. Protection of property has been on a particular basis, in some cases reducing the threat of erosion against property and in other cases not, but the reduction in the recreational capacity of the beach has not been addressed.

The shoreline preservation strategy

The need of a shoreline preservation strategy is critical to maintain a desirable beach width that provides property protection and recreational capacity of the beaches. A preservation strategy would provide a list of solutions for the shoreline erosion, giving consideration to the coastal processes, sand supplies, infrastructure, property protection, recreational demand, financial resources, regulatory constrains and local community interests on the shore.

To develop the shoreline preservation strategy it is needed to address each segment of the beach to its particular problems and interests, working in the context of the coastal processes of the entire beach system.

The primary land uses on the Rosarito Beach shoreline are:

- Industrial
- Residential
- Mass tourism
- Local population housing

Considering these uses, the Rosarito Beach shoreline can be divided in three segment :

North zone	Central zone	South zone
Residential	Industrial Local pop. Housing	Residential Mass tourism Local pop. Housing

The north zone is characterized by sandy beaches backed with low cliffs. This zone has experienced accelerated residential development in the last decade. Erosion problems are present and because the lack of an erosion management program, homeowners have protected themselves, armoring the shoreline with seawalls. This may be affecting downdrift beaches in the south, adjacent property, and also the beaches in the same zone, reducing its recreational capacity.

The central zone is the transition between the sandy beach backed by low cliffs and the sandy beach backed with sand dunes, although dunes are no longer present due to the urbanization. This zone is where the thermal-electric power plant and the Mexican oil company Petroleos Mexicanos (PEMEX) are situated. South of these industries some local population houses have suffered severe erosion, as a direct effect of the water intake structures of the thermal-electric power plant.

The south zone is characterized by sandy beaches that where used to be backed by dunes, but now the dune fields have been replaced with infrastructure. In the north part of this zone is where most local population housing is located, and severe erosion problems and flooding during storms occurs. Consequently local residents have abandon houses or protected themselves with poor degree of success. In the south part of this zone is where the mass tourism is located, as well as some residential housing. In this area the erosion problems are not as critical as in the other areas, but narrowing of the beach has been observed. As a result the Rosarito Beach Hotel have try to restore the sand dunes as a protective measure to storm waves.

From this zonation we can appreciate that the erosion problems and the response to those problems differ from zone to zone. The consideration of this zones in the shoreline preservation strategy is indispensable to address the erosion, considering the needs and problems in each segment of the shoreline, but thinking of the segments as part of one integrated system.

As presented by Shoreline Solutions (1995) basics phases towards the development of the shoreline preservation strategy rely on the elaboration of inventories, analysis of data and implementing actions. This approach facilitate the study of particular segments of beach according to its nature of development, that is, to avoid hazards or to protect, by some engineering technique, the beach segment under consideration.

The major steps towards the shoreline preservation strategies are the following:

Coastal processes

The understanding of the coastal processes is critical to provide the framework to develop strategies. The needs and priorities are:

- Historic evaluation of shoreline changes
- Geomorphologic characterization
- Sediment budget
- Human impacts on the sediment budget

Actors Involvement

The actors involvement during the implementation of the strategy is indispensable for its success. The principal actors to consider in the strategy are the local government, hotel owners association, property owners, PEMEX and the thermal-electric power plant. The principal points to consider are:

- Education and public awareness
- Regulatory regimes
- Local development plans

Management strategies

Working with the frame of coastal processes, and considering the actors interests on the shore, a set of management tactics can be developed. From this tactics, the most adequate are to be selected to provide the shoreline preservation strategy

Conclusions

Rosarito Beach is a tourism oriented town that relies primary on its beaches. Erosion has been evident in recent years and is a critical problem that has not been address. The development of a shoreline preservation strategy is critical for the economy of the area.

In Mexico there is no prior experience in these kind of management planning. The development and implementation of this kind of plans are consider by the authors as a major step towards the coastal zone management in Mexico.

References

Espacio Urbano y Arquitectura, S.C. (EURA). (1992). "Programa regional de desarrollo urbano del corredor Tijuana-Ensenada, B.C. Segunda etapa." EURA unpublished report. Tijuana, B.C.

Piñeda-Ramirez, D., Ortiz-Figueroa, J. (1989). "Historia de Tijuana. Edición conmemorativa del centenario de su fundación." Centro de Investigaciones Históricas UNAM-UABC, Universidad Autónoma de Baja California, Tijuana, B.C., Mexico.

Rosarito Tourism Trust (RTT). Rosarito Beach County Brochure.

Secretaria de Turismo (SECTUR) and Gobierno del Estado de Baja California. Baja California, Mexico Brochure.

Shoreline Solutions. (1995). "Littoral cell management planning along the Oregon coast." Oregon Department of Land Conservation and Development, Salem, OR.

CHAPTER 199

RESTORING THE STATE'S ERODING BEACHES AN INNOVATIVE CASE STUDY - SAND REPLENISHMENT FROM A RAILROAD GRADE SEPARATION IN THE CITY OF SOLANA BEACH

Councilmember Joe Kellejian
Director of Community Development Steven Apple
City of Solana Beach

Since 1994, two public agencies have coordinated a multi-phased, 18 million dollar, railroad grade separation project designed to lower North County Transit District existing railroad tracks below the City of Solana Beach's main east-west vehicular arterial, Lomas Santa Fe Drive. Beach sand quality deposits were identified during test borings and an opportunistic beach sand replenishment program was developed.

During the past 30 months, these two public agencies have pursued this opportunistic beach sand undertaking, with Federal and State Agency permits obtained with varying degrees of success. Expensive chemical and physical properties testing were undertaken and extensive elected official, staff and community resources expended. Lessons learned include the importance of persistence, being focused, knowing when to compromise and when not to, and obtaining permit processing time-line commitments. As of the preparation date of this abstract, all permits have been obtained for a minority portion of the opportunistic Grade Separation material. An open-ended blank check mitigation measure has jeopardized the majority of material available for beach sand replenishment.

CHAPTER 200

Governing a Community at the Edge of the Ocean

James Bond and Patrick Murphy[1]

Governing a city at the edge of the ocean has revealed that conflicts between those living at the beach, using the beach and governing the beach are more difficult to resolve than the natural conflicts between the ocean and land. It is generally recognized that science and technology exist to resolve nearly all issues that arise from natural damage and erosion, however, getting all of the stakeholders involved in, and in alignment with, an overall solution is, indeed, most challenging.

In an attempt to address these concerns in a way that gained the involvement and alignment of all of stakeholders, the City of Encinitas prepared a Comprehensive Coastal Bluff and Shoreline Plan. The purpose of this long-range planning document is to address shoreline erosion and bluff recession in a comprehensive and inclusive way.

What follows is our experiences, issues and setbacks encountered along with some perspective of local policy development while considering this very diverse group of stakeholders.

Introduction - Perspective From A Policy-Maker, Councilmember James Bond

A little over four years ago when campaigning for local office, I became aware of groups of alarmed and angry homeowners, surfers and beach-users. Each group with generally opposing positions, e.g. homeowners desperately wanted to protect

[1] James Bond, City Council Member, City of Encinitas, 505 S. Vulcan Avenue, Encinitas, CA 92024-3633
Patrick Murphy, Planning Consultant, P.O. Box 230367, Encinitas, CA 92023-0367

their homes and property with rip-rap, while beach-users and surfers didn't want any part of a solution that would limit public access along our narrow to non-existent beaches. In time, one thing became abundantly clear .... all of these groups were in basic alignment around the notion that sand on our beaches was both good and necessary for all concerned.

This experience, coupled with new understanding of our shoreline, how sand is naturally transported up and down the coast and the general causes of shoreline erosion, led to the inescapable conclusion that science, politics and policy-making, though distinctly different, are inextricably linked and must collaborate and cooperate in the ultimate solution to shoreline retreat and erosion.

As a policy-maker for the City of Encinitas, I have little scientific, engineering or geological expertise in the area of bluff or shoreline protection. In spite of that, I am heavily involved in long-range plans and setting standards that address bluff and shoreline erosion. Likewise, the scientific community along with beach-goers and surfers now find themselves involved with some very interesting and challenging policy issues. This is as it should be. No one of us can possibly know all there is to know about everything. It is possible, however, for each group to "add value" to the others' thinking and capability in the pursuit of a common, worthy goal. Such diverse expertise, properly focused and aligned, can create endless new possibilities that not only lead to a solution, but it will be a long-term, durable solution that is acceptable to all stakeholders.

As can be seen, governing a City at the edge of the ocean is not only involved with resolving the natural conflicts between the ocean and the land, but between those using the beach and the myriad of regulatory bodies also attempting to save and restore the shoreline.

Accordingly, the City of Encinitas is now deeply involved in addressing social, natural and political conflicts. Much of this, through the development of a Comprehensive Coastal Bluff and Shoreline Plan. Our aim is to address coastal bluff recession and shoreline erosion in a comprehensive way.

Encinitas, The Setting

The City of Encinitas is located in the north San Diego County Area, California, and has approximately 6.1 miles of shoreline extending between two lagoons, Batiquitos Lagoon to the north and San Elijo Lagoon to the south. Much of the City's shoreline is backed by steep bluffs, ranging in height from 70 feet to 90 feet, with residential development along most of its length.

The residential development along the top of the bluff is primarily single-family with some multi-family buildings. The typical single-family lot is relatively small with lot depths ranging from 80 feet to 130 feet measured from the top of bluff to the street right-of-way. The typical lot width is approximately 50 feet. The existing building setback from the bluff edge ranges from 5 feet to 50 feet, with the majority of homes having a setback of about 30 feet.

Most of the Encinitas' coastline consists of narrow to nonexistent sand or cobble beaches. Under natural conditions the beaches are fairly narrow. Human intervention has, however, affected the shoreline in Encinitas during the last 50 years by the construction of Oceanside Harbor, located to the north. As a result, the naturally narrow beaches have become narrower and coastal bluff failures have occurred in the recent past.

The combination of narrow beaches and unstable bluffs has created an undesirable situation. Residential development has historically been allowed on the top of the bluffs with minimum setbacks from the bluff edge. As that is the case, many coastal homeowners have implemented beach and bluff stabilization measures to protect their homes from unstable bluff conditions.

The narrow to nonexistent beaches, steep and unstable bluffs, and homes in close proximity to the bluff's edge have led alarmed property owners and politically sensitive regulatory bodies to some reactive solutions through the emergency permitting process.

Such protective measures and other improvements have been incrementally done without comprehensive policies and standards.

A reactive permitting process, responding only to emergency situations, has created visually unpleasant conditions, gaps between seawalls, and a variety of protective devices that differ in appearance, size and effectiveness, many of which, are massive seawalls and upper bluff walls.

City's First Attempt to Reconcile The Issue

With several homes being threatened, property owners requested that the City step-in and attempt to circumvent the Coastal Commission emergency permitting process. The blufftop property owners' desperate cry's for help persuaded the City to declare the deteriorating bluffs a dangerous "public nuisance." The theory being that if the City declared the bluffs a public nuisance, then the City would have the ability to circumvent State law and the Coastal Commission. Initially, the Council thought that this approach had merit, but later discovered that the risks were far greater than the rewards to the few it would benefit.

In our next effort, the City worked with various blufftop property owners to establish a Geological Hazard Abatement District (GHAD), which is authorized by the State of California legislation. The purpose of a GHAD is to provide a mechanism for people to protect their property from known geologic hazards, or in our case, coastal bluff erosion, landslides, etc. State law requires that a GHAD establish a "Plan of Control" and allows for the establishment of a funding mechanism for construction purposes and long-term maintenance effort.

This seemed like a viable solution. Consequently, both the City and California Coastal Commission supported this concept. The establishment of a GHAD and companion comprehensive plan left the California Coastal Commission with the feeling that a more comprehensive approach was, indeed, underway, and they authorized a number of seawalls along many of the properties located in the GHAD.

The City assisted the GHAD Board, which was composed of property owners located within the District, with loans and staffing during its first two years to get it up and running. In particular, the City was interested in the District being able to financially self-support itself. However, the decision to assess the property owners for the construction of bluff protective devices and long-term maintenance belonged to the GHAD, or property owners, and this is where things began to unravel.

Property owners did not want to contribute through an assessment district to maintain or construct seawalls unless it was their own seawall. Hence, the GHAD had become economically infeasible .... no funding.

It also became apparent that property owners didn't want to be part of the GHAD unless they were experiencing a problem. It was difficult for property owners to see the advantages and financially contribute through an assessment district if "today" they didn't have a problem, even though the odds are pretty good that "tomorrow" they will.

With no funding source the Geologic Hazard Abatement District was dissolved.

A More Comprehensive Approach Is Needed

Every coastal community within the State of California is required to prepare and adopt a Local Coastal Program which helps to protect the natural and scenic resources of the California coastal area. The City's Local Coastal Program was approved by the California Coastal Commission in May of 1995. As part of the Local Coastal Program, the Coastal Commission required the City to prepare a Comprehensive Plan to address coastal bluff recession and shoreline erosion problems within the City.

The Comprehensive Plan is to address: minimum blufftop setback requirements; alternatives to shore/bluff protection such as beach sand replenishment; removal of threatened portions of a residence or the entire residence or underpinning existing structures; addressing bluff stability and the need for protective measures over the entire bluff (lower, mid and upper); impacts of shoreline structures on beach and sand areas as well as mitigation for such impacts; impacts of groundwater and irrigation on bluff stability; and visual impacts of necessary/required protective structures.

It has been said that the Encinitas shoreline is the most studied coastline in the State. This study along with other previously prepared studies were used as the basis for recommendations within the Comprehensive Plan.

Studies of Encinitas Coastline

- ✓ US Army Corps of Engineers Reconnaissance Report for the Encinitas Shoreline, March 1996.
- ✓ Moffatt and Nichol Technical Report for the City of Encinitas Comprehensive Plan, 1996.
- ✓ Beach Bluff Erosion Technical Report for the City of Encinitas, Zeiser Kling Report, January 1994.
- ✓ Shoreline erosion Assessment and Atlas of the San Diego Region, Volume I and II, California Department of Boating and Waterways and SANDAG, December 1994.

Several reports have been prepared addressing the existing conditions of the Encinitas shoreline. The above reports generally describe the Encinitas shoreline in segments and typically identify whether the existing conditions represent a high, moderate or low risk. High risk is generally defined as a segment of a bluff where a high probability exists of structural damage unless bluff protection is provided in a timely manner. The reports conclude that much of the Encinitas shoreline is considered to be in a high risk category requiring bluff protection.

The US Army Corps of Engineers Reconnaissance Report for the Encinitas Shoreline concluded that for a relatively large area of our coastline the bluff has already been so compromised that more than 100 homes are jeopardized and in danger of experiencing undermining and damage, resulting in an estimated annual costs of $598,000. Even with an aggressive sand replenishment program, the report concluded that the upper bluff would continue to retreat, resulting in an annual structural damage of $465,000.

To address this and other issues, the City has prepared a Comprehensive Coastal Bluff and Shoreline Plan. As a long-range planning document that addresses the coastal bluff recession and shoreline erosion problems in the City, the Plan takes a comprehensive look at the bluff and shoreline issues within the City of Encinitas and establishes goals, policies, standards and strategies that the City will pursue.

Components of the Comprehensive Plan

- ✓ *Sand Replenishment and Retention*
- ✓ *Recreation*
- ✓ *Public Access*
- ✓ *Groundwater*
- ✓ *Cobble Management*
- ✓ *Shoreline and Bluff Protective Devices/Measures*
- ✓ *Blufftop Development Standards*
  (Setbacks, Private Stairways, Irrigation, Landscaping, Surface Runoff/Drainage, and Small-Scale Groundwater Control)

The Comprehensive Plan addresses a variety of issues, including: sand replenishment and retention, recreation, public access, groundwater, cobble management, shoreline and bluff protective devices, and blufftop development standards, relating to setbacks, private stairways, irrigation and landscaping, surface drainage runoff, and small-scale groundwater control.

The overall goal of the Plan relates to sand replenishment and the desire to establish a wide sandy beach along the City's shoreline to provide increased recreational opportunities and bluff protection, benefiting both the City through tourism and economic incentives, and property owners through bluff protection.

The Comprehensive Plan has received wide public support on a local level. One of the challenges facing the City will be working with other agencies to broaden that support.

The Comprehensive Plan is not without controversy. Much of the controversy and disagreement is between agencies. These issues relate to 1) financial resources needed to implement the Plan, 2) sand mitigation fees to address the impacts caused by seawalls, 3) planned retreat - the concept of moving your home back as the bluff retreats, 4) installation of bluff protective devices - seawalls and bluff walls, and 5) blufftop lateral access.

Although significant differences exists between agencies on what should be contained in a comprehensive plan and the supporting documentation, the bottom line for the City is that we want to stop being reactive in addressing the needs of the our coastline .... from sand replenishment to property protection .... the City is developing strategies to guide both property owners and decision-makers for the long-term benefit of everyone.

Concluding Observations:

The City does not want reactive solutions that result in massive seawalls and equally large upper bluff walls. Since bluff protection is authorized for the protection of one's structure and bluff failures will continue to happen, the City wants to be able to react before a bluff failure occurs through the proactive installation of seawalls that match the existing bluffs and halt the undercutting of the bluffs that eventually leads to bluff failures, structural damage, and massive seawalls. Coupled with this, the City needs to pursue other programs relating to sand replenishment, groundwater programs, and cobble management, along with private property owners implementing various standards and strategies for their individual properties.

From a local agency's perspective, governing a city at the edge of an ocean is not easy and requires not only local support, but regional, state and federal support.

That is why it is important to gain the cooperation and collaboration of all of the local, regional, state and federal governments and agencies in the pursuit of the common goal to maintain a safe, sandy beaches

As a local leader of a beach community on the edge of the ocean, I can tell you that at times the challenge is literally overwhelming. I have to regularly look over my shoulder to see if these diverse groups are following. If not, it's a good bet we are not leading the way to a realistic, long-term solution.

CHAPTER 201

Carlsbad Opportunistic Beach Fill Program

Steven Jantz[1] and Christopher K. Webb,[2] Secretary of the California Shore and Beach Preservation Association

ABSTRACT

The Opportunistic Beach Fill Program is being established by the City of Carlsbad to streamline the permit process for beach replenishment. Opportunistic beach fill is beach-suitable material generated from construction projects and made available at low or no cost. The program will: 1) renourish the littoral cell, 2) maintain a federal replenishment project, and 3) improve shore protection and recreation. The City proposes to create a general permit for beach fill projects assuming sand properties and placement operations meet certain criteria. Carlsbad would be granted discretion to approve beach fills, while notifying resource agencies of pending projects. Material criteria pertain to chemistry, gradation, color, compactibility, etc. A checklist will be used by City staff in assessing each project. A greater level of agency scrutiny could occur when beach fill sources are marginally acceptable (consist of more than 15% fine-grains) or are large quantities (greater than 80,000 cubic meters).

INTRODUCTION

Beaches in the City of Carlsbad are typically narrow and covered with cobble, and beach replenishment is considered a long-term solution. The City of Carlsbad

---

[1] Steven Jantz, Associate Engineer, City of Carlsbad, 2075 Las Palmas Drive, Carlsbad, CA 92009-4859

[2] Christopher K. Webb, Coastal Scientist, Moffatt & Nichol Engineers, 250 W. Wardlow Road, P.O. Box 7707, Long Beach, CA 90807

seeks to implement an Opportunistic Beach fill Program, in which suitable beach fill material generated from terrestrial sources are considered for beach disposal. Opportunistic beach fill is material which becomes available from upland construction projects, and is therefore available at no cost or low cost compared to material used primarily for beach enhancement.

The program purpose is to streamline the permit process for implementing beach fill projects for purposes of 1) renourishing the Oceanside Littoral Cell; 2) providing on-going maintenance of the U.S. Navy Homeporting Project, and 3) improving protection to coastal structures and beach recreation opportunities in the City.

The program proposes to obtain general permits from agencies to allow for potential opportunistic beach fill projects in the future, rather than obtaining individual permits for each project. Pre-approval would be based on a pre-determined set of criteria that each project would have to meet, including material characteristics and placement operations.

The City proposes transport of the material by truck, and placement on two receiving beaches. The beaches include the north end of the City adjacent to Buena Vista lagoon and immediately south of warm water jetty at the San Diego Gas & Electric (SDG&E) Encina Power Plant as shown in Figure 1. A temporary stockpile location is proposed at a vacant lot south of Buena Vista lagoon.

AGENCY CONCERNS

The U.S. Army Corps of Engineers (USACE) and the Environmental Protection Agency (EPA) consider beach fill to be a beneficial use of dredged material (Draft Inland Testing Manual, or DITM, 1994). Local EPA staff recognizes the same benefit for opportunistic sand (S. John, Personal Communication, 1996). The Coastal Commission concurs. Agencies are concerned with effects of beach fills on public health and the environment, as summarized below.

Chemical Compatibility - Chemical suitability of beach fill is required to minimize adverse effects to the environment. Chemistry is evaluated through testing. The need for chemical testing is based on whether the resource agencies have "reason to believe" that the material is contaminated (DITM, 1994). Material is most likely to be free of contaminants if it consists of sand, gravel, or other inert material, is found in areas of high current or wave energy, or is isolated from sources of contamination.

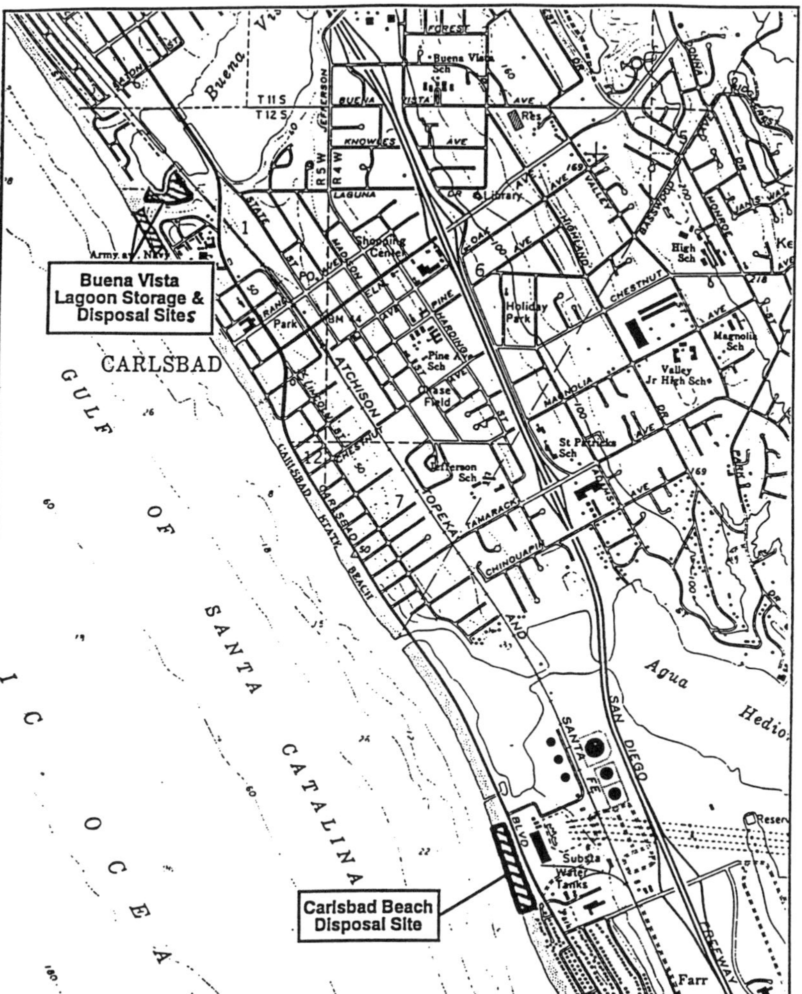

Project Vicinity Map
USGS 7.5' San Luis Rey Quadrangle

FIGURE 1

NO SCALE

Sediment Grain Size - Sediment grain size ranges indicate potential impacts on the environment. Reef habitat, bottom-dwelling organisms, and kelp may become smothered when fine sediment (defined as clays and silts less than 0.074 millimeters in diameter) settles offshore from suspension. Other impacts may be contribution of contaminants which adhere to fine grains, and interference with bird feeding as turbidity obscures ocean visibility.

Color - Color has been an insignificant issue in the past due to extensive use of dredged material for beach fill which eventually "bleaches out" under exposure to the marine environment. The result is fill similar in appearance to the receiving beach. Terrestrial sand is sometimes a reddish tone. The only agency criteria for color is to match the color of the receiving beach for aesthetic reasons.

Particle Shape - Natural sand consists of rounded particles rather than sharp or angular particles. The City is concerned that sharp or jagged manufactured particles may cause beach user discomfort.

Debris - Debris, trash and organic material are concerns due to possible health hazards, nuisance odors and visual impacts associated with their presence.

Compactability/Moldability - Certain types of fill may form a hard pan when exposed to wetting and drying conditions and compaction. This property is not acceptable for beach fill. Materials possessing this property are to be mixed with existing beach sand or buried beneath a layer of sand to minimize hardening (Zoutendyk, 1996).

Placement Site and Timing - Placement operations are a concern due to effects to sensitive resources, grunion, and beach use during high use times (U.S. EPA, Section 404(B)(1) Guidelines).

Placement Rate - Placement rates are a concern due to turbidity levels. Extending the placement period reduces turbidity but extends the period of turbid conditions.

REQUIRED SEDIMENT GRADATION ANALYSIS

Specifying acceptable grain sizes is done to define the maximum acceptable fines content of opportunistic beach fill. The DITM (1994) emphasizes the "reference sediment" concept, which is interpreted as follows:

1. The quantity of fines introduced into the littoral zone should not exceed the quantity of fines at beaches at the point of discharge.

2. The site of fines discharge should not differ greatly from sites of natural fines discharge, such as stream and ravine outlets, eroding beaches, shore platforms and seacliffs. Effects to biota would be minimized below significant if this were incorporated into the placement plan (PSBS, 1995).

3. The time of year when fines are mobilized by waves and suspended in the water column should not differ greatly from natural processes.

Sediments will be sampled at each proposed opportunistic beach fill borrow site, and material gradation will be compared to that of the receiving beach and shoreface (littoral zone). Sampling, testing, and data analysis methods are described below.

Sampling and Testing Methods

Borrow Sites - The USACE (1989) specifies the number of locations to be sampled at each borrow site (for dredging projects), N, as

$$N = 0.85\, A^{1/2} / 50 \qquad 1$$

in which A = plan area of the borrow site in square meters (about 3.5 sample stations per hectare). A minimum of three sampling locations are required. This procedure will also be applied to land sources. Samples are to be taken down to the proposed depth of dredging plus 1 meter.

Sampling of opportunistic beach fill will be consistent with the USACE Equation 1, and one sample at each location be obtained below the soil horizon (to the vertical extent of the excavation). All samples will be sieved in accordance with American Society for Testing and Methods (ASTM) MD 422-63 (Standard Method of Particle-Size Analysis of Soils). This procedure should be reconsidered for specific borrow site conditions, with additional samples taken as necessary.

Receiving Sites - Samples were collected to characterize beach gradation. Two transects each were established for the Buena Vista and SDG&E Encina Power Plant (referred to as the Agua Hedionda site) beaches. Sampling sites were at 2 meter elevation intervals referenced to MLLW (+4, +2, 0, -2, -4, -6, -8 and -10 meters).

Data Analysis

Borrow Sites - A composite grain size distribution curve of potential opportunistic beach fill samples will be prepared. A composite gradation is the mean gradation of all material sampled in the borrow area. If individual samples appear dissimilar in size and character, each distinctive area will be analyzed independently.

Receiving Sites - The USACE (1989) requires grain size distribution curves for each beach sample collected (8 samples per transect, and 2 transects per receiving beach, for 32 samples total for Carlsbad sites). These curves were used to define the envelope of particle sizes within the littoral zone.

The composite curve(s) for the opportunistic beach fill should fall largely within the envelope of curves for the receiving beach. Figure 2 shows the composite curves for the Buena Vista Lagoon receiving beach. The beach fill gradation should fall within the shaded area of the chart.

The coarse-grained portion (fine gravel, anything caught by the #4 sieve or greater than 4.76 mm in diameter) of the composite curve may fall outside the envelope, but the opportunistic beach fill must not contain more than 10%, by volume, gravel and larger, rounded particles (USACE, 1989). The fine-grained portion may also fall outside of the envelope by 10% of the total quantity of fines at the beach (USACE, 1989).

Test Results of Carlsbad Beach Sites

Grain sizes vary between the beach sites, and between transects at the same site. Everts (1995) reached the following conclusions regarding the receiving beaches:

1. Fines increase with depth everywhere;

2. Fine percentages remain small (less than 5%) to a depth between -2 and -4 meters at the Buena Vista site and to a depth between -4 and -6 meters at Agua Hedionda, but increase to between 14 and 15% at the next sampling depth at both sites;

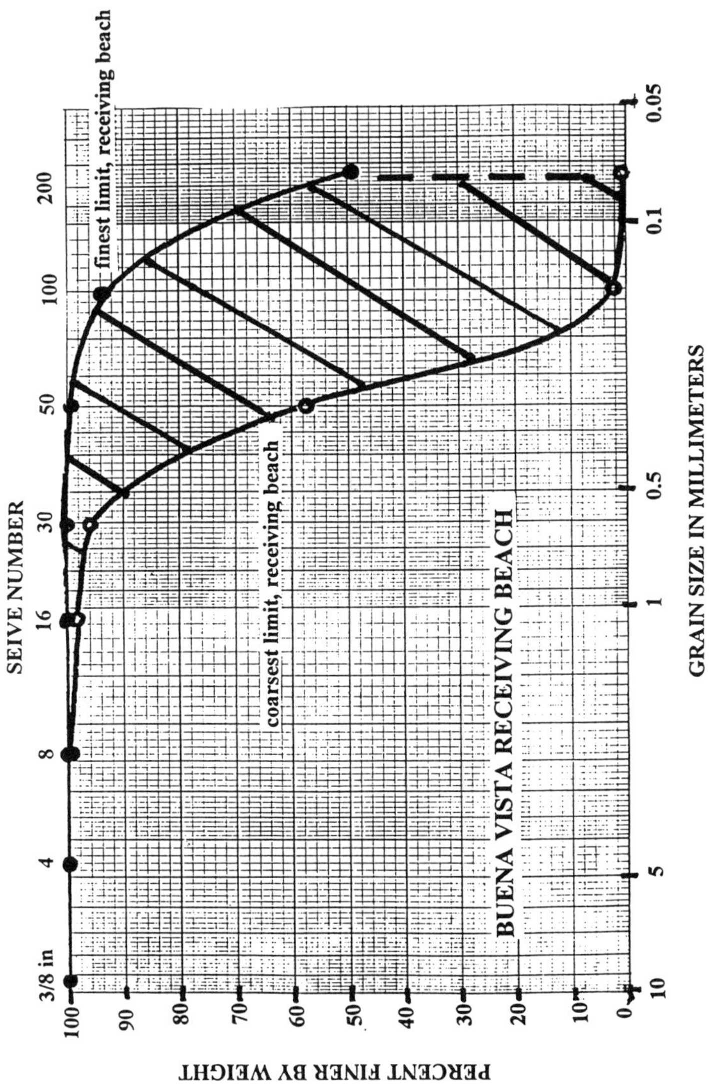

FIGURE 2 COMPOSITE SEDIMENT GRADATION CURVES FOR BUENA VISTA RECEIVING BEACH

3. Fine percentages are twice as large at a depth of -10 meters at the Buena Vista site as the Agua Hedionda site, which is reasonable given its proximity to the closest lagoon mouth and Santa Margarita and San Luis Rey Rivers, and

4. Fines reach 41% at -10 meters at the Buena Vista Lagoon (BVL) north transect and 49% at the (BVL) south transect, and 25% at the Agua Hedionda (AH) north transect and 19% at the (AH) south transect.

PROPOSED CRITERIA FOR BEACH FILL CHARACTERISTICS

Criteria proposed by the City for beach fills are in accordance with environmental guidelines established by the Federal government as discussed below. The City proposes to use a checklist of these items to assess material suitability.

Chemical Compatibility - Material with unhealthful chemical contents is not acceptable. The City will forego chemical testing to minimize costs. Tier I testing, consisting of initial evaluation of existing information is proposed. If Tier I testing suggests the need for chemical testing, it may occur if the benefits of the project outweigh the costs of testing.

Sediment Gradation - The following criteria are proposed for gradation:

1. Material containing 15% fines or less is a candidate as opportunistic beach fill. Beach fills with more than 15% fines may also be candidates, but may require more scrutiny by regulatory agencies.

2. Opportunistic beach fill will be delivered using trucks in most situations. Placement will be on the beach and/or in shallow water (less than 1 meter depth).

3. Opportunistic beach fill of less than 12,000 cm with less than 15% fines may be placed on the beach at any time of year, except during grunion runs. Carlsbad beaches are often cobble with little or no sand, so predicted grunion runs may not actually occur.

4. Opportunistic beach fill of more than 12,000 cm and 15% fines shall be placed mainly from 15 November to 15 April (the wet season). This period also coincides with the storm season when beaches and seacliffs erode, shoreface sediment is

reworked, and turbidity caused by suspended sediment is highest.

5. The level of scrutiny of the beach fill material will follow the guidelines expressed in Table 1. Under the umbrella of the general permits, the City would be able to approve relatively small projects with low levels of fines. Projects of larger quantities and higher percentages of fines will possibly fall out of the umbrella of general permits.

Table 1
Required Tier Levels of City Review for the Carlsbad Opportunistic Beach Fill Program

Quantity of Proposed Beach Fill (cm)	Percentage of Fines Content in Beach Fill			
	0-4%	5-9%	10-15%	more than 15%
0-12,000	Tier I	Tier I	Tier I	Tiers II or III
12,000-40,000	Tier I	Tier I	Tier II	Tiers II or III
40,000-80,000	Tier I	Tier I	Tier II	Tier III
>80,000 (depending on quantity)	Tier II	Tier II	Tier III	Tier III

LEGEND: Tier I - Preliminary Level (existing data) - EPA Tier I

Tier II - Secondary Level (supplemental data and agency review suggested) - EPA Tier II

Tier III -Tertiary Level (supplemental data and resource agency review required) - EPA Tier II

6. Beach fill material with a high percentage of coarse grains is only acceptable if the particles are no larger than cobble-sized, and do not constitute greater than 10% of the volume of the beach fill.

Color - Opportunistic beach fill should become consistent in color with existing beach sand after the material has been washed and reworked by waves, and bleached under exposure to the marine environment. The

USACE Shore Protection Manual (SPM, 1984) states that "...fill material... "reddened" by oxidized clay minerals (Imperial Beach, California) will be bleached quickly by the sun to achieve a more natural beach color."

Color will be addressed by: 1) wash and bleach testing of the material, 2) burial of the material beneath a layer of existing beach sand, and 3) implementation of a public information program. Color incompatibility presents no adverse physical or chemical effects to the coastal environment.

Particle Shape - Angular particles are not acceptable for beach fill because sharp points and edges may cause user discomfort. Manufactured sands will not constitute more than 10%, or a maximum of 8,000 cm of beach fill.

Debris - Opportunistic beach fill will be generated from undisturbed sites and contain no debris (trash, wood, etc.). Material obtained from stream beds or rivers may require screening of debris. Screening is necessary if the debris content appears significant.

Compactability/Moldability - Material will be considered acceptable if it does not form a crust or hardpan. Material tending to form a hardpan can be rejected as beach fill, or mixed with other suitable material or buried beneath a layer of beach sand to preclude hardpan formation. Burial will reduce hardpan formation due to less variable moisture and temperature conditions.

Placement Sites and Timing - The project will utilize the proposed discharge sites. The Buena Vista site is optimum for replicating natural conditions due to its proximity to sediment discharges from Buena Vista lagoon and the San Luis Rey and Santa Margarita Rivers. Beach placement will occur during all seasons, but stockpiling may be required.

Placement Rate - A guideline for the placement of beach fill by season is shown in Table 2. A combination of material stockpiling at the lagoon site, burial beneath the existing beach and diking along the bluff toe can be utilized to control turbidity.

MONITORING

Monitoring is proposed to document the effectiveness of the program. Measurement of the horizontal berm width will occur to identify beach fill changes. These data can be analyzed to improve beach fill design in the future. The City will also document each beach fill and record the date, volume and location of each project.

Table 2
Proposed Quantities of Fines Allowable for Beach Placement during Fall/Winter and Spring/Summer Seasons

	SEASONS		
Maximum Quantities (cm)	Fall/ Winter	Spring/ Summer	Yearly Total
Fines per Season	8,000	4,000	12,000
Fines per Week	2,000	800	12,000
Total Beach Fill Quantity per Season (Assuming 15% Fines)	53,000	26,000	80,000 (approx.)
Total Beach Fill Quantity per Week (Assuming 15% Fines)	13,000	5,250	80,000

References

Everts, Craig, Everts Coastal, *Sediment Size Compatibility: Proposed Opportunistic Receiving Beaches, Carlsbad, California*, July 29, 1995.

John, Stephen, U.S. Environmental Protection Agency, Personal Communication, July 31, 1996

Pacific Southwest Biological Services, *Potential Biological Impacts from the Placement of Opportunistic Beach Fill Materials in Storage or Disposal Sites Adjacent to Buena Vista Lagoon and at Carlsbad Beach, Carlsbad, California*, February 9, 1995.

U.S. Army Corps of Engineers, Waterways Experiment Station, *Shore Protection Manual*, 1984.

U.S. Army Corps of Engineers, Los Angeles District, 1989, *Requirements for Sampling, Testing and Data Analysis of Dredged Material*, unpublished dated report appended to the June 1989 COE report "San Gabriel River to Newport Beach, Beach Replenishment at Surfside-Sunset Beach, Geotechnical Report."

U.S. Environmental Protection Agency, *Guidelines for Specification of Disposal Sites for Dredged or Fill*

Material, 404(B)(1) Guidelines), Federal Register, Volume 45, No. 249, 1980.

U. S. Environmental Protection Agency, and Department of the Army, U. S. Army Corps of Engineers, *Draft Inland Testing Manual (DITM), Evaluation of Dredged Material Proposed for Discharge in Waters of the U. S. - Testing Manual (Draft)*, EPA reference 823-B-94-002, COE Office of Water, June 1994.

Zoutendyk, David, U.S. Army Corps of Engineers, Personal Communication, July 1, 1996.

CHAPTER 202

BEACH RESTORATION AS A NATIONAL POLICY

Must We Go It Alone?

Shared Cost - Shared Benefit

by

Don Hadley, Director of Harbor and Beaches, City of Oceanside
Dana Whitson, Assistant City Manager, City of Oceanside

Abstract

Tourism is one of the nation's major industries. This country's beaches are the number one destination of choice for the traveler. Yet, there is no focused national strategy for protecting and enhancing this valuable asset.

If we are to maintain this beach attraction, the local, state and federal governments must make every effort to assure that beaches continue to viably exist. Very few areas of the world have coastline, especially ones with moderate weather conducive to rest, relaxation and recreation.

Competition for the tourist and the ensuing revenues increase day-by-day. Nature is also with us day-by-day, bombarding our coastline with erosion taking the sand we depend on to protect the shoreline and provide a recreational venue for the local, regional, statewide, national and international users.

Why should beach restoration be more than a simple local effort?

First and foremost, most of the nation's beaches are of regional, state or even national significance and are used by more than our local populations.

Secondly, local agencies simply do not have the fiscal or technical resources to respond to beach erosion problems.

National and sate public responsibility, along with benefits, must be identified and be brought to bear. The responsibilities are far-reaching, but so are the benefits.

In 1995, travel and tourism contributed $746 billion to the U.S. Gross Domestic Product (10% of the total GDP), $47 billion in annual tax revenue, and 343,000 new jobs.

The number one concern of people who visit beaches is beach erosion. After beach nourishment occurred in the 1970's at Miami Beach, beach attendance increased from 8 million in 1978 to 21 million in 1983.

Beach nourishment efforts must be taken on by the full spectrum of governmental agencies. No local community is capable of accomplishing such a feat. Every level of the spectrum stands to benefit far beyond their responsibility.

CHAPTER 203

The Santa Barbara Channel - Santa Maria Basin Circulation Study

Dr. Myrl C. Hendershott[1]

The goals of this study are (i) to describe and understand the flow with emphasis on the surface circulation, and (ii) to summarize the description and understanding for Minerals Management Service oil spill risk analysts.

The salient features of the surface circulation are: (i) Poleward monthly mean subtidal flow through the SBC except in wintertime, with flow reversals lasting days. (ii) Counterclockwise recirculation in the western SBC, strongest spring through autumn in monthly mean, with flow past Pt Conception almost always westward and flow past San Miguel Island almost always eastward. (iii) Transient episodes of vigorous counterclockwise eddying lasting weeks and recurring remarkably regularly during spring and summer in the western half of the SBC. (iv) Strong upwelling near Pts. Conception and Arguello year round, with upwelled water frequently entering the SBC eastward along the Channel Islands. (v) Occasional penetration of eddies from the open ocean into the far-western SBC, sometimes bringing California Current water to the Channel Islands region. (vi) Frequent concentration of flow away from the Channel Islands into a jet flowing northeastward from Santa Rosa Island and bifurcating into westward and eastward flow between Santa Barbara and Pt. Hueneme. (vii) Poleward flow through the SBC often following the coast northward past Pt. Arguello, but during episodes of strong upwelling continuing westward as a concentrated jet. (viii) Equatorward flow through the SBC generally turning clockwise to ultimately flow westward along the south side of the Channel Islands.

Direct flow correlation with local winds is observed in the southwest SBC but is elsewhere much weaker than in e.g. CODE. Transient counterclockwise eddying in the western SBC is not clearly correlated with local winds. Much of the remaining flow variance reflects pressure fluctuations observed over spatial scales much larger than the size of the SBC. Pressure fluctuations associated with density variations of upwelled water and water in the Southern California Bight may be a third forcing agent.

[1]Center for Coastal Studies, Scripps Institution of Oceanography, University of California, San Diego, 9500 Gilman Dr., La Jolla, CA 92093-0209

CHAPTER 204

Wind Forcing of the Ocean Circulation in the Santa Barbara Channel and the Santa Maria Basin

Dr. Clinton D. Winant[1]

Abstract

The boundary between the Santa Barbara Channel and the Santa Maria Basin is the site of mixing between warm, saline waters from the Southern California Bight, cold, relatively fresh waters upwelled to the north and water flowing south in the California Current. Prevailing winds vary strongly here; north of Point Conception they are equatorward and strong in spring and summer, while the channel is relatively sheltered. The resulting circulation is more complex than elsewhere along the west coast of the U.S.

The Minerals Management Service has sponsored a 4 year field study of the ocean circulation in this area. Measurements include time series of currents, temperature, conductivity and pressure at depths and locations throughout the channel, drifter paths and ship surveys. These observations are used to document the response of the ocean to a wind event, and the subsequent response to the relaxation from strong wind forcing, with particular emphasis on two periods: 11-20 August and 15-25 October 1994.

Upwelling favorable winds begin to blow in a pattern where the wind direction is basically parallel to the coastline with maximum amplitudes near the western entrance to the channel and minimum values in the eastern portion. The current response consists of eastward flow of cold water north of the channel islands; in some cases extending the eastern entrance, while the circulation along the coast is weak. After several days a significant horizontal surface temperature gradient builds, separating the cold recently upwelled waters in the southern and western portion of the channel from the warmer waters in the northeastern area. That temperature gradient and the related pressure gradient oppose the wind stress, and the circulation changes, resulting in a strong cyclonic eddy in the western half of the channel. After the wind stress relaxes, the remaining pressure gradient drives warm water westward along the northern boundary of the channel, redistributing it from the northeast portion. Once the jet reaches Point Conception, it either veers south and becomes entrained into the California Current, mixes locally, or follows the coast north towards Morro Bay. At the same time, as the horizontal temperature distribution becomes more uniform, the associated pressure gradient diminishes, and velocities weaken.

---

[1]Center for Coastal Studies, Scripps Institution of Oceanography, University of California, San Diego, 9500 Gilman Dr., La Jolla, CA 92093-0209

Introduction

The Santa Barbara Channel - Santa Maria Basin (SBC-SMB) Circulation Study is sponsored by the Minerals Management Service (MMS) with the dual goals of: (a) improving the description of surface circulation in these areas, and connecting the circulation patterns with existing forcing agents, the wind stress and the pressure difference between the Southern California Bight and the coast of Central California; and (b) to summarize the circulation in a way which is useful to analysts charged with assessing the impact of oil spill development and exploration.

The Observations

The experimental program consisted of five components: (a) an array of meteorological sensors deployed on buoys, oil platforms, and coastal stations; (b) an array of current meters, temperature and conductivity sensors and bottom pressure sensors deployed at the locations illustrated in Figure 1; (c) program of regular deployment of drifters drogued at 1 m. depth and released at 14 locations within the channel every two months; (d) current and water mass property surveys conducted twice each year; (e) acquisition of remotely sensed (AVHRR) sea surface temperature several times each day.

The experimental program was focused on the Santa Barbara Channel between 1993 and 1995. Currently most of the observational resources have been moved to the Santa Maria Basin, north of Pt. Arguello.

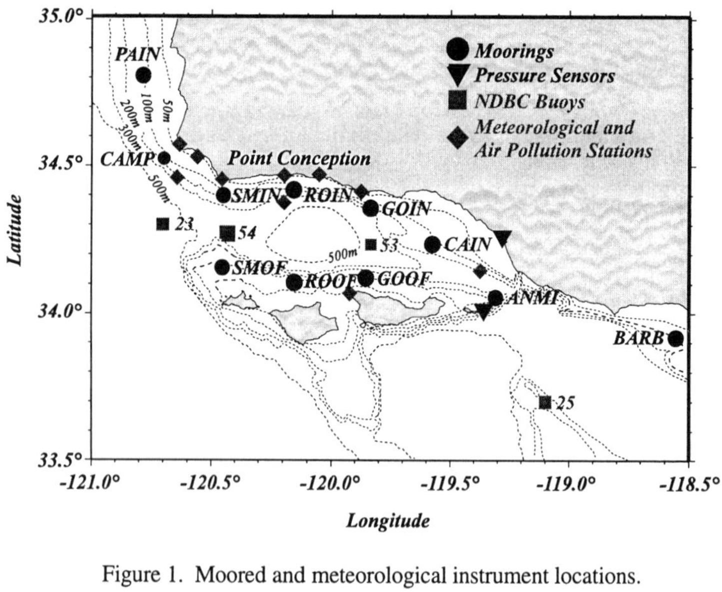

Figure 1. Moored and meteorological instrument locations.

The Circulation

The surface circulation is illustrated in this AVHRR image (Figure 2). The shading of the image represents the temperature of the sea surface ranging between 20°C in the SBC to 13°C in the upwelling area off Pt. Conception.

At the same time we have represented average currents for the day, as measured by the 5m current meters deployed at each of the mooring sites. This particular day is chosen because it illustrates the two most salient features of the correlation seen to date, that is: the average surface circulation in the Santa Barbara Channel consists of a cyclonic gyre centered north of Santa Rosa Island, superposed on a larger scale flow, confined near the coast, extending from the Santa Maria Basin to the Southern California Bight, and alternating direction.

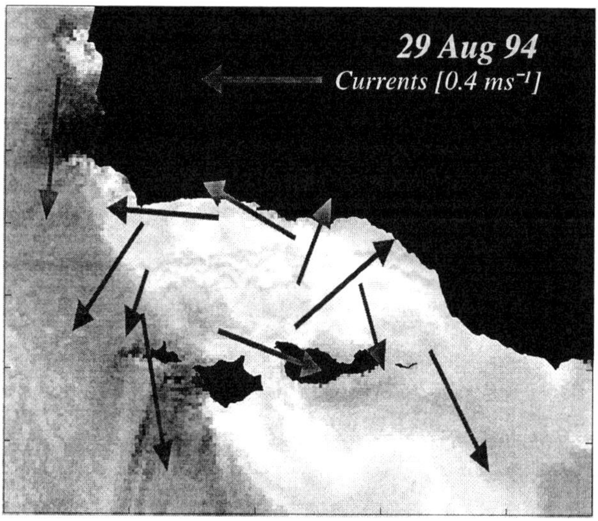

Figure 2. AVHRR satellite image of sea surface temperature (SST) on 29 August 1994. Warmer SSTs are shown in dark gray, colder SSTs are shown in light gray. Arrows represent averaged 5-m currents at each mooring location for the same day.

The Relationship Between *Fluctuating* Bottom Pressure (BP), Baroclinic Pressure and Sub-Surface Pressure (SSP)

The hydrostatic equation:

$$\frac{\partial \rho}{\partial z} = \rho \times g$$

Integrate from the bottom to the surface:

$$p_{z=-h} = p_{z=\eta} + g \int_{z=-h}^{z=\eta} \rho \, dz$$

The term on the left hand side is the bottom pressure, which is directly observed. The first term on the right hand side is the pressure at the surface, or sub-surface (SSP), while the second term on the right hand side represents contributions to the bottom pressure which arise due to density variations in the water column, the baroclinic pressure.

If that the equations of state can be locally represented by a linear relationship between density and temperature:

$$\frac{\Delta \rho}{\rho_0} = \alpha \, \Delta T$$

(Note that α is negative, so that increasing temperatures correspond to decreasing densities) then

$$\rho_{z=-h} = \rho_{z=\eta} + g \, \alpha \int_{z=-h}^{z=\eta} (T - \overline{T}) \, dz$$

where the p's are now understood to refer to fluctuations from the mean over the period of the measurements, and $\overline{T}$ represents the average temperature over the period of observations, at any location and depth. Since the bottom pressure and temperatures are directly observed, the fluctuating SSP can be estimated at each mooring.

These relationships are illustrated with the observations presented in Figure 3, where time series of temperature at several depths at the mooring PAIN are illustrated for the year 1994. The baroclinic pressure computed as described above is shown to vary seasonally with an amplitude of ± 1 kPa. The pressure near the surface is estimated by adding the baroclinic pressure to the bottom pressure. Note that since the datum reference for the bottom pressure is not known, this method cannot be used to estimate the long term average of the pressure.

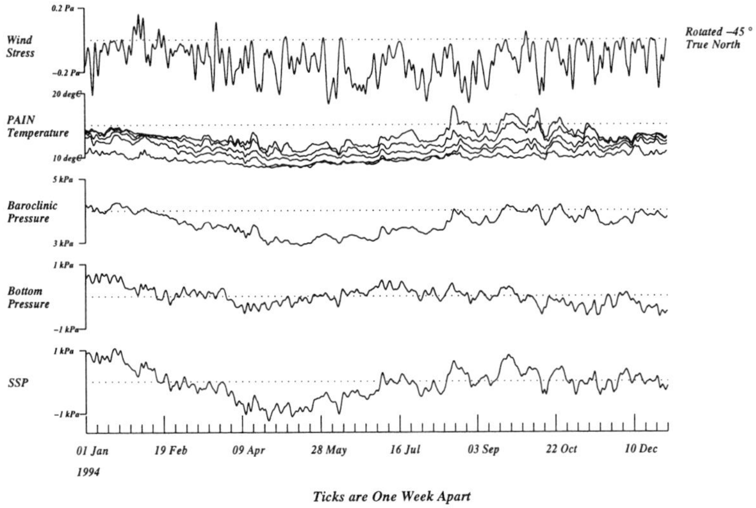

Figure 3. Time series of wind stress at NDBC buoy 54 (top panel), temperatures at 5, 25, 45, 65, and 100 m depth at PAIN (second panel), baroclinic pressure (third panel), bottom pressure (fourth panel), and synthetic subsurface pressure (SSP, bottom panel), all at PAIN.

The pressure difference across the Santa Barbara Channel (Barb-Pain) is illustrated in Figure 4. During the early part of the yearly cycle, wind stress and pressure gradients act in the same direction driving the current equatorward, but after mid-April, the pressure gradient begins to reverse, corresponding to warmer water in the Southern California Bight and colder waters off the Upwelling areas in Central California. During the late summer, the pressure difference reaches maximum amplitude, in the direction opposite to the wind stress.

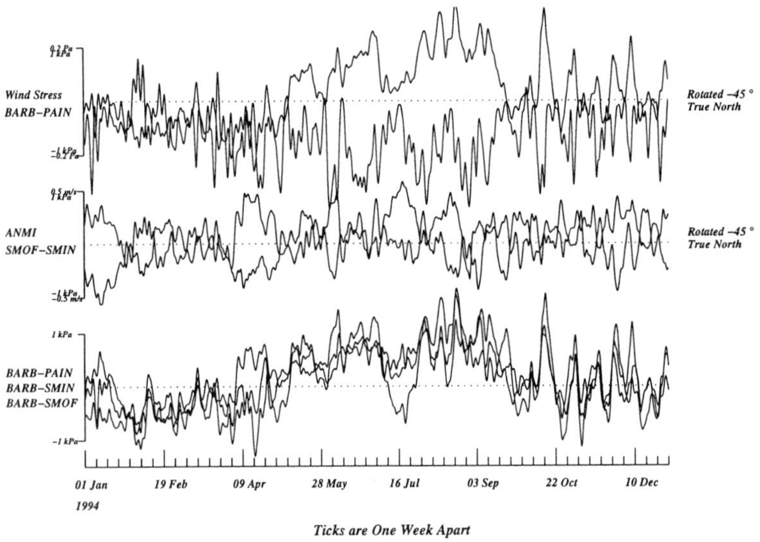

Figure 4. Top panel: Time series of wind stress at NDBC buoy 54 and the surface pressure difference between BARB and PAIN. Center panel: Time series of alongshelf currents at ANMI and the surface pressure difference between SMOF and SMIN. Bottom panel: Time series of the pressure difference between BARB and PAIN, BARB and SMIN, BARB and SMOF.

Conclusions

• The strength of the wind varies, from very vigorous near the western entrance to the channel and north of Point Conception, to weak values in the eastern portion of the channel.

• The spatial variation in upwelling due to the wind results in strong temperature gradients at the surface of the water.

• Warmer temperatures and lower densities result in high pressure at the surface of the ocean, so that temperature gradients are associated with surface pressure gradients.

• The direction of these gradients is in the sense to oppose the wind stress. The circulation pattern in the area thus results from the competition between these two forcing mechanisms.

CHAPTER 205

Propagating Eddies in the Santa Barbara Channel

Dr. Sabine Harms[1]

Abstract

In summer and fall vigorous cyclonic eddies frequently form in the eastern and central Santa Barbara Channel and slowly drift westward. Empirical orthogonal function (EOF) and correlation analyses applied to moored observations of wind stress, currents, temperatures, and surface pressures from the Santa Barbara Channel (SBC) - Santa Maria Basin (SMB) Coastal Circulation Study are used to describe these structures.

In summer and fall the flow in the SBC switches between three energetic states. The *Upwelling* circulation is dominated by strong equatorward flows near Points Arguello and Conception and eastward flow on the southern shelf of the SBC. The *Cyclonic* pattern describes a condition which consists of currents oriented in opposite direction with comparable speeds on alternate sides of the channel. The *Relaxation* pattern describes a circulation which consists of a strong and continuous poleward jet along the northern shelf of the SBC between Port Hueneme and Point Conception. During the transition from the *Upwelling* to the *Cyclonic* regime, the shear between alongshelf flows on opposite shelves is largest in the eastern SBC. When the *Cyclonic* circulation is strongest, maximum shear flows are observed in the central channel. When the *Cyclonic* circulation turns into the *Relaxation* regime, maximum shear occurs in the western channel. The shear between alongshelf currents on opposite shelves may give rise to unstable conditions and cyclonic eddies, about 20 to 30 km in diameter, can form. Results of the lagged correlation analysis suggest that the eddies travel westward with an average speed of 6 $km\ day^{-1}$.

Introduction

The Santa Barbara Channel (SBC), located on the northern edge of the Southern California Bight, is bounded to the north by the south central California coast and to the south by four of the channel islands; San Miguel, Santa Rosa, Santa Cruz, and Anacapa (Figure 1). The channel is directed east-west

[1]Center for Coastal Studies, Scripps Institution of Oceanography, University of California, San Diego, 9500 Gilman Dr., La Jolla, CA 92093-0209

and is approximately 100 km long and 40 km wide. The central basin depth is about 500 m, the sills at the eastern and western entrances of the channel are 220 m and 430 m deep, respectively, and the passages between the islands are about 40 m deep.

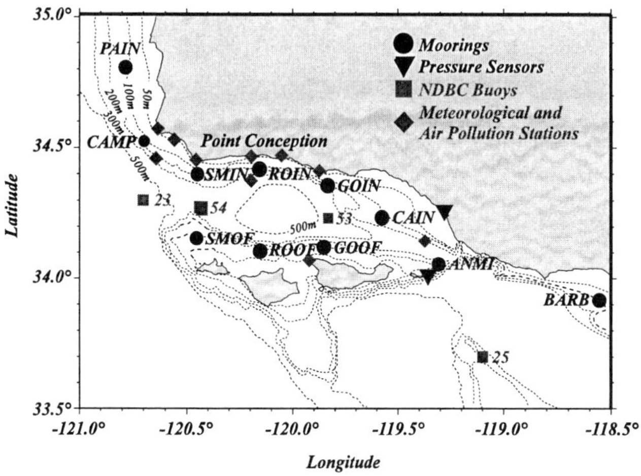

Figure 1: Santa Barbara Channel - Santa Maria Basin moored and meteorological instrument locations.

In early 1992 an extensive multi-year-long observational program, sponsored by the Minerals Management Service (MMS) of the U.S. Department of the Interior, was initiated in the SBC to yield a detailed description of the circulation in this area. The moored array (Figure 1) consists of a total of ten moorings deployed over the shelf along the 100-m isobath and at 200 m depth in the center of the eastern entrance of the SBC, supplemented by a pair of shallow pressure sensors deployed on either side of the eastern entrance. NDBC buoys 53 and 54 are equipped with a downward looking Acoustic Doppler Current Profiler (ADCP). Also included are observations obtained during the California Monitoring Program (CAMP) off Point Arguello [*Hamilton et al.*, 1994]. Instruments deployed at each mooring include current meters at 5 and 45 m depth, temperature sensors at 1, 5, 25, 45, 65 and 100 m depth, conductivity sensors at 5, 45 and 100 m depth, and a pressure sensor at the bottom. Meteorological observations are available from NDBC buoys, from Scripps Institution of Oceanography operated meteorological stations and from air pollution stations maintained by the Santa Barbara County Air Pollution Control District. Other observations obtained as part of this program are described, for example, by *Hendershott and Winant* [1996]. Results presented in this report are based on the analysis of the moored and meteorological observations obtained between 1993 and 1995. All time series have been hourly averaged and lowpass filtered by convolution with a symmetric filter with a half power period of 38 hours [*Limeburner*, 1985].

Large-scale and Local-scale

The circulation in the SBC results from a superposition of a larger-than-SBC scale flow and a cyclonic circulation of variable strength inside the channel [*Harms*, 1996]. Current fluctuations at ANMI (Figure 2a), located over the sill which separates the SBC from the Southern California Bight, are significantly correlated with the flow along the central California coast and represent the "large-scale signal". Within the SBC alongshelf currents on the northern shelf are often in opposition to currents on the southern shelf (Figures 2b and 2c), both in terms of the mean and the fluctuations, a situation first described by *Brink and Muench* [1986] and by *Gunn et al.* [1987]. The strengths of these anti-correlated events give an indication of the intensity of the cyclonic circulation within the channel. Anti-correlated events are strongest in summer and fall, coincidental with strongest poleward flow through the eastern entrance, and weakest in winter when the flow ANMI is directed equatorward.

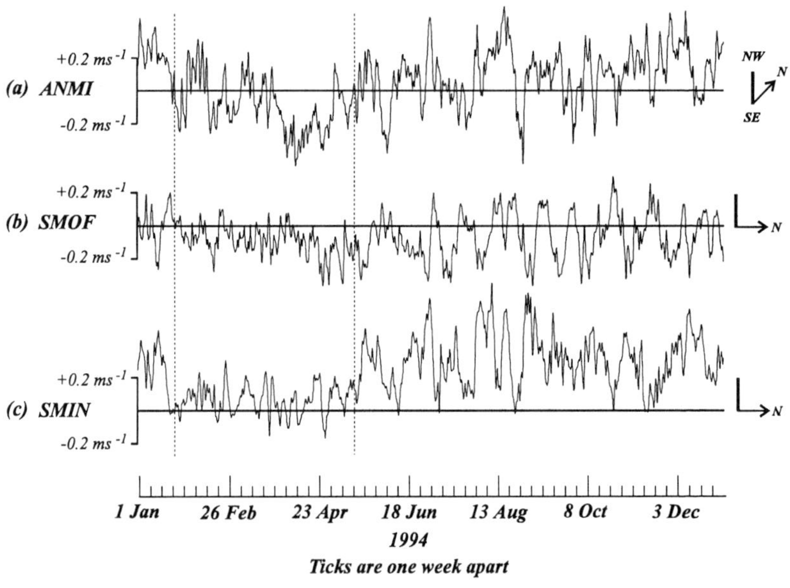

Figure 2: (a) Time series of principal axis 5-m currents at ANMI. (b),(c) Time series of 5-m currents at SMOF and SMIN along the shelf, a direction closely resembling that of the principal axes.

Circulation Patterns

The most frequently-occurring combinations of various features of the surface flow in the SBC can be summarized in terms of a few synoptic patterns. Empirical orthogonal functions (EOFs) are used to synthesize the current measurements, and to establish the principal patterns of variability. The time-averaged currents are removed from the time series before the analysis is performed. The time-averaged circulation consists mainly of a cyclonic eddy centered in the western portion of the SBC. The first three EOF modes account for 50% of the total variance over the 1994 year. The first mode (23%) describes currents which are parallel to the coast, with large amplitude and similar direction at each end of the SBC (Figure 3a). The second mode (Figure 3b) accounts for 16% of the variability. This mode consists of currents which are in opposite direction on alternate sides of the channel. In the middle of the channel the amplitude and direction of this mode are such as to close the circulation pattern. The spatial pattern of the second mode resembles closely that of the mean currents. The third mode (Figure 3c) accounts for 11% of the variance. This mode describes circulation features with smaller spatial scales than the first two modes. The third mode consists of current fluctuations which, across any given north-south section, are opposite in direction on alternate shelves. Current fluctuations at nearby stations on the same shelf tend to oppose each other.

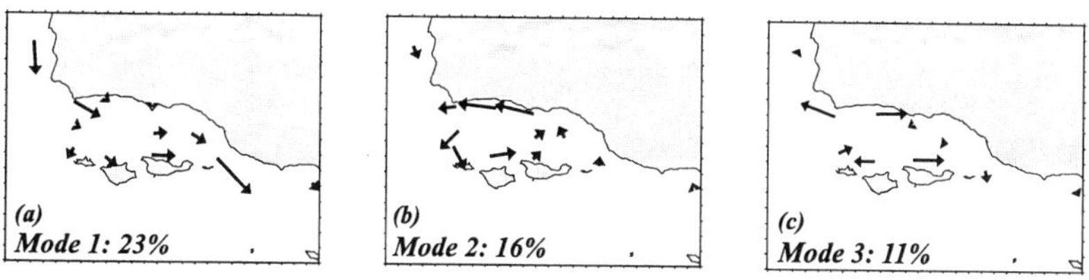

Figure 3: Spatial structures of the first three empirical orthogonal functions (EOFs) of the low-frequency 5-m currents during 1994. The first three modes describe 23%, 16% and 11% of the total low-frequency variance in 1994.

Recurring Flow Patterns

During the summer and fall seasons, the first two EOF modes have a maximum correlation of 0.58, when the second mode lags the first by 4 days, suggesting that in combination with the mean flow these modes describe a repeating pattern of circulation with an approximate period of sixteen days (Figure 4). Beginning when mode 1 has maximum temporal values, the circulation is composed of large equatorward (alongshore) currents at either end of the channel and along its southern boundary, while the flow along the northern boundary is weakly towards the west. This flow pattern is labelled *Upwelling*. About four days later mode 2 reaches maximum positive weighting. The westward flow along the northern boundary intensifies, while the current north of the SBC and north of the channel islands is still directed equatorward. The resulting flow pattern, labelled *Cyclonic*, consists of currents oriented in opposite direction with comparable speeds on alternate sides of the channel. After four more days, when the first mode reaches maximum negative weighting, the flow is everywhere weak, with the exception of a strong poleward jet which begins at the eastern entrance and continues along the northern side of the channel to Point Conception. This pattern is labelled *Relaxation*. Another four days later, when the second mode reaches maximum negative weighting, the flow is everywhere weak. After about sixteen days this circulation pattern repeats itself.

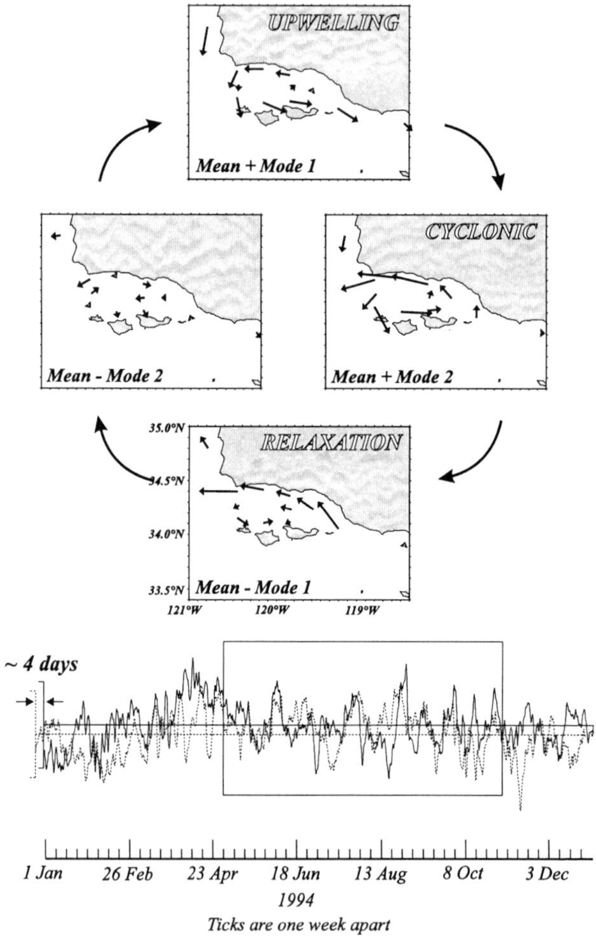

Figure 4: Spatial structures of the two largest 5-m current eigenfunctions when combined with the mean circulation (upper 4 panels) and their temporal amplitudes (bottom panel). The time series of the second mode is shifted relative to the time series of the first by about 4 days. Between late spring and fall the maximum lagged correlation between both modes is 0.58, when the second mode lags the first by about 4 days.

Propagating Cyclones

When combined with the mean field, mode 3 of the 5-m current field describes cyclonic eddies of limited extent at different locations in the SBC (Figure 5, left panels).

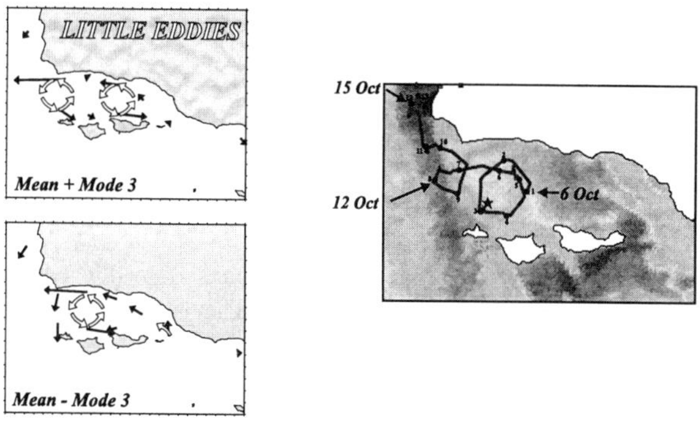

Figure 5: Left panels: Spatial structures of the third 5-m current eigenfunction when combined with the mean circulation. Schematic eddies are superposed. Right panel: AVHRR satellite image taken on 8 October 1993 (indicated by the star) and the track of a drifter released on 6 October in the central channel.

The right panel in Figure 5 shows an AVHRR satellite image taken on 8 October 1993 and the track of a drifter released in the central channel on 6 October 1993. On October 8 both the drifter track and the surface temperature field show a small cyclonic eddy of about 20 to 30 km in diameter. Over the next few days the drifter, caught in the eddy, slowly moves towards the west. In terms of synoptic patterns these structures are labelled *Propagating Cyclones*. Examination of the lagged correlations between 5-m current records in the western channel (Table 1) shows quantitative evidence for such westward drift, at a speed of about 6 $km\ day^{-1}$, of the eddy field during summer and fall.

Stationpairs	C_{max}	R_{max}	*Lag [days]*
Across-shelf			
GOIN-GOOF	-0.63	-0.73	2.8
ROIN-ROOF	-0.79	-0.71	1.3
SMIN-SMOF	-0.74	-0.74	1.3
Along the northern shelf			
GOIN-ROIN	0.64	0.49	4.0
ROIN-SMIN	0.46	0.43	4.5
Along the southern shelf			
GOOF-ROOF	0.70	0.73	4.8
ROOF-SMOF	0.64	0.54	5.3

Table 1: Maximum lagged correlation (C_{max}), regression coefficient (R_{max}) and lag between alongshelf 5-m currents during summer and fall. Displayed are results for the bandpass filtered time series ($10d < T < 25d$).

Conclusions

• The circulation in the Santa Barbara Channel tends to be cyclonic most of the time. Cyclonic flow is strongest in summer and weakest in winter.

• The most frequently-occurring combinations of various features of the surface flow in the SBC can be summarized in terms of a few synoptic patterns. Four of these patterns are described in this report.

• The *Upwelling*, *Relaxation* and *Cyclonic* patterns occur predominantly between spring and fall. In summer and fall these flow regimes describe a repeating pattern of circulation in which the sequence of states *Upwelling*, *Cyclonic*, *Relaxation*, *Weak Flow* is traversed roughly every 16 days.

• The fourth pattern, labelled *Propagating Cyclones*, describes recirculating flow trajectories involving smaller cyclonic eddies which slowly drift from east to west with a speed of about 6 $km\ day^{-1}$. The eddies tend to form in regions where the shear flow between alongshelf currents on opposite shelves is strongest, suggesting that these eddies result when the shear flow becomes unstable. Once formed the eddies appear to be advected westward with the mean flow.

References

Brink, K. H., and R. D. Muench, Circulation in the Point Conception-Santa Barbara Channel region, *J. Geophys. Res.*, 91(C1), 877-895, 1986.

Gunn, J. T., P. Hamilton, H. J. Herring, K. L. Kantha, G. S. Lagerloef, G. L. Mellor, R. D. Muench, and G. R. Stegen, Santa Barbara Channel circulation model and field study, Final Report, *Dynalisis of Princeton*, Rep. 92.1 and 92.2, 1987.

Hamilton, P., T. J. Berger, and E. Waddell, Long term monitoring of currents off Point Arguello, California (abstract), Fall Meeting suppl., *EOS Trans. AGU*, 75(44), 337, 1994.

Harms, S., Circulation induced by winds and pressure gradients in the Santa Barbara Channel, Ph.D. thesis, Scripps Institution of Oceanography, University of California at San Diego, La Jolla, CA, 161pp, 1996.

Hendershott, M. C. and C. D. Winant, The circulation in the Santa Barbara Channel, *Oceanography*, 9(2), 114-121, 1996.

Limeburner, R. (Ed.), CODE-2 moored array and large-scale data report, *WHOI Tech. Rep. 85-35, CODE Tech. Rep. 38*, Woods Hole Oceanogr. Inst., Woods Hole, Mass., 1985.

CHAPTER 206

The Surface Circulation in the Santa Barbara Channel Observed with Drifters and Current Meters

Dr. Edward P. Dever[1]

To learn more about circulation within the Santa Barbara Channel and better assess oil spill risks there, the Minerals Management Service has funded an extensive oceanographic field program, the Santa Barbara Channel - Santa Maria Basin Coastal Circulation Study. This study included an extensive moored current meter array with near-surface Vector-Measuring Current Meters (VMCMs) at 5 m depth. It also included over 200 surface drifters deployed at 12 locations in a grid pattern within the Santa Barbara Channel (Figure 1).

Both current meters and drifters can be used to give estimates of surface circulation within the channel. Current meters have the advantage of providing continuous sampling in time. However, the entire array samples only a limited number of fixed locations limiting spatial resolution. They also sample at 5 m depth, which, while well within the surface boundary layer, may differ quantitatively from the velocity just at the surface. Surface drifters provide direct estimates of parcel trajectories at 1 m depth and hence are of utmost importance in oil spill risk analysis. However, they sample intermittently in time and space so that the possibility of spatial or temporal bias is present. For this reason, comparisons of drifter estimates of velocity with those of moored current meters are essential to an assessment of drifter-derived maps of surface circulation.

Maps of near-surface circulation deduced from drifters are estimated by time differencing surface drifter positions to form velocity estimates and averaging these velocity estimates within latitude and longitude bins of 12.5×12.5 km^2 (Figure 2). Mean velocities derived from drifter tracks show westward flow along the northern boundary of the Santa Barbara Channel and eastward flow of similar magnitude along the southern boundary. In the center of the channel where mean velocities are lower, a mean southern flow is observed at the west end of the Santa Barbara Channel and a weak mean northwestward flow is observed at the eastern end of the channel. The spatial structure of the mean flow and its magnitude agree in general with those estimated from moored velocity data. This comparison indicates drifter sampling was adequate.

---

[1]Center for Coastal Studies, Scripps Institution of Oceanography, University of California, San Diego, 9500 Gilman Dr., La Jolla, CA 92093-0209

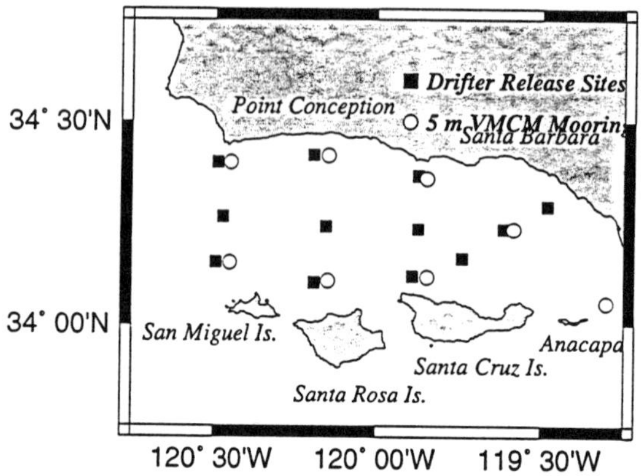

Figure 1. Common drifter deployment sites and VMCM mooring locations

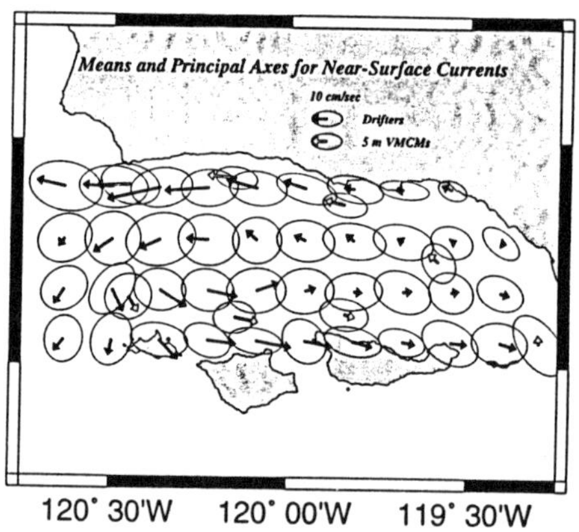

Figure 2. Means of VMCM and drifter observations. Standard deviations along principal axes for 6 hr-averaged velocities are indicated by the ellipses.

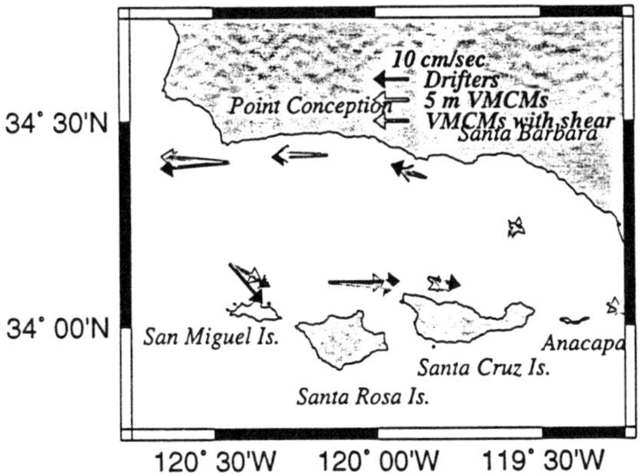

Figure 3. Means of simultaneous VMCM and drifter observations. The effect of wind-induced shear is also shown

Differences between drifter and moored velocities are due in part to the action of wind and waves within the upper 5 meters of the water column. One model of the vertical structure of near-surface velocity is the logarithmic boundary layer. The theoretical estimate of the log layer shear is given by:

$$\Delta \mathbf{u}_{\mathbf{log}} = \mathbf{u}_{*} \kappa^{-1} ln(z_1/z_0) \tag{1}$$

where $\mathbf{u}_{*} = \tau/\rho_{air}$ is the shear velocity induced by the wind stress, κ is von Karman's constant and $z_1 = 5$m and $z_0 = 0.5$m are the depths of the VMCM velocity and drifter velocity estimates respectively. Here $\mathbf{u}_{*}$ and $\Delta \mathbf{u}_{\mathbf{log}}$ are in the direction of wind velocity. Simultaneous drifter and VMCM velocity estimates agree more closely within the southern portions of the Santa Barbara Channel when the theoretical shear induced by wind forcing is accounted for (Figure 3).

Flow direction and speed in the channel may be described in terms of synoptic patterns which last several weeks, and different synoptic patterns prevail seasonally. Both drifter and VMCM seasonal mean velocities agree qualitatively (Figures 4,5,6,7).

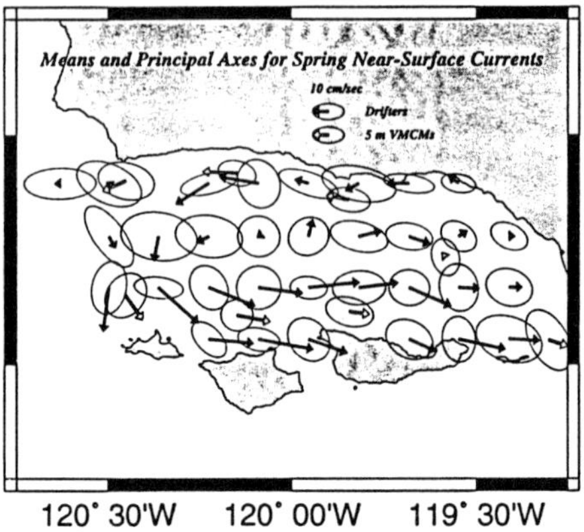

Figure 4. In spring (March, April, and May), both the drifters and VMCMs indicate primarily southward flow at the western Channel entrance and relatively strong eastward flow along the southern boundary of the Channel.

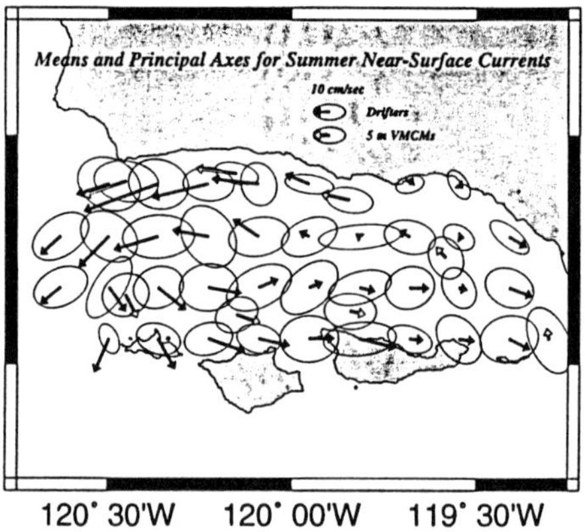

Figure 5. Means are quite strong in the summer and are often larger principal axis standard deviations, particularly in the western half of the Channel. The means in this region also show a well developed cyclonic circulation.

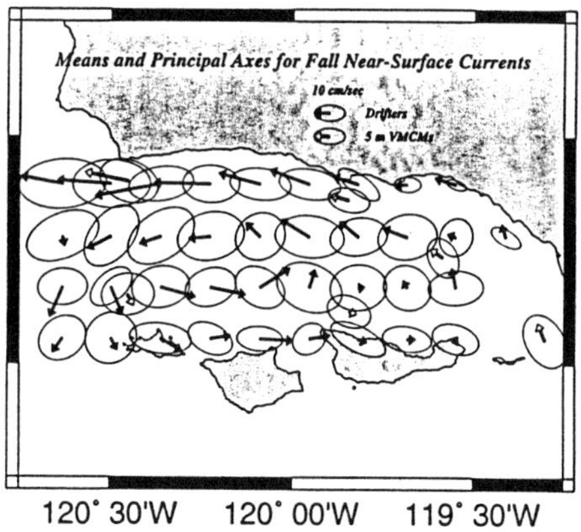

Figure 6. Fall means are quite similar to those in summer.

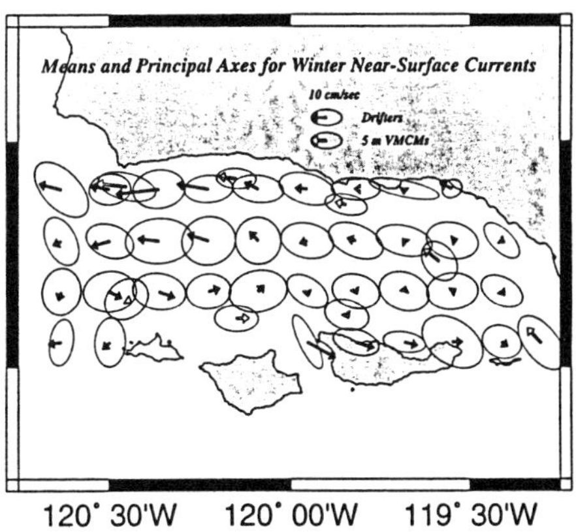

Figure 7. The winter seasonal circulation has weak means relative to other seasons. Principal axes standard deviations exceed means at all locations for both 5 m VMCMs and drifters.

CHAPTER 207

Results of the Coastal Waves Meteorological Project: Analysis of the Santa Barbara Channel Data

Kathleen Edwards[1]

The talk presented analysis of low-level flight data in the Santa Barbara Channel region. The data were obtained during the NSF-funded Coastal Waves 96 meteorological project, which investigated high-speed flow in the marine atmospheric boundary layer along the central California coast. In the Santa Barbara Channel region, two flights were performed using a C-130 from the National Center for Atmospheric Research. The aircraft, capable of extended flights, was instrumented for mean and turbulent measurements. We flew same-altitude legs at a number of heights as well as vertical profiles; our intent was to capture the along and cross-shore variation of the marine atmospheric boundary layer in the region. Our flight data is a significant addition to the long-term meteorological observations from the National Data Buoy Center, coastal air pollution stations, and the oil platform and island stations which the Minerals Management Service supports. The strengths of this station array–its long and consistent coverage in time–is supplemented by the advantages of aircraft measurement: high-resolution spatial coverage over the long flight tracks. Thus questions such as the evolution of the marine atmospheric boundary layer offshore and the extent of the coast's influence on the flow can be addressed.

A better knowledge of the area's marine atmospheric boundary layer will have two significant consequences for its coastal resource managers. First, we expect our insights into local atmospheric effects to improve marine and aviation forecasts in the area. In addition, analysis of data from the Coastal Waves project in combination with data from the Santa Barbara Channel monitoring effort funded by the Minerals Management Service will permit better understanding of the air-sea interaction in the coastal zone.

[1]Center for Coastal Studies, Scripps Institution of Oceanography, University of California, San Diego, 9500 Gilman Dr., La Jolla, CA 92093-0209

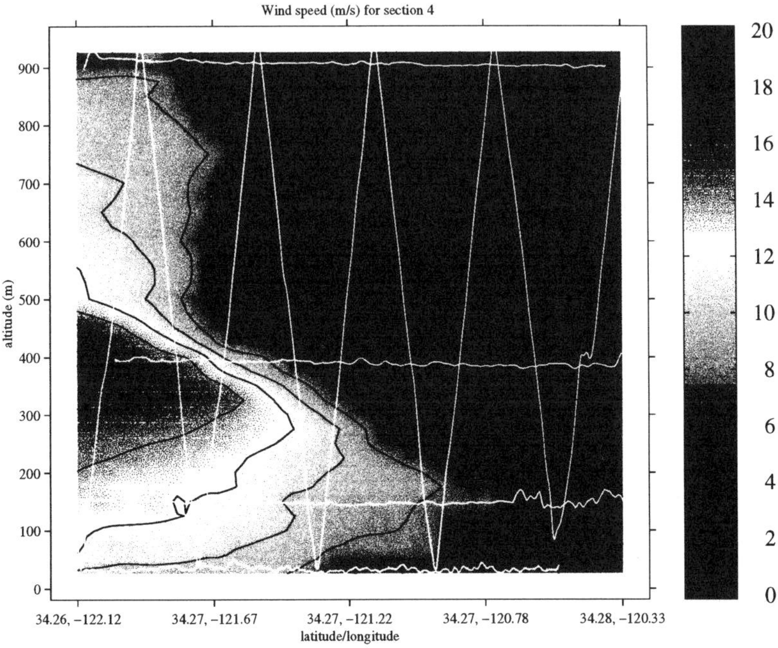

Figure 1: A vertical cross section of wind speeds from June 19, 1996 is shown. Winds are contoured in color with the scale on the right, in (m/s). The track the plane flew is shown in white. The section begins offshore on the left and transits onshore into the Santa Barbara Channel. Note the core of high speed winds at the top of the marine atmospheric boundary layerwhich decreases in altitude towards the shore, and a decrease in wind strength inside the Channel (east of about -120.78 west).

CHAPTER 208

THE CALIFORNIA MARINE MAMMAL RESEARCH PROGRAM OF THE ACOUSTIC THERMOMETRY OF OCEAN CLIMATE EXPERIMENT

POTENTIAL EFFECTS OF LOW FREQUENCY SOUND ON DISTRIBUTION AND BEHAVIOR OF MARINE MAMMALS

Daniel P. Costa, Daniel E. Crocker, Danielle M. Waples, Paul M. Webb, Jason Gedamke, Dorian Houser, P. Dawn Goley, Burney J. Le Boeuf
Department of Biology & Institute of Marine Sciences
University of California
Santa Cruz, California 95064
and
John Calambokidis
Cascadia Research Collective
218 1/2 W 4th Ave
Olympia, WA 98501

INTRODUCTION

The overall rationale behind the combination of approaches included here is to determine, using established scientific methods, as well to develop new techniques to measure, what species were in the area where they were exposed to the operational ATOC source, how many animals were in this area throughout the year, and what species would most likely be prone to being affected by the ATOC source. To be affected, it is presumed that an animal would have to be able to hear the sound, and respond to it in some meaningful way. The research methods focused on observations of behavioral effects; e.g., changes in distribution and behavior during or after ATOC transmissions. The protocol was designed around the following assumptions 1) reactions from marine animals are unlikely at ATOC received levels <120 dB or at distances >20 km; 2) marine mammal distribution and behavior were similar during the 1-4 days after the start of a transmission period, and during the 5-7 days after the start of a control (off) period. Aerial surveys were conducted in order to determine which species are present in the study area, and their sighting distributions, densities, and seasonality and how these change in response to ATOC transmissions. Data were collected to evaluate changes in behavior and distribution for free-ranging animals in the study area, using a combination of aerial surveys/observations. Potential changes in diving behavior of free-ranging northern elephant seals in the vicinity of the proposed ATOC source site were measured using satellite tags and acoustic dive recorders attached to the animals. The highest priority research efforts concentrated on the study of species believed to be the most likely to react to the sound transmissions mysticetes (blue, humpback, whales); odontocetes (sperm whales); pinnipeds (elephant seals).

The intent of the study was to determine if the ATOC sound source when optimized for climate studies has an adverse affect on marine mammals. Therefore, it was necessary for the source to operate on the same duty cycle anticipated for ATOC climate studies. The ATOC climate study duty cycle consists of 20 minutes of continuous transmissions every 4 hours for a 24 hour period. These 20 minute transmissions are preceded by a 5 minute ramp up period. The ATOC transmission is a M-coded pseudo-random signal with a center frequency of 75 Hz and a sound pressure level of 195 dB re 1 μPa at 1 meter. The ATOC MMRP protocol (4 days on, 7 days off) allowed an initial 24 hr. for animals to respond to the transmissions, followed by 3 days of transmissions to complete surveys and observations. The 7 day control interval was designed to allow a 4 day period for the study area to return to steady-state, followed by 3 days to complete control surveys and observations. Given typical swimming speeds of marine mammals (1-2 m/sec), 3-4 days of no transmissions should have allowed sufficient time for exposed animals to leave (if they desired) or for new animals to enter the study area.

Ideally, the sampling protocol was designed to consist of control surveys and observations were conducted for 1-2 days before transmissions begin, followed by surveys 1-4 days after transmissions begin. Fluctuations in the duration of the on (transmission) and off (control) periods were primarily due to weather preventing aerial surveys. If surveys were completed early during each period, the duration of the sampling periods was reduced; whereas, when weather prevented sampling, the duration was extended. In this manner, the study design was basically a paired sample (one off period paired with an on period).

The primary objectives of the California MMRP were to: 1) use aerial surveys determine if ATOC sound transmissions change the relative distribution and abundance of marine mammals within the 120 dB sound field (modeled at 100 m depth), 2) use tags that record the acoustic environment and animal behavior, determine if elephant seals modify their behavior in response to the ATOC sound source and 3) record and catalog the occurrence and ultimate fate of individual humpback, blue and sperm whales using photo-identification methods.

MATERIALS AND METHODS

Aerial Surveys

Aerial surveys were the primary responsibility of the Cascadia Research Collective. The aims of the aerial surveys were to identify which species were present in the study area; to provide estimates of relative density and relative abundance of the more common species in the study area; and to evaluate whether low frequency sound (from ATOC) alters the habitat utilization (occurrence, distribution), or behavior of marine animals in the study area. Aerial surveys were designed to test whether there was a difference in sighting rate, distribution, orientation, general activity, or group size of different species (or groups of species) between surveys conducted when the source was on and when it was off, and as a function of distance from the source.

Measurements of marine mammal densities, distributions, and behavior in the vicinity of the Pioneer Seamount area during both control and transmission periods were initiated in December 1995. Specific methodologies included aerial surveys/observations and shipboard photo identification 1996. From December 1995 to January 1997 a minimum of two replicate (control: experimental) aerial surveys were flown every month. The timing of these surveys was based on putting the greatest effort when we expected the greatest number of mysticete whales. Typically during a period of intensive aerial surveys we were able to complete up to 4 replicate surveys a month. One replicate aerial survey refers to one control aerial survey prior to ATOC transmissions, followed by one experimental aerial survey during the 4 day period while the source is on (i.e., transmitting 6 times/day). Surveys/observations were conducted using standard line-transect methodology. The survey area was an 80 km x 80 km square centered at the ATOC source site on Pioneer Seamount. Transect lines were flown at a nominal altitude of 700ft at a nominal 100 kts airspeed.

Coverage was 3x as high in the inner box (40 x 40 km) vs. the outer box (80 x 80 km). Due to the larger size of the outer box, this yielded the same total effort in each box. Biological data collected was: species, number, perpendicular distance from the transect, direction of travel, and general behavior. Date, time, and position (using a GPS system) are recorded each minute. Beaufort state, cloud cover, sighting conditions, and glare were noted at the beginning and end of each transect and when significant changes occurred. The data are still being analyzed and will be used to determine the distribution and abundance (and density) of marine animals in the study area, and test whether abundance and distribution of marine animals changed when the ATOC sound source operated.

The surveys were flown with a twin-engine Partenavia P-68 aircraft specifically outfitted for aerial surveys. It was equipped with bubble windows on each side of the aircraft that allow full downward visibility and an opening in the bottom allowing photography or a dedicated downward-viewing observer. An onboard global positioning system (GPS) was used for accurate position information and navigation. Five personnel were used on each survey: pilot, recorder/navigator, two side observers, and a center observer. The use of an observer looking through a downward hatch was beneficial because it: 1) is similar to the methodology used by the South West Fisheries Science Center (SWFSC) during aerial surveys and was recommended for these surveys, 2) would provide critical near-transect sightings for small species that occur in small groups, especially Pacific white-sided dolphins, and sea turtles and 3) were used as an independent observer to evaluate the rate at which sightings are missed by the side observers.

When a sighting was made, the aircraft continued on transect until the animal(s) were abeam and the observer determined the angle to the sighting using a hand-held clinometer. The belly observer used calibrated graduations on the belly opening to determine angle to sightings. Confirmation of species and group size, was accomplished for most sighting, except the most common (such as California sea lions) by circling with the aircraft. Data were recorded using an onboard Data Acquisition System (DAS) developed by Cascadia and widely used by Cascadia and SWFSC.

Aerial surveys were scheduled to coincide with the 2-4 day on, 2-7 day off transmission schedule. Both the exact transmission schedule and the timing of the aerial surveys were set by the following considerations: 1) maximize the number of replicates; 2) flexibility to take advantage of good weather windows; 3) ability to maintain a pattern that keeps the observers blind to whether the sound source is on or off. Regardless of the above considerations, the sound source never transmitted continuously longer than 4 days, and the total duty cycle average less than 3%. The actual ATOC transmission schedule is reported in Table 1.

Photo-Identification

A separate effort to photographically identify large whales in the vicinity of the Pioneer Seamount was carried out by Cascadia Research Collective. This was in addition to ongoing photographic identification efforts carried out by Cascadia in southern and central California. Photographic identification research on humpback and blue whales has been successfully conducted off California since 1986. Recognition of individuals allows long-term assessments of stock structure, population size, movements (migrations), birth and death rates, survival, and growth

Table 1. ATOC transmission schedule through 18 February 1997.

Cycle	Survey	Start date	time	End date	time	Intensity
1	1	2-Dec	1450	3-Dec	1600	185 dB
2	2	21-Dec	1200	23-Dec	1600	185 dB
3	3	28-Dec	1200	1-Jan	1200	195 dB
4	4	4-Jan	1630	6-Jan	1600	195 dB
5	5	10-Jan	1245	12-Jan	1200	195 dB
6	6	28-Jan	1230	31-Jan	1200	195 dB
7	7	10-Feb	1600	12-Feb	1600	195 dB
8	8	1-Mar	1700	3-Mar	1200	195 dB
9	9	7-Mar	1200	11-Mar	800	195 dB
10	10	8-Apr	1400	11-Apr	1300	195 dB
11	11	14-Apr	900	17-Apr	100	195 dB
12	12	21-Apr	900	25-Apr	500	195 dB
13		29-Apr	900	3-May	100	195 dB
14	13	12-May	1700	16-May	1300	195 dB
15	14	19-May	1700	23-May	1300	195 dB
16	15	17-Jun	1300	19-Jun	1700	195 dB
17	16	23-Jun	1300	25-Jun	500	195 dB
AST-1	17	30-Jun	1045	4-Jul	900	195 dB
AST-2	18	5-Jul	2100	8-Jul	2100	195 dB
20	19	18-Jul	1700	20-Jul	2100	195 dB
21	20	2-Aug	1700	5-Aug	1700	195 dB
22	21	11-Aug	2100	15-Aug	1300	195 dB
23	22	18-Aug	1445	20-Aug	900	195 dB
24	23	26-Aug	1300	28-Aug	900	195 dB
25	24	2-Sep	1700	4-Sep	1700	195 dB
26	25	11-Sep	2100	14-Sep	900	195 dB
27	26	23-Sep	1700	26-Sep	1700	195 dB
28	27	1-Oct	1700	5-Oct	1300	195 dB
29	27	7-Oct	930	9-Oct	1300	195 dB
30	28	17-Oct	2100	21-Oct	900	195 dB
31	29	29-Oct	400	30-Oct	1600	195 dB
32	30	4-Nov	800	7-Nov	1600	195 dB
33	31	12-Nov	800	15-Nov	1600	195 dB
34	31	18-Nov	800	21-Nov	1200	195 dB
35	32	23-Nov	2400	27-Nov	2000	195 dB
36	33	3-Dec	2000	6-Dec	1200	195 dB
37	34	12-Dec	2400	16-Dec	2000	195 dB
38	35	8-Jan	1200	10-Jan	1200	195 dB
39	36	16-Jan	1600	20-Jan	400	195 dB
40	37	23-Jan	2000	28-Jan	1600	195 dB
41	38	1-Feb	2000	3-Feb	2400	195 dB
42	39	7-Feb	400	10-Feb	400	195 dB
43	40	15-Feb	400	18-Feb	2400	195 dB

(Calambokidis *et al.*, 1990a, 1990b, 1993). The purpose of the research under the aegis of the MMRP was to document specific individual cetaceans (particularly blue, humpback and sperm whales), that were present in the area around Pioneer Seamount. Virtually, all of the humpback whales that feed off the California coast have been individually recognized; a total of 645 through 1993 (Calambokidis and Steiger, 1995). Further, 642 blue whales have been identified through 1993, out of an estimated population off California of 2000 individuals (Calambokidis and Steiger, 1995).

Blue, humpback, and killer (when available) whales were approached to within 30 m and photographed using a 35 mm SLR camera with a 70-300 mm lens (Calambokidis *et al.*, 1993). Photographs were then compared to others collected and catalogued by Cascadia Research Collective.

Northern Elephant Seals

Elephant seal studies were the primary responsibility of researchers from UCSC. Satellite and recoverable tag-based behavioral studies of northern elephant seals were designed to test whether there was a detectable change in their behavior when the source was on and when it was off. Northern elephant seals were chosen because they are known to have the best low frequency hearing capability of any pinniped measured to date (Kastak 1996), are relatively abundant and naturally migrate past the Pioneer Seamount, are deep divers (up to 1500 m depth), and make good research subjects (Leboeuf and Laws 1994). Two types of instruments were used in conjunction with an ARGOS satellite tag. The ARGOS satellite tag provides information on the animals location while at-sea, whereas a archival tag upon recovery provides information in the animals swim-speed, time and depth (STDR), and an acoustic data logger (ADL) provides information on the animals ambient acoustic environment as well as the actual sound pressure level of the ATOC source at the animal. A combination of these devices allowed us to relate received sound levels to changes in dive depth or duration, and any avoidance of the source by monitoring changes in swim speed and location. Tagging individuals has advantages for obtaining accurate and detailed information on movements, dive depth and duration, duration at surface, swimming speed and acoustic exposure (Costa, 1994; Le Boeuf and Laws, 1994; Fletcher *et al.* 1996, Burgess *et al.* 1997). Tracking an individual with an attached satellite tag provides individual-specific data that usually is of greater detail and duration than can be obtained using visual techniques. Satellite tracking allows animals to be monitored at great distances (decreasing biases associated with potential disturbance), at nighttime; and greater amounts of data can be gathered than with visual techniques alone.

These instruments were deployed on animals to study the response of naturally migrating adult male elephant seals that were expected to swim over or near the source site on both departure and return, and upon juvenile animals that were captured translocated and released beyond the source site, with a relatively high probability that they would pass over or near the source (i.e., within the 120 dB sound field) in returning to their Año Nuevo rookeries.

Adult male elephant seals that were deemed to be near departure were instrumented with ARGOS satellite tags and TDRs. Fifteen adult males were instrumented in August/September 1995, 2 in August 1996, and 12 in March/April 1997 at Año Nuevo mainland. Adult males are a good choice for this part of the study as their 3 month migration allows a single instrument deployment to obtain data on two natural passes in the vicinity of the source. Previous studies have demonstrated that satellite tracks on migrating animals are based on intermittent position fixes accurate to within 1-2 km, so that the diving behavior can be analyzed from a period in which the animal was in the vicinity of the sound source. Adult male elephant seals exhibit a highly regular continuous diving pattern while in the initial transit periods of their migrations (Le Boeuf *et al.*, 1993). Potential changes in diving behavior due to proximity to the source site were most evident during this period.

A total of 29 juvenile elephant seals were instrumented with ARGOS satellite tags and separate data loggers (3 during Fall 1995, 11 during Spring 1996, 11 during Fall 1996, 4 during spring 1997) and translocated to the vicinity of Pioneer Seamount. These animals were transported from their rookery to Long Marine Laboratory, where they were instrumented, then transported by vessel to an area due west of the Pioneer Seamount source site. This protocol increased the potential that released seals would swim through the source site area (or actively avoid it) in returning, while being tracked by satellite. Upon their return (3-5 days) the instruments were recovered and data downloaded for analysis. The recorders measured depth, swim velocity ambient acoustic environment.

The value of the translocation paradigm is that it makes it possible to conduct acoustic monitoring during diving in a matter of days. This is important because studies of the free-ranging dive pattern must be conducted over a period of months (a minimum of 3.5 months and a maximum of 8 months), the periods that the animals are at sea (Le Boeuf, 1994). Moreover, two year-old juveniles dive to the same mean depths and exhibit the same mean dive duration's as adult males and adult females (Le Boeuf and Laws, 1994). By the end of the fourth trip to sea, when juveniles average 270 ± 26 kg, the adult diving pattern and migratory pattern is set. Being able to conduct short-term acoustic-diving studies on translocated seals allows a flexible, efficient approach by reducing costly long-term experiments.

RESULTS

Aerial surveys

Aerial surveys through the end of 1996 included 15,687 nmi of on-effort aerial survey coverage of the inner box and 15,872 nmi of coverage of the outer box. Coverage was equally divided between control and experimental periods (7,792 to 7,999 nmi in each of the four strata). A total of 2,270 sightings of 41,002 marine mammals has been made through the end of 1996; this is roughly equally split between control and experimental periods.

Nineteen species were sighted during the surveys. Most commonly sighted species included by category were: 1) mystecetes: humpback whales, 2) large odontocetes: sperm whales, 3) small odontocetes: Pacific white sided dolphins, and 4) pinnipeds: California sea lions. There were clear seasonal patterns in occurrence of the mystecetes and some of the odontocetes. There were also pronounced patterns in habitat preferences for many species including those seen close to shore or on the continental shelf (i.e. gray whales), other along the shelf edge (humpback whales), along the steep slope waters (Pacific white sided dolphins) or in deeper waters and around the Seamount (sperm whales).

Though there were not significant differences in the number of sightings when the source was on or off (Figure 1), there were significant differences in how two species were distributed in relation to the sound source. Both humpback and sperm whales were generally seen farther from the sound source when it was transmitting than when it was off ($p<0.01$). Additional analyses of these differences

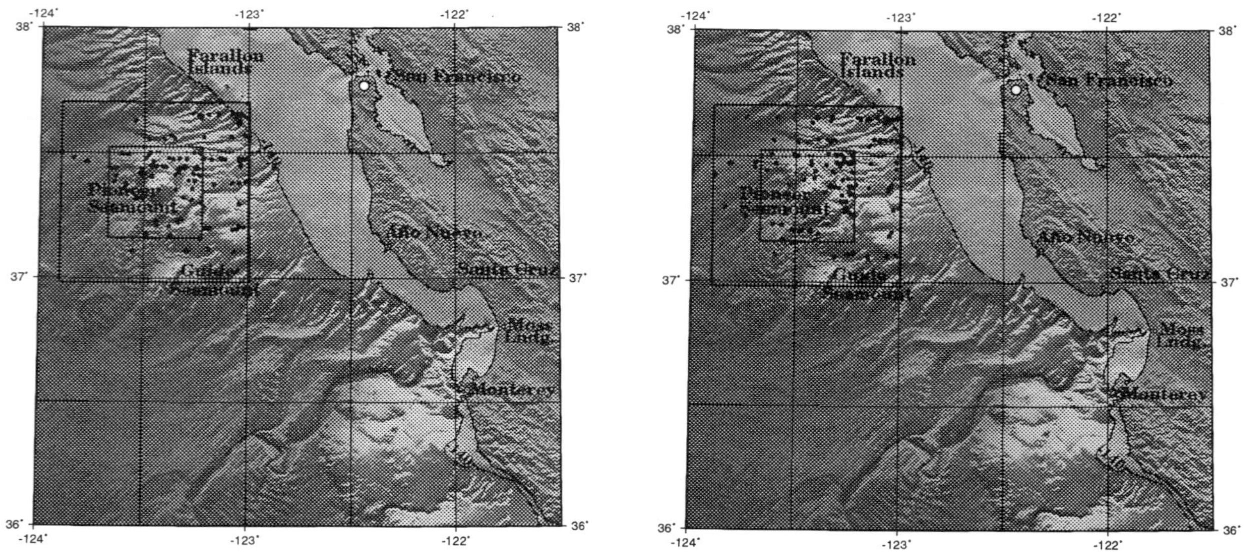

Control Period

Experimental Period

Fig. 1. Summary Sightings of Humpback Whales During Control and Experimental Periods Through Sept. 13, 1996

is currently underway.

Photo-ID

Small boat photographic identification surveys were completed through 12 November 1996. In total 32, vessel surveys were conducted in and around the study area in 1996. A total of 25 identifications representing 23 different humpback whales were made in 1996. Fifteen of these individuals were identified from 0.5 to 12.6 nmi from the sound source during periods when it was transmitting. These identified whales will provide a key group of individuals than can be tracked in future years to evaluate any potential long-term injury from exposure to the ATOC signal. Five of the 15 individuals identified during transmissions have already been resighted again later in 1996. These animals were seen from one to more than 6 months later primarily around the Farallon Islands.

Individual identifications were also obtained near the ATOC sound source of a number of other species, including blue whales, fin whales, and sperm whales. Fewer identifications of these species were obtained compared to humpback whales because they were less common during the periods of boat effort.

Elephant Seal Research

Three juvenile elephant seals were translocated to the Pioneer seamount in October 1995 to establish the time course and feasibility of translocations and to collect control data before initiation of the ATOC transmissions. Eleven juvenile elephant seals were translocated in April 1996 and 11 in October-November 1996 while the ATOC source was in operation. Four juvenile elephant seals were translocated in April of 1997, when the source was not operational. Of the 22 deployments, 19 animals returned to Año Nuevo, where the instruments were recovered.

Measured mean intensity of sound exposure during transmission for analyzed acoustic records ranged from 120 - 135 dB for 60-90 Hz ($n = 6$) compared to ambient levels of 100 - 107 dB (60-90 Hz). Animals did not alter return track, diving pattern, or swim speed, did not go to the surface, and often continued to dive closer to the sound source if on the descending segment of a dive. Analysis is underway to quantify these results, utilizing appropriate statistics. An example of a record of elephant seal diving behavior (depth of dive and swim speed) coupled with measurements of the ATOC transmission are shown in Figure 2.

The migration study subjects were 14 male elephant seals that were tagged in August and September 1995, went to sea on a natural migration, and then returned during the months of December 1995 and January 1996 to the Año Nuevo breeding ground. Two male elephant seals were tagged with satellite tags and swim velocity/time-depth recorders in September 1996. These instruments were recovered in January 1997. Twelve adult males were instrumented with satellite tags and time depth recorders in March 1997. Six of these subjects also carried acoustic data loggers. As the ATOC source became non-operational just prior to their departure, these migrations will serve as additional control data when recovered in July 1997. The paths of male elephant seal as they passed by the Pioneer Sea mount on their return migration is hown in Figure 3. A paired t-test comparison of track distance from the Pioneer Seamount utilizing animals that departed when the source was off and returned when the source was in operation ($n = 8$) showed no significant difference ($t = 1.89$, $df = 7$, $p > 0.05$).

The diving pattern of one adult male elephant seal for which a complete TDR record was obtained and analyzed for differences when the sound source was

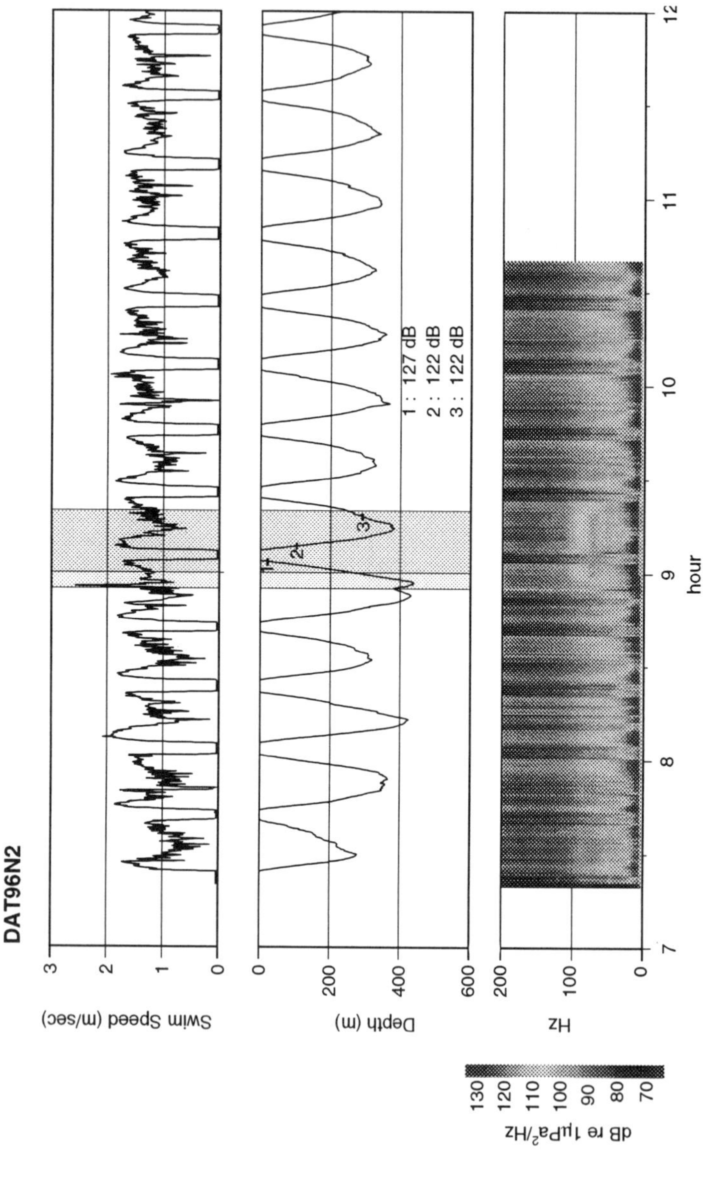

Figure 2. Swimming Speed, Diving Depth and Spectrogram of Received Sound Intensities. Shaded area area is period of ATOC transmission.

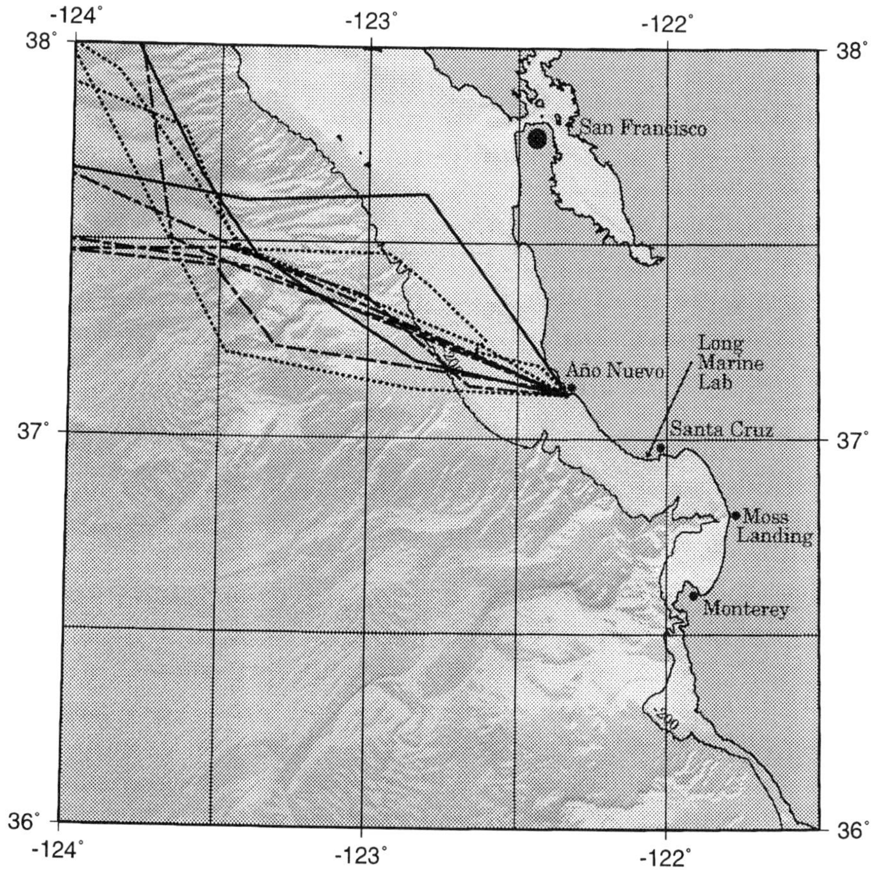

Figure 3. ARGOS satellite migration tracks of adult male elephant seals returning from the North Pacific Ocean to Ano Nuevo Rookery during December 1995 to January 1996. Solid tracks are for animals returning when the source was not operating. Dotted tracks are for animals returning while the source was transitting at 185 dB per re 1 μ Pa. Dashed tracks are for animals returning while the source was transmitting at 195 dB re 1 μ Pa.

transmitting. This animal yielded a class 1 (error < 1 km) satellite location ~9 km from the source approximately 40 min before a 195 dB transmission began. The diving pattern for 10 hrs following the initiation of transmission was compared to the 10 hrs of data surrounding a departure location that was ~8.5 km from the source. Statistical comparison was attempted using a students t-test. When conditions of normality or variance failed a Mann-Whitney rank sum test was used. There was no significant difference in diving depth, surface interval, descent rate, and ascent rate during exposure to the sound source (rank sum test; $n = 23,28$: $T = 510.5, 556.5, 688.5, 512.5$; $p > 0.05$). Dive duration was significantly longer during exposure (t test; $t = 8.80$, $df = 49$; $p < 0.05$). However, mean dive often increases across the foraging migration. Analysis of this dive record reveals an increase in dive duration across the record and prior to sound exposure.

CONCLUSIONS

At the present stage of analysis it would be premature to make any statements regarding the biological significance of the differences in the distribution of humpback and sperm whales observed in relation to the operation of the ATOC sound source. We are currently assessing the distribution of the animals relative to the ATOC sound field. We hope to increase our sample size when the source begins operation later this year (October 1997).

The northern elephant seal translocation paradigm has allowed conclusive measurement of exposure to the ATOC source in translocated individuals. This exposure was not correlated with any gross changes in return track or diving behavior. A finer statistical comparison of diving patterns and the incorporation of physiological measures such as heart rate are forthcoming. Application of acoustic data loggers on natural migrations can potentially provide information on exposure levels and behavior during natural movements.

LITERATURE CITED

Burgess, W. P. Tyack, B.J. LeBoeuf, D.P. Costa. An Intelligent Acoustic Recording Tag First Results from Free-Ranging Northern elephant Seals. Deep Sea Research (in press).

Calambokidis, J., and G.H. Steiger. 1995. Population estimates of humpback and blue whales through photo-identification from 1193 surveys off California. Report to southwest Fisheries Science Center, National Marine Fisheries Service, La Jolla, California. 36 pp.

Calambokidis, J., J.C. Cubbage, G.H. Steiger, K.C. Balcomb, P. Bloedel. 1990a. Population estimates of humpback whales in the Gulf of the Farralones, California. In: Individual recognition of cetaceans: Use of photo identification techniques to estimate population parameters. P.S. Hammond, S.A. Mizroch, G.P. Donovan, eds. Rep. Int. Whal. Comm, Special Issue. 12:325-333.

Calambokidis, J., G.H. Steiger, J.C. Cubbage, K.C. Balcomb, C. Ewald, S. Kruse, R. Wells, and R. Sears. 1990b. Sightings and movements of blue whales off central California 1986-88 from photo-identification of individuals. Rep. Int. Whal. Comm, Special Issue. 12:343-348.

Calambokidis, J., G.H. Steiger, and J.R. Evenson. 1993. Photographic identification and abundance estimates of humpback and blue whales off California in 1991-92. Rept. to Southwest Fisheries Center, National Marine Fisheries Service, La Jolla, California. 67 pp.

Costa, D.P. 1994. The Secret Life of Marine Mammals: New tools for the study of their biology and ecology. Oceanography: 6(3):120-128.

Fletcher, S., B.J. Le Boeuf, D.P. Costa, P.L.Tyack, and S.B. Blackwell. 1996. Onboard acoustic recording from diving northern elephant seals. J. Acoust. Soc. Am. 100(4):2531-2539.

Le Boeuf, B.J. 1994. Variation in the diving pattern of Northern elephant seals with age, mass, sex, and reproductive condition. In: B.J. Le Boeuf and R.M. Laws (eds.), Elephant seals: Population ecology, behavior and Press, Los Angeles.

Le Boeuf, B.J. and R.M. Laws (eds.) 1994. Elephant seals: Population ecology, behavior and physiology. University of California Press Berkeley, CA. 414 pp.

CHAPTER 209

The Effect of El Niño on Coastal New Production

Frances P Wilkerson and Richard C Dugdale[1]

Abstract

El Niño events have been shown to strongly reduce the primary production of coastal upwelling systems and have devastating consequences on the coastal fisheries that are supported by "new" production. New production is production based on the input of new nutrients (e.g. upwelled nutrients) and sets an upper limit to the biomass yield at higher trophic levels and the export of biomass from the euphotic zone. El Niño's reduce the rate of new production as a consequence of the decrease in concentration of upwelled nutrients as the thermocline deepens. This climatological phenomenon and its consequences were originally thought to be restricted to the equatorial Pacific and the associated coastal fisheries there (e.g. the anchovy fisheries of Peru), but more recently, elevated temperatures and reduced nutrients have been shown to reach California coastal waters. This communication compares new production measured in El Niño and non-El Niño years in Monterey Bay, California during the upwelling season. Small changes in surface temperature resulting from El Niño events, are shown to have large effects on surface nutrients and biological new production at this coastal upwelling location.

Introduction

A major focus of contemporary oceanographic research is the study of the flux of biogenic elements between the upper and deep ocean and between the ocean and atmosphere, both on recent and geologic time scales. Although major emphasis in these biogeochemical studies has been on open ocean systems, the biogeochemistry of the coastal ocean deserves serious attention, since a

[1] Senior Research Scientists, Romberg Tiburon Centers, San Francisco State University, PO Box 855, Tiburon, CA 94920

and Barber, 1987). In these productive waters, the biological pump serves to drawdown CO_2 at the air/sea interface and export carbon to depth as sinking particles produced as a result of photosynthesis. The efficiency of the biological pump depends on the fraction of photosynthetic carbon fixation that escapes recycling within the mixed layer (Berger et al., 1988) - i.e. new production (Dugdale and Goering, 1967). New production can be interpreted as the production supported by the influx of nutrients (e.g. nitrate, NO_3 or silicate, $Si(OH)_4$), that sets an upper limit to the biomass yield at higher trophic levels and the export of biomass from the euphotic zone (Eppley and Peterson, 1979). Nitrogen was originally used as a tracer element to study new production because of the ability to distinguish between new (i.e. NO_3) and regenerated (ammonia and urea) forms; the implication being that autotrophic NO_3 uptake can be used as an estimate of new production when NO_3 is the significant nitrogenous form in the system. In many coastal situations with high NO_3 (> 5 μM) this is the case. Typically ecosystems with high levels of NO_3 have high new production in which phytoplankton blooms are common, often have diatom dominated communities and conventional food chains that result in coastal fisheries (e.g. Ryther, 1969). Monterey Bay, California, an eastern boundary system, has the characteristics of this type of coastal ecosystem. The vertical motion of the water near the coast feeds high concentrations of NO_3, $Si(OH)_4$ and total CO_2 to the illuminated surface waters enabling phytoplankton growth, in particular diatom blooms (Garrison, 1976; 1979).

ENSO (El Niño Southern Oscillation) is an interannual climatic condition that results in sea-surface warming, and a deeper thermocline and nutricline in eastern boundary currents (Philander, 1990), especially in the tropical Pacific. In these upwelling areas (e.g. off Peru) the devastating consequences of reduced nutrients and productivity is well documented (e.g. Barber and Chavez, 1986), in particular the impact on local fisheries resulting from nutrient-rich water being placed below the euphotic zone and upwelling source water. The onset of the El Niño condition can also be observed at higher latitudes, in California (Simpson, 1984) and Oregon (Huyer and Smith, 1985). Apparently, oceanic Kelvin waves link the atmospheric events in the western equatorial Pacific and warming anomalies in central California (Chavez, 1996) with associated chemical and biological consequences (Lenarz et al., 1995). Chavez (1996) reported ocean warming in Monterey Bay, California during the 1992 El Niño and demonstrated how the high productivity, diatom-dominated ecosystem normally found in Spring was practically eliminated and replaced by small solitary picoplanktonic populations, characteristic of open ocean ecosystems that subsist on recycled nutrients. The most dramatic effects of the 1992 El Niño reported by Chavez (1996) were manifested in the reduced supply of new nutrients, i.e. NO_3 with the implication that new production must have been reduced significantly. Roemmich and McGowan (1995) in their paper describing how climatic warming may be linked to the decline of zooplankton in the California Current also implicate reduced new production, resulting from less nutrients as waters are upwelled from a shallower source. Reduced new production during El Niño warming events has been documented for the coastal upwelling areas of 15°S Peru, Cap Blanc northwest Africa and Baja California (e.g. Wilkerson et al., 1987; Dugdale and Wilkerson, 1992). This contribution describes data collected in the upwelling ecosytem of Monterey Bay, California and shows how the 1992 El Niño event resulted in both reduced ambient nutrients and new production in coastal upwelling areas.

Methods

Surface mapping of ambient nutrient concentrations and nutrient uptake rates was conducted during the upwelling season aboard the R/V's New Horizon and Point Sur in Monterey Bay, California, as part of a series of seasonal cruises made in 1992 and 1993 (designated SU92 and SU93). The results presented here were collected in 23 May - 3 June, 1992 (SU92-2) and 7 - 13 May, 1993 (SU93-2). Typically maps were made at the start and end of each week-long cruise. Data from the first maps (called Map 1) are shown here (Figs. 1, 2). Seawater was collected from 5m depth using 10-L PVC Nisken bottles equipped with Teflon coated springs and fittings mounted on a rosette sampler. NO_3 concentrations and uptake rates were determined for all stations. Nutrient samples were collected and stored frozen in acid-washed 30-mL polypropylene bottles until analysis for NO_3 plus nitrite using a Technicon AutoAnalyzer II according to the procedures of Whitledge et al. (1981). Combined NO_3 and nitrite concentrations are reported as NO_3. ^{15}N incubations to measure NO_3 uptake were conducted using additions of $K^{15}NO_3$ (99 atom%, Cambridge Isotope Laboratories) at approximately 10% of ambient NO_3 to 280-mL seawater samples. Incubations for nutrient uptake determination were conducted in 280-mL polycarbonate bottles, shaded by 50% neutral density light screening, for 24 h in on-deck Plexiglas incubators, cooled by flowing surface seawater. Following incubation, all samples were collected on 25 mm Whatman GF/F glass microfiber filters and stored frozen. Filters were dried (<60°C for 24 h) and analyzed for ^{15}N enrichment and total particulate nitrogen content with a Europa Scientific Roboprep-mass spectrometer system. The transport rate ρNO_3 was calculated according to Dugdale and Wilkerson (1986).

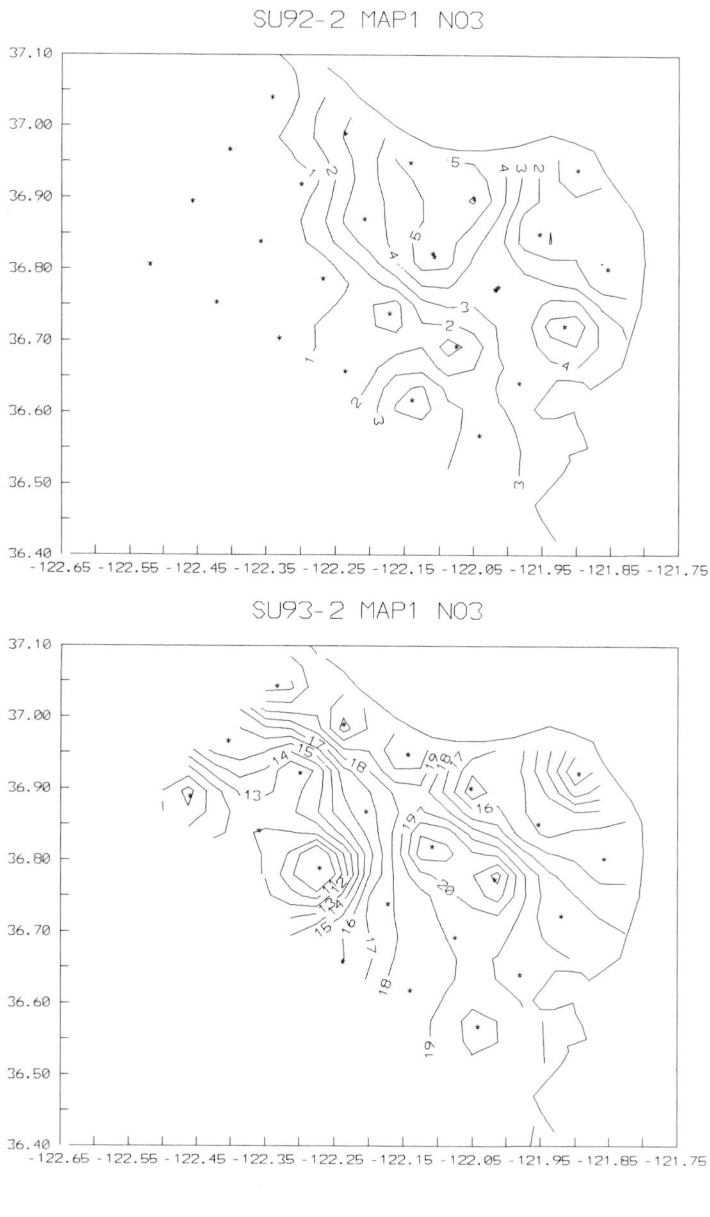

Figure 1. Map of Monterey Bay, California showing near surface (5m) contours of NO_3, (μM) obtained during May 1992 (upper panel, SU92-2, El Niño conditions) and May 1993 (lower panel, SU93-2, typical conditions).

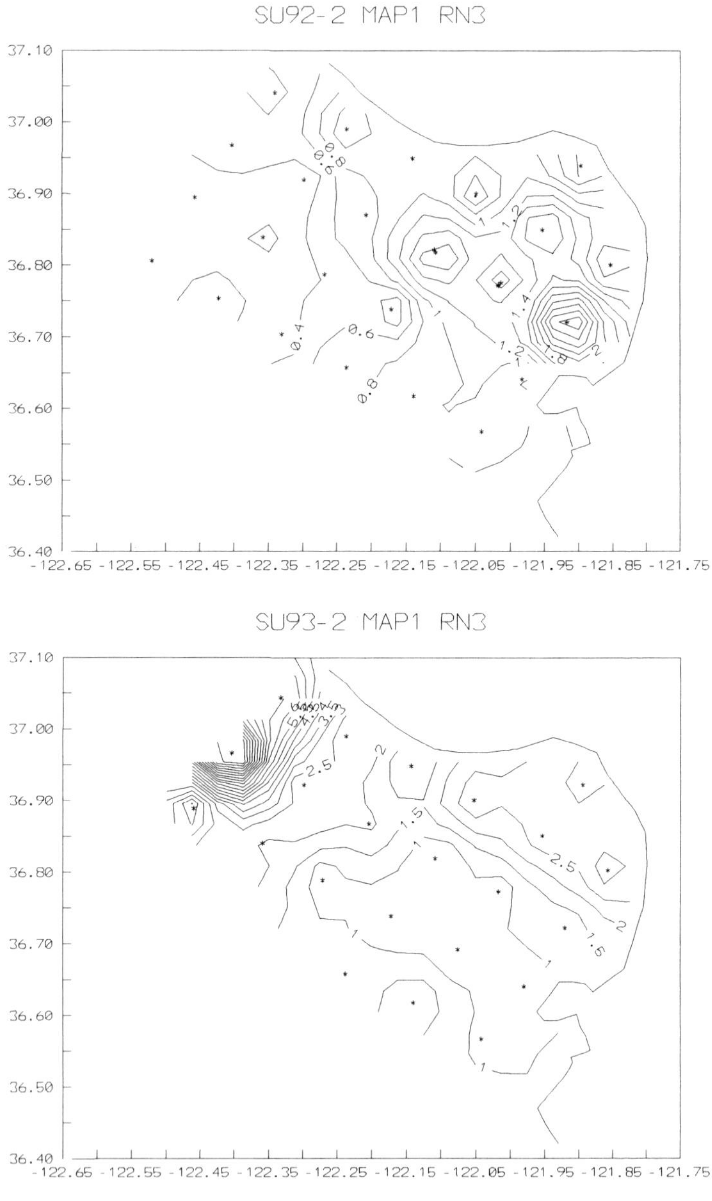

Figure 2. Map of Monterey Bay, California showing near surface (5m) contours of ρNO_3, ($\mu mol\ l^{-1}\ d^{-1}$) obtained during May 1992 (upper panel, SU92-2, El Niño conditions) and May 1993 (lower panel, SU93-2, typical conditions).

Results

Both surface maps (SU92-2 map 1 and SU93-2 map 1) were sampled during upwelling favorable conditions (i.e. northwesterly winds). The upwelling center to the north of the Bay, near Davenport is shown clearly by the higher NO_3 concentrations there. Rosenfeld et al. (1994) previously identified this area as the upwelling center using a variety of hydrographic parameters. However the highest surface NO_3 concentration observed in the map made in May 1992 is only 5 μM compared to 22.4 μM measured on the map made in 1993 (Fig. 1). This reduction in surface NO_3 during the 1992 El Niño conditions was also illustrated by Chavez (1996), with concentrations in 1992 always less than the mean for the "normal" years 1989-1991. NO_3 uptake maps also show the same contrast with lower values of ρNO_3 during the May 1992 cruise compared to the 1993 cruise. As implicated by both Chavez (1996) and Roemmich and McGowan (1995) the reduction in new production was significant. The highest value of ρNO_3 measured in the El Niño cruise was 3.2 μmol l^{-1} d^{-1} compared with 13.2 μmol l^{-1} d^{-1} during the more normal condition in 1993 (Fig. 2). In addition to the peak in values at the upwelling center, another peak in NO_3 and uptake was observed just to the north of Monterey in the map made during El Niño conditions (SU92-2), but a peak at this location was not seen in the more normal year (SU93-2). Both peaks were much lower that those measured in SU93-2.

Discussion

This paper provides the observations to support the suggestions of both Roemmich and McGowan (1995) and Chavez (1996), that El Niño warming events result in reduced new production rates, along the coast of California, and that this phenomenon is not limited to the equatorial pacific (e.g. Wilkerson et al., 1987, McCarthy et al., 1996).

The spatial patterns of ρNO_3 uptake observed in Monterey Bay matched those reported by Kudela and Dugdale (1996) and White and Dugdale (1997); and also those for the upwelling center at Point Conception (Dugdale and Wilkerson, 1989) with maximum rates coinciding with the NO_3 maxima near the coast. White and Dugdale (1997) observed similar patterns for $Si(OH)_4$ and $Si(OH)_4$ uptake. Kudela and Dugdale (1996) used the strong correlation between sea-surface temperature and surface NO_3 concentrations in the Monterey Bay upwelling region, along with physiological models of NO_3 uptake (Dugdale et al., 1990), to estimate new production for Monterey Bay using remotely sensed data, as was also carried out for the upwelling areas of Cap Blanc (Dugdale et al., 1989) and Point Conception, California (Dugdale et al., 1997). The model uses remotely sensed sea-surface temperature data from AVHRR (Advanced Very High Resolution Radiometer) to predict new production from temperature changes.

Consequently it could be employed to predict the productivity effects of surface warming from El Niño events. Another satellite sensor could also be used to forecast El Niño related productivity changes along the California coast. The TOPEX/POSEIDON altimeter provides information of sea-surface height changes (Fu et al. 1996). Typically during El Niño events the sea-surface level is raised, the thermocline deepened with warmer and nutrient-poor surface waters. Similar physiological models could be linked with data of the changing sea-surface height to evaluate effects on coastal new production. A new avenue for research would be to use a combination of data from these two sensors to aid in predicting the effects of El Niño events on new production, and the associated cascade of impacts on higher trophic levels.

References

Barber, R.T. and F.P. Chavez. 1986. Ocean variability in relation to living resources during the 1982/83 El Niño. Nature 319: 279-285.

Berger, W.H., K. Fischer, C. Lai and G. Wu. 1988. Ocean carbon flux: global maps of primary production and export production. In: C.R. Agegian (ed.), Biogeochemical Cycling and Fluxes between the Deep Euphotic Zone and other Oceanic Realms, pp. 131-176.

Chavez, F.P. 1996. Forcing and biological impact of onset of the 1992 El Niño in central California. Geophysical Research Letters 23: 265-268.

Chavez, F.P. and R.T. Barber. 1987. An estimate of new production in the equatorial Pacific, Deep-Sea Research 34: 1229-1243.

Dugdale, R.C., C.O. Davis and F.P. Wilkerson. 1997. Assessment of new production at the upwelling center at Point Conception, California using nitrate estimated from remotely sensed sea surface temperature. Journal of Geophysical Research 102: 8753-8585.

Dugdale, R.C. and J.J. Goering. 1967. Uptake of new and regenerated forms of nitrogen in primary productivity. Limnology and Oceanography 12: 196-206.

Dugdale, R.C., A. Morel, A. Bricaud and F.P. Wilkerson. 1989. Modeling new production in upwelling centers: II A case-study of modeling new production from remotely sensed temperature and color. Journal of Geophysical Research 94: 18119-18133.

Dugdale, R.C. and F.P. Wilkerson. 1986. The use of ^{15}N to measure nitrogen uptake in eutrophic oceans; experimental considerations. Limnology and Oceanography 31: 673-689.

Dugdale, R.C. and F.P. Wilkerson. 1989. New production in the upwelling center at Point Conception, California: temporal and spatial patterns. Deep-Sea Research 36: 985-1007.

Dugdale, R.C. and F.P. Wilkerson. 1992. Nutrient limitation of new production. In: P.G. Falkowski and A.D. Woodhead (eds.), Primary Productivity and Biogeochemical Cycles in the Sea, Plenum Press, New York, pp. 107-122.

Dugdale, R.C., F.P. Wilkerson and A. Morel. 1990. Modeling new production in upwelling centers: I Applications of a shift-up model to 3 upwelling areas. Limnology and Oceanography 35: 822-829.

Eppley, R.W. and B.J. Peterson. 1979. Particulate organic matter flux and planktonic new production in the deep ocean. Nature 282: 677-680.

Fu, L-L, C.J. Koblinsky, J-F Minster and J. Picaut. 1996. Reflecting on the first three years of TOPEX/POSEIDON. Eos Trans., 77: 109, 111, 117.

Garrison, D.L. 1976. Contribution of the net plankton and nannoplankton to the standing stocks and primary productivity in Monterey Bay, California during the upwelling season. Fisheries Bulletin, 74: 183-194.

Garrison, D.L. 1979. Monterey Bay phytoplankton I. Seasonal cycles of phytoplankton assemblages. Journal of Plankton Research 1: 241-265.

Huyer, A. and R.L. Smith. 1985. The signature of El Niño off Oregon, 1982-1983. Journal of Geophysical Research 90: 7133-7142.

Kudela, R.K. and R.C. Dugdale. 1996. Estimation of new production from remotely-sensed data in a coastal upwelling regime. Advances in Space Research 18: 91-97.

Lenarz, W.H., D.A. Ventresca, W.M. Graham, F.B. Schwing and F.P. Chavez. 1995. Expolorations of El Niño events and associated biological population dynamics off central California. CalCOFI Report 36: 106-119.

McCarthy, J.J., C. Garside, J.L. Nevins and R.T. Barber. 1996. New production along 140°W in the equatorial Pacific during and following the 1992 El Niño. Deep-Sea Research 43, Part II: 1065-1095.

Philander, S.G. 1990. El Niño, La Niña and the Southern Oscillation. Academic Press, 293 pp.

Roemmich, D. and J. McGowan. 1995 Climatic warming and the decline of zooplankton in the California Current. Science 267: 1324-1326.

Rosenfeld, L.K., F.B. Schwing, N. Garfield and D.E. Tracy. 1994. The origin of upwelled water in Monterey Bay: an example of offshore and equatorward advected tongues from an upwelling center. Continental Shelf Research 14: 931-964.

Ryther, J.H. 1969. Photosynthesis and fish production in the sea. Science 166: 72-76.

Simpson, J.J. 1984. El Niño induced onshore transport in the California Current during 1982-1983. Geophysical Research Letters 11: 241-242.

Walsh, J.J. 1991. Importance of continental margins in the marine biogeochemical cycling of carbon and nitrogen. Nature 350: 53-55.

White, K.K. and R.C. Dugdale. 1997. Silicate and nitrate uptake in the Monterey Bay upwelling center. Continental Shelf Research 17: 455-472.

Whitledge, T.E., S.C. Malloy, C.J. Patton and C.D. Wirick. 1981. Automated nutrient analyses in seawater, Report BNL 51398, 216 pp., Brookhaven National Laboratory, Upton NY.

Wilkerson, F.P., R.C. Dugdale and R.T. Barber. 1987. Effects of El Niño on new, regenerated and total production in eastern boundary upwelling systems. Journal of Geophysical Research 92: 14347-14355.

Acknowledgements

This research was supported by the National Science Foundation (OCE 91-15929, OCE 94-10445). We thank Raphael M. Kudela for his contributions at sea and in the lab, and the crews of the R/V Point Sur and the R/V New Horizon for their assistance at sea.

CHAPTER 210

Will global change affect planktonic productivity?

Lynn J. Rothschild[1]

Abstract

Phytoplankton are a major source of marine primary productivity. Factors that control algal productivity range from the physical to the chemical to the biological. Currently, levels of atmospheric carbon dioxide and surficial fluxes of ultraviolet radiation are rising. Both of these environmental variables can have a profound effect on algal productivity. Atmospheric CO_2 may increase levels of dissolved inorganic carbon in the ocean surface. Laboratory and field studies of algal mats and phytoplankton cultures under ambient and elevated levels of pCO_2 show that elevated levels of inorganic carbon can cause an increase in photosynthetic rates. The increase in photosynthetic rates can result in higher population density. There may be an increase in the excretion of fixed carbon, which in turn may enhance bacterial productivity. Alternatively, in analogy with studies on the effect of elevated pCO_2 on plants, the phytoplankton could change their carbon to nitrogen ratios, which will affect the feeding of the planktonic grazers. The seasonal depletion of stratospheric ozone has resulted in elevated fluxes of UVB radiation superimposed on the normal seasonal variation. Present surface UV fluxes have a significant direct impact on algal photosynthesis, as well as indirect effects on cell physiology and aquatic chemistry. While not disputing the importance of other factors in controlling marine phytoplankton productivity, it is argued here that the global change variables of elevated pCO_2 and UVB radiation should not be ignored.

Introduction

Algae are at the base of oceanic food webs, providing new organic carbon for the remaining members of the community. Thus, the rate that they produce organic carbon to a large extent controls the productivity of the entire ecosystem. Factors that control algal productivity range from the physical (e.g., temperature, light), chemical (e.g., nutrient levels) to the biological (e.g., grazing). The three main biochemical

[1]Research Scientist, MS 239-12, NASA/Ames Research Center, Moffett Field, CA 94035

controls on photosynthesis are the availability of carbon dioxide, energy in the form of ATP, and energy in the form of radiation. Two of these variables, the availability of CO_2 and the relative flux of ultraviolet radiation, are changing. Atmospheric levels of pCO_2 are rising and there have been seasonal increases in UVB radiation due to depletion of stratospheric ozone. Workers on terrestrial and freshwater systems are acutely aware of the repercussions of changes in these variables for primary productivity in their ecosystems. The oceanographic community has been slower in its response. The reasons for this include a perception that elements other than carbon, for example, iron, nitrogen and phosphorus, are the rate limiting factors in the marine environment (e.g., Schlesinger, 1991; Coale et al., 1996). Similarly, UV penetration is not seen as important. Recently there has been an increase in awareness of the importance of UV radiation on marine ecosystems. For example, the Inter-American Institute for Global Change Research sponsored a workshop on UV effects on aquatic ecosystems in September 1996 in Ensenada, Mexico.

The purpose of this paper is to help make oceanographers aware of the potential importance of rising levels of atmospheric pCO_2 and UVB radiation on phytoplankton productivity. This will be accomplished through a review of the literature on the effects of rising CO_2 and UVB radiation on photosynthetic organisms, including marine phytoplankton.

Rising levels of atmospheric CO_2

There is much evidence that atmospheric levels of pCO_2 are rising (Boden et al., 1993). Atmospheric CO_2 is a greenhouse gas, and thus increases in atmospheric CO_2 causes an increase surficial temperatures. Atmospheric CO_2 dissolves in water as function of its concentration in the overlying atmosphere, as well as temperature (CO_2 is more soluble in colder waters), surface mixing and salinity. Theoretically, the concentration of dissolved inorganic carbon in the surface ocean is in equilibrium the atmosphere. However, because of the uptake of this compound for photosynthesis, it remains undersaturated with respect to the atmosphere (Schlesinger, 1991).

The total pool of dissolved inorganic carbon (ΣCO_2) is the sum of all the species of inorganic carbon present. These species are carbon dioxide (CO_2), bicarbonate (HCO_3^-) and carbonate (CO_3^{--}). The addition of CO_2 to an aqueous solution affects the carbonate equilibrium (Fig. 1). From the first equation, it is seen that the addition of CO_2 to an aqueous system decreases the pH of the solution as a byproduct of the production of the HCO_3^-. Of importance is the fact that many species of phytoplankton are unable to utilize bicarbonate directly, and must first expend energy converting it to CO_2 before fixation (Riebesell et al., 1993).

$$CO_2 + H_2O \leftrightarrow H^+ + HCO_3^-$$
$$HCO_3^- \leftrightarrow H^+ + CO_3^{--}$$

Figure 1. The carbonate system.

Seawater in the euphotic zone is approximately pH 8.2. At that pH ~99% of the dissolved inorganic carbon is in the form of HCO_3^-, while only ~1% is in the form of free CO_2. Seawater is, however, far better buffered than freshwater systems, so the effect of relatively small amounts of additional CO_2 in the next few decades on pH is unlikely to be as significant. But, this situation could change at higher atmospheric concentrations of CO_2. For example, 21 million years ago atmospheric concentrations of CO_2 are to have been 4.5 times their present levels, suggesting that levels of dissolved inorganic carbon were higher (Rothschild, 1994). At that time, surface sea water in the equatorial Pacific was pH 7.4 (Spivack et al., 1993).

Elevated levels of CO_2 are known to increase carbon fixation rates in a diversity of photosynthetic organisms, and chemoautotrophs including the nitrifiers. This is because CO_2 is both an activator of, and substrate for, the carbon fixation enzyme ribulose-1,5-bisphosphate carboxylase (RuBisCO). Vascular plants almost always increase photosynthetic carbon fixation in response to increased levels of CO_2, with a resulting increase in growth and development (rev. in Kimball, 1983; Idso, 1989). Likewise, laboratory experiments on cyanobacteria (e.g., Badger and Andrews, 1982), phytoplankton (cited in Sikes and Fabry, 1993) and marine macroalgae (e.g., Brechignac and Andre, 1984; Beer and Shragge, 1987; Holbrook et al., 1988; Reiskind et al., 1989; Gao et al., 1991; Gao et al., 1992; Levasseur et al., 1991) indicate that many photosynthetic protists may be carbon limited at present. Carbon limitation might be particularly severe under microalgal bloom conditions where surface levels of dissolved inorganic carbon decrease drastically (Riebesell et al., 1993). Also suggestive of carbon limitation in nature is that an array of photosynthetic organisms, from cyanobacteria to angiosperms, concentrate CO_2 prior to fixation by RuBisCO (rev. in Raven, 1985; Badger and Price, 1992).

Both short and long-term field studies on selected ecosystems have tended to support the concept of carbon limitation to photosynthesis in nature. Our short-term and longer term studies on carbon fixation in a taxonomic and ecological diversity of algal mats have shown a dramatic increase in carbon fixation rates with the addition of inorganic carbon (Rothschild and Mancinelli, 1990; Rothschild, 1991; Rothschild, 1994; Bartley and Rothschild, 1996). We have measured diurnal changes in carbon fixation in several microbial mat communities in situ in the presence and absence of elevated levels of inorganic carbon (Rothschild and Mancinelli, 1990; Rothschild, 1991; Rothschild and Giver, 1993; Rothschild, 1994b). Three of the mats were marine with a blue-green algal autotrophic component and included a hypersaline (5-10% NaCl) Microcoleus mat and two marine intertidal mats dominated by Lyngbya aestuarii. The other two mats were acidic freshwater eukaryotic algal mats found in Yellowstone National Park. One was composed primarily of the unicellular red alga Cyanidium caldarium and found in a hotspring that flows into Nymph Lake, and the other was primarily composed of the filamentous green alga Zygogonium and was found at ambient temperatures near Norris Annex. We estimated gross daily fixation from carbon fixation data taken over a 24-h period using a 15-20 min incubation in the presence of ^{14}C-bicarbonate (Rothschild, 1991). We tested the immediate impact of elevated levels of dissolved inorganic carbon (DIC) on carbon fixation in microbial mats in field settings, and used the results to hypothesize rates of carbon fixation during the last 3.5 Gyr (Rothschild and Mancinelli, 1990). Our results demonstrate that, in spite of the wide taxonomic and ecological diversity, carbon fixation rates in these mat communities are limited, in the short term, by the availability of DIC.

Most long-term field studies on the effects of elevated atmospheric CO_2 on plants support these results (rev. in Drake, 1992a). For example, wetland grasses growing by the Chesapeake Bay showed an increase in the biomass of C_3 plants under elevated CO_2 over a four-year period (Drake, 1992b). But in the Arctic tundra an immediate response to elevated levels of CO_2 is modified over time so that eventually there is no difference between low and high CO_2 treatments (e.g., Oechel et al., 1994). We have tested the effect of longer-term exposure to elevated pCO_2 on Lyngbya mats in a controlled environmental chamber (Bartley and Rothschild, 1996). After a month-long pre-acclimation period, gross carbon fixation increased almost four-fold at elevated levels of dissolved inorganic carbon.

There are several fates for fixed carbon: excretion into the environment, remineralization during metabolism, storage for later use, or use as an organic carbon skeleton upon which to build other cellular constituents. The last two possibilities result in increased organismal size, distribution of resources (e.g., increased root to shoot material), or reproductive rate. The fate of the fixed carbon affects both the individual organism as well as ecosystem dynamics because interactions among species have been altered (e.g., Mooney et al., 1991; Bazzaz et al., 1990). For example, if the extra fixed carbon is retained altering the C:N ratio of the organism, this changes its nutritional quality which can affect potential herbivores (Lincoln et al., 1993). Other changes in plant quality from elevated CO_2, including carbon-based allelochemical production and increased leaf toughness, are known to have a deleterious effect on insect herbivores (Lincoln et al., 1993). If the photosynthate is excreted[2], then the increased availability of dissolved organic carbon could support the development of a larger or more complex heterotrophic community. Alternatively, the released material could have an inhibitory effect on bacterial growth (e.g., Jones and Cannon, 1986; Murray et al., 1986). Excreted photosynthate can be utilized by other algae (e.g., Rivkin and Putt, 1987; Tenaud et al., 1989). Algal excretion of photosynthate can be linked to bacterial uptake of the excreted products in nature (e.g., Bauld and Brock, 1974). In two angiosperm communities, doubled CO_2 concentrations did not change the total above-ground biomass, but did result in an increased efflux of fixed carbon from the roots to the soil (Díaz et al., 1993). Díaz and co-workers (1993) hypothesized that the increased excretion of photosynthate enhanced the population growth of soil microbes. Zak et al. (1993) showed that for a Populus grandidentata system that under elevated CO_2, total above and below ground

---

[2]Mechanisms of excretion include release from healthy cells, autolysis or attack by other organisms.

biomass increased, labile C in the rhizosphere increased, microbial biomass C was greater in the rhizosphere and bulk soil, and N availability was greater in the bulk soil. Although there are methodological difficulties with estimating release of photosynthate, estimates range from 0-62% of photosynthate in marine phytoplankton (Jones and Cannon, 1986) and from 2-12% for benthic algal mats in thermal springs (Belly et al., 1973; Bauld and Brock, 1974). Thus, any change in carbon fixation rates has ramifications throughout the ecosystem.

We traced the fate of fixed carbon in mats grown at high and low concentrations of DIC in order to determine the significance of the higher photosynthetic rates in the algal mats (Rothschild and Bartley, 1996). In the Lyngbya mats pre-incubated at high [DIC] for one month, total excreted carbon increased almost six-fold at high [DIC] and comprised 25% of total carbon fixed, as opposed to 14% at low [DIC]. Net carbon fixation (^{14}C remaining as organic carbon after 48 hours) increased from 3% of gross fixation at low [DIC] to 16% at high [DIC]. Total remineralization of fixed carbon (gross C fixed - net C fixed = C returned to inorganic form over 48 hours) increased, but the proportion of fixed carbon remineralized declined from 77% at low [DIC] to 59% at high [DIC]. This result is expected because RuBisCO is both an carboxylase and an oxygenase. Higher concentrations of CO_2 effectively decrease the relative concentration of O_2, thus decreasing the oxygenase function of RuBisCO, which is a primary pathway for remineralization. The results indicate that biomass turnover was slower at high [DIC], and that standing biomass increased. This increase might be reflected in increased reproductive rates or in per-cell increases in organic carbon content (Thompson et al., 1991). The density of photoautotrophs in a microbial mat is light- and space-limited, making it unlikely that cyanobacterial cell density increased. Thus, the reproductive rate of cyanobacteria in these mats probably did not increase, unless turnover rates also increased. Cell size was assessed visually with phase-contrast microscopy, and differences in trichome width were undetectable. It is possible that mat organisms increased production of C-rich storage compounds, which would suggest increased C:N and/or C:P ratios at high [DIC]. Cyanobacteria produce carbon-based reserve materials such as cyanophycin, phycobiliprotein, poly-β-hydroxybutyrate, and glycogen that are higher in stationary phase cultures than in exponential ones, suggesting an inverse correlation between cell storage and reproductive rate (Smith, 1982; Stal, 1989). Although samples were not retained to analyze the elemental composition of the mats, numerous studies have demonstrated that C:N ratios in plants often increases when grown under elevated pCO_2 (Mooney et al., 1991; Pitelka, 1994; Strain and Cure, 1991) and starch can accumulate (Körner and Arnone III, 1992). In analogy, it is possible that the C:N ratio of the Lyngbya mat increased here.

Rising surficial fluxes of UVB radiation

On present day earth, UVA (320 to 400 nm) and UVB (280 to 320 nm) reach the surface (Fig. 2). Stratospheric ozone attenuates much of the solar UVB, but because it is highly damaging, small changes in UVB flux can be significant

(Crawford, 1987, Frederick and Snell, 1988, Kerr and McElroy, 1993). Both UVA and photosynthetically-active radiation (PAR) are nearly independent of ozone concentration. During the last two decades, regional and seasonal stratospheric ozone depletion has been documented (Lloyd, 1993). Thus, even though there are daily and seasonal variations in UV flux as the result of the earth's rotation and orbit, and weather conditions (Németh et al., 1996), average surficial fluxes of UVB radiation have risen. Present surficial UV fluxes have a significant impact on organismal physiology, including inhibition of the light and dark reactions of photosynthesis (David et al., 1988; Vincent and Roy, 1993; Cockell and Rothschild, unpublished).

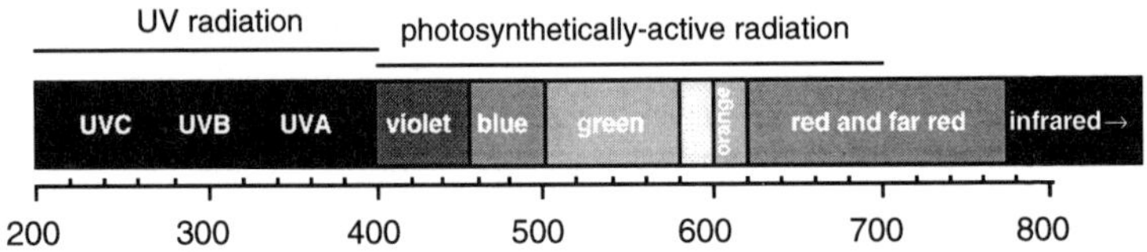

Figure 2. Solar spectrum relevant to primary production.

UV effects depend on the wavelength (Cullen et al., 1992). UVB effects on primary production have received the greatest attention (Häder and Worrest, 1991, Jokiel and York, 1984, Cullen and Neale, 1994, Prézelin et al., 1994a, b, Steeman Nielsen, 1964, Lorenzen, 1979, Smith et al., 1980; Smith et al., 1992, Neale et al., 1994, Vincent and Roy, 1993). UVA is less energetic than UVB, but the greater region of the spectrum that it occupies and its higher flux to the earth's surface makes its total contribution to photosynthesis inhibition significant. It is only weakly screened by atmospheric ozone. UVA can affect primary production (Buhlmann et al., 1987, Helbling et al., 1992, Kim and Watanabe, 1994, Prézelin et al., 1994b). In some Antarctic species of diatoms, UVA has been implicated as the major cause of UV-induced mortality (Davidson and Marchant, 1994). UVB also inhibits primary production. In a three-season study of the loblolly pine seedlings from seven sources (Sullivan and Teramura, 1992), an elevated level of UVB radiation meant to simulate 25% ozone depletion over Maryland reduced biomass by 12-20%. Thus, stratospheric ozone depletion could significantly reduce the growth of loblolly pine over its lifetime (Sullivan and Teramura, 1992). Therefore in understanding the effects of UV radiation on primary production and other cellular processes, and to understand the effect of stratospheric ozone depletion, it is important to dissect the effects of UVA and UVB (Bothwell et al. 1993; Bothwell et al., 1994).

Some organisms are thought to be well adapted to coping with UV radiation fluxes. One putative mechanism for UV resistance is the production of UV-screening compounds. For example, some organisms contain mycosporine-like amino acids (MAA's) which screen in the UVB portion of the spectrum (Karentz et al. 1991a, Marchant et al. 1991, Donker and Häder, 1995), or pigments such as scytonemin, present in a range of cyanobacteria, which may also screen UVA (Garcia-Pichel and

Castenholz, 1991; 1993; Garcia-Pichel et al., 1992).

Effect of UV radiation on carbon fixation

Because both photosynthesis and UV fluxes are subject to diurnal variations, we determined whether there is diurnal variability in the effect of UVB on photosynthesis (Cockell and Rothschild, unpublished). We assessed the effect of UV radiation on in situ carbon fixation rates in Lyngbya, Cyanidium and Zygogonium over a diurnal cycle under three UV attenuation regimes. The results showed that despite the ecological and taxonomic diversity of these mats, the screening of UVB was always found to increase total primary production averaged over the day with a further increase when both UVA and UVB were screened. The effects on primary production varied at different times of day in the different organisms. They were more pronounced in the morning than in the afternoon, demonstrating that measuring diurnal change is important in assessing UV impact on organisms. In contrast, the typical midday dip in photosynthesis is not significantly accentuated by UV radiation.

It is important to note that the effect of UV radiation on ecosystems cannot be simply extracted from known data about the primary producer. For example, Bothwell and collaborators (Bothwell et al. 1993; Bothwell et al., 1994) showed that although UV inhibits growth in diatom communities, UV had an even more deleterious effect on the grazers. For this reason, there were fewer diatoms than in the presence of UV than in its absence.

Other examples of the indirect effect of UVB on primary productivity exist. Besides effects on grazers as in Bothwell's studies, UVB radiation can affect the chemical composition of the water. For example, previously unavailable nutrients can become refractory after UV-induced chemical changes. Bushaw et al. (1996) demonstrated that when humic substances are exposed to the solar spectrum there is release of a biologically-refractory form of nitrogen, ammonium. The wavelengths responsible for the photochemical ammonium formation were primarily in the UV portion of the spectrum, from 295 to 360 nm. UV radiation affects many other cellular processes that themselves may inhibit primary productivity. These include inhibition of nitrogen metabolism (Behrenfeld et al., 1995), inhibition of nitrogenase activity (Newton et al., 1979; Sinha et al., 1996), inhibition of heterocyst formation (Sinha et al., 1996), reduction in motility (e.g., Donkor et al., 1993a; 1993b; Donkor and Häder, 1991), and increased synthesis of the UV-screening pigment scytonemin (Garcia-Pichel and Castenholz, 1991). Of these inhibition of nitrogen fixation may be the most critical if the availability of N is the factor limiting primary productivity. This effect may be mitigated by the photochemical production of biologically-refractory N, as discussed above. However, the relative contributions of the two UV-induced reactions are unknown.

UVB can induce DNA damage (e.g., Harm, 1980; Karentz et al., 1991b; Karentz, 1994). Even if the damage is repaired, repair siphons off energy that could be used for photosynthesis. Some of these phenomena are being studied by our lab as well as others in terms of ecological impact.

What does rising pCO_2 and UVB mean for marine phytoplankton?

Increased atmospheric levels of pCO_2 can affect marine phytoplankton productivity by changing surface temperature, pH and the availability of inorganic carbon for photosynthesis. In general, phytoplankton productivity may not be enhanced by the relatively small increases in available pCO_2 predicted for the next few decades. However, there by the time a phytoplankton bloom collapses, there is severe carbon limitation. Thus, any increase in available inorganic carbon in surface waters should serve to prolong phytoplankton blooms.

In laboratory studies and field experiments on shallow water ecosystems it is clear that elevated fluxes of UVB radiation affects primary productivity both directly and indirectly through inhibition of other cellular processes. But solar UVB will only have a direct effect on aquatic organisms if it is not attenuated by the overlying water column. Although UVB radiation is strongly absorbed by water, penetration still may be significant. For example, in clear oceanic waters UVB may penetrate dozens of meters, and even in coastal waters with high turbidity UVB may penetrate as much as several meters (Häder, 1993). Studies of Antarctic phytoplankton demonstrated a significant UVB inhibition of photosynthesis depending on when the measurements were made within or outside of the ozone hole (Häder, 1993). UV-induced photochemical reactions in the marine environment are thought to affect primary productivity significantly . For example, Bushaw et al. (1996) calculated that ~0.44×10^4 tons of ammonium can be produced in this manner on the southeastern United States shelf annually, thus increasing the terrestrial N available to microorganisms on the shelf by 20%.

In conclusion, while not disputing the importance N, P, iron, etc. in controlling phytoplankton primary productivity, it is argued here that the global change variables of elevated pCO_2 and UVB radiation must be considered as well.

References

Badger, M.R., and G.D. Price. 1992. The CO_2 concentrating mechanism in cyanobacteria and microalgae. Physiologia Plantarum 84: 606-615.

Badger, M.R., and T.J. Andrews. 1982. Photosynthesis and inorganic carbon usage by the marine cyanobacterium, Synechococcus sp. Plant Physiol. 70: 517-523.

Bauld, J., and T.D. Brock. 1974. Algal excretion and bacterial assimilation in hot spring algal mats. J. Phycol. 10: 101-106.

Bazzaz, F.A. 1990. The response of natural ecosystems to the rising global CO_2 levels. Ann. Rev. Ecol. Syst. 21: 167-196.

Beer, S., and B. Shragge. 1987. Photosynthetic carbon metabolism in Enteromorpha compressa (Chlorophyta). J. Phycol. 23: 580-584.

Behrenfeld, M. J., D.R.S. Lean, and H. Lee II. 1995. Ultraviolet-B radiation effects on inorganic nitrogen uptake by natural assemblages of oceanic plankton. J. Phycol. 31 25-36.

Belly, R.T., M.R. Tansey, and T.D. Brock. 1973. Algal excetion of ^{14}C-labeled compounds and microbial interactions in Cyanidium caldarium mats. J. Phycol. 9: 123-7.

Boden, T.A., D.P. Kaiser, R.J. Sepanski, and F.W. Stoss. 1993. Trends '93: A Compendium of Data on Global Change. Carbon Dioxide Information Analysis Center, Oak Ridge, Tennessee. Publication No. ORNL/CDIAC-65, ESD Publication No. 4195, U.S. Department of Energy.

Bothwell, M.L., D. Sherbot, A.C. Roberge and R.J. Daley. 1993. Influence of natural ultraviolet radiation on periphytic diatom community growth, biomass accrual, and species composition : short-term versus long-term effects. J.Phycol. 29 : 24-35.

Bothwell, M.L., D.M.J. Sherbot and C.M. Pollock. 1994. Ecosystem response to solar ultraviolet-B radiation: influence of trophic-level interactions. Science 265: 97-100.

Brechignac, F., and M. Andre. 1984. Oxygen uptake and photosynthesis of the red macroalga, Chondrus crispus, in seawater. Plant Physiol. 75: 919-923.

Buhlmann, B., P. Bossard, and U. Uehlinger. 1987. The influence of longwave ultraviolet radiation (UV A) on the photosynthetic activity (^{14}C assimilation) of phytoplankton. J. Plankton Res. 9: 935-43.

Bushaw, K.L., R.G. Zepp, M.A. Tarr, D. Schulz-Jander, R.A. Bourbonniere, R.E. Hodson, W.L. Miller, D.A. Bronk, and M.A. Moran. 1996. Photochemical release of biologically available nitrogen from aquatic dissolved organic matter. Nature 381: 404-407.

Coale, K.H., K.S. Johnson, S.E. Fitzwater, R.M. Gordon, S.Tanner, F.P. Chavez, L.Ferioli, C. Sakamoto, P. Rogers, F. Millero, P. Steinberg, P. Nightingale, D. Cooper, W.P. Cochlan, M.R. Landry, J. Constantinou, G. Rollwagen, A. Trasvina, and R. Kudela. 1996. A massive phytoplankton bloom induced by an ecosystem-scale iron fertilization experiment in the equatorial Pacific Ocean. Nature 383: 495-517.

Crawford, M. 1987. Landmark Ozone treaty negotiated Science 237: 1557.

Cullen, J.C., and P.J. Neale. 1994. Ultraviolet radiation, ozone depletion, and marine photosynthesis. Photosynth. Res. 39: 303-320.

Cullen, J.C., P.J. Neale, and M.P Lesser. 1992. Biological weighting function for the inhibition of phytoplankton photosynthesis by ultraviolet radiation. Science 258: 646-650.

David, K.A.V., S.S. Rane, M. Mathus, and S.K. Bhattacharjee. 1988. Herbicide mediated UV-resistance in cyanobacteria on the role of photosynthetic electron transport system rather than replicative DNA as lethal target determining dark-survival of Anacystis nidulans. Photochem. Photobiol. 47: 107-114.

Davidson, A.T., and Marchant, H.J. 1994. Comparative impact of in situ UV exposure on productivity, growth and survival of Antarctic Phaeocystis and diatoms. Proc. NIPR Symp. Polar Biol. 7 : 53-69.

Díaz, S., J.P. Grime, J. Harris, and E. McPherson. 1993. Evidence of a feedback mechanism limiting plant response to elevated carbon dioxide. Nature 364: 616-617.

Donkor, V.A., and D.-P. Häder. 1991. Effects of solar and ultraviolet radiation on motility photomovement and pigmentation in filamentous gliding cyanobacteria. FEMS Microbiol. Ecol. 86: 159-168.

Donkor, V.A., and Häder, D. 1995. Protective strategies of several cyanobacteria against solar radiation. J. Plant Physiol. 145 : 750-755.

Donkor, V.A., D.H.A.K. Amewowor, and D.-P. Häder. 1993a. Effects of tropical solar radiation on the velocity and photophobic behavior of filamentous gliding cyanobacteria. Acta Protozool. 32: 67-72.

Donkor, V.A., D.H.A.K. Amewowor and D.-P. Häder. 1993b. Effects of tropical solar radiation on the motility of filamentous cyanobacteria. FEMS Microbiol. Ecol. 12: 143-148.

Drake, B.G. 1992a. The impact of rising on CO_2 ecosystem production. Water, Air and Soil Pollution 64: 25-44.

Drake, B.G. 1992b. A field study of the effects of elevated CO_2 on ecosystem processes in a Chesapeake Bay wetland. Aust. J. Bot. 40: 579-595.

Frederick, J.E., and H.E. Snell. 1988. Ultraviolet radiation levels during the Antarctic spring. Science 241 : 438-440.

Gao, K., Y. Aruga, K. Asada, T. Ishihara, T. Akano, and M. Kiyohara. 1991. Enhanced growth of the red alga Porphyra yezoensis Ueda in high CO_2 concentrations. J. Appl. Phycol. 3: 355-362.

Gao, K., Y. Aruga, K. Asada, T. Ishihara, T. Akano, and M. Kiyohara. 1992. Photorespiration and CO_2 fixation in the red alga Porphyra yezoensis Ueda. Jpn. J. Phycol. (Sôrui) 40: 373-377.

Garcia-Pichel, F., and R.W. Castenholz. 1991. Characterization and biological implications of scytonemin, a cyanobacterial sheath pigment. J. Phycol. 27: 395-409.

Garcia-Pichel, F., and R.W. Castenholz. 1993. Occurence of UV-absorbing mycosporine-like compounds among cyanobacterial isolates and an estimate of their screening capacity. Appl. Environ. Microbiol. 59: 163-169.

Garcia-Pichel, F., N.D. Sherry and R.W. Castenholz. 1992. Evidence for an ultraviolet sunscreen role of the extracellular pigment scytonemin in the terrestrial cyanobacterium Chlorogloeopsis sp. Photochem. Photobiol. 56: 17-23.

Gieskes, W.C., and G.W. Kraay. 1990. Transmission of ultraviolet light in the Weddell Sea. Report on the first measurements made in Antarctic. Biomass Neswletter 12: 12-14.

Häder, D-P., and Worrest, R.C. 1991. Effects of enhanced solar ultraviolet radiation on aquatic ecosystems. Photochem. Photobiol. 53 : 717-725.

Häder, D.-P. 1993. Effects of enhanced solar ultraviolet radiation on aquatic ecosystems. In: UV-B Radiation and Ozone Depletion (M. Tevini, ed) Lewis Publishers, Boca Raton. pp. 155-192.

Harm, W. 1980. Biological Effects of Ultraviolet Radiation. IUPAB Biophysics Series I. Cambridge : Cambridge University Press, 216 pp.

Hebling, E.W., V. Villafane, M. Ferrario, and O. Holm-Hansen. 1992. Impact of natural ultraviolet radiation on rates of photosynthesis and on specific marine phytoplankton species. Mar. Ecol. Prog Ser. 80 : 89-100.

Holbrook, G.P., S. Beer, W.E. Spencer, J.B. Reiskind, J.S. Davis, and G. Bowes. 1988. Photosynthesis in marine macroalgae: evidence for carbon limitation. Can. J. Bot. 66: 577-82.

Idso, S.B. 1989. Carbon Dioxide and Global Change: Earth in Transition. IBR Press, Tempe, Arizona. ch. 7.

Jeffrey, S. W., and Humphrey, G. F. 1975. New spectrophotometric equations for determining chlorophylls a, b, c1, and c2 in higher plants, algae, and natural phytoplankton. Biochem. Physiol. Pflanzen. 167: 191.

Jokiel, P.L., and York, R.H.Jr. 1984. Importance of ultraviolet radiation in photoinhibition of microalgal growth. Limnol. Oceanogr. 29 : 192-199.

Jones, A.K., and R.C. Cannon. 1986. The release of micro-algal photosynthate and ssociated bacterial uptake and hetertrophic growth. Br. Phycol. J. 21: 341-358.

Karentz, D. 1994. Ultraviolet tolerance mechanisms in Antarctic marine organisms. In: Ultraviolet Radiation in Antarctica: Measurements and Biological Effects. Antarctic Research Series, vol. 62, pp. 93-110.

Karentz, D., F.S. McEuen, M.C. Land, and W.C. Dunlap. 1991a. Survey of mycosporine-like amino acids in Antarctic marine organisms: potential protection from ultraviolet exposure. Mar. Biol. 108 : 157-166.

Karentz, D., J.E. Cleaver, and D.L. Mitchell. 1991b. Cell survival characteristics and molecular responses of Antarctic phytoplankton to ultraviolet-B radiation. J. Phycol. 27 : 326-341.

Kerr, J.B., and C.T. McElroy. 1993. Evidence for large upward trends of ultraviolet-B radiation linked top ozone depletion. Science 262 : 1032-1034.

Kim, D., and Y. Watanabe. 1994. Inhibition of growth and photosynthesis of freshwater phytoplankton by ultraviolet A (UVA) irradiation and subsequent recovery from stress. J. Plankton Res. 16 : 1645-1654.

Kimball, B.A. 1983. Carbon dioxide and agricultural yield: an assemblage and analysis of 430 prior observations. Agronomy J. 75: 779-788.

Körner, C., and J.A. Arnone III. 1992. Responses to elevated carbon dioxide in artificial tropical ecosystems. Science 257: 1672-1675.

Levavasseur, G., G.E. Edwards, C.B. Osmond, and J. Ramus. 1991. Inorganic carbon limitation of photosynthesis in Ulva rotundata (Chlorophyta). J. Phycol. 27: 667-72.

Lincoln, D.E., E.D. Fajer, and R.H. Johnson. 1993. Plant-insect herbivore interactions in elevated CO_2 environments. Trends Ecol. Evol. 8: 64-68.

Lloyd, S.A. 1993. Stratospheric ozone depletion. The Lancet 342: 1156-8.

Lorenzen, C.J. 1979. Ultraviolet radiation and phytoplankton photosynthesis. Limnl. Oceanogr. 24, 1117-20.

Marchant, H.J., A.T. Davidson, and G.J. Kelly. 1991. UVB protecting compounds in the marine alga Pheocystis pouchetii from Antarctica. Mar. Biol. 109 : 391-395.

Mooney, H.A., B.G. Drake, R.J. Luxmoore, W.C. Oechel, and L.F. Pitelka. 1991. Predicting ecosystem responses to elevated CO_2 concentrations. BioScience 41:

96-104.

Murray, R.E., K.E. Cooksey, and J.C. Priscu. 1986. Stimulation of bacterial DNA synthesis by algal exudates in attached algal-bacterial consortia. Appl. Env. Microbiol. 52: 1177-1182.

Németh, P., Z. Tóth, and Z. Nagy. 1996. Effect of weather conditions on UV-B radiation reaching the earth's surface. J. Photochem. Photobiol. B: Biology. 32: 177-181.

Newton, J.W., D.D. Tyler, and M.E. Slodki. 1979. Effect of ultraviolet-B (280-320 nm) radiation on blue-green algae (Cyanobacteria), possible biological indicators of stratospheric ozone depletion. Appl. Environ. Microbiol. 37: 1137-1141.

Oechel, W.C., S. Cowles, N. Grulke, S.J. Hastings, B. Lawrence, T. Prudhomme, G. Riechers, B. Strain, D. Tissue, and G. Vourlltis. 1994. Transient nature of CO_2 fertilization in Arctic tundra. Nature 371: 500-503.

Pitelka, L. 1994. Ecosystem response to elevated CO_2. Trends Ecol. Evol. 9: 204-207.

Prézelin, B.B., N.P. Boucher, and O. Schofield. 1994a. Evaluation of field studies of UVB radiation effects on Antarctic marine primary productivity. In: Stratopheric Ozone Deletion/ UV-B Radiaton in the Biosphere. (R.H. Biggs and M.E.B. Joyner, eds) NATO ASI Series, Vol. 118. Springer-Verlag, Berlin. pp. 181-194.

Prézelin, B.B., N.P. Boucher, and R.C. Smith. 1994b. Marine primary production under the influence of the Antarctic Ozone hole: Icecolors '90. In: Ultraviolet Radiation In Antarctica: Measurements and Biological Effects. Antarctic Research Series. 62: 159-186.

Raven, J.A. 1985. The CO_2 concentrating mechanism. In: Inorganic Carbon Uptake by Aquatic Photosynthetic Organisms. (W.J. Lucas and J.A. Berry , eds) American Society of Plant Physiologists, pp. 67-82.

Reiskind, J.B., P.T. Seamon, and G. Bowes. 1989. Photosynthetic responses and anatomical features of two marine macroalgae with different carbon dioxide compensation points. Aquat. Bot. 33: 71-86.

Riebesell, U., D.A. Wolf-Gladrow, and V. Smetacek. 1993. Carbon dioxide limitation of marine phytoplankton growth rates. Nature 361: 249-251.

Rivkin, R.B., and M. Putt. 1987. Heterotrophy and photoheterotrophy by Antarctic microalgae: Light-dependent incorporation of amino acids and glucose. J. Phycol. 23: 442-452.

Rothschild, L.J. 1991. A model for diurnal patterns of carbon fixation in a Precambrian microbial mat based on a modern analog. BioSystems 25: 13-23.

Rothschild, L. 1994a. CO_2 and diatom mats. Nature 368: 817.

Rothschild, L.J. 1994b. Elevated CO_2: Impact on diurnal patterns of photosynthesis in natural microbial ecosystems. Adv. Space Res. 14(11): 285-289.

Rothschild, L.J., and J. Bartley. 1996. High pCO_2 results in low $\delta^{13}C_{org}$ and high carbon fixation rates: implications for ancient biogeochemistry and modern carbon limitation. J. Phycol. 32: 41.

Rothschild, L.J., and L.J. Giver. 1993. Diurnal patterns of carbon fixation and DNA synthesis in a Cyanidium caldarium mat community. J. Protozoology 40: 18A.

Rothschild, L.J., and R.L. Mancinelli. 1990. A model for the evolution of carbon fixation in microbial mats, 3500 million years ago to the present. Nature 345: 710-712.

Rowan, K.S. 1989. Photosynthetic Pigments of Algae. Cambridge University Press, Cambridge. 334 pp.

Schlesinger, W.H. 1991. Biogeochemistry. An analysis of global change. Academic Press, San Diego. 443 pp.

Sikes, C.S., and V. J. Fabry. 1993. Photosynthesis, $CaCO_3$ deposition, coccolithophorids, and the global carbon cycle. In: Photosynthetic Carbon Metabolism and Regulation of Atmospheric CO_2 and O_2. (N. E. Tolbert and J. Preiss, eds) Oxford Press, Oxford.

Sinha, R.P., N. Singh, A. Kumar, H.D. Kumar, M. Häder ,and D.-P. Häder. 1996. Effects of UV irradiation on certain physiological and biochemical processes in cyanobacteria. J. Photochem. Photobiol. B: Biology. 32: 107-113.

Smith, A.J. 1982. Models of cyanobacterial carbon metabolism. In: The Biology of Cyanobacteria Botanical Monographs vol. 19 (Carr, N.G. and Whitton, B.A., eds) University of California Press, Berkeley. pp. 47-85.

Smith, R.C., K.S. Baker, O. Holm-Hansen, and R. Olsen. 1980. Photoinhibition of photosynthesis in natural waters. Photochem. Photobiol. 31 : 585-92.

Spivack, A.J., C.-F. You, and H.J. Smith. 1993. Foraminiferal boran isotope ratios as a proxy for surface ocean pH over the past 21 Myr. Nature 363: 149-151.

Stal, L.J., H. Heyer, S. Bekker, M. Villbrandt, and W.E. Krumbein. 1989. Aerobic-anaerobic metabolism in the cyanobacterium Oscillatoria limosa In: Microbial Mats (Cohen, Y., and Rosenberg, E., eds.) American Society for Microbiology, Washington, D.C. pp. 255-276.

Steemann Nielsen, E. 1964. On a complication in marine productivity work due to the influence of ultraviolet light. J. Cons. Cons. Int. Explor. Mer. 29 : 130-5.

Strain, B.R., and J.D. Cure, eds. 1991. Direct effects of atmospheric CO_2 enrichment on plants and ecosystems: An updated bibliographic data base. Carbon Dioxide Information Analysis Center, Oak Ridge National Laboratory, TN, Environmental Sciences Division Publication No. 4253. 287 pp.

Sullivan, J.H., and A.H. Teramura. 1992. The effects of UV-B radiation on loblolly pine. 2. Growth of field-grown seedlings. Trees 6: 115-120.

Tenaud, M., M. Ohmori, and S. Miyachi. 1989. Inorganic carbon and acetate assimilation in Botryococcus braunii (Chlorophyta). J. Phycol. 25: 662-667.

Thompson, P.A., P.J. Harrison, and J.S. Parslow. 1991. Influence of irradiance on cell volume and carbon quota for ten species of marine phytoplankton. J. Phycol. 27: 351-360.

Vincent, W.F., and S. Roy. 1993. Solar ultraviolet-B radiation and aquatic primary production: damage, protection, and recovery. Environ. Rev. 1: 1-12.

Zak, D.R., K.S. Pregitzer, P.S. Curtis, J.A. Teeri, R. Fogel and D.L. Randlett. 1993. Elevate atmopheric CO_2 and feedback between carbon and nitrogen cyles. Plant and Soil 151: 105-117.

CHAPTER 211

CLIMATE CHANGE AND THE CALIFORNIA CURRENT ECOSYSTEM

John A. McGowan[1]
Thomas L. Hayward, Ph.D.

Abstract

There has been a very large change in the biology and physics of the California Current. Forty five years of study by the California Cooperative Oceanic Fisheries Investigations (CalCOFI) has shown that plankton, the base of the food web, has declined by over 70%, the catch rate of commercially important fish has been greatly depressed and the abundance of plankton feeding oceanic sea-birds is down by about 90%. This great ecosystem deterioration is closely associated with the observed ocean warming brought about by climatic change. An anomalously warm surface layer makes it much more difficult for the winds to stir, mix or upwell deeper waters rich in plant nutrients, hence a decline in biological production.

This major oceanographic change began about 1977 but has only become clear in the past two years. It is continuing. The ocean warming may be part of a natural climate cycle of long duration or it may be due to man-caused global warming. In either case this change in our ocean is of importance to the management of California's ocean and coastal resources. Our ability to separate anthropogenic from natural changes is the main issue.

---

[1] Scripps Institution of Oceanography, University of California, San Diego, La Jolla, California 92093-0227

CHAPTER 212

Bio-enrichment by Ocean Upwelling along the West Coast of North America

John T. Howe[1] and Lynn J. Rothschild[2]

Abstract

Primary biological productivity in the coastal ocean is affected by physical oceanography, which in turn is coupled to weather; particularly wind stress, solar heating, and illumination related to photosynthesis. In this paper we begin to put these phenomena together to examine their interrelationships and to estimate the extent to which upwelling enhances productivity along that part of the west coast of North America where it is prominent for a season. Accordingly, the long-term temporal and spatial features of upwelling in ocean waters along the west coast are summarized. In regions of particularly strong upwelling, illustrative ocean observations, measurements, and computations are performed at selected locations spanning more than a decade in time. Results are compared with the results of other studies. Upwelling is shown to enrich biomass as much as fiftyfold. An analytic solution to an approximate set of the rigorous coupled hydrodynamic, energy, radiative transfer, and species conservation equations is presented and described. This includes the effects of weather, wind induced shear stress and its variation (which affect the upwelling formation), illumination (which affects the production of a biomass in surface waters), solar heating, and other physical, chemical, photochemical, and biological events in coastal waters. An approximate analogy (or index) which relates local sea surface temperature and primary productivity parameters in the upwelling formation is discussed. Computed results are compared to NOAA coastal ocean observations.

Acknowledgment; The authors gratefully acknowledge the contribution of D. Gibson, T. Evans, P. Evans, and F. Watters of the F. V. "Bountiful" for assistance at sea; J. Bartley for help in chlorophyll sampling; and E. P. Daghir, NOAA Oceanographer, for Sea Surface Thermal Analysis maps.

Introduction

Coastal ocean regions are valuable far beyond their portion of ocean area. It has been estimated that half of the world's marketable fish catch is obtained from 0.1% of the ocean area (Ryther 1969) To a considerable extent, these regions include the shallow and coastal oceans. It is indeed important to understand coastal upwelling as it relates to the production, distribution temporal variation, and eventual demise of a biomass. The

[1]Chief Scientist, NASA Ames Research Center, Mail Stop 233-20, Moffett Field, CA 94035-1000.
[2]Research Scientist, JCWS, NASA Ames Research Center, Mail Stop 239-12, Moffett Field, CA 94035-1000.

effects of climate, oceanography, biology, local solar heating, illumination, and other physical and chemical processes should be examined together.

Along much of the west coast of North America the ocean tends to be biologically rich because of the coupling of these effects. In the coastal ocean, organic matter from both land mass effluents and ocean origins undergoes microbial remineralization. Atmospheric wind stress at the ocean surface promotes ocean upwelling which brings these mineral nutrients from the ocean floor to sunlit surface waters, where a biomass can be produced by photosynthesis. This primary productivity, in turn, supports a rich fauna.

Accordingly, this paper presents a brief long-term overview of coastal ocean upwelling. In an extensive region of strong upwelling, ocean observations and measurements performed over more than a decade in time are presented to illustrate the process and estimate the influence of upwelling on biomass production. A mathematical framework which models the behavior of the system is described and presented. Calculated results are obtained and compared with observations.

General Features

Figure 1 is a typical NOAA satellite image of cool upwelled coastal waters (the light tones in fig. 1) from Cape Mendocino to Point Conception, California, in July. Dominant upwelling along the west coast of North America occurs in the spring and summer. Long-term climate data show that the wind tends to have a component directed toward the equator during that time. Application of an early upwelling model (Ekman 1905) indicates that the wind stress and the Coriolis effect of the rotation of the earth combine to move surface waters offshore and cause cooler nutrient-rich water to be upwelled from the depths into sunlit surface layers of the ocean.

The temporal variation of the intensity of upwelling at Point Sal over a year is characterized by the solid line in figure 2. It is interpolated from NOAA data in which the 20-year mean monthly "upwelling index" (Bakun 1973) was computed by an Ekman-like model. The upwelling index is defined as the calculated component of the local transport of upwelled waters directed offshore and has the units, volume-per-second-per-unit length of shoreline as defined by Bakun. Derived information from this index will be used for illustrative purposes to indicate the overall distribution, behavior, and intensity of upwelling in this first part of the paper. Figure 2 shows that at Point Sal the upwelling is intense and persistent at midyear; subsiding around the turn of the year. In contrast, northward at Cape Blanco in southern Oregon, upwelling is less intense, and the ocean tends to downwell for four months around the turn of the year.

The peak of the mean monthly upwelling index derived from the Bakun computations is shown as a function of latitude from Baja California to the Gulf of Alaska in figure 3. It is prominent from south of Baja California to near the California/Oregon border. North of Oregon, the monthly index peak is small. Temporally, the local peak moves northward from April to July as shown on the upper abscissa. The maximum monthly index occurs between San Diego and Point Conception in June. The maximum 20-year mean weekly index occurs in August at about 39°, near Point Arena (Bakun 1975). The dashed line in figure 3 shows the long-term mean of the curl of the wind stress vector (Nelson 1977). Positive curl relates to increased wind stress offshore, which tends to move upwelled water farther offshore; negative curl corresponds to diminished wind stress offshore which inhibits the offshore movement of upwelled water

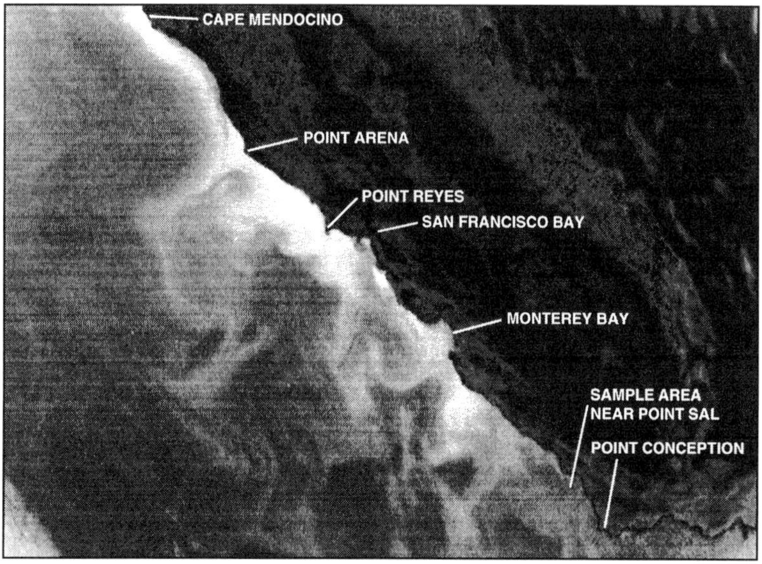

Figure 1. Strong upwelling along the west coast of North America in July.

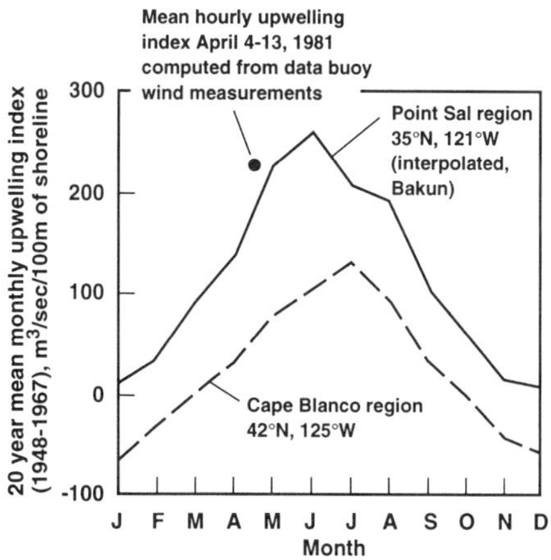

Figure 2. Upwelling index comparison.

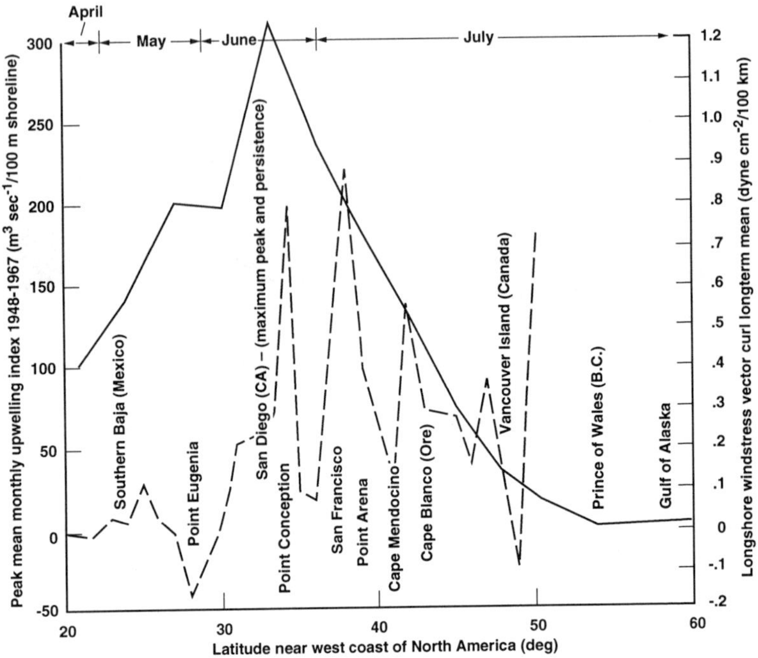

Figure 3. Latitudinal variation of long-term mean-monthly upwelling index peak, and associated wind stress vector curl.

and is conducive to the formation of biologically rich offshore fronts that terminate the surface flow of upwelled water. Importantly, the curl of the wind stress vector is predominantly positive in the mean. However, as an episode of strong upwelling wind subsides, positive curl can reverse itself and become negative on a transient or a quasi-steady basis. Thus, biologically rich offshore downwelling fronts can develop locally anywhere in the upwelling region. These two variations of the model adapted from Nelson are depicted schematically in figure 4.

Illustrative Observations

Point Sal: Near Point Sal, both the wind velocity locally parallel to the coast and the sea surface temperature record were obtained from a local NOAA data buoy for an upwelling episode lasting about 10 days in April 1981. Figure 5 shows that the sea surface temperature (dotted line) rises and falls in response to the variable wind (with about a one-day lag). Peak wind is about 12 m/sec. The surface temperature tends to decrease until the wind subsides late in the upwelling episode. The corresponding calculated mean hourly upwelling index for the episode is shown by the solid symbol in figure 2. The upwelling index was near its peak 20-year mean monthly value. As the upwelling wind diminished sufficiently (April 13, fig. 5) to allow trawl operations and

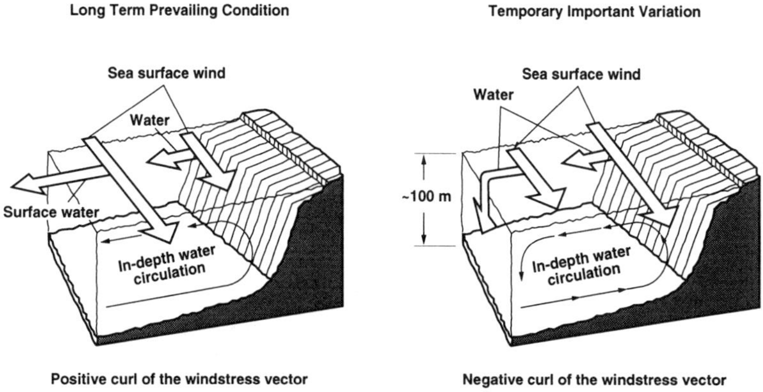

Figure 4. Model of effect of curl of the wind stress vector.

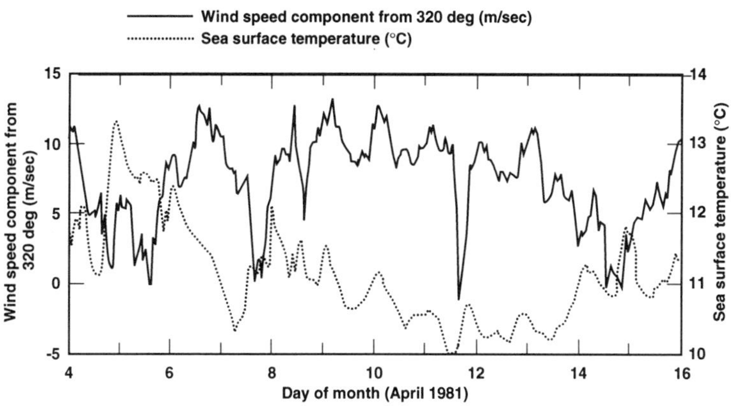

Figure 5. Point Sal wind speed and sea temperature; NOAA buoy data.

shipboard observations, Secchi disk observations were made by the first author (verified by an associate) in five locations in the sample area offshore from Point Sal shown in figure 1. The Secchi depth ranged from 4 m to 6 m, which corresponded to a mean chlorophyll a concentration of approximately 12–5 mg/m^3 (as can be seen in figure 9). Thus, the Secchi measurements indicated high chlorophyll concentration (indicative of a phytoplankton bloom), as would be expected for the intense upwelling episode.

The corresponding shipboard echo-sounder trace (fig. 6) shows schools of fish labeled by the captain of the trawler. The zooplankton at the ocean floor rise at night, perhaps as fast as 7 m/min (Starkey 1981), graze on the surface phytoplankton, and, in turn, are grazed upon by larger species.

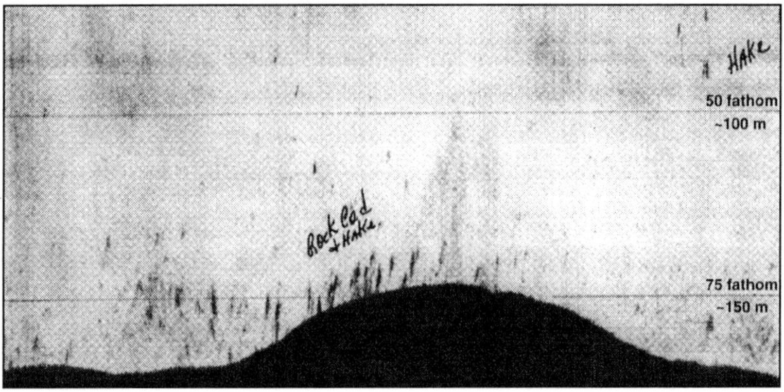

Trace A

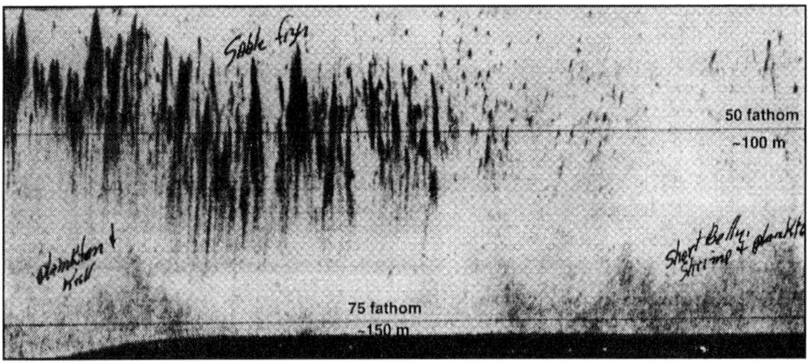

Trace B

Figure 6. Echo-sounder traces in lower part of water column with species noted.

Monterey Bay: Two sites were sampled on April 6, 1994 (site 1 at 36° 47' 28" N, 122° 00' 51" W is about 10 n.mi. west of Moss Landing, in the vicinity of the confluence of the Monterey Canyon and the Soquel Canyon; site 2 at 36° 48' 21" N, 121° 55' 90" W is about 6 n.mi. west of Moss Landing). One site was sampled on November 29, 1994 (site 3 at 36° 50' 23" N, 122° 09' 50" W). The sites were tested for light extinction using Secchi disk measurements and a vertical assessment of photosynthetic photon flux (400–700 nm) as determined using a LiCor model LI-185B quantum radiometer photometer with an atmospheric/underwater quantum sensor (LiCor, Lincoln, Nebraska). Chlorophyll a content was determined using standard spectrophotometric (Lorenzen 1967) and fluorometric (Arar and Collins 1992) methods. Secchi depth was determined, after which water samples were collected with 1 liter Niskin bottles at 1 m depth intervals down to the Secchi depth. The samples were immediately stored on ice in the dark and were returned to the lab for analysis in that state. Within 24 hours the samples were filtered onto Whatman GF/F glass fiber filters (particle retention size 0.7 μm) wetted with a saturated solution of $MgCO_3$. All samples were extracted in acetone on ice.

Upwelling along the north bay and an inflow of upwelled water from outside the bay occurred immediately prior to and on April 6, 1994. At site 1, light extinction and Secchi disk measurements were performed at 0800 hours. The Secchi depth was 6 m. Similarly, at site 2 at midmorning, the Secchi depth was 4 m. The light for wavelengths from 400–700 nm is shown as a function of depth for both sites in figure 7. A semi-log scale shows that at sites 1 and 2 the light variation is essentially linear to depths of 4–5 m, and attenuation (the slope) is similar. The attenuation coefficients calculate to about 4.9×10^{-3} cm^{-1}, and 3.9×10^{-3} cm^{-1}, respectively.

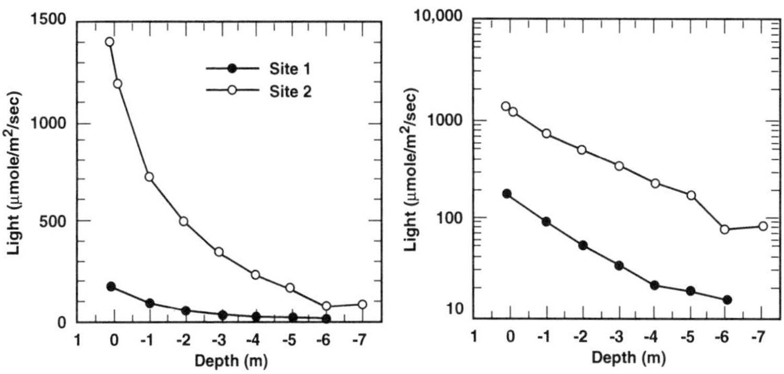

Figure 7. Light attenuation with depth in Monterey Bay, April 6, 1994.

At site 2, light, temperature, and chlorophyll profiles are shown in figure 8. Temperature varied little until it dropped sharply at about 6 m, suggesting that the surface waters were well mixed to that depth. Although the chlorophyll a concentration measurements in the upper layers are rather irregular, they are generally high, averaging 15.2 mg/m^3. The Secchi and the chlorophyll measurements for site 2 are shown by the designated upper symbol in figure 9. The figure shows our correlation of chlorophyll a versus Secchi depth for a NASA/NOAA data base (Nolton 1980). During this strong upwelling season our measurements are consistently among the high measured values, whether the measurements were estimated by Secchi disk or measured in the laboratory. The measurements span a decade in time, and are physically separated by a significant distance within the strong upwelling region (as shown by figs. 1 and 3). Biologically, these are very rich upwelled waters.

In the absence of upwelling (site 3, November 1994), the Secchi depth was 11 m. The data point for site 3 averaged over samples taken from the surface to the Secchi depth (like site 2) is 0.25 mg/m^3 and is shown in the lower symbol in figure 9. Importantly, in this area, chlorophyll a concentration is about fiftyfold, higher in April when upwelling was prominent, than in November when upwelling was not evident.

Point Arena, Gulf of the Farallones, analysis and computational result: We have described upwelling episodes, the formation of major upwelling features, and their temporal and spatial variation along the west coast of North America in terms of a particular model. We have presented some physical and biological observations.

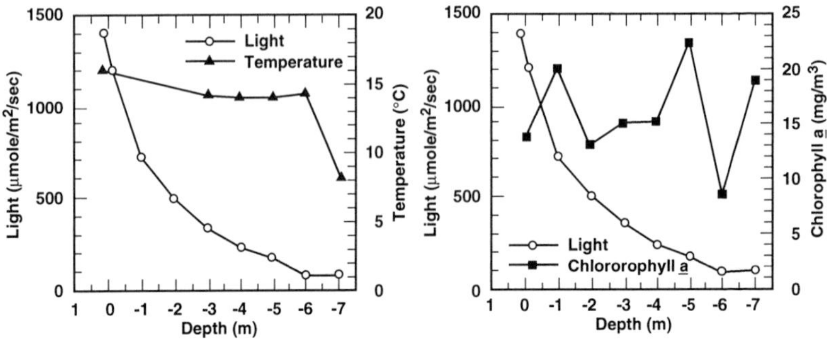

Figure 8. Light, temperature, and chlorophyll a depth profiles in Monterey Bay, April 6, 1994.

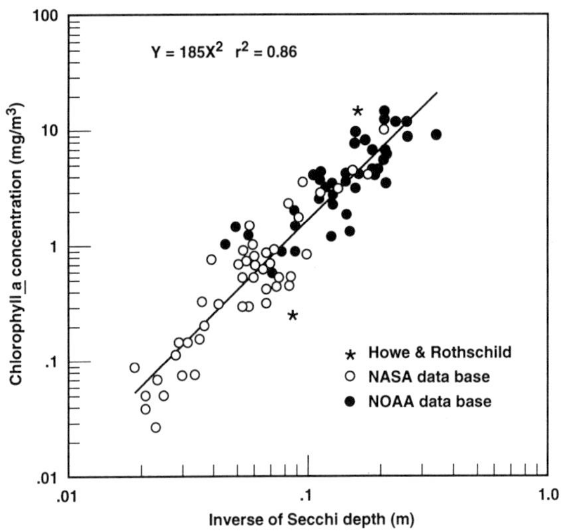

Figure 9. Chlorophyll-Secchi data and correlation.

Finally, we couple physical oceanography and biological productivity deterministically. We describe the formulation of a detailed mathematical model of wind-driven upwelling in the ocean. The formulation (presented in the Appendix) includes the coupled effects of the conservation of mass; the momentum equations in the three coordinate directions (including viscous, gravity, and Coriolis effects); the energy equation (including the divergence of the radiative flux); the spectral radiative transfer equations in the ocean (including effects of absorption, scattering, and emission of radiation) transformed (Howe and Yang 1993) from the Schuster-Schwarzchild form (Schuster 1905) to the two-flux form of incident and backscattered radiation (Kubelka-Munk 1931); and species conservation equations (which include chemical, photochemical, biochemical, and predation processes that relate to the formation or depletion of each species). Boundary conditions include wind stress, solar heating, and illumination.

An approximation to the mathematical formulation described above yields an analytical solution to the equations. In the approximation, we have assumed that the motion of the surface water is in the direction of the wind (which is probably realistic for shallow coastal waters). Correspondingly, the horizontal momentum equation is solved by matching the boundary layer in the water to that in the air. A comparison of our momentum result with that of the USGS oil spill trajectory equation result (Samuels and Huang 1982) shows remarkable agreement. Our analysis reveals an approximate analogy between the local temperature of the well-mixed surface layer of an upwelling formation and the biomass concentration to a definable level of approximation. The computed surface thermal variation for an upwelling episode just south of 39° latitude is shown in figure 10. The figure inset, obtained from the NOAA Sea Surface Thermal Map for July 6, 1984, is configured like the Point Arena upwelling in figure 1. The detailed inset includes surface isotherms, and an offshore convergent front that terminates the upwelling formation (the first author has seen such fronts many times in the Gulf of the Farallones, just to the south). In the figure, the theory is represented by the curve, the

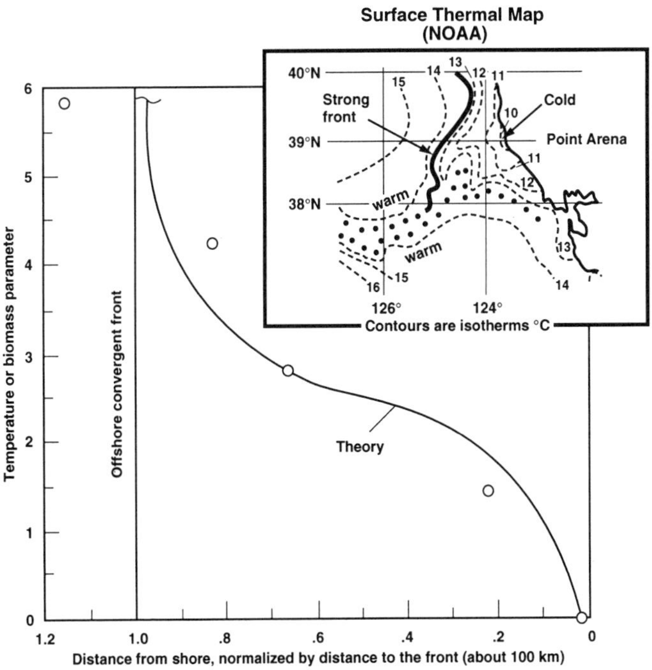

Figure 10. Surface temperature/biomass profile near Point Arena computation and NOAA map. The data are scaled from NOAA SSTA near Point Arena July 6, 1984 (~39° N).

offshore front by the vertical line, and the isotherms by the open circular symbols. The abscissa is the normalized distance from shore. The ordinate is the "temperature parameter"; the surface temperature increase from shore normalized by a factor containing the ratio of nearshore wind velocity to the product of radiative heating and local distance from shore to the front at the local latitude, as defined in the Appendix. Alternatively, because of the approximate analogy, the ordinate is the "biomass parameter"; the biomass increase from shore normalized by a similar factor where the radiative heating is replaced by a biological term (Appendix). Both nearshore and near the front, that part of the solution pertaining to the ocean dynamics results in very low surface velocity and, therefore, large residence times for upwelled surface waters. Correspondingly, the large residence time is manifested in a locally abrupt increase in the temperature curve and biomass concentration in accord with the approximate analogy in both locations. There is a singularity in the theory at the front, and the theoretical curve is arbitrarily terminated at the level of the isotherm just outside the front. Thus, we can form some notion of the development of the biomass in an upwelling formation as the surface water flows seaward.

Concluding Remarks

Major spatial and temporal features of coastal upwelling along the west coast of North America have been presented and discussed. At various locations and times, physical and biological information obtained from a data buoy and in situ observations (including the entire water column) have been presented as well as large scale imagery obtained by satellite remote sensing. Comparisons with upwelling parameters have been made. Biological and physical measurements have been presented. Importantly, during the strong upwelling season, our biological measurements of chlorophyll a are consistently among the high reported values (whether estimated by Secchi disk or measured in the laboratory). They span more than a decade in time and are physically separated by a significant distance (33°–39° north latitude) within the strong upwelling region along the Central California coastal ocean. Biologically, these are very rich waters. In the absence of upwelling, our chlorophyll a measurements are lower by as much as fiftyfold. Thus, upwelling has a large effect on the biomass production. Relationships among weather, oceanography, illumination, solar heating, physical, and biological phenomena that have been formulated deterministically have been described. Reasonable simplifying assumptions employed to solve these coupled relationships have yielded analytical solutions to the equations. Results have been compared to NOAA Sea Surface Thermal Analysis information. A useful analogy between sea surface thermal and biological parameters was identified to a defined level of approximation in the energy and biomass conservation equations and applied to the thermal results as an index to biological productivity.

References

Arar, E. J. and G. B. Collins. 1992. Method 445.0 in vitro determination of chlorophyll a and pheophytin a in marine and freshwater phytoplankton by fluorescence, pp. 445.0-1–445.0-9. In Methods for the determination of chemical substances in marine and estuarine environmental samples. EPA/600/R-92/121.

Bakun, A. 1973. Coastal upwelling indices, west coast of North America 1946–1971. NOAA Tech. Rep. NMFS-SSRF-671.

Bakun, A. 1975. Daily and weekly upwelling indices, west coast of North America 1967–1973. NOAA Tech. Rep. NMFS-SSRF-693.

Codespotti, L. A. 1983. Nitrogen in Upwelling Systems, Nitrogen in the Marine Environment, Carpenter, E. J. and Capone, D. G., eds., Academic Press, New York, p. 540.

Denman, K. L., and Platt, T. 1975. Biological Prediction in the Sea. Contribution to NATO Advanced Study Institute on Modeling and Prediction of the Upper Layers of the Ocean, Urbino, Italy.

Ekman, V. W. 1905. On the influence of the earth's rotation on ocean currents. Ark. f. Mat. Astr. och. Fysik. K. Vet. Ak. (Stockholm.) 2(11).

Howe, J. T., and Yang, L. 1993. Earth atmosphere entry thermal protection by radiation backscattering ablating materials. J. Therm. Heat Trans. 7(1): 74–81.

Hsueh, Y., and Kenny, R. N. III. 1972. Steady Coastal Upwelling in a Continuously Stratified Ocean, Jour. Physical Oceanography, Vol. 2, no. 1, pp. 27–33.

Kubelka, P., and Munk, R. 1931. Z. Tech. Phys. 12: 593.

List, R. J. 1963. American Institute of Physics Handbook, Second Ed., McGraw-Hill, New York, pp. 2–142.

Lorenzen, C. J. 1967. Determination of chlorophyll and pheo-pigments: spectrophotometric equations. Limnol. Oceanogr. 12: 343–346.5.

McEwen, G. F. 1912. The Distribution of Ocean Temperatures Along the West Coast of North America Deduced from Ekman's Theory of the Upwelling of Cold Water from the Adjacent Ocean Depths, Intern. Revue ges Hydrobiol. u Hydrogr., Bd. 5, pp. 243–286.

Nelson, C. S. 1977. Wind stress and wind stress curl over the California current. NOAA Tech. Rep. NMFS-SSRF-714.

Nolton, J. W. 1980. In situ optical methods for chlorophyll estimation in the sea. Thesis, San Jose State University.

Pedlosky, J. 1974. On Coastal Jets and Upwelling in Bounded Basins, Journal of Physical Oceanography, Vol. 4. no. 1 pp. 3–18.

Ryther, J. H. 1969. Photosynthesis and fish production in the sea. Sci. 166:71–76.

Samuels, W. B., and Huang, N. E. 1982. An oilspill trajectory analysis model with a variable wind deflection angle. Ocean Eng. 9(4): 347–360.

Schuster, A. 1905. Radiation through a foggy atmosphere. Astrophys. J. 21(1): 1–22.

Sverdrup, H. U. 1942. Oceanography for Meteorologists, Prentice-Hall, New York.

Sverdrup, H. U., Johnson, M. W., and Fleming, R. H. 1942. The Oceans, Prentice-Hall, New Jersey.

Starkey, R. J. Jr. 1981. Antisubmarine warfare oceanography. Part II: military electronics/countermeasures, p. 65.

Appendix; Analytical Formulation; John T. Howe

Nomenclature

B	biomass concentration
c	mass fraction of species, coefficient for surface velocity
c_p	specific heat at constant pressure
d_e	depth of euphotic layer
D	Secchi depth
g	gravitational constant
I_R	intensity of radiation backscattered upward in the sea
I_T	intensity of radiation transmitted downward in the sea
i	small distance from coast where properties are known
J_i	depletion rate of species i per unit mass
K_ν	Kubelka-Munk absorption coefficient at frequency ν
k	eddy conductivity
m	slope of the linear variation of the wind
n	index of refraction, compatibility condition fraction
p	pressure
P_m^B	specific biomass production rate at optimum light intensity
q_r	radiative flux
R	component of frictional force per unit volume
r_m	maximum ration for zooplankton grazing
S_ν	Kubelka-Munk scattering coefficient at frequency ν
t	time
U	air velocity component in x direction
u	water velocity component in x direction
V	air velocity component in the y direction
$\overline{V}$	horizontal air velocity
v	water velocity component in y direction
v_e	water velocity component in the eastward direction
w	water velocity component in z direction
x,y	sea surface coordinates normal and parallel to the coast
z	coordinate measured downward from sea surface
α	initial slope of specific photosynthesis rate at irradiance near zero
δ	distance offshore where convergent downwelling front is located
ε	distance offshore where weather data are known
μ	radiative absorption coefficient
ρ	density, reflectivity
σ	zooplankton grazing rate constant
ϕ	latitude
ω_i	production rate of species i per unit volume
Ω	angular velocity of rotation of Earth

Subscripts

B	blackbody
i	species i
o	value at, or near coast; value at ocean surface
r	radiative
s	surface value, scattering
x	component in x direction
y	component in y direction
z	component in z direction
δ	value at offshore front
ν	spectral frequency

Analysis

In the ocean, the equations of change in a left hand coordinate system with the coordinate, z, directed downward include the conservation of mass,

$$\frac{\partial\rho}{\partial t}+\frac{\partial(\rho u)}{\partial x}+\frac{\partial(\rho v)}{\partial y}+\frac{\partial(\rho w)}{\partial z}=0 \tag{1}$$

the momentum equations in the three coordinate directions

$$\frac{Du}{Dt}=\frac{\partial u}{\partial t}+u\frac{\partial u}{\partial x}+v\frac{\partial u}{\partial y}+w\frac{\partial u}{\partial z}=2v\Omega\sin\phi-\frac{1}{\rho}\frac{\partial p}{\partial x}+\frac{1}{\rho}R_x \tag{2}$$

$$\frac{Dv}{Dt}=\frac{\partial v}{\partial t}+u\frac{\partial v}{\partial x}+v\frac{\partial v}{\partial y}+w\frac{\partial v}{\partial z}=-2u\Omega\sin\phi-\frac{1}{\rho}\frac{\partial p}{\partial y}+\frac{1}{\rho}R_y \tag{3}$$

$$\frac{Dw}{Dt}=\frac{\partial w}{\partial t}+u\frac{\partial w}{\partial x}+v\frac{\partial w}{\partial y}+w\frac{\partial w}{\partial z}=-2v_e\Omega\sin\phi-\frac{1}{\rho}\frac{\partial p}{\partial z}+g+\frac{1}{\rho}R_z \tag{4}$$

an energy equation,

$$\rho c_p\frac{\partial T}{\partial t}+\rho u c_p\frac{\partial T}{\partial x}+\rho v c_p\frac{\partial T}{\partial y}+\rho w c_p\frac{\partial T}{\partial z}=\frac{\partial}{\partial z}\left(k\frac{\partial T}{\partial z}-q_{r_z}\right) \tag{5}$$

and a species conservation equation,

$$\frac{\partial c_i}{\partial t}+u\frac{\partial c_i}{\partial x}+v\frac{\partial c_i}{\partial y}+w\frac{\partial c_i}{\partial z}=J_i+\frac{\omega_i}{\rho} \tag{6}$$

The divergence of the radiative flux is (Howe and Yang 1993, Schuster 1905)

$$-\frac{\partial}{\partial z}q_{r_z}=\pi\sum_{\nu=h}^{h'}K_\nu\left(I_{R_\nu}+I_{T_\nu}-2n^2 I_{B_\nu}\right) \tag{7}$$

where the incident and backscattered radiative intensity in the ocean are

$$\frac{dI_{T_v}}{dz} = -(K_v + S_v)I_{T_v} + S_v I_{R_v} + K_v n^2 I_{B_v} \tag{8}$$

$$\frac{dI_{R_v}}{dz} = (K_v + S_v)I_{R_v} - S_v I_{T_v} - K_v n^2 I_{B_v} \tag{9}$$

The right side of equation (6) contains the chemical, photochemical, biochemical, and predation processes that relate to the formation or depletion of each specie, i. Boundary conditions for these equations will be discussed subsequently.

For wind-driven currents it is commonly assumed (Ekman 1905, McEwen 1912, Sverdrup 1942, Pedlosky 1974, and Hsueh and Kenny 1972) that the pressure and acceleration (actually the entire substantial derivative) terms can be neglected in the momentum equations (2) and (3). The result is a balance between the viscous and Coriolis terms (the first and last terms on the right sides of the equations). The simplified equations were originally integrated by Ekman, and subsequently by many others (as discussed by Sverdrup et al. 1942). For present purposes, for a shallow ocean, the surface current is aligned with the wind.

Example: In the Gulf of the Farallones, the broad ocean floor slopes gently to the edge of the continental shelf. Depths are of the order of 100 m or less, which allows upwelling of bottom nutrients into sunlit surface waters where photosynthesis can occur to produce a biomass. A coordinate system is aligned with the coast, and the curl of the windstress vector is allowed to be either positive or negative. For illustrative purposes, it is assumed that the wind velocity is known near shore and vanishes offshore. Thus the curl is negative, which promotes the formation of a biologically rich convergent downwelling front offshore. The author has observed offshore convergent fronts which terminated upwelling structures in this region. Illustrative ocean analysis and computations will be made in the euphotic layer to the north of the gulf of the Farallones near 39° N latitude, where results will be compared with an existing NOAA sea surface thermal map. The approximate solution of the equations of change can be obtained as follows.

Simplifications: Assume that real time weather data are known at two locations, nearshore at x = i, and offshore at x= ε (e.g. the NOAA weather radio gives wind speed and direction at the San Francisco buoy and at the Southeast Farallone Island). Further assume a quasi-steady state; the wind varies linearly with distance offshore; and any alongshore current, v, is allowed which varies negligibly with y locally. The solution of the continuity equation (1) in nearshore water is

$$u = ax, \quad w = -az \tag{10}$$

respectively. (Similarly, if a convergent front exists offshore at x= δ, the solution of equation (1) in that vicinity will be of the form

$$u = (\delta - x), \quad w = bz \tag{11}$$

The wind can be described by

$$\vec{V} = \vec{V}_0(1 + mx) \tag{12}$$

with the components

$$U(x) = U_0(1+mx), \quad V(x) = V_0(1+mx) \tag{13}$$

These are the boundary conditions on the horizontal momentum equations, which are simplified as follows. The x component of the surface velocity of the water is

$$u(x) = cU = cU_0(1+mx) \tag{14}$$

where the constant, c, is evaluated by matching the boundary layer shear stress at the air/water interface. The constant is essentially the cube root of the ratio of the density viscosity product in the air to that in the water. For laminar flow it varies from 0.025 to 0.03 for surface temperatures between 10° C to 20° C. For turbulent flow, 0.035 has been used in the USGS oilspill trajectory model (Samuels and Huang 1982).

At $x = n\varepsilon$, the u velocity component given by equations (10) and (14) are equal and

$$u(n\varepsilon) = cU_0(1+mn\varepsilon) = an\varepsilon \tag{15}$$

Thus the coefficient a is determined to be

$$a = \frac{cU_0}{n\varepsilon} = cU_0 m\left(1+\frac{1}{mn\varepsilon}\right) \tag{16}$$

where n will be determined subsequently by compatibility conditions. Thus for a given wind condition, the motion of the surface waters can be approximated simply for present purposes.

Solar Radiation in the Sea: Although upwelling may arise from depths of the order of 100 m, the euphotic zone in coastal waters where sunlight interacts with organic matter is of the order of 10 m. It is estimated that 90% of all sunlight is absorbed in this surface layer.

Briefly, three conditions are employed concerning the euphotic zone in what follows: it is a surface layer, is well mixed, and the residence time of living phytoplankton is large compared to the life of the front, as estimated by the Stokes settling rate.

For present purposes, at depth z the unreflected local intensity of partially absorbed incident light of frequency ν is

$$I_{T_\nu} = \frac{q_{r_{\nu 0}}}{\pi}\left(1-\rho_{s_\nu}\right)e^{-K_\nu z} \tag{17}$$

and the divergence of the incident spectral radiative flux from equations (17) and (7) is

$$-\frac{\partial}{\partial z}q_{r_{z_\nu}} = q_{r_{\nu 0}}K_\nu\left(1-\rho_{s_\nu}\right)e^{-K_\nu z} \tag{18}$$

To obtain the radiant energy absorbed in the euphotic layer, we sum over the entire spectrum, and average the radiation absorbed by integrating through the layer of thickness d_e. The result is substituted for the right hand side of the energy equation (5), which by use of assumptions already made can be written for the quasi-steady state

$$u\frac{\partial T}{\partial x}=\left(\frac{1}{\rho c_p d_e}\right)\sum_{\nu=0}^{\infty} q_{r_{\nu 0}}\left(1-\rho_{s_\nu}\right)\left(e^{-\mu_\nu d_e}-1\right)\equiv Q_r \tag{19}$$

The solution describes the temperature variation in the euphotic layer as a function of distance from shore. Importantly, the growth rate of phytoplankton increases with temperature (Codespotti 1983); the rate of metabolism of ocean organisms is approximately doubled by a 10° C increase in temperature (Sverdrup et al 1942). Normalize x by i, the location of the nearshore weather data ($x = i\bar{x}$), and equation (19) becomes (where the left side is the temperature parameter in fig. 10)

$$\frac{T - T_0}{Q_r}=\int_1^{\bar{x}}\frac{d\bar{x}}{u} \tag{20}$$

Equation (20) can be integrated to obtain

$$\left(\frac{T-T_0}{Q_r}\right)=\frac{1}{a}\ell n\frac{x}{i}, \quad x\le n\varepsilon \tag{21}$$

Similarly from weather data at a second location, $x = \varepsilon$

$$\left(\frac{T_\varepsilon-T}{Q_r}\right)=\frac{1}{cU_0 m}\ell n\left(\frac{1+m\varepsilon}{1+mx}\right), \quad x\ge n\varepsilon \tag{22}$$

At $x = n\varepsilon$, the temperature is continuous, and from equations (21) and (22)

$$\left(\frac{T_\varepsilon-T_0}{Q_r}\right)cU_0 m=\ell n\left(\frac{1+m\varepsilon}{1+mn\varepsilon}\right)+\left(\frac{mn\varepsilon}{1+mn\varepsilon}\right)\ell n\frac{n\varepsilon}{i} \tag{23}$$

This is the transcendental compatibility equation whose solution is n, the match point in surface water temperature and velocity mentioned in connection with equations (15) and (16).

Convergent Front Offshore: Conveniently in the right hand diagram of figure 4, the weather is known at the front; that is, the driving wind goes to zero. Thus $\delta = \varepsilon$, and $m = -1/\delta$. In the vicinity of the front

$$u=cU_0\left(1-\frac{x}{\delta}\right)=\frac{cU_0}{\delta}(\delta-x)=b(\delta-x) \tag{24}$$

which defines b.

In order to avoid dealing with a mathematical singularity at the front, allow T_δ to occur near the front at $x = (\delta - i)$ with the constraint $\bar{\delta} = \delta/i >> 1$. The compatibility equation (23) can be written

$$\left(\frac{T_\delta-T_0}{Q_r}\right)\frac{cU_0}{\delta}=\ell n\bar{\delta}(1-n)+\frac{n}{n-1}\ell nn\bar{\delta} \tag{25}$$

Correspondingly, equation (20) can be integrated to yield the surface temperature distribution offshore from the matchpoint n, and written

$$\left(\frac{T-T_0}{Q_r}\right)\frac{cU_0}{\delta}=\left(\frac{T_\delta-T_0}{Q_r}\right)\frac{cU_0}{\delta}-\ell n\,\bar{\delta}\left[\left(1-\frac{x}{\delta}\right)\right],\quad \frac{x}{\delta}\geq n \tag{26}$$

(while equation (21) gives the temperature distribution inshore of n).

To determine n, Q_r is taken to be 0.296×10^{-5} °C/sec (derived from the adjusted solar flux absorbed by water averaged over a 24 hour day of typical cloudiness along the west coast in July (List 1963)). Other representative numbers are that $T_\delta - T_0$ is 4° C, U_0 is 12 m/sec, and δ is approximately 100 km. If T_0 is known at x = i = 2 km; n is found to be 0.44 from equation (25). The corresponding computed surface temperature as a function of distance from shore is shown by the curve in figure 10. The abrupt temperature rise nearshore and near the front caused by low surface velocity (eqs. (10) and (11)), and long residence time was mentioned in the discussion of figure 10 in the text.

Analog: Let us examine the energy equation and the species equation to inquire, "is there an analogy between surface temperature and biomass to any definable level of approximation in this upwelling formation?" A mathematical model of the phytoplankton biomass as it grows by nourishment in sunlit surface layers and is diminished by predation was developed by Denman and Platt (1975). It can be shown to be of the form of the species equation (6). For the quasi-steady state, not nutrient limited, with smoothly varying predation, locally well mixed surface water, where along-shore derivatives are negligible, the model equation can be written (in their terminology)

$$u\frac{dB}{dx}=P_m^B B\tanh\left(\frac{\alpha\bar{I}_V}{P_m^B}\right)-r_m\left(1-e^{-\sigma B}\right) \tag{27}$$

where for convenience, their "threshold biomass," B, has been set to zero, the next to the last term corresponds to the species production rate as affected by light intensity (and their $I_Z(z)$ has been replaced by $\bar{I}_V$, our unreflected portion of incident photons which take part in photosynthesis), and the last term accounts for predation. Under the limiting conditions where tanh g ≈ g, and

$$\left|\frac{g^2}{2!}\right|<<1,\quad e^{-g^2}\approx 1-g^2 \tag{28}$$

equation (27) becomes

$$u\frac{dB}{dx}=B\alpha\bar{I}_V-r_m\sigma B \tag{29}$$

The solution yields the phytoplankton biomass variation as a function of distance offshore

$$B(x)=B_0\,e^{\left(\alpha\bar{I}_V-r_m\sigma\right)\int_i^x dx/u} \tag{30}$$

Again if the exponent of the exponential is sufficiently small, this becomes

$$\frac{B(x)-B_0}{B_0\left(\alpha\bar{I}_v - r_m\sigma\right)} = \int_1^{\bar{x}} \frac{d\bar{x}}{u} \tag{31}$$

which is the same form as the energy equation solution (20); the approximate solutions for temperature and biomass are identical in form. Correspondingly, the lefthand ordinate is also labeled the "biomass parameter" in figure 10. The temperature parameter is defined as the left hand side of equation (26). The biomass parameter is defined as the left hand side of equation (31) multiplied by cU_0/δ. To this extent the temperature and biomass parameters are related, and temperature is an index to biomass in the upwelling structure. It should be noted with respect to equation (31) that the integral in the exponential in equation (30) is the time required to move surface waters offshore from x = i. Thus for large x, the exponential may be sufficiently large that equation (30) is more appropriate than equation (31). Thus the analogy is not general, although a local analogy would be easy to define. All things considered, it seems to the author that any deviation between theory and temperature data is probably attributable to the data, and that a thermal parameter can be formulated that is related approximately to a localized biomass parameter in the well behaved upwelling formation. A beginning that includes physical and biological features has been made to quantify the relationship between upwelling and biomass production.

Practical Inferences and Personal Observations

As a practical matter, the quasi steady existence of the convergent front offshore can be predicted by local weather information (strong wind nearshore and light wind, which terminates in the vicinity of the front, offshore). Moreover, a vessel can locate the front: if the temperature of the engine cooling water inlet is rising, the vessel is headed in the direction of the front. This simple notion has proved to be useful for commercial fishing. As mentioned previously, frontal features have been observed by the author just south of the region in figure 10. That is, after several days, strong winds nearshore diminished and offshore vanished completely in the vicinity of a sharp front that extended as far as the eye could see. Significantly, the front was very rich biologically, both at the surface and for the depth of water downwelling at the front. Salmon and other species congregate there. Fish may be able to follow rising temperature or biological gradients to locate the fronts.

Generally, both fish and commercial fishermen tend to follow the upwelling as the peak moves northward during the season as shown in figure 3. Interestingly, the author found that in Monterey Bay, commercial fish processing facilities were strained to capacity one day during a strong upwelling episode. The upwelling terminated abruptly, and the next day the facilities were idle. Both the fish and the fishing fleet followed the upwelling elsewhere.

CHAPTER 213

Demersal Fish and Invertebrate Population Variability

M. James Allen[1]
Shelly L. Moore[2]

Abstract

Demersal fishes and megabenthic invertebrates are the most characteristic larger organisms of the soft-bottom habitat, the most extensive habitat of the mainland shelf of Southern California. Populations of these organisms have been extensively monitored during the past 25 years to assess impacts resulting from wastewater discharge on the shelf. Although much is known about their condition in local areas, less is known about their spatial variability throughout Southern California. This study briefly describes some results of a regional survey of demersal fish and invertebrate populations on the mainland shelf of Southern California in summer 1994 and discusses variability in these populations in the context of previous studies. Spatially, these populations vary more consistently by depth than by region or outfall area. In particular, the inner shelf (< 25 m in depth) is generally much lower in abundance, biomass, and numbers of species than are deeper zones on the shelf. Outfall monitoring areas had enhanced populations although areas immediately near some outfalls may have depressed populations. Changes in demersal fish and invertebrate populations in the Southern California Bight during the past two decades partly reflect a general warming of the area and an improvement in environmental conditions as wastewater treatment has improved.

Introduction

Demersal fish and invertebrate populations are important indicators of human impact to the soft-bottom habitat. Populations have been regularly monitored near

----

[1] Principal Scientist, Southern California Coastal Water Research Project, 7171 Fenwick Lane, Westminster, CA 92683-5218
[2] Research Technician, Southern California Coastal Water Research Project, 7171 Fenwick Lane, Westminster, CA 92683-5218

ocean outfalls for many years but not throughout Southern California. Some past studies accumulated information from a variety of studies to provide a regional description of the populations (Allen and Voglin 1976, Thompson et al. 1993a). However, these studies are not recent nor were designed to examine spatial variability in the populations. Knowledge of this variability is essential for assessing impacts of human activities to this fauna.

This study summarizes variation in fish and invertebrate populations from a regional survey conducted in Southern California in 1994 and places the findings in context of previous studies to help understand general patterns in their variability. The study focuses on spatial (e.g., depth, region, outfall) and temporal (long-term) variability of the populations.

Regional Survey of 1994

A regional trawl survey was conducted on the mainland shelf of Southern California in summer 1994 to assess the extent of environmental change in the Southern California Bight (SCB). One of the objectives of the study was to assess the extent of change in demersal fish and invertebrate populations and assemblages. Only a brief summary of a part of that study is given here. Allen and Moore (1996) provide details to the population analyses summarized here.

Methods. Trawl samples were collected from 114 stations from Point Conception, California to the United States-Mexico border at depths of 10 to 200 m in July to August 1994 (Fig. 1). A stratified random design was used with the following depth strata: 1) inner shelf -- 10-25 m; middle shelf -- 26-100 m; outer shelf -- 101-200 m (Bergen 1996). Samples were collected using a 7.6-m (head-rope) otter trawl with 1.25-cm cod-end mesh. Trawls were towed for 10 min at 1.0 m/sec along isobaths. Fish and invertebrates were identified to species, counted, examined for anomalies, and weighed to the nearest 0.1 kg by species. Fish lengths (maximum standard length) were measured and reported by centimeter size-class.

Data analysis focused on fish and invertebrate population metrics (e.g., abundance, biomass, number of species, and Shannon-Wiener diversity, all expressed per haul) and assemblages. In additional to depth strata defined by the survey design, regional and outfall status strata were defined for data analysis. Regional strata had the following ranges: 1) north -- Point Conception to Point Dume; 2) central -- Point Dume to Dana Point; and 3) Dana Point to the U. S.-Mexico border. Outfalls status categories were 1) outfall -- areas on the middle shelf that are monitored around outfalls of the major public-owned treatment works (POTWs) of Southern California and 2) nonoutfall -- areas on the middle shelf that do not occur in the POTW monitoring areas. The largest POTWs discharge at depths of 60 m near Los Angeles (Santa Monica Bay, Palos Verdes Shelf, San Pedro Bay) or 100 m at San Diego.

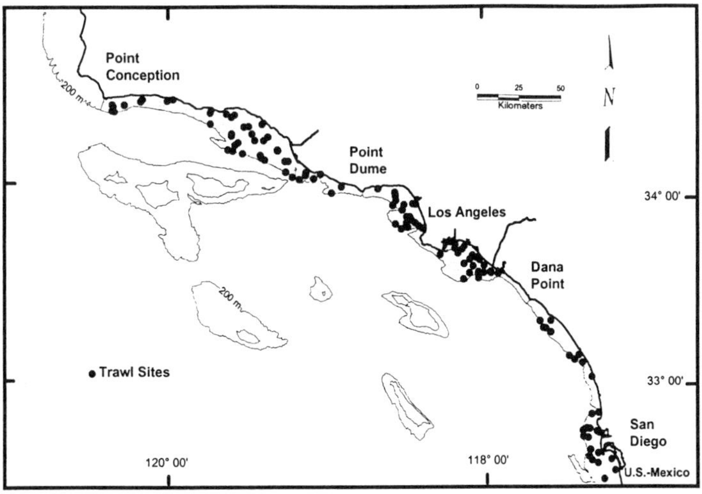

Figure 1. Stations sampled by trawl on the mainland shelf of Southern California at depths of 10-200 m, July-August 1994.

Summary statistics of population metrics were area-weighted and analyzed by procedures described in Allen and Moore (1996). G-tests (Sokal and Rohlf 1995) were used to test for differences between subpopulations in the amount of area lying below the bight-wide area-weighted median for each metric.

Results. A total of 87 demersal fish species, representing 34 families, and 3 classes were collected. Bony fishes were the most diverse class while rockfishes (Scorpaenidae) were the most diverse family. Twelve species occurred in at least 45% of the survey area. Pacific sanddab (*Citharichthys sordidus*), Dover sole (*Microstomus pacificus*), and plainfin midshipman (*Porichthys notatus*) were the most common species.

An average (median) trawl haul in the SCB caught 103 fish weighing 3.4 kg, 11.4 species, and had a diversity of 1.58. Fish catch metrics varied with region, depth, and outfall status (Table 1). Fish abundance and numbers of species were particularly low on the inner shelf; only 6% of the inner shelf had values above the bight medians. In contrast, more than 75% of the area near outfalls had high fish abundance and diversity. Demersal fish populations showed significant regional, depth, and outfall variability (Table 2). Fish abundance was significantly higher in the south, increased with increasing depth, and was higher in outfall areas. Fish

TABLE 1. Percent area of Southern California Bight with values above Bight median for demersal fish and invertebrate population metrics in Southern California regional survey, July-August 1994.

Category	Number of Stations	Percent Area above Bight Median			
		Abundance	Biomass	Species	Diversity
FISH					
Region					
Northern	45	46	37	58	65
Central	41	40	64	37	49
Southern	28	66	61	47	68
Depth Zone					
Inner Shelf	30	6	45	6	35
Middle Shelf	53	57	42	61	54
Outer Shelf	31	74	73	65	52
Outfall Status					
Outfall	23	78	56	71	79
Nonoutfall	30	52	37	59	53
Total	114	50	50	50	50
INVERTEBRATES					
Region					
Northern	45	58	54	52	44
Central	41	35	34	50	70
Southern	28	43	59	45	33
Depth Zone					
Inner Shelf	30	0	0	7	65
Middle Shelf	53	60	60	59	44
Outer Shelf	31	72	87	64	44
Outfall Status					
Outfall	23	62	72	68	58
Nonoutfall	30	72	51	57	40
Total	114	50	50	50	50

Bight medians are as follows:
FISH: abundance -- 103 individuals/haul; biomass -- 3.4 kg/haul; numbers of species -- 11.4 species/haul; diversity-- 1.58 bits/individual/haul.
INVERTEBRATES: abundance -- 171 individuals/haul; biomass -- 2.9 kg/haul; numbers of species -- 11.7 species/haul; diversity-- 1.05 bits/individual/haul.

TABLE 2. Results of significance tests of trawl catch metrics by subpopulation of Southern California regional survey, July to August 1994.

Metric	Regions	Depth Zones	Middle Shelf Outfall Status
Fish			
Abundance	N=C<S	I<M<O	OF>NOF
Biomass	N<C=S	I=M<O	OF>NOF
Species	N>C, N=S, C=S	I<M=O	OF=NOF
Diversity	N=S>C	I<M=O	OF>NOF
Invertebrates			
Abundance	N>C=S	I<M=O	OF=NOF
Biomass	N=S>C	I<M<O	OF>NOF
Species	N=S=C	I<M=O	OF=NOF
Diversity	N=S<C	I>M=O	OF>NOF

Null hypothesis was that there was no difference between subpopulations in amount of area below the median for the Southern California Bight. Significance tests used were 2 x 2 G-tests for two subpopulations and R x C G-tests for three populations (Sokal and Rohlf 1995). Tests were significant at $p<0.05$.
N = north; C = central; S = south; I = inner shelf; M = middle shelf; O = outer shelf; OF = outfall, middle shelf; NOF = nonoutfall, middle shelf.

biomass was lower in north, higher on outer shelf, and higher in outfall areas. The number of fish species per haul was higher in the north and lower on the inner shelf. Fish diversity was lower in the central region and on the middle shelf and higher in outfall areas.

A total of 222 species of megabenthic invertebrates representing 114 families, 20 classes, and 8 phyla. Mollusks were the most diverse phylum, spider crabs (Majidae) the most diverse family. Six species occurred in more than 45% of the area. California sand star (*Astropecten verrilli*), ridgeback rock shrimp (*Sicyonia ingentis*), and gray sand star (*Luidia foliolata*) were the most common species. An average (median) trawl haul caught 171 invertebrates weighing 2.9 kg, 11.7 species, and had a diversity of 1.12. Invertebrate abundance, biomass, and numbers of species were particularly low on the inner shelf (Table 1). None of the area of the inner shelf had invertebrate abundance or biomass values above the bight medians; only 7% of the area had numbers of species above the bight median. In contrast, 87% of the outer shelf had biomass values above the bight median. Demersal invertebrate populations showed significant regional, depth, and outfall variability (Table 2). Invertebrate abundance was significantly higher in north and lower on inner shelf. Biomass was lower in the central region, increased with increasing depth, and higher in outfall areas. Numbers of invertebrate species were lower on inner shelf. whereas invertebrate diversity was higher in central region, on the inner shelf, and near outfalls.

Discussion

Depth Variability. The 1994 regional survey showed significant depth variability in median values of all fish and invertebrate catch metrics and there was considerable similarity of pattern among metrics (Table 2). The inner shelf was significantly lower in all but fish biomass and invertebrate diversity; it was significantly higher in the latter. The outer shelf was significantly higher in fish abundance and biomass, and invertebrate biomass. Fish abundance and invertebrate biomass increased significantly between each depth zone with increasing depth.

Variability in fish catch metrics over depths from 10 to 200 m have been summarized in more limited surveys (Allen and Mearns 1977, Word and Mearns 1977, Allen 1982, Thompson et al. 1987, Thompson et al. 1993a,b). Similar patterns of increasing fish abundance and lower numbers of fish species on the inner shelf and higher fish biomass on the outer shelf have been shown as trends in the past (Allen and Mearns 1977, Allen 1982, Thompson et al. 1987, Thompson et al. 1993b) but these had not been tested for significance. As in this study, Thompson et al. (1987) found a distinct increase in invertebrate biomass with depth but that study found fewer individuals on the middle shelf and more species on the inner shelf. The low catch metrics in the inner shelf zone may be related to a more variable environment (e.g., of temperature, salinity, turbulence, and food availability). Higher daytime light levels there may also select for more cryptic invertebrate species and facilitate net avoidance by fish.

Regional Variability. The 1994 regional survey showed significant regional variability in median values of all fish and invertebrate catch metrics except for number of invertebrate species (Table 2). However, unlike depth, there was little similarity of regional pattern among the metrics. The north had significantly lower fish biomass but significantly higher numbers of fish species (than the central region) and invertebrate abundance. The central region had significantly lower fish diversity and invertebrate biomass but significantly higher invertebrate diversity. The south had significantly higher fish abundance.

Comparisons between mean catch metrics of historical and 1994 regional data can be made for fish but not for invertebrates. Allen and Voglin (1976) provide some information on regional variability of fish catch metrics but the regions were divided more finely than in this study. Comparison of 1994 data to this data base indicates that the highest fish abundances have shifted from the central to the southern region during the last 20 years. However, the general regional pattern of fish biomass was similar between the two time periods.

The highest numbers of fish species have shifted from the central to the northern region during the past two decades (Allen and Voglin 1976, Allen and Moore 1996). However, numbers of fish species appeared to be higher in the south in 1994 than in the past, perhaps due to the warming and consequent recruitment changes since the early 1980s. Fish diversity was also higher in the north in 1994 than it had been in the past.

Along the 60-m isobath, Word et al. (1977) found fish abundance higher in the north and diversity higher in the south but invertebrate abundance and number of fish and invertebrate species did not show any trend. Thus, compared with Allen and Voglin (1976) and Word et al. (1977), fish abundance appear to have shifted to the south and diversity to the north since the late 1970s. This is perhaps due to the general warming of the Bight since the early 1980s and the consequent reduced recruitment of northern and increased recruitment of southern species (CSDLAC 1990, MBC 1990, Stull and Tang 1996).

Outfall Variability. In this study, outfall areas on the middle shelf had significantly higher fish abundance, biomass and diversity, and invertebrate biomass and diversity than nonoutfall areas in the same depth zone. The two areas did not differ significantly in numbers of fish and invertebrate species and invertebrate abundance. Because the shelf areas in most of the outfall areas (the Palos Verdes Shelf excepted) is broad whereas most of the shelf of the SCB is narrow, the shelf breadth may influence the higher catch metrics in the outfall areas. Further analysis is needed to resolve whether the higher catches are due to stimulation from wastewater discharge or shelf breadth.

Before the 1994 regional survey, there was no regional comparison of trawl catch metrics from all of the major POTW outfall areas with those from nonoutfall areas. Mearns et al. (1976) compared fish catch metrics at each outfall (in the monitoring program station areas, which is much smaller than the outfall areas of the 1994 regional survey) and showed differences among the outfall areas. With one exception, fish catch metrics on the Palos Verdes Shelf near the White Point outfall (County Sanitation Districts of Los Angeles County) in 1985-1986 did not differ significantly from those of reference areas in 1985 (MBC 1987, Thompson et al. 1987, CSDLAC 1988). However, fish biomass was significantly higher on the middle shelf of the Palos Verdes Peninsula than in reference areas. Although broad areas near POTW outfalls generally had higher catch metrics in 1994, metrics can be depressed in the immediate vicinity of outfalls (Mearns et al. 1976, Allen 1977, CSDLAC 1988).

Long-term Variability in the Southern California Bight. Many trawl studies have been conducted in Southern California during the past 40 years, but most are focused in local areas rather than the SCB as a whole. However, two studies, Allen and Voglin (1976) and Thompson et al. (1993a) provide information on catch metrics of the SCB for fish and invertebrates, respectively.

Allen and Voglin (1976) summarized regional and local variability in fish catch metrics from 2,237 trawl samples collected on the mainland shelf and slope throughout the SCB from 1957 to 1975 at depths from 5-659 m. Of these, 2,220 were collected over the mainland shelf (5-200 m). Similarly, Thompson et al. (1993a) summarized information on demersal (megabenthic) invertebrates from 1203 trawl samples in Southern California from 1971 to 1985 over a depth range of 10 to 915 m. Of these, 658 were collected over the mainland shelf (10-137 m). The 1994 regional survey collected trawl samples from 10 to 200 m.

In general, the Bight-wide means of fish and invertebrate catch metrics in the 1994 regional survey were similar to the historical values (Allen and Moore 1996). However, a number of differences between the two studies may affect these comparisons. Although the historical fish data were predominantly collected at depths of 10-200 m, the overall depth range was greater than in the 1994 regional survey. In contrast, the invertebrate data used for comparison were collected over a slightly more limited depth range. In addition, the 1994 regional survey means were weighted in accordance to the sampling design whereas those of the historical databases were not; it is not known yet how this might affect this comparison.

Although overall fish and invertebrate population metrics have not changed much during the past few decades, considerable changes have occurred in certain species and at certain sites. With a general oceanic warming following the El Niños of the early 1980s, many individual species locally increased or decreased in abundance (MBC 1990, Stull and Tang 1996). At the same time, as these population shifts were occurring, wastewater treatment improved dramatically, with subsequent shifts in local populations (Stull 1995, Stull and Tang 1996). However, the impact of these changes on individual species populations and on species assemblages throughout the mainland shelf of Southern California during this period has not yet been described.

Conclusions

In conclusion, demersal fish and invertebrate populations showed significant regional, depth, and outfall variability. Trawl catch metrics for demersal fish and invertebrates varied more consistently by depth than by region; generally catch metrics were lower on the inner shelf and higher on the outer shelf. Regional variability occurred in almost all fish and invertebrate catch metrics but the pattern of variability differed greatly by catch metric. Outfall areas generally had higher catch metric values than nonoutfall areas; however, although these may be due to stimulation from wastewater discharge, they also may be related to the broader shelf in the vicinity of most outfalls relative to that of the SCB. In general, mean fish and invertebrate catch metrics were similar between the 1994 regional survey of 1994 and historical surveys for the mainland shelf of the SCB as a whole. There appears to have been a shift in increased fish abundance and numbers of species to the south and increased fish diversity to the north since the late 1970s.

Acknowledgments

The authors thank Southern California Coastal Water Research Project staff for their help with this project, especially Larry Cooper, and Dario Diehl. They also thank the members of the Southern California Bight Pilot Project Steering Committee and the Trawl Report Group for input into the study.

Appendix I. References

Allen, M. J. (1977). Pollution-related alterations of demersal fish communities. Am. Fish. Soc., Cal-Neva Wildl. Trans. 1977, 103-107.

Allen, M. J. (1982). Functional structure of soft-bottom fish communities of the Southern California shelf. Ph.D. dissertation. University of California, San Diego, La Jolla, Calif. 577 p.

Allen, M. J., and Mearns, M. J. (1977). Bottom fish populations below 200 meters. Coastal Water Research Project annual report for the year ended 30 June 1977. Southern California Coastal Water Research Project, El Segundo, Calif. 109-115.

Allen, M. J., and Moore, S. L. (1996). Spatial variability in Southern California demersal fish and invertebrate catch parameters in 1994. M. J. Allen, C. Francisco, and D. Hallock, eds., Southern California Coastal Water Research Project annual report 1994-1995. Southern California Coastal Water Research Project, Westminster, Calif. 114-127.

Allen, M. J., and Voglin, R. (1976). Regional and local variability of bottom fish and invertebrate populations. Southern California Coastal Water Research Project, Annual report for the year ended 30 June 1976. Southern California Coastal Water Research Project, El Segundo, Calif. 217-221.

Bergen, M. (1996). The Southern California Bight Pilot: sampling design. M. J. Allen, C. Francisco, and D. Hallock, eds., Southern California Coastal Water Research Project Annual Report 1994-95. Southern California Coastal Water Research Project, Westminster, Calif. 109-113.

CSDLAC (County Sanitation Districts of Los Angeles County). (1988). Joint Water Pollution Control Plant 1988 revised application for modification of secondary treatment requirements for discharges into marine waters, Appendix F-12: Organization of demersal fish assemblages of the Palos Verdes Shelf during 1985-1986. County Sanitation Districts of Los Angeles County, Whittier, Calif.

CSDLAC (County Sanitation Districts of Los Angeles County). (1990). Palos Verdes ocean monitoring annual report, 1989, Chapter 4: Trawls. County Sanitation Districts of Los Angeles County, Whittier, Calif.

MBC Applied Environmental Sciences. (1987). Organization of demersal fish assemblages of the Palos Verdes Shelf during 1985-1986. Prepared for County Sanitation Districts of Los Angeles County, Whittier, Calif. Prepared by MBC Applied Environmental Sciences, Costa Mesa, Calif. 140 p.

MBC Applied Environmental Sciences. (1990). LACSD monitoring and reporting program, 1989: trawl surveys. Prepared for County Sanitation Districts of Los Angeles County, Whittier, Calif. Prepared by MBC Applied Environmental Sciences, Costa Mesa, Calif. 186 p.

Mearns, A. J., Allen, M. J., Word, L. S., Word, J. Q., Greene, C. S., Sherwood, M. J., and Myers, B. (1976). Quantitative responses of demersal fish and benthic invertebrate communities to coastal municipal wastewater discharges. Grant R801152. Prepared for United States Environmental Protection Agency, National Marine Water Quality Laboratory, Corvallis, Ore. Prepared by Southern California Coastal Water Research Project, El Segundo, Calif. 246 p.

Sokal, R. R., and Rohlf, F. J.. 1995. Biometry. 3rd edition. W. H. Freeman and Co., New York, NY. 887 p.

Stull, J. K. (1995). Two decades of marine biological monitoring, Palos Verdes, California, 1972 to 1992. Bull. So. Calif. Acad. Sci. 94(1), 21-45.

Stull, J. K., and Tang, C.-L. (1996). Demersal fish trawls off Palos Verdes, Southern California, 1973-1993. Calif, Coop. Oceanic Fish. Invest. Rep. 37, 211-240.

Thompson, B. E., Laughlin, J. D., and Tsukada, D. T. (1987). 1985 reference site survey. Southern California Coastal Water Research Project, Long Beach, Calif. Tech. Rep. 221. 50 p.

Thompson, B. [E.], Tsukada, D. [T.], and Laughlin, J. [D]. (1993a). Megabenthic assemblages of coastal shelves, slopes, and basins off Southern California. Bull. So. Calif. Acad. Sci. 92(1), 25-42.

Thompson, B. [E.], Tsukada, D. [T.], and O'Donohue, D. (1993b). 1990 reference site survey. Southern California Coastal Water Research Project, Westminster, Calif. Tech. Rep. 269. 105 p.

Word, J. Q., and Mearns, A. J. (1977). Bottom invertebrate populations below 200 meters. Coastal Water Research Project annual report for the year ended 30 June 1977. Southern California Coastal Water Research Project, El Segundo, Calif. 117-120.

Word, J. Q., Mearns, A. J., and Allen, M. J. (1977). Better control stations: the 60-meter survey. Southern California Coastal Water Research Project, Annual report for the year ended 30 June 1977. Southern California Coastal Water Research Project, El Segundo, Calif. 89-97.

CHAPTER 214

San Francisco Ocean Monitoring Program: Analysis of Long Term Data.

Michael G. Kellogg[1,2], Laura Riege[3], Arleen Navarret[4], and Robert W. Smith[3]

Abstract

The City and County of San Francisco discharges an average of 18 MGD (dry weather flow) of treated municipal wastewater into the Pacific Ocean approximately 6 km offshore. Receiving water environmental monitoring data have been collected from 1982 to present, spanning periods of pre-discharge, discharge of primary treated effluent, and discharge of secondary treated effluent. Using ordination and BACI (before-after-control-impact) analysis techniques, we examine shifts in the benthic infauna community to determine if the discharge is having an impact on the receiving water environment. The primary community gradients showed shifts likely due to large scale oceanographic conditions, including several El Niño (ENSO) events. The BACI analyses revealed that one of the gradients also captures a shift in biota at the impact stations, but not at the reference station, during the period of primary effluent discharge. After the commencement of secondary effluent discharge, the community structure shifted back to resemble the reference station again. Multiple regression analysis showed that this biological shift corresponds to temporal changes in effluent mass loading of solids. Although this study showed some evidence of an outfall effect, the sampling design is inadequate to answer a number of management questions relative to effluent discharge, and it cannot address confounding factors, including impact of outflow from San Francisco Bay. A new sampling design that addresses these problems is discussed.

---

[1] Marine Biologist, City and County of San Francisco, Public Utilities Commission. Oceanside Laboratory, 3500 Great Highway, San Francisco, CA 94132
[2] Research Associate, Departments of Invertebrate Zoology and Geology, California Academy of Sciences
[3] Senior Environmental Scientist, EcoAnalysis, Inc., Ojai, California
[4] Senior Marine Biologist, City and County of San Francisco, Public Utilities Commission.

Introduction

The City and County of San Francisco (the City) discharges treated municipal wastewater into the Pacific Ocean and thereby incurs the responsibility of providing an assessment of the impact of that discharge on the receiving water environment. Municipal wastewater contains solids, nutrients, bacteria, and other contaminants that have the potential to adversely affect the environment. The nature and extent of an environmental impact (i.e. outfall effect) from municipal wastewater will be determined by the volume and quality of the effluent, the energy regime of the receiving waters, the design of the outfall, and other factors (Diener and Fuller 1995). The ability to detect such an impact is closely linked to the sampling design implemented to monitor the receiving water environment. An inadequate sampling design may not detect subtle environmental effects from a low volume discharge into strong currents that rapidly mix the effluent plume with receiving waters. It is therefore important to consider sampling design when evaluating environmental effects due to wastewater discharges.

The service area for the Oceanside Water Pollution Control Plant, which discharges through the ocean outfall, is the largely residential western side of the City. Discharge of primary treated effluent began in the fall of 1986 and treatment was upgraded to secondary in the fall of 1993. The average dry weather discharge is 18 million gallons per day (MGD). Because the City has a combined sewer system, wet weather flows are substantially higher, up to 175 MGD. However, these high flows are tied to rainfall and rarely last more than a few hours. The degree of effluent treatment varies with flow volume such that flows up to 43 MGD receive full secondary treatment, flows reaching treatment plant capacity of 65 MGD are a combination of primary and secondary treated effluent, and flows in excess of 65 MGD include a component which receives the equivalent of primary treatment in holding boxes (the City's combined sewer system and west-side treatment regimes are discussed in detail by Foreé and McGregor in this volume).

Yearly analyses of monitoring data (BWPC 1984, 1985, 1988, 1989, 1990, 1992a,b, 1993, 1994, 1995, in prep.) did not show a consistent outfall effect in the benthic infauna community. A pattern suggestive of such an effect during a single sampling period might be absent or even reversed the next sampling period. We have identified several factors that may act in combination to account for this apparent lack of an outfall effect: the relatively low volume and high quality of the effluent, the high energy regime of the receiving water environment, natural seasonal fluctuations and perturbations due to storms, and outflow of San Francisco Bay and Delta waters through the nearby mouth of the Golden Gate. Given these factors and the lack of a clear outfall effect in short-term analyses, we turn to long-term trends and patterns in the benthic infauna and sediment quality data to detect an environmental effect from our wastewater discharge.

Methods and Materials

Study Site and Sampling

Stations 1 and 25 are within the zone of initial dilution, stations 2 and 28 are nearfield, and stations 4 and 31 are farfield. Station 6, 4.8 km south of the outfall and at the same depth, is the reference site (Figure 1). The outfall terminates in a diffuser section 6 km offshore in 24 m water depth (MLLW). This site was chosen for the outfall so that strong currents would effect rapid mixture of the effluent plume with the receiving waters. Results of current, density, and dye studies (CH2M-Hill 1989) demonstrate currents in the vicinity of the diffuser average 15 cm/sec, maximum speeds exceed 40 cm/sec, and the current speed never appears to reach zero. Minimum dilution was calculated at 100:1 and generally observed to exceed 200:1.

Separate samples for benthic infauna and sediment quality were collected with a Smith-McIntyre grab (0.1 m^2 surface area). Samples for benthic infauna were sieved through a 0.5 mm mesh screen and identified to the lowest possible taxon. Sediment samples were collected from the top 2 cm of the grab. Three seasons (winter, summer, fall) were sampled from 1983 through 1989 and two seasons (winter and fall) were sampled from 1990 through 1995.

Statistical Analyses

Using the methods of Smith, *et al.* (1988), the Bray-Curtis dissimilarity index was used to characterize the infauna communities for each year at each station. These index values were input to principal coordinates analysis to identify unique biological community gradients, or "ordination scores". Diversity measures, including Shannon-Wiener (H') (Shannon and Weaver 1949), Simpson's dominance (D) (Pool 1974) and Pielou's evenness (J') (Pielou 1966) were calculated for each sample. In order to see long term patterns that might be masked by the strong seasonality, the data were also averaged over the fall and winter seasons. Before-After-Control-Impact (BACI) analyses (Bernstein and Zalinski 1983, Stewart-Oaten and Murdoch 1986) on the ordination scores, Bray-Curtis dissimilarities, and individual species abundances tested for possible outfall effects between the impact stations and the reference station. Multiple regression analyses examined relationships among the biological community gradients and a suite of environmental measurements. These measures included rainfall, wave energy measurements, grain size, effluent TOC, TKN, and TVS, volume of San Francisco Bay-Delta outflow, and effluent mass loading.

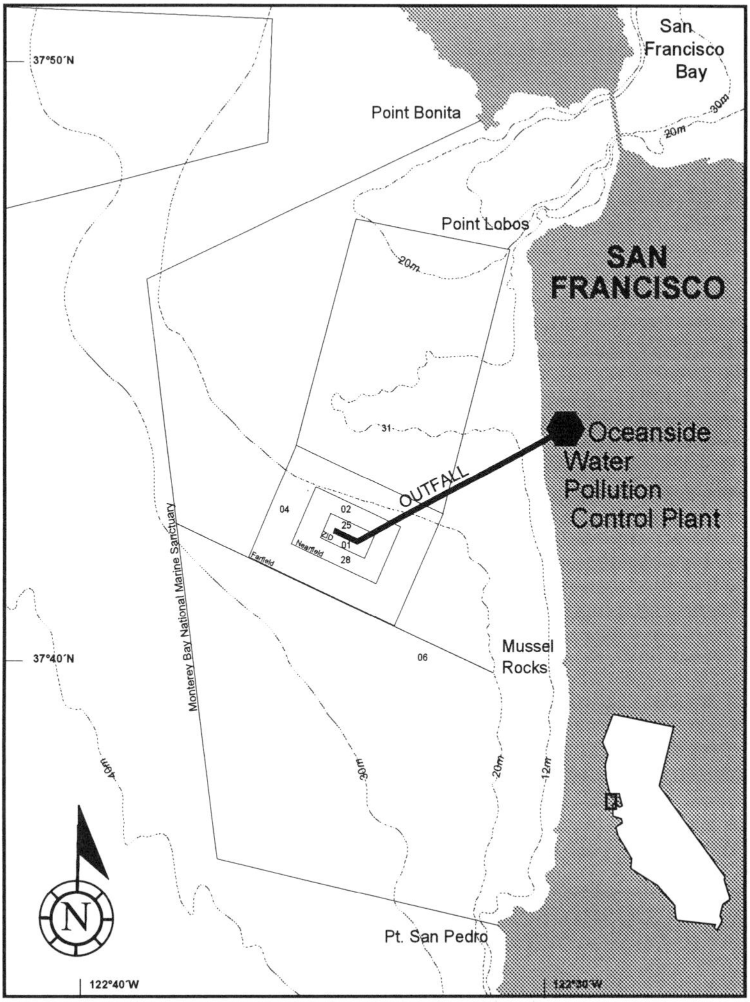

Figure 1.
Study area and sampling stations. Demarcated areas are part of a new sample design (see text for discussion).

Results and Discussion

Oceanographic Effects

One of the strongest patterns in the data is that of seasonality. Higher species diversity typically occurs in the fall (or summer 1983-1989) than in the winter (Figure 2). High diversity in the fall results from spring and summer recruitment (*e.g.* McCormick 1984, Silberstein 1987, Hannan, *et al.* 1979). Low diversity in the winter most likely reflects disturbance of bottom sediments by storms. The pattern of seasonality is strongest during wet (*i.e.* stormy) years (*e.g.* 1982-1984, 1993, 1995) and less pronounced during dry years (*e.g.* 1987-1992).

The strongest community gradient identified in the ordination analyses reflects differences in community composition due to the influence of oceanographic effects in the study area (Figure 3). The study area was profoundly affected by a series of winter storms in early 1983. Prior to those storms the polychaete worm *Owenia collaris* was the numerically dominant organism with densities often exceeding 10,000 per m^2 (McCormick 1984, BWPC 1984). The worm was essentially wiped out by the storms and has never regained its former abundance. Other infaunal organisms also showed reduced abundances after the winter

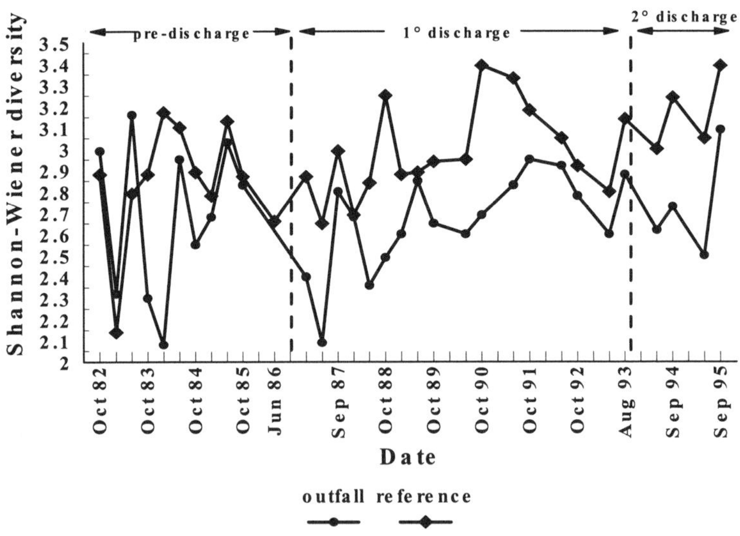

Figure 2.
Shannon-Wiener diversity values for station 1 (outfall) and station 6 (reference) demonstrating seasonality in the benthic infauna data.

1983 storms (BWPC 1984). High community index scores in Figure 3 during the period of primary effluent discharge also coincide with a five-year drought from 1987 to 1992. The higher scores are due to higher abundances of tube-dwelling polychaetes (*e.g. Spiophanes berkeleyorum, Prionospio cirrifera, Spiochaetopterus costarum*, and *Pherusa neopapillata*), and are coupled with reduced seasonality during the same period (Figure 2). This pattern is suggestive of less bottom disturbance during the drought. Lower index scores, in contrast, are due to higher abundances of Crustacea and errant polychaetes (*e.g. Eohaustorius* spp., *Rhepoxynius* spp., *Lissocrangon stylirostris, Nephtys* spp., and *Ophiodromus pugettensis*). Oliver, *et al.* (1980) describe a wave-disturbance related zonation pattern of soft-bottom benthic invertebrate communities where Crustacea and errant polychaetes are indicative of shallower, wave disturbed bottoms while tube-dwelling polychaetes are characteristic of deeper, less disturbed bottoms. The data presented here indicate a similar pattern on a temporal rather than spatial scale.

The approximate durations of ENSO (El Niño-Southern Oscillation) events (Dr. Jerry Bell, Climate Analysis Center, pers. comm.) are also shown on Figure 3. The 1982-83 event is the strongest of the century and marked the beginning of three consecutive wet winters. The 1986-87

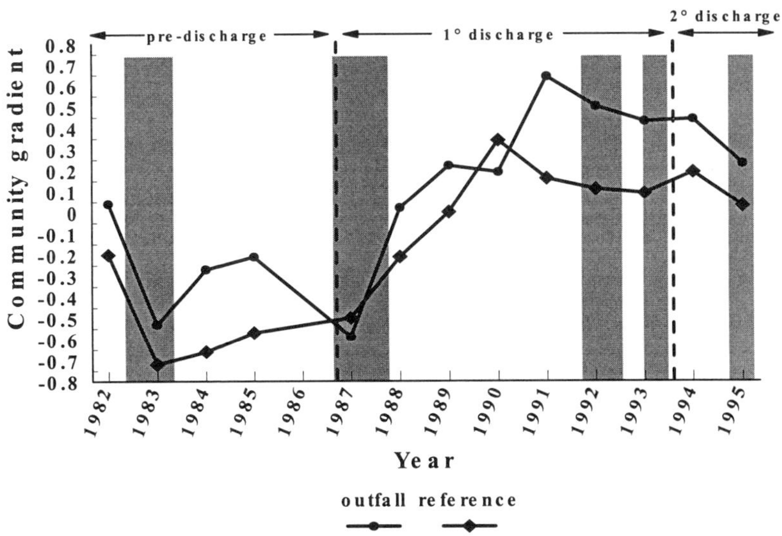

Figure 3.
Station 1 (outfall) and station 6 (reference) plotted along the primary ordination axis. The shaded bars indicate the approximate duration of El Niño (ENSO) events.

event in contrast was a dry El Niño and marked the beginning of a five year drought. It is interesting to note the response in the study area is the same (reduction of index scores) regardless of whether the ENSO is a wet or dry event. Oceanographically, the primary effects of ENSO are intensified storms during wet events, and sustained southwest winds which reduce upwelling and result in higher sea surface temperatures. During the 1982-83 event, for example, the sea level offshore of San Francisco was 10 cm higher than predicted and sea surface temperatures were 1 - 3 °C warmer than normal (NOAA 1997).

Outfall Effects

The second strongest community gradient identified in the ordination analyses also reflects shifts in community composition associated with oceanographic cycles (Figure 4). In addition to the large-scale impacts, this gradient distinguishes between the infauna communities at the impact and reference stations during the period of primary treated effluent discharge. BACI analysis results are significant ($p < 0.05$) at all stations for this gradient. The communities at the reference

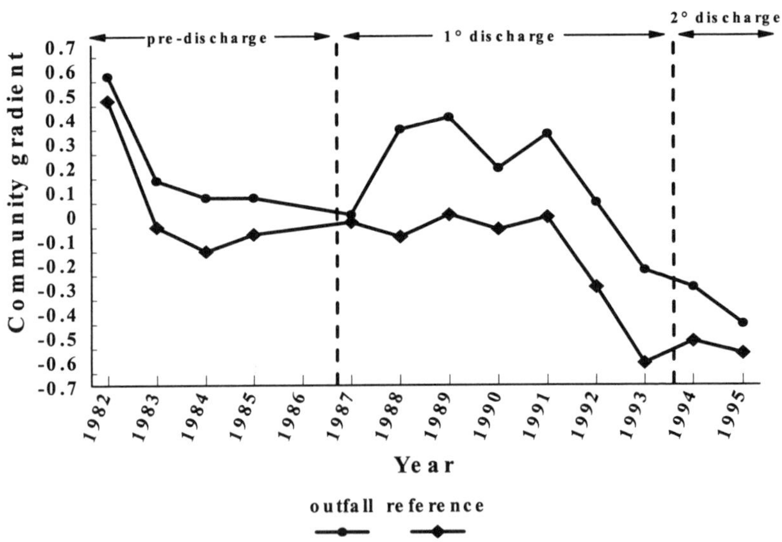

Figure 4.
Station 1 (outfall) and station 6 (reference) plotted along the second ordination axis. This gradient captures a significant divergence in community structure during the period of primary effluent discharge.

and outfall stations are similar prior to the onset of effluent discharge. Shortly after the onset of primary treated effluent discharge, there is divergence in the communities at the impact and reference stations along this gradient, which continues until after the change to secondary treatment. This would appear to be a classic demonstration of an outfall effect during the period of primary treatment with recovery after the onset of secondary treatment.

Biologically, separation of the impact and reference sites is due to higher total abundances of Mollusca and Crustacea at the outfall, and several individual species that had significantly ($p < 0.05$) higher abundances at the outfall or reference site (Table 1). Multiple regression analysis revealed that effluent mass loading of solids correlates with the observed separation of the impact stations from the reference station during the period of primary treatment. Suspended solids in the effluent averaged 5974 Kg/day during primary treatment. During secondary treatment the average fell to 1227 Kg/day (Figure 5). This correlation is suggestive of enrichment, however, none of the species in Table 1 with higher abundances at the outfall are indicative of enrichment, pollution, or other environmental degradation.

Taxon	Outfall	Reference
Bathycopea daltonae (C)	X	
Modiolus rectus (M)	X	
Hesionella mccullochae (P)	X	
Podarkeopsis glabra (P)	X	
Cooperella subdiaphana (M)		X
Glycera convoluta (P)		X
Typosyllis nr. *farallonensis* (P)		X
Aedicira sp. A (P)		X
Protothaca spp. (M)		X
Magelona spp. (P)		X
Euclymeninae (P)		X

Table 1.
Taxa with distributions significantly higher ($p < 0.05$) at either the outfall (Station 1) or reference (Station 6) during the period of primary treated effluent discharge. C = crustacean, M = mollusk, P = polychaete

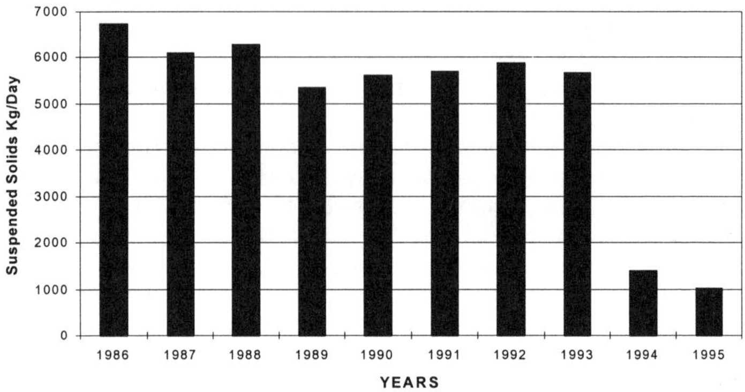

Figure 5.
Average suspended solids in Oceanside Water Pollution Control Plant effluent. Conversion from primary to secondary treatment occurred in September 1993.

Sample Design

The interpretation of an outfall effect from these data is confounded by outflow of San Francisco Bay-Delta water through the Golden Gate. Our analysis relies on a single reference station south of the Bay and the outfall. The Bay-Delta system drains 40% of the land area of California (Conomos 1979) and nearly half of the state's total runoff (SFEP 1993). The Delta Outflow Index, calculated on a daily basis by the California Department of Water Resources (DWR 1997), provides an estimate of these flows (90% of the Bay's water enters through the Delta [Conomos 1979]). During the study period, 1982 to 1995, the highest Delta outflow was 385 billion gallons per day (BGD) on February 20, 1986 and the lowest outflow was 12.9 MGD on September 1, 1989. The average daily outflow during the five year drought from 1987 to 1992 was 3.9 BGD, so that even during dry years the outflow is substantial. Thus, an alternative explanation, that the observed pattern is due to suspended particulates in the Bay-Delta outflow that affected the closer outfall stations more than the southern reference station, cannot be eliminated.

A new sampling scheme (Figure 1) designed to address some of these concerns has been incorporated into the City's NPDES (national pollution discharge elimination system) permit for the Oceanside Water Pollution Control Plant. The challenge in designing the new scheme was to do so while maintaining a similar level of effort. It is one thing to design a sampling program given unlimited equipment, money, and manpower, but quite another when given practical limits to those resources. The new design achieves this balance by adding more stations, while reducing the sample frequency to once per year and reducing the number of station replicates. Samples will be collected in the fall when biological diversity is highest and the sediments are least disturbed. The scheme is a stratified-random design with areas designated as zone of initial dilution, nearfield, farfield, and a southern reference area. Two areas to the north were added to help identify effects of San Francisco Bay-Delta outflow on the study area. Stations with historical data will be retained and additional stations added randomly through the application of an EMAP grid by the U.S. Environmental Protection Agency (Overton, *et al.* 1990, White, *et al.* 1992). An important aspect of the new design is that it will be evaluated and modified as necessary. In addition to added perspective, the new scheme is more regional in nature providing information of value to the manager(s) of the Gulf of the Farallones, Cordell Bank, and Monterey Bay National Marine Sanctuaries, as well as complimenting studies inside San Francisco Bay.

Acknowledgments

Few long term environmental monitoring programs can be accomplished without the hard work and dedication of many people. To all the field crew and laboratory support personnel we offer our sincere appreciation. Dale Davis, Maggie Dutch, Dale Hopkins, Caroline Karp, Kathy Langan-Cranford, Susan McCormick, Patricia McGregor, Cheryl Niemi, Brian Sak, Marilyn Smith, John Trone, and two of us (AMN and MGK) have been responsible at various times for invertebrate taxonomy. Jim Salerno, Mike Carlin, Mariana Flemmings, Kathy Langan-Cranford, and Patricia McGregor have been responsible at various times for the sediment analyses. We are grateful to the following people who contributed materially to the completion of this paper: Brian Sak, Susan McCormick, Kathy Langan-Cranford, Laura Targgart, Sylvia Rostami, Patricia McGregor, Pat Conroy, Marilene Harvey, and Dena Sherick.

References

Bernstein, B.B. and J. Zalinski 1983. An optimum sampling design and power tests for environmental biologists. Journal of Environmental Management 16:35-43.

Bureau of Water Pollution Control (BWPC) 1984. Ocean outfall monitoring program 1982-83 annual report. City and County of San Francisco, Department of Public Works, submitted to U.S.

EPA, Region 9 and S.F. Bay Regional Water Quality Control Board.

Bureau of Water Pollution Control (BWPC) 1985. Ocean outfall monitoring program 1983-84 annual report. City and County of San Francisco, Department of Public Works, submitted to U.S. EPA, Region 9 and S.F. Bay Regional Water Quality Control Board.

Bureau of Water Pollution Control (BWPC) 1988. Ocean outfall monitoring program 1987 annual report. City and County of San Francisco, Department of Public Works, submitted to U.S. EPA, Region 9 and S.F. Bay Regional Water Quality Control Board.

Bureau of Water Pollution Control (BWPC) 1989. Ocean outfall monitoring program 1988 annual report. City and County of San Francisco, Department of Public Works, submitted to U.S. EPA, Region 9 and S.F. Bay Regional Water Quality Control Board.

Bureau of Water Pollution Control (BWPC) 1990. Ocean outfall monitoring program 1989 annual report. City and County of San Francisco, Department of Public Works, submitted to U.S. EPA, Region 9 and S.F. Bay Regional Water Quality Control Board.

Bureau of Water Pollution Control (BWPC) 1992a. Ocean outfall monitoring program 1990 annual report. City and County of San Francisco, Department of Public Works, submitted to U.S. EPA, Region 9 and S.F. Bay Regional Water Quality Control Board.

Bureau of Water Pollution Control (BWPC) 1992b. Ocean outfall monitoring program 1991 annual report. City and County of San Francisco, Department of Public Works, submitted to U.S. EPA, Region 9 and S.F. Bay Regional Water Quality Control Board.

Bureau of Water Pollution Control (BWPC) 1993. Ocean outfall monitoring program 1992 annual report. City and County of San Francisco, Department of Public Works, submitted to U.S. EPA, Region 9 and S.F. Bay Regional Water Quality Control Board.

Bureau of Water Pollution Control (BWPC) 1994. Ocean outfall monitoring program 1993 annual report. City and County of San Francisco, Department of Public Works, submitted to U.S. EPA, Region 9 and S.F. Bay Regional Water Quality Control Board.

Bureau of Water Pollution Control (BWPC) 1995. Ocean outfall monitoring program 1994 annual report. City and County of San Francisco, Department of Public Works, submitted to U.S. EPA, Region 9 and S.F. Bay Regional Water Quality Control Board.

Bureau of Water Pollution Control (BWPC) in prep. Ocean outfall monitoring program 1995 annual report. City and County of San Francisco, Public Utilities Commission.

BWPC see Bureau of Water Pollution Control

California Department of Water Resources (DWR) 1997. "Interagency Ecological Program for the Sacramento-San Joaquin Estuary: Dayflow." [http://www.iep.water.ca.gov/dayflow.html]

CH2M-Hill 1989. Wastefield transport and bacteriological compliance studies of the San Francisco ocean outfall. Report prepared for the City and County of San Francisco and submitted to U.S. EPA, Region 9 and S.F. Bay Regional Water Quality Control Board.

Conomos, T.J. 1979. Properties and circulation of San Francisco Bay waters. pp. 47-84 *in:* T.J. Conomos, Ed. San Francisco Bay: The urbanized estuary. American Association for the Advancement of Science, Pacific Division; California Academy of Sciences, San Francisco, CA; 493 pp.

Diener, D.R. and S.C. Fuller 1995. Infaunal patterns in the vicinity of a small coastal wastewater outfall and lack of infaunal community response to secondary treatment. Bulletin of the Southern California Academy of Sciences 94(1):5-20.

DWR see California Department of Water Resources

Hannan, C.A., L.W. Hulberg, K.M. Mawn, M.G. Kellogg, and J.W. Nybakken 1979. A study to develop standard procedures for life-history analyses of benthic invertebrates for biological monitoring in marine and estuarine environments. California State Water Resources Control Board, Publication No. 63; 398 pp.

McCormick, S.J. 1984. The life history of *Owenia collaris* Hartman in the Gulf of the Farallones, California. M.A. Thesis, San Francisco State University; 101 pp.

National Oceanographic and Atmospheric Administration (NOAA) 1997. "An El Nino Theme Page: Accessing Distributed information related to El Nino." [http://www.pmel.noaa.gov/toga-tao/el-nino/]

NOAA see National Oceanographic and Atmospheric Administration

Overton, W.S., D. White and D.L. Stevens, Jr. 1990. Design report for EMAP environmental monitoring and assessment program. U.S. Environmental Protection Agency, EPA/600/3-91/053; 41 pp.

Pielou, E.C. 1966. Species diversity and pattern diversity in the study of ecological succession. Journal of Theoretical Biology 10: 370-383.

Pool, R.W. 1974. An introduction to quantitative ecology. McGraw-Hill, Inc., New York, NY. 532 pp.

San Francisco Estuary Project (SFEP) 1993. San Francisco Bay-Delta Estuary. Information Brochure, San Francisco Estuary Project, P.O. Box 2050, Oakland, CA 94604; 4 pp.

SFEP see San Francisco Estuary Project

Shannon, C.E. and W. Weaver 1949. The mathematical theory of communication. University of Illinois Press, Urbana.

Silberstein, M. 1987. Feeding ecology of the polychaete worm, *Glycinde polygnatha* Hartman 1950, an infaunal predator, with notes on life history. M.A. Thesis, San Jose State University; 56 pp.

Smith, R.W., B.B. Bernstein, and R.L. Cimberg 1988. Community-environmental relationships in the benthos: Applications of multivariate techniques. pp. 247-326 *in:* Soule, D.F. and G.S. Kleppel, Eds. Marine organisms as indicators. Springer-Verlag, New York, NY.

Stewart-Oaten, A. and W.W. Murdoch 1986. Environmental impact assessment: "pseudoreplication" in time? Ecology 67(4):929-940.

White, D., A.J. Kimerling, and W.S. Overton 1992. Cartographic and geometric components of a global sampling design for environmental monitoring. Cartography and Geographic Information Systems 19(1):5-22.

CHAPTER 215

Contaminant Levels in Fish Tissue from San Francisco Bay.

[1]Ross Clark
[2]Karen Taberski

Abstract

Edible fish species were collected from thirteen locations throughout San Francisco Bay, during the spring of 1994, for determination of contaminant levels in muscle tissue. Species collected included white croaker, surfperch, leopard and brown smoothhound sharks, striped bass, white sturgeon and halibut. Sixty six composite tissue samples were analyzed for the presence of PAHs, PCBs, pesticides, trace elements and dioxin/furans. The U.S. EPA approach to assessing chemical contaminant data for fish tissue consumption was used for identifying the primary chemicals of concern. Six chemicals or chemical groups were found to exceed screening levels established using the U.S. EPA approach. PCBs (as total Aroclors) exceeded the screening level of 3 ng/g in all sixty six tissue samples, with the greatest concentrations (638 ng/g) found near San Francisco's industrial areas. Mercury was elevated (>0.14 mg/g) in forty of the sixty-six samples with the highest levels (1.26 mg/g) occurring in shark muscle tissues. Concentrations of the organochlorine pesticides dieldrin, total chlordane and total DDT exceeded screening levels in a number of samples. Dioxin/Furans (as TEQ's) were elevated (above 0.15 pg/g) in 16 of the 19 samples analyzed. Fish with high lipid content (croaker and surfperch) in their muscle tissue generally exhibited higher contaminant levels while fish with low lipid levels (halibut and shark) exhibited lower organic contaminant levels. Tissue samples taken from North Bay stations most often exhibited high levels of chemical contamination. The California Office of Health Hazard Assessment is currently evaluating the results of this study and has issued an interim Health Advisory concerning the human consumption of fish tissue from San Francisco Bay.

[1]Moss Landing Marine Laboratories/California Dept. of Fish and Game, Moss Landing, CA
[2]San Francisco Bay Regional Water Quality Control Board, Oakland, CA.

CHAPTER 216

A Ten-Year Summary of Ocean Monitoring on the San Pedro Shelf, California

George Robertson[1], Don Maurer and Irwin Haydock

Abstract

Presented are ten years of selected results from the County Sanitation Districts of Orange County, California (Districts) ocean monitoring program. Spatial and temporal trends of sediment contaminant concentrations and benthic infauna showed that discharge effects have decreased over time and the magnitude of the observed changes has declined. Moreover, contaminant–biological relationships were only weakly associated and were mainly coincidental, not causal.

Introduction

From 1985 to 1996 the County Sanitation Districts of Orange County, California (Districts) discharged an average of 8.8×10^8 L (232 million gallons) of treated wastewater per day onto the San Pedro Shelf. The effluent is discharged ~8.2 km (5.1 mi) offshore, at a depth of ~60 m, through a ~1.8 km (6000 ft) 500-port outfall diffuser (Figure 1). The Districts' 1985 301(h) modified National Pollution Discharge Elimination System (NPDES) discharge permit requires a multifaceted ocean monitoring program (i.e., water quality, sediment geochemistry and infauna, epibenthic fish and macroinvertebrates) to study the effects of this discharge. Complete results of 10 years of monitoring are presented in CSDOC (1996). This account presents selected results from that more detailed and extensive report.

---

[1]County Sanitation Districts of Orange County; P.O. Box 8127, Fountain Valley, CA 91728-8127

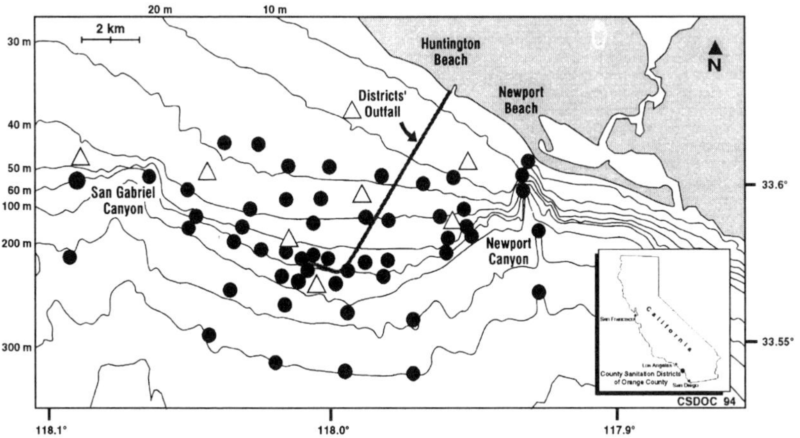

Figure 1. Districts' outfall, OMP study area and benthic (●) and trawl (Δ) station locations.

Discharged treated wastewater represent a source of particulate organic carbon metals, and organic contaminants to the marine environment that may then accumulate in bottom sediments (Young, et al., 1978; Gossett, et al, 1982). The deposition and accumulation of effluent particles are expected to result in changes in the benthic environment. Benthic infauna can exhibit a wide range of responses to gradients in organic and contaminant enrichment (Pearson and Rosenberg, 1978; Gray, 1979, 1989, Maurer et al., 1993). The 10–year program provided a unique opportunity to evaluate the effect of the Districts discharge to the benthic environment.

Sediment Geochemistry

Sediments in the Districts' study area are predominantly sand (70–80%) with smaller fractions of silt plus clay. The median grain size ranged from 50–150 microns (Figure 2). Little temporal change occurred in grain size at stations along the 60 m depth contour. Long-term reductions in mass emissions of metals from the outfall were reflected in sediment concentrations (Table 1). However, the absence of significant changes in the sediment metal concentrations of the Newport Canyon, a functional sediment trap, showed that the magnitude of regional, multiple, non-point sources of contaminants did not vary whereas those attributed to the outfall dramatically reduced. Even with the reduction in most of the sediment trace metals, there still was a signature or imprint of a few elevated trace metals associated with the Districts' discharge.

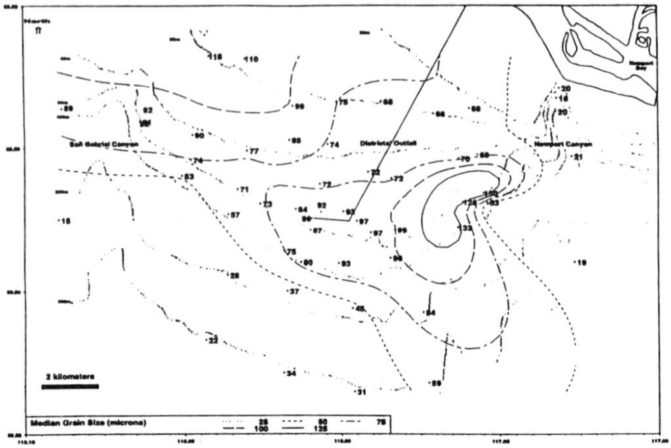

Figure 2. Spatial distribution of median grain size (microns) averaged by station from 1985-1995 (n=10).

Table 1. Mean (n=156) annual sediment metal concentrations (µg/g dry wt.) at the Districts' 60-m stations; n=133 for 1994-95.

	85-86	86-87	87-88	88-89	89-90	90-91	91-92	92-93	93-94	94-95
Arsenic	8.9	4.4	4.6	4.1	3.4	3.2	3.4	3.6	3.0	3.3
Beryllium	0.4	0.3	0.3	0.3	0.3	0.3	0.3	0.3	0.3	0.3
Cadmium	0.9	0.7	0.6	0.6	0.5	0.6	0.5	0.4	0.8	0.5
Chromium	23.0	21.0	21.0	20.0	20.0	20.0	19.0	19.0	21.0	20.0
Copper	21.0	18.0	18.0	16.0	15.0	16.0	14.0	14.0	16.0	15.0
Mercury	0.1	0.1	0.1	0.1	0.1	0.1	0.1	0.1	0.0	0.1
Nickel	10.0	8.5	8.3	8.4	8.6	8.4	8.6	8.9	10.0	11.0
Lead	12.0	11.0	11.0	10.0	10.0	9.1	8.4	9.0	7.9	9.2
Silver	0.5	0.5	0.5	0.5	0.4	0.4	0.3	0.4	0.6	0.3
Zinc	59.0	53.0	51.0	49.0	49.0	46.0	50.0	43.0	50.0	50.0

Cadmium (Figure 3),silver and copper provide an imprint. While sediment cadmium and copper concentrations declined over the ten-year period, silver remained constant (Table 1). These metal concentrations were comparable with other regional studies (Table 2).

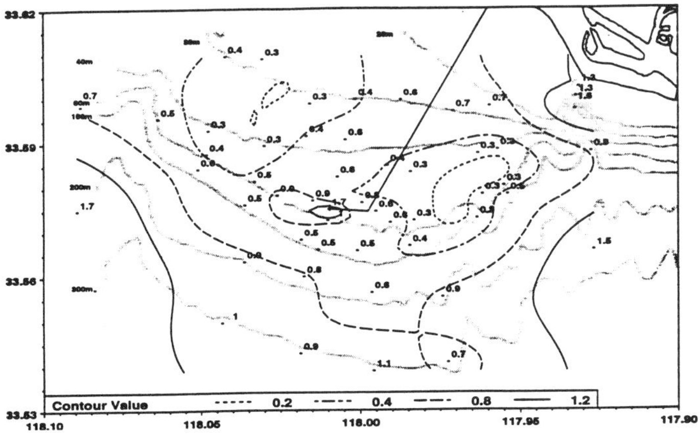

Figure 3. Spatial pattern of cadmium concentrations (mg/kg) during July 1993.

Table 2. Median (Ranges) of sediment metal concentrations (µg/g) during the Southern California Bight Pilot Project (July/August 1994).

	San Pedro Shelf (n=21)	Santa Monica Bay (n=19)	Santa Barbara Channel (n=32)
Aluminum	4,680 (1,730–9,360)	5,050 (1,540–6,940)	4,060 (1,300–6,150)
Silver	0.05 (0.05–0.28)	0.34 (0.05–0.98)	0.05 (0.05–0.11)
Arsenic	1.6 (1.0–5.0)	5.5 (1.8–6.7)	2.4 (1.3–8.2)
Cadmium	0.10 (0.05–0.28)	0.28 (0.07–1.07)	0.16 (0.07–0.4)
Chromium	12 (5.3–18)	21 (5.9–80)	12 (4.7–18)
Copper	6.0 (2.0–14)	9.0 (2.0–49)	5.0 (2.0–11)
Iron	8,930 (4,150–13,500)	11,300 (3,810–20,800)	11,150 (3,550–39,400)
Mercury	0.03 (0.02–0.11)	0.10 (0.04–1.17)	0.03 (0.02–0.09)
Nickel	6.0 (3.0–9.0)	12 (5.0–19)	10 (4.0–18)
Lead	4.0 (0.9–7.9)	9.0 (4.3–18.8)	4.0 (1.6–7.1)
Zinc	25 (12–39)	33 (12–54)	28 (9.0–53)

With most of organic contaminants below detection limits temporal and spatial trends for sediment organic contaminants were not as discernable as with trace metals. However, Figure 4 illustrates the general declines during 1985-1995 and occasional spikes for total DDT (tDDT) and total PCBs (tPCB). The report concluded that tDDT in sediments near the outfall were comparable or lower than those at the reference site. In contrast, tPCB and tPAH were elevated near the outfall. The Newport Canyon also had elevated sediment concentrations of organic sediments which was consistent with the picture of the canyon serving as a regional sediment-contaminant trap. Sediment bioassay tests conducted on selected organisms (brittlestar, worm, microcrustacean), maintained in sediment from the outfall and reference stations, showed no significant differences in mortality or growth. The outfall sediment with its complete complement of chemical constituents (copper, cadmium, silver, et al.; tDDT, tPCB, tPAH, et al.) did not produce toxic responses under test conditions.

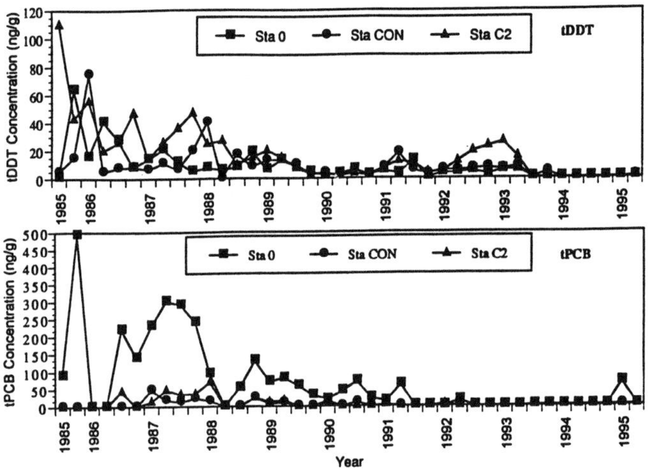

Figure 4. Concentrations of tDDT and TPCB in sediments at the outfall (0), reference (CON) and Newport Canyon (C2) stations from 1985-1995.

Infaunal Communities

In terms of species diversity the Southern California Bight (SCB) is one of the richest in the world containing at least 5000 invertebrate species (Dailey, et al., 1993). Because of their value in monitoring studies the ten-year program placed great emphasis on infaunal communities collecting ~1.7 million organisms, consisting of ~1150 species with a mean abundance and number of species per sample of $5776/m^2$ and 100 species, respectively. While these numbers were impressive, however, what is critical is to find out whether the change caused by the outfall has produced a degraded community.

Based on the Pearson-Rosenberg Model (PRM) (Pearson and Rosenberg, 1978) the number of species, abundance and biomass of infaunal species did not indicate a stressed community (Table 3). Abundance was enhanced near the outfall (Figure 5) was probably due to bioenhancement of a few species, exploiting the organic load from the outfall; this bioenhancement occurred without a decrease in the number of species found at the outfall (Figure 5). This is an important finding since, typically, under stressed conditions, a few, pollution tolerant species characteristically dominate and exclude other species, eventually producing low density.

Table 3. Results of ANOVA-SNK analyses for selected infaunal community measures and species along the nominal 60 m depth for quarterly surveys (n=48), July 1994 Through April 1995.

Parameter	**Probability**	**Mean Value (untransformed) SNK Station Separation**											
Num. of Species[1] (spp./sample)	0.101	114	113	111	110	109	107	106	104	100	92.3	90.7	90.4
		37	5	1	4	0	CON	9	12	ZB2	13	ZB	C
Total Abundance[2] (no./sample)	<0.001	1089	863	853	712	653	554	527	479	426	393	390	329
		0	1	ZB2	4	ZB	5	9	12	CON	37	13	C
Total Biomass[2] (kg/sample)	<0.001	14.4	11.6	18.0	8.1	8.3	7.9	7.8	6.8	5.9	6.0	4.7	3.5
		0	ZB	ZB2	13	1	5	C	4	9	CON	37	12
Infaunal Trophic[2] Index	<0.001	87.2	87.1	82.8	81.2	79.7	76.8	75.9	73.3	70.9	68.2	65.3	60.3
		C	13	CON	37	5	12	9	4	1	ZB	ZB2	0
Amphiodia spp.[2]	<0.001	75.2	65.9	41.5	36.3	8.2	6.1	3.5	3.6	1.7	2.6	1.1	0.6
		13	C	CON	5	37	9	1	4	ZB2	12	ZB	0
Capitella capitata[2]	<0.001	80.5	49.6	17.6	4.3	3.3	1.6	1.5	0.8	1.2	0.3	0.10	0.10
		0	ZB2	ZB	4	1	9	5	12	37	CON	C	13

Transformations: 1 = none, 2 = rank
Stations: CON=upcoast reference; 37=downcoast reference; 0 and 4=outfall; ZB and ZB2=ZID boundry; 1, 5, 13 and C=upcoast transition; 9 and 12=downcoast transition

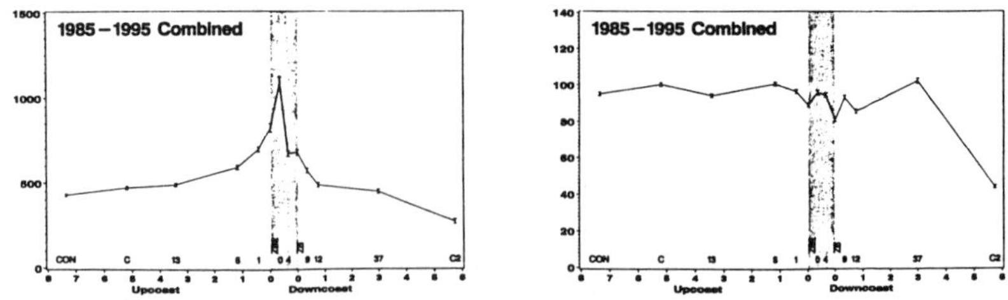

Abundance Number of Species

Figure 5. Mean infaunal abundance per sample (a) and number of species from 1985-1995).

During the ten-year period abundance of a dominant pollution–tolerant, species, *Parvilucina tenusiculpta*, declined and a pollution–sensitive, brittlestar species, *Amphiodia urtica*, increased (Figure 6). Major variability in population dynamics was statistically shown to be associated with water depth, the uniqueness of the canyons, sediment type, and annual and seasonal cycles. Outfall effects accounted for only 6.3% of the infaunal distribution, due primarily to the bioenhancement previously mentioned.

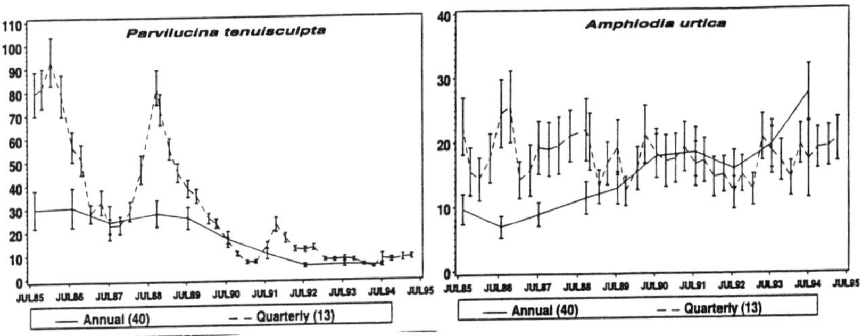

Figure 6. Temporal trends for *Parvilucina tenuisculpta* and *Amphiodia urtica*.

The outfall represents a hard–bottom structure in a natural soft–bottom regime. The outfall as a reef structure altered sediment type which affected infaunal organisms and by providing a substratum, affected larger, attached epifaunal organisms. With extensive development of epifauna on the reef-like outfall there was increased opportunity for predatory fish and invertebrates. Increased predation may be partially responsible for low populations of the *A. urtica* near the outfall. From 1985 through 1995 the spatial extent of outfall effects has reduced and stabilized which corroborates the position that the outfall area was not degraded but supported a balanced indigenous population in critically (legal) defined areas.

Fish and Macroinvertebrate Communities

About 109 species of fish were sampled in the monitoring program. Similar to benthic infauna, water depth was the most important factor affecting the composition of demersal fish and macroinvertebrates rather than any outfall effects. Despite the year class phenomenon of several fish species (yellowchin sculpin and Pacific sanddab), there was a decline in abundance encompassing almost 25 years (Figure 7). The decline was associated with regional water temperature warming (~2°C) from 1951 to the present (Roemmich and McGowan, 1995). Since independent surveys have documented declines throughout the SCB, increased water temperature, over fishing, reduced nutrient loads from dammed rivers are considered causative agents rather than the effects of a single outfall.

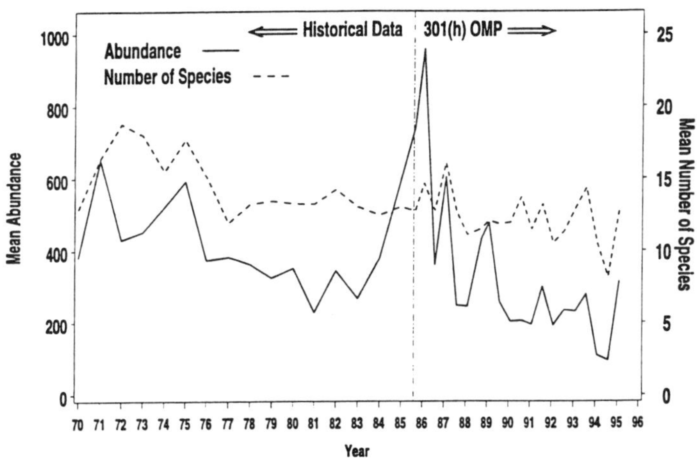

Figure 7. Mean abundance and number of fish species per trawl from 1970-1995.

Macroinvertebrate community structure was variable among stations. One macroinvertebrate (seastar) thrived near the outfall whereas abundances of two macroinvertebrates (ridgeback prawn and sea cucumber) were low there (Table 4). The latter species have been commercially harvested. Spatial variability in abundances was greatly influenced by the white sea urchin which occurred in thousands per fishing trawl. The distribution and abundance of this species controlled all analyses and comparisons of macroinvertebrates.

Fish species composition and abundance were not negatively affected by the outfall. The regional decline in SCB fisheries coincides with reduced harvests reported from other United States and international fisheries' regions. Major changes in regional, national and global fishing practices with revised watershed management should be made to realize sustainable harvests or to begin to resuscitate some fisheries regions.

Table 4. Results of ANOVA-SNK analysis on abundances of selected species for 55-m trawl stations, July 1991–January 1994.

Parameter	Probability		Mean Value (untransformed) SNK Station Separation				
Pacific sanddab (*Citharichthys sordidus*) 2	0.009	summer	298 TC	221 T11	145 TB	88 T1	79 T3
Speckled sanddab (Citharichthys stigmaeus)1	<0.001	summer	42 T1	3.4 TB	1.3 T11	0.3 TC	0 T3
Longfin sanddab (*Citharichthys xanthostigma*)2	<0.001	summer	53 T3	51 T1	48 TB	29 TC	16 T11
Bigmouth sole (*Hyppoglossina stomata*)1	<0.001	summer	7.3 T11	4.9 T1	6.5 T3	3.1 TB	1.0 TC
Hornyhead turbot (*Pleuronichthys verticalis*)1	<0.001	summer	6.2 T3	4.0 T1	0.9 TB	0.7 T11	0.3 TC
Sand star (*Astropecten verrilli*) 1	<0.001	summer	179 T1	18 TB	31 T11	7.7 TC	4.3 T3
White urchin (*Lytechinus pictus*) 1	<0.001	summer	2730 T3	46 T1	183 TB	1.7 T11	1.4 TC
California sea cucumber (*Parastichopus californicus*)1	0.114	summer	22 TB	16 T3	12 T11	12 TC	7.1 T1
Ridgeback prawn (*Sicyonia ingentis*)2	<0.001	summer	45 TB	37 TC	19 T11	13 T1	11 T3

Transformations: 1 = rank; 2 = log(x+1)
Stations: T11=upcoast reference; T1=outfall; T3=downcoast reference; TC and TB = upcoast transition

Discussion

Sediment geochemistry results showed that concentrations of total organic carbon (TOC), trace metals and organic contaminants in sediments were primarily related to depth and sediment grain size. Grain size was unaffected by the outfall while sulfides, TOC and some trace metals (e.g., Ag, Cd, Cu) were higher near the outfall. The magnitudes of the elevated contaminant concentrations at the outfall were small and generally within a factor of ten above background concentrations found in the SCB. Sediment contaminant concentrations have declined since this monitoring program began 1985, in particularly at sample sites near the outfall. Finally, based upon laboratory bioassay analyses of bottom sediments collected at the outfall that showed no toxicity, present levels of sediment contaminants are below levels expected to affect bottom-dwelling organisms adversely.

The magnitude and types of alterations to benthic infauna attributed to the outfall were mixed. Among those seen were: (a) increased abundances and biomass of infaunal organisms near the outfall; (b) species richness at the outfall that was comparable to reference conditions; and (c) decreases in diversity and dominance indices (i.e., H′, J′, Infaunal Trophic Index [ITI]). In addition, selected indicator species (i.e., *Amphiodia urtica*, *Capitella capitata*) showed abundance gradients attributed to the outfall. Since 1985, the observed changes have improved and become more subtle. It was concluded that a balanced indigenous population exists at the outfall.

The 10-year summary report (CSDOC, 1996) described achievements and limitations of an ocean monitoring program that, initially, focused on the effects of the outfall, but eventually encompassed a suite of natural and anthropogenically induced agents operating outside the area and interacting with local processes and events. The report made it clearer which effects were associated with the outfall and which effects were associated with regional inputs. As such the report was able to urge the recognition of regional practices that need to be assessed for a more cogent understanding of the health of coastal waters. Finally, while the summary report brought the 10–year results to a culmination, it did not represent the final chapter in the Districts' monitoring efforts. Rather, the 10 years of monitoring have formed the foundation of a new proposed monitoring program designed to provide better management information for protection of the natural resources on the San Pedro Shelf.

References

County Sanitation Districts of Orange County (CSDOC). 1996. Marine Monitoring Annual Report: Science Report — Ten Year Synthesis, 1985-1995. 377 pages.

Dailey, M.D., J.W. Anderson, D.J. Reish, and D.S. Gorsline. 1993. The California Bight: Background and setting. In: *Ecology of the Southern California Bight*. eds. M.D. Dailey, D.J. Reish, and J.W. Anderson. 1-18. Berkeley, CA. University of California Press.

Gossett, R.W., D.A. Brown, and D.R. Young. 1982. Predicting the bioaccumulation and toxicity of organic compounds. In: *SCCWRP Biennial Report, 1981-1982*, ed. W. Bascom, 149-156. Long Beach, CA: SCCWRP.

Gray, J.S. 1989. Effects of environmental stress on species rich assemblages. *Biol. J. Linnean. Soc.* 37:19-32.

Maurer, D., G. Robertson, and T. Gerlinger. 1993. San Pedro Shelf, California: Testing the Pearson–Rosenberg Model (PRM). *Mar. Environ. Res.* 35:303-321.

Pearson, T.H., and R. Rosenberg. 1978. Macrobenthic succession in relation to organic enrichment and pollution of the marine environment. *Oceanogr. Mar. Biol. Ann. Rev.* 16:229-311.

Roemmich, D., and J. McGowan. 1995. Climatic warming and the decline of zooplankton in the California Current. *Science* 267:1324-1326.

Young, D.R., T.K. Jan, and T.C. Heesen. 1978. Cycling of trace metal and chlorinated hydrocarbon wastes in the Southern California Bight. In: *Estuarine Interactions*, ed. M.L. Wiley, 481-496. New York, NY: Academic Press.

CHAPTER 217

City and County of San Francisco Combined Sewer System: Reduction of Overflows and High Bacteria Counts

Sonya F. Foree[1] and Patricia McGregor[2]

Abstract

The City and County of San Francisco has a combined sewer system (CSS) which collects all sanitary, industrial and stormwater flow. A concern of the public is how these systems operate during periods of heavy rainfall when all flow cannot be contained. San Francisco developed a master construction plan to control the number of CSS overflows to the shoreline. Recent historical data (1986 to 1997) associated with the City and County of San Francisco CSS including rainfall, number of overflows, shoreline bacteria counts and beach postings are summarized. Results since the completion of the CSS control structures indicate a reduction in the number of overflows to the shoreline, and an associated reduction in the elevated bacteria counts and days of beach posting for public health concerns.

Introduction

The City and County of San Francisco recently completed a large portion of the Master Plan for Wastewater Management developed in 1971. This includes completion of Westside, Northshore and Southeast Core sections and Oceanside Water Pollution Control Plant. The effect completion of the west side has had on the number of overflows, the number of high bacteria counts and the number of beach postings before and after these sections were completed is summarized.

Background

There are two main types of sewer systems: combined sewer systems and separate sewer systems. A CSS is defined as "a wastewater collection system owned by a state or municipality which conveys sanitary wastewater (domestic, commercial, and industrial wastewater) and storm water through a single pipe to a POTW (publicly

---

[1] Water Quality Chemist, City and County of San Francisco, Public Utilities Commission
[2] Marine Biologist, City and County of San Francisco, Public Utilities Commission

owned treatment works) treatment plant" (EPA 1994). Conventional systems separate industrial and sanitary flow from stormwater runoff and have distinct pipelines to retain this separation. Combined sewer systems are able to treat all wastewater and runoff during dry weather; this can include sewage, pesticides, herbicides, urban and agricultural runoff, automobile leaks, leacheate from toxic waste, air pollution and industrial waste. In effect, a CSS treats all non-point source pollution during dry weather. A CSS also treats a large percentage of stormwater flows during wet weather that would otherwise go untreated into receiving waters. In separate systems, this flow is often diverted along with wastewater into local receiving water sources, such as lakes, rivers, oceans, etc. Combined sewer systems contain all dry weather flow and as much wet weather flow as possible. However, when stormwater flow is excessive the capacity of the system is exceeded and overflows occur to the receiving waters.

The City and County of San Francisco (the City) has a combined sewer system which consists of a single network of sewer pipes, catch basins and transport systems. Historically, whenever it rained as little as 0.02 inches per hour wet weather induced overflows exceeded treatment plant capacity (Eng.-Sci., 1967). This would result in 80% of the combined wastewater and rainwater flow to be discharged untreated into the Pacific Ocean and San Francisco Bay. Frequently, beaches were posted during the first rain in October or November and remained posted until May or April, the end of the rainy season.

To address public health concerns along the City's shoreline and alleviate the general dissatisfaction of the public to combined sewer system overflows (CSO), the City developed a Master Plan for Wastewater Management in 1971 to control CSOs and improve wastewater treatment. The plan was subsequently updated in 1974 to address the Federal Water Pollution Act of 1972 which required wastewater treatment plants be upgraded to provide full secondary treatment.

The City rejected the possibility of a separate sewer system in the early 1970s because of the high costs and disruption associated with replumbing existing sewer lines (necessitating the excavation of every street in the city) and the failure of a separate system to treat pollution caused by stormwater runoff (EPA 1997). Instead, the City chose a combination of increased treatment capacity and large volume storage as the most cost-effective and environmentally-protective means of controlling the City's CSO problems.

Construction and upgrading of facilities began in 1977 under the direction of the Department of Public Works. Prior to 1982, collected wastewater was treated at three primary treatment plants having a combined wet weather capacity of 225 million gallons per day (MGD). On the east side of the City, construction of new treatment facilities in the Bayside Core System included the Southeast Water Pollution Control Plant (WPCP) and the Channel and Northshore Pump Stations, completed in 1982 (Figure 1). Other Southeast area improvements such as the Mariposa, Yosemite and Sunnydale facilities were completed in 1991 and improvements to Islais Creek will be completed In 1997. On the west side of the City, the Westside Core System encompasses the

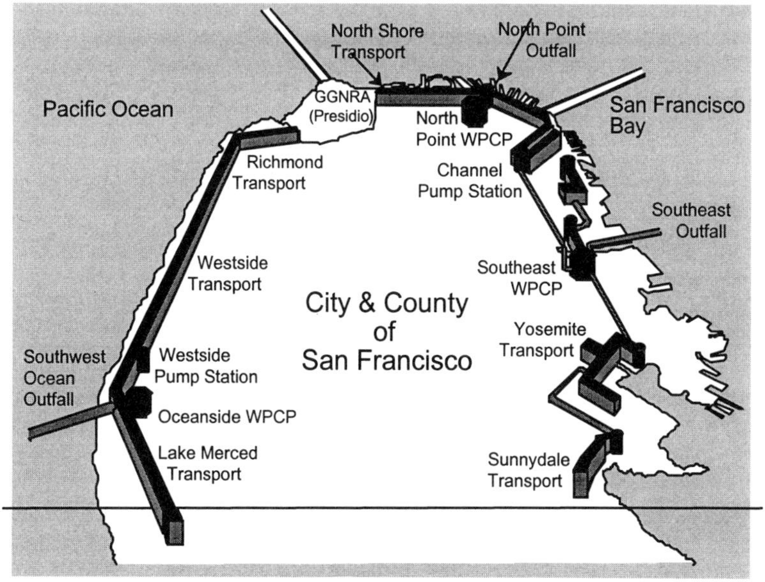

Figure 1. CCSF Master Plan for Wastewater Management.

Westside Pump Station, the Westside Transport and the Southwest Ocean Outfall; all became operational in 1986 (BWPC1988,1990,1994,1995). The Lake Merced Transport was completed in 1993 and the Richmond Transport in 1997. The total wet weather capacity for the City today is 585 MGD.

One of the major components of the plan included upgrading the level of effluent treatment from primary to secondary. Three water pollution control plants facilitate wastewater storage and treatment around the City currently: the Southeast WPCP, a secondary plant on the east side with an outfall structure in the Bay and a capacity of 150 MGD secondary treatment and an additional 100 MGD primary treatment; the North Point WPCP; a 150 MGD primary wet weather facility to the north with an outfall structure in the Bay; and on the west side replacement of an old primary plant with a new secondary treatment plant, the Oceanside

WPCP. The capacity of Oceanside WPCP is 43 MGD secondary treatment and an additional 22 MGD primary treatment for a total wet weather capacity of 65 MGD.

A second component of the plan included the construction of transport and storage collection systems to prevent overflows into receiving waters. These transport/storage collection systems are massive underground boxes which lie ten to fifteen feet beneath the roadways; they extend approximately 2.5 miles on the west side and 3.0 miles on the east side. The three transport/storage collection systems most recently constructed on the west side are the Westside Transport completed in 1986 with a storage capacity of 47.6 MGD, the Lake Merced Transport completed in 1993 with a storage capacity of 10 MGD and the Richmond Transport completed in 1997 with a storage capacity of 9.7 MGD (BWPC 1994).

The Westside Transport is structurally different from the Bayside in that there are two underground boxes which run side by side; sewage flows into the east box, and during dry weather is transported to the treatment plant. In a storm when the treatment capacity of Oceanside WPCP is exceeded (greater than 65 MGD), the east box fills to capacity, solids are settled and floatables are baffled before flow decants to the west box and is pumped directly out the Southwest Ocean Outfall. When the capacity of the west box is exceeded a treated overflow occurs through an overflow structure. This treatment is approximately equivalent to primary treatment. There are seven overflow structures along the west side of the City. The five main overflow sites on the west side are Lincoln and Vicente along Ocean beach, Mile Rock near Lands End, Lake Merced at the southeastern end of the City and Sea Cliff II at the northwestern edge of the City.

Storage capacity becomes crucial when rainfall occurs because capacity determines the point at which an overflow will occur. The process to determine the long term average number of overflows allowed each year by the State and the Regional Water Quality Control Board was based on the beneficial uses and sensitivity of the receiving waters, storage capacity, and cost. Overflows are prohibited unless all storage, pumping and treatment plant facilities are operating at maximum potential (EPA 1997).

A third component of the plan was to provide for the eventual discharge of all wastewater into the Pacific Ocean via an offshore outfall. The Southwest Ocean Outfall was built in 1986 and extends 3.75 miles offshore to a depth of 80 feet. Diffusers distribute the flow along a length of approximately 3000 feet. Oceanside WPCP and the Westside Transport System connect to this pipe.

The wet weather treatment design goal for the west side of the City is to achieve a combination of secondary and primary treatment for 50% of all flow treated at the Oceanside WPCP, approximate primary treatment for 37% of flow in the Westside Transport discharged via the ocean outfall, and approximate primary treatment for 13% of all flow discharged along the shoreline (EPA 1997). This design estimates the annual treated discharge time along the shoreline at 0.4%.

Methods

Currently bacteriological samples are taken three times per week at nine locations along the westside, chosen for their proximity to overflow sites. Shoreline bacteriological samples have been taken for the past fifteen years at 45 sites with a frequency of 8 to 12 times per month. Samples were analyzed for total, fecal and enterococcus bacteria using either multiple tube fermentation or membrane filtration techniques according to Standard Methods, 17th. ed. Total coliform bacteria are used as an indicator species for recreational water contact standards. The California Ocean Plan water contact total coliform standard requires that "not more than 20% of the samples in any 30 day period at any given station exceed 1000 colony forming units/100 mL" (State of California 1993). San Francisco beaches are posted whenever an overflow occurs and/or whenever the California Ocean Plan standard is exceeded. Beaches remain posted until coliform counts return to ambient conditions.

Data from overflow structures is summarized from the Wet Weather Discharge Summary monthly report submitted to the Regional Water Quality Control Board.

A rain gauge at the Oceanside WPCP is an official weather station in San Francisco for the Department of Commerce, Weather Bureau. This data is used to determine rainfall on the west side. On the Bayside of the City new and improved rain gauges will be installed to better estimate when overflows will occur.

Telemetry in the transport system indicates when an overflow is occurring; this uses a capacitance probe which sits in a trough inside the storage/transport boxes at the overflow structures. The probe sends an electrical charge when liquid (or flow) reaches the level of the sensor. When flow in the box reaches this overflow level the probe sends a signal (or alarm) to the treatment plant control room (Lanny Beach, CCSF, pers. comm.). The telemetry indicates immediately the beginning of an overflow and the end of an overflow enabling personnel to post the beach in a timely period.

Results

Overflows vs. Rainfall

January is one the rainiest months in San Francisco and hence a useful month for comparing overflows. Figure 2 shows data from the Lake

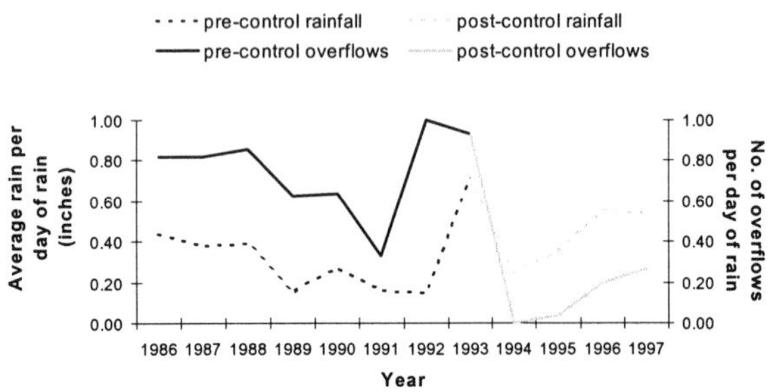

FIGURE 2. Lake Merced Overflows vs. Rainfall

Merced overflow structure for the month of January from 1986 to 1997 summarizing the rainfall and the number of overflows. The number of overflows that occur is always rain dependent. Taking that into consideration Figure 2 is standardized for the number of rain days, making drought and excessive rain periods more comparable. As might be expected the number of overflows follow rainfall rather closely for the time period before the CSO controls were put in. After the Lake Merced Transport section was completed in 1993, the number of overflows continued to be rain dependent but the number of overflows decreased while the average rainfall per number of days of rain was still fairly normal.

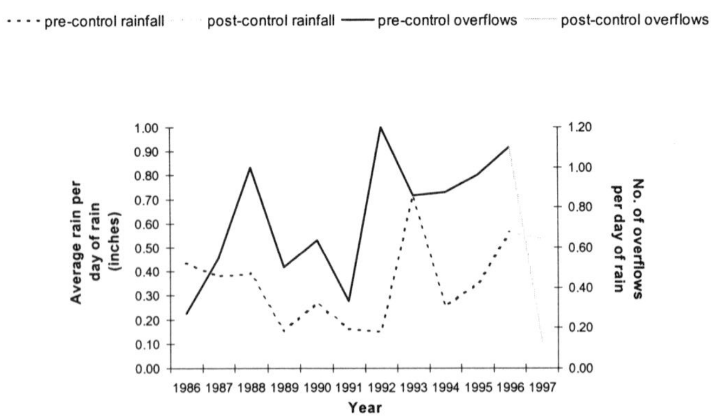

FIGURE 3. Sea Cliff II Overflows vs. Rainfall

Figure 3 shows the same reduction at the Sea Cliff II overflow structure which was completed in January 1997. The completed CSO structure provides a substantial reduction in the number of overflows per days of rain, even with a fairly wet January in 1997.

The transport and storage structures along the west side of the City have led to a reduction in the total number of overflows each rainy season. With the completion of the Master Plan, there is a long term average of 8 overflows per year along the west side. Figure 4 shows the overall reduction in overflows during the rainy season over the last eight years. The number of overflows occurring from each of the transport structures on the west side compared with total yearly rainfall is shown: Vicente from the Westside Transport (completed in 1986), Lake Merced Transport (completed in 1993) and Sea Cliff II from the Richmond Transport (completed in 1997). In each case, the number of overflows has been substantially reduced once the controls were implemented. The pattern displayed in 1997 with a fairly low number of overflows is expected to be the long term trend, showing a minimal amount of increases or decreases which are dependent on rainfall.

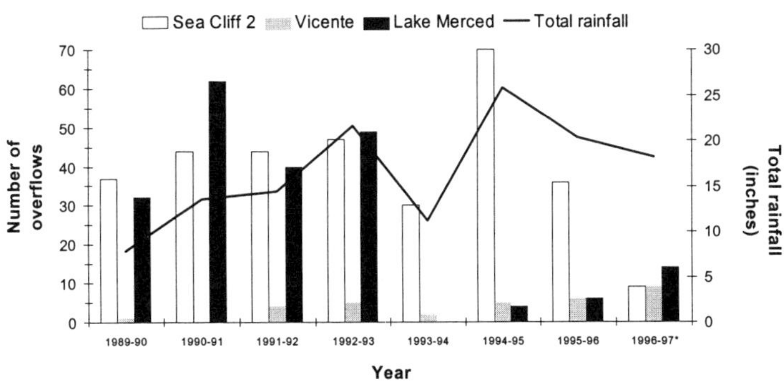

FIGURE 4. Wet Season November to April

Bacteria Counts vs. Overflows

Water quality samples show that bacteria counts are elevated at shoreline stations during and shortly after overflow events, but generally return to background levels within one or two tidal cycles following the end of the overflow. Sampling indicates that bacteria counts are higher at those sites in close proximity to the overflow structures. Figure 5 shows data from the Lake Merced Transport structure for the month of January from 1986 to 1997 with the percent coliform counts greater than 1000 along with rainfall. Completion of the Lake Merced Transport in 1993 reduced the number of overflows caused by rainfall and resulted in a decrease in the number of times water contact limits were exceeded.

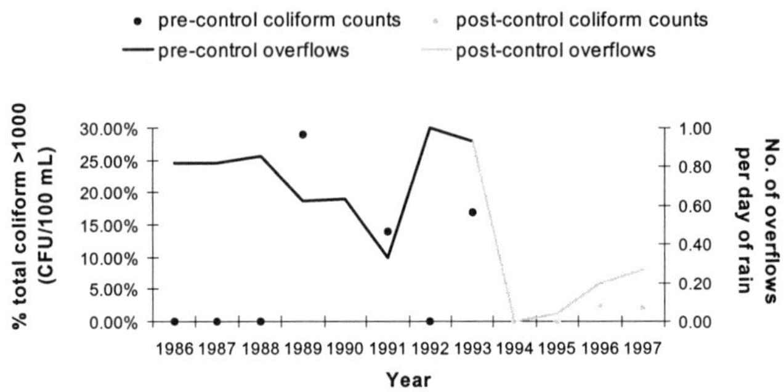

FIGURE 5. Lake Merced January Overflows and High Bacteria Counts

Figure 6 shows the decrease in the number of high bacteria counts

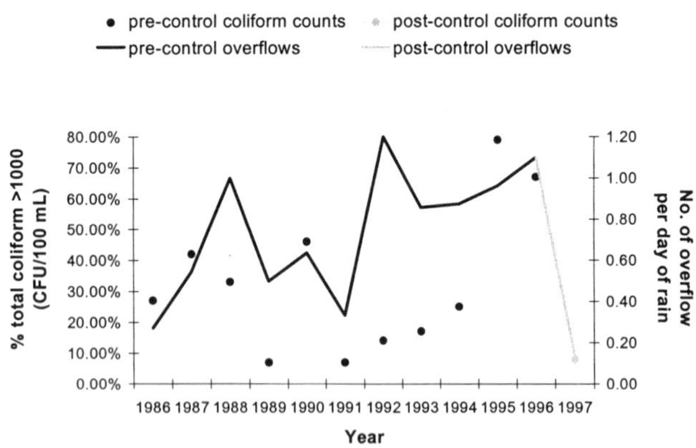

FIGURE 6. Sea Cliff II January Overflows and High Bacteria Counts

associated with a drop in overflows when the Sea Cliff II structure, part of the Richmond Transport system, was completed in 1997.

A reduction in both the number of overflows and the number of high bacteria counts is evident in Figure 7; here the two graphs compare January 1996 data before control structures were completed to January

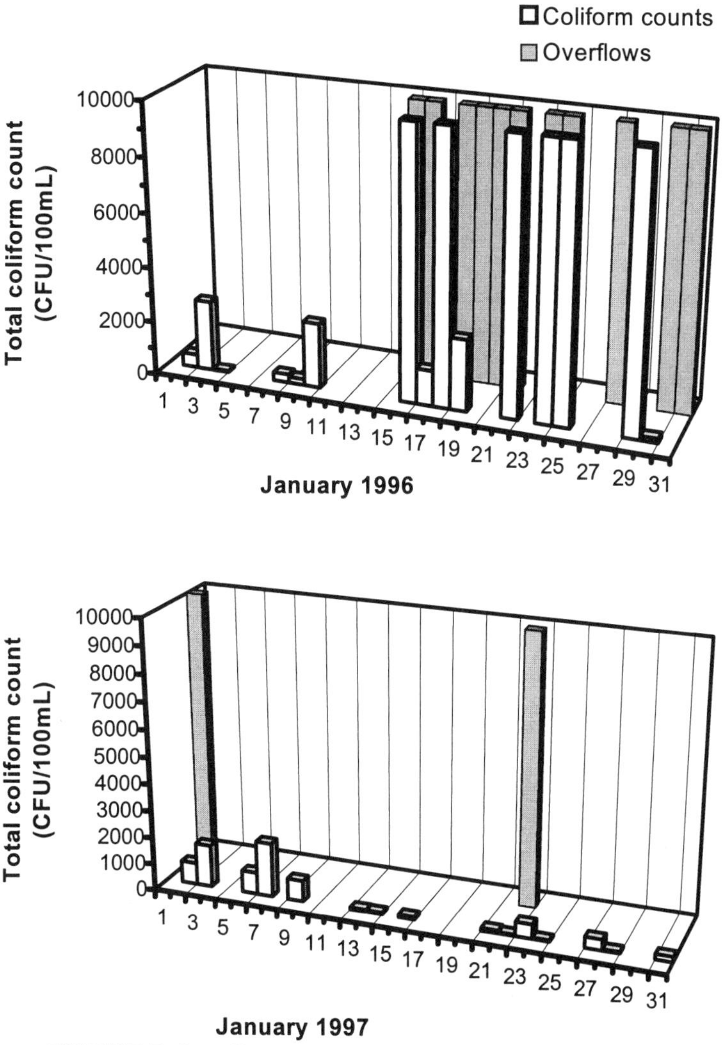

FIGURE 7. Overflows and Bacteria Counts

1997 data after controls were in place. The bacteria counts are in the foreground and the overflow event is the background column. Prior to the control structures being completed in 1997, (1) there were a significant number of overflows and (2) bacteria levels exceeded recreational water contact standards. After controls were completed, there are fewer overflows and substantially lower bacteria counts.

A reduction in the number of high bacteria counts also reduces the number of times the beaches are posted to warn users that waters do not meet recreational water contact standards. Figure 8 shows the cumulative number of postings during the last five years.

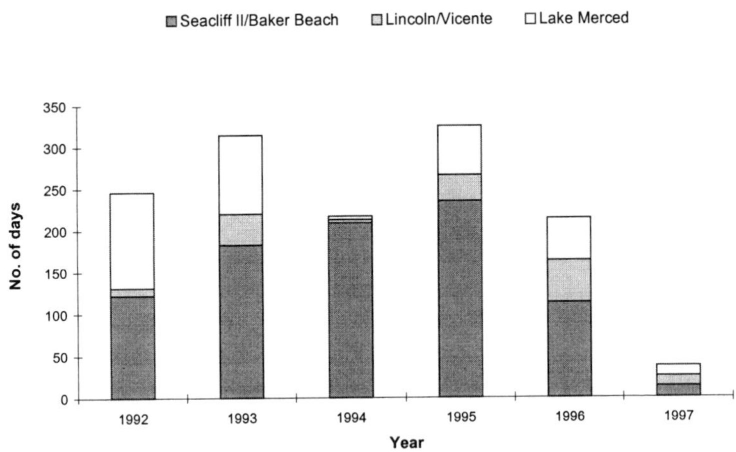

FIGURE 8. Beach Postings Total Number of Days

Conclusion

To achieve the current overflow number at a long term average of eight per year (for the west side) has cost approximately 1.5 billion dollars for the entire system. When the City went forward with the Master Plan in 1971, justification for the construction was based on "the most cost effective and environmentally protective strategy for addressing" the City's CSO needs (NPDES 1997). But to further reduce the number of overflows by increasing capacity is unlikely in the near future because of cost and the difficulty of increasing storage capacity.

The U.S. Environmental Protection Agency (EPA) recently performed a Best Professional Judgment analysis for the City's CSS discharge from the Southwest Ocean Outfall and concluded that (1) the system currently in place provides effluent reduction at a cost in excess of that which would be required by Best Practical Technology (BPT)/Best Conventional Pollutant Control Technology (BCT) /Best Available Technology Economically Available (BAT) ; and (2) no additional treatment facilities can be justified on a BPT/BCT/BAT cost basis (EPA 1997).

Recently there has been much discussion on findings of stormwater runoff consisting of a variety of pollutants, especially from highly populated areas. The EPA has issued stormwater runoff permits to detect and control this runoff to local receiving waters. In CSS the responsibility of locating and controlling point-source pollutants, before these pollutants are discharged to the sewer system, lies much more heavily on city or county governments to avoid any disruption these pollutants might cause to the treatment process. In separate systems, "stormwater runoff spills" could go undetected because the effect of the spill occurs in large bodies of receiving waters which are not monitored daily. As part of the City's pretreatment requirements the Bureau of Environmental Regulation and Management administers a program which includes enforcement inspections, pretreatment monitoring, and permitting of Significant Industrial Users.

Because the combined system can encompass and virtually surround the city and the small size of San Francisco, the City has been able to successfully complete the sewer system, treat the large majority of runoff and decrease a previously high number of overflows. Ocean beach in San Francisco has been attracting more and surfers who consider the surf conditions desirable, especially in the winter months. The city has initiated a Recreational Water Quality Hotline whereby ocean users can call to find out the water quality each day.

To make the City's CSS as efficient as possible the City is working on further reducing the hydraulic loading of the CSS and reducing the decant rate to the outfall.

References

American Public Health Association, 1987. Standard Methods for the Examination of Water and Wastewater, 17th ed. APHA.

Bureau of Water Pollution Control (BWPC) 1988. Ocean outfall monitoring program 1987 annual report. City and County of San Francisco, Department of Public Works, submitted to U.S. EPA, Region 9 and S.F. Bay Regional Water Quality Control Board.

Bureau of Water Pollution Control (BWPC) 1990. Ocean outfall monitoring program 1989 annual report. City and County of San Francisco, Department of Public Works, submitted to U.S. EPA, Region 9 and S.F. Bay Regional Water Quality Control Board.

Bureau of Water Pollution Control (BWPC) 1994. Ocean outfall monitoring program 1993 annual report. City and County of San Francisco, Department of Public Works, submitted to U.S. EPA, Region 9 and S.F. Bay Regional Water Quality Control Board.

Bureau of Water Pollution Control (BWPC) 1995. Ocean outfall monitoring program 1994 annual report. City and County of San Francisco, Department of Public Works, submitted to U.S. EPA, Region 9 and S.F. Bay Regional Water Quality Control Board.

City and County of San Francisco, 1994 NPDES Permit Program, Attachment 1

City and County of San Francisco, Department of Public Works. September 1991. Op. Cit.

Engineering-Science, Inc.1967. Characterization and Treatment of Combined Sewer Overflows. City and County of San Francisco, Department of Public Works, submitted to Federal Water Pollution Control Association.

EPA Region IX and the California Regional Water Quality Control Board. Fact Sheet, NPDES No. CA 0037681.

EPA Section IA EPA. 1994 CSO Control Policy

State of California, 1993. California Ocean Plan.

CHAPTER 218

Stepping Beyond Public/Private Partnerships for Consensus on Environmental Policy

Janet J. Dillon, Ph.D., Esq.[1]

Abstract

A new successful approach to environmental protection is achieved through public participation. While this approach encourages environmental goals it allows flexibility for economic viability to industry. A strong infrastructure of process is essential to reach any type of consensus. This new process involves the coordination of the participants and stakeholders, both public and private, by a trusted, truly neutral third-party mediator. An effective public/private process can lead to effective partnerships and creative solutions that balance community concerns to the benefit of all.

Introduction

Dialogue and consensus are playing a critical role in the move away from prescriptive command and control regulatory measures and yielding a new approach to environmental protection. This new approach, rooted in public/private partnerships, is prompting private experimentation with technology and pollution prevention that achieves environmental goals while allowing industry the flexibility for economic viability. I have participated in many of these forums in water and air quality regulation, and I see a need for standing back and spending time considering the structure for these processes. Too often inattention to the structure for public/private partnering has led to aborted efforts or more limited success than might otherwise have been achieved.

Successful, productive partnerships are rooted in a strong infrastructure of process. Even before the formal public participation begins, stakeholders must be involved in developing the process whereby public and private entities will contribute

[1] Executive Director and Founder, Eco-Partners nonprofit foundation, 2275 Huntington Drive, #225, San Marino, CA 91108

their expertise and express their goals and concerns. For a legitimate process, this involves coordination by a truly neutral third-party who is trusted by all participants and potential stakeholders.

Over and over this point of a neutral third-party who is not a principal or employee or contractor of any of the participating public or private entities bears repeating. As I have watched groups around me in the last several years who are eager to tackle environmental issues by public participation, it is clear to me that this is the single most important lack of understanding of effective public process. Too often one or more of the larger participants will attempt to play the coordination role or the day-to-day facilitation role. One organization even articulated a desire to fulfill the coordination role by "tricking" participants into thinking it was neutral. Most often, this misapprehension of neutrality is formed in an attempt of the party to be helpful and listen to all sides. However, attempts to assume a neutral role by an interested party will not work.

Participants might profit from standing back and taking a close look at their ultimate needs. Granted many times those needs include resolution of issues with a broad base of participants. If they attempt to jump in and fill the void themselves, one of two outcomes may occur:

1. Their own individual viewpoints, needs, and goals may not be effectively communicated in the attempt to meet group needs: or
2. Their own needs are communicated, and in the process other participants may drop out because it is perceived that the process is "slanted" toward a particular outcome.

If sufficient attention is paid to a truly neutral process, participants may then concentrate on effectively getting their individual viewpoints, needs, and goals across without fear of "tainting" the process and jeopardizing its outcome.

Key factors that help ensure a legitimate process to achieve results include:

1. Identifying individuals and organizations that have been overlooked and participants who have resources to contribute;
2. Among the participants, using stakeholders and a facilitator who have technical expertise and knowledge of the specific subject, and knowledge of other participants;
3. Understanding the administrative and legal context of the issue(s) and the task at hand (for example, to rewrite an environmental regulation); and
4. Understanding the legal consequences of not reaching consensus.

Beyond these fundamentals and other protocol for public participation, information must be free-flowing and in clear language that is understood by technical and non-technical participants. Further, information must be complete, not anecdotal, and from every source possible.

An effective public/private process is what leads to effective partnerships and creative solutions that balance community concerns to the benefit of all.

Case in point: As a severe nonattainment area for federal air quality standards, the County of Ventura was mandated to comply with a Federal Implementation Plan (FIP). In response to this FIP, a coalition of community organizations was formed including the Ventura County Economic Development Association, the Council on Economic Vitality, Citizens to Preserve Ojai, and the Ventura County Air Pollution Control District, among other organizations.

Coordinating public and private input involved liaison with the US EPA and other regulatory agencies, industries and businesses, municipalities, special districts, environmental organizations and citizen groups. Comments and concerns were drafted into an alternative plan, endorsed by the community. The plan, "The Ventura County Alternative Federal Implementation Plan," was heralded as a national model of partnership among industry, regulators and the community and recently won approval by the federal Environmental Protection Agency.

I have a suggestion for applying the concept of a truly neutral consensus-building process to watershed management planning using the example of the Los Angeles Regional Water Quality Control Board Pilot Program on Calleguas Creek in Ventura County.

The Calleguas Creek watershed extends approximately 35 miles in a 343 square mile area of Ventura County from the City of Simi Valley through the cities of Moorpark, Thousand Oaks, and Camarillo to the Mugu Lagoon. Its flow varies seasonally and consists substantially of treated water from several municipal sewer plants, pumped groundwater, and urban and agricultural run-off. Calleguas Creek recharges several of the county's ground-water basins, and farmers use the creek water for crops and for return irrigation flows. When it empties into Mugu Lagoon it flows through sensitive habitat including several endangered species. Mugu Lagoon is one of the most important estuaries in Southern California.

The Los Angeles Regional Water Quality Control Board is synchronizing consideration of all its permits in the Calleguas Creek watershed and expects to reissue them all from a new template. In conjunction with permit renewals is a new pilot watershed management planning program on the Calleguas Creek watershed. This planning effort represents a new direction: coordination of land uses and agency decisions. The process is expected to take several years and requires collection of data on creek pollutants, which may include excess sedimentation, nitrogen compounds, pesticides and metals, and identification of the most environmentally beneficial and cost-effective practices to reduce pollution. Community-based watershed planning is recognized by the US EPA as the most effective way to control both point sources, such as wastewater treatment plants and industrial centers, and "nonpoint sources" coming from urban run-off and farms.

In Ventura County, more than 70 groups have been meeting for more than a year to discuss data collection and coordination. The planning committee includes representatives of cities and the county, water agencies, the Navy, environmentalists, developers, business people, farmers, private landowners, Native Americans, and individual residents. A vision statement and list of issues to be addressed have been drafted as a group effort.

The steering committee has wrestled with a structure for the planning process, staffing and funding. The biggest hurdles appear to be recognition of the importance of neutral coordination and willingness to contribute the level of funding required for professional coordination that will produce the level of results needed in a reasonable time frame. A similar effort for the river to the north of the Calleguas, spearheaded by the County, used only in-kind contributions from the interested public agencies and discrete funding for specific pieces of data collection. Because neutral coordinators knowledgeable of the participants as well as administrative and technical requirements was not available, the effort failed to galvanize the broad spectrum of support needed for such an effort to be maximally effective. Many participants lost interest before the planning effort came to fruition. Adequate funding and professional neutral staffing can shave years off the time required and lead to an amazingly fruitful effort that will not only have success far greater than dreamed by the local participants but stand as a model for tackling other local, regional and national environmental issues.

CHAPTER 219

Using Consensus Building Techniques to Implement Integrated Coastal Management

Eugenia Laychak

Abstract

Integrated Coastal Management (ICM) concepts call for achieving natural resource sustainability to safeguard the ability of future generations of people and other species to meet their needs. Putting this theory into practice can be challenging to project proponents, agency regulators, resource managers, affected communities and other interested parties. One of the biggest challenges is implementing a project in an environmentally sustainable manner while adhering to very real budget, political and time constraints and ensuring mutually satisfying benefits.

This paper compares similarities and differences between two consensus building processes used to manage coastal and bay-delta conflicts. One case study involves a focused conflict between a small commercial fishing fleet and a scientific research firm which was charged to conduct a study sponsored by the Department of Interior's Mineral Management Service. The other process is the much more complex CALFED Bay-Delta Program, headed by a consortium of 10 state and federal agencies which is trying to reach agreement with about 50 major private and public organizations and individuals. The goal of both projects is to achieve project success, including satisfying the interests and concerns of the stakeholders. The approaches used to achieve consensus are different, due to vast differences in complexity and roles of the stakeholders.

The author compares the two case study processes to conflict resolution models, tools, and principles often used to resolve public disputes. She describes the techniques used to address concerns raised by stakeholders and discuss resulting outcomes. Highlights of lessons learned include: 1) leaders must be identified, 2) clearly state the objectives of and limitations to the collaborative process, 3) partnerships must include commitments of time, effort, and resources, 4) stakeholders need a common vision, but they and the process must be flexible to accommodate unforeseen changes in circumstances, and 5) plan for implementation of the agreement.

CHAPTER 220

Cooperative Research Efforts as a Tool for Bridging Information Gaps And Resolving Conflicts: The COOGER Study

Frederick L. White[1]

Abstract

The Minerals Management Service (MMS), the State of California, and local governments are using a unique partnership to acquire data needed for planning and resolution of longstanding conflicts associated with oil and gas development in federal waters off California. This partnership is a useful model for groups that are working out disagreements that would benefit from a common information base that each party could use to make decisions. The partnership involves joint funding and oversight by key oil and gas industry and government stakeholders of a cooperative research project entitled *California Offshore Oil and Gas Energy Resources: a Joint Study of the Development Scenarios and Onshore Constraints in the Tri-county Area of San Luis Obispo, Santa Barbara, and Ventura.*

Introduction

This paper explores the use of cooperative research efforts and their role in conflict resolution. While the paper focuses on the *California Offshore Oil and Gas Energy Resources: a Joint Study of the Development Scenarios and Onshore Constraints in the Tri-county Area of San Luis Obispo, Santa Barbara, and Ventura* (COOGER Study) the general principles and philosophy behind the study are applicable to a wide range of environmental and social issues where information gaps and communication failures can be effectively addressed by developing common information and language.

---

[1]Social Scientist, Minerals Management Service, Pacific OCS Region, Office of Environmental Evaluation, Environmental Studies Section, 770 Paseo Camarillo, Camarillo, CA 93010

COOGER Description

COOGER is jointly funded by the Minerals Management Service and the oil and gas industry and conducted under the auspices of the MMS Environmental Studies Program. Dames & Moore is the primary contractor for the study. Unique to this study is the management structure of COOGER and the extent of public involvement, participation and consultation incorporated in the study. The purpose of COOGER is to develop a common information base on the relationship between offshore oil and gas lease development and various onshore constraints and resources. COOGER addresses only existing Federal and state leases in the California offshore environment. COOGER is not intended to be nor is it, a National Environmental Policy Act or California Environmental Quality Act document or process. COOGER is specifically designed to look not generally addressed by MMS in its NEPA documents.

Background

COOGER is the MMS response to many years of comments from the State of California and its agencies, local governments, and others on specific lease sales and Five Year Leasing Programs, for a more comprehensive look at the potential onshore effects from development of existing undeveloped leases. COOGER was conceived of by local and State government interests together with MMS in the MMS/Tri-County Forum, a cooperative problem-solving entity in the Pacific OCS Region. The Tri-County Forum provided the initial opportunity for the interested parties to express their vision of the study as well as the opportunity to actually frame the study and the management of the effort. The holders of Federal undeveloped leases were approached in 1992 by MMS to help frame the Study and to share in the funding and management of the study. In February 1995, MMS issued a $1.4 million contract to study the onshore constraints to the potential development of existing Federal and state leases offshore San Luis Obispo, Santa Barbara and Ventura Counties California to Dames & Moore and associated subcontractors. A brief chronology of the COOGER Study is in Table 1.

Table 1
COOGER Chronology

November 1992 - COOGER Concept Introduced
January 1993 - Ad Hoc Committee Formed to Write Scope for Study, Evolves into Technical Management Team
September 1993 - Bidders List Prepared
October 1993 - Statement of Work Developed

May 1994 - Request for Proposal Issued
August 1994 - Proposals Received and Reviewed
February 1995 - Dames & Moore Selected as Prime Contractor
October 1995 - Public Workshops Held on Scope & Purpose of Study
October 1996 - Environmental and Business Representatives Added to the Steering Committee

Management Structure

A unique tri-parte management structure is used to guide the COOGER study and the prime contractor Dames & Moore. The management structure is composed of a Steering Committee, Technical Management Team, and MMS (see Figure 1). The Steering Committee provides policy direction for the study. The Technical Management Team provides routine guidance in implementing the policy, and MMS carries out the day to day management of the study. The Steering Committee and Technical Management Team strive for a consensus in all decisions. A consensus for the Steering Committee and Technical Management Team is defined as a decision everyone can live with, not necessarily everyone's preferred choice. When a consensus cannot be reached, MMS has been designated, the impasse decision maker.

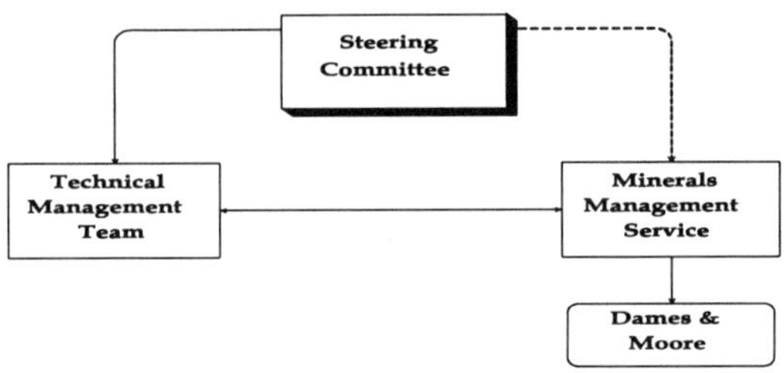

Figure 1 COOGER Management Structure

Membership of the Steering Committee and Technical Management Team are shown in the Tables 2 and 3. The Steering Committee had a significant membership revision when representatives from the environmental and business communities from each county were added to the Steering Committee. The Steering Committee meetings are public meetings and are advertised through Press Releases, paid display ads, public service announcements, mailings, and on the Pacific OCS Region homepage.

Table 2
Steering Committee Membership

Representatives of San Luis Obispo, Santa Barbara, and Ventura County
Representatives of the California Coastal and State Lands Commissions, and Department of Natural Resources
All Lessee and Operators of the undeveloped Federal Leases
Representative of the Western States Petroleum Association
Representatives of the Non-oil and gas business communities of San Luis Obispo, Santa Barbara, and Ventura County
Representatives of the environmental communities of San Luis Obispo, Santa Barbara, and Ventura County
MMS

Table 3
Technical Management Team

Representatives of San Luis Obispo, Santa Barbara, and Ventura County
Representatives of the California Coastal and State Lands Commissions, and Department of Natural Resources
Representatives of CalResources LLC, Nuevo Energy Company, and Samedan
Representative of the Western States Petroleum Association
MMS

Minerals Management Service manages the day to day work of the contract through a Contracting Officers Technical Representative. The Contracting Officers Technical Representative operates within the confines of the appropriate Federal laws and regulations. Any directions or changes which may affect cost must be given by the MMS Contracting Officer.

The interaction of the three parts of the COOGER Management Structure allows MMS to give guidance to Dames & Moore the prime contractor for the study. While Federal Law requires MMS to give direct guidance and direction to Dames & Moore, MMS gives direction and guidance based on the instructions of the Steering Committee and Technical Management Team.

Public Involvement

The COOGER public involvement process is one of the most comprehensive efforts that MMS has used in an environmental study. The public involvement process is built on a three pronged approach, 1) Public Workshops at key points in the Study, 2) regular mailings about progress on COOGER, and 3) open and advertised Steering Committee Meetings. The intent of the COOGER public involvement process is to reach out beyond the usual constituents of special interest groups and to build a process that allows MMS to directly involve the largest number of individuals in the study.

Why use Cooperative Research Efforts

Why should we use cooperative research efforts? Cooperative research efforts have the possibility to improve communication by creating a working environment where issues are raised, discussed, and ideally openly addressed. Trust can be built, because improved communication and more frequent contact will allow different groups to see each other as human beings, not merely as someone we are in conflict with. Resource allocation is improved because time can be spent working on the real conflicts and concerns and less time responding to artificial or trial balloon concerns. Additionally synergies can be accomplished by pooling funds and efforts.

Decision making is invariably improved with more complete and diverse information. Diversity of opinion allows decision makers the opportunity to see the full spectrum of issues before making a decision Once decisions are made, they generally are more effective because all parties will have had an opportunity to help shape the decision. Cooperative research efforts achieve synergy through share decision making, improved resource allocation, increased buy-in, and more effective decision making. The costs of cooperative research efforts are the perceived loss of control, and an expanded time frame for decision making.

When to Use Cooperative Research Efforts

Cooperative research efforts should be used in most cases involving complex, diverse, or controversial decisions. The use of cooperative research efforts can generally be viewed as tool for the effective resolution of conflicts, misunderstandings, distrust, or other commendations failures. In general, cooperative research efforts are best applied to problems and issues, where conflict exists, when filling information gaps can help resolve conflicts, where positions have become entrenched, when communication is ineffective or non-existent, or where multiple parties are required to implement the decision. The key to deciding to use a cooperative research effort is the willingness of all appropriate parties to participate in good faith.

Structuring a Cooperative Research Effort

When developing a cooperative research effort, it is essential as a first step to insure that the appropriate parties have been identified and the research problem is clearly identified. The more accurately described and defined is the research problem the more timely can effective results be expected. Fundamental to structuring cooperative research efforts are a clear and concise descriptions of the roles and responsibilities of those involved in the research effort. No promises or agreements should be made which are outside the ability of those involved in the research effort to fulfill.

When the appropriate parties have been identified and invited to participate in a good faith effort to work on the cooperative research effort sufficient startup time for the research effort must be allocated to build trust and cooperation in the process. Trust and cooperation are best built through a series of successes, beginning with small successes too larger and more complex successes. The most effective way to build trust and cooperation is through the use of a consensus decision making process, where every participant can see their hand in shaping the result. A consensus does not mean that everyone gets their way, but rather, that we have arrived at the other best answer. At a minimum, the consensus arrived should be one where all parties can accept and support, even if it is not their preference, when necessary a mechanism for resolving may be developed.

After the appropriate parties have been invited to participate, the scope of the project identified, a decision on sharing the financial burden of the effort is necessary. The sharing of the financial burden does not necessarily require equal or even proportionate sharing, but it does mean that all parties should have some direct or indirect financial interest in the research effort. Table 4 outlines the basic structure for cooperative research efforts.

Table 4
Basic Structure Requirements
Cooperative Research Efforts

Clear and Concise Problem Statement
Identification and early Invitation to Participate of all appropriate parties
Clear and Concise Descriptions of Roles and Responsibilities
Sufficient Time to Build Trust and Cooperation
Good Faith Effort
Consensus Decision Making
Shared Financial Burden

Conclusions

Through our efforts in COOGER we have developed a set of Principles for Cooperative Research Efforts, these Principles are itemized in Table 5. The Principles for Cooperative Research Efforts can be characterized by one over arching consideration or philosophy that is flexibility. To have an effective and productive cooperative research effort participants must recognize these projects take more time to complete. Most of the literature addressing consensus decision making stress the need for more time early in the decision process. The payoff for more time early in the research or decision process is more effective and timely implementation of decisions. To insure an effective cooperative research effort management must allow employees to fully represent the agency/company and support the employee in his/her efforts at representing the agency/company.

Table 5
Principles of Cooperative Research Efforts

Everything Takes More Time
Empowered Employees
Supportive Management
Early Invitation for Participation
Early and Small Successes Build Trust
Follow the Spirit of the Law
Do Not Say We "Cannot Do That"
Accommodate Even the Most Offbeat Request
Everything Takes More Time

Flexibility must be demonstrated in a variety of ways including, following the spirit of the law and adopting an attitude that accommodates the most offbeat and unusual request or idea. The accommodation of unusual or offbeat request may be the route to arriving at the other best answer, one to which all can agree. Even though something has not been done before, it is not a reason not to try it in the cooperative research effort. At the core of all cooperative research efforts must be the concept that everything is on the table and all discussions are open and honest. To insure the most effective result of the research effort those embarking on cooperative research efforts must remember that everything takes more time.

CHAPTER 221

Mediating Conflicts Between Fish and Oil: A Prototype for Joint Problem Solving of Space-Use and Resource Conflicts

Alana S. Knaster, Craig Fusaro, John Richards[1]

Abstract

On April 9, 1983, negotiations between the offshore oil industry and commercial fishermen were convened to address a number of problems that resulted from the increase in oil and gas permitting activities along the Central California coast. Both industries were experiencing crisis after crisis in their attempts to co-exist in a constrained work space and without any formal mechanism for communication and coordination. Representatives of the two industries invited the State Lands Commission, Coastal Commission and Minerals Management Service to participate in these initial talks. At the behest of the Sea Grant Program Marine Advisor, who had been working closely with the industries and agencies for several years, a team of mediators was retained to formalize the dispute resolution process.

This meeting set a course for future problem-solving that remains operational to the present and that has already been replicated as an approach for resolving a variety of issues among disputants in the region. Although the specific solutions utilized by the parties may not be appropriate in other contexts, the lessons learned regarding the effective elements of a dispute resolution system and consensus-building process are applicable to a wide variety of ecosystem and resource management conflicts.

---

[1]Alana S. Knaster, The Mediation Institute: 4508 Park Cordero, Calabasas, CA 91302 Craig Fusaro, Joint Oil Fisheries Liaison Officer: 121 Gray Avenue, Santa Barbara, CA 93101, John Richards, UC Santa Barbara, 5266A Hollister Avenue, Santa Barbara, CA, 93111

was built in 100 feet of water off the town of Summerland in 1958, leasing and offshore development continued within the three mile state boundary. Oil industry conflicts with the fishing industry increased again when leasing of federal offshore lands began in 1963 and offshore oil exploration and development accelerated.

Santa Barbara Channel trawl fisherman actively sought relief in the public arena in 1966 to solve gear-loss problems attributed to wellhead obstructions and seafloor debris related to offshore oil activities. Testimony from the fishing industry was heard and acknowledged by the California State Assembly's Subcommittee on Marine Resources in February of 1966, but there was no resolution Although there was some relief for the fishing industry when restrictions were placed on oil activity following the 1969 oil spill in Santa Barbara, these same conflicts, primarily involving trawlers and seafloor obstructions, emerged in the late 1970's and intensified when federal offshore leasing was renewed with vigor in the early 1980's.

The fishing industry also lobbied extensively against the use of explosives in seismic exploration. Their claims that the explosives killed significant numbers of fish went unheeded until the late 70's when pressure from a number of interest groups resulted in a ban on explosives and the conversion in California to air and water acoustic devices.

Sea Grant Extension Involvement. Early efforts to resolve the seafloor obstruction problem began in 1976 when they newly established Sea Grant Extension Program (SGEP) was approached by the trawlers and asked for help in facilitating communication between the fishermen, the offshore oil industry, and the regulatory agencies. Initially, the U.S. Geological Survey and the California State Lands Commission were the agencies responsible for seafloor issues. The a Marine Advisor was able to serve as a neutral intervener/facilitator in finding solutions to several specific problems, mainly involving the Santa Barbara Channel trawl fleet and two or three established offshore oil companies. The major issues were related to abandoned wellheads, seafloor debris, snags on submerged pipelines, and a particularly difficult problem of anchor-scars left by pipe-laying barge in a productive shrimp ground. The latter effectively eliminated trawl activity near the pipeline for over three years.

Though not all of these conflicts were totally resolved, lines of communication were established during meetings held at the Marine Advisor's offices; , each industries operations and problems were beginning to be clarified and understood by the participating parties, and the regulatory agencies, and offshore industries were actively cooperating in find solutions.

Rise of Multiple Conflicts Between the Industries. The initiation of federal leasing in the early 1980's brought intensive geophysical exploration again as well as exploratory drilling and supply boat activity. Now, besides the small group of trawlers originally involved, trappers, gill-gilnetters, hook and line fishermen, and trollers began to have space-use conflicts with seismic vessels. Also, a previously unaffected fishing fleet of various gear-types, operating in the Santa Maria Basin off San Luis Obispo County, began encountering the geophysical vessels. The oil and gas industry and their geophysical contractors had foreseen this problem and had requested SGEP and regulatory agency assistance in developing a broader communication network with the fishing industry in 1981. Following several months of dialogue and two meetings facilitated by the Marine Advisor, a prototype seismic notification system was set up using industry and agency volunteers. However, offshore interactions and gear losses were increasing and better communications alone could not relieve the escalating problems. Additional local, state, and federal agencies, such as the California Coastal Commission, Department of Fish and Game, U.S. Coast Guard, and the National Marine Fisheries Service also intervened in a number of the offshore conflicts, but were unsuccessful in grappling with the ever expanding scope of problems.

The story of the Gus-D vs ARCO , a drill rig scheduled for placement in a traditional shrimp/prawn trawl area in the Santa Maria Basin, was indicative of the need for a coordinated effort to facilitate communications and solve potential problems. Fishermen brought a complaint before the Coastal Commission claiming that the project would seriously disrupt fishing in the area. The Commission delayed the project for several weeks while it conferred with the fishermen, company and several sets of agency staff. The company lost thousands of dollars by putting the drill-rig on stand-by status; in addition, the Commission process required extensive amounts of staff time from other agencies plus travel costs. This drain of public and private money and time, which might have been avoided if a mechanism for expeditious industry to industry communication had been in place. suggested that an alternative course of action was necessary.

Confrontation versus negotiation. Bad weather and silting in of the harbor only increased the frustrations of the fishing community in the winter of 1982-83. Various factions within the fishing industry in Santa Barbara began to discuss both political and legal action as the only approach that would get the attention of the oil industry and the government. Plans were set in motion for a large media event to expose all of the problems that the fishermen believed could only get worse. The Marine Advisor was asked to assist the fishermen in organizing the technical components of the program. However, believing that the media event would not provide any long lasting solutions, the Marine Advisor began suggesting the possibility of bringing in professional mediators to determine if face- to -face negotiations to resolve some of the space-use conflicts was feasible. A majority of the representatives of both industries agreed to consider this approach.

Tensions at the meeting were high. The press was excluded in order to enable the representatives of each of the industries and agencies to speak openly and directly. The oil industry acknowledged that they needed to keep fishermen informed on what projects were proposed and to work out solutions for avoiding impacts. Fishermen indicated that they had traditionally been supportive of new oil and gas development, but could not alone bear the cost with respect to damage to gear and loss of fishing grounds. The parties present decided that space use conflicts cold best be resolved by the two industries negotiating on their own. If a government regulation or procedure appeared necessary, the two industries would strive to present a joint proposal to the agencies. A second set of issues, potential damage to fishery resources from cumulative exposure to high energy seismic sources, was assigned to a multi-agency/industry group that would operate in parallel.

Formation of the Joint Oil/ Fisheries Committee. Three fishermen and three oil industry representatives met again several weeks later to begin to grapple with solutions to the problems that had been identified on April 9th. The central issues defined by group included 1) improving notification regarding impending activities, 2) improving coordination of fishing and oil vessel traffic 3) compensation for gear damage and short-term areal exclusion and 4) compensation or mitigation to fishermen for perceived accumulated impacts over the previous decades. With the assistance of the mediators, the parties negotiated a package of process groundrules for a "Joint Oil/Fisheries Committee. Decisions are made by consensus, i.e., a position, program, or issue would only be implemented or released by the Joint Committee if every individual representative could support it. It is presumed that when the representatives vote in favor of an item, they are representing the broader group of constitutes, not just themselves. If an aspect of a negotiation item is troublesome to one of the representatives in either caucus, it was modified, rewritten and re-reviewed, until it was in a form that all could support. This sometimes resembled dentistry without Novocaine, but was the only way to ensure compliance with policies, guidelines, programs, and negotiation results by all parties. Most of the programs developed by the committee have been voluntary and on a goodwill basis. Without up-front agreement, implementation would be difficult if not impossible, hence the need for consensus.

Over the course of subsequent years and due to various issues, each caucus added two representatives, for a total of five on each of the fishing and oil caucuses. The Joint Committee met monthly for the first four or five years to develop and establish programs to address each of the identified issue areas. The mediator worked closely with the group often undertaking "shuttle diplomacy" between the respect interest caucuses, assisting in clarifying and presenting interests in a way that facilitated solutions to problems. Notes are kept of each session to keep track of consensus decisions ; formal agreements on major topics are negotiated periodically and signed by the members.

Improving Inter-industry Communications. The first issue addressed by the Joint Committee was the mutually perceived need for improving communication between industries. Neither side really knew much about how the other worked, nor where, nor when. The Committee spent the first six months working out the detailed structure and function of both the Committee itself, and a staffed communications office, the Liaison Office. The Joint Committee members jointly interviewed and hired a Liaison Officer who is based in Santa Barbara but is expected to cover the region stretching from Ventura in the south to Morro Bay in the north. The oil industry contributes the cost for operations of the Liaison Office, but the contract for services with the Liaison Officer clearly establishes that he serves at the behest of both industries.

In October, 1983, the Liaison Office began to develop key contacts in both industries for efficient inter-industry communication, and developed other information about both industries as directed by the Joint Committee. In general, the Liaison Officer has the day-to-day role of facilitating inter-industry communications, assisting in the filing of claims of gear damage or areal preclusion (depending on circumstances), keeping records of at-sea conflicts and claims, and hosting short-notice interindustry meetings on an as-needed basis.

One thread of the discussion on inter-industry communications by the Joint Committee focused on advance notice to fishermen of upcoming oil industry projects. Seismic survey, exploratory drilling and offshore development and/or construction activities all have the potential to interact with commercial fishing gear and vessels. The Joint Committee took up how much advance notice and specificity would be useful for pre-planning, and settled on a compromise of about 30 days advance notice to fishermen. This "gentlemen's agreement" is not hard and fast, but a guideline which accounts for several factors. First, the farther in advance notice is given, the less specific as to exact dates and times of operations it must be. Conversely, the closer to actual operations the notice is given, the more realistic the dates, times, precise vessels, etc. are in the advance notice. Further, for fishermen who stay out days to weeks at a time, 30 days would give sufficient time to respond, and attend meetings as necessary to discuss space-use sharing.

To further facilitate timely inter-industry communication and advance notice of oil industry plans, the Marine Advisor and his staff developed a monthly newsletter entitled "Oil and Gas Project Newsletter for Fishermen and Offshore Operators." This newsletter was a monthly forum for the publication of oil industry proposed activities, as well as a review of the previous month's fisheries activities. It proved useful for a wide variety of fishermen, oil industry representatives, government entities and consultants. At its peak, the newsletter circulated about 700 copies.

Seismic Survey Notification Procedures. The Liaison Office and Joint

Committee developed an information handbook to assist fishermen in understanding seismic survey operations and to assist geophysical contractors in understanding the operations of the commercial fishing industry in this region. The manual, patterned after a similar effort in Alaska, is entitled "A Manual for Geophysical Operations in Fishing Areas of South/Central California." It provides non-binding guidelines in a step-by-step fashion for oil companies and their survey contractors to become informed about which fisheries are likely to be active in any area of concern, key contacts in the fishing industry, and helpful hints about communications with fishermen. It provided for fishermen an overview of the technology and operating procedures of survey contractors, photos and key information about seismic vessels , and key company contacts.

A corollary issue emerged as seismic survey vessels did their work. Fishermen 1) felt that the fish left the area, and/or 2) were very concerned that the 2 mile long survey cable would cut their set gear off so they couldn't find it. When gear is lost in that way, the gear still "fishes" until trap self-destruct clips give way or the net "balls up". In response, the Joint Committee developed guidelines covering advanced notification and documentation of claims for preclusion.

Vessel Traffic/ Right-of-Way. To address the concerns of the "set gear" fisheries (crab and lobster trap fleets, set gillnet fleet) about oil industry crew, supply, and tanker vessels "cutting off their gear," the Joint Committee and Liaison Office developed a "Vessel Traffic Corridor Program." After considerable discussion, the oil industry agreed to voluntarily abide by a set of corridors or traffic "lanes" in nearshore waters of the central coast counties when transiting to and from shore crew and supply bases and tanker terminals to offshore platforms and exploratory drilling rigs. The purpose of this was to reduce potential damage to trap and net buoy lines from the propellers of the relatively deep-draft vessels. The corridors were routed initially in 1984 after comments were received by the Joint Committee from their constituent groups regarding the precise alignment of the corridors to provide convenience for the oil industry while avoiding prime fishing areas. The corridor program is not intended to supersede U.S. Coast Guard rules of the road, nor common-sense safety considerations for crew and vessel. Revisions to the corridor program have been made several times as necessary and require the concurrence of both industries in accordance with the Committee's groundrules.

To address situations in which a contractor's vessel strays outside the corridors and damages fishing gear, the Joint Committee developed generic guidelines for negotiations on gear damage reimbursements, similar to the guidelines for geophysical surveys. These guidelines succeeded in reducing the numbers of claims and alleged violations of the corridors in non-emergency situations.

Compensation/Mitigation of Impacts. For many reasons, the most difficult component to resolve was the issue of the perceived accumulated impacts of oil industry activities on commercial fisheries in the Santa Barbara Channel and Santa Maria Basin for prior decades. Initially, the Joint Committee retained a neutral economist to assist them identifying and quantifying impacts from the various off-shore oil activities. Although the economist was able to demonstrate that there were real economic impacts that resulted from oil and fishing interaction, there was a significant difference of opinion about the type of compensation mechanism that would be appropriate and equitable. To address this issue, the Joint Committee worked with the Governor's office and the California Legislature to enact legislation calling for federal-state shared oil revenues [Outer Continental Shelf Lands Act Sec. 8(g)] to fund an impact mitigation program (AB2605, 1987, Seastrand). In 1988, the Local Marine Fisheries Impact Program (LMFIP) began implementation through the Governor's Office of Environmental Affairs. The LMFIP had a number of elements: Area, Gear, and Technology R&D contracts; Vessel Safety Gear Grants, Seafloor Debris Cleanup Program, Purse Seine Fleet Assistance, Marketing Assistance Program, and a Fisheries Development Corporation. The development corporation was later folded into the marketing assistance program, which resulted in the formation of the California Seafood Council, now in its sixth year. Overall, and during the course of six years, the program disbursed about $6.45 million in 8(g) revenues among the various program elements.

Potential Resource Damage Issues. Per the agreements reached at the April 9th session, issues pertaining to potential damage to fishery resources was delegated to an ad hoc committee that included representatives of several key state and federal agencies as well as delegates from the Joint Committee. The fishermen contended that during a geophysical survey, fish were observed to leave an area and not return for an extensive time period. Other fishermen expressed concern that repeated exposure to sound could do more than temporarily affect fish behavior; they questioned whether there could be permanent damage to the resource especially in the earlier life stages of fish and shellfish of commercial significance. Agency scientists felt that there was enough evidence to warrant conducting studies to determine whether the concerns of the fishermen would prove correct.

Each industry and agency prepared to bring in their best scientists to justify their particular point of view. Sensing that this approach would only fuel the controversy and never result in any conclusions that would be acceptable to all of the parties, the mediator suggested an alternative approach to the "battle of the experts". The committee jointly identified neutral experts in the fields of acoustics and fish behavior. Any expert who had worked extensively for either the oil or fishing industry or who was perceived to be closely aligned with either industry or any of the agencies was eliminated from consideration. The mediator and Marine

Advisor were asked to contact each of the experts, collect resume information and determine their interest in working with the group. The final list of experts was agreed upon by a consensus of all of the members. In addition, the committee developed questions for the panel members and an agenda for the session. It was agreed upon in advance, and formalized in a "Memorandum of Understanding" that , if the panel reached a consensus (unanimity) on its answers to the questions posed and if the panel was able to reach a consensus on a proposed design for an ocean research study, then all of the members would support such a study and not challenge the integrity of the study once it was completed.

The "Fish Dispersal Panel" was able to reach a consensus and the committee conducted a pilot study in accordance with the panel recommendations. The approach used for selecting a panel and obtaining expert guidance on critical resource issues was replicated again several years later by a subsequent industry/agency group composed of representatives of the same organizations and agencies. This time the group focussed on determining whether there were lethal and/or sublethal effects of such acoustic energy pulses on the Northern Anchovy, and on the larval stages of Dungeness Crab. These studies were co-funded by the oil industry with state and federal agencies and foundation support. Fishermen provided in-kind services, for example volunteering their vessels and crew for several of the at-sea experiments.

Most importantly, the process that evolved from this joint industry/agency effort has withstood the test of time. The studies have been defended against detractors by each of the individuals and their respective organizations as well as by the scientists who participated on the consensus panels and who were retained by the committee to conduct this jointly sponsored research.

Model for Resolution of Other Coastal Disputes. The specific solutions generated by the deliberations of the Joint Committee and the parallel Fish Dispersal Committee are important examples of creative approaches for addressing coastal resource disputes. More significantly, however, the dispute resolution mechanisms and problem-solving processes that were developed by the parties themselves have served as the prototype for subsequent deliberations and are applicable to a wide variety of potential conflicts in the areas of ecosystem and coastal resource management. The concluding section of this paper highlights some of the generic principles of a dispute resolution/consensus-building model that are worthy of consideration in other contexts. These principles are based upon both the "lessons learned" from the 14 year fishing-oil mediation process and the experience of each of the neutrals, the Marine Advisor, the mediator and the Liaison Officer, who have played key roles in the design and implementation of this unique, durable experiment in joint problem-solving.

1. Neutral roles. There has been a critical distinction among the roles of the three neutrals who have interfaced with the participants in the Joint Committee and fish dispersal processes. The **Marine Advisor** served as the neutral provider of technical information to each of the constituent groups. In the years leading up to the mediation process, he assisted the parties in gaining a better understanding of their respective industries, and facilitated meetings to prevent conflicts from escalating , but all in the context of the role of "technical" advisor. After the Joint Committee was convened, the Marine Advisor provided valuable information on fishing seasons, and gear adaptation that was utilized in generating at-sea conflict avoidance strategies. The " Oil and Gas Project Newsletter" provided a *neutral* source of information on the operations of both industries and served as an important tool for keeping individuals who were not directly involved on the negotiation committees informed.

The **Liaison Officer** is charged with ensuring that all of the key parties are informed of impending activities in a timely manner and is proactive in bringing stakeholders together to discuss individual projects and problems. The Liaison Officer is responsible for maintaining the institution of early problem-solving on a case by case basis that is a critical function of the Liaison Office. As is true of the Marine Advisor, he also provides technical information to the parties regarding operations of both industries to Joint Committee member organizations and the numerous state and federal agencies that have come to rely on information provided by the Liaison Office in carrying out their own mandates with respect to oil and gas exploration and development.

The **mediator** plays a key role in the crafting system -wide procedures that govern the relationships among the stakeholders. In addition to chairing meetings of the group, the mediator employs shuttle diplomacy to help break an impasse on the proposal under consideration, helps the parties evaluating the reasonableness of their positions and assists in the crafting of formal written agreements. For the Fish Dispersal Committee process, the mediator served as the contact between the negotiators and the scientific panel and facilitated the panel proceedings to ensure the neutrality and fairness of the process.

Each of these neutral roles has been critical to the success of the overall process and suggests that there is a range of neutral tasks that can be provided in the context of problem-solving and dispute resolution of natural resource disputes. It is important for stakeholders who are considering a dispute resolution or consensus-building effort to identify what role they envision for a neutral and to set the expectations of participants accordingly, especially since the needs of the parties may change over time. As happened in the Santa Barbara case, the Marine Advisor recognized that the parties needed a different set of neutral services than could be provided by his office. Accordingly, he recommended that they transition from his informal dispute resolution role and technical consultation to that of a formalized, traditional mediation approach.

2. Representation of Stakeholder Interests. The Joint Committee has been in existence 14 years. Those years saw activity cycles peak and wane in both industries. Interest in the process was intensive during the formative years of the process; as oil activity declined and there was less interaction between the two industries, there was less focus on the Joint Committee process. Committee membership and representation has, but for a single individual, turned over in each of the ten seats, often several times. It has been noted throughout these continual personnel changes that the knowledge level of Joint Committee programs, agreements, and ongoing negotiations among rank-and-file members of both industries is a function of individual members of the Committee. When and as individual committee members have taken an active role in ensuring their constituents' commitment to programs and agreements, and have actively and regularly disseminated information on the Joint Committee actions and sought constituent input to the process, the Joint Committee has been able to provide meaningful solutions to inter-industry problems. When a more passive role has been taken by individual members of either industry, communication breakdowns have occurred, and implementation of agreed-upon programs and actions has been less consistent, and more difficult. Thus, the extent to which selected representatives remain committed to the goals and programs of the committee, and are active agents of those goals and programs with their constituents, is perhaps the most important determinant of success in this type of problem-solving effort.

It is essential for each interest group to such negotiations to have reached internal agreement with respect to goals and "bottom line" negotiation positions prior to meeting with other differing interest groups. This may be a particularly acute problem for non-government organizations or loosely-knit coalitions. Internal dissention regarding goals and bottom-lines can prevent effective inter-group negotiations by setting up conditions where clear directions to appointed negotiators are lacking. The individuals representing an organization or agency do not have to include upper management or the chair of the group, but the negotiators must have access to key decision-makers on a timely basis. When a representative indicates support for a particular proposal, the other representatives need to have assurance that the individual delegates are negotiating on behalf of his or her constituents and can "deliver" their support if a consensus is reached.

3. Importance of process groundrules and written agreements

Although initially time-consuming to craft, formal process groundrules enable the parties to gain experience in working on process issues as a prelude for deciding whether deliberations on substantive issues are feasible. Groundrules also provide a structure that can guide the talks, especially those that extend over a several month-year timeframe. Groundrules should address the issue of constituent representation, confidentiality of data, expected workproducts, and decision-making (consensus, super-majority...). In the case of the Joint Committee, formal groundrules addressed the roles and responsibilities of the Liaison Officer. From time to time, one side or another might attempt to dictate a different set of duties or relationship to the parties. This was especially true given the turnover in representatives. The groundrules became key in ensuring the integrity of the office as it had been originally intended and without jeopardizing the relationship of the Liaison Officer with the new Joint Committee member or his group.

The Santa Barbara experience as well as the authors' experience in other contexts indicates the importance of capturing what has been agreed upon by multiple stakeholder groups in writing. Often parties have left a meeting, each thinking that they understand what was agreed upon, only to learn at a subsequent meeting that several individuals had a different memory of what was stated or a different interpretation of an agreement. Written agreements can be in the form of a meeting summary that everyone approves at a subsequent session or, as appropriate, the parties may wish to craft a more formal, legalistic document that has to be ratified before it can be signed by the process representatives.

4. Involvement of all stakeholder groups. The decision of the Joint Committee not to include either representatives of government agencies or the general public has been criticized on a number of occasions. Their exclusion has led critics to conjecture on the nature of the talks and the scope of issues covered. For the most part, the oil and fishing industry representatives have negotiated on issues that affect their own interaction and their own economic impacts. Accordingly, they continue to believe that industry to industry talks are appropriate and have not chosen to expand participation on these matters. However, as they gained experience working together, and when the focus of their talks did involve a broader public interest, other groups were included in the deliberations.

The forums established for resolution of seismic impact issues were more inclusive of key stakeholder groups and had proceedings that were open to the public. As new scientific issues have been identified, the agencies have chosen to adopt the mediation negotiations and consensus science panel process. The Minerals Management Service, for example convened a broad-based stakeholder group to consider the potential effects of disposing drill muds and cuttings on hard bottom substrate. More recently, MMS initiated a consensus process, again inclusive of all of the key stakeholder interests, to develop a recommended process for reviewing high energy seismic survey permit applications. That group has sponsored a scientific workshop to consider potential effects on marine mammals and strategies for mitigation effects that are identified.

New Frontiers. Stakeholder groups in California who have been involved on many of the consensus-building deliberative process that were an outgrowth of the fishing/oil talks have gained valuable experience in joint problem-solving. This success indicates that there may be additional issue areas which are now ripe for mediated resolution in U.S. coastal and offshore areas. These issues might include (1) the perceived decline in fisheries stocks overall as well as in specific regions, 2) fish and shellfish resource allocation disputes among various ocean resource user groups including commercial fishers, sport fishers, and non-consumptive users, and 3) reconciliation of marine-focused environmental groups interested in ocean resource sustainability (sustainable yield estimates, fish and non-fish bycatch issues) with a commercial fisheries short-term cash flow needs, commercial sport boat use of local areas, and or sport boat use of local areas, and/or sportfishers' recreations interests.).

Conflict can be a productive force for creating meaningful solutions to complex economic and resource problems. The processes described in this paper did not avoid conflict; they channeled the energy of the disputing parties into productive problem-solving activities that addressed all of the interests of the participants. No individual, organization or government entity was required to compromise on a basic principle. In each situation, the goal of the mediated/facilitated talks was to reach an accommodation that maximized the ability to address the particular needs of all the interests, whether that interest was cost, operational constraints, or the critical biological needs of a marine species. When the parties lacked information on how to proceed, they engaged in a creative process for obtaining mutually acceptable advice from recognized experts, rather than pitting the advocates of their institutional position against one another. The most important lesson of the Santa Barbara case is that confronting conflict, engaging in honest, direct talks and working towards building a consensus can result in positive, sustainable solutions to seemingly unresolvable conflicts.

CHAPTER 222

Redefining the Federal Review Process of High Energy Seismic Exploration for Oil and Gas on Southern California's Outer Continental Shelf: An Example of Federal, State, Local, Industry, and Public Cooperation

Drew Mayerson[1], John Lane, James Grant, and Maurice Hill

Abstract

The Federal review process of offshore geophysical operations on the Pacific Outer Continental Shelf (OCS) has become increasingly complex and controversial since its inception during the late 1950's and early 1960's. Prior to about 1980, the Federal review process consisted of little more than a record keeping exercise and a means of procuring data collected by industry, Today's review process includes a recognition of other users of the OCS and the need to work with all parties that monitor the impacts from OCS activities on the marine and coastal environment. Though commercial fishers, the military, and other Federal and State agencies have been involved in the review process since the early 1980's, some local government agencies and environmental interest groups have not been included in the process.

In 1995, Exxon proposed conducting a high energy seismic survey over their leases in the Santa Barbara Channel. Although the federal review process was even broader than it had been in the 1980's, advocates for the protection of marine mammals argued successfully that their concerns were not being properly addressed for the proposed operation. Recognizing that future exploration of existing offshore leases using high energy survey techniques depends on the ability to reach agreement with all affected parties, the Minerals Management Service (MMS) proposed the formation of a multi-disciplinary team to redefine the Federal review process.

In June, 1996 the MMS convened a meeting to bring other Federal and State agencies, local governments, the oil and geophysical exploration industries, fisher, and environmental interest groups together with the goal of forming a team to build a model for the future review of high energy seismic survey operations pertaining

[1]Geologist, U.S. Dept. of the Interior, Minerals Management Service, POCS Region, Camarillo, CA

to oil and gas exploration on the OCS off of southern California. Additionally, the team was charged with defining the scope of environmental review necessary to assess the impacts from future geophysical exploration projects, and recommending mitigation measures that will help minimize those impacts.

The team began meeting on a quasi-monthly basis in September, 1996 and hired a facilitator to manage the meetings. The team has met five times to date. The initial meetings resulted in a team statement of purpose, an acceptable definition of high energy seismic surveying, educational primers regarding sound transmission in the sea and its impacts on marine mammals and fish, and formation of two subcommittees to examine permit process issues and technical issues, respectively. The latter subcommittee is responsible for planning a workshop next June. At the workshop experts in bio-acoustics, marine mammals, and seismic data acquisition will attempt to answer questions posed by the team relating to the impacts of high energy seismic surveys on marine mammals.

This paper provides a brief history of the Federal review process of high energy seismic surveys, why the process needed to be changed, the steps that have been taken to bring all parties together during and after the Exxon survey in 1995, and the results to date of the team approach for developing the new review process.

CHAPTER 223

BREAKING THE IMPASSE ON COASTAL PROTECTION

JOHN J. O'CONNELL*

Mr. Chairman, thank you for your kind introduction.

It's a pleasure to be with you today. Like you (I hope), I am enjoying the conference enormously -- a worthwhile purpose; congenial company; a beautiful setting; excellent talks (at least until now). Who could ask for more?

Yet, despite all those good things, I suspect that most of us will leave here today with a certain uneasiness about the future.

Why is that?

Let me suggest the following as an answer: despite our brave words about confronting and solving the coastal erosion problems that plague this state and country, the reality is that we cannot seem to get our act together. The problems really are not being addressed. When you regularly come to symposia like this, you begin to feel like (to borrow Yogi Berra's famous phrase) "it's deja vu all over again." Each year we hear that the problems are about to be dealt with. Each year we hear of projects that are just about to be funded or just about to begin. Each year we hear about how the public perception of the value of shore protection

---

*** Mr. O'Connell is a trial attorney with the New York City law firm of Solomon, Zauderer, Ellenhorn, Frischer & Sharp. He specializes in coastal environmental and regulatory law.**

projects is beginning to change for the better, which should itself result in more and better projects. Yet, when we come back the next year we hear that the problems still are there, the projects are not yet funded or are not yet started, that there is political or public opposition to a particular project. In other words, the more things change, the more they stay the same. Here we are in 1997 and we still do not have a national coastal policy.

And added to that is the policy of the Clinton administration to take the U.S. Army Corps of Engineers out of the shore protection business. Sadly, we may come to look on the past decade or so as the golden age of shore protection projects. If that happens, God help us, and God help our coastline.

Let me pose this question: why, despite all our efforts, have we achieved so little?

I submit to you -- and I believe many of you will agree -- that the reason for our scant success is that the coastal zone has become a battleground for conflicting interests -- all well intentioned, but nevertheless destructive. The very people who should be working together to protect the nation's shorelines instead are battling with each other in an effort to achieve their own particular goals to the exclusion of all others.

The property owners want to keep the shoreline in front of their homes private -- protected at public expense, mind you, but off limits to the public that has paid to protect it. They also doesn't want to hear about environmentalists' concerns about the protection of endangered or threatened species in or around their property -- God knows, it might interfere with their lifestyle, and that would be unconstitutional. I've seen projects killed by that type of thinking.

Then there are the environmentalists. Many of them haven't seen a developed coastal area they like. The shoreline should be kept natural or returned to nature. Their goal is to force coastal zone property owners into the hinterlands -- to so-called strategic retreat (or should I say "strategic relocation"). Trouble is, we have billions upon billions of dollars of developed real estate in the coastal zone, and there's really nothing that can be done about that now. Nevertheless, these environmentalists will try to block any measure that helps the coastal property owner -- even if (as is often the case) failure to adopt the measure could lead to the destruction of irreplaceable habitat for the very endangered species the environmentalists are trying to save.

Then there are what I call the "populists." They believe that not one thin dime of public money should be spent on coastal protection projects unless there is unimpeded public access to the coastline covered by the projects. They want public beaches. They want facilities -- restaurants, rest rooms, showers, etc. They want parking lots. And, if at all possible, they would like the houses removed from the beach area.

Local government also is part of the problem. On the one hand, local government officials love the tax revenues that coastal properties generate. They also love that the shorefront owners (at least the summer residents among them) don't send their children to local schools or make much of a demand on local government resources. Also (and quite significantly) many coastal property owners (being largely non-residents) don't vote.

Compounding the problem at the local level is the media. They often oppose shore protection projects. Editorial after editorial will rant and rave about throwing good money after bad -- "After all, the sand will all wash away someday" -- or "That project is nothing more than welfare for the wealthy." Many residents (especially elderly voters on fixed income) agree with such views and write supportive letters to the media and the politicians.

All of this leads the local politician to oppose the project. He may appreciate the value of the shorefront property to the local economy or ecology. In his heart, he may even know a restoration and protection project would be the right thing to do. But the trouble is it would be political suicide for him to support a project that would involve a major outlay of cash (and an increase of taxes) at the local level.

The same problem exists at the state and federal level. State legislators and U.S. congressmen are boxed in. Most of them know what should be done, but do not have the political courage to do it. Their constituents (the local voters) would run them out of office if they voted for a project. So they do nothing but make sympathetic sounds.

Their counterparts in various agencies of the executive branches of the state and federal governments with responsibilities for the coastal zone are in the same dilemma. They also know that the shoreline should be protected, but are unable to easily implement any projects to achieve that goal. Compounding the problem is that these state and federal agencies contain within them individuals who fit into one of the other interest groups, specifically the environmentalists or the populists. They often attempt to include into any project either public access or environmental requirements that the property owners will find unacceptable and that will doom the project.

Could you ask for a recipe better suited for paralysis? These groups will go at each other for years, arguing that their vision -- and only their vision -- is the acceptable basis for a project -- or for no project. Meanwhile, like the fabled city of Atlantis, the eroding shoreline disappears into the sea, and everyone loses.

It's my opinion that the fundamental reason we have been unable to solve our coastal problems is the inability of the interested parties to resolve their differences and to move forward in a unified fashion. And, judging by the looks on your faces, I can see that many of you agree with me.

So what's to be done?

There's no easy answer. The mutual suspicions and antagonisms among the interested parties took years to develop, and, as we all know, bad habits die hard.

But there is, I urge, an answer that would work in many situations -- at least where the cause of the erosion is man-made (or to be more accurate, government-made) -- and the dirty little secret is that many, if not most, of our erosion hot spots are man-made.

I have been asked to share with you my experiences as the attorney for the property owners in bringing about the successful resolution of the infamous erosion problems on the barrier island in Long Island's West Hampton. They provide, I believe, a way out of the dilemma many of you now face.

The root of the West Hampton erosion problem was the stabilization of an inlet known as Shinnecock Inlet in the early 1950s. Upon its stabilization, the inlet began to trap enormous quantities of sand flowing westward in the littoral drift. The result was severely accelerated shoreline recession west of the inlet, with erosion rates jumping from approximately 1.25 ft per yr. to in excess of 7 ft. per yr.

In the 1960s and 1970s, the federal, state and local governments undertook an erosion control and hurricane protection project in the area west of the inlet in part to mitigate the erosion. In a series of events that can only be described as bizarre, the project was never completed and was constructed backwards -- yes, I said backwards. That led to even more accelerated erosion in the area downdrift of the uncompleted project -- about 15 ft per year. In that area, which is now known as the Village of West Hampton Dunes, there stood well in excess of 300 homes as of the early 1980's. By the early 1990's, over 250 of them had been destroyed, the island had been breached, and a half-mile wide inlet had been created, isolating the remaining houses and a large public park. Over the years, well in excess of $20 million was spent on emergency projects to close the breaches.

Despite all this, absolutely nothing was done to solve the underlying man-made causes of the erosion. There were endless discussions between federal, state and local governments concerning a how to resolve it. The property owners petitioned their elected representatives. Environmentalists and populists weighed in with their views. All to no avail. No one could agree on what to do. It looked as if the area was simply going to sink beneath the sea.

In early 1993, just after the island had been breached again, the property owners and I developed a four- pronged strategy to break the impasse.

The first and most important prong of the strategy was to reactivate the litigation against the federal, state, and county governments and announce that we would press forward as fast as possible to judgment, unless a firm binding and acceptable settlement proposal was placed on the table. Our thought was that without what we came to call "judgment day" (*i.e.*, a trial) the parties would continue to bicker among themselves forever. With judgment day looming in the not-to-distant future, the parties would be forced to identify their fundamental concerns and attempt to negotiate an appropriate resolution of the problem. Otherwise, a judge or a jury would do it for them.

It would force the parties to say themselves: "What is it we really want here? Because if we don't put that on the table and then try to accommodate the legitimate interests of the other parties, we could lose everything." The populist might not get public access; the environmentalist might lose the very habitat he's trying to save; the property owners might lose their lawsuit and watch their property continue to wash away; or the governments might lose the lawsuit, and wind up paying considerably more in damages than the costs of a restoration project. In the West Hampton litigation, the property owners asserted claims of in excess of $200 million. Those were legitimate claims, not lawyers' hyperbole. A project to save the island was projected to cost $75 million spread out over 30 years. When one compared the present value of the $75 million project against a near certain loss in excess of $200 million in damages, there simply was no comparison.

The second prong of the strategy was an aggressive public relations campaign to demonstrate the benefits of a restoration project to the local community. We identified every possible benefit from maintenance of the tax base to restored food and liquor purchases by the property owners whose homes had been destroyed. You name it, we identified it. We estimated that each of our 300 (plus) homes generated between $35,000 and $45,000 per year between taxes, purchases, etc. -- for an annual total of at least $10,500,000 to $13,500,000. In essence, we said to the local community, your share of the project costs -- $7.5 million over 30 years -- will be dwarfed by the direct economic benefits you will receive -- at least $300 million over the life of the project.

We also worked hard to get TV, radio, and press coverage and were quite successful. Whenever there was a good story -- particularly photo opportunities of homes falling into the ocean -- we'd call the local TV stations. We provided background material and property owners to be interviewed. We met with editorial boards to discuss our plight. We showed ourselves as ordinary people -- all the residents we presented to the media looked more like John Goodman or Roseanne than the Duke and Duchess of Windsor.

We also harped on the fact that the erosion problem we were suffering was not of our own making. Neither God, nature or the property owners was the cause of the erosion problems -- the governments were, by stabilizing Shinnecock Inlet and starting, but not completing, the West Hampton hurricane protection and erosion control project.

That was our "fairness" point. Understand that we never attacked the stabilization of Shinnecock Inlet. The project clearly had benefitted Shinnecock Bay and the adjacent mainland property both environmentally and economically, and that was just fine with us. However, those benefits should not be purchased at our expense. We were literally losing our property, watching it sacrificed for the benefit of the mainland people. In fairness (we argued), shouldn't our property be protected?

The P.R. campaign was long and hard but ultimately successful. By the time the matter was resolved, the local media were on our side and hailed the settlement as being to the benefit of all Long Island and a masterpiece of inter-governmental cooperation. Quite a change.

The third prong of our strategy was political. Any time two or more politicians gathered together in our area of Long Island, we were there. We supported politicians who supported us. We were one issue people. We persuaded our people to register to vote locally, even if they were from out of state. We became the swing votes in tight elections.

We also wanted the politicians to see our erosion problem in human terms. We wanted them to realize that a "no" vote on a restoration and maintenance project meant that the homes of people they knew personally were going to go into the ocean. I can assure you, even the most hardened politician is affected when a young couple with their children confront him and ask if he's going to let their home be destroyed.

Again, as with the P.R. campaign, it was long, hard, but ultimately successful work. To show you how effective, when our project came before the county legislature, it was approved unanimously.

The final prong of the strategy was to develop a plan that achieved the essential objectives of each of the interest groups I mentioned earlier. Note that I said <u>essential</u> objectives -- not each and every objective.

So, for example, the property owners had to have their property restored and protected and had to have the right to rebuild and reoccupy their homes and get to the ocean. The environmentalists had to have the right to protect endangered and threatened species -- consistent with the rights of the property owners to rebuild and occupy their homes and with access to the beach and ocean. The populists had to be given access to the beach and ocean -- not unimpeded access, but access consistent both with the rights of the adjacent property owners and the concerns of environmentalists for the fragile coastal ecosystem. (I called this "environmentally sensitive public access.") The local governments had to have a fully restored tax base and the assurance that the per property expenditures by the owners, both in paying their taxes and in purchasing goods and services, would at least equal and probably surpass (the local government's share of the project's costs. Finally, the plan had to be consistent with federal and state regulations concerning the coastal zone.

There is a saying that nothing focuses a man's thinking like the setting of a date for his hanging. Well, the experience at West Hampton proved the truth of that saying. We reactivated the litigation in January 1993, telling all the involved interest groups that we weren't going to stop until we had a judgment (which we were confident would be in our favor) unless a firm and acceptable settlement offer were placed on the table. We also told them we were willing to address their essential concerns in any such settlement. Document production took place in the spring and summer and depositions began in the fall.

Guess what happened?

On December 16, 1993, a scant eleven months later, we had a firm and very attractive settlement proposal in hand. We refined the offer over the next few months, and arrived at an agreement that consisted of the following:

1. Modifications to the existing, uncompleted hurricane protection and erosion control project to eliminate (or at least significantly reduce) its erosion-producing aspects;

2. construction of a 2.5 mile long, 400' deep beach, culminating in a large duneline;

3. Thirty years guaranteed maintenance of the project, pending a larger protection project for the bulk of the south shore of Long Island;

4. Recognition of the property owners' right to rebuild their homes and restoration of the property to the tax rolls;

5. $2,000,000 for infrastructure replacement and development;

6. Reimbursement of all the property owners' legal expenses;

7. Access to the beach on the seaward side of the duneline, with full public access (with parking and other facilities) on to two sites; and

8. A comprehensive and unprecedented plan for the protection of endangered and threatened species, while recognizing the rights of the property owners to the use of and enjoyment of their homes and to access to the beach and ocean.

9. A favorable cost sharing formula was agreed-upon -- 70% federal; 21% state; and 9% local.

In my judgment, that agreement never could have been reached were it not for the litigation (and the clock ticking toward trial). It concentrated everyone's thinking. Without it, we'd all still be bickering. And I'm not the only one who believes that -- a number of government officials and government attorneys have expressed the same view to me.

Will the same strategy work for you?

Let me answer (and conclude) this way. If your beach is falling apart because of man-made erosion, you've got only three options:

1. Do nothing and watch your shoreline sink beneath the sea -- which I assume is unacceptable, or you wouldn't be here;

2. Continue to lobby and hope that you'll get a project started before it's too late -- an option which in these times of leaner and meaner government budgets might at best (*i.e.*, most charitably) be characterized as wishful thinking; or

3. Sue the parties responsible for your erosion mess and tell them that *they* have two options:

   a. Fix the mess, in return for which you'll allow reasonable and responsible public access and wildlife habitat protection and enhancement; or

   b. Pay the property owners money damages of three, four, or five times the amount of the cost of the project you are willing to settle for -- and run the risk of losing forever an opportunity to preserve an irreplaceable natural asset.

Let me suggest to you that the hearts and minds of those responsible for the erosion quickly will begin to wrap around the first option.

Thank you.

CHAPTER 224

Chemistry, toxicity and benthic community conditions of sediments from selected southern California bays and estuaries

Brian S. Anderson[1], John W. Hunt[1], Russel Fairey[2], H. Max Puckett[3], Craig J. Wilson[4], and Edward R. Long[5]

Abstract

Chemicals discharged into California watersheds through point and non-point sources often become associated with sediments and may pose significant risk to beneficial uses of freshwater and marine systems. A cooperative research project to assess sediment contamination was conducted jointly by the California State Water Resources Control Board, the National Ocean and Atmospheric Administration, and the Environmental Protection Agency through the State Board'. Bay Protection and Toxic Cleanup Program. Using a weight-of-evidence approacl based on the Sediment Quality Triad, measures of chemical contamination, toxicity, and benthic community structure were completed at 43 stations to determent relative degradation in selected Southern California bays, estuaries and lagoons. Degree of chemical contamination was assessed using sediment quality guidelines developed by NOAA (Long et al., 1995), the State of Florida (MacDonald, 1994), and EP (1995). Relative to these guidelines, DDT, Chlordane, Copper, Mercury, and Zinc were found to be the chemicals or chemical groups of greatest concern. Chemical contamination was considered to be moderate relative to other more highly industrialized areas. Based on toxicity information from randomly selected stations, 58% of the total study area was found to be significantly toxic to amphipods (*Rhepoxynius abronius*). Using sea urchin development as an indicator of toxicity, 91, 83, and 51% of the randomly-sampled study area was found to be significantly toxic in 100, 50, and 25% concentrations of sediment interstitial water, respectively. Forty-three percent of the randomly-sampled study area was toxic using a test which

[1]Institute of Marine Sciences, University of California, Santa Cruz
[2]San Jose State University, Moss Landing Marine Laboratories
[3]Marine Pollution Studies Laboratory, California Department of Fish and Game
[4]Bays and Estuaries Unit, State Water Resources Control Board
[5]ORCA, National Oceanic and Atmospheric Administration

assesses sea urchin sperm viability as a measure of contaminant toxicity. There were weak negative associations between toxicity test results and some chemical compounds measured in bulk-phase samples. Survival of the amphipod (*Rhepoxynius abronius*) was negatively associated with DDE, PCB52, un-ionized ammonia, two metals, and fine-grained sediment. Survival of the amphipod *Ampelosca abdita* was negatively associated with PCBs and several metals. Sea urchin embryo development was negatively associated with two metals, chlordanes, and DDT compounds. There was a strong negative correlation between sea urchin embryo development and interstitial water un-ionized ammonia concentrations. Benthic community structure was assessed using a Benthic Index, calculated based on measures of the total number of fauna, number of crustacean species, and numbers of positive and negative indicator species. Using this index, 15 of the 43 stations sampled (35%) were considered to be significantly degraded; 10 of the 15 degraded stations were located in 4 of the coastal lagoons sampled. Benthic community degradation was not strongly associated with measured bulk-phase chemicals. The Benthic Index was negatively correlated with interstitial water hydrogen sulfide concentrations, indicating that sediment anoxia influenced benthic community structure, particularly in the coastal lagoons. All measures of sediment contamination and degradation proved useful in this study. Stations recommended for further investigation were prioritized to help direct future investigations by State and Regional Water Board staff. Each station receiving a high, moderate or low priority ranking met one or more of the criteria under evaluation for determining hotspot status in the Bay Protection Toxic Cleanup Program.

CHAPTER 225

Modeling Longshore Sediment Transport and Shoaling Assessment

Christian M. Appendini[1] , Román Lizárraga-Arciniega[2], Rafael García-Krasovsky[3]

Abstract

Along the beach of Rosarito, Baja California, Mexico, a thermal-electric power plant was built, presenting shoaling problems at the water intake since its construction. Sand transport rates were calculated with the GENESIS model to estimate the potential volume of sand contributing to the creation of shoals. The lack of on-site data made necessary to use alternative data to perform model simulations. The model was first calibrated and verified using the corresponding wave data. Using a 20 year simulation with an annual wave climate, the model was calibrated to perform a simulation considered representative of a given year. From these results, sediment budgets were performed, allowing estimates of the monthly volumes of sand that enter the water intake. In general, the shoaling problem is throughout the entire year, but it is greater from November to April (35,800 m^3) and low from May to October (16,200 m^3). Site data on beach evolution, dredging and wave parameters is spare or non-existent, making more difficult the model usage and validity, therefore the results are considered a first approximation.

Introduction

It is recognized that proper planning and management of any coastal development requires a good understanding of the complex coastal processes. The use of numerical models for simulating coastal processes

---

[1]Graduate student, Facultad de Ciencias Marinas, Universidad Autónoma de Baja California. Suite ESE-1306, 4492 Camino de la Plaza, San Ysidro, CA 92173-3097.
[2]Researcher, Instituto de Investigaciones Oceanológicas, Universidad Autónoma de Baja California, P.O. Box 453, adm. 1, Ensenada, BC, Mexico.
[3]Professor, Facultad de Ciencias Marinas, Universidad Autónoma de Baja California, km 107 carretera Tijuana-Ensenada, Ensenada, BC, Mexico

has been proven to be a powerful tool to improve coastal structure design and increase the ability for the development of proper management programs, e.g., dredging and sand nourishment (Davidson et. al. 1992; Gravens 1989; Ludwick et. al. 1987). One of the most frequently used models for simulating shoreline change in the United States is the one-line GENESIS model developed by the U.S. Army Corps of Engineers (Hanson and Kraus 1989; Gravens et al. 1991). The model is designed to describe the long-term trend of the shoreline in response to breaking wave action, coastal structures and given boundary conditions, assuming that all depth contours remain parallel to one another throughout the simulation (Work and Dean 1995).

In the early sixties a thermal-electric power plant was built in the coastal area of Rosarito, Baja California, Mexico (figure 1). The required sea-water intake for the cooling system was generated through the construction of coastal structures on a near straight sandy beach, modifying the erosion-deposition patterns and resulting in shoaling at the intake. This paper describes the use of the GENESIS model to assess the effect of the longshore sediment transport (LST) on the development of shoals at the water intake of the thermal-electric power plant. The lack of on-site data made necessary the use of alternative data to performed simulations, therefore the results are consider a first approximation to the problem.

Study area

Rosarito is located 20 km south of the U.S.-Mexican border on the northwest coast of Baja California. Its main activity is tourism although Petroleos Mexicanos's (PEMEX) oil tanks and Comisión Federal de Electricidad (CFE) thermal-electric power plant are present. The straight sandy beach has an extension of 7.5 km, it is gently sloping and exposed to the ocean, with only two small streams reaching the ocean during rainy seasons. The main sources of sand to the coastal area are the Tijuana River in the north, longshore sediment transport (LST) from adjacent beaches and cross-shore sediment transport (CST) from offshore areas (Ureña-Portales 1986). Waves arriving at the coast are similar to those of the Southern California Bight, showing a typical climate in height and direction of approach associated with winter and summer seasons, as described by Moffatt & Nichol Engineers (1989), Wiegel (1993), and observed from the data obtained by the Coastal Data Information Program (CDIP, annual reports).

The sea-water intake structures of the thermal-electric power plant were built in the early 1960´s. The structures consist of two breakwaters and two groins. The breakwaters provide the access channel to the pump house; the north groin is to prevent littoral material from entering the intake, and the

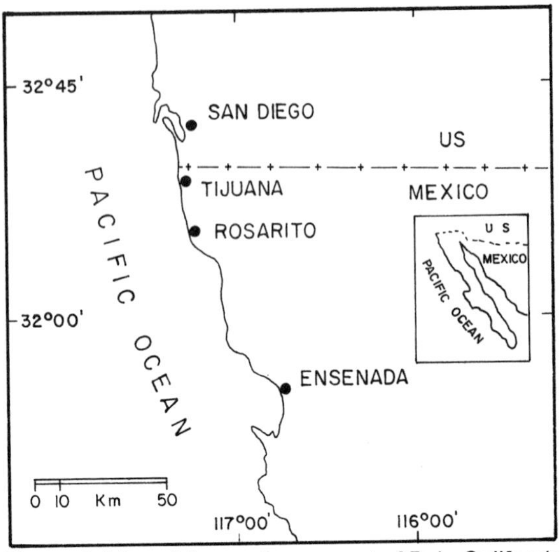

Figure 1. Map of the northern coast of Baja California

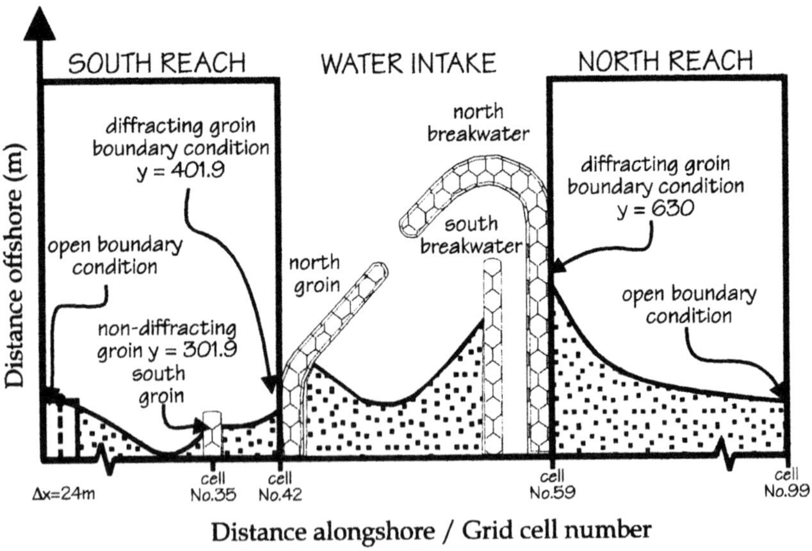

Figure 2. Water intake structures of the power plant and definition sketch for modeling conditions

south groin stabilizes the discharge channel (figure 2). Rapid economic growth in Baja California required an expansion of the plant in the mid-eighties. To prevent increased shoaling problems at the intake, lengthening of the north groin was undertaken, but the shoaling processes continued. Today, a permanent dredging program is operated by one fixed dredge and two movable dredges.

Previous studies of LST in Playas de Tijuana and Punta Los Buenos (20 and 10 km north of the area of study) have shown a northward transport pattern (Ureña-Portales 1986; Elizarrarras-Vargas 1986). In the study area the littoral circulation pattern directed to the intake was from both the north and south, with a net southward sediment transport of 150,000 m^3/year (Marmolejo-Lara 1985). Unfortunately, no data of dredged sediment volumes are available.

Shoreline boundary conditions and coastal structures

Shoreline change simulations with GENESIS were performed individually for the north and south beach segments adjacent to the intake structures (figure 2). For the north reach an open boundary condition was considered as its northern limit whereas the southern boundary was defined by a diffracting groin representing the north breakwater of the intake. For the south reach the northern boundary was defined by a diffracting groin representing the north groin of the intake, with an open boundary condition considered as the southern limit (figure 2).

The open boundary condition allows the sand to freely enter or exit the grid, whereas the diffracting groin boundary condition controls the sand entering or leaving the grid. The amount of sand passing the groin boundary condition is determined from the distance to the shoreline on either side of the groin tip and the specified groin permeability. The shoreline grid for both the north and the south model reaches consisted of forty one cells with 24 m of longshore length for a total model reach of 984 m on each side of the power plant. The water discharge groin located 200 m south of the north end of the south reach was defined as a non-diffracting groin. The location of the structures were defined by a CFE survey in 1992 and their location were translated into corresponding cell numbers of the GENESIS reference grid (figure 2).

The seaward boundary is defined as the maximum depth of active LST known in literature as closure depth (D_c). The changing nature of wave characteristics makes difficult its evaluation. It is possible to estimate this value from bathymetric surveys where available, but as sounding and positioning errors increase with depth and distance from shore (Hands

1983) closure depth was calculated from the Hallermeir (1981) equation. Utilizing deep water data for a period of six years (1979 to 1981 and 1991) a closure depth of 5 m was calculated.

Wave data and model simulations

For calibration and verification, shallow water wave data were taken from the Imperial Beach array, located north of the Rosarito area, and operated under the Coastal Data Information Program (CDIP) run by the Center for Coastal Studies at Scripps Institute of Oceanography and the Coastal Engineering Research Center (CERC). The data selected comprised significant wave height (Hs), period (T) and direction of approach to the coast at a depth of 10 m and analyzed every six hours. In order to model the variety of wave patterns, the wave data file consisted of all information provided every six hours. Calibration was performed for a whole year (March 1991 - March 1992) minimizing cyclical seasonal displacements of the shoreline (Hanson and Kraus 1989). Verification of proper model predictions was performed for wave and shoreline position data from October 1990 to March 1992.

A 20-year simulation was performed in order to calibrate a wave climate to be used with the model. The wave climate was obtained from deep water significant height, average period and significant direction, taken from a five-year record of wave data based on ship observations measured in the marsden square 28 from 1979 through 1984. These data were complemented with an additional year of wave data from the NOAA buoy 46048 during 1991. The data were considered representative for a wide spectrum of wave climate patterns from which a representative single year of wave data could be extracted. Using the REDSEA model (Worthington and Herbich 1970), the six year data set was transformed to the 10 m isobath shoreward of the thermal-electric power plant. Percent occurrence of significant wave height ranges (calms, 1-1.49, 1.5-1.99, 2-2.49, 2.5-2.99, 3-3.49, 3.5-3.99, 4-4.49 and >4.5m), period bands (<6, 6-7.9, 8-9.9, 10-11.9, 12-13.9 and >14s) and wave direction (S, SSW, SW, WSW, W, WNW, NW, NNW and N) was obtained to elaborate the wave climate. Based on this analysis, a wave data file was compiled on a 1 hour interval for one year. The authors consider six years of data acceptable for a 20 year simulation since Gravens (1989) uses a three years of wave data for 5 and 20 year simulations, and obtained favorable results.

Calibration and verification

Calibration was performed from March 1991 to March 1992. The values of the calibration coefficients were varied until a good visual fit was observed between the measured and the calculated shorelines as proposed by CCSTWS (1991). The k1 and k2 values corresponded to 0.1 and 0.5 respectively for the north reach, and 0.25 and 0.15 for the south reach, where k1 controls the time scale of the simulated shoreline change, as well as the magnitude of the LST rate; and k2 controls the LST produced by the longshore gradients in breaking wave height. An adjustment of the wave incidence angle was performed setting the wave angle amount (ZCNGA) to 20 and 10 on the north and south reach, respectively. This amount is added to the wave angles in the wave data files and is attributed to the variation induced by waves traveling from the measurement location to the Rosarito area ($\approx$ 30 km). Similar criteria was applied by Simpson and Cox (1993) while using wave data from a source outside of the study area.

Maximum differences of 15 m were obtained between measured and calculated shorelines in both model reaches (figure 3), except in the south reach between the groins where a difference of 76 m was found (figure 3). The reason for this disagreement is explained if it is considered that the dredged sand of the fixed dredge is discharged in this area. Net LST for the calibration period was calculated to be 100,000 m^3 to the south considering both reaches. Calibration was accomplished by the agreement between measured and calculated shorelines, as well as by the agreement between the calculated net LST and the value observed by Marmolejo-Lara (1985).

Verification was performed from October 1990 to March 1992 where coefficients k1 and k2 were kept unchanged, obtaining a good visual fit for both reaches (figure 3). The exception was along the stretch of beach between the groins where 50 m differences were observed (figure 3). As pointed out for the calibration procedure, dredged material is brought to this area, and even though sediment supplied by dredges to this area could be modeled, data on dredged sediment is not available.

20 year simulation

Once the model was calibrated and verified it was considered independent of the simulation interval, so a 20-year simulation (1972-1992) could be performed with the one-year, hourly-wave data file. The k1 and k2 coefficient values were kept unchanged, but it was found that better results were obtained if the values of ZCNGA were set to 15 in both the north and the south reaches. This was considered by the authors to account for the variability of the wave climate in the 20 years of the simulation, whereas the wave climate used was elaborated with 6 years of wave data.

Greater differences were found in the stretch of beach between groins, with 85 m of difference (figure 3) as a result of what has already being discussed

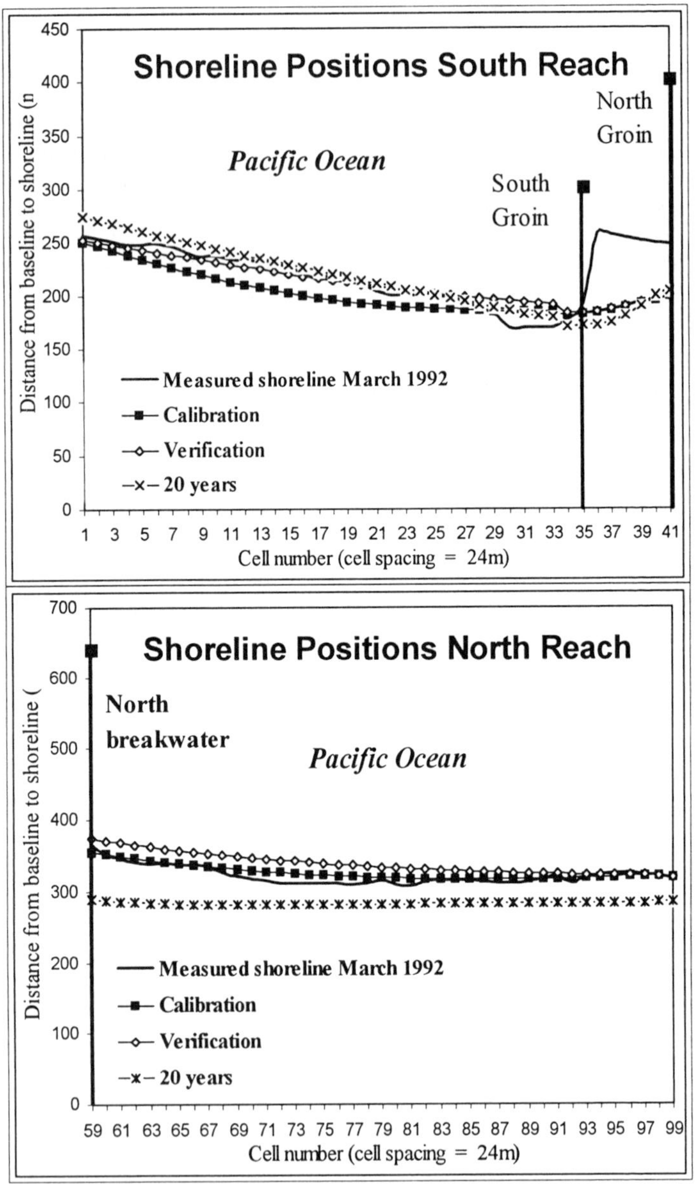

Figure 3. Calculated shoreline positions during model simulations

for the calibration and verification simulations. Differences of 30 m were found near the north breakwater where material from the movable dredges' is discharged. Here it is noticed that the movable dredges discharge is important for the long-term shoreline change, while the fixed dredge discharge is important in the short-term. The one year, hourly-wave data file was considered as representative of the long term wave climate in the area, since the results obtained in this simulation were satisfactory.

Sediment budget

Simulations were performed for each month of the year using the wave statistics. Since the model was calibrated to represent long term conditions using the wave statistics, the values obtained for LST in each month were considered reliable and representative of long term conditions. Sediment budgets were constructed for both seasons representing the LST to the right and to the left hand on each of the grid boundaries, as well as for the volumes of accretion and erosion, and the volume of sediment that enters the water intake (figure 4).

The net LST has a northerly direction in the south reach and a southerly direction in the north reach, so that the sediment is directed to the intake during the whole year. The south reach has shoreline erosion during the winter season and accretion during the summer season, where erosion dominates over accretion by 7,300 m^3/year. This amount represents the net annual erosion to be expected in the long term. In the north reach accretion was found all of the time due to the impact of the north breakwater. The results obtained do not mean that erosion does not occur within the north reach (as observed in the calibration and verification periods), but that in the long term, the sedimentary processes are predominant.

On an annual basis an erosion rate of 7,300 m^3 was found in the south reach and an accretion rate of 132,200 m^3 for the north reach, denoting a non-equilibrium condition. On the other hand, equilibrium conditions were observed from the calibration, verification and 20-year simulations, where the shoreline maintained a defined interval. The discrepancy between these two findings are accredited to the transport mechanisms that are not modeled by GENESIS, e.g., cross-shore transport and rip current induced transport; and the dredges' discharge.

The sediment volumes estimated to enter the intake each month are presented in table 1. These values represent the volume of sediment that could to enter the intake each month of the year due to long term wave and

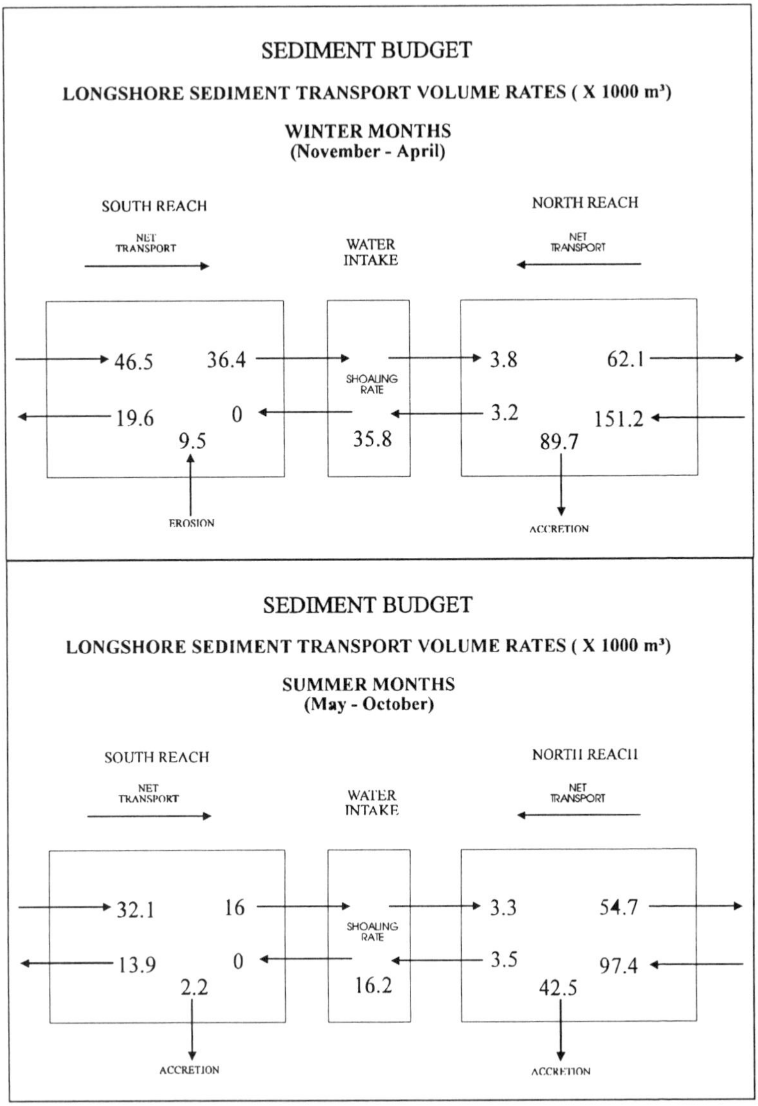

Figure 4. Sediment budgets

Table 1. Volume of potential sediment

Month	J	F	M	A	M	J	J	A	S	O	N	D
Sediment volume (m^3x1000)	8.4	7.1	6.2	3.6	2.9	3.4	2.9	0.9	3	3.1	5.6	4.9

beach conditions at the intake of the plant. It was found that greater volumes of sediment would be transported to the intake during winter months, because of greater wave energy. Volumes may vary because even though winter waves have more energy, the summer waves produce a cross-shore transport with a net direction to the shore, causing accretion. These results may lead to somewhat different estimates for the volumes of sediment that enters the water intake.

Conclusion

The shoreline change model GENESIS was successfully calibrated and verified, giving reasonable good simulations ; the 20-year simulation showed that selection of the wave climate from existing data is adequate. Thus, the results of a one year simulation can be considered as representative of the long term LST rates at the power plant beach. It was found that in the winter season 35,800 m^3 of sediment are potentially able to enter the water intake, while in the summer season this was reduced to 16,200 m^3. The sediment budgets only represent the volume of sediment moved by the LST, so other important factors such as the cross-shore transport, river discharge, wind induced transport, dredge discharge, etc., are not included. More detailed studies must be performed to identify and integrate these factors in order to quantify their sand supply. This work provides valuable information on the longshore sand contributions to the water intake; however, recommendation is made to quantify dredged volumes of sand since it will provide valuable information on sediment distribution as well as for verifying modeling results. Along with studies on other sand contributions, the corresponding agency can use this information to elaborate a dredging program to reduce costs and elevate the power plant efficiency. Because of the lack of on-site data, the results obtained are consider as a first approximation. In this regard, recommendations is made on the gathering of systematic information related to wave parameters, beach evolution and dredged volumes.

Acknowledgments

The authors would like to thank Mark B. Gravens for providing the GENESIS model; William O'Reilly and Don Collins for providing the wave data; Ing. José Antonio Maza, Ing. Carlos Sanchez Linares and Mario Marmolejo-Lara from the Comisón Federal de Electricidad for the information supplied; George W. Domurat and Reinhard Flick for their valuable advice; and Gregory W. Stone and David W. Fischer for their comments on the manuscript.

References

Coastal Data Information Program. Monthly Reports. Center for Coastal Studies Scripps Institute of Oceanography, Coastal Engineering Research Center and the Department of Boating and Waterways.

Coast of California Storm and Tidal Wave Study (CCSTWS). (1991). "State of the coast report: San Diego region." U.S. Army Corps of Engineers, Los Angeles District, Vol. 1, main report.

Davidson, A.T., Nicholls, R.J., and Leatherman, S.P. (1992). "Beach nourishment as a coastal management tool : An annotated bibliography." Journal of Coastal Research, (8)4, 984-1022.

Elizarrarras-Vargas, S.R. (1986). "Determinación de la dirección del transporte littoral predominante por medio de minerales trazadores entre Punta Los Buenos y Playas de Rosarito, BC." B.Sc. thesis, Escuela Superior de Ciencias Marinas, Universidad Autónoma de Baja California, Ensenada, Baja California.

Gravens, M.B. (1989). "A new ocean-entrance system at Bolsa Chica Bay, California: preconstruction assessment of potential shoreline impacts." Proceedings Coastal Zone '89, ASCE, New York, N.Y., 583-594.

Gravens, M.B.; Kraus, N.C., and Hanson, H. (1991). "GENESIS: generalized model for simulating shoreline change. Report 2: workbook and system user's manual." Technical Report CERC-89-19, Coastal Engineering Research Center, U.S. Army Corps of Engineers, Vicksburg, Mississippi.

Hallermeier, R.J. (1981). "A profile zonation for the seasonal sand beaches from wave climate." Coastal Engineering, 4(3), 253-277.

Hands, E.B. (1983). "The Great Lakes as a test model for profile responses to sea level changes." Handbook of Coastal Processes and Erosion, P. D. Komar, editor, CRC Press Inc., Boca Raton, Florida, 167-189.

Hanson, H.; Kraus, N.C. (1989). "GENESIS: generalized model for simulating shoreline change. Report 1: technical reference." Technical Report CERC-89-19, Coastal Engineering Research Center, U.S. Army Corps of Engineers, Vicksburg, Mississippi.

Ludwick, J.C., Kang, H.J., and Reynolds, R.N. (1987). "Loss of filled sand from an Estuarine Groin System." Proceedings Coastal Sediments '87, ASCE, New York, N.Y., 16-26.

Marmolejo-Lara, M. (1985). "Control de azolvamiento en una obra de toma de agua marina." B.Sc. thesis, Escuela Superior de Ciencias Marinas, Universidad Autónoma de Baja California, Ensenada, Baja California.

Moffatt & Nichol, Engineers. (1989). "Historic wave and sea level data report, San Diego region." Coast of California Storm and Tidal Wave Study. U.S. Army Corps of Engineers, Los Angeles District, 88-6.

Simpson, D.P., and Cox, J.C. (1993). "Incremental influence of shore structures determined using a shoreline change model." Coastal Engineering Considerations in Coastal Zone Management, S. A. Hughes, editor, ASCE, 16-26.

Ureña-Portales, E. (1986). "Transporte litoral paralelo a la costa en Playas de Tijuana." B.Sc. thesis, Escuela Superior de Ciencias Marinas, Universidad Autónoma de Baja California, Ensenada, Baja California.

Wiegel, R.L. (1993). "Dana Point Harbor." Shore & Beach, 61(3), 37-55.

Work, P.D., and Dean, R.G. (1995). "Assessment and prediction of beach-nourishment evolution." Journal of Waterway, Port, Coastal and Ocean Engineering, ASCE, 121(3) 182-189.

Worthington, H.W., and Herbich, J.B. (1970). "A computer program to estimate the combined effect of refraction and diffraction of water waves." Coastal and Ocean Engineering Division, Sea Grant publication 219 Report 127, Texas A&M University.

CHAPTER 226

OCEAN BENEFITS OF HEALTHY RIPARIAN SYSTEMS

Scott Clemons[1]

ABSTRACT

Healthy riparian systems (that is, riparian systems that are functioning properly) can provide a vital connection between land and ocean. Resource experts estimate that up to 90 percent of this habitat has been removed and that the remaining portions are fragmented and not functioning properly. This report summarizes the Wildlife Conservation Board's California Riparian Habitat Conservation Program, (CRHCP) and describes how this program relates to the management of California's ocean resources.

CRHCP was authorized by the passage of Senate Bill 906 (Hill) in 1991 and signed into law by Governor Wilson. The enabling legislation firmly established riparian conservation as a high priority for all state agencies whose activities affect riparian habitats. Utilizing several new authorities provided by the statute, the CRHCP's goal is to protect, preserve, and restore riparian habitats throughout the state. A program plan was developed including the following priority objectives:
1. Assess the current amount and status of riparian habitat throughout the state. 2. Identify those areas which are critical to the maintenance of California's riparian ecosystems. 3. Identify those areas which are in imminent danger of destruction or significant degradation. 4. Prioritize protection needs based on the significance of the site and the potential for loss or degradation of habitat. 5. Develop and fund project specific strategies to protect, enhance or restore significant riparian habitat. 6. Develop, fund, and administer a grants program for riparian habitat conservation.

The CRHCP is coordinating development of the statewide riparian habitat inventory and assessment which will be arrayed on a Geographic Information System developed by the University of California, Davis as an element of the multi-agency California Rivers Assessment. Information accumulated by this approach is being used in an interactive statewide riparian conservation decision support model. The program has also utilized the experience and technical expertise of the Department of Fish and Game to recommend available state funding for Wildlife Conservation Board approval and subsequent acquisition and protection of lands which support significant remaining examples of riparian habitats in the coastal zone and throughout the state. The Wildlife Conservation Board has also approved the use of state funds for habitat restoration grants to a wide collection of conservation agencies and nonprofit organizations. The program also coordinates with and assists several major habitat conservation organizations, including the Santa Clara River Enhancement and Management Plan Steering Committee, the National Park Service's Rivers and Trails Conservation Program, the California Partners in Flight Riparian Habitat Joint Venture, the Upper Sacramento River Advisory Council, and the recently formed National Riparian Team.

This report presents several examples of successful coastal zone riparian habitat acquisition and restoration projects and links the projects to the benefits they can provide for the coastal zone and ocean resources. Several California Rivers Assessment habitat assessment products are shown as examples of available information technology. This information can assist riparian conservation strategies in the coastal zone and elsewhere in the state.

---

[1]Riparian Program Manager, Wildlife Conservation Board, 801 K Street, Suite 806, Sacramento, California 95814

Artificial Attachment of Surf Grass (*Phyllospadix* sp.) Within A Nearshore, Rocky Habitat in Southern California

Leray A. de Wit[1], Ronald C. Phillips[2], Robert R. Ware[3], and Leland E. Fausak[4]

Abstract

In July 1993, a total of 60 surf grass planting units, ranging in size from 225 to over 300 cm^2 each, were attached to 14 to 18 kg boulders with epoxy adhesives and placed onto armor rock covering a constructed trench in three water depths (0.9, 1.8, and 3.7 m, MLLW) in Santa Barbara Channel. A four-year monitoring program to record the development, health, and growth of the transplants was initiated. The results of the first two years' monitoring indicate that approximately 50% of the initial 1.3 m^2 of transplanted material and 20 of the 60 planting units had been lost within one year, however, the transplants that remained increased in areal cover by an average of over 57%. Despite extensive epibiota fouling of the armor rock and on the exposed portions of the planting unit boulders, increases of up to 94% in mean areal cover per planting unit was recorded at the -0.9 m station. Kelp shaded much of the transplanted surf grass at the -3.7 m station, however, mean areal cover there increased by 51%. Urchin grazing of the artificially attached surf grass at the -1.8 m station resulted in a net decrease in mean area cover per planting unit there. Annual monitoring is continuing through July 1997.

The methods were effective in securing the surf grass, however the relatively small boulders, which were not attached to the armor rock, were somewhat unstable in the wave-exposed environment. Shading of the transplants by algae which developed at the 0.9 m and 3.7 m stations may have affected growth of surf grass, and urchins grazed heavily on the 1.8 m station transplants. Larger scale transplant efforts should consider the possible negative effects of removing extensive amounts of healthy surf grass from existing beds. Monitoring of the transplants should be should be at least quarterly in order to allow the removal of algae from around the transplants.

- - - - - - - - - - - - - - - - - - -

[1]Leray A. de Wit, Consultant, 2054 Bluerock Circle, Concord, California 94521; phone: (510) 685-9441 Fax (510) 685-9401 e-mail: ladewit@aol.com

[2]Battelle, Pacific Northwest Laboratories, P.O. Box 999, MSIN: K6-05, Richland, Washington 99352; phone: (509) 372-3691 Fax (509) 376-6663

[3]Coastal Resources Management, 3334 E. Coast Highway, Corona del Mar, California 92625; phone: (714) 673-3076 Fax (714) 673-8652 e-mail: crmrrw@aol.com

[4]Marine Resource Consultants, P.O. Box 2057, Santa Barbara, California 93120; phone (805) 898-0401 Fax (805) 898-0501 e-mail: lefausak@msn.com

Introduction

Seagrasses include approximately 48 species of grass-like flowering plants found in the shallow-water coastal areas of the world between the Arctic and Antarctic Circles. Most form dense meadows over sedimentary bottom, while at least five species, including two within the genus *Phyllospadix*, are associated with solid substrate. *Phyllospadix* occurs exclusively on rocky substrate (Phillips and Menez, 1988). Along the west coast of the United States, three species of *Phyllospadix* are found attached to boulders and rock benches within the intertidal and shallow subtidal zones along exposed coastlines. The preferred habitat of this genus has resulted in species being collectively called "surfgrass". *P. scouleri* (Hooker) occurs from Sitka, Alaska to Baja California while *P. torreyi* (S. Watson) occurs from the northern tip of Vancouver Island to Baja California; the later is most abundant south of Monterey, California. *P. serrulatus* is found on mud and solid substrates from the Gulf of Alaska to the Puget Sound, Washington (R. Phillips, pers. comm.).

Phyllospadix is long-lived, slow growing, and forms dense beds that have been found to preclude macroalgal development below the canopy. Blade length of up to 2 m has been documented by Phillips and Menez (1988) and Stewart and Meyers (1980). Attachment to the substrate is by rhizomes and roots. Rhizome growth ranges from 2.5 to 6.0 cm/year in the intertidal zone (Turner, 1985). Reproduction is vegetative (i.e., rhizome growth onto uninhabited substrates) and sexual, with plants in southern California releasing seeds in the spring and fall (Phillips, pers. comm.). Previous experiments have found that *Phyllospadix* is slow to recover from disturbance and that large (*e.g.* 0.25 m^2 or greater) disturbances can "...drastically alter the community" (Turner, 1985). The surfgrass beds support an abundant and diverse algal and invertebrate community; many of the associated species are unique to that habitat (Stewart and Meyers, 1980). Extrapolating from observations by several researchers on subtidal seagrass meadows of that those meadows provide shelter and refuge for resident and transient adult and juvenile animals, many of which are of commercial and recreational importance (Phillips and Menez, 1988), it has been postulated that surfgrass beds serve a similar function. This expectation, coupled with the slow recovery rate of disturbed beds, was the primary reason for attempting to mitigate the damage to an existing *Phyllospadix* bed offshore of the mouth of Las Flores Canyon within Santa Barbara Channel, approximately 48 km. west of Santa Barbara, California. As a result of agency decisions, a series of studies were conducted to ascertain the possibility of artificially attaching surfgrass to various solid substrates and transplanting those substrates into the affected area.

As documented in Phillips, de Wit and Fausak (1992), a major oil company funded a study designed to test the efficacy and feasibility of artificially attaching and transplanting *Phyllospadix* rhizomes and mats to replace vegetation lost during construction within a nearshore area. The initial study, which focused on identifying methods of attaching surfgrass rhizomes to artificial substrates, concluded that 1) 13.6 to 18.1 kg (30- to 40-pound) boulders could be handled by divers and were apparently stable within the nearshore habitat to be utilized as substrate for attachment of surfgrass rhizomes; 2) incremental sand burial did affect blade growth and therefore future transplants should be placed in rocky habitats; 3) patches of surfgrass, nominally 10 cm by 10 cm each, should be used as planting units instead of individual rhizomes; 4) monitoring of transplants should occur for at least one year; and 5) the epoxy adhesives used did not appear to be acutely toxic to the surfgrass.

Following the initial study, the authors prepared a work plan for removing and transplanting 1.3 m^2 of naturally attached surfgrass from a nearby bed to the affected area. Upon approval from several county, state, and federal resource agencies and following preliminary surveys to locate donor surfgrass beds, the transplanting was initiated in July 1993.

The study had three distinct tasks: to ascertain the effectiveness of plastic mesh to collect surfgrass seeds; to document what, if any, effects clearing algae around naturally attached surfgrass had on increasing rhizome growth, and to monitor growth of artificially attached surfgrass.

Methods

In an effort to remove existing surfgrass from within the proposed construction zone prior to trenching, a preliminary survey to locate 14 to 18 kg boulders with naturally attached surfgrass within the intertidal and shallow subtidal habitats near the mouth of Las Flores Canyon/Corral Creek was completed. Finding only one boulder of that size with attached surfgrass, a second survey to document existing conditions within the trench corridor, locate the required number of "clean" boulders to be used as the transplant substrate, and to locate proximal "donor beds" of naturally attached surfgrass was conducted following construction. The results of the later survey indicated that the trench corridor, which had been covered with armor rock, was suitable to support the transplanted surfgrass boulders; a sufficient number of boulders within the weight range existed within the upper intertidal zone of the site; and naturally attached surfgrass was abundant within the intertidal and shallow subtidal habitats immediately adjacent to the trench corridor. The transplanting commenced on July 26, 1993.

The first task in the transplant study was the collection of 105, 14- to 18-kg boulders. Epoxy was used to attach a uniquely numbered aluminum tag to each boulder, and a photograph was taken of each boulder. Wide-blade paint scrapers were used to remove approximately 50 mats of naturally attached surfgrass from existing beds. The removed mats were stored in buckets of seawater prior to attachment to the boulders. Individual planting units, averaging 225 cm^2 each, were cut from the mats and attached to the boulders by one of two epoxy adhesives (SuperGlu and Z-Spar Splash Zone Compound), which, due to their short setup/drying time, had been found to be the most effective (Phillips, *et al.*, 1992). A total of 61 transplants (32 attached with SuperGlu and 29 with Z-Spar) were attached to boulders, transported to one of three subtidal areas (20 each in water depths of 0.9, 1.8, and 3.7 m [MLLW]) within the trench corridor, and placed onto the armor rock. In addition to the adhesive, several of the transplants were also secured with elastic cord to enhance attachment of the surfgrass.

Underwater photographs of each area were taken following placement of all transplants and the location of each station was surveyed from shore markers. A small, subsurface buoy was placed at each station to facilitate divers finding them during subsequent monitoring surveys.

Monitoring surveys have been conducted over two years consisting of biweekly observations for the first month, then quarterly for next 23 months. Annual monitoring is continuing through July 1997. Monitoring surveys consist of diver-biologists observing and recording the changes in the surfgrass and surrounding habitat. Areal cover of the individual transplants was measured by a hand-held ruler placed along two sides of the surfgrass mats. Measurements were recorded to the nearest cm. Observations included documentation of the general health of the transplanted surfgrass, fouling of the substrate, and the number and areal cover of the transplanted and naturally attached surfgrass at each station and within surrounding control areas. Control areas, consisting of pre-mapped portions of the exposed armor rock around the transplant area, ranged from 19 to 25 m^2 each. Observations on the success of the plastic mesh and algal clearing were also completed during each survey and overturned boulders were uprighted and returned to the transplant area.

During the first two monitoring surveys, attempts were made to reattach surfgrass mats that had been detached from the boulders. However, turbulence reduced the effectiveness of the adhesive underwater. Further reattachment efforts were not attempted.

Results

Approximately 40 hours of diver observations have been completed during the nine monitoring surveys conducted from August 1993 through July 1995. The January/February 1995 monitoring was cancelled due to poor water clarity and sea conditions. Although some blade discoloration and increased attached epibiota were observed during the August and October 1993 (one and three months after the initial transplant), new blades and rhizome growth were first observed three months after the initial transplant (October 1993). Seed pods were observed within the naturally attached surfgrass in July, August, and October of each year, while the first seed pods on the transplants appeared in July 1994, 12 months after the initial placement. Abundant macroalgae developed on the armor rock throughout the survey area during the spring and fall surveys, complicating the divers' ability to occupy the stations. Table 1 compares the total and mean cover of the transplanted surfgrass of the initial planting (July 1993) to that recorded in the two-year (July 1995) survey.

Table 1
Summary of Changes in Total Area Cover (cm^2) and Mean Area (cm^2) of Transplanted Surfgrass
(July 1993 through July 1995)

Total Area Cover (cm^2)

STATION	JULY 1993	JULY 1995	AREA CHANGE (% CHANGE)
-0.9 m.	4,158	3,225	-933 (24%)
-1.8 m.	4,063	1,000	-3,063 (-75%)
-3.7 m.	5,296	2,800	-2,496 (-47%)
All Stations	13,517	7,425	-6,092 (-45%)

Mean Area Cover (cm^2)

STATION	JULY 1993	JULY 1995	AREA CHANGE (% CHANGE)
-0.9 m.	207.9 (20)[a]	403.1 (8)[b]	+195.2 (+94%)
-1.8 m.	203.2 (20)	200.0 (5)	-3.2 (-2%)
-3.7 m.	264.8 (20)	400.0 (7)	+135.2 (+51%)
All Stations	225.3 (60)	353.6 (20)	+128.3 (+57%)

a Number of surfgrass mats initially transplanted (July 1993)
b Number of surfgrass mats remaining (July 1995)

As this table indicates, after two years, the total areal cover of the transplants decreased by approximately 45%, and 66% of the transplants have been lost. The mean area for the remaining transplants has, however, increased by an average of 57% (225.3 cm^2 to over 353 cm^2). The transplanted mats at the -0.9 m station have had the largest increase in mean cover (+94%), while those at the -3.7 m station have increased by approximately 51%. Twenty of the initial 60 transplants remained after two years.

Urchin grazing on the transplants at the -1.8 m station, apparently resulted in the mean area for transplants there decreasing by 2% over the two year period. Urchin densities of up to 10/m^2 have been observed there. In addition, the reduction in the number of transplants observed at the -0.9 m station may be attributable to the coalescing of adjacent transplants.

Table 2 provides a survey-by-survey summary of the areal cover of transplants at each station and the total for all stations through the July 1995 monitoring. The total areal cover of the transplanted surfgrass declined through the first eight months of monitoring (October 1993), followed by minimal increases in January, August, and October 1994. No survey was conducted in January/February 1995 due to poor water clarity resulting from almost continuous storm activity.

Table 2
Summary Area Cover, Number of Transplants, Mean Cover, and Changes in Areal Cover Per Survey (July 1993 through July 1995)

All Stations

	7-93	8-93[a]	10-93	1-94	4-94	8-94	10-94	4-95[b]	7-95
Total Area (m^2)	1.35	1.34 1.26	0.82	0.84	0.66	0.77	0.92	0.24	0.74
Total Trans-plants	60	60 59	51	48	36	50	37	22	20[c]
Mean Area per trans-plant (cm^2)	225.3	219.0 212.8	160.6	175.9	167.0	153.0	247.3	109.1	353.6
Change in Total Area Cover from Initial (cm^2)	---	-105 -962	-5,326	-5,074	-7,491	-5,868	-4,367	-11,117	-6,092
Change in Mean Area Cover from Initial (cm^2)	---	-6.3 -12.5	-64.7	-49.4	-58.3	-72.3	+22.0	-116.2	+128.3

a Two surveys completed in August 1993; top number is two weeks after initial transplant, bottom number is after four weeks after initial transplant.
b No survey in January/February 1995 due to storm conditions.
c Number does not include one apparently naturally attached surfgrass patch that was observed at the -0.9 m. station during this survey.

While only 10 boulders with attached surfgrass were lost over the first 12 months (through August 1994), only 22 boulders were found during the April 1995 monitoring, the first survey following substantial storm activity during the winter of 1994-95. Poor water clarity contributed to the divers' inability to locate all of the transplants during that survey.

Discounting the April surveys as aberrations of overall trend, the stability of the boulders, both in number and total area, is evident after approximately six months in-place (through January 1994). Although sloughing of the transplanted surfgrass was evident during the first three surveys, blade growth was observed at all of the stations during the third survey (October 1993); new blade production and rhizome growth has continued through July 1995.

Discussion

The primary objective of the study was to monitor the success of transplanting artificially attached surfgrass into an area affected by construction. In addition, this project must also be considered an experiment to ascertain whether relatively large scale artificial attachment of living surfgrass would be a viable method of restoring an impacted resource. Previous studies had limited success in maintaining viable plants from the attachment of individual rhizomes. Using mats of surfgrass was expected to increase the chances of survival and provide researchers the opportunity to observe growth on larger vegetative transplants.

The testing of several adhesives during earlier experiments resulted in the identification of two epoxy glues that were not toxic to the transplants and were relatively easy to use in the field. Set up time for the adhesives was short and relatively small amounts were sufficient to adhere the vegetative material to the substrata.

The observations to date indicate that the method utilized is effective in securing the material to the substrates and artificially attached surfgrass does grow and produce seed pods. If, as has been documented in previous studies, the natural recovery of *Phyllospadix* is a relatively slow process, the transplanting of artificially attached material may decrease the recovery time of the habitat. Through July 1995, we have found only four naturally attached surfgrass patches within the approximately 75 m^2 control area around the transplant locations. Three plants (total area approximately 50 cm^2) first appeared in October 1994 at the -0.9 m. station. By April 1995 a fourth was recorded. Combined, those plants covered approximately 100 cm^2 in April 1995, but they were not found in the July 1995 survey. Despite the extensive algal growth around and within the -0.9 m station, the transplanted surfgrass there was the most robust of all three locations. A limiting factor to surfgrass growth at the deeper station appears to be reduced light, due both to water depth and from shading by kelp and other algae.

Since the use of mechanical equipment for placing the boulders was not permitted within the study area, we were limited to using boulders that could be manually handled on the beach and in the water. The relatively small size and weight of the boulders likely contributed to the loss of some of the boulders from wave activity. We would suggest that larger boulders be utilized in subsequent transplants and, if possible, those boulders be attached to the surrounding natural substrate by fast-setting concrete or other adhesive. We have also found that surge, poor water clarity, and/or heavy algal growth at the two shallow stations complicated diver observations. The subsurface buoys placed during the initial transplant were lost by the third survey, and diver-biologists had to rely on subsurface and shoreline features to locate the stations.

This method of introducing surfgrass into existing habitats should be viewed as a way of reducing the time for natural reestablishment of surfgrass, however its use in larger scale efforts must consider the potential effects of removal of extensive amounts of surfgrass from healthy beds. A 50% loss of boulders and transplants following initial placement should be expected, particularly if the boulders are not secured to the natural substrate.

Acknowledgements

We wish to thank Exxon Company, U.S.A., Santa Barbara County Energy Division, California Department of Fish and Game, California State Lands Commission, California Coastal Commission, and the National Marine Fisheries Service for support and guidance during this study.

References

Phillips, R.C. and E.G. Menez (1988) Seagrasses Smithsonian Contrib. Mar. Sci., No. 34. 104 pp.

Phillips, R.C., L.A. de Wit, and L.E. Fausak (1992) Transplantation of surfgrass (*Phyllospadix torreyi*) into a high energy shallow coastal zone near Santa Barbara, California. Presented at National Society of Wetland Scientists, New Orleans, LA., June.

Phillips, R.C. (personal communication). Associate, Battelle Northwest Laboratories, Richland, Washington.

Stewart, J.G. and B. Meyers (1980) Assemblages of algae and invertebrates in southern California *Phyllospadix*-dominated intertidal habitats. Aquatic Botany, 9, pp. 73-94.

Turner, T. (1985) Stability of rocky intertidal surfgrass beds: persistence, preemption, and recovery. Ecology Vol. 66, No. 1 pp. 83-92.

CHAPTER 228

A CASE STUDY ON THE STORM SURGE NORTHEAST OF TAIWAN

H.S. Hou, Y.H. Wang* and S.H. Chen**

Inst. of Transportation, Ministry of Trans. & Communication, Taiwan, ROC

*Institute of Oceanography, National Taiwan University, Taipei

**Union-Tech Engineering Consultants Company, Taipei

ABSTRACT: The impact of storm surge is the single most serious disaster to the coastal area of Taiwan during typhoon seasons. For example, in some lower coastal areas, lots of properties, or even lives were lost each year due to storm induced surges and flooding. Many coastal structures, seawalls and breakwaters were damaged or destroyed by typhoon waves in associated with storm surges. In August 1994, the Typhoon Doug caused serious damage to the breakwaters and docking facilities at the Lungtung harbor, a newly constructed yacht harbor at the Northeast of Taiwan. The controlling factors of this storm surge are studied by analyzing field data including tide and wave measurements and meteorological records collected in several nearby land-base stations. The results show that the serious impacts are due to large waves associated with swells generated by the typhoon. There was an extraordinary sea level rise lasted for about 12 hours at the peak of typhoon. Under such circumstance, the wave forces can reach a higher level of the coast line to damage the docking facilities. This investigation points out the importance of the local geometry to the pile-up of wind driven currents and the change of near-shore bottom topography to the modification and deformation of deep water waves. The enhanced force due to the interaction of swells and local wind waves is believed to be important too. It is suggested that the number of waves and their duration, in addition to the significant wave height and period, are important in the designing and construction of coastal structures and facilities.

1. INTRODUCTION

As located at the Northwest of sub-tropical Pacific Ocean, Taiwan experiences a lot of typhoons in the summer every year. The typhoon associated rain, flooding, waves and surges always cause serious damages to the coastal area. For example, in August 1994, the Typhoon Doug caused serious damages to the breakwater of Hualient and Suao Harbors in the eastern Taiwan (Figure 1). This typhoon also destroyed many harbor facilities and coastal structures at the Lungtung yacht harbor northeast of Taiwan (Chang and Tseng, 1996). On the other hand, the Typhoon Tim, passing through middle of Taiwan, did not cause big problems at the northeast corner of Taiwan. The motivation of this paper is to study what are the factors that cause the unusual forces during the Typhoon Doug.

The study area, Lungtung, is located at the Northeast of Taiwan (Figure 1). In this about 20 km coastal line is a bay-shape geometry with Sandiao Cape at the south end and Pitou Cape at the north end. To the East is the Pacific Ocean that Lungtung receives strong northeast monsoon during the winter. The coastal line around the harbor site is rather straight which make it lack of good sheltering from the northeasterly. The water depth is less than 10 m mostly within 1 km offshore from the harbor site and then increases sharply to over 100 m at the bay mouth.

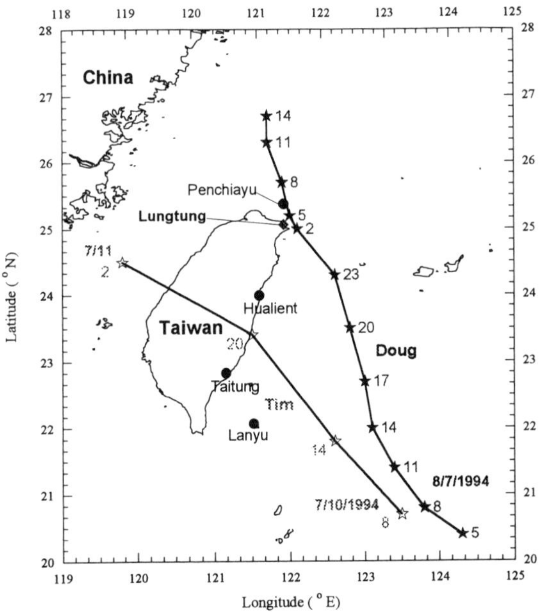

Figure 1. The map of study area indicating the passages of two typical Typhoons -Doug and Tim. The stations of wave(Taitung) and winds(Lanyu, Penchiayu) are also indicated.

2. DATA ANALYSIS AND RESULTS

Data used in this study include sea level measurements, waves and meteorological records. The sea level data are obtained from Yenliao station, about 4 km south of Lungtung. The wind data are from 3 stations including Yenliao, Lanyu and Penchiayu (see Figure 1). The wave data are measured at Taitung, located in eastern Taiwan, since there is no measurement available at the study site. Although a total of one year data are analyzed, the presentation is focus on the summer typhoon season, July and August of 1994.

2.1 Typhoons

During 1994, there were six major Typhoons (Figure 2) passing Taiwan: Tim(7/9), Caitlin(8/3), Doug (8/8), Fred (8/20), Gladys(9/1) and Seth(10/10). The Typhoon Doug was originated at 17.5N, 127.4E on August 6 with a center pressure of 930mb (Figure1). This typhoon was heading northeast on the route to the Northeast of Taiwan. Doug stayed at about 20 km off Lungtung for over 3 hours on early morning of August 8. This typhoon then left Taiwan slowly heading to the North. Wind speed reached over 40 m/s during the passage of Doug while the year round wind speed at the area is about 5 to 10 m/s (Wang and Chuang, 1996).

2.2 Sea levels

Figure 2 shows the sea levels recorded at Yenliao station. The tidal range is about 1 m with the high water line at about 60 cm above mean sea level. A sharp rise of sea level occurred at 8/8, which is about twice of the regular tidal variations. To distinguish the storm induced sea level rise, a tidal harmonic analysis (Foreman, 1979) is applied to the hourly sea level records of 1994. The results show that the major tidal components are O1, K1, M2 and S2. The form ratio is 1.1 indicating that the area is belong to the mixed-tide with semi-diurnal tide dominated. Comparisons of tidal predictions with measurements show that the storm induced sea level rise is 1.37 meters during the passage of Doug (Figure 3). The winds were from the northeast with a speed of about 5 m/s on August 7. A small sea level rise can be identified. At the evening of 8/7, the continuing northeast wind increased to a maximum speed of 40 m/s. This 8-hour on-shore strong wind is believed to be the major factor caused the peak of sea level rise.

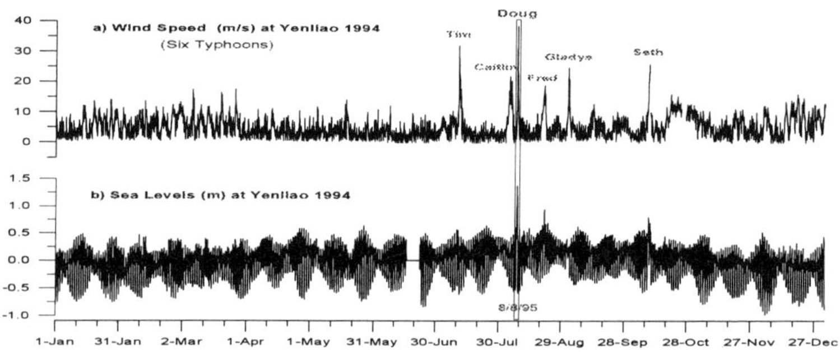

Figure 2. Time series of a) winds (m/s) and b) sea levels (m) at Yenliao during 1994.

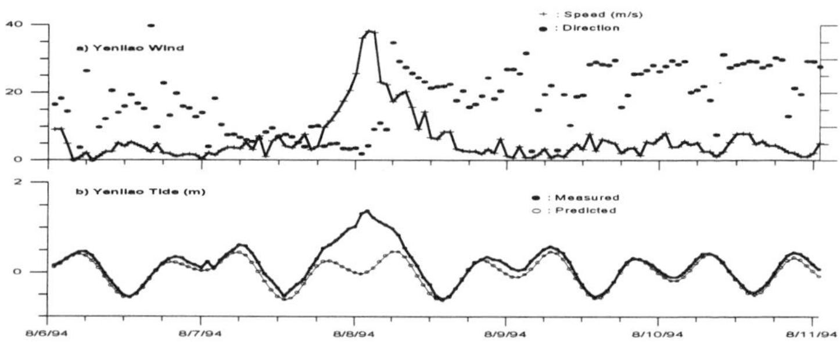

Figure 3. Time series of a) wind speed (solid line) and direction (dots), b) sea levels of measured and prediction at Yenliao during the passage of the Typhoon Doug.

In contrast to Doug, during typhoon Tim the sea level did not show large increase at the Lungtung area (Figure 4). The winds were from the Northeast with a speed of 5 m/s on 7/9. This on-shore wind caused a small sea level rise. However, at the evening of 7/9 when the wind speed started to increase, the wind direction also turned to the southeasterly. Only 40 cm of sea level rise occurred at the peak of the Typhoon Tim. The continuing southerly wind (off-shore wind) even caused a small drop of sea level (about 20cm) at the early morning of July 10, 1994.

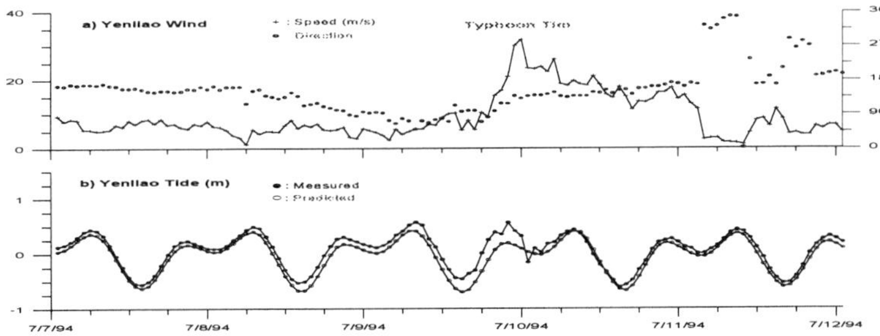

Figure 4. Time series of a) wind speed (solid line) and direction (dots), b) sea levels of measured and prediction at Yenliao during the passage of the Typhoon Tim.

2.3 <u>Wind and waves</u>

Since there are no local wave measurements available, data from Taitung are used to indicate the wave conditions during typhoons of 1994. The wave records at Taitung (Figure 5) indicate that the significant wave height was about 10 m during Tim and 5 m during Doug. The significant wave period was about 15 seconds for most of the typhoons in 1994. A special note is that the duration of waves during Doug was much longer. This indicates that the number of large waves during typhoon Doug should be more than other typhoons. In order to imply the wave conditions at Lungtung area, wind data from Lanyu (an island near Taitung) and Penchiayu (an islet near Lungtung) are link to the wave records of Taitung.

Figure 6 shows the time series of wind vectors at the Lanyu and Penchiayu stations from July to October of 1994. At the Lanyu station, the wind speeds during Tim were larger than the other typhoons of 1994. This agrees with the bigger wave records as shown in Figure 5. Figure 6 also shows the variation of atmospheric pressure (low center) during the passages of typhoons. It is clear that there are bursts of northeasterly wind occurred right before the low center arrived. This is because typhoon is an anti-cyclone and most of the typhoons are coming from the southeast of the weather stations in the study area.

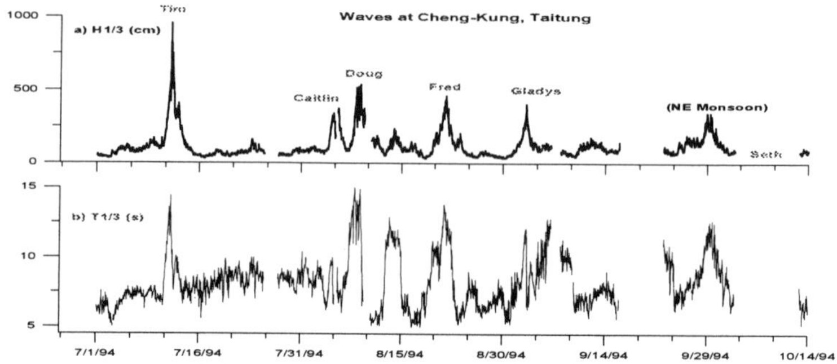

Figure 5. Time series of significant a) wave height b) wave period at Taitung during the summer typhoon season of 1994.

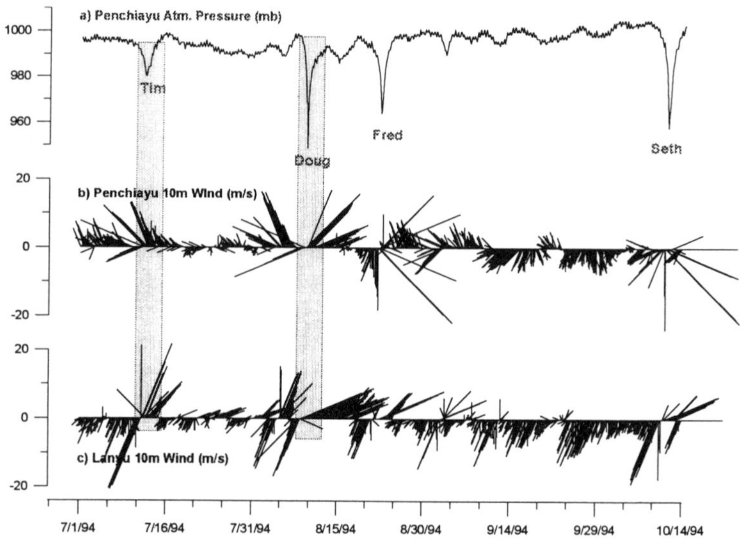

Figure 6. Time series of a) atmosphere pressure b) wind speed vector at Penchiayu and c) wind speed vector at Lanyu during the passage of typhoons Tim and Doug.

At the Penchiayu station, although most of the winds are from the southeast during the passage of Tim and Doug, the northeasterly wind is responsible to the waves at the sea northeast of Taiwan. The magnitude of the northeasterly during Doug is much stronger than that of Tim. Previous reports suggest that waves generate by Tim reached 16 m at Hualient harbor (Chang and Tseng, 1996). Assuming the wave height is proportional to the wind speed and duration, Doug should generate waves over 20 m at the sea northeast of Taiwan during the passage of typhoon (the estimation of maximum wave height can be find in, such as, Shore Protection Manual of U.S. Army Corps Engrg., 1984). The large wind induced waves, in additional to the swells, during the high water line must have strong impact to the coastal structure and docking facilities at the Lungtung harbor.

The water mark left on the wall of building at Lungtung harbor suggests that there was a maximum of 3 m of sea level rise during the passage of typhoon Doug. If this is real, the actual sea level rise is double that of measured at the near by tidal station - Yenliao. The reason for this additional water pile-up can be very complex. One possibility is due to the shoaling and breaking of large waves and than accumulated at the coastal line. The platform like shallow bathmetry near the coastal line may play an important role in this process. The detail of wave pile-up and associated effect will require further investigations.

3. SUMMARY

The passage of typhoon Doug over the northeastern Taiwan was used as a typical example to study the storm induced surges and its consequence effects. The results show that the sea level variations are related to the local wind speed, direction and duration. The local geometry is a critical factor to the pile-up of sea levels in storm surges. It is suggested that the combined effects of wind wave and swell are important to the destruction of coastal structures during storm surges. The bottom topography variations and the number of waves, in additional to the significant wave high and period, are important too.

ACKNOWLEDGEMENTS

The authors greatly appreciate the Taiwan Power Company and the Central Weather Bureau of ROC provide valuable tide, wave and weather records.

REFERENCES

[1] Chang, C.K. and H.M. Tseng (1996). "Investigation of the Hwa-Lain Harbor Instability," The 17th Ocean Engineering Conference, Tainan, Taiwan, 131-143.

[2] Foreman, M.G.G. (1979). *Manual for Tidal Heights Analysis and Prediction*, Pacific Marine Science Report No.77-10, Institute of Ocean Sciences, Patricia Bay, Victoria, B.C.

[3] U.S. Army Corps Engrg. (1984). *Shore Protection Manual*, Government Printing office, Washington, D.C.

[4] Wang, Y.H. and W.S. Chuang (1996). "Surface wind measurements around Taiwan," Meteorological Bulletin, Central Weather Bureau of ROC, V41, #2, 81-89.

CHAPTER 229

Carpinteria Offshore Field Study
Offshore Southern California

Jeff Kennedy
Minerals Management Service, Pacific OCS Region

The California State Lands Commission (CSLC), Minerals Management Service (MMS), Pacific Operators Offshore, Inc. (POOI), and Los Alamos National Laboratory (LANL) are working together to build a reservoir simulation of the Carpinteria Offshore Field. The field located in the eastern Santa Barbara Channel, is the only offshore field in California straddling the State/Federal boundary which has produced both from Federal (POOI and Torch Energy) and State (Chevron and Exxon) leases. Chevron Oil Company started to abandoned the State portion of the field in 1992 when they considered the field to be no longer economical . In August 1996, Chevron removed the State platforms (Hope and Heidi). CSLC reassign the mineral rights to POOI to the State portion of the field in November 1996.

The CSLC approached MMS in 1993 to conduct a joint study of the field to look for opportunities ro recover bypassed oil within both the State and Federal portions of the field. The study group was expanded to include the Federal lessee, POOI and LANL. At this time, a reservoir simulation is currently being developed by the group to identify areas of bypassed oil within the field. POOI has received funding from the Department of Energy to drill a tri-lateral horizontal well within the field. Results from the reservoir simulations, which will be finished later this year (1997), will be used to identify the best location for the well or wells if the field warrants full secondary recovery operations. With the removal of State platforms, the remaining reserves within the State portion of the field can only be developed from the existing Federal platforms. The CSLC has expressed interest in a joint agreement with MMS in recovering any royalty allocation (revenue) from future production within the State portion of the field. This session, describes the findings of the our cooperative study.

CHAPTER 230

THE SOUTHERN CALIFORNIA MARINE INSTITUTE (SCMI)- A RECENTLY FORMED CONSORTIUM OF SOUTHERN CALIFORNIA UNIVERSITIES

Lon L. McClanahan and Don Newman

Abstract

The Southern California Marine Institute is a recently (1193) formed consortium representing seven campuses form the California State University System (Dominguez Hills, Fullerton, Long Beach, Los Angeles, Northridge, Pomaona, San Marcos), the University of Southern California, and Occidental College. SCMI is a strategic alliance of marine groups that maximizes the resources of participation entities, and combines the collective knowledge of the members. SCMI provides modern facilities located at Fish Harbor, Terminal Island, Calif. The facilities include laboratories, classrooms, storage space, and a machine shop. SCMI also provides a state certified laboratory with state of the art equipment for marine environmental chemistry. SCMI maintains a fleet of four modern seagoing research vessels operated by affiliate institutions. SCMI has 10,000 sq. ft. of deep harbor space that accommodates the vessels.

The mission of SCMI is to provide marine support in teaching and research for all the member campuses as well as other educational institutions throughout Southern California. SCMI has also initiated outreach programs to K - 12 schools in the Los Angeles and Long Beach Unified School Districts providing student and teacher training in marine sciences.

Lon L. McClanahan, Don Newman, Southern California Marine Institute, 820 South Seaside Avenue, Terminal Island, CA 90731

CHAPTER 231

Review of California's Marine Bacterial Standards

Linda P. O'Connell[1] and Francis H. Palmer, D. Env[2]

Abstract

The California Ocean Plan contains bacterial standards for protecting two beneficial uses: water contact recreation and shellfish harvesting. These bacteriological standards are based on coliforms as indicator organisms. Indicator organisms are used as surrogates for disease-causing organisms, with the assumption that high levels of the indicators imply the presence of fecal contamination. The reliability of using bacteriological indicators to predict the presence of non-bacterial pathogens, such as enteric viruses or protozoans, has been questioned. Research is being conducted to determine if there is a more appropriate indicator organism for use in marine waters. Two questions regarding microbiological near-shore ocean water quality are frequently raised:

1) In 1986, U.S. EPA recommended that states adopt an enterococcus standard for marine waters. Is enterococcus a more appropriate indicator of viral contamination than total and/or fecal coliform?
2) Should fecal coliform or another indicator organism be added to the total coliform Ocean Plan standard currently in place for shellfish harvesting areas?

The State Water Resources Control Board (State Board) has been working cooperatively with various public and governmental agencies to answer these questions. With respect to enterococcus, the State Board staff convened the Microbiological Advisory Committee (MAC), representing academia, ocean dischargers, and the State Department of Health Services (DHS). The MAC designed a study to review monitoring data from two large ocean dischargers and to review recent epidemiological studies to determine if one indicator organism is superior for California use. This study report is in draft form and

----

[1]Environmental Specialist, Ocean Standards Unit, State Water Resources Control Board
[2]Chief, Ocean Standards Unit, State Water Resources Control Board
901 P Street, Sacramento, California, 95814

should be final by mid-1997.

Several studies have been initiated to address shellfish harvesting waters. The State Water Resources Control Board (State Board), working cooperatively with Regional Water Quality Control Boards (Regional Board), DHS, Department of Fish and Game, and California State University at Hayward, conducted a study to identify pollution sources impacting shellfish growing areas in Tomales Bay. Results of this study should be available by September 1997. A similar study began in 1996 to identify pollution sources impacting Morro Bay. The Morro Bay study is a cooperative effort involving State Board, Regional Board, DHS, Morro Bay National Estuary Program, and interested citizens.

Introduction

The State Board is the public agency with primary responsibility for protecting California's water quality. The State Board, headed by a five-member Board appointed by the Governor, works in conjunction with nine Regional Boards located in each of the State's major watersheds. The Regional Boards are charged with maintaining the quality of water bodies within their region. Six of the nine Regional Boards are located along California's coastline.

The California Ocean Plan establishes water quality objectives for California's ocean waters and is the basis for regulation of wastes discharged into the State's coastal waters. The Ocean Plan, first adopted in 1972, describes beneficial use designations, water quality objectives, requirements for management of wastes, effluent and receiving water requirements, discharge prohibitions, and general provisions for exceptions and monitoring programs. The State Board adopts the Ocean Plan, and both the State Board and the six coastal Regional Boards interpret and implement the Ocean Plan.

Federal and State laws require that the Ocean Plan be reviewed at least once every three years. During the last Triennial Review, comments were received concerning the adequacy of the bacterial standards contained in the Ocean Plan, both for the protection of water-contact recreation and for shellfish harvesting areas. These comments raised valid questions that need to be investigated. Staff from the State Board's Ocean Standards Unit have been working with staff from Regional Boards, local governments, environmental groups, private industry, and other interested parties to investigate these questions.

Water-Contact Standards

Discussion:

The Ocean Plan contains the following bacteriological standards for the protection of water-contact recreation: "Within a zone bounded by the shoreline and a distance of 1,000 feet from the shoreline or the 30-foot depth contour, whichever is further from the shoreline, and in areas outside this zone used for water contact sports, as determined by the Regional Board, but including all kelp beds, the following bacterial objectives shall be maintained throughout the water column:

a. Samples of water from each sampling station shall have a density of total coliform organisms less than 1,000 per 100 ml (10 per ml); provided that not more than 20 percent of the samples at any sampling station, in any 30-day period, may exceed 1,000 per 100 ml (10 per ml), and provided further that no single sample when verified by a repeat sample taken within 48 hours shall exceed 10,000 per 100 ml (100 per ml).

b. The fecal coliform density based on a minimum of not less that five samples for any 30-day period, shall not exceed a geometric mean of 200 per 100 ml nor shall more than 10 percent of the total samples during any 60-day period exceed 400 per 100 ml.

The adequacy of total and fecal coliform bacteria as indicators of human disease-causing organisms has been questioned for a number of years. Indicator organisms are used as surrogates for disease-causing organisms, with the assumption that high levels of the indicators imply the presence of fecal contamination. However, bacterial indicator organisms may not be reliable predictors of non-bacterial pathogens, such as enteric viruses or protozoans. In 1986, EPA recommended that states adopt an enterococcus standard for marine waters, based on epidemiological studies conducted in east coast waters. These studies showed enterococcus to be better correlated with human illness than were total and fecal coliform bacteria. Like the coliform bacteria, enterococcus bacteria are a group of bacteria that are normally found in the gastrointestinal tract of warm-blooded animals. However, there is concern that the correlations developed in the EPA studies may not be applicable to the cooler California waters. To resolve the issue of which bacterial group would be a better indicator organism, the Ocean Plan was amended in 1990 to require dischargers to measure enterococcus density at all stations where total and fecal coliform monitoring are required. Also, if a shore station consistently exceeded a coliform objective or exceeded a geometric mean enterococcus density of 24 organisms per 100 ml for a 30-day period or 12 organisms per 100 ml for a six-month period, the Regional Board was to require the appropriate discharger to conduct sanitary surveys. The intent was that these surveys would illustrate which indicator organism, along with its associated numerical level, was a superior indicator of bacterial contamination for California use. This approach has resulted in some controversy because discharges are being required to bear the expense of monitoring for an additional indicator organism. Also, since sampling for total, fecal, and enterococcus bacteria has been performed for several years, the regulated community has requested that the State Board take action to resolve this issue.

State Board Investigation of Water Contact Standards:

An independent technical group, the Microbiological Advisory Committee (MAC) was formed in 1992 to advise State Board staff on how to investigate this issue. A study design was agreed upon, and a contract was initiated with the University of California, Berkeley in 1993. The contract allowed for data from two major ocean dischargers to be analyzed using two approaches:

1. at each monitoring station, for each month and for each individual indicator organism, the number of times the measured level exceeded the allowable value contained in the Ocean Plan was determined; and
2. for each monitoring station, the density of indicator organisms were compared against each other and to physical parameters measured at the same time (water temperature, salinity, dissolved oxygen, etc.).

The contract also required that recent epidemiological studies be reviewed, summarized and related (if possible) to the discharger data analyses. Based on review of both discharger monitoring data and results of recent epidemiological studies, UC Berkeley was: (1) to make recommendations for possible revision of the Ocean Plan water contact bacterial standards and (2) to identify areas in which additional research is necessary.

Because there was interest in the environmental fate of indicator organisms based on monitoring data taken over a time course of several years and under diverse environmental conditions, data from the City of San Diego and the City and County of San Francisco were analyzed. Locations of the discharges and sampling locations are provided in Figures 1 and 2.

Data analysis is complete, and the draft report is under review by the MAC. Appropriate changes to the Ocean Plan's water-contact bacterial standards should be proposed for public review in late 1997 or early 1998.

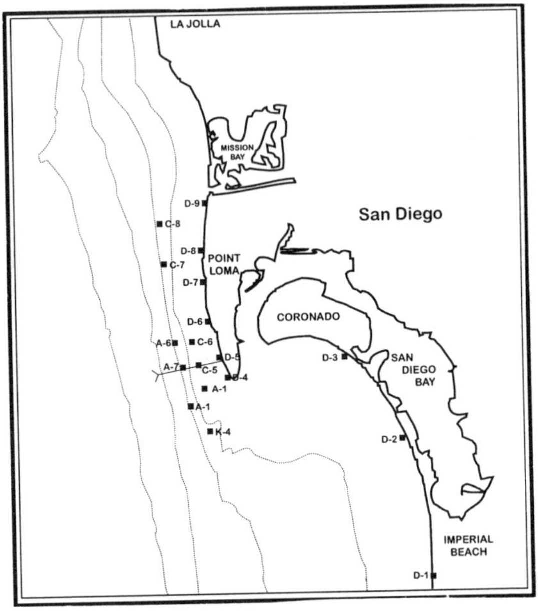

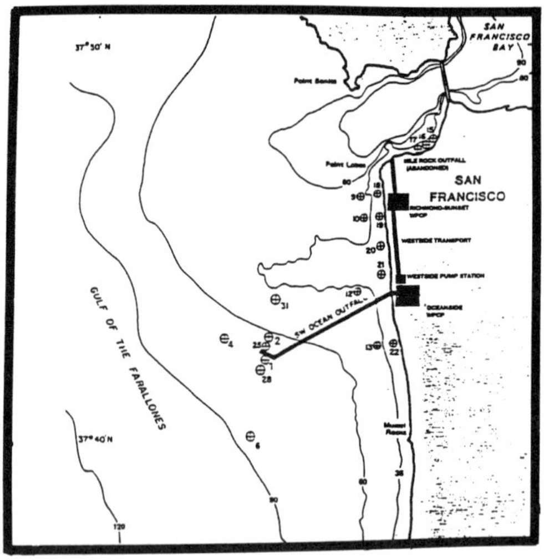

Figure 1. (above) City of San Diego Ocean Monitoring Program

Figure 2. (below) City and County of San Francisco Ocean Monitoring Program

Shellfish Harvesting Standards

Discussion:

The Ocean Plan contains the following bacterial standard for shellfish harvesting waters:

> "At all areas where shellfish may be harvested for human consumption, as determined by the Regional Board, the following bacterial objectives shall be maintained throughout the water column:
> The median total coliform density shall not exceed 70 per 100 ml, and not more that 10 percent of the samples shall exceed 230 per 100 ml."

The six commercial growing areas in California (Figure 3) are under the regulatory control of the Department of Health Services, which is a participant of the National Shellfish Sanitation Program (NSSP). The objective of the program is to prevent human illness associated with the consumption of fresh and fresh/frozen molluscan shellfish by developing sanitary controls over all phases of the growing, harvesting, shucking, packing, and distribution of shellfish. The NSSP developed a manual containing the minimum criteria and guidelines for the safe growing and harvesting of shellfish. A related organization, the Interstate Shellfish Sanitation Conference (ISSC) was formed to provide a formal structure by which NSSP guidelines can be revised. The NSSP standard for commercial shellfish harvesting waters shall meet one of the following:

1. The total coliform median or geometric mean MPN of the water does not exceed 70 per 100 ml water and no more than 10 percent of the samples exceed an MPN of 230 per 100 ml for a 5-tube decimal dilution test.

2. The fecal coliform median or geometric mean MPN of the water does not exceed 14 per 100 ml and not more than 10 percent of the samples exceed an MPN of 43 per 100 ml for a 5-tube decimal dilution test.

Similarly to using indicator organisms for water-contact standards, the adequacy of current coliform standards for shellfish harvesting waters has been questioned, particularly with respect to viral pathogens.

In California, interest in shellfish harvesting waters has received increased attention since the passage of the Shellfish Protection Act of 1993. This Act acknowledges shellfish harvesting as a beneficial use of the State's waters and notes that pollution from point and nonpoint sources is currently threatening many of the State's commercial shellfish growing areas. Whenever a commercial shellfish growing area is identified as threatened under the terms of the Act, the appropriate Regional Board is required to form a technical advisory committee (TAC) to investigate the problem and suggest remedial action. Bacterial contamination of shellfish growing waters most often occurs during the winter months, when the rain runoff carries fecal material from the watersheds into the waterbodies.

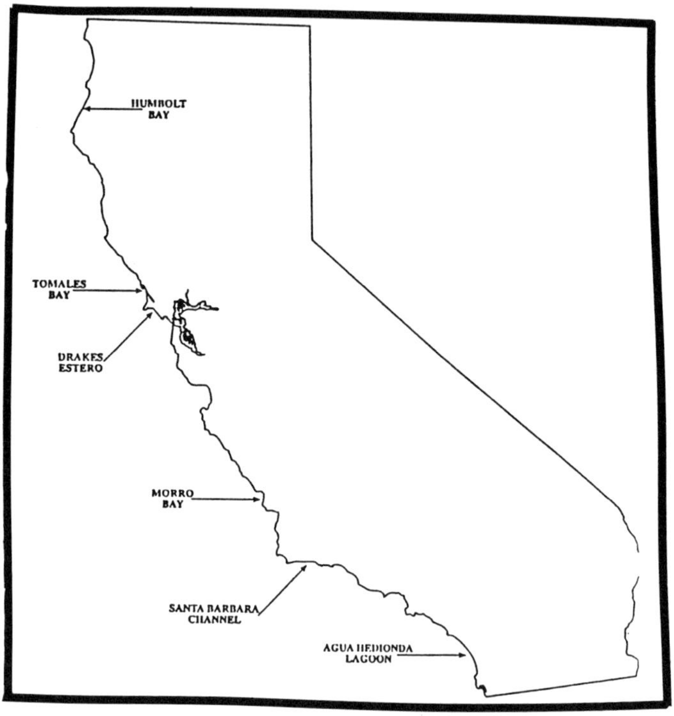

Figure 3. Commercial Shellfish Growing Areas

Regional Board Advisory Committees:

Current effort to identify and remediate fecal contamination to shellfish growing waters is directed toward commercial operations. Four of the six coastal commercial shellfish harvesting areas have been identified as threatened (as defined by the Shellfish Protection Act) and have subsequently formed technical advisory committees:

Humboldt Bay TAC (North Coast Regional Board) - Currently there are two commercial growers operating three leases in Humboldt Bay. The TAC has reviewed available data, and has identified several potential sources of fecal contamination to the Bay: dairy and beef cattle operations, horse operations, septic systems, and timber operations. No mitigation plans have begun at this time.

Morro Bay TAC (Central Coast Regional Board) - There is one commercial grower operating two leases in Morro Bay. The TAC, working in conjunction with the National Estuary Program, has a two phase plan to identify pollution sources to the Bay. The first phase requires compiling existing information and implementing studies necessary to fill identified data gaps, provide baseline data on pollutant loadings, and develop estimates of spacial and temporal variability. The second phase of technical activity is to develop a long term environmental monitoring program that will identify current and emerging problems, assess effectiveness of actions and programs implemented under the plan, and provide essential information to direct and refocus the management plan as needed over time. Phase one, which began in 1996, is a cooperative effort involving the State and Regional Board, DHS, Morro Bay National Estuary Program, and interested citizens.

Aqua Hedionda TAC (San Diego Regional Board) - One commercial shellfish operation is located in this lagoon. Because the growing waters are classified as restricted, all shellfish from this operation must be depurated before they can be sold. The Agua Hedionda TAC has conducted studies in 1995 and 1996 to identify fecal coliform pollution sources in the shellfish growing waters. These studies are continuing into 1997 to identify the pollution sources and the TAC will recommend mitigation measures to the Regional Board.

Tomales Bay TAC (San Francisco Regional Board) - There are eight shellfish growers currently operating a total of 14 leases in Tomales Bay. The TAC determined that a study should be performed to determine the various point and non-point sources of fecal contamination that are impacting shellfish growing areas. In addition, water samples should be included which show potential for being more human-specific than coliform bacteria.

A pilot monitoring study was conducted during the winter of 1994-1995. The goals of the pilot study were to :

1. evaluate enterococcus, Clostridium perfringens and Bacteroides species as potential indicators;
2. test sample volume, collection, sample transport logistics, and laboratory analysis in preparation for the full-scale study; and
3. finalize site selection of watershed sampling stations.

Using the results of the pilot study, the study protocol was finalized for the full-scale study. C. perfringens analysis was not included in the full-scale study. C. perfringens forms endospores, which are very resistant to adverse environmental conditions. Endospores can remain dormant many years, but can be triggered to convert back into a vegetative cell when placed in appropriate environmental conditions. Because of this condition, it is not possible to determine if a samples positive for C. perfringens is the result of a recent fecal contamination or is from past contamination.

The full-scale study was conducted during the winter of 1995-1996. The study consisted of six sampling periods (one pre-rain season sampling, four rainfall samplings, and one post rain season sampling) at 40 sampling stations (Figure 4). Sampling stations included both watershed and bay sites. All samples were analyzed for four standard indicators of microbiological water quality: total coliform, fecal coliform, enterococcus, and Escherichia coli. Four of these primary stations, plus freshwater and seawater control sites, were analyzed for Bacteriodes vulgatus and bacteriophage. Midway through the study, testing for viruses and for an E. coli toxin gene were added. In addition to water samples, oyster samples were taken from bay water sample stations. Station locations were selected to:

1. monitor levels of contamination in the outer, middle, and inner sections of the bay; and
2. measure the impact of birds on the shellfish growing areas.

Final samples were collected in July 1996, and sample analysis was completed in early August. The TAC anticipates a draft report of this study in mid 1997. A public meeting was held in February 1997 to discuss the results of the study. A second meeting in May 1997 will form working groups to suggest ways to control the identified pollution sources.

FIG. 4 - SAMPLING STATIONS FOR TOMALES BAY STUDY

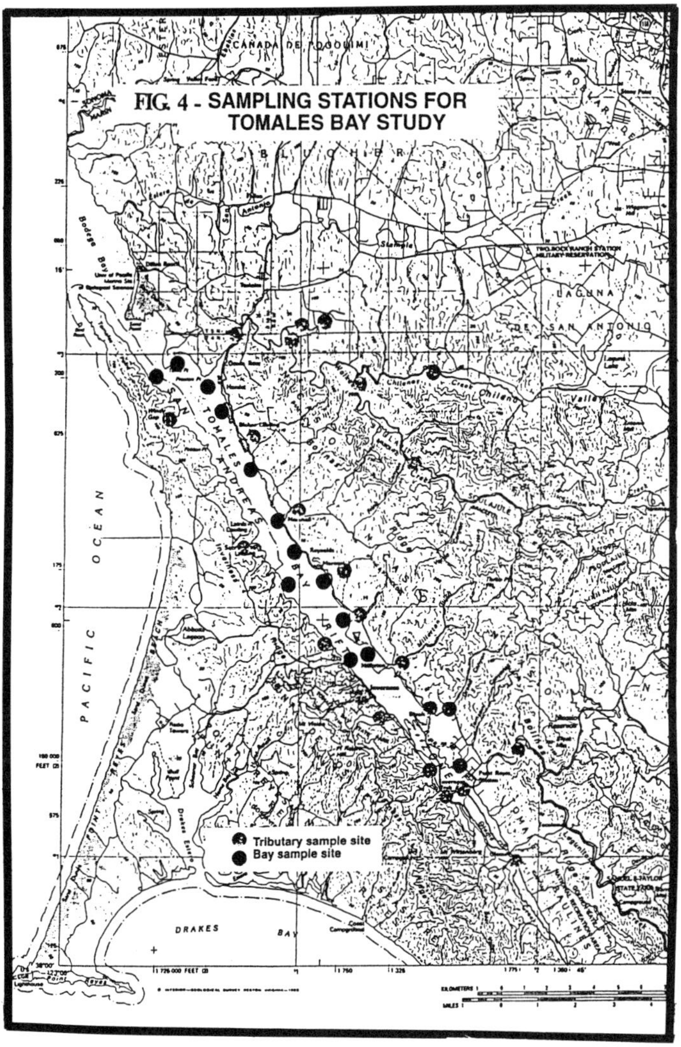

Summary and Conclusions

Because routine monitoring for all possible human disease-causing agents is impractical, indicator bacteria are used as an alternative to the measurement of pathogens. Total and fecal coliform are the bacteria most commonly used to indicate fecal contamination. These indicators are not human specific; total coliform bacteria can exist on soil particles and plant surfaces, and fecal coliform bacteria are normally found in the gastrointestinal tract of warm-blooded animals. Enterococcus, another group of bacteria used as indicator organisms, is also common to all warm-blooded animals. EPA epidemiological studies suggest that enterococcus is a better indicator of human health effects than either total or fecal coliform. However, none of the current indicator organisms are thought to be reliable predictors of non-bacterial pathogens, such as enteric viruses or protozoans.

Staff from the State Board's Ocean Standards Unit have been working with staff from Regional Boards, local governments, environmental groups, private industry, and other interested parties to investigate questions raised about the adequacy of bacterial standards for water contact and for shellfish harvesting waters:

- The State Board contracted with the University of California, Berkeley to analyze data collected by two ocean dischargers, to review recent epidemiological studies, to make recommendations for possible revision of the Ocean Plan water contact bacterial standards, and to identify areas in which additional research is necessary. Data analysis is complete, and the draft report is under review by the MAC. Appropriate changes to the Ocean Plan's water-contact bacterial standards should be proposed for public review in late 1997 or early 1998.

- State Board staff are participants on several shellfish TACs, attempting to identify fecal pollution sources to shellfish harvesting waters. A study to identify sources of fecal contamination into Tomales Bay has been completed. The Tomales Bay TAC anticipates a draft report of this study in mid 1997. A public meeting was held in February 1997 to discuss the results of the study. A second meeting, not yet scheduled, will form working groups to suggest ways to control the identified pollution sources.

References

State Water Resources Control Board. 1990. Water Quality Control Plan for Ocean Waters of California (California Ocean Plan).

U.S. Department of Health and Human Services. Public Health Service. Food and Drug Administration. 1995. National Shellfish Sanitation Program Manual of Operations. Part I - Sanitation of Shellfish Growing Areas.

CHAPTER 232

The State Water Resources Control Board's Marine Bioassay Project

Matthew W. Reeve, M.S.[1] and Francis H. Palmer, D.Env.[2]

Abstract

The Marine Bioassay Project (MBP) was initiated in 1984 by the State Water Resources Control Board (SWRCB) to design and develop sensitive test protocols to measure toxicity of discharges to marine waters. Many toxicity tests previously in use were relatively insensitive because the adverse effect measured was lethality to adult organisms. A newer generation of tests developed by the MBP has been designed to estimate more subtle long-term adverse effects of waste discharges that ultimately may damage populations of important marine species.

The work of the MBP was mandated in 1986 by addition of Section 13170.2 to the California Water Code. This Section directed the SWRCB to "develop bioassay protocols to evaluate the effect of municipal and industrial discharges on the marine environment". The project consists of four different entities, the SWRCB, the Department of Fish and Game (DFG), the University of California at Santa Cruz, and the Scientific Review Committee (SRC), an independent group of advisers representing industry, academia, and government. The contribution of each group is as follows: (1) the SWRCB plans, administers, and funds the MBP, (2) the DFG provides research facilities for the MBP at the Granite Canyon Marine Pollution Studies Laboratory, (3) the University of California at Santa Cruz provides the principal investigators and necessary staff to conduct the research, and (4) the SRC provides MBP the scientific guidance and focus to successfully conduct their research.

The MBP has developed and implemented four toxicity test methods using species indigenous to California's coastal waters. These tests reflect recent advancements in the field of aquatic toxicology and have a broad taxonomic

----

[1]Environmental Specialist, Ocean Standards Unit, State Water Resources Control Board
[2]Chief, Ocean Standards Unit, State Water Resources Control Board
901 P Street, Sacramento, California, 95814

representation. The test methods use an alga (giant kelp), a fish (topsmelt), and two invertebrates (red abalone and Pacific mysid shrimp). The giant kelp (*Macrocystis pyrifera*) and red abalone (*Haliotis rufescens*) are currently on the 1990 Ocean Plan approved list of critical life stage protocols for use in National Pollutant Discharge Elimination System compliance monitoring. The mysid (*Holmesimysis costata*) and topsmelt (*Atherinops affinis)* protocols will be added to the list if the proposed 1996-1997 Ocean Plan amendments are adopted by the SWRCB in March 1997.

All four of the methods developed by the MBP have been adopted by the U.S.EPA for use in compliance monitoring of waste discharges along the entire west coast of the United States (California, Oregon, and Washington). They constitute the majority of the tests published in U.S.EPA's West Coast Procedures Manual and have been pivotal in evaluating the potential effect of municipal and industrial ocean discharges on California marine life.

Introduction

The goal of the SWRCB's Marine Bioassay Project is to protect California's ocean resources by determining the impacts of waste discharges on marine waters. The Project's primary objective is development of critical life stage tests using indigenous organisms to measure toxicity of these discharges. A second objective is the actual use of these tests for regulatory purposes, specifically for compliance monitoring of chronic toxicity in complex effluents discharged into the Pacific Ocean.

Consistent with these two objectives, the MBP has developed four protocols that use an alga (giant kelp), a fish (topsmelt), and two invertebrates (red abalone and Pacific mysid crustacean) native to California's waters. The giant kelp (*Macrocystis pyrifera*) and red abalone (*Haliotis rufescens)* protocols are currently on the 1990 Ocean Plan approved list of critical life stage protocols for use in NPDES compliance monitoring. The mysid (*Holmesimysis costata*) and topsmelt (*Atherinops affinis*) protocols were added to the list with the adoption of the 1996-1997 Ocean Plan amendments by the SWRCB. The four protocols developed by the MBP reflect recent advancements in the field of aquatic toxicology.

Aquatic Toxicity Testing

The use of aquatic toxicity tests is dependent upon the testing objective (e.g., determining toxicity of effluents, receiving water, or known chemical compounds and elements). Toxicity tests are conducted on: (1) effluents to assess compliance with discharge permit limits and the water quality toxicity objective, (2) receiving waters to determine if toxicity has resulted from a discharge or if ambient toxicity is present, and (3) specific chemical compounds (e.g., sodium azide) and elements (e.g., copper, zinc, cadmium) to determine the concentration at which adverse effects are detected and the sensitivity of the test organism.

The endpoints measured in tests to determine the adverse effects of toxicants (i.e., the measured response) include survival, reproduction, growth, locomotor activity and other behavioral responses, gill ventilation rate, heart rate, blood chemistry, histopathology, enzyme activity, terata, and genetic abnormalities, among others. Because it is not possible to detect and/or quantify all of these effects on a routine basis, toxicity test measurements are usually limited to a few endpoints such as survival, growth, reproduction, and developmental abnormalities.

There are three general categories of toxicity tests: acute, chronic, and critical life stage. The terms "acute" and "chronic" are sometimes confusing because they may refer to either the duration of exposure or to the adverse effect produced by exposure to a toxicant. An acute exposure is a short time period, usually 96 hours or less for toxicity tests. An acute effect generally refers to mortality. For example, when an acute test is conducted on larval fish with an endpoint of mortality and duration of 96 hours, acute describes both duration of exposure and toxic effect.

Chronic refers to a long term exposure; a chronic test may involve exposing the test organism for its entire reproductive cycle. For fish, the duration may exceed twelve months. Chronic toxicity tests are inherently more sensitive to toxicants than acute tests; that is, adverse effects usually will be detected at lower concentrations of the toxicant. While a chronic effect can be lethal or sublethal, chronic is frequently interpreted to mean a sublethal effect.

A third type of toxicity test, the critical life stage or early life stage test, is intermediate to acute and chronic tests in duration and sensitivity to toxicants. These tests generally focus on early periods of an organism's life cycle when it is most sensitive to toxicants; however this variety of test can also refer to a sensitive adult life stage, such as during gametogenesis. When properly designed, a critical life stage test serves as a "short-term estimate of chronic toxicity".

Regulatory History

Development of toxicity test protocols to estimate long-term effects of waste discharges is consistent with both federal and state requirements. In 1984, the U.S.EPA issued a national "Policy for the Development of Water Quality-Based Limitations for Toxic Pollutants"(49 CFR, No. 48, March 9, 1984). This policy outlined a technical approach for controlling discharge of toxic substances through the federal system of discharge permits. In addition to meeting numerical standards for individual chemicals, the policy requires U.S.EPA and the States to use biological testing to complement chemical testing. Biological testing also provides information not available from chemical testing. For example, it incorporates bioavailability and interactions in complicated mixtures of toxic materials.

In 1986, the California Legislature (AB 3500) added Section 13170.2 to the California Water Code. Section 13170.2 specifically refers to the California Ocean Plan and directs the SWRCB to "develop bioassay protocols to evaluate the effect of municipal and industrial discharges on the marine environment." This section also required the SWRCB to adopt a list of critical life stage tests for use in monitoring ocean discharges by 1990. Ocean discharges exceeding 100 million gallons per day were required to use these tests as part of their permit monitoring commencing January 1, 1991, with the requirement extended to smaller dischargers on January 1, 1992.

To implement section 13170.2, as part of the March 1990 Ocean Plan amendments, the SWRCB added a chronic toxicity water quality objective for protection of marine aquatic life. The SWRCB approved an initial list of seven toxicity test protocols deemed sufficiently developed for measuring compliance with the new chronic toxicity objective. The list included two MBP tests, the giant kelp and red abalone protocols, three tests developed by the U.S.EPA, and two protocols developed elsewhere.

During the current triennial review, SWRCB staff were asked to consider updating the existing critical life stage protocol list. To perform this review, staff convened a 10 member external advisory group called the Protocol Review Committee (PRC). The PRC is an assemblage of aquatic toxicity experts representing industry, academia, and government.

In October 1994, the PRC recommended to SWRCB staff a revised list of critical life stage protocols acceptable for use in measuring compliance. The list is a culmination of four additional years of test method refinement and development since the use of specific toxicity tests were first included in the 1990 Ocean Plan. Included on the list are four tests (the giant kelp, red abalone, mysid shrimp, and topsmelt protocols) that have been developed by the State Board's MBP. These four are also included in the EPA's Methods Manuals for estimating chronic toxicity to marine organisms.

The Use of Native Species for Toxicity Testing

With the adoption of 1996-1997 Ocean Plan amendments, all four toxicity test methods developed by the MBP are on the Ocean Plan list of critical life stage protocols. The adoption of the mysid (crustacean) and topsmelt (fish) tests, deletion of the *Champia* (alga) test, and potential de-listing of the east coast mysid (shrimp) and *Menidia* (fish) tests could result in a list consisting entirely of indigenous species.

The DFG has been a major supporter of MBP research and development of toxicity test methods using indigenous species. The DFG expressed a number of concerns regarding the use of non-native species required by the U.S.EPA for toxicity testing. These concerns were described in a letter dated February 1, 1990, from Peter Bonadelli, then director of DFG, to James J. Lichtenberg of EPA's Environmental Monitoring Systems Laboratory in Cincinnati:

"The DFG is concerned about the importation of non-native, east-coast species. The widespread use of certain species as a bioassay animal on the west coast could eventually lead to an accidental establishment in coastal areas. All live aquatic life shipped into California, other than pets used in closed systems, are subject to Fish and Game commission importation regulation..."

"As with acute organisms, the DFG is concerned with the use in California of non-native species for toxicity testing. All species recommended, except *Menidia beryllina* in a restricted location, do not occur in California. The use of these animals as standard bioassay species in California is not commensurate with the threats they pose to our native estuarine fishes, should they be accidentally released or escape and become established in California. The DFG will oppose the implementation of monitoring programs which include toxicity tests using non-native species, which may threaten native fishes or wildlife."

Organization of the MBP

Two State agencies and one university work cooperatively in the operation of the MBP. The DFG's Marine Pollution Studies Laboratory serves as the research facility. U.C. Santa Cruz provides the principal investigators and staff to conduct the research. The SWRCB oversees and provides funding for the Project by contracting with DFG to develop and conduct toxicity tests, provide technical expertise, and complete related tasks specified in the contract's agreement. DFG in turn subcontracts with U.C. Santa Cruz to provide staff and operate the Project. Partial funding for the Project was provided by U.S.EPA 205(j) funds during the years 1986 to 1989.

The MBP staff has also consulted closely with the U.S.EPA in the completion of the U.S.EPA West Coast Methods Manual. John Hunt and Brian Anderson of the MBP worked with Debra Denton (U.S.EPA Region 9), and Gary Chapman and Jim Lazorchak (U.S.EPA Office of Research and Development) in adapting MBP protocols for inclusion into the U.S.EPA West Coast Methods Manual. The U.S.EPA manual contains seven critical life stage protocols, four developed by the MBP, one developed by the staff of the Southern California Water Research Project, and two older protocols revised by U.S.EPA. Completion of the MBP Procedures Manual and the U.S.EPA West Coast Methods Manual has occurred as a result of collaboration by numerous agencies, groups, and individuals.

Results and Discussion

The MBP was initiated in 1984 by the SWRCB to develop and design sensitive measures for testing the toxicity of discharges to marine waters. The program's success has been demonstrated by development of critical life stage test methods using indigenous species. All four of these toxicity protocols have been adopted by the SWRCB to the Ocean Plan list for monitoring compliance of ocean waste discharges. The MBP developed methods comprise the majority of toxicity tests published in U.S.EPA's West Coast Procedures Manual.

In its 13 years of existence, the Project has become essential to staff of government agencies, the waste discharger community, and consulting laboratories. The two principal investigators, John Hunt and Brian Anderson, are recognized nationally in the field of aquatic toxicology and are often called upon for their technical expertise. They have also conducted workshops on test methods developed by the MBP for representatives of the waste discharger community and SWRCB and Regional Board staff. The MBP provides the vital link in the transfer of knowledge from the research phase (the development of the toxicity test methods) to the implementation phase (use of the protocols for compliance monitoring of ocean waste discharges).

The three accompanying figures illustrate the four organisms used in the marine aquatic toxicity tests developed by the MBP (Figure 1) and typical dose response curves one would observe when testing with a treated wastewater discharge (Figure 2) and a reference toxicant (Figure 3).

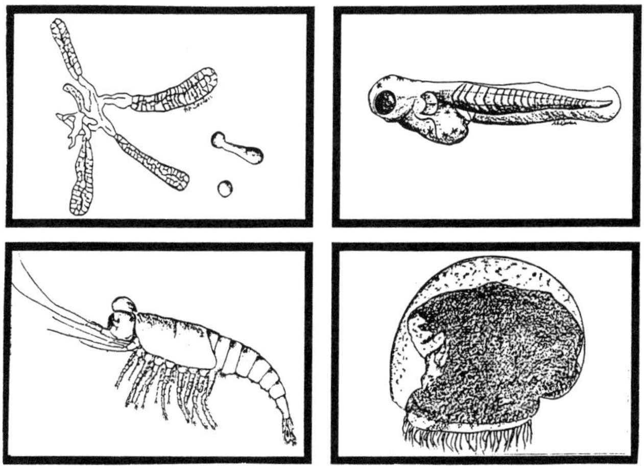

Figure 1. microscopic kelp sporophytes and germinated and un-germinated kelp spores (upper left); nine day old topsmelt larva (upper right); three day old juvenile (lower left); 48 hour old abalone veliger larva (lower right)

Figure 2. Dose Response Curve - Treated Wastewater Discharge Topsmelt (*Atherinops affinis*)

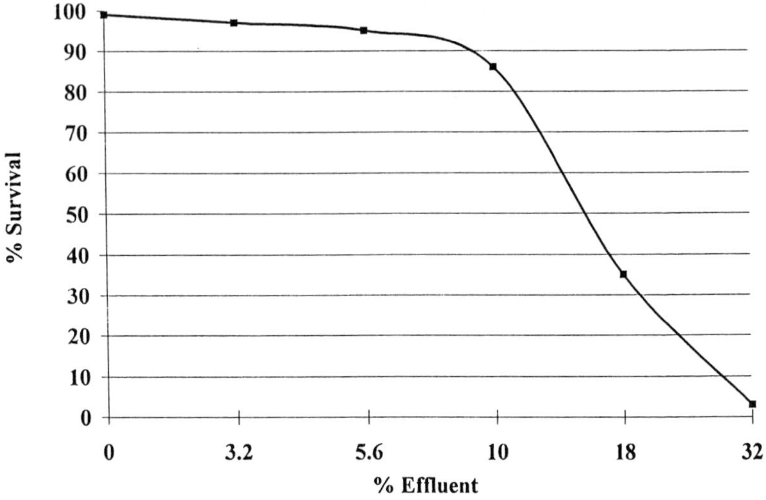

Figure 3. Dose Response Curve - Reference Toxicant (Cu) Giant Kelp (*Macrocystis pyrifera*)

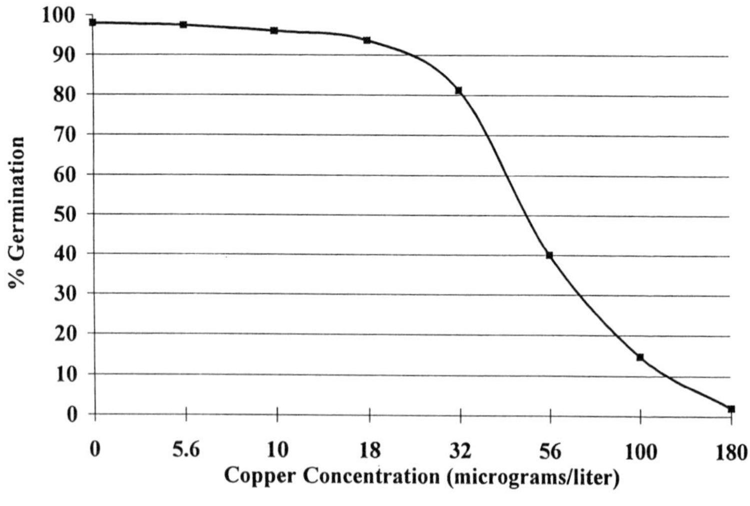

References

SWRCB. 1996. Procedures Manual for Conducting Toxicity Tests Developed by the Marine Bioassay Project, 96-1WQ.

U.S.EPA. 1994. Short-term Methods for Estimating the Chronic Toxicity of Effluents and Receiving Water to Marine and Estuarine Organisms. Klemm, D.J., G.E. Morrison, T.J. Norberg-King, W.J. Peltier, and M.A. Heber. EPA/600/4-91-003

U.S.EPA. 1995. Short-term Methods for Estimating the Chronic Toxicity of Effluents and Receiving Waters to West Coast Marine and Estuarine Organisms. Chapman, G.A., D.L. Denton, and J.M. Lazorchak. EPA/600/R-95-136

CHAPTER 233

Bioeconomics of Harvest Refguia

James N. Sanchirico
James E. Wilen[1]

Abstract

This paper develops a simple model of harvest refugia formation. A two patch biological model is used to depict a system the dynamics of which depend upon patch-specific parameters as well as dispersal between patches. We model a special case with one-way flows, commonly referred to as a source/sink model. A harvesting sector is added by appending a model of a fishing industry whose entry/exit dynamics respond to rents in the fashion suggested by Gordon and V. Smith. The questions addressed with this framework center on the implications of shutting one fishery down in order to create a refuge free from exploitation. We examine conditions under which biomass, harvests, and rents increase under reserve formation. Conditions are derived which characterize the configurations of biological and economic parameters that result in increased biomass, harvests and rents. Using both the conceptual model and a simulation exercise, we show that refugia can be directly beneficial even to the industry that is displaced from the reserve.

---

[1]The authors are Sea Grant Trainee/Research Assistant and Professor in the Department of Agricultural and Resource Economics at the University of California at Davis.

CHAPTER 234

Southern California Educational Initiative and Coastal Marine Institute: Long-term Environmental, Social and Economic Consequences of Offshore Oil and Gas Production

Russell J. Schmitt[1 2], Mark H. Carr[2], and Bonnie Williamson[2]

Abstract

During the past decade, the Environmental Studies Program of the Minerals Management Service recognized the need for more and better information on long-term environmental and social effects of OCS oil and gas production. Additionally, the agency has sought mechanisms to foster innovative, peer-reviewed science, and to more effectively engage academic scientists to word on questions relevant to MMS. The agency implemented two programs in the Pacific Region to accomplish these goals, which are entitled the Southern California Educational Initiative (SCEI) and the Coastal Marine Institute (CMI). The overall goal of the 5-year programs is to develop new information on long-term social and environmental effects of oil and gas production in the Pacific Region, Outer continental Shelf of the United States. This will be achieved primarily through research studies that focus on broad, unresolved issues of regional and national importance, the results of which will be published in the peer-reviewed literature. A complementary goal is to develop a "pool" of scientific expertise capable of resolving current and future problems arising form such coastally-dependent enterprises as offshore oil and gas development. This will be realized by attracting academic scientists to conduct original research on issues appropriate to MMS, and by a variety of mechanisms to involve undergraduate, graduate and post-graduate students in a broad array of research activities. Program objectives include:

- respond to MMS, State and local information needs and interests with UC expertise of national caliber in relevant disciplines;
- broaden recognition and comprehension of study results through

[1]Department of Ecology, Evolution an Marine Biology, University of Califonia, Santa Barbara, CA 93106

[2]Coastal Research Center, Marine sciemce Institute, University of California, Santa Barbara, CA 93106

performance and presentation / disseminating findings in the peer-reviewed literature;

- improve existing local capabilities and facilities for innovative scientific research relevant to OCS resource management issues;
- use the interdisciplinary environment of the University of California to foster process oriented studies, needed technologies and concepts, and syntheses of information that will benefit environmental and resource management;
- develop human resources to address MMS information needs, including attracting high caliber faculty scientists to pursue new research endeavors on identified issues and training students in research applicable to decision-making;
- achieve consensus between MMS and California regarding the most important environmental research needs relevant to the OCS program.

CHAPTER 235

STATE AND LOCAL GOVERNMENT INVOLVEMENT IN COASTAL MANAGEMENT, QUEENSLAND, AUSTRALIA

Mr L.A. (Angus)JACKSON [1] Mr A.W.(Sam) SMITH [2]

In 1967 after a number of cyclones along Queensland's Gold Coast, there was resultant severe erosion which in turn had a significant adverse economic impact on the City. The value of the beaches to the local and State economies was evident and consequently the State Government commissioned the Delft Hydraulics Laboratory to produce a detailed and definitive report. To expedite restoration works, the Gold Coast City Council set up its own coastal section to design works and run contracts, staffed with experienced engineers, technicians and surveyors. Meanwhile the Queensland Government set up the Beach protection Authority and adopted a Management Scheme for the Gold Coast and eventually other areas.

To obtain approvals to carry out works in the Coastal Zone approval is now required from at least that Authority as well as the Department of Ernvironment. Often several other departments are also involved. As the time frames and objectives are by nature different It has been been very difficult at times to attain effective co-operation and harmony. Works in the coastal zone often have some risk and the State departments often adopt rigid policies whilst the City has tended to be more project orientated. As the City is responsible for implementation and 75% of funding, the relationship between the City which has a degree of independance through its monetary resources and local expertise and the controlling State has been difficult..

The local City Councillors have always seen beach protection works as vote catchers. Large dredges and sand pumping are highly visible and make good TV news items. The State Government politicians have also come to see the benefits of successful beach projects. With the rising cost of coastal protection works and with restructuring of the Council several years ago, the Council has made greater requests from the State Government and the State has become more involved. As a result in recent times, the State Government has been more and more willing to project manage and fund the majority of large projects. This has lead to a greater understanding of the practical problems at a local level and at times availability of resources. However the costs of project administration have risen..

[1] Managing Director, International Coastal Management, P.O.Box 7196, Gold Coast MC 9726 AUSTRALIA PH/FAX 61-7-5592 9147 EMAIL; jackson@on the net com.au
[2] Consulting Engineer, 5 Ilkinia Ave, Florida Gardens, Gold Coast 4217 AUSTRALIA Ph 61-55381856

CHAPTER 236

Support for Coastal Watershed Activities Using the CERES Watershed Information Technical System

Janine Stenback and Rebecca Fawver[1]

Abstract

The California Resources Agency will debut in April 1997 a new web site in support of local watershed assessment, planning, management, restoration, monitoring and education statewide. The Watershed Information Technical System (WITS) is a focused application of the California Environmental Resources Evaluation System (CERES) -- an on-line information service developed by the California Resources Agency to facilitate access to a variety of data describing California's diverse environments (*http://ceres.ca.gov)*.

The initial product development is focused on the coastal watersheds of the Russian River and Mendocino Coast. In these hydrologic units, WITS will provide customized access to a digital collection of watershed related documents (e.g., planning guidelines, restoration procedures, basin plans and assessments); watershed maps and spatial data; a database of all known ecological information and field collected data; and local points of contact, such as local watershed coordinators and watershed groups. WITS has been developed in partnership with the local coastal communities, State and federal government, and the University of California.

---

[1] Special Assistants for the California Resources Agency, 1416 Ninth Street, Suite 1311, Sacramento, CA 95814

CHAPTER 237

Mass Loading and Subsequent Baywide Transport of Sediment Resuspension During Tug Assisted Ship Movements at the Naval Station San Diego.

Dr. Don Sutton, San Diego Supercomputer Center

Bart Chadwick, Dr. Ken Richter, Naval Research and Development

Abstract

Large ship, propeller induced, resuspension of bottom sediments is potentially a viable mechanism for sediment and contaminant transport in harbor areas. A field survey was conducted to quantify the water column mass loading of resuspended sediments during tug assisted ship movements occurring in the pier areas of the Naval Station in San Diego Bay. A towed transmissometer was used to assess total suspended solids in the water column before and after ship movement events. The average mass of resuspended material attributed to a single tug assisted ship movement event was found to be 8.6 x 10^3 kg. For the Naval Station, this amounts to an average loading of 41.7 x 10^3 kg/day. Once resuspended, the transport of the material is primarily governed by tidal current flows and particle settling rates. This loading data, together with grain-size analysis data, was inputted into a validated, high resolution hydrodynamic computer model to predict baywide transport and fate of the resuspended sediments.

Field measurements

Five field surveys were conducted for this study. Field surveys were carried out using a NRaD 40 ft survey craft, Ecos, outfitted with differential GPS, a towed transmissometer, and a water sample collection system. The transmissometer generates an analog voltage proportional to light transmission over a 25 cm distance. Total suspended solids concentrations (TSS) are calculated using an exponential curve fit of transmissometer data and TSS values obtained by weighing the solid fraction from water samples taken during the surveys. Background field measurements were made by towing the transmissometer 2 meters below the surface in the pier areas and main channel prior to the ship launching or docking. Once the ship and associated tugs had stopped or cleared, a duplicate survey was made. The transmissometer data together with the differential GPS information enable suspended solid concentrations to be GIS mapped.

Average daily mass loading analysis

The background and resuspension transmissometer data sets are interpolated onto 10,000 node grids. Land areas and piers within the grid, along with wetted areas not effected are masked off. The total mass loading per ship movement event is calculated by subtracting the background from the resuspension interpolated concentrations and multiplying by grid cell depths. Average TSS concentrations and total mass values are shown in **Table I** for background, resuspension, and their difference from the interpolations from each of the surveys. The total mass loading per ship movement event ranges from 3,800 to 20,400 kg with an average of 11,200 kg and a standard deviation of 6,500 kg.

Daily ship movement records over a 6 month period, obtained from the Fleet Support Office, are used to estimate the average daily mass loading. An average of 4.8 ships using 1.7 tugs per ship move either in or out of the Naval Station on a daily basis, **Table II**. Ship movement events include assistance by 2 to 4 tugs. A weighting scheme is used in which, the ship is weighted twice that of a tug, in estimating source of resuspension. For instance, Survey2, which sampled the movement of *USS Tarawa* and 3 tugs, accounted for 5 unit movements or from **Table I**, 20.4 x 10^3 kg divided 5 unit movements, or 4.1 x 10^3 kg/unit movement.

Using this scheme the average daily movements in terms of units from the Fleet Support Office records is 17.5 units/day, **Table II**. Average daily total mass loading is estimated by calculating the mass loading per unit movement from the surveys found in **Table I**, and multiplying this by 17.5 units/day. These estimates range from 15.3 to 71.4 x 10^3 kg/day with an average of 41.7 x 10^3 kg/day, **Table III**.

Table I: Survey TSS data

Surv1 (2tugs)	**Background**	**Resuspension**	**Difference**
Ave conc, mg/L	2.02	6.94	4.9, 244%
Total Mass kg	4.52 x 10^3	15.54 x 10^3	11.0 x 10^3
Area m^2	0.245 x 10^6	0.245 x 10^6	
Surv2 (3tugs)			
Ave con., mg/L	1.68	7.23	5.5, 324%
Total Mass kg	6.19 x 10^3	26.62 x 10^3	20.4 x 10^3
Area m^2	0.402 x 10^6	0.402 x 10^6	
Surv3 (2 tugs)			
Ave conc, mg/L	2.26	3.20	0.9, 41%
Total Mass kg	9.11 x 10^3	12.9 x 10^3	3.8 x 10^3
Area m^2	0.442 x 10^6	0.442 x 10^6	
Surv4 (3 tugs)			
Ave conc, mg/L	6.61	12.07	5.5, 83%
Total Mass kg	19.59 x 10^3	35.78 x 10^3	16.2 x 10^3
Area m^2	0.324 x 10^6	0.324 x 10^6	
Surv5 (3 tugs)			
Ave conc, mg/L	5.46	6.81	1.35, 25%
Total Mass kg	17.66 x 10^3	22.01 x 10^3	4.4 x 10^3
Area m^2	0.353 x 10^6	0.353 x 10^6	
Average, SDev.			
Ave conc, mg/L	**3.61 +/- 2.02**	**7.25 +/- 2.83**	**3.3 +/- 1.6**
Total Mass kg	**11.4 +/- 6.1 x 10^3**	**22.6 +/- 8.2 x 10^3**	**11.2 +/- 6.5 x 10^3**

Table II: average number of daily ship movements for NAVSTA

Month	Ships moved	tugs used	tugs/ship	ships/day	units/day
Oct'95	150	265	1.77	4.84	18.25
Nov '95	134	229	1.71	4.47	16.58
Jan '96	132	236	1.79	4.26	16.14
Feb '96	127	219	1.72	4.38	16.29
Mar '96	165	267	1.62	5.50	19.91
Apr '96	165	270	1.64	5.50	20.02
May '96	136	215	1.58	4.39	15.71
Average	**144**	**244**	**1.7+/- .07**	**4.8+/-.5**	**17.5+/-1.7**

Table III: Average daily total mass of resuspended sediment for NAVSTA

	resuspension, kg/unit	Ave daily resuspension, kg
Survey1	2.76 x 10^3 kg/unit	48.44 x 10^3 kg/day
Survey2	4.07 x 10^3 kg/unit	71.42 x 10^3 kg/day
Survey3	0.95 x 10^3 kg/unit	16.67 x 10^3 kg/day
Survey4	3.24 x 10^3 kg/unit	56.86 x 10^3 kg/day
Survey5	0.87 x 10^3 kg/unit	15.27 x 10^3 kg/day
Average, SDev	2.4 +/- 1.3 x 10^3 kg/unit	41.7 +/- 22.3 x 10^3 kg/day

Model transport and fate predictions

TRIM2D, a hydrodynamic model developed by R.Cheng at US Geological Survey, is used to model the transport and fate of the resuspended sediments. The model uses an estimate for the average total mass resuspended for a single ship movement event. From Table II, the average ship movement is 3.64 vessel units. The average resuspended mass from a single ship movement using the value in Table III is 8.6 x 10^3 kg. In the model simulations this amount is released 4.8 times per day over three locations spanning the Naval Station. Additionally, grain size is an important parameter in this modeling study since transport processes (settling rate) and adsorption of contaminants depend on grain size distribution of the resuspended sediments. Model prediction show that particles larger than 100 microns remain in the Naval Station area. Smaller particles settle over larger areas in a non linear fashion. The smaller 20 micron particles can be transported over a 5 km radius during strong tidal surges.

Discussion

Possible significant errors may arise from the assumption that the suspended solids are well mixed vertically or that the values measured 2 meters below the surface are equal to depth averaged TSS concentrations. Although a 3-dimensional map is desired, the dynamics of the sediment transport are rapid compared to the time it takes to depth profile with the transmissometer over the effected area. At present a horizontal mapping at a constant 2 m depth below the surface is taken and for analysis purposes the water column is assumed to be well mixed. Since the sediment is resuspended from the bottom it is possible that a significant gradient exists. A number of depth profiles taken, however, show fairly uniform concentrations from top to bottom of the water column. Along with the spatial limitations of the interpolation, errors can be introduced in the temporal assumption that the survey mapping is a snapshot in time. The resuspension process is dynamic and visible horizontal spreading occurs on the order of minutes, while a survey mapping can take up to an hour. Surveying during slack tides helps significantly to reduce this possible error. All of the ship movements surveyed were within 45 minutes of slack tide.

It is likely that a significant fraction of the four to five fold difference in TSS concentration and mass loading values are actual and not due to assumptions in the analysis. Several factors contribute to sediment resuspension. Pier site or bottom bathymetry, ship type or draft, propeller speeds, variability in technique of operators of tugs and of ship captain (the degree to which the ship uses its own propellers can vary widely) may all play a role.

To put the resuspension estimates in perspective, the Naval Station including the shipping channel directly in front of it has a combined water volume of about 40 x 10^6 m^3. Using the average background measured TSS concentration, 3.61 mg/L, this volume has a total background suspended sediment mass of about 144.4 x 10^3 kg. The average daily total resuspended mass estimate from this report, 41.7 x 10^3 kg/day, represents 29 % of the background value.

The average daily total resuspended mass estimate can be compared to creek loadings during storm events. 1994-96 data from Kinetics Inc. for Chollas Creek (north end of the Naval Station) and Paleta Creek (center of the Naval Station) show TSS loadings of 230 x 10^3 kg/storm and 130 x 10^3 kg/storm, respectively, for a single average storm event which last about a day. The daily resuspension estimates from this study are 11.6 % of this average storm event value. On a yearly TSS mass balance basis, where roughly 15 average storm events occur, or a total combined storm loading of 5.4 x 10^6 kg/yr. from Chollas and Paleta Creeks, the ship induced resuspension is 15.0 x 10^6 kg/yr., or roughly three times the creek loading.

Summary

A reasonable estimate is made of the daily resuspension of sediments due to ship movements in the Naval Station area. This estimate ranges from 16.7 x 10^3 to 71.4 x 10^3 kg/day with an average value of 41.7 x 10^3 kg/day. Although significant experimental error may exist some variability can be attributed to differing conditions during ship movements. This daily input value represents 29 % of the background mass of suspended sediment for the Naval Station and adjacent shipping channel. In comparison to TSS loading from Chollas and Paleta Creeks, which drain into the Naval Station, the yearly estimated total sediment resuspension from tug assisted ship movements is roughly 300 % of the storm estimated total mass coming from the creeks. From this data together with visual observation of the plumes around the ships and tugs during these movements, it is clear that this loading is a significant source of sediment loading in San Diego Bay and must be considered in mass balance analysis of sediment.

SUBJECT INDEX

Page number refers to the first page of paper

AUTHOR INDEX

Page number refers to the first page of paper